Commercial Renovation Costs with RSMeans data

Thomas Lane, Senior Editor

2021
42nd annual edition

Vice President, Data
Tim Duggan

Vice President, Product
Ted Kail

Product Manager
Jason Jordan

Principal Engineer
Bob Mewis

Engineers: Architectural Divisions
1, 3, 4, 5, 6, 7, 8, 9, 10, 11, 12, 13, 41
Matthew Doheny, Manager
Sam Babbitt
Sergeleen Edouard
Richard Goldin

Scott Keller
Thomas Lane

Engineers: Civil Divisions and Wages
2, 31, 32, 33, 34, 35, 44, 46
Derrick Hale, PE, Manager
Christopher Babbitt
Stephen Bell
Michael Lynch
Elisa Mello
David Yazbek

Engineers: Mechanical, Electrical, Plumbing & Conveying Divisions
14, 21, 22, 23, 25, 26, 27, 28, 48
Joseph Kelble, Manager
Brian Adams

Michelle Curran
Antonio D'Aulerio
Thomas Lyons
Jake MacDonald

Contributing Engineers
John Melin, PE, Manager
Paul Cowan
Barry Hutchinson
Gerard Lafond, PE
Matthew Sorrentino

Technical Project Lead
Kevin Souza

Production
Debra Panarelli, Manager
Jonathan Forgit
Sharon Larsen
Sheryl Rose
Janice Thalin

Data Quality
Joseph Ingargiola, Manager
David Byars
Audrey Considine
Ellen D'Amico

Cover Design
Blaire Collins

Innovation
Ray Diwakar, Vice President
Kedar Gaikwad
Todd Glowac
Srini Narla
Joseph Woughter

RSMeans data from Gordian
Construction Publishers & Consultants
1099 Hingham Street, Suite 201
Rockland, MA 02370
United States of America
1.800.448.8182
RSMeans.com

Copyright 2020 by The Gordian Group Inc.
All rights reserved.
Cover photo © iStock.com/Pulvret80

Printed in the United States of America
ISSN 2161-4628
ISBN 978-1-950656-52-3

0041 $319.00 per copy (in United States)
Price is subject to change without prior notice.

Related Data and Services

Our engineers recommend the following products and services to complement *Commercial Renovation Costs with RSMeans data:*

Annual Cost Data Books
2021 Facilities Construction Costs with RSMeans data
2021 Facilities Maintenance & Repair Costs with RSMeans data

Reference Books
Estimating Building Costs
RSMeans Estimating Handbook
Green Building: Project Planning & Estimating
How to Estimate with RSMeans data
Plan Reading & Material Takeoff
Project Scheduling & Management for Construction
Universal Design for Style, Comfort & Safety

Virtual, Instructor-led & On-site Training Offerings
Facilities Construction Estimating
Training for our online estimating solution
Practical Project Management for Construction Professionals
Mechanical & Electrical Estimating
Maintenance & Repair Estimating for Facilities

RSMeans data
For access to the latest cost data, an intuitive search, and an easy-to-use estimate builder, take advantage of the time savings available from our online application.

To learn more visit: **RSMeans.com/online**

Enterprise Solutions
Building owners, facility managers, building product manufacturers, and attorneys across the public and private sectors engage with RSMeans data Enterprise to solve unique challenges where trusted construction cost data is critical.

To learn more visit: **RSMeans.com/Enterprise**

Custom Built Data Sets
Building and Space Models: Quickly plan construction costs across multiple locations based on geography, project size, building system component, product options, and other variables for precise budgeting and cost control.

Predictive Analytics: Accurately plan future builds with custom graphical interactive dashboards, negotiate future costs of tenant build-outs, and identify and compare national account pricing.

Consulting
Building Product Manufacturing Analytics: Validate your claims and assist with new product launches.

Third-Party Legal Resources: Used in cases of construction cost or estimate disputes, construction product failure vs. installation failure, eminent domain, class action construction product liability, and more.

API
For resellers or internal application integration, RSMeans data is offered via API. Deliver Unit, Assembly, and Square Foot Model data within your interface. To learn more about how you can provide your customers with the latest in localized construction cost data visit:
RSMeans.com/API

Table of Contents

Foreword

The Value of RSMeans data from Gordian

Since 1942, RSMeans data has been the industry-standard materials, labor, and equipment cost information database for contractors, facility owners and managers, architects, engineers, and anyone else that requires the latest localized construction cost information. More than 75 years later, the objective remains the same: to provide facility and construction professionals with the most current and comprehensive construction cost database possible.

With the constant influx of new construction methods and materials, in addition to ever-changing labor and material costs, last year's cost data is not reliable for today's designs, estimates, or budgets. Gordian's cost engineers apply real-world construction experience to identify and quantify new building products and methodologies, adjust productivity rates, and adjust costs to local market conditions across the nation. This adds up to more than 22,000 hours in cost research annually. This unparalleled construction cost expertise is why so many facility and construction professionals rely on RSMeans data year over year.

About Gordian

Gordian originated in the spirit of innovation and a strong commitment to helping clients reach and exceed their construction goals. In 1982, Gordian's chairman and founder, Harry H. Mellon, created Job Order Contracting while serving as chief engineer at the Supreme Headquarters Allied Powers Europe. Job Order Contracting is a unique indefinite delivery/indefinite quantity (IDIQ) process, which enables facility owners to complete a substantial number of repair, maintenance, and construction projects with a single, competitively awarded contract. Realizing facility and infrastructure owners across various industries could greatly benefit from the time and cost saving advantages of this innovative construction procurement solution, he established Gordian in 1990.

Continuing the commitment to provide the most relevant and accurate facility and construction data, software, and expertise in the industry, Gordian enhanced the fortitude of its data with the acquisition of RSMeans in 2014. And in an effort to expand its facility management capabilities, Gordian acquired Sightlines, the leading provider of facilities benchmarking data and analysis, in 2015.

Our Offerings

Gordian is the leader in facility and construction cost data, software, and expertise for all phases of the building life cycle. From planning to design, procurement, construction, and operations, Gordian's solutions help clients maximize efficiency, optimize cost savings, and increase building quality with its highly specialized data engineers, software, and unique proprietary data sets.

Our Commitment

At Gordian, we do more than talk about the quality of our data and the usefulness of its application. We stand behind all of our RSMeans data—from historical cost indexes to construction materials and techniques—to craft current costs and predict future trends. If you have any questions about our products or services, please call us toll-free at 800.448.8182 or visit our website at gordian.com.

How the Cost Data Is Built: An Overview

Unit Prices*

All cost data have been divided into 50 divisions according to the MasterFormat® system of classification and numbering.

Assemblies*

The cost data in this section have been organized in an "Assemblies" format. These assemblies are the functional elements of a building and are arranged according to the 7 elements of the UNIFORMAT II classification system. For a complete explanation of a typical "Assembly", see "RSMeans data: Assemblies—How They Work."

*Residential Models**

Model buildings for four classes of construction—economy, average, custom, and luxury—are developed and shown with complete costs per square foot.

*Commercial/Industrial/ Institutional Models**

This section contains complete costs for 77 typical model buildings expressed as costs per square foot.

*Green Commercial/Industrial/ Institutional Models**

This section contains complete costs for 25 green model buildings expressed as costs per square foot.

*References**

This section includes information on Equipment Rental Costs, Crew Listings, Historical Cost Indexes, City Cost Indexes, Location Factors, Reference Tables, and Change Orders, as well as a listing of abbreviations.

- **Equipment Rental Costs:** Included are the average costs to rent and operate hundreds of pieces of construction equipment.
- **Crew Listings:** This section lists all the crews referenced in the cost data. A crew is composed of more than one trade classification and/or the addition of power equipment to any trade classification. Power equipment is included in the cost of the crew. Costs are shown both with bare labor rates and with the installing contractor's overhead and profit added. For each, the total crew cost per eight-hour day and the composite cost per labor-hour are listed.

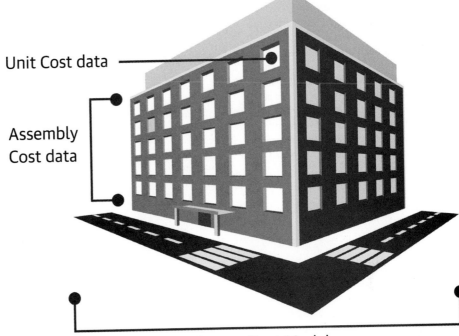

Unit Cost data

Assembly Cost data

Square Foot Models

- **Historical Cost Indexes**: These indexes provide you with data to adjust construction costs over time.
- **City Cost Indexes**: All costs in this data set are U.S. national averages. Costs vary by region. You can adjust for this by CSI Division to over 730 cities in 900+ 3-digit zip codes throughout the U.S. and Canada by using this data.
- **Location Factors**: You can adjust total project costs to over 730 cities in 900+ 3-digit zip codes throughout the U.S. and Canada by using the weighted number, which applies across all divisions.
- **Reference Tables**: At the beginning of selected major classifications in the Unit Prices are reference numbers indicators. These numbers refer you to related information in the Reference Section. In this section, you'll find reference tables, explanations, and estimating information that support how we develop the unit price data, technical data, and estimating procedures.
- **Change Orders**: This section includes information on the factors that influence the pricing of change orders.

- **Abbreviations**: A listing of abbreviations used throughout this information, along with the terms they represent, is included.

Index (printed versions only)

A comprehensive listing of all terms and subjects will help you quickly find what you need when you are not sure where it occurs in MasterFormat®.

Conclusion

This information is designed to be as comprehensive and easy to use as possible.

The Construction Specifications Institute (CSI) and Construction Specifications Canada (CSC) have produced the 2018 edition of MasterFormat®, a system of titles and numbers used extensively to organize construction information.

All unit prices in the RSMeans cost data are now arranged in the 50-division MasterFormat® 2018 system.

---

*   Not all information is available in all data sets

Note: The material prices in RSMeans cost data are "contractor's prices." They are the prices that contractors can expect to pay at the lumberyards, suppliers'/distributors' warehouses, etc. Small orders of specialty items would be higher than the costs shown, while very large orders, such as truckload lots, would be less. The variation would depend on the size, timing, and negotiating power of the contractor. The labor costs are primarily for new construction or major renovation rather than repairs or minor alterations. With reasonable exercise of judgment, the figures can be used for any building work.

Estimating with RSMeans data: Unit Prices

Following these steps will allow you to complete an accurate estimate using RSMeans data: Unit Prices.

1. Scope Out the Project

- Think through the project and identify the CSI divisions needed in your estimate.
- Identify the individual work tasks that will need to be covered in your estimate.
- The Unit Price data have been divided into 50 divisions according to CSI MasterFormat® 2018.
- In printed versions, the Unit Price Section Table of Contents on page 1 may also be helpful when scoping out your project.
- Experienced estimators find it helpful to begin with Division 2 and continue through completion. Division 1 can be estimated after the full project scope is known.

2. Quantify

- Determine the number of units required for each work task that you identified.
- Experienced estimators include an allowance for waste in their quantities. (Waste is not included in our Unit Price line items unless otherwise stated.)

3. Price the Quantities

- Use the search tools available to locate individual Unit Price line items for your estimate.
- Reference Numbers indicated within a Unit Price section refer to additional information that you may find useful.
- The crew indicates who is performing the work for that task. Crew codes are expanded in the Crew Listings in the Reference Section to include all trades and equipment that comprise the crew.
- The Daily Output is the amount of work the crew is expected to complete in one day.
- The Labor-Hours value is the amount of time it will take for the crew to install one unit of work.
- The abbreviated Unit designation indicates the unit of measure upon which the crew, productivity, and prices are based.
- Bare Costs are shown for materials, labor, and equipment needed to complete the Unit Price line item. Bare costs do not include waste, project overhead, payroll insurance, payroll taxes, main office overhead, or profit.
- The Total Incl O&P cost is the billing rate or invoice amount of the installing contractor or subcontractor who performs the work for the Unit Price line item.

4. Multiply

- Multiply the total number of units needed for your project by the Total Incl O&P cost for each Unit Price line item.
- Be careful that your take off unit of measure matches the unit of measure in the Unit column.
- The price you calculate is an estimate for a completed item of work.
- Keep scoping individual tasks, determining the number of units required for those tasks, matching each task with individual Unit Price line items, and multiplying quantities by Total Incl O&P costs.
- An estimate completed in this manner is priced as if a subcontractor, or set of subcontractors, is performing the work. The estimate does not yet include Project Overhead or Estimate Summary components such as general contractor markups on subcontracted work, general contractor office overhead and profit, contingency, and location factors.

5. Project Overhead

- Include project overhead items from Division 1-General Requirements.
- These items are needed to make the job run. They are typically, but not always, provided by the general contractor. Items include, but are not limited to, field personnel, insurance, performance bond, permits, testing, temporary utilities, field office and storage facilities, temporary scaffolding and platforms, equipment mobilization and demobilization, temporary roads and sidewalks, winter protection, temporary barricades and fencing, temporary security, temporary signs, field engineering and layout, final cleaning, and commissioning.
- Each item should be quantified and matched to individual Unit Price line items in Division 1, then priced and added to your estimate.
- An alternate method of estimating project overhead costs is to apply a percentage of the total project cost—usually 5% to 15% with an average of 10% (see General Conditions).
- Include other project related expenses in your estimate such as:
  - Rented equipment not itemized in the Crew Listings
  - Rubbish handling throughout the project (see section 02 41 19.19)

6. Estimate Summary

- Include sales tax as required by laws of your state or county.
- Include the general contractor's markup on self-performed work, usually 5% to 15% with an average of 10%.
- Include the general contractor's markup on subcontracted work, usually 5% to 15% with an average of 10%.
- Include the general contractor's main office overhead and profit:
  - RSMeans data provides general guidelines on the general contractor's main office overhead (see section 01 31 13.60 and Reference Number R013113-50).
  - Markups will depend on the size of the general contractor's operations, projected annual revenue, the level of risk, and the level of competition in the local area and for this project in particular.
- Include a contingency, usually 3% to 5%, if appropriate.
- Adjust your estimate to the project's location by using the City Cost Indexes or the Location Factors in the Reference Section:
  - Look at the rules in "How to Use the City Cost Indexes" to see how to apply the Indexes for your location.
  - When the proper Index or Factor has been identified for the project's location, convert it to a multiplier by dividing it by 100, then multiply that multiplier by your estimated total cost. The original estimated total cost will now be adjusted up or down from the national average to a total that is appropriate for your location.

How to Use the Cost Data: The Details

What's Behind the Numbers? The Development of Cost Data

RSMeans data engineers continually monitor developments in the construction industry in order to ensure reliable, thorough, and up-to-date cost information. While overall construction costs may vary relative to general economic conditions, price fluctuations within the industry are dependent upon many factors. Individual price variations may, in fact, be opposite to overall economic trends. Therefore, costs are constantly tracked and complete updates are performed yearly. Also, new items are frequently added in response to changes in materials and methods.

Costs in U.S. Dollars

All costs represent U.S. national averages and are given in U.S. dollars. The City Cost Index (CCI) with RSMeans data can be used to adjust costs to a particular location. The CCI for Canada can be used to adjust U.S. national averages to local costs in Canadian dollars. No exchange rate conversion is necessary because it has already been factored in.

G The processes or products identified by the green symbol in our publications have been determined to be environmentally responsible and/or resource-efficient solely by RSMeans data engineering staff. The inclusion of the green symbol does not represent compliance with any specific industry association or standard.

Material Costs

RSMeans data engineers contact manufacturers, dealers, distributors, and contractors all across the U.S. and Canada to determine national average material costs. If you have access to current material costs for your specific location, you may wish to make adjustments to reflect differences from the national average. Included within material costs are fasteners for a normal installation. RSMeans data engineers use manufacturers' recommendations, written specifications, and/or standard construction practices for the sizing and spacing of fasteners. Adjustments to material costs may be required for your specific application or location. The manufacturer's warranty is assumed. Extended warranties are not included in the material costs. **Material costs do not include sales tax.**

Labor Costs

Labor costs are based upon a mathematical average of trade-specific wages in 30 major U.S. cities. The type of wage (union, open shop, or residential) is identified on the inside back cover of printed publications or selected by the estimator when using the electronic products. Markups for the wages can also be found on the inside back cover of printed publications and/or under the labor references found in the electronic products.

- If wage rates in your area vary from those used, or if rate increases are expected within a given year, labor costs should be adjusted accordingly.

Labor costs reflect productivity based on actual working conditions. In addition to actual installation, these figures include time spent during a normal weekday on tasks, such as material receiving and handling, mobilization at the site, site movement, breaks, and cleanup.

Productivity data is developed over an extended period so as not to be influenced by abnormal variations and reflects a typical average.

Equipment Costs

Equipment costs include not only rental but also operating costs for equipment under normal use. The operating costs include parts and labor for routine servicing, such as the repair and replacement of pumps, filters, and worn lines. Normal operating expendables, such as fuel, lubricants, tires, and electricity (where applicable), are also included. Extraordinary operating expendables with highly variable wear patterns, such as diamond bits and blades, are excluded. These costs are included under materials. Equipment rental rates are obtained from industry sources throughout North America—contractors, suppliers, dealers, manufacturers, and distributors.

Rental rates can also be treated as reimbursement costs for contractor-owned equipment. Owned equipment costs include depreciation, loan payments, interest, taxes, insurance, storage, and major repairs.

Equipment costs do not include operators' wages.

Equipment Cost/Day—The cost of equipment required for each crew is included in the Crew Listings in the Reference Section (small tools that are considered essential everyday tools are not listed out separately). The Crew Listings itemize specialized tools and heavy equipment along with labor trades. The daily cost of itemized equipment included in a crew is based on dividing the weekly bare rental rate by 5 (number of working days per week), then adding the hourly operating cost times 8 (the number of hours per day). This Equipment Cost/Day is shown in the last column of the Equipment Rental Costs in the Reference Section.

Mobilization, Demobilization—The cost to move construction equipment from an equipment yard or rental company to the job site and back again is not included in equipment costs. Mobilization (to the site) and demobilization (from the site) costs can be found in the Unit Price Section. If a piece of equipment is already at the job site, it is not appropriate to utilize mobilization or demobilization costs again in an estimate.

Overhead and Profit

Total Cost including O&P for the installing contractor is shown in the last column of the Unit Price and/or Assemblies. This figure is the sum of the bare material cost plus 10% for profit, the bare labor cost plus total overhead and profit, and the bare equipment cost plus 10% for profit. Details for the calculation of overhead and profit on labor are shown on the inside back cover of the printed product and in the Reference Section of the electronic product.

General Conditions

Cost data in this data set are presented in two ways: Bare Costs and Total Cost including O&P (Overhead and Profit). General Conditions, or General Requirements, of the contract should also be added to the Total Cost including O&P when applicable. Costs for General Conditions are listed in Division 1 of the Unit Price Section and in the Reference Section.

General Conditions for the installing contractor may range from 0% to 10% of the Total Cost including O&P. For the general or prime contractor, costs for General Conditions may range from 5% to 15% of the Total Cost including O&P, with a figure of 10% as the most typical allowance. If applicable, the Assemblies and Models sections use costs that include the installing contractor's overhead and profit (O&P).

Factors Affecting Costs

Costs can vary depending upon a number of variables. Here's a listing of some factors that affect costs and points to consider.

Quality—The prices for materials and the workmanship upon which productivity is based represent sound construction work. They are also in line with industry standard and manufacturer specifications and are frequently used by federal, state, and local governments.

Overtime—We have made no allowance for overtime. If you anticipate premium time or work beyond normal working hours, be sure to make an appropriate adjustment to your labor costs.

Productivity—The productivity, daily output, and labor-hour figures for each line item are based on an eight-hour work day in daylight hours in moderate temperatures and up to a 14' working height unless otherwise indicated. For work that extends beyond normal work hours or is performed under adverse conditions, productivity may decrease.

Size of Project—The size, scope of work, and type of construction project will have a significant impact on cost. Economies of scale can reduce costs for large projects. Unit costs can often run higher for small projects.

Location—Material prices are for metropolitan areas. However, in dense urban areas, traffic and site storage limitations may increase costs. Beyond a 20-mile radius of metropolitan areas, extra trucking or transportation charges may also increase the material costs slightly. On the other hand, lower wage rates may be in effect. Be sure to consider both of these factors when preparing an estimate, particularly if the job site is located in a central city or remote rural location. In addition, highly specialized subcontract items may require travel and per-diem expenses for mechanics.

Other Factors—

- season of year
- contractor management
- weather conditions
- local union restrictions
- building code requirements
- availability of:
  - adequate energy
  - skilled labor
  - building materials
- owner's special requirements/restrictions
- safety requirements
- environmental considerations
- access

Unpredictable Factors—General business conditions influence "in-place" costs of all items. Substitute materials and construction methods may have to be employed. These may affect the installed cost and/or life cycle costs. Such factors may be difficult to evaluate and cannot necessarily be predicted on the basis of the job's location in a particular section of the country. Thus, where these factors apply, you may find significant but unavoidable cost variations for which you will have to apply a measure of judgment to your estimate.

Rounding of Costs

In printed publications only, all unit prices in excess of $5.00 have been rounded to make them easier to use and still maintain adequate precision of the results.

How Subcontracted Items Affect Costs

A considerable portion of all large construction jobs is usually subcontracted. In fact, the percentage done by subcontractors is constantly increasing and may run over 90%. Since the workers employed by these companies do nothing else but install their particular products, they soon become experts in that line. As a result, installation by these firms is accomplished so efficiently that the total in-place cost, even with the general contractor's overhead and profit, is no more, and often less, than if the principal contractor had handled the installation. Companies that deal with construction specialties are anxious to have their products perform well and, consequently, the installation will be the best possible.

Contingencies

The allowance for contingencies generally provides for unforeseen construction difficulties. On alterations or repair jobs, 20% is not too much. If drawings are final and only field contingencies are being considered, 2% or 3% is probably sufficient and often nothing needs to be added. Contractually, changes in plans will be covered by extras. The contractor should consider inflationary price trends and possible material shortages during the course of the job. These escalation factors are dependent upon both economic conditions and the anticipated time between the estimate and actual construction. If drawings are not complete or approved, or a budget cost is wanted, it is wise to add 5% to 10%. Contingencies, then, are a matter of judgment.

Important Estimating Considerations

The productivity, or daily output, of each craftsman or crew assumes a well-managed job where tradesmen with the proper tools and equipment, along with the appropriate construction materials, are present. Included are daily set-up and cleanup time, break time, and plan layout time. Unless otherwise indicated, time for material movement on site (for items that can be transported by hand) of up to 200' into the building and to the first or second floor is also included. If material has to be transported by other means, over greater distances, or to higher floors, an additional allowance should be considered by the estimator.

While horizontal movement is typically a sole function of distances, vertical transport introduces other variables that can significantly impact productivity. In an occupied building, the use of elevators (assuming access, size, and required protective measures are acceptable) must be understood at the time of the estimate. For new construction, hoist wait and cycle times can easily be 15 minutes and may result in scheduled access extending beyond the normal work day. Finally, all vertical transport will impose strict weight limits likely to preclude the use of any motorized material handling.

The productivity, or daily output, also assumes installation that meets manufacturer/designer/ standard specifications. A time allowance for quality control checks, minor adjustments, and any task required to ensure proper function or operation is also included. For items that require connections to services, time is included for positioning, leveling, securing the unit, and making all the necessary connections (and start up where applicable) to ensure a complete installation. Estimating of the services themselves (electrical, plumbing, water, steam, hydraulics, dust collection, etc.) is separate.

In some cases, the estimator must consider the use of a crane and an appropriate crew for the installation of large or heavy items. For those situations where a crane is not included in the assigned crew and as part of the line item cost,

then equipment rental costs, mobilization and demobilization costs, and operator and support personnel costs must be considered.

Labor-Hours

The labor-hours expressed in this publication are derived by dividing the total daily labor-hours for the crew by the daily output. Based on average installation time and the assumptions listed above, the labor-hours include: direct labor, indirect labor, and nonproductive time. A typical day for a craftsman might include but is not limited to:

- Direct Work
  - ☐ Measuring and layout
  - ☐ Preparing materials
  - ☐ Actual installation
  - ☐ Quality assurance/quality control
- Indirect Work
  - ☐ Reading plans or specifications
  - ☐ Preparing space
  - ☐ Receiving materials
  - ☐ Material movement
  - ☐ Giving or receiving instruction
  - ☐ Miscellaneous
- Non-Work
  - ☐ Chatting
  - ☐ Personal issues
  - ☐ Breaks
  - ☐ Interruptions (i.e., sickness, weather, material or equipment shortages, etc.)

If any of the items for a typical day do not apply to the particular work or project situation, the estimator should make any necessary adjustments.

Final Checklist

Estimating can be a straightforward process provided you remember the basics. Here's a checklist of some of the steps you should remember to complete before finalizing your estimate.

Did you remember to:

- factor in the City Cost Index for your locale?
- take into consideration which items have been marked up and by how much?
- mark up the entire estimate sufficiently for your purposes?
- read the background information on techniques and technical matters that could impact your project time span and cost?
- include all components of your project in the final estimate?
- double check your figures for accuracy?
- call RSMeans data engineers if you have any questions about your estimate or the data you've used? Remember, Gordian stands behind all of our products, including our extensive RSMeans data solutions. If you have any questions about your estimate, about the costs you've used from our data, or even about the technical aspects of the job that may affect your estimate, feel free to call the Gordian RSMeans editors at 1.800.448.8182.

Using Minimum Labor/Equipment Charges for Small Quantities

Estimating small construction or repair tasks often creates situations in which the quantity of work to be performed is very small. When this occurs, the labor and/or equipment costs to perform the work may be too low to allow for the crew to get to the job, receive instructions, find materials, get set up, perform the work, clean up, and get to the next job. In these situations, the estimator should compare the developed labor and/or equipment costs for performing the work (e.g., quantity × labor and/or equipment costs) with the *minimum labor/equipment charge* within the Unit Price section of the data set. (These minimum labor/equipment charge line items appear only in *Facilities Construction Costs with RSMeans data* and *Commercial Renovation Costs with RSMeans data*.)

If the labor and/or equipment costs developed by the estimator are LOWER THAN the *minimum labor/equipment charge* listed at the bottom of specific sections of Unit Price costs, the estimator should adjust the developed costs upward to the *minimum labor/equipment charge*. The proper use of a *minimum labor/equipment charge* results in having enough money in the estimate to cover the contractor's higher cost of performing a very small amount of work during a partial workday.

A *minimum labor/equipment charge* should be used only when the task being estimated is the only task the crew will perform at the job site that day. **If, however, the crew will be able to perform other tasks at the job site that day, the use of a *minimum labor/equipment charge* is not appropriate.**

08 52 Wood Windows

08 52 10 – Plain Wood Windows

08 52 10.40 Casement Window		Crew	Daily Output	Labor-Hours	Unit	Material	2021 Bare Costs Labor	Equipment	Total	Total Incl O&P
0010 **CASEMENT WINDOW**, including frame, screen and grilles										
0100 2'-0" x 3'-0" H, double insulated glass	G	1 Carp	10	.800	Ea.	278	44		322	375
0150 Low E glass	G		10	.800		288	44		332	385
0200 2'-0" x 4'-6" high, double insulated glass	G		9	.889		390	48.50		438.50	510
0250 Low E glass	G		9	.889		440	48.50		488.50	565
9000 Minimum labor/equipment charge		1 Carp	3	2.667	Job		146		146	232

Example:

Establish the bid price to install two casement windows. Assume installation of 2' × 4'-6" metal clad windows with insulating glass [Unit Price line number 08 52 10.40 0250] and that this is the only task this crew will perform at the job site that day.

Solution:

Step One — Develop the Bare Labor Cost for this task:
Bare Labor Cost = 2 windows @ $48.50 each = $97.00

Step Two — Evaluate the *minimum labor/equipment charge* for this Unit Price section against the developed Bare Labor Cost for this task:
Minimum labor/equipment charge = $146.00 (compare with $97.00)

Step Three — Choose to adjust the developed labor cost upward to the *minimum labor/equipment charge*.

Step Four — Develop the bid price for this task (including O&P):
Add together the marked-up Bare Material Cost for this task and the marked-up *minimum labor/equipment charge* for this Unit Price section.

2 × ($440.00 + 10%) + ($146.00 + 59.2%)

= 2 × ($440.00 + $44.00) + ($146.00 + $86.43)

= 2 × ($484.00) + $232.43

= $968.00 + $232.43

= $1,200.43

ANSWER: $1,200.43 is the correct bid price to use. This sum takes into consideration the material cost (with 10% for profit) for these two windows, plus the *minimum labor/equipment charge* (with O&P included) for this section of the Unit Price.

Unit Price Section

Table of Contents

Table of Contents (cont.)

Table of Contents (cont.)

3

RSMeans data: Unit Prices— How They Work

All RSMeans data: Unit Prices are organized in the same way.

03 30 Cast-In-Place Concrete

03 30 53 – Miscellaneous Cast-In-Place Concrete

03 30 53.40 Concrete In Place		Crew	Daily Output	Labor-Hours	Unit	Material	2021 Bare Costs Labor	2021 Bare Costs Equipment	Total	Total Incl O&P
0010	**CONCRETE IN PLACE**									
0020	Including forms (4 uses), Grade 60 rebar, concrete (Portland cement	R033105-10								
0050	Type I), placement and finishing unless otherwise indicated	R033105-20								
0300	Beams (3500 psi), 5 kip/L.F., 10' span	R033105-70 / R033105-85	C-14A	15.62	12.804	C.Y.	420	700	29	1,149	1,600
0350	25' span		"	18.55	10.782		440	590	24.50	1,054.50	1,450
0500	Chimney foundations (5000 psi), over 5 C.Y.		C-14C	32.22	3.476		187	182	.83	369.83	495
0510	(3500 psi), under 5 C.Y.		"	23.71	4.724		223	247	1.13	471.13	635
0700	Columns, square (4000 psi), 12" x 12", up to 1% reinforcing by area		C-14A	11.96	16.722		475	915	37.50	1,427.50	2,025
3540	Equipment pad (3000 psi), 3' x 3' x 6" thick		C-14H	45	1.067	Ea.	55.50	57	.61	113.11	152
3550	4' x 4' x 6" thick			30	1.600		84	85.50	.91	170.41	229
3560	5' x 5' x 8" thick			18	2.667		150	143	1.52	294.52	395
3570	6' x 6' x 8" thick			14	3.429		204	184	1.95	389.95	520
3580	8' x 8' x 10" thick			8	6		415	320	3.42	738.42	975
3590	10' x 10' x 12" thick			5	9.600		720	515	5.45	1,240.45	1,600

It is important to understand the structure of RSMeans data: Unit Prices so that you can find information easily and use it correctly.

1 Line Numbers

Line Numbers consist of 12 characters, which identify a unique location in the database for each task. The first 6 or 8 digits conform to the Construction Specifications Institute MasterFormat® 2018. The remainder of the digits are a further breakdown in order to arrange items in understandable groups of similar tasks. Line numbers are consistent across all of our publications, so a line number in any of our products will always refer to the same item of work.

2 Descriptions

Descriptions are shown in a hierarchical structure to make them readable. In order to read a complete description, read up through the indents to the top of the section. Include everything that is above and to the left that is not contradicted by information below. For instance, the complete description for line 03 30 53.40 3550 is "Concrete in place, including forms (4 uses), Grade 60 rebar, concrete (Portland cement Type 1), placement and finishing unless otherwise indicated; Equipment pad (3,000 psi), 4' × 4' × 6" thick."

3 RSMeans data

When using **RSMeans data**, it is important to read through an entire section to ensure that you use the data that most closely matches your work. Note that sometimes there is additional information shown in the section that may improve your price. There are frequently lines that further describe, add to, or adjust data for specific situations.

4 Reference Information

Gordian's RSMeans engineers have created **reference** information to assist you in your estimate. If there is information that applies to a section, it will be indicated at the start of the section. The Reference Section is located in the back of the data set.

5 Crews

Crews include labor and/or equipment necessary to accomplish each task. In this case, Crew C-14H is used. Gordian's RSMeans staff selects a crew to represent the workers and equipment that are typically used for that task. In this case, Crew C-14H consists of one carpenter foreman (outside), two carpenters, one rodman, one laborer, one cement finisher, and one gas engine vibrator. Details of all crews can be found in the Reference Section.

Crews - Renovation

Crew No.	Bare Costs		Incl. Subs O & P		Cost Per Labor-Hour	
Crew C-14H	Hr.	Daily	Hr.	Daily	Bare Costs	Incl. O&P
1 Carpenter Foreman (outside)	$56.70	$453.60	$90.30	$722.40	$53.53	$84.75
2 Carpenters	54.70	875.20	87.10	1393.60		
1 Rodman (reinf.)	58.90	471.20	92.20	737.60		
1 Laborer	44.40	355.20	70.70	565.60		
1 Cement Finisher	51.80	414.40	81.10	648.80		
1 Gas Engine Vibrator		27.15		29.86	.57	.62
48 L.H., Daily Totals		$2596.75		$4097.86	$54.10	$85.37

6 Daily Output

The **Daily Output** is the amount of work that the crew can do in a normal 8-hour workday, including mobilization, layout, movement of materials, and cleanup. In this case, crew C-14H can install thirty 4' × 4' × 6" thick concrete pads in a day. Daily output is variable and based on many factors, including the size of the job, location, and environmental conditions. RSMeans data represents work done in daylight (or adequate lighting) and temperate conditions.

7 Labor-Hours

The figure in the **Labor-Hours** column is the amount of labor required to perform one unit of work–in this case the amount of labor required to construct one 4' × 4' equipment pad. This figure is calculated by dividing the number of hours of labor in the crew by the daily output (48 labor-hours divided by 30 pads = 1.6 hours of labor per pad). Multiply 1.6 times 60 to see the value in minutes: 60 × 1.6 = 96

minutes. Note: the labor-hour figure is not dependent on the crew size. A change in crew size will result in a corresponding change in daily output, but the labor-hours per unit of work will not change.

8 Unit of Measure

All RSMeans data: Unit Prices include the typical **Unit of Measure** used for estimating that item. For concrete-in-place the typical unit is cubic yards (C.Y.) or each (Ea.). For installing broadloom carpet it is square yard and for gypsum board it is square foot. The estimator needs to take special care that the unit in the data matches the unit in the take-off. Unit conversions may be found in the Reference Section.

9 Bare Costs

Bare Costs are the costs of materials, labor, and equipment that the installing contractor pays. They represent the cost, in U.S. dollars, for one unit of work. They do not include any markups for profit or labor burden.

10 Bare Total

The **Total column** represents the total bare cost for the installing contractor in U.S. dollars. In this case, the sum of $84 for material + $85.50 for labor + $.91 for equipment is $170.41.

11 Total Incl O&P

The **Total Incl O&P column** is the total cost, including overhead and profit, that the installing contractor will charge the customer. This represents the cost of materials plus 10% profit, the cost of labor plus labor burden and 10% profit, and the cost of equipment plus 10% profit. It does not include the general contractor's overhead and profit. Note: See the inside back cover of the printed product or the Reference Section of the electronic product for details on how the labor burden is calculated.

National Average

The RSMeans data in our print publications represent a "national average" cost. This data should be modified to the project location using the **City Cost Indexes** *or* **Location Factors** *tables found in the Reference Section. Use the Location Factors to adjust estimate totals if the project covers multiple trades. Use the City Cost Indexes (CCI) for single trade projects or projects where a more detailed analysis is required. All figures in the two tables are derived from the same research. The last row of data in the CCI—the weighted average—is the same as the numbers reported for each location in the location factor table.*

RSMeans data: Unit Prices— How They Work (Continued)

Project Name: Pre-Engineered Steel Building			Architect: As Shown					
Location:	Anywhere, USA							
							01/01/21	R+R
Line Number	**Description**	**Qty**	**Unit**	**Material**	**Labor**	**Equipment**	**SubContract**	**Estimate Total**
03 30 53.40 3940	Strip footing, 12" x 24", reinforced	15	C.Y.	$2,625.00	$1,830.00	$8.40	$0.00	
03 30 53.40 3950	Strip footing, 12" x 36", reinforced	34	C.Y.	$5,644.00	$3,315.00	$15.30	$0.00	
03 11 13.65 3000	Concrete slab edge forms	500	L.F.	$225.00	$1,390.00	$0.00	$0.00	
03 22 11.10 0200	Welded wire fabric reinforcing	150	C.S.F.	$3,150.00	$4,575.00	$0.00	$0.00	
03 31 13.35 0300	Ready mix concrete, 4000 psi for slab on grade	278	C.Y.	$35,862.00	$0.00	$0.00	$0.00	
03 31 13.70 4300	Place, strike off & consolidate concrete slab	278	C.Y.	$0.00	$5,560.00	$136.22	$0.00	
03 35 13.30 0250	Machine float & trowel concrete slab	15,000	S.F.	$0.00	$10,350.00	$750.00	$0.00	
03 15 16.20 0140	Cut control joints in concrete slab	950	L.F.	$38.00	$437.00	$57.00	$0.00	
03 39 23.13 0300	Sprayed concrete curing membrane	150	C.S.F.	$1,830.00	$1,125.00	$0.00	$0.00	
Division 03	**Subtotal**			**$49,374.00**	**$28,582.00**	**$966.92**	**$0.00**	**$78,922.92**
08 36 13.10 2650	Manual 10' x 10' steel sectional overhead door	8	Ea.	$9,800.00	$3,880.00	$0.00	$0.00	
08 36 13.10 2860	Insulation and steel back panel for OH door	800	S.F.	$4,760.00	$0.00	$0.00	$0.00	
Division 08	**Subtotal**			**$14,560.00**	**$3,880.00**	**$0.00**	**$0.00**	**$18,440.00**
13 34 19.50 1100	Pre-Engineered Steel Building, 100' x 150' x 24'	15,000	SF Flr.	$0.00	$0.00	$0.00	$375,000.00	
13 34 19.50 6050	Framing for PESB door opening, 3' x 7'	4	Opng.	$0.00	$0.00	$0.00	$2,360.00	
13 34 19.50 6100	Framing for PESB door opening, 10' x 10'	8	Opng.	$0.00	$0.00	$0.00	$9,400.00	
13 34 19.50 6200	Framing for PESB window opening, 4' x 3'	6	Opng.	$0.00	$0.00	$0.00	$3,420.00	
13 34 19.50 5750	PESB door, 3' x 7', single leaf	4	Opng.	$2,740.00	$772.00	$0.00	$0.00	
13 34 19.50 7750	PESB sliding window, 4' x 3' with screen	6	Opng.	$2,940.00	$660.00	$68.70	$0.00	
13 34 19.50 6550	PESB gutter, eave type, 26 ga., painted	300	L.F.	$2,595.00	$906.00	$0.00	$0.00	
13 34 19.50 8650	PESB roof vent, 12" wide x 10' long	15	Ea.	$705.00	$3,615.00	$0.00	$0.00	
13 34 19.50 6900	PESB insulation, vinyl faced, 4" thick	27,400	S.F.	$14,522.00	$10,412.00	$0.00	$0.00	
Division 13	**Subtotal**			**$23,502.00**	**$16,365.00**	**$68.70**	**$390,180.00**	**$430,115.70**
			Subtotal	$87,436.00	$48,827.00	$1,035.62	$390,180.00	$527,478.62
Division 01	**General Requirements @ 7%**			6,120.52	3,417.89	72.49	27,312.60	
			Estimate Subtotal	$93,556.52	$52,244.89	$1,108.11	$417,492.60	$527,478.62
			Sales Tax @ 5%	4,677.83		55.41	10,437.32	
			Subtotal A	98,234.35	52,244.89	1,163.52	427,929.92	
			GC O & P	9,823.43	30,563.26	116.35	42,792.99	
			Subtotal B	108,057.78	82,808.15	1,279.87	470,722.91	$662,868.71
			Contingency @ 5%					33,143.44
			Subtotal C					$696,012.14
			Bond @ $12/1000 +10% O&P					9,187.36
			Subtotal D					$705,199.50
			Location Adjustment Factor		114.20			100,138.33
			Grand Total					**$805,337.83**

This estimate is based on an interactive spreadsheet. You are free to download it and adjust it to your methodology.
A copy of this spreadsheet is available at **RSMeans.com/2021books.**

Sample Estimate

This sample demonstrates the elements of an estimate, including a tally of the RSMeans data lines and a summary of the markups on a contractor's work to arrive at a total cost to the owner. The Location Factor with RSMeans data is added at the bottom of the estimate to adjust the cost of the work to a specific location.

1 Work Performed

The body of the estimate shows the RSMeans data selected, including the line number, a brief description of each item, its take-off unit and quantity, and the bare costs of materials, labor, and equipment. This estimate also includes a column titled "SubContract." This data is taken from the column "Total Incl O&P" and represents the total that a subcontractor would charge a general contractor for the work, including the sub's markup for overhead and profit.

2 Division 1, General Requirements

This is the first division numerically but the last division estimated. Division 1 includes project-wide needs provided by the general contractor. These requirements vary by project but may include temporary facilities and utilities, security, testing, project cleanup, etc. For small projects a percentage can be used—typically between 5% and 15% of project cost. For large projects the costs may be itemized and priced individually.

3 Sales Tax

If the work is subject to state or local sales taxes, the amount must be added to the estimate. Sales tax may be added to material costs, equipment costs, and subcontracted work. In this case, sales tax was added in all three categories. It was assumed that approximately half the subcontracted work would be material cost, so the tax was applied to 50% of the subcontract total.

4 GC O&P

This entry represents the general contractor's markup on material, labor, equipment, and subcontractor costs. Our standard markup on materials, equipment, and subcontracted work is 10%. In this estimate, the markup on the labor performed by the GC's workers uses "Skilled Workers Average" shown in Column F on the table "Installing Contractor's Overhead & Profit," which can be found on the inside back cover of the printed product or in the Reference Section of the electronic product.

5 Contingency

A factor for contingency may be added to any estimate to represent the cost of unknowns that may occur between the time that the estimate is performed and the time the project is constructed. The amount of the allowance will depend on the stage of design at which the estimate is done and the contractor's assessment of the risk involved. Refer to section 01 21 16.50 for contingency allowances.

6 Bonds

Bond costs should be added to the estimate. The figures here represent a typical performance bond, ensuring the owner that if the general contractor does not complete the obligations in the construction contract the bonding company will pay the cost for completion of the work.

7 Location Adjustment

Published prices are based on national average costs. If necessary, adjust the total cost of the project using a location factor from the "Location Factor" table or the "City Cost Index" table. Use location factors if the work is general, covering multiple trades. If the work is by a single trade (e.g., masonry) use the more specific data found in the "City Cost Indexes."

Estimating Tips
01 20 00 Price and Payment Procedures

- Allowances that should be added to estimates to cover contingencies and job conditions that are not included in the national average material and labor costs are shown in Section 01 21.
- When estimating historic preservation projects (depending on the condition of the existing structure and the owner's requirements), a 15–20% contingency or allowance is recommended, regardless of the stage of the drawings.

01 30 00 Administrative Requirements

- Before determining a final cost estimate, it is good practice to review all the items listed in Subdivisions 01 31 and 01 32 to make final adjustments for items that may need customizing to specific job conditions.
- Requirements for initial and periodic submittals can represent a significant cost to the General Requirements of a job. Thoroughly check the submittal specifications when estimating a project to determine any costs that should be included.

01 40 00 Quality Requirements

- All projects will require some degree of quality control. This cost is not included in the unit cost of construction listed in each division. Depending upon the terms of the contract, the various costs of inspection and testing can be the responsibility of either the owner or the contractor. Be sure to include the required costs in your estimate.

01 50 00 Temporary Facilities and Controls

- Barricades, access roads, safety nets, scaffolding, security, and many more requirements for the execution of a safe project are elements of direct cost. These costs can easily be overlooked when preparing an estimate. When looking through the major classifications of this subdivision, determine which items apply to each division in your estimate.
- Construction equipment rental costs can be found in the Reference Section in Section 01 54 33. Operators' wages are not included in equipment rental costs.
- Equipment mobilization and demobilization costs are not included in equipment rental costs and must be considered separately.

- The cost of small tools provided by the installing contractor for his workers is covered in the "Overhead" column on the "Installing Contractor's Overhead and Profit" table that lists labor trades, base rates, and markups. Therefore, it is included in the "Total Incl. O&P" cost of any unit price line item.

01 70 00 Execution and Closeout Requirements

- When preparing an estimate, thoroughly read the specifications to determine the requirements for Contract Closeout. Final cleaning, record documentation, operation and maintenance data, warranties and bonds, and spare parts and maintenance materials can all be elements of cost for the completion of a contract. Do not overlook these in your estimate.

Reference Numbers

Reference numbers are shown at the beginning of some major classifications. These numbers refer to related items in the Reference Section. The reference information may be an estimating procedure, an alternate pricing method, or technical information.

Note: Not all subdivisions listed here necessarily appear. ■

Same Data. Simplified.

Enjoy the convenience and efficiency of accessing your costs anywhere:

- **Skip the multiplier** by setting your location
- **Quickly search,** edit, favorite and share costs
- **Stay on top of price changes** with automatic updates

Discover more at rsmeans.com/online

01 11 Summary of Work

01 11 31 – Professional Consultants

01 11 31.10 Architectural Fees

		Crew	Daily Output	Labor-Hours	Unit	Material	2021 Bare Costs Labor	Equipment	Total	Total Incl O&P
0010	**ARCHITECTURAL FEES** R011110-10									
0020	For new construction									
0060	Minimum				Project				4.90%	4.90%
0090	Maximum								16%	16%
0100	For alteration work, to $500,000, minimum								7.35%	7.35%
0125	Maximum								24%	24%
0150	For alteration work, over $500,000, minumum								6.13%	6.13%
0175	Maximum								20%	20%

01 11 31.20 Construction Management Fees

		Crew	Daily Output	Labor-Hours	Unit	Material	Labor	Equipment	Total	Total Incl O&P
0010	**CONSTRUCTION MANAGEMENT FEES**									
0060	For work to $100,000				Project				10%	10%
0070	To $250,000								9%	9%
0090	To $1,000,000								6%	6%
0100	To $5,000,000								5%	5%
0110	To $10,000,000								4%	4%

01 11 31.30 Engineering Fees

		Crew	Daily Output	Labor-Hours	Unit	Material	Labor	Equipment	Total	Total Incl O&P
0010	**ENGINEERING FEES** R011110-30									
0020	Educational planning consultant, minimum				Project				.50%	.50%
0100	Maximum				"				2.50%	2.50%
0400	Elevator & conveying systems, minimum				Contrct				2.50%	2.50%
0500	Maximum								5%	5%
1000	Mechanical (plumbing & HVAC), minimum								4.10%	4.10%
1100	Maximum								10.10%	10.10%
1200	Structural, minimum				Project				1%	1%
1300	Maximum				"				2.50%	2.50%

01 11 31.75 Renderings

		Crew	Daily Output	Labor-Hours	Unit	Material	Labor	Equipment	Total	Total Incl O&P
0010	**RENDERINGS** Color, matted, 20" x 30", eye level,									
0020	1 building, minimum				Ea.	2,375			2,375	2,600
0050	Average					3,425			3,425	3,775
0100	Maximum					5,600			5,600	6,150
1000	5 buildings, minimum					4,125			4,125	4,550
1100	Maximum					8,250			8,250	9,075

01 21 Allowances

01 21 16 – Contingency Allowances

01 21 16.50 Contingencies

		Crew	Daily Output	Labor-Hours	Unit	Material	Labor	Equipment	Total	Total Incl O&P
0010	**CONTINGENCIES**, Add to estimate									
0020	Conceptual stage				Project				20%	20%
0050	Schematic stage								15%	15%
0100	Preliminary working drawing stage (Design Dev.)								10%	10%
0150	Final working drawing stage								3%	3%

01 21 53 – Factors Allowance

01 21 53.50 Factors

		Crew	Daily Output	Labor-Hours	Unit	Material	Labor	Equipment	Total	Total Incl O&P
0010	**FACTORS** Cost adjustments R012153-10									
0100	Add to construction costs for particular job requirements									
0500	Cut & patch to match existing construction, add, minimum				Costs	2%	3%			
0550	Maximum					5%	9%			
0800	Dust protection, add, minimum					1%	2%			
0850	Maximum					4%	11%			
1100	Equipment usage curtailment, add, minimum					1%	1%			

10

01 21 53 – Factors Allowance

01 21 53.50 Factors

		Crew	Daily Output	Labor-Hours	Unit	Material	2021 Bare Costs Labor	Equipment	Total	Total Incl O&P
1150	Maximum				Costs	3%	10%			
1400	Material handling & storage limitation, add, minimum					1%	1%			
1450	Maximum					6%	7%			
1700	Protection of existing work, add, minimum					2%	2%			
1750	Maximum					5%	7%			
2000	Shift work requirements, add, minimum						5%			
2050	Maximum						30%			
2300	Temporary shoring and bracing, add, minimum					2%	5%			
2350	Maximum					5%	12%			

01 21 53.60 Security Factors

		Crew	Daily Output	Labor-Hours	Unit	Material	2021 Bare Costs Labor	Equipment	Total	Total Incl O&P
0010	SECURITY FACTORS R012153-60									
0100	Additional costs due to security requirements									
0110	Daily search of personnel, supplies, equipment and vehicles									
0120	Physical search, inventory and doc of assets, at entry				Costs		30%			
0130	At entry and exit						50%			
0140	Physical search, at entry						6.25%			
0150	At entry and exit						12.50%			
0160	Electronic scan search, at entry						2%			
0170	At entry and exit						4%			
0180	Visual inspection only, at entry						.25%			
0190	At entry and exit						.50%			
0200	ID card or display sticker only, at entry						.12%			
0210	At entry and exit						.25%			
0220	Day 1 as described below, then visual only for up to 5 day job duration									
0230	Physical search, inventory and doc of assets, at entry				Costs		5%			
0240	At entry and exit						10%			
0250	Physical search, at entry						1.25%			
0260	At entry and exit						2.50%			
0270	Electronic scan search, at entry						.42%			
0280	At entry and exit						.83%			
0290	Day 1 as described below, then visual only for 6-10 day job duration									
0300	Physical search, inventory and doc of assets, at entry				Costs		2.50%			
0310	At entry and exit				Costs		5%			
0320	Physical search, at entry						.63%			
0330	At entry and exit						1.25%			
0340	Electronic scan search, at entry						.21%			
0350	At entry and exit						.42%			
0360	Day 1 as described below, then visual only for 11-20 day job duration									
0370	Physical search, inventory and doc of assets, at entry				Costs		1.25%			
0380	At entry and exit						2.50%			
0390	Physical search, at entry						.31%			
0400	At entry and exit						.63%			
0410	Electronic scan search, at entry						.10%			
0420	At entry and exit						.21%			
0430	Beyond 20 days, costs are negligible									
0440	Escort required to be with tradesperson during work effort				Costs		6.25%			

01 21 53.65 Infectious Disease Precautions

		Crew	Daily Output	Labor-Hours	Unit	Material	2021 Bare Costs Labor	Equipment	Total	Total Incl O&P
0010	INFECTIOUS DISEASE PRECAUTIONS R012153-65									
0100	Additional costs due to infectious disease precautions									
0120	Daily temperature checks				Costs		1%			
0130	Donning & doffing masks and gloves						1%			
0140	Washing hands						1%			

01 21 Allowances

01 21 53 – Factors Allowance

01 21 53.65 Infectious Disease Precautions	Crew	Daily Output	Labor-Hours	Unit	Material	2021 Bare Costs Labor	Equipment	Total	Total Incl O&P
0150 Informational meetings				Costs		2%			
0160 Maintaining social distance						1%			
0170 Disinfecting tools or equipment				↓		3%			
0200 N95 rated masks				Ea.	1.40			1.40	1.54
0210 Surgical masks				"	.16			.16	.18
0220 Black nitrile disposable gloves				Pair	.18			.18	.20
0250 Hand washing station & 2 x weekly service				Week				158	174

01 21 55 – Job Conditions Allowance

01 21 55.50 Job Conditions	Crew	Daily Output	Labor-Hours	Unit	Material	2021 Bare Costs Labor	Equipment	Total	Total Incl O&P
0010 JOB CONDITIONS Modifications to applicable									
0020 cost summaries									
0100 Economic conditions, favorable, deduct				Project				2%	2%
0200 Unfavorable, add								5%	5%
0300 Hoisting conditions, favorable, deduct								2%	2%
0400 Unfavorable, add								5%	5%
0500 General contractor management, experienced, deduct								2%	2%
0600 Inexperienced, add								10%	10%
0700 Labor availability, surplus, deduct								1%	1%
0800 Shortage, add								10%	10%
0900 Material storage area, available, deduct								1%	1%
1000 Not available, add								2%	2%
1100 Subcontractor availability, surplus, deduct				Project				5%	5%
1200 Shortage, add								12%	12%
1300 Work space, available, deduct								2%	2%
1400 Not available, add				↓				5%	5%

01 21 63 – Taxes

01 21 63.10 Taxes	Crew	Daily Output	Labor-Hours	Unit	Material	2021 Bare Costs Labor	Equipment	Total	Total Incl O&P
0010 TAXES									
0020 Sales tax, State, average R012909-80				%	5.08%				
0050 Maximum R012909-85					7.50%				
0200 Social Security, on first $118,500 of wages						7.65%			
0300 Unemployment, combined Federal and State, minimum R012909-86						.60%			
0350 Average						9.60%			
0400 Maximum				↓		12%			

01 31 Project Management and Coordination

01 31 13 – Project Coordination

01 31 13.20 Field Personnel	Crew	Daily Output	Labor-Hours	Unit	Material	2021 Bare Costs Labor	Equipment	Total	Total Incl O&P
0010 FIELD PERSONNEL									
0020 Clerk, average				Week		495		495	790
0160 General purpose laborer, average	1 Clab	.20	40			1,775		1,775	2,825
0180 Project manager, minimum						2,175		2,175	3,475
0200 Average						2,500		2,500	4,000
0220 Maximum						2,850		2,850	4,575
0240 Superintendent, minimum						2,125		2,125	3,400
0260 Average						2,325		2,325	3,725
0280 Maximum				↓		2,650		2,650	4,225

01 31 Project Management and Coordination

01 31 13 – Project Coordination

01 31 13.30 Insurance

		Crew	Daily Output	Labor-Hours	Unit	Material	Labor	Equipment	Total	Total Incl O&P
0010	**INSURANCE** R013113-40								.24%	.24%
0020	Builders risk, standard, minimum				Job				.80%	.80%
0050	Maximum R013113-50								.25%	.25%
0200	All-risk type, minimum								.62%	.62%
0250	Maximum R013113-60				Value				.50%	.50%
0400	Contractor's equipment floater, minimum				"				1.50%	1.50%
0450	Maximum				"				2.02%	2.02%
0600	Public liability, average				Job					
0810	Workers' compensation & employer's liability									
2000	Range of 35 trades in 50 states, excl. wrecking & clerical, min.				Payroll		1.37%			
2100	Average						10.60%			
2200	Maximum						120.29%			

01 31 13.40 Main Office Expense

		Crew	Daily Output	Labor-Hours	Unit	Material	Labor	Equipment	Total	Total Incl O&P
0010	**MAIN OFFICE EXPENSE** Average for General Contractors									
0020	As a percentage of their annual volume									
0030	Annual volume to $300,000, minimum				% Vol.				20%	
0040	Maximum				"				30%	
0060	To $500,000, minimum				% Vol.				17%	
0070	Maximum								22%	
0080	To $1,000,000, minimum								16%	
0090	Maximum								19%	
0110	To $3,000,000, minimum								14%	
0120	Maximum								16%	
0130	To $5,000,000, minimum								8%	
0140	Maximum								10%	

01 31 13.50 General Contractor's Mark-Up

		Crew	Daily Output	Labor-Hours	Unit	Material	Labor	Equipment	Total	Total Incl O&P
0010	**GENERAL CONTRACTOR'S MARK-UP** on Change Orders				%				10%	10%
0200	Extra work, by subcontractors, add								15%	15%
0250	By General Contractor, add								5%	5%
0400	Omitted work, by subcontractors, deduct all but								7.50%	7.50%
0450	By General Contractor, deduct all but								15%	15%
0600	Overtime work, by subcontractors, add								10%	10%
0650	By General Contractor, add									

01 31 13.80 Overhead and Profit

		Crew	Daily Output	Labor-Hours	Unit	Material	Labor	Equipment	Total	Total Incl O&P
0010	**OVERHEAD & PROFIT** Allowance to add to items in this R013113-50				%				25%	
0020	book that do not include Subs O&P, average									
0100	Allowance to add to items in this book that R013113-55				%				5%	5%
0110	do include Subs O&P, minimum								10%	10%
0150	Average								15%	15%
0200	Maximum								40%	
0290	Typical, by size of project, under $50,000								35%	
0310	$50,000 to $100,000								25%	
0320	$100,000 to $500,000								20%	
0330	$500,000 to $1,000,000									

01 31 13.90 Performance Bond

		Crew	Daily Output	Labor-Hours	Unit	Material	Labor	Equipment	Total	Total Incl O&P
0010	**PERFORMANCE BOND**								.60%	.60%
0020	For buildings, minimum				Job				2.50%	2.50%
0100	Maximum				"					

01 31 Project Management and Coordination

01 31 14 – Facilities Services Coordination

01 31 14.20 Lock Out/Tag Out

	01 31 14.20 Lock Out/Tag Out	Crew	Daily Output	Labor-Hours	Unit	Material	2021 Bare Costs Labor	Equipment	Total	Total Incl O&P
0010	**LOCK OUT/TAG OUT**									
0020	Miniature circuit breaker lock out device	1 Elec	220	.036	Ea.	24	2.32		26.32	30
0030	Miniature pin circuit breaker lock out device		220	.036		19.90	2.32		22.22	25.50
0040	Single circuit breaker lock out device		220	.036		21	2.32		23.32	26.50
0050	Multi-pole circuit breaker lock out device (15 to 225 Amp)		210	.038		19.90	2.43		22.33	25.50
0060	Large 3 pole circuit breaker lock out device (over 225 Amp)		210	.038		21	2.43		23.43	27
0080	Square D I-Line circuit breaker lock out device		210	.038		33	2.43		35.43	39.50
0090	Lock out disconnect switch, 30 to 100 Amp		330	.024		11	1.54		12.54	14.45
0100	100 to 400 Amp		330	.024		11	1.54		12.54	14.45
0110	Over 400 Amp	1 Elec	330	.024	Ea.	11	1.54		12.54	14.45
0120	Lock out hasp for multiple lockout tags		200	.040		5.60	2.55		8.15	10.05
0130	Electrical cord plug lock out device		220	.036		10.50	2.32		12.82	15.10
0140	Electrical plug prong lock out device (3-wire grounding plug)		220	.036		6.95	2.32		9.27	11.20
0150	Wall switch lock out		200	.040		25	2.55		27.55	32
0160	Fire alarm pull station lock out	1 Stpi	200	.040		18.75	2.73		21.48	24.50
0170	Sprinkler valve tamper and flow switch lock out device	1 Skwk	220	.036		16.60	2.08		18.68	21.50
0180	Lock out sign		330	.024		18.90	1.38		20.28	23
0190	Lock out tag		440	.018		5	1.04		6.04	7.15

01 41 Regulatory Requirements

01 41 26 – Permit Requirements

01 41 26.50 Permits

		Crew	Daily Output	Labor-Hours	Unit	Material	2021 Bare Costs Labor	Equipment	Total	Total Incl O&P
0010	**PERMITS**									
0020	Rule of thumb, most cities, minimum				Job				.50%	.50%
0100	Maximum				"				2%	2%

01 45 Quality Control

01 45 23 – Testing and Inspecting Services

01 45 23.50 Testing

		Crew	Daily Output	Labor-Hours	Unit	Material	2021 Bare Costs Labor	Equipment	Total	Total Incl O&P
0010	**TESTING** and Inspecting Services									
1800	Compressive test, cylinder, delivered to lab, ASTM C 39				Ea.				36	40
1900	Picked up by lab, 30 minute round trip, add to above	1 Skwk	16	.500			28.50		28.50	45.50
1950	1 hour round trip, add to above		8	1			57		57	90.50
2000	2 hour round trip, add to above		4	2			114		114	181
2200	Compressive strength, cores (not incl. drilling), ASTM C 42								95	105
2300	Patching core holes, to 12" diameter	1 Cefi	22	.364		35	18.85		53.85	68
4730	Soil testing									
4750	Moisture content, ASTM D 2216				Ea.				18	20
4780	Permeability test, double ring infiltrometer								500	550
4800	Permeability, var. or constant head, undist., ASTM D 2434								264	290
4850	Recompacted								250	275
4900	Proctor compaction, 4" standard mold, ASTM D 698								230	253
4950	6" modified mold								110	120
5100	Shear tests, triaxial, minimum								410	450
5150	Maximum								545	600
5550	Technician for inspection, per day, earthwork								556	612
5650	Bolting								556	612
5750	Roofing								556	612
5790	Welding								556	612

01 45 Quality Control

01 45 23 – Testing and Inspecting Services

01 45 23.50 Testing	Crew	Daily Output	Labor-Hours	Unit	Material	2021 Bare Costs Labor	Equipment	Total	Total Incl O&P
5820 Non-destructive metal testing, dye penetrant				Day				556	612
5840 Magnetic particle				"				556	612
5860 Radiography				Day				556	612
5880 Ultrasonic				"				556	612
6000 Welding certification, minimum				Ea.				91	100
6100 Maximum				"				364	400
7000 Underground storage tank									
7520 Volumetric tightness, <= 12,000 gal.				Ea.	470			470	515
7530 12,000 – 29,999 gal.					585			585	645
7540 >= 30,000 gal.					665			665	730
7600 Vadose zone (soil gas) sampling, 10-40 samples, min.				Day				1,375	1,500
7610 Maximum				"				2,275	2,500
7700 Ground water monitoring incl. drilling 3 wells, min.				Total				4,550	5,000
7710 Maximum				"				6,375	7,000
8000 X-ray concrete slabs				Ea.				182	200

01 51 Temporary Utilities

01 51 13 – Temporary Electricity

01 51 13.80 Temporary Utilities	Crew	Daily Output	Labor-Hours	Unit	Material	2021 Bare Costs Labor	Equipment	Total	Total Incl O&P
0010 **TEMPORARY UTILITIES**									
0350 Lighting, lamps, wiring, outlets, 40,000 S.F. building, 8 strings	1 Elec	34	.235	CSF Flr	3.55	15		18.55	27
0360 16 strings	"	17	.471		7.10	30		37.10	54
0400 Power for temp lighting only, 6.6 KWH, per month								.92	1.01
0430 11.8 KWH, per month								1.65	1.82
0450 23.6 KWH, per month								3.30	3.63
0600 Power for job duration incl. elevator, etc., minimum								53	58
0650 Maximum								110	121
0675 Temporary cooling				Ea.	1,050			1,050	1,150

01 52 Construction Facilities

01 52 13 – Field Offices and Sheds

01 52 13.20 Office and Storage Space	Crew	Daily Output	Labor-Hours	Unit	Material	2021 Bare Costs Labor	Equipment	Total	Total Incl O&P
0010 **OFFICE AND STORAGE SPACE**									
0020 Office trailer, furnished, no hookups, 20' x 8', buy	2 Skwk	1	16	Ea.	10,000	915		10,915	12,500
0250 Rent per month					192			192	211
0300 32' x 8', buy	2 Skwk	.70	22.857		15,000	1,300		16,300	18,600
0350 Rent per month					243			243	267
0400 50' x 10', buy	2 Skwk	.60	26.667		30,500	1,525		32,025	35,900
0450 Rent per month					355			355	390
0500 50' x 12', buy	2 Skwk	.50	32		25,800	1,825		27,625	31,300
0550 Rent per month					450			450	500
0700 For air conditioning, rent per month, add					53			53	58.50
0800 For delivery, add per mile				Mile	11.95			11.95	13.15
1200 Storage boxes, 20' x 8', buy	2 Skwk	1.80	8.889	Ea.	3,100	510		3,610	4,225
1250 Rent per month				Ea.	88			88	96.50
1300 40' x 8', buy	2 Skwk	1.40	11.429		4,600	655		5,255	6,100
1350 Rent per month					135			135	148

For customer support on your Commercial Renovation Costs with RSMeans Data, call 800.448.8182.

01 54 Construction Aids

01 54 16 – Temporary Hoists

01 54 16.50 Weekly Forklift Crew	Crew	Daily Output	Labor-Hours	Unit	Material	2021 Bare Costs Labor	Equipment	Total	Total Incl O&P
0010 **WEEKLY FORKLIFT CREW**									
0100 All-terrain forklift, 45' lift, 35' reach, 9000 lb. capacity	A-3P	.20	40	Week		2,225	1,725	3,950	5,375

01 54 19 – Temporary Cranes

01 54 19.50 Daily Crane Crews

	Crew	Daily Output	Labor-Hours	Unit	Material	Labor	Equipment	Total	Total Incl O&P
0010 **DAILY CRANE CREWS** for small jobs, portal to portal									
0100 12-ton truck-mounted hydraulic crane	A-3H	1	8	Day		490	735	1,225	1,575
0200 25-ton	A-3I	1	8			490	810	1,300	1,650
0300 40-ton	A-3J	1	8			490	1,275	1,765	2,200
0400 55-ton	A-3K	1	16			910	1,525	2,435	3,100
0500 80-ton	A-3L	1	16			910	2,200	3,110	3,850
0900 If crane is needed on a Saturday, Sunday or Holiday									
0910 At time-and-a-half, add				Day		50%			
0920 At double time, add				"		100%			

01 54 19.60 Monthly Tower Crane Crew

	Crew	Daily Output	Labor-Hours	Unit	Material	Labor	Equipment	Total	Total Incl O&P
0010 **MONTHLY TOWER CRANE CREW**, excludes concrete footing									
0100 Static tower crane, 130' high, 106' jib, 6200 lb. capacity	A-3N	.05	176	Month		10,800	38,200	49,000	59,000

01 54 23 – Temporary Scaffolding and Platforms

01 54 23.60 Pump Staging

	Crew	Daily Output	Labor-Hours	Unit	Material	Labor	Equipment	Total	Total Incl O&P
0010 **PUMP STAGING**, Aluminum R015423-20									
0200 24' long pole section, buy				Ea.	380			380	420
0300 18' long pole section, buy					310			310	345
0400 12' long pole section, buy					199			199	219
0500 6' long pole section, buy					131			131	145
0600 6' long splice joint section, buy					78			78	85.50
0700 Pump jack, buy					147			147	162
0900 Foldable brace, buy					54.50			54.50	60
1000 Workbench/back safety rail support, buy					75			75	82.50
1100 Scaffolding planks/workbench, 14" wide x 24' long, buy					675			675	740
1200 Plank end safety rail, buy					241			241	265
1250 Safety net, 22' long, buy					350			350	385
1300 System in place, 50' working height, per use based on 50 uses	2 Carp	84.80	.189	C.S.F.	6	10.30		16.30	23
1400 100 uses		84.80	.189		3.01	10.30		13.31	19.75
1500 150 uses		84.80	.189		2.01	10.30		12.31	18.65

01 54 23.70 Scaffolding

	Crew	Daily Output	Labor-Hours	Unit	Material	Labor	Equipment	Total	Total Incl O&P
0010 **SCAFFOLDING** R015423-10									
0015 Steel tube, regular, no plank, labor only to erect & dismantle									
0090 Building exterior, wall face, 1 to 5 stories, 6'-4" x 5' frames	3 Carp	8	3	C.S.F.		164		164	261
0200 6 to 12 stories	4 Carp	8	4	C.S.F.		219		219	350
0301 13 to 20 stories	5 Clab	8	5			222		222	355
0460 Building interior, wall face area, up to 16' high	3 Carp	12	2			109		109	174
0560 16' to 40' high		10	2.400			131		131	209
0800 Building interior floor area, up to 30' high		150	.160	C.C.F.		8.75		8.75	13.95
0900 Over 30' high	4 Carp	160	.200	"		10.95		10.95	17.40
0906 Complete system for face of walls, no plank, material only rent/mo				C.S.F.	81.50			81.50	90
0908 Interior spaces, no plank, material only rent/mo				C.C.F.	5.60			5.60	6.15
0910 Steel tubular, heavy duty shoring, buy									
0920 Frames 5' high 2' wide				Ea.	92.50			92.50	102
0925 5' high 4' wide					118			118	130
0930 6' high 2' wide					120			120	132
0935 6' high 4' wide					119			119	131
0940 Accessories									

16

01 54 Construction Aids

01 54 23 – Temporary Scaffolding and Platforms

01 54 23.70 Scaffolding	Crew	Daily Output	Labor-Hours	Unit	Material	2021 Bare Costs Labor	Equipment	Total	Total Incl O&P
0945 Cross braces				Ea.	19.20			19.20	21
0950 U-head, 8" x 8"					23			23	25
0955 J-head, 4" x 8"					16.70			16.70	18.35
0960 Base plate, 8" x 8"					18.05			18.05	19.90
0965 Leveling jack					36			36	39.50
1000 Steel tubular, regular, buy									
1100 Frames 3' high 5' wide				Ea.	54			54	59.50
1150 5' high 5' wide					71			71	78
1200 6'-4" high 5' wide					83.50			83.50	92
1350 7'-6" high 6' wide					156			156	171
1500 Accessories, cross braces					19.20			19.20	21
1550 Guardrail post					24			24	26.50
1600 Guardrail 7' section					8.15			8.15	9
1650 Screw jacks & plates					26			26	28.50
1700 Sidearm brackets					26			26	29
1750 8" casters					37.50			37.50	41
1800 Plank 2" x 10" x 16'-0"					66.50			66.50	73.50
1900 Stairway section					298			298	330
1910 Stairway starter bar					32.50			32.50	35.50
1920 Stairway inside handrail					57.50			57.50	63.50
1930 Stairway outside handrail					91.50			91.50	101
1940 Walk-thru frame guardrail					42			42	46.50
2000 Steel tubular, regular, rent/mo.									
2100 Frames 3' high 5' wide				Ea.	9.90			9.90	10.90
2150 5' high 5' wide				"	9.90			9.90	10.90
2200 6'-4" high 5' wide				Ea.	5.75			5.75	6.35
2250 7'-6" high 6' wide					9.90			9.90	10.90
2500 Accessories, cross braces					3.35			3.35	3.69
2550 Guardrail post					3.25			3.25	3.58
2600 Guardrail 7' section					11.10			11.10	12.20
2650 Screw jacks & plates					4.35			4.35	4.79
2700 Sidearm brackets					5.20			5.20	5.70
2750 8" casters					7.85			7.85	8.65
2800 Outrigger for rolling tower					9.10			9.10	10
2850 Plank 2" x 10" x 16'-0"					9.90			9.90	10.90
2900 Stairway section					32			32	35.50
2940 Walk-thru frame guardrail					8.25			8.25	9.10
3000 Steel tubular, heavy duty shoring, rent/mo.									
3250 5' high 2' & 4' wide				Ea.	8.45			8.45	9.25
3300 6' high 2' & 4' wide					8.45			8.45	9.25
3500 Accessories, cross braces					.91			.91	1
3600 U-head, 8" x 8"					2.49			2.49	2.74
3650 J-head, 4" x 8"					2.49			2.49	2.74
3700 Base plate, 8" x 8"					.91			.91	1
3750 Leveling jack					2.47			2.47	2.72
5700 Planks, 2" x 10" x 16'-0", labor only to erect & remove to 50' H	3 Carp	72	.333			18.25		18.25	29
5800 Over 50' high	4 Carp	80	.400			22		22	35

17

01 54 Construction Aids

01 54 23.75 Scaffolding Specialties

		Crew	Daily Output	Labor-Hours	Unit	Material	2021 Bare Costs Labor	Equipment	Total	Total Incl O&P
0010	**SCAFFOLDING SPECIALTIES**									
1200	Sidewalk bridge, heavy duty steel posts & beams, including									
1210	parapet protection & waterproofing (material cost is rent/month)									
1220	8' to 10' wide, 2 posts	3 Carp	15	1.600	L.F.	65.50	87.50		153	211
1230	3 posts	"	10	2.400	"	96.50	131		227.50	315
1500	Sidewalk bridge using tubular steel scaffold frames including									
1510	planking (material cost is rent/month)	3 Carp	45	.533	L.F.	9.80	29		38.80	57.50
1600	For 2 uses per month, deduct from all above					50%				
1700	For 1 use every 2 months, add to all above					100%				
1900	Catwalks, 20" wide, no guardrails, 7' span, buy				Ea.	148			148	163
2000	10' span, buy					207			207	228
3720	Putlog, standard, 8' span, with hangers, buy					78.50			78.50	86
3730	Rent per month					16.30			16.30	17.90
3750	12' span, buy					100			100	110
3755	Rent per month					20.50			20.50	22.50
3760	Trussed type, 16' span, buy					254			254	280
3770	Rent per month					24.50			24.50	27
3790	22' span, buy				Ea.	287			287	315
3795	Rent per month					32.50			32.50	35.50
4000	7 step					1,125			1,125	1,225
4100	Rolling towers, buy, 5' wide, 7' long, 10' high					1,250			1,250	1,375
4200	For additional 5' high sections, to buy					180			180	198
4300	Complete incl. wheels, railings, outriggers,									
4350	21' high, to buy				Ea.	2,050			2,050	2,250
4400	Rent/month = 5% of purchase cost				"	103			103	113
5000	Motorized work platform, mast climber									
5050	Base unit, 50' W, less than 100' tall, rent/mo				Ea.	3,700			3,700	4,075
5100	Less than 200' tall, rent/mo					3,725			3,725	4,100
5150	Less than 300' tall, rent/mo					4,575			4,575	5,025
5200	Less than 400' tall, rent/mo					5,250			5,250	5,775
5250	Set up and demob, per unit, less than 100' tall	B-68F	16.60	1.446	C.S.F.		83.50	17.05	100.55	151
5300	Less than 200' tall		25	.960			55.50	11.35	66.85	100
5350	Less than 300' tall		30	.800			46	9.45	55.45	83.50
5400	Less than 400' tall		33	.727			42	8.60	50.60	76
5500	Mobilization (price includes freight in and out) per unit	B-34N	1	16	Ea.		880	690	1,570	2,150

01 54 23.80 Staging Aids

		Crew	Daily Output	Labor-Hours	Unit	Material	2021 Bare Costs Labor	Equipment	Total	Total Incl O&P
0010	**STAGING AIDS** and fall protection equipment									
0100	Sidewall staging bracket, tubular, buy				Ea.	60.50			60.50	67
0110	Cost each per day, based on 250 days use				Day	.24			.24	.27
0200	Guard post, buy				Ea.	51			51	56.50
0210	Cost each per day, based on 250 days use				Day	.20			.20	.23
0300	End guard chains, buy per pair				Pair	49			49	54
0310	Cost per set per day, based on 250 days use				Day	.24			.24	.27
1010	Cost each per day, based on 250 days use				"	.04			.04	.05
1100	Wood bracket, buy				Ea.	32.50			32.50	35.50
1110	Cost each per day, based on 250 days use				Day	.13			.13	.14
2010	Cost per pair per day, based on 250 days use					.57			.57	.63
3010	Cost each per day, based on 250 days use					.21			.21	.23
3100	Aluminum scaffolding plank, 20" wide x 24' long, buy				Ea.	705			705	775
3110	Cost each per day, based on 250 days use				Day	2.81			2.81	3.10
4000	Nylon full body harness, lanyard and rope grab				Ea.	175			175	193
4010	Cost each per day, based on 250 days use				Day	.70			.70	.77

18

01 54 Construction Aids

01 54 23 – Temporary Scaffolding and Platforms

01 54 23.80 Staging Aids

		Crew	Daily Output	Labor-Hours	Unit	Material	2021 Bare Costs Labor	Equipment	Total	Total Incl O&P
4100	Rope for safety line, 5/8" x 100' nylon, buy				Ea.	61			61	67
4110	Cost each per day, based on 250 days use				Day	.24			.24	.27
4200	Permanent U-Bolt roof anchor, buy				Ea.	78.50			78.50	86
4300	Temporary (one use) roof ridge anchor, buy				Ea.	6.25			6.25	6.90
5000	Installation (setup and removal) of staging aids									
5010	Sidewall staging bracket	2 Carp	64	.250	Ea.		13.70		13.70	22
5020	Guard post with 2 wood rails	"	64	.250			13.70		13.70	22
5030	End guard chains, set	1 Carp	64	.125			6.85		6.85	10.90
5100	Roof shingling bracket		96	.083			4.56		4.56	7.25
5200	Ladder jack		64	.125			6.85		6.85	10.90
5300	Wood plank, 2" x 10" x 16'	2 Carp	80	.200			10.95		10.95	17.40
5310	Aluminum scaffold plank, 20" x 24'	"	40	.400			22		22	35
5410	Safety rope	1 Carp	40	.200			10.95		10.95	17.40
5420	Permanent U-Bolt roof anchor (install only)	2 Carp	40	.400			22		22	35
5430	Temporary roof ridge anchor (install only)	1 Carp	64	.125			6.85		6.85	10.90

01 54 26 – Temporary Swing Staging

01 54 26.50 Swing Staging

		Crew	Daily Output	Labor-Hours	Unit	Material	2021 Bare Costs Labor	Equipment	Total	Total Incl O&P
0010	SWING STAGING, 500 lb. cap., 2' wide to 24' long, hand operated									
0020	Steel cable type, with 60' cables, buy				Ea.	6,425			6,425	7,075
0030	Rent per month				"	645			645	705
0600	Lightweight (not for masons) 24' long for 150' height,									
0610	manual type, buy				Ea.	13,200			13,200	14,500
0620	Rent per month					1,325			1,325	1,450
0700	Powered, electric or air, to 150' high, buy					34,200			34,200	37,600
0710	Rent per month					2,400			2,400	2,625
0780	To 300' high, buy					34,700			34,700	38,200
0800	Rent per month					2,425			2,425	2,675
1000	Bosun's chair or work basket 3' x 3'6", to 300' high, electric, buy					13,600			13,600	15,000
1010	Rent per month					950			950	1,050
2200	Move swing staging (setup and remove)	E-4	2	16	Move		975	74.50	1,049.50	1,650

01 54 36 – Equipment Mobilization

01 54 36.50 Mobilization

		Crew	Daily Output	Labor-Hours	Unit	Material	2021 Bare Costs Labor	Equipment	Total	Total Incl O&P
0010	MOBILIZATION (Use line item again for demobilization) R015436-50									
0015	Up to 25 mi. haul dist. (50 mi. RT for mob/demob crew)									
1200	Small equipment, placed in rear of, or towed by pickup truck	A-3A	4	2	Ea.		111	44	155	222
1300	Equipment hauled on 3-ton capacity towed trailer	A-3Q	2.67	3			167	93	260	360
1400	20-ton capacity	B-34U	2	8			425	224	649	920
1500	40-ton capacity	B-34N	2	8			440	345	785	1,075
1600	50-ton capacity	B-34V	1	24			1,350	995	2,345	3,225
1700	Crane, truck-mounted, up to 75 ton (driver only)	1 Eqhv	4	2			123		123	192
1800	Over 75 ton (with chase vehicle)	A-3E	2.50	6.400			360	70.50	430.50	650
2400	Crane, large lattice boom, requiring assembly	B-34W	.50	144			7,775	7,075	14,850	20,100
2500	For each additional 5 miles haul distance, add						10%	10%		
3000	For large pieces of equipment, allow for assembly/knockdown									
3100	For mob/demob of micro-tunneling equip, see Section 33 05 07.36									
3200	For mob/demob of pile driving equip, see Section 31 62 19.10									

01 55 Vehicular Access and Parking

01 55 23 – Temporary Roads

01 55 23.50 Roads and Sidewalks	Crew	Daily Output	Labor-Hours	Unit	Material	2021 Bare Costs Labor	Equipment	Total	Total Incl O&P
0010 **ROADS AND SIDEWALKS** Temporary									
0050 Roads, gravel fill, no surfacing, 4" gravel depth	B-14	715	.067	S.Y.	5.35	3.13	.30	8.78	11.15
0100 8" gravel depth	"	615	.078	"	10.65	3.64	.35	14.64	17.90

01 56 Temporary Barriers and Enclosures

01 56 13 – Temporary Air Barriers

01 56 13.60 Tarpaulins

	Crew	Daily Output	Labor-Hours	Unit	Material	2021 Bare Costs Labor	Equipment	Total	Total Incl O&P
0010 **TARPAULINS**									
0020 Cotton duck, 10-13.13 oz./S.Y., 6' x 8'				S.F.	.74			.74	.81
0050 30' x 30'					.62			.62	.68
0100 Polyvinyl coated nylon, 14-18 oz., minimum					1.46			1.46	1.61
0150 Maximum					1.46			1.46	1.61
0200 Reinforced polyethylene 3 mils thick, white					.06			.06	.07
0300 4 mils thick, white, clear or black					.17			.17	.19
0400 5.5 mils thick, clear					.16			.16	.18
0500 White, fire retardant					.56			.56	.62
0600 12 mils, oil resistant, fire retardant					.51			.51	.56
0700 8.5 mils, black					.31			.31	.34
0720 Steel reinforced polyethylene, 20 mils thick					.56			.56	.62
0730 Polyester reinforced w/integral fastening system, 11 mils thick					.23			.23	.25
0740 Polyethylene, reflective, 23 mils thick					1.34			1.34	1.47

01 56 13.90 Winter Protection

	Crew	Daily Output	Labor-Hours	Unit	Material	2021 Bare Costs Labor	Equipment	Total	Total Incl O&P
0010 **WINTER PROTECTION**									
0100 Framing to close openings	2 Clab	500	.032	S.F.	.77	1.42		2.19	3.11
0200 Tarpaulins hung over scaffolding, 8 uses, not incl. scaffolding		1500	.011		.25	.47		.72	1.03
0250 Tarpaulin polyester reinf. w/integral fastening system, 11 mils thick		1600	.010		.25	.44		.69	.99
0300 Prefab fiberglass panels, steel frame, 8 uses		1200	.013		3.06	.59		3.65	4.31

01 56 16 – Temporary Dust Barriers

01 56 16.10 Dust Barriers, Temporary

	Crew	Daily Output	Labor-Hours	Unit	Material	2021 Bare Costs Labor	Equipment	Total	Total Incl O&P
0010 **DUST BARRIERS, TEMPORARY**, erect and dismantle									
0020 Spring loaded telescoping pole & head, to 12', erect and dismantle	1 Clab	240	.033	Ea.		1.48		1.48	2.36
0025 Cost per day (based upon 250 days)				Day	.22			.22	.25
0030 To 21', erect and dismantle	1 Clab	240	.033	Ea.		1.48		1.48	2.36
0035 Cost per day (based upon 250 days)				Day	.73			.73	.81
0040 Accessories, caution tape reel, erect and dismantle	1 Clab	480	.017	Ea.		.74		.74	1.18
0045 Cost per day (based upon 250 days)				Day	.34			.34	.38
0060 Foam rail and connector, erect and dismantle	1 Clab	240	.033	Ea.		1.48		1.48	2.36
0065 Cost per day (based upon 250 days)				Day	.10			.10	.11
0070 Caution tape	1 Clab	384	.021	C.L.F.	2.55	.93		3.48	4.27
0080 Zipper, standard duty		60	.133	Ea.	10.75	5.90		16.65	21.50
0090 Heavy duty		48	.167	"	10.85	7.40		18.25	24
0100 Polyethylene sheet, 4 mil		37	.216	Sq.	2.70	9.60		12.30	18.25
0110 6 mil		37	.216	"	7.65	9.60		17.25	24
1000 Dust partition, 6 mil polyethylene, 1" x 3" frame	2 Carp	2000	.008	S.F.	.49	.44		.93	1.24
1080 2" x 4" frame	"	2000	.008	"	.60	.44		1.04	1.36
1085 Negative air machine, 1800 CFM				Ea.	900			900	990
1090 Adhesive strip application, 2" width	1 Clab	192	.042	C.L.F.	10.65	1.85		12.50	14.70
4000 Dust & infectious control partition, adj. to 10' high, obscured, 4' panel	2 Carp	90	.178	Ea.	580	9.70		589.70	655
4010 3' panel		90	.178		550	9.70		559.70	620
4020 2' panel		90	.178		430	9.70		439.70	490

01 56 Temporary Barriers and Enclosures

01 56 16 – Temporary Dust Barriers

01 56 16.10 Dust Barriers, Temporary	Crew	Daily Output	Labor-Hours	Unit	Material	2021 Bare Costs Labor	Equipment	Total	Total Incl O&P
4030	1' panel	1 Carp	90	.089	Ea.	285	4.86		289.86	325
4040	6" panel	"	90	.089		265	4.86		269.86	300
4050	2' panel with HEPA filtered discharge port	2 Carp	90	.178		500	9.70		509.70	565
4060	3' panel with 32" door		90	.178		895	9.70		904.70	1,000
4070	4' panel with 36" door		90	.178		995	9.70		1,004.70	1,125
4080	4'-6" panel with 44" door		90	.178		1,200	9.70		1,209.70	1,350
4090	Hinged corner		80	.200		185	10.95		195.95	221
4100	Outside corner		80	.200		150	10.95		160.95	182
4110	T post		80	.200		150	10.95		160.95	182
4120	Accessories, ceiling grid clip		360	.044		7.45	2.43		9.88	12.05
4130	Panel locking clip		360	.044		5.15	2.43		7.58	9.50
4140	Panel joint closure strip		360	.044		8	2.43		10.43	12.65
4150	Screw jack		360	.044		6.65	2.43		9.08	11.15
4160	Digital pressure difference gauge					275			275	305
4180	Combination lockset	1 Carp	13	.615		200	33.50		233.50	274
4185	Sealant tape, 2" wide	1 Clab	192	.042	C.L.F.	10.65	1.85		12.50	14.70
4190	System in place, including door and accessories				L.F.					
4200	Based upon 25 uses	2 Carp	51	.314		9.85	17.15		27	38.50
4210	Based upon 50 uses		51	.314		4.94	17.15		22.09	33
4230	Based upon 100 uses		51	.314		2.47	17.15		19.62	30

01 56 23 – Temporary Barricades

01 56 23.10 Barricades	Crew	Daily Output	Labor-Hours	Unit	Material	2021 Bare Costs Labor	Equipment	Total	Total Incl O&P
0010	**BARRICADES**									
0020	5' high, 3 rail @ 2" x 8", fixed	2 Carp	20	.800	L.F.	9.45	44		53.45	80
0150	Movable	"	30	.533	"	8.15	29		37.15	55.50
0300	Stock units, 58' high, 8' wide, reflective, buy				Ea.	211			211	232
0350	With reflective tape, buy				Ea.	410			410	455
0400	Break-a-way 3" PVC pipe barricade									
0410	with 3 ea. 1' x 4' reflectorized panels, buy				Ea.	146			146	161
0500	Barricades, plastic, 8" x 24" wide, foldable					57.50			57.50	63.50
0800	Traffic cones, PVC, 18" high					11.60			11.60	12.75
0850	28" high					26			26	28.50
0900	Barrels, 55 gal., with flasher	1 Clab	96	.083		99	3.70		102.70	115
1000	Guardrail, wooden, 3' high, 1" x 6" on 2" x 4" posts	2 Carp	200	.080	L.F.	2.05	4.38		6.43	9.20
1100	2" x 6" on 4" x 4" posts	"	165	.097		3.46	5.30		8.76	12.25
1200	Portable metal with base pads, buy					14.35			14.35	15.80
1250	Typical installation, assume 10 reuses	2 Carp	600	.027		2.19	1.46		3.65	4.73
1300	Barricade tape, polyethylene, 7 mil, 3" wide x 300' long roll	1 Clab	128	.063	Ea.	7.65	2.78		10.43	12.80
3000	Detour signs, set up and remove									
3010	Reflective aluminum, MUTCD, 24" x 24", post mounted	1 Clab	20	.400	Ea.	3.46	17.75		21.21	32.50
4000	Roof edge portable barrier stands and warning flags, 50 uses	1 Rohe	9100	.001	L.F.	.08	.03		.11	.14
4010	100 uses	"	9100	.001	"	.04	.03		.07	.09

01 56 26 – Temporary Fencing

01 56 26.50 Temporary Fencing	Crew	Daily Output	Labor-Hours	Unit	Material	2021 Bare Costs Labor	Equipment	Total	Total Incl O&P
0010	**TEMPORARY FENCING**									
0020	Chain link, 11 ga., 4' high	2 Clab	400	.040	L.F.	3.05	1.78		4.83	6.20
0100	6' high		300	.053		5.80	2.37		8.17	10.10
0200	Rented chain link, 6' high, to 1000' (up to 12 mo.)		400	.040		3.44	1.78		5.22	6.60
0250	Over 1000' (up to 12 mo.)		300	.053		4.07	2.37		6.44	8.25
0350	Plywood, painted, 2" x 4" frame, 4' high	A-4	135	.178		9.45	9.25		18.70	25
0400	4" x 4" frame, 8' high	"	110	.218		17.05	11.35		28.40	37
0500	Wire mesh on 4" x 4" posts, 4' high	2 Carp	100	.160		12.25	8.75		21	27.50

01 56 Temporary Barriers and Enclosures

01 56 26 – Temporary Fencing

01 56 26.50 Temporary Fencing	Crew	Daily Output	Labor-Hours	Unit	Material	2021 Bare Costs Labor	Equipment	Total	Total Incl O&P
0550 8' high	2 Carp	80	.200	L.F.	18	10.95		28.95	37

01 56 29 – Temporary Protective Walkways

01 56 29.50 Protection

		Crew	Daily Output	Labor-Hours	Unit	Material	2021 Bare Costs Labor	Equipment	Total	Total Incl O&P
0010	**PROTECTION**									
0020	Stair tread, 2" x 12" planks, 1 use	1 Carp	75	.107	Tread	7.50	5.85		13.35	17.55
0100	Exterior plywood, 1/2" thick, 1 use		65	.123		3.01	6.75		9.76	14
0200	3/4" thick, 1 use		60	.133	↓	4.15	7.30		11.45	16.15
2200	Sidewalks, 2" x 12" planks, 2 uses		350	.023	S.F.	1.25	1.25		2.50	3.36
2300	Exterior plywood, 2 uses, 1/2" thick		750	.011		.50	.58		1.08	1.48
2400	5/8" thick		650	.012		.58	.67		1.25	1.71
2500	3/4" thick	↓	600	.013	↓	.69	.73		1.42	1.92

01 57 Temporary Controls

01 57 33 – Temporary Security

01 57 33.50 Watchman

		Crew	Daily Output	Labor-Hours	Unit	Material	2021 Bare Costs Labor	Equipment	Total	Total Incl O&P
0010	**WATCHMAN**									
0020	Service, monthly basis, uniformed person, minimum				Hr.				27.65	30.40
0100	Maximum								56	61.62
0200	Person and command dog, minimum								31	34
0300	Maximum				↓				60	65

01 58 Project Identification

01 58 13 – Temporary Project Signage

01 58 13.50 Signs

		Crew	Daily Output	Labor-Hours	Unit	Material	2021 Bare Costs Labor	Equipment	Total	Total Incl O&P
0010	**SIGNS**									
0020	High intensity reflectorized, no posts, buy				Ea.	22			22	24

01 66 Product Storage and Handling Requirements

01 66 19 – Material Handling

01 66 19.10 Material Handling

		Crew	Daily Output	Labor-Hours	Unit	Material	2021 Bare Costs Labor	Equipment	Total	Total Incl O&P
0010	**MATERIAL HANDLING**									
0020	Above 2nd story, via stairs, per C.Y. of material per floor	2 Clab	145	.110	C.Y.		4.90		4.90	7.80
0030	Via elevator, per C.Y. of material		240	.067			2.96		2.96	4.71
0050	Distances greater than 200', per C.Y. of material per each addl 200'	↓	300	.053	↓		2.37		2.37	3.77

01 71 Examination and Preparation

01 71 23 – Field Engineering

01 71 23.13 Construction Layout

		Crew	Daily Output	Labor-Hours	Unit	Material	2021 Bare Costs Labor	Equipment	Total	Total Incl O&P
0010	**CONSTRUCTION LAYOUT**									
1100	Crew for layout of building, trenching or pipe laying, 2 person crew	A-6	1	16	Day		880	34.50	914.50	1,450
1200	3 person crew	A-7	1	24			1,425	34.50	1,459.50	2,300
1400	Crew for roadway layout, 4 person crew	A-8	1	32	↓		1,850	34.50	1,884.50	2,950

01 74 Cleaning and Waste Management

01 74 13 – Progress Cleaning

01 74 13.20 Cleaning Up

	Crew	Daily Output	Labor-Hours	Unit	Material	2021 Bare Costs Labor	Equipment	Total	Total Incl O&P
0010 **CLEANING UP**									
0020 After job completion, allow, minimum				Job				.30%	.30%
0040 Maximum				"				1%	1%
0052 Cleanup of floor area, continuous, per day, during const.	A-5	16	1.125	M.S.F.	2.77	50.50	3.11	56.38	87
0100 Final by GC at end of job	"	11.50	1.565	"	2.93	70.50	4.32	77.75	120

01 76 Protecting Installed Construction

01 76 13 – Temporary Protection of Installed Construction

01 76 13.20 Temporary Protection

	Crew	Daily Output	Labor-Hours	Unit	Material	2021 Bare Costs Labor	Equipment	Total	Total Incl O&P
0010 **TEMPORARY PROTECTION**									
0020 Flooring, 1/8" tempered hardboard, taped seams	2 Carp	1500	.011	S.F.	.51	.58		1.09	1.49
0030 Peel away carpet protection	1 Clab	3200	.003	"	.11	.11		.22	.30

01 91 Commissioning

01 91 13 – General Commissioning Requirements

01 91 13.50 Building Commissioning

	Crew	Daily Output	Labor-Hours	Unit	Material	2021 Bare Costs Labor	Equipment	Total	Total Incl O&P
0010 **BUILDING COMMISSIONING**									
0100 Systems operation and verification during turnover				%				.25%	.25%
0150 Including all systems subcontractors								.50%	.50%
0200 Systems design assistance, operation, verification and training								.50%	.50%
0250 Including all systems subcontractors				%				1%	1%

01 93 Facility Maintenance

01 93 09 – Facility Equipment

01 93 09.50 Moving Equipment

	Crew	Daily Output	Labor-Hours	Unit	Material	2021 Bare Costs Labor	Equipment	Total	Total Incl O&P
0010 **MOVING EQUIPMENT**, Remove and reset, 100' distance,									
0020 No obstructions, no assembly or leveling unless noted									
0100 Annealing furnace, 24' overall	B-67	4	4	Ea.		229	68	297	430
0200 Annealing oven, small		14	1.143			65.50	19.50	85	123
0240 Very large		1	16			915	273	1,188	1,725
0400 Band saw, small		12	1.333			76	22.50	98.50	143
0440 Large		8	2			114	34	148	215
0500 Blue print copy machine		7	2.286			131	39	170	246
0600 Bonding mill, 6"		7	2.286			131	39	170	246
0620 12"		6	2.667			152	45.50	197.50	286
0640 18"		4	4			229	68	297	430
0660 24"		2	8			455	136	591	860
0700 Boring machine (jig)	B-68	7	3.429			198	39	237	350
0800 Bridgeport mill, standard	B-67	14	1.143			65.50	19.50	85	123
1000 Calibrator, 6 unit	"	14	1.143			65.50	19.50	85	123
1100 Comparitor, bench top	2 Clab	14	1.143			50.50		50.50	81
1140 Floor mounted	B-67	7	2.286			131	39	170	246
1200 Computer, desk top	2 Clab	25	.640			28.50		28.50	45.50
1300 Copy machine	"	25	.640			28.50		28.50	45.50
1500 Deflasher	B-67	14	1.143			65.50	19.50	85	123
1600 Degreaser, small		14	1.143			65.50	19.50	85	123
1640 Large 24' overall		1	16			915	273	1,188	1,725

23

01 93 09.50 Moving Equipment	Crew	Daily Output	Labor-Hours	Unit	Material	2021 Bare Costs Labor	Equipment	Total	Total Incl O&P
1700 Desk with chair	2 Clab	25	.640	Ea.		28.50		28.50	45.50
1725 Desk only	"	22	.727			32.50		32.50	51.50
1750 Chair only	1 Clab	300	.027			1.18		1.18	1.89
1800 Dial press	B-67	7	2.286			131	39	170	246
1900 Drafting table	2 Clab	14	1.143			50.50		50.50	81
2000 Drill press, bench top	"	14	1.143			50.50		50.50	81
2040 Floor mounted	B-67	14	1.143			65.50	19.50	85	123
2080 Industrial radial	"	7	2.286			131	39	170	246
2100 Dust collector, portable	2 Clab	25	.640			28.50		28.50	45.50
2140 Stationary, small	B-67	7	2.286			131	39	170	246
2180 Stationary, large	"	2	8			455	136	591	860
2300 Electric discharge machine	B-68	7	3.429			198	39	237	350
2400 Environmental chamber walls, including assembly	4 Clab	18	1.778	L.F.		79		79	126
2600 File cabinet	2 Clab	25	.640	Ea.		28.50		28.50	45.50
2800 Grinder/sander, pedestal mount	B-67	14	1.143			65.50	19.50	85	123
3000 Hack saw, power	2 Clab	24	.667			29.50		29.50	47
3100 Hydraulic press	B-67	14	1.143			65.50	19.50	85	123
3500 Laminar flow tables	"	14	1.143			65.50	19.50	85	123
3600 Lathe, bench	2 Clab	14	1.143			50.50		50.50	81
3640 6"	B-67	14	1.143			65.50	19.50	85	123
3680 10"		13	1.231			70.50	21	91.50	132
3720 12"		12	1.333			76	22.50	98.50	143
4000 Milling machine		8	2			114	34	148	215
4100 Molding press, 25 ton		5	3.200			183	54.50	237.50	345
4140 60 ton		4	4			229	68	297	430
4180 100 ton		2	8			455	136	591	860
4220 150 ton		1.50	10.667			610	182	792	1,150
4260 200 ton		1	16			915	273	1,188	1,725
4300 300 ton		.75	21.333			1,225	365	1,590	2,300
4700 Oil pot stand		14	1.143			65.50	19.50	85	123
5000 Press, 10 ton		14	1.143			65.50	19.50	85	123
5040 15 ton		12	1.333			76	22.50	98.50	143
5080 20 ton		10	1.600			91.50	27.50	119	172
5120 30 ton		8	2			114	34	148	215
5160 45 ton		6	2.667			152	45.50	197.50	286
5200 60 ton		4	4			229	68	297	430
5240 75 ton		2.50	6.400			365	109	474	690
5280 100 ton		2	8			455	136	591	860
5500 Raised floor, including assembly	2 Carp	250	.064	S.F.		3.50		3.50	5.55
5600 Rolling mill, 6"	B-67	7	2.286	Ea.		131	39	170	246
5640 9"		6	2.667			152	45.50	197.50	286
5680 12"		4	4			229	68	297	430
5720 13"		3.50	4.571			261	78	339	490
5760 18"		2	8			455	136	591	860
5800 25"		1	16			915	273	1,188	1,725
6000 Sander, floor stand		14	1.143			65.50	19.50	85	123
6100 Screw machine		7	2.286			131	39	170	246
6200 Shaper, 16"		14	1.143			65.50	19.50	85	123
6300 Shear, power assist		4	4			229	68	297	430
6400 Slitter, 6"		14	1.143			65.50	19.50	85	123
6440 8"		13	1.231			70.50	21	91.50	132
6480 10"		12	1.333			76	22.50	98.50	143
6520 12"		11	1.455			83	25	108	157

01 93 Facility Maintenance

01 93 09 – Facility Equipment

01 93 09.50 Moving Equipment		Crew	Daily Output	Labor-Hours	Unit	Material	2021 Bare Costs Labor	Equipment	Total	Total Incl O&P
6560	16"	B-67	10	1.600	Ea.		91.50	27.50	119	172
6600	20"		8	2			114	34	148	215
6640	24"		6	2.667			152	45.50	197.50	286
6800	Snag and tap machine		7	2.286			131	39	170	246
6900	Solder machine (auto)		7	2.286			131	39	170	246
7000	Storage cabinet metal, small	2 Clab	36	.444			19.75		19.75	31.50
7040	Large		25	.640			28.50		28.50	45.50
7100	Storage rack open, small		14	1.143			50.50		50.50	81
7140	Large		7	2.286			101		101	162
7200	Surface bench, small	B-67	14	1.143			65.50	19.50	85	123
7240	Large	"	5	3.200			183	54.50	237.50	345
7300	Surface grinder, large wet	B-68	5	4.800			277	54.50	331.50	490
7500	Time check machine	2 Clab	14	1.143			50.50		50.50	81
8000	Welder, 30 kVA (bench)		14	1.143			50.50		50.50	81
8100	Work bench with chair		25	.640			28.50		28.50	45.50
8900	For moving distance to 200', add						25%			

Division Notes

	CREW	DAILY OUTPUT	LABOR-HOURS	UNIT	BARE COSTS				TOTAL INCL O&P
					MAT.	LABOR	EQUIP.	TOTAL	

Estimating Tips

02 30 00 Subsurface Investigation

In preparing estimates on structures involving earthwork or foundations, all information concerning soil characteristics should be obtained. Look particularly for hazardous waste, evidence of prior dumping of debris, and previous stream beds.

02 40 00 Demolition and Structure Moving

The costs shown for selective demolition do not include rubbish handling or disposal. These items should be estimated separately using RSMeans data or other sources.

- Historic preservation often requires that the contractor remove materials from the existing structure, rehab them, and replace them. The estimator must be aware of any related measures and precautions that must be taken when doing selective demolition and cutting and patching. Requirements may include special handling and storage, as well as security.

- In addition to Subdivision 02 41 00, you can find selective demolition items in each division. Example: Roofing demolition is in Division 7.
- Absent of any other specific reference, an approximate demolish-in-place cost can be obtained by halving the new-install labor cost. To remove for reuse, allow the entire new-install labor figure.

02 40 00 Building Deconstruction

This section provides costs for the careful dismantling and recycling of most low-rise building materials.

02 50 00 Containment of Hazardous Waste

This section addresses on-site hazardous waste disposal costs.

02 80 00 Hazardous Material Disposal/Remediation

This subdivision includes information on hazardous waste handling, asbestos remediation, lead remediation, and mold remediation. See reference numbers

RO28213-20 and RO28319-60 for further guidance in using these unit price lines.

02 90 00 Monitoring Chemical Sampling, Testing Analysis

This section provides costs for on-site sampling and testing hazardous waste.

Reference Numbers

Reference numbers are shown at the beginning of some major classifications. These numbers refer to related items in the Reference Section. The reference information may be an estimating procedure, an alternate pricing method, or technical information.

Note: Not all subdivisions listed here necessarily appear. ■

Same Data. Simplified.

Enjoy the convenience and efficiency of accessing your costs anywhere:

- **Skip the multiplier** by setting your location
- **Quickly search,** edit, favorite and share costs
- **Stay on top of price changes** with automatic updates

Discover more at rsmeans.com/online

02 21 Surveys

02 21 13 – Site Surveys

02 21 13.09 Topographical Surveys

		Crew	Daily Output	Labor-Hours	Unit	Material	2021 Bare Costs Labor	2021 Bare Costs Equipment	Total	Total Incl O&P
0010	**TOPOGRAPHICAL SURVEYS**									
0020	Topographical surveying, conventional, minimum	A-7	3.30	7.273	Acre	35	430	10.40	475.40	730
0050	Average	"	1.95	12.308		52.50	730	17.60	800.10	1,225
0100	Maximum	A-8	.60	53.333	↓	70	3,075	57.50	3,202.50	5,000

02 21 13.13 Boundary and Survey Markers

		Crew	Daily Output	Labor-Hours	Unit	Material	2021 Bare Costs Labor	2021 Bare Costs Equipment	Total	Total Incl O&P
0010	**BOUNDARY AND SURVEY MARKERS**									
0300	Lot location and lines, large quantities, minimum	A-7	2	12	Acre	34.50	715	17.15	766.65	1,175
0320	Average	"	1.25	19.200		61	1,150	27.50	1,238.50	1,900
0400	Small quantities, maximum	A-8	1	32	↓	77	1,850	34.50	1,961.50	3,025

02 31 Geophysical Investigations

02 31 23 – Electromagnetic Investigations

02 31 23.10 Ground Penetrating Radar

		Crew	Daily Output	Labor-Hours	Unit	Material	2021 Bare Costs Labor	2021 Bare Costs Equipment	Total	Total Incl O&P
0010	**GROUND PENETRATING RADAR**									
0100	Ground Penetrating Radar				Hr.		210		210	235

02 32 Geotechnical Investigations

02 32 13 – Subsurface Drilling and Sampling

02 32 13.10 Boring and Exploratory Drilling

		Crew	Daily Output	Labor-Hours	Unit	Material	2021 Bare Costs Labor	2021 Bare Costs Equipment	Total	Total Incl O&P
0010	**BORING AND EXPLORATORY DRILLING**									
0020	Borings, initial field stake out & determination of elevations	A-6	1	16	Day		880	34.50	914.50	1,450
0100	Drawings showing boring details				Total		335		335	425
0200	Report and recommendations from P.E.						775		775	970
0300	Mobilization and demobilization	B-55	4	6	↓		275	310	585	780
0350	For over 100 miles, per added mile		450	.053	Mile		2.45	2.76	5.21	6.95
0600	Auger holes in earth, no samples, 2-1/2" diameter		78.60	.305	L.F.		14	15.85	29.85	40
0650	4" diameter		67.50	.356			16.30	18.45	34.75	46.50
0800	Cased borings in earth, with samples, 2-1/2" diameter		55.50	.432		25	19.85	22.50	67.35	83.50
0850	4" diameter		32.60	.736		28.50	34	38	100.50	128
1000	Drilling in rock, "BX" core, no sampling	B-56	34.90	.458			23	45.50	68.50	86
1050	With casing & sampling	"	31.70	.505		25	25	50	100	122
1200	"NX" core, no sampling	B-56	25.92	.617	L.F.		31	61.50	92.50	116
1250	With casing and sampling	"	25	.640	"	26	32	63.50	121.50	149
1400	Borings, earth, drill rig and crew with truck mounted auger	B-55	1	24	Day		1,100	1,250	2,350	3,125
1450	Rock using crawler type drill	B-56	1	16	"		800	1,600	2,400	3,000
1500	For inner city borings, add, minimum								10%	10%
1510	Maximum								20%	20%

02 41 Demolition

02 41 13 – Selective Site Demolition

	Crew	Daily Output	Labor-Hours	Unit	Material	2021 Bare Costs Labor	2021 Bare Costs Equipment	Total	Total Incl O&P

02 41 13.15 Hydrodemolition

		Crew	Daily Output	Labor-Hours	Unit	Material	Labor	Equipment	Total	Total Incl O&P
0010	**HYDRODEMOLITION**									
0015	Hydrodemolition, concrete pavement									
0120	Includes removal but excludes disposal of concrete						3.15	2.75	5.90	8
0130	2" depth	B-5E	1000	.064	S.F.		3.15	2.75	5.90	8
0410	4" depth		800	.080			3.93	3.43	7.36	10.05
0420	6" depth		600	.107			5.25	4.58	9.83	13.35
0520	8" depth		300	.213			10.50	9.15	19.65	26.50

02 41 13.16 Hydroexcavation

		Crew	Daily Output	Labor-Hours	Unit	Material	Labor	Equipment	Total	Total Incl O&P
0010	**HYDROEXCAVATION**									
0015	Hydroexcavation									
0105	Assumes onsite disposal									
0110	Mobilization or Demobilization	B-6D	8	2.500	Ea.		127	160	287	375
0130	Normal Conditions		48	.417	B.C.Y.		21	26.50	47.50	63
0140	Adverse Conditions		30	.667	"		34	42.50	76.50	101
0160	Minimum labor/equipment charge		4	5	Ea.		253	320	573	750

02 41 13.17 Demolish, Remove Pavement and Curb

		Crew	Daily Output	Labor-Hours	Unit	Material	Labor	Equipment	Total	Total Incl O&P
0010	**DEMOLISH, REMOVE PAVEMENT AND CURB** R024119-10									
5010	Pavement removal, bituminous roads, up to 3" thick	B-38	690	.058	S.Y.		2.90	1.76	4.66	6.50
5050	4"-6" thick		420	.095			4.76	2.90	7.66	10.70
5100	Bituminous driveways		640	.063			3.12	1.90	5.02	7
5200	Concrete to 6" thick, hydraulic hammer, mesh reinforced		255	.157			7.85	4.77	12.62	17.60
5300	Rod reinforced		200	.200			10	6.10	16.10	22.50
5400	Concrete, 7"-24" thick, plain		33	1.212	C.Y.		60.50	37	97.50	136
5500	Reinforced		24	1.667	"		83	50.50	133.50	187
5590	Minimum labor/equipment charge		6	6.667	Job		335	203	538	750
5600	With hand held air equipment, bituminous, to 6" thick	B-39	1900	.025	S.F.		1.18	.19	1.37	2.08
5700	Concrete to 6" thick, no reinforcing		1600	.030			1.40	.22	1.62	2.46
5800	Mesh reinforced		1400	.034			1.60	.25	1.85	2.81
5900	Rod reinforced		765	.063			2.92	.46	3.38	5.15
5990	Minimum labor/equipment charge	B-38	6	6.667	Job		335	203	538	750
6000	Curbs, concrete, plain	B-6	360	.067	L.F.		3.21	.60	3.81	5.70
6100	Reinforced	"	275	.087	"		4.20	.79	4.99	7.50
6200	Granite	B-6	360	.067	L.F.		3.21	.60	3.81	5.70
6300	Bituminous		528	.045	"		2.19	.41	2.60	3.90
6390	Minimum labor/equipment charge		6	4	Job		192	36	228	345

02 41 13.23 Utility Line Removal

		Crew	Daily Output	Labor-Hours	Unit	Material	Labor	Equipment	Total	Total Incl O&P
0010	**UTILITY LINE REMOVAL**	B-6	7	3.429	Ea.		165	31	196	295
0015	No hauling, abandon catch basin or manhole		4	6			289	54	343	515
0020	Remove existing catch basin or manhole, masonry		13	1.846			89	16.65	105.65	158
0030	Catch basin or manhole frames and covers, stored		7	3.429			165	31	196	295
0040	Remove and reset	B-21A	5	8			440	86.50	526.50	780
0900	Hydrants, fire, remove only	"	2	20			1,100	216	1,316	1,975
0950	Remove and reset	2 Plum	2	8	Job		540		540	835
0990	Minimum labor/equipment charge									
2900	Pipe removal, sewer/water, no excavation, 12" diameter	B-6	175	.137	L.F.		6.60	1.24	7.84	11.75
2930	15"-18" diameter	B-12Z	150	.160			8	10.45	18.45	24
2960	21"-24" diameter		120	.200			10	13.05	23.05	30
3000	27"-36" diameter		90	.267			13.35	17.40	30.75	40
3200	Steel, welded connections, 4" diameter	B-6	160	.150			7.20	1.35	8.55	12.90
3300	10" diameter		80	.300			14.45	2.70	17.15	26
3390	Minimum labor/equipment charge		3	8	Job		385	72	457	690

02 41 Demolition

02 41 13 – Selective Site Demolition

02 41 13.30 Minor Site Demolition

	02 41 13.30 Minor Site Demolition	Crew	Daily Output	Labor-Hours	Unit	Material	2021 Bare Costs Labor	Equipment	Total	Total Incl O&P
0010	**MINOR SITE DEMOLITION** R024119-10									
0100	Roadside delineators, remove only	B-80	175	.183	Ea.		8.90	6	14.90	21
0110	Remove and reset	"	100	.320	"		15.60	10.50	26.10	36
0400	Minimum labor/equipment charge	B-6	4	6	Job		289	54	343	515
0800	Guiderail, corrugated steel, remove only	B-80A	100	.240	L.F.		10.65	8.50	19.15	26.50
0850	Remove and reset	"	40	.600	"		26.50	21.50	48	66
0860	Guide posts, remove only	B-80B	120	.267	Ea.		12.60	1.99	14.59	22
0870	Remove and reset	B-55	50	.480	"		22	25	47	62.50
0890	Minimum labor/equipment charge	2 Clab	4	4	Job		178		178	283
1000	Masonry walls, block, solid	B-5	1800	.031	C.F.		1.52	.83	2.35	3.32
1200	Brick, solid		900	.062			3.04	1.67	4.71	6.65
1400	Stone, with mortar		900	.062			3.04	1.67	4.71	6.65
1500	Dry set	↓	1500	.037	↓		1.82	1	2.82	3.98
1600	Median barrier, precast concrete, remove and store	B-3	430	.112	L.F.		5.50	5.35	10.85	14.65
1610	Remove and reset	"	390	.123	"		6.10	5.90	12	16.15
1650	Minimum labor/equipment charge	A-1	4	2	Job		89	28.50	117.50	173
4000	Sidewalk removal, bituminous, 2" thick	B-6	350	.069	S.Y.		3.30	.62	3.92	5.90
4010	2-1/2" thick		325	.074			3.55	.67	4.22	6.35
4050	Brick, set in mortar		185	.130			6.25	1.17	7.42	11.15
4100	Concrete, plain, 4"	↓	160	.150	↓		7.20	1.35	8.55	12.90
4110	Plain, 5"	B-6	140	.171	S.Y.		8.25	1.54	9.79	14.75
4120	Plain, 6"		120	.200			9.60	1.80	11.40	17.20
4200	Mesh reinforced, concrete, 4"		150	.160			7.70	1.44	9.14	13.75
4210	5" thick		131	.183			8.80	1.65	10.45	15.70
4220	6" thick	↓	112	.214	↓		10.30	1.93	12.23	18.40
4290	Minimum labor/equipment charge	B-39	12	4	Job		186	29.50	215.50	330

02 41 13.33 Railtrack Removal

	02 41 13.33 Railtrack Removal	Crew	Daily Output	Labor-Hours	Unit	Material	Labor	Equipment	Total	Total Incl O&P
0010	**RAILTRACK REMOVAL**									
3500	Railroad track removal, ties and track	B-13	330	.170	L.F.		8.20	1.78	9.98	14.90
3600	Ballast	B-14	500	.096	C.Y.		4.47	.43	4.90	7.60
3700	Remove and re-install, ties & track using new bolts & spikes		50	.960	L.F.		44.50	4.32	48.82	76
3800	Turnouts using new bolts and spikes		1	48	Ea.	2,225	216		2,441	3,800
3890	Minimum labor/equipment charge	↓	5	9.600	Job		445	43	488	760

02 41 13.60 Selective Demolition Fencing

	02 41 13.60 Selective Demolition Fencing	Crew	Daily Output	Labor-Hours	Unit	Material	Labor	Equipment	Total	Total Incl O&P
0010	**SELECTIVE DEMOLITION FENCING** R024119-10									
1600	Fencing, barbed wire, 3 strand	2 Clab	430	.037	L.F.		1.65		1.65	2.63
1650	5 strand	"	280	.057			2.54		2.54	4.04
1700	Chain link, posts & fabric, 8'-10' high, remove only	B-6	445	.054	↓		2.59	.49	3.08	4.63

02 41 16 – Structure Demolition

02 41 16.13 Building Demolition

	02 41 16.13 Building Demolition	Crew	Daily Output	Labor-Hours	Unit	Material	Labor	Equipment	Total	Total Incl O&P
0010	**BUILDING DEMOLITION** Large urban projects, incl. 20 mi. haul R024119-10									
0011	No foundation or dump fees, C.F. is vol. of building standing									
0020	Steel	B-8	21500	.003	C.F.		.15	.13	.28	.39
0050	Concrete		15300	.004			.21	.19	.40	.55
0080	Masonry		20100	.003			.16	.14	.30	.42
0100	Mixture of types	↓	20100	.003			.16	.14	.30	.42
0500	Small bldgs, or single bldgs, no salvage included, steel	B-3	14800	.003			.16	.16	.32	.42
0600	Concrete		11300	.004			.21	.20	.41	.55
0650	Masonry		14800	.003			.16	.16	.32	.42
0700	Wood	↓	14800	.003			.16	.16	.32	.42
0750	For buildings with no interior walls, deduct				↓				30%	30%

02 41 Demolition

02 41 16 – Structure Demolition

		Crew	Daily Output	Labor-Hours	Unit	Material	2021 Bare Costs Labor	2021 Bare Costs Equipment	Total	Total Incl O&P
02 41 16.13	**Building Demolition**									
1000	Demolition single family house, one story, wood 1600 S.F.	B-3	1	48	Ea.		2,375	2,300	4,675	6,300
1020	3200 S.F.		.50	96			4,750	4,600	9,350	12,600
1200	Demolition two family house, two story, wood 2400 S.F.		.67	71.964			3,550	3,450	7,000	9,450
1220	4200 S.F.		.38	128			6,325	6,150	12,475	16,800
1300	Demolition three family house, three story, wood 3200 S.F.		.50	96			4,750	4,600	9,350	12,600
1320	5400 S.F.	↓	.30	160	↓		7,925	7,675	15,600	21,100
5000	For buildings with no interior walls, deduct								30%	30%

02 41 16.17 Building Demolition Footings and Foundations

		Crew	Daily Output	Labor-Hours	Unit	Material	2021 Bare Costs Labor	2021 Bare Costs Equipment	Total	Total Incl O&P
0010	**BUILDING DEMOLITION FOOTINGS AND FOUNDATIONS** R024119-10									
0200	Floors, concrete slab on grade,									
0240	4" thick, plain concrete	B-13L	5000	.003	S.F.		.20	.42	.62	.77
0280	Reinforced, wire mesh	B-13L	4000	.004	S.F.		.25	.53	.78	.96
0300	Rods		4500	.004			.22	.47	.69	.85
0400	6" thick, plain concrete		4000	.004			.25	.53	.78	.96
0420	Reinforced, wire mesh		3200	.005			.31	.66	.97	1.20
0440	Rods		3600	.004	↓		.27	.58	.85	1.07
1000	Footings, concrete, 1' thick, 2' wide		300	.053	L.F.		3.28	7	10.28	12.80
1080	1'-6" thick, 2' wide		250	.064			3.93	8.40	12.33	15.40
1120	3' wide		200	.080			4.92	10.50	15.42	19.20
1140	2' thick, 3' wide	↓	175	.091			5.60	12	17.60	22
1200	Average reinforcing, add								10%	10%
1220	Heavy reinforcing, add								20%	20%
2000	Walls, block, 4" thick	B-13L	8000	.002	S.F.		.12	.26	.38	.48
2040	6" thick		6000	.003			.16	.35	.51	.65
2080	8" thick		4000	.004			.25	.53	.78	.96
2100	12" thick	↓	3000	.005			.33	.70	1.03	1.28
2200	For horizontal reinforcing, add								10%	10%
2220	For vertical reinforcing, add				↓				20%	20%
2400	Concrete, plain concrete, 6" thick	B-13L	4000	.004			.25	.53	.78	.96
2420	8" thick		3500	.005			.28	.60	.88	1.10
2440	10" thick		3000	.005			.33	.70	1.03	1.28
2500	12" thick	↓	2500	.006			.39	.84	1.23	1.54
2600	For average reinforcing, add								10%	10%
2620	For heavy reinforcing, add				↓				20%	20%
9000	Minimum labor/equipment charge	A-1	2	4	Job		178	57	235	345

02 41 16.19 Minor Building Deconstruction

			Crew	Daily Output	Labor-Hours	Unit	Material	2021 Bare Costs Labor	2021 Bare Costs Equipment	Total	Total Incl O&P
0010	**MINOR BUILDING DECONSTRUCTION** R024119-10										
0011	For salvage, avg house										
0225	2 story, pre 1970 house, 1400 S.F., labor for all salv mat, min	G	6 Clab	25	1.920	SF Flr.		85.50		85.50	136
0230	Maximum	G	"	15	3.200	"		142		142	226
0235	Salvage of carpet, tackless	G	2 Clab	2400	.007	S.F.		.30		.30	.47
0240	Wood floors, incl. denailing and packaging	G	3 Clab	270	.089	"		3.95		3.95	6.30
0260	Wood doors and trim, standard	G	1 Clab	16	.500	Ea.		22		22	35.50
0280	Base or cove mouldings, incl. denailing and packaging	G		500	.016	L.F.		.71		.71	1.13
0320	Closet shelving and trim, incl. denailing and packaging	G		18	.444	Set		19.75		19.75	31.50
0340	Kitchen cabinets, uppers and lowers, prefab type	G	2 Clab	24	.667	L.F.		29.50		29.50	47
0360	Kitchen cabinets, uppers and lowers, built-in	G		12	1.333	"		59		59	94.50
0370	Bath fixt, incl. toilet, tub, vanity, and med cabinet	G	↓	3	5.333	Set		237		237	375

31

02 41 19.13 Selective Building Demolition

	Crew	Daily Output	Labor-Hours	Unit	Material	2021 Bare Costs Labor	Equipment	Total	Total Incl O&P
0010 **SELECTIVE BUILDING DEMOLITION**									
0020 Costs related to selective demolition of specific building components									
0025 are included under Common Work Results (XX 05)									
0030 in the component's appropriate division.									

02 41 19.16 Selective Demolition, Cutout

		Crew	Daily Output	Labor-Hours	Unit	Material	2021 Bare Costs Labor	Equipment	Total	Total Incl O&P
0010	**SELECTIVE DEMOLITION, CUTOUT** R024119-10									
0020	Concrete, elev. slab, light reinforcement, under 6 C.F.	B-9	65	.615	C.F.		27.50	5.45	32.95	50
0050	Light reinforcing, over 6 C.F.		75	.533	"		24	4.74	28.74	43
0200	Slab on grade to 6" thick, not reinforced, under 8 S.F.		85	.471	S.F.		21	4.18	25.18	38
0250	8-16 S.F.	↓	175	.229	"		10.25	2.03	12.28	18.55
0255	For over 16 S.F. see Line 02 41 16.17 0400									
0600	Walls, not reinforced, under 6 C.F.	B-9	60	.667	C.F.		30	5.95	35.95	54
0650	6-12 C.F.	"	80	.500	"		22.50	4.45	26.95	40.50
0655	For over 12 C.F. see Line 02 41 16.17 2500									
1000	Concrete, elevated slab, bar reinforced, under 6 C.F.	B-9	45	.889	C.F.		40	7.90	47.90	72
1050	Bar reinforced, over 6 C.F.		50	.800	"		36	7.10	43.10	65
1200	Slab on grade to 6" thick, bar reinforced, under 8 S.F.		75	.533	S.F.		24	4.74	28.74	43
1250	8-16 S.F.	↓	150	.267	"		11.95	2.37	14.32	21.50
1255	For over 16 S.F. see Line 02 41 16.17 0440									
1400	Walls, bar reinforced, under 6 C.F.	B-9	50	.800	C.F.		36	7.10	43.10	65
1450	6-12 C.F.	"	70	.571	"		25.50	5.10	30.60	46.50
1455	For over 12 C.F. see Lines 02 41 16.17 2500 and 2600									
2000	Brick, to 4 S.F. opening, not including toothing									
2040	4" thick	B-9	30	1.333	Ea.		59.50	11.85	71.35	108
2060	8" thick		18	2.222			99.50	19.75	119.25	181
2080	12" thick		10	4			179	35.50	214.50	325
2400	Concrete block, to 4 S.F. opening, 2" thick		35	1.143			51	10.15	61.15	92.50
2420	4" thick		30	1.333			59.50	11.85	71.35	108
2440	8" thick		27	1.481			66.50	13.15	79.65	121
2460	12" thick		24	1.667			74.50	14.80	89.30	135
2600	Gypsum block, to 4 S.F. opening, 2" thick		80	.500			22.50	4.45	26.95	40.50
2620	4" thick		70	.571			25.50	5.10	30.60	46.50
2640	8" thick		55	.727			32.50	6.45	38.95	59
2800	Terra cotta, to 4 S.F. opening, 4" thick		70	.571			25.50	5.10	30.60	46.50
2840	8" thick		65	.615			27.50	5.45	32.95	50
2880	12" thick	↓	50	.800	↓		36	7.10	43.10	65
4000	For toothing masonry, see Section 04 01 20.50									
6000	Walls, interior, not including re-framing,									
6010	openings to 5 S.F.									
6100	Drywall to 5/8" thick	1 Clab	24	.333	Ea.		14.80		14.80	23.50
6200	Paneling to 3/4" thick	"	20	.400	"		17.75		17.75	28.50
6300	Plaster, on gypsum lath	1 Clab	20	.400	Ea.		17.75		17.75	28.50
6340	On wire lath	"	14	.571	"		25.50		25.50	40.50
7000	Wood frame, not including re-framing, openings to 5 S.F.									
7200	Floors, sheathing and flooring to 2" thick	1 Clab	5	1.600	Ea.		71		71	113
7310	Roofs, sheathing to 1" thick, not including roofing		6	1.333			59		59	94.50
7410	Walls, sheathing to 1" thick, not including siding		7	1.143	↓		50.50		50.50	81
8500	Minimum labor/equipment charge	↓	4	2	Job		89		89	141

For customer support on your Commercial Renovation Costs with RSMeans Data, call 800.448.8182.

02 41 Demolition

02 41 19 – Selective Demolition

02 41 19.18 Selective Demolition, Disposal Only

	Crew	Daily Output	Labor-Hours	Unit	Material	2021 Bare Costs Labor	Equipment	Total	Total Incl O&P
0010	**SELECTIVE DEMOLITION, DISPOSAL ONLY** R024119-10								
0015	Urban bldg w/salvage value allowed								
0020	Including loading and 5 mile haul to dump								
0200	Steel frame	B-3	430	.112	C.Y.		5.50	5.35	10.85	14.65
0300	Concrete frame		365	.132			6.50	6.30	12.80	17.25
0400	Masonry construction		445	.108			5.35	5.20	10.55	14.15
0500	Wood frame		247	.194			9.60	9.35	18.95	25.50

02 41 19.19 Selective Demolition

	Crew	Daily Output	Labor-Hours	Unit	Material	2021 Bare Costs Labor	Equipment	Total	Total Incl O&P
0010	**SELECTIVE DEMOLITION,** Rubbish Handling R024119-10								
0020	The following are to be added to the demolition prices								
0050	The following are components for a complete chute system								
0100	Top chute circular steel, 4' long, 18" diameter	B-1C	15	1.600	Ea.	272	72	19.50	363.50	435
0102	23" diameter		15	1.600		295	72	19.50	386.50	460
0104	27" diameter		15	1.600		320	72	19.50	411.50	485
0106	30" diameter		15	1.600		340	72	19.50	431.50	510
0108	33" diameter		15	1.600		365	72	19.50	456.50	535
0110	36" diameter		15	1.600		385	72	19.50	476.50	560
0112	Regular chute, 18" diameter		15	1.600		204	72	19.50	295.50	360
0114	23" diameter		15	1.600		227	72	19.50	318.50	385
0116	27" diameter		15	1.600		249	72	19.50	340.50	410
0118	30" diameter		15	1.600		261	72	19.50	352.50	425
0120	33" diameter		15	1.600		295	72	19.50	386.50	460
0122	36" diameter		15	1.600		320	72	19.50	411.50	485
0124	Control door chute, 18" diameter		15	1.600		385	72	19.50	476.50	560
0126	23" diameter		15	1.600		410	72	19.50	501.50	585
0128	27" diameter		15	1.600		430	72	19.50	521.50	610
0130	30" diameter		15	1.600		455	72	19.50	546.50	635
0132	33" diameter		15	1.600		475	72	19.50	566.50	660
0134	36" diameter		15	1.600		500	72	19.50	591.50	685
0136	Chute liners, 14 ga., 18"-30" diameter		15	1.600		209	72	19.50	300.50	365
0138	33"-36" diameter		15	1.600		261	72	19.50	352.50	425
0140	17% thinner chute, 30" diameter		15	1.600		227	72	19.50	318.50	385
0142	33% thinner chute, 30" diameter		15	1.600		170	72	19.50	261.50	325
0144	Top chute cover	1 Clab	24	.333		143	14.80		157.80	181
0146	Door chute cover	"	24	.333		143	14.80		157.80	181
0148	Top chute trough	2 Clab	12	1.333		475	59		534	620
0150	Bolt down frame & counter weights, 250 lb.	B-1	4	6		4,175	270		4,445	5,025
0152	500 lb.		4	6		6,175	270		6,445	7,225
0154	750 lb.		4	6		10,500	270		10,770	12,000
0156	1000 lb.		2.67	8.989		11,500	405		11,905	13,300
0158	1500 lb.		2.67	8.989		14,800	405		15,205	16,800
0160	Chute warning light system, 5 stories	B-1C	4	6		7,975	270	73	8,318	9,275
0162	10 stories	"	2	12		14,800	540	146	15,486	17,300
0164	Dust control device for dumpsters	1 Clab	8	1		166	44.50		210.50	254
0166	Install or replace breakaway cord		8	1		27	44.50		71.50	100
0168	Install or replace warning sign		16	.500		10	22		32	46.50
0600	Dumpster, weekly rental, 1 dump/week, 6 C.Y. capacity (2 tons)				Week	415			415	455
0700	10 C.Y. capacity (3 tons)					480			480	530
0725	20 C.Y. capacity (5 tons) R024119-20					565			565	625
0800	30 C.Y. capacity (7 tons)					730			730	800
0840	40 C.Y. capacity (10 tons)					775			775	850
0900	Alternate pricing for dumpsters									

02 41 19.19 Selective Demolition

02 41 19.19 Selective Demolition	Crew	Daily Output	Labor-Hours	Unit	Material	2021 Bare Costs Labor	Equipment	Total	Total Incl O&P
0910	Delivery, average for all sizes				Ea.	75			75	82.50
0920	Haul, average for all sizes					235			235	259
0930	Rent per day, average for all sizes					20			20	22
0940	Rent per month, average for all sizes					80			80	88
0950	Disposal fee per ton, average for all sizes				Ton	88			88	97
2000	Load, haul, dump and return, 0'-50' haul, hand carried	2 Clab	24	.667	C.Y.		29.50		29.50	47
2005	Wheeled		37	.432			19.20		19.20	30.50
2040	0'-100' haul, hand carried		16.50	.970			43		43	68.50
2045	Wheeled		25	.640			28.50		28.50	45.50
2050	Forklift	A-3R	25	.320			17.75	11.35	29.10	40
2080	Haul and return, add per each extra 100' haul, hand carried	2 Clab	35.50	.451			20		20	32
2085	Wheeled		54	.296			13.15		13.15	21
2120	For travel in elevators, up to 10 floors, add		140	.114			5.05		5.05	8.10
2130	0'-50' haul, incl. up to 5 riser stairs, hand carried		23	.696			31		31	49
2135	Wheeled		35	.457			20.50		20.50	32.50
2140	6-10 riser stairs, hand carried		22	.727			32.50		32.50	51.50
2145	Wheeled		34	.471			21		21	33.50
2150	11-20 riser stairs, hand carried		20	.800			35.50		35.50	56.50
2155	Wheeled		31	.516			23		23	36.50
2160	21-40 riser stairs, hand carried		16	1			44.50		44.50	70.50
2165	Wheeled	2 Clab	24	.667	C.Y.		29.50		29.50	47
2170	0'-100' haul, incl. 5 riser stairs, hand carried		15	1.067			47.50		47.50	75.50
2175	Wheeled		23	.696			31		31	49
2180	6-10 riser stairs, hand carried		14	1.143			50.50		50.50	81
2185	Wheeled		21	.762			34		34	54
2190	11-20 riser stairs, hand carried		12	1.333			59		59	94.50
2195	Wheeled		18	.889			39.50		39.50	63
2200	21-40 riser stairs, hand carried		8	2			89		89	141
2205	Wheeled		12	1.333			59		59	94.50
2210	Haul and return, add per each extra 100' haul, hand carried		35.50	.451			20		20	32
2215	Wheeled		54	.296			13.15		13.15	21
2220	For each additional flight of stairs, up to 5 risers, add		550	.029	Flight		1.29		1.29	2.06
2225	6-10 risers, add		275	.058			2.58		2.58	4.11
2230	11-20 risers, add		138	.116			5.15		5.15	8.20
2235	21-40 risers, add		69	.232			10.30		10.30	16.40
3000	Loading & trucking, including 2 mile haul, chute loaded	B-16	45	.711	C.Y.		33	12.85	45.85	67
3040	Hand loading truck, 50' haul	"	48	.667			31	12.05	43.05	63
3080	Machine loading truck	B-17	120	.267			13.05	5.20	18.25	26
3120	Wheeled 50' and ramp dump loaded	2 Clab	24	.667			29.50		29.50	47
5000	Haul, per mile, up to 8 C.Y. truck	B-34B	1165	.007			.35	.50	.85	1.11
5100	Over 8 C.Y. truck	"	1550	.005			.26	.37	.63	.83

02 41 19.20 Selective Demolition, Dump Charges

		Crew	Daily Output	Labor-Hours	Unit	Material	2021 Bare Costs Labor	Equipment	Total	Total Incl O&P
0010	SELECTIVE DEMOLITION, DUMP CHARGES R024119-10									
0020	Dump charges, typical urban city, tipping fees only									
0100	Building construction materials				Ton	74			74	81
0200	Trees, brush, lumber					63			63	69.50
0300	Rubbish only					63			63	69.50
0500	Reclamation station, usual charge					74			74	81

34

For customer support on your Commercial Renovation Costs with RSMeans Data, call 800.448.8182.

02 41 Demolition

02 41 19 – Selective Demolition

02 41 19.21 Selective Demolition, Gutting

		Crew	Daily Output	Labor-Hours	Unit	Material	2021 Bare Costs Labor	Equipment	Total	Total Incl O&P
0010	**SELECTIVE DEMOLITION, GUTTING** R024119-10									
0020	Building interior, including disposal, dumpster fees not included									
0500	Residential building									
0560	Minimum	B-16	400	.080	SF Flr.		3.73	1.45	5.18	7.55
0580	Maximum	"	360	.089	"		4.14	1.61	5.75	8.35
0900	Commercial building									
1000	Minimum	B-16	350	.091	SF Flr.		4.26	1.65	5.91	8.60
1020	Maximum		250	.128	"		5.95	2.32	8.27	12.05
3000	Minimum labor/equipment charge		4	8	Job		375	145	520	755

02 41 19.25 Selective Demolition, Saw Cutting

		Crew	Daily Output	Labor-Hours	Unit	Material	2021 Bare Costs Labor	Equipment	Total	Total Incl O&P
0010	**SELECTIVE DEMOLITION, SAW CUTTING** R024119-10									
0012	For concrete saw cutting, see Section 03 81									
0015	Asphalt, up to 3" deep	B-89	1050	.015	L.F.	.11	.79	1.02	1.92	2.49
0020	Each additional inch of depth	B-89	1800	.009	L.F.	.04	.46	.59	1.09	1.42
1200	Masonry walls, hydraulic saw, brick, per inch of depth	B-89B	300	.053		.04	2.78	5.20	8.02	10.15
1220	Block walls, solid, per inch of depth	"	250	.064		.04	3.34	6.25	9.63	12.15
2000	Brick or masonry w/hand held saw, per inch of depth	A-1	125	.064		.05	2.84	.91	3.80	5.55
5000	Wood sheathing to 1" thick, on walls	1 Carp	200	.040			2.19		2.19	3.48
5020	On roof	"	250	.032			1.75		1.75	2.79
9000	Minimum labor/equipment charge	A-1	2	4	Job		178	57	235	345

02 41 19.27 Selective Demolition, Torch Cutting

		Crew	Daily Output	Labor-Hours	Unit	Material	2021 Bare Costs Labor	Equipment	Total	Total Incl O&P
0010	**SELECTIVE DEMOLITION, TORCH CUTTING** R024119-10									
0020	Steel, 1" thick plate	E-25	333	.024	L.F.	2.23	1.50	.04	3.77	4.91
0040	1" diameter bar	"	600	.013	Ea.	.37	.83	.02	1.22	1.77
1000	Oxygen lance cutting, reinforced concrete walls									
1040	12"-16" thick walls	1 Clab	10	.800	L.F.		35.50		35.50	56.50
1080	24" thick walls	"	6	1.333	"		59		59	94.50
1090	Minimum labor/equipment charge	E-25	2	4	Job		249	6.50	255.50	410
1100	See Section 05 05 21.10									

02 42 Removal and Diversion of Construction Materials

02 42 10 – Building Deconstruction

02 42 10.10 Estimated Salvage Value or Savings

			Crew	Daily Output	Labor-Hours	Unit	Material	2021 Bare Costs Labor	Equipment	Total	Total Incl O&P
0010	**ESTIMATED SALVAGE VALUE OR SAVINGS**										
0015	Excludes material handling, packaging, container costs and										
0020	transportation for salvage or disposal										
0050	All items in Section 02 42 10.10 are credit deducts and not costs										
0100	Copper wire salvage value	G				Lb.	1.70			1.70	1.87
0200	Brass salvage value	G					1.25			1.25	1.38
0250	Bronze						1.45			1.45	1.60
0300	Steel	G					.05			.05	.06
0304	304 Stainless Steel						.30			.30	.33
0316	316 Stainless Steel						.45			.45	.50
0400	Cast Iron	G					.06			.06	.07
0600	Aluminum siding	G					.33			.33	.36
0610	Aluminum wire						.47			.47	.52

02 42 10.20 Deconstruction of Building Components		Crew	Daily Output	Labor-Hours	Unit	Material	2021 Bare Costs Labor	2021 Bare Costs Equipment	Total	Total Incl O&P
0010	**DECONSTRUCTION OF BUILDING COMPONENTS**									
0012	Buildings one or two stories only									
0015	Excludes material handling, packaging, container costs and									
0020	transportation for salvage or disposal									
0050	Deconstruction of plumbing fixtures									
0100	Wall hung or countertop lavatory	G	2 Clab	16	1	Ea.		44.50		44.50	70.50
0110	Single or double compartment kitchen sink	G		14	1.143			50.50		50.50	81
0120	Wall hung urinal	G		14	1.143			50.50		50.50	81
0130	Floor mounted	G	2 Clab	8	2	Ea.		89		89	141
0140	Floor mounted water closet	G		16	1			44.50		44.50	70.50
0150	Wall hung	G		14	1.143			50.50		50.50	81
0160	Water fountain, free standing	G		16	1			44.50		44.50	70.50
0170	Wall hung or deck mounted	G		12	1.333			59		59	94.50
0180	Bathtub, steel or fiberglass	G		10	1.600			71		71	113
0190	Cast iron	G		8	2			89		89	141
0200	Shower, single	G		6	2.667			118		118	189
0210	Group	G		7	2.286			101		101	162
0300	Deconstruction of electrical fixtures									
0310	Surface mounted incandescent fixtures	G	2 Clab	48	.333	Ea.		14.80		14.80	23.50
0320	Fluorescent, 2 lamp	G		32	.500			22		22	35.50
0330	4 lamp	G		24	.667			29.50		29.50	47
0340	Strip fluorescent, 1 lamp	G		40	.400			17.75		17.75	28.50
0350	2 lamp	G		32	.500			22		22	35.50
0400	Recessed drop-in fluorescent fixture, 2 lamp	G		27	.593			26.50		26.50	42
0410	4 lamp	G		18	.889			39.50		39.50	63
0500	Deconstruction of appliances									
0510	Cooking stoves	G	2 Clab	26	.615	Ea.		27.50		27.50	43.50
0520	Dishwashers	G	"	26	.615	"		27.50		27.50	43.50
0600	Deconstruction of millwork and trim									
0610	Cabinets, wood	G	2 Carp	40	.400	L.F.		22		22	35
0620	Countertops	G		100	.160	"		8.75		8.75	13.95
0630	Wall paneling, 1" thick	G		500	.032	S.F.		1.75		1.75	2.79
0640	Ceiling trim	G		500	.032	L.F.		1.75		1.75	2.79
0650	Wainscoting	G		500	.032	S.F.		1.75		1.75	2.79
0660	Base, 3/4"-1" thick	G		600	.027	L.F.		1.46		1.46	2.32
0700	Deconstruction of doors and windows									
0710	Doors, wrap, interior, wood, single, no closers	G	2 Carp	21	.762	Ea.	4	41.50		45.50	71
0720	Double	G		13	1.231		8	67.50		75.50	116
0730	Solid core, single, exterior or interior	G		10	1.600		4	87.50		91.50	143
0740	Double	G		8	2		8	109		117	183
0810	Windows, wrap, wood, single									
0812	with no casement or cladding	G	2 Carp	21	.762	Ea.	4	41.50		45.50	71
0820	with casement and/or cladding	G	"	18	.889	"	4	48.50		52.50	82
0900	Deconstruction of interior finishes									
0910	Drywall for recycling	G	2 Clab	1775	.009	S.F.		.40		.40	.64
0920	Plaster wall, first floor	G		1775	.009			.40		.40	.64
0930	Second floor	G		1330	.012			.53		.53	.85
1000	Deconstruction of roofing and accessories									
1010	Built-up roofs	G	2 Clab	570	.028	S.F.		1.25		1.25	1.98
1020	Gutters, fascia and rakes	G	"	1140	.014	L.F.		.62		.62	.99
2000	Deconstruction of wood components									
2010	Roof sheeting	G	2 Clab	570	.028	S.F.		1.25		1.25	1.98

36

02 42 10 – Building Deconstruction

02 42 10.20 Deconstruction of Building Components		Crew	Daily Output	Labor-Hours	Unit	Material	2021 Bare Costs Labor	Equipment	Total	Total Incl O&P
2020	Main roof framing	G	2 Clab	760	.021	L.F.		.93		.93	1.49
2030	Porch roof framing	G	↓	445	.036			1.60		1.60	2.54
2040	Beams 4" x 8"	G	B-1	375	.064			2.88		2.88	4.59
2050	4" x 10"	G		300	.080			3.61		3.61	5.75
2055	4" x 12"	G		250	.096			4.33		4.33	6.90
2060	6" x 8"	G		250	.096			4.33		4.33	6.90
2065	6" x 10"	G		200	.120			5.40		5.40	8.60
2070	6" x 12"	G		170	.141			6.35		6.35	10.15
2075	8" x 12"	G		126	.190			8.60		8.60	13.65
2080	10" x 12"	G	↓	100	.240			10.80		10.80	17.20
2100	Ceiling joists	G	2 Clab	800	.020			.89		.89	1.41
2150	Wall framing, interior	G		1230	.013	↓		.58		.58	.92
2160	Sub-floor	G		2000	.008	S.F.		.36		.36	.57
2170	Floor joists	G		2000	.008	L.F.		.36		.36	.57
2200	Wood siding (no lead or asbestos)	G		1300	.012	S.F.		.55		.55	.87
2300	Wall framing, exterior	G		1600	.010	L.F.		.44		.44	.71
2400	Stair risers	G		53	.302	Ea.		13.40		13.40	21.50
2500	Posts	G	↓	800	.020	L.F.		.89		.89	1.41
3000	Deconstruction of exterior brick walls										
3010	Exterior brick walls, first floor	G	2 Clab	200	.080	S.F.		3.55		3.55	5.65
3020	Second floor	G		64	.250	"		11.10		11.10	17.70
3030	Brick chimney	G	↓	100	.160	C.F.		7.10		7.10	11.30
4000	Deconstruction of concrete										
4010	Slab on grade, 4" thick, plain concrete	G	B-9	500	.080	S.F.		3.58	.71	4.29	6.50
4020	Wire mesh reinforced	G		470	.085			3.81	.76	4.57	6.90
4030	Rod reinforced	G		400	.100			4.48	.89	5.37	8.15
4110	Foundation wall, 6" thick, plain concrete	G		160	.250			11.20	2.22	13.42	20.50
4120	8" thick	G		140	.286			12.80	2.54	15.34	23.50
4130	10" thick	G	↓	120	.333	↓		14.95	2.96	17.91	27.50
9000	Deconstruction process, support equipment as needed										
9010	Daily use, portal to portal, 12-ton truck-mounted hydraulic crane crew	G	A-3H	1	8	Day		490	735	1,225	1,575
9020	Daily use, skid steer and operator	G	A-3C	1	8			445	445	890	1,175
9030	Daily use, backhoe 48 HP, operator and labor	G	"	1	8	↓		445	445	890	1,175

02 42 10.30 Deconstruction Material Handling

			Crew	Daily Output	Labor-Hours	Unit	Material	Labor	Equipment	Total	Total Incl O&P
0010	**DECONSTRUCTION MATERIAL HANDLING**										
0012	Buildings one or two stories only										
0100	Clean and stack brick on pallet	G	2 Clab	1200	.013	Ea.		.59		.59	.94
0200	Haul 50' and load rough lumber up to 2" x 8" size	G	2 Clab	2000	.008	Ea.		.36		.36	.57
0210	Lumber larger than 2" x 8"	G		3200	.005	B.F.		.22		.22	.35
0300	Finish wood for recycling stack and wrap per pallet	G		8	2	Ea.	32	89		121	176
0350	Light fixtures			6	2.667		57.50	118		175.50	253
0375	Windows			6	2.667		54.50	118		172.50	249
0400	Miscellaneous materials		↓	8	2	↓	16	89		105	159
1000	See Section 02 41 19.19 for bulk material handling										

37

02 43 Structure Moving

02 43 13 – Structure Relocation

02 43 13.13 Building Relocation		Crew	Daily Output	Labor-Hours	Unit	Material	2021 Bare Costs Labor	Equipment	Total	Total Incl O&P
0010	**BUILDING RELOCATION**									
0011	One day move, up to 24' wide									
0020	Reset on existing foundation				Total				11,500	11,500
0040	Wood or steel frame bldg., based on ground floor area	G	B-4	185	.259	S.F.		11.90	2.70	14.60	22
0060	Masonry bldg., based on ground floor area	G	"	137	.350			16.05	3.64	19.69	29.50
0200	For 24'-42' wide, add							15%		15%

02 58 Snow Control

02 58 13 – Snow Fencing

02 58 13.10 Snow Fencing System

		Crew	Daily Output	Labor-Hours	Unit	Material	Labor	Equipment	Total	Total Incl O&P
0010	**SNOW FENCING SYSTEM**									
7001	Snow fence on steel posts 10' OC, 4' high	B-1	500	.048	L.F.	.39	2.16		2.55	3.87

02 65 Underground Storage Tank Removal

02 65 10 – Underground Tank and Contaminated Soil Removal

02 65 10.30 Removal of Underground Storage Tanks

			Crew	Daily Output	Labor-Hours	Unit	Material	2021 Bare Costs Labor	Equipment	Total	Total Incl O&P
0010	**REMOVAL OF UNDERGROUND STORAGE TANKS**	R026510-20									
0011	Petroleum storage tanks, non-leaking										
0100	Excavate & load onto trailer										
0110	3,000 gal. to 5,000 gal. tank	G	B-14	4	12	Ea.		560	54	614	945
0120	6,000 gal. to 8,000 gal. tank	G	B-3A	3	13.333			630	232	862	1,250
0130	9,000 gal. to 12,000 gal. tank	G	"	2	20			945	350	1,295	1,875
0190	Known leaking tank, add					%				100%	100%
0200	Remove sludge, water and remaining product from bottom										
0201	of tank with vacuum truck										
0300	3,000 gal. to 5,000 gal. tank	G	A-13	5	1.600	Ea.		89	149	238	305
0310	6,000 gal. to 8,000 gal. tank	G		4	2			111	187	298	380
0320	9,000 gal. to 12,000 gal. tank	G		3	2.667			148	249	397	505
0390	Dispose of sludge off-site, average					Gal.				6.25	6.80
0400	Insert inert solid CO_2 "dry ice" into tank										
0401	For cleaning/transporting tanks (1.5 lb./100 gal. cap)	G	1 Clab	500	.016	Lb.	1	.71		1.71	2.23
0503	Disconnect and remove piping	G	1 Plum	160	.050	L.F.		3.39		3.39	5.20
0603	Transfer liquids, 10% of volume	G	"	1600	.005	Gal.		.34		.34	.52
0703	Cut accessway into underground storage tank	G	1 Clab	5.33	1.501	Ea.		66.50		66.50	106
0813	Remove sludge, wash and wipe tank, 500 gal.	G	1 Plum	8	1			67.50		67.50	104
0823	3,000 gal.	G		6.67	1.199			81		81	125
0833	5,000 gal.	G		6.15	1.301			88		88	136
0843	8,000 gal.	G		5.33	1.501			102		102	157
0853	10,000 gal.	G		4.57	1.751			119		119	183
0863	12,000 gal.	G		4.21	1.900			129		129	198
1020	Haul tank to certified salvage dump, 100 miles round trip										
1023	3,000 gal. to 5,000 gal. tank					Ea.				760	830
1026	6,000 gal. to 8,000 gal. tank									880	960
1029	9,000 gal. to 12,000 gal. tank									1,050	1,150
1100	Disposal of contaminated soil to landfill										
1110	Minimum					C.Y.				145	160
1111	Maximum					"				400	440
1120	Disposal of contaminated soil to										
1121	bituminous concrete batch plant										

02 65 Underground Storage Tank Removal

02 65 10 – Underground Tank and Contaminated Soil Removal

02 65 10.30 Removal of Underground Storage Tanks		Crew	Daily Output	Labor-Hours	Unit	Material	2021 Bare Costs Labor	2021 Bare Costs Equipment	Total	Total Incl O&P
1130	Minimum				C.Y.				80	88
1131	Maximum				"				115	125
1203	Excavate, pull & load tank, backfill hole, 8,000 gal. +	G	B-12C	.50	32	Ea.		1,700	1,875	3,575	4,750
1213	Haul tank to certified dump, 100 miles rt, 8,000 gal. +	G	B-34K	1	8			410	865	1,275	1,600
1223	Excavate, pull & load tank, backfill hole, 500 gal.	G	B-11C	1	16			825	216	1,041	1,550
1233	Excavate, pull & load tank, backfill hole, 3,000-5,000 gal.	G	B-11M	.50	32			1,650	470	2,120	3,125
1243	Haul tank to certified dump, 100 miles rt, 500 gal.	G	B-34L	1	8			445	198	643	915
1253	Haul tank to certified dump, 100 miles rt, 3,000-5,000 gal.	G	B-34M	1	8			445	850	1,295	1,625
2010	Decontamination of soil on site incl poly tarp on top/bottom										
2011	Soil containment berm and chemical treatment										
2020	Minimum	G	B-11C	100	.160	C.Y.	6.25	8.25	2.16	16.66	22.50
2021	Maximum	G	"	100	.160		8.10	8.25	2.16	18.51	24.50
2050	Disposal of decontaminated soil, minimum									135	150
2055	Maximum									400	440

02 81 Transportation and Disposal of Hazardous Materials

02 81 20 – Hazardous Waste Handling

02 81 20.10 Hazardous Waste Cleanup/Pickup/Disposal

		Crew	Daily Output	Labor-Hours	Unit	Material	Labor	Equipment	Total	Total Incl O&P
0010	**HAZARDOUS WASTE CLEANUP/PICKUP/DISPOSAL**									
0100	For contractor rental equipment, i.e., dozer,									
0110	Front end loader, dump truck, etc., see 01 54 33 Reference Section									
1000	Solid pickup									
1100	55 gal. drums				Ea.				240	265
1120	Bulk material, minimum				Ton				190	210
1130	Maximum				Ton				595	655
1200	Transportation to disposal site									
1220	Truckload = 80 drums or 25 C.Y. or 18 tons									
1260	Minimum				Mile				3.95	4.45
1270	Maximum				"				7.25	7.98
3000	Liquid pickup, vacuum truck, stainless steel tank									
3100	Minimum charge, 4 hours				Hr.				140	155
3110	1 compartment, 2200 gallon				"				200	225
3120	2 compartment, 5000 gallon				Mile				7.95	8.75
3400	Transportation in 6900 gallon bulk truck				"				10.20	11.25
3410	In teflon lined truck				Hr.				140	155
5000	Heavy sludge or dry vacuumable material				Ton				140	155
6000	Dumpsite disposal charge, minimum				"				415	455
6020	Maximum									

02 82 Asbestos Remediation

02 82 13 – Asbestos Abatement

02 82 13.39 Asbestos Remediation Plans and Methods

		Crew	Daily Output	Labor-Hours	Unit	Material	Labor	Equipment	Total	Total Incl O&P
0010	**ASBESTOS REMEDIATION PLANS AND METHODS**									
0100	Building Survey-Commercial Building				Ea.				2,200	2,400
0200	Asbestos Abatement Remediation Plan				"				1,350	1,475

For customer support on your Commercial Renovation Costs with RSMeans Data, call 800.448.8182.

39

02 82 13.41 Asbestos Abatement Equipment

	Crew	Daily Output	Labor-Hours	Unit	Material	2021 Bare Costs Labor	2021 Bare Costs Equipment	Total	Total Incl O&P
0010	**ASBESTOS ABATEMENT EQUIPMENT** R028213-20								
0011	Equipment and supplies, buy								
0200	Air filtration device, 2000 CFM			Ea.	965			965	1,050
0250	Large volume air sampling pump, minimum				350			350	385
0260	Maximum				320			320	350
0300	Airless sprayer unit, 2 gun				2,100			2,100	2,325
0350	Light stand, 500 watt				34			34	37.50
0400	Personal respirators								
0410	Negative pressure, 1/2 face, dual operation, minimum			Ea.	28.50			28.50	31.50
0420	Maximum				26.50			26.50	29
0450	P.A.P.R., full face, minimum				680			680	745
0460	Maximum				1,225			1,225	1,325
0470	Supplied air, full face, including air line, minimum				289			289	320
0480	Maximum				470			470	515
0500	Personnel sampling pump				210			210	231
1500	Power panel, 20 unit, including GFI				500			500	550
1600	Shower unit, including pump and filters				1,225			1,225	1,350
1700	Supplied air system (type C)				4,125			4,125	4,525
1750	Vacuum cleaner, HEPA, 16 gal., stainless steel, wet/dry				1,300			1,300	1,450
1760	55 gallon			Ea.	1,500			1,500	1,650
1800	Vacuum loader, 9-18 ton/hr.				100,000			100,000	110,000
1900	Water atomizer unit, including 55 gal. drum				310			310	340
2000	Worker protection, whole body, foot, head cover & gloves, plastic				8.90			8.90	9.80
2500	Respirator, single use				27.50			27.50	30.50
2550	Cartridge for respirator				4.20			4.20	4.62
2570	Glove bag, 7 mil, 50" x 64"				8.75			8.75	9.60
2580	10 mil, 44" x 60"				6.85			6.85	7.50
6000	Disposable polyethylene bags, 6 mil, 3 C.F.				1.23			1.23	1.35
6300	Disposable fiber drums, 3 C.F.				19.80			19.80	22
6400	Pressure sensitive caution labels, 3" x 5"				3.46			3.46	3.81
6450	11" x 17"				7.80			7.80	8.55
6500	Negative air machine, 1800 CFM				900			900	990

02 82 13.42 Preparation of Asbestos Containment Area

	Crew	Daily Output	Labor-Hours	Unit	Material	2021 Bare Costs Labor	2021 Bare Costs Equipment	Total	Total Incl O&P
0010	**PREPARATION OF ASBESTOS CONTAINMENT AREA**								
0100	Pre-cleaning, HEPA vacuum and wet wipe, flat surfaces	A-9	12000	.005	S.F.	.01	.33		.34	.53
0200	Protect carpeted area, 2 layers 6 mil poly on 3/4" plywood	"	1000	.064		2.16	3.90		6.06	8.55
0300	Separation barrier, 2" x 4" @ 16", 1/2" plywood ea. side, 8' high	2 Carp	400	.040		4.07	2.19		6.26	7.95
0310	12' high		320	.050		3.98	2.74		6.72	8.75
0320	16' high		200	.080		3.94	4.38		8.32	11.30
0400	Personnel decontam. chamber, 2" x 4" @ 16", 3/4" ply ea. side		280	.057		4.74	3.13		7.87	10.20
0450	Waste decontam. chamber, 2" x 4" studs @ 16", 3/4" ply ea. side		360	.044		4.74	2.43		7.17	9.05
0500	Cover surfaces with polyethylene sheeting								
0501	Including glue and tape								
0550	Floors, each layer, 6 mil	A-9	8000	.008	S.F.	.08	.49		.57	.86
0551	4 mil		9000	.007		.03	.43		.46	.72
0560	Walls, each layer, 6 mil		6000	.011		.08	.65		.73	1.12
0561	4 mil		7000	.009		.03	.56		.59	.91
0570	For heights above 14', add						20%			
0575	For heights above 20', add						30%			
0580	For fire retardant poly, add					100%				
0590	For large open areas, deduct					10%	20%			
0600	Seal floor penetrations with foam firestop to 36 sq. in.	2 Carp	200	.080	Ea.	13.45	4.38		17.83	22

For customer support on your Commercial Renovation Costs with RSMeans Data, call 800.448.8182.

02 82 13 – Asbestos Abatement

02 82 13.42 Preparation of Asbestos Containment Area

		Crew	Daily Output	Labor-Hours	Unit	Material	2021 Bare Costs Labor	2021 Bare Costs Equipment	Total	Total Incl O&P
	02 82 13.42 Preparation of Asbestos Containment Area	2 Carp	125	.128	Ea.	27	7		34	40.50
0610	36 sq. in. to 72 sq. in.		80	.200		54	10.95		64.95	76.50
0615	72 sq. in. to 144 sq. in.		180	.089		13.45	4.86		18.31	22.50
0620	Wall penetrations, to 36 sq. in.		100	.160		27	8.75		35.75	43.50
0630	36 sq. in. to 72 sq. in.		60	.267		54	14.60		68.60	82
0640	72 sq. in. to 144 sq. in.									
0800	Caulk seams with latex	1 Carp	230	.035	L.F.	.19	1.90		2.09	3.24
0900	Set up neg. air machine, 1-2k CFM/25 M.C.F. volume	1 Asbe	4.30	1.860	Ea.		113		113	179
0950	Set up and remove portable shower unit	2 Asbe	4	4	"		244		244	385

02 82 13.43 Bulk Asbestos Removal

		Crew	Daily Output	Labor-Hours	Unit	Material	2021 Bare Costs Labor	2021 Bare Costs Equipment	Total	Total Incl O&P
0010	**BULK ASBESTOS REMOVAL**									
0020	Includes disposable tools and 2 suits and 1 respirator filter/day/worker									
0100	Beams, W 10 x 19	A-9	235	.272	L.F.	.75	16.60		17.35	27.50
0110	W 12 x 22		210	.305		.84	18.60		19.44	30.50
0120	W 14 x 26		180	.356		.98	21.50		22.48	35.50
0130	W 16 x 31		160	.400		1.10	24.50		25.60	39.50
0140	W 18 x 40		140	.457		1.26	28		29.26	45.50
0150	W 24 x 55		110	.582		1.60	35.50		37.10	58
0160	W 30 x 108		85	.753		2.07	46		48.07	75
0170	W 36 x 150		72	.889		2.44	54		56.44	88
0200	Boiler insulation		480	.133	S.F.	.43	8.15		8.58	13.30
0210	With metal lath, add				%				50%	50%
0300	Boiler breeching or flue insulation	A-9	520	.123	S.F.	.34	7.50		7.84	12.20
0310	For active boiler, add				%				100%	100%
0400	Duct or AHU insulation	A-10B	440	.073	S.F.	.20	4.44		4.64	7.20
0500	Duct vibration isolation joints, up to 24 sq. in. duct	A-9	56	1.143	Ea.	3.14	69.50		72.64	113
0520	25 sq. in. to 48 sq. in. duct		48	1.333		3.67	81.50		85.17	133
0530	49 sq. in. to 76 sq. in. duct		40	1.600		4.40	97.50		101.90	159
0600	Pipe insulation, air cell type, up to 4" diameter pipe		900	.071	L.F.	.20	4.34		4.54	7.05
0610	4" to 8" diameter pipe		800	.080		.22	4.88		5.10	7.95
0620	10" to 12" diameter pipe		700	.091		.25	5.60		5.85	9.10
0630	14" to 16" diameter pipe		550	.116		.32	7.10		7.42	11.55
0650	Over 16" diameter pipe		650	.098	S.F.	.27	6		6.27	9.80
0700	With glove bag up to 3" diameter pipe		200	.320	L.F.	9.95	19.50		29.45	42
1000	Pipe fitting insulation up to 4" diameter pipe		320	.200	Ea.	.55	12.20		12.75	19.90
1100	6" to 8" diameter pipe		304	.211		.58	12.85		13.43	21
1110	10" to 12" diameter pipe		192	.333		.92	20.50		21.42	33
1120	14" to 16" diameter pipe		128	.500		1.38	30.50		31.88	49.50
1130	Over 16" diameter pipe		176	.364	S.F.	1	22		23	36
1200	With glove bag, up to 8" diameter pipe		75	.853	L.F.	6.55	52		58.55	89.50
2000	Scrape foam fireproofing from flat surface		2400	.027	S.F.	.07	1.63		1.70	2.65
2100	Irregular surfaces		1200	.053		.15	3.25		3.40	5.30
3000	Remove cementitious material from flat surface		1800	.036		.10	2.17		2.27	3.54
3100	Irregular surface		1000	.064		.13	3.90		4.03	6.30
4000	Scrape acoustical coating/fireproofing, from ceiling		3200	.020		.06	1.22		1.28	1.99
5000	Remove VAT and mastic from floor by hand		2400	.027		.07	1.63		1.70	2.65
5100	By machine	A-11	4800	.013		.04	.81	.01	.86	1.35
5150	For 2 layers, add				%				50%	50%
6000	Remove contaminated soil from crawl space by hand	A-9	400	.160	C.F.	.44	9.75		10.19	15.90
6100	With large production vacuum loader	A-12	700	.091	"	.25	5.60	1.07	6.92	10.25
7000	Radiator backing, not including radiator removal	A-9	1200	.053	S.F.	.15	3.25		3.40	5.30
8000	Cement-asbestos transite board and cement wall board	2 Asbe	1000	.016		.16	.98		1.14	1.72
8100	Transite shingle siding	A-10B	750	.043		.24	2.61		2.85	4.38

41

02 82 Asbestos Remediation

02 82 13 – Asbestos Abatement

02 82 13.43 Bulk Asbestos Removal	Crew	Daily Output	Labor-Hours	Unit	Material	2021 Bare Costs Labor	2021 Bare Costs Equipment	Total	Total Incl O&P
8200	Shingle roofing	A-10B	2000	.016	S.F.	.07	.98		1.05	1.62
8250	Built-up, no gravel, non-friable	B-2	1400	.029		.07	1.28		1.35	2.12
8260	Bituminous flashing	1 Rofc	300	.027		.07	1.29		1.36	2.29
8300	Asbestos millboard, flat board and VAT contaminated plywood	2 Asbe	1000	.016	↓	.07	.98		1.05	1.62
9000	For type B (supplied air) respirator equipment, add				%				10%	10%

02 82 13.44 Demolition In Asbestos Contaminated Area

	02 82 13.44 Demolition In Asbestos Contaminated Area	Crew	Daily Output	Labor-Hours	Unit	Material	2021 Bare Costs Labor	Equipment	Total	Total Incl O&P
0010	**DEMOLITION IN ASBESTOS CONTAMINATED AREA**									
0200	Ceiling, including suspension system, plaster and lath	A-9	2100	.030	S.F.	.08	1.86		1.94	3.03
0210	Finished plaster, leaving wire lath		585	.109		.30	6.65		6.95	10.90
0220	Suspended acoustical tile		3500	.018		.05	1.12		1.17	1.82
0230	Concealed tile grid system		3000	.021		.06	1.30		1.36	2.12
0240	Metal pan grid system		1500	.043		.12	2.60		2.72	4.24
0250	Gypsum board		2500	.026	↓	.07	1.56		1.63	2.55
0260	Lighting fixtures up to 2' x 4'	↓	72	.889	Ea.	2.44	54		56.44	88
0400	Partitions, non load bearing									
0410	Plaster, lath, and studs	A-9	690	.093	S.F.	.86	5.65		6.51	9.90
0450	Gypsum board and studs	"	1390	.046	"	.13	2.81		2.94	4.58
9000	For type B (supplied air) respirator equipment, add				%				10%	10%

02 82 13.45 OSHA Testing

	02 82 13.45 OSHA Testing	Crew	Daily Output	Labor-Hours	Unit	Material	2021 Bare Costs Labor	Equipment	Total	Total Incl O&P
0010	**OSHA TESTING**									
0100	Certified technician, minimum				Day		340		340	340
0110	Maximum						670		670	670
0121	Industrial hygienist, minimum						385		385	385
0130	Maximum				↓		755		755	755
0200	Asbestos sampling and PCM analysis, NIOSH 7400, minimum	1 Asbe	8	1	Ea.	15.85	61		76.85	114
0210	Maximum		4	2		27	122		149	223
1000	Cleaned area samples		8	1		185	61		246	300
1100	PCM air sample analysis, NIOSH 7400, minimum		8	1		17.05	61		78.05	115
1110	Maximum		4	2		20	122		142	215
1200	TEM air sample analysis, NIOSH 7402, minimum	↓							80	106
1210	Maximum				↓				360	450

02 82 13.46 Decontamination of Asbestos Containment Area

	02 82 13.46 Decontamination of Asbestos Containment Area	Crew	Daily Output	Labor-Hours	Unit	Material	2021 Bare Costs Labor	Equipment	Total	Total Incl O&P
0010	**DECONTAMINATION OF ASBESTOS CONTAINMENT AREA**									
0100	Spray exposed substrate with surfactant (bridging)									
0200	Flat surfaces	A-9	6000	.011	S.F.	.42	.65		1.07	1.49
0250	Irregular surfaces	A-9	4000	.016	S.F.	.45	.98		1.43	2.04
0300	Pipes, beams, and columns		2000	.032	L.F.	.61	1.95		2.56	3.75
1000	Spray encapsulate polyethylene sheeting		8000	.008	S.F.	.42	.49		.91	1.23
1100	Roll down polyethylene sheeting		8000	.008	"		.49		.49	.77
1500	Bag polyethylene sheeting		400	.160	Ea.	1.05	9.75		10.80	16.55
2000	Fine clean exposed substrate, with nylon brush		2400	.027	S.F.		1.63		1.63	2.57
2500	Wet wipe substrate		4800	.013			.81		.81	1.29
2600	Vacuum surfaces, fine brush		6400	.010	↓		.61		.61	.96
3000	Structural demolition									
3100	Wood stud walls	A-9	2800	.023	S.F.		1.39		1.39	2.20
3500	Window manifolds, not incl. window replacement		4200	.015			.93		.93	1.47
3600	Plywood carpet protection	↓	2000	.032	↓		1.95		1.95	3.08
4000	Remove custom decontamination facility	A-10A	8	3	Ea.	15.45	183		198.45	305
4100	Remove portable decontamination facility	3 Asbe	12	2	"	13.25	122		135.25	208
5000	HEPA vacuum, shampoo carpeting	A-9	4800	.013	S.F.	.11	.81		.92	1.41
9000	Final cleaning of protected surfaces	A-10A	8000	.003	"		.18		.18	.29

For customer support on your Commercial Renovation Costs with RSMeans Data, call 800.448.8182.

02 82 Asbestos Remediation

02 82 13 – Asbestos Abatement

02 82 13.47 Asbestos Waste Pkg., Handling, and Disp.

		Crew	Daily Output	Labor-Hours	Unit	Material	2021 Bare Costs Labor	Equipment	Total	Total Incl O&P
0010	**ASBESTOS WASTE PACKAGING, HANDLING, AND DISPOSAL**									
0100	Collect and bag bulk material, 3 C.F. bags, by hand	A-9	400	.160	Ea.	1.23	9.75		10.98	16.75
0200	Large production vacuum loader	A-12	880	.073		1.04	4.44	.85	6.33	9.05
1000	Double bag and decontaminate	A-9	960	.067		1.23	4.07		5.30	7.80
2000	Containerize bagged material in drums, per 3 C.F. drum	"	800	.080		19.80	4.88		24.68	29.50
3000	Cart bags 50' to dumpster	2 Asbe	400	.040	↓		2.44		2.44	3.85
5000	Disposal charges, not including haul, minimum				C.Y.				61	67
5020	Maximum				"				355	395
9000	For type B (supplied air) respirator equipment, add				%				10%	10%

02 82 13.48 Asbestos Encapsulation With Sealants

		Crew	Daily Output	Labor-Hours	Unit	Material	2021 Bare Costs Labor	Equipment	Total	Total Incl O&P
0010	**ASBESTOS ENCAPSULATION WITH SEALANTS**									
0100	Ceilings and walls, minimum	A-9	21000	.003	S.F.	.44	.19		.63	.77
0110	Maximum		10600	.006		.53	.37		.90	1.16
0200	Columns and beams, minimum		13300	.005		.44	.29		.73	.94
0210	Maximum		5325	.012	↓	.65	.73		1.38	1.88
0300	Pipes to 12" diameter including minor repairs, minimum		800	.080	L.F.	.59	4.88		5.47	8.35
0310	Maximum	↓	400	.160	"	1.27	9.75		11.02	16.80

02 83 Lead Remediation

02 83 19 – Lead-Based Paint Remediation

02 83 19.21 Lead Paint Remediation Plans and Methods

		Crew	Daily Output	Labor-Hours	Unit	Material	2021 Bare Costs Labor	Equipment	Total	Total Incl O&P
0010	**LEAD PAINT REMEDIATION PLANS AND METHODS**									
0100	Building Survey-Commercial Building				Ea.				2,050	2,250
0200	Lead Abatement Remediation Plan								1,225	1,350
0300	Lead Paint Testing, AAS Analysis				↓				51	56
0400	Lead Paint Testing, X-Ray Fluorescence				Ea.				51	56

02 83 19.22 Preparation of Lead Containment Area

		Crew	Daily Output	Labor-Hours	Unit	Material	2021 Bare Costs Labor	Equipment	Total	Total Incl O&P
0010	**PREPARATION OF LEAD CONTAINMENT AREA**									
0020	Lead abatement work area, test kit, per swab	1 Skwk	16	.500	Ea.	3.40	28.50		31.90	49
0025	For dust barriers see Section 01 56 16.10									
0050	Caution sign	1 Skwk	48	.167	Ea.	7.30	9.50		16.80	23
0100	Pre-cleaning, HEPA vacuum and wet wipe, floor and wall surfaces	3 Skwk	5000	.005	S.F.	.01	.27		.28	.44
0105	Ceiling, 6'-11' high		4100	.006		.08	.33		.41	.61
0108	12'-15' high		3550	.007		.07	.39		.46	.69
0115	Over 15' high	↓	3000	.008	↓	.11	.46		.57	.84
0500	Cover surfaces with polyethylene sheeting									
0550	Floors, each layer, 6 mil	3 Skwk	8000	.003	S.F.	.08	.17		.25	.36
0560	Walls, each layer, 6 mil	"	6000	.004	"	.08	.23		.31	.45
0570	For heights above 14', add						20%			
2400	Post abatement cleaning of protective sheeting, HEPA vacuum & wet wipe	3 Skwk	5000	.005	S.F.	.01	.27		.28	.44
2450	Doff, bag and seal protective sheeting		12000	.002		.01	.11		.12	.19
2500	Post abatement cleaning, HEPA vacuum & wet wipe	↓	5000	.005	↓	.01	.27		.28	.44

02 83 19.23 Encapsulation of Lead-Based Paint

		Crew	Daily Output	Labor-Hours	Unit	Material	2021 Bare Costs Labor	Equipment	Total	Total Incl O&P
0010	**ENCAPSULATION OF LEAD-BASED PAINT**									
0020	Interior, brushwork, trim, under 6"	1 Pord	240	.033	L.F.	2.42	1.55		3.97	5.10
0030	6"-12" wide		180	.044		3.15	2.06		5.21	6.75
0040	Balustrades		300	.027		2.14	1.24		3.38	4.31
0050	Pipe to 4" diameter		500	.016		1.31	.74		2.05	2.62
0060	To 8" diameter		375	.021		1.54	.99		2.53	3.26
0070	To 12" diameter	↓	250	.032	↓	2.33	1.49		3.82	4.91

43

02 83 Lead Remediation

02 83 19 – Lead-Based Paint Remediation

02 83 19.23 Encapsulation of Lead-Based Paint

		Crew	Daily Output	Labor-Hours	Unit	Material	2021 Bare Costs Labor	Equipment	Total	Total Incl O&P
0080	To 16" diameter	1 Pord	170	.047	L.F.	3.45	2.19		5.64	7.25
0090	Cabinets, ornate design		200	.040	S.F.	3.29	1.86		5.15	6.55
0100	Simple design		250	.032	"	2.54	1.49		4.03	5.15
0110	Doors, 3' x 7', both sides, incl. frame & trim									
0120	Flush	1 Pord	6	1.333	Ea.	29.50	62		91.50	131
0130	French, 10-15 lite		3	2.667		6.65	124		130.65	203
0140	Panel		4	2		35	93		128	186
0150	Louvered		2.75	2.909		33	135		168	250
0160	Windows, per interior side, per 15 S.F.									
0170	1-6 lite	1 Pord	14	.571	Ea.	23	26.50		49.50	67
0180	7-10 lite		7.50	1.067		23	49.50		72.50	104
0190	12 lite		5.75	1.391		34	64.50		98.50	140
0200	Radiators		8	1		73.50	46.50		120	154
0210	Grilles, vents		275	.029	S.F.	2.09	1.35		3.44	4.44
0220	Walls, roller, drywall or plaster		1000	.008		.65	.37		1.02	1.31
0230	With spunbonded reinforcing fabric		720	.011		.76	.52		1.28	1.66
0240	Wood	1 Pord	800	.010	S.F.	.71	.46		1.17	1.52
0250	Ceilings, roller, drywall or plaster		900	.009		.76	.41		1.17	1.49
0260	Wood		700	.011		.93	.53		1.46	1.86
0270	Exterior, brushwork, gutters and downspouts		300	.027	L.F.	2.16	1.24		3.40	4.34
0280	Columns		400	.020	S.F.	1.46	.93		2.39	3.08
0290	Spray, siding		600	.013	"	1.09	.62		1.71	2.18
0300	Miscellaneous									
0310	Electrical conduit, brushwork, to 2" diameter	1 Pord	500	.016	L.F.	1.31	.74		2.05	2.62
0320	Brick, block or concrete, spray		500	.016	S.F.	1.31	.74		2.05	2.62
0330	Steel, flat surfaces and tanks to 12"		500	.016		1.31	.74		2.05	2.62
0340	Beams, brushwork		400	.020		1.46	.93		2.39	3.08
0350	Trusses		400	.020		1.46	.93		2.39	3.08

02 83 19.26 Removal of Lead-Based Paint

		Crew	Daily Output	Labor-Hours	Unit	Material	2021 Bare Costs Labor	Equipment	Total	Total Incl O&P
0010	**REMOVAL OF LEAD-BASED PAINT** R028319-60									
0011	By chemicals, per application									
0050	Baseboard, to 6" wide	1 Pord	64	.125	L.F.	.86	5.80		6.66	10.15
0070	To 12" wide		32	.250	"	1.54	11.60		13.14	20
0200	Balustrades, one side		28	.286	S.F.	1.50	13.25		14.75	22.50
1400	Cabinets, simple design		32	.250		1.42	11.60		13.02	19.95
1420	Ornate design		25	.320		1.76	14.85		16.61	25.50
1600	Cornice, simple design		60	.133		1.58	6.20		7.78	11.55
1620	Ornate design		20	.400		5.55	18.60		24.15	35.50
2800	Doors, one side, flush		84	.095		1.88	4.42		6.30	9.05
2820	Two panel		80	.100		1.42	4.65		6.07	8.90
2840	Four panel		45	.178		1.41	8.25		9.66	14.65
2880	For trim, one side, add		64	.125	L.F.	.82	5.80		6.62	10.10
3000	Fence, picket, one side		30	.267	S.F.	1.42	12.40		13.82	21
3200	Grilles, one side, simple design		30	.267		1.42	12.40		13.82	21
3220	Ornate design		25	.320		1.47	14.85		16.32	25
3240	Handrails		90	.089	L.F.	1.42	4.13		5.55	8.10
4400	Pipes, to 4" diameter		90	.089		1.88	4.13		6.01	8.60
4420	To 8" diameter		50	.160		3.77	7.45		11.22	15.90
4440	To 12" diameter		36	.222		5.65	10.30		15.95	22.50
4460	To 16" diameter		20	.400		7.55	18.60		26.15	38
4500	For hangers, add		40	.200	Ea.	2.54	9.30		11.84	17.50
4800	Siding		90	.089	S.F.	1.41	4.13		5.54	8.10

44

02 83 Lead Remediation

02 83 19 – Lead-Based Paint Remediation

02 83 19.26 Removal of Lead-Based Paint	Crew	Daily Output	Labor-Hours	Unit	Material	2021 Bare Costs Labor	Equipment	Total	Total Incl O&P
5000 Trusses, open	1 Pord	55	.145	SF Face	2.04	6.75		8.79	12.95
6200 Windows, one side only, double-hung, 1/1 light, 24" x 48" high		4	2	Ea.	25	93		118	175
6220 30" x 60" high		3	2.667		33	124		157	233
6240 36" x 72" high		2.50	3.200		40.50	149		189.50	280
6280 40" x 80" high		2	4		50	186		236	350
6400 Colonial window, 6/6 light, 24" x 48" high	1 Pord	2	4	Ea.	50.50	186		236.50	350
6420 30" x 60" high		1.50	5.333		66	248		314	465
6440 36" x 72" high		1	8		99.50	370		469.50	700
6480 40" x 80" high		1	8		99.50	370		469.50	700
6600 8/8 light, 24" x 48" high		2	4		50	186		236	350
6620 40" x 80" high		1	8		99.50	370		469.50	700
6800 12/12 light, 24" x 48" high		1	8		99.50	370		469.50	700
6820 40" x 80" high		.75	10.667		134	495		629	930
6840 Window frame & trim items, included in pricing above									
7000 Hand scraping and HEPA vacuum, less than 4 S.F.	1 Pord	8	1	Ea.	1.23	46.50		47.73	75
8000 Collect and bag bulk material, 3 C.F. bags, by hand		30	.267	"	1.23	12.40		13.63	21
9000 Minimum labor/equipment charge		3	2.667	Job		124		124	196

02 87 Biohazard Remediation

02 87 13 – Mold Remediation

02 87 13.16 Mold Remediation Preparation and Containment

	Crew	Daily Output	Labor-Hours	Unit	Material	2021 Bare Costs Labor	Equipment	Total	Total Incl O&P
0010 **MOLD REMEDIATION PREPARATION AND CONTAINMENT**									
0015 Mold remediation plans and methods									
0020 Initial inspection, areas to 2,500 S.F.				Total				268	295
0030 Areas to 5,000 S.F.								440	485
0032 Areas to 10,000 S.F.								440	485
0040 Testing, air sample each								126	139
0050 Swab sample								115	127
0060 Tape sample								125	140
0070 Post remediation air test								126	139
0080 Mold abatement plan, area to 2,500 S.F.								1,225	1,325
0090 Areas to 5,000 S.F.								1,625	1,775
0095 Areas to 10,000 S.F.								2,550	2,800
0100 Packup & removal of contents, average 3 bedroom home, excl. storage								8,150	8,975
0110 Average 5 bedroom home, excl. storage								15,300	16,800
0600 For demolition in mold contaminated areas, see Section 02 87 13.33									
0610 For personal protection equipment, see Section 02 82 13.41									
6010 Preparation of mold containment area									
6100 Pre-cleaning, HEPA vacuum and wet wipe, flat surfaces	A-9	12000	.005	S.F.	.01	.33		.34	.53
6300 Separation barrier, 2" x 4" @ 16", 1/2" plywood ea. side, 8' high	2 Carp	400	.040		3.11	2.19		5.30	6.90
6310 12' high		320	.050		3.52	2.74		6.26	8.25
6320 16' high		200	.080		2.64	4.38		7.02	9.85
6400 Personnel decontam. chamber, 2" x 4" @ 16", 3/4" ply ea. side		280	.057		5.40	3.13		8.53	10.90
6450 Waste decontam. chamber, 2" x 4" studs @ 16", 3/4" ply each side		360	.044		5.30	2.43		7.73	9.70
6500 Cover surfaces with polyethylene sheeting									
6501 Including glue and tape									
6550 Floors, each layer, 6 mil	A-9	8000	.008	S.F.	.08	.49		.57	.86
6551 4 mil		9000	.007		.03	.43		.46	.72
6560 Walls, each layer, 6 mil		6000	.011		.08	.65		.73	1.12
6561 4 mil		7000	.009		.03	.56		.59	.91
6570 For heights above 14', add						20%			

45

For customer support on your Commercial Renovation Costs with RSMeans Data, call 800.448.8182.

02 87 Biohazard Remediation

02 87 13 – Mold Remediation

02 87 13.16 Mold Remediation Preparation and Containment

		Crew	Daily Output	Labor-Hours	Unit	Material	2021 Bare Costs Labor	2021 Bare Costs Equipment	Total	Total Incl O&P
6575	For heights above 20', add						30%			
6580	For fire retardant poly, add					100%				
6590	For large open areas, deduct					10%	20%			
6600	Seal floor penetrations with foam firestop to 36 sq. in.	2 Carp	200	.080	Ea.	13.45	4.38		17.83	22
6610	36 sq. in. to 72 sq. in.		125	.128		27	7		34	40.50
6615	72 sq. in. to 144 sq. in.		80	.200		54	10.95		64.95	76.50
6620	Wall penetrations, to 36 sq. in.		180	.089		13.45	4.86		18.31	22.50
6630	36 sq. in. to 72 sq. in.		100	.160		27	8.75		35.75	43.50
6640	72 sq. in. to 144 sq. in.		60	.267		54	14.60		68.60	82
6800	Caulk seams with latex caulk	1 Carp	230	.035	L.F.	.19	1.90		2.09	3.24
6900	Set up neg. air machine, 1-2k CFM/25 M.C.F. volume	1 Asbe	4.30	1.860	Ea.		113		113	179

02 87 13.33 Removal and Disposal of Materials With Mold

		Crew	Daily Output	Labor-Hours	Unit	Material	2021 Bare Costs Labor	2021 Bare Costs Equipment	Total	Total Incl O&P
0010	**REMOVAL AND DISPOSAL OF MATERIALS WITH MOLD**									
0015	Demolition in mold contaminated area									
0200	Ceiling, including suspension system, plaster and lath	A-9	2100	.030	S.F.	.08	1.86		1.94	3.03
0210	Finished plaster, leaving wire lath		585	.109		.30	6.65		6.95	10.90
0220	Suspended acoustical tile		3500	.018		.05	1.12		1.17	1.82
0230	Concealed tile grid system		3000	.021		.06	1.30		1.36	2.12
0240	Metal pan grid system		1500	.043		.12	2.60		2.72	4.24
0250	Gypsum board		2500	.026		.07	1.56		1.63	2.55
0255	Plywood		2500	.026		.07	1.56		1.63	2.55
0260	Lighting fixtures up to 2' x 4'		72	.889	Ea.	2.44	54		56.44	88
0400	Partitions, non load bearing									
0410	Plaster, lath, and studs	A-9	690	.093	S.F.	.86	5.65		6.51	9.90
0450	Gypsum board and studs		1390	.046		.13	2.81		2.94	4.58
0465	Carpet & pad		1390	.046		.13	2.81		2.94	4.58
0600	Pipe insulation, air cell type, up to 4" diameter pipe		900	.071	L.F.	.20	4.34		4.54	7.05
0610	4" to 8" diameter pipe		800	.080		.22	4.88		5.10	7.95
0620	10" to 12" diameter pipe		700	.091		.25	5.60		5.85	9.10
0630	14" to 16" diameter pipe		550	.116		.32	7.10		7.42	11.55
0650	Over 16" diameter pipe		650	.098	S.F.	.27	6		6.27	9.80
9000	For type B (supplied air) respirator equipment, add				%		10%			10%

02 87 19 – Bio-Hazard and Infectious Disease Control Remediation

02 87 19.10 Temporary Barriers, Including PPE

		Crew	Daily Output	Labor-Hours	Unit	Material	2021 Bare Costs Labor	2021 Bare Costs Equipment	Total	Total Incl O&P
0010	**TEMPORARY BARRIERS, including PPE**									
0100	Infectious control partition, adjustable to 10' high, obscured, 4' panel	2 Carp	75	.213	Ea.	580	11.65		591.65	660
0110	3' panel		75	.213		550	11.65		561.65	625
0120	2' panel		75	.213		430	11.65		441.65	495
0130	1' panel	1 Carp	75	.107		285	5.85		290.85	325
0140	6" panel	"	75	.107		265	5.85		270.85	300
0150	2' panel with HEPA filtered discharge port	2 Carp	75	.213		500	11.65		511.65	570
0160	3' panel with 32" door		75	.213		895	11.65		906.65	1,000
0170	4' panel with 36" door		75	.213		995	11.65		1,006.65	1,125
0180	4'-6" panel with 44" door		75	.213		1,200	11.65		1,211.65	1,350
0190	Hinged corner		68	.235		186	12.85		198.85	225
0200	Outside corner		68	.235		151	12.85		163.85	187
0210	T post		68	.235		151	12.85		163.85	187
0220	Accessories, ceiling grid clip		300	.053		7.60	2.92		10.52	13.05
0230	Panel locking clip		300	.053		5.30	2.92		8.22	10.50
0240	Panel joint closure strip		300	.053		8.15	2.92		11.07	13.65
0250	Screw jack		300	.053		6.80	2.92		9.72	12.15
0260	Digital pressure difference guage					275			275	305

For customer support on your Commercial Renovation Costs with RSMeans Data, call 800.448.8182.

02 87 Biohazard Remediation

02 87 19 – Bio-Hazard and Infectious Disease Control Remediation

		Crew	Daily Output	Labor-Hours	Unit	Material	2021 Bare Costs Labor	Equipment	Total	Total Incl O&P
02 87 19.10	**Temporary Barriers, Including PPE**									
0280	Combination lockset	1 Carp	11	.727	Ea.	202	40		242	287
0285	Sealant tape, 2" wide	1 Clab	165	.048	C.L.F.	10.85	2.15		13	15.35
0290	System in place, including door and accessories	2 Carp	43	.372	L.F.	11.10	20.50		31.60	44.50
0300	Based upon 25 uses		43	.372		11.10	20.50		31.60	44.50
0310	Based upon 50 uses		43	.372		6.15	20.50		26.65	39.50
0330	Based upon 100 uses		43	.372		3.69	20.50		24.19	36.50
0400	Separation barrier, 2" x 4" @ 16" O.C., 1/2" plywood each side, 8' high		340	.047	S.F.	4.22	2.57		6.79	8.75
0410	12' high		275	.058		4.17	3.18		7.35	9.65
0420	16' high		170	.094		4.25	5.15		9.40	12.85
0430	5/8" FR GWB each side, 8' high		290	.055		2.68	3.02		5.70	7.75
0440	12' high		235	.068		2.60	3.72		6.32	8.80
0450	16' high		144	.111		2.55	6.10		8.65	12.50
02 87 19.20	**Temporary Protection, Including PPE**									
0010	**TEMPORARY PROTECTION, including PPE**									
0100	Infectious control protection, 6 mil plastic sheeting, applied to floors	1 Skwk	1900	.004	S.F.	.08	.24		.32	.46
0110	Applied to walls		1400	.006		.08	.33		.41	.60
0120	Applied to ceilings		1000	.008		.08	.46		.54	.80
0130	Sealant tape, 2" wide	1 Clab	165	.048	C.L.F.	10.85	2.15		13	15.35
0140	Floor covering, 1/8" tempered hardboard	2 Carp	878	.018	S.F.	.69	1		1.69	2.35
0150	1/2" CDX		878	.018		1.18	1		2.18	2.89
0160	3/4" CDX		755	.021		1.56	1.16		2.72	3.57
02 87 19.30	**Infectious Control Equipment and Supplies**									
0010	**INFECTIOUS CONTROL EQUIPMENT AND SUPPLIES**									
0100	Entry/exit decontamination chamber, portable, 10 uses	A-10C	12	2	Ea.	228	122	16.55	366.55	460
0105	Negative air machine, 1800 CFM					900			900	990
0110	Disposable polyethylene bags, 6 mil, 3 C.F.					1.23			1.23	1.35
0130	Paper towels, carton of 12, 350' rolls					65.50			65.50	72.50
0140	Heavy duty wipes, 160 per box					38			38	42
0200	Tent, including frame, weights, roof, walls, 5' x 5'	2 Skwk	32	.500		2,150	28.50		2,178.50	2,425
0205	5' x 10'		24	.667		2,500	38		2,538	2,800
0210	10' x 10'		24	.667		3,125	38		3,163	3,475
0220	13' x 26'		12	1.333		9,350	76		9,426	10,400
0230	13' x 26', flat roof infirmary		12	1.333		7,600	76		7,676	8,475
0240	13' x 26', drive through screening		12	1.333		8,800	76		8,876	9,800
0242	LED light kit for tents, 15 watt	1 Elec	48	.167		199	10.60		209.60	235
0244	Halogen heating kit, 1500 watt	"	48	.167		500	10.60		510.60	565
0250	Medical partition system, including frame, drapes, 5' x 5'	2 Skwk	48	.333		1,300	19.05		1,319.05	1,450
02 87 19.40	**Infectious Control Cleaning, Includes PPE**									
0010	**INFECTIOUS CONTROL CLEANING, Includes PPE**									
0100	Infectious control cleaning, including PPE and antimicrobial spray, floors	1 Skwk	3600	.002	S.F.	.01	.13		.14	.21
0110	Walls		3000	.003		.01	.15		.16	.25
0120	Ceilings		2700	.003		.01	.17		.18	.28
0130	Countertops		3300	.002		.01	.14		.15	.23
0140	Other surfaces		2600	.003		.01	.18		.19	.29
0150	Plumbing fixtures		14	.571	Ea.	1.87	32.50		34.37	53.50
0160	Bathroom accessories		30	.267		.87	15.25		16.12	25
0170	Kitchen appliance		9	.889		2.92	51		53.92	83.50
0180	Electrical switches and receptacles		140	.057		.19	3.26		3.45	5.35
0190	Work station		9	.889		2.92	51		53.92	83.50
0200	Door and window hardware		88	.091		.30	5.20		5.50	8.60
0250	Supply / return ductwork, up to 2 S.F. cross section		240	.033	L.F.	.11	1.90		2.01	3.14
0260	2 to 4 S.F. cross section		192	.042		.14	2.38		2.52	3.92

47

02 87 Biohazard Remediation

02 87 19 – Bio-Hazard and Infectious Disease Control Remediation

02 87 19.40 Infectious Control Cleaning, Includes PPE	Crew	Daily Output	Labor-Hours	Unit	Material	2021 Bare Costs Labor	Equipment	Total	Total Incl O&P
0270	4 to 8 S.F. cross section	1 Skwk	144	.056	L.F.	.19	3.17		3.36	5.25
0280	8 SF cross section		102	.078	↓	.26	4.48		4.74	7.40
0300	Duct coil		5	1.600	Ea.	5.25	91.50		96.75	151
0310	Duct damper and vane set		77	.104		.35	5.95		6.30	9.80
0320	Fan coil unit or VAV box		8	1		3.28	57		60.28	94
0330	Air handling unit and coil		2	4		13.10	228		241.10	375
0340	Grill / diffuser or register	↓	32	.250	↓	.82	14.30		15.12	23.50

For customer support on your Commercial Renovation Costs with RSMeans Data, call 800.448.8182.

Estimating Tips
General

- Carefully check all the plans and specifications. Concrete often appears on drawings other than structural drawings, including mechanical and electrical drawings for equipment pads. The cost of cutting and patching is often difficult to estimate. See Subdivision 03 81 for Concrete Cutting, Subdivision 02 41 19.16 for Cutout Demolition, Subdivision 03 05 05.10 for Concrete Demolition, and Subdivision 02 41 19.19 for Rubbish Handling (handling, loading, and hauling of debris).

- Always obtain concrete prices from suppliers near the job site. A volume discount can often be negotiated, depending upon competition in the area. Remember to add for waste, particularly for slabs and footings on grade.

03 10 00 Concrete Forming and Accessories

- A primary cost for concrete construction is forming. Most jobs today are constructed with prefabricated forms. The selection of the forms best suited for the job and the total square feet of forms required for efficient concrete forming and placing are key elements in estimating concrete construction. Enough forms must be available for erection to make efficient use of the concrete placing equipment and crew.

- Concrete accessories for forming and placing depend upon the systems used. Study the plans and specifications to ensure that all special accessory requirements have been included in the cost estimate, such as anchor bolts, inserts, and hangers.

- Included within costs for forms-in-place are all necessary bracing and shoring.

03 20 00 Concrete Reinforcing

- Ascertain that the reinforcing steel supplier has included all accessories, cutting, bending, and an allowance for lapping, splicing, and waste. A good rule of thumb is 10% for lapping, splicing, and waste. Also, 10% waste should be allowed for welded wire fabric.

- The unit price items in the subdivisions for Reinforcing In Place, Glass Fiber Reinforcing, and Welded Wire Fabric include the labor to install accessories such as beam and slab bolsters, high chairs, and bar ties and tie wire. The material cost for these accessories is not included; they may be obtained from the Accessories Subdivisions.

03 30 00 Cast-In-Place Concrete

- When estimating structural concrete, pay particular attention to requirements for concrete additives, curing methods, and surface treatments. Special consideration for climate, hot or cold, must be included in your estimate. Be sure to include requirements for concrete placing equipment and concrete finishing.

- For accurate concrete estimating, the estimator must consider each of the following major components individually: forms, reinforcing steel, ready-mix concrete, placement of the concrete, and finishing of the top surface. For faster estimating, Subdivision 03 30 53.40 for Concrete-In-Place can be used; here, various items of concrete work are presented that include the costs of all five major components (unless specifically stated otherwise).

03 40 00 Precast Concrete
03 50 00 Cast Decks and Underlayment

- The cost of hauling precast concrete structural members is often an important factor. For this reason, it is important to get a quote from the nearest supplier. It may become economically feasible to set up precasting beds on the site if the hauling costs are prohibitive.

Reference Numbers

Reference numbers are shown at the beginning of some major classifications. These numbers refer to related items in the Reference Section. The reference information may be an estimating procedure, an alternate pricing method, or technical information.

Note: Not all subdivisions listed here necessarily appear. ∎

Same Data. Simplified.

Enjoy the convenience and efficiency of accessing your costs anywhere:

- **Skip the multiplier** by setting your location
- **Quickly search**, edit, favorite and share costs
- **Stay on top of price changes** with automatic updates

Discover more at rsmeans.com/online

03 01 30.62 Concrete Patching

		Crew	Daily Output	Labor-Hours	Unit	Material	2021 Bare Costs Labor	Equipment	Total	Total Incl O&P
0010	**CONCRETE PATCHING**									
0100	Floors, 1/4" thick, small areas, regular grout	1 Cefi	170	.047	S.F.	1.53	2.44		3.97	5.50
0150	Epoxy grout	1 Cefi	100	.080	S.F.	11.15	4.14		15.29	18.75
2000	Walls, including chipping, cleaning and epoxy grout									
2100	1/4" deep	1 Cefi	65	.123	S.F.	6.55	6.40		12.95	17.20
2150	1/2" deep		50	.160		13.10	8.30		21.40	27.50
2200	3/4" deep		40	.200		19.65	10.35		30	37.50
9000	Minimum labor/equipment charge		4.50	1.778	Job		92		92	144

03 01 30.71 Concrete Crack Repair

		Crew	Daily Output	Labor-Hours	Unit	Material	2021 Bare Costs Labor	Equipment	Total	Total Incl O&P
0010	**CONCRETE CRACK REPAIR**									
1000	Structural repair of concrete cracks by epoxy injection (ACI RAP-1)									
1001	suitable for horizontal, vertical and overhead repairs									
1010	Clean/grind concrete surface(s) free of contaminants	1 Cefi	400	.020	L.F.		1.04		1.04	1.62
1015	Rout crack with v-notch crack chaser, if needed	C-32	600	.027		.08	1.28	.25	1.61	2.38
1020	Blow out crack with oil-free dry compressed air (1 pass)	C-28	3000	.003			.14	.01	.15	.23
1030	Install surface-mounted entry ports (spacing = concrete depth)	1 Cefi	400	.020	Ea.	.62	1.04		1.66	2.30
1040	Cap crack at surface with epoxy gel (per side/face)		400	.020	L.F.	.69	1.04		1.73	2.38
1050	Snap off ports, grind off epoxy cap residue after injection		200	.040	"		2.07		2.07	3.24
1100	Manual injection with 2-part epoxy cartridge, excludes prep									
1110	Up to 1/32" (0.03125") wide x 4" deep	1 Cefi	160	.050	L.F.	.35	2.59		2.94	4.44
1120	6" deep		120	.067		.52	3.45		3.97	5.95
1130	8" deep		107	.075		.69	3.87		4.56	6.80
1140	10" deep		100	.080		.87	4.14		5.01	7.45
1150	12" deep		96	.083		1.04	4.32		5.36	7.90
1210	Up to 1/16" (0.0625") wide x 4" deep		160	.050		.69	2.59		3.28	4.82
1220	6" deep		120	.067		1.04	3.45		4.49	6.55
1230	8" deep		107	.075		1.39	3.87		5.26	7.60
1240	10" deep		100	.080		1.74	4.14		5.88	8.40
1250	12" deep		96	.083		2.08	4.32		6.40	9.05
1310	Up to 3/32" (0.09375") wide x 4" deep		160	.050		1.04	2.59		3.63	5.20
1320	6" deep		120	.067		1.56	3.45		5.01	7.10
1330	8" deep		107	.075		2.08	3.87		5.95	8.35
1340	10" deep		100	.080		2.60	4.14		6.74	9.35
1350	12" deep		96	.083		3.13	4.32		7.45	10.20
1410	Up to 1/8" (0.125") wide x 4" deep		160	.050		1.39	2.59		3.98	5.60
1420	6" deep		120	.067		2.08	3.45		5.53	7.70
1430	8" deep		107	.075		2.78	3.87		6.65	9.10
1440	10" deep		100	.080		3.47	4.14		7.61	10.30
1450	12" deep		96	.083		4.17	4.32		8.49	11.35
1500	Pneumatic injection with 2-part bulk epoxy, excludes prep									
1510	Up to 5/32" (0.15625") wide x 4" deep	C-31	240	.033	L.F.	.32	1.73	1.34	3.39	4.52
1520	6" deep	"	180	.044	"	.48	2.30	1.79	4.57	6.10
1530	8" deep	C-31	160	.050	L.F.	.63	2.59	2.01	5.23	6.95
1540	10" deep		150	.053		.79	2.76	2.15	5.70	7.55
1550	12" deep		144	.056		.95	2.88	2.23	6.06	8
1610	Up to 3/16" (0.1875") wide x 4" deep		240	.033		.38	1.73	1.34	3.45	4.59
1620	6" deep		180	.044		.57	2.30	1.79	4.66	6.20
1630	8" deep		160	.050		.76	2.59	2.01	5.36	7.10
1640	10" deep		150	.053		.95	2.76	2.15	5.86	7.75
1650	12" deep		144	.056		1.14	2.88	2.23	6.25	8.25
1710	Up to 1/4" (0.25") wide x 4" deep		240	.033		.51	1.73	1.34	3.58	4.73
1720	6" deep		180	.044		.76	2.30	1.79	4.85	6.40

50

03 01 30 – Maintenance of Cast-In-Place Concrete

03 01 30.71 Concrete Crack Repair

		Crew	Daily Output	Labor-Hours	Unit	Material	2021 Bare Costs Labor	Equipment	Total	Total Incl O&P
1730	8" deep	C-31	160	.050	L.F.	1.02	2.59	2.01	5.62	7.40
1740	10" deep		150	.053		1.27	2.76	2.15	6.18	8.10
1750	12" deep	↓	144	.056	↓	1.52	2.88	2.23	6.63	8.65
2000	Non-structural filling of concrete cracks by gravity-fed resin (ACI RAP-2)									
2001	suitable for individual cracks in stable horizontal surfaces only									
2010	Clean/grind concrete surface(s) free of contaminants	1 Cefi	400	.020	L.F.		1.04		1.04	1.62
2020	Rout crack with v-notch crack chaser, if needed	C-32	600	.027		.08	1.28	.25	1.61	2.38
2030	Blow out crack with oil-free dry compressed air (1 pass)	C-28	3000	.003			.14	.01	.15	.23
2040	Cap crack with epoxy gel at underside of elevated slabs, if needed	1 Cefi	400	.020		.69	1.04		1.73	2.38
2050	Insert backer rod into crack, if needed		400	.020		.03	1.04		1.07	1.65
2060	Partially fill crack with fine dry sand, if needed		800	.010		.02	.52		.54	.83
2070	Apply two beads of sealant alongside crack to form reservoir, if needed	↓	400	.020	↓	.27	1.04		1.31	1.91
2100	Manual filling with squeeze bottle of 2-part epoxy resin									
2110	Full depth crack up to 1/16" (0.0625") wide x 4" deep	1 Cefi	300	.027	L.F.	.07	1.38		1.45	2.24
2120	6" deep		240	.033		.11	1.73		1.84	2.82
2130	8" deep		200	.040		.14	2.07		2.21	3.39
2140	10" deep		185	.043		.18	2.24		2.42	3.70
2150	12" deep		175	.046		.21	2.37		2.58	3.94
2210	Full depth crack up to 1/8" (0.125") wide x 4" deep		270	.030		.14	1.53		1.67	2.55
2220	6" deep		215	.037		.21	1.93		2.14	3.25
2230	8" deep		180	.044		.28	2.30		2.58	3.91
2240	10" deep		165	.048		.35	2.51		2.86	4.32
2250	12" deep		155	.052		.42	2.67		3.09	4.65
2310	Partial depth crack up to 3/16" (0.1875") wide x 1" deep		300	.027		.05	1.38		1.43	2.22
2410	Up to 1/4" (0.250") wide x 1" deep		270	.030		.07	1.53		1.60	2.48
2510	Up to 5/16" (0.3125") wide x 1" deep		250	.032		.09	1.66		1.75	2.70
2610	Up to 3/8" (0.375") wide x 1" deep	↓	240	.033	↓	.11	1.73		1.84	2.82

03 01 30.72 Concrete Surface Repairs

		Crew	Daily Output	Labor-Hours	Unit	Material	2021 Bare Costs Labor	Equipment	Total	Total Incl O&P
0010	**CONCRETE SURFACE REPAIRS**									
1000	Spall repairs using low-pressure spraying (ACI RAP-3)									
1001	Suitable for vertical and overhead repairs up to 3" thick									
1010	Sound the concrete surface to locate delaminated areas	1 Cefi	2000	.004	S.F.		.21		.21	.32
1020	Remove concrete in repair areas to fully expose reinforcing bars	B-9E	160	.100			4.81	.21	5.02	7.85
1030	Mark the perimeter of each repair area	1 Cefi	2000	.004	↓		.21		.21	.32
1040	Saw cut the perimeter of each repair area down to reinforcing bars	B-89C	160	.050	L.F.	.03	2.59	.36	2.98	4.50
1050	If corroded rebar is exposed, remove concrete to 3/4" below corroded rebars									
1051	Single layer of #4 rebar	B-9E	64	.250	S.F.		12.05	.51	12.56	19.55
1052	#5 rebar		53.30	.300			14.45	.62	15.07	23.50
1053	#6 rebar		45.70	.350			16.85	.72	17.57	27.50
1054	#7 rebar		40	.400			19.25	.82	20.07	31.50
1055	#8 rebar		35.60	.449			21.50	.92	22.42	35
1056	Double layer of #4 rebar		35.60	.449			21.50	.92	22.42	35
1057	#5 rebar		32	.500			24	1.03	25.03	39
1058	#6 rebar		29.10	.550			26.50	1.13	27.63	42.50
1059	#7 rebar		26.70	.599			29	1.23	30.23	47
1060	#8 rebar	↓	24.60	.650	↓		31.50	1.33	32.83	51
1100	See section 03 21 11.60 for placement of supplemental reinforcement bars									
1104	See section 02 41 19.19 for handling & removal of debris									
1106	See section 02 41 19.20 for disposal of debris									
1110	Final cleaning by sandblasting	E-11	1100	.029	S.F.	.52	1.41	.27	2.20	3.19
1120	By high pressure water	C-29	500	.016			.71	.19	.90	1.34
1130	Blow off dust and debris with oil-free dry compressed air	C-28	16000	.001	↓		.03		.03	.04

03 01 30.72 Concrete Surface Repairs		Crew	Daily Output	Labor-Hours	Unit	Material	2021 Bare Costs Labor	Equipment	Total	Total Incl O&P
1140	Mix up repair material with concrete mixer	C-30	200	.040	C.F.	12.50	1.78	.74	15.02	17.40
1150	Place repair material by low-pressure spray in lifts up to 3" thick	C-9	200	.160	"		7.85	2.38	10.23	14.95
1160	Screed and initial float of repair material by hand	C-10	2000	.012	S.F.		.59		.59	.93
1170	Float and steel trowel finish repair material by hand	"	1265	.019			.94		.94	1.47
1180	Cure with sprayed membrane curing compound	2 Cefi	6200	.003	↓	.12	.13		.25	.34
2000	Surface repairs using form-and-pour techniques (ACI RAP-4)									
2001	Suitable for walls, columns, beam sides/bottoms, and slab soffits									
2010	Sound the concrete surface to locate delaminated areas	1 Cefi	2000	.004	S.F.		.21		.21	.32
2020	Remove concrete in repair areas to fully expose reinforcing bars	B-9E	160	.100			4.81	.21	5.02	7.85
2030	Mark the perimeter of each repair area	1 Cefi	2000	.004	↓		.21		.21	.32
2040	Saw cut the perimeter of each repair area down to reinforcing bars	B-89C	160	.050	L.F.	.03	2.59	.36	2.98	4.50
2050	If corroded rebar is exposed, remove concrete to 3/4" below corroded rebars									
2051	Single layer of #4 rebar	B-9E	64	.250	S.F.		12.05	.51	12.56	19.55
2052	#5 rebar	"	53.30	.300	"		14.45	.62	15.07	23.50
2053	#6 rebar	B-9E	45.70	.350	S.F.		16.85	.72	17.57	27.50
2054	#7 rebar		40	.400			19.25	.82	20.07	31.50
2055	#8 rebar		35.60	.449			21.50	.92	22.42	35
2056	Double layer of #4 rebar		35.60	.449			21.50	.92	22.42	35
2057	#5 rebar		32	.500			24	1.03	25.03	39
2058	#6 rebar		29.10	.550			26.50	1.13	27.63	42.50
2059	#7 rebar		26.70	.599			29	1.23	30.23	47
2060	#8 rebar	↓	24.60	.650	↓		31.50	1.33	32.83	51
2100	See section 03 21 11.60 for placement of supplemental reinforcement bars									
2105	See section 02 41 19.19 for handling & removal of debris									
2110	See section 02 41 19.20 for disposal of debris									
2120	For slab soffit repairs, drill 2" diam. holes to 6" deep for conc placement	B-89A	16.50	.970	Ea.	.27	49	7.35	56.62	86.50
2130	For each additional inch of slab thickness in same hole, add	"	1080	.015		.04	.75	.11	.90	1.36
2140	Drill holes for expansion shields, 3/4" diameter x 4" deep	1 Carp	45	.178	↓		9.70		9.70	15.50
2150	Final cleaning by sandblasting	E-11	1100	.029	S.F.	.52	1.41	.27	2.20	3.19
2160	By high pressure water	C-29	500	.016			.71	.19	.90	1.34
2170	Blow off dust and debris with oil-free dry compressed air	C-28	16000	.001	↓		.03		.03	.04
2180	Build wood forms including chutes where needed for concrete placement	C-2	960	.050	SFCA	4.63	2.67		7.30	9.35
2190	Insert expansion shield, coil rod, coil tie and plastic cone in each hole	1 Carp	32	.250	Ea.	5.50	13.70		19.20	28
2200	Install wood forms and fasten with coil rod and coil nut	C-2	960	.050	SFCA	.58	2.67		3.25	4.89
2210	Mix up bagged repair material by hand with a mixing paddle	1 Cefi	30	.267	C.F.	12.50	13.80		26.30	35.50
2220	By concrete mixer	C-30	200	.040		12.50	1.78	.74	15.02	17.40
2230	Place repair material by hand and consolidate	C-6A	200	.080	↓		4.14	.23	4.37	6.75
2240	Remove wood forms	C-2	1600	.030	SFCA		1.60		1.60	2.55
2250	Chip/grind off extra concrete at fill chutes	B-9E	16	1	Ea.		48	2.05	50.17	78.50
2260	Break ties and patch voids	1 Cefi	270	.030	"	.08	1.53		1.61	2.49
2270	Cure with sprayed membrane curing compound	2 Cefi	6200	.003	S.F.	.12	.13		.25	.34
3000	Surface repairs using form-and-pump techniques (ACI RAP-5)									
3001	Suitable for walls, columns, beam sides/bottoms, and slab soffits									
3010	Sound the concrete surface to locate delaminated areas	1 Cefi	2000	.004	S.F.		.21		.21	.32
3020	Remove concrete in repair areas to fully expose reinforcing bars	B-9E	160	.100	↓		4.81	.21	5.02	7.85
3030	Mark the perimeter of each repair area	1 Cefi	2000	.004	↓		.21		.21	.32
3040	Saw cut the perimeter of each repair area down to reinforcing bars	B-89C	160	.050	L.F.	.03	2.59	.36	2.98	4.50
3050	If corroded rebar is exposed, remove concrete to 3/4" below corroded rebars									
3051	Single layer of #4 rebar	B-9E	64	.250	S.F.		12.05	.51	12.56	19.55
3052	#5 rebar		53.30	.300			14.45	.62	15.07	23.50
3053	#6 rebar	↓	45.70	.350	↓		16.85	.72	17.57	27.50
3054	#7 rebar	B-9E	40	.400	S.F.		19.25	.82	20.07	31.50
3055	#8 rebar	↓	35.60	.449	↓		21.50	.92	22.42	35

52

03 01 30 – Maintenance of Cast-In-Place Concrete

03 01 30.72 Concrete Surface Repairs		Crew	Daily Output	Labor-Hours	Unit	Material	2021 Bare Costs Labor	Equipment	Total	Total Incl O&P
3056	Double layer of #4 rebar	B-9E	35.60	.449	S.F.		21.50	.92	22.42	35
3057	#5 rebar		32	.500			24	1.03	25.03	39
3058	#6 rebar		29.10	.550			26.50	1.13	27.63	42.50
3059	#7 rebar		26.70	.599			29	1.23	30.23	47
3060	#8 rebar		24.60	.650			31.50	1.33	32.83	51
3100	See section 03 21 11.60 for placement of supplemental reinforcement bars									
3104	See section 02 41 19.19 for handling & removal of debris									
3106	See section 02 41 19.20 for disposal of debris									
3110	Drill holes for expansion shields, 3/4" diameter x 4" deep	1 Carp	45	.178	Ea.		9.70		9.70	15.50
3120	Final cleaning by sandblasting	E-11	1100	.029	S.F.	.52	1.41	.27	2.20	3.19
3130	By high pressure water	C-29	500	.016			.71	.19	.90	1.34
3140	Blow off dust and debris with oil-free dry compressed air	C-28	16000	.001			.03		.03	.04
3150	Build wood forms and drill holes for injection ports	C-2	960	.050	SFCA	4.63	2.67		7.30	9.35
3160	Assemble PVC injection ports with ball valves and install in forms	1 Plum	8	1	Ea.	90.50	67.50		158	204
3170	Insert expansion shield, coil rod, coil tie and plastic cone in each hole	1 Carp	32	.250	"	5.50	13.70		19.20	28
3180	Install wood forms and fasten with coil rod and coil nut	C-2	960	.050	SFCA	.58	2.67		3.25	4.89
3190	Mix up self-consolidating repair material with concrete mixer	C-30	200	.040	C.F.	12.75	1.78	.74	15.27	17.65
3200	Place repair material by pump and pressurize	C-9	200	.160	"		7.85	2.38	10.23	14.95
3210	Remove wood forms	C-2	1600	.030	SFCA		1.60		1.60	2.55
3220	Chip/grind off extra concrete at injection ports	B-9E	64	.250	Ea.	.12	12.05	.51	12.68	19.70
3230	Break ties and patch voids	1 Cefi	270	.030	"	.08	1.53		1.61	2.49
3240	Cure with sprayed membrane curing compound	2 Cefi	6200	.003	S.F.	.12	.13		.25	.34
4000	Vertical & overhead spall repair by hand application (ACI RAP-6)									
4001	Generally recommended for thin repairs that are cosmetic in nature									
4010	Remove concrete in repair areas to fully expose reinforcing bars	B-9E	160	.100	S.F.		4.81	.21	5.02	7.85
4020	Saw cut the perimeter of each repair area to 1/2" depth	B-89C	160	.050	L.F.	.02	2.59	.36	2.97	4.48
4030	If corroded rebar is exposed, remove concrete to 3/4" below corroded rebars	B-9E	64	.250	S.F.		12.05	.51	12.56	19.55
4031	Single layer of #4 rebar		53.30	.300			14.45	.62	15.07	23.50
4032	#5 rebar		45.70	.350			16.85	.72	17.57	27.50
4033	#6 rebar		40	.400			19.25	.82	20.07	31.50
4034	#7 rebar		35.60	.449			21.50	.92	22.42	35
4035	#8 rebar		35.60	.449			21.50	.92	22.42	35
4036	Double layer of #4 rebar		32	.500			24	1.03	25.03	39
4037	#5 rebar		29.10	.550			26.50	1.13	27.63	42.50
4038	#6 rebar		26.70	.599			29	1.23	30.23	47
4039	#7 rebar		24.60	.650			31.50	1.33	32.83	51
4040	#8 rebar	B-9E	24.60	.650	S.F.		31.50	1.33	32.83	51
4080	See section 03 21 11.60 for placement of supplemental reinforcement bars									
4090	See section 02 41 19.19 for handling & removal of debris									
4095	See section 02 41 19.20 for disposal of debris									
4100	Final cleaning by sandblasting	E-11	1100	.029	S.F.	.52	1.41	.27	2.20	3.19
4110	By high pressure water	C-29	500	.016			.71	.19	.90	1.34
4120	Blow off dust and debris with oil-free dry compressed air	C-28	16000	.001			.03		.03	.04
4130	Brush on bonding material	1 Cefi	320	.025		.11	1.30		1.41	2.15
4140	Mix up bagged repair material by hand with a mixing paddle	"	30	.267	C.F.	73.50	13.80		87.30	102
4150	By concrete mixer	C-30	200	.040		73.50	1.78	.74	76.02	84
4160	Place repair material by hand in lifts up to 2" deep for vertical repairs	C-10	48	.500			24.50		24.50	39
4170	Up to 1" deep for overhead repairs		36	.667			33		33	52
4180	Screed and initial float of repair material by hand		2000	.012	S.F.		.59		.59	.93
4190	Float and steel trowel finish repair material by hand		1265	.019			.94		.94	1.47
4200	Cure with sprayed membrane curing compound	2 Cefi	6200	.003		.12	.13		.25	.34
5000	Spall repair of horizontal concrete surfaces (ACI RAP-7)									
5001	Suitable for structural slabs, slabs-on-grade, balconies, interior floors									

53

03 01 30.72 Concrete Surface Repairs	Crew	Daily Output	Labor-Hours	Unit	Material	2021 Bare Costs Labor	Equipment	Total	Total Incl O&P
5010	Sound the concrete surface to locate delaminated areas	1 Cefi	2000	.004	SF Flr.		.21		.21	.32
5020	Mark the perimeter of each repair area	"	2000	.004	"		.21		.21	.32
5030	Saw cut the perimeter of each repair area down to reinforcing bars	B-89	1060	.015	L.F.	.03	.79	1.01	1.83	2.39
5040	Remove concrete in repair areas to fully expose reinforcing bars	B-9E	160	.100	S.F.		4.81	.21	5.02	7.85
5050	If corroded rebar is exposed, remove concrete to 3/4" below corroded rebars									
5051	Single layer of #4 rebar	B-9E	64	.250	S.F.		12.05	.51	12.56	19.55
5052	#5 rebar		53.30	.300			14.45	.62	15.07	23.50
5053	#6 rebar		45.70	.350			16.85	.72	17.57	27.50
5054	#7 rebar		40	.400			19.25	.82	20.07	31.50
5055	#8 rebar		35.60	.449			21.50	.92	22.42	35
5056	Double layer of #4 rebar		35.60	.449			21.50	.92	22.42	35
5057	#5 rebar		32	.500			24	1.03	25.03	39
5058	#6 rebar		29.10	.550			26.50	1.13	27.63	42.50
5059	#7 rebar		26.70	.599			29	1.23	30.23	47
5060	#8 rebar		24.60	.650			31.50	1.33	32.83	51
5100	See section 03 21 11.60 for placement of supplemental reinforcement bars									
5110	See section 02 41 19.19 for handling & removal of debris									
5115	See section 02 41 19.20 for disposal of debris									
5120	Final cleaning by sandblasting	E-11	1100	.029	S.F.	.52	1.41	.27	2.20	3.19
5130	By high pressure water	C-29	500	.016	"		.71	.19	.90	1.34
5140	Blow off dust and debris with oil-free dry compressed air	C-28	16000	.001	S.F.		.03		.03	.04
5160	Brush on bonding material	1 Cefi	320	.025	"	.11	1.30		1.41	2.15
5170	Mix up bagged repair material by hand with a mixing paddle	"	30	.267	C.F.	73.50	13.80		87.30	102
5180	By concrete mixer	C-30	200	.040		73.50	1.78	.74	76.02	84
5190	Place repair material by hand, consolidate and screed	C-6A	200	.080			4.14	.23	4.37	6.75
5200	Bull float, manual float & broom finish	C-10	1850	.013	S.F.		.64		.64	1.01
5210	Bull float, manual float & trowel finish	"	1265	.019			.94		.94	1.47
5220	Cure with sprayed membrane curing compound	2 Cefi	6200	.003		.12	.13		.25	.34
6000	Spall repairs by the Preplaced Aggregate method (ACI RAP-9)									
6001	Suitable for vertical, overhead, and underwater repairs									
6010	Sound the concrete surface to locate delaminated areas	1 Cefi	2000	.004	S.F.		.21		.21	.32
6020	Remove concrete in repair areas to fully expose reinforcing bars	B-9E	160	.100			4.81	.21	5.02	7.85
6030	Mark the perimeter of each repair area	1 Cefi	2000	.004			.21		.21	.32
6040	Saw cut the perimeter of each repair area down to reinforcing bars	B-89C	160	.050	L.F.	.03	2.59	.36	2.98	4.50
6050	If corroded rebar is exposed, remove concrete to 3/4" below corroded rebars									
6051	Single layer of #4 rebar	B-9E	64	.250	S.F.		12.05	.51	12.56	19.55
6052	#5 rebar		53.30	.300			14.45	.62	15.07	23.50
6053	#6 rebar		45.70	.350			16.85	.72	17.57	27.50
6054	#7 rebar		40	.400			19.25	.82	20.07	31.50
6055	#8 rebar		35.60	.449			21.50	.92	22.42	35
6056	Double layer of #4 rebar		35.60	.449			21.50	.92	22.42	35
6057	#5 rebar		32	.500			24	1.03	25.03	39
6058	#6 rebar		29.10	.550			26.50	1.13	27.63	42.50
6059	#7 rebar		26.70	.599			29	1.23	30.23	47
6060	#8 rebar		24.60	.650			31.50	1.33	32.83	51
6100	See section 03 21 11.60 for placement of supplemental reinforcement bars									
6104	See section 02 41 19.19 for handling & removal of debris									
6106	See section 02 41 19.20 for disposal of debris									
6110	Drill holes for expansion shields, 3/4" diameter x 4" deep	1 Carp	45	.178	Ea.		9.70		9.70	15.50
6120	Final cleaning by sandblasting	E-11	1100	.029	S.F.	.52	1.41	.27	2.20	3.19
6130	By high pressure water	C-29	500	.016			.71	.19	.90	1.34
6140	Blow off dust and debris with oil-free dry compressed air	C-28	16000	.001			.03		.03	.04
6150	Build wood forms and drill holes for injection ports	C-2	960	.050	SFCA	4.63	2.67		7.30	9.35

03 01 30 – Maintenance of Cast-In-Place Concrete

03 01 30.72 Concrete Surface Repairs

		Crew	Daily Output	Labor-Hours	Unit	Material	2021 Bare Costs Labor	Equipment	Total	Total Incl O&P
6160	Assemble PVC injection ports with ball valves and install in forms	1 Plum	8	1	Ea.	90.50	67.50		158	204
6170	Insert expansion shield, coil rod, coil tie and plastic cone in each hole	1 Carp	32	.250	"	5.50	13.70		19.20	28
6180	Install wood forms and fasten with coil rod and coil nut	C-2	960	.050	SFCA	.58	2.67		3.25	4.89
6185	Install pre-washed gap-graded aggregate as forms are installed	B-24	160	.150	C.F.	1.11	7.55		8.66	13.15
6190	Mix up pre-bagged grout material with concrete mixer	C-30	200	.040	C.F.	15.40	1.78	.74	17.92	20.50
6200	Place grout by pump from bottom up	C-9	200	.160	"		7.85	2.38	10.23	14.95
6210	Remove wood forms	C-2	1600	.030	SFCA		1.60		1.60	2.55
6220	Chip/grind off extra concrete at injection ports	B-9E	64	.250	Ea.	.12	12.05	.51	12.68	19.70
6230	Break ties and patch voids	1 Cefi	270	.030	"	.08	1.53		1.61	2.49
6240	Cure with sprayed membrane curing compound	2 Cefi	6200	.003	S.F.	.12	.13		.25	.34
7000	Leveling & reprofiling of vertical & overhead surfaces (ACI RAP-10)									
7001	Used to provide an acceptable surface for aesthetic or protective coatings									
7010	Clean the surface by sandblasting, leaving an open profile	E-11	1100	.029	S.F.	.52	1.41	.27	2.20	3.19
7020	By high pressure water	C-29	500	.016			.71	.19	.90	1.34
7030	Blow off dust and debris with oil-free dry compressed air	C-28	16000	.001	↓		.03		.03	.04
7040	Mix up bagged repair material by hand with a mixing paddle	1 Cefi	30	.267	C.F.	73.50	13.80		87.30	102
7050	Place repair material by hand with steel trowel to fill pores & honeycombs	C-10	1265	.019	S.F.		.94		.94	1.47
7060	Cure with sprayed membrane curing compound	2 Cefi	6200	.003	"	.12	.13		.25	.34
8000	Concrete repair by shotcrete application (ACI RAP-12)									
8001	Suitable for vertical and overhead surfaces									
8010	Sound the concrete surface to locate delaminated areas	1 Cefi	2000	.004	S.F.		.21		.21	.32
8020	Remove concrete in repair areas to fully expose reinforcing bars	B-9E	160	.100			4.81	.21	5.02	7.85
8030	Mark the perimeter of each repair area	1 Cefi	2000	.004	↓		.21		.21	.32
8040	Saw cut the perimeter of each repair area down to reinforcing bars	B-89C	160	.050	L.F.	.03	2.59	.36	2.98	4.50
8050	If corroded rebar is exposed, remove concrete to 3/4" below corroded rebars									
8051	Single layer of #4 rebar	B-9E	64	.250	S.F.		12.05	.51	12.56	19.55
8052	#5 rebar		53.30	.300			14.45	.62	15.07	23.50
8053	#6 rebar		45.70	.350			16.85	.72	17.57	27.50
8054	#7 rebar		40	.400			19.25	.82	20.07	31.50
8055	#8 rebar		35.60	.449			21.50	.92	22.42	35
8056	Double layer of #4 rebar		35.60	.449			21.50	.92	22.42	35
8057	#5 rebar		32	.500			24	1.03	25.03	39
8058	#6 rebar		29.10	.550			26.50	1.13	27.63	42.50
8059	#7 rebar		26.70	.599			29	1.23	30.23	47
8060	#8 rebar	↓	24.60	.650	↓		31.50	1.33	32.83	51
8100	See section 03 21 11.60 for placement of supplemental reinforcement bars									
8110	Final cleaning by sandblasting	E-11	1100	.029	S.F.	.52	1.41	.27	2.20	3.19
8120	By high pressure water	C-29	500	.016			.71	.19	.90	1.34
8130	Blow off dust and debris with oil-free dry compressed air	C-28	16000	.001	↓		.03		.03	.04
8140	Mix up repair material with concrete mixer	C-30	200	.040	C.F.	12.50	1.78	.74	15.02	17.40
8150	Place repair material by wet shotcrete process to full depth of repair	C-8C	200	.240	"		11.60	3.15	14.75	22
8160	Carefully trim excess material and steel trowel finish without overworking	C-10	1265	.019	S.F.		.94		.94	1.47
8170	Cure with sprayed membrane curing compound	2 Cefi	6200	.003	"	.12	.13		.25	.34
9000	Surface repair by Methacrylate flood coat (ACI RAP-13)									
9010	Suitable for healing and sealing horizontal surfaces only									
9020	Large cracks must previously have been repaired or filled									
9100	Shotblast entire surface to remove contaminants	A-1A	4000	.002	S.F.		.11	.05	.16	.24
9200	Blow off dust and debris with oil-free dry compressed air	C-28	16000	.001			.03		.03	.04
9300	Flood coat surface w/Methacrylate, distribute w/broom/squeegee, no prep.	3 Cefi	8000	.003		.97	.16		1.13	1.31
9400	Lightly broadcast even coat of dry silica sand while sealer coat is tacky	1 Cefi	8000	.001	↓	.01	.05		.06	.10

55

For customer support on your Commercial Renovation Costs with RSMeans Data, call 800.448.8182.

03 05 Common Work Results for Concrete

03 05 05 – Selective Demolition for Concrete

03 05 05.10 Selective Demolition, Concrete

		Crew	Daily Output	Labor-Hours	Unit	Material	2021 Bare Costs Labor	Equipment	Total	Total Incl O&P
0010	**SELECTIVE DEMOLITION, CONCRETE** R024119-10									
0012	Excludes saw cutting, torch cutting, loading or hauling									
0050	Break into small pieces, reinf. less than 1% of cross-sectional area	B-9	24	1.667	C.Y.		74.50	14.80	89.30	135
0060	Reinforcing 1% to 2% of cross-sectional area		16	2.500			112	22	134	203
0070	Reinforcing more than 2% of cross-sectional area		8	5			224	44.50	268.50	405
0150	Remove whole pieces, up to 2 tons per piece	E-18	36	1.111	Ea.		67	42.50	109.50	155
0160	2-5 tons per piece		30	1.333			80.50	51	131.50	185
0170	5-10 tons per piece		24	1.667			101	63.50	164.50	232
0180	10-15 tons per piece		18	2.222			134	85	219	310
0250	Precast unit embedded in masonry, up to 1 C.F.	D-1	16	1			48.50		48.50	78.50
0260	1-2 C.F.		12	1.333			65		65	104
0270	2-5 C.F.		10	1.600			78		78	125
0280	5-10 C.F.		8	2			97.50		97.50	157
0990	For hydrodemolition see Section 02 41 13.15									
1910	Minimum labor/equipment charge	B-9	2	20	Job		895	178	1,073	1,625

03 05 13 – Basic Concrete Materials

03 05 13.20 Concrete Admixtures and Surface Treatments

		Crew	Daily Output	Labor-Hours	Unit	Material	2021 Bare Costs Labor	Equipment	Total	Total Incl O&P
0010	**CONCRETE ADMIXTURES AND SURFACE TREATMENTS**									
0040	Abrasives, aluminum oxide, over 20 tons				Lb.	1.90			1.90	2.09
0050	1 to 20 tons					1.98			1.98	2.17
0070	Under 1 ton					2.05			2.05	2.26
0100	Silicon carbide, black, over 20 tons					2.97			2.97	3.27
0110	1 to 20 tons					3.09			3.09	3.40
0120	Under 1 ton					3.21			3.21	3.53
0200	Air entraining agent, .7 to 1.5 oz. per bag, 55 gallon drum				Gal.	22			22	24.50
0220	5 gallon pail				"	34.50			34.50	38
0300	Bonding agent, acrylic latex, 250 S.F./gallon, 5 gallon pail				Gal.	27			27	29.50
0320	Epoxy resin, 80 S.F./gallon, 4 gallon case				"	62.50			62.50	69
0400	Calcium chloride, 50 lb. bags, T.L. lots				Ton	1,450			1,450	1,575
0420	Less than truckload lots				Bag	24			24	26.50
0500	Carbon black, liquid, 2 to 8 lb. per bag of cement				Lb.	16.05			16.05	17.65
0600	Concrete admixture, integral colors, dry pigment, 5 lb. bag				Ea.	31			31	34
0610	10 lb. bag					47			47	51.50
0620	25 lb. bag					82			82	90.50
0920	Dustproofing compound, 250 S.F./gal., 5 gallon pail				Gal.	8			8	8.80
1010	Epoxy based, 125 S.F./gal., 5 gallon pail				"	65.50			65.50	72
1100	Hardeners, metallic, 55 lb. bags, natural (grey)				Lb.	1.04			1.04	1.15
1200	Colors					1.53			1.53	1.68
1300	Non-metallic, 55 lb. bags, natural grey					.60			.60	.66
1320	Colors					.85			.85	.93
1550	Release agent, for tilt slabs, 5 gallon pail				Gal.	18.55			18.55	20.50
1570	For forms, 5 gallon pail					13.90			13.90	15.30
1590	Concrete release agent for forms, 100% biodegradable, zero VOC, 5 gal. pail G					21.50			21.50	24
1595	55 gallon drum G					19.90			19.90	22
1600	Sealer, hardener and dustproofer, epoxy-based, 125 S.F./gal., 5 gallon unit					65.50			65.50	72
1620	3 gallon unit					72			72	79
1630	Sealer, solvent-based, 250 S.F./gal., 55 gallon drum					21			21	23
1640	5 gallon pail					41			41	45
1650	Sealer, water based, 350 S.F., 55 gallon drum					26.50			26.50	29
1660	5 gallon pail					36			36	39.50
1900	Set retarder, 100 S.F./gal., 1 gallon pail					6.35			6.35	6.95
2000	Waterproofing, integral 1 lb. per bag of cement				Lb.	1.04			1.04	1.14

56

03 05 Common Work Results for Concrete

03 05 13 – Basic Concrete Materials

03 05 13.20 Concrete Admixtures and Surface Treatments		Crew	Daily Output	Labor-Hours	Unit	Material	2021 Bare Costs Labor	Equipment	Total	Total Incl O&P
2100	Powdered metallic, 40 lbs. per 100 S.F., standard colors				Lb.	4.50			4.50	4.95
2120	Premium colors				▼	6.30			6.30	6.95
3000	For colored ready-mix concrete, add to prices in section 03 31 13.35									
3100	Subtle shades, 5 lb. dry pigment per C.Y., add				C.Y.	31			31	34
3400	Medium shades, 10 lb. dry pigment per C.Y., add					47			47	51.50
3700	Deep shades, 25 lb. dry pigment per C.Y., add				▼	82			82	90.50

03 05 13.85 Winter Protection

		Crew	Daily Output	Labor-Hours	Unit	Material	2021 Bare Costs Labor	Equipment	Total	Total Incl O&P
0010	**WINTER PROTECTION**									
0012	For heated ready mix, add				C.Y.	5.85			5.85	6.40
0100	Temporary heat to protect concrete, 24 hours	2 Clab	50	.320	M.S.F.	181	14.20		195.20	222
0200	Temporary shelter for slab on grade, wood frame/polyethylene sheeting									
0201	Build or remove, light framing for short spans	2 Carp	10	1.600	M.S.F.	530	87.50		617.50	720
0210	Large framing for long spans	"	3	5.333	"	620	292		912	1,150
0710	Electrically heated pads, 15 watts/S.F., 20 uses				S.F.	.72			.72	.80

03 11 Concrete Forming

03 11 13 – Structural Cast-In-Place Concrete Forming

03 11 13.20 Forms In Place, Beams and Girders

			Crew	Daily Output	Labor-Hours	Unit	Material	2021 Bare Costs Labor	Equipment	Total	Total Incl O&P
0010	**FORMS IN PLACE, BEAMS AND GIRDERS**	R031113-40									
0500	Exterior spandrel, job-built plywood, 12" wide, 1 use	R031113-60	C-2	225	.213	SFCA	4.66	11.35		16.01	23
0650	4 use			310	.155		1.51	8.25		9.76	14.80
1000	18" wide, 1 use			250	.192		4.22	10.25		14.47	21
1150	4 use			315	.152		1.39	8.10		9.49	14.50
1500	24" wide, 1 use			265	.181		3.86	9.65		13.51	19.65
1650	4 use			325	.148		1.25	7.85		9.10	13.95
2000	Interior beam, job-built plywood, 12" wide, 1 use			300	.160		5.45	8.55		14	19.60
2150	4 use			377	.127		1.78	6.80		8.58	12.75
2500	24" wide, 1 use			320	.150		3.96	8		11.96	17.10
2650	4 use			395	.122		1.28	6.50		7.78	11.70
3000	Encasing steel beam, hung, job-built plywood, 1 use			325	.148		4.48	7.85		12.33	17.50
3150	4 use			430	.112	▼	1.46	5.95		7.41	11.10
9000	Minimum labor/equipment charge		2 Carp	2	8	Job		440		440	695

03 11 13.25 Forms In Place, Columns

			Crew	Daily Output	Labor-Hours	Unit	Material	2021 Bare Costs Labor	Equipment	Total	Total Incl O&P
0010	**FORMS IN PLACE, COLUMNS**	R031113-40									
0500	Round fiberglass, 4 use per mo., rent, 12" diameter		C-1	160	.200	L.F.	22.50	10.45		32.95	41.50
0550	16" diameter	R031113-60		150	.213		26	11.10		37.10	46
0600	18" diameter			140	.229		28.50	11.90		40.40	50.50
0650	24" diameter			135	.237		35.50	12.35		47.85	58.50
0700	28" diameter			130	.246		40	12.85		52.85	64.50
0800	30" diameter			125	.256		42	13.35		55.35	67.50
0850	36" diameter			120	.267		55	13.90		68.90	82.50
1500	Round fiber tube, recycled paper, 1 use, 8" diameter	G		155	.206		2.94	10.75		13.69	20.50
1550	10" diameter	G		155	.206		3.51	10.75		14.26	21
1600	12" diameter	G		150	.213		4.14	11.10		15.24	22.50
1650	14" diameter	G		145	.221		5.85	11.50		17.35	25
1700	16" diameter	G		140	.229		6.30	11.90		18.20	26
1720	18" diameter	G		140	.229		7.40	11.90		19.30	27
1750	20" diameter	G		135	.237		7.65	12.35		20	28
1800	24" diameter	G		130	.246		12.70	12.85		25.55	34.50
1850	30" diameter	G		125	.256		17.60	13.35		30.95	41

03 11 13.25 Forms In Place, Columns

			Crew	Daily Output	Labor-Hours	Unit	Material	2021 Bare Costs Labor	Equipment	Total	Total Incl O&P
1900	36" diameter	G	C-1	115	.278	L.F.	21.50	14.50		36	46.50
1950	42" diameter	G		100	.320		43	16.70		59.70	74
2000	48" diameter	G		85	.376		89	19.65		108.65	130
2200	For seamless type, add						15%				
3000	Round, steel, 4 use per mo., rent, regular duty, 14" diameter	G	C-1	145	.221	L.F.	18.50	11.50		30	39
3050	16" diameter	G		125	.256		18.85	13.35		32.20	42.50
3100	Heavy duty, 20" diameter	G		105	.305		20.50	15.90		36.40	48
3150	24" diameter	G		85	.376		22.50	19.65		42.15	56
3200	30" diameter	G		70	.457		25.50	24		49.50	66
3250	36" diameter	G		60	.533		27.50	28		55.50	74.50
3300	48" diameter	G		50	.640		41.50	33.50		75	99
3350	60" diameter	G		45	.711		50.50	37		87.50	115
4500	For second and succeeding months, deduct						50%				
5000	Job-built plywood, 8" x 8" columns, 1 use		C-1	165	.194	SFCA	3.73	10.10		13.83	20
5050	2 use			195	.164		2.14	8.55		10.69	15.95
5100	3 use			210	.152		1.49	7.95		9.44	14.30
5150	4 use			215	.149		1.22	7.75		8.97	13.70
5500	12" x 12" columns, 1 use			180	.178		3.76	9.25		13.01	18.90
5550	2 use			210	.152		2.07	7.95		10.02	14.95
5600	3 use			220	.145		1.50	7.60		9.10	13.70
5650	4 use			225	.142		1.22	7.40		8.62	13.15
6000	16" x 16" columns, 1 use			185	.173		3.85	9		12.85	18.60
6050	2 use			215	.149		2.04	7.75		9.79	14.60
6100	3 use			230	.139		1.55	7.25		8.80	13.25
6150	4 use			235	.136		1.26	7.10		8.36	12.70
6500	24" x 24" columns, 1 use			190	.168		4.43	8.80		13.23	18.85
6550	2 use			216	.148		2.44	7.70		10.14	15
6600	3 use			230	.139		1.77	7.25		9.02	13.50
6650	4 use			238	.134		1.44	7		8.44	12.75
7000	36" x 36" columns, 1 use			200	.160		3.55	8.35		11.90	17.20
7050	2 use			230	.139		1.99	7.25		9.24	13.75
7100	3 use			245	.131		1.42	6.80		8.22	12.40
7150	4 use			250	.128		1.15	6.65		7.80	11.85
7400	Steel framed plywood, based on 50 uses of purchased										
7420	forms, and 4 uses of bracing lumber										
7500	8" x 8" column		C-1	340	.094	SFCA	2.29	4.91		7.20	10.30
7550	10" x 10"			350	.091		2.03	4.77		6.80	9.85
7600	12" x 12"			370	.086		1.72	4.51		6.23	9.10
7650	16" x 16"			400	.080		1.33	4.17		5.50	8.10
7700	20" x 20"			420	.076		1.18	3.97		5.15	7.60
7750	24" x 24"			440	.073		.84	3.79		4.63	6.95
7755	30" x 30"			440	.073		1.04	3.79		4.83	7.20
7760	36" x 36"			460	.070		.92	3.63		4.55	6.75
9000	Minimum labor/equipment charge		2 Carp	2	8	Job		440		440	695

03 11 13.35 Forms In Place, Elevated Slabs

			Crew	Daily Output	Labor-Hours	Unit	Material	2021 Bare Costs Labor	Equipment	Total	Total Incl O&P
0010	**FORMS IN PLACE, ELEVATED SLABS**	R031113-40									
1000	Flat plate, job-built plywood, to 15' high, 1 use		C-2	470	.102	S.F.	5.75	5.45		11.20	15
1150	4 use	R031113-60		560	.086		1.87	4.57		6.44	9.35
1500	15' to 20' high ceilings, 4 use			495	.097		1.90	5.15		7.05	10.35
2000	Flat slab, drop panels, job-built plywood, to 15' high, 1 use			449	.107		6.75	5.70		12.45	16.55
2150	4 use			544	.088		2.20	4.70		6.90	9.90
2250	15' to 20' high ceilings, 4 use			480	.100		3.15	5.35		8.50	11.95

03 11 13 – Structural Cast-In-Place Concrete Forming

03 11 13.35 Forms In Place, Elevated Slabs

		Crew	Daily Output	Labor-Hours	Unit	Material	2021 Bare Costs Labor	Equipment	Total	Total Incl O&P
3500	Floor slab, with 1-way joist pans, 1 use	C-2	415	.116	S.F.	10.20	6.15		16.35	21
3650	4 use		500	.096		5.65	5.10		10.75	14.35
4500	With 2-way waffle domes, 1 use		405	.119		10.75	6.30		17.05	22
4550	4 use		470	.102		6.15	5.45		11.60	15.45
5000	Box out for slab openings, over 16" deep, 1 use		190	.253	SFCA	4.83	13.45		18.28	27
5050	2 use		240	.200	"	2.66	10.65		13.31	19.90
5500	Shallow slab box outs, to 10 S.F.		42	1.143	Ea.	19.20	61		80.20	118
5550	Over 10 S.F. (use perimeter)		600	.080	L.F.	2.56	4.27		6.83	9.60
6000	Bulkhead forms for slab, with keyway, 1 use, 2 piece		500	.096		2.35	5.10		7.45	10.75
6100	3 piece (see also edge forms)		460	.104		2.68	5.55		8.23	11.80
6200	Slab bulkhead form, 4-1/2" high, exp metal, w/keyway & stakes [G]	C-1	1200	.027		.94	1.39		2.33	3.24
6210	5-1/2" high [G]		1100	.029		1.19	1.52		2.71	3.72
6215	7-1/2" high [G]		960	.033		1.40	1.74		3.14	4.31
6220	9-1/2" high [G]		840	.038		1.56	1.99		3.55	4.88
6500	Curb forms, wood, 6" to 12" high, on elevated slabs, 1 use		180	.178	SFCA	2.63	9.25		11.88	17.65
6550	2 use		205	.156		1.45	8.15		9.60	14.55
6600	3 use		220	.145		1.05	7.60		8.65	13.20
6650	4 use		225	.142		.86	7.40		8.26	12.75
7000	Edge forms to 6" high, on elevated slab, 4 use		500	.064	L.F.	.34	3.34		3.68	5.70
7070	7" to 12" high, 1 use		162	.198	SFCA	1.94	10.30		12.24	18.55
7080	2 use		198	.162		1.07	8.45		9.52	14.55
7090	3 use		222	.144		.78	7.50		8.28	12.80
7101	4 use		350	.091		.63	4.77		5.40	8.30
7500	Depressed area forms to 12" high, 4 use		300	.107	L.F.	1.07	5.55		6.62	10.05
7550	12" to 24" high, 4 use		175	.183		1.46	9.55		11.01	16.80
8000	Perimeter deck and rail for elevated slabs, straight		90	.356		18.85	18.55		37.40	50
8050	Curved		65	.492		26	25.50		51.50	69.50
8500	Void forms, round plastic, 8" high x 3" diameter [G]		450	.071	Ea.	1.62	3.71		5.33	7.70
8550	4" diameter [G]		425	.075		2.41	3.93		6.34	8.90
8600	6" diameter [G]		400	.080		4	4.17		8.17	11.05
8650	8" diameter [G]		375	.085		7.10	4.45		11.55	14.90
9000	Minimum labor/equipment charge	2 Carp	2	8	Job		440		440	695

03 11 13.40 Forms In Place, Equipment Foundations

		Crew	Daily Output	Labor-Hours	Unit	Material	2021 Bare Costs Labor	Equipment	Total	Total Incl O&P
0010	**FORMS IN PLACE, EQUIPMENT FOUNDATIONS** R031113-40									
0020	1 use	C-2	160	.300	SFCA	3.96	16		19.96	30
0050	2 use R031113-60		190	.253		2.18	13.45		15.63	24
0100	3 use		200	.240		1.59	12.80		14.39	22
0150	4 use		205	.234		1.30	12.50		13.80	21.50
9000	Minimum labor/equipment charge	1 Carp	3	2.667	Job		146		146	232

03 11 13.45 Forms In Place, Footings

		Crew	Daily Output	Labor-Hours	Unit	Material	2021 Bare Costs Labor	Equipment	Total	Total Incl O&P
0010	**FORMS IN PLACE, FOOTINGS** R031113-40									
0020	Continuous wall, plywood, 1 use	C-1	375	.085	SFCA	7.85	4.45		12.30	15.75
0150	4 use R031113-60	"	485	.066	"	2.56	3.44		6	8.30
1500	Keyway, 4 use, tapered wood, 2" x 4"	1 Carp	530	.015	L.F.	.37	.83		1.20	1.72
1550	2" x 6"	"	500	.016	"	.53	.88		1.41	1.97
3000	Pile cap, square or rectangular, job-built plywood, 1 use	C-1	290	.110	SFCA	4.48	5.75		10.23	14.10
3150	4 use		383	.084		1.46	4.36		5.82	8.55
5000	Spread footings, job-built lumber, 1 use		305	.105		3.28	5.45		8.73	12.30
5150	4 use		414	.077		1.06	4.03		5.09	7.55
9000	Minimum labor/equipment charge	1 Carp	3	2.667	Job		146		146	232

For customer support on your Commercial Renovation Costs with RSMeans Data, call 800.448.8182.

59

03 11 Concrete Forming

03 11 13 – Structural Cast-In-Place Concrete Forming

03 11 13.47 Forms In Place, Gas Station Forms

		Crew	Daily Output	Labor-Hours	Unit	Material	2021 Bare Costs Labor	Equipment	Total	Total Incl O&P
0010	**FORMS IN PLACE, GAS STATION FORMS**									
0050	Curb fascia, with template, 12 ga. steel, left in place, 9" high [G]	1 Carp	50	.160	L.F.	13.80	8.75		22.55	29
1000	Sign or light bases, 18" diameter, 9" high [G]		9	.889	Ea.	87	48.50		135.50	174
1050	30" diameter, 13" high [G]		8	1	"	138	54.50		192.50	239
1990	Minimum labor/equipment charge		2	4	Job		219		219	350
2000	Island forms, 10' long, 9" high, 3'-6" wide [G]	C-1	10	3.200	Ea.	385	167		552	690
2050	4' wide [G]		9	3.556		400	185		585	735
2500	20' long, 9" high, 4' wide [G]		6	5.333		640	278		918	1,150
2550	5' wide [G]		5	6.400		670	335		1,005	1,275
9000	Minimum labor/equipment charge	1 Carp	3	2.667	Job		146		146	232

03 11 13.50 Forms In Place, Grade Beam

		Crew	Daily Output	Labor-Hours	Unit	Material	2021 Bare Costs Labor	Equipment	Total	Total Incl O&P
0010	**FORMS IN PLACE, GRADE BEAM** R031113-40									
0020	Job-built plywood, 1 use	C-2	530	.091	SFCA	3.97	4.83		8.80	12.05
0150	4 use R031113-60	"	605	.079	"	1.29	4.23		5.52	8.15
9000	Minimum labor/equipment charge	2 Carp	2	8	Job		440		440	695

03 11 13.65 Forms In Place, Slab On Grade

		Crew	Daily Output	Labor-Hours	Unit	Material	2021 Bare Costs Labor	Equipment	Total	Total Incl O&P
0010	**FORMS IN PLACE, SLAB ON GRADE** R031113-40									
1000	Bulkhead forms w/keyway, wood, 6" high, 1 use	C-1	510	.063	L.F.	1.65	3.27		4.92	7
1400	Bulkhead form for slab, 4-1/2" high, exp metal, incl. keyway & stakes R031113-60		1200	.027		.94	1.39		2.33	3.24
1410	5-1/2" high [G]		1100	.029		1.19	1.52		2.71	3.72
1420	7-1/2" high [G]		960	.033		1.40	1.74		3.14	4.31
1430	9-1/2" high [G]		840	.038		1.56	1.99		3.55	4.88
2000	Curb forms, wood, 6" to 12" high, on grade, 1 use		215	.149	SFCA	2.99	7.75		10.74	15.65
2150	4 use	C-1	275	.116	SFCA	.97	6.05		7.02	10.70
3000	Edge forms, wood, 4 use, on grade, to 6" high		600	.053	L.F.	.45	2.78		3.23	4.93
3050	7" to 12" high		435	.074	SFCA	1.14	3.83		4.97	7.35
3060	Over 12"		350	.091	"	1.43	4.77		6.20	9.15
3500	For depressed slabs, 4 use, to 12" high		300	.107	L.F.	1.18	5.55		6.73	10.15
3550	To 24" high		175	.183		1.53	9.55		11.08	16.90
4000	For slab blockouts, to 12" high, 1 use		200	.160		1.20	8.35		9.55	14.60
4050	To 24" high, 1 use		120	.267		1.51	13.90		15.41	23.50
4100	Plastic (extruded), to 6" high, multiple use, on grade		800	.040		6.50	2.09		8.59	10.45
5000	Screed, 24 ga. metal key joint, see Section 03 15 16.30									
5020	Wood, incl. wood stakes, 1" x 3"	C-1	900	.036	L.F.	1.23	1.85		3.08	4.31
5050	2" x 4"		900	.036	"	1.31	1.85		3.16	4.39
6000	Trench forms in floor, wood, 1 use		160	.200	SFCA	3.15	10.45		13.60	20
6150	4 use		185	.173	"	1.02	9		10.02	15.50
8760	Void form, corrugated fiberboard, 4" x 12", 4' long [G]		3000	.011	S.F.	3.85	.56		4.41	5.15
8770	6" x 12", 4' long		3000	.011		4.64	.56		5.20	6
8780	1/4" thick hardboard protective cover for void form	2 Carp	1500	.011		.94	.58		1.52	1.96
9000	Minimum labor/equipment charge	1 Carp	2	4	Job		219		219	350

03 11 13.85 Forms In Place, Walls

		Crew	Daily Output	Labor-Hours	Unit	Material	2021 Bare Costs Labor	Equipment	Total	Total Incl O&P
0010	**FORMS IN PLACE, WALLS** R031113-10									
0100	Box out for wall openings, to 16" thick, to 10 S.F.	C-2	24	2	Ea.	42	107		149	216
0150	Over 10 S.F. (use perimeter) R031113-40	"	280	.171	L.F.	3.66	9.15		12.81	18.55
0250	Brick shelf, 4" W, add to wall forms, use wall area above shelf									
0260	1 use R031113-60	C-2	240	.200	SFCA	4.02	10.65		14.67	21.50
0350	4 use		300	.160	"	1.61	8.55		10.16	15.35
0500	Bulkhead, wood with keyway, 1 use, 2 piece		265	.181	L.F.	3.19	9.65		12.84	18.90
0600	Bulkhead forms with keyway, 1 piece expanded metal, 8" wall [G]	C-1	1000	.032		1.40	1.67		3.07	4.20
0610	10" wall [G]		800	.040		1.56	2.09		3.65	5.05

03 11 Concrete Forming

03 11 13 – Structural Cast-In-Place Concrete Forming

03 11 13.85 Forms In Place, Walls

			Crew	Daily Output	Labor-Hours	Unit	Material	2021 Bare Costs Labor	Equipment	Total	Total Incl O&P
0620	12" wall	G	C-1	525	.061	L.F.	1.87	3.18		5.05	7.10
0700	Buttress, to 8' high, 1 use		C-2	350	.137	SFCA	5.60	7.30		12.90	17.80
0850	4 use			480	.100	"	1.84	5.35		7.19	10.55
1000	Corbel or haunch, to 12" wide, add to wall forms, 1 use			150	.320	L.F.	3.75	17.05		20.80	31
1150	4 use			180	.267	"	1.22	14.20		15.42	24
2000	Wall, job-built plywood, to 8' high, 1 use			370	.130	SFCA	4.22	6.90		11.12	15.65
2050	2 use			435	.110		2.72	5.90		8.62	12.35
2100	3 use			495	.097		1.98	5.15		7.13	10.45
2150	4 use			505	.095		1.61	5.05		6.66	9.80
2400	Over 8' to 16' high, 1 use			280	.171		4.66	9.15		13.81	19.65
2450	2 use			345	.139		2.03	7.40		9.43	14.05
2500	3 use		C-2	375	.128	SFCA	1.45	6.80		8.25	12.45
2550	4 use			395	.122		1.19	6.50		7.69	11.60
2700	Over 16' high, 1 use			235	.204		4.14	10.90		15.04	22
2750	2 use			290	.166		2.28	8.85		11.13	16.55
2800	3 use			315	.152		1.66	8.10		9.76	14.75
2850	4 use			330	.145		1.35	7.75		9.10	13.85
4000	Radial, smooth curved, job-built plywood, 1 use			245	.196		3.77	10.45		14.22	21
4150	4 use			335	.143		1.23	7.65		8.88	13.50
4200	Below grade, job-built plywood, 1 use			225	.213		4.23	11.35		15.58	23
4210	2 use			225	.213		2.33	11.35		13.68	20.50
4220	3 use			225	.213		1.96	11.35		13.31	20.50
4230	4 use			225	.213		1.38	11.35		12.73	19.60
4600	Retaining wall, battered, job-built plyw'd, to 8' high, 1 use			300	.160		3.08	8.55		11.63	17
4750	4 use			390	.123		1	6.55		7.55	11.55
4900	Over 8' to 16' high, 1 use			240	.200		3.39	10.65		14.04	20.50
5050	4 use			320	.150		1.10	8		9.10	13.95
7500	Lintel or sill forms, 1 use		1 Carp	30	.267		5.15	14.60		19.75	28.50
7560	4 use		"	37	.216		1.68	11.85		13.53	20.50
7800	Modular prefabricated plywood, based on 20 uses of purchased										
7820	forms, and 4 uses of bracing lumber										
7860	To 8' high		C-2	800	.060	SFCA	1.31	3.20		4.51	6.55
8060	Over 8' to 16' high			600	.080		1.40	4.27		5.67	8.35
8600	Pilasters, 1 use			270	.178		4.92	9.50		14.42	20.50
8660	4 use			385	.125		1.60	6.65		8.25	12.35
9010	Steel framed plywood, based on 50 uses of purchased										
9020	forms, and 4 uses of bracing lumber										
9060	To 8' high		C-2	600	.080	SFCA	.74	4.27		5.01	7.60
9260	Over 8' to 16' high			450	.107		.74	5.70		6.44	9.85
9460	Over 16' to 20' high			400	.120		.74	6.40		7.14	11
9475	For elevated walls, add							10%			
9480	For battered walls, 1 side battered, add							10%	10%		
9485	For battered walls, 2 sides battered, add							15%	15%		
9900	Minimum labor/equipment charge		2 Carp	2	8	Job		440		440	695

03 11 16 – Architectural Cast-in-Place Concrete Forming

03 11 16.13 Concrete Form Liners

			Crew	Daily Output	Labor-Hours	Unit	Material	2021 Bare Costs Labor	Equipment	Total	Total Incl O&P
0010	**CONCRETE FORM LINERS**										
5750	Liners for forms (add to wall forms), ABS plastic										
5800	Aged wood, 4" wide, 1 use		1 Carp	256	.031	SFCA	3.11	1.71		4.82	6.15
5820	2 use			256	.031		1.71	1.71		3.42	4.60
5830	3 use			256	.031		1.24	1.71		2.95	4.09
5840	4 use		1 Carp	256	.031	SFCA	1.01	1.71		2.72	3.83

03 11 Concrete Forming

03 11 16 – Architectural Cast-in-Place Concrete Forming

03 11 16.13 Concrete Form Liners

		Crew	Daily Output	Labor-Hours	Unit	Material	2021 Bare Costs Labor	Equipment	Total	Total Incl O&P
5900	Fractured rope rib, 1 use	1 Carp	192	.042	SFCA	4.57	2.28		6.85	8.70
5925	2 use		192	.042		2.51	2.28		4.79	6.40
5950	3 use		192	.042		1.83	2.28		4.11	5.65
6000	4 use		192	.042		1.49	2.28		3.77	5.25
6100	Ribbed, 3/4" deep x 1-1/2" OC, 1 use		224	.036		5.05	1.95		7	8.65
6125	2 use		224	.036		2.77	1.95		4.72	6.15
6150	3 use		224	.036		2.01	1.95		3.96	5.30
6200	4 use		224	.036		1.63	1.95		3.58	4.91
6300	Rustic brick pattern, 1 use		224	.036		3.23	1.95		5.18	6.65
6325	2 use		224	.036		1.78	1.95		3.73	5.05
6350	3 use		224	.036		1.29	1.95		3.24	4.53
6400	4 use		224	.036		1.05	1.95		3	4.26
6500	3/8" striated, random, 1 use		224	.036		3.52	1.95		5.47	7
6525	2 use		224	.036		1.94	1.95		3.89	5.25
6550	3 use		224	.036		1.41	1.95		3.36	4.66
6600	4 use		224	.036		1.14	1.95		3.09	4.37

03 15 Concrete Accessories

03 15 05 – Concrete Forming Accessories

03 15 05.70 Shores

			Crew	Daily Output	Labor-Hours	Unit	Material	2021 Bare Costs Labor	Equipment	Total	Total Incl O&P
0010	**SHORES**										
0020	Erect and strip, by hand, horizontal members										
0500	Aluminum joists and stringers	G	2 Carp	60	.267	Ea.		14.60		14.60	23
0600	Steel, adjustable beams	G		45	.356			19.45		19.45	31
0700	Wood joists			50	.320			17.50		17.50	28
0800	Wood stringers			30	.533			29		29	46.50
1000	Vertical members to 10' high	G		55	.291			15.90		15.90	25.50
1050	To 13' high	G		50	.320			17.50		17.50	28
1100	To 16' high	G		45	.356			19.45		19.45	31
1500	Reshoring	G		1400	.011	S.F.	.62	.63		1.25	1.68
1600	Flying truss system	G	C-17D	9600	.009	SFCA		.50	.08	.58	.88
1760	Horizontal, aluminum joists, 6-1/4" high x 5' to 21' span, buy	G				L.F.	16.05			16.05	17.65
1770	Beams, 7-1/4" high x 4' to 30' span	G				"	19.25			19.25	21
1810	Horizontal, steel beam, W8x10, 7' span, buy	G				Ea.	63			63	69
1830	10' span	G					73			73	80
1920	15' span	G					125			125	138
1940	20' span	G					176			176	194
1970	Steel stringer, W8x10, 4' to 16' span, buy	G				L.F.	7.30			7.30	8
3000	Rent for job duration, aluminum joist @ 2' OC, per mo.	G				SF Flr.	.40			.40	.44
3050	Steel W8x10	G					.18			.18	.20
3060	Steel adjustable	G					.18			.18	.20
3500	#1 post shore, steel, 5'-7" to 9'-6" high, 10,000# cap., buy	G				Ea.	158			158	174
3550	#2 post shore, 7'-3" to 12'-10" high, 7800# capacity	G					182			182	200
3600	#3 post shore, 8'-10" to 16'-1" high, 3800# capacity	G					199			199	219
5010	Frame shoring systems, steel, 12,000#/leg, buy										
5040	Frame, 2' wide x 6' high	G				Ea.	120			120	132
5250	X-brace	G					19.20			19.20	21
5550	Base plate	G					18.05			18.05	19.90
5600	Screw jack	G					36			36	39.50
5650	U-head, 8" x 8"	G					23			23	25

03 15 Concrete Accessories

03 15 05 – Concrete Forming Accessories

03 15 05.85 Stair Tread Inserts

		Crew	Daily Output	Labor-Hours	Unit	Material	2021 Bare Costs Labor	Equipment	Total	Total Incl O&P
0010	**STAIR TREAD INSERTS**									
0105	Cast nosing insert, abrasive surface, pre-drilled, includes screws									
0110	Aluminum, 3" wide x 3' long	1 Cefi	32	.250	Ea.	57.50	12.95		70.45	84
0120	4' long		31	.258		74.50	13.35		87.85	103
0130	5' long		30	.267		89.50	13.80		103.30	120
0135	Extruded nosing insert, black abrasive strips, continuous anchor									
0140	Aluminum, 3" wide x 3' long	1 Cefi	64	.125	Ea.	35.50	6.50		42	49
0150	4' long		60	.133		48	6.90		54.90	64
0160	5' long		56	.143		62.50	7.40		69.90	80
0165	Extruded nosing insert, black abrasive strips, pre-drilled, incl. screws									
0170	Aluminum, 3" wide x 3' long	1 Cefi	32	.250	Ea.	45	12.95		57.95	70
0180	4' long		31	.258		59	13.35		72.35	86
0190	5' long		30	.267		83.50	13.80		97.30	113
9000	Minimum labor/equipment charge	1 Carp	4	2	Job		109		109	174

03 15 13 – Waterstops

03 15 13.50 Waterstops

		Crew	Daily Output	Labor-Hours	Unit	Material	2021 Bare Costs Labor	Equipment	Total	Total Incl O&P
0010	**WATERSTOPS**, PVC and Rubber									
0020	PVC, ribbed 3/16" thick, 4" wide	1 Carp	155	.052	L.F.	1.66	2.82		4.48	6.35
0050	6" wide		145	.055		2.62	3.02		5.64	7.70
0500	With center bulb, 6" wide, 3/16" thick		135	.059		2.73	3.24		5.97	8.15
0550	3/8" thick		130	.062		3.16	3.37		6.53	8.85
0600	9" wide x 3/8" thick		125	.064		4.17	3.50		7.67	10.15

03 15 16 – Concrete Construction Joints

03 15 16.20 Control Joints, Saw Cut

		Crew	Daily Output	Labor-Hours	Unit	Material	2021 Bare Costs Labor	Equipment	Total	Total Incl O&P
0010	**CONTROL JOINTS, SAW CUT**									
0100	Sawcut control joints in green concrete									
0120	1" depth	C-27	2000	.008	L.F.	.03	.41	.06	.50	.74
0140	1-1/2" depth		1800	.009		.04	.46	.06	.56	.84
0160	2" depth		1600	.010		.06	.52	.07	.65	.95
0180	Sawcut joint reservoir in cured concrete									
0182	3/8" wide x 3/4" deep, with single saw blade	C-27	1000	.016	L.F.	.04	.83	.11	.98	1.47
0184	1/2" wide x 1" deep, with double saw blades	"	900	.018	"	.08	.92	.13	1.13	1.67
0186	3/4" wide x 1-1/2" deep, with double saw blades	C-27	800	.020	L.F.	.17	1.04	.14	1.35	1.97
0190	Water blast joint to wash away laitance, 2 passes	C-29	2500	.003			.14	.04	.18	.27
0200	Air blast joint to blow out debris and air dry, 2 passes	C-28	2000	.004			.21	.02	.23	.34
0300	For backer rod, see Section 07 91 23.10									
0340	For joint sealant, see Section 03 15 16.30 or 07 92 13.20									
0900	For replacement of joint sealant, see Section 07 01 90.81									

03 15 16.30 Expansion Joints

			Crew	Daily Output	Labor-Hours	Unit	Material	2021 Bare Costs Labor	Equipment	Total	Total Incl O&P
0010	**EXPANSION JOINTS**										
0020	Keyed, cold, 24 ga., incl. stakes, 3-1/2" high	G	1 Carp	200	.040	L.F.	.89	2.19		3.08	4.46
0050	4-1/2" high	G		200	.040		.94	2.19		3.13	4.51
0100	5-1/2" high	G		195	.041		1.19	2.24		3.43	4.88
0150	7-1/2" high	G		190	.042		1.40	2.30		3.70	5.20
0160	9-1/2" high	G		185	.043		1.56	2.37		3.93	5.50
0300	Poured asphalt, plain, 1/2" x 1"		1 Clab	450	.018		.94	.79		1.73	2.29
0350	1" x 2"			400	.020		3.77	.89		4.66	5.55
0500	Neoprene, liquid, cold applied, 1/2" x 1"			450	.018		2.72	.79		3.51	4.25
0550	1" x 2"			400	.020		10.75	.89		11.64	13.20
0700	Polyurethane, poured, 2 part, 1/2" x 1"			400	.020		1.42	.89		2.31	2.97
0750	1" x 2"			350	.023		5.65	1.01		6.66	7.85

63

For customer support on your Commercial Renovation Costs with RSMeans Data, call 800.448.8182.

03 15 16.30 Expansion Joints

		Crew	Daily Output	Labor-Hours	Unit	Material	2021 Bare Costs Labor	Equipment	Total	Total Incl O&P
0900	Rubberized asphalt, hot or cold applied, 1/2" x 1"	1 Clab	450	.018	L.F.	.34	.79		1.13	1.63
0950	1" x 2"		400	.020		1.29	.89		2.18	2.83
1100	Hot applied, fuel resistant, 1/2" x 1"		450	.018		.51	.79		1.30	1.82
1150	1" x 2"		400	.020		1.94	.89		2.83	3.54
2000	Premolded, bituminous fiber, 1/2" x 6"	1 Carp	375	.021		.41	1.17		1.58	2.31
2050	1" x 12"		300	.027		1.63	1.46		3.09	4.11
2140	Concrete expansion joint, recycled paper and fiber, 1/2" x 6" G		390	.021		.43	1.12		1.55	2.26
2150	1/2" x 12" G		360	.022		.86	1.22		2.08	2.89
2250	Cork with resin binder, 1/2" x 6"		375	.021		1.49	1.17		2.66	3.50
2300	1" x 12"		300	.027		3.65	1.46		5.11	6.35
2500	Neoprene sponge, closed cell, 1/2" x 6"		375	.021		2.41	1.17		3.58	4.51
2550	1" x 12"		300	.027		7.85	1.46		9.31	10.90
2750	Polyethylene foam, 1/2" x 6"		375	.021		.65	1.17		1.82	2.58
2800	1" x 12"		300	.027		3.44	1.46		4.90	6.10
3000	Polyethylene backer rod, 3/8" diameter		460	.017		.16	.95		1.11	1.69
3050	3/4" diameter		460	.017		.08	.95		1.03	1.59
3100	1" diameter		460	.017		.15	.95		1.10	1.68
3500	Polyurethane foam, with polybutylene, 1/2" x 1/2"		475	.017		1.25	.92		2.17	2.85
3550	1" x 1"		450	.018		3.18	.97		4.15	5.05
3750	Polyurethane foam, regular, closed cell, 1/2" x 6"		375	.021		.93	1.17		2.10	2.88
3800	1" x 12"		300	.027		3.31	1.46		4.77	5.95
4000	Polyvinyl chloride foam, closed cell, 1/2" x 6"	1 Carp	375	.021	L.F.	2.52	1.17		3.69	4.63
4050	1" x 12"		300	.027		8.70	1.46		10.16	11.85
4250	Rubber, gray sponge, 1/2" x 6"		375	.021		2.12	1.17		3.29	4.19
4300	1" x 12"		300	.027		7.60	1.46		9.06	10.70
4400	Redwood heartwood, 1" x 4"		400	.020		2.97	1.09		4.06	5
4450	1" x 6"		375	.021		2.70	1.17		3.87	4.83
5000	For installation in walls, add						75%			
5250	For installation in boxouts, add						25%			

03 15 19 – Cast-In Concrete Anchors

03 15 19.05 Anchor Bolt Accessories

		Crew	Daily Output	Labor-Hours	Unit	Material	2021 Bare Costs Labor	Equipment	Total	Total Incl O&P
0010	**ANCHOR BOLT ACCESSORIES**									
0015	For anchor bolts set in fresh concrete, see Section 03 15 19.10									
8150	Anchor bolt sleeve, plastic, 1" diameter bolts	1 Carp	60	.133	Ea.	10.45	7.30		17.75	23
8500	1-1/2" diameter		28	.286		13.10	15.65		28.75	39.50
8600	2" diameter		24	.333		15.05	18.25		33.30	45.50
8650	3" diameter		20	.400		27.50	22		49.50	65.50

03 15 19.10 Anchor Bolts

		Crew	Daily Output	Labor-Hours	Unit	Material	2021 Bare Costs Labor	Equipment	Total	Total Incl O&P
0010	**ANCHOR BOLTS**									
0015	Made from recycled materials									
0025	Single bolts installed in fresh concrete, no templates									
0030	Hooked w/nut and washer, 1/2" diameter, 8" long G	1 Carp	132	.061	Ea.	1.55	3.32		4.87	7
0040	12" long G		131	.061		1.72	3.34		5.06	7.20
0050	5/8" diameter, 8" long G		129	.062		3.40	3.39		6.79	9.15
0060	12" long G		127	.063		4.18	3.45		7.63	10.10
0070	3/4" diameter, 8" long G		127	.063		4.18	3.45		7.63	10.10
0080	12" long G		125	.064		5.25	3.50		8.75	11.30
0090	2-bolt pattern, including job-built 2-hole template, per set									
0100	J-type, incl. hex nut & washer, 1/2" diameter x 6" long G	1 Carp	21	.381	Set	7.20	21		28.20	41
0110	12" long G		21	.381		7.90	21		28.90	41.50
0120	18" long G		21	.381		8.90	21		29.90	43
0130	3/4" diameter x 8" long G		20	.400		12.80	22		34.80	49

For customer support on your Commercial Renovation Costs with RSMeans Data, call 800.448.8182.

03 15 Concrete Accessories

03 15 19 – Cast-In Concrete Anchors

03 15 19.10 Anchor Bolts		Crew	Daily Output	Labor-Hours	Unit	Material	2021 Bare Costs Labor	Equipment	Total	Total Incl O&P
0140	12" long	G 1 Carp	20	.400	Set	14.90	22		36.90	51.50
0150	18" long	G	20	.400		18.05	22		40.05	55
0160	1" diameter x 12" long	G	19	.421		28	23		51	67.50
0170	18" long	G	19	.421		33	23		56	72.50
0180	24" long	G	19	.421		39	23		62	79.50
0190	36" long	G	18	.444		52	24.50		76.50	95.50
0200	1-1/2" diameter x 18" long	G	17	.471		49	25.50		74.50	95
0210	24" long	G	16	.500		57.50	27.50		85	107
0300	L-type, incl. hex nut & washer, 3/4" diameter x 12" long	G	20	.400		21.50	22		43.50	58.50
0310	18" long	G	20	.400		26.50	22		48.50	64
0320	24" long	G	20	.400		31	22		53	69.50
0330	30" long	G 1 Carp	20	.400	Set	38.50	22		60.50	77
0340	36" long	G	20	.400		43	22		65	82.50
0350	1" diameter x 12" long	G	19	.421		34	23		57	74
0360	18" long	G	19	.421		41.50	23		64.50	82
0370	24" long	G	19	.421		50	23		73	91.50
0380	30" long	G	19	.421		58.50	23		81.50	101
0390	36" long	G	18	.444		66	24.50		90.50	111
0400	42" long	G	18	.444		79	24.50		103.50	126
0410	48" long	G	18	.444		88.50	24.50		113	136
0420	1-1/4" diameter x 18" long	G	18	.444		43	24.50		67.50	86
0430	24" long	G	18	.444		50	24.50		74.50	94
0440	30" long	G	17	.471		57.50	25.50		83	104
0450	36" long	G	17	.471		64.50	25.50		90	112
1000	4-bolt pattern, including job-built 4-hole template, per set									
1100	J-type, incl. hex nut & washer, 1/2" diameter x 6" long	G 1 Carp	19	.421	Set	9.95	23		32.95	47.50
1110	12" long	G	19	.421		11.30	23		34.30	49
1120	18" long	G	18	.444		13.40	24.50		37.90	53
1130	3/4" diameter x 8" long	G	17	.471		21	25.50		46.50	64.50
1140	12" long	G	17	.471		25.50	25.50		51	69
1150	18" long	G	17	.471		31.50	25.50		57	76
1160	1" diameter x 12" long	G	16	.500		51.50	27.50		79	100
1170	18" long	G	15	.533		61	29		90	114
1180	24" long	G	15	.533		73.50	29		102.50	128
1190	36" long	G	15	.533		99	29		128	156
1200	1-1/2" diameter x 18" long	G	13	.615		94	33.50		127.50	157
1210	24" long	G	12	.667		111	36.50		147.50	180
1300	L-type, incl. hex nut & washer, 3/4" diameter x 12" long	G	17	.471		38.50	25.50		64	83.50
1310	18" long	G	17	.471		48.50	25.50		74	94
1320	24" long	G	17	.471		58	25.50		83.50	105
1330	30" long	G	16	.500		72.50	27.50		100	123
1340	36" long	G	16	.500		82	27.50		109.50	134
1350	1" diameter x 12" long	G	16	.500		63.50	27.50		91	114
1360	18" long	G	15	.533		78.50	29		107.50	133
1370	24" long	G	15	.533		95.50	29		124.50	152
1380	30" long	G	15	.533		112	29		141	170
1390	36" long	G	15	.533		128	29		157	187
1400	42" long	G	14	.571		154	31.50		185.50	219
1410	48" long	G	14	.571		172	31.50		203.50	239
1420	1-1/4" diameter x 18" long	G	14	.571		81.50	31.50		113	140
1430	24" long	G	14	.571		96	31.50		127.50	156
1440	30" long	G	13	.615		110	33.50		143.50	176
1450	36" long	G	13	.615		125	33.50		158.50	191

03 15 Concrete Accessories

03 15 19 – Cast-In Concrete Anchors

03 15 19.45 Machinery Anchors

		Crew	Daily Output	Labor-Hours	Unit	Material	2021 Bare Costs Labor	Equipment	Total	Total Incl O&P
0010	**MACHINERY ANCHORS,** heavy duty, incl. sleeve, floating base nut,									
0020	lower stud & coupling nut, fiber plug, connecting stud, washer & nut.									
0030	For flush mounted embedment in poured concrete heavy equip. pads.									
0200	Stud & bolt, 1/2" diameter	G E-16	40	.400	Ea.	60.50	24.50	3.72	88.72	110
0300	5/8" diameter	G	35	.457		67	28	4.25	99.25	124
0500	3/4" diameter	G	30	.533		80.50	32.50	4.96	117.96	147
0600	7/8" diameter	G	25	.640		95	39	5.95	139.95	174
0800	1" diameter	G	20	.800		105	49	7.45	161.45	203
0900	1-1/4" diameter	G	15	1.067		135	65.50	9.90	210.40	265

03 21 Reinforcement Bars

03 21 11 – Plain Steel Reinforcement Bars

03 21 11.60 Reinforcing In Place

		Crew	Daily Output	Labor-Hours	Unit	Material	2021 Bare Costs Labor	Equipment	Total	Total Incl O&P
0010	**REINFORCING IN PLACE,** 50-60 ton lots, A615 Grade 60 R032110-10									
0020	Includes labor, but not material cost, to install accessories									
0030	Made from recycled materials									
0102	Beams & Girders, #3 to #7	G 4 Rodm	3200	.010	Lb.	.62	.59		1.21	1.60
0152	#8 to #18	G	5400	.006		.62	.35		.97	1.23
0202	Columns, #3 to #7	G	3000	.011		.62	.63		1.25	1.66
0252	#8 to #18	G	4600	.007		.62	.41		1.03	1.32
0402	Elevated slabs, #4 to #7	G	5800	.006		.62	.32		.94	1.19
0502	Footings, #4 to #7	G	4200	.008		.62	.45		1.07	1.38
0552	#8 to #18	G	7200	.004		.62	.26		.88	1.09
0602	Slab on grade, #3 to #7	G	4200	.008		.62	.45		1.07	1.38
0702	Walls, #3 to #7	G	6000	.005		.62	.31		.93	1.17
0752	#8 to #18	G	8000	.004		.62	.24		.86	1.05
0900	For other than 50-60 ton lots									
1000	Under 10 ton job, #3 to #7, add					25%	10%			
1010	#8 to #18, add					20%	10%			
1050	10-50 ton job, #3 to #7, add					10%				
1060	#8 to #18, add					5%				
1100	60-100 ton job, #3 to #7, deduct					5%				
1110	#8 to #18, deduct					10%				
1150	Over 100 ton job, #3 to #7, deduct					10%				
1160	#8 to #18, deduct					15%				
2000	Unloading & sorting, add to above	C-5	100	.560	Ton		33	5.85	38.85	58
2200	Crane cost for handling, 90 picks/day, up to 1.5 tons/bundle, add to above		135	.415			24.50	4.35	28.85	43
2210	1.0 ton/bundle		92	.609			35.50	6.40	41.90	63
2220	0.5 ton/bundle		35	1.600			94	16.75	110.75	165
2400	Dowels, 2 feet long, deformed, #3	G 2 Rodm	520	.031	Ea.	.51	1.81		2.32	3.40
2410	#4	G 2 Rodm	480	.033	Ea.	.91	1.96		2.87	4.07
2420	#5	G	435	.037		1.42	2.17		3.59	4.96
2430	#6	G	360	.044		2.05	2.62		4.67	6.35
2450	Longer and heavier dowels, add	G	725	.022	Lb.	.68	1.30		1.98	2.78
2500	Smooth dowels, 12" long, 1/4" or 3/8" diameter	G	140	.114	Ea.	.87	6.75		7.62	11.50
2520	5/8" diameter	G	125	.128		1.53	7.55		9.08	13.50
2530	3/4" diameter	G	110	.145		1.89	8.55		10.44	15.50
2600	Dowel sleeves for CIP concrete, 2-part system									
2610	Sleeve base, plastic, for 5/8" smooth dowel sleeve, fasten to edge form	1 Rodm	200	.040	Ea.	.53	2.36		2.89	4.27
2615	Sleeve, plastic, 12" long, for 5/8" smooth dowel, snap onto base		400	.020		1.54	1.18		2.72	3.53

66

03 21 Reinforcement Bars

03 21 11 – Plain Steel Reinforcement Bars

03 21 11.60 Reinforcing In Place

		Crew	Daily Output	Labor-Hours	Unit	Material	2021 Bare Costs Labor	Equipment	Total	Total Incl O&P
2620	Sleeve base, for 3/4" smooth dowel sleeve	1 Rodm	175	.046	Ea.	.64	2.69		3.33	4.91
2625	Sleeve, 12" long, for 3/4" smooth dowel		350	.023		1.42	1.35		2.77	3.67
2630	Sleeve base, for 1" smooth dowel sleeve		150	.053		.68	3.14		3.82	5.65
2635	Sleeve, 12" long, for 1" smooth dowel		300	.027		1.37	1.57		2.94	3.97
2700	Dowel caps, visual warning only, plastic, #3 to #8	2 Rodm	800	.020		.39	1.18		1.57	2.27
2720	#8 to #18		750	.021		.89	1.26		2.15	2.94
2750	Impalement protective, plastic, #4 to #9		800	.020		1.22	1.18		2.40	3.18
9000	Minimum labor/equipment charge	1 Rodm	4	2	Job		118		118	184

03 21 21 – Composite Reinforcement Bars

03 21 21.11 Glass Fiber-Reinforced Polymer Reinf. Bars

		Crew	Daily Output	Labor-Hours	Unit	Material	2021 Bare Costs Labor	Equipment	Total	Total Incl O&P
0010	**GLASS FIBER-REINFORCED POLYMER REINFORCEMENT BARS**									
0020	Includes labor, but not material cost, to install accessories									
0050	#2 bar, .043 lb./L.F.	4 Rodm	9500	.003	L.F.	.41	.20		.61	.76
0100	#3 bar, .092 lb./L.F.		9300	.003		.66	.20		.86	1.05
0150	#4 bar, .160 lb./L.F.		9100	.004		.95	.21		1.16	1.37
0200	#5 bar, .258 lb./L.F.		8700	.004		1.17	.22		1.39	1.63
0250	#6 bar, .372 lb./L.F.		8300	.004		2.04	.23		2.27	2.60
0300	#7 bar, .497 lb./L.F.		7900	.004		2.52	.24		2.76	3.14
0350	#8 bar, .620 lb./L.F.		7400	.004		2.29	.25		2.54	2.92
0400	#9 bar, .800 lb./L.F.		6800	.005		4.15	.28		4.43	5
0450	#10 bar, 1.08 lb./L.F.		5800	.006		3.77	.32		4.09	4.66
0500	For bends, add per bend				Ea.	1.59			1.59	1.75

03 22 Fabric and Grid Reinforcing

03 22 11 – Plain Welded Wire Fabric Reinforcing

03 22 11.10 Plain Welded Wire Fabric

			Crew	Daily Output	Labor-Hours	Unit	Material	2021 Bare Costs Labor	Equipment	Total	Total Incl O&P
0010	**PLAIN WELDED WIRE FABRIC** ASTM A185	R032205-30									
0020	Includes labor, but not material cost, to install accessories										
0030	Made from recycled materials										
0050	Sheets										
0100	6 x 6 - W1.4 x W1.4 (10 x 10) 21 lb./C.S.F.	G	2 Rodm	35	.457	C.S.F.	14.95	27		41.95	58.50
0200	6 x 6 - W2.1 x W2.1 (8 x 8) 30 lb./C.S.F.	G	2 Rodm	31	.516	C.S.F.	21	30.50		51.50	71
0300	6 x 6 - W2.9 x W2.9 (6 x 6) 42 lb./C.S.F.	G		29	.552		33.50	32.50		66	88
0400	6 x 6 - W4 x W4 (4 x 4) 58 lb./C.S.F.	G		27	.593		39	35		74	97.50
0500	4 x 4 - W1.4 x W1.4 (10 x 10) 31 lb./C.S.F.	G		31	.516		24.50	30.50		55	74.50
0600	4 x 4 - W2.1 x W2.1 (8 x 8) 44 lb./C.S.F.	G		29	.552		30.50	32.50		63	84.50
0650	4 x 4 - W2.9 x W2.9 (6 x 6) 61 lb./C.S.F.	G		27	.593		49	35		84	108
0700	4 x 4 - W4 x W4 (4 x 4) 85 lb./C.S.F.	G		25	.640		61	37.50		98.50	126
0750	Rolls										
0800	2 x 2 - #14 galv., 21 lb./C.S.F., beam & column wrap	G	2 Rodm	6.50	2.462	C.S.F.	45.50	145		190.50	277
0900	2 x 2 - #12 galv. for gunite reinforcing	G	"	6.50	2.462	"	63	145		208	297
9000	Minimum labor/equipment charge		1 Rodm	4	2	Job		118		118	184

03 22 13 – Galvanized Welded Wire Fabric Reinforcing

03 22 13.10 Galvanized Welded Wire Fabric

		Crew	Daily Output	Labor-Hours	Unit	Material	2021 Bare Costs Labor	Equipment	Total	Total Incl O&P
0010	**GALVANIZED WELDED WIRE FABRIC**									
0100	Add to plain welded wire pricing for galvanized welded wire				Lb.	.24			.24	.27

03 22 Fabric and Grid Reinforcing

03 22 16 – Epoxy-Coated Welded Wire Fabric Reinforcing

03 22 16.10 Epoxy-Coated Welded Wire Fabric	Crew	Daily Output	Labor-Hours	Unit	Material	2021 Bare Costs Labor	Equipment	Total	Total Incl O&P
0010 **EPOXY-COATED WELDED WIRE FABRIC**									
0100 Add to plain welded wire pricing for epoxy-coated welded wire				Lb.	.49			.49	.53

03 24 Fibrous Reinforcing

03 24 05 – Reinforcing Fibers

03 24 05.30 Synthetic Fibers

	Crew	Daily Output	Labor-Hours	Unit	Material	2021 Bare Costs Labor	Equipment	Total	Total Incl O&P
0010 **SYNTHETIC FIBERS**									
0100 Synthetic fibers, add to concrete				Lb.	6.15			6.15	6.75
0110 1-1/2 lb./C.Y.				C.Y.	9.50			9.50	10.45

03 24 05.70 Steel Fibers

		Crew	Daily Output	Labor-Hours	Unit	Material	2021 Bare Costs Labor	Equipment	Total	Total Incl O&P
0010 **STEEL FIBERS**										
0140 ASTM A850, Type V, continuously deformed, 1-1/2" long x 0.045" diam.										
0150 Add to price of ready mix concrete	G				Lb.	1.20			1.20	1.32
0205 Alternate pricing, dosing at 5 lb./C.Y., add to price of RMC	G				C.Y.	6			6	6.60
0210 10 lb./C.Y.	G					12			12	13.20
0215 15 lb./C.Y.	G					18			18	19.80
0220 20 lb./C.Y.	G					24			24	26.50
0225 25 lb./C.Y.	G					30			30	33
0230 30 lb./C.Y.	G					36			36	39.50
0235 35 lb./C.Y.	G					42			42	46
0240 40 lb./C.Y.	G					48			48	53
0250 50 lb./C.Y.	G					60			60	66
0275 75 lb./C.Y.	G					90			90	99
0300 100 lb./C.Y.	G					120			120	132

03 30 Cast-In-Place Concrete

03 30 53 – Miscellaneous Cast-In-Place Concrete

03 30 53.40 Concrete In Place

		Crew	Daily Output	Labor-Hours	Unit	Material	2021 Bare Costs Labor	Equipment	Total	Total Incl O&P
0010 **CONCRETE IN PLACE**	R033105-10									
0020 Including forms (4 uses), Grade 60 rebar, concrete (Portland cement	R033105-20									
0050 Type I), placement and finishing unless otherwise indicated	R033105-70									
0300 Beams (3500 psi), 5 kip/L.F., 10' span	R033105-85	C-14A	15.62	12.804	C.Y.	420	700	29	1,149	1,600
0350 25' span		"	18.55	10.782		440	590	24.50	1,054.50	1,450
0500 Chimney foundations (5000 psi), over 5 C.Y.		C-14C	32.22	3.476		187	182	.83	369.83	495
0510 (3500 psi), under 5 C.Y.		"	23.71	4.724		223	247	1.13	471.13	635
0700 Columns, square (4000 psi), 12" x 12", up to 1% reinforcing by area		C-14A	11.96	16.722		475	915	37.50	1,427.50	2,025
0720 Up to 2% reinforcing by area			10.13	19.743		735	1,075	44.50	1,854.50	2,550
0740 Up to 3% reinforcing by area			9.03	22.148		1,075	1,200	50	2,325	3,175
0800 16" x 16", up to 1% reinforcing by area			16.22	12.330		375	675	27.50	1,077.50	1,525
0820 Up to 2% reinforcing by area			12.57	15.911		620	870	36	1,526	2,100
0840 Up to 3% reinforcing by area			10.25	19.512		945	1,075	44	2,064	2,800
0900 24" x 24", up to 1% reinforcing by area			23.66	8.453		310	460	19	789	1,100
0920 Up to 2% reinforcing by area			17.71	11.293		545	620	25.50	1,190.50	1,600
0940 Up to 3% reinforcing by area			14.15	14.134		860	775	32	1,667	2,200
1000 36" x 36", up to 1% reinforcing by area			33.69	5.936		268	325	13.35	606.35	825
1020 Up to 2% reinforcing by area			23.32	8.576		475	470	19.30	964.30	1,275
1040 Up to 3% reinforcing by area			17.82	11.223		795	615	25.50	1,435.50	1,875
1100 Columns, round (4000 psi), tied, 12" diameter, up to 1% reinforcing by area			20.97	9.537		420	520	21.50	961.50	1,325

03 30 53.40 Concrete In Place		Crew	Daily Output	Labor-Hours	Unit	Material	2021 Bare Costs Labor	Equipment	Total	Total Incl O&P
1120	Up to 2% reinforcing by area	C-14A	15.27	13.098	C.Y.	675	715	29.50	1,419.50	1,900
1140	Up to 3% reinforcing by area		12.11	16.515		1,025	905	37	1,967	2,600
1200	16" diameter, up to 1% reinforcing by area		31.49	6.351		365	345	14.30	724.30	965
1220	Up to 2% reinforcing by area		19.12	10.460		625	570	23.50	1,218.50	1,625
1240	Up to 3% reinforcing by area		13.77	14.524		935	795	32.50	1,762.50	2,300
1300	20" diameter, up to 1% reinforcing by area		41.04	4.873		335	266	10.95	611.95	800
1320	Up to 2% reinforcing by area		24.05	8.316		575	455	18.70	1,048.70	1,375
1340	Up to 3% reinforcing by area		17.01	11.758		905	645	26.50	1,576.50	2,050
1400	24" diameter, up to 1% reinforcing by area		51.85	3.857		340	211	8.70	559.70	720
1420	Up to 2% reinforcing by area		27.06	7.391		600	405	16.65	1,021.65	1,325
1440	Up to 3% reinforcing by area		18.29	10.935		915	600	24.50	1,539.50	1,975
1500	36" diameter, up to 1% reinforcing by area		75.04	2.665		330	146	6	482	605
1520	Up to 2% reinforcing by area		37.49	5.335		560	292	12	864	1,100
1540	Up to 3% reinforcing by area	▼	22.84	8.757		880	480	19.70	1,379.70	1,750
1900	Elevated slab (4000 psi), flat slab with drops, 125 psf Sup. Load, 20' span	C-14B	38.45	5.410		345	295	11.75	651.75	865
1950	30' span		50.99	4.079		360	223	8.85	591.85	760
2100	Flat plate, 125 psf Sup. Load, 15' span		30.24	6.878		310	375	14.95	699.95	955
2150	25' span		49.60	4.194		320	229	9.10	558.10	725
2300	Waffle const., 30" domes, 125 psf Sup. Load, 20' span		37.07	5.611		340	305	12.20	657.20	875
2350	30' span		44.07	4.720		315	258	10.25	583.25	765
2500	One way joists, 30" pans, 125 psf Sup. Load, 15' span		27.38	7.597		440	415	16.50	871.50	1,150
2550	25' span		31.15	6.677		405	365	14.50	784.50	1,025
2700	One way beam & slab, 125 psf Sup. Load, 15' span		20.59	10.102		340	550	22	912	1,275
2750	25' span		28.36	7.334		315	400	15.90	730.90	1,000
2900	Two way beam & slab, 125 psf Sup. Load, 15' span		24.04	8.652		325	470	18.80	813.80	1,125
2950	25' span	▼	35.87	5.799	▼	276	315	12.60	603.60	820
3100	Elevated slabs, flat plate, including finish, not									
3110	including forms or reinforcing									
3150	Regular concrete (4000 psi), 4" slab	C-8	2613	.021	S.F.	1.72	1.05	.16	2.93	3.72
3200	6" slab		2585	.022		2.52	1.06	.16	3.74	4.62
3250	2-1/2" thick floor fill		2685	.021		1.12	1.02	.16	2.30	3.01
3300	Lightweight, 110 #/C.F., 2-1/2" thick floor fill		2585	.022		1.31	1.06	.16	2.53	3.29
3400	Cellular concrete, 1-5/8" fill, under 5000 S.F.		2000	.028		.89	1.37	.21	2.47	3.37
3450	Over 10,000 S.F.		2200	.025		.85	1.24	.19	2.28	3.11
3500	Add per floor for 3 to 6 stories high		31800	.002			.09	.01	.10	.15
3520	For 7 to 20 stories high	▼	21200	.003	▼		.13	.02	.15	.22
3540	Equipment pad (3000 psi), 3' x 3' x 6" thick	C-14H	45	1.067	Ea.	55.50	57	.61	113.11	152
3550	4' x 4' x 6" thick		30	1.600		84	85.50	.91	170.41	229
3560	5' x 5' x 8" thick		18	2.667		150	143	1.52	294.52	395
3570	6' x 6' x 8" thick		14	3.429		204	184	1.95	389.95	520
3580	8' x 8' x 10" thick		8	6		415	320	3.42	738.42	975
3590	10' x 10' x 12" thick	▼	5	9.600	▼	720	515	5.45	1,240.45	1,600
3800	Footings (3000 psi), spread under 1 C.Y.	C-14C	28	4	C.Y.	208	209	.96	417.96	560
3825	1 C.Y. to 5 C.Y.		43	2.605		266	136	.63	402.63	510
3850	Over 5 C.Y.	▼	75	1.493		245	78	.36	323.36	395
3900	Footings, strip (3000 psi), 18" x 9", unreinforced	C-14L	40	2.400		159	123	.67	282.67	370
3920	18" x 9", reinforced	C-14C	35	3.200		189	167	.77	356.77	475
3925	20" x 10", unreinforced	C-14L	45	2.133		153	109	.60	262.60	345
3930	20" x 10", reinforced	C-14C	40	2.800		178	146	.67	324.67	430
3935	24" x 12", unreinforced	C-14L	55	1.745		150	89.50	.49	239.99	310
3940	24" x 12", reinforced	C-14C	48	2.333		175	122	.56	297.56	385
3945	36" x 12", unreinforced	C-14L	70	1.371		144	70	.38	214.38	270
3950	36" x 12", reinforced	C-14C	60	1.867	▼	166	97.50	.45	263.95	340

For customer support on your Commercial Renovation Costs with RSMeans Data, call 800.448.8182.

69

03 30 53.40 Concrete In Place	Crew	Daily Output	Labor-Hours	Unit	Material	2021 Bare Costs Labor	Equipment	Total	Total Incl O&P
4000	Foundation mat (3000 psi), under 10 C.Y.	C-14C	38.67	2.896	C.Y.	259	151	.70	410.70	525
4050	Over 20 C.Y.	↓	56.40	1.986		225	104	.48	329.48	415
4200	Wall, free-standing (3000 psi), 8" thick, 8' high	C-14D	45.83	4.364		199	237	9.80	445.80	605
4250	14' high		27.26	7.337		237	400	16.50	653.50	915
4260	12" thick, 8' high		64.32	3.109		179	169	7	355	475
4270	14' high		40.01	4.999		191	272	11.25	474.25	655
4300	15" thick, 8' high		80.02	2.499		172	136	5.60	313.60	410
4350	12' high		51.26	3.902		172	212	8.80	392.80	535
4500	18' high	↓	48.85	4.094	↓	195	223	9.20	427.20	580
4520	Handicap access ramp (4000 psi), railing both sides, 3' wide	C-14H	14.58	3.292	L.F.	405	176	1.88	582.88	725
4525	5' wide		12.22	3.928		420	210	2.24	632.24	795
4530	With 6" curb and rails both sides, 3' wide		8.55	5.614		415	300	3.20	718.20	940
4535	5' wide	↓	7.31	6.566	↓	425	350	3.74	778.74	1,025
4650	Slab on grade (3500 psi), not including finish, 4" thick	C-14E	60.75	1.449	C.Y.	149	77.50	.45	226.95	286
4700	6" thick	"	92	.957	"	144	51	.30	195.30	239
4701	Thickened slab edge (3500 psi), for slab on grade poured									
4702	monolithically with slab; depth is in addition to slab thickness;									
4703	formed vertical outside edge, earthen bottom and inside slope									
4705	8" deep x 8" wide bottom, unreinforced	C-14L	2190	.044	L.F.	4.58	2.24	.01	6.83	8.65
4710	8" x 8", reinforced	C-14C	1670	.067		7.50	3.51	.02	11.03	13.80
4715	12" deep x 12" wide bottom, unreinforced	C-14L	1800	.053		9.05	2.73	.01	11.79	14.30
4720	12" x 12", reinforced	C-14C	1310	.086		14.40	4.47	.02	18.89	23
4725	16" deep x 16" wide bottom, unreinforced	C-14L	1440	.067		15	3.41	.02	18.43	22
4730	16" x 16", reinforced	C-14C	1120	.100		21.50	5.25	.02	26.77	32
4735	20" deep x 20" wide bottom, unreinforced	C-14L	1150	.083		22.50	4.27	.02	26.79	31.50
4740	20" x 20", reinforced	C-14C	920	.122		30.50	6.35	.03	36.88	43.50
4745	24" deep x 24" wide bottom, unreinforced	C-14L	930	.103		31.50	5.30	.03	36.83	43
4750	24" x 24", reinforced	C-14C	740	.151	↓	42	7.90	.04	49.94	58.50
4751	Slab on grade (3500 psi), incl. troweled finish, not incl. forms									
4760	or reinforcing, over 10,000 S.F., 4" thick	C-14F	3425	.021	S.F.	1.65	1.04	.01	2.70	3.46
4820	6" thick		3350	.021		2.41	1.07	.01	3.49	4.34
4840	8" thick		3184	.023		3.29	1.12	.01	4.42	5.40
4900	12" thick		2734	.026		4.94	1.31	.01	6.26	7.50
4950	15" thick	↓	2505	.029	↓	6.20	1.42	.01	7.63	9.10
5000	Slab on grade (3000 psi), incl. broom finish, not incl. forms									
5001	or reinforcing, 4" thick	C-14G	2873	.019	S.F.	1.62	.95	.01	2.58	3.29
5010	6" thick		2590	.022		2.53	1.06	.01	3.60	4.46
5020	8" thick	↓	2320	.024	↓	3.30	1.18	.01	4.49	5.50
5200	Lift slab in place above the foundation, incl. forms, reinforcing,									
5210	concrete (4000 psi) and columns, over 20,000 S.F./floor	C-14B	2113	.098	S.F.	7.50	5.35	.21	13.06	17
5250	10,000 S.F. to 20,000 S.F./floor		1650	.126		8.20	6.90	.27	15.37	20
5300	Under 10,000 S.F./floor	↓	1500	.139	↓	8.85	7.55	.30	16.70	22
5500	Lightweight, ready mix, including screed finish only,									
5510	not including forms or reinforcing									
5550	1:4 (2500 psi) for structural roof decks	C-14B	260	.800	C.Y.	126	43.50	1.74	171.24	209
5600	1:6 (3000 psi) for ground slab with radiant heat	C-14F	92	.783		118	39	.30	157.30	191
5650	1:3:2 (2000 psi) with sand aggregate, roof deck	C-14B	260	.800		112	43.50	1.74	157.24	194
5700	Ground slab (2000 psi)	C-14F	107	.673		112	33.50	.26	145.76	176
5900	Pile caps (3000 psi), incl. forms and reinf., sq. or rect., under 10 C.Y.	C-14C	54.14	2.069		221	108	.50	329.50	415
5950	Over 10 C.Y.		75	1.493		202	78	.36	280.36	345
6000	Triangular or hexagonal, under 10 C.Y.		53	2.113		153	111	.51	264.51	345
6050	Over 10 C.Y.	↓	85	1.318		173	69	.32	242.32	299
6200	Retaining walls (3000 psi), gravity, 4' high see Section 32 32	C-14D	66.20	3.021	↓	176	164	6.80	346.80	465

03 30 Cast-In-Place Concrete

03 30 53 – Miscellaneous Cast-In-Place Concrete

03 30 53.40 Concrete In Place		Crew	Daily Output	Labor-Hours	Unit	Material	2021 Bare Costs Labor	Equipment	Total	Total Incl O&P
6250	10' high	C-14D	125	1.600	C.Y.	166	87	3.60	256.60	325
6300	Cantilever, level backfill loading, 8' high		70	2.857		185	155	6.45	346.45	455
6350	16' high		91	2.198		179	119	4.95	302.95	390
6800	Stairs (3500 psi), not including safety treads, free standing, 3'-6" wide	C-14H	83	.578	LF Nose	7.60	31	.33	38.93	57.50
6850	Cast on ground		125	.384	"	6.05	20.50	.22	26.77	39.50
7000	Stair landings, free standing		200	.240	S.F.	6.10	12.85	.14	19.09	27.50
7050	Cast on ground		475	.101	"	4.50	5.40	.06	9.96	13.55
9000	Minimum labor/equipment charge	2 Carp	1	16	Job		875		875	1,400

03 31 Structural Concrete

03 31 13 – Heavyweight Structural Concrete

03 31 13.25 Concrete, Hand Mix

03 31 13.25 Concrete, Hand Mix		Crew	Daily Output	Labor-Hours	Unit	Material	2021 Bare Costs Labor	Equipment	Total	Total Incl O&P
0010	**CONCRETE, HAND MIX** for small quantities or remote areas									
0050	Includes bulk local aggregate, bulk sand, bagged Portland									
0060	cement (Type I) and water, using gas powered cement mixer									
0125	2500 psi	C-30	135	.059	C.F.	4.35	2.63	1.09	8.07	10.20
0130	3000 psi		135	.059		4.71	2.63	1.09	8.43	10.60
0135	3500 psi		135	.059		4.92	2.63	1.09	8.64	10.80
0140	4000 psi		135	.059		5.15	2.63	1.09	8.87	11.10
0145	4500 psi		135	.059		5.45	2.63	1.09	9.17	11.40
0150	5000 psi		135	.059		5.85	2.63	1.09	9.57	11.80
0300	Using pre-bagged dry mix and wheelbarrow (80-lb. bag = 0.6 C.F.)									
0340	4000 psi	1 Clab	48	.167	C.F.	13.90	7.40		21.30	27

03 31 13.30 Concrete, Volumetric Site-Mixed

03 31 13.30 Concrete, Volumetric Site-Mixed		Crew	Daily Output	Labor-Hours	Unit	Material	2021 Bare Costs Labor	Equipment	Total	Total Incl O&P
0010	**CONCRETE, VOLUMETRIC SITE-MIXED**									
0015	Mixed on-site in volumetric truck									
0020	Includes local aggregate, sand, Portland cement (Type I) and water									
0025	Excludes all additives and treatments									
0100	3000 psi, 1 C.Y. mixed and discharged				C.Y.	205			205	226
0110	2 C.Y.					160			160	176
0120	3 C.Y.					145			145	160
0130	4 C.Y.					128			128	140
0140	5 C.Y.					125			125	138
0200	For truck holding/waiting time past first 2 on-site hours, add				Hr.	93			93	102
0210	For trip charge beyond first 20 miles, each way, add				Mile	3.68			3.68	4.05
0220	For each additional increase of 500 psi, add				Ea.	4.72			4.72	5.20

03 31 13.35 Heavyweight Concrete, Ready Mix

03 31 13.35 Heavyweight Concrete, Ready Mix		Crew	Daily Output	Labor-Hours	Unit	Material	2021 Bare Costs Labor	Equipment	Total	Total Incl O&P
0010	**HEAVYWEIGHT CONCRETE, READY MIX**, delivered									
0012	Includes local aggregate, sand, Portland cement (Type I) and water									
0015	Excludes all additives and treatments	R033105-20								
0020	2000 psi				C.Y.	109			109	120
0100	2500 psi					112			112	124
0150	3000 psi					124			124	137
0200	3500 psi					127			127	139
0300	4000 psi					129			129	142
0350	4500 psi					133			133	146
0400	5000 psi					137			137	151
0411	6000 psi					141			141	155
0412	8000 psi					149			149	163
0413	10,000 psi					156			156	172

03 31 13 – Heavyweight Structural Concrete

03 31 13.35 Heavyweight Concrete, Ready Mix	Crew	Daily Output	Labor-Hours	Unit	Material	2021 Bare Costs Labor	Equipment	Total	Total Incl O&P
0414	12,000 psi				C.Y.	164			164	180
1000	For high early strength (Portland cement Type III), add					10%				
1300	For winter concrete (hot water), add					5.85			5.85	6.40
1410	For mid-range water reducer, add					4.22			4.22	4.64
1420	For high-range water reducer/superplasticizer, add					6.65			6.65	7.30
1430	For retarder, add					4.22			4.22	4.64
1440	For non-Chloride accelerator, add					8.70			8.70	9.60
1450	For Chloride accelerator, per 1%, add					3.69			3.69	4.06
1460	For fiber reinforcing, synthetic (1 lb./C.Y.), add					8.30			8.30	9.15
1500	For Saturday delivery, add					7.70			7.70	8.45
1510	For truck holding/waiting time past 1st hour per load, add				Hr.	114			114	125
1520	For short load (less than 4 C.Y.), add per load				Ea.	97			97	107
2000	For all lightweight aggregate, add				C.Y.	45%				

03 31 13.70 Placing Concrete

		Crew	Daily Output	Labor-Hours	Unit	Material	2021 Bare Costs Labor	Equipment	Total	Total Incl O&P
0010	**PLACING CONCRETE** R033105-70									
0020	Includes labor and equipment to place, level (strike off) and consolidate									
0050	Beams, elevated, small beams, pumped	C-20	60	1.067	C.Y.		50.50	7.95	58.45	89
0100	With crane and bucket	C-7	45	1.600	C.Y.		77	24	101	148
0200	Large beams, pumped	C-20	90	.711			33.50	5.30	38.80	59.50
0250	With crane and bucket	C-7	65	1.108			53	16.75	69.75	102
0400	Columns, square or round, 12" thick, pumped	C-20	60	1.067			50.50	7.95	58.45	89
0450	With crane and bucket	C-7	40	1.800			86.50	27.50	114	166
0600	18" thick, pumped	C-20	90	.711			33.50	5.30	38.80	59.50
0650	With crane and bucket	C-7	55	1.309			63	19.80	82.80	122
0800	24" thick, pumped	C-20	92	.696			33	5.20	38.20	57.50
0850	With crane and bucket	C-7	70	1.029			49.50	15.55	65.05	95
1000	36" thick, pumped	C-20	140	.457			21.50	3.41	24.91	38.50
1050	With crane and bucket	C-7	100	.720			34.50	10.90	45.40	66.50
1200	Duct bank, direct chute	C-6	155	.310			14.25	.35	14.60	23
1400	Elevated slabs, less than 6" thick, pumped	C-20	140	.457			21.50	3.41	24.91	38.50
1450	With crane and bucket	C-7	95	.758			36.50	11.45	47.95	70
1500	6" to 10" thick, pumped	C-20	160	.400			18.95	2.99	21.94	33.50
1550	With crane and bucket	C-7	110	.655			31.50	9.90	41.40	60.50
1600	Slabs over 10" thick, pumped	C-20	180	.356			16.85	2.66	19.51	29.50
1650	With crane and bucket	C-7	130	.554			26.50	8.40	34.90	51.50
1900	Footings, continuous, shallow, direct chute	C-6	120	.400			18.40	.45	18.85	29.50
1950	Pumped	C-20	150	.427			20	3.19	23.19	35.50
2000	With crane and bucket	C-7	90	.800			38.50	12.10	50.60	74
2100	Footings, continuous, deep, direct chute	C-6	140	.343			15.75	.39	16.14	25.50
2150	Pumped	C-20	160	.400			18.95	2.99	21.94	33.50
2200	With crane and bucket	C-7	110	.655			31.50	9.90	41.40	60.50
2400	Footings, spread, under 1 C.Y., direct chute	C-6	55	.873			40	.99	40.99	64.50
2450	Pumped	C-20	65	.985			46.50	7.35	53.85	82
2500	With crane and bucket	C-7	45	1.600			77	24	101	148
2600	Over 5 C.Y., direct chute	C-6	120	.400			18.40	.45	18.85	29.50
2650	Pumped	C-20	150	.427			20	3.19	23.19	35.50
2700	With crane and bucket	C-7	100	.720			34.50	10.90	45.40	66.50
2900	Foundation mats, over 20 C.Y., direct chute	C-6	350	.137			6.30	.15	6.45	10.15
2950	Pumped	C-20	400	.160			7.60	1.20	8.80	13.30
3000	With crane and bucket	C-7	300	.240			11.50	3.63	15.13	22
3200	Grade beams, direct chute	C-6	150	.320			14.70	.36	15.06	24
3250	Pumped	C-20	180	.356			16.85	2.66	19.51	29.50

03 31 13.70 Placing Concrete

	Crew	Daily Output	Labor-Hours	Unit	Material	2021 Bare Costs Labor	2021 Bare Costs Equipment	Total	Total Incl O&P
3300 With crane and bucket	C-7	120	.600	C.Y.		29	9.10	38.10	55.50
3500 High rise, for more than 5 stories, pumped, add per story	C-20	2100	.030			1.44	.23	1.67	2.54
3510 With crane and bucket, add per story	C-7	2100	.034			1.64	.52	2.16	3.17
3700 Pile caps, under 5 C.Y., direct chute	C-6	90	.533			24.50	.60	25.10	39.50
3750 Pumped	C-20	110	.582	C.Y.		27.50	4.35	31.85	48.50
3800 With crane and bucket	C-7	80	.900			43	13.65	56.65	83.50
3850 Pile cap, 5 C.Y. to 10 C.Y., direct chute	C-6	175	.274			12.60	.31	12.91	20.50
3900 Pumped	C-20	200	.320			15.15	2.39	17.54	26.50
3950 With crane and bucket	C-7	150	.480			23	7.25	30.25	44.50
4000 Over 10 C.Y., direct chute	C-6	215	.223			10.25	.25	10.50	16.60
4050 Pumped	C-20	240	.267			12.65	1.99	14.64	22
4100 With crane and bucket	C-7	185	.389			18.65	5.90	24.55	36
4300 Slab on grade, up to 6" thick, direct chute	C-6	110	.436			20	.49	20.49	32.50
4350 Pumped	C-20	130	.492			23.50	3.68	27.18	41
4400 With crane and bucket	C-7	110	.655			31.50	9.90	41.40	60.50
4600 Over 6" thick, direct chute	C-6	165	.291			13.35	.33	13.68	21.50
4650 Pumped	C-20	185	.346			16.40	2.58	18.98	29
4700 With crane and bucket	C-7	145	.497			24	7.50	31.50	46
4900 Walls, 8" thick, direct chute	C-6	90	.533			24.50	.60	25.10	39.50
4950 Pumped	C-20	100	.640			30.50	4.78	35.28	53.50
5000 With crane and bucket	C-7	80	.900			43	13.65	56.65	83.50
5050 12" thick, direct chute	C-6	100	.480			22	.54	22.54	35.50
5100 Pumped	C-20	110	.582			27.50	4.35	31.85	48.50
5200 With crane and bucket	C-7	90	.800			38.50	12.10	50.60	74
5300 15" thick, direct chute	C-6	105	.457			21	.52	21.52	34
5350 Pumped	C-20	120	.533			25.50	3.98	29.48	44.50
5400 With crane and bucket	C-7	95	.758			36.50	11.45	47.95	70
5600 Wheeled concrete dumping, add to placing costs above									
5610 Walking cart, 50' haul, add	C-18	32	.281	C.Y.		12.55	3.65	16.20	24
5620 150' haul, add		24	.375			16.75	4.87	21.62	32
5700 250' haul, add		18	.500			22.50	6.50	29	42.50
5800 Riding cart, 50' haul, add	C-19	80	.113			5	1.74	6.74	9.90
5810 150' haul, add		60	.150			6.70	2.32	9.02	13.20
5900 250' haul, add		45	.200			8.90	3.09	11.99	17.60
6000 Concrete in-fill for pan-type metal stairs and landings. Manual placement									
6010 includes up to 50' horizontal haul from point of concrete discharge.									
6100 Stair pan treads, 2" deep									
6110 Flights in 1st floor level up/down from discharge point	C-8A	3200	.015	S.F.		.71		.71	1.12
6120 2nd floor level		2500	.019			.91		.91	1.43
6130 3rd floor level		2000	.024			1.13		1.13	1.79
6140 4th floor level		1800	.027			1.26		1.26	1.99
6200 Intermediate stair landings, pan-type 4" deep									
6210 Flights in 1st floor level up/down from discharge point	C-8A	2000	.024	S.F.		1.13		1.13	1.79
6220 2nd floor level	C-8A	1500	.032	S.F.		1.51		1.51	2.39
6230 3rd floor level		1200	.040			1.89		1.89	2.99
6240 4th floor level		1000	.048			2.27		2.27	3.59
9000 Minimum labor/equipment charge	C-6	2	24	Job		1,100	27	1,127	1,775

For customer support on your Commercial Renovation Costs with RSMeans Data, call 800.448.8182.

03 35 13 – High-Tolerance Concrete Floor Finishing

03 35 13.30 Finishing Floors, High Tolerance

	Crew	Daily Output	Labor-Hours	Unit	Material	2021 Bare Costs Labor	Equipment	Total	Total Incl O&P
0010 **FINISHING FLOORS, HIGH TOLERANCE**									
0012 Finishing of fresh concrete flatwork requires that concrete									
0013 first be placed, struck off & consolidated									
0015 Basic finishing for various unspecified flatwork									
0100 Bull float only	C-10	4000	.006	S.F.		.30		.30	.47
0125 Bull float & manual float		2000	.012			.59		.59	.93
0150 Bull float, manual float & broom finish, w/edging & joints		1850	.013			.64		.64	1.01
0200 Bull float, manual float & manual steel trowel		1265	.019			.94		.94	1.47
0210 For specified Random Access Floors in ACI Classes 1, 2, 3 and 4 to achieve									
0215 Composite Overall Floor Flatness and Levelness values up to FF35/FL25									
0250 Bull float, machine float & machine trowel (walk-behind)	C-10C	1715	.014	S.F.		.69	.05	.74	1.15
0300 Power screed, bull float, machine float & trowel (walk-behind)	C-10D	2400	.010			.49	.08	.57	.86
0350 Power screed, bull float, machine float & trowel (ride-on)	C-10E	4000	.006			.30	.06	.36	.54
0352 For specified Random Access Floors in ACI Classes 5, 6, 7 and 8 to achieve									
0354 Composite Overall Floor Flatness and Levelness values up to FF50/FL50									
0356 Add for two-dimensional restraightening after power float	C-10	6000	.004	S.F.		.20		.20	.31
0358 For specified Random or Defined Access Floors in ACI Class 9 to achieve									
0360 Composite Overall Floor Flatness and Levelness values up to FF100/FL100									
0362 Add for two-dimensional restraightening after bull float & power float	C-10	3000	.008	S.F.		.39		.39	.62
0364 For specified Superflat Defined Access Floors in ACI Class 9 to achieve									
0366 Minimum Floor Flatness and Levelness values of FF100/FL100									
0368 Add for 2-dim'l restraightening after bull float, power float, power trowel	C-10	2000	.012	S.F.		.59		.59	.93
9100 Minimum labor/equipment charge	"	2	12	Job		590		590	930

03 35 29 – Tooled Concrete Finishing

03 35 29.60 Finishing Walls

	Crew	Daily Output	Labor-Hours	Unit	Material	2021 Bare Costs Labor	Equipment	Total	Total Incl O&P
0010 **FINISHING WALLS**									
0020 Break ties and patch voids	1 Cefi	540	.015	S.F.	.04	.77		.81	1.25
0050 Burlap rub with grout	"	450	.018		.04	.92		.96	1.49
0300 Bush hammer, green concrete	B-39	1000	.048			2.24	.36	2.60	3.94
0350 Cured concrete	"	650	.074			3.44	.55	3.99	6.05
0500 Acid etch	1 Cefi	575	.014	S.F.	.13	.72		.85	1.28
0700 Sandblast, light penetration	E-11	1100	.029		.52	1.41	.27	2.20	3.19
0750 Heavy penetration	"	375	.085		1.04	4.15	.81	6	8.85
0850 Grind form fins flush	1 Clab	700	.011	L.F.		.51		.51	.81
9000 Minimum labor/equipment charge	C-10	2	12	Job		590		590	930

03 35 33 – Stamped Concrete Finishing

03 35 33.50 Slab Texture Stamping

	Crew	Daily Output	Labor-Hours	Unit	Material	2021 Bare Costs Labor	Equipment	Total	Total Incl O&P
0010 **SLAB TEXTURE STAMPING**									
0050 Stamping requires that concrete first be placed, struck off, consolidated,									
0060 bull floated and free of bleed water. Decorative stamping tasks include:									
0100 Step 1 - first application of dry shake colored hardener	1 Cefi	6400	.001	S.F.	.43	.06		.49	.57
0110 Step 2 - bull float		6400	.001			.06		.06	.10
0130 Step 3 - second application of dry shake colored hardener		6400	.001		.21	.06		.27	.33
0140 Step 4 - bull float, manual float & steel trowel	3 Cefi	1280	.019			.97		.97	1.52
0150 Step 5 - application of dry shake colored release agent	1 Cefi	6400	.001		.10	.06		.16	.21
0160 Step 6 - place, tamp & remove mats	3 Cefi	2400	.010		.85	.52		1.37	1.74
0170 Step 7 - touch up edges, mat joints & simulated grout lines	1 Cefi	1280	.006			.32		.32	.51
0300 Alternate stamping estimating method includes all tasks above	4 Cefi	800	.040		1.58	2.07		3.65	4.98
0400 Step 8 - pressure wash @ 3000 psi after 24 hours	1 Cefi	1600	.005			.26		.26	.41
0500 Step 9 - roll 2 coats cure/seal compound when dry	"	800	.010		.81	.52		1.33	1.70

For customer support on your Commercial Renovation Costs with RSMeans Data, call 800.448.8182.

03 35 Concrete Finishing

03 35 43 – Polished Concrete Finishing

03 35 43.10 Polished Concrete Floors	Crew	Daily Output	Labor-Hours	Unit	Material	2021 Bare Costs Labor	2021 Bare Costs Equipment	Total	Total Incl O&P
0010 **POLISHED CONCRETE FLOORS**									
0015	Processing of cured concrete to include grinding, honing,								
0020	and polishing of interior floors with 22" segmented diamond								
0025	planetary floor grinder (2 passes in different directions per grit)								
0100	Removal of pre-existing coatings, dry, with carbide discs using								
0105	dry vacuum pick-up system, final hand sweeping	J-4	1.60	15	M.S.F.	23	740	134	897	1,350
0110	Glue, adhesive or tar		3.60	6.667		23	330	59.50	412.50	610
0120	Paint, epoxy, 1 coat		1.80	13.333		23	660	119	802	1,175
0130	2 coats									
0200	Grinding and edging, wet, including wet vac pick-up and auto								
0205	scrubbing between grit changes	J-4A	1.60	20	M.S.F.	42.50	960	247	1,249.50	1,850
0210	40-grit diamond/metal matrix		2	16		42.50	770	197	1,009.50	1,500
0220	80-grit diamond/metal matrix		2.40	13.333		42.50	640	164	846.50	1,225
0230	120-grit diamond/metal matrix		2.80	11.429		42.50	550	141	733.50	1,075
0240	200-grit diamond/metal matrix	1 Cefi	16	.500		215	26		241	278
0300	Spray on dye or stain (1 coat)	1 Cefi	8	1	M.S.F.	305	52		357	415
0400	Spray on densifier/hardener (2 coats)	J-4B	16	.500	"		22	11.20	33.20	48
0410	Auto scrubbing after 2nd coat, when dry									
0500	Honing and edging, wet, including wet vac pick-up and auto								
0505	scrubbing between grit changes	J-4A	2.80	11.429	M.S.F.	42.50	550	141	733.50	1,075
0510	100-grit diamond/resin matrix	"	2.80	11.429	"	42.50	550	141	733.50	1,075
0520	200-grit diamond/resin matrix									
0530	Dry, including dry vacuum pick-up system, final hand sweeping	J-4A	2.80	11.429	M.S.F.	42.50	550	141	733.50	1,075
0540	400-grit diamond/resin matrix									
0600	Polishing and edging, dry, including dry vac pick-up and hand								
0605	sweeping between grit changes	J-4A	2.80	11.429	M.S.F.	42.50	550	141	733.50	1,075
0610	800-grit diamond/resin matrix		2.80	11.429		42.50	550	141	733.50	1,075
0620	1500-grit diamond/resin matrix		2.80	11.429		42.50	550	141	733.50	1,075
0630	3000-grit diamond/resin matrix									
0700	Auto scrubbing after final polishing step	J-4B	16	.500			22	11.20	33.20	48

03 37 Specialty Placed Concrete

03 37 13 – Shotcrete

03 37 13.30 Gunite (Dry-Mix)

		Crew	Daily Output	Labor-Hours	Unit	Material	2021 Bare Costs Labor	2021 Bare Costs Equipment	Total	Total Incl O&P
0010	**GUNITE (DRY-MIX)**									
0020	Typical in place, 1" layers, no mesh included	C-16	2000	.028	S.F.	.52	1.37	.20	2.09	2.95
0300	Typical in place, including mesh, 2" thick, flat surfaces		1000	.056		1.66	2.74	.39	4.79	6.60
0350	Curved surfaces		500	.112		1.66	5.50	.79	7.95	11.35
0500	4" thick, flat surfaces		750	.075		2.70	3.65	.52	6.87	9.30
0550	Curved surfaces		350	.160		2.70	7.80	1.12	11.62	16.55
0900	Prepare old walls, no scaffolding, good condition	C-10	1000	.024			1.18		1.18	1.86
0950	Poor condition	"	275	.087			4.31		4.31	6.75
1100	For high finish requirement or close tolerance, add						50%			
1150	Very high						110%			
9000	Minimum labor/equipment charge	C-10	1	24	Job		1,175		1,175	1,875

03 37 13.60 Shotcrete (Wet-Mix)

		Crew	Daily Output	Labor-Hours	Unit	Material	2021 Bare Costs Labor	2021 Bare Costs Equipment	Total	Total Incl O&P
0010	**SHOTCRETE (WET-MIX)**									
0020	Wet mix, placed @ up to 12 C.Y./hour, 3000 psi	C-8C	80	.600	C.Y.	167	29	7.85	203.85	239
0100	Up to 35 C.Y./hour	C-8E	240	.200	"	151	9.55	2.72	163.27	184
1010	Fiber reinforced, 1" thick	C-8C	1740	.028	S.F.	1.23	1.34	.36	2.93	3.86

03 37 Specialty Placed Concrete

03 37 13 – Shotcrete

03 37 13.60 Shotcrete (Wet-Mix)	Crew	Daily Output	Labor-Hours	Unit	Material	2021 Bare Costs Labor	Equipment	Total	Total Incl O&P
1020	2" thick	C-8C	900	.053	S.F.	2.45	2.58	.70	5.73	7.55
1030	3" thick		825	.058		3.68	2.82	.76	7.26	9.35
1040	4" thick	↓	750	.064	↓	4.91	3.10	.84	8.85	11.20

03 39 Concrete Curing

03 39 13 – Water Concrete Curing

03 39 13.50 Water Curing

		Crew	Daily Output	Labor-Hours	Unit	Material	2021 Bare Costs Labor	Equipment	Total	Total Incl O&P
0010	**WATER CURING**									
0015	With burlap, 4 uses assumed, 7.5 oz.	2 Clab	55	.291	C.S.F.	13.75	12.90		26.65	35.50
0100	10 oz.	2 Clab	55	.291	C.S.F.	25	12.90		37.90	47.50
0400	Curing blankets, 1" to 2" thick, buy				S.F.	.64			.64	.70
9000	Minimum labor/equipment charge	1 Clab	5	1.600	Job		71		71	113

03 39 23 – Membrane Concrete Curing

03 39 23.13 Chemical Compound Membrane Concrete Curing

		Crew	Daily Output	Labor-Hours	Unit	Material	2021 Bare Costs Labor	Equipment	Total	Total Incl O&P
0010	**CHEMICAL COMPOUND MEMBRANE CONCRETE CURING**									
0300	Sprayed membrane curing compound	2 Clab	95	.168	C.S.F.	12.20	7.50		19.70	25.50
0700	Curing compound, solvent based, 400 S.F./gal., 55 gallon lots				Gal.	21			21	23
0720	5 gallon lots					41			41	45
0800	Curing compound, water based, 250 S.F./gal., 55 gallon lots					26.50			26.50	29
0820	5 gallon lots				↓	36			36	39.50

03 39 23.23 Sheet Membrane Concrete Curing

		Crew	Daily Output	Labor-Hours	Unit	Material	2021 Bare Costs Labor	Equipment	Total	Total Incl O&P
0010	**SHEET MEMBRANE CONCRETE CURING**									
0200	Curing blanket, burlap/poly, 2-ply	2 Clab	70	.229	C.S.F.	20.50	10.15		30.65	38.50

03 41 Precast Structural Concrete

03 41 13 – Precast Concrete Hollow Core Planks

03 41 13.50 Precast Slab Planks

			Crew	Daily Output	Labor-Hours	Unit	Material	2021 Bare Costs Labor	Equipment	Total	Total Incl O&P
0010	**PRECAST SLAB PLANKS**	R034105-30									
0020	Prestressed roof/floor members, grouted, solid, 4" thick		C-11	2400	.030	S.F.	8.40	1.79	.97	11.16	13.20
0050	6" thick			2800	.026		8.55	1.54	.83	10.92	12.80
0100	Hollow, 8" thick			3200	.023		12.20	1.35	.73	14.28	16.40
0150	10" thick			3600	.020		9.90	1.20	.65	11.75	13.55
0200	12" thick		↓	4000	.018	↓	11.70	1.08	.58	13.36	15.20

03 41 23 – Precast Concrete Stairs

03 41 23.50 Precast Stairs

		Crew	Daily Output	Labor-Hours	Unit	Material	2021 Bare Costs Labor	Equipment	Total	Total Incl O&P
0010	**PRECAST STAIRS**									
0020	Precast concrete treads on steel stringers, 3' wide	C-12	75	.640	Riser	135	35	6.35	176.35	211
0300	Front entrance, 5' wide with 48" platform, 2 risers		16	3	Flight	690	163	29.50	882.50	1,050
0350	5 risers		12	4		1,275	218	39.50	1,532.50	1,800
0500	6' wide, 2 risers		15	3.200		735	174	31.50	940.50	1,125
0550	5 risers	↓	11	4.364		1,400	238	43	1,681	1,975
1200	Basement entrance stairwell, 6 steps, incl. steel bulkhead door	B-51	22	2.182		1,900	99	9.05	2,008.05	2,275
1250	14 steps	"	11	4.364	↓	3,425	198	18.05	3,641.05	4,075

03 41 Precast Structural Concrete

03 41 33 – Precast Structural Pretensioned Concrete

03 41 33.10 Precast Beams

		Crew	Daily Output	Labor-Hours	Unit	Material	2021 Bare Costs Labor	Equipment	Total	Total Incl O&P
0010	**PRECAST BEAMS** R034105-30	C-11	32	2.250	Ea.	4,800	135	72.50	5,007.50	5,575
0011	L-shaped, 20' span, 12" x 20"		24	3		6,625	179	97	6,901	7,675
0060	18" x 36"		22	3.273		7,925	196	106	8,227	9,150
0100	24" x 44"									
0150	30' span, 12" x 36"		24	3		9,350	179	97	9,626	10,700
0200	18" x 44"		20	3.600		11,100	215	116	11,431	12,800
0250	24" x 52"		16	4.500		13,400	269	145	13,814	15,400
0400	40' span, 12" x 52"		20	3.600		14,100	215	116	14,431	16,000
0450	18" x 52"	C-11	16	4.500	Ea.	15,800	269	145	16,214	18,000
0500	24" x 52"		12	6		17,900	360	194	18,454	20,500
1200	Rectangular, 20' span, 12" x 20"		32	2.250		4,150	135	72.50	4,357.50	4,875
1250	18" x 36"		24	3		6,050	179	97	6,326	7,050
1300	24" x 44"		22	3.273		7,050	196	106	7,352	8,175
1400	30' span, 12" x 36"		24	3		6,875	179	97	7,151	7,975
1450	18" x 44"		20	3.600		9,550	215	116	9,881	11,000
1500	24" x 52"		16	4.500		11,600	269	145	12,014	13,400
1600	40' span, 12" x 52"		20	3.600		11,400	215	116	11,731	13,100
1650	18" x 52"		16	4.500		13,500	269	145	13,914	15,500
1700	24" x 52"		12	6		15,500	360	194	16,054	17,900
2000	"T" shaped, 20' span, 12" x 20"		32	2.250		5,775	135	72.50	5,982.50	6,650
2050	18" x 36"		24	3		7,600	179	97	7,876	8,775
2100	24" x 44"		22	3.273		7,825	196	106	8,127	9,025
2200	30' span, 12" x 36"		24	3		10,400	179	97	10,676	11,800
2250	18" x 44"		20	3.600		13,100	215	116	13,431	14,900
2300	24" x 52"		16	4.500		12,800	269	145	13,214	14,600
2500	40' span, 12" x 52"		20	3.600		16,800	215	116	17,131	19,000
2550	18" x 52"		16	4.500		15,900	269	145	16,314	18,100
2600	24" x 52"		12	6		17,000	360	194	17,554	19,500

03 41 33.15 Precast Columns

		Crew	Daily Output	Labor-Hours	Unit	Material	2021 Bare Costs Labor	Equipment	Total	Total Incl O&P
0010	**PRECAST COLUMNS** R034105-30	C-11	120	.600	L.F.	272	36	19.35	327.35	380
0020	Rectangular to 12' high, 16" x 16"									
0050	24" x 24"	"	96	.750	"	370	45	24	439	510

03 45 Precast Architectural Concrete

03 45 13 – Faced Architectural Precast Concrete

03 45 13.50 Precast Wall Panels

		Crew	Daily Output	Labor-Hours	Unit	Material	2021 Bare Costs Labor	Equipment	Total	Total Incl O&P
0010	**PRECAST WALL PANELS**									
2200	Fiberglass reinforced cement with urethane core									
2210	R20, 8' x 8', 5" plain finish	E-2	750	.075	S.F.	30	4.45	2.28	36.73	42.50
2220	Exposed aggregate or brick finish	"	600	.093	"	49.50	5.55	2.86	57.91	66.50

03 48 Precast Concrete Specialties

03 48 43 – Precast Concrete Trim

03 48 43.40 Precast Lintels

		Crew	Daily Output	Labor-Hours	Unit	Material	2021 Bare Costs		Total	Total Incl O&P
							Labor	Equipment		
0010	**PRECAST LINTELS**, smooth gray, prestressed, stock units only									
0800	4" wide x 8" high x 4' long	D-10	28	1.143	Ea.	35.50	61.50	15.45	112.45	154
0850	8' long		24	1.333		81	71.50	18.05	170.55	223
1000	6" wide x 8" high x 4' long		26	1.231		60	66	16.65	142.65	189
1050	10' long		22	1.455		156	78	19.65	253.65	320
1200	8" wide x 8" high x 4' long		24	1.333		58	71.50	18.05	147.55	198
1250	12' long	↓	20	1.600	↓	201	86	21.50	308.50	380
1275	For custom sizes, types, colors, or finishes of precast lintels, add					150%				

03 48 43.90 Precast Window Sills

		Crew	Daily Output	Labor-Hours	Unit	Material	2021 Bare Costs		Total	Total Incl O&P
							Labor	Equipment		
0010	**PRECAST WINDOW SILLS**									
0600	Precast concrete, 4" tapers to 3", 9" wide	D-1	70	.229	L.F.	22.50	11.15		33.65	42.50
0650	11" wide	"	60	.267	"	40.50	13		53.50	65.50

03 51 Cast Roof Decks

03 51 13 – Cementitious Wood Fiber Decks

03 51 13.50 Cementitious/Wood Fiber Planks

		Crew	Daily Output	Labor-Hours	Unit	Material	2021 Bare Costs		Total	Total Incl O&P
							Labor	Equipment		
0010	**CEMENTITIOUS/WOOD FIBER PLANKS**									
0050	Plank, beveled edge, 1" thick	2 Carp	1000	.016	S.F.	3.66	.88		4.54	5.40
0100	1-1/2" thick		975	.016		5.55	.90		6.45	7.55
0150	T&G, 2" thick		950	.017		3.83	.92		4.75	5.70
0200	2-1/2" thick	↓	925	.017		4.34	.95		5.29	6.30
1000	Bulb tee, sub-purlin and grout, 6' span, add	E-1	5000	.005		2.20	.28	.03	2.51	2.91
1100	8' span	"	4200	.006	↓	2.20	.34	.04	2.58	3

03 52 Lightweight Concrete Roof Insulation

03 52 16 – Lightweight Insulating Concrete

03 52 16.13 Lightweight Cellular Insulating Concrete

		Crew	Daily Output	Labor-Hours	Unit	Material	2021 Bare Costs		Total	Total Incl O&P
							Labor	Equipment		
0010	**LIGHTWEIGHT CELLULAR INSULATING CONCRETE** R035216-10									
0020	Portland cement and foaming agent [G]	C-8	50	1.120	C.Y.	134	55	8.50	197.50	244

03 52 16.16 Lightweight Aggregate Insulating Concrete

		Crew	Daily Output	Labor-Hours	Unit	Material	2021 Bare Costs		Total	Total Incl O&P
							Labor	Equipment		
0010	**LIGHTWEIGHT AGGREGATE INSULATING CONCRETE** R035216-10									
0100	Poured vermiculite or perlite, field mix,									
0110	1:6 field mix [G]	C-8	50	1.120	C.Y.	330	55	8.50	393.50	460
0200	Ready mix, 1:6 mix, roof fill, 2" thick [G]		10000	.006	S.F.	1.85	.27	.04	2.16	2.51
0250	3" thick [G]	↓	7700	.007	"	2.77	.36	.06	3.19	3.67

03 53 Concrete Topping

03 53 16 – Iron-Aggregate Concrete Topping

03 53 16.50 Floor Topping

		Crew	Daily Output	Labor-Hours	Unit	Material	2021 Bare Costs		Total	Total Incl O&P
							Labor	Equipment		
0010	**FLOOR TOPPING**									
0400	Integral topping/finish, on fresh concrete, using 1:1:2 mix, 3/16" thick	C-10B	1000	.040	S.F.	.17	1.89	.34	2.40	3.54
0450	1/2" thick		950	.042		.44	1.99	.35	2.78	4.02
0500	3/4" thick		850	.047		.66	2.23	.39	3.28	4.68
0600	1" thick		750	.053		.88	2.53	.45	3.86	5.45
0800	Granolithic topping, on fresh or cured concrete, 1:1:1-1/2 mix, 1/2" thick		590	.068		.49	3.21	.57	4.27	6.25
0820	3/4" thick		580	.069		.74	3.27	.58	4.59	6.60
0850	1" thick	↓	575	.070		.99	3.29	.58	4.86	6.90

03 53 Concrete Topping

03 53 16 – Iron-Aggregate Concrete Topping

03 53 16.50 Floor Topping	Crew	Daily Output	Labor-Hours	Unit	Material	2021 Bare Costs Labor	Equipment	Total	Total Incl O&P
0950	2" thick	C-10B	500	.080	S.F.	1.97	3.79	.67	6.43	8.90

03 54 Cast Underlayment

03 54 13 – Gypsum Cement Underlayment

03 54 13.50 Poured Gypsum Underlayment

		Crew	Daily Output	Labor-Hours	Unit	Material	2021 Bare Costs Labor	Equipment	Total	Total Incl O&P
0010	**POURED GYPSUM UNDERLAYMENT**									
0400	Underlayment, gypsum based, self-leveling 2500 psi, pumped, 1/2" thick	C-8	24000	.002	S.F.	.20	.11	.02	.33	.42
0500	3/4" thick		20000	.003		.30	.14	.02	.46	.57
0600	1" thick		16000	.004		.40	.17	.03	.60	.74
1400	Hand placed, 1/2" thick	C-18	450	.020		.20	.89	.26	1.35	1.93
1500	3/4" thick	C-18	300	.030	S.F.	.30	1.34	.39	2.03	2.89

03 54 16 – Hydraulic Cement Underlayment

03 54 16.50 Cement Underlayment

		Crew	Daily Output	Labor-Hours	Unit	Material	2021 Bare Costs Labor	Equipment	Total	Total Incl O&P
0010	**CEMENT UNDERLAYMENT**									
2510	Underlayment, P.C. based, self-leveling, 4100 psi, pumped, 1/4" thick	C-8	20000	.003	S.F.	2.09	.14	.02	2.25	2.54
2520	1/2" thick		19000	.003		4.19	.14	.02	4.35	4.86
2530	3/4" thick		18000	.003		6.30	.15	.02	6.47	7.15
2540	1" thick		17000	.003		8.35	.16	.02	8.53	9.50
2550	1-1/2" thick		15000	.004		12.55	.18	.03	12.76	14.10
2560	Hand placed, 1/2" thick	C-18	450	.020		4.19	.89	.26	5.34	6.30
2610	Topping, P.C. based, self-leveling, 6100 psi, pumped, 1/4" thick	C-8	20000	.003		3.40	.14	.02	3.56	3.98
2620	1/2" thick		19000	.003		6.80	.14	.02	6.96	7.75
2630	3/4" thick		18000	.003		10.20	.15	.02	10.37	11.50
2660	1" thick		17000	.003		13.60	.16	.02	13.78	15.25
2670	1-1/2" thick		15000	.004		20.50	.18	.03	20.71	23
2680	Hand placed, 1/2" thick	C-18	450	.020		6.80	.89	.26	7.95	9.20

03 63 Epoxy Grouting

03 63 05 – Grouting of Dowels and Fasteners

03 63 05.10 Epoxy Only

		Crew	Daily Output	Labor-Hours	Unit	Material	2021 Bare Costs Labor	Equipment	Total	Total Incl O&P
0010	**EPOXY ONLY**									
1500	Chemical anchoring, epoxy cartridge, excludes layout, drilling, fastener									
1530	For fastener 3/4" diam. x 6" embedment	2 Skwk	72	.222	Ea.	4.89	12.70		17.59	25.50
1535	1" diam. x 8" embedment		66	.242		7.35	13.85		21.20	30
1540	1-1/4" diam. x 10" embedment		60	.267		14.70	15.25		29.95	40
1545	1-3/4" diam. x 12" embedment		54	.296		24.50	16.90		41.40	54
1550	14" embedment		48	.333		29.50	19.05		48.55	62.50
1555	2" diam. x 12" embedment		42	.381		39	22		61	77.50
1560	18" embedment		32	.500		49	28.50		77.50	99.50

03 81 Concrete Cutting

03 81 13 – Flat Concrete Sawing

03 81 13.50 Concrete Floor/Slab Cutting

03 81 13.50 Concrete Floor/Slab Cutting	Crew	Daily Output	Labor-Hours	Unit	Material	2021 Bare Costs Labor	Equipment	Total	Total Incl O&P
0010 **CONCRETE FLOOR/SLAB CUTTING**									
0050 Includes blade cost, layout and set-up time									
0300 Saw cut concrete slabs, plain, up to 3" deep	B-89	1060	.015	L.F.	.10	.79	1.01	1.90	2.46
0320 Each additional inch of depth		3180	.005		.03	.26	.34	.63	.82
0400 Mesh reinforced, up to 3" deep		980	.016		.12	.85	1.09	2.06	2.67
0420 Each additional inch of depth		2940	.005		.04	.28	.36	.68	.89
0500 Rod reinforced, up to 3" deep		800	.020		.14	1.04	1.33	2.51	3.27
0520 Each additional inch of depth		2400	.007		.05	.35	.44	.84	1.09
0590 Minimum labor/equipment charge		2	8	Job		415	535	950	1,250

03 81 13.75 Concrete Saw Blades

03 81 13.75 Concrete Saw Blades	Crew	Daily Output	Labor-Hours	Unit	Material	2021 Bare Costs Labor	Equipment	Total	Total Incl O&P
0010 **CONCRETE SAW BLADES**									
3000 Blades for saw cutting, included in cutting line items									
3020 Diamond, 12" diameter				Ea.	219			219	241
3040 18" diameter					345			345	380
3080 24" diameter					490			490	535
3120 30" diameter					865			865	950
3160 36" diameter					1,225			1,225	1,350
3200 42" diameter					1,850			1,850	2,050

03 81 16 – Track Mounted Concrete Wall Sawing

03 81 16.50 Concrete Wall Cutting

03 81 16.50 Concrete Wall Cutting	Crew	Daily Output	Labor-Hours	Unit	Material	2021 Bare Costs Labor	Equipment	Total	Total Incl O&P
0010 **CONCRETE WALL CUTTING**									
0750 Includes blade cost, layout and set-up time									
0800 Concrete walls, hydraulic saw, plain, per inch of depth	B-89B	250	.064	L.F.	.03	3.34	6.25	9.62	12.15
0820 Rod reinforcing, per inch of depth		150	.107	"	.05	5.55	10.40	16	20.50
0890 Minimum labor/equipment charge		2	8	Job		415	780	1,195	1,525

03 82 Concrete Boring

03 82 13 – Concrete Core Drilling

03 82 13.10 Core Drilling

03 82 13.10 Core Drilling	Crew	Daily Output	Labor-Hours	Unit	Material	2021 Bare Costs Labor	Equipment	Total	Total Incl O&P
0010 **CORE DRILLING**									
0015 Includes bit cost, layout and set-up time									
0020 Reinforced concrete slab, up to 6" thick									
0100 1" diameter core	B-89A	17	.941	Ea.	.20	48	7.15	55.35	84
0150 For each additional inch of slab thickness in same hole, add		1440	.011		.03	.56	.08	.67	1.03
0200 2" diameter core		16.50	.970		.27	49	7.35	56.62	86.50
0250 For each additional inch of slab thickness in same hole, add		1080	.015		.04	.75	.11	.90	1.36
0300 3" diameter core		16	1		.41	51	7.60	59.01	89.50
0350 For each additional inch of slab thickness in same hole, add		720	.022		.07	1.13	.17	1.37	2.06
0500 4" diameter core		15	1.067		.50	54	8.10	62.60	95.50
0550 For each additional inch of slab thickness in same hole, add		480	.033		.08	1.69	.25	2.02	3.06
0700 6" diameter core		14	1.143		.80	58	8.70	67.50	102
0750 For each additional inch of slab thickness in same hole, add		360	.044		.13	2.26	.34	2.73	4.10
0900 8" diameter core		13	1.231		1.16	62.50	9.35	73.01	111
0950 For each additional inch of slab thickness in same hole, add		288	.056		.19	2.82	.42	3.43	5.15
1100 10" diameter core		12	1.333		1.43	67.50	10.15	79.08	120
1150 For each additional inch of slab thickness in same hole, add		240	.067		.24	3.38	.51	4.13	6.15
1300 12" diameter core		11	1.455		1.63	74	11.05	86.68	131
1350 For each additional inch of slab thickness in same hole, add		206	.078		.27	3.94	.59	4.80	7.20
1500 14" diameter core		10	1.600		2.02	81	12.15	95.17	145
1550 For each additional inch of slab thickness in same hole, add		180	.089		.34	4.51	.68	5.53	8.25

03 82 13 – Concrete Core Drilling

03 82 13.10 Core Drilling

		Crew	Daily Output	Labor-Hours	Unit	Material	2021 Bare Costs Labor	Equipment	Total	Total Incl O&P
1700	18" diameter core	B-89A	9	1.778	Ea.	2.80	90	13.50	106.30	161
1750	For each additional inch of slab thickness in same hole, add		144	.111		.47	5.65	.84	6.96	10.40
1754	24" diameter core		8	2		4.07	102	15.20	121.27	182
1756	For each additional inch of slab thickness in same hole, add	B-89A	120	.133	Ea.	.68	6.75	1.01	8.44	12.60
1760	For horizontal holes, add to above				"		20%	20%		
1770	Prestressed hollow core plank, 8" thick									
1780	1" diameter core	B-89A	17.50	.914	Ea.	.26	46.50	6.95	53.71	81.50
1790	For each additional inch of plank thickness in same hole, add		3840	.004		.03	.21	.03	.27	.41
1794	2" diameter core		17.25	.928		.36	47	7.05	54.41	83
1796	For each additional inch of plank thickness in same hole, add		2880	.006		.04	.28	.04	.36	.55
1800	3" diameter core		17	.941		.55	48	7.15	55.70	84.50
1810	For each additional inch of plank thickness in same hole, add		1920	.008		.07	.42	.06	.55	.82
1820	4" diameter core		16.50	.970		.67	49	7.35	57.02	87
1830	For each additional inch of plank thickness in same hole, add		1280	.013		.08	.63	.10	.81	1.20
1840	6" diameter core		15.50	1.032		1.06	52.50	7.85	61.41	93
1850	For each additional inch of plank thickness in same hole, add		960	.017		.13	.85	.13	1.11	1.63
1860	8" diameter core		15	1.067		1.55	54	8.10	63.65	96.50
1870	For each additional inch of plank thickness in same hole, add		768	.021		.19	1.06	.16	1.41	2.06
1880	10" diameter core		14	1.143		1.90	58	8.70	68.60	104
1890	For each additional inch of plank thickness in same hole, add		640	.025		.24	1.27	.19	1.70	2.49
1900	12" diameter core		13.50	1.185		2.18	60	9	71.18	108
1910	For each additional inch of plank thickness in same hole, add		548	.029		.27	1.48	.22	1.97	2.89
1999	Drilling, core, minimum labor/equipment charge		2	8	Job		405	61	466	710
3000	Bits for core drilling, included in drilling line items									
3010	Diamond, premium, 1" diameter				Ea.	78			78	86
3020	2" diameter					107			107	118
3030	3" diameter					165			165	182
3040	4" diameter					202			202	222
3060	6" diameter					320			320	350
3080	8" diameter					465			465	510
3110	10" diameter					570			570	630
3120	12" diameter					655			655	720
3140	14" diameter					805			805	890
3180	18" diameter					1,125			1,125	1,225
3240	24" diameter					1,625			1,625	1,800

03 82 16 – Concrete Drilling

03 82 16.10 Concrete Impact Drilling

		Crew	Daily Output	Labor-Hours	Unit	Material	2021 Bare Costs Labor	Equipment	Total	Total Incl O&P
0010	**CONCRETE IMPACT DRILLING**									
0020	Includes bit cost, layout and set-up time, no anchors									
0050	Up to 4" deep in concrete/brick floors/walls									
0100	Holes, 1/4" diameter	1 Carp	75	.107	Ea.	.05	5.85		5.90	9.35
0150	For each additional inch of depth in same hole, add	"	430	.019	"	.01	1.02		1.03	1.63
0200	3/8" diameter	1 Carp	63	.127	Ea.	.05	6.95		7	11.10
0250	For each additional inch of depth in same hole, add		340	.024		.01	1.29		1.30	2.06
0300	1/2" diameter		50	.160		.05	8.75		8.80	14
0350	For each additional inch of depth in same hole, add		250	.032		.01	1.75		1.76	2.80
0400	5/8" diameter		48	.167		.08	9.10		9.18	14.60
0450	For each additional inch of depth in same hole, add		240	.033		.02	1.82		1.84	2.92
0500	3/4" diameter		45	.178		.11	9.70		9.81	15.60
0550	For each additional inch of depth in same hole, add		220	.036		.03	1.99		2.02	3.20
0600	7/8" diameter		43	.186		.13	10.20		10.33	16.35
0650	For each additional inch of depth in same hole, add		210	.038		.03	2.08		2.11	3.36

03 82 Concrete Boring

03 82 16 – Concrete Drilling

03 82 16.10 Concrete Impact Drilling	Crew	Daily Output	Labor-Hours	Unit	Material	2021 Bare Costs Labor	Equipment	Total	Total Incl O&P
0700	1" diameter	1 Carp	40	.200	Ea.	.21	10.95		11.16	17.65
0750	For each additional inch of depth in same hole, add		190	.042		.05	2.30		2.35	3.73
0800	1-1/4" diameter		38	.211		.35	11.50		11.85	18.75
0850	For each additional inch of depth in same hole, add		180	.044		.09	2.43		2.52	3.97
0900	1-1/2" diameter		35	.229		.37	12.50		12.87	20.50
0950	For each additional inch of depth in same hole, add	↓	165	.048	↓	.09	2.65		2.74	4.32
1000	For ceiling installations, add						40%			

82

Estimating Tips

04 05 00 Common Work Results for Masonry

- The terms mortar and grout are often used interchangeably—and incorrectly. Mortar is used to bed masonry units, seal the entry of air and moisture, provide architectural appearance, and allow for size variations in the units. Grout is used primarily in reinforced masonry construction and to bond the masonry to the reinforcing steel. Common mortar types are M (2500 psi), S (1800 psi), N (750 psi), and O (350 psi), and they conform to ASTM C270. Grout is either fine or coarse and conforms to ASTM C476, and in-place strengths generally exceed 2500 psi. Mortar and grout are different components of masonry construction and are placed by entirely different methods. An estimator should be aware of their unique uses and costs.

- Mortar is included in all assembled masonry line items. The mortar cost, part of the assembled masonry material cost, includes all ingredients, all labor, and all equipment required. Please see reference number R040513-10.

- Waste, specifically the loss/droppings of mortar and the breakage of brick and block, is included in all unit cost lines that include mortar and masonry units in this division. A factor of 25% is added for mortar and 3% for brick and concrete masonry units.

- Scaffolding or staging is not included in any of the Division 4 costs. Refer to Subdivision 01 54 23 for scaffolding and staging costs.

04 20 00 Unit Masonry

- The most common types of unit masonry are brick and concrete masonry. The major classifications of brick are building brick (ASTM C62), facing brick (ASTM C216), glazed brick, fire brick, and pavers. Many varieties of texture and appearance can exist within these classifications, and the estimator would be wise to check local custom and availability within the project area. For repair and remodeling jobs, matching the existing brick may be the most important criteria.

- Brick and concrete block are priced by the piece and then converted into a price per square foot of wall. Openings less than two square feet are generally ignored by the estimator because any savings in units used are offset by the cutting and trimming required.

- It is often difficult and expensive to find and purchase small lots of historic brick. Costs can vary widely. Many design issues affect costs, selection of mortar mix, and repairs or replacement of masonry materials. Cleaning techniques must be reflected in the estimate.

- All masonry walls, whether interior or exterior, require bracing. The cost of bracing walls during construction should be included by the estimator, and this bracing must remain in place until permanent bracing is complete. Permanent bracing of masonry walls is accomplished by masonry itself, in the form of pilasters or abutting wall corners, or by anchoring the walls to the structural frame. Accessories in the form of anchors, anchor slots, and ties are used, but their supply and installation can be by different trades. For instance, anchor slots on spandrel beams and columns are supplied and welded in place by the steel fabricator, but the ties from the slots into the masonry are installed by the bricklayer. Regardless of the installation method, the estimator must be certain that these accessories are accounted for in pricing.

Reference Numbers

Reference numbers are shown at the beginning of some major classifications. These numbers refer to related items in the Reference Section. The reference information may be an estimating procedure, an alternate pricing method, or technical information.

Note: Not all subdivisions listed here necessarily appear. ■

Same Data. Simplified.

Enjoy the convenience and efficiency of accessing your costs anywhere:

- **Skip the multiplier** by setting your location
- **Quickly search,** edit, favorite and share costs
- **Stay on top of price changes** with automatic updates

Discover more at rsmeans.com/online

04 01 20.20 Pointing Masonry

		Crew	Daily Output	Labor-Hours	Unit	Material	2021 Bare Costs Labor	2021 Bare Costs Equipment	Total	Total Incl O&P
0010	**POINTING MASONRY**									
0300	Cut and repoint brick, hard mortar, running bond	1 Bric	80	.100	S.F.	.57	5.35		5.92	9.30
0320	Common bond		77	.104		.57	5.60		6.17	9.60
0360	Flemish bond		70	.114		.60	6.15		6.75	10.50
0400	English bond		65	.123		.60	6.60		7.20	11.25
0600	Soft old mortar, running bond		100	.080		.57	4.30		4.87	7.55
0620	Common bond		96	.083		.57	4.48		5.05	7.85
0640	Flemish bond		90	.089		.60	4.77		5.37	8.30
0680	English bond		82	.098		.60	5.25		5.85	9.05
0700	Stonework, hard mortar		140	.057	L.F.	.76	3.07		3.83	5.75
0720	Soft old mortar		160	.050	"	.76	2.69		3.45	5.15
1000	Repoint, mask and grout method, running bond		95	.084	S.F.	.76	4.52		5.28	8.10
1020	Common bond		90	.089		.76	4.77		5.53	8.50
1040	Flemish bond		86	.093		.80	5		5.80	8.95
1060	English bond		77	.104		.80	5.60		6.40	9.85
2000	Scrub coat, sand grout on walls, thin mix, brushed		120	.067		3.89	3.58		7.47	10
2020	Troweled		98	.082		5.40	4.38		9.78	12.95
9000	Minimum labor/equipment charge		3	2.667	Job		143		143	230

04 01 20.30 Pointing CMU

		Crew	Daily Output	Labor-Hours	Unit	Material	2021 Bare Costs Labor	2021 Bare Costs Equipment	Total	Total Incl O&P
0010	**POINTING CMU**									
0300	Cut and repoint block, hard mortar, running bond	1 Bric	190	.042	S.F.	.23	2.26		2.49	3.89
0310	Stacked bond	1 Bric	200	.040	S.F.	.23	2.15		2.38	3.71
0600	Soft old mortar, running bond		230	.035		.23	1.87		2.10	3.26
0610	Stacked bond		245	.033		.23	1.75		1.98	3.08

04 01 20.41 Unit Masonry Stabilization

		Crew	Daily Output	Labor-Hours	Unit	Material	2021 Bare Costs Labor	2021 Bare Costs Equipment	Total	Total Incl O&P
0010	**UNIT MASONRY STABILIZATION**									
0100	Structural repointing method									
0110	Cut/grind mortar joint	1 Bric	240	.033	L.F.		1.79		1.79	2.88
0120	Clean and mask joint		2500	.003		.13	.17		.30	.42
0130	Epoxy paste and 1/4" FRP rod		240	.033		2.06	1.79		3.85	5.15
0132	3/8" FRP rod		160	.050		2.92	2.69		5.61	7.55
0140	Remove masking		14400	.001			.03		.03	.05
0300	Structural fabric method									
0310	Primer	1 Bric	600	.013	S.F.	1.01	.72		1.73	2.26
0320	Apply filling/leveling paste		720	.011		.88	.60		1.48	1.92
0330	Epoxy, glass fiber fabric		720	.011		9.25	.60		9.85	11.15
0340	Carbon fiber fabric		720	.011		19.70	.60		20.30	22.50

04 01 20.50 Toothing Masonry

		Crew	Daily Output	Labor-Hours	Unit	Material	2021 Bare Costs Labor	2021 Bare Costs Equipment	Total	Total Incl O&P
0010	**TOOTHING MASONRY**									
0500	Brickwork, soft old mortar	1 Clab	40	.200	V.L.F.		8.90		8.90	14.15
0520	Hard mortar		30	.267			11.85		11.85	18.85
0700	Blockwork, soft old mortar		70	.114			5.05		5.05	8.10
0720	Hard mortar		50	.160			7.10		7.10	11.30
9000	Minimum labor/equipment charge		4	2	Job		89		89	141

04 01 20.52 Cleaning Masonry

		Crew	Daily Output	Labor-Hours	Unit	Material	2021 Bare Costs Labor	2021 Bare Costs Equipment	Total	Total Incl O&P
0010	**CLEANING MASONRY**									
0200	By chemical, brush and rinse, new work, light construction dust	D-1	1000	.016	S.F.	.07	.78		.85	1.33
0220	Medium construction dust		800	.020		.10	.97		1.07	1.68
0240	Heavy construction dust, drips or stains		600	.027		.14	1.30		1.44	2.24
0260	Low pressure wash and rinse, light restoration, light soil		800	.020		.16	.97		1.13	1.75
0270	Average soil, biological staining		400	.040		.24	1.95		2.19	3.40
0280	Heavy soil, biological and mineral staining, paint		330	.048		.32	2.36		2.68	4.15

For customer support on your Commercial Renovation Costs with RSMeans Data, call 800.448.8182.

04 01 Maintenance of Masonry

04 01 20 – Maintenance of Unit Masonry

04 01 20.52 Cleaning Masonry

		Crew	Daily Output	Labor-Hours	Unit	Material	2021 Bare Costs Labor	Equipment	Total	Total Incl O&P
0300	High pressure wash and rinse, heavy restoration, light soil	D-1	600	.027	S.F.	.09	1.30		1.39	2.19
0310	Average soil, biological staining		400	.040		.14	1.95		2.09	3.28
0320	Heavy soil, biological and mineral staining, paint		250	.064		.19	3.12		3.31	5.20
0400	High pressure wash, water only, light soil	C-29	500	.016			.71	.19	.90	1.34
0420	Average soil, biological staining		375	.021			.95	.26	1.21	1.80
0440	Heavy soil, biological and mineral staining, paint		250	.032			1.42	.39	1.81	2.69
0800	High pressure water and chemical, light soil		450	.018		.19	.79	.22	1.20	1.71
0820	Average soil, biological staining		300	.027		.29	1.18	.32	1.79	2.57
0840	Heavy soil, biological and mineral staining, paint		200	.040		.39	1.78	.49	2.66	3.80
1200	Sandblast, wet system, light soil	J-6	1750	.018		.35	.88	.17	1.40	1.96
1220	Average soil, biological staining	J-6	1100	.029	S.F.	.52	1.40	.27	2.19	3.08
1240	Heavy soil, biological and mineral staining, paint		700	.046		.69	2.20	.43	3.32	4.71
1400	Dry system, light soil		2500	.013		.35	.62	.12	1.09	1.48
1420	Average soil, biological staining		1750	.018		.52	.88	.17	1.57	2.15
1440	Heavy soil, biological and mineral staining, paint		1000	.032		.69	1.54	.30	2.53	3.53
1800	For walnut shells, add					.95			.95	1.05
1820	For corn chips, add					.99			.99	1.09
2000	Steam cleaning, light soil	A-1H	750	.011			.47	.10	.57	.86
2020	Average soil, biological staining		625	.013			.57	.12	.69	1.04
2040	Heavy soil, biological and mineral staining		375	.021			.95	.21	1.16	1.74
4000	Add for masking doors and windows	1 Clab	800	.010		.19	.44		.63	.92
4200	Add for pedestrian protection				Job				10%	10%
9000	Minimum labor/equipment charge	D-4	2	16	"		785	95	880	1,350

04 01 20.70 Brick Washing

		Crew	Daily Output	Labor-Hours	Unit	Material	2021 Bare Costs Labor	Equipment	Total	Total Incl O&P
0010	**BRICK WASHING**									
0012	Acid cleanser, smooth brick surface	1 Bric	560	.014	S.F.	.06	.77		.83	1.29
0050	Rough brick		400	.020		.07	1.07		1.14	1.81
0060	Stone, acid wash		600	.013		.09	.72		.81	1.25
1000	Muriatic acid, price per gallon in 5 gallon lots				Gal.	11.15			11.15	12.30

04 05 Common Work Results for Masonry

04 05 05 – Selective Demolition for Masonry

04 05 05.10 Selective Demolition

		Crew	Daily Output	Labor-Hours	Unit	Material	2021 Bare Costs Labor	Equipment	Total	Total Incl O&P
0010	**SELECTIVE DEMOLITION** R024119-10									
0200	Bond beams, 8" block with #4 bar	2 Clab	32	.500	L.F.		22		22	35.50
0300	Concrete block walls, unreinforced, 2" thick		1200	.013	S.F.		.59		.59	.94
0310	4" thick		1150	.014			.62		.62	.98
0320	6" thick		1100	.015			.65		.65	1.03
0330	8" thick		1050	.015			.68		.68	1.08
0340	10" thick		1000	.016			.71		.71	1.13
0360	12" thick		950	.017			.75		.75	1.19
0380	Reinforced alternate courses, 2" thick		1130	.014			.63		.63	1
0390	4" thick		1080	.015			.66		.66	1.05
0400	6" thick		1035	.015			.69		.69	1.09
0410	8" thick		990	.016			.72		.72	1.14
0420	10" thick		940	.017			.76		.76	1.20
0430	12" thick		890	.018			.80		.80	1.27
0440	Reinforced alternate courses & vertically 48" OC, 4" thick		900	.018			.79		.79	1.26
0450	6" thick		850	.019			.84		.84	1.33
0460	8" thick		800	.020			.89		.89	1.41
0480	10" thick		750	.021			.95		.95	1.51

04 05 05.10 Selective Demolition	Crew	Daily Output	Labor-Hours	Unit	Material	2021 Bare Costs Labor	2021 Bare Costs Equipment	Total	Total Incl O&P
0490	12" thick	2 Clab	700	.023	S.F.		1.01		1.01	1.62
1000	Chimney, 16" x 16", soft old mortar	1 Clab	55	.145	C.F.		6.45		6.45	10.30
1020	Hard mortar		40	.200			8.90		8.90	14.15
1030	16" x 20", soft old mortar		55	.145			6.45		6.45	10.30
1040	Hard mortar		40	.200			8.90		8.90	14.15
1050	16" x 24", soft old mortar		55	.145			6.45		6.45	10.30
1060	Hard mortar		40	.200			8.90		8.90	14.15
1080	20" x 20", soft old mortar		55	.145			6.45		6.45	10.30
1100	Hard mortar		40	.200			8.90		8.90	14.15
1110	20" x 24", soft old mortar		55	.145			6.45		6.45	10.30
1120	Hard mortar		40	.200			8.90		8.90	14.15
1140	20" x 32", soft old mortar		55	.145			6.45		6.45	10.30
1160	Hard mortar		40	.200			8.90		8.90	14.15
1200	48" x 48", soft old mortar		55	.145			6.45		6.45	10.30
1220	Hard mortar	↓	40	.200	↓		8.90		8.90	14.15
1250	Metal, high temp steel jacket, 24" diameter	E-2	130	.431	V.L.F.		25.50	13.20	38.70	55.50
1260	60" diameter	"	60	.933			55.50	28.50	84	121
1280	Flue lining, up to 12" x 12"	1 Clab	200	.040			1.78		1.78	2.83
1282	Up to 24" x 24"		150	.053			2.37		2.37	3.77
2000	Columns, 8" x 8", soft old mortar		48	.167			7.40		7.40	11.80
2020	Hard mortar		40	.200			8.90		8.90	14.15
2060	16" x 16", soft old mortar		16	.500			22		22	35.50
2100	Hard mortar		14	.571			25.50		25.50	40.50
2140	24" x 24", soft old mortar		8	1			44.50		44.50	70.50
2160	Hard mortar		6	1.333			59		59	94.50
2200	36" x 36", soft old mortar		4	2			89		89	141
2220	Hard mortar		3	2.667	↓		118		118	189
2230	Alternate pricing method, soft old mortar		30	.267	C.F.		11.85		11.85	18.85
2240	Hard mortar	↓	23	.348	"		15.45		15.45	24.50
3000	Copings, precast or masonry, to 8" wide									
3020	Soft old mortar	1 Clab	180	.044	L.F.		1.97		1.97	3.14
3040	Hard mortar	"	160	.050	"		2.22		2.22	3.54
3100	To 12" wide									
3120	Soft old mortar	1 Clab	160	.050	L.F.		2.22		2.22	3.54
3140	Hard mortar	"	140	.057	"		2.54		2.54	4.04
4000	Fireplace, brick, 30" x 24" opening									
4020	Soft old mortar	1 Clab	2	4	Ea.		178		178	283
4040	Hard mortar		1.25	6.400			284		284	450
4100	Stone, soft old mortar		1.50	5.333			237		237	375
4120	Hard mortar		1	8	↓		355		355	565
5000	Veneers, brick, soft old mortar	↓	140	.057	S.F.		2.54		2.54	4.04
5020	Hard mortar	1 Clab	125	.064	S.F.		2.84		2.84	4.52
5050	Glass block, up to 4" thick		500	.016			.71		.71	1.13
5100	Granite and marble, 2" thick		180	.044			1.97		1.97	3.14
5120	4" thick		170	.047			2.09		2.09	3.33
5140	Stone, 4" thick		180	.044			1.97		1.97	3.14
5160	8" thick		175	.046	↓		2.03		2.03	3.23
5400	Alternate pricing method, stone, 4" thick		60	.133	C.F.		5.90		5.90	9.45
5420	8" thick		85	.094	"		4.18		4.18	6.65
9000	Minimum labor/equipment charge	↓	2	4	Job		178		178	283

04 05 Common Work Results for Masonry

04 05 13 – Masonry Mortaring

04 05 13.10 Cement

		Crew	Daily Output	Labor-Hours	Unit	Material	Labor	Equipment	Total	Total Incl O&P
0010	**CEMENT**				Bag	11.75			11.75	12.90
0100	Masonry, 70 lb. bag, T.L. lots					12.45			12.45	13.70
0150	L.T.L. lots					20.50			20.50	22.50
0200	White, 70 lb. bag, T.L. lots					17.10			17.10	18.80
0250	L.T.L. lots				↓					

04 05 13.20 Lime

		Crew	Daily Output	Labor-Hours	Unit	Material	Labor	Equipment	Total	Total Incl O&P
0010	**LIME**				Bag	11.05			11.05	12.15
0020	Masons, hydrated, 50 lb. bag, T.L. lots					12.15			12.15	13.40
0050	L.T.L. lots					12.05			12.05	13.25
0200	Finish, double hydrated, 50 lb. bag, T.L. lots					13.25			13.25	14.55
0250	L.T.L. lots				↓					

04 05 13.23 Surface Bonding Masonry Mortaring

		Crew	Daily Output	Labor-Hours	Unit	Material	Labor	Equipment	Total	Total Incl O&P
0010	**SURFACE BONDING MASONRY MORTARING**									
0020	Gray or white colors, not incl. block work	1 Bric	540	.015	S.F.	.20	.80		1	1.50

04 05 13.30 Mortar

		Crew	Daily Output	Labor-Hours	Unit	Material	Labor	Equipment	Total	Total Incl O&P
0010	**MORTAR**									
0020	With masonry cement									
0100	Type M, 1:1:6 mix	1 Brhe	143	.056	C.F.	6.85	2.44		9.29	11.50
0200	Type N, 1:3 mix		143	.056		5.45	2.44		7.89	9.95
0300	Type O, 1:3 mix		143	.056		4.82	2.44		7.26	9.25
0400	Type PM, 1:1:6 mix, 2500 psi		143	.056		6.80	2.44		9.24	11.40
0500	Type S, 1/2:1:4 mix	↓	143	.056	↓	6.60	2.44		9.04	11.20
2000	With Portland cement and lime									
2100	Type M, 1:1/4:3 mix	1 Brhe	143	.056	C.F.	9.45	2.44		11.89	14.35
2200	Type N, 1:1:6 mix, 750 psi		143	.056		7.30	2.44		9.74	12
2300	Type O, 1:2:9 mix (Pointing Mortar)		143	.056		8.85	2.44		11.29	13.65
2400	Type PL, 1:1/2:4 mix, 2500 psi		143	.056		6.10	2.44		8.54	10.65
2600	Type S, 1:1/2:4 mix, 1800 psi	↓	143	.056		9.50	2.44		11.94	14.40
2650	Pre-mixed, type S or N				↓	6.40			6.40	7.05
2700	Mortar for glass block	1 Brhe	143	.056	↓	13	2.44		15.44	18.25
2900	Mortar for fire brick, dry mix, 10 lb. pail				Ea.	25			25	27.50

04 05 13.91 Masonry Restoration Mortaring

		Crew	Daily Output	Labor-Hours	Unit	Material	Labor	Equipment	Total	Total Incl O&P
0010	**MASONRY RESTORATION MORTARING**									
0020	Masonry restoration mix				Lb.	.73			.73	.80
0050	White				"	1.20			1.20	1.32

04 05 13.93 Mortar Pigments

		Crew	Daily Output	Labor-Hours	Unit	Material	Labor	Equipment	Total	Total Incl O&P
0010	**MORTAR PIGMENTS**, 50 lb. bags (2 bags per M bricks)									
0020	Color admixture, range 2 to 10 lb. per bag of cement, light colors				Lb.	6.65			6.65	7.30
0050	Medium colors					7.65			7.65	8.40
0100	Dark colors				↓	15.85			15.85	17.40

04 05 13.95 Sand

		Crew	Daily Output	Labor-Hours	Unit	Material	Labor	Equipment	Total	Total Incl O&P
0010	**SAND**, screened and washed at pit				Ton	21.50			21.50	23.50
0020	For mortar, per ton					43			43	47
0050	With 10 mile haul					71.50			71.50	78.50
0100	With 30 mile haul									
0200	Screened and washed, at the pit				C.Y.	30			30	33
0250	With 10 mile haul					59.50			59.50	65.50
0300	With 30 mile haul				↓	99			99	109

87

For customer support on your Commercial Renovation Costs with RSMeans Data, call 800.448.8182.

04 05 13 – Masonry Mortaring

04 05 13.98 Mortar Admixtures	Crew	Daily Output	Labor-Hours	Unit	Material	2021 Bare Costs Labor	Equipment	Total	Total Incl O&P
0010 **MORTAR ADMIXTURES**									
0020 Waterproofing admixture, per quart (1 qt. to 2 bags of masonry cement)				Qt.	3.66			3.66	4.03

04 05 16 – Masonry Grouting

04 05 16.30 Grouting	Crew	Daily Output	Labor-Hours	Unit	Material	2021 Bare Costs Labor	Equipment	Total	Total Incl O&P
0010 **GROUTING**									
0011 Bond beams & lintels, 8" deep, 6" thick, 0.15 C.F./L.F.	D-4	1480	.022	L.F.	.83	1.06	.13	2.02	2.74
0020 8" thick, 0.2 C.F./L.F.		1400	.023		1.33	1.12	.14	2.59	3.40
0050 10" thick, 0.25 C.F./L.F.		1200	.027		1.39	1.31	.16	2.59	3.40
0060 12" thick, 0.3 C.F./L.F.		1040	.031		1.66	1.51	.18	3.35	4.44
0200 Concrete block cores, solid, 4" thk., by hand, 0.067 C.F./S.F. of wall	D-8	1100	.036	S.F.	.37	1.81		2.18	3.31
0210 6" thick, pumped, 0.175 C.F./S.F.	D-4	720	.044		.97	2.18	.26	3.41	4.84
0250 8" thick, pumped, 0.258 C.F./S.F.		680	.047		1.43	2.31	.28	4.02	5.55
0300 10" thick, pumped, 0.340 C.F./S.F.		660	.048		1.88	2.38	.29	4.55	6.20
0350 12" thick, pumped, 0.422 C.F./S.F.		640	.050		2.34	2.46	.30	5.10	6.80
0500 Cavity walls, 2" space, pumped, 0.167 C.F./S.F. of wall		1700	.019		.93	.93	.11	1.97	2.61
0550 3" space, 0.250 C.F./S.F.		1200	.027		1.39	1.31	.16	2.86	3.78
0600 4" space, 0.333 C.F./S.F.		1150	.028		1.85	1.37	.17	3.39	4.39
0700 6" space, 0.500 C.F./S.F.		800	.040		2.77	1.97	.24	4.98	6.45
0800 Door frames, 3' x 7' opening, 2.5 C.F. per opening		60	.533	Opng.	13.85	26	3.17	43.02	60.50
0850 6' x 7' opening, 3.5 C.F. per opening		45	.711	"	19.40	35	4.23	58.63	81.50
2000 Grout, C476, for bond beams, lintels and CMU cores		350	.091	C.F.	5.55	4.49	.54	10.58	13.85
9000 Minimum labor/equipment charge	1 Bric	2	4	Job		215		215	345

04 05 19 – Masonry Anchorage and Reinforcing

04 05 19.05 Anchor Bolts	Crew	Daily Output	Labor-Hours	Unit	Material	2021 Bare Costs Labor	Equipment	Total	Total Incl O&P
0010 **ANCHOR BOLTS**									
0015 Installed in fresh grout in CMU bond beams or filled cores, no templates									
0020 Hooked, with nut and washer, 1/2" diameter, 8" long	1 Bric	132	.061	Ea.	1.55	3.25		4.80	6.95
0030 12" long		131	.061		1.72	3.28		5	7.15
0040 5/8" diameter, 8" long		129	.062		3.40	3.33		6.73	9.10
0050 12" long		127	.063		4.18	3.38		7.56	10.05
0060 3/4" diameter, 8" long		127	.063		4.18	3.38		7.56	10.05
0070 12" long		125	.064		5.25	3.44		8.69	11.25

04 05 19.16 Masonry Anchors	Crew	Daily Output	Labor-Hours	Unit	Material	2021 Bare Costs Labor	Equipment	Total	Total Incl O&P
0010 **MASONRY ANCHORS**									
0020 For brick veneer, galv., corrugated, 7/8" x 7", 22 ga.	1 Bric	10.50	.762	C	15.95	41		56.95	83.50
0100 24 ga.		10.50	.762		10.40	41		51.40	77.50
0150 16 ga.		10.50	.762		33	41		74	102
0200 Buck anchors, galv., corrugated, 16 ga., 2" bend, 8" x 2"		10.50	.762		58.50	41		99.50	131
0250 8" x 3"		10.50	.762		63.50	41		104.50	136
0300 Adjustable, rectangular, 4-1/8" wide									
0350 Anchor and tie, 3/16" wire, mill galv.									
0400 2-3/4" eye, 3-1/4" tie	1 Bric	1.05	7.619	M	480	410		890	1,200
0500 4-3/4" tie		1.05	7.619		520	410		930	1,225
0520 5-1/2" tie		1.05	7.619		565	410		975	1,275
0550 4-3/4" eye, 3-1/4" tie		1.05	7.619		525	410		935	1,250
0570 4-3/4" tie		1.05	7.619		580	410		990	1,300
0580 5-1/2" tie		1.05	7.619		680	410		1,090	1,400
0660 Cavity wall, Z-type, galvanized, 6" long, 1/8" diam.		10.50	.762	C	24.50	41		65.50	93
0670 3/16" diameter		10.50	.762		30.50	41		71.50	99.50
0680 1/4" diameter		10.50	.762		40	41		81	110
0850 8" long, 3/16" diameter		10.50	.762		28	41		69	96.50

04 05 19 – Masonry Anchorage and Reinforcing

04 05 19.16 Masonry Anchors	Crew	Daily Output	Labor-Hours	Unit	Material	2021 Bare Costs Labor	Equipment	Total	Total Incl O&P
0855	1/4" diameter	1 Bric	10.50	.762	C	50.50	41		91.50	122
1000	Rectangular type, galvanized, 1/4" diameter, 2" x 6"		10.50	.762		80	41		121	154
1050	4" x 6"		10.50	.762		91	41		132	166
1100	3/16" diameter, 2" x 6"		10.50	.762		54	41		95	126
1150	4" x 6"		10.50	.762		55.50	41		96.50	127
1200	Mesh wall tie, 1/2" mesh, hot dip galvanized									
1400	16 ga., 12" long, 3" wide	1 Bric	9	.889	C	91.50	47.50		139	178
1420	6" wide		9	.889		134	47.50		181.50	225
1440	12" wide		8.50	.941		212	50.50		262.50	315
1500	Rigid partition anchors, plain, 8" long, 1" x 1/8"		10.50	.762		191	41		232	277
1550	1" x 1/4"		10.50	.762		325	41		366	425
1580	1-1/2" x 1/8"		10.50	.762		227	41		268	315
1600	1-1/2" x 1/4"		10.50	.762		360	41		401	460
1650	2" x 1/8"		10.50	.762		335	41		376	430
1700	2" x 1/4"		10.50	.762		395	41		436	500
2000	Column flange ties, wire, galvanized									
2300	3/16" diameter, up to 3" wide	1 Bric	10.50	.762	C	80.50	41		121.50	155
2350	To 5" wide		10.50	.762		87.50	41		128.50	163
2400	To 7" wide		10.50	.762		94.50	41		135.50	170
2600	To 9" wide		10.50	.762		100	41		141	176
2650	1/4" diameter, up to 3" wide		10.50	.762		112	41		153	189
2700	To 5" wide		10.50	.762		133	41		174	212
2800	To 7" wide		10.50	.762		149	41		190	230
2850	To 9" wide		10.50	.762		162	41		203	244
2900	For hot dip galvanized, add					35%				
4000	Channel slots, 1-3/8" x 1/2" x 8"									
4100	12 ga., plain	1 Bric	10.50	.762	C	254	41		295	345
4150	16 ga., galvanized	"	10.50	.762	"	143	41		184	223
4200	Channel slot anchors									
4300	16 ga., galvanized, 1-1/4" x 3-1/2"				C	64			64	70.50
4350	1-1/4" x 5-1/2"					74.50			74.50	82
4400	1-1/4" x 7-1/2"					68.50			68.50	75.50
4500	1/8" plain, 1-1/4" x 3-1/2"					143			143	158
4550	1-1/4" x 5-1/2"					153			153	168
4600	1-1/4" x 7-1/2"					167			167	184
4700	For corrugation, add					77			77	85
4750	For hot dip galvanized, add					35%				
5000	Dowels									
5100	Plain, 1/4" diameter, 3" long				C	49			49	54
5150	4" long					54.50			54.50	59.50
5200	6" long					66.50			66.50	73
5300	3/8" diameter, 3" long					65			65	71.50
5350	4" long					80.50			80.50	88.50
5400	6" long					93			93	102
5500	1/2" diameter, 3" long					94.50			94.50	104
5550	4" long					112			112	123
5600	6" long					146			146	161
5700	5/8" diameter, 3" long					130			130	143
5750	4" long					159			159	175
5800	6" long					219			219	241
6000	3/4" diameter, 3" long					161			161	177
6100	4" long					205			205	225
6150	6" long					293			293	320

For customer support on your Commercial Renovation Costs with RSMeans Data, call 800.448.8182.

89

04 05 19.16 Masonry Anchors	Crew	Daily Output	Labor-Hours	Unit	Material	2021 Bare Costs Labor	Equipment	Total	Total Incl O&P
6300	For hot dip galvanized, add				C	35%				

04 05 19.26 Masonry Reinforcing Bars	Crew	Daily Output	Labor-Hours	Unit	Material	2021 Bare Costs Labor	Equipment	Total	Total Incl O&P
0010	**MASONRY REINFORCING BARS**									
0015	Steel bars A615, placed horiz., #3 & #4 bars	1 Bric	450	.018	Lb.	.62	.95		1.57	2.21
0020	#5 & #6 bars		800	.010		.62	.54		1.16	1.54
0050	Placed vertical, #3 & #4 bars		350	.023		.62	1.23		1.85	2.65
0060	#5 & #6 bars		650	.012		.62	.66		1.28	1.74
0200	Joint reinforcing, regular truss, to 6" wide, mill std galvanized		30	.267	C.L.F.	23	14.30		37.30	48.50
0250	12" wide		20	.400		29.50	21.50		51	67
0400	Cavity truss with drip section, to 6" wide		30	.267		23.50	14.30		37.80	48.50
0450	12" wide		20	.400		26.50	21.50		48	64
0500	Joint reinforcing, ladder type, mill std galvanized									
0600	9 ga. sides, 9 ga. ties, 4" wall	1 Bric	30	.267	C.L.F.	22	14.30		36.30	47
0650	6" wall		30	.267		22	14.30		36.30	47
0700	8" wall		25	.320		26.50	17.20		43.70	56.50
0750	10" wall		20	.400		26	21.50		47.50	63
0800	12" wall		20	.400		27.50	21.50		49	64.50
1000	Truss type									
1100	9 ga. sides, 9 ga. ties, 4" wall	1 Bric	30	.267	C.L.F.	24.50	14.30		38.80	50
1150	6" wall		30	.267		26	14.30		40.30	51.50
1200	8" wall		25	.320		27.50	17.20		44.70	58
1250	10" wall		20	.400		23.50	21.50		45	60.50
1300	12" wall		20	.400		25	21.50		46.50	62
1500	3/16" sides, 9 ga. ties, 4" wall		30	.267		28.50	14.30		42.80	54.50
1550	6" wall		30	.267		36	14.30		50.30	63
1600	8" wall		25	.320		32.50	17.20		49.70	63.50
1650	10" wall		20	.400		39	21.50		60.50	77.50
1700	12" wall		20	.400		38	21.50		59.50	76.50
2000	3/16" sides, 3/16" ties, 4" wall		30	.267		40.50	14.30		54.80	67.50
2050	6" wall		30	.267		41.50	14.30		55.80	69
2100	8" wall		25	.320		40.50	17.20		57.70	72
2150	10" wall		20	.400		45	21.50		66.50	84
2200	12" wall		20	.400		46.50	21.50		68	86
2500	Cavity truss type, galvanized									
2600	9 ga. sides, 9 ga. ties, 4" wall	1 Bric	25	.320	C.L.F.	35.50	17.20		52.70	66.50
2650	6" wall		25	.320		40.50	17.20		57.70	72
2700	8" wall		20	.400		48	21.50		69.50	87
2750	10" wall		15	.533		43	28.50		71.50	93.50
2800	12" wall		15	.533		52	28.50		80.50	103
3000	3/16" sides, 9 ga. ties, 4" wall		25	.320		45	17.20		62.20	77
3050	6" wall		25	.320		33	17.20		50.20	63.50
3100	8" wall		20	.400		60.50	21.50		82	101
3150	10" wall		15	.533		43	28.50		71.50	93
3200	12" wall		15	.533		43	28.50		71.50	93
3500	For hot dip galvanizing, add				Ton	485			485	535

04 05 23.13 Masonry Control and Expansion Joints	Crew	Daily Output	Labor-Hours	Unit	Material	2021 Bare Costs Labor	Equipment	Total	Total Incl O&P
0010	**MASONRY CONTROL AND EXPANSION JOINTS**									
0020	Rubber, for double wythe 8" minimum wall (Brick/CMU)	1 Bric	400	.020	L.F.	2.55	1.07		3.62	4.54
0025	"T" shaped		320	.025		1.37	1.34		2.71	3.67
0030	Cross-shaped for CMU units		280	.029		1.68	1.53		3.21	4.32
0050	PVC, for double wythe 8" minimum wall (Brick/CMU)		400	.020		1.64	1.07		2.71	3.53

04 05 Common Work Results for Masonry

04 05 23 – Masonry Accessories

04 05 23.13 Masonry Control and Expansion Joints

		Crew	Daily Output	Labor-Hours	Unit	Material	2021 Bare Costs Labor	2021 Bare Costs Equipment	Total	Total Incl O&P
0120	"T" shaped	1 Bric	320	.025	L.F.	1.08	1.34		2.42	3.35
0160	Cross-shaped for CMU units	↓	280	.029	↓	1.28	1.53		2.81	3.88

04 05 23.19 Masonry Cavity Drainage, Weepholes, and Vents

		Crew	Daily Output	Labor-Hours	Unit	Material	Labor	Equipment	Total	Total Incl O&P
0010	**MASONRY CAVITY DRAINAGE, WEEPHOLES, AND VENTS**									
0020	Extruded aluminum, 4" deep, 2-3/8" x 8-1/8"	1 Bric	30	.267	Ea.	41	14.30		55.30	68.50
0050	5" x 8-1/8"		25	.320		50	17.20		67.20	82.50
0100	2-1/4" x 25"		25	.320		84.50	17.20		101.70	121
0150	5" x 16-1/2"		22	.364		62	19.55		81.55	100
0175	5" x 24"		22	.364		94	19.55		113.55	135
0200	6" x 16-1/2"		22	.364		101	19.55		120.55	143
0250	7-3/4" x 16-1/2"	↓	20	.400		93.50	21.50		115	138
0400	For baked enamel finish, add					35%				
0500	For cast aluminum, painted, add					60%				
1000	Stainless steel ventilators, 6" x 6"	1 Bric	25	.320		227	17.20		244.20	278
1050	8" x 8"		24	.333		254	17.90		271.90	310
1100	12" x 12"		23	.348		274	18.70		292.70	330
1150	12" x 6"		24	.333		275	17.90		292.90	330
1200	Foundation block vent, galv., 1-1/4" thk, 8" high, 16" long, no damper	↓	30	.267	↓	18.55	14.30		32.85	43.50
1250	For damper, add					3.74			3.74	4.11
1450	Drainage and ventilation fabric	2 Bric	1450	.011	S.F.	.82	.59		1.41	1.85

04 05 23.95 Wall Plugs

		Crew	Daily Output	Labor-Hours	Unit	Material	Labor	Equipment	Total	Total Incl O&P
0010	**WALL PLUGS** (for nailing to brickwork)									
0020	25 ga., galvanized, plain	1 Bric	10.50	.762	C	52	41		93	124
0050	Wood filled	"	10.50	.762	"	121	41		162	200

04 21 Clay Unit Masonry

04 21 13 – Brick Masonry

04 21 13.13 Brick Veneer Masonry

		Crew	Daily Output	Labor-Hours	Unit	Material	Labor	Equipment	Total	Total Incl O&P
0010	**BRICK VENEER MASONRY**, T.L. lots, excl. scaff., grout & reinforcing R042110-20									
0015	Material costs incl. 3% brick and 25% mortar waste									
2000	Standard, sel. common, 4" x 2-2/3" x 8" (6.75/S.F.) R042110-50	D-8	230	.174	S.F.	4.75	8.65		13.40	19.10
2020	Red, 4" x 2-2/3" x 8", running bond		220	.182		5.05	9.05		14.10	20
2050	Full header every 6th course (7.88/S.F.)		185	.216		5.90	10.75		16.65	24
2100	English, full header every 2nd course (10.13/S.F.)		140	.286		7.55	14.20		21.75	31.50
2150	Flemish, alternate header every course (9.00/S.F.)	↓	150	.267	↓	6.75	13.25		20	29
2200	Flemish, alt. header every 6th course (7.13/S.F.)	D-8	205	.195	S.F.	5.35	9.70		15.05	21.50
2250	Full headers throughout (13.50/S.F.)		105	.381		10.05	18.95		29	41.50
2300	Rowlock course (13.50/S.F.)		100	.400		10.05	19.90		29.95	43
2350	Rowlock stretcher (4.50/S.F.)		310	.129		3.41	6.40		9.81	14.05
2400	Soldier course (6.75/S.F.)		200	.200		5.05	9.95		15	21.50
2450	Sailor course (4.50/S.F.)		290	.138		3.41	6.85		10.26	14.75
2600	Buff or gray face, running bond (6.75/S.F.)		220	.182		5.35	9.05		14.40	20.50
2700	Glazed face brick, running bond		210	.190		16	9.45		25.45	33
2750	Full header every 6th course (7.88/S.F.)		170	.235		18.70	11.70		30.40	39.50
3000	Jumbo, 6" x 4" x 12" running bond (3.00/S.F.)		435	.092		5.40	4.57		9.97	13.30
3050	Norman, 4" x 2-2/3" x 12" running bond (4.5/S.F.)		320	.125		6.70	6.20		12.90	17.40
3100	Norwegian, 4" x 3-1/5" x 12" (3.75/S.F.)		375	.107		7.70	5.30		13	16.95
3150	Economy, 4" x 4" x 8" (4.50/S.F.)		310	.129		4.72	6.40		11.12	15.50
3200	Engineer, 4" x 3-1/5" x 8" (5.63/S.F.)		260	.154		3.99	7.65		11.64	16.70
3250	Roman, 4" x 2" x 12" (6.00/S.F.)	↓	250	.160	↓	8.10	7.95		16.05	22

04 21 13 – Brick Masonry

04 21 13.13 Brick Veneer Masonry

		Crew	Daily Output	Labor-Hours	Unit	Material	2021 Bare Costs Labor	Equipment	Total	Total Incl O&P
3300	S.C.R., 6" x 2-2/3" x 12" (4.50/S.F.)	D-8	310	.129	S.F.	6.15	6.40		12.55	17.05
3350	Utility, 4" x 4" x 12" (3.00/S.F.)	↓	360	.111		5.55	5.50		11.05	14.95
3360	For less than truck load lots, add					.10%				
3400	For cavity wall construction, add						15%			
3450	For stacked bond, add						10%			
3500	For interior veneer construction, add						15%			
3550	For curved walls, add				↓		30%			
9000	Minimum labor/equipment charge	D-1	2	8	Job		390		390	625

04 21 13.14 Thin Brick Veneer

		Crew	Daily Output	Labor-Hours	Unit	Material	2021 Bare Costs Labor	Equipment	Total	Total Incl O&P
0010	**THIN BRICK VENEER**									
0015	Material costs incl. 3% brick and 25% mortar waste									
0020	On & incl. metal panel support sys, modular, 2-2/3" x 5/8" x 8", red	D-7	92	.174	S.F.	13.85	8.10		21.95	28
0100	Closure, 4" x 5/8" x 8"		110	.145		9.45	6.80		16.25	21
0110	Norman, 2-2/3" x 5/8" x 12"		110	.145		9.85	6.80		16.65	21.50
0120	Utility, 4" x 5/8" x 12"		125	.128		9.20	5.95		15.15	19.40
0130	Emperor, 4" x 3/4" x 16"		175	.091		10.40	4.27		14.67	18.10
0140	Super emperor, 8" x 3/4" x 16"	↓	195	.082	↓	11.40	3.83		15.23	18.55
0150	For L shaped corners with 4" return, add				L.F.	9.25			9.25	10.20
0200	On masonry/plaster back-up, modular, 2-2/3" x 5/8" x 8", red	D-7	137	.117	S.F.	8.80	5.45		14.25	18.20
0210	Closure, 4" x 5/8" x 8"		165	.097		4.40	4.52		8.92	11.90
0220	Norman, 2-2/3" x 5/8" x 12"		165	.097		4.77	4.52		9.29	12.30
0230	Utility, 4" x 5/8" x 12"		185	.086		4.12	4.03		8.15	10.85
0240	Emperor, 4" x 3/4" x 16"		260	.062		5.30	2.87		8.17	10.35
0250	Super emperor, 8" x 3/4" x 16"	↓	285	.056	↓	6.35	2.62		8.97	11.05
0260	For L shaped corners with 4" return, add				L.F.	10			10	11
0270	For embedment into pre-cast concrete panels, add				S.F.	14.40			14.40	15.85

04 21 13.15 Chimney

		Crew	Daily Output	Labor-Hours	Unit	Material	2021 Bare Costs Labor	Equipment	Total	Total Incl O&P
0010	**CHIMNEY**, excludes foundation, scaffolding, grout and reinforcing									
0100	Brick, 16" x 16", 8" flue	D-1	18.20	.879	V.L.F.	26	43		69	98
0150	16" x 20" with one 8" x 12" flue		16	1		44	48.50		92.50	127
0200	16" x 24" with two 8" x 8" flues		14	1.143		65	55.50		120.50	161
0250	20" x 20" with one 12" x 12" flue		13.70	1.168		51.50	57		108.50	149
0300	20" x 24" with two 8" x 12" flues		12	1.333		73	65		138	184
0350	20" x 32" with two 12" x 12" flues	↓	10	1.600	↓	91	78		169	225

04 21 13.18 Columns

		Crew	Daily Output	Labor-Hours	Unit	Material	2021 Bare Costs Labor	Equipment	Total	Total Incl O&P
0010	**COLUMNS**, solid, excludes scaffolding, grout and reinforcing									
0050	Brick, 8" x 8", 9 brick/V.L.F.	D-1	56	.286	V.L.F.	6.50	13.90		20.40	29.50
0100	12" x 8", 13.5 brick/V.L.F.		37	.432		9.75	21		30.75	45
0200	12" x 12", 20 brick/V.L.F.		25	.640		14.50	31		45.50	66
0300	16" x 12", 27 brick/V.L.F.		19	.842		19.55	41		60.55	87.50
0400	16" x 16", 36 brick/V.L.F.		14	1.143		26	55.50		81.50	118
0500	20" x 16", 45 brick/V.L.F.		11	1.455		32.50	71		103.50	150
0600	20" x 20", 56 brick/V.L.F.		9	1.778		40.50	86.50		127	184
0700	24" x 20", 68 brick/V.L.F.		7	2.286		49	111		160	233
0800	24" x 24", 81 brick/V.L.F.		6	2.667		58.50	130		188.50	274
1000	36" x 36", 182 brick/V.L.F.		3	5.333	↓	132	260		392	560
9000	Minimum labor/equipment charge		2	8	Job		390		390	625

04 21 13.30 Oversized Brick

		Crew	Daily Output	Labor-Hours	Unit	Material	2021 Bare Costs Labor	Equipment	Total	Total Incl O&P
0010	**OVERSIZED BRICK**, excludes scaffolding, grout and reinforcing									
0100	Veneer, 4" x 2.25" x 16"	D-8	387	.103	S.F.	4.90	5.15		10.05	13.65
0102	8" x 2.25" x 16", multicell		265	.151		17.75	7.50		25.25	31.50
0105	4" x 2.75" x 16"	↓	412	.097		5.85	4.83		10.68	14.20

04 21 Clay Unit Masonry

04 21 13 – Brick Masonry

	04 21 13.30 Oversized Brick	Crew	Daily Output	Labor-Hours	Unit	Material	2021 Bare Costs Labor	Equipment	Total	Total Incl O&P
0107	8" x 2.75" x 16", multicell	D-8	295	.136	S.F.	17.75	6.75		24.50	30.50
0110	4" x 4" x 16"		460	.087		3.90	4.32		8.22	11.25
0120	4" x 8" x 16"		533	.075		5.60	3.73		9.33	12.20
0122	4" x 8" x 16" multicell		327	.122		15.90	6.10		22	27.50
0125	Loadbearing, 6" x 4" x 16", grouted and reinforced		387	.103		12.05	5.15		17.20	21.50
0130	8" x 4" x 16", grouted and reinforced		327	.122		12.20	6.10		18.30	23
0132	10" x 4" x 16", grouted and reinforced		327	.122		28.50	6.10		34.60	41.50
0135	6" x 8" x 16", grouted and reinforced		440	.091		15.60	4.52		20.12	24.50
0140	8" x 8" x 16", grouted and reinforced		400	.100		16.45	4.97		21.42	26
0145	Curtainwall/reinforced veneer, 6" x 4" x 16"		387	.103		17.45	5.15		22.60	27.50
0150	8" x 4" x 16"		327	.122		20.50	6.10		26.60	32.50
0152	10" x 4" x 16"		327	.122		31.50	6.10		37.60	45
0155	6" x 8" x 16"	↓	440	.091	↓	21.50	4.52		26.02	31.50
0160	8" x 8" x 16"	D-8	400	.100	S.F.	29	4.97		33.97	40
0200	For 1 to 3 slots in face, add					15%				
0210	For 4 to 7 slots in face, add					25%				
0220	For bond beams, add					20%				
0230	For bullnose shapes, add					20%				
0240	For open end knockout, add					10%				
0250	For white or gray color group, add					10%				
0260	For 135 degree corner, add					250%				

04 21 13.35 Common Building Brick

		Crew	Daily Output	Labor-Hours	Unit	Material	2021 Bare Costs Labor	Equipment	Total	Total Incl O&P
0010	COMMON BUILDING BRICK, C62, T.L. lots, material only R042110-20									
0020	Standard				M	610			610	670
0050	Select				"	580			580	640

04 21 13.40 Structural Brick

		Crew	Daily Output	Labor-Hours	Unit	Material	2021 Bare Costs Labor	Equipment	Total	Total Incl O&P
0010	STRUCTURAL BRICK C652, Grade SW, incl. mortar, scaffolding not incl.									
0100	Standard unit, 4-5/8" x 2-3/4" x 9-5/8"	D-8	245	.163	S.F.	4.33	8.10		12.43	17.80
0120	Bond beam		225	.178		4.45	8.85		13.30	19.10
0140	V cut bond beam		225	.178		4.34	8.85		13.19	19
0160	Stretcher quoin, 5-5/8" x 2-3/4" x 9-5/8"		245	.163		7.35	8.10		15.45	21
0180	Corner quoin		245	.163		8	8.10		16.10	22
0200	Corner, 45 degree, 4-5/8" x 2-3/4" x 10-7/16"	↓	235	.170	↓	9.05	8.45		17.50	23.50

04 21 13.45 Face Brick

		Crew	Daily Output	Labor-Hours	Unit	Material	2021 Bare Costs Labor	Equipment	Total	Total Incl O&P
0010	FACE BRICK Material Only, C216, T.L. lots R042110-20									
0300	Standard modular, 4" x 2-2/3" x 8"				M	625			625	690
0450	Economy, 4" x 4" x 8"					910			910	1,000
0510	Economy, 4" x 4" x 12"					1,450			1,450	1,575
0550	Jumbo, 6" x 4" x 12"					1,625			1,625	1,800
0610	Jumbo, 8" x 4" x 12"					1,625			1,625	1,800
0650	Norwegian, 4" x 3-1/5" x 12"					1,875			1,875	2,075
0710	Norwegian, 6" x 3-1/5" x 12"					1,575			1,575	1,750
0850	Standard glazed, plain colors, 4" x 2-2/3" x 8"					2,200			2,200	2,425
1000	Deep trim shades, 4" x 2-2/3" x 8"					2,375			2,375	2,625
1080	Jumbo utility, 4" x 4" x 12"					1,675			1,675	1,850
1120	4" x 8" x 8"					1,975			1,975	2,150
1140	4" x 8" x 16"					6,875			6,875	7,550
1260	Engineer, 4" x 3-1/5" x 8"					580			580	640
1350	King, 4" x 2-3/4" x 10"					575			575	635
1770	Standard modular, double glazed, 4" x 2-2/3" x 8"					3,000			3,000	3,300
1850	Jumbo, colored glazed ceramic, 6" x 4" x 12"					2,800			2,800	3,075
2050	Jumbo utility, glazed, 4" x 4" x 12"					5,750			5,750	6,325

04 21 Clay Unit Masonry

04 21 13 – Brick Masonry

04 21 13.45 Face Brick

		Crew	Daily Output	Labor-Hours	Unit	Material	2021 Bare Costs Labor	Equipment	Total	Total Incl O&P
2100	4" x 8" x 8"				M	6,775			6,775	7,450
2150	4" x 16" x 8"				↓	7,225			7,225	7,950
3050	Used brick				M	495			495	540
3150	Add for brick to match existing work, minimum					5%				
3200	Maximum				↓	50%				

04 21 26 – Glazed Structural Clay Tile Masonry

04 21 26.10 Structural Facing Tile

		Crew	Daily Output	Labor-Hours	Unit	Material	2021 Bare Costs Labor	Equipment	Total	Total Incl O&P
0010	**STRUCTURAL FACING TILE**, std. colors, excl. scaffolding, grout, reinforcing									
0020	6T series, 5-1/3" x 12", 2.3 pieces per S.F., glazed 1 side, 2" thick	D-8	225	.178	S.F.	9.90	8.85		18.75	25
0100	4" thick		220	.182		13.10	9.05		22.15	29
0150	Glazed 2 sides		195	.205		17.45	10.20		27.65	35.50
0250	6" thick		210	.190		18.70	9.45		28.15	35.50
0300	Glazed 2 sides		185	.216		22.50	10.75		33.25	42
0400	8" thick		180	.222	↓	23	11.05		34.05	43
0500	Special shapes, group 1		400	.100	Ea.	7.55	4.97		12.52	16.30
0550	Group 2		375	.107		13.55	5.30		18.85	23.50
0600	Group 3		350	.114		17.05	5.70		22.75	28
0650	Group 4		325	.123		35.50	6.10		41.60	49
0700	Group 5		300	.133		42	6.65		48.65	57
0750	Group 6		275	.145	↓	62.50	7.25		69.75	80.50
1000	Fire rated, 4" thick, 1 hr. rating		210	.190	S.F.	18.45	9.45		27.90	35.50
1050	2 hr. rating		170	.235		28	11.70		39.70	50
1300	Acoustic, 4" thick		210	.190		40	9.45		49.45	58.50
1310	6" thick		170	.235		41	11.70		52.70	64.50
1320	8" thick		145	.276		43.50	13.70		57.20	70
1350	8T acoustic, 4" thick		265	.151		19.80	7.50		27.30	34
1360	6" thick		256	.156		29	7.75		36.75	44.50
1370	8" thick	↓	208	.192	↓	31.50	9.55		41.05	50
2000	8W series, 8" x 16", 1.125 pieces per S.F.									
2050	2" thick, glazed 1 side	D-8	360	.111	S.F.	9.55	5.50		15.05	19.35
2100	4" thick, glazed 1 side		345	.116		13.50	5.75		19.25	24
2150	Glazed 2 sides		325	.123		14.40	6.10		20.50	25.50
2200	6" thick, glazed 1 side		330	.121		31	6		37	44
2210	Glazed 2 sides		250	.160		31.50	7.95		39.45	47.50
2250	8" thick, glazed 1 side		310	.129		29.50	6.40		35.90	43
2260	Glazed 2 sides		220	.182	↓	35	9.05		44.05	53
2500	Special shapes, group 1		300	.133	Ea.	15.90	6.65		22.55	28
2550	Group 2		280	.143		21	7.10		28.10	34.50
2600	Group 3		260	.154		22.50	7.65		30.15	37.50
2650	Group 4		250	.160		54	7.95		61.95	72
2700	Group 5		240	.167		47	8.30		55.30	65
2750	Group 6		230	.174	↓	103	8.65		111.65	127
2900	Fire rated, 2" thick, glazed 1 side		360	.111	S.F.	12.70	5.50		18.20	23
3000	4" thick, glazed 1 side	D-8	345	.116	S.F.	15.80	5.75		21.55	26.50
3020	Glazed 2 sides		345	.116		24	5.75		29.75	36
3040	6" thick, glazed 1 side		330	.121		24	6		30	35.50
3045	Glazed 2 sides		250	.160		25	7.95		32.95	40.50
3047	8" thick, glazed 1 side		310	.129		30	6.40		36.40	43.50
3048	Glazed 2 sides		220	.182		31.50	9.05		40.55	49.50
3050	Special shapes, group 1, fire rated		300	.133		20.50	6.65		27.15	33.50
3060	Group 2		280	.143		29.50	7.10		36.60	44
3070	Group 3	↓	260	.154	↓	30	7.65		37.65	45.50

04 21 26 – Glazed Structural Clay Tile Masonry

04 21 26.10 Structural Facing Tile	Crew	Daily Output	Labor-Hours	Unit	Material	2021 Bare Costs Labor	Equipment	Total	Total Incl O&P
3080	Group 4	D-8	250	.160	S.F.	53	7.95		60.95	71.50
3090	Group 5		240	.167		52.50	8.30		60.80	71
3100	Acoustic, 4" thick		345	.116		22.50	5.75		28.25	34
3120	4W series, 8" x 8", 2.25 pieces per S.F.									
3125	2" thick, glazed 1 side	D-8	360	.111	S.F.	9.20	5.50		14.70	18.95
3130	4" thick, glazed 1 side		345	.116		12.65	5.75		18.40	23
3135	Glazed 2 sides		325	.123		15.75	6.10		21.85	27
3140	6" thick, glazed 1 side		330	.121		16.15	6		22.15	27.50
3145	Glazed 2 sides		250	.160		22	7.95		29.95	37
3150	8" thick, glazed 1 side		310	.129		24.50	6.40		30.90	37.50
3155	Special shapes, group 1		300	.133	Ea.	7.60	6.65		14.25	19.05
3160	Group 2		280	.143	"	8.85	7.10		15.95	21
3200	For designer colors, add					25%				
3300	For epoxy mortar joints, add				S.F.	1.88			1.88	2.07
9000	Minimum labor/equipment charge	D-1	2	8	Job		390		390	625

04 21 29 – Terra Cotta Masonry

04 21 29.10 Terra Cotta Masonry Components

04 21 29.10 Terra Cotta Masonry Components	Crew	Daily Output	Labor-Hours	Unit	Material	2021 Bare Costs Labor	Equipment	Total	Total Incl O&P
0010	**TERRA COTTA MASONRY COMPONENTS**									
0020	Coping, split type, not glazed, 9" wide	D-1	90	.178	L.F.	11.60	8.65		20.25	26.50
0100	13" wide		80	.200		16.70	9.75		26.45	34
0200	Coping, split type, glazed, 9" wide		90	.178		20.50	8.65		29.15	36.50
0250	13" wide		80	.200		24.50	9.75		34.25	42.50
0500	Partition or back-up blocks, scored, in C.L. lots									
0700	Non-load bearing 12" x 12", 3" thick, special order	D-8	550	.073	S.F.	20.50	3.61		24.11	28.50
0750	4" thick, standard		500	.080		6.95	3.98		10.93	14.05
0800	6" thick		450	.089		8.40	4.42		12.82	16.35
0850	8" thick		400	.100		10.35	4.97		15.32	19.35
1000	Load bearing, 12" x 12", 4" thick, in walls		500	.080		6.45	3.98		10.43	13.50
1050	In floors		750	.053		6.45	2.65		9.10	11.35
1200	6" thick, in walls		450	.089		9.80	4.42		14.22	17.85
1250	In floors		675	.059		9.80	2.95		12.75	15.50
1400	8" thick, in walls		400	.100		10.85	4.97		15.82	19.95
1450	In floors	D-8	575	.070	S.F.	10.85	3.46		14.31	17.50
1600	10" thick, in walls, special order		350	.114		28.50	5.70		34.20	40
1650	In floors, special order		500	.080		28.50	3.98		32.48	37.50
1800	12" thick, in walls, special order		300	.133		26.50	6.65		33.15	39.50
1850	In floors, special order		450	.089		26.50	4.42		30.92	36
2000	For reinforcing with steel rods, add to above					15%	5%			
2025	Placed vertical, #3 & #4 bars	1 Bric	350	.023	Lb.	.62	1.23		1.85	2.65
2100	For smooth tile instead of scored, add				S.F.	2.94			2.94	3.23
2200	For L.C.L. quantities, add				"	10%	10%			
9000	Minimum labor/equipment charge	D-1	2	8	Job		390		390	625

04 21 29.20 Terra Cotta Tile

04 21 29.20 Terra Cotta Tile	Crew	Daily Output	Labor-Hours	Unit	Material	2021 Bare Costs Labor	Equipment	Total	Total Incl O&P
0010	**TERRA COTTA TILE**, on walls, dry set, 1/2" thick									
0100	Square, hexagonal or lattice shapes, unglazed	1 Tilf	135	.059	S.F.	5.20	3.06		8.26	10.50
0300	Glazed, plain colors		130	.062		8.10	3.18		11.28	13.85
0400	Intense colors		125	.064		9.45	3.31		12.76	15.55

For customer support on your Commercial Renovation Costs with RSMeans Data, call 800.448.8182.

95

04 22 Concrete Unit Masonry

04 22 10 – Concrete Masonry Units

04 22 10.11 Autoclave Aerated Concrete Block

		Crew	Daily Output	Labor-Hours	Unit	Material	2021 Bare Costs Labor	Equipment	Total	Total Incl O&P
0010	**AUTOCLAVE AERATED CONCRETE BLOCK**, excl. scaffolding, grout & reinforcing									
0050	Solid, 4" x 8" x 24", incl. mortar **G**	D-8	600	.067	S.F.	1.59	3.31		4.90	7.05
0060	6" x 8" x 24" **G**		600	.067		2.56	3.31		5.87	8.10
0070	8" x 8" x 24" **G**		575	.070		3.40	3.46		6.86	9.30
0080	10" x 8" x 24" **G**		575	.070		4.08	3.46		7.54	10.05
0090	12" x 8" x 24" **G**	↓	550	.073	↓	4.89	3.61		8.50	11.20

04 22 10.14 Concrete Block, Back-Up

		Crew	Daily Output	Labor-Hours	Unit	Material	2021 Bare Costs Labor	Equipment	Total	Total Incl O&P
0010	**CONCRETE BLOCK, BACK-UP**, C90, 2000 psi									
0020	Normal weight, 8" x 16" units, tooled joint 1 side									
0050	Not-reinforced, 2000 psi, 2" thick	D-8	475	.084	S.F.	1.72	4.19		5.91	8.65
0200	4" thick		460	.087		2.20	4.32		6.52	9.35
0300	6" thick		440	.091		2.81	4.52		7.33	10.35
0350	8" thick		400	.100		2.74	4.97		7.71	11
0400	10" thick	↓	330	.121		3.29	6		9.29	13.30
0450	12" thick	D-9	310	.155		4.74	7.55		12.29	17.30
1000	Reinforced, alternate courses, 4" thick	D-8	450	.089		2.38	4.42		6.80	9.70
1100	6" thick		430	.093		3.01	4.62		7.63	10.75
1150	8" thick		395	.101		2.95	5.05		8	11.35
1200	10" thick	↓	320	.125		3.47	6.20		9.67	13.80
1250	12" thick	D-9	300	.160	↓	4.93	7.80		12.73	17.95
9000	Minimum labor/equipment charge	D-1	2	8	Job		390		390	625

04 22 10.16 Concrete Block, Bond Beam

		Crew	Daily Output	Labor-Hours	Unit	Material	2021 Bare Costs Labor	Equipment	Total	Total Incl O&P
0010	**CONCRETE BLOCK, BOND BEAM**, C90, 2000 psi									
0020	Not including grout or reinforcing									
0125	Regular block, 6" thick	D-8	584	.068	L.F.	2.76	3.40		6.16	8.50
0130	8" high, 8" thick	"	565	.071		2.93	3.52		6.45	8.90
0150	12" thick	D-9	510	.094		4.35	4.58		8.93	12.15
0525	Lightweight, 6" thick	D-8	592	.068		3.11	3.36		6.47	8.80
0530	8" high, 8" thick	"	575	.070		3.66	3.46		7.12	9.55
0550	12" thick	D-9	520	.092	↓	4.91	4.50		9.41	12.60
2000	Including grout and 2 #5 bars									
2100	Regular block, 8" high, 8" thick	D-8	300	.133	L.F.	5.55	6.65		12.20	16.75
2150	12" thick	D-9	250	.192		7.70	9.35		17.05	23.50
2500	Lightweight, 8" high, 8" thick	D-8	305	.131		6.30	6.50		12.80	17.35
2550	12" thick	D-9	255	.188	↓	8.25	9.15		17.40	24
9000	Minimum labor/equipment charge	D-1	2	8	Job		390		390	625

04 22 10.19 Concrete Block, Insulation Inserts

		Crew	Daily Output	Labor-Hours	Unit	Material	2021 Bare Costs Labor	Equipment	Total	Total Incl O&P
0010	**CONCRETE BLOCK, INSULATION INSERTS**									
0100	Styrofoam, plant installed, add to block prices									
0200	8" x 16" units, 6" thick				S.F.	1.28			1.28	1.41
0250	8" thick					1.65			1.65	1.82
0300	10" thick					1.34			1.34	1.47
0350	12" thick					1.71			1.71	1.88
0500	8" x 8" units, 8" thick					1.63			1.63	1.79
0550	12" thick				↓	1.62			1.62	1.78

04 22 10.23 Concrete Block, Decorative

		Crew	Daily Output	Labor-Hours	Unit	Material	2021 Bare Costs Labor	Equipment	Total	Total Incl O&P
0010	**CONCRETE BLOCK, DECORATIVE**, C90, 2000 psi									
0020	Embossed, simulated brick face									
0100	8" x 16" units, 4" thick	D-8	400	.100	S.F.	3.17	4.97		8.14	11.50
0200	8" thick		340	.118		3.29	5.85		9.14	13
0250	12" thick	↓	300	.133	↓	4.82	6.65		11.47	15.95
0400	Embossed both sides									

96

04 22 10.23 **Concrete Block, Decorative**	Crew	Daily Output	Labor-Hours	Unit	Material	2021 Bare Costs Labor	Equipment	Total	Total Incl O&P
0500	8" thick	D-8	300	.133	S.F.	5.55	6.65		12.20	16.75
0550	12" thick	"	275	.145	"	6.75	7.25		14	19
1000	Fluted high strength								9.92	13.85
1100	8" x 16" x 4" thick, flutes 1 side,	D-8	345	.116	S.F.	4.17	5.75		9.92	13.85
1150	Flutes 2 sides		335	.119		4.72	5.95		10.67	14.75
1200	8" thick	↓	300	.133		5.30	6.65		11.95	16.45
1250	For special colors, add				↓	.66			.66	.72
1400	Deep grooved, smooth face									
1450	8" x 16" x 4" thick	D-8	345	.116	S.F.	2.64	5.75		8.39	12.15
1500	8" thick	"	300	.133	"	4.54	6.65		11.19	15.65
2000	Formblock, incl. inserts & reinforcing									
2100	8" x 16" x 8" thick	D-8	345	.116	S.F.	4.49	5.75		10.24	14.20
2150	12" thick	"	310	.129	"	5.85	6.40		12.25	16.70
2500	Ground face									
2600	8" x 16" x 4" thick	D-8	345	.116	S.F.	4.30	5.75		10.05	14
2650	6" thick		325	.123		4.92	6.10		11.02	15.25
2700	8" thick		300	.133		5.70	6.65		12.35	16.90
2725	10" thick	↓	280	.143		5.60	7.10		12.70	17.60
2750	12" thick	D-9	265	.181	↓	7.30	8.80		16.10	22
2900	For special colors, add, minimum					15%				
2950	For special colors, add, maximum					45%				
4000	Slump block									
4100	4" face height x 16" x 4" thick	D-1	165	.097	S.F.	5.40	4.72		10.12	13.50
4150	6" thick		160	.100		6.65	4.87		11.52	15.15
4200	8" thick		155	.103		9.50	5.05		14.55	18.55
4250	10" thick		140	.114		12.15	5.55		17.70	22.50
4300	12" thick		130	.123		13.45	6		19.45	24.50
4400	6" face height x 16" x 6" thick		155	.103		6.20	5.05		11.25	14.90
4450	8" thick		150	.107		7.65	5.20		12.85	16.75
4500	10" thick		130	.123		14.55	6		20.55	25.50
4550	12" thick	↓	120	.133	↓	14.55	6.50		21.05	26.50
5000	Split rib profile units, 1" deep ribs, 8 ribs									
5100	8" x 16" x 4" thick	D-8	345	.116	S.F.	4.90	5.75		10.65	14.65
5150	6" thick		325	.123		5.65	6.10		11.75	16.05
5200	8" thick		300	.133		6.85	6.65		13.50	18.20
5225	10" thick	↓	300	.133		6.95	6.65		13.60	18.30
5250	12" thick	D-9	275	.175		7.85	8.50		16.35	22.50
5400	For special deeper colors, 4" thick, add					1.42			1.42	1.57
5450	12" thick, add					1.38			1.38	1.52
5600	For white, 4" thick, add					1.42			1.42	1.57
5650	6" thick, add					1.38			1.38	1.52
5700	8" thick, add					1.48			1.48	1.63
5750	12" thick, add					1.42			1.42	1.57
6000	Split face									
6100	8" x 16" x 4" thick	D-8	350	.114	S.F.	3.86	5.70		9.56	13.40
6150	6" thick		325	.123		4.38	6.10		10.48	14.65
6200	8" thick		300	.133		5.60	6.65		12.25	16.80
6225	10" thick	↓	275	.145		5.85	7.25		13.10	18.05
6250	12" thick	D-9	270	.178		6.25	8.65		14.90	21
6300	For scored, add					.39			.39	.43
6400	For special deeper colors, 4" thick, add					.66			.66	.72
6450	6" thick, add					.84			.84	.92
6500	8" thick, add					.71			.71	.78

97

04 22 Concrete Unit Masonry

04 22 10 – Concrete Masonry Units

04 22 10.23 Concrete Block, Decorative

		Crew	Daily Output	Labor-Hours	Unit	Material	2021 Bare Costs Labor	Equipment	Total	Total Incl O&P
6550	12" thick, add				S.F.	.84			.84	.92
6650	For white, 4" thick, add				S.F.	1.42			1.42	1.57
6700	6" thick, add					1.29			1.29	1.42
6750	8" thick, add					1.41			1.41	1.55
6800	12" thick, add					1.38			1.38	1.52
7000	Scored ground face, 2 to 5 scores									
7100	8" x 16" x 4" thick	D-8	340	.118	S.F.	8.85	5.85		14.70	19.15
7150	6" thick		310	.129		10.15	6.40		16.55	21.50
7200	8" thick		290	.138		11.50	6.85		18.35	23.50
7250	12" thick	D-9	265	.181		17.10	8.80		25.90	33
8000	Hexagonal face profile units, 8" x 16" units									
8100	4" thick, hollow	D-8	340	.118	S.F.	4.18	5.85		10.03	14
8200	Solid		340	.118		5.30	5.85		11.15	15.20
8300	6" thick, hollow		310	.129		4.19	6.40		10.59	14.90
8350	8" thick, hollow		290	.138		4.17	6.85		11.02	15.60
8500	For stacked bond, add						26%			
8550	For high rise construction, add per story	D-8	67.80	.590	M.S.F.		29.50		29.50	47
8600	For scored block, add					10%				
8650	For honed or ground face, per face, add				Ea.	1.19			1.19	1.31
8700	For honed or ground end, per end, add				"	1.54			1.54	1.69
8750	For bullnose block, add					10%				
8800	For special color, add					13%				
9000	Minimum labor/equipment charge	D-1	2	8	Job		390		390	625

04 22 10.24 Concrete Block, Exterior

		Crew	Daily Output	Labor-Hours	Unit	Material	2021 Bare Costs Labor	Equipment	Total	Total Incl O&P
0010	**CONCRETE BLOCK, EXTERIOR**, C90, 2000 psi									
0020	Reinforced alt courses, tooled joints 2 sides									
0100	Normal weight, 8" x 16" x 6" thick	D-8	395	.101	S.F.	2.56	5.05		7.61	10.90
0200	8" thick		360	.111		4.25	5.50		9.75	13.50
0250	10" thick		290	.138		4.58	6.85		11.43	16.05
0300	12" thick	D-9	250	.192		5.35	9.35		14.70	21
9000	Minimum labor/equipment charge	D-1	2	8	Job		390		390	625

04 22 10.26 Concrete Block Foundation Wall

		Crew	Daily Output	Labor-Hours	Unit	Material	2021 Bare Costs Labor	Equipment	Total	Total Incl O&P
0010	**CONCRETE BLOCK FOUNDATION WALL**, C90/C145									
0050	Normal-weight, cut joints, horiz joint reinf, no vert reinf.									
0200	Hollow, 8" x 16" x 6" thick	D-8	455	.088	S.F.	3.38	4.37		7.75	10.70
0250	8" thick		425	.094		3.33	4.68		8.01	11.15
0300	10" thick		350	.114		3.87	5.70		9.57	13.40
0350	12" thick	D-9	300	.160		5.35	7.80		13.15	18.40
0500	Solid, 8" x 16" block, 6" thick	D-8	440	.091		4.22	4.52		8.74	11.90
0550	8" thick	"	415	.096		4.50	4.79		9.29	12.65
0600	12" thick	D-9	350	.137		6.35	6.70		13.05	17.70
1000	Reinforced, #4 vert @ 48"									
1100	Hollow, 8" x 16" block, 4" thick	D-8	455	.088	S.F.	3.42	4.37		7.79	10.75
1125	6" thick		445	.090		4.47	4.47		8.94	12.10
1150	8" thick		415	.096		4.88	4.79		9.67	13.05
1200	10" thick		340	.118		5.90	5.85		11.75	15.85
1250	12" thick	D-9	290	.166		7.80	8.05		15.85	21.50
1500	Solid, 8" x 16" block, 6" thick	D-8	430	.093		4.22	4.62		8.84	12.10
1600	8" thick	"	405	.099		4.50	4.91		9.41	12.85
1650	12" thick	D-9	340	.141		6.35	6.90		13.25	18
9000	Minimum labor/equipment charge	D-1	2	8	Job		390		390	625

04 22 10.30 Concrete Block, Interlocking

		Crew	Daily Output	Labor-Hours	Unit	Material	2021 Bare Costs Labor	2021 Bare Costs Equipment	Total	Total Incl O&P
0010	**CONCRETE BLOCK, INTERLOCKING**									
0100	Not including grout or reinforcing									
0200	8" x 16" units, 2,000 psi, 8" thick	D-1	245	.065	S.F.	3.31	3.18		6.49	8.75
0300	12" thick	"	220	.073		4.90	3.54		8.44	11.10
0400	Including grout & reinforcing, 8" thick	D-4	245	.131		9.35	6.40	.78	16.53	21.50
0450	12" thick		220	.145		11.25	7.15	.87	19.27	24.50
0500	16" thick	↓	185	.173	↓	13.75	8.50	1.03	23.28	30
9000	Minimum labor/equipment charge	D-1	2	8	Job		390		390	625

04 22 10.32 Concrete Block, Lintels

		Crew	Daily Output	Labor-Hours	Unit	Material	2021 Bare Costs Labor	2021 Bare Costs Equipment	Total	Total Incl O&P
0010	**CONCRETE BLOCK, LINTELS**, C90, normal weight									
0100	Including grout and horizontal reinforcing									
0200	8" x 8" x 8", 1 #4 bar	D-4	300	.107	L.F.	4.90	5.25	.63	10.78	14.45
0250	2 #4 bars		295	.108		5.20	5.35	.65	11.20	14.90
0400	8" x 16" x 8", 1 #4 bar		275	.116		4.31	5.70	.69	10.70	14.60
0450	2 #4 bars		270	.119		4.59	5.85	.71	11.15	15.15
1000	12" x 8" x 8", 1 #4 bar		275	.116		6.45	5.70	.69	12.84	16.95
1100	2 #4 bars		270	.119		6.75	5.85	.71	13.31	17.50
1150	2 #5 bars		270	.119		7.05	5.85	.71	13.61	17.85
1200	2 #6 bars		265	.121		7.40	5.95	.72	14.07	18.40
1500	12" x 16" x 8", 1 #4 bar		250	.128		7.60	6.30	.76	14.66	19.25
1600	2 #3 bars		245	.131		7.60	6.40	.78	14.78	19.50
1650	2 #4 bars		245	.131		7.85	6.40	.78	15.03	19.75
1700	2 #5 bars	↓	240	.133	↓	8.20	6.55	.79	15.54	20.50
9000	Minimum labor/equipment charge	D-1	2	8	Job		390		390	625

04 22 10.33 Lintel Block

		Crew	Daily Output	Labor-Hours	Unit	Material	2021 Bare Costs Labor	2021 Bare Costs Equipment	Total	Total Incl O&P
0010	**LINTEL BLOCK**									
3481	Lintel block 6" x 8" x 8"	D-1	300	.053	Ea.	1.45	2.60		4.05	5.75
3501	6" x 16" x 8"		275	.058		2.02	2.83		4.85	6.75
3521	8" x 8" x 8"		275	.058		1.40	2.83		4.23	6.10
3561	8" x 16" x 8"	↓	250	.064	↓	1.55	3.12		4.67	6.70

04 22 10.34 Concrete Block, Partitions

		Crew	Daily Output	Labor-Hours	Unit	Material	2021 Bare Costs Labor	2021 Bare Costs Equipment	Total	Total Incl O&P
0010	**CONCRETE BLOCK, PARTITIONS**, excludes scaffolding									
1000	Lightweight block, tooled joints, 2 sides, hollow									
1100	Not reinforced, 8" x 16" x 4" thick	D-8	440	.091	S.F.	2	4.52		6.52	9.45
1150	6" thick		410	.098		3.10	4.85		7.95	11.20
1200	8" thick		385	.104		3.97	5.15		9.12	12.65
1250	10" thick	↓	370	.108		4.22	5.35		9.57	13.30
1300	12" thick	D-9	350	.137		4.02	6.70		10.72	15.15
1500	Reinforced alternate courses, 4" thick	D-8	435	.092		2.17	4.57		6.74	9.75
1600	6" thick		405	.099		3.26	4.91		8.17	11.50
1650	8" thick		380	.105		4.17	5.25		9.42	13
1700	10" thick	↓	365	.110		4.41	5.45		9.86	13.60
1750	12" thick	D-9	345	.139	↓	4.23	6.80		11.03	15.55
4000	Regular block, tooled joints, 2 sides, hollow									
4100	Not reinforced, 8" x 16" x 4" thick	D-8	430	.093	S.F.	2.09	4.62		6.71	9.75
4150	6" thick		400	.100		2.71	4.97		7.68	11
4200	8" thick		375	.107		2.64	5.30		7.94	11.40
4250	10" thick	↓	360	.111		3.19	5.50		8.69	12.35
4300	12" thick	D-9	340	.141		4.64	6.90		11.54	16.15
4500	Reinforced alternate courses, 8" x 16" x 4" thick	D-8	425	.094		2.28	4.68		6.96	10
4550	6" thick	↓	395	.101		2.91	5.05		7.96	11.30

04 22 Concrete Unit Masonry

04 22 10 – Concrete Masonry Units

04 22 10.34 Concrete Block, Partitions

		Crew	Daily Output	Labor-Hours	Unit	Material	2021 Bare Costs Labor	Equipment	Total	Total Incl O&P
4600	8" thick	D-8	370	.108	S.F.	2.85	5.35		8.20	11.80
4650	10" thick	↓	355	.113		4.68	5.60		10.28	14.15
4700	12" thick	D-9	335	.143	↓	4.83	7		11.83	16.50
9000	Minimum labor/equipment charge	D-1	2	8	Job		390		390	625

04 22 10.38 Concrete Brick

		Crew	Daily Output	Labor-Hours	Unit	Material	2021 Bare Costs Labor	Equipment	Total	Total Incl O&P
0010	**CONCRETE BRICK**, C55, grade N, type 1									
0100	Regular, 4" x 2-1/4" x 8"	D-8	660	.061	Ea.	.66	3.01		3.67	5.55
0125	Rusticated, 4" x 2-1/4" x 8"		660	.061		.74	3.01		3.75	5.65
0150	Frog, 4" x 2-1/4" x 8"		660	.061		.71	3.01		3.72	5.60
0200	Double, 4" x 4-7/8" x 8"	↓	535	.075	↓	1.16	3.72		4.88	7.20

04 22 10.44 Glazed Concrete Block

		Crew	Daily Output	Labor-Hours	Unit	Material	2021 Bare Costs Labor	Equipment	Total	Total Incl O&P
0010	**GLAZED CONCRETE BLOCK** C744									
0100	Single face, 8" x 16" units, 2" thick	D-8	360	.111	S.F.	12.25	5.50		17.75	22.50
0200	4" thick		345	.116		12.60	5.75		18.35	23
0250	6" thick		330	.121		14.85	6		20.85	26
0300	8" thick		310	.129		16.50	6.40		22.90	28.50
0350	10" thick	↓	295	.136		16.95	6.75		23.70	29.50
0400	12" thick	D-9	280	.171		18.20	8.35		26.55	33.50
0700	Double face, 8" x 16" units, 4" thick	D-8	340	.118		17.10	5.85		22.95	28
0750	6" thick		320	.125		21.50	6.20		27.70	34
0800	8" thick		300	.133	↓	23	6.65		29.65	36
1000	Jambs, bullnose or square, single face, 8" x 16", 2" thick		315	.127	Ea.	20.50	6.30		26.80	32.50
1050	4" thick		285	.140	"	21.50	7		28.50	34.50
1200	Caps, bullnose or square, 8" x 16", 2" thick	↓	420	.095	L.F.	20.50	4.73		25.23	30
1250	4" thick	D-8	380	.105	L.F.	23.50	5.25		28.75	34
1256	Corner, bullnose or square, 2" thick		280	.143	Ea.	23.50	7.10		30.60	37.50
1258	4" thick		270	.148		27	7.35		34.35	42
1260	6" thick		260	.154		30	7.65		37.65	45.50
1270	8" thick		250	.160		39.50	7.95		47.45	56.50
1280	10" thick		240	.167		32	8.30		40.30	49
1290	12" thick		230	.174	↓	31.50	8.65		40.15	48.50
1500	Cove base, 8" x 16", 2" thick		315	.127	L.F.	10.80	6.30		17.10	22
1550	4" thick		285	.140		10.90	7		17.90	23
1600	6" thick		265	.151		11.60	7.50		19.10	25
1650	8" thick	↓	245	.163	↓	12.15	8.10		20.25	26.50
9000	Minimum labor/equipment charge	D-1	2	8	Job		390		390	625

04 23 Glass Unit Masonry

04 23 13 – Vertical Glass Unit Masonry

04 23 13.10 Glass Block

		Crew	Daily Output	Labor-Hours	Unit	Material	2021 Bare Costs Labor	Equipment	Total	Total Incl O&P
0010	**GLASS BLOCK**									
0100	Plain, 4" thick, under 1,000 S.F., 6" x 6"	D-8	115	.348	S.F.	26.50	17.30		43.80	57
0150	8" x 8"		160	.250		16.15	12.45		28.60	37.50
0160	end block		160	.250		54	12.45		66.45	79.50
0170	90 degree corner		160	.250		68	12.45		80.45	95
0180	45 degree corner		160	.250		56	12.45		68.45	82
0200	12" x 12"		175	.229		27.50	11.35		38.85	48.50
0210	4" x 8"		160	.250		35.50	12.45		47.95	59
0220	6" x 8"		160	.250		21	12.45		33.45	43
0300	1,000 to 5,000 S.F., 6" x 6"	↓	135	.296		26	14.75		40.75	52

For customer support on your Commercial Renovation Costs with RSMeans Data, call 800.448.8182.

04 23 Glass Unit Masonry

04 23 13 – Vertical Glass Unit Masonry

04 23 13.10 Glass Block

		Crew	Daily Output	Labor-Hours	Unit	Material	2021 Bare Costs Labor	Equipment	Total	Total Incl O&P
0350	8" x 8"	D-8	190	.211	S.F.	15.85	10.45		26.30	34
0400	12" x 12"		215	.186		27	9.25		36.25	44.50
0410	4" x 8"		215	.186		34.50	9.25		43.75	53
0420	6" x 8"		215	.186		20.50	9.25		29.75	37.50
0500	Over 5,000 S.F., 6" x 6"	↓	145	.276	↓	25	13.70		38.70	49.50
0700	For solar reflective blocks, add					100%				
1000	Thinline, plain, 3-1/8" thick, under 1,000 S.F., 6" x 6"	D-8	115	.348	S.F.	24.50	17.30		41.80	55
1050	8" x 8"		160	.250		15.75	12.45		28.20	37.50
1250	8" x 8"		215	.186		15.80	9.25		25.05	32
1400	For cleaning block after installation (both sides), add	↓	1000	.040	↓	.16	1.99		2.15	3.37
9000	Minimum labor/equipment charge	D-1	2	8	Job		390		390	625

04 27 Multiple-Wythe Unit Masonry

04 27 10 – Multiple-Wythe Masonry

04 27 10.10 Cornices

		Crew	Daily Output	Labor-Hours	Unit	Material	2021 Bare Costs Labor	Equipment	Total	Total Incl O&P
0010	**CORNICES**									
0110	Face bricks, 12 brick/S.F.	D-1	30	.533	SF Face	5.20	26		31.20	47.50
0150	15 brick/S.F.	↓	23	.696	"	6.25	34		40.25	61.50
9000	Minimum labor/equipment charge		1.50	10.667	Job		520		520	835

04 27 10.30 Brick Walls

		Crew	Daily Output	Labor-Hours	Unit	Material	2021 Bare Costs Labor	Equipment	Total	Total Incl O&P
0010	**BRICK WALLS**, including mortar, excludes scaffolding									
0020	Estimating by number of brick									
0140	Face brick, 4" thick wall, 6.75 brick/S.F.	D-8	1.45	27.586	M	740	1,375		2,115	3,025
0150	Common brick, 4" thick wall, 6.75 brick/S.F.		1.60	25		720	1,250		1,970	2,800
0204	8" thick, 13.50 brick/S.F.		1.80	22.222		740	1,100		1,840	2,600
0250	12" thick, 20.25 brick/S.F.		1.90	21.053		745	1,050		1,795	2,500
0304	16" thick, 27.00 brick/S.F.		2	20		755	995		1,750	2,425
0500	Reinforced, face brick, 4" thick wall, 6.75 brick/S.F.		1.40	28.571		760	1,425		2,185	3,125
0520	Common brick, 4" thick wall, 6.75 brick/S.F.		1.55	25.806		745	1,275		2,020	2,875
0550	8" thick, 13.50 brick/S.F.		1.75	22.857		765	1,125		1,890	2,675
0600	12" thick, 20.25 brick/S.F.		1.85	21.622		770	1,075		1,845	2,575
0650	16" thick, 27.00 brick/S.F.	↓	1.95	20.513	↓	775	1,025		1,800	2,500
0790	Alternate method of figuring by square foot									
0800	Face brick, 4" thick wall, 6.75 brick/S.F.	D-8	215	.186	S.F.	4.99	9.25		14.24	20.50
0850	Common brick, 4" thick wall, 6.75 brick/S.F. R042110-50		240	.167		4.87	8.30		13.17	18.65
0900	8" thick, 13.50 brick/S.F.		135	.296		10	14.75		24.75	34.50
1000	12" thick, 20.25 brick/S.F.		95	.421		15.05	21		36.05	50
1050	16" thick, 27.00 brick/S.F.		75	.533		20.50	26.50		47	65
1200	Reinforced, face brick, 4" thick wall, 6.75 brick/S.F.		210	.190		5.15	9.45		14.60	21
1220	Common brick, 4" thick wall, 6.75 brick/S.F.		235	.170		5	8.45		13.45	19.10
1250	8" thick, 13.50 brick/S.F.		130	.308		10.30	15.30		25.60	36
1260	8" thick, 2.25 brick/S.F.		130	.308		1.89	15.30		17.19	26.50
1300	12" thick, 20.25 brick/S.F.		90	.444		15.55	22		37.55	52.50
1350	16" thick, 27.00 brick/S.F.	↓	70	.571	↓	21	28.50		49.50	68.50
9000	Minimum labor/equipment charge	D-1	2	8	Job		390		390	625

04 27 10.40 Steps

		Crew	Daily Output	Labor-Hours	Unit	Material	2021 Bare Costs Labor	Equipment	Total	Total Incl O&P
0010	**STEPS**									
0012	Entry steps, select common brick	D-1	.30	53.333	M	580	2,600		3,180	4,825

04 41 Dry-Placed Stone

04 41 10 – Dry Placed Stone

04 41 10.10 Rough Stone Wall

	04 41 10.10 Rough Stone Wall		Crew	Daily Output	Labor-Hours	Unit	Material	2021 Bare Costs Labor	Equipment	Total	Total Incl O&P
0011	**ROUGH STONE WALL**, Dry										
0012	Dry laid (no mortar), under 18" thick	G	D-1	60	.267	C.F.	14.75	13		27.75	37.50
0100	Random fieldstone, under 18" thick	G	D-12	60	.533		14.75	26		40.75	58.50
0150	Over 18" thick	G	"	63	.508		17.70	25		42.70	59.50
0500	Field stone veneer	G	D-8	120	.333	S.F.	14	16.55		30.55	42
0510	Valley stone veneer	G		120	.333		14	16.55		30.55	42
0520	River stone veneer	G		120	.333		14	16.55		30.55	42
0600	Rubble stone walls, in mortar bed, up to 18" thick	G	D-11	75	.320	C.F.	17.80	16.35		34.15	45.50
9000	Minimum labor/equipment charge		D-1	2	8	Job		390		390	625

04 43 Stone Masonry

04 43 10 – Masonry with Natural and Processed Stone

04 43 10.05 Ashlar Veneer

	04 43 10.05 Ashlar Veneer	Crew	Daily Output	Labor-Hours	Unit	Material	2021 Bare Costs Labor	Equipment	Total	Total Incl O&P
0011	**ASHLAR VENEER** +/- 4" thk, random or random rectangular									
0150	Sawn face, split joints, low priced stone	D-8	140	.286	S.F.	13	14.20		27.20	37.50
0200	Medium priced stone		130	.308		15.55	15.30		30.85	41.50
0300	High priced stone		120	.333		18.75	16.55		35.30	47
0600	Seam face, split joints, medium price stone		125	.320		18.45	15.90		34.35	46
0700	High price stone		120	.333		17.95	16.55		34.50	46.50
1000	Split or rock face, split joints, medium price stone		125	.320		10.45	15.90		26.35	37
1100	High price stone		120	.333		18.30	16.55		34.85	46.50

04 43 10.10 Bluestone

	04 43 10.10 Bluestone	Crew	Daily Output	Labor-Hours	Unit	Material	2021 Bare Costs Labor	Equipment	Total	Total Incl O&P
0010	**BLUESTONE**, cut to size									
0500	Sills, natural cleft, 10" wide to 6' long, 1-1/2" thick	D-11	70	.343	L.F.	13.50	17.50		31	43
0550	2" thick		63	.381		14.75	19.45		34.20	47.50
0600	Smooth finish, 1-1/2" thick		70	.343		14.45	17.50		31.95	44
0650	2" thick		63	.381		17.10	19.45		36.55	50
0800	Thermal finish, 1-1/2" thick		70	.343		12.75	17.50		30.25	42
0850	2" thick		63	.381		10.95	19.45		30.40	43
1000	Stair treads, natural cleft, 12" wide, 6' long, 1-1/2" thick	D-10	115	.278		15.60	14.95	3.76	34.31	45.50
1050	2" thick		105	.305		17	16.35	4.12	37.47	49
1100	Smooth finish, 1-1/2" thick		115	.278		14.60	14.95	3.76	33.31	44
1150	2" thick		105	.305		15.45	16.35	4.12	35.92	47.50
1300	Thermal finish, 1-1/2" thick		115	.278		15	14.95	3.76	33.71	44.50
1350	2" thick		105	.305		14.30	16.35	4.12	34.77	46.50
9000	Minimum labor/equipment charge	D-1	2.50	6.400	Job		310		310	500

04 43 10.45 Granite

	04 43 10.45 Granite	Crew	Daily Output	Labor-Hours	Unit	Material	2021 Bare Costs Labor	Equipment	Total	Total Incl O&P
0010	**GRANITE**, cut to size									
0050	Veneer, polished face, 3/4" to 1-1/2" thick									
0150	Low price, gray, light gray, etc.	D-10	130	.246	S.F.	25	13.20	3.33	41.53	52
0180	Medium price, pink, brown, etc.		130	.246		28.50	13.20	3.33	45.03	56
0220	High price, red, black, etc.		130	.246		43	13.20	3.33	59.53	71.50
0300	1-1/2" to 2-1/2" thick, veneer									
0350	Low price, gray, light gray, etc.	D-10	130	.246	S.F.	29	13.20	3.33	45.53	56.50
0500	Medium price, pink, brown, etc.	"	130	.246	"	34	13.20	3.33	50.53	62
2500	Steps, copings, etc., finished on more than one surface									
2550	Low price, gray, light gray, etc.	D-10	50	.640	C.F.	92	34.50	8.65	135.15	165
2575	Medium price, pink, brown, etc.		50	.640		120	34.50	8.65	163.15	196
2600	High price, red, black, etc.		50	.640		147	34.50	8.65	190.15	226
9000	Minimum labor/equipment charge	D-1	2	8	Job		390		390	625

102

04 43 10 – Masonry with Natural and Processed Stone

04 43 10.55 Limestone

	Crew	Daily Output	Labor-Hours	Unit	Material	2021 Bare Costs Labor	2021 Bare Costs Equipment	Total	Total Incl O&P
0010 **LIMESTONE**, cut to size									
0020 Veneer facing panels									
0500 Texture finish, light stick, 4-1/2" thick, 5' x 12'	D-4	300	.107	S.F.	26	5.25	.63	31.88	38
0750 5" thick, 5' x 14' panels	D-10	275	.116		28	6.25	1.57	35.82	42
1000 Sugarcube finish, 2" thick, 3' x 5' panels		275	.116		31	6.25	1.57	38.82	45.50
1050 3" thick, 4' x 9' panels		275	.116		32	6.25	1.57	39.82	46.50
1200 4" thick, 5' x 11' panels		275	.116		35	6.25	1.57	42.82	50
1400 Sugarcube, textured finish, 4-1/2" thick, 5' x 12'		275	.116		38.50	6.25	1.57	46.32	53.50
1450 5" thick, 5' x 14' panels		275	.116		39	6.25	1.57	46.82	54.50
2000 Coping, sugarcube finish, top & 2 sides		30	1.067	C.F.	66.50	57	14.40	137.90	180
2100 Sills, lintels, jambs, trim, stops, sugarcube finish, simple		20	1.600		66.50	86	21.50	174	234
2150 Detailed		20	1.600		66.50	86	21.50	174	234
2300 Steps, extra hard, 14" wide, 6" rise		50	.640	L.F.	29	34.50	8.65	72.15	95.50
3000 Quoins, plain finish, 6" x 12" x 12"	D-12	25	1.280	Ea.	54.50	63		117.50	161
3050 6" x 16" x 24"	"	25	1.280	"	73	63		136	181
9000 Minimum labor/equipment charge	D-1	2	8	Job		390		390	625

04 43 10.60 Marble

	Crew	Daily Output	Labor-Hours	Unit	Material	2021 Bare Costs Labor	2021 Bare Costs Equipment	Total	Total Incl O&P
0011 **MARBLE**, ashlar, split face, +/- 4" thick, random									
0040 Lengths 1' to 4' & heights 2" to 7-1/2", average	D-8	175	.229	S.F.	16.85	11.35		28.20	37
0100 Base, polished, 3/4" or 7/8" thick, polished, 6" high	D-10	65	.492	L.F.	11.25	26.50	6.65	44.40	61.50
0300 Carvings or bas-relief, from templates, simple design		80	.400	S.F.	157	21.50	5.40	183.90	212
0350 Intricate design		80	.400	"	345	21.50	5.40	371.90	420
1000 Facing, polished finish, cut to size, 3/4" to 7/8" thick									
1050 Carrara or equal	D-10	130	.246	S.F.	35	13.20	3.33	51.53	63
1100 Arabescato or equal		130	.246		37	13.20	3.33	53.53	65
1300 1-1/4" thick, Botticino Classico or equal		125	.256		22.50	13.75	3.46	39.71	51
1350 Statuarietto or equal		125	.256		37.50	13.75	3.46	54.71	67.50
1500 2" thick, Crema Marfil or equal		120	.267		48	14.30	3.61	65.91	80
1550 Cafe Pinta or equal		120	.267		66.50	14.30	3.61	84.41	100
1700 Rubbed finish, cut to size, 4" thick									
1740 Average	D-10	100	.320	S.F.	41.50	17.15	4.33	62.98	78
1780 Maximum	"	100	.320	"	77.50	17.15	4.33	98.98	118
2200 Window sills, 6" x 3/4" thick	D-1	85	.188	L.F.	13.10	9.15		22.25	29
2500 Flooring, polished tiles, 12" x 12" x 3/8" thick									
2510 Thin set, Giallo Solare or equal	D-11	90	.267	S.F.	14.75	13.60		28.35	38
2600 Sky Blue or equal		90	.267		11.35	13.60		24.95	34.50
2700 Mortar bed, Giallo Solare or equal		65	.369		14.85	18.85		33.70	47
2740 Sky Blue or equal		65	.369		11.35	18.85		30.20	43
2780 Travertine, 3/8" thick, Sierra or equal	D-10	130	.246		11.50	13.20	3.33	28.03	37.50
2790 Silver or equal	"	130	.246		30.50	13.20	3.33	47.03	58
2800 Patio tile, non-slip, 1/2" thick, flame finish	D-11	75	.320		13.45	16.35		29.80	41
2900 Shower or toilet partitions, 7/8" thick partitions									
3050 3/4" or 1-1/4" thick stiles, polished 2 sides, average	D-11	75	.320	S.F.	48.50	16.35		64.85	79
3201 Soffits, add to above prices				"	20%	100%			
3210 Stairs, risers, 7/8" thick x 6" high	D-10	115	.278	L.F.	14.95	14.95	3.76	33.66	44.50
3360 Treads, 12" wide x 1-1/4" thick	"	115	.278	"	46	14.95	3.76	64.71	78.50
3500 Thresholds, 3' long, 7/8" thick, 4" to 5" wide, plain	D-12	24	1.333	Ea.	34.50	65.50		100	143
3550 Beveled		24	1.333	"	76	65.50		141.50	189
3700 Window stools, polished, 7/8" thick, 5" wide		85	.376	L.F.	19.65	18.50		38.15	51.50
9000 Minimum labor/equipment charge	D-1	2	8	Job		390		390	625

For customer support on your Commercial Renovation Costs with RSMeans Data, call 800.448.8182.

103

04 43 Stone Masonry

04 43 10 – Masonry with Natural and Processed Stone

04 43 10.75 Sandstone or Brownstone

		Crew	Daily Output	Labor-Hours	Unit	Material	2021 Bare Costs Labor	Equipment	Total	Total Incl O&P
0011	**SANDSTONE OR BROWNSTONE**									
0100	Sawed face veneer, 2-1/2" thick, to 2' x 4' panels	D-10	130	.246	S.F.	21.50	13.20	3.33	38.03	48.50
0150	4" thick, to 3'-6" x 8' panels		100	.320		21.50	17.15	4.33	42.98	56.50
0300	Split face, random sizes		100	.320		15	17.15	4.33	36.48	49
9000	Minimum labor/equipment charge	D-1	2.50	6.400	Job		310		310	500

04 43 10.80 Slate

		Crew	Daily Output	Labor-Hours	Unit	Material	2021 Bare Costs Labor	Equipment	Total	Total Incl O&P
0010	**SLATE**									
0040	Pennsylvania - blue gray to black									
0050	Vermont - unfading green, mottled green & purple, gray & purple									
0100	Virginia - blue black									
3100	Stair landings, 1" thick, black, clear	D-1	65	.246	S.F.	21	12		33	42.50
3200	Ribbon	"	65	.246	"	23	12		35	44.50
3500	Stair treads, sand finish, 1" thick x 12" wide									
3550	Under 3 L.F.	D-10	85	.376	L.F.	23	20	5.10	48.10	62.50
3600	3 L.F. to 6 L.F.	"	120	.267	"	25	14.30	3.61	42.91	54
3700	Ribbon, sand finish, 1" thick x 12" wide									
3750	To 6 L.F.	D-10	120	.267	L.F.	21	14.30	3.61	38.91	50
4000	Stools or sills, sand finish, 1" thick, 6" wide	D-12	160	.200		12	9.85		21.85	29
4100	Honed finish		160	.200		11.45	9.85		21.30	28.50
4200	10" wide		90	.356		18.50	17.50		36	48.50
4250	Honed finish		90	.356		17.20	17.50		34.70	47
4400	2" thick, 6" wide		140	.229		19.30	11.25		30.55	39
4450	Honed finish		140	.229		18.40	11.25		29.65	38.50
4600	10" wide		90	.356		30	17.50		47.50	61
4650	Honed finish		90	.356		28.50	17.50		46	59.50
4800	For lengths over 3', add					25%				
9000	Minimum labor/equipment charge	D-1	2.50	6.400	Job		310		310	500

04 43 10.85 Window Sill

		Crew	Daily Output	Labor-Hours	Unit	Material	2021 Bare Costs Labor	Equipment	Total	Total Incl O&P
0010	**WINDOW SILL**									
0020	Bluestone, thermal top, 10" wide, 1-1/2" thick	D-1	85	.188	S.F.	8.40	9.15		17.55	24
0050	2" thick		75	.213	"	12.95	10.40		23.35	31
0100	Cut stone, 5" x 8" plain		48	.333	L.F.	12.80	16.25		29.05	40
0200	Face brick on edge, brick, 8" wide		80	.200		3.53	9.75		13.28	19.55
0400	Marble, 9" wide, 1" thick		85	.188		9.10	9.15		18.25	25
0900	Slate, colored, unfading, honed, 12" wide, 1" thick		85	.188		9.30	9.15		18.45	25
0950	2" thick		70	.229		11.30	11.15		22.45	30.50
9000	Minimum labor/equipment charge	1 Bric	2	4	Job		215		215	345

04 54 Refractory Brick Masonry

04 54 10 – Refractory Brick Work

04 54 10.10 Fire Brick

		Crew	Daily Output	Labor-Hours	Unit	Material	2021 Bare Costs Labor	Equipment	Total	Total Incl O&P
0010	**FIRE BRICK**									
0012	Low duty, 2000°F, 9" x 2-1/2" x 4-1/2"	D-1	.60	26.667	M	1,750	1,300		3,050	4,000
0050	High duty, 3000°F	"	.60	26.667	"	3,150	1,300		4,450	5,525

04 54 10.20 Fire Clay

		Crew	Daily Output	Labor-Hours	Unit	Material	2021 Bare Costs Labor	Equipment	Total	Total Incl O&P
0010	**FIRE CLAY**									
0020	Gray, high duty, 100 lb. bag				Bag	28.50			28.50	31
0050	100 lb. drum, premixed (400 brick per drum)				Drum	44			44	48.50

04 57 Masonry Fireplaces

04 57 10 – Brick or Stone Fireplaces

04 57 10.10 Fireplace

		Crew	Daily Output	Labor-Hours	Unit	Material	2021 Bare Costs Labor	2021 Bare Costs Equipment	Total	Total Incl O&P
0010	**FIREPLACE**									
0100	Brick fireplace, not incl. foundations or chimneys	D-1	.40	40	Ea.	630	1,950		2,580	3,825
0110	30" x 29" opening, incl. chamber, plain brickwork	"	2	8	"	162	390		552	805
0200	Fireplace box only (110 brick)									
0300	For elaborate brickwork and details, add					35%	35%			
0400	For hearth, brick & stone, add	D-1	2	8	Ea.	214	390		604	860
0410	For steel, damper, cleanouts, add		4	4		18.10	195		213.10	335
0600	Plain brickwork, incl. metal circulator		.50	32		1,075	1,550		2,625	3,675
0800	Face brick only, standard size, 8" x 2-2/3" x 4"		.30	53.333	M	615	2,600		3,215	4,850
0900	Stone fireplace, fieldstone, add				SF Face	9.10			9.10	10
1000	Cut stone, add				"	9.95			9.95	10.95
9000	Minimum labor/equipment charge	D-1	2	8	Job		390		390	625

04 71 Manufactured Brick Masonry

04 71 10 – Simulated or Manufactured Brick

04 71 10.10 Simulated Brick

		Crew	Daily Output	Labor-Hours	Unit	Material	2021 Bare Costs Labor	2021 Bare Costs Equipment	Total	Total Incl O&P
0010	**SIMULATED BRICK**									
0020	Aluminum, baked on colors	1 Carp	200	.040	S.F.	4.53	2.19		6.72	8.45
0050	Fiberglass panels		200	.040		9.90	2.19		12.09	14.40
0100	Urethane pieces cemented in mastic		150	.053		8.30	2.92		11.22	13.75
0150	Vinyl siding panels		200	.040		11.10	2.19		13.29	15.70

04 72 Cast Stone Masonry

04 72 10 – Cast Stone Masonry Features

04 72 10.10 Coping

		Crew	Daily Output	Labor-Hours	Unit	Material	2021 Bare Costs Labor	2021 Bare Costs Equipment	Total	Total Incl O&P
0010	**COPING**, stock units									
0050	Precast concrete, 10" wide, 4" tapers to 3-1/2", 8" wall	D-1	75	.213	L.F.	25	10.40		35.40	44
0100	12" wide, 3-1/2" tapers to 3", 10" wall		70	.229		27	11.15		38.15	47.50
0110	14" wide, 4" tapers to 3-1/2", 12" wall		65	.246		36.50	12		48.50	59.50
0150	16" wide, 4" tapers to 3-1/2", 14" wall		60	.267		39	13		52	64
0300	Limestone for 12" wall, 4" thick		90	.178		16.90	8.65		25.55	32.50
0350	6" thick		80	.200		23.50	9.75		33.25	41.50
0500	Marble, to 4" thick, no wash, 9" wide		90	.178		10.40	8.65		19.05	25.50
0550	12" wide		80	.200		18.50	9.75		28.25	36
0700	Terra cotta, 9" wide		90	.178		8.10	8.65		16.75	23
0750	12" wide		80	.200		8.20	9.75		17.95	24.50
0800	Aluminum, for 12" wall		80	.200		9.45	9.75		19.20	26
9000	Minimum labor/equipment charge		2	8	Job		390		390	625

04 72 20 – Cultured Stone Veneer

04 72 20.10 Cultured Stone Veneer Components

		Crew	Daily Output	Labor-Hours	Unit	Material	2021 Bare Costs Labor	2021 Bare Costs Equipment	Total	Total Incl O&P
0010	**CULTURED STONE VENEER COMPONENTS**									
0110	On wood frame and sheathing substrate, random sized cobbles, corner stones	D-8	70	.571	V.L.F.	8.45	28.50		36.95	55
0120	Field stones		140	.286	S.F.	5.90	14.20		20.10	29.50
0130	Random sized flats, corner stones		70	.571	V.L.F.	8.65	28.50		37.15	55
0140	Field stones		140	.286	S.F.	9.20	14.20		23.40	33
0150	Horizontal lined ledgestones, corner stones		75	.533	V.L.F.	8.35	26.50		34.85	51.50
0160	Field stones		150	.267	S.F.	6.30	13.25		19.55	28.50
0170	Random shaped flats, corner stones		65	.615	V.L.F.	8.40	30.50		38.90	58
0180	Field stones		150	.267	S.F.	6.15	13.25		19.40	28.50

105

04 72 Cast Stone Masonry

04 72 20 – Cultured Stone Veneer

04 72 20.10 Cultured Stone Veneer Components	Crew	Daily Output	Labor-Hours	Unit	Material	2021 Bare Costs Labor	Equipment	Total	Total Incl O&P
0190	Random shaped/textured face, corner stones	D-8	65	.615	V.L.F.	8.45	30.50		38.95	58.50
0200	Field stones		130	.308	S.F.	6.10	15.30		21.40	31.50
0210	Random shaped river rock, corner stones		65	.615	V.L.F.	8.45	30.50		38.95	58.50
0220	Field stones		130	.308	S.F.	6.10	15.30		21.40	31.50
0240	On concrete or CMU substrate, random sized cobbles, corner stones		70	.571	V.L.F.	7.60	28.50		36.10	54
0250	Field stones		140	.286	S.F.	5.45	14.20		19.65	29
0260	Random sized flats, corner stones		70	.571	V.L.F.	7.80	28.50		36.30	54
0270	Field stones		140	.286	S.F.	8.75	14.20		22.95	32.50
0280	Horizontal lined ledgestones, corner stones		75	.533	V.L.F.	7.50	26.50		34	51
0290	Field stones		150	.267	S.F.	5.85	13.25		19.10	28
0300	Random shaped flats, corner stones		70	.571	V.L.F.	7.50	28.50		36	54
0310	Field stones		140	.286	S.F.	5.75	14.20		19.95	29.50
0320	Random shaped/textured face, corner stones		65	.615	V.L.F.	7.60	30.50		38.10	57.50
0330	Field stones		130	.308	S.F.	5.70	15.30		21	31
0340	Random shaped river rock, corner stones		65	.615	V.L.F.	7.60	30.50		38.10	57.50
0350	Field stones		130	.308	S.F.	5.70	15.30		21	31
0360	Cultured stone veneer, #15 felt weather resistant barrier	1 Clab	3700	.002	Sq.	5.70	.10		5.80	6.40
0370	Expanded metal lath, diamond, 2.5 lb./S.Y., galvanized	1 Lath	85	.094	S.Y.	3.87	5.10		8.97	12.20
0390	Water table or window sill, 18" long	1 Bric	80	.100	Ea.	9.25	5.35		14.60	18.80

04 73 Manufactured Stone Masonry

04 73 20 – Simulated or Manufactured Stone

04 73 20.10 Simulated Stone

		Crew	Daily Output	Labor-Hours	Unit	Material	2021 Bare Costs Labor	Equipment	Total	Total Incl O&P
0010	**SIMULATED STONE**									
0100	Insulated fiberglass panels, 5/8" ply backer	L-4	200	.120	S.F.	11.60	6.25		17.85	22.50

Estimating Tips
05 05 00 Common Work Results for Metals

- Nuts, bolts, washers, connection angles, and plates can add a significant amount to both the tonnage of a structural steel job and the estimated cost. As a rule of thumb, add 10% to the total weight to account for these accessories.

- Type 2 steel construction, commonly referred to as "simple construction," consists generally of field-bolted connections with lateral bracing supplied by other elements of the building, such as masonry walls or x-bracing. The estimator should be aware, however, that shop connections may be accomplished by welding or bolting. The method may be particular to the fabrication shop and may have an impact on the estimated cost.

05 10 00 Structural Steel

- Steel items can be obtained from two sources: a fabrication shop or a metals service center. Fabrication shops can fabricate items under more controlled conditions than crews in the field can. They are also more efficient and can produce items more economically. Metal service centers serve as a source of long mill shapes to both fabrication shops and contractors.

- Most line items in this structural steel subdivision, and most items in 05 50 00 Metal Fabrications, are indicated as being shop fabricated. The bare material cost for these shop fabricated items is the "Invoice Cost" from the shop and includes the mill base price of steel plus mill extras, transportation to the shop, shop drawings and detailing where warranted, shop fabrication and handling, sandblasting and a shop coat of primer paint, all necessary structural bolts, and delivery to the job site. The bare labor cost and bare equipment cost for these shop fabricated items are for field installation or erection.

- Line items in Subdivision 05 12 23.40 Lightweight Framing, and other items scattered in Division 5, are indicated as being field fabricated. The bare material cost for these field fabricated items is the "Invoice Cost" from the metals service center and includes the mill base price of steel plus mill extras, transportation to the metals service center, material handling, and delivery of long lengths of mill shapes to the job site. Material costs for structural bolts and welding rods should be added to the estimate. The bare labor cost and bare equipment cost for these items are for both field fabrication and field installation or erection, and include time for cutting, welding, and drilling in the fabricated metal items. Drilling into concrete and fasteners to fasten field fabricated items to other work is not included and should be added to the estimate.

05 20 00 Steel Joist Framing

- In any given project the total weight of open web steel joists is determined by the loads to be supported and the design. However, economies can be realized in minimizing the amount of labor used to place the joists. This is done by maximizing the joist spacing and therefore minimizing the number of joists required to be installed on the job. Certain spacings and locations may be required by the design, but in other cases maximizing the spacing and keeping it as uniform as possible will keep the costs down.

05 30 00 Steel Decking

- The takeoff and estimating of a metal deck involve more than the area of the floor or roof and the type of deck specified or shown on the drawings. Many different sizes and types of openings may exist. Small openings for individual pipes or conduits may be drilled after the floor/roof is installed, but larger openings may require special deck lengths as well as reinforcing or structural support. The estimator should determine who will be supplying this reinforcing. Additionally, some deck terminations are part of the deck package, such as screed angles and pour stops, and others will be part of the steel contract, such as angles attached to structural members and cast-in-place angles and plates. The estimator must ensure that all pieces are accounted for in the complete estimate.

05 50 00 Metal Fabrications

- The most economical steel stairs are those that use common materials, standard details, and most importantly, a uniform and relatively simple method of field assembly. Commonly available A36/A992 channels and plates are very good choices for the main stringers of the stairs, as are angles and tees for the carrier members. Risers and treads are usually made by specialty shops, and it is most economical to use a typical detail in as many places as possible. The stairs should be pre-assembled and shipped directly to the site. The field connections should be simple and straightforward enough to be accomplished efficiently, and with minimum equipment and labor.

Reference Numbers

Reference numbers are shown at the beginning of some major classifications. These numbers refer to related items in the Reference Section. The reference information may be an estimating procedure, an alternate pricing method, or technical information.

Note: Not all subdivisions listed here necessarily appear. ∎

Same Data. Simplified.

Enjoy the convenience and efficiency of accessing your costs anywhere:

- **Skip the multiplier** by setting your location
- **Quickly search,** edit, favorite and share costs
- **Stay on top of price changes** with automatic updates

Discover more at rsmeans.com/online

---

05 05 Common Work Results for Metals

05 05 05 – Selective Demolition for Metals

05 05 05.10 Selective Demolition, Metals

		Crew	Daily Output	Labor-Hours	Unit	Material	2021 Bare Costs Labor	2021 Bare Costs Equipment	Total	Total Incl O&P
0010	**SELECTIVE DEMOLITION, METALS** R024119-10									
0015	Excludes shores, bracing, cutting, loading, hauling, dumping									
0020	Remove nuts only up to 3/4" diameter	1 Sswk	480	.017	Ea.		1.01		1.01	1.63
0030	7/8" to 1-1/4" diameter		240	.033			2.01		2.01	3.25
0040	1-3/8" to 2" diameter		160	.050			3.02		3.02	4.88
0060	Unbolt and remove structural bolts up to 3/4" diameter		240	.033			2.01		2.01	3.25
0070	7/8" to 2" diameter		160	.050			3.02		3.02	4.88
0140	Light weight framing members, remove whole or cut up, up to 20 lb.	↓	240	.033			2.01		2.01	3.25
0150	21-40 lb.	2 Sswk	210	.076			4.59		4.59	7.45
0160	41-80 lb.	3 Sswk	180	.133			8.05		8.05	13
0170	81-120 lb.	4 Sswk	150	.213			12.85		12.85	21
0230	Structural members, remove whole or cut up, up to 500 lb.	E-19	48	.500			29.50	32	61.50	82.50
0240	1/4-2 tons	E-18	36	1.111			67	42.50	109.50	155
0250	2-5 tons	E-24	30	1.067			64	19.55	83.55	125
0260	5-10 tons	E-20	24	2.667			159	88.50	247.50	355
0270	10-15 tons	E-2	18	3.111			186	95	281	400
0340	Fabricated item, remove whole or cut up, up to 20 lb.	1 Sswk	96	.083			5.05		5.05	8.15
0350	21-40 lb.	2 Sswk	84	.190			11.50		11.50	18.60
0360	41-80 lb.	3 Sswk	72	.333			20		20	32.50
0370	81-120 lb.	4 Sswk	60	.533			32		32	52
0380	121-500 lb.	E-19	48	.500			29.50	32	61.50	82.50
0390	501-1000 lb.	"	36	.667	↓		39.50	42.50	82	110
0500	Steel roof decking, uncovered, bare	B-2	5000	.008	S.F.		.36		.36	.57
2950	Minimum labor/equipment charge	E-19	2	12	Job		710	765	1,475	2,000

05 05 19 – Post-Installed Concrete Anchors

05 05 19.10 Chemical Anchors

		Crew	Daily Output	Labor-Hours	Unit	Material	2021 Bare Costs Labor	2021 Bare Costs Equipment	Total	Total Incl O&P
0010	**CHEMICAL ANCHORS**									
0020	Includes layout & drilling									
1430	Chemical anchor, w/rod & epoxy cartridge, 3/4" diameter x 9-1/2" long	B-89A	27	.593	Ea.	11.90	30	4.50	46.40	66
1435	1" diameter x 11-3/4" long		24	.667		24	34	5.05	63.05	85.50
1440	1-1/4" diameter x 14" long		21	.762		38.50	38.50	5.80	82.80	110
1445	1-3/4" diameter x 15" long		20	.800		70.50	40.50	6.10	117.10	149
1450	18" long		17	.941		84.50	48	7.15	139.65	177
1455	2" diameter x 18" long		16	1		148	51	7.60	206.60	252
1460	24" long	↓	15	1.067	↓	194	54	8.10	256.10	310

05 05 19.20 Expansion Anchors

			Crew	Daily Output	Labor-Hours	Unit	Material	2021 Bare Costs Labor	2021 Bare Costs Equipment	Total	Total Incl O&P
0010	**EXPANSION ANCHORS**										
0100	Anchors for concrete, brick or stone, no layout and drilling										
0200	Expansion shields, zinc, 1/4" diameter, 1-5/16" long, single	G	1 Carp	90	.089	Ea.	.31	4.86		5.17	8.10
0300	1-3/8" long, double	G		85	.094		.42	5.15		5.57	8.65
0400	3/8" diameter, 1-1/2" long, single	G		85	.094		.51	5.15		5.66	8.75
0500	2" long, double	G		80	.100		1.22	5.45		6.67	10.05
0600	1/2" diameter, 2-1/16" long, single	G		80	.100		1.23	5.45		6.68	10.05
0700	2-1/2" long, double	G		75	.107		2.25	5.85		8.10	11.80
0800	5/8" diameter, 2-5/8" long, single	G		75	.107		1.97	5.85		7.82	11.45
0900	2-3/4" long, double	G		70	.114		3.80	6.25		10.05	14.15
1000	3/4" diameter, 2-3/4" long, single	G		70	.114		3.75	6.25		10	14.10
1100	3-15/16" long, double	G		65	.123		5.55	6.75		12.30	16.80
2100	Hollow wall anchors for gypsum wall board, plaster or tile										
2300	1/8" diameter, short	G	1 Carp	160	.050	Ea.	.31	2.74		3.05	4.70
2400	Long	G		150	.053		.30	2.92		3.22	4.98
2500	3/16" diameter, short	G	↓	150	.053		.61	2.92		3.53	5.30

108

05 05 Common Work Results for Metals

05 05 19 – Post-Installed Concrete Anchors

05 05 19.20 Expansion Anchors

		Crew	Daily Output	Labor-Hours	Unit	Material	2021 Bare Costs Labor	Equipment	Total	Total Incl O&P
2600	Long **G**	1 Carp	140	.057	Ea.	.71	3.13		3.84	5.75
2700	1/4" diameter, short **G**		140	.057		.58	3.13		3.71	5.60
2800	Long **G**		130	.062		1.04	3.37		4.41	6.50
3000	Toggle bolts, bright steel, 1/8" diameter, 2" long **G**		85	.094		.33	5.15		5.48	8.55
3100	4" long **G**		80	.100		.38	5.45		5.83	9.10
3200	3/16" diameter, 3" long **G**		80	.100		.40	5.45		5.85	9.15
3300	6" long **G**		75	.107		.62	5.85		6.47	10
3400	1/4" diameter, 3" long **G**		75	.107		.44	5.85		6.29	9.80
3500	6" long **G**		70	.114		.59	6.25		6.84	10.60
3600	3/8" diameter, 3" long **G**		70	.114		.73	6.25		6.98	10.75
3700	6" long **G**		60	.133		1.62	7.30		8.92	13.40
3800	1/2" diameter, 4" long **G**		60	.133		2.66	7.30		9.96	14.55
3900	6" long **G**	▼	50	.160	▼	3.05	8.75		11.80	17.30
4000	Nailing anchors									
4100	Nylon nailing anchor, 1/4" diameter, 1" long	1 Carp	3.20	2.500	C	19.40	137		156.40	240
4200	1-1/2" long		2.80	2.857		22	156		178	274
4300	2" long	▼	2.40	3.333	▼	24	182		206	315
4400	Metal nailing anchor, 1/4" diameter, 1" long **G**	1 Carp	3.20	2.500	C	20.50	137		157.50	241
4500	1-1/2" long **G**		2.80	2.857		21	156		177	272
4600	2" long **G**	▼	2.40	3.333	▼	44	182		226	340
8000	Wedge anchors, not including layout or drilling									
8050	Carbon steel, 1/4" diameter, 1-3/4" long **G**	1 Carp	150	.053	Ea.	.76	2.92		3.68	5.50
8100	3-1/4" long **G**		140	.057		1	3.13		4.13	6.10
8150	3/8" diameter, 2-1/4" long **G**		145	.055		.43	3.02		3.45	5.30
8200	5" long **G**		140	.057		.76	3.13		3.89	5.80
8250	1/2" diameter, 2-3/4" long **G**		140	.057		1.14	3.13		4.27	6.25
8300	7" long **G**		125	.064		1.96	3.50		5.46	7.70
8350	5/8" diameter, 3-1/2" long **G**		130	.062		2.13	3.37		5.50	7.70
8400	8-1/2" long **G**		115	.070		4.54	3.81		8.35	11.05
8450	3/4" diameter, 4-1/4" long **G**		115	.070		3.45	3.81		7.26	9.85
8500	10" long **G**		95	.084		7.85	4.61		12.46	16
8550	1" diameter, 6" long **G**		100	.080		5.65	4.38		10.03	13.15
8575	9" long **G**		85	.094		7.35	5.15		12.50	16.25
8600	12" long **G**		75	.107		7.95	5.85		13.80	18
8650	1-1/4" diameter, 9" long **G**		70	.114		33.50	6.25		39.75	47
8700	12" long **G**	▼	60	.133	▼	43	7.30		50.30	59
8750	For type 303 stainless steel, add					350%				
8800	For type 316 stainless steel, add					450%				
8950	Self-drilling concrete screw, hex washer head, 3/16" diam. x 1-3/4" long **G**	1 Carp	300	.027	Ea.	.21	1.46		1.67	2.55
8960	2-1/4" long **G**		250	.032		.29	1.75		2.04	3.11
8970	Phillips flat head, 3/16" diam. x 1-3/4" long **G**		300	.027		.20	1.46		1.66	2.54
8980	2-1/4" long **G**	▼	250	.032	▼	.31	1.75		2.06	3.13
9000	Minimum labor/equipment charge		4	2	Job		109		109	174

05 05 21 – Fastening Methods for Metal

05 05 21.10 Cutting Steel

		Crew	Daily Output	Labor-Hours	Unit	Material	2021 Bare Costs Labor	Equipment	Total	Total Incl O&P
0010	**CUTTING STEEL**									
0020	Hand burning, incl. preparation, torch cutting & grinding, no staging									
0050	Steel to 1/4" thick	E-25	400	.020	L.F.	.64	1.25	.03	1.92	2.77
0100	1/2" thick		320	.025		1.04	1.56	.04	2.64	3.71
0150	3/4" thick		260	.031		1.61	1.92	.05	3.58	4.93
0200	1" thick	▼	200	.040	▼	2.23	2.49	.06	4.78	6.55
9000	Minimum labor/equipment charge		2	4	Job		249	6.50	255.50	410

05 05 21.15 Drilling Steel

		Crew	Daily Output	Labor-Hours	Unit	Material	2021 Bare Costs Labor	Equipment	Total	Total Incl O&P
0010	**DRILLING STEEL**									
1910	Drilling & layout for steel, up to 1/4" deep, no anchor									
1920	Holes, 1/4" diameter	1 Sswk	112	.071	Ea.	.06	4.31		4.37	7
1925	For each additional 1/4" depth, add		336	.024		.06	1.44		1.50	2.39
1930	3/8" diameter	↓	104	.077	↓	.08	4.64		4.72	7.60
1935	For each additional 1/4" depth, add	1 Sswk	312	.026	Ea.	.08	1.55		1.63	2.59
1940	1/2" diameter		96	.083		.08	5.05		5.13	8.25
1945	For each additional 1/4" depth, add		288	.028		.08	1.68		1.76	2.80
1950	5/8" diameter		88	.091		.12	5.50		5.62	9
1955	For each additional 1/4" depth, add		264	.030		.12	1.83		1.95	3.09
1960	3/4" diameter		80	.100		.16	6.05		6.21	9.95
1965	For each additional 1/4" depth, add		240	.033		.16	2.01		2.17	3.43
1970	7/8" diameter		72	.111		.19	6.70		6.89	11.05
1975	For each additional 1/4" depth, add		216	.037		.19	2.23		2.42	3.82
1980	1" diameter		64	.125		.30	7.55		7.85	12.55
1985	For each additional 1/4" depth, add	↓	192	.042		.30	2.51		2.81	4.39
1990	For drilling up, add				↓		40%			
2000	Minimum labor/equipment charge	1 Carp	4	2	Job		109		109	174

05 05 21.90 Welding Steel

		Crew	Daily Output	Labor-Hours	Unit	Material	2021 Bare Costs Labor	Equipment	Total	Total Incl O&P
0010	**WELDING STEEL**, Structural R050521-20									
0020	Field welding, 1/8" E6011, cost per welder, no operating engineer	E-14	8	1	Hr.	5.45	62.50	18.60	86.55	128
0200	With 1/2 operating engineer	E-13	8	1.500		5.45	90	18.60	114.05	171
0300	With 1 operating engineer	E-12	8	2		5.45	118	18.60	142.05	214
0500	With no operating engineer, 2# weld rod per ton	E-14	8	1	Ton	5.45	62.50	18.60	86.55	128
0600	8# E6011 per ton	"	2	4		22	249	74.50	345.50	510
0800	With one operating engineer per welder, 2# E6011 per ton	E-12	8	2		5.45	118	18.60	142.05	214
0900	8# E6011 per ton	"	2	8	↓	22	470	74.50	566.50	855
1200	Continuous fillet, down welding									
1300	Single pass, 1/8" thick, 0.1#/L.F.	E-14	150	.053	L.F.	.27	3.32	.99	4.58	6.75
1400	3/16" thick, 0.2#/L.F.		75	.107		.54	6.65	1.98	9.17	13.55
1500	1/4" thick, 0.3#/L.F.		50	.160		.82	9.95	2.97	13.74	20.50
1610	5/16" thick, 0.4#/L.F.		38	.211		1.09	13.10	3.91	18.10	26.50
1800	3 passes, 3/8" thick, 0.5#/L.F.		30	.267		1.36	16.60	4.96	22.92	34
2010	4 passes, 1/2" thick, 0.7#/L.F.		22	.364		1.90	22.50	6.75	31.15	46
2200	5 to 6 passes, 3/4" thick, 1.3#/L.F.		12	.667		3.54	41.50	12.40	57.44	84.50
2400	8 to 11 passes, 1" thick, 2.4#/L.F.	↓	6	1.333		6.55	83	25	114.55	169
2600	For vertical joint welding, add						20%			
2700	Overhead joint welding, add						300%			
2900	For semi-automatic welding, obstructed joints, deduct						5%			
3000	Exposed joints, deduct				↓		15%			
4000	Cleaning and welding plates, bars, or rods									
4010	to existing beams, columns, or trusses	E-14	12	.667	L.F.	1.36	41.50	12.40	55.26	82
9000	Minimum labor/equipment charge	"	4	2	Job		125	37	162	243

05 05 23 – Metal Fastenings

05 05 23.30 Lag Screws

			Crew	Daily Output	Labor-Hours	Unit	Material	2021 Bare Costs Labor	Equipment	Total	Total Incl O&P
0010	**LAG SCREWS**										
0020	Steel, 1/4" diameter, 2" long	G	1 Carp	200	.040	Ea.	.13	2.19		2.32	3.62
0200	1/2" diameter, 3" long	G		130	.062		1.08	3.37		4.45	6.55
0300	5/8" diameter, 3" long	G	↓	120	.067		1.72	3.65		5.37	7.70

05 05 Common Work Results for Metals

05 05 23 – Metal Fastenings

05 05 23.50 Powder Actuated Tools and Fasteners

		Crew	Daily Output	Labor-Hours	Unit	Material	2021 Bare Costs Labor	Equipment	Total	Total Incl O&P
0010	**POWDER ACTUATED TOOLS & FASTENERS**				Ea.	146			146	160
0020	Stud driver, .22 caliber, single shot				"	625			625	685
0100	.27 caliber, semi automatic, strip				C	7.85			7.85	8.60
0300	Powder load, single shot, .22 cal, power level 2, brown					12			12	13.20
0400	Strip, .27 cal, power level 4, red					16.95	91		107.95	164
0600	Drive pin, .300 x 3/4" long	G	1 Carp	4.80	1.667					
0700	.300 x 3" long with washer	G	"	4	2		22	109	131	198

05 05 23.70 Structural Blind Bolts

		Crew	Daily Output	Labor-Hours	Unit	Material	Labor	Equipment	Total	Total Incl O&P
0010	**STRUCTURAL BLIND BOLTS**									
0100	1/4" diameter x 1/4" grip	G	1 Sswk	240	.033	Ea.	1.27	2.01		3.28	4.65
0150	1/2" grip	G		216	.037		1.01	2.23		3.24	4.72
0200	3/8" diameter x 1/2" grip	G		232	.034		1.69	2.08		3.77	5.20
0250	3/4" grip	G		208	.038		1.96	2.32		4.28	5.90
0300	1/2" diameter x 1/2" grip	G		224	.036		3.82	2.15		5.97	7.70
0350	3/4" grip	G		200	.040		3.82	2.41		6.23	8.10
0400	5/8" diameter x 3/4" grip	G		216	.037		6	2.23		8.23	10.20
0450	1" grip	G		192	.042		6	2.51		8.51	10.65

05 05 23.87 Weld Studs

		Crew	Daily Output	Labor-Hours	Unit	Material	Labor	Equipment	Total	Total Incl O&P
0010	**WELD STUDS**									
0020	1/4" diameter, 2-11/16" long	G	E-10	1120	.014	Ea.	.39	.88	.89	2.16	2.83
0100	4-1/8" long	G		1080	.015		.37	.91	.92	2.20	2.90
0200	3/8" diameter, 4-1/8" long	G		1080	.015		.42	.91	.92	2.25	2.95
0300	6-1/8" long	G		1040	.015		.55	.94	.96	2.45	3.19
9000	Minimum labor/equipment charge		1 Sswk	2	4	Job		241		241	390

05 12 Structural Steel Framing

05 12 23 – Structural Steel for Buildings

05 12 23.05 Canopy Framing

		Crew	Daily Output	Labor-Hours	Unit	Material	Labor	Equipment	Total	Total Incl O&P
0010	**CANOPY FRAMING**									
0020	6" and 8" members, shop fabricated	G	E-4	3000	.011	Lb.	1.58	.65	.05	2.28	2.84
9000	Minimum labor/equipment charge		1 Sswk	1	8	Job		480		480	780

05 12 23.10 Ceiling Supports

		Crew	Daily Output	Labor-Hours	Unit	Material	Labor	Equipment	Total	Total Incl O&P
0010	**CEILING SUPPORTS**									
1000	Entrance door/folding partition supports, shop fabricated	G	E-4	60	.533	L.F.	26.50	32.50	2.48	61.48	84
1100	Linear accelerator door supports	G		14	2.286		120	139	10.65	269.65	370
1200	Lintels or shelf angles, hung, exterior hot dipped galv.	G		267	.120		18	7.30	.56	25.86	32
1250	Two coats primer paint instead of galv.	G		267	.120		15.60	7.30	.56	23.46	29.50
1400	Monitor support, ceiling hung, expansion bolted	G		4	8	Ea.	420	485	37	942	1,275
1450	Hung from pre-set inserts	G		6	5.333		450	325	25	800	1,050
1600	Motor supports for overhead doors	G		4	8		213	485	37	735	1,050
1700	Partition support for heavy folding partitions, without pocket	G	E-4	24	1.333	L.F.	60	81	6.20	147.20	204
1750	Supports at pocket only	G		12	2.667		120	162	12.40	294.40	410
2000	Rolling grilles & fire door supports	G		34	.941		51.50	57	4.38	112.88	154
2100	Spider-leg light supports, expansion bolted to ceiling slab	G		8	4	Ea.	171	243	18.60	432.60	605
2150	Hung from pre-set inserts	G		12	2.667	"	185	162	12.40	359.40	480
2400	Toilet partition support	G		36	.889	L.F.	60	54	4.13	118.13	158
2500	X-ray travel gantry support	G		12	2.667	"	206	162	12.40	380.40	500

111

05 12 23 – Structural Steel for Buildings

05 12 23.17 Columns, Structural		Crew	Daily Output	Labor-Hours	Unit	Material	2021 Bare Costs Labor	Equipment	Total	Total Incl O&P
0010	**COLUMNS, STRUCTURAL**	R051223-10								
0015	Made from recycled materials									
0020	Shop fab'd for 100-ton, 1-2 story project, bolted connections									
0800	Steel, concrete filled, extra strong pipe, 3-1/2" diameter	E-2	660	.085	L.F.	43.50	5.05	2.60	51.15	59
0830	4" diameter		780	.072		48.50	4.28	2.20	54.98	63
0890	5" diameter		1020	.055		58	3.27	1.68	62.95	70.50
0930	6" diameter		1200	.047		76.50	2.78	1.43	80.71	90
0940	8" diameter		1100	.051		76.50	3.04	1.56	81.10	90.50
1500	Steel pipe, extra strong, no concrete, 3" to 5" diameter	G	16000	.004	Lb.	1.32	.21	.11	1.64	1.90
3300	Structural tubing, square, A500GrB, 4" to 6" square, light section	G	11270	.005	"	1.32	.30	.15	1.77	2.09
8090	For projects 75 to 99 tons, add				%	10%				
8092	50 to 74 tons, add					20%				
8094	25 to 49 tons, add					30%	10%			
8096	10 to 24 tons, add					50%	25%			
8098	2 to 9 tons, add					75%	50%			
8099	Less than 2 tons, add					100%	100%			
9000	Minimum labor/equipment charge	1 Sswk	1	8	Job		480		480	780

05 12 23.18 Corner Guards

		Crew	Daily Output	Labor-Hours	Unit	Material	Labor	Equipment	Total	Total Incl O&P
0010	**CORNER GUARDS**									
0020	Steel angle w/anchors, 1" x 1" x 1/4", 1.5#/L.F.	2 Carp	160	.100	L.F.	7.45	5.45		12.90	16.90
0100	2" x 2" x 1/4" angles, 3.2#/L.F.	"	150	.107	"	11.50	5.85		17.35	22
9000	Minimum labor/equipment charge	1 Carp	2	4	Job		219		219	350

05 12 23.20 Curb Edging

		Crew	Daily Output	Labor-Hours	Unit	Material	Labor	Equipment	Total	Total Incl O&P
0010	**CURB EDGING**									
0020	Steel angle w/anchors, shop fabricated, on forms, 1" x 1", 0.8#/L.F.	G E-4	350	.091	L.F.	1.69	5.55	.43	7.67	11.35
0300	4" x 4" angles, 8.2#/L.F.	G	275	.116	"	14.20	7.05	.54	21.79	27.50
9000	Minimum labor/equipment charge		4	8	Job		485	37	522	825

05 12 23.40 Lightweight Framing

		Crew	Daily Output	Labor-Hours	Unit	Material	Labor	Equipment	Total	Total Incl O&P
0010	**LIGHTWEIGHT FRAMING**									
0015	Made from recycled materials									
0200	For load-bearing steel studs see Section 05 41 13.30									
0400	Angle framing, field fabricated, 4" and larger	G E-3	440	.055	Lb.	.77	3.33	.34	4.44	6.60
0450	Less than 4" angles	G "	265	.091	"	.79	5.50	.56	6.85	10.45
0600	Channel framing, field fabricated, 8" and larger	G E-3	500	.048	Lb.	.79	2.93	.30	4.02	5.95
0650	Less than 8" channels	G "	335	.072		.79	4.37	.44	5.60	8.40
1000	Continuous slotted channel framing system, shop fab, simple framing	G 2 Sswk	2400	.007		4.09	.40		4.49	5.15
1200	Complex framing	G "	1600	.010		4.62	.60		5.22	6.10
1250	Plate & bar stock for reinforcing beams and trusses	G				1.45			1.45	1.60
1300	Cross bracing, rods, shop fabricated, 3/4" diameter	G E-3	700	.034		1.58	2.09	.21	3.88	5.35
1310	7/8" diameter	G	850	.028		1.58	1.72	.18	3.48	4.71
1320	1" diameter	G	1000	.024		1.58	1.46	.15	3.19	4.27
1330	Angle, 5" x 5" x 3/8"	G	2800	.009		1.58	.52	.05	2.15	2.65
1350	Hanging lintels, shop fabricated	G	850	.028		1.58	1.72	.18	3.48	4.71
1380	Roof frames, shop fabricated, 3'-0" square, 5' span	G E-2	4200	.013		1.58	.80	.41	2.79	3.46
1400	Tie rod, not upset, 1-1/2" to 4" diameter, with turnbuckle	G 2 Sswk	800	.020		1.71	1.21		2.92	3.84
1420	No turnbuckle	G	700	.023		1.65	1.38		3.03	4.04
1500	Upset, 1-3/4" to 4" diameter, with turnbuckle	G	800	.020		1.71	1.21		2.92	3.84
1520	No turnbuckle	G	700	.023		1.65	1.38		3.03	4.04
9000	Minimum labor/equipment charge		2	8	Job		480		480	780

05 12 Structural Steel Framing

05 12 23 – Structural Steel for Buildings

05 12 23.45 Lintels

			Crew	Daily Output	Labor-Hours	Unit	Material	2021 Bare Costs Labor	Equipment	Total	Total Incl O&P
0010	**LINTELS**										
0015	Made from recycled materials										
0020	Plain steel angles, shop fabricated, under 500 lb.	G	1 Bric	550	.015	Lb.	1.02	.78		1.80	2.38
0100	500 to 1,000 lb.	G		640	.013		.99	.67		1.66	2.17
0200	1,000 to 2,000 lb.	G		640	.013		.96	.67		1.63	2.14
0300	2,000 to 4,000 lb.	G		640	.013		.94	.67		1.61	2.11
0500	For built-up angles and plates, add to above	G					1.32			1.32	1.45
0700	For engineering, add to above						.13			.13	.15
0900	For galvanizing, add to above, under 500 lb.						.39			.39	.42
0950	500 to 2,000 lb.						.35			.35	.39
1000	Over 2,000 lb.						.33			.33	.37
2000	Steel angles, 3-1/2" x 3", 1/4" thick, 2'-6" long	G	1 Bric	47	.170	Ea.	14.25	9.15		23.40	30.50
2100	4'-6" long	G		26	.308		25.50	16.50		42	54.50
2500	3-1/2" x 3-1/2" x 5/16", 5'-0" long	G		18	.444		38	24		62	80.50
2600	4" x 3-1/2", 1/4" thick, 5'-0" long	G		21	.381		32.50	20.50		53	69
2700	9'-0" long	G		12	.667		59	36		95	123
2800	4" x 3-1/2" x 5/16", 7'-0" long	G		12	.667		57	36		93	120
2900	5" x 3-1/2" x 5/16", 10'-0" long	G		8	1		92	53.50		145.50	188
9000	Minimum labor/equipment charge			4	2	Job		107		107	173

05 12 23.60 Pipe Support Framing

			Crew	Daily Output	Labor-Hours	Unit	Material	2021 Bare Costs Labor	Equipment	Total	Total Incl O&P
0010	**PIPE SUPPORT FRAMING**										
0020	Under 10#/L.F., shop fabricated	G	E-4	3900	.008	Lb.	1.77	.50	.04	2.31	2.79
0200	10.1 to 15#/L.F.	G	"	4300	.007	"	1.74	.45	.03	2.22	2.69
0400	15.1 to 20#/L.F.	G	E-4	4800	.007	Lb.	1.71	.41	.03	2.15	2.58
0600	Over 20#/L.F.	G	"	5400	.006	"	1.69	.36	.03	2.08	2.47

05 12 23.65 Plates

			Crew	Daily Output	Labor-Hours	Unit	Material	2021 Bare Costs Labor	Equipment	Total	Total Incl O&P
0010	**PLATES**										
0015	Made from recycled materials										
0020	For connections & stiffener plates, shop fabricated										
0050	1/8" thick (5.1 lb./S.F.)	G				S.F.	6.75			6.75	7.40
0100	1/4" thick (10.2 lb./S.F.)	G					13.45			13.45	14.80
0300	3/8" thick (15.3 lb./S.F.)	G					20			20	22
0400	1/2" thick (20.4 lb./S.F.)	G					27			27	29.50
0450	3/4" thick (30.6 lb./S.F.)	G					40.50			40.50	44.50
0500	1" thick (40.8 lb./S.F.)	G					54			54	59
2000	Steel plate, warehouse prices, no shop fabrication										
2100	1/4" thick (10.2 lb./S.F.)	G				S.F.	8.10			8.10	8.90
2210	1/4" steel plate, welded in place		E-18	528	.076		8.10	4.58	2.89	15.57	19.45
2220	1/2" steel plate, welded in place			480	.083		16.15	5.05	3.18	24.38	29.50
2230	3/4" steel plate, welded in place			384	.104		24	6.30	3.97	34.27	41
2240	1" steel plate, welded in place			320	.125		32.50	7.55	4.77	44.82	53
2250	1-1/2" steel plate, welded in place			256	.156		48.50	9.45	5.95	63.90	75
2260	2" steel plate, welded in place			192	.208		64.50	12.60	7.95	85.05	100

05 12 23.75 Structural Steel Members

			Crew	Daily Output	Labor-Hours	Unit	Material	2021 Bare Costs Labor	Equipment	Total	Total Incl O&P
0010	**STRUCTURAL STEEL MEMBERS** R051223-10										
0015	Made from recycled materials										
0020	Shop fab'd for 100-ton, 1-2 story project, bolted connections										
0102	Beam or girder, W 6 x 9 R051223-15	G	E-2	600	.093	L.F.	13.05	5.55	2.86	21.46	26.50
0302	W 8 x 10	G		600	.093		14.50	5.55	2.86	22.91	28
0502	x 31	G		550	.102		45	6.05	3.11	54.16	62.50
0702	W 10 x 22	G		600	.093		32	5.55	2.86	40.41	47

113

05 12 23.75 Structural Steel Members

			Crew	Daily Output	Labor-Hours	Unit	Material	2021 Bare Costs Labor	Equipment	Total	Total Incl O&P
0902	x 49	G	E-2	550	.102	L.F.	71	6.05	3.11	80.16	91
1102	W 12 x 16	G		880	.064		23	3.80	1.95	28.75	33.50
1302	x 22	G		880	.064		32	3.80	1.95	37.75	43
1502	x 26	G		880	.064		37.50	3.80	1.95	43.25	49.50
1902	W 14 x 26	G		990	.057		37.50	3.37	1.73	42.60	49
2102	x 30	G		900	.062		43.50	3.71	1.90	49.11	56
2702	W 16 x 26	G		1000	.056		37.50	3.34	1.71	42.55	48.50
2902	x 31	G		900	.062		45	3.71	1.90	50.61	57.50
8490	For projects 75 to 99 tons, add						10%				
8492	50 to 74 tons, add						20%				
8494	25 to 49 tons, add						30%	10%			
8496	10 to 24 tons, add						50%	25%			
8498	2 to 9 tons, add						75%	50%			
8499	Less than 2 tons, add						100%	100%			
9000	Minimum labor/equipment charge		E-2	2	28	Job		1,675	855	2,530	3,625

05 12 23.77 Structural Steel Projects

			Crew	Daily Output	Labor-Hours	Unit	Material	2021 Bare Costs Labor	Equipment	Total	Total Incl O&P
0010	**STRUCTURAL STEEL PROJECTS**										
0015	Made from recycled materials										
0020	Shop fab'd for 100-ton, 1-2 story project, bolted connections										
0700	Offices, hospitals, etc., steel bearing, 1 to 2 stories R050521-20	G	E-5	10.30	7.767	Ton	2,650	465	181	3,296	3,850
1300	Industrial bldgs., 1 story, beams & girders, steel bearing	G		12.90	6.202		2,650	370	144	3,164	3,650
1400	Masonry bearing R051223-10	G		10	8		2,650	480	186	3,316	3,875
1600	1 story with roof trusses, steel bearing	G		10.60	7.547		3,125	455	176	3,756	4,350
1700	Masonry bearing R051223-15	G		8.30	9.639		3,125	580	224	3,929	4,600
5390	For projects 75 to 99 tons, add						10%				
5392	50 to 74 tons, add						20%				
5394	25 to 49 tons, add						30%	10%			
5396	10 to 24 tons, add						50%	25%			
5398	2 to 9 tons, add						75%	50%			
5399	Less than 2 tons, add						100%	100%			

05 12 23.78 Structural Steel Secondary Members

			Crew	Daily Output	Labor-Hours	Unit	Material	2021 Bare Costs Labor	Equipment	Total	Total Incl O&P
0010	**STRUCTURAL STEEL SECONDARY MEMBERS**										
0015	Made from recycled materials										
0020	Shop fabricated for 20-ton girt/purlin framing package, materials only										
0100	Girts/purlins, C/Z-shapes, includes clips and bolts										
0110	6" x 2-1/2" x 2-1/2", 16 ga., 3.0 lb./L.F.					L.F.	3.56			3.56	3.92
0115	14 ga., 3.5 lb./L.F.						4.16			4.16	4.57
0120	8" x 2-3/4" x 2-3/4", 16 ga., 3.4 lb./L.F.						4.04			4.04	4.44
0125	14 ga., 4.1 lb./L.F.						4.87			4.87	5.35
0130	12 ga., 5.6 lb./L.F.						6.65			6.65	7.30
0135	10" x 3-1/2" x 3-1/2", 14 ga., 4.7 lb./L.F.						5.60			5.60	6.15
0140	12 ga., 6.7 lb./L.F.						7.95			7.95	8.75
0145	12" x 3-1/2" x 3-1/2", 14 ga., 5.3 lb./L.F.						6.30			6.30	6.90
0150	12 ga., 7.4 lb./L.F.						8.80			8.80	9.65
0200	Eave struts, C-shape, includes clips and bolts										
0210	6" x 4" x 3", 16 ga., 3.1 lb./L.F.					L.F.	3.68			3.68	4.05
0215	14 ga., 3.9 lb./L.F.						4.63			4.63	5.10
0220	8" x 4" x 3", 16 ga., 3.5 lb./L.F.						4.16			4.16	4.57
0225	14 ga., 4.4 lb./L.F.						5.20			5.20	5.75
0230	12 ga., 6.2 lb./L.F.						7.35			7.35	8.10
0235	10" x 5" x 3", 14 ga., 5.2 lb./L.F.						6.15			6.15	6.80
0240	12 ga., 7.3 lb./L.F.						8.65			8.65	9.55

05 12 Structural Steel Framing

05 12 23 – Structural Steel for Buildings

05 12 23.78 Structural Steel Secondary Members	Crew	Daily Output	Labor-Hours	Unit	Material	2021 Bare Costs Labor	Equipment	Total	Total Incl O&P
0245 12" x 5" x 4", 14 ga., 6.0 lb./L.F.				L.F.	7.10			7.10	7.85
0250 12 ga., 8.4 lb./L.F.				↓	9.95			9.95	10.95
0300 Rake/base angle, excludes concrete drilling and expansion anchors									
0310 2" x 2", 14 ga., 1.0 lb./L.F.	2 Sswk	640	.025	L.F.	1.19	1.51		2.70	3.75
0315 3" x 2", 14 ga., 1.3 lb./L.F.		535	.030		1.54	1.80		3.34	4.62
0320 3" x 3", 14 ga., 1.6 lb./L.F.		500	.032		1.90	1.93		3.83	5.20
0325 4" x 3", 14 ga., 1.8 lb./L.F.	↓	480	.033	↓	2.14	2.01		4.15	5.60
0600 Installation of secondary members, erection only									
0610 Girts, purlins, eave struts, 16 ga., 6" deep	E-18	100	.400	Ea.		24	15.25	39.25	56
0615 8" deep		80	.500			30	19.10	49.10	69.50
0620 14 ga., 6" deep		80	.500			30	19.10	49.10	69.50
0625 8" deep		65	.615			37	23.50	60.50	86
0630 10" deep		55	.727			44	28	72	101
0635 12" deep		50	.800			48.50	30.50	79	111
0640 12 ga., 8" deep		50	.800			48.50	30.50	79	111
0645 10" deep		45	.889			53.50	34	87.50	124
0650 12" deep	↓	40	1	↓		60.50	38	98.50	139
0900 For less than 20-ton job lots									
0905 For 15 to 19 tons, add				%	10%				
0910 For 10 to 14 tons, add					25%				
0915 For 5 to 9 tons, add					50%	50%	50%		
0920 For 1 to 4 tons, add					75%	75%	75%		
0925 For less than 1 ton, add				↓	100%	100%	100%		
0990 Minimum labor/equipment charge	E-18	1	40	Job		2,425	1,525	3,950	5,550

05 15 Wire Rope Assemblies

05 15 16 – Steel Wire Rope Assemblies

05 15 16.70 Temporary Cable Safety Railing

	Crew	Daily Output	Labor-Hours	Unit	Material	2021 Bare Costs Labor	Equipment	Total	Total Incl O&P
0010 **TEMPORARY CABLE SAFETY RAILING**, Each 100' strand incl.									
0020 2 eyebolts, 1 turnbuckle, 100' cable, 2 thimbles, 6 clips									
0025 Made from recycled materials									
0100 One strand using 1/4" cable & accessories [G]	2 Sswk	4	4	C.L.F.	79.50	241		320.50	475
0200 1/2" cable & accessories [G]	"	2	8	"	157	480		637	955

05 21 Steel Joist Framing

05 21 13 – Deep Longspan Steel Joist Framing

05 21 13.50 Deep Longspan Joists

	Crew	Daily Output	Labor-Hours	Unit	Material	2021 Bare Costs Labor	Equipment	Total	Total Incl O&P
0010 **DEEP LONGSPAN JOISTS**									
3010 DLH series, 40-ton job lots, bolted cross bridging, shop primer									
3015 Made from recycled materials									
3040 Spans to 144' (shipped in 2 pieces) [G]	E-7	13	6.154	Ton	2,475	370	155	3,000	3,500
3500 For less than 40-ton job lots									
3502 For 30 to 39 tons, add				%	10%				
3504 20 to 29 tons, add					20%				
3506 10 to 19 tons, add				↓	30%				
3507 5 to 9 tons, add				%	50%	25%			
3508 1 to 4 tons, add					75%	50%			
3509 Less than 1 ton, add				↓	100%	100%			
4010 SLH series, 40-ton job lots, bolted cross bridging, shop primer									

For customer support on your Commercial Renovation Costs with RSMeans Data, call 800.448.8182.

115

05 21 Steel Joist Framing

05 21 13 – Deep Longspan Steel Joist Framing

05 21 13.50 Deep Longspan Joists		Crew	Daily Output	Labor-Hours	Unit	Material	2021 Bare Costs Labor	Equipment	Total	Total Incl O&P
4040	Spans to 200' (shipped in 3 pieces) [G]	E-7	13	6.154	Ton	2,550	370	155	3,075	3,575
6100	For less than 40-ton job lots									
6102	For 30 to 39 tons, add				%	10%				
6104	20 to 29 tons, add					20%				
6106	10 to 19 tons, add					30%				
6107	5 to 9 tons, add					50%	25%			
6108	1 to 4 tons, add					75%	50%			
6109	Less than 1 ton, add					100%	100%			

05 21 16 – Longspan Steel Joist Framing

05 21 16.50 Longspan Joists

		Crew	Daily Output	Labor-Hours	Unit	Material	2021 Bare Costs Labor	Equipment	Total	Total Incl O&P
0010	**LONGSPAN JOISTS**									
2000	LH series, 40-ton job lots, bolted cross bridging, shop primer									
2015	Made from recycled materials									
2040	Longspan joists, LH series, up to 96' [G]	E-7	13	6.154	Ton	2,150	370	155	2,675	3,150
2600	For less than 40-ton job lots									
2602	For 30 to 39 tons, add				%	10%				
2604	20 to 29 tons, add					20%				
2606	10 to 19 tons, add					30%				
2607	5 to 9 tons, add					50%	25%			
2608	1 to 4 tons, add					75%	50%			
2609	Less than 1 ton, add					100%	100%			

05 21 19 – Open Web Steel Joist Framing

05 21 19.10 Open Web Joists

		Crew	Daily Output	Labor-Hours	Unit	Material	2021 Bare Costs Labor	Equipment	Total	Total Incl O&P
0010	**OPEN WEB JOISTS**									
0015	Made from recycled materials									
0050	K series, 40-ton lots, horiz. bridging, spans to 30', shop primer [G]	E-7	12	6.667	Ton	1,900	400	168	2,468	2,925
0440	K series, 30' to 50' spans [G]	"	17	4.706	"	1,950	283	118	2,351	2,725
0800	For less than 40-ton job lots									
0802	For 30 to 39 tons, add				%	10%				
0804	20 to 29 tons, add					20%				
0806	10 to 19 tons, add					30%				
0807	5 to 9 tons, add					50%	25%			
0808	1 to 4 tons, add					75%	50%			
0809	Less than 1 ton, add					100%	100%			
1010	CS series, 40-ton job lots, horizontal bridging, shop primer									
1040	Spans to 30' [G]	E-7	12	6.667	Ton	1,975	400	168	2,543	3,000
1500	For less than 40-ton job lots									
1502	For 30 to 39 tons, add				%	10%				
1504	20 to 29 tons, add					20%				
1506	10 to 19 tons, add					30%				
1507	5 to 9 tons, add					50%	25%			
1508	1 to 4 tons, add					75%	50%			
1509	Less than 1 ton, add					100%	100%			
6400	Individual steel bearing plate, 6" x 6" x 1/4" with J-hook [G]	1 Bric	160	.050	Ea.	7.90	2.69		10.59	13
9000	Minimum labor and equipment, bar joists	F-6	1	40	Job		2,075	475	2,550	3,825

05 31 Steel Decking

05 31 13 – Steel Floor Decking

05 31 13.50 Floor Decking

		Crew	Daily Output	Labor-Hours	Unit	Material	2021 Bare Costs Labor	2021 Bare Costs Equipment	Total	Total Incl O&P
0010	**FLOOR DECKING** R053100-10									
0015	Made from recycled materials									
5100	Non-cellular composite decking, galvanized, 1-1/2" deep, 16 ga. G	E-4	3500	.009	S.F.	4.90	.56	.04	5.50	6.35
5120	18 ga. G		3650	.009		3.02	.53	.04	3.59	4.23
5140	20 ga. G		3800	.008		3.57	.51	.04	4.12	4.79
5200	2" deep, 22 ga. G		3860	.008		2.41	.50	.04	2.95	3.51
5300	20 ga. G		3600	.009		3.43	.54	.04	4.01	4.69
5400	18 ga. G		3380	.009		3.15	.58	.04	3.77	4.45
5500	16 ga. G		3200	.010		4.84	.61	.05	5.50	6.35
5700	3" deep, 22 ga. G		3200	.010		2.63	.61	.05	3.29	3.92
5800	20 ga. G		3000	.011		3.77	.65	.05	4.47	5.25
5900	18 ga. G		2850	.011		3.15	.68	.05	3.88	4.63
6000	16 ga. G		2700	.012		5.50	.72	.06	6.28	7.30
9000	Minimum labor/equipment charge	1 Sswk	1	8	Job		480		480	780

05 31 23 – Steel Roof Decking

05 31 23.50 Roof Decking

		Crew	Daily Output	Labor-Hours	Unit	Material	2021 Bare Costs Labor	2021 Bare Costs Equipment	Total	Total Incl O&P
0010	**ROOF DECKING**									
0015	Made from recycled materials									
2100	Open type, 1-1/2" deep, Type B, wide rib, galv., 22 ga., under 50 sq. G	E-4	4500	.007	S.F.	2.72	.43	.03	3.18	3.73
2400	Over 500 squares G	"	5100	.006	"	1.96	.38	.03	2.37	2.80

05 31 33 – Steel Form Decking

05 31 33.50 Form Decking

		Crew	Daily Output	Labor-Hours	Unit	Material	2021 Bare Costs Labor	2021 Bare Costs Equipment	Total	Total Incl O&P
0010	**FORM DECKING**									
0015	Made from recycled materials									
6100	Slab form, steel, 28 ga., 9/16" deep, Type UFS, uncoated G	E-4	4000	.008	S.F.	1.91	.49	.04	2.44	2.93
6200	Galvanized G	"	4000	.008	"	1.69	.49	.04	2.22	2.69

05 35 Raceway Decking Assemblies

05 35 13 – Steel Cellular Decking

05 35 13.50 Cellular Decking

		Crew	Daily Output	Labor-Hours	Unit	Material	2021 Bare Costs Labor	2021 Bare Costs Equipment	Total	Total Incl O&P
0010	**CELLULAR DECKING**									
0015	Made from recycled materials									
0200	Cellular units, galv, 1-1/2" deep, Type BC, 20-20 ga., over 15 squares G	E-4	1460	.022	S.F.	11.70	1.33	.10	13.13	15.15
0400	3" deep, Type NC, galvanized, 20-20 ga. G	E-4	1375	.023	S.F.	12.90	1.41	.11	14.42	16.55
1000	4-1/2" deep, Type JC, galvanized, 18-20 ga. G	"	1100	.029	"	12.40	1.77	.14	14.31	16.65
1500	For acoustical deck, add					15%				
1900	For multi-story or congested site, add						50%			

05 41 Structural Metal Stud Framing

05 41 13 – Load-Bearing Metal Stud Framing

05 41 13.05 Bracing

		Crew	Daily Output	Labor-Hours	Unit	Material	2021 Bare Costs Labor	2021 Bare Costs Equipment	Total	Total Incl O&P
0010	**BRACING**, shear wall X-bracing, per 10' x 10' bay, one face									
0015	Made of recycled materials									
0120	Metal strap, 20 ga. x 4" wide G	2 Carp	18	.889	Ea.	20.50	48.50		69	101
0130	6" wide G		18	.889		33	48.50		81.50	114
0160	18 ga. x 4" wide G		16	1		34	54.50		88.50	124
0170	6" wide G		16	1		49.50	54.50		104	142
0410	Continuous strap bracing, per horizontal row on both faces									

117

05 41 Structural Metal Stud Framing

05 41 13 – Load-Bearing Metal Stud Framing

05 41 13.05 Bracing

		Crew	Daily Output	Labor-Hours	Unit	Material	2021 Bare Costs Labor	Equipment	Total	Total Incl O&P
0420	Metal strap, 20 ga. x 2" wide, studs 12" OC	G 1 Carp	7	1.143	C.L.F.	59.50	62.50		122	165
0430	16" OC	G	8	1		59.50	54.50		114	153
0440	24" OC	G	10	.800		59.50	44		103.50	135
0450	18 ga. x 2" wide, studs 12" OC	G	6	1.333		81.50	73		154.50	206
0460	16" OC	G	7	1.143		81.50	62.50		144	189
0470	24" OC	G	8	1		81.50	54.50		136	177

05 41 13.10 Bridging

		Crew	Daily Output	Labor-Hours	Unit	Material	2021 Bare Costs Labor	Equipment	Total	Total Incl O&P
0010	**BRIDGING**, solid between studs w/1-1/4" leg track, per stud bay									
0015	Made from recycled materials									
0200	Studs 12" OC, 18 ga. x 2-1/2" wide	G 1 Carp	125	.064	Ea.	.98	3.50		4.48	6.65
0210	3-5/8" wide	G	120	.067		1.17	3.65		4.82	7.10
0220	4" wide	G	120	.067		1.11	3.65		4.76	7
0230	6" wide	G	115	.070		1.62	3.81		5.43	7.85
0240	8" wide	G	110	.073		1.99	3.98		5.97	8.55
0300	16 ga. x 2-1/2" wide	G	115	.070		1.21	3.81		5.02	7.40
0310	3-5/8" wide	G	110	.073		1.45	3.98		5.43	7.95
0320	4" wide	G	110	.073		1.61	3.98		5.59	8.10
0330	6" wide	G	105	.076		2.02	4.17		6.19	8.85
0340	8" wide	G	100	.080		2.48	4.38		6.86	9.70
1200	Studs 16" OC, 18 ga. x 2-1/2" wide	G	125	.064		1.25	3.50		4.75	6.95
1210	3-5/8" wide	G	120	.067		1.50	3.65		5.15	7.45
1220	4" wide	G	120	.067		1.42	3.65		5.07	7.35
1230	6" wide	G	115	.070		2.08	3.81		5.89	8.35
1240	8" wide	G	110	.073		2.56	3.98		6.54	9.15
1300	16 ga. x 2-1/2" wide	G	115	.070		1.55	3.81		5.36	7.75
1310	3-5/8" wide	G	110	.073		1.86	3.98		5.84	8.40
1320	4" wide	G	110	.073		2.06	3.98		6.04	8.60
1330	6" wide	G	105	.076		2.59	4.17		6.76	9.50
1340	8" wide	G 1 Carp	100	.080	Ea.	3.18	4.38		7.56	10.45
2200	Studs 24" OC, 18 ga. x 2-1/2" wide	G	125	.064		1.81	3.50		5.31	7.55
2210	3-5/8" wide	G	120	.067		2.17	3.65		5.82	8.20
2220	4" wide	G	120	.067		2.05	3.65		5.70	8.05
2230	6" wide	G	115	.070		3.01	3.81		6.82	9.35
2240	8" wide	G	110	.073		3.70	3.98		7.68	10.40
2300	16 ga. x 2-1/2" wide	G	115	.070		2.24	3.81		6.05	8.50
2310	3-5/8" wide	G	110	.073		2.70	3.98		6.68	9.30
2320	4" wide	G	110	.073		2.98	3.98		6.96	9.65
2330	6" wide	G	105	.076		3.75	4.17		7.92	10.75
2340	8" wide	G	100	.080		4.61	4.38		8.99	12
3000	Continuous bridging, per row									
3100	16 ga. x 1-1/2" channel thru studs 12" OC	G 1 Carp	6	1.333	C.L.F.	55	73		128	177
3110	16" OC	G	7	1.143		55	62.50		117.50	160
3120	24" OC	G	8.80	.909		55	49.50		104.50	140
4100	2" x 2" angle x 18 ga., studs 12" OC	G	7	1.143		82.50	62.50		145	191
4110	16" OC	G	9	.889		82.50	48.50		131	169
4120	24" OC	G	12	.667		82.50	36.50		119	149
4200	16 ga., studs 12" OC	G	5	1.600		103	87.50		190.50	253
4210	16" OC	G	7	1.143		103	62.50		165.50	214
4220	24" OC	G	10	.800		103	44		147	184

05 41 13 – Load-Bearing Metal Stud Framing

05 41 13.25 Framing, Boxed Headers/Beams		Crew	Daily Output	Labor-Hours	Unit	Material	2021 Bare Costs Labor	Equipment	Total	Total Incl O&P
0010	**FRAMING, BOXED HEADERS/BEAMS**									
0015	Made from recycled materials									
0200	Double, 18 ga. x 6" deep	G	2 Carp	220	.073	L.F.	5.80	3.98		9.78	12.75
0210	8" deep	G		210	.076		6	4.17		10.17	13.30
0220	10" deep	G		200	.080		7.70	4.38		12.08	15.40
0230	12" deep	G		190	.084		8.40	4.61		13.01	16.60
0300	16 ga. x 8" deep	G		180	.089		7	4.86		11.86	15.45
0310	10" deep	G		170	.094		8.65	5.15		13.80	17.70
0320	12" deep	G		160	.100		9.40	5.45		14.85	19
0400	14 ga. x 10" deep	G		140	.114		10.15	6.25		16.40	21
0410	12" deep	G		130	.123		11.15	6.75		17.90	23
1210	Triple, 18 ga. x 8" deep	G		170	.094		9.10	5.15		14.25	18.20
1220	10" deep	G		165	.097		11	5.30		16.30	20.50
1230	12" deep	G		160	.100		12.05	5.45		17.50	22
1300	16 ga. x 8" deep	G		145	.110		10.55	6.05		16.60	21
1310	10" deep	G		140	.114		12.40	6.25		18.65	23.50
1320	12" deep	G		135	.119		13.50	6.50		20	25
1400	14 ga. x 10" deep	G		115	.139		13.85	7.60		21.45	27.50
1410	12" deep	G		110	.145		15.35	7.95		23.30	29.50

05 41 13.30 Framing, Stud Walls

		Crew	Daily Output	Labor-Hours	Unit	Material	2021 Bare Costs Labor	Equipment	Total	Total Incl O&P
0010	**FRAMING, STUD WALLS** w/top & bottom track, no openings,									
0020	Headers, beams, bridging or bracing									
0025	Made from recycled materials									
4100	8' high walls, 18 ga. x 2-1/2" wide, studs 12" OC	G	2 Carp	54	.296	L.F.	8.50	16.20		24.70	35.50
4110	16" OC	G		77	.208		6.85	11.35		18.20	25.50
4120	24" OC	G		107	.150		5.20	8.20		13.40	18.70
4130	3-5/8" wide, studs 12" OC	G		53	.302		10.15	16.50		26.65	37.50
4140	16" OC	G		76	.211		8.15	11.50		19.65	27.50
4150	24" OC	G		105	.152		6.20	8.35		14.55	20
4160	4" wide, studs 12" OC	G		52	.308		10.35	16.85		27.20	38.50
4170	16" OC	G		74	.216		8.30	11.85		20.15	28
4180	24" OC	G		103	.155		6.25	8.50		14.75	20.50
4190	6" wide, studs 12" OC	G		51	.314		13.50	17.15		30.65	42.50
4200	16" OC	G		73	.219		10.90	12		22.90	31
4210	24" OC	G		101	.158		8.30	8.65		16.95	23
4220	8" wide, studs 12" OC	G		50	.320		17.65	17.50		35.15	47.50
4230	16" OC	G		72	.222		14.20	12.15		26.35	35
4240	24" OC	G		100	.160		10.70	8.75		19.45	26
4300	16 ga. x 2-1/2" wide, studs 12" OC	G		47	.340		11.15	18.60		29.75	42
4310	16" OC	G		68	.235		8.80	12.85		21.65	30
4320	24" OC	G		94	.170		6.50	9.30		15.80	22
4330	3-5/8" wide, studs 12" OC	G		46	.348		12.80	19.05		31.85	44.50
4340	16" OC	G		66	.242		10.15	13.25		23.40	32
4350	24" OC	G		92	.174		7.50	9.50		17	23.50
4360	4" wide, studs 12" OC	G		45	.356		13.80	19.45		33.25	46
4370	16" OC	G		65	.246		10.85	13.45		24.30	33.50
4380	24" OC	G		90	.178		7.95	9.70		17.65	24.50
4390	6" wide, studs 12" OC	G		44	.364		17.40	19.90		37.30	50.50
4400	16" OC	G		64	.250		13.85	13.70		27.55	37
4410	24" OC	G		88	.182		10.25	9.95		20.20	27
4420	8" wide, studs 12" OC	G		43	.372		21	20.50		41.50	55.50
4430	16" OC	G		63	.254		16.70	13.90		30.60	40.50

05 41 13.30 Framing, Stud Walls		Crew	Daily Output	Labor-Hours	Unit	Material	2021 Bare Costs Labor	Equipment	Total	Total Incl O&P
4440	24" OC	G 2 Carp	86	.186	L.F.	12.40	10.20		22.60	30
5100	10' high walls, 18 ga. x 2-1/2" wide, studs 12" OC	G	54	.296		10.15	16.20		26.35	37
5110	16" OC	G	77	.208		8.10	11.35		19.45	27
5120	24" OC	G	107	.150		6	8.20		14.20	19.60
5130	3-5/8" wide, studs 12" OC	G	53	.302		12.15	16.50		28.65	40
5140	16" OC	G	76	.211		9.65	11.50		21.15	29
5150	24" OC	G	105	.152		7.20	8.35		15.55	21
5160	4" wide, studs 12" OC	G	52	.308		12.40	16.85		29.25	40.50
5170	16" OC	G	74	.216		9.85	11.85		21.70	29.50
5180	24" OC	G	103	.155		7.25	8.50		15.75	21.50
5190	6" wide, studs 12" OC	G 2 Carp	51	.314	L.F.	16.10	17.15		33.25	45
5200	16" OC	G	73	.219		12.85	12		24.85	33
5210	24" OC	G	101	.158		9.60	8.65		18.25	24.50
5220	8" wide, studs 12" OC	G	50	.320		21	17.50		38.50	51
5230	16" OC	G	72	.222		16.75	12.15		28.90	38
5240	24" OC	G	100	.160		12.45	8.75		21.20	27.50
5300	16 ga. x 2-1/2" wide, studs 12" OC	G	47	.340		13.45	18.60		32.05	44.50
5310	16" OC	G	68	.235		10.55	12.85		23.40	32
5320	24" OC	G	94	.170		7.65	9.30		16.95	23.50
5330	3-5/8" wide, studs 12" OC	G	46	.348		15.45	19.05		34.50	47.50
5340	16" OC	G	66	.242		12.15	13.25		25.40	34.50
5350	24" OC	G	92	.174		8.85	9.50		18.35	25
5360	4" wide, studs 12" OC	G	45	.356		16.70	19.45		36.15	49.50
5370	16" OC	G	65	.246		13.05	13.45		26.50	36
5380	24" OC	G	90	.178		9.40	9.70		19.10	26
5390	6" wide, studs 12" OC	G	44	.364		21	19.90		40.90	54.50
5400	16" OC	G	64	.250		16.50	13.70		30.20	40
5410	24" OC	G	88	.182		12.05	9.95		22	29
5420	8" wide, studs 12" OC	G	43	.372		25.50	20.50		46	60.50
5430	16" OC	G	63	.254		19.90	13.90		33.80	44
5440	24" OC	G	86	.186		14.55	10.20		24.75	32
6190	12' high walls, 18 ga. x 6" wide, studs 12" OC	G	41	.390		18.70	21.50		40.20	54.50
6200	16" OC	G	58	.276		14.80	15.10		29.90	40.50
6210	24" OC	G	81	.198		10.90	10.80		21.70	29
6220	8" wide, studs 12" OC	G	40	.400		24.50	22		46.50	62
6230	16" OC	G	57	.281		19.35	15.35		34.70	46
6240	24" OC	G	80	.200		14.20	10.95		25.15	33
6390	16 ga. x 6" wide, studs 12" OC	G	35	.457		24.50	25		49.50	67
6400	16" OC	G	51	.314		19.20	17.15		36.35	48.50
6410	24" OC	G	70	.229		13.85	12.50		26.35	35
6420	8" wide, studs 12" OC	G	34	.471		29.50	25.50		55	73.50
6430	16" OC	G	50	.320		23	17.50		40.50	53.50
6440	24" OC	G	69	.232		16.70	12.70		29.40	38.50
6530	14 ga. x 3-5/8" wide, studs 12" OC	G	34	.471		23.50	25.50		49	67
6540	16" OC	G	48	.333		18.45	18.25		36.70	49.50
6550	24" OC	G	65	.246		13.20	13.45		26.65	36
6560	4" wide, studs 12" OC	G	33	.485		25	26.50		51.50	69.50
6570	16" OC	G	47	.340		19.45	18.60		38.05	51
6580	24" OC	G	64	.250		14	13.70		27.70	37.50
6730	12 ga. x 3-5/8" wide, studs 12" OC	G	31	.516		32.50	28		60.50	81
6740	16" OC	G	43	.372		25	20.50		45.50	60
6750	24" OC	G	59	.271		17.70	14.85		32.55	43
6760	4" wide, studs 12" OC	G 2 Carp	30	.533	L.F.	34	29		63	84

05 41 13 – Load-Bearing Metal Stud Framing

05 41 13.30 Framing, Stud Walls		Crew	Daily Output	Labor-Hours	Unit	Material	2021 Bare Costs Labor	Equipment	Total	Total Incl O&P
6770	16" OC	G 2 Carp	42	.381	L.F.	26.50	21		47.50	62
6780	24" OC	G	58	.276		18.60	15.10		33.70	44.50
7390	16' high walls, 16 ga. x 6" wide, studs 12" OC	G	33	.485		31.50	26.50		58	77
7400	16" OC	G	48	.333		24.50	18.25		42.75	56
7410	24" OC	G	67	.239		17.40	13.05		30.45	40
7420	8" wide, studs 12" OC	G	32	.500		38	27.50		65.50	85.50
7430	16" OC	G	47	.340		29.50	18.60		48.10	62
7440	24" OC	G	66	.242		21	13.25		34.25	44
7560	14 ga. x 4" wide, studs 12" OC	G	31	.516		32	28		60	80.50
7570	16" OC	G	45	.356		25	19.45		44.45	58.50
7580	24" OC	G	61	.262		17.60	14.35		31.95	42.50
7590	6" wide, studs 12" OC	G	30	.533		40.50	29		69.50	91
7600	16" OC	G	44	.364		31.50	19.90		51.40	66
7610	24" OC	G	60	.267		22.50	14.60		37.10	47.50
7760	12 ga. x 4" wide, studs 12" OC	G	29	.552		44.50	30		74.50	97
7770	16" OC	G	40	.400		34	22		56	72.50
7780	24" OC	G	55	.291		24	15.90		39.90	51.50
7790	6" wide, studs 12" OC	G	28	.571		57.50	31.50		89	114
7800	16" OC	G	39	.410		44	22.50		66.50	84
7810	24" OC	G	54	.296		30.50	16.20		46.70	60
8590	20' high walls, 14 ga. x 6" wide, studs 12" OC	G	29	.552		50	30		80	103
8600	16" OC	G	42	.381		38.50	21		59.50	75
8610	24" OC	G	57	.281		27	15.35		42.35	54
8620	8" wide, studs 12" OC	G	28	.571		55.50	31.50		87	111
8630	16" OC	G	41	.390		43	21.50		64.50	81
8640	24" OC	G	56	.286		30	15.65		45.65	58
8790	12 ga. x 6" wide, studs 12" OC	G	27	.593		71	32.50		103.50	130
8800	16" OC	G	37	.432		54.50	23.50		78	97
8810	24" OC	G	51	.314		37.50	17.15		54.65	68.50
8820	8" wide, studs 12" OC	G	26	.615		83.50	33.50		117	146
8830	16" OC	G	36	.444		64	24.50		88.50	109
8840	24" OC	G	50	.320	Job	44	17.50		61.50	76.50
9000	Minimum labor/equipment charge		4	4			219		219	350

05 42 13 – Cold-Formed Metal Floor Joist Framing

05 42 13.05 Bracing

			Crew	Daily Output	Labor-Hours	Unit	Material	2021 Bare Costs Labor	Equipment	Total	Total Incl O&P
0010	**BRACING**, continuous, per row, top & bottom										
0015	Made from recycled materials										
0120	Flat strap, 20 ga. x 2" wide, joists at 12" OC	G	1 Carp	4.67	1.713	C.L.F.	62	93.50		155.50	218
0130	16" OC	G		5.33	1.501		60	82		142	197
0140	24" OC	G		6.66	1.201		58	65.50		123.50	169
0150	18 ga. x 2" wide, joists at 12" OC	G	1 Carp	4	2	C.L.F.	80.50	109		189.50	263
0160	16" OC	G		4.67	1.713		79	93.50		172.50	236
0170	24" OC	G		5.33	1.501		77.50	82		159.50	217

05 42 13.10 Bridging

			Crew	Daily Output	Labor-Hours	Unit	Material	2021 Bare Costs Labor	Equipment	Total	Total Incl O&P
0010	**BRIDGING**, solid between joists w/1-1/4" leg track, per joist bay										
0015	Made from recycled materials										
0230	Joists 12" OC, 18 ga. track x 6" wide	G	1 Carp	80	.100	Ea.	1.62	5.45		7.07	10.50
0240	8" wide	G		75	.107		1.99	5.85		7.84	11.50
0250	10" wide	G		70	.114		2.55	6.25		8.80	12.75

For customer support on your Commercial Renovation Costs with RSMeans Data, call 800.448.8182.

121

05 42 Cold-Formed Metal Joist Framing

05 42 13 – Cold-Formed Metal Floor Joist Framing

05 42 13.10 Bridging

			Crew	Daily Output	Labor-Hours	Unit	Material	2021 Bare Costs Labor	Equipment	Total	Total Incl O&P
0260	12" wide	G	1 Carp	65	.123	Ea.	2.83	6.75		9.58	13.80
0330	16 ga. track x 6" wide	G		70	.114		2.02	6.25		8.27	12.15
0340	8" wide	G		65	.123		2.48	6.75		9.23	13.45
0350	10" wide	G		60	.133		3.22	7.30		10.52	15.15
0360	12" wide	G		55	.145		3.69	7.95		11.64	16.70
0440	14 ga. track x 8" wide	G		60	.133		3.24	7.30		10.54	15.15
0450	10" wide	G		55	.145		3.94	7.95		11.89	17
0460	12" wide	G		50	.160		4.56	8.75		13.31	18.95
0550	12 ga. track x 10" wide	G		45	.178		5.75	9.70		15.45	22
0560	12" wide	G		40	.200		6.10	10.95		17.05	24
1230	16" OC, 18 ga. track x 6" wide	G		80	.100		2.08	5.45		7.53	11
1240	8" wide	G		75	.107		2.56	5.85		8.41	12.10
1250	10" wide	G		70	.114		3.27	6.25		9.52	13.55
1260	12" wide	G		65	.123		3.63	6.75		10.38	14.70
1330	16 ga. track x 6" wide	G		70	.114		2.59	6.25		8.84	12.80
1340	8" wide	G		65	.123		3.18	6.75		9.93	14.20
1350	10" wide	G		60	.133		4.13	7.30		11.43	16.15
1360	12" wide	G		55	.145		4.74	7.95		12.69	17.85
1440	14 ga. track x 8" wide	G		60	.133		4.16	7.30		11.46	16.15
1450	10" wide	G		55	.145		5.05	7.95		13	18.20
1460	12" wide	G		50	.160		5.85	8.75		14.60	20.50
1550	12 ga. track x 10" wide	G		45	.178		7.40	9.70		17.10	23.50
1560	12" wide	G		40	.200		7.80	10.95		18.75	26
2230	24" OC, 18 ga. track x 6" wide	G		80	.100		3.01	5.45		8.46	12
2240	8" wide	G		75	.107		3.70	5.85		9.55	13.35
2250	10" wide	G		70	.114		4.73	6.25		10.98	15.15
2260	12" wide	G		65	.123		5.25	6.75		12	16.50
2330	16 ga. track x 6" wide	G		70	.114		3.75	6.25		10	14.05
2340	8" wide	G		65	.123		4.61	6.75		11.36	15.75
2350	10" wide	G		60	.133		5.95	7.30		13.25	18.15
2360	12" wide	G		55	.145		6.85	7.95		14.80	20
2440	14 ga. track x 8" wide	G		60	.133		6	7.30		13.30	18.20
2450	10" wide	G		55	.145		7.30	7.95		15.25	20.50
2460	12" wide	G		50	.160		8.45	8.75		17.20	23.50
2550	12 ga. track x 10" wide	G	1 Carp	45	.178	Ea.	10.70	9.70		20.40	27.50
2560	12" wide	G	"	40	.200	"	11.30	10.95		22.25	30

05 42 13.25 Framing, Band Joist

			Crew	Daily Output	Labor-Hours	Unit	Material	2021 Bare Costs Labor	Equipment	Total	Total Incl O&P
0010	**FRAMING, BAND JOIST** (track) fastened to bearing wall										
0015	Made from recycled materials										
0220	18 ga. track x 6" deep	G	2 Carp	1000	.016	L.F.	1.32	.88		2.20	2.85
0230	8" deep	G		920	.017		1.63	.95		2.58	3.30
0240	10" deep	G		860	.019		2.08	1.02		3.10	3.91
0320	16 ga. track x 6" deep	G		900	.018		1.65	.97		2.62	3.36
0330	8" deep	G		840	.019		2.03	1.04		3.07	3.89
0340	10" deep	G		780	.021		2.63	1.12		3.75	4.68
0350	12" deep	G		740	.022		3.01	1.18		4.19	5.20
0430	14 ga. track x 8" deep	G		750	.021		2.65	1.17		3.82	4.77
0440	10" deep	G		720	.022		3.21	1.22		4.43	5.45
0450	12" deep	G		700	.023		3.72	1.25		4.97	6.10
0540	12 ga. track x 10" deep	G		670	.024		4.70	1.31		6.01	7.25
0550	12" deep	G		650	.025		4.98	1.35		6.33	7.60

05 42 13 – Cold-Formed Metal Floor Joist Framing

05 42 13.30 Framing, Boxed Headers/Beams

05 42 13.30 Framing, Boxed Headers/Beams		Crew	Daily Output	Labor-Hours	Unit	Material	2021 Bare Costs Labor	Equipment	Total	Total Incl O&P
0010	**FRAMING, BOXED HEADERS/BEAMS**									
0015	Made from recycled materials									
0200	Double, 18 ga. x 6" deep	G	2 Carp	220	.073	L.F.	5.80	3.98		9.78	12.75
0210	8" deep	G		210	.076		6	4.17		10.17	13.30
0220	10" deep	G		200	.080		7.70	4.38		12.08	15.40
0230	12" deep	G		190	.084		8.40	4.61		13.01	16.60
0300	16 ga. x 8" deep	G		180	.089		7	4.86		11.86	15.45
0310	10" deep	G		170	.094		8.65	5.15		13.80	17.70
0320	12" deep	G		160	.100		9.40	5.45		14.85	19
0400	14 ga. x 10" deep	G		140	.114		10.15	6.25		16.40	21
0410	12" deep	G		130	.123		11.15	6.75		17.90	23
0500	12 ga. x 10" deep	G		110	.145		13.40	7.95		21.35	27.50
0510	12" deep	G		100	.160		14.85	8.75		23.60	30.50
1210	Triple, 18 ga. x 8" deep	G		170	.094		9.10	5.15		14.25	18.20
1220	10" deep	G		165	.097		11	5.30		16.30	20.50
1230	12" deep	G		160	.100		12.05	5.45		17.50	22
1300	16 ga. x 8" deep	G		145	.110		10.55	6.05		16.60	21
1310	10" deep	G		140	.114		12.40	6.25		18.65	23.50
1320	12" deep	G		135	.119		13.50	6.50		20	25
1400	14 ga. x 10" deep	G		115	.139		14.65	7.60		22.25	28.50
1410	12" deep	G		110	.145		16.15	7.95		24.10	30.50
1500	12 ga. x 10" deep	G		90	.178		19.55	9.70		29.25	37
1510	12" deep	G		85	.188		21.50	10.30		31.80	40.50

05 42 13.40 Framing, Joists

05 42 13.40 Framing, Joists		Crew	Daily Output	Labor-Hours	Unit	Material	2021 Bare Costs Labor	Equipment	Total	Total Incl O&P
0010	**FRAMING, JOISTS**, no band joists (track), web stiffeners, headers,									
0020	Beams, bridging or bracing									
0025	Made from recycled materials									
0030	Joists (2" flange) and fasteners, materials only									
0220	18 ga. x 6" deep	G				L.F.	1.96			1.96	2.16
0230	8" deep	G					2.08			2.08	2.29
0240	10" deep	G					2.47			2.47	2.71
0320	16 ga. x 6" deep	G					2.12			2.12	2.33
0330	8" deep	G					2.59			2.59	2.85
0340	10" deep	G					2.96			2.96	3.26
0350	12" deep	G					3.35			3.35	3.68
0430	14 ga. x 8" deep	G					3.27			3.27	3.59
0440	10" deep	G					3.76			3.76	4.13
0450	12" deep	G					4.27			4.27	4.70
0540	12 ga. x 10" deep	G					5.45			5.45	6
0550	12" deep	G					6.20			6.20	6.85
1010	Installation of joists to band joists, beams & headers, labor only									
1220	18 ga. x 6" deep		2 Carp	110	.145	Ea.		7.95		7.95	12.65
1230	8" deep			90	.178			9.70		9.70	15.50
1240	10" deep			80	.200			10.95		10.95	17.40
1320	16 ga. x 6" deep			95	.168			9.20		9.20	14.65
1330	8" deep			70	.229			12.50		12.50	19.90
1340	10" deep			60	.267			14.60		14.60	23
1350	12" deep			55	.291			15.90		15.90	25.50
1430	14 ga. x 8" deep			65	.246			13.45		13.45	21.50
1440	10" deep			45	.356			19.45		19.45	31
1450	12" deep			35	.457			25		25	40
1540	12 ga. x 10" deep			40	.400			22		22	35

For customer support on your **Commercial Renovation Costs** with RSMeans Data, call 800.448.8182.

123

05 42 Cold-Formed Metal Joist Framing

05 42 13 – Cold-Formed Metal Floor Joist Framing

05 42 13.40 Framing, Joists		Crew	Daily Output	Labor-Hours	Unit	Material	2021 Bare Costs		Total	Total Incl O&P
							Labor	Equipment		
1550	12" deep	2 Carp	30	.533	Ea.		29		29	46.50
9000	Minimum labor/equipment charge	↓	4	4	Job		219		219	350

05 42 13.45 Framing, Web Stiffeners			Crew	Daily Output	Labor-Hours	Unit	Material	Labor	Equipment	Total	Total Incl O&P
0010	**FRAMING, WEB STIFFENERS** at joist bearing, fabricated from										
0020	Stud piece (1-5/8" flange) to stiffen joist (2" flange)										
0025	Made from recycled materials										
2120	For 6" deep joist, with 18 ga. x 2-1/2" stud	G	1 Carp	120	.067	Ea.	.83	3.65		4.48	6.70
2130	3-5/8" stud	G		110	.073		.99	3.98		4.97	7.45
2140	4" stud	G		105	.076		1.03	4.17		5.20	7.80
2150	6" stud	G		100	.080		1.30	4.38		5.68	8.40
2160	8" stud	G		95	.084		1.73	4.61		6.34	9.25
2220	8" deep joist, with 2-1/2" stud	G		120	.067		1.11	3.65		4.76	7
2230	3-5/8" stud	G		110	.073		1.33	3.98		5.31	7.80
2240	4" stud	G		105	.076		1.38	4.17		5.55	8.15
2250	6" stud	G		100	.080		1.74	4.38		6.12	8.85
2260	8" stud	G	↓	95	.084	↓	2.32	4.61		6.93	9.90
2320	10" deep joist, with 2-1/2" stud	G	1 Carp	110	.073	Ea.	1.38	3.98		5.36	7.85
2330	3-5/8" stud	G		100	.080		1.64	4.38		6.02	8.75
2340	4" stud	G		95	.084		1.71	4.61		6.32	9.25
2350	6" stud	G		90	.089		2.16	4.86		7.02	10.10
2360	8" stud	G		85	.094		2.87	5.15		8.02	11.35
2420	12" deep joist, with 2-1/2" stud	G		110	.073		1.66	3.98		5.64	8.20
2430	3-5/8" stud	G		100	.080		1.98	4.38		6.36	9.15
2440	4" stud	G		95	.084		2.06	4.61		6.67	9.60
2450	6" stud	G		90	.089		2.60	4.86		7.46	10.60
2460	8" stud	G		85	.094		3.46	5.15		8.61	12
3130	For 6" deep joist, with 16 ga. x 3-5/8" stud	G		100	.080		1.32	4.38		5.70	8.40
3140	4" stud	G		95	.084		1.46	4.61		6.07	8.95
3150	6" stud	G		90	.089		1.79	4.86		6.65	9.70
3160	8" stud	G		85	.094		2.15	5.15		7.30	10.55
3230	8" deep joist, with 3-5/8" stud	G		100	.080		1.77	4.38		6.15	8.90
3240	4" stud	G		95	.084		1.96	4.61		6.57	9.50
3250	6" stud	G		90	.089		2.40	4.86		7.26	10.40
3260	8" stud	G		85	.094		2.88	5.15		8.03	11.35
3330	10" deep joist, with 3-5/8" stud	G		85	.094		2.19	5.15		7.34	10.60
3340	4" stud	G		80	.100		2.42	5.45		7.87	11.35
3350	6" stud	G		75	.107		2.97	5.85		8.82	12.55
3360	8" stud	G		70	.114		3.57	6.25		9.82	13.90
3430	12" deep joist, with 3-5/8" stud	G		85	.094		2.64	5.15		7.79	11.10
3440	4" stud	G		80	.100		2.92	5.45		8.37	11.90
3450	6" stud	G		75	.107		3.58	5.85		9.43	13.25
3460	8" stud	G		70	.114		4.30	6.25		10.55	14.70
4230	For 8" deep joist, with 14 ga. x 3-5/8" stud	G		90	.089		2.33	4.86		7.19	10.30
4240	4" stud	G		85	.094		2.44	5.15		7.59	10.90
4250	6" stud	G		80	.100		3.08	5.45		8.53	12.10
4260	8" stud	G		75	.107		3.40	5.85		9.25	13.05
4330	10" deep joist, with 3-5/8" stud	G		75	.107		2.89	5.85		8.74	12.50
4340	4" stud	G		70	.114		3.02	6.25		9.27	13.25
4350	6" stud	G		65	.123		3.82	6.75		10.57	14.90
4360	8" stud	G		60	.133		4.22	7.30		11.52	16.25
4430	12" deep joist, with 3-5/8" stud	G		75	.107		3.48	5.85		9.33	13.15
4440	4" stud	G	↓	70	.114	↓	3.64	6.25		9.89	13.95

05 42 Cold-Formed Metal Joist Framing

05 42 13 – Cold-Formed Metal Floor Joist Framing

05 42 13.45 Framing, Web Stiffeners

		Crew	Daily Output	Labor-Hours	Unit	Material	2021 Bare Costs Labor	Equipment	Total	Total Incl O&P
4450	6" stud	G 1 Carp	65	.123	Ea.	4.60	6.75		11.35	15.75
4460	8" stud	G	60	.133		5.10	7.30		12.40	17.20
5330	For 10" deep joist, with 12 ga. x 3-5/8" stud	G	65	.123		4.13	6.75		10.88	15.25
5340	4" stud	G	60	.133		4.30	7.30		11.60	16.35
5350	6" stud	G	55	.145		5.60	7.95		13.55	18.80
5360	8" stud	G	50	.160		6.55	8.75		15.30	21
5430	12" deep joist, with 3-5/8" stud	G 1 Carp	65	.123	Ea.	4.98	6.75		11.73	16.20
5440	4" stud	G	60	.133		5.20	7.30		12.50	17.30
5450	6" stud	G	55	.145		6.70	7.95		14.65	20
5460	8" stud	G	50	.160		7.90	8.75		16.65	22.50

05 42 23 – Cold-Formed Metal Roof Joist Framing

05 42 23.05 Framing, Bracing

		Crew	Daily Output	Labor-Hours	Unit	Material	2021 Bare Costs Labor	Equipment	Total	Total Incl O&P
0010	**FRAMING, BRACING**									
0015	Made from recycled materials									
0020	Continuous bracing, per row									
0100	16 ga. x 1-1/2" channel thru rafters/trusses @ 16" OC	G 1 Carp	4.50	1.778	C.L.F.	55	97		152	216
0120	24" OC	G	6	1.333		55	73		128	177
0300	2" x 2" angle x 18 ga., rafters/trusses @ 16" OC	G	6	1.333		82.50	73		155.50	207
0320	24" OC	G	8	1		82.50	54.50		137	178
0400	16 ga., rafters/trusses @ 16" OC	G	4.50	1.778		103	97		200	269
0420	24" OC	G	6.50	1.231		103	67.50		170.50	221

05 42 23.10 Framing, Bridging

		Crew	Daily Output	Labor-Hours	Unit	Material	2021 Bare Costs Labor	Equipment	Total	Total Incl O&P
0010	**FRAMING, BRIDGING**									
0015	Made from recycled materials									
0020	Solid, between rafters w/1-1/4" leg track, per rafter bay									
1200	Rafters 16" OC, 18 ga. x 4" deep	G 1 Carp	60	.133	Ea.	1.42	7.30		8.72	13.15
1210	6" deep	G	57	.140		2.08	7.70		9.78	14.50
1220	8" deep	G	55	.145		2.56	7.95		10.51	15.45
1230	10" deep	G	52	.154		3.27	8.40		11.67	17
1240	12" deep	G	50	.160		3.63	8.75		12.38	17.95
2200	24" OC, 18 ga. x 4" deep	G	60	.133		2.05	7.30		9.35	13.85
2210	6" deep	G	57	.140		3.01	7.70		10.71	15.50
2220	8" deep	G	55	.145		3.70	7.95		11.65	16.70
2230	10" deep	G	52	.154		4.73	8.40		13.13	18.60
2240	12" deep	G	50	.160		5.25	8.75		14	19.75

05 42 23.50 Framing, Parapets

		Crew	Daily Output	Labor-Hours	Unit	Material	2021 Bare Costs Labor	Equipment	Total	Total Incl O&P
0010	**FRAMING, PARAPETS**									
0015	Made from recycled materials									
0100	3' high installed on 1st story, 18 ga. x 4" wide studs, 12" OC	G 2 Carp	100	.160	L.F.	5.20	8.75		13.95	19.65
0110	16" OC	G	150	.107		4.42	5.85		10.27	14.15
0120	24" OC	G	200	.080		3.65	4.38		8.03	10.95
0200	6" wide studs, 12" OC	G	100	.160		7	8.75		15.75	21.50
0210	16" OC	G	150	.107		6	5.85		11.85	15.90
0220	24" OC	G	200	.080		5.05	4.38		9.43	12.50
1100	Installed on 2nd story, 18 ga. x 4" wide studs, 12" OC	G	95	.168		5.20	9.20		14.40	20.50
1110	16" OC	G	145	.110		4.42	6.05		10.47	14.45
1120	24" OC	G	190	.084		3.65	4.61		8.26	11.35
1200	6" wide studs, 12" OC	G	95	.168		7	9.20		16.20	22.50
1210	16" OC	G	145	.110		6	6.05		12.05	16.20
1220	24" OC	G 2 Carp	190	.084	L.F.	5.05	4.61		9.66	12.90
2100	Installed on gable, 18 ga. x 4" wide studs, 12" OC	G	85	.188		5.20	10.30		15.50	22
2110	16" OC	G	130	.123		4.42	6.75		11.17	15.55

05 42 23 – Cold-Formed Metal Roof Joist Framing

05 42 23.50 Framing, Parapets		Crew	Daily Output	Labor-Hours	Unit	Material	2021 Bare Costs Labor	Equipment	Total	Total Incl O&P
2120	24" OC	2 Carp	170	.094	L.F.	3.65	5.15		8.80	12.20
2200	6" wide studs, 12" OC	G	85	.188		7	10.30		17.30	24
2210	16" OC	G	130	.123		6	6.75		12.75	17.30
2220	24" OC	G	170	.094		5.05	5.15		10.20	13.75

05 42 23.60 Framing, Roof Rafters

		Crew	Daily Output	Labor-Hours	Unit	Material	2021 Bare Costs Labor	Equipment	Total	Total Incl O&P
0010	**FRAMING, ROOF RAFTERS**									
0015	Made from recycled materials									
0100	Boxed ridge beam, double, 18 ga. x 6" deep	G 2 Carp	160	.100	L.F.	5.80	5.45		11.25	15.10
0110	8" deep	G	150	.107		6	5.85		11.85	15.95
0120	10" deep	G	140	.114		7.70	6.25		13.95	18.40
0130	12" deep	G	130	.123		8.40	6.75		15.15	19.95
0200	16 ga. x 6" deep	G	150	.107		6.10	5.85		11.95	16
0210	8" deep	G	140	.114		7	6.25		13.25	17.65
0220	10" deep	G	130	.123		8.65	6.75		15.40	20
0230	12" deep	G	120	.133		9.40	7.30		16.70	22
1100	Rafters, 2" flange, material only, 18 ga. x 6" deep	G				1.96			1.96	2.16
1110	8" deep	G				2.08			2.08	2.29
1120	10" deep	G				2.47			2.47	2.71
1130	12" deep	G				2.85			2.85	3.13
1200	16 ga. x 6" deep	G				2.12			2.12	2.33
1210	8" deep	G				2.59			2.59	2.85
1220	10" deep	G				2.96			2.96	3.26
1230	12" deep	G				3.35			3.35	3.68
2100	Installation only, ordinary rafter to 4:12 pitch, 18 ga. x 6" deep	2 Carp	35	.457	Ea.		25		25	40
2110	8" deep		30	.533			29		29	46.50
2120	10" deep		25	.640			35		35	55.50
2130	12" deep		20	.800			44		44	69.50
2200	16 ga. x 6" deep		30	.533			29		29	46.50
2210	8" deep		25	.640			35		35	55.50
2220	10" deep		20	.800			44		44	69.50
2230	12" deep		15	1.067			58.50		58.50	93
8100	Add to labor, ordinary rafters on steep roofs						25%			
8110	Dormers & complex roofs						50%			
8200	Hip & valley rafters to 4:12 pitch						25%			
8210	Steep roofs						50%			
8220	Dormers & complex roofs						75%			
8300	Hip & valley jack rafters to 4:12 pitch						50%			
8310	Steep roofs						75%			
8320	Dormers & complex roofs						100%			
9000	Minimum labor/equipment charge	2 Carp	4	4	Job		219		219	350

05 42 23.70 Framing, Soffits and Canopies

		Crew	Daily Output	Labor-Hours	Unit	Material	2021 Bare Costs Labor	Equipment	Total	Total Incl O&P
0010	**FRAMING, SOFFITS & CANOPIES**									
0015	Made from recycled materials									
0130	Continuous ledger track @ wall, studs @ 16" OC, 18 ga. x 4" wide	G 2 Carp	535	.030	L.F.	.95	1.64		2.59	3.64
0140	6" wide	G	500	.032		1.39	1.75		3.14	4.31
0150	8" wide	G	465	.034		1.71	1.88		3.59	4.88
0160	10" wide	G	430	.037		2.18	2.04		4.22	5.65
0230	Studs @ 24" OC, 18 ga. x 4" wide	G	800	.020		.90	1.09		1.99	2.73
0240	6" wide	G	750	.021		1.32	1.17		2.49	3.32
0250	8" wide	G	700	.023		1.63	1.25		2.88	3.78
0260	10" wide	G	650	.025		2.08	1.35		3.43	4.43
1000	Horizontal soffit and canopy members, material only									

05 42 Cold-Formed Metal Joist Framing

05 42 23 – Cold-Formed Metal Roof Joist Framing

05 42 23.70 Framing, Soffits and Canopies

			Crew	Daily Output	Labor-Hours	Unit	Material	2021 Bare Costs Labor	Equipment	Total	Total Incl O&P
1030	1-5/8" flange studs, 18 ga. x 4" deep	G				L.F.	1.24			1.24	1.36
1040	6" deep	G					1.56			1.56	1.72
1050	8" deep	G					2.08			2.08	2.28
1140	2" flange joists, 18 ga. x 6" deep	G					2.24			2.24	2.47
1150	8" deep	G					2.38			2.38	2.61
1160	10" deep	G					2.82			2.82	3.10
4030	Installation only, 18 ga., 1-5/8" flange x 4" deep		2 Carp	130	.123	Ea.		6.75		6.75	10.70
4040	6" deep			110	.145			7.95		7.95	12.65
4050	8" deep			90	.178			9.70		9.70	15.50
4140	2" flange, 18 ga. x 6" deep			110	.145			7.95		7.95	12.65
4150	8" deep			90	.178			9.70		9.70	15.50
4160	10" deep			80	.200			10.95		10.95	17.40
6010	Clips to attach fascia to rafter tails, 2" x 2" x 18 ga. angle	G	1 Carp	120	.067		.98	3.65		4.63	6.85
6020	16 ga. angle	G	"	100	.080		1.22	4.38		5.60	8.30
9000	Minimum labor/equipment charge		2 Carp	4	4	Job		219		219	350

05 44 Cold-Formed Metal Trusses

05 44 13 – Cold-Formed Metal Roof Trusses

05 44 13.60 Framing, Roof Trusses

			Crew	Daily Output	Labor-Hours	Unit	Material	2021 Bare Costs Labor	Equipment	Total	Total Incl O&P
0010	**FRAMING, ROOF TRUSSES**										
0015	Made from recycled materials										
0020	Fabrication of trusses on ground, Fink (W) or King Post, to 4:12 pitch										
0120	18 ga. x 4" chords, 16' span	G	2 Carp	12	1.333	Ea.	57.50	73		130.50	180
0130	20' span	G		11	1.455		72	79.50		151.50	207
0140	24' span	G		11	1.455		86.50	79.50		166	222
0150	28' span	G		10	1.600		101	87.50		188.50	250
0160	32' span	G		10	1.600		115	87.50		202.50	266
0250	6" chords, 28' span	G		9	1.778		127	97		224	295
0260	32' span	G		9	1.778		146	97		243	315
0270	36' span	G		8	2		164	109		273	355
0280	40' span	G	2 Carp	8	2	Ea.	182	109		291	375
1120	5:12 to 8:12 pitch, 18 ga. x 4" chords, 16' span	G		10	1.600		66	87.50		153.50	212
1130	20' span	G		9	1.778		82.50	97		179.50	246
1140	24' span	G		9	1.778		99	97		196	264
1150	28' span	G		8	2		115	109		224	300
1160	32' span	G		8	2		132	109		241	320
1250	6" chords, 28' span	G		7	2.286		146	125		271	360
1260	32' span	G		7	2.286		166	125		291	380
1270	36' span	G		6	2.667		187	146		333	440
1280	40' span	G		6	2.667		208	146		354	460
2120	9:12 to 12:12 pitch, 18 ga. x 4" chords, 16' span	G		8	2		82.50	109		191.50	265
2130	20' span	G		7	2.286		103	125		228	310
2140	24' span	G		7	2.286		124	125		249	335
2150	28' span	G		6	2.667		144	146		290	390
2160	32' span	G		6	2.667		165	146		311	415
2250	6" chords, 28' span	G		5	3.200		182	175		357	480
2260	32' span	G		5	3.200		208	175		383	510
2270	36' span	G		4	4		234	219		453	605
2280	40' span	G		4	4		260	219		479	635
4900	Minimum labor/equipment charge			4	4	Job		219		219	350
5120	Erection only of roof trusses, to 4:12 pitch, 16' span		F-6	48	.833	Ea.		43.50	9.90	53.40	79.50

05 44 Cold-Formed Metal Trusses

05 44 13 – Cold-Formed Metal Roof Trusses

05 44 13.60 Framing, Roof Trusses

		Crew	Daily Output	Labor-Hours	Unit	Material	2021 Bare Costs Labor	Equipment	Total	Total Incl O&P
5130	20' span	F-6	46	.870	Ea.		45	10.35	55.35	83
5140	24' span		44	.909			47	10.80	57.80	87
5150	28' span		42	.952			49.50	11.35	60.85	91
5160	32' span		40	1			52	11.90	63.90	95.50
5170	36' span		38	1.053			54.50	12.55	67.05	100
5180	40' span		36	1.111			57.50	13.20	70.70	106
5220	5:12 to 8:12 pitch, 16' span		42	.952			49.50	11.35	60.85	91
5230	20' span		40	1			52	11.90	63.90	95.50
5240	24' span		38	1.053			54.50	12.55	67.05	100
5250	28' span		36	1.111			57.50	13.20	70.70	106
5260	32' span		34	1.176			61	14	75	112
5270	36' span		32	1.250			65	14.90	79.90	119
5280	40' span		30	1.333			69	15.85	84.85	127
5320	9:12 to 12:12 pitch, 16' span		36	1.111			57.50	13.20	70.70	106
5330	20' span		34	1.176			61	14	75	112
5340	24' span		32	1.250			65	14.90	79.90	119
5350	28' span		30	1.333			69	15.85	84.85	127
5360	32' span		28	1.429			74	17	91	137
5370	36' span		26	1.538			80	18.30	98.30	147
5380	40' span		24	1.667			86.50	19.85	106.35	159
9000	Minimum labor/equipment charge		2	20	Job		1,050	238	1,288	1,900

05 51 Metal Stairs

05 51 13 – Metal Pan Stairs

05 51 13.50 Pan Stairs

			Crew	Daily Output	Labor-Hours	Unit	Material	2021 Bare Costs Labor	Equipment	Total	Total Incl O&P
0010	**PAN STAIRS**, shop fabricated, steel stringers										
0015	Made from recycled materials										
0200	Metal pan tread for concrete in-fill, picket rail, 3'-6" wide	G	E-4	35	.914	Riser	600	55.50	4.25	659.75	755
0300	4'-0" wide	G		30	1.067		590	65	4.96	659.96	760
0350	Wall rail, both sides, 3'-6" wide	G		53	.604		515	36.50	2.81	554.31	630
1500	Landing, steel pan, conventional	G		160	.200	S.F.	79.50	12.15	.93	92.58	108

05 51 16 – Metal Floor Plate Stairs

05 51 16.50 Floor Plate Stairs

			Crew	Daily Output	Labor-Hours	Unit	Material	2021 Bare Costs Labor	Equipment	Total	Total Incl O&P
0010	**FLOOR PLATE STAIRS**, shop fabricated, steel stringers										
0015	Made from recycled materials										
0400	Cast iron tread and pipe rail, 3'-6" wide	G	E-4	35	.914	Riser	565	55.50	4.25	624.75	720
0450	4'-0" wide	G		30	1.067		595	65	4.96	664.96	765
0475	5'-0" wide	G		28	1.143		655	69.50	5.30	729.80	840
0500	Checkered plate tread, industrial, 3'-6" wide	G		28	1.143		325	69.50	5.30	399.80	480
0550	Circular, for tanks, 3'-0" wide	G		33	.970		370	59	4.51	433.51	505
0600	For isolated stairs, add							100%			
0800	Custom steel stairs, 3'-6" wide, economy	G	E-4	35	.914		495	55.50	4.25	554.75	640
0810	Medium priced	G		30	1.067		655	65	4.96	724.96	830
0900	Deluxe	G		20	1.600		820	97.50	7.45	924.95	1,075
1100	For 4' wide stairs, add							5%	5%		
1300	For 5' wide stairs, add							10%	10%		

05 51 Metal Stairs

05 51 19 – Metal Grating Stairs

05 51 19.50 Grating Stairs

		Crew	Daily Output	Labor-Hours	Unit	Material	2021 Bare Costs Labor	Equipment	Total	Total Incl O&P
0010	**GRATING STAIRS**, shop fabricated, steel stringers, safety nosing on treads									
0015	Made from recycled materials									
0020	Grating tread and pipe railing, 3'-6" wide	G	E-4	35	.914	Riser	390	55.50	4.25	449.75	525
0100	4'-0" wide	G		30	1.067	"	385	65	4.96	454.96	535
9000	Minimum labor/equipment charge			2	16	Job		975	74.50	1,049.50	1,650

05 51 23 – Metal Fire Escapes

05 51 23.25 Fire Escapes

		Crew	Daily Output	Labor-Hours	Unit	Material	Labor	Equipment	Total	Total Incl O&P
0010	**FIRE ESCAPES**, shop fabricated									
0200	2' wide balcony, 1" x 1/4" bars 1-1/2" OC, with railing	G	2 Sswk	10	1.600	L.F.	58	96.50		154.50	220
0400	1st story cantilevered stair, standard, with railing	G		.50	32	Ea.	2,350	1,925		4,275	5,700
0500	Cable counterweighted, with railing	G		.40	40	"	2,250	2,400		4,650	6,375
0700	36" x 40" platform & fixed stair, with railing	G		.40	40	Flight	1,025	2,400		3,425	5,050
0900	For 3'-6" wide escapes, add to above						100%	150%			

05 51 23.50 Fire Escape Stairs

		Crew	Daily Output	Labor-Hours	Unit	Material	Labor	Equipment	Total	Total Incl O&P
0010	**FIRE ESCAPE STAIRS**, portable									
0100	Portable ladder					Ea.	119			119	131

05 51 33 – Metal Ladders

05 51 33.13 Vertical Metal Ladders

		Crew	Daily Output	Labor-Hours	Unit	Material	Labor	Equipment	Total	Total Incl O&P
0010	**VERTICAL METAL LADDERS**, shop fabricated									
0015	Made from recycled materials									
0020	Steel, 20" wide, bolted to concrete, with cage	G	E-4	50	.640	V.L.F.	107	39	2.98	148.98	183
0100	Without cage	G		85	.376		42	23	1.75	66.75	85
0300	Aluminum, bolted to concrete, with cage	G		50	.640		106	39	2.98	147.98	183
0400	Without cage	G		85	.376		51	23	1.75	75.75	95
9000	Minimum labor/equipment charge			2	16	Job		975	74.50	1,049.50	1,650

05 51 33.16 Inclined Metal Ladders

		Crew	Daily Output	Labor-Hours	Unit	Material	Labor	Equipment	Total	Total Incl O&P
0010	**INCLINED METAL LADDERS**, shop fabricated									
0015	Made from recycled materials									
3900	Industrial ships ladder, steel, 24" W, grating treads, 2 line pipe rail	G	E-4	30	1.067	Riser	181	65	4.96	250.96	310
4000	Aluminum	G	"	30	1.067	"	310	65	4.96	379.96	450

05 51 33.23 Alternating Tread Ladders

		Crew	Daily Output	Labor-Hours	Unit	Material	Labor	Equipment	Total	Total Incl O&P
0010	**ALTERNATING TREAD LADDERS**, shop fabricated									
0015	Made from recycled materials									
0800	Alternating tread ladders, 68° angle of incline									
0810	8' vertical rise, steel, 149 lb., standard paint color	B-68G	3	5.333	Ea.	1,875	320	94.50	2,289.50	2,675
0820	Non-standard paint color		3	5.333		2,150	320	94.50	2,564.50	3,000
0830	Galvanized		3	5.333		2,175	320	94.50	2,589.50	3,025
0840	Stainless		3	5.333		3,175	320	94.50	3,589.50	4,100
0850	Aluminum, 87 lb.		3	5.333		2,325	320	94.50	2,739.50	3,175
1010	10' vertical rise, steel, 181 lb., standard paint color		2.75	5.818		2,275	350	103	2,728	3,175
1020	Non-standard paint color		2.75	5.818		2,575	350	103	3,028	3,525
1030	Galvanized		2.75	5.818		2,650	350	103	3,103	3,575
1040	Stainless		2.75	5.818		3,825	350	103	4,278	4,875
1050	Aluminum, 103 lb.		2.75	5.818		2,800	350	103	3,253	3,750
1210	12' vertical rise, steel, 245 lb., standard paint color		2.50	6.400		2,650	385	113	3,148	3,675
1220	Non-standard paint color		2.50	6.400		3,000	385	113	3,498	4,050
1230	Galvanized		2.50	6.400		3,100	385	113	3,598	4,150
1240	Stainless		2.50	6.400		4,450	385	113	4,948	5,650
1250	Aluminum, 103 lb.		2.50	6.400		3,275	385	113	3,773	4,350
1410	14' vertical rise, steel, 281 lb., standard paint color		2.25	7.111		3,050	430	126	3,606	4,175

05 51 Metal Stairs

05 51 33 – Metal Ladders

05 51 33.23 Alternating Tread Ladders

		Crew	Daily Output	Labor-Hours	Unit	Material	2021 Bare Costs Labor	Equipment	Total	Total Incl O&P
1420	Non-standard paint color	B-68G	2.25	7.111	Ea.	3,425	430	126	3,981	4,600
1430	Galvanized		2.25	7.111		3,550	430	126	4,106	4,750
1440	Stainless		2.25	7.111		5,100	430	126	5,656	6,450
1450	Aluminum, 136 lb.		2.25	7.111		3,750	430	126	4,306	4,925
1610	16' vertical rise, steel, 317 lb., standard paint color		2	8		3,450	480	142	4,072	4,725
1620	Non-standard paint color		2	8		3,850	480	142	4,472	5,175
1630	Galvanized		2	8		4,025	480	142	4,647	5,350
1640	Stainless		2	8		5,750	480	142	6,372	7,250
1650	Aluminum, 153 lb.		2	8		4,225	480	142	4,847	5,550

05 52 Metal Railings

05 52 13 – Pipe and Tube Railings

05 52 13.50 Railings, Pipe

			Crew	Daily Output	Labor-Hours	Unit	Material	2021 Bare Costs Labor	Equipment	Total	Total Incl O&P
0010	**RAILINGS, PIPE**, shop fab'd, 3'-6" high, posts @ 5' OC										
0015	Made from recycled materials										
0020	Aluminum, 2 rail, satin finish, 1-1/4" diameter	G	E-4	160	.200	L.F.	50.50	12.15	.93	63.58	76.50
0030	Clear anodized	G		160	.200		60	12.15	.93	73.08	86.50
0040	Dark anodized	G		160	.200		67.50	12.15	.93	80.58	95
0080	1-1/2" diameter, satin finish	G		160	.200		57.50	12.15	.93	70.58	83.50
0090	Clear anodized	G		160	.200		64.50	12.15	.93	77.58	91.50
0100	Dark anodized	G		160	.200		71	12.15	.93	84.08	98.50
0140	Aluminum, 3 rail, 1-1/4" diam., satin finish	G		137	.234		65	14.20	1.09	80.29	95.50
0150	Clear anodized	G		137	.234		90	14.20	1.09	105.29	123
0160	Dark anodized	G		137	.234		89	14.20	1.09	104.29	122
0200	1-1/2" diameter, satin finish	G		137	.234		85.50	14.20	1.09	100.79	118
0210	Clear anodized	G		137	.234		87.50	14.20	1.09	102.79	121
0220	Dark anodized	G		137	.234		107	14.20	1.09	122.29	142
0500	Steel, 2 rail, on stairs, primed, 1-1/4" diameter	G		160	.200		36	12.15	.93	49.08	60.50
0520	1-1/2" diameter	G		160	.200		31	12.15	.93	44.08	54.50
0540	Galvanized, 1-1/4" diameter	G		160	.200		46.50	12.15	.93	59.58	71.50
0560	1-1/2" diameter	G		160	.200		55.50	12.15	.93	68.58	81.50
0580	Steel, 3 rail, primed, 1-1/4" diameter	G		137	.234		51.50	14.20	1.09	66.79	80.50
0600	1-1/2" diameter	G		137	.234		56.50	14.20	1.09	71.79	86.50
0620	Galvanized, 1-1/4" diameter	G		137	.234		72	14.20	1.09	87.29	104
0640	1-1/2" diameter	G		137	.234		87	14.20	1.09	102.29	120
0700	Stainless steel, 2 rail, 1-1/4" diam., #4 finish	G		137	.234		146	14.20	1.09	161.29	185
0720	High polish	G		137	.234		236	14.20	1.09	251.29	284
0740	Mirror polish	G		137	.234		295	14.20	1.09	310.29	350
0760	Stainless steel, 3 rail, 1-1/2" diam., #4 finish	G		120	.267		219	16.20	1.24	236.44	268
0770	High polish	G		120	.267		365	16.20	1.24	382.44	425
0780	Mirror finish	G		120	.267		445	16.20	1.24	462.44	515
0900	Wall rail, alum. pipe, 1-1/4" diam., satin finish	G		213	.150		24	9.15	.70	33.85	42
0905	Clear anodized	G		213	.150		33.50	9.15	.70	43.35	52.50
0910	Dark anodized	G		213	.150		39	9.15	.70	48.85	58.50
0915	1-1/2" diameter, satin finish	G		213	.150		29.50	9.15	.70	39.35	48
0920	Clear anodized	G		213	.150		37	9.15	.70	46.85	56.50
0925	Dark anodized	G		213	.150		47	9.15	.70	56.85	67
0930	Steel pipe, 1-1/4" diameter, primed	G		213	.150		21	9.15	.70	30.85	38.50
0935	Galvanized	G		213	.150		30	9.15	.70	39.85	48.50
0940	1-1/2" diameter	G		176	.182		17.60	11.05	.85	29.50	38
0945	Galvanized	G	E-4	213	.150	L.F.	30	9.15	.70	39.85	48.50

130

05 52 Metal Railings

05 52 13 – Pipe and Tube Railings

05 52 13.50 Railings, Pipe

		Crew	Daily Output	Labor-Hours	Unit	Material	2021 Bare Costs Labor	Equipment	Total	Total Incl O&P
0955	Stainless steel pipe, 1-1/2" diam., #4 finish	G E-4	107	.299	L.F.	117	18.20	1.39	136.59	160
0960	High polish	G	107	.299		238	18.20	1.39	257.59	292
0965	Mirror polish	G	107	.299		281	18.20	1.39	300.59	340
2000	2-line pipe rail (1-1/2" T&B) with 1/2" pickets @ 4-1/2" OC,									
2005	attached handrail on brackets									
2010	42" high aluminum, satin finish, straight & level	G E-4	120	.267	L.F.	390	16.20	1.24	407.44	455
2050	42" high steel, primed, straight & level	G "	120	.267		148	16.20	1.24	165.44	189
4000	For curved and level rails, add						10%	10%		
4100	For sloped rails for stairs, add						30%	30%		
9000	Minimum labor/equipment charge	1 Sswk	2	4	Job		241		241	390

05 52 16 – Industrial Railings

05 52 16.50 Railings, Industrial

		Crew	Daily Output	Labor-Hours	Unit	Material	2021 Bare Costs Labor	Equipment	Total	Total Incl O&P
0010	**RAILINGS, INDUSTRIAL**, shop fab'd, 3'-6" high, posts @ 5' OC									
0020	2 rail, 3'-6" high, 1-1/2" pipe	G E-4	255	.125	L.F.	40.50	7.65	.58	48.73	57.50
0200	For 4" high kick plate, 10 ga., add	G				5.70			5.70	6.30
0500	For curved level rails, add					10%	10%			
0550	For sloped rails for stairs, add					30%	30%			
9000	Minimum labor/equipment charge	1 Sswk	2	4	Job		241		241	390

05 55 Metal Stair Treads and Nosings

05 55 13 – Metal Stair Treads

05 55 13.50 Stair Treads

		Crew	Daily Output	Labor-Hours	Unit	Material	2021 Bare Costs Labor	Equipment	Total	Total Incl O&P
0010	**STAIR TREADS**, stringers and bolts not included									
3000	Diamond plate treads, steel, 1/8" thick									
3005	Open riser, black enamel									
3010	9" deep x 36" long	G 2 Sswk	48	.333	Ea.	114	20		134	159
3020	42" long	G	48	.333		121	20		141	166
3030	48" long	G	48	.333		126	20		146	171
3040	11" deep x 36" long	G	44	.364		121	22		143	169
3050	42" long	G	44	.364		128	22		150	176
3060	48" long	G	44	.364		136	22		158	185
3110	Galvanized, 9" deep x 36" long	G	48	.333		181	20		201	232
3120	42" long	G	48	.333		192	20		212	244
3130	48" long	G	48	.333		204	20		224	257
3140	11" deep x 36" long	G	44	.364		187	22		209	242
3150	42" long	G	44	.364		204	22		226	260
3160	48" long	G	44	.364		212	22		234	269
3200	Closed riser, black enamel									
3210	12" deep x 36" long	G 2 Sswk	40	.400	Ea.	138	24		162	191
3220	42" long	G	40	.400		150	24		174	204
3230	48" long	G	40	.400		158	24		182	213
3240	Galvanized, 12" deep x 36" long	G 2 Sswk	40	.400	Ea.	218	24		242	279
3250	42" long	G	40	.400		238	24		262	300
3260	48" long	G	40	.400		249	24		273	315
4000	Bar grating treads									
4005	Steel, 1-1/4" x 3/16" bars, anti-skid nosing, black enamel									
4010	8-5/8" deep x 30" long	G 2 Sswk	48	.333	Ea.	56	20		76	94
4020	36" long	G	48	.333		65.50	20		85.50	105
4030	48" long	G	48	.333		101	20		121	144
4040	10-15/16" deep x 36" long	G	44	.364		73	22		95	116

05 55 13 – Metal Stair Treads

05 55 13.50 Stair Treads		Crew	Daily Output	Labor-Hours	Unit	Material	2021 Bare Costs Labor	2021 Bare Costs Equipment	Total	Total Incl O&P
4050	48" long	G	2 Sswk	44	.364	Ea.	103	22		125	149
4060	Galvanized, 8-5/8" deep x 30" long	G		48	.333		64.50	20		84.50	104
4070	36" long	G		48	.333		76.50	20		96.50	117
4080	48" long	G		48	.333		112	20		132	156
4090	10-15/16" deep x 36" long	G		44	.364		89.50	22		111.50	134
4100	48" long	G		44	.364		116	22		138	163
4200	Aluminum, 1-1/4" x 3/16" bars, serrated, with nosing										
4210	7-5/8" deep x 18" long	G	2 Sswk	52	.308	Ea.	54	18.55		72.55	89.50
4220	24" long	G		52	.308		62	18.55		80.55	98.50
4230	30" long	G		52	.308		72	18.55		90.55	109
4240	36" long	G		52	.308		186	18.55		204.55	235
4250	8-13/16" deep x 18" long	G		48	.333		71	20		91	111
4260	24" long	G		48	.333		117	20		137	162
4270	30" long	G		48	.333		117	20		137	162
4280	36" long	G		48	.333		205	20		225	258
4290	10" deep x 18" long	G		44	.364		137	22		159	187
4300	30" long	G		44	.364		182	22		204	236
4310	36" long	G		44	.364		221	22		243	279
5000	Channel grating treads										
5005	Steel, 14 ga., 2-1/2" thick, galvanized										
5010	9" deep x 36" long	G	2 Sswk	48	.333	Ea.	103	20		123	146
5020	48" long	G		48	.333	"	137	20		157	183
9000	Minimum labor/equipment charge			4	4	Job		241		241	390

05 55 19 – Metal Stair Tread Covers

05 55 19.50 Stair Tread Covers for Renovation

		Crew	Daily Output	Labor-Hours	Unit	Material	2021 Bare Costs Labor	2021 Bare Costs Equipment	Total	Total Incl O&P
0010	**STAIR TREAD COVERS FOR RENOVATION**										
0205	Extruded tread cover with nosing, pre-drilled, includes screws										
0210	Aluminum with black abrasive strips, 9" wide x 3' long	1 Carp	24	.333	Ea.	105	18.25		123.25	144
0220	4' long		22	.364		137	19.90		156.90	183
0230	5' long		20	.400		186	22		208	240
0240	11" wide x 3' long		24	.333		147	18.25		165.25	191
0250	4' long		22	.364		201	19.90		220.90	253
0260	5' long	1 Carp	20	.400	Ea.	240	22		262	299
0305	Black abrasive strips with yellow front strips										
0310	Aluminum, 9" wide x 3' long	1 Carp	24	.333	Ea.	127	18.25		145.25	169
0320	4' long		22	.364		170	19.90		189.90	219
0330	5' long		20	.400		218	22		240	274
0340	11" wide x 3' long		24	.333		162	18.25		180.25	207
0350	4' long		22	.364		206	19.90		225.90	259
0360	5' long		20	.400		269	22		291	330
0405	Black abrasive strips with photoluminescent front strips										
0410	Aluminum, 9" wide x 3' long	1 Carp	24	.333	Ea.	161	18.25		179.25	206
0420	4' long		22	.364		182	19.90		201.90	233
0430	5' long		20	.400		256	22		278	315
0440	11" wide x 3' long		24	.333		182	18.25		200.25	230
0450	4' long		22	.364		243	19.90		262.90	299
0460	5' long		20	.400		305	22		327	370
9000	Minimum labor/equipment charge		4	2	Job		109		109	174

05 56 Metal Castings

05 56 13 – Metal Construction Castings

05 56 13.50 Construction Castings		Crew	Daily Output	Labor-Hours	Unit	Material	2021 Bare Costs Labor	Equipment	Total	Total Incl O&P
0010	**CONSTRUCTION CASTINGS**									
0020	Manhole covers and frames, see Section 33 44 13.13									
0100	Column bases, cast iron, 16" x 16", approx. 65 lb.	G	E-4	46	.696	Ea.	139	42.50	3.23	184.73	225
0200	32" x 32", approx. 256 lb.	G		23	1.391	"	505	84.50	6.45	595.95	700
0600	Miscellaneous C.I. castings, light sections, less than 150 lb.	G		3200	.010	Lb.	9.05	.61	.05	9.71	11.05
1300	Special low volume items	G		3200	.010	"	11.25	.61	.05	11.91	13.45

05 58 Formed Metal Fabrications

05 58 13 – Column Covers

05 58 13.05 Column Covers

			Crew	Daily Output	Labor-Hours	Unit	Material	2021 Bare Costs Labor	Equipment	Total	Total Incl O&P
0010	**COLUMN COVERS**										
0015	Made from recycled materials										
0020	Excludes structural steel, light ga. metal framing, misc. metals, sealants										
0100	Round covers, 2 halves with 2 vertical joints for backer rod and sealant										
0110	Up to 12' high, no horizontal joints										
0120	12" diameter, 0.125" aluminum, anodized/painted finish	G	2 Sswk	32	.500	V.L.F.	33	30		63	85
0130	Type 304 stainless steel, 16 gauge, #4 brushed finish	G		32	.500		48	30		78	102
0140	Type 316 stainless steel, 16 gauge, #4 brushed finish	G		32	.500		57.50	30		87.50	113
0150	18" diameter, aluminum	G		32	.500		49.50	30		79.50	104
0160	Type 304 stainless steel	G		32	.500		71.50	30		101.50	128
0170	Type 316 stainless steel	G		32	.500		86.50	30		116.50	144
0180	24" diameter, aluminum	G		32	.500		66	30		96	122
0190	Type 304 stainless steel	G	2 Sswk	32	.500	V.L.F.	95.50	30		125.50	154
0200	Type 316 stainless steel	G		32	.500		115	30		145	176
0210	30" diameter, aluminum	G		30	.533		82.50	32		114.50	143
0220	Type 304 stainless steel	G		30	.533		120	32		152	183
0230	Type 316 stainless steel	G		30	.533		144	32		176	211
0240	36" diameter, aluminum	G		30	.533		99	32		131	161
0250	Type 304 stainless steel	G		30	.533		143	32		175	210
0260	Type 316 stainless steel	G		30	.533		173	32		205	242
0400	Up to 24' high, 2 stacked sections with 1 horizontal joint										
0410	18" diameter, aluminum	G	2 Sswk	28	.571	V.L.F.	52	34.50		86.50	113
0450	Type 304 stainless steel	G		28	.571		75.50	34.50		110	139
0460	Type 316 stainless steel	G		28	.571		91	34.50		125.50	156
0470	24" diameter, aluminum	G		28	.571		69	34.50		103.50	132
0480	Type 304 stainless steel	G		28	.571		100	34.50		134.50	166
0490	Type 316 stainless steel	G		28	.571		121	34.50		155.50	189
0500	30" diameter, aluminum	G		24	.667		86.50	40		126.50	160
0510	Type 304 stainless steel	G		24	.667		126	40		166	203
0520	Type 316 stainless steel	G		24	.667		151	40		191	232
0530	36" diameter, aluminum	G		24	.667		104	40		144	179
0540	Type 304 stainless steel	G		24	.667		151	40		191	231
0550	Type 316 stainless steel	G		24	.667		182	40		222	265

05 58 23 – Formed Metal Guards

05 58 23.90 Window Guards

			Crew	Daily Output	Labor-Hours	Unit	Material	2021 Bare Costs Labor	Equipment	Total	Total Incl O&P
0010	**WINDOW GUARDS**, shop fabricated										
0015	Expanded metal, steel angle frame, permanent	G	E-4	350	.091	S.F.	23	5.55	.43	28.98	34.50
0025	Steel bars, 1/2" x 1/2", spaced 5" OC	G	"	290	.110	"	15.90	6.70	.51	23.11	29
0030	Hinge mounted, add	G				Opng.	46			46	50.50
0040	Removable type, add	G				"	29			29	32

05 58 Formed Metal Fabrications

05 58 23 – Formed Metal Guards

05 58 23.90 Window Guards

		Crew	Daily Output	Labor-Hours	Unit	Material	2021 Bare Costs Labor	Equipment	Total	Total Incl O&P
0050	For galvanized guards, add				S.F.	35%				
0070	For pivoted or projected type, add					105%	40%			
0100	Mild steel, stock units, economy	G	E-4	405	.079		6.25	4.80	.37	11.42	15.05
0200	Deluxe	G		405	.079		12.70	4.80	.37	17.87	22
0400	Woven wire, stock units, 3/8" channel frame, 3' x 5' opening	G		40	.800	Opng.	169	48.50	3.72	221.22	269
0500	4' x 6' opening	G		38	.842		270	51	3.92	324.92	385
0800	Basket guards for above, add	G					241			241	266
1000	Swinging guards for above, add	G					82.50			82.50	90.50
9000	Minimum labor/equipment charge		1 Sswk	2	4	Job		241		241	390

05 58 25 – Formed Lamp Posts

05 58 25.40 Lamp Posts

		Crew	Daily Output	Labor-Hours	Unit	Material	2021 Bare Costs Labor	Equipment	Total	Total Incl O&P
0010	**LAMP POSTS**									
0020	Aluminum, 7' high, stock units, post only	G	1 Carp	16	.500	Ea.	70.50	27.50		98	121
0100	Mild steel, plain	G	1 Carp	16	.500	Ea.	100	27.50		127.50	154
9000	Minimum labor/equipment charge		"	4	2	Job		109		109	174

05 71 Decorative Metal Stairs

05 71 13 – Fabricated Metal Spiral Stairs

05 71 13.50 Spiral Stairs

		Crew	Daily Output	Labor-Hours	Unit	Material	2021 Bare Costs Labor	Equipment	Total	Total Incl O&P
0010	**SPIRAL STAIRS**									
1805	Shop fabricated, custom ordered									
1810	Aluminum, 5'-0" diameter, plain units	G	E-4	45	.711	Riser	640	43	3.31	686.31	780
1820	Fancy units	G		45	.711		1,050	43	3.31	1,096.31	1,250
1900	Cast iron, 4'-0" diameter, plain units	G		45	.711		770	43	3.31	816.31	920
1920	Fancy units	G		25	1.280		1,650	78	5.95	1,733.95	1,950
2000	Steel, industrial checkered plate, 4' diameter	G		45	.711		610	43	3.31	656.31	745
2200	6' diameter	G		40	.800		655	48.50	3.72	707.22	805
3100	Spiral stair kits, 12 stacking risers to fit exact floor height									
3110	Steel, flat metal treads, primed, 3'-6" diameter	G	2 Carp	1.60	10	Flight	1,300	545		1,845	2,300
3120	4'-0" diameter	G		1.45	11.034		1,400	605		2,005	2,500
3130	4'-6" diameter	G		1.35	11.852		1,675	650		2,325	2,875
3140	5'-0" diameter			1.25	12.800		1,875	700		2,575	3,200
3210	Galvanized, 3'-6" diameter			1.60	10		2,125	545		2,670	3,200
3220	4'-0" diameter			1.45	11.034		2,225	605		2,830	3,400
3230	4'-6" diameter			1.35	11.852		2,500	650		3,150	3,775
3240	5'-0" diameter			1.25	12.800		2,800	700		3,500	4,225
3310	Checkered plate tread, primed, 3'-6" diameter			1.45	11.034		1,550	605		2,155	2,675
3320	4'-0" diameter			1.35	11.852		1,675	650		2,325	2,850
3330	4'-6" diameter			1.25	12.800		1,950	700		2,650	3,275
3340	5'-0" diameter			1.15	13.913		2,150	760		2,910	3,575
3410	Galvanized, 3'-6" diameter			1.45	11.034		2,375	605		2,980	3,575
3420	4'-0" diameter			1.35	11.852		2,475	650		3,125	3,750
3430	4'-6" diameter			1.25	12.800		2,675	700		3,375	4,075
3440	5'-0" diameter	G		1.15	13.913		3,125	760		3,885	4,650
3510	Red oak covers on flat metal treads, 3'-6" diameter			1.35	11.852		1,800	650		2,450	3,000
3520	4'-0" diameter			1.25	12.800		1,950	700		2,650	3,275
3530	4'-6" diameter			1.15	13.913		2,225	760		2,985	3,625
3540	5'-0" diameter			1.05	15.238		2,625	835		3,460	4,225

05 73 Decorative Metal Railings

05 73 16 – Wire Rope Decorative Metal Railings

05 73 16.10 Cable Railings		Crew	Daily Output	Labor-Hours	Unit	Material	2021 Bare Costs Labor	Equipment	Total	Total Incl O&P
0010	**CABLE RAILINGS**, with 316 stainless steel 1 x 19 cable, 3/16" diameter									
0015	Made from recycled materials									
0100	1-3/4" diameter stainless steel posts x 42" high, cables 4" OC	G	2 Sswk	25	.640	L.F.	39	38.50		77.50	106

05 73 23 – Ornamental Railings

05 73 23.50 Railings, Ornamental		Crew	Daily Output	Labor-Hours	Unit	Material	2021 Bare Costs Labor	Equipment	Total	Total Incl O&P
0010	**RAILINGS, ORNAMENTAL**, 3'-6" high, posts @ 6' OC										
0020	Bronze or stainless, hand forged, plain	G	2 Sswk	24	.667	L.F.	265	40		305	355
0100	Fancy	G		18	.889		530	53.50		583.50	670
0200	Aluminum, panelized, plain	G		24	.667		13.60	40		53.60	80
0300	Fancy	G		18	.889		26	53.50		79.50	115
0400	Wrought iron, hand forged, plain	G		24	.667		75	40		115	148
0500	Fancy	G		18	.889		233	53.50		286.50	345
0550	Steel, panelized, plain	G		24	.667		21.50	40		61.50	89
0560	Fancy	G		18	.889		34	53.50		87.50	124
0600	Composite metal/wood/glass, plain			18	.889		154	53.50		207.50	256
0700	Fancy			12	1.333		310	80.50		390.50	470
9000	Minimum labor/equipment charge		1 Sswk	2	4	Job		241		241	390

05 75 Decorative Formed Metal

05 75 13 – Columns

05 75 13.10 Aluminum Columns		Crew	Daily Output	Labor-Hours	Unit	Material	2021 Bare Costs Labor	Equipment	Total	Total Incl O&P
0010	**ALUMINUM COLUMNS**										
0015	Made from recycled materials										
0020	Aluminum, extruded, stock units, no cap or base, 6" diameter	G	E-4	240	.133	L.F.	18	8.10	.62	26.72	33.50
0100	8" diameter	G	"	170	.188	"	19.80	11.45	.88	32.13	41.50
0460	Caps, ornamental, plain	G				Set	250			250	275
0470	Fancy	G				"	1,400			1,400	1,525
0500	For square columns, add to column prices above					L.F.	50%				

135

For customer support on your Commercial Renovation Costs with RSMeans Data, call 800.448.8182.

Division Notes

	CREW	DAILY OUTPUT	LABOR-HOURS	UNIT	BARE COSTS				TOTAL INCL O&P
					MAT.	LABOR	EQUIP.	TOTAL	

Estimating Tips
06 05 00 Common Work Results for Wood, Plastics, and Composites

- Common to any wood-framed structure are the accessory connector items such as screws, nails, adhesives, hangers, connector plates, straps, angles, and hold-downs. For typical wood-framed buildings, such as residential projects, the aggregate total for these items can be significant, especially in areas where seismic loading is a concern. For floor and wall framing, the material cost is based on 10 to 25 lbs. of accessory connectors per MBF. Hold-downs, hangers, and other connectors should be taken off by the piece.

  Included with material costs are fasteners for a normal installation. Gordian's RSMeans engineers use manufacturers' recommendations, written specifications, and/or standard construction practice for the sizing and spacing of fasteners. Prices for various fasteners are shown for informational purposes only. Adjustments should be made if unusual fastening conditions exist.

06 10 00 Carpentry

- Lumber is a traded commodity and therefore sensitive to supply and demand in the marketplace. Even with "budgetary" estimating of wood-framed projects, it is advisable to call local suppliers for the latest market pricing.

- The common quantity unit for wood-framed projects is "thousand board feet" (MBF). A board foot is a volume of wood—1" x 1' x 1' or 144 cubic inches. Board-foot quantities are generally calculated using nominal material dimensions—dressed sizes are ignored. Board foot per lineal foot of any stick of lumber can be calculated by dividing the nominal cross-sectional area by 12. As an example, 2,000 lineal feet of 2 x 12 equates to 4 MBF by dividing the nominal area, 2 x 12, by 12, which equals 2, and multiplying that by 2,000 to give 4,000 board feet. This simple rule applies to all nominal dimensioned lumber.

- Waste is an issue of concern at the quantity takeoff for any area of construction. Framing lumber is sold in even foot lengths, i.e., 8', 10', 12', 14', 16', and depending on spans, wall heights, and the grade of lumber, waste is inevitable. A rule of thumb for lumber waste is 5–10% depending on material quality and the complexity of the framing.

- Wood in various forms and shapes is used in many projects, even where the main structural framing is steel, concrete, or masonry. Plywood as a back-up partition material and 2x boards used as blocking and cant strips around roof edges are two common examples. The estimator should ensure that the costs of all wood materials are included in the final estimate.

06 20 00 Finish Carpentry

- It is necessary to consider the grade of workmanship when estimating labor costs for erecting millwork and an interior finish. In practice, there are three grades: premium, custom, and economy. The RSMeans daily output for base and case moldings is in the range of 200 to 250 L.F. per carpenter per day. This is appropriate for most average custom-grade projects. For premium projects, an adjustment to productivity of 25–50% should be made, depending on the complexity of the job.

Reference Numbers

Reference numbers are shown at the beginning of some major classifications. These numbers refer to related items in the Reference Section. The reference information may be an estimating procedure, an alternate pricing method, or technical information.

Note: Not all subdivisions listed here necessarily appear. ■

---

06 05 05.10 Selective Demolition Wood Framing	Crew	Daily Output	Labor-Hours	Unit	Material	2021 Bare Costs Labor	2021 Bare Costs Equipment	Total	Total Incl O&P
0010	**SELECTIVE DEMOLITION WOOD FRAMING** R024119-10								
0100	Timber connector, nailed, small	1 Clab	96	.083	Ea.		3.70		3.70	5.90
0110	Medium		60	.133			5.90		5.90	9.45
0120	Large		48	.167			7.40		7.40	11.80
0130	Bolted, small		48	.167			7.40		7.40	11.80
0140	Medium		32	.250			11.10		11.10	17.70
0150	Large		24	.333			14.80		14.80	23.50
2958	Beams, 2" x 6"	2 Clab	1100	.015	L.F.		.65		.65	1.03
2960	2" x 8"		825	.019			.86		.86	1.37
2965	2" x 10"		665	.024			1.07		1.07	1.70
2970	2" x 12"		550	.029			1.29		1.29	2.06
2972	2" x 14"		470	.034			1.51		1.51	2.41
2975	4" x 8"	B-1	413	.058			2.62		2.62	4.17
2980	4" x 10"		330	.073			3.28		3.28	5.20
2985	4" x 12"		275	.087			3.93		3.93	6.25
3000	6" x 8"		275	.087			3.93		3.93	6.25
3040	6" x 10"		220	.109			4.92		4.92	7.85
3080	6" x 12"	B-1	185	.130	L.F.		5.85		5.85	9.30
3120	8" x 12"		140	.171			7.75		7.75	12.30
3160	10" x 12"		110	.218			9.85		9.85	15.65
3162	Alternate pricing method		1.10	21.818	M.B.F.		985		985	1,575
3170	Blocking, in 16" OC wall framing, 2" x 4"	1 Clab	600	.013	L.F.		.59		.59	.94
3172	2" x 6"		400	.020			.89		.89	1.41
3174	In 24" OC wall framing, 2" x 4"		600	.013			.59		.59	.94
3176	2" x 6"		400	.020			.89		.89	1.41
3178	Alt method, wood blocking removal from wood framing		.40	20	M.B.F.		890		890	1,425
3179	Wood blocking removal from steel framing		.36	22.222	"		985		985	1,575
3180	Bracing, let in, 1" x 3", studs 16" OC		1050	.008	L.F.		.34		.34	.54
3181	Studs 24" OC		1080	.007			.33		.33	.52
3182	1" x 4", studs 16" OC		1050	.008			.34		.34	.54
3183	Studs 24" OC		1080	.007			.33		.33	.52
3184	1" x 6", studs 16" OC		1050	.008			.34		.34	.54
3185	Studs 24" OC		1080	.007			.33		.33	.52
3186	2" x 3", studs 16" OC		800	.010			.44		.44	.71
3187	Studs 24" OC		830	.010			.43		.43	.68
3188	2" x 4", studs 16" OC		800	.010			.44		.44	.71
3189	Studs 24" OC		830	.010			.43		.43	.68
3190	2" x 6", studs 16" OC		800	.010			.44		.44	.71
3191	Studs 24" OC		830	.010			.43		.43	.68
3192	2" x 8", studs 16" OC		800	.010			.44		.44	.71
3193	Studs 24" OC		830	.010			.43		.43	.68
3194	"T" shaped metal bracing, studs at 16" OC		1060	.008			.34		.34	.53
3195	Studs at 24" OC		1200	.007			.30		.30	.47
3196	Metal straps, studs at 16" OC		1200	.007			.30		.30	.47
3197	Studs at 24" OC		1240	.006			.29		.29	.46
3200	Columns, round, 8' to 14' tall		40	.200	Ea.		8.90		8.90	14.15
3202	Dimensional lumber sizes	2 Clab	1.10	14.545	M.B.F.		645		645	1,025
3250	Blocking, between joists	1 Clab	320	.025	Ea.		1.11		1.11	1.77
3252	Bridging, metal strap, between joists		320	.025	Pr.		1.11		1.11	1.77
3254	Wood, between joists		320	.025	"		1.11		1.11	1.77
3260	Door buck, studs, header & access., 8' high 2" x 4" wall, 3' wide		32	.250	Ea.		11.10		11.10	17.70
3261	4' wide		32	.250			11.10		11.10	17.70
3262	5' wide		32	.250			11.10		11.10	17.70

138

06 05 05 – Selective Demolition for Wood, Plastics, and Composites

06 05 05.10 Selective Demolition Wood Framing	Crew	Daily Output	Labor-Hours	Unit	Material	2021 Bare Costs Labor	2021 Bare Costs Equipment	Total	Total Incl O&P
3263	6' wide	1 Clab	32	.250	Ea.		11.10		11.10	17.70
3264	8' wide		30	.267			11.85		11.85	18.85
3265	10' wide		30	.267			11.85		11.85	18.85
3266	12' wide		30	.267			11.85		11.85	18.85
3267	2" x 6" wall, 3' wide		32	.250			11.10		11.10	17.70
3268	4' wide	1 Clab	32	.250	Ea.		11.10		11.10	17.70
3269	5' wide		32	.250			11.10		11.10	17.70
3270	6' wide		32	.250			11.10		11.10	17.70
3271	8' wide		30	.267			11.85		11.85	18.85
3272	10' wide		30	.267			11.85		11.85	18.85
3273	12' wide		30	.267			11.85		11.85	18.85
3274	Window buck, studs, header & access., 8' high 2" x 4" wall, 2' wide		24	.333			14.80		14.80	23.50
3275	3' wide		24	.333			14.80		14.80	23.50
3276	4' wide		24	.333			14.80		14.80	23.50
3277	5' wide		24	.333			14.80		14.80	23.50
3278	6' wide		24	.333			14.80		14.80	23.50
3279	7' wide		24	.333			14.80		14.80	23.50
3280	8' wide		22	.364			16.15		16.15	25.50
3281	10' wide		22	.364			16.15		16.15	25.50
3282	12' wide		22	.364			16.15		16.15	25.50
3283	2" x 6" wall, 2' wide		24	.333			14.80		14.80	23.50
3284	3' wide		24	.333			14.80		14.80	23.50
3285	4' wide		24	.333			14.80		14.80	23.50
3286	5' wide		24	.333			14.80		14.80	23.50
3287	6' wide		24	.333			14.80		14.80	23.50
3288	7' wide		24	.333			14.80		14.80	23.50
3289	8' wide		22	.364			16.15		16.15	25.50
3290	10' wide		22	.364			16.15		16.15	25.50
3291	12' wide		22	.364			16.15		16.15	25.50
3360	Deck or porch decking		825	.010	L.F.		.43		.43	.69
3400	Fascia boards, 1" x 6"		500	.016			.71		.71	1.13
3440	1" x 8"		450	.018			.79		.79	1.26
3480	1" x 10"		400	.020			.89		.89	1.41
3490	2" x 6"		450	.018			.79		.79	1.26
3500	2" x 8"		400	.020			.89		.89	1.41
3510	2" x 10"		350	.023			1.01		1.01	1.62
3610	Furring, on wood walls or ceiling		4000	.002	S.F.		.09		.09	.14
3620	On masonry or concrete walls or ceiling		1200	.007	"		.30		.30	.47
3800	Headers over openings, 2 @ 2" x 6"		110	.073	L.F.		3.23		3.23	5.15
3840	2 @ 2" x 8"		100	.080			3.55		3.55	5.65
3880	2 @ 2" x 10"		90	.089			3.95		3.95	6.30
3885	Alternate pricing method		.26	30.651	M.B.F.		1,350		1,350	2,175
3920	Joists, 1" x 4"		1250	.006	L.F.		.28		.28	.45
3930	1" x 6"		1135	.007			.31		.31	.50
3940	1" x 8"		1000	.008			.36		.36	.57
3950	1" x 10"		895	.009			.40		.40	.63
3960	1" x 12"		765	.010			.46		.46	.74
4200	2" x 4"	2 Clab	1000	.016	L.F.		.71		.71	1.13
4230	2" x 6"		970	.016			.73		.73	1.17
4240	2" x 8"		940	.017			.76		.76	1.20
4250	2" x 10"		910	.018			.78		.78	1.24
4280	2" x 12"		880	.018			.81		.81	1.29
4281	2" x 14"		850	.019			.84		.84	1.33

06 05 05.10 Selective Demolition Wood Framing		Crew	Daily Output	Labor-Hours	Unit	Material	2021 Bare Costs Labor	2021 Bare Costs Equipment	Total	Total Incl O&P
4282	Composite joists, 9-1/2"	2 Clab	960	.017	L.F.		.74		.74	1.18
4283	11-7/8"		930	.017			.76		.76	1.22
4284	14"		897	.018			.79		.79	1.26
4285	16"		865	.019			.82		.82	1.31
4290	Wood joists, alternate pricing method		1.50	10.667	M.B.F.		475		475	755
4500	Open web joist, 12" deep		500	.032	L.F.		1.42		1.42	2.26
4505	14" deep		475	.034			1.50		1.50	2.38
4510	16" deep		450	.036			1.58		1.58	2.51
4520	18" deep		425	.038			1.67		1.67	2.66
4530	24" deep		400	.040			1.78		1.78	2.83
4550	Ledger strips, 1" x 2"	1 Clab	1200	.007			.30		.30	.47
4560	1" x 3"		1200	.007			.30		.30	.47
4570	1" x 4"		1200	.007			.30		.30	.47
4580	2" x 2"		1100	.007			.32		.32	.51
4590	2" x 4"		1000	.008			.36		.36	.57
4600	2" x 6"		1000	.008			.36		.36	.57
4601	2" x 8" or 2" x 10"		800	.010			.44		.44	.71
4602	4" x 6"		600	.013			.59		.59	.94
4604	4" x 8"		450	.018			.79		.79	1.26
5400	Posts, 4" x 4"	2 Clab	800	.020			.89		.89	1.41
5405	4" x 6"		550	.029			1.29		1.29	2.06
5410	4" x 8"		440	.036			1.61		1.61	2.57
5425	4" x 10"		390	.041			1.82		1.82	2.90
5430	4" x 12"		350	.046			2.03		2.03	3.23
5440	6" x 6"		400	.040			1.78		1.78	2.83
5445	6" x 8"		350	.046			2.03		2.03	3.23
5450	6" x 10"		320	.050			2.22		2.22	3.54
5455	6" x 12"		290	.055			2.45		2.45	3.90
5480	8" x 8"		300	.053			2.37		2.37	3.77
5500	10" x 10"		240	.067			2.96		2.96	4.71
5660	T&G floor planks		2	8	M.B.F.		355		355	565
5682	Rafters, ordinary, 16" OC, 2" x 4"		880	.018	S.F.		.81		.81	1.29
5683	2" x 6"		840	.019			.85		.85	1.35
5684	2" x 8"		820	.020			.87		.87	1.38
5685	2" x 10"		820	.020			.87		.87	1.38
5686	2" x 12"		810	.020			.88		.88	1.40
5687	24" OC, 2" x 4"	2 Clab	1170	.014	S.F.		.61		.61	.97
5688	2" x 6"		1117	.014			.64		.64	1.01
5689	2" x 8"		1091	.015			.65		.65	1.04
5690	2" x 10"		1091	.015			.65		.65	1.04
5691	2" x 12"		1077	.015			.66		.66	1.05
5795	Rafters, ordinary, 2" x 4" (alternate method)		862	.019	L.F.		.82		.82	1.31
5800	2" x 6" (alternate method)		850	.019			.84		.84	1.33
5840	2" x 8" (alternate method)		837	.019			.85		.85	1.35
5855	2" x 10" (alternate method)		825	.019			.86		.86	1.37
5865	2" x 12" (alternate method)		812	.020			.87		.87	1.39
5870	Sill plate, 2" x 4"	1 Clab	1170	.007			.30		.30	.48
5871	2" x 6"		780	.010			.46		.46	.73
5872	2" x 8"		586	.014			.61		.61	.97
5873	Alternate pricing method		.78	10.256	M.B.F.		455		455	725
5885	Ridge board, 1" x 4"	2 Clab	900	.018	L.F.		.79		.79	1.26
5886	1" x 6"		875	.018			.81		.81	1.29
5887	1" x 8"		850	.019			.84		.84	1.33

For customer support on your Commercial Renovation Costs with RSMeans Data, call 800.448.8182.

06 05 05 – Selective Demolition for Wood, Plastics, and Composites

	06 05 05.10 Selective Demolition Wood Framing	Crew	Daily Output	Labor-Hours	Unit	Material	2021 Bare Costs Labor	Equipment	Total	Total Incl O&P
5888	1" x 10"	2 Clab	825	.019	L.F.		.86		.86	1.37
5889	1" x 12"		800	.020			.89		.89	1.41
5890	2" x 4"		900	.018			.79		.79	1.26
5892	2" x 6"		875	.018			.81		.81	1.29
5894	2" x 8"		850	.019			.84		.84	1.33
5896	2" x 10"		825	.019			.86		.86	1.37
5898	2" x 12"		800	.020			.89		.89	1.41
5900	Hip & valley rafters, 2" x 6"		500	.032			1.42		1.42	2.26
5940	2" x 8"		420	.038			1.69		1.69	2.69
6050	Rafter tie, 1" x 4"		1250	.013			.57		.57	.90
6052	1" x 6"		1135	.014			.63		.63	1
6054	2" x 4"		1000	.016			.71		.71	1.13
6056	2" x 6"		970	.016			.73		.73	1.17
6070	Sleepers, on concrete, 1" x 2"	1 Clab	4700	.002			.08		.08	.12
6075	1" x 3"		4000	.002			.09		.09	.14
6080	2" x 4"		3000	.003			.12		.12	.19
6085	2" x 6"		2600	.003			.14		.14	.22
6086	Sheathing from roof, 5/16"	2 Clab	1600	.010	S.F.		.44		.44	.71
6088	3/8"		1525	.010			.47		.47	.74
6090	1/2"		1400	.011			.51		.51	.81
6092	5/8"		1300	.012			.55		.55	.87
6094	3/4"		1200	.013			.59		.59	.94
6096	Board sheathing from roof		1400	.011			.51		.51	.81
6100	Sheathing, from walls, 1/4"		1200	.013			.59		.59	.94
6110	5/16"		1175	.014			.60		.60	.96
6120	3/8"	2 Clab	1150	.014	S.F.		.62		.62	.98
6130	1/2"		1125	.014			.63		.63	1.01
6140	5/8"		1100	.015			.65		.65	1.03
6150	3/4"		1075	.015			.66		.66	1.05
6152	Board sheathing from walls		1500	.011			.47		.47	.75
6158	Subfloor/roof deck, with boards		2200	.007			.32		.32	.51
6159	Subfloor/roof deck, with tongue & groove boards		2000	.008			.36		.36	.57
6160	Plywood, 1/2" thick		768	.021			.93		.93	1.47
6162	5/8" thick		760	.021			.93		.93	1.49
6164	3/4" thick		750	.021			.95		.95	1.51
6165	1-1/8" thick		720	.022			.99		.99	1.57
6166	Underlayment, particle board, 3/8" thick	1 Clab	780	.010			.46		.46	.73
6168	1/2" thick		768	.010			.46		.46	.74
6170	5/8" thick		760	.011			.47		.47	.74
6172	3/4" thick		750	.011			.47		.47	.75
6200	Stairs and stringers, straight run	2 Clab	40	.400	Riser		17.75		17.75	28.50
6240	With platforms, winders or curves	"	26	.615	"		27.50		27.50	43.50
6300	Components, tread	1 Clab	110	.073	Ea.		3.23		3.23	5.15
6320	Riser		80	.100	"		4.44		4.44	7.05
6390	Stringer, 2" x 10"		260	.031	L.F.		1.37		1.37	2.18
6400	2" x 12"		260	.031			1.37		1.37	2.18
6410	3" x 10"		250	.032			1.42		1.42	2.26
6420	3" x 12"		250	.032			1.42		1.42	2.26
6590	Wood studs, 2" x 3"	2 Clab	3076	.005			.23		.23	.37
6600	2" x 4"		2000	.008			.36		.36	.57
6640	2" x 6"		1600	.010			.44		.44	.71
6720	Wall framing, including studs, plates and blocking, 2" x 4"	1 Clab	600	.013	S.F.		.59		.59	.94
6740	2" x 6"		480	.017	"		.74		.74	1.18

06 05 05.10 Selective Demolition Wood Framing

		Crew	Daily Output	Labor-Hours	Unit	Material	2021 Bare Costs Labor	Equipment	Total	Total Incl O&P
6750	Headers, 2" x 4"	1 Clab	1125	.007	L.F.		.32		.32	.50
6755	2" x 6"		1125	.007			.32		.32	.50
6760	2" x 8"		1050	.008			.34		.34	.54
6765	2" x 10"		1050	.008			.34		.34	.54
6770	2" x 12"		1000	.008			.36		.36	.57
6780	4" x 10"		525	.015			.68		.68	1.08
6785	4" x 12"		500	.016			.71		.71	1.13
6790	6" x 8"		560	.014			.63		.63	1.01
6795	6" x 10"		525	.015			.68		.68	1.08
6797	6" x 12"		500	.016			.71		.71	1.13
7000	Trusses									
7050	12' span	2 Clab	74	.216	Ea.		9.60		9.60	15.30
7150	24' span	F-3	66	.606			34	7.20	41.20	62
7200	26' span	"	64	.625			35	7.45	42.45	63.50
7250	28' span	F-3	62	.645	Ea.		36	7.70	43.70	66
7300	30' span		58	.690			38.50	8.20	46.70	70.50
7350	32' span		56	.714			40	8.50	48.50	73
7400	34' span		54	.741			41.50	8.80	50.30	75.50
7450	36' span		52	.769			43	9.15	52.15	78.50
8000	Soffit, T&G wood	1 Clab	520	.015	S.F.		.68		.68	1.09
8010	Hardboard, vinyl or aluminum	"	640	.013			.56		.56	.88
8030	Plywood	2 Carp	315	.051			2.78		2.78	4.42
9000	Minimum labor/equipment charge	1 Clab	4	2	Job		89		89	141
9500	See Section 02 41 19.19 for rubbish handling									

06 05 05.20 Selective Demolition Millwork and Trim

		Crew	Daily Output	Labor-Hours	Unit	Material	2021 Bare Costs Labor	Equipment	Total	Total Incl O&P
0010	**SELECTIVE DEMOLITION MILLWORK AND TRIM** R024119-10									
1000	Cabinets, wood, base cabinets, per L.F.	2 Clab	80	.200	L.F.		8.90		8.90	14.15
1020	Wall cabinets, per L.F.	"	80	.200	"		8.90		8.90	14.15
1060	Remove and reset, base cabinets	2 Carp	18	.889	Ea.		48.50		48.50	77.50
1070	Wall cabinets		20	.800			44		44	69.50
1072	Oven cabinet, 7' high		11	1.455			79.50		79.50	127
1074	Cabinet door, up to 2' high	1 Clab	66	.121			5.40		5.40	8.55
1076	2' - 4' high	"	46	.174			7.70		7.70	12.30
1100	Steel, painted, base cabinets	2 Clab	60	.267	L.F.		11.85		11.85	18.85
1120	Wall cabinets		60	.267	"		11.85		11.85	18.85
1200	Casework, large area		320	.050	S.F.		2.22		2.22	3.54
1220	Selective		200	.080	"		3.55		3.55	5.65
1500	Counter top, straight runs		200	.080	L.F.		3.55		3.55	5.65
1510	L, U or C shapes		120	.133			5.90		5.90	9.45
1550	Remove and reset, straight runs	2 Carp	50	.320			17.50		17.50	28
1560	L, U or C shape	"	40	.400			22		22	35
2000	Paneling, 4' x 8' sheets	2 Clab	2000	.008	S.F.		.36		.36	.57
2100	Boards, 1" x 4"		700	.023			1.01		1.01	1.62
2120	1" x 6"		750	.021			.95		.95	1.51
2140	1" x 8"		800	.020			.89		.89	1.41
3000	Trim, baseboard, to 6" wide		1200	.013	L.F.		.59		.59	.94
3040	Greater than 6" and up to 12" wide		1000	.016			.71		.71	1.13
3080	Remove and reset, minimum	2 Carp	400	.040			2.19		2.19	3.48
3090	Maximum	"	300	.053			2.92		2.92	4.65
3100	Ceiling trim	2 Clab	1000	.016			.71		.71	1.13
3120	Chair rail		1200	.013			.59		.59	.94
3140	Railings with balusters		240	.067			2.96		2.96	4.71

For customer support on your Commercial Renovation Costs with RSMeans Data, call 800.448.8182.

06 05 Common Work Results for Wood, Plastics, and Composites

06 05 05 – Selective Demolition for Wood, Plastics, and Composites

06 05 05.20 Selective Demolition Millwork and Trim	Crew	Daily Output	Labor-Hours	Unit	Material	2021 Bare Costs Labor	Equipment	Total	Total Incl O&P
3160	Wainscoting	2 Clab	700	.023	S.F.		1.01		1.01	1.62
9000	Minimum labor/equipment charge	1 Clab	4	2	Job		89		89	141

06 05 23 – Wood, Plastic, and Composite Fastenings

06 05 23.10 Nails

		Crew	Daily Output	Labor-Hours	Unit	Material	Labor	Equipment	Total	Total Incl O&P
0010	**NAILS**, material only, based upon 50# box purchase									
0020	Copper nails, plain				Lb.	11.25			11.25	12.40
0400	Stainless steel, plain					9.20			9.20	10.10
0500	Box, 3d to 20d, bright					1.67			1.67	1.84
0520	Galvanized					2.48			2.48	2.73
0600	Common, 3d to 60d, plain					1.15			1.15	1.27
0700	Galvanized					1.78			1.78	1.96
0800	Aluminum					11.05			11.05	12.20
1000	Annular or spiral thread, 4d to 60d, plain					2.88			2.88	3.17
1200	Galvanized					3.22			3.22	3.54
1400	Drywall nails, plain					2.11			2.11	2.32
1600	Galvanized					1.93			1.93	2.12
1800	Finish nails, 4d to 10d, plain					1.55			1.55	1.71
2000	Galvanized					2.21			2.21	2.43
2100	Aluminum					10.70			10.70	11.75
2300	Flooring nails, hardened steel, 2d to 10d, plain					3.04			3.04	3.34
2400	Galvanized					5.10			5.10	5.60
2500	Gypsum lath nails, 1-1/8", 13 ga. flathead, blued					2.80			2.80	3.08
2600	Masonry nails, hardened steel, 3/4" to 3" long, plain					3.02			3.02	3.32
2700	Galvanized					3.94			3.94	4.33
2900	Roofing nails, threaded, galvanized					3.06			3.06	3.37
3100	Aluminum					7.50			7.50	8.25
3300	Compressed lead head, threaded, galvanized					3.63			3.63	3.99
3600	Siding nails, plain shank, galvanized					2.19			2.19	2.41
3800	Aluminum					6.85			6.85	7.55
5000	Add to prices above for cement coating					.16			.16	.18
5200	Zinc or tin plating					.21			.21	.23
5500	Vinyl coated sinkers, 8d to 16d					2.28			2.28	2.51

06 05 23.50 Wood Screws

		Crew	Daily Output	Labor-Hours	Unit	Material	Labor	Equipment	Total	Total Incl O&P
0010	**WOOD SCREWS**									
0020	#8, 1" long, steel				C	4.42			4.42	4.86
0100	Brass					15.70			15.70	17.30
0200	#8, 2" long, steel					7.60			7.60	8.35
0300	Brass					26			26	28.50
0400	#10, 1" long, steel					7.35			7.35	8.10
0500	Brass					18.10			18.10	19.90
0600	#10, 2" long, steel					11.85			11.85	13
0700	Brass					25			25	27.50
0800	#10, 3" long, steel					18.85			18.85	20.50
1000	#12, 2" long, steel					12			12	13.20
1100	Brass				C	40			40	44.50
1500	#12, 3" long, steel					15.50			15.50	17.05
2000	#12, 4" long, steel					28.50			28.50	31.50
2050	Composite trim screws									
2100	Stainless steel, trim head, #7 x 1-5/8" long				C	219			219	240
2200	#7 x 2" long					216			216	238
2300	#7 x 2-1/2" long					243			243	268
2400	#7 x 3" long					2,275			2,275	2,500

For customer support on your Commercial Renovation Costs with RSMeans Data, call 800.448.8182.

143

06 05 23.50 Wood Screws		Crew	Daily Output	Labor-Hours	Unit	Material	2021 Bare Costs Labor	Equipment	Total	Total Incl O&P
2500	#9 x 3-1/2" long				C	76,500			76,500	84,500
2600	#9 x 4" long					90,000			90,000	99,000
3000	Stainless steel, flat head, #6 x 1-5/8" long					203			203	223
3100	#8 x 2" long					144			144	159
3200	#8 x 2-1/2" long					172			172	189
3300	#8 x 3" long					200			200	220
3400	#10 x 2" long					163			163	180
3500	#10 x 3-1/2" long					2,625			2,625	2,900
3600	#12 x 3" long					3,025			3,025	3,325
3700	#12 x 4" long					72,500			72,500	80,000
3800	#12 x 6" long					124,500			124,500	137,000

06 05 23.60 Timber Connectors		Crew	Daily Output	Labor-Hours	Unit	Material	2021 Bare Costs Labor	Equipment	Total	Total Incl O&P
0010	**TIMBER CONNECTORS**									
0020	Add up cost of each part for total cost of connection									
0100	Connector plates, steel, with bolts, straight	2 Carp	75	.213	Ea.	31	11.65		42.65	52.50
0110	Tee, 7 ga.		50	.320		45.50	17.50		63	78
0120	T- Strap, 14 ga., 12" x 8" x 2"		50	.320		45.50	17.50		63	78
0150	Anchor plates, 7 ga., 9" x 7"		75	.213		31	11.65		42.65	52.50
0200	Bolts, machine, sq. hd. with nut & washer, 1/2" diameter, 4" long	1 Carp	140	.057		.74	3.13		3.87	5.80
0300	7-1/2" long		130	.062		1.37	3.37		4.74	6.85
0500	3/4" diameter, 7-1/2" long		130	.062		3.37	3.37		6.74	9.05
0610	Machine bolts, w/nut, washer, 3/4" diameter, 15" L, HD's & beam hangers		95	.084		6.35	4.61		10.96	14.30
0800	Drilling bolt holes in timber, 1/2" diameter		450	.018	Inch		.97		.97	1.55
0900	1" diameter		350	.023	"		1.25		1.25	1.99
1100	Framing anchor, angle, 3" x 3" x 1-1/2", 12 ga.		175	.046	Ea.	2.29	2.50		4.79	6.50
1150	Framing anchors, 18 ga., 4-1/2" x 2-3/4"		175	.046		2.29	2.50		4.79	6.50
1160	Framing anchors, 18 ga., 4-1/2" x 3"		175	.046		2.29	2.50		4.79	6.50
1170	Clip anchors plates, 18 ga., 12" x 1-1/8"		175	.046		2.29	2.50		4.79	6.50
1250	Holdowns, 3 ga. base, 10 ga. body		8	1		50	54.50		104.50	142
1260	Holdowns, 7 ga. 11-1/16" x 3-1/4"		8	1		50	54.50		104.50	142
1270	Holdowns, 7 ga. 14-3/8" x 3-1/8"		8	1		50	54.50		104.50	142
1275	Holdowns, 12 ga. 8" x 2-1/2"	1 Carp	8	1	Ea.	50	54.50		104.50	142
1300	Joist and beam hangers, 18 ga. galv., for 2" x 4" joist		175	.046		.79	2.50		3.29	4.85
1400	2" x 6" to 2" x 10" joist		165	.048		1.41	2.65		4.06	5.75
1600	16 ga. galv., 3" x 6" to 3" x 10" joist		160	.050		5.20	2.74		7.94	10.10
1700	3" x 10" to 3" x 14" joist		160	.050		5.50	2.74		8.24	10.40
1800	4" x 6" to 4" x 10" joist		155	.052		4.41	2.82		7.23	9.35
1900	4" x 10" to 4" x 14" joist		155	.052		5.90	2.82		8.72	11
2000	Two-2" x 6" to two-2" x 10" joists		150	.053		4.66	2.92		7.58	9.80
2100	Two-2" x 10" to two-2" x 14" joists		150	.053		4.18	2.92		7.10	9.25
2300	3/16" thick, 6" x 8" joist		145	.055		58	3.02		61.02	68.50
2400	6" x 10" joist		140	.057		57.50	3.13		60.63	68.50
2500	6" x 12" joist		135	.059		68	3.24		71.24	79.50
2700	1/4" thick, 6" x 14" joist		130	.062		78	3.37		81.37	91
2900	Plywood clips, extruded aluminum H clip, for 3/4" panels					.25			.25	.28
3000	Galvanized 18 ga. back-up clip					.11			.11	.12
3200	Post framing, 16 ga. galv. for 4" x 4" base, 2 piece	1 Carp	130	.062		12	3.37		15.37	18.55
3300	Cap		130	.062		35.50	3.37		38.87	44.50
3500	Rafter anchors, 18 ga. galv., 1-1/2" wide, 5-1/4" long		145	.055		.43	3.02		3.45	5.30
3600	10-3/4" long		145	.055		1.36	3.02		4.38	6.30
3800	Shear plates, 2-5/8" diameter		120	.067		4.15	3.65		7.80	10.35
3900	4" diameter		115	.070		3.48	3.81		7.29	9.90

06 05 Common Work Results for Wood, Plastics, and Composites

06 05 23 – Wood, Plastic, and Composite Fastenings

06 05 23.60 Timber Connectors

		Crew	Daily Output	Labor-Hours	Unit	Material	2021 Bare Costs Labor	Equipment	Total	Total Incl O&P
4000	Sill anchors, embedded in concrete or block, 25-1/2" long	1 Carp	115	.070	Ea.	2.31	3.81		6.12	8.60
4100	Spike grids, 3" x 6"		120	.067		1.65	3.65		5.30	7.60
4400	Split rings, 2-1/2" diameter		120	.067		4.29	3.65		7.94	10.50
4500	4" diameter		110	.073		4.52	3.98		8.50	11.30
4550	Tie plate, 20 ga., 7" x 3-1/8"		110	.073		4.52	3.98		8.50	11.30
4560	5" x 4-1/8"		110	.073		4.52	3.98		8.50	11.30
4575	Twist straps, 18 ga., 12" x 1-1/4"		110	.073		4.52	3.98		8.50	11.30
4580	16" x 1-1/4"		110	.073		4.52	3.98		8.50	11.30
4600	Strap ties, 20 ga., 2-1/16" wide, 12-13/16" long		180	.044		.97	2.43		3.40	4.94
4700	16 ga., 1-3/8" wide, 12" long		180	.044		.97	2.43		3.40	4.94
4800	1-1/4" wide, 21-5/8" long		160	.050		1.83	2.74		4.57	6.35
5000	Toothed rings, 2-5/8" or 4" diameter		90	.089		2.15	4.86		7.01	10.10
5200	Truss plates, nailed, 20 ga., up to 32' span		17	.471	Truss	15.15	25.50		40.65	57.50
5400	Washers, 2" x 2" x 1/8"				Ea.	.44			.44	.48
5500	3" x 3" x 3/16"				"	1.18			1.18	1.30
9000	Minimum labor/equipment charge	1 Carp	4	2	Job		109		109	174

06 05 23.80 Metal Bracing

		Crew	Daily Output	Labor-Hours	Unit	Material	2021 Bare Costs Labor	Equipment	Total	Total Incl O&P
0010	**METAL BRACING**									
0302	Let-in, "T" shaped, 22 ga. galv. steel, studs at 16" OC	1 Carp	580	.014	L.F.	.84	.75		1.59	2.12
0402	Studs at 24" OC		600	.013		.84	.73		1.57	2.08
0502	Steel straps, 16 ga. galv. steel, studs at 16" OC		600	.013		1.09	.73		1.82	2.36
0602	Studs at 24" OC		620	.013		1.09	.71		1.80	2.32

06 11 Wood Framing

06 11 10 – Framing with Dimensional, Engineered or Composite Lumber

06 11 10.01 Forest Stewardship Council Certification

		Crew	Daily Output	Labor-Hours	Unit	Material	2021 Bare Costs Labor	Equipment	Total	Total Incl O&P
0010	**FOREST STEWARDSHIP COUNCIL CERTIFICATION**									
0020	For Forest Stewardship Council (FSC) cert dimension lumber, add [G]					65%				

06 11 10.02 Blocking

		Crew	Daily Output	Labor-Hours	Unit	Material	2021 Bare Costs Labor	Equipment	Total	Total Incl O&P
0010	**BLOCKING**									
2600	Miscellaneous, to wood construction									
2620	2" x 4"	1 Carp	.17	47.059	M.B.F.	1,050	2,575		3,625	5,250
2625	Pneumatic nailed		.21	38.095		1,050	2,075		3,125	4,500
2660	2" x 8"		.27	29.630		1,200	1,625		2,825	3,900
2665	Pneumatic nailed		.33	24.242		1,225	1,325		2,550	3,450
2720	To steel construction									
2740	2" x 4"	1 Carp	.14	57.143	M.B.F.	1,050	3,125		4,175	6,125
2780	2" x 8"		.21	38.095	"	1,200	2,075		3,275	4,650
9000	Minimum labor/equipment charge		4	2	Job		109		109	174

06 11 10.04 Wood Bracing

		Crew	Daily Output	Labor-Hours	Unit	Material	2021 Bare Costs Labor	Equipment	Total	Total Incl O&P
0010	**WOOD BRACING**									
0012	Let-in, with 1" x 6" boards, studs @ 16" OC	1 Carp	150	.053	L.F.	1.15	2.92		4.07	5.90
0202	Studs @ 24" OC	"	230	.035	"	1.15	1.90		3.05	4.29

06 11 10.06 Bridging

		Crew	Daily Output	Labor-Hours	Unit	Material	2021 Bare Costs Labor	Equipment	Total	Total Incl O&P
0010	**BRIDGING**									
0012	Wood, for joists 16" OC, 1" x 3"	1 Carp	130	.062	Pr.	.98	3.37		4.35	6.40
0017	Pneumatic nailed		170	.047		1.06	2.57		3.63	5.25
0102	2" x 3" bridging		130	.062		.98	3.37		4.35	6.45
0107	Pneumatic nailed		170	.047		1.02	2.57		3.59	5.25
0302	Steel, galvanized, 18 ga., for 2" x 10" joists at 12" OC		130	.062		1.77	3.37		5.14	7.30

145

06 11 10 – Framing with Dimensional, Engineered or Composite Lumber

06 11 10.06 Bridging		Crew	Daily Output	Labor-Hours	Unit	Material	2021 Bare Costs Labor	Equipment	Total	Total Incl O&P
0352	16" OC	1 Carp	135	.059	Pr.	.03	3.24		3.27	5.20
0402	24" OC		140	.057		2.62	3.13		5.75	7.85
0602	For 2" x 14" joists at 16" OC		130	.062		2.26	3.37		5.63	7.85
0902	Compression type, 16" OC, 2" x 8" joists		200	.040		1.34	2.19		3.53	4.96
1002	2" x 12" joists		200	.040		1.34	2.19		3.53	4.95

06 11 10.10 Beam and Girder Framing

		Crew	Daily Output	Labor-Hours	Unit	Material	2021 Bare Costs Labor	Equipment	Total	Total Incl O&P
0010	**BEAM AND GIRDER FRAMING** R061110-30									
1000	Single, 2" x 6"	2 Carp	700	.023	L.F.	1.06	1.25		2.31	3.16
1005	Pneumatic nailed		812	.020		1.07	1.08		2.15	2.90
1020	2" x 8"		650	.025		1.62	1.35		2.97	3.92
1025	Pneumatic nailed		754	.021		1.63	1.16		2.79	3.64
1040	2" x 10"		600	.027		2.03	1.46		3.49	4.56
1045	Pneumatic nailed		696	.023		2.05	1.26		3.31	4.25
1060	2" x 12"	2 Carp	550	.029	L.F.	2.51	1.59		4.10	5.30
1065	Pneumatic nailed		638	.025		2.53	1.37		3.90	4.96
1080	2" x 14"		500	.032		3.56	1.75		5.31	6.70
1085	Pneumatic nailed		580	.028		3.58	1.51		5.09	6.35
1100	3" x 8"		550	.029		4.10	1.59		5.69	7.05
1120	3" x 10"		500	.032		4.96	1.75		6.71	8.25
1140	3" x 12"		450	.036		7.05	1.94		8.99	10.85
1160	3" x 14"		400	.040		6.45	2.19		8.64	10.60
1170	4" x 6"	F-3	1100	.036		2.89	2.04	.43	5.36	6.90
1180	4" x 8"		1000	.040		6.50	2.24	.48	9.22	11.20
1200	4" x 10"		950	.042		8	2.36	.50	10.86	13.10
1220	4" x 12"		900	.044		7.30	2.49	.53	10.32	12.55
1240	4" x 14"		850	.047		9.35	2.64	.56	12.55	15.10
1250	6" x 8"		525	.076		9.65	4.27	.91	14.83	18.40
1260	6" x 10"		500	.080		9.90	4.48	.95	15.33	19.05
1290	8" x 12"		300	.133		29.50	7.45	1.59	38.54	46
2000	Double, 2" x 6"	2 Carp	625	.026		2.13	1.40		3.53	4.57
2005	Pneumatic nailed		725	.022		2.15	1.21		3.36	4.28
2020	2" x 8"		575	.028		3.23	1.52		4.75	5.95
2025	Pneumatic nailed		667	.024		3.25	1.31		4.56	5.65
2040	2" x 10"		550	.029		4.07	1.59		5.66	7
2045	Pneumatic nailed		638	.025		4.10	1.37		5.47	6.70
2060	2" x 12"		525	.030		5.05	1.67		6.72	8.20
2065	Pneumatic nailed		610	.026		5.05	1.43		6.48	7.85
2080	2" x 14"		475	.034		7.15	1.84		8.99	10.80
2085	Pneumatic nailed		551	.029		7.15	1.59		8.74	10.45
3000	Triple, 2" x 6"		550	.029		3.19	1.59		4.78	6.05
3005	Pneumatic nailed		638	.025		3.22	1.37		4.59	5.70
3020	2" x 8"		525	.030		4.85	1.67		6.52	8
3025	Pneumatic nailed		609	.026		4.88	1.44		6.32	7.65
3040	2" x 10"		500	.032		6.10	1.75		7.85	9.50
3045	Pneumatic nailed		580	.028		6.15	1.51		7.66	9.15
3060	2" x 12"		475	.034		7.55	1.84		9.39	11.25
3065	Pneumatic nailed		551	.029		7.60	1.59		9.19	10.90
3080	2" x 14"		450	.036		11.25	1.94		13.19	15.50
3085	Pneumatic nailed		522	.031		10.75	1.68		12.43	14.45
9000	Minimum labor/equipment charge	1 Carp	2	4	Job		219		219	350

146

06 11 Wood Framing

06 11 10 – Framing with Dimensional, Engineered or Composite Lumber

06 11 10.12 Ceiling Framing

		Crew	Daily Output	Labor-Hours	Unit	Material	2021 Bare Costs Labor	Equipment	Total	Total Incl O&P
0010	**CEILING FRAMING**									
6000	Suspended, 2" x 3"	2 Carp	1000	.016	L.F.	.60	.88		1.48	2.05
6050	2" x 4"		900	.018		.70	.97		1.67	2.32
6100	2" x 6"		800	.020		1.06	1.09		2.15	2.91
6150	2" x 8"		650	.025		1.62	1.35		2.97	3.92
9000	Minimum labor/equipment charge	1 Carp	4	2	Job		109		109	174

06 11 10.14 Posts and Columns

		Crew	Daily Output	Labor-Hours	Unit	Material	2021 Bare Costs Labor	Equipment	Total	Total Incl O&P
0010	**POSTS AND COLUMNS**									
0100	4" x 4"	2 Carp	390	.041	L.F.	2.31	2.24		4.55	6.10
0150	4" x 6"		275	.058		2.89	3.18		6.07	8.25
0200	4" x 8"		220	.073		6.50	3.98		10.48	13.50
0250	6" x 6"		215	.074		6.85	4.07		10.92	14.05
0300	6" x 8"		175	.091		9.65	5		14.65	18.60
0350	6" x 10"		150	.107		9.90	5.85		15.75	20
9000	Minimum labor/equipment charge	1 Carp	2	4	Job		219		219	350

06 11 10.18 Joist Framing

		Crew	Daily Output	Labor-Hours	Unit	Material	2021 Bare Costs Labor	Equipment	Total	Total Incl O&P
0010	**JOIST FRAMING** R061110-30									
2000	Joists, 2" x 4"	2 Carp	1250	.013	L.F.	.70	.70		1.40	1.88
2005	Pneumatic nailed		1438	.011		.71	.61		1.32	1.75
2100	2" x 6"		1250	.013		1.06	.70		1.76	2.28
2105	Pneumatic nailed		1438	.011		1.07	.61		1.68	2.15
2150	2" x 8"		1100	.015		1.62	.80		2.42	3.05
2155	Pneumatic nailed		1265	.013		1.63	.69		2.32	2.89
2200	2" x 10"		900	.018		2.03	.97		3	3.79
2205	Pneumatic nailed		1035	.015		2.05	.85		2.90	3.60
2250	2" x 12"		875	.018		2.51	1		3.51	4.36
2255	Pneumatic nailed		1006	.016		2.53	.87		3.40	4.17
2300	2" x 14"		770	.021		3.56	1.14		4.70	5.75
2305	Pneumatic nailed		886	.018		3.58	.99		4.57	5.50
2350	3" x 6"		925	.017		3.05	.95		4	4.86
2400	3" x 10"		780	.021		4.96	1.12		6.08	7.25
2450	3" x 12"		600	.027		7.05	1.46		8.51	10.05
2500	4" x 6"		800	.020		2.89	1.09		3.98	4.92
2550	4" x 10"		600	.027		8	1.46		9.46	11.10
2600	4" x 12"		450	.036		7.30	1.94		9.24	11.10
2605	Sister joist, 2" x 6"		800	.020		1.06	1.09		2.15	2.91
2606	Pneumatic nailed		960	.017		1.07	.91		1.98	2.63
2610	2" x 8"		640	.025		1.62	1.37		2.99	3.96
2611	Pneumatic nailed		768	.021		1.63	1.14		2.77	3.60
2615	2" x 10"		535	.030		2.03	1.64		3.67	4.84
2616	Pneumatic nailed		642	.025		2.05	1.36		3.41	4.42
2620	2" x 12"		455	.035		2.51	1.92		4.43	5.85
2625	Pneumatic nailed		546	.029		2.53	1.60		4.13	5.35
3000	Composite wood joist 9-1/2" deep		.90	17.778	M.L.F.	2,275	970		3,245	4,075
3010	11-1/2" deep		.88	18.182		2,625	995		3,620	4,450
3020	14" deep		.82	19.512		3,275	1,075		4,350	5,300
3030	16" deep		.78	20.513		5,275	1,125		6,400	7,575
4000	Open web joist 12" deep		.88	18.182		5,625	995		6,620	7,750
4002	Per linear foot		880	.018	L.F.	5.60	.99		6.59	7.80
4004	Treated, per linear foot	2 Carp	880	.018	L.F.	6.55	.99		7.54	8.80
4010	14" deep		.82	19.512	M.L.F.	6,350	1,075		7,425	8,675
4012	Per linear foot		820	.020	L.F.	6.35	1.07		7.42	8.65

06 11 Wood Framing

06 11 10 – Framing with Dimensional, Engineered or Composite Lumber

06 11 10.18 Joist Framing		Crew	Daily Output	Labor-Hours	Unit	Material	2021 Bare Costs Labor	Equipment	Total	Total Incl O&P
4014	Treated, per linear foot	2 Carp	820	.020	L.F.	7.40	1.07		8.47	9.85
4020	16" deep		.78	20.513	M.L.F.	5,950	1,125		7,075	8,325
4022	Per linear foot		780	.021	L.F.	5.95	1.12		7.07	8.35
4024	Treated, per linear foot		780	.021	"	7.15	1.12		8.27	9.70
4030	18" deep		.74	21.622	M.L.F.	5,150	1,175		6,325	7,525
4032	Per linear foot		740	.022	L.F.	5.15	1.18		6.33	7.55
4034	Treated, per linear foot		740	.022	"	6.50	1.18		7.68	9.05
6000	Composite rim joist, 1-1/4" x 9-1/2"		.90	17.778	M.L.F.	2,075	970		3,045	3,825
6010	1-1/4" x 11-1/2"		.88	18.182		2,575	995		3,570	4,400
6020	1-1/4" x 14-1/2"		.82	19.512		4,525	1,075		5,600	6,675
6030	1-1/4" x 16-1/2"		.78	20.513		4,100	1,125		5,225	6,275
9000	Minimum labor/equipment charge	1 Carp	4	2	Job		109		109	174

06 11 10.24 Miscellaneous Framing

		Crew	Daily Output	Labor-Hours	Unit	Material	2021 Bare Costs Labor	Equipment	Total	Total Incl O&P
0010	**MISCELLANEOUS FRAMING**									
2000	Firestops, 2" x 4"	2 Carp	780	.021	L.F.	.70	1.12		1.82	2.56
2005	Pneumatic nailed		952	.017		.71	.92		1.63	2.24
2100	2" x 6"		600	.027		1.06	1.46		2.52	3.49
2105	Pneumatic nailed		732	.022		1.07	1.20		2.27	3.08
5000	Nailers, treated, wood construction, 2" x 4"		800	.020		.90	1.09		1.99	2.73
5005	Pneumatic nailed		960	.017		.91	.91		1.82	2.45
5100	2" x 6"		750	.021		1.26	1.17		2.43	3.25
5105	Pneumatic nailed		900	.018		1.27	.97		2.24	2.95
5120	2" x 8"		700	.023		1.91	1.25		3.16	4.09
5125	Pneumatic nailed		840	.019		1.92	1.04		2.96	3.77
5200	Steel construction, 2" x 4"		750	.021		.90	1.17		2.07	2.85
5220	2" x 6"		700	.023		1.26	1.25		2.51	3.38
5240	2" x 8"		650	.025		1.91	1.35		3.26	4.24
7000	Rough bucks, treated, for doors or windows, 2" x 6"		400	.040		1.26	2.19		3.45	4.87
7005	Pneumatic nailed		480	.033		1.27	1.82		3.09	4.30
7100	2" x 8"		380	.042		1.91	2.30		4.21	5.75
7105	Pneumatic nailed		456	.035		1.92	1.92		3.84	5.15
8000	Stair stringers, 2" x 10"		130	.123		2.03	6.75		8.78	12.95
8100	2" x 12"		130	.123		2.51	6.75		9.26	13.45
8150	3" x 10"		125	.128		4.96	7		11.96	16.60
8200	3" x 12"		125	.128		7.05	7		14.05	18.90
8870	Laminated structural lumber, 1-1/4" x 11-1/2"		130	.123		2.57	6.75		9.32	13.50
8880	1-1/4" x 14-1/2"		130	.123		4.50	6.75		11.25	15.65
9000	Minimum labor/equipment charge	1 Carp	4	2	Job		109		109	174

06 11 10.26 Partitions

		Crew	Daily Output	Labor-Hours	Unit	Material	2021 Bare Costs Labor	Equipment	Total	Total Incl O&P
0010	**PARTITIONS**									
0020	Single bottom and double top plate, no waste, std. & better lumber									
0180	2" x 4" studs, 8' high, studs 12" OC	2 Carp	80	.200	L.F.	7.70	10.95		18.65	26
0185	12" OC, pneumatic nailed		96	.167		7.75	9.10		16.85	23
0200	16" OC		100	.160		6.30	8.75		15.05	21
0205	16" OC, pneumatic nailed		120	.133		6.35	7.30		13.65	18.60
0300	24" OC		125	.128		4.90	7		11.90	16.55
0305	24" OC, pneumatic nailed		150	.107		4.94	5.85		10.79	14.75
0380	10' high, studs 12" OC		80	.200		9.10	10.95		20.05	27.50
0385	12" OC, pneumatic nailed		96	.167		9.15	9.10		18.25	24.50
0400	16" OC		100	.160		7.35	8.75		16.10	22
0405	16" OC, pneumatic nailed		120	.133		7.40	7.30		14.70	19.75
0500	24" OC		125	.128		5.60	7		12.60	17.30

148

06 11 Wood Framing

06 11 10 – Framing with Dimensional, Engineered or Composite Lumber

06 11 10.26 Partitions

		Crew	Daily Output	Labor-Hours	Unit	Material	2021 Bare Costs Labor	Equipment	Total	Total Incl O&P
0505	24" OC, pneumatic nailed	2 Carp	150	.107	L.F.	5.65	5.85		11.50	15.50
0580	12' high, studs 12" OC		65	.246		10.50	13.45		23.95	33
0585	12" OC, pneumatic nailed		78	.205		10.60	11.20		21.80	29.50
0600	16" OC		80	.200		8.40	10.95		19.35	26.50
0605	16" OC, pneumatic nailed		96	.167		8.45	9.10		17.55	24
0700	24" OC		100	.160		6.30	8.75		15.05	21
0705	24" OC, pneumatic nailed		120	.133		6.35	7.30		13.65	18.60
0780	2" x 6" studs, 8' high, studs 12" OC		70	.229		11.70	12.50		24.20	33
0785	12" OC, pneumatic nailed		84	.190		11.80	10.40		22.20	29.50
0800	16" OC		90	.178		9.60	9.70		19.30	26
0805	16" OC, pneumatic nailed		108	.148		9.65	8.10		17.75	23.50
0900	24" OC		115	.139		7.45	7.60		15.05	20.50
0905	24" OC, pneumatic nailed		138	.116		7.50	6.35		13.85	18.35
0980	10' high, studs 12" OC		70	.229		13.85	12.50		26.35	35
0985	12" OC, pneumatic nailed		84	.190		13.95	10.40		24.35	32
1000	16" OC		90	.178		11.20	9.70		20.90	28
1005	16" OC, pneumatic nailed		108	.148		11.25	8.10		19.35	25.50
1100	24" OC		115	.139		8.50	7.60		16.10	21.50
1105	24" OC, pneumatic nailed		138	.116		8.60	6.35		14.95	19.55
1180	12' high, studs 12" OC		55	.291		15.95	15.90		31.85	43
1185	12" OC, pneumatic nailed		66	.242		16.10	13.25		29.35	38.50
1200	16" OC		70	.229		12.80	12.50		25.30	34
1205	16" OC, pneumatic nailed		84	.190		12.90	10.40		23.30	31
1300	24" OC		90	.178		9.60	9.70		19.30	26
1305	24" OC, pneumatic nailed		108	.148		9.65	8.10		17.75	23.50
1400	For horizontal blocking, 2" x 4", add		600	.027		.70	1.46		2.16	3.09
1500	2" x 6", add		600	.027		1.06	1.46		2.52	3.49
1600	For openings, add	2 Carp	250	.064	L.F.		3.50		3.50	5.55
1702	Headers for above openings, material only, add				B.F.	1.34			1.34	1.47
9000	Minimum labor/equipment charge	1 Carp	4	2	Job		109		109	174

06 11 10.28 Porch or Deck Framing

		Crew	Daily Output	Labor-Hours	Unit	Material	2021 Bare Costs Labor	Equipment	Total	Total Incl O&P
0010	**PORCH OR DECK FRAMING**									
0100	Treated lumber, posts or columns, 4" x 4"	2 Carp	390	.041	L.F.	2.26	2.24		4.50	6.05
0110	4" x 6"		275	.058		2.97	3.18		6.15	8.30
0120	4" x 8"		220	.073		6.25	3.98		10.23	13.20
0130	Girder, single, 4" x 4"		675	.024		2.26	1.30		3.56	4.55
0140	4" x 6"		600	.027		2.97	1.46		4.43	5.60
0150	4" x 8"		525	.030		6.25	1.67		7.92	9.50
0160	Double, 2" x 4"		625	.026		1.82	1.40		3.22	4.24
0170	2" x 6"		600	.027		2.56	1.46		4.02	5.15
0180	2" x 8"		575	.028		3.86	1.52		5.38	6.65
0190	2" x 10"		550	.029		4.12	1.59		5.71	7.05
0200	2" x 12"		525	.030		7.45	1.67		9.12	10.85
0210	Triple, 2" x 4"		575	.028		2.73	1.52		4.25	5.45
0220	2" x 6"		550	.029		3.84	1.59		5.43	6.75
0230	2" x 8"		525	.030		5.80	1.67		7.47	9.05
0240	2" x 10"		500	.032		6.20	1.75		7.95	9.60
0250	2" x 12"		475	.034		11.15	1.84		12.99	15.20
0260	Ledger, bolted 4' OC, 2" x 4"		400	.040		1.16	2.19		3.35	4.76
0270	2" x 6"		395	.041		1.52	2.22		3.74	5.20
0280	2" x 8"		390	.041		2.17	2.24		4.41	5.95
0290	2" x 10"		385	.042		2.29	2.27		4.56	6.15

06 11 Wood Framing

06 11 10 – Framing with Dimensional, Engineered or Composite Lumber

06 11 10.28 Porch or Deck Framing

		Crew	Daily Output	Labor-Hours	Unit	Material	2021 Bare Costs Labor	Equipment	Total	Total Incl O&P
0300	2" x 12"	2 Carp	380	.042	L.F.	3.94	2.30		6.24	8
0310	Joists, 2" x 4"		1250	.013		.92	.70		1.62	2.12
0320	2" x 6"		1250	.013		1.29	.70		1.99	2.53
0330	2" x 8"		1100	.015		1.95	.80		2.75	3.41
0340	2" x 10"		900	.018		2.08	.97		3.05	3.84
0350	2" x 12"		875	.018		2.50	1		3.50	4.33
0360	Railings and trim, 1" x 4"	1 Carp	300	.027		.83	1.46		2.29	3.24
0370	2" x 2"		300	.027		.64	1.46		2.10	3.02
0380	2" x 4"		300	.027		.90	1.46		2.36	3.31
0390	2" x 6"		300	.027		1.27	1.46		2.73	3.71
0400	Decking, 1" x 4"		275	.029	S.F.	2.05	1.59		3.64	4.79
0410	2" x 4"		300	.027		3.07	1.46		4.53	5.70
0420	2" x 6"		320	.025		2.75	1.37		4.12	5.20
0430	5/4" x 6"		320	.025		3.15	1.37		4.52	5.65
0440	Balusters, square, 2" x 2"	2 Carp	660	.024	L.F.	.64	1.33		1.97	2.82
0450	Turned, 2" x 2"		420	.038		.85	2.08		2.93	4.26
0460	Stair stringer, 2" x 10"		130	.123		2.08	6.75		8.83	13
0470	2" x 12"		130	.123		2.50	6.75		9.25	13.45
0480	Stair treads, 1" x 4"	2 Carp	140	.114	L.F.	2.06	6.25		8.31	12.20
0490	2" x 4"		140	.114		.92	6.25		7.17	10.95
0500	2" x 6"		160	.100		1.38	5.45		6.83	10.20
0510	5/4" x 6"		160	.100		1.47	5.45		6.92	10.30
0520	Turned handrail post, 4" x 4"		64	.250	Ea.	16.05	13.70		29.75	39.50
0530	Lattice panel, 4' x 8', 1/2"		1600	.010	S.F.	.83	.55		1.38	1.78
0535	3/4"		1600	.010	"	1.23	.55		1.78	2.23
0540	Cedar, posts or columns, 4" x 4"		390	.041	L.F.	5.30	2.24		7.54	9.35
0550	4" x 6"		275	.058		8.45	3.18		11.63	14.35
0560	4" x 8"		220	.073		14.55	3.98		18.53	22.50
0800	Decking, 1" x 4"		550	.029		3.85	1.59		5.44	6.75
0810	2" x 4"		600	.027		7.40	1.46		8.86	10.45
0820	2" x 6"		640	.025		14.30	1.37		15.67	17.90
0830	5/4" x 6"		640	.025		10.35	1.37		11.72	13.60
0840	Railings and trim, 1" x 4"		600	.027		3.85	1.46		5.31	6.55
0860	2" x 4"		600	.027		7.40	1.46		8.86	10.45
0870	2" x 6"		600	.027		14.30	1.46		15.76	18
0920	Stair treads, 1" x 4"		140	.114		3.85	6.25		10.10	14.20
0930	2" x 4"		140	.114		7.40	6.25		13.65	18.10
0940	2" x 6"		160	.100		14.30	5.45		19.75	24.50
0950	5/4" x 6"		160	.100		10.35	5.45		15.80	20
0980	Redwood, posts or columns, 4" x 4"		390	.041		11.30	2.24		13.54	15.95
0990	4" x 6"		275	.058		17.95	3.18		21.13	25
1000	4" x 8"		220	.073		33.50	3.98		37.48	43.50
1240	Decking, 1" x 4"	1 Carp	275	.029	S.F.	10.15	1.59		11.74	13.75
1260	2" x 6"		340	.024		20.50	1.29		21.79	24.50
1270	5/4" x 6"		320	.025		7.35	1.37		8.72	10.30
1280	Railings and trim, 1" x 4"	2 Carp	600	.027	L.F.	2.98	1.46		4.44	5.60
1310	2" x 6"		600	.027		20.50	1.46		21.96	25
1420	Alternative decking, wood/plastic composite, 5/4" x 6" [G]		640	.025		4.44	1.37		5.81	7.05
1440	1" x 4" square edge fir		550	.029		2.07	1.59		3.66	4.80
1450	1" x 4" tongue and groove fir		450	.036		2.04	1.94		3.98	5.35
1460	1" x 4" mahogany		550	.029		2.28	1.59		3.87	5.05
1462	5/4" x 6" PVC		550	.029		3.55	1.59		5.14	6.45
1465	Framing, porch or deck, alt deck fastening, screws, add	1 Carp	240	.033	S.F.		1.82		1.82	2.90

150

06 11 10 – Framing with Dimensional, Engineered or Composite Lumber

06 11 10.28 Porch or Deck Framing

		Crew	Daily Output	Labor-Hours	Unit	Material	2021 Bare Costs Labor	2021 Bare Costs Equipment	Total	Total Incl O&P
1470	Accessories, joist hangers, 2" x 4"	1 Carp	160	.050	Ea.	.79	2.74		3.53	5.25
1480	2" x 6" through 2" x 12"		150	.053		1.41	2.92		4.33	6.20
1530	Post footing, incl excav, backfill, tube form & concrete, 4' deep, 8" diam.	F-7	12	2.667		18	132		150	230
1540	10" diameter		11	2.909		24	144		168	257
1550	12" diameter		10	3.200		31.50	159		190.50	287

06 11 10.30 Roof Framing

		Crew	Daily Output	Labor-Hours	Unit	Material	2021 Bare Costs Labor	2021 Bare Costs Equipment	Total	Total Incl O&P
0010	**ROOF FRAMING** R061110-30									
1900	Rough fascia, 2" x 6"	2 Carp	250	.064	L.F.	1.06	3.50		4.56	6.70
2000	2" x 8"		225	.071		1.62	3.89		5.51	8
2100	2" x 10"		180	.089		2.03	4.86		6.89	10
2200	2" x 12"		180	.089		2.51	4.86		7.37	10.50
5000	Rafters, to 4 in 12 pitch, 2" x 6", ordinary		1000	.016		1.06	.88		1.94	2.56
5060	2" x 8", ordinary		950	.017		1.62	.92		2.54	3.25
5120	2" x 10", ordinary		630	.025		2.03	1.39		3.42	4.45
5180	2" x 12", ordinary		575	.028		2.51	1.52		4.03	5.20
5250	Composite rafter, 9-1/2" deep		575	.028		2.29	1.52		3.81	4.93
5260	11-1/2" deep		575	.028		2.61	1.52		4.13	5.30
5300	Hip and valley rafters, 2" x 6", ordinary		760	.021		1.06	1.15		2.21	3
5360	2" x 8", ordinary		720	.022		1.62	1.22		2.84	3.72
5420	2" x 10", ordinary		570	.028		2.03	1.54		3.57	4.68
5480	2" x 12", ordinary		525	.030		2.51	1.67		4.18	5.40
5540	Hip and valley jacks, 2" x 6", ordinary		600	.027		1.06	1.46		2.52	3.49
5600	2" x 8", ordinary		490	.033		1.62	1.79		3.41	4.62
5660	2" x 10", ordinary		450	.036		2.03	1.94		3.97	5.35
5720	2" x 12", ordinary		375	.043		2.51	2.33		4.84	6.50
5761	For slopes steeper than 4 in 12, add						30%			
5780	Rafter tie, 1" x 4", #3	2 Carp	800	.020	L.F.	.83	1.09		1.92	2.65
5790	2" x 4", #3		800	.020		.70	1.09		1.79	2.51
5800	Ridge board, #2 or better, 1" x 6"		600	.027		1.15	1.46		2.61	3.58
5820	1" x 8"		550	.029		1.69	1.59		3.28	4.39
5840	1" x 10"		500	.032		2.85	1.75		4.60	5.90
5860	2" x 6"		500	.032		1.06	1.75		2.81	3.96
5880	2" x 8"		450	.036		1.62	1.94		3.56	4.88
5900	2" x 10"		400	.040		2.03	2.19		4.22	5.70
5920	Roof cants, split, 4" x 4"		650	.025		2.31	1.35		3.66	4.68
5940	6" x 6"		600	.027		6.85	1.46		8.31	9.85
5960	Roof curbs, untreated, 2" x 6"		520	.031		1.06	1.68		2.74	3.85
5980	2" x 12"		400	.040		2.51	2.19		4.70	6.25
6000	Sister rafters, 2" x 6"		800	.020		1.06	1.09		2.15	2.91
6020	2" x 8"		640	.025		1.62	1.37		2.99	3.96
6040	2" x 10"		535	.030		2.03	1.64		3.67	4.84
6060	2" x 12"		455	.035		2.51	1.92		4.43	5.85
9000	Minimum labor/equipment charge	1 Carp	4	2	Job		109		109	174

06 11 10.32 Sill and Ledger Framing

		Crew	Daily Output	Labor-Hours	Unit	Material	2021 Bare Costs Labor	2021 Bare Costs Equipment	Total	Total Incl O&P
0010	**SILL AND LEDGER FRAMING**									
0020	Extruded polystyrene sill sealer, 5-1/2" wide	1 Carp	1600	.005	L.F.	.14	.27		.41	.59
2002	Ledgers, nailed, 2" x 4"	2 Carp	755	.021		.70	1.16		1.86	2.62
2052	2" x 6"	"	600	.027		1.06	1.46		2.52	3.49
2102	Bolted, not including bolts, 3" x 6"	2 Carp	325	.049	L.F.	3.03	2.69		5.72	7.65
2152	3" x 12"		233	.069		7	3.76		10.76	13.70
2602	Mud sills, redwood, construction grade, 2" x 4"		895	.018		5.45	.98		6.43	7.55
2622	2" x 6"		780	.021		9.35	1.12		10.47	12.10

06 11 Wood Framing

06 11 10 – Framing with Dimensional, Engineered or Composite Lumber

06 11 10.32 Sill and Ledger Framing

		Crew	Daily Output	Labor-Hours	Unit	Material	2021 Bare Costs Labor	2021 Bare Costs Equipment	Total	Total Incl O&P
4002	Sills, 2" x 4"	2 Carp	600	.027	L.F.	.69	1.46		2.15	3.08
4052	2" x 6"		550	.029		1.06	1.59		2.65	3.69
4082	2" x 8"		500	.032		1.60	1.75		3.35	4.55
4101	2" x 10"		450	.036		2.02	1.94		3.96	5.30
4121	2" x 12"		400	.040		2.50	2.19		4.69	6.25
4202	Treated, 2" x 4"		550	.029		.89	1.59		2.48	3.51
4222	2" x 6"		500	.032		1.25	1.75		3	4.17
4242	2" x 8"		450	.036		1.90	1.94		3.84	5.20
4261	2" x 10"		400	.040		2.02	2.19		4.21	5.70
4281	2" x 12"		350	.046		3.67	2.50		6.17	8
4402	4" x 4"		450	.036		2.23	1.94		4.17	5.55
4422	4" x 6"		350	.046		2.92	2.50		5.42	7.20
4462	4" x 8"		300	.053		6.15	2.92		9.07	11.45
4481	4" x 10"		260	.062		7.15	3.37		10.52	13.20
9000	Minimum labor/equipment charge	1 Carp	4	2	Job		109		109	174

06 11 10.34 Sleepers

		Crew	Daily Output	Labor-Hours	Unit	Material	2021 Bare Costs Labor	2021 Bare Costs Equipment	Total	Total Incl O&P
0010	**SLEEPERS**									
0100	On concrete, treated, 1" x 2"	2 Carp	2350	.007	L.F.	.45	.37		.82	1.08
0150	1" x 3"		2000	.008		.73	.44		1.17	1.51
0200	2" x 4"		1500	.011		.97	.58		1.55	1.99
0250	2" x 6"		1300	.012		1.40	.67		2.07	2.61
9000	Minimum labor/equipment charge	1 Carp	4	2	Job		109		109	174

06 11 10.36 Soffit and Canopy Framing

		Crew	Daily Output	Labor-Hours	Unit	Material	2021 Bare Costs Labor	2021 Bare Costs Equipment	Total	Total Incl O&P
0010	**SOFFIT AND CANOPY FRAMING**									
1002	Canopy or soffit framing, 1" x 4"	2 Carp	900	.018	L.F.	.83	.97		1.80	2.46
1021	1" x 6"		850	.019		1.15	1.03		2.18	2.90
1042	1" x 8"		750	.021		1.69	1.17		2.86	3.72
1102	2" x 4"		620	.026		.70	1.41		2.11	3.02
1121	2" x 6"		560	.029		1.06	1.56		2.62	3.66
1142	2" x 8"		500	.032		1.62	1.75		3.37	4.57
1202	3" x 4"		500	.032		2.13	1.75		3.88	5.15
1221	3" x 6"		400	.040		3.05	2.19		5.24	6.85
1242	3" x 10"		300	.053		4.96	2.92		7.88	10.10
9000	Minimum labor/equipment charge	1 Carp	4	2	Job		109		109	174

06 11 10.38 Treated Lumber Framing Material

		Crew	Daily Output	Labor-Hours	Unit	Material	2021 Bare Costs Labor	2021 Bare Costs Equipment	Total	Total Incl O&P
0010	**TREATED LUMBER FRAMING MATERIAL**									
0100	2" x 4"				M.B.F.	1,350			1,350	1,475
0110	2" x 6"					1,250			1,250	1,375
0120	2" x 8"					1,425			1,425	1,575
0130	2" x 10"					1,200			1,200	1,325
0140	2" x 12"				M.B.F.	1,825			1,825	2,025
0200	4" x 4"					1,675			1,675	1,850
0210	4" x 6"					1,450			1,450	1,600
0220	4" x 8"					2,300			2,300	2,550

06 11 10.40 Wall Framing

		Crew	Daily Output	Labor-Hours	Unit	Material	2021 Bare Costs Labor	2021 Bare Costs Equipment	Total	Total Incl O&P
0010	**WALL FRAMING** R061110-30									
0100	Door buck, studs, header, access, 8' high, 2" x 4" wall, 3' wide	1 Carp	32	.250	Ea.	29.50	13.70		43.20	54
0110	4' wide		32	.250		31.50	13.70		45.20	56.50
0120	5' wide		32	.250		39.50	13.70		53.20	65
0130	6' wide		32	.250		42.50	13.70		56.20	69
0140	8' wide		30	.267		56	14.60		70.60	84.50
0150	10' wide		30	.267		73.50	14.60		88.10	104

For customer support on your Commercial Renovation Costs with RSMeans Data, call 800.448.8182.

06 11 Wood Framing

06 11 10 – Framing with Dimensional, Engineered or Composite Lumber

06 11 10.40 Wall Framing	Crew	Daily Output	Labor-Hours	Unit	Material	2021 Bare Costs Labor	Equipment	Total	Total Incl O&P
0160	12' wide	1 Carp	30	.267	Ea.	109	14.60		123.60	143
0170	2" x 6" wall, 3' wide		32	.250		41	13.70		54.70	67
0180	4' wide		32	.250		43	13.70		56.70	69.50
0190	5' wide		32	.250		51	13.70		64.70	78
0200	6' wide		32	.250		54	13.70		67.70	81.50
0210	8' wide		30	.267		67.50	14.60		82.10	97.50
0220	10' wide		30	.267		85.50	14.60		100.10	117
0230	12' wide		30	.267		121	14.60		135.60	156
0240	Window buck, studs, header & access, 8' high 2" x 4" wall, 2' wide		24	.333		31	18.25		49.25	63
0250	3' wide		24	.333		36.50	18.25		54.75	69
0260	4' wide		24	.333		40	18.25		58.25	73
0270	5' wide		24	.333		47.50	18.25		65.75	81.50
0280	6' wide		24	.333		53	18.25		71.25	87.50
0290	7' wide		24	.333		64	18.25		82.25	99.50
0300	8' wide		22	.364		70.50	19.90		90.40	109
0310	10' wide		22	.364		90.50	19.90		110.40	131
0320	12' wide		22	.364		130	19.90		149.90	175
0330	2" x 6" wall, 2' wide		24	.333		46	18.25		64.25	79.50
0340	3' wide		24	.333		51	18.25		69.25	85.50
0350	4' wide		24	.333		55.50	18.25		73.75	90
0360	5' wide		24	.333		64	18.25		82.25	99
0370	6' wide		24	.333		70	18.25		88.25	106
0380	7' wide		24	.333		82.50	18.25		100.75	120
0390	8' wide		22	.364		90	19.90		109.90	131
0400	10' wide		22	.364		111	19.90		130.90	154
0410	12' wide	↓	22	.364	↓	153	19.90		172.90	200
2002	Headers over openings, 2" x 6"	2 Carp	360	.044	L.F.	1.06	2.43		3.49	5.05
2007	2" x 6", pneumatic nailed		432	.037		1.07	2.03		3.10	4.41
2052	2" x 8"		340	.047		1.62	2.57		4.19	5.90
2057	2" x 8", pneumatic nailed		408	.039		1.63	2.15		3.78	5.20
2101	2" x 10"	↓	320	.050	↓	2.03	2.74		4.77	6.60
2106	2" x 10", pneumatic nailed	2 Carp	384	.042	L.F.	2.05	2.28		4.33	5.90
2152	2" x 12"		300	.053		2.51	2.92		5.43	7.40
2157	2" x 12", pneumatic nailed		360	.044		2.53	2.43		4.96	6.65
2180	4" x 8"		260	.062		6.50	3.37		9.87	12.50
2185	4" x 8", pneumatic nailed		312	.051		6.50	2.81		9.31	11.60
2191	4" x 10"		240	.067		8	3.65		11.65	14.60
2196	4" x 10", pneumatic nailed		288	.056		8.05	3.04		11.09	13.70
2202	4" x 12"		190	.084		7.30	4.61		11.91	15.35
2207	4" x 12", pneumatic nailed		228	.070		7.30	3.84		11.14	14.15
2241	6" x 10"		165	.097		9.90	5.30		15.20	19.35
2246	6" x 10", pneumatic nailed		198	.081		9.95	4.42		14.37	18
2251	6" x 12"		140	.114		13.70	6.25		19.95	25
2256	6" x 12", pneumatic nailed		168	.095		13.75	5.20		18.95	23.50
5002	Plates, untreated, 2" x 3"		850	.019		.60	1.03		1.63	2.30
5007	2" x 3", pneumatic nailed		1020	.016		.60	.86		1.46	2.03
5022	2" x 4"		800	.020		.70	1.09		1.79	2.51
5027	2" x 4", pneumatic nailed		960	.017		.71	.91		1.62	2.23
5041	2" x 6"		750	.021		1.06	1.17		2.23	3.03
5046	2" x 6", pneumatic nailed		900	.018		1.07	.97		2.04	2.73
5122	Studs, 8' high wall, 2" x 3"		1200	.013		.60	.73		1.33	1.82
5127	2" x 3", pneumatic nailed		1440	.011		.60	.61		1.21	1.63
5142	2" x 4"	↓	1100	.015	↓	.69	.80		1.49	2.03

For customer support on your Commercial Renovation Costs with RSMeans Data, call 800.448.8182.

153

06 11 Wood Framing

06 11 10 – Framing with Dimensional, Engineered or Composite Lumber

06 11 10.40 Wall Framing

		Crew	Daily Output	Labor-Hours	Unit	Material	2021 Bare Costs Labor	Equipment	Total	Total Incl O&P
5147	2" x 4", pneumatic nailed	2 Carp	1320	.012	L.F.	.71	.66		1.37	1.84
5162	2" x 6"		1000	.016		1.06	.88		1.94	2.56
5167	2" x 6", pneumatic nailed		1200	.013		1.07	.73		1.80	2.34
5182	3" x 4"		800	.020		2.13	1.09		3.22	4.08
5187	3" x 4", pneumatic nailed		960	.017		2.14	.91		3.05	3.80
8200	For 12' high walls, deduct						5%			
8220	For stub wall, 6' high, add						20%			
8240	3' high, add						40%			
8250	For second story & above, add						5%			
8300	For dormer & gable, add						15%			
9000	Minimum labor/equipment charge	1 Carp	4	2	Job		109		109	174

06 11 10.42 Furring

		Crew	Daily Output	Labor-Hours	Unit	Material	2021 Bare Costs Labor	Equipment	Total	Total Incl O&P
0010	**FURRING**									
0012	Wood strips, 1" x 2", on walls, on wood	1 Carp	550	.015	L.F.	.42	.80		1.22	1.73
0015	On wood, pneumatic nailed		710	.011		.42	.62		1.04	1.44
0300	On masonry		495	.016		.45	.88		1.33	1.91
0400	On concrete		260	.031		.45	1.68		2.13	3.18
0600	1" x 3", on walls, on wood		550	.015		.62	.80		1.42	1.95
0605	On wood, pneumatic nailed		710	.011		.62	.62		1.24	1.66
0700	On masonry	1 Carp	495	.016	L.F.	.67	.88		1.55	2.15
0800	On concrete		260	.031		.67	1.68		2.35	3.42
0850	On ceilings, on wood		350	.023		.62	1.25		1.87	2.67
0855	On wood, pneumatic nailed		450	.018		.62	.97		1.59	2.23
0900	On masonry		320	.025		.67	1.37		2.04	2.92
0950	On concrete		210	.038		.67	2.08		2.75	4.06
9000	Minimum labor/equipment charge		4	2	Job		109		109	174

06 11 10.44 Grounds

		Crew	Daily Output	Labor-Hours	Unit	Material	2021 Bare Costs Labor	Equipment	Total	Total Incl O&P
0010	**GROUNDS**									
0020	For casework, 1" x 2" wood strips, on wood	1 Carp	330	.024	L.F.	.42	1.33		1.75	2.57
0100	On masonry		285	.028		.45	1.54		1.99	2.94
0200	On concrete		250	.032		.45	1.75		2.20	3.29
0400	For plaster, 3/4" deep, on wood		450	.018		.42	.97		1.39	2.01
0500	On masonry		225	.036		.45	1.94		2.39	3.60
0600	On concrete		175	.046		.45	2.50		2.95	4.48
0700	On metal lath		200	.040		.45	2.19		2.64	3.98
9000	Minimum labor/equipment charge		4	2	Job		109		109	174

06 12 Structural Panels

06 12 10 – Structural Insulated Panels

06 12 10.10 OSB Faced Panels

			Crew	Daily Output	Labor-Hours	Unit	Material	2021 Bare Costs Labor	Equipment	Total	Total Incl O&P
0010	**OSB FACED PANELS**										
0100	Structural insul. panels, 7/16" OSB both faces, EPS insul., 3-5/8" T	G	F-3	2075	.019	S.F.	4.29	1.08	.23	5.60	6.70
0110	5-5/8" thick	G		1725	.023		4.36	1.30	.28	5.94	7.15
0120	7-3/8" thick	G		1425	.028		5.45	1.57	.33	7.35	8.85
0130	9-3/8" thick	G		1125	.036		5.90	1.99	.42	8.31	10.15
0140	7/16" OSB one face, EPS insul., 3-5/8" thick	G		2175	.018		5.05	1.03	.22	6.30	7.40
0150	5-5/8" thick	G		1825	.022		5.20	1.23	.26	6.69	8
0160	7-3/8" thick	G		1525	.026		6	1.47	.31	7.78	9.25
0170	9-3/8" thick	G		1225	.033		6.40	1.83	.39	8.62	10.40
0190	7/16" OSB - 1/2" GWB faces, EPS insul., 3-5/8" T	G		2075	.019		3.66	1.08	.23	4.97	6

154

06 12 Structural Panels

06 12 10 – Structural Insulated Panels

06 12 10.10 OSB Faced Panels

		Crew	Daily Output	Labor-Hours	Unit	Material	2021 Bare Costs Labor	2021 Bare Costs Equipment	Total	Total Incl O&P
0200	5-5/8" thick [G]	F-3	1725	.023	S.F.	4.74	1.30	.28	6.32	7.55
0210	7-3/8" thick [G]		1425	.028		5.45	1.57	.33	7.35	8.85
0220	9-3/8" thick [G]		1125	.036		5.20	1.99	.42	7.61	9.40
0240	7/16" OSB - 1/2" MRGWB faces, EPS insul., 3-5/8" T [G]		2075	.019		4.49	1.08	.23	5.80	6.90
0250	5-5/8" thick [G]		1725	.023		5.45	1.30	.28	7.03	8.35
0260	7-3/8" thick [G]		1425	.028		5.60	1.57	.33	7.50	9
0270	9-3/8" thick [G]	▼	1125	.036		5.40	1.99	.42	7.81	9.60
0300	For 1/2" GWB added to OSB skin, add [G]					1.54			1.54	1.69
0310	For 1/2" MRGWB added to OSB skin, add [G]					1.70			1.70	1.87
0320	For one T1-11 skin, add to OSB-OSB [G]					2.01			2.01	2.21
0330	For one 19/32" CDX skin, add to OSB-OSB [G]				▼	1.65			1.65	1.82
0500	Structural insulated panel, 7/16" OSB both sides, straw core									
0510	4-3/8" T, walls (w/sill, splines, plates) [G]	F-6	2400	.017	S.F.	8.30	.87	.20	9.37	10.75
0520	Floors (w/splines) [G]		2400	.017		8.30	.87	.20	9.37	10.75
0530	Roof (w/splines) [G]		2400	.017		8.30	.87	.20	9.37	10.75
0550	7-7/8" T, walls (w/sill, splines, plates) [G]		2400	.017		11.75	.87	.20	12.82	14.55
0560	Floors (w/splines) [G]		2400	.017		11.75	.87	.20	12.82	14.55
0570	Roof (w/splines) [G]	▼	2400	.017	▼	11.75	.87	.20	12.82	14.55

06 12 19 – Composite Shearwall Panels

06 12 19.10 Steel and Wood Composite Shearwall Panels

		Crew	Daily Output	Labor-Hours	Unit	Material	2021 Bare Costs Labor	2021 Bare Costs Equipment	Total	Total Incl O&P
0010	**STEEL & WOOD COMPOSITE SHEARWALL PANELS**									
0020	Anchor bolts, 36" long (must be placed in wet concrete)	1 Carp	150	.053	Ea.	43	2.92		45.92	52
0030	On concrete, 2" x 4" & 2" x 6" walls, 7'-10' high, 360 lb. shear, 12" wide	2 Carp	8	2		510	109		619	735
0040	715 lb. shear, 15" wide		8	2		635	109		744	870
0050	1860 lb. shear, 18" wide		8	2		690	109		799	935
0060	2780 lb. shear, 21" wide		8	2		810	109		919	1,075
0070	3790 lb. shear, 24" wide		8	2		800	109		909	1,050
0080	2" x 6" walls, 11'-13' high, 1180 lb. shear, 18" wide		6	2.667		765	146		911	1,075
0090	1555 lb. shear, 21" wide		6	2.667		1,175	146		1,321	1,500
0100	2280 lb. shear, 24" wide	▼	6	2.667	▼	910	146		1,056	1,225
0110	For installing above on wood floor frame, add									
0120	Coupler nuts, threaded rods, bolts, shear transfer plate kit	1 Carp	16	.500	Ea.	64.50	27.50		92	115
0130	Framing anchors, angle (2 required)	"	96	.083	"	2.29	4.56		6.85	9.75
0140	For blocking see Section 06 11 10.02									
0150	For installing above, first floor to second floor, wood floor frame, add									
0160	Add stack option to first floor wall panel				Ea.	71			71	78
0170	Threaded rods, bolts, shear transfer plate kit	1 Carp	16	.500		79.50	27.50		107	131
0180	Framing anchors, angle (2 required)	"	96	.083	▼	2.29	4.56		6.85	9.75
0190	For blocking see section 06 11 10.02									
0200	For installing stacked panels, balloon framing									
0210	Add stack option to first floor wall panel				Ea.	71			71	78
0220	Threaded rods, bolts kit	1 Carp	16	.500	"	45	27.50		72.50	93

06 13 Heavy Timber Construction

06 13 23 – Heavy Timber Framing

06 13 23.10 Heavy Framing

		Crew	Daily Output	Labor-Hours	Unit	Material	2021 Bare Costs Labor	Equipment	Total	Total Incl O&P
0010	**HEAVY FRAMING**									
0020	Beams, single 6" x 10"	2 Carp	1.10	14.545	M.B.F.	2,150	795		2,945	3,650
0100	Single 8" x 16"		1.20	13.333	"	2,675	730		3,405	4,075
0202	Built from 2" lumber, multiple 2" x 14"		900	.018	B.F.	1.52	.97		2.49	3.22
0212	Built from 3" lumber, multiple 3" x 6"		700	.023	"	2.02	1.25		3.27	4.21
0222	Multiple 3" x 8"	2 Carp	800	.020	B.F.	2.04	1.09		3.13	3.98
0232	Multiple 3" x 10"		900	.018		1.98	.97		2.95	3.72
0242	Multiple 3" x 12"		1000	.016		2.33	.88		3.21	3.96
0252	Built from 4" lumber, multiple 4" x 6"		800	.020		1.44	1.09		2.53	3.32
0262	Multiple 4" x 8"		900	.018		2.43	.97		3.40	4.22
0272	Multiple 4" x 10"		1000	.016		2.40	.88		3.28	4.03
0282	Multiple 4" x 12"		1100	.015		1.81	.80		2.61	3.26
0292	Columns, structural grade, 1500 fb, 4" x 4"		450	.036	L.F.	3.23	1.94		5.17	6.65
0302	6" x 6"		225	.071		7	3.89		10.89	13.85
0402	8" x 8"		240	.067		10.90	3.65		14.55	17.80
0502	10" x 10"		90	.178		20.50	9.70		30.20	38
0602	12" x 12"		70	.229		26	12.50		38.50	48.50
0802	Floor planks, 2" thick, T&G, 2" x 6"		1050	.015	B.F.	1.90	.83		2.73	3.42
0902	2" x 10"		1100	.015		1.92	.80		2.72	3.38
1102	3" thick, 3" x 6"		1050	.015		2.27	.83		3.10	3.82
1202	3" x 10"		1100	.015		2.31	.80		3.11	3.81
1402	Girders, structural grade, 12" x 12"		800	.020		2.16	1.09		3.25	4.11
1502	10" x 16"		1000	.016		3.60	.88		4.48	5.35
2302	Roof purlins, 4" thick, structural grade		1050	.015		2.43	.83		3.26	4
9000	Minimum labor/equipment charge	1 Carp	2	4	Job		219		219	350

06 15 Wood Decking

06 15 16 – Wood Roof Decking

06 15 16.10 Solid Wood Roof Decking

		Crew	Daily Output	Labor-Hours	Unit	Material	2021 Bare Costs Labor	Equipment	Total	Total Incl O&P
0010	**SOLID WOOD ROOF DECKING**									
0350	Cedar planks, 2" thick	2 Carp	350	.046	S.F.	8.40	2.50		10.90	13.25
0400	3" thick		320	.050		12.60	2.74		15.34	18.20
0500	4" thick		250	.064		16.80	3.50		20.30	24
0550	6" thick		200	.080		25	4.38		29.38	34.50
0650	Douglas fir, 2" thick		350	.046		2.74	2.50		5.24	7
0700	3" thick		320	.050		4.11	2.74		6.85	8.90
0800	4" thick		250	.064		5.50	3.50		9	11.60
0850	6" thick		200	.080		8.20	4.38		12.58	16
0950	Hemlock, 2" thick		350	.046		2.79	2.50		5.29	7.05
1000	3" thick		320	.050		4.18	2.74		6.92	8.95
1100	4" thick		250	.064		5.60	3.50		9.10	11.70
1150	6" thick		200	.080		8.35	4.38		12.73	16.15
1250	Western white spruce, 2" thick		350	.046		1.78	2.50		4.28	5.95
1300	3" thick		320	.050		2.67	2.74		5.41	7.30
1400	4" thick		250	.064		3.56	3.50		7.06	9.45
1450	6" thick		200	.080		5.35	4.38		9.73	12.85
9000	Minimum labor/equipment charge	1 Carp	2	4	Job		219		219	350

06 15 Wood Decking

06 15 23 – Laminated Wood Decking

06 15 23.10 Laminated Roof Deck

		Crew	Daily Output	Labor-Hours	Unit	Material	2021 Bare Costs Labor	Equipment	Total	Total Incl O&P
0010	**LAMINATED ROOF DECK**									
0020	Pine or hemlock, 3" thick	2 Carp	425	.038	S.F.	7.95	2.06		10.01	12.05
0100	4" thick		325	.049		10.10	2.69		12.79	15.40
0300	Cedar, 3" thick		425	.038		11.50	2.06		13.56	15.95
0400	4" thick		325	.049		15.50	2.69		18.19	21.50
0600	Fir, 3" thick		425	.038		9	2.06		11.06	13.20
0700	4" thick		325	.049		11.45	2.69		14.14	16.90
9000	Minimum labor/equipment charge	1 Carp	3	2.667	Job		146		146	232

06 16 Sheathing

06 16 13 – Insulating Sheathing

06 16 13.10 Insulating Sheathing

			Crew	Daily Output	Labor-Hours	Unit	Material	2021 Bare Costs Labor	Equipment	Total	Total Incl O&P
0010	**INSULATING SHEATHING**										
0020	Expanded polystyrene, 1#/C.F. density, 3/4" thick, R2.89	G	2 Carp	1400	.011	S.F.	.40	.63		1.03	1.44
0030	1" thick, R3.85	G		1300	.012		.48	.67		1.15	1.60
0040	2" thick, R7.69	G		1200	.013		.80	.73		1.53	2.04
0050	Extruded polystyrene, 15 psi compressive strength, 1" thick, R5	G		1300	.012		.68	.67		1.35	1.82
0060	2" thick, R10	G		1200	.013		.83	.73		1.56	2.08
0070	Polyisocyanurate, 2#/C.F. density, 3/4" thick	G		1400	.011		.65	.63		1.28	1.71
0080	1" thick	G		1300	.012		.62	.67		1.29	1.75
0090	1-1/2" thick	G		1250	.013		1.06	.70		1.76	2.27
0100	2" thick	G		1200	.013		1.11	.73		1.84	2.38

06 16 23 – Subflooring

06 16 23.10 Subfloor

			Crew	Daily Output	Labor-Hours	Unit	Material	2021 Bare Costs Labor	Equipment	Total	Total Incl O&P
0010	**SUBFLOOR**	R061636-20									
0011	Plywood, CDX, 1/2" thick		2 Carp	1500	.011	SF Flr.	1	.58		1.58	2.03
0015	Pneumatic nailed			1860	.009		1	.47		1.47	1.85
0102	5/8" thick			1350	.012		1.16	.65		1.81	2.30
0107	Pneumatic nailed			1674	.010		1.16	.52		1.68	2.10
0202	3/4" thick			1250	.013		1.38	.70		2.08	2.63
0207	Pneumatic nailed			1550	.010		1.38	.56		1.94	2.42
0302	1-1/8" thick, 2-4-1 including underlayment			1050	.015		2.88	.83		3.71	4.50
0440	With boards, 1" x 6", S4S, laid regular			900	.018		2.49	.97		3.46	4.29
0452	1" x 8", laid regular			1000	.016		2.69	.88		3.57	4.35
0462	Laid diagonal			850	.019		2.69	1.03		3.72	4.60
0502	1" x 10", laid regular			1100	.015		3.58	.80		4.38	5.20
0602	Laid diagonal			900	.018		3.58	.97		4.55	5.50
8990	Subfloor adhesive, 3/8" bead		1 Carp	2300	.003	L.F.	.10	.19		.29	.41
9000	Minimum labor/equipment charge		"	4	2	Job		109		109	174

06 16 26 – Underlayment

06 16 26.10 Wood Product Underlayment

			Crew	Daily Output	Labor-Hours	Unit	Material	2021 Bare Costs Labor	Equipment	Total	Total Incl O&P
0010	**WOOD PRODUCT UNDERLAYMENT**	R061636-20									
0015	Plywood, underlayment grade, 1/4" thick		2 Carp	1500	.011	S.F.	1.34	.58		1.92	2.40
0018	Pneumatic nailed		2 Carp	1860	.009	S.F.	1.34	.47		1.81	2.22
0030	3/8" thick			1500	.011		1.48	.58		2.06	2.56
0070	Pneumatic nailed			1860	.009		1.48	.47		1.95	2.38
0102	1/2" thick			1450	.011		1.83	.60		2.43	2.98
0107	Pneumatic nailed			1798	.009		1.83	.49		2.32	2.80
0202	5/8" thick			1400	.011		2.16	.63		2.79	3.37

06 16 26 – Underlayment

06 16 26.10 Wood Product Underlayment		Crew	Daily Output	Labor-Hours	Unit	Material	2021 Bare Costs Labor	Equipment	Total	Total Incl O&P
0207	Pneumatic nailed	2 Carp	1736	.009	S.F.	2.16	.50		2.66	3.17
0302	3/4" thick		1300	.012		2.21	.67		2.88	3.50
0306	Pneumatic nailed		1612	.010		2.21	.54		2.75	3.29
0502	Particle board, 3/8" thick		1500	.011		.45	.58		1.03	1.42
0507	Pneumatic nailed		1860	.009		.45	.47		.92	1.24
0602	1/2" thick		1450	.011		.71	.60		1.31	1.75
0607	Pneumatic nailed		1798	.009		.71	.49		1.20	1.57
0802	5/8" thick		1400	.011		.98	.63		1.61	2.08
0807	Pneumatic nailed		1736	.009		.98	.50		1.48	1.88
0902	3/4" thick		1300	.012		1.06	.67		1.73	2.24
0907	Pneumatic nailed		1612	.010		1.06	.54		1.60	2.03
1100	Hardboard, underlayment grade, 4' x 4', .215" thick G		1500	.011		.94	.58		1.52	1.96
9000	Minimum labor/equipment charge	1 Carp	4	2	Job		109		109	174

06 16 33 – Wood Board Sheathing

06 16 33.10 Board Sheathing		Crew	Daily Output	Labor-Hours	Unit	Material	2021 Bare Costs Labor	Equipment	Total	Total Incl O&P
0009	**BOARD SHEATHING**									
0010	Roof, 1" x 6" boards, laid horizontal	2 Carp	725	.022	S.F.	2.49	1.21		3.70	4.66
0020	On steep roof		520	.031		2.49	1.68		4.17	5.40
0040	On dormers, hips, & valleys		480	.033		2.49	1.82		4.31	5.65
0050	Laid diagonal		650	.025		2.49	1.35		3.84	4.88
0070	1" x 8" boards, laid horizontal		875	.018		2.69	1		3.69	4.55
0080	On steep roof		635	.025		2.69	1.38		4.07	5.15
0090	On dormers, hips, & valleys		580	.028		2.75	1.51		4.26	5.45
0100	Laid diagonal		725	.022		2.69	1.21		3.90	4.88
0110	Skip sheathing, 1" x 4", 7" OC	1 Carp	1200	.007		1.03	.36		1.39	1.72
0120	1" x 6", 9" OC		1450	.006		1.14	.30		1.44	1.74
0180	T&G sheathing/decking, 1" x 6"		1000	.008		2.44	.44		2.88	3.38
0190	2" x 6"		1000	.008		5.60	.44		6.04	6.90
0200	Walls, 1" x 6" boards, laid regular	2 Carp	650	.025		2.49	1.35		3.84	4.88
0210	Laid diagonal		585	.027		2.49	1.50		3.99	5.10
0220	1" x 8" boards, laid regular		765	.021		2.69	1.14		3.83	4.78
0230	Laid diagonal		650	.025		2.69	1.35		4.04	5.10

06 16 36 – Wood Panel Product Sheathing

06 16 36.10 Sheathing		Crew	Daily Output	Labor-Hours	Unit	Material	2021 Bare Costs Labor	Equipment	Total	Total Incl O&P
0010	**SHEATHING** R061110-30									
0012	Plywood on roofs, CDX									
0032	5/16" thick	2 Carp	1600	.010	S.F.	.90	.55		1.45	1.86
0037	Pneumatic nailed R061636-20		1952	.008		.90	.45		1.35	1.70
0052	3/8" thick		1525	.010		.78	.57		1.35	1.77
0057	Pneumatic nailed		1860	.009		.78	.47		1.25	1.61
0102	1/2" thick		1400	.011		1	.63		1.63	2.10
0103	Pneumatic nailed		1708	.009		1	.51		1.51	1.92
0202	5/8" thick		1300	.012		1.16	.67		1.83	2.34
0207	Pneumatic nailed		1586	.010		1.16	.55		1.71	2.15
0302	3/4" thick		1200	.013		1.38	.73		2.11	2.68
0307	Pneumatic nailed		1464	.011		1.38	.60		1.98	2.47
0502	Plywood on walls, with exterior CDX, 3/8" thick		1200	.013		.78	.73		1.51	2.02
0507	Pneumatic nailed		1488	.011		.78	.59		1.37	1.80
0602	1/2" thick		1125	.014		1	.78		1.78	2.34
0607	Pneumatic nailed		1395	.011		1	.63		1.63	2.10
0702	5/8" thick		1050	.015		1.16	.83		1.99	2.60
0707	Pneumatic nailed		1302	.012		1.16	.67		1.83	2.34

For customer support on your Commercial Renovation Costs with RSMeans Data, call 800.448.8182.

06 16 Sheathing

06 16 36 – Wood Panel Product Sheathing

06 16 36.10 Sheathing

		Crew	Daily Output	Labor-Hours	Unit	Material	2021 Bare Costs Labor	2021 Bare Costs Equipment	Total	Total Incl O&P
0802	3/4" thick	2 Carp	975	.016	S.F.	1.38	.90		2.28	2.95
0807	Pneumatic nailed	↓	1209	.013	↓	1.38	.72		2.10	2.67
1000	For shear wall construction, add						20%			
1200	For structural 1 exterior plywood, add				S.F.	10%				
3000	Wood fiber, regular, no vapor barrier, 1/2" thick	2 Carp	1200	.013		.72	.73		1.45	1.95
3100	5/8" thick		1200	.013		.68	.73		1.41	1.91
3300	No vapor barrier, in colors, 1/2" thick		1200	.013		.78	.73		1.51	2.02
3400	5/8" thick		1200	.013		.82	.73		1.55	2.06
3600	With vapor barrier one side, white, 1/2" thick		1200	.013		.80	.73		1.53	2.04
3700	Vapor barrier 2 sides, 1/2" thick		1200	.013		.95	.73		1.68	2.21
3800	Asphalt impregnated, 25/32" thick		1200	.013		.43	.73		1.16	1.64
3850	Intermediate, 1/2" thick		1200	.013		.45	.73		1.18	1.65
4500	Oriented strand board, on roof, 7/16" thick Ⓖ		1460	.011		.41	.60		1.01	1.40
4505	Pneumatic nailed Ⓖ		1780	.009		.41	.49		.90	1.23
4550	1/2" thick Ⓖ		1400	.011		.41	.63		1.04	1.45
4555	Pneumatic nailed Ⓖ		1736	.009		.41	.50		.91	1.25
4600	5/8" thick Ⓖ		1300	.012		1.05	.67		1.72	2.23
4605	Pneumatic nailed Ⓖ		1586	.010		1.05	.55		1.60	2.04
4610	On walls, 7/16" thick Ⓖ		1200	.013		.41	.73		1.14	1.61
4615	Pneumatic nailed Ⓖ		1488	.011		.41	.59		1	1.39
4620	1/2" thick Ⓖ		1195	.013		.41	.73		1.14	1.62
4625	Pneumatic nailed Ⓖ		1325	.012		.41	.66		1.07	1.50
4630	5/8" thick Ⓖ		1050	.015		1.05	.83		1.88	2.49
4635	Pneumatic nailed Ⓖ	↓	1302	.012	↓	1.05	.67		1.72	2.23
4700	Oriented strand board, factory laminated W.R. barrier, on roof, 1/2" thick Ⓖ	2 Carp	1400	.011	S.F.	1.15	.63		1.78	2.26
4705	Pneumatic nailed Ⓖ		1736	.009		1.15	.50		1.65	2.06
4720	5/8" thick Ⓖ		1300	.012		1.40	.67		2.07	2.61
4725	Pneumatic nailed Ⓖ		1586	.010		1.40	.55		1.95	2.42
4730	5/8" thick, T&G Ⓖ		1150	.014		1.51	.76		2.27	2.87
4735	Pneumatic nailed, T&G Ⓖ		1400	.011		1.51	.63		2.14	2.66
4740	On walls, 7/16" thick Ⓖ		1200	.013		1.04	.73		1.77	2.30
4745	Pneumatic nailed Ⓖ		1488	.011		1.04	.59		1.63	2.08
4750	1/2" thick Ⓖ		1195	.013		1.15	.73		1.88	2.43
4755	Pneumatic nailed Ⓖ		1325	.012	↓	1.15	.66		1.81	2.31
4800	Joint sealant tape, 3-1/2"		7600	.002	L.F.	.37	.12		.49	.59
4810	Joint sealant tape, 6"	↓	7600	.002	"	.56	.12		.68	.80
9000	Minimum labor/equipment charge	1 Carp	2	4	Job		219		219	350

06 16 43 – Gypsum Sheathing

06 16 43.10 Gypsum Sheathing

		Crew	Daily Output	Labor-Hours	Unit	Material	2021 Bare Costs Labor	2021 Bare Costs Equipment	Total	Total Incl O&P
0010	**GYPSUM SHEATHING**									
0020	Gypsum, weatherproof, 1/2" thick	2 Carp	1125	.014	S.F.	.46	.78		1.24	1.75
0040	With embedded glass mats	"	1100	.015	"	.56	.80		1.36	1.89

06 17 Shop-Fabricated Structural Wood

06 17 33 – Wood I-Joists

06 17 33.10 Wood and Composite I-Joists

		Crew	Daily Output	Labor-Hours	Unit	Material	2021 Bare Costs Labor	Equipment	Total	Total Incl O&P
0010	**WOOD AND COMPOSITE I-JOISTS**									
0100	Plywood webs, incl. bridging & blocking, panels 24" OC									
1200	15' to 24' span, 50 psf live load	F-5	2400	.013	SF Flr.	2.59	.74		3.33	4.01
1300	55 psf live load		2250	.014		2.96	.79		3.75	4.50
1400	24' to 30' span, 45 psf live load		2600	.012		3.70	.68		4.38	5.15
1500	55 psf live load		2400	.013		6	.74		6.74	7.75

06 17 53 – Shop-Fabricated Wood Trusses

06 17 53.10 Roof Trusses

		Crew	Daily Output	Labor-Hours	Unit	Material	2021 Bare Costs Labor	Equipment	Total	Total Incl O&P
0010	**ROOF TRUSSES**									
0100	Fink (W) or King post type, 2'-0" OC									
0200	Metal plate connected, 4 in 12 slope									
0210	24' to 29' span	F-3	3000	.013	SF Flr.	2.62	.75	.16	3.53	4.23
0300	30' to 43' span		3000	.013		3.50	.75	.16	4.41	5.20
0400	44' to 60' span		3000	.013		3.37	.75	.16	4.28	5.05
0700	Glued and nailed, add					50%				

06 18 Glued-Laminated Construction

06 18 13 – Glued-Laminated Beams

06 18 13.20 Laminated Framing

		Crew	Daily Output	Labor-Hours	Unit	Material	2021 Bare Costs Labor	Equipment	Total	Total Incl O&P
0010	**LAMINATED FRAMING**									
0020	30 lb., short term live load, 15 lb. dead load									
0200	Straight roof beams, 20' clear span, beams 8' OC	F-3	2560	.016	SF Flr.	3.25	.88	.19	4.32	5.15
0300	Beams 16' OC	"	3200	.013	"	2.38	.70	.15	3.23	3.89
0500	40' clear span, beams 8' OC	F-3	3200	.013	SF Flr.	6.15	.70	.15	7	8.05
0600	Beams 16' OC	"	3840	.010		5.10	.58	.12	5.80	6.65
0800	60' clear span, beams 8' OC	F-4	2880	.017		10.55	.92	.34	11.81	13.50
0900	Beams 16' OC	"	3840	.013		7.90	.69	.26	8.85	10.10
1100	Tudor arches, 30' to 40' clear span, frames 8' OC	F-3	1680	.024		13.80	1.33	.28	15.41	17.60
1200	Frames 16' OC	"	2240	.018		10.75	1	.21	11.96	13.65
1400	50' to 60' clear span, frames 8' OC	F-4	2200	.022		14.80	1.21	.45	16.46	18.70
1500	Frames 16' OC		2640	.018		12.65	1.01	.38	14.04	15.90
1700	Radial arches, 60' clear span, frames 8' OC		1920	.025		13.85	1.39	.52	15.76	18
1800	Frames 16' OC		2880	.017		11.05	.92	.34	12.31	14.05
2000	100' clear span, frames 8' OC		1600	.030		14.30	1.66	.62	16.58	19
2100	Frames 16' OC		2400	.020		12.60	1.11	.41	14.12	16.10
2300	120' clear span, frames 8' OC		1440	.033		19	1.85	.69	21.54	24.50
2400	Frames 16' OC		1920	.025		17.35	1.39	.52	19.26	22
2600	Bowstring trusses, 20' OC, 40' clear span	F-3	2400	.017		8.60	.93	.20	9.73	11.15
2700	60' clear span	F-4	3600	.013		7.75	.74	.28	8.77	9.95
2800	100' clear span		4000	.012		10.95	.67	.25	11.87	13.35
2900	120' clear span		3600	.013		11.60	.74	.28	12.62	14.25
3000	For less than 1000 B.F., add					20%				
3100	For premium appearance, add to S.F. prices					5%				
3300	For industrial type, deduct					15%				
3500	For stain and varnish, add					5%				
3900	For 3/4" laminations, add to straight					25%				
4100	Add to curved					15%				
4300	Alternate pricing method: (use nominal footage of									
4310	components). Straight beams, camber less than 6"	F-3	3.50	11.429	M.B.F.	4,050	640	136	4,826	5,650
4400	Columns, including hardware		2	20		4,350	1,125	238	5,713	6,825

160

06 18 13 – Glued-Laminated Beams

06 18 13.20 Laminated Framing	Crew	Daily Output	Labor-Hours	Unit	Material	2021 Bare Costs Labor	Equipment	Total	Total Incl O&P
4600 Curved members, radius over 32'	F-3	2.50	16	M.B.F.	4,450	895	190	5,535	6,525
4700 Radius 10' to 32'	↓	3	13.333		4,425	745	159	5,329	6,200
4900 For complicated shapes, add maximum					100%				
5100 For pressure treating, add to straight					35%				
5200 Add to curved				↓	45%				
6000 Laminated veneer members, southern pine or western species									
6050 1-3/4" wide x 5-1/2" deep	2 Carp	480	.033	L.F.	6.10	1.82		7.92	9.60
6100 9-1/2" deep		480	.033		5.20	1.82		7.02	8.60
6150 14" deep		450	.036		8.15	1.94		10.09	12.10
6200 18" deep	↓	450	.036	↓	11.30	1.94		13.24	15.50
6300 Parallel strand members, southern pine or western species									
6350 1-3/4" wide x 9-1/4" deep	2 Carp	480	.033	L.F.	5.40	1.82		7.22	8.85
6400 11-1/4" deep		450	.036		6.45	1.94		8.39	10.20
6450 14" deep		400	.040		9.50	2.19		11.69	13.95
6500 3-1/2" wide x 9-1/4" deep		480	.033		26	1.82		27.82	31.50
6550 11-1/4" deep		450	.036		28.50	1.94		30.44	34.50
6600 14" deep		400	.040		26	2.19		28.19	32
6650 7" wide x 9-1/4" deep		450	.036		48	1.94		49.94	56
6700 11-1/4" deep		420	.038		58.50	2.08		60.58	68
6750 14" deep	↓	400	.040	↓	70	2.19		72.19	80.50
9000 Minimum labor/equipment charge	F-3	2.50	16	Job		895	190	1,085	1,625

06 22 13 – Standard Pattern Wood Trim

06 22 13.15 Moldings, Base

06 22 13.15 Moldings, Base	Crew	Daily Output	Labor-Hours	Unit	Material	2021 Bare Costs Labor	Equipment	Total	Total Incl O&P
0010 **MOLDINGS, BASE**									
5100 Classic profile, 5/8" x 5-1/2", finger jointed and primed	1 Carp	250	.032	L.F.	1.66	1.75		3.41	4.62
5105 Poplar		240	.033		1.72	1.82		3.54	4.79
5110 Red oak		220	.036		2.64	1.99		4.63	6.05
5115 Maple		220	.036		3.98	1.99		5.97	7.55
5120 Cherry		220	.036		4.72	1.99		6.71	8.35
5125 3/4" x 7-1/2", finger jointed and primed		250	.032		2.10	1.75		3.85	5.10
5130 Poplar		240	.033		2.70	1.82		4.52	5.85
5135 Red oak		220	.036		3.74	1.99		5.73	7.30
5140 Maple		220	.036		5.20	1.99		7.19	8.90
5145 Cherry		220	.036		6.20	1.99		8.19	10
5150 Modern profile, 5/8" x 3-1/2", finger jointed and primed		250	.032		.99	1.75		2.74	3.88
5155 Poplar		240	.033		1.07	1.82		2.89	4.08
5160 Red oak		220	.036		1.76	1.99		3.75	5.10
5165 Maple		220	.036		2.74	1.99		4.73	6.20
5170 Cherry		220	.036		2.99	1.99		4.98	6.45
5175 Ogee profile, 7/16" x 3", finger jointed and primed		250	.032		.67	1.75		2.42	3.53
5180 Poplar		240	.033		.78	1.82		2.60	3.76
5185 Red oak		220	.036		1.02	1.99		3.01	4.29
5200 9/16" x 3-1/2", finger jointed and primed		250	.032		1.07	1.75		2.82	3.97
5205 Pine		240	.033		2.80	1.82		4.62	6
5210 Red oak		220	.036		2.96	1.99		4.95	6.45
5215 9/16" x 4-1/2", red oak		220	.036		4.44	1.99		6.43	8.05
5220 5/8" x 3-1/2", finger jointed and primed		250	.032		1.01	1.75		2.76	3.90
5225 Poplar		240	.033		1.07	1.82		2.89	4.08
5230 Red oak	↓	220	.036	↓	1.76	1.99		3.75	5.10

06 22 13.15 Moldings, Base

		Crew	Daily Output	Labor-Hours	Unit	Material	2021 Bare Costs Labor	Equipment	Total	Total Incl O&P
5235	Maple	1 Carp	220	.036	L.F.	2.74	1.99		4.73	6.20
5240	Cherry		220	.036		2.97	1.99		4.96	6.45
5245	5/8" x 4", finger jointed and primed		250	.032		1.30	1.75		3.05	4.22
5250	Poplar	1 Carp	240	.033	L.F.	1.39	1.82		3.21	4.43
5255	Red oak		220	.036		1.98	1.99		3.97	5.35
5260	Maple		220	.036		3.06	1.99		5.05	6.55
5265	Cherry		220	.036		3.22	1.99		5.21	6.70
5270	Rectangular profile, oak, 3/8" x 1-1/4"		260	.031		1.56	1.68		3.24	4.39
5275	1/2" x 2-1/2"		255	.031		2.28	1.72		4	5.25
5280	1/2" x 3-1/2"		250	.032		2.89	1.75		4.64	5.95
5285	1" x 6"		240	.033		2.84	1.82		4.66	6
5290	1" x 8"		240	.033		3.97	1.82		5.79	7.25
5295	Pine, 3/8" x 1-3/4"		260	.031		.50	1.68		2.18	3.23
5300	7/16" x 2-1/2"		255	.031		1.01	1.72		2.73	3.84
5305	1" x 6"		240	.033		1.90	1.82		3.72	4.99
5310	1" x 8"		240	.033		2.86	1.82		4.68	6.05
5315	Shoe, 1/2" x 3/4", primed		260	.031		.53	1.68		2.21	3.26
5320	Pine		240	.033		.78	1.82		2.60	3.76
5325	Poplar		240	.033		.44	1.82		2.26	3.38
5330	Red oak		220	.036		.59	1.99		2.58	3.82
5335	Maple		220	.036		.84	1.99		2.83	4.09
5340	Cherry		220	.036		.88	1.99		2.87	4.14
5345	11/16" x 1-1/2", pine		240	.033		.69	1.82		2.51	3.66
5350	Caps, 11/16" x 1-3/8", pine		240	.033		.68	1.82		2.50	3.65
5355	3/4" x 1-3/4", finger jointed and primed		260	.031		.86	1.68		2.54	3.63
5360	Poplar		240	.033		1.08	1.82		2.90	4.09
5365	Red oak		220	.036		1.37	1.99		3.36	4.68
5370	Maple		220	.036		1.71	1.99		3.70	5.05
5375	Cherry		220	.036		3.70	1.99		5.69	7.25
5380	Combination base & shoe, 9/16" x 3-1/2" & 1/2" x 3/4", pine		125	.064		3.58	3.50		7.08	9.50
5385	Three piece oak, 6" high		80	.100		4.80	5.45		10.25	14
5390	Including 3/4" x 1" base shoe		70	.114		5.40	6.25		11.65	15.90
5395	Flooring cant strip, 3/4" x 3/4", pre-finished pine		260	.031		.39	1.68		2.07	3.11
5400	For pre-finished, stain and clear coat, add					1			1	1.10
5405	Clear coat only, add					.85			.85	.94
9000	Minimum labor/equipment charge	1 Carp	4	2	Job		109		109	174

06 22 13.30 Moldings, Casings

		Crew	Daily Output	Labor-Hours	Unit	Material	2021 Bare Costs Labor	Equipment	Total	Total Incl O&P
0010	**MOLDINGS, CASINGS**									
0085	Apron, 9/16" x 2-1/2", pine	1 Carp	250	.032	L.F.	1.09	1.75		2.84	3.99
0090	5/8" x 2-1/2", pine		250	.032		1.90	1.75		3.65	4.88
0110	5/8" x 3-1/2", pine		220	.036		1.97	1.99		3.96	5.35
0300	Band, 11/16" x 1-1/8", pine		270	.030		1.85	1.62		3.47	4.62
0310	11/16" x 1-1/2", finger jointed and primed		270	.030		1.69	1.62		3.31	4.44
0320	Pine		270	.030		2.44	1.62		4.06	5.25
0330	11/16" x 1-3/4", finger jointed and primed		270	.030		1.69	1.62		3.31	4.44
0350	Pine	1 Carp	270	.030	L.F.	2.44	1.62		4.06	5.25
0355	Beaded, 3/4" x 3-1/2", finger jointed and primed		220	.036		1.01	1.99		3	4.28
0360	Poplar		220	.036		1.22	1.99		3.21	4.51
0365	Red oak		220	.036		1.81	1.99		3.80	5.15
0370	Maple		220	.036		2.80	1.99		4.79	6.25
0375	Cherry		220	.036		3.38	1.99		5.37	6.90
0380	3/4" x 4", finger jointed and primed		220	.036		1.82	1.99		3.81	5.15

06 22 13 – Standard Pattern Wood Trim

06 22 13.30 Moldings, Casings	Crew	Daily Output	Labor-Hours	Unit	Material	2021 Bare Costs Labor	Equipment	Total	Total Incl O&P
0385 Poplar	1 Carp	220	.036	L.F.	2.14	1.99		4.13	5.50
0390 Red oak		220	.036		2.54	1.99		4.53	5.95
0395 Maple		220	.036		4.38	1.99		6.37	8
0400 Cherry		220	.036		4.14	1.99		6.13	7.70
0405 3/4" x 5-1/2", finger jointed and primed		200	.040		1.66	2.19		3.85	5.30
0410 Poplar		200	.040		2.12	2.19		4.31	5.80
0415 Red oak		200	.040		3.12	2.19		5.31	6.90
0420 Maple		200	.040		4.09	2.19		6.28	8
0425 Cherry		200	.040		4.70	2.19		6.89	8.65
0430 Classic profile, 3/4" x 2-3/4", finger jointed and primed		250	.032		.88	1.75		2.63	3.76
0435 Poplar		250	.032		1.10	1.75		2.85	4
0440 Red oak		250	.032		1.61	1.75		3.36	4.56
0445 Maple		250	.032		2.29	1.75		4.04	5.30
0450 Cherry		250	.032		2.60	1.75		4.35	5.65
0455 Fluted, 3/4" x 3-1/2", poplar		220	.036		1.22	1.99		3.21	4.51
0460 Red oak		220	.036		1.81	1.99		3.80	5.15
0465 Maple		220	.036		2.80	1.99		4.79	6.25
0470 Cherry		220	.036		3.38	1.99		5.37	6.90
0475 3/4" x 4", poplar		220	.036		1.99	1.99		3.98	5.35
0480 Red oak		220	.036		2.38	1.99		4.37	5.80
0485 Maple		220	.036		4.38	1.99		6.37	8
0490 Cherry		220	.036		4.58	1.99		6.57	8.20
0495 3/4" x 5-1/2", poplar		200	.040		2.12	2.19		4.31	5.80
0500 Red oak		200	.040		3.12	2.19		5.31	6.90
0505 Maple		200	.040		4.09	2.19		6.28	8
0510 Cherry		200	.040		4.70	2.19		6.89	8.65
0515 3/4" x 7-1/2", poplar		190	.042		1.58	2.30		3.88	5.40
0520 Red oak		190	.042		5.40	2.30		7.70	9.55
0525 Maple		190	.042		7.40	2.30		9.70	11.80
0530 Cherry		190	.042		8.85	2.30		11.15	13.40
0535 3/4" x 9-1/2", poplar		180	.044		5.30	2.43		7.73	9.65
0540 Red oak		180	.044		8.35	2.43		10.78	13
0545 Maple		180	.044		11.50	2.43		13.93	16.50
0550 Cherry		180	.044		12.50	2.43		14.93	17.60
0555 Modern profile, 9/16" x 2-1/4", poplar		250	.032		1.36	1.75		3.11	4.29
0560 Red oak	1 Carp	250	.032	L.F.	1.83	1.75		3.58	4.80
0565 11/16" x 2-1/2", finger jointed & primed		250	.032		1.11	1.75		2.86	4.01
0570 Pine		250	.032		2.77	1.75		4.52	5.85
0575 3/4" x 2-1/2", poplar		250	.032		.97	1.75		2.72	3.86
0580 Red oak		250	.032		1.31	1.75		3.06	4.23
0585 Maple		250	.032		1.98	1.75		3.73	4.97
0590 Cherry		250	.032		2.49	1.75		4.24	5.55
0595 Mullion, 5/16" x 2", pine		270	.030		1.05	1.62		2.67	3.74
0600 9/16" x 2-1/2", finger jointed and primed		250	.032		1.11	1.75		2.86	4.01
0605 Pine		250	.032		1.35	1.75		3.10	4.28
0610 Red oak		250	.032		3.45	1.75		5.20	6.60
0615 1-1/16" x 3-3/4", red oak		220	.036		7.10	1.99		9.09	11
0620 Ogee, 7/16" x 2-1/2", poplar		250	.032		1.36	1.75		3.11	4.29
0625 Red oak		250	.032		.96	1.75		2.71	3.85
0630 9/16" x 2-1/4", finger jointed and primed		250	.032		.56	1.75		2.31	3.41
0635 Poplar		250	.032		.63	1.75		2.38	3.48
0640 Red oak		250	.032		.86	1.75		2.61	3.74
0645 11/16" x 2-1/2", finger jointed and primed		250	.032		1.11	1.75		2.86	4.01

06 22 13.30 Moldings, Casings

		Crew	Daily Output	Labor-Hours	Unit	Material	2021 Bare Costs Labor	Equipment	Total	Total Incl O&P
0700	Pine	1 Carp	250	.032	L.F.	1.38	1.75		3.13	4.31
0701	Red oak		250	.032		3.29	1.75		5.04	6.40
0730	11/16" x 3-1/2", finger jointed and primed		220	.036		1.01	1.99		3	4.28
0750	Pine		220	.036		1.70	1.99		3.69	5.05
0755	3/4" x 2-1/2", finger jointed and primed		250	.032		1.69	1.75		3.44	4.65
0760	Poplar		250	.032		1.46	1.75		3.21	4.40
0765	Red oak		250	.032		1.31	1.75		3.06	4.23
0770	Maple		250	.032		1.98	1.75		3.73	4.97
0775	Cherry		250	.032		2.49	1.75		4.24	5.55
0780	3/4" x 3-1/2", finger jointed and primed		220	.036		1.01	1.99		3	4.28
0785	Poplar		220	.036		1.22	1.99		3.21	4.51
0790	Red oak		220	.036		1.81	1.99		3.80	5.15
0795	Maple		220	.036		2.80	1.99		4.79	6.25
0800	Cherry		220	.036		3.38	1.99		5.37	6.90
4700	Square profile, 1" x 1", teak		215	.037		1.99	2.04		4.03	5.45
4800	Rectangular profile, 1" x 3", teak		200	.040		5.95	2.19		8.14	10.05
9000	Minimum labor/equipment charge		4	2	Job		109		109	174

06 22 13.35 Moldings, Ceilings

		Crew	Daily Output	Labor-Hours	Unit	Material	2021 Bare Costs Labor	Equipment	Total	Total Incl O&P
0010	**MOLDINGS, CEILINGS**									
0600	Bed, 9/16" x 1-3/4", pine	1 Carp	270	.030	L.F.	.79	1.62		2.41	3.45
0650	9/16" x 2", pine		270	.030		1.26	1.62		2.88	3.97
0710	9/16" x 1-3/4", oak		270	.030		2.61	1.62		4.23	5.45
1200	Cornice, 9/16" x 1-3/4", pine		270	.030		1.06	1.62		2.68	3.75
1300	9/16" x 2-1/4", pine		265	.030		1.47	1.65		3.12	4.25
1350	Cove, 1/2" x 2-1/4", poplar	1 Carp	265	.030	L.F.	1.37	1.65		3.02	4.14
1360	Red oak		265	.030		1.84	1.65		3.49	4.65
1370	Hard maple		265	.030		2.78	1.65		4.43	5.70
1380	Cherry		265	.030		3.50	1.65		5.15	6.50
2400	9/16" x 1-3/4", pine		270	.030		.76	1.62		2.38	3.42
2500	11/16" x 2-3/4", pine		265	.030		2.50	1.65		4.15	5.40
2510	Crown, 5/8" x 5/8", poplar		300	.027		.58	1.46		2.04	2.96
2520	Red oak		300	.027		.58	1.46		2.04	2.96
2530	Hard maple		300	.027		.78	1.46		2.24	3.18
2540	Cherry		300	.027		.82	1.46		2.28	3.22
2600	9/16" x 3-5/8", pine		250	.032		1.53	1.75		3.28	4.47
2700	11/16" x 4-1/4", pine		250	.032		2.44	1.75		4.19	5.45
2705	Oak		250	.032		3.17	1.75		4.92	6.30
2710	3/4" x 1-3/4", poplar		270	.030		.79	1.62		2.41	3.45
2720	Red oak		270	.030		1.16	1.62		2.78	3.86
2730	Hard maple		270	.030		1.53	1.62		3.15	4.26
2740	Cherry		270	.030		1.88	1.62		3.50	4.65
2750	3/4" x 2", poplar		270	.030		1.03	1.62		2.65	3.71
2760	Red oak		270	.030		1.38	1.62		3	4.10
2770	Hard maple		270	.030		1.85	1.62		3.47	4.62
2780	Cherry		270	.030		2.06	1.62		3.68	4.85
2790	3/4" x 2-3/4", poplar		265	.030		1.12	1.65		2.77	3.86
2800	Red oak		265	.030		1.72	1.65		3.37	4.52
2810	Hard maple		265	.030		2.26	1.65		3.91	5.10
2820	Cherry		265	.030		2.61	1.65		4.26	5.50
2830	3/4" x 3-1/2", poplar		250	.032		1.43	1.75		3.18	4.36
2840	Red oak		250	.032		2.15	1.75		3.90	5.15
2850	Hard maple		250	.032		2.81	1.75		4.56	5.90

06 22 Millwork

06 22 13 – Standard Pattern Wood Trim

06 22 13.35 Moldings, Ceilings

		Crew	Daily Output	Labor-Hours	Unit	Material	2021 Bare Costs Labor	Equipment	Total	Total Incl O&P
2860	Cherry	1 Carp	250	.032	L.F.	3.28	1.75		5.03	6.40
2870	FJP poplar		250	.032		1.02	1.75		2.77	3.91
2880	3/4" x 5", poplar		245	.033		2.12	1.79		3.91	5.15
2890	Red oak		245	.033		3.13	1.79		4.92	6.30
2900	Hard maple		245	.033		4.10	1.79		5.89	7.35
2910	Cherry		245	.033		4.72	1.79		6.51	8.05
2920	FJP poplar		245	.033		1.47	1.79		3.26	4.46
2930	3/4" x 6-1/4", poplar		240	.033		2.52	1.82		4.34	5.65
2940	Red oak		240	.033		3.80	1.82		5.62	7.10
2950	Hard maple		240	.033		4.96	1.82		6.78	8.35
2960	Cherry		240	.033		5.90	1.82		7.72	9.40
2970	7/8" x 8-3/4", poplar		220	.036		5.05	1.99		7.04	8.70
2980	Red oak		220	.036		8.35	1.99		10.34	12.35
2990	Hard maple		220	.036		11.50	1.99		13.49	15.80
3000	Cherry	1 Carp	220	.036	L.F.	8.05	1.99		10.04	12.05
3010	1" x 7-1/4", poplar		220	.036		4.14	1.99		6.13	7.75
3020	Red oak		220	.036		6.45	1.99		8.44	10.20
3030	Hard maple		220	.036		8.90	1.99		10.89	12.95
3040	Cherry		220	.036		10.35	1.99		12.34	14.55
3050	1-1/16" x 4-1/4", poplar		250	.032		2.51	1.75		4.26	5.55
3060	Red oak		250	.032		2.95	1.75		4.70	6.05
3070	Hard maple		250	.032		3.90	1.75		5.65	7.10
3080	Cherry		250	.032		5.55	1.75		7.30	8.95
3090	Dentil crown, 3/4" x 5", poplar		250	.032		2.04	1.75		3.79	5.05
3100	Red oak		250	.032		3.13	1.75		4.88	6.25
3110	Hard maple		250	.032		4.02	1.75		5.77	7.20
3120	Cherry		250	.032		4.72	1.75		6.47	8
3130	Dentil piece for above, 1/2" x 1/2", poplar		300	.027		2.99	1.46		4.45	5.60
3140	Red oak		300	.027		3.44	1.46		4.90	6.10
3150	Hard maple		300	.027		3.91	1.46		5.37	6.60
3160	Cherry		300	.027		4.26	1.46		5.72	7
9000	Minimum labor/equipment charge		4	2	Job		109		109	174

06 22 13.40 Moldings, Exterior

		Crew	Daily Output	Labor-Hours	Unit	Material	2021 Bare Costs Labor	Equipment	Total	Total Incl O&P
0010	**MOLDINGS, EXTERIOR**									
0100	Band board, cedar, rough sawn, 1" x 2"	1 Carp	300	.027	L.F.	.68	1.46		2.14	3.07
0110	1" x 3"		300	.027		1.01	1.46		2.47	3.43
0120	1" x 4"		250	.032		1.35	1.75		3.10	4.27
0130	1" x 6"		250	.032		2.02	1.75		3.77	5
0140	1" x 8"		225	.036		2.70	1.94		4.64	6.05
0150	1" x 10"		225	.036		3.35	1.94		5.29	6.80
0160	1" x 12"		200	.040		4.03	2.19		6.22	7.90
0240	STK, 1" x 2"		300	.027		.39	1.46		1.85	2.75
0250	1" x 3"		300	.027		.43	1.46		1.89	2.79
0260	1" x 4"		250	.032		.86	1.75		2.61	3.74
0270	1" x 6"		250	.032		1.40	1.75		3.15	4.33
0280	1" x 8"		225	.036		2.47	1.94		4.41	5.80
0290	1" x 10"		225	.036		3.40	1.94		5.34	6.85
0300	1" x 12"		200	.040		4.69	2.19		6.88	8.65
0310	Pine, #2, 1" x 2"		300	.027		.44	1.46		1.90	2.80
0320	1" x 3"		300	.027		.64	1.46		2.10	3.02
0330	1" x 4"		250	.032		.87	1.75		2.62	3.74
0340	1" x 6"		250	.032		1.18	1.75		2.93	4.09

For customer support on your Commercial Renovation Costs with RSMeans Data, call 800.448.8182.

165

06 22 13.40 Moldings, Exterior		Crew	Daily Output	Labor-Hours	Unit	Material	2021 Bare Costs Labor	Equipment	Total	Total Incl O&P
0350	1" x 8"	1 Carp	225	.036	L.F.	1.72	1.94		3.66	4.99
0360	1" x 10"		225	.036		2.90	1.94		4.84	6.30
0370	1" x 12"		200	.040		3.79	2.19		5.98	7.65
0380	D & better, 1" x 2"		300	.027		.57	1.46		2.03	2.95
0390	1" x 3"		300	.027		.86	1.46		2.32	3.27
0400	1" x 4"	1 Carp	250	.032	L.F.	1.10	1.75		2.85	4
0410	1" x 6"		250	.032		1.58	1.75		3.33	4.53
0420	1" x 8"		225	.036		2.02	1.94		3.96	5.30
0430	1" x 10"		225	.036		2.70	1.94		4.64	6.05
0440	1" x 12"		200	.040		3.43	2.19		5.62	7.25
0450	Redwood, clear all heart, 1" x 2"		300	.027		.91	1.46		2.37	3.32
0460	1" x 3"		300	.027		1.36	1.46		2.82	3.81
0470	1" x 4"		250	.032		3.01	1.75		4.76	6.10
0480	1" x 6"		252	.032		2.74	1.74		4.48	5.80
0490	1" x 8"		225	.036		7.45	1.94		9.39	11.30
0500	1" x 10"		225	.036		15	1.94		16.94	19.60
0510	1" x 12"		200	.040		6.30	2.19		8.49	10.45
0530	Corner board, cedar, rough sawn, 1" x 2"		225	.036		.68	1.94		2.62	3.85
0540	1" x 3"		225	.036		1.01	1.94		2.95	4.21
0550	1" x 4"		200	.040		1.35	2.19		3.54	4.96
0560	1" x 6"		200	.040		2.02	2.19		4.21	5.70
0570	1" x 8"		200	.040		2.70	2.19		4.89	6.45
0580	1" x 10"		175	.046		3.35	2.50		5.85	7.65
0590	1" x 12"		175	.046		4.03	2.50		6.53	8.40
0670	STK, 1" x 2"		225	.036		.39	1.94		2.33	3.53
0680	1" x 3"		225	.036		.43	1.94		2.37	3.57
0690	1" x 4"		200	.040		.84	2.19		3.03	4.40
0700	1" x 6"		200	.040		1.40	2.19		3.59	5
0710	1" x 8"		200	.040		2.47	2.19		4.66	6.20
0720	1" x 10"		175	.046		3.40	2.50		5.90	7.70
0730	1" x 12"		175	.046		4.69	2.50		7.19	9.15
0740	Pine, #2, 1" x 2"		225	.036		.44	1.94		2.38	3.58
0750	1" x 3"		225	.036		.64	1.94		2.58	3.80
0760	1" x 4"		200	.040		.87	2.19		3.06	4.43
0770	1" x 6"		200	.040		1.18	2.19		3.37	4.78
0780	1" x 8"		200	.040		1.72	2.19		3.91	5.35
0790	1" x 10"		175	.046		2.90	2.50		5.40	7.15
0800	1" x 12"		175	.046		3.79	2.50		6.29	8.15
0810	D & better, 1" x 2"		225	.036		.57	1.94		2.51	3.73
0820	1" x 3"		225	.036		.86	1.94		2.80	4.05
0830	1" x 4"		200	.040		1.10	2.19		3.29	4.69
0840	1" x 6"		200	.040		1.58	2.19		3.77	5.20
0850	1" x 8"		200	.040		2.02	2.19		4.21	5.70
0860	1" x 10"		175	.046		2.70	2.50		5.20	6.95
0870	1" x 12"		175	.046		3.43	2.50		5.93	7.75
0880	Redwood, clear all heart, 1" x 2"		225	.036		.91	1.94		2.85	4.10
0890	1" x 3"		225	.036		1.36	1.94		3.30	4.59
0900	1" x 4"		200	.040		3.01	2.19		5.20	6.80
0910	1" x 6"	1 Carp	200	.040	L.F.	2.74	2.19		4.93	6.50
0920	1" x 8"		200	.040		7.45	2.19		9.64	11.70
0930	1" x 10"		175	.046		15	2.50		17.50	20.50
0940	1" x 12"		175	.046		6.30	2.50		8.80	10.95
0950	Cornice board, cedar, rough sawn, 1" x 2"		330	.024		.68	1.33		2.01	2.86

06 22 13 – Standard Pattern Wood Trim

06 22 13.40 Moldings, Exterior		Crew	Daily Output	Labor-Hours	Unit	Material	2021 Bare Costs Labor	Equipment	Total	Total Incl O&P
0960	1" x 3"	1 Carp	290	.028	L.F.	1.01	1.51		2.52	3.51
0970	1" x 4"		250	.032		1.35	1.75		3.10	4.27
0980	1" x 6"		250	.032		2.02	1.75		3.77	5
0990	1" x 8"		200	.040		2.70	2.19		4.89	6.45
1000	1" x 10"		180	.044		3.35	2.43		5.78	7.55
1010	1" x 12"		180	.044		4.03	2.43		6.46	8.30
1020	STK, 1" x 2"		330	.024		.39	1.33		1.72	2.54
1030	1" x 3"		290	.028		.43	1.51		1.94	2.87
1040	1" x 4"		250	.032		.86	1.75		2.61	3.74
1050	1" x 6"		250	.032		1.40	1.75		3.15	4.33
1060	1" x 8"		200	.040		2.47	2.19		4.66	6.20
1070	1" x 10"		180	.044		3.40	2.43		5.83	7.60
1080	1" x 12"		180	.044		4.69	2.43		7.12	9
1500	Pine, #2, 1" x 2"		330	.024		.44	1.33		1.77	2.59
1510	1" x 3"		290	.028		.43	1.51		1.94	2.87
1600	1" x 4"		250	.032		.87	1.75		2.62	3.74
1700	1" x 6"		250	.032		1.18	1.75		2.93	4.09
1800	1" x 8"		200	.040		1.72	2.19		3.91	5.35
1900	1" x 10"		180	.044		2.90	2.43		5.33	7.05
2000	1" x 12"		180	.044		3.79	2.43		6.22	8.05
2020	D & better, 1" x 2"		330	.024		.57	1.33		1.90	2.74
2030	1" x 3"		290	.028		.86	1.51		2.37	3.35
2040	1" x 4"		250	.032		1.10	1.75		2.85	4
2050	1" x 6"		250	.032		1.58	1.75		3.33	4.53
2060	1" x 8"		200	.040		2.02	2.19		4.21	5.70
2070	1" x 10"		180	.044		2.70	2.43		5.13	6.85
2080	1" x 12"		180	.044		3.43	2.43		5.86	7.65
2090	Redwood, clear all heart, 1" x 2"		330	.024		.91	1.33		2.24	3.11
2100	1" x 3"		290	.028		1.36	1.51		2.87	3.89
2110	1" x 4"		250	.032		3.01	1.75		4.76	6.10
2120	1" x 6"		250	.032		2.74	1.75		4.49	5.80
2130	1" x 8"		200	.040		7.45	2.19		9.64	11.70
2140	1" x 10"		180	.044		15	2.43		17.43	20.50
2150	1" x 12"		180	.044		6.30	2.43		8.73	10.80
2160	3 piece, 1" x 2", 1" x 4", 1" x 6", rough sawn cedar		80	.100		4.07	5.45		9.52	13.20
2180	STK cedar		80	.100		2.65	5.45		8.10	11.60
2200	#2 pine		80	.100		2.49	5.45		7.94	11.45
2210	D & better pine	1 Carp	80	.100	L.F.	3.25	5.45		8.70	12.25
2220	Clear all heart redwood		80	.100		6.65	5.45		12.10	16.05
2230	1" x 8", 1" x 10", 1" x 12", rough sawn cedar		65	.123		10.05	6.75		16.80	22
2240	STK cedar		65	.123		10.50	6.75		17.25	22.50
2300	#2 pine		65	.123		8.35	6.75		15.10	19.90
2320	D & better pine		65	.123		8.10	6.75		14.85	19.60
2330	Clear all heart redwood		65	.123		28.50	6.75		35.25	42
2340	Door/window casing, cedar, rough sawn, 1" x 2"		275	.029		.68	1.59		2.27	3.28
2350	1" x 3"		275	.029		1.01	1.59		2.60	3.64
2360	1" x 4"		250	.032		1.35	1.75		3.10	4.27
2370	1" x 6"		250	.032		2.02	1.75		3.77	5
2380	1" x 8"		230	.035		2.70	1.90		4.60	6
2390	1" x 10"		230	.035		3.35	1.90		5.25	6.70
2395	1" x 12"		210	.038		4.03	2.08		6.11	7.75
2410	STK, 1" x 2"		275	.029		.39	1.59		1.98	2.96
2420	1" x 3"		275	.029		.43	1.59		2.02	3

06 22 13.40 Moldings, Exterior		Crew	Daily Output	Labor-Hours	Unit	Material	2021 Bare Costs Labor	Equipment	Total	Total Incl O&P
2430	1" x 4"	1 Carp	250	.032	L.F.	.86	1.75		2.61	3.74
2440	1" x 6"		250	.032		1.40	1.75		3.15	4.33
2450	1" x 8"		230	.035		2.47	1.90		4.37	5.75
2460	1" x 10"		230	.035		3.40	1.90		5.30	6.75
2470	1" x 12"		210	.038		4.69	2.08		6.77	8.45
2550	Pine, #2, 1" x 2"		275	.029		.44	1.59		2.03	3.01
2560	1" x 3"		275	.029		.64	1.59		2.23	3.23
2570	1" x 4"		250	.032		.87	1.75		2.62	3.74
2580	1" x 6"		250	.032		1.18	1.75		2.93	4.09
2590	1" x 8"		230	.035		1.72	1.90		3.62	4.92
2600	1" x 10"		230	.035		2.90	1.90		4.80	6.20
2610	1" x 12"		210	.038		3.79	2.08		5.87	7.50
2620	Pine, D & better, 1" x 2"		275	.029		.57	1.59		2.16	3.16
2630	1" x 3"		275	.029		.86	1.59		2.45	3.48
2640	1" x 4"		250	.032		1.10	1.75		2.85	4
2650	1" x 6"		250	.032		1.58	1.75		3.33	4.53
2660	1" x 8"		230	.035		2.02	1.90		3.92	5.25
2670	1" x 10"		230	.035		2.70	1.90		4.60	6
2680	1" x 12"		210	.038		3.43	2.08		5.51	7.10
2690	Redwood, clear all heart, 1" x 2"		275	.029		.91	1.59		2.50	3.53
2695	1" x 3"		275	.029		1.36	1.59		2.95	4.02
2710	1" x 4"		250	.032		3.01	1.75		4.76	6.10
2715	1" x 6"		250	.032		2.74	1.75		4.49	5.80
2730	1" x 8"		230	.035		7.45	1.90		9.35	11.25
2740	1" x 10"		230	.035		15	1.90		16.90	19.55
2750	1" x 12"		210	.038		6.30	2.08		8.38	10.25
3500	Bellyband, pine, 11/16" x 4-1/4"	1 Carp	250	.032	L.F.	2.83	1.75		4.58	5.90
3610	Brickmold, pine, 1-1/4" x 2"		200	.040		2.09	2.19		4.28	5.80
3620	FJP, 1-1/4" x 2"		200	.040		1.11	2.19		3.30	4.70
5100	Fascia, cedar, rough sawn, 1" x 2"		275	.029		.68	1.59		2.27	3.28
5110	1" x 3"		275	.029		1.01	1.59		2.60	3.64
5120	1" x 4"		250	.032		1.35	1.75		3.10	4.27
5200	1" x 6"		250	.032		2.02	1.75		3.77	5
5300	1" x 8"		230	.035		2.70	1.90		4.60	6
5310	1" x 10"		230	.035		3.35	1.90		5.25	6.70
5320	1" x 12"		210	.038		4.03	2.08		6.11	7.75
5400	2" x 4"		220	.036		1.38	1.99		3.37	4.69
5500	2" x 6"		220	.036		2.07	1.99		4.06	5.45
5600	2" x 8"		200	.040		2.76	2.19		4.95	6.50
5700	2" x 10"		180	.044		3.43	2.43		5.86	7.65
5800	2" x 12"		170	.047		8.20	2.57		10.77	13.10
6120	STK, 1" x 2"		275	.029		.39	1.59		1.98	2.96
6130	1" x 3"		275	.029		.43	1.59		2.02	3
6140	1" x 4"		250	.032		.86	1.75		2.61	3.74
6150	1" x 6"		250	.032		1.40	1.75		3.15	4.33
6160	1" x 8"		230	.035		2.47	1.90		4.37	5.75
6170	1" x 10"		230	.035		3.40	1.90		5.30	6.75
6180	1" x 12"		210	.038		4.69	2.08		6.77	8.45
6185	2" x 2"		260	.031		.86	1.68		2.54	3.63
6190	Pine, #2, 1" x 2"		275	.029		.44	1.59		2.03	3.01
6200	1" x 3"		275	.029		.64	1.59		2.23	3.23
6210	1" x 4"		250	.032		.87	1.75		2.62	3.74
6220	1" x 6"		250	.032		1.18	1.75		2.93	4.09

06 22 13 – Standard Pattern Wood Trim

06 22 13.40 Moldings, Exterior	Crew	Daily Output	Labor-Hours	Unit	Material	2021 Bare Costs Labor	Equipment	Total	Total Incl O&P
6230	1" x 8"	1 Carp	230	.035	L.F.	1.72	1.90		3.62	4.92
6240	1" x 10"		230	.035		2.90	1.90		4.80	6.20
6250	1" x 12"		210	.038		3.79	2.08		5.87	7.50
6260	D & better, 1" x 2"		275	.029		.57	1.59		2.16	3.16
6270	1" x 3"		275	.029		.86	1.59		2.45	3.48
6280	1" x 4"		250	.032		1.10	1.75		2.85	4
6290	1" x 6"		250	.032		1.58	1.75		3.33	4.53
6300	1" x 8"		230	.035		2.02	1.90		3.92	5.25
6310	1" x 10"		230	.035		2.70	1.90		4.60	6
6312	1" x 12"		210	.038		3.43	2.08		5.51	7.10
6330	Southern yellow, 1-1/4" x 5"		240	.033		3.09	1.82		4.91	6.30
6340	1-1/4" x 6"		240	.033		2.51	1.82		4.33	5.65
6350	1-1/4" x 8"		215	.037		3.35	2.04		5.39	6.90
6360	1-1/4" x 12"		190	.042		5.40	2.30		7.70	9.60
6370	Redwood, clear all heart, 1" x 2"		275	.029		.91	1.59		2.50	3.53
6380	1" x 3"		275	.029		3.01	1.59		4.60	5.85
6390	1" x 4"		250	.032		3.01	1.75		4.76	6.10
6400	1" x 6"		250	.032		2.74	1.75		4.49	5.80
6410	1" x 8"		230	.035		7.45	1.90		9.35	11.25
6420	1" x 10"		230	.035		15	1.90		16.90	19.55
6430	1" x 12"		210	.038		6.30	2.08		8.38	10.25
6440	1-1/4" x 5"		240	.033		2.86	1.82		4.68	6.05
6450	1-1/4" x 6"		240	.033		3.43	1.82		5.25	6.65
6460	1-1/4" x 8"		215	.037		11.20	2.04		13.24	15.55
6470	1-1/4" x 12"		190	.042		9.45	2.30		11.75	14.05
6580	Frieze, cedar, rough sawn, 1" x 2"		275	.029		.68	1.59		2.27	3.28
6590	1" x 3"		275	.029		1.01	1.59		2.60	3.64
6600	1" x 4"		250	.032		1.35	1.75		3.10	4.27
6610	1" x 6"		250	.032		2.02	1.75		3.77	5
6620	1" x 8"		250	.032		2.70	1.75		4.45	5.75
6630	1" x 10"		225	.036		3.35	1.94		5.29	6.80
6640	1" x 12"		200	.040		3.99	2.19		6.18	7.85
6650	STK, 1" x 2"		275	.029		.39	1.59		1.98	2.96
6660	1" x 3"		275	.029		.43	1.59		2.02	3
6670	1" x 4"		250	.032		.86	1.75		2.61	3.74
6680	1" x 6"		250	.032		1.40	1.75		3.15	4.33
6690	1" x 8"		250	.032		2.47	1.75		4.22	5.50
6700	1" x 10"		225	.036		3.40	1.94		5.34	6.85
6710	1" x 12"		200	.040		4.69	2.19		6.88	8.65
6790	Pine, #2, 1" x 2"		275	.029		.44	1.59		2.03	3.01
6800	1" x 3"		275	.029		.64	1.59		2.23	3.23
6810	1" x 4"		250	.032		.87	1.75		2.62	3.74
6820	1" x 6"		250	.032		1.18	1.75		2.93	4.09
6830	1" x 8"		250	.032		1.72	1.75		3.47	4.68
6840	1" x 10"		225	.036		2.90	1.94		4.84	6.30
6850	1" x 12"		200	.040		3.79	2.19		5.98	7.65
6860	D & better, 1" x 2"		275	.029		.57	1.59		2.16	3.16
6870	1" x 3"		275	.029		.86	1.59		2.45	3.48
6880	1" x 4"		250	.032		1.10	1.75		2.85	4
6890	1" x 6"		250	.032		1.58	1.75		3.33	4.53
6900	1" x 8"		250	.032		2.02	1.75		3.77	5
6910	1" x 10"		225	.036		2.70	1.94		4.64	6.05
6920	1" x 12"		200	.040		3.43	2.19		5.62	7.25

06 22 13.40 Moldings, Exterior		Crew	Daily Output	Labor-Hours	Unit	Material	2021 Bare Costs Labor	Equipment	Total	Total Incl O&P
6930	Redwood, clear all heart, 1" x 2"	1 Carp	275	.029	L.F.	.91	1.59		2.50	3.53
6940	1" x 3"		275	.029		1.36	1.59		2.95	4.02
6950	1" x 4"		250	.032		3.01	1.75		4.76	6.10
6960	1" x 6"		250	.032		2.74	1.75		4.49	5.80
6970	1" x 8"		250	.032		7.45	1.75		9.20	11
6980	1" x 10"		225	.036		15	1.94		16.94	19.60
6990	1" x 12"		200	.040		6.30	2.19		8.49	10.45
7000	Grounds, 1" x 1", cedar, rough sawn		300	.027		.35	1.46		1.81	2.70
7010	STK		300	.027		.24	1.46		1.70	2.59
7020	Pine, #2		300	.027		.27	1.46		1.73	2.62
7030	D & better		300	.027		.35	1.46		1.81	2.70
7050	Redwood		300	.027		.55	1.46		2.01	2.93
7060	Rake/verge board, cedar, rough sawn, 1" x 2"		225	.036		.68	1.94		2.62	3.85
7070	1" x 3"		225	.036		1.01	1.94		2.95	4.21
7080	1" x 4"		200	.040		1.35	2.19		3.54	4.96
7090	1" x 6"		200	.040		2.02	2.19		4.21	5.70
7100	1" x 8"		190	.042		2.70	2.30		5	6.65
7110	1" x 10"		190	.042		3.35	2.30		5.65	7.35
7120	1" x 12"		180	.044		4.03	2.43		6.46	8.30
7130	STK, 1" x 2"		225	.036		.39	1.94		2.33	3.53
7140	1" x 3"		225	.036		.43	1.94		2.37	3.57
7150	1" x 4"		200	.040		.86	2.19		3.05	4.43
7160	1" x 6"		200	.040		1.40	2.19		3.59	5
7170	1" x 8"		190	.042		2.47	2.30		4.77	6.40
7180	1" x 10"		190	.042		3.40	2.30		5.70	7.40
7190	1" x 12"		180	.044		4.69	2.43		7.12	9
7200	Pine, #2, 1" x 2"		225	.036		.44	1.94		2.38	3.58
7210	1" x 3"		225	.036		.64	1.94		2.58	3.80
7220	1" x 4"		200	.040		.87	2.19		3.06	4.43
7230	1" x 6"		200	.040		1.18	2.19		3.37	4.78
7240	1" x 8"		190	.042		1.72	2.30		4.02	5.55
7250	1" x 10"		190	.042		2.90	2.30		5.20	6.85
7260	1" x 12"		180	.044		3.79	2.43		6.22	8.05
7340	D & better, 1" x 2"		225	.036		.57	1.94		2.51	3.73
7350	1" x 3"		225	.036		.86	1.94		2.80	4.05
7360	1" x 4"		200	.040		1.10	2.19		3.29	4.69
7370	1" x 6"		200	.040		1.58	2.19		3.77	5.20
7380	1" x 8"		190	.042		2.02	2.30		4.32	5.90
7390	1" x 10"		190	.042		2.70	2.30		5	6.65
7400	1" x 12"		180	.044		3.43	2.43		5.86	7.65
7410	Redwood, clear all heart, 1" x 2"		225	.036		.91	1.94		2.85	4.10
7420	1" x 3"		225	.036		1.36	1.94		3.30	4.59
7430	1" x 4"		200	.040		3.01	2.19		5.20	6.80
7440	1" x 6"		200	.040		2.74	2.19		4.93	6.50
7450	1" x 8"		190	.042		7.45	2.30		9.75	11.85
7460	1" x 10"		190	.042		15	2.30		17.30	20
7470	1" x 12"		180	.044		6.30	2.43		8.73	10.80
7480	2" x 4"		200	.040		5.50	2.19		7.69	9.55
7490	2" x 6"		182	.044		9.40	2.40		11.80	14.20
7500	2" x 8"		165	.048		7.35	2.65		10	12.25
7630	Soffit, cedar, rough sawn, 1" x 2"	2 Carp	440	.036		.68	1.99		2.67	3.92
7640	1" x 3"		440	.036		1.01	1.99		3	4.28
7650	1" x 4"		420	.038		1.35	2.08		3.43	4.80

06 22 Millwork

06 22 13 – Standard Pattern Wood Trim

06 22 13.40 Moldings, Exterior		Crew	Daily Output	Labor-Hours	Unit	Material	2021 Bare Costs Labor	Equipment	Total	Total Incl O&P
7660	1" x 6"	2 Carp	420	.038	L.F.	2.02	2.08		4.10	5.55
7670	1" x 8"		420	.038		2.70	2.08		4.78	6.30
7680	1" x 10"		400	.040		3.35	2.19		5.54	7.15
7690	1" x 12"		400	.040		4.03	2.19		6.22	7.90
7700	STK, 1" x 2"		440	.036		.39	1.99		2.38	3.60
7710	1" x 3"		440	.036		.43	1.99		2.42	3.64
7720	1" x 4"		420	.038		.86	2.08		2.94	4.27
7730	1" x 6"		420	.038		1.40	2.08		3.48	4.86
7740	1" x 8"		420	.038		2.47	2.08		4.55	6.05
7750	1" x 10"		400	.040		3.40	2.19		5.59	7.20
7760	1" x 12"		400	.040		4.69	2.19		6.88	8.65
7770	Pine, #2, 1" x 2"		440	.036		.44	1.99		2.43	3.65
7780	1" x 3"		440	.036		.64	1.99		2.63	3.87
7790	1" x 4"		420	.038		.87	2.08		2.95	4.27
7800	1" x 6"		420	.038		1.18	2.08		3.26	4.62
7810	1" x 8"		420	.038		1.72	2.08		3.80	5.20
7820	1" x 10"		400	.040		2.90	2.19		5.09	6.65
7830	1" x 12"		400	.040		3.79	2.19		5.98	7.65
7840	D & better, 1" x 2"		440	.036		.57	1.99		2.56	3.80
7850	1" x 3"		440	.036		.86	1.99		2.85	4.12
7860	1" x 4"		420	.038		1.10	2.08		3.18	4.53
7870	1" x 6"		420	.038		1.58	2.08		3.66	5.05
7880	1" x 8"		420	.038		2.02	2.08		4.10	5.55
7890	1" x 10"		400	.040		2.70	2.19		4.89	6.45
7900	1" x 12"		400	.040		3.43	2.19		5.62	7.25
7910	Redwood, clear all heart, 1" x 2"		440	.036		.91	1.99		2.90	4.17
7920	1" x 3"		440	.036		1.36	1.99		3.35	4.66
7930	1" x 4"		420	.038		3.01	2.08		5.09	6.65
7940	1" x 6"		420	.038		2.74	2.08		4.82	6.35
7950	1" x 8"		420	.038		7.45	2.08		9.53	11.50
7960	1" x 10"		400	.040		15	2.19		17.19	20
7970	1" x 12"		400	.040		6.30	2.19		8.49	10.45
8050	Trim, crown molding, pine, 11/16" x 4-1/4"	1 Carp	250	.032		4.46	1.75		6.21	7.70
8060	Back band, 11/16" x 1-1/16"		250	.032		.88	1.75		2.63	3.76
8070	Insect screen frame stock, 1-1/16" x 1-3/4"		395	.020		.42	1.11		1.53	2.22
8080	Dentils, 2-1/2" x 2-1/2" x 4", 6" OC		30	.267		1.17	14.60		15.77	24.50
8100	Fluted, 5-1/2"		165	.048		1.74	2.65		4.39	6.15
8110	Stucco bead, 1-3/8" x 1-5/8"		250	.032		2.30	1.75		4.05	5.30
9000	Minimum labor/equipment charge		4	2	Job		109		109	174

06 22 13.45 Moldings, Trim		Crew	Daily Output	Labor-Hours	Unit	Material	2021 Bare Costs Labor	Equipment	Total	Total Incl O&P
0010	**MOLDINGS, TRIM**									
0200	Astragal, stock pine, 11/16" x 1-3/4"	1 Carp	255	.031	L.F.	1.15	1.72		2.87	4
0250	1-5/16" x 2-3/16"		240	.033		2.77	1.82		4.59	5.95
0800	Chair rail, stock pine, 5/8" x 2-1/2"		270	.030		1.65	1.62		3.27	4.40
0900	5/8" x 3-1/2"		240	.033		3.91	1.82		5.73	7.20
1000	Closet pole, stock pine, 1-1/8" diameter		200	.040		4.27	2.19		6.46	8.20
1100	Fir, 1-5/8" diameter		200	.040		1.99	2.19		4.18	5.65
3300	Half round, stock pine, 1/4" x 1/2"		270	.030		.40	1.62		2.02	3.02
3350	1/2" x 1"		255	.031		.54	1.72		2.26	3.32
3400	Handrail, fir, single piece, stock, hardware not included									
3450	1-1/2" x 1-3/4"	1 Carp	80	.100	L.F.	2.53	5.45		7.98	11.50
3470	Pine, 1-1/2" x 1-3/4"		80	.100		1.87	5.45		7.32	10.75

06 22 13 – Standard Pattern Wood Trim

06 22 13.45 Moldings, Trim

		Crew	Daily Output	Labor-Hours	Unit	Material	2021 Bare Costs Labor	Equipment	Total	Total Incl O&P
3500	1-1/2" x 2-1/2"	1 Carp	76	.105	L.F.	2.52	5.75		8.27	11.90
3600	Lattice, stock pine, 1/4" x 1-1/8"		270	.030		.38	1.62		2	3
3700	1/4" x 1-3/4"		250	.032		.87	1.75		2.62	3.75
3800	Miscellaneous, custom, pine, 1" x 1"		270	.030		.55	1.62		2.17	3.18
3850	1" x 2"		265	.030		1.09	1.65		2.74	3.83
3900	1" x 3"		240	.033		1.64	1.82		3.46	4.71
4100	Birch or oak, nominal 1" x 1"		240	.033		1.26	1.82		3.08	4.29
4200	Nominal 1" x 3"		215	.037		3.79	2.04		5.83	7.40
4400	Walnut, nominal 1" x 1"		215	.037		.94	2.04		2.98	4.28
4500	Nominal 1" x 3"		200	.040		2.82	2.19		5.01	6.60
4700	Teak, nominal 1" x 1"		215	.037		5.20	2.04		7.24	9
4800	Nominal 1" x 3"		200	.040		15.60	2.19		17.79	20.50
4900	Quarter round, stock pine, 1/4" x 1/4"		275	.029		.33	1.59		1.92	2.89
4950	3/4" x 3/4"		255	.031		.54	1.72		2.26	3.32
5600	Wainscot moldings, 1-1/8" x 9/16", 2' high, minimum		76	.105	S.F.	11.85	5.75		17.60	22
5700	Maximum		65	.123	"	16.80	6.75		23.55	29
9000	Minimum labor/equipment charge		4	2	Job		109		109	174

06 22 13.50 Moldings, Window and Door

		Crew	Daily Output	Labor-Hours	Unit	Material	2021 Bare Costs Labor	Equipment	Total	Total Incl O&P
0010	**MOLDINGS, WINDOW AND DOOR**									
2800	Door moldings, stock, decorative, 1-1/8" wide, plain	1 Carp	17	.471	Set	46.50	25.50		72	92
2900	Detailed		17	.471	"	112	25.50		137.50	164
2960	Clear pine door jamb, no stops, 11/16" x 4-9/16"		240	.033	L.F.	5.60	1.82		7.42	9.10
3150	Door trim set, 1 head and 2 sides, pine, 2-1/2" wide		12	.667	Opng.	23.50	36.50		60	83.50
3170	3-1/2" wide		11	.727	"	28.50	40		68.50	95
3250	Glass beads, stock pine, 3/8" x 1/2"		275	.029	L.F.	.77	1.59		2.36	3.38
3270	3/8" x 7/8"		270	.030		.48	1.62		2.10	3.11
4850	Parting bead, stock pine, 3/8" x 3/4"		275	.029		.47	1.59		2.06	3.05
4870	1/2" x 3/4"		255	.031		.44	1.72		2.16	3.21
5000	Stool caps, stock pine, 11/16" x 3-1/2"		200	.040		2.47	2.19		4.66	6.20
5100	1-1/16" x 3-1/4"		150	.053		3.66	2.92		6.58	8.70
5300	Threshold, oak, 3' long, inside, 5/8" x 3-5/8"		32	.250	Ea.	12.20	13.70		25.90	35.50
5400	Outside, 1-1/2" x 7-5/8"		16	.500	"	48.50	27.50		76	97
5900	Window trim sets, including casings, header, stops,									
5910	stool and apron, 2-1/2" wide, FJP	1 Carp	13	.615	Opng.	32	33.50		65.50	89
5950	Pine		10	.800		38	44		82	111
6000	Oak		6	1.333		79.50	73		152.50	204
9000	Minimum labor/equipment charge		4	2	Job		109		109	174

06 22 13.60 Moldings, Soffits

		Crew	Daily Output	Labor-Hours	Unit	Material	2021 Bare Costs Labor	Equipment	Total	Total Incl O&P
0010	**MOLDINGS, SOFFITS** R061110-30									
0200	Soffits, pine, 1" x 4"	2 Carp	420	.038	L.F.	.83	2.08		2.91	4.23
0210	1" x 6"		420	.038		1.14	2.08		3.22	4.58
0220	1" x 8"		420	.038		1.68	2.08		3.76	5.15
0230	1" x 10"		400	.040		2.84	2.19		5.03	6.60
0240	1" x 12"		400	.040		3.73	2.19		5.92	7.60
0250	STK cedar, 1" x 4"		420	.038		.82	2.08		2.90	4.22
0260	1" x 6"		420	.038		1.36	2.08		3.44	4.82
0270	1" x 8"		420	.038		2.43	2.08		4.51	6
0280	1" x 10"		400	.040		3.34	2.19		5.53	7.15
0290	1" x 12"		400	.040		4.63	2.19		6.82	8.60
1000	Exterior AC plywood, 1/4" thick		400	.040	S.F.	1.37	2.19		3.56	4.99
1050	3/8" thick		400	.040		1.48	2.19		3.67	5.10
1100	1/2" thick		400	.040		1.83	2.19		4.02	5.50

06 22 Millwork

06 22 13 – Standard Pattern Wood Trim

06 22 13.60 Moldings, Soffits		Crew	Daily Output	Labor-Hours	Unit	Material	2021 Bare Costs Labor	2021 Bare Costs Equipment	Total	Total Incl O&P
1150	Polyvinyl chloride, white, solid	1 Carp	230	.035	S.F.	2.15	1.90		4.05	5.40
1160	Perforated	"	230	.035	↓	2.15	1.90		4.05	5.40
1170	Accessories, "J" channel 5/8"	2 Carp	700	.023	L.F.	.63	1.25		1.88	2.68
9000	Minimum labor/equipment charge	"	5	3.200	Job		175		175	279

06 25 Prefinished Paneling

06 25 13 – Prefinished Hardboard Paneling

06 25 13.10 Paneling, Hardboard

			Crew	Daily Output	Labor-Hours	Unit	Material	Labor	Equipment	Total	Total Incl O&P
0010	**PANELING, HARDBOARD**										
0050	Not incl. furring or trim, hardboard, tempered, 1/8" thick	G	2 Carp	500	.032	S.F.	.51	1.75		2.26	3.35
0100	1/4" thick	G		500	.032		.55	1.75		2.30	3.40
0300	Tempered pegboard, 1/8" thick	G		500	.032		.42	1.75		2.17	3.25
0400	1/4" thick	G		500	.032		.76	1.75		2.51	3.63
0600	Untempered hardboard, natural finish, 1/8" thick	G		500	.032		.28	1.75		2.03	3.10
0700	1/4" thick	G		500	.032		.47	1.75		2.22	3.31
0900	Untempered pegboard, 1/8" thick	G		500	.032		.50	1.75		2.25	3.34
1000	1/4" thick	G		500	.032		.56	1.75		2.31	3.41
1200	Plastic faced hardboard, 1/8" thick	G		500	.032		.43	1.75		2.18	3.26
1300	1/4" thick	G		500	.032		.83	1.75		2.58	3.70
1500	Plastic faced pegboard, 1/8" thick	G		500	.032		.64	1.75		2.39	3.49
1600	1/4" thick	G		500	.032		.86	1.75		2.61	3.74
1800	Wood grained, plain or grooved, 1/8" thick	G		500	.032		.71	1.75		2.46	3.57
1900	1/4" thick	G		425	.038	↓	1.34	2.06		3.40	4.75
2100	Moldings, wood grained MDF			500	.032	L.F.	.41	1.75		2.16	3.24
2200	Pine		↓	425	.038	"	1.37	2.06		3.43	4.79
9000	Minimum labor/equipment charge		1 Carp	2	4	Job		219		219	350

06 25 16 – Prefinished Plywood Paneling

06 25 16.10 Paneling, Plywood

			Crew	Daily Output	Labor-Hours	Unit	Material	Labor	Equipment	Total	Total Incl O&P
0010	**PANELING, PLYWOOD**	R061636-20									
2400	Plywood, prefinished, 1/4" thick, 4' x 8' sheets										
2410	with vertical grooves. Birch faced, economy		2 Carp	500	.032	S.F.	1.97	1.75		3.72	4.96
2420	Average			420	.038		1.40	2.08		3.48	4.86
2430	Custom			350	.046		1.03	2.50		3.53	5.10
2600	Mahogany, African			400	.040		2.56	2.19		4.75	6.30
2700	Philippine (Lauan)			500	.032		.42	1.75		2.17	3.25
2900	Oak			500	.032		1.66	1.75		3.41	4.62
3000	Cherry			400	.040		1.94	2.19		4.13	5.60
3200	Rosewood			320	.050		3.10	2.74		5.84	7.75
3400	Teak			400	.040		3.29	2.19		5.48	7.10
3600	Chestnut			375	.043		5.45	2.33		7.78	9.65
3800	Pecan			400	.040		2.53	2.19		4.72	6.25
3900	Walnut, average			500	.032		2.67	1.75		4.42	5.75
3950	Custom			400	.040		2.47	2.19		4.66	6.20
4000	Plywood, prefinished, 3/4" thick, stock grades, economy			320	.050		1.50	2.74		4.24	6
4100	Average			224	.071		5.05	3.91		8.96	11.75
4300	Architectural grade, custom			224	.071		5.45	3.91		9.36	12.20
4400	Luxury			160	.100		5.60	5.45		11.05	14.90
4600	Plywood, "A" face, birch, VC, 1/2" thick, natural			450	.036		1.43	1.94		3.37	4.67
4700	Select			450	.036		1.91	1.94		3.85	5.20
4900	Veneer core, 3/4" thick, natural		↓	320	.050	↓	2.28	2.74		5.02	6.85

173

06 25 Prefinished Paneling

06 25 16 – Prefinished Plywood Paneling

06 25 16.10 Paneling, Plywood		Crew	Daily Output	Labor-Hours	Unit	Material	2021 Bare Costs Labor	Equipment	Total	Total Incl O&P
5000	Select	2 Carp	320	.050	S.F.	2.69	2.74		5.43	7.30
5200	Lumber core, 3/4" thick, natural		320	.050		3.01	2.74		5.75	7.65
5500	Plywood, knotty pine, 1/4" thick, A2 grade		450	.036		1.64	1.94		3.58	4.90
5600	A3 grade		450	.036		1.90	1.94		3.84	5.20
5800	3/4" thick, veneer core, A2 grade		320	.050		2.44	2.74		5.18	7.05
5900	A3 grade		320	.050		2.22	2.74		4.96	6.80
6100	Aromatic cedar, 1/4" thick, plywood		400	.040		2.32	2.19		4.51	6.05
6200	1/4" thick, particle board		400	.040		1.17	2.19		3.36	4.77
9000	Minimum labor/equipment charge	1 Carp	2	4	Job		219		219	350

06 25 26 – Panel System

06 25 26.10 Panel Systems		Crew	Daily Output	Labor-Hours	Unit	Material	2021 Bare Costs Labor	Equipment	Total	Total Incl O&P
0010	**PANEL SYSTEMS**									
0100	Raised panel, eng. wood core w/wood veneer, std., paint grade	2 Carp	300	.053	S.F.	13.20	2.92		16.12	19.15
0110	Oak veneer		300	.053		23.50	2.92		26.42	30.50
0120	Maple veneer		300	.053		33	2.92		35.92	40.50
0130	Cherry veneer		300	.053		35.50	2.92		38.42	43.50
0300	Class I fire rated, paint grade		300	.053		14.10	2.92		17.02	20
0310	Oak veneer		300	.053		31.50	2.92		34.42	39
0320	Maple veneer		300	.053		35.50	2.92		38.42	43.50
0330	Cherry veneer		300	.053		51.50	2.92		54.42	61
0510	Beadboard, 5/8" MDF, standard, primed		300	.053		8.05	2.92		10.97	13.55
0520	Oak veneer, unfinished		300	.053		12.45	2.92		15.37	18.35
0530	Maple veneer, unfinished		300	.053		20.50	2.92		23.42	27
0610	Rustic paneling, 5/8" MDF, standard, maple veneer, unfinished		300	.053		17.50	2.92		20.42	24

06 26 Board Paneling

06 26 13 – Profile Board Paneling

06 26 13.10 Paneling, Boards		Crew	Daily Output	Labor-Hours	Unit	Material	2021 Bare Costs Labor	Equipment	Total	Total Incl O&P
0010	**PANELING, BOARDS**									
6400	Wood board paneling, 3/4" thick, knotty pine	2 Carp	300	.053	S.F.	2.40	2.92		5.32	7.30
6500	Rough sawn cedar		300	.053		4.45	2.92		7.37	9.55
6700	Redwood, clear, 1" x 4" boards		300	.053		12.70	2.92		15.62	18.65
6900	Aromatic cedar, closet lining, boards		275	.058		2.33	3.18		5.51	7.60
9000	Minimum labor/equipment charge	1 Carp	2	4	Job		219		219	350

06 43 Wood Stairs and Railings

06 43 13 – Wood Stairs

06 43 13.20 Prefabricated Wood Stairs		Crew	Daily Output	Labor-Hours	Unit	Material	2021 Bare Costs Labor	Equipment	Total	Total Incl O&P
0010	**PREFABRICATED WOOD STAIRS**									
0100	Box stairs, prefabricated, 3'-0" wide									
0110	Oak treads, up to 14 risers	2 Carp	39	.410	Riser	97	22.50		119.50	143
0600	With pine treads for carpet, up to 14 risers	"	39	.410	"	62.50	22.50		85	105
1100	For 4' wide stairs, add				Flight	25%				
1550	Stairs, prefabricated stair handrail with balusters	1 Carp	30	.267	L.F.	72.50	14.60		87.10	103
1700	Basement stairs, prefabricated, pine treads									
1710	Pine risers, 3' wide, up to 14 risers	2 Carp	52	.308	Riser	62.50	16.85		79.35	96
4000	Residential, wood, oak treads, prefabricated		1.50	10.667	Flight	1,250	585		1,835	2,300
4200	Built in place		.44	36.364	"	2,175	2,000		4,175	5,575
4400	Spiral, oak, 4'-6" diameter, unfinished, prefabricated,									

174

06 43 13 – Wood Stairs

06 43 13.20 Prefabricated Wood Stairs	Crew	Daily Output	Labor-Hours	Unit	Material	2021 Bare Costs Labor	Equipment	Total	Total Incl O&P
4500	incl. railing, 9' high	2 Carp	1.50	10.667	Flight	2,700	585		3,285	3,900
9000	Minimum labor/equipment charge	"	3	5.333	Job		292		292	465

06 43 13.40 Wood Stair Parts

		Crew	Daily Output	Labor-Hours	Unit	Material	Labor	Equipment	Total	Total Incl O&P
0010	**WOOD STAIR PARTS**									
0020	Pin top balusters, 1-1/4", oak, 34"	1 Carp	96	.083	Ea.	6.70	4.56		11.26	14.60
0030	38"		96	.083		7.35	4.56		11.91	15.30
0040	42"		96	.083		7.70	4.56		12.26	15.70
0050	Poplar, 34"		96	.083		4.99	4.56		9.55	12.75
0060	38"		96	.083		5.35	4.56		9.91	13.15
0070	42"		96	.083		5.60	4.56		10.16	13.40
0080	Maple, 34"		96	.083		15.45	4.56		20.01	24.50
0090	38"		96	.083		17.75	4.56		22.31	27
0100	42"		96	.083		19.10	4.56		23.66	28.50
0130	Primed, 34"		96	.083		7.25	4.56		11.81	15.25
0140	38"		96	.083		8.15	4.56		12.71	16.20
0150	42"		96	.083		9.05	4.56		13.61	17.20
0180	Box top balusters, 1-1/4", oak, 34"		60	.133		7.85	7.30		15.15	20.50
0190	38"		60	.133		8.05	7.30		15.35	20.50
0200	42"		60	.133		8.45	7.30		15.75	21
0210	Poplar, 34"		60	.133		5.90	7.30		13.20	18.10
0220	38"		60	.133		5.90	7.30		13.20	18.05
0230	42"		60	.133		6.15	7.30		13.45	18.35
0240	Maple, 34"		60	.133		17.20	7.30		24.50	30.50
0250	38"		60	.133		18.30	7.30		25.60	31.50
0260	42"		60	.133		16.25	7.30		23.55	29.50
0290	Primed, 34"		60	.133		7.65	7.30		14.95	20
0300	38"		60	.133		8.35	7.30		15.65	21
0310	42"		60	.133		9.30	7.30		16.60	22
0340	Square balusters, cut from lineal stock, pine, 1-1/16" x 1-1/16"		180	.044	L.F.	6	2.43		8.43	10.45
0350	1-5/16" x 1-5/16"		180	.044		2.02	2.43		4.45	6.10
0360	1-5/8" x 1-5/8"		180	.044		8.35	2.43		10.78	13.05
0370	Turned newel, oak, 3-1/2" square, 48" high		8	1	Ea.	86.50	54.50		141	182
0380	62" high		8	1		132	54.50		186.50	232
0390	Poplar, 3-1/2" square, 48" high		8	1		55	54.50		109.50	148
0400	62" high		8	1		90.50	54.50		145	187
0410	Maple, 3-1/2" square, 48" high		8	1		55	54.50		109.50	148
0420	62" high		8	1		144	54.50		198.50	245
0430	Square newel, oak, 3-1/2" square, 48" high		8	1		86.50	54.50		141	182
0440	58" high		8	1		108	54.50		162.50	206
0450	Poplar, 3-1/2" square, 48" high		8	1		55	54.50		109.50	148
0460	58" high		8	1		69	54.50		123.50	163
0470	Maple, 3" square, 48" high		8	1		117	54.50		171.50	216
0480	58" high		8	1		147	54.50		201.50	248
0490	Railings, oak, economy		96	.083	L.F.	10.65	4.56		15.21	18.95
0500	Average		96	.083		12.30	4.56		16.86	21
0510	Custom		96	.083		19.80	4.56		24.36	29.50
0520	Maple, economy		96	.083		10.60	4.56		15.16	18.90
0530	Average		96	.083		12.90	4.56		17.46	21.50
0540	Custom		96	.083		22	4.56		26.56	31.50
0550	Oak, for bending rail, economy		48	.167		26	9.10		35.10	43
0560	Average		48	.167		25.50	9.10		34.60	42.50
0570	Custom		48	.167		33	9.10		42.10	51

175

06 43 13.40 Wood Stair Parts	Crew	Daily Output	Labor-Hours	Unit	Material	2021 Bare Costs Labor	Equipment	Total	Total Incl O&P
0580	Maple, for bending rail, economy	1 Carp	48	.167	L.F.	32.50	9.10		41.60	50.50
0590	Average		48	.167		30.50	9.10		39.60	48
0600	Custom		48	.167		32.50	9.10		41.60	50.50
0610	Risers, oak, 3/4" x 8", 36" long		80	.100	Ea.	23	5.45		28.45	33.50
0620	42" long		70	.114		26.50	6.25		32.75	39
0630	48" long		63	.127		30.50	6.95		37.45	44.50
0640	54" long		56	.143		34	7.80		41.80	50
0650	60" long		50	.160		38	8.75		46.75	56
0660	72" long		42	.190		45.50	10.40		55.90	66.50
0670	Poplar, 3/4" x 8", 36" long		80	.100		16.65	5.45		22.10	27
0680	42" long		71	.113		19.45	6.15		25.60	31.50
0690	48" long		63	.127		22	6.95		28.95	35.50
0700	54" long		56	.143		25	7.80		32.80	40
0710	60" long		50	.160		28	8.75		36.75	44.50
0720	72" long		42	.190		33.50	10.40		43.90	53
0730	Pine, 1" x 8", 36" long		80	.100		5.05	5.45		10.50	14.25
0740	42" long		70	.114		5.90	6.25		12.15	16.45
0750	48" long		63	.127		6.75	6.95		13.70	18.45
0760	54" long		56	.143		7.55	7.80		15.35	21
0770	60" long		50	.160		8.40	8.75		17.15	23
0780	72" long		42	.190		10.10	10.40		20.50	27.50
0790	Treads, oak, no returns, 1-1/32" x 11-1/2" x 36" long		32	.250		46	13.70		59.70	72.50
0800	42" long		32	.250		54	13.70		67.70	81
0810	48" long		32	.250		61.50	13.70		75.20	89.50
0820	54" long		32	.250		69	13.70		82.70	98
0830	60" long		32	.250		77	13.70		90.70	107
0840	72" long		32	.250		92	13.70		105.70	123
0850	Mitred return one end, 1-1/32" x 11-1/2" x 36" long		24	.333		51.50	18.25		69.75	86
0860	42" long		24	.333		60	18.25		78.25	95
0870	48" long		24	.333		69	18.25		87.25	105
0880	54" long		24	.333		77.50	18.25		95.75	114
0890	60" long		24	.333		86	18.25		104.25	124
0900	72" long		24	.333		103	18.25		121.25	143
0910	Mitred return two ends, 1-1/32" x 11-1/2" x 36" long		12	.667		65.50	36.50		102	130
0920	42" long		12	.667		76.50	36.50		113	142
0930	48" long		12	.667		87	36.50		123.50	154
0940	54" long		12	.667		98	36.50		134.50	166
0950	60" long		12	.667		109	36.50		145.50	178
0960	72" long		12	.667		131	36.50		167.50	202
0970	Starting step, oak, 48", bullnose		8	1		172	54.50		226.50	276
0980	Double end bullnose		8	1		241	54.50		295.50	350
1030	Skirt board, pine, 1" x 10"		55	.145	L.F.	2.84	7.95		10.79	15.75
1040	1" x 12"		52	.154	"	3.73	8.40		12.13	17.50
1050	Oak landing tread, 1-1/16" thick		54	.148	S.F.	8.35	8.10		16.45	22
1060	Oak cove molding		96	.083	L.F.	1.82	4.56		6.38	9.25
1070	Oak stringer molding		96	.083	"	3.53	4.56		8.09	11.15
1090	Rail bolt, 5/16" x 3-1/2"		48	.167	Ea.	3.27	9.10		12.37	18.10
1100	5/16" x 4-1/2"		48	.167		2.95	9.10		12.05	17.75
1120	Newel post anchor		16	.500		29	27.50		56.50	75
1130	Tapered plug, 1/2"		240	.033		.88	1.82		2.70	3.87
1140	1"		240	.033		.36	1.82		2.18	3.30
9000	Minimum labor/equipment charge		3	2.667	Job		146		146	232

176

06 43 Wood Stairs and Railings

06 43 16 – Wood Railings

06 43 16.10 Wood Handrails and Railings		Crew	Daily Output	Labor-Hours	Unit	Material	2021 Bare Costs Labor	Equipment	Total	Total Incl O&P
0010	**WOOD HANDRAILS AND RAILINGS**									
0020	Custom design, architectural grade, hardwood, plain	1 Carp	38	.211	L.F.	12	11.50		23.50	31.50
0100	Shaped		30	.267		59.50	14.60		74.10	88.50
0300	Stock interior railing with spindles 4" OC, 4' long		40	.200		45	10.95		55.95	67
0400	8' long		48	.167		45	9.10		54.10	64
9000	Minimum labor/equipment charge		3	2.667	Job		146		146	232

06 44 Ornamental Woodwork

06 44 19 – Wood Grilles

06 44 19.10 Grilles

		Crew	Daily Output	Labor-Hours	Unit	Material	2021 Bare Costs Labor	Equipment	Total	Total Incl O&P
0010	**GRILLES** and panels, hardwood, sanded									
0020	2' x 4' to 4' x 8', custom designs, unfinished, economy	1 Carp	38	.211	S.F.	54.50	11.50		66	78.50
0050	Average		30	.267		62.50	14.60		77.10	92
0100	Custom		19	.421		73	23		96	117
9000	Minimum labor/equipment charge		2	4	Job		219		219	350

06 44 33 – Wood Mantels

06 44 33.10 Fireplace Mantels

		Crew	Daily Output	Labor-Hours	Unit	Material	2021 Bare Costs Labor	Equipment	Total	Total Incl O&P
0010	**FIREPLACE MANTELS**									
0015	6" molding, 6' x 3'-6" opening, plain, paint grade	1 Carp	5	1.600	Opng.	460	87.50		547.50	645
0100	Ornate, oak		5	1.600		640	87.50		727.50	845
0300	Prefabricated pine, colonial type, stock, deluxe		2	4		2,150	219		2,369	2,700
0400	Economy		3	2.667		755	146		901	1,050
9000	Minimum labor/equipment charge		3	2.667	Job		146		146	232

06 44 33.20 Fireplace Mantel Beam

		Crew	Daily Output	Labor-Hours	Unit	Material	2021 Bare Costs Labor	Equipment	Total	Total Incl O&P
0010	**FIREPLACE MANTEL BEAM**									
0020	Rough texture wood, 4" x 8"	1 Carp	36	.222	L.F.	11	12.15		23.15	31.50
0100	4" x 10"		35	.229	"	12.35	12.50		24.85	33.50
0300	Laminated hardwood, 2-1/4" x 10-1/2" wide, 6' long		5	1.600	Ea.	109	87.50		196.50	258
0400	8' long		5	1.600	"	182	87.50		269.50	340
0600	Brackets for above, rough sawn		12	.667	Pr.	11.85	36.50		48.35	71
0700	Laminated		12	.667	"	89.50	36.50		126	157
9000	Minimum labor/equipment charge		4	2	Job		109		109	174

06 44 39 – Wood Posts and Columns

06 44 39.10 Decorative Beams

		Crew	Daily Output	Labor-Hours	Unit	Material	2021 Bare Costs Labor	Equipment	Total	Total Incl O&P
0010	**DECORATIVE BEAMS**									
0020	Rough sawn cedar, non-load bearing, 4" x 4"	2 Carp	180	.089	L.F.	1.89	4.86		6.75	9.85
0100	4" x 6"		170	.094		1.70	5.15		6.85	10.05
0200	4" x 8"		160	.100		2.26	5.45		7.71	11.20
0300	4" x 10"		150	.107		4.15	5.85		10	13.85
0400	4" x 12"		140	.114		4.84	6.25		11.09	15.25
0500	8" x 8"		130	.123		4.65	6.75		11.40	15.80
9000	Minimum labor/equipment charge	1 Carp	3	2.667	Job		146		146	232

06 44 39.20 Columns

		Crew	Daily Output	Labor-Hours	Unit	Material	2021 Bare Costs Labor	Equipment	Total	Total Incl O&P
0010	**COLUMNS**									
0050	Aluminum, round colonial, 6" diameter	2 Carp	80	.200	V.L.F.	11.15	10.95		22.10	29.50
0100	8" diameter		62.25	.257		16.90	14.05		30.95	41
0200	10" diameter		55	.291		26.50	15.90		42.40	54.50
0250	Fir, stock units, hollow round, 6" diameter		80	.200		34.50	10.95		45.45	55.50
0300	8" diameter		80	.200		31	10.95		41.95	51.50

06 44 Ornamental Woodwork

06 44 39 – Wood Posts and Columns

06 44 39.20 Columns

		Crew	Daily Output	Labor-Hours	Unit	Material	2021 Bare Costs Labor	2021 Bare Costs Equipment	Total	Total Incl O&P
0350	10" diameter	2 Carp	70	.229	V.L.F.	48	12.50		60.50	73
0360	12" diameter		65	.246		70	13.45		83.45	98.50
0400	Solid turned, to 8' high, 3-1/2" diameter		80	.200		11.50	10.95		22.45	30
0500	4-1/2" diameter		75	.213		14.40	11.65		26.05	34.50
0600	5-1/2" diameter		70	.229		19.35	12.50		31.85	41.50
0800	Square columns, built-up, 5" x 5"		65	.246		33.50	13.45		46.95	58.50
0900	Solid, 3-1/2" x 3-1/2"		130	.123		11.35	6.75		18.10	23
1600	Hemlock, tapered, T&G, 12" diam., 10' high		100	.160		47	8.75		55.75	66
1700	16' high		65	.246		81.50	13.45		94.95	111
1900	14" diameter, 10' high		100	.160		111	8.75		119.75	136
2000	18' high		65	.246		97	13.45		110.45	129
2200	18" diameter, 12' high		65	.246		170	13.45		183.45	209
2300	20' high		50	.320		124	17.50		141.50	164
2500	20" diameter, 14' high		40	.400		233	22		255	292
2600	20' high		35	.457		189	25		214	247
2800	For flat pilasters, deduct					33%				
3000	For splitting into halves, add				Ea.	132			132	145
4000	Rough sawn cedar posts, 4" x 4"	2 Carp	250	.064	V.L.F.	4.12	3.50		7.62	10.10
4100	4" x 6"		235	.068		6.85	3.72		10.57	13.50
4200	6" x 6"		220	.073		13.60	3.98		17.58	21.50
4300	8" x 8"		200	.080		21.50	4.38		25.88	30.50
9000	Minimum labor/equipment charge	1 Carp	3	2.667	Job		146		146	232

06 48 Wood Frames

06 48 13 – Exterior Wood Door Frames

06 48 13.10 Exterior Wood Door Frames and Accessories

		Crew	Daily Output	Labor-Hours	Unit	Material	2021 Bare Costs Labor	2021 Bare Costs Equipment	Total	Total Incl O&P
0010	**EXTERIOR WOOD DOOR FRAMES AND ACCESSORIES**									
0400	Exterior frame, incl. ext. trim, pine, 5/4 x 4-9/16" deep	2 Carp	375	.043	L.F.	6.90	2.33		9.23	11.30
0420	5-3/16" deep		375	.043		7.80	2.33		10.13	12.30
0440	6-9/16" deep		375	.043		8.35	2.33		10.68	12.85
0600	Oak, 5/4 x 4-9/16" deep		350	.046		13.45	2.50		15.95	18.80
0620	5-3/16" deep		350	.046		14.80	2.50		17.30	20
0640	6-9/16" deep		350	.046		23.50	2.50		26	29.50
1000	Sills, 8/4 x 8" deep, oak, no horns		100	.160		10.80	8.75		19.55	26
1020	2" horns		100	.160		21	8.75		29.75	37.50
1040	3" horns		100	.160		21	8.75		29.75	37.50
1100	8/4 x 10" deep, oak, no horns		90	.178		7.15	9.70		16.85	23.50
1120	2" horns		90	.178		27.50	9.70		37.20	46
1140	3" horns		90	.178		27.50	9.70		37.20	46
2000	Wood frame & trim, ext., colonial, 3' opng., fluted pilasters, flat head		22	.727	Ea.	500	40		540	615
2010	Dentil head		21	.762		630	41.50		671.50	755
2020	Ram's head		20	.800		605	44		649	735
2100	5'-4" opening, in-swing, fluted pilasters, flat head		17	.941		535	51.50		586.50	670
2120	Ram's head		15	1.067		1,325	58.50		1,383.50	1,550
2140	Out-swing, fluted pilasters, flat head		17	.941		475	51.50		526.50	605
2160	Ram's head		15	1.067		1,350	58.50		1,408.50	1,575
2400	6'-0" opening, in-swing, fluted pilasters, flat head		16	1		480	54.50		534.50	615
2420	Ram's head		10	1.600		1,500	87.50		1,587.50	1,800
2460	Out-swing, fluted pilasters, flat head		16	1		480	54.50		534.50	615
2480	Ram's head		10	1.600		1,525	87.50		1,612.50	1,825
2600	For two sidelights, flat head, add		30	.533	Opng.	340	29		369	420

For customer support on your Commercial Renovation Costs with RSMeans Data, call 800.448.8182.

06 48 Wood Frames

06 48 13 – Exterior Wood Door Frames

06 48 13.10 Exterior Wood Door Frames and Accessories	Crew	Daily Output	Labor-Hours	Unit	Material	2021 Bare Costs Labor	2021 Bare Costs Equipment	Total	Total Incl O&P
2620	Ram's head, add	2 Carp	20	.800	Opng.	845	44		889	1,000
2700	Custom birch frame, 3'-0" opening		16	1	Ea.	257	54.50		311.50	370
2750	6'-0" opng.		16	1		415	54.50		469.50	540
2900	Exterior, modern, plain trim, 3' opng., in-swing, FJP		26	.615		52.50	33.50		86	111
2920	Fir		24	.667		60.50	36.50		97	125
2940	Oak		22	.727		69.50	40		109.50	140

06 48 16 – Interior Wood Door Frames

06 48 16.10 Interior Wood Door Jamb and Frames

		Crew	Daily Output	Labor-Hours	Unit	Material	2021 Bare Costs Labor	2021 Bare Costs Equipment	Total	Total Incl O&P
0010	**INTERIOR WOOD DOOR JAMB AND FRAMES**									
3000	Interior frame, pine, 11/16" x 3-5/8" deep	2 Carp	375	.043	L.F.	5.55	2.33		7.88	9.80
3020	4-9/16" deep		375	.043		6.45	2.33		8.78	10.75
3200	Oak, 11/16" x 3-5/8" deep		350	.046		5.35	2.50		7.85	9.90
3220	4-9/16" deep		350	.046		13.40	2.50		15.90	18.75
3240	5-3/16" deep		350	.046		18.40	2.50		20.90	24
3400	Walnut, 11/16" x 3-5/8" deep		350	.046		14.50	2.50		17	19.95
3420	4-9/16" deep		350	.046		11.95	2.50		14.45	17.15
3440	5-3/16" deep		350	.046		11.60	2.50		14.10	16.75
3600	Pocket door frame		16	1	Ea.	69	54.50		123.50	163
3800	Threshold, oak, 5/8" x 3-5/8" deep		200	.080	L.F.	2.63	4.38		7.01	9.85
3820	4-5/8" deep		190	.084		4.31	4.61		8.92	12.10
3840	5-5/8" deep		180	.089		6.85	4.86		11.71	15.30
9000	Minimum labor/equipment charge	1 Carp	4	2	Job		109		109	174

06 49 Wood Screens and Exterior Wood Shutters

06 49 19 – Exterior Wood Shutters

06 49 19.10 Shutters, Exterior

		Crew	Daily Output	Labor-Hours	Unit	Material	2021 Bare Costs Labor	2021 Bare Costs Equipment	Total	Total Incl O&P
0010	**SHUTTERS, EXTERIOR**									
0012	Aluminum, louvered, 1'-4" wide, 3'-0" long	1 Carp	10	.800	Pr.	224	44		268	315
0200	4'-0" long		10	.800		271	44		315	370
0300	5'-4" long		10	.800		284	44		328	380
0400	6'-8" long		9	.889		350	48.50		398.50	465
1000	Pine, louvered, primed, each 1'-2" wide, 3'-3" long		10	.800		229	44		273	320
1100	4'-7" long		10	.800		265	44		309	360
1250	Each 1'-4" wide, 3'-0" long		10	.800		254	44		298	350
1350	5'-3" long		10	.800		480	44		524	600
1500	Each 1'-6" wide, 3'-3" long		10	.800		350	44		394	455
1600	4'-7" long		10	.800		475	44		519	595
1620	Cedar, louvered, 1'-2" wide, 5'-7" long		10	.800		315	44		359	415
1630	Each 1'-4" wide, 2'-2" long		10	.800		205	44		249	296
1640	3'-0" long		10	.800		219	44		263	310
1650	3'-3" long		10	.800		231	44		275	325
1660	3'-11" long		10	.800		263	44		307	360
1670	4'-3" long		10	.800		415	44		459	525
1680	5'-3" long		10	.800		340	44		384	440
1690	5'-11" long		10	.800		375	44		419	485
1700	Door blinds, 6'-9" long, each 1'-3" wide		9	.889		600	48.50		648.50	740
1710	1'-6" wide		9	.889		565	48.50		613.50	700
1720	Cedar, solid raised panel, each 1'-4" wide, 3'-3" long		10	.800		325	44		369	425
1730	3'-11" long		10	.800		330	44		374	430
1740	4'-3" long		10	.800		325	44		369	425

06 49 Wood Screens and Exterior Wood Shutters

06 49 19 – Exterior Wood Shutters

06 49 19.10 Shutters, Exterior

		Crew	Daily Output	Labor-Hours	Unit	Material	2021 Bare Costs Labor	Equipment	Total	Total Incl O&P
1750	4'-7" long	1 Carp	10	.800	Pr.	340	44		384	445
1760	4'-11" long		10	.800		370	44		414	475
1770	5'-11" long		10	.800		515	44		559	640
1800	Door blinds, 6'-9" long, each 1'-3" wide		9	.889		485	48.50		533.50	610
1900	1'-6" wide		9	.889		725	48.50		773.50	880
2500	Polystyrene, solid raised panel, each 1'-4" wide, 3'-3" long		10	.800		88.50	44		132.50	167
2600	3'-11" long		10	.800		94.50	44		138.50	174
2700	4'-7" long		10	.800		107	44		151	188
2800	5'-3" long		10	.800		121	44		165	203
2900	6'-8" long		9	.889		155	48.50		203.50	249
4500	Polystyrene, louvered, each 1'-2" wide, 3'-3" long		10	.800		30.50	44		74.50	103
4600	4'-7" long		10	.800		61	44		105	137
4750	5'-3" long		10	.800		71	44		115	148
4850	6'-8" long		9	.889		90	48.50		138.50	177
6000	Vinyl, louvered, each 1'-2" x 4'-7" long		10	.800		81.50	44		125.50	160
6200	Each 1'-4" x 6'-8" long		9	.889		117	48.50		165.50	206
9000	Minimum labor/equipment charge		4	2	Job		109		109	174

06 51 Structural Plastic Shapes and Plates

06 51 13 – Plastic Lumber

06 51 13.10 Recycled Plastic Lumber

			Crew	Daily Output	Labor-Hours	Unit	Material	2021 Bare Costs Labor	Equipment	Total	Total Incl O&P
0010	**RECYCLED PLASTIC LUMBER**										
4000	Sheeting, recycled plastic, black or white, 4' x 8' x 1/8"	G	2 Carp	1100	.015	S.F.	1.07	.80		1.87	2.45
4010	4' x 8' x 3/16"	G		1100	.015		1.39	.80		2.19	2.80
4020	4' x 8' x 1/4"	G		950	.017		1.48	.92		2.40	3.10
4030	4' x 8' x 3/8"	G		950	.017		2.72	.92		3.64	4.46
4040	4' x 8' x 1/2"	G		900	.018		3.61	.97		4.58	5.50
4050	4' x 8' x 5/8"	G		900	.018		6.30	.97		7.27	8.45
4060	4' x 8' x 3/4"	G		850	.019		7.05	1.03		8.08	9.40
4070	Add for colors	G				Ea.	5%				
8500	100% recycled plastic, var colors, NLB, 2" x 2"	G				L.F.	1.99			1.99	2.19
8510	2" x 4"	G					4.11			4.11	4.52
8520	2" x 6"	G					6.20			6.20	6.80
8530	2" x 8"	G					9.10			9.10	10
8540	2" x 10"	G					12.90			12.90	14.20
8550	5/4" x 4"	G					4.33			4.33	4.76
8560	5/4" x 6"	G					6.05			6.05	6.65
8570	1" x 6"	G					2.97			2.97	3.27
8580	1/2" x 8"	G					3.50			3.50	3.85
8590	2" x 10" T&G	G					12.90			12.90	14.20
8600	3" x 10" T&G	G					22			22	24.50
8610	Add for premium colors	G					20%				

06 51 13.12 Structural Plastic Lumber

		Crew	Daily Output	Labor-Hours	Unit	Material	2021 Bare Costs Labor	Equipment	Total	Total Incl O&P
0010	**STRUCTURAL PLASTIC LUMBER**									
1320	Plastic lumber, posts or columns, 4" x 4"	2 Carp	390	.041	L.F.	11.20	2.24		13.44	15.85
1325	4" x 6"		275	.058		14.80	3.18		17.98	21.50
1330	4" x 8"		220	.073		21.50	3.98		25.48	30.50
1340	Girder, single, 4" x 4"		675	.024		11.20	1.30		12.50	14.35
1345	4" x 6"		600	.027		14.80	1.46		16.26	18.60
1350	4" x 8"		525	.030		21.50	1.67		23.17	26.50
1352	Double, 2" x 4"		625	.026		14.55	1.40		15.95	18.25

For customer support on your Commercial Renovation Costs with RSMeans Data, call 800.448.8182.

06 51 Structural Plastic Shapes and Plates

06 51 13 – Plastic Lumber

06 51 13.12 Structural Plastic Lumber

		Crew	Daily Output	Labor-Hours	Unit	Material	2021 Bare Costs Labor	Equipment	Total	Total Incl O&P
1354	2" x 6"	2 Carp	600	.027	L.F.	10.70	1.46		12.16	14.10
1356	2" x 8"		575	.028		18.75	1.52		20.27	23
1358	2" x 10"		550	.029		22.50	1.59		24.09	27
1360	2" x 12"		525	.030		42	1.67		43.67	48.50
1362	Triple, 2" x 4"		575	.028		22	1.52		23.52	26.50
1364	2" x 6"		550	.029		16.05	1.59		17.64	20
1366	2" x 8"		525	.030		28	1.67		29.67	33.50
1368	2" x 10"		500	.032		33.50	1.75		35.25	39.50
1370	2" x 12"		475	.034		62.50	1.84		64.34	72
1372	Ledger, bolted 4' OC, 2" x 4"		400	.040		7.50	2.19		9.69	11.75
1374	2" x 6"		550	.029		5.40	1.59		6.99	8.50
1376	2" x 8"		390	.041		9.60	2.24		11.84	14.10
1378	2" x 10"		385	.042		11.35	2.27		13.62	16.10
1380	2" x 12"		380	.042		21	2.30		23.30	26.50
1382	Joists, 2" x 4"		1250	.013		7.25	.70		7.95	9.10
1384	2" x 6"		1250	.013		5.35	.70		6.05	7
1386	2" x 8"		1100	.015		9.40	.80		10.20	11.60
1388	2" x 10"		500	.032		11.20	1.75		12.95	15.15
1390	2" x 12"		875	.018		21	1		22	24.50
1392	Railings and trim, 5/4" x 4"	1 Carp	300	.027		5.05	1.46		6.51	7.85
1394	2" x 2"		300	.027		3.84	1.46		5.30	6.55
1396	2" x 4"		300	.027		7.25	1.46		8.71	10.30
1398	2" x 6"		300	.027		5.35	1.46		6.81	8.15

06 52 Plastic Structural Assemblies

06 52 10 – Fiberglass Structural Assemblies

06 52 10.30 Fiberglass Grating

		Crew	Daily Output	Labor-Hours	Unit	Material	2021 Bare Costs Labor	Equipment	Total	Total Incl O&P
0010	**FIBERGLASS GRATING**									
0100	Molded, green (for mod. corrosive environment)									
0140	1" x 4" mesh, 1" thick	2 Sswk	400	.040	S.F.	7.65	2.41		10.06	12.30
0180	1-1/2" square mesh, 1" thick		400	.040		16.85	2.41		19.26	22.50
0220	1-1/4" thick		400	.040		7.95	2.41		10.36	12.65
0260	1-1/2" thick		400	.040		24.50	2.41		26.91	31
0300	2" square mesh, 2" thick		320	.050		24	3.02		27.02	31.50
1000	Orange (for highly corrosive environment)									
1040	1" x 4" mesh, 1" thick	2 Sswk	400	.040	S.F.	15.45	2.41		17.86	21
1080	1-1/2" square mesh, 1" thick		400	.040		21.50	2.41		23.91	27.50
1120	1-1/4" thick		400	.040		17.95	2.41		20.36	23.50
1160	1-1/2" thick		400	.040		21	2.41		23.41	27.50
1200	2" square mesh, 2" thick		320	.050		18.85	3.02		21.87	26
3000	Pultruded, green (for mod. corrosive environment)									
3040	1" OC bar spacing, 1" thick	2 Sswk	400	.040	S.F.	14.45	2.41		16.86	19.80
3080	1-1/2" thick		320	.050		16.65	3.02		19.67	23
3120	1-1/2" OC bar spacing, 1" thick		400	.040		15	2.41		17.41	20.50
3160	1-1/2" thick		400	.040		21.50	2.41		23.91	27.50
4000	Grating support legs, fixed height, no base				Ea.	46			46	50.50
4040	With base					37			37	41
4080	Adjustable to 60"					55.50			55.50	61.50

181

06 52 Plastic Structural Assemblies

06 52 10 – Fiberglass Structural Assemblies

06 52 10.40 Fiberglass Floor Grating	Crew	Daily Output	Labor-Hours	Unit	Material	2021 Bare Costs Labor	Equipment	Total	Total Incl O&P
0010 **FIBERGLASS FLOOR GRATING**									
0100 Reinforced polyester, fire retardant, 1" x 4" grid, 1" thick	E-4	510	.063	S.F.	13.10	3.81	.29	17.20	21
0200 1-1/2" x 6" mesh, 1-1/2" thick		500	.064		15.15	3.89	.30	19.34	23.50
0300 With grit surface, 1-1/2" x 6" grid, 1-1/2" thick	↓	500	.064	↓	15.45	3.89	.30	19.64	23.50

06 63 Plastic Railings

06 63 10 – Plastic (PVC) Railings

06 63 10.10 Plastic Railings

	Crew	Daily Output	Labor-Hours	Unit	Material	2021 Bare Costs Labor	Equipment	Total	Total Incl O&P
0010 **PLASTIC RAILINGS**									
0100 Horizontal PVC handrail with balusters, 3-1/2" wide, 36" high	1 Carp	96	.083	L.F.	31	4.56		35.56	41.50
0150 42" high		96	.083		29.50	4.56		34.06	40
0200 Angled PVC handrail with balusters, 3-1/2" wide, 36" high		72	.111		21	6.10		27.10	33
0250 42" high		72	.111		31	6.10		37.10	44
0300 Post sleeve for 4 x 4 post		96	.083	↓	18.65	4.56		23.21	28
0400 Post cap for 4 x 4 post, flat profile		48	.167	Ea.	13.30	9.10		22.40	29
0450 Newel post style profile		48	.167		35	9.10		44.10	53
0500 Raised corbeled profile		48	.167		33.50	9.10		42.60	51.50
0550 Post base trim for 4 x 4 post	↓	96	.083	↓	17.60	4.56		22.16	26.50

06 65 Plastic Trim

06 65 10 – PVC Trim

06 65 10.10 PVC Trim, Exterior

	Crew	Daily Output	Labor-Hours	Unit	Material	2021 Bare Costs Labor	Equipment	Total	Total Incl O&P
0010 **PVC TRIM, EXTERIOR**									
0100 Cornerboards, 5/4" x 6" x 6"	1 Carp	240	.033	L.F.	16.20	1.82		18.02	20.50
0110 Door/window casing, 1" x 4"		200	.040		5.50	2.19		7.69	9.55
0120 1" x 6"		200	.040		6.05	2.19		8.24	10.15
0130 1" x 8"		195	.041		6.85	2.24		9.09	11.10
0140 1" x 10"		195	.041		9.40	2.24		11.64	13.90
0150 1" x 12"		190	.042		10.10	2.30		12.40	14.75
0160 5/4" x 4"		195	.041		6.30	2.24		8.54	10.50
0170 5/4" x 6"		195	.041		7.55	2.24		9.79	11.85
0180 5/4" x 8"		190	.042		8.55	2.30		10.85	13.05
0190 5/4" x 10"		190	.042		11.45	2.30		13.75	16.25
0200 5/4" x 12"		185	.043		13.30	2.37		15.67	18.35
0210 Fascia, 1" x 4"		250	.032		5.50	1.75		7.25	8.85
0220 1" x 6"		250	.032		6.05	1.75		7.80	9.45
0230 1" x 8"		225	.036		6.85	1.94		8.79	10.65
0240 1" x 10"		225	.036		9.40	1.94		11.34	13.45
0250 1" x 12"		200	.040		10.10	2.19		12.29	14.60
0260 5/4" x 4"		240	.033		6.30	1.82		8.12	9.85
0270 5/4" x 6"		240	.033		7.55	1.82		9.37	11.20
0280 5/4" x 8"		215	.037		8.55	2.04		10.59	12.65
0290 5/4" x 10"		215	.037		11.45	2.04		13.49	15.85
0300 5/4" x 12"		190	.042		13.30	2.30		15.60	18.25
0310 Frieze, 1" x 4"		250	.032		5.50	1.75		7.25	8.85
0320 1" x 6"		250	.032		6.05	1.75		7.80	9.45
0330 1" x 8"		225	.036		6.85	1.94		8.79	10.65
0340 1" x 10"		225	.036		9.40	1.94		11.34	13.45
0350 1" x 12"	↓	200	.040	↓	10.10	2.19		12.29	14.60

182

06 65 Plastic Trim

06 65 10 – PVC Trim

06 65 10.10 PVC Trim, Exterior

		Crew	Daily Output	Labor-Hours	Unit	Material	2021 Bare Costs Labor	Equipment	Total	Total Incl O&P
0360	5/4" x 4"	1 Carp	240	.033	L.F.	6.30	1.82		8.12	9.85
0370	5/4" x 6"		240	.033		7.55	1.82		9.37	11.20
0380	5/4" x 8"		215	.037		8.55	2.04		10.59	12.65
0390	5/4" x 10"		215	.037		11.45	2.04		13.49	15.85
0400	5/4" x 12"		190	.042		13.30	2.30		15.60	18.25
0410	Rake, 1" x 4"		200	.040		5.50	2.19		7.69	9.55
0420	1" x 6"		200	.040		6.05	2.19		8.24	10.15
0430	1" x 8"		190	.042		6.85	2.30		9.15	11.20
0440	1" x 10"		190	.042		9.40	2.30		11.70	14
0450	1" x 12"		180	.044		10.10	2.43		12.53	14.95
0460	5/4" x 4"		195	.041		6.30	2.24		8.54	10.50
0470	5/4" x 6"		195	.041		7.55	2.24		9.79	11.85
0480	5/4" x 8"		185	.043		8.55	2.37		10.92	13.15
0490	5/4" x 10"		185	.043		11.45	2.37		13.82	16.35
0500	5/4" x 12"		175	.046		13.30	2.50		15.80	18.60
0510	Rake trim, 1" x 4"		225	.036		5.50	1.94		7.44	9.15
0520	1" x 6"		225	.036		6.05	1.94		7.99	9.75
0560	5/4" x 4"		220	.036		6.30	1.99		8.29	10.10
0570	5/4" x 6"		220	.036		7.55	1.99		9.54	11.45
0610	Soffit, 1" x 4"	2 Carp	420	.038		5.50	2.08		7.58	9.35
0620	1" x 6"		420	.038		6.05	2.08		8.13	9.95
0630	1" x 8"		420	.038		6.85	2.08		8.93	10.85
0640	1" x 10"		400	.040		9.40	2.19		11.59	13.85
0650	1" x 12"		400	.040		10.10	2.19		12.29	14.60
0660	5/4" x 4"		410	.039		6.30	2.13		8.43	10.35
0670	5/4" x 6"		410	.039		7.55	2.13		9.68	11.70
0680	5/4" x 8"		410	.039		8.55	2.13		10.68	12.80
0690	5/4" x 10"		390	.041		11.45	2.24		13.69	16.15
0700	5/4" x 12"		390	.041		13.30	2.24		15.54	18.15

06 73 Composite Decking

06 73 13 – Composite Decking

06 73 13.10 Woodgrained Composite Decking

		Crew	Daily Output	Labor-Hours	Unit	Material	2021 Bare Costs Labor	Equipment	Total	Total Incl O&P
0010	**WOODGRAINED COMPOSITE DECKING**									
0100	Woodgrained composite decking, 1" x 6"	2 Carp	640	.025	L.F.	4.31	1.37		5.68	6.90
0110	Grooved edge		660	.024		3.03	1.33		4.36	5.45
0120	2" x 6"		640	.025		4.62	1.37		5.99	7.30
0130	Encased, 1" x 6"		640	.025		4.72	1.37		6.09	7.40
0140	Grooved edge		660	.024		4.95	1.33		6.28	7.55
0150	2" x 6"		640	.025		5.55	1.37		6.92	8.30

183

06 81 Composite Railings

06 81 10 – Encased Railings

06 81 10.10 Encased Composite Railings	Crew	Daily Output	Labor-Hours	Unit	Material	2021 Bare Costs Labor	Equipment	Total	Total Incl O&P
0010 **ENCASED COMPOSITE RAILINGS**									
0100 Encased composite railing, 6' long, 36" high, incl. balusters	1 Carp	16	.500	Ea.	144	27.50		171.50	202
0110 42" high, incl. balusters		16	.500		222	27.50		249.50	288
0120 8' long, 36" high, incl. balusters		12	.667		247	36.50		283.50	330
0130 42" high, incl. balusters		12	.667		310	36.50		346.50	400
0140 Accessories, post sleeve, 4" x 4", 39" long		32	.250		30	13.70		43.70	55
0150 96" long		24	.333		75.50	18.25		93.75	112
0160 6" x 6", 39" long		32	.250		55	13.70		68.70	82.50
0170 96" long		24	.333		153	18.25		171.25	197
0180 Accessories, post skirt, 4" x 4"		96	.083		6.95	4.56		11.51	14.90
0190 6" x 6"		96	.083		7.95	4.56		12.51	16
0200 Post cap, 4" x 4", flat		48	.167		9	9.10		18.10	24.50
0210 Pyramid		48	.167		7.85	9.10		16.95	23
0220 Post cap, 6" x 6", flat		48	.167		14.95	9.10		24.05	31
0230 Pyramid		48	.167		15	9.10		24.10	31

Estimating Tips
07 10 00 Dampproofing and Waterproofing

- Be sure of the job specifications before pricing this subdivision. The difference in cost between waterproofing and dampproofing can be great. Waterproofing will hold back standing water. Dampproofing prevents the transmission of water vapor. Also included in this section are vapor retarding membranes.

07 20 00 Thermal Protection

- Insulation and fireproofing products are measured by area, thickness, volume, or R-value. Specifications may give only what the specific R-value should be in a certain situation. The estimator may need to choose the type of insulation to meet that R-value.

07 30 00 Steep Slope Roofing
07 40 00 Roofing and Siding Panels

- Many roofing and siding products are bought and sold by the square. One square is equal to an area that measures 100 square feet.

  This simple change in unit of measure could create a large error if the estimator is not observant. Accessories necessary for a complete installation must be figured into any calculations for both material and labor.

07 50 00 Membrane Roofing
07 60 00 Flashing and Sheet Metal
07 70 00 Roofing and Wall Specialties and Accessories

- The items in these subdivisions compose a roofing system. No one component completes the installation, and all must be estimated. Built-up or single-ply membrane roofing systems are made up of many products and installation trades. Wood blocking at roof perimeters or penetrations, parapet coverings, reglets, roof drains, gutters, downspouts, sheet metal flashing, skylights, smoke vents, and roof hatches all need to be considered along with the roofing material. Several different installation trades will need to work together on the roofing system. Inherent difficulties in the scheduling and coordination of various trades must be accounted for when estimating labor costs.

07 90 00 Joint Protection

- To complete the weather-tight shell, the sealants and caulkings must be estimated. Where different materials meet—at expansion joints, at flashing penetrations, and at hundreds of other locations throughout a construction project—caulking and sealants provide another line of defense against water penetration. Often, an entire system is based on the proper location and placement of caulking or sealants. The detailed drawings that are included as part of a set of architectural plans show typical locations for these materials. When caulking or sealants are shown at typical locations, this means the estimator must include them for all the locations where this detail is applicable. Be careful to keep different types of sealants separate, and remember to consider backer rods and primers if necessary.

Reference Numbers

Reference numbers are shown at the beginning of some major classifications. These numbers refer to related items in the Reference Section. The reference information may be an estimating procedure, an alternate pricing method, or technical information.

Note: Not all subdivisions listed here necessarily appear. ■

Same Data. Simplified.

Enjoy the convenience and efficiency of accessing your costs anywhere:

- **Skip the multiplier** by setting your location
- **Quickly search,** edit, favorite and share costs
- **Stay on top of price changes** with automatic updates

Discover more at rsmeans.com/online

07 01 50 – Maintenance of Membrane Roofing

07 01 50.10 Roof Coatings

		Crew	Daily Output	Labor-Hours	Unit	Material	2021 Bare Costs Labor	2021 Bare Costs Equipment	Total	Total Incl O&P
0010	**ROOF COATINGS**									
0012	Asphalt, brush grade, material only				Gal.	8.80			8.80	9.65
0200	Asphalt base, fibered aluminum coating [G]					9.55			9.55	10.50
0300	Asphalt primer, 5 gal.					5.50			5.50	6.05
0600	Coal tar pitch, 200 lb. barrels				Ton	1,200			1,200	1,325
0700	Tar roof cement, 5 gal. lots				Gal.	14.35			14.35	15.80
0800	Glass fibered roof & patching cement, 5 gal.				"	9.55			9.55	10.50
0900	Reinforcing glass membrane, 450 S.F./roll				Ea.	59.50			59.50	65.50
1000	Neoprene roof coating, 5 gal., 2 gal./sq.				Gal.	30.50			30.50	33.50
1100	Roof patch & flashing cement, 5 gal.					9.75			9.75	10.70
1200	Roof resaturant, glass fibered, 3 gal./sq.					8.85			8.85	9.75
1600	Reflective roof coating, white, elastomeric, approx. 50 S.F./gal. [G]					17.20			17.20	18.90

07 01 90 – Maintenance of Joint Protection

07 01 90.81 Joint Sealant Replacement

		Crew	Daily Output	Labor-Hours	Unit	Material	2021 Bare Costs Labor	2021 Bare Costs Equipment	Total	Total Incl O&P
0010	**JOINT SEALANT REPLACEMENT**									
0050	Control joints in concrete floors/slabs									
0100	Option 1 for joints with hard dry sealant									
0110	Step 1: Sawcut to remove 95% of old sealant									
0112	1/4" wide x 1/2" deep, with single saw blade	C-27	4800	.003	L.F.	.01	.17	.02	.20	.31
0114	3/8" wide x 3/4" deep, with single saw blade		4000	.004		.02	.21	.03	.26	.38
0116	1/2" wide x 1" deep, with double saw blades		3600	.004		.05	.23	.03	.31	.44
0118	3/4" wide x 1-1/2" deep, with double saw blades		3200	.005		.10	.26	.04	.40	.56
0120	Step 2: Water blast joint faces and edges	C-29	2500	.003			.14	.04	.18	.27
0130	Step 3: Air blast joint faces and edges	C-28	2000	.004			.21	.02	.23	.34
0140	Step 4: Sand blast joint faces and edges	E-11	2000	.016			.78	.15	.93	1.44
0150	Step 5: Air blast joint faces and edges	C-28	2000	.004			.21	.02	.23	.34
0200	Option 2 for joints with soft pliable sealant									
0210	Step 1: Plow joint with rectangular blade	B-62	2600	.009	L.F.		.44	.07	.51	.78
0220	Step 2: Sawcut to re-face joint faces									
0222	1/4" wide x 1/2" deep, with single saw blade	C-27	2400	.007	L.F.	.02	.35	.05	.42	.61
0224	3/8" wide x 3/4" deep, with single saw blade		2000	.008		.03	.41	.06	.50	.75
0226	1/2" wide x 1" deep, with double saw blades		1800	.009		.06	.46	.06	.58	.86
0228	3/4" wide x 1-1/2" deep, with double saw blades		1600	.010		.13	.52	.07	.72	1.03
0230	Step 3: Water blast joint faces and edges	C-29	2500	.003			.14	.04	.18	.27
0240	Step 4: Air blast joint faces and edges	C-28	2000	.004			.21	.02	.23	.34
0250	Step 5: Sand blast joint faces and edges	E-11	2000	.016			.78	.15	.93	1.44
0260	Step 6: Air blast joint faces and edges	C-28	2000	.004			.21	.02	.23	.34
0290	For saw cutting new control joints, see Section 03 15 16.20									
8910	For backer rod, see Section 07 91 23.10									
8920	For joint sealant, see Section 03 15 16.30 or 07 92 13.20									

07 05 05 – Selective Demolition for Thermal and Moisture Protection

07 05 05.10 Selective Demo., Thermal and Moist. Protection	Crew	Daily Output	Labor-Hours	Unit	Material	2021 Bare Costs Labor	Equipment	Total	Total Incl O&P
0010 SELECTIVE DEMO., THERMAL AND MOISTURE PROTECTION									
0020 Caulking/sealant, to 1" x 1" joint R024119-10	1 Clab	600	.013	L.F.		.59		.59	.94
0120 Downspouts, including hangers		350	.023	"		1.01		1.01	1.62
0220 Flashing, sheet metal		290	.028	S.F.		1.22		1.22	1.95
0420 Gutters, aluminum or wood, edge hung		240	.033	L.F.		1.48		1.48	2.36
0520 Built-in		100	.080	"		3.55		3.55	5.65
0620 Insulation, air/vapor barrier		3500	.002	S.F.		.10		.10	.16
0670 Batts or blankets		1400	.006	C.F.		.25		.25	.40
0720 Foamed or sprayed in place	2 Clab	1000	.016	B.F.		.71		.71	1.13
0770 Loose fitting	1 Clab	3000	.003	C.F.		.12		.12	.19
0870 Rigid board		3450	.002	B.F.		.10		.10	.16
1120 Roll roofing, cold adhesive		12	.667	Sq.		29.50		29.50	47
1170 Roof accessories, adjustable metal chimney flashing		9	.889	Ea.		39.50		39.50	63
1325 Plumbing vent flashing		32	.250	"		11.10		11.10	17.70
1375 Ridge vent strip, aluminum		310	.026	L.F.		1.15		1.15	1.82
1620 Skylight to 10 S.F.		8	1	Ea.		44.50		44.50	70.50
2120 Roof edge, aluminum soffit and fascia		570	.014	L.F.		.62		.62	.99
2170 Concrete coping, up to 12" wide	2 Clab	160	.100			4.44		4.44	7.05
2220 Drip edge	1 Clab	1000	.008			.36		.36	.57
2270 Gravel stop		950	.008			.37		.37	.60
2370 Sheet metal coping, up to 12" wide		240	.033			1.48		1.48	2.36
2470 Roof insulation board, over 2" thick	B-2	7800	.005	B.F.		.23		.23	.37
2520 Up to 2" thick	"	3900	.010	S.F.		.46		.46	.73
2620 Roof ventilation, louvered gable vent	1 Clab	16	.500	Ea.		22		22	35.50
2670 Remove, roof hatch	G-3	15	2.133			117		117	184
2675 Rafter vents	1 Clab	960	.008			.37		.37	.59
2720 Soffit vent and/or fascia vent		575	.014	L.F.		.62		.62	.98
2775 Soffit vent strip, aluminum, 3" to 4" wide		160	.050			2.22		2.22	3.54
2820 Roofing accessories, shingle moulding, to 1" x 4"		1600	.005			.22		.22	.35
2870 Cant strip	B-2	2000	.020			.90		.90	1.43
2920 Concrete block walkway	1 Clab	230	.035			1.54		1.54	2.46
3070 Roofing, felt paper, #15		70	.114	Sq.		5.05		5.05	8.10
3125 #30 felt		30	.267	"		11.85		11.85	18.85
3170 Asphalt shingles, 1 layer	B-2	3500	.011	S.F.		.51		.51	.82
3180 2 layers		1750	.023	"		1.02		1.02	1.63
3370 Modified bitumen		26	1.538	Sq.		69		69	110
3420 Built-up, no gravel, 3 ply		25	1.600			71.50		71.50	114
3470 4 ply		21	1.905			85.50		85.50	136
3620 5 ply		1600	.025	S.F.		1.12		1.12	1.78
3720 5 ply, with gravel		890	.045			2.01		2.01	3.21
3725 Loose gravel removal		5000	.008			.36		.36	.57
3730 Embedded gravel removal		2000	.020			.90		.90	1.43
3870 Fiberglass sheet		1200	.033			1.49		1.49	2.38
4120 Slate shingles		1900	.021			.94		.94	1.50
4170 Ridge shingles, clay or slate		2000	.020	L.F.		.90		.90	1.43
4320 Single ply membrane, attached at seams		52	.769	Sq.		34.50		34.50	55
4370 Ballasted		75	.533			24		24	38
4420 Fully adhered		39	1.026			46		46	73
4550 Roof hatch, 2'-6" x 3'-0"	1 Clab	10	.800	Ea.		35.50		35.50	56.50
4670 Wood shingles	B-2	2200	.018	S.F.		.81		.81	1.30
4820 Sheet metal roofing	"	2150	.019			.83		.83	1.33
4970 Siding, horizontal wood clapboards	1 Clab	380	.021			.93		.93	1.49
5025 Exterior insulation finish system	"	120	.067			2.96		2.96	4.71

187

07 05 Common Work Results for Thermal and Moisture Protection

07 05 05 – Selective Demolition for Thermal and Moisture Protection

07 05 05.10 Selective Demo., Thermal and Moist. Protection	Crew	Daily Output	Labor-Hours	Unit	Material	2021 Bare Costs Labor	Equipment	Total	Total Incl O&P
5070	Tempered hardboard, remove and reset	1 Carp	380	.021	S.F.		1.15		1.15	1.83
5120	Tempered hardboard sheet siding	"	375	.021			1.17		1.17	1.86
5170	Metal, corner strips	1 Clab	850	.009	L.F.		.42		.42	.67
5225	Horizontal strips		444	.018	S.F.		.80		.80	1.27
5320	Vertical strips		400	.020			.89		.89	1.41
5520	Wood shingles		350	.023			1.01		1.01	1.62
5620	Stucco siding		360	.022			.99		.99	1.57
5670	Textured plywood		725	.011			.49		.49	.78
5720	Vinyl siding		510	.016			.70		.70	1.11
5770	Corner strips		900	.009	L.F.		.39		.39	.63
5870	Wood, boards, vertical		400	.020	S.F.		.89		.89	1.41
5880	Steel siding, corrugated/ribbed		402.50	.020	"		.88		.88	1.41
5920	Waterproofing, protection/drain board	2 Clab	3900	.004	B.F.		.18		.18	.29
5970	Over 1/2" thick		1750	.009	S.F.		.41		.41	.65
6020	To 1/2" thick		2000	.008	"		.36		.36	.57
9000	Minimum labor/equipment charge	1 Clab	2	4	Job		178		178	283

07 11 Dampproofing

07 11 13 – Bituminous Dampproofing

07 11 13.10 Bituminous Asphalt Coating

		Crew	Daily Output	Labor-Hours	Unit	Material	2021 Bare Costs Labor	Equipment	Total	Total Incl O&P
0010	**BITUMINOUS ASPHALT COATING**									
0030	Brushed on, below grade, 1 coat	1 Rofc	665	.012	S.F.	.22	.58		.80	1.24
0100	2 coat		500	.016		.44	.77		1.21	1.81
0300	Sprayed on, below grade, 1 coat		830	.010		.22	.46		.68	1.04
0400	2 coat		500	.016		.43	.77		1.20	1.80
0500	Asphalt coating, with fibers				Gal.	9.55			9.55	10.50
0600	Troweled on, asphalt with fibers, 1/16" thick	1 Rofc	500	.016	S.F.	.42	.77		1.19	1.79
0700	1/8" thick		400	.020		.73	.96		1.69	2.47
1000	1/2" thick		350	.023		2.39	1.10		3.49	4.53
9000	Minimum labor/equipment charge		3	2.667	Job		129		129	221

07 11 16 – Cementitious Dampproofing

07 11 16.20 Cementitious Parging

		Crew	Daily Output	Labor-Hours	Unit	Material	2021 Bare Costs Labor	Equipment	Total	Total Incl O&P
0010	**CEMENTITIOUS PARGING**									
0020	Portland cement, 2 coats, 1/2" thick	D-1	250	.064	S.F.	.49	3.12		3.61	5.55
0100	Waterproofed Portland cement, 1/2" thick, 2 coats	"	250	.064	"	2.26	3.12		5.38	7.50

07 12 Built-up Bituminous Waterproofing

07 12 13 – Built-Up Asphalt Waterproofing

07 12 13.20 Membrane Waterproofing

		Crew	Daily Output	Labor-Hours	Unit	Material	2021 Bare Costs Labor	Equipment	Total	Total Incl O&P
0010	**MEMBRANE WATERPROOFING**									
0012	On slabs, 1 ply, felt, mopped	G-1	3000	.019	S.F.	.48	.84	.19	1.51	2.18
0015	On walls, 1 ply, felt, mopped		3000	.019		.48	.84	.19	1.51	2.18
0100	On slabs, 1 ply, glass fiber fabric, mopped		2100	.027		.53	1.20	.27	2	2.94
0105	On walls, 1 ply, glass fiber fabric, mopped		2100	.027		.53	1.20	.27	2	2.94
0300	On slabs, 2 ply, felt, mopped		2500	.022		.96	1.01	.23	2.20	3.03
0305	On walls, 2 ply, felt, mopped		2500	.022		.96	1.01	.23	2.20	3.03
0400	On slabs, 2 ply, glass fiber fabric, mopped		1650	.034		1.13	1.53	.35	3.01	4.25
0405	On walls, 2 ply, glass fiber fabric, mopped		1650	.034		1.13	1.53	.35	3.01	4.25
0600	On slabs, 3 ply, felt, mopped		2100	.027		1.44	1.20	.27	2.91	3.94

07 12 Built-up Bituminous Waterproofing

07 12 13 – Built-Up Asphalt Waterproofing

07 12 13.20 Membrane Waterproofing	Crew	Daily Output	Labor-Hours	Unit	Material	2021 Bare Costs Labor	Equipment	Total	Total Incl O&P
0605	On walls, 3 ply, felt, mopped	G-1	2100	.027	S.F.	1.44	1.20	.27	2.91	3.94
0700	On slabs, 3 ply, glass fiber fabric, mopped		1550	.036		1.59	1.63	.37	3.59	4.96
0705	On walls, 3 ply, glass fiber fabric, mopped		1550	.036		1.59	1.63	.37	3.59	4.96
0710	Asphaltic hardboard protection board, 1/8" thick	2 Rofc	500	.032		.77	1.54		2.31	3.50
0715	1/4" thick		450	.036		1.40	1.71		3.11	4.49
1000	EPS membrane protection board, 1/4" thick		3500	.005		.34	.22		.56	.75
1050	3/8" thick		3500	.005		.38	.22		.60	.79
1060	1/2" thick		3500	.005		.42	.22		.64	.84
1070	Fiberglass fabric, black, 20/10 mesh		116	.138	Sq.	15.45	6.65		22.10	28.50
9000	Minimum labor/equipment charge		2	8	Job		385		385	665

07 13 Sheet Waterproofing

07 13 53 – Elastomeric Sheet Waterproofing

07 13 53.10 Elastomeric Sheet Waterproofing and Access.

		Crew	Daily Output	Labor-Hours	Unit	Material	2021 Bare Costs Labor	Equipment	Total	Total Incl O&P
0010	**ELASTOMERIC SHEET WATERPROOFING AND ACCESS.**									
0090	EPDM, plain, 45 mils thick	2 Rofc	580	.028	S.F.	.74	1.33		2.07	3.11
0100	60 mils thick		570	.028		.95	1.35		2.30	3.37
0300	Nylon reinforced sheets, 45 mils thick		580	.028		.87	1.33		2.20	3.24
0400	60 mils thick		570	.028		1.01	1.35		2.36	3.44
0600	Vulcanizing splicing tape for above, 2" wide				C.L.F.	61.50			61.50	67.50
0700	4" wide				"	320			320	350
0900	Adhesive, bonding, 60 S.F./gal.				Gal.	30.50			30.50	33.50
1000	Splicing, 75 S.F./gal.				"	37.50			37.50	41.50
1200	Neoprene sheets, plain, 45 mils thick	2 Rofc	580	.028	S.F.	2.25	1.33		3.58	4.77
1300	60 mils thick		570	.028		1.59	1.35		2.94	4.08
1500	Nylon reinforced, 45 mils thick		580	.028		2.36	1.33		3.69	4.89
1600	60 mils thick		570	.028		2.64	1.35		3.99	5.25
1800	120 mils thick		500	.032		5.70	1.54		7.24	8.95
1900	Adhesive, splicing, 150 S.F./gal. per coat				Gal.	37.50			37.50	41.50
2100	Fiberglass reinforced, fluid applied, 1/8" thick	2 Rofc	500	.032	S.F.	1.63	1.54		3.17	4.44
2200	Polyethylene and rubberized asphalt sheets, 60 mils thick		550	.029		1.26	1.40		2.66	3.80
2400	Polyvinyl chloride sheets, plain, 10 mils thick		580	.028		.15	1.33		1.48	2.46
2500	20 mils thick		570	.028		.26	1.35		1.61	2.62
2700	30 mils thick		560	.029		.21	1.38		1.59	2.60
3000	Adhesives, trowel grade, 40-100 S.F./gal.				Gal.	25.50			25.50	28
3100	Brush grade, 100-250 S.F./gal.				"	23			23	25.50
3300	Bitumen modified polyurethane, fluid applied, 55 mils thick	2 Rofc	665	.024	S.F.	1.18	1.16		2.34	3.30
9000	Minimum labor/equipment charge	"	2	8	Job		385		385	665

07 16 Cementitious and Reactive Waterproofing

07 16 16 – Crystalline Waterproofing

07 16 16.20 Cementitious Waterproofing

		Crew	Daily Output	Labor-Hours	Unit	Material	2021 Bare Costs Labor	Equipment	Total	Total Incl O&P
0010	**CEMENTITIOUS WATERPROOFING**									
0020	1/8" application, sprayed on	G-2A	1000	.024	S.F.	.58	1.03	.64	2.25	3.07
0050	4 coat cementitious metallic slurry	1 Cefi	1.20	6.667	C.S.F.	72	345		417	620

For customer support on your Commercial Renovation Costs with RSMeans Data, call 800.448.8182.

189

07 19 Water Repellents

07 19 19 – Silicone Water Repellents

07 19 19.10 Silicone Based Water Repellents		Crew	Daily Output	Labor-Hours	Unit	Material	2021 Bare Costs Labor	Equipment	Total	Total Incl O&P
0010	**SILICONE BASED WATER REPELLENTS**									
0020	Water base liquid, roller applied	2 Rofc	7000	.002	S.F.	.43	.11		.54	.66
0200	Silicone or stearate, sprayed on CMU, 1 coat	1 Rofc	4000	.002		.40	.10		.50	.62
0300	2 coats		3000	.003	↓	.81	.13		.94	1.11
9000	Minimum labor/equipment charge	↓	3	2.667	Job		129		129	221

07 21 Thermal Insulation

07 21 13 – Board Insulation

07 21 13.10 Rigid Insulation

07 21 13.10 Rigid Insulation			Crew	Daily Output	Labor-Hours	Unit	Material	2021 Bare Costs Labor	Equipment	Total	Total Incl O&P
0010	**RIGID INSULATION**, for walls										
0040	Fiberglass, 1.5#/C.F., unfaced, 1" thick, R4.1	G	1 Carp	1000	.008	S.F.	.36	.44		.80	1.10
0060	1-1/2" thick, R6.2	G		1000	.008		.44	.44		.88	1.18
0080	2" thick, R8.3	G		1000	.008		.49	.44		.93	1.24
0120	3" thick, R12.4	G		800	.010		.59	.55		1.14	1.52
0370	3#/C.F., unfaced, 1" thick, R4.3	G		1000	.008		.57	.44		1.01	1.33
0390	1-1/2" thick, R6.5	G		1000	.008		.73	.44		1.17	1.50
0400	2" thick, R8.7	G		890	.009		1.25	.49		1.74	2.16
0420	2-1/2" thick, R10.9	G		800	.010		1.04	.55		1.59	2.01
0440	3" thick, R13	G		800	.010		1.63	.55		2.18	2.66
0520	Foil faced, 1" thick, R4.3	G		1000	.008		.80	.44		1.24	1.58
0540	1-1/2" thick, R6.5	G		1000	.008		1.18	.44		1.62	2
0560	2" thick, R8.7	G		890	.009		1.49	.49		1.98	2.42
0580	2-1/2" thick, R10.9	G		800	.010		1.75	.55		2.30	2.80
0600	3" thick, R13	G	↓	800	.010	↓	1.85	.55		2.40	2.91
1600	Isocyanurate, 4' x 8' sheet, foil faced, both sides										
1610	1/2" thick	G	1 Carp	800	.010	S.F.	.29	.55		.84	1.19
1620	5/8" thick	G		800	.010		.62	.55		1.17	1.55
1630	3/4" thick	G		800	.010		.44	.55		.99	1.35
1640	1" thick	G		800	.010		.62	.55		1.17	1.55
1650	1-1/2" thick	G		730	.011		.76	.60		1.36	1.79
1660	2" thick	G		730	.011		.78	.60		1.38	1.81
1670	3" thick	G		730	.011		3.13	.60		3.73	4.39
1680	4" thick	G		730	.011		2.60	.60		3.20	3.81
1700	Perlite, 1" thick, R2.77	G		800	.010		.46	.55		1.01	1.38
1750	2" thick, R5.55	G		730	.011		.83	.60		1.43	1.86
1900	Extruded polystyrene, 25 psi compressive strength, 1" thick, R5	G		800	.010		.61	.55		1.16	1.54
1940	2" thick, R10	G		730	.011		1.23	.60		1.83	2.30
1960	3" thick, R15	G		730	.011		1.75	.60		2.35	2.88
2100	Expanded polystyrene, 1" thick, R3.85	G		800	.010		.32	.55		.87	1.22
2120	2" thick, R7.69	G		730	.011		.64	.60		1.24	1.65
2140	3" thick, R11.49	G		730	.011	↓	.96	.60		1.56	2.01
9000	Minimum labor/equipment charge		↓	4	2	Job		109		109	174

07 21 13.13 Foam Board Insulation

07 21 13.13 Foam Board Insulation			Crew	Daily Output	Labor-Hours	Unit	Material	2021 Bare Costs Labor	Equipment	Total	Total Incl O&P
0010	**FOAM BOARD INSULATION**										
0600	Polystyrene, expanded, 1" thick, R4	G	1 Carp	680	.012	S.F.	.32	.64		.96	1.37
0700	2" thick, R8	G		675	.012	"	.64	.65		1.29	1.73
9000	Minimum labor/equipment charge		↓	4	2	Job		109		109	174

07 21 16 – Blanket Insulation

07 21 16.10 Blanket Insulation for Floors/Ceilings

		Crew	Daily Output	Labor-Hours	Unit	Material	2021 Bare Costs Labor	Equipment	Total	Total Incl O&P
0010	**BLANKET INSULATION FOR FLOORS/CEILINGS**									
0020	Including spring type wire fasteners									
2000	Fiberglass, blankets or batts, paper or foil backing									
2100	3-1/2" thick, R13	G 1 Carp	700	.011	S.F.	.49	.63		1.12	1.54
2150	6-1/4" thick, R19	G	600	.013		.60	.73		1.33	1.82
2210	9-1/2" thick, R30	G	500	.016		.91	.88		1.79	2.39
2220	12" thick, R38	G	475	.017		1.13	.92		2.05	2.71
3000	Unfaced, 3-1/2" thick, R13	G 1 Carp	600	.013	S.F.	.33	.73		1.06	1.52
3010	6-1/4" thick, R19	G	500	.016		.56	.88		1.44	2.01
3020	9-1/2" thick, R30	G	450	.018		.56	.97		1.53	2.17
3030	12" thick, R38	G	425	.019		.78	1.03		1.81	2.50
9000	Minimum labor/equipment charge		4	2	Job		109		109	174

07 21 16.20 Blanket Insulation for Walls

		Crew	Daily Output	Labor-Hours	Unit	Material	2021 Bare Costs Labor	Equipment	Total	Total Incl O&P
0010	**BLANKET INSULATION FOR WALLS**									
0020	Kraft faced fiberglass, 3-1/2" thick, R11, 15" wide	G 1 Carp	1350	.006	S.F.	.32	.32		.64	.87
0030	23" wide	G	1600	.005		.32	.27		.59	.79
0060	R13, 11" wide	G	1150	.007		.35	.38		.73	1
0080	15" wide	G	1350	.006		.35	.32		.67	.91
0100	23" wide	G	1600	.005		.35	.27		.62	.83
0110	R15, 11" wide	G	1150	.007		.55	.38		.93	1.22
0120	15" wide	G	1350	.006		.55	.32		.87	1.13
0130	23" wide	G	1600	.005		.55	.27		.82	1.05
0140	6" thick, R19, 11" wide	G	1150	.007		.46	.38		.84	1.12
0160	15" wide	G	1350	.006		.46	.32		.78	1.03
0180	23" wide	G	1600	.005		.46	.27		.73	.95
0182	R21, 11" wide	G	1150	.007		.85	.38		1.23	1.55
0184	15" wide	G	1350	.006		.85	.32		1.17	1.46
0186	23" wide	G	1600	.005		.85	.27		1.12	1.38
0188	9" thick, R30, 11" wide	G	985	.008		.91	.44		1.35	1.71
0200	15" wide	G	1150	.007		.91	.38		1.29	1.61
0220	23" wide	G	1350	.006		.91	.32		1.23	1.52
0230	12" thick, R38, 11" wide	G	985	.008		1.13	.44		1.57	1.95
0240	15" wide	G	1150	.007		1.13	.38		1.51	1.85
0260	23" wide	G	1350	.006		1.13	.32		1.45	1.76
0410	Foil faced fiberglass, 3-1/2" thick, R13, 11" wide	G	1150	.007		.55	.38		.93	1.22
0420	15" wide	G	1350	.006		.55	.32		.87	1.13
0440	23" wide	G	1600	.005		.55	.27		.82	1.05
0442	R15, 11" wide	G	1150	.007		.47	.38		.85	1.13
0444	15" wide	G	1350	.006		.47	.32		.79	1.04
0446	23" wide	G	1600	.005		.47	.27		.74	.96
0448	6" thick, R19, 11" wide	G	1150	.007		.76	.38		1.14	1.45
0460	15" wide	G	1350	.006		.76	.32		1.08	1.36
0480	23" wide	G	1600	.005		.76	.27		1.03	1.28
0482	R21, 11" wide	G	1150	.007		.66	.38		1.04	1.34
0484	15" wide	G	1350	.006		.66	.32		.98	1.25
0486	23" wide	G	1600	.005		.66	.27		.93	1.17
0488	9" thick, R30, 11" wide	G	985	.008		1.19	.44		1.63	2.02
0500	15" wide	G	1150	.007		1.19	.38		1.57	1.92
0550	23" wide	G	1350	.006		1.19	.32		1.51	1.83
0560	12" thick, R38, 11" wide	G 1 Carp	985	.008	S.F.	1.13	.44		1.57	1.95
0570	15" wide	G	1150	.007		1.13	.38		1.51	1.85
0580	23" wide	G	1350	.006		1.13	.32		1.45	1.76

07 21 Thermal Insulation

07 21 16 – Blanket Insulation

07 21 16.20 Blanket Insulation for Walls

		Crew	Daily Output	Labor-Hours	Unit	Material	2021 Bare Costs Labor	Equipment	Total	Total Incl O&P
0620	Unfaced fiberglass, 3-1/2" thick, R13, 11" wide [G]	1 Carp	1150	.007	S.F.	.33	.38		.71	.97
0820	15" wide [G]		1350	.006		.33	.32		.65	.88
0830	23" wide [G]		1600	.005		.33	.27		.60	.80
0832	R15, 11" wide [G]		1150	.007		.41	.38		.79	1.06
0836	23" wide [G]		1600	.005		.41	.27		.68	.89
0838	6" thick, R19, 11" wide [G]		1150	.007		.56	.38		.94	1.23
0860	15" wide [G]		1150	.007		.56	.38		.94	1.23
0880	23" wide [G]		1350	.006		.56	.32		.88	1.14
0882	R21, 11" wide [G]		1150	.007		.68	.38		1.06	1.36
0886	15" wide [G]		1350	.006		.68	.32		1	1.27
0888	23" wide [G]		1600	.005		.68	.27		.95	1.19
0890	9" thick, R30, 11" wide [G]		985	.008		.56	.44		1	1.33
0900	15" wide [G]		1150	.007		.56	.38		.94	1.23
0920	23" wide [G]		1350	.006		.56	.32		.88	1.14
0930	12" thick, R38, 11" wide [G]		985	.008		.78	.44		1.22	1.57
0940	15" wide [G]		1000	.008		.78	.44		1.22	1.56
0960	23" wide [G]		1150	.007		.78	.38		1.16	1.47
1300	Wall or ceiling insulation, mineral wool batts									
1320	3-1/2" thick, R15 [G]	1 Carp	1600	.005	S.F.	.84	.27		1.11	1.36
1340	5-1/2" thick, R23 [G]		1600	.005		1.32	.27		1.59	1.89
1380	7-1/4" thick, R30 [G]		1350	.006		1.74	.32		2.06	2.43
1700	Non-rigid insul., recycled blue cotton fiber, unfaced batts, R13, 16" wide [G]		1600	.005		1.23	.27		1.50	1.79
1710	R19, 16" wide [G]		1600	.005		1.39	.27		1.66	1.97
1850	Friction fit wire insulation supports, 16" OC		960	.008	Ea.	.14	.46		.60	.88
9000	Minimum labor/equipment charge		4	2	Job		109		109	174

07 21 19 – Foamed In Place Insulation

07 21 19.10 Masonry Foamed In Place Insulation

		Crew	Daily Output	Labor-Hours	Unit	Material	2021 Bare Costs Labor	Equipment	Total	Total Incl O&P
0010	**MASONRY FOAMED IN PLACE INSULATION**									
0100	Amino-plast foam, injected into block core, 6" block [G]	G-2A	6000	.004	Ea.	.17	.17	.11	.45	.60
0110	8" block [G]		5000	.005		.21	.21	.13	.55	.71
0120	10" block [G]		4000	.006		.26	.26	.16	.68	.89
0130	12" block [G]		3000	.008		.35	.34	.21	.90	1.19
0140	Injected into cavity wall [G]		13000	.002	B.F.	.06	.08	.05	.19	.25
0150	Preparation, drill holes into mortar joint every 4 V.L.F., 5/8" diameter	1 Clab	960	.008	Ea.		.37		.37	.59
0160	7/8" diameter		680	.012			.52		.52	.83
0170	Patch drilled holes, 5/8" diameter		1800	.004		.04	.20		.24	.35
0180	7/8" diameter		1200	.007		.05	.30		.35	.53
9000	Minimum labor/equipment charge	G-2A	2	12	Job		515	320	835	1,225
9010	Minimum labor/equipment charge	1 Clab	4	2	"		89		89	141

07 21 23 – Loose-Fill Insulation

07 21 23.10 Poured Loose-Fill Insulation

		Crew	Daily Output	Labor-Hours	Unit	Material	2021 Bare Costs Labor	Equipment	Total	Total Incl O&P
0010	**POURED LOOSE-FILL INSULATION**									
0020	Cellulose fiber, R3.8 per inch [G]	1 Carp	200	.040	C.F.	.70	2.19		2.89	4.25
0021	4" thick [G]		1000	.008	S.F.	.17	.44		.61	.88
0022	6" thick [G]		800	.010	"	.28	.55		.83	1.18
0080	Fiberglass wool, R4 per inch [G]		200	.040	C.F.	.55	2.19		2.74	4.09
0081	4" thick [G]		600	.013	S.F.	.19	.73		.92	1.37
0082	6" thick [G]		400	.020	"	.27	1.09		1.36	2.03
0100	Mineral wool, R3 per inch [G]		200	.040	C.F.	.56	2.19		2.75	4.10
0101	4" thick [G]		600	.013	S.F.	.18	.73		.91	1.36
0102	6" thick [G]		400	.020	"	.28	1.09		1.37	2.05
0300	Polystyrene, R4 per inch [G]		200	.040	C.F.	1.56	2.19		3.75	5.20

07 21 Thermal Insulation

07 21 23 – Loose-Fill Insulation

07 21 23.10 Poured Loose-Fill Insulation

		Crew	Daily Output	Labor-Hours	Unit	Material	2021 Bare Costs Labor	Equipment	Total	Total Incl O&P
0301	4" thick	G 1 Carp	600	.013	S.F.	.51	.73		1.24	1.73
0302	6" thick	G	400	.020	"	.78	1.09		1.87	2.60
0400	Perlite, R2.78 per inch	G	200	.040	C.F.	6.55	2.19		8.74	10.75
0401	4" thick	G	1000	.008	S.F.	2.19	.44		2.63	3.11
0402	6" thick	G	800	.010	"	3.29	.55		3.84	4.48
9000	Minimum labor/equipment charge		4	2	Job		109		109	174

07 21 23.20 Masonry Loose-Fill Insulation

		Crew	Daily Output	Labor-Hours	Unit	Material	2021 Bare Costs Labor	Equipment	Total	Total Incl O&P
0010	**MASONRY LOOSE-FILL INSULATION,** vermiculite or perlite									
0100	In cores of concrete block, 4" thick wall, .115 C.F./S.F.	G D-1	4800	.003	S.F.	.76	.16		.92	1.09
0200	6" thick wall, .175 C.F./S.F.	G	3000	.005		1.15	.26		1.41	1.68
0300	8" thick wall, .258 C.F./S.F.	G	2400	.007		1.70	.32		2.02	2.38
0400	10" thick wall, .340 C.F./S.F.	G	1850	.009		2.23	.42		2.65	3.14
0500	12" thick wall, .422 C.F./S.F.	G	1200	.013		2.77	.65		3.42	4.09
0600	Poured cavity wall, vermiculite or perlite, water repellent	G	250	.064	C.F.	6.55	3.12		9.67	12.25
0700	Foamed in place, urethane in 2-5/8" cavity	G G-2A	1035	.023	S.F.	1.45	.99	.62	3.06	3.94
0800	For each 1" added thickness, add	G "	2372	.010	"	.55	.43	.27	1.25	1.64

07 21 26 – Blown Insulation

07 21 26.10 Blown Insulation

		Crew	Daily Output	Labor-Hours	Unit	Material	2021 Bare Costs Labor	Equipment	Total	Total Incl O&P
0010	**BLOWN INSULATION** Ceilings, with open access									
0020	Cellulose, 3-1/2" thick, R13	G G-4	5000	.005	S.F.	.24	.22	.08	.54	.70
0030	5-3/16" thick, R19	G	3800	.006		.36	.28	.11	.75	.96
0050	6-1/2" thick, R22	G	3000	.008		.46	.36	.14	.96	1.22
0100	8-11/16" thick, R30	G	2600	.009		.62	.42	.16	1.20	1.51
0120	10-7/8" thick, R38	G	1800	.013		.79	.60	.23	1.62	2.08
1000	Fiberglass, 5.5" thick, R11	G	3800	.006		.19	.28	.11	.58	.78
1050	6" thick, R12	G	3000	.008		.27	.36	.14	.77	1.01
1100	8.8" thick, R19	G	2200	.011		.34	.49	.19	1.02	1.36
1200	10" thick, R22	G G-4	1800	.013	S.F.	.39	.60	.23	1.22	1.64
1300	11.5" thick, R26	G	1500	.016		.46	.72	.27	1.45	1.96
1350	13" thick, R30	G	1400	.017		.54	.77	.29	1.60	2.14
1450	16" thick, R38	G	1145	.021		.69	.94	.36	1.99	2.65
1500	20" thick, R49	G	920	.026		.91	1.18	.45	2.54	3.36
9000	Minimum labor/equipment charge		4	6	Job		270	103	373	545

07 21 29 – Sprayed Insulation

07 21 29.10 Sprayed-On Insulation

		Crew	Daily Output	Labor-Hours	Unit	Material	2021 Bare Costs Labor	Equipment	Total	Total Incl O&P
0010	**SPRAYED-ON INSULATION**									
0020	Fibrous/cementitious, finished wall, 1" thick, R3.7	G G-2	2050	.012	S.F.	.36	.54	.09	.99	1.36
0100	Attic, 5.2" thick, R19	G	1550	.015	"	.42	.72	.12	1.26	1.74
0200	Fiberglass, R4 per inch, vertical	G	1600	.015	B.F.	.20	.69	.12	1.01	1.45
0210	Horizontal	G	1200	.020	"	.20	.93	.16	1.29	1.86
0300	Closed cell, spray polyurethane foam, 2 lb./C.F. density									
0310	1" thick	G G-2A	6000	.004	S.F.	.55	.17	.11	.83	1.02
0320	2" thick	G	3000	.008		1.10	.34	.21	1.65	2.02
0330	3" thick	G	2000	.012		1.65	.51	.32	2.48	3.03
0335	3-1/2" thick	G	1715	.014		1.93	.60	.37	2.90	3.54
0340	4" thick	G	1500	.016		2.21	.69	.43	3.33	4.05
0350	5" thick	G	1200	.020		2.76	.86	.54	4.16	5.05
0355	5-1/2" thick	G	1090	.022		3.03	.94	.59	4.56	5.55
0360	6" thick	G	1000	.024		3.31	1.03	.64	4.98	6.05
9000	Minimum labor/equipment charge	G-2	2	12	Job		555	95	650	990

For customer support on your Commercial Renovation Costs with RSMeans Data, call 800.448.8182.

193

07 21 Thermal Insulation

07 21 53 – Reflective Insulation

07 21 53.10 Reflective Insulation Options		Crew	Daily Output	Labor-Hours	Unit	Material	2021 Bare Costs Labor	Equipment	Total	Total Incl O&P
0010	**REFLECTIVE INSULATION OPTIONS**									
0020	Aluminum foil on reinforced scrim	G 1 Carp	19	.421	C.S.F.	14.95	23		37.95	53
0100	Reinforced with woven polyolefin	G	19	.421		23.50	23		46.50	62.50
0500	With single bubble air space, R8.8	G	15	.533		26.50	29		55.50	75.50
0600	With double bubble air space, R9.8	G	15	.533		31.50	29		60.50	81.50
9000	Minimum labor/equipment charge		4	2	Job		109		109	174

07 22 Roof and Deck Insulation

07 22 16 – Roof Board Insulation

07 22 16.10 Roof Deck Insulation

07 22 16.10 Roof Deck Insulation		Crew	Daily Output	Labor-Hours	Unit	Material	2021 Bare Costs Labor	Equipment	Total	Total Incl O&P
0010	**ROOF DECK INSULATION**, fastening excluded									
0016	Asphaltic cover board, fiberglass lined, 1/8" thick	1 Rofc	1400	.006	S.F.	.49	.28		.77	1.01
0018	1/4" thick		1400	.006		.99	.28		1.27	1.56
0020	Fiberboard low density, 1/2" thick, R1.39	G	1300	.006		.36	.30		.66	.91
0030	1" thick, R2.78	G	1040	.008		.76	.37		1.13	1.48
0080	1-1/2" thick, R4.17	G	1040	.008		1.15	.37		1.52	1.91
0100	2" thick, R5.56	G	1040	.008		1.10	.37		1.47	1.85
0110	Fiberboard high density, 1/2" thick, R1.3	G	1300	.006		.32	.30		.62	.86
0120	1" thick, R2.5	G 1 Rofc	1040	.008	S.F.	.65	.37		1.02	1.36
0130	1-1/2" thick, R3.8	G	1040	.008		1.30	.37		1.67	2.06
0200	Fiberglass, 3/4" thick, R2.78	G	1300	.006		.62	.30		.92	1.19
0400	15/16" thick, R3.70	G	1300	.006		.86	.30		1.16	1.46
0460	1-1/16" thick, R4.17	G	1300	.006		1.20	.30		1.50	1.83
0600	1-5/16" thick, R5.26	G	1300	.006		1.49	.30		1.79	2.15
0650	2-1/16" thick, R8.33	G	1040	.008		1.67	.37		2.04	2.48
0700	2-7/16" thick, R10	G	1040	.008		1.79	.37		2.16	2.61
0800	Gypsum cover board, fiberglass mat facer, 1/4" thick		1400	.006		.51	.28		.79	1.03
0810	1/2" thick		1300	.006		.67	.30		.97	1.25
0820	5/8" thick		1200	.007		.66	.32		.98	1.28
0830	Primed fiberglass mat facer, 1/4" thick		1400	.006		.61	.28		.89	1.14
0840	1/2" thick		1300	.006		.69	.30		.99	1.27
0850	5/8" thick		1200	.007		.71	.32		1.03	1.33
1650	Perlite, 1/2" thick, R1.32	G	1365	.006		.30	.28		.58	.82
1655	3/4" thick, R2.08	G	1040	.008		.39	.37		.76	1.07
1660	1" thick, R2.78	G	1040	.008		.60	.37		.97	1.30
1670	1-1/2" thick, R4.17	G	1040	.008		.86	.37		1.23	1.59
1680	2" thick, R5.56	G	910	.009		1.07	.42		1.49	1.91
1685	2-1/2" thick, R6.67	G	910	.009		1.37	.42		1.79	2.24
1690	Tapered for drainage	G	1040	.008	B.F.	1.18	.37		1.55	1.94
1700	Polyisocyanurate, 2#/C.F. density, 3/4" thick	G	1950	.004	S.F.	.49	.20		.69	.88
1705	1" thick	G	1820	.004		.46	.21		.67	.87
1715	1-1/2" thick	G	1625	.005		.90	.24		1.14	1.40
1725	2" thick	G	1430	.006		.95	.27		1.22	1.51
1735	2-1/2" thick	G	1365	.006		1	.28		1.28	1.59
1745	3" thick	G	1300	.006		1.10	.30		1.40	1.72
1755	3-1/2" thick	G	1300	.006		1.69	.30		1.99	2.37
1765	Tapered for drainage	G	1820	.004	B.F.	.65	.21		.86	1.08
1900	Extruded polystyrene									
1910	15 psi compressive strength, 1" thick, R5	G 1 Rofc	1950	.004	S.F.	.52	.20		.72	.91
1920	2" thick, R10	G	1625	.005		.68	.24		.92	1.15

07 22 Roof and Deck Insulation

07 22 16 – Roof Board Insulation

07 22 16.10 Roof Deck Insulation		Crew	Daily Output	Labor-Hours	Unit	Material	2021 Bare Costs Labor	Equipment	Total	Total Incl O&P
1930	3" thick, R15	G 1 Rofc	1300	.006	S.F.	1.35	.30		1.65	2
1932	4" thick, R20	G	1300	.006	↓	1.82	.30		2.12	2.51
1934	Tapered for drainage	G	1950	.004	B.F.	.51	.20		.71	.90
1940	25 psi compressive strength, 1" thick, R5	G	1950	.004	S.F.	1.05	.20		1.25	1.50
1942	2" thick, R10	G	1625	.005		2	.24		2.24	2.60
1944	3" thick, R15	G	1300	.006		3.05	.30		3.35	3.86
1946	4" thick, R20	G	1300	.006	↓	4.20	.30		4.50	5.15
1948	Tapered for drainage	G	1950	.004	B.F.	.57	.20		.77	.97
1950	40 psi compressive strength, 1" thick, R5	G	1950	.004	S.F.	.86	.20		1.06	1.29
1952	2" thick, R10	G	1625	.005		1.63	.24		1.87	2.21
1954	3" thick, R15	G	1300	.006		2.37	.30		2.67	3.11
1956	4" thick, R20	G	1300	.006	↓	3.10	.30		3.40	3.92
1958	Tapered for drainage	G	1820	.004	B.F.	.87	.21		1.08	1.32
1960	60 psi compressive strength, 1" thick, R5	G	1885	.004	S.F.	1.07	.20		1.27	1.53
1962	2" thick, R10	G	1560	.005		2.03	.25		2.28	2.67
1964	3" thick, R15	G	1270	.006		3.32	.30		3.62	4.17
1966	4" thick, R20	G	1235	.006	↓	4.12	.31		4.43	5.05
1968	Tapered for drainage	G	1820	.004	B.F.	.98	.21		1.19	1.44
2010	Expanded polystyrene, 1#/C.F. density, 3/4" thick, R2.89	G	1950	.004	S.F.	.24	.20		.44	.60
2020	1" thick, R3.85	G	1950	.004		.32	.20		.52	.69
2100	2" thick, R7.69	G	1625	.005		.64	.24		.88	1.11
2110	3" thick, R11.49	G	1625	.005		.96	.24		1.20	1.47
2120	4" thick, R15.38	G	1625	.005		1.28	.24		1.52	1.82
2130	5" thick, R19.23	G	1495	.005		1.60	.26		1.86	2.20
2140	6" thick, R23.26	G	1495	.005	↓	1.92	.26		2.18	2.55
2150	Tapered for drainage	G	1950	.004	B.F.	.45	.20		.65	.84
2400	Composites with 2" EPS									
2410	1" fiberboard	G 1 Rofc	1325	.006	S.F.	1.55	.29		1.84	2.21
2420	7/16" oriented strand board	G	1040	.008		1.37	.37		1.74	2.15
2430	1/2" plywood	G	1040	.008		3.19	.37		3.56	4.15
2440	1" perlite	G	1040	.008	↓	1.04	.37		1.41	1.78
2450	Composites with 1-1/2" polyisocyanurate									
2460	1" fiberboard	G 1 Rofc	1040	.008	S.F.	1.20	.37		1.57	1.96
2470	1" perlite	G	1105	.007		.99	.35		1.34	1.69
2480	7/16" oriented strand board	G	1040	.008	↓	1.18	.37		1.55	1.94
3000	Fastening alternatives, coated screws, 2" long		3744	.002	Ea.	.05	.10		.15	.24
3010	4" long		3120	.003		.11	.12		.23	.33
3020	6" long		2675	.003		.20	.14		.34	.47
3030	8" long		2340	.003		.28	.16		.44	.59
3040	10" long		1872	.004		.52	.21		.73	.92
3050	Pre-drill and drive wedge spike, 2-1/2"		1248	.006		.62	.31		.93	1.21
3060	3-1/2"		1101	.007		1.38	.35		1.73	2.12
3070	4-1/2"		936	.009		1.93	.41		2.34	2.83
3075	3" galvanized deck plates		7488	.001		.10	.05		.15	.20
3080	Spot mop asphalt	G-1	295	.190	Sq.	6.45	8.55	1.94	16.94	24
3090	Full mop asphalt	"	192	.292		12.95	13.10	2.98	29.03	40
3110	Low-rise polyurethane adhesive, 12" OC beads	1 Rofc	45	.178		41.50	8.55		50.05	60.50
3120	6" OC beads		32	.250		83	12.05		95.05	112
3130	4" OC beads		30	.267	↓	126	12.85		138.85	160
9000	Minimum labor/equipment charge		3.25	2.462	Job		119		119	204

07 24 Exterior Insulation and Finish Systems

07 24 13 – Polymer-Based Exterior Insulation and Finish System

07 24 13.10 Exterior Insulation and Finish Systems		Crew	Daily Output	Labor-Hours	Unit	Material	2021 Bare Costs Labor	Equipment	Total	Total Incl O&P
0010	**EXTERIOR INSULATION AND FINISH SYSTEMS**									
0095	Field applied, 1" EPS insulation	G	J-1	390	.103	S.F.	.58	4.90	.29	5.77	8.75
0100	With 1/2" cement board sheathing	G		268	.149		1.30	7.10	.42	8.82	13.25
0105	2" EPS insulation	G		390	.103		.90	4.90	.29	6.09	9.10
0110	With 1/2" cement board sheathing	G		268	.149		1.62	7.10	.42	9.14	13.60
0115	3" EPS insulation	G		390	.103		1.22	4.90	.29	6.41	9.45
0120	With 1/2" cement board sheathing	G		268	.149		1.94	7.10	.42	9.46	13.95
0125	4" EPS insulation	G		390	.103		1.54	4.90	.29	6.73	9.80
0130	With 1/2" cement board sheathing	G		268	.149		2.98	7.10	.42	10.50	15.10
0140	Premium finish add			1265	.032		.44	1.51	.09	2.04	2.98
0145	Drainage and ventilation cavity, add		2 Plas	1450	.011		.82	.55		1.37	1.77
0150	Heavy duty reinforcement add		J-1	914	.044		.58	2.09	.12	2.79	4.10
0160	2.5#/S.Y. metal lath substrate add		1 Lath	75	.107	S.Y.	3.87	5.75		9.62	13.25
0170	3.4#/S.Y. metal lath substrate add		"	75	.107	"	4.99	5.75		10.74	14.50
0180	Color or texture change		J-1	1265	.032	S.F.	.74	1.51	.09	2.34	3.31
0190	With substrate leveling base coat		1 Plas	530	.015		.85	.75		1.60	2.14
0210	With substrate sealing base coat		1 Pord	1224	.007		.14	.30		.44	.63
0370	V groove shape in panel face					L.F.	.72			.72	.79
0380	U groove shape in panel face					"	.87			.87	.96
0433	Crack repair, acrylic rubber, fluid applied, 20 mils thick		1 Plas	350	.023	S.F.	1.56	1.14		2.70	3.53
0437	50 mils thick, reinforced		"	200	.040	"	2.77	1.99		4.76	6.20
0440	For higher than one story, add							25%			

07 25 Weather Barriers

07 25 10 – Weather Barriers or Wraps

07 25 10.10 Weather Barriers

		Crew	Daily Output	Labor-Hours	Unit	Material	2021 Bare Costs Labor	Equipment	Total	Total Incl O&P
0010	**WEATHER BARRIERS**									
0400	Asphalt felt paper, #15	1 Carp	37	.216	Sq.	5.70	11.85		17.55	25
0401	Per square foot	"	3700	.002	S.F.	.06	.12		.18	.25
0450	Housewrap, exterior, spun bonded polypropylene									
0470	Small roll	1 Carp	3800	.002	S.F.	.15	.12		.27	.34
0480	Large roll	"	4000	.002	"	.18	.11		.29	.36
2100	Asphalt felt roof deck vapor barrier, class 1 metal decks	1 Rofc	37	.216	Sq.	24.50	10.40		34.90	45
2200	For all other decks	"	37	.216		18.65	10.40		29.05	38.50
2800	Asphalt felt, 50% recycled content, 15 lb., 4 sq./roll	1 Carp	36	.222		5.95	12.15		18.10	26
2810	30 lb., 2 sq./roll	"	36	.222		10.20	12.15		22.35	30.50
3000	Building wrap, spun bonded polyethylene	2 Carp	8000	.002	S.F.	.14	.11		.25	.32
9960	Minimum labor/equipment charge	1 Carp	2	4	Job		219		219	350

07 26 Vapor Retarders

07 26 13 – Above-Grade Vapor Retarders

07 26 13.10 Vapor Retarders

			Crew	Daily Output	Labor-Hours	Unit	Material	2021 Bare Costs Labor	Equipment	Total	Total Incl O&P
0010	**VAPOR RETARDERS**										
0020	Aluminum and kraft laminated, foil 1 side	G	1 Carp	37	.216	Sq.	17.65	11.85		29.50	38.50
0100	Foil 2 sides	G		37	.216		18.40	11.85		30.25	39
0600	Polyethylene vapor barrier, standard, 2 mil	G		37	.216		3.03	11.85		14.88	22
0700	4 mil	G		37	.216		2.70	11.85		14.55	22
0900	6 mil	G		37	.216		7.65	11.85		19.50	27.50
1200	10 mil	G		37	.216		8.15	11.85		20	28

196

07 26 Vapor Retarders

07 26 13 – Above-Grade Vapor Retarders

07 26 13.10 Vapor Retarders

		Crew	Daily Output	Labor-Hours	Unit	Material	2021 Bare Costs Labor	Equipment	Total	Total Incl O&P
1300	Clear reinforced, fire retardant, 8 mil	1 Carp	37	.216	Sq.	26.50	11.85		38.35	48
1350	Cross laminated type, 3 mil	G	37	.216		12.85	11.85		24.70	33
1400	4 mil	G	37	.216		14.25	11.85		26.10	34.50
1800	Reinf. waterproof, 2 mil polyethylene backing, 1 side		37	.216		12	11.85		23.85	32
1900	2 sides		37	.216		15.15	11.85		27	35.50
2400	Waterproofed kraft with sisal or fiberglass fibers		37	.216		18.90	11.85		30.75	40
9950	Minimum labor/equipment charge		4	2	Job		109		109	174

07 27 Air Barriers

07 27 13 – Modified Bituminous Sheet Air Barriers

07 27 13.10 Modified Bituminous Sheet Air Barrier

		Crew	Daily Output	Labor-Hours	Unit	Material	2021 Bare Costs Labor	Equipment	Total	Total Incl O&P
0010	**MODIFIED BITUMINOUS SHEET AIR BARRIER**									
0100	SBS modified sheet laminated to polyethylene sheet, 40 mils, 4" wide	1 Carp	1200	.007	L.F.	.15	.36		.51	.75
0120	6" wide		1100	.007		.21	.40		.61	.86
0140	9" wide		1000	.008		.29	.44		.73	1.02
0160	12" wide		900	.009		.37	.49		.86	1.18
0180	18" wide	2 Carp	1700	.009	S.F.	.35	.51		.86	1.20
0200	36" wide	"	1800	.009		.34	.49		.83	1.14
0220	Adhesive for above	1 Carp	1400	.006			.31		.31	.50

07 27 26 – Fluid-Applied Membrane Air Barriers

07 27 26.10 Fluid Applied Membrane Air Barrier

		Crew	Daily Output	Labor-Hours	Unit	Material	2021 Bare Costs Labor	Equipment	Total	Total Incl O&P
0010	**FLUID APPLIED MEMBRANE AIR BARRIER**									
0100	Spray applied vapor barrier, 25 S.F./gallon	1 Pord	1375	.006	S.F.	.02	.27		.29	.45

07 31 Shingles and Shakes

07 31 13 – Asphalt Shingles

07 31 13.10 Asphalt Roof Shingles

		Crew	Daily Output	Labor-Hours	Unit	Material	2021 Bare Costs Labor	Equipment	Total	Total Incl O&P
0010	**ASPHALT ROOF SHINGLES**									
0100	Standard strip shingles									
0150	Inorganic, class A, 25 year	1 Rofc	5.50	1.455	Sq.	75	70		145	204
0155	Pneumatic nailed		7	1.143		75	55		130	178
0200	30 year		5	1.600		115	77		192	259
0205	Pneumatic nailed		6.25	1.280		115	61.50		176.50	232
0250	Standard laminated multi-layered shingles									
0300	Class A, 240-260 lb./square	1 Rofc	4.50	1.778	Sq.	113	85.50		198.50	272
0305	Pneumatic nailed		5.63	1.422		113	68.50		181.50	243
0350	Class A, 250-270 lb./square		4	2		113	96.50		209.50	291
0355	Pneumatic nailed		5	1.600		113	77		190	258
0400	Premium, laminated multi-layered shingles									
0450	Class A, 260-300 lb./square	1 Rofc	3.50	2.286	Sq.	143	110		253	345
0455	Pneumatic nailed		4.37	1.831		143	88		231	310
0500	Class A, 300-385 lb./square		3	2.667		370	129		499	630
0505	Pneumatic nailed		3.75	2.133		370	103		473	585
0800	#15 felt underlayment		64	.125		5.70	6.05		11.75	16.60
0825	#30 felt underlayment		58	.138		9.60	6.65		16.25	22
0850	Self adhering polyethylene and rubberized asphalt underlayment		22	.364		68.50	17.55		86.05	106
0900	Ridge shingles		330	.024	L.F.	2.44	1.17		3.61	4.69
0905	Pneumatic nailed		412.50	.019	"	2.44	.93		3.37	4.29
1000	For steep roofs (7 to 12 pitch or greater), add						50%			

07 31 Shingles and Shakes

07 31 13 – Asphalt Shingles

07 31 13.10 Asphalt Roof Shingles

	07 31 13.10 Asphalt Roof Shingles	Crew	Daily Output	Labor-Hours	Unit	Material	2021 Bare Costs Labor	Equipment	Total	Total Incl O&P
9000	Minimum labor/equipment charge	1 Rofc	3	2.667	Job		129		129	221

07 31 16 – Metal Shingles

07 31 16.10 Aluminum Shingles

	07 31 16.10 Aluminum Shingles	Crew	Daily Output	Labor-Hours	Unit	Material	2021 Bare Costs Labor	Equipment	Total	Total Incl O&P
0010	ALUMINUM SHINGLES									
0020	Mill finish, .019" thick	1 Carp	5	1.600	Sq.	234	87.50		321.50	395
0100	.020" thick	"	5	1.600		280	87.50		367.50	450
0300	For colors, add					21			21	23.50
0600	Ridge cap, .024" thick	1 Carp	170	.047	L.F.	3.95	2.57		6.52	8.45
0700	End wall flashing, .024" thick		170	.047		2.70	2.57		5.27	7.05
0900	Valley section, .024" thick		170	.047		4.14	2.57		6.71	8.65
1000	Starter strip, .024" thick		400	.020		2.23	1.09		3.32	4.19
1200	Side wall flashing, .024" thick		170	.047		2.62	2.57		5.19	7
1500	Gable flashing, .024" thick		400	.020		2.01	1.09		3.10	3.95
9000	Minimum labor/equipment charge		3	2.667	Job		146		146	232

07 31 16.20 Steel Shingles

	07 31 16.20 Steel Shingles	Crew	Daily Output	Labor-Hours	Unit	Material	2021 Bare Costs Labor	Equipment	Total	Total Incl O&P
0010	STEEL SHINGLES									
0012	Galvanized, 26 ga.	1 Rots	2.20	3.636	Sq.	395	175		570	735
0200	24 ga.	"	2.20	3.636		385	175		560	725
0300	For colored galvanized shingles, add					54.50			54.50	60
9000	Minimum labor/equipment charge	1 Rots	3	2.667	Job		129		129	221

07 31 26 – Slate Shingles

07 31 26.10 Slate Roof Shingles

	07 31 26.10 Slate Roof Shingles		Crew	Daily Output	Labor-Hours	Unit	Material	2021 Bare Costs Labor	Equipment	Total	Total Incl O&P
0010	SLATE ROOF SHINGLES										
0100	Buckingham Virginia black, 3/16" - 1/4" thick	G	1 Rots	1.75	4.571	Sq.	555	220		775	990
0200	1/4" thick	G		1.75	4.571		555	220		775	990
0900	Pennsylvania black, Bangor, #1 clear	G		1.75	4.571		500	220		720	930
1200	Vermont, unfading, green, mottled green	G		1.75	4.571		505	220		725	935
1300	Semi-weathering green & gray	G		1.75	4.571		405	220		625	825
1400	Purple	G	1 Rots	1.75	4.571		450	220		670	870
1500	Black or gray	G		1.75	4.571	Sq.	485	220		705	915
2500	Slate roof repair, extensive replacement			1	8		605	385		990	1,325
2600	Repair individual pieces, scattered			19	.421	Ea.	7.15	20.50		27.65	43
2700	Ridge shingles, slate			200	.040	L.F.	10.20	1.93		12.13	14.50
9000	Minimum labor/equipment charge			3	2.667	Job		129		129	221

07 31 29 – Wood Shingles and Shakes

07 31 29.13 Wood Shingles

	07 31 29.13 Wood Shingles		Crew	Daily Output	Labor-Hours	Unit	Material	2021 Bare Costs Labor	Equipment	Total	Total Incl O&P
0010	WOOD SHINGLES	R061110-30									
0012	16" No. 1 red cedar shingles, 5" exposure, on roof		1 Carp	2.50	3.200	Sq.	330	175		505	640
0015	Pneumatic nailed			3.25	2.462		330	135		465	575
0200	7-1/2" exposure, on walls			2.05	3.902		219	213		432	580
0205	Pneumatic nailed			2.67	2.996		219	164		383	500
0300	18" No. 1 red cedar perfections, 5-1/2" exposure, on roof			2.75	2.909		297	159		456	580
0305	Pneumatic nailed			3.57	2.241		297	123		420	520
0500	7-1/2" exposure, on walls			2.25	3.556		219	194		413	550
0505	Pneumatic nailed			2.92	2.740		219	150		369	480
0600	Resquared and rebutted, 5-1/2" exposure, on roof			3	2.667		300	146		446	560
0605	Pneumatic nailed			3.90	2.051		300	112		412	510
0900	7-1/2" exposure, on walls			2.45	3.265		221	179		400	525
0905	Pneumatic nailed			3.18	2.516		221	138		359	460
1000	Add to above for fire retardant shingles						60			60	66
1060	Preformed ridge shingles		1 Carp	400	.020	L.F.	4.06	1.09		5.15	6.20

198

07 31 Shingles and Shakes

07 31 29 – Wood Shingles and Shakes

07 31 29.13 Wood Shingles

		Crew	Daily Output	Labor-Hours	Unit	Material	2021 Bare Costs Labor	2021 Bare Costs Equipment	Total	Total Incl O&P
2000	White cedar shingles, 16" long, extras, 5" exposure, on roof	1 Carp	2.40	3.333	Sq.	170	182		352	475
2005	Pneumatic nailed		3.12	2.564		170	140		310	410
2050	5" exposure on walls		2	4		170	219		389	535
2055	Pneumatic nailed		2.60	3.077		170	168		338	455
2100	7-1/2" exposure, on walls		2	4		121	219		340	485
2105	Pneumatic nailed		2.60	3.077		121	168		289	400
2150	"B" grade, 5" exposure on walls		2	4		151	219		370	515
2155	Pneumatic nailed		2.60	3.077		151	168		319	435
2300	For #15 organic felt underlayment on roof, 1 layer, add		64	.125		5.70	6.85		12.55	17.15
2400	2 layers, add		32	.250		11.40	13.70		25.10	34.50
2600	For steep roofs (7/12 pitch or greater), add to above						50%			
2700	Panelized systems, No.1 cedar shingles on 5/16" CDX plywood									
2800	On walls, 8' strips, 7" or 14" exposure	2 Carp	700	.023	S.F.	7.75	1.25		9	10.55
3500	On roofs, 8' strips, 7" or 14" exposure	1 Carp	3	2.667	Sq.	590	146		736	880
3505	Pneumatic nailed		4	2	"	590	109		699	825
9000	Minimum labor/equipment charge		3	2.667	Job		146		146	232

07 31 29.16 Wood Shakes

		Crew	Daily Output	Labor-Hours	Unit	Material	2021 Bare Costs Labor	2021 Bare Costs Equipment	Total	Total Incl O&P
0010	**WOOD SHAKES**									
1100	Hand-split red cedar shakes, 1/2" thick x 24" long, 10" exp. on roof	1 Carp	2.50	3.200	Sq.	380	175		555	695
1105	Pneumatic nailed		3.25	2.462		380	135		515	630
1110	3/4" thick x 24" long, 10" exp. on roof		2.25	3.556		380	194		574	725
1115	Pneumatic nailed		2.92	2.740		380	150		530	655
1200	1/2" thick, 18" long, 8-1/2" exp. on roof		2	4		249	219		468	625
1205	Pneumatic nailed		2.60	3.077		249	168		417	540
1210	3/4" thick x 18" long, 8-1/2" exp. on roof		1.80	4.444		249	243		492	660
1215	Pneumatic nailed		2.34	3.419		249	187		436	570
1255	10" exposure on walls		2	4		241	219		460	615
1260	10" exposure on walls, pneumatic nailed		2.60	3.077		241	168		409	535
1700	Add to above for fire retardant shakes, 24" long					60			60	66
1800	18" long					60			60	66
1810	Ridge shakes	1 Carp	350	.023	L.F.	5.95	1.25		7.20	8.55

07 32 Roof Tiles

07 32 13 – Clay Roof Tiles

07 32 13.10 Clay Tiles

		Crew	Daily Output	Labor-Hours	Unit	Material	2021 Bare Costs Labor	2021 Bare Costs Equipment	Total	Total Incl O&P
0010	**CLAY TILES**, including accessories									
0300	Flat shingle, interlocking, 15", 166 pcs./sq., fireflashed blend	3 Rots	6	4	Sq.	555	193		748	940
0500	Terra cotta red		6	4		655	193		848	1,050
0600	Roman pan and top, 18", 102 pcs./sq., fireflashed blend		5.50	4.364		515	210		725	925
0640	Terra cotta red	1 Rots	2.40	3.333		595	161		756	930
1100	Barrel mission tile, 18", 166 pcs./sq., fireflashed blend	3 Rots	5.50	4.364		430	210		640	830
1140	Terra cotta red		5.50	4.364		435	210		645	840
1700	Scalloped edge flat shingle, 14", 145 pcs./sq., fireflashed blend		6	4		1,125	193		1,318	1,550
1800	Terra cotta red		6	4		1,100	193		1,293	1,550
3010	#15 felt underlayment	1 Rofc	64	.125		5.70	6.05		11.75	16.60
3020	#30 felt underlayment		58	.138		9.60	6.65		16.25	22
3040	Polyethylene and rubberized asph. underlayment		22	.364		68.50	17.55		86.05	106
9010	Minimum labor/equipment charge	1 Rots	2	4	Job		193		193	330

199

07 32 16 – Concrete Roof Tiles

07 32 16.10 Concrete Tiles

		Crew	Daily Output	Labor-Hours	Unit	Material	2021 Bare Costs Labor	Equipment	Total	Total Incl O&P
0010	**CONCRETE TILES**									
0020	Corrugated, 13" x 16-1/2", 90 per sq., 950 lb./sq.									
0050	Earthtone colors, nailed to wood deck	1 Rots	1.35	5.926	Sq.	118	286		404	620
0150	Blues		1.35	5.926		106	286		392	605
0200	Greens		1.35	5.926		117	286		403	620
0250	Premium colors		1.35	5.926		117	286		403	620
0500	Shakes, 13" x 16-1/2", 90 per sq., 950 lb./sq.									
0600	All colors, nailed to wood deck	1 Rots	1.50	5.333	Sq.	172	257		429	630
1500	Accessory pieces, ridge & hip, 10" x 16-1/2", 8 lb. each	"	120	.067	Ea.	3.74	3.21		6.95	9.65
1700	Rake, 6-1/2" x 16-3/4", 9 lb. each				Ea.	3.74			3.74	4.11
1800	Mansard hip, 10" x 16-1/2", 9.2 lb. each					3.74			3.74	4.11
1900	Hip starter, 10" x 16-1/2", 10.5 lb. each					10.35			10.35	11.40
2000	3 or 4 way apex, 10" each side, 11.5 lb. each					11.85			11.85	13.05
9000	Minimum labor/equipment charge	1 Rots	3	2.667	Job		129		129	221

07 32 19 – Metal Roof Tiles

07 32 19.10 Metal Roof Tiles

		Crew	Daily Output	Labor-Hours	Unit	Material	2021 Bare Costs Labor	Equipment	Total	Total Incl O&P
0010	**METAL ROOF TILES**									
0020	Accessories included, .032" thick aluminum, mission tile	1 Carp	2.50	3.200	Sq.	795	175		970	1,150
0200	Spanish tiles		3	2.667	"	565	146		711	850
9000	Minimum labor/equipment charge		3	2.667	Job		146		146	232

07 33 63 – Vegetated Roofing

07 33 63.10 Green Roof Systems

			Crew	Daily Output	Labor-Hours	Unit	Material	2021 Bare Costs Labor	Equipment	Total	Total Incl O&P
0010	**GREEN ROOF SYSTEMS**										
0020	Soil mixture for green roof 30% sand, 55% gravel, 15% soil										
0100	Hoist and spread soil mixture 4" depth up to 5 stories tall roof	G	B-13B	4000	.014	S.F.	.30	.68	.25	1.23	1.68
0150	6" depth	G		2667	.021		.46	1.01	.37	1.84	2.51
0200	8" depth	G		2000	.028		.61	1.35	.50	2.46	3.35
0250	10" depth	G		1600	.035		.76	1.69	.62	3.07	4.19
0300	12" depth	G		1335	.042		.91	2.03	.74	3.68	5.05
0310	Alt. man-made soil mix, hoist & spread, 4" deep up to 5 stories tall roof	G		4000	.014		1.85	.68	.25	2.78	3.38
0350	Mobilization 55 ton crane to site	G	1 Eqhv	3.60	2.222	Ea.		137		137	213
0355	Hoisting cost to 5 stories per day (Avg. 28 picks per day)	G	B-13B	1	56	Day		2,700	990	3,690	5,375
0360	Mobilization or demobilization, 100 ton crane to site driver & escort	G	A-3E	2.50	6.400	Ea.		360	70.50	430.50	650
0365	Hoisting cost 6-10 stories per day (Avg. 21 picks per day)	G	B-13C	1	56	Day		2,700	2,300	5,000	6,825
0370	Hoist and spread soil mixture 4" depth 6-10 stories tall roof	G		4000	.014	S.F.	.30	.68	.58	1.56	2.05
0375	6" depth	G		2667	.021		.46	1.01	.87	2.34	3.05
0380	8" depth	G		2000	.028		.61	1.35	1.16	3.12	4.08
0385	10" depth	G		1600	.035		.76	1.69	1.44	3.89	5.10
0390	12" depth	G		1335	.042		.91	2.03	1.73	4.67	6.10
0400	Green roof edging treated lumber 4" x 4", no hoisting included	G	2 Carp	400	.040	L.F.	2.26	2.19		4.45	5.95
0410	4" x 6"	G		400	.040		2.97	2.19		5.16	6.75
0420	4" x 8"	G		360	.044		6.25	2.43		8.68	10.70
0430	4" x 6" double stacked	G		300	.053		5.95	2.92		8.87	11.20
0500	Green roof edging redwood lumber 4" x 4", no hoisting included	G		400	.040		11.30	2.19		13.49	15.90
0510	4" x 6"	G		400	.040		17.95	2.19		20.14	23
0520	4" x 8"	G		360	.044		33.50	2.43		35.93	41
0530	4" x 6" double stacked	G		300	.053		36	2.92		38.92	44
0550	Components, not including membrane or insulation:										

07 33 Natural Roof Coverings

07 33 63 – Vegetated Roofing

07 33 63.10 Green Roof Systems

		Crew	Daily Output	Labor-Hours	Unit	Material	2021 Bare Costs Labor	Equipment	Total	Total Incl O&P
0560	Fluid applied rubber membrane, reinforced, 215 mil thick	G · G-5	350	.114	S.F.	.28	4.99	.56	5.83	9.50
0570	Root barrier	G · 2 Rofc	775	.021		1.01	1		2.01	2.82
0580	Moisture retention barrier and reservoir	G · "	900	.018		4.55	.86		5.41	6.45
0600	Planting sedum, light soil, potted, 2-1/4" diameter, 2 per S.F.	G · 1 Clab	420	.019		9.95	.85		10.80	12.30
0610	1 per S.F.	G · "	840	.010		4.98	.42		5.40	6.10
0630	Planting sedum mat per S.F. including shipping (4000 S.F. min)	G · 4 Clab	4000	.008		11.95	.36		12.31	13.70
0640	Installation sedum mat system (no soil required) per S.F. (4000 S.F. min)	G · "	4000	.008	▼	16.60	.36		16.96	18.85
0645	Note: pricing of sedum mats shipped in full truck loads (4000-5000 S.F.)									

07 41 Roof Panels

07 41 13 – Metal Roof Panels

07 41 13.10 Aluminum Roof Panels

		Crew	Daily Output	Labor-Hours	Unit	Material	2021 Bare Costs Labor	Equipment	Total	Total Incl O&P
0010	**ALUMINUM ROOF PANELS**									
0020	Corrugated or ribbed, .0155" thick, natural	G-3	1200	.027	S.F.	.77	1.46		2.23	3.15
0300	Painted		1200	.027		1.13	1.46		2.59	3.54
0400	Corrugated, .018" thick, on steel frame, natural finish		1200	.027		1.33	1.46		2.79	3.76
0600	Painted		1200	.027		1.66	1.46		3.12	4.13
0700	Corrugated, on steel frame, natural, .024" thick		1200	.027		1.93	1.46		3.39	4.42
0800	Painted		1200	.027		2.20	1.46		3.66	4.72
0900	.032" thick, natural		1200	.027		3.14	1.46		4.60	5.75
1200	Painted	▼	1200	.027	▼	3.41	1.46		4.87	6.05
9000	Minimum labor/equipment charge	1 Rofc	3	2.667	Job		129		129	221

07 41 13.20 Steel Roofing Panels

		Crew	Daily Output	Labor-Hours	Unit	Material	2021 Bare Costs Labor	Equipment	Total	Total Incl O&P
0010	**STEEL ROOFING PANELS**									
0012	Corrugated or ribbed, on steel framing, 30 ga. galv	G-3	1100	.029	S.F.	1.89	1.60		3.49	4.59
0100	28 ga.		1050	.030		1.54	1.67		3.21	4.32
0300	26 ga.		1000	.032		1.95	1.76		3.71	4.91
0400	24 ga.		950	.034		3.07	1.85		4.92	6.30
0600	Colored, 28 ga.		1050	.030		1.92	1.67		3.59	4.74
0700	26 ga.		1000	.032		2.17	1.76		3.93	5.15
0710	Flat profile, 1-3/4" standing seams, 10" wide, standard finish, 26 ga.		1000	.032		3.99	1.76		5.75	7.15
0715	24 ga.		950	.034		4.65	1.85		6.50	8
0720	22 ga.		900	.036		5.75	1.95		7.70	9.40
0725	Zinc aluminum alloy finish, 26 ga.		1000	.032		3.18	1.76		4.94	6.25
0730	24 ga.		950	.034		3.76	1.85		5.61	7.05
0735	22 ga.		900	.036		4.29	1.95		6.24	7.80
0740	12" wide, standard finish, 26 ga.		1000	.032		4.26	1.76		6.02	7.45
0745	24 ga.		950	.034		5.25	1.85		7.10	8.65
0750	Zinc aluminum alloy finish, 26 ga.		1000	.032		4.52	1.76		6.28	7.75
0755	24 ga.		950	.034		3.98	1.85		5.83	7.30
0840	Flat profile, 1" x 3/8" batten, 12" wide, standard finish, 26 ga.		1000	.032		3.51	1.76		5.27	6.60
0845	24 ga.	G-3	950	.034	S.F.	4.13	1.85		5.98	7.45
0850	22 ga.		900	.036		4.96	1.95		6.91	8.50
0855	Zinc aluminum alloy finish, 26 ga.		1000	.032		3.38	1.76		5.14	6.50
0860	24 ga.		950	.034		3.82	1.85		5.67	7.10
0865	22 ga.		900	.036		4.33	1.95		6.28	7.85
0870	16-1/2" wide, standard finish, 24 ga.		950	.034		4.07	1.85		5.92	7.40
0875	22 ga.		900	.036		4.58	1.95		6.53	8.10
0880	Zinc aluminum alloy finish, 24 ga.		950	.034		3.56	1.85		5.41	6.85
0885	22 ga.		900	.036		4.03	1.95		5.98	7.50
0890	Flat profile, 2" x 2" batten, 12" wide, standard finish, 26 ga.	▼	1000	.032	▼	4.03	1.76		5.79	7.20

07 41 Roof Panels

07 41 13 – Metal Roof Panels

07 41 13.20 Steel Roofing Panels

		Crew	Daily Output	Labor-Hours	Unit	Material	2021 Bare Costs Labor	2021 Bare Costs Equipment	Total	Total Incl O&P
0895	24 ga.	G-3	950	.034	S.F.	5.15	1.85		7	8.60
0900	22 ga.		900	.036		6.30	1.95		8.25	10
0905	Zinc aluminum alloy finish, 26 ga.		1000	.032		4.08	1.76		5.84	7.25
0910	24 ga.		950	.034		4.32	1.85		6.17	7.65
0915	22 ga.		900	.036		5	1.95		6.95	8.55
0920	16-1/2" wide, standard finish, 24 ga.		950	.034		4.48	1.85		6.33	7.85
0925	22 ga.		900	.036		5.20	1.95		7.15	8.75
0930	Zinc aluminum alloy finish, 24 ga.		950	.034		4.03	1.85		5.88	7.35
0935	22 ga.		900	.036		4.60	1.95		6.55	8.10
9000	Minimum labor/equipment charge	1 Rofc	2	4	Job		193		193	330

07 41 33 – Plastic Roof Panels

07 41 33.10 Fiberglass Panels

		Crew	Daily Output	Labor-Hours	Unit	Material	2021 Bare Costs Labor	2021 Bare Costs Equipment	Total	Total Incl O&P
0010	**FIBERGLASS PANELS**									
0012	Corrugated panels, roofing, 8 oz./S.F.	G-3	1000	.032	S.F.	2.57	1.76		4.33	5.60
0100	12 oz./S.F.		1000	.032		4.65	1.76		6.41	7.85
0300	Corrugated siding, 6 oz./S.F.		880	.036		2.53	2		4.53	5.90
0400	8 oz./S.F.		880	.036		2.57	2		4.57	5.95
0500	Fire retardant		880	.036		4	2		6	7.55
0600	12 oz. siding, textured		880	.036		3.78	2		5.78	7.30
0700	Fire retardant		880	.036		4.85	2		6.85	8.50
0900	Flat panels, 6 oz./S.F., clear or colors		880	.036		3.06	2		5.06	6.50
1100	Fire retardant, class A		880	.036		3.63	2		5.63	7.15
1300	8 oz./S.F., clear or colors		880	.036		2.41	2		4.41	5.80
9000	Minimum labor/equipment charge	1 Rofc	2	4	Job		193		193	330

07 42 Wall Panels

07 42 13 – Metal Wall Panels

07 42 13.10 Mansard Panels

		Crew	Daily Output	Labor-Hours	Unit	Material	2021 Bare Costs Labor	2021 Bare Costs Equipment	Total	Total Incl O&P
0010	**MANSARD PANELS**									
0600	Aluminum, stock units, straight surfaces	1 Shee	115	.070	S.F.	4.72	4.55		9.27	12.30
0700	Concave or convex surfaces, add		75	.107	"	2.56	7		9.56	13.70
0800	For framing, to 5' high, add		115	.070	L.F.	3.86	4.55		8.41	11.35
0900	Soffits, to 1' wide		125	.064	S.F.	2.97	4.19		7.16	9.80
9000	Minimum labor/equipment charge	1 Shee	2.50	3.200	Job		209		209	325

07 42 13.20 Aluminum Siding Panels

		Crew	Daily Output	Labor-Hours	Unit	Material	2021 Bare Costs Labor	2021 Bare Costs Equipment	Total	Total Incl O&P
0010	**ALUMINUM SIDING PANELS**									
0012	Corrugated, on steel framing, .019" thick, natural finish	G-3	775	.041	S.F.	1.92	2.27		4.19	5.70
0100	Painted		775	.041		1.85	2.27		4.12	5.60
0400	Farm type, .021" thick on steel frame, natural		775	.041		1.75	2.27		4.02	5.50
0600	Painted		775	.041		1.85	2.27		4.12	5.60
0700	Industrial type, corrugated, on steel, .024" thick, mill		775	.041		2.41	2.27		4.68	6.20
0900	Painted		775	.041		2.56	2.27		4.83	6.40
1000	.032" thick, mill		775	.041		2.06	2.27		4.33	5.85
1200	Painted		775	.041		3.12	2.27		5.39	7
1300	V-Beam, on steel frame, .032" thick, mill		775	.041		2.79	2.27		5.06	6.65
1500	Painted		775	.041		3.44	2.27		5.71	7.35
1600	.040" thick, mill		775	.041		2.89	2.27		5.16	6.75
1800	Painted		775	.041		4.10	2.27		6.37	8.10
3800	Horizontal, colored clapboard, 8" wide, plain	2 Carp	515	.031		3.13	1.70		4.83	6.15
3810	Insulated		515	.031		3.02	1.70		4.72	6.05

07 42 Wall Panels

07 42 13 – Metal Wall Panels

07 42 13.20 Aluminum Siding Panels

		Crew	Daily Output	Labor-Hours	Unit	Material	2021 Bare Costs Labor	Equipment	Total	Total Incl O&P
3830	8" embossed, painted	2 Carp	515	.031	S.F.	3.76	1.70		5.46	6.85
3840	Insulated		515	.031		3.29	1.70		4.99	6.35
3860	12" painted, smooth		600	.027		3.55	1.46		5.01	6.25
3870	Insulated		600	.027		3.22	1.46		4.68	5.85
3890	12" embossed, painted		600	.027		3.09	1.46		4.55	5.70
3900	Insulated		515	.031		3.15	1.70		4.85	6.20
4000	Vertical board & batten, colored, non-insulated		515	.031		2.63	1.70		4.33	5.60
4200	For simulated wood design, add					.18			.18	.20
4300	Corners for above, outside	2 Carp	515	.031	V.L.F.	4.02	1.70		5.72	7.15
4500	Inside corners	"	515	.031	"	1.80	1.70		3.50	4.69
9000	Minimum labor/equipment charge	1 Carp	3	2.667	Job		146		146	232

07 42 13.30 Steel Siding

		Crew	Daily Output	Labor-Hours	Unit	Material	2021 Bare Costs Labor	Equipment	Total	Total Incl O&P
0010	**STEEL SIDING**									
0020	Beveled, vinyl coated, 8" wide	1 Carp	265	.030	S.F.	1.90	1.65		3.55	4.72
0050	10" wide	"	275	.029		1.86	1.59		3.45	4.58
0080	Galv, corrugated or ribbed, on steel frame, 30 ga.	G-3	800	.040		1.18	2.20		3.38	4.75
0100	28 ga.		795	.040		1.28	2.21		3.49	4.89
0300	26 ga.		790	.041		1.66	2.23		3.89	5.35
0400	24 ga.		785	.041		1.67	2.24		3.91	5.35
0600	22 ga.		770	.042		1.81	2.28		4.09	5.60
0700	Colored, corrugated/ribbed, on steel frame, 10 yr. finish, 28 ga.		800	.040		2.32	2.20		4.52	6
0900	26 ga.		795	.040		2.04	2.21		4.25	5.70
1000	24 ga.		790	.041		2.28	2.23		4.51	6
1020	20 ga.		785	.041		2.90	2.24		5.14	6.70
9000	Minimum labor/equipment charge	1 Carp	3	2.667	Job		146		146	232

07 46 Siding

07 46 23 – Wood Siding

07 46 23.10 Wood Board Siding

		Crew	Daily Output	Labor-Hours	Unit	Material	2021 Bare Costs Labor	Equipment	Total	Total Incl O&P
0010	**WOOD BOARD SIDING**									
3200	Wood, cedar bevel, A grade, 1/2" x 6"	1 Carp	295	.027	S.F.	4.42	1.48		5.90	7.20
3300	1/2" x 8"		330	.024		7.50	1.33		8.83	10.35
3500	3/4" x 10", clear grade		375	.021		7.35	1.17		8.52	9.90
3600	"B" grade		375	.021		4.38	1.17		5.55	6.70
3800	Cedar, rough sawn, 1" x 4", A grade, natural		220	.036		8.80	1.99		10.79	12.80
3900	Stained		220	.036		8.75	1.99		10.74	12.75
4100	1" x 12", board & batten, #3 & Btr., natural		420	.019		4.83	1.04		5.87	6.95
4200	Stained		420	.019		5.25	1.04		6.29	7.45
4400	1" x 8" channel siding, #3 & Btr., natural		330	.024		5.25	1.33		6.58	7.90
4500	Stained		330	.024		6	1.33		7.33	8.70
4700	Redwood, clear, beveled, vertical grain, 1/2" x 4"		220	.036		4.99	1.99		6.98	8.65
4750	1/2" x 6"		295	.027		5.40	1.48		6.88	8.30
4800	1/2" x 8"		330	.024		5.55	1.33		6.88	8.20
5000	3/4" x 10"		375	.021		5.35	1.17		6.52	7.75
5200	Channel siding, 1" x 10", B grade		375	.021		4.90	1.17		6.07	7.25
5250	Redwood, T&G boards, B grade, 1" x 4"		220	.036		7	1.99		8.99	10.85
5270	1" x 8"		330	.024		9	1.33		10.33	12
5400	White pine, rough sawn, 1" x 8", natural		330	.024		2.41	1.33		3.74	4.76
5500	Stained		330	.024		2.41	1.33		3.74	4.76
9000	Minimum labor/equipment charge		2	4	Job		219		219	350

For customer support on your Commercial Renovation Costs with RSMeans Data, call 800.448.8182.

203

07 46 Siding

07 46 29 – Plywood Siding

07 46 29.10 Plywood Siding Options	Crew	Daily Output	Labor-Hours	Unit	Material	2021 Bare Costs Labor	Equipment	Total	Total Incl O&P
0010 **PLYWOOD SIDING OPTIONS**									
0900 Plywood, medium density overlaid, 3/8" thick	2 Carp	750	.021	S.F.	1.20	1.17		2.37	3.18
1000 1/2" thick		700	.023		1.59	1.25		2.84	3.74
1100 3/4" thick		650	.025		2.62	1.35		3.97	5
1600 Texture 1-11, cedar, 5/8" thick, natural		675	.024		2.58	1.30		3.88	4.90
1700 Factory stained		675	.024		2.86	1.30		4.16	5.20
1900 Texture 1-11, fir, 5/8" thick, natural		675	.024		1.38	1.30		2.68	3.58
2000 Factory stained		675	.024		2.02	1.30		3.32	4.28
2050 Texture 1-11, S.Y.P., 5/8" thick, natural		675	.024		1.44	1.30		2.74	3.64
2100 Factory stained		675	.024		1.54	1.30		2.84	3.75
2200 Rough sawn cedar, 3/8" thick, natural		675	.024		1.28	1.30		2.58	3.47
2300 Factory stained		675	.024		1.54	1.30		2.84	3.75
2500 Rough sawn fir, 3/8" thick, natural		675	.024		1.02	1.30		2.32	3.18
2600 Factory stained		675	.024		1.13	1.30		2.43	3.30
2800 Redwood, textured siding, 5/8" thick		675	.024		2.05	1.30		3.35	4.32
3000 Polyvinyl chloride coated, 3/8" thick		750	.021		1.18	1.17		2.35	3.16
9000 Minimum labor/equipment charge	1 Carp	2	4	Job		219		219	350

07 46 33 – Plastic Siding

07 46 33.10 Vinyl Siding

	Crew	Daily Output	Labor-Hours	Unit	Material	2021 Bare Costs Labor	Equipment	Total	Total Incl O&P
0010 **VINYL SIDING**									
3995 Clapboard profile, woodgrain texture, .048 thick, double 4	2 Carp	495	.032	S.F.	1.10	1.77		2.87	4.03
4000 Double 5		550	.029		1.10	1.59		2.69	3.74
4005 Single 8		495	.032		1.87	1.77		3.64	4.87
4010 Single 10		550	.029		2.24	1.59		3.83	4.99
4015 .044 thick, double 4		495	.032		1.09	1.77		2.86	4.02
4020 Double 5		550	.029		1.11	1.59		2.70	3.75
4025 .042 thick, double 4		495	.032		1.19	1.77		2.96	4.13
4030 Double 5		550	.029		1.19	1.59		2.78	3.84
4035 Cross sawn texture, .040 thick, double 4		495	.032		.76	1.77		2.53	3.65
4040 Double 5		550	.029		.73	1.59		2.32	3.33
4045 Smooth texture, .042 thick, double 4		495	.032		.97	1.77		2.74	3.88
4050 Double 5		550	.029		.81	1.59		2.40	3.42
4055 Single 8		495	.032		.73	1.77		2.50	3.62
4060 Cedar texture, .044 thick, double 4		495	.032		1.15	1.77		2.92	4.08
4065 Double 6		600	.027		1.49	1.46		2.95	3.96
4070 Dutch lap profile, woodgrain texture, .048 thick, double 5		550	.029		1.22	1.59		2.81	3.87
4075 .044 thick, double 4.5		525	.030		1.10	1.67		2.77	3.86
4080 .042 thick, double 4.5		525	.030		.91	1.67		2.58	3.65
4085 .040 thick, double 4.5		525	.030		.75	1.67		2.42	3.47
4100 Shake profile, 10" wide		400	.040		3.76	2.19		5.95	7.60
4105 Vertical pattern, .046 thick, double 5		550	.029		2.01	1.59		3.60	4.74
4110 .044 thick, triple 3		550	.029		1.72	1.59		3.31	4.42
4115 .040 thick, triple 4		550	.029		1.74	1.59		3.33	4.44
4120 .040 thick, triple 2.66		550	.029		1.92	1.59		3.51	4.64
4125 Insulation, fan folded extruded polystyrene, 1/4"		2000	.008		.30	.44		.74	1.03
4130 3/8"		2000	.008		.33	.44		.77	1.06
4135 Accessories, J channel, 5/8" pocket		700	.023	L.F.	.64	1.25		1.89	2.69
4140 3/4" pocket		695	.023		.72	1.26		1.98	2.80
4145 1-1/4" pocket		680	.024		1.22	1.29		2.51	3.39
4150 Flexible, 3/4" pocket		600	.027		3.35	1.46		4.81	6
4155 Under sill finish trim		500	.032		.69	1.75		2.44	3.55
4160 Vinyl starter strip		700	.023		.75	1.25		2	2.81

07 46 Siding

07 46 33 – Plastic Siding

07 46 33.10 Vinyl Siding

		Crew	Daily Output	Labor-Hours	Unit	Material	2021 Bare Costs Labor	Equipment	Total	Total Incl O&P
4165	Aluminum starter strip	2 Carp	700	.023	L.F.	.30	1.25		1.55	2.32
4170	Window casing, 2-1/2" wide, 3/4" pocket		510	.031		1.95	1.72		3.67	4.87
4175	Outside corner, woodgrain finish, 4" face, 3/4" pocket		700	.023		2.62	1.25		3.87	4.87
4180	5/8" pocket	▼	700	.023	▼	2.02	1.25		3.27	4.21
4185	Smooth finish, 4" face, 3/4" pocket	2 Carp	700	.023	L.F.	2.66	1.25		3.91	4.91
4190	7/8" pocket		690	.023		1.92	1.27		3.19	4.13
4195	1-1/4" pocket		700	.023		1.52	1.25		2.77	3.66
4200	Soffit and fascia, 1' overhang, solid		120	.133		5.55	7.30		12.85	17.70
4205	Vented		120	.133		5.55	7.30		12.85	17.70
4207	18" overhang, solid		110	.145		6.55	7.95		14.50	19.85
4208	Vented		110	.145		6.55	7.95		14.50	19.85
4210	2' overhang, solid		100	.160		7.55	8.75		16.30	22.50
4215	Vented		100	.160		7.55	8.75		16.30	22.50
4217	3' overhang, solid		100	.160		9.55	8.75		18.30	24.50
4218	Vented	▼	100	.160	▼	9.55	8.75		18.30	24.50
4220	Colors for siding and soffits, add				S.F.	.15			.15	.17
4225	Colors for accessories and trim, add				L.F.	.29			.29	.32
9000	Minimum labor/equipment charge	1 Carp	3	2.667	Job		146		146	232

07 46 33.20 Polypropylene Siding

		Crew	Daily Output	Labor-Hours	Unit	Material	2021 Bare Costs Labor	Equipment	Total	Total Incl O&P
0010	**POLYPROPYLENE SIDING**									
4090	Shingle profile, random grooves, double 7	2 Carp	400	.040	S.F.	3.54	2.19		5.73	7.35
4092	Cornerpost for above	1 Carp	365	.022	L.F.	15.65	1.20		16.85	19.10
4095	Triple 5	2 Carp	400	.040	S.F.	3.72	2.19		5.91	7.55
4097	Cornerpost for above	1 Carp	365	.022	L.F.	12.50	1.20		13.70	15.65
5000	Staggered butt, double 7"	2 Carp	400	.040	S.F.	4.15	2.19		6.34	8.05
5002	Cornerpost for above	1 Carp	365	.022	L.F.	14.50	1.20		15.70	17.85
5010	Half round, double 6-1/4"	2 Carp	360	.044	S.F.	4.31	2.43		6.74	8.60
5020	Shake profile, staggered butt, double 9"	"	510	.031	"	3.84	1.72		5.56	6.95
5022	Cornerpost for above	1 Carp	365	.022	L.F.	11.65	1.20		12.85	14.75
5030	Straight butt, double 7"	2 Carp	400	.040	S.F.	4.31	2.19		6.50	8.20
5032	Cornerpost for above	1 Carp	365	.022	L.F.	14.65	1.20		15.85	18.05
6000	Accessories, J channel, 5/8" pocket	2 Carp	700	.023		.64	1.25		1.89	2.69
6010	3/4" pocket		695	.023		.72	1.26		1.98	2.80
6020	1-1/4" pocket		680	.024		1.22	1.29		2.51	3.39
6030	Aluminum starter strip	▼	700	.023	▼	.30	1.25		1.55	2.32

07 46 46 – Fiber Cement Siding

07 46 46.10 Fiber Cement Siding

		Crew	Daily Output	Labor-Hours	Unit	Material	2021 Bare Costs Labor	Equipment	Total	Total Incl O&P
0010	**FIBER CEMENT SIDING**									
0020	Lap siding, 5/16" thick, 6" wide, 4-3/4" exposure, smooth texture	2 Carp	415	.039	S.F.	1.39	2.11		3.50	4.88
0025	Woodgrain texture		415	.039		1.39	2.11		3.50	4.88
0030	7-1/2" wide, 6-1/4" exposure, smooth texture		425	.038		2.05	2.06		4.11	5.55
0035	Woodgrain texture		425	.038		2.05	2.06		4.11	5.55
0040	8" wide, 6-3/4" exposure, smooth texture		425	.038		1.10	2.06		3.16	4.49
0045	Rough sawn texture		425	.038		1.10	2.06		3.16	4.49
0050	9-1/2" wide, 8-1/4" exposure, smooth texture		440	.036		1.08	1.99		3.07	4.35
0055	Woodgrain texture	▼	440	.036	▼	1.08	1.99		3.07	4.35
0060	12" wide, 10-3/8" exposure, smooth texture	2 Carp	455	.035	S.F.	2.14	1.92		4.06	5.40
0065	Woodgrain texture		455	.035		2.14	1.92		4.06	5.40
0070	Panel siding, 5/16" thick, smooth texture		750	.021		1.06	1.17		2.23	3.02
0075	Stucco texture		750	.021		1.06	1.17		2.23	3.02
0080	Grooved woodgrain texture		750	.021		1.06	1.17		2.23	3.02
0088	Shingle siding, 48" x 15-1/4" panels, 7" exposure	▼	700	.023	▼	1.99	1.25		3.24	4.18

07 46 Siding

07 46 46 – Fiber Cement Siding

07 46 46.10 Fiber Cement Siding	Crew	Daily Output	Labor-Hours	Unit	Material	2021 Bare Costs Labor	Equipment	Total	Total Incl O&P
0090 Wood starter strip	2 Carp	400	.040	L.F.	.68	2.19		2.87	4.23

07 46 73 – Soffit

07 46 73.10 Soffit Options

	Crew	Daily Output	Labor-Hours	Unit	Material	2021 Bare Costs Labor	Equipment	Total	Total Incl O&P
0010 **SOFFIT OPTIONS**									
0012 Aluminum, residential, .020" thick	1 Carp	210	.038	S.F.	2.23	2.08		4.31	5.75
0100 Baked enamel on steel, 16 or 18 ga.		105	.076		6.50	4.17		10.67	13.80
0300 Polyvinyl chloride, white, solid		230	.035		2.15	1.90		4.05	5.40
0400 Perforated		230	.035		2.15	1.90		4.05	5.40
0500 For colors, add					.15			.15	.17
9000 Minimum labor/equipment charge	1 Carp	3	2.667	Job		146		146	232

07 51 Built-Up Bituminous Roofing

07 51 13 – Built-Up Asphalt Roofing

07 51 13.10 Built-Up Roofing Components

	Crew	Daily Output	Labor-Hours	Unit	Material	2021 Bare Costs Labor	Equipment	Total	Total Incl O&P
0010 **BUILT-UP ROOFING COMPONENTS**									
0012 Asphalt saturated felt, #30, 2 sq./roll	1 Rofc	58	.138	Sq.	9.60	6.65		16.25	22
0200 #15, 4 sq./roll, plain or perforated, not mopped		58	.138		5.70	6.65		12.35	17.70
0300 Roll roofing, smooth, #65		15	.533		10	25.50		35.50	55
0500 #90		12	.667		35	32		67	94
0520 Mineralized		12	.667		36	32		68	95
0540 D.C. (double coverage), 19" selvage edge		10	.800		48	38.50		86.50	120
0580 Adhesive (lap cement)				Gal.	7.85			7.85	8.65
0800 Steep, flat or dead level asphalt, 10 ton lots, packaged				Ton	1,075			1,075	1,175
9000 Minimum labor/equipment charge	1 Rofc	4	2	Job		96.50		96.50	166

07 51 13.13 Cold-Applied Built-Up Asphalt Roofing

	Crew	Daily Output	Labor-Hours	Unit	Material	2021 Bare Costs Labor	Equipment	Total	Total Incl O&P
0010 **COLD-APPLIED BUILT-UP ASPHALT ROOFING**									
0020 3 ply system, installation only (components listed below)	G-5	50	.800	Sq.		35	3.90	38.90	64.50
0100 Spunbond poly. fabric, 1.35 oz./S.Y., 36"W, 10.8 sq./roll				Ea.	129			129	142
0500 Base & finish coat, 3 gal./sq., 5 gal./can				Gal.	7.80			7.80	8.60
0600 Coating, ceramic granules, 1/2 sq./bag				Ea.	22.50			22.50	25
0700 Aluminum, 2 gal./sq.				Gal.	12.25			12.25	13.50
0800 Emulsion, fibered or non-fibered, 4 gal./sq.				"	5.65			5.65	6.25

07 51 13.20 Built-Up Roofing Systems

	Crew	Daily Output	Labor-Hours	Unit	Material	2021 Bare Costs Labor	Equipment	Total	Total Incl O&P
0010 **BUILT-UP ROOFING SYSTEMS**									
0120 Asphalt flood coat with gravel/slag surfacing, not including									
0140 Insulation, flashing or wood nailers									
0200 Asphalt base sheet, 3 plies #15 asphalt felt, mopped	G-1	22	2.545	Sq.	127	115	26	268	365
0350 On nailable decks		21	2.667		118	120	27.50	265.50	365
0500 4 plies #15 asphalt felt, mopped		20	2.800		157	126	28.50	311.50	420
0550 On nailable decks		19	2.947		138	133	30	301	415
0700 Coated glass base sheet, 2 plies glass (type IV), mopped		22	2.545		124	115	26	265	360
0850 3 plies glass, mopped		20	2.800		150	126	28.50	304.50	415
0950 On nailable decks		19	2.947		140	133	30	303	415
1100 4 plies glass fiber felt (type IV), mopped		20	2.800		185	126	28.50	339.50	450
1150 On nailable decks		19	2.947		166	133	30	329	445
1200 Coated & saturated base sheet, 3 plies #15 asph. felt, mopped		20	2.800		128	126	28.50	282.50	390
1250 On nailable decks		19	2.947		119	133	30	282	390
1300 4 plies #15 asphalt felt, mopped		22	2.545		148	115	26	289	390
2000 Asphalt flood coat, smooth surface									
2200 Asphalt base sheet & 3 plies #15 asphalt felt, mopped	G-1	24	2.333	Sq.	120	105	24	249	340

07 51 13 – Built-Up Asphalt Roofing

07 51 13.20 Built-Up Roofing Systems

		Crew	Daily Output	Labor-Hours	Unit	Material	2021 Bare Costs Labor	Equipment	Total	Total Incl O&P
2400	On nailable decks	G-1	23	2.435	Sq.	111	110	25	246	340
2600	4 plies #15 asphalt felt, mopped		24	2.333		141	105	24	270	365
2700	On nailable decks		23	2.435		132	110	25	267	360
2900	Coated glass fiber base sheet, mopped, and 2 plies of									
2910	glass fiber felt (type IV)	G-1	25	2.240	Sq.	117	101	23	241	325
3100	On nailable decks		24	2.333		111	105	24	240	330
3200	3 plies, mopped		23	2.435		143	110	25	278	375
3300	On nailable decks		22	2.545		134	115	26	275	375
3800	4 plies glass fiber felt (type IV), mopped		23	2.435		169	110	25	304	400
3900	On nailable decks		22	2.545		160	115	26	301	400
4000	Coated & saturated base sheet, 3 plies #15 asph. felt, mopped		24	2.333		121	105	24	250	340
4200	On nailable decks		23	2.435		112	110	25	247	340
4300	4 plies #15 organic felt, mopped		22	2.545		142	115	26	283	380
4500	Coal tar pitch with gravel/slag surfacing									
4600	4 plies #15 tarred felt, mopped	G-1	21	2.667	Sq.	206	120	27.50	353.50	460
4800	3 plies glass fiber felt (type IV), mopped	"	19	2.947	"	169	133	30	332	445
5000	Coated glass fiber base sheet, and 2 plies of									
5010	glass fiber felt (type IV), mopped	G-1	19	2.947	Sq.	172	133	30	335	450
5300	On nailable decks		18	3.111		151	140	32	323	440
5600	4 plies glass fiber felt (type IV), mopped		21	2.667		235	120	27.50	382.50	495
5800	On nailable decks		20	2.800		214	126	28.50	368.50	485

07 51 13.30 Cants

		Crew	Daily Output	Labor-Hours	Unit	Material	2021 Bare Costs Labor	Equipment	Total	Total Incl O&P
0010	**CANTS**									
0012	Lumber, treated, 4" x 4" cut diagonally	1 Rofc	325	.025	L.F.	1.89	1.19		3.08	4.12
0300	Mineral or fiber, trapezoidal, 1" x 4" x 48"		325	.025		.36	1.19		1.55	2.44
0400	1-1/2" x 5-5/8" x 48"		325	.025		.47	1.19		1.66	2.56
9000	Minimum labor/equipment charge		4	2	Job		96.50		96.50	166

07 51 13.40 Felts

		Crew	Daily Output	Labor-Hours	Unit	Material	2021 Bare Costs Labor	Equipment	Total	Total Incl O&P
0010	**FELTS**									
0012	Glass fibered roofing felt, #15, not mopped	1 Rofc	58	.138	Sq.	10.75	6.65		17.40	23.50
0300	Base sheet, #80, channel vented		58	.138		46.50	6.65		53.15	63
0400	#70, coated		58	.138		18.60	6.65		25.25	32
0500	Cap, #87, mineral surfaced		58	.138		71	6.65		77.65	89.50
0600	Flashing membrane, #65		16	.500		10	24		34	52.50
0800	Coal tar fibered, #15, no mopping		58	.138		18.50	6.65		25.15	32
0900	Asphalt felt, #15, 4 sq./roll, no mopping		58	.138		5.70	6.65		12.35	17.70
1100	#30, 2 sq./roll		58	.138		9.60	6.65		16.25	22
1200	Double coated, #33		58	.138		11.95	6.65		18.60	24.50
1400	#40, base sheet		58	.138		10.95	6.65		17.60	23.50
1450	Coated and saturated		58	.138		12.05	6.65		18.70	24.50
1500	Tarred felt, organic, #15, 4 sq. rolls		58	.138		16	6.65		22.65	29
1550	#30, 2 sq. roll		58	.138		26.50	6.65		33.15	40.50
1700	Add for mopping above felts, per ply, asphalt, 24 lb./sq.	G-1	192	.292		12.95	13.10	2.98	29.03	40
1800	Coal tar mopping, 30 lb./sq.		186	.301		17.90	13.55	3.08	34.53	46.50
1900	Flood coat, with asphalt, 60 lb./sq.		60	.933		32.50	42	9.55	84.05	119
2000	With coal tar, 75 lb./sq.		56	1		45	45	10.25	100.25	138
9000	Minimum labor/equipment charge	1 Rofc	4	2	Job		96.50		96.50	166

07 51 13.50 Walkways for Built-Up Roofs

		Crew	Daily Output	Labor-Hours	Unit	Material	2021 Bare Costs Labor	Equipment	Total	Total Incl O&P
0010	**WALKWAYS FOR BUILT-UP ROOFS**									
0020	Asphalt impregnated, 3' x 6' x 1/2" thick	1 Rofc	400	.020	S.F.	2.11	.96		3.07	3.98
0100	3' x 3' x 3/4" thick		400	.020	"	6.10	.96		7.06	8.35
0600	100% recycled rubber, 3' x 4' x 3/8" [G]		400	.020	L.F.	6.95	.96		7.91	9.30

For customer support on your Commercial Renovation Costs with RSMeans Data, call 800.448.8182.

207

07 51 Built-Up Bituminous Roofing

07 51 13 – Built-Up Asphalt Roofing

07 51 13.50 Walkways for Built-Up Roofs		Crew	Daily Output	Labor-Hours	Unit	Material	2021 Bare Costs Labor	Equipment	Total	Total Incl O&P
0610	3' x 4' x 1/2"	G 1 Rofc	400	.020	L.F.	7.80	.96		8.76	10.20
0620	3' x 4' x 3/4"	G	400	.020		8.15	.96		9.11	10.60
9000	Minimum labor/equipment charge		2.75	2.909	Job		140		140	241

07 52 Modified Bituminous Membrane Roofing

07 52 13 – Atactic-Polypropylene-Modified Bituminous Membrane Roofing

07 52 13.10 APP Modified Bituminous Membrane

		Crew	Daily Output	Labor-Hours	Unit	Material	2021 Bare Costs Labor	Equipment	Total	Total Incl O&P
0010	**APP MODIFIED BITUMINOUS MEMBRANE** R075213-30									
0020	Base sheet, #15 glass fiber felt, nailed to deck	1 Rofc	58	.138	Sq.	12.50	6.65		19.15	25
0030	Spot mopped to deck	G-1	295	.190		17.25	8.55	1.94	27.74	36
0040	Fully mopped to deck	"	192	.292		23.50	13.10	2.98	39.58	52
0050	#15 organic felt, nailed to deck	1 Rofc	58	.138		7.40	6.65		14.05	19.60
0060	Spot mopped to deck	G-1	295	.190		12.15	8.55	1.94	22.64	30
0070	Fully mopped to deck	"	192	.292		18.65	13.10	2.98	34.73	46.50
2100	APP mod., smooth surf. cap sheet, poly. reinf., torched, 160 mils	G-5	2100	.019	S.F.	.85	.83	.09	1.77	2.47
2150	170 mils	G-5	2100	.019	S.F.	.75	.83	.09	1.67	2.36
2200	Granule surface cap sheet, poly. reinf., torched, 180 mils		2000	.020		.95	.87	.10	1.92	2.66
2250	Smooth surface flashing, torched, 160 mils		1260	.032		.85	1.39	.15	2.39	3.50
2300	170 mils		1260	.032		.75	1.39	.15	2.29	3.39
2350	Granule surface flashing, torched, 180 mils		1260	.032		.95	1.39	.15	2.49	3.61
2400	Fibrated aluminum coating	1 Rofc	3800	.002		.10	.10		.20	.28
2450	Seam heat welding	"	205	.039	L.F.	.10	1.88		1.98	3.35

07 52 16 – Styrene-Butadiene-Styrene Modified Bituminous Membrane Roofing

07 52 16.10 SBS Modified Bituminous Membrane

		Crew	Daily Output	Labor-Hours	Unit	Material	2021 Bare Costs Labor	Equipment	Total	Total Incl O&P
0010	**SBS MODIFIED BITUMINOUS MEMBRANE**									
0080	Mod. bit. rfng., SBS mod, gran surf. cap sheet, poly. reinf.									
0650	120 to 149 mils thick	G-1	2000	.028	S.F.	1.34	1.26	.29	2.89	3.96
0750	150 to 160 mils		2000	.028		1.78	1.26	.29	3.33	4.45
1600	Smooth surface cap sheet, mopped, 145 mils		2100	.027		.86	1.20	.27	2.33	3.31
1620	Lightweight base sheet, fiberglass reinforced, 35 to 47 mil		2100	.027		.29	1.20	.27	1.76	2.68
1625	Heavyweight base/ply sheet, reinforced, 87 to 120 mil thick		2100	.027		.92	1.20	.27	2.39	3.37
1650	Granulated walkpad, 180 to 220 mils	1 Rofc	400	.020		1.84	.96		2.80	3.68
1700	Smooth surface flashing, 145 mils	G-1	1260	.044		.86	2	.45	3.31	4.89
1800	150 mils		1260	.044		.52	2	.45	2.97	4.51
1900	Granular surface flashing, 150 mils		1260	.044		.69	2	.45	3.14	4.70
2000	160 mils		1260	.044		.77	2	.45	3.22	4.79
2010	Elastomeric asphalt primer	1 Rofc	2600	.003		.17	.15		.32	.44
2015	Roofing asphalt, 30 lb./square	G-1	19000	.003		.16	.13	.03	.32	.44
2020	Cold process adhesive, 20 to 30 mils thick	1 Rofc	750	.011		.26	.51		.77	1.17
2025	Self adhering vapor retarder, 30 to 45 mils thick	G-5	2150	.019		1.06	.81	.09	1.96	2.67
2050	Seam heat welding	1 Rofc	205	.039	L.F.	.10	1.88		1.98	3.35

07 53 Elastomeric Membrane Roofing

07 53 16 – Chlorosulfonate-Polyethylene Roofing

07 53 16.10 Chlorosulfonated Polyethylene Roofing

	Crew	Daily Output	Labor-Hours	Unit	Material	2021 Bare Costs Labor	Equipment	Total	Total Incl O&P
0010 **CHLOROSULFONATED POLYETHYLENE ROOFING**									
0800 Chlorosulfonated polyethylene (CSPE)									
0900 45 mils, heat welded seams, plate attachment	G-5	35	1.143	Sq.	258	50	5.55	313.55	375
1100 Heat welded seams, plate attachment and ballasted		26	1.538		272	67	7.50	346.50	425
1200 60 mils, heat welded seams, plate attachment		35	1.143		355	50	5.55	410.55	480
1300 Heat welded seams, plate attachment and ballasted		26	1.538		365	67	7.50	439.50	530

07 53 23 – Ethylene-Propylene-Diene-Monomer Roofing

07 53 23.20 Ethylene-Propylene-Diene-Monomer Roofing

	Crew	Daily Output	Labor-Hours	Unit	Material	2021 Bare Costs Labor	Equipment	Total	Total Incl O&P
0010 **ETHYLENE-PROPYLENE-DIENE-MONOMER ROOFING (EPDM)**									
3500 Ethylene-propylene-diene-monomer (EPDM), 45 mils, 0.28 psf									
3600 Loose-laid & ballasted with stone (10 psf)	G-5	51	.784	Sq.	105	34.50	3.82	143.32	178
3700 Mechanically attached	G-5	35	1.143	Sq.	95.50	50	5.55	151.05	197
3800 Fully adhered with adhesive	"	26	1.538	"	93	67	7.50	167.50	226
4500 60 mils, 0.40 psf									
4600 Loose-laid & ballasted with stone (10 psf)	G-5	51	.784	Sq.	125	34.50	3.82	163.32	200
4700 Mechanically attached		35	1.143		114	50	5.55	169.55	218
4800 Fully adhered with adhesive		26	1.538		111	67	7.50	185.50	247
4810 45 mil, 0.28 psf, membrane only					57.50			57.50	63.50
4820 60 mil, 0.40 psf, membrane only					74.50			74.50	82
4850 Seam tape for membrane, 3" x 100' roll				Ea.	87			87	95.50
4900 Batten strips, 10' sections					3.61			3.61	3.97
4910 Cover tape for batten strips, 6" x 100' roll					188			188	207
4930 Plate anchors				M	103			103	113
4970 Adhesive for fully adhered systems, 60 S.F./gal.				Gal.	.68			.68	.75

07 53 29 – Polyisobutylene Roofing

07 53 29.10 Polyisobutylene Roofing

	Crew	Daily Output	Labor-Hours	Unit	Material	2021 Bare Costs Labor	Equipment	Total	Total Incl O&P
0010 **POLYISOBUTYLENE ROOFING**									
7500 Polyisobutylene (PIB), 100 mils, 0.57 psf									
7600 Loose-laid & ballasted with stone/gravel (10 psf)	G-5	51	.784	Sq.	215	34.50	3.82	253.32	299
7700 Partially adhered with adhesive		35	1.143		258	50	5.55	313.55	375
7800 Hot asphalt attachment		35	1.143		246	50	5.55	301.55	365
7900 Fully adhered with contact cement		26	1.538		272	67	7.50	346.50	425

07 54 Thermoplastic Membrane Roofing

07 54 16 – Ketone Ethylene Ester Roofing

07 54 16.10 Ketone Ethylene Ester Roofing

	Crew	Daily Output	Labor-Hours	Unit	Material	2021 Bare Costs Labor	Equipment	Total	Total Incl O&P
0010 **KETONE ETHYLENE ESTER ROOFING**									
0100 Ketone ethylene ester roofing, 50 mil, fully adhered	G-5	26	1.538	Sq.	212	67	7.50	286.50	360
0120 Mechanically attached		35	1.143		137	50	5.55	192.55	243
0140 Ballasted with stone		51	.784		148	34.50	3.82	186.32	226
0160 50 mil, fleece backed, adhered w/hot asphalt	G-1	26	2.154		160	97	22	279	365
0180 Accessories, pipe boot	1 Rofc	32	.250	Ea.	25.50	12.05		37.55	48.50
0200 Pre-formed corners		32	.250	"	9.60	12.05		21.65	31
0220 Ketone clad metal, including up to 4 bends		330	.024	S.F.	4.05	1.17		5.22	6.45
0240 Walkway pad	2 Rofc	800	.020	"	4.21	.96		5.17	6.30
0260 Stripping material	1 Rofc	310	.026	L.F.	.96	1.24		2.20	3.20

For customer support on your Commercial Renovation Costs with RSMeans Data, call 800.448.8182.

209

07 54 Thermoplastic Membrane Roofing

07 54 19 – Polyvinyl-Chloride Roofing

07 54 19.10 Polyvinyl-Chloride Roofing (PVC)	Crew	Daily Output	Labor-Hours	Unit	Material	2021 Bare Costs Labor	Equipment	Total	Total Incl O&P
0010 **POLYVINYL-CHLORIDE ROOFING (PVC)**									
8200 Heat welded seams									
8700 Reinforced, 48 mils, 0.33 psf									
8750 Loose-laid & ballasted with stone/gravel (12 psf)	G-5	51	.784	Sq.	131	34.50	3.82	169.32	208
8800 Mechanically attached	"	35	1.143	"	121	50	5.55	176.55	226
8850 Fully adhered with adhesive	G-5	26	1.538	Sq.	116	67	7.50	190.50	252
8860 Reinforced, 60 mils, 0.40 psf									
8870 Loose-laid & ballasted with stone/gravel (12 psf)	G-5	51	.784	Sq.	132	34.50	3.82	170.32	208
8880 Mechanically attached		35	1.143		122	50	5.55	177.55	226
8890 Fully adhered with adhesive		26	1.538		117	67	7.50	191.50	252

07 54 23 – Thermoplastic-Polyolefin Roofing

07 54 23.10 Thermoplastic Polyolefin Roofing (T.P.O.)	Crew	Daily Output	Labor-Hours	Unit	Material	2021 Bare Costs Labor	Equipment	Total	Total Incl O&P
0010 **THERMOPLASTIC POLYOLEFIN ROOFING (T.P.O.)**									
0100 45 mil, loose laid & ballasted with stone (1/2 ton/sq.)	G-5	51	.784	Sq.	90	34.50	3.82	128.32	162
0120 Fully adhered		25	1.600		77	70	7.80	154.80	213
0140 Mechanically attached		34	1.176		76.50	51.50	5.75	133.75	179
0160 Self adhered		35	1.143		79	50	5.55	134.55	179
0180 60 mil membrane, heat welded seams, ballasted		50	.800		105	35	3.90	143.90	180
0200 Fully adhered		25	1.600		92	70	7.80	169.80	230
0220 Mechanically attached		34	1.176		95.50	51.50	5.75	152.75	200
0240 Self adhered		35	1.143		106	50	5.55	161.55	209

07 55 Protected Membrane Roofing

07 55 10 – Protected Membrane Roofing Components

07 55 10.10 Protected Membrane Roofing Components	Crew	Daily Output	Labor-Hours	Unit	Material	2021 Bare Costs Labor	Equipment	Total	Total Incl O&P
0010 **PROTECTED MEMBRANE ROOFING COMPONENTS**									
0100 Choose roofing membrane from 07 50									
0120 Then choose roof deck insulation from 07 22									
0130 Filter fabric	2 Rofc	10000	.002	S.F.	.09	.08		.17	.22
0140 Ballast, 3/8" - 1/2" in place	G-1	36	1.556	Ton	26.50	70	15.90	112.40	167
0150 3/4" - 1-1/2" in place	"	36	1.556	"	26.50	70	15.90	112.40	167
0200 2" concrete blocks, natural	1 Clab	115	.070	S.F.	3.57	3.09		6.66	8.85
0210 Colors	"	115	.070	"	3.56	3.09		6.65	8.85

07 56 Fluid-Applied Roofing

07 56 10 – Fluid-Applied Roofing Elastomers

07 56 10.10 Elastomeric Roofing	Crew	Daily Output	Labor-Hours	Unit	Material	2021 Bare Costs Labor	Equipment	Total	Total Incl O&P
0010 **ELASTOMERIC ROOFING**									
0020 Acrylic, 44% solids, 2 coats, on corrugated metal	2 Rofc	2400	.007	S.F.	.56	.32		.88	1.17
0025 On smooth metal		3000	.005		.45	.26		.71	.93
0030 On foam or modified bitumen		1500	.011		.90	.51		1.41	1.87
0035 On concrete		1500	.011		.90	.51		1.41	1.87
0040 On tar and gravel		1500	.011		.90	.51		1.41	1.87
0045 36% solids, 2 coats, on corrugated metal		2400	.007		.48	.32		.80	1.08
0050 On smooth metal		3000	.005		.38	.26		.64	.86
0055 On foam or modified bitumen		1500	.011		.77	.51		1.28	1.72
0060 On concrete		1500	.011		.77	.51		1.28	1.72
0065 On tar and gravel		1500	.011		.77	.51		1.28	1.72

07 56 Fluid-Applied Roofing

07 56 10 – Fluid-Applied Roofing Elastomers

07 56 10.10 Elastomeric Roofing	Crew	Daily Output	Labor-Hours	Unit	Material	2021 Bare Costs Labor	Equipment	Total	Total Incl O&P
0070	Primer if required, 2 coats on corrugated metal	2 Rofc	2400	.007	S.F.	1.26	.32		1.58	1.94
0075	On smooth metal	2 Rofc	3000	.005	S.F.	1.26	.26		1.52	1.83
0080	On foam or modified bitumen		1500	.011		.70	.51		1.21	1.65
0085	On concrete		1500	.011		.52	.51		1.03	1.46
0090	On tar & gravel/rolled roof		1500	.011		.70	.51		1.21	1.65
0110	Acrylic rubber, fluid applied, 20 mils thick	G-5	2000	.020		2.20	.87	.10	3.17	4.03
0120	50 mils, reinforced		1200	.033		.53	1.46	.16	2.15	3.27
0130	For walking surface, add		900	.044		1.41	1.94	.22	3.57	5.15
0300	Neoprene, fluid applied, 20 mil thick, not reinforced	G-1	1135	.049		1.32	2.22	.50	4.04	5.85
0600	Non-woven polyester, reinforced		960	.058		1.44	2.62	.60	4.66	6.75
0700	5 coat neoprene deck, 60 mil thick, under 10,000 S.F.		325	.172		4.40	7.75	1.76	13.91	20
0900	Over 10,000 S.F.		625	.090		4.40	4.03	.92	9.35	12.80
9000	Minimum labor/equipment charge	1 Rofc	2	4	Job		193		193	330

07 57 Coated Foamed Roofing

07 57 13 – Sprayed Polyurethane Foam Roofing

07 57 13.10 Sprayed Polyurethane Foam Roofing (S.P.F.)

		Crew	Daily Output	Labor-Hours	Unit	Material	2021 Bare Costs Labor	Equipment	Total	Total Incl O&P
0010	**SPRAYED POLYURETHANE FOAM ROOFING (S.P.F.)**									
0100	Primer for metal substrate (when required)	G-2A	3000	.008	S.F.	.49	.34	.21	1.04	1.35
0200	Primer for non-metal substrate (when required)		3000	.008		.18	.34	.21	.73	1.01
0300	Closed cell spray, polyurethane foam, 3 lb./C.F. density, 1", R6.7		15000	.002		.65	.07	.04	.76	.88
0400	2", R13.4		13125	.002		1.30	.08	.05	1.43	1.61
0500	3", R18.6		11485	.002		1.95	.09	.06	2.10	2.36
0550	4", R24.8		10080	.002		2.60	.10	.06	2.76	3.10
0700	Spray-on silicone coating		2500	.010		.02	.41	.26	.69	.99

07 58 Roll Roofing

07 58 10 – Asphalt Roll Roofing

07 58 10.10 Roll Roofing

		Crew	Daily Output	Labor-Hours	Unit	Material	2021 Bare Costs Labor	Equipment	Total	Total Incl O&P
0010	**ROLL ROOFING**									
0100	Asphalt, mineral surface									
0200	1 ply #15 organic felt, 1 ply mineral surfaced									
0300	Selvage roofing, lap 19", nailed & mopped	G-1	27	2.074	Sq.	74	93.50	21	188.50	266
0400	3 plies glass fiber felt (type IV), 1 ply mineral surfaced									
0500	Selvage roofing, lapped 19", mopped	G-1	25	2.240	Sq.	132	101	23	256	345
0600	Coated glass fiber base sheet, 2 plies of glass fiber									
0700	Felt (type IV), 1 ply mineral surfaced selvage									
0800	Roofing, lapped 19", mopped	G-1	25	2.240	Sq.	140	101	23	264	350
0900	On nailable decks	"	24	2.333	"	127	105	24	256	350
1000	3 plies glass fiber felt (type III), 1 ply mineral surfaced									
1100	Selvage roofing, lapped 19", mopped	G-1	25	2.240	Sq.	132	101	23	256	345

For customer support on your Commercial Renovation Costs with RSMeans Data, call 800.448.8182.

211

07 61 Sheet Metal Roofing

07 61 13 – Standing Seam Sheet Metal Roofing

07 61 13.10 Standing Seam Sheet Metal Roofing, Field Fab.	Crew	Daily Output	Labor-Hours	Unit	Material	2021 Bare Costs Labor	Equipment	Total	Total Incl O&P
0010 **STANDING SEAM SHEET METAL ROOFING, FIELD FABRICATED**									
0400 Copper standing seam roofing, over 10 squares, 16 oz., 125 lb./sq.	1 Shee	1.30	6.154	Sq.	990	405		1,395	1,700
0600 18 oz., 140 lb./sq.		1.20	6.667		1,100	435		1,535	1,900
0700 20 oz., 150 lb./sq.		1.10	7.273		1,400	475		1,875	2,300
1200 For abnormal conditions or small areas, add					25%	100%			
1300 For lead-coated copper, add					25%				
1400 Lead standing seam roofing, 5 lb./S.F.	1 Shee	1.30	6.154		1,600	405		2,005	2,400
1500 Zinc standing seam roofing, .020" thick		1.20	6.667		1,175	435		1,610	1,975
1510 .027" thick		1.15	6.957		1,575	455		2,030	2,425
1520 .032" thick		1.10	7.273		1,850	475		2,325	2,800
1530 .040" thick		1.05	7.619		2,700	500		3,200	3,750
9000 Minimum labor/equipment charge		2	4	Job		262		262	410

07 61 16 – Batten Seam Sheet Metal Roofing

07 61 16.10 Batten Seam Sheet Metal Roofing, Field Fab.

	Crew	Daily Output	Labor-Hours	Unit	Material	2021 Bare Costs Labor	Equipment	Total	Total Incl O&P
0010 **BATTEN SEAM SHEET METAL ROOFING, FIELD FABRICATED**									
0012 Copper batten seam roofing, over 10 sq., 16 oz., 130 lb./sq.	1 Shee	1.10	7.273	Sq.	1,250	475		1,725	2,125
0020 Lead batten seam roofing, 5 lb./S.F.		1.20	6.667		1,675	435		2,110	2,525
0100 Zinc/copper alloy batten seam roofing, .020" thick		1.20	6.667		1,225	435		1,660	2,025
0200 Copper roofing, batten seam, over 10 sq., 18 oz., 145 lb./sq.		1	8		1,400	525		1,925	2,375
0300 20 oz., 160 lb./sq.		1	8		1,775	525		2,300	2,775
0500 Stainless steel batten seam roofing, type 304, 28 ga.		1.20	6.667		650	435		1,085	1,400
0600 26 ga.		1.15	6.957		805	455		1,260	1,600
0800 Zinc, copper alloy roofing, batten seam, .027" thick		1.15	6.957		1,625	455		2,080	2,500
0900 .032" thick		1.10	7.273		1,950	475		2,425	2,875
1000 .040" thick		1.05	7.619		2,825	500		3,325	3,875
9000 Minimum labor/equipment charge		2	4	Job		262		262	410

07 61 19 – Flat Seam Sheet Metal Roofing

07 61 19.10 Flat Seam Sheet Metal Roofing, Field Fabricated

	Crew	Daily Output	Labor-Hours	Unit	Material	2021 Bare Costs Labor	Equipment	Total	Total Incl O&P
0010 **FLAT SEAM SHEET METAL ROOFING, FIELD FABRICATED**									
0900 Copper flat seam roofing, over 10 squares, 16 oz., 115 lb./sq.	1 Shee	1.20	6.667	Sq.	915	435		1,350	1,675
0950 18 oz., 130 lb./sq.		1.15	6.957		1,025	455		1,480	1,825
1000 20 oz., 145 lb./sq.		1.10	7.273		1,300	475		1,775	2,175
1008 Zinc flat seam roofing, .020" thick		1.20	6.667		1,050	435		1,485	1,825
1010 .027" thick		1.15	6.957		1,400	455		1,855	2,225
1020 .032" thick		1.12	7.143		1,650	470		2,120	2,550
1030 .040" thick		1.05	7.619		2,425	500		2,925	3,425
1100 Lead flat seam roofing, 5 lb./S.F.		1.30	6.154		1,425	405		1,830	2,200
9000 Minimum labor/equipment charge		2.75	2.909	Job		190		190	297

07 62 Sheet Metal Flashing and Trim

07 62 10 – Sheet Metal Trim

07 62 10.10 Sheet Metal Cladding

	Crew	Daily Output	Labor-Hours	Unit	Material	2021 Bare Costs Labor	Equipment	Total	Total Incl O&P
0010 **SHEET METAL CLADDING**									
0100 Aluminum, up to 6 bends, .032" thick, window casing	1 Carp	180	.044	S.F.	2.06	2.43		4.49	6.15
0200 Window sill		72	.111	L.F.	2.06	6.10		8.16	11.95
0300 Door casing		180	.044	S.F.	2.06	2.43		4.49	6.15
0400 Fascia		250	.032		2.06	1.75		3.81	5.05
0500 Rake trim		225	.036		2.06	1.94		4	5.35
0700 .024" thick, window casing		180	.044		1.50	2.43		3.93	5.50
0800 Window sill		72	.111	L.F.	1.50	6.10		7.60	11.35

07 62 Sheet Metal Flashing and Trim

07 62 10 – Sheet Metal Trim

07 62 10.10 Sheet Metal Cladding

		Crew	Daily Output	Labor-Hours	Unit	Material	2021 Bare Costs Labor	2021 Bare Costs Equipment	Total	Total Incl O&P
0900	Door casing	1 Carp	180	.044	S.F.	1.50	2.43		3.93	5.50
1000	Fascia		250	.032		1.50	1.75		3.25	4.44
1100	Rake trim		225	.036		1.50	1.94		3.44	4.75
1200	Vinyl coated aluminum, up to 6 bends, window casing		180	.044		1.92	2.43		4.35	6
1300	Window sill		72	.111	L.F.	1.92	6.10		8.02	11.80
1400	Door casing		180	.044	S.F.	1.92	2.43		4.35	6
1500	Fascia		250	.032		1.92	1.75		3.67	4.90
1600	Rake trim		225	.036		1.92	1.94		3.86	5.20

07 65 Flexible Flashing

07 65 10 – Sheet Metal Flashing

07 65 10.10 Sheet Metal Flashing and Counter Flashing

		Crew	Daily Output	Labor-Hours	Unit	Material	2021 Bare Costs Labor	2021 Bare Costs Equipment	Total	Total Incl O&P
0010	**SHEET METAL FLASHING AND COUNTER FLASHING**									
0011	Including up to 4 bends									
0020	Aluminum, mill finish, .013" thick	1 Rofc	145	.055	S.F.	1.30	2.66		3.96	6
0030	.016" thick		145	.055		1.16	2.66		3.82	5.85
0060	.019" thick		145	.055		1.56	2.66		4.22	6.30
0100	.032" thick		145	.055		1.53	2.66		4.19	6.25
0200	.040" thick		145	.055		2.38	2.66		5.04	7.20
0300	.050" thick		145	.055		3.25	2.66		5.91	8.15
0325	Mill finish 5" x 7" step flashing, .016" thick		1920	.004	Ea.	.17	.20		.37	.54
0350	Mill finish 12" x 12" step flashing, .016" thick		1600	.005	"	.61	.24		.85	1.08
0400	Painted finish, add				S.F.	.31			.31	.34
1000	Mastic-coated 2 sides, .005" thick	1 Rofc	330	.024		1.87	1.17		3.04	4.07
1100	.016" thick		330	.024		2.06	1.17		3.23	4.28
1600	Copper, 16 oz. sheets, under 1000 lb.		115	.070		7.90	3.35		11.25	14.45
1700	Over 4000 lb.		155	.052		7.90	2.49		10.39	13
1900	20 oz. sheets, under 1000 lb.		110	.073		11.20	3.51		14.71	18.40
2000	Over 4000 lb.		145	.055		10.65	2.66		13.31	16.30
2200	24 oz. sheets, under 1000 lb.		105	.076		15.60	3.67		19.27	23.50
2300	Over 4000 lb.		135	.059		14.80	2.86		17.66	21
2500	32 oz. sheets, under 1000 lb.		100	.080		22.50	3.86		26.36	31.50
2600	Over 4000 lb.		130	.062		21.50	2.97		24.47	29
2700	W shape for valleys, 16 oz., 24" wide	1 Rofc	100	.080	L.F.	15.75	3.86		19.61	24
5800	Lead, 2.5 lb./S.F., up to 12" wide		135	.059	S.F.	6.30	2.86		9.16	11.85
5900	Over 12" wide		135	.059		4.34	2.86		7.20	9.70
8900	Stainless steel sheets, 32 ga.		155	.052		3.55	2.49		6.04	8.20
9000	28 ga.		155	.052		5.10	2.49		7.59	9.90
9100	26 ga.		155	.052		5.20	2.49		7.69	10.05
9200	24 ga.		155	.052		5.95	2.49		8.44	10.85
9290	For mechanically keyed flashing, add					40%				
9320	Steel sheets, galvanized, 20 ga.	1 Rofc	130	.062	S.F.	1.24	2.97		4.21	6.45
9322	22 ga.		135	.059		1.65	2.86		4.51	6.75
9324	24 ga.		140	.057		.97	2.75		3.72	5.80
9326	26 ga.		148	.054		.85	2.61		3.46	5.40
9328	28 ga.		155	.052		.73	2.49		3.22	5.10
9340	30 ga.		160	.050		.62	2.41		3.03	4.83
9400	Terne coated stainless steel, .015" thick, 28 ga.		155	.052		8.15	2.49		10.64	13.25
9500	.018" thick, 26 ga.		155	.052		9.10	2.49		11.59	14.30
9600	Zinc and copper alloy (brass), .020" thick		155	.052		9.05	2.49		11.54	14.25
9700	.027" thick		155	.052		10.65	2.49		13.14	16

07 65 Flexible Flashing

07 65 10 – Sheet Metal Flashing

07 65 10.10 Sheet Metal Flashing and Counter Flashing		Crew	Daily Output	Labor-Hours	Unit	Material	2021 Bare Costs Labor	Equipment	Total	Total Incl O&P
9800	.032" thick	1 Rofc	155	.052	S.F.	13.30	2.49		15.79	18.95
9900	.040" thick		155	.052		19.45	2.49		21.94	26
9950	Minimum labor/equipment charge		3	2.667	Job		129		129	221

07 65 12 – Fabric and Mastic Flashings

07 65 12.10 Fabric and Mastic Flashing and Counter Flashing

		Crew	Daily Output	Labor-Hours	Unit	Material	2021 Bare Costs Labor	Equipment	Total	Total Incl O&P
0010	**FABRIC AND MASTIC FLASHING AND COUNTER FLASHING**									
1300	Asphalt flashing cement, 5 gallon				Gal.	9.60			9.60	10.55
4900	Fabric, asphalt-saturated cotton, specification grade	1 Rofc	35	.229	S.Y.	3.32	11		14.32	22.50
5000	Utility grade		35	.229		1.45	11		12.45	20.50
5300	Close-mesh fabric, saturated, 17 oz./S.Y.		35	.229		2.21	11		13.21	21.50
5500	Fiberglass, resin-coated		35	.229		1.14	11		12.14	20
8500	Shower pan, bituminous membrane, 7 oz.		155	.052	S.F.	1.80	2.49		4.29	6.25

07 65 13 – Laminated Sheet Flashing

07 65 13.10 Laminated Sheet Flashing

		Crew	Daily Output	Labor-Hours	Unit	Material	2021 Bare Costs Labor	Equipment	Total	Total Incl O&P
0010	**LAMINATED SHEET FLASHING**, Including up to 4 bends									
0500	Aluminum, fabric-backed 2 sides, mill finish, .004" thick	1 Rofc	330	.024	S.F.	1.58	1.17		2.75	3.75
0700	.005" thick		330	.024		1.86	1.17		3.03	4.06
0750	Mastic-backed, self adhesive		460	.017		3.41	.84		4.25	5.20
0800	Mastic-coated 2 sides, .004" thick		330	.024		1.58	1.17		2.75	3.75
2800	Copper, paperbacked 1 side, 2 oz.		330	.024		2.15	1.17		3.32	4.38
2900	3 oz.		330	.024		3.14	1.17		4.31	5.45
3100	Paperbacked 2 sides, 2 oz.		330	.024		2.38	1.17		3.55	4.63
3150	3 oz.		330	.024		2.23	1.17		3.40	4.46
3200	5 oz.	1 Rofc	330	.024	S.F.	3.88	1.17		5.05	6.30
3250	7 oz.		330	.024		6.10	1.17		7.27	8.75
3400	Mastic-backed 2 sides, copper, 2 oz.		330	.024		2.27	1.17		3.44	4.51
3500	3 oz.		330	.024		2.57	1.17		3.74	4.84
3700	5 oz.		330	.024		3.91	1.17		5.08	6.30
3800	Fabric-backed 2 sides, copper, 2 oz.		330	.024		2.04	1.17		3.21	4.25
4000	3 oz.		330	.024		2.88	1.17		4.05	5.20
4100	5 oz.		330	.024		4.04	1.17		5.21	6.45
4300	Copper-clad stainless steel, .015" thick, under 500 lb.		115	.070		6.90	3.35		10.25	13.35
4400	Over 2000 lb.		155	.052		6.80	2.49		9.29	11.75
4600	.018" thick, under 500 lb.		100	.080		7.90	3.86		11.76	15.35
4700	Over 2000 lb.		145	.055		8.15	2.66		10.81	13.55
8550	Shower pan, 3 ply copper and fabric, 3 oz.		155	.052		4.12	2.49		6.61	8.80
8600	7 oz.		155	.052		4.98	2.49		7.47	9.80
9300	Stainless steel, paperbacked 2 sides, .005" thick		330	.024		4.05	1.17		5.22	6.45

07 65 19 – Plastic Sheet Flashing

07 65 19.10 Plastic Sheet Flashing and Counter Flashing

		Crew	Daily Output	Labor-Hours	Unit	Material	2021 Bare Costs Labor	Equipment	Total	Total Incl O&P
0010	**PLASTIC SHEET FLASHING AND COUNTER FLASHING**									
7300	Polyvinyl chloride, black, 10 mil	1 Rofc	285	.028	S.F.	.27	1.35		1.62	2.63
7400	20 mil		285	.028		.26	1.35		1.61	2.62
7600	30 mil		285	.028		.33	1.35		1.68	2.69
7700	60 mil		285	.028		.88	1.35		2.23	3.30
7900	Black or white for exposed roofs, 60 mil		285	.028		4.78	1.35		6.13	7.60
8060	PVC tape, 5" x 45 mils, for joint covers, 100 L.F./roll				Ea.	185			185	203
8850	Polyvinyl chloride, 30 mil	1 Rofc	160	.050	S.F.	1.90	2.41		4.31	6.25

214

07 65 Flexible Flashing

07 65 23 – Rubber Sheet Flashing

07 65 23.10 Rubber Sheet Flashing and Counter Flashing	Crew	Daily Output	Labor-Hours	Unit	Material	2021 Bare Costs Labor	Equipment	Total	Total Incl O&P
0010 **RUBBER SHEET FLASHING AND COUNTER FLASHING**									
4810 EPDM 90 mils, 1" diameter pipe flashing	1 Rofc	32	.250	Ea.	21.50	12.05		33.55	44
4820 2" diameter		30	.267		21.50	12.85		34.35	45.50
4830 3" diameter		28	.286		24	13.75		37.75	50
4840 4" diameter		24	.333		31	16.05		47.05	61.50
4850 6" diameter		22	.364		31	17.55		48.55	64
8100 Rubber, butyl, 1/32" thick		285	.028	S.F.	2.30	1.35		3.65	4.86
8200 1/16" thick		285	.028		4.80	1.35		6.15	7.65
8300 Neoprene, cured, 1/16" thick		285	.028		2.30	1.35		3.65	4.86
8400 1/8" thick		285	.028		6.05	1.35		7.40	9

07 65 26 – Self-Adhering Sheet Flashing

07 65 26.10 Self-Adhering Sheet or Roll Flashing	Crew	Daily Output	Labor-Hours	Unit	Material	2021 Bare Costs Labor	Equipment	Total	Total Incl O&P
0010 **SELF-ADHERING SHEET OR ROLL FLASHING**									
0020 Self-adhered flashing, 25 mil cross laminated HDPE, 4" wide	1 Rofc	960	.008	L.F.	.25	.40		.65	.96
0040 6" wide	"	896	.009	"	.37	.43		.80	1.15
0060 9" wide	1 Rofc	832	.010	L.F.	.56	.46		1.02	1.41
0080 12" wide	"	768	.010	"	.74	.50		1.24	1.67

07 71 Roof Specialties

07 71 16 – Manufactured Counterflashing Systems

07 71 16.10 Roof Drain Boot	Crew	Daily Output	Labor-Hours	Unit	Material	2021 Bare Costs Labor	Equipment	Total	Total Incl O&P
0010 **ROOF DRAIN BOOT**									
0100 Cast iron, 4" diameter	1 Shee	125	.064	L.F.	46	4.19		50.19	57
0300 4" x 3"		125	.064		68	4.19		72.19	81
0400 5" x 4"		125	.064		102	4.19		106.19	120

07 71 16.20 Pitch Pockets, Variable Sizes	Crew	Daily Output	Labor-Hours	Unit	Material	2021 Bare Costs Labor	Equipment	Total	Total Incl O&P
0010 **PITCH POCKETS, VARIABLE SIZES**									
0100 Adjustable, 4" to 7", welded corners, 4" deep	1 Rofc	48	.167	Ea.	23	8.05		31.05	39.50
0200 Side extenders, 6"	"	240	.033	"	3.16	1.61		4.77	6.25

07 71 19 – Manufactured Gravel Stops and Fasciae

07 71 19.10 Gravel Stop	Crew	Daily Output	Labor-Hours	Unit	Material	2021 Bare Costs Labor	Equipment	Total	Total Incl O&P
0010 **GRAVEL STOP**									
0020 Aluminum, .050" thick, 4" face height, mill finish	1 Shee	145	.055	L.F.	6.95	3.61		10.56	13.30
0080 Duranodic finish		145	.055		8.40	3.61		12.01	14.90
0100 Painted		145	.055		6.60	3.61		10.21	12.90
0300 6" face height		135	.059		7.85	3.88		11.73	14.70
0350 Duranodic finish		135	.059		8.55	3.88		12.43	15.50
0400 Painted		135	.059		8.85	3.88		12.73	15.75
0600 8" face height		125	.064		7.50	4.19		11.69	14.80
0650 Duranodic finish		125	.064		9.10	4.19		13.29	16.55
0700 Painted		125	.064		10.15	4.19		14.34	17.70
0900 12" face height, .080" thick, 2 piece		100	.080		11.05	5.25		16.30	20.50
0950 Duranodic finish		100	.080		10.75	5.25		16	19.95
1000 Painted		100	.080		14.40	5.25		19.65	24
1200 Copper, 16 oz., 3" face height		145	.055		35	3.61		38.61	44
1300 6" face height		135	.059		42	3.88		45.88	52.50
1350 Galv steel, 24 ga., 4" leg, plain, with continuous cleat, 4" face		145	.055		7.20	3.61		10.81	13.55
1360 6" face height		145	.055		6.60	3.61		10.21	12.90
1500 Polyvinyl chloride, 6" face height		135	.059		5.80	3.88		9.68	12.45

For customer support on your Commercial Renovation Costs with RSMeans Data, call 800.448.8182.

215

07 71 Roof Specialties

07 71 19 – Manufactured Gravel Stops and Fasciae

07 71 19.10 Gravel Stop

		Crew	Daily Output	Labor-Hours	Unit	Material	2021 Bare Costs Labor	Equipment	Total	Total Incl O&P
1600	9" face height	1 Shee	125	.064	L.F.	7	4.19		11.19	14.25
1800	Stainless steel, 24 ga., 6" face height		135	.059		17.10	3.88		20.98	25
1900	12" face height		100	.080		22.50	5.25		27.75	32.50
2100	20 ga., 6" face height		135	.059		21	3.88		24.88	29
2200	12" face height		100	.080		26.50	5.25		31.75	37
9000	Minimum labor/equipment charge		3.50	2.286	Job		150		150	233

07 71 19.30 Fascia

		Crew	Daily Output	Labor-Hours	Unit	Material	2021 Bare Costs Labor	Equipment	Total	Total Incl O&P
0010	**FASCIA**									
0100	Aluminum, reverse board and batten, .032" thick, colored, no furring incl.	1 Shee	145	.055	S.F.	7.50	3.61		11.11	13.90
0200	Residential type, aluminum	1 Carp	200	.040	L.F.	2.20	2.19		4.39	5.90
0220	Vinyl	1 Carp	200	.040	L.F.	2.91	2.19		5.10	6.70
0300	Steel, galv and enameled, stock, no furring, long panels	1 Shee	145	.055	S.F.	5.85	3.61		9.46	12.05
0600	Short panels		115	.070	"	5.50	4.55		10.05	13.15
9000	Minimum labor/equipment charge		4	2	Job		131		131	204

07 71 23 – Manufactured Gutters and Downspouts

07 71 23.10 Downspouts

		Crew	Daily Output	Labor-Hours	Unit	Material	2021 Bare Costs Labor	Equipment	Total	Total Incl O&P
0010	**DOWNSPOUTS**									
0020	Aluminum, embossed, .020" thick, 2" x 3"	1 Shee	190	.042	L.F.	.98	2.76		3.74	5.35
0100	Enameled		190	.042		1.40	2.76		4.16	5.85
0300	.024" thick, 2" x 3"		180	.044		2.23	2.91		5.14	7
0400	3" x 4"		140	.057		2.30	3.74		6.04	8.40
0600	Round, corrugated aluminum, 3" diameter, .020" thick		190	.042		1.93	2.76		4.69	6.40
0700	4" diameter, .025" thick		140	.057		6.85	3.74		10.59	13.40
0900	Wire strainer, round, 2" diameter		155	.052	Ea.	1.92	3.38		5.30	7.35
1000	4" diameter		155	.052		2.45	3.38		5.83	7.95
1200	Rectangular, perforated, 2" x 3"		145	.055		2.36	3.61		5.97	8.25
1300	3" x 4"		145	.055		3.57	3.61		7.18	9.60
1500	Copper, round, 16 oz., stock, 2" diameter		190	.042	L.F.	7.60	2.76		10.36	12.65
1600	3" diameter		190	.042		10.40	2.76		13.16	15.75
1800	4" diameter		145	.055		10.90	3.61		14.51	17.60
1900	5" diameter		130	.062		14.40	4.03		18.43	22
2100	Rectangular, corrugated copper, stock, 2" x 3"		190	.042		9.95	2.76		12.71	15.25
2200	3" x 4"		145	.055		11.75	3.61		15.36	18.55
2400	Rectangular, plain copper, stock, 2" x 3"		190	.042		10.40	2.76		13.16	15.75
2500	3" x 4"		145	.055		15.65	3.61		19.26	23
2700	Wire strainers, rectangular, 2" x 3"		145	.055	Ea.	17.15	3.61		20.76	24.50
2800	3" x 4"		145	.055		20	3.61		23.61	27.50
3000	Round, 2" diameter		145	.055		9.40	3.61		13.01	16
3100	3" diameter		145	.055		10.90	3.61		14.51	17.65
3300	4" diameter		145	.055		12.60	3.61		16.21	19.50
3400	5" diameter		115	.070		25.50	4.55		30.05	35
3600	Lead-coated copper, round, stock, 2" diameter		190	.042	L.F.	27	2.76		29.76	34
3700	3" diameter		190	.042		26	2.76		28.76	33
3900	4" diameter		145	.055		28.50	3.61		32.11	37
4000	5" diameter, corrugated		130	.062		26.50	4.03		30.53	35.50
4200	6" diameter, corrugated		105	.076		31	4.99		35.99	42
4300	Rectangular, corrugated, stock, 2" x 3"		190	.042		15.25	2.76		18.01	21
4500	Plain, stock, 2" x 3"		190	.042		26.50	2.76		29.26	33.50
4600	3" x 4"		145	.055		42.50	3.61		46.11	52
4800	Steel, galvanized, round, corrugated, 2" or 3" diameter, 28 ga.		190	.042		2.82	2.76		5.58	7.40
4900	4" diameter, 28 ga.	1 Shee	145	.055	L.F.	2.09	3.61		5.70	7.95
5100	5" diameter, 26 ga.		130	.062		3.87	4.03		7.90	10.55

07 71 Roof Specialties

07 71 23 – Manufactured Gutters and Downspouts

07 71 23.10 Downspouts	Crew	Daily Output	Labor-Hours	Unit	Material	2021 Bare Costs Labor	2021 Bare Costs Equipment	Total	Total Incl O&P
5400 6" diameter, 28 ga.	1 Shee	105	.076	L.F.	3.74	4.99		8.73	11.85
5500 26 ga.		105	.076		3.38	4.99		8.37	11.45
5700 Rectangular, corrugated, 28 ga., 2" x 3"		190	.042		2.43	2.76		5.19	6.95
5800 3" x 4"		145	.055		1.98	3.61		5.59	7.85
6000 Rectangular, plain, 28 ga., galvanized, 2" x 3"		190	.042		3.98	2.76		6.74	8.65
6100 3" x 4"		145	.055		5.10	3.61		8.71	11.25
6300 Epoxy painted, 24 ga., corrugated, 2" x 3"		190	.042		2.54	2.76		5.30	7.10
6400 3" x 4"		145	.055		2.70	3.61		6.31	8.60
6600 Wire strainers, rectangular, 2" x 3"		145	.055	Ea.	19.25	3.61		22.86	26.50
6700 3" x 4"		145	.055		19.75	3.61		23.36	27
6900 Round strainers, 2" or 3" diameter		145	.055		4.82	3.61		8.43	10.95
7000 4" diameter		145	.055		8.40	3.61		12.01	14.90
9000 Minimum labor/equipment charge		4	2	Job		131		131	204

07 71 23.20 Downspout Elbows

	Crew	Daily Output	Labor-Hours	Unit	Material	2021 Bare Costs Labor	2021 Bare Costs Equipment	Total	Total Incl O&P
0010 **DOWNSPOUT ELBOWS**									
0020 Aluminum, embossed, 2" x 3", .020" thick	1 Shee	100	.080	Ea.	1.08	5.25		6.33	9.35
0100 Enameled		100	.080		2.64	5.25		7.89	11.05
0200 Embossed, 3" x 4", .025" thick		100	.080		2.92	5.25		8.17	11.35
0300 Enameled		100	.080		3.84	5.25		9.09	12.35
0400 Embossed, corrugated, 3" diameter, .020" thick		100	.080		2.97	5.25		8.22	11.40
0500 4" diameter, .025" thick		100	.080		7.20	5.25		12.45	16.10
0600 Copper, 16 oz., 2" diameter		100	.080		10.85	5.25		16.10	20
0700 3" diameter		100	.080		12.35	5.25		17.60	21.50
0800 4" diameter		100	.080		13.10	5.25		18.35	22.50
1000 Rectangular, 2" x 3" corrugated		100	.080		11.90	5.25		17.15	21.50
1100 3" x 4" corrugated		100	.080		15.30	5.25		20.55	25
1300 Vinyl, 2-1/2" diameter, 45 or 75 degree bend		100	.080		4.05	5.25		9.30	12.60
1400 Tee Y junction		75	.107		12.85	7		19.85	25
9000 Minimum labor/equipment charge		4	2	Job		131		131	204

07 71 23.30 Gutters

	Crew	Daily Output	Labor-Hours	Unit	Material	2021 Bare Costs Labor	2021 Bare Costs Equipment	Total	Total Incl O&P
0010 **GUTTERS**									
0012 Aluminum, stock units, 5" K type, .027" thick, plain	1 Shee	125	.064	L.F.	2.95	4.19		7.14	9.80
0100 Enameled		125	.064		2.97	4.19		7.16	9.80
0300 5" K type, .032" thick, plain		125	.064		3.45	4.19		7.64	10.35
0400 Enameled		125	.064		3.57	4.19		7.76	10.50
0700 Copper, half round, 16 oz., stock units, 4" wide		125	.064		9.60	4.19		13.79	17.10
0900 5" wide		125	.064		7.80	4.19		11.99	15.15
1000 6" wide		118	.068		10.40	4.44		14.84	18.35
1200 K type, 16 oz., stock, 5" wide		125	.064		8.15	4.19		12.34	15.50
1300 6" wide		125	.064		8.70	4.19		12.89	16.15
1500 Lead coated copper, 16 oz., half round, stock, 4" wide	1 Shee	125	.064	L.F.	17.70	4.19		21.89	26
1600 6" wide		118	.068		17.80	4.44		22.24	26.50
1800 K type, stock, 5" wide		125	.064		18.65	4.19		22.84	27
1900 6" wide		125	.064		21	4.19		25.19	30
2100 Copper clad stainless steel, K type, 5" wide		125	.064		7.45	4.19		11.64	14.70
2200 6" wide		125	.064		10.25	4.19		14.44	17.85
2400 Steel, galv, half round or box, 28 ga., 5" wide, plain		125	.064		2.61	4.19		6.80	9.40
2500 Enameled		125	.064		2.23	4.19		6.42	9
2700 26 ga., stock, 5" wide		125	.064		2.60	4.19		6.79	9.40
2800 6" wide		125	.064		2.65	4.19		6.84	9.45
3000 Vinyl, O.G., 4" wide	1 Carp	115	.070		1.43	3.81		5.24	7.60
3100 5" wide		115	.070		1.74	3.81		5.55	7.95

07 71 23.30 Gutters	Crew	Daily Output	Labor-Hours	Unit	Material	2021 Bare Costs Labor	Equipment	Total	Total Incl O&P
3200	4" half round, stock units	1 Carp	115	.070	L.F.	1.40	3.81		5.21	7.60
3250	Joint connectors				Ea.	3.24			3.24	3.56
3300	Wood, clear treated cedar, fir or hemlock, 3" x 4"	1 Carp	100	.080	L.F.	10.85	4.38		15.23	18.90
3400	4" x 5"	"	100	.080	"	24	4.38		28.38	33.50
5000	Accessories, end cap, K type, aluminum 5"	1 Shee	625	.013	Ea.	.74	.84		1.58	2.12
5010	6"		625	.013		1.50	.84		2.34	2.96
5020	Copper, 5"		625	.013		3.44	.84		4.28	5.10
5030	6"		625	.013		3.62	.84		4.46	5.30
5040	Lead coated copper, 5"		625	.013		13.20	.84		14.04	15.80
5050	6"		625	.013		14.05	.84		14.89	16.80
5060	Copper clad stainless steel, 5"		625	.013		3.58	.84		4.42	5.25
5070	6"		625	.013		3.58	.84		4.42	5.25
5080	Galvanized steel, 5"		625	.013		1.58	.84		2.42	3.05
5090	6"		625	.013		2.74	.84		3.58	4.32
5100	Vinyl, 4"	1 Carp	625	.013		6.55	.70		7.25	8.30
5110	5"	"	625	.013		2.99	.70		3.69	4.40
5120	Half round, copper, 4"	1 Shee	625	.013		4.66	.84		5.50	6.45
5130	5"		625	.013		5.30	.84		6.14	7.10
5140	6"		625	.013		8.55	.84		9.39	10.75
5150	Lead coated copper, 5"		625	.013		14.85	.84		15.69	17.65
5160	6"		625	.013		22	.84		22.84	26
5170	Copper clad stainless steel, 5"		625	.013		5.05	.84		5.89	6.85
5180	6"		625	.013		4.62	.84		5.46	6.40
5190	Galvanized steel, 5"		625	.013		2.82	.84		3.66	4.41
5200	6"		625	.013		3.50	.84		4.34	5.15
5210	Outlet, aluminum, 2" x 3"		420	.019		.73	1.25		1.98	2.74
5220	3" x 4"		420	.019		1.19	1.25		2.44	3.25
5230	2-3/8" round		420	.019		.67	1.25		1.92	2.68
5240	Copper, 2" x 3"		420	.019		7.80	1.25		9.05	10.55
5250	3" x 4"	1 Shee	420	.019	Ea.	8.80	1.25		10.05	11.60
5260	2-3/8" round		420	.019		4.91	1.25		6.16	7.35
5270	Lead coated copper, 2" x 3"		420	.019		26	1.25		27.25	31
5280	3" x 4"		420	.019		30	1.25		31.25	35
5290	2-3/8" round		420	.019		26.50	1.25		27.75	31
5300	Copper clad stainless steel, 2" x 3"		420	.019		7.30	1.25		8.55	10
5310	3" x 4"		420	.019		8.35	1.25		9.60	11.10
5320	2-3/8" round		420	.019		4.91	1.25		6.16	7.35
5330	Galvanized steel, 2" x 3"		420	.019		3.81	1.25		5.06	6.15
5340	3" x 4"		420	.019		6.15	1.25		7.40	8.70
5350	2-3/8" round		420	.019		4.99	1.25		6.24	7.45
5360	K type mitres, aluminum		65	.123		4.94	8.05		12.99	18
5370	Copper		65	.123		17.60	8.05		25.65	32
5380	Lead coated copper		65	.123		56	8.05		64.05	74
5390	Copper clad stainless steel		65	.123		28	8.05		36.05	43
5400	Galvanized steel		65	.123		28	8.05		36.05	43.50
5420	Half round mitres, copper		65	.123		71	8.05		79.05	91
5430	Lead coated copper		65	.123		91.50	8.05		99.55	114
5440	Copper clad stainless steel		65	.123		57.50	8.05		65.55	75.50
5450	Galvanized steel		65	.123		34	8.05		42.05	50
5460	Vinyl mitres and outlets		65	.123		10.95	8.05		19	24.50
5470	Sealant		940	.009	L.F.	.01	.56		.57	.88
5480	Soldering		96	.083	"	.34	5.45		5.79	8.85
9000	Minimum labor/equipment charge		3.75	2.133	Job		140		140	218

07 71 Roof Specialties

07 71 23 – Manufactured Gutters and Downspouts

07 71 23.35 Gutter Guard

		Crew	Daily Output	Labor-Hours	Unit	Material	2021 Bare Costs Labor	2021 Bare Costs Equipment	Total	Total Incl O&P
0010	**GUTTER GUARD**									
0020	6" wide strip, aluminum mesh	1 Carp	500	.016	L.F.	2.47	.88		3.35	4.11
0100	Vinyl mesh		500	.016	"	3.08	.88		3.96	4.78
9000	Minimum labor/equipment charge		4	2	Job		109		109	174

07 71 26 – Reglets

07 71 26.10 Reglets and Accessories

		Crew	Daily Output	Labor-Hours	Unit	Material	2021 Bare Costs Labor	2021 Bare Costs Equipment	Total	Total Incl O&P
0010	**REGLETS AND ACCESSORIES**									
0020	Reglet, aluminum, .025" thick, in parapet	1 Carp	225	.036	L.F.	1.90	1.94		3.84	5.20
0300	16 oz. copper		225	.036		7.30	1.94		9.24	11.15
0400	Galvanized steel, 24 ga.		225	.036		1.51	1.94		3.45	4.76
0600	Stainless steel, .020" thick		225	.036		4.12	1.94		6.06	7.65
0900	Counter flashing for above, 12" wide, .032" aluminum	1 Shee	150	.053		2.01	3.49		5.50	7.65
1200	16 oz. copper		150	.053		7.30	3.49		10.79	13.50
1300	Galvanized steel, 26 ga.		150	.053		1.40	3.49		4.89	7
1500	Stainless steel, .020" thick		150	.053		6.85	3.49		10.34	13
9000	Minimum labor/equipment charge	1 Carp	3	2.667	Job		146		146	232

07 71 29 – Manufactured Roof Expansion Joints

07 71 29.10 Expansion Joints

		Crew	Daily Output	Labor-Hours	Unit	Material	2021 Bare Costs Labor	2021 Bare Costs Equipment	Total	Total Incl O&P
0010	**EXPANSION JOINTS**									
0300	Butyl or neoprene center with foam insulation, metal flanges									
0400	Aluminum, .032" thick for openings to 2-1/2"	1 Rofc	165	.048	L.F.	14.90	2.34		17.24	20.50
0600	For joint openings to 3-1/2"		165	.048		12.25	2.34		14.59	17.50
0610	For joint openings to 5"		165	.048		14.60	2.34		16.94	20
0620	For joint openings to 8"		165	.048		17.85	2.34		20.19	23.50
0700	Copper, 16 oz. for openings to 2-1/2"		165	.048		25	2.34		27.34	31.50
0900	For joint openings to 3-1/2"		165	.048		19.45	2.34		21.79	25.50
0910	For joint openings to 5"		165	.048		24.50	2.34		26.84	31
0920	For joint openings to 8"		165	.048		26.50	2.34		28.84	33
1000	Galvanized steel, 26 ga. for openings to 2-1/2"		165	.048		12.35	2.34		14.69	17.60
1200	For joint openings to 3-1/2"		165	.048		10.40	2.34		12.74	15.45
1210	For joint openings to 5"		165	.048		15.05	2.34		17.39	20.50
1220	For joint openings to 8"		165	.048		16.20	2.34		18.54	22
1300	Lead-coated copper, 16 oz. for openings to 2-1/2"		165	.048		36	2.34		38.34	43.50
1500	For joint openings to 3-1/2"		165	.048		39.50	2.34		41.84	47
1600	Stainless steel, .018", for openings to 2-1/2"		165	.048		12.30	2.34		14.64	17.50
1800	For joint openings to 3-1/2"		165	.048		14.15	2.34		16.49	19.55
1810	For joint openings to 5"		165	.048		13.85	2.34		16.19	19.20
1820	For joint openings to 8"		165	.048		22	2.34		24.34	28.50
1900	Neoprene, double-seal type with thick center, 4-1/2" wide		125	.064		14.45	3.08		17.53	21
1950	Polyethylene bellows, with galv steel flat flanges		100	.080		7.75	3.86		11.61	15.15
1960	With galvanized angle flanges		100	.080		7.80	3.86		11.66	15.25
2000	Roof joint with extruded aluminum cover, 2"	1 Shee	115	.070		33.50	4.55		38.05	44
2100	Expansion, stainless flange, foam center, standard	1 Rofc	100	.080		13.75	3.86		17.61	22
2200	Large	"	100	.080		18.50	3.86		22.36	27
2500	Roof to wall joint with extruded aluminum cover	1 Shee	115	.070		30.50	4.55		35.05	40.50
2700	Wall joint, closed cell foam on PVC cover, 9" wide	1 Rofc	125	.064		6.25	3.08		9.33	12.15
2800	12" wide	"	115	.070		7.05	3.35		10.40	13.50
9000	Minimum labor/equipment charge	1 Shee	3	2.667	Job		175		175	272

07 71 Roof Specialties

07 71 43 – Drip Edge

07 71 43.10 Drip Edge, Rake Edge, Ice Belts	Crew	Daily Output	Labor-Hours	Unit	Material	2021 Bare Costs Labor	Equipment	Total	Total Incl O&P
0010 **DRIP EDGE, RAKE EDGE, ICE BELTS**									
0020 Aluminum, .016" thick, 5" wide, mill finish	1 Carp	400	.020	L.F.	.80	1.09		1.89	2.62
0100 White finish		400	.020		.64	1.09		1.73	2.44
0200 8" wide, mill finish		400	.020		1.40	1.09		2.49	3.28
0300 Ice belt, 28" wide, mill finish		100	.080		7.95	4.38		12.33	15.70
0310 Vented, mill finish		400	.020		2.25	1.09		3.34	4.22
0320 Painted finish		400	.020		2.63	1.09		3.72	4.63
0400 Galvanized, 5" wide	1 Carp	400	.020	L.F.	.69	1.09		1.78	2.50
0500 8" wide, mill finish		400	.020		.89	1.09		1.98	2.72
0510 Rake edge, aluminum, 1-1/2" x 1-1/2"		400	.020		.34	1.09		1.43	2.11
0520 3-1/2" x 1-1/2"		400	.020		.49	1.09		1.58	2.28
9000 Minimum labor/equipment charge		4	2	Job		109		109	174

07 72 Roof Accessories

07 72 23 – Relief Vents

07 72 23.10 Roof Vents

	Crew	Daily Output	Labor-Hours	Unit	Material	2021 Bare Costs Labor	Equipment	Total	Total Incl O&P
0010 **ROOF VENTS**									
0020 Mushroom shape, for built-up roofs, aluminum	1 Rofc	30	.267	Ea.	70.50	12.85		83.35	99.50
0100 PVC, 6" high		30	.267	"	28	12.85		40.85	52.50
9000 Minimum labor/equipment charge		2.75	2.909	Job		140		140	241

07 72 23.20 Vents

	Crew	Daily Output	Labor-Hours	Unit	Material	2021 Bare Costs Labor	Equipment	Total	Total Incl O&P
0010 **VENTS**									
0100 Soffit or eave, aluminum, mill finish, strips, 2-1/2" wide	1 Carp	200	.040	L.F.	.50	2.19		2.69	4.03
0200 3" wide		200	.040		.48	2.19		2.67	4.01
0300 Enamel finish, 3" wide		200	.040		.54	2.19		2.73	4.07
0400 Mill finish, rectangular, 4" x 16"		72	.111	Ea.	1.77	6.10		7.87	11.65
0500 8" x 16"		72	.111		2.58	6.10		8.68	12.55
2420 Roof ventilator	Q-9	16	1		55	59		114	153
2500 Vent, roof vent	1 Rofc	24	.333		27.50	16.05		43.55	57.50

07 72 26 – Ridge Vents

07 72 26.10 Ridge Vents and Accessories

	Crew	Daily Output	Labor-Hours	Unit	Material	2021 Bare Costs Labor	Equipment	Total	Total Incl O&P
0010 **RIDGE VENTS AND ACCESSORIES**									
0100 Aluminum strips, mill finish	1 Rofc	160	.050	L.F.	2.72	2.41		5.13	7.15
0150 Painted finish		160	.050	"	3.42	2.41		5.83	7.90
0200 Connectors		48	.167	Ea.	5.20	8.05		13.25	19.55
0300 End caps		48	.167	"	2.49	8.05		10.54	16.60
0400 Galvanized strips		160	.050	L.F.	4.68	2.41		7.09	9.30
0430 Molded polyethylene, shingles not included		160	.050	"	2.84	2.41		5.25	7.25
0440 End plugs		48	.167	Ea.	2.49	8.05		10.54	16.60
0450 Flexible roll, shingles not included		160	.050	L.F.	3.09	2.41		5.50	7.55
2300 Ridge vent strip, mill finish	1 Shee	155	.052	"	4.17	3.38		7.55	9.85

07 72 33 – Roof Hatches

07 72 33.10 Roof Hatch Options

	Crew	Daily Output	Labor-Hours	Unit	Material	2021 Bare Costs Labor	Equipment	Total	Total Incl O&P
0010 **ROOF HATCH OPTIONS**									
0500 2'-6" x 3', aluminum curb and cover	G-3	10	3.200	Ea.	895	176		1,071	1,250
0520 Galvanized steel curb and aluminum cover		10	3.200		1,000	176		1,176	1,375
0540 Galvanized steel curb and cover		10	3.200		690	176		866	1,025
0600 2'-6" x 4'-6", aluminum curb and cover		9	3.556		1,150	195		1,345	1,575
0800 Galvanized steel curb and aluminum cover		9	3.556		990	195		1,185	1,400

220

07 72 Roof Accessories

07 72 33 – Roof Hatches

07 72 33.10 Roof Hatch Options

	07 72 33.10 Roof Hatch Options	Crew	Daily Output	Labor-Hours	Unit	Material	2021 Bare Costs Labor	2021 Bare Costs Equipment	Total	Total Incl O&P
0900	Galvanized steel curb and cover	G-3	9	3.556	Ea.	985	195		1,180	1,375
1100	4' x 4' aluminum curb and cover	↓	8	4	↓	1,750	220		1,970	2,275
1120	Galvanized steel curb and aluminum cover	G-3	8	4	Ea.	1,800	220		2,020	2,325
1140	Galvanized steel curb and cover		8	4		950	220		1,170	1,400
1200	2'-6" x 8'-0", aluminum curb and cover		6.60	4.848		2,400	266		2,666	3,075
1400	Galvanized steel curb and aluminum cover		6.60	4.848		1,875	266		2,141	2,475
1500	Galvanized steel curb and cover	↓	6.60	4.848	↓	1,200	266		1,466	1,750
1800	For plexiglass panels, 2'-6" x 3'-0", add to above					485			485	535
9000	Minimum labor/equipment charge	2 Carp	2	8	Job		440		440	695

07 72 36 – Smoke Vents

07 72 36.10 Smoke Hatches

	07 72 36.10 Smoke Hatches	Crew	Daily Output	Labor-Hours	Unit	Material	2021 Bare Costs Labor	2021 Bare Costs Equipment	Total	Total Incl O&P
0010	**SMOKE HATCHES**									
0200	For 3'-0" long, add to roof hatches from Section 07 72 33.10				Ea.	25%	5%			
0250	For 4'-0" long, add to roof hatches from Section 07 72 33.10					20%	5%			
0300	For 8'-0" long, add to roof hatches from Section 07 72 33.10				↓	10%	5%			

07 72 36.20 Smoke Vent Options

	07 72 36.20 Smoke Vent Options	Crew	Daily Output	Labor-Hours	Unit	Material	2021 Bare Costs Labor	2021 Bare Costs Equipment	Total	Total Incl O&P
0010	**SMOKE VENT OPTIONS**									
0100	4' x 4' aluminum cover and frame	G-3	13	2.462	Ea.	2,175	135		2,310	2,625
0200	Galvanized steel cover and frame		13	2.462		1,700	135		1,835	2,075
0300	4' x 8' aluminum cover and frame		8	4		3,300	220		3,520	3,975
0400	Galvanized steel cover and frame	↓	8	4	↓	2,725	220		2,945	3,350
9000	Minimum labor/equipment charge	2 Carp	2	8	Job		440		440	695

07 72 53 – Snow Guards

07 72 53.10 Snow Guard Options

	07 72 53.10 Snow Guard Options	Crew	Daily Output	Labor-Hours	Unit	Material	2021 Bare Costs Labor	2021 Bare Costs Equipment	Total	Total Incl O&P
0010	**SNOW GUARD OPTIONS**									
0100	Slate & asphalt shingle roofs, fastened with nails	1 Rofc	160	.050	Ea.	12.45	2.41		14.86	17.85
0200	Standing seam metal roofs, fastened with set screws		48	.167		17.80	8.05		25.85	33.50
0300	Surface mount for metal roofs, fastened with solder		48	.167	↓	7.75	8.05		15.80	22.50
0400	Double rail pipe type, including pipe	↓	130	.062	L.F.	36	2.97		38.97	44.50

07 72 80 – Vent Options

07 72 80.30 Vent Options

	07 72 80.30 Vent Options	Crew	Daily Output	Labor-Hours	Unit	Material	2021 Bare Costs Labor	2021 Bare Costs Equipment	Total	Total Incl O&P
0010	**VENT OPTIONS**									
0020	Plastic, for insulated decks, 1 per M.S.F.	1 Rofc	40	.200	Ea.	25	9.65		34.65	44
0100	Heavy duty		20	.400		61.50	19.30		80.80	101
0300	Aluminum	↓	30	.267		23.50	12.85		36.35	48
0800	Polystyrene baffles, 12" wide for 16" OC rafter spacing	1 Carp	90	.089		.52	4.86		5.38	8.30
0900	For 24" OC rafter spacing		110	.073	↓	.90	3.98		4.88	7.35
9000	Minimum labor/equipment charge	↓	3	2.667	Job		146		146	232

07 76 Roof Pavers

07 76 16 – Roof Decking Pavers

07 76 16.10 Roof Pavers and Supports

07 76 16.10 Roof Pavers and Supports	Crew	Daily Output	Labor-Hours	Unit	Material	2021 Bare Costs Labor	Equipment	Total	Total Incl O&P
0010 ROOF PAVERS AND SUPPORTS									
1000 Roof decking pavers, concrete blocks, 2" thick, natural	1 Clab	115	.070	S.F.	3.57	3.09		6.66	8.85
1100 Colors		115	.070	"	3.56	3.09		6.65	8.85
1200 Support pedestal, bottom cap	↓	960	.008	Ea.	2.65	.37		3.02	3.51
1300 Top cap	1 Clab	960	.008	Ea.	4.86	.37		5.23	5.95
1400 Leveling shims, 1/16"		1920	.004		1.21	.19		1.40	1.62
1500 1/8"		1920	.004		1.99	.19		2.18	2.48
1600 Buffer pad		960	.008	↓	4.50	.37		4.87	5.55
1700 PVC legs (4" SDR 35)		2880	.003	Inch	.18	.12		.30	.40
2000 Alternate pricing method, system in place	↓	101	.079	S.F.	8	3.52		11.52	14.35

07 81 Applied Fireproofing

07 81 16 – Cementitious Fireproofing

07 81 16.10 Sprayed Cementitious Fireproofing

07 81 16.10 Sprayed Cementitious Fireproofing	Crew	Daily Output	Labor-Hours	Unit	Material	2021 Bare Costs Labor	Equipment	Total	Total Incl O&P
0010 SPRAYED CEMENTITIOUS FIREPROOFING									
0050 Not including canvas protection, normal density									
0100 Per 1" thick, on flat plate steel	G-2	3000	.008	S.F.	.59	.37	.06	1.02	1.31
0200 Flat decking		2400	.010		.59	.46	.08	1.13	1.48
0400 Beams		1500	.016		.59	.74	.13	1.46	1.97
0500 Corrugated or fluted decks		1250	.019		.88	.89	.15	1.92	2.55
0700 Columns, 1-1/8" thick		1100	.022		.66	1.01	.17	1.84	2.53
0800 2-3/16" thick		700	.034		1.63	1.59	.27	3.49	4.61
0900 For canvas protection, add	↓	5000	.005	↓	.11	.22	.04	.37	.51
1000 Not including canvas protection, high density									
1100 Per 1" thick, on flat plate steel	G-2	3000	.008	S.F.	2.30	.37	.06	2.73	3.19
1110 On flat decking		2400	.010		2.30	.46	.08	2.84	3.36
1120 On beams		1500	.016		2.30	.74	.13	3.17	3.85
1130 Corrugated or fluted decks		1250	.019		2.30	.89	.15	3.34	4.11
1140 Columns, 1-1/8" thick		1100	.022		2.59	1.01	.17	3.77	4.65
1150 2-3/16" thick		1100	.022		5.20	1.01	.17	6.38	7.50
1170 For canvas protection, add	↓	5000	.005	↓	.11	.22	.04	.37	.51
1200 Not including canvas protection, retrofitting									
1210 Per 1" thick, on flat plate steel	G-2	1500	.016	S.F.	.52	.74	.13	1.39	1.89
1220 On flat decking		1200	.020		.52	.93	.16	1.61	2.21
1230 On beams		750	.032		.52	1.48	.25	2.25	3.21
1240 Corrugated or fluted decks		625	.038		.78	1.78	.30	2.86	4.02
1250 Columns, 1-1/8" thick		550	.044		.59	2.02	.35	2.96	4.24
1260 2-3/16" thick		500	.048		1.17	2.22	.38	3.77	5.25
1400 Accessories, preliminary spattered texture coat	↓	4500	.005		.05	.25	.04	.34	.49
1410 Bonding agent	1 Plas	1000	.008		.11	.40		.51	.75
9000 Minimum labor/equipment charge	G-2	3	8	Job		370	63.50	433.50	660

07 84 13 – Penetration Firestopping

07 84 13.10 Firestopping		Crew	Daily Output	Labor-Hours	Unit	Material	2021 Bare Costs Labor	Equipment	Total	Total Incl O&P
0010	**FIRESTOPPING** R078413-30									
0100	Metallic piping, non insulated									
0110	Through walls, 2" diameter	1 Carp	16	.500	Ea.	3.70	27.50		31.20	47.50
0120	4" diameter		14	.571		8.30	31.50		39.80	59
0130	6" diameter		12	.667		12.70	36.50		49.20	72
0140	12" diameter		10	.800		25	44		69	97
0150	Through floors, 2" diameter		32	.250		1.88	13.70		15.58	24
0160	4" diameter		28	.286		3.31	15.65		18.96	28.50
0170	6" diameter		24	.333		6.50	18.25		24.75	36
0180	12" diameter		20	.400		12.90	22		34.90	49
0190	Metallic piping, insulated									
0200	Through walls, 2" diameter	1 Carp	16	.500	Ea.	8.65	27.50		36.15	53
0210	4" diameter		14	.571		12.65	31.50		44.15	64
0220	6" diameter		12	.667		16.90	36.50		53.40	76.50
0230	12" diameter		10	.800		28.50	44		72.50	101
0240	Through floors, 2" diameter		32	.250		3.30	13.70		17	25.50
0250	4" diameter		28	.286		6.35	15.65		22	32
0260	6" diameter		24	.333		8.75	18.25		27	38.50
0270	12" diameter		20	.400		13.05	22		35.05	49.50
0280	Non metallic piping, non insulated									
0290	Through walls, 2" diameter	1 Carp	12	.667	Ea.	54.50	36.50		91	118
0300	4" diameter		10	.800		109	44		153	190
0310	6" diameter		8	1		156	54.50		210.50	259
0330	Through floors, 2" diameter		16	.500		28	27.50		55.50	74.50
0340	4" diameter		6	1.333		64	73		137	187
0350	6" diameter		6	1.333		120	73		193	248
0370	Ductwork, insulated & non insulated, round									
0380	Through walls, 6" diameter	1 Carp	12	.667	Ea.	16.65	36.50		53.15	76.50
0390	12" diameter		10	.800		29	44		73	101
0400	18" diameter		8	1		34	54.50		88.50	124
0410	Through floors, 6" diameter		16	.500		8.55	27.50		36.05	53
0420	12" diameter		14	.571		14.70	31.50		46.20	66
0430	18" diameter		12	.667		17.75	36.50		54.25	77.50
0440	Ductwork, insulated & non insulated, rectangular									
0450	With stiffener/closure angle, through walls, 6" x 12"	1 Carp	8	1	Ea.	22	54.50		76.50	112
0460	12" x 24"		6	1.333		37	73		110	157
0470	24" x 48"		4	2		93	109		202	276
0480	With stiffener/closure angle, through floors, 6" x 12"		10	.800		10.50	44		54.50	81
0490	12" x 24"		8	1		15.85	54.50		70.35	104
0500	24" x 48"		6	1.333		44	73		117	164
0510	Multi trade openings									
0520	Through walls, 6" x 12"	1 Carp	2	4	Ea.	56	219		275	410
0530	12" x 24"	"	1	8	"	210	440		650	925
0540	24" x 48"	2 Carp	1	16	Ea.	675	875		1,550	2,150
0550	48" x 96"	"	.75	21.333		2,550	1,175		3,725	4,650
0560	Through floors, 6" x 12"	1 Carp	2	4		45	219		264	400
0570	12" x 24"	"	1	8		64	440		504	765
0580	24" x 48"	2 Carp	.75	21.333		138	1,175		1,313	2,000
0590	48" x 96"	"	.50	32		325	1,750		2,075	3,125
0600	Structural penetrations, through walls									
0610	Steel beams, W8 x 10	1 Carp	8	1	Ea.	74	54.50		128.50	169
0620	W12 x 14		6	1.333		108	73		181	235
0630	W21 x 44		5	1.600		165	87.50		252.50	320

223

07 84 Firestopping

07 84 13 – Penetration Firestopping

07 84 13.10 Firestopping	Crew	Daily Output	Labor-Hours	Unit	Material	2021 Bare Costs Labor	Equipment	Total	Total Incl O&P
0640	W36 x 135	1 Carp	3	2.667	Ea.	234	146		380	490
0650	Bar joists, 18" deep		6	1.333		43	73		116	164
0660	24" deep		6	1.333		61	73		134	183
0670	36" deep		5	1.600		91.50	87.50		179	240
0680	48" deep	↓	4	2	↓	132	109		241	320
0690	Construction joints, floor slab at exterior wall									
0700	Precast, brick, block or drywall exterior									
0710	2" wide joint	1 Carp	125	.064	L.F.	11.05	3.50		14.55	17.70
0720	4" wide joint	"	75	.107	"	17.05	5.85		22.90	28
0730	Metal panel, glass or curtain wall exterior									
0740	2" wide joint	1 Carp	40	.200	L.F.	12.55	10.95		23.50	31
0750	4" wide joint	"	25	.320	"	14.25	17.50		31.75	43.50
0760	Floor slab to drywall partition									
0770	Flat joint	1 Carp	100	.080	L.F.	14.45	4.38		18.83	23
0780	Fluted joint		50	.160		16.80	8.75		25.55	32.50
0790	Etched fluted joint	↓	75	.107	↓	20	5.85		25.85	31.50
0800	Floor slab to concrete/masonry partition									
0810	Flat joint	1 Carp	75	.107	L.F.	16.85	5.85		22.70	28
0820	Fluted joint	"	50	.160	"	18.35	8.75		27.10	34
0830	Concrete/CMU wall joints									
0840	1" wide	1 Carp	100	.080	L.F.	20.50	4.38		24.88	29.50
0850	2" wide		75	.107		32	5.85		37.85	44.50
0860	4" wide	↓	50	.160	↓	36	8.75		44.75	53.50
0870	Concrete/CMU floor joints									
0880	1" wide	1 Carp	200	.040	L.F.	21.50	2.19		23.69	27
0890	2" wide		150	.053		24.50	2.92		27.42	31.50
0900	4" wide	↓	100	.080	↓	36	4.38		40.38	46.50

07 91 Preformed Joint Seals

07 91 13 – Compression Seals

07 91 13.10 Compression Seals

		Crew	Daily Output	Labor-Hours	Unit	Material	2021 Bare Costs Labor	Equipment	Total	Total Incl O&P
0010	**COMPRESSION SEALS**									
4900	O-ring type cord, 1/4"	1 Bric	472	.017	L.F.	.42	.91		1.33	1.92
4910	1/2"	1 Bric	440	.018	L.F.	1.11	.98		2.09	2.79
4920	3/4"		424	.019		2.23	1.01		3.24	4.08
4930	1"		408	.020		4.48	1.05		5.53	6.60
4940	1-1/4"		384	.021		10.80	1.12		11.92	13.65
4950	1-1/2"		368	.022		13.25	1.17		14.42	16.45
4960	1-3/4"		352	.023		15.45	1.22		16.67	18.95
4970	2"		344	.023	↓	27	1.25		28.25	31.50

07 91 16 – Joint Gaskets

07 91 16.10 Joint Gaskets

		Crew	Daily Output	Labor-Hours	Unit	Material	2021 Bare Costs Labor	Equipment	Total	Total Incl O&P
0010	**JOINT GASKETS**									
4400	Joint gaskets, neoprene, closed cell w/adh, 1/8" x 3/8"	1 Bric	240	.033	L.F.	.35	1.79		2.14	3.27
4500	1/4" x 3/4"		215	.037		.74	2		2.74	4.02
4700	1/2" x 1"		200	.040		1.79	2.15		3.94	5.40
4800	3/4" x 1-1/2"	↓	165	.048	↓	1.88	2.60		4.48	6.25

07 91 Preformed Joint Seals

07 91 23 – Backer Rods

07 91 23.10 Backer Rods

		Crew	Daily Output	Labor-Hours	Unit	Material	2021 Bare Costs Labor	2021 Bare Costs Equipment	Total	Total Incl O&P
0010	**BACKER RODS**									
0032	Backer rod, polyethylene, 1/4" diameter	1 Bric	460	.017	L.F.	.03	.93		.96	1.53
0052	1/2" diameter		460	.017		.20	.93		1.13	1.72
0072	3/4" diameter		460	.017		.08	.93		1.01	1.58
0092	1" diameter	↓	460	.017	↓	.15	.93		1.08	1.67

07 91 26 – Joint Fillers

07 91 26.10 Joint Fillers

		Crew	Daily Output	Labor-Hours	Unit	Material	2021 Bare Costs Labor	2021 Bare Costs Equipment	Total	Total Incl O&P
0010	**JOINT FILLERS**									
4360	Butyl rubber filler, 1/4" x 1/4"	1 Bric	290	.028	L.F.	.23	1.48		1.71	2.63
4365	1/2" x 1/2"		250	.032		.92	1.72		2.64	3.77
4370	1/2" x 3/4"		210	.038		1.38	2.05		3.43	4.81
4375	3/4" x 3/4"		230	.035		2.07	1.87		3.94	5.30
4380	1" x 1"	↓	180	.044	↓	2.76	2.39		5.15	6.90
4390	For coloring, add					12%				
4980	Polyethylene joint backing, 1/4" x 2"	1 Bric	2.08	3.846	C.L.F.	14.85	207		221.85	345
4990	1/4" x 6"		1.28	6.250	"	33.50	335		368.50	575
5600	Silicone, room temp vulcanizing foam seal, 1/4" x 1/2"		1312	.006	L.F.	.45	.33		.78	1.03
5610	1/2" x 1/2"		656	.012		.91	.65		1.56	2.05
5620	1/2" x 3/4"		442	.018		1.36	.97		2.33	3.05
5630	3/4" x 3/4"		328	.024		2.04	1.31		3.35	4.34
5640	1/8" x 1"		1312	.006		.45	.33		.78	1.03
5650	1/8" x 3"		442	.018		1.36	.97		2.33	3.05
5670	1/4" x 3"		295	.027		2.72	1.46		4.18	5.35
5680	1/4" x 6"		148	.054		5.45	2.90		8.35	10.65
5690	1/2" x 6"		82	.098		10.85	5.25		16.10	20.50
5700	1/2" x 9"		52.50	.152		16.30	8.20		24.50	31
5710	1/2" x 12"	↓	33	.242	↓	21.50	13		34.50	45

07 92 Joint Sealants

07 92 13 – Elastomeric Joint Sealants

07 92 13.10 Masonry Joint Sealants

		Crew	Daily Output	Labor-Hours	Unit	Material	2021 Bare Costs Labor	2021 Bare Costs Equipment	Total	Total Incl O&P
0010	**MASONRY JOINT SEALANTS**, 1/2" x 1/2" joint									
0050	Re-caulk only, oil base	1 Bric	225	.036	L.F.	.53	1.91		2.44	3.65
0100	Acrylic latex		205	.039		.30	2.10		2.40	3.70
0200	Polyurethane		200	.040		.73	2.15		2.88	4.25
0300	Silicone		195	.041		.59	2.20		2.79	4.19
1000	Cut out and re-caulk, oil base		145	.055		.53	2.96		3.49	5.35
1050	Acrylic latex		130	.062		.30	3.30		3.60	5.65
1100	Polyurethane		125	.064		.73	3.44		4.17	6.30
1150	Silicone		120	.067	↓	.59	3.58		4.17	6.40
9000	Minimum labor/equipment charge	↓	4	2	Job		107		107	173

07 92 13.20 Caulking and Sealant Options

		Crew	Daily Output	Labor-Hours	Unit	Material	2021 Bare Costs Labor	2021 Bare Costs Equipment	Total	Total Incl O&P
0010	**CAULKING AND SEALANT OPTIONS**									
0050	Latex acrylic based, bulk				Gal.	30.50			30.50	33.50
0055	Bulk in place 1/4" x 1/4" bead	1 Bric	300	.027	L.F.	.10	1.43		1.53	2.41
0060	1/4" x 3/8"		294	.027		.16	1.46		1.62	2.53
0065	1/4" x 1/2"		288	.028		.21	1.49		1.70	2.63
0075	3/8" x 3/8"		284	.028		.24	1.51		1.75	2.69
0080	3/8" x 1/2"		280	.029		.32	1.53		1.85	2.82
0085	3/8" x 5/8"	↓	276	.029	↓	.40	1.56		1.96	2.94

225

07 92 Joint Sealants

07 92 13 – Elastomeric Joint Sealants

07 92 13.20 Caulking and Sealant Options	Crew	Daily Output	Labor-Hours	Unit	Material	2021 Bare Costs Labor	Equipment	Total	Total Incl O&P
0095	3/8" x 3/4"	1 Bric	272	.029	L.F.	.48	1.58		2.06	3.06
0100	1/2" x 1/2"		275	.029		.42	1.56		1.98	2.98
0105	1/2" x 5/8"		269	.030		.53	1.60		2.13	3.15
0110	1/2" x 3/4"		263	.030		.64	1.63		2.27	3.33
0115	1/2" x 7/8"		256	.031		.74	1.68		2.42	3.52
0120	1/2" x 1"		250	.032		.85	1.72		2.57	3.69
0125	3/4" x 3/4"		244	.033		.96	1.76		2.72	3.88
0130	3/4" x 1"		225	.036		1.28	1.91		3.19	4.47
0135	1" x 1"		200	.040		1.70	2.15		3.85	5.30
0190	Cartridges				Gal.	41.50			41.50	45.50
0200	11 fl. oz. cartridge				Ea.	3.57			3.57	3.93
0500	1/4" x 1/2"	1 Bric	288	.028	L.F.	.29	1.49		1.78	2.72
0600	1/2" x 1/2"		275	.029		.58	1.56		2.14	3.15
0800	3/4" x 3/4"		244	.033		1.31	1.76		3.07	4.27
0900	3/4" x 1"		225	.036		1.75	1.91		3.66	5
1000	1" x 1"		200	.040		2.19	2.15		4.34	5.85
1400	Butyl based, bulk				Gal.	41			41	45.50
1500	Cartridges				"	41			41	45.50
1700	1/4" x 1/2", 154 L.F./gal.	1 Bric	288	.028	L.F.	.27	1.49		1.76	2.69
1800	1/2" x 1/2", 77 L.F./gal.	"	275	.029	"	.53	1.56		2.09	3.10
2300	Polysulfide compounds, 1 component, bulk				Gal.	84			84	92
2400	Cartridges				"	170			170	187
2600	1 or 2 component, in place, 1/4" x 1/4", 308 L.F./gal.	1 Bric	300	.027	L.F.	.27	1.43		1.70	2.60
2700	1/2" x 1/4", 154 L.F./gal.		288	.028		.54	1.49		2.03	3
2900	3/4" x 3/8", 68 L.F./gal.		272	.029		1.23	1.58		2.81	3.90
3000	1" x 1/2", 38 L.F./gal.		250	.032		2.21	1.72		3.93	5.20
3200	Polyurethane, 1 or 2 component				Gal.	54.50			54.50	60
3300	Cartridges				"	65			65	71.50
3500	Bulk, in place, 1/4" x 1/4"	1 Bric	300	.027	L.F.	.18	1.43		1.61	2.50
3655	1/2" x 1/4"		288	.028		.35	1.49		1.84	2.79
3800	3/4" x 3/8"		272	.029		.80	1.58		2.38	3.42
3900	1" x 1/2"		250	.032		1.42	1.72		3.14	4.32
4100	Silicone rubber, bulk				Gal.	65.50			65.50	72.50
4200	Cartridges				"	44.50			44.50	49
4300	Bulk in place, 1/4" x 1/2", 154 L.F./gal.	1 Bric	235	.034	L.F.	.43	1.83		2.26	3.41

07 92 16 – Rigid Joint Sealants

07 92 16.10 Rigid Joint Sealants

0010	**RIGID JOINT SEALANTS**									
5802	Tapes, sealant, PVC foam adhesive, 1/16" x 1/4"				L.F.	.09			.09	.10
5902	1/16" x 1/2"					.09			.09	.10
5952	1/16" x 1"					.15			.15	.17
6002	1/8" x 1/2"					.09			.09	.10

07 92 19 – Acoustical Joint Sealants

07 92 19.10 Acoustical Sealant

0010	**ACOUSTICAL SEALANT**									
0020	Acoustical sealant, elastomeric, cartridges				Ea.	8.05			8.05	8.85
0025	In place, 1/4" x 1/4"	1 Bric	300	.027	L.F.	.33	1.43		1.76	2.66
0030	1/4" x 1/2"		288	.028		.66	1.49		2.15	3.12
0035	1/2" x 1/2"		275	.029		1.31	1.56		2.87	3.96
0040	1/2" x 3/4"		263	.030		1.97	1.63		3.60	4.80
0045	3/4" x 3/4"		244	.033		2.96	1.76		4.72	6.10
0050	1" x 1"		200	.040		5.25	2.15		7.40	9.25

226

07 95 Expansion Control

07 95 13 – Expansion Joint Cover Assemblies

07 95 13.50 Expansion Joint Assemblies	Crew	Daily Output	Labor-Hours	Unit	Material	2021 Bare Costs Labor	Equipment	Total	Total Incl O&P
0010 **EXPANSION JOINT ASSEMBLIES**									
0200 Floor cover assemblies, 1" space, aluminum	1 Sswk	38	.211	L.F.	22.50	12.70		35.20	45.50
0300 Bronze		38	.211		53	12.70		65.70	79
0500 2" space, aluminum		38	.211		14.90	12.70		27.60	37
0600 Bronze		38	.211		61.50	12.70		74.20	88.50
0800 Wall and ceiling assemblies, 1" space, aluminum		38	.211		17.20	12.70		29.90	39.50
0900 Bronze		38	.211		54.50	12.70		67.20	80
1100 2" space, aluminum		38	.211		18.05	12.70		30.75	40.50
1200 Bronze		38	.211		52	12.70		64.70	78
1400 Floor to wall assemblies, 1" space, aluminum		38	.211		21	12.70		33.70	43.50
1500 Bronze or stainless	1 Sswk	38	.211	L.F.	63.50	12.70		76.20	90.50
1700 Gym floor angle covers, aluminum, 3" x 3" angle		46	.174		21	10.50		31.50	40
1800 3" x 4" angle		46	.174		25	10.50		35.50	44.50
2000 Roof closures, aluminum, flat roof, low profile, 1" space		57	.140		18.65	8.45		27.10	34
2100 High profile		57	.140		25.50	8.45		33.95	41.50
2300 Roof to wall, low profile, 1" space		57	.140		25	8.45		33.45	40.50
2400 High profile		57	.140		32.50	8.45		40.95	49
9000 Minimum labor/equipment charge		2	4	Job		241		241	390

Division Notes

	CREW	DAILY OUTPUT	LABOR-HOURS	UNIT	BARE COSTS				TOTAL INCL O&P
					MAT.	LABOR	EQUIP.	TOTAL	

Estimating Tips
08 10 00 Doors and Frames
All exterior doors should be addressed for their energy conservation (insulation and seals).

- Most metal doors and frames look alike, but there may be significant differences among them. When estimating these items, be sure to choose the line item that most closely compares to the specification or door schedule requirements regarding:
  - ☐ type of metal
  - ☐ metal gauge
  - ☐ door core material
  - ☐ fire rating
  - ☐ finish
- Wood and plastic doors vary considerably in price. The primary determinant is the veneer material. Lauan, birch, and oak are the most common veneers. Other variables include the following:
  - ☐ hollow or solid core
  - ☐ fire rating
  - ☐ flush or raised panel
  - ☐ finish
- Door pricing includes bore for cylindrical locksets and mortise for hinges.

08 30 00 Specialty Doors and Frames
- There are many varieties of special doors, and they are usually priced per each. Add frames, hardware, or operators required for a complete installation.

08 40 00 Entrances, Storefronts, and Curtain Walls
- Glazed curtain walls consist of the metal tube framing and the glazing material. The cost data in this subdivision is presented for the metal tube framing alone or the composite wall. If your estimate requires a detailed takeoff of the framing, be sure to add the glazing cost and any tints.

08 50 00 Windows
- Steel windows are unglazed and aluminum can be glazed or unglazed. Some metal windows are priced without glass. Refer to 08 80 00 Glazing for glass pricing. The grade C indicates commercial grade windows, usually ASTM C-35.
- All wood windows and vinyl are priced preglazed. The glazing is insulating glass. Add the cost of screens and grills if required and not already included.

08 70 00 Hardware
- Hardware costs add considerably to the cost of a door. The most efficient method to determine the hardware requirements for a project is to review the door and hardware schedule together. One type of door may have different hardware, depending on the door usage.
- Door hinges are priced by the pair, with most doors requiring 1-1/2 pairs per door. The hinge prices do not include installation labor because it is included in door installation.

Hinges are classified according to the frequency of use, base material, and finish.

08 80 00 Glazing
- Different openings require different types of glass. The most common types are:
  - ☐ float
  - ☐ tempered
  - ☐ insulating
  - ☐ impact-resistant
  - ☐ ballistic-resistant
- Most exterior windows are glazed with insulating glass. Entrance doors and window walls, where the glass is less than 18" from the floor, are generally glazed with tempered glass. Interior windows and some residential windows are glazed with float glass.
- Coastal communities require the use of impact-resistant glass, dependent on wind speed.
- The insulation or 'u' value is a strong consideration, along with solar heat gain, to determine total energy efficiency.

Reference Numbers
Reference numbers are shown at the beginning of some major classifications. These numbers refer to related items in the Reference Section. The reference information may be an estimating procedure, an alternate pricing method, or technical information.

Note: Not all subdivisions listed here necessarily appear. ■

Same Data. Simplified.

Enjoy the convenience and efficiency of accessing your costs anywhere:

- **Skip the multiplier** by setting your location
- **Quickly search,** edit, favorite and share costs
- **Stay on top of price changes** with automatic updates

Discover more at rsmeans.com/online

08 01 Operation and Maintenance of Openings

08 01 11 – Operation and Maintenance of Metal Doors and Frames

08 01 11.10 Door and Window Maintenance

08 01 11.10 Door and Window Maintenance		Crew	Daily Output	Labor-Hours	Unit	Material	2021 Bare Costs Labor	2021 Bare Costs Equipment	Total	Total Incl O&P
0010	**DOOR & WINDOW MAINTENANCE**									
1130	Install door	2 Carp	13.07	1.224	Ea.		67		67	107
1140	Remove deadbolt	1 Carp	13.85	.578			31.50		31.50	50.50
1150	Remove panic bar		7.69	1.040			57		57	90.50
1420	Remove door, plane to fit, rail, reinstall door		9.60	.833			45.50		45.50	72.50

08 01 14 – Operation and Maintenance of Wood Doors

08 01 14.15 Door and Window Maintenance

08 01 14.15		Crew	Daily Output	Labor-Hours	Unit	Material	Labor	Equipment	Total	Total Incl O&P
0010	**DOOR & WINDOW MAINTENANCE**									
0050	Rehang single door	1 Carp	32	.250	Ea.		13.70		13.70	22
0100	Rehang double door		16	.500	Pr.		27.50		27.50	43.50
0350	Install window sash cord		16	.500	Ea.	4.89	27.50		32.39	49

08 01 53 – Operation and Maintenance of Plastic Windows

08 01 53.81 Solid Vinyl Replacement Windows

08 01 53.81			Crew	Daily Output	Labor-Hours	Unit	Material	Labor	Equipment	Total	Total Incl O&P
0010	**SOLID VINYL REPLACEMENT WINDOWS**	R085313-20									
0020	Double-hung, insulated glass, up to 83 united inches	G	2 Carp	8	2	Ea.	181	109		290	375
0040	84 to 93	G		8	2		197	109		306	390
0060	94 to 101	G		6	2.667		197	146		343	450
0080	102 to 111	G		6	2.667		207	146		353	460
0100	112 to 120	G		6	2.667		222	146		368	475
0120	For each united inch over 120, add	G		800	.020	Inch	2.90	1.09		3.99	4.93
0140	Casement windows, one operating sash, 42 to 60 united inches	G		8	2	Ea.	284	109		393	485
0160	61 to 70	G		8	2		310	109		419	515
0180	71 to 80	G		8	2		365	109		474	575
0200	81 to 96	G		8	2		360	109		469	570
0220	Two operating sash, 58 to 78 united inches	G		8	2		655	109		764	900
0240	79 to 88	G		8	2		740	109		849	990
0260	89 to 98	G		8	2		960	109		1,069	1,225
0280	99 to 108	G		6	2.667		785	146		931	1,100
0300	109 to 121	G		6	2.667		860	146		1,006	1,175
0320	Two operating, one fixed sash, 73 to 108 united inches	G		8	2		625	109		734	860
0340	109 to 118	G		8	2		680	109		789	920
0360	119 to 128	G	2 Carp	6	2.667	Ea.	775	146		921	1,075
0380	129 to 138	G		6	2.667		775	146		921	1,075
0400	139 to 156	G		6	2.667		955	146		1,101	1,275
0420	Four operating sash, 98 to 118 united inches	G		8	2		1,325	109		1,434	1,625
0440	119 to 128	G		8	2		1,375	109		1,484	1,700
0460	129 to 138	G		6	2.667		1,425	146		1,571	1,800
0480	139 to 148	G		6	2.667		1,450	146		1,596	1,825
0500	149 to 168	G		6	2.667		1,575	146		1,721	1,950
0520	169 to 178	G		6	2.667		1,625	146		1,771	2,000
0560	Fixed picture window, up to 63 united inches	G		8	2		181	109		290	375
0580	64 to 83	G		8	2		194	109		303	390
0600	84 to 101	G		8	2		236	109		345	435
0620	For each united inch over 101, add	G		900	.018	Inch	3.39	.97		4.36	5.30
0800	Cellulose fiber insulation, poured into sash balance cavity	G	1 Carp	36	.222	C.F.	.70	12.15		12.85	20
0820	Silicone caulking at perimeter	G	"	800	.010	L.F.	.14	.55		.69	1.03
2000	Impact resistant replacement windows										
2005	Laminated glass, 120 MPH rating, measure in united inches										
2010	Installation labor does not cover any rework of the window opening										
2020	Double-hung, insulated glass, up to 101 united inches		2 Carp	8	2	Ea.	420	109		529	635
2025	For each united inch over 101, add			80	.200	Inch	3.49	10.95		14.44	21
2100	Casement windows, impact resistant, up to 60 united inches			8	2	Ea.	345	109		454	555

230

08 01 Operation and Maintenance of Openings

08 01 53 – Operation and Maintenance of Plastic Windows

08 01 53.81 Solid Vinyl Replacement Windows	Crew	Daily Output	Labor-Hours	Unit	Material	2021 Bare Costs Labor	Equipment	Total	Total Incl O&P
2120	61 to 70	2 Carp	8	2	Ea.	385	109		494	595
2130	71 to 80		8	2		420	109		529	640
2140	81 to 100		8	2		440	109		549	655
2150	For each united inch over 100, add		80	.200	Inch	5.30	10.95		16.25	23.50
2200	Awning windows, impact resistant, up to 60 united inches		8	2	Ea.	340	109		449	545
2220	61 to 70		8	2		380	109		489	590
2230	71 to 80		8	2		420	109		529	635
2240	For each united inch over 80, add		80	.200	Inch	5	10.95		15.95	23
2300	Picture windows, impact resistant, up to 63 united inches		8	2	Ea.	305	109		414	510
2320	63 to 83		8	2		350	109		459	560
2330	84 to 101		8	2		420	109		529	635
2340	For each united inch over 101, add		80	.200	Inch	3.36	10.95		14.31	21

08 05 Common Work Results for Openings

08 05 05 – Selective Demolition for Openings

08 05 05.10 Selective Demolition Doors

		Crew	Daily Output	Labor-Hours	Unit	Material	2021 Bare Costs Labor	Equipment	Total	Total Incl O&P
0010	**SELECTIVE DEMOLITION DOORS** R024119-10									
0200	Doors, exterior, 1-3/4" thick, single, 3' x 7' high	1 Clab	16	.500	Ea.		22		22	35.50
0202	Doors, exterior, 3' - 6' wide x 7' high	"	12	.667	"		29.50		29.50	47
0210	3' x 8' high	1 Clab	10	.800	Ea.		35.50		35.50	56.50
0215	Double, 3' x 8' high		6	1.333			59		59	94.50
0220	Double, 6' x 7' high		12	.667			29.50		29.50	47
0500	Interior, 1-3/8" thick, single, 3' x 7' high		20	.400			17.75		17.75	28.50
0520	Double, 6' x 7' high		16	.500			22		22	35.50
0700	Bi-folding, 3' x 6'-8" high		20	.400			17.75		17.75	28.50
0720	6' x 6'-8" high		18	.444			19.75		19.75	31.50
0900	Bi-passing, 3' x 6'-8" high		16	.500			22		22	35.50
0940	6' x 6'-8" high		14	.571			25.50		25.50	40.50
0960	Interior metal door 1-3/4" thick, 3'-0" x 6'-8"		18	.444			19.75		19.75	31.50
0980	Interior metal door 1-3/4" thick, 3'-0" x 7'-0"		18	.444			19.75		19.75	31.50
1000	Interior wood door 1-3/4" thick, 3'-0" x 6'-8"		20	.400			17.75		17.75	28.50
1020	Interior wood door 1-3/4" thick, 3'-0" x 7'-0"		20	.400			17.75		17.75	28.50
1100	Door demo, floor door	2 Sswk	5	3.200			193		193	310
1500	Remove and reset, hollow core	1 Carp	8	1			54.50		54.50	87
1520	Solid		6	1.333			73		73	116
2000	Frames, including trim, metal		8	1			54.50		54.50	87
2200	Wood	2 Carp	32	.500			27.50		27.50	43.50
2201	Alternate pricing method	1 Carp	200	.040	L.F.		2.19		2.19	3.48
2205	Remove door hardware	"	45.70	.175	Ea.		9.60		9.60	15.25
3000	Special doors, counter doors	2 Carp	6	2.667			146		146	232
3100	Double acting		10	1.600			87.50		87.50	139
3200	Floor door (trap type), or access type		8	2			109		109	174
3300	Glass, sliding, including frames		12	1.333			73		73	116
3400	Overhead, commercial, 12' x 12' high		4	4			219		219	350
3440	up to 20' x 16' high		3	5.333			292		292	465
3445	up to 35' x 30' high		1	16			875		875	1,400
3500	Residential, 9' x 7' high		8	2			109		109	174
3540	16' x 7' high		7	2.286			125		125	199
3600	Remove and reset, small		4	4			219		219	350
3620	Large		2.50	6.400			350		350	555
3700	Roll-up grille		5	3.200			175		175	279

08 05 05 – Selective Demolition for Openings

08 05 05.10 Selective Demolition Doors

		Crew	Daily Output	Labor-Hours	Unit	Material	2021 Bare Costs Labor	Equipment	Total	Total Incl O&P
3800	Revolving door, no wire or connections	2 Carp	2	8	Ea.		440		440	695
3900	Storefront swing door	↓	3	5.333			292		292	465
3902	Cafe/bar swing door	2 Clab	8	2			89		89	141
4000	Residential lockset, exterior	1 Carp	28	.286			15.65		15.65	25
4224	Pocket door, no frame		8	1			54.50		54.50	87
4300	Remove and reset deadbolt		8	1			54.50		54.50	87
5500	Remove door closer	1 Clab	10	.800			35.50		35.50	56.50
5520	Remove push/pull plate		20	.400			17.75		17.75	28.50
5540	Remove door bolt		30	.267			11.85		11.85	18.85
5560	Remove kick/mop plate		20	.400			17.75		17.75	28.50
5585	Remove panic device		10	.800			35.50		35.50	56.50
5590	Remove mail slot		45	.178			7.90		7.90	12.55
5595	Remove peep hole	↓	45	.178			7.90		7.90	12.55
5604	Remove 9' x 7' garage door	2 Carp	9.50	1.684			92		92	147
5624	16' x 7' garage door		8	2			109		109	174
5626	9' x 7' swing up garage door		7	2.286			125		125	199
5628	16' x 7' swing up garage door	↓	7	2.286			125		125	199
5630	Garage door demolition, remove garage door track	1 Carp	6	1.333			73		73	116
5644	Remove overhead door opener	1 Clab	5	1.600			71		71	113
5774	Remove heavy gage sectional door, 20 ga., 8' x 8'	2 Carp	7	2.286			125		125	199
5784	10' x 10'		6	2.667			146		146	232
5794	12' x 12'		5	3.200			175		175	279
5805	14' x 14'	↓	4	4			219		219	350
6384	Remove French door unit	1 Clab	8	1			44.50		44.50	70.50
7100	Remove double swing pneumatic doors, openers and sensors	2 Skwk	.50	32	Opng.		1,825		1,825	2,900
7570	Remove shock absorbing door	2 Sswk	1.90	8.421	"		510		510	820
9000	Minimum labor/equipment charge	1 Carp	4	2	Job		109		109	174

08 05 05.20 Selective Demolition of Windows

		Crew	Daily Output	Labor-Hours	Unit	Material	2021 Bare Costs Labor	Equipment	Total	Total Incl O&P
0010	**SELECTIVE DEMOLITION OF WINDOWS** R024119-10									
0200	Aluminum, including trim, to 12 S.F.	1 Clab	16	.500	Ea.		22		22	35.50
0240	To 25 S.F.		11	.727			32.50		32.50	51.50
0280	To 50 S.F.		5	1.600			71		71	113
0320	Storm windows/screens, to 12 S.F.		27	.296			13.15		13.15	21
0360	To 25 S.F.		21	.381			16.90		16.90	27
0400	To 50 S.F.		16	.500	↓		22		22	35.50
0600	Glass, up to 10 S.F./window		200	.040	S.F.		1.78		1.78	2.83
0620	Over 10 S.F./window		150	.053	"		2.37		2.37	3.77
1000	Steel, including trim, to 12 S.F.		13	.615	Ea.		27.50		27.50	43.50
1020	To 25 S.F.		9	.889			39.50		39.50	63
1040	To 50 S.F.		4	2			89		89	141
2000	Wood, including trim, to 12 S.F.		22	.364			16.15		16.15	25.50
2020	To 25 S.F.		18	.444			19.75		19.75	31.50
2060	To 50 S.F.		13	.615			27.50		27.50	43.50
2065	To 180 S.F.	↓	8	1	↓		44.50		44.50	70.50
4400	Remove skylight, prefabricated glass block with metal frame	G-3	180	.178	S.F.		9.75		9.75	15.35
4410	Remove skylight, plstc domes, flush/curb mtd	2 Carp	395	.041	"		2.22		2.22	3.53
4420	Remove skylight, plstc/glass up to 2' x 3'	1 Carp	15	.533	Ea.		29		29	46.50
4440	Remove skylight, plstc/glass up to 4' x 6'	2 Carp	10	1.600			87.50		87.50	139
4480	Remove roof window up to 3' x 4'	1 Carp	8	1			54.50		54.50	87
4500	Remove roof window up to 4' x 6'	2 Carp	6	2.667			146		146	232
5020	Remove and reset window, up to a 2' x 2' window	1 Carp	6	1.333			73		73	116
5040	Up to a 3' x 3' window	↓	4	2	↓		109		109	174

08 05 Common Work Results for Openings

08 05 05 – Selective Demolition for Openings

08 05 05.20 Selective Demolition of Windows	Crew	Daily Output	Labor-Hours	Unit	Material	Labor	Equipment	Total	Total Incl O&P
5080	Up to a 4' x 5' window	1 Carp	2	4	Ea.		219		219	350
6000	Screening only	1 Clab	4000	.002	S.F.		.09		.09	.14
9000	Minimum labor/equipment charge	"	4	2	Job		89		89	141

08 11 Metal Doors and Frames

08 11 16 – Aluminum Doors and Frames

08 11 16.10 Entrance Doors

	08 11 16.10 Entrance Doors	Crew	Daily Output	Labor-Hours	Unit	Material	Labor	Equipment	Total	Total Incl O&P
0010	**ENTRANCE DOORS** and frame, aluminum, narrow stile									
0011	Including standard hardware, clear finish, no glass									
0012	Top and bottom offset pivots, 1/4" beveled glass stops, threshold									
0013	Dead bolt lock with inside thumb screw, standard push pull									
0020	3'-0" x 7'-0" opening	2 Sswk	2	8	Ea.	1,100	480		1,580	1,975
0025	Anodizing aluminum entr. door & frame, add					107			107	118
0030	3'-6" x 7'-0" opening	2 Sswk	2	8		860	480		1,340	1,725
0100	3'-0" x 10'-0" opening, 3' high transom		1.80	8.889		1,325	535		1,860	2,350
0200	3'-6" x 10'-0" opening, 3' high transom		1.80	8.889		1,450	535		1,985	2,475
0280	5'-0" x 7'-0" opening		2	8		1,575	480		2,055	2,500
0300	6'-0" x 7'-0" opening		1.30	12.308		1,075	740		1,815	2,375
0301	6'-0" x 7'-0" opening		1.30	12.308	Pr.	1,075	740		1,815	2,375
0400	6'-0" x 10'-0" opening, 3' high transom		1.10	14.545	"	1,600	875		2,475	3,175
0520	3'-0" x 7'-0" opening, wide stile		2	8	Ea.	905	480		1,385	1,775
0540	3'-6" x 7'-0" opening		2	8		990	480		1,470	1,875
0560	5'-0" x 7'-0" opening		2	8		1,500	480		1,980	2,425
0580	6'-0" x 7'-0" opening		1.30	12.308	Pr.	1,625	740		2,365	3,000
0600	7'-0" x 7'-0" opening	1	16	"	2,300	965		3,265	4,075
1200	For non-standard size, add				Leaf	80%				
1250	For installation of non-standard size, add						20%			
1300	Light bronze finish, add				Leaf	36%				
1400	Dark bronze finish, add					25%				
1500	For black finish, add					40%				
1600	Concealed panic device, add					1,075			1,075	1,200
1700	Electric striker release, add				Opng.	325			325	360
1800	Floor check, add				Leaf	700			700	770
1900	Concealed closer, add				"	560			560	615
9000	Minimum labor/equipment charge	2 Carp	4	4	Job		219		219	350

08 11 63 – Metal Screen and Storm Doors and Frames

08 11 63.23 Aluminum Screen and Storm Doors and Frames

	08 11 63.23 Aluminum Screen and Storm Doors and Frames	Crew	Daily Output	Labor-Hours	Unit	Material	Labor	Equipment	Total	Total Incl O&P
0010	**ALUMINUM SCREEN AND STORM DOORS AND FRAMES**									
0020	Combination storm and screen									
0420	Clear anodic coating, 2'-8" wide	2 Carp	14	1.143	Ea.	237	62.50		299.50	360
0440	3'-0" wide	"	14	1.143	"	191	62.50		253.50	310
0500	For 7'-0" door height, add					8%				
1020	Mill finish, 2'-8" wide	2 Carp	14	1.143	Ea.	248	62.50		310.50	370
1040	3'-0" wide	"	14	1.143		267	62.50		329.50	395
1100	For 7'-0" door, add					8%				
1520	White painted, 2'-8" wide	2 Carp	14	1.143		259	62.50		321.50	385
1540	3'-0" wide	"	14	1.143		293	62.50		355.50	420
1600	For 7'-0" door, add					8%				
2000	Wood door & screen, see Section 08 14 33.20									
9000	Minimum labor/equipment charge	1 Carp	4	2	Job		109		109	174

08 12 13.13 Standard Hollow Metal Frames		Crew	Daily Output	Labor-Hours	Unit	Material	2021 Bare Costs Labor	Equipment	Total	Total Incl O&P
0010	**STANDARD HOLLOW METAL FRAMES**									
0020	16 ga., up to 5-3/4" jamb depth									
0025	3'-0" x 6'-8" single	G	2 Carp	16	1	Ea.	228	54.50		282.50	340
0028	3'-6" wide, single	G		16	1		269	54.50		323.50	385
0030	4'-0" wide, single	G		16	1		305	54.50		359.50	420
0040	6'-0" wide, double	G		14	1.143		299	62.50		361.50	430
0045	8'-0" wide, double	G		14	1.143		245	62.50		307.50	370
0100	3'-0" x 7'-0" single	G		16	1		217	54.50		271.50	325
0110	3'-6" wide, single			16	1		228	54.50		282.50	340
0112	4'-0" wide, single			16	1		253	54.50		307.50	365
0140	6'-0" wide, double			14	1.143		296	62.50		358.50	425
0145	8'-0" wide, double			14	1.143		231	62.50		293.50	355
1000	16 ga., up to 4-7/8" deep, 3'-0" x 7'-0" single	G		16	1		174	54.50		228.50	278
1140	6'-0" wide, double	G		14	1.143		212	62.50		274.50	335
1200	16 ga., 8-3/4" deep, 3'-0" x 7'-0" single	G		16	1		231	54.50		285.50	340
1240	6'-0" wide, double	G		14	1.143		275	62.50		337.50	400
2800	14 ga., up to 3-7/8" deep, 3'-0" x 7'-0" single	G		16	1		230	54.50		284.50	340
2840	6'-0" wide, double	G		14	1.143		350	62.50		412.50	485
3000	14 ga., up to 5-3/4" deep, 3'-0" x 6'-8" single	G		16	1		156	54.50		210.50	258
3002	3'-6" wide, single	G		16	1		246	54.50		300.50	360
3005	4'-0" wide, single	G		16	1		133	54.50		187.50	233
3600	up to 5-3/4" jamb depth, 4'-0" x 7'-0" single			15	1.067		216	58.50		274.50	330
3620	6'-0" wide, double			12	1.333		209	73		282	345
3640	8'-0" wide, double			12	1.333		315	73		388	465
3700	8'-0" high, 4'-0" wide, single		2 Carp	15	1.067	Ea.	298	58.50		356.50	425
3740	8'-0" wide, double			12	1.333		345	73		418	495
4000	6-3/4" deep, 4'-0" x 7'-0" single	G		15	1.067		288	58.50		346.50	410
4020	6'-0" wide, double	G		12	1.333		325	73		398	475
4040	8'-0" wide, double	G		12	1.333		243	73		316	385
4100	8'-0" high, 4'-0" wide, single	G		15	1.067		160	58.50		218.50	269
4140	8'-0" wide, double	G		12	1.333		450	73		523	610
4400	8-3/4" deep, 4'-0" x 7'-0", single			15	1.067		460	58.50		518.50	600
4440	8'-0" wide, double			12	1.333		440	73		513	595
4500	4'-0" x 8'-0", single			15	1.067		480	58.50		538.50	625
4540	8'-0" wide, double			12	1.333		475	73		548	635
4900	For welded frames, add						69.50			69.50	76.50
5380	Steel frames, KD, 14 ga., "B" label, to 5-3/4" throat, to 3'-0" x 7'-0"		2 Carp	15	1.067		239	58.50		297.50	355
5400	14 ga., "B" label, up to 5-3/4" deep, 4'-0" x 7'-0" single	G		15	1.067		183	58.50		241.50	294
5440	8'-0" wide, double	G		12	1.333		230	73		303	370
5800	6-3/4" deep, 7'-0" high, 4'-0" wide, single	G		15	1.067		166	58.50		224.50	275
5840	8'-0" wide, double	G		12	1.333		345	73		418	495
6200	8-3/4" deep, 4'-0" x 7'-0" single	G		15	1.067		271	58.50		329.50	390
6240	8'-0" wide, double	G		12	1.333		395	73		468	550
6300	For "A" label use same price as "B" label										
6400	For baked enamel finish, add						30%	15%			
6500	For galvanizing, add						20%				
6600	For hospital stop, add					Ea.	253			253	278
6620	For hospital stop, stainless steel, add					"	139			139	153
7900	Transom lite frames, fixed, add		2 Carp	155	.103	S.F.	53	5.65		58.65	67.50
8000	Movable, add		"	130	.123	"	71.50	6.75		78.25	89
9000	Minimum labor/equipment charge		1 Carp	4	2	Job		109		109	174

234

08 12 Metal Frames

08 12 13 – Hollow Metal Frames

08 12 13.15 Borrowed Lites	Crew	Daily Output	Labor-Hours	Unit	Material	2021 Bare Costs Labor	Equipment	Total	Total Incl O&P
0010	**BORROWED LITES**									
0100	Hollow metal section 20 ga., 3-1/2" with glass stop	2 Carp	100	.160	L.F.	21	8.75		29.75	37
0110	3-3/4" with glass stop		100	.160		7.55	8.75		16.30	22.50
0120	4" with glass stop		100	.160		9.70	8.75		18.45	24.50
0130	4-5/8" with glass stop		100	.160		10.05	8.75		18.80	25
0140	4-7/8" with glass stop		100	.160		10.25	8.75		19	25.50
0150	5" with glass stop		100	.160		11.15	8.75		19.90	26
0160	5-3/8" with glass stop		100	.160		13.15	8.75		21.90	28.50
0300	Hollow metal section 18 ga., 3-1/2" with glass stop		80	.200		10.35	10.95		21.30	29
0310	3-3/4" with glass stop		80	.200		10.55	10.95		21.50	29
0320	4" with glass stop		80	.200		14.65	10.95		25.60	33.50
0330	4-5/8" with glass stop		80	.200		11.25	10.95		22.20	30
0340	4-7/8" with glass stop		80	.200		11.50	10.95		22.45	30
0350	5" with glass stop		80	.200		11.65	10.95		22.60	30
0360	5-3/8" with glass stop		80	.200		11.60	10.95		22.55	30
0370	6-5/8" with glass stop		80	.200		12.95	10.95		23.90	31.50
0380	6-7/8" with glass stop		80	.200		13.15	10.95		24.10	32
0390	7-1/4" with glass stop		80	.200		13.70	10.95		24.65	32.50
0500	Mullion section 18 ga., 3-1/2" with double stop		50	.320		24	17.50		41.50	54.50
0510	3-3/4" with double stop		50	.320		24.50	17.50		42	55
0530	4-5/8" with double stop		50	.320		26.50	17.50		44	57
0540	4-7/8" with double stop		50	.320		27	17.50		44.50	57.50
0550	5" with double stop		50	.320		27	17.50		44.50	57.50
0560	5-3/8" with double stop		50	.320		27.50	17.50		45	58.50
0570	6-5/8" with double stop		50	.320		30	17.50		47.50	61
0580	6-7/8" with double stop		50	.320		30.50	17.50		48	62
0590	7-1/4" with double stop		50	.320		31	17.50		48.50	62.50
1000	Assembled frame 2'-0" x 2'-0" 16 ga., 3-3/4" with glass stop		10	1.600	Ea.	227	87.50		314.50	390
1010	3'-0" x 2'-0"		10	1.600		227	87.50		314.50	390
1020	4'-0" x 2'-0"		10	1.600		228	87.50		315.50	390
1030	5'-0" x 2'-0"		10	1.600		300	87.50		387.50	470
1040	6'-0" x 2'-0"		8	2		340	109		449	550
1050	7'-0" x 2'-0"		8	2		340	109		449	550
1060	8'-0" x 2'-0"		8	2		340	109		449	550
1100	2'-0" x 7'-0"		8	2		340	109		449	550
1110	3'-0" x 7'-0"		8	2		340	109		449	550
1120	4'-0" x 7'-0"		8	2		340	109		449	550
1130	5'-0" x 7'-0"		8	2		340	109		449	550
1140	6'-0" x 7'-0"		6	2.667		370	146		516	635
1150	7'-0" x 7'-0"		6	2.667		370	146		516	635
1160	8'-0" x 7'-0"		6	2.667		370	146		516	635
1200	Assembled frame 2'-0" x 2'-0" 16 ga., 4-7/8" with glass stop		10	1.600		228	87.50		315.50	390
1210	3'-0" x 2'-0"		10	1.600		228	87.50		315.50	390
1220	4'-0" x 2'-0"		8	2		228	109		337	425
1230	5'-0" x 2'-0"		8	2		340	109		449	550
1240	6'-0" x 2'-0"		8	2		340	109		449	550
1250	7'-0" x 2'-0"		8	2		340	109		449	550
1260	8'-0" x 2'-0"		8	2		340	109		449	550
1300	2'-0" x 7'-0"		8	2		340	109		449	550
1310	3'-0" x 7'-0"		8	2		340	109		449	550
1320	4'-0" x 7'-0"		8	2		340	109		449	550
1330	5'-0" x 7'-0"		6	2.667		340	146		486	605

08 12 13 – Hollow Metal Frames

08 12 13.15 Borrowed Lites

		Crew	Daily Output	Labor-Hours	Unit	Material	2021 Bare Costs Labor	Equipment	Total	Total Incl O&P
1340	6'-0" x 7'-0"	2 Carp	6	2.667	Ea.	370	146		516	635
1350	7'-0" x 7'-0"		6	2.667		325	146		471	590
1360	8'-0" x 7'-0"		6	2.667		325	146		471	590
1400	Assembled frame 2'-0" x 3'-0" 16 ga., 6-1/8" with glass stop		6	2.667		229	146		375	485
1410	3'-0" x 3'-0"		8	2		228	109		337	425
1420	4'-0" x 3'-0"		8	2		228	109		337	425
1430	5'-0" x 3'-0"		6	2.667		340	146		486	605
1440	6'-0" x 3'-0"		6	2.667		330	146		476	590
1450	7'-0" x 3'-0"		6	2.667		340	146		486	605
1460	8'-0" x 3'-0"		10	1.600		340	87.50		427.50	515

08 12 13.25 Channel Metal Frames

			Crew	Daily Output	Labor-Hours	Unit	Material	2021 Bare Costs Labor	Equipment	Total	Total Incl O&P
0010	**CHANNEL METAL FRAMES**										
0020	Steel channels with anchors and bar stops										
0100	6" channel @ 8.2#/L.F., 3' x 7' door, weighs 150#	G	E-4	13	2.462	Ea.	237	150	11.45	398.45	515
0200	8" channel @ 11.5#/L.F., 6' x 8' door, weighs 275#	G		9	3.556		435	216	16.55	667.55	850
0300	8' x 12' door, weighs 400#	G		6.50	4.923		635	299	23	957	1,200
0400	10" channel @ 15.3#/L.F., 10' x 10' door, weighs 500#	G		6	5.333		790	325	25	1,140	1,425
0500	12' x 12' door, weighs 600#	G		5.50	5.818		950	355	27	1,332	1,650
0600	12" channel @ 20.7#/L.F., 12' x 12' door, weighs 825#	G		4.50	7.111		1,300	430	33	1,763	2,150
0700	12' x 16' door, weighs 1000#	G		4	8		1,575	485	37	2,097	2,575
0800	For frames without bar stops, light sections, deduct						15%				
0900	Heavy sections, deduct						10%				
9000	Minimum labor/equipment charge		E-4	4	8	Job		485	37	522	825

08 13 13 – Hollow Metal Doors

08 13 13.13 Standard Hollow Metal Doors

			Crew	Daily Output	Labor-Hours	Unit	Material	2021 Bare Costs Labor	Equipment	Total	Total Incl O&P
0010	**STANDARD HOLLOW METAL DOORS**	R081313-20									
0015	Flush, full panel, hollow core										
0017	When noted doors are prepared but do not include glass or louvers										
0020	1-3/8" thick, 20 ga., 2'-0" x 6'-8"	G	2 Carp	20	.800	Ea.	390	44		434	500
0040	2'-8" x 6'-8"	G		18	.889		360	48.50		408.50	480
0060	3'-0" x 6'-8"	G		17	.941		400	51.50		451.50	520
0100	3'-0" x 7'-0"			17	.941		405	51.50		456.50	525
0120	For vision lite, add						108			108	119
0140	For narrow lite, add						114			114	126
0320	Half glass, 20 ga., 2'-0" x 6'-8"	G	2 Carp	20	.800		775	44		819	920
0340	2'-8" x 6'-8"	G		18	.889		670	48.50		718.50	820
0360	3'-0" x 6'-8"	G		17	.941		610	51.50		661.50	750
0400	3'-0" x 7'-0"	G		17	.941		475	51.50		526.50	605
0410	1-3/8" thick, 18 ga., 2'-0" x 6'-8"	G		20	.800		525	44		569	645
0420	3'-0" x 6'-8"	G		17	.941		530	51.50		581.50	660
0425	3'-0" x 7'-0"	G		17	.941		470	51.50		521.50	595
0450	For vision lite, add					Ea.	108			108	119
0452	For narrow lite, add						114			114	126
0460	Half glass, 18 ga., 2'-0" x 6'-8"	G	2 Carp	20	.800		710	44		754	850
0465	2'-8" x 6'-8"	G		18	.889		735	48.50		783.50	890
0470	3'-0" x 6'-8"	G		17	.941		720	51.50		771.50	870
0475	3'-0" x 7'-0"	G		17	.941		595	51.50		646.50	735
0500	Hollow core, 1-3/4" thick, full panel, 20 ga., 2'-8" x 6'-8"			18	.889		450	48.50		498.50	570
0520	3'-0" x 6'-8"	G		17	.941		435	51.50		486.50	560

08 13 13 – Hollow Metal Doors

	08 13 13.13 Standard Hollow Metal Doors		Crew	Daily Output	Labor-Hours	Unit	Material	2021 Bare Costs Labor	Equipment	Total	Total Incl O&P
0640	3'-0" x 7'-0"	G	2 Carp	17	.941	Ea.	470	51.50		521.50	595
0680	4'-0" x 7'-0"	G		15	1.067		565	58.50		623.50	715
0700	4'-0" x 8'-0"	G		13	1.231		680	67.50		747.50	855
1000	18 ga., 2'-8" x 6'-8"	G		17	.941		485	51.50		536.50	610
1020	3'-0" x 6'-8"	G		16	1		495	54.50		549.50	630
1120	3'-0" x 7'-0"	G		17	.941		495	51.50		546.50	625
1150	3'-6" x 7'-0"		1 Carp	14	.571		615	31.50		646.50	730
1180	4'-0" x 7'-0"	G	2 Carp	14	1.143		695	62.50		757.50	865
1200	4'-0" x 8'-0"	G	"	17	.941		850	51.50		901.50	1,025
1212	For vision lite, add						108			108	119
1214	For narrow lite, add						114			114	126
1230	Half glass, 20 ga., 2'-8" x 6'-8"	G	2 Carp	20	.800		575	44		619	705
1240	3'-0" x 6'-8"	G		18	.889		545	48.50		593.50	680
1260	3'-0" x 7'-0"	G		18	.889		570	48.50		618.50	710
1280	Embossed panel, 1-3/4" thick, poly core, 20 ga., 3'-0" x 7'-0"	G		18	.889		730	48.50		778.50	880
1290	Half glass, 1-3/4" thick, poly core, 20 ga., 3'-0" x 7'-0"	G		18	.889		665	48.50		713.50	810
1320	18 ga., 2'-8" x 6'-8"	G		18	.889		615	48.50		663.50	760
1340	3'-0" x 6'-8"	G		17	.941		610	51.50		661.50	750
1360	3'-0" x 7'-0"	G		17	.941		620	51.50		671.50	765
1380	4'-0" x 7'-0"	G		15	1.067		785	58.50		843.50	960
1400	4'-0" x 8'-0"	G		14	1.143		845	62.50		907.50	1,025
1500	Flush full panel, 16 ga., steel hollow core										
1520	2'-0" x 6'-8"	G	2 Carp	20	.800	Ea.	645	44		689	780
1530	2'-8" x 6'-8"	G		20	.800		645	44		689	780
1540	3'-0" x 6'-8"	G		20	.800		645	44		689	780
1560	2'-8" x 7'-0"	G		18	.889		680	48.50		728.50	825
1570	3'-0" x 7'-0"	G		18	.889		680	48.50		728.50	825
1580	3'-6" x 7'-0"	G		18	.889		785	48.50		833.50	940
1590	4'-0" x 7'-0"	G		18	.889		880	48.50		928.50	1,050
1600	2'-8" x 8'-0"	G		18	.889		815	48.50		863.50	975
1620	3'-0" x 8'-0"	G		18	.889		890	48.50		938.50	1,050
1630	3'-6" x 8'-0"	G		18	.889		955	48.50		1,003.50	1,125
1640	4'-0" x 8'-0"	G		18	.889		1,025	48.50		1,073.50	1,200
1650	1-13/16", 14 ga., 2'-8" x 7'-0"	G	2 Carp	10	1.600	Ea.	725	87.50		812.50	940
1670	3'-0" x 7'-0"	G		10	1.600		725	87.50		812.50	940
1690	3'-6" x 7'-0"	G		10	1.600		810	87.50		897.50	1,025
1700	4'-0" x 7'-0"	G		10	1.600		875	87.50		962.50	1,100
1720	Insulated, 1-3/4" thick, full panel, 18 ga., 3'-0" x 6'-8"	G		15	1.067		535	58.50		593.50	685
1740	2'-8" x 7'-0"	G		16	1		490	54.50		544.50	625
1760	3'-0" x 7'-0"	G		15	1.067		560	58.50		618.50	710
1800	4'-0" x 8'-0"	G		13	1.231		785	67.50		852.50	965
1820	Half glass, 18 ga., 3'-0" x 6'-8"	G		16	1		625	54.50		679.50	770
1840	2'-8" x 7'-0"	G		17	.941		645	51.50		696.50	790
1860	3'-0" x 7'-0"	G		16	1		690	54.50		744.50	845
1900	4'-0" x 8'-0"	G		14	1.143		595	62.50		657.50	755
2000	For vision lite, add						108			108	119
2010	For narrow lite, add						114			114	126
8100	For bottom louver, add						243			243	267
8110	For baked enamel finish, add						30%	15%			
8120	For galvanizing, add						20%				
8190	Commercial door, flush steel, fire door see section 08 13 13 15										
8270	For dutch door shelf, add to standard door					Ea.	300			300	330
9000	Minimum labor/equipment charge		1 Carp	4	2	Job		109		109	174

237

08 13 13 – Hollow Metal Doors

08 13 13.15 Metal Fire Doors

		Crew	Daily Output	Labor-Hours	Unit	Material	2021 Bare Costs Labor	2021 Bare Costs Equipment	Total	Total Incl O&P
0010	**METAL FIRE DOORS** R081313-20									
0015	Steel, flush, "B" label, 90 minutes									
0020	Full panel, 20 ga., 2'-0" x 6'-8"	2 Carp	20	.800	Ea.	420	44		464	535
0040	2'-8" x 6'-8"		18	.889		420	48.50		468.50	545
0060	3'-0" x 6'-8"		17	.941		425	51.50		476.50	545
0080	3'-0" x 7'-0"		17	.941		440	51.50		491.50	560
0140	18 ga., 3'-0" x 6'-8"		16	1		585	54.50		639.50	725
0160	2'-8" x 7'-0"		17	.941		615	51.50		666.50	760
0180	3'-0" x 7'-0"		16	1		510	54.50		564.50	650
0200	4'-0" x 7'-0"		15	1.067		630	58.50		688.50	785
0220	For "A" label, 3 hour, 18 ga., use same price as "B" label									
0240	For vision lite, add				Ea.	165			165	182
0300	Full panel, 16 ga., 2'-0" x 6'-8"	2 Carp	20	.800		615	44		659	750
0310	2'-8" x 6'-8"		18	.889		635	48.50		683.50	780
0320	3'-0" x 6'-8"		17	.941		460	51.50		511.50	585
0350	2'-8" x 7'-0"		17	.941		660	51.50		711.50	805
0360	3'-0" x 7'-0"		16	1		660	54.50		714.50	810
0370	4'-0" x 7'-0"		15	1.067		835	58.50		893.50	1,000
0520	Flush, "B" label, 90 minutes, egress core, 20 ga., 2'-0" x 6'-8"		18	.889		685	48.50		733.50	830
0540	2'-8" x 6'-8"		17	.941		685	51.50		736.50	835
0560	3'-0" x 6'-8"	2 Carp	16	1	Ea.	690	54.50		744.50	840
0580	3'-0" x 7'-0"		16	1		715	54.50		769.50	875
0640	Flush, "A" label, 3 hour, egress core, 18 ga., 3'-0" x 6'-8"		15	1.067		745	58.50		803.50	915
0660	2'-8" x 7'-0"		16	1		780	54.50		834.50	945
0680	3'-0" x 7'-0"		15	1.067		415	58.50		473.50	550
0700	4'-0" x 7'-0"		14	1.143		950	62.50		1,012.50	1,150
9000	Minimum labor/equipment charge	1 Carp	4	2	Job		109		109	174

08 13 13.20 Residential Steel Doors

			Crew	Daily Output	Labor-Hours	Unit	Material	2021 Bare Costs Labor	2021 Bare Costs Equipment	Total	Total Incl O&P
0010	**RESIDENTIAL STEEL DOORS**										
0020	Prehung, insulated, exterior										
0030	Embossed, full panel, 2'-8" x 6'-8"	G	2 Carp	17	.941	Ea.	410	51.50		461.50	530
0040	3'-0" x 6'-8"	G		15	1.067		272	58.50		330.50	390
0060	3'-0" x 7'-0"	G		15	1.067		360	58.50		418.50	490
0070	5'-4" x 6'-8", double	G		8	2		615	109		724	855
0220	Half glass, 2'-8" x 6'-8"	G		17	.941		340	51.50		391.50	455
0240	3'-0" x 6'-8"	G		16	1		340	54.50		394.50	460
0260	3'-0" x 7'-0"	G		16	1		410	54.50		464.50	535
0270	5'-4" x 6'-8", double	G		8	2		975	109		1,084	1,250
1320	Flush face, full panel, 2'-8" x 6'-8"	G		16	1		350	54.50		404.50	475
1340	3'-0" x 6'-8"	G		15	1.067		350	58.50		408.50	485
1360	3'-0" x 7'-0"	G		15	1.067		315	58.50		373.50	440
1380	5'-4" x 6'-8", double	G		8	2		840	109		949	1,100
1420	Half glass, 2'-8" x 6'-8"	G		17	.941		350	51.50		401.50	465
1440	3'-0" x 6'-8"	G		16	1		350	54.50		404.50	470
1460	3'-0" x 7'-0"	G		16	1		470	54.50		524.50	605
1480	5'-4" x 6'-8", double	G		8	2		680	109		789	925
1500	Sidelight, full lite, 1'-0" x 6'-8" with grille	G					410			410	450
1510	1'-0" x 6'-8", low E	G					410			410	450
1520	1'-0" x 6'-8", half lite	G					310			310	340
1530	1'-0" x 6'-8", half lite, low E	G					300			300	330
2300	Interior, residential, closet, bi-fold, 2'-0" x 6'-8"	G	2 Carp	16	1		273	54.50		327.50	385
2330	3'-0" wide	G		16	1		284	54.50		338.50	400

08 13 Metal Doors

08 13 13 – Hollow Metal Doors

08 13 13.20 Residential Steel Doors

		Crew	Daily Output	Labor-Hours	Unit	Material	2021 Bare Costs Labor	Equipment	Total	Total Incl O&P
2360	4'-0" wide **G**	2 Carp	15	1.067	Ea.	295	58.50		353.50	420
2400	5'-0" wide **G**		14	1.143		375	62.50		437.50	510
2420	6'-0" wide **G**		13	1.231		297	67.50		364.50	430
9000	Minimum labor/equipment charge	1 Carp	4	2	Job		109		109	174

08 13 13.25 Doors Hollow Metal

		Crew	Daily Output	Labor-Hours	Unit	Material	2021 Bare Costs Labor	Equipment	Total	Total Incl O&P
0010	**DOORS HOLLOW METAL**									
0500	Exterior, commercial, flush, 20 ga., 1-3/4" x 7'-0" x 2'-6" wide **G**	2 Carp	15	1.067	Ea.	505	58.50		563.50	650
0530	2'-8" wide **G**		15	1.067		505	58.50		563.50	650
0560	3'-0" wide **G**		14	1.143		505	62.50		567.50	655
0590	3'-6" wide **G**	2 Carp	14	1.143	Ea.	645	62.50		707.50	805
1000	18 ga., 1-3/4" x 7'-0" x 2'-6" wide **G**		15	1.067		575	58.50		633.50	725
1030	2'-8" wide **G**		15	1.067		575	58.50		633.50	725
1060	3'-0" wide **G**		14	1.143		575	62.50		637.50	730
1500	16 ga., 1-3/4" x 7'-0" x 2'-6" wide **G**		15	1.067		680	58.50		738.50	840
1530	2'-8" wide **G**		15	1.067		680	58.50		738.50	840
1560	3'-0" wide **G**		14	1.143		680	62.50		742.50	845
1590	3'-6" wide **G**		14	1.143		785	62.50		847.50	960
2900	Fire door, "A" label, 18 gauge, 1-3/4" x 2'-6" x 7'-0" **G**		15	1.067		620	58.50		678.50	780
2930	2'-8" wide **G**		15	1.067		620	58.50		678.50	780
2960	3'-0" wide **G**		14	1.143		620	62.50		682.50	785
2990	3'-6" wide **G**		14	1.143		755	62.50		817.50	930
3100	"B" label, 2'-6" wide **G**		15	1.067		550	58.50		608.50	700
3130	2'-8" wide **G**		15	1.067		550	58.50		608.50	700
3160	3'-0" wide **G**		14	1.143		550	62.50		612.50	705

08 13 16 – Aluminum Doors

08 13 16.10 Commercial Aluminum Doors

		Crew	Daily Output	Labor-Hours	Unit	Material	2021 Bare Costs Labor	Equipment	Total	Total Incl O&P
0010	**COMMERCIAL ALUMINUM DOORS**, flush, no glazing									
5000	Flush panel doors, pair of 2'-6" x 7'-0"	2 Sswk	2	8	Pr.	1,250	480		1,730	2,150
5050	3'-0" x 7'-0", single		2.50	6.400	Ea.	650	385		1,035	1,350
5100	Pair of 3'-0" x 7'-0"		2	8	Pr.	1,400	480		1,880	2,325
5150	3'-6" x 7'-0", single		2.50	6.400	Ea.	855	385		1,240	1,575

08 14 Wood Doors

08 14 13 – Carved Wood Doors

08 14 13.10 Types of Wood Doors, Carved

		Crew	Daily Output	Labor-Hours	Unit	Material	2021 Bare Costs Labor	Equipment	Total	Total Incl O&P
0010	**TYPES OF WOOD DOORS, CARVED**									
3000	Solid wood, 1-3/4" thick stile and rail									
3020	Mahogany, 3'-0" x 7'-0", six panel	2 Carp	14	1.143	Ea.	1,275	62.50		1,337.50	1,500
3030	With two lites		10	1.600		3,525	87.50		3,612.50	4,025
3040	3'-6" x 8'-0", six panel		10	1.600		2,125	87.50		2,212.50	2,500
3050	With two lites		8	2		3,050	109		3,159	3,525
3100	Pine, 3'-0" x 7'-0", six panel		14	1.143		680	62.50		742.50	850
3110	With two lites		10	1.600		990	87.50		1,077.50	1,250
3120	3'-6" x 8'-0", six panel		10	1.600		1,150	87.50		1,237.50	1,400
3130	With two lites		8	2		1,750	109		1,859	2,100
3200	Red oak, 3'-0" x 7'-0", six panel		14	1.143		1,350	62.50		1,412.50	1,600
3210	With two lites		10	1.600		2,650	87.50		2,737.50	3,075
3220	3'-6" x 8'-0", six panel		10	1.600		2,775	87.50		2,862.50	3,200
3230	With two lites		8	2		3,600	109		3,709	4,125
4000	Hand carved door, mahogany									

08 14 Wood Doors

08 14 13 – Carved Wood Doors

08 14 13.10 Types of Wood Doors, Carved	Crew	Daily Output	Labor-Hours	Unit	Material	2021 Bare Costs Labor	Equipment	Total	Total Incl O&P
4020	3'-0" x 7'-0", simple design	2 Carp	14	1.143	Ea.	1,875	62.50		1,937.50	2,175
4030	Intricate design		11	1.455		3,725	79.50		3,804.50	4,225
4040	3'-6" x 8'-0", simple design		10	1.600		3,400	87.50		3,487.50	3,900
4050	Intricate design		8	2		3,950	109		4,059	4,525
4400	For custom finish, add					585			585	645
4600	Side light, mahogany, 7'-0" x 1'-6" wide, 4 lites	2 Carp	18	.889		1,200	48.50		1,248.50	1,400
4610	6 lites		14	1.143		3,150	62.50		3,212.50	3,575
4620	8'-0" x 1'-6" wide, 4 lites		14	1.143		2,225	62.50		2,287.50	2,550
4630	6 lites		10	1.600		2,450	87.50		2,537.50	2,850
4640	Side light, oak, 7'-0" x 1'-6" wide, 4 lites		18	.889		1,550	48.50		1,598.50	1,775
4650	6 lites		14	1.143		2,550	62.50		2,612.50	2,900
4660	8'-0" x 1'-6" wide, 4 lites		14	1.143		1,450	62.50		1,512.50	1,700
4670	6 lites		10	1.600		2,550	87.50		2,637.50	2,950

08 14 16 – Flush Wood Doors

08 14 16.09 Smooth Wood Doors

		Crew	Daily Output	Labor-Hours	Unit	Material	2021 Bare Costs Labor	Equipment	Total	Total Incl O&P
0010	**SMOOTH WOOD DOORS**									
0015	Flush, interior, hollow core									
0025	Lauan face, 1-3/8", 3'-0" x 6'-8"	2 Carp	17	.941	Ea.	46.50	51.50		98	133
0030	4'-0" x 6'-8"		16	1		125	54.50		179.50	224
0080	1-3/4", 2'-0" x 6'-8"		17	.941		60	51.50		111.50	148
0085	3'-0" x 6'-8"	1 Carp	16	.500		46.50	27.50		74	94.50
0108	3'-0" x 7'-0"	2 Carp	16	1		168	54.50		222.50	271
0112	Pair of 3'-0" x 7'-0"		9	1.778	Pr.	150	97		247	320
0140	Birch face, 1-3/8", 2'-6" x 6'-8"		17	.941	Ea.	87	51.50		138.50	178
0180	3'-0" x 6'-8"		17	.941		97.50	51.50		149	189
0200	4'-0" x 6'-8"		16	1		158	54.50		212.50	261
0202	1-3/4", 2'-0" x 6'-8"		17	.941		69	51.50		120.50	158
0204	2'-4" x 7'-0"		16	1		108	54.50		162.50	206
0206	2'-6" x 7'-0"		16	1		108	54.50		162.50	206
0210	3'-0" x 7'-0"		16	1		140	54.50		194.50	241
0214	Pair of 3'-0" x 7'-0"		9	1.778	Pr.	273	97		370	455
0220	Oak face, 1-3/8", 2'-0" x 6'-8"		17	.941	Ea.	123	51.50		174.50	217
0280	3'-0" x 6'-8"		17	.941		161	51.50		212.50	259
0300	4'-0" x 6'-8"		16	1		163	54.50		217.50	266
0305	1-3/4", 2'-6" x 6'-8"		17	.941		148	51.50		199.50	245
0310	3'-0" x 7'-0"		16	1		262	54.50		316.50	375
0320	Walnut face, 1-3/8", 2'-0" x 6'-8"		17	.941		229	51.50		280.50	335
0340	2'-6" x 6'-8"		17	.941		214	51.50		265.50	315
0380	3'-0" x 6'-8"		17	.941		223	51.50		274.50	325
0400	4'-0" x 6'-8"		16	1		231	54.50		285.50	340
0430	For 7'-0" high, add					30			30	33
0440	For 8'-0" high, add					49.50			49.50	54.50
0480	For prefinishing, clear, add					63			63	69.50
0500	For prefinishing, stain, add				Ea.	69.50			69.50	76.50
1320	M.D. overlay on hardboard, 1-3/8", 2'-0" x 6'-8"	2 Carp	17	.941		133	51.50		184.50	228
1340	2'-6" x 6'-8"		17	.941		147	51.50		198.50	244
1380	3'-0" x 6'-8"		17	.941		135	51.50		186.50	230
1400	4'-0" x 6'-8"		16	1		224	54.50		278.50	335
1420	For 7'-0" high, add					20.50			20.50	22.50
1440	For 8'-0" high, add					46			46	50.50
1720	H.P. plastic laminate, 1-3/8", 2'-0" x 6'-8"	2 Carp	16	1		298	54.50		352.50	415
1740	2'-6" x 6'-8"		16	1		293	54.50		347.50	410

08 14 16.09 Smooth Wood Doors	Crew	Daily Output	Labor-Hours	Unit	Material	2021 Bare Costs Labor	Equipment	Total	Total Incl O&P
1780	3'-0" x 6'-8"	2 Carp	15	1.067	Ea.	310	58.50		368.50	435
1800	4'-0" x 6'-8"	↓	14	1.143		410	62.50		472.50	550
1820	For 7'-0" high, add					20			20	22.50
1840	For 8'-0" high, add					27.50			27.50	30
2020	Particle core, lauan face, 1-3/8", 2'-6" x 6'-8"	2 Carp	15	1.067		105	58.50		163.50	208
2040	3'-0" x 6'-8"		14	1.143		123	62.50		185.50	235
2080	3'-0" x 7'-0"		13	1.231		133	67.50		200.50	253
2085	4'-0" x 7'-0"		12	1.333		160	73		233	293
2110	1-3/4", 3'-0" x 7'-0"		13	1.231		168	67.50		235.50	291
2120	Birch face, 1-3/8", 2'-6" x 6'-8"		15	1.067		135	58.50		193.50	241
2140	3'-0" x 6'-8"		14	1.143		147	62.50		209.50	262
2180	3'-0" x 7'-0"		13	1.231		154	67.50		221.50	276
2200	4'-0" x 7'-0"		12	1.333		175	73		248	310
2205	1-3/4", 3'-0" x 7'-0"		13	1.231		181	67.50		248.50	305
2220	Oak face, 1-3/8", 2'-6" x 6'-8"		15	1.067		154	58.50		212.50	263
2240	3'-0" x 6'-8"		14	1.143		145	62.50		207.50	259
2280	3'-0" x 7'-0"		13	1.231		161	67.50		228.50	284
2300	4'-0" x 7'-0"		12	1.333		198	73		271	335
2305	1-3/4", 3'-0" x 7'-0"		13	1.231		145	67.50		212.50	266
2320	Walnut face, 1-3/8", 2'-0" x 6'-8"		15	1.067		136	58.50		194.50	243
2340	2'-6" x 6'-8"		14	1.143		153	62.50		215.50	268
2380	3'-0" x 6'-8"		13	1.231		177	67.50		244.50	300
2400	4'-0" x 6'-8"	↓	12	1.333		227	73		300	365
2440	For 8'-0" high, add					49			49	54
2460	For 8'-0" high walnut, add					40			40	44
2720	For prefinishing, clear, add					46.50			46.50	51
2740	For prefinishing, stain, add					59			59	65
3320	M.D. overlay on hardboard, 1-3/8", 2'-6" x 6'-8"	2 Carp	14	1.143		189	62.50		251.50	310
3340	3'-0" x 6'-8"		13	1.231		214	67.50		281.50	340
3380	3'-0" x 7'-0"		12	1.333		216	73		289	355
3400	4'-0" x 7'-0"	↓	10	1.600	↓	292	87.50		379.50	460
3440	For 8'-0" height, add				Ea.	45			45	49.50
3460	For solid wood core, add					76.50			76.50	84
3720	H.P. plastic laminate, 1-3/8", 2'-6" x 6'-8"	2 Carp	13	1.231		186	67.50		253.50	310
3740	3'-0" x 6'-8"		12	1.333		198	73		271	335
3780	3'-0" x 7'-0"		11	1.455		211	79.50		290.50	360
3800	4'-0" x 7'-0"	↓	8	2		475	109		584	700
3840	For 8'-0" height, add					45			45	49.50
3860	For solid wood core, add					45.50			45.50	50
4000	Exterior, flush, solid core, birch, 1-3/4" x 2'-6" x 7'-0"	2 Carp	15	1.067		159	58.50		217.50	268
4020	2'-8" wide		15	1.067		231	58.50		289.50	345
4040	3'-0" wide	↓	14	1.143		181	62.50		243.50	299
4045	3'-0" x 8'-0"	1 Carp	8	1		292	54.50		346.50	405
4100	Oak faced 1-3/4" x 2'-6" x 7'-0"	2 Carp	15	1.067		251	58.50		309.50	370
4120	2'-8" wide		15	1.067		278	58.50		336.50	400
4140	3'-0" wide	↓	14	1.143		264	62.50		326.50	390
4160	Walnut faced, 1-3/4" x 3'-0" x 6'-8"	1 Carp	17	.471		535	25.50		560.50	630
4180	3'-6" wide	"	17	.471		610	25.50		635.50	710
4200	Walnut faced, 1-3/4" x 2'-6" x 7'-0"	2 Carp	15	1.067		355	58.50		413.50	485
4220	2'-8" wide		15	1.067		390	58.50		448.50	520
4240	3'-0" wide	↓	14	1.143		325	62.50		387.50	455
4300	For 6'-8" high door, deduct from 7'-0" door					18.95			18.95	21
5000	Wood doors, for vision lite, add					108			108	119

08 14 Wood Doors

08 14 16 – Flush Wood Doors

08 14 16.09 Smooth Wood Doors	Crew	Daily Output	Labor-Hours	Unit	Material	2021 Bare Costs Labor	2021 Bare Costs Equipment	Total	Total Incl O&P
5010 Wood doors, for narrow lite, add				Ea.	114			114	126
5015 Wood doors, for bottom (or top) louver, add				↓	243			243	267
9000 Minimum labor/equipment charge	1 Carp	4	2	Job		109		109	174

08 14 16.10 Wood Doors Decorator	Crew	Daily Output	Labor-Hours	Unit	Material	2021 Bare Costs Labor	2021 Bare Costs Equipment	Total	Total Incl O&P
0010 **WOOD DOORS DECORATOR**									
1800 Exterior, flush, solid wood core, birch 1-3/4" x 2'-6" x 7'-0"	2 Carp	15	1.067	Ea.	350	58.50		408.50	480
1820 2'-8" wide		15	1.067		300	58.50		358.50	425
1840 3'-0" wide		14	1.143		320	62.50		382.50	450
1900 Oak faced, 1-3/4" x 2'-6" x 7'-0"		15	1.067		232	58.50		290.50	350
1920 2'-8" wide		15	1.067		585	58.50		643.50	740
1940 3'-0" wide		14	1.143		550	62.50		612.50	705
2100 Walnut faced, 1-3/4" x 2'-6" x 7'-0"		15	1.067		475	58.50		533.50	615
2120 2'-8" wide		15	1.067		465	58.50		523.50	605
2140 3'-0" wide	↓	14	1.143	↓	425	62.50		487.50	565

08 14 16.20 Wood Fire Doors	Crew	Daily Output	Labor-Hours	Unit	Material	2021 Bare Costs Labor	2021 Bare Costs Equipment	Total	Total Incl O&P
0010 **WOOD FIRE DOORS**									
0020 Particle core, 7 face plys, "B" label,									
0040 1 hour, birch face, 1-3/4" x 2'-6" x 6'-8"	2 Carp	14	1.143	Ea.	470	62.50		532.50	620
0080 3'-0" x 6'-8"		13	1.231		535	67.50		602.50	695
0090 3'-0" x 7'-0"	↓	12	1.333	↓	500	73		573	665
0100 4'-0" x 7'-0"	2 Carp	12	1.333	Ea.	735	73		808	920
0140 Oak face, 2'-6" x 6'-8"		14	1.143		580	62.50		642.50	740
0180 3'-0" x 6'-8"		13	1.231		530	67.50		597.50	690
0190 3'-0" x 7'-0"		12	1.333		550	73		623	720
0200 4'-0" x 7'-0"		12	1.333		685	73		758	870
0240 Walnut face, 2'-6" x 6'-8"		14	1.143		550	62.50		612.50	705
0280 3'-0" x 6'-8"		13	1.231		555	67.50		622.50	715
0290 3'-0" x 7'-0"		12	1.333		545	73		618	715
0300 4'-0" x 7'-0"		12	1.333		805	73		878	1,000
0440 M.D. overlay on hardboard, 2'-6" x 6'-8"		15	1.067		375	58.50		433.50	505
0480 3'-0" x 6'-8"		14	1.143		415	62.50		477.50	555
0490 3'-0" x 7'-0"		13	1.231		440	67.50		507.50	590
0500 4'-0" x 7'-0"		12	1.333		480	73		553	640
0740 90 minutes, birch face, 1-3/4" x 2'-6" x 6'-8"		14	1.143		345	62.50		407.50	480
0780 3'-0" x 6'-8"		13	1.231		440	67.50		507.50	590
0790 3'-0" x 7'-0"		12	1.333		460	73		533	620
0800 4'-0" x 7'-0"		12	1.333		605	73		678	785
0840 Oak face, 2'-6" x 6'-8"		14	1.143		410	62.50		472.50	550
0880 3'-0" x 6'-8"		13	1.231		455	67.50		522.50	605
0890 3'-0" x 7'-0"		12	1.333		490	73		563	650
0900 4'-0" x 7'-0"		12	1.333		615	73		688	790
0940 Walnut face, 2'-6" x 6'-8"		14	1.143		525	62.50		587.50	675
0980 3'-0" x 6'-8"		13	1.231		420	67.50		487.50	565
0990 3'-0" x 7'-0"		12	1.333		580	73		653	755
1000 4'-0" x 7'-0"		12	1.333		705	73		778	890
1140 M.D. overlay on hardboard, 2'-6" x 6'-8"		15	1.067		460	58.50		518.50	600
1180 3'-0" x 6'-8"		14	1.143		450	62.50		512.50	595
1190 3'-0" x 7'-0"		13	1.231		485	67.50		552.50	640
1200 4'-0" x 7'-0"	↓	12	1.333	↓	565	73		638	735
1240 For 8'-0" height, add					87.50			87.50	96.50
1260 For 8'-0" height walnut, add					97			97	107
2200 Custom architectural "B" label, flush, 1-3/4" thick, birch,									

08 14 Wood Doors

08 14 16 – Flush Wood Doors

08 14 16.20 Wood Fire Doors

		Crew	Daily Output	Labor-Hours	Unit	Material	2021 Bare Costs Labor	Equipment	Total	Total Incl O&P
2210	Solid core									
2220	2'-6" x 7'-0"	2 Carp	15	1.067	Ea.	430	58.50		488.50	570
2260	3'-0" x 7'-0"		14	1.143		355	62.50		417.50	490
2300	4'-0" x 7'-0"		13	1.231		570	67.50		637.50	730
2420	4'-0" x 8'-0"		11	1.455		570	79.50		649.50	750
2480	For oak veneer, add					50%				
2500	For walnut veneer, add					75%				
9000	Minimum labor/equipment charge	1 Carp	4	2	Job		109		109	174

08 14 33 – Stile and Rail Wood Doors

08 14 33.10 Wood Doors Paneled

		Crew	Daily Output	Labor-Hours	Unit	Material	2021 Bare Costs Labor	Equipment	Total	Total Incl O&P
0010	**WOOD DOORS PANELED**									
0020	Interior, six panel, hollow core, 1-3/8" thick									
0040	Molded hardboard, 2'-0" x 6'-8"	2 Carp	17	.941	Ea.	71	51.50		122.50	161
0060	2'-6" x 6'-8"		17	.941		70	51.50		121.50	159
0070	2'-8" x 6'-8"		17	.941		73	51.50		124.50	163
0080	3'-0" x 6'-8"		17	.941		84.50	51.50		136	175
0140	Embossed print, molded hardboard, 2'-0" x 6'-8"		17	.941		70	51.50		121.50	159
0160	2'-6" x 6'-8"		17	.941		70	51.50		121.50	159
0180	3'-0" x 6'-8"		17	.941		84.50	51.50		136	175
0540	Six panel, solid, 1-3/8" thick, pine, 2'-0" x 6'-8"		15	1.067		169	58.50		227.50	279
0560	2'-6" x 6'-8"		14	1.143		177	62.50		239.50	295
0580	3'-0" x 6'-8"		13	1.231		146	67.50		213.50	268
1020	Two panel, bored rail, solid, 1-3/8" thick, pine, 1'-6" x 6'-8"		16	1		258	54.50		312.50	370
1040	2'-0" x 6'-8"		15	1.067		350	58.50		408.50	475
1060	2'-6" x 6'-8"		14	1.143		385	62.50		447.50	520
1340	Two panel, solid, 1-3/8" thick, fir, 2'-0" x 6'-8"		15	1.067		171	58.50		229.50	281
1360	2'-6" x 6'-8"		14	1.143		230	62.50		292.50	355
1380	3'-0" x 6'-8"		13	1.231		415	67.50		482.50	565
1740	Five panel, solid, 1-3/8" thick, fir, 2'-0" x 6'-8"		15	1.067		310	58.50		368.50	435
1760	2'-6" x 6'-8"		14	1.143		400	62.50		462.50	545
1780	3'-0" x 6'-8"		13	1.231		400	67.50		467.50	550
4190	Exterior, Knotty pine, paneled, 1-3/4", 3'-0" x 6'-8"		16	1		590	54.50		644.50	730
4195	Double 1-3/4", 3'-0" x 6'-8"		16	1		1,175	54.50		1,229.50	1,375
4200	Ash, paneled, 1-3/4", 3'-0" x 6'-8"		16	1		445	54.50		499.50	575
4205	Double 1-3/4", 3'-0" x 6'-8"		16	1		890	54.50		944.50	1,075
4210	Cherry, paneled, 1-3/4", 3'-0" x 6'-8"		16	1		1,200	54.50		1,254.50	1,375
4215	Double 1-3/4", 3'-0" x 6'-8"		16	1		2,375	54.50		2,429.50	2,700
4230	Ash, paneled, 1-3/4", 3'-0" x 8'-0"		16	1		700	54.50		754.50	855
4235	Double 1-3/4", 3'-0" x 8'-0"		16	1		1,400	54.50		1,454.50	1,625
4240	Hard maple, paneled, 1-3/4", 3'-0" x 8'-0"		16	1		820	54.50		874.50	985
4245	Double 1-3/4", 3'-0" x 8'-0"		16	1		1,650	54.50		1,704.50	1,875
4250	Cherry, paneled, 1-3/4", 3'-0" x 8'-0"		16	1		735	54.50		789.50	890
4255	Double 1-3/4", 3'-0" x 8'-0"		16	1		1,475	54.50		1,529.50	1,675
9000	Minimum labor/equipment charge	1 Carp	4	2	Job		109		109	174

08 14 33.20 Wood Doors Residential

		Crew	Daily Output	Labor-Hours	Unit	Material	2021 Bare Costs Labor	Equipment	Total	Total Incl O&P
0010	**WOOD DOORS RESIDENTIAL**									
0200	Exterior, combination storm & screen, pine									
0260	2'-8" wide	2 Carp	10	1.600	Ea.	325	87.50		412.50	500
0280	3'-0" wide	2 Carp	9	1.778	Ea.	345	97		442	530
0300	7'-1" x 3'-0" wide		9	1.778		365	97		462	555
0400	Full lite, 6'-9" x 2'-6" wide		11	1.455		350	79.50		429.50	510
0420	2'-8" wide		10	1.600		340	87.50		427.50	515

08 14 33.20 Wood Doors Residential	Crew	Daily Output	Labor-Hours	Unit	Material	2021 Bare Costs Labor	Equipment	Total	Total Incl O&P
0440	3'-0" wide	2 Carp	9	1.778	Ea.	350	97		447	540
0500	7'-1" x 3'-0" wide		9	1.778		380	97		477	570
0700	Dutch door, pine, 1-3/4" x 2'-8" x 6'-8", 6 panel		12	1.333		620	73		693	800
0720	Half glass		10	1.600		1,050	87.50		1,137.50	1,300
0800	3'-0" wide, 6 panel		12	1.333		655	73		728	835
0820	Half glass		10	1.600		1,025	87.50		1,112.50	1,300
1000	Entrance door, colonial, 1-3/4" x 6'-8" x 2'-8" wide		16	1		635	54.50		689.50	780
1020	6 panel pine, 3'-0" wide		15	1.067		615	58.50		673.50	775
1100	8 panel pine, 2'-8" wide		16	1		740	54.50		794.50	900
1120	3'-0" wide		15	1.067		680	58.50		738.50	845
1200	For tempered safety glass lites (min. of 2), add					85.50			85.50	94
1300	Flush, birch, solid core, 1-3/4" x 6'-8" x 2'-8" wide	2 Carp	16	1		261	54.50		315.50	375
1320	3'-0" wide		15	1.067		154	58.50		212.50	263
1350	7'-0" x 2'-8" wide		16	1		141	54.50		195.50	242
1360	3'-0" wide		15	1.067		171	58.50		229.50	281
1380	For tempered safety glass lites, add					120			120	132
1420	6'-8" x 3'-0" wide, fir	2 Carp	16	1		485	54.50		539.50	620
1720	Carved mahogany 3'-0" x 6'-8"		15	1.067		1,700	58.50		1,758.50	1,975
1800	Lauan, solid core, 1-3/4" x 7'-0" x 2'-4" wide		16	1		147	54.50		201.50	249
1810	2'-6" wide		15	1.067		155	58.50		213.50	263
1820	2'-8" wide		9	1.778		192	97		289	365
1830	3'-0" wide		16	1		209	54.50		263.50	315
1840	3'-4" wide		16	1		277	54.50		331.50	390
1850	Pair of 3'-0" wide		15	1.067	Pr.	420	58.50		478.50	555
2700	Interior, closet, bi-fold, w/hardware, no frame or trim incl.									
2720	Flush, birch, 2'-6" x 6'-8"	2 Carp	13	1.231	Ea.	82	67.50		149.50	197
2740	3'-0" wide		13	1.231		77	67.50		144.50	192
2760	4'-0" wide		12	1.333		110	73		183	237
2780	5'-0" wide		11	1.455		122	79.50		201.50	261
2800	6'-0" wide		10	1.600		138	87.50		225.50	291
2820	Flush, hardboard, primed, 6'-8" x 2'-6" wide		13	1.231		77	67.50		144.50	192
2840	3'-0" wide		13	1.231		95.50	67.50		163	212
2860	4'-0" wide		12	1.333		153	73		226	284
2880	5'-0" wide		11	1.455		197	79.50		276.50	345
2900	6'-0" wide		10	1.600		166	87.50		253.50	320
3000	Raised panel pine, 6'-6" or 6'-8" x 2'-6" wide		13	1.231		182	67.50		249.50	305
3020	3'-0" wide		13	1.231		305	67.50		372.50	440
3040	4'-0" wide		12	1.333		330	73		403	480
3060	5'-0" wide		11	1.455		425	79.50		504.50	590
3080	6'-0" wide		10	1.600		465	87.50		552.50	655
3200	Louvered, pine, 6'-6" or 6'-8" x 2'-6" wide		13	1.231		171	67.50		238.50	295
3220	3'-0" wide		13	1.231		295	67.50		362.50	430
3240	4'-0" wide		12	1.333		256	73		329	395
3260	5'-0" wide		11	1.455		277	79.50		356.50	430
3280	6'-0" wide		10	1.600		340	87.50		427.50	515
4400	Bi-passing closet, incl. hardware and frame, no trim incl.									
4420	Flush, lauan, 6'-8" x 4'-0" wide	2 Carp	12	1.333	Opng.	178	73		251	310
4440	5'-0" wide		11	1.455		196	79.50		275.50	345
4460	6'-0" wide		10	1.600		177	87.50		264.50	335
4600	Flush, birch, 6'-8" x 4'-0" wide		12	1.333		264	73		337	405
4620	5'-0" wide		11	1.455		226	79.50		305.50	375
4640	6'-0" wide		10	1.600		380	87.50		467.50	560
4800	Louvered, pine, 6'-8" x 4'-0" wide		12	1.333		530	73		603	695

244

08 14 33 – Stile and Rail Wood Doors

08 14 33.20 Wood Doors Residential	Crew	Daily Output	Labor-Hours	Unit	Material	2021 Bare Costs Labor	Equipment	Total	Total Incl O&P
4820	5'-0" wide	2 Carp	11	1.455	Opng.	775	79.50		854.50	980
4840	6'-0" wide		10	1.600		790	87.50		877.50	1,000
5000	Paneled, pine, 6'-8" x 4'-0" wide		12	1.333		505	73		578	670
5020	5'-0" wide		11	1.455		795	79.50		874.50	1,000
5040	6'-0" wide		10	1.600		920	87.50		1,007.50	1,175
5042	8'-0" wide		12	1.333		1,000	73		1,073	1,225
6100	Folding accordion, closet, including track and frame									
6120	Vinyl, 2 layer, stock	2 Carp	10	1.600	Ea.	89	87.50		176.50	237
6200	Rigid PVC	"	10	1.600	"	128	87.50		215.50	279
7310	Passage doors, flush, no frame included									
7320	Hardboard, hollow core, 1-3/8" x 6'-8" x 1'-6" wide	2 Carp	18	.889	Ea.	43	48.50		91.50	125
7330	2'-0" wide		18	.889		48	48.50		96.50	131
7340	2'-6" wide		18	.889		54.50	48.50		103	138
7350	2'-8" wide		18	.889		57	48.50		105.50	140
7360	3'-0" wide		17	.941		57	51.50		108.50	145
7420	Lauan, hollow core, 1-3/8" x 6'-8" x 1'-6" wide		18	.889		68	48.50		116.50	153
7440	2'-0" wide		18	.889		59	48.50		107.50	143
7450	2'-4" wide		18	.889		67.50	48.50		116	152
7460	2'-6" wide		18	.889		67.50	48.50		116	152
7480	2'-8" wide		18	.889		70	48.50		118.50	155
7500	3'-0" wide		17	.941		85	51.50		136.50	176
7700	Birch, hollow core, 1-3/8" x 6'-8" x 1'-6" wide		18	.889		81.50	48.50		130	167
7720	2'-0" wide		18	.889		92.50	48.50		141	180
7740	2'-6" wide		18	.889		88	48.50		136.50	174
7760	2'-8" wide		18	.889		99.50	48.50		148	188
7780	3'-0" wide		17	.941		94	51.50		145.50	186
8000	Pine louvered, 1-3/8" x 6'-8" x 1'-6" wide		19	.842		161	46		207	251
8020	2'-0" wide		18	.889		179	48.50		227.50	274
8040	2'-6" wide		18	.889		192	48.50		240.50	289
8060	2'-8" wide		18	.889		219	48.50		267.50	320
8080	3'-0" wide		17	.941		233	51.50		284.50	340
8300	Pine paneled, 1-3/8" x 6'-8" x 1'-6" wide		19	.842		178	46		224	270
8320	2'-0" wide		18	.889		256	48.50		304.50	360
8330	2'-4" wide		18	.889		257	48.50		305.50	360
8340	2'-6" wide		18	.889		278	48.50		326.50	385
8360	2'-8" wide		18	.889		275	48.50		323.50	380
8380	3'-0" wide		17	.941		297	51.50		348.50	405
8804	Pocket door, 6 panel pine, 2'-6" x 6'-8" with frame		10.50	1.524		410	83.50		493.50	585
8814	2'-8" x 6'-8"		10.50	1.524		405	83.50		488.50	580
8824	3'-0" x 6'-8"		10.50	1.524		385	83.50		468.50	560
9000	Passage doors, flush, no frame, birch, solid core, 1-3/8" x 2'-4" x 7'-0"		16	1		133	54.50		187.50	233
9020	2'-8" wide		16	1		140	54.50		194.50	241
9040	3'-0" wide		16	1		167	54.50		221.50	270
9060	3'-4" wide		15	1.067		298	58.50		356.50	420
9080	Pair of 3'-0" wide		9	1.778	Pr.	295	97		392	480
9100	Lauan, solid core, 1-3/8" x 7'-0" x 2'-4" wide		16	1	Ea.	178	54.50		232.50	283
9120	2'-8" wide		16	1		159	54.50		213.50	262
9140	3'-0" wide		16	1		210	54.50		264.50	320
9160	3'-4" wide		15	1.067		217	58.50		275.50	330
9180	Pair of 3'-0" wide		9	1.778	Pr.	425	97		522	625
9200	Hardboard, solid core, 1-3/8" x 7'-0" x 2'-4" wide		16	1	Ea.	176	54.50		230.50	281
9220	2'-8" wide		16	1		180	54.50		234.50	285
9240	3'-0" wide		16	1		185	54.50		239.50	291

08 14 Wood Doors

08 14 33 - Stile and Rail Wood Doors

08 14 33.20 Wood Doors Residential

	Crew	Daily Output	Labor-Hours	Unit	Material	2021 Bare Costs Labor	2021 Bare Costs Equipment	Total	Total Incl O&P
9260	3'-4" wide	2 Carp	15	1.067	Ea.	385	58.50		443.50	520
9900	Minimum labor/equipment charge	1 Carp	4	2	Job		109		109	174

08 14 35 - Torrified Doors

08 14 35.10 Torrified Exterior Doors

	Crew	Daily Output	Labor-Hours	Unit	Material	2021 Bare Costs Labor	2021 Bare Costs Equipment	Total	Total Incl O&P
0010	**TORRIFIED EXTERIOR DOORS**									
0020	Wood doors made from torrified wood, exterior									
0030	All doors require a finish be applied, all glass is insulated									
0040	All doors require pilot holes for all fasteners									
0100	6 panel, paint grade poplar, 1-3/4" x 3'-0" x 6'-8"	2 Carp	12	1.333	Ea.	1,325	73		1,398	1,575
0120	Half glass 3'-0" x 6'-8"	"	12	1.333	"	1,425	73		1,498	1,675
0200	Side lite, full glass, 1-3/4" x 1'-2" x 6'-8"				Ea.	785			785	865
0220	Side lite, half glass, 1-3/4" x 1'-2" x 6'-8"					765			765	840
0300	Raised face, 2 panel, paint grade poplar, 1-3/4" x 3'-0" x 7'-0"	2 Carp	12	1.333		1,550	73		1,623	1,850
0320	Side lite, raised face, half glass, 1-3/4" x 1'-2" x 7'-0"					890			890	980
0500	6 panel, Fir, 1-3/4" x 3'-0" x 6'-8"	2 Carp	12	1.333		1,875	73		1,948	2,175
0520	Half glass 3'-0" x 6'-8"	"	12	1.333		1,700	73		1,773	2,000
0600	Side lite, full glass, 1-3/4" x 1'-2" x 6'-8"					795			795	875
0620	Side lite, half glass, 1-3/4" x 1'-2" x 6'-8"					825			825	905
0700	6 panel, Mahogany, 1-3/4" x 3'-0" x 6'-8"	2 Carp	12	1.333		2,475	73		2,548	2,850
0800	Side lite, full glass, 1-3/4" x 1'-2" x 6'-8"					900			900	990
0820	Side lite, half glass, 1-3/4" x 1'-2" x 6'-8"					910			910	1,000

08 14 40 - Interior Cafe Doors

08 14 40.10 Cafe Style Doors

	Crew	Daily Output	Labor-Hours	Unit	Material	2021 Bare Costs Labor	2021 Bare Costs Equipment	Total	Total Incl O&P
0010	**CAFE STYLE DOORS**									
6520	Interior cafe doors, 2'-6" opening, stock, panel pine	2 Carp	16	1	Ea.	435	54.50		489.50	560
6540	3'-0" opening	"	16	1	"	485	54.50		539.50	620
6550	Louvered pine									
6560	2'-6" opening	2 Carp	16	1	Ea.	345	54.50		399.50	465
8000	3'-0" opening		16	1		310	54.50		364.50	430
8010	2'-6" opening, hardwood		16	1		385	54.50		439.50	505
8020	3'-0" opening		16	1		430	54.50		484.50	560
9000	Minimum labor/equipment charge	1 Carp	4	2	Job		109		109	174

08 16 Composite Doors

08 16 13 - Fiberglass Doors

08 16 13.10 Entrance Doors, Fibrous Glass

		Crew	Daily Output	Labor-Hours	Unit	Material	2021 Bare Costs Labor	2021 Bare Costs Equipment	Total	Total Incl O&P
0010	**ENTRANCE DOORS, FIBROUS GLASS**										
0020	Exterior, fiberglass, door, 2'-8" wide x 6'-8" high	G	2 Carp	15	1.067	Ea.	274	58.50		332.50	395
0040	3'-0" wide x 6'-8" high	G		15	1.067		278	58.50		336.50	400
0060	3'-0" wide x 7'-0" high	G		15	1.067		520	58.50		578.50	670
0080	3'-0" wide x 6'-8" high, with two lites	G		15	1.067		345	58.50		403.50	475
0100	3'-0" wide x 8'-0" high, with two lites	G		15	1.067		545	58.50		603.50	690
0110	Half glass, 3'-0" wide x 6'-8" high	G		15	1.067		480	58.50		538.50	625
0120	3'-0" wide x 6'-8" high, low E	G		15	1.067		525	58.50		583.50	670
0130	3'-0" wide x 8'-0" high	G		15	1.067		610	58.50		668.50	765
0140	3'-0" wide x 8'-0" high, low E	G		15	1.067		685	58.50		743.50	850
0150	Side lights, 1'-0" wide x 6'-8" high	G					282			282	310
0160	1'-0" wide x 6'-8" high, low E	G					278			278	305
0180	1'-0" wide x 6'-8" high, full glass	G					370			370	405
0190	1'-0" wide x 6'-8" high, low E	G					390			390	430

246

08 16 Composite Doors

08 16 14 – French Doors

08 16 14.10 Exterior Doors With Glass Lites

		Crew	Daily Output	Labor-Hours	Unit	Material	2021 Bare Costs Labor	2021 Bare Costs Equipment	Total	Total Incl O&P
0010	**EXTERIOR DOORS WITH GLASS LITES**									
0020	French, Fir, 1-3/4", 3'-0" wide x 6'-8" high	2 Carp	12	1.333	Ea.	510	73		583	675
0025	Double		12	1.333		1,025	73		1,098	1,250
0030	Maple, 1-3/4", 3'-0" wide x 6'-8" high		12	1.333		735	73		808	925
0035	Double		12	1.333		1,475	73		1,548	1,750
0040	Cherry, 1-3/4", 3'-0" wide x 6'-8" high		12	1.333		860	73		933	1,050
0045	Double		12	1.333		1,725	73		1,798	2,025
0100	Mahogany, 1-3/4", 3'-0" wide x 8'-0" high		10	1.600		1,250	87.50		1,337.50	1,525
0105	Double		10	1.600		2,500	87.50		2,587.50	2,900
0110	Fir, 1-3/4", 3'-0" wide x 8'-0" high		10	1.600		550	87.50		637.50	745
0115	Double		10	1.600		1,100	87.50		1,187.50	1,350
0120	Oak, 1-3/4", 3'-0" wide x 8'-0" high		10	1.600		595	87.50		682.50	795
0125	Double	↓	10	1.600	↓	1,200	87.50		1,287.50	1,450

08 17 Integrated Door Opening Assemblies

08 17 13 – Integrated Metal Door Opening Assemblies

08 17 13.10 Hollow Metal Doors and Frames

		Crew	Daily Output	Labor-Hours	Unit	Material	2021 Bare Costs Labor	2021 Bare Costs Equipment	Total	Total Incl O&P
0010	**HOLLOW METAL DOORS AND FRAMES**									
0100	Prehung, flush, 18 ga., 1-3/4" x 6'-8" x 2'-8" wide	1 Carp	16	.500	Ea.	580	27.50		607.50	685
0120	3'-0" wide		15	.533		680	29		709	790
0130	3'-6" wide		13	.615		710	33.50		743.50	840
0140	4'-0" wide		10	.800		785	44		829	930
0300	Double, 1-3/4" x 3'-0" x 6'-8"		5	1.600		870	87.50		957.50	1,100
0350	3'-0" x 8'-0"	↓	4	2	↓	1,875	109		1,984	2,225

08 17 13.20 Stainless Steel Doors and Frames

			Crew	Daily Output	Labor-Hours	Unit	Material	2021 Bare Costs Labor	2021 Bare Costs Equipment	Total	Total Incl O&P
0010	**STAINLESS STEEL DOORS AND FRAMES**										
0020	Stainless steel (304) prehung 24 ga. 2'-6" x 6'-8" door w/16 g frame	G	2 Carp	6	2.667	Ea.	1,925	146		2,071	2,350
0025	2'-8" x 6'-8"	G		6	2.667		1,950	146		2,096	2,375
0030	3'-0" x 6'-8"	G		6	2.667		1,925	146		2,071	2,350
0040	3'-0" x 7'-0"	G		6	2.667		2,000	146		2,146	2,425
0050	4'-0" x 7'-0"	G		5	3.200		3,550	175		3,725	4,200
0100	Stainless steel (316) prehung 24 ga. 2'-6" x 6'-8" door w/16 g frame	G		6	2.667		4,250	146		4,396	4,900
0110	2'-8" x 6'-8"	G		6	2.667		2,300	146		2,446	2,750
0120	3'-0" x 6'-8"	G		6	2.667		4,350	146		4,496	5,025
0150	3'-0" x 7'-0"	G		6	2.667		3,825	146		3,971	4,425
0160	4'-0" x 7'-0"	G		6	2.667		4,575	146		4,721	5,275
0300	Stainless steel (304) prehung 18 ga. 2'-6" x 6'-8" door w/16 g frame	G		6	2.667		3,250	146		3,396	3,800
0310	2'-8" x 6'-8"	G		6	2.667		3,300	146		3,446	3,850
0320	3'-0" x 6'-8"	G		6	2.667		2,250	146		2,396	2,700
0350	3'-0" x 7'-0"	G		6	2.667		2,350	146		2,496	2,825
0360	4'-0" x 7'-0"	G		5	3.200		3,250	175		3,425	3,875
0500	Stainless steel, prehung door, foam core, 14 ga., 3'-0" x 7'-0"	G		5	3.200		4,000	175		4,175	4,675
0600	Stainless steel, prehung double door, foam core, 14 ga., 3'-0" x 7'-0"	G	↓	4	4	↓	6,600	219		6,819	7,625

08 17 23 – Integrated Wood Door Opening Assemblies

08 17 23.10 Pre-Hung Doors

		Crew	Daily Output	Labor-Hours	Unit	Material	2021 Bare Costs Labor	2021 Bare Costs Equipment	Total	Total Incl O&P
0010	**PRE-HUNG DOORS**									
0300	Exterior, wood, comb. storm & screen, 6'-9" x 2'-6" wide	2 Carp	15	1.067	Ea.	263	58.50		321.50	380
0320	2'-8" wide		15	1.067		350	58.50		408.50	480
0340	3'-0" wide		15	1.067		330	58.50		388.50	460
0360	For 7'-0" high door, add	↓			↓	46			46	50.50

08 17 23.10 Pre-Hung Doors		Crew	Daily Output	Labor-Hours	Unit	Material	2021 Bare Costs Labor	2021 Bare Costs Equipment	Total	Total Incl O&P
1600	Entrance door, flush, birch, solid core									
1620	4-5/8" solid jamb, 1-3/4" x 6'-8" x 2'-8" wide	2 Carp	16	1	Ea.	300	54.50		354.50	415
1640	3'-0" wide		16	1		410	54.50		464.50	535
1642	5-5/8" jamb		16	1		345	54.50		399.50	465
1680	For 7'-0" high door, add					23			23	25
2000	Entrance door, colonial, 6 panel pine									
2020	4-5/8" solid jamb, 1-3/4" x 6'-8" x 2'-8" wide	2 Carp	16	1	Ea.	675	54.50		729.50	825
2040	3'-0" wide	"	16	1		710	54.50		764.50	865
2060	For 7'-0" high door, add					57.50			57.50	63
2200	For 5-5/8" solid jamb, add					44			44	48.50
2230	French style, exterior, 1 lite, 1-3/4" x 3'-0" x 6'-8"	1 Carp	14	.571		715	31.50		746.50	840
2235	9 lites	"	14	.571		725	31.50		756.50	850
2245	15 lites	2 Carp	14	1.143		800	62.50		862.50	980
2250	Double, 15 lites, 2'-0" x 6'-8", 4'-0" opening		7	2.286	Pr.	1,300	125		1,425	1,625
2260	2'-6" x 6'-8", 5'-0" opening		7	2.286		1,450	125		1,575	1,800
2280	3'-0" x 6'-8", 6'-0" opening		7	2.286		1,675	125		1,800	2,025
2430	3'-0" x 7'-0", 15 lites		14	1.143	Ea.	800	62.50		862.50	980
2432	Two 3'-0" x 7'-0"		7	2.286	Pr.	1,675	125		1,800	2,025
2435	3'-0" x 8'-0"		14	1.143	Ea.	800	62.50		862.50	980
2437	Two, 3'-0" x 8'-0"		7	2.286	Pr.	1,675	125		1,800	2,025
4000	Interior, passage door, 4-5/8" solid jamb									
4350	Paneled, primed, hollow core, 2'-8" wide	2 Carp	17	.941	Ea.	134	51.50		185.50	229
4360	3'-0" wide		17	.941		192	51.50		243.50	294
4400	Lauan, flush, solid core, 1-3/8" x 6'-8" x 2'-6" wide		17	.941		199	51.50		250.50	300
4420	2'-8" wide		17	.941		199	51.50		250.50	300
4440	3'-0" wide		16	1		220	54.50		274.50	330
4600	Hollow core, 1-3/8" x 6'-8" x 2'-6" wide		17	.941		133	51.50		184.50	228
4620	2'-8" wide		17	.941		134	51.50		185.50	230
4640	3'-0" wide		16	1		140	54.50		194.50	241
4700	For 7'-0" high door, add					43.50			43.50	48
5000	Birch, flush, solid core, 1-3/8" x 6'-8" x 2'-6" wide	2 Carp	17	.941		360	51.50		411.50	480
5020	2'-8" wide	2 Carp	17	.941	Ea.	218	51.50		269.50	320
5040	3'-0" wide		16	1		355	54.50		409.50	480
5200	Hollow core, 1-3/8" x 6'-8" x 2'-6" wide		17	.941		315	51.50		366.50	425
5220	2'-8" wide		17	.941		269	51.50		320.50	380
5240	3'-0" wide		16	1		290	54.50		344.50	405
5280	For 7'-0" high door, add					37			37	40.50
5500	Hardboard paneled, 1-3/8" x 6'-8" x 2'-6" wide	2 Carp	17	.941		156	51.50		207.50	254
5520	2'-8" wide		17	.941		164	51.50		215.50	262
5540	3'-0" wide		16	1		165	54.50		219.50	269
6000	Pine paneled, 1-3/8" x 6'-8" x 2'-6" wide		17	.941		279	51.50		330.50	385
6020	2'-8" wide		17	.941		295	51.50		346.50	405
6040	3'-0" wide		16	1		305	54.50		359.50	420
8200	Birch, flush, solid core, 1-3/4" x 6'-8" x 2'-4" wide	1 Carp	17	.471		217	25.50		242.50	280
8220	2'-6" wide		17	.471		209	25.50		234.50	271
8240	2'-8" wide		17	.471		196	25.50		221.50	257
8260	3'-0" wide		16	.500		204	27.50		231.50	269
8280	3'-6" wide		15	.533		360	29		389	440
8500	Pocket door frame with lauan, flush, hollow core, 1-3/8" x 3'-0" x 6'-8"		17	.471		275	25.50		300.50	340
9000	Minimum labor/equipment charge		4	2	Job		109		109	174

248

08 31 13 – Access Doors and Frames

08 31 13.10 Types of Framed Access Doors	Crew	Daily Output	Labor-Hours	Unit	Material	2021 Bare Costs Labor	2021 Bare Costs Equipment	Total	Total Incl O&P
0010 **TYPES OF FRAMED ACCESS DOORS**									
1000 Fire rated door with lock									
1100 Metal, 12" x 12"	1 Carp	10	.800	Ea.	158	44		202	243
1150 18" x 18"		9	.889		222	48.50		270.50	320
1200 24" x 24"		9	.889		345	48.50		393.50	460
1250 24" x 36"		8	1		345	54.50		399.50	465
1300 24" x 48"		8	1		410	54.50		464.50	535
1350 36" x 36"		7.50	1.067		515	58.50		573.50	660
1400 48" x 48"		7.50	1.067		665	58.50		723.50	830
1600 Stainless steel, 12" x 12"		10	.800		335	44		379	440
1650 18" x 18"		9	.889		370	48.50		418.50	490
1700 24" x 24"		9	.889		545	48.50		593.50	680
1750 24" x 36"		8	1		650	54.50		704.50	800
2000 Flush door for finishing									
2100 Metal 8" x 8"	1 Carp	10	.800	Ea.	38	44		82	112
2150 12" x 12"	"	10	.800	"	38.50	44		82.50	112
3000 Recessed door for acoustic tile									
3100 Metal, 12" x 12"	1 Carp	4.50	1.778	Ea.	48.50	97		145.50	209
3150 12" x 24"	"	4.50	1.778	"	111	97		208	277
3200 24" x 24"	1 Carp	4	2	Ea.	118	109		227	305
3250 24" x 36"	"	4	2	"	163	109		272	355
4000 Recessed door for drywall									
4100 Metal 12" x 12"	1 Carp	6	1.333	Ea.	75.50	73		148.50	199
4150 12" x 24"		5.50	1.455		118	79.50		197.50	257
4200 24" x 36"		5	1.600		189	87.50		276.50	345
6000 Standard door									
6100 Metal, 8" x 8"	1 Carp	10	.800	Ea.	31.50	44		75.50	105
6150 12" x 12"		10	.800		33.50	44		77.50	107
6200 18" x 18"		9	.889		35	48.50		83.50	116
6250 24" x 24"		9	.889		50	48.50		98.50	133
6300 24" x 36"		8	1		91	54.50		145.50	187
6350 36" x 36"		8	1		99	54.50		153.50	196
6500 Stainless steel, 8" x 8"		10	.800		80.50	44		124.50	158
6550 12" x 12"		10	.800		94.50	44		138.50	174
6600 18" x 18"		9	.889		171	48.50		219.50	266
6650 24" x 24"		9	.889		260	48.50		308.50	365
9000 Minimum labor/equipment charge		4	2	Job		109		109	174

08 31 13.20 Bulkhead/Cellar Doors

	Crew	Daily Output	Labor-Hours	Unit	Material	2021 Bare Costs Labor	2021 Bare Costs Equipment	Total	Total Incl O&P
0010 **BULKHEAD/CELLAR DOORS**									
0020 Steel, not incl. sides, 44" x 62"	1 Carp	5.50	1.455	Ea.	720	79.50		799.50	915
0100 52" x 73"		5.10	1.569		900	86		986	1,125
0500 With sides and foundation plates, 57" x 45" x 24"		4.70	1.702		895	93		988	1,125
0600 42" x 49" x 51"		4.30	1.860		610	102		712	830
9000 Minimum labor/equipment charge		2	4	Job		219		219	350

08 31 13.30 Commercial Floor Doors

	Crew	Daily Output	Labor-Hours	Unit	Material	2021 Bare Costs Labor	2021 Bare Costs Equipment	Total	Total Incl O&P
0010 **COMMERCIAL FLOOR DOORS**									
0020 Aluminum tile, steel frame, one leaf, 2' x 2' opng.	2 Sswk	3.50	4.571	Opng.	740	276		1,016	1,250
0050 3'-6" x 3'-6" opening		3.50	4.571		1,575	276		1,851	2,175
0500 Double leaf, 4' x 4' opening		3	5.333		1,600	320		1,920	2,300
0550 5' x 5' opening		3	5.333		2,675	320		2,995	3,450
9000 Minimum labor/equipment charge		2	8	Job		480		480	780

For customer support on your Commercial Renovation Costs with RSMeans Data, call 800.448.8182.

249

08 31 Access Doors and Panels

08 31 13 – Access Doors and Frames

08 31 13.35 Industrial Floor Doors

		Crew	Daily Output	Labor-Hours	Unit	Material	2021 Bare Costs Labor	2021 Bare Costs Equipment	Total	Total Incl O&P
0010	**INDUSTRIAL FLOOR DOORS**									
0020	Steel 300 psf L.L., single leaf, 2' x 2', 175#	2 Sswk	6	2.667	Opng.	835	161		996	1,175
0050	3' x 3' opening, 300#		5.50	2.909		1,400	175		1,575	1,825
0300	Double leaf, 4' x 4' opening, 455#		5	3.200		2,675	193		2,868	3,250
0350	5' x 5' opening, 645#		4.50	3.556		3,175	214		3,389	3,825
1000	Aluminum, 300 psf L.L., single leaf, 2' x 2', 60#		6	2.667		735	161		896	1,075
1050	3' x 3' opening, 100#		5.50	2.909		890	175		1,065	1,275
1500	Double leaf, 4' x 4' opening, 160#		5	3.200		3,200	193		3,393	3,825
1550	5' x 5' opening, 235#		4.50	3.556		4,125	214		4,339	4,900
9000	Minimum labor/equipment charge	2 Sswk	2	8	Job		480		480	780

08 32 Sliding Glass Doors

08 32 13 – Sliding Aluminum-Framed Glass Doors

08 32 13.10 Sliding Aluminum Doors

		Crew	Daily Output	Labor-Hours	Unit	Material	2021 Bare Costs Labor	2021 Bare Costs Equipment	Total	Total Incl O&P
0010	**SLIDING ALUMINUM DOORS**									
0350	Aluminum, 5/8" tempered insulated glass, 6' wide									
0400	Premium	2 Carp	4	4	Ea.	1,550	219		1,769	2,050
0450	Economy		4	4		885	219		1,104	1,325
0500	8' wide, premium		3	5.333		1,775	292		2,067	2,425
0550	Economy		3	5.333		1,575	292		1,867	2,200
0600	12' wide, premium		2.50	6.400		3,100	350		3,450	3,950
0650	Economy		2.50	6.400		1,575	350		1,925	2,275
4000	Aluminum, baked on enamel, temp glass, 6'-8" x 10'-0" wide		4	4		1,200	219		1,419	1,650
4020	Insulating glass, 6'-8" x 6'-0" wide		4	4		980	219		1,199	1,425
4040	8'-0" wide		3	5.333		1,175	292		1,467	1,750
4060	10'-0" wide		2	8		1,325	440		1,765	2,175
4080	Anodized, temp glass, 6'-8" x 6'-0" wide		4	4		480	219		699	875
4100	8'-0" wide		3	5.333		605	292		897	1,125
4120	10'-0" wide		2	8		710	440		1,150	1,475
5000	Aluminum sliding glass door system									
5010	Sliding door 4' wide opening single side	2 Carp	2	8	Ea.	6,550	440		6,990	7,900
5015	8' wide opening single side		2	8		1,750	440		2,190	2,625
5020	Telescoping glass door system, 4' wide opening biparting		2	8		5,625	440		6,065	6,875
5025	8' wide opening biparting		2	8		6,550	440		6,990	7,900
5030	Folding glass door, 4' wide opening biparting		2	8		8,725	440		9,165	10,300
5035	8' wide opening biparting		2	8		5,225	440		5,665	6,450
5040	ICU-CCU sliding telescoping glass door, 4' x 7', single side opening		2	8		3,350	440		3,790	4,375
5045	8' x 7', single side opening		2	8		5,075	440		5,515	6,275
7000	Electric swing door operator and control, single door w/sensors	1 Carp	4	2		2,975	109		3,084	3,425
7005	Double door w/sensors		2	4		5,975	219		6,194	6,900
7010	Electric folding door operator and control, single door w/sensors		4	2		7,025	109		7,134	7,900
7015	Bi-folding door		4	2		4,850	109		4,959	5,500
7020	Electric swing door operator and control, single door		4	2		4,025	109		4,134	4,600

08 32 19 – Sliding Wood-Framed Glass Doors

08 32 19.10 Sliding Wood Doors

		Crew	Daily Output	Labor-Hours	Unit	Material	2021 Bare Costs Labor	2021 Bare Costs Equipment	Total	Total Incl O&P
0010	**SLIDING WOOD DOORS**									
0020	Wood, tempered insul. glass, 6' wide, premium	2 Carp	4	4	Ea.	1,550	219		1,769	2,050
0100	Economy		4	4		1,225	219		1,444	1,700
0150	8' wide, wood, premium		3	5.333		1,975	292		2,267	2,650
0200	Economy		3	5.333		1,450	292		1,742	2,075

08 32 Sliding Glass Doors

08 32 19 – Sliding Wood-Framed Glass Doors

08 32 19.10 Sliding Wood Doors

		Crew	Daily Output	Labor-Hours	Unit	Material	2021 Bare Costs Labor	Equipment	Total	Total Incl O&P
0235	10' wide, wood, premium	2 Carp	2.50	6.400	Ea.	2,900	350		3,250	3,750
0240	Economy		2.50	6.400		2,475	350		2,825	3,275
0250	12' wide, wood, premium		2.50	6.400		3,200	350		3,550	4,075
0300	Economy		2.50	6.400		2,650	350		3,000	3,475
9000	Minimum labor/equipment charge	1 Carp	2	4	Job		219		219	350

08 32 19.15 Sliding Glass Vinyl-Clad Wood Doors

			Crew	Daily Output	Labor-Hours	Unit	Material	2021 Bare Costs Labor	Equipment	Total	Total Incl O&P
0010	**SLIDING GLASS VINYL-CLAD WOOD DOORS**										
0020	Glass, sliding vinyl-clad, insul. glass, 6'-0" x 6'-8"	G	2 Carp	4	4	Opng.	1,625	219		1,844	2,125
0025	6'-0" x 6'-10" high	G		4	4		1,775	219		1,994	2,300
0030	6'-0" x 8'-0" high	G		4	4		2,175	219		2,394	2,750
0050	5'-0" x 6'-8" high	G		4	4		1,700	219		1,919	2,225
0100	8'-0" x 6'-10" high	G		4	4		2,325	219		2,544	2,900
0150	8'-0" x 8'-0" high	G		4	4		2,500	219		2,719	3,100
0500	4 leaf, 9'-0" x 6'-10" high	G		3	5.333		3,350	292		3,642	4,175
0550	9'-0" x 8'-0" high	G		3	5.333		4,025	292		4,317	4,900
0600	12'-0" x 6'-10" high	G		3	5.333		4,375	292		4,667	5,275
0650	12'-0" x 8'-0" high	G		3	5.333		4,375	292		4,667	5,275
9000	Minimum labor/equipment charge		1 Carp	4	2	Job		109		109	174

08 33 Coiling Doors and Grilles

08 33 13 – Coiling Counter Doors

08 33 13.10 Counter Doors, Coiling Type

		Crew	Daily Output	Labor-Hours	Unit	Material	2021 Bare Costs Labor	Equipment	Total	Total Incl O&P
0010	**COUNTER DOORS, COILING TYPE**									
0020	Manual, incl. frame and hardware, galv. stl., 4' roll-up, 6' long	2 Carp	2	8	Opng.	1,300	440		1,740	2,125
0300	Galvanized steel, UL label		1.80	8.889		1,300	485		1,785	2,225
0600	Stainless steel, 4' high roll-up, 6' long		2	8		2,200	440		2,640	3,100
0700	10' long		1.80	8.889		2,625	485		3,110	3,675
2000	Aluminum, 4' high, 4' long		2.20	7.273		1,825	400		2,225	2,625
2020	6' long		2	8		2,450	440		2,890	3,400
2040	8' long		1.90	8.421		2,325	460		2,785	3,275
2060	10' long		1.80	8.889		2,300	485		2,785	3,300
2080	14' long		1.40	11.429		3,550	625		4,175	4,900
2100	6' high, 4' long		2	8		2,500	440		2,940	3,450
2120	6' long		1.60	10		1,825	545		2,370	2,875
2140	10' long		1.40	11.429		2,400	625		3,025	3,625
9000	Minimum labor/equipment charge	1 Carp	2	4	Job		219		219	350

08 33 16 – Coiling Counter Grilles

08 33 16.10 Coiling Grilles

		Crew	Daily Output	Labor-Hours	Unit	Material	2021 Bare Costs Labor	Equipment	Total	Total Incl O&P
0010	**COILING GRILLES**									
2020	Aluminum, manual operated, mill finish	2 Sswk	82	.195	S.F.	30	11.75		41.75	52
2040	Bronze anodized		82	.195	"	55	11.75		66.75	79.50
2060	Steel, manual operated, 10' x 10' high		1	16	Opng.	2,650	965		3,615	4,475
2080	15' x 8' high		.80	20	"	3,525	1,200		4,725	5,825
3000	For safety edge bottom bar, electric, add				L.F.	49.50			49.50	54.50
8000	For motor operation, add	2 Sswk	5	3.200	Opng.	1,525	193		1,718	2,000
9000	Minimum labor/equipment charge	"	1	16	Job		965		965	1,550

For customer support on your Commercial Renovation Costs with RSMeans Data, call 800.448.8182.

251

08 33 Coiling Doors and Grilles

08 33 23 – Overhead Coiling Doors

08 33 23.10 Coiling Service Doors	Crew	Daily Output	Labor-Hours	Unit	Material	2021 Bare Costs Labor	Equipment	Total	Total Incl O&P
0010 **COILING SERVICE DOORS** Steel, manual, 20 ga., incl. hardware									
0050 8' x 8' high	2 Sswk	1.60	10	Ea.	720	605		1,325	1,775
0130 12' x 12' high, standard		1.20	13.333		1,325	805		2,130	2,775
0160 10' x 20' high, standard		.50	32		1,575	1,925		3,500	4,850
2000 Class A fire doors, manual, 20 ga., 8' x 8' high		1.40	11.429		1,825	690		2,515	3,150
2100 10' x 10' high		1.10	14.545		2,525	875		3,400	4,200
2200 20' x 10' high		.80	20		5,150	1,200		6,350	7,625
2300 12' x 12' high		1	16		3,975	965		4,940	5,925
2400 20' x 12' high		.80	20		5,550	1,200		6,750	8,050
2500 14' x 14' high		.60	26.667		4,350	1,600		5,950	7,375
2600 20' x 16' high		.50	32		6,000	1,925		7,925	9,750
2700 10' x 20' high		.40	40		5,475	2,400		7,875	9,925
3000 For 18 ga. doors, add				S.F.	1.55			1.55	1.71
3300 For enamel finish, add				"	2.89			2.89	3.18
3600 For safety edge bottom bar, pneumatic, add				L.F.	28.50			28.50	31.50
4000 For weatherstripping, extruded rubber, jambs, add					19.45			19.45	21.50
4100 Hood, add					9.25			9.25	10.15
4200 Sill, add					9.40			9.40	10.30
4500 Motor operators, to 14' x 14' opening	2 Sswk	5	3.200	Ea.	1,150	193		1,343	1,550
4700 For fire door, additional fusible link, add				"	78.50			78.50	86.50
9000 Minimum labor/equipment charge	2 Sswk	1	16	Job		965		965	1,550

08 34 Special Function Doors

08 34 13 – Cold Storage Doors

08 34 13.10 Doors for Cold Area Storage

	Crew	Daily Output	Labor-Hours	Unit	Material	2021 Bare Costs Labor	Equipment	Total	Total Incl O&P
0010 **DOORS FOR COLD AREA STORAGE**									
0020 Single, 20 ga. galvanized steel									
0300 Horizontal sliding, 5' x 7', manual operation, 3.5" thick	2 Carp	2	8	Ea.	3,050	440		3,490	4,050
0400 4" thick		2	8		3,550	440		3,990	4,600
0500 6" thick		2	8		3,500	440		3,940	4,550
0800 5' x 7', power operation, 2" thick		1.90	8.421		5,050	460		5,510	6,275
0900 4" thick		1.90	8.421		5,825	460		6,285	7,125
1000 6" thick		1.90	8.421		6,625	460		7,085	8,000
1300 9' x 10', manual operation, 2" insulation		1.70	9.412		4,400	515		4,915	5,675
1400 4" insulation		1.70	9.412		4,975	515		5,490	6,300
1500 6" insulation		1.70	9.412		5,025	515		5,540	6,375
1800 Power operation, 2" insulation		1.60	10		6,925	545		7,470	8,500
1900 4" insulation	2 Carp	1.60	10	Ea.	7,900	545		8,445	9,575
2000 6" insulation	"	1.70	9.412	"	8,375	515		8,890	10,000
2300 For stainless steel face, add					25%				
3000 Hinged, lightweight, 3' x 7'-0", 2" thick	2 Carp	2	8	Ea.	1,275	440		1,715	2,100
3050 4" thick		1.90	8.421		1,600	460		2,060	2,475
3300 Polymer doors, 3' x 7'-0"		1.90	8.421		1,375	460		1,835	2,225
3350 6" thick		1.40	11.429		2,375	625		3,000	3,625
3600 Stainless steel, 3' x 7'-0", 4" thick		1.90	8.421		1,775	460		2,235	2,675
3650 6" thick		1.40	11.429		2,925	625		3,550	4,225
3900 Painted, 3' x 7'-0", 4" thick		1.90	8.421		1,350	460		1,810	2,200
3950 6" thick		1.40	11.429		2,400	625		3,025	3,625
5000 Bi-parting, electric operated									
5010 6' x 8' opening, galv. faces, 4" thick for cooler	2 Carp	.80	20	Opng.	7,375	1,100		8,475	9,850

252

08 34 Special Function Doors

08 34 13 – Cold Storage Doors

08 34 13.10 Doors for Cold Area Storage

		Crew	Daily Output	Labor-Hours	Unit	Material	2021 Bare Costs Labor	2021 Bare Costs Equipment	Total	Total Incl O&P
5050	For freezer, 4" thick	2 Carp	.80	20	Opng.	8,100	1,100		9,200	10,700
5300	For door buck framing and door protection, add		2.50	6.400		655	350		1,005	1,275
6000	Galvanized batten door, galvanized hinges, 4' x 7'		2	8		1,925	440		2,365	2,825
6050	6' x 8'		1.80	8.889		2,525	485		3,010	3,550
6500	Fire door, 3 hr., 6' x 8', single slide		.80	20		8,725	1,100		9,825	11,400
6550	Double, bi-parting		.70	22.857		11,200	1,250		12,450	14,300
9000	Minimum labor/equipment charge	1 Carp	2	4	Job		219		219	350

08 34 36 – Darkroom Doors

08 34 36.10 Various Types of Darkroom Doors

		Crew	Daily Output	Labor-Hours	Unit	Material	2021 Bare Costs Labor	2021 Bare Costs Equipment	Total	Total Incl O&P
0010	**VARIOUS TYPES OF DARKROOM DOORS**									
0015	Revolving, standard, 2 way, 36" diameter	2 Carp	3.10	5.161	Opng.	3,500	282		3,782	4,300
0020	41" diameter		3.10	5.161		3,250	282		3,532	4,025
0050	3 way, 51" diameter		1.40	11.429		4,950	625		5,575	6,450
1000	4 way, 49" diameter		1.40	11.429		3,775	625		4,400	5,150
2000	Hinged safety, 2 way, 41" diameter		2.30	6.957		3,925	380		4,305	4,925
2500	3 way, 51" diameter		1.40	11.429		5,150	625		5,775	6,650
3000	Pop out safety, 2 way, 41" diameter		3.10	5.161		3,975	282		4,257	4,825
4000	3 way, 51" diameter		1.40	11.429		5,775	625		6,400	7,350
5000	Wheelchair-type, pop out, 51" diameter		1.40	11.429		5,775	625		6,400	7,350
5020	72" diameter		.90	17.778		10,900	970		11,870	13,600

08 34 53 – Security Doors and Frames

08 34 53.20 Steel Door

		Crew	Daily Output	Labor-Hours	Unit	Material	2021 Bare Costs Labor	2021 Bare Costs Equipment	Total	Total Incl O&P
0010	**STEEL DOOR** flush with ballistic core and welded frame both 14 ga.									
0120	UL 752 Level 3, 1-3/4", 3'-0" x 7'-0"	2 Carp	1.50	10.667	Opng.	2,875	585		3,460	4,100
0125	1-3/4", 3'-6" x 7'-0"		1.50	10.667		2,725	585		3,310	3,900
0130	1-3/4", 4'-0" x 7'-0"		1.20	13.333		2,725	730		3,455	4,125
0150	UL 752 Level 8, 1-3/4", 3'-0" x 7'-0"		1.50	10.667		9,125	585		9,710	10,900
0155	1-3/4", 3'-6" x 7'-0"		1.50	10.667		9,400	585		9,985	11,200
0160	1-3/4", 4'-0" x 7'-0"		1.20	13.333		15,300	730		16,030	18,000
1000	Safe room sliding door and hardware, 1-3/4", 3'-0" x 7'-0" UL 752 Level 3		.50	32		24,100	1,750		25,850	29,400
1050	Safe room swinging door and hardware, 1-3/4", 3'-0" x 7'-0" UL 752 Level 3		.50	32		29,000	1,750		30,750	34,700

08 34 53.30 Wood Ballistic Doors

		Crew	Daily Output	Labor-Hours	Unit	Material	2021 Bare Costs Labor	2021 Bare Costs Equipment	Total	Total Incl O&P
0010	**WOOD BALLISTIC DOORS** with frames and hardware									
0050	Wood, 1-3/4", 3'-0" x 7'-0" UL 752 Level 3	2 Carp	1.50	10.667	Opng.	4,350	585		4,935	5,700

08 34 56 – Security Gates

08 34 56.10 Gates

		Crew	Daily Output	Labor-Hours	Unit	Material	2021 Bare Costs Labor	2021 Bare Costs Equipment	Total	Total Incl O&P
0010	**GATES**									
0015	Driveway gates include mounting hardware									
0500	Wood, security gate, driveway, dual, 10' wide	H-4	.80	25	Opng.	3,300	1,275		4,575	5,700
0505	12' wide		.80	25		3,800	1,275		5,075	6,225
0510	15' wide		.80	25		4,600	1,275		5,875	7,100
0600	Steel, security gate, driveway, single, 10' wide		.80	25		1,825	1,275		3,100	4,050
0605	12' wide		.80	25		1,700	1,275		2,975	3,925
0620	Steel, security gate, driveway, dual, 12' wide		.80	25		1,700	1,275		2,975	3,925
0625	14' wide		.80	25		1,800	1,275		3,075	4,025
0630	16' wide		.80	25		2,925	1,275		4,200	5,275
0700	Aluminum, security gate, driveway, dual, 10' wide		.80	25		1,000	1,275		2,275	3,150
0705	12' wide		.80	25		1,025	1,275		2,300	3,200
0710	16' wide		.80	25		3,400	1,275		4,675	5,800
1000	Security gate, driveway, opener 12 VDC				Ea.	1,800			1,800	1,975
1010	Wireless					1,475			1,475	1,625

08 34 Special Function Doors

08 34 56 – Security Gates

08 34 56.10 Gates

		Crew	Daily Output	Labor-Hours	Unit	Material	2021 Bare Costs Labor	Equipment	Total	Total Incl O&P
1020	Security gate, driveway, opener 24 VDC				Ea.	1,725			1,725	1,900
1030	Wireless					1,650			1,650	1,800
1040	Security gate, driveway, opener 12 VDC, solar panel 10 watt	1 Elec	2	4		1,775	255		2,030	2,350
1050	20 watt	"	2	4		945	255		1,200	1,450

08 34 73 – Sound Control Door Assemblies

08 34 73.10 Acoustical Doors

		Crew	Daily Output	Labor-Hours	Unit	Material	2021 Bare Costs Labor	Equipment	Total	Total Incl O&P
0010	**ACOUSTICAL DOORS**									
0020	Including framed seals, 3' x 7', wood, 40 STC rating	2 Carp	1.50	10.667	Ea.	3,200	585		3,785	4,425
0100	Steel, 41 STC rating		1.50	10.667		3,900	585		4,485	5,225
0200	45 STC rating		1.50	10.667		1,650	585		2,235	2,750
0300	48 STC rating		1.50	10.667		4,900	585		5,485	6,300
0400	52 STC rating		1.50	10.667		5,600	585		6,185	7,075
9000	Minimum labor/equipment charge	1 Carp	4	2	Job		109		109	174

08 36 Panel Doors

08 36 13 – Sectional Doors

08 36 13.10 Overhead Commercial Doors

		Crew	Daily Output	Labor-Hours	Unit	Material	2021 Bare Costs Labor	Equipment	Total	Total Incl O&P
0010	**OVERHEAD COMMERCIAL DOORS**									
1000	Stock, sectional, heavy duty, wood, 1-3/4" thick, 8' x 8' high	2 Carp	2	8	Ea.	1,225	440		1,665	2,025
1200	12' x 12' high	2 Carp	1.50	10.667	Ea.	2,800	585		3,385	4,000
1300	Chain hoist, 14' x 14' high		1.30	12.308		3,800	675		4,475	5,250
1600	20' x 16' high		.65	24.615		7,175	1,350		8,525	10,100
2100	For medium duty custom door, deduct					5%	5%			
2150	For medium duty stock doors, deduct					10%	5%			
2300	Fiberglass and aluminum, heavy duty, sectional, 12' x 12' high	2 Carp	1.50	10.667	Ea.	3,400	585		3,985	4,675
2450	Chain hoist, 20' x 20' high	"	.50	32		7,800	1,750		9,550	11,400
2900	For electric trolley operator, 1/3 HP, to 12' x 12', add	1 Carp	2	4		1,150	219		1,369	1,600
2950	Over 12' x 12', 1/2 HP, add		1	8		1,250	440		1,690	2,075
2980	Overhead, for row of clear lites, add		1	8		206	440		646	920
9000	Minimum labor/equipment charge	2 Carp	1.50	10.667	Job		585		585	930

08 36 13.20 Residential Garage Doors

		Crew	Daily Output	Labor-Hours	Unit	Material	2021 Bare Costs Labor	Equipment	Total	Total Incl O&P
0010	**RESIDENTIAL GARAGE DOORS**									
0050	Hinged, wood, custom, double door, 9' x 7'	2 Carp	4	4	Ea.	2,300	219		2,519	2,875
0070	16' x 7'		3	5.333		1,425	292		1,717	2,050
0200	Overhead, sectional, incl. hardware, fiberglass, 9' x 7', standard		5	3.200		965	175		1,140	1,325
0220	Deluxe		5	3.200		1,175	175		1,350	1,575
0300	16' x 7', standard		6	2.667		2,200	146		2,346	2,650
0320	Deluxe		6	2.667		2,425	146		2,571	2,900
0500	Hardboard, 9' x 7', standard		8	2		735	109		844	980
0520	Deluxe		8	2		935	109		1,044	1,200
0600	16' x 7', standard		6	2.667		1,350	146		1,496	1,725
0620	Deluxe		6	2.667		1,675	146		1,821	2,075
0700	Metal, 9' x 7', standard		8	2		1,150	109		1,259	1,450
0720	Deluxe		6	2.667		1,150	146		1,296	1,500
0800	16' x 7', standard		6	2.667		1,250	146		1,396	1,600
0820	Deluxe		5	3.200		1,475	175		1,650	1,900
0900	Wood, 9' x 7', standard		8	2		1,050	109		1,159	1,325
0920	Deluxe		8	2		2,375	109		2,484	2,800
1000	16' x 7', standard		6	2.667		3,525	146		3,671	4,100
1020	Deluxe		6	2.667		3,175	146		3,321	3,725

08 36 Panel Doors

08 36 13 – Sectional Doors

08 36 13.20 Residential Garage Doors

		Crew	Daily Output	Labor-Hours	Unit	Material	2021 Bare Costs Labor	2021 Bare Costs Equipment	Total	Total Incl O&P
1800	Door hardware, sectional	1 Carp	4	2	Ea.	375	109		484	585
1810	Door tracks only		4	2		175	109		284	365
1820	One side only		7	1.143		124	62.50		186.50	236
4000	For electric operator, economy, add		8	1		445	54.50		499.50	570
4100	Deluxe, including remote control		8	1		645	54.50		699.50	795
4500	For transmitter/receiver control, add to operator				Total	118			118	130
4600	Transmitters, additional				"	69			69	76
6000	Replace section, on sectional door, fiberglass, 9' x 7'	1 Carp	4	2	Ea.	720	109		829	970
6020	16' x 7'	1 Carp	3.50	2.286	Ea.	760	125		885	1,025
6200	Hardboard, 9' x 7'		4	2		190	109		299	385
6220	16' x 7'		3.50	2.286		249	125		374	475
6300	Metal, 9' x 7'		4	2		227	109		336	425
6320	16' x 7'		3.50	2.286		335	125		460	570
6500	Wood, 9' x 7'		4	2		124	109		233	310
6520	16' x 7'		3.50	2.286		244	125		369	470
7010	Garage doors, row of lites					228			228	251
9000	Minimum labor/equipment charge	1 Carp	2.50	3.200	Job		175		175	279

08 36 19 – Multi-Leaf Vertical Lift Doors

08 36 19.10 Sectional Vertical Lift Doors

		Crew	Daily Output	Labor-Hours	Unit	Material	2021 Bare Costs Labor	2021 Bare Costs Equipment	Total	Total Incl O&P
0010	**SECTIONAL VERTICAL LIFT DOORS**									
0020	Motorized, 14 ga. steel, incl. frame and control panel									
0050	16' x 16' high	L-10	.50	48	Ea.	22,600	2,950	950	26,500	30,600
0100	10' x 20' high		1.30	18.462		39,100	1,125	365	40,590	45,200
0120	15' x 20' high		1.30	18.462		47,600	1,125	365	49,090	54,500
0140	20' x 20' high		1	24		56,000	1,475	475	57,950	64,500
0160	25' x 20' high		1	24		56,500	1,475	475	58,450	65,000
0170	32' x 24' high		.75	32		52,500	1,975	635	55,110	61,500
0180	20' x 25' high		1	24		58,500	1,475	475	60,450	67,000
0200	25' x 25' high		.70	34.286		81,000	2,100	680	83,780	93,000
0220	25' x 30' high		.70	34.286		85,000	2,100	680	87,780	97,500
0240	30' x 30' high		.70	34.286		93,500	2,100	680	96,280	107,000
0260	35' x 30' high		.70	34.286		102,000	2,100	680	104,780	116,000

08 38 Traffic Doors

08 38 19 – Rigid Traffic Doors

08 38 19.20 Double Acting Swing Doors

		Crew	Daily Output	Labor-Hours	Unit	Material	2021 Bare Costs Labor	2021 Bare Costs Equipment	Total	Total Incl O&P
0010	**DOUBLE ACTING SWING DOORS**									
0020	Including frame, closer, hardware and vision panel									
1000	Polymer, 7'-0" high, 4'-0" wide	2 Carp	4.20	3.810	Pr.	2,075	208		2,283	2,600
1025	6'-0" wide		4	4		2,550	219		2,769	3,150
1050	6'-8" wide		4	4		2,700	219		2,919	3,325
2000	3/4" thick, stainless steel									
2010	Stainless steel, 7' high opening, 4' wide	2 Carp	4	4	Pr.	2,550	219		2,769	3,150
2050	7' wide		3.80	4.211	"	2,750	230		2,980	3,400
9000	Minimum labor/equipment charge		2	8	Job		440		440	695

08 38 19.30 Shock Absorbing Doors

		Crew	Daily Output	Labor-Hours	Unit	Material	2021 Bare Costs Labor	2021 Bare Costs Equipment	Total	Total Incl O&P
0010	**SHOCK ABSORBING DOORS**									
0020	Rigid, no frame, 1-1/2" thick, 5' x 7'	2 Sswk	1.90	8.421	Opng.	1,550	510		2,060	2,525
0100	8' x 8'		1.80	8.889		3,150	535		3,685	4,350
0500	Flexible, no frame, insulated, .16" thick, economy, 5' x 7'		2	8		2,900	480		3,380	3,975

08 38 Traffic Doors

08 38 19 – Rigid Traffic Doors

08 38 19.30 Shock Absorbing Doors	Crew	Daily Output	Labor-Hours	Unit	Material	2021 Bare Costs Labor	Equipment	Total	Total Incl O&P
0600	Deluxe	2 Sswk	1.90	8.421	Opng.	3,625	510		4,135	4,800
1000	8' x 8' opening, economy	↓	2	8	↓	3,150	480		3,630	4,250
1100	Deluxe	2 Sswk	1.90	8.421	Opng.	3,575	510		4,085	4,750
9000	Minimum labor/equipment charge	"	2	8	Job		480		480	780

08 41 Entrances and Storefronts

08 41 13 – Aluminum-Framed Entrances and Storefronts

08 41 13.20 Tube Framing

		Crew	Daily Output	Labor-Hours	Unit	Material	2021 Bare Costs Labor	Equipment	Total	Total Incl O&P
0010	**TUBE FRAMING**, For window walls and storefronts, aluminum stock									
0050	Plain tube frame, mill finish, 1-3/4" x 1-3/4"	2 Glaz	103	.155	L.F.	10.80	8.20		19	25
0150	1-3/4" x 4"		98	.163		15.35	8.60		23.95	30.50
0200	1-3/4" x 4-1/2"		95	.168		16.90	8.85		25.75	32.50
0250	2" x 6"		89	.180		30	9.45		39.45	48
0350	4" x 4"		87	.184		24.50	9.70		34.20	42.50
0400	4-1/2" x 4-1/2"		85	.188		31	9.90		40.90	50
0450	Glass bead		240	.067		3.39	3.51		6.90	9.35
1000	Flush tube frame, mill finish, 1/4" glass, 1-3/4" x 4", open header		80	.200		15.40	10.55		25.95	33.50
1050	Open sill		82	.195		13.75	10.25		24	31.50
1100	Closed back header		83	.193		24	10.15		34.15	42.50
1150	Closed back sill	↓	85	.188	↓	23	9.90		32.90	41.50
1160	Tube frmg., spandrel cover both sides, alum 1" wide	1 Sswk	85	.094	S.F.	98.50	5.70		104.20	117
1170	Tube frmg., spandrel cover both sides, alum 2" wide	"	85	.094	"	40.50	5.70		46.20	53.50
1200	Vertical mullion, one piece	2 Glaz	75	.213	L.F.	21.50	11.25		32.75	41.50
1250	Two piece		73	.219		23	11.55		34.55	44
1300	90° or 180° vertical corner post		75	.213		36	11.25		47.25	57.50
1400	1-3/4" x 4-1/2", open header		80	.200		18.10	10.55		28.65	36.50
1450	Open sill		82	.195		15	10.25		25.25	33
1500	Closed back header		83	.193		21	10.15		31.15	39
1550	Closed back sill		85	.188		21	9.90		30.90	39
1600	Vertical mullion, one piece		75	.213		23.50	11.25		34.75	44
1650	Two piece		73	.219		25	11.55		36.55	46
1700	90° or 180° vertical corner post		75	.213		25.50	11.25		36.75	46
2000	Flush tube frame, mil fin.,ins. glass w/thml brk, 2" x 4-1/2", open header		75	.213		15.85	11.25		27.10	35.50
2050	Open sill		77	.208		13.35	10.95		24.30	32
2100	Closed back header		78	.205		14.75	10.80		25.55	33.50
2150	Closed back sill		80	.200		14.30	10.55		24.85	32.50
2200	Vertical mullion, one piece		70	.229		16.85	12.05		28.90	37.50
2250	Two piece		68	.235		18.60	12.40		31	40
2300	90° or 180° vertical corner post		70	.229		19.15	12.05		31.20	40
5000	Flush tube frame, mill fin., thermal brk., 2-1/4" x 4-1/2", open header		74	.216		16.40	11.40		27.80	36
5050	Open sill		75	.213		14.20	11.25		25.45	33.50
5100	Vertical mullion, one piece		69	.232		18.85	12.20		31.05	40.50
5150	Two piece		67	.239		21	12.55		33.55	43
5200	90° or 180° vertical corner post	2 Glaz	69	.232	L.F.	19.10	12.20		31.30	40.50
5295	Flush tube frame, mill finish, 4-1/2" x 4-1/2"		85	.188		31	9.90		40.90	50
5300	Plain tube frame, mill finish, 2" x 3" jamb		115	.139		11.35	7.35		18.70	24
5310	2" x 4"		115	.139		16.25	7.35		23.60	29.50
5320	3" x 5-1/2"		107	.150		25.50	7.85		33.35	40.50
5330	Head, 2" x 3"		200	.080		21	4.21		25.21	29.50
5340	Mullion, 2" x 3"		200	.080		21.50	4.21		25.71	30
5350	2" x 4"	↓	115	.139	↓	27.50	7.35		34.85	41.50

256

08 41 Entrances and Storefronts

08 41 13 – Aluminum-Framed Entrances and Storefronts

08 41 13.20 Tube Framing		Crew	Daily Output	Labor-Hours	Unit	Material	2021 Bare Costs Labor	Equipment	Total	Total Incl O&P
5360	3" x 5-1/2"	2 Glaz	107	.150	L.F.	34.50	7.85		42.35	50.50
5370	4" corner mullion		90	.178		30.50	9.35		39.85	48.50
5380	Horizontal, 2" x 3"		115	.139		25	7.35		32.35	39
5390	3" x 5-1/2"		107	.150		29	7.85		36.85	44.50
5430	Sill section, 1/8" x 6"		133	.120		36	6.35		42.35	49.50
5440	1/8" x 7"		133	.120		43	6.35		49.35	57.50
5450	1/8" x 8-1/2"		133	.120		50.50	6.35		56.85	65.50
5460	Column covers, aluminum, 1/8" x 26"		57	.281		69.50	14.80		84.30	100
5470	1/8" x 34"		53	.302		99	15.90		114.90	135
5480	1/8" x 38"		53	.302		100	15.90		115.90	136
5700	Vertical mullions, clear finish, 1/4" thick glass		528	.030		14.60	1.60		16.20	18.65
5720	3/8" thick		445	.036		13.60	1.89		15.49	17.95
5740	1/2" thick		400	.040		14.40	2.11		16.51	19.20
5760	3/4" thick		344	.047		15.65	2.45		18.10	21
5780	1" thick		304	.053		15.25	2.77		18.02	21
6980	Door stop (snap in)		380	.042		3.20	2.22		5.42	7.05
7000	For joints, 90°, clip type, add				Ea.	27			27	30
7050	Screw spline joint, add					24.50			24.50	26.50
7100	For joint other than 90°, add					51			51	56.50
8000	For bronze anodized aluminum, add					15%				
8020	For black finish, add					30%				
8050	For stainless steel materials, add					350%				
8100	For monumental grade, add					53%				
8150	For steel stiffener, add	2 Glaz	200	.080	L.F.	11.50	4.21		15.71	19.35
8200	For 2 to 5 stories, add per story				Story		8%			
9000	Minimum labor/equipment charge	2 Glaz	2	8	Job		420		420	670

08 41 19 – Stainless-Steel-Framed Entrances and Storefronts

08 41 19.10 Stainless-Steel and Glass Entrance Unit

		Crew	Daily Output	Labor-Hours	Unit	Material	2021 Bare Costs Labor	Equipment	Total	Total Incl O&P
0010	**STAINLESS-STEEL AND GLASS ENTRANCE UNIT**, narrow stiles									
0020	3' x 7' opening, including hardware, minimum	2 Sswk	1.60	10	Opng.	7,775	605		8,380	9,525
0050	Average		1.40	11.429		8,250	690		8,940	10,200
0100	Maximum		1.20	13.333		8,850	805		9,655	11,100
1000	For solid bronze entrance units, statuary finish, add					64%				
1100	Without statuary finish, add				Opng.	45%				
2000	Balanced doors, 3' x 7', economy	2 Sswk	.90	17.778	Ea.	10,300	1,075		11,375	13,100
2100	Premium		.70	22.857	"	16,800	1,375		18,175	20,700
9000	Minimum labor/equipment charge		2	8	Job		480		480	780

08 41 26 – All-Glass Entrances and Storefronts

08 41 26.10 Window Walls Aluminum, Stock

		Crew	Daily Output	Labor-Hours	Unit	Material	2021 Bare Costs Labor	Equipment	Total	Total Incl O&P
0010	**WINDOW WALLS ALUMINUM, STOCK**, including glazing									
0020	Minimum	H-2	160	.150	S.F.	58.50	7.50		66	76
0050	Average		140	.171		80.50	8.55		89.05	102
0100	Maximum		110	.218		190	10.90		200.90	226
0500	For translucent sandwich wall systems, see Section 07 41 33.10									
0850	Cost of the above walls depends on material,									
0860	finish, repetition, and size of units.									
0870	The larger the opening, the lower the S.F. cost									

For customer support on your Commercial Renovation Costs with RSMeans Data, call 800.448.8182.

257

08 42 Entrances

08 42 26 – All-Glass Entrances

08 42 26.10 Swinging Glass Doors

		Crew	Daily Output	Labor-Hours	Unit	Material	2021 Bare Costs Labor	Equipment	Total	Total Incl O&P
0010	**SWINGING GLASS DOORS**									
0020	Including hardware, 1/2" thick, tempered, 3' x 7' opening	2 Glaz	2	8	Opng.	2,425	420		2,845	3,325
0100	6' x 7' opening		1.40	11.429	"	4,625	600		5,225	6,050
9000	Minimum labor/equipment charge		2	8	Job		420		420	670

08 42 36 – Balanced Door Entrances

08 42 36.10 Balanced Entrance Doors

		Crew	Daily Output	Labor-Hours	Unit	Material	2021 Bare Costs Labor	Equipment	Total	Total Incl O&P
0010	**BALANCED ENTRANCE DOORS**									
0020	Hardware & frame, alum. & glass, 3' x 7', econ.	2 Sswk	.90	17.778	Ea.	7,325	1,075		8,400	9,775
0150	Premium		.70	22.857	"	8,725	1,375		10,100	11,800
9000	Minimum labor/equipment charge		1	16	Job		965		965	1,550

08 43 Storefronts

08 43 13 – Aluminum-Framed Storefronts

08 43 13.10 Aluminum-Framed Entrance Doors and Frames

		Crew	Daily Output	Labor-Hours	Unit	Material	2021 Bare Costs Labor	Equipment	Total	Total Incl O&P
0010	**ALUMINUM-FRAMED ENTRANCE DOORS AND FRAMES**									
0015	Standard hardware and glass stops but no glass									
0020	Entrance door, 3' x 7' opening, clear anodized finish	2 Sswk	7	2.286	Opng.	700	138		838	995
0040	Bronze finish		7	2.286		875	138		1,013	1,175
0060	Black finish		7	2.286		850	138		988	1,150
0200	3'-6" x 7'-0", mill finish		7	2.286		945	138		1,083	1,275
0220	Bronze finish		7	2.286		750	138		888	1,050
0240	Black finish		7	2.286		760	138		898	1,050
0500	6' x 7' opening, clear finish		6	2.667		940	161		1,101	1,275
0520	Bronze finish		6	2.667		1,300	161		1,461	1,675
0600	Door frame for above doors 3'-0" x 7'-0", mill finish		6	2.667		595	161		756	910
0620	Bronze finish		6	2.667		630	161		791	955
0640	Black finish	2 Sswk	6	2.667	Opng.	710	161		871	1,050
0700	3'-6" x 7'-0", mill finish		6	2.667		425	161		586	730
0720	Bronze finish		6	2.667		425	161		586	730
0740	Black finish		6	2.667		425	161		586	730
0800	6'-0" x 7'-0", mill finish		6	2.667		425	161		586	730
0820	Bronze finish		6	2.667		430	161		591	730
0840	Black finish		6	2.667		465	161		626	770
1000	With 3' high transom above, 3' x 7' opening, clear finish		5.50	2.909		660	175		835	1,000
1050	Bronze finish		5.50	2.909		655	175		830	1,000
1100	Black finish		5.50	2.909		640	175		815	990
1300	3'-6" x 7'-0" opening, clear finish		5.50	2.909		425	175		600	750
1320	Bronze finish		5.50	2.909		450	175		625	780
1340	Black finish		5.50	2.909		455	175		630	785
1500	6' x 7' opening, clear finish		5.50	2.909		695	175		870	1,050
1550	Bronze finish		5.50	2.909		815	175		990	1,175
1600	Black finish		5.50	2.909		905	175		1,080	1,275
8000	For 8' high doors, add				Ea.	174			174	174
9000	Minimum labor/equipment charge	2 Sswk	4	4	Job		241		241	390

08 43 13.20 Storefront Systems

		Crew	Daily Output	Labor-Hours	Unit	Material	2021 Bare Costs Labor	Equipment	Total	Total Incl O&P
0010	**STOREFRONT SYSTEMS**, aluminum frame clear 3/8" plate glass									
0020	incl. 3' x 7' door with hardware (400 sq. ft. max. wall)									
0500	Wall height to 12' high, commercial grade	2 Glaz	150	.107	S.F.	24	5.60		29.60	35.50
0600	Institutional grade		130	.123		28	6.50		34.50	41.50
0700	Monumental grade		115	.139		48.50	7.35		55.85	64.50

08 43 Storefronts

08 43 13 – Aluminum-Framed Storefronts

08 43 13.20 Storefront Systems

		Crew	Daily Output	Labor-Hours	Unit	Material	2021 Bare Costs Labor	Equipment	Total	Total Incl O&P
1000	6' x 7' door with hardware, commercial grade	2 Glaz	135	.119	S.F.	73.50	6.25		79.75	91
1100	Institutional grade		115	.139		42	7.35		49.35	57.50
1200	Monumental grade		100	.160		118	8.40		126.40	143
1500	For bronze anodized finish, add					15%				
1600	For black anodized finish, add					36%				
1700	For stainless steel framing, add to monumental					78%				
9000	Minimum labor/equipment charge	2 Glaz	1	16	Job		840		840	1,350

08 43 29 – Sliding Storefronts

08 43 29.10 Sliding Panels

		Crew	Daily Output	Labor-Hours	Unit	Material	2021 Bare Costs Labor	Equipment	Total	Total Incl O&P
0010	**SLIDING PANELS**									
0020	Mall fronts, aluminum & glass, 15' x 9' high	2 Glaz	1.30	12.308	Opng.	4,475	650		5,125	5,950
0100	24' x 9' high		.70	22.857		6,275	1,200		7,475	8,825
0200	48' x 9' high, with fixed panels		.90	17.778		11,200	935		12,135	13,800
0500	For bronze finish, add					17%				
9000	Minimum labor/equipment charge	2 Glaz	1	16	Job		840		840	1,350

08 45 Translucent Wall and Roof Assemblies

08 45 10 – Translucent Roof Assemblies

08 45 10.10 Skyroofs

		Crew	Daily Output	Labor-Hours	Unit	Material	2021 Bare Costs Labor	Equipment	Total	Total Incl O&P
0010	**SKYROOFS**									
1200	Skylights, circular, clear, double glazed acrylic									
1230	30" diameter	2 Carp	3	5.333	Ea.	1,725	292		2,017	2,375
1250	60" diameter		3	5.333		2,850	292		3,142	3,600
1290	96" diameter		2	8		5,900	440		6,340	7,200
1300	Skylight Barrel Vault, clear, double glazed, acrylic									
1330	3'-0" x 12'-0"	G-3	3	10.667	Ea.	4,275	585		4,860	5,625
1350	4'-0" x 12'-0"		3	10.667		4,725	585		5,310	6,125
1390	5'-0" x 12'-0"		2	16		6,225	880		7,105	8,225
1400	Skylight Pyramid, aluminum frame, clear low E laminated glass									
1410	The glass is installed in the frame except where noted									
1430	Square, 3' x 3'	G-3	3	10.667	Ea.	4,150	585		4,735	5,475
1440	4' x 4'		3	10.667		4,700	585		5,285	6,100
1450	5' x 5', glass must be field installed		3	10.667		6,625	585		7,210	8,200
1460	6' x 6', glass must be field installed		2	16		9,175	880		10,055	11,500
1550	Install pre-cut laminated glass in aluminum frame on a flat roof	2 Glaz	55	.291	SF Surf		15.30		15.30	24.50
1560	Install pre-cut laminated glass in aluminum frame on a sloped roof	"	40	.400	"		21		21	33.50
9000	Minimum labor/equipment charge	2 Carp	8	2	Job		109		109	174

08 51 Metal Windows

08 51 13 – Aluminum Windows

08 51 13.10 Aluminum Sash

		Crew	Daily Output	Labor-Hours	Unit	Material	2021 Bare Costs Labor	Equipment	Total	Total Incl O&P
0010	**ALUMINUM SASH**									
0020	Stock, grade C, glaze & trim not incl., casement	2 Sswk	200	.080	S.F.	43.50	4.82		48.32	55.50
0050	Double-hung		200	.080		42	4.82		46.82	54.50
0100	Fixed casement		200	.080		18.60	4.82		23.42	28.50
0150	Picture window		200	.080		19.95	4.82		24.77	30
0200	Projected window		200	.080		38.50	4.82		43.32	50.50
0250	Single-hung		200	.080		17.95	4.82		22.77	27.50
0300	Sliding		200	.080		24	4.82		28.82	34

For customer support on your Commercial Renovation Costs with RSMeans Data, call 800.448.8182.

259

08 51 13 – Aluminum Windows

08 51 13.10 Aluminum Sash

		Crew	Daily Output	Labor-Hours	Unit	Material	2021 Bare Costs Labor	Equipment	Total	Total Incl O&P
1000	Mullions for above, tubular	2 Sswk	240	.067	L.F.	6.60	4.02		10.62	13.75
9000	Minimum labor/equipment charge	1 Sswk	2	4	Job		241		241	390

08 51 13.20 Aluminum Windows

		Crew	Daily Output	Labor-Hours	Unit	Material	2021 Bare Costs Labor	Equipment	Total	Total Incl O&P
0010	**ALUMINUM WINDOWS**, incl. frame and glazing, commercial grade									
1000	Stock units, casement, 3'-1" x 3'-2" opening	2 Sswk	10	1.600	Ea.	405	96.50		501.50	600
1050	Add for storms					130			130	143
1600	Projected, with screen, 3'-1" x 3'-2" opening	2 Sswk	10	1.600		385	96.50		481.50	575
1700	Add for storms					126			126	139
2000	4'-5" x 5'-3" opening	2 Sswk	8	2		425	121		546	660
2100	Add for storms					129			129	142
2500	Enamel finish windows, 3'-1" x 3'-2"	2 Sswk	10	1.600		390	96.50		486.50	585
2600	4'-5" x 5'-3"	2 Sswk	8	2	Ea.	440	121		561	680
3000	Single-hung, 2' x 3' opening, enameled, standard glazed		10	1.600		225	96.50		321.50	405
3100	Insulating glass		10	1.600		273	96.50		369.50	455
3300	2'-8" x 6'-8" opening, standard glazed		8	2		380	121		501	615
3400	Insulating glass		8	2		500	121		621	745
3700	3'-4" x 5'-0" opening, standard glazed		9	1.778		325	107		432	535
3800	Insulating glass		9	1.778		355	107		462	565
4000	Sliding aluminum, 3' x 2' opening, standard glazed		10	1.600		246	96.50		342.50	425
4100	Insulating glass		10	1.600		243	96.50		339.50	425
4300	5' x 3' opening, standard glazed		9	1.778		345	107		452	555
4400	Insulating glass		9	1.778		435	107		542	655
4600	8' x 4' opening, standard glazed		6	2.667		370	161		531	665
4700	Insulating glass		6	2.667		595	161		756	910
5000	9' x 5' opening, standard glazed		4	4		555	241		796	1,000
5100	Insulating glass		4	4		850	241		1,091	1,325
5500	Sliding, with thermal barrier and screen, 6' x 4', 2 track		8	2		725	121		846	990
5700	4 track		8	2		915	121		1,036	1,200
6000	For above units with bronze finish, add					15%				
6200	For installation in concrete openings, add					8%				
7000	Double-hung, insulating glass, 2'-0" x 2'-0"	2 Sswk	11	1.455		119	87.50		206.50	273
7020	2'-0" x 2'-6"		11	1.455		134	87.50		221.50	289
7040	3'-0" x 1'-6"		10	1.600		273	96.50		369.50	455
7060	3'-0" x 2'-0"		10	1.600		315	96.50		411.50	505
7080	3'-0" x 2'-6"		10	1.600		380	96.50		476.50	575
7100	3'-0" x 3'-0"		10	1.600		425	96.50		521.50	620
7120	3'-0" x 3'-6"		10	1.600		445	96.50		541.50	645
7140	3'-0" x 4'-0"		10	1.600		485	96.50		581.50	685
7160	3'-0" x 5'-0"		10	1.600		530	96.50		626.50	740
7180	3'-0" x 6'-0"		9	1.778		670	107		777	915

08 51 13.30 Impact Resistant Aluminum Windows

		Crew	Daily Output	Labor-Hours	Unit	Material	2021 Bare Costs Labor	Equipment	Total	Total Incl O&P
0010	**IMPACT RESISTANT ALUMINUM WINDOWS**, incl. frame and glazing									
0100	Single-hung, impact resistant, 2'-8" x 5'-0"	2 Carp	9	1.778	Ea.	1,125	97		1,222	1,400
0120	3'-0" x 5'-0"		9	1.778		1,350	97		1,447	1,650
0130	4'-0" x 5'-0"		9	1.778		1,900	97		1,997	2,250
0250	Horizontal slider, impact resistant, 5'-5" x 5'-2"		9	1.778		2,200	97		2,297	2,575

08 51 23 – Steel Windows

08 51 23.10 Steel Sash

		Crew	Daily Output	Labor-Hours	Unit	Material	2021 Bare Costs Labor	Equipment	Total	Total Incl O&P
0010	**STEEL SASH** Custom units, glazing and trim not included									
0100	Casement, 100% vented	2 Sswk	200	.080	S.F.	70	4.82		74.82	85
0200	50% vented		200	.080		57	4.82		61.82	71
0300	Fixed		200	.080		29.50	4.82		34.32	40.50

08 51 Metal Windows

08 51 23 – Steel Windows

08 51 23.10 Steel Sash

		Crew	Daily Output	Labor-Hours	Unit	Material	2021 Bare Costs Labor	Equipment	Total	Total Incl O&P
1000	Projected, commercial, 40% vented	2 Sswk	200	.080	S.F.	54	4.82		58.82	67.50
1100	Intermediate, 50% vented		200	.080		64.50	4.82		69.32	79
1500	Industrial, horizontally pivoted		200	.080		56.50	4.82		61.32	70
1600	Fixed		200	.080		32.50	4.82		37.32	44
2000	Industrial security sash, 50% vented		200	.080		62	4.82		66.82	76
2100	Fixed		200	.080		49.50	4.82		54.32	62.50
2500	Picture window		200	.080		33	4.82		37.82	44
3000	Double-hung		200	.080		62.50	4.82		67.32	77
5000	Mullions for above, open interior face		240	.067	L.F.	11	4.02		15.02	18.60
5100	With interior cover		240	.067	"	18.25	4.02		22.27	26.50
5200	Single glazing for above, add	2 Glaz	200	.080	S.F.	7.40	4.21		11.61	14.85
6000	Double glazing for above, add		200	.080		13.40	4.21		17.61	21.50
6100	Triple glazing for above, add		85	.188		12.80	9.90		22.70	30
9000	Minimum labor/equipment charge	1 Sswk	2	4	Job		241		241	390

08 51 23.20 Steel Windows

		Crew	Daily Output	Labor-Hours	Unit	Material	2021 Bare Costs Labor	Equipment	Total	Total Incl O&P
0010	**STEEL WINDOWS** Stock, including frame, trim and insul. glass R085123-10									
1000	Custom units, double-hung, 2'-8" x 4'-6" opening	2 Sswk	12	1.333	Ea.	725	80.50		805.50	925
1100	2'-4" x 3'-9" opening		12	1.333		595	80.50		675.50	785
1500	Commercial projected, 3'-9" x 5'-5" opening		10	1.600		1,250	96.50		1,346.50	1,550
1600	6'-9" x 4'-1" opening		7	2.286		1,675	138		1,813	2,050
2000	Intermediate projected, 2'-9" x 4'-1" opening		12	1.333		710	80.50		790.50	910
2100	4'-1" x 5'-5" opening		10	1.600		1,450	96.50		1,546.50	1,725
9000	Minimum labor/equipment charge	1 Sswk	3	2.667	Job		161		161	260

08 51 66 – Metal Window Screens

08 51 66.10 Screens

		Crew	Daily Output	Labor-Hours	Unit	Material	2021 Bare Costs Labor	Equipment	Total	Total Incl O&P
0010	**SCREENS**									
0020	For metal sash, aluminum or bronze mesh, flat screen	2 Sswk	1200	.013	S.F.	4.74	.80		5.54	6.50
0500	Wicket screen, inside window		1000	.016		7.05	.96		8.01	9.30
0800	Security screen, aluminum frame with stainless steel cloth		1200	.013		25	.80		25.80	29
0900	Steel grate, painted, on steel frame		1600	.010		14.10	.60		14.70	16.50
1000	Screens for solar louvers		160	.100		26	6.05		32.05	38.50
4000	See Section 05 58 23.90 for window guards									

08 52 Wood Windows

08 52 10 – Plain Wood Windows

08 52 10.20 Awning Window

		Crew	Daily Output	Labor-Hours	Unit	Material	2021 Bare Costs Labor	Equipment	Total	Total Incl O&P
0010	**AWNING WINDOW**, Including frame, screens and grilles									
0100	34" x 22", insulated glass	1 Carp	10	.800	Ea.	297	44		341	395
0200	Low E glass		10	.800		330	44		374	430
0300	40" x 28", insulated glass		9	.889		315	48.50		363.50	430
0400	Low E glass		9	.889		350	48.50		398.50	465
0500	48" x 36", insulated glass		8	1		480	54.50		534.50	615
0600	Low E glass	1 Carp	8	1	Ea.	505	54.50		559.50	640
4000	Impact windows, minimum, add					60%				
4010	Impact windows, maximum, add					160%				
9000	Minimum labor/equipment charge	1 Carp	4	2	Job		109		109	174

08 52 10.30 Wood Windows

		Crew	Daily Output	Labor-Hours	Unit	Material	2021 Bare Costs Labor	Equipment	Total	Total Incl O&P
0010	**WOOD WINDOWS**, double-hung									
0020	Including frame, double insulated glass, screens and grilles									
0040	Double-hung, 2'-2" x 3'-4" high	2 Carp	15	1.067	Ea.	222	58.50		280.50	335
0060	2'-2" x 4'-4"		14	1.143		236	62.50		298.50	360
0080	2'-6" x 3'-4"		13	1.231		233	67.50		300.50	365
0100	2'-6" x 4'-0"		12	1.333		238	73		311	380
0120	2'-6" x 4'-8"		12	1.333		251	73		324	395
0140	2'-10" x 3'-4"		10	1.600		229	87.50		316.50	390
0160	2'-10" x 4'-0"		10	1.600		255	87.50		342.50	420
0180	3'-7" x 3'-4"		9	1.778		261	97		358	440
0200	3'-7" x 5'-4"		9	1.778		305	97		402	490
0220	3'-10" x 5'-4"		8	2		520	109		629	750
3800	Triple glazing for above, add					25%				

08 52 10.40 Casement Window

			Crew	Daily Output	Labor-Hours	Unit	Material	2021 Bare Costs Labor	Equipment	Total	Total Incl O&P
0010	**CASEMENT WINDOW**, including frame, screen and grilles										
0100	2'-0" x 3'-0" H, double insulated glass	G	1 Carp	10	.800	Ea.	278	44		322	375
0150	Low E glass	G		10	.800		288	44		332	385
0200	2'-0" x 4'-6" high, double insulated glass	G		9	.889		390	48.50		438.50	510
0250	Low E glass	G		9	.889		440	48.50		488.50	565
0260	Casement 4'-2" x 4'-2" double insulated glass	G		11	.727		945	40		985	1,125
0270	4'-0" x 4'-0" Low E glass	G		11	.727		620	40		660	745
0290	6'-4" x 5'-7" Low E glass	G		9	.889		1,200	48.50		1,248.50	1,400
0300	2'-4" x 6'-0" high, double insulated glass	G		8	1		450	54.50		504.50	580
0350	Low E glass	G		8	1		465	54.50		519.50	595
0522	Vinyl-clad, premium, double insulated glass, 2'-0" x 3'-0"	G		10	.800		266	44		310	365
0524	2'-0" x 4'-0"	G		9	.889		335	48.50		383.50	450
0525	2'-0" x 5'-0"	G		8	1		355	54.50		409.50	475
0528	2'-0" x 6'-0"	G		8	1		515	54.50		569.50	655
0600	3'-0" x 5'-0"	G		8	1		650	54.50		704.50	800
0700	4'-0" x 3'-0"	G		8	1		775	54.50		829.50	935
0710	4'-0" x 4'-0"	G		8	1		615	54.50		669.50	760
0720	4'-8" x 4'-0"	G		8	1		675	54.50		729.50	830
0730	4'-8" x 5'-0"	G		6	1.333		835	73		908	1,025
0740	4'-8" x 6'-0"	G		6	1.333		910	73		983	1,125
0750	6'-0" x 4'-0"	G		6	1.333		790	73		863	985
0800	6'-0" x 5'-0"	G		6	1.333		1,075	73		1,148	1,325
0900	5'-6" x 5'-6"	G	2 Carp	15	1.067		1,425	58.50		1,483.50	1,675
8190	For installation, add per leaf					Ea.		15%			
8200	For multiple leaf units, deduct for stationary sash										
8220	2' high					Ea.	25			25	27.50
8240	4'-6" high						28			28	31
8260	6' high						37.50			37.50	41
8300	Impact windows, minimum, add						60%				
8310	Impact windows, maximum, add						160%				
9000	Minimum labor/equipment charge		1 Carp	3	2.667	Job		146		146	232

08 52 10.50 Double-Hung

			Crew	Daily Output	Labor-Hours	Unit	Material	2021 Bare Costs Labor	Equipment	Total	Total Incl O&P
0010	**DOUBLE-HUNG**, Including frame, screens and grilles										
0100	2'-0" x 3'-0" high, low E insul. glass	G	1 Carp	10	.800	Ea.	192	44		236	281
0200	3'-0" x 4'-0" high, double insulated glass	G		9	.889		285	48.50		333.50	395
0300	4'-0" x 4'-6" high, low E insulated glass	G		8	1		320	54.50		374.50	440
8000	Impact windows, minimum, add						60%				
8010	Impact windows, maximum, add						160%				

08 52 Wood Windows

08 52 10 – Plain Wood Windows

08 52 10.50 Double-Hung		Crew	Daily Output	Labor-Hours	Unit	Material	2021 Bare Costs Labor	Equipment	Total	Total Incl O&P
9000	Minimum labor/equipment charge	1 Carp	3	2.667	Job		146		146	232

08 52 10.55 Picture Window

		Crew	Daily Output	Labor-Hours	Unit	Material	Labor	Equipment	Total	Total Incl O&P
0010	**PICTURE WINDOW**, Including frame and grilles									
0100	3'-6" x 4'-0" high, double insulated glass	2 Carp	12	1.333	Ea.	440	73		513	600
0150	Low E glass		12	1.333		450	73		523	610
0200	4'-0" x 4'-6" high, double insulated glass		11	1.455		540	79.50		619.50	720
0250	Low E glass		11	1.455		610	79.50		689.50	800
0300	5'-0" x 4'-0" high, double insulated glass		11	1.455		615	79.50		694.50	800
0350	Low E glass		11	1.455		610	79.50		689.50	795
0400	6'-0" x 4'-6" high, double insulated glass		10	1.600		655	87.50		742.50	865
0450	Low E glass		10	1.600		635	87.50		722.50	840

08 52 10.65 Wood Sash

		Crew	Daily Output	Labor-Hours	Unit	Material	Labor	Equipment	Total	Total Incl O&P
0010	**WOOD SASH**, Including glazing but not trim									
0050	Custom, 5'-0" x 4'-0", 1" double glazed, 3/16" thick lites	2 Carp	3.20	5	Ea.	245	274		519	705
0100	1/4" thick lites		5	3.200		355	175		530	670
0200	1" thick, triple glazed		5	3.200		425	175		600	750
0300	7'-0" x 4'-6" high, 1" double glazed, 3/16" thick lites		4.30	3.721		455	204		659	830
0400	1/4" thick lites		4.30	3.721		490	204		694	865
0500	1" thick, triple glazed		4.30	3.721		600	204		804	985
0600	8'-6" x 5'-0" high, 1" double glazed, 3/16" thick lites		3.50	4.571		625	250		875	1,075
0700	1/4" thick lites		3.50	4.571		685	250		935	1,150
0800	1" thick, triple glazed		3.50	4.571		640	250		890	1,100
0900	Window frames only, based on perimeter length				L.F.	3.88			3.88	4.27
1200	Window sill, stock, per lineal foot					9.05			9.05	9.95
1250	Casing, stock					3.30			3.30	3.63

08 52 10.70 Sliding Windows

			Crew	Daily Output	Labor-Hours	Unit	Material	Labor	Equipment	Total	Total Incl O&P
0010	**SLIDING WINDOWS**										
0100	3'-0" x 3'-0" high, double insulated	G	1 Carp	10	.800	Ea.	292	44		336	390
0120	Low E glass	G		10	.800		310	44		354	415
0200	4'-0" x 3'-6" high, double insulated	G		9	.889		425	48.50		473.50	545
0220	Low E glass	G		9	.889		430	48.50		478.50	555
0300	6'-0" x 5'-0" high, double insulated	G		8	1		490	54.50		544.50	625
0320	Low E glass	G		8	1		600	54.50		654.50	745
9000	Minimum labor/equipment charge			3	2.667	Job		146		146	232

08 52 13 – Metal-Clad Wood Windows

08 52 13.10 Awning Windows, Metal-Clad

		Crew	Daily Output	Labor-Hours	Unit	Material	Labor	Equipment	Total	Total Incl O&P
0010	**AWNING WINDOWS, METAL-CLAD**									
2000	Metal-clad, awning deluxe, double insulated glass, 34" x 22"	1 Carp	9	.889	Ea.	251	48.50		299.50	355
2050	36" x 25"		9	.889		281	48.50		329.50	390
2100	40" x 22"		9	.889		300	48.50		348.50	410
2150	40" x 30"		9	.889		350	48.50		398.50	465
2200	48" x 28"		8	1		355	54.50		409.50	480
2250	60" x 36"		8	1		380	54.50		434.50	505

08 52 13.20 Casement Windows, Metal-Clad

			Crew	Daily Output	Labor-Hours	Unit	Material	Labor	Equipment	Total	Total Incl O&P
0010	**CASEMENT WINDOWS, METAL-CLAD**										
0100	Metal-clad, deluxe, dbl. insul. glass, 2'-0" x 3'-0" high	G	1 Carp	10	.800	Ea.	305	44		349	405
0120	2'-0" x 4'-0" high	G		9	.889		345	48.50		393.50	460
0130	2'-0" x 5'-0" high	G		8	1		365	54.50		419.50	490
0140	2'-0" x 6'-0" high	G		8	1		395	54.50		449.50	515
0300	Metal-clad, casement, bldrs mdl, 6'-0" x 4'-0", dbl. insul. glass, 3 panels		2 Carp	10	1.600		1,250	87.50		1,337.50	1,525
0310	9'-0" x 4'-0", 4 panels			8	2		1,575	109		1,684	1,900

08 52 Wood Windows

08 52 13 – Metal-Clad Wood Windows

08 52 13.20 Casement Windows, Metal-Clad		Crew	Daily Output	Labor-Hours	Unit	Material	2021 Bare Costs Labor	Equipment	Total	Total Incl O&P
0320	10'-0" x 5'-0", 5 panels	2 Carp	7	2.286	Ea.	2,150	125		2,275	2,575
0330	12'-0" x 6'-0", 6 panels	▼	6	2.667	▼	2,725	146		2,871	3,225

08 52 13.30 Double-Hung Windows, Metal-Clad

0010	**DOUBLE-HUNG WINDOWS, METAL-CLAD**									
0100	Metal-clad, deluxe, dbl. insul. glass, 2'-6" x 3'-0" high [G]	1 Carp	10	.800	Ea.	296	44		340	395
0120	3'-0" x 3'-6" high [G]		10	.800		330	44		374	430
0140	3'-0" x 4'-0" high [G]		9	.889		335	48.50		383.50	450
0160	3'-0" x 4'-6" high [G]		9	.889		360	48.50		408.50	475
0180	3'-0" x 5'-0" high [G]		8	1		395	54.50		449.50	520
0200	3'-6" x 6'-0" high [G]	▼	8	1	▼	480	54.50		534.50	610

08 52 13.35 Picture and Sliding Windows Metal-Clad

0010	**PICTURE AND SLIDING WINDOWS METAL-CLAD**									
2000	Metal-clad, dlx picture, dbl. insul. glass, 4'-0" x 4'-0" high	2 Carp	12	1.333	Ea.	385	73		458	540
2100	4'-0" x 6'-0" high		11	1.455		575	79.50		654.50	760
2200	5'-0" x 6'-0" high		10	1.600		640	87.50		727.50	840
2300	6'-0" x 6'-0" high	▼	10	1.600		730	87.50		817.50	945
2400	Metal-clad, dlx sliding, dbl. insul. glass, 3'-0" x 3'-0" high [G]	1 Carp	10	.800		330	44		374	430
2420	4'-0" x 3'-6" high [G]		9	.889		400	48.50		448.50	520
2440	5'-0" x 4'-0" high [G]	▼	9	.889	▼	480	48.50		528.50	605
2460	6'-0" x 5'-0" high [G]	1 Carp	8	1	Ea.	775	54.50		829.50	940
9000	Minimum labor/equipment charge	2 Carp	2.75	5.818	Job		320		320	505

08 52 13.40 Bow and Bay Windows, Metal-Clad

0010	**BOW AND BAY WINDOWS, METAL-CLAD**									
0100	Metal-clad, deluxe, dbl. insul. glass, 8'-0" x 5'-0" high, 4 panels	2 Carp	10	1.600	Ea.	1,750	87.50		1,837.50	2,075
0120	10'-0" x 5'-0" high, 5 panels		8	2		1,875	109		1,984	2,250
0140	10'-0" x 6'-0" high, 5 panels		7	2.286		2,200	125		2,325	2,625
0160	12'-0" x 6'-0" high, 6 panels		6	2.667		3,525	146		3,671	4,100
0400	Double-hung, bldrs. model, bay, 8' x 4' high, dbl. insul. glass		10	1.600		1,375	87.50		1,462.50	1,675
0440	Low E glass		10	1.600		1,500	87.50		1,587.50	1,800
0460	9'-0" x 5'-0" high, dbl. insul. glass		6	2.667		1,500	146		1,646	1,875
0480	Low E glass		6	2.667		1,575	146		1,721	1,950
0500	Metal-clad, deluxe, dbl. insul. glass, 7'-0" x 4'-0" high		10	1.600		1,325	87.50		1,412.50	1,600
0520	8'-0" x 4'-0" high		8	2		1,375	109		1,484	1,700
0540	8'-0" x 5'-0" high		7	2.286		1,400	125		1,525	1,725
0560	9'-0" x 5'-0" high	▼	6	2.667	▼	1,525	146		1,671	1,900

08 52 16 – Plastic-Clad Wood Windows

08 52 16.10 Bow Window

0010	**BOW WINDOW** including frames, screens, and grilles									
0020	End panels operable									
1000	Bow type, casement, wood, bldrs. mdl., 8' x 5' dbl. insul. glass, 4 panel	2 Carp	10	1.600	Ea.	1,575	87.50		1,662.50	1,900
1050	Low E glass		10	1.600		1,325	87.50		1,412.50	1,600
1100	10'-0" x 5'-0", dbl. insul. glass, 6 panels		6	2.667		1,400	146		1,546	1,750
1200	Low E glass, 6 panels		6	2.667		1,500	146		1,646	1,850
1300	Vinyl-clad, bldrs. model, dbl. insul. glass, 6'-0" x 4'-0", 3 panel		10	1.600		1,025	87.50		1,112.50	1,275
1340	9'-0" x 4'-0", 4 panel		8	2		1,450	109		1,559	1,775
1380	10'-0" x 6'-0", 5 panels		7	2.286		2,350	125		2,475	2,800
1420	12'-0" x 6'-0", 6 panels		6	2.667		3,100	146		3,246	3,625
2000	Bay window, 8' x 5', dbl. insul. glass		10	1.600		1,975	87.50		2,062.50	2,325
2050	Low E glass		10	1.600		2,375	87.50		2,462.50	2,775
2100	12'-0" x 6'-0", dbl. insul. glass, 6 panels		6	2.667		2,450	146		2,596	2,925
2200	Low E glass	▼	6	2.667	▼	3,450	146		3,596	4,025

08 52 Wood Windows

08 52 16 – Plastic-Clad Wood Windows

08 52 16.10 Bow Window		Crew	Daily Output	Labor-Hours	Unit	Material	2021 Bare Costs Labor	Equipment	Total	Total Incl O&P
2300	Vinyl-clad, premium, dbl. insul. glass, 8'-0" x 5'-0"	2 Carp	10	1.600	Ea.	1,675	87.50		1,762.50	1,975
2340	10'-0" x 5'-0"		8	2		2,350	109		2,459	2,775
2380	10'-0" x 6'-0"		7	2.286		3,100	125		3,225	3,600
2420	12'-0" x 6'-0"		6	2.667		2,975	146		3,121	3,500
3300	Vinyl-clad, premium, dbl. insul. glass, 7'-0" x 4'-6"		10	1.600		1,375	87.50		1,462.50	1,675
3340	8'-0" x 4'-6"		8	2		1,450	109		1,559	1,775
3380	8'-0" x 5'-0"		7	2.286		1,525	125		1,650	1,900
3420	9'-0" x 5'-0"		6	2.667		1,575	146		1,721	1,950
9000	Minimum labor/equipment charge		2.50	6.400	Job		350		350	555

08 52 16.15 Awning Window Vinyl-Clad		Crew	Daily Output	Labor-Hours	Unit	Material	2021 Bare Costs Labor	Equipment	Total	Total Incl O&P
0010	**AWNING WINDOW VINYL-CLAD** including frames, screens, and grilles									
0240	Vinyl-clad, 34" x 22"	1 Carp	10	.800	Ea.	271	44		315	370
0280	36" x 28"		9	.889		310	48.50		358.50	420
0300	36" x 36"		9	.889		355	48.50		403.50	470
0340	40" x 22"		10	.800		300	44		344	400
0360	48" x 28"		8	1		375	54.50		429.50	495
0380	60" x 36"		8	1		515	54.50		569.50	650

08 52 16.20 Half Round, Vinyl-Clad		Crew	Daily Output	Labor-Hours	Unit	Material	2021 Bare Costs Labor	Equipment	Total	Total Incl O&P
0010	**HALF ROUND, VINYL-CLAD**, double insulated glass, incl. grille									
0800	14" height x 24" base	2 Carp	9	1.778	Ea.	410	97		507	605
1040	15" height x 25" base		8	2		375	109		484	585
1060	16" height x 28" base		7	2.286		430	125		555	675
1080	17" height x 29" base		7	2.286		445	125		570	690
2000	19" height x 33" base	1 Carp	6	1.333		455	73		528	615
2100	20" height x 35" base		6	1.333		455	73		528	615
2200	21" height x 37" base		6	1.333		480	73		553	640
2250	23" height x 41" base	2 Carp	6	2.667		535	146		681	820
2300	26" height x 48" base		6	2.667		560	146		706	845
2350	30" height x 56" base		6	2.667		650	146		796	945
3000	36" height x 67" base	1 Carp	4	2		1,150	109		1,259	1,425
3040	38" height x 71" base	2 Carp	5	3.200		1,000	175		1,175	1,400
3050	40" height x 75" base	"	5	3.200		1,325	175		1,500	1,725
5000	Elliptical, 71" x 16"	1 Carp	11	.727		1,000	40		1,040	1,175
5100	95" x 21"	"	10	.800		1,425	44		1,469	1,650

08 52 16.35 Double-Hung Window			Crew	Daily Output	Labor-Hours	Unit	Material	2021 Bare Costs Labor	Equipment	Total	Total Incl O&P
0010	**DOUBLE-HUNG WINDOW** including frames, screens, and grilles										
0300	Vinyl-clad, premium, double insulated glass, 2'-6" x 3'-0"	G	1 Carp	10	.800	Ea.	430	44		474	545
0305	2'-6" x 4'-0"	G		10	.800		440	44		484	555
0400	3'-0" x 3'-6"	G		10	.800		360	44		404	465
0500	3'-0" x 4'-0"	G		9	.889		405	48.50		453.50	525
0600	3'-0" x 4'-6"	G		9	.889		450	48.50		498.50	570
0700	3'-0" x 5'-0"	G		8	1		460	54.50		514.50	590
0790	3'-4" x 5'-0"	G		8	1		470	54.50		524.50	605
0800	3'-6" x 6'-0"	G		8	1		530	54.50		584.50	665
0820	4'-0" x 5'-0"	G		7	1.143		555	62.50		617.50	710
0830	4'-0" x 6'-0"	G		7	1.143		675	62.50		737.50	845

08 52 16.40 Transom Windows		Crew	Daily Output	Labor-Hours	Unit	Material	2021 Bare Costs Labor	Equipment	Total	Total Incl O&P
0010	**TRANSOM WINDOWS**									
0050	Vinyl-clad, premium, dbl. insul. glass, 32" x 8"	1 Carp	16	.500	Ea.	232	27.50		259.50	299
0100	36" x 8"		16	.500		209	27.50		236.50	274
0110	36" x 12"		16	.500		224	27.50		251.50	291
2000	Custom sizes, up to 350 sq. in.		12	.667		289	36.50		325.50	380

08 52 16 – Plastic-Clad Wood Windows

08 52 16.40 Transom Windows		Crew	Daily Output	Labor-Hours	Unit	Material	2021 Bare Costs Labor	Equipment	Total	Total Incl O&P
2100	351 to 750 sq. in.	1 Carp	12	.667	Ea.	395	36.50		431.50	495
2200	751 to 1150 sq. in.		11	.727		430	40		470	535
2300	1151 to 1450 sq. in.		11	.727		505	40		545	620
2400	1451 to 1850 sq. in.	2 Carp	12	1.333		650	73		723	830
2500	1851 to 2250 sq. in.		12	1.333		655	73		728	835
2600	2251 to 2650 sq. in.		11	1.455		865	79.50		944.50	1,075
2700	2651 to 3050 sq. in.		11	1.455		765	79.50		844.50	965
2800	3051 to 3450 sq. in.		11	1.455		820	79.50		899.50	1,025
2900	3451 to 3850 sq. in.		10	1.600		885	87.50		972.50	1,100
3000	3851 to 4250 sq. in.		10	1.600		1,050	87.50		1,137.50	1,300
3100	4251 to 4650 sq. in.		10	1.600		1,075	87.50		1,162.50	1,325
3200	4651 to 5050 sq. in.		10	1.600		1,175	87.50		1,262.50	1,450
3300	5051 to 5450 sq. in.		9	1.778		1,150	97		1,247	1,400
3400	5451 to 5850 sq. in.		9	1.778		1,000	97		1,097	1,275
3600	6251 to 6650 sq. in.		8	2		1,250	109		1,359	1,550
3700	6651 to 7050 sq. in.		8	2		1,575	109		1,684	1,925

08 52 16.45 Trapezoid Windows

		Crew	Daily Output	Labor-Hours	Unit	Material	2021 Bare Costs Labor	Equipment	Total	Total Incl O&P
0010	**TRAPEZOID WINDOWS**									
0900	20" base x 44" leg x 53" leg	2 Carp	13	1.231	Ea.	410	67.50		477.50	555
1000	24" base x 90" leg x 102" leg		8	2		740	109		849	990
3000	36" base x 40" leg x 22" leg		12	1.333		460	73		533	625
3010	36" base x 44" leg x 25" leg		13	1.231		490	67.50		557.50	640
3050	36" base x 26" leg x 48" leg		9	1.778		490	97		587	690
3100	36" base x 42" legs, 50" peak		9	1.778		575	97		672	785
3200	36" base x 60" leg x 81" leg		11	1.455		750	79.50		829.50	950
4320	44" base x 23" leg x 56" leg		11	1.455		580	79.50		659.50	765
4350	44" base x 59" leg x 92" leg		10	1.600		940	87.50		1,027.50	1,175
4500	46" base x 15" leg x 46" leg		8	2		465	109		574	685
4550	46" base x 16" leg x 48" leg		8	2		490	109		599	715
4600	46" base x 50" leg x 80" leg		7	2.286		735	125		860	1,000
6600	66" base x 12" leg x 42" leg		8	2		635	109		744	870
6650	66" base x 12" legs, 28" peak		9	1.778		490	97		587	695
6700	68" base x 23" legs, 31" peak		8	2		635	109		744	870

08 52 16.70 Vinyl-Clad, Premium, DBL. Insul. Glass

			Crew	Daily Output	Labor-Hours	Unit	Material	2021 Bare Costs Labor	Equipment	Total	Total Incl O&P
0010	**VINYL-CLAD, PREMIUM, DBL. INSUL. GLASS**										
1000	Sliding, 3'-0" x 3'-0"	G	1 Carp	10	.800	Ea.	655	44		699	790
1050	4'-0" x 3'-6"	G		9	.889		895	48.50		943.50	1,075
1100	5'-0" x 4'-0"	G		9	.889		945	48.50		993.50	1,125
1150	6'-0" x 5'-0"	G		8	1		1,225	54.50		1,279.50	1,425

08 52 50 – Window Accessories

08 52 50.10 Window Grille or Muntin

		Crew	Daily Output	Labor-Hours	Unit	Material	2021 Bare Costs Labor	Equipment	Total	Total Incl O&P
0010	**WINDOW GRILLE OR MUNTIN**, snap in type									
0020	Standard pattern interior grilles									
2000	Wood, awning window, glass size, 28" x 16" high	1 Carp	30	.267	Ea.	31	14.60		45.60	57.50
2060	44" x 24" high		32	.250		45	13.70		58.70	71.50
2100	Casement, glass size, 20" x 36" high		30	.267		34.50	14.60		49.10	61
2180	20" x 56" high		32	.250		47	13.70		60.70	73.50
2200	Double-hung, glass size, 16" x 24" high		24	.333	Set	60.50	18.25		78.75	95.50
2280	32" x 32" high		34	.235	"	140	12.85		152.85	175
2500	Picture, glass size, 48" x 48" high		30	.267	Ea.	130	14.60		144.60	166
2580	60" x 68" high		28	.286	"	205	15.65		220.65	251
2600	Sliding, glass size, 14" x 36" high		24	.333	Set	47	18.25		65.25	81

For customer support on your Commercial Renovation Costs with RSMeans Data, call 800.448.8182.

08 52 Wood Windows

08 52 50 – Window Accessories

08 52 50.10 Window Grille or Muntin		Crew	Daily Output	Labor-Hours	Unit	Material	2021 Bare Costs Labor	Equipment	Total	Total Incl O&P
2680	36" x 36" high	1 Carp	22	.364	Set	46.50	19.90		66.40	82.50
9000	Minimum labor/equipment charge	↓	5	1.600	Job		87.50		87.50	139

08 52 66 – Wood Window Screens

08 52 66.10 Wood Screens

	08 52 66.10 Wood Screens	Crew	Daily Output	Labor-Hours	Unit	Material	Labor	Equipment	Total	Total Incl O&P
0010	**WOOD SCREENS**									
0020	Over 3 S.F., 3/4" frames	2 Carp	375	.043	S.F.	4.84	2.33		7.17	9
0100	1-1/8" frames		375	.043		8.65	2.33		10.98	13.20
0200	Rescreen wood frame	↓	500	.032	↓	1.70	1.75		3.45	4.66
9000	Minimum labor/equipment charge	1 Carp	4	2	Job		109		109	174

08 52 69 – Wood Storm Windows

08 52 69.10 Storm Windows

	08 52 69.10 Storm Windows		Crew	Daily Output	Labor-Hours	Unit	Material	Labor	Equipment	Total	Total Incl O&P
0010	**STORM WINDOWS**, aluminum residential										
0300	Basement, mill finish, incl. fiberglass screen										
0320	1'-10" x 1'-0" high	G	2 Carp	30	.533	Ea.	37.50	29		66.50	88
0340	2'-9" x 1'-6" high	G		30	.533		41	29		70	91.50
0360	3'-4" x 2'-0" high	G	↓	30	.533	↓	53	29		82	105
1600	Double-hung, combination, storm & screen										
1700	Custom, clear anodic coating, 2'-0" x 3'-5" high		2 Carp	30	.533	Ea.	94	29		123	150
1720	2'-6" x 5'-0" high			28	.571		137	31.50		168.50	200
1740	4'-0" x 6'-0" high			25	.640		242	35		277	320
1800	White painted, 2'-0" x 3'-5" high			30	.533		115	29		144	173
1820	2'-6" x 5'-0" high			28	.571		194	31.50		225.50	264
1840	4'-0" x 6'-0" high			25	.640		297	35		332	380
2000	Clear anodic coating, 2'-0" x 3'-5" high	G		30	.533		104	29		133	161
2020	2'-6" x 5'-0" high	G		28	.571		129	31.50		160.50	192
2040	4'-0" x 6'-0" high	G		25	.640		133	35		168	202
2400	White painted, 2'-0" x 3'-5" high	G		30	.533		107	29		136	165
2420	2'-6" x 5'-0" high	G		28	.571		182	31.50		213.50	251
2440	4'-0" x 6'-0" high	G		25	.640		121	35		156	189
2600	Mill finish, 2'-0" x 3'-5" high	G		30	.533		86.50	29		115.50	142
2620	2'-6" x 5'-0" high	G		28	.571		109	31.50		140.50	169
2640	4'-0" x 6'-8" high	G	↓	25	.640	↓	137	35		172	207
4000	Picture window, storm, 1 lite, white or bronze finish										
4020	4'-6" x 4'-6" high		2 Carp	25	.640	Ea.	148	35		183	219
4040	5'-8" x 4'-6" high			20	.800		154	44		198	239
4400	Mill finish, 4'-6" x 4'-6" high			25	.640		140	35		175	209
4420	5'-8" x 4'-6" high		↓	20	.800	↓	178	44		222	266
4600	3 lite, white or bronze finish										
4620	4'-6" x 4'-6" high		2 Carp	25	.640	Ea.	180	35		215	254
4640	5'-8" x 4'-6" high			20	.800		180	44		224	268
4800	Mill finish, 4'-6" x 4'-6" high			25	.640		155	35		190	226
4820	5'-8" x 4'-6" high		↓	20	.800	↓	180	44		224	268
6000	Sliding window, storm, 2 lite, white or bronze finish										
6020	3'-4" x 2'-7" high		2 Carp	28	.571	Ea.	150	31.50		181.50	215
6040	4'-4" x 3'-3" high			25	.640		161	35		196	233
6060	5'-4" x 6'-0" high		↓	20	.800	↓	262	44		306	360
8000	PVC framed										
8100	Double-hung, combination, storm & screen										
8120	2'-6" x 3'-5"		2 Carp	15	1.067	Ea.	133	58.50		191.50	239
8140	4' x 6'		"	12.50	1.280	"	186	70		256	315
8600	Single lite picture storm										
8620	4'-6" x 4'-6"		2 Carp	12.50	1.280	Ea.	180	70		250	310

267

08 52 Wood Windows

08 52 69 – Wood Storm Windows

08 52 69.10 Storm Windows		Crew	Daily Output	Labor-Hours	Unit	Material	2021 Bare Costs Labor	Equipment	Total	Total Incl O&P
8640	5'-8" x 4'-6"	2 Carp	10	1.600	Ea.	199	87.50		286.50	360
9000	Interior storm window									
9100	Storm window interior glass	1 Glaz	107	.075	S.F.	5.70	3.94		9.64	12.50
9410	Minimum labor/equipment charge	1 Carp	4	2	Job		109		109	174

08 53 Plastic Windows

08 53 13 – Vinyl Windows

08 53 13.20 Vinyl Single-Hung Windows

			Crew	Daily Output	Labor-Hours	Unit	Material	2021 Bare Costs Labor	Equipment	Total	Total Incl O&P
0010	**VINYL SINGLE-HUNG WINDOWS**, insulated glass										
0020	Grids, low E, J fin, extension jambs										
0130	25" x 41"	G	2 Carp	20	.800	Ea.	205	44		249	296
0140	25" x 49"	G		18	.889		212	48.50		260.50	310
0150	25" x 57"	G		17	.941		232	51.50		283.50	335
0160	25" x 65"	G		16	1		260	54.50		314.50	375
0170	29" x 41"	G		18	.889		206	48.50		254.50	305
0180	29" x 53"	G		18	.889		224	48.50		272.50	325
0190	29" x 57"	G		17	.941		242	51.50		293.50	350
0200	29" x 65"	G		16	1		288	54.50		342.50	400
0210	33" x 41"	G		20	.800		223	44		267	315
0220	33" x 53"	G		18	.889		238	48.50		286.50	340
0230	33" x 57"	G		17	.941		288	51.50		339.50	395
0240	33" x 65"	G		16	1		295	54.50		349.50	410
0250	37" x 41"	G		20	.800		226	44		270	320
0260	37" x 53"	G	2 Carp	18	.889	Ea.	259	48.50		307.50	360
0270	37" x 57"	G		17	.941		270	51.50		321.50	380
0280	37" x 65"	G		16	1		310	54.50		364.50	425

08 53 13.30 Vinyl Double-Hung Windows

			Crew	Daily Output	Labor-Hours	Unit	Material	2021 Bare Costs Labor	Equipment	Total	Total Incl O&P
0010	**VINYL DOUBLE-HUNG WINDOWS**, insulated glass										
0100	Grids, low E, J fin, ext. jambs, 21" x 53"	G	2 Carp	18	.889	Ea.	191	48.50		239.50	288
0102	21" x 37"	G		18	.889		310	48.50		358.50	420
0104	21" x 41"	G		18	.889		325	48.50		373.50	435
0106	21" x 49"	G		18	.889		287	48.50		335.50	395
0110	21" x 57"	G		17	.941		355	51.50		406.50	475
0120	21" x 65"	G		16	1		385	54.50		439.50	505
0128	25" x 37"	G		20	.800		305	44		349	410
0130	25" x 41"	G		20	.800		320	44		364	425
0140	25" x 49"	G		18	.889		325	48.50		373.50	440
0145	25" x 53"	G		18	.889		360	48.50		408.50	475
0150	25" x 57"	G		17	.941		289	51.50		340.50	400
0160	25" x 65"	G		16	1		266	54.50		320.50	380
0162	25" x 69"	G		16	1		315	54.50		369.50	430
0164	25" x 77"	G		16	1		320	54.50		374.50	440
0168	29" x 37"	G		18	.889		320	48.50		368.50	430
0170	29" x 41"	G		18	.889		345	48.50		393.50	460
0172	29" x 49"	G		18	.889		355	48.50		403.50	470
0180	29" x 53"	G		18	.889		365	48.50		413.50	480
0190	29" x 57"	G		17	.941		247	51.50		298.50	355
0200	29" x 65"	G		16	1		370	54.50		424.50	490
0202	29" x 69"	G		16	1		340	54.50		394.50	460
0205	29" x 77"	G		16	1		305	54.50		359.50	420
0208	33" x 37"	G		20	.800		340	44		384	445

08 53 Plastic Windows

08 53 13 – Vinyl Windows

08 53 13.30 Vinyl Double-Hung Windows		Crew	Daily Output	Labor-Hours	Unit	Material	2021 Bare Costs Labor	Equipment	Total	Total Incl O&P
0210	33" x 41" G	2 Carp	20	.800	Ea.	350	44		394	455
0215	33" x 49" G		20	.800		291	44		335	390
0220	33" x 53" G		18	.889		250	48.50		298.50	355
0230	33" x 57" G		17	.941		283	51.50		334.50	390
0240	33" x 65" G		16	1		325	54.50		379.50	445
0242	33" x 69" G		16	1		315	54.50		369.50	435
0246	33" x 77" G		16	1		295	54.50		349.50	410
0250	37" x 41" G		20	.800		299	44		343	400
0255	37" x 49" G		20	.800		320	44		364	420
0260	37" x 53" G		18	.889		325	48.50		373.50	440
0270	37" x 57" G		17	.941		350	51.50		401.50	465
0280	37" x 65" G		16	1		375	54.50		429.50	500
0282	37" x 69" G		16	1		390	54.50		444.50	515
0286	37" x 77" G		16	1		385	54.50		439.50	510
0300	Solid vinyl, average quality, double insulated glass, 2'-0" x 3'-0" G	1 Carp	10	.800		299	44		343	400
0310	3'-0" x 4'-0" G	1 Carp	9	.889	Ea.	239	48.50		287.50	340
0330	Premium, double insulated glass, 2'-6" x 3'-0" G		10	.800		320	44		364	425
0340	3'-0" x 3'-6" G		9	.889		340	48.50		388.50	450
0350	3'-0" x 4'-0" G		9	.889		345	48.50		393.50	460
0360	3'-0" x 4'-6" G		9	.889		375	48.50		423.50	495
0370	3'-0" x 5'-0" G		8	1		425	54.50		479.50	550
0380	3'-6" x 6'-0" G		8	1		420	54.50		474.50	550

08 53 13.40 Vinyl Casement Windows

08 53 13.40 Vinyl Casement Windows		Crew	Daily Output	Labor-Hours	Unit	Material	2021 Bare Costs Labor	Equipment	Total	Total Incl O&P
0010	**VINYL CASEMENT WINDOWS**, insulated glass									
0015	Grids, low E, J fin, extension jambs, screens									
0100	One lite, 21" x 41" G	2 Carp	20	.800	Ea.	415	44		459	525
0110	21" x 47" G		20	.800		425	44		469	535
0120	21" x 53" G		20	.800		455	44		499	575
0128	24" x 35" G		19	.842		360	46		406	470
0130	24" x 41" G		19	.842		420	46		466	535
0140	24" x 47" G		19	.842		445	46		491	565
0150	24" x 53" G		19	.842		465	46		511	590
0158	28" x 35" G		19	.842		415	46		461	530
0160	28" x 41" G		19	.842		370	46		416	485
0170	28" x 47" G		19	.842		465	46		511	590
0180	28" x 53" G		19	.842		305	46		351	410
0184	28" x 59" G		19	.842		355	46		401	465
0188	Two lites, 33" x 35" G		18	.889		425	48.50		473.50	545
0190	33" x 41" G		18	.889		410	48.50		458.50	535
0200	33" x 47" G		18	.889		435	48.50		483.50	560
0210	33" x 53" G		18	.889		460	48.50		508.50	585
0212	33" x 59" G		18	.889		480	48.50		528.50	610
0215	33" x 72" G		18	.889		520	48.50		568.50	650
0220	41" x 41" G		18	.889		510	48.50		558.50	640
0230	41" x 47" G		18	.889		510	48.50		558.50	640
0240	41" x 53" G		17	.941		610	51.50		661.50	750
0242	41" x 59" G		17	.941		590	51.50		641.50	730
0246	41" x 72" G		17	.941		610	51.50		661.50	750
0250	47" x 41" G		17	.941		660	51.50		711.50	810
0260	47" x 47" G		17	.941		500	51.50		551.50	630
0270	47" x 53" G		17	.941		550	51.50		601.50	685
0272	47" x 59" G		17	.941		745	51.50		796.50	900

08 53 Plastic Windows

08 53 13 – Vinyl Windows

08 53 13.40 Vinyl Casement Windows

			Crew	Daily Output	Labor-Hours	Unit	Material	2021 Bare Costs Labor	Equipment	Total	Total Incl O&P
0280	56" x 41"	G	2 Carp	15	1.067	Ea.	550	58.50		608.50	700
0290	56" x 47"	G		15	1.067		550	58.50		608.50	700
0300	56" x 53"	G		15	1.067		605	58.50		663.50	760
0302	56" x 59"	G		15	1.067		605	58.50		663.50	760
0310	56" x 72"	G		15	1.067		780	58.50		838.50	950
0340	Solid vinyl, premium, double insulated glass, 2'-0" x 3'-0" high	G	1 Carp	10	.800		274	44		318	370
0360	2'-0" x 4'-0" high	G	1 Carp	9	.889	Ea.	300	48.50		348.50	410
0380	2'-0" x 5'-0" high	G	"	8	1	"	340	54.50		394.50	460

08 53 13.50 Vinyl Picture Windows

			Crew	Daily Output	Labor-Hours	Unit	Material	2021 Bare Costs Labor	Equipment	Total	Total Incl O&P
0010	**VINYL PICTURE WINDOWS**, insulated glass										
0120	Grids, low E, J fin, ext. jambs, 47" x 35"		2 Carp	12	1.333	Ea.	305	73		378	450
0130	47" x 41"			12	1.333		430	73		503	585
0140	47" x 47"			12	1.333		315	73		388	460
0150	47" x 53"			11	1.455		405	79.50		484.50	575
0160	71" x 35"			11	1.455		450	79.50		529.50	620
0170	71" x 41"			11	1.455		475	79.50		554.50	645
0180	71" x 47"			11	1.455		510	79.50		589.50	685

08 53 13.60 Vinyl Half Round Windows

			Crew	Daily Output	Labor-Hours	Unit	Material	2021 Bare Costs Labor	Equipment	Total	Total Incl O&P
0010	**VINYL HALF ROUND WINDOWS**, Including grille, J fin, low E, ext. jambs										
0100	10" height x 20" base		2 Carp	8	2	Ea.	340	109		449	550
0110	15" height x 30" base			8	2		315	109		424	525
0120	17" height x 34" base			7	2.286		410	125		535	650
0130	19" height x 38" base			7	2.286		625	125		750	885
0140	19" height x 33" base			7	2.286		545	125		670	800
0150	24" height x 48" base		1 Carp	6	1.333		455	73		528	615
0160	25" height x 50" base		"	6	1.333		560	73		633	730
0170	30" height x 60" base		2 Carp	6	2.667		620	146		766	915

08 54 Composite Windows

08 54 13 – Fiberglass Windows

08 54 13.10 Fiberglass Single-Hung Windows

			Crew	Daily Output	Labor-Hours	Unit	Material	2021 Bare Costs Labor	Equipment	Total	Total Incl O&P
0010	**FIBERGLASS SINGLE-HUNG WINDOWS**										
0100	Grids, low E, 18" x 24"	G	2 Carp	18	.889	Ea.	375	48.50		423.50	495
0110	18" x 40"	G		17	.941		370	51.50		421.50	485
0130	24" x 40"	G		20	.800		400	44		444	510
0230	36" x 36"	G		17	.941		405	51.50		456.50	530
0250	36" x 48"	G		20	.800		445	44		489	560
0260	36" x 60"	G		18	.889		490	48.50		538.50	620
0280	36" x 72"	G		16	1		530	54.50		584.50	665
0290	48" x 40"	G		16	1		530	54.50		584.50	665

08 54 13.30 Fiberglass Slider Windows

			Crew	Daily Output	Labor-Hours	Unit	Material	2021 Bare Costs Labor	Equipment	Total	Total Incl O&P
0010	**FIBERGLASS SLIDER WINDOWS**										
0100	Grids, low E, 36" x 24"	G	2 Carp	20	.800	Ea.	380	44		424	485
0110	36" x 36"	G	"	20	.800	"	380	44		424	490

08 54 13.50 Fiberglass Bay Windows

			Crew	Daily Output	Labor-Hours	Unit	Material	2021 Bare Costs Labor	Equipment	Total	Total Incl O&P
0010	**FIBERGLASS BAY WINDOWS**										
0150	48" x 36"	G	2 Carp	11	1.455	Ea.	1,125	79.50		1,204.50	1,350

08 56 Special Function Windows

08 56 46 – Radio-Frequency-Interference Shielding Windows

08 56 46.10 Radio-Frequency-Interference Mesh	Crew	Daily Output	Labor-Hours	Unit	Material	2021 Bare Costs Labor	Equipment	Total	Total Incl O&P
0010 **RADIO-FREQUENCY-INTERFERENCE MESH**									
0100 16 mesh copper 0.011" wire				S.F.	1.68			1.68	1.85
0150 22 mesh copper 0.015" wire				"	6			6	6.60
0200 100 mesh copper 0.022" wire				S.F.	2.88			2.88	3.17
0250 100 mesh stainless steel 0.0012" wire				"	6.90			6.90	7.55

08 61 Roof Windows

08 61 13 – Metal Roof Windows

08 61 13.10 Roof Windows

	Crew	Daily Output	Labor-Hours	Unit	Material	2021 Bare Costs Labor	Equipment	Total	Total Incl O&P
0010 **ROOF WINDOWS**, fixed high perf tmpd glazing, metallic framed									
0020 46" x 21-1/2", Flashed for shingled roof	1 Carp	8	1	Ea.	297	54.50		351.50	410
0100 46" x 28"		8	1		330	54.50		384.50	450
0125 57" x 44"		6	1.333		395	73		468	550
0130 72" x 28"		7	1.143		395	62.50		457.50	535
0150 Fixed, laminated tempered glazing, 46" x 21-1/2"		8	1		480	54.50		534.50	610
0175 46" x 28"		8	1		530	54.50		584.50	665
0200 57" x 44"		6	1.333		505	73		578	670
0500 Vented flashing set for shingled roof, 46" x 21-1/2"		7	1.143		455	62.50		517.50	600
0525 46" x 28"		6	1.333		570	73		643	745
0550 57" x 44"		5	1.600		945	87.50		1,032.50	1,200
0560 72" x 28"		5	1.600		945	87.50		1,032.50	1,200
0575 Flashing set for low pitched roof, 46" x 21-1/2"		7	1.143		475	62.50		537.50	620
0600 46" x 28"		7	1.143		510	62.50		572.50	660
0625 57" x 44"		5	1.600		615	87.50		702.50	815
0650 Flashing set for curb, 46" x 21-1/2"		7	1.143		670	62.50		732.50	835
0675 46" x 28"		7	1.143		730	62.50		792.50	900
0700 57" x 44"		5	1.600		900	87.50		987.50	1,125

08 62 Unit Skylights

08 62 13 – Domed Unit Skylights

08 62 13.10 Domed Skylights

		Crew	Daily Output	Labor-Hours	Unit	Material	2021 Bare Costs Labor	Equipment	Total	Total Incl O&P
0010 **DOMED SKYLIGHTS**										
0020 Skylight, fixed dome type, 22" x 22"	G	G-3	12	2.667	Ea.	231	146		377	485
0030 22" x 46"	G		10	3.200		273	176		449	575
0040 30" x 30"	G		12	2.667		289	146		435	550
0050 30" x 46"	G		10	3.200		375	176		551	685
0110 Fixed, double glazed, 22" x 27"	G		12	2.667		273	146		419	530
0120 22" x 46"	G		10	3.200		315	176		491	625
0130 44" x 46"	G		10	3.200		430	176		606	745
0210 Operable, double glazed, 22" x 27"	G		12	2.667		400	146		546	670
0220 22" x 46"	G		10	3.200		485	176		661	805
0230 44" x 46"	G		10	3.200		835	176		1,011	1,200
9000 Minimum labor/equipment charge			2	16	Job		880		880	1,375

08 62 13.20 Skylights

		Crew	Daily Output	Labor-Hours	Unit	Material	2021 Bare Costs Labor	Equipment	Total	Total Incl O&P
0010 **SKYLIGHTS**, flush or curb mounted										
2120 Ventilating insulated plexiglass dome with										
2130 curb mounting, 36" x 36"	G	G-3	12	2.667	Ea.	460	146		606	735
2150 52" x 52"	G	"	12	2.667	"	645	146		791	940
2160 28" x 52"	G	G-3	10	3.200	Ea.	535	176		711	865

08 62 Unit Skylights

08 62 13 – Domed Unit Skylights

08 62 13.20 Skylights

		Crew	Daily Output	Labor-Hours	Unit	Material	2021 Bare Costs Labor	2021 Bare Costs Equipment	Total	Total Incl O&P
2170	36" x 52"	G	G-3	10	3.200	Ea.	590	176		766	925
2180	For electric opening system, add	G					345			345	375
2210	Operating skylight, with thermopane glass, 24" x 48"	G	G-3	10	3.200		595	176		771	930
2220	32" x 48"	G		9	3.556		625	195		820	990
2300	Insulated safety glass with aluminum frame	G		160	.200	S.F.	95	11		106	122
4000	Skylight, solar tube kit, incl. dome, flashing, diffuser, 1 pipe, 10" diam.	G	1 Carp	2	4	Ea.	305	219		524	685
4010	14" diam.	G		2	4		375	219		594	760
4020	21" diam.	G		2	4		465	219		684	865
4030	Accessories for, 1' long x 9" diam. pipe	G		24	.333		58.50	18.25		76.75	93
4040	2' long x 9" diam. pipe	G		24	.333		43.50	18.25		61.75	76.50
4050	4' long x 9" diam. pipe	G		20	.400		73.50	22		95.50	116
4060	1' long x 13" diam. pipe	G		24	.333		66.50	18.25		84.75	102
4070	2' long x 13" diam. pipe	G		24	.333		55	18.25		73.25	89.50
4080	4' long x 13" diam. pipe	G		20	.400		78.50	22		100.50	121
4090	6.5" turret ext. for 21" diam. pipe	G		16	.500		133	27.50		160.50	191
4100	12' long x 21" diam. flexible pipe	G		12	.667		99	36.50		135.50	167
4110	45 degree elbow, 10"	G		16	.500		190	27.50		217.50	253
4120	14"	G		16	.500		100	27.50		127.50	154
4130	Interior decorative ring, 9"	G		20	.400		53.50	22		75.50	94
4140	13"	G		20	.400		54.50	22		76.50	95

08 63 Metal-Framed Skylights

08 63 13 – Domed Metal-Framed Skylights

08 63 13.20 Skylight Rigid Metal-Framed

		Crew	Daily Output	Labor-Hours	Unit	Material	2021 Bare Costs Labor	2021 Bare Costs Equipment	Total	Total Incl O&P
0010	**SKYLIGHT RIGID METAL-FRAMED** skylight framing is aluminum									
0050	Fixed acrylic double domes, curb mount, 25-1/2" x 25-1/2"	G-3	10	3.200	Ea.	219	176		395	515
0060	25-1/2" x 33-1/2"		10	3.200		264	176		440	565
0070	25-1/2" x 49-1/2"		6	5.333		290	293		583	780
0080	33-1/2" x 33-1/2"		8	4		325	220		545	700
0090	37-1/2" x 25-1/2"		8	4		282	220		502	655
0100	37-1/2" x 37-1/2"		8	4		385	220		605	770
0110	37-1/2" x 49-1/2"		6	5.333		460	293		753	965
0120	49-1/2" x 33-1/2"		6	5.333		435	293		728	940
0130	49-1/2" x 49-1/2"		6	5.333		485	293		778	995
1000	Fixed tempered glass, curb mount, 17-1/2" x 33-1/2"		10	3.200		161	176		337	455
1020	17-1/2" x 49-1/2"		6	5.333		183	293		476	660
1030	25-1/2" x 25-1/2"		10	3.200		161	176		337	455
1040	25-1/2" x 33-1/2"		10	3.200		192	176		368	485
1050	25-1/2" x 37-1/2"		8	4		204	220		424	570
1060	25-1/2" x 49-1/2"		6	5.333		540	293		833	1,050
1070	25-1/2" x 73-1/2"	G-3	6	5.333	Ea.	370	293		663	865
1080	33-1/2" x 33-1/2"		8	4		224	220		444	590
2000	Manual vent tempered glass & screen, curb, 25-1/2" x 25-1/2"		10	3.200		420	176		596	735
2020	25-1/2" x 37-1/2"		10	3.200		480	176		656	800
2030	25-1/2" x 49-1/2"		8	4		515	220		735	915
2040	33-1/2" x 33-1/2"		8	4		550	220		770	950
2050	33-1/2" x 49-1/2"		6	5.333		695	293		988	1,225
2060	37-1/2" x 37-1/2"		6	5.333		635	293		928	1,150
2070	49-1/2" x 49-1/2"		6	5.333		815	293		1,108	1,350
3000	Electric vent tempered glass, curb mount, 25-1/2" x 25-1/2"		10	3.200		1,075	176		1,251	1,475
3020	25-1/2" x 37-1/2"		10	3.200		1,175	176		1,351	1,575

08 63 Metal-Framed Skylights

08 63 13 – Domed Metal-Framed Skylights

	08 63 13.20 Skylight Rigid Metal-Framed	Crew	Daily Output	Labor-Hours	Unit	Material	2021 Bare Costs Labor	Equipment	Total	Total Incl O&P
3030	25-1/2" x 49-1/2"	G-3	8	4	Ea.	1,200	220		1,420	1,675
3040	33-1/2" x 33-1/2"		8	4		1,225	220		1,445	1,700
3050	33-1/2" x 49-1/2"		6	5.333		1,300	293		1,593	1,875
3060	37-1/2" x 37-1/2"		6	5.333		1,275	293		1,568	1,850
3070	49-1/2" x 49-1/2"		6	5.333		1,400	293		1,693	2,000

08 71 Door Hardware

08 71 13 – Automatic Door Operators

08 71 13.10 Automatic Openers Commercial

		Crew	Daily Output	Labor-Hours	Unit	Material	2021 Bare Costs Labor	Equipment	Total	Total Incl O&P
0010	**AUTOMATIC OPENERS COMMERCIAL**									
0020	Pneumatic door opener incl. motion sens, control box, tubing, compressor									
0050	For single swing door, per opening	2 Skwk	.80	20	Ea.	5,525	1,150		6,675	7,900
0100	Pair, per opening		.50	32	Opng.	7,900	1,825		9,725	11,600
1000	For single sliding door, per opening		.60	26.667		5,900	1,525		7,425	8,925
1300	Bi-parting pair		.50	32		7,900	1,825		9,725	11,600
1420	Electronic door opener incl. motion sens, 12 V control box, motor									
1450	For single swing door, per opening	2 Skwk	.80	20	Opng.	3,975	1,150		5,125	6,175
1500	Pair, per opening		.50	32		7,950	1,825		9,775	11,600
1600	For single sliding door, per opening		.60	26.667		5,200	1,525		6,725	8,150
1700	Bi-parting pair		.50	32		5,825	1,825		7,650	9,300
1750	Handicap actuator buttons, 2, incl. 12 V DC wiring, add	1 Carp	1.50	5.333	Pr.	480	292		772	990
2000	Electric panic button for door	2 Skwk	.50	32	Opng.	242	1,825		2,067	3,175

08 71 20.15 Hardware

		Crew	Daily Output	Labor-Hours	Unit	Material	2021 Bare Costs Labor	Equipment	Total	Total Incl O&P
0010	**HARDWARE**									
0020	Average hardware cost									
1000	Door hardware, apartment, interior	1 Carp	4	2	Door	585	109		694	815
1300	Average, door hardware, motel/hotel interior, with access card		4	2		710	109		819	960
1500	Hospital bedroom, average quality		4	2		650	109		759	890
2000	High quality		3	2.667		695	146		841	995
2250	School, single exterior, incl. lever, incl. panic device		3	2.667		1,450	146		1,596	1,825
2500	Single interior, regular use, lever included	1 Carp	3	2.667	Door	620	146		766	910
2600	Average, door hdwe. set, school, classroom, ANSI F88, incl. lever		3	2.667		950	146		1,096	1,275
2850	Stairway, single interior		3	2.667		670	146		816	970
3100	Double exterior, with panic device		2	4	Pr.	2,750	219		2,969	3,375
6020	Add for fire alarm door holder, electro-magnetic	1 Elec	4	2	Ea.	92	127		219	297

08 71 20.30 Door Closers

		Crew	Daily Output	Labor-Hours	Unit	Material	2021 Bare Costs Labor	Equipment	Total	Total Incl O&P
0010	**DOOR CLOSERS** adjustable backcheck, multiple mounting									
0015	and rack and pinion									
0020	Standard Regular Arm	1 Carp	6	1.333	Ea.	128	73		201	257
0040	Hold open arm		6	1.333		102	73		175	228
0100	Fusible link		6.50	1.231		247	67.50		314.50	380
1500	Concealed closers, normal use, head, pivot hung, interior		5.50	1.455		380	79.50		459.50	540
1510	Exterior		5.50	1.455		445	79.50		524.50	615
1520	Overhead concealed, all sizes, regular arm		5.50	1.455		199	79.50		278.50	345
1525	Concealed arm		5	1.600		340	87.50		427.50	515
1530	Concealed in door, all sizes, regular arm		5.50	1.455		435	79.50		514.50	600
1535	Concealed arm		5	1.600		270	87.50		357.50	435
1560	Floor concealed, all sizes, single acting		2.20	3.636		595	199		794	965
1565	Double acting		2.20	3.636		510	199		709	880
1570	Interior, floor, offset pivot, single acting		3.50	2.286		1,050	125		1,175	1,350

273

08 71 Door Hardware

08 71 13 – Automatic Door Operators

08 71 20.30 Door Closers

		Crew	Daily Output	Labor-Hours	Unit	Material	2021 Bare Costs Labor	2021 Bare Costs Equipment	Total	Total Incl O&P
1590	Exterior	1 Carp	3.50	2.286	Ea.	1,150	125		1,275	1,475
1610	Hold open arm		6	1.333		450	73		523	610
1620	Double acting, standard arm		6	1.333		1,000	73		1,073	1,225
1630	Hold open arm		6	1.333		1,050	73		1,123	1,275
1640	Floor, center hung, single acting, bottom arm		6	1.333		420	73		493	580
1650	Double acting		6	1.333		900	73		973	1,100
1660	Offset hung, single acting, bottom arm		6	1.333		710	73		783	895
2000	Backcheck and adjustable power, hinge face mount									
5000	For cast aluminum cylinder, deduct				Ea.	41.50			41.50	45.50
5040	For delayed action, add					51			51	56
5080	For fusible link arm, add					61			61	67
5120	For shock absorbing arm, add					54			54	59.50
5160	For spring power adjustment, add					39.50			39.50	43.50
6000	Closer-holder, hinge face mount, all sizes, exposed arm	1 Carp	6.50	1.231		236	67.50		303.50	365
6500	Electro magnetic closer/holder									
6510	Single point, no detector	1 Carp	4	2	Ea.	645	109		754	885
6515	Including detector		4	2		740	109		849	990
6520	Multi-point, no detector		4	2		985	109		1,094	1,250
6524	Including detector		4	2		1,525	109		1,634	1,850
6550	Electric automatic operators									
6555	Operator	1 Carp	4	2	Ea.	2,625	109		2,734	3,075
6570	Wall plate actuator		4	2		266	109		375	465
7000	Electronic closer-holder, hinge facemount, concealed arm		5	1.600		490	87.50		577.50	680
7400	With built-in detector		5	1.600		620	87.50		707.50	820
9000	Minimum labor/equipment charge		4	2	Job		109		109	174

08 71 20.36 Panic Devices

		Crew	Daily Output	Labor-Hours	Unit	Material	2021 Bare Costs Labor	2021 Bare Costs Equipment	Total	Total Incl O&P
0010	**PANIC DEVICES** R087110-10									
0015	Touch bars various styles									
0040	Single door exit only, rim device, wide stile									
0050	Economy US28	1 Carp	5	1.600	Ea.	280	87.50		367.50	450
0060	Standard duty US28		5	1.600		820	87.50		907.50	1,050
0065	US26D		5	1.600		825	87.50		912.50	1,050
0070	US10		5	1.600		880	87.50		967.50	1,100
0075	US3		5	1.600		1,000	87.50		1,087.50	1,250
0080	Night latch, economy US28		5	1.600		370	87.50		457.50	545
0085	Standard duty, night latch US28		5	1.600		920	87.50		1,007.50	1,150
0090	US26D		5	1.600		1,025	87.50		1,112.50	1,275
0095	US10		5	1.600		1,075	87.50		1,162.50	1,325
0100	US3		5	1.600		1,225	87.50		1,312.50	1,500
0500	Single door exit only, rim device, narrow stile									
0540	Economy US28	1 Carp	5	1.600	Ea.	475	87.50		562.50	665
0550	Standard duty US28		5	1.600		885	87.50		972.50	1,125
0565	US26D		5	1.600		1,150	87.50		1,237.50	1,400
0570	US10		5	1.600		900	87.50		987.50	1,125
0575	US3		5	1.600		1,125	87.50		1,212.50	1,400
0580	Economy with night latch US28		5	1.600		360	87.50		447.50	535
0585	Standard duty US28		5	1.600		995	87.50		1,082.50	1,250
0590	US26D		5	1.600		1,150	87.50		1,237.50	1,400
0595	US10		5	1.600		920	87.50		1,007.50	1,175
0600	US32D		5	1.600		1,300	87.50		1,387.50	1,575
1040	Single door exit only, surface vertical rod									
1050	Surface vertical rod, economy US28	1 Carp	4	2	Ea.	545	109		654	775

08 71 Door Hardware

08 71 13 – Automatic Door Operators

08 71 20.36 Panic Devices

		Crew	Daily Output	Labor-Hours	Unit	Material	2021 Bare Costs Labor	Equipment	Total	Total Incl O&P
1060	Surface vertical rod, standard duty US28	1 Carp	4	2	Ea.	1,100	109		1,209	1,400
1065	US26D		4	2		1,200	109		1,309	1,500
1070	US10		4	2		1,200	109		1,309	1,500
1075	US3		4	2		1,250	109		1,359	1,550
1080	US32D		4	2		1,400	109		1,509	1,700
1500	Single door exit, rim, narrow stile, economy, surface vertical rod									
1540	Surface vertical rod, narrow, economy US28	1 Carp	4	2	Ea.	545	109		654	775
2040	Single door exit only, concealed vertical rod									
2042	These devices will require separate and extra door preparation									
2060	Concealed rod, standard duty US28	1 Carp	4	2	Ea.	1,100	109		1,209	1,400
2065	US26D		4	2		1,250	109		1,359	1,550
2070	US10		4	2		1,350	109		1,459	1,675
2560	Concealed rod, narrow, standard duty US28		4	2		1,275	109		1,384	1,575
2565	US26D		4	2		1,500	109		1,609	1,825
2567	US26		4	2		1,575	109		1,684	1,900
2570	US10		4	2		1,325	109		1,434	1,625
2575	US3		4	2		1,550	109		1,659	1,875
3040	Single door exit only, mortise device, wide stile									
3042	These devices must be combined with a mortise lock									
3060	Mortise, standard duty US28	1 Carp	4	2	Ea.	1,175	109		1,284	1,475
3065	US26D		4	2		1,175	109		1,284	1,450
3067	US26		4	2		1,400	109		1,509	1,725
3070	US10		4	2		1,275	109		1,384	1,575
3075	US3		4	2		1,225	109		1,334	1,525
3080	US32D		4	2		1,225	109		1,334	1,525
3500	Single door exit only, mortise device, narrow stile									
3510	These devices must be combined with a mortise lock									
3540	Mortise, economy US28	1 Carp	4	2	Ea.	440	109		549	660

08 71 20.40 Lockset

		Crew	Daily Output	Labor-Hours	Unit	Material	2021 Bare Costs Labor	Equipment	Total	Total Incl O&P
0010	**LOCKSET**, Standard duty									
0020	Non-keyed, passage, w/sect. trim	1 Carp	12	.667	Ea.	84	36.50		120.50	151
0100	Privacy		12	.667		75.50	36.50		112	141
0400	Keyed, single cylinder function		10	.800		196	44		240	285
0420	Hotel (see also Section 08 71 20.15)		8	1		208	54.50		262.50	315
0500	Lever handled, keyed, single cylinder function		10	.800		162	44		206	248
1000	Heavy duty with sectional trim, non-keyed, passages		12	.667		119	36.50		155.50	189
1100	Privacy		12	.667		154	36.50		190.50	228
1400	Keyed, single cylinder function		10	.800		184	44		228	273
1420	Hotel		8	1		650	54.50		704.50	800
1600	Communicating		10	.800		276	44		320	375
1690	For re-core cylinder, add					66.50			66.50	73
3800	Cipher lockset w/key pad (security item)	1 Carp	13	.615		1,075	33.50		1,108.50	1,225
3900	Cipher lockset with dial for swinging doors (security item)		13	.615		1,900	33.50		1,933.50	2,125
3920	with dial for swinging doors & drill resistant plate (security item)		12	.667		2,300	36.50		2,336.50	2,575
3950	Cipher lockset with dial for safe/vault door (security item)		12	.667		1,650	36.50		1,686.50	1,875
3980	Keyless, pushbutton type									
4000	Residential/light commercial, deadbolt, standard	1 Carp	9	.889	Ea.	158	48.50		206.50	252
4010	Heavy duty	"	9	.889	"	305	48.50		353.50	415
4020	Industrial, heavy duty, with deadbolt	1 Carp	9	.889	Ea.	525	48.50		573.50	660
4030	Key override		9	.889		515	48.50		563.50	650
4040	Lever activated handle		9	.889		500	48.50		548.50	630
4050	Key override		9	.889		495	48.50		543.50	625

For customer support on your Commercial Renovation Costs with RSMeans Data, call 800.448.8182.

275

08 71 20.40 Lockset

		Crew	Daily Output	Labor-Hours	Unit	Material	2021 Bare Costs Labor	2021 Bare Costs Equipment	Total	Total Incl O&P
4060	Double sided pushbutton type	1 Carp	8	1	Ea.	855	54.50		909.50	1,025
4070	Key override		8	1		945	54.50		999.50	1,125
9000	Minimum labor/equipment charge		6	1.333	Job		73		73	116

08 71 20.41 Dead Locks

		Crew	Daily Output	Labor-Hours	Unit	Material	2021 Bare Costs Labor	2021 Bare Costs Equipment	Total	Total Incl O&P
0010	**DEAD LOCKS**									
0011	Mortise heavy duty outside key (security item)	1 Carp	9	.889	Ea.	193	48.50		241.50	290
0020	Double cylinder		9	.889		192	48.50		240.50	289
0100	Medium duty, outside key		10	.800		133	44		177	216
0110	Double cylinder		10	.800		169	44		213	256
1000	Tubular, standard duty, outside key		10	.800		56	44		100	132
1010	Double cylinder		10	.800		71.50	44		115.50	148
1200	Night latch, outside key		10	.800		53	44		97	128
1203	Deadlock night latch		7.70	1.039		53	57		110	149
1420	Deadbolt lock, single cylinder		10	.800		45.50	44		89.50	120
1440	Double cylinder		10	.800		69	44		113	146

08 71 20.42 Mortise Locksets

		Crew	Daily Output	Labor-Hours	Unit	Material	2021 Bare Costs Labor	2021 Bare Costs Equipment	Total	Total Incl O&P
0010	**MORTISE LOCKSETS**, Comm., wrought knobs & full escutcheon trim									
0015	Assumes mortise is cut									
0020	Non-keyed, passage, Grade 3	1 Carp	9	.889	Ea.	157	48.50		205.50	251
0030	Grade 1		8	1		450	54.50		504.50	575
0040	Privacy set, Grade 3		9	.889		171	48.50		219.50	266
0050	Grade 1		8	1		495	54.50		549.50	630
0100	Keyed, office/entrance/apartment, Grade 2		8	1		200	54.50		254.50	305
0110	Grade 1		7	1.143		515	62.50		577.50	670
0120	Single cylinder, typical, Grade 3		8	1		196	54.50		250.50	305
0130	Grade 1		7	1.143		535	62.50		597.50	690
0200	Hotel, room, Grade 3		7	1.143		192	62.50		254.50	310
0210	Grade 1 (see also Section 08 71 20.15)		6	1.333		555	73		628	730
0300	Double cylinder, Grade 3		8	1		240	54.50		294.50	350
0310	Grade 1		7	1.143		585	62.50		647.50	740
1000	Wrought knobs and sectional trim, non-keyed, passage, Grade 3		10	.800		139	44		183	223
1010	Grade 1		9	.889		430	48.50		478.50	550
1040	Privacy, Grade 3		10	.800		162	44		206	248
1050	Grade 1		9	.889		480	48.50		528.50	610
1100	Keyed, entrance, office/apartment, Grade 3		9	.889		234	48.50		282.50	335
1103	Install lockset		6.92	1.156		234	63		297	360
1110	Grade 1		8	1		550	54.50		604.50	690
1120	Single cylinder, Grade 3	1 Carp	9	.889	Ea.	239	48.50		287.50	340
1130	Grade 1	"	8	1	"	510	54.50		564.50	645
2000	Cast knobs and full escutcheon trim									
2010	Non-keyed, passage, Grade 3	1 Carp	9	.889	Ea.	277	48.50		325.50	385
2020	Grade 1		8	1		455	54.50		509.50	585
2040	Privacy, Grade 3		9	.889		345	48.50		393.50	460
2050	Grade 1		8	1		445	54.50		499.50	575
2120	Keyed, single cylinder, Grade 3		8	1		340	54.50		394.50	455
2123	Mortise lock		6.15	1.301		340	71		411	485
2130	Grade 1		7	1.143		545	62.50		607.50	700
3000	Cast knob and sectional trim, non-keyed, passage, Grade 3		10	.800		221	44		265	315
3010	Grade 1		10	.800		430	44		474	540
3040	Privacy, Grade 3		10	.800		230	44		274	325
3050	Grade 1		10	.800		475	44		519	595
3100	Keyed, office/entrance/apartment, Grade 3		9	.889		261	48.50		309.50	365

For customer support on your Commercial Renovation Costs with RSMeans Data, call 800.448.8182.

08 71 Door Hardware

08 71 13 – Automatic Door Operators

08 71 20.42 Mortise Locksets

		Crew	Daily Output	Labor-Hours	Unit	Material	2021 Bare Costs Labor	Equipment	Total	Total Incl O&P
3110	Grade 1	1 Carp	9	.889	Ea.	485	48.50		533.50	610
3120	Single cylinder, Grade 3		9	.889		273	48.50		321.50	380
3130	Grade 1		9	.889		555	48.50		603.50	690
3190	For re-core cylinder, add					73.50			73.50	80.50

08 71 20.44 Anti-Ligature Locksets Grade 1

		Crew	Daily Output	Labor-Hours	Unit	Material	2021 Bare Costs Labor	Equipment	Total	Total Incl O&P
0010	**ANTI-LIGATURE LOCKSETS GRADE 1**									
0100	Anti-ligature cylindrical locksets									
0500	Arch handle passage set, US32D	1 Carp	8	1	Ea.	545	54.50		599.50	685
0510	privacy set		8	1		595	54.50		649.50	740
0520	classroom set		8	1		615	54.50		669.50	760
0530	storeroom set		8	1		615	54.50		669.50	760
0550	Lever handle passage set US32D		8	1		420	54.50		474.50	550
0560	Hospital/privacy set		8	1		440	54.50		494.50	565
0570	Office set		8	1		440	54.50		494.50	570
0580	Classroom set		8	1		455	54.50		509.50	585
0590	Storeroom set		8	1		440	54.50		494.50	565
0600	Exit set		8	1		520	54.50		574.50	655
0610	Entry set		8	1		465	54.50		519.50	595
0620	Asylum set		8	1		500	54.50		554.50	635
0630	Back to back dummy set		8	1		320	54.50		374.50	435
0690	Anti-ligature mortise locksets									
0700	Knob handle privacy set, US32D	1 Carp	8	1	Ea.	625	54.50		679.50	775
0710	Office set		8	1		615	54.50		669.50	760
0720	Classroom set		8	1		615	54.50		669.50	760
0730	Hotel set		8	1		630	54.50		684.50	780
0740	Apartment set		8	1		630	54.50		684.50	780
0750	Institutional privacy set		8	1		675	54.50		729.50	830
0760	Lever handle office set	1 Carp	8	1	Ea.	605	54.50		659.50	750
0770	Classroom set sectional trim		8	1		605	54.50		659.50	750
0780	Hotel set		8	1		620	54.50		674.50	770
0790	Apartment set		8	1		620	54.50		674.50	770
0800	Institutional privacy set		8	1		670	54.50		724.50	825
0810	Office set US32D		8	1		655	54.50		709.50	805
0820	Classroom set escutcheon trim		8	1		770	54.50		824.50	935
0830	Hotel set escutcheon trim		8	1		790	54.50		844.50	955
0840	Apartment set escutcheon trim		8	1		790	54.50		844.50	955
0850	Institutional privacy set escutcheon trim		8	1		790	54.50		844.50	955

08 71 20.45 Peepholes

		Crew	Daily Output	Labor-Hours	Unit	Material	2021 Bare Costs Labor	Equipment	Total	Total Incl O&P
0010	**PEEPHOLES**									
2010	Peephole	1 Carp	32	.250	Ea.	15.85	13.70		29.55	39.50
2020	Peephole, wide view	"	32	.250	"	11.70	13.70		25.40	35

08 71 20.50 Door Stops

		Crew	Daily Output	Labor-Hours	Unit	Material	2021 Bare Costs Labor	Equipment	Total	Total Incl O&P
0010	**DOOR STOPS**									
0020	Holder & bumper, floor or wall	1 Carp	32	.250	Ea.	36	13.70		49.70	61.50
1300	Wall bumper, 4" diameter, with rubber pad, aluminum		32	.250		13.80	13.70		27.50	37
1600	Door bumper, floor type, aluminum		32	.250		4.48	13.70		18.18	27
1620	Brass		32	.250		13.65	13.70		27.35	37
1630	Bronze		32	.250		20.50	13.70		34.20	44.50
1900	Plunger type, door mounted		32	.250		33	13.70		46.70	58.50
2500	Holder, floor type, aluminum		32	.250		22.50	13.70		36.20	46.50
2520	Wall type, aluminum		32	.250		35.50	13.70		49.20	61
2530	Overhead type, bronze		32	.250		115	13.70		128.70	148

08 71 Door Hardware

08 71 13 – Automatic Door Operators

08 71 20.50 Door Stops

		Crew	Daily Output	Labor-Hours	Unit	Material	2021 Bare Costs Labor	2021 Bare Costs Equipment	Total	Total Incl O&P
2540	Plunger type, aluminum	1 Carp	32	.250	Ea.	33	13.70		46.70	58.50
2560	Brass		32	.250		47	13.70		60.70	73.50
3000	Electromagnetic, wall mounted, US3		3	2.667		291	146		437	550
3020	Floor mounted, US3		3	2.667		340	146		486	605
4000	Doorstop, ceiling mounted		3	2.667		19.50	146		165.50	254
4030	Doorstop, header		3	2.667		45.50	146		191.50	283
9000	Minimum labor/equipment charge		6	1.333	Job		73		73	116

08 71 20.55 Push-Pull Plates

		Crew	Daily Output	Labor-Hours	Unit	Material	2021 Bare Costs Labor	2021 Bare Costs Equipment	Total	Total Incl O&P
0010	**PUSH-PULL PLATES**									
0090	Push plate, 0.050 thick, 3" x 12", aluminum	1 Carp	12	.667	Ea.	6.25	36.50		42.75	65
0100	4" x 16"		12	.667		9.40	36.50		45.90	68.50
0110	6" x 16"		12	.667		8.40	36.50		44.90	67.50
0120	8" x 16"		12	.667		9.90	36.50		46.40	69
0200	Push plate, 0.050 thick, 3" x 12", brass		12	.667		14.40	36.50		50.90	74
0210	4" x 16"		12	.667		18	36.50		54.50	78
0220	6" x 16"		12	.667		27.50	36.50		64	88.50
0230	8" x 16"		12	.667		36	36.50		72.50	98
0250	Push plate, 0.050 thick, 3" x 12", satin brass		12	.667		14.60	36.50		51.10	74
0260	4" x 16"	1 Carp	12	.667	Ea.	18.20	36.50		54.70	78
0270	6" x 16"		12	.667		28	36.50		64.50	88.50
0280	8" x 16"		12	.667		36.50	36.50		73	98
0490	Push plate, 0.050 thick, 3" x 12", bronze		12	.667		17.35	36.50		53.85	77
0500	4" x 16"		12	.667		26	36.50		62.50	86.50
0510	6" x 16"		12	.667		33	36.50		69.50	94.50
0520	8" x 16"		12	.667		40	36.50		76.50	102
0600	Push plate, antimicrobial copper alloy finish, 3.5" x 15"		13	.615		56.50	33.50		90	116
0610	4" x 16"		13	.615		60.50	33.50		94	120
0620	6" x 16"		13	.615		24	33.50		57.50	80
0630	6" x 20"		13	.615		43	33.50		76.50	101
0740	Push plate, 0.050 thick, 3" x 12", stainless steel		12	.667		12.25	36.50		48.75	71.50
0750	4" x 16"		12	.667		11.60	36.50		48.10	71
0760	6" x 16"		12	.667		19.25	36.50		55.75	79
0780	8" x 16"		12	.667		24.50	36.50		61	85
0790	Push plate, 0.050 thick, 3" x 12", satin stainless steel		12	.667		6.95	36.50		43.45	65.50
0810	4" x 16"		12	.667		8.50	36.50		45	67.50
0820	6" x 16"		12	.667		12.10	36.50		48.60	71.50
0830	8" x 16"		12	.667		16.25	36.50		52.75	76
0980	Pull plate, 0.050 thick, 3" x 12", aluminum		12	.667		26.50	36.50		63	87
1000	4" x 16"		12	.667		30.50	36.50		67	91.50
1050	Pull plate, 0.050 thick, 3" x 12", brass		12	.667		41	36.50		77.50	103
1060	4" x 16"		12	.667		65.50	36.50		102	130
1080	Pull plate, 0.050 thick, 3" x 12", bronze		12	.667		52	36.50		88.50	115
1100	4" x 16"		12	.667		60.50	36.50		97	125
1180	Pull plate, 0.050 thick, 3" x 12", stainless steel		12	.667		47.50	36.50		84	110
1200	4" x 16"		12	.667		62.50	36.50		99	127
1250	Pull plate, 0.050 thick, 3" x 12", chrome		12	.667		47.50	36.50		84	110
1270	4" x 16"		12	.667		49.50	36.50		86	113
1500	Pull handle and push bar, aluminum		11	.727		126	40		166	202
2000	Bronze		10	.800		173	44		217	260
9800	Minimum labor/equipment charge		5	1.600	Job		87.50		87.50	139

08 71 13 – Automatic Door Operators

08 71 20.60 Entrance Locks		Crew	Daily Output	Labor-Hours	Unit	Material	2021 Bare Costs Labor	Equipment	Total	Total Incl O&P
0010	**ENTRANCE LOCKS**									
0015	Cylinder, grip handle deadlocking latch	1 Carp	9	.889	Ea.	179	48.50		227.50	275
0020	Deadbolt		8	1		231	54.50		285.50	340
0100	Push and pull plate, dead bolt		8	1		247	54.50		301.50	360
0900	For handicapped lever, add					175			175	192

08 71 20.65 Thresholds										
0010	**THRESHOLDS**									
0011	Threshold 3' long saddles aluminum	1 Carp	48	.167	L.F.	19.55	9.10		28.65	36
0100	Aluminum, 8" wide, 1/2" thick	"	12	.667	Ea.	64	36.50		100.50	129
0500	Bronze	1 Carp	60	.133	L.F.	42.50	7.30		49.80	58
0600	Bronze, panic threshold, 5" wide, 1/2" thick		12	.667	Ea.	159	36.50		195.50	232
0700	Rubber, 1/2" thick, 5-1/2" wide		20	.400		41	22		63	80
0800	2-3/4" wide		20	.400		49.50	22		71.50	89.50
1950	ADA compliant thresholds									
2000	Threshold, wood oak 3-1/2" wide x 24" long	1 Carp	12	.667	Ea.	27	36.50		63.50	87.50
2010	3-1/2" wide x 36" long		12	.667		37	36.50		73.50	98.50
2020	3-1/2" wide x 48" long		12	.667		41	36.50		77.50	103
2030	4-1/2" wide x 24" long		12	.667		14.70	36.50		51.20	74
2040	4-1/2" wide x 36" long		12	.667		38	36.50		74.50	100
2050	4-1/2" wide x 48" long		12	.667		44	36.50		80.50	107
2060	6-1/2" wide x 24" long		12	.667		31	36.50		67.50	92
2070	6-1/2" wide x 36" long		12	.667		46	36.50		82.50	109
2080	6-1/2" wide x 48" long		12	.667		59	36.50		95.50	123
2090	Threshold, wood cherry 3-1/2" wide x 24" long		12	.667		27	36.50		63.50	87.50
2100	3-1/2" wide x 36" long		12	.667		37	36.50		73.50	98.50
2110	3-1/2" wide x 48" long		12	.667		40.50	36.50		77	103
2120	4-1/2" wide x 24" long		12	.667		29	36.50		65.50	90
2130	4-1/2" wide x 36" long		12	.667		39.50	36.50		76	101
2140	4-1/2" wide x 48" long		12	.667		44	36.50		80.50	107
2150	6-1/2" wide x 24" long		12	.667		28.50	36.50		65	89.50
2160	6-1/2" wide x 36" long		12	.667		46	36.50		82.50	109
2170	6-1/2" wide x 48" long		12	.667		65	36.50		101.50	130
2180	Threshold, wood walnut 3-1/2" wide x 24" long		12	.667		27	36.50		63.50	87.5
2190	3-1/2" wide x 36" long		12	.667		36.50	36.50		73	98
2200	3-1/2" wide x 48" long		12	.667		41.50	36.50		78	104
2210	4-1/2" wide x 24" long		12	.667		29	36.50		65.50	90
2220	4-1/2" wide x 36" long		12	.667		38	36.50		74.50	100
2230	4-1/2" wide x 48" long		12	.667		44	36.50		80.50	107
2240	6-1/2" wide x 24" long		12	.667		31	36.50		67.50	92
2250	6-1/2" wide x 36" long		12	.667		49	36.50		85.50	112
2260	6-1/2" wide x 48" long		12	.667		65	36.50		101.50	130
2300	Threshold, aluminum 4" wide x 36" long		12	.667		30	36.50		66.50	91
2310	4" wide x 48" long		12	.667		45.50	36.50		82	108
2320	4" wide x 72" long		12	.667		7.40	36.50		43.90	66
2330	5" wide x 36" long		12	.667		32	36.50		68.50	93.50
2340	5" wide x 48" long		12	.667		24	36.50		60.50	84.50
2350	5" wide x 72" long		12	.667		56.50	36.50		93	121
2360	6" wide x 36" long		12	.667		36	36.50		72.50	98
2370	6" wide x 48" long	1 Carp	12	.667	Ea.	46.50	36.50		83	110
2380	6" wide x 72" long		12	.667		72.50	36.50		109	138
2390	7" wide x 36" long		12	.667		48	36.50		84.50	111
2400	7" wide x 48" long		12	.667		60.50	36.50		97	125

08 71 20.65 Thresholds	Crew	Daily Output	Labor-Hours	Unit	Material	2021 Bare Costs Labor	2021 Bare Costs Equipment	Total	Total Incl O&P
2410	7" wide x 72" long	1 Carp	12	.667	Ea.	91	36.50		127.50	158
2500	Threshold, ramp, aluminum or rubber 24" x 24"		12	.667	↓	180	36.50		216.50	256
9000	Minimum labor/equipment charge	↓	4	2	Job		109		109	174

08 71 20.75 Door Hardware Accessories	Crew	Daily Output	Labor-Hours	Unit	Material	Labor	Equipment	Total	Total Incl O&P
0010	**DOOR HARDWARE ACCESSORIES**									
0050	Door closing coordinator, 36" (for paired openings up to 56")	1 Carp	8	1	Ea.	133	54.50		187.50	234
0060	48" (for paired openings up to 84")		8	1		141	54.50		195.50	243
0070	56" (for paired openings up to 96")		8	1		151	54.50		205.50	253
1000	Knockers, brass, standard		16	.500		67.50	27.50		95	118
1100	Deluxe	↓	10	.800	↓	97.50	44		141.50	177
2000	Torsion springs for overhead doors									
2050	1-3/4" diam., 32" long, 0.243" spring	1 Carp	8	1	Ea.	59.50	54.50		114	152
2060	1-3/4" diam., 32" long, 0.250" spring		8	1		44.50	54.50		99	136
2070	2" diam., 32" long, 0.243" spring		8	1		45.50	54.50		100	137
2080	2" diam., 32" long, 0.253" spring		8	1		46.50	54.50		101	138
4500	Rubber door silencers		540	.015	↓	.37	.81		1.18	1.70
9000	Minimum labor/equipment charge	↓	6	1.333	Job		73		73	116

08 71 20.80 Hasps	Crew	Daily Output	Labor-Hours	Unit	Material	Labor	Equipment	Total	Total Incl O&P
0010	**HASPS**, steel assembly									
0015	3"	1 Carp	26	.308	Ea.	5.90	16.85		22.75	33.50
0020	4-1/2"		13	.615		7.25	33.50		40.75	61.50
0040	6"	↓	12.50	.640	↓	10.45	35		45.45	67

08 71 20.90 Hinges	Crew	Daily Output	Labor-Hours	Unit	Material	Labor	Equipment	Total	Total Incl O&P
0010	**HINGES**									
0012	Full mortise, avg. freq., steel base, USP, 4-1/2" x 4-1/2"				Pr.	48			48	53
0100	5" x 5", USP					73			73	80.50
0200	6" x 6", USP					133			133	146
0400	Brass base, 4-1/2" x 4-1/2", US10					90.50			90.50	99.50
0500	5" x 5", US10					58.50			58.50	64.50
0600	6" x 6", US10					180			180	198
0800	Stainless steel base, 4-1/2" x 4-1/2", US32				↓	72			72	79.50
0900	For non removable pin, add (security item)				Ea.	2.46			2.46	2.71
0910	For floating pin, driven tips, add					3.32			3.32	3.65
0930	For hospital type tip on pin, add					15			15	16.50
0940	For steeple type tip on pin, add				↓	22.50			22.50	25
0950	Full mortise, high frequency, steel base, 3-1/2" x 3-1/2", US26D				Pr.	31.50			31.50	34.50
1000	4-1/2" x 4-1/2", USP					65			65	71.50
1100	5" x 5", USP					55			55	60.50
1200	6" x 6", USP					112			112	123
1400	Brass base, 3-1/2" x 3-1/2", US4					47.50			47.50	52.50
1430	4-1/2" x 4-1/2", US10					60.50			60.50	66.50
1500	5" x 5", US10					123			123	136
1600	6" x 6", US10					143			143	157
1800	Stainless steel base, 4-1/2" x 4-1/2", US32					107			107	118
1810	5" x 4-1/2", US32				↓	142			142	156
1930	For hospital type tip on pin, add				Ea.	14.60			14.60	16.05
1950	Full mortise, low frequency, steel base, 3-1/2" x 3-1/2", US26D				Pr.	33			33	36
2000	4-1/2" x 4-1/2", USP					25			25	27.50
2100	5" x 5", USP					55			55	60.50
2200	6" x 6", USP					95.50			95.50	105
2300	4-1/2" x 4-1/2", US3					20.50			20.50	22.50
2310	5" x 5", US3				↓	41			41	45

08 71 13 – Automatic Door Operators

08 71 20.90 Hinges	Crew	Daily Output	Labor-Hours	Unit	Material	2021 Bare Costs Labor	Equipment	Total	Total Incl O&P
2400 Brass bass, 4-1/2" x 4-1/2", US10				Pr.	93.50			93.50	103
2500 5" x 5", US10					120			120	132
2800 Stainless steel base, 4-1/2" x 4-1/2", US32					80			80	88
8000 Install hinge	1 Carp	34	.235			12.85		12.85	20.50

08 71 20.91 Special Hinges	Crew	Daily Output	Labor-Hours	Unit	Material	2021 Bare Costs Labor	Equipment	Total	Total Incl O&P
0010 **SPECIAL HINGES**									
0015 Paumelle, high frequency									
0020 Steel base, 6" x 4-1/2", US10				Pr.	138			138	152
0100 Brass base, 5" x 4-1/2", US10				Ea.	235			235	259
0200 Paumelle, average frequency, steel base, 4-1/2" x 3-1/2", US10				Pr.	93			93	103
0400 Olive knuckle, low frequency, brass base, 6" x 4-1/2", US10				Ea.	153			153	169
1000 Electric hinge with concealed conductor, average frequency									
1010 Steel base, 4-1/2" x 4-1/2", US26D				Pr.	345			345	380
1100 Bronze base, 4-1/2" x 4-1/2", US26D				"	350			350	385
1200 Electric hinge with concealed conductor, high frequency									
1210 Steel base, 4-1/2" x 4-1/2", US26D				Pr.	284			284	315
1600 Double weight, 800 lb., steel base, removable pin, 5" x 6", USP					625			625	690
1700 Steel base-welded pin, 5" x 6", USP					174			174	191
1800 Triple weight, 2000 lb., steel base, welded pin, 5" x 6", USP					850			850	935
2000 Pivot reinf., high frequency, steel base, 7-3/4" door plate, USP					170			170	187
2200 Bronze base, 7-3/4" door plate, US10					256			256	282
3000 Swing clear, full mortise, full or half surface, high frequency,									
3010 Steel base, 5" high, USP				Pr.	150			150	165
3200 Swing clear, full mortise, average frequency									
3210 Steel base, 4-1/2" high, USP				Pr.	133			133	146
4000 Wide throw, average frequency, steel base, 4-1/2" x 6", USP					94			94	103
4200 High frequency, steel base, 4-1/2" x 6", USP					128			128	141
4440 US26D					94			94	103
4600 Spring hinge, single acting, 6" flange, steel				Ea.	50			50	55
4700 Brass					96.50			96.50	106
4900 Double acting, 6" flange, steel					61			61	67
4950 Brass					134			134	147
5000 T-strap, galvanized, 4"				Pr.	22.50			22.50	25
5010 6"					31			31	34
5020 8"					61.50			61.50	68
8000 Continuous hinges									
8010 Steel, piano, 2" x 72"	1 Carp	20	.400	Ea.	28	22		50	66
8020 Brass, piano, 1-1/16" x 30"		30	.267		12.70	14.60		27.30	37
8030 Acrylic, piano, 1-3/4" x 12"		40	.200		15.80	10.95		26.75	35
8040 Aluminum, door, standard duty, 7'		3	2.667		125	146		271	370
8050 Heavy duty, 7'		3	2.667		149	146		295	395
8060 8'		3	2.667		177	146		323	425
8070 Steel, door, heavy duty, 7'		3	2.667		213	146		359	465
8080 8'		3	2.667		425	146		571	695
8090 Stainless steel, door, heavy duty, 7'		3	2.667		271	146		417	530
8100 8'		3	2.667		305	146		451	565
9000 Continuous hinge, steel, full mortise, heavy duty, 96"		2	4		755	219		974	1,175
9200 Continuous geared hinge, aluminum, full mortise, standard duty, 83"		3	2.667		159	146		305	405
9250 Continuous geared hinge, aluminum, full mortise, heavy duty, 83"		3	2.667		223	146		369	475

08 71 20.95 Kick Plates		Crew	Daily Output	Labor-Hours	Unit	Material	2021 Bare Costs Labor	Equipment	Total	Total Incl O&P
0010	**KICK PLATES**									
0020	Stainless steel, .050", 16 ga., 8" x 28", US32	1 Carp	15	.533	Ea.	49.50	29		78.50	101
0030	8" x 30"		15	.533		50.50	29		79.50	103
0040	8" x 34"		15	.533		57.50	29		86.50	110
0050	10" x 28"		15	.533		62	29		91	115
0060	10" x 30"		15	.533		66	29		95	120
0070	10" x 34"		15	.533		75	29		104	129
0080	Mop/Kick, 4" x 28"		15	.533		39	29		68	89.50
0090	4" x 30"		15	.533		41	29		70	91.50
0100	4" x 34"		15	.533		47	29		76	98
0110	6" x 28"		15	.533		37	29		66	87.50
0120	6" x 30"		15	.533		39.50	29		68.50	90
0130	6" x 34"		15	.533		45	29		74	96
0500	Bronze, .050", 8" x 28"		15	.533		69.50	29		98.50	123
0510	8" x 30"	1 Carp	15	.533	Ea.	75	29		104	129
0520	8" x 34"		15	.533		85	29		114	141
0530	10" x 28"		15	.533		87.50	29		116.50	143
0540	10" x 30"		15	.533		94	29		123	150
0550	10" x 34"		15	.533		106	29		135	164
0560	Mop/Kick, 4" x 28"		15	.533		37.50	29		66.50	87.50
0570	4" x 30"		15	.533		41.50	29		70.50	92
0580	4" x 34"		15	.533		41.50	29		70.50	92.50
0590	6" x 28"		15	.533		55	29		84	107
0600	6" x 30"		15	.533		59	29		88	112
0610	6" x 34"		15	.533		67	29		96	121
1000	Acrylic, .125", 8" x 26"		15	.533		28.50	29		57.50	77.50
1010	8" x 36"		15	.533		39.50	29		68.50	90
1020	8" x 42"		15	.533		46	29		75	97
1030	10" x 26"		15	.533		35.50	29		64.50	86
1040	10" x 36"		15	.533		50	29		79	102
1050	10" x 42"		15	.533		58	29		87	111
1060	Mop/Kick, 4" x 26"		15	.533		17	29		46	65
1070	4" x 36"		15	.533		23.50	29		52.50	72.50
1080	4" x 42"		15	.533		27.50	29		56.50	76.50
1090	6" x 26"		15	.533		25.50	29		54.50	74.50
1100	6" x 36"		15	.533		34.50	29		63.50	84.50
1110	6" x 42"		15	.533		41	29		70	91.50
1220	Brass, .050", 8" x 26"		15	.533		73	29		102	127
1230	8" x 36"		15	.533		71.50	29		100.50	126
1240	8" x 42"		15	.533		89	29		118	145
1250	10" x 26"		15	.533		74	29		103	128
1260	10" x 36"		15	.533		90	29		119	146
1270	10" x 42"		15	.533		108	29		137	166
1320	Mop/Kick, 4" x 26"		15	.533		27	29		56	76.50
1330	4" x 36"		15	.533		37.50	29		66.50	88
1340	4" x 42"		15	.533		44	29		73	95
1350	6" x 26"		15	.533		40.50	29		69.50	91
1360	6" x 36"		15	.533		49.50	29		78.50	101
1370	6" x 42"		15	.533		149	29		178	211
1800	Aluminum, .050", 8" x 26"		15	.533		37	29		66	87.50
1810	8" x 36"		15	.533		44.50	29		73.50	95.50
1820	8" x 42"		15	.533		57	29		86	110

08 71 Door Hardware

08 71 13 – Automatic Door Operators

		Crew	Daily Output	Labor-Hours	Unit	Material	2021 Bare Costs Labor	Equipment	Total	Total Incl O&P
08 71 20.95 Kick Plates										
1830	10" x 26"	1 Carp	15	.533	Ea.	34.50	29		63.50	84.50
1840	10" x 36"		15	.533		44.50	29		73.50	95.50
1850	10" x 42"		15	.533		49.50	29		78.50	101
1860	Mop/Kick, 4" x 26"		15	.533		26.50	29		55.50	75.50
1870	4" x 36"		15	.533		22	29		51	70.50
1880	4" x 42"	1 Carp	15	.533	Ea.	32.50	29		61.50	82
1890	6" x 26"		15	.533		24.50	29		53.50	73.50
1900	6" x 36"		15	.533		37	29		66	87
1910	6" x 42"		15	.533		40	29		69	90.50
9000	Minimum labor/equipment charge		6	1.333	Job		73		73	116

08 71 21 – Astragals

08 71 21.10 Exterior Mouldings, Astragals

		Crew	Daily Output	Labor-Hours	Unit	Material	2021 Bare Costs Labor	Equipment	Total	Total Incl O&P
0010	**EXTERIOR MOULDINGS, ASTRAGALS**									
0400	One piece, overlapping cadmium plated steel, flat, 3/16" x 2"	1 Carp	90	.089	L.F.	5.10	4.86		9.96	13.35
0600	Prime coated steel, flat, 1/8" x 3"		90	.089		6.10	4.86		10.96	14.45
0800	Stainless steel, flat, 3/32" x 1-5/8"		90	.089		15.70	4.86		20.56	25
1000	Aluminum, flat, 1/8" x 2"		90	.089		4.53	4.86		9.39	12.75
1200	Nail on, "T" extrusion		120	.067		2.16	3.65		5.81	8.20
1300	Vinyl bulb insert		105	.076		2.64	4.17		6.81	9.55
1600	Screw on, "T" extrusion		90	.089		4.66	4.86		9.52	12.90
1700	Vinyl insert		75	.107		4.08	5.85		9.93	13.80
2000	"L" extrusion, neoprene bulbs		75	.107		4.37	5.85		10.22	14.10
2100	Neoprene sponge insert		75	.107		6.80	5.85		12.65	16.75
2200	Magnetic		75	.107		10.35	5.85		16.20	20.50
2400	Spring hinged security seal, with cam		75	.107		6.65	5.85		12.50	16.60
2600	Spring loaded locking bolt, vinyl insert		45	.178		9	9.70		18.70	25.50
2800	Neoprene sponge strip, "Z" shaped, aluminum		60	.133		8	7.30		15.30	20.50
2900	Solid neoprene strip, nail on aluminum strip		90	.089		3.76	4.86		8.62	11.90
3000	One piece stile protection									
3020	Neoprene fabric loop, nail on aluminum strips	1 Carp	60	.133	L.F.	1.23	7.30		8.53	12.95
3110	Flush mounted aluminum extrusion, 1/2" x 1-1/4"		60	.133		6.60	7.30		13.90	18.85
3140	3/4" x 1-3/8"		60	.133		3.91	7.30		11.21	15.90
3160	1-1/8" x 1-3/4"		60	.133		5.35	7.30		12.65	17.50
3300	Mortise, 9/16" x 3/4"		60	.133		4.45	7.30		11.75	16.50
3320	13/16" x 1-3/8"		60	.133		4.14	7.30		11.44	16.15
3600	Spring bronze strip, nail on type		105	.076		2.35	4.17		6.52	9.25
3620	Screw on, with retainer		75	.107		2.61	5.85		8.46	12.15
3800	Flexible stainless steel housing, pile insert, 1/2" door		105	.076		6.95	4.17		11.12	14.25
3820	3/4" door		105	.076		7.95	4.17		12.12	15.35
4000	Extruded aluminum retainer, flush mount, pile insert		105	.076		2.72	4.17		6.89	9.65
4080	Mortise, felt insert		90	.089		4.53	4.86		9.39	12.75
4160	Mortise with spring, pile insert		90	.089		3.73	4.86		8.59	11.85
4400	Rigid vinyl retainer, mortise, pile insert		105	.076		3.05	4.17		7.22	10
4600	Wool pile filler strip, aluminum backing		105	.076		3.01	4.17		7.18	9.95
5000	Two piece overlapping astragal, extruded aluminum retainer									
5010	Pile insert	1 Carp	60	.133	L.F.	3.16	7.30		10.46	15.10
5020	Vinyl bulb insert		60	.133		1.71	7.30		9.01	13.50
5040	Vinyl flap insert		60	.133		4.31	7.30		11.61	16.35
5060	Solid neoprene flap insert		60	.133		6.40	7.30		13.70	18.65
5080	Hypalon rubber flap insert		60	.133		6.35	7.30		13.65	18.60
5090	Snap on cover, pile insert		60	.133		9.60	7.30		16.90	22
5400	Magnetic aluminum, surface mounted		60	.133		27	7.30		34.30	41

08 71 Door Hardware

08 71 21 – Astragals

08 71 21.10 Exterior Mouldings, Astragals

		Crew	Daily Output	Labor-Hours	Unit	Material	2021 Bare Costs Labor	Equipment	Total	Total Incl O&P
5500	Interlocking aluminum, 5/8" x 1" neoprene bulb insert	1 Carp	45	.178	L.F.	4.29	9.70		13.99	20
5600	Adjustable aluminum, 9/16" x 21/32", pile insert		45	.178		17.65	9.70		27.35	35
5800	Magnetic, adjustable, 9/16" x 21/32"		45	.178		23	9.70		32.70	41
6000	Two piece stile protection									
6010	Cloth backed rubber loop, 1" gap, nail on aluminum strips	1 Carp	45	.178	L.F.	4.40	9.70		14.10	20.50
6040	Screw on aluminum strips		45	.178		6.40	9.70		16.10	22.50
6100	1-1/2" gap, screw on aluminum extrusion		45	.178		5.70	9.70		15.40	22
6240	Vinyl fabric loop, slotted aluminum extrusion, 1" gap		45	.178		2.33	9.70		12.03	18.05
6300	1-1/4" gap		45	.178		6.05	9.70		15.75	22

08 71 25 – Door Weatherstripping

08 71 25.10 Mechanical Seals, Weatherstripping

		Crew	Daily Output	Labor-Hours	Unit	Material	2021 Bare Costs Labor	Equipment	Total	Total Incl O&P
0010	**MECHANICAL SEALS, WEATHERSTRIPPING**									
1000	Doors, wood frame, interlocking, for 3' x 7' door, zinc	1 Carp	3	2.667	Opng.	53.50	146		199.50	291
1100	Bronze		3	2.667		69	146		215	310
1300	6' x 7' opening, zinc		2	4		56.50	219		275.50	410
1400	Bronze		2	4		69.50	219		288.50	425
1700	Wood frame, spring type, bronze									
1800	3' x 7' door	1 Carp	7.60	1.053	Opng.	22.50	57.50		80	116
1900	6' x 7' door	"	7	1.143	"	31.50	62.50		94	134
2200	Metal frame, spring type, bronze									
2300	3' x 7' door	1 Carp	3	2.667	Opng.	51	146		197	288
2400	6' x 7' door	"	2.50	3.200	"	52	175		227	335
2500	For stainless steel, spring type, add					133%				
2700	Metal frame, extruded sections, 3' x 7' door, aluminum	1 Carp	3	2.667	Opng.	33	146		179	269
2800	Bronze		3	2.667		95.50	146		241.50	335
3100	6' x 7' door, aluminum		1.50	5.333		41	292		333	510
3200	Bronze		1.50	5.333		147	292		439	625
3500	Threshold weatherstripping									
3650	Door sweep, flush mounted, aluminum	1 Carp	25	.320	Ea.	22	17.50		39.50	52
3700	Vinyl		25	.320		19.65	17.50		37.15	49.50
5000	Garage door bottom weatherstrip, 12' aluminum, clear		14	.571		24	31.50		55.50	76.50
5010	Bronze	1 Carp	14	.571	Ea.	94	31.50		125.50	153
5050	Bottom protection, rubber		14	.571		55.50	31.50		87	111
5100	Threshold		14	.571		68.50	31.50		100	126
9000	Minimum labor/equipment charge		3	2.667	Job		146		146	232

08 75 Window Hardware

08 75 10 – Window Handles and Latches

08 75 10.10 Handles and Latches

		Crew	Daily Output	Labor-Hours	Unit	Material	2021 Bare Costs Labor	Equipment	Total	Total Incl O&P
0010	**HANDLES AND LATCHES**									
1000	Handles, surface mounted, aluminum	1 Carp	24	.333	Ea.	6.15	18.25		24.40	36
1020	Brass		24	.333		5.45	18.25		23.70	35
1040	Chrome		24	.333		8.60	18.25		26.85	38.50
1200	Window handles window crank ADA		24	.333		17.40	18.25		35.65	48
1500	Recessed, aluminum		12	.667		3.81	36.50		40.31	62
1520	Brass		12	.667		5.65	36.50		42.15	64
1540	Chrome		12	.667		4.85	36.50		41.35	63.50
2000	Latches, aluminum		20	.400		4.27	22		26.27	39.50
2020	Brass		20	.400		5.30	22		27.30	41
2040	Chrome		20	.400		5.70	22		27.70	41.50

08 75 Window Hardware

08 75 10 – Window Handles and Latches

08 75 10.10 Handles and Latches	Crew	Daily Output	Labor-Hours	Unit	Material	2021 Bare Costs Labor	Equipment	Total	Total Incl O&P
9000 Minimum labor/equipment charge	1 Carp	6	1.333	Job		73		73	116

08 75 10.15 Window Opening Control

		Crew	Daily Output	Labor-Hours	Unit	Material	Labor	Equipment	Total	Total Incl O&P
0010	**WINDOW OPENING CONTROL**									
0015	Window stops									
0020	For double-hung window	1 Carp	35	.229	Ea.	41	12.50		53.50	65
0030	Cam action for sliding window		35	.229		4.61	12.50		17.11	25
0040	Thumb screw for sliding window	↓	35	.229	↓	2.72	12.50		15.22	23
0100	Window guards, child safety bars for single or double-hung									
0110	14" to 17" wide max vert opening 26"	1 Carp	16	.500	Ea.	69.50	27.50		97	120
0120	17" to 23" wide max vert opening 26"		16	.500		79	27.50		106.50	131
0130	23" to 36" wide max vert opening 26"		16	.500		90	27.50		117.50	143
0140	35" to 58" wide max vert opening 26"		16	.500		104	27.50		131.50	159
0150	58" to 90" wide max vert opening 26"		16	.500		141	27.50		168.50	199
0160	73" to 120" wide max vert opening 26"	↓	16	.500	↓	246	27.50		273.50	315

08 75 30 – Window Weatherstripping

08 75 30.10 Mechanical Weather Seals

		Crew	Daily Output	Labor-Hours	Unit	Material	Labor	Equipment	Total	Total Incl O&P
0010	**MECHANICAL WEATHER SEALS**, Window, double-hung, 3' x 5'									
0020	Zinc	1 Carp	7.20	1.111	Opng.	19.55	61		80.55	119
0100	Bronze		7.20	1.111		39.50	61		100.50	141
0200	Vinyl V strip		7	1.143		4.59	62.50		67.09	105
0500	As above but heavy duty, zinc		4.60	1.739		20	95		115	173
0600	Bronze	↓	4.60	1.739	↓	77	95		172	236
9000	Minimum labor/equipment charge	1 Clab	4.60	1.739	Job		77		77	123

08 79 Hardware Accessories

08 79 20 – Door Accessories

08 79 20.10 Door Hardware Accessories

		Crew	Daily Output	Labor-Hours	Unit	Material	Labor	Equipment	Total	Total Incl O&P
0010	**DOOR HARDWARE ACCESSORIES**									
0140	Door bolt, surface, 4"	1 Carp	32	.250	Ea.	19.65	13.70		33.35	43.50
0160	Door latch	"	12	.667	"	8.90	36.50		45.40	68
0200	Sliding closet door									
0220	Track and hanger, single	1 Carp	10	.800	Ea.	66	44		110	142
0240	Double		8	1		84.50	54.50		139	180
0260	Door guide, single		48	.167		33	9.10		42.10	50.50
0280	Double		48	.167		42	9.10		51.10	60.50
0600	Deadbolt and lock cover plate, brass or stainless steel		30	.267		8.45	14.60		23.05	32.50
0620	Hole cover plate, brass or chrome		35	.229		9.10	12.50		21.60	30
2240	Mortise lockset, passage, lever handle		9	.889		182	48.50		230.50	278
4000	Security chain, standard		18	.444		15.75	24.50		40.25	56
4100	Deluxe	↓	18	.444	↓	62	24.50		86.50	107

For customer support on your Commercial Renovation Costs with RSMeans Data, call 800.448.8182.

285

08 81 Glass Glazing

08 81 10 – Float Glass

08 81 10.10 Various Types and Thickness of Float Glass

	Crew	Daily Output	Labor-Hours	Unit	Material	2021 Bare Costs Labor	Equipment	Total	Total Incl O&P
0010 **VARIOUS TYPES AND THICKNESS OF FLOAT GLASS**									
0020 3/16" plain	2 Glaz	130	.123	S.F.	7.10	6.50		13.60	18.10
0200 Tempered, clear		130	.123		21	6.50		27.50	34
0300 Tinted		130	.123		7.65	6.50		14.15	18.70
0600 1/4" thick, clear, plain		120	.133		8.35	7		15.35	20.50
0700 Tinted		120	.133		9.45	7		16.45	21.50
0800 Tempered, clear		120	.133		7.70	7		14.70	19.65
0900 Tinted		120	.133		25.50	7		32.50	39
1600 3/8" thick, clear, plain		75	.213		14.85	11.25		26.10	34
1700 Tinted		75	.213		20	11.25		31.25	40
1800 Tempered, clear		75	.213		27.50	11.25		38.75	48.50
1900 Tinted		75	.213		40.50	11.25		51.75	62.50
2200 1/2" thick, clear, plain		55	.291		19.20	15.30		34.50	45.50
2300 Tinted		55	.291		35	15.30		50.30	63
2400 Tempered, clear		55	.291		32	15.30		47.30	59.50
2500 Tinted		55	.291		28	15.30		43.30	55.50
2800 5/8" thick, clear, plain		45	.356		35	18.70		53.70	68.50
2900 Tempered, clear		45	.356		40	18.70		58.70	74.50
3200 3/4" thick, clear, plain		35	.457		46.50	24		70.50	90
3300 Tempered, clear		35	.457		56	24		80	100
3600 1" thick, clear, plain	↓	30	.533		75.50	28		103.50	128
8900 For low emissivity coating for 3/16" & 1/4" only, add to above				↓	18%				
9000 Minimum labor/equipment charge	1 Glaz	2	4	Job		211		211	335

08 81 17 – Fire Glass

08 81 17.10 Fire Resistant Glass

	Crew	Daily Output	Labor-Hours	Unit	Material	2021 Bare Costs Labor	Equipment	Total	Total Incl O&P
0010 **FIRE RESISTANT GLASS**									
0020 Fire glass minimum	2 Glaz	40	.400	S.F.	177	21		198	229
0030 Mid range		40	.400		234	21		255	291
0050 High end	↓	40	.400		234	21		255	291

08 81 20 – Vision Panels

08 81 20.10 Full Vision

	Crew	Daily Output	Labor-Hours	Unit	Material	2021 Bare Costs Labor	Equipment	Total	Total Incl O&P
0010 **FULL VISION**, window system with 3/4" glass mullions									
0020 Up to 10' high	H-2	130	.185	S.F.	69.50	9.20		78.70	91
0100 10' to 20' high, minimum		110	.218		76.50	10.90		87.40	102
0150 Average		100	.240		79.50	12		91.50	107
0200 Maximum	↓	80	.300	↓	92.50	14.95		107.45	126
9000 Minimum labor/equipment charge	1 Glaz	2	4	Job		211		211	335

08 81 25 – Glazing Variables

08 81 25.10 Applications of Glazing

	Crew	Daily Output	Labor-Hours	Unit	Material	2021 Bare Costs Labor	Equipment	Total	Total Incl O&P
0010 **APPLICATIONS OF GLAZING** R088110-10									
0600 For glass replacement, add				S.F.		100%			
0700 For gasket settings, add				L.F.	6			6	6.60
0900 For sloped glazing, add				S.F.		26%			
2000 Fabrication, polished edges, 1/4" thick				Inch	.59			.59	.65
2100 1/2" thick					1.45			1.45	1.60
2500 Mitered edges, 1/4" thick					1.45			1.45	1.60
2600 1/2" thick				↓	2.40			2.40	2.64

08 81 Glass Glazing

08 81 30 – Insulating Glass

08 81 30.10 Reduce Heat Transfer Glass

		Crew	Daily Output	Labor-Hours	Unit	Material	2021 Bare Costs Labor	Equipment	Total	Total Incl O&P
0010	**REDUCE HEAT TRANSFER GLASS**									
0015	2 lites 1/8" float, 1/2" thk under 15 S.F.									
0020	Clear R088110-10	G 2 Glaz	95	.168	S.F.	27.50	8.85		36.35	44.50
0100	Tinted	G	95	.168		30.50	8.85		39.35	47.50
0200	2 lites 3/16" float, for 5/8" thk unit, 15 to 30 S.F., clear	G	90	.178		17	9.35		26.35	33.50
0400	1" thk, dbl. glazed, 1/4" float, 30 to 70 S.F., clear	G	75	.213		17.35	11.25		28.60	37
0500	Tinted	G	75	.213		29	11.25		40.25	50
2000	Both lites, light & heat reflective	G	85	.188		39	9.90		48.90	59
2500	Heat reflective, film inside, 1" thick unit, clear	G	85	.188		34	9.90		43.90	53.50
2600	Tinted	G	85	.188		32.50	9.90		42.40	51.50
3000	Film on weatherside, clear, 1/2" thick unit	G	95	.168		24.50	8.85		33.35	41
3100	5/8" thick unit	G	90	.178		21.50	9.35		30.85	38.50
3200	1" thick unit	G	85	.188		34	9.90		43.90	53.50
3350	Clear heat reflective film on inside	G 1 Glaz	50	.160		13.75	8.40		22.15	28.50
3360	Heat reflective film on inside, tinted	G	25	.320		14.40	16.85		31.25	43
3370	Heat reflective film on the inside, metalized	G	20	.400		21.50	21		42.50	57
5000	Spectrally selective film, on ext., blocks solar gain/allows 70% of light	G 2 Glaz	95	.168		15.10	8.85		23.95	31
9000	Minimum labor/equipment charge	1 Glaz	2	4	Job		211		211	335

08 81 40 – Plate Glass

08 81 40.10 Plate Glass

		Crew	Daily Output	Labor-Hours	Unit	Material	Labor	Equipment	Total	Total Incl O&P
0010	**PLATE GLASS** Twin ground, polished,									
0020	3/16" thick	2 Glaz	100	.160	S.F.	6.10	8.40		14.50	20
0100	1/4" thick		94	.170		8.35	8.95		17.30	23.50
0200	3/8" thick		60	.267		14.15	14.05		28.20	38
0300	1/2" thick		40	.400		27.50	21		48.50	64

08 81 55 – Window Glass

08 81 55.10 Sheet Glass

		Crew	Daily Output	Labor-Hours	Unit	Material	Labor	Equipment	Total	Total Incl O&P
0010	**SHEET GLASS** (window), clear float, stops, putty bed									
0015	1/8" thick, clear float	2 Glaz	480	.033	S.F.	4.32	1.76		6.08	7.55
0500	3/16" thick, clear		480	.033		6.75	1.76		8.51	10.25
0600	Tinted		480	.033		7.65	1.76		9.41	11.25
0700	Tempered		480	.033		12.55	1.76		14.31	16.60
9000	Minimum labor/equipment charge		5	3.200	Job		168		168	268

08 81 65 – Wire Glass

08 81 65.10 Glass Reinforced With Wire

		Crew	Daily Output	Labor-Hours	Unit	Material	Labor	Equipment	Total	Total Incl O&P
0010	**GLASS REINFORCED WITH WIRE**									
0012	1/4" thick rough obscure	2 Glaz	135	.119	S.F.	25.50	6.25		31.75	38
1000	Polished wire, 1/4" thick, diamond, clear		135	.119		33.50	6.25		39.75	46.50
1500	Pinstripe, obscure		135	.119		54	6.25		60.25	69

08 83 Mirrors

08 83 13 – Mirrored Glass Glazing

08 83 13.10 Mirrors		Crew	Daily Output	Labor-Hours	Unit	Material	2021 Bare Costs Labor	Equipment	Total	Total Incl O&P
0010	**MIRRORS**, No frames, wall type, 1/4" plate glass, polished edge									
0100	Up to 5 S.F.	2 Glaz	125	.128	S.F.	9.80	6.75		16.55	21.50
0200	Over 5 S.F.		160	.100		9.40	5.25		14.65	18.70
0500	Door type, 1/4" plate glass, up to 12 S.F.		160	.100		9.75	5.25		15	19.05
1000	Float glass, up to 10 S.F., 1/8" thick		160	.100		6.85	5.25		12.10	15.90
1100	3/16" thick		150	.107		8.85	5.60		14.45	18.70
1500	12" x 12" wall tiles, square edge, clear		195	.082		2.92	4.32		7.24	10.05
1600	Veined		195	.082		6.50	4.32		10.82	14
2000	1/4" thick, stock sizes, one way transparent		125	.128		23.50	6.75		30.25	36
2010	Bathroom, unframed, laminated		160	.100		16.80	5.25		22.05	27
2500	Tempered		160	.100		19	5.25		24.25	29.50

08 83 13.15 Reflective Glass

			Crew	Daily Output	Labor-Hours	Unit	Material	Labor	Equipment	Total	Total Incl O&P
0010	**REFLECTIVE GLASS**										
0100	1/4" float with fused metallic oxide fixed	G	2 Glaz	115	.139	S.F.	18.65	7.35		26	32
0500	1/4" float glass with reflective applied coating	G	"	115	.139	"	40	7.35		47.35	56

08 84 Plastic Glazing

08 84 10 – Plexiglass Glazing

08 84 10.10 Plexiglass Acrylic

		Crew	Daily Output	Labor-Hours	Unit	Material	Labor	Equipment	Total	Total Incl O&P
0010	**PLEXIGLASS ACRYLIC**, clear, masked,									
0020	1/8" thick, cut sheets	2 Glaz	170	.094	S.F.	16.70	4.96		21.66	26.50
0200	Full sheets		195	.082		3.42	4.32		7.74	10.60
0500	1/4" thick, cut sheets		165	.097		12.50	5.10		17.60	22
0600	Full sheets	2 Glaz	185	.086	S.F.	6.75	4.55		11.30	14.65
0900	3/8" thick, cut sheets		155	.103		11	5.45		16.45	21
1000	Full sheets		180	.089		18.90	4.68		23.58	28.50
1300	1/2" thick, cut sheets		135	.119		37	6.25		43.25	50.50
1400	Full sheets		150	.107		27	5.60		32.60	39
1700	3/4" thick, cut sheets		115	.139		32	7.35		39.35	46.50
1800	Full sheets		130	.123		18.35	6.50		24.85	30.50
2100	1" thick, cut sheets		105	.152		36	8		44	52.50
2200	Full sheets		125	.128		22	6.75		28.75	35
3000	Colored, 1/8" thick, cut sheets		170	.094		21	4.96		25.96	31
3200	Full sheets		195	.082		12.15	4.32		16.47	20.50
3500	1/4" thick, cut sheets		165	.097		23.50	5.10		28.60	33.50
3600	Full sheets		185	.086		17.35	4.55		21.90	26.50
4000	Mirrors, untinted, cut sheets, 1/8" thick		185	.086		17	4.55		21.55	26
4200	1/4" thick		180	.089		20.50	4.68		25.18	30

08 84 20 – Polycarbonate

08 84 20.10 Thermoplastic

		Crew	Daily Output	Labor-Hours	Unit	Material	Labor	Equipment	Total	Total Incl O&P
0010	**THERMOPLASTIC**, clear, masked, cut sheets									
0020	1/8" thick	2 Glaz	170	.094	S.F.	15.70	4.96		20.66	25
0500	3/16" thick		165	.097		6.80	5.10		11.90	15.55
1000	1/4" thick		155	.103		18.90	5.45		24.35	29.50
1500	3/8" thick		150	.107		35	5.60		40.60	47.50
9000	Minimum labor/equipment charge	1 Glaz	2	4	Job		211		211	335

288

For customer support on your Commercial Renovation Costs with RSMeans Data, call 800.448.8182.

08 85 Glazing Accessories

08 85 10 - Miscellaneous Glazing Accessories

08 85 10.10 Glazing Gaskets

		Crew	Daily Output	Labor-Hours	Unit	Material	2021 Bare Costs Labor	Equipment	Total	Total Incl O&P
0010	**GLAZING GASKETS**, Neoprene for glass tongued mullion									
0015	1/4" glass	2 Glaz	200	.080	L.F.	3.96	4.21		8.17	11.05
0020	3/8"		200	.080		1.13	4.21		5.34	7.95
0040	1/2"		200	.080		1.44	4.21		5.65	8.30
0060	3/4"		180	.089		1.84	4.68		6.52	9.45
0080	1"		180	.089		1.72	4.68		6.40	9.35
1000	Glazing compound, wood	1 Glaz	58	.138			7.25		7.25	11.55
1005	Glazing compound, per window, up to 30 L.F.									
1006	Glazing compound, per window, up to 30 L.F.				Ea.	.12			.12	.13
1020	Metal	1 Glaz	58	.138	L.F.		7.25		7.25	11.55

08 85 20 - Structural Glazing Sealants

08 85 20.10 Structural Glazing Adhesives

		Crew	Daily Output	Labor-Hours	Unit	Material	2021 Bare Costs Labor	Equipment	Total	Total Incl O&P
0010	**STRUCTURAL GLAZING ADHESIVES**									
0050	Structural glazing adhesive, 1/4" x 1/4" joint seal	1 Glaz	200	.040	L.F.	.20	2.11		2.31	3.57

08 87 Glazing Surface Films

08 87 13 - Solar Control Films

08 87 13.10 Solar Films On Glass

		Crew	Daily Output	Labor-Hours	Unit	Material	2021 Bare Costs Labor	Equipment	Total	Total Incl O&P
0010	**SOLAR FILMS ON GLASS** (glass not included)									
1000	Bronze, 20% VLT	H-2	180	.133	S.F.	1.59	6.65		8.24	12.35
1020	50% VLT		180	.133		.95	6.65		7.60	11.65
1050	Neutral, 20% VLT		180	.133		.63	6.65		7.28	11.30
1100	Silver, 15% VLT		180	.133		.75	6.65		7.40	11.45
1120	35% VLT		180	.133		1.04	6.65		7.69	11.75
1150	68% VLT		180	.133		.41	6.65		7.06	11.05
3000	One way mirror with night vision 5% in and 15% out VLT		180	.133		3.04	6.65		9.69	13.95

08 87 16 - Glass Safety Films

08 87 16.10 Safety Films On Glass

		Crew	Daily Output	Labor-Hours	Unit	Material	2021 Bare Costs Labor	Equipment	Total	Total Incl O&P
0010	**SAFETY FILMS ON GLASS** (glass not included)									
0015	Safety film helps hold glass together when broken									
0050	Safety film, clear, 2 mil	H-2	180	.133	S.F.	1.80	6.65		8.45	12.60
0100	Safety film, clear, 4 mil		180	.133		1.60	6.65		8.25	12.35
0110	Safety film, tinted, 4 mil		180	.133		1.20	6.65		7.85	11.90
0150	Safety film, black tint, 8 mil		180	.133		2.07	6.65		8.72	12.90

08 87 26 - Bird Control Film

08 87 26.10 Bird Control Film

		Crew	Daily Output	Labor-Hours	Unit	Material	2021 Bare Costs Labor	Equipment	Total	Total Incl O&P
0010	**BIRD CONTROL FILM**									
0050	Patterned, adhered to glass	2 Glaz	180	.089	S.F.	6.15	4.68		10.83	14.20
0200	Decals small, adhered to glass	1 Glaz	50	.160	Ea.	1.74	8.40		10.14	15.30
0250	Decals large, adhered to glass		50	.160	"	8.10	8.40		16.50	22.50
0300	Decals set of 4, adhered to glass		25	.320	Set	7.45	16.85		24.30	35
0400	Bird control film tape		10	.800	Roll	4.98	42		46.98	72.50

08 87 33 - Electrically Tinted Window Film

08 87 33.20 Window Film

		Crew	Daily Output	Labor-Hours	Unit	Material	2021 Bare Costs Labor	Equipment	Total	Total Incl O&P
0010	**WINDOW FILM** adhered on glass (glass not included)									
0015	Film is pre-wired and can be trimmed									
0100	Window film 4 S.F. adhered on glass	1 Glaz	200	.040	S.F.	222	2.11		224.11	247
0120	6 S.F.		180	.044		330	2.34		332.34	370
0160	9 S.F.		170	.047		500	2.48		502.48	555

08 87 Glazing Surface Films

08 87 33 - Electrically Tinted Window Film

08 87 33.20 Window Film	Crew	Daily Output	Labor-Hours	Unit	Material	2021 Bare Costs Labor	Equipment	Total	Total Incl O&P
0180	10 S.F.	1 Glaz	150	.053	S.F.	138	2.81		140.81	156
0200	12 S.F.		144	.056		665	2.93		667.93	735
0220	14 S.F.		140	.057		775	3.01		778.01	855
0240	16 S.F.		128	.063		885	3.29		888.29	980
0260	18 S.F.		126	.063		995	3.34		998.34	1,100
0280	20 S.F.		120	.067		1,100	3.51		1,103.51	1,225

08 87 53 - Security Films On Glass

08 87 53.10 Security Film Adhered On Glass

		Crew	Daily Output	Labor-Hours	Unit	Material	2021 Bare Costs Labor	Equipment	Total	Total Incl O&P
0010	**SECURITY FILM ADHERED ON GLASS** (glass not included)									
0020	Security film, clear, 7 mil	1 Glaz	200	.040	S.F.	2.11	2.11		4.22	5.65
0030	8 mil		200	.040		2.31	2.11		4.42	5.90
0040	9 mil		200	.040		1.11	2.11		3.22	4.57
0050	10 mil		200	.040		1.91	2.11		4.02	5.45
0060	12 mil		200	.040		2.80	2.11		4.91	6.45
0075	15 mil	1 Glaz	200	.040	S.F.	2.26	2.11		4.37	5.85
0100	Security film, sealed with structural adhesive, 7 mil		180	.044		2.12	2.34		4.46	6.05
0110	8 mil		180	.044		2.32	2.34		4.66	6.25
0140	14 mil		180	.044		3.84	2.34		6.18	7.95

08 88 Special Function Glazing

08 88 52 - Prefabricated Glass Block Windows

08 88 52.10 Prefabricated Glass Block Windows

		Crew	Daily Output	Labor-Hours	Unit	Material	2021 Bare Costs Labor	Equipment	Total	Total Incl O&P
0010	**PREFABRICATED GLASS BLOCK WINDOWS**									
0015	Includes frame and silicone seal									
0020	Glass Block Window 16" x 8"	2 Glaz	6	2.667	Ea.	153	140		293	390
0025	16" x 16"		6	2.667		192	140		332	435
0030	16" x 24"		6	2.667		225	140		365	470
0050	16" x 48"		4	4		330	211		541	695
0100	Glass Block Window 24" x 8"		6	2.667		181	140		321	420
0110	24" x 16"		6	2.667		232	140		372	480
0120	24" x 24"		6	2.667		292	140		432	545
0150	24" x 48"		6	2.667		430	140		570	695
0200	Glass Block Window 32" x 8"		6	2.667		181	140		321	420
0210	32" x 16"		6	2.667		265	140		405	515
0220	32" x 24"		4	4		335	211		546	700
0250	32" x 48"		4	4		525	211		736	910
0300	Glass Block Window 40" x 8"		6	2.667		231	140		371	475
0310	40" x 16"		6	2.667		305	140		445	560
0320	40" x 24"		6	2.667		375	140		515	635
0350	40" x 48"		6	2.667		615	140		755	900
0400	Glass Block Window 48" x 8"		6	2.667		273	140		413	525
0410	48" x 16"		6	2.667		360	140		500	620
0420	48" x 24"		6	2.667		430	140		570	695
0450	48" x 48"		6	2.667		700	140		840	995
0500	Glass Block Window 56" x 8"		6	2.667		275	140		415	525
0510	56" x 16"		4	4		405	211		616	785
0520	56" x 24"		4	4		500	211		711	890
0550	56" x 48"		4	4		795	211		1,006	1,200

For customer support on your Commercial Renovation Costs with RSMeans Data, call 800.448.8182.

08 88 Special Function Glazing

08 88 52 – Prefabricated Glass Block Windows

08 88 52.20 Prefab. Glass Block Windows Hurricane Resistant	Crew	Daily Output	Labor-Hours	Unit	Material	2021 Bare Costs Labor	Equipment	Total	Total Incl O&P
0010 **PREFABRICATED GLASS BLOCK WINDOWS HURRICANE RESISTANT**									
0015 Includes frame and silicone seal									
3020 Glass Block Window 16" x 16"	2 Glaz	6	2.667	Ea.	430	140		570	700
3030 16" x 24"		6	2.667		505	140		645	780
3040 16" x 32"		4	4		595	211		806	990
3050 16" x 40"		4	4		630	211		841	1,025
3060 16" x 48"		4	4		665	211		876	1,075
3070 16" x 56"		4	4		825	211		1,036	1,250
3080 Glass Block Window 24" x 24"		6	2.667		585	140		725	870
3090 24" x 32"		4	4		665	211		876	1,075
3100 24" x 40"		4	4		835	211		1,046	1,250
3110 24" x 48"		4	4		870	211		1,081	1,300
3120 24" x 56"		4	4		920	211		1,131	1,325
3130 Glass Block Window 32" x 32"		4	4		745	211		956	1,150
3140 32" x 40"		4	4		570	211		781	960
3150 32" x 48"		4	4		985	211		1,196	1,400
3160 32" x 56"		4	4		870	211		1,081	1,300
3200 Glass Block Window 40" x 40"		4	4		985	211		1,196	1,400
3210 40" x 48"		3	5.333		1,200	281		1,481	1,750
3220 40" x 56"		3	5.333		1,325	281		1,606	1,900
3260 Glass Block Window 48" x 48"		3	5.333		1,375	281		1,656	1,975
3270 48" x 56"		3	5.333		1,550	281		1,831	2,150
3310 Glass Block Window 56" x 56"		3	5.333		1,700	281		1,981	2,325

08 88 52.30 Prefabricated Acrylic Block Casement Windows

	Crew	Daily Output	Labor-Hours	Unit	Material	2021 Bare Costs Labor	Equipment	Total	Total Incl O&P
0010 **PREFABRICATED ACRYLIC BLOCK CASEMENT WINDOWS**									
0015 Windows include frame and nail fin.									
3050 6" x 6" x 1-3/4" Block Casement Window 2 x 2 Pattern	2 Glaz	6	2.667	Ea.	450	140		590	720
3055 2 x 3		6	2.667		470	140		610	740
3060 2 x 4		6	2.667		510	140		650	790
3065 2 x 5		6	2.667		545	140		685	825
3070 2 x 6		6	2.667		600	140		740	885
3075 2 x 7		4	4		715	211		926	1,125
3080 2 x 8		4	4		745	211		956	1,150
3085 2 x 9		4	4		820	211		1,031	1,225
3090 3 x 3		4	4		495	211		706	880
3095 3 x 4		4	4		545	211		756	935
3100 3 x 5		4	4		610	211		821	1,000
3110 3 x 6		4	4		650	211		861	1,050
3115 3 x 7		4	4		740	211		951	1,150
3120 3 x 8		4	4		815	211		1,026	1,225
3125 3 x 9		3	5.333		885	281		1,166	1,425
3130 4 x 4		3	5.333		640	281		921	1,150
3135 4 x 5		3	5.333		680	281		961	1,200
3140 4 x 6		3	5.333		730	281		1,011	1,250
3145 4 x 7		3	5.333		770	281		1,051	1,300
3150 4 x 8		3	5.333		885	281		1,166	1,425
3500 8" x 8" x 1-3/4" Block Casement Window 2 x 2 Pattern		6	2.667		415	140		555	685
3510 2 x 3		6	2.667		460	140		600	735
3515 2 x 4		6	2.667		520	140		660	795
3520 2 x 5		6	2.667		595	140		735	875
3525 2 x 6		4	4		660	211		871	1,075
3530 2 x 7		4	4		695	211		906	1,100

For customer support on your Commercial Renovation Costs with RSMeans Data, call 800.448.8182.

291

08 88 Special Function Glazing

08 88 52 – Prefabricated Glass Block Windows

08 88 52.30 Prefabricated Acrylic Block Casement Windows	Crew	Daily Output	Labor-Hours	Unit	Material	2021 Bare Costs Labor	Equipment	Total	Total Incl O&P
3535	2 x 8	2 Glaz	3	5.333	Ea.	755	281		1,036	1,275
3540	2 x 9		3	5.333		800	281		1,081	1,325
3545	2 x 10		3	5.333		800	281		1,081	1,325
3550	2 x 11		3	5.333		825	281		1,106	1,350
3555	3 x 3		4	4		515	211		726	905
3560	3 x 4		4	4		575	211		786	965
3565	3 x 5		4	4		685	211		896	1,100
3570	3 x 6		4	4		750	211		961	1,150
3575	3 x 7		4	4		860	211		1,071	1,275
3580	3 x 8		3	5.333		955	281		1,236	1,500
3585	3 x 9		3	5.333		880	281		1,161	1,400
3590	4 x 4		3	5.333		710	281		991	1,225
3595	4 x 5		3	5.333		865	281		1,146	1,400
3600	4 x 6		3	5.333		930	281		1,211	1,475

08 88 56 – Ballistics-Resistant Glazing

08 88 56.10 Laminated Glass

		Crew	Daily Output	Labor-Hours	Unit	Material	2021 Bare Costs Labor	Equipment	Total	Total Incl O&P
0010	**LAMINATED GLASS**									
0020	Clear float .03" vinyl 1/4" thick	2 Glaz	90	.178	S.F.	13.45	9.35		22.80	29.50
0100	3/8" thick		78	.205		32	10.80		42.80	52
0200	.06" vinyl, 1/2" thick		65	.246		28	12.95		40.95	51.50
1000	5/8" thick		90	.178		39.50	9.35		48.85	58.50
2000	Bullet-resisting, 1-3/16" thick, to 15 S.F.		16	1		133	52.50		185.50	230
2100	Over 15 S.F.		16	1		132	52.50		184.50	229
2200	2" thick, to 15 S.F.		10	1.600		151	84		235	300
2300	Over 15 S.F.		10	1.600		134	84		218	281
2500	2-1/4" thick, to 15 S.F.		12	1.333		207	70		277	340
2600	Over 15 S.F.		12	1.333		195	70		265	325
2700	Level 2 (.357 magnum), NIJ and UL		12	1.333		56.50	70		126.50	174
2750	Level 3A (.44 magnum) NIJ, UL 3		12	1.333		60.50	70		130.50	179
2800	Level 4 (AK-47) NIJ, UL 7 & 8		12	1.333		83	70		153	204
2850	Level 5 (M-16) UL		12	1.333		82	70		152	202
2900	Level 3 (7.62 Armor Piercing) NIJ, UL 4 & 5		12	1.333		97	70		167	218

08 91 Louvers

08 91 16 – Operable Wall Louvers

08 91 16.10 Movable Blade Louvers

		Crew	Daily Output	Labor-Hours	Unit	Material	2021 Bare Costs Labor	Equipment	Total	Total Incl O&P
0010	**MOVABLE BLADE LOUVERS**									
0100	PVC, commercial grade, 12" x 12"	1 Shee	20	.400	Ea.	83	26		109	133
0110	16" x 16"		14	.571		97	37.50		134.50	165
0120	18" x 18"		14	.571		99.50	37.50		137	169
0130	24" x 24"		14	.571		142	37.50		179.50	216
0140	30" x 30"		12	.667		171	43.50		214.50	256
0150	36" x 36"		12	.667		201	43.50		244.50	290
0300	Stainless steel, commercial grade, 12" x 12"		20	.400		254	26		280	320
0310	16" x 16"	1 Shee	14	.571	Ea.	315	37.50		352.50	405
0320	18" x 18"		14	.571		325	37.50		362.50	420
0330	20" x 20"		14	.571		375	37.50		412.50	475
0340	24" x 24"		14	.571		445	37.50		482.50	550
0350	30" x 30"		12	.667		500	43.50		543.50	620
0360	36" x 36"		12	.667		605	43.50		648.50	740

08 91 Louvers

08 91 19 – Fixed Louvers

08 91 19.10 Aluminum Louvers

		Crew	Daily Output	Labor-Hours	Unit	Material	2021 Bare Costs Labor	Equipment	Total	Total Incl O&P
0010	**ALUMINUM LOUVERS**									
0020	Aluminum with screen, residential, 8" x 8"	1 Carp	38	.211	Ea.	19.60	11.50		31.10	40
0100	12" x 12"		38	.211		18.80	11.50		30.30	39
0200	12" x 18"		35	.229		22	12.50		34.50	44
0250	14" x 24"		30	.267		34	14.60		48.60	60
0300	18" x 24"		27	.296		26.50	16.20		42.70	55.50
0500	24" x 30"		24	.333		72.50	18.25		90.75	109
0700	Triangle, adjustable, small		20	.400		75	22		97	118
0800	Large		15	.533		104	29		133	162
2100	Round, aluminum, 3/4" deep, 1" diameter		85	.094		1.08	5.15		6.23	9.40
2150	3" diameter		60	.133		2.55	7.30		9.85	14.40
2200	4" diameter		50	.160		5.55	8.75		14.30	20
2250	6" diameter	▼	30	.267		4.03	14.60		18.63	27.50
3000	PVC, commercial grade, 12" x 12"	1 Shee	20	.400		98	26		124	149
3010	12" x 18"		20	.400		110	26		136	163
3020	12" x 24"		20	.400		148	26		174	204
3030	14" x 24"		20	.400		155	26		181	211
3100	Aluminum, commercial grade, 12" x 12"		20	.400		170	26		196	228
3110	24" x 24"		16	.500		234	32.50		266.50	310
3120	24" x 36"		16	.500		335	32.50		367.50	415
3130	36" x 24"		16	.500		345	32.50		377.50	430
3140	36" x 36"		14	.571		430	37.50		467.50	535
3150	36" x 48"		10	.800		475	52.50		527.50	600
3160	48" x 36"		10	.800		500	52.50		552.50	630
3170	48" x 48"		10	.800		505	52.50		557.50	635
3180	60" x 48"		8	1		1,150	65.50		1,215.50	1,350
3190	60" x 60"	▼	8	1	▼	750	65.50		815.50	925

08 91 19.20 Steel Louvers

		Crew	Daily Output	Labor-Hours	Unit	Material	2021 Bare Costs Labor	Equipment	Total	Total Incl O&P
0010	**STEEL LOUVERS**									
3300	Galvanized steel, fixed blades, commercial grade, 18" x 18"	1 Shee	20	.400	Ea.	163	26		189	220
3310	24" x 24"		20	.400		218	26		244	281
3320	24" x 36"		16	.500		284	32.50		316.50	360
3330	36" x 24"		16	.500		290	32.50		322.50	370
3340	36" x 36"		14	.571		355	37.50		392.50	455
3350	36" x 48"	▼	10	.800	▼	420	52.50		472.50	540
3360	48" x 36"	1 Shee	10	.800	Ea.	410	52.50		462.50	530
3370	48" x 48"		10	.800		510	52.50		562.50	640
3380	60" x 48"		10	.800		555	52.50		607.50	690
3390	60" x 60"	▼	10	.800	▼	665	52.50		717.50	810

08 91 26 – Door Louvers

08 91 26.10 Steel Louvers, 18 Gauge, Fixed Blade

		Crew	Daily Output	Labor-Hours	Unit	Material	2021 Bare Costs Labor	Equipment	Total	Total Incl O&P
0010	**STEEL LOUVERS, 18 GAUGE, FIXED BLADE**									
0050	12" x 12", with powder coat	1 Carp	20	.400	Ea.	102	22		124	147
0055	18" x 12"		20	.400		106	22		128	152
0060	18" x 18"		20	.400		112	22		134	158
0065	24" x 12"		20	.400		122	22		144	170
0070	24" x 18"		20	.400		129	22		151	177
0075	24" x 24"		20	.400		164	22		186	215
0100	12" x 12", galvanized		20	.400		269	22		291	330
0105	18" x 12"		20	.400		285	22		307	350
0115	24" x 12"	▼	20	.400	▼	112	22		134	158

For customer support on your Commercial Renovation Costs with RSMeans Data, call 800.448.8182.

293

08 91 Louvers

08 91 26 – Door Louvers

08 91 26.10 Steel Louvers, 18 Gauge, Fixed Blade		Crew	Daily Output	Labor-Hours	Unit	Material	2021 Bare Costs Labor	Equipment	Total	Total Incl O&P
0125	24" x 24"	1 Carp	20	.400	Ea.	400	22		422	475
0300	6" x 12", painted		20	.400		46.50	22		68.50	86.50
0320	10" x 16"		20	.400		73	22		95	115
0340	12" x 22"		20	.400		108	22		130	153
0360	14" x 14"		20	.400		109	22		131	154
0400	18" x 18"		20	.400		128	22		150	176
0420	20" x 26"		20	.400		147	22		169	197
0440	22" x 22"		20	.400		139	22		161	188
0460	26" x 26"		20	.400		226	22		248	283
3000	Fire rated									
3010	12" x 12"	1 Carp	10	.800	Ea.	183	44		227	271
3050	18" x 18"		10	.800		268	44		312	365
3100	24" x 12"		10	.800		280	44		324	380
3120	24" x 18"		10	.800		300	44		344	400
3130	24" x 24"		10	.800		300	44		344	400

08 91 26.20 Aluminum Louvers

		Crew	Daily Output	Labor-Hours	Unit	Material	2021 Bare Costs Labor	Equipment	Total	Total Incl O&P
0010	**ALUMINUM LOUVERS**, 18 ga., fixed blade, clear anodized									
0050	6" x 12"	1 Carp	20	.400	Ea.	54.50	22		76.50	95
0060	10" x 10"		20	.400		63	22		85	105
0065	10" x 14"		20	.400		71.50	22		93.50	114
0080	12" x 22"		20	.400		100	22		122	145
0090	14" x 14"		20	.400		90	22		112	134
0100	18" x 10"		20	.400		74.50	22		96.50	117
0110	18" x 18"		20	.400		131	22		153	179
0120	22" x 22"		20	.400		163	22		185	214
0130	26" x 26"		20	.400		208	22		230	264

08 95 Vents

08 95 13 – Soffit Vents

08 95 13.10 Wall Louvers

		Crew	Daily Output	Labor-Hours	Unit	Material	2021 Bare Costs Labor	Equipment	Total	Total Incl O&P
0010	**WALL LOUVERS**									
2340	Baked enamel finish	1 Carp	200	.040	L.F.	4.92	2.19		7.11	8.90
2400	Under eaves vent, aluminum, mill finish, 16" x 4"		48	.167	Ea.	2.08	9.10		11.18	16.80
2500	16" x 8"		48	.167	"	2.77	9.10		11.87	17.55

08 95 16 – Wall Vents

08 95 16.10 Louvers

		Crew	Daily Output	Labor-Hours	Unit	Material	2021 Bare Costs Labor	Equipment	Total	Total Incl O&P
0010	**LOUVERS**									
0020	Redwood, 2'-0" diameter, full circle	1 Carp	16	.500	Ea.	95	27.50		122.50	148
0100	Half circle		16	.500		230	27.50		257.50	297
0200	Octagonal		16	.500		216	27.50		243.50	282
0300	Triangular, 5/12 pitch, 5'-0" at base		16	.500		660	27.50		687.50	770
1000	Rectangular, 1'-4" x 1'-3"		16	.500		28.50	27.50		56	75
1100	Rectangular, 1'-4" x 1'-8"		16	.500		44.50	27.50		72	92.50
1200	1'-4" x 2'-2"		15	.533		51	29		80	103
1300	1'-9" x 2'-2"		15	.533		89.50	29		118.50	145
1400	2'-3" x 2'-2"		14	.571		101	31.50		132.50	161
1700	2'-4" x 2'-11"		13	.615		101	33.50		134.50	165
2000	Aluminum, 12" x 16"		25	.320		24.50	17.50		42	55
2010	16" x 20"		25	.320		36	17.50		53.50	68
2020	24" x 30"		25	.320		64	17.50		81.50	98.50

08 95 16.10 Louvers		Crew	Daily Output	Labor-Hours	Unit	Material	2021 Bare Costs Labor	Equipment	Total	Total Incl O&P
2100	6' triangle	1 Carp	12	.667	Ea.	196	36.50		232.50	273
3100	Round, 2'-2" diameter		16	.500		198	27.50		225.50	262
7000	Vinyl gable vent, 8" x 8"		38	.211		17.40	11.50		28.90	37.50
7020	12" x 12"		38	.211		32	11.50		43.50	53.50
7080	12" x 18"		35	.229		41	12.50		53.50	65
7200	18" x 24"		30	.267		57.50	14.60		72.10	86.50
9000	Minimum labor/equipment charge		3.50	2.286	Job		125		125	199

Division Notes

	CREW	DAILY OUTPUT	LABOR-HOURS	UNIT	BARE COSTS				TOTAL INCL O&P
					MAT.	LABOR	EQUIP.	TOTAL	

Estimating Tips
General

- Room Finish Schedule: A complete set of plans should contain a room finish schedule. If one is not available, it would be well worth the time and effort to obtain one.

09 20 00 Plaster and Gypsum Board

- Lath is estimated by the square yard plus a 5% allowance for waste. Furring, channels, and accessories are measured by the linear foot. An extra foot should be allowed for each accessory miter or stop.

- Plaster is also estimated by the square yard. Deductions for openings vary by preference, from zero deduction to 50% of all openings over 2 feet in width. The estimator should allow one extra square foot for each linear foot of horizontal interior or exterior angle located below the ceiling level. Also, double the areas of small radius work.

- Drywall accessories, studs, track, and acoustical caulking are all measured by the linear foot. Drywall taping is figured by the square foot. Gypsum wallboard is estimated by the square foot. No material deductions should be made for door or window openings under 32 S.F.

09 60 00 Flooring

- Tile and terrazzo areas are taken off on a square foot basis. Trim and base materials are measured by the linear foot. Accent tiles are listed per each. Two basic methods of installation are used. Mud set is approximately 30% more expensive than thin set.

The cost of grout is included with tile unit price lines unless otherwise noted. In terrazzo work, be sure to include the linear footage of embedded decorative strips, grounds, machine rubbing, and power cleanup.

- Wood flooring is available in strip, parquet, or block configuration. The latter two types are set in adhesives with quantities estimated by the square foot. The laying pattern will influence labor costs and material waste. In addition to the material and labor for laying wood floors, the estimator must make allowances for sanding and finishing these areas, unless the flooring is prefinished.

- Sheet flooring is measured by the square yard. Roll widths vary, so consideration should be given to use the most economical width, as waste must be figured into the total quantity. Consider also the installation methods available—direct glue down or stretched. Direct glue-down installation is assumed with sheet carpet unit price lines unless otherwise noted.

09 70 00 Wall Finishes

- Wall coverings are estimated by the square foot. The area to be covered is measured—length by height of the wall above the baseboards—to calculate the square footage of each wall. This figure is divided by the number of square feet in the single roll which is being used. Deduct, in full, the areas of openings such as doors and windows. Where a pattern match is required allow 25–30% waste.

09 80 00 Acoustic Treatment

- Acoustical systems fall into several categories. The takeoff of these materials should be by the square foot of area with a 5% allowance for waste. Do not forget about scaffolding, if applicable, when estimating these systems.

09 90 00 Painting and Coating

- New line items created for cut-ins with reference diagram.

- A major portion of the work in painting involves surface preparation. Be sure to include cleaning, sanding, filling, and masking costs in the estimate.

- Protection of adjacent surfaces is not included in painting costs. When considering the method of paint application, an important factor is the amount of protection and masking required. These must be estimated separately and may be the determining factor in choosing the method of application.

Reference Numbers

Reference numbers are shown at the beginning of some major classifications. These numbers refer to related items in the Reference Section. The reference information may be an estimating procedure, an alternate pricing method, or technical information.

Note: Not all subdivisions listed here necessarily appear. ■

Same Data. Simplified.

Enjoy the convenience and efficiency of accessing your costs anywhere:

- **Skip the multiplier** by setting your location
- **Quickly search,** edit, favorite and share costs
- **Stay on top of price changes** with automatic updates

Discover more at rsmeans.com/online

09 01 70 – Maintenance of Wall Finishes

09 01 70.10 Gypsum Wallboard Repairs

		Crew	Daily Output	Labor-Hours	Unit	Material	2021 Bare Costs Labor	Equipment	Total	Total Incl O&P
0010	**GYPSUM WALLBOARD REPAIRS**									
0100	Fill and sand, pin/nail holes	1 Carp	960	.008	Ea.		.46		.46	.73
0110	Screw head pops		480	.017			.91		.91	1.45
0120	Dents, up to 2" square		48	.167		.01	9.10		9.11	14.50
0130	2" to 4" square		24	.333		.04	18.25		18.29	29
0140	Cut square, patch, sand and finish, holes, up to 2" square		12	.667		.03	36.50		36.53	58
0150	2" to 4" square		11	.727		.08	40		40.08	63.50
0160	4" to 8" square		10	.800		.23	44		44.23	70
0170	8" to 12" square		8	1		.48	54.50		54.98	87.50
0180	12" to 32" square		6	1.333		1.59	73		74.59	118
0210	16" by 48"		5	1.600		2.75	87.50		90.25	142
0220	32" by 48"		4	2		4.31	109		113.31	179
0230	48" square	1 Carp	3.50	2.286	Ea.	6.10	125		131.10	206
0240	60" square		3.20	2.500	"	10.05	137		147.05	229
0500	Skim coat surface with joint compound		1600	.005	S.F.	.03	.27		.30	.48
0510	Prepare, retape and refinish joints		60	.133	L.F.	.67	7.30		7.97	12.35
9000	Minimum labor/equipment charge		2	4	Job		219		219	350

09 01 90 – Maintenance of Painting and Coating

09 01 90.92 Sanding

		Crew	Daily Output	Labor-Hours	Unit	Material	2021 Bare Costs Labor	Equipment	Total	Total Incl O&P
0010	**SANDING** and puttying interior trim, compared to									
0100	Painting 1 coat, on quality work				L.F.		100%			
0300	Medium work						50%			
0400	Industrial grade						25%			
0500	Surface protection, placement and removal									
0510	Surface protection, placement and removal, basic drop cloths	1 Pord	6400	.001	S.F.		.06		.06	.09
0520	Masking with paper		800	.010		.19	.46		.65	.95
0530	Volume cover up (using plastic sheathing or building paper)		16000	.001			.02		.02	.04

09 01 90.93 Exterior Surface Preparation

		Crew	Daily Output	Labor-Hours	Unit	Material	2021 Bare Costs Labor	Equipment	Total	Total Incl O&P
0010	**EXTERIOR SURFACE PREPARATION**									
0015	Doors, per side, not incl. frames or trim									
0020	Scrape & sand									
0030	Wood, flush	1 Pord	616	.013	S.F.		.60		.60	.96
0040	Wood, detail		496	.016			.75		.75	1.19
0050	Wood, louvered		280	.029			1.33		1.33	2.10
0060	Wood, overhead		616	.013			.60		.60	.96
0070	Wire brush									
0080	Metal, flush	1 Pord	640	.013	S.F.		.58		.58	.92
0090	Metal, detail		520	.015			.71		.71	1.13
0100	Metal, louvered		360	.022			1.03		1.03	1.63
0110	Metal or fibr., overhead		640	.013			.58		.58	.92
0120	Metal, roll up		560	.014			.66		.66	1.05
0130	Metal, bulkhead		640	.013			.58		.58	.92
0140	Power wash, based on 2500 lb. operating pressure									
0150	Metal, flush	A-1H	2240	.004	S.F.		.16	.03	.19	.29
0160	Metal, detail		2120	.004			.17	.04	.21	.31
0170	Metal, louvered		2000	.004			.18	.04	.22	.32
0180	Metal or fibr., overhead		2400	.003			.15	.03	.18	.28
0190	Metal, roll up		2400	.003			.15	.03	.18	.28
0200	Metal, bulkhead		2200	.004			.16	.04	.20	.30
0400	Windows, per side, not incl. trim									
0410	Scrape & sand									
0420	Wood, 1-2 lite	1 Pord	320	.025	S.F.		1.16		1.16	1.84

09 01 90 – Maintenance of Painting and Coating

09 01 90.93 Exterior Surface Preparation

	Crew	Daily Output	Labor-Hours	Unit	Material	2021 Bare Costs Labor	Equipment	Total	Total Incl O&P
0430	Wood, 3-6 lite	1 Pord	280	.029	S.F.		1.33		1.33	2.10
0440	Wood, 7-10 lite	↓	240	.033	↓		1.55		1.55	2.45
0450	Wood, 12 lite	1 Pord	200	.040	S.F.		1.86		1.86	2.94
0460	Wood, Bay/Bow	"	320	.025	"		1.16		1.16	1.84
0470	Wire brush									
0480	Metal, 1-2 lite	1 Pord	480	.017	S.F.		.77		.77	1.23
0490	Metal, 3-6 lite		400	.020			.93		.93	1.47
0500	Metal, Bay/Bow	↓	480	.017	↓		.77		.77	1.23
0510	Power wash, based on 2500 lb. operating pressure									
0520	1-2 lite	A-1H	4400	.002	S.F.		.08	.02	.10	.15
0530	3-6 lite		4320	.002			.08	.02	.10	.15
0540	7-10 lite		4240	.002			.08	.02	.10	.15
0550	12 lite		4160	.002			.09	.02	.11	.16
0560	Bay/Bow	↓	4400	.002	↓		.08	.02	.10	.15
0600	Siding, scrape and sand, light=10-30%, med.=30-70%									
0610	Heavy=70-100% of surface to sand									
0650	Texture 1-11, light	1 Pord	480	.017	S.F.		.77		.77	1.23
0660	Med.		440	.018			.84		.84	1.34
0670	Heavy		360	.022			1.03		1.03	1.63
0680	Wood shingles, shakes, light		440	.018			.84		.84	1.34
0690	Med.		360	.022			1.03		1.03	1.63
0700	Heavy		280	.029			1.33		1.33	2.10
0710	Clapboard, light		520	.015			.71		.71	1.13
0720	Med.		480	.017			.77		.77	1.23
0730	Heavy	↓	400	.020	↓		.93		.93	1.47
0740	Wire brush									
0750	Aluminum, light	1 Pord	600	.013	S.F.		.62		.62	.98
0760	Med.		520	.015			.71		.71	1.13
0770	Heavy	↓	440	.018	↓		.84		.84	1.34
0780	Pressure wash, based on 2500 lb. operating pressure									
0790	Stucco	A-1H	3080	.003	S.F.		.12	.03	.15	.21
0800	Aluminum or vinyl		3200	.003			.11	.02	.13	.21
0810	Siding, masonry, brick & block	↓	2400	.003	↓		.15	.03	.18	.28
1300	Miscellaneous, wire brush									
1310	Metal, pedestrian gate	1 Pord	100	.080	S.F.		3.72		3.72	5.90
1320	Aluminum chain link, both sides		250	.032			1.49		1.49	2.35
1400	Existing galvanized surface, clean and prime, prep for painting	↓	380	.021	↓	.13	.98		1.11	1.70
8000	For chemical washing, see Section 04 01 30									
8010	For steam cleaning, see Section 04 01 30.20									

09 01 90.94 Interior Surface Preparation

	Crew	Daily Output	Labor-Hours	Unit	Material	2021 Bare Costs Labor	Equipment	Total	Total Incl O&P
0010	**INTERIOR SURFACE PREPARATION**									
0020	Doors, per side, not incl. frames or trim									
0030	Scrape & sand									
0040	Wood, flush	1 Pord	616	.013	S.F.		.60		.60	.96
0050	Wood, detail		496	.016			.75		.75	1.19
0060	Wood, louvered	↓	280	.029	↓		1.33		1.33	2.10
0070	Wire brush									
0080	Metal, flush	1 Pord	640	.013	S.F.		.58		.58	.92
0090	Metal, detail		520	.015			.71		.71	1.13
0100	Metal, louvered		360	.022			1.03		1.03	1.63
0110	Hand wash									
0120	Wood, flush	1 Pord	2160	.004	S.F.		.17		.17	.27

09 01 Maintenance of Finishes

09 01 90 – Maintenance of Painting and Coating

09 01 90.94 Interior Surface Preparation

		Crew	Daily Output	Labor-Hours	Unit	Material	2021 Bare Costs Labor	Equipment	Total	Total Incl O&P
0130	Wood, detail	1 Pord	2000	.004	S.F.		.19		.19	.29
0140	Wood, louvered		1360	.006			.27		.27	.43
0150	Metal, flush		2160	.004			.17		.17	.27
0160	Metal, detail		2000	.004			.19		.19	.29
0170	Metal, louvered		1360	.006			.27		.27	.43
0400	Windows, per side, not incl. trim									
0410	Scrape & sand									
0420	Wood, 1-2 lite	1 Pord	360	.022	S.F.		1.03		1.03	1.63
0430	Wood, 3-6 lite		320	.025			1.16		1.16	1.84
0440	Wood, 7-10 lite		280	.029			1.33		1.33	2.10
0450	Wood, 12 lite		240	.033			1.55		1.55	2.45
0460	Wood, Bay/Bow		360	.022			1.03		1.03	1.63
0470	Wire brush									
0480	Metal, 1-2 lite	1 Pord	520	.015	S.F.		.71		.71	1.13
0490	Metal, 3-6 lite		440	.018			.84		.84	1.34
0500	Metal, Bay/Bow		520	.015			.71		.71	1.13
0600	Walls, sanding, light=10-30%, medium=30-70%,									
0610	heavy=70-100% of surface to sand									
0650	Walls, sand									
0660	Gypsum board or plaster, light	1 Pord	3077	.003	S.F.		.12		.12	.19
0670	Gypsum board or plaster, medium		2160	.004			.17		.17	.27
0680	Gypsum board or plaster, heavy		923	.009			.40		.40	.64
0690	Wood, T&G, light		2400	.003			.15		.15	.25
0700	Wood, T&G, medium		1600	.005			.23		.23	.37
0710	Wood, T&G, heavy		800	.010			.46		.46	.74
0720	Walls, wash									
0730	Gypsum board or plaster	1 Pord	3200	.003	S.F.		.12		.12	.18
0740	Wood, T&G		3200	.003			.12		.12	.18
0750	Masonry, brick & block, smooth		2800	.003			.13		.13	.21
0760	Masonry, brick & block, coarse		2000	.004			.19		.19	.29
8000	For chemical washing, see Section 04 01 30									
8010	For steam cleaning, see Section 04 01 30.20									
9010	Minimum labor/equipment charge	1 Pord	3	2.667	Job		124		124	196

09 01 90.95 Scrape After Fire Damage

		Crew	Daily Output	Labor-Hours	Unit	Material	2021 Bare Costs Labor	Equipment	Total	Total Incl O&P
0010	**SCRAPE AFTER FIRE DAMAGE**									
0050	Boards, 1" x 4"	1 Pord	336	.024	L.F.		1.11		1.11	1.75
0060	1" x 6"		260	.031			1.43		1.43	2.26
0070	1" x 8"		207	.039			1.80		1.80	2.84
0080	1" x 10"		174	.046			2.14		2.14	3.38
0500	Framing, 2" x 4"		265	.030			1.40		1.40	2.22
0510	2" x 6"		221	.036			1.68		1.68	2.66
0520	2" x 8"		190	.042			1.96		1.96	3.10
0530	2" x 10"		165	.048			2.25		2.25	3.57
0540	2" x 12"		144	.056			2.58		2.58	4.09
1000	Heavy framing, 3" x 4"		226	.035			1.64		1.64	2.60
1010	4" x 4"		210	.038			1.77		1.77	2.80
1020	4" x 6"		191	.042			1.95		1.95	3.08
1030	4" x 8"		165	.048			2.25		2.25	3.57
1040	4" x 10"		144	.056			2.58		2.58	4.09
1060	4" x 12"		131	.061			2.84		2.84	4.49
2900	For sealing, light damage		825	.010	S.F.	.15	.45		.60	.88
2920	Heavy damage		460	.017	"	.38	.81		1.19	1.70

300

For customer support on your Commercial Renovation Costs with RSMeans Data, call 800.448.8182.

09 01 Maintenance of Finishes

09 01 90 – Maintenance of Painting and Coating

09 01 90.95 Scrape After Fire Damage	Crew	Daily Output	Labor-Hours	Unit	Material	2021 Bare Costs Labor	2021 Bare Costs Equipment	Total	Total Incl O&P
3000	For sandblasting, see Section 04 01 30.20									
9000	Minimum labor/equipment charge	1 Pord	3	2.667	Job		124		124	196

09 05 Common Work Results for Finishes

09 05 05 – Selective Demolition for Finishes

09 05 05.10 Selective Demolition, Ceilings

		Crew	Daily Output	Labor-Hours	Unit	Material	Labor	Equipment	Total	Total Incl O&P
0010	**SELECTIVE DEMOLITION, CEILINGS** R024119-10									
0200	Ceiling, gypsum wall board, furred and nailed or screwed	2 Clab	800	.020	S.F.		.89		.89	1.41
0220	On metal frame		760	.021			.93		.93	1.49
0240	On suspension system, including system		720	.022			.99		.99	1.57
1000	Plaster, lime and horse hair, on wood lath, incl. lath		700	.023			1.01		1.01	1.62
1020	On metal lath		570	.028			1.25		1.25	1.98
1100	Gypsum, on gypsum lath		720	.022			.99		.99	1.57
1120	On metal lath		500	.032			1.42		1.42	2.26
1200	Suspended ceiling, mineral fiber, 2' x 2' or 2' x 4'		1500	.011			.47		.47	.75
1250	On suspension system, incl. system		1200	.013			.59		.59	.94
1500	Tile, wood fiber, 12" x 12", glued		900	.018			.79		.79	1.26
1540	Stapled		1500	.011			.47		.47	.75
1580	On suspension system, incl. system		760	.021			.93		.93	1.49
2000	Wood, tongue and groove, 1" x 4"		1000	.016			.71		.71	1.13
2040	1" x 8"		1100	.015			.65		.65	1.03
2400	Plywood or wood fiberboard, 4' x 8' sheets		1200	.013			.59		.59	.94
2500	Remove & refinish textured ceiling	1 Plas	222	.036	S.F.	.03	1.80		1.83	2.90
9000	Minimum labor/equipment charge	1 Clab	2	4	Job		178		178	283

09 05 05.20 Selective Demolition, Flooring

		Crew	Daily Output	Labor-Hours	Unit	Material	Labor	Equipment	Total	Total Incl O&P
0010	**SELECTIVE DEMOLITION, FLOORING** R024119-10									
0200	Brick with mortar	2 Clab	475	.034	S.F.		1.50		1.50	2.38
0400	Carpet, bonded, including surface scraping		2000	.008			.36		.36	.57
0480	Tackless		9000	.002			.08		.08	.13
0550	Carpet tile, releasable adhesive		5000	.003			.14		.14	.23
0560	Permanent adhesive		1850	.009			.38		.38	.61
0600	Composition, acrylic or epoxy		400	.040			1.78		1.78	2.83
0700	Concrete, scarify skin	A-1A	225	.036			2.03	.93	2.96	4.24
0800	Resilient, sheet goods	2 Clab	1400	.011			.51		.51	.81
0820	For gym floors	"	900	.018			.79		.79	1.26
0850	Vinyl or rubber cove base	1 Clab	1000	.008	L.F.		.36		.36	.57
0860	Vinyl or rubber cove base, molded corner	"	1000	.008	Ea.		.36		.36	.57
0870	For glued and caulked installation, add to labor						50%			
0900	Vinyl composition tile, 12" x 12"	2 Clab	1000	.016	S.F.		.71		.71	1.13
2000	Tile, ceramic, thin set		675	.024			1.05		1.05	1.68
2020	Mud set		625	.026			1.14		1.14	1.81
2200	Marble, slate, thin set		675	.024			1.05		1.05	1.68
2220	Mud set		625	.026			1.14		1.14	1.81
2600	Terrazzo, thin set		450	.036			1.58		1.58	2.51
2620	Mud set		425	.038			1.67		1.67	2.66
2640	Terrazzo, cast in place		300	.053			2.37		2.37	3.77
3000	Wood, block, on end	1 Carp	400	.020			1.09		1.09	1.74
3200	Parquet		450	.018			.97		.97	1.55
3400	Strip flooring, interior, 2-1/4" x 25/32" thick		325	.025			1.35		1.35	2.14
3500	Exterior, porch flooring, 1" x 4"		220	.036			1.99		1.99	3.17
3800	Subfloor, tongue and groove, 1" x 6"		325	.025			1.35		1.35	2.14

For customer support on your Commercial Renovation Costs with RSMeans Data, call 800.448.8182.

301

09 05 05 – Selective Demolition for Finishes

09 05 05.20 Selective Demolition, Flooring

		Crew	Daily Output	Labor-Hours	Unit	Material	2021 Bare Costs Labor	Equipment	Total	Total Incl O&P
3820	1" x 8"	1 Carp	430	.019	S.F.		1.02		1.02	1.62
3840	1" x 10"		520	.015			.84		.84	1.34
4000	Plywood, nailed		600	.013			.73		.73	1.16
4100	Glued and nailed		400	.020			1.09		1.09	1.74
4200	Hardboard, 1/4" thick		760	.011			.58		.58	.92
8000	Remove flooring, bead blast, simple floor plan	A-1A	1000	.008		1.68	.46	.21	2.35	2.80
8100	Complex floor plan		400	.020		1.68	1.14	.52	3.34	4.23
8150	Mastic only		1500	.005		1.68	.30	.14	2.12	2.48
9000	Minimum labor/equipment charge	1 Clab	4	2	Job		89		89	141
9050	For grinding concrete floors, see Section 03 35 43.10									

09 05 05.30 Selective Demolition, Walls and Partitions

		Crew	Daily Output	Labor-Hours	Unit	Material	2021 Bare Costs Labor	Equipment	Total	Total Incl O&P
0010	**SELECTIVE DEMOLITION, WALLS AND PARTITIONS** R024119-10									
0020	Walls, concrete, reinforced	B-39	120	.400	C.F.		18.65	2.96	21.61	33
0025	Plain	"	160	.300			13.95	2.22	16.17	24.50
0100	Brick, 4" to 12" thick	B-9	220	.182			8.15	1.62	9.77	14.75
0200	Concrete block, 4" thick		1150	.035	S.F.		1.56	.31	1.87	2.82
0280	8" thick		1050	.038			1.71	.34	2.05	3.09
0300	Exterior stucco 1" thick over mesh		3200	.013			.56	.11	.67	1.01
1000	Gypsum wallboard, nailed or screwed	1 Clab	1000	.008			.36		.36	.57
1010	2 layers		400	.020			.89		.89	1.41
1020	Glued and nailed		900	.009			.39		.39	.63
1500	Fiberboard, nailed		900	.009			.39		.39	.63
1520	Glued and nailed		800	.010			.44		.44	.71
1568	Plenum barrier, sheet lead		300	.027			1.18		1.18	1.89
1600	Glass block		65	.123			5.45		5.45	8.70
2000	Movable walls, metal, 5' high		300	.027			1.18		1.18	1.89
2020	8' high		400	.020			.89		.89	1.41
2200	Metal or wood studs, finish 2 sides, fiberboard	B-1	520	.046			2.08		2.08	3.31
2250	Lath and plaster		260	.092			4.16		4.16	6.60
2300	Gypsum wallboard		520	.046			2.08		2.08	3.31
2350	Plywood		450	.053			2.40		2.40	3.83
2800	Paneling, 4' x 8' sheets	1 Clab	475	.017			.75		.75	1.19
3000	Plaster, lime and horsehair, on wood lath		400	.020			.89		.89	1.41
3020	On metal lath		335	.024			1.06		1.06	1.69
3400	Gypsum or perlite, on gypsum lath		410	.020			.87		.87	1.38
3420	On metal lath		300	.027			1.18		1.18	1.89
3450	Plaster, interior gypsum, acoustic, or cement		60	.133	S.Y.		5.90		5.90	9.45
3500	Stucco, on masonry		145	.055			2.45		2.45	3.90
3510	Commercial 3-coat		80	.100			4.44		4.44	7.05
3520	Interior stucco		25	.320			14.20		14.20	22.50
3753	Remove damaged glass block	B-1	134.62	.178	S.F.		8.05		8.05	12.80
3760	Tile, ceramic, on walls, thin set	1 Clab	300	.027			1.18		1.18	1.89
3765	Mud set		250	.032			1.42		1.42	2.26
3800	Toilet partitions, slate or marble		5	1.600	Ea.		71		71	113
3820	Metal or plastic		8	1	"		44.50		44.50	70.50
5000	Wallcovering, vinyl	1 Pape	700	.011	S.F.		.53		.53	.84
5010	With release agent		1500	.005			.25		.25	.39
5025	Wallpaper, 2 layers or less, by hand		250	.032			1.48		1.48	2.35
5035	3 layers or more		165	.048			2.24		2.24	3.55
9000	Minimum labor/equipment charge	1 Clab	4	2	Job		89		89	141

09 05 Common Work Results for Finishes

09 05 71 – Acoustic Underlayment

09 05 71.10 Acoustical Underlayment	Crew	Daily Output	Labor-Hours	Unit	Material	2021 Bare Costs Labor	2021 Bare Costs Equipment	Total	Total Incl O&P
0010 **ACOUSTICAL UNDERLAYMENT**									
0100 Rubber underlayment, 5/64"	1 Tilf	275	.029	S.F.	.41	1.50		1.91	2.80
0110 1/8" thk.		275	.029		.58	1.50		2.08	2.99
0120 13/64" thk.		270	.030		.81	1.53		2.34	3.28
0130 15/64" thk.		270	.030		1.20	1.53		2.73	3.71
0140 23/64" thk.		265	.030		1.47	1.56		3.03	4.06
0150 15/32" thk.		265	.030		1.53	1.56		3.09	4.12
0400 Cork underlayment, 3/32"		275	.029		.62	1.50		2.12	3.03
0410 1/8" thk.		275	.029		.56	1.50		2.06	2.97
0420 5/32" thk.		275	.029		.51	1.50		2.01	2.91
0430 15/64" thk.		270	.030		1.12	1.53		2.65	3.62
0440 15/32" thk.		265	.030		1.69	1.56		3.25	4.30
0600 Rubber cork underlayment, 3/64"		275	.029		.38	1.50		1.88	2.77
0610 13/64" thk.		270	.030		1.63	1.53		3.16	4.18
0620 3/8" thk.		270	.030		2.71	1.53		4.24	5.35
0630 25/64" thk.		270	.030		2.84	1.53		4.37	5.50
0800 Foam underlayment, 5/64"		275	.029		.44	1.50		1.94	2.83
0810 1/8" thk.		275	.029		.67	1.50		2.17	3.09
0820 15/32" thk.	↓	265	.030	↓	.54	1.56		2.10	3.03
4000 Nylon matting 0.4" thick, with carbon black spinerette									
4010 plus polyester fabric, on floor	D-7	1600	.010	S.F.	1.51	.47		1.98	2.39

09 21 Plaster and Gypsum Board Assemblies

09 21 16 – Gypsum Board Assemblies

09 21 16.23 Gypsum Board Shaft Wall Assemblies

	Crew	Daily Output	Labor-Hours	Unit	Material	2021 Bare Costs Labor	2021 Bare Costs Equipment	Total	Total Incl O&P
0010 **GYPSUM BOARD SHAFT WALL ASSEMBLIES**									
0020 Cavity type on 25 ga. J-track & C-H studs, 24" OC									
0030 1" thick coreboard wall liner on shaft side									
0040 2-hour assembly with double layer									
0060 5/8" f.r. gyp bd on rm side, 2-1/2" J-track & C-H studs	2 Carp	220	.073	S.F.	3.05	3.98		7.03	9.70
0065 4" J-track & C-H studs	↓	220	.073		3.21	3.98		7.19	9.90
0070 6" J-track & C-H studs	↓	220	.073	↓	3.42	3.98		7.40	10.10
0100 3-hour assembly with triple layer									
0300 5/8" f.r. gyp bd on rm side, 2-1/2" J-track & C-H studs	2 Carp	180	.089	S.F.	2.69	4.86		7.55	10.70
0305 4" J-track & C-H studs		180	.089		2.85	4.86		7.71	10.90
0310 6" J-track & C-H studs	↓	180	.089	↓	3.06	4.86		7.92	11.10
0400 4-hour assembly, 1" coreboard, 5/8" fire rated gypsum board									
0600 and 3/4" galv. metal furring channels, 24" OC, with									
0700 Dbl. layer 5/8" f.r. gyp bd on rm side, 2-1/2" trk. & C-H studs	2 Carp	110	.145	S.F.	3.42	7.95		11.37	16.40
0705 4" J-track & C-H studs		110	.145		3.21	7.95		11.16	16.20
0710 6" J-track & C-H studs	↓	110	.145		3.42	7.95		11.37	16.40
0900 For taping & finishing, add per side	1 Carp	1050	.008	↓	.05	.42		.47	.71
1000 For insulation, see Section 07 21									
5200 For work over 8' high, add	2 Carp	3060	.005	S.F.		.29		.29	.46
5300 For distribution cost 3 stories and above, add per story	"	6100	.003	"		.14		.14	.23

09 21 16.33 Partition Wall

	Crew	Daily Output	Labor-Hours	Unit	Material	2021 Bare Costs Labor	2021 Bare Costs Equipment	Total	Total Incl O&P
0010 **PARTITION WALL** Stud wall, 8' to 12' high									
0050 1/2", interior, gypsum board, std. tape & finish 2 sides									
0500 Installed on and incl., 2" x 4" wood studs, 16" OC	2 Carp	310	.052	S.F.	1.53	2.82		4.35	6.20
1000 Metal studs, NLB, 25 ga., 16" OC, 3-5/8" wide	↓	350	.046	↓	1.23	2.50		3.73	5.35

For customer support on your Commercial Renovation Costs with RSMeans Data, call 800.448.8182.

303

09 21 16.33 Partition Wall		Crew	Daily Output	Labor-Hours	Unit	Material	2021 Bare Costs Labor	Equipment	Total	Total Incl O&P
1200	6" wide	2 Carp	330	.048	S.F.	1.32	2.65		3.97	5.65
1400	Water resistant, on 2" x 4" wood studs, 16" OC		310	.052		1.61	2.82		4.43	6.30
1600	Metal studs, NLB, 25 ga., 16" OC, 3-5/8" wide		350	.046		1.31	2.50		3.81	5.40
1800	6" wide		330	.048		1.40	2.65		4.05	5.75
2000	Fire res., 2 layers, 1-1/2 hr., on 2" x 4" wood studs, 16" OC		210	.076		2.35	4.17		6.52	9.25
2200	Metal studs, NLB, 25 ga., 16" OC, 3-5/8" wide		250	.064		2.05	3.50		5.55	7.80
2400	6" wide		230	.070		2.14	3.81		5.95	8.40
2600	Fire & water res., 2 layers, 1-1/2 hr., 2" x 4" studs, 16" OC		210	.076		2.35	4.17		6.52	9.25
2800	Metal studs, NLB, 25 ga., 16" OC, 3-5/8" wide		250	.064		2.05	3.50		5.55	7.80
3000	6" wide		230	.070		2.14	3.81		5.95	8.40
3200	5/8", interior, gypsum board, standard, tape & finish 2 sides									
3400	Installed on and including 2" x 4" wood studs, 16" OC	2 Carp	300	.053	S.F.	1.59	2.92		4.51	6.40
3600	24" OC		330	.048		1.42	2.65		4.07	5.80
3800	Metal studs, NLB, 25 ga., 16" OC, 3-5/8" wide		340	.047		1.29	2.57		3.86	5.50
4000	6" wide		320	.050		1.38	2.74		4.12	5.90
4200	24" OC, 3-5/8" wide		360	.044		1.17	2.43		3.60	5.15
4400	6" wide		340	.047		1.23	2.57		3.80	5.45
4800	Water resistant, on 2" x 4" wood studs, 16" OC		300	.053		1.73	2.92		4.65	6.55
5000	24" OC		330	.048		1.56	2.65		4.21	5.95
5200	Metal studs, NLB, 25 ga. 16" OC, 3-5/8" wide		340	.047		1.43	2.57		4	5.65
5400	6" wide		320	.050		1.52	2.74		4.26	6.05
5600	24" OC, 3-5/8" wide		360	.044		1.31	2.43		3.74	5.30
5800	6" wide		340	.047		1.37	2.57		3.94	5.60
6000	Fire resistant, 2 layers, 2 hr., on 2" x 4" wood studs, 16" OC		205	.078		2.39	4.27		6.66	9.45
6200	24" OC		235	.068		2.22	3.72		5.94	8.40
6400	Metal studs, NLB, 25 ga., 16" OC, 3-5/8" wide		245	.065		2.09	3.57		5.66	8
6600	6" wide		225	.071		2.18	3.89		6.07	8.60
6800	24" OC, 3-5/8" wide		265	.060		1.97	3.30		5.27	7.40
7000	6" wide		245	.065		2.03	3.57		5.60	7.95
7200	Fire & water resistant, 2 layers, 2 hr., 2" x 4" studs, 16" OC		205	.078		2.39	4.27		6.66	9.45
7400	24" OC		235	.068		2.22	3.72		5.94	8.40
7600	Metal studs, NLB, 25 ga., 16" OC, 3-5/8" wide		245	.065		2.09	3.57		5.66	8
7800	6" wide		225	.071		2.18	3.89		6.07	8.60
8000	24" OC, 3-5/8" wide		265	.060		1.97	3.30		5.27	7.40
8200	6" wide		245	.065		2.03	3.57		5.60	7.95
8600	1/2" blueboard, mesh tape both sides									
8620	Installed on and including 2" x 4" wood studs, 16" OC	2 Carp	300	.053	S.F.	1.61	2.92		4.53	6.45
8640	Metal studs, NLB, 25 ga., 16" OC, 3-5/8" wide		340	.047		1.31	2.57		3.88	5.55
8660	6" wide		320	.050		1.40	2.74		4.14	5.90
8800	Hospital security partition, 5/8" fiber reinf. high abuse gyp. bd.									
8810	Mtl. studs, NLB, 20 ga., 16" OC, 3-5/8" wide, w/sec. mesh, gyp. bd.	2 Carp	208	.077	S.F.	4.74	4.21		8.95	11.90
9000	Exterior, 1/2" gypsum sheathing, 1/2" gypsum finished, interior,									
9100	including foil faced insulation, metal studs, 20 ga.									
9200	16" OC, 3-5/8" wide	2 Carp	290	.055	S.F.	2.03	3.02		5.05	7.05
9400	6" wide		270	.059		2.25	3.24		5.49	7.60
9600	Partitions, for work over 8' high, add		1530	.010			.57		.57	.91

09 22 Supports for Plaster and Gypsum Board

09 22 03 – Fastening Methods for Finishes

09 22 03.20 Drilling Plaster/Drywall	Crew	Daily Output	Labor-Hours	Unit	Material	2021 Bare Costs Labor	Equipment	Total	Total Incl O&P
0010 **DRILLING PLASTER/DRYWALL**									
1100 Drilling & layout for drywall/plaster walls, up to 1" deep, no anchor									
1200 Holes, 1/4" diameter	1 Carp	150	.053	Ea.	.01	2.92		2.93	4.66
1300 3/8" diameter		140	.057		.01	3.13		3.14	4.99
1400 1/2" diameter		130	.062		.01	3.37		3.38	5.35
1500 3/4" diameter		120	.067		.01	3.65		3.66	5.80
1600 1" diameter		110	.073		.03	3.98		4.01	6.40
1700 1-1/4" diameter		100	.080		.04	4.38		4.42	7
1800 1-1/2" diameter		90	.089		.05	4.86		4.91	7.80
1900 For ceiling installations, add						40%			

09 22 13 – Metal Furring

09 22 13.13 Metal Channel Furring

09 22 13.13 Metal Channel Furring	Crew	Daily Output	Labor-Hours	Unit	Material	2021 Bare Costs Labor	Equipment	Total	Total Incl O&P
0010 **METAL CHANNEL FURRING**									
0030 Beams and columns, 7/8" hat channels, galvanized, 12" OC	1 Lath	155	.052	S.F.	.47	2.79		3.26	4.87
0050 16" OC		170	.047		.38	2.55		2.93	4.39
0070 24" OC		185	.043		.25	2.34		2.59	3.93
0100 Ceilings, on steel, 7/8" hat channels, galvanized, 12" OC		210	.038		.42	2.06		2.48	3.69
0300 16" OC		290	.028		.38	1.49		1.87	2.75
0400 24" OC		420	.019		.25	1.03		1.28	1.89
0600 1-5/8" hat channels, galvanized, 12" OC		190	.042		.57	2.28		2.85	4.19
0700 16" OC		260	.031		.51	1.66		2.17	3.16
0900 24" OC		390	.021		.34	1.11		1.45	2.11
0930 7/8" hat channels with sound isolation clips, 12" OC		120	.067		2.26	3.61		5.87	8.15
0940 16" OC		100	.080		1.70	4.33		6.03	8.60
0950 24" OC		165	.048		1.13	2.62		3.75	5.35
0960 1-5/8" hat channels, galvanized, 12" OC		110	.073		2.41	3.93		6.34	8.80
0970 16" OC		100	.080		1.81	4.33		6.14	8.75
0980 24" OC		155	.052		1.20	2.79		3.99	5.70
1000 Walls, 7/8" hat channels, galvanized, 12" OC		235	.034		.42	1.84		2.26	3.34
1200 16" OC		265	.030		.38	1.63		2.01	2.97
1300 24" OC		350	.023		.25	1.24		1.49	2.21
1500 1-5/8" hat channels, galvanized, 12" OC		210	.038		.57	2.06		2.63	3.85
1600 16" OC		240	.033		.51	1.80		2.31	3.38
1800 24" OC		305	.026		.34	1.42		1.76	2.60
1920 7/8" hat channels with sound isolation clips, 12" OC		125	.064		2.26	3.46		5.72	7.90
1940 16" OC		100	.080		1.70	4.33		6.03	8.60
1950 24" OC		150	.053		1.13	2.89		4.02	5.75
1960 1-5/8" hat channels, galvanized, 12" OC		115	.070		2.41	3.76		6.17	8.50
1970 16" OC		95	.084		1.81	4.56		6.37	9.10
1980 24" OC		140	.057		1.20	3.09		4.29	6.15
3000 Z Furring, walls, 1" deep, 25 ga., 24" OC		350	.023		.34	1.24		1.58	2.30
3010 48" OC		700	.011		.17	.62		.79	1.15
3020 1-1/2" deep, 24" OC		345	.023		1.61	1.25		2.86	3.73
3030 48" OC		695	.012		.80	.62		1.42	1.85
3040 2" deep, 24" OC		340	.024		1.93	1.27		3.20	4.11
3050 48" OC		690	.012		.97	.63		1.60	2.04
3060 1" deep, 20 ga., 24" OC		350	.023		2.28	1.24		3.52	4.44
3070 48" OC		700	.011		1.14	.62		1.76	2.22
3080 1-1/2" deep, 24" OC		345	.023		2.62	1.25		3.87	4.84
3090 48" OC		695	.012		1.31	.62		1.93	2.41
4000 2" deep, 24" OC		340	.024		3.22	1.27		4.49	5.55
4010 48" OC		690	.012		1.61	.63		2.24	2.75

09 22 Supports for Plaster and Gypsum Board

09 22 13 – Metal Furring

09 22 13.13 Metal Channel Furring	Crew	Daily Output	Labor-Hours	Unit	Material	Labor	2021 Bare Costs Equipment	Total	Total Incl O&P
9000 Minimum labor/equipment charge	1 Lath	4	2	Job		108		108	169

09 22 16 – Non-Structural Metal Framing

09 22 16.13 Non-Structural Metal Stud Framing

		Crew	Daily Output	Labor-Hours	Unit	Material	Labor	2021 Bare Costs Equipment	Total	Total Incl O&P
0010	**NON-STRUCTURAL METAL STUD FRAMING**									
1600	Non-load bearing, galv., 8' high, 25 ga. 1-5/8" wide, 16" OC	1 Carp	619	.013	S.F.	.49	.71		1.20	1.67
1610	24" OC		950	.008		.35	.46		.81	1.12
1620	2-1/2" wide, 16" OC		613	.013		.42	.71		1.13	1.60
1630	24" OC		938	.009		.32	.47		.79	1.09
1640	3-5/8" wide, 16" OC		600	.013		.51	.73		1.24	1.72
1650	24" OC		925	.009		.39	.47		.86	1.18
1660	4" wide, 16" OC		594	.013		.46	.74		1.20	1.67
1670	24" OC		925	.009		.34	.47		.81	1.13
1680	6" wide, 16" OC		588	.014		.60	.74		1.34	1.85
1690	24" OC		906	.009		.45	.48		.93	1.26
1700	20 ga. studs, 1-5/8" wide, 16" OC		494	.016		.36	.89		1.25	1.80
1710	24" OC		763	.010		.27	.57		.84	1.21
1720	2-1/2" wide, 16" OC		488	.016		.39	.90		1.29	1.86
1730	24" OC		750	.011		.30	.58		.88	1.26
1740	3-5/8" wide, 16" OC		481	.017		.69	.91		1.60	2.21
1750	24" OC		738	.011		.53	.59		1.12	1.53
1760	4" wide, 16" OC		475	.017		.75	.92		1.67	2.30
1770	24" OC		738	.011		.57	.59		1.16	1.56
1780	6" wide, 16" OC		469	.017		.91	.93		1.84	2.49
1790	24" OC		725	.011		.68	.60		1.28	1.71
2000	Non-load bearing, galv., 10' high, 25 ga. 1-5/8" wide, 16" OC		495	.016		.47	.88		1.35	1.92
2100	24" OC		760	.011		.33	.58		.91	1.29
2200	2-1/2" wide, 16" OC		490	.016		.39	.89		1.28	1.85
2250	24" OC		750	.011		.29	.58		.87	1.25
2300	3-5/8" wide, 16" OC		480	.017		.47	.91		1.38	1.97
2350	24" OC		740	.011		.36	.59		.95	1.33
2400	4" wide, 16" OC		475	.017		.43	.92		1.35	1.94
2450	24" OC		740	.011		.32	.59		.91	1.29
2500	6" wide, 16" OC		470	.017		.56	.93		1.49	2.10
2550	24" OC		725	.011		.42	.60		1.02	1.42
2600	20 ga. studs, 1-5/8" wide, 16" OC		395	.020		.34	1.11		1.45	2.13
2650	24" OC		610	.013		.25	.72		.97	1.41
2700	2-1/2" wide, 16" OC		390	.021		.36	1.12		1.48	2.19
2750	24" OC		600	.013		.27	.73		1	1.46
2800	3-5/8" wide, 16" OC		385	.021		.65	1.14		1.79	2.52
2850	24" OC		590	.014		.49	.74		1.23	1.72
2900	4" wide, 16" OC		380	.021		.71	1.15		1.86	2.61
2950	24" OC		590	.014		.53	.74		1.27	1.76
3000	6" wide, 16" OC		375	.021		.86	1.17		2.03	2.81
3050	24" OC		580	.014		.64	.75		1.39	1.90
3060	Non-load bearing, galv., 12' high, 25 ga. 1-5/8" wide, 16" OC		413	.019		.45	1.06		1.51	2.19
3070	24" OC		633	.013		.32	.69		1.01	1.45
3080	2-1/2" wide, 16" OC		408	.020		.37	1.07		1.44	2.11
3090	24" OC		625	.013		.27	.70		.97	1.41
3100	3-5/8" wide, 16" OC		400	.020		.44	1.09		1.53	2.23
3110	24" OC		617	.013		.33	.71		1.04	1.50
3120	4" wide, 16" OC		396	.020		.41	1.11		1.52	2.21
3130	24" OC		617	.013		.30	.71		1.01	1.46

09 22 16 – Non-Structural Metal Framing

09 22 16.13 Non-Structural Metal Stud Framing

		Crew	Daily Output	Labor-Hours	Unit	Material	2021 Bare Costs Labor	Equipment	Total	Total Incl O&P
3140	6" wide, 16" OC	1 Carp	392	.020	S.F.	.54	1.12		1.66	2.37
3150	24" OC		604	.013		.39	.72		1.11	1.58
3160	20 ga. studs, 1-5/8" wide, 16" OC		329	.024		.32	1.33		1.65	2.47
3170	24" OC		508	.016		.24	.86		1.10	1.63
3180	2-1/2" wide, 16" OC		325	.025		.34	1.35		1.69	2.52
3190	24" OC		500	.016		.26	.88		1.14	1.67
3200	3-5/8" wide, 16" OC		321	.025		.61	1.36		1.97	2.84
3210	24" OC		492	.016		.46	.89		1.35	1.92
3220	4" wide, 16" OC		317	.025		.68	1.38		2.06	2.95
3230	24" OC		492	.016		.50	.89		1.39	1.97
3240	6" wide, 16" OC		313	.026		.82	1.40		2.22	3.14
3250	24" OC		483	.017		.60	.91		1.51	2.10
3260	Non-load bearing, galv., 16' high, 25 ga. 4" wide, 12" OC		195	.041		.53	2.24		2.77	4.15
3270	16" OC		275	.029		.42	1.59		2.01	2.99
3280	24" OC		400	.020		.31	1.09		1.40	2.08
3290	6" wide, 12" OC		190	.042		.70	2.30		3	4.44
3300	16" OC		280	.029		.55	1.56		2.11	3.09
3310	24" OC		400	.020		.40	1.09		1.49	2.18
3320	20 ga. studs, 2-1/2" wide, 12" OC		180	.044		.43	2.43		2.86	4.35
3330	16" OC		254	.032		.34	1.72		2.06	3.12
3340	24" OC		390	.021		.26	1.12		1.38	2.07
3350	3-5/8" wide, 12" OC		170	.047		.76	2.57		3.33	4.94
3360	16" OC		251	.032		.61	1.74		2.35	3.45
3370	24" OC		384	.021		.46	1.14		1.60	2.31
3380	4" wide, 12" OC		170	.047		.87	2.57		3.44	5.05
3390	16" OC		247	.032		.69	1.77		2.46	3.58
3400	24" OC		384	.021		.51	1.14		1.65	2.37
3410	6" wide, 12" OC		175	.046		1.06	2.50		3.56	5.15
3420	16" OC		245	.033		.84	1.79		2.63	3.76
3430	24" OC		400	.020		.61	1.09		1.70	2.41
3440	Non-load bearing, galv., 20' high, 25 ga. 6" wide, 12" OC		125	.064		.68	3.50		4.18	6.30
3450	16" OC		220	.036		.54	1.99		2.53	3.76
3460	24" OC		360	.022		.39	1.22		1.61	2.37
3470	20 ga. studs, 4" wide, 12" OC		120	.067		.86	3.65		4.51	6.75
3480	16" OC		215	.037		.69	2.04		2.73	4
3490	6" wide, 12" OC		115	.070		1.04	3.81		4.85	7.20
3500	16" OC		215	.037		.81	2.04		2.85	4.14
3510	24" OC		331	.024		.59	1.32		1.91	2.76
5000	For load bearing studs, see Section 05 41 13.30									
9000	Minimum labor/equipment charge	1 Carp	4	2	Job		109		109	174

09 22 26 – Suspension Systems

09 22 26.13 Ceiling Suspension Systems

		Crew	Daily Output	Labor-Hours	Unit	Material	2021 Bare Costs Labor	Equipment	Total	Total Incl O&P
0010	**CEILING SUSPENSION SYSTEMS** for gypsum board or plaster									
8000	Suspended ceilings, including carriers									
8200	1-1/2" carriers, 24" OC with:									
8300	7/8" channels, 16" OC	1 Lath	275	.029	S.F.	.62	1.57		2.19	3.14
8320	24" OC		310	.026		.49	1.40		1.89	2.72
8400	1-5/8" channels, 16" OC		205	.039		.75	2.11		2.86	4.12
8420	24" OC		250	.032		.58	1.73		2.31	3.33
8600	2" carriers, 24" OC with:									
8700	7/8" channels, 16" OC	1 Lath	250	.032	S.F.	.65	1.73		2.38	3.42
8720	24" OC		285	.028		.52	1.52		2.04	2.95

For customer support on your Commercial Renovation Costs with RSMeans Data, call 800.448.8182.

307

09 22 Supports for Plaster and Gypsum Board

09 22 26 – Suspension Systems

09 22 26.13 Ceiling Suspension Systems

		Crew	Daily Output	Labor-Hours	Unit	Material	2021 Bare Costs Labor	Equipment	Total	Total Incl O&P
8800	1-5/8" channels, 16" OC	1 Lath	190	.042	S.F.	.78	2.28		3.06	4.42
8820	24" OC	↓	225	.036	↓	.61	1.92		2.53	3.67

09 22 36 – Lath

09 22 36.13 Gypsum Lath

		Crew	Daily Output	Labor-Hours	Unit	Material	2021 Bare Costs Labor	Equipment	Total	Total Incl O&P
0010	**GYPSUM LATH** R092000-50									
0020	Plain or perforated, nailed, 3/8" thick	1 Lath	85	.094	S.Y.	3.33	5.10		8.43	11.60
0100	1/2" thick		80	.100		2.88	5.40		8.28	11.60
0300	Clipped to steel studs, 3/8" thick		75	.107		3.33	5.75		9.08	12.65
0400	1/2" thick		70	.114		2.88	6.20		9.08	12.80
1500	For ceiling installations, add		216	.037			2		2	3.13
1600	For columns and beams, add		170	.047	↓		2.55		2.55	3.97
9000	Minimum labor/equipment charge	↓	4.25	1.882	Job		102		102	159

09 22 36.23 Metal Lath

		Crew	Daily Output	Labor-Hours	Unit	Material	2021 Bare Costs Labor	Equipment	Total	Total Incl O&P
0010	**METAL LATH** R092000-50									
0020	Diamond, expanded, 2.5 lb./S.Y., painted				S.Y.	5.45			5.45	6
0100	Galvanized					3.87			3.87	4.26
0300	3.4 lb./S.Y., painted					4.72			4.72	5.20
0400	Galvanized					4.99			4.99	5.50
0600	For #15 asphalt sheathing paper, add					.51			.51	.56
0900	Flat rib, 1/8" high, 2.75 lb., painted					3.87			3.87	4.26
1000	Foil backed					3.63			3.63	3.99
1200	3.4 lb./S.Y., painted					4.07			4.07	4.48
1300	Galvanized					4.24			4.24	4.66
1500	For #15 asphalt sheathing paper, add					.51			.51	.56
1800	High rib, 3/8" high, 3.4 lb./S.Y., painted					4.82			4.82	5.30
1900	Galvanized				↓	4			4	4.40
2400	3/4" high, painted, .60 lb./S.F.				S.F.	.82			.82	.90
2500	.75 lb./S.F.				"	1.72			1.72	1.89
2800	Stucco mesh, painted, 3.6 lb.				S.Y.	4.09			4.09	4.50
3000	K-lath, perforated, absorbent paper, regular					4.39			4.39	4.83
3100	Heavy duty					5.20			5.20	5.70
3300	Waterproof, heavy duty, grade B backing					5.05			5.05	5.55
3400	Fire resistant backing					5.60			5.60	6.15
3600	2.5 lb. diamond painted, on wood framing, on walls	1 Lath	85	.094		5.45	5.10		10.55	13.95
3700	On ceilings		75	.107		5.45	5.75		11.20	15
3900	3.4 lb. diamond painted, on wood framing, on walls		80	.100		4.07	5.40		9.47	12.95
4000	On ceilings		70	.114		4.07	6.20		10.27	14.15
4200	3.4 lb. diamond painted, wired to steel framing		75	.107		4.07	5.75		9.82	13.50
4300	On ceilings		60	.133		4.07	7.20		11.27	15.75
4600	Cornices, wired to steel		35	.229		4.07	12.35		16.42	24
4800	Screwed to steel studs, 2.5 lb.		80	.100		5.45	5.40		10.85	14.45
4900	3.4 lb.		75	.107		4.72	5.75		10.47	14.20
5100	Rib lath, painted, wired to steel, on walls, 2.5 lb.		75	.107		3.87	5.75		9.62	13.25
5200	3.4 lb.		70	.114		4.82	6.20		11.02	14.95
5400	4.0 lb.	↓	65	.123		5.75	6.65		12.40	16.70
5500	For self-furring lath, add					.12			.12	.13
5700	Suspended ceiling system, incl. 3.4 lb. diamond lath, painted	1 Lath	15	.533		4.20	29		33.20	49.50
5800	Galvanized	"	15	.533	↓	3.84	29		32.84	49
6000	Hollow metal stud partitions, 3.4 lb. painted lath both sides									
6010	Non-load bearing, 25 ga., w/rib lath, 2-1/2" studs, 12" OC	1 Lath	20.30	.394	S.Y.	13.95	21.50		35.45	49
6300	16" OC		21.10	.379		13.05	20.50		33.55	46.50
6350	24" OC	↓	22.70	.352	↓	12.30	19.05		31.35	43.50

09 22 Supports for Plaster and Gypsum Board

09 22 36 – Lath

09 22 36.23 Metal Lath

		Crew	Daily Output	Labor-Hours	Unit	Material	2021 Bare Costs Labor	2021 Bare Costs Equipment	Total	Total Incl O&P
6400	3-5/8" studs, 16" OC	1 Lath	19.50	.410	S.Y.	13.85	22		35.85	50
6600	24" OC		20.40	.392		12.85	21		33.85	47
6700	4" studs, 16" OC		20.40	.392		13.50	21		34.50	48
6900	24" OC		21.60	.370		12.50	20		32.50	45.50
7000	6" studs, 16" OC		19.50	.410		14.70	22		36.70	50.50
7100	24" OC		21.10	.379		13.40	20.50		33.90	46.50
7200	L.B. partitions, 16 ga., w/rib lath, 2-1/2" studs, 16" OC		20	.400		13.65	21.50		35.15	49
7300	3-5/8" studs, 16 ga.		19.70	.406		15.10	22		37.10	51
7500	4" studs, 16 ga.		19.50	.410		16.25	22		38.25	52.50
7600	6" studs, 16 ga.		18.70	.428		19.05	23		42.05	57
9000	Minimum labor/equipment charge		4.25	1.882	Job		102		102	159

09 23 Gypsum Plastering

09 23 13 – Acoustical Gypsum Plastering

09 23 13.10 Perlite or Vermiculite Plaster

		Crew	Daily Output	Labor-Hours	Unit	Material	2021 Bare Costs Labor	2021 Bare Costs Equipment	Total	Total Incl O&P
0010	**PERLITE OR VERMICULITE PLASTER** R092000-50									
0020	In 100 lb. bags, under 200 bags				Bag	19.25			19.25	21
0300	2 coats, no lath included, on walls	J-1	92	.435	S.Y.	5.95	21	1.23	28.18	41
0400	On ceilings		79	.506		5.95	24	1.43	31.38	46.50
0900	3 coats, no lath included, on walls		74	.541		6.65	26	1.53	34.18	50
1000	On ceilings		63	.635		6.65	30.50	1.80	38.95	57.50
1700	For irregular or curved surfaces, add to above						30%			
1800	For columns and beams, add to above						50%			
1900	For soffits, add to ceiling prices						40%			
9000	Minimum labor/equipment charge	1 Plas	1	8	Job		400		400	635

09 23 20 – Gypsum Plaster

09 23 20.10 Gypsum Plaster On Walls and Ceilings

		Crew	Daily Output	Labor-Hours	Unit	Material	2021 Bare Costs Labor	2021 Bare Costs Equipment	Total	Total Incl O&P
0010	**GYPSUM PLASTER ON WALLS AND CEILINGS** R092000-50									
0020	80# bag, less than 1 ton				Bag	16.20			16.20	17.85
0300	2 coats, no lath included, on walls	J-1	105	.381	S.Y.	4.01	18.20	1.08	23.29	34.50
0400	On ceilings		92	.435		4.01	21	1.23	26.24	39
0900	3 coats, no lath included, on walls		87	.460		5.70	22	1.30	29	42.50
1000	On ceilings		78	.513		5.70	24.50	1.45	31.65	47
1600	For irregular or curved surfaces, add						30%			
1800	For columns & beams, add						50%			
9000	Minimum labor/equipment charge	1 Plas	1	8	Job		400		400	635

09 24 Cement Plastering

09 24 23 – Cement Stucco

09 24 23.40 Stucco

		Crew	Daily Output	Labor-Hours	Unit	Material	2021 Bare Costs Labor	2021 Bare Costs Equipment	Total	Total Incl O&P
0010	**STUCCO** R092000-50									
0015	3 coats 7/8" thick, float finish, with mesh, on wood frame	J-2	63	.762	S.Y.	7.70	37	1.80	46.50	69.50
0100	On masonry construction, no mesh incl.	J-1	67	.597		3.10	28.50	1.69	33.29	51
0300	For trowel finish, add	1 Plas	170	.047			2.35		2.35	3.73
0600	For coloring, add	J-1	685	.058		.43	2.79	.17	3.39	5.10
0700	For special texture, add	"	200	.200		1.50	9.55	.57	11.62	17.45
0900	For soffits, add	J-2	155	.310		2.29	15.10	.73	18.12	27.50
1000	Stucco, with bonding agent, 3 coats, on walls, no mesh incl.	J-1	200	.200		4.31	9.55	.57	14.43	20.50
1200	Ceilings		180	.222		3.83	10.60	.63	15.06	22

For customer support on your Commercial Renovation Costs with RSMeans Data, call 800.448.8182.

309

09 24 Cement Plastering

09 24 23 – Cement Stucco

09 24 23.40 Stucco

		Crew	Daily Output	Labor-Hours	Unit	Material	2021 Bare Costs Labor	Equipment	Total	Total Incl O&P
1300	Beams	J-1	80	.500	S.Y.	3.83	24	1.42	29.25	44
1500	Columns	↓	100	.400	↓	3.83	19.10	1.13	24.06	36
1550	Minimum labor/equipment charge	1 Plas	1	8	Job		400		400	635
1600	Mesh, galvanized, nailed to wood, 1.8 lb.	1 Lath	60	.133	S.Y.	8	7.20		15.20	20
1800	3.6 lb.		55	.145		4.09	7.85		11.94	16.80
1900	Wired to steel, galvanized, 1.8 lb.		53	.151		8	8.15		16.15	21.50
2100	3.6 lb.		50	.160	↓	4.09	8.65		12.74	18
9000	Minimum labor/equipment charge	↓	4	2	Job		108		108	169

09 25 Other Plastering

09 25 23 – Lime Based Plastering

09 25 23.10 Venetian Plaster

		Crew	Daily Output	Labor-Hours	Unit	Material	2021 Bare Costs Labor	Equipment	Total	Total Incl O&P
0010	**VENETIAN PLASTER**									
0100	Walls, 1 coat primer, roller applied	1 Plas	950	.008	S.F.	.15	.42		.57	.83
0200	Plaster, 3 coats, incl. sanding	2 Plas	700	.023	"	.41	1.14		1.55	2.26
0210	For pigment, light colors add per S.F. plaster				S.F.	.02			.02	.02
0220	For pigment, dark colors add per S.F. plaster					.04			.04	.04
0300	For sealer/wax coat incl. burnishing, add	1 Plas	300	.027	↓	.36	1.33		1.69	2.50

09 26 Veneer Plastering

09 26 13 – Gypsum Veneer Plastering

09 26 13.20 Blueboard

		Crew	Daily Output	Labor-Hours	Unit	Material	2021 Bare Costs Labor	Equipment	Total	Total Incl O&P
0010	**BLUEBOARD** For use with thin coat									
0100	plaster application see Section 09 26 13.80									
1000	3/8" thick, on walls or ceilings, standard, no finish included	2 Carp	1900	.008	S.F.	.34	.46		.80	1.10
1100	With thin coat plaster finish		875	.018		.44	1		1.44	2.08
1400	On beams, columns, or soffits, standard, no finish included		675	.024		.39	1.30		1.69	2.49
1450	With thin coat plaster finish		475	.034		.49	1.84		2.33	3.47
3000	1/2" thick, on walls or ceilings, standard, no finish included		1900	.008		.36	.46		.82	1.13
3100	With thin coat plaster finish		875	.018		.46	1		1.46	2.10
3300	Fire resistant, no finish included		1900	.008		.36	.46		.82	1.13
3400	With thin coat plaster finish		875	.018		.46	1		1.46	2.10
3450	On beams, columns, or soffits, standard, no finish included		675	.024		.41	1.30		1.71	2.52
3500	With thin coat plaster finish		475	.034		.52	1.84		2.36	3.50
3700	Fire resistant, no finish included		675	.024		.41	1.30		1.71	2.52
3800	With thin coat plaster finish		475	.034		.52	1.84		2.36	3.50
5000	5/8" thick, on walls or ceilings, fire resistant, no finish included		1900	.008		.39	.46		.85	1.16
5100	With thin coat plaster finish		875	.018		.49	1		1.49	2.13
5500	On beams, columns, or soffits, no finish included		675	.024		.45	1.30		1.75	2.55
5600	With thin coat plaster finish		475	.034		.55	1.84		2.39	3.53
6000	For high ceilings, over 8' high, add		3060	.005			.29		.29	.46
6500	For distribution costs 3 stories and above, add per story		6100	.003			.14		.14	.23
9000	Minimum labor/equipment charge	1 Carp	2	4	Job		219		219	350

09 26 13.80 Thin Coat Plaster

		Crew	Daily Output	Labor-Hours	Unit	Material	2021 Bare Costs Labor	Equipment	Total	Total Incl O&P
0010	**THIN COAT PLASTER** R092000-50									
0012	1 coat veneer, not incl. lath	J-1	3600	.011	S.F.	.10	.53	.03	.66	.98
1000	In 50 lb. bags				Bag	13.65			13.65	15.05

09 28 Backing Boards and Underlayments

09 28 13 – Cementitious Backing Boards

09 28 13.10 Cementitious Backerboard	Crew	Daily Output	Labor-Hours	Unit	Material	2021 Bare Costs Labor	Equipment	Total	Total Incl O&P
0010 **CEMENTITIOUS BACKERBOARD**									
0070 Cementitious backerboard, on floor, 3' x 4' x 1/2" sheets	2 Carp	525	.030	S.F.	.99	1.67		2.66	3.74
0080 3' x 5' x 1/2" sheets		525	.030		.81	1.67		2.48	3.54
0090 3' x 6' x 1/2" sheets		525	.030		.72	1.67		2.39	3.44
0100 3' x 4' x 5/8" sheets		525	.030		.92	1.67		2.59	3.66
0110 3' x 5' x 5/8" sheets		525	.030		.99	1.67		2.66	3.74
0120 3' x 6' x 5/8" sheets	2 Carp	525	.030	S.F.	.98	1.67		2.65	3.72
0150 On wall, 3' x 4' x 1/2" sheets		350	.046		.99	2.50		3.49	5.05
0160 3' x 5' x 1/2" sheets		350	.046		.81	2.50		3.31	4.87
0170 3' x 6' x 1/2" sheets		350	.046		.72	2.50		3.22	4.77
0180 3' x 4' x 5/8" sheets		350	.046		.92	2.50		3.42	4.99
0190 3' x 5' x 5/8" sheets		350	.046		.99	2.50		3.49	5.05
0200 3' x 6' x 5/8" sheets		350	.046		.98	2.50		3.48	5.05
0250 On counter, 3' x 4' x 1/2" sheets		180	.089		.99	4.86		5.85	8.85
0260 3' x 5' x 1/2" sheets		180	.089		.81	4.86		5.67	8.65
0270 3' x 6' x 1/2" sheets		180	.089		.72	4.86		5.58	8.55
0300 3' x 4' x 5/8" sheets		180	.089		.92	4.86		5.78	8.75
0310 3' x 5' x 5/8" sheets		180	.089		.99	4.86		5.85	8.85
0320 3' x 6' x 5/8" sheets		180	.089		.98	4.86		5.84	8.80

09 29 Gypsum Board

09 29 10 – Gypsum Board Panels

09 29 10.30 Gypsum Board	Crew	Daily Output	Labor-Hours	Unit	Material	2021 Bare Costs Labor	Equipment	Total	Total Incl O&P
0010 **GYPSUM BOARD** on walls & ceilings R092910-10									
0100 Nailed or screwed to studs unless otherwise noted									
0110 1/4" thick, on walls or ceilings, standard, no finish included	2 Carp	1330	.012	S.F.	.44	.66		1.10	1.53
0115 1/4" thick, on walls or ceilings, flexible, no finish included		1050	.015		.32	.83		1.15	1.68
0117 1/4" thick, on columns or soffits, flexible, no finish included		1050	.015		.32	.83		1.15	1.68
0130 1/4" thick, standard, no finish included, less than 800 S.F.		510	.031		.44	1.72		2.16	3.21
0150 3/8" thick, on walls, standard, no finish included		2000	.008		.35	.44		.79	1.09
0200 On ceilings, standard, no finish included		1800	.009		.35	.49		.84	1.16
0250 On beams, columns, or soffits, no finish included		675	.024		.35	1.30		1.65	2.45
0300 1/2" thick, on walls, standard, no finish included		2000	.008		.33	.44		.77	1.06
0350 Taped and finished (level 4 finish)		965	.017		.38	.91		1.29	1.86
0390 With compound skim coat (level 5 finish)		775	.021		.43	1.13		1.56	2.28
0400 Fire resistant, no finish included		2000	.008		.37	.44		.81	1.11
0450 Taped and finished (level 4 finish)		965	.017		.42	.91		1.33	1.90
0490 With compound skim coat (level 5 finish)		775	.021		.47	1.13		1.60	2.32
0500 Water resistant, no finish included		2000	.008		.37	.44		.81	1.11
0550 Taped and finished (level 4 finish)		965	.017		.42	.91		1.33	1.90
0590 With compound skim coat (level 5 finish)		775	.021		.47	1.13		1.60	2.32
0600 Prefinished, vinyl, clipped to studs		900	.018		.53	.97		1.50	2.13
0700 Mold resistant, no finish included		2000	.008		.44	.44		.88	1.18
0710 Taped and finished (level 4 finish)		965	.017		.49	.91		1.40	1.98
0720 With compound skim coat (level 5 finish)		775	.021		.54	1.13		1.67	2.40
1000 On ceilings, standard, no finish included	2 Carp	1800	.009	S.F.	.33	.49		.82	1.13
1050 Taped and finished (level 4 finish)		765	.021		.38	1.14		1.52	2.24
1090 With compound skim coat (level 5 finish)		610	.026		.43	1.43		1.86	2.76
1100 Fire resistant, no finish included		1800	.009		.37	.49		.86	1.18
1150 Taped and finished (level 4 finish)		765	.021		.42	1.14		1.56	2.28
1195 With compound skim coat (level 5 finish)		610	.026		.47	1.43		1.90	2.80

09 29 10.30 Gypsum Board		Crew	Daily Output	Labor-Hours	Unit	Material	2021 Bare Costs Labor	Equipment	Total	Total Incl O&P
1200	Water resistant, no finish included	2 Carp	1800	.009	S.F.	.37	.49		.86	1.18
1250	Taped and finished (level 4 finish)		765	.021		.42	1.14		1.56	2.28
1290	With compound skim coat (level 5 finish)		610	.026		.47	1.43		1.90	2.80
1310	Mold resistant, no finish included		1800	.009		.44	.49		.93	1.25
1320	Taped and finished (level 4 finish)		765	.021		.49	1.14		1.63	2.36
1330	With compound skim coat (level 5 finish)		610	.026		.54	1.43		1.97	2.88
1350	Sag resistant, no finish included		1600	.010		.35	.55		.90	1.26
1360	Taped and finished (level 4 finish)		765	.021		.40	1.14		1.54	2.26
1370	With compound skim coat (level 5 finish)		610	.026		.45	1.43		1.88	2.78
1500	On beams, columns, or soffits, standard, no finish included		675	.024		.38	1.30		1.68	2.48
1550	Taped and finished (level 4 finish)		540	.030		.38	1.62		2	3
1590	With compound skim coat (level 5 finish)		475	.034		.43	1.84		2.27	3.41
1600	Fire resistant, no finish included		675	.024		.37	1.30		1.67	2.47
1650	Taped and finished (level 4 finish)		540	.030		.42	1.62		2.04	3.04
1690	With compound skim coat (level 5 finish)		475	.034		.47	1.84		2.31	3.45
1700	Water resistant, no finish included		675	.024		.43	1.30		1.73	2.53
1750	Taped and finished (level 4 finish)		540	.030		.42	1.62		2.04	3.04
1790	With compound skim coat (level 5 finish)		475	.034		.47	1.84		2.31	3.45
1800	Mold resistant, no finish included		675	.024		.51	1.30		1.81	2.62
1810	Taped and finished (level 4 finish)		540	.030		.49	1.62		2.11	3.12
1820	With compound skim coat (level 5 finish)		475	.034		.54	1.84		2.38	3.53
1850	Sag resistant, no finish included		675	.024		.40	1.30		1.70	2.50
1860	Taped and finished (level 4 finish)		540	.030		.40	1.62		2.02	3.02
1870	With compound skim coat (level 5 finish)		475	.034		.45	1.84		2.29	3.43
2000	5/8" thick, on walls, standard, no finish included		2000	.008		.36	.44		.80	1.10
2050	Taped and finished (level 4 finish)		965	.017		.41	.91		1.32	1.89
2090	With compound skim coat (level 5 finish)		775	.021		.46	1.13		1.59	2.31
2100	Fire resistant, no finish included		2000	.008		.38	.44		.82	1.12
2150	Taped and finished (level 4 finish)		965	.017		.43	.91		1.34	1.91
2195	With compound skim coat (level 5 finish)		775	.021		.48	1.13		1.61	2.33
2200	Water resistant, no finish included		2000	.008		.43	.44		.87	1.17
2250	Taped and finished (level 4 finish)	2 Carp	965	.017	S.F.	.48	.91		1.39	1.97
2290	With compound skim coat (level 5 finish)		775	.021		.53	1.13		1.66	2.39
2300	Prefinished, vinyl, clipped to studs		900	.018		1.11	.97		2.08	2.77
2510	Mold resistant, no finish included		2000	.008		.45	.44		.89	1.20
2520	Taped and finished (level 4 finish)		965	.017		.50	.91		1.41	1.99
2530	With compound skim coat (level 5 finish)		775	.021		.55	1.13		1.68	2.41
3000	On ceilings, standard, no finish included		1800	.009		.36	.49		.85	1.17
3050	Taped and finished (level 4 finish)		765	.021		.41	1.14		1.55	2.27
3090	With compound skim coat (level 5 finish)		615	.026		.46	1.42		1.88	2.78
3100	Fire resistant, no finish included		1800	.009		.38	.49		.87	1.19
3150	Taped and finished (level 4 finish)		765	.021		.43	1.14		1.57	2.29
3190	With compound skim coat (level 5 finish)		615	.026		.48	1.42		1.90	2.80
3200	Water resistant, no finish included		1800	.009		.43	.49		.92	1.24
3250	Taped and finished (level 4 finish)		765	.021		.48	1.14		1.62	2.35
3290	With compound skim coat (level 5 finish)		615	.026		.53	1.42		1.95	2.86
3300	Mold resistant, no finish included		1800	.009		.45	.49		.94	1.27
3310	Taped and finished (level 4 finish)		765	.021		.50	1.14		1.64	2.37
3320	With compound skim coat (level 5 finish)		615	.026		.55	1.42		1.97	2.88
3500	On beams, columns, or soffits, no finish included		675	.024		.41	1.30		1.71	2.52
3550	Taped and finished (level 4 finish)		475	.034		.47	1.84		2.31	3.45
3590	With compound skim coat (level 5 finish)		380	.042		.53	2.30		2.83	4.26
3600	Fire resistant, no finish included		675	.024		.44	1.30		1.74	2.54

312

09 29 10 – Gypsum Board Panels

09 29 10.30 Gypsum Board	Crew	Daily Output	Labor-Hours	Unit	Material	2021 Bare Costs Labor	Equipment	Total	Total Incl O&P
3650	Taped and finished (level 4 finish)	2 Carp	475	.034	S.F.	.49	1.84		2.33	3.47
3690	With compound skim coat (level 5 finish)		380	.042		.48	2.30		2.78	4.20
3700	Water resistant, no finish included		675	.024		.49	1.30		1.79	2.60
3750	Taped and finished (level 4 finish)		475	.034		.53	1.84		2.37	3.52
3790	With compound skim coat (level 5 finish)		380	.042		.55	2.30		2.85	4.27
3800	Mold resistant, no finish included		675	.024		.52	1.30		1.82	2.63
3810	Taped and finished (level 4 finish)		475	.034		.55	1.84		2.39	3.54
3820	With compound skim coat (level 5 finish)		380	.042		.57	2.30		2.87	4.30
4000	Fireproofing, beams or columns, 2 layers, 1/2" thick, incl finish		330	.048		.83	2.65		3.48	5.15
4010	Mold resistant		330	.048		.97	2.65		3.62	5.30
4050	5/8" thick		300	.053		.85	2.92		3.77	5.60
4060	Mold resistant		300	.053		.99	2.92		3.91	5.75
4100	3 layers, 1/2" thick		225	.071		1.25	3.89		5.14	7.60
4110	Mold resistant		225	.071		1.46	3.89		5.35	7.80
4150	5/8" thick		210	.076		1.28	4.17		5.45	8.05
4160	Mold resistant		210	.076		1.49	4.17		5.66	8.30
5050	For 1" thick coreboard on columns	2 Carp	480	.033	S.F.	.82	1.82		2.64	3.80
5100	For foil-backed board, add					.22			.22	.24
5200	For work over 8' high, add	2 Carp	3060	.005			.29		.29	.46
5270	For textured spray, add	2 Lath	1600	.010		.04	.54		.58	.88
5300	For distribution cost 3 stories and above, add per story	2 Carp	6100	.003			.14		.14	.23
5350	For finishing inner corners, add		950	.017	L.F.	.10	.92		1.02	1.58
5355	For finishing outer corners, add		1250	.013		.24	.70		.94	1.37
5500	For acoustical sealant, add per bead	1 Carp	500	.016		.04	.88		.92	1.44
5550	Sealant, 1 quart tube				Ea.	7.15			7.15	7.90
6000	Gypsum sound dampening panels									
6010	1/2" thick on walls, multi-layer, lightweight, no finish included	2 Carp	1500	.011	S.F.	1.66	.58		2.24	2.76
6015	Taped and finished (level 4 finish)		725	.022		1.71	1.21		2.92	3.80
6020	With compound skim coat (level 5 finish)		580	.028		1.76	1.51		3.27	4.34
6025	5/8" thick on walls, for wood studs, no finish included		1500	.011		2.81	.58		3.39	4.02
6030	Taped and finished (level 4 finish)		725	.022		2.86	1.21		4.07	5.05
6035	With compound skim coat (level 5 finish)		580	.028		2.91	1.51		4.42	5.60
6040	For metal stud, no finish included		1500	.011		1.86	.58		2.44	2.98
6045	Taped and finished (level 4 finish)		725	.022		1.91	1.21		3.12	4.02
6050	With compound skim coat (level 5 finish)		580	.028		1.96	1.51		3.47	4.56
6055	Abuse resist, no finish included		1500	.011		1.62	.58		2.20	2.71
6060	Taped and finished (level 4 finish)		725	.022		1.67	1.21		2.88	3.75
6065	With compound skim coat (level 5 finish)		580	.028		1.72	1.51		3.23	4.30
6070	Shear rated, no finish included		1500	.011		1.63	.58		2.21	2.72
6075	Taped and finished (level 4 finish)		725	.022		1.68	1.21		2.89	3.77
6080	With compound skim coat (level 5 finish)		580	.028		1.73	1.51		3.24	4.31
6085	For SCIF applications, no finish included		1500	.011		1.75	.58		2.33	2.86
6090	Taped and finished (level 4 finish)		725	.022		1.80	1.21		3.01	3.90
6095	With compound skim coat (level 5 finish)		580	.028		1.85	1.51		3.36	4.44
6100	1-3/8" thick on walls, THX certified, no finish included		1500	.011		9.50	.58		10.08	11.40
6105	Taped and finished (level 4 finish)		725	.022		9.55	1.21		10.76	12.40
6110	With compound skim coat (level 5 finish)		580	.028		9.60	1.51		11.11	13
6115	5/8" thick on walls, score & snap installation, no finish included		2000	.008		1.86	.44		2.30	2.75
6120	Taped and finished (level 4 finish)		965	.017		1.91	.91		2.82	3.54
6125	With compound skim coat (level 5 finish)		775	.021		1.96	1.13		3.09	3.96
7020	5/8" thick on ceilings, for wood joists, no finish included		1200	.013		2.81	.73		3.54	4.25
7025	Taped and finished (level 4 finish)		510	.031		2.86	1.72		4.58	5.85
7030	With compound skim coat (level 5 finish)		410	.039		2.91	2.13		5.04	6.60

09 29 10.30 Gypsum Board

		Crew	Daily Output	Labor-Hours	Unit	Material	2021 Bare Costs Labor	Equipment	Total	Total Incl O&P
7035	For metal joists, no finish included	2 Carp	1200	.013	S.F.	1.86	.73		2.59	3.21
7040	Taped and finished (level 4 finish)		510	.031		1.91	1.72		3.63	4.83
7045	With compound skim coat (level 5 finish)		410	.039		1.96	2.13		4.09	5.55
7050	Abuse resist, no finish included		1200	.013		1.62	.73		2.35	2.94
7055	Taped and finished (level 4 finish)		510	.031		1.67	1.72		3.39	4.56
7060	With compound skim coat (level 5 finish)		410	.039		1.72	2.13		3.85	5.30
7065	Shear rated, no finish included		1200	.013		1.63	.73		2.36	2.95
7070	Taped and finished (level 4 finish)		510	.031		1.68	1.72		3.40	4.58
7075	With compound skim coat (level 5 finish)		410	.039		1.73	2.13		3.86	5.30
7080	For SCIF applications, no finish included		1200	.013		1.75	.73		2.48	3.09
7085	Taped and finished (level 4 finish)		510	.031		1.80	1.72		3.52	4.71
7090	With compound skim coat (level 5 finish)		410	.039		1.85	2.13		3.98	5.45
8010	5/8" thick on ceilings, score & snap installation, no finish included		1600	.010		1.86	.55		2.41	2.92
8015	Taped and finished (level 4 finish)		680	.024		1.91	1.29		3.20	4.15
8020	With compound skim coat (level 5 finish)		545	.029		1.96	1.61		3.57	4.72
9000	Minimum labor/equipment charge	1 Carp	2	4	Job		219		219	350

09 29 10.50 High Abuse Gypsum Board

		Crew	Daily Output	Labor-Hours	Unit	Material	2021 Bare Costs Labor	Equipment	Total	Total Incl O&P
0010	**HIGH ABUSE GYPSUM BOARD**, fiber reinforced, nailed or									
0100	screwed to studs unless otherwise noted									
0110	1/2" thick, on walls, no finish included	2 Carp	1800	.009	S.F.	.83	.49		1.32	1.68
0120	Taped and finished (level 4 finish)		870	.018		.88	1.01		1.89	2.57
0130	With compound skim coat (level 5 finish)		700	.023		.93	1.25		2.18	3.02
0150	On ceilings, no finish included		1620	.010		.83	.54		1.37	1.77
0160	Taped and finished (level 4 finish)		690	.023		.88	1.27		2.15	2.99
0170	With compound skim coat (level 5 finish)		550	.029		.93	1.59		2.52	3.56
0210	5/8" thick, on walls, no finish included		1800	.009		.80	.49		1.29	1.65
0220	Taped and finished (level 4 finish)		870	.018		.85	1.01		1.86	2.53
0230	With compound skim coat (level 5 finish)		700	.023		.90	1.25		2.15	2.98
0250	On ceilings, no finish included		1620	.010		.80	.54		1.34	1.74
0260	Taped and finished (level 4 finish)		690	.023		.85	1.27		2.12	2.95
0270	With compound skim coat (level 5 finish)		550	.029		.90	1.59		2.49	3.52
0272	5/8" thick, on roof sheathing for 1-hour rating		1620	.010		.80	.54		1.34	1.74
0310	5/8" thick, on walls, very high impact, no finish included		1800	.009		.95	.49		1.44	1.82
0320	Taped and finished (level 4 finish)		870	.018		1	1.01		2.01	2.70
0330	With compound skim coat (level 5 finish)		700	.023		1.05	1.25		2.30	3.15
0350	On ceilings, no finish included		1620	.010		.95	.54		1.49	1.91
0360	Taped and finished (level 4 finish)		690	.023		1	1.27		2.27	3.12
0370	With compound skim coat (level 5 finish)		550	.029		1.05	1.59		2.64	3.69
0400	High abuse, gypsum core, paper face									
0410	1/2" thick, on walls, no finish included	2 Carp	1800	.009	S.F.	.80	.49		1.29	1.65
0420	Taped and finished (level 4 finish)		870	.018		.85	1.01		1.86	2.53
0430	With compound skim coat (level 5 finish)		700	.023		.90	1.25		2.15	2.98
0450	On ceilings, no finish included		1620	.010		.80	.54		1.34	1.74
0460	Taped and finished (level 4 finish)		690	.023		.85	1.27		2.12	2.95
0470	With compound skim coat (level 5 finish)		550	.029		.90	1.59		2.49	3.52
0510	5/8" thick, on walls, no finish included		1800	.009		.82	.49		1.31	1.67
0520	Taped and finished (level 4 finish)		870	.018		.87	1.01		1.88	2.55
0530	With compound skim coat (level 5 finish)		700	.023		.92	1.25		2.17	3.01
0550	On ceilings, no finish included		1620	.010		.82	.54		1.36	1.76
0560	Taped and finished (level 4 finish)		690	.023		.87	1.27		2.14	2.97
0570	With compound skim coat (level 5 finish)		550	.029		.92	1.59		2.51	3.55
1000	For high ceilings, over 8' high, add		2750	.006			.32		.32	.51

09 29 Gypsum Board

09 29 10 – Gypsum Board Panels

09 29 10.50 High Abuse Gypsum Board	Crew	Daily Output	Labor-Hours	Unit	Material	2021 Bare Costs Labor	2021 Bare Costs Equipment	Total	Total Incl O&P
1010	For distribution cost 3 stories and above, add per story	2 Carp	5500	.003	S.F.		.16		.16	.25

09 29 15 – Gypsum Board Accessories

09 29 15.10 Accessories, Gypsum Board

		Crew	Daily Output	Labor-Hours	Unit	Material	Labor	Equipment	Total	Total Incl O&P
0010	**ACCESSORIES, GYPSUM BOARD**									
0020	Casing bead, galvanized steel	1 Carp	2.90	2.759	C.L.F.	22.50	151		173.50	265
0100	Vinyl		3	2.667		24	146		170	259
0300	Corner bead, galvanized steel, 1" x 1"		4	2		15.95	109		124.95	192
0400	1-1/4" x 1-1/4"		3.50	2.286		17.35	125		142.35	218
0600	Vinyl		4	2		19.60	109		128.60	196
0900	Furring channel, galv. steel, 7/8" deep, standard		2.60	3.077		34.50	168		202.50	305
1000	Resilient		2.55	3.137		26	172		198	300
1100	J trim, galvanized steel, 1/2" wide		3	2.667		23	146		169	258
1120	5/8" wide		2.95	2.712		30	148		178	269
1500	Z stud, galvanized steel, 1-1/2" wide		2.60	3.077		44	168		212	315
9000	Minimum labor/equipment charge		3	2.667	Job		146		146	232

09 30 Tiling

09 30 13 – Ceramic Tiling

09 30 13.20 Ceramic Tile Repairs

		Crew	Daily Output	Labor-Hours	Unit	Material	Labor	Equipment	Total	Total Incl O&P
0010	**CERAMIC TILE REPAIRS**									
1000	Grout removal, carbide tipped, rotary grinder	1 Clab	240	.033	L.F.		1.48		1.48	2.36

09 30 13.45 Ceramic Tile Accessories

		Crew	Daily Output	Labor-Hours	Unit	Material	Labor	Equipment	Total	Total Incl O&P
0010	**CERAMIC TILE ACCESSORIES**									
0100	Spacers, 1/8"				C	2.41			2.41	2.65
1310	Sealer for natural stone tile, installed	1 Tilf	650	.012	S.F.	.05	.64		.69	1.05

09 30 29 – Metal Tiling

09 30 29.10 Metal Tile

		Crew	Daily Output	Labor-Hours	Unit	Material	Labor	Equipment	Total	Total Incl O&P
0010	**METAL TILE** 4' x 4' sheet, 24 ga., tile pattern, nailed									
0200	Stainless steel	2 Carp	512	.031	S.F.	28	1.71		29.71	33.50
0400	Aluminized steel	2 Carp	512	.031	S.F.	20.50	1.71		22.21	25
9000	Minimum labor/equipment charge	1 Carp	4	2	Job		109		109	174

09 30 95 – Tile & Stone Setting Materials and Specialties

09 30 95.10 Moisture Resistant, Anti-Fracture Membrane

		Crew	Daily Output	Labor-Hours	Unit	Material	Labor	Equipment	Total	Total Incl O&P
0010	**MOISTURE RESISTANT, ANTI-FRACTURE MEMBRANE**									
0200	Elastomeric membrane, 1/16" thick	D-7	275	.058	S.F.	1.53	2.71		4.24	5.90

09 31 Thin-Set Tiling

09 31 13 – Thin-Set Ceramic Tiling

09 31 13.10 Thin-Set Ceramic Tile

		Crew	Daily Output	Labor-Hours	Unit	Material	Labor	Equipment	Total	Total Incl O&P
0010	**THIN-SET CERAMIC TILE**									
0700	Cove base, 4-1/4" x 4-1/4"	D-7	128	.125	L.F.	4.09	5.85		9.94	13.60
1000	6" x 4-1/4" high		137	.117		4.24	5.45		9.69	13.15
1300	Sanitary cove base, 6" x 4-1/4" high		124	.129		4.86	6		10.86	14.75
1600	6" x 6" high		117	.137		4.92	6.40		11.32	15.35
2500	Bullnose trim, 4-1/4" x 4-1/4"		128	.125		4.18	5.85		10.03	13.70
2800	2" x 6"		124	.129		4	6		10	13.80
3300	Ceramic tile, porcelain type, 1 color, color group 2, 1" x 1"		183	.087	S.F.	5.75	4.08		9.83	12.70

09 31 Thin-Set Tiling

09 31 13 – Thin-Set Ceramic Tiling

09 31 13.10 Thin-Set Ceramic Tile

		Crew	Daily Output	Labor-Hours	Unit	Material	2021 Bare Costs Labor	Equipment	Total	Total Incl O&P
3310	2" x 2" or 2" x 1"	D-7	190	.084	S.F.	4.04	3.93		7.97	10.60
3350	For random blend, 2 colors, add					1			1	1.10
3360	4 colors, add					1.43			1.43	1.57
4300	Specialty tile, 4-1/4" x 4-1/4" x 1/2", decorator finish	D-7	183	.087		13.75	4.08		17.83	21.50
4500	Add for epoxy grout, 1/16" joint, 1" x 1" tile		800	.020		.66	.93		1.59	2.19
4600	2" x 2" tile		820	.020		.62	.91		1.53	2.10
4610	Add for epoxy grout, 1/8" joint, 8" x 8" x 3/8" tile, add		900	.018		1.56	.83		2.39	3.01
4800	Pregrouted sheets, walls, 4-1/4" x 4-1/4", 6" x 4-1/4"									
4810	and 8-1/2" x 4-1/4", 4 S.F. sheets, silicone grout	D-7	240	.067	S.F.	5.80	3.11		8.91	11.25
5100	Floors, unglazed, 2 S.F. sheets,									
5110	urethane adhesive	D-7	180	.089	S.F.	2.07	4.15		6.22	8.75
5400	Walls, interior, 4-1/4" x 4-1/4" tile		190	.084		3.02	3.93		6.95	9.45
5500	6" x 4-1/4" tile		190	.084		3.17	3.93		7.10	9.65
5700	8-1/2" x 4-1/4" tile		190	.084		6.40	3.93		10.33	13.15
5800	6" x 6" tile		175	.091		4.22	4.27		8.49	11.30
5810	8" x 8" tile		170	.094		4.86	4.39		9.25	12.20
5820	12" x 12" tile		160	.100		4.82	4.67		9.49	12.60
5830	16" x 16" tile		150	.107		5.50	4.98		10.48	13.80
6000	Decorated wall tile, 4-1/4" x 4-1/4", color group 1		270	.059		3.73	2.76		6.49	8.40
6100	Color group 4		180	.089		69	4.15		73.15	82
9300	Ceramic tiles, recycled glass, standard colors, 2" x 2" thru 6" x 6" ⒢		190	.084		22.50	3.93		26.43	30.50
9310	6" x 6" ⒢		175	.091		22.50	4.27		26.77	31
9320	8" x 8" ⒢		170	.094		23	4.39		27.39	32.50
9330	12" x 12" ⒢	D-7	160	.100	S.F.	23	4.67		27.67	33
9340	Earthtones, 2" x 2" to 4" x 8" ⒢		190	.084		27.50	3.93		31.43	36
9350	6" x 6" ⒢		175	.091		27.50	4.27		31.77	36.50
9360	8" x 8" ⒢		170	.094		26.50	4.39		30.89	36
9370	12" x 12" ⒢		160	.100		26.50	4.67		31.17	36.50
9380	Deep colors, 2" x 2" to 4" x 8" ⒢		190	.084		31.50	3.93		35.43	40.50
9390	6" x 6" ⒢		175	.091		31.50	4.27		35.77	41
9400	8" x 8" ⒢		170	.094		32.50	4.39		36.89	43
9410	12" x 12" ⒢		160	.100		32.50	4.67		37.17	43.50

09 31 33 – Thin-Set Stone Tiling

09 31 33.10 Tiling, Thin-Set Stone

		Crew	Daily Output	Labor-Hours	Unit	Material	2021 Bare Costs Labor	Equipment	Total	Total Incl O&P
0010	**TILING, THIN-SET STONE**									
3000	Floors, natural clay, random or uniform, color group 1	D-7	183	.087	S.F.	4.98	4.08		9.06	11.80
3100	Color group 2	"	183	.087	"	5.70	4.08		9.78	12.60

09 32 Mortar-Bed Tiling

09 32 13 – Mortar-Bed Ceramic Tiling

09 32 13.10 Ceramic Tile

		Crew	Daily Output	Labor-Hours	Unit	Material	2021 Bare Costs Labor	Equipment	Total	Total Incl O&P
0010	**CERAMIC TILE**									
0600	Cove base, 4-1/4" x 4-1/4" high	D-7	91	.176	L.F.	4.28	8.20		12.48	17.50
0900	6" x 4-1/4" high		100	.160		4.43	7.45		11.88	16.50
1200	Sanitary cove base, 6" x 4-1/4" high		93	.172		5.05	8.05		13.10	18.10
1500	6" x 6" high		84	.190		5.10	8.90		14	19.45
2400	Bullnose trim, 4-1/4" x 4-1/4"		82	.195		4.31	9.10		13.41	18.95
2700	2" x 6" bullnose trim		84	.190		4.09	8.90		12.99	18.35
6210	Wall tile, 4-1/4" x 4-1/4", better grade	1 Tilf	50	.160	S.F.	9.60	8.25		17.85	23.50
6240	2" x 2"		50	.160		7.85	8.25		16.10	21.50

09 32 Mortar-Bed Tiling

09 32 13 – Mortar-Bed Ceramic Tiling

09 32 13.10 Ceramic Tile

		Crew	Daily Output	Labor-Hours	Unit	Material	2021 Bare Costs Labor	2021 Bare Costs Equipment	Total	Total Incl O&P
6250	6" x 6"	1 Tilf	55	.145	S.F.	9.95	7.50		17.45	22.50
6260	8" x 8"	↓	60	.133		10.25	6.90		17.15	22
6600	Crystalline glazed, 4-1/4" x 4-1/4", plain	D-7	100	.160		4.86	7.45		12.31	17
6700	4-1/4" x 4-1/4", scored tile		100	.160		6.70	7.45		14.15	19
6900	6" x 6" plain		93	.172		5.25	8.05		13.30	18.35
7000	For epoxy grout, 1/16" joints, 4-1/4" tile, add		800	.020		.43	.93		1.36	1.93
7200	For tile set in dry mortar, add		1735	.009			.43		.43	.67
7300	For tile set in Portland cement mortar, add		290	.055	↓	.22	2.57		2.79	4.26
9500	Minimum labor/equipment charge	↓	3.25	4.923	Job		230		230	360

09 32 16 – Mortar-Bed Quarry Tiling

09 32 16.10 Quarry Tile

		Crew	Daily Output	Labor-Hours	Unit	Material	2021 Bare Costs Labor	2021 Bare Costs Equipment	Total	Total Incl O&P
0010	**QUARRY TILE**									
0100	Base, cove or sanitary, to 5" high, 1/2" thick	D-7	110	.145	L.F.	6.65	6.80		13.45	17.90
0300	Bullnose trim, red, 6" x 6" x 1/2" thick		120	.133		5.20	6.20		11.40	15.45
0400	4" x 4" x 1/2" thick		110	.145		4.23	6.80		11.03	15.25
0600	4" x 8" x 1/2" thick, using 8" as edge	↓	130	.123	↓	5.45	5.75		11.20	14.95
0700	Floors, 1,000 S.F. lots, red, 4" x 4" x 1/2" thick	D-7	120	.133	S.F.	9.40	6.20		15.60	20
0900	6" x 6" x 1/2" thick		140	.114		10.15	5.35		15.50	19.45
1000	4" x 8" x 1/2" thick	↓	130	.123		6.65	5.75		12.40	16.25
1300	For waxed coating, add					.76			.76	.84
1500	For non-standard colors, add					.50			.50	.55
1600	For abrasive surface, add					.54			.54	.59
1800	Brown tile, imported, 6" x 6" x 3/4"	D-7	120	.133		7.80	6.20		14	18.25
1900	8" x 8" x 1"		110	.145		10.25	6.80		17.05	22
2100	For thin set mortar application, deduct		700	.023			1.07		1.07	1.66
2700	Stair tread, 6" x 6" x 3/4", plain		50	.320		8.55	14.95		23.50	33
2800	Abrasive		47	.340		8.75	15.90		24.65	34.50
3000	Wainscot, 6" x 6" x 1/2", thin set, red		105	.152		3.31	7.10		10.41	14.75
3100	Non-standard colors		105	.152	↓	3.79	7.10		10.89	15.25
3300	Window sill, 6" wide, 3/4" thick		90	.178	L.F.	8.50	8.30		16.80	22.50
3400	Corners		80	.200	Ea.	6.20	9.35		15.55	21.50
9000	Minimum labor/equipment charge	↓	3.25	4.923	Job		230		230	360

09 32 23 – Mortar-Bed Glass Mosaic Tiling

09 32 23.10 Glass Mosaics

		Crew	Daily Output	Labor-Hours	Unit	Material	2021 Bare Costs Labor	2021 Bare Costs Equipment	Total	Total Incl O&P
0010	**GLASS MOSAICS** 3/4" tile on 12" sheets, standard grout									
1020	1" tile on 12" sheets, opalescent finish	D-7	73	.219	S.F.	18.85	10.20		29.05	36.50
1040	1" x 2" tile on 12" sheet, blend		73	.219		21	10.20		31.20	39
1060	2" tile on 12" sheet, blend		73	.219		15.55	10.20		25.75	33
1080	5/8" x random tile, linear, on 12" sheet, blend		73	.219		35	10.20		45.20	54.50
1600	Dots on 12" sheet		73	.219		35	10.20		45.20	54.50
1700	For glass mosaic tiles set in dry mortar, add		290	.055		.45	2.57		3.02	4.52
1720	For glass mosaic tiles set in Portland cement mortar, add		290	.055	↓	.01	2.57		2.58	4.03
1730	For polyblend sanded tile grout	↓	96.15	.166	Lb.	2.19	7.75		9.94	14.50

09 34 Waterproofing-Membrane Tiling

09 34 13 – Waterproofing-Membrane Ceramic Tiling

09 34 13.10 Ceramic Tile Waterproofing Membrane	Crew	Daily Output	Labor-Hours	Unit	Material	2021 Bare Costs Labor	Equipment	Total	Total Incl O&P
0010 **CERAMIC TILE WATERPROOFING MEMBRANE**									
0020 On floors, including thinset									
0030 Fleece laminated polyethylene grid, 1/8" thick	D-7	250	.064	S.F.	2.13	2.99		5.12	7
0040 5/16" thick	"	250	.064	"	2.63	2.99		5.62	7.55
0050 On walls, including thinset									
0060 Fleece laminated polyethylene sheet, 8 mil thick	D-7	480	.033	S.F.	2.43	1.56		3.99	5.10
0070 Accessories, including thinset									
0080 Joint and corner sheet, 4 mils thick, 5" wide	1 Tilf	240	.033	L.F.	1.34	1.72		3.06	4.17
0090 7-1/4" wide	"	180	.044	"	1.72	2.30		4.02	5.50
0100 10" wide	1 Tilf	120	.067	L.F.	2.09	3.45		5.54	7.70
0110 Pre-formed corners, inside		32	.250	Ea.	9.15	12.95		22.10	30
0120 Outside		32	.250		7.65	12.95		20.60	28.50
0130 2" flanged floor drain with 6" stainless steel grate		16	.500		390	26		416	470
0140 EPS, sloped shower floor		480	.017	S.F.	11.60	.86		12.46	14.10
0150 Curb		32	.250	L.F.	17.55	12.95		30.50	39.50

09 35 Chemical-Resistant Tiling

09 35 13 – Chemical-Resistant Ceramic Tiling

09 35 13.10 Chemical-Resistant Ceramic Tiling

	Crew	Daily Output	Labor-Hours	Unit	Material	2021 Bare Costs Labor	Equipment	Total	Total Incl O&P
0010 **CHEMICAL-RESISTANT CERAMIC TILING**									
0100 4-1/4" x 4-1/4" x 1/4", 1/8" joint	D-7	130	.123	S.F.	10.80	5.75		16.55	21
0200 6" x 6" x 1/2" thick		120	.133		8.40	6.20		14.60	18.95
0300 8" x 8" x 1/2" thick		110	.145		9.95	6.80		16.75	21.50
0400 4-1/4" x 4-1/4" x 1/4", 1/4" joint		130	.123		11.60	5.75		17.35	21.50
0500 6" x 6" x 1/2" thick		120	.133		9.60	6.20		15.80	20.50
0600 8" x 8" x 1/2" thick		110	.145		10.65	6.80		17.45	22.50
0700 4-1/4" x 4-1/4" x 1/4", 3/8" joint		130	.123		12.35	5.75		18.10	22.50
0800 6" x 6" x 1/2" thick		120	.133		10.60	6.20		16.80	21.50
0900 8" x 8" x 1/2" thick		110	.145		11.90	6.80		18.70	23.50

09 35 16 – Chemical-Resistant Quarry Tiling

09 35 16.10 Chemical-Resistant Quarry Tiling

	Crew	Daily Output	Labor-Hours	Unit	Material	2021 Bare Costs Labor	Equipment	Total	Total Incl O&P
0010 **CHEMICAL-RESISTANT QUARRY TILING**									
0100 4" x 8" x 1/2" thick, 1/8" joint	D-7	130	.123	S.F.	10.10	5.75		15.85	20
0200 6" x 6" x 1/2" thick		120	.133		10.20	6.20		16.40	21
0300 8" x 8" x 1/2" thick		110	.145		7.60	6.80		14.40	19
0400 4" x 8" x 1/2" thick, 1/4" joint		130	.123		11.40	5.75		17.15	21.50
0500 6" x 6" x 1/2" thick		120	.133		11.35	6.20		17.55	22
0600 8" x 8" x 1/2" thick		110	.145		8.30	6.80		15.10	19.75
0700 4" x 8" x 1/2" thick, 3/8" joint		130	.123		12.55	5.75		18.30	23
0800 6" x 6" x 1/2" thick		120	.133		12.40	6.20		18.60	23.50
0900 8" x 8" x 1/2" thick		110	.145		9.55	6.80		16.35	21

09 51 13 – Acoustical Panel Ceilings

09 51 13.10 Ceiling, Acoustical Panel

	09 51 13.10 Ceiling, Acoustical Panel	Crew	Daily Output	Labor-Hours	Unit	Material	2021 Bare Costs Labor	Equipment	Total	Total Incl O&P
0010	**CEILING, ACOUSTICAL PANEL**									
0100	Fiberglass boards, film faced, 2' x 2' or 2' x 4', 5/8" thick	1 Carp	625	.013	S.F.	2.27	.70		2.97	3.61
0120	3/4" thick		600	.013		4.91	.73		5.64	6.55
0130	3" thick, thermal, R11		450	.018		3.99	.97		4.96	5.95

09 51 14 – Acoustical Fabric-Faced Panel Ceilings

09 51 14.10 Ceiling, Acoustical Fabric-Faced Panel

	09 51 14.10 Ceiling, Acoustical Fabric-Faced Panel	Crew	Daily Output	Labor-Hours	Unit	Material	Labor	Equipment	Total	Total Incl O&P
0010	**CEILING, ACOUSTICAL FABRIC-FACED PANEL**									
0100	Glass cloth faced fiberglass, 3/4" thick	1 Carp	500	.016	S.F.	3.35	.88		4.23	5.10
0120	1" thick	1 Carp	485	.016	S.F.	3.37	.90		4.27	5.15
0130	1-1/2" thick, nubby face	"	475	.017	"	2.82	.92		3.74	4.57

09 51 23 – Acoustical Tile Ceilings

09 51 23.10 Suspended Acoustic Ceiling Tiles

	09 51 23.10 Suspended Acoustic Ceiling Tiles	Crew	Daily Output	Labor-Hours	Unit	Material	Labor	Equipment	Total	Total Incl O&P
0010	**SUSPENDED ACOUSTIC CEILING TILES**, not including									
0100	suspension system									
1110	Mineral fiber tile, lay-in, 2' x 2' or 2' x 4', 5/8" thick, fine texture	1 Carp	625	.013	S.F.	.82	.70		1.52	2.01
1115	Rough textured		625	.013		.64	.70		1.34	1.81
1125	3/4" thick, fine textured		600	.013		2.14	.73		2.87	3.51
1130	Rough textured		600	.013		2.05	.73		2.78	3.42
1135	Fissured		600	.013		2.26	.73		2.99	3.65
1150	Tegular, 5/8" thick, fine textured		470	.017		1.08	.93		2.01	2.67
1155	Rough textured		470	.017		1.41	.93		2.34	3.03
1165	3/4" thick, fine textured		450	.018		2.35	.97		3.32	4.14
1170	Rough textured		450	.018		1.41	.97		2.38	3.10
1175	Fissured		450	.018		3.25	.97		4.22	5.15
1185	For plastic film face, add					.75			.75	.83
1190	For fire rating, add					.54			.54	.59
3720	Mineral fiber, 24" x 24" or 48", reveal edge, painted, 5/8" thick	1 Carp	600	.013		1.15	.73		1.88	2.43
3740	3/4" thick		575	.014		1.72	.76		2.48	3.10
5020	66-78% recycled content, 3/4" thick [G]		600	.013		6.85	.73		7.58	8.70
5040	Mylar, 42% recycled content, 3/4" thick [G]		600	.013		4.80	.73		5.53	6.45
6000	Remove & install new ceiling tiles, min fiber, 2' x 2' or 2' x 4', 5/8"thk.		335	.024		.82	1.31		2.13	2.98
9000	Minimum labor/equipment charge		4	2	Job		109		109	174

09 51 23.30 Suspended Ceilings, Complete

	09 51 23.30 Suspended Ceilings, Complete	Crew	Daily Output	Labor-Hours	Unit	Material	Labor	Equipment	Total	Total Incl O&P
0010	**SUSPENDED CEILINGS, COMPLETE**, incl. standard									
0100	suspension system but not incl. 1-1/2" carrier channels									
0600	Fiberglass ceiling board, 2' x 4' x 5/8", plain faced	1 Carp	500	.016	S.F.	3.57	.88		4.45	5.30
0700	Offices, 2' x 4' x 3/4"		380	.021		6.20	1.15		7.35	8.70
0800	Mineral fiber, on 15/16" T bar susp. 2' x 2' x 3/4" lay-in board		345	.023		3.73	1.27		5	6.10
0810	2' x 4' x 5/8" tile		380	.021		2.82	1.15		3.97	4.93
0820	Tegular, 2' x 2' x 5/8" tile on 9/16" grid		250	.032		3.08	1.75		4.83	6.20
0830	2' x 4' x 3/4" tile		275	.029		3.32	1.59		4.91	6.20
0900	Luminous panels, prismatic, acrylic		255	.031		4.48	1.72		6.20	7.65
1200	Metal pan with acoustic pad, steel		75	.107		5.60	5.85		11.45	15.50
1300	Painted aluminum		75	.107		4.13	5.85		9.98	13.85
1500	Aluminum, degreased finish		75	.107		6.40	5.85		12.25	16.35
1600	Stainless steel		75	.107		10.15	5.85		16	20.50
1800	Tile, Z bar suspension, 5/8" mineral fiber tile		150	.053		2.62	2.92		5.54	7.55
1900	3/4" mineral fiber tile		150	.053		2.37	2.92		5.29	7.25
2402	For strip lighting, see Section 26 51 13.50									
2500	For rooms under 500 S.F., add				S.F.	25%				
9000	Minimum labor/equipment charge	1 Carp	2	4	Job		219		219	350

09 51 Acoustical Ceilings

09 51 33 – Acoustical Metal Pan Ceilings

09 51 33.10 Ceiling, Acoustical Metal Pan

		Crew	Daily Output	Labor-Hours	Unit	Material	2021 Bare Costs Labor	Equipment	Total	Total Incl O&P
0010	**CEILING, ACOUSTICAL METAL PAN**									
0100	Metal panel, lay-in, 2' x 2', sq. edge	1 Carp	500	.016	S.F.	11.20	.88		12.08	13.70
0110	Tegular edge		500	.016		14.40	.88		15.28	17.25
0120	2' x 4', sq. edge		500	.016		14.15	.88		15.03	16.95
0130	Tegular edge		500	.016		14.25	.88		15.13	17.10
0140	Perforated alum. clip-in, 2' x 2'		500	.016		14.40	.88		15.28	17.25
0150	2' x 4'		500	.016		11.75	.88		12.63	14.35

09 51 53 – Direct-Applied Acoustical Ceilings

09 51 53.10 Ceiling Tile

		Crew	Daily Output	Labor-Hours	Unit	Material	2021 Bare Costs Labor	Equipment	Total	Total Incl O&P
0010	**CEILING TILE**, stapled or cemented									
0100	12" x 12" or 12" x 24", not including furring									
0600	Mineral fiber, vinyl coated, 5/8" thick	1 Carp	300	.027	S.F.	2.20	1.46		3.66	4.74
0700	3/4" thick		300	.027		3.77	1.46		5.23	6.45
0900	Fire rated, 3/4" thick, plain faced		300	.027		1.53	1.46		2.99	4
1000	Plastic coated face		300	.027		1.73	1.46		3.19	4.22
1200	Aluminum faced, 5/8" thick, plain		300	.027		1.95	1.46		3.41	4.47
3000	Wood fiber tile, 1/2" thick		400	.020		1.33	1.09		2.42	3.20
3100	3/4" thick		400	.020		1.30	1.09		2.39	3.17
3300	For flameproofing, add					.09			.09	.10
3400	For sculptured 3 dimensional, add					.34			.34	.37
3900	For ceiling primer, add					.12			.12	.13
4000	For ceiling cement, add					.40			.40	.44
9000	Minimum labor/equipment charge	1 Carp	4	2	Job		109		109	174

09 53 Acoustical Ceiling Suspension Assemblies

09 53 23 – Metal Acoustical Ceiling Suspension Assemblies

09 53 23.30 Ceiling Suspension Systems

		Crew	Daily Output	Labor-Hours	Unit	Material	2021 Bare Costs Labor	Equipment	Total	Total Incl O&P
0010	**CEILING SUSPENSION SYSTEMS** for boards and tile									
0050	Class A suspension system, 15/16" T bar, 2' x 4' grid	1 Carp	800	.010	S.F.	1.30	.55		1.85	2.30
0300	2' x 2' grid		650	.012		1.59	.67		2.26	2.82
0310	25% recycled steel, 2' x 4' grid [G]		800	.010		1	.55		1.55	1.97
0320	2' x 2' grid [G]		650	.012		1.40	.67		2.07	2.61
0350	For 9/16" grid, add					.14			.14	.15
0360	For fire rated grid, add					.08			.08	.09
0370	For colored grid, add					.20			.20	.22
0400	Concealed Z bar suspension system, 12" module	1 Carp	520	.015		.95	.84		1.79	2.39
0600	1-1/2" carrier channels, 4' OC, add	"	470	.017		.13	.93		1.06	1.62
0700	Carrier channels for ceilings with									
0900	recessed lighting fixtures, add	1 Carp	460	.017	S.F.	.23	.95		1.18	1.77
3000	Seismic ceiling bracing, IBC Site Class D, Occupancy Category II									
3050	For ceilings less than 2500 S.F.									
3060	Seismic clips at attached walls	1 Carp	180	.044	Ea.	.96	2.43		3.39	4.92
3100	For ceilings greater than 2500 S.F., add									
3120	Seismic clips, joints at cross tees	1 Carp	120	.067	Ea.	.89	3.65		4.54	6.80
3140	At cross tees and mains, mains field cut	"	60	.133	"	.89	7.30		8.19	12.60
3200	Compression posts, telescopic, attached to structure above									
3210	To 30" high	1 Carp	26	.308	Ea.	36.50	16.85		53.35	67.50
3220	30" to 48" high		25.50	.314		37.50	17.15		54.65	69
3230	48" to 84" high		25	.320		45.50	17.50		63	78
3240	84" to 102" high		24.50	.327		51.50	17.85		69.35	85.50

09 53 Acoustical Ceiling Suspension Assemblies

09 53 23 – Metal Acoustical Ceiling Suspension Assemblies

09 53 23.30 Ceiling Suspension Systems	Crew	Daily Output	Labor-Hours	Unit	Material	2021 Bare Costs Labor	Equipment	Total	Total Incl O&P
3250	102" to 120" high	1 Carp	24	.333	Ea.	74	18.25		92.25	111
3260	120" to 144" high	↓	24	.333	↓	82.50	18.25		100.75	120
3300	Stabilizer bars									
3310	12" long	1 Carp	240	.033	Ea.	1.02	1.82		2.84	4.02
3320	24" long		235	.034		.94	1.86		2.80	4
3330	36" long		230	.035		.94	1.90		2.84	4.06
3340	48" long	↓	220	.036	↓	.75	1.99		2.74	4
3400	Wire support for light fixtures, per L.F. height to structure above									
3410	Less than 10 lb.	1 Carp	400	.020	L.F.	.40	1.09		1.49	2.18
3420	10 lb. to 56 lb.	"	240	.033	"	.81	1.82		2.63	3.79
3500	Retrofit existing suspended ceiling grid to current code									
3510	Less than 2500 S.F. (using clips @ perimeter)	1 Carp	1455	.006	S.F.	.06	.30		.36	.54
3520	Greater than 2500 S.F. (using compression posts and clips)		550	.015	↓	.38	.80		1.18	1.69
4000	Remove and replace ceiling grid, class A, 15/16" T bar, 2' x 2' grid		435	.018		1.59	1.01		2.60	3.35
4010	2' x 4' grid	↓	535	.015	↓	1.30	.82		2.12	2.73

09 54 Specialty Ceilings

09 54 16 – Luminous Ceilings

09 54 16.10 Ceiling, Luminous

		Crew	Daily Output	Labor-Hours	Unit	Material	2021 Bare Costs Labor	Equipment	Total	Total Incl O&P
0010	**CEILING, LUMINOUS**									
0020	Translucent lay-in panels, 2' x 2'	1 Carp	500	.016	S.F.	1.74	.88		2.62	3.30
0030	2' x 6'	"	500	.016	"	18.30	.88		19.18	21.50

09 54 23 – Linear Metal Ceilings

09 54 23.10 Metal Ceilings

		Crew	Daily Output	Labor-Hours	Unit	Material	2021 Bare Costs Labor	Equipment	Total	Total Incl O&P
0010	**METAL CEILINGS**									
0015	Solid alum. planks, 3-1/4" x 12', open reveal	1 Carp	500	.016	S.F.	2.42	.88		3.30	4.05
0020	Closed reveal		500	.016		3.10	.88		3.98	4.80
0030	7-1/4" x 12', open reveal		500	.016		4.13	.88		5.01	5.95
0040	Closed reveal	↓	500	.016	↓	5.25	.88		6.13	7.20
0050	Metal, open cell, 2' x 2', 6" cell	1 Carp	500	.016	S.F.	8.95	.88		9.83	11.25
0060	8" cell		500	.016		9.90	.88		10.78	12.30
0070	2' x 4', 6" cell		500	.016		5.90	.88		6.78	7.85
0080	8" cell		500	.016		5.90	.88		6.78	7.85

09 54 26 – Suspended Wood Ceilings

09 54 26.10 Wood Ceilings

		Crew	Daily Output	Labor-Hours	Unit	Material	2021 Bare Costs Labor	Equipment	Total	Total Incl O&P
0010	**WOOD CEILINGS**									
1000	4"-6" wood slats on heavy duty 15/16" T-bar grid	2 Carp	250	.064	S.F.	26.50	3.50		30	34.50

09 54 33 – Decorative Panel Ceilings

09 54 33.20 Metal Panel Ceilings

		Crew	Daily Output	Labor-Hours	Unit	Material	2021 Bare Costs Labor	Equipment	Total	Total Incl O&P
0010	**METAL PANEL CEILINGS**									
0020	Lay-in or screwed to furring, not including grid									
0100	Tin ceilings, 2' x 2' or 2' x 4', bare steel finish	2 Carp	300	.053	S.F.	4.90	2.92		7.82	10.05
0120	Painted white finish		300	.053	"	3.59	2.92		6.51	8.60
0140	Copper, chrome or brass finish		300	.053	L.F.	4.47	2.92		7.39	9.55
0200	Cornice molding, 2-1/2" to 3-1/2" wide, 4' long, bare steel finish		200	.080	S.F.	2.28	4.38		6.66	9.45
0220	Painted white finish		200	.080		2.98	4.38		7.36	10.25
0240	Copper, chrome or brass finish		200	.080		4.07	4.38		8.45	11.45
0320	5" to 6-1/2" wide, 4' long, bare steel finish		150	.107		5.30	5.85		11.15	15.15
0340	Painted white finish	↓	150	.107	↓	7	5.85		12.85	17

For customer support on your Commercial Renovation Costs with RSMeans Data, call 800.448.8182.

321

09 54 Specialty Ceilings

09 54 33 – Decorative Panel Ceilings

09 54 33.20 Metal Panel Ceilings	Crew	Daily Output	Labor-Hours	Unit	Material	2021 Bare Costs Labor	Equipment	Total	Total Incl O&P
0360 Copper, chrome or brass finish	2 Carp	150	.107	S.F.	4.87	5.85		10.72	14.65
0420 Flat molding, 3-1/2" to 5" wide, 4' long, bare steel finish		250	.064		3.99	3.50		7.49	9.95
0440 Painted white finish		250	.064		3.45	3.50		6.95	9.35
0460 Copper, chrome or brass finish		250	.064		3.58	3.50		7.08	9.50

09 61 Flooring Treatment

09 61 19 – Concrete Floor Staining

09 61 19.40 Floors, Interior

	Crew	Daily Output	Labor-Hours	Unit	Material	2021 Bare Costs Labor	Equipment	Total	Total Incl O&P
0010 **FLOORS, INTERIOR**									
0300 Acid stain and sealer									
0310 Stain, one coat	1 Pord	650	.012	S.F.	.14	.57		.71	1.07
0320 Two coats		570	.014		.28	.65		.93	1.34
0330 Acrylic sealer, one coat		2600	.003		.24	.14		.38	.50
0340 Two coats		1400	.006		.49	.27		.76	.96

09 62 Specialty Flooring

09 62 19 – Laminate Flooring

09 62 19.10 Floating Floor

	Crew	Daily Output	Labor-Hours	Unit	Material	2021 Bare Costs Labor	Equipment	Total	Total Incl O&P
0010 **FLOATING FLOOR**									
8300 Floating floor, laminate, wood pattern strip, complete	1 Clab	133	.060	S.F.	4.80	2.67		7.47	9.55
8310 Components, T&G wood composite strips					4.15			4.15	4.56
8320 Film					.21			.21	.23
8330 Foam					.32			.32	.35
8340 Adhesive					.43			.43	.47
8350 Installation kit					.24			.24	.26
8360 Trim, 2" wide x 3' long				L.F.	5.90			5.90	6.50
8370 Reducer moulding				"	6.70			6.70	7.35

09 62 29 – Cork Flooring

09 62 29.10 Cork Tile Flooring

		Crew	Daily Output	Labor-Hours	Unit	Material	2021 Bare Costs Labor	Equipment	Total	Total Incl O&P
0010 **CORK TILE FLOORING**										
2200 Cork tile, standard finish, 1/8" thick	G	1 Tilf	315	.025	S.F.	4.86	1.31		6.17	7.40
2250 3/16" thick	G		315	.025		6.20	1.31		7.51	8.85
2300 5/16" thick	G		315	.025		7.50	1.31		8.81	10.30
2350 1/2" thick	G		315	.025		4.52	1.31		5.83	7
2500 Urethane finish, 1/8" thick	G		315	.025		6.25	1.31		7.56	8.95
2550 3/16" thick	G		315	.025		9.20	1.31		10.51	12.15
2600 5/16" thick	G		315	.025		7.55	1.31		8.86	10.40
2650 1/2" thick	G		315	.025		7.85	1.31		9.16	10.65

09 63 Masonry Flooring

09 63 13 – Brick Flooring

09 63 13.10 Miscellaneous Brick Flooring

		Crew	Daily Output	Labor-Hours	Unit	Material	2021 Bare Costs Labor	Equipment	Total	Total Incl O&P
0010	**MISCELLANEOUS BRICK FLOORING**									
0020	Acid-proof shales, red, 8" x 3-3/4" x 1-1/4" thick	D-7	.43	37.209	M	615	1,725		2,340	3,375
0050	2-1/4" thick	D-1	.40	40		1,000	1,950		2,950	4,225
0200	Acid-proof clay brick, 8" x 3-3/4" x 2-1/4" thick G		.40	40		1,025	1,950		2,975	4,250
0250	9" x 4-1/2" x 3" G		95	.168	S.F.	4.14	8.20		12.34	17.75
0260	Cast ceramic, pressed, 4" x 8" x 1/2", unglazed	D-7	100	.160		7.40	7.45		14.85	19.80
0270	Glazed		100	.160		9.85	7.45		17.30	22.50
0280	Hand molded flooring, 4" x 8" x 3/4", unglazed		95	.168		9.75	7.85		17.60	23
0290	Glazed		95	.168		10.75	7.85		18.60	24
0300	8" hexagonal, 3/4" thick, unglazed		85	.188		9.35	8.80		18.15	24
0310	Glazed		85	.188		16.90	8.80		25.70	32.50
0400	Heavy duty industrial, cement mortar bed, 2" thick, not incl. brick	D-1	80	.200		.90	9.75		10.65	16.65
0450	Acid-proof joints, 1/4" wide	"	65	.246		1.70	12		13.70	21
0500	Pavers, 8" x 4", 1" to 1-1/4" thick, red	D-7	95	.168		4.30	7.85		12.15	17
0510	Ironspot	"	95	.168		5.30	7.85		13.15	18.10
0540	1-3/8" to 1-3/4" thick, red	D-1	95	.168		4.15	8.20		12.35	17.75
0560	Ironspot		95	.168		5.25	8.20		13.45	19
0580	2-1/4" thick, red		90	.178		4.23	8.65		12.88	18.55
0590	Ironspot		90	.178		6.55	8.65		15.20	21
0700	Paver, adobe brick, 6" x 12", 1/2" joint G		42	.381		1.42	18.55		19.97	31.50
0710	Mexican red, 12" x 12" G	1 Tilf	48	.167		2.08	8.60		10.68	15.75
0720	Saltillo, 12" x 12" G	"	48	.167		1.73	8.60		10.33	15.35
0800	For sidewalks and patios with pavers, see Section 32 14 16.10									
0870	For epoxy joints, add	D-1	600	.027	S.F.	3.22	1.30		4.52	5.65
0880	For Furan underlayment, add	"	600	.027	"	2.66	1.30		3.96	5
0890	For waxed surface, steam cleaned, add	A-1H	1000	.008	S.F.	.21	.36	.08	.65	.88
9000	Minimum labor/equipment charge	1 Bric	2	4	Job		215		215	345

09 63 40 – Stone Flooring

09 63 40.10 Marble

		Crew	Daily Output	Labor-Hours	Unit	Material	2021 Bare Costs Labor	Equipment	Total	Total Incl O&P
0010	**MARBLE**									
0020	Thin gauge tile, 12" x 6", 3/8", white Carara	D-7	60	.267	S.F.	16.90	12.45		29.35	38
0100	Travertine		60	.267		9.55	12.45		22	30
0200	12" x 12" x 3/8", thin set, floors		60	.267		12.40	12.45		24.85	33
0300	On walls		52	.308		9.95	14.35		24.30	33.50
1000	Marble threshold, 4" wide x 36" long x 5/8" thick, white		60	.267	Ea.	12.40	12.45		24.85	33
9000	Minimum labor/equipment charge		3	5.333	Job		249		249	390

09 63 40.20 Slate Tile

		Crew	Daily Output	Labor-Hours	Unit	Material	2021 Bare Costs Labor	Equipment	Total	Total Incl O&P
0010	**SLATE TILE**									
0020	Vermont, 6" x 6" x 1/4" thick, thin set	D-7	180	.089	S.F.	7.65	4.15		11.80	14.90
9000	Minimum labor/equipment charge	"	3	5.333	Job		249		249	390

323

09 64 Wood Flooring

09 64 16 – Wood Block Flooring

09 64 16.10 End Grain Block Flooring	Crew	Daily Output	Labor-Hours	Unit	Material	2021 Bare Costs Labor	Equipment	Total	Total Incl O&P
0010 **END GRAIN BLOCK FLOORING**									
0020 End grain flooring, coated, 2" thick	1 Carp	295	.027	S.F.	3.80	1.48		5.28	6.55
0400 Natural finish, 1" thick, fir		125	.064		3.93	3.50		7.43	9.85
0600 1-1/2" thick, pine		125	.064		3.85	3.50		7.35	9.80
0700 2" thick, pine		125	.064		4.46	3.50		7.96	10.45
9000 Minimum labor/equipment charge		2	4	Job		219		219	350

09 64 23 – Wood Parquet Flooring

09 64 23.10 Wood Parquet	Crew	Daily Output	Labor-Hours	Unit	Material	2021 Bare Costs Labor	Equipment	Total	Total Incl O&P
0010 **WOOD PARQUET** flooring									
5200 Parquetry, 5/16" thk, no finish, oak, plain pattern	1 Carp	160	.050	S.F.	5.95	2.74		8.69	10.90
5300 Intricate pattern		100	.080		11.85	4.38		16.23	19.95
5500 Teak, plain pattern		160	.050		5.90	2.74		8.64	10.85
5600 Intricate pattern		100	.080		10.10	4.38		14.48	18.05
5650 13/16" thick, select grade oak, plain pattern		160	.050		11.40	2.74		14.14	16.90
5700 Intricate pattern		100	.080		17.05	4.38		21.43	26
5800 Custom parquetry, including finish, plain pattern		100	.080		18	4.38		22.38	27
5900 Intricate pattern		50	.160		24.50	8.75		33.25	41
6700 Parquetry, prefinished white oak, 5/16" thick, plain pattern		160	.050		8.55	2.74		11.29	13.75
6800 Intricate pattern		100	.080		7.85	4.38		12.23	15.60
7000 Walnut or teak, parquetry, plain pattern		160	.050		9.35	2.74		12.09	14.65
7100 Intricate pattern		100	.080		16.50	4.38		20.88	25
7200 Acrylic wood parquet blocks, 12" x 12" x 5/16",									
7210 Irradiated, set in epoxy	1 Carp	160	.050	S.F.	11	2.74		13.74	16.45

09 64 29 – Wood Strip and Plank Flooring

09 64 29.10 Wood	Crew	Daily Output	Labor-Hours	Unit	Material	2021 Bare Costs Labor	Equipment	Total	Total Incl O&P
0010 **WOOD** R061110-30									
0020 Fir, vertical grain, 1" x 4", not incl. finish, grade B & better	1 Carp	255	.031	S.F.	3.34	1.72		5.06	6.40
0100 Grade C & better		255	.031		3.14	1.72		4.86	6.20
0300 Flat grain, 1" x 4", not incl. finish, grade B & better		255	.031		3.80	1.72		5.52	6.90
0400 Grade C & better		255	.031		3.67	1.72		5.39	6.75
4000 Maple, strip, 25/32" x 2-1/4", not incl. finish, select		170	.047		4.64	2.57		7.21	9.20
4100 #2 & better		170	.047		6.15	2.57		8.72	10.85
4300 33/32" x 3-1/4", not incl. finish, #1 grade		170	.047		6.20	2.57		8.77	10.95
4400 #2 & better		170	.047		5.35	2.57		7.92	10
4600 Oak, white or red, 25/32" x 2-1/4", not incl. finish									
4700 #1 common	1 Carp	170	.047	S.F.	4.23	2.57		6.80	8.75
4900 Select quartered, 2-1/4" wide		170	.047		4.66	2.57		7.23	9.25
5000 Clear		170	.047		4.24	2.57		6.81	8.75
6100 Prefinished, white oak, prime grade, 2-1/4" wide		170	.047		6.65	2.57		9.22	11.45
6200 3-1/4" wide		185	.043		6.65	2.37		9.02	11.10
6400 Ranch plank		145	.055		6.90	3.02		9.92	12.40
6500 Hardwood blocks, 9" x 9", 25/32" thick		160	.050		7.95	2.74		10.69	13.10
7400 Yellow pine, 3/4" x 3-1/8", T&G, C & better, not incl. finish		200	.040		2.02	2.19		4.21	5.70
7500 Refinish wood floor, sand, 2 coats poly, wax, soft wood	1 Clab	400	.020		.23	.89		1.12	1.66
7600 Hardwood		130	.062		.23	2.73		2.96	4.60
7800 Sanding and finishing, 2 coats polyurethane		295	.027		.23	1.20		1.43	2.17
7900 Subfloor and underlayment, see Section 06 16									
8015 Transition molding, 2-1/4" wide, 5' long	1 Carp	19.20	.417	Ea.	20.50	23		43.50	59
9000 Minimum labor/equipment charge	"	2	4	Job		219		219	350

09 64 Wood Flooring

09 64 36 – Bamboo Flooring

09 64 36.10 Flooring, Bamboo

		Crew	Daily Output	Labor-Hours	Unit	Material	2021 Bare Costs Labor	2021 Bare Costs Equipment	Total	Total Incl O&P
0010	**FLOORING, BAMBOO**									
8600	Flooring, wood, bamboo strips, unfinished, 5/8" x 4" x 3' **G**	1 Carp	255	.031	S.F.	6.05	1.72		7.77	9.45
8610	5/8" x 4" x 4' **G**		275	.029		6.10	1.59		7.69	9.25
8620	5/8" x 4" x 6' **G**		295	.027		6.90	1.48		8.38	9.95
8630	Finished, 5/8" x 4" x 3' **G**		255	.031		6.45	1.72		8.17	9.85
8640	5/8" x 4" x 4' **G**		275	.029		7	1.59		8.59	10.25
8650	5/8" x 4" x 6' **G**		295	.027		5.10	1.48		6.58	8
8660	Stair treads, unfinished, 1-1/16" x 11-1/2" x 4' **G**		18	.444	Ea.	56.50	24.50		81	101
8670	Finished, 1-1/16" x 11-1/2" x 4' **G**		18	.444		90.50	24.50		115	138
8680	Stair risers, unfinished, 5/8" x 7-1/2" x 4' **G**		18	.444		21	24.50		45.50	61.50
8690	Finished, 5/8" x 7-1/2" x 4' **G**		18	.444		41	24.50		65.50	83.50
8700	Stair nosing, unfinished, 6' long **G**		16	.500		46	27.50		73.50	94
8710	Finished, 6' long **G**		16	.500		46	27.50		73.50	94.50

09 65 Resilient Flooring

09 65 10 – Resilient Tile Underlayment

09 65 10.10 Latex Underlayment

		Crew	Daily Output	Labor-Hours	Unit	Material	2021 Bare Costs Labor	2021 Bare Costs Equipment	Total	Total Incl O&P
0010	**LATEX UNDERLAYMENT**									
3600	Latex underlayment, 1/8" thk., cementitious for resilient flooring	1 Tilf	160	.050	S.F.	1.29	2.59		3.88	5.45
4000	Liquid, fortified				Gal.	32			32	35.50

09 65 13 – Resilient Base and Accessories

09 65 13.13 Resilient Base

		Crew	Daily Output	Labor-Hours	Unit	Material	2021 Bare Costs Labor	2021 Bare Costs Equipment	Total	Total Incl O&P
0010	**RESILIENT BASE**									
0690	1/8" vinyl base, 2-1/2" H, straight or cove, standard colors	1 Tilf	315	.025	L.F.	.78	1.31		2.09	2.91
0700	4" high		315	.025		1.23	1.31		2.54	3.40
0710	6" high		315	.025		1.72	1.31		3.03	3.94
0720	Corners, 2-1/2" high		315	.025	Ea.	2.73	1.31		4.04	5.05
0730	4" high		315	.025		3.16	1.31		4.47	5.55
0740	6" high		315	.025		5.65	1.31		6.96	8.25
0800	1/8" rubber base, 2-1/2" H, straight or cove, standard colors		315	.025	L.F.	1.13	1.31		2.44	3.29
1100	4" high		315	.025		1.58	1.31		2.89	3.79
1110	6" high		315	.025		2.01	1.31		3.32	4.26
1150	Corners, 2-1/2" high		315	.025	Ea.	3.58	1.31		4.89	6
1153	4" high		315	.025		2.47	1.31		3.78	4.77
1155	6" high		315	.025		3.86	1.31		5.17	6.30
1450	For premium color/finish add					50%				
1500	Millwork profile	1 Tilf	315	.025	L.F.	4.16	1.31		5.47	6.65

09 65 13.23 Resilient Stair Treads and Risers

		Crew	Daily Output	Labor-Hours	Unit	Material	2021 Bare Costs Labor	2021 Bare Costs Equipment	Total	Total Incl O&P
0010	**RESILIENT STAIR TREADS AND RISERS**									
0300	Rubber, molded tread, 12" wide, 5/16" thick, black	1 Tilf	115	.070	L.F.	14.70	3.60		18.30	22
0400	Colors		115	.070		16.10	3.60		19.70	23.50
0600	1/4" thick, black		115	.070		13.50	3.60		17.10	20.50
0700	Colors		115	.070		16.05	3.60		19.65	23.50
0900	Grip strip safety tread, colors, 5/16" thick		115	.070		17.20	3.60		20.80	24.50
1000	3/16" thick		120	.067		18.90	3.45		22.35	26.50
1200	Landings, smooth sheet rubber, 1/8" thick		120	.067	S.F.	4.57	3.45		8.02	10.45
1300	3/16" thick		120	.067	"	6	3.45		9.45	12
1500	Nosings, 3" wide, 3/16" thick, black		140	.057	L.F.	4.77	2.95		7.72	9.85
1600	Colors		140	.057		6.80	2.95		9.75	12.05
1800	Risers, 7" high, 1/8" thick, flat		250	.032		8.75	1.65		10.40	12.25

325

09 65 Resilient Flooring

09 65 13 – Resilient Base and Accessories

09 65 13.23 Resilient Stair Treads and Risers

		Crew	Daily Output	Labor-Hours	Unit	Material	2021 Bare Costs Labor	Equipment	Total	Total Incl O&P
1900	Coved	1 Tilf	250	.032	L.F.	9.75	1.65		11.40	13.30
2100	Vinyl, molded tread, 12" wide, colors, 1/8" thick		115	.070		5.95	3.60		9.55	12.10
2200	1/4" thick		115	.070	↓	7.20	3.60		10.80	13.55
2300	Landing material, 1/8" thick		200	.040	S.F.	6.65	2.07		8.72	10.55
2400	Riser, 7" high, 1/8" thick, coved		175	.046	L.F.	2.92	2.36		5.28	6.90
2500	Tread and riser combined, 1/8" thick	↓	80	.100	"	10.20	5.15		15.35	19.30
9000	Minimum labor/equipment charge	1 Tilf	3	2.667	Job		138		138	215

09 65 13.37 Vinyl Transition Strips

		Crew	Daily Output	Labor-Hours	Unit	Material	2021 Bare Costs Labor	Equipment	Total	Total Incl O&P
0010	**VINYL TRANSITION STRIPS**									
0100	Various mats. to various mats., adhesive applied, 1/4" to 1/8"	1 Tilf	315	.025	L.F.	3.31	1.31		4.62	5.70
0105	0.08" to 1/8"		315	.025		3.37	1.31		4.68	5.75
0110	0.08" to 1/4"		315	.025		4.20	1.31		5.51	6.65
0115	1/4" to 3/8"		315	.025		5.15	1.31		6.46	7.75
0120	1/4" to 1/2"		315	.025		1.15	1.31		2.46	3.32
0125	1/4" to 0.08"		315	.025		1.21	1.31		2.52	3.38
0200	Vinyl wheeled trans. strips, carpet to var. mats., 1/4" to 1/8" x 2-1/2"		315	.025		5.30	1.31		6.61	7.85
0205	1/4" to 1/8" x 4"		315	.025		5.80	1.31		7.11	8.40
0210	Various mats. to various mats. 1/4" to 0.08" x 2-1/2"		315	.025		4.20	1.31		5.51	6.65
0215	Carpet to various materials, 1/4" to flush x 2-1/2"		315	.025		4.40	1.31		5.71	6.90
0220	1/4" to flush x 4"		315	.025		6.80	1.31		8.11	9.50
0225	Various materials to resilient, 3/8" to 1/8" x 2-1/2"		315	.025		4.40	1.31		5.71	6.90
0230	Carpet to various materials, 3/8" to 1/4" x 2-1/2"		315	.025		6.10	1.31		7.41	8.75
0235	1/4" to 1/4" x 2-1/2"		315	.025		6.80	1.31		8.11	9.50
0240	Various materials to resilient, 1/8" to 1/8" x 2-1/2"		315	.025		5.15	1.31		6.46	7.70
0245	Various materials to var. mats., 1/8" to flush x 2-1/2"		315	.025		3.44	1.31		4.75	5.85
0250	3/8" to flush x 4"		315	.025		6.90	1.31		8.21	9.65
0255	1/2" to flush x 4"		315	.025		9.30	1.31		10.61	12.25
0260	Various materials to resilient, 1/8" to 0.08" x 2-1/2"		315	.025		3.98	1.31		5.29	6.45
0265	0.08" to 0.08" x 2-1/2"		315	.025		3.78	1.31		5.09	6.20
0270	3/8" to 0.08" x 2-1/2"	↓	315	.025	↓	3.78	1.31		5.09	6.20

09 65 16 – Resilient Sheet Flooring

09 65 16.10 Rubber and Vinyl Sheet Flooring

		Crew	Daily Output	Labor-Hours	Unit	Material	2021 Bare Costs Labor	Equipment	Total	Total Incl O&P
0010	**RUBBER AND VINYL SHEET FLOORING**									
5500	Linoleum, sheet goods **G**	1 Tilf	360	.022	S.F.	3.34	1.15		4.49	5.45
5900	Rubber, sheet goods, 36" wide, 1/8" thick		120	.067		10.05	3.45		13.50	16.45
5950	3/16" thick		100	.080		10.05	4.14		14.19	17.50
6000	1/4" thick		90	.089		11.85	4.60		16.45	20
8000	Vinyl sheet goods, backed, .065" thick, plain pattern/colors		250	.032		4.61	1.65		6.26	7.65
8050	Intricate pattern/colors		200	.040		4.96	2.07		7.03	8.70
8100	.080" thick, plain pattern/colors		230	.035		4.17	1.80		5.97	7.40
8150	Intricate pattern/colors		200	.040		5.70	2.07		7.77	9.55
8200	.125" thick, plain pattern/colors		230	.035		3.80	1.80		5.60	7
8250	Intricate pattern/colors		200	.040	↓	7.45	2.07		9.52	11.40
8400	For welding seams, add		100	.080	L.F.	.14	4.14		4.28	6.60
8450	For integral cove base, add	↓	175	.046	"	9.55	2.36		11.91	14.20
8700	Adhesive cement, 1 gallon per 200 to 300 S.F.				Gal.	33			33	36
8800	Asphalt primer, 1 gallon per 300 S.F.				Gal.	15.20			15.20	16.70
8900	Emulsion, 1 gallon per 140 S.F.				"	21			21	23

09 65 Resilient Flooring

09 65 19 – Resilient Tile Flooring

09 65 19.19 Vinyl Composition Tile Flooring

		Crew	Daily Output	Labor-Hours	Unit	Material	2021 Bare Costs Labor	2021 Bare Costs Equipment	Total	Total Incl O&P
0010	**VINYL COMPOSITION TILE FLOORING**									
7000	Vinyl composition tile, 12" x 12", 1/16" thick	1 Tilf	500	.016	S.F.	.88	.83		1.71	2.26
7050	Embossed		500	.016		2.78	.83		3.61	4.35
7100	Marbleized		500	.016		2.78	.83		3.61	4.35
7150	Solid		500	.016		3.59	.83		4.42	5.25
7200	3/32" thick, embossed		500	.016		1.52	.83		2.35	2.96
7250	Marbleized		500	.016		3.20	.83		4.03	4.81
7300	Solid		500	.016		1.33	.83		2.16	2.75
7350	1/8" thick, marbleized		500	.016		2.15	.83		2.98	3.66
7400	Solid		500	.016		1.81	.83		2.64	3.28
7450	Conductive		500	.016		5.90	.83		6.73	7.80

09 65 19.23 Vinyl Tile Flooring

		Crew	Daily Output	Labor-Hours	Unit	Material	2021 Bare Costs Labor	2021 Bare Costs Equipment	Total	Total Incl O&P
0010	**VINYL TILE FLOORING**									
7500	Vinyl tile, 12" x 12", 3/32" thick, standard colors/patterns	1 Tilf	500	.016	S.F.	3.51	.83		4.34	5.15
7550	1/8" thick, standard colors/patterns		500	.016		1.28	.83		2.11	2.70
7600	1/8" thick, premium colors/patterns		500	.016		6.70	.83		7.53	8.65
7650	Solid colors		500	.016		3.22	.83		4.05	4.83
7700	Marbleized or Travertine pattern		500	.016		6.35	.83		7.18	8.30
7750	Florentine pattern		500	.016		6.65	.83		7.48	8.65
7800	Premium colors/patterns		500	.016		8.25	.83		9.08	10.40
9500	Minimum labor/equipment charge		4	2	Job		103		103	161

09 65 19.33 Rubber Tile Flooring

		Crew	Daily Output	Labor-Hours	Unit	Material	2021 Bare Costs Labor	2021 Bare Costs Equipment	Total	Total Incl O&P
0010	**RUBBER TILE FLOORING**									
6050	Rubber tile, marbleized colors, 12" x 12", 1/8" thick	1 Tilf	400	.020	S.F.	7	1.03		8.03	9.30
6100	3/16" thick		400	.020		8.65	1.03		9.68	11.10
6300	Special tile, plain colors, 1/8" thick		400	.020		8.60	1.03		9.63	11.10
6350	3/16" thick		400	.020		10.15	1.03		11.18	12.80

09 65 33 – Conductive Resilient Flooring

09 65 33.10 Conductive Rubber and Vinyl Flooring

		Crew	Daily Output	Labor-Hours	Unit	Material	2021 Bare Costs Labor	2021 Bare Costs Equipment	Total	Total Incl O&P
0010	**CONDUCTIVE RUBBER AND VINYL FLOORING**									
1700	Conductive flooring, rubber tile, 1/8" thick	1 Tilf	315	.025	S.F.	7.60	1.31		8.91	10.45
1800	Homogeneous vinyl tile, 1/8" thick	"	315	.025	"	7.15	1.31		8.46	9.90

09 65 66 – Resilient Athletic Flooring

09 65 66.10 Resilient Athletic Flooring

		Crew	Daily Output	Labor-Hours	Unit	Material	2021 Bare Costs Labor	2021 Bare Costs Equipment	Total	Total Incl O&P
0010	**RESILIENT ATHLETIC FLOORING**									
1000	Recycled rubber rolled goods, for weight rooms, 3/8" thk.	1 Tilf	315	.025	S.F.	2.89	1.31		4.20	5.25
1050	Interlocking 2' x 2' squares, rubber, 1/4" thk.		310	.026		12.75	1.33		14.08	16.15
1055	5/16" thk.		310	.026		2.54	1.33		3.87	4.87
1060	3/8" thk.		300	.027		3.07	1.38		4.45	5.55
1065	1/2" thk.	1 Tilf	310	.026	S.F.	3.83	1.33		5.16	6.30
2000	Vinyl sheet flooring, 1/4" thk.	"	315	.025	"	5	1.31		6.31	7.55

09 66 13 – Portland Cement Terrazzo Flooring

09 66 13.10 Portland Cement Terrazzo	Crew	Daily Output	Labor-Hours	Unit	Material	2021 Bare Costs Labor	Equipment	Total	Total Incl O&P
0010 **PORTLAND CEMENT TERRAZZO**, cast-in-place									
0020 Cove base, 6" high, 16 ga. zinc	1 Mstz	20	.400	L.F.	3.85	20.50		24.35	36.50
0100 Curb, 6" high and 6" wide		6	1.333		6.35	69		75.35	115
0300 Divider strip for floors, 14 ga., 1-1/4" deep, zinc		375	.021		1.40	1.10		2.50	3.26
0400 Brass		375	.021		2.82	1.10		3.92	4.82
0600 Heavy top strip 1/4" thick, 1-1/4" deep, zinc		300	.027		2.14	1.38		3.52	4.50
1200 For thin set floors, 16 ga., 1/2" x 1/2", zinc		350	.023		1.46	1.18		2.64	3.46
1500 Floor, bonded to concrete, 1-3/4" thick, gray cement	J-3	75	.213	S.F.	3.86	10.15	3.45	17.46	24
1600 White cement, mud set		75	.213		4.55	10.15	3.45	18.15	24.50
1800 Not bonded, 3" total thickness, gray cement		70	.229		4.75	10.90	3.70	19.35	26.50
1900 White cement, mud set		70	.229		5.50	10.90	3.70	20.10	27
9000 Minimum labor/equipment charge	1 Mstz	1	8	Job		415		415	645

09 66 16 – Terrazzo Floor Tile

09 66 16.10 Tile or Terrazzo Base

	Crew	Daily Output	Labor-Hours	Unit	Material	2021 Bare Costs Labor	Equipment	Total	Total Incl O&P
0010 **TILE OR TERRAZZO BASE**									
0020 Scratch coat only	1 Mstz	150	.053	S.F.	.48	2.76		3.24	4.84
0500 Scratch and brown coat only		75	.107	"	.96	5.50		6.46	9.65
9000 Minimum labor/equipment charge		1	8	Job		415		415	645

09 66 16.13 Portland Cement Terrazzo Floor Tile

	Crew	Daily Output	Labor-Hours	Unit	Material	2021 Bare Costs Labor	Equipment	Total	Total Incl O&P
0010 **PORTLAND CEMENT TERRAZZO FLOOR TILE**									
1200 Floor tiles, non-slip, 1" thick, 12" x 12"	D-1	60	.267	S.F.	27.50	13		40.50	51
1300 1-1/4" thick, 12" x 12"		60	.267		25.50	13		38.50	49
1500 16" x 16"		50	.320		28	15.60		43.60	56
1600 1-1/2" thick, 16" x 16"		45	.356		26	17.30		43.30	56.50
1800 For Venetian terrazzo, add					7.75			7.75	8.55
1900 For white cement, add					.71			.71	.78

09 66 16.30 Terrazzo, Precast

	Crew	Daily Output	Labor-Hours	Unit	Material	2021 Bare Costs Labor	Equipment	Total	Total Incl O&P
0010 **TERRAZZO, PRECAST**									
0020 Base, 6" high, straight	1 Mstz	70	.114	L.F.	12.50	5.90		18.40	23
0100 Cove		60	.133		16.95	6.90		23.85	29.50
0300 8" high, straight		60	.133		15.95	6.90		22.85	28.50
0400 Cove		50	.160		24.50	8.30		32.80	40
0600 For white cement, add					.60			.60	.66
0700 For 16 ga. zinc toe strip, add					2.35			2.35	2.59
0900 Curbs, 4" x 4" high	1 Mstz	40	.200		43.50	10.35		53.85	63.50
1000 8" x 8" high	"	30	.267		47.50	13.80		61.30	74
2400 Stair treads, 1-1/2" thick, non-slip, three line pattern	2 Mstz	70	.229		54	11.85		65.85	78
2500 Nosing and two lines	2 Mstz	70	.229	L.F.	54	11.85		65.85	78
2700 2" thick treads, straight		60	.267		62.50	13.80		76.30	90.50
2800 Curved		50	.320		84.50	16.55		101.05	119
3000 Stair risers, 1" thick, to 6" high, straight sections		60	.267		16.25	13.80		30.05	39.50
3100 Cove		50	.320		20.50	16.55		37.05	48.50
3300 Curved, 1" thick, to 6" high, vertical		48	.333		28.50	17.25		45.75	58
3400 Cove		38	.421		42	22		64	80
3600 Stair tread and riser, single piece, straight, smooth surface		60	.267		66.50	13.80		80.30	95
3700 Non skid surface		40	.400		86	20.50		106.50	128
3900 Curved tread and riser, smooth surface		40	.400		95.50	20.50		116	138
4000 Non skid surface		32	.500		117	26		143	170
4200 Stair stringers, notched, 1" thick		25	.640		38	33		71	93.50
4300 2" thick		22	.727		45	37.50		82.50	109
4500 Stair landings, structural, non-slip, 1-1/2" thick		85	.188	S.F.	42.50	9.75		52.25	62
4600 3" thick		75	.213		59	11.05		70.05	82.50

09 66 Terrazzo Flooring

09 66 16 – Terrazzo Floor Tile

09 66 16.30 Terrazzo, Precast

		Crew	Daily Output	Labor-Hours	Unit	Material	2021 Bare Costs Labor	Equipment	Total	Total Incl O&P
4800	Wainscot, 12" x 12" x 1" tiles	1 Mstz	12	.667	S.F.	9.50	34.50		44	64.50
4900	16" x 16" x 1-1/2" tiles	"	8	1	↓	18.75	52		70.75	102
9500	Minimum labor/equipment charge	1 Tilf	2	4	Job		207		207	325

09 66 23 – Resinous Matrix Terrazzo Flooring

09 66 23.16 Epoxy-Resin Terrazzo Flooring

		Crew	Daily Output	Labor-Hours	Unit	Material	2021 Bare Costs Labor	Equipment	Total	Total Incl O&P
0010	**EPOXY-RESIN TERRAZZO FLOORING**									
1800	Epoxy terrazzo, 1/4" thick, chemical resistant, granite chips	J-3	200	.080	S.F.	6.05	3.81	1.29	11.15	14
1900	Recycled porcelain	"	150	.107	"	10.25	5.10	1.73	17.08	21

09 67 Fluid-Applied Flooring

09 67 13 – Elastomeric Liquid Flooring

09 67 13.13 Elastomeric Liquid Flooring

		Crew	Daily Output	Labor-Hours	Unit	Material	2021 Bare Costs Labor	Equipment	Total	Total Incl O&P
0010	**ELASTOMERIC LIQUID FLOORING**									
0020	Cementitious acrylic, 1/4" thick	C-6	520	.092	S.F.	1.79	4.24	.10	6.13	8.85
0200	Methyl methacrylate, 1/4" thick	C-8A	3000	.016	"	7.40	.76		8.16	9.35

09 67 23 – Resinous Flooring

09 67 23.23 Resinous Flooring

		Crew	Daily Output	Labor-Hours	Unit	Material	2021 Bare Costs Labor	Equipment	Total	Total Incl O&P
0010	**RESINOUS FLOORING**									
1200	Heavy duty epoxy topping, 1/4" thick,									
1300	500 to 1,000 S.F.	C-6	420	.114	S.F.	6.25	5.25	.13	11.63	15.35
1500	1,000 to 2,000 S.F.		450	.107		5.30	4.90	.12	10.32	13.80
1600	Over 10,000 S.F.	↓	480	.100	↓	5	4.60	.11	9.71	12.90

09 67 26 – Quartz Flooring

09 67 26.26 Quartz Flooring

		Crew	Daily Output	Labor-Hours	Unit	Material	2021 Bare Costs Labor	Equipment	Total	Total Incl O&P
0010	**QUARTZ FLOORING**									
0600	Epoxy, with colored quartz chips, broadcast, 3/8" thick	C-6	675	.071	S.F.	3.32	3.27	.08	6.67	8.95
0700	1/2" thick		490	.098		4.45	4.50	.11	9.06	12.15
0900	Troweled, minimum	↓	560	.086	↓	3.56	3.94	.10	7.60	10.30
1000	Maximum	C-6	480	.100	S.F.	6.15	4.60	.11	10.86	14.20

09 68 Carpeting

09 68 05 – Carpet Accessories

09 68 05.11 Flooring Transition Strip

		Crew	Daily Output	Labor-Hours	Unit	Material	2021 Bare Costs Labor	Equipment	Total	Total Incl O&P
0010	**FLOORING TRANSITION STRIP**									
0107	Clamp down brass divider, 12' strip, vinyl to carpet	1 Tilf	31.25	.256	Ea.	14.25	13.25		27.50	36
0117	Vinyl to hard surface	"	31.25	.256	"	14.25	13.25		27.50	36

09 68 10 – Carpet Pad

09 68 10.10 Commercial Grade Carpet Pad

		Crew	Daily Output	Labor-Hours	Unit	Material	2021 Bare Costs Labor	Equipment	Total	Total Incl O&P
0010	**COMMERCIAL GRADE CARPET PAD**									
9000	Sponge rubber pad, 20 oz./sq. yd.	1 Tilf	150	.053	S.Y.	4.90	2.76		7.66	9.70
9100	40 to 62 oz./sq. yd.		150	.053		8.15	2.76		10.91	13.30
9200	Felt pad, 20 oz./sq. yd.		150	.053		6.40	2.76		9.16	11.30
9300	32 to 56 oz./sq. yd.		150	.053		12.25	2.76		15.01	17.80
9400	Bonded urethane pad, 2.7 density		150	.053		6	2.76		8.76	10.90
9500	13.0 density		150	.053		8	2.76		10.76	13.10
9600	Prime urethane pad, 2.7 density		150	.053		3.67	2.76		6.43	8.35
9700	13.0 density	↓	150	.053		8.15	2.76		10.91	13.30

09 68 13 – Tile Carpeting

09 68 13.10 Carpet Tile	Crew	Daily Output	Labor-Hours	Unit	Material	2021 Bare Costs Labor	Equipment	Total	Total Incl O&P
0010 **CARPET TILE**									
0100 Tufted nylon, 18" x 18", hard back, 20 oz.	1 Tilf	80	.100	S.Y.	26.50	5.15		31.65	37
0110 26 oz.		80	.100		25	5.15		30.15	35.50
0200 Cushion back, 20 oz.		80	.100		11.60	5.15		16.75	21
0210 26 oz.		80	.100		37	5.15		42.15	48.50
1100 Tufted, 24" x 24", hard back, 24 oz. nylon		80	.100		32.50	5.15		37.65	43.50
1180 35 oz.		80	.100		22.50	5.15		27.65	32.50
5060 42 oz.		80	.100		44	5.15		49.15	56.50
6000 Electrostatic dissapative carpet tile, 24" x 24", 24 oz.		80	.100		39	5.15		44.15	50.50
6100 Electrostatic dissapative carpet tile for access floors, 24" x 24", 24 oz.	▼	80	.100	▼	48.50	5.15		53.65	61.50

09 68 16 – Sheet Carpeting

09 68 16.10 Sheet Carpet

09 68 16.10 Sheet Carpet	Crew	Daily Output	Labor-Hours	Unit	Material	2021 Bare Costs Labor	Equipment	Total	Total Incl O&P
0010 **SHEET CARPET**									
0700 Nylon, level loop, 26 oz., light to medium traffic	1 Tilf	75	.107	S.Y.	16	5.50		21.50	26
0720 28 oz., light to medium traffic		75	.107		32.50	5.50		38	44
0900 32 oz., medium traffic		75	.107		41.50	5.50		47	54
1100 40 oz., medium to heavy traffic		75	.107		56.50	5.50		62	70.50
2920 Nylon plush, 30 oz., medium traffic		75	.107		15.20	5.50		20.70	25.50
3000 36 oz., medium traffic		75	.107		39	5.50		44.50	51
3100 42 oz., medium to heavy traffic		70	.114		20.50	5.90		26.40	31.50
3200 46 oz., medium to heavy traffic		70	.114		54	5.90		59.90	68.50
3300 54 oz., heavy traffic		70	.114		32	5.90		37.90	44.50
4110 Wool, level loop, 40 oz., medium traffic	▼	70	.114	▼	118	5.90		123.90	138
4500 50 oz., medium to heavy traffic	1 Tilf	70	.114	S.Y.	110	5.90		115.90	130
4700 Patterned, 32 oz., medium to heavy traffic		70	.114		100	5.90		105.90	119
4900 48 oz., heavy traffic	▼	70	.114	▼	110	5.90		115.90	130
5000 For less than full roll (approx. 1500 S.F.), add					25%				
5100 For small rooms, less than 12' wide, add						25%			
5200 For large open areas (no cuts), deduct						25%			
5600 For bound carpet baseboard, add	1 Tilf	300	.027	L.F.	2.04	1.38		3.42	4.39
5610 For stairs, not incl. price of carpet, add	"	30	.267	Riser		13.80		13.80	21.50
5620 For borders and patterns, add to labor						18%			
8950 For tackless, stretched installation, add padding from 09 68 10.10 to above									
9850 For brand-named specific fiber, add				S.Y.	25%				
9900 Carpet cleaning machine, rent				Day				38	42
9910 Minimum labor/equipment charge	1 Tilf	3	2.667	Job		138		138	215

09 68 20 – Athletic Carpet

09 68 20.10 Indoor Athletic Carpet

09 68 20.10 Indoor Athletic Carpet	Crew	Daily Output	Labor-Hours	Unit	Material	2021 Bare Costs Labor	Equipment	Total	Total Incl O&P
0010 **INDOOR ATHLETIC CARPET**									
3700 Polyethylene, in rolls, no base incl., landscape surfaces	1 Tilf	275	.029	S.F.	5.25	1.50		6.75	8.15
3800 Nylon action surface, 1/8" thick		275	.029		4.44	1.50		5.94	7.25
3900 1/4" thick		275	.029		6.40	1.50		7.90	9.40
4000 3/8" thick		275	.029		8.05	1.50		9.55	11.20
5500 Polyvinyl chloride, sheet goods for gyms, 1/4" thick		80	.100		6.70	5.15		11.85	15.45
5600 3/8" thick	▼	60	.133	▼	9.80	6.90		16.70	21.50

09 69 Access Flooring

09 69 13 – Rigid-Grid Access Flooring

09 69 13.10 Access Floors

		Crew	Daily Output	Labor-Hours	Unit	Material	2021 Bare Costs Labor	Equipment	Total	Total Incl O&P
0010	**ACCESS FLOORS**									
0015	Access floor pkg. including panel, pedestal, & stringers 1500 lbs. load									
0100	Package pricing, conc. fill panels, no fin., 6" ht.	4 Carp	750	.043	S.F.	15	2.33		17.33	20
0105	12" ht.		750	.043		15.20	2.33		17.53	20.50
0110	18" ht.		750	.043		15.40	2.33		17.73	20.50
0115	24" ht.		750	.043		15.55	2.33		17.88	21
0120	Package pricing, steel panels, no fin., 6" ht.		750	.043		18	2.33		20.33	23.50
0125	12" ht.		750	.043		18.05	2.33		20.38	23.50
0130	18" ht.		750	.043		18.25	2.33		20.58	23.50
0135	24" ht.		750	.043		18.45	2.33		20.78	24
0140	Package pricing, wood core panels, no fin., 6" ht.		750	.043		12.20	2.33		14.53	17.10
0145	12" ht.		750	.043		12.40	2.33		14.73	17.30
0150	18" ht.		750	.043		12.55	2.33		14.88	17.50
0155	24" ht.		750	.043		12.75	2.33		15.08	17.75
0160	Pkg. pricing, conc. fill pnls, no stringers, no fin., 6" ht, 1250 lbs. load		700	.046		17.85	2.50		20.35	23.50
0165	12" ht.	4 Carp	700	.046	S.F.	18.10	2.50		20.60	24
0170	18" ht.		700	.046		18.30	2.50		20.80	24
0175	24" ht.		700	.046		18.50	2.50		21	24.50
0250	Panels, 2' x 2' conc. fill, no fin.	2 Carp	500	.032		23	1.75		24.75	28
0255	With 1/8" high pressure laminate		500	.032		56.50	1.75		58.25	65
0260	Metal panels with 1/8" high pressure laminate		500	.032		66.50	1.75		68.25	76.50
0265	Wood core panels with 1/8" high pressure laminate		500	.032		44	1.75		45.75	51
0400	Aluminum panels, no fin.		500	.032		35	1.75		36.75	41.50
0600	For carpet covering, add					9.20			9.20	10.15
0700	For vinyl floor covering, add					9.80			9.80	10.80
0900	For high pressure laminate covering, add					8.10			8.10	8.90
0910	For snap on stringer system, add	2 Carp	1000	.016		1.45	.88		2.33	2.99
1000	Machine cutouts after initial installation	1 Carp	50	.160	Ea.	19.75	8.75		28.50	35.50
1050	Pedestals, 6" to 12"	2 Carp	85	.188		9.20	10.30		19.50	26.50
1100	Air conditioning grilles, 4" x 12"	1 Carp	17	.471		72.50	25.50		98	121
1150	4" x 18"	"	14	.571		99	31.50		130.50	159
1200	Approach ramps, steel	2 Carp	60	.267	S.F.	29.50	14.60		44.10	55.50
1300	Aluminum	"	40	.400	"	35	22		57	73.50
1500	Handrail, 2 rail, aluminum	1 Carp	15	.533	L.F.	126	29		155	185

09 72 Wall Coverings

09 72 13 – Cork Wall Coverings

09 72 13.10 Covering, Cork Wall

		Crew	Daily Output	Labor-Hours	Unit	Material	2021 Bare Costs Labor	Equipment	Total	Total Incl O&P
0010	**COVERING, CORK WALL**									
0600	Cork tiles, light or dark, 12" x 12" x 3/16"	1 Pape	240	.033	S.F.	4.04	1.54		5.58	6.90
0700	5/16" thick		235	.034		3.38	1.58		4.96	6.20
0900	1/4" basket weave		240	.033		3.12	1.54		4.66	5.85
1000	1/2" natural, non-directional pattern		240	.033		6.60	1.54		8.14	9.70
1100	3/4" natural, non-directional pattern		240	.033		12.35	1.54		13.89	16
1200	Granular surface, 12" x 36", 1/2" thick		385	.021		1.35	.96		2.31	3.01
1300	1" thick		370	.022		1.76	1		2.76	3.52
1500	Polyurethane coated, 12" x 12" x 3/16" thick		240	.033		4.26	1.54		5.80	7.15
1600	5/16" thick		235	.034		6.10	1.58		7.68	9.20
1800	Cork wallpaper, paperbacked, natural		480	.017		1.65	.77		2.42	3.04
1900	Colors		480	.017		2.14	.77		2.91	3.57

09 72 16 – Vinyl-Coated Fabric Wall Coverings

09 72 16.13 Flexible Vinyl Wall Coverings		Crew	Daily Output	Labor-Hours	Unit	Material	2021 Bare Costs Labor	Equipment	Total	Total Incl O&P
0010	**FLEXIBLE VINYL WALL COVERINGS**									
3000	Vinyl wall covering, fabric-backed, lightweight, type 1 (12-15 oz./S.Y.)	1 Pape	640	.013	S.F.	1.57	.58		2.15	2.65
3300	Medium weight, type 2 (20-24 oz./S.Y.)		480	.017		1.23	.77		2	2.57
3400	Heavy weight, type 3 (28 oz./S.Y.)		435	.018		1.87	.85		2.72	3.41
3600	Adhesive, 5 gal. lots (18 S.Y./gal.)				Gal.	8.30			8.30	9.15

09 72 16.16 Rigid-Sheet Vinyl Wall Coverings

		Crew	Daily Output	Labor-Hours	Unit	Material	Labor	Equipment	Total	Total Incl O&P
0010	**RIGID-SHEET VINYL WALL COVERINGS**									
0100	Acrylic, modified, semi-rigid PVC, .028" thick	2 Carp	330	.048	S.F.	1.54	2.65		4.19	5.90
0110	.040" thick	"	320	.050	"	2.04	2.74		4.78	6.60

09 72 19 – Textile Wall Coverings

09 72 19.10 Textile Wall Covering

		Crew	Daily Output	Labor-Hours	Unit	Material	Labor	Equipment	Total	Total Incl O&P
0010	**TEXTILE WALL COVERING**, including sizing; add 10-30% waste @ takeoff									
0020	Silk	1 Pape	640	.013	S.F.	6.10	.58		6.68	7.60
0030	Cotton		640	.013		4.75	.58		5.33	6.15
0040	Linen		640	.013		4.36	.58		4.94	5.70
0050	Blend		640	.013		3.28	.58		3.86	4.53
0090	Grass cloth, natural fabric ☐G		400	.020		1.82	.93		2.75	3.47
0100	Grass cloths with lining paper ☐G		400	.020		1.26	.93		2.19	2.86
0110	Premium texture/color ☐G		350	.023		3.47	1.06		4.53	5.50

09 72 20 – Natural Fiber Wall Covering

09 72 20.10 Natural Fiber Wall Covering

		Crew	Daily Output	Labor-Hours	Unit	Material	Labor	Equipment	Total	Total Incl O&P
0010	**NATURAL FIBER WALL COVERING**, including sizing; add 10-30% waste @ takeoff									
0015	Bamboo	1 Pape	640	.013	S.F.	2.72	.58		3.30	3.91
0030	Burlap		640	.013		2.72	.58		3.30	3.91
0045	Jute		640	.013		2.36	.58		2.94	3.52
0060	Sisal		640	.013		1.74	.58		2.32	2.83

09 72 23 – Wallpapering

09 72 23.10 Wallpaper

		Crew	Daily Output	Labor-Hours	Unit	Material	Labor	Equipment	Total	Total Incl O&P
0010	**WALLPAPER** including sizing; add 10-30% waste @ takeoff									
0050	Aluminum foil	1 Pape	275	.029	S.F.	1.07	1.35		2.42	3.31
0100	Copper sheets, .025" thick, vinyl backing		240	.033		5.70	1.54		7.24	8.70
0300	Phenolic backing		240	.033		7.40	1.54		8.94	10.60
2400	Gypsum-based, fabric-backed, fire resistant									
2500	for masonry walls, 21 oz./S.Y.	1 Pape	800	.010	S.F.	.89	.46		1.35	1.71
2600	Average		720	.011		1.31	.51		1.82	2.25
2700	Small quantities		640	.013		.89	.58		1.47	1.90
3700	Wallpaper, average workmanship, solid pattern, low cost paper		640	.013		.60	.58		1.18	1.58
3900	Basic patterns (matching required), avg. cost paper		535	.015		1.35	.69		2.04	2.59
4000	Paper at $85 per double roll, quality workmanship		435	.018		2.34	.85		3.19	3.92
5990	Wallpaper removal, 1 layer		800	.010		.02	.46		.48	.75
6000	Wallpaper removal, 3 layer		400	.020		.07	.93		1	1.54
9000	Minimum labor/equipment charge		2	4	Job		185		185	293

09 74 Flexible Wood Sheets

09 74 16 – Flexible Wood Veneers

09 74 16.10 Veneer, Flexible Wood	Crew	Daily Output	Labor-Hours	Unit	Material	2021 Bare Costs Labor	Equipment	Total	Total Incl O&P
0010 **VENEER, FLEXIBLE WOOD**									
0100 Flexible wood veneer, 1/32" thick, plain woods	1 Pape	100	.080	S.F.	2.55	3.70		6.25	8.65
0110 Exotic woods	1 Pape	95	.084	S.F.	3.85	3.90		7.75	10.40

09 77 Special Wall Surfacing

09 77 30 – Fiberglass Reinforced Panels

09 77 30.10 Fiberglass Reinforced Plastic Panels

	Crew	Daily Output	Labor-Hours	Unit	Material	2021 Bare Costs Labor	Equipment	Total	Total Incl O&P
0010 **FIBERGLASS REINFORCED PLASTIC PANELS**, .090" thick									
0020 On walls, adhesive mounted, embossed surface	2 Carp	640	.025	S.F.	1.20	1.37		2.57	3.50
0030 Smooth surface		640	.025		1.48	1.37		2.85	3.81
0040 Fire rated, embossed surface		640	.025		2.11	1.37		3.48	4.50
0050 Nylon rivet mounted, on drywall, embossed surface		480	.033		1.19	1.82		3.01	4.21
0060 Smooth surface		480	.033		1.49	1.82		3.31	4.54
0070 Fire rated, embossed surface		480	.033		2.11	1.82		3.93	5.20
0080 On masonry, embossed surface		320	.050		1.19	2.74		3.93	5.65
0090 Smooth surface		320	.050		1.49	2.74		4.23	6
0100 Fire rated, embossed surface		320	.050		2.11	2.74		4.85	6.70
0110 Nylon rivet and adhesive mounted, on drywall, embossed surface		240	.067		1.37	3.65		5.02	7.30
0120 Smooth surface		240	.067		1.46	3.65		5.11	7.40
0130 Fire rated, embossed surface		240	.067		2.37	3.65		6.02	8.40
0140 On masonry, embossed surface		190	.084		1.37	4.61		5.98	8.85
0150 Smooth surface		190	.084		1.46	4.61		6.07	8.95
0160 Fire rated, embossed surface		190	.084		2.37	4.61		6.98	9.95
0170 For moldings, add	1 Carp	250	.032	L.F.	.25	1.75		2	3.07
0180 On ceilings, for lay in grid system, embossed surface		400	.020	S.F.	1.20	1.09		2.29	3.06
0190 Smooth surface		400	.020		1.48	1.09		2.57	3.37
0200 Fire rated, embossed surface		400	.020		2.11	1.09		3.20	4.06

09 77 43 – Panel Systems

09 77 43.20 Slatwall Panels and Accessories

	Crew	Daily Output	Labor-Hours	Unit	Material	2021 Bare Costs Labor	Equipment	Total	Total Incl O&P
0010 **SLATWALL PANELS AND ACCESSORIES**									
0100 Slatwall panel, 4' x 8' x 3/4" T, MDF, paint grade	1 Carp	500	.016	S.F.	1.58	.88		2.46	3.13
0110 Melamine finish		500	.016		2.13	.88		3.01	3.73
0120 High pressure plastic laminate finish		500	.016		3.49	.88		4.37	5.25
0125 Wood veneer		500	.016		4.10	.88		4.98	5.90
0130 Aluminum channel inserts, add					2.66			2.66	2.93
0200 Accessories, corner forms, 8' L				L.F.	6.20			6.20	6.85
0210 T-connector, 8' L					7.90			7.90	8.70
0220 J-mold, 8' L					1.23			1.23	1.35
0230 Edge cap, 8' L					1.48			1.48	1.63
0240 Finish end cap, 8' L					3.39			3.39	3.73
0300 Display hook, metal, 4" L				Ea.	.43			.43	.47
0310 6" L					.48			.48	.53
0320 8" L					.52			.52	.57
0330 10" L					.50			.50	.55
0340 12" L				Ea.	.47			.47	.52
0350 Acrylic, 4" L					.15			.15	.17
0360 6" L					.17			.17	.19
0400 Waterfall hanger, metal, 12"-16"					3.82			3.82	4.20
0410 Acrylic					10.30			10.30	11.35
0500 Shelf bracket, metal, 8"					1.83			1.83	2.01

09 77 Special Wall Surfacing

09 77 43 – Panel Systems

09 77 43.20 Slatwall Panels and Accessories	Crew	Daily Output	Labor-Hours	Unit	Material	2021 Bare Costs Labor	Equipment	Total	Total Incl O&P
0510	10"				Ea.	1.91			1.91	2.10
0520	12"					2.12			2.12	2.33
0530	14"					2.34			2.34	2.57
0540	16"					2.99			2.99	3.29
0550	Acrylic, 8"					4.02			4.02	4.42
0560	10"					4.26			4.26	4.69
0570	12"					4.64			4.64	5.10
0580	14"					5.30			5.30	5.85
0600	Shelf, acrylic, 12" x 16" x 1/4"					23			23	25
0610	12" x 24" x 1/4"					36.50			36.50	40

09 81 Acoustic Insulation

09 81 13 – Acoustic Board Insulation

09 81 13.10 Acoustic Board Insulation

		Crew	Daily Output	Labor-Hours	Unit	Material	Labor	Equipment	Total	Total Incl O&P
0010	**ACOUSTIC BOARD INSULATION**									
0020	Cellulose fiber board, 1/2" thk.	1 Carp	800	.010	S.F.	.84	.55		1.39	1.79

09 81 16 – Acoustic Blanket Insulation

09 81 16.10 Sound Attenuation Blanket

		Crew	Daily Output	Labor-Hours	Unit	Material	Labor	Equipment	Total	Total Incl O&P
0010	**SOUND ATTENUATION BLANKET**									
0020	Blanket, 1" thick	1 Carp	925	.009	S.F.	.28	.47		.75	1.06
0500	1-1/2" thick		920	.009		.36	.48		.84	1.16
1000	2" thick		915	.009		.38	.48		.86	1.18
1500	3" thick		910	.009		.57	.48		1.05	1.40
2000	Wall hung, STC 18-21, 1" thick, 4' x 20'	2 Carp	22	.727	Ea.	110	40		150	185
2010	10' x 20'	"	19	.842		276	46		322	380
2020	Wall hung, STC 27-28, 3" thick, 4' x 20'	3 Carp	12	2		580	109		689	815
2030	10' x 20'	"	9	2.667		1,450	146		1,596	1,825
3400	Urethane plastic foam, open cell, on wall, 2" thick	2 Carp	2050	.008	S.F.	3.12	.43		3.55	4.11
3500	3" thick		1550	.010		4.15	.56		4.71	5.45
3600	4" thick		1050	.015		5.80	.83		6.63	7.75
3700	On ceiling, 2" thick		1700	.009		3.11	.51		3.62	4.24
3800	3" thick		1300	.012		4.15	.67		4.82	5.65
3900	4" thick		900	.018		5.80	.97		6.77	7.95
9000	Minimum labor/equipment charge	1 Carp	5	1.600	Job		87.50		87.50	139

09 84 Acoustic Room Components

09 84 13 – Fixed Sound-Absorptive Panels

09 84 13.10 Fixed Panels

		Crew	Daily Output	Labor-Hours	Unit	Material	Labor	Equipment	Total	Total Incl O&P
0010	**FIXED PANELS** Perforated steel facing, painted with									
0100	Fiberglass or mineral filler, no backs, 2-1/4" thick, modular									
0200	space units, ceiling or wall hung, white or colored	1 Carp	100	.080	S.F.	11.85	4.38		16.23	20
0300	Fiberboard sound deadening panels, 1/2" thick	"	600	.013	"	.30	.73		1.03	1.49
0500	Fiberglass panels, 4' x 8' x 1" thick, with									
0600	glass cloth face for walls, cemented	1 Carp	155	.052	S.F.	9.15	2.82		11.97	14.55
0700	1-1/2" thick, dacron covered, inner aluminum frame,									
0710	wall mounted	1 Carp	300	.027	S.F.	8.40	1.46		9.86	11.55
0900	Mineral fiberboard panels, fabric covered, 30" x 108",									
1000	3/4" thick, concealed spline, wall mounted	1 Carp	150	.053	S.F.	6.05	2.92		8.97	11.30
9000	Minimum labor/equipment charge	"	4	2	Job		109		109	174

09 91 Painting

09 91 13 – Exterior Painting

09 91 13.30 Fences

		Crew	Daily Output	Labor-Hours	Unit	Material	2021 Bare Costs Labor	Equipment	Total	Total Incl O&P
0010	**FENCES**									
0100	Chain link or wire metal, one side, water base									
0110	Roll & brush, first coat	1 Pord	960	.008	S.F.	.07	.39		.46	.69
0120	Second coat		1280	.006		.07	.29		.36	.53
0130	Spray, first coat		2275	.004		.07	.16		.23	.34
0140	Second coat		2600	.003		.07	.14		.21	.31
0150	Picket, water base									
0160	Roll & brush, first coat	1 Pord	865	.009	S.F.	.08	.43		.51	.76
0170	Second coat		1050	.008		.08	.35		.43	.64
0180	Spray, first coat		2275	.004		.08	.16		.24	.34
0190	Second coat		2600	.003		.08	.14		.22	.31
0200	Stockade, water base									
0210	Roll & brush, first coat	1 Pord	1040	.008	S.F.	.08	.36		.44	.65
0220	Second coat		1200	.007		.08	.31		.39	.57
0230	Spray, first coat		2275	.004		.08	.16		.24	.34
0240	Second coat		2600	.003		.08	.14		.22	.31
9000	Minimum labor/equipment charge		2	4	Job		186		186	294

09 91 13.42 Miscellaneous, Exterior

		Crew	Daily Output	Labor-Hours	Unit	Material	2021 Bare Costs Labor	Equipment	Total	Total Incl O&P
0010	**MISCELLANEOUS, EXTERIOR**									
0100	Railing, ext., decorative wood, incl. cap & baluster									
0110	Newels & spindles @ 12" OC									
0120	Brushwork, stain, sand, seal & varnish									
0130	First coat	1 Pord	90	.089	L.F.	.87	4.13		5	7.50
0140	Second coat	"	120	.067	"	.87	3.10		3.97	5.85
0150	Rough sawn wood, 42" high, 2" x 2" verticals, 6" OC									
0160	Brushwork, stain, each coat	1 Pord	90	.089	L.F.	.29	4.13		4.42	6.85
0170	Wrought iron, 1" rail, 1/2" sq. verticals									
0180	Brushwork, zinc chromate, 60" high, bars 6" OC									
0190	Primer	1 Pord	130	.062	L.F.	.80	2.86		3.66	5.40
0200	Finish coat		130	.062		1.22	2.86		4.08	5.85
0210	Additional coat		190	.042		1.42	1.96		3.38	4.66
0220	Shutters or blinds, single panel, 2' x 4', paint all sides									
0230	Brushwork, primer	1 Pord	20	.400	Ea.	.67	18.60		19.27	30
0240	Finish coat, exterior latex		20	.400		.57	18.60		19.17	30
0250	Primer & 1 coat, exterior latex		13	.615		1.09	28.50		29.59	46.50
0260	Spray, primer		35	.229		.97	10.60		11.57	17.85
0270	Finish coat, exterior latex		35	.229		1.20	10.60		11.80	18.10
0280	Primer & 1 coat, exterior latex		20	.400		1.04	18.60		19.64	30.50
0290	For louvered shutters, add				S.F.	10%				
0300	Stair stringers, exterior, metal									
0310	Roll & brush, zinc chromate, to 14", each coat	1 Pord	320	.025	L.F.	.41	1.16		1.57	2.29
0320	Rough sawn wood, 4" x 12"									
0330	Roll & brush, exterior latex, each coat	1 Pord	215	.037	L.F.	.08	1.73		1.81	2.83
0340	Trellis/lattice, 2" x 2" @ 3" OC with 2" x 8" supports									
0350	Spray, latex, per side, each coat	1 Pord	475	.017	S.F.	.08	.78		.86	1.33
0450	Decking, ext., sealer, alkyd, brushwork, sealer coat		1140	.007		.10	.33		.43	.63
0460	1st coat		1140	.007		.11	.33		.44	.64
0470	2nd coat		1300	.006		.08	.29		.37	.54
0500	Paint, alkyd, brushwork, primer coat		1140	.007		.10	.33		.43	.64
0510	1st coat		1140	.007		.13	.33		.46	.66
0520	2nd coat		1300	.006		.09	.29		.38	.55
0600	Sand paint, alkyd, brushwork, 1 coat		150	.053		.14	2.48		2.62	4.07

09 91 13.42 Miscellaneous, Exterior

		Crew	Daily Output	Labor-Hours	Unit	Material	2021 Bare Costs Labor	2021 Bare Costs Equipment	Total	Total Incl O&P
9000	Minimum labor/equipment charge	1 Pord	2	4	Job		186		186	294

09 91 13.60 Siding Exterior

		Crew	Daily Output	Labor-Hours	Unit	Material	Labor	Equipment	Total	Total Incl O&P
0010	**SIDING EXTERIOR**, Alkyd (oil base)									
0450	Steel siding, oil base, paint 1 coat, brushwork	2 Pord	2015	.008	S.F.	.10	.37		.47	.69
0500	Spray		4550	.004		.15	.16		.31	.43
0800	Paint 2 coats, brushwork		1300	.012		.20	.57		.77	1.13
1000	Spray		2750	.006		.15	.27		.42	.60
1200	Stucco, rough, oil base, paint 2 coats, brushwork		1300	.012		.20	.57		.77	1.13
1400	Roller		1625	.010		.21	.46		.67	.96
1600	Spray		2925	.005		.23	.25		.48	.65
1800	Texture 1-11 or clapboard, oil base, primer coat, brushwork		1300	.012		.14	.57		.71	1.06
2000	Spray		4550	.004		.14	.16		.30	.41
2400	Paint 2 coats, brushwork		810	.020		.30	.92		1.22	1.78
2600	Spray		2600	.006		.33	.29		.62	.81
3400	Stain 2 coats, brushwork	2 Pord	950	.017	S.F.	.19	.78		.97	1.45
4000	Spray		3050	.005		.21	.24		.45	.62
4200	Wood shingles, oil base primer coat, brushwork		1300	.012		.13	.57		.70	1.05
4400	Spray		3900	.004		.12	.19		.31	.43
5000	Paint 2 coats, brushwork		810	.020		.25	.92		1.17	1.72
5200	Spray		2275	.007		.23	.33		.56	.78
6500	Stain 2 coats, brushwork		950	.017		.19	.78		.97	1.45
7000	Spray		2660	.006		.26	.28		.54	.73
8000	For latex paint, deduct					10%				
8100	For work over 12' H, from pipe scaffolding, add						15%			
8200	For work over 12' H, from extension ladder, add						25%			
8300	For work over 12' H, from swing staging, add						35%			
9000	Minimum labor/equipment charge	1 Pord	2	4	Job		186		186	294

09 91 13.62 Siding, Misc.

		Crew	Daily Output	Labor-Hours	Unit	Material	Labor	Equipment	Total	Total Incl O&P
0010	**SIDING, MISC.**, latex paint									
0100	Aluminum siding									
0110	Brushwork, primer	2 Pord	2275	.007	S.F.	.06	.33		.39	.59
0120	Finish coat, exterior latex		2275	.007		.06	.33		.39	.59
0130	Primer & 1 coat exterior latex		1300	.012		.14	.57		.71	1.06
0140	Primer & 2 coats exterior latex		975	.016		.20	.76		.96	1.43
0150	Mineral fiber shingles									
0160	Brushwork, primer	2 Pord	1495	.011	S.F.	.14	.50		.64	.94
0170	Finish coat, industrial enamel		1495	.011		.20	.50		.70	1.01
0180	Primer & 1 coat enamel		810	.020		.34	.92		1.26	1.82
0190	Primer & 2 coats enamel		540	.030		.54	1.38		1.92	2.77
0200	Roll, primer		1625	.010		.15	.46		.61	.89
0210	Finish coat, industrial enamel		1625	.010		.22	.46		.68	.96
0220	Primer & 1 coat enamel		975	.016		.37	.76		1.13	1.62
0230	Primer & 2 coats enamel		650	.025		.59	1.14		1.73	2.45
0240	Spray, primer		3900	.004		.12	.19		.31	.43
0250	Finish coat, industrial enamel		3900	.004		.18	.19		.37	.50
0260	Primer & 1 coat enamel		2275	.007		.30	.33		.63	.85
0270	Primer & 2 coats enamel		1625	.010		.48	.46		.94	1.25
0280	Waterproof sealer, first coat		4485	.004		.12	.17		.29	.39
0290	Second coat		5235	.003		.12	.14		.26	.35
0300	Rough wood incl. shingles, shakes or rough sawn siding									
0310	Brushwork, primer	2 Pord	1280	.013	S.F.	.13	.58		.71	1.07
0320	Finish coat, exterior latex		1280	.013		.11	.58		.69	1.04

09 91 13 – Exterior Painting

09 91 13.62 Siding, Misc.	Crew	Daily Output	Labor-Hours	Unit	Material	2021 Bare Costs Labor	Equipment	Total	Total Incl O&P
0330 Primer & 1 coat exterior latex	2 Pord	960	.017	S.F.	.25	.77		1.02	1.50
0340 Primer & 2 coats exterior latex		700	.023		.36	1.06		1.42	2.07
0350 Roll, primer	2 Pord	2925	.005	S.F.	.18	.25		.43	.60
0360 Finish coat, exterior latex		2925	.005		.13	.25		.38	.55
0370 Primer & 1 coat exterior latex		1790	.009		.31	.42		.73	1
0380 Primer & 2 coats exterior latex		1300	.012		.45	.57		1.02	1.40
0390 Spray, primer		3900	.004		.15	.19		.34	.47
0400 Finish coat, exterior latex		3900	.004		.10	.19		.29	.41
0410 Primer & 1 coat exterior latex		2600	.006		.25	.29		.54	.73
0420 Primer & 2 coats exterior latex		2080	.008		.36	.36		.72	.96
0430 Waterproof sealer, first coat		4485	.004		.22	.17		.39	.50
0440 Second coat		4485	.004		.12	.17		.29	.39
0450 Smooth wood incl. butt, T&G, beveled, drop or B&B siding									
0460 Brushwork, primer	2 Pord	2325	.007	S.F.	.10	.32		.42	.62
0470 Finish coat, exterior latex		1280	.013		.11	.58		.69	1.04
0480 Primer & 1 coat exterior latex		800	.020		.21	.93		1.14	1.70
0490 Primer & 2 coats exterior latex		630	.025		.32	1.18		1.50	2.22
0500 Roll, primer		2275	.007		.11	.33		.44	.64
0510 Finish coat, exterior latex		2275	.007		.12	.33		.45	.65
0520 Primer & 1 coat exterior latex		1300	.012		.23	.57		.80	1.16
0530 Primer & 2 coats exterior latex		975	.016		.35	.76		1.11	1.59
0540 Spray, primer		4550	.004		.08	.16		.24	.35
0550 Finish coat, exterior latex		4550	.004		.10	.16		.26	.37
0560 Primer & 1 coat exterior latex		2600	.006		.19	.29		.48	.66
0570 Primer & 2 coats exterior latex		1950	.008		.29	.38		.67	.92
0580 Waterproof sealer, first coat		5230	.003		.12	.14		.26	.36
0590 Second coat		5980	.003		.12	.12		.24	.33
0600 For oil base paint, add					10%				
9000 Minimum labor/equipment charge	1 Pord	2	4	Job		186		186	294

09 91 13.70 Doors and Windows, Exterior

	Crew	Daily Output	Labor-Hours	Unit	Material	2021 Bare Costs Labor	Equipment	Total	Total Incl O&P
0010 **DOORS AND WINDOWS, EXTERIOR**									
0100 Door frames & trim, only									
0110 Brushwork, primer	1 Pord	512	.016	L.F.	.06	.73		.79	1.22
0120 Finish coat, exterior latex		512	.016		.07	.73		.80	1.23
0130 Primer & 1 coat, exterior latex		300	.027		.13	1.24		1.37	2.10
0135 2 coats, exterior latex, both sides		15	.533	Ea.	6.30	25		31.30	46
0140 Primer & 2 coats, exterior latex		265	.030	L.F.	.20	1.40		1.60	2.44
0150 Doors, flush, both sides, incl. frame & trim									
0160 Roll & brush, primer	1 Pord	10	.800	Ea.	4.60	37		41.60	64
0170 Finish coat, exterior latex		10	.800		5.40	37		42.40	65
0180 Primer & 1 coat, exterior latex		7	1.143		10	53		63	95
0190 Primer & 2 coats, exterior latex		5	1.600		15.35	74.50		89.85	135
0200 Brushwork, stain, sealer & 2 coats polyurethane	1 Pord	4	2	Ea.	31	93		124	181
0210 Doors, French, both sides, 10-15 lite, incl. frame & trim									
0220 Brushwork, primer	1 Pord	6	1.333	Ea.	2.30	62		64.30	101
0230 Finish coat, exterior latex		6	1.333		2.69	62		64.69	101
0240 Primer & 1 coat, exterior latex		3	2.667		4.99	124		128.99	202
0250 Primer & 2 coats, exterior latex		2	4		7.55	186		193.55	300
0260 Brushwork, stain, sealer & 2 coats polyurethane		2.50	3.200		11.05	149		160.05	247
0270 Doors, louvered, both sides, incl. frame & trim									
0280 Brushwork, primer	1 Pord	7	1.143	Ea.	4.60	53		57.60	89
0290 Finish coat, exterior latex		7	1.143		5.40	53		58.40	90

09 91 13.70 Doors and Windows, Exterior

		Crew	Daily Output	Labor-Hours	Unit	Material	2021 Bare Costs Labor	Equipment	Total	Total Incl O&P
0300	Primer & 1 coat, exterior latex	1 Pord	4	2	Ea.	10	93		103	158
0310	Primer & 2 coats, exterior latex		3	2.667		15.05	124		139.05	213
0320	Brushwork, stain, sealer & 2 coats polyurethane		4.50	1.778		31	82.50		113.50	165
0330	Doors, panel, both sides, incl. frame & trim									
0340	Roll & brush, primer	1 Pord	6	1.333	Ea.	4.60	62		66.60	103
0350	Finish coat, exterior latex		6	1.333		5.40	62		67.40	104
0360	Primer & 1 coat, exterior latex		3	2.667		10	124		134	207
0370	Primer & 2 coats, exterior latex		2.50	3.200		15.05	149		164.05	252
0380	Brushwork, stain, sealer & 2 coats polyurethane		3	2.667		31	124		155	230
0400	Windows, per ext. side, based on 15 S.F.									
0410	1 to 6 lite									
0420	Brushwork, primer	1 Pord	13	.615	Ea.	.91	28.50		29.41	46.50
0430	Finish coat, exterior latex		13	.615		1.06	28.50		29.56	46.50
0440	Primer & 1 coat, exterior latex		8	1		1.97	46.50		48.47	75.50
0450	Primer & 2 coats, exterior latex		6	1.333		2.97	62		64.97	101
0460	Stain, sealer & 1 coat varnish		7	1.143		4.36	53		57.36	89
0470	7 to 10 lite									
0480	Brushwork, primer	1 Pord	11	.727	Ea.	.91	34		34.91	54.50
0490	Finish coat, exterior latex		11	.727		1.06	34		35.06	54.50
0500	Primer & 1 coat, exterior latex		7	1.143		1.97	53		54.97	86
0510	Primer & 2 coats, exterior latex		5	1.600		2.97	74.50		77.47	121
0520	Stain, sealer & 1 coat varnish		6	1.333		4.36	62		66.36	103
0530	12 lite									
0540	Brushwork, primer	1 Pord	10	.800	Ea.	.91	37		37.91	60
0550	Finish coat, exterior latex		10	.800		1.06	37		38.06	60
0560	Primer & 1 coat, exterior latex		6	1.333		1.97	62		63.97	100
0570	Primer & 2 coats, exterior latex		5	1.600		2.97	74.50		77.47	121
0580	Stain, sealer & 1 coat varnish		6	1.333		4.64	62		66.64	103
0590	For oil base paint, add					10%				
9000	Minimum labor/equipment charge	1 Pord	2	4	Job		186		186	294

09 91 13.80 Trim, Exterior

		Crew	Daily Output	Labor-Hours	Unit	Material	2021 Bare Costs Labor	Equipment	Total	Total Incl O&P
0010	**TRIM, EXTERIOR**									
0100	Door frames & trim (see Doors, interior or exterior)									
0110	Fascia, latex paint, one coat coverage									
0120	1" x 4", brushwork	1 Pord	640	.013	L.F.	.02	.58		.60	.94
0130	Roll		1280	.006		.02	.29		.31	.49
0140	Spray		2080	.004		.02	.18		.20	.30
0150	1" x 6" to 1" x 10", brushwork		640	.013		.08	.58		.66	1
0160	Roll		1230	.007		.08	.30		.38	.57
0170	Spray		2100	.004		.06	.18		.24	.35
0180	1" x 12", brushwork		640	.013		.08	.58		.66	1
0190	Roll		1050	.008		.08	.35		.43	.65
0200	Spray		2200	.004		.06	.17		.23	.34
0210	Gutters & downspouts, metal, zinc chromate paint									
0220	Brushwork, gutters, 5", first coat	1 Pord	640	.013	L.F.	.43	.58		1.01	1.39
0230	Second coat		960	.008		.41	.39		.80	1.06
0240	Third coat		1280	.006		.33	.29		.62	.82
0250	Downspouts, 4", first coat		640	.013		.43	.58		1.01	1.39
0260	Second coat		960	.008		.41	.39		.80	1.06
0270	Third coat		1280	.006		.33	.29		.62	.82
0280	Gutters & downspouts, wood									
0290	Brushwork, gutters, 5", primer	1 Pord	640	.013	L.F.	.06	.58		.64	.99

09 91 13 – Exterior Painting

09 91 13.80 Trim, Exterior

		Crew	Daily Output	Labor-Hours	Unit	Material	2021 Bare Costs Labor	Equipment	Total	Total Incl O&P
0300	Finish coat, exterior latex	1 Pord	640	.013	L.F.	.06	.58		.64	.99
0310	Primer & 1 coat exterior latex		400	.020		.13	.93		1.06	1.61
0320	Primer & 2 coats exterior latex		325	.025		.20	1.14		1.34	2.03
0330	Downspouts, 4", primer		640	.013		.06	.58		.64	.99
0340	Finish coat, exterior latex		640	.013		.06	.58		.64	.99
0350	Primer & 1 coat exterior latex		400	.020		.13	.93		1.06	1.61
0360	Primer & 2 coats exterior latex		325	.025		.10	1.14		1.24	1.92
0370	Molding, exterior, up to 14" wide									
0380	Brushwork, primer	1 Pord	640	.013	L.F.	.07	.58		.65	1
0390	Finish coat, exterior latex		640	.013		.08	.58		.66	1
0400	Primer & 1 coat exterior latex		400	.020		.16	.93		1.09	1.64
0410	Primer & 2 coats exterior latex		315	.025		.16	1.18		1.34	2.04
0420	Stain & fill		1050	.008		.12	.35		.47	.69
0430	Shellac		1850	.004		.15	.20		.35	.49
0440	Varnish		1275	.006		.12	.29		.41	.59
9000	Minimum labor/equipment charge		2	4	Job		186		186	294

09 91 13.90 Walls, Masonry (CMU), Exterior

		Crew	Daily Output	Labor-Hours	Unit	Material	2021 Bare Costs Labor	Equipment	Total	Total Incl O&P
0010	**WALLS, MASONRY (CMU), EXTERIOR**									
0360	Concrete masonry units (CMU), smooth surface									
0370	Brushwork, latex, first coat	1 Pord	640	.013	S.F.	.06	.58		.64	.99
0380	Second coat		960	.008		.05	.39		.44	.66
0390	Waterproof sealer, first coat		736	.011		.27	.50		.77	1.10
0400	Second coat		1104	.007		.27	.34		.61	.83
0410	Roll, latex, paint, first coat		1465	.005		.07	.25		.32	.48
0420	Second coat		1790	.004		.06	.21		.27	.39
0430	Waterproof sealer, first coat		1680	.005		.27	.22		.49	.65
0440	Second coat		2060	.004		.27	.18		.45	.59
0450	Spray, latex, paint, first coat		1950	.004		.06	.19		.25	.36
0460	Second coat		2600	.003		.05	.14		.19	.28
0470	Waterproof sealer, first coat		2245	.004		.27	.17		.44	.56
0480	Second coat		2990	.003		.27	.12		.39	.50
0490	Concrete masonry unit (CMU), porous									
0500	Brushwork, latex, first coat	1 Pord	640	.013	S.F.	.12	.58		.70	1.05
0510	Second coat		960	.008		.06	.39		.45	.68
0520	Waterproof sealer, first coat		736	.011		.27	.50		.77	1.10
0530	Second coat		1104	.007		.27	.34		.61	.83
0540	Roll latex, first coat		1465	.005		.09	.25		.34	.50
0550	Second coat		1790	.004		.06	.21		.27	.39
0560	Waterproof sealer, first coat		1680	.005		.27	.22		.49	.65
0570	Second coat		2060	.004		.27	.18		.45	.59
0580	Spray latex, first coat		1950	.004		.07	.19		.26	.37
0590	Second coat		2600	.003		.05	.14		.19	.28
0600	Waterproof sealer, first coat		2245	.004		.27	.17		.44	.56
0610	Second coat		2990	.003		.27	.12		.39	.50
9000	Minimum labor/equipment charge		2	4	Job		186		186	294

09 91 23 – Interior Painting

09 91 23.20 Cabinets and Casework

		Crew	Daily Output	Labor-Hours	Unit	Material	2021 Bare Costs Labor	Equipment	Total	Total Incl O&P
0010	**CABINETS AND CASEWORK**									
1000	Primer coat, oil base, brushwork	1 Pord	650	.012	S.F.	.08	.57		.65	.99
2000	Paint, oil base, brushwork, 1 coat		650	.012		.12	.57		.69	1.04
3000	Stain, brushwork, wipe off		650	.012		.10	.57		.67	1.02
4000	Shellac, 1 coat, brushwork		650	.012		.13	.57		.70	1.05

09 91 23.20 Cabinets and Casework

		Crew	Daily Output	Labor-Hours	Unit	Material	2021 Bare Costs Labor	2021 Bare Costs Equipment	Total	Total Incl O&P
4500	Varnish, 3 coats, brushwork, sand after 1st coat	1 Pord	325	.025	S.F.	.30	1.14		1.44	2.14
5000	For latex paint, deduct				↓	10%				
6300	Strip, prep and refinish wood furniture									
6310	Remove paint using chemicals, wood furniture	1 Pord	28	.286	S.F.	1.76	13.25		15.01	23
6320	Prep for painting, sanding		75	.107		.22	4.95		5.17	8.10
6350	Stain and wipe, brushwork		600	.013		.10	.62		.72	1.09
6355	Spray applied	↓	900	.009	↓	.09	.41		.50	.74
6360	Sealer or varnish, brushwork	1 Pord	1080	.007	S.F.	.10	.34		.44	.65
6365	Spray applied		2100	.004		.10	.18		.28	.39
6370	Paint, primer, brushwork		720	.011		.08	.52		.60	.90
6375	Spray applied		2100	.004		.07	.18		.25	.36
6380	Finish coat, brushwork		810	.010		.12	.46		.58	.86
6385	Spray applied		2100	.004	↓	.11	.18		.29	.41
9010	Minimum labor/equipment charge	↓	2	4	Job		186		186	294

09 91 23.33 Doors and Windows, Interior Alkyd (Oil Base)

		Crew	Daily Output	Labor-Hours	Unit	Material	Labor	Equipment	Total	Total Incl O&P
0010	**DOORS AND WINDOWS, INTERIOR ALKYD (OIL BASE)**									
0500	Flush door & frame, 3' x 7', oil, primer, brushwork	1 Pord	10	.800	Ea.	3.47	37		40.47	63
1000	Paint, 1 coat		10	.800		4.79	37		41.79	64.50
1400	Stain, brushwork, wipe off		18	.444		2.01	20.50		22.51	34.50
1600	Shellac, 1 coat, brushwork		25	.320		2.67	14.85		17.52	26.50
1800	Varnish, 3 coats, brushwork, sand after 1st coat		9	.889		6.35	41.50		47.85	72.50
2000	Panel door & frame, 3' x 7', oil, primer, brushwork		6	1.333		2.89	62		64.89	101
2200	Paint, 1 coat		6	1.333		4.79	62		66.79	103
2600	Stain, brushwork, panel door, 3' x 7', not incl. frame		16	.500		2.01	23		25.01	39
2800	Shellac, 1 coat, brushwork		22	.364		2.67	16.90		19.57	30
3000	Varnish, 3 coats, brushwork, sand after 1st coat	↓	7.50	1.067	↓	6.35	49.50		55.85	85.50
4400	Windows, including frame and trim, per side									
4600	Colonial type, 6/6 lites, 2' x 3', oil, primer, brushwork	1 Pord	14	.571	Ea.	.46	26.50		26.96	42.50
5800	Paint, 1 coat		14	.571		.76	26.50		27.26	43
6200	3' x 5' opening, 6/6 lites, primer coat, brushwork		12	.667		1.14	31		32.14	50.50
6400	Paint, 1 coat		12	.667		1.89	31		32.89	51
6800	4' x 8' opening, 6/6 lites, primer coat, brushwork		8	1		2.44	46.50		48.94	76
7000	Paint, 1 coat		8	1		4.03	46.50		50.53	78
8000	Single lite type, 2' x 3', oil base, primer coat, brushwork		33	.242		.46	11.25		11.71	18.35
8200	Paint, 1 coat		33	.242		.76	11.25		12.01	18.70
8600	3' x 5' opening, primer coat, brushwork		20	.400		1.14	18.60		19.74	31
8800	Paint, 1 coat		20	.400		1.89	18.60		20.49	31.50
9200	4' x 8' opening, primer coat, brushwork		14	.571		2.44	26.50		28.94	44.50
9400	Paint, 1 coat		14	.571	↓	4.03	26.50		30.53	46.50
9900	Minimum labor/equipment charge	↓	2	4	Job		186		186	294

09 91 23.35 Doors and Windows, Interior Latex

		Crew	Daily Output	Labor-Hours	Unit	Material	Labor	Equipment	Total	Total Incl O&P
0010	**DOORS & WINDOWS, INTERIOR LATEX**									
0100	Doors, flush, both sides, incl. frame & trim									
0110	Roll & brush, primer	1 Pord	10	.800	Ea.	4.41	37		41.41	64
0120	Finish coat, latex		10	.800		5.70	37		42.70	65.50
0130	Primer & 1 coat latex		7	1.143		10.10	53		63.10	95
0140	Primer & 2 coats latex	↓	5	1.600	↓	15.50	74.50		90	135
0160	Spray, both sides, primer	1 Pord	20	.400	Ea.	4.64	18.60		23.24	34.50
0170	Finish coat, latex		20	.400		6	18.60		24.60	36
0180	Primer & 1 coat latex		11	.727		10.70	34		44.70	65.50
0190	Primer & 2 coats latex	↓	8	1		16.40	46.50		62.90	91.50
0200	Doors, French, both sides, 10-15 lite, incl. frame & trim									

For customer support on your Commercial Renovation Costs with RSMeans Data, call 800.448.8182.

09 91 23.35 Doors and Windows, Interior Latex

		Crew	Daily Output	Labor-Hours	Unit	Material	2021 Bare Costs Labor	Equipment	Total	Total Incl O&P
0210	Roll & brush, primer	1 Pord	6	1.333	Ea.	2.20	62		64.20	100
0220	Finish coat, latex		6	1.333		2.85	62		64.85	101
0230	Primer & 1 coat latex		3	2.667		5.05	124		129.05	202
0240	Primer & 2 coats latex		2	4		7.75	186		193.75	305
0260	Doors, louvered, both sides, incl. frame & trim									
0270	Roll & brush, primer	1 Pord	7	1.143	Ea.	4.41	53		57.41	89
0280	Finish coat, latex		7	1.143		5.70	53		58.70	90.50
0290	Primer & 1 coat, latex		4	2		9.85	93		102.85	158
0300	Primer & 2 coats, latex		3	2.667		15.80	124		139.80	213
0320	Spray, both sides, primer		20	.400		4.64	18.60		23.24	34.50
0330	Finish coat, latex		20	.400		6	18.60		24.60	36
0340	Primer & 1 coat, latex		11	.727		10.70	34		44.70	65.50
0350	Primer & 2 coats, latex		8	1		16.75	46.50		63.25	92
0360	Doors, panel, both sides, incl. frame & trim									
0370	Roll & brush, primer	1 Pord	6	1.333	Ea.	4.64	62		66.64	103
0380	Finish coat, latex		6	1.333		5.70	62		67.70	104
0390	Primer & 1 coat, latex		3	2.667		10.10	124		134.10	207
0400	Primer & 2 coats, latex		2.50	3.200		15.80	149		164.80	252
0420	Spray, both sides, primer		10	.800		4.64	37		41.64	64
0430	Finish coat, latex		10	.800		6	37		43	65.50
0440	Primer & 1 coat, latex		5	1.600		10.70	74.50		85.20	130
0450	Primer & 2 coats, latex		4	2		16.75	93		109.75	165
0460	Windows, per interior side, based on 15 S.F.									
0470	1 to 6 lite									
0480	Brushwork, primer	1 Pord	13	.615	Ea.	.87	28.50		29.37	46.50
0490	Finish coat, enamel		13	.615		1.13	28.50		29.63	46.50
0500	Primer & 1 coat enamel		8	1		2	46.50		48.50	75.50
0510	Primer & 2 coats enamel		6	1.333		3.12	62		65.12	101
0530	7 to 10 lite									
0540	Brushwork, primer	1 Pord	11	.727	Ea.	.87	34		34.87	54.50
0550	Finish coat, enamel		11	.727		1.13	34		35.13	54.50
0560	Primer & 1 coat enamel		7	1.143		2	53		55	86
0570	Primer & 2 coats enamel		5	1.600		3.12	74.50		77.62	121
0590	12 lite									
0600	Brushwork, primer	1 Pord	10	.800	Ea.	.87	37		37.87	60
0610	Finish coat, enamel	1 Pord	10	.800	Ea.	1.13	37		38.13	60
0620	Primer & 1 coat enamel		6	1.333		2	62		64	100
0630	Primer & 2 coats enamel		5	1.600		3.12	74.50		77.62	121
0650	For oil base paint, add					10%				
9000	Minimum labor/equipment charge	1 Pord	2	4	Job		186		186	294

09 91 23.39 Doors and Windows, Interior Latex, Zero Voc

			Crew	Daily Output	Labor-Hours	Unit	Material	2021 Bare Costs Labor	Equipment	Total	Total Incl O&P
0010	**DOORS & WINDOWS, INTERIOR LATEX, ZERO VOC**										
0100	Doors flush, both sides, incl. frame & trim										
0110	Roll & brush, primer	G	1 Pord	10	.800	Ea.	5.75	37		42.75	65.50
0120	Finish coat, latex	G		10	.800		5.70	37		42.70	65.50
0130	Primer & 1 coat latex	G		7	1.143		11.45	53		64.45	96.50
0140	Primer & 2 coats latex	G		5	1.600		16.85	74.50		91.35	137
0160	Spray, both sides, primer	G		20	.400		6.05	18.60		24.65	36
0170	Finish coat, latex	G		20	.400		6	18.60		24.60	36
0180	Primer & 1 coat latex	G		11	.727		12.15	34		46.15	67
0190	Primer & 2 coats latex	G		8	1		17.85	46.50		64.35	93
0200	Doors, French, both sides, 10-15 lite, incl. frame & trim										

For customer support on your Commercial Renovation Costs with RSMeans Data, call 800.448.8182.

341

09 91 23.39 Doors and Windows, Interior Latex, Zero Voc

			Crew	Daily Output	Labor-Hours	Unit	Material	2021 Bare Costs Labor	Equipment	Total	Total Incl O&P
0210	Roll & brush, primer	G	1 Pord	6	1.333	Ea.	2.88	62		64.88	101
0220	Finish coat, latex	G		6	1.333		2.86	62		64.86	101
0230	Primer & 1 coat latex	G		3	2.667		5.75	124		129.75	202
0240	Primer & 2 coats latex	G		2	4		8.45	186		194.45	305
0360	Doors, panel, both sides, incl. frame & trim										
0370	Roll & brush, primer	G	1 Pord	6	1.333	Ea.	6.05	62		68.05	105
0380	Finish coat, latex	G		6	1.333		5.70	62		67.70	104
0390	Primer & 1 coat, latex	G		3	2.667		11.45	124		135.45	209
0400	Primer & 2 coats, latex	G		2.50	3.200		17.20	149		166.20	254
0420	Spray, both sides, primer	G		10	.800		6.05	37		43.05	65.50
0430	Finish coat, latex	G		10	.800		6	37		43	65.50
0440	Primer & 1 coat, latex	G		5	1.600		12.15	74.50		86.65	131
0450	Primer & 2 coats, latex	G		4	2		18.20	93		111.20	167
0460	Windows, per interior side, based on 15 S.F.										
0470	1 to 6 lite										
0480	Brushwork, primer	G	1 Pord	13	.615	Ea.	1.14	28.50		29.64	47
0490	Finish coat, enamel	G		13	.615		1.13	28.50		29.63	46.50
0500	Primer & 1 coat enamel	G		8	1		2.26	46.50		48.76	76
0510	Primer & 2 coats enamel	G		6	1.333		3.39	62		65.39	102
9000	Minimum labor/equipment charge			2	4	Job		186		186	294

09 91 23.40 Floors, Interior

		Crew	Daily Output	Labor-Hours	Unit	Material	2021 Bare Costs Labor	Equipment	Total	Total Incl O&P
0010	**FLOORS, INTERIOR**									
0100	Concrete paint, latex									
0110	Brushwork									
0120	1st coat	1 Pord	975	.008	S.F.	.15	.38		.53	.77
0130	2nd coat		1150	.007		.10	.32		.42	.62
0140	3rd coat		1300	.006		.08	.29		.37	.54
0150	Roll									
0160	1st coat	1 Pord	2600	.003	S.F.	.20	.14		.34	.45
0170	2nd coat		3250	.002		.12	.11		.23	.31
0180	3rd coat		3900	.002		.09	.10		.19	.25
0190	Spray									
0200	1st coat	1 Pord	2600	.003	S.F.	.18	.14		.32	.42
0210	2nd coat		3250	.002		.10	.11		.21	.29
0220	3rd coat		3900	.002		.08	.10		.18	.23

09 91 23.44 Anti-Slip Floor Treatments

		Crew	Daily Output	Labor-Hours	Unit	Material	2021 Bare Costs Labor	Equipment	Total	Total Incl O&P
0010	**ANTI-SLIP FLOOR TREATMENTS**									
1000	Walking surface treatment, ADA compliant, mop on and rinse									
1100	For tile, terrazzo, stone or smooth concrete	1 Pord	4000	.002	S.F.	.14	.09		.23	.30
1110	For marble		4000	.002		.19	.09		.28	.36
1120	For wood		4000	.002		.18	.09		.27	.35
1130	For baths and showers		500	.016		.15	.74		.89	1.35
2000	Granular additive for paint or sealer, add to paint cost					.01			.01	.02

09 91 23.52 Miscellaneous, Interior

		Crew	Daily Output	Labor-Hours	Unit	Material	2021 Bare Costs Labor	Equipment	Total	Total Incl O&P
0010	**MISCELLANEOUS, INTERIOR**									
2400	Floors, conc./wood, oil base, primer/sealer coat, brushwork	2 Pord	1950	.008	S.F.	.09	.38		.47	.69
2450	Roller		5200	.003		.09	.14		.23	.33
2600	Spray		6000	.003		.09	.12		.21	.30
2650	Paint 1 coat, brushwork		1950	.008		.08	.38		.46	.69
2800	Roller		5200	.003		.09	.14		.23	.33
2850	Spray		6000	.003		.09	.12		.21	.30
3000	Stain, wood floor, brushwork, 1 coat		4550	.004		.10	.16		.26	.37

09 91 23 – Interior Painting

09 91 23.52 Miscellaneous, Interior	Crew	Daily Output	Labor-Hours	Unit	Material	2021 Bare Costs Labor	Equipment	Total	Total Incl O&P
3200	Roller	2 Pord	5200	.003	S.F.	.10	.14		.24	.34
3250	Spray		6000	.003		.10	.12		.22	.31
3400	Varnish, wood floor, brushwork		4550	.004		.10	.16		.26	.37
3450	Roller		5200	.003		.11	.14		.25	.35
3600	Spray		6000	.003		.11	.12		.23	.32
3800	Grilles, per side, oil base, primer coat, brushwork	1 Pord	520	.015		.15	.71		.86	1.30
3850	Spray		1140	.007		.16	.33		.49	.70
3920	Paint 2 coats, brushwork		325	.025		.49	1.14		1.63	2.35
3940	Spray		650	.012		.56	.57		1.13	1.53
4600	Miscellaneous surfaces, metallic paint, spray applied									
4610	Water based, non-tintable, warm silver	1 Pord	1140	.007	S.F.	.61	.33		.94	1.19
4620	Rusted iron		1140	.007		1.05	.33		1.38	1.68
4630	Low VOC, tintable		1140	.007		.41	.33		.74	.97
5000	Pipe, 1"-4" diameter, primer or sealer coat, oil base, brushwork	2 Pord	1250	.013	L.F.	.09	.59		.68	1.04
5100	Spray		2165	.007		.08	.34		.42	.63
5350	Paint 2 coats, brushwork		775	.021		.27	.96		1.23	1.81
5400	Spray		1240	.013		.30	.60		.90	1.28
6300	13"-16" diameter, primer or sealer coat, brushwork		310	.052		.36	2.40		2.76	4.19
6450	Spray		540	.030		.61	1.38		1.99	2.85
6500	Paint 2 coats, brushwork		195	.082		1.07	3.81		4.88	7.25
6550	Spray		310	.052		1.19	2.40		3.59	5.10
7000	Trim, wood, incl. puttying, under 6" wide									
7200	Primer coat, oil base, brushwork	1 Pord	650	.012	L.F.	.04	.57		.61	.95
7250	Paint, 1 coat, brushwork		650	.012		.06	.57		.63	.98
7450	3 coats		325	.025		.18	1.14		1.32	2.01
7500	Over 6" wide, primer coat, brushwork		650	.012		.08	.57		.65	.99
7550	Paint, 1 coat, brushwork		650	.012		.13	.57		.70	1.05
7650	3 coats		325	.025		.36	1.14		1.50	2.21
8000	Cornice, simple design, primer coat, oil base, brushwork		650	.012	S.F.	.08	.57		.65	.99
8250	Paint, 1 coat		650	.012		.13	.57		.70	1.05
8350	Ornate design, primer coat		350	.023		.08	1.06		1.14	1.76
8400	Paint, 1 coat		350	.023		.13	1.06		1.19	1.82
8600	Balustrades, primer coat, oil base, brushwork		520	.015		.08	.71		.79	1.21
8900	Trusses and wood frames, primer coat, oil base, brushwork		800	.010		.08	.46		.54	.82
8950	Spray		1200	.007		.08	.31		.39	.58
9220	Paint 2 coats, brushwork		500	.016		.24	.74		.98	1.45
9240	Spray		600	.013		.27	.62		.89	1.28
9260	Stain, brushwork, wipe off		600	.013		.10	.62		.72	1.09
9280	Varnish, 3 coats, brushwork		275	.029		.30	1.35		1.65	2.47
9350	For latex paint, deduct					10%				
9900	Minimum labor/equipment charge	1 Pord	2	4	Job		186		186	294

09 91 23.62 Electrostatic Painting

		Crew	Daily Output	Labor-Hours	Unit	Material	2021 Bare Costs Labor	Equipment	Total	Total Incl O&P
0010	**ELECTROSTATIC PAINTING**									
0100	In shop									
0200	Flat surfaces (lockers, casework, elevator doors, etc.)									
0300	One coat	1 Pord	200	.040	S.F.	.22	1.86		2.08	3.19
0400	Two coats	"	120	.067	"	.33	3.10		3.43	5.25
0500	Irregular surfaces (furniture, door frames, etc.)									
0600	One coat	1 Pord	150	.053	S.F.	.22	2.48		2.70	4.17
0700	Two coats	"	100	.080	"	.33	3.72		4.05	6.25
0800	On site									
0900	Flat surfaces (lockers, casework, elevator doors, etc.)									

09 91 23.62 Electrostatic Painting		Crew	Daily Output	Labor-Hours	Unit	Material	2021 Bare Costs Labor	Equipment	Total	Total Incl O&P
1000	One coat	1 Pord	150	.053	S.F.	.22	2.48		2.70	4.17
1100	Two coats	"	100	.080	"	.33	3.72		4.05	6.25
1200	Irregular surfaces (furniture, door frames, etc.)									
1300	One coat	1 Pord	115	.070	S.F.	.22	3.23		3.45	5.35
1400	Two coats	↓	70	.114		.33	5.30		5.63	8.75
2000	Anti-microbial coating, hospital application		150	.053	↓	.42	2.48		2.90	4.38

09 91 23.72 Walls and Ceilings, Interior

		Crew	Daily Output	Labor-Hours	Unit	Material	2021 Bare Costs Labor	Equipment	Total	Total Incl O&P
0010	**WALLS AND CEILINGS, INTERIOR**									
0100	Concrete, drywall or plaster, latex, primer or sealer coat									
0150	Smooth finish, cut-in by brush	1 Pord	1150	.007	L.F.	.02	.32		.34	.53
0200	Brushwork		1150	.007	S.F.	.07	.32		.39	.58
0240	Roller		1350	.006		.07	.28		.35	.51
0280	Spray		2750	.003	↓	.06	.14		.20	.27
0290	Sand finish, cut-in by brush		975	.008	L.F.	.02	.38		.40	.62
0300	Brushwork		975	.008	S.F.	.07	.38		.45	.67
0340	Roller		1150	.007		.07	.32		.39	.58
0380	Spray		2275	.004	↓	.06	.16		.22	.32
0590	Paint 2 coats, smooth finish, cut-in by brush		680	.012	L.F.	.04	.55		.59	.91
0800	Brushwork		680	.012	S.F.	.16	.55		.71	1.05
0840	Roller		800	.010		.16	.46		.62	.92
0880	Spray		1625	.005	↓	.15	.23		.38	.53
0890	Sand finish, cut-in by brush		605	.013	L.F.	.04	.61		.65	1.01
0900	Brushwork		605	.013	S.F.	.16	.61		.77	1.15
0940	Roller		1020	.008		.16	.36		.52	.76
0980	Spray		1700	.005		.15	.22		.37	.52
1600	Glaze coating, 2 coats, spray, clear		1200	.007		.56	.31		.87	1.11
1640	Multicolor		1200	.007	↓	.92	.31		1.23	1.50
1660	Painting walls, complete, including surface prep, primer &									
1670	2 coats finish, on drywall or plaster, with roller	1 Pord	325	.025	S.F.	.24	1.14		1.38	2.07
1700	For oil base paint, add					10%				
1800	For ceiling installations, add				↓		25%			
2000	Masonry or concrete block, primer/sealer, latex paint									
2090	Primer, smooth finish, cut-in by brush	1 Pord	1000	.008	L.F.	.04	.37		.41	.63
2100	Brushwork		1000	.008	S.F.	.15	.37		.52	.76
2110	Roller		1150	.007		.12	.32		.44	.64
2180	Spray		2400	.003	↓	.10	.15		.25	.36
2190	Sand finish, cut-in by brush		850	.009	L.F.	.03	.44		.47	.72
2200	Brushwork		850	.009	S.F.	.12	.44		.56	.82
2210	Roller		975	.008		.12	.38		.50	.73
2280	Spray		2050	.004	↓	.10	.18		.28	.40
2590	Primer plus one finish coat, smooth cut-in by brush		525	.015	L.F.	.08	.71		.79	1.21
2800	Brushwork		525	.015	S.F.	.32	.71		1.03	1.47
2810	Roller		615	.013		.20	.60		.80	1.18
2880	Spray		1200	.007	↓	.18	.31		.49	.68
2890	Sand finish, cut-in by brush		450	.018	L.F.	.05	.83		.88	1.36
2900	Brushwork		450	.018	S.F.	.20	.83		1.03	1.53
2910	Roller		515	.016		.20	.72		.92	1.36
2980	Spray		1025	.008		.18	.36		.54	.76
3600	Glaze coating, 3 coats, spray, clear		900	.009		.80	.41		1.21	1.53
3620	Multicolor		900	.009		1.12	.41		1.53	1.88
4000	Block filler, 1 coat, brushwork		425	.019		.14	.87		1.01	1.53
4100	Silicone, water repellent, 2 coats, spray	↓	2000	.004	↓	.47	.19		.66	.80

09 91 23.72 Walls and Ceilings, Interior

		Crew	Daily Output	Labor-Hours	Unit	Material	2021 Bare Costs Labor	2021 Bare Costs Equipment	Total	Total Incl O&P
4120	For oil base paint, add				S.F.	10%				
8200	For work 8'-15' H, add						10%			
8300	For work over 15' H, add						20%			
8400	For light textured surfaces, add						10%			
8410	Heavy textured, add						25%			
9900	Minimum labor/equipment charge	1 Pord	2	4	Job		186		186	294

09 91 23.74 Walls and Ceilings, Interior, Zero VOC Latex

		Crew	Daily Output	Labor-Hours	Unit	Material	2021 Bare Costs Labor	2021 Bare Costs Equipment	Total	Total Incl O&P
0010	**WALLS AND CEILINGS, INTERIOR, ZERO VOC LATEX**									
0100	Concrete, dry wall or plaster, latex, primer or sealer coat									
0190	Smooth finish, cut-in by brush	1 Pord	1150	.007	L.F.	.02	.32		.34	.53
0200	Brushwork [G]		1150	.007	S.F.	.08	.32		.40	.59
0240	Roller [G]		1350	.006		.07	.28		.35	.52
0280	Spray [G]		2750	.003		.06	.14		.20	.27
0290	Sand finish, cut-in by brush		975	.008	L.F.	.02	.38		.40	.62
0300	Brushwork [G]		975	.008	S.F.	.07	.38		.45	.68
0340	Roller [G]		1150	.007		.08	.32		.40	.60
0380	Spray [G]		2275	.004		.06	.16		.22	.33
0790	Paint 2 coats, smooth finish, cut-in by brush		680	.012	L.F.	.04	.55		.59	.91
0800	Brushwork [G]		680	.012	S.F.	.04	.55		.59	.91
0840	Roller [G]		800	.010		.16	.46		.62	.91
0880	Spray [G]		1625	.005		.14	.23		.37	.51
0890	Sand finish, cut-in by brush		605	.013	L.F.	.04	.61		.65	1.01
0900	Brushwork [G]		605	.013	S.F.	.15	.61		.76	1.13
0940	Roller [G]		1020	.008		.16	.36		.52	.75
0980	Spray [G]		1700	.005		.14	.22		.36	.50
1800	For ceiling installations, add						25%			
8200	For work 8' - 15' H, add						10%			
8300	For work over 15' H, add						20%			
9900	Minimum labor/equipment charge	1 Pord	2	4	Job		186		186	294

09 91 23.75 Dry Fall Painting

		Crew	Daily Output	Labor-Hours	Unit	Material	2021 Bare Costs Labor	2021 Bare Costs Equipment	Total	Total Incl O&P
0010	**DRY FALL PAINTING**									
0100	Sprayed on walls, gypsum board or plaster									
0220	One coat	1 Pord	2600	.003	S.F.	.09	.14		.23	.32
0250	Two coats		1560	.005		.17	.24		.41	.57
0280	Concrete or textured plaster, one coat		1560	.005		.09	.24		.33	.47
0310	Two coats		1300	.006		.17	.29		.46	.64
0340	Concrete block, one coat		1560	.005		.09	.24		.33	.47
0370	Two coats		1300	.006		.17	.29		.46	.64
0400	Wood, one coat		877	.009		.09	.42		.51	.76
0430	Two coats		650	.012		.17	.57		.74	1.10
0440	On ceilings, gypsum board or plaster									
0470	One coat	1 Pord	1560	.005	S.F.	.09	.24		.33	.47
0500	Two coats		1300	.006		.17	.29		.46	.64
0530	Concrete or textured plaster, one coat		1560	.005		.09	.24		.33	.47
0560	Two coats		1300	.006		.17	.29		.46	.64
0570	Structural steel, bar joists or metal deck, one coat		1560	.005		.09	.24		.33	.47
0580	Two coats		1040	.008		.17	.36		.53	.76
9900	Minimum labor/equipment charge		2	4	Job		186		186	294

09 93 Staining and Transparent Finishing

09 93 23 – Interior Staining and Finishing

09 93 23.10 Varnish	Crew	Daily Output	Labor-Hours	Unit	Material	2021 Bare Costs Labor	Equipment	Total	Total Incl O&P
0010 **VARNISH**									
0012 1 coat + sealer, on wood trim, brush, no sanding included	1 Pord	400	.020	S.F.	.07	.93		1	1.55
0020 1 coat + sealer, on wood trim, brush, no sanding included, no VOC		400	.020		.22	.93		1.15	1.71
0100 Hardwood floors, 2 coats, no sanding included, roller		1890	.004		.15	.20		.35	.48
9000 Minimum labor/equipment charge		4	2	Job		93		93	147

09 96 High-Performance Coatings

09 96 23 – Graffiti-Resistant Coatings

09 96 23.10 Graffiti-Resistant Treatments

09 96 23.10 Graffiti-Resistant Treatments	Crew	Daily Output	Labor-Hours	Unit	Material	2021 Bare Costs Labor	Equipment	Total	Total Incl O&P
0010 **GRAFFITI-RESISTANT TREATMENTS**, sprayed on walls									
0100 Non-sacrificial, permanent non-stick coating, clear, on metals	1 Pord	2000	.004	S.F.	1.98	.19		2.17	2.47
0200 Concrete		2000	.004		2.25	.19		2.44	2.77
0300 Concrete block		2000	.004		2.91	.19		3.10	3.49
0400 Brick		2000	.004		3.30	.19		3.49	3.92
0500 Stone		2000	.004		3.30	.19		3.49	3.92
0600 Unpainted wood		2000	.004		3.81	.19		4	4.48
2000 Semi-permanent cross linking polymer primer, on metals		2000	.004		.61	.19		.80	.96
2100 Concrete		2000	.004		.73	.19		.92	1.09
2200 Concrete block		2000	.004		.91	.19		1.10	1.29
2300 Brick		2000	.004		.73	.19		.92	1.09
2400 Stone		2000	.004		.73	.19		.92	1.09
2500 Unpainted wood		2000	.004		1.01	.19		1.20	1.40
3000 Top coat, on metals		2000	.004		.56	.19		.75	.90
3100 Concrete		2000	.004		.64	.19		.83	.99
3200 Concrete block		2000	.004		.89	.19		1.08	1.27
3300 Brick		2000	.004		.74	.19		.93	1.11
3400 Stone		2000	.004		.74	.19		.93	1.11
3500 Unpainted wood		2000	.004		.89	.19		1.08	1.27
5000 Sacrificial, water based, on metal		2000	.004		.32	.19		.51	.65
5100 Concrete		2000	.004		.32	.19		.51	.65
5200 Concrete block		2000	.004		.32	.19		.51	.65
5300 Brick		2000	.004		.32	.19		.51	.65
5400 Stone		2000	.004		.32	.19		.51	.65
5500 Unpainted wood		2000	.004		.32	.19		.51	.65
8000 Cleaner for use after treatment									
8100 Towels or wipes, per package of 30				Ea.	.72			.72	.80
8200 Aerosol spray, 24 oz. can				"	19			19	21

09 96 46 – Intumescent Painting

09 96 46.10 Coatings, Intumescent

09 96 46.10 Coatings, Intumescent	Crew	Daily Output	Labor-Hours	Unit	Material	2021 Bare Costs Labor	Equipment	Total	Total Incl O&P
0010 **COATINGS, INTUMESCENT**, spray applied									
0100 On exterior structural steel, 0.25" d.f.t.	1 Pord	475	.017	S.F.	.50	.78		1.28	1.80
0150 0.51" d.f.t.		350	.023		.50	1.06		1.56	2.24
0200 0.98" d.f.t.		280	.029		.50	1.33		1.83	2.66
0300 On interior structural steel, 0.108" d.f.t.		300	.027		.49	1.24		1.73	2.49
0350 0.310" d.f.t.		150	.053		.49	2.48		2.97	4.45
0400 0.670" d.f.t.		100	.080		.49	3.72		4.21	6.45

09 96 High-Performance Coatings

09 96 53 – Elastomeric Coatings

09 96 53.10 Coatings, Elastomeric	Crew	Daily Output	Labor-Hours	Unit	Material	2021 Bare Costs Labor	Equipment	Total	Total Incl O&P
0010 **COATINGS, ELASTOMERIC**									
0020 High build, water proof, one coat system									
0100 Concrete, brush	1 Pord	650	.012	S.F.	.20	.57		.77	1.13

09 96 56 – Epoxy Coatings

09 96 56.20 Wall Coatings

	Crew	Daily Output	Labor-Hours	Unit	Material	2021 Bare Costs Labor	Equipment	Total	Total Incl O&P
0010 **WALL COATINGS**									
0100 Acrylic glazed coatings, matte	1 Pord	525	.015	S.F.	.38	.71		1.09	1.54
0200 Gloss		305	.026		.77	1.22		1.99	2.78
0300 Epoxy coatings, solvent based		525	.015		.48	.71		1.19	1.65
0400 Water based		170	.047		.38	2.19		2.57	3.88
2400 Sprayed perlite or vermiculite, 1/16" thick, solvent based		2935	.003		.28	.13		.41	.51
2500 Water based		640	.013		.88	.58		1.46	1.89
2700 Vinyl plastic wall coating, solvent based		735	.011		.41	.51		.92	1.25
2800 Water based		240	.033		1.03	1.55		2.58	3.58
3000 Urethane on smooth surface, 2 coats, solvent based		1135	.007		.35	.33		.68	.91
3100 Water based		665	.012		.63	.56		1.19	1.57

09 97 Special Coatings

09 97 35 – Dry Erase Coatings

09 97 35.10 Dry Erase Coatings

	Crew	Daily Output	Labor-Hours	Unit	Material	2021 Bare Costs Labor	Equipment	Total	Total Incl O&P
0010 **DRY ERASE COATINGS**									
0020 Dry erase coatings, clear, roller applied	1 Pord	1325	.006	S.F.	2.16	.28		2.44	2.82

Division Notes

	CREW	DAILY OUTPUT	LABOR-HOURS	UNIT	BARE COSTS				TOTAL INCL O&P
					MAT.	LABOR	EQUIP.	TOTAL	

Estimating Tips
General
- The items in this division are usually priced per square foot or each.
- Many items in Division 10 require some type of support system or special anchors that are not usually furnished with the item. The required anchors must be added to the estimate in the appropriate division.
- Some items in Division 10, such as lockers, may require assembly before installation. Verify the amount of assembly required. Assembly can often exceed installation time.

10 20 00 Interior Specialties
- Support angles and blocking are not included in the installation of toilet compartments, shower/dressing compartments, or cubicles. Appropriate line items from Division 5 or 6 may need to be added to support the installations.
- Toilet partitions are priced by the stall. A stall consists of a side wall, pilaster, and door with hardware. Toilet tissue holders and grab bars are extra.
- The required acoustical rating of a folding partition can have a significant impact on costs. Verify the sound transmission coefficient rating of the panel priced against the specification requirements.

- Grab bar installation does not include supplemental blocking or backing to support the required load. When grab bars are installed at an existing facility, provisions must be made to attach the grab bars to a solid structure.

Reference Numbers
Reference numbers are shown at the beginning of some major classifications. These numbers refer to related items in the Reference Section. The reference information may be an estimating procedure, an alternate pricing method, or technical information.

Note: Not all subdivisions listed here necessarily appear. ■

Same Data. Simplified.

Enjoy the convenience and efficiency of accessing your costs anywhere:

- **Skip the multiplier** by setting your location
- **Quickly search,** edit, favorite and share costs
- **Stay on top of price changes** with automatic updates

Discover more at rsmeans.com/online

10 11 13.13 Fixed Chalkboards

		Crew	Daily Output	Labor-Hours	Unit	Material	2021 Bare Costs Labor	Equipment	Total	Total Incl O&P
0010	**FIXED CHALKBOARDS** Porcelain enamel steel									
3900	Wall hung									
4000	Aluminum frame and chalktrough									
4300	3' x 5'	2 Carp	15	1.067	Ea.	267	58.50		325.50	385
4600	4' x 12'	"	13	1.231	"	510	67.50		577.50	665
4700	Wood frame and chalktrough									
4800	3' x 4'	2 Carp	16	1	Ea.	166	54.50		220.50	269
5300	4' x 8'	"	13	1.231	"	320	67.50		387.50	455
5400	Liquid chalk, white porcelain enamel, wall hung									
5420	Deluxe units, aluminum trim and chalktrough									
5450	4' x 4'	2 Carp	16	1	Ea.	320	54.50		374.50	440
5550	4' x 12'	"	12	1.333	"	630	73		703	810
5700	Wood trim and chalktrough									
5900	4' x 4'	2 Carp	16	1	Ea.	635	54.50		689.50	785
6200	4' x 8'	↓	14	1.143	"	875	62.50		937.50	1,050
9000	Minimum labor/equipment charge		3	5.333	Job		292		292	465

10 11 13.43 Portable Chalkboards

		Crew	Daily Output	Labor-Hours	Unit	Material	2021 Bare Costs Labor	Equipment	Total	Total Incl O&P
0010	**PORTABLE CHALKBOARDS**									
0100	Freestanding, reversible									
0120	Economy, wood frame, 4' x 6'									
0140	Chalkboard both sides				Ea.	625			625	685
0160	Chalkboard one side, cork other side				"	665			665	730
0200	Standard, lightweight satin finished aluminum, 4' x 6'									
0220	Chalkboard both sides				Ea.	535			535	590
0240	Chalkboard one side, cork other side				"	660			660	725
0300	Deluxe, heavy duty extruded aluminum, 4' x 6'									
0320	Chalkboard both sides				Ea.	715			715	790
0340	Chalkboard one side, cork other side				"	1,400			1,400	1,525

10 11 16 – Markerboards

10 11 16.53 Electronic Markerboards

		Crew	Daily Output	Labor-Hours	Unit	Material	2021 Bare Costs Labor	Equipment	Total	Total Incl O&P
0010	**ELECTRONIC MARKERBOARDS**									
0100	Wall hung or free standing, 3' x 4' to 4' x 6'	2 Carp	8	2	S.F.	365	109		474	575
0150	5' x 6' to 4' x 8'	↓	8	2	"	460	109		569	680
0500	Interactive projection module for existing whiteboards		8	2	Ea.	1,325	109		1,434	1,650

10 11 23 – Tackboards

10 11 23.10 Fixed Tackboards

		Crew	Daily Output	Labor-Hours	Unit	Material	2021 Bare Costs Labor	Equipment	Total	Total Incl O&P
0010	**FIXED TACKBOARDS**									
2120	Prefabricated, 1/4" cork, 3' x 5' with aluminum frame	2 Carp	16	1	Ea.	139	54.50		193.50	239
2140	Wood frame	"	16	1	"	170	54.50		224.50	274
2300	Glass enclosed cabinets, alum., cork panel, hinged doors									
2600	4' x 7', 3 door	2 Carp	10	1.600	Ea.	2,700	87.50		2,787.50	3,100
9000	Minimum labor/equipment charge	"	4	4	Job		219		219	350

10 11 23.20 Control Boards

		Crew	Daily Output	Labor-Hours	Unit	Material	2021 Bare Costs Labor	Equipment	Total	Total Incl O&P
0010	**CONTROL BOARDS**									
1000	Hospital patient display board, 4-color custom design									
1010	Porcelain steel dry erase board, 36" x 24"	2 Carp	7.50	2.133	Ea.	239	117		356	450

10 13 Directories

10 13 10 – Building Directories

10 13 10.10 Directory Boards	Crew	Daily Output	Labor-Hours	Unit	Material	2021 Bare Costs Labor	Equipment	Total	Total Incl O&P
0010 **DIRECTORY BOARDS**									
0050 Plastic, glass covered, 30" x 20"	2 Carp	3	5.333	Ea.	183	292		475	665
0100 36" x 48"		2	8		565	440		1,005	1,325
0900 Outdoor, weatherproof, black plastic, 36" x 24"		2	8		685	440		1,125	1,450
1000 36" x 36"	↓	1.50	10.667	↓	263	585		848	1,225
9000 Minimum labor/equipment charge	1 Carp	1	8	Job		440		440	695

10 14 Signage

10 14 19 – Dimensional Letter Signage

10 14 19.10 Exterior Signs

	Crew	Daily Output	Labor-Hours	Unit	Material	2021 Bare Costs Labor	Equipment	Total	Total Incl O&P
0010 **EXTERIOR SIGNS**									
3900 Plaques, custom, 20" x 30", for up to 450 letters, cast aluminum	2 Carp	4	4	Ea.	1,825	219		2,044	2,375
4000 Cast bronze		4	4		1,975	219		2,194	2,500
4200 30" x 36", up to 900 letters, cast aluminum		3	5.333		2,975	292		3,267	3,750
4300 Cast bronze	↓	3	5.333		4,175	292		4,467	5,050
5100 Exit signs, 24 ga. alum., 14" x 12" surface mounted	1 Carp	30	.267		52.50	14.60		67.10	80.50
5200 10" x 7"	"	20	.400		28.50	22		50.50	66.50
6400 Replacement sign faces, 6" or 8"	1 Clab	50	.160	↓	64	7.10		71.10	81.50
9000 Minimum labor/equipment charge	1 Carp	4	2	Job		109		109	174

10 14 23 – Panel Signage

10 14 23.13 Engraved Panel Signage

	Crew	Daily Output	Labor-Hours	Unit	Material	2021 Bare Costs Labor	Equipment	Total	Total Incl O&P
0010 **ENGRAVED PANEL SIGNAGE**, interior									
1010 Flexible door sign, adhesive back, w/Braille, 5/8" letters, 4" x 4"	1 Clab	32	.250	Ea.	33.50	11.10		44.60	54
1050 6" x 6"		32	.250		45	11.10		56.10	67
1100 8" x 2"		32	.250		36.50	11.10		47.60	57.50
1150 8" x 4"		32	.250		43	11.10		54.10	64.50
1200 8" x 8"		32	.250		72	11.10		83.10	96.50
1250 12" x 2"		32	.250		35	11.10		46.10	56
1300 12" x 6"		32	.250		45	11.10		56.10	67
1350 12" x 12"		32	.250		129	11.10		140.10	160
1500 Graphic symbols, 2" x 2"		32	.250		18.50	11.10		29.60	38
1550 6" x 6"		32	.250		31	11.10		42.10	51.50
1600 8" x 8"	↓	32	.250		39	11.10		50.10	60
2010 Corridor, stock acrylic, 2-sided, with mounting bracket, 2" x 8"	1 Carp	24	.333		8	18.25		26.25	38
2020 2" x 10"		24	.333		35	18.25		53.25	67
2050 3" x 8"		24	.333		18.20	18.25		36.45	49
2060 3" x 10"		24	.333		40	18.25		58.25	73
2070 3" x 12"		24	.333		47.50	18.25		65.75	81.50
2100 4" x 8"		24	.333		24	18.25		42.25	55
2110 4" x 10"		24	.333		40.50	18.25		58.75	73.50
2120 4" x 12"	↓	24	.333	↓	49.50	18.25		67.75	83.50

10 14 26 – Post and Panel/Pylon Signage

10 14 26.10 Post and Panel Signage

	Crew	Daily Output	Labor-Hours	Unit	Material	2021 Bare Costs Labor	Equipment	Total	Total Incl O&P
0010 **POST AND PANEL SIGNAGE**, Alum., incl. 2 posts,									
0011 2-sided panel (blank), concrete footings per post									
0025 24" W x 18" H	B-1	25	.960	Ea.	123	43.50		166.50	205
0050 36" W x 24" H		25	.960		975	43.50		1,018.50	1,150
0100 57" W x 36" H		25	.960		2,125	43.50		2,168.50	2,400
0150 81" W x 46" H	↓	25	.960	↓	475	43.50		518.50	590

10 14 Signage

10 14 43 – Photoluminescent Signage

10 14 43.10 Photoluminescent Signage	Crew	Daily Output	Labor-Hours	Unit	Material	2021 Bare Costs Labor	Equipment	Total	Total Incl O&P
0010 **PHOTOLUMINESCENT SIGNAGE**									
0011 Photoluminescent exit sign, plastic, 7"x10"	1 Carp	32	.250	Ea.	23.50	13.70		37.20	48
0022 Polyester		32	.250		15.05	13.70		28.75	38.50
0033 Aluminum		32	.250		52.50	13.70		66.20	80
0044 Plastic, 10"x14"		32	.250		53	13.70		66.70	80.50
0055 Polyester		32	.250		55	13.70		68.70	82.50
0066 Aluminum		32	.250		121	13.70		134.70	155
0111 Directional exit sign, plastic, 10"x14"		32	.250		55	13.70		68.70	82.50
0122 Polyester		32	.250		55	13.70		68.70	82.50
0133 Aluminum		32	.250		67.50	13.70		81.20	96
0153 Kick plate exit sign, 10"x34"		28	.286		305	15.65		320.65	365
0173 Photoluminescent tape, 1"x60'		300	.027	L.F.	.53	1.46		1.99	2.91
0183 2"x60'		290	.028	"	1.05	1.51		2.56	3.55
0211 Photoluminescent directional arrows		42	.190	Ea.	12.75	10.40		23.15	30.50

10 14 53 – Traffic Signage

10 14 53.20 Traffic Signs

	Crew	Daily Output	Labor-Hours	Unit	Material	2021 Bare Costs Labor	Equipment	Total	Total Incl O&P
0010 **TRAFFIC SIGNS**									
0012 Stock, 24" x 24", no posts, .080" alum. reflectorized	B-80	70	.457	Ea.	97	22.50	15.05	134.55	159
0100 High intensity		70	.457		28.50	22.50	15.05	66.05	83.50
0300 30" x 30", reflectorized		70	.457		68.50	22.50	15.05	106.05	127
0400 High intensity		70	.457		182	22.50	15.05	219.55	252
0600 Guide and directional signs, 12" x 18", reflectorized		70	.457		42.50	22.50	15.05	80.05	98.50
0700 High intensity		70	.457		66.50	22.50	15.05	104.05	125
0900 18" x 24", stock signs, reflectorized		70	.457		32.50	22.50	15.05	70.05	88
1000 High intensity		70	.457		48.50	22.50	15.05	86.05	106
1200 24" x 24", stock signs, reflectorized	B-80	70	.457	Ea.	38.50	22.50	15.05	76.05	94.50
1300 High intensity		70	.457		54.50	22.50	15.05	92.05	112
1500 Add to above for steel posts, galvanized, 10'-0" upright, bolted		200	.160		44.50	7.80	5.25	57.55	67
1600 12'-0" upright, bolted		140	.229		39.50	11.15	7.50	58.15	69.50
1800 Highway road signs, aluminum, over 20 S.F., reflectorized		350	.091	S.F.	13	4.46	3.01	20.47	24.50
2000 High intensity		350	.091		13	4.46	3.01	20.47	24.50
2200 Highway, suspended over road, 80 S.F. min., reflectorized		165	.194		18.75	9.45	6.40	34.60	42.50
2300 High intensity		165	.194		18.60	9.45	6.40	34.45	42.50
9000 Replace directional sign		6	5.333	Ea.	182	260	175	617	805

10 17 Telephone Specialties

10 17 16 – Telephone Enclosures

10 17 16.10 Commercial Telephone Enclosures

	Crew	Daily Output	Labor-Hours	Unit	Material	2021 Bare Costs Labor	Equipment	Total	Total Incl O&P
0010 **COMMERCIAL TELEPHONE ENCLOSURES**									
0300 Shelf type, wall hung, recessed	2 Carp	5	3.200	Ea.	1,025	175		1,200	1,400
0400 Surface mount		5	3.200	"	1,725	175		1,900	2,175
9000 Minimum labor/equipment charge		4	4	Job		219		219	350

10 21 13 – Toilet Compartments

10 21 13.13 Metal Toilet Compartments

		Crew	Daily Output	Labor-Hours	Unit	Material	2021 Bare Costs Labor	2021 Bare Costs Equipment	Total	Total Incl O&P
0010	**METAL TOILET COMPARTMENTS**									
0110	Cubicles, ceiling hung									
0200	Powder coated steel	2 Carp	4	4	Ea.	540	219		759	945
0500	Stainless steel	"	4	4		890	219		1,109	1,325
0600	For handicap units, add					480			480	530
0900	Floor and ceiling anchored									
1000	Powder coated steel	2 Carp	5	3.200	Ea.	710	175		885	1,075
1300	Stainless steel	"	5	3.200		1	175		176	280
1400	For handicap units, add					380			380	415
1610	Floor anchored									
1700	Powder coated steel	2 Carp	7	2.286	Ea.	920	125		1,045	1,225
2000	Stainless steel	"	7	2.286		1,425	125		1,550	1,750
2100	For handicap units, add					360			360	395
2200	For juvenile units, deduct					45			45	49.50
2450	Floor anchored, headrail braced									
2500	Powder coated steel	2 Carp	6	2.667	Ea.	350	146		496	615
2804	Stainless steel	"	6	2.667		1,200	146		1,346	1,525
2900	For handicap units, add					330			330	360
3000	Wall hung partitions, powder coated steel	2 Carp	7	2.286		690	125		815	960
3300	Stainless steel	"	7	2.286		1,800	125		1,925	2,175
3400	For handicap units, add					330			330	360
4000	Screens, entrance, floor mounted, 58" high, 48" wide									
4200	Powder coated steel	2 Carp	15	1.067	Ea.	243	58.50		301.50	360
4500	Stainless steel	"	15	1.067	"	990	58.50		1,048.50	1,200
4650	Urinal screen, 18" wide									
4704	Powder coated steel	2 Carp	6.15	2.602	Ea.	229	142		371	480
5004	Stainless steel	"	6.15	2.602	"	530	142		672	810
5100	Floor mounted, headrail braced									
5300	Powder coated steel	2 Carp	8	2	Ea.	240	109		349	440
5600	Stainless steel	"	8	2	"	615	109		724	855
5750	Pilaster, flush									
5800	Powder coated steel	2 Carp	10	1.600	Ea.	268	87.50		355.50	435
6100	Stainless steel		10	1.600		615	87.50		702.50	815
6300	Post braced, powder coated steel		10	1.600		169	87.50		256.50	325
6800	Powder coated steel		10	1.600		104	87.50		191.50	253
7800	Wedge type, powder coated steel		10	1.600		137	87.50		224.50	290
8100	Stainless steel		10	1.600		625	87.50		712.50	830
9000	Minimum labor/equipment charge	1 Carp	2.50	3.200	Job		175		175	279

10 21 13.16 Plastic-Laminate-Clad Toilet Compartments

		Crew	Daily Output	Labor-Hours	Unit	Material	2021 Bare Costs Labor	2021 Bare Costs Equipment	Total	Total Incl O&P
0010	**PLASTIC-LAMINATE-CLAD TOILET COMPARTMENTS**									
0110	Cubicles, ceiling hung									
0300	Plastic laminate on particle board	2 Carp	4	4	Ea.	745	219		964	1,175
0600	For handicap units, add				"	480			480	530
0900	Floor and ceiling anchored									
1100	Plastic laminate on particle board	2 Carp	5	3.200	Ea.	810	175		985	1,175
1400	For handicap units, add				"	380			380	415
1610	Floor mounted									
1800	Plastic laminate on particle board	2 Carp	7	2.286	Ea.	540	125		665	795
2450	Floor mounted, headrail braced									
2600	Plastic laminate on particle board	2 Carp	6	2.667	Ea.	805	146		951	1,125
3400	For handicap units, add					330			330	360
4300	Entrance screen, floor mtd., plas. lam., 58" high, 48" wide	2 Carp	15	1.067		525	58.50		583.50	675

For customer support on your Commercial Renovation Costs with RSMeans Data, call 800.448.8182.

353

10 21 Compartments and Cubicles

10 21 13 – Toilet Compartments

10 21 13.16 Plastic-Laminate-Clad Toilet Compartments

		Crew	Daily Output	Labor-Hours	Unit	Material	2021 Bare Costs Labor	Equipment	Total	Total Incl O&P
4800	Urinal screen, 18" wide, ceiling braced, plastic laminate	2 Carp	8	2	Ea.	210	109		319	405
5400	Floor mounted, headrail braced		8	2		212	109		321	405
5900	Pilaster, flush, plastic laminate		10	1.600		435	87.50		522.50	620
6400	Post braced, plastic laminate	▼	10	1.600	▼	235	87.50		322.50	400
6700	Wall hung, bracket supported									
6900	Plastic laminate on particle board	2 Carp	10	1.600	Ea.	97	87.50		184.50	245
7450	Flange supported									
7500	Plastic laminate on particle board	2 Carp	10	1.600	Ea.	263	87.50		350.50	430
9000	Minimum labor/equipment charge	1 Carp	2.50	3.200	Job		175		175	279

10 21 13.19 Plastic Toilet Compartments

		Crew	Daily Output	Labor-Hours	Unit	Material	2021 Bare Costs Labor	Equipment	Total	Total Incl O&P
0010	**PLASTIC TOILET COMPARTMENTS**									
0110	Cubicles, ceiling hung									
0250	Phenolic	2 Carp	4	4	Ea.	920	219		1,139	1,350
0260	Polymer plastic	"	4	4		1,025	219		1,244	1,475
0600	For handicap units, add				▼	480			480	530
0900	Floor and ceiling anchored									
1050	Phenolic	2 Carp	5	3.200	Ea.	970	175		1,145	1,350
1060	Polymer plastic	"	5	3.200		1,200	175		1,375	1,600
1400	For handicap units, add				▼	380			380	415
1610	Floor mounted									
1750	Phenolic	2 Carp	7	2.286	Ea.	935	125		1,060	1,225
1760	Polymer plastic	"	7	2.286		845	125		970	1,125
2100	For handicap units, add					360			360	395
2200	For juvenile units, deduct				▼	45			45	49.50
2450	Floor mounted, headrail braced									
2550	Phenolic	2 Carp	6	2.667	Ea.	865	146		1,011	1,175
3600	Polymer plastic		6	2.667		890	146		1,036	1,200
3810	Entrance screen, polymer plastic, flr. mtd., 48"x58"		6	2.667		252	146		398	510
3820	Entrance screen, polymer plastic, flr. to clg pilaster, 48"x58"		6	2.667		252	146		398	510
6110	Urinal screen, polymer plastic, pilaster flush, 18" w		6	2.667		244	146		390	500
7110	Wall hung		6	2.667		166	146		312	415
7710	Flange mounted	▼	6	2.667	▼	198	146		344	450
9000	Minimum labor/equipment charge	1 Carp	2.50	3.200	Job		175		175	279

10 21 13.40 Stone Toilet Compartments

		Crew	Daily Output	Labor-Hours	Unit	Material	2021 Bare Costs Labor	Equipment	Total	Total Incl O&P
0010	**STONE TOILET COMPARTMENTS**									
0100	Cubicles, ceiling hung, marble	2 Marb	2	8	Ea.	1,650	425		2,075	2,500
0600	For handicap units, add					480			480	530
0800	Floor & ceiling anchored, marble	2 Marb	2.50	6.400		1,675	340		2,015	2,400
1400	For handicap units, add					380			380	415
1600	Floor mounted, marble	2 Marb	3	5.333		1,050	284		1,334	1,600
2400	Floor mounted, headrail braced, marble	"	3	5.333		1,200	284		1,484	1,775
2900	For handicap units, add					330			330	360
4100	Entrance screen, floor mounted marble, 58" high, 48" wide	2 Marb	9	1.778		725	94.50		819.50	945
4600	Urinal screen, 18" wide, ceiling braced, marble	D-1	6	2.667	▼	760	130		890	1,050
5100	Floor mounted, headrail braced									
5200	Marble	D-1	6	2.667	Ea.	650	130		780	925
5700	Pilaster, flush, marble		9	1.778		850	86.50		936.50	1,075
6200	Post braced, marble	▼	9	1.778	▼	830	86.50		916.50	1,050
9000	Minimum labor/equipment charge	1 Carp	2.50	3.200	Job		175		175	279

10 21 14 – Toilet Compartment Components

10 21 14.13 Metal Toilet Compartment Components

		Crew	Daily Output	Labor-Hours	Unit	Material	2021 Bare Costs Labor	Equipment	Total	Total Incl O&P
0010	**METAL TOILET COMPARTMENT COMPONENTS**									
0100	Pilasters									
0110	Overhead braced, powder coated steel, 7" wide x 82" high	2 Carp	22.20	.721	Ea.	77.50	39.50		117	149
0120	Stainless steel		22.20	.721		141	39.50		180.50	218
0130	Floor braced, powder coated steel, 7" wide x 70" high		23.30	.687		105	37.50		142.50	175
0140	Stainless steel		23.30	.687		234	37.50		271.50	315
0150	Ceiling hung, powder coated steel, 7" wide x 83" high		13.30	1.203		132	66		198	250
0160	Stainless steel		13.30	1.203		330	66		396	465
0170	Wall hung, powder coated steel, 3" wide x 58" high		18.90	.847		135	46.50		181.50	223
0180	Stainless steel		18.90	.847		222	46.50		268.50	320
0200	Panels									
0210	Powder coated steel, 31" wide x 58" high	2 Carp	18.90	.847	Ea.	127	46.50		173.50	213
0220	Stainless steel		18.90	.847		320	46.50		366.50	425
0230	Powder coated steel, 53" wide x 58" high		18.90	.847		187	46.50		233.50	280
0240	Stainless steel		18.90	.847		435	46.50		481.50	555
0250	Powder coated steel, 63" wide x 58" high		18.90	.847		153	46.50		199.50	242
0260	Stainless steel		18.90	.847		535	46.50		581.50	665
0300	Doors									
0310	Powder coated steel, 24" wide x 58" high	2 Carp	14.10	1.135	Ea.	130	62		192	242
0320	Stainless steel		14.10	1.135		420	62		482	560
0330	Powder coated steel, 26" wide x 58" high		14.10	1.135		162	62		224	278
0340	Stainless steel		14.10	1.135		430	62		492	570
0350	Powder coated steel, 28" wide x 58" high		14.10	1.135		176	62		238	293
0360	Stainless steel		14.10	1.135		445	62		507	590
0370	Powder coated steel, 36" wide x 58" high		14.10	1.135		183	62		245	300
0380	Stainless steel		14.10	1.135		495	62		557	645
0400	Headrails									
0410	For powder coated steel, 62" long	2 Carp	65	.246	Ea.	35	13.45		48.45	60
0420	Stainless steel		65	.246		35	13.45		48.45	60
0430	For powder coated steel, 84" long		50	.320		29.50	17.50		47	60.50
0440	Stainless steel		50	.320		30.50	17.50		48	61.50
0450	For powder coated steel, 120" long		30	.533		40.50	29		69.50	91
0460	Stainless steel		30	.533		40.50	29		69.50	91
9000	Minimum labor/equipment charge	1 Carp	4	2	Job		109		109	174

10 21 14.16 Plastic-Laminate Clad Toilet Compartment Components

		Crew	Daily Output	Labor-Hours	Unit	Material	2021 Bare Costs Labor	Equipment	Total	Total Incl O&P
0010	**PLASTIC-LAMINATE CLAD TOILET COMPARTMENT COMPONENTS**									
0100	Pilasters									
0110	Overhead braced, 7" wide x 82" high	2 Carp	22.20	.721	Ea.	102	39.50		141.50	176
0130	Floor anchored, 7" wide x 70" high		23.30	.687		108	37.50		145.50	178
0150	Ceiling hung, 7" wide x 83" high		13.30	1.203		127	66		193	245
0180	Wall hung, 3" wide x 58" high		18.90	.847		96	46.50		142.50	180
0200	Panels									
0210	31" wide x 58" high	2 Carp	18.90	.847	Ea.	171	46.50		217.50	262
0230	51" wide x 58" high		18.90	.847		210	46.50		256.50	305
0250	63" wide x 58" high		18.90	.847		265	46.50		311.50	365
0300	Doors									
0310	24" wide x 58" high	2 Carp	14.10	1.135	Ea.	156	62		218	271
0330	26" wide x 58" high		14.10	1.135		161	62		223	276
0350	28" wide x 58" high		14.10	1.135		166	62		228	282
0370	36" wide x 58" high		14.10	1.135		177	62		239	294
0400	Headrails									
0410	62" long	2 Carp	65	.246	Ea.	24	13.45		37.45	48

10 21 Compartments and Cubicles

10 21 14 – Toilet Compartment Components

10 21 14.16 Plastic-Laminate Clad Toilet Compartment Components

		Crew	Daily Output	Labor-Hours	Unit	Material	2021 Bare Costs Labor	2021 Bare Costs Equipment	Total	Total Incl O&P
0430	84" long	2 Carp	60	.267	Ea.	33.50	14.60		48.10	59.50
0450	120" long	↓	30	.533	↓	41.50	29		70.50	92.50
9000	Minimum labor/equipment charge	1 Carp	4	2	Job		109		109	174

10 21 14.19 Plastic Toilet Compartment Components

		Crew	Daily Output	Labor-Hours	Unit	Material	2021 Bare Costs Labor	2021 Bare Costs Equipment	Total	Total Incl O&P
0010	**PLASTIC TOILET COMPARTMENT COMPONENTS**									
0100	Pilasters									
0110	Overhead braced, polymer plastic, 7" wide x 82" high	2 Carp	22.20	.721	Ea.	104	39.50		143.50	177
0120	Phenolic		22.20	.721		139	39.50		178.50	216
0130	Floor braced, polymer plastic, 7" wide x 70" high		23.30	.687		192	37.50		229.50	271
0140	Phenolic		23.30	.687		104	37.50		141.50	174
0150	Ceiling hung, polymer plastic, 7" wide x 83" high		13.30	1.203		177	66		243	300
0160	Phenolic		13.30	1.203		127	66		193	245
0180	Wall hung, phenolic, 3" wide x 58" high	↓	18.90	.847		109	46.50		155.50	193
0200	Panels									
0203	Polymer plastic, 18" wide x 55" high	2 Carp	18.90	.847	Ea.	310	46.50		356.50	415
0206	Phenolic, 18" wide x 58" high		18.90	.847		305	46.50		351.50	410
0210	Polymer plastic, 31" wide x 55" high		18.90	.847		350	46.50		396.50	460
0220	Phenolic, 31" wide x 58" high		18.90	.847		284	46.50		330.50	385
0223	Polymer plastic, 48" wide x 55" high		18.90	.847		535	46.50		581.50	665
0226	Phenolic, 48" wide x 58" high		18.90	.847		470	46.50		516.50	590
0230	Polymer plastic, 51" wide x 55" high		18.90	.847		490	46.50		536.50	615
0240	Phenolic, 51" wide x 58" high		18.90	.847		435	46.50		481.50	555
0250	Polymer plastic, 63" wide x 55" high		18.90	.847		680	46.50		726.50	825
0260	Phenolic, 63" wide x 58" high	↓	18.90	.847	↓	560	46.50		606.50	690
0300	Doors									
0310	Polymer plastic, 24" wide x 55" high	2 Carp	14.10	1.135	Ea.	263	62		325	390
0320	Phenolic, 24" wide x 58" high		14.10	1.135		350	62		412	485
0330	Polymer plastic, 26" wide x 55" high		14.10	1.135		273	62		335	400
0340	Phenolic, 26" wide x 58" high		14.10	1.135		370	62		432	505
0350	Polymer plastic, 28" wide x 55" high		14.10	1.135		293	62		355	420
0360	Phenolic, 28" wide x 58" high		14.10	1.135		375	62		437	515
0370	Polymer plastic, 36" wide x 55" high		14.10	1.135		325	62		387	460
0380	Phenolic, 36" wide x 58" high	↓	14.10	1.135	↓	520	62		582	670
0400	Headrails									
0410	For polymer plastic, 62" long	2 Carp	65	.246	Ea.	35	13.45		48.45	60
0420	Phenolic		65	.246		35	13.45		48.45	60
0430	For polymer plastic, 84" long		50	.320		35	17.50		52.50	66.50
0440	Phenolic		50	.320		36.50	17.50		54	68
0450	For polymer plastic, 120" long		30	.533		50.50	29		79.50	103
0460	Phenolic	↓	30	.533	↓	40	29		69	91
9000	Minimum labor/equipment charge	1 Carp	4	2	Job		109		109	174

10 21 23 – Cubicle Curtains and Track

10 21 23.16 Cubicle Track and Hardware

		Crew	Daily Output	Labor-Hours	Unit	Material	2021 Bare Costs Labor	2021 Bare Costs Equipment	Total	Total Incl O&P
0010	**CUBICLE TRACK AND HARDWARE**									
0020	Curtain track, box channel, ceiling mounted	1 Carp	135	.059	L.F.	6.90	3.24		10.14	12.75
0100	Suspended		100	.080		8.60	4.38		12.98	16.45
0550	Polyester, antimicrobial, 9' ceiling height		425	.019		19.45	1.03		20.48	23
0560	8' ceiling height	↓	425	.019	↓	17.35	1.03		18.38	20.50

10 22 Partitions

10 22 16 – Folding Gates

10 22 16.10 Security Gates

	Crew	Daily Output	Labor-Hours	Unit	Material	2021 Bare Costs Labor	Equipment	Total	Total Incl O&P
0010 **SECURITY GATES**									
0015 For roll up type, see Section 08 33 13.10									
0300 Scissors type folding gate, ptd. steel, single, 6-1/2' high, 5-1/2' wide	2 Sswk	4	4	Opng.	170	241		411	575
0350 6-1/2' wide		4	4		250	241		491	665
0400 7-1/2' wide		4	4		264	241		505	680
0600 Double gate, 8' high, 8' wide		2.50	6.400		375	385		760	1,050
0650 10' wide		2.50	6.400		450	385		835	1,125
0700 12' wide		2	8		640	480		1,120	1,475
0750 14' wide		2	8		385	480		865	1,200
0900 Door gate, folding steel, 4' wide, 61" high		4	4		152	241		393	555
1000 71" high		4	4		145	241		386	550
1200 81" high		4	4		157	241		398	565
1300 Window gates, 2' to 4' wide, 31" high		4	4		82.50	241		323.50	480
1500 55" high		3.75	4.267		122	257		379	550
1600 79" high		3.50	4.571		144	276		420	605

10 22 19 – Demountable Partitions

10 22 19.43 Demountable Composite Partitions

	Crew	Daily Output	Labor-Hours	Unit	Material	2021 Bare Costs Labor	Equipment	Total	Total Incl O&P
0010 **DEMOUNTABLE COMPOSITE PARTITIONS**, add for doors									
0100 Do not deduct door openings from total L.F.									
0900 Demountable gypsum system on 2" to 2-1/2"									
1000 Steel studs, 9' high, 3" to 3-3/4" thick									
1200 Vinyl-clad gypsum	2 Carp	48	.333	L.F.	61	18.25		79.25	96.50
1300 Fabric clad gypsum		44	.364		153	19.90		172.90	200
1500 Steel clad gypsum		40	.400		171	22		193	223
1600 1.75 system, aluminum framing, vinyl-clad hardboard,									
1800 Paper honeycomb core panel, 1-3/4" to 2-1/2" thick									
1900 9' high	2 Carp	48	.333	L.F.	103	18.25		121.25	142
2100 7' high		60	.267		92	14.60		106.60	124
2200 5' high		80	.200		78	10.95		88.95	103
2250 Unitized gypsum system									
2300 Unitized panel, 9' high, 2" to 2-1/2" thick									
2350 Vinyl-clad gypsum	2 Carp	48	.333	L.F.	132	18.25		150.25	175
2400 Fabric clad gypsum	"	44	.364	"	218	19.90		237.90	272
2500 Unitized mineral fiber system									
2510 Unitized panel, 9' high, 2-1/4" thick, aluminum frame									
2550 Vinyl-clad mineral fiber	2 Carp	48	.333	L.F.	132	18.25		150.25	174
2600 Fabric clad mineral fiber	"	44	.364	"	196	19.90		215.90	248
2800 Movable steel walls, modular system									
2900 Unitized panels, 9' high, 48" wide									
3100 Baked enamel, pre-finished	2 Carp	60	.267	L.F.	149	14.60		163.60	187
3200 Fabric clad steel	"	56	.286	"	216	15.65		231.65	262
5500 For acoustical partitions, add, unsealed				S.F.	2.38			2.38	2.62
5550 Sealed				"	11.10			11.10	12.20
5700 For doors, see Section 08 16									
5800 For door hardware, see Section 08 71									
6100 In-plant modular office system, w/prehung hollow core door									
6200 3" thick polystyrene core panels									
6250 12' x 12', 2 wall	2 Clab	3.80	4.211	Ea.	4,575	187		4,762	5,325
6300 4 wall		1.90	8.421		6,400	375		6,775	7,625
6350 16' x 16', 2 wall		3.60	4.444		6,925	197		7,122	7,950
6400 4 wall		1.80	8.889		11,100	395		11,495	12,800
9000 Minimum labor/equipment charge	2 Carp	3	5.333	Job		292		292	465

For customer support on your Commercial Renovation Costs with RSMeans Data, call 800.448.8182.

357

10 22 23 – Portable Partitions, Screens, and Panels

10 22 23.13 Wall Screens

		Crew	Daily Output	Labor-Hours	Unit	Material	2021 Bare Costs Labor	Equipment	Total	Total Incl O&P
0010	**WALL SCREENS**, divider panels, free standing, fiber core									
0020	Fabric face straight									
0100	3'-0" long, 4'-0" high	2 Carp	100	.160	L.F.	137	8.75		145.75	164
0200	5'-0" high		90	.178		118	9.70		127.70	146
0500	6'-0" high		75	.213		119	11.65		130.65	150
0900	5'-0" long, 4'-0" high		175	.091		61	5		66	75
1000	5'-0" high		150	.107		83	5.85		88.85	101
1500	6'-0" high		125	.128		112	7		119	134
1600	6'-0" long, 5'-0" high		162	.099		112	5.40		117.40	132
3200	Economical panels, fabric face, 4'-0" long, 5'-0" high		132	.121		59	6.65		65.65	75
3250	6'-0" high		112	.143		62.50	7.80		70.30	81.50
3300	5'-0" long, 5'-0" high		150	.107		40.50	5.85		46.35	54
3350	6'-0" high		125	.128		48	7		55	64
3450	Acoustical panels, 60 to 90 NRC, 3'-0" long, 5'-0" high		90	.178		91	9.70		100.70	116
3550	6'-0" high		75	.213		86.50	11.65		98.15	114
3600	5'-0" long, 5'-0" high		150	.107		52.50	5.85		58.35	67
3650	6'-0" high		125	.128		66	7		73	83.50
3700	6'-0" long, 5'-0" high		162	.099		46.50	5.40		51.90	59.50
3750	6'-0" high		138	.116		83.50	6.35		89.85	102
3800	Economy acoustical panels, 40 NRC, 4'-0" long, 5'-0" high		132	.121		59	6.65		65.65	75
3850	6'-0" high		112	.143		62.50	7.80		70.30	81.50
3900	5'-0" long, 6'-0" high		125	.128		48	7		55	64
3950	6'-0" long, 5'-0" high		162	.099		43.50	5.40		48.90	56.50
9000	Minimum labor/equipment charge		3	5.333	Job		292		292	465

10 22 33 – Accordion Folding Partitions

10 22 33.10 Partitions, Accordion Folding

		Crew	Daily Output	Labor-Hours	Unit	Material	2021 Bare Costs Labor	Equipment	Total	Total Incl O&P
0010	**PARTITIONS, ACCORDION FOLDING**									
0100	Vinyl covered, over 150 S.F., frame not included									
0300	Residential, 1.25 lb./S.F., 8' maximum height	2 Carp	300	.053	S.F.	22.50	2.92		25.42	29
0400	Commercial, 1.75 lb./S.F., 8' maximum height		225	.071		25.50	3.89		29.39	34
0900	Acoustical, 3 lb./S.F., 17' maximum height		100	.160		27.50	8.75		36.25	44.50
1200	5 lb./S.F., 20' maximum height		95	.168		44.50	9.20		53.70	63.50
1500	Vinyl-clad wood or steel, electric operation, 5.0 psf		160	.100		55.50	5.45		60.95	69.50
1900	Wood, non-acoustic, birch or mahogany, to 10' high		300	.053		30	2.92		32.92	37.50
9000	Minimum labor/equipment charge		4	4	Job		219		219	350

10 22 39 – Folding Panel Partitions

10 22 39.10 Partitions, Folding Panel

		Crew	Daily Output	Labor-Hours	Unit	Material	2021 Bare Costs Labor	Equipment	Total	Total Incl O&P
0010	**PARTITIONS, FOLDING PANEL**, acoustic, wood									
0100	Vinyl faced, to 18' high, 6 psf, economy trim	2 Carp	60	.267	S.F.	74.50	14.60		89.10	105
0150	Standard trim	2 Carp	45	.356	S.F.	89	19.45		108.45	129
0200	Premium trim		30	.533		115	29		144	173
0400	Plastic laminate or hardwood finish, standard trim		60	.267		76.50	14.60		91.10	107
0500	Premium trim		30	.533		81.50	29		110.50	137
0600	Wood, low acoustical type, 4.5 psf, to 14' high		50	.320		55.50	17.50		73	89.50
9000	Minimum labor/equipment charge		4	4	Job		219		219	350

10 26 Wall and Door Protection

10 26 23 – Protective Wall Covering

10 26 23.10 Wall Covering, Protective

	Crew	Daily Output	Labor-Hours	Unit	Material	2021 Bare Costs Labor	Equipment	Total	Total Incl O&P
0010 **WALL COVERING, PROTECTIVE**									
1775 Wall end guard, stainless steel, 16 ga., 36" tall, screwed to studs	1 Skwk	30	.267	Ea.	34	15.25		49.25	61.50

10 26 23.13 Impact Resistant Wall Protection

	Crew	Daily Output	Labor-Hours	Unit	Material	2021 Bare Costs Labor	Equipment	Total	Total Incl O&P
0010 **IMPACT RESISTANT WALL PROTECTION**									
0100 Vinyl wall protection, complete instl. incl. panels, trim, adhesive									
0110 .040" thk., std. colors	2 Carp	320	.050	S.F.	11.70	2.74		14.44	17.20
0120 .040" thk., element patterns		300	.053		16.55	2.92		19.47	23
0130 .060" thk., std. colors		300	.053		16.50	2.92		19.42	23
0140 .060" thk., element patterns		280	.057		19.10	3.13		22.23	26
1775 Wall protection, stainless steel, 16 ga., 48" x 36" tall, screwed to studs	2 Skwk	500	.032		8.25	1.83		10.08	12

10 26 33 – Door and Frame Protection

10 26 33.10 Protection, Door and Frame

	Crew	Daily Output	Labor-Hours	Unit	Material	2021 Bare Costs Labor	Equipment	Total	Total Incl O&P
0010 **PROTECTION, DOOR AND FRAME**									
0100 Door frame guard, vinyl, 3" x 3" x 4'	1 Carp	30	.267	Ea.	93	14.60		107.60	125
0110 Door frame guard, vinyl, 3" x 3" x 8'		26	.308		135	16.85		151.85	176
0120 Door frame guard, stainless steel, 3" x 3" x 4'		30	.267		440	14.60		454.60	510
0500 Steel door track/wheel guards, 4'-0" high	E-4	22	1.455		113	88.50	6.75	208.25	274

10 28 Toilet, Bath, and Laundry Accessories

10 28 13 – Toilet Accessories

10 28 13.13 Commercial Toilet Accessories

	Crew	Daily Output	Labor-Hours	Unit	Material	2021 Bare Costs Labor	Equipment	Total	Total Incl O&P
0010 **COMMERCIAL TOILET ACCESSORIES**									
0200 Curtain rod, stainless steel, 5' long, 1" diameter	1 Carp	13	.615	Ea.	26.50	33.50		60	82.50
0300 1-1/4" diameter		13	.615		18.30	33.50		51.80	73.50
0350 Chrome, 1" diameter		13	.615		28	33.50		61.50	84.50
0360 For vinyl curtain, add		1950	.004	S.F.	131	.22		131.22	144
0400 Diaper changing station, horizontal, wall mounted, plastic		10	.800	Ea.	315	44		359	415
0420 Vertical		10	.800		221	44		265	315
0430 Oval shaped		10	.800		305	44		349	405
0440 Recessed, with stainless steel flange		6	1.333		800	73		873	995
0500 Dispenser units, combined soap & towel dispensers,									
0510 Mirror and shelf, flush mounted	1 Carp	10	.800	Ea.	355	44		399	460
0600 Towel dispenser and waste receptacle,									
0610 18 gallon capacity	1 Carp	10	.800	Ea.	385	44		429	495
0800 Grab bar, straight, 1-1/4" diameter, stainless steel, 18" long		24	.333		10.90	18.25		29.15	41
0900 24" long		23	.348		13.30	19.05		32.35	45
1000 30" long		22	.364		27.50	19.90		47.40	61.50
1100 36" long		20	.400		13.30	22		35.30	49.50
1105 42" long		20	.400		41.50	22		63.50	80.50
1120 Corner, 36" long		20	.400		95	22		117	139
1200 1-1/2" diameter, 24" long		23	.348		37.50	19.05		56.55	72
1300 36" long		20	.400		43	22		65	82
1310 42" long		18	.444		34	24.50		58.50	76
1500 Tub bar, 1-1/4" diameter, 24" x 36"		14	.571		110	31.50		141.50	171
1600 Plus vertical arm		12	.667		108	36.50		144.50	177
1900 End tub bar, 1" diameter, 90° angle, 16" x 32"		12	.667		102	36.50		138.50	170
2010 Tub/shower/toilet, 2-wall, 36" x 24"		12	.667		88	36.50		124.50	155
2300 Hand dryer, surface mounted, electric, 115 volt, 20 amp		4	2		425	109		534	640
2400 230 volt, 10 amp		4	2		585	109		694	820
2450 Hand dryer, touch free, 1400 watt, 81,000 rpm		4	2		1,350	109		1,459	1,675

For customer support on your Commercial Renovation Costs with RSMeans Data, call 800.448.8182.

359

10 28 13.13 Commercial Toilet Accessories

		Crew	Daily Output	Labor-Hours	Unit	Material	2021 Bare Costs Labor	2021 Bare Costs Equipment	Total	Total Incl O&P
2600	Hat and coat strip, stainless steel, 4 hook, 36" long	1 Carp	24	.333	Ea.	36.50	18.25		54.75	69
2700	6 hook, 60" long		20	.400		106	22		128	151
3000	Mirror, with stainless steel 3/4" square frame, 18" x 24"		20	.400		50	22		72	90
3100	36" x 24"		15	.533		63.50	29		92.50	117
3200	48" x 24"		10	.800		168	44		212	255
3300	72" x 24"		6	1.333		385	73		458	540
3500	With 5" stainless steel shelf, 18" x 24"		20	.400		83	22		105	126
3600	36" x 24"		15	.533		290	29		319	365
3700	48" x 24"		10	.800		285	44		329	385
3800	72" x 24"		6	1.333		241	73		314	380
4100	Mop holder strip, stainless steel, 5 holders, 48" long		20	.400		72	22		94	114
4200	Napkin/tampon dispenser, recessed		15	.533		635	29		664	745
4220	Semi-recessed		6.50	1.231		345	67.50		412.50	485
4250	Napkin receptacle, recessed		6.50	1.231		170	67.50		237.50	294
4300	Robe hook, single, regular		96	.083		15.30	4.56		19.86	24
4400	Heavy duty, concealed mounting		56	.143		25.50	7.80		33.30	40.50
4600	Soap dispenser, chrome, surface mounted, liquid		20	.400		67.50	22		89.50	110
5000	Recessed stainless steel, liquid		10	.800		172	44		216	259
5600	Shelf, stainless steel, 5" wide, 18 ga., 24" long		24	.333		66.50	18.25		84.75	102
5700	48" long		16	.500		212	27.50		239.50	277
5800	8" wide shelf, 18 ga., 24" long		22	.364		85	19.90		104.90	125
5900	48" long	1 Carp	14	.571	Ea.	146	31.50		177.50	210
6000	Toilet seat cover dispenser, stainless steel, recessed		20	.400		181	22		203	234
6050	Surface mounted		15	.533		34	29		63	84
6100	Toilet tissue dispenser, surface mounted, SS, single roll		30	.267		16	14.60		30.60	40.50
6200	Double roll		24	.333		23.50	18.25		41.75	55
6240	Plastic, twin/jumbo dbl. roll		24	.333		30.50	18.25		48.75	62.50
6290	Toilet seat	1 Plum	40	.200		26	13.55		39.55	49.50
6400	Towel bar, stainless steel, 18" long	1 Carp	23	.348		38.50	19.05		57.55	72.50
6500	30" long		21	.381		46	21		67	84
6700	Towel dispenser, stainless steel, surface mounted		16	.500		38.50	27.50		66	86
6800	Flush mounted, recessed		10	.800		67.50	44		111.50	144
6900	Plastic, touchless, battery operated		16	.500		79	27.50		106.50	131
7000	Towel holder, hotel type, 2 guest size		20	.400		54	22		76	94.50
7200	Towel shelf, stainless steel, 24" long, 8" wide		20	.400		85	22		107	129
7400	Tumbler holder, for tumbler only		30	.267		37.50	14.60		52.10	64
7410	Tumbler holder, recessed		20	.400		8.25	22		30.25	44
7500	Soap, tumbler & toothbrush		30	.267		17.40	14.60		32	42
7510	Tumbler & toothbrush holder		20	.400		11.15	22		33.15	47.50
8000	Waste receptacles, stainless steel, with top, 13 gallon		10	.800		315	44		359	420
8100	36 gallon		8	1		535	54.50		589.50	675
9000	Minimum labor/equipment charge		5	1.600	Job		87.50		87.50	139
9996	Bathroom access., grab bar, straight, 1-1/2" diam., SS, 42" L install only		18	.444	Ea.		24.50		24.50	38.50

10 28 16.20 Medicine Cabinets

		Crew	Daily Output	Labor-Hours	Unit	Material	2021 Bare Costs Labor	2021 Bare Costs Equipment	Total	Total Incl O&P
0010	**MEDICINE CABINETS**									
0020	With mirror, sst frame, 16" x 22", unlighted	1 Carp	14	.571	Ea.	102	31.50		133.50	162
0100	Wood frame		14	.571		128	31.50		159.50	191
0300	Sliding mirror doors, 20" x 16" x 4-3/4", unlighted		7	1.143		136	62.50		198.50	250
0400	24" x 19" x 8-1/2", lighted		5	1.600		220	87.50		307.50	380
0600	Triple door, 30" x 32", unlighted, plywood body		7	1.143		271	62.50		333.50	400
0700	Steel body		7	1.143		355	62.50		417.50	490

For customer support on your Commercial Renovation Costs with RSMeans Data, call 800.448.8182.

10 28 Toilet, Bath, and Laundry Accessories

10 28 16 – Bath Accessories

10 28 16.20 Medicine Cabinets	Crew	Daily Output	Labor-Hours	Unit	Material	2021 Bare Costs Labor	Equipment	Total	Total Incl O&P
0900	Oak door, wood body, beveled mirror, single door	1 Carp	7	1.143	Ea.	119	62.50		181.50	231
1000	Double door		6	1.333		375	73		448	525
1200	Hotel cabinets, stainless, with lower shelf, unlighted		10	.800		296	44		340	395
1300	Lighted		5	1.600		310	87.50		397.50	485
9000	Minimum labor/equipment charge		4	2	Job		109		109	174

10 28 19 – Tub and Shower Enclosures

10 28 19.10 Partitions, Shower

		Crew	Daily Output	Labor-Hours	Unit	Material	2021 Bare Costs Labor	Equipment	Total	Total Incl O&P
0010	**PARTITIONS, SHOWER** floor mounted, no plumbing									
0400	Cabinet, one piece, fiberglass, 32" x 32"	2 Carp	5	3.200	Ea.	625	175		800	965
0420	36" x 36"		5	3.200		545	175		720	880
0440	36" x 48"		5	3.200		1,375	175		1,550	1,800
0460	Acrylic, 32" x 32"		5	3.200		225	175		400	525
0480	36" x 36"		5	3.200		1,050	175		1,225	1,425
0500	36" x 48"		5	3.200		815	175		990	1,175
0520	Shower door for above, clear plastic, 24" wide	1 Carp	8	1		150	54.50		204.50	252
0540	28" wide		8	1		190	54.50		244.50	296
0560	Tempered glass, 24" wide		8	1		135	54.50		189.50	236
0580	28" wide		8	1		160	54.50		214.50	263
4100	Shower doors, economy plastic, 24" wide	1 Shee	9	.889		133	58		191	237
4200	Tempered glass door, economy		8	1		278	65.50		343.50	405
4700	Deluxe, tempered glass, chrome on brass frame, 42" to 44"		8	1		320	65.50		385.50	450
4800	39" to 48" wide		1	8		725	525		1,250	1,600
9990	Minimum labor/equipment charge		2.50	3.200	Job		209		209	325

10 31 Manufactured Fireplaces

10 31 13 – Manufactured Fireplace Chimneys

10 31 13.10 Fireplace Chimneys

		Crew	Daily Output	Labor-Hours	Unit	Material	2021 Bare Costs Labor	Equipment	Total	Total Incl O&P
0010	**FIREPLACE CHIMNEYS**									
0500	Chimney dbl. wall, all stainless, over 8'-6", 7" diam., add to fireplace	1 Carp	33	.242	V.L.F.	97	13.25		110.25	128
0600	10" diameter, add to fireplace		32	.250		102	13.70		115.70	135
0700	12" diameter, add to fireplace		31	.258		204	14.10		218.10	247
0800	14" diameter, add to fireplace		30	.267		212	14.60		226.60	256
1000	Simulated brick chimney top, 4' high, 16" x 16"		10	.800	Ea.	485	44		529	605
1100	24" x 24"		7	1.143	"	575	62.50		637.50	735

10 31 13.20 Chimney Accessories

		Crew	Daily Output	Labor-Hours	Unit	Material	2021 Bare Costs Labor	Equipment	Total	Total Incl O&P
0010	**CHIMNEY ACCESSORIES**									
0020	Chimney screens, galv., 13" x 13" flue	1 Bric	8	1	Ea.	58	53.50		111.50	151
0050	24" x 24" flue		5	1.600		124	86		210	275
0200	Stainless steel, 13" x 13" flue		8	1		106	53.50		159.50	203
0250	20" x 20" flue		5	1.600		148	86		234	300
2400	Squirrel and bird screens, galvanized, 8" x 8" flue		16	.500		57	27		84	106
2450	13" x 13" flue		12	.667		148	36		184	221
9000	Minimum labor/equipment charge		3.50	2.286	Job		123		123	197

10 31 16 – Manufactured Fireplace Forms

10 31 16.10 Fireplace Forms

		Crew	Daily Output	Labor-Hours	Unit	Material	2021 Bare Costs Labor	Equipment	Total	Total Incl O&P
0010	**FIREPLACE FORMS**									
1800	Fireplace forms, no accessories, 32" opening	1 Bric	3	2.667	Ea.	700	143		843	995
1900	36" opening		2.50	3.200		890	172		1,062	1,250
2000	40" opening		2	4		1,175	215		1,390	1,650
2100	78" opening		1.50	5.333		1,725	286		2,011	2,350

361

10 31 Manufactured Fireplaces

10 31 23 – Prefabricated Fireplaces

10 31 23.10 Fireplace, Prefabricated	Crew	Daily Output	Labor-Hours	Unit	Material	2021 Bare Costs Labor	2021 Bare Costs Equipment	Total	Total Incl O&P
0010 **FIREPLACE, PREFABRICATED**, free standing or wall hung									
0100 With hood & screen, painted	1 Carp	1.30	6.154	Ea.	1,650	335		1,985	2,325
0150 Average		1	8		1,525	440		1,965	2,375
0200 Stainless steel		.90	8.889		2,800	485		3,285	3,850
1500 Simulated logs, gas fired, 40,000 BTU, 2' long, manual safety pilot		7	1.143	Set	575	62.50		637.50	730
1600 Adjustable flame remote pilot		6	1.333		2,425	73		2,498	2,775
1700 Electric, 1,500 BTU, 1'-6" long, incandescent flame		7	1.143		284	62.50		346.50	410
1800 1,500 BTU, LED flame		6	1.333		345	73		418	490
2000 Fireplace, built-in, 36" hearth, radiant		1.30	6.154	Ea.	760	335		1,095	1,375
2100 Recirculating, small fan		1	8		910	440		1,350	1,700
2150 Large fan		.90	8.889		2,450	485		2,935	3,475
2200 42" hearth, radiant		1.20	6.667		1,075	365		1,440	1,750
2300 Recirculating, small fan		.90	8.889		1,225	485		1,710	2,125
2350 Large fan		.80	10		1,325	545		1,870	2,325
2400 48" hearth, radiant		1.10	7.273		2,100	400		2,500	2,950
2500 Recirculating, small fan		.80	10		2,675	545		3,220	3,825
2550 Large fan		.70	11.429		2,675	625		3,300	3,925
3000 See through, including doors		.80	10		2,125	545		2,670	3,200
3200 Corner (2 wall)		1	8		3,550	440		3,990	4,600

10 32 Fireplace Specialties

10 32 13 – Fireplace Dampers

10 32 13.10 Dampers

10 32 13.10 Dampers	Crew	Daily Output	Labor-Hours	Unit	Material	2021 Bare Costs Labor	2021 Bare Costs Equipment	Total	Total Incl O&P
0010 **DAMPERS**									
0800 Damper, rotary control, steel, 30" opening	1 Bric	6	1.333	Ea.	125	71.50		196.50	253
0850 Cast iron, 30" opening		6	1.333		124	71.50		195.50	252
0880 36" opening		6	1.333		128	71.50		199.50	255
0900 48" opening		6	1.333		150	71.50		221.50	280
0920 60" opening		6	1.333		360	71.50		431.50	510
1000 84" opening, special order		5	1.600		925	86		1,011	1,175
1050 96" opening, special order		4	2		940	107		1,047	1,200
1200 Steel plate, poker control, 60" opening		8	1		330	53.50		383.50	445
1250 84" opening, special order		5	1.600		600	86		686	800
1400 "Universal" type, chain operated, 32" x 20" opening		8	1		255	53.50		308.50	370
1450 48" x 24" opening		5	1.600		380	86		466	560

10 32 23 – Fireplace Doors

10 32 23.10 Doors

10 32 23.10 Doors	Crew	Daily Output	Labor-Hours	Unit	Material	2021 Bare Costs Labor	2021 Bare Costs Equipment	Total	Total Incl O&P
0010 **DOORS**									
0400 Cleanout doors and frames, cast iron, 8" x 8"	1 Bric	12	.667	Ea.	53	36		89	116
0450 12" x 12"		10	.800		103	43		146	182
0500 18" x 24"		8	1		153	53.50		206.50	255
0550 Cast iron frame, steel door, 24" x 30"		5	1.600		380	86		466	560
1600 Dutch oven door and frame, cast iron, 12" x 15" opening		13	.615		134	33		167	200
1650 Copper plated, 12" x 15" opening		13	.615		262	33		295	340

10 35 Stoves

10 35 13 – Heating Stoves

10 35 13.10 Wood Burning Stoves		Crew	Daily Output	Labor-Hours	Unit	Material	2021 Bare Costs Labor	Equipment	Total	Total Incl O&P
0010	**WOOD BURNING STOVES**									
0015	Cast iron, less than 1,500 S.F.	2 Carp	1.30	12.308	Ea.	1,450	675		2,125	2,675
0020	1,500 to 2,000 S.F.		1	16		2,550	875		3,425	4,200
0030	greater than 2,000 S.F.		.80	20		3,200	1,100		4,300	5,250
0050	For gas log lighter, add					53.50			53.50	59

10 43 Emergency Aid Specialties

10 43 13 – Defibrillator Cabinets

10 43 13.05 Defibrillator Cabinets

		Crew	Daily Output	Labor-Hours	Unit	Material	2021 Bare Costs Labor	Equipment	Total	Total Incl O&P
0010	**DEFIBRILLATOR CABINETS**, not equipped, stainless steel									
0050	Defibrillator cabinet, stainless steel with strobe & alarm 12" x 27"	1 Carp	10	.800	Ea.	415	44		459	525
0100	Automatic External Defibrillator	"	30	.267	"	1,450	14.60		1,464.60	1,625

10 44 Fire Protection Specialties

10 44 13 – Fire Protection Cabinets

10 44 13.53 Fire Equipment Cabinets

		Crew	Daily Output	Labor-Hours	Unit	Material	2021 Bare Costs Labor	Equipment	Total	Total Incl O&P
0010	**FIRE EQUIPMENT CABINETS**, not equipped, 20 ga. steel box									
0040	Recessed, D.S. glass in door, box size given									
1000	Portable extinguisher, single, 8" x 12" x 27", alum. door & frame	Q-12	8	2	Ea.	155	120		275	355
1100	Steel door and frame	"	8	2	"	183	120		303	385
3000	Hose rack assy., 1-1/2" valve & 100' hose, 24" x 40" x 5-1/2"									
3200	Steel door and frame	Q-12	6	2.667	Ea.	229	160		389	500
4000	Hose rack assy., 2-1/2" x 1-1/2" valve, 100' hose, 24" x 40" x 8"									
4200	Steel door and frame	Q-12	6	2.667	Ea.	390	160		550	675
5000	Hose rack assy., 2-1/2" x 1-1/2" valve, 100' hose									
5010	and extinguisher, 30" x 40" x 8"									
5200	Steel door and frame	Q-12	5	3.200	Ea.	282	192		474	605

10 44 16 – Fire Extinguishers

10 44 16.13 Portable Fire Extinguishers

		Crew	Daily Output	Labor-Hours	Unit	Material	2021 Bare Costs Labor	Equipment	Total	Total Incl O&P
0010	**PORTABLE FIRE EXTINGUISHERS**									
0140	CO_2, with hose and "H" horn, 10 lb.				Ea.	335			335	370
1000	Dry chemical, pressurized									
1040	Standard type, portable, painted, 2-1/2 lb.				Ea.	43.50			43.50	48
1080	10 lb.					98.50			98.50	109
1100	20 lb.					230			230	253
1120	30 lb.					430			430	475
2000	ABC all purpose type, portable, 2-1/2 lb.					40.50			40.50	44.50
2080	9-1/2 lb.				Ea.	73			73	80.50
3500	Halotron 1, 2-1/2 lb.					87			87	95.50
3600	5 lb.					211			211	232
3700	60 lb.					435			435	480

10 51 13.10 Lockers	Crew	Daily Output	Labor-Hours	Unit	Material	2021 Bare Costs Labor	Equipment	Total	Total Incl O&P
0011	**LOCKERS** steel, baked enamel, knock down construction								
0012	Body- 24ga, door- 16 or 18ga								
0110	1-tier locker, 12" x 15" x 72"	1 Shee	20	.400	Ea.	273	26		299	340
0120	18" x 15" x 72"		20	.400		335	26		361	410
0130	12" x 18" x 72"		20	.400		251	26		277	315
0140	18" x 18" x 72"		20	.400		272	26		298	340
0410	2- tier, 12" x 15" x 36"		30	.267		287	17.45		304.45	340
0420	18" x 15" x 36"		30	.267		238	17.45		255.45	289
0430	12" x 18" x 36"		30	.267		405	17.45		422.45	470
0440	18" x 18" x 36"		30	.267		269	17.45		286.45	325
0450	3- tier 12" x 15" x 24"		30	.267		280	17.45		297.45	335
0455	12" x 18" x 24"		30	.267		310	17.45		327.45	365
0480	4- tier 12" x 15" x 18"		30	.267		129	17.45		146.45	169
0485	12" x 18" x 18"		30	.267		400	17.45		417.45	465
0500	Two person, 18" x 15" x 72"		20	.400		245	26		271	310
0510	18" x 18" x 72"		20	.400		310	26		336	380
0520	Duplex, 15" x 15" x 72"		20	.400		217	26		243	280
0530	15" x 21" x 72"		20	.400		355	26		381	430
1098	All welded, ventilated athletic lockers, body- 16 ga., door- 14 ga.								
1100	1-tier, 12" x 18" x 72"	1 Shee	7.50	1.067	Ea.	320	70		390	465
1110	18" x 15" x 72"		7.50	1.067		244	70		314	375
1115	18" x 18" x 72"		7.50	1.067		580	70		650	750
1130	2-tier, 12" x 15" x 36"		7.50	1.067		595	70		665	765
1135	12" x 18" x 36"		7.50	1.067		635	70		705	810
1140	18" x 15" x 36"		7.50	1.067		690	70		760	870
1145	18" x 18" x 36"		7.50	1.067		705	70		775	885
1160	3-tier, 12" x 18" x 24"		7.50	1.067		505	70		575	665
1180	4-tier, 12" x 18" x 18"		7.50	1.067		249	70		319	385
1190	5H box, 12" x 18" x 14.4"		7.50	1.067		485	70		555	645
1195	18" x 18" x 14.4"		7.50	1.067		815	70		885	1,000
1196	6H box, 12" x 15" x 12"		7.50	1.067		545	70		615	710
1197	12" x 18" x 12"		7.50	1.067		620	70		690	790
1198	18" x 18" x 12"		7.50	1.067		740	70		810	925
2399	Standard duty lockers, body- 24 ga., doors- 16-18 ga.								
2400	16-person locker unit with clothing rack								
2500	72" wide x 15" deep x 72" high	1 Shee	15	.533	Ea.	585	35		620	700
2550	18" deep		15	.533	"	495	35		530	600
3250	Rack w/24 wire mesh baskets		1.50	5.333	Set	470	350		820	1,050
3260	30 baskets		1.25	6.400		395	420		815	1,100
3270	36 baskets		.95	8.421		465	550		1,015	1,375
3280	42 baskets		.80	10		475	655		1,130	1,550
3600	For hanger rods, add				Ea.	2.53			2.53	2.78
3650	For number plate kit, 100 plates #1 - #100, add	1 Shee	4	2		87.50	131		218.50	300
3700	For locker base, closed front panel		90	.089		4.24	5.80		10.04	13.70
3710	End panel, bolted		36	.222		6.90	14.55		21.45	30
3800	For sloping top, 12" wide		24	.333		32	22		54	69.50
3810	15" wide		24	.333		32.50	22		54.50	69.50
3820	18" wide		24	.333		34	22		56	71.50
3850	Sloping top end panel, 12" deep		72	.111		16.40	7.25		23.65	29.50
3860	15" deep		72	.111		17.20	7.25		24.45	30.50
3870	18" deep		72	.111		20	7.25		27.25	33.50
3900	For finish end panels, steel, 60" high, 15" deep		12	.667		85	43.50		128.50	162
3910	72" high, 12" deep		12	.667		27.50	43.50		71	98.50

10 51 Lockers

10 51 13 – Metal Lockers

10 51 13.10 Lockers

		Crew	Daily Output	Labor-Hours	Unit	Material	2021 Bare Costs Labor	Equipment	Total	Total Incl O&P
3920	18" deep	1 Shee	12	.667	Ea.	44	43.50		87.50	117
5000	For "ready to assemble" lockers,									
5010	Add to labor						75%			
5020	Deduct from material					20%				
6000	Heavy duty for detention facility, tamper proof, 14 ga. welded steel, solid									
6100	24" W x 24" D x 74" H, single tier	1 Shee	18	.444	Ea.	610	29		639	715
6110	Double tier		18	.444		575	29		604	675
6120	Triple tier		18	.444		615	29		644	720
9000	Minimum labor/equipment charge		2.50	3.200	Job		209		209	325

10 55 Postal Specialties

10 55 23 – Mail Boxes

10 55 23.10 Commercial Mail Boxes

		Crew	Daily Output	Labor-Hours	Unit	Material	2021 Bare Costs Labor	Equipment	Total	Total Incl O&P
0010	**COMMERCIAL MAIL BOXES**									
0020	Horiz., key lock, 5"H x 6"W x 15"D, alum., rear load	1 Carp	34	.235	Ea.	40.50	12.85		53.35	65
0100	Front loading		34	.235		39.50	12.85		52.35	64
0200	Double, 5"H x 12"W x 15"D, rear loading		26	.308		73	16.85		89.85	107
0300	Front loading		26	.308		77.50	16.85		94.35	113
0500	Quadruple, 10"H x 12"W x 15"D, rear loading		20	.400		155	22		177	206
0600	Front loading		20	.400		137	22		159	186
1600	Vault type, horizontal, for apartments, 4" x 5"		34	.235		31.50	12.85		44.35	55
1700	Alphabetical directories, 120 names		10	.800		103	44		147	183
1900	Letter slot, residential	1 Carp	20	.400	Ea.	78.50	22		100.50	122
2000	Post office type		8	1	"	32	54.50		86.50	122
9000	Minimum labor/equipment charge		5	1.600	Job		87.50		87.50	139

10 56 Storage Assemblies

10 56 13 – Metal Storage Shelving

10 56 13.10 Shelving

		Crew	Daily Output	Labor-Hours	Unit	Material	2021 Bare Costs Labor	Equipment	Total	Total Incl O&P
0010	**SHELVING**									
0020	Metal, industrial, cross-braced, 3' W, 12" D	1 Sswk	175	.046	SF Shlf	7.55	2.76		10.31	12.75
0100	24" D		330	.024		5.65	1.46		7.11	8.55
2200	Wide span, 1600 lb. capacity per shelf, 6' W, 24" D		380	.021		7.35	1.27		8.62	10.10
2400	36" D		440	.018		6.55	1.10		7.65	9
9000	Minimum labor/equipment charge	1 Carp	4	2	Job		109		109	174

10 57 Wardrobe and Closet Specialties

10 57 23 – Closet and Utility Shelving

10 57 23.19 Wood Closet and Utility Shelving

		Crew	Daily Output	Labor-Hours	Unit	Material	2021 Bare Costs Labor	Equipment	Total	Total Incl O&P
0010	**WOOD CLOSET AND UTILITY SHELVING**									
0020	Pine, clear grade, no edge band, 1" x 8"	1 Carp	115	.070	L.F.	4.38	3.81		8.19	10.85
0100	1" x 10"		110	.073		5.45	3.98		9.43	12.35
0200	1" x 12"		105	.076		6.55	4.17		10.72	13.85
0600	Plywood, 3/4" thick with lumber edge, 12" wide		75	.107		2.76	5.85		8.61	12.35
0700	24" wide		70	.114		4.97	6.25		11.22	15.40
0900	Bookcase, clear grade pine, shelves 12" OC, 8" deep, per S.F. shelf		70	.114	S.F.	14.25	6.25		20.50	25.50
1000	12" deep shelves		65	.123	"	21.50	6.75		28.25	34

10 57 Wardrobe and Closet Specialties

10 57 23 – Closet and Utility Shelving

10 57 23.19 Wood Closet and Utility Shelving	Crew	Daily Output	Labor-Hours	Unit	Material	2021 Bare Costs Labor	Equipment	Total	Total Incl O&P
1200 Adjustable closet rod and shelf, 12" wide, 3' long	1 Carp	20	.400	Ea.	14.50	22		36.50	51
1300　　　　　8' long		15	.533	"	28.50	29		57.50	78
1500 Prefinished shelves with supports, stock, 8" wide		75	.107	L.F.	4.50	5.85		10.35	14.25
1600　　　　10" wide		70	.114	"	5.60	6.25		11.85	16.15
9000 Minimum labor/equipment charge	▼	4	2	Job		109		109	174

10 73 Protective Covers

10 73 13 – Awnings

10 73 13.10 Awnings, Fabric

		Crew	Daily Output	Labor-Hours	Unit	Material	2021 Bare Costs Labor	Equipment	Total	Total Incl O&P
0010	**AWNINGS, FABRIC**									
0020	Including acrylic canvas and frame, standard design									
0100	Door and window, slope, 3' high, 4' wide	1 Carp	4.50	1.778	Ea.	755	97		852	985
0110	6' wide		3.50	2.286		970	125		1,095	1,275
0120	8' wide		3	2.667		1,175	146		1,321	1,525
0200	Quarter round convex, 4' wide		3	2.667		1,175	146		1,321	1,525
0210	6' wide		2.25	3.556		1,525	194		1,719	1,975
0220	8' wide		1.80	4.444		1,850	243		2,093	2,425
0300	Dome, 4' wide		7.50	1.067		450	58.50		508.50	595
0310	6' wide		3.50	2.286		1,025	125		1,150	1,325
0320	8' wide		2	4		1,800	219		2,019	2,325
0350	Elongated dome, 4' wide	1 Carp	1.33	6.015	Ea.	1,700	330		2,030	2,400
0360	6' wide		1.11	7.207		2,025	395		2,420	2,850
0370	8' wide	▼	1	8		2,375	440		2,815	3,300
1000	Entry or walkway, peak, 12' long, 4' wide	2 Carp	.90	17.778		5,325	970		6,295	7,400
1010	6' wide		.60	26.667		8,200	1,450		9,650	11,400
1020	8' wide		.40	40		11,300	2,200		13,500	15,900
1100	Radius with dome end, 4' wide		1.10	14.545		4,025	795		4,820	5,700
1110	6' wide		.70	22.857		6,475	1,250		7,725	9,125
1120	8' wide	▼	.50	32	▼	9,225	1,750		10,975	12,900
2000	Retractable lateral arm awning, manual									
2010	To 12' wide, 8'-6" projection	2 Carp	1.70	9.412	Ea.	1,200	515		1,715	2,150
2020	To 14' wide, 8'-6" projection		1.10	14.545		1,400	795		2,195	2,825
2030	To 19' wide, 8'-6" projection		.85	18.824		1,900	1,025		2,925	3,750
2040	To 24' wide, 8'-6" projection	▼	.67	23.881		2,400	1,300		3,700	4,725
2050	Motor for above, add	1 Carp	2.67	3	▼	1,025	164		1,189	1,375
3000	Patio/deck canopy with frame									
3010	12' wide, 12' projection	2 Carp	2	8	Ea.	1,700	440		2,140	2,575
3020	16' wide, 14' projection	"	1.20	13.333		2,650	730		3,380	4,050
9000	For fire retardant canvas, add					7%				
9010	For lettering or graphics, add					35%				
9020	For painted or coated acrylic canvas, deduct					8%				
9030	For translucent or opaque vinyl canvas, add					10%				
9040	For 6 or more units, deduct					20%	15%			

10 73 16 – Canopies

10 73 16.10 Canopies, Residential

		Crew	Daily Output	Labor-Hours	Unit	Material	2021 Bare Costs Labor	Equipment	Total	Total Incl O&P
0010	**CANOPIES, RESIDENTIAL** Prefabricated									
0500	Carport, free standing, baked enamel, alum., .032", 40 psf									
0520	16' x 8', 4 posts	2 Carp	3	5.333	Ea.	3,875	292		4,167	4,725
0600	20' x 10', 6 posts		2	8		4,875	440		5,315	6,075
0605	30' x 10', 8 posts	▼	2	8	▼	7,325	440		7,765	8,750

10 73 16 – Canopies

10 73 16.10 Canopies, Residential

		Crew	Daily Output	Labor-Hours	Unit	Material	2021 Bare Costs Labor	Equipment	Total	Total Incl O&P
1000	Door canopies, extruded alum., .032", 42" projection, 4' wide	1 Carp	8	1	Ea.	195	54.50		249.50	300
1020	6' wide	"	6	1.333		288	73		361	430
1040	8' wide	2 Carp	9	1.778		495	97		592	700
1060	10' wide		7	2.286		620	125		745	880
1080	12' wide		5	3.200		595	175		770	935
1200	54" projection, 4' wide	1 Carp	8	1		236	54.50		290.50	345
1220	6' wide	"	6	1.333		297	73		370	440
1240	8' wide	2 Carp	9	1.778		395	97		492	590
1260	10' wide		7	2.286		825	125		950	1,100
1280	12' wide		5	3.200	Ea.	890	175		1,065	1,250
1300	Painted, add					20%				
1310	Bronze anodized, add					50%				
3000	Window awnings, aluminum, window 3' high, 4' wide	1 Carp	10	.800		115	44		159	196
3020	6' wide	"	8	1		187	54.50		241.50	293
3040	9' wide	2 Carp	9	1.778		420	97		517	615
3060	12' wide	"	5	3.200		530	175		705	865
3100	Window, 4' high, 4' wide	1 Carp	10	.800		283	44		327	380
3120	6' wide	"	8	1		335	54.50		389.50	450
3140	9' wide	2 Carp	9	1.778		710	97		807	940
3160	12' wide	"	5	3.200		765	175		940	1,125
3200	Window, 6' high, 4' wide	1 Carp	10	.800		375	44		419	485
3220	6' wide	"	8	1		390	54.50		444.50	515
3240	9' wide	2 Carp	9	1.778		965	97		1,062	1,225
3260	12' wide	"	5	3.200		1,475	175		1,650	1,900
3400	Roll-up aluminum, 2'-6" wide	1 Carp	14	.571		210	31.50		241.50	281
3420	3' wide		12	.667		259	36.50		295.50	345
3440	4' wide		10	.800		279	44		323	375
3460	6' wide		8	1		335	54.50		389.50	450
3480	9' wide	2 Carp	9	1.778		395	97		492	590
3500	12' wide	"	5	3.200		535	175		710	870
3600	Window awnings, canvas, 24" drop, 3' wide	1 Carp	30	.267	L.F.	61.50	14.60		76.10	90.50
3620	4' wide		40	.200		46.50	10.95		57.45	68.50
3700	30" drop, 3' wide		30	.267		62	14.60		76.60	91.50
3720	4' wide		40	.200		49	10.95		59.95	71.50
3740	5' wide		45	.178		47.50	9.70		57.20	68
3760	6' wide		48	.167		44	9.10		53.10	63
3780	8' wide		48	.167		39.50	9.10		48.60	57.50
3800	10' wide		50	.160		43	8.75		51.75	61
3900	Repair canvas, minimum		8	1	Ea.		54.50		54.50	87
3920	Maximum		4	2	"		109		109	174
9000	Minimum labor/equipment charge		3	2.667	Job		146		146	232

10 73 16.20 Metal Canopies

		Crew	Daily Output	Labor-Hours	Unit	Material	2021 Bare Costs Labor	Equipment	Total	Total Incl O&P
0010	**METAL CANOPIES**									
0020	Wall hung, .032", aluminum, prefinished, 8' x 10'	K-2	1.30	18.462	Ea.	3,100	1,050	655	4,805	5,825
0300	8' x 20'		1.10	21.818		4,225	1,250	775	6,250	7,500
1000	12' x 20'		1	24		6,000	1,375	850	8,225	9,725
2300	Aluminum entrance canopies, flat soffit, .032"									
2500	3'-6" x 4'-0", clear anodized	2 Carp	4	4	Ea.	1,050	219		1,269	1,500
9000	Minimum labor/equipment charge	"	2	8	Job		440		440	695

10 74 Manufactured Exterior Specialties

10 74 23 – Cupolas

10 74 23.10 Wood Cupolas

10 74 23.10 Wood Cupolas	Crew	Daily Output	Labor-Hours	Unit	Material	2021 Bare Costs Labor	Equipment	Total	Total Incl O&P
0010 **WOOD CUPOLAS**									
0020 Stock units, pine, painted, 18" sq., 28" high, alum. roof	1 Carp	4.10	1.951	Ea.	340	107		447	545
0100 Copper roof		3.80	2.105		275	115		390	490
0300 23" square, 33" high, aluminum roof		3.70	2.162		445	118		563	680
0400 Copper roof		3.30	2.424		705	133		838	985
0600 30" square, 37" high, aluminum roof		3.70	2.162		680	118		798	935
0700 Copper roof		3.30	2.424		795	133		928	1,075
0900 Hexagonal, 31" wide, 46" high, copper roof		4	2		1,125	109		1,234	1,400
1000 36" wide, 50" high, copper roof		3.50	2.286		1,800	125		1,925	2,175
1200 For deluxe stock units, add to above					25%				
1400 For custom built units, add to above					50%	50%			
9000 Minimum labor/equipment charge	1 Carp	2.75	2.909	Job		159		159	253

10 74 29 – Steeples

10 74 29.10 Prefabricated Steeples

	Crew	Daily Output	Labor-Hours	Unit	Material	2021 Bare Costs Labor	Equipment	Total	Total Incl O&P
0010 **PREFABRICATED STEEPLES**									
4000 Steeples, translucent fiberglass, 30" square, 15' high	F-3	2	20	Ea.	10,300	1,125	238	11,663	13,300
4150 25' high	"	1.80	22.222		11,900	1,250	264	13,414	15,300
4600 Aluminum, baked finish, 16" square, 14' high					6,350			6,350	7,000
4640 35' high, 8' base					39,600			39,600	43,500
4680 152' high, custom					741,500			741,500	815,500

10 74 33 – Weathervanes

10 74 33.10 Residential Weathervanes

	Crew	Daily Output	Labor-Hours	Unit	Material	2021 Bare Costs Labor	Equipment	Total	Total Incl O&P
0010 **RESIDENTIAL WEATHERVANES**									
0020 Residential types, 18" to 24"	1 Carp	8	1	Ea.	177	54.50		231.50	282
0100 24" to 48"		2	4	"	210	219		429	580
9000 Minimum labor/equipment charge		4	2	Job		109		109	174

10 74 46 – Window Wells

10 74 46.10 Area Window Wells

	Crew	Daily Output	Labor-Hours	Unit	Material	2021 Bare Costs Labor	Equipment	Total	Total Incl O&P
0010 **AREA WINDOW WELLS**, Galvanized steel									
0020 20 ga., 3'-2" wide, 1' deep	1 Sswk	29	.276	Ea.	17.95	16.65		34.60	46.50
0100 2' deep		23	.348		29.50	21		50.50	66.50
0300 16 ga., 3'-2" wide, 1' deep		29	.276		27.50	16.65		44.15	57
0400 3' deep		23	.348		48	21		69	87
0600 Welded grating for above, 15 lb., painted		45	.178		103	10.70		113.70	130
0700 Galvanized		45	.178		126	10.70		136.70	155
0900 Translucent plastic cap for above		60	.133		20.50	8.05		28.55	35.50

10 86 Security Mirrors and Domes

10 86 10 – Security Mirrors

10 86 10.10 Exterior Traffic Control Mirrors

	Crew	Daily Output	Labor-Hours	Unit	Material	2021 Bare Costs Labor	Equipment	Total	Total Incl O&P
0010 **EXTERIOR TRAFFIC CONTROL MIRRORS**									
0100 Convex, stainless steel, 20 ga., 26" diameter	1 Carp	12	.667	Ea.	315	36.50		351.50	405

10 86 20 – Security Domes

10 86 20.10 Domes

	Crew	Daily Output	Labor-Hours	Unit	Material	2021 Bare Costs Labor	Equipment	Total	Total Incl O&P
0010 **DOMES** for security cameras (CCTV)									
0100 Ceiling mounted, 10" diameter	1 Carp	30	.267	Ea.	56	14.60		70.60	84.50
0110 12" diameter	"	30	.267	"	56	14.60		70.60	84.50

Estimating Tips
General

- The items in this division are usually priced per square foot or each. Many of these items are purchased by the owner for installation by the contractor. Check the specifications for responsibilities and include time for receiving, storage, installation, and mechanical and electrical hookups in the appropriate divisions.

- Many items in Division 11 require some type of support system that is not usually furnished with the item. Examples of these systems include blocking for the attachment of casework and support angles for ceiling-hung projection screens. The required blocking or supports must be added to the estimate in the appropriate division.

- Some items in Division 11 may require assembly or electrical hookups. Verify the amount of assembly required or the need for a hard electrical connection and add the appropriate costs.

Reference Numbers

Reference numbers are shown at the beginning of some major classifications. These numbers refer to related items in the Reference Section. The reference information may be an estimating procedure, an alternate pricing method, or technical information.

Same Data. Simplified.

Enjoy the convenience and efficiency of accessing your costs anywhere:

- **Skip the multiplier** by setting your location
- **Quickly search**, edit, favorite and share costs
- **Stay on top of price changes** with automatic updates

Discover more at rsmeans.com/online

11 11 Vehicle Service Equipment

11 11 13 – Compressed-Air Vehicle Service Equipment

11 11 13.10 Compressed Air Equipment	Crew	Daily Output	Labor-Hours	Unit	Material	2021 Bare Costs Labor	Equipment	Total	Total Incl O&P
0010 **COMPRESSED AIR EQUIPMENT**									
0030 Compressors, electric, 1-1/2 HP, standard controls	L-4	1.50	16	Ea.	1,125	835		1,960	2,550
0550 Dual controls		1.50	16		1,050	835		1,885	2,475
0600 5 HP, 115/230 volt, standard controls		1	24		1,000	1,250		2,250	3,100
0650 Dual controls		1	24		1,325	1,250		2,575	3,450

11 11 19 – Vehicle Lubrication Equipment

11 11 19.10 Lubrication Equipment

	Crew	Daily Output	Labor-Hours	Unit	Material	Labor	Equipment	Total	Total Incl O&P
0010 **LUBRICATION EQUIPMENT**									
3000 Lube equipment, 3 reel type, with pumps, not including piping	L-4	.50	48	Set	10,400	2,500		12,900	15,500
3100 Hose reel, including hose, oil/lube, 1000 psi	2 Sswk	2	8	Ea.	295	480		775	1,100
3200 Grease, 5000 psi		2	8		455	480		935	1,275
3300 Air, 50', 160 psi		2	8		90	480		570	880
3350 25', 160 psi		2	8		54	480		534	840

11 11 33 – Vehicle Spray Painting Equipment

11 11 33.10 Spray Painting Equipment

	Crew	Daily Output	Labor-Hours	Unit	Material	Labor	Equipment	Total	Total Incl O&P
0010 **SPRAY PAINTING EQUIPMENT**									
4000 Spray painting booth, 26' long, complete	L-4	.40	60	Ea.	18,100	3,125		21,225	24,900

11 12 Parking Control Equipment

11 12 13 – Parking Key and Card Control Units

11 12 13.10 Parking Control Units

	Crew	Daily Output	Labor-Hours	Unit	Material	Labor	Equipment	Total	Total Incl O&P
0010 **PARKING CONTROL UNITS**									
5100 Card reader	1 Elec	2	4	Ea.	895	255		1,150	1,375
5120 Proximity with customer display	2 Elec	1	16		6,100	1,025		7,125	8,275
6000 Parking control software, basic functionality	1 Elec	.50	16		23,000	1,025		24,025	26,900
6020 Multi-function	"	.20	40		90,000	2,550		92,550	103,000

11 12 16 – Parking Ticket Dispensers

11 12 16.10 Ticket Dispensers

	Crew	Daily Output	Labor-Hours	Unit	Material	Labor	Equipment	Total	Total Incl O&P
0010 **TICKET DISPENSERS**									
5900 Ticket spitter with time/date stamp, standard	2 Elec	2	8	Ea.	6,525	510		7,035	7,950
5920 Mag stripe encoding	"	2	8	"	18,900	510		19,410	21,600

11 12 26 – Parking Fee Collection Equipment

11 12 26.13 Parking Fee Coin Collection Equipment

	Crew	Daily Output	Labor-Hours	Unit	Material	Labor	Equipment	Total	Total Incl O&P
0010 **PARKING FEE COIN COLLECTION EQUIPMENT**									
5200 Cashier booth, average	B-22	1	30	Ea.	10,400	1,550	287	12,237	14,300
5300 Collector station, pay on foot	2 Elec	.20	80		111,500	5,100		116,600	131,000
5320 Credit card only	"	.50	32		20,600	2,050		22,650	25,800

11 12 26.23 Fee Equipment

	Crew	Daily Output	Labor-Hours	Unit	Material	Labor	Equipment	Total	Total Incl O&P
0010 **FEE EQUIPMENT**									
5600 Fee computer	1 Elec	1.50	5.333	Ea.	13,000	340		13,340	14,800

11 12 33 – Parking Gates

11 12 33.13 Lift Arm Parking Gates

	Crew	Daily Output	Labor-Hours	Unit	Material	Labor	Equipment	Total	Total Incl O&P
0010 **LIFT ARM PARKING GATES**									
5000 Barrier gate with programmable controller	2 Elec	3	5.333	Ea.	3,325	340		3,665	4,200
5020 Industrial		3	5.333		5,175	340		5,515	6,200
5050 Non-programmable, with reader and 12' arm		3	5.333		1,725	340		2,065	2,425
5500 Exit verifier		1	16		19,800	1,025		20,825	23,400
5700 Full sign, 4" letters	1 Elec	2	4		1,225	255		1,480	1,750

11 12 Parking Control Equipment

11 12 33 – Parking Gates

11 12 33.13 Lift Arm Parking Gates

		Crew	Daily Output	Labor-Hours	Unit	Material	2021 Bare Costs Labor	Equipment	Total	Total Incl O&P
5800	Inductive loop	2 Elec	4	4	Ea.	208	255		463	620
5950	Vehicle detector, microprocessor based	1 Elec	3	2.667		465	170		635	775
7100	Traffic spike unit, flush mount, spring loaded, 72" L	B-89	4	4		1,325	209	266	1,800	2,075
7200	Surface mount, 72" L	2 Skwk	10	1.600		2,450	91.50		2,541.50	2,850

11 13 Loading Dock Equipment

11 13 13 – Loading Dock Bumpers

11 13 13.10 Dock Bumpers

		Crew	Daily Output	Labor-Hours	Unit	Material	2021 Bare Costs Labor	Equipment	Total	Total Incl O&P
0010	**DOCK BUMPERS** Bolts not included									
0012	2" x 6" to 4" x 8", average	1 Carp	300	.027	B.F.	.74	1.46		2.20	3.13
0050	Bumpers, lam. rubber blocks 4-1/2" thick, 10" high, 14" long		26	.308	Ea.	70	16.85		86.85	105
0200	24" long		22	.364		103	19.90		122.90	145
0300	36" long		17	.471		365	25.50		390.50	440
0500	12" high, 14" long		25	.320		69.50	17.50		87	105

11 13 16 – Loading Dock Seals and Shelters

11 13 16.10 Dock Seals and Shelters

		Crew	Daily Output	Labor-Hours	Unit	Material	2021 Bare Costs Labor	Equipment	Total	Total Incl O&P
0010	**DOCK SEALS AND SHELTERS**									
6200	Shelters, fabric, for truck or train, scissor arms, minimum	1 Carp	1	8	Ea.	2,400	440		2,840	3,350
6300	Maximum	"	.50	16	"	2,300	875		3,175	3,950

11 13 19 – Stationary Loading Dock Equipment

11 13 19.10 Dock Equipment

		Crew	Daily Output	Labor-Hours	Unit	Material	2021 Bare Costs Labor	Equipment	Total	Total Incl O&P
0010	**DOCK EQUIPMENT**									
2200	Dock boards, heavy duty, 60" x 60", aluminum, 5,000 lb. capacity				Ea.	1,050			1,050	1,150
4200	Platform lifter, 6' x 6', portable, 3,000 lb. capacity				"	9,775			9,775	10,700
6000	Dock leveler, 15 ton capacity									
6100	I-beam construction, mechanical, 6' x 8'	E-16	.50	32	Ea.	5,650	1,950	298	7,898	9,700
6150	Hydraulic		.50	32		6,625	1,950	298	8,873	10,800
6200	Formed beam deck construction, mechanical, 6' x 8'		.50	32		3,800	1,950	298	6,048	7,700
6250	Hydraulic		.50	32		5,225	1,950	298	7,473	9,225
6300	Edge of dock leveler, mechanical, 15 ton capacity		2	8		2,350	490	74.50	2,914.50	3,450
6301	Edge of dock leveler, mechanical, 10 ton capacity		2	8		2,350	490	74.50	2,914.50	3,450
7000	22.5 ton capacity									
7100	Vertical storing dock leveler, hydraulic, 6' x 6'	E-16	.40	40	Ea.	8,625	2,450	370	11,445	13,900

11 21 Retail and Service Equipment

11 21 13 – Cash Registers and Checking Equipment

11 21 13.10 Checkout Counter

		Crew	Daily Output	Labor-Hours	Unit	Material	2021 Bare Costs Labor	Equipment	Total	Total Incl O&P
0010	**CHECKOUT COUNTER**									
0020	Supermarket conveyor, single belt	2 Clab	10	1.600	Ea.	3,850	71		3,921	4,350
0100	Double belt, power take-away		9	1.778		5,325	79		5,404	5,975
0400	Double belt, power take-away, incl. side scanning		7	2.286		5,850	101		5,951	6,575
0800	Warehouse or bulk type		6	2.667		6,800	118		6,918	7,675
1000	Scanning system, 2 lanes, w/registers, scan gun & memory				System	19,300			19,300	21,300
1100	10 lanes, single processor, full scan, with scales				"	184,000			184,000	202,000
2000	Register, restaurant, minimum				Ea.	810			810	895
2100	Maximum					3,100			3,100	3,425
2150	Store, minimum					810			810	895
2200	Maximum					3,100			3,100	3,425

371

11 21 Retail and Service Equipment

11 21 33 – Checkroom Equipment

11 21 33.10 Clothes Check Equipment

		Crew	Daily Output	Labor-Hours	Unit	Material	2021 Bare Costs Labor	2021 Bare Costs Equipment	Total	Total Incl O&P
0010	**CLOTHES CHECK EQUIPMENT**									
0030	Clothes check rack, free standing, st. stl., 2-tier, 90 bag capacity				Ea.	1,575			1,575	1,725
0050	Wall mounted, 45 bag capacity	L-2	8	2		925	97		1,022	1,175
0100	Garment checking bag, green mesh fabric, 21" H x 17" W with 4.5" hook					28.50			28.50	31.50

11 21 53 – Barber and Beauty Shop Equipment

11 21 53.10 Barber Equipment

		Crew	Daily Output	Labor-Hours	Unit	Material	2021 Bare Costs Labor	2021 Bare Costs Equipment	Total	Total Incl O&P
0010	**BARBER EQUIPMENT**									
0020	Chair, hydraulic, movable, minimum	1 Carp	24	.333	Ea.	635	18.25		653.25	730
0050	Maximum	"	16	.500		3,950	27.50		3,977.50	4,375
0200	Wall hung styling station with mirrors, minimum	L-2	8	2		650	97		747	870
0300	Maximum	"	4	4		2,625	194		2,819	3,200
0500	Sink, hair washing basin, rough plumbing not incl.	1 Plum	8	1		505	67.50		572.50	660
1000	Sterilizer, liquid solution for tools					163			163	179
1100	Total equipment, rule of thumb, per chair, minimum	L-8	1	20		1,275	1,150		2,425	3,225
1150	Maximum	"	1	20		5,825	1,150		6,975	8,225

11 21 73 – Commercial Laundry and Dry Cleaning Equipment

11 21 73.13 Dry Cleaning Equipment

		Crew	Daily Output	Labor-Hours	Unit	Material	2021 Bare Costs Labor	2021 Bare Costs Equipment	Total	Total Incl O&P
0010	**DRY CLEANING EQUIPMENT**									
2000	Dry cleaners, electric, 20 lb. capacity, not incl. rough-in	L-1	.20	80	Ea.	35,000	5,250		40,250	46,600
2050	25 lb. capacity		.17	94.118		52,500	6,175		58,675	67,000
2100	30 lb. capacity		.15	107		55,500	7,000		62,500	72,000
2150	60 lb. capacity		.09	178		78,500	11,700		90,200	104,500

11 21 73.16 Drying and Conditioning Equipment

		Crew	Daily Output	Labor-Hours	Unit	Material	2021 Bare Costs Labor	2021 Bare Costs Equipment	Total	Total Incl O&P
0010	**DRYING AND CONDITIONING EQUIPMENT**									
0100	Dryers, not including rough-in									
1500	Industrial, 30 lb. capacity	1 Plum	2	4	Ea.	1,625	271		1,896	2,200
1600	50 lb. capacity	"	1.70	4.706		4,575	320		4,895	5,525
4700	Lint collector, ductwork not included, 8,000 to 10,000 CFM	Q-10	.30	80		11,200	4,875		16,075	20,000

11 21 73.19 Finishing Equipment

		Crew	Daily Output	Labor-Hours	Unit	Material	2021 Bare Costs Labor	2021 Bare Costs Equipment	Total	Total Incl O&P
0010	**FINISHING EQUIPMENT**									
3500	Folders, blankets & sheets, minimum	1 Elec	.17	47.059	Ea.	35,900	3,000		38,900	44,100
3700	King size with automatic stacker		.10	80		65,000	5,100		70,100	79,500
3800	For conveyor delivery, add		.45	17.778		16,500	1,125		17,625	20,000

11 21 73.23 Commercial Ironing Equipment

		Crew	Daily Output	Labor-Hours	Unit	Material	2021 Bare Costs Labor	2021 Bare Costs Equipment	Total	Total Incl O&P
0010	**COMMERCIAL IRONING EQUIPMENT**									
4500	Ironers, institutional, 110", single roll	1 Elec	.20	40	Ea.	34,800	2,550		37,350	42,100

11 21 73.26 Commercial Washers and Extractors

		Crew	Daily Output	Labor-Hours	Unit	Material	2021 Bare Costs Labor	2021 Bare Costs Equipment	Total	Total Incl O&P
0010	**COMMERCIAL WASHERS AND EXTRACTORS**, not including rough-in									
6000	Combination washer/extractor, 20 lb. capacity	L-6	1.50	8	Ea.	7,175	530		7,705	8,725
6100	30 lb. capacity		.80	15		10,900	995		11,895	13,500
6200	50 lb. capacity		.68	17.647		13,800	1,175		14,975	17,000
6300	75 lb. capacity		.30	40		19,200	2,650		21,850	25,200
6350	125 lb. capacity		.16	75		30,100	4,975		35,075	40,800

11 21 73.33 Coin-Operated Laundry Equipment

		Crew	Daily Output	Labor-Hours	Unit	Material	2021 Bare Costs Labor	2021 Bare Costs Equipment	Total	Total Incl O&P
0010	**COIN-OPERATED LAUNDRY EQUIPMENT**									
0990	Dryer, gas fired									
1000	Commercial, 30 lb. capacity, coin operated, single	1 Plum	3	2.667	Ea.	1,150	181		1,331	1,525
1100	Double stacked	"	2	4	"	2,925	271		3,196	3,625
5290	Clothes washer									
5300	Commercial, coin operated, average	1 Plum	3	2.667	Ea.	1,350	181		1,531	1,775

11 21 Retail and Service Equipment

11 21 83 – Photo Processing Equipment

11 21 83.13 Darkroom Equipment	Crew	Daily Output	Labor-Hours	Unit	Material	2021 Bare Costs Labor	Equipment	Total	Total Incl O&P
0010 **DARKROOM EQUIPMENT**									
0020 Developing sink, 5" deep, 24" x 48"	Q-1	2	8	Ea.	630	485		1,115	1,450
0050 48" x 52"		1.70	9.412		1,500	575		2,075	2,525
0200 10" deep, 24" x 48"		1.70	9.412		1,650	575		2,225	2,700
0250 24" x 108"		1.50	10.667		4,050	650		4,700	5,450
0500 Dryers, dehumidified filtered air, 36" x 25" x 68" high	L-7	6	4.667		2,250	248		2,498	2,850
3000 Viewing lights, 20" x 24"		6	4.667		277	248		525	695
3100 20" x 24" with color correction		6	4.667		415	248		663	845

11 22 Banking Equipment

11 22 13 – Vault Equipment

11 22 13.16 Safes

	Crew	Daily Output	Labor-Hours	Unit	Material	2021 Bare Costs Labor	Equipment	Total	Total Incl O&P
0010 **SAFES**									
0800 Money, "B" label, 9" x 14" x 14"				Ea.	625			625	685
0900 Tool resistive, 24" x 24" x 20"					4,175			4,175	4,600
1050 Tool and torch resistive, 24" x 24" x 20"					8,825			8,825	9,700

11 22 16 – Teller and Service Equipment

11 22 16.13 Teller Equipment Systems

	Crew	Daily Output	Labor-Hours	Unit	Material	2021 Bare Costs Labor	Equipment	Total	Total Incl O&P
0010 **TELLER EQUIPMENT SYSTEMS**									
0020 Alarm system, police	2 Elec	1.60	10	Ea.	5,100	635		5,735	6,600
0400 Bullet resistant teller window, 44" x 60"	1 Glaz	.60	13.333		4,675	700		5,375	6,250
0500 48" x 60"	"	.60	13.333		7,350	700		8,050	9,225
3000 Counters for banks, frontal only	2 Carp	1	16	Station	2,125	875		3,000	3,750
3100 Complete with steel undercounter	2 Carp	.50	32	Station	3,725	1,750		5,475	6,875
5400 Partitions, bullet-resistant, 1-3/16" glass, 8' high	"	10	1.600	L.F.	206	87.50		293.50	365
5600 Pass thru, bullet-res. window, painted steel, 24" x 36"	2 Sswk	1.60	10	Ea.	2,875	605		3,480	4,125

11 30 Residential Equipment

11 30 13 – Residential Appliances

11 30 13.15 Cooking Equipment

	Crew	Daily Output	Labor-Hours	Unit	Material	2021 Bare Costs Labor	Equipment	Total	Total Incl O&P
0010 **COOKING EQUIPMENT**									
0020 Cooking range, 30" free standing, 1 oven, minimum	2 Clab	10	1.600	Ea.	460	71		531	620
0050 Maximum	"	4	4		2,300	178		2,478	2,800
0350 Built-in, 30" wide, 1 oven, minimum	1 Elec	6	1.333		865	85		950	1,075
0400 Maximum	2 Carp	2	8		2,000	440		2,440	2,900
0900 Countertop cooktops, 4 burner, standard, minimum	1 Elec	6	1.333		335	85		420	500
0950 Maximum		3	2.667		1,850	170		2,020	2,300
1250 Microwave oven, minimum		4	2		107	127		234	315
1300 Maximum		2	4		485	255		740	925
5380 Oven, built-in, standard		4	2		800	127		927	1,075
5390 Deluxe		2	4		3,000	255		3,255	3,700

11 30 13.16 Refrigeration Equipment

	Crew	Daily Output	Labor-Hours	Unit	Material	2021 Bare Costs Labor	Equipment	Total	Total Incl O&P
0010 **REFRIGERATION EQUIPMENT**									
5450 Refrigerator, no frost, 6 C.F.	2 Clab	15	1.067	Ea.	365	47.50		412.50	475
5500 Refrigerator, no frost, 10 C.F. to 12 C.F., minimum		10	1.600		455	71		526	615
5600 Maximum		6	2.667		560	118		678	810
6790 Energy-star qualified, 18 C.F., minimum [G]	2 Carp	4	4		550	219		769	955
6795 Maximum [G]		2	8		2,000	440		2,440	2,900

373

11 30 Residential Equipment

11 30 13 - Residential Appliances

11 30 13.16 Refrigeration Equipment

		Crew	Daily Output	Labor-Hours	Unit	Material	2021 Bare Costs Labor	2021 Bare Costs Equipment	Total	Total Incl O&P
6797	21.7 C.F., minimum	**G** 2 Carp	4	4	Ea.	1,450	219		1,669	1,950
6799	Maximum	**G** ↓	4	4	↓	1,975	219		2,194	2,525

11 30 13.17 Kitchen Cleaning Equipment

		Crew	Daily Output	Labor-Hours	Unit	Material	Labor	Equipment	Total	Total Incl O&P
0010	**KITCHEN CLEANING EQUIPMENT**									
2750	Dishwasher, built-in, 2 cycles, minimum	L-1	4	4	Ea.	310	263		573	745
2800	Maximum		2	8		445	525		970	1,300
3100	Energy-star qualified, minimum	**G**	4	4		350	263		613	790
3110	Maximum	**G** ↓	2	8	↓	1,725	525		2,250	2,700

11 30 13.18 Waste Disposal Equipment

		Crew	Daily Output	Labor-Hours	Unit	Material	Labor	Equipment	Total	Total Incl O&P
0010	**WASTE DISPOSAL EQUIPMENT**									
1750	Compactor, residential size, 4 to 1 compaction, minimum	1 Carp	5	1.600	Ea.	705	87.50		792.50	915
1800	Maximum	"	3	2.667		1,150	146		1,296	1,475
3300	Garbage disposal, sink type, minimum	L-1	10	1.600		109	105		214	282
3350	Maximum	"	10	1.600	↓	242	105		347	430

11 30 13.19 Kitchen Ventilation Equipment

		Crew	Daily Output	Labor-Hours	Unit	Material	Labor	Equipment	Total	Total Incl O&P
0010	**KITCHEN VENTILATION EQUIPMENT**									
4150	Hood for range, 2 speed, vented, 30" wide, minimum	L-3	5	3.200	Ea.	107	191		298	415
4200	Maximum		3	5.333		1,050	320		1,370	1,650
4300	42" wide, minimum		5	3.200		152	191		343	465
4330	Custom		5	3.200		1,975	191		2,166	2,475
4350	Maximum	↓	3	5.333		2,400	320		2,720	3,150
4400	Ventless hood, 2 speed, 30" wide, minimum	1 Elec	8	1		820	63.50		883.50	1,000
4450	Maximum		7	1.143		575	73		648	740
4510	36" wide, minimum		7	1.143		78	73		151	198
4550	Maximum		6	1.333		800	85		885	1,000
4580	42" wide, minimum		6	1.333		99	85		184	239
4600	Maximum	↓	5	1.600		625	102		727	845
4650	For vented 1 speed, deduct from maximum				↓	70			70	77

11 30 13.24 Washers

		Crew	Daily Output	Labor-Hours	Unit	Material	Labor	Equipment	Total	Total Incl O&P
0010	**WASHERS**									
5000	Residential, 4 cycle, average	1 Plum	3	2.667	Ea.	995	181		1,176	1,375
6750	Energy star qualified, front loading, minimum	**G**	3	2.667		910	181		1,091	1,275
6760	Maximum	**G**	1	8		2,275	540		2,815	3,325
6764	Top loading, minimum	**G**	3	2.667		735	181		916	1,100
6766	Maximum	**G** ↓	3	2.667	↓	1,175	181		1,356	1,550

11 30 13.25 Dryers

		Crew	Daily Output	Labor-Hours	Unit	Material	Labor	Equipment	Total	Total Incl O&P
0010	**DRYERS**									
0500	Gas fired residential, 16 lb. capacity, average	1 Plum	3	2.667	Ea.	950	181		1,131	1,325
6770	Electric, front loading, energy-star qualified, minimum	**G** L-2	3	5.333		510	258		768	975
6780	Maximum	**G** "	2	8	↓	1,575	385		1,960	2,350

11 30 15 - Miscellaneous Residential Appliances

11 30 15.13 Sump Pumps

		Crew	Daily Output	Labor-Hours	Unit	Material	Labor	Equipment	Total	Total Incl O&P
0010	**SUMP PUMPS**									
6400	Cellar drainer, pedestal, 1/3 HP, molded PVC base	1 Plum	3	2.667	Ea.	138	181		319	430
6450	Solid brass	"	2	4	"	250	271		521	695
6460	Sump pump, see also Section 22 14 29.16									

11 30 Residential Equipment

11 30 33 – Retractable Stairs

11 30 33.10 Disappearing Stairway	Crew	Daily Output	Labor-Hours	Unit	Material	2021 Bare Costs Labor	Equipment	Total	Total Incl O&P
0010 **DISAPPEARING STAIRWAY** No trim included									
0020 One piece, yellow pine, 8'-0" ceiling	2 Carp	4	4	Ea.	262	219		481	640
0030 9'-0" ceiling		4	4		350	219		569	740
0040 10'-0" ceiling		3	5.333		249	292		541	740
0050 11'-0" ceiling		3	5.333		400	292		692	905
0060 12'-0" ceiling		3	5.333		385	292		677	890
0100 Custom grade, pine, 8'-6" ceiling, minimum	1 Carp	4	2		134	109		243	320
0150 Average		3.50	2.286		282	125		407	510
0200 Maximum		3	2.667		264	146		410	520
0500 Heavy duty, pivoted, from 7'-7" to 12'-10" floor to floor		3	2.667		1,325	146		1,471	1,700
0600 16'-0" ceiling		2	4		1,650	219		1,869	2,175
0800 Economy folding, pine, 8'-6" ceiling		4	2		216	109		325	410
0900 9'-6" ceiling		4	2		262	109		371	460
1100 Automatic electric, aluminum, floor to floor height, 8' to 9'	2 Carp	1	16		8,500	875		9,375	10,800
1400 11' to 12'	"	.90	17.778		9,725	970		10,695	12,300
1700 14' to 15'	2 Carp	.70	22.857	Ea.	9,650	1,250		10,900	12,600
9000 Minimum labor/equipment charge	1 Carp	2	4	Job		219		219	350

11 32 Unit Kitchens

11 32 13 – Metal Unit Kitchens

11 32 13.10 Commercial Unit Kitchens

	Crew	Daily Output	Labor-Hours	Unit	Material	2021 Bare Costs Labor	Equipment	Total	Total Incl O&P
0010 **COMMERCIAL UNIT KITCHENS**									
1500 Combination range, refrigerator and sink, 30" wide, minimum	L-1	2	8	Ea.	635	525		1,160	1,500
1550 Maximum	"	1	16	"	1,125	1,050		2,175	2,875
1640 Combination range, refrigerator, sink, microwave									
1660 Oven and ice maker	L-1	.80	20	Ea.	4,875	1,325		6,200	7,375

11 41 Foodservice Storage Equipment

11 41 13 – Refrigerated Food Storage Cases

11 41 13.10 Refrigerated Food Cases

	Crew	Daily Output	Labor-Hours	Unit	Material	2021 Bare Costs Labor	Equipment	Total	Total Incl O&P
0010 **REFRIGERATED FOOD CASES**									
0030 Dairy, multi-deck, 12' long	Q-5	3	5.333	Ea.	12,700	330		13,030	14,500
0100 For rear sliding doors, add					1,975			1,975	2,175
0200 Delicatessen case, service deli, 12' long, single deck	Q-5	3.90	4.103		9,800	252		10,052	11,200
0300 Multi-deck, 18 S.F. shelf display		3	5.333		8,350	330		8,680	9,675
0400 Freezer, self-contained, chest-type, 30 C.F.		3.90	4.103		5,025	252		5,277	5,925
0500 Glass door, upright, 78 C.F.		3.30	4.848		9,300	298		9,598	10,700
0600 Frozen food, chest type, 12' long		3.30	4.848		6,875	298		7,173	8,000
0700 Glass door, reach-in, 5 door		3	5.333		10,600	330		10,930	12,200
0800 Island case, 12' long, single deck		3.30	4.848		8,775	298		9,073	10,100
0900 Multi-deck		3	5.333		10,600	330		10,930	12,200
1000 Meat case, 12' long, single deck		3.30	4.848		7,450	298		7,748	8,650
1050 Multi-deck		3.10	5.161		10,200	320		10,520	11,700
1100 Produce, 12' long, single deck		3.30	4.848		6,600	298		6,898	7,700
1200 Multi-deck		3.10	5.161		8,525	320		8,845	9,900

11 41 Foodservice Storage Equipment

11 41 13 – Refrigerated Food Storage Cases

11 41 13.20 Refrigerated Food Storage Equipment		Crew	Daily Output	Labor-Hours	Unit	Material	2021 Bare Costs Labor	2021 Bare Costs Equipment	Total	Total Incl O&P
0010	**REFRIGERATED FOOD STORAGE EQUIPMENT**									
2350	Cooler, reach-in, beverage, 6' long	Q-1	6	2.667	Ea.	2,875	162		3,037	3,425
4300	Freezers, reach-in, 44 C.F.		4	4		4,425	244		4,669	5,250
4500	68 C.F.		3	5.333		5,450	325		5,775	6,475
4600	Freezer, pre-fab, 8' x 8' w/refrigeration	2 Carp	.45	35.556		9,250	1,950		11,200	13,300
4620	8' x 12'		.35	45.714		11,200	2,500		13,700	16,300
4640	8' x 16'		.25	64		15,000	3,500		18,500	22,100
4660	8' x 20'		.17	94.118		21,100	5,150		26,250	31,400
4680	Reach-in, 1 compartment	Q-1	4	4		3,975	244		4,219	4,750
4685	Energy star rated [G]	R-18	7.80	3.333		3,625	186		3,811	4,275
4700	2 compartment	Q-1	3	5.333		4,975	325		5,300	5,975
4705	Energy star rated [G]	R-18	6.20	4.194		3,850	234		4,084	4,575
4710	3 compartment	Q-1	3	5.333		6,375	325		6,700	7,525
4715	Energy star rated [G]	R-18	5.60	4.643		5,325	259		5,584	6,250
8320	Refrigerator, reach-in, 1 compartment		7.80	3.333		3,700	186		3,886	4,350
8325	Energy star rated [G]		7.80	3.333		3,625	186		3,811	4,275
8330	2 compartment		6.20	4.194		3,475	234		3,709	4,175
8335	Energy star rated [G]		6.20	4.194		3,850	234		4,084	4,575
8340	3 compartment		5.60	4.643		4,950	259		5,209	5,850
8345	Energy star rated [G]		5.60	4.643		5,325	259		5,584	6,250
8350	Pre-fab, with refrigeration, 8' x 8'	2 Carp	.45	35.556		6,400	1,950		8,350	10,200
8360	8' x 12'		.35	45.714		9,825	2,500		12,325	14,800
8370	8' x 16'		.25	64		10,400	3,500		13,900	17,100
8380	8' x 20'		.17	94.118		13,400	5,150		18,550	22,900
8390	Pass-thru/roll-in, 1 compartment	R-18	7.80	3.333		4,300	186		4,486	5,000
8400	2 compartment		6.24	4.167		5,525	233		5,758	6,425
8410	3 compartment		5.60	4.643		23,700	259		23,959	26,500
8420	Walk-in, alum, door & floor only, no refrig, 6' x 6' x 7'-6"	2 Carp	1.40	11.429		6,975	625		7,600	8,650
8430	10' x 6' x 7'-6"		.55	29.091		5,600	1,600		7,200	8,675
8440	12' x 14' x 7'-6"		.25	64		13,400	3,500		16,900	20,400
8450	12' x 20' x 7'-6"		.17	94.118		12,600	5,150		17,750	22,100
8460	Refrigerated cabinets, mobile					5,900			5,900	6,500
8470	Refrigerator/freezer, reach-in, 1 compartment	R-18	5.60	4.643		7,025	259		7,284	8,125
8480	2 compartment	"	4.80	5.417		7,775	305		8,080	9,025

11 41 13.30 Wine Cellar

		Crew	Daily Output	Labor-Hours	Unit	Material	2021 Bare Costs Labor	2021 Bare Costs Equipment	Total	Total Incl O&P
0010	**WINE CELLAR**, refrigerated, Redwood interior, carpeted, walk-in type									
0020	6'-8" high, including racks									
0200	80" W x 48" D for 900 bottles	2 Carp	1.50	10.667	Ea.	4,200	585		4,785	5,550
0250	80" W x 72" D for 1300 bottles		1.33	12.030		5,875	660		6,535	7,500
0300	80" W x 94" D for 1900 bottles		1.17	13.675		6,875	750		7,625	8,775
0400	80" W x 124" D for 2500 bottles		1	16		7,950	875		8,825	10,200

11 41 33 – Foodservice Shelving

11 41 33.20 Metal Food Storage Shelving

		Crew	Daily Output	Labor-Hours	Unit	Material	2021 Bare Costs Labor	2021 Bare Costs Equipment	Total	Total Incl O&P
0010	**METAL FOOD STORAGE SHELVING**									
8600	Stainless steel shelving, louvered 4-tier, 20" x 3'	1 Clab	6	1.333	Ea.	1,350	59		1,409	1,575
8605	20" x 4'		6	1.333		1,450	59		1,509	1,700
8610	20" x 6'		6	1.333		1,750	59		1,809	2,025
8615	24" x 3'		6	1.333		2,075	59		2,134	2,375
8620	24" x 4'		6	1.333		2,600	59		2,659	2,950
8625	24" x 6'		6	1.333		3,525	59		3,584	3,975
8630	Flat 4-tier, 20" x 3'		6	1.333		1,350	59		1,409	1,575

11 41 Foodservice Storage Equipment

11 41 33 – Foodservice Shelving

11 41 33.20 Metal Food Storage Shelving	Crew	Daily Output	Labor-Hours	Unit	Material	2021 Bare Costs Labor	Equipment	Total	Total Incl O&P
8635	20" x 4'	1 Clab	6	1.333	Ea.	1,450	59		1,509	1,675
8640	20" x 5'		6	1.333		1,725	59		1,784	2,000
8645	24" x 3'		6	1.333		910	59		969	1,100
8650	24" x 4'		6	1.333		2,350	59		2,409	2,675
8655	24" x 6'		6	1.333		2,875	59		2,934	3,275
8700	Galvanized shelving, louvered 4-tier, 20" x 3'		6	1.333		805	59		864	980
8705	20" x 4'		6	1.333		890	59		949	1,075
8710	20" x 6'		6	1.333		940	59		999	1,125
8715	24" x 3'		6	1.333		915	59		974	1,100
8720	24" x 4'		6	1.333		960	59		1,019	1,150
8725	24" x 6'		6	1.333		1,275	59		1,334	1,500
8730	Flat 4-tier, 20" x 3'		6	1.333		635	59		694	795
8735	20" x 4'		6	1.333		695	59		754	860
8740	20" x 6'		6	1.333		1,000	59		1,059	1,200
8745	24" x 3'		6	1.333		810	59		869	985
8750	24" x 4'		6	1.333		560	59		619	715
8755	24" x 6'		6	1.333		995	59		1,054	1,200
8760	Stainless steel dunnage rack, 24" x 3'		8	1		325	44.50		369.50	430
8765	24" x 4'		8	1		375	44.50		419.50	480
8770	Galvanized dunnage rack, 24" x 3'		8	1		142	44.50		186.50	227
8775	24" x 4'		8	1		195	44.50		239.50	286

11 42 Food Preparation Equipment

11 42 10 – Commercial Food Preparation Equipment

11 42 10.10 Choppers, Mixers and Misc. Equipment	Crew	Daily Output	Labor-Hours	Unit	Material	2021 Bare Costs Labor	Equipment	Total	Total Incl O&P
0010	**CHOPPERS, MIXERS AND MISC. EQUIPMENT**									
1700	Choppers, 5 pounds	R-18	7	3.714	Ea.	3,500	208		3,708	4,175
1720	16 pounds		5	5.200		2,275	291		2,566	2,950
1740	35 to 40 pounds		4	6.500		3,550	365		3,915	4,450
1840	Coffee brewer, 5 burners	1 Plum	3	2.667		935	181		1,116	1,300
1850	Coffee urn, twin 6 gallon urns		2	4		2,625	271		2,896	3,300
1860	Single, 3 gallon		3	2.667		1,900	181		2,081	2,375
3000	Fast food equipment, total package, minimum	6 Skwk	.08	600		226,000	34,300		260,300	303,000
3100	Maximum	"	.07	686		290,500	39,200		329,700	381,500
3800	Food mixers, bench type, 20 quarts	L-7	7	4		3,075	212		3,287	3,700
3850	40 quarts		5.40	5.185		8,325	275		8,600	9,575
3900	60 quarts		5	5.600		11,900	297		12,197	13,600
4040	80 quarts		3.90	7.179		20,900	380		21,280	23,600
4100	Floor type, 20 quarts		15	1.867		3,800	99		3,899	4,350
4120	60 quarts		14	2		10,800	106		10,906	12,100
4140	80 quarts		12	2.333		18,900	124		19,024	21,000
4160	140 quarts		8.60	3.256		26,600	173		26,773	29,600
6700	Peelers, small	R-18	8	3.250		1,875	182		2,057	2,350
6720	Large	"	6	4.333		3,225	242		3,467	3,925
6800	Pulper/extractor, close coupled, 5 HP	1 Plum	1.90	4.211		2,500	285		2,785	3,225
8580	Slicer with table	R-18	9	2.889		4,500	161		4,661	5,200

For customer support on your Commercial Renovation Costs with RSMeans Data, call 800.448.8182.

377

11 43 13.10 Mobile Carts, Racks and Trays		Crew	Daily Output	Labor-Hours	Unit	Material	2021 Bare Costs Labor	Equipment	Total	Total Incl O&P
0010	**MOBILE CARTS, RACKS AND TRAYS**									
1650	Cabinet, heated, 1 compartment, reach-in	R-18	5.60	4.643	Ea.	12,000	259		12,259	13,600
1655	Pass-thru roll-in		5.60	4.643		2,050	259		2,309	2,650
1660	2 compartment, reach-in	↓	4.80	5.417		11,500	305		11,805	13,100
1670	Mobile					4,300			4,300	4,725
2000	Hospital food cart, hot and cold service, 20 tray capacity					15,500			15,500	17,000
6850	Mobile rack w/pan slide					510			510	560
9180	Tray and silver dispenser, mobile	1 Clab	16	.500	↓	695	22		717	800

11 44 13.10 Cooking Equipment		Crew	Daily Output	Labor-Hours	Unit	Material	2021 Bare Costs Labor	Equipment	Total	Total Incl O&P
0010	**COOKING EQUIPMENT**									
0020	Bake oven, gas, one section	Q-1	8	2	Ea.	6,250	122		6,372	7,075
0300	Two sections		7	2.286		10,000	139		10,139	11,200
0600	Three sections	↓	6	2.667		12,400	162		12,562	13,900
0900	Electric convection, single deck	L-7	4	7		5,500	370		5,870	6,650
1300	Broiler, without oven, standard	Q-1	8	2		3,750	122		3,872	4,325
1550	Infrared	L-7	4	7		5,300	370		5,670	6,425
4750	Fryer, with twin baskets, modular model	Q-1	7	2.286		1,575	139		1,714	1,975
5000	Floor model, on 6" legs	"	5	3.200		3,100	195		3,295	3,725
5100	Extra single basket, large					51			51	56.50
5170	Energy star rated, 50 lb. capacity [G]	R-18	4	6.500		5,025	365		5,390	6,075
5175	85 lb. capacity [G]	"	4	6.500		1,525	365		1,890	2,225
5300	Griddle, SS, 24" plate, w/4" legs, elec, 208 V, 3 phase, 3' long	Q-1	7	2.286		4,100	139		4,239	4,750
5550	4' long	"	6	2.667		2,375	162		2,537	2,875
6200	Iced tea brewer	1 Plum	3.44	2.326		695	157		852	1,000
6350	Kettle, w/steam jacket, tilting, w/positive lock, SS, 20 gallons	L-7	7	4		10,800	212		11,012	12,200
6600	60 gallons	"	6	4.667		14,600	248		14,848	16,500
6900	Range, restaurant type, 6 burners and 1 standard oven, 36" wide	Q-1	7	2.286		2,700	139		2,839	3,175
6950	Convection		7	2.286		5,150	139		5,289	5,900
7150	2 standard ovens, 24" griddle, 60" wide		6	2.667		4,400	162		4,562	5,075
7200	1 standard, 1 convection oven		6	2.667		9,125	162		9,287	10,300
7450	Heavy duty, single 34" standard oven, open top		5	3.200		4,900	195		5,095	5,700
7500	Convection oven		5	3.200		5,175	195		5,370	6,000
7700	Griddle top		6	2.667		2,600	162		2,762	3,125
7750	Convection oven	↓	6	2.667		3,425	162		3,587	4,025
7760	Induction cooker, electric	L-7	7	4	↓	1,950	212		2,162	2,475
8850	Steamer, electric 27 KW	L-7	7	4	Ea.	16,600	212		16,812	18,600
9100	Electric, 10 KW or gas 100,000 BTU	"	5	5.600		7,850	297		8,147	9,100
9150	Toaster, conveyor type, 16-22 slices/minute					1,100			1,100	1,200
9160	Pop-up, 2 slot				↓	89.50			89.50	98.50
9200	For deluxe models of above equipment, add					75%				
9400	Rule of thumb: Equipment cost based									
9410	on kitchen work area									
9420	Office buildings, minimum	L-7	77	.364	S.F.	98.50	19.30		117.80	139
9450	Maximum		58	.483		160	25.50		185.50	217
9550	Public eating facilities, minimum		77	.364		131	19.30		150.30	175
9600	Maximum		46	.609		210	32.50		242.50	282
9750	Hospitals, minimum		58	.483		127	25.50		152.50	181
9800	Maximum	↓	39	.718	↓	235	38		273	320

11 46 Food Dispensing Equipment

11 46 16 – Service Line Equipment

11 46 16.10 Commercial Food Dispensing Equipment		Crew	Daily Output	Labor-Hours	Unit	Material	2021 Bare Costs Labor	2021 Bare Costs Equipment	Total	Total Incl O&P
0010	COMMERCIAL FOOD DISPENSING EQUIPMENT									
1050	Butter pat dispenser	1 Clab	13	.615	Ea.	780	27.50		807.50	900
1100	Bread dispenser, counter top		13	.615		940	27.50		967.50	1,075
1900	Cup and glass dispenser, drop in		4	2		610	89		699	810
1920	Disposable cup, drop in		16	.500		765	22		787	880
2650	Dish dispenser, drop in, 12"		11	.727		2,950	32.50		2,982.50	3,300
2660	Mobile		10	.800		3,550	35.50		3,585.50	3,950
3300	Food warmer, counter, 1.2 KW					630			630	695
3550	1.6 KW					2,275			2,275	2,500
3600	Well, hot food, built-in, rectangular, 12" x 20"	R-30	10	2.600		330	135		465	575
3610	Circular, 7 qt.		10	2.600		395	135		530	645
3620	Refrigerated, 2 compartments		10	2.600		2,900	135		3,035	3,400
3630	3 compartments		9	2.889		3,700	150		3,850	4,300
3640	4 compartments		8	3.250		5,000	169		5,169	5,775
4720	Frost cold plate		9	2.889		23,300	150		23,450	25,800
5700	Hot chocolate dispenser	1 Plum	4	2		1,175	135		1,310	1,500
5750	Ice dispenser 567 pound	Q-1	6	2.667		6,000	162		6,162	6,850
6250	Jet spray dispenser	R-18	4.50	5.778		2,225	325		2,550	2,950
6300	Juice dispenser, concentrate	"	4.50	5.778		2,000	325		2,325	2,700
6690	Milk dispenser, bulk, 2 flavor	R-30	8	3.250		2,400	169		2,569	2,925
6695	3 flavor	"	8	3.250		2,975	169		3,144	3,550
8800	Serving counter, straight	1 Carp	40	.200	L.F.	1,750	10.95		1,760.95	1,950
8820	Curved section	"	30	.267	"	2,250	14.60		2,264.60	2,500
8825	Solid surface, see Section 12 36 61.16									
8860	Sneeze guard with lights, 60" L	1 Clab	16	.500	Ea.	335	22		357	400
8900	Sneeze guard, stainless steel and glass, single sided									
8910	Portable, 48" W				Ea.	400			400	440
8920	Portable, 72" W					455			455	500
8930	Adjustable, 36" W	1 Carp	24	.333		350	18.25		368.25	415
8940	Adjustable, 48" W	"	20	.400		320	22		342	385
9100	Soft serve ice cream machine, medium	R-18	11	2.364		8,350	132		8,482	9,400
9110	Large	"	9	2.889		22,000	161		22,161	24,400

11 46 83 – Ice Machines

11 46 83.10 Commercial Ice Equipment

11 46 83.10 Commercial Ice Equipment		Crew	Daily Output	Labor-Hours	Unit	Material	2021 Bare Costs Labor	2021 Bare Costs Equipment	Total	Total Incl O&P
0010	COMMERCIAL ICE EQUIPMENT									
5800	Ice cube maker, 50 lbs./day	Q-1	6	2.667	Ea.	1,825	162		1,987	2,250
5810	65 lbs./day, energy star rated		6	2.667		1,675	162		1,837	2,100
5900	250 lbs./day		1.20	13.333		2,175	810		2,985	3,650
5950	300 lbs./day, remote condensing		1.20	13.333		2,525	810		3,335	4,025
6050	500 lbs./day		4	4		2,625	244		2,869	3,250
6060	With bin		1.20	13.333		3,975	810		4,785	5,625
6070	Modular, with bin and condenser		1.20	13.333		3,825	810		4,635	5,450
6090	1000 lbs./day, with bin		1	16		5,575	975		6,550	7,650
6100	Ice flakers, 300 lbs./day		1.60	10		3,750	610		4,360	5,075
6120	600 lbs./day		.95	16.842		4,375	1,025		5,400	6,375
6130	1000 lbs./day		.75	21.333		5,550	1,300		6,850	8,125
6140	2000 lbs./day		.65	24.615		22,500	1,500		24,000	27,100
6160	Ice storage bin, 500 pound capacity	Q-5	1	16		1,025	985		2,010	2,650
6180	1000 pound	"	.56	28.571		3,150	1,750		4,900	6,150

11 48 Foodservice Cleaning and Disposal Equipment

11 48 13 – Commercial Dishwashers

11 48 13.10 Dishwashers

		Crew	Daily Output	Labor-Hours	Unit	Material	2021 Bare Costs Labor	2021 Bare Costs Equipment	Total	Total Incl O&P
0010	**DISHWASHERS**									
2700	Dishwasher, commercial, rack type									
2720	10 to 12 racks/hour	Q-1	3.20	5	Ea.	3,500	305		3,805	4,325
2730	Energy star rated, 35 to 40 racks/hour [G]		1.30	12.308		5,650	750		6,400	7,350
2740	50 to 60 racks/hour [G]	↓	1.30	12.308		10,800	750		11,550	13,100
2800	Automatic, 190 to 230 racks/hour	L-6	.35	34.286		16,600	2,275		18,875	21,800
2820	235 to 275 racks/hour		.25	48		29,000	3,175		32,175	36,800
2840	8,750 to 12,500 dishes/hour	↓	.10	120	↓	56,500	7,975		64,475	74,500
2950	Dishwasher hood, canopy type	L-3A	10	1.200	L.F.	1,200	71.50		1,271.50	1,425
2960	Pant leg type	"	2.50	4.800	Ea.	8,050	286		8,336	9,300
5200	Garbage disposal 1.5 HP, 100 GPH	L-1	4.80	3.333		2,200	219		2,419	2,725
5210	3 HP, 120 GPH		4.60	3.478		2,550	229		2,779	3,175
5220	5 HP, 250 GPH	↓	4.50	3.556	↓	3,050	234		3,284	3,725
6750	Pot sink, 3 compartment	1 Plum	7.25	1.103	L.F.	870	74.50		944.50	1,075
6760	Pot washer, low temp wash/rinse		1.60	5	Ea.	5,550	340		5,890	6,625
6770	High pressure wash, high temperature rinse	↓	1.20	6.667		17,100	450		17,550	19,500
9170	Trash compactor, small, up to 125 lb. compacted weight	L-4	4	6		20,900	310		21,210	23,500
9175	Large, up to 175 lb. compacted weight	"	3	8	↓	29,100	415		29,515	32,700

11 52 Audio-Visual Equipment

11 52 13 – Projection Screens

11 52 13.10 Projection Screens, Wall or Ceiling Hung

		Crew	Daily Output	Labor-Hours	Unit	Material	2021 Bare Costs Labor	2021 Bare Costs Equipment	Total	Total Incl O&P
0010	**PROJECTION SCREENS, WALL OR CEILING HUNG**, matte white									
0100	Manually operated, economy	2 Carp	500	.032	S.F.	6.60	1.75		8.35	10.05
0300	Intermediate		450	.036		7.90	1.94		9.84	11.75
0400	Deluxe		400	.040	↓	10	2.19		12.19	14.50
9000	Minimum labor/equipment charge	↓	3	5.333	Job		292		292	465

11 52 16 – Projectors

11 52 16.10 Movie Equipment

		Crew	Daily Output	Labor-Hours	Unit	Material	2021 Bare Costs Labor	2021 Bare Costs Equipment	Total	Total Incl O&P
0011	**MOVIE EQUIPMENT**									
0020	Changeover, minimum				Ea.	520			520	570
3000	Projection screens, rigid, in wall, acrylic, 1/4" thick	2 Glaz	195	.082	S.F.	49	4.32		53.32	61
3100	1/2" thick	"	130	.123	"	56	6.50		62.50	72
3700	Sound systems, incl. amplifier, mono, minimum	1 Elec	.90	8.889	Ea.	3,875	565		4,440	5,125
3800	Dolby/Super Sound, maximum		.40	20		20,800	1,275		22,075	24,900
4100	Dual system, 2 channel, front surround, minimum		.70	11.429		5,325	730		6,055	7,000
4200	Dolby/Super Sound, 4 channel, maximum	↓	.40	20		20,000	1,275		21,275	24,000
5700	Seating, painted steel, upholstered, minimum	2 Carp	35	.457		174	25		199	231
5800	Maximum	"	28	.571	↓	560	31.50		591.50	665

11 53 Laboratory Equipment

11 53 33 – Emergency Safety Appliances

11 53 33.13 Emergency Equipment

		Crew	Daily Output	Labor-Hours	Unit	Material	2021 Bare Costs Labor	Equipment	Total	Total Incl O&P
0010	**EMERGENCY EQUIPMENT**									
1400	Safety equipment, eye wash, hand held				Ea.	370			370	410
1450	Deluge shower				"	825			825	905

11 53 43 – Service Fittings and Accessories

11 53 43.13 Fittings

		Crew	Daily Output	Labor-Hours	Unit	Material	2021 Bare Costs Labor	Equipment	Total	Total Incl O&P
0010	**FITTINGS**									
1600	Sink, one piece plastic, flask wash, hose, free standing	1 Plum	1.60	5	Ea.	2,075	340		2,415	2,800
1610	Epoxy resin sink, 25" x 16" x 10"	"	2	4	"	228	271		499	670
1630	Steel table, open underneath				L.F.	595			595	655
1640	Doors underneath				"	975			975	1,075
1900	Utility cabinet, stainless steel				Ea.	1,300			1,300	1,425
1950	Utility table, acid resistant top with drawers	2 Carp	30	.533	L.F.	185	29		214	250
8000	Alternate pricing method: as percent of lab furniture									
8050	Installation, not incl. plumbing & duct work				% Furn.				22%	22%
8100	Plumbing, final connections, simple system				% Furn.				10%	10%
8110	Moderately complex system								15%	15%
8120	Complex system								20%	20%
8150	Electrical, simple system								10%	10%
8160	Moderately complex system								20%	20%
8170	Complex system								35%	35%

11 53 53 – Biological Safety Cabinets

11 53 53.10 Pharmacy Cabinets

		Crew	Daily Output	Labor-Hours	Unit	Material	2021 Bare Costs Labor	Equipment	Total	Total Incl O&P
0010	**PHARMACY CABINETS**, vertical flow									
0100	Class II, type B2, 6' L	2 Carp	1.50	10.667	Ea.	15,800	585		16,385	18,200

11 57 Vocational Shop Equipment

11 57 10 – Shop Equipment

11 57 10.10 Vocational School Shop Equipment

		Crew	Daily Output	Labor-Hours	Unit	Material	2021 Bare Costs Labor	Equipment	Total	Total Incl O&P
0010	**VOCATIONAL SCHOOL SHOP EQUIPMENT**									
0020	Benches, work, wood, average	2 Carp	5	3.200	Ea.	780	175		955	1,125
0100	Metal, average		5	3.200		450	175		625	775
0400	Combination belt & disc sander, 6"		4	4		1,775	219		1,994	2,325
0700	Drill press, floor mounted, 12", 1/2 HP		4	4		455	219		674	850
0800	Dust collector, not incl. ductwork, 6" diameter	1 Shee	1.10	7.273		5,775	475		6,250	7,100
0810	Dust collector bag, 20" diameter	"	5	1.600		605	105		710	830
1000	Grinders, double wheel, 1/2 HP	2 Carp	5	3.200		247	175		422	550
1300	Jointer, 4", 3/4 HP		4	4		1,550	219		1,769	2,050
1600	Kilns, 16 C.F., to 2000°		4	4		1,975	219		2,194	2,525
1900	Lathe, woodworking, 10", 1/2 HP		4	4		580	219		799	985
2200	Planer, 13" x 6"		4	4		1,350	219		1,569	1,825
2500	Potter's wheel, motorized		4	4		1,275	219		1,494	1,750
2800	Saws, band, 14", 3/4 HP		4	4		1,275	219		1,494	1,750
3100	Metal cutting band saw, 14"		4	4		2,450	219		2,669	3,050
3400	Radial arm saw, 10", 2 HP		4	4		1,350	219		1,569	1,850
3700	Scroll saw, 24"		4	4		625	219		844	1,050
4000	Table saw, 10", 3 HP		4	4		1,050	219		1,269	1,500
4300	Welder AC arc, 30 amp capacity		4	4		3,825	219		4,044	4,550

For customer support on your Commercial Renovation Costs with RSMeans Data, call 800.448.8182.

381

11 61 Broadcast, Theater, and Stage Equipment

11 61 23 – Folding and Portable Stages

11 61 23.10 Portable Stages	Crew	Daily Output	Labor-Hours	Unit	Material	2021 Bare Costs Labor	2021 Bare Costs Equipment	Total	Total Incl O&P
0010 **PORTABLE STAGES**									
5000 Stages, portable with steps, folding legs, stock, 8" high				SF Stg.	47.50			47.50	52
5100 16" high					59			59	65
5200 32" high					60			60	66
5300 40" high					65.50			65.50	72

11 61 33 – Rigging Systems and Controls

11 61 33.10 Controls

	Crew	Daily Output	Labor-Hours	Unit	Material	Labor	Equipment	Total	Total Incl O&P
0010 **CONTROLS**									
0050 Control boards with dimmers and breakers, minimum	1 Elec	1	8	Ea.	15,500	510		16,010	17,900
0150 Maximum	"	.20	40	"	124,500	2,550		127,050	141,000

11 61 43 – Stage Curtains

11 61 43.10 Curtains

	Crew	Daily Output	Labor-Hours	Unit	Material	Labor	Equipment	Total	Total Incl O&P
0010 **CURTAINS**									
0500 Curtain track, straight, light duty	2 Carp	20	.800	L.F.	31	44		75	104
0700 Curved sections		12	1.333	"	202	73		275	340
1000 Curtains, velour, medium weight		600	.027	S.F.	8.75	1.46		10.21	11.90

11 66 Athletic Equipment

11 66 13 – Exercise Equipment

11 66 13.10 Physical Training Equipment

	Crew	Daily Output	Labor-Hours	Unit	Material	Labor	Equipment	Total	Total Incl O&P
0010 **PHYSICAL TRAINING EQUIPMENT**									
0020 Abdominal rack, 2 board capacity				Ea.	525			525	575
0050 Abdominal board, upholstered					780			780	860
0200 Bicycle trainer, minimum					515			515	570
0300 Deluxe, electric					5,075			5,075	5,575
0400 Barbell set, chrome plated steel, 25 lb.					310			310	345
0420 100 lb.					590			590	650
0450 200 lb.					790			790	865
0500 Weight plates, cast iron, per lb.				Lb.	3.26			3.26	3.59
0520 Storage rack, 10 station				Ea.	1,225			1,225	1,350
0600 Circuit training apparatus, 12 machines minimum	2 Clab	1.25	12.800	Set	34,700	570		35,270	39,000
0700 Average		1	16		42,100	710		42,810	47,400
0800 Maximum		.75	21.333		50,500	945		51,445	57,000
0820 Dumbbell set, cast iron, with rack and 5 pair					510			510	560
0900 Squat racks	2 Clab	5	3.200	Ea.	835	142		977	1,150
4150 Exercise equipment, bicycle trainer					1,200			1,200	1,325
4180 Chinning bar, adjustable, wall mounted	1 Carp	5	1.600		240	87.50		327.50	405
4200 Exercise ladder, 16' x 1'-7", suspended	L-2	3	5.333		1,400	258		1,658	1,975
4210 High bar, floor plate attached	1 Carp	4	2		2,775	109		2,884	3,225
4240 Parallel bars, adjustable		4	2		1,625	109		1,734	1,975
4270 Uneven parallel bars, adjustable		4	2		4,025	109		4,134	4,625
4280 Wall mounted, adjustable	L-2	1.50	10.667	Set	970	515		1,485	1,900
4300 Rope, ceiling mounted, 18' long	1 Carp	3.66	2.186	Ea.	207	120		327	420
4330 Side horse, vaulting		5	1.600		1,700	87.50		1,787.50	2,025
4360 Treadmill, motorized, deluxe, training type		5	1.600		4,150	87.50		4,237.50	4,700
4390 Weight lifting multi-station, minimum	2 Clab	1	16		310	710		1,020	1,475

11 66 Athletic Equipment

11 66 23 – Gymnasium Equipment

11 66 23.13 Basketball Equipment

		Crew	Daily Output	Labor-Hours	Unit	Material	2021 Bare Costs Labor	2021 Bare Costs Equipment	Total	Total Incl O&P
0010	**BASKETBALL EQUIPMENT**									
1000	Backstops, wall mtd., 6' extended, fixed, minimum	L-2	1	16	Ea.	1,825	775		2,600	3,275
1100	Maximum		1	16		1,875	775		2,650	3,300
1200	Swing up, minimum		1	16		1,575	775		2,350	2,975
1250	Maximum		1	16		3,650	775		4,425	5,275
1300	Portable, manual, heavy duty, spring operated		1.90	8.421		14,100	405		14,505	16,200
1400	Ceiling suspended, stationary, minimum		.78	20.513		3,900	990		4,890	5,900
1450	Fold up, with accessories, maximum		.40	40		6,200	1,925		8,125	9,925
1600	For electrically operated, add	1 Elec	1	8	▼	2,600	510		3,110	3,625
5800	Wall pads, 1-1/2" thick, standard (not fire rated)	2 Carp	640	.025	S.F.	6.45	1.37		7.82	9.30

11 66 23.19 Boxing Ring

		Crew	Daily Output	Labor-Hours	Unit	Material	2021 Bare Costs Labor	2021 Bare Costs Equipment	Total	Total Incl O&P
0010	**BOXING RING**									
4100	Elevated, 22' x 22'	L-4	.10	240	Ea.	6,350	12,500		18,850	26,900
4110	For cellular plastic foam padding, add		.10	240		1,275	12,500		13,775	21,300
4120	Floor level, including posts and ropes only, 20' x 20'		.80	30		4,975	1,550		6,525	7,975
4130	Canvas, 30' x 30'	▼	5	4.800	▼	1,525	250		1,775	2,075

11 66 23.47 Gym Mats

		Crew	Daily Output	Labor-Hours	Unit	Material	2021 Bare Costs Labor	2021 Bare Costs Equipment	Total	Total Incl O&P
0010	**GYM MATS**									
5500	2" thick, naugahyde covered				S.F.	4.37			4.37	4.81
5600	Vinyl/nylon covered					9			9	9.90
6000	Wrestling mats, 1" thick, heavy duty			▼		4.92			4.92	5.40

11 66 43 – Interior Scoreboards

11 66 43.10 Scoreboards

		Crew	Daily Output	Labor-Hours	Unit	Material	2021 Bare Costs Labor	2021 Bare Costs Equipment	Total	Total Incl O&P
0010	**SCOREBOARDS**									
7000	Baseball, minimum	R-3	1.30	15.385	Ea.	4,200	975	147	5,322	6,275
7200	Maximum		.05	400		18,000	25,400	3,825	47,225	63,000
7300	Football, minimum		.86	23.256		5,175	1,475	222	6,872	8,225
7400	Maximum		.20	100		25,600	6,350	955	32,905	38,900
7500	Basketball (one side), minimum		2.07	9.662		2,400	615	92.50	3,107.50	3,675
7600	Maximum		.30	66.667		9,250	4,225	635	14,110	17,400
7700	Hockey-basketball (four sides), minimum		.25	80		10,300	5,075	765	16,140	20,000
7800	Maximum	▼	.15	133	▼	17,500	8,450	1,275	27,225	33,600

11 66 53 – Gymnasium Dividers

11 66 53.10 Divider Curtains

		Crew	Daily Output	Labor-Hours	Unit	Material	2021 Bare Costs Labor	2021 Bare Costs Equipment	Total	Total Incl O&P
0010	**DIVIDER CURTAINS**									
4500	Gym divider curtain, mesh top, vinyl bottom, manual	L-4	500	.048	S.F.	9.85	2.50		12.35	14.80
4700	Electric roll up	L-7	400	.070	"	14.45	3.71		18.16	22

11 71 Medical Sterilizing Equipment

11 71 10 – Medical Sterilizers & Distillers

11 71 10.10 Sterilizers and Distillers

		Crew	Daily Output	Labor-Hours	Unit	Material	2021 Bare Costs Labor	2021 Bare Costs Equipment	Total	Total Incl O&P
0010	**STERILIZERS AND DISTILLERS**									
3010	Portable, top loading, 105-135 degree C, 3 to 30 psi, 50 L chamber				Ea.	9,800			9,800	10,800
3020	Stainless steel basket, 10.7" diam. x 11.8" H					216			216	238
3025	Stainless steel pail, 10.7" diam. x 10.7" H				▼	320			320	355
3050	85 L chamber				Ea.	18,200			18,200	20,000
3060	Stainless steel basket, 15.3" diam. x 11.5" H					470			470	515
3065	Stainless steel pail, 15.3" diam. x 11" H			▼		685			685	750

For customer support on your Commercial Renovation Costs with RSMeans Data, call 800.448.8182.

383

11 76 Operating Room Equipment

11 76 10 – Equipment for Operating Rooms

11 76 10.10 Surgical Equipment	Crew	Daily Output	Labor-Hours	Unit	Material	2021 Bare Costs Labor	Equipment	Total	Total Incl O&P
0010 **SURGICAL EQUIPMENT**									
6550 Major surgery table, minimum	1 Sswk	.50	16	Ea.	23,300	965		24,265	27,200
6570 Maximum		.50	16		44,700	965		45,665	51,000
6600 Hydraulic, hand-held control, general surgery		.60	13.333		43,900	805		44,705	49,600
6650 Stationary, universal	↓	.50	16		55,500	965		56,465	62,500
6800 Surgical lights, major operating room, dual head, minimum	2 Elec	1	16		6,775	1,025		7,800	9,025
6850 Maximum		1	16		41,600	1,025		42,625	47,400
6900 Ceiling mount articulation, single arm	↓	1	16	↓	5,025	1,025		6,050	7,100

11 81 Facility Maintenance Equipment

11 81 19 – Vacuum Cleaning Systems

11 81 19.10 Vacuum Cleaning

	Crew	Daily Output	Labor-Hours	Unit	Material	2021 Bare Costs Labor	Equipment	Total	Total Incl O&P
0010 **VACUUM CLEANING**									
0020 Central, 3 inlet, residential	1 Skwk	.90	8.889	Total	1,175	510		1,685	2,100
0200 Commercial		.70	11.429		1,375	655		2,030	2,550
0400 5 inlet system, residential		.50	16		1,850	915		2,765	3,475
0600 7 inlet system, commercial	↓	.40	20	↓	2,400	1,150		3,550	4,450

11 82 Facility Solid Waste Handling Equipment

11 82 26 – Facility Waste Compactors

11 82 26.10 Compactors

	Crew	Daily Output	Labor-Hours	Unit	Material	2021 Bare Costs Labor	Equipment	Total	Total Incl O&P
0010 **COMPACTORS**									
0020 Compactors, 115 volt, 250 lbs./hr., chute fed	L-4	1	24	Ea.	12,500	1,250		13,750	15,700
0100 Hand fed		2.40	10		16,800	520		17,320	19,200
1000 Heavy duty industrial compactor, 0.5 C.Y. capacity		1	24		11,100	1,250		12,350	14,200
1050 1.0 C.Y. capacity	↓	1	24		16,900	1,250		18,150	20,600
1400 For handling hazardous waste materials, 55 gallon drum packer, std.					21,800			21,800	24,000
1410 55 gallon drum packer w/HEPA filter					26,200			26,200	28,800
1420 55 gallon drum packer w/charcoal & HEPA filter					34,900			34,900	38,400
1430 All of the above made explosion proof, add					2,025			2,025	2,225

11 82 39 – Medical Waste Disposal Systems

11 82 39.10 Off-Site Disposal

	Crew	Daily Output	Labor-Hours	Unit	Material	2021 Bare Costs Labor	Equipment	Total	Total Incl O&P
0010 **OFF-SITE DISPOSAL**									
0100 Medical waste disposal, Red Bag system, pick up & treat, 200 lbs./week				Week	206			206	227
0110 Per month				Month	745			745	820
0150 Red bags, 7-10 gal., 1.2 mil, pkg of 500				Ea.	65.50			65.50	72
0200 15 gal., package of 250				"	64			64	70.50
0250 33 gal., package of 250				Ea.	72.50			72.50	79.50
0300 45 gal., package of 100				"	59.50			59.50	65.50

11 82 39.20 Disposal Carts

	Crew	Daily Output	Labor-Hours	Unit	Material	2021 Bare Costs Labor	Equipment	Total	Total Incl O&P
0010 **DISPOSAL CARTS**									
2010 Medical waste disposal cart, HDPE, w/lid, 28 gal. capacity				Ea.	291			291	320
2020 96 gal. capacity					340			340	375
2030 150 gal. capacity, low profile					490			490	540
2040 200 gal. capacity				↓	1,025			1,025	1,125

For customer support on your Commercial Renovation Costs with RSMeans Data, call 800.448.8182.

11 82 Facility Solid Waste Handling Equipment

11 82 39 - Medical Waste Disposal Systems

11 82 39.30 Medical Waste Sanitizers	Crew	Daily Output	Labor-Hours	Unit	Material	2021 Bare Costs Labor	Equipment	Total	Total Incl O&P
0010 **MEDICAL WASTE SANITIZERS**									
2010 Small, hand loaded, 1.5 C.Y., 225 lb. capacity				Ea.	83,500			83,500	91,500
2020 Medium, cart loaded, 6.25 C.Y., 938 lb. capacity					111,000			111,000	122,500
2030 Large, cart loaded, 15 C.Y., 2250 lb. capacity					134,000			134,000	147,500
3010 Cart, aluminum, 75 lb. capacity					2,225			2,225	2,450
3020 95 lb. capacity					2,400			2,400	2,650
4010 Stainless steel, 173 lb. capacity					3,150			3,150	3,475
4020 232 lb. capacity					3,525			3,525	3,900
4030 Cart lift, hydraulic scissor type					6,250			6,250	6,875
4040 Portable aluminum ramp					1,775			1,775	1,950
4050 Fold-down steel tracks					1,425			1,425	1,575
4060 Pull-out drawer, small					6,750			6,750	7,425
4070 Medium					9,750			9,750	10,700
4080 Large					13,700			13,700	15,100
5000 Medical waste treatment, sanitize, on-site									
5010 Less than 15,000 lbs./month				Lb.	.20			.20	.22
5020 Over 15,000 lbs./month				"	.16			.16	.18

11 91 Religious Equipment

11 91 13 - Baptisteries

11 91 13.10 Baptistry

	Crew	Daily Output	Labor-Hours	Unit	Material	2021 Bare Costs Labor	Equipment	Total	Total Incl O&P
0010 **BAPTISTRY**									
0150 Fiberglass, 3'-6" deep, x 13'-7" long,									
0160 steps at both ends, incl. plumbing, minimum	L-8	1	20	Ea.	7,575	1,150		8,725	10,100
0200 Maximum	"	.70	28.571		12,400	1,625		14,025	16,200
0250 Add for filter, heater and lights					1,700			1,700	1,875

11 91 23 - Sanctuary Equipment

11 91 23.10 Sanctuary Furnishings

	Crew	Daily Output	Labor-Hours	Unit	Material	2021 Bare Costs Labor	Equipment	Total	Total Incl O&P
0010 **SANCTUARY FURNISHINGS**									
0020 Altar, wood, custom design, plain	1 Carp	1.40	5.714	Ea.	2,775	315		3,090	3,575
0050 Deluxe		.20	40	"	13,400	2,200		15,600	18,200
5000 Wall cross, aluminum, extruded, 2" x 2" section		34	.235	L.F.	250	12.85		262.85	296
5150 4" x 4" section		29	.276		360	15.10		375.10	420
5300 Bronze, extruded, 1" x 2" section		31	.258		495	14.10		509.10	570
5350 2-1/2" x 2-1/2" section		34	.235		745	12.85		757.85	840

11 97 Security Equipment

11 97 30 - Security Drawers

11 97 30.10 Pass Through Drawer

	Crew	Daily Output	Labor-Hours	Unit	Material	2021 Bare Costs Labor	Equipment	Total	Total Incl O&P
0010 **PASS THROUGH DRAWER**									
0100 Pass-thru drawer for personal items, 18" x 15" x 24"	1 Skwk	2	4	Ea.	2,725	228		2,953	3,350
0110 Including speakers	"	1.50	5.333	"	3,075	305		3,380	3,850

For customer support on your Commercial Renovation Costs with RSMeans Data, call 800.448.8182.

385

11 98 Detention Equipment

11 98 21 – Detention Windows

11 98 21.13 Visitor Cubicle Windows	Crew	Daily Output	Labor-Hours	Unit	Material	2021 Bare Costs Labor	Equipment	Total	Total Incl O&P
0010 **VISITOR CUBICLE WINDOWS**									
4000 Visitor cubicle, vision panel, no intercom	E-4	2	16	Ea.	3,675	975	74.50	4,724.50	5,675

11 98 30 – Detention Cell Equipment

11 98 30.10 Cell Equipment

	Crew	Daily Output	Labor-Hours	Unit	Material	Labor	Equipment	Total	Total Incl O&P
0010 **CELL EQUIPMENT**									
3000 Toilet apparatus including wash basin, average	L-8	1.50	13.333	Ea.	3,525	765		4,290	5,075

Estimating Tips
General

- The items in this division are usually priced per square foot or each. Most of these items are purchased by the owner and installed by the contractor. Do not assume the items in Division 12 will be purchased and installed by the contractor. Check the specifications for responsibilities and include receiving, storage, installation, and mechanical and electrical hookups in the appropriate divisions.

- Some items in this division require some type of support system that is not usually furnished with the item. Examples of these systems include blocking for the attachment of casework and heavy drapery rods. The required blocking must be added to the estimate in the appropriate division.

Reference Numbers

Reference numbers are shown at the beginning of some major classifications. These numbers refer to related items in the Reference Section. The reference information may be an estimating procedure, an alternate pricing method, or technical information.

Same Data. Simplified.

Enjoy the convenience and efficiency of accessing your costs anywhere:

- **Skip the multiplier** by setting your location
- **Quickly search,** edit, favorite and share costs
- **Stay on top of price changes** with automatic updates

Discover more at rsmeans.com/online

12 21 Window Blinds

12 21 13 – Horizontal Louver Blinds

12 21 13.13 Metal Horizontal Louver Blinds

		Crew	Daily Output	Labor-Hours	Unit	Material	2021 Bare Costs Labor	2021 Bare Costs Equipment	Total	Total Incl O&P
0010	**METAL HORIZONTAL LOUVER BLINDS**									
0020	Horizontal, 1" aluminum slats, solid color, stock	1 Carp	590	.014	S.F.	5.60	.74		6.34	7.40

12 21 13.33 Vinyl Horizontal Louver Blinds

		Crew	Daily Output	Labor-Hours	Unit	Material	2021 Bare Costs Labor	2021 Bare Costs Equipment	Total	Total Incl O&P
0010	**VINYL HORIZONTAL LOUVER BLINDS**									
0100	2" composite, 48" wide, 48" high	1 Carp	30	.267	Ea.	119	14.60		133.60	154
0120	72" high		29	.276		161	15.10		176.10	201
0140	96" high		28	.286		219	15.65		234.65	266
0200	60" wide, 60" high		27	.296		204	16.20		220.20	251
0220	72" high		25	.320		235	17.50		252.50	286
0240	96" high		24	.333		350	18.25		368.25	415
0300	72" wide, 72" high		25	.320		292	17.50		309.50	350
0320	96" high		23	.348		405	19.05		424.05	475
0400	96" wide, 96" high		20	.400		850	22		872	970
1000	2" faux wood, 48" wide, 48" high		30	.267		69.50	14.60		84.10	99
1020	72" high		29	.276		90	15.10		105.10	123
1040	96" high		28	.286		107	15.65		122.65	143
1300	72" wide, 72" high		25	.320		118	17.50		135.50	158
1320	96" high		23	.348		178	19.05		197.05	226
1400	96" wide, 96" high		20	.400		360	22		382	435

12 23 Interior Shutters

12 23 13 – Wood Interior Shutters

12 23 13.10 Wood Interior Shutters

		Crew	Daily Output	Labor-Hours	Unit	Material	2021 Bare Costs Labor	2021 Bare Costs Equipment	Total	Total Incl O&P
0010	**WOOD INTERIOR SHUTTERS**, louvered									
0200	Two panel, 27" wide, 36" high	1 Carp	5	1.600	Set	175	87.50		262.50	330
0300	33" wide, 36" high		5	1.600		226	87.50		313.50	390
0500	47" wide, 36" high		5	1.600		305	87.50		392.50	475
1000	Four panel, 27" wide, 36" high		5	1.600		216	87.50		303.50	375
1100	33" wide, 36" high	1 Carp	5	1.600	Set	276	87.50		363.50	445
1300	47" wide, 36" high		5	1.600	"	370	87.50		457.50	545
1400	Plantation shutters, 16" x 48"		5	1.600	Ea.	141	87.50		228.50	294
1450	16" x 96"		4	2		232	109		341	430
1460	36" x 96"		3	2.667		510	146		656	790

12 23 13.13 Wood Panels

		Crew	Daily Output	Labor-Hours	Unit	Material	2021 Bare Costs Labor	2021 Bare Costs Equipment	Total	Total Incl O&P
0010	**WOOD PANELS**									
3000	Wood folding panels with movable louvers, 7" x 20" each	1 Carp	17	.471	Pr.	92	25.50		117.50	142
4000	Fixed louver type, stock units, 8" x 20" each		17	.471		121	25.50		146.50	175
4450	18" x 40" each		17	.471		174	25.50		199.50	232

12 24 Window Shades

12 24 13 – Roller Window Shades

12 24 13.10 Shades

		Crew	Daily Output	Labor-Hours	Unit	Material	2021 Bare Costs Labor	Equipment	Total	Total Incl O&P
0010	**SHADES**									
0020	Basswood, roll-up, stain finish, 3/8" slats	1 Carp	300	.027	S.F.	23	1.46		24.46	27.50
0200	7/8" slats		300	.027		21	1.46		22.46	25.50
0900	Mylar, single layer, non-heat reflective		685	.012		3.28	.64		3.92	4.63
0910	Mylar, single layer, heat reflective		685	.012		2.95	.64		3.59	4.27
1000	Double layered, heat reflective		685	.012		6.85	.64		7.49	8.55
1100	Triple layered, heat reflective		685	.012		10.35	.64		10.99	12.35
2000	Polyester, room darkening, with continuous cord, GEI									
2010	36" x 72"	1 Carp	38	.211	Ea.	100	11.50		111.50	128
2020	48" x 72"		28	.286		151	15.65		166.65	191
2030	60" x 72"		23	.348		146	19.05		165.05	191
2040	72" x 72"		19	.421		153	23		176	205
5011	Insulative shades		125	.064	S.F.	17.85	3.50		21.35	25
5030	R10.7		44	.182	"	21	9.95		30.95	39
5050	Magnetic clips, set of 20				Set	39.50			39.50	43.50
6011	Solar screening, fiberglass [G]	1 Carp	85	.094	S.F.	8.15	5.15		13.30	17.15

12 32 Manufactured Wood Casework

12 32 23 – Hardwood Casework

12 32 23.10 Manufactured Wood Casework, Stock Units

		Crew	Daily Output	Labor-Hours	Unit	Material	2021 Bare Costs Labor	Equipment	Total	Total Incl O&P
0010	**MANUFACTURED WOOD CASEWORK, STOCK UNITS**									
0300	Built-in drawer units, pine, 18" deep, 32" high, unfinished									
0400	Minimum	2 Carp	53	.302	L.F.	135	16.50		151.50	176
0500	Maximum	"	40	.400	"	160	22		182	211
0700	Kitchen base cabinets, hardwood, not incl. counter tops,									
0710	24" deep, 35" high, prefinished									
0800	One top drawer, one door below, 12" wide	2 Carp	24.80	.645	Ea.	320	35.50		355.50	405
0820	15" wide		24	.667		370	36.50		406.50	465
0840	18" wide		23.30	.687		215	37.50		252.50	297
0860	21" wide		22.70	.705		305	38.50		343.50	395
0880	24" wide	2 Carp	22.30	.717	Ea.	415	39.50		454.50	525
1000	Four drawers, 12" wide		24.80	.645		296	35.50		331.50	380
1020	15" wide		24	.667		294	36.50		330.50	385
1040	18" wide		23.30	.687		330	37.50		367.50	420
1060	24" wide		22.30	.717		375	39.50		414.50	480
1200	Two top drawers, two doors below, 27" wide		22	.727		440	40		480	545
1220	30" wide		21.40	.748		420	41		461	530
1240	33" wide		20.90	.766		285	42		327	380
1260	36" wide		20.30	.788		485	43		528	605
1280	42" wide		19.80	.808		470	44		514	590
1300	48" wide		18.90	.847		585	46.50		631.50	715
1500	Range or sink base, two doors below, 30" wide		21.40	.748		360	41		401	465
1520	33" wide		20.90	.766		410	42		452	515
1540	36" wide		20.30	.788		420	43		463	530
1560	42" wide		19.80	.808		455	44		499	575
1580	48" wide		18.90	.847		465	46.50		511.50	585
1800	For sink front units, deduct					172			172	190
2000	Corner base cabinets, 36" wide, standard	2 Carp	18	.889		565	48.50		613.50	705
2100	Lazy Susan with revolving door	"	16.50	.970		425	53		478	550
4000	Kitchen wall cabinets, hardwood, 12" deep with two doors									
4050	12" high, 30" wide	2 Carp	24.80	.645	Ea.	216	35.50		251.50	293

12 32 Manufactured Wood Casework

12 32 23 – Hardwood Casework

12 32 23.10 Manufactured Wood Casework, Stock Units	Crew	Daily Output	Labor-Hours	Unit	Material	2021 Bare Costs Labor	Equipment	Total	Total Incl O&P
4400	15" high, 30" wide	2 Carp	24	.667	Ea.	320	36.50		356.50	410
4420	33" wide		23.30	.687		144	37.50		181.50	218
4440	36" wide		22.70	.705		151	38.50		189.50	228
4450	42" wide		22.70	.705		350	38.50		388.50	445
4700	24" high, 30" wide		23.30	.687		161	37.50		198.50	237
4720	36" wide		22.70	.705		465	38.50		503.50	570
4740	42" wide		22.30	.717		292	39.50		331.50	385
5000	30" high, one door, 12" wide		22	.727		105	40		145	180
5020	15" wide		21.40	.748		239	41		280	325
5040	18" wide		20.90	.766		263	42		305	355
5060	24" wide		20.30	.788		305	43		348	405
5300	Two doors, 27" wide		19.80	.808		176	44		220	265
5320	30" wide		19.30	.829		390	45.50		435.50	500
5340	36" wide		18.80	.851		375	46.50		421.50	485
5360	42" wide		18.50	.865		440	47.50		487.50	560
5380	48" wide		18.40	.870		755	47.50		802.50	905
6000	Corner wall, 30" high, 24" wide		18	.889		195	48.50		243.50	293
6050	30" wide		17.20	.930		515	51		566	645
6100	36" wide		16.50	.970		455	53		508	585
6500	Revolving Lazy Susan		15.20	1.053		142	57.50		199.50	248
7000	Broom cabinet, 84" high, 24" deep, 18" wide	2 Carp	10	1.600	Ea.	790	87.50		877.50	1,000
7500	Oven cabinets, 84" high, 24" deep, 27" wide		8	2	"	1,025	109		1,134	1,300
7750	Valance board trim		396	.040	L.F.	24	2.21		26.21	29.50
7780	Toe kick trim	1 Carp	256	.031	"	3.67	1.71		5.38	6.75
7790	Base cabinet corner filler		16	.500	Ea.	51.50	27.50		79	100
7800	Cabinet filler, 3" x 24"		20	.400		19.95	22		41.95	57
7810	3" x 30"		20	.400		25	22		47	62.50
7820	3" x 42"		18	.444		35	24.50		59.50	77
7830	3" x 80"		16	.500		66.50	27.50		94	117
7850	Cabinet panel		50	.160	S.F.	11.70	8.75		20.45	27
9000	For deluxe models of all cabinets, add					40%				
9500	For custom built in place, add					25%	10%			
9558	Rule of thumb, kitchen cabinets not including									
9560	appliances & counter top, minimum	2 Carp	30	.533	L.F.	202	29		231	270
9600	Maximum	"	25	.640	"	450	35		485	550
9700	Minimum labor/equipment charge	1 Carp	3	2.667	Job		146		146	232

12 32 23.30 Manufactured Wood Casework Vanities

		Crew	Daily Output	Labor-Hours	Unit	Material	2021 Bare Costs Labor	Equipment	Total	Total Incl O&P
0010	**MANUFACTURED WOOD CASEWORK VANITIES**									
8000	Vanity bases, 2 doors, 30" high, 21" deep, 24" wide	2 Carp	20	.800	Ea.	440	44		484	555
8050	30" wide		16	1		465	54.50		519.50	595
8100	36" wide		13.33	1.200		395	65.50		460.50	540
8150	48" wide		11.43	1.400		515	76.50		591.50	685
9000	For deluxe models of all vanities, add to above					40%				
9500	For custom built in place, add to above					25%	10%			

12 32 23.35 Manufactured Wood Casework Hardware

		Crew	Daily Output	Labor-Hours	Unit	Material	2021 Bare Costs Labor	Equipment	Total	Total Incl O&P
0010	**MANUFACTURED WOOD CASEWORK HARDWARE**									
1000	Catches, minimum	1 Carp	235	.034	Ea.	1.58	1.86		3.44	4.71
1040	Maximum	"	80	.100	"	9.15	5.45		14.60	18.75
2000	Door/drawer pulls, handles									
2200	Handles and pulls, projecting, metal, minimum	1 Carp	48	.167	Ea.	1.10	9.10		10.20	15.70
2240	Maximum		36	.222		11.65	12.15		23.80	32
2300	Wood, minimum		48	.167		5.90	9.10		15	21

12 32 Manufactured Wood Casework

12 32 23 – Hardwood Casework

12 32 23.35 Manufactured Wood Casework Hardware	Crew	Daily Output	Labor- Hours	Unit	Material	2021 Bare Costs Labor	Equipment	Total	Total Incl O&P
2340	Maximum	1 Carp	36	.222	Ea.	10.90	12.15		23.05	31.50
2400	Drawer pulls, antimicrobial copper alloy finish		50	.160		13.35	8.75		22.10	28.50
2600	Flush, metal, minimum		48	.167		5.90	9.10		15	21
2640	Maximum		36	.222		11.05	12.15		23.20	31.50
2900	Drawer knobs, antimicrobial copper alloy finish		50	.160		7.65	8.75		16.40	22.50
3000	Drawer tracks/glides, minimum		48	.167	Pr.	9.80	9.10		18.90	25.50
3040	Maximum		24	.333		28.50	18.25		46.75	60.50
4000	Cabinet hinges, minimum		160	.050		3.32	2.74		6.06	8
4040	Maximum		68	.118		14.25	6.45		20.70	26
7000	Appliance pulls, antimicrobial copper alloy finish	1 Carp	50	.160	L.F.	75	8.75		83.75	96.50

12 34 Manufactured Plastic Casework

12 34 16 – Manufactured Solid-Plastic Casework

12 34 16.10 Outdoor Casework

		Crew	Daily Output	Labor- Hours	Unit	Material	Labor	Equipment	Total	Total Incl O&P
0010	**OUTDOOR CASEWORK**									
0020	Cabinet, base, sink/range, 36"	2 Carp	20.30	.788	Ea.	1,825	43		1,868	2,100
0100	Base, 36"		20.30	.788		1,825	43		1,868	2,100
0200	Filler strip, 1" x 30"		158	.101		32	5.55		37.55	44
0210	Filler strip, 2" x 30"		158	.101		53	5.55		58.55	67.50

12 35 Specialty Casework

12 35 53 – Laboratory Casework

12 35 53.13 Metal Laboratory Casework

		Crew	Daily Output	Labor- Hours	Unit	Material	Labor	Equipment	Total	Total Incl O&P
0010	**METAL LABORATORY CASEWORK**									
0020	Cabinets, base, door units, metal	2 Carp	18	.889	L.F.	286	48.50		334.50	395
0300	Drawer units		18	.889		630	48.50		678.50	775
0700	Tall storage cabinets, open, 7' high		20	.800		605	44		649	740
0900	With glazed doors		20	.800		1,125	44		1,169	1,300
1300	Wall cabinets, metal, 12-1/2" deep, open		20	.800		209	44		253	300
1500	With doors		20	.800		465	44		509	580

12 35 70 – Healthcare Casework

12 35 70.13 Hospital Casework

		Crew	Daily Output	Labor- Hours	Unit	Material	Labor	Equipment	Total	Total Incl O&P
0010	**HOSPITAL CASEWORK**									
3000	Hospital cabinets, stainless steel with glass door(s), lockable									
3010	One door, 24" W x 18" D x 60" H	2 Clab	18	.889	Ea.	3,250	39.50		3,289.50	3,650
3020	Two doors, 36" W x 18" D x 60" H		15	1.067		3,750	47.50		3,797.50	4,200
3030	36" W x 24" D x 67" H		15	1.067		4,825	47.50		4,872.50	5,375
3040	48" W x 24" D x 66" H		12	1.333		5,975	59		6,034	6,675
3050	48" W x 24" D x 72" H		12	1.333		4,900	59		4,959	5,475
3060	60" W x 24" D x 72" H		9	1.778		7,400	79		7,479	8,275

12 36 Countertops

12 36 16 – Metal Countertops

12 36 16.10 Stainless Steel Countertops	Crew	Daily Output	Labor-Hours	Unit	Material	2021 Bare Costs Labor	Equipment	Total	Total Incl O&P
0010 **STAINLESS STEEL COUNTERTOPS**									
3200 Stainless steel, custom	1 Carp	24	.333	S.F.	190	18.25		208.25	238

12 36 19 – Wood Countertops

12 36 19.10 Maple Countertops

	Crew	Daily Output	Labor-Hours	Unit	Material	Labor	Equipment	Total	Total Incl O&P
0010 **MAPLE COUNTERTOPS**									
2900 Solid, laminated, 1-1/2" thick, no splash	1 Carp	28	.286	L.F.	89	15.65		104.65	123
3000 With square splash		28	.286	"	106	15.65		121.65	141
3400 Recessed cutting block with trim, 16" x 20" x 1"		8	1	Ea.	137	54.50		191.50	238

12 36 23 – Plastic Countertops

12 36 23.13 Plastic-Laminate-Clad Countertops

	Crew	Daily Output	Labor-Hours	Unit	Material	Labor	Equipment	Total	Total Incl O&P
0010 **PLASTIC-LAMINATE-CLAD COUNTERTOPS**									
0020 Stock, 24" wide w/backsplash, minimum	1 Carp	30	.267	L.F.	16.80	14.60		31.40	41.50
0100 Maximum	1 Carp	25	.320	L.F.	40.50	17.50		58	72.50
0300 Custom plastic, 7/8" thick, aluminum molding, no splash		30	.267		37	14.60		51.60	64
0400 Cove splash		30	.267		47.50	14.60		62.10	75
0600 1-1/4" thick, no splash		28	.286		41	15.65		56.65	70
0700 Square splash		28	.286		45	15.65		60.65	75
0900 Square edge, plastic face, 7/8" thick, no splash		30	.267		35.50	14.60		50.10	62
1000 With splash		30	.267		54.50	14.60		69.10	83
1200 For stainless channel edge, 7/8" thick, add					3.86			3.86	4.25
1300 1-1/4" thick, add					4.43			4.43	4.87
1500 For solid color suede finish, add					5.80			5.80	6.35
1700 For end splash, add				Ea.	26.50			26.50	29
1900 For cut outs, standard, add, minimum	1 Carp	32	.250		16.85	13.70		30.55	40.50
2000 Maximum		8	1		8.75	54.50		63.25	96.50
2100 Postformed, including backsplash and front edge		30	.267	L.F.	12.65	14.60		27.25	37
2110 Mitred, add		12	.667	Ea.		36.50		36.50	58
2200 Built-in place, 25" wide, plastic laminate		25	.320	L.F.	66.50	17.50		84	101
9000 Minimum labor/equipment charge		3.75	2.133	Job		117		117	186

12 36 33 – Tile Countertops

12 36 33.10 Ceramic Tile Countertops

	Crew	Daily Output	Labor-Hours	Unit	Material	Labor	Equipment	Total	Total Incl O&P
0010 **CERAMIC TILE COUNTERTOPS**									
2300 Ceramic tile mosaic	1 Carp	25	.320	L.F.	44.50	17.50		62	77

12 36 40 – Stone Countertops

12 36 40.10 Natural Stone Countertops

	Crew	Daily Output	Labor-Hours	Unit	Material	Labor	Equipment	Total	Total Incl O&P
0010 **NATURAL STONE COUNTERTOPS**									
2500 Marble, stock, with splash, 1/2" thick, minimum	1 Bric	17	.471	L.F.	53.50	25.50		79	99.50
2700 3/4" thick, maximum	"	13	.615	"	130	33		163	196

12 36 53 – Laboratory Countertops

12 36 53.10 Laboratory Countertops and Sinks

	Crew	Daily Output	Labor-Hours	Unit	Material	Labor	Equipment	Total	Total Incl O&P
0010 **LABORATORY COUNTERTOPS AND SINKS**									
0020 Countertops, epoxy resin, not incl. base cabinets, acid-proof, minimum	2 Carp	82	.195	S.F.	52.50	10.65		63.15	74.50
0030 Maximum		70	.229		56	12.50		68.50	82
0040 Stainless steel		82	.195		226	10.65		236.65	266

12 36 61 – Simulated Stone Countertops

12 36 61.16 Solid Surface Countertops

	Crew	Daily Output	Labor-Hours	Unit	Material	Labor	Equipment	Total	Total Incl O&P
0010 **SOLID SURFACE COUNTERTOPS**, Acrylic polymer									
0020 Pricing for orders of 100 L.F. or greater									
0100 25" wide, solid colors	2 Carp	28	.571	L.F.	55.50	31.50		87	111

392

12 36 Countertops

12 36 61 – Simulated Stone Countertops

12 36 61.16 Solid Surface Countertops

		Crew	Daily Output	Labor-Hours	Unit	Material	2021 Bare Costs Labor	2021 Bare Costs Equipment	Total	Total Incl O&P
0200	Patterned colors	2 Carp	28	.571	L.F.	70.50	31.50		102	128
0300	Premium patterned colors		28	.571		94	31.50		125.50	154
0400	With silicone attached 4" backsplash, solid colors		27	.593		65	32.50		97.50	123
0500	Patterned colors		27	.593		82.50	32.50		115	143
0600	Premium patterned colors		27	.593		103	32.50		135.50	165
0700	With hard seam attached 4" backsplash, solid colors	2 Carp	23	.696	L.F.	65	38		103	132
0800	Patterned colors		23	.696		82.50	38		120.50	152
0900	Premium patterned colors		23	.696		103	38		141	174
1000	Pricing for order of 51-99 L.F.									
1100	25" wide, solid colors	2 Carp	24	.667	L.F.	64	36.50		100.50	129
1200	Patterned colors		24	.667		81.50	36.50		118	148
1300	Premium patterned colors		24	.667		108	36.50		144.50	177
1400	With silicone attached 4" backsplash, solid colors		23	.696		75	38		113	143
1500	Patterned colors		23	.696		95	38		133	166
1600	Premium patterned colors		23	.696		118	38		156	191
1700	With hard seam attached 4" backsplash, solid colors		20	.800		75	44		119	152
1800	Patterned colors		20	.800		95	44		139	175
1900	Premium patterned colors		20	.800		118	44		162	200
2000	Pricing for order of 1-50 L.F.									
2100	25" wide, solid colors	2 Carp	20	.800	L.F.	75	44		119	152
2200	Patterned colors		20	.800		95.50	44		139.50	175
2300	Premium patterned colors		20	.800		127	44		171	210
2400	With silicone attached 4" backsplash, solid colors		19	.842		88	46		134	171
2500	Patterned colors		19	.842		112	46		158	197
2600	Premium patterned colors		19	.842		139	46		185	227
2700	With hard seam attached 4" backsplash, solid colors		15	1.067		88	58.50		146.50	190
2800	Patterned colors		15	1.067		112	58.50		170.50	216
2900	Premium patterned colors		15	1.067		139	58.50		197.50	246
3000	Sinks, pricing for order of 100 or greater units									
3100	Single bowl, hard seamed, solid colors, 13" x 17"	1 Carp	3	2.667	Ea.	400	146		546	670
3200	10" x 15"		7	1.143		186	62.50		248.50	305
3300	Cutouts for sinks		8	1			54.50		54.50	87
3400	Sinks, pricing for order of 51-99 units									
3500	Single bowl, hard seamed, solid colors, 13" x 17"	1 Carp	2.55	3.137	Ea.	460	172		632	785
3600	10" x 15"		6	1.333		213	73		286	350
3700	Cutouts for sinks		7	1.143			62.50		62.50	99.50
3800	Sinks, pricing for order of 1-50 units									
3900	Single bowl, hard seamed, solid colors, 13" x 17"	1 Carp	2	4	Ea.	540	219		759	945
4000	10" x 15"		4.55	1.758		250	96		346	430
4100	Cutouts for sinks		5.25	1.524			83.50		83.50	133
4200	Cooktop cutouts, pricing for 100 or greater units		4	2		27	109		136	204
4300	51-99 units		3.40	2.353		31	129		160	240
4400	1-50 units		3	2.667		36.50	146		182.50	273

12 36 61.17 Solid Surface Vanity Tops

		Crew	Daily Output	Labor-Hours	Unit	Material	2021 Bare Costs Labor	2021 Bare Costs Equipment	Total	Total Incl O&P
0010	**SOLID SURFACE VANITY TOPS**									
0015	Solid surface, center bowl, 17" x 19"	1 Carp	12	.667	Ea.	171	36.50		207.50	246
0020	19" x 25"		12	.667		201	36.50		237.50	279
0030	19" x 31"		12	.667		250	36.50		286.50	335
0040	19" x 37"		12	.667		292	36.50		328.50	380
0050	22" x 25"		10	.800		410	44		454	525
0060	22" x 31"		10	.800		545	44		589	665
0070	22" x 37"		10	.800		560	44		604	685

12 36 Countertops

12 36 61 – Simulated Stone Countertops

12 36 61.17 Solid Surface Vanity Tops		Crew	Daily Output	Labor-Hours	Unit	Material	2021 Bare Costs Labor	Equipment	Total	Total Incl O&P
0080	22" x 43"	1 Carp	10	.800	Ea.	645	44		689	775
0090	22" x 49"		10	.800		700	44		744	840
0110	22" x 55"		8	1		520	54.50		574.50	655
0120	22" x 61"		8	1		585	54.50		639.50	730
0220	Double bowl, 22" x 61"		8	1		655	54.50		709.50	805
0230	Double bowl, 22" x 73"	↓	8	1	↓	1,050	54.50		1,104.50	1,225
0240	For aggregate colors, add					35%				
0250	For faucets and fittings, see Section 22 41 39.10									

12 36 61.19 Quartz Agglomerate Countertops		Crew	Daily Output	Labor-Hours	Unit	Material	2021 Bare Costs Labor	Equipment	Total	Total Incl O&P
0010	**QUARTZ AGGLOMERATE COUNTERTOPS**									
0100	25" wide, 4" backsplash, color group A, minimum	2 Carp	15	1.067	L.F.	61	58.50		119.50	160
0110	Maximum		15	1.067		100	58.50		158.50	203
0120	Color group B, minimum		15	1.067		71	58.50		129.50	171
0130	Maximum		15	1.067		112	58.50		170.50	217
0140	Color group C, minimum		15	1.067		88	58.50		146.50	190
0150	Maximum		15	1.067		159	58.50		217.50	267
0160	Color group D, minimum		15	1.067		94.50	58.50		153	197
0170	Maximum	↓	15	1.067	↓	122	58.50		180.50	227

12 48 Rugs and Mats

12 48 13 – Entrance Floor Mats and Frames

12 48 13.13 Entrance Floor Mats			Crew	Daily Output	Labor-Hours	Unit	Material	2021 Bare Costs Labor	Equipment	Total	Total Incl O&P
0010	**ENTRANCE FLOOR MATS**										
2000	Recycled rubber tire tile, 12" x 12" x 3/8" thick	G	1 Clab	125	.064	S.F.	10.70	2.84		13.54	16.25
2510	Natural cocoa fiber, 1/2" thick	G		125	.064		7.95	2.84		10.79	13.25
2520	3/4" thick	G		125	.064		7.30	2.84		10.14	12.55
2530	1" thick	G	↓	125	.064	↓	7.95	2.84		10.79	13.25
3000	Hospital tacky mats, package of 30 with frame					Ea.	72			72	79
3010	4 packages of 30					"	71			71	78

12 51 Office Furniture

12 51 16 – Case Goods

12 51 16.13 Metal Case Goods		Crew	Daily Output	Labor-Hours	Unit	Material	2021 Bare Costs Labor	Equipment	Total	Total Incl O&P
0010	**METAL CASE GOODS**									
0020	Desks, 29" high, double pedestal, 30" x 60", metal, minimum				Ea.	715			715	785
0030	Maximum				"	815			815	895

12 54 Hospitality Furniture

12 54 13 – Hotel and Motel Furniture

12 54 13.10 Hotel Furniture		Crew	Daily Output	Labor-Hours	Unit	Material	2021 Bare Costs Labor	Equipment	Total	Total Incl O&P
0010	**HOTEL FURNITURE**									
0020	Standard quality set, minimum				Room	2,675			2,675	2,950
0200	Maximum				"	8,500			8,500	9,350

12 54 16 – Restaurant Furniture

12 54 16.20 Furniture, Restaurant		Crew	Daily Output	Labor-Hours	Unit	Material	2021 Bare Costs Labor	Equipment	Total	Total Incl O&P
0010	**FURNITURE, RESTAURANT**									
0020	Bars, built-in, front bar	1 Carp	5	1.600	L.F.	320	87.50		407.50	495

12 54 Hospitality Furniture

12 54 16 – Restaurant Furniture

12 54 16.20 Furniture, Restaurant	Crew	Daily Output	Labor-Hours	Unit	Material	2021 Bare Costs Labor	Equipment	Total	Total Incl O&P
0200	Back bar	1 Carp	5	1.600	L.F.	234	87.50		321.50	395
0500	Booth unit, molded plastic, stub wall and 2 seats, minimum		2	4	Set	355	219		574	745
0600	Maximum		1.50	5.333	"	1,775	292		2,067	2,425
0800	Booth seat, upholstered, foursome, single (end) minimum		5	1.600	Ea.	930	87.50		1,017.50	1,175
0900	Maximum		4	2		1,250	109		1,359	1,550
1000	Foursome, double, minimum		4	2		1,200	109		1,309	1,500
1100	Maximum		3	2.667		2,100	146		2,246	2,550
1300	Circle booth, upholstered, 1/4 circle, minimum		3	2.667		1,700	146		1,846	2,100
1400	Maximum		2	4		2,025	219		2,244	2,575
1500	3/4 circle, minimum		1.50	5.333		7,550	292		7,842	8,775
1600	Maximum	↓	1	8	↓	5,375	440		5,815	6,625

12 55 Detention Furniture

12 55 13 – Detention Bunks

12 55 13.13 Cots

		Crew	Daily Output	Labor-Hours	Unit	Material	2021 Bare Costs Labor	Equipment	Total	Total Incl O&P
0010	**COTS**									
2500	Bolted, single, painted steel	E-4	20	1.600	Ea.	350	97.50	7.45	454.95	550
2700	Stainless steel	"	20	1.600	"	1,000	97.50	7.45	1,104.95	1,275

12 56 Institutional Furniture

12 56 43 – Dormitory Furniture

12 56 43.10 Dormitory Furnishings

		Crew	Daily Output	Labor-Hours	Unit	Material	2021 Bare Costs Labor	Equipment	Total	Total Incl O&P
0010	**DORMITORY FURNISHINGS**									
0300	Bunkable bed, twin, minimum				Ea.	400			400	440
0320	Maximum				"	460			460	505

12 56 51 – Library Furniture

12 56 51.10 Library Furnishings

		Crew	Daily Output	Labor-Hours	Unit	Material	2021 Bare Costs Labor	Equipment	Total	Total Incl O&P
0010	**LIBRARY FURNISHINGS**									
1710	Carrels, hardwood, 36" x 24", minimum	1 Carp	5	1.600	Ea.	660	87.50		747.50	865
1720	Maximum		4	2		1,900	109		2,009	2,250
1730	Metal, minimum		5	1.600		182	87.50		269.50	340
1740	Maximum		4	2	↓	460	109		569	680
6010	Bookshelf, metal, 90" high, 10" shelf, double face		11.50	.696	L.F.	208	38		246	290
6020	Single face	↓	12	.667	"	116	36.50		152.50	186
6050	For 8" shelving, subtract from above					10%				
6070	For 42" high with countertop, subtract from above					20%				

12 56 70 – Healthcare Furniture

12 56 70.10 Furniture, Hospital

		Crew	Daily Output	Labor-Hours	Unit	Material	2021 Bare Costs Labor	Equipment	Total	Total Incl O&P
0010	**FURNITURE, HOSPITAL**									
0020	Beds, manual, minimum				Ea.	540			540	595
0100	Maximum				"	3,000			3,000	3,300
1100	Patient wall systems, not incl. plumbing, minimum				Room	1,325			1,325	1,475
1200	Maximum				"	2,400			2,400	2,650

12 63 Stadium and Arena Seating

12 63 13 – Stadium and Arena Bench Seating

12 63 13.13 Bleachers

		Crew	Daily Output	Labor-Hours	Unit	Material	2021 Bare Costs Labor	Equipment	Total	Total Incl O&P
0010	**BLEACHERS**									
3000	Telescoping, manual to 15 tier, minimum	F-5	65	.492	Seat	98	27		125	152
3100	Maximum		60	.533		210	29.50		239.50	278
3300	16 to 20 tier, minimum		60	.533		212	29.50		241.50	280
3400	Maximum		55	.582		415	32		447	505
3600	21 to 30 tier, minimum		50	.640		278	35.50		313.50	360
3700	Maximum		40	.800		445	44		489	560
3900	For integral power operation, add, minimum	2 Elec	300	.053		67	3.40		70.40	78.50
4000	Maximum	"	250	.064		110	4.08		114.08	127
5000	Benches, folding, in wall, 14' table, 2 benches	L-4	2	12	Set	745	625		1,370	1,825

12 67 Pews and Benches

12 67 13 – Pews

12 67 13.13 Sanctuary Pews

		Crew	Daily Output	Labor-Hours	Unit	Material	2021 Bare Costs Labor	Equipment	Total	Total Incl O&P
0010	**SANCTUARY PEWS**									
1500	Bench type, hardwood, minimum	1 Carp	20	.400	L.F.	125	22		147	173
1550	Maximum	"	15	.533		186	29		215	252
1570	For kneeler, add					21			21	23.50

12 93 Interior Public Space Furnishings

12 93 13 – Bicycle Racks

12 93 13.10 Bicycle Racks

		Crew	Daily Output	Labor-Hours	Unit	Material	2021 Bare Costs Labor	Equipment	Total	Total Incl O&P
0010	**BICYCLE RACKS**									
0020	Single side, grid, 1-5/8" OD stl. pipe, w/1/2" bars, galv, 5 bike cap	2 Clab	10	1.600	Ea.	199	71		270	330
0025	Powder coat finish		10	1.600		109	71		180	233
0030	Single side, grid, 1-5/8" OD stl. pipe, w/1/2" bars, galv, 9 bike cap		8	2		299	89		388	470
0035	Powder coat finish		8	2		141	89		230	296
0040	Single side, grid, 1-5/8" OD stl. pipe, w/1/2" bars, galv, 18 bike cap		4	4		350	178		528	670
0045	Powder coat finish		4	4		380	178		558	700
0050	S curve, 1-7/8" OD stl. pipe, 11 ga., galv, 5 bike cap		10	1.600		147	71		218	275
0055	Powder coat finish		10	1.600		147	71		218	275
0060	S curve, 1-7/8" OD stl. pipe, 11 ga., galv, 7 bike cap		9	1.778		262	79		341	415
0065	Powder coat finish		9	1.778		206	79		285	355
0070	S curve, 1-7/8" OD stl. pipe, 11 ga., galv, 9 bike cap		8	2		335	89		424	510
0075	Powder coat finish		8	2		269	89		358	435
0080	S curve, 1-7/8" OD stl. pipe, 11 ga., galv, 11 bike cap		6	2.667		425	118		543	655
0085	Powder coat finish		6	2.667		425	118		543	655

12 93 23 – Trash and Litter Receptacles

12 93 23.10 Trash Receptacles

			Crew	Daily Output	Labor-Hours	Unit	Material	2021 Bare Costs Labor	Equipment	Total	Total Incl O&P
0010	**TRASH RECEPTACLES**										
0500	Recycled plastic, var colors, round, 32 gal., 28" x 38" high	G	2 Clab	5	3.200	Ea.	745	142		887	1,050
0510	32 gal., 31" x 32" high	G		5	3.200		725	142		867	1,025
9110	Plastic, with dome lid, 32 gal. capacity			35	.457		61.50	20.50		82	100
9120	Recycled plastic slats, plastic dome lid, 32 gal. capacity			35	.457		305	20.50		325.50	370

Estimating Tips
General

- The items and systems in this division are usually estimated, purchased, supplied, and installed as a unit by one or more subcontractors. The estimator must ensure that all parties are operating from the same set of specifications and assumptions, and that all necessary items are estimated and will be provided. Many times the complex items and systems are covered, but the more common ones, such as excavation or a crane, are overlooked for the very reason that everyone assumes nobody could miss them. The estimator should be the central focus and be able to ensure that all systems are complete.

- It is important to consider factors such as site conditions, weather, shape and size of building, as well as labor availability as they may impact the overall cost of erecting special structures and systems included in this division.

- Another area where problems can develop in this division is at the interface between systems.

The estimator must ensure, for instance, that anchor bolts, nuts, and washers are estimated and included for the air-supported structures and pre-engineered buildings to be bolted to their foundations. Utility supply is a common area where essential items or pieces of equipment can be missed or overlooked because each subcontractor may feel it is another's responsibility. The estimator should also be aware of certain items which may be supplied as part of a package but installed by others, and ensure that the installing contractor's estimate includes the cost of installation. Conversely, the estimator must also ensure that items are not costed by two different subcontractors, resulting in an inflated overall estimate.

13 30 00 Special Structures

- The foundations and floor slab, as well as rough mechanical and electrical, should be estimated, as this work is required for the assembly and erection of the structure. Generally, as noted in the data set, the pre-engineered building comes

as a shell. Pricing is based on the size and structural design parameters stated in the reference section. Additional features, such as windows and doors with their related structural framing, must also be included by the estimator. Here again, the estimator must have a clear understanding of the scope of each portion of the work and all the necessary interfaces.

Reference Numbers

Reference numbers are shown at the beginning of some major classifications. These numbers refer to related items in the Reference Section. The reference information may be an estimating procedure, an alternate pricing method, or technical information.

Note: Not all subdivisions listed here necessarily appear. ■

Same Data. Simplified.

Enjoy the convenience and efficiency of accessing your costs anywhere:

- **Skip the multiplier** by setting your location
- **Quickly search**, edit, favorite and share costs
- **Stay on top of price changes** with automatic updates

Discover more at rsmeans.com/online

13 05 05 – Selective Demolition for Special Construction

13 05 05.10 Selective Demolition, Air Supported Structures	Crew	Daily Output	Labor-Hours	Unit	Material	2021 Bare Costs Labor	Equipment	Total	Total Incl O&P
0010 **SELECTIVE DEMOLITION, AIR SUPPORTED STRUCTURES**									
0020 Tank covers, scrim, dbl. layer, vinyl poly w/hdwe., blower & controls									
0050 Round and rectangular R024119-10	B-2	9000	.004	S.F.		.20		.20	.32
0100 Warehouse structures									
0120 Poly/vinyl fabric, 28 oz., incl. tension cables & inflation system	4 Clab	9000	.004	SF Flr.		.16		.16	.25
0150 Reinforced vinyl, 12 oz., 3,000 S.F.	"	5000	.006			.28		.28	.45
0200 12,000 to 24,000 S.F.	8 Clab	20000	.003			.14		.14	.23
0250 Tedlar vinyl fabric, 28 oz. w/liner, to 3,000 S.F.	4 Clab	5000	.006			.28		.28	.45
0300 12,000 to 24,000 S.F.	8 Clab	20000	.003	↓		.14		.14	.23
0350 Greenhouse/shelter, woven polyethylene with liner									
0400 3,000 S.F.	4 Clab	5000	.006	SF Flr.		.28		.28	.45
0450 12,000 to 24,000 S.F.	8 Clab	20000	.003			.14		.14	.23
0500 Tennis/gymnasium, poly/vinyl fabric, 28 oz., incl. thermal liner	4 Clab	9000	.004			.16		.16	.25
0600 Stadium/convention center, teflon coated fiberglass, incl. thermal liner	9 Clab	40000	.002	↓		.08		.08	.13
0700 Doors, air lock, 15' long, 10' x 10'	2 Carp	1.50	10.667	Ea.		585		585	930
0720 15' x 15'		.80	20			1,100		1,100	1,750
0750 Revolving personnel door, 6' diam. x 6'-6" high	↓	1.50	10.667	↓		585		585	930

13 05 05.20 Selective Demolition, Garden Houses

	Crew	Daily Output	Labor-Hours	Unit	Material	Labor	Equipment	Total	Total Incl O&P
0010 **SELECTIVE DEMOLITION, GARDEN HOUSES**									
0020 Prefab, wood, excl. foundation, average	2 Clab	400	.040	SF Flr.		1.78		1.78	2.83

13 05 05.25 Selective Demolition, Geodesic Domes

	Crew	Daily Output	Labor-Hours	Unit	Material	Labor	Equipment	Total	Total Incl O&P
0010 **SELECTIVE DEMOLITION, GEODESIC DOMES**									
0050 Shell only, interlocking plywood panels, 30' diameter	F-5	3.20	10	Ea.		550		550	880
0060 34' diameter		2.30	13.913			770		770	1,225
0070 39' diameter	↓	2	16			885		885	1,400
0080 45' diameter	F-3	2.20	18.182			1,025	216	1,241	1,875
0090 55' diameter		2	20			1,125	238	1,363	2,025
0100 60' diameter		2	20			1,125	238	1,363	2,025
0110 65' diameter	↓	1.60	25	↓		1,400	298	1,698	2,550

13 05 05.30 Selective Demolition, Greenhouses

	Crew	Daily Output	Labor-Hours	Unit	Material	Labor	Equipment	Total	Total Incl O&P
0010 **SELECTIVE DEMOLITION, GREENHOUSES**									
0020 Resi-type, free standing, excl. foundations, 9' long x 8' wide	2 Clab	160	.100	SF Flr.		4.44		4.44	7.05
0030 9' long x 11' wide	2 Clab	170	.094	SF Flr.		4.18		4.18	6.65
0040 9' long x 14' wide		220	.073			3.23		3.23	5.15
0050 9' long x 17' wide		320	.050			2.22		2.22	3.54
0060 Lean-to type, 4' wide		64	.250			11.10		11.10	17.70
0070 7' wide		120	.133	↓		5.90		5.90	9.45
0080 Geodesic hemisphere, 1/8" plexiglass glazing, 8' diam.		4	4	Ea.		178		178	283
0090 24' diam.		.80	20			890		890	1,425
0100 48' diam.	↓	.40	40	↓		1,775		1,775	2,825

13 05 05.35 Selective Demolition, Hangars

	Crew	Daily Output	Labor-Hours	Unit	Material	Labor	Equipment	Total	Total Incl O&P
0010 **SELECTIVE DEMOLITION, HANGARS**									
0020 T type hangars, prefab, steel, galv roof & walls, incl doors, excl fndtn	E-2	2550	.022	SF Flr.		1.31	.67	1.98	2.84
0030 Circular type, prefab, steel frame, plastic skin, incl foundation, 80' diam	"	.50	112	Total		6,675	3,425	10,100	14,500

13 05 05.45 Selective Demolition, Lightning Protection

	Crew	Daily Output	Labor-Hours	Unit	Material	Labor	Equipment	Total	Total Incl O&P
0010 **SELECTIVE DEMOLITION, LIGHTNING PROTECTION**									
0020 Air terminal & base, copper, 3/8" diam. x 10", to 75' H	1 Clab	16	.500	Ea.		22		22	35.50
0030 1/2" diam. x 12", over 75' H		16	.500			22		22	35.50
0050 Aluminum, 1/2" diam. x 12", to 75' H		16	.500			22		22	35.50
0060 5/8" diam. x 12", over 75' H		16	.500	↓		22		22	35.50
0070 Cable, copper, 220 lb. per thousand feet, to 75' H	↓	640	.013	L.F.		.56		.56	.88

13 05 Common Work Results for Special Construction

13 05 05 – Selective Demolition for Special Construction

13 05 05.45 Selective Demolition, Lightning Protection	Crew	Daily Output	Labor-Hours	Unit	Material	2021 Bare Costs Labor	Equipment	Total	Total Incl O&P
0080	375 lb. per thousand feet, over 75' H	1 Clab	460	.017	L.F.		.77		.77	1.23
0090	Aluminum, 101 lb. per thousand feet, to 75' H		560	.014			.63		.63	1.01
0100	199 lb. per thousand feet, over 75' H		480	.017	▼		.74		.74	1.18
0110	Arrester, 175 V AC, to ground		16	.500	Ea.		22		22	35.50
0120	650 V AC, to ground	▼	13	.615	"		27.50		27.50	43.50

13 05 05.50 Selective Demolition, Pre-Engineered Steel Buildings

		Crew	Daily Output	Labor-Hours	Unit	Material	Labor	Equipment	Total	Total Incl O&P
0010	**SELECTIVE DEMOLITION, PRE-ENGINEERED STEEL BUILDINGS**									
0500	Pre-engd. steel bldgs., rigid frame, clear span & multi post, excl. salvage									
0550	3,500 to 7,500 S.F.	L-10	1000	.024	SF Flr.		1.47	.48	1.95	2.87
0600	7,501 to 12,500 S.F.		1500	.016			.98	.32	1.30	1.92
0650	12,501 S.F. or greater	▼	1650	.015	▼		.89	.29	1.18	1.75
0700	Pre-engd. steel building components									
0710	Entrance canopy, including frame 4' x 4'	E-24	8	4	Ea.		240	73.50	313.50	465
0720	4' x 8'	"	7	4.571			274	84	358	530
0730	HM doors, self framing, single leaf	2 Skwk	8	2			114		114	181
0740	Double leaf		5	3.200	▼		183		183	290
0760	Gutter, eave type		600	.027	L.F.		1.52		1.52	2.41
0770	Sash, single slide, double slide or fixed		24	.667	Ea.		38		38	60.50
0780	Skylight, fiberglass, to 30 S.F.		16	1			57		57	90.50
0785	Roof vents, circular, 12" to 24" diameter		12	1.333			76		76	121
0790	Continuous, 10' long	▼	8	2	▼		114		114	181
0900	Shelters, aluminum frame									
0910	Acrylic glazing, 3' x 9' x 8' high	2 Skwk	2	8	Ea.		455		455	725
0920	9' x 12' x 8' high	"	1.50	10.667	"		610		610	965

13 05 05.60 Selective Demolition, Silos

		Crew	Daily Output	Labor-Hours	Unit	Material	Labor	Equipment	Total	Total Incl O&P
0010	**SELECTIVE DEMOLITION, SILOS**									
0020	Conc stave, indstrl, conical/sloping bott, excl fndtn, 12' diam., 35' H	E-24	.18	178	Ea.		10,700	3,250	13,950	20,700
0030	16' diam., 45' H		.12	267			16,000	4,900	20,900	31,000
0040	25' diam., 75' H	▼	.08	400			24,000	7,325	31,325	46,600
0050	Steel, factory fabricated, 30,000 gal. cap, painted or epoxy lined	L-5	2	28	▼		1,700	293	1,993	3,050

13 05 05.65 Selective Demolition, Sound Control

		Crew	Daily Output	Labor-Hours	Unit	Material	Labor	Equipment	Total	Total Incl O&P
0010	**SELECTIVE DEMOLITION, SOUND CONTROL**									
0120	Acoustical enclosure, 4" thick walls & ceiling panels, 8 lb./S.F.	3 Carp	144	.167	SF Surf		9.10		9.10	14.50
0130	10.5 lb./S.F.		128	.188			10.25		10.25	16.35
0140	Reverb chamber, parallel walls, 4" thick		120	.200			10.95		10.95	17.40
0150	Skewed walls, parallel roof, 4" thick		110	.218			11.95		11.95	19
0160	Skewed walls/roof, 4" layer/air space		96	.250			13.70		13.70	22
0170	Sound-absorbing panels, painted metal, 2'-6" x 8', under 1,000 S.F.		430	.056			3.05		3.05	4.86
0180	Over 1,000 S.F.	▼	480	.050			2.74		2.74	4.36
0190	Flexible transparent curtain, clear	3 Shee	430	.056			3.65		3.65	5.70
0192	50% clear, 50% foam		430	.056			3.65		3.65	5.70
0194	25% clear, 75% foam		430	.056			3.65		3.65	5.70
0196	100% foam	▼	430	.056	▼		3.65		3.65	5.70
0200	Audio-masking sys., incl. speakers, amplfr., signal gnrtr.									
0205	Ceiling mounted, 5,000 S.F.	2 Elec	4800	.003	S.F.		.21		.21	.33
0210	10,000 S.F.		5600	.003			.18		.18	.28
0220	Plenum mounted, 5,000 S.F.		7600	.002			.13		.13	.21
0230	10,000 S.F.	▼	8800	.002	▼		.12		.12	.18

13 05 05.70 Selective Demolition, Special Purpose Rooms

		Crew	Daily Output	Labor-Hours	Unit	Material	Labor	Equipment	Total	Total Incl O&P
0010	**SELECTIVE DEMOLITION, SPECIAL PURPOSE ROOMS**									
0100	Audiometric rooms, under 500 S.F. surface	4 Carp	200	.160	SF Surf		8.75		8.75	13.95
0110	Over 500 S.F. surface	"	240	.133	"		7.30		7.30	11.60

13 05 05.70 Selective Demolition, Special Purpose Rooms

		Crew	Daily Output	Labor-Hours	Unit	Material	2021 Bare Costs Labor	2021 Bare Costs Equipment	Total	Total Incl O&P
0200	Clean rooms, 12' x 12' soft wall, class 100	1 Carp	.30	26.667	Ea.		1,450		1,450	2,325
0210	Class 1,000		.30	26.667			1,450		1,450	2,325
0220	Class 10,000		.35	22.857			1,250		1,250	2,000
0230	Class 100,000		.35	22.857			1,250		1,250	2,000
0300	Darkrooms, shell complete, 8' high	2 Carp	220	.073	SF Flr.		3.98		3.98	6.35
0310	12' high		110	.145	"		7.95		7.95	12.65
0350	Darkrooms doors, mini-cylindrical, revolving		4	4	Ea.		219		219	350
0400	Music room, practice modular		140	.114	SF Surf		6.25		6.25	9.95
0500	Refrigeration structures and finishes									
0510	Wall finish, 2 coat Portland cement plaster, 1/2" thick	1 Clab	200	.040	S.F.		1.78		1.78	2.83
0520	Fiberglass panels, 1/8" thick	"	400	.020	"		.89		.89	1.41
0530	Ceiling finish, polystyrene plastic, 1" to 2" thick	1 Clab	500	.016	S.F.		.71		.71	1.13
0540	4" thick	"	450	.018	"		.79		.79	1.26
0550	Refrigerator, prefab aluminum walk-in, 7'-6" high, 6' x 6' OD	2 Carp	100	.160	SF Flr.		8.75		8.75	13.95
0560	10' x 10' OD		160	.100			5.45		5.45	8.70
0570	Over 150 S.F.		200	.080			4.38		4.38	6.95
0600	Sauna, prefabricated, including heater & controls, 7' high, to 30 S.F.		120	.133			7.30		7.30	11.60
0610	To 40 S.F.		140	.114			6.25		6.25	9.95
0620	To 60 S.F.		175	.091			5		5	7.95
0630	To 100 S.F.		220	.073			3.98		3.98	6.35
0640	To 130 S.F.		250	.064			3.50		3.50	5.55
0650	Steam bath, heater, timer, head, single, to 140 C.F.	1 Plum	2.20	3.636	Ea.		246		246	380
0660	To 300 C.F.		2.20	3.636			246		246	380
0670	Steam bath, comm. size, w/blow-down assembly, to 800 C.F.		1.80	4.444			300		300	465
0680	To 2,500 C.F.		1.60	5			340		340	520
0690	Steam bath, comm. size, multiple, for motels, apts, 500 C.F., 2 baths		2	4			271		271	420
0700	1,000 C.F., 4 baths		1.40	5.714			385		385	595

13 05 05.75 Selective Demolition, Storage Tanks

		Crew	Daily Output	Labor-Hours	Unit	Material	2021 Bare Costs Labor	2021 Bare Costs Equipment	Total	Total Incl O&P
0010	**SELECTIVE DEMOLITION, STORAGE TANKS**									
0500	Steel tank, single wall, above ground, not incl. fdn., pumps or piping									
0510	Single wall, 275 gallon	Q-1	3	5.333	Ea.		325		325	500
0520	550 thru 2,000 gallon	B-34P	2	12			705	535	1,240	1,675
0530	5,000 thru 10,000 gallon	B-34Q	2	12			715	565	1,280	1,750
0540	15,000 thru 30,000 gallon	B-34S	2	16			1,000	1,550	2,550	3,250
0600	Steel tank, double wall, above ground not incl. fdn., pumps & piping									
0620	500 thru 2,000 gallon	B-34P	2	12	Ea.		705	535	1,240	1,675

13 05 05.85 Selective Demolition, Swimming Pool Equip

		Crew	Daily Output	Labor-Hours	Unit	Material	2021 Bare Costs Labor	2021 Bare Costs Equipment	Total	Total Incl O&P
0010	**SELECTIVE DEMOLITION, SWIMMING POOL EQUIP**									
0020	Diving stand, stainless steel, 3 meter	2 Clab	3	5.333	Ea.		237		237	375
0030	1 meter		5	3.200			142		142	226
0040	Diving board, 16' long, aluminum		5.40	2.963			132		132	209
0050	Fiberglass		5.40	2.963			132		132	209
0070	Ladders, heavy duty, stainless steel, 2 tread		14	1.143			50.50		50.50	81
0080	4 tread		12	1.333			59		59	94.50
0090	Lifeguard chair, stainless steel, fixed		5	3.200			142		142	226
0100	Slide, tubular, fiberglass, aluminum handrails & ladder, 5', straight		4	4			178		178	283
0110	8', curved		6	2.667			118		118	189
0120	10', curved		3	5.333			237		237	375
0130	12' straight, with platform		2.50	6.400			284		284	450
0140	Removable access ramp, stainless steel		4	4			178		178	283
0150	Removable stairs, stainless steel, collapsible		4	4			178		178	283

13 05 Common Work Results for Special Construction

13 05 05 – Selective Demolition for Special Construction

13 05 05.90 Selective Demolition, Tension Structures

		Crew	Daily Output	Labor-Hours	Unit	Material	2021 Bare Costs Labor	Equipment	Total	Total Incl O&P
0010	**SELECTIVE DEMOLITION, TENSION STRUCTURES**									
0020	Steel/alum. frame, fabric shell, 60' clear span, 6,000 S.F.	B-41	2000	.022	SF Flr.		1.01	.15	1.16	1.77
0030	12,000 S.F.		2200	.020			.92	.14	1.06	1.61
0040	80' clear span, 20,800 S.F.		2440	.018			.83	.13	.96	1.46
0050	100' clear span, 10,000 S.F.	L-5	4350	.013			.78	.13	.91	1.41
0060	26,000 S.F.		4600	.012			.74	.13	.87	1.33
0070	36,000 S.F.		5000	.011			.68	.12	.80	1.22

13 05 05.95 Selective Demo, X-Ray/Radio Freq Protection

		Crew	Daily Output	Labor-Hours	Unit	Material	2021 Bare Costs Labor	Equipment	Total	Total Incl O&P
0010	**SELECTIVE DEMO, X-RAY/RADIO FREQ PROTECTION**									
0020	Shielding lead, lined door frame, excl. hdwe., 1/16" thick	1 Clab	4.80	1.667	Ea.		74		74	118
0030	Lead sheets, 1/16" thick	2 Clab	270	.059	S.F.		2.63		2.63	4.19
0040	1/8" thick		240	.067			2.96		2.96	4.71
0050	Lead shielding, 1/4" thick		270	.059			2.63		2.63	4.19
0060	1/2" thick		240	.067			2.96		2.96	4.71
0070	Lead glass, 1/4" thick, 2.0 mm LE, 12" x 16"	2 Glaz	16	1	Ea.		52.50		52.50	83.50
0080	24" x 36"		8	2			105		105	167
0090	36" x 60"		4	4			211		211	335
0100	Lead glass window frame, with 1/16" lead & voice passage, 36" x 60"		4	4			211		211	335
0110	Lead glass window frame, 24" x 36"		8	2			105		105	167
0120	Lead gypsum board, 5/8" thick with 1/16" lead	2 Clab	320	.050	S.F.		2.22		2.22	3.54
0130	1/8" lead		280	.057			2.54		2.54	4.04
0140	1/32" lead		400	.040			1.78		1.78	2.83
0150	Butt joints, 1/8" lead or thicker, 2" x 7' long batten strip		480	.033	Ea.		1.48		1.48	2.36
0160	X-ray protection, average radiography room, up to 300 S.F., 1/16" lead, min		.50	32	Total		1,425		1,425	2,250
0170	Maximum		.30	53.333			2,375		2,375	3,775
0180	Deep therapy X-ray room, 250 kV cap, up to 300 S.F., 1/4" lead, min		.20	80			3,550		3,550	5,650
0190	Maximum		.12	133			5,925		5,925	9,425
0880	Radio frequency shielding, prefab or screen-type copper or steel, minimum		360	.044	SF Surf		1.97		1.97	3.14
0890	Average		310	.052			2.29		2.29	3.65
0895	Maximum		290	.055			2.45		2.45	3.90

13 12 Fountains

13 12 13 – Exterior Fountains

13 12 13.10 Outdoor Fountains

		Crew	Daily Output	Labor-Hours	Unit	Material	2021 Bare Costs Labor	Equipment	Total	Total Incl O&P
0010	**OUTDOOR FOUNTAINS**									
0100	Outdoor fountain, 48" high with bowl and figures	2 Clab	2	8	Ea.	385	355		740	990
0200	Commercial, concrete or cast stone, 40-60" H, simple		2	8		975	355		1,330	1,650
0220	Average		2	8		1,900	355		2,255	2,650
0240	Ornate		2	8		4,300	355		4,655	5,325
0260	Metal, 72" high		2	8		1,525	355		1,880	2,250
0280	90" high		2	8		2,200	355		2,555	3,000
0300	120" high		2	8		5,275	355		5,630	6,375
0320	Resin or fiberglass, 40-60" H, wall type	2 Clab	2	8	Ea.	690	355		1,045	1,325
0340	Waterfall type	"	2	8	"	1,175	355		1,530	1,850

13 12 23 – Interior Fountains

13 12 23.10 Indoor Fountains

		Crew	Daily Output	Labor-Hours	Unit	Material	2021 Bare Costs Labor	Equipment	Total	Total Incl O&P
0010	**INDOOR FOUNTAINS**									
0100	Commercial, floor type, resin or fiberglass, lighted, cascade type	2 Clab	2	8	Ea.	435	355		790	1,050
0120	Tiered type		2	8		680	355		1,035	1,325
0140	Waterfall type		2	8		485	355		840	1,100

13 17 Tubs and Pools

13 17 13 – Hot Tubs

13 17 13.10 Redwood Hot Tub System	Crew	Daily Output	Labor-Hours	Unit	Material	2021 Bare Costs Labor	2021 Bare Costs Equipment	Total	Total Incl O&P
0010 **REDWOOD HOT TUB SYSTEM**									
7050 4' diameter x 4' deep	Q-1	1	16	Ea.	4,650	975		5,625	6,625
7150 6' diameter x 4' deep		.80	20		6,300	1,225		7,525	8,800
7200 8' diameter x 4' deep	↓	.80	20	↓	9,000	1,225		10,225	11,800

13 17 33 – Whirlpool Tubs

13 17 33.10 Whirlpool Bath	Crew	Daily Output	Labor-Hours	Unit	Material	2021 Bare Costs Labor	2021 Bare Costs Equipment	Total	Total Incl O&P
0010 **WHIRLPOOL BATH**									
6000 Whirlpool, bath with vented overflow, molded fiberglass									
6100 66" x 36" x 24"	Q-1	1	16	Ea.	1,075	975		2,050	2,675
6400 72" x 36" x 21"		1	16		1,550	975		2,525	3,200
6500 60" x 34" x 21"		1	16		1,150	975		2,125	2,775
6600 72" x 42" x 23"	↓	1	16		1,450	975		2,425	3,100
6710 For color, add					10%				
6711 For designer colors and trim, add				↓	25%				

13 21 Controlled Environment Rooms

13 21 26 – Cold Storage Rooms

13 21 26.50 Refrigeration	Crew	Daily Output	Labor-Hours	Unit	Material	2021 Bare Costs Labor	2021 Bare Costs Equipment	Total	Total Incl O&P
0010 **REFRIGERATION**									
0020 Curbs, 12" high, 4" thick, concrete	2 Carp	58	.276	L.F.	3.32	15.10		18.42	27.50
6300 Rule of thumb for complete units, w/o doors & refrigeration, cooler		146	.110	SF Flr.	177	6		183	204
6400 Freezer	↓	109.60	.146	"	113	8		121	137

13 24 Special Activity Rooms

13 24 16 – Saunas

13 24 16.50 Saunas and Heaters	Crew	Daily Output	Labor-Hours	Unit	Material	2021 Bare Costs Labor	2021 Bare Costs Equipment	Total	Total Incl O&P
0010 **SAUNAS AND HEATERS**									
0020 Prefabricated, incl. heater & controls, 7' high, 6' x 4', C/C	L-7	2.20	12.727	Ea.	6,425	675		7,100	8,125
1700 Door only, cedar, 2' x 6', w/ 1' x 4' tempered insulated glass window	2 Carp	3.40	4.706		590	257		847	1,050
1800 Prehung, incl. jambs, pulls & hardware	"	12	1.333		935	73		1,008	1,150
2500 Heaters only (incl. above), wall mounted, to 200 C.F.				↓	1,025			1,025	1,125
4480 For additional equipment, see Section 11 66 13.10									

13 24 26 – Steam Baths

13 24 26.50 Steam Baths and Components	Crew	Daily Output	Labor-Hours	Unit	Material	2021 Bare Costs Labor	2021 Bare Costs Equipment	Total	Total Incl O&P
0010 **STEAM BATHS AND COMPONENTS**									
0020 Heater, timer & head, single, to 140 C.F.	1 Plum	1.20	6.667	Ea.	2,150	450		2,600	3,050
0500 To 300 C.F.	"	1.10	7.273		2,425	490		2,915	3,425
2000 Multiple, motels, apts., 2 baths, w/blow-down assm., 500 C.F.	Q-1	1.30	12.308		6,750	750		7,500	8,575
2500 4 baths	"	.70	22.857	↓	10,200	1,400		11,600	13,400

13 28 Athletic and Recreational Special Construction

13 28 33 – Athletic and Recreational Court Walls

13 28 33.50 Sport Court	Crew	Daily Output	Labor-Hours	Unit	Material	2021 Bare Costs Labor	Equipment	Total	Total Incl O&P
0010 **SPORT COURT**									
0020 Sport court floors, no. 2 & better maple, 25/32" thick, see 09 64 66.10				SF Flr.					
0300 Squash, regulation court in existing building, minimum				Court	43,900			43,900	48,300
0400 Maximum				"	48,900			48,900	54,000
0450 Rule of thumb for components:									
0470 Walls	3 Carp	.15	160	Court	13,400	8,750		22,150	28,600
0500 Floor	"	.25	96		10,600	5,250		15,850	20,000
0550 Lighting	2 Elec	.60	26.667	↓	2,500	1,700		4,200	5,350

13 34 Fabricated Engineered Structures

13 34 13 – Glazed Structures

13 34 13.13 Greenhouses

	Crew	Daily Output	Labor-Hours	Unit	Material	2021 Bare Costs Labor	Equipment	Total	Total Incl O&P
0010 **GREENHOUSES**, Shell only, stock units, not incl. 2' stub walls,									
0020 foundation, floors, heat or compartments									
0300 Residential type, free standing, 8'-6" long x 7'-6" wide	2 Carp	59	.271	SF Flr.	24	14.85		38.85	49.50
0400 10'-6" wide		85	.188		43.50	10.30		53.80	64.50
0600 13'-6" wide		108	.148		46	8.10		54.10	63.50
0700 17'-0" wide		160	.100		42.50	5.45		47.95	55
0900 Lean-to type, 3'-10" wide		34	.471		42.50	25.50		68	88
1000 6'-10" wide		58	.276		26	15.10		41.10	52.50
1050 8'-0" wide	↓	60	.267	↓	75	14.60		89.60	106
1100 Wall mounted to existing window, 3' x 3'	1 Carp	4	2	Ea.	2,625	109		2,734	3,075
1120 4' x 5'	"	3	2.667		2,425	146		2,571	2,875
3900 Cooling, 1,200 CFM exhaust fan, add					325			325	355
4000 7,850 CFM					850			850	935
4200 For heaters, 10 MBH, add					245			245	270
4300 60 MBH, add					730			730	805
4500 For benches, 2' x 8', add					188			188	207
4600 4' x 10', add				↓	495			495	545
4800 For ventilation & humidity control w/4 integrated outlets, add				Total	288			288	315
4900 For environmental controls and automation, 8 outputs, 9 stages, add				"	1,325			1,325	1,450
5100 For humidification equipment, add				Ea.	325			325	360
5200 For vinyl shading, add				S.F.	.39			.39	.43

13 34 13.19 Swimming Pool Enclosures

	Crew	Daily Output	Labor-Hours	Unit	Material	2021 Bare Costs Labor	Equipment	Total	Total Incl O&P
0010 **SWIMMING POOL ENCLOSURES** Translucent, free standing									
0020 not including foundations, heat or light									
0200 Economy	2 Carp	200	.080	SF Hor.	42	4.38		46.38	53
0600 Deluxe	"	70	.229	"	115	12.50		127.50	146

13 34 23 – Fabricated Structures

13 34 23.16 Fabricated Control Booths

	Crew	Daily Output	Labor-Hours	Unit	Material	2021 Bare Costs Labor	Equipment	Total	Total Incl O&P
0010 **FABRICATED CONTROL BOOTHS**									
0100 Guard House, prefab conc. w/bullet resistant doors & windows, roof & wiring									
0110 8' x 8', Level III	L-10	1	24	Ea.	59,500	1,475	475	61,450	68,500
0120 8' x 8', Level IV	"	1	24	"	77,000	1,475	475	78,950	87,500

13 34 23.45 Kiosks

	Crew	Daily Output	Labor-Hours	Unit	Material	2021 Bare Costs Labor	Equipment	Total	Total Incl O&P
0010 **KIOSKS**									
0020 Round, advertising type, 5' diameter, 7' high, aluminum wall, illuminated				Ea.	22,600			22,600	24,800
0100 Aluminum wall, non-illuminated					21,800			21,800	24,000
0500 Rectangular, 5' x 9', 7'-6" high, aluminum wall, illuminated					24,900			24,900	27,300
0600 Aluminum wall, non-illuminated				↓	25,500			25,500	28,100

13 34 Fabricated Engineered Structures

13 34 63 – Natural Fiber Construction

13 34 63.50 Straw Bale Construction		Crew	Daily Output	Labor-Hours	Unit	Material	2021 Bare Costs Labor	Equipment	Total	Total Incl O&P
0010	**STRAW BALE CONSTRUCTION**									
2020	Straw bales in walls w/modified post and beam frame	G	2 Carp	320	.050	S.F.	6.70	2.74		9.44	11.75

13 42 Building Modules

13 42 63 – Detention Cell Modules

13 42 63.16 Steel Detention Cell Modules

		Crew	Daily Output	Labor-Hours	Unit	Material	Labor	Equipment	Total	Total Incl O&P
0010	**STEEL DETENTION CELL MODULES**									
2000	Cells, prefab., 5' to 6' wide, 7' to 8' high, 7' to 8' deep,									
2010	bar front, cot, not incl. plumbing	E-4	1.50	21.333	Ea.	9,650	1,300	99	11,049	12,800

13 48 Sound, Vibration, and Seismic Control

13 48 13 – Manufactured Sound and Vibration Control Components

13 48 13.50 Audio Masking

		Crew	Daily Output	Labor-Hours	Unit	Material	Labor	Equipment	Total	Total Incl O&P
0010	**AUDIO MASKING**, acoustical enclosure, 4" thick wall and ceiling									
0020	8 lb./S.F., up to 12' span	3 Carp	72	.333	SF Surf	32.50	18.25		50.75	64.50
0300	Better quality panels, 10.5 #/S.F		64	.375		36.50	20.50		57	73
0400	Reverb-chamber, 4" thick, parallel walls	↓	60	.400	↓	46	22		68	85.50

13 49 Radiation Protection

13 49 13 – Integrated X-Ray Shielding Assemblies

13 49 13.50 Lead Sheets

		Crew	Daily Output	Labor-Hours	Unit	Material	Labor	Equipment	Total	Total Incl O&P
0010	**LEAD SHEETS**									
0300	Lead sheets, 1/16" thick	2 Lath	135	.119	S.F.	11.10	6.40		17.50	22
0400	1/8" thick		120	.133		25	7.20		32.20	39
0500	Lead shielding, 1/4" thick		135	.119		47	6.40		53.40	61.50
0550	1/2" thick	↓	120	.133	↓	87.50	7.20		94.70	107
1200	X-ray protection, average radiography or fluoroscopy									
1210	room, up to 300 S.F. floor, 1/16" lead, economy	2 Lath	.25	64	Total	13,300	3,450		16,750	20,000
1500	7'-0" walls, deluxe	"	.15	107	"	15,900	5,775		21,675	26,500
1600	Deep therapy X-ray room, 250 kV capacity,									
1800	up to 300 S.F. floor, 1/4" lead, economy	2 Lath	.08	200	Total	37,000	10,800		47,800	57,500
1900	7'-0" walls, deluxe		.06	267	"	45,700	14,400		60,100	72,500
1999	Minimum labor/equipment charge	↓	4.50	3.556	Job		192		192	300

13 49 21 – Lead Glazing

13 49 21.50 Lead Glazing

		Crew	Daily Output	Labor-Hours	Unit	Material	Labor	Equipment	Total	Total Incl O&P
0010	**LEAD GLAZING**									
0600	Lead glass, 1/4" thick, 2.0 mm LE, 12" x 16"	2 Glaz	13	1.231	Ea.	495	65		560	645
0700	24" x 36"	"	8	2	"	1,675	105		1,780	2,000
2000	X-ray viewing panels, clear lead plastic									
2010	7 mm thick, 0.3 mm LE, 2.3 lb./S.F.	H-3	139	.115	S.F.	256	5.45		261.45	291
2020	12 mm thick, 0.5 mm LE, 3.9 lb./S.F.		82	.195		345	9.25		354.25	395
2030	18 mm thick, 0.8 mm LE, 5.9 lb./S.F.		54	.296		420	14.05		434.05	485
2040	22 mm thick, 1.0 mm LE, 7.2 lb./S.F.		44	.364		690	17.20		707.20	790
2050	35 mm thick, 1.5 mm LE, 11.5 lb./S.F.		28	.571		915	27		942	1,050
2060	46 mm thick, 2.0 mm LE, 15.0 lb./S.F.	↓	21	.762	↓	1,200	36		1,236	1,350

13 49 Radiation Protection

13 49 23 – Integrated RFI/EMI Shielding Assemblies

13 49 23.50 Modular Shielding Partitions	Crew	Daily Output	Labor-Hours	Unit	Material	2021 Bare Costs Labor	Equipment	Total	Total Incl O&P
0010 **MODULAR SHIELDING PARTITIONS**									
4000 X-ray barriers, modular, panels mounted within framework for									
4002 attaching to floor, wall or ceiling, upper portion is clear lead									
4005 plastic window panels 48"H, lower portion is opaque leaded									
4008 steel panels 36"H, structural supports not incl.									
4010 1-section barrier, 36"W x 84"H overall									
4020 0.5 mm LE panels	H-3	6.40	2.500	Ea.	9,250	118		9,368	10,400
4030 0.8 mm LE panels		6.40	2.500		9,975	118		10,093	11,200
4040 1.0 mm LE panels		5.33	3.002		13,800	142		13,942	15,400
4050 1.5 mm LE panels	↓	5.33	3.002	↓	16,300	142		16,442	18,100
4060 2-section barrier, 72"W x 84"H overall									
4070 0.5 mm LE panels	H-3	4	4	Ea.	14,000	189		14,189	15,700
4080 0.8 mm LE panels		4	4		14,400	189		14,589	16,200
4090 1.0 mm LE panels		3.56	4.494		15,700	213		15,913	17,600
5000 1.5 mm LE panels	↓	3.20	5	↓	28,100	237		28,337	31,400
5010 3-section barrier, 108"W x 84"H overall									
5020 0.5 mm LE panels	H-3	3.20	5	Ea.	21,000	237		21,237	23,500
5030 0.8 mm LE panels		3.20	5		21,700	237		21,937	24,200
5040 1.0 mm LE panels		2.67	5.993		23,600	284		23,884	26,400
5050 1.5 mm LE panels	↓	2.46	6.504	↓	42,700	310		43,010	47,400
7000 X-ray barriers, mobile, mounted within framework w/casters on									
7005 bottom, clear lead plastic window panels on upper portion,									
7010 opaque on lower, 30"W x 75"H overall, incl. framework									
7020 24"H upper w/0.5 mm LE, 48"H lower w/0.8 mm LE	1 Carp	16	.500	Ea.	4,000	27.50		4,027.50	4,475
7030 48"W x 75"H overall, incl. framework									
7040 36"H upper w/0.5 mm LE, 36"H lower w/0.8 mm LE	1 Carp	16	.500	Ea.	6,325	27.50		6,352.50	7,000
7050 36"H upper w/1.0 mm LE, 36"H lower w/1.5 mm LE	"	16	.500	"	7,500	27.50		7,527.50	8,275
7060 72"W x 75"H overall, incl. framework									
7070 36"H upper w/0.5 mm LE, 36"H lower w/0.8 mm LE	1 Carp	16	.500	Ea.	7,500	27.50		7,527.50	8,275
7080 36"H upper w/1.0 mm LE, 36"H lower w/1.5 mm LE	"	16	.500	"	9,375	27.50		9,402.50	10,300

For customer support on your Commercial Renovation Costs with RSMeans Data, call 800.448.8182.

405

Division Notes

	CREW	DAILY OUTPUT	LABOR-HOURS	UNIT	BARE COSTS				TOTAL INCL O&P
					MAT.	LABOR	EQUIP.	TOTAL	

Estimating Tips
General
- Many products in Division 14 will require some type of support or blocking for installation not included with the item itself. Examples are supports for conveyors or tube systems, attachment points for lifts, and footings for hoists or cranes. Add these supports in the appropriate division.

14 10 00 Dumbwaiters
14 20 00 Elevators
- Dumbwaiters and elevators are estimated and purchased in a method similar to buying a car. The manufacturer has a base unit with standard features. Added to this base unit price will be whatever options the owner or specifications require. Increased load capacity, additional vertical travel, additional stops, higher speed, and cab finish options are items to be considered. When developing an estimate for dumbwaiters and elevators, remember that some items needed by the installers may have to be included as part of the general contract.

Examples are:
- ☐ shaftway
- ☐ rail support brackets
- ☐ machine room
- ☐ electrical supply
- ☐ sill angles
- ☐ electrical connections
- ☐ pits
- ☐ roof penthouses
- ☐ pit ladders

Check the job specifications and drawings before pricing.

- Installation of elevators and handicapped lifts in historic structures can require significant additional costs. The associated structural requirements may involve cutting into and repairing finishes, moldings, flooring, etc. The estimator must account for these special conditions.

14 30 00 Escalators and Moving Walks
- Escalators and moving walks are specialty items installed by specialty contractors. There are numerous options associated with these items. For specific options, contact a manufacturer or contractor. In a method similar to estimating dumbwaiters and elevators, you should verify the extent of general contract work and add items as necessary.

14 40 00 Lifts
14 90 00 Other Conveying Equipment
- Products such as correspondence lifts, chutes, and pneumatic tube systems, as well as other items specified in this subdivision, may require trained installers. The general contractor might not have any choice as to who will perform the installation or when it will be performed. Long lead times are often required for these products, making early decisions in scheduling necessary.

Reference Numbers
Reference numbers are shown at the beginning of some major classifications. These numbers refer to related items in the Reference Section. The reference information may be an estimating procedure, an alternate pricing method, or technical information.

Note: Not all subdivisions listed here necessarily appear. ■

Same Data. Simplified.

Enjoy the convenience and efficiency of accessing your costs anywhere:

- **Skip the multiplier** by setting your location
- **Quickly search,** edit, favorite and share costs
- **Stay on top of price changes** with automatic updates

Discover more at rsmeans.com/online

14 11 Manual Dumbwaiters

14 11 10 – Hand Operated Dumbwaiters

14 11 10.20 Manual Dumbwaiters	Crew	Daily Output	Labor-Hours	Unit	Material	2021 Bare Costs Labor	Equipment	Total	Total Incl O&P
0010 **MANUAL DUMBWAITERS**									
0020 2 stop, hand powered, up to 75 lb. capacity	2 Elev	.75	21.333	Ea.	3,400	1,925		5,325	6,675
0100 76 lb. capacity and up		.50	32	"	7,125	2,900		10,025	12,300
0300 For each additional stop, add	↓	.75	21.333	Stop	1,125	1,925		3,050	4,200

14 12 Electric Dumbwaiters

14 12 10 – Dumbwaiters

14 12 10.10 Electric Dumbwaiters

	Crew	Daily Output	Labor-Hours	Unit	Material	2021 Bare Costs Labor	Equipment	Total	Total Incl O&P
0010 **ELECTRIC DUMBWAITERS**									
0020 2 stop, up to 75 lb. capacity	2 Elev	.13	123	Ea.	3,500	11,100		14,600	20,900
0100 76 lb. capacity and up		.11	145	"	13,200	13,100		26,300	34,600
0600 For each additional stop, add	↓	.54	29.630	Stop	555	2,675		3,230	4,700

14 21 Electric Traction Elevators

14 21 13 – Electric Traction Freight Elevators

14 21 13.10 Electric Traction Freight Elevators and Options

		Crew	Daily Output	Labor-Hours	Unit	Material	2021 Bare Costs Labor	Equipment	Total	Total Incl O&P
0010 **ELECTRIC TRACTION FREIGHT ELEVATORS AND OPTIONS**	R142000-10									
0425 Electric freight, base unit, 4000 lb., 200 fpm, 4 stop, std. fin.	R142000-20	2 Elev	.05	320	Ea.	120,500	28,900		149,400	177,000
0450 For 5000 lb. capacity, add	R142000-30					6,150			6,150	6,775
0500 For 6000 lb. capacity, add	R142000-40					14,400			14,400	15,900
0525 For 7000 lb. capacity, add						19,800			19,800	21,800
0550 For 8000 lb. capacity, add						23,100			23,100	25,400
0575 For 10000 lb. capacity, add						32,200			32,200	35,400
0600 For 12000 lb. capacity, add						38,900			38,900	42,800
0625 For 16000 lb. capacity, add						47,400			47,400	52,000
0650 For 20000 lb. capacity, add						55,000			55,000	60,500
0675 For increased speed, 250 fpm, add						14,300			14,300	15,800
0700 300 fpm, geared electric, add						17,900			17,900	19,700
0725 350 fpm, geared electric, add						21,900			21,900	24,100
0750 400 fpm, geared electric, add						25,700			25,700	28,200
0775 500 fpm, gearless electric, add						35,700			35,700	39,300
0800 600 fpm, gearless electric, add					↓	39,700			39,700	43,700
0825 700 fpm, gearless electric, add					Ea.	47,000			47,000	51,500
0850 800 fpm, gearless electric, add						52,000			52,000	57,500
0875 For class "B" loading, add						2,900			2,900	3,200
0900 For class "C-1" loading, add						7,175			7,175	7,875
0925 For class "C-2" loading, add						8,425			8,425	9,275
0950 For class "C-3" loading, add					↓	11,800			11,800	13,000
0975 For travel over 40 V.L.F., add		2 Elev	7.25	2.207	V.L.F.	600	199		799	965
1000 For number of stops over 4, add		"	.27	59.259	Stop	4,550	5,350		9,900	13,200

14 21 23 – Electric Traction Passenger Elevators

14 21 23.10 Electric Traction Passenger Elevators and Options

		Crew	Daily Output	Labor-Hours	Unit	Material	2021 Bare Costs Labor	Equipment	Total	Total Incl O&P
0010 **ELECTRIC TRACTION PASSENGER ELEVATORS AND OPTIONS**	R142000-10									
1625 Electric pass., base unit, 2,000 lb., 200 fpm, 4 stop, std. fin.	R142000-20	2 Elev	.05	320	Ea.	101,500	28,900		130,400	156,000
1650 For 2,500 lb. capacity, add	R142000-30					4,325			4,325	4,775
1675 For 3,000 lb. capacity, add	R142000-40					4,750			4,750	5,225
1700 For 3,500 lb. capacity, add						6,025			6,025	6,650
1725 For 4,000 lb. capacity, add						7,100			7,100	7,800
1750 For 4,500 lb. capacity, add						9,775			9,775	10,800

14 21 Electric Traction Elevators

14 21 23 – Electric Traction Passenger Elevators

14 21 23.10 Electric Traction Passenger Elevators and Options		Crew	Daily Output	Labor-Hours	Unit	Material	2021 Bare Costs Labor	Equipment	Total	Total Incl O&P
1775	For 5000 lb. capacity, add				Ea.	12,600			12,600	13,800
1800	For increased speed, 250 fpm, geared electric, add					3,450			3,450	3,775
1825	300 fpm, geared electric, add					6,650			6,650	7,300
1850	350 fpm, geared electric, add					8,525			8,525	9,400
1875	400 fpm, geared electric, add					12,500			12,500	13,700
1900	500 fpm, gearless electric, add					30,000			30,000	33,000
1925	600 fpm, gearless electric, add					48,600			48,600	53,500
1950	700 fpm, gearless electric, add					54,500			54,500	60,000
1975	800 fpm, gearless electric, add					60,500			60,500	66,500
2000	For travel over 40 V.L.F., add	2 Elev	7.25	2.207	V.L.F.	720	199		919	1,100
2025	For number of stops over 4, add		.27	59.259	Stop	3,225	5,350		8,575	11,800
2400	Electric hospital, base unit, 4,000 lb., 200 fpm, 4 stop, std fin.		.05	320	Ea.	94,000	28,900		122,900	148,000
2425	For 4,500 lb. capacity, add					6,025			6,025	6,625
2450	For 5,000 lb. capacity, add					7,875			7,875	8,675
2475	For increased speed, 250 fpm, geared electric, add					3,825			3,825	4,225
2500	300 fpm, geared electric, add					7,350			7,350	8,075
2525	350 fpm, geared electric, add					8,850			8,850	9,725
2550	400 fpm, geared electric, add					12,800			12,800	14,100
2575	500 fpm, gearless electric, add					35,200			35,200	38,800
2600	600 fpm, gearless electric, add					53,000			53,000	58,000
2625	700 fpm, gearless electric, add					59,000			59,000	64,500
2650	800 fpm, gearless electric, add					66,000			66,000	72,500
2675	For travel over 40 V.L.F., add	2 Elev	7.25	2.207	V.L.F.	190	199		389	515
2700	For number of stops over 4, add	"	.27	59.259	Stop	4,775	5,350		10,125	13,500

14 21 33 – Electric Traction Residential Elevators

14 21 33.20 Residential Elevators

		Crew	Daily Output	Labor-Hours	Unit	Material	2021 Bare Costs Labor	Equipment	Total	Total Incl O&P
0010	**RESIDENTIAL ELEVATORS**									
7000	Residential, cab type, 1 floor, 2 stop, economy model	2 Elev	.20	80	Ea.	10,500	7,225		17,725	22,600
7100	Custom model		.10	160		17,700	14,400		32,100	41,500
7200	2 floor, 3 stop, economy model		.12	133		15,500	12,000		27,500	35,500
7300	Custom model		.06	267		25,400	24,100		49,500	65,000

14 24 Hydraulic Elevators

14 24 13 – Hydraulic Freight Elevators

14 24 13.10 Hydraulic Freight Elevators and Options

			Crew	Daily Output	Labor-Hours	Unit	Material	2021 Bare Costs Labor	Equipment	Total	Total Incl O&P
0010	**HYDRAULIC FREIGHT ELEVATORS AND OPTIONS**	R142000-10									
1025	Hydraulic freight, base unit, 2,000 lb., 50 fpm, 2 stop, std. fin.	R142000-20	2 Elev	.10	160	Ea.	83,500	14,400		97,900	114,000
1050	For 2,500 lb. capacity, add	R142000-30					4,225			4,225	4,625
1075	For 3,000 lb. capacity, add	R142000-40					5,575			5,575	6,125
1100	For 3,500 lb. capacity, add						9,100			9,100	10,000
1125	For 4,000 lb. capacity, add						10,400			10,400	11,500
1150	For 4,500 lb. capacity, add						12,100			12,100	13,300
1175	For 5,000 lb. capacity, add						16,400			16,400	18,000
1200	For 6,000 lb. capacity, add						16,700			16,700	18,400
1225	For 7,000 lb. capacity, add						25,700			25,700	28,200
1250	For 8,000 lb. capacity, add						31,100			31,100	34,200
1275	For 10,000 lb. capacity, add						32,900			32,900	36,200
1300	For 12,000 lb. capacity, add						40,000			40,000	44,000
1325	For 16,000 lb. capacity, add						53,500			53,500	59,000
1350	For 20,000 lb. capacity, add						59,500			59,500	65,500

14 24 Hydraulic Elevators

14 24 13 – Hydraulic Freight Elevators

14 24 13.10 Hydraulic Freight Elevators and Options	Crew	Daily Output	Labor-Hours	Unit	Material	2021 Bare Costs Labor	Equipment	Total	Total Incl O&P
1375	For increased speed, 100 fpm, add				Ea.	1,200			1,200	1,325
1400	125 fpm, add					3,150			3,150	3,450
1425	150 fpm, add					4,950			4,950	5,450
1450	175 fpm, add					7,100			7,100	7,800
1475	For class "B" loading, add					2,825			2,825	3,100
1500	For class "C-1" loading, add					7,125			7,125	7,825
1525	For class "C-2" loading, add					8,550			8,550	9,400
1550	For class "C-3" loading, add					11,700			11,700	12,900
1575	For travel over 20 V.L.F., add	2 Elev	7.25	2.207	V.L.F.	895	199		1,094	1,275
1600	For number of stops over 2, add	"	.27	59.259	Stop	2,250	5,350		7,600	10,700

14 24 23 – Hydraulic Passenger Elevators

14 24 23.10 Hydraulic Passenger Elevators and Options

		Crew	Daily Output	Labor-Hours	Unit	Material	2021 Bare Costs Labor	Equipment	Total	Total Incl O&P
0010	**HYDRAULIC PASSENGER ELEVATORS AND OPTIONS** R142000-10									
2050	Hyd. pass., base unit, 1,500 lb., 100 fpm, 2 stop, std. fin. R142000-20	2 Elev	.10	160	Ea.	40,800	14,400		55,200	67,000
2075	For 2,000 lb. capacity, add R142000-30				"	845			845	930
2100	For 2,500 lb. capacity, add R142000-40				Ea.	3,175			3,175	3,500
2125	For 3,000 lb. capacity, add					4,550			4,550	5,025
2150	For 3,500 lb. capacity, add					7,625			7,625	8,375
2175	For 4,000 lb. capacity, add					9,175			9,175	10,100
2200	For 4,500 lb. capacity, add					12,100			12,100	13,300
2225	For 5,000 lb. capacity, add					17,000			17,000	18,700
2250	For increased speed, 125 fpm, add					1,225			1,225	1,350
2275	150 fpm, add					2,750			2,750	3,025
2300	175 fpm, add					5,350			5,350	5,875
2325	200 fpm, add					10,300			10,300	11,400
2350	For travel over 12 V.L.F., add	2 Elev	7.25	2.207	V.L.F.	730	199		929	1,100
2375	For number of stops over 2, add		.27	59.259	Stop	1,000	5,350		6,350	9,325
2725	Hydraulic hospital, base unit, 4,000 lb., 100 fpm, 2 stop, std. fin.		.10	160	Ea.	67,000	14,400		81,400	95,500
2775	For 4,500 lb. capacity, add					7,525			7,525	8,275
2800	For 5,000 lb. capacity, add					10,900			10,900	12,000
2825	For increased speed, 125 fpm, add					2,250			2,250	2,475
2850	150 fpm, add					3,750			3,750	4,125
2875	175 fpm, add					6,325			6,325	6,950
2900	200 fpm, add					9,300			9,300	10,200
2925	For travel over 12 V.L.F., add	2 Elev	7.25	2.207	V.L.F.	410	199		609	755
2950	For number of stops over 2, add	"	.27	59.259	Stop	4,675	5,350		10,025	13,400

14 27 Custom Elevator Cabs and Doors

14 27 13 – Custom Elevator Cab Finishes

14 27 13.10 Cab Finishes

		Crew	Daily Output	Labor-Hours	Unit	Material	2021 Bare Costs Labor	Equipment	Total	Total Incl O&P
0010	**CAB FINISHES**									
3325	Passenger elevator cab finishes (based on 3,500 lb. cab size)									
3350	Acrylic panel ceiling				Ea.	795			795	875
3375	Aluminum eggcrate ceiling					715			715	790
3400	Stainless steel doors					4,150			4,150	4,550
3425	Carpet flooring					635			635	695
3450	Epoxy flooring					480			480	525
3475	Quarry tile flooring					920			920	1,025
3500	Slate flooring					1,650			1,650	1,825
3525	Textured rubber flooring					665			665	735

14 27 Custom Elevator Cabs and Doors

14 27 13 – Custom Elevator Cab Finishes

14 27 13.10 Cab Finishes

		Crew	Daily Output	Labor-Hours	Unit	Material	2021 Bare Costs Labor	Equipment	Total	Total Incl O&P
3550	Stainless steel walls				Ea.	4,150			4,150	4,575
3575	Stainless steel returns at door				↓	1,200			1,200	1,300
4450	Hospital elevator cab finishes (based on 3,500 lb. cab size)									
4475	Aluminum eggcrate ceiling				Ea.	725			725	795
4500	Stainless steel doors					4,125			4,125	4,550
4525	Epoxy flooring				↓	475			475	525
4550	Quarry tile flooring				Ea.	920			920	1,025
4575	Textured rubber flooring					660			660	730
4600	Stainless steel walls					4,700			4,700	5,175
4625	Stainless steel returns at door				↓	930			930	1,025

14 28 Elevator Equipment and Controls

14 28 10 – Elevator Equipment and Control Options

14 28 10.10 Elevator Controls and Doors

		Crew	Daily Output	Labor-Hours	Unit	Material	2021 Bare Costs Labor	Equipment	Total	Total Incl O&P
0010	**ELEVATOR CONTROLS AND DOORS**									
2975	Passenger elevator options									
3000	2 car group automatic controls	2 Elev	.66	24.242	Ea.	3,625	2,200		5,825	7,325
3025	3 car group automatic controls		.44	36.364		5,525	3,275		8,800	11,100
3050	4 car group automatic controls		.33	48.485		9,650	4,375		14,025	17,300
3075	5 car group automatic controls		.26	61.538		14,500	5,550		20,050	24,500
3100	6 car group automatic controls		.22	72.727		22,100	6,575		28,675	34,400
3125	Intercom service		3	5.333		580	480		1,060	1,375
3150	Duplex car selective collective		.66	24.242		4,175	2,200		6,375	7,950
3175	Center opening 1 speed doors		2	8		2,075	720		2,795	3,400
3200	Center opening 2 speed doors		2	8		2,925	720		3,645	4,325
3225	Rear opening doors (opposite front)		2	8		4,475	720		5,195	6,025
3250	Side opening 2 speed doors		2	8		7,650	720		8,370	9,525
3275	Automatic emergency power switching		.66	24.242		1,275	2,200		3,475	4,750
3300	Manual emergency power switching	↓	8	2		540	181		721	870
3625	Hall finishes, stainless steel doors					1,475			1,475	1,625
3650	Stainless steel frames					1,550			1,550	1,700
3675	12 month maintenance contract								4,425	4,425
3700	Signal devices, hall lanterns	2 Elev	8	2		545	181		726	875
3725	Position indicators, up to 3		9.40	1.702		350	154		504	620
3750	Position indicators, per each over 3	↓	32	.500		94.50	45		139.50	173
3775	High speed heavy duty door opener					2,775			2,775	3,050
3800	Variable voltage, O.H. gearless machine, min.	2 Elev	.16	100		34,600	9,025		43,625	52,000
3815	Maximum		.07	229		77,500	20,600		98,100	117,000
3825	Basement installed geared machine	↓	.33	48.485	↓	13,600	4,375		17,975	21,700
3850	Freight elevator options									
3875	Doors, bi-parting	2 Elev	.66	24.242	Ea.	6,250	2,200		8,450	10,200
3900	Power operated door and gate	"	.66	24.242		25,100	2,200		27,300	31,000
3925	Finishes, steel plate floor					1,125			1,125	1,250
3950	14 ga. 1/4" x 4' steel plate walls					2,075			2,075	2,275
3975	12 month maintenance contract								3,400	3,400
4000	Signal devices, hall lanterns	2 Elev	8	2		530	181		711	860
4025	Position indicators, up to 3	"	9.40	1.702	↓	375	154		529	650
4050	Position indicators, per each over 3	2 Elev	32	.500	Ea.	103	45		148	182
4075	Variable voltage basement installed geared machine	"	.66	24.242	"	21,000	2,200		23,200	26,500
4100	Hospital elevator options									
4125	2 car group automatic controls	2 Elev	.66	24.242	Ea.	3,625	2,200		5,825	7,325

14 28 Elevator Equipment and Controls

14 28 10 – Elevator Equipment and Control Options

14 28 10.10 Elevator Controls and Doors

14 28 10.10 Elevator Controls and Doors		Crew	Daily Output	Labor-Hours	Unit	Material	2021 Bare Costs Labor	Equipment	Total	Total Incl O&P
4150	3 car group automatic controls	2 Elev	.44	36.364	Ea.	5,500	3,275		8,775	11,100
4175	4 car group automatic controls		.33	48.485		13,100	4,375		17,475	21,100
4200	5 car group automatic controls		.26	61.538		13,100	5,550		18,650	22,900
4225	6 car group automatic controls		.22	72.727		20,000	6,575		26,575	32,100
4250	Intercom service		3	5.333		550	480		1,030	1,350
4275	Duplex car selective collective		.66	24.242		4,075	2,200		6,275	7,850
4300	Center opening 1 speed doors		2	8		2,075	720		2,795	3,375
4325	Center opening 2 speed doors		2	8		2,750	720		3,470	4,100
4350	Rear opening doors (opposite front)		2	8		4,475	720		5,195	6,025
4375	Side opening 2 speed doors		2	8		6,650	720		7,370	8,400
4400	Automatic emergency power switching		.66	24.242		1,250	2,200		3,450	4,725
4425	Manual emergency power switching	▼	8	2		525	181		706	850
4675	Hall finishes, stainless steel doors					1,550			1,550	1,700
4700	Stainless steel frames					1,550			1,550	1,725
4725	12 month maintenance contract								6,125	6,125
4750	Signal devices, hall lanterns	2 Elev	8	2		520	181		701	845
4775	Position indicators, up to 3		9.40	1.702		355	154		509	625
4800	Position indicators, per each over 3	▼	32	.500		95	45		140	174
4825	High speed heavy duty door opener					2,800			2,800	3,075
4850	Variable voltage, O.H. gearless machine, min.	2 Elev	.16	100		36,400	9,025		45,425	54,000
4865	Maximum		.07	229		79,000	20,600		99,600	118,500
4875	Basement installed geared machine	▼	.33	48.485	▼	18,200	4,375		22,575	26,800
5000	Drilling for piston, casing included, 18" diameter	B-48	80	.700	V.L.F.	66	35	13.10	114.10	142

14 42 Wheelchair Lifts

14 42 13 – Inclined Wheelchair Lifts

14 42 13.10 Inclined Wheelchair Lifts and Stairclimbers

		Crew	Daily Output	Labor-Hours	Unit	Material	2021 Bare Costs Labor	Equipment	Total	Total Incl O&P
0010	**INCLINED WHEELCHAIR LIFTS AND STAIRCLIMBERS**									
7700	Stair climber (chair lift), single seat, minimum	2 Elev	1	16	Ea.	4,975	1,450		6,425	7,700
7800	Maximum		.20	80		6,825	7,225		14,050	18,600
8700	Stair lift, minimum		1	16		13,500	1,450		14,950	17,000
8900	Maximum	▼	.20	80	▼	21,400	7,225		28,625	34,600

14 42 16 – Vertical Wheelchair Lifts

14 42 16.10 Wheelchair Lifts

		Crew	Daily Output	Labor-Hours	Unit	Material	2021 Bare Costs Labor	Equipment	Total	Total Incl O&P
0010	**WHEELCHAIR LIFTS**									
8000	Wheelchair lift, minimum	2 Elev	1	16	Ea.	6,800	1,450		8,250	9,725
8500	Maximum	"	.50	32	"	16,100	2,900		19,000	22,200

14 45 Vehicle Lifts

14 45 10 – Hydraulic Vehicle Lifts

14 45 10.10 Hydraulic Lifts

		Crew	Daily Output	Labor-Hours	Unit	Material	2021 Bare Costs Labor	Equipment	Total	Total Incl O&P
0010	**HYDRAULIC LIFTS**									
2200	Single post, 8,000 lb. capacity	L-4	.40	60	Ea.	7,600	3,125		10,725	13,300
2810	Double post, 6,000 lb. capacity		2.67	8.989		2,375	470		2,845	3,375
2815	9,000 lb. capacity		2.29	10.480		3,125	545		3,670	4,325
2820	15,000 lb. capacity		2	12		6,525	625		7,150	8,175
2822	Four post, 26,000 lb. capacity		1.80	13.333		15,900	695		16,595	18,600
2825	30,000 lb. capacity		1.60	15		62,000	780		62,780	69,500
2830	Ramp style, 4 post, 25,000 lb. capacity	▼	2	12	▼	15,400	625		16,025	18,000

14 45 Vehicle Lifts

14 45 10 – Hydraulic Vehicle Lifts

14 45 10.10 Hydraulic Lifts

14 45 10.10 Hydraulic Lifts	Crew	Daily Output	Labor-Hours	Unit	Material	2021 Bare Costs Labor	2021 Bare Costs Equipment	Total	Total Incl O&P
2835	35,000 lb. capacity	L-4	1	24	Ea.	19,400	1,250		20,650	23,400
2840	50,000 lb. capacity		1	24		129,500	1,250		130,750	144,500
2845	75,000 lb. capacity	↓	1	24		150,500	1,250		151,750	167,500
2850	For drive thru tracks, add, minimum					1,100			1,100	1,200
2855	Maximum					1,875			1,875	2,075
2860	Ramp extensions, 3' (set of 2)					445			445	490
2865	Rolling jack platform					1,575			1,575	1,725
2870	Electric/hydraulic jacking beam					12,200			12,200	13,400
2880	Scissor lift, portable, 6,000 lb. capacity					1,700			1,700	1,875

14 91 Facility Chutes

14 91 33 – Laundry and Linen Chutes

14 91 33.10 Chutes

14 91 33.10 Chutes	Crew	Daily Output	Labor-Hours	Unit	Material	2021 Bare Costs Labor	2021 Bare Costs Equipment	Total	Total Incl O&P
0011	CHUTES, linen, trash or refuse									
0050	Aluminized steel, 16 ga., 18" diameter	2 Shee	3.50	4.571	Floor	1,800	299		2,099	2,475
0100	24" diameter		3.20	5		2,000	325		2,325	2,700
0400	Galvanized steel, 16 ga., 18" diameter		3.50	4.571		1,075	299		1,374	1,675
0500	24" diameter		3.20	5		1,225	325		1,550	1,850
0800	Stainless steel, 18" diameter		3.50	4.571		3,475	299		3,774	4,300
0900	24" diameter	↓	3.20	5	↓	3,050	325		3,375	3,850

14 91 82 – Trash Chutes

14 91 82.10 Trash Chutes and Accessories

14 91 82.10 Trash Chutes and Accessories	Crew	Daily Output	Labor-Hours	Unit	Material	2021 Bare Costs Labor	2021 Bare Costs Equipment	Total	Total Incl O&P
0010	TRASH CHUTES AND ACCESSORIES									
2900	Package chutes, spiral type, minimum	2 Shee	4.50	3.556	Floor	2,575	233		2,808	3,225
3000	Maximum	"	1.50	10.667	"	6,750	700		7,450	8,525
9000	Minimum labor/equipment charge	1 Shee	1	8	Job		525		525	815

14 92 Pneumatic Tube Systems

14 92 10 – Conventional, Automatic and Computer Controlled Pneumatic Tube Systems

14 92 10.10 Pneumatic Tube Systems

14 92 10.10 Pneumatic Tube Systems	Crew	Daily Output	Labor-Hours	Unit	Material	2021 Bare Costs Labor	2021 Bare Costs Equipment	Total	Total Incl O&P
0010	PNEUMATIC TUBE SYSTEMS									
0020	100' long, single tube, 2 stations, stock									
0100	3" diameter	2 Stpi	.12	133	Total	4,175	9,125		13,300	18,700
0300	4" diameter	"	.09	178	"	5,125	12,200		17,325	24,400
0400	Twin tube, two stations or more, conventional system									
0700	3" round	2 Stpi	46	.348	L.F.	39	24		63	79.50
1050	Add for blower		2	8	System	5,025	545		5,570	6,375
1110	Plus for each round station, add		7.50	2.133	Ea.	1,350	146		1,496	1,700
1200	Alternate pricing method: base cost, economy model		.75	21.333	Total	6,000	1,450		7,450	8,850
1300	Custom model		.25	64	"	12,000	4,375		16,375	20,000
1500	Plus total system length, add, for economy model		93.40	.171	L.F.	7.95	11.70		19.65	27
1600	For custom model	↓	37.60	.426	"	25	29		54	72.50

For customer support on your Commercial Renovation Costs with RSMeans Data, call 800.448.8182.

413

Division Notes

		CREW	DAILY OUTPUT	LABOR-HOURS	UNIT	BARE COSTS				TOTAL INCL O&P
						MAT.	LABOR	EQUIP.	TOTAL	

Estimating Tips

Pipe for fire protection and all uses is located in Subdivisions 21 11 13 and 22 11 13.

The labor adjustment factors listed in Subdivision 22 01 02.20 also apply to Division 21.

Many, but not all, areas in the U.S. require backflow protection in the fire system. Insurance underwriters may have specific requirements for the type of materials to be installed or design requirements based on the hazard to be protected. Local jurisdictions may have requirements not covered by code. It is advisable to be aware of any special conditions.

For your reference, the following is a list of the most applicable Fire Codes and Standards, which may be purchased from the NFPA, 1 Batterymarch Park, Quincy, MA 02169-7471.

- NFPA 1: Uniform Fire Code
- NFPA 10: Portable Fire Extinguishers
- NFPA 11: Low-, Medium-, and High-Expansion Foam
- NFPA 12: Carbon Dioxide Extinguishing Systems (Also companion 12A)
- NFPA 13: Installation of Sprinkler Systems (Also companion 13D, 13E, and 13R)
- NFPA 14: Installation of Standpipe and Hose Systems
- NFPA 15: Water Spray Fixed Systems for Fire Protection
- NFPA 16: Installation of Foam-Water Sprinkler and Foam-Water Spray Systems
- NFPA 17: Dry Chemical Extinguishing Systems (Also companion 17A)
- NFPA 18: Wetting Agents
- NFPA 20: Installation of Stationary Pumps for Fire Protection
- NFPA 22: Water Tanks for Private Fire Protection
- NFPA 24: Installation of Private Fire Service Mains and their Appurtenances
- NFPA 25: Inspection, Testing and Maintenance of Water-Based Fire Protection

Reference Numbers

Reference numbers are shown at the beginning of some major classifications. These numbers refer to related items in the Reference Section. The reference information may be an estimating procedure, an alternate pricing method, or technical information.

Same Data. Simplified.

Enjoy the convenience and efficiency of accessing your costs anywhere:

- **Skip the multiplier** by setting your location
- **Quickly search,** edit, favorite and share costs
- **Stay on top of price changes** with automatic updates

Discover more at rsmeans.com/online

21 05 Common Work Results for Fire Suppression

21 05 23 – General-Duty Valves for Water-Based Fire-Suppression Piping

21 05 23.50 General-Duty Valves	Crew	Daily Output	Labor-Hours	Unit	Material	2021 Bare Costs Labor	Equipment	Total	Total Incl O&P
0010 **GENERAL-DUTY VALVES,** for water-based fire suppression									
6500 Check, swing, C.I. body, brass fittings, auto. ball drip									
6520 4" size	Q-12	3	5.333	Ea.	435	320		755	975
9990 Minimum labor/equipment charge	1 Spri	3	2.667	Job		177		177	274

21 11 Facility Fire-Suppression Water-Service Piping

21 11 16 – Facility Fire Hydrants

21 11 16.50 Fire Hydrants for Buildings

	Crew	Daily Output	Labor-Hours	Unit	Material	Labor	Equipment	Total	Total Incl O&P
0010 **FIRE HYDRANTS FOR BUILDINGS**									
3750 Hydrants, wall, w/caps, single, flush, polished brass									
3800 2-1/2" x 2-1/2"	Q-12	5	3.200	Ea.	233	192		425	555
4350 Double, projecting, polished brass									
4400 2-1/2" x 2-1/2" x 4"	Q-12	5	3.200	Ea.	248	192		440	570

21 11 19 – Fire-Department Connections

21 11 19.50 Connections for the Fire-Department

	Crew	Daily Output	Labor-Hours	Unit	Material	Labor	Equipment	Total	Total Incl O&P
0010 **CONNECTIONS FOR THE FIRE-DEPARTMENT**									
7140 Standpipe connections, wall, w/plugs & chains									
7280 Double, flush, polished brass									
7300 2-1/2" x 2-1/2" x 4"	Q-12	5	3.200	Ea.	585	192		777	940
9990 Minimum labor/equipment charge	1 Spri	4	2	Job		133		133	205

21 12 Fire-Suppression Standpipes

21 12 13 – Fire-Suppression Hoses and Nozzles

21 12 13.50 Fire Hoses and Nozzles

	Crew	Daily Output	Labor-Hours	Unit	Material	Labor	Equipment	Total	Total Incl O&P
0010 **FIRE HOSES AND NOZZLES**									
0200 Adapters, rough brass, straight hose threads									
0220 One piece, female to male, rocker lugs									
0240 1" x 1"				Ea.	47			47	52
0320 2" x 2"					54.50			54.50	60
0380 2-1/2" x 2-1/2"					25			25	27.50
2200 Hose, less couplings									
2260 Synthetic jacket, lined, 300 lb. test, 1-1/2" diameter	Q-12	2600	.006	L.F.	3.31	.37		3.68	4.21
2270 2" diameter		2200	.007		3.61	.44		4.05	4.64
2280 2-1/2" diameter		2200	.007		5.80	.44		6.24	7.05
2290 3" diameter	Q-12	2200	.007	L.F.	7.10	.44		7.54	8.45
2360 High strength, 500 lb. test, 1-1/2" diameter		2600	.006		3.62	.37		3.99	4.55
2380 2-1/2" diameter		2200	.007		6.10	.44		6.54	7.35
5600 Nozzles, brass									
5620 Adjustable fog, 3/4" booster line				Ea.	154			154	170
5640 1-1/2" leader line					120			120	132
5660 2-1/2" direct connection					280			280	310
5680 2-1/2" playpipe nozzle					261			261	287
5780 For chrome plated, add					8%				
9900 Minimum labor/equipment charge	1 Plum	2	4	Job		271		271	420

21 12 Fire-Suppression Standpipes

21 12 19 – Fire-Suppression Hose Racks

21 12 19.50 Fire Hose Racks	Crew	Daily Output	Labor-Hours	Unit	Material	2021 Bare Costs Labor	Equipment	Total	Total Incl O&P
0010 **FIRE HOSE RACKS**									
2600 Hose rack, swinging, for 1-1/2" diameter hose,									
2620 Enameled steel, 50' and 75' lengths of hose	Q-12	20	.800	Ea.	132	48		180	219

21 12 23 – Fire-Suppression Hose Valves

21 12 23.70 Fire Hose Valves

	Crew	Daily Output	Labor-Hours	Unit	Material	Labor	Equipment	Total	Total Incl O&P
0010 **FIRE HOSE VALVES**									
3000 Gate, hose, wheel handle, N.R.S., rough brass, 1-1/2"	1 Spri	12	.667	Ea.	144	44.50		188.50	227
3040 2-1/2", 300 lb.	"	7	1.143		184	76		260	320
3080 For polished brass, add					40%				
3090 For polished chrome, add				↓	50%				

21 13 Fire-Suppression Sprinkler Systems

21 13 13 – Wet-Pipe Sprinkler Systems

21 13 13.50 Wet-Pipe Sprinkler System Components

	Crew	Daily Output	Labor-Hours	Unit	Material	Labor	Equipment	Total	Total Incl O&P
0010 **WET-PIPE SPRINKLER SYSTEM COMPONENTS**									
1220 Water motor gong	1 Spri	4	2	Ea.	325	133		458	560
1800 Firecycle system, controls, includes panel									
1900 Flexible sprinkler head connectors									
1910 Braided stainless steel hose with mounting bracket									
1920 1/2" and 3/4" outlet size									
1940 40" length	1 Spri	30	.267	Ea.	62	17.75		79.75	95.50
1960 60" length	"	22	.364	"	72	24		96	117
1982 May replace hard-pipe armovers									
1984 for wet and pre-action systems.									
2600 Sprinkler heads, not including supply piping									
3700 Standard spray, pendent or upright, brass, 135°F to 286°F									
3740 1/2" NPT, K5.6	1 Spri	16	.500	Ea.	14.90	33.50		48.40	68
3790 For riser & feeder piping, see Section 22 11 13.00									

21 13 16 – Dry-Pipe Sprinkler Systems

21 13 16.50 Dry-Pipe Sprinkler System Components

	Crew	Daily Output	Labor-Hours	Unit	Material	Labor	Equipment	Total	Total Incl O&P
0010 **DRY-PIPE SPRINKLER SYSTEM COMPONENTS**									
0600 Accelerator	1 Spri	8	1	Ea.	955	66.50		1,021.50	1,150
0800 Air compressor for dry pipe system, automatic, complete									
0820 30 gal. system capacity, 3/4 HP	1 Spri	1.30	6.154	Ea.	1,400	410		1,810	2,175
2600 Sprinkler heads, not including supply piping									
2640 Dry, pendent, 1/2" orifice, 3/4" or 1" NPT									
2660 3" to 6" length	1 Spri	14	.571	Ea.	197	38		235	275
2700 15-1/4" to 18" length	"	14	.571	"	260	38		298	345
6330 Valves and components									
8200 Dry pipe valve, incl. trim and gauges, 3" size	Q-12	2	8	Ea.	3,200	480		3,680	4,275
8220 4" size	"	1	16	"	3,425	960		4,385	5,250

21 13 39 – Foam-Water Systems

21 13 39.50 Foam-Water System Components

	Crew	Daily Output	Labor-Hours	Unit	Material	Labor	Equipment	Total	Total Incl O&P
0010 **FOAM-WATER SYSTEM COMPONENTS**									
2600 Sprinkler heads, not including supply piping									
3600 Foam-water, pendent or upright, 1/2" NPT	1 Spri	12	.667	Ea.	248	44.50		292.50	340

Division Notes

		CREW	DAILY OUTPUT	LABOR-HOURS	UNIT	BARE COSTS				TOTAL INCL O&P
						MAT.	LABOR	EQUIP.	TOTAL	

Estimating Tips
22 10 00 Plumbing Piping and Pumps

This subdivision is primarily basic pipe and related materials. The pipe may be used by any of the mechanical disciplines, i.e., plumbing, fire protection, heating, and air conditioning.

Note: CPVC plastic piping approved for fire protection is located in 21 11 13.

- The labor adjustment factors listed in Subdivision 22 01 02.20 apply throughout Divisions 21, 22, and 23. CAUTION: the correct percentage may vary for the same items. For example, the percentage add for the basic pipe installation should be based on the maximum height that the installer must install for that particular section. If the pipe is to be located 14' above the floor but it is suspended on threaded rod from beams, the bottom flange of which is 18' high (4' rods), then the height is actually 18' and the add is 20%. The pipe cover, however, does not have to go above the 14' and so the add should be 10%.

- Most pipe is priced first as straight pipe with a joint (coupling, weld, etc.) every 10' and a hanger usually every 10'. There are exceptions with hanger spacing such as for cast iron pipe (5')

and plastic pipe (3 per 10'). Following each type of pipe there are several lines listing sizes and the amount to be subtracted to delete couplings and hangers. This is for pipe that is to be buried or supported together on trapeze hangers. The reason that the couplings are deleted is that these runs are usually long, and frequently longer lengths of pipe are used. By deleting the couplings, the estimator is expected to look up and add back the correct reduced number of couplings.

- When preparing an estimate, it may be necessary to approximate the fittings. Fittings usually run between 25% and 50% of the cost of the pipe. The lower percentage is for simpler runs, and the higher number is for complex areas, such as mechanical rooms.

- For historic restoration projects, the systems must be as invisible as possible, and pathways must be sought for pipes, conduit, and ductwork. While installations in accessible spaces (such as basements and attics) are relatively straightforward to estimate, labor costs may be more difficult to determine when delivery systems must be concealed.

22 40 00 Plumbing Fixtures

- Plumbing fixture costs usually require two lines: the fixture itself and its "rough-in, supply, and waste."

- In the Assemblies Section (Plumbing D2010) for the desired fixture, the System Components Group at the center of the page shows the fixture on the first line. The rest of the list (fittings, pipe, tubing, etc.) will total up to what we refer to in the Unit Price section as "Rough-in, supply, waste, and vent." Note that for most fixtures we allow a nominal 5' of tubing to reach from the fixture to a main or riser.

- Remember that gas- and oil-fired units need venting.

Reference Numbers

Reference numbers are shown at the beginning of some major classifications. These numbers refer to related items in the Reference Section. The reference information may be an estimating procedure, an alternate pricing method, or technical information.

Note: Not all subdivisions listed here necessarily appear. ■

Same Data. Simplified.

Enjoy the convenience and efficiency of accessing your costs anywhere:

- **Skip the multiplier** by setting your location
- **Quickly search,** edit, favorite and share costs
- **Stay on top of price changes** with automatic updates

Discover more at rsmeans.com/online

---

22 01 02 – Labor Adjustments

22 01 02.20 Labor Adjustment Factors	Crew	Daily Output	Labor-Hours	Unit	Material	2021 Bare Costs Labor	Equipment	Total	Total Incl O&P
0010 **LABOR ADJUSTMENT FACTORS** (For Div. 21, 22 and 23) R220102-20									
0100 Labor factors: The below are reasonable suggestions, but									
0110 each project must be evaluated for its own peculiarities, and									
0120 the adjustments be increased or decreased depending on the									
0130 severity of the special conditions.									
1000 Add to labor for elevated installation (Above floor level)									
1080 10' to 14.5' high						10%			
1100 15' to 19.5' high						20%			
1120 20' to 24.5' high						25%			
1140 25' to 29.5' high						35%			
1160 30' to 34.5' high						40%			
1180 35' to 39.5' high						50%			
1200 40' and higher						55%			
2000 Add to labor for crawl space									
2100 3' high						40%			
2140 4' high						30%			
3000 Add to labor for multi-story building									
3010 For new construction (No elevator available)									
3100 Add for floors 3 thru 10						5%			
3110 Add for floors 11 thru 15						10%			
3120 Add for floors 16 thru 20						15%			
3130 Add for floors 21 thru 30						20%			
3140 Add for floors 31 and up						30%			
3170 For existing structure (Elevator available)									
3180 Add for work on floor 3 and above						2%			
4000 Add to labor for working in existing occupied buildings									
4100 Hospital						35%			
4140 Office building						25%			
4180 School						20%			
4220 Factory or warehouse						15%			
4260 Multi dwelling						15%			
5000 Add to labor, miscellaneous									
5100 Cramped shaft						35%			
5140 Congested area						15%			
5180 Excessive heat or cold						30%			
9000 Labor factors: The above are reasonable suggestions, but									
9010 each project should be evaluated for its own peculiarities.									
9100 Other factors to be considered are:									
9140 Movement of material and equipment through finished areas									
9180 Equipment room									
9220 Attic space									
9260 No service road									
9300 Poor unloading/storage area									
9340 Congested site area/heavy traffic									

22 05 Common Work Results for Plumbing

22 05 05 – Selective Demolition for Plumbing

22 05 05.10 Plumbing Demolition

		Crew	Daily Output	Labor-Hours	Unit	Material	2021 Bare Costs Labor	2021 Bare Costs Equipment	Total	Total Incl O&P
0010	**PLUMBING DEMOLITION** R220105-10									
1020	Fixtures, including 10' piping									
1100	Bathtubs, cast iron R024119-10	1 Plum	4	2	Ea.		135		135	209
1120	Fiberglass		6	1.333			90.50		90.50	139
1140	Steel		5	1.600			108		108	167
1200	Lavatory, wall hung		10	.800			54		54	83.50
1220	Counter top		8	1			67.50		67.50	104
1300	Sink, single compartment		8	1			67.50		67.50	104
1320	Double compartment		7	1.143			77.50		77.50	119
1400	Water closet, floor mounted		8	1			67.50		67.50	104
1420	Wall mounted		7	1.143			77.50		77.50	119
1500	Urinal, floor mounted		4	2			135		135	209
1520	Wall mounted		7	1.143			77.50		77.50	119
1600	Water fountains, free standing		8	1			67.50		67.50	104
1620	Wall or deck mounted	1 Plum	6	1.333	Ea.		90.50		90.50	139
2000	Piping, metal, up thru 1-1/2" diameter		200	.040	L.F.		2.71		2.71	4.18
2050	2" thru 3-1/2" diameter		150	.053			3.61		3.61	5.55
2100	4" thru 6" diameter	2 Plum	100	.160			10.85		10.85	16.70
2150	8" thru 14" diameter	"	60	.267			18.05		18.05	28
2153	16" thru 20" diameter	Q-18	70	.343			22	1.54	23.54	35
2155	24" thru 26" diameter		55	.436			28	1.95	29.95	45
2156	30" thru 36" diameter		40	.600			38.50	2.69	41.19	62
2160	Plastic pipe with fittings, up thru 1-1/2" diameter	1 Plum	250	.032			2.17		2.17	3.34
2162	2" thru 3" diameter	"	200	.040			2.71		2.71	4.18
2164	4" thru 6" diameter	Q-1	200	.080			4.87		4.87	7.50
2166	8" thru 14" diameter		150	.107			6.50		6.50	10
2168	16" diameter		100	.160			9.75		9.75	15.05
2250	Water heater, 40 gal.	1 Plum	6	1.333	Ea.		90.50		90.50	139
6000	Remove and reset fixtures, easy access		6	1.333			90.50		90.50	139
6100	Difficult access		4	2			135		135	209
9000	Minimum labor/equipment charge		2	4	Job		271		271	420

22 05 23 – General-Duty Valves for Plumbing Piping

22 05 23.20 Valves, Bronze

		Crew	Daily Output	Labor-Hours	Unit	Material	2021 Bare Costs Labor	2021 Bare Costs Equipment	Total	Total Incl O&P
0010	**VALVES, BRONZE**									
1020	Angle, 150 lb., rising stem, threaded									
1070	3/4"	1 Plum	20	.400	Ea.	235	27		262	300
1080	1"		19	.421		325	28.50		353.50	405
1100	1-1/2"		13	.615		550	41.50		591.50	670
1102	Soldered same price as threaded									
1110	2"	1 Plum	11	.727	Ea.	865	49		914	1,025
1300	Ball									
1398	Threaded, 150 psi									
1460	3/4"	1 Plum	20	.400	Ea.	30	27		57	75
1522	Solder the same price as threaded									
1750	Check, swing, class 150, regrinding disc, threaded									
1850	1/2"	1 Plum	24	.333	Ea.	89.50	22.50		112	134
1860	3/4"		20	.400		124	27		151	178
1870	1"		19	.421		176	28.50		204.50	238
1890	1-1/2"		13	.615		294	41.50		335.50	390
1900	2"		11	.727		425	49		474	540
2850	Gate, N.R.S., soldered, 125 psi									
2920	1/2"	1 Plum	24	.333	Ea.	76.50	22.50		99	119

For customer support on your Commercial Renovation Costs with RSMeans Data, call 800.448.8182.

421

22 05 23 – General-Duty Valves for Plumbing Piping

22 05 23.20 Valves, Bronze		Crew	Daily Output	Labor-Hours	Unit	Material	2021 Bare Costs Labor	Equipment	Total	Total Incl O&P
2940	3/4"	1 Plum	20	.400	Ea.	84.50	27		111.50	135
2950	1"		19	.421		122	28.50		150.50	178
2970	1-1/2"		13	.615		207	41.50		248.50	293
2980	2"	1 Plum	11	.727	Ea.	293	49		342	395
4850	Globe, class 150, rising stem, threaded									
4950	1/2"	1 Plum	24	.333	Ea.	129	22.50		151.50	177
4960	3/4"	"	20	.400	"	172	27		199	232
5600	Relief, pressure & temperature, self-closing, ASME, threaded									
5640	3/4"	1 Plum	28	.286	Ea.	267	19.35		286.35	325
5650	1"		24	.333		425	22.50		447.50	505
5660	1-1/4"		20	.400		855	27		882	985
5670	1-1/2"		18	.444		1,650	30		1,680	1,850
5680	2"		16	.500		1,800	34		1,834	2,025
6400	Pressure, water, ASME, threaded									
6440	3/4"	1 Plum	28	.286	Ea.	58	19.35		77.35	94
6450	1"		24	.333		360	22.50		382.50	430
6470	1-1/2"		18	.444		780	30		810	905
6900	Reducing, water pressure									
6920	300 psi to 25-75 psi, threaded or sweat									
6940	1/2"	1 Plum	24	.333	Ea.	560	22.50		582.50	650
6950	3/4"		20	.400		585	27		612	685
6960	1"		19	.421		905	28.50		933.50	1,050
8350	Tempering, water, sweat connections									
8400	1/2"	1 Plum	24	.333	Ea.	119	22.50		141.50	166
8440	3/4"	"	20	.400	"	175	27		202	235
8650	Threaded connections									
8700	1/2"	1 Plum	24	.333	Ea.	167	22.50		189.50	219
8740	3/4"	"	20	.400	"	1,050	27		1,077	1,200
8800	Water heater water & gas safety shut off									
8810	Protection against a leaking water heater									
8814	Shut off valve	1 Plum	16	.500	Ea.	196	34		230	267
8818	Water heater dam		32	.250		52.50	16.95		69.45	84
8822	Gas control wiring harness		32	.250		25.50	16.95		42.45	54
8830	Whole house flood safety shut off									
8834	Connections									
8838	3/4" NPT	1 Plum	12	.667	Ea.	1,075	45		1,120	1,250
8842	1" NPT		11	.727		1,100	49		1,149	1,300
8846	1-1/4" NPT		10	.800		1,150	54		1,204	1,325
9000	Minimum labor/equipment charge		4	2	Job		135		135	209

22 05 23.60 Valves, Plastic

		Crew	Daily Output	Labor-Hours	Unit	Material	Labor	Equipment	Total	Total Incl O&P
0010	**VALVES, PLASTIC**									
1150	Ball, PVC, socket or threaded, true union									
1230	1/2"	1 Plum	26	.308	Ea.	38	21		59	74
1240	3/4"	"	25	.320	"	45	21.50		66.50	83
1250	1"	1 Plum	23	.348	Ea.	54	23.50		77.50	95.50
1260	1-1/4"		21	.381		85	26		111	134
1270	1-1/2"		20	.400		85	27		112	136
1280	2"		17	.471		134	32		166	197
1650	CPVC, socket or threaded, single union									
1700	1/2"	1 Plum	26	.308	Ea.	59	21		80	97
1720	3/4"		25	.320		74.50	21.50		96	116
1730	1"		23	.348		89.50	23.50		113	135

22 05 Common Work Results for Plumbing

22 05 23 – General-Duty Valves for Plumbing Piping

	22 05 23.60 Valves, Plastic	Crew	Daily Output	Labor-Hours	Unit	Material	2021 Bare Costs Labor	Equipment	Total	Total Incl O&P
1750	1-1/4"	1 Plum	21	.381	Ea.	142	26		168	196
1760	1-1/2"		20	.400		142	27		169	198
1770	2"	↓	17	.471	↓	196	32		228	265
3150	Ball check, PVC, socket or threaded									
3290	2"	1 Plum	17	.471	Ea.	149	32		181	212
9000	Minimum labor/equipment charge	"	3.50	2.286	Job		155		155	239

22 05 76 – Facility Drainage Piping Cleanouts

22 05 76.10 Cleanouts

		Crew	Daily Output	Labor-Hours	Unit	Material	Labor	Equipment	Total	Total Incl O&P
0010	**CLEANOUTS**									
0060	Floor type									
0080	Round or square, scoriated nickel bronze top									
0100	2" pipe size	1 Plum	10	.800	Ea.	440	54		494	565
0120	3" pipe size		8	1		460	67.50		527.50	610
0140	4" pipe size	↓	6	1.333	↓	615	90.50		705.50	820
0980	Round top, recessed for terrazzo									
1000	2" pipe size	1 Plum	9	.889	Ea.	630	60		690	790
1080	3" pipe size		6	1.333		675	90.50		765.50	880
1100	4" pipe size	↓	4	2		805	135		940	1,100
1120	5" pipe size	Q-1	6	2.667	↓	1,275	162		1,437	1,650
9000	Minimum labor/equipment charge	1 Plum	3	2.667	Job		181		181	278

22 05 76.20 Cleanout Tees

		Crew	Daily Output	Labor-Hours	Unit	Material	Labor	Equipment	Total	Total Incl O&P
0010	**CLEANOUT TEES**									
0100	Cast iron, B&S, with countersunk plug									
0220	3" pipe size	1 Plum	3.60	2.222	Ea.	205	150		355	455
0240	4" pipe size	"	3.30	2.424	"	310	164		474	595
0500	For round smooth access cover, same price									
4000	Plastic, tees and adapters. Add plugs									
4010	ABS, DWV									
4020	Cleanout tee, 1-1/2" pipe size	1 Plum	15	.533	Ea.	17.30	36		53.30	74.50
9000	Minimum labor/equipment charge	"	2.75	2.909	Job		197		197	305

22 07 Plumbing Insulation

22 07 16 – Plumbing Equipment Insulation

22 07 16.10 Insulation for Plumbing Equipment

			Crew	Daily Output	Labor-Hours	Unit	Material	Labor	Equipment	Total	Total Incl O&P
0010	**INSULATION FOR PLUMBING EQUIPMENT**										
2900	Domestic water heater wrap kit										
2920	1-1/2" with vinyl jacket, 20 to 60 gal.	G	1 Plum	8	1	Ea.	19.70	67.50		87.20	126
2925	50 to 80 gal.	G	"	8	1	"	30.50	67.50		98	138

22 07 19 – Plumbing Piping Insulation

22 07 19.10 Piping Insulation

			Crew	Daily Output	Labor-Hours	Unit	Material	Labor	Equipment	Total	Total Incl O&P
0010	**PIPING INSULATION**										
0110	Insulation req'd. is based on the surface size/area to be covered										
0230	Insulated protectors (ADA)										
0235	For exposed piping under sinks or lavatories	♿									
0240	Vinyl coated foam, velcro tabs										
0245	P Trap, 1-1/4" or 1-1/2"		1 Plum	32	.250	Ea.	17.10	16.95		34.05	45
0260	Valve and supply cover										
0265	1/2", 3/8", and 7/16" pipe size		1 Plum	32	.250	Ea.	16.80	16.95		33.75	44.50
0280	Tailpiece offset (wheelchair)	♿									
0285	1-1/4" pipe size		1 Plum	32	.250	Ea.	13.95	16.95		30.90	41.50

For customer support on your Commercial Renovation Costs with RSMeans Data, call 800.448.8182.

423

22 07 Plumbing Insulation

22 07 19 – Plumbing Piping Insulation

22 07 19.10 Piping Insulation		Crew	Daily Output	Labor-Hours	Unit	Material	2021 Bare Costs Labor	Equipment	Total	Total Incl O&P
0600	Pipe covering (price copper tube one size less than IPS)									
6120	2" wall, 1/2" iron pipe size	G	Q-14	145	.110	L.F.	7.15	6.05		13.20	17.45
6210	4" iron pipe size	G	"	125	.128	"	13.05	7		20.05	25.50
6600	Fiberglass, with all service jacket									
6840	1" wall, 1/2" iron pipe size	G	Q-14	240	.067	L.F.	.96	3.66		4.62	6.85
6860	3/4" iron pipe size	G		230	.070		1.03	3.82		4.85	7.20
6870	1" iron pipe size	G		220	.073		1.14	3.99		5.13	7.55
6880	1-1/4" iron pipe size	G		210	.076		1.22	4.18		5.40	7.95
6890	1-1/2" iron pipe size	G		210	.076		1.32	4.18		5.50	8.05
6900	2" iron pipe size	G		200	.080		2.19	4.39		6.58	9.35
7080	1-1/2" wall, 1/2" iron pipe size	G		230	.070		2.55	3.82		6.37	8.85
7140	2" iron pipe size	G		190	.084		3.48	4.62		8.10	11.15
7160	3" iron pipe size	G		170	.094		3.87	5.15		9.02	12.40
7879	Rubber tubing, flexible closed cell foam									
8100	1/2" wall, 1/4" iron pipe size	G	1 Asbe	90	.089	L.F.	1.14	5.40		6.54	9.80
8120	3/8" iron pipe size	G		90	.089		1.13	5.40		6.53	9.80
8130	1/2" iron pipe size	G		89	.090		1.25	5.50		6.75	10.05
8140	3/4" iron pipe size	G		89	.090		1.29	5.50		6.79	10.05
8150	1" iron pipe size	G		88	.091		1.56	5.55		7.11	10.45
8170	1-1/2" iron pipe size	G		87	.092		1.97	5.60		7.57	11
8180	2" iron pipe size	G		86	.093		2.52	5.65		8.17	11.70
8200	3" iron pipe size	G		85	.094		3.51	5.75		9.26	12.90
8220	4" iron pipe size	G		80	.100		5.25	6.10		11.35	15.45
8300	3/4" wall, 1/4" iron pipe size	G		90	.089		1.60	5.40		7	10.30
8330	1/2" iron pipe size	G		89	.090		1.90	5.50		7.40	10.75
8340	3/4" iron pipe size	G		89	.090		2.28	5.50		7.78	11.15
8350	1" iron pipe size	G	1 Asbe	88	.091	L.F.	2.77	5.55		8.32	11.80
8360	1-1/4" iron pipe size	G		87	.092		3.17	5.60		8.77	12.35
8370	1-1/2" iron pipe size	G		87	.092		4.21	5.60		9.81	13.50
8380	2" iron pipe size	G		86	.093		5.10	5.65		10.75	14.60
8444	1" wall, 1/2" iron pipe size	G		86	.093		3.40	5.65		9.05	12.70
8445	3/4" iron pipe size	G		84	.095		4.19	5.80		9.99	13.75
8446	1" iron pipe size	G		84	.095		5.10	5.80		10.90	14.75
8447	1-1/4" iron pipe size	G		82	.098		5.90	5.95		11.85	15.90
8448	1-1/2" iron pipe size	G		82	.098		6.55	5.95		12.50	16.60
8449	2" iron pipe size	G		80	.100		7.85	6.10		13.95	18.30
8450	2-1/2" iron pipe size	G		80	.100		10.65	6.10		16.75	21.50
8456	Rubber insulation tape, 1/8" x 2" x 30'	G				Ea.	25			25	27.50
9600	Minimum labor/equipment charge		1 Plum	4	2	Job		135		135	209

22 11 Facility Water Distribution

22 11 13 – Facility Water Distribution Piping

22 11 13.14 Pipe, Brass

22 11 13.14		Crew	Daily Output	Labor-Hours	Unit	Material	2021 Bare Costs Labor	Equipment	Total	Total Incl O&P
0010	**PIPE, BRASS**, Plain end									
0900	Field threaded, coupling & clevis hanger assembly 10' OC									
0920	Regular weight									
1120	1/2" diameter		1 Plum	48	.167	L.F.	8.30	11.30		19.60	26.50
1140	3/4" diameter			46	.174		10.85	11.75		22.60	30
1160	1" diameter			43	.186		5.30	12.60		17.90	25.50
1180	1-1/4" diameter		Q-1	72	.222		22	13.55		35.55	45
1200	1-1/2" diameter			65	.246		26.50	15		41.50	52

22 11 Facility Water Distribution

22 11 13 – Facility Water Distribution Piping

22 11 13.14 Pipe, Brass	Crew	Daily Output	Labor-Hours	Unit	Material	2021 Bare Costs Labor	Equipment	Total	Total Incl O&P
1220	2" diameter	Q-1	53	.302	L.F.	17.60	18.40		36	48
9000	Minimum labor/equipment charge	1 Plum	4	2	Job		135		135	209

22 11 13.16 Pipe Fittings, Brass

		Crew	Daily Output	Labor-Hours	Unit	Material	2021 Bare Costs Labor	Equipment	Total	Total Incl O&P
0010	**PIPE FITTINGS, BRASS**, Rough bronze, threaded, lead free.									
1000	Standard wt., 90° elbow									
1100	1/2"	1 Plum	12	.667	Ea.	23.50	45		68.50	95.50
1120	3/4"		11	.727		31.50	49		80.50	111
1140	1"	↓	10	.800		51	54		105	140
1160	1-1/4"	Q-1	17	.941		82.50	57.50		140	180
1180	1-1/2"	"	16	1		104	61		165	209
1500	45° elbow, 1/8"	1 Plum	13	.615		29	41.50		70.50	96.50
1580	1/2"		12	.667		29	45		74	102
1600	3/4"		11	.727		41	49		90	121
1620	1"	↓	10	.800		70	54		124	161
1640	1-1/4"	Q-1	17	.941		111	57.50		168.50	211
1660	1-1/2"	"	16	1		140	61		201	248
2000	Tee, 1/8"	1 Plum	9	.889		27.50	60		87.50	124
2080	1/2"		8	1		28	67.50		95.50	135
2100	3/4"	↓	7	1.143	↓	39	77.50		116.50	162
2120	1"	1 Plum	6	1.333	Ea.	75.50	90.50		166	223
2140	1-1/4"	Q-1	10	1.600		122	97.50		219.50	284
2160	1-1/2"	"	9	1.778		138	108		246	320
2500	Coupling, 1/8"	1 Plum	26	.308		19.65	21		40.65	53.50
2580	1/2"		15	.533		19.65	36		55.65	77
2600	3/4"		14	.571		27.50	38.50		66	90
2620	1"	↓	13	.615		47	41.50		88.50	117
2640	1-1/4"	Q-1	22	.727		78.50	44.50		123	155
2660	1-1/2"		20	.800		102	48.50		150.50	187
2680	2"	↓	18	.889	↓	169	54		223	270
9000	Minimum labor/equipment charge	1 Plum	4	2	Job		135		135	209

22 11 13.23 Pipe/Tube, Copper

		Crew	Daily Output	Labor-Hours	Unit	Material	2021 Bare Costs Labor	Equipment	Total	Total Incl O&P
0010	**PIPE/TUBE, COPPER**, Solder joints									
1000	Type K tubing, couplings & clevis hanger assemblies 10' OC									
1100	1/4" diameter	1 Plum	84	.095	L.F.	5	6.45		11.45	15.45
1200	1" diameter		66	.121		12.20	8.20		20.40	26
1260	2" diameter	↓	40	.200	↓	29.50	13.55		43.05	53.50
2000	Type L tubing, couplings & clevis hanger assemblies 10' OC									
2100	1/4" diameter	1 Plum	88	.091	L.F.	1.83	6.15		7.98	11.50
2120	3/8" diameter		84	.095		2.13	6.45		8.58	12.30
2140	1/2" diameter		81	.099		2.51	6.70		9.21	13.05
2160	5/8" diameter		79	.101		6.05	6.85		12.90	17.20
2180	3/4" diameter		76	.105		3.04	7.15		10.19	14.35
2200	1" diameter		68	.118		3.55	7.95		11.50	16.20
2220	1-1/4" diameter		58	.138		10.70	9.35		20.05	26
2240	1-1/2" diameter		52	.154		6.35	10.40		16.75	23
2260	2" diameter	↓	42	.190		11.50	12.90		24.40	32.50
2280	2-1/2" diameter	Q-1	62	.258		15.85	15.70		31.55	42
2300	3" diameter		56	.286		46	17.40		63.40	78
2320	3-1/2" diameter		43	.372		64	22.50		86.50	106
2340	4" diameter		39	.410		41	25		66	83.50
2360	5" diameter	↓	34	.471		178	28.50		206.50	240
2380	6" diameter	Q-2	40	.600	↓	114	38		152	184

For customer support on your Commercial Renovation Costs with RSMeans Data, call 800.448.8182.

425

22 11 13 – Facility Water Distribution Piping

22 11 13.23 Pipe/Tube, Copper

		Crew	Daily Output	Labor-Hours	Unit	Material	2021 Bare Costs Labor	2021 Bare Costs Equipment	Total	Total Incl O&P
2410	For other than full hard temper, add				L.F.	21%				
2590	For silver solder, add						15%			
3000	Type M tubing, couplings & clevis hanger assemblies 10' OC									
3140	1/2" diameter	1 Plum	84	.095	L.F.	3.75	6.45		10.20	14.10
3180	3/4" diameter		78	.103		5.40	6.95		12.35	16.65
3200	1" diameter		70	.114		7.35	7.75		15.10	20
4000	Type DWV tubing, couplings & clevis hanger assemblies 10' OC									
4100	1-1/4" diameter	1 Plum	60	.133	L.F.	10.90	9.05		19.95	26
4120	1-1/2" diameter		54	.148		6.65	10.05		16.70	23
4140	2" diameter		44	.182		9.25	12.30		21.55	29
4160	3" diameter	Q-1	58	.276		18.95	16.80		35.75	47
4180	4" diameter		40	.400		68	24.50		92.50	112
4200	5" diameter		36	.444		136	27		163	192
4220	6" diameter	Q-2	42	.571		203	36		239	279
9000	Minimum labor/equipment charge	1 Plum	4	2	Job		135		135	209

22 11 13.25 Pipe/Tube Fittings, Copper

		Crew	Daily Output	Labor-Hours	Unit	Material	2021 Bare Costs Labor	2021 Bare Costs Equipment	Total	Total Incl O&P
0010	**PIPE/TUBE FITTINGS, COPPER**, Wrought unless otherwise noted									
0020	For silver solder, add						15%			
0040	Solder joints, copper x copper									
0070	90° elbow, 1/4"	1 Plum	22	.364	Ea.	5.05	24.50		29.55	43.50
0100	1/2"		20	.400		1.77	27		28.77	44
0120	3/4"		19	.421		3.89	28.50		32.39	48.50
0130	1"		16	.500		9.55	34		43.55	62.50
0250	45° elbow, 1/4"		22	.364		9	24.50		33.50	48
0270	3/8"		22	.364		7.70	24.50		32.20	46.50
0280	1/2"		20	.400		3.18	27		30.18	45.50
0290	5/8"		19	.421		13.50	28.50		42	59
0300	3/4"		19	.421		5.40	28.50		33.90	50
0310	1"		16	.500		13.55	34		47.55	67
0320	1-1/4"		15	.533		18.20	36		54.20	75.50
0330	1-1/2"		13	.615		22.50	41.50		64	89
0340	2"		11	.727		37.50	49		86.50	117
0350	2-1/2"	Q-1	13	1.231		72.50	75		147.50	196
0360	3"		13	1.231		101	75		176	227
0370	3-1/2"		10	1.600		174	97.50		271.50	340
0380	4"		9	1.778		209	108		317	395
0390	5"		6	2.667		755	162		917	1,075
0400	6"	Q-2	9	2.667		1,200	168		1,368	1,550
0450	Tee, 1/4"	1 Plum	14	.571		10.10	38.50		48.60	70.50
0470	3/8"		14	.571		8.10	38.50		46.60	68.50
0480	1/2"		13	.615		2.96	41.50		44.46	68
0490	5/8"		12	.667		18.50	45		63.50	90
0500	3/4"		12	.667		7.15	45		52.15	77.50
0510	1"		10	.800		11.50	54		65.50	96
0520	1-1/4"		9	.889		29.50	60		89.50	126
0530	1-1/2"		8	1		44.50	67.50		112	153
0540	2"		7	1.143		71	77.50		148.50	197
0550	2-1/2"	Q-1	8	2		131	122		253	330
0560	3"		7	2.286		188	139		327	420
0570	3-1/2"		6	2.667		545	162		707	850
0580	4"		5	3.200		440	195		635	785
0590	5"		4	4		1,350	244		1,594	1,850

22 11 13.25 Pipe/Tube Fittings, Copper		Crew	Daily Output	Labor-Hours	Unit	Material	2021 Bare Costs Labor	Equipment	Total	Total Incl O&P
0600	6"	Q-2	6	4	Ea.	1,850	253		2,103	2,425
0612	Tee, reducing on the outlet, 1/4"	1 Plum	15	.533		27.50	36		63.50	86
0613	3/8"		15	.533		23.50	36		59.50	81.50
0614	1/2"		14	.571		21.50	38.50		60	83.50
0615	5/8"		13	.615		43.50	41.50		85	113
0616	3/4"		12	.667		9.95	45		54.95	80.50
0617	1"		11	.727		47.50	49		96.50	128
0618	1-1/4"		10	.800		46.50	54		100.50	135
0619	1-1/2"		9	.889		49.50	60		109.50	148
0620	2"		8	1		81	67.50		148.50	193
0621	2-1/2"	Q-1	9	1.778		223	108		331	410
0622	3"		8	2		253	122		375	465
0623	4"		6	2.667		455	162		617	750
0624	5"		5	3.200		1,975	195		2,170	2,475
0625	6"	Q-2	7	3.429		3,475	217		3,692	4,125
0626	8"	"	6	4		13,400	253		13,653	15,100
0630	Tee, reducing on the run, 1/4"	1 Plum	15	.533		31	36		67	90
0631	3/8"		15	.533		35.50	36		71.50	94.50
0632	1/2"		14	.571		30	38.50		68.50	92.50
0633	5/8"		13	.615		12.25	41.50		53.75	78
0634	3/4"		12	.667		24	45		69	96
0635	1"		11	.727		38	49		87	118
0636	1-1/4"		10	.800		61.50	54		115.50	151
0637	1-1/2"		9	.889		106	60		166	209
0638	2"		8	1		135	67.50		202.50	253
0639	2-1/2"	Q-1	9	1.778		288	108		396	480
0640	3"		8	2		400	122		522	630
0641	4"		6	2.667		860	162		1,022	1,200
0642	5"		5	3.200		2,400	195		2,595	2,950
0643	6"	Q-2	7	3.429		3,650	217		3,867	4,350
0644	8"	"	6	4		12,600	253		12,853	14,200
0650	Coupling, 1/4"	1 Plum	24	.333		1.78	22.50		24.28	37
0670	3/8"		24	.333		2.31	22.50		24.81	37.50
0680	1/2"		22	.364		1.92	24.50		26.42	40
0690	5/8"		21	.381		5.05	26		31.05	45.50
0700	3/4"		21	.381		3.85	26		29.85	44
0710	1"		18	.444		7.65	30		37.65	55
0715	1-1/4"		17	.471		13.40	32		45.40	64
0716	1-1/2"		15	.533		17.70	36		53.70	75
0718	2"		13	.615		29.50	41.50		71	97
0721	2-1/2"	Q-1	15	1.067		60.50	65		125.50	167
0722	3"		13	1.231		82.50	75		157.50	207
0724	3-1/2"		8	2		155	122		277	360
0726	4"		7	2.286		182	139		321	415
0728	5"		6	2.667		380	162		542	665
0731	6"	Q-2	8	3		630	190		820	980
2000	DWV, solder joints, copper x copper									
2030	90° elbow, 1-1/4"	1 Plum	13	.615	Ea.	27	41.50		68.50	94.50
2050	1-1/2"		12	.667		36	45		81	109
2070	2"		10	.800		57.50	54		111.50	147
2090	3"	Q-1	10	1.600		139	97.50		236.50	305
2100	4"	"	9	1.778		675	108		783	910
2250	Tee, sanitary, 1-1/4"	1 Plum	9	.889		47.50	60		107.50	145

22 11 Facility Water Distribution

22 11 13 – Facility Water Distribution Piping

22 11 13.25 Pipe/Tube Fittings, Copper

		Crew	Daily Output	Labor-Hours	Unit	Material	2021 Bare Costs Labor	2021 Bare Costs Equipment	Total	Total Incl O&P
2270	1-1/2"	1 Plum	8	1	Ea.	59	67.50		126.50	169
2290	2"	↓	7	1.143		81	77.50		158.50	208
2310	3"	Q-1	7	2.286		340	139		479	590
2330	4"	"	6	2.667		750	162		912	1,075
2400	Coupling, 1-1/4"	1 Plum	14	.571		11.30	38.50		49.80	72
2420	1-1/2"		13	.615		13.95	41.50		55.45	80
2440	2"	↓	11	.727		19.40	49		68.40	97.50
2460	3"	Q-1	11	1.455		45	88.50		133.50	187
2480	4"	"	10	1.600	↓	99.50	97.50		197	259
9000	Minimum labor/equipment charge	1 Plum	4	2	Job		135		135	209

22 11 13.44 Pipe, Steel

		Crew	Daily Output	Labor-Hours	Unit	Material	2021 Bare Costs Labor	2021 Bare Costs Equipment	Total	Total Incl O&P
0010	**PIPE, STEEL** R221113-50									
0020	All pipe sizes are to Spec. A-53 unless noted otherwise									
0050	Schedule 40, threaded, with couplings, and clevis hanger									
0060	assemblies sized for covering, 10' OC									
0540	Black, 1/4" diameter	1 Plum	66	.121	L.F.	7.85	8.20		16.05	21.50
0560	1/2" diameter		63	.127		2.69	8.60		11.29	16.20
0570	3/4" diameter		61	.131		3.27	8.90		12.17	17.30
0580	1" diameter	↓	53	.151		3.73	10.20		13.93	19.85
0590	1-1/4" diameter	Q-1	89	.180		4.82	10.95		15.77	22
0600	1-1/2" diameter		80	.200		5.65	12.20		17.85	25
0610	2" diameter		64	.250		7.60	15.25		22.85	32
0620	2-1/2" diameter		50	.320		14.30	19.50		33.80	45.50
0630	3" diameter		43	.372		18.65	22.50		41.15	55.50
0640	3-1/2" diameter		40	.400		28	24.50		52.50	68
0650	4" diameter	↓	36	.444	↓	28	27		55	72.50
1280	All pipe sizes are to Spec. A-53 unless noted otherwise									
1281	Schedule 40, threaded, with couplings and clevis hanger									
1282	assemblies sized for covering, 10' OC									
1290	Galvanized, 1/4" diameter	1 Plum	66	.121	L.F.	10.85	8.20		19.05	24.50
1310	1/2" diameter		63	.127		3.07	8.60		11.67	16.60
1320	3/4" diameter		61	.131		3.76	8.90		12.66	17.85
1330	1" diameter	↓	53	.151		4.45	10.20		14.65	20.50
1350	1-1/2" diameter	Q-1	80	.200		6.80	12.20		19	26.50
1360	2" diameter		64	.250		9.20	15.25		24.45	33.50
1370	2-1/2" diameter		50	.320		16.10	19.50		35.60	47.50
1380	3" diameter		43	.372		21	22.50		43.50	58
1400	4" diameter	↓	36	.444	↓	32.50	27		59.50	78
2000	Welded, sch. 40, on yoke & roll hanger assy's, sized for covering, 10' OC									
2040	Black, 1" diameter	Q-15	93	.172	L.F.	4.27	10.50	1.16	15.93	22
2070	2" diameter		61	.262		7.40	16	1.76	25.16	34.50
2090	3" diameter		43	.372		15.50	22.50	2.50	40.50	55
2110	4" diameter		37	.432		22	26.50	2.91	51.41	67.50
2120	5" diameter	↓	32	.500	↓	24.50	30.50	3.36	58.36	77.50
2130	6" diameter	Q-16	36	.667	↓	26	42	2.99	70.99	97
9990	Minimum labor/equipment charge	1 Plum	3	2.667	Job		181		181	278

22 11 13.45 Pipe Fittings, Steel, Threaded

		Crew	Daily Output	Labor-Hours	Unit	Material	2021 Bare Costs Labor	2021 Bare Costs Equipment	Total	Total Incl O&P
0010	**PIPE FITTINGS, STEEL, THREADED** R221113-50									
0020	Cast iron									
0040	Standard weight, black									
0060	90° elbow, straight									
0070	1/4"	1 Plum	16	.500	Ea.	12.65	34		46.65	66

22 11 13.45 Pipe Fittings, Steel, Threaded	Crew	Daily Output	Labor-Hours	Unit	Material	2021 Bare Costs Labor	Equipment	Total	Total Incl O&P
0080	3/8"	1 Plum	16	.500	Ea.	18.35	34		52.35	72
0090	1/2"		15	.533		8.05	36		44.05	64.50
0100	3/4"		14	.571		8.35	38.50		46.85	68.50
0110	1"		13	.615		9.95	41.50		51.45	75.50
0130	1-1/2"	Q-1	20	.800		19.50	48.50		68	96.50
0140	2"		18	.889		30.50	54		84.50	117
0150	2-1/2"		14	1.143		73	69.50		142.50	187
0160	3"		10	1.600		119	97.50		216.50	281
0180	4"		6	2.667		221	162		383	495
0500	Tee, straight									
0510	1/4"	1 Plum	10	.800	Ea.	21	54		75	107
0520	3/8"		10	.800		19.30	54		73.30	105
0540	3/4"		9	.889		14.60	60		74.60	109
0550	1"		8	1		13	67.50		80.50	118
0570	1-1/2"	Q-1	13	1.231		31	75		106	150
0580	2"		11	1.455		43	88.50		131.50	184
0590	2-1/2"		9	1.778		112	108		220	290
0600	3"		6	2.667		171	162		333	440
0620	4"		4	4		335	244		579	740
0700	Standard weight, galvanized cast iron									
0720	90° elbow, straight									
0730	1/4"	1 Plum	16	.500	Ea.	16.80	34		50.80	70.50
0740	3/8"		16	.500		16.80	34		50.80	70.50
0750	1/2"		15	.533		21	36		57	79
0760	3/4"		14	.571		16.35	38.50		54.85	77.50
0770	1"		13	.615		21.50	41.50		63	88.50
0790	1-1/2"	Q-1	20	.800		46.50	48.50		95	126
0800	2"		18	.889		59	54		113	149
0810	2-1/2"		14	1.143		122	69.50		191.50	241
0820	3"		10	1.600		185	97.50		282.50	355
0840	4"		6	2.667		340	162		502	625
1100	Tee, straight									
1110	1/4"	1 Plum	10	.800	Ea.	20.50	54		74.50	106
1120	3/8"		10	.800		17.70	54		71.70	103
1140	3/4"		9	.889		27	60		87	123
1150	1"		8	1		25.50	67.50		93	132
1170	1-1/2"	Q-1	13	1.231		58.50	75		133.50	181
1180	2"		11	1.455		73	88.50		161.50	218
1190	2-1/2"		9	1.778		150	108		258	330
1200	3"		6	2.667		490	162		652	790
1220	4"		4	4		580	244		824	1,000
5000	Malleable iron, 150 lb.									
5020	Black									
5040	90° elbow, straight									
5060	1/4"	1 Plum	16	.500	Ea.	6.25	34		40.25	59
5070	3/8"		16	.500		6.25	34		40.25	59
5090	3/4"		14	.571		5.20	38.50		43.70	65.50
5100	1"		13	.615		9.05	41.50		50.55	74.50
5120	1-1/2"	Q-1	20	.800		19.65	48.50		68.15	96.50
5130	2"		18	.889		34	54		88	121
5140	2-1/2"		14	1.143		75.50	69.50		145	190
5150	3"		10	1.600		110	97.50		207.50	272
5170	4"		6	2.667		211	162		373	485

22 11 13.45 Pipe Fittings, Steel, Threaded

		Crew	Daily Output	Labor-Hours	Unit	Material	2021 Bare Costs Labor	Equipment	Total	Total Incl O&P
5450	Tee, straight									
5470	1/4"	1 Plum	10	.800	Ea.	9.05	54		63.05	93.50
5480	3/8"		10	.800		9.05	54		63.05	93.50
5500	3/4"		9	.889		8.30	60		68.30	102
5510	1"		8	1		14.20	67.50		81.70	120
5520	1-1/4"	Q-1	14	1.143		23	69.50		92.50	133
5530	1-1/2"		13	1.231		28.50	75		103.50	148
5540	2"		11	1.455		49	88.50		137.50	191
5550	2-1/2"		9	1.778		105	108		213	283
5560	3"		6	2.667		155	162		317	420
5570	3-1/2"		5	3.200		375	195		570	715
5580	4"		4	4		375	244		619	785
5650	Coupling									
5670	1/4"	1 Plum	19	.421	Ea.	7.75	28.50		36.25	52.50
5680	3/8"		19	.421		7.75	28.50		36.25	52.50
5690	1/2"		19	.421		6	28.50		34.50	50.50
5700	3/4"		18	.444		7	30		37	54
5710	1"		15	.533		10.50	36		46.50	67
5720	1-1/4"	Q-1	26	.615		12.05	37.50		49.55	71.50
5730	1-1/2"		24	.667		16.30	40.50		56.80	80.50
5740	2"		21	.762		24	46.50		70.50	98
5750	2-1/2"		18	.889		66.50	54		120.50	157
5760	3"		14	1.143		90	69.50		159.50	206
5780	4"		10	1.600		181	97.50		278.50	350
6000	For galvanized elbows, tees, and couplings, add					20%				
9990	Minimum labor/equipment charge	1 Plum	4	2	Job		135		135	209

22 11 13.47 Pipe Fittings, Steel

		Crew	Daily Output	Labor-Hours	Unit	Material	2021 Bare Costs Labor	Equipment	Total	Total Incl O&P
0010	**PIPE FITTINGS, STEEL**, flanged, welded & special									
3000	Weld joint, butt, carbon steel, standard weight									
3040	90° elbow, long radius									
3050	1/2" pipe size	Q-15	16	1	Ea.	41	61	6.70	108.70	146
3060	3/4" pipe size		16	1		41	61	6.70	108.70	146
3070	1" pipe size		16	1		24.50	61	6.70	92.20	128
3080	1-1/4" pipe size		14	1.143		24.50	69.50	7.70	101.70	142
3090	1-1/2" pipe size		13	1.231		24.50	75	8.25	107.75	152
3100	2" pipe size		10	1.600		21	97.50	10.75	129.25	185
3110	2-1/2" pipe size		8	2		32.50	122	13.45	167.95	239
3120	3" pipe size		7	2.286		23	139	15.35	177.35	257
3130	4" pipe size		5	3.200		54	195	21.50	270.50	385
3136	5" pipe size		4	4		110	244	27	381	525
3140	6" pipe size	Q-16	5	4.800		119	305	21.50	445.50	625
3350	Tee, straight									
3360	1/2" pipe size	Q-15	10	1.600	Ea.	108	97.50	10.75	216.25	280
3370	3/4" pipe size		10	1.600		108	97.50	10.75	216.25	280
3380	1" pipe size		10	1.600		53.50	97.50	10.75	161.75	221
3390	1-1/4" pipe size		9	1.778		67	108	11.95	186.95	254
3400	1-1/2" pipe size		8	2		67	122	13.45	202.45	276
3410	2" pipe size		6	2.667		53.50	162	17.90	233.40	330
3420	2-1/2" pipe size		5	3.200		73.50	195	21.50	290	405
3430	3" pipe size		4	4		82	244	27	353	495
3440	4" pipe size		3	5.333		115	325	36	476	665
3446	5" pipe size		2.50	6.400		190	390	43	623	855

22 11 13 – Facility Water Distribution Piping

22 11 13.47 Pipe Fittings, Steel		Crew	Daily Output	Labor-Hours	Unit	Material	2021 Bare Costs Labor	Equipment	Total	Total Incl O&P
3450	6" pipe size	Q-16	3	8	Ea.	199	505	36	740	1,050
9805	2" x 1"	1 Plum	50	.160		106	10.85		116.85	134
9810	2" x 1-1/4"		50	.160		124	10.85		134.85	154
9815	2" x 1-1/2"	▼	50	.160		129	10.85		139.85	159
9820	2-1/2" x 1"	Q-1	80	.200		111	12.20		123.20	141
9825	2-1/2" x 1-1/4"		80	.200		163	12.20		175.20	198
9830	2-1/2" x 1-1/2"		80	.200		163	12.20		175.20	198
9835	2-1/2" x 2"		80	.200		111	12.20		123.20	141
9840	3" x 1"		80	.200		120	12.20		132.20	151
9845	3" x 1-1/4"		80	.200		156	12.20		168.20	191
9850	3" x 1-1/2"		80	.200		156	12.20		168.20	191
9852	3" x 2"		80	.200		190	12.20		202.20	229
9856	4" x 1"		67	.239		149	14.55		163.55	187
9858	4" x 1-1/4"		67	.239		195	14.55		209.55	238
9860	4" x 1-1/2"		67	.239		197	14.55		211.55	240
9862	4" x 2"		67	.239		204	14.55		218.55	248
9864	4" x 2-1/2"		67	.239		211	14.55		225.55	255
9866	4" x 3"	▼	67	.239		211	14.55		225.55	255
9870	6" x 1-1/2"	Q-2	50	.480		247	30.50		277.50	320
9872	6" x 2"		50	.480		251	30.50		281.50	325
9874	6" x 2-1/2"		50	.480		249	30.50		279.50	320
9876	6" x 3"		50	.480		284	30.50		314.50	355
9878	6" x 4"		50	.480		315	30.50		345.50	395
9880	8" x 2"		42	.571		425	36		461	525
9882	8" x 2-1/2"		42	.571		425	36		461	525
9884	8" x 3"		42	.571		450	36		486	550
9886	8" x 4"	▼	42	.571	▼	915	36		951	1,050
9990	Minimum labor/equipment charge	Q-15	3	5.333	Job		325	36	361	540

22 11 13.48 Pipe, Fittings and Valves, Steel, Grooved-Joint

		Crew	Daily Output	Labor-Hours	Unit	Material	2021 Bare Costs Labor	Equipment	Total	Total Incl O&P
0010	**PIPE, FITTINGS AND VALVES, STEEL, GROOVED-JOINT**									
0012	Fittings are ductile iron. Steel fittings noted.									
0020	Pipe includes coupling & clevis type hanger assemblies, 10' OC									
1000	Schedule 40, black									
1040	3/4" diameter	1 Plum	71	.113	L.F.	5.85	7.65		13.50	18.20
1050	1" diameter		63	.127		5.40	8.60		14	19.20
1060	1-1/4" diameter		58	.138		6.95	9.35		16.30	22
1070	1-1/2" diameter		51	.157		7.85	10.60		18.45	25
1080	2" diameter	▼	40	.200		9.30	13.55		22.85	31
1090	2-1/2" diameter	Q-1	57	.281		14.90	17.10		32	43
1100	3" diameter		50	.320		18.35	19.50		37.85	50
1110	4" diameter		45	.356		26	21.50		47.50	62.50
1120	5" diameter	▼	37	.432	▼	31.50	26.50		58	75
3990	Fittings: coupling material required at joints not incl. in fitting price.									
3994	Add 1 selected coupling, material only, per joint for installed price.									
4000	Elbow, 90° or 45°, painted									
4030	3/4" diameter	1 Plum	50	.160	Ea.	95	10.85		105.85	122
4040	1" diameter		50	.160		51	10.85		61.85	72.50
4050	1-1/4" diameter		40	.200		51	13.55		64.55	77
4060	1-1/2" diameter		33	.242		51	16.40		67.40	81.50
4070	2" diameter	▼	25	.320		51	21.50		72.50	89.50
4080	2-1/2" diameter	Q-1	40	.400		51	24.50		75.50	93.50
4090	3" diameter	↓	33	.485	▼	89.50	29.50		119	144

22 11 Facility Water Distribution

22 11 13 – Facility Water Distribution Piping

22 11 13.48 Pipe, Fittings and Valves, Steel, Grooved-Joint	Crew	Daily Output	Labor-Hours	Unit	Material	2021 Bare Costs Labor	Equipment	Total	Total Incl O&P
4100 4" diameter	Q-1	25	.640	Ea.	97	39		136	167
4110 5" diameter	▼	20	.800		230	48.50		278.50	330
4250 For galvanized elbows, add				▼	26%				
4690 Tee, painted									
4700 3/4" diameter	1 Plum	38	.211	Ea.	102	14.25		116.25	134
4740 1" diameter		33	.242		78.50	16.40		94.90	112
4750 1-1/4" diameter		27	.296		78.50	20		98.50	118
4760 1-1/2" diameter		22	.364		78.50	24.50		103	125
4770 2" diameter	▼	17	.471		78.50	32		110.50	136
4780 2-1/2" diameter	Q-1	27	.593		78.50	36		114.50	142
4790 3" diameter		22	.727		107	44.50		151.50	187
4800 4" diameter		17	.941		163	57.50		220.50	268
4810 5" diameter	▼	13	1.231		380	75		455	530
4900 For galvanized tees, add					24%				
4933 20" diameter	Q-2	16	1.500	▼	835	95		930	1,050
4940 Flexible, standard, painted									
4950 3/4" diameter	1 Plum	100	.080	Ea.	29	5.40		34.40	40.50
4960 1" diameter		100	.080		29	5.40		34.40	40.50
4970 1-1/4" diameter		80	.100		37.50	6.75		44.25	51.50
4980 1-1/2" diameter		67	.119		40.50	8.10		48.60	57
4990 2" diameter	▼	50	.160		43.50	10.85		54.35	64.50
5000 2-1/2" diameter	Q-1	80	.200		50.50	12.20		62.70	74.50
5010 3" diameter		67	.239		56	14.55		70.55	84
5020 3-1/2" diameter		57	.281		80	17.10		97.10	115
5030 4" diameter		50	.320		80.50	19.50		100	119
5040 5" diameter	▼	40	.400		121	24.50		145.50	171
5050 6" diameter	Q-2	50	.480		143	30.50		173.50	204
5200 For galvanized couplings, add				▼	33%				
7790 Valves: coupling material required at joints not incl. in valve price.									
7794 Add 1 selected coupling, material only, per joint for installed price.									
9990 Minimum labor/equipment charge	1 Plum	4	2	Job		135		135	209

22 11 13.74 Pipe, Plastic

	Crew	Daily Output	Labor-Hours	Unit	Material	2021 Bare Costs Labor	Equipment	Total	Total Incl O&P
0010 **PIPE, PLASTIC**									
1800 PVC, couplings 10' OC, clevis hanger assemblies, 3 per 10'									
1820 Schedule 40									
1860 1/2" diameter	1 Plum	54	.148	L.F.	4.17	10.05		14.22	20
1870 3/4" diameter		51	.157		4.77	10.60		15.37	21.50
1880 1" diameter		46	.174		3.35	11.75		15.10	22
1890 1-1/4" diameter		42	.190		3.36	12.90		16.26	23.50
1900 1-1/2" diameter	▼	36	.222		4.90	15.05		19.95	28.50
1910 2" diameter	Q-1	59	.271		4.85	16.50		21.35	31
1920 2-1/2" diameter		56	.286		7.85	17.40		25.25	35.50
1930 3" diameter		53	.302		10.05	18.40		28.45	39.50
1940 4" diameter		48	.333		13	20.50		33.50	46
1950 5" diameter		43	.372		25	22.50		47.50	62.50
1960 6" diameter	▼	39	.410	▼	25	25		50	66
4100 DWV type, schedule 40, couplings 10' OC, clevis hanger assy's, 3 per 10'									
4210 ABS, schedule 40, foam core type									
4212 Plain end black									
4214 1-1/2" diameter	1 Plum	39	.205	L.F.	2.79	13.90		16.69	24.50
4216 2" diameter	Q-1	62	.258		3.27	15.70		18.97	28
4218 3" diameter	▼	56	.286	▼	5.65	17.40		23.05	33

22 11 13 – Facility Water Distribution Piping

22 11 13.74 Pipe, Plastic

		Crew	Daily Output	Labor-Hours	Unit	Material	2021 Bare Costs Labor	2021 Bare Costs Equipment	Total	Total Incl O&P
4220	4" diameter	Q-1	51	.314	L.F.	8.05	19.10		27.15	38.50
4222	6" diameter	↓	42	.381	↓	19.75	23		42.75	57.50
4240	To delete coupling & hangers, subtract									
4244	1-1/2" diam. to 6" diam.					43%	48%			
4400	PVC									
4410	1-1/4" diameter	1 Plum	42	.190	L.F.	3.42	12.90		16.32	23.50
4460	2" diameter	Q-1	59	.271		3.13	16.50		19.63	29
4470	3" diameter		53	.302		5.35	18.40		23.75	34.50
4480	4" diameter		48	.333		7.40	20.50		27.90	39.50
4490	6" diameter	↓	39	.410	↓	15.75	25		40.75	56
5300	CPVC, socket joint, couplings 10' OC, clevis hanger assemblies, 3 per 10'									
5302	Schedule 40									
5304	1/2" diameter	1 Plum	54	.148	L.F.	5.30	10.05		15.35	21.50
5305	3/4" diameter		51	.157		6.45	10.60		17.05	23.50
5306	1" diameter		46	.174		5.30	11.75		17.05	24
5307	1-1/4" diameter		42	.190		6.50	12.90		19.40	27
5308	1-1/2" diameter	↓	36	.222		6.35	15.05		21.40	30
5309	2" diameter	Q-1	59	.271		9.35	16.50		25.85	36
5310	2-1/2" diameter		56	.286		15.20	17.40		32.60	43.50
5311	3" diameter	↓	53	.302	↓	19.80	18.40		38.20	50.50
5360	CPVC, threaded, couplings 10' OC, clevis hanger assemblies, 3 per 10'									
5380	Schedule 40									
5460	1/2" diameter	1 Plum	54	.148	L.F.	6.15	10.05		16.20	22
5470	3/4" diameter		51	.157		7.90	10.60		18.50	25
5480	1" diameter		46	.174		6.85	11.75		18.60	25.50
5490	1-1/4" diameter		42	.190		7.65	12.90		20.55	28.50
5500	1-1/2" diameter	↓	36	.222		7.35	15.05		22.40	31
5510	2" diameter	Q-1	59	.271		10.55	16.50		27.05	37
5520	2-1/2" diameter		56	.286		16.50	17.40		33.90	45
5530	3" diameter	↓	53	.302	↓	21.50	18.40		39.90	52.50
9900	Minimum labor/equipment charge	1 Plum	4	2	Job		135		135	209

22 11 13.76 Pipe Fittings, Plastic

		Crew	Daily Output	Labor-Hours	Unit	Material	2021 Bare Costs Labor	2021 Bare Costs Equipment	Total	Total Incl O&P
0010	**PIPE FITTINGS, PLASTIC**									
2700	PVC (white), schedule 40, socket joints									
2760	90° elbow, 1/2"	1 Plum	33.30	.240	Ea.	.55	16.25		16.80	25.50
2810	2"	Q-1	36.40	.440	"	3.29	27		30.29	45
4500	DWV, ABS, non pressure, socket joints									
4540	1/4 bend, 1-1/4"	1 Plum	20.20	.396	Ea.	8.15	27		35.15	50.50
4560	1-1/2"	"	18.20	.440		6.35	30		36.35	53
4570	2"	Q-1	33.10	.483	↓	6.05	29.50		35.55	52
4650	1/8 bend, same as 1/4 bend									
4800	Tee, sanitary									
4820	1-1/4"	1 Plum	13.50	.593	Ea.	10.85	40		50.85	74
4830	1-1/2"	"	12.10	.661		9.40	45		54.40	79.50
4840	2"	Q-1	20	.800	↓	14.45	48.50		62.95	91
5000	DWV, PVC, schedule 40, socket joints									
5040	1/4 bend, 1-1/4"	1 Plum	20.20	.396	Ea.	16.55	27		43.55	59.50
5060	1-1/2"	"	18.20	.440		4.75	30		34.75	51.50
5070	2"	Q-1	33.10	.483		7.45	29.50		36.95	53.50
5080	3"		20.80	.769		22	47		69	96.50
5090	4"		16.50	.970		43.50	59		102.50	139
5100	6"	↓	10.10	1.584		152	96.50		248.50	315

For customer support on your Commercial Renovation Costs with RSMeans Data, call 800.448.8182.

433

22 11 13.76 Pipe Fittings, Plastic		Crew	Daily Output	Labor-Hours	Unit	Material	2021 Bare Costs Labor	Equipment	Total	Total Incl O&P
5105	8"	Q-2	9.30	2.581	Ea.	197	163		360	465
5110	1/4 bend, long sweep, 1-1/2"	1 Plum	18.20	.440		11	30		41	58
5112	2"	Q-1	33.10	.483		12.25	29.50		41.75	59
5114	3"		20.80	.769		28.50	47		75.50	104
5116	4"	▼	16.50	.970		53.50	59		112.50	150
5150	1/8 bend, 1-1/4"	1 Plum	20.20	.396		11.30	27		38.30	54
5215	8"	Q-2	9.30	2.581		87	163		250	345
5250	Tee, sanitary 1-1/4"	1 Plum	13.50	.593		17.75	40		57.75	81.50
5254	1-1/2"	"	12.10	.661		8.30	45		53.30	78
5255	2"	Q-1	20	.800		12.20	48.50		60.70	88.50
5256	3"		13.90	1.151		32	70		102	144
5257	4"		11	1.455		58.50	88.50		147	202
5259	6"	▼	6.70	2.388		235	146		381	485
5261	8"	Q-2	6.20	3.871		520	245		765	950
5264	2" x 1-1/2"	Q-1	22	.727		10.75	44.50		55.25	80.50
5266	3" x 1-1/2"		15.50	1.032		23.50	63		86.50	123
5268	4" x 3"		12.10	1.322		68.50	80.50		149	200
5271	6" x 4"	▼	6.90	2.319		228	141		369	470
5314	Combination Y & 1/8 bend, 1-1/2"	1 Plum	12.10	.661		20	45		65	91
5315	2"	Q-1	20	.800		25	48.50		73.50	103
5317	3"		13.90	1.151		55	70		125	169
5318	4"	▼	11	1.455	▼	109	88.50		197.50	257
5324	Combination Y & 1/8 bend, reducing									
5325	2" x 2" x 1-1/2"	Q-1	22	.727	Ea.	28	44.50		72.50	99.50
5327	3" x 3" x 1-1/2"		15.50	1.032		50.50	63		113.50	153
5328	3" x 3" x 2"		15.30	1.046		37.50	63.50		101	140
5329	4" x 4" x 2"	▼	12.20	1.311		57	80		137	186
5331	Wye, 1-1/4"	1 Plum	13.50	.593		22.50	40		62.50	87
5332	1-1/2"	"	12.10	.661		15.10	45		60.10	85.50
5333	2"	Q-1	20	.800		14.85	48.50		63.35	91.50
5334	3"		13.90	1.151		40	70		110	152
5335	4"		11	1.455		73	88.50		161.50	217
5336	6"	▼	6.70	2.388		210	146		356	455
5337	8"	Q-2	6.20	3.871		370	245		615	785
5341	2" x 1-1/2"	Q-1	22	.727		10.75	44.50		55.25	80.50
5342	3" x 1-1/2"		15.50	1.032		15.90	63		78.90	115
5343	4" x 3"		12.10	1.322		35	80.50		115.50	163
5344	6" x 4"	▼	6.90	2.319		94.50	141		235.50	320
5345	8" x 6"	Q-2	6.40	3.750		205	237		442	590
5347	Double wye, 1-1/2"	1 Plum	9.10	.879		34	59.50		93.50	130
5348	2"	Q-1	16.60	.964		40.50	58.50		99	135
5349	3"		10.40	1.538		79	93.50		172.50	232
5350	4"	▼	8.25	1.939	▼	170	118		288	370
5353	Double wye, reducing									
5354	2" x 2" x 1-1/2" x 1-1/2"	Q-1	16.80	.952	Ea.	34.50	58		92.50	128
5355	3" x 3" x 2" x 2"		10.60	1.509		58.50	92		150.50	207
5356	4" x 4" x 3" x 3"		8.45	1.893		127	115		242	315
5357	6" x 6" x 4" x 4"	▼	7.25	2.207		445	134		579	695
5374	Coupling, 1-1/4"	1 Plum	20.20	.396		10.70	27		37.70	53.50
5376	1-1/2"	"	18.20	.440		2.23	30		32.23	48.50
5378	2"	Q-1	33.10	.483		3.06	29.50		32.56	49
5380	3"		20.80	.769		10.65	47		57.65	84.50
5390	4"	▼	16.50	.970	▼	18.10	59		77.10	111

22 11 13.76 Pipe Fittings, Plastic		Crew	Daily Output	Labor-Hours	Unit	Material	2021 Bare Costs Labor	Equipment	Total	Total Incl O&P
5410	Reducer bushing, 2" x 1-1/4"	Q-1	36.50	.438	Ea.	3.73	26.50		30.23	45
5412	3" x 1-1/2"		27.30	.586		11	35.50		46.50	67
5414	4" x 2"		18.20	.879		32.50	53.50		86	118
5416	6" x 4"		11.10	1.441		83	88		171	227
5418	8" x 6"	Q-2	10.20	2.353		163	149		312	410
5500	CPVC, Schedule 80, threaded joints									
5540	90° elbow, 1/4"	1 Plum	32	.250	Ea.	12.80	16.95		29.75	40
5560	1/2"		30.30	.264		7.40	17.85		25.25	35.50
5570	3/4"		26	.308		11.10	21		32.10	44
5580	1"		22.70	.352		15.60	24		39.60	54
5590	1-1/4"		20.20	.396		30	27		57	74.50
5600	1-1/2"		18.20	.440		32.50	30		62.50	81.50
5610	2"	Q-1	33.10	.483		43.50	29.50		73	93
5620	2-1/2"		24.20	.661		134	40.50		174.50	210
5630	3"		20.80	.769		144	47		191	231
5730	Coupling, 1/4"	1 Plum	32	.250		16.30	16.95		33.25	44
5732	1/2"		30.30	.264		13.40	17.85		31.25	42.50
5734	3/4"		26	.308		21.50	21		42.50	56
5736	1"		22.70	.352		24.50	24		48.50	64
5738	1-1/4"		20.20	.396		26	27		53	70
5740	1-1/2"		18.20	.440		28	30		58	77
5742	2"	Q-1	33.10	.483		33	29.50		62.50	82
5744	2-1/2"		24.20	.661		59	40.50		99.50	127
5746	3"		20.80	.769		69	47		116	149
5900	CPVC, Schedule 80, socket joints									
5904	90° elbow, 1/4"	1 Plum	32	.250	Ea.	12.15	16.95		29.10	39.50
5906	1/2"		30.30	.264		4.77	17.85		22.62	33
5908	3/4"		26	.308		6.10	21		27.10	38.50
5910	1"		22.70	.352		9.65	24		33.65	47.50
5912	1-1/4"		20.20	.396		21	27		48	64.50
5914	1-1/2"		18.20	.440		23.50	30		53.50	71.50
5916	2"	Q-1	33.10	.483		28	29.50		57.50	76.50
5918	2-1/2"		24.20	.661		65	40.50		105.50	134
5920	3"		20.80	.769		73.50	47		120.50	153
5930	45° elbow, 1/4"	1 Plum	32	.250		18.10	16.95		35.05	46
5932	1/2"		30.30	.264		5.85	17.85		23.70	34
5934	3/4"		26	.308		8.40	21		29.40	41.50
5936	1"		22.70	.352		13.40	24		37.40	52
5938	1-1/4"		20.20	.396		26.50	27		53.50	70.50
5940	1-1/2"		18.20	.440		27	30		57	75.50
5942	2"	Q-1	33.10	.483		30.50	29.50		60	79
5944	2-1/2"		24.20	.661		62	40.50		102.50	130
5946	3"		20.80	.769		79.50	47		126.50	160
5990	Coupling, 1/4"	1 Plum	32	.250		12.95	16.95		29.90	40.50
5992	1/2"		30.30	.264		5.05	17.85		22.90	33
5994	3/4"		26	.308		7.05	21		28.05	40
5996	1"		22.70	.352		9.50	24		33.50	47.50
5998	1-1/4"		20.20	.396		14.20	27		41.20	57
6000	1-1/2"		18.20	.440		17.85	30		47.85	65.50
6002	2"	Q-1	33.10	.483		21	29.50		50.50	68.50
6004	2-1/2"		24.20	.661		46	40.50		86.50	113
6006	3"		20.80	.769		50	47		97	128
9990	Minimum labor/equipment charge	1 Plum	4	2	Job		135		135	209

22 11 19 – Domestic Water Piping Specialties

22 11 19.10 Flexible Connectors

		Crew	Daily Output	Labor-Hours	Unit	Material	2021 Bare Costs Labor	Equipment	Total	Total Incl O&P
0010	**FLEXIBLE CONNECTORS**, Corrugated, 5/8" OD, 3/4" ID									
0050	Gas, seamless brass, steel fittings									
0200	12" long	1 Plum	36	.222	Ea.	20.50	15.05		35.55	46
0220	18" long		36	.222		25.50	15.05		40.55	51.50
0240	24" long		34	.235		30.50	15.95		46.45	58
0260	30" long		34	.235		33	15.95		48.95	60.50
9000	Minimum labor/equipment charge		4	2	Job		135		135	209

22 11 19.18 Mixing Valve

		Crew	Daily Output	Labor-Hours	Unit	Material	2021 Bare Costs Labor	Equipment	Total	Total Incl O&P
0010	**MIXING VALVE**, Automatic, water tempering.									
0040	1/2" size	1 Stpi	19	.421	Ea.	655	29		684	765
0050	3/4" size		18	.444	"	565	30.50		595.50	665
9000	Minimum labor/equipment charge		5	1.600	Job		109		109	169

22 11 19.38 Water Supply Meters

		Crew	Daily Output	Labor-Hours	Unit	Material	2021 Bare Costs Labor	Equipment	Total	Total Incl O&P
0010	**WATER SUPPLY METERS**									
2000	Domestic/commercial, bronze									
2020	Threaded									
2060	5/8" diameter, to 20 GPM	1 Plum	16	.500	Ea.	56.50	34		90.50	115
2080	3/4" diameter, to 30 GPM		14	.571		103	38.50		141.50	174
2100	1" diameter, to 50 GPM		12	.667		157	45		202	243
2300	Threaded/flanged									
2340	1-1/2" diameter, to 100 GPM	1 Plum	8	1	Ea.	385	67.50		452.50	525
2600	Flanged, compound									
9000	Minimum labor/equipment charge	1 Plum	3.25	2.462	Job		167		167	257

22 11 19.42 Backflow Preventers

		Crew	Daily Output	Labor-Hours	Unit	Material	2021 Bare Costs Labor	Equipment	Total	Total Incl O&P
0010	**BACKFLOW PREVENTERS**, Includes valves									
0020	and four test cocks, corrosion resistant, automatic operation									
1000	Double check principle									
1010	Threaded, with ball valves									
1020	3/4" pipe size	1 Plum	16	.500	Ea.	240	34		274	315
1030	1" pipe size		14	.571		335	38.50		373.50	430
1040	1-1/2" pipe size		10	.800		640	54		694	790
1050	2" pipe size		7	1.143		690	77.50		767.50	880
1080	Threaded, with gate valves									
1100	3/4" pipe size	1 Plum	16	.500	Ea.	1,275	34		1,309	1,450
1120	1" pipe size		14	.571		1,300	38.50		1,338.50	1,475
1140	1-1/2" pipe size		10	.800		1,150	54		1,204	1,350
1160	2" pipe size		7	1.143		2,050	77.50		2,127.50	2,375
4000	Reduced pressure principle									
4100	Threaded, bronze, valves are ball									
4120	3/4" pipe size	1 Plum	16	.500	Ea.	450	34		484	545
4140	1" pipe size		14	.571		480	38.50		518.50	590
4150	1-1/4" pipe size		12	.667		1,175	45		1,220	1,375
4160	1-1/2" pipe size		10	.800		970	54		1,024	1,150
4180	2" pipe size		7	1.143		1,550	77.50		1,627.50	1,825
5000	Flanged, bronze, valves are OS&Y									
5060	2-1/2" pipe size	Q-1	5	3.200	Ea.	5,225	195		5,420	6,050
5080	3" pipe size		4.50	3.556		5,725	217		5,942	6,625
5100	4" pipe size		3	5.333		7,350	325		7,675	8,600
5120	6" pipe size	Q-2	3	8		10,400	505		10,905	12,200
5600	Flanged, iron, valves are OS&Y									
5660	2-1/2" pipe size	Q-1	5	3.200	Ea.	3,825	195		4,020	4,525

22 11 19 – Domestic Water Piping Specialties

22 11 19.42 Backflow Preventers

		Crew	Daily Output	Labor-Hours	Unit	Material	2021 Bare Costs Labor	Equipment	Total	Total Incl O&P
5680	3" pipe size	Q-1	4.50	3.556	Ea.	3,250	217		3,467	3,900
5700	4" pipe size	↓	3	5.333		4,725	325		5,050	5,700
5720	6" pipe size	Q-2	3	8		5,650	505		6,155	6,975
5800	Rebuild 4" diameter reduced pressure	1 Plum	4	2		405	135		540	655
5810	6" diameter		2.66	3.008		425	204		629	785
5820	8" diameter		2	4		530	271		801	1,000
5830	10" diameter		1.60	5	↓	620	340		960	1,200
9010	Minimum labor/equipment charge	↓	2	4	Job		271		271	420

22 11 19.50 Vacuum Breakers

		Crew	Daily Output	Labor-Hours	Unit	Material	2021 Bare Costs Labor	Equipment	Total	Total Incl O&P
0010	**VACUUM BREAKERS**									
0013	See also backflow preventers Section 22 11 19.42									
1000	Anti-siphon continuous pressure type									
1010	Max. 150 psi - 210°F									
1020	Bronze body									
1030	1/2" size	1 Stpi	24	.333	Ea.	221	23		244	278
1040	3/4" size		20	.400		221	27.50		248.50	285
1050	1" size		19	.421		226	29		255	293
1060	1-1/4" size		15	.533		450	36.50		486.50	550
1070	1-1/2" size		13	.615		545	42		587	665
1080	2" size	↓	11	.727	↓	560	49.50		609.50	690
1200	Max. 125 psi with atmospheric vent									
1210	Brass, in-line construction									
1220	1/4" size	1 Stpi	24	.333	Ea.	151	23		174	201
1230	3/8" size	"	24	.333		151	23		174	201
1260	For polished chrome finish, add				↓	13%				
2000	Anti-siphon, non-continuous pressure type									
2010	Hot or cold water 125 psi - 210°F									
2020	Bronze body									
2030	1/4" size	1 Stpi	24	.333	Ea.	92	23		115	136
2040	3/8" size		24	.333		92	23		115	136
2050	1/2" size		24	.333		103	23		126	149
2060	3/4" size		20	.400		124	27.50		151.50	178
2070	1" size		19	.421		191	29		220	255
2080	1-1/4" size		15	.533		335	36.50		371.50	420
2090	1-1/2" size		13	.615		370	42		412	470
2100	2" size		11	.727		610	49.50		659.50	745
2110	2-1/2" size		8	1		1,750	68.50		1,818.50	2,025
2120	3" size	↓	6	1.333	↓	2,325	91		2,416	2,700
2150	For polished chrome finish, add					50%				
3000	Air gap fitting									
3020	1/2" NPT size	1 Plum	19	.421	Ea.	62	28.50		90.50	112
3030	1" NPT size	"	15	.533		71	36		107	134
3040	2" NPT size	Q-1	21	.762		139	46.50		185.50	225
3050	3" NPT size		14	1.143		275	69.50		344.50	410
3060	4" NPT size	↓	10	1.600	↓	275	97.50		372.50	455

22 11 19.54 Water Hammer Arresters/Shock Absorbers

		Crew	Daily Output	Labor-Hours	Unit	Material	2021 Bare Costs Labor	Equipment	Total	Total Incl O&P
0010	**WATER HAMMER ARRESTERS/SHOCK ABSORBERS**									
0490	Copper									
0500	3/4" male IPS for 1 to 11 fixtures	1 Plum	12	.667	Ea.	31	45		76	104
0600	1" male IPS for 12 to 32 fixtures		8	1		52	67.50		119.50	161
0700	1-1/4" male IPS for 33 to 60 fixtures		8	1		52.50	67.50		120	162
0800	1-1/2" male IPS for 61 to 113 fixtures	↓	8	1		77	67.50		144.50	189

22 11 19 – Domestic Water Piping Specialties

22 11 19.54 Water Hammer Arresters/Shock Absorbers		Crew	Daily Output	Labor-Hours	Unit	Material	2021 Bare Costs Labor	Equipment	Total	Total Incl O&P
0900	2" male IPS for 114 to 154 fixtures	1 Plum	8	1	Ea.	118	67.50		185.50	234
1000	2-1/2" male IPS for 155 to 330 fixtures		4	2		330	135		465	575
9000	Minimum labor/equipment charge		3.50	2.286	Job		155		155	239

22 11 19.64 Hydrants

		Crew	Daily Output	Labor-Hours	Unit	Material	2021 Bare Costs Labor	Equipment	Total	Total Incl O&P
0010	**HYDRANTS**									
0050	Wall type, moderate climate, bronze, encased									
0200	3/4" IPS connection	1 Plum	16	.500	Ea.	930	34		964	1,075
1000	Non-freeze, bronze, exposed									
1100	3/4" IPS connection, 4" to 9" thick wall	1 Plum	14	.571	Ea.	530	38.50		568.50	640
1120	10" to 14" thick wall	"	12	.667		465	45		510	580
1280	For anti-siphon type, add					198			198	218
9000	Minimum labor/equipment charge	1 Plum	3	2.667	Job		181		181	278

22 13 16 – Sanitary Waste and Vent Piping

22 13 16.20 Pipe, Cast Iron

		Crew	Daily Output	Labor-Hours	Unit	Material	2021 Bare Costs Labor	Equipment	Total	Total Incl O&P
0010	**PIPE, CAST IRON**, Soil, on clevis hanger assemblies, 5' OC R221113-50									
0020	Single hub, service wt., lead & oakum joints 10' OC									
2120	2" diameter	Q-1	63	.254	L.F.	13.90	15.45		29.35	39.50
2140	3" diameter		60	.267		18.65	16.25		34.90	45.50
2160	4" diameter		55	.291		24.50	17.75		42.25	54
2180	5" diameter	Q-2	76	.316		29	19.95		48.95	63
2200	6" diameter	"	73	.329		41.50	21		62.50	77.50
2220	8" diameter	Q-3	59	.542		66	35		101	127
4000	No hub, couplings 10' OC									
4100	1-1/2" diameter	Q-1	71	.225	L.F.	9.95	13.75		23.70	32
4120	2" diameter		67	.239		10.35	14.55		24.90	34
4140	3" diameter		64	.250		13.55	15.25		28.80	38.50
4160	4" diameter		58	.276		17.85	16.80		34.65	45.50
4180	5" diameter	Q-2	83	.289		21.50	18.25		39.75	51.50
9000	Minimum labor/equipment charge	1 Plum	4	2	Job		135		135	209

22 13 16.30 Pipe Fittings, Cast Iron

		Crew	Daily Output	Labor-Hours	Unit	Material	2021 Bare Costs Labor	Equipment	Total	Total Incl O&P
0010	**PIPE FITTINGS, CAST IRON**, Soil R221113-50									
0040	Hub and spigot, service weight, lead & oakum joints									
0080	1/4 bend, 2"	Q-1	16	1	Ea.	25.50	61		86.50	122
0120	3"		14	1.143		34	69.50		103.50	145
0140	4"		13	1.231		53.50	75		128.50	175
0160	5"	Q-2	18	1.333		74.50	84		158.50	212
0180	6"	"	17	1.412		93	89		182	240
0200	8"	Q-3	11	2.909		280	187		467	600
0340	1/8 bend, 2"	Q-1	16	1		18.15	61		79.15	114
0350	3"		14	1.143		28.50	69.50		98	139
0360	4"		13	1.231		41.50	75		116.50	162
0380	5"	Q-2	18	1.333		58.50	84		142.50	195
0400	6"	"	17	1.412		70.50	89		159.50	216
0420	8"	Q-3	11	2.909		211	187		398	520
0500	Sanitary tee, 2"	Q-1	10	1.600		35.50	97.50		133	189
0540	3"		9	1.778		57.50	108		165.50	231
0620	4"		8	2		70.50	122		192.50	266
0700	5"	Q-2	12	2		141	126		267	350

438

22 13 16.30 Pipe Fittings, Cast Iron		Crew	Daily Output	Labor- Hours	Unit	Material	2021 Bare Costs Labor	Equipment	Total	Total Incl O&P
0800	6"	Q-2	11	2.182	Ea.	159	138		297	390
0880	8"	Q-3	7	4.571	↓	420	295		715	920
5990	No hub									
6000	Cplg. & labor required at joints not incl. in fitting									
6010	price. Add 1 coupling per joint for installed price									
6020	1/4 bend, 1-1/2"				Ea.	13.15			13.15	14.50
6060	2"					14.30			14.30	15.70
6080	3"					19.95			19.95	22
6120	4"					29.50			29.50	32.50
6184	1/4 bend, long sweep, 1-1/2"					33.50			33.50	37
6186	2"					31.50			31.50	34.50
6188	3"					38			38	41.50
6189	4"					60			60	66
6190	5"					117			117	128
6191	6"					133			133	146
6192	8"					345			345	380
6193	10"					670			670	735
6200	1/8 bend, 1-1/2"					11			11	12.10
6210	2"					12.30			12.30	13.55
6212	3"					16.50			16.50	18.15
6214	4"					21.50			21.50	24
6380	Sanitary tee, tapped, 1-1/2"					26			26	28.50
6382	2" x 1-1/2"					23			23	25.50
6384	2"					25			25	27.50
6386	3" x 2"					37			37	40.50
6388	3"					63.50			63.50	70
6390	4" x 1-1/2"					32.50			32.50	36
6392	4" x 2"					37			37	41
6393	4"					37			37	41
6394	6" x 1-1/2"					85.50			85.50	94
6396	6" x 2"					90			90	99.50
6459	Sanitary tee, 1-1/2"					18.40			18.40	20.50
6460	2"					19.75			19.75	21.50
6470	3"					24.50			24.50	27
6472	4"				↓	46			46	50.50
8000	Coupling, standard (by CISPI Mfrs.)									
8020	1-1/2"	Q-1	48	.333	Ea.	13.90	20.50		34.40	47
8040	2"		44	.364		15.50	22		37.50	51
8080	3"		38	.421		18.20	25.50		43.70	59.50
8120	4"	↓	33	.485	↓	24	29.50		53.50	72
8300	Coupling, cast iron clamp & neoprene gasket (by MG)									
8310	1-1/2"	Q-1	48	.333	Ea.	9.70	20.50		30.20	42
8320	2"		44	.364		13.05	22		35.05	48.50
8330	3"		38	.421		12.80	25.50		38.30	53.50
8340	4"	↓	33	.485	↓	19.45	29.50		48.95	67
8600	Coupling, stainless steel, heavy duty									
8620	1-1/2"	Q-1	48	.333	Ea.	5.35	20.50		25.85	37.50
8630	2"		44	.364		5.80	22		27.80	40.50
8640	2" x 1-1/2"		44	.364		13.05	22		35.05	48.50
8650	3"		38	.421		6.05	25.50		31.55	46
8660	4"	↓	33	.485	↓	6.80	29.50		36.30	53
9000	Minimum labor/equipment charge	1 Plum	4	2	Job		135		135	209

22 13 16.60 Traps

		Crew	Daily Output	Labor-Hours	Unit	Material	2021 Bare Costs Labor	Equipment	Total	Total Incl O&P
0010	**TRAPS**									
0030	Cast iron, service weight									
0050	Running P trap, without vent									
1100	2"	Q-1	16	1	Ea.	184	61		245	297
1150	4"	"	13	1.231		184	75		259	320
1160	6"	Q-2	17	1.412		815	89		904	1,025
3000	P trap, B&S, 2" pipe size	Q-1	16	1		38.50	61		99.50	136
3040	3" pipe size	"	14	1.143		57.50	69.50		127	170
4700	Copper, drainage, drum trap									
4840	3" x 6" swivel, 1-1/2" pipe size	1 Plum	16	.500	Ea.	790	34		824	915
5100	P trap, standard pattern									
5200	1-1/4" pipe size	1 Plum	18	.444	Ea.	234	30		264	305
5240	1-1/2" pipe size		17	.471		178	32		210	245
5260	2" pipe size		15	.533		275	36		311	355
5280	3" pipe size		11	.727		785	49		834	940
9000	Minimum labor/equipment charge		3	2.667	Job		181		181	278

22 13 16.80 Vent Flashing and Caps

		Crew	Daily Output	Labor-Hours	Unit	Material	2021 Bare Costs Labor	Equipment	Total	Total Incl O&P
0010	**VENT FLASHING AND CAPS**									
0120	Vent caps									
0140	Cast iron									
0180	2-1/2" to 3-5/8" pipe	1 Plum	21	.381	Ea.	51.50	26		77.50	96.50
0190	4" to 4-1/8" pipe	"	19	.421	"	74	28.50		102.50	126
0900	Vent flashing									
1000	Aluminum with lead ring									
1050	3" pipe	1 Plum	17	.471	Ea.	5.85	32		37.85	55.50
1060	4" pipe	"	16	.500	"	7.05	34		41.05	60
1350	Copper with neoprene ring									
1440	2" pipe	1 Plum	18	.444	Ea.	78.50	30		108.50	133
1450	3" pipe		17	.471		94.50	32		126.50	153
1460	4" pipe		16	.500		94.50	34		128.50	156
9000	Minimum labor/equipment charge		4	2	Job		135		135	209

22 13 19 – Sanitary Waste Piping Specialties

22 13 19.13 Sanitary Drains

		Crew	Daily Output	Labor-Hours	Unit	Material	2021 Bare Costs Labor	Equipment	Total	Total Incl O&P
0010	**SANITARY DRAINS**									
0400	Deck, auto park, CI, 13" top									
0440	3", 4", 5", and 6" pipe size	Q-1	8	2	Ea.	2,100	122		2,222	2,525
0480	For galvanized body, add				"	1,300			1,300	1,425
2000	Floor, medium duty, CI, deep flange, 7" diam. top									
2040	2" and 3" pipe size	Q-1	12	1.333	Ea.	355	81		436	515
2080	For galvanized body, add					162			162	178
2120	With polished bronze top					440			440	485
2500	Heavy duty, cleanout & trap w/bucket, CI, 15" top									
2540	2", 3", and 4" pipe size	Q-1	6	2.667	Ea.	6,950	162		7,112	7,875
2560	For galvanized body, add					2,800			2,800	3,100
2580	With polished bronze top					11,200			11,200	12,300

22 13 19.14 Floor Receptors

		Crew	Daily Output	Labor-Hours	Unit	Material	2021 Bare Costs Labor	Equipment	Total	Total Incl O&P
0010	**FLOOR RECEPTORS**, For connection to 2", 3" & 4" diameter pipe									
0200	12-1/2" square top, 25 sq. in. open area	Q-1	10	1.600	Ea.	1,750	97.50		1,847.50	2,075
0300	For grate with 4" diameter x 3-3/4" high funnel, add					1,875			1,875	2,050
0400	For grate with 6" diameter x 6" high funnel, add					245			245	269
0700	For acid-resisting bucket, add					310			310	345

22 13 19 – Sanitary Waste Piping Specialties

22 13 19.14 Floor Receptors

		Crew	Daily Output	Labor-Hours	Unit	Material	2021 Bare Costs Labor	Equipment	Total	Total Incl O&P
0900	For stainless steel mesh bucket liner, add				Ea.	315			315	350
2000	12-5/8" diameter top, 40 sq. in. open area	Q-1	10	1.600	↓	815	97.50		912.50	1,050
2100	For options, add same prices as square top									
3000	8" x 4" rectangular top, 7.5 sq. in. open area	Q-1	14	1.143	Ea.	990	69.50		1,059.50	1,200
3100	For trap primer connections, add				"	98.50			98.50	108
9000	Minimum labor/equipment charge	Q-1	3	5.333	Job		325		325	500

22 13 19.15 Sink Waste Treatment

		Crew	Daily Output	Labor-Hours	Unit	Material	2021 Bare Costs Labor	Equipment	Total	Total Incl O&P
0010	**SINK WASTE TREATMENT**, System for commercial kitchens									
0100	includes clock timer & fittings									
0200	System less chemical, wall mounted cabinet	1 Plum	16	.500	Ea.	480	34		514	580
2000	Chemical, 1 gallon, add					42.50			42.50	46.50
2100	6 gallons, add					191			191	210
2200	15 gallons, add					520			520	570
2300	30 gallons, add					975			975	1,075
2400	55 gallons, add					1,650			1,650	1,825

22 13 23 – Sanitary Waste Interceptors

22 13 23.10 Interceptors

		Crew	Daily Output	Labor-Hours	Unit	Material	2021 Bare Costs Labor	Equipment	Total	Total Incl O&P
0010	**INTERCEPTORS**									
0150	Grease, fabricated steel, 4 GPM, 8 lb. fat capacity	1 Plum	4	2	Ea.	1,750	135		1,885	2,125
0200	7 GPM, 14 lb. fat capacity		4	2		2,300	135		2,435	2,725
1000	10 GPM, 20 lb. fat capacity	↓	4	2		2,875	135		3,010	3,350
1120	50 GPM, 100 lb. fat capacity	Q-1	2	8		9,050	485		9,535	10,700
1160	100 GPM, 200 lb. fat capacity		2	8		21,700	485		22,185	24,700
1240	300 GPM, 600 lb. fat capacity	↓	1	16	↓	45,300	975		46,275	51,500
9000	Minimum labor/equipment charge	1 Plum	3	2.667	Job		181		181	278

22 13 29 – Sanitary Sewerage Pumps

22 13 29.14 Sewage Ejector Pumps

		Crew	Daily Output	Labor-Hours	Unit	Material	2021 Bare Costs Labor	Equipment	Total	Total Incl O&P
0010	**SEWAGE EJECTOR PUMPS**, With operating and level controls									
0100	Simplex system incl. tank, cover, pump 15' head									
0500	37 gal. PE tank, 12 GPM, 1/2 HP, 2" discharge	Q-1	3.20	5	Ea.	515	305		820	1,025
0510	3" discharge		3.10	5.161		550	315		865	1,075
0530	87 GPM, .7 HP, 2" discharge		3.20	5		795	305		1,100	1,350
0540	3" discharge		3.10	5.161		860	315		1,175	1,425
0600	45 gal. coated stl. tank, 12 GPM, 1/2 HP, 2" discharge		3	5.333		925	325		1,250	1,525
0610	3" discharge		2.90	5.517		960	335		1,295	1,575
0630	87 GPM, .7 HP, 2" discharge		3	5.333		1,175	325		1,500	1,800
0640	3" discharge		2.90	5.517		1,250	335		1,585	1,900
0660	134 GPM, 1 HP, 2" discharge		2.80	5.714		1,275	350		1,625	1,925
0680	3" discharge		2.70	5.926		1,350	360		1,710	2,025
0700	70 gal. PE tank, 12 GPM, 1/2 HP, 2" discharge		2.60	6.154		995	375		1,370	1,675
0710	3" discharge		2.40	6.667		1,075	405		1,480	1,800
0730	87 GPM, .7 HP, 2" discharge		2.50	6.400		1,325	390		1,715	2,050
0740	3" discharge		2.30	6.957		1,375	425		1,800	2,150
0760	134 GPM, 1 HP, 2" discharge		2.20	7.273		1,400	445		1,845	2,225
0770	3" discharge		2	8	↓	1,500	485		1,985	2,400
9000	Minimum labor/equipment charge	↓	2.50	6.400	Job		390		390	600

For customer support on your Commercial Renovation Costs with RSMeans Data, call 800.448.8182.

441

22 14 Facility Storm Drainage

22 14 26 – Facility Storm Drains

22 14 26.13 Roof Drains

	Crew	Daily Output	Labor-Hours	Unit	Material	2021 Bare Costs Labor	2021 Bare Costs Equipment	Total	Total Incl O&P
0010 ROOF DRAINS									
0140 Cornice, CI, 45° or 90° outlet									
0200 3" and 4" pipe size	Q-1	12	1.333	Ea.	400	81		481	560
0260 For galvanized body, add					128			128	141
0280 For polished bronze dome, add				↓	95.50			95.50	105
3860 Roof, flat metal deck, CI body, 12" CI dome									
3890 3" pipe size	Q-1	14	1.143	Ea.	470	69.50		539.50	620
3900 4" pipe size	"	13	1.231	"	565	75		640	735
4620 Main, all aluminum, 12" low profile dome									
4640 2", 3" and 4" pipe size	Q-1	14	1.143	Ea.	665	69.50		734.50	835
9000 Minimum labor/equipment charge	1 Plum	4	2	Job		135		135	209

22 14 29 – Sump Pumps

22 14 29.16 Submersible Sump Pumps

	Crew	Daily Output	Labor-Hours	Unit	Material	2021 Bare Costs Labor	2021 Bare Costs Equipment	Total	Total Incl O&P
0010 SUBMERSIBLE SUMP PUMPS									
1000 Elevator sump pumps, automatic									
1010 Complete systems, pump, oil detector, controls and alarm									
1020 1-1/2" discharge, does not include the sump pit/tank									
1040 1/3 HP, 115 V	1 Plum	4.40	1.818	Ea.	1,900	123		2,023	2,300
1050 1/2 HP, 115 V		4	2		2,025	135		2,160	2,425
1060 1/2 HP, 230 V		4	2		2,000	135		2,135	2,400
1070 3/4 HP, 115 V		3.60	2.222		2,125	150		2,275	2,550
1080 3/4 HP, 230 V	↓	3.60	2.222	↓	2,100	150		2,250	2,525
1100 Sump pump only									
1110 1/3 HP, 115 V	1 Plum	6.40	1.250	Ea.	177	84.50		261.50	325
1120 1/2 HP, 115 V		5.80	1.379		253	93.50		346.50	425
1130 1/2 HP, 230 V		5.80	1.379		294	93.50		387.50	470
1140 3/4 HP, 115 V		5.40	1.481		335	100		435	525
1150 3/4 HP, 230 V	↓	5.40	1.481	↓	380	100		480	575
1200 Oil detector, control and alarm only									
1210 115 V	1 Plum	8	1	Ea.	1,825	67.50		1,892.50	2,125
1220 230 V	"	8	1	"	1,825	67.50		1,892.50	2,125
7000 Sump pump, automatic									
7100 Plastic, 1-1/4" discharge, 1/4 HP	1 Plum	6.40	1.250	Ea.	173	84.50		257.50	320
7140 1/3 HP		6	1.333		229	90.50		319.50	390
7180 1-1/2" discharge, 1/2 HP		5.20	1.538		350	104		454	545
7500 Cast iron, 1-1/4" discharge, 1/4 HP		6	1.333		243	90.50		333.50	405
7540 1/3 HP		6	1.333		285	90.50		375.50	455
7560 1/2 HP		5	1.600	↓	345	108		453	545
9000 Minimum labor/equipment charge	↓	4	2	Job		135		135	209

22 15 General Service Compressed-Air Systems

22 15 19 – General Service Packaged Air Compressors and Receivers

22 15 19.10 Air Compressors

	Crew	Daily Output	Labor-Hours	Unit	Material	2021 Bare Costs Labor	2021 Bare Costs Equipment	Total	Total Incl O&P
0010 AIR COMPRESSORS									
5250 Air, reciprocating air cooled, splash lubricated, tank mounted									
5300 Single stage, 1 phase, 140 psi									
5303 1/2 HP, 17 gal. tank	1 Stpi	3	2.667	Ea.	1,975	182		2,157	2,425
5305 3/4 HP, 30 gal. tank		2.60	3.077		1,625	210		1,835	2,100
5307 1 HP, 30 gal. tank	↓	2.20	3.636	↓	2,375	249		2,624	3,000

22 31 Domestic Water Softeners

22 31 13 – Residential Domestic Water Softeners

22 31 13.10 Residential Water Softeners		Crew	Daily Output	Labor-Hours	Unit	Material	2021 Bare Costs Labor	Equipment	Total	Total Incl O&P
0010	**RESIDENTIAL WATER SOFTENERS**									
7350	Water softener, automatic, to 30 grains per gallon	2 Plum	5	3.200	Ea.	405	217		622	780
7400	To 100 grains per gallon	"	4	4	"	865	271		1,136	1,375

22 33 Electric Domestic Water Heaters

22 33 13 – Instantaneous Electric Domestic Water Heaters

22 33 13.20 Instantaneous Elec. Point-Of-Use Water Heaters

			Crew	Daily Output	Labor-Hours	Unit	Material	2021 Bare Costs Labor	Equipment	Total	Total Incl O&P
0010	**INSTANTANEOUS ELECTRIC POINT-OF-USE WATER HEATERS**										
8965	Point of use, electric, glass lined										
8969	Energy saver										
8970	2.5 gal. single element	G	1 Plum	2.80	2.857	Ea.	315	193		508	645
8971	4 gal. single element	G		2.80	2.857		360	193		553	695
8974	6 gal. single element	G		2.50	3.200		465	217		682	850
8975	10 gal. single element	G		2.50	3.200		465	217		682	845
8976	15 gal. single element	G		2.40	3.333		545	226		771	945
8977	20 gal. single element	G		2.40	3.333		685	226		911	1,100
8978	30 gal. single element	G		2.30	3.478		700	235		935	1,125
8979	40 gal. single element	G		2.20	3.636		1,375	246		1,621	1,900
8988	Commercial (ASHRAE energy std. 90)										
8989	6 gallon	G	1 Plum	2.50	3.200	Ea.	1,025	217		1,242	1,475
8990	10 gallon	G		2.50	3.200		1,200	217		1,417	1,650
8991	15 gallon	G		2.40	3.333		1,200	226		1,426	1,675
8992	20 gallon	G		2.40	3.333		1,375	226		1,601	1,875
8993	30 gallon	G		2.30	3.478		2,925	235		3,160	3,575
8995	Under the sink, copper, w/bracket										
8996	2.5 gallon	G	1 Plum	4	2	Ea.	410	135		545	665
9000	Minimum labor/equipment charge		"	1.75	4.571	Job		310		310	475

22 33 30 – Residential, Electric Domestic Water Heaters

22 33 30.13 Residential, Small-Capacity Elec. Water Heaters

		Crew	Daily Output	Labor-Hours	Unit	Material	2021 Bare Costs Labor	Equipment	Total	Total Incl O&P
0010	**RESIDENTIAL, SMALL-CAPACITY ELECTRIC DOMESTIC WATER HEATERS**									
1000	Residential, electric, glass lined tank, 5 yr., 10 gal., single element	1 Plum	2.30	3.478	Ea.	465	235		700	875
1060	30 gallon, double element		2.20	3.636		1,225	246		1,471	1,725
1080	40 gallon, double element		2	4		1,400	271		1,671	1,950
1100	52 gallon, double element		2	4		1,575	271		1,846	2,175
1120	66 gallon, double element		1.80	4.444		2,150	300		2,450	2,825
1140	80 gallon, double element		1.60	5		2,425	340		2,765	3,200

22 33 33 – Light-Commercial Electric Domestic Water Heaters

22 33 33.10 Commercial Electric Water Heaters

		Crew	Daily Output	Labor-Hours	Unit	Material	2021 Bare Costs Labor	Equipment	Total	Total Incl O&P
0010	**COMMERCIAL ELECTRIC WATER HEATERS**									
4000	Commercial, 100° rise. NOTE: for each size tank, a range of									
4010	heaters between the ones shown is available									
4020	Electric									
4100	5 gal., 3 kW, 12 GPH, 208 volt	1 Plum	2	4	Ea.	4,200	271		4,471	5,025
4160	50 gal., 36 kW, 148 GPH, 208 volt	"	1.80	4.444		9,700	300		10,000	11,200
4480	400 gal., 210 kW, 860 GPH, 480 volt	Q-1	1	16		84,000	975		84,975	94,000

443

For customer support on your Commercial Renovation Costs with RSMeans Data, call 800.448.8182.

22 34 Fuel-Fired Domestic Water Heaters

22 34 13 – Instantaneous, Tankless, Gas Domestic Water Heaters

22 34 13.10 Instantaneous, Tankless, Gas Water Heaters		Crew	Daily Output	Labor-Hours	Unit	Material	2021 Bare Costs Labor	2021 Bare Costs Equipment	Total	Total Incl O&P
0010	**INSTANTANEOUS, TANKLESS, GAS WATER HEATERS**									
9410	Natural gas/propane, 3.2 GPM G	1 Plum	2	4	Ea.	605	271		876	1,075
9420	6.4 GPM G		1.90	4.211		770	285		1,055	1,300
9430	8.4 GPM G		1.80	4.444		865	300		1,165	1,425
9440	9.5 GPM G		1.60	5		1,025	340		1,365	1,650

22 34 30 – Residential Gas Domestic Water Heaters

22 34 30.13 Residential, Atmos, Gas Domestic Wtr Heaters

22 34 30.13		Crew	Daily Output	Labor-Hours	Unit	Material	2021 Bare Costs Labor	2021 Bare Costs Equipment	Total	Total Incl O&P
0010	**RESIDENTIAL, ATMOSPHERIC, GAS DOMESTIC WATER HEATERS**									
2000	Gas fired, foam lined tank, 10 yr., vent not incl.									
2040	30 gallon	1 Plum	2	4	Ea.	2,225	271		2,496	2,850
2060	40 gallon		1.90	4.211		1,850	285		2,135	2,475
2100	75 gallon		1.50	5.333		2,575	360		2,935	3,375
3000	Tank leak safety, water & gas shut off see 22 05 23.20 8800									

22 34 36 – Commercial Gas Domestic Water Heaters

22 34 36.13 Commercial, Atmos., Gas Domestic Water Htrs.

22 34 36.13		Crew	Daily Output	Labor-Hours	Unit	Material	2021 Bare Costs Labor	2021 Bare Costs Equipment	Total	Total Incl O&P
0010	**COMMERCIAL, ATMOSPHERIC, GAS DOMESTIC WATER HEATERS**									
6000	Gas fired, flush jacket, std. controls, vent not incl.									
6040	75 MBH input, 73 GPH	1 Plum	1.40	5.714	Ea.	3,700	385		4,085	4,675
6060	98 MBH input, 95 GPH		1.40	5.714		9,075	385		9,460	10,600
6180	200 MBH input, 192 GPH		.60	13.333		11,700	905		12,605	14,300
6900	For low water cutoff, add		8	1		405	67.50		472.50	550
6960	For bronze body hot water circulator, add		4	2		2,300	135		2,435	2,725

22 34 46 – Oil-Fired Domestic Water Heaters

22 34 46.10 Residential Oil-Fired Water Heaters

22 34 46.10		Crew	Daily Output	Labor-Hours	Unit	Material	2021 Bare Costs Labor	2021 Bare Costs Equipment	Total	Total Incl O&P
0010	**RESIDENTIAL OIL-FIRED WATER HEATERS**									
3000	Oil fired, glass lined tank, 5 yr., vent not included, 30 gallon	1 Plum	2	4	Ea.	1,675	271		1,946	2,275
3040	50 gallon	"	1.80	4.444	"	1,525	300		1,825	2,150

22 34 46.20 Commercial Oil-Fired Water Heaters

22 34 46.20		Crew	Daily Output	Labor-Hours	Unit	Material	2021 Bare Costs Labor	2021 Bare Costs Equipment	Total	Total Incl O&P
0010	**COMMERCIAL OIL-FIRED WATER HEATERS**									
8000	Oil fired, glass lined, UL listed, std. controls, vent not incl.									
8060	140 gal., 140 MBH input, 134 GPH	Q-1	2.13	7.512	Ea.	24,900	460		25,360	28,100
8080	140 gal., 199 MBH input, 191 GPH		2	8		25,700	485		26,185	29,100
8160	140 gal., 540 MBH input, 519 GPH		.96	16.667		35,100	1,025		36,125	40,200
8280	201 gal., 1,250 MBH input, 1,200 GPH	Q-2	1.22	19.672		55,000	1,250		56,250	62,500

22 41 Residential Plumbing Fixtures

22 41 13 – Residential Water Closets, Urinals, and Bidets

22 41 13.13 Water Closets

22 41 13.13		Crew	Daily Output	Labor-Hours	Unit	Material	2021 Bare Costs Labor	2021 Bare Costs Equipment	Total	Total Incl O&P
0010	**WATER CLOSETS**									
0032	For automatic flush, see Line 22 42 39.10 0972									
0150	Tank type, vitreous china, incl. seat, supply pipe w/stop, 1.6 gpf or noted									
0200	Wall hung									
0400	Two piece, close coupled	Q-1	5.30	3.019	Ea.	440	184		624	765
0960	For rough-in, supply, waste, vent and carrier	"	2.73	5.861	"	1,425	355		1,780	2,125
0999	Floor mounted									
1100	Two piece, close coupled	Q-1	5.30	3.019	Ea.	191	184		375	495
1102	Economy		5.30	3.019		106	184		290	400
1110	Two piece, close coupled, dual flush		5.30	3.019		284	184		468	595
1140	Two piece, close coupled, 1.28 gpf, ADA G		5.30	3.019		268	184		452	580

22 41 Residential Plumbing Fixtures

22 41 13 – Residential Water Closets, Urinals, and Bidets

22 41 13.13 Water Closets

		Crew	Daily Output	Labor-Hours	Unit	Material	2021 Bare Costs Labor	2021 Bare Costs Equipment	Total	Total Incl O&P
1960	For color, add					30%				
1980	For rough-in, supply, waste and vent	Q-1	3.05	5.246	Ea.	345	320		665	875

22 41 16 – Residential Lavatories and Sinks

22 41 16.13 Lavatories

		Crew	Daily Output	Labor-Hours	Unit	Material	2021 Bare Costs Labor	2021 Bare Costs Equipment	Total	Total Incl O&P
0010	**LAVATORIES**, With trim, white unless noted otherwise									
0500	Vanity top, porcelain enamel on cast iron									
0600	20" x 18"	Q-1	6.40	2.500	Ea.	400	152		552	675
0640	33" x 19" oval	"	6.40	2.500	"	560	152		712	850
0860	For color, add					25%				
1000	Cultured marble, 19" x 17", single bowl	Q-1	6.40	2.500	Ea.	159	152		311	410
1120	25" x 22", single bowl	"	6.40	2.500	"	179	152		331	430
1580	For color, same price									
1900	Stainless steel, self-rimming, 25" x 22", single bowl, ledge	Q-1	6.40	2.500	Ea.	345	152		497	615
1960	17" x 22", single bowl		6.40	2.500		335	152		487	605
2600	Steel, enameled, 20" x 17", single bowl		5.80	2.759		169	168		337	445
2900	Vitreous china, 20" x 16", single bowl		5.40	2.963		242	181		423	545
2960	20" x 17", single bowl		5.40	2.963		145	181		326	440
3580	Rough-in, supply, waste and vent for all above lavatories		2.30	6.957		405	425		830	1,100
4000	Wall hung									
4040	Porcelain enamel on cast iron, 16" x 14", single bowl	Q-1	8	2	Ea.	510	122		632	750
4180	20" x 18", single bowl	"	8	2	"	305	122		427	525
4580	For color, add					30%				
6000	Vitreous china, 18" x 15", single bowl with backsplash	Q-1	7	2.286	Ea.	192	139		331	425
6960	Rough-in, supply, waste and vent for above lavatories	"	1.66	9.639	"	665	585		1,250	1,625
7000	Pedestal type									
7600	Vitreous china, 27" x 21", white	Q-1	6.60	2.424	Ea.	730	148		878	1,025
7610	27" x 21", colored		6.60	2.424		920	148		1,068	1,250
7620	27" x 21", premium color		6.60	2.424		1,050	148		1,198	1,375
7660	26" x 20", white		6.60	2.424		710	148		858	1,000
7670	26" x 20", colored		6.60	2.424		895	148		1,043	1,225
7680	26" x 20", premium color		6.60	2.424		1,075	148		1,223	1,425
7700	24" x 20", white		6.60	2.424		490	148		638	765
7710	24" x 20", colored		6.60	2.424		590	148		738	880
7720	24" x 20", premium color		6.60	2.424		650	148		798	945
7760	21" x 18", white		6.60	2.424		293	148		441	550
7770	21" x 18", colored		6.60	2.424		335	148		483	595
7990	Rough-in, supply, waste and vent for pedestal lavatories		1.66	9.639		665	585		1,250	1,625
9000	Minimum labor/equipment charge	1 Plum	3	2.667	Job		181		181	278

22 41 16.16 Sinks

		Crew	Daily Output	Labor-Hours	Unit	Material	2021 Bare Costs Labor	2021 Bare Costs Equipment	Total	Total Incl O&P
0010	**SINKS**, With faucets and drain									
2000	Kitchen, counter top style, PE on CI, 24" x 21" single bowl	Q-1	5.60	2.857	Ea.	305	174		479	605
2100	31" x 22" single bowl		5.60	2.857		910	174		1,084	1,275
2200	32" x 21" double bowl		4.80	3.333		390	203		593	745
3000	Stainless steel, self rimming, 19" x 18" single bowl		5.60	2.857		630	174		804	960
3100	25" x 22" single bowl		5.60	2.857		695	174		869	1,025
3200	33" x 22" double bowl		4.80	3.333		1,000	203		1,203	1,425
3300	43" x 22" double bowl		4.80	3.333		1,175	203		1,378	1,600
4000	Steel, enameled, with ledge, 24" x 21" single bowl		5.60	2.857		480	174		654	800
4100	32" x 21" double bowl		4.80	3.333		480	203		683	845
4960	For color sinks except stainless steel, add					10%				
4980	For rough-in, supply, waste and vent, counter top sinks	Q-1	2.14	7.477		355	455		810	1,100
5790	For rough-in, supply, waste & vent, sinks	"	1.85	8.649		355	525		880	1,200

22 41 19 – Residential Bathtubs

22 41 19.10 Baths

		Crew	Daily Output	Labor-Hours	Unit	Material	2021 Bare Costs Labor	Equipment	Total	Total Incl O&P
0010	**BATHS**									
0100	Tubs, recessed porcelain enamel on cast iron, with trim									
0180	48" x 42"	Q-1	4	4	Ea.	3,225	244		3,469	3,925
0220	72" x 36"	"	3	5.333	"	3,150	325		3,475	3,975
0300	Mat bottom									
0380	5' long	Q-1	4.40	3.636	Ea.	1,125	222		1,347	1,600
0560	Corner 48" x 44"		4.40	3.636		2,925	222		3,147	3,575
2000	Enameled formed steel, 4'-6" long		5.80	2.759		540	168		708	855
4600	Module tub & showerwall surround, molded fiberglass									
4610	5' long x 34" wide x 76" high	Q-1	4	4	Ea.	830	244		1,074	1,275
4750	ADA compliant with 1-1/2" OD grab bar, antiskid bottom									
4760	60" x 32-3/4" x 72" high	Q-1	4	4	Ea.	595	244		839	1,025
4770	60" x 30" x 71" high with molded seat		3.50	4.571		750	279		1,029	1,250
9600	Rough-in, supply, waste and vent, for all above tubs, add		2.07	7.729		460	470		930	1,225
9900	Minimum labor/equipment charge		3	5.333	Job		325		325	500

22 41 23 – Residential Showers

22 41 23.20 Showers

		Crew	Daily Output	Labor-Hours	Unit	Material	2021 Bare Costs Labor	Equipment	Total	Total Incl O&P
0011	**SHOWERS**, Stall, with drain only									
1520	32" square	Q-1	5	3.200	Ea.	1,150	195		1,345	1,575
1530	36" square		4.80	3.333		3,075	203		3,278	3,725
1540	Terrazzo receptor, 32" square		5	3.200		1,350	195		1,545	1,775
1580	36" corner angle		4.80	3.333		2,200	203		2,403	2,750
3000	Fiberglass, one piece, with 3 walls, 32" x 32" square		5.50	2.909		370	177		547	685
3100	36" x 36" square		5.50	2.909		415	177		592	735
4200	Rough-in, supply, waste and vent for above showers		2.05	7.805		445	475		920	1,225

22 41 23.40 Shower System Components

		Crew	Daily Output	Labor-Hours	Unit	Material	2021 Bare Costs Labor	Equipment	Total	Total Incl O&P
0010	**SHOWER SYSTEM COMPONENTS**									
5500	Head, water economizer, 1.6 GPM [G]	1 Plum	24	.333	Ea.	53.50	22.50		76	94

22 41 36 – Residential Laundry Trays

22 41 36.10 Laundry Sinks

		Crew	Daily Output	Labor-Hours	Unit	Material	2021 Bare Costs Labor	Equipment	Total	Total Incl O&P
0010	**LAUNDRY SINKS**, With trim									
0020	Porcelain enamel on cast iron, black iron frame									
0050	24" x 21", single compartment	Q-1	6	2.667	Ea.	620	162		782	930
0100	26" x 21", single compartment	"	6	2.667	"	680	162		842	995
2000	Molded stone, on wall hanger or legs									
2020	22" x 23", single compartment	Q-1	6	2.667	Ea.	179	162		341	450
2100	45" x 21", double compartment	"	5	3.200	"	365	195		560	700
3000	Plastic, on wall hanger or legs									
3020	18" x 23", single compartment	Q-1	6.50	2.462	Ea.	146	150		296	390
3100	20" x 24", single compartment		6.50	2.462		163	150		313	410
3200	36" x 23", double compartment		5.50	2.909		216	177		393	510
3300	40" x 24", double compartment		5.50	2.909		284	177		461	585
5000	Stainless steel, counter top, 22" x 17" single compartment		6	2.667		77	162		239	335
5200	33" x 22", double compartment		5	3.200		92.50	195		287.50	400
9600	Rough-in, supply, waste and vent, for all laundry sinks		2.14	7.477		355	455		810	1,100
9810	Minimum labor/equipment charge	1 Plum	3	2.667	Job		181		181	278

22 41 39 – Residential Faucets, Supplies and Trim

22 41 39.10 Faucets and Fittings

		Crew	Daily Output	Labor-Hours	Unit	Material	2021 Bare Costs Labor	Equipment	Total	Total Incl O&P
0010	**FAUCETS AND FITTINGS**									
0150	Bath, faucets, diverter spout combination, sweat	1 Plum	8	1	Ea.	86	67.50		153.50	199

22 41 39 – Residential Faucets, Supplies and Trim

22 41 39.10 Faucets and Fittings	Crew	Daily Output	Labor-Hours	Unit	Material	2021 Bare Costs Labor	Equipment	Total	Total Incl O&P
0200 For integral stops, IPS unions, add				Ea.	111			111	123
0420 Bath, press-bal mix valve w/diverter, spout, shower head, arm/flange	1 Plum	8	1		200	67.50		267.50	325
0500 Drain, central lift, 1-1/2" IPS male		20	.400		53	27		80	101
0600 Trip lever, 1-1/2" IPS male	↓	20	.400	↓	60	27		87	108
0810 Bidet									
0812 Fitting, over the rim, swivel spray/pop-up drain	1 Plum	8	1	Ea.	231	67.50		298.50	360
1000 Kitchen sink faucets, top mount, cast spout		10	.800		83.50	54		137.50	175
1100 For spray, add		24	.333		17.20	22.50		39.70	54
1200 Wall type, swing tube spout	↓	10	.800	↓	73.50	54		127.50	165
1300 Single control lever handle									
1310 With pull out spray									
1320 Polished chrome	1 Plum	10	.800	Ea.	225	54		279	330
2000 Laundry faucets, shelf type, IPS or copper unions		12	.667		61.50	45		106.50	137
2100 Lavatory faucet, centerset, without drain		10	.800		67	54		121	158
2120 With pop-up drain	↓	6.66	1.201	↓	86.50	81.50		168	220
2210 Porcelain cross handles and pop-up drain									
2220 Polished chrome	1 Plum	6.66	1.201	Ea.	214	81.50		295.50	360
2230 Polished brass	"	6.66	1.201	"	283	81.50		364.50	435
2260 Single lever handle and pop-up drain									
2280 Satin nickel	1 Plum	6.66	1.201	Ea.	300	81.50		381.50	455
2290 Polished chrome		6.66	1.201		215	81.50		296.50	360
2800 Self-closing, center set		10	.800		149	54		203	248
2810 Automatic sensor and operator, with faucet head [G]		6.15	1.301		590	88		678	785
4000 Shower by-pass valve with union		18	.444		57	30		87	110
4100 Shower arm with flange and head		22	.364		19.10	24.50		43.60	59
4140 Shower, hand held, pin mount, massage action, chrome		22	.364		73.50	24.50		98	119
4142 Polished brass		22	.364		160	24.50		184.50	214
4144 Shower, hand held, wall mtd, adj. spray, 2 wall mounts, chrome		20	.400		115	27		142	168
4146 Polished brass		20	.400		245	27		272	310
4148 Shower, hand held head, bar mounted 24", adj. spray, chrome		20	.400		184	27		211	244
4150 Polished brass		20	.400		380	27		407	455
4200 Shower thermostatic mixing valve, concealed, with shower head trim kit	↓	8	1	↓	365	67.50		432.50	505
4220 Shower pressure balancing mixing valve									
4230 With shower head, arm, flange and diverter tub spout									
4240 Chrome	1 Plum	6.14	1.303	Ea.	420	88		508	595
4250 Satin nickel		6.14	1.303		550	88		638	740
4260 Polished graphite		6.14	1.303		535	88		623	725
5000 Sillcock, compact, brass, IPS or copper to hose	↓	24	.333	↓	12.90	22.50		35.40	49
6000 Stop and waste valves, bronze									
6100 Angle, solder end 1/2"	1 Plum	24	.333	Ea.	42.50	22.50		65	81.50
6110 3/4"		20	.400		49	27		76	96
6300 Straightway, solder end 3/8"		24	.333		19.30	22.50		41.80	56.50
6310 1/2"		24	.333		21.50	22.50		44	58.50
6410 Straightway, threaded 1/2"		24	.333		29	22.50		51.50	67
6420 3/4"		20	.400		32.50	27		59.50	77.50
6430 1"		19	.421		22	28.50		50.50	68
7800 Water closet, wax gasket		96	.083		1.25	5.65		6.90	10.10
7820 Gasket toilet tank to bowl		32	.250		2.89	16.95		19.84	29
7830 Replacement diaphragm washer assy for ballcock valve		12	.667		2.89	45		47.89	72.50
7850 Dual flush valve	↓	12	.667		111	45		156	192
8000 Water supply stops, polished chrome plate									
8200 Angle, 3/8"	1 Plum	24	.333	Ea.	11.30	22.50		33.80	47.50
8300 1/2"	↓	22	.364	↓	12.30	24.50		36.80	51.50

22 41 Residential Plumbing Fixtures

22 41 39 – Residential Faucets, Supplies and Trim

22 41 39.10 Faucets and Fittings	Crew	Daily Output	Labor-Hours	Unit	Material	2021 Bare Costs Labor	Equipment	Total	Total Incl O&P
8400	Straight, 3/8"	1 Plum	26	.308	Ea.	14.65	21		35.65	48
8500	1/2"		24	.333		11.85	22.50		34.35	48
8600	Water closet, angle, w/flex riser, 3/8"		24	.333		40	22.50		62.50	79
9000	Minimum labor/equipment charge		4	2	Job		135		135	209

22 42 Commercial Plumbing Fixtures

22 42 13 – Commercial Water Closets, Urinals, and Bidets

22 42 13.13 Water Closets

		Crew	Daily Output	Labor-Hours	Unit	Material	2021 Bare Costs Labor	Equipment	Total	Total Incl O&P
0010	**WATER CLOSETS**									
3000	Bowl only, with flush valve, seat, 1.6 gpf unless noted									
3100	Wall hung	Q-1	5.80	2.759	Ea.	945	168		1,113	1,300
3200	For rough-in, supply, waste and vent, single WC		2.56	6.250		1,450	380		1,830	2,175
3360	With floor outlet, 1.28 gpf [G]		5.80	2.759		550	168		718	865
3362	With floor outlet, 1.28 gpf, ADA [G]		5.80	2.759		575	168		743	890
3370	For rough-in, supply, waste and vent, single WC		2.84	5.634		375	345		720	940
3390	Floor mounted children's size, 10-3/4" high									
3392	With automatic flush sensor, 1.6 gpf	Q-1	6.20	2.581	Ea.	660	157		817	965
3396	With automatic flush sensor, 1.28 gpf		6.20	2.581	"	610	157		767	910
9000	Minimum labor/equipment charge		4	4	Job		244		244	375

22 42 13.16 Urinals

		Crew	Daily Output	Labor-Hours	Unit	Material	2021 Bare Costs Labor	Equipment	Total	Total Incl O&P
0010	**URINALS**									
3000	Wall hung, vitreous china, with self-closing valve									
3100	Siphon jet type	Q-1	3	5.333	Ea.	335	325		660	870
3120	Blowout type		3	5.333		495	325		820	1,050
3140	Water saving .5 gpf [G]		3	5.333		575	325		900	1,125
3300	Rough-in, supply, waste & vent		2.83	5.654		835	345		1,180	1,450
5000	Stall type, vitreous china, includes valve		2.50	6.400		765	390		1,155	1,450
6980	Rough-in, supply, waste and vent		1.99	8.040		500	490		990	1,300
9000	Minimum labor/equipment charge		4	4	Job		244		244	375

22 42 16 – Commercial Lavatories and Sinks

22 42 16.13 Lavatories

0010	**LAVATORIES**, With trim, white unless noted otherwise									
0020	Commercial lavatories same as residential. See Section 22 41 16									

22 42 16.40 Service Sinks

		Crew	Daily Output	Labor-Hours	Unit	Material	2021 Bare Costs Labor	Equipment	Total	Total Incl O&P
0010	**SERVICE SINKS**									
6650	Service, floor, corner, PE on CI, 28" x 28"	Q-1	4.40	3.636	Ea.	1,350	222		1,572	1,850
6790	For rough-in, supply, waste & vent, floor service sinks		1.64	9.756	"	1,350	595		1,945	2,425
9000	Minimum labor/equipment charge		4	4	Job		244		244	375

22 42 23 – Commercial Showers

22 42 23.30 Group Showers

		Crew	Daily Output	Labor-Hours	Unit	Material	2021 Bare Costs Labor	Equipment	Total	Total Incl O&P
0010	**GROUP SHOWERS**									
6000	Group, w/pressure balancing valve, rough-in and rigging not included									
6800	Column, 6 heads, no receptors, less partitions	Q-1	3	5.333	Ea.	9,675	325		10,000	11,100
6900	With stainless steel partitions		1	16		12,800	975		13,775	15,600
7600	5 heads, no receptors, less partitions		3	5.333		6,650	325		6,975	7,800
7620	4 heads (1 ADA compliant) no receptors, less partitions		3	5.333		6,175	325		6,500	7,275
7700	With stainless steel partitions		1	16		6,650	975		7,625	8,825
9000	Minimum labor/equipment charge		4	4	Job		244		244	375

448

22 42 Commercial Plumbing Fixtures

22 42 33 – Wash Fountains

22 42 33.20 Commercial Wash Fountains

	22 42 33.20 Commercial Wash Fountains	Crew	Daily Output	Labor-Hours	Unit	Material	2021 Bare Costs Labor	2021 Bare Costs Equipment	Total	Total Incl O&P
0010	**COMMERCIAL WASH FOUNTAINS**									
1900	Group, foot control									
2000	Precast terrazzo, circular, 36" diam., 5 or 6 persons	Q-2	3	8	Ea.	7,225	505		7,730	8,725
2100	54" diam. for 8 or 10 persons		2.50	9.600		10,900	605		11,505	12,900
2400	Semi-circular, 36" diam. for 3 persons		3	8		6,100	505		6,605	7,475
2500	54" diam. for 4 or 5 persons		2.50	9.600		11,200	605		11,805	13,200
5610	Group, infrared control, barrier free									
5614	Precast terrazzo									
5620	Semi-circular 36" diam. for 3 persons	Q-2	3	8	Ea.	7,500	505		8,005	9,025
5630	46" diam. for 4 persons		2.80	8.571		8,125	540		8,665	9,750
5640	Circular, 54" diam. for 8 persons, button control		2.50	9.600		9,750	605		10,355	11,600
5700	Rough-in, supply, waste and vent for above wash fountains	Q-1	1.82	8.791		510	535		1,045	1,375
9000	Minimum labor/equipment charge	Q-2	3	8	Job		505		505	780

22 42 39 – Commercial Faucets, Supplies, and Trim

22 42 39.10 Faucets and Fittings

	22 42 39.10 Faucets and Fittings	Crew	Daily Output	Labor-Hours	Unit	Material	2021 Bare Costs Labor	2021 Bare Costs Equipment	Total	Total Incl O&P
0010	**FAUCETS AND FITTINGS**									
0840	Flush valves, with vacuum breaker									
0850	Water closet									
0860	Exposed, rear spud	1 Plum	8	1	Ea.	152	67.50		219.50	271
0870	Top spud		8	1		198	67.50		265.50	320
0880	Concealed, rear spud		8	1		212	67.50		279.50	335
0890	Top spud		8	1		172	67.50		239.50	293
0900	Wall hung		8	1		223	67.50		290.50	350
0910	Dual flush flushometer		12	.667		237	45		282	330
0912	Flushometer retrofit kit		18	.444		17.70	30		47.70	66
0920	Urinal									
0930	Exposed, stall	1 Plum	8	1	Ea.	198	67.50		265.50	320
0940	Wall (washout)		8	1		163	67.50		230.50	284
0950	Pedestal, top spud		8	1		142	67.50		209.50	261
0960	Concealed, stall		8	1		169	67.50		236.50	290
0970	Wall (washout)		8	1		182	67.50		249.50	305
0971	Automatic flush sensor and operator for									
0972	urinals or water closets, standard [G]	1 Plum	8	1	Ea.	525	67.50		592.50	680
0980	High efficiency water saving									
0984	Water closets, 1.28 gpf [G]	1 Plum	8	1	Ea.	435	67.50		502.50	585
0988	Urinals, .5 gpf [G]	"	8	1	"	435	67.50		502.50	585
2790	Faucets for lavatories									
2800	Self-closing, center set	1 Plum	10	.800	Ea.	149	54		203	248
2810	Automatic sensor and operator, with faucet head		6.15	1.301		590	88		678	785
3000	Service sink faucet, cast spout, pail hook, hose end		14	.571		76	38.50		114.50	143

22 42 39.30 Carriers and Supports

	22 42 39.30 Carriers and Supports	Crew	Daily Output	Labor-Hours	Unit	Material	2021 Bare Costs Labor	2021 Bare Costs Equipment	Total	Total Incl O&P
0010	**CARRIERS AND SUPPORTS**, For plumbing fixtures									
3000	Lavatory, concealed arm									
3050	Floor mounted, single									
3100	High back fixture	1 Plum	6	1.333	Ea.	800	90.50		890.50	1,025
3200	Flat slab fixture		6	1.333		935	90.50		1,025.50	1,175
3220	ADA compliant		6	1.333		770	90.50		860.50	985
6980	Water closet, siphon jet									
7000	Horizontal, adjustable, caulk									
7040	Single, 4" pipe size	1 Plum	5.33	1.501	Ea.	1,275	102		1,377	1,575
7060	5" pipe size		5.33	1.501		1,125	102		1,227	1,400

22 42 Commercial Plumbing Fixtures

22 42 39 – Commercial Faucets, Supplies, and Trim

22 42 39.30 Carriers and Supports		Crew	Daily Output	Labor-Hours	Unit	Material	2021 Bare Costs Labor	Equipment	Total	Total Incl O&P
7100	Double, 4" pipe size	1 Plum	5	1.600	Ea.	2,400	108		2,508	2,825
7120	5" pipe size	↓	5	1.600	↓	1,625	108		1,733	1,975
8200	Water closet, residential									
8220	Vertical centerline, floor mount									
8240	Single, 3" caulk, 2" or 3" vent	1 Plum	6	1.333	Ea.	925	90.50		1,015.50	1,175
8260	4" caulk, 2" or 4" vent		6	1.333	"	1,200	90.50		1,290.50	1,450
9990	Minimum labor/equipment charge	↓	3.50	2.286	Job		155		155	239

22 43 Healthcare Plumbing Fixtures

22 43 39 – Healthcare Faucets

22 43 39.10 Faucets and Fittings

		Crew	Daily Output	Labor-Hours	Unit	Material	2021 Bare Costs Labor	Equipment	Total	Total Incl O&P
0010	**FAUCETS AND FITTINGS**									
2850	Medical, bedpan cleanser, with pedal valve,	1 Plum	12	.667	Ea.	870	45		915	1,025
2860	With screwdriver stop valve		12	.667		450	45		495	565
2870	With self-closing spray valve		12	.667		278	45		323	375
2900	Faucet, gooseneck spout, wrist handles, grid drain		10	.800		145	54		199	243
2940	Mixing valve, knee action, screwdriver stops	↓	4	2	↓	485	135		620	745

22 45 Emergency Plumbing Fixtures

22 45 13 – Emergency Showers

22 45 13.10 Emergency Showers

		Crew	Daily Output	Labor-Hours	Unit	Material	2021 Bare Costs Labor	Equipment	Total	Total Incl O&P
0010	**EMERGENCY SHOWERS**, Rough-in not included									
5000	Shower, single head, drench, ball valve, pull, freestanding	Q-1	4	4	Ea.	385	244		629	795
5200	Horizontal or vertical supply		4	4		630	244		874	1,075
6000	Multi-nozzle, eye/face wash combination		4	4		840	244		1,084	1,300
6400	Multi-nozzle, 12 spray, shower only		4	4		2,050	244		2,294	2,625
6600	For freeze-proof, add		6	2.667	↓	520	162		682	820
9000	Minimum labor/equipment charge	↓	3	5.333	Job		325		325	500

22 45 16 – Eyewash Equipment

22 45 16.10 Eyewash Safety Equipment

		Crew	Daily Output	Labor-Hours	Unit	Material	2021 Bare Costs Labor	Equipment	Total	Total Incl O&P
0010	**EYEWASH SAFETY EQUIPMENT**, Rough-in not included									
1000	Eye wash fountain									
1400	Plastic bowl, pedestal mounted	Q-1	4	4	Ea.	340	244		584	750

22 47 Drinking Fountains and Water Coolers

22 47 13 – Drinking Fountains

22 47 13.10 Drinking Water Fountains

		Crew	Daily Output	Labor-Hours	Unit	Material	2021 Bare Costs Labor	Equipment	Total	Total Incl O&P
0010	**DRINKING WATER FOUNTAINS**, For connection to cold water supply									
1000	Wall mounted, non-recessed									
2700	Stainless steel, single bubbler, no back	1 Plum	4	2	Ea.	870	135		1,005	1,175
2740	With back		4	2		1,125	135		1,260	1,450
2780	Dual handle, ADA compliant		4	2		785	135		920	1,075
2820	Dual level, ADA compliant		3.20	2.500		1,450	169		1,619	1,850
3980	For rough-in, supply and waste, add	↓	2.21	3.620	↓	370	245		615	785
4000	Wall mounted, semi-recessed									
4200	Poly-marble, single bubbler	1 Plum	4	2	Ea.	950	135		1,085	1,250
4600	Stainless steel, satin finish, single bubbler	"	4	2	"	1,700	135		1,835	2,075

22 47 Drinking Fountains and Water Coolers

22 47 13 – Drinking Fountains

22 47 13.10 Drinking Water Fountains	Crew	Daily Output	Labor-Hours	Unit	Material	2021 Bare Costs Labor	Equipment	Total	Total Incl O&P
6000	Wall mounted, fully recessed									
6400	Poly-marble, single bubbler	1 Plum	4	2	Ea.	1,750	135		1,885	2,125
6800	Stainless steel, single bubbler		4	2		1,350	135		1,485	1,675
7580	For rough-in, supply and waste, add	↓	1.83	4.372	↓	370	296		666	860
7600	Floor mounted, pedestal type									
8600	Enameled iron, heavy duty service, 2 bubblers	1 Plum	2	4	Ea.	3,475	271		3,746	4,250
8880	For freeze-proof valve system, add		2	4		680	271		951	1,175
8900	For rough-in, supply and waste, add		1.83	4.372	↓	370	296		666	860
9000	Minimum labor/equipment charge	↓	2	4	Job		271		271	420

22 47 16 – Pressure Water Coolers

22 47 16.10 Electric Water Coolers

		Crew	Daily Output	Labor-Hours	Unit	Material	2021 Bare Costs Labor	Equipment	Total	Total Incl O&P
0010	**ELECTRIC WATER COOLERS**									
0100	Wall mounted, non-recessed									
0140	4 GPH	Q-1	4	4	Ea.	690	244		934	1,125
0160	8 GPH, barrier free, sensor operated		4	4		1,750	244		1,994	2,300
1000	Dual height, 8.2 GPH		3.80	4.211		1,775	257		2,032	2,350
1040	14.3 GPH		3.80	4.211		2,025	257		2,282	2,625
3300	Semi-recessed, 8.1 GPH	↓	4	4	↓	1,025	244		1,269	1,500
4600	Floor mounted, flush-to-wall									
4640	4 GPH	1 Plum	3	2.667	Ea.	980	181		1,161	1,350
4980	For stainless steel cabinet, add					141			141	155
5000	Dual height, 8.2 GPH	1 Plum	2	4	↓	1,200	271		1,471	1,750
9000	Minimum labor/equipment charge	"	2	4	Job		271		271	420

22 66 Chemical-Waste Systems for Lab. and Healthcare Facilities

22 66 53 – Laboratory Chemical-Waste and Vent Piping

22 66 53.30 Glass Pipe

		Crew	Daily Output	Labor-Hours	Unit	Material	2021 Bare Costs Labor	Equipment	Total	Total Incl O&P
0010	**GLASS PIPE**, Borosilicate, couplings & clevis hanger assemblies, 10' OC									
0020	Drainage									
1100	1-1/2" diameter	Q-1	52	.308	L.F.	13.60	18.75		32.35	44
1120	2" diameter		44	.364		17.90	22		39.90	53.50
1140	3" diameter		39	.410		22.50	25		47.50	63.50
1160	4" diameter		30	.533		43	32.50		75.50	97
1180	6" diameter	↓	26	.615	↓	77	37.50		114.50	143
9000	Minimum labor/equipment charge	1 Plum	4	2	Job		135		135	209

22 66 53.40 Pipe Fittings, Glass

		Crew	Daily Output	Labor-Hours	Unit	Material	2021 Bare Costs Labor	Equipment	Total	Total Incl O&P
0010	**PIPE FITTINGS, GLASS**									
0020	Drainage, beaded ends									
0040	Coupling & labor required at joints not incl. in fitting									
0050	price. Add 1 per joint for installed price									
0070	90° bend or sweep, 1-1/2"				Ea.	35.50			35.50	39
0090	2"					45.50			45.50	50
0100	3"					74			74	81.50
0110	4"					118			118	130
0120	6" (sweep only)					320			320	350
0350	Tee, single sanitary, 1-1/2"					57.50			57.50	63.50
0370	2"					57.50			57.50	63.50
0380	3"					83			83	91
0390	4"					153			153	168
0400	6"				↓	410			410	450

451

22 66 53.40 Pipe Fittings, Glass

		Crew	Daily Output	Labor-Hours	Unit	Material	2021 Bare Costs Labor	Equipment	Total	Total Incl O&P
0500	Coupling, stainless steel, TFE seal ring									
0520	1-1/2"	Q-1	32	.500	Ea.	27	30.50		57.50	76.50
0530	2"		30	.533		34	32.50		66.50	87.50
0540	3"		25	.640		32.50	39		71.50	96
0550	4"		23	.696		78.50	42.50		121	152
0560	6"		20	.800		177	48.50		225.50	269
9000	Minimum labor/equipment charge	1 Plum	4	2	Job		135		135	209

22 66 53.60 Corrosion Resistant Pipe

		Crew	Daily Output	Labor-Hours	Unit	Material	2021 Bare Costs Labor	Equipment	Total	Total Incl O&P
0010	**CORROSION RESISTANT PIPE**, No couplings or hangers									
0020	Iron alloy, drain, mechanical joint									
1000	1-1/2" diameter	Q-1	70	.229	L.F.	75.50	13.95		89.45	105
1100	2" diameter		66	.242		77.50	14.75		92.25	109
1120	3" diameter		60	.267		85	16.25		101.25	119
1140	4" diameter		52	.308		108	18.75		126.75	148
2980	Plastic, epoxy, fiberglass filament wound, B&S joint									
3000	2" diameter	Q-1	62	.258	L.F.	12	15.70		27.70	37.50
3100	3" diameter		51	.314		14.05	19.10		33.15	45
3120	4" diameter		45	.356		20	21.50		41.50	55.50
3160	8" diameter	Q-2	38	.632		44	40		84	110
3200	12" diameter	"	28	.857		73.50	54		127.50	164
9800	Minimum labor/equipment charge	1 Plum	4	2	Job		135		135	209

22 66 53.70 Pipe Fittings, Corrosion Resistant

		Crew	Daily Output	Labor-Hours	Unit	Material	2021 Bare Costs Labor	Equipment	Total	Total Incl O&P
0010	**PIPE FITTINGS, CORROSION RESISTANT**									
0030	Iron alloy									
0050	Mechanical joint									
0060	1/4 bend, 1-1/2"	Q-1	12	1.333	Ea.	128	81		209	266
0080	2"		10	1.600		209	97.50		306.50	380
0090	3"		9	1.778		251	108		359	445
0100	4"		8	2		288	122		410	505
0160	Tee and Y, sanitary, straight									
0170	1-1/2"	Q-1	8	2	Ea.	139	122		261	340
0180	2"		7	2.286		186	139		325	420
0190	3"		6	2.667		288	162		450	565
0200	4"		5	3.200		530	195		725	885
0360	Coupling, 1-1/2"		14	1.143		67	69.50		136.50	181
0380	2"		12	1.333		76	81		157	209
0390	3"		11	1.455		79.50	88.50		168	225
0400	4"		10	1.600		90	97.50		187.50	249
3000	Epoxy, filament wound									
3030	Quick-lock joint									
3040	90° elbow, 2"	Q-1	28	.571	Ea.	99	35		134	163
3060	3"		16	1		113	61		174	218
3070	4"		13	1.231		154	75		229	286
3190	Tee, 2"		19	.842		234	51.50		285.50	335
3200	3"		11	1.455		282	88.50		370.50	445
3210	4"		9	1.778		340	108		448	540
9000	Minimum labor/equipment charge	1 Plum	4	2	Job		135		135	209

Estimating Tips

The labor adjustment factors listed in Subdivision 22 01 02.20 also apply to Division 23.

23 10 00 Facility Fuel Systems

- The prices in this subdivision for above- and below-ground storage tanks do not include foundations or hold-down slabs, unless noted. The estimator should refer to Divisions 3 and 31 for foundation system pricing. In addition to the foundations, required tank accessories, such as tank gauges, leak detection devices, and additional manholes and piping, must be added to the tank prices.

23 50 00 Central Heating Equipment

- When estimating the cost of an HVAC system, check to see who is responsible for providing and installing the temperature control system. It is possible to overlook controls, assuming that they would be included in the electrical estimate.

- When looking up a boiler, be careful on specified capacity. Some manufacturers rate their products on output while others use input.

- Include HVAC insulation for pipe, boiler, and duct (wrap and liner).

- Be careful when looking up mechanical items to get the correct pressure rating and connection type (thread, weld, flange).

23 70 00 Central HVAC Equipment

- Combination heating and cooling units are sized by the air conditioning requirements. (See Reference No. R236000-20 for the preliminary sizing guide.)

- A ton of air conditioning is nominally 400 CFM.

- Rectangular duct is taken off by the linear foot for each size, but its cost is usually estimated by the pound. Remember that SMACNA standards now base duct on internal pressure.

- Prefabricated duct is estimated and purchased like pipe: straight sections and fittings.

- Note that cranes or other lifting equipment are not included on any lines in Division 23. For example, if a crane is required to lift a heavy piece of pipe into place high above a gym floor, or to put a rooftop unit on the roof of a four-story building, etc., it must be added. Due to the potential for extreme variation—from nothing additional required to a major crane or helicopter—we feel that including a nominal amount for "lifting contingency" would be useless and detract from the accuracy of the estimate. When using equipment rental cost data from RSMeans, do not forget to include the cost of the operator(s).

Reference Numbers

Reference numbers are shown at the beginning of some major classifications. These numbers refer to related items in the Reference Section. The reference information may be an estimating procedure, an alternate pricing method, or technical information.

Note: Not all subdivisions listed here necessarily appear. ■

Same Data. Simplified.

Enjoy the convenience and efficiency of accessing your costs anywhere:

- **Skip the multiplier** by setting your location
- **Quickly search,** edit, favorite and share costs
- **Stay on top of price changes** with automatic updates

Discover more at rsmeans.com/online

---

23 05 05.10 HVAC Demolition

		Crew	Daily Output	Labor-Hours	Unit	Material	2021 Bare Costs Labor	2021 Bare Costs Equipment	Total	Total Incl O&P
0010	**HVAC DEMOLITION** R220105-10									
0100	Air conditioner, split unit, 3 ton	Q-5	2	8	Ea.		490		490	760
0150	Package unit, 3 ton R024119-10	Q-6	3	8	"		510		510	785
0298	Boilers									
0300	Electric, up thru 148 kW	Q-19	2	12	Ea.		745		745	1,150
0310	150 thru 518 kW	"	1	24			1,500		1,500	2,300
0320	550 thru 2,000 kW	Q-21	.40	80			5,100		5,100	7,850
0330	2,070 kW and up	"	.30	107			6,800		6,800	10,500
0340	Gas and/or oil, up thru 150 MBH	Q-7	2.20	14.545			945		945	1,450
0350	160 thru 2,000 MBH		.80	40			2,600		2,600	4,025
0360	2,100 thru 4,500 MBH		.50	64			4,175		4,175	6,425
0370	4,600 thru 7,000 MBH		.30	107			6,950		6,950	10,700
0380	7,100 thru 12,000 MBH		.16	200			13,000		13,000	20,100
0390	12,200 thru 25,000 MBH		.12	267			17,300		17,300	26,800
1000	Ductwork, 4" high, 8" wide	1 Clab	200	.040	L.F.		1.78		1.78	2.83
1020	10" wide		190	.042			1.87		1.87	2.98
1040	14" wide		180	.044			1.97		1.97	3.14
1100	6" high, 8" wide		165	.048			2.15		2.15	3.43
1120	12" wide		150	.053			2.37		2.37	3.77
1140	18" wide		135	.059			2.63		2.63	4.19
1200	10" high, 12" wide		125	.064			2.84		2.84	4.52
1220	18" wide		115	.070			3.09		3.09	4.92
1240	24" wide		110	.073			3.23		3.23	5.15
1300	12"-14" high, 16"-18" wide		85	.094			4.18		4.18	6.65
1320	24" wide		75	.107			4.74		4.74	7.55
1340	48" wide		71	.113			5		5	7.95
1400	18" high, 24" wide		67	.119			5.30		5.30	8.45
1420	36" wide		63	.127			5.65		5.65	9
1440	48" wide		59	.136			6		6	9.60
1500	30" high, 36" wide		56	.143			6.35		6.35	10.10
1520	48" wide		53	.151			6.70		6.70	10.65
1540	72" wide		50	.160			7.10		7.10	11.30
1550	Duct heater, electric strip	1 Elec	8	1	Ea.		63.50		63.50	98
1850	Minimum labor/equipment charge	1 Clab	3	2.667	Job		118		118	189
2200	Furnace, electric	Q-20	2	10	Ea.		600		600	930
2300	Gas or oil, under 120 MBH	Q-9	4	4			236		236	365
2340	Over 120 MBH	"	3	5.333			315		315	490
2800	Heat pump, package unit, 3 ton	Q-5	2.40	6.667			410		410	635
2840	Split unit, 3 ton		2	8			490		490	760
3000	Mechanical equipment, light items. Unit is weight, not cooling.		.90	17.778	Ton		1,100		1,100	1,675
3600	Heavy items		1.10	14.545	"		895		895	1,375
5090	Remove refrigerant from system	1 Stpi	40	.200	Lb.		13.65		13.65	21
9000	Minimum labor/equipment charge	Q-6	3	8	Job		510		510	785

23 05 23.30 Valves, Iron Body

		Crew	Daily Output	Labor-Hours	Unit	Material	2021 Bare Costs Labor	2021 Bare Costs Equipment	Total	Total Incl O&P
0010	**VALVES, IRON BODY**									
1020	Butterfly, wafer type, gear actuator, 200 lb.									
1030	2"	1 Plum	14	.571	Ea.	118	38.50		156.50	190
1060	4"	Q-1	5	3.200	"	138	195		333	450
1650	Gate, 125 lb., N.R.S.									
2150	Flanged									
2240	2-1/2"	Q-1	5	3.200	Ea.	935	195		1,130	1,325

23 05 23 – General-Duty Valves for HVAC Piping

23 05 23.30 Valves, Iron Body

		Crew	Daily Output	Labor-Hours	Unit	Material	2021 Bare Costs Labor	Equipment	Total	Total Incl O&P
2260	3"	Q-1	4.50	3.556	Ea.	1,050	217		1,267	1,475
2280	4"	↓	3	5.333	↓	1,500	325		1,825	2,150
3550	OS&Y, 125 lb., flanged									
3680	4"	Q-1	3	5.333	Ea.	1,000	325		1,325	1,600
3690	5"	Q-2	3.40	7.059		1,650	445		2,095	2,525
3700	6"	"	3	8	↓	1,650	505		2,155	2,600
9000	Minimum labor/equipment charge	1 Plum	3	2.667	Job		181		181	278

23 05 93 – Testing, Adjusting, and Balancing for HVAC

23 05 93.10 Balancing, Air

		Crew	Daily Output	Labor-Hours	Unit	Material	2021 Bare Costs Labor	Equipment	Total	Total Incl O&P
0010	**BALANCING, AIR** (Subcontractor's quote incl. material and labor)									
0900	Heating and ventilating equipment									
1000	Centrifugal fans, utility sets				Ea.				420	420
1100	Heating and ventilating unit								630	630
1200	In-line fan								630	630
1300	Propeller and wall fan								119	119
1400	Roof exhaust fan								280	280
2000	Air conditioning equipment, central station								910	910
2100	Built-up low pressure unit								840	840
2200	Built-up high pressure unit								980	980
2300	Built-up high pressure dual duct								1,550	1,550
2400	Built-up variable volume								1,825	1,825
2500	Multi-zone A.C. and heating unit								630	630
2600	For each zone over one, add								140	140
2700	Package A.C. unit								350	350
2800	Rooftop heating and cooling unit								490	490
3000	Supply, return, exhaust, registers & diffusers, avg. height ceiling								84	84
3100	High ceiling								126	126
3200	Floor height								70	70
3300	Off mixing box								56	56
3500	Induction unit								91	91
3600	Lab fume hood								420	420
3700	Linear supply								210	210
3800	Linear supply high								245	245
4000	Linear return								70	70
4100	Light troffers								84	84
4200	Moduline - master								84	84
4300	Moduline - slaves								42	42
4400	Regenerators								560	560
4500	Taps into ceiling plenums								105	105
4600	Variable volume boxes				↓				84	84

23 05 93.20 Balancing, Water

		Crew	Daily Output	Labor-Hours	Unit	Material	2021 Bare Costs Labor	Equipment	Total	Total Incl O&P
0010	**BALANCING, WATER** (Subcontractor's quote incl. material and labor)									
0050	Air cooled condenser				Ea.				256	256
0080	Boiler								515	515
0100	Cabinet unit heater								88	88
0200	Chiller								620	620
0300	Convector								73	73
0400	Converter								365	365
0500	Cooling tower								475	475
0600	Fan coil unit, unit ventilator								132	132
0700	Fin tube and radiant panels								146	146
0800	Main and duct re-heat coils				↓				135	135

For customer support on your Commercial Renovation Costs with RSMeans Data, call 800.448.8182.

455

23 05 Common Work Results for HVAC

23 05 93 – Testing, Adjusting, and Balancing for HVAC

23 05 93.20 Balancing, Water		Crew	Daily Output	Labor-Hours	Unit	Material	2021 Bare Costs Labor	Equipment	Total	Total Incl O&P
0810	Heat exchanger				Ea.				135	135
0900	Main balancing cocks								110	110
1000	Pumps								320	320
1100	Unit heater				▼				102	102

23 07 HVAC Insulation

23 07 13 – Duct Insulation

23 07 13.10 Duct Thermal Insulation

			Crew	Daily Output	Labor-Hours	Unit	Material	2021 Bare Costs Labor	Equipment	Total	Total Incl O&P
0010	**DUCT THERMAL INSULATION**										
0110	Insulation req'd. is based on the surface size/area to be covered										
3000	Ductwork										
3020	Blanket type, fiberglass, flexible										
3030	Fire rated for grease and hazardous exhaust ducts										
3060	1-1/2" thick		Q-14	84	.190	S.F.	4.45	10.45		14.90	21.50
3090	Fire rated for plenums										
3100	1/2" x 24" x 25'		Q-14	1.94	8.247	Roll	180	450		630	915
3110	1/2" x 24" x 25'			98	.163	S.F.	3.60	8.95		12.55	18.10
3120	1/2" x 48" x 25'			1.04	15.385	Roll	355	845		1,200	1,725
3126	1/2" x 48" x 25'		▼	104	.154	S.F.	3.56	8.45		12.01	17.25
3140	FSK vapor barrier wrap, .75 lb. density										
3160	1" thick	G	Q-14	350	.046	S.F.	.21	2.51		2.72	4.19
3170	1-1/2" thick	G		320	.050		.27	2.74		3.01	4.63
3180	2" thick	G		300	.053		.34	2.93		3.27	4.99
3190	3" thick	G		260	.062		.63	3.38		4.01	6.05
3200	4" thick	G	▼	242	.066	▼	.66	3.63		4.29	6.50
3210	Vinyl jacket, same as FSK										
9600	Minimum labor/equipment charge		1 Stpi	4	2	Job		137		137	211

23 07 16 – HVAC Equipment Insulation

23 07 16.10 HVAC Equipment Thermal Insulation

			Crew	Daily Output	Labor-Hours	Unit	Material	2021 Bare Costs Labor	Equipment	Total	Total Incl O&P
0010	**HVAC EQUIPMENT THERMAL INSULATION**										
0110	Insulation req'd. is based on the surface size/area to be covered										
9610	Minimum labor/equipment charge		1 Stpi	4	2	Job		137		137	211

23 09 Instrumentation and Control for HVAC

23 09 23.10 Control Components/DDC Systems

			Crew	Daily Output	Labor-Hours	Unit	Material	2021 Bare Costs Labor	Equipment	Total	Total Incl O&P
0010	**CONTROL COMPONENTS/DDC SYSTEMS** (Sub's quote incl. M & L)										
0100	Analog inputs										
0110	Sensors (avg. 50' run in 1/2" EMT)										
0120	Duct temperature					Ea.				465	465
0130	Space temperature									665	665
0140	Duct humidity, +/- 3%									595	595
0150	Space humidity, +/- 2%									1,150	1,150
0160	Duct static pressure									680	680
0170	CFM/transducer									920	920
0172	Water temperature									940	940
0174	Water flow									3,450	3,450
0176	Water pressure differential									975	975
0177	Steam flow									2,400	2,400
0178	Steam pressure					▼				1,025	1,025

23 07 16 – HVAC Equipment Insulation

23 09 23.10 Control Components/DDC Systems	Crew	Daily Output	Labor-Hours	Unit	Material	2021 Bare Costs Labor	Equipment	Total	Total Incl O&P
0180	KW/transducer				Ea.				1,350	1,350
0182	KWH totalization (not incl. elec. meter pulse xmtr.)								625	625
0190	Space static pressure				▼				1,075	1,075
1000	Analog outputs (avg. 50' run in 1/2" EMT)									
1010	P/I transducer				Ea.				635	635
1020	Analog output, matl. in MUX								305	305
1030	Pneumatic (not incl. control device)								645	645
1040	Electric (not incl. control device)				▼				380	380
2000	Status (alarms)									
2100	Digital inputs (avg. 50' run in 1/2" EMT)									
2110	Freeze				Ea.				435	435
2120	Fire								395	395
2130	Differential pressure (air)	2 Elec	3	5.333		283	340		623	830
2140	Differential pressure (water)								975	975
2150	Current sensor								435	435
2160	Duct high temperature thermostat								570	570
2170	Duct smoke detector				▼				705	705
2200	Digital output (avg. 50' run in 1/2" EMT)									
2210	Start/stop				Ea.				340	340
2220	On/off (maintained contact)				"				585	585
3000	Controller MUX panel, incl. function boards									
3100	48 point				Ea.				5,275	5,275
3110	128 point				"				7,225	7,225
3200	DDC controller (avg. 50' run in conduit)									
3210	Mechanical room									
3214	16 point controller (incl. 120 volt/1 phase power supply)				Ea.				3,275	3,275
3229	32 point controller (incl. 120 volt/1 phase power supply)				"				5,425	5,425
3230	Includes software programming and checkout									
3260	Space									
3266	VAV terminal box (incl. space temp. sensor)				Ea.				840	840
3280	Host computer (avg. 50' run in conduit)									
3281	Package complete with PC, keyboard,									
3282	printer, monitor, basic software				Ea.				3,150	3,150
4000	Front end costs									
4100	Computer (P.C.) with software program				Ea.				6,350	6,350
4200	Color graphics software								3,925	3,925
4300	Color graphics slides								490	490
4350	Additional printer				▼				980	980
4400	Communications trunk cable				L.F.				3.80	3.80
4500	Engineering labor (not incl. dftg.)				Point				94	94
4600	Calibration labor								120	120
4700	Start-up, checkout labor				▼				120	120
4800	Programming labor, as req'd									
5000	Communications bus (data transmission cable)									
5010	#18 twisted shielded pair in 1/2" EMT conduit				C.L.F.				380	380
8000	Applications software									
8050	Basic maintenance manager software (not incl. data base entry)				Ea.				1,950	1,950
8100	Time program				Point				6.85	6.85
8120	Duty cycle								13.65	13.65
8140	Optimum start/stop								41.50	41.50
8160	Demand limiting								20.50	20.50
8180	Enthalpy program				▼				41.50	41.50
8200	Boiler optimization				Ea.				1,225	1,225

23 09 Instrumentation and Control for HVAC

23 07 16 – HVAC Equipment Insulation

23 09 23.10 Control Components/DDC Systems		Crew	Daily Output	Labor-Hours	Unit	Material	2021 Bare Costs Labor	Equipment	Total	Total Incl O&P
8220	Chiller optimization				Ea.				1,625	1,625
8240	Custom applications									
8260	Cost varies with complexity									

23 09 53 – Pneumatic and Electric Control System for HVAC

23 09 53.10 Control Components

		Crew	Daily Output	Labor-Hours	Unit	Material	2021 Bare Costs Labor	Equipment	Total	Total Incl O&P
0010	**CONTROL COMPONENTS**									
5000	Thermostats									
5030	Manual	1 Stpi	8	1	Ea.	44.50	68.50		113	154
5040	1 set back, electric, timed Ⓖ		8	1		99.50	68.50		168	214
5050	2 set back, electric, timed Ⓖ	↓	8	1		247	68.50		315.50	375
5200	24 hour, automatic, clock Ⓖ	1 Shee	8	1	↓	186	65.50		251.50	305
6000	Valves, motorized zone									
6100	Sweat connections, 1/2" C x C	1 Stpi	20	.400	Ea.	139	27.50		166.50	195
6110	3/4" C x C		20	.400		78.50	27.50		106	128
6120	1" C x C	↓	19	.421		183	29		212	246
9000	Minimum labor/equipment charge	1 Plum	4	2	Job		135		135	209

23 21 Hydronic Piping and Pumps

23 21 20 – Hydronic HVAC Piping Specialties

23 21 20.34 Dielectric Unions

		Crew	Daily Output	Labor-Hours	Unit	Material	2021 Bare Costs Labor	Equipment	Total	Total Incl O&P
0010	**DIELECTRIC UNIONS**, Standard gaskets for water and air									
0020	250 psi maximum pressure									
0280	Female IPT to sweat, straight									
0340	3/4" pipe size	1 Plum	20	.400	Ea.	9.80	27		36.80	53
0780	Female IPT to female IPT, straight									
0800	1/2" pipe size	1 Plum	24	.333	Ea.	16.25	22.50		38.75	53
0840	3/4" pipe size		20	.400	"	18.55	27		45.55	62.50
9000	Minimum labor/equipment charge	↓	4	2	Job		135		135	209

23 21 20.46 Expansion Tanks

		Crew	Daily Output	Labor-Hours	Unit	Material	2021 Bare Costs Labor	Equipment	Total	Total Incl O&P
0010	**EXPANSION TANKS**									
2000	Steel, liquid expansion, ASME, painted, 15 gallon capacity	Q-5	17	.941	Ea.	880	58		938	1,050
2040	30 gallon capacity		12	1.333		1,025	82		1,107	1,250
2080	60 gallon capacity		8	2		1,425	123		1,548	1,775
2120	100 gallon capacity		6	2.667		2,075	164		2,239	2,550
3000	Steel ASME expansion, rubber diaphragm, 19 gal. cap. accept.		12	1.333		3,350	82		3,432	3,825
3020	31 gallon capacity		8	2		3,725	123		3,848	4,300
3040	61 gallon capacity		6	2.667		5,450	164		5,614	6,225
3080	119 gallon capacity		4	4	↓	5,650	246		5,896	6,600
9000	Minimum labor/equipment charge	↓	4	4	Job		246		246	380

23 21 20.70 Steam Traps

		Crew	Daily Output	Labor-Hours	Unit	Material	2021 Bare Costs Labor	Equipment	Total	Total Incl O&P
0010	**STEAM TRAPS**									
0030	Cast iron body, threaded									
0040	Inverted bucket									
0050	1/2" pipe size	1 Stpi	12	.667	Ea.	170	45.50		215.50	258
0100	1" pipe size		9	.889		455	61		516	595
0120	1-1/4" pipe size	↓	8	1	↓	690	68.50		758.50	865
1000	Float & thermostatic, 15 psi									
1010	3/4" pipe size	1 Stpi	16	.500	Ea.	191	34		225	263
1020	1" pipe size		15	.533		194	36.50		230.50	270
1040	1-1/2" pipe size	↓	9	.889	↓	340	61		401	470

23 21 Hydronic Piping and Pumps

23 21 20 – Hydronic HVAC Piping Specialties

23 21 20.70 Steam Traps

		Crew	Daily Output	Labor-Hours	Unit	Material	2021 Bare Costs Labor	Equipment	Total	Total Incl O&P
1060	2" pipe size	1 Stpi	6	1.333	Ea.	975	91		1,066	1,225
9000	Minimum labor/equipment charge		4	2	Job		137		137	211

23 21 20.76 Strainers, Y Type, Bronze Body

		Crew	Daily Output	Labor-Hours	Unit	Material	2021 Bare Costs Labor	Equipment	Total	Total Incl O&P
0010	**STRAINERS, Y TYPE, BRONZE BODY**									
0050	Screwed, 125 lb., 1/4" pipe size	1 Stpi	24	.333	Ea.	45.50	23		68.50	85
0100	1/2" pipe size		20	.400		45.50	27.50		73	92
0120	3/4" pipe size		19	.421		56.50	29		85.50	107
0140	1" pipe size		17	.471		83	32		115	141
0160	1-1/2" pipe size		14	.571		178	39		217	256
0180	2" pipe size		13	.615		234	42		276	325
0182	3" pipe size		12	.667		1,575	45.50		1,620.50	1,800
0200	300 lb., 2-1/2" pipe size	Q-5	17	.941		890	58		948	1,075
0220	3" pipe size		16	1		1,325	61.50		1,386.50	1,550
0240	4" pipe size		15	1.067		2,425	65.50		2,490.50	2,775
1000	Flanged, 150 lb., 1-1/2" pipe size	1 Stpi	11	.727		565	49.50		614.50	695
1020	2" pipe size	"	8	1		690	68.50		758.50	865
1030	2-1/2" pipe size	Q-5	5	3.200		1,150	197		1,347	1,575
1040	3" pipe size		4.50	3.556		1,425	219		1,644	1,900
1060	4" pipe size		3	5.333		2,200	330		2,530	2,900
1100	6" pipe size	Q-6	3	8		4,225	510		4,735	5,425
1106	8" pipe size	"	2.60	9.231		4,600	590		5,190	5,950
1500	For 300 lb. rating, add					40%				
9000	Minimum labor/equipment charge	1 Stpi	3.75	2.133	Job		146		146	225

23 21 23 – Hydronic Pumps

23 21 23.13 In-Line Centrifugal Hydronic Pumps

		Crew	Daily Output	Labor-Hours	Unit	Material	2021 Bare Costs Labor	Equipment	Total	Total Incl O&P
0010	**IN-LINE CENTRIFUGAL HYDRONIC PUMPS**									
0600	Bronze, sweat connections, 1/40 HP, in line									
0640	3/4" size	Q-1	16	1	Ea.	273	61		334	395
1000	Flange connection, 3/4" to 1-1/2" size									
1040	1/12 HP	Q-1	6	2.667	Ea.	680	162		842	995
1060	1/8 HP		6	2.667		1,250	162		1,412	1,600
1100	1/3 HP		6	2.667		1,375	162		1,537	1,775
1140	2" size, 1/6 HP		5	3.200		1,775	195		1,970	2,250
1180	2-1/2" size, 1/4 HP		5	3.200		2,275	195		2,470	2,800
1220	3" size, 1/4 HP		4	4		2,450	244		2,694	3,075
9000	Minimum labor/equipment charge		3.25	4.923	Job		300		300	465

23 23 Refrigerant Piping

23 23 16 – Refrigerant Piping Specialties

23 23 16.16 Refrigerant Line Sets

		Crew	Daily Output	Labor-Hours	Unit	Material	2021 Bare Costs Labor	Equipment	Total	Total Incl O&P
0010	**REFRIGERANT LINE SETS**, Standard									
0100	Copper tube									
0110	1/2" insulation, both tubes									
0120	Combination 1/4" and 1/2" tubes									
0130	10' set	Q-5	42	.381	Ea.	50.50	23.50		74	91.50
0135	15' set		42	.381		72.50	23.50		96	116
0140	20' set		40	.400		78.50	24.50		103	124
0150	30' set		37	.432		111	26.50		137.50	163
0160	40' set		35	.457		138	28		166	196
0170	50' set		32	.500		172	31		203	237

23 23 16.16 Refrigerant Line Sets	Crew	Daily Output	Labor-Hours	Unit	Material	2021 Bare Costs Labor	Equipment	Total	Total Incl O&P
0180	100' set	Q-5	22	.727	Ea.	375	45		420	480
0300	Combination 1/4" and 3/4" tubes									
0310	10' set	Q-5	40	.400	Ea.	56	24.50		80.50	100
0320	20' set		38	.421		99.50	26		125.50	149
0330	30' set		35	.457		149	28		177	208
0340	40' set		33	.485		195	30		225	260
0350	50' set		30	.533		245	33		278	320
0380	100' set		20	.800		555	49		604	690
0500	Combination 3/8" & 3/4" tubes									
0510	10' set	Q-5	28	.571	Ea.	64.50	35		99.50	125
0520	20' set		36	.444		101	27.50		128.50	153
0530	30' set		34	.471		137	29		166	196
0540	40' set		31	.516		180	32		212	247
0550	50' set		28	.571		211	35		246	286
0580	100' set		18	.889		625	54.50		679.50	770
0700	Combination 3/8" & 1-1/8" tubes									
0710	10' set	Q-5	36	.444	Ea.	113	27.50		140.50	166
0720	20' set		33	.485		189	30		219	254
0730	30' set		31	.516		254	32		286	330
0740	40' set		28	.571		380	35		415	475
0750	50' set		26	.615		370	38		408	465
0900	Combination 1/2" & 3/4" tubes									
0910	10' set	Q-5	37	.432	Ea.	68	26.50		94.50	116
0920	20' set		35	.457		121	28		149	177
0930	30' set		33	.485		182	30		212	246
0940	40' set		30	.533		239	33		272	315
0950	50' set		27	.593		300	36.50		336.50	385
0980	100' set		17	.941		680	58		738	840
2100	Combination 1/2" & 1-1/8" tubes									
2110	10' set	Q-5	35	.457	Ea.	117	28		145	172
2120	20' set		31	.516		200	32		232	268
2130	30' set		29	.552		300	34		334	385
2140	40' set		25	.640		400	39.50		439.50	505
2150	50' set		14	1.143		490	70.50		560.50	650
2300	For 1" thick insulation add					30%	15%			
3000	Refrigerant line sets, min-split, flared									
3100	Combination 1/4" & 3/8" tubes									
3120	15' set	Q-5	41	.390	Ea.	72	24		96	116
3140	25' set		38.50	.416		102	25.50		127.50	152
3160	35' set		37	.432		128	26.50		154.50	182
3180	50' set		30	.533		175	33		208	243
3200	Combination 1/4" & 1/2" tubes									
3220	15' set	Q-5	41	.390	Ea.	77	24		101	122
3240	25' set		38.50	.416		105	25.50		130.50	155
3260	35' set		37	.432		134	26.50		160.50	189
3280	50' set		30	.533		182	33		215	251

23 31 HVAC Ducts and Casings

23 31 13 – Metal Ducts

23 31 13.13 Rectangular Metal Ducts

		Crew	Daily Output	Labor-Hours	Unit	Material	2021 Bare Costs Labor	Equipment	Total	Total Incl O&P
0010	**RECTANGULAR METAL DUCTS**									
0020	Fabricated rectangular, includes fittings, joints, supports,									
0021	allowance for flexible connections and field sketches.									
0030	Does not include "as-built dwgs." or insulation.									
0031	NOTE: Fabrication and installation are combined									
0040	as LABOR cost. Approx. 25% fittings assumed.									
0042	Fabrication/Inst. is to commercial quality standards									
0043	(SMACNA or equiv.) for structure, sealing, leak testing, etc.									
0100	Aluminum, alloy 3003-H14, under 100 lb.	Q-10	75	.320	Lb.	2.97	19.55		22.52	34
0110	100 to 500 lb.		80	.300		1.68	18.30		19.98	30.50
0120	500 to 1,000 lb.		95	.253		1.90	15.45		17.35	26
0160	Over 5,000 lb.		145	.166		1.85	10.10		11.95	17.80
0500	Galvanized steel, under 200 lb.		235	.102		.59	6.25		6.84	10.35
0520	200 to 500 lb.		245	.098		.65	6		6.65	10.05
0540	500 to 1,000 lb.		255	.094		.62	5.75		6.37	9.65
0560	1,000 to 2,000 lb.		265	.091		.63	5.55		6.18	9.30
0580	Over 5,000 lb.		285	.084		.63	5.15		5.78	8.70
1000	Stainless steel, type 304, under 100 lb.		165	.145		4.84	8.90		13.74	19.15
1020	100 to 500 lb.		175	.137		3.89	8.40		12.29	17.35
1030	500 to 1,000 lb.		190	.126		2.83	7.70		10.53	15.15

23 31 13.16 Round and Flat-Oval Spiral Ducts

		Crew	Daily Output	Labor-Hours	Unit	Material	2021 Bare Costs Labor	Equipment	Total	Total Incl O&P
0010	**ROUND AND FLAT-OVAL SPIRAL DUCTS**									
0020	Fabricated round and flat oval spiral, includes hangers,									
0021	supports and field sketches.									
5400	Spiral preformed, steel, galv., straight lengths, max 10" spwg.									
5410	4" diameter, 26 ga.	Q-9	360	.044	L.F.	2.01	2.62		4.63	6.30
5416	5" diameter, 26 ga.		320	.050		1.77	2.95		4.72	6.55
5420	6" diameter, 26 ga.		280	.057		1.74	3.37		5.11	7.15
5425	7" diameter, 26 ga.		240	.067		2.37	3.93		6.30	8.70
5430	8" diameter, 26 ga.		200	.080		2.86	4.71		7.57	10.50
5440	10" diameter, 26 ga.		160	.100		3.06	5.90		8.96	12.55
9990	Minimum labor/equipment charge	1 Shee	3	2.667	Job		175		175	272

23 31 16 – Nonmetal Ducts

23 31 16.13 Fibrous-Glass Ducts

		Crew	Daily Output	Labor-Hours	Unit	Material	2021 Bare Costs Labor	Equipment	Total	Total Incl O&P
0010	**FIBROUS-GLASS DUCTS**									
3490	Rigid fiberglass duct board, foil reinf. kraft facing									
3500	Rectangular, 1" thick, alum. faced, (FRK), std. weight	Q-10	350	.069	SF Surf	.87	4.19		5.06	7.50
9990	Minimum labor/equipment charge	1 Shee	3	2.667	Job		175		175	272

23 31 16.19 PVC Ducts

		Crew	Daily Output	Labor-Hours	Unit	Material	2021 Bare Costs Labor	Equipment	Total	Total Incl O&P
0010	**PVC DUCTS**									
4000	Rigid plastic, corrosive fume resistant PVC									
4020	Straight, 6" diameter	Q-9	220	.073	L.F.	14.05	4.28		18.33	22
4040	8" diameter		160	.100		20.50	5.90		26.40	31.50
4060	10" diameter		120	.133		25.50	7.85		33.35	40.50
4070	12" diameter		100	.160		31.50	9.40		40.90	49
4080	14" diameter		70	.229		38	13.45		51.45	63
4090	16" diameter		62	.258		44	15.20		59.20	72
4100	18" diameter		58	.276		50	16.25		66.25	80.50
4110	20" diameter	Q-10	75	.320		47.50	19.55		67.05	83
4130	24" diameter	"	55	.436		68	26.50		94.50	117
4250	Coupling, 6" diameter	Q-9	88	.182	Ea.	33	10.70		43.70	53
4270	8" diameter		78	.205		34	12.10		46.10	56.50

461

23 31 16.19 PVC Ducts

		Crew	Daily Output	Labor-Hours	Unit	Material	2021 Bare Costs Labor	Equipment	Total	Total Incl O&P
4290	10" diameter	Q-9	70	.229	Ea.	40.50	13.45		53.95	65.50
4300	12" diameter		55	.291		44	17.15		61.15	75
4310	14" diameter		44	.364		55.50	21.50		77	94.50
4320	16" diameter		40	.400		59.50	23.50		83	102
4330	18" diameter	↓	38	.421		76.50	25		101.50	123
4340	20" diameter	Q-10	55	.436		94.50	26.50		121	146
4360	24" diameter	"	45	.533		138	32.50		170.50	203
4470	Elbow, 90°, 6" diameter	Q-9	44	.364		139	21.50		160.50	187
4490	8" diameter		28	.571		152	33.50		185.50	220
4510	10" diameter		18	.889		150	52.50		202.50	247
4520	12" diameter		15	1.067		185	63		248	300
4540	16" diameter	↓	10	1.600		325	94		419	500
4550	18" diameter	Q-10	15	1.600		395	97.50		492.50	585
4560	20" diameter		14	1.714		590	105		695	815
4580	24" diameter	↓	12	2	↓	830	122		952	1,100
4750	Elbow 45°, use 90° and deduct					25%				
9990	Minimum labor/equipment charge	1 Shee	3	2.667	Job		175		175	272

23 31 16.22 Duct Board

0010	**DUCT BOARD**									
3800	Rigid, resin bonded fibrous glass, FSK									
3810	Temperature, bacteria and fungi resistant									
3820	1" thick	Q-14	150	.107	S.F.	1.71	5.85		7.56	11.15
3830	1-1/2" thick		130	.123		2.26	6.75		9.01	13.15
3840	2" thick	↓	120	.133	↓	3.89	7.30		11.19	15.85

23 33 Air Duct Accessories

23 33 13 – Dampers

23 33 13.13 Volume-Control Dampers

0010	**VOLUME-CONTROL DAMPERS**									
5990	Multi-blade dampers, opposed blade, 8" x 6"	1 Shee	24	.333	Ea.	38.50	22		60.50	76.50
5994	8" x 8"		22	.364		128	24		152	178
5996	10" x 10"		21	.381		124	25		149	175
6000	12" x 12"		21	.381		124	25		149	175
6020	12" x 18"		18	.444		189	29		218	254
6030	14" x 10"		20	.400		129	26		155	183
6031	14" x 14"		17	.471		148	31		179	210
6033	16" x 12"		17	.471		186	31		217	252
6035	16" x 16"		16	.500		148	32.50		180.50	213
6037	18" x 16"		15	.533		156	35		191	226
6038	18" x 18"		15	.533		186	35		221	260
6070	20" x 16"		14	.571		224	37.50		261.50	305
6072	20" x 20"		13	.615		186	40.50		226.50	268
6074	22" x 18"		14	.571		94.50	37.50		132	163
6076	24" x 16"		11	.727		233	47.50		280.50	330
6078	24" x 20"		8	1		198	65.50		263.50	320
6080	24" x 24"		8	1		214	65.50		279.50	335
6110	26" x 26"	↓	6	1.333		248	87.50		335.50	410
6133	30" x 30"	Q-9	6.60	2.424		287	143		430	540
6135	32" x 32"		6.40	2.500		287	147		434	545
6180	48" x 36"	↓	5.60	2.857	↓	530	168		698	845
7500	Variable volume modulating motorized damper, incl. elect. mtr.									

23 33 Air Duct Accessories

23 33 13 – Dampers

23 33 13.13 Volume-Control Dampers

		Crew	Daily Output	Labor-Hours	Unit	Material	2021 Bare Costs Labor	Equipment	Total	Total Incl O&P
7504	8" x 6"	1 Shee	15	.533	Ea.	200	35		235	274
7506	10" x 6"		14	.571		200	37.50		237.50	278
7510	10" x 10"		13	.615		208	40.50		248.50	292
7520	12" x 12"		12	.667		215	43.50		258.50	305
7522	12" x 16"		11	.727		216	47.50		263.50	310
7524	16" x 10"		12	.667		212	43.50		255.50	300
7526	16" x 14"		10	.800		222	52.50		274.50	325
7528	16" x 18"		9	.889		229	58		287	345
7542	18" x 18"		8	1		232	65.50		297.50	360
7544	20" x 14"		8	1		241	65.50		306.50	365
7546	20" x 18"		7	1.143		248	75		323	390
7560	24" x 12"		8	1		243	65.50		308.50	370
7562	24" x 18"		7	1.143		262	75		337	405
7564	24" x 24"		6	1.333		273	87.50		360.50	435
7568	28" x 10"		7	1.143		243	75		318	385
7590	30" x 14"		5	1.600		249	105		354	435
7600	30" x 18"		4	2		320	131		451	560
7610	30" x 24"		3.80	2.105		430	138		568	690
7700	For thermostat, add	↓	8	1	↓	47.50	65.50		113	155
8000	Multi-blade dampers, parallel blade									
8100	8" x 8"	1 Shee	24	.333	Ea.	96	22		118	140
8140	16" x 10"		20	.400		128	26		154	182
8160	18" x 12"		18	.444		163	29		192	225
8220	28" x 16"	↓	10	.800	↓	176	52.50		228.50	275

23 33 13.16 Fire Dampers

		Crew	Daily Output	Labor-Hours	Unit	Material	2021 Bare Costs Labor	Equipment	Total	Total Incl O&P
0010	**FIRE DAMPERS**									
3000	Fire damper, curtain type, 1-1/2 hr. rated, vertical, 6" x 6"	1 Shee	24	.333	Ea.	28.50	22		50.50	65
3020	8" x 6"		22	.364		28.50	24		52.50	68
3240	16" x 14"		18	.444		56.50	29		85.50	108
3400	24" x 20"	↓	8	1	↓	67.50	65.50		133	177

23 33 19 – Duct Silencers

23 33 19.10 Duct Silencers

		Crew	Daily Output	Labor-Hours	Unit	Material	2021 Bare Costs Labor	Equipment	Total	Total Incl O&P
0010	**DUCT SILENCERS**									
9000	Silencers, noise control for air flow, duct				MCFM	83.50			83.50	91.50

23 33 23 – Turning Vanes

23 33 23.13 Air Turning Vanes

		Crew	Daily Output	Labor-Hours	Unit	Material	2021 Bare Costs Labor	Equipment	Total	Total Incl O&P
0010	**AIR TURNING VANES**									
9400	Turning vane components									
9410	Turning vane rail	1 Shee	160	.050	L.F.	.85	3.27		4.12	6.05
9420	Double thick, factory fab. vane		300	.027		1.24	1.75		2.99	4.08
9428	12" high set		170	.047		1.92	3.08		5	6.90
9432	14" high set		160	.050	↓	2.24	3.27		5.51	7.55
9900	Minimum labor/equipment charge	↓	4	2	Job		131		131	204

23 33 46 – Flexible Ducts

23 33 46.10 Flexible Air Ducts

		Crew	Daily Output	Labor-Hours	Unit	Material	2021 Bare Costs Labor	Equipment	Total	Total Incl O&P
0010	**FLEXIBLE AIR DUCTS**									
1280	Add to labor for elevated installation									
1282	of prefabricated (purchased) ductwork									
1283	10' to 15' high						10%			
1284	15' to 20' high						20%			
1285	20' to 25' high						25%			

463

23 33 46 – Flexible Ducts

23 33 46.10 Flexible Air Ducts		Crew	Daily Output	Labor-Hours	Unit	Material	2021 Bare Costs Labor	Equipment	Total	Total Incl O&P
1286	25' to 30' high						35%			
1287	30' to 35' high						40%			
1288	35' to 40' high						50%			
1289	Over 40' high						55%			
1300	Flexible, coated fiberglass fabric on corr. resist. metal helix									
1400	pressure to 12" (WG) UL-181									
1500	Noninsulated, 3" diameter	Q-9	400	.040	L.F.	1.42	2.36		3.78	5.25
1540	5" diameter		320	.050		1.65	2.95		4.60	6.40
1560	6" diameter		280	.057		1.94	3.37		5.31	7.40
1580	7" diameter		240	.067		2.19	3.93		6.12	8.50
1600	8" diameter		200	.080		2.45	4.71		7.16	10.05
1640	10" diameter		160	.100		3.12	5.90		9.02	12.65
1660	12" diameter		120	.133		3.67	7.85		11.52	16.30
1900	Insulated, 1" thick, PE jacket, 3" diameter [G]		380	.042		2.73	2.48		5.21	6.85
1910	4" diameter [G]		340	.047		3.44	2.77		6.21	8.10
1920	5" diameter [G]		300	.053		3.60	3.14		6.74	8.85
1940	6" diameter [G]		260	.062		3.81	3.62		7.43	9.85
1960	7" diameter [G]		220	.073		4.41	4.28		8.69	11.55
1980	8" diameter [G]		180	.089		2.71	5.25		7.96	11.15
2020	10" diameter [G]		140	.114		5.40	6.75		12.15	16.45
2040	12" diameter [G]		100	.160		6.30	9.40		15.70	21.50
2060	14" diameter [G]		80	.200		7.65	11.80		19.45	27
9990	Minimum labor/equipment charge	1 Shee	3	2.667	Job		175		175	272

23 33 53 – Duct Liners

23 33 53.10 Duct Liner Board

			Crew	Daily Output	Labor-Hours	Unit	Material	Labor	Equipment	Total	Total Incl O&P
0010	**DUCT LINER BOARD**										
3490	Board type, fiberglass liner, 3 lb. density										
3680	No finish										
3700	1" thick	[G]	Q-14	170	.094	S.F.	.73	5.15		5.88	8.95
3710	1-1/2" thick	[G]		140	.114		1.09	6.25		7.34	11.10
3720	2" thick	[G]		130	.123		1.43	6.75		8.18	12.20
3940	Board type, non-fibrous foam										
3950	Temperature, bacteria and fungi resistant										
3960	1" thick	[G]	Q-14	150	.107	S.F.	2.88	5.85		8.73	12.40
3970	1-1/2" thick	[G]		130	.123		5	6.75		11.75	16.15
3980	2" thick	[G]		120	.133		4.02	7.30		11.32	15.95

23 34 HVAC Fans

23 34 13 – Axial HVAC Fans

23 34 13.10 Axial Flow HVAC Fans

0010	**AXIAL FLOW HVAC FANS**										
0020	Air conditioning and process air handling										

23 34 14 – Blower HVAC Fans

23 34 14.10 Blower Type HVAC Fans

			Crew	Daily Output	Labor-Hours	Unit	Material	Labor	Equipment	Total	Total Incl O&P
0010	**BLOWER TYPE HVAC FANS**										
2500	Ceiling fan, right angle, extra quiet, 0.10" S.P.										
2540	210 CFM		Q-20	19	1.053	Ea.	325	63		388	460
2580	885 CFM			16	1.250		815	75		890	1,000
2620	2,960 CFM			11	1.818		1,500	109		1,609	1,825
2640	For wall or roof cap, add		1 Shee	16	.500		276	32.50		308.50	355

23 34 HVAC Fans

23 34 16 – Centrifugal HVAC Fans

23 34 16.10 Centrifugal Type HVAC Fans

	23 34 16.10 Centrifugal Type HVAC Fans	Crew	Daily Output	Labor-Hours	Unit	Material	2021 Bare Costs Labor	Equipment	Total	Total Incl O&P
0010	**CENTRIFUGAL TYPE HVAC FANS**									
0200	In-line centrifugal, supply/exhaust booster									
0220	aluminum wheel/hub, disconnect switch, 1/4" S.P.									
0240	500 CFM, 10" diameter connection	Q-20	3	6.667	Ea.	1,575	400		1,975	2,350
0280	1,520 CFM, 16" diameter connection		2	10		1,700	600		2,300	2,800
0320	3,480 CFM, 20" diameter connection		.80	25		2,175	1,500		3,675	4,725
0326	5,080 CFM, 20" diameter connection		.75	26.667		2,350	1,600		3,950	5,075
5000	Utility set, centrifugal, V belt drive, motor									
5020	1/4" S.P., 1,200 CFM, 1/4 HP	Q-20	6	3.333	Ea.	1,975	200		2,175	2,475
5040	1,520 CFM, 1/3 HP		5	4		2,025	239		2,264	2,600
5060	1,850 CFM, 1/2 HP		4	5		2,025	299		2,324	2,700
5080	2,180 CFM, 3/4 HP		3	6.667		2,400	400		2,800	3,250
7000	Roof exhauster, centrifugal, aluminum housing, 12" galvanized									
7020	curb, bird screen, back draft damper, 1/4" S.P.									
7100	Direct drive, 320 CFM, 11" sq. damper	Q-20	7	2.857	Ea.	725	171		896	1,050
7120	600 CFM, 11" sq. damper	"	6	3.333	"	935	200		1,135	1,325
9900	Minimum labor/equipment charge	1 Elec	4	2	Job		127		127	196

23 34 23 – HVAC Power Ventilators

23 34 23.10 HVAC Power Circulators and Ventilators

	23 34 23.10 HVAC Power Circulators and Ventilators	Crew	Daily Output	Labor-Hours	Unit	Material	2021 Bare Costs Labor	Equipment	Total	Total Incl O&P
0010	**HVAC POWER CIRCULATORS AND VENTILATORS**									
6650	Residential, bath exhaust, grille, back draft damper									
6660	50 CFM	Q-20	24	.833	Ea.	48	50		98	130
6670	110 CFM		22	.909		96	54.50		150.50	191
6680	Light combination, squirrel cage, 100 watt, 70 CFM		24	.833		112	50		162	201
6700	Light/heater combination, ceiling mounted									
6710	70 CFM, 1,450 watt	Q-20	24	.833	Ea.	239	50		289	340
6800	Heater combination, recessed, 70 CFM		24	.833		62.50	50		112.50	147
6820	With 2 infrared bulbs		23	.870		94	52		146	184
6900	Kitchen exhaust, grille, complete, 160 CFM		22	.909		129	54.50		183.50	226
6910	180 CFM		20	1		103	60		163	206
6920	270 CFM		18	1.111		203	66.50		269.50	325
6940	Residential roof jacks and wall caps									
6944	Wall cap with back draft damper									
6946	3" & 4" diam. round duct	1 Shee	11	.727	Ea.	18.75	47.50		66.25	94.50
6948	6" diam. round duct	"	11	.727	"	54.50	47.50		102	134
6958	Roof jack with bird screen and back draft damper									
6960	3" & 4" diam. round duct	1 Shee	11	.727	Ea.	24	47.50		71.50	101
6962	3-1/4" x 10" rectangular duct	"	10	.800	"	38.50	52.50		91	124
6980	Transition									
6982	3-1/4" x 10" to 6" diam. round	1 Shee	20	.400	Ea.	38	26		64	83
8020	Attic, roof type									
8030	Aluminum dome, damper & curb									
8080	12" diameter, 1,000 CFM (gravity)	1 Elec	10	.800	Ea.	640	51		691	785
8090	16" diameter, 1,500 CFM (gravity)		9	.889		775	56.50		831.50	935
8100	20" diameter, 2,500 CFM (gravity)		8	1		950	63.50		1,013.50	1,150
8110	26" diameter, 4,000 CFM (gravity)		7	1.143		1,150	73		1,223	1,375
8120	32" diameter, 6,500 CFM (gravity)		6	1.333		1,575	85		1,660	1,850
8130	38" diameter, 8,000 CFM (gravity)		5	1.600		2,350	102		2,452	2,725
8140	50" diameter, 13,000 CFM (gravity)		4	2		3,400	127		3,527	3,925
8160	Plastic, ABS dome									
8180	1,050 CFM	1 Elec	14	.571	Ea.	189	36.50		225.50	263
8200	1,600 CFM	"	12	.667	"	282	42.50		324.50	375

For customer support on your Commercial Renovation Costs with RSMeans Data, call 800.448.8182.

465

23 34 HVAC Fans

23 34 23 – HVAC Power Ventilators

23 34 23.10 HVAC Power Circulators and Ventilators	Crew	Daily Output	Labor-Hours	Unit	Material	2021 Bare Costs Labor	Equipment	Total	Total Incl O&P
8240	Attic, wall type, with shutter, one speed									
8250	12" diameter, 1,000 CFM	1 Elec	14	.571	Ea.	470	36.50		506.50	570
8260	14" diameter, 1,500 CFM		12	.667		505	42.50		547.50	625
8270	16" diameter, 2,000 CFM		9	.889		575	56.50		631.50	715
8290	Whole house, wall type, with shutter, one speed									
8300	30" diameter, 4,800 CFM	1 Elec	7	1.143	Ea.	1,225	73		1,298	1,450
8310	36" diameter, 7,000 CFM		6	1.333		1,325	85		1,410	1,600
8320	42" diameter, 10,000 CFM		5	1.600		1,500	102		1,602	1,800
8330	48" diameter, 16,000 CFM		4	2		1,850	127		1,977	2,250
8340	For two speed, add					112			112	123
8350	Whole house, lay-down type, with shutter, one speed									
8360	30" diameter, 4,500 CFM	1 Elec	8	1	Ea.	1,300	63.50		1,363.50	1,550
8370	36" diameter, 6,500 CFM		7	1.143		1,400	73		1,473	1,650
8380	42" diameter, 9,000 CFM		6	1.333		1,550	85		1,635	1,825
8390	48" diameter, 12,000 CFM		5	1.600		1,750	102		1,852	2,075
8440	For two speed, add					84			84	92.50
8450	For 12 hour timer switch, add	1 Elec	32	.250		84	15.95		99.95	117

23 36 Air Terminal Units

23 36 13 – Constant-Air-Volume Units

23 36 13.10 Constant Volume Mixing Boxes

		Crew	Daily Output	Labor-Hours	Unit	Material	2021 Bare Costs Labor	Equipment	Total	Total Incl O&P
0010	**CONSTANT VOLUME MIXING BOXES**									
5180	Mixing box, includes electric or pneumatic motor									
5200	Constant volume, 150 to 270 CFM	Q-9	12	1.333	Ea.	1,125	78.50		1,203.50	1,350
5210	270 to 600 CFM	"	11	1.455	"	1,025	85.50		1,110.50	1,250

23 36 16 – Variable-Air-Volume Units

23 36 16.10 Variable Volume Mixing Boxes

		Crew	Daily Output	Labor-Hours	Unit	Material	2021 Bare Costs Labor	Equipment	Total	Total Incl O&P
0010	**VARIABLE VOLUME MIXING BOXES**									
5180	Mixing box, includes electric or pneumatic motor									
5500	VAV cool only, pneumatic, pressure independent 300 to 600 CFM	Q-9	11	1.455	Ea.	810	85.50		895.50	1,025
5510	500 to 1,000 CFM		9	1.778		835	105		940	1,075
5520	800 to 1,600 CFM		9	1.778		860	105		965	1,125
5530	1,100 to 2,000 CFM		8	2		880	118		998	1,150
5540	1,500 to 3,000 CFM		7	2.286		940	135		1,075	1,225
5550	2,000 to 4,000 CFM		6	2.667		945	157		1,102	1,300

23 37 Air Outlets and Inlets

23 37 13 – Diffusers, Registers, and Grilles

23 37 13.10 Diffusers

		Crew	Daily Output	Labor-Hours	Unit	Material	2021 Bare Costs Labor	Equipment	Total	Total Incl O&P
0010	**DIFFUSERS**, Aluminum, opposed blade damper unless noted									
0100	Ceiling, linear, also for sidewall									
0120	2" wide	1 Shee	32	.250	L.F.	26.50	16.35		42.85	54.50
0160	4" wide		26	.308		22.50	20		42.50	56.50
0180	6" wide		24	.333		28	22		50	64.50
0200	8" wide		22	.364		50.50	24		74.50	93
0500	Perforated, 24" x 24" lay-in panel size, 6" x 6"		16	.500	Ea.	184	32.50		216.50	253
0520	8" x 8"		15	.533		95	35		130	159
0530	9" x 9"		14	.571		196	37.50		233.50	274
0540	10" x 10"		14	.571		195	37.50		232.50	274

23 37 Air Outlets and Inlets

23 37 13 – Diffusers, Registers, and Grilles

23 37 13.10 Diffusers

		Crew	Daily Output	Labor-Hours	Unit	Material	2021 Bare Costs Labor	Equipment	Total	Total Incl O&P
0560	12" x 12"	1 Shee	12	.667	Ea.	204	43.50		247.50	293
0590	16" x 16"		11	.727		217	47.50		264.50	315
0600	18" x 18"		10	.800		233	52.50		285.50	340
0610	20" x 20"		10	.800		250	52.50		302.50	355
0620	24" x 24"		9	.889		293	58		351	410
1000	Rectangular, 1 to 4 way blow, 6" x 6"		16	.500		44.50	32.50		77	100
1010	8" x 8"		15	.533		61.50	35		96.50	122
1014	9" x 9"		15	.533		52	35		87	112
1016	10" x 10"		15	.533		90.50	35		125.50	154
1020	12" x 6"		15	.533		75.50	35		110.50	138
1040	12" x 9"		14	.571		80.50	37.50		118	147
1060	12" x 12"		12	.667		75.50	43.50		119	151
1070	14" x 6"		13	.615		82	40.50		122.50	154
1074	14" x 14"		12	.667		134	43.50		177.50	215
1150	18" x 18"		9	.889		129	58		187	233
1170	24" x 12"		10	.800		179	52.50		231.50	279
2000	T-bar mounting, 24" x 24" lay-in frame, 6" x 6"		16	.500		72	32.50		104.50	131
2020	8" x 8"		14	.571		72.50	37.50		110	138
2040	12" x 12"		12	.667		88.50	43.50		132	166
2060	16" x 16"		11	.727		112	47.50		159.50	197
2080	18" x 18"	▼	10	.800		110	52.50		162.50	203
6000	For steel diffusers instead of aluminum, deduct				▼	10%				
9000	Minimum labor/equipment charge	1 Shee	4	2	Job		131		131	204

23 37 13.30 Grilles

		Crew	Daily Output	Labor-Hours	Unit	Material	2021 Bare Costs Labor	Equipment	Total	Total Incl O&P
0010	**GRILLES**									
0020	Aluminum, unless noted otherwise									
1000	Air return, steel, 6" x 6"	1 Shee	26	.308	Ea.	24	20		44	58
1020	10" x 6"		24	.333		24	22		46	60.50
1080	16" x 8"		22	.364		35	24		59	75.50
1100	12" x 12"		22	.364		35	24		59	75.50
1120	24" x 12"		18	.444		45.50	29		74.50	95.50
1300	48" x 24"		12	.667	▼	164	43.50		207.50	249
9000	Minimum labor/equipment charge		4	2	Job		131		131	204

23 37 13.60 Registers

		Crew	Daily Output	Labor-Hours	Unit	Material	2021 Bare Costs Labor	Equipment	Total	Total Incl O&P
0010	**REGISTERS**									
0980	Air supply									
1000	Ceiling/wall, O.B. damper, anodized aluminum									
1010	One or two way deflection, adj. curved face bars									
1120	12" x 12"	1 Shee	18	.444	Ea.	53.50	29		82.50	105
1140	14" x 8"		17	.471		28.50	31		59.50	79
1240	20" x 6"		18	.444		48	29		77	98
1340	24" x 8"		13	.615		64.50	40.50		105	134
1350	24" x 18"		12	.667		119	43.50		162.50	199
2700	Above registers in steel instead of aluminum, deduct	▼			▼	10%				
3000	Baseboard, hand adj. damper, enameled steel									
3012	8" x 6"	1 Shee	26	.308	Ea.	10.40	20		30.40	43
3020	10" x 6"		24	.333		11.30	22		33.30	46.50
3040	12" x 5"		23	.348		9.70	23		32.70	46
3060	12" x 6"		23	.348		12.30	23		35.30	49
3080	12" x 8"		22	.364		17.80	24		41.80	56.50
3100	14" x 6"		20	.400	▼	13.30	26		39.30	55.50
9000	Minimum labor/equipment charge		4	2	Job		131		131	204

For customer support on your Commercial Renovation Costs with RSMeans Data, call 800.448.8182.

467

23 37 Air Outlets and Inlets

23 37 23 – HVAC Gravity Ventilators

23 37 23.10 HVAC Gravity Air Ventilators	Crew	Daily Output	Labor-Hours	Unit	Material	2021 Bare Costs Labor	Equipment	Total	Total Incl O&P
0010 **HVAC GRAVITY AIR VENTILATORS**, Includes base									
1280 Rotary ventilators, wind driven, galvanized									
1300 4" neck diameter	Q-9	20	.800	Ea.	69	47		116	150
1400 12" neck diameter		10	1.600		97.50	94		191.50	254
1500 24" neck diameter		8	2		435	118		553	660
1600 For aluminum, add					30%				
9000 Minimum labor/equipment charge	1 Plum	2	4	Job		271		271	420

23 38 Ventilation Hoods

23 38 13 – Commercial-Kitchen Hoods

23 38 13.10 Hood and Ventilation Equipment

	Crew	Daily Output	Labor-Hours	Unit	Material	Labor	Equipment	Total	Total Incl O&P
0010 **HOOD AND VENTILATION EQUIPMENT**									
2970 Exhaust hood, sst, gutter on all sides, 4' x 4' x 2'	1 Carp	1.80	4.444	Ea.	1,675	243		1,918	2,225
2980 4' x 4' x 7'	"	1.60	5	"	6,850	274		7,124	7,950

23 41 Particulate Air Filtration

23 41 13 – Panel Air Filters

23 41 13.10 Panel Type Air Filters

	Crew	Daily Output	Labor-Hours	Unit	Material	Labor	Equipment	Total	Total Incl O&P
0010 **PANEL TYPE AIR FILTERS**									
2950 Mechanical media filtration units									
3000 High efficiency type, with frame, non-supported [G]				MCFM	33.50			33.50	37
3100 Supported type [G]				"	48			48	53
5500 Throwaway glass or paper media type, 12" x 36" x 1"				Ea.	2.34			2.34	2.57

23 41 16 – Renewable-Media Air Filters

23 41 16.10 Disposable Media Air Filters

	Crew	Daily Output	Labor-Hours	Unit	Material	Labor	Equipment	Total	Total Incl O&P
0010 **DISPOSABLE MEDIA AIR FILTERS**									
5000 Renewable disposable roll				C.S.F.	5.20			5.20	5.70

23 41 19 – Washable Air Filters

23 41 19.10 Permanent Air Filters

	Crew	Daily Output	Labor-Hours	Unit	Material	Labor	Equipment	Total	Total Incl O&P
0010 **PERMANENT AIR FILTERS**									
4500 Permanent washable [G]				MCFM	25			25	27.50

23 41 23 – Extended Surface Filters

23 41 23.10 Expanded Surface Filters

	Crew	Daily Output	Labor-Hours	Unit	Material	Labor	Equipment	Total	Total Incl O&P
0010 **EXPANDED SURFACE FILTERS**									
4000 Medium efficiency, extended surface [G]				MCFM	5.85			5.85	6.40

23 42 Gas-Phase Air Filtration

23 42 13 – Activated-Carbon Air Filtration

23 42 13.10 Charcoal Type Air Filtration

		Crew	Daily Output	Labor-Hours	Unit	Material	2021 Bare Costs Labor	Equipment	Total	Total Incl O&P
0010	**CHARCOAL TYPE AIR FILTRATION**									
0050	Activated charcoal type, full flow				MCFM	650			650	715
0060	Full flow, impregnated media 12" deep					224			224	246
0070	HEPA filter & frame for field erection					500			500	550
0080	HEPA filter-diffuser, ceiling install.				↓	330			330	360

23 43 Electronic Air Cleaners

23 43 13 – Washable Electronic Air Cleaners

23 43 13.10 Electronic Air Cleaners

		Crew	Daily Output	Labor-Hours	Unit	Material	2021 Bare Costs Labor	Equipment	Total	Total Incl O&P
0010	**ELECTRONIC AIR CLEANERS**									
2000	Electronic air cleaner, duct mounted									
2150	1,000 CFM	1 Shee	4	2	Ea.	525	131		656	785
2200	1,200 CFM		3.80	2.105		495	138		633	755
2250	1,400 CFM	↓	3.60	2.222	↓	640	145		785	930

23 51 Breechings, Chimneys, and Stacks

23 51 13 – Draft Control Devices

23 51 13.16 Vent Dampers

		Crew	Daily Output	Labor-Hours	Unit	Material	2021 Bare Costs Labor	Equipment	Total	Total Incl O&P
0010	**VENT DAMPERS**									
5000	Vent damper, bi-metal, gas, 3" diameter	Q-9	24	.667	Ea.	74.50	39.50		114	143
5010	4" diameter		24	.667		73.50	39.50		113	142
5020	5" diameter		23	.696		74.50	41		115.50	146
5030	6" diameter		22	.727		73.50	43		116.50	148
5040	7" diameter		21	.762		75.50	45		120.50	153
5050	8" diameter	↓	20	.800	↓	81.50	47		128.50	163
9000	Minimum labor/equipment charge	1 Shee	4	2	Job		131		131	204

23 51 13.19 Barometric Dampers

		Crew	Daily Output	Labor-Hours	Unit	Material	2021 Bare Costs Labor	Equipment	Total	Total Incl O&P
0010	**BAROMETRIC DAMPERS**									
1000	Barometric, gas fired system only, 6" size for 5" and 6" pipes	1 Shee	20	.400	Ea.	103	26		129	154
1040	8" size, for 7" and 8" pipes	"	18	.444	"	144	29		173	204
2000	All fuel, oil, oil/gas, coal									
2020	10" for 9" and 10" pipes	1 Shee	15	.533	Ea.	249	35		284	330
3260	For thermal switch for above, add	"	24	.333	"	102	22		124	147

23 51 23 – Gas Vents

23 51 23.10 Gas Chimney Vents

		Crew	Daily Output	Labor-Hours	Unit	Material	2021 Bare Costs Labor	Equipment	Total	Total Incl O&P
0010	**GAS CHIMNEY VENTS**, Prefab metal, UL listed									
0020	Gas, double wall, galvanized steel									
0080	3" diameter	Q-9	72	.222	V.L.F.	7.95	13.10		21.05	29.50
0100	4" diameter		68	.235		9.65	13.85		23.50	32
0120	5" diameter		64	.250		10.40	14.75		25.15	34.50
0140	6" diameter		60	.267		13.85	15.70		29.55	40
0160	7" diameter		56	.286		25.50	16.85		42.35	54
0180	8" diameter		52	.308		27	18.10		45.10	58.50
0200	10" diameter		48	.333		52	19.65		71.65	87.50
0220	12" diameter		44	.364		62	21.50		83.50	102
0260	16" diameter		40	.400		148	23.50		171.50	200
0300	20" diameter	Q-10	36	.667		215	40.50		255.50	300
0340	24" diameter	"	32	.750	↓	335	46		381	440

For customer support on your Commercial Renovation Costs with RSMeans Data, call 800.448.8182.

469

23 51 26.30 All-Fuel Vent Chimneys, Double Wall, St. Stl.	Crew	Daily Output	Labor-Hours	Unit	Material	2021 Bare Costs Labor	Equipment	Total	Total Incl O&P
0010 **ALL-FUEL VENT CHIMNEYS, DOUBLE WALL, STAINLESS STEEL**									
7780 All fuel, pressure tight, double wall, 4" insulation, UL listed, 1,400°F.									
7790 304 stainless steel liner, aluminized steel outer jacket									
7800 6" diameter	Q-9	60	.267	V.L.F.	54	15.70		69.70	83.50
7804 8" diameter		52	.308		78	18.10		96.10	115
7806 10" diameter		48	.333		89.50	19.65		109.15	129
7808 12" diameter		44	.364		104	21.50		125.50	149
7810 14" diameter		42	.381		117	22.50		139.50	163
7880 For 316 stainless steel liner, add				L.F.	30%				
8000 All fuel, double wall, stainless steel fittings									
8010 Roof support, 6" diameter	Q-9	30	.533	Ea.	121	31.50		152.50	182
8030 8" diameter		26	.615		127	36.50		163.50	197
8040 10" diameter		24	.667		151	39.50		190.50	227
8050 12" diameter		22	.727		159	43		202	242
8060 14" diameter		21	.762		171	45		216	258
8100 Elbow 45°, 6" diameter		30	.533		246	31.50		277.50	320
8140 8" diameter		26	.615		281	36.50		317.50	365
8160 10" diameter		24	.667		340	39.50		379.50	435
8180 12" diameter		22	.727		365	43		408	465
8200 14" diameter		21	.762		420	45		465	535
8300 Insulated tee, 6" diameter		30	.533		310	31.50		341.50	390
8360 8" diameter		26	.615		325	36.50		361.50	415
8380 10" diameter		24	.667		350	39.50		389.50	445
8400 12" diameter		22	.727		425	43		468	530
8420 14" diameter		20	.800		500	47		547	625
8500 Boot tee, 6" diameter		28	.571		705	33.50		738.50	830
8520 8" diameter		24	.667		750	39.50		789.50	885
8530 10" diameter		22	.727		850	43		893	1,000
8540 12" diameter		20	.800		990	47		1,037	1,175
8550 14" diameter		18	.889		1,025	52.50		1,077.50	1,200
8600 Rain cap with bird screen, 6" diameter		30	.533		315	31.50		346.50	400
8640 8" diameter		26	.615		345	36.50		381.50	435
8660 10" diameter		24	.667		350	39.50		389.50	445
8680 12" diameter		22	.727		495	43		538	610
8700 14" diameter		21	.762		520	45		565	640
8800 Flat roof flashing, 6" diameter		30	.533		116	31.50		147.50	176
8840 8" diameter		26	.615		118	36.50		154.50	187
8860 10" diameter		24	.667		127	39.50		166.50	200
8880 12" diameter		22	.727		141	43		184	222
8900 14" diameter		21	.762		149	45		194	234

23 52 Heating Boilers

23 52 13 – Electric Boilers

23 52 13.10 Electric Boilers, ASME

		Crew	Daily Output	Labor-Hours	Unit	Material	2021 Bare Costs Labor	2021 Bare Costs Equipment	Total	Total Incl O&P
0010	**ELECTRIC BOILERS, ASME**, Standard controls and trim									
1000	Steam, 6 KW, 20.5 MBH	Q-19	1.20	20	Ea.	4,500	1,250		5,750	6,875
1160	60 KW, 205 MBH		1	24		7,125	1,500		8,625	10,200
1300	296 KW, 1,010 MBH	↓	.45	53.333		26,300	3,325		29,625	34,100
1400	592 KW, 2,020 MBH	Q-21	.34	94.118		35,700	6,000		41,700	48,600
1540	1,110 KW, 3,788 MBH	"	.19	168		50,000	10,700		60,700	71,500
2000	Hot water, 7.5 KW, 25.6 MBH	Q-19	1.30	18.462		5,150	1,150		6,300	7,425
2040	30 KW, 102 MBH		1.20	20		5,525	1,250		6,775	8,000
2060	45 KW, 164 MBH		1.20	20		5,400	1,250		6,650	7,850
2070	60 KW, 205 MBH		1.20	20		5,775	1,250		7,025	8,275
2080	75 KW, 256 MBH		1.10	21.818		6,100	1,350		7,450	8,800
2100	90 KW, 307 MBH	↓	1.10	21.818	↓	6,125	1,350		7,475	8,850
9000	Minimum labor/equipment charge	Q-20	1	20	Job		1,200		1,200	1,850

23 52 23 – Cast-Iron Boilers

23 52 23.20 Gas-Fired Boilers

		Crew	Daily Output	Labor-Hours	Unit	Material	2021 Bare Costs Labor	2021 Bare Costs Equipment	Total	Total Incl O&P
0010	**GAS-FIRED BOILERS**, Natural or propane, standard controls, packaged									
1000	Cast iron, with insulated jacket									
2000	Steam, gross output, 81 MBH	Q-7	1.40	22.857	Ea.	3,125	1,475		4,600	5,750
2060	163 MBH		.90	35.556		3,425	2,325		5,750	7,350
2200	440 MBH		.51	62.500		6,225	4,075		10,300	13,100
2320	1,875 MBH		.30	107		29,000	6,950		35,950	42,600
2400	3,570 MBH		.18	182		40,200	11,800		52,000	62,500
2540	6,970 MBH		.10	320		96,500	20,800		117,300	138,000
3000	Hot water, gross output, 80 MBH		1.46	21.918		2,850	1,425		4,275	5,350
3020	100 MBH		1.35	23.704		2,800	1,550		4,350	5,450
3040	122 MBH		1.10	29.091		3,025	1,900		4,925	6,250
3060	163 MBH		1	32		3,550	2,075		5,625	7,125
3200	440 MBH		.58	54.983		5,525	3,575		9,100	11,600
3320	2,000 MBH		.26	125		25,200	8,125		33,325	40,200
3540	6,970 MBH	↓	.09	360		128,500	23,400		151,900	177,500
7000	For tankless water heater, add					10%				
7050	For additional zone valves up to 312 MBH, add				↓	199			199	219
9900	Minimum labor/equipment charge	Q-6	1	24	Job		1,525		1,525	2,350

23 52 23.30 Gas/Oil Fired Boilers

		Crew	Daily Output	Labor-Hours	Unit	Material	2021 Bare Costs Labor	2021 Bare Costs Equipment	Total	Total Incl O&P
0010	**GAS/OIL FIRED BOILERS**, Combination with burners and controls, packaged									
1000	Cast iron with insulated jacket									
2000	Steam, gross output, 720 MBH	Q-7	.43	74.074	Ea.	16,100	4,825		20,925	25,100
2140	2,700 MBH		.19	166		26,200	10,800		37,000	45,400
2380	6,970 MBH	↓	.09	372	↓	112,000	24,200		136,200	161,000
2900	Hot water, gross output									
2910	200 MBH	Q-6	.62	39.024	Ea.	11,700	2,500		14,200	16,800
2920	300 MBH		.49	49.080		11,700	3,125		14,825	17,700
2930	400 MBH		.41	57.971		13,700	3,700		17,400	20,800
2940	500 MBH	↓	.36	67.039		14,800	4,275		19,075	22,800
3000	584 MBH	Q-7	.44	72.072		14,800	4,700		19,500	23,500
3060	1,460 MBH		.28	113		30,800	7,350		38,150	45,200
3300	13,500 MBH, 403.3 BHP	↓	.04	727	↓	212,000	47,300		259,300	306,500

23 52 23.40 Oil-Fired Boilers

		Crew	Daily Output	Labor-Hours	Unit	Material	2021 Bare Costs Labor	2021 Bare Costs Equipment	Total	Total Incl O&P
0010	**OIL-FIRED BOILERS**, Standard controls, flame retention burner, packaged									
1000	Cast iron, with insulated flush jacket									
2000	Steam, gross output, 109 MBH	Q-7	1.20	26.667	Ea.	2,475	1,725		4,200	5,400
2060	207 MBH	↓	.90	35.556	↓	3,375	2,325		5,700	7,300

471

23 52 Heating Boilers

23 52 23 – Cast-Iron Boilers

23 52 23.40 Oil-Fired Boilers	Crew	Daily Output	Labor-Hours	Unit	Material	2021 Bare Costs Labor	2021 Bare Costs Equipment	Total	Total Incl O&P
2180	1,084 MBH	Q-7	.38	85.106	Ea.	11,700	5,525		17,225	21,500
2200	1,360 MBH		.33	98.160		14,300	6,375		20,675	25,700
2240	2,175 MBH		.24	134		17,400	8,700		26,100	32,500
2340	4,360 MBH		.15	215		39,300	14,000		53,300	64,500
2460	6,970 MBH		.09	364		108,500	23,700		132,200	156,000
3000	Hot water, same price as steam									
4000	For tankless coil in smaller sizes, add				Ea.	15%				

23 52 26 – Steel Boilers

23 52 26.40 Oil-Fired Boilers

		Crew	Daily Output	Labor-Hours	Unit	Material	Labor	Equipment	Total	Total Incl O&P
0010	**OIL-FIRED BOILERS**, Standard controls, flame retention burner									
5000	Steel, with insulated flush jacket									
7000	Hot water, gross output, 103 MBH	Q-6	1.60	15	Ea.	2,225	955		3,180	3,925
7020	122 MBH		1.45	16.506		2,100	1,050		3,150	3,950
7040	137 MBH		1.36	17.595		2,850	1,125		3,975	4,850
7060	168 MBH		1.30	18.405		2,800	1,175		3,975	4,875
7080	225 MBH		1.22	19.704		4,300	1,250		5,550	6,675
7100	315 MBH		.96	25.105		5,625	1,600		7,225	8,675
7120	420 MBH		.70	34.483		5,300	2,200		7,500	9,250
9000	Minimum labor/equipment charge		1.75	13.714	Job		875		875	1,350

23 52 88 – Burners

23 52 88.10 Replacement Type Burners

		Crew	Daily Output	Labor-Hours	Unit	Material	Labor	Equipment	Total	Total Incl O&P
0010	**REPLACEMENT TYPE BURNERS**									
0990	Residential, conversion, gas fired, LP or natural									
1000	Gun type, atmospheric input 50 to 225 MBH	Q-1	2.50	6.400	Ea.	1,725	390		2,115	2,500
1020	100 to 400 MBH		2	8		2,875	485		3,360	3,900
1040	300 to 1,000 MBH		1.70	9.412		8,700	575		9,275	10,500
3000	Flame retention oil fired assembly, input									
3020	.50 to 2.25 GPH	Q-1	2.40	6.667	Ea.	360	405		765	1,025
3040	2.0 to 5.0 GPH		2	8		335	485		820	1,125
3060	3.0 to 7.0 GPH		1.80	8.889		630	540		1,170	1,525
3080	6.0 to 12.0 GPH		1.60	10		1,025	610		1,635	2,075

23 54 Furnaces

23 54 13 – Electric-Resistance Furnaces

23 54 13.10 Electric Furnaces

		Crew	Daily Output	Labor-Hours	Unit	Material	Labor	Equipment	Total	Total Incl O&P
0010	**ELECTRIC FURNACES**, Hot air, blowers, std. controls									
0011	not including gas, oil or flue piping									
1000	Electric, UL listed									
1070	10.2 MBH	Q-20	4.40	4.545	Ea.	345	272		617	805
1080	17.1 MBH		4.60	4.348		425	260		685	870
1100	34.1 MBH		4.40	4.545		530	272		802	1,000

23 54 16 – Fuel-Fired Furnaces

23 54 16.13 Gas-Fired Furnaces

		Crew	Daily Output	Labor-Hours	Unit	Material	Labor	Equipment	Total	Total Incl O&P
0010	**GAS-FIRED FURNACES**									
3000	Gas, AGA certified, upflow, direct drive models									
3020	45 MBH input	Q-9	4	4	Ea.	715	236		951	1,150
3040	60 MBH input		3.80	4.211		645	248		893	1,100
3060	75 MBH input		3.60	4.444		880	262		1,142	1,375
3100	100 MBH input		3.20	5		840	295		1,135	1,375

23 54 Furnaces

23 54 16 – Fuel-Fired Furnaces

23 54 16.13 Gas-Fired Furnaces

		Crew	Daily Output	Labor-Hours	Unit	Material	2021 Bare Costs Labor	Equipment	Total	Total Incl O&P
3120	125 MBH input	Q-9	3	5.333	Ea.	875	315		1,190	1,450
3130	150 MBH input		2.80	5.714		1,750	335		2,085	2,450
3140	200 MBH input		2.60	6.154		3,425	360		3,785	4,325

23 54 16.16 Oil-Fired Furnaces

		Crew	Daily Output	Labor-Hours	Unit	Material	2021 Bare Costs Labor	Equipment	Total	Total Incl O&P
0010	**OIL-FIRED FURNACES**									
6000	Oil, UL listed, atomizing gun type burner									
6020	56 MBH output	Q-9	3.60	4.444	Ea.	3,375	262		3,637	4,125
6040	95 MBH output		3.40	4.706		3,600	277		3,877	4,375
6060	134 MBH output		3.20	5		3,500	295		3,795	4,300
6080	151 MBH output		3	5.333		3,900	315		4,215	4,775
6100	200 MBH input		2.60	6.154		4,800	360		5,160	5,850
9000	Minimum labor/equipment charge		2.75	5.818	Job		345		345	535

23 54 24 – Furnace Components for Cooling

23 54 24.10 Furnace Components and Combinations

		Crew	Daily Output	Labor-Hours	Unit	Material	2021 Bare Costs Labor	Equipment	Total	Total Incl O&P
0010	**FURNACE COMPONENTS AND COMBINATIONS**									
0080	Coils, A.C. evaporator, for gas or oil furnaces									
0090	Add-on, with holding charge									
0100	Upflow									
0120	1-1/2 ton cooling	Q-5	4	4	Ea.	220	246		466	620
0130	2 ton cooling		3.70	4.324		253	266		519	690
0140	3 ton cooling		3.30	4.848		385	298		683	885
0150	4 ton cooling		3	5.333		470	330		800	1,025
0160	5 ton cooling		2.70	5.926		530	365		895	1,150
0300	Downflow									
0330	2-1/2 ton cooling	Q-5	3	5.333	Ea.	297	330		627	830
0340	3-1/2 ton cooling		2.60	6.154		445	380		825	1,075
0350	5 ton cooling		2.20	7.273		530	445		975	1,275
0600	Horizontal									
0630	2 ton cooling	Q-5	3.90	4.103	Ea.	320	252		572	745
0640	3 ton cooling		3.50	4.571		370	281		651	840
0650	4 ton cooling		3.20	5		520	310		830	1,050
0660	5 ton cooling		2.90	5.517		565	340		905	1,150
2000	Cased evaporator coils for air handlers									
2100	1-1/2 ton cooling	Q-5	4.40	3.636	Ea.	263	224		487	635
2110	2 ton cooling		4.10	3.902		305	240		545	705
2120	2-1/2 ton cooling		3.90	4.103		320	252		572	740
2130	3 ton cooling		3.70	4.324		360	266		626	805
2140	3-1/2 ton cooling		3.50	4.571		490	281		771	975
2150	4 ton cooling		3.20	5		520	310		830	1,050
2160	5 ton cooling		2.90	5.517		515	340		855	1,100
3010	Air handler, modular									
3100	With cased evaporator cooling coil									
3120	1-1/2 ton cooling	Q-5	3.80	4.211	Ea.	600	259		859	1,050
3130	2 ton cooling		3.50	4.571		605	281		886	1,100
3140	2-1/2 ton cooling		3.30	4.848		710	298		1,008	1,250
3150	3 ton cooling		3.10	5.161		825	320		1,145	1,400
3160	3-1/2 ton cooling		2.90	5.517		830	340		1,170	1,450
3170	4 ton cooling		2.50	6.400		1,050	395		1,445	1,775
3180	5 ton cooling		2.10	7.619		1,125	470		1,595	1,950
3500	With no cooling coil									
3520	1-1/2 ton coil size	Q-5	12	1.333	Ea.	620	82		702	805
3530	2 ton coil size		10	1.600		1,025	98.50		1,123.50	1,275

23 54 Furnaces

23 54 24 – Furnace Components for Cooling

23 54 24.10 Furnace Components and Combinations	Crew	Daily Output	Labor-Hours	Unit	Material	2021 Bare Costs Labor	Equipment	Total	Total Incl O&P
3540	2-1/2 ton coil size	Q-5	10	1.600	Ea.	1,125	98.50		1,223.50	1,400
3554	3 ton coil size		9	1.778		1,175	109		1,284	1,475
3560	3-1/2 ton coil size		9	1.778		1,275	109		1,384	1,575
3570	4 ton coil size		8.50	1.882		1,225	116		1,341	1,525
3580	5 ton coil size		8	2		1,425	123		1,548	1,750
4000	With heater									
4120	5 kW, 17.1 MBH	Q-5	16	1	Ea.	970	61.50		1,031.50	1,175
4130	7.5 kW, 25.6 MBH		15.60	1.026		860	63		923	1,050
4140	10 kW, 34.2 MBH		15.20	1.053		1,025	65		1,090	1,250
4150	12.5 kW, 42.7 MBH		14.80	1.081		1,225	66.50		1,291.50	1,450

23 55 Fuel-Fired Heaters

23 55 13 – Fuel-Fired Duct Heaters

23 55 13.16 Gas-Fired Duct Heaters

23 55 13.16 Gas-Fired Duct Heaters	Crew	Daily Output	Labor-Hours	Unit	Material	Labor	Equipment	Total	Total Incl O&P
0010	**GAS-FIRED DUCT HEATERS**, Includes burner, controls, stainless steel									
0020	heat exchanger. Gas fired, electric ignition									
1000	Outdoor installation, with power venter									
1120	225 MBH output	Q-5	2.50	6.400	Ea.	5,875	395		6,270	7,050
1160	375 MBH output		1.60	10		6,675	615		7,290	8,275
1180	450 MBH output		1.40	11.429		7,550	705		8,255	9,375

23 55 33 – Fuel-Fired Unit Heaters

23 55 33.16 Gas-Fired Unit Heaters

23 55 33.16 Gas-Fired Unit Heaters	Crew	Daily Output	Labor-Hours	Unit	Material	Labor	Equipment	Total	Total Incl O&P
0010	**GAS-FIRED UNIT HEATERS**, Cabinet, grilles, fan, ctrls., burner, no piping									
0022	thermostat, no piping. For flue see Section 23 51 23.10									
1000	Gas fired, floor mounted									
1100	60 MBH output	Q-5	10	1.600	Ea.	755	98.50		853.50	985
1180	180 MBH output		6	2.667		1,200	164		1,364	1,575
2000	Suspension mounted, propeller fan, 20 MBH output		8.50	1.882		775	116		891	1,025
2040	60 MBH output		7	2.286		1,875	141		2,016	2,275
2100	130 MBH output		5	3.200		1,300	197		1,497	1,750
2240	320 MBH output		2	8		2,550	490		3,040	3,550
2500	For powered venter and adapter, add					400			400	440
5000	Wall furnace, 17.5 MBH output	Q-5	6	2.667		670	164		834	995
5020	24 MBH output		5	3.200		670	197		867	1,050
5040	35 MBH output		4	4		725	246		971	1,175
9000	Minimum labor/equipment charge		3.50	4.571	Job		281		281	435

23 56 Solar Energy Heating Equipment

23 56 16 – Packaged Solar Heating Equipment

23 56 16.40 Solar Heating Systems

23 56 16.40 Solar Heating Systems	Crew	Daily Output	Labor-Hours	Unit	Material	Labor	Equipment	Total	Total Incl O&P
0010	**SOLAR HEATING SYSTEMS**									
0020	System/package prices, not including connecting									
0030	pipe, insulation, or special heating/plumbing fixtures									
0500	Hot water, standard package, low temperature									
0580	2 collectors, circulator, fittings, 120 gal. tank [G]	Q-1	.40	40	Ea.	5,475	2,425		7,900	9,775

23 56 19 – Solar Heating Components

23 56 19.50 Solar Heating Ancillary		Crew	Daily Output	Labor-Hours	Unit	Material	2021 Bare Costs Labor	Equipment	Total	Total Incl O&P
0010	**SOLAR HEATING ANCILLARY**									
2300	Circulators, air									
2310	Blowers									
2550	Booster fan 6" diameter, 120 CFM	G	Q-9	16	1	Ea.	38	59		97	134
2580	8" diameter, 150 CFM	G		16	1		42.50	59		101.50	139
2660	Shutter/damper	G		12	1.333		59	78.50		137.50	187
2670	Shutter motor	G		16	1		147	59		206	254
2800	Circulators, liquid, 1/25 HP, 5.3 GPM	G	Q-1	14	1.143		208	69.50		277.50	335
2820	1/20 HP, 17 GPM	G	"	12	1.333		310	81		391	465
3000	Collector panels, air with aluminum absorber plate									
3010	Wall or roof mount									
3040	Flat black, plastic glazing									
3080	4' x 8'	G	Q-9	6	2.667	Ea.	685	157		842	1,000
3300	Collector panels, liquid with copper absorber plate									
3320	Black chrome, tempered glass glazing									
3330	Alum. frame, 4' x 8', 5/32" single glazing	G	Q-1	9.50	1.684	Ea.	1,050	103		1,153	1,300
3450	Flat black, alum. frame, 3.5' x 7.5'	G		9	1.778		765	108		873	1,000
3500	4' x 8'	G		5.50	2.909		930	177		1,107	1,300
3600	Liquid, full wetted, plastic, alum. frame, 4' x 10'	G		5	3.200		330	195		525	660
3650	Collector panel mounting, flat roof or ground rack	G		7	2.286		267	139		406	510
3670	Roof clamps	G		70	.229	Set	3.57	13.95		17.52	25.50
3700	Roof strap, teflon	G	1 Plum	205	.039	L.F.	25	2.64		27.64	31.50
3900	Differential controller with two sensors									
3930	Thermostat, hard wired	G	1 Plum	8	1	Ea.	115	67.50		182.50	231
4050	Pool valve system	G		2.50	3.200		149	217		366	500
4070	With 12 VAC actuator	G		2	4		330	271		601	785
4150	Sensors									
4220	Freeze prevention	G	1 Plum	32	.250	Ea.	26	16.95		42.95	54.50
4300	Heat exchanger									
4315	includes coil, blower, circulator									
4316	and controller for DHW and space hot air									
4330	Fluid to air coil, up flow, 45 MBH	G	Q-1	4	4	Ea.	335	244		579	745
4380	70 MBH	G		3.50	4.571		380	279		659	845
4400	80 MBH	G		3	5.333		500	325		825	1,050
4580	Fluid to fluid package includes two circulating pumps									
4590	expansion tank, check valve, relief valve									
4600	controller, high temperature cutoff and sensors	G	Q-1	2.50	6.400	Ea.	785	390		1,175	1,450
4650	Heat transfer fluid									
4700	Propylene glycol, inhibited anti-freeze	G	1 Plum	28	.286	Gal.	23	19.35		42.35	55
4800	Solar storage tanks, knocked down									
5190	6'-3" high, 7' x 7' = 306 C.F./2,000 gallons	G	Q-10	1.20	20	Ea.	14,000	1,225		15,225	17,300
5210	7' x 14' = 613 C.F./4,000 gallons	G		.60	40		22,100	2,450		24,550	28,100
5230	10'-6" x 14' = 919 C.F./6,000 gallons	G		.40	60		24,300	3,675		27,975	32,400
5250	14' x 17'-6" = 1,531 C.F./10,000 gallons	G	Q-11	.30	107		31,500	6,650		38,150	45,000
7000	Solar control valves and vents									
7070	Air eliminator, automatic 3/4" size	G	1 Plum	32	.250	Ea.	31	16.95		47.95	60
7090	Air vent, automatic, 1/8" fitting	G		32	.250		15.95	16.95		32.90	43.50
7100	Manual, 1/8" NPT	G		32	.250		3.31	16.95		20.26	29.50
7120	Backflow preventer, 1/2" pipe size	G		16	.500		89	34		123	150
7130	3/4" pipe size	G		16	.500		74	34		108	134
7150	Balancing valve, 3/4" pipe size	G		20	.400		56	27		83	104
7180	Draindown valve, 1/2" copper tube	G		9	.889		220	60		280	335

23 56 Solar Energy Heating Equipment

23 56 19 – Solar Heating Components

23 56 19.50 Solar Heating Ancillary

		Crew	Daily Output	Labor-Hours	Unit	Material	2021 Bare Costs Labor	Equipment	Total	Total Incl O&P
7200	Flow control valve, 1/2" pipe size	G 1 Plum	22	.364	Ea.	143	24.50		167.50	195
7220	Expansion tank, up to 5 gal.	G	32	.250		65.50	16.95		82.45	98
7400	Pressure gauge, 2" dial	G	32	.250		25.50	16.95		42.45	54
7450	Relief valve, temp. and pressure 3/4" pipe size	G	30	.267		26.50	18.05		44.55	57
7500	Solenoid valve, normally closed									
7520	Brass, 3/4" NPT, 24V	G 1 Plum	9	.889	Ea.	161	60		221	270
7530	1" NPT, 24V	G	9	.889		257	60		317	375
7750	Vacuum relief valve, 3/4" pipe size	G	32	.250		31	16.95		47.95	60
7800	Thermometers									
7820	Digital temperature monitoring, 4 locations	G 1 Plum	2.50	3.200	Ea.	167	217		384	520
7900	Upright, 1/2" NPT	G "	8	1	"	59.50	67.50		127	169
8250	Water storage tank with heat exchanger and electric element									
8270	66 gal. with 2" x 2 lb. density insulation	G 1 Plum	1.60	5	Ea.	2,075	340		2,415	2,825
8300	80 gal. with 2" x 2 lb. density insulation	G "	1.60	5	"	1,800	340		2,140	2,500
8500	Water storage module, plastic									
8600	Tubular, 12" diameter, 4' high	G 1 Carp	48	.167	Ea.	101	9.10		110.10	126
8610	12" diameter, 8' high	G "	40	.200		159	10.95		169.95	192
8650	Cap, 12" diameter	G				20			20	22
9000	Minimum labor/equipment charge	1 Plum	2	4	Job		271		271	420

23 61 Refrigerant Compressors

23 61 16 – Reciprocating Refrigerant Compressors

23 61 16.10 Reciprocating Compressors

		Crew	Daily Output	Labor-Hours	Unit	Material	2021 Bare Costs Labor	Equipment	Total	Total Incl O&P
0010	**RECIPROCATING COMPRESSORS**									
0990	Refrigeration, recip. hermetic, switches & protective devices									
1000	10 ton	Q-5	1	16	Ea.	17,200	985		18,185	20,500
1100	20 ton	Q-6	.72	33.333		23,000	2,125		25,125	28,500
1400	50 ton	"	.20	120		34,000	7,650		41,650	49,200
1600	130 ton	Q-7	.21	152		42,900	9,925		52,825	62,500

23 62 Packaged Compressor and Condenser Units

23 62 13 – Packaged Air-Cooled Refrigerant Compressor and Condenser Units

23 62 13.10 Packaged Air-Cooled Refrig. Condensing Units

		Crew	Daily Output	Labor-Hours	Unit	Material	2021 Bare Costs Labor	Equipment	Total	Total Incl O&P
0010	**PACKAGED AIR-COOLED REFRIGERANT CONDENSING UNITS**									
0020	Condensing unit									
0030	Air cooled, compressor, standard controls									
0050	1.5 ton	Q-5	2.50	6.400	Ea.	945	395		1,340	1,650
0200	2.5 ton		1.70	9.412		1,225	580		1,805	2,250
0300	3 ton		1.30	12.308		1,250	755		2,005	2,525
0400	4 ton		.90	17.778		1,675	1,100		2,775	3,500
0500	5 ton		.60	26.667		2,000	1,650		3,650	4,725

23 64 Packaged Water Chillers

23 64 19 – Reciprocating Water Chillers

23 64 19.10 Reciprocating Type Water Chillers

		Crew	Daily Output	Labor-Hours	Unit	Material	2021 Bare Costs Labor	Equipment	Total	Total Incl O&P
0010	**RECIPROCATING TYPE WATER CHILLERS**, With standard controls									
0494	Water chillers, integral air cooled condenser									
0980	Water cooled, multiple compressor, semi-hermetic, tower not incl.									
1090	45 ton cooling	Q-7	.29	111	Ea.	28,100	7,225		35,325	42,000
1160	100 ton cooling		.18	180		47,700	11,700		59,400	70,500
1200	145 ton cooling		.16	203		60,000	13,200		73,200	86,000

23 64 23 – Scroll Water Chillers

23 64 23.10 Scroll Water Chillers

		Crew	Daily Output	Labor-Hours	Unit	Material	2021 Bare Costs Labor	Equipment	Total	Total Incl O&P
0010	**SCROLL WATER CHILLERS**, With standard controls									
0480	Packaged w/integral air cooled condenser									
0482	10 ton cooling	Q-7	.34	94.118	Ea.	15,400	6,125		21,525	26,400
0490	15 ton cooling		.37	86.486		18,400	5,625		24,025	29,000
0500	20 ton cooling		.34	94.118		23,400	6,125		29,525	35,200
0520	40 ton cooling		.30	108		38,000	7,025		45,025	52,500
0680	Scroll water cooled, single compressor, hermetic, tower not incl.									
0760	10 ton cooling	Q-6	.36	67.039	Ea.	7,425	4,275		11,700	14,800
0900	Scroll water cooled, multiple compressor, hermetic, tower not incl.									

23 64 26 – Rotary-Screw Water Chillers

23 64 26.10 Rotary-Screw Type Water Chillers

		Crew	Daily Output	Labor-Hours	Unit	Material	2021 Bare Costs Labor	Equipment	Total	Total Incl O&P
0010	**ROTARY-SCREW TYPE WATER CHILLERS**, With standard controls									
0110	Screw, liquid chiller, air cooled, insulated evaporator									
0120	130 ton	Q-7	.14	229	Ea.	100,500	14,900		115,400	133,500
0124	160 ton	"	.13	246	"	129,000	16,000		145,000	166,500

23 64 33 – Modular Water Chillers

23 64 33.10 Direct Expansion Type Water Chillers

		Crew	Daily Output	Labor-Hours	Unit	Material	2021 Bare Costs Labor	Equipment	Total	Total Incl O&P
0010	**DIRECT EXPANSION TYPE WATER CHILLERS**, With standard controls									
8000	Direct expansion, shell and tube type, for built up systems									
8020	1 ton	Q-5	2	8	Ea.	8,375	490		8,865	9,975
8030	5 ton		1.90	8.421		13,900	520		14,420	16,100
8040	10 ton		1.70	9.412		17,300	580		17,880	19,900
9000	Minimum labor/equipment charge	Q-6	1	24	Job		1,525		1,525	2,350

23 65 Cooling Towers

23 65 13 – Forced-Draft Cooling Towers

23 65 13.10 Forced-Draft Type Cooling Towers

		Crew	Daily Output	Labor-Hours	Unit	Material	2021 Bare Costs Labor	Equipment	Total	Total Incl O&P
0010	**FORCED-DRAFT TYPE COOLING TOWERS**, Packaged units									
0070	Galvanized steel									
0080	Induced draft, crossflow									
0100	Vertical, belt drive, 61 tons	Q-6	90	.267	TonAC	208	17		225	255
0150	100 ton		100	.240		201	15.30		216.30	245
0200	115 ton		109	.220		175	14.05		189.05	214
0250	131 ton		120	.200		210	12.75		222.75	251
0260	162 ton		132	.182		170	11.60		181.60	205
1000	For higher capacities, use multiples									
1500	Induced air, double flow									
1900	Vertical, gear drive, 167 ton	Q-6	126	.190	TonAC	165	12.15		177.15	200
2000	297 ton		129	.186		101	11.85		112.85	130
2100	582 ton		132	.182		56.50	11.60		68.10	80
2200	1,016 ton		150	.160		76.50	10.20		86.70	100

23 65 Cooling Towers

23 65 13 – Forced-Draft Cooling Towers

23 65 13.10 Forced-Draft Type Cooling Towers	Crew	Daily Output	Labor-Hours	Unit	Material	2021 Bare Costs Labor	Equipment	Total	Total Incl O&P
3000	For higher capacities, use multiples									
3500	For pumps and piping, add	Q-6	38	.632	TonAC	104	40.50		144.50	177
4000	For absorption systems, add				"	75%	75%			
4100	Cooling water chemical feeder	Q-5	3	5.333	Ea.	440	330		770	990
5000	Fiberglass tower on galvanized steel support structure									
5010	Draw thru									
5100	100 ton	Q-6	1.40	17.143	Ea.	13,900	1,100		15,000	17,000
5120	120 ton		1.20	20		16,700	1,275		17,975	20,300
5140	140 ton		1	24		17,000	1,525		18,525	21,100
5160	160 ton		.80	30		18,400	1,925		20,325	23,200
5180	180 ton		.65	36.923		21,500	2,350		23,850	27,200
5251	600 ton		.16	150		78,500	9,575		88,075	101,500
5252	1,000 ton		.10	250		131,000	16,000		147,000	168,500
5300	For stainless steel support structure, add					30%				
5360	For higher capacities, use multiples of each size									
6000	Stainless steel									
6010	Induced draft, crossflow, horizontal, belt drive									
6100	57 ton	Q-6	1.50	16	Ea.	27,700	1,025		28,725	32,100
6120	91 ton		.99	24.242		34,000	1,550		35,550	39,800
6140	111 ton		.43	55.814		51,500	3,550		55,050	62,000
6160	126 ton		.22	109		53,000	6,950		59,950	68,500
6170	Induced draft, crossflow, vertical, gear drive									
6172	167 ton	Q-6	.75	32	Ea.	62,000	2,050		64,050	71,500
6174	297 ton		.43	55.814		68,500	3,550		72,050	81,000
6176	582 ton		.23	104		110,000	6,650		116,650	131,500
6180	1,016 ton		.15	160		194,000	10,200		204,200	229,000
9000	Minimum labor/equipment charge		1	24	Job		1,525		1,525	2,350

23 72 Air-to-Air Energy Recovery Equipment

23 72 13 – Heat-Wheel Air-to-Air Energy-Recovery Equipment

23 72 13.10 Heat-Wheel Air-to-Air Energy Recovery Equip.

		Crew	Daily Output	Labor-Hours	Unit	Material	2021 Bare Costs Labor	Equipment	Total	Total Incl O&P
0010	**HEAT-WHEEL AIR-TO-AIR ENERGY RECOVERY EQUIPMENT**									
0100	Air to air									
9000	Minimum labor/equipment charge	1 Shee	4	2	Job		131		131	204

23 73 Indoor Central-Station Air-Handling Units

23 73 39 – Indoor, Direct Gas-Fired Heating and Ventilating Units

23 73 39.10 Make-Up Air Unit

		Crew	Daily Output	Labor-Hours	Unit	Material	2021 Bare Costs Labor	Equipment	Total	Total Incl O&P
0010	**MAKE-UP AIR UNIT**									
0020	Indoor suspension, natural/LP gas, direct fired,									
0032	standard control. For flue see Section 23 51 23.10									
0040	70°F temperature rise, MBH is input									
0100	75 MBH input	Q-6	3.60	6.667	Ea.	4,325	425		4,750	5,400
0220	225 MBH input		2.40	10		5,850	640		6,490	7,400
0300	400 MBH input		1.60	15		14,500	955		15,455	17,400

23 74 Packaged Outdoor HVAC Equipment

23 74 23 – Packaged, Outdoor, Heating-Only Makeup-Air Units

23 74 23.16 Packaged, Ind.-Fired, In/Outdoor

		Crew	Daily Output	Labor-Hours	Unit	Material	2021 Bare Costs Labor	2021 Bare Costs Equipment	Total	Total Incl O&P
0010	**PACKAGED, IND.-FIRED, IN/OUTDOOR,** Heat-Only Makeup-Air Units									
1000	Rooftop unit, natural gas, gravity vent, S.S. exchanger									
1010	70°F temperature rise, MBH is input									
1020	250 MBH	Q-6	4	6	Ea.	16,500	385		16,885	18,800
1040	400 MBH		3.60	6.667		18,900	425		19,325	21,400
1060	550 MBH	↓	3.30	7.273		20,600	465		21,065	23,300
1600	For cleanable filters, add					5%				
1700	For electric modulating gas control, add				↓	10%				
9000	Minimum labor/equipment charge	Q-6	2.75	8.727	Job		555		555	860

23 74 33 – Dedicated Outdoor-Air Units

23 74 33.10 Rooftop Air Conditioners

		Crew	Daily Output	Labor-Hours	Unit	Material	2021 Bare Costs Labor	2021 Bare Costs Equipment	Total	Total Incl O&P
0010	**ROOFTOP AIR CONDITIONERS,** Standard controls, curb, economizer									
1000	Single zone, electric cool, gas heat									
1100	3 ton cooling, 60 MBH heating	Q-5	.70	22.857	Ea.	2,650	1,400		4,050	5,075
1120	4 ton cooling, 95 MBH heating		.61	26.403		3,425	1,625		5,050	6,275
1150	7.5 ton cooling, 170 MBH heating		.50	32.258		5,825	1,975		7,800	9,450
1156	8.5 ton cooling, 170 MBH heating	↓	.46	34.783		7,300	2,150		9,450	11,300
1160	10 ton cooling, 200 MBH heating	Q-6	.67	35.982		9,500	2,300		11,800	14,000
1220	30 ton cooling, 540 MBH heating	Q-7	.47	68.376		36,800	4,450		41,250	47,300
1240	40 ton cooling, 675 MBH heating	"	.35	91.168	↓	44,800	5,925		50,725	58,500
2000	Multizone, electric cool, gas heat, economizer									
2120	20 ton cooling, 360 MBH heating	Q-7	.53	60.038	Ea.	63,000	3,900		66,900	75,500
2180	30 ton cooling, 540 MBH heating		.37	85.562		96,000	5,575		101,575	114,000
2210	50 ton cooling, 540 MBH heating		.23	142		140,500	9,250		149,750	169,000
2220	70 ton cooling, 1,500 MBH heating		.16	199		149,500	12,900		162,400	184,500
2240	80 ton cooling, 1,500 MBH heating		.14	229		171,000	14,900		185,900	211,000
2280	105 ton cooling, 1,500 MBH heating	↓	.11	291		225,000	18,900		243,900	276,500
2400	For hot water heat coil, deduct					5%				
2500	For steam heat coil, deduct					2%				
2600	For electric heat, deduct				↓	3%	5%			
9000	Minimum labor/equipment charge	Q-5	1.50	10.667	Job		655		655	1,000

23 81 Decentralized Unitary HVAC Equipment

23 81 19 – Self-Contained Air-Conditioners

23 81 19.10 Window Unit Air Conditioners

		Crew	Daily Output	Labor-Hours	Unit	Material	2021 Bare Costs Labor	2021 Bare Costs Equipment	Total	Total Incl O&P
0010	**WINDOW UNIT AIR CONDITIONERS**									
4000	Portable/window, 15 amp, 125 V grounded receptacle required									
4500	10,000 BTUH	1 Carp	6	1.333	Ea.	365	73		438	520
4520	12,000 BTUH	L-2	8	2	"	1,875	97		1,972	2,200
9000	Minimum labor/equipment charge	1 Carp	2	4	Job		219		219	350

23 81 19.20 Self-Contained Single Package

		Crew	Daily Output	Labor-Hours	Unit	Material	2021 Bare Costs Labor	2021 Bare Costs Equipment	Total	Total Incl O&P
0010	**SELF-CONTAINED SINGLE PACKAGE**									
0100	Air cooled, for free blow or duct, not incl. remote condenser									
0110	Constant volume									
0200	3 ton cooling	Q-5	1	16	Ea.	4,025	985		5,010	5,950
0210	4 ton cooling	"	.80	20		4,375	1,225		5,600	6,725
0240	10 ton cooling	Q-7	1	32	↓	8,275	2,075		10,350	12,300
0280	30 ton cooling	"	.80	40	↓	26,900	2,600		29,500	33,600
1000	Water cooled for free blow or duct, not including tower									
1010	Constant volume									

For customer support on your Commercial Renovation Costs with RSMeans Data, call 800.448.8182.

479

23 81 Decentralized Unitary HVAC Equipment

23 81 19 – Self-Contained Air-Conditioners

23 81 19.20 Self-Contained Single Package

23 81 19.20 Self-Contained Single Package	Crew	Daily Output	Labor-Hours	Unit	Material	2021 Bare Costs Labor	Equipment	Total	Total Incl O&P
1100	3 ton cooling	Q-6	1	24	Ea.	3,725	1,525		5,250	6,425
1120	5 ton cooling	"	1	24		4,875	1,525		6,400	7,725
1140	10 ton cooling	Q-7	.90	35.556		8,675	2,325		11,000	13,100
1180	30 ton cooling	"	.70	45.714		35,200	2,975		38,175	43,400
1240	60 ton cooling	Q-8	.30	107		56,000	6,950	360	63,310	72,500
1300	For hot water or steam heat coils, add					12%	10%			
9000	Minimum labor/equipment charge	Q-5	1	16	Job		985		985	1,525

23 81 23 – Computer-Room Air-Conditioners

23 81 23.10 Computer Room Units

		Crew	Daily Output	Labor-Hours	Unit	Material	Labor	Equipment	Total	Total Incl O&P
0010	**COMPUTER ROOM UNITS**									
1000	Air cooled, includes remote condenser but not									
1020	interconnecting tubing or refrigerant									
1160	6 ton	Q-5	.30	53.333	Ea.	56,500	3,275		59,775	67,000
1240	10 ton	"	.25	64		61,500	3,950		65,450	73,500
1300	20 ton	Q-6	.26	92.308		80,500	5,900		86,400	98,000
1320	22 ton	"	.24	100		81,500	6,375		87,875	100,000

23 81 26 – Split-System Air-Conditioners

23 81 26.10 Split Ductless Systems

		Crew	Daily Output	Labor-Hours	Unit	Material	Labor	Equipment	Total	Total Incl O&P
0010	**SPLIT DUCTLESS SYSTEMS**									
0100	Cooling only, single zone									
0110	Wall mount									
0120	3/4 ton cooling	Q-5	2	8	Ea.	1,100	490		1,590	1,975
0130	1 ton cooling		1.80	8.889		1,250	545		1,795	2,250
0140	1-1/2 ton cooling		1.60	10		1,975	615		2,590	3,125
0150	2 ton cooling		1.40	11.429		2,250	705		2,955	3,550
0160	3 ton cooling		1.20	13.333		2,250	820		3,070	3,750
1000	Ceiling mount									
1020	2 ton cooling	Q-5	1.40	11.429	Ea.	2,000	705		2,705	3,275
1030	3 ton cooling	"	1.20	13.333	"	2,550	820		3,370	4,075
3000	Multizone									
3010	Wall mount									
3020	2 @ 3/4 ton cooling	Q-5	1.80	8.889	Ea.	3,375	545		3,920	4,550
5000	Cooling/Heating									
5010	Wall mount									
5110	1 ton cooling	Q-5	1.70	9.412	Ea.	1,300	580		1,880	2,325
5120	1-1/2 ton cooling	"	1.50	10.667	"	1,975	655		2,630	3,175
7000	Accessories for all split ductless systems									
7010	Add for ambient frost control	Q-5	8	2	Ea.	259	123		382	475
7020	Add for tube/wiring kit (line sets)									
7030	15' kit	Q-5	32	.500	Ea.	63.50	31		94.50	117
7036	25' kit		28	.571		53	35		88	113
7040	35' kit		24	.667		108	41		149	183
7050	50' kit		20	.800		144	49		193	235

23 81 29 – Variable Refrigerant Flow HVAC Systems

23 81 29.10 Heat Pump, Gas Driven

		Crew	Daily Output	Labor-Hours	Unit	Material	Labor	Equipment	Total	Total Incl O&P
0010	**HEAT PUMP, GAS DRIVEN**, Variable refrigerant volume (VRV) type									
0020	Not including interconnecting tubing or multi-zone controls									
1000	For indoor fan VRV type AHU see 23 82 19.40									
1010	Multi-zone split									
1020	Outdoor unit									
1100	8 tons cooling, for up to 17 zones	Q-5	1.30	12.308	Ea.	30,600	755		31,355	34,900

23 81 Decentralized Unitary HVAC Equipment

23 81 29 – Variable Refrigerant Flow HVAC Systems

23 81 29.10 Heat Pump, Gas Driven	Crew	Daily Output	Labor-Hours	Unit	Material	2021 Bare Costs Labor	Equipment	Total	Total Incl O&P
1110	Isolation rails	Q-5	2.60	6.154	Pair	1,075	380		1,455	1,750
1160	15 tons cooling, for up to 33 zones		1	16	Ea.	40,400	985		41,385	45,900
1170	Isolation rails	↓	2.60	6.154	Pair	1,275	380		1,655	1,975
2000	Packaged unit									
2020	Outdoor unit									
2200	11 tons cooling	Q-5	1.30	12.308	Ea.	39,500	755		40,255	44,600
2210	Roof curb adapter	"	2.60	6.154	↓	685	380		1,065	1,325
2220	Thermostat	1 Stpi	1.30	6.154	↓	410	420		830	1,100

23 81 43 – Air-Source Unitary Heat Pumps

23 81 43.10 Air-Source Heat Pumps

		Crew	Daily Output	Labor-Hours	Unit	Material	2021 Bare Costs Labor	Equipment	Total	Total Incl O&P
0010	**AIR-SOURCE HEAT PUMPS**, Not including interconnecting tubing									
1000	Air to air, split system, not including curbs, pads, fan coil and ductwork									
1012	Outside condensing unit only, for fan coil see Section 23 82 19.10									
1020	2 ton cooling, 8.5 MBH heat @ 0°F	Q-5	2	8	Ea.	1,600	490		2,090	2,500
1040	3 ton cooling, 13 MBH heat @ 0°F		1.20	13.333		1,950	820		2,770	3,425
1080	7.5 ton cooling, 33 MBH heat @ 0°F	↓	.45	35.556		3,650	2,200		5,850	7,400
1120	15 ton cooling, 64 MBH heat @ 0°F	Q-6	.50	48		10,500	3,050		13,550	16,200
1130	20 ton cooling, 85 MBH heat @ 0°F		.35	68.571		19,100	4,375		23,475	27,800
1140	25 ton cooling, 119 MBH heat @ 0°F	↓	.25	96	↓	21,900	6,125		28,025	33,600
1500	Single package, not including curbs, pads, or plenums									
1520	2 ton cooling, 6.5 MBH heat @ 0°F	Q-5	1.50	10.667	Ea.	3,325	655		3,980	4,675
1560	3 ton cooling, 10 MBH heat @ 0°F	"	1.20	13.333	"	3,725	820		4,545	5,375
6000	Air to water, single package, excluding storage tank and ductwork									
6010	Includes circulating water pump, air duct connections, digital temperature									
6020	controller with remote tank temp. probe and sensor for storage tank.									
6040	Water heating - air cooling capacity									
6110	35.5 MBH heat water, 2.3 ton cool air	Q-5	1.60	10	Ea.	17,100	615		17,715	19,800
6120	58 MBH heat water, 3.8 ton cool air		1.10	14.545		19,500	895		20,395	22,900
6130	76 MBH heat water, 4.9 ton cool air		.87	18.391		23,300	1,125		24,425	27,400
6140	98 MBH heat water, 6.5 ton cool air		.62	25.806		30,100	1,600		31,700	35,600
6150	113 MBH heat water, 7.4 ton cool air		.59	27.119		33,000	1,675		34,675	38,900
6160	142 MBH heat water, 9.2 ton cool air		.52	30.769		40,000	1,900		41,900	46,900
6170	171 MBH heat water, 11.1 ton cool air	↓	.49	32.653	↓	47,100	2,000		49,100	55,000

23 81 46 – Water-Source Unitary Heat Pumps

23 81 46.10 Water Source Heat Pumps

		Crew	Daily Output	Labor-Hours	Unit	Material	2021 Bare Costs Labor	Equipment	Total	Total Incl O&P
0010	**WATER SOURCE HEAT PUMPS**, Not incl. connecting tubing or water source									
2000	Water source to air, single package									
2100	1 ton cooling, 13 MBH heat @ 75°F	Q-5	2	8	Ea.	2,075	490		2,565	3,050
2140	2 ton cooling, 19 MBH heat @ 75°F		1.70	9.412		2,300	580		2,880	3,425
2220	5 ton cooling, 29 MBH heat @ 75°F		.90	17.778	↓	3,350	1,100		4,450	5,375
9000	Minimum labor/equipment charge	↓	1.75	9.143	Job		565		565	865

23 82 16 – Air Coils

23 82 16.10 Flanged Coils		Crew	Daily Output	Labor-Hours	Unit	Material	2021 Bare Costs Labor	Equipment	Total	Total Incl O&P
0010	**FLANGED COILS**									
1500	Hot water heating, 1 row, 24" x 48"	Q-5	4	4	Ea.	1,850	246		2,096	2,425
9000	Minimum labor/equipment charge	1 Plum	1	8	Job		540		540	835

23 82 16.20 Duct Heaters

		Crew	Daily Output	Labor-Hours	Unit	Material	Labor	Equipment	Total	Total Incl O&P
0010	**DUCT HEATERS**, Electric, 480 V, 3 Ph.									
0020	Finned tubular insert, 500°F									
0100	8" wide x 6" high, 4.0 kW	Q-20	16	1.250	Ea.	820	75		895	1,025
0120	12" high, 8.0 kW		15	1.333		1,350	80		1,430	1,625
0140	18" high, 12.0 kW		14	1.429		1,900	85.50		1,985.50	2,225
0160	24" high, 16.0 kW		13	1.538		2,450	92		2,542	2,850
0180	30" high, 20.0 kW		12	1.667		3,000	100		3,100	3,450
0300	12" wide x 6" high, 6.7 kW		15	1.333		870	80		950	1,075
0320	12" high, 13.3 kW		14	1.429		1,400	85.50		1,485.50	1,675
0340	18" high, 20.0 kW		13	1.538		1,975	92		2,067	2,325
8000	To obtain BTU multiply kW by 3413									
9900	Minimum labor/equipment charge	1 Shee	4	2	Job		131		131	204

23 82 19 – Fan Coil Units

23 82 19.10 Fan Coil Air Conditioning

		Crew	Daily Output	Labor-Hours	Unit	Material	Labor	Equipment	Total	Total Incl O&P
0010	**FAN COIL AIR CONDITIONING**									
0030	Fan coil AC, cabinet mounted, filters and controls									
0100	Chilled water, 1/2 ton cooling	Q-5	8	2	Ea.	680	123		803	940
0120	1 ton cooling		6	2.667		785	164		949	1,125
0160	2-1/2 ton cooling		5	3.200		2,175	197		2,372	2,700
0180	3 ton cooling		4	4		2,400	246		2,646	3,000
0262	For hot water coil, add					40%	10%			
0320	1 ton cooling	Q-5	6	2.667	Ea.	1,825	164		1,989	2,250
0940	Direct expansion, for use w/air cooled condensing unit, 1-1/2 ton cooling		5	3.200		700	197		897	1,075
1000	5 ton cooling		3	5.333		1,250	330		1,580	1,900
1040	10 ton cooling	Q-6	2.60	9.231		2,175	590		2,765	3,300
1060	20 ton cooling	"	.70	34.286		4,325	2,175		6,500	8,125
1500	For hot water coil, add					40%	10%			
1512	For condensing unit add see Section 23 62									
3000	Chilled water, horizontal unit, housing, 2 pipe, fan control, no valves									
3100	1/2 ton cooling	Q-5	8	2	Ea.	1,475	123		1,598	1,800
3110	1 ton cooling		6	2.667		1,700	164		1,864	2,125
3120	1-1/2 ton cooling		5.50	2.909		1,925	179		2,104	2,400
3130	2 ton cooling		5.25	3.048		2,275	188		2,463	2,800
3140	3 ton cooling		4	4		2,825	246		3,071	3,475
3150	3-1/2 ton cooling		3.80	4.211		3,175	259		3,434	3,875
3160	4 ton cooling		3.80	4.211		3,175	259		3,434	3,875
4000	With electric heat, 2 pipe									
4100	1/2 ton cooling	Q-5	8	2	Ea.	1,700	123		1,823	2,075
4105	3/4 ton cooling		7	2.286		1,900	141		2,041	2,300
4110	1 ton cooling		6	2.667		2,125	164		2,289	2,575
4120	1-1/2 ton cooling		5.50	2.909		2,350	179		2,529	2,875
4130	2 ton cooling		5.25	3.048		2,900	188		3,088	3,500
4135	2-1/2 ton cooling		4.80	3.333		4,100	205		4,305	4,825
4140	3 ton cooling		4	4		5,300	246		5,546	6,200
4150	3-1/2 ton cooling		3.80	4.211		6,325	259		6,584	7,375
4160	4 ton cooling		3.60	4.444		7,275	273		7,548	8,425
9000	Minimum labor/equipment charge		3.75	4.267	Job		263		263	405

23 82 19 – Fan Coil Units

23 82 19.40 Fan Coil Air Conditioning		Crew	Daily Output	Labor-Hours	Unit	Material	2021 Bare Costs Labor	Equipment	Total	Total Incl O&P
0010	**FAN COIL AIR CONDITIONING**, Variable refrigerant volume (VRV) type									
0020	Not including interconnecting tubing or multi-zone controls									
0030	For VRV condensing unit see section 23 81 29.10									
0050	Indoor type, ducted									
0100	Vertical concealed									
0130	1 ton cooling	G	Q-5	2.97	5.387	Ea.	2,675	330		3,005	3,450
0140	1.5 ton cooling	G		2.77	5.776		2,725	355		3,080	3,550
0150	2 ton cooling	G		2.70	5.926		2,925	365		3,290	3,775
0160	2.5 ton cooling	G		2.60	6.154		3,075	380		3,455	3,950
0170	3 ton cooling	G		2.50	6.400		3,125	395		3,520	4,025
0180	3.5 ton cooling	G		2.30	6.957		3,275	430		3,705	4,250
0190	4 ton cooling	G		2.19	7.306		3,300	450		3,750	4,325
0200	4.5 ton cooling	G	↓	2.08	7.692	↓	3,625	475		4,100	4,725
0230	Outside air connection possible									
1100	Ceiling concealed									
1130	0.6 ton cooling	G	Q-5	2.60	6.154	Ea.	1,750	380		2,130	2,500
1140	0.75 ton cooling	G		2.60	6.154		1,800	380		2,180	2,550
1150	1 ton cooling	G	↓	2.60	6.154		1,900	380		2,280	2,675
1152	Screening door	G	1 Stpi	5.20	1.538		105	105		210	278
1160	1.5 ton cooling	G	Q-5	2.60	6.154		1,975	380		2,355	2,750
1162	Screening door	G	1 Stpi	5.20	1.538		122	105		227	296
1170	2 ton cooling	G	Q-5	2.60	6.154		2,325	380		2,705	3,125
1172	Screening door	G	1 Stpi	5.20	1.538		105	105		210	278
1180	2.5 ton cooling	G	Q-5	2.60	6.154		2,675	380		3,055	3,525
1182	Screening door	G	1 Stpi	5.20	1.538		183	105		288	365
1190	3 ton cooling	G	Q-5	2.60	6.154		2,850	380		3,230	3,700
1192	Screening door	G	1 Stpi	5.20	1.538		183	105		288	365
1200	4 ton cooling	G	Q-5	2.30	6.957		2,800	430		3,230	3,725
1202	Screening door	G	1 Stpi	5.20	1.538		183	105		288	365
1220	6 ton cooling	G	Q-5	2.08	7.692		4,950	475		5,425	6,150
1230	8 ton cooling	G	"	1.89	8.466	↓	5,550	520		6,070	6,900
1260	Outside air connection possible									
4050	Indoor type, duct-free									
4100	Ceiling mounted cassette									
4130	0.75 ton cooling	G	Q-5	3.46	4.624	Ea.	2,075	285		2,360	2,725
4140	1 ton cooling	G		2.97	5.387		2,200	330		2,530	2,925
4150	1.5 ton cooling	G		2.77	5.776		2,275	355		2,630	3,050
4160	2 ton cooling	G		2.60	6.154		2,400	380		2,780	3,225
4170	2.5 ton cooling	G		2.50	6.400		2,475	395		2,870	3,325
4180	3 ton cooling	G		2.30	6.957		2,650	430		3,080	3,575
4190	4 ton cooling	G	↓	2.08	7.692		2,975	475		3,450	3,975
4198	For ceiling cassette decoration panel, add	G	1 Stpi	5.20	1.538	↓	315	105		420	505
4230	Outside air connection possible									
4500	Wall mounted									
4530	0.6 ton cooling	G	Q-5	5.20	3.077	Ea.	1,050	189		1,239	1,475
4540	0.75 ton cooling	G		5.20	3.077		1,100	189		1,289	1,525
4550	1 ton cooling	G		4.16	3.846		1,275	237		1,512	1,775
4560	1.5 ton cooling	G		3.77	4.244		1,400	261		1,661	1,925
4570	2 ton cooling	G	↓	3.46	4.624		1,500	285		1,785	2,100
4590	Condensate pump for the above air handlers	G	1 Stpi	6.46	1.238	↓	227	84.50		311.50	380
4700	Floor standing unit									
4730	1 ton cooling	G	Q-5	2.60	6.154	Ea.	1,775	380		2,155	2,550

23 82 19 – Fan Coil Units

23 82 19.40 Fan Coil Air Conditioning		Crew	Daily Output	Labor-Hours	Unit	Material	2021 Bare Costs Labor	2021 Bare Costs Equipment	Total	Total Incl O&P
4740	1.5 ton cooling ⒢	Q-5	2.60	6.154	Ea.	2,025	380		2,405	2,800
4750	2 ton cooling ⒢	↓	2.60	6.154	↓	2,175	380		2,555	2,975
4800	Floor standing unit, concealed									
4830	1 ton cooling ⒢	Q-5	2.60	6.154	Ea.	1,725	380		2,105	2,475
4840	1.5 ton cooling ⒢		2.60	6.154		1,950	380		2,330	2,700
4850	2 ton cooling ⒢	↓	2.60	6.154	↓	2,100	380		2,480	2,900
4880	Outside air connection possible									
8000	Accessories									
8100	Branch divergence pipe fitting									
8110	Capacity under 76 MBH	1 Stpi	2.50	3.200	Ea.	171	219		390	525
8120	76 MBH to 112 MBH		2.50	3.200		190	219		409	545
8130	112 MBH to 234 MBH	↓	2.50	3.200	↓	535	219		754	925
8200	Header pipe fitting									
8210	Max 4 branches									
8220	Capacity under 76 MBH	1 Stpi	1.70	4.706	Ea.	235	320		555	755
8260	Max 8 branches									
8270	76 MBH to 112 MBH	1 Stpi	1.70	4.706	Ea.	480	320		800	1,025
8280	112 MBH to 234 MBH	"	1.30	6.154	"	760	420		1,180	1,500

23 82 29 – Radiators

23 82 29.10 Hydronic Heating

		Crew	Daily Output	Labor-Hours	Unit	Material	2021 Bare Costs Labor	2021 Bare Costs Equipment	Total	Total Incl O&P
0010	**HYDRONIC HEATING,** Terminal units, not incl. main supply pipe									
1000	Radiation									
1100	Panel, baseboard, C.I., including supports, no covers	Q-5	46	.348	L.F.	53.50	21.50		75	92
3000	Radiators, cast iron									
3100	Free standing or wall hung, 6 tube, 25" high	Q-5	96	.167	Section	70.50	10.25		80.75	93.50
9000	Minimum labor/equipment charge	"	3.20	5	Job		310		310	475

23 82 36 – Finned-Tube Radiation Heaters

23 82 36.10 Finned Tube Radiation

		Crew	Daily Output	Labor-Hours	Unit	Material	2021 Bare Costs Labor	2021 Bare Costs Equipment	Total	Total Incl O&P
0010	**FINNED TUBE RADIATION,** Terminal units, not incl. main supply pipe									
1150	Fin tube, wall hung, 14" slope top cover, with damper									
1200	1-1/4" copper tube, 4-1/4" alum. fin	Q-5	38	.421	L.F.	49	26		75	94
1250	1-1/4" steel tube, 4-1/4" steel fin		36	.444		43	27.50		70.50	89.50
1255	2" steel tube, 4-1/4" steel fin		32	.500		47	31		78	99.50
1310	Baseboard, pkgd, 1/2" copper tube, alum. fin, 7" high		60	.267		11.75	16.40		28.15	38.50
1320	3/4" copper tube, alum. fin, 7" high		58	.276		7.85	16.95		24.80	34.50
1340	1" copper tube, alum. fin, 8-7/8" high		56	.286		20	17.60		37.60	49
1360	1-1/4" copper tube, alum. fin, 8-7/8" high	↓	54	.296	↓	30	18.25		48.25	61
1500	Note: fin tube may also require corners, caps, etc.									

23 82 39 – Unit Heaters

23 82 39.16 Propeller Unit Heaters

		Crew	Daily Output	Labor-Hours	Unit	Material	2021 Bare Costs Labor	2021 Bare Costs Equipment	Total	Total Incl O&P
0010	**PROPELLER UNIT HEATERS**									
3950	Unit heaters, propeller, 115 V 2 psi steam, 60°F entering air									
4000	Horizontal, 12 MBH	Q-5	12	1.333	Ea.	430	82		512	600
4060	43.9 MBH		8	2		640	123		763	895
4240	286.9 MBH		2	8		1,825	490		2,315	2,775
4260	364 MBH		1.80	8.889		2,300	545		2,845	3,400
4270	404 MBH	↓	1.60	10	↓	2,400	615		3,015	3,600
4300	Vertical diffuser same price									
4310	Vertical flow, 40 MBH	Q-5	11	1.455	Ea.	655	89.50		744.50	860
4326	131 MBH	"	4	4		975	246		1,221	1,450
4354	420 MBH (460 V)	Q-6	1.80	13.333	↓	2,600	850		3,450	4,175

484

23 82 Convection Heating and Cooling Units

23 82 39 – Unit Heaters

23 82 39.16 Propeller Unit Heaters	Crew	Daily Output	Labor-Hours	Unit	Material	2021 Bare Costs Labor	Equipment	Total	Total Incl O&P
4358	500 MBH (460 V)	Q-6	1.71	14.035	Ea.	3,325	895		4,220	5,025
4362	570 MBH (460 V)		1.40	17.143		4,125	1,100		5,225	6,225
4366	620 MBH (460 V)		1.30	18.462		4,950	1,175		6,125	7,250
4370	960 MBH (460 V)		1.10	21.818		8,450	1,400		9,850	11,500

23 83 Radiant Heating Units

23 83 33 – Electric Radiant Heaters

23 83 33.10 Electric Heating

		Crew	Daily Output	Labor-Hours	Unit	Material	2021 Bare Costs Labor	Equipment	Total	Total Incl O&P
0010	**ELECTRIC HEATING**, not incl. conduit or feed wiring									
1100	Rule of thumb: Baseboard units, including control	1 Elec	4.40	1.818	kW	119	116		235	310
1300	Baseboard heaters, 2' long, 350 watt		8	1	Ea.	31.50	63.50		95	133
1400	3' long, 750 watt		8	1		36.50	63.50		100	139
1600	4' long, 1,000 watt		6.70	1.194		38.50	76		114.50	160
1800	5' long, 935 watt		5.70	1.404		52.50	89.50		142	195
2000	6' long, 1,500 watt		5	1.600		53	102		155	216
2400	8' long, 2,000 watt		4	2		73	127		200	277
2600	9' long, 1,680 watt		3.60	2.222		85.50	142		227.50	310
2800	10' long, 1,875 watt		3.30	2.424		97.50	154		251.50	345
2950	Wall heaters with fan, 120 to 277 volt									
3170	1,000 watt	1 Elec	6	1.333	Ea.	335	85		420	495
3180	1,250 watt		5	1.600		113	102		215	282
3190	1,500 watt		4	2		360	127		487	590
3230	2,500 watt		3.50	2.286		365	146		511	625
3250	4,000 watt		2.70	2.963		330	189		519	650
3600	Thermostats, integral		16	.500		29.50	32		61.50	81
3800	Line voltage, 1 pole		8	1		16.40	63.50		79.90	116
3810	2 pole		8	1		23	63.50		86.50	124
9990	Minimum labor/equipment charge		4	2	Job		127		127	196

23 84 Humidity Control Equipment

23 84 13 – Humidifiers

23 84 13.10 Humidifier Units

		Crew	Daily Output	Labor-Hours	Unit	Material	2021 Bare Costs Labor	Equipment	Total	Total Incl O&P
0010	**HUMIDIFIER UNITS**									
0520	Steam, room or duct, filter, regulators, auto. controls, 220 V									
0560	22 lb./hr.	Q-5	5	3.200	Ea.	3,075	197		3,272	3,700
0580	33 lb./hr.		4	4		3,175	246		3,421	3,875
0620	100 lb./hr.		3	5.333		5,225	330		5,555	6,225
9000	Minimum labor/equipment charge		3.50	4.571	Job		281		281	435

For customer support on your Commercial Renovation Costs with RSMeans Data, call 800.448.8182.

485

23 91 10.10 Prefabricated Pads and Stands	Crew	Daily Output	Labor-Hours	Unit	Material	2021 Bare Costs Labor	Equipment	Total	Total Incl O&P
0010 **PREFABRICATED PADS AND STANDS**									
6000 Pad, fiberglass reinforced concrete with polystyrene foam core									
6090 24" x 36"	1 Shee	12	.667	Ea.	72	43.50		115.50	147
6280 36" x 36"	Q-9	8	2		104	118		222	298
6300 36" x 40"	↓	7	2.286	↓	118	135		253	340
6320 36" x 48"		7	2.286		138	135		273	360

486

For customer support on your Commercial Renovation Costs with RSMeans Data, call 800.448.8182.

Estimating Tips

- When estimating material costs for special systems, it is always prudent to obtain manufacturers' quotations for equipment prices and special installation requirements that may affect the total cost.

- For cost modifications for elevated tray installation, add the percentages to labor according to the height of the installation and only to the quantities exceeding the different height levels, not to the total tray quantities. Refer to subdivision 26 01 02.20 for labor adjustment factors.

- Do not overlook the costs for equipment used in the installation.

Reference Numbers

Reference numbers are shown at the beginning of some major classifications. These numbers refer to related items in the Reference Section. The reference information may be an estimating procedure, an alternate pricing method, or technical information.

Same Data. Simplified.

Enjoy the convenience and efficiency of accessing your costs anywhere:

- **Skip the multiplier** by setting your location
- **Quickly search,** edit, favorite and share costs
- **Stay on top of price changes** with automatic updates

Discover more at rsmeans.com/online

Note: Trade Service, in part, has been used as a reference source for some of the material prices used in Division 25.

25 05 Common Work Results for Integrated Automation

25 05 28 – Pathways for Integrated Automation

25 05 28.39 Surface Raceways for Integrated Automation	Crew	Daily Output	Labor-Hours	Unit	Material	2021 Bare Costs Labor	2021 Bare Costs Equipment	Total	Total Incl O&P
0010 **SURFACE RACEWAYS FOR INTEGRATED AUTOMATION**									
1000 Din rail, G type, 15 mm, steel, galvanized, 3', for termination blocks	1 Elec	48	.167	Ea.	12.55	10.60		23.15	30
1010 32 mm, steel, galvanized, 3', for termination blocks		48	.167		16.55	10.60		27.15	34.50
1020 35 mm, steel, galvanized, 3', for termination blocks		48	.167		17.75	10.60		28.35	36
1030 Din rail, rail mounted, 35 mm, aluminum, 3', for termination blocks		48	.167		26	10.60		36.60	45
1040 35 mm, aluminum, 6', for termination blocks		48	.167		34.50	10.60		45.10	54.50
2000 Terminal block, Din mounted, 35 mm, medium screw, connector end bracket		48	.167		2.92	10.60		13.52	19.50
2010 35 mm, large screw, connector end bracket		48	.167		4.94	10.60		15.54	22
2020 Large screw, connector end bracket		48	.167		5.10	10.60		15.70	22
3000 Terminal block, feed through, 35 mm, Din mounted, 600V, 20A		48	.167		1.01	10.60		11.61	17.40
3010 35 mm, Din mounted, 600V, 30A		48	.167		1.66	10.60		12.26	18.15
3020 Terminal block, fuse holder, LED indicator, Din mounted, 110V, 6A		48	.167		9.15	10.60		19.75	26.50
3030 Din mounted, 220V, 6A		48	.167		8	10.60		18.60	25
3040 Din mounted, 300V, 6A		48	.167		9.45	10.60		20.05	26.50

25 36 Integrated Automation Instrumentation & Terminal Devices

25 36 19 – Integrated Automation Current Sensors

25 36 19.10 Integrated Automation Current Sensors

	Crew	Daily Output	Labor-Hours	Unit	Material	Labor	Equipment	Total	Total Incl O&P
0010 **INTEGRATED AUTOMATION CURRENT SENSORS**									
1000 Current limiter inrush, 100 to 480 VAC, 10A, power supply	1 Elec	48	.167	Ea.	88.50	10.60		99.10	114

Estimating Tips

26 05 00 Common Work Results for Electrical

- Conduit should be taken off in three main categories—power distribution, branch power, and branch lighting—so the estimator can concentrate on systems and components, therefore making it easier to ensure all items have been accounted for.

- For cost modifications for elevated conduit installation, add the percentages to labor according to the height of installation and only to the quantities exceeding the different height levels, not to the total conduit quantities. Refer to subdivision 26 01 02.20 for labor adjustment factors.

- Remember that aluminum wiring of equal ampacity is larger in diameter than copper and may require larger conduit.

- If more than three wires at a time are being pulled, deduct percentages from the labor hours of that grouping of wires.

- When taking off grounding systems, identify separately the type and size of wire, and list each unique type of ground connection.

- The estimator should take the weights of materials into consideration when completing a takeoff. Topics to consider include: How will the materials be supported? What methods of support are available? How high will the support structure have to reach? Will the final support structure be able to withstand the total burden? Is the support material included or separate from the fixture, equipment, and material specified?

- Do not overlook the costs for equipment used in the installation. If scaffolding or highlifts are available in the field, contractors may use them in lieu of the proposed ladders and rolling staging.

26 20 00 Low-Voltage Electrical Transmission

- Supports and concrete pads may be shown on drawings for the larger equipment, or the support system may be only a piece of plywood for the back of a panelboard. In either case, they must be included in the costs.

26 40 00 Electrical and Cathodic Protection

- When taking off cathodic protection systems, identify the type and size of cable, and list each unique type of anode connection.

26 50 00 Lighting

- Fixtures should be taken off room by room using the fixture schedule, specifications, and the ceiling plan. For large concentrations of lighting fixtures in the same area, deduct the percentages from labor hours.

Reference Numbers

Reference numbers are shown at the beginning of some major classifications. These numbers refer to related items in the Reference Section. The reference information may be an estimating procedure, an alternate pricing method, or technical information.

Note: Not all subdivisions listed here necessarily appear. ■

Same Data. Simplified.

Enjoy the convenience and efficiency of accessing your costs anywhere:

- **Skip the multiplier** by setting your location
- **Quickly search,** edit, favorite and share costs
- **Stay on top of price changes** with automatic updates

Discover more at rsmeans.com/online

26 01 02.20 Labor Adjustment Factors	Crew	Daily Output	Labor-Hours	Unit	Material	2021 Bare Costs Labor	Equipment	Total	Total Incl O&P
0010 LABOR ADJUSTMENT FACTORS (For Div. 26, 27 and 28) R260519-90									
0100 Subtract from labor for Economy of Scale for Wire									
0110 4-5 wires						25%			
0120 6-10 wires						30%			
0130 11-15 wires						35%			
0140 over 15 wires						40%			
0150 Labor adjustment factors (For Div. 26, 27, 28 and 48)									
0200 Labor factors: The below are reasonable suggestions, but									
0210 each project must be evaluated for its own peculiarities, and									
0220 the adjustments be increased or decreased depending on the									
0230 severity of the special conditions.									
1000 Add to labor for elevated installation (above floor level)									
1010 10' to 14.5' high						10%			
1020 15' to 19.5' high						20%			
1030 20' to 24.5' high						25%			
1040 25' to 29.5' high						35%			
1050 30' to 34.5' high						40%			
1060 35' to 39.5' high						50%			
1070 40' and higher						55%			
2000 Add to labor for crawl space									
2010 3' high						40%			
2020 4' high						30%			
3000 Add to labor for multi-story building									
3100 For new construction (No elevator available)									
3110 Add for floors 3 thru 10						5%			
3120 Add for floors 11 thru 15						10%			
3130 Add for floors 16 thru 20						15%			
3140 Add for floors 21 thru 30						20%			
3150 Add for floors 31 and up						30%			
3200 For existing structure (Elevator available)									
3210 Add for work on floor 3 and above						2%			
4000 Add to labor for working in existing occupied buildings									
4010 Hospital						35%			
4020 Office building						25%			
4030 School						20%			
4040 Factory or warehouse						15%			
4050 Multi-dwelling						15%			
5000 Add to labor, miscellaneous									
5010 Cramped shaft						35%			
5020 Congested area						15%			
5030 Excessive heat or cold						30%			
6000 Labor factors: the above are reasonable suggestions, but									
6100 each project should be evaluated for its own peculiarities									
6200 Other factors to be considered are:									
6210 Movement of material and equipment through finished areas						10%			
6220 Equipment room min security direct access w/authorization						15%			
6230 Attic space						25%			
6240 No service road						25%			
6250 Poor unloading/storage area, no hydraulic lifts or jacks						20%			
6260 Congested site area/heavy traffic						20%			
7000 Correctional facilities (no compounding division 1 adjustment factors)									
7010 Minimum security w/facilities escort						30%			
7020 Medium security w/facilities and correctional officer escort						40%			

26 01 Operation and Maintenance of Electrical Systems

26 01 02 – Labor Adjustment

26 01 02.20 Labor Adjustment Factors	Crew	Daily Output	Labor-Hours	Unit	Material	2021 Bare Costs Labor	2021 Bare Costs Equipment	Total	Total Incl O&P
7030	Max security w/facilities & correctional officer escort (no inmate contact)						50%			

26 01 50 – Operation and Maintenance of Lighting

26 01 50.81 Luminaire Replacement

		Crew	Daily Output	Labor-Hours	Unit	Material	2021 Bare Costs Labor	2021 Bare Costs Equipment	Total	Total Incl O&P
0010	**LUMINAIRE REPLACEMENT**									
0962	Remove and install new replacement lens, 2' x 2' troffer	1 Elec	16	.500	Ea.	18.25	32		50.25	69
0964	2' x 4' troffer		16	.500		14.95	32		46.95	65.50
3200	Remove and replace (reinstall), lighting fixture		4	2			127		127	196

26 05 Common Work Results for Electrical

26 05 05 – Selective Demolition for Electrical

26 05 05.10 Electrical Demolition

			Crew	Daily Output	Labor-Hours	Unit	Material	2021 Bare Costs Labor	2021 Bare Costs Equipment	Total	Total Incl O&P
0010	**ELECTRICAL DEMOLITION**	R260105-30									
0020	Electrical demolition, conduit to 10' high, incl. fittings & hangers										
0100	Rigid galvanized steel, 1/2" to 1" diameter	R024119-10	1 Elec	242	.033	L.F.		2.11		2.11	3.23
0120	1-1/4" to 2"		"	200	.040			2.55		2.55	3.91
0140	2-1/2" to 3-1/2"		2 Elec	302	.053			3.37		3.37	5.20
0200	Electric metallic tubing (EMT), 1/2" to 1"		1 Elec	394	.020			1.29		1.29	1.99
0220	1-1/4" to 1-1/2"			326	.025			1.56		1.56	2.40
0240	2" to 3"			236	.034			2.16		2.16	3.32
0400	Wiremold raceway, including fittings & hangers										
0420	No. 3000		1 Elec	250	.032	L.F.		2.04		2.04	3.13
0440	No. 4000		"	217	.037	"		2.35		2.35	3.61
0500	Channels, steel, including fittings & hangers										
0520	3/4" x 1-1/2"		1 Elec	308	.026	L.F.		1.65		1.65	2.54
0540	1-1/2" x 1-1/2"			269	.030			1.89		1.89	2.91
0560	1-1/2" x 1-7/8"			229	.035			2.23		2.23	3.42
0600	Copper bus duct, indoor, 3 phase										
0610	Including hangers & supports										
0620	225 amp		2 Elec	135	.119	L.F.		7.55		7.55	11.60
0640	400 amp			106	.151			9.60		9.60	14.75
0660	600 amp			86	.186			11.85		11.85	18.20
0680	1,000 amp			60	.267			17		17	26
0700	1,600 amp			40	.400			25.50		25.50	39
0720	3,000 amp			10	1.600			102		102	157
0800	Plug-in switches, 600 V 3 ph., incl. disconnecting										
0820	wire, conduit terminations, 30 amp		1 Elec	15.50	.516	Ea.		33		33	50.50
0840	60 amp			13.90	.576			36.50		36.50	56.50
0850	100 amp			10.40	.769			49		49	75.50
0860	200 amp			6.20	1.290			82		82	126
0880	400 amp		2 Elec	5.40	2.963			189		189	290
0900	600 amp			3.40	4.706			300		300	460
0920	800 amp			2.60	6.154			390		390	600
0940	1,200 amp			2	8			510		510	785
0960	1,600 amp			1.70	9.412			600		600	920
1010	Safety switches, 250 or 600 V, incl. disconnection										
1050	of wire & conduit terminations										
1100	30 amp		1 Elec	12.30	.650	Ea.		41.50		41.50	63.50
1120	60 amp			8.80	.909			58		58	89
1140	100 amp			7.30	1.096			70		70	107
1160	200 amp			5	1.600			102		102	157

26 05 05.10 Electrical Demolition	Crew	Daily Output	Labor-Hours	Unit	Material	2021 Bare Costs Labor	Equipment	Total	Total Incl O&P
1210	Panel boards, incl. removal of all breakers,									
1220	conduit terminations & wire connections									
1230	3 wire, 120/240 V, 100A, to 20 circuits	1 Elec	2.60	3.077	Ea.		196		196	300
1240	200 amps, to 42 circuits	2 Elec	2.60	6.154			390		390	600
1241	225 amps, to 42 circuits		2.40	6.667			425		425	650
1250	400 amps, to 42 circuits	↓	2.20	7.273			465		465	710
1260	4 wire, 120/208 V, 125A, to 20 circuits	1 Elec	2.40	3.333			212		212	325
1270	200 amps, to 42 circuits	2 Elec	2.40	6.667	↓		425		425	650
1300	Transformer, dry type, 1 phase, incl. removal of									
1320	supports, wire & conduit terminations									
1340	1 kVA	1 Elec	7.70	1.039	Ea.		66		66	102
1360	5 kVA		4.70	1.702			108		108	167
1380	10 kVA	↓	3.60	2.222			142		142	217
1400	37.5 kVA	2 Elec	3	5.333			340		340	520
1420	75 kVA	"	2.50	6.400	↓		410		410	625
1440	3 phase to 600 V, primary									
1460	3 kVA	1 Elec	3.87	2.067	Ea.		132		132	202
1480	15 kVA	2 Elec	3.67	4.360			278		278	425
1490	25 kVA		3.51	4.558			290		290	445
1500	30 kVA	↓	3.42	4.678			298		298	460
1530	112.5 kVA	R-3	2.90	6.897			440	66	506	750
1560	500 kVA		1.40	14.286			905	136	1,041	1,550
1570	750 kVA	↓	1.10	18.182	↓		1,150	174	1,324	1,975
1600	Pull boxes & cabinets, sheet metal, incl. removal									
1620	of supports and conduit terminations									
1640	6" x 6" x 4"	1 Elec	31.10	.257	Ea.		16.40		16.40	25
1660	12" x 12" x 4"		23.30	.343			22		22	33.50
1720	Junction boxes, 4" sq. & oct.		80	.100			6.35		6.35	9.80
1740	Handy box		107	.075			4.76		4.76	7.30
1760	Switch box		107	.075			4.76		4.76	7.30
1780	Receptacle & switch plates	↓	257	.031	↓		1.98		1.98	3.05
1800	Wire, THW-THWN-THHN, removed from									
1810	in place conduit, to 10' high									
1830	#14	1 Elec	65	.123	C.L.F.		7.85		7.85	12.05
1840	#12		55	.145			9.25		9.25	14.25
1850	#10	↓	45.50	.176			11.20		11.20	17.20
1880	#4	2 Elec	53	.302			19.25		19.25	29.50
1890	#3		50	.320			20.50		20.50	31.50
1910	1/0		33.20	.482			30.50		30.50	47
1920	2/0		29.20	.548			35		35	53.50
1930	3/0		25	.640			41		41	62.50
1980	400 kcmil		17	.941			60		60	92
1990	500 kcmil	↓	16.20	.988	↓		63		63	96.50
2000	Interior fluorescent fixtures, incl. supports									
2010	& whips, to 10' high									
2100	Recessed drop-in 2' x 2', 2 lamp	2 Elec	35	.457	Ea.		29		29	44.50
2110	2' x 2', 4 lamp		30	.533			34		34	52
2120	2' x 4', 2 lamp		33	.485			31		31	47.50
2140	2' x 4', 4 lamp		30	.533			34		34	52
2160	4' x 4', 4 lamp		20	.800			51		51	78.50
2180	Surface mount, acrylic lens & hinged frame									
2200	1' x 4', 2 lamp	2 Elec	44	.364	Ea.		23		23	35.50
2210	6" x 4', 2 lamp	↓	44	.364			23		23	35.50

26 05 05 – Selective Demolition for Electrical

26 05 05.10 Electrical Demolition	Crew	Daily Output	Labor-Hours	Unit	Material	2021 Bare Costs Labor	Equipment	Total	Total Incl O&P
2220	2' x 2', 2 lamp	2 Elec	44	.364	Ea.		23		23	35.50
2260	2' x 4', 4 lamp		33	.485			31		31	47.50
2280	4' x 4', 4 lamp		23	.696			44.50		44.50	68
2281	4' x 4', 6 lamp		23	.696			44.50		44.50	68
2300	Strip fixtures, surface mount									
2320	4' long, 1 lamp	2 Elec	53	.302	Ea.		19.25		19.25	29.50
2340	4' long, 2 lamp		50	.320			20.50		20.50	31.50
2360	8' long, 1 lamp		42	.381			24.50		24.50	37.50
2380	8' long, 2 lamp		40	.400			25.50		25.50	39
2400	Pendant mount, industrial, incl. removal									
2410	of chain or rod hangers, to 10' high									
2420	4' long, 2 lamp	2 Elec	35	.457	Ea.		29		29	44.50
2421	4' long, 4 lamp		35	.457			29		29	44.50
2440	8' long, 2 lamp		27	.593			38		38	58
2460	Interior incandescent, surface, ceiling									
2470	or wall mount, to 10' high									
2480	Metal cylinder type, 75 Watt	2 Elec	62	.258	Ea.		16.45		16.45	25.50
2500	150 Watt	"	62	.258	"		16.45		16.45	25.50
2520	Metal halide, high bay									
2540	400 Watt	2 Elec	15	1.067	Ea.		68		68	104
2560	1,000 Watt		12	1.333			85		85	130
2580	150 Watt, low bay		20	.800			51		51	78.50
2600	Exterior fixtures, incandescent, wall mount									
2620	100 Watt	2 Elec	50	.320	Ea.		20.50		20.50	31.50
2640	Quartz, 500 Watt		33	.485			31		31	47.50
2660	1,500 Watt		27	.593			38		38	58
2680	Wall pack, mercury vapor									
2700	175 Watt	2 Elec	25	.640	Ea.		41		41	62.50
2720	250 Watt	"	25	.640	"		41		41	62.50
9000	Minimum labor/equipment charge	1 Elec	4	2	Job		127		127	196
9900	Add to labor for higher elevated installation									
9905	10' to 14.5' high, add						10%			
9910	15' to 20' high, add						20%			
9920	20' to 25' high, add						25%			
9930	25' to 30' high, add						35%			
9940	30' to 35' high, add						40%			
9950	35' to 40' high, add						50%			
9960	Over 40' high, add						55%			

26 05 19 – Low-Voltage Electrical Power Conductors and Cables

26 05 19.20 Armored Cable

		Crew	Daily Output	Labor-Hours	Unit	Material	2021 Bare Costs Labor	Equipment	Total	Total Incl O&P
0010	**ARMORED CABLE**									
0050	600 volt, copper (BX), #14, 2 conductor, solid	1 Elec	2.40	3.333	C.L.F.	38.50	212		250.50	370
0100	3 conductor, solid		2.20	3.636		63.50	232		295.50	425
0150	#12, 2 conductor, solid		2.30	3.478		40.50	222		262.50	385
0200	3 conductor, solid		2	4		65.50	255		320.50	460
0250	#10, 2 conductor, solid		2	4		79.50	255		334.50	480
0300	3 conductor, solid		1.60	5		107	320		427	610
0340	#8, 2 conductor, stranded		1.50	5.333		156	340		496	690
0350	3 conductor, stranded		1.30	6.154		156	390		546	770
9010	600 volt, copper (MC) steel clad, #14, 2 wire	R-1A	8.16	1.961		64.50	112		176.50	244
9020	3 wire		7.21	2.219		71.50	127		198.50	274
9030	4 wire		6.69	2.392		185	137		322	415

For customer support on your Commercial Renovation Costs with RSMeans Data, call 800.448.8182.

493

26 05 19 – Low-Voltage Electrical Power Conductors and Cables

26 05 19.20 Armored Cable		Crew	Daily Output	Labor-Hours	Unit	Material	2021 Bare Costs Labor	Equipment	Total	Total Incl O&P
9040	#12, 2 wire	R-1A	7.21	2.219	C.L.F.	45.50	127		172.50	246
9050	3 wire		6.69	2.392		110	137		247	330
9070	#10, 2 wire		6.04	2.649		155	152		307	405
9080	3 wire		5.65	2.832		190	162		352	460
9100	#8, 2 wire, stranded		3.94	4.061		146	233		379	520
9110	3 wire, stranded	▼	3.40	4.706	▼	194	270		464	630
9900	Minimum labor/equipment charge	R-1C	4	8	Job		460	25.50	485.50	735

26 05 19.50 Mineral Insulated Cable

		Crew	Daily Output	Labor-Hours	Unit	Material	2021 Bare Costs Labor	Equipment	Total	Total Incl O&P
0010	**MINERAL INSULATED CABLE** 600 volt									
0100	1 conductor, #12	1 Elec	1.60	5	C.L.F.	410	320		730	940
0200	#10		1.60	5		615	320		935	1,175
0400	#8		1.50	5.333		690	340		1,030	1,275
0500	#6	▼	1.40	5.714		825	365		1,190	1,475
0600	#4	2 Elec	2.40	6.667		1,100	425		1,525	1,875
0800	#2		2.20	7.273		1,575	465		2,040	2,425
0900	#1		2.10	7.619		1,825	485		2,310	2,750
1000	1/0		2	8		2,125	510		2,635	3,100
1100	2/0		1.90	8.421		2,525	535		3,060	3,625
1200	3/0		1.80	8.889		3,050	565		3,615	4,225
1400	4/0	▼	1.60	10		3,500	635		4,135	4,825
1410	250 kcmil	3 Elec	2.40	10		3,950	635		4,585	5,325
1420	350 kcmil		1.95	12.308		4,650	785		5,435	6,300
1430	500 kcmil	▼	1.95	12.308		5,800	785		6,585	7,600
1500	2 conductor, #12	1 Elec	1.40	5.714		1,000	365		1,365	1,650
1600	#10		1.20	6.667		1,250	425		1,675	2,025
1800	#8		1.10	7.273		1,325	465		1,790	2,175
2000	#6	▼	1.05	7.619		1,975	485		2,460	2,900
2100	#4	2 Elec	2	8		2,650	510		3,160	3,700
2200	3 conductor, #12	1 Elec	1.20	6.667		1,250	425		1,675	2,025
2400	#10		1.10	7.273		1,425	465		1,890	2,275
2600	#8		1.05	7.619		1,800	485		2,285	2,725
2800	#6	▼	1	8		2,375	510		2,885	3,375
3000	#4	2 Elec	1.80	8.889		2,950	565		3,515	4,125
3100	4 conductor, #12	1 Elec	1.20	6.667		1,325	425		1,750	2,100
3200	#10		1.10	7.273		1,600	465		2,065	2,450
3400	#8		1	8		2,100	510		2,610	3,075
3600	#6		.90	8.889		2,725	565		3,290	3,850
3620	7 conductor, #12		1.10	7.273		1,700	465		2,165	2,575
3640	#10		1	8	▼	2,250	510		2,760	3,250
3800	Terminations, 600 volt, 1 conductor, #12		8	1	Ea.	21.50	63.50		85	122
5500	2 conductor, #12		6.70	1.194		21.50	76		97.50	141
5600	#10		6.40	1.250		32	79.50		111.50	157
5800	#8		6.20	1.290		32	82		114	161
6000	#6		5.70	1.404		32	89.50		121.50	172
6200	#4		5.30	1.509		71	96		167	226
6400	3 conductor, #12		5.70	1.404		32	89.50		121.50	172
6500	#10		5.50	1.455		32	92.50		124.50	177
6600	#8		5.20	1.538		32	98		130	186
6800	#6		4.80	1.667		32	106		138	198
7200	#4		4.60	1.739		71	111		182	248
7400	4 conductor, #12		4.60	1.739		35.50	111		146.50	209
7500	#10		4.40	1.818		35.50	116		151.50	217

26 05 Common Work Results for Electrical

26 05 19 – Low-Voltage Electrical Power Conductors and Cables

26 05 19.50 Mineral Insulated Cable

		Crew	Daily Output	Labor-Hours	Unit	Material	2021 Bare Costs Labor	Equipment	Total	Total Incl O&P
7600	#8	1 Elec	4.20	1.905	Ea.	35.50	121		156.50	225
8400	#6		4	2		74.50	127		201.50	278
8500	7 conductor, #12		3.50	2.286		35.50	146		181.50	263
8600	#10		3	2.667		77	170		247	345
8800	Crimping tool, plier type					78.50			78.50	86
9000	Stripping tool					296			296	325
9200	Hand vise					88			88	97
9500	Minimum labor/equipment charge	1 Elec	4	2	Job		127		127	196

26 05 19.55 Non-Metallic Sheathed Cable

		Crew	Daily Output	Labor-Hours	Unit	Material	2021 Bare Costs Labor	Equipment	Total	Total Incl O&P
0010	**NON-METALLIC SHEATHED CABLE** 600 volt									
0100	Copper with ground wire (Romex)									
0150	#14, 2 conductor	1 Elec	2.70	2.963	C.L.F.	17.40	189		206.40	310
0200	3 conductor		2.40	3.333		24.50	212		236.50	350
0220	4 conductor		2.20	3.636		41	232		273	400
0250	#12, 2 conductor		2.50	3.200		23.50	204		227.50	340
0300	3 conductor		2.20	3.636		37	232		269	395
0320	4 conductor		2	4		60.50	255		315.50	455
0350	#10, 2 conductor		2.20	3.636		45.50	232		277.50	405
0400	3 conductor		1.80	4.444		57.50	283		340.50	500
0430	#8, 2 conductor		1.60	5		77.50	320		397.50	575
0450	3 conductor		1.50	5.333		110	340		450	640
0500	#6, 3 conductor		1.40	5.714		385	365		750	980
0550	SE type SER aluminum cable, 3 RHW and									
0600	1 bare neutral, 3 #8 & 1 #8	1 Elec	1.60	5	C.L.F.	71	320		391	570
0650	3 #6 & 1 #6	"	1.40	5.714		74.50	365		439.50	640
0700	3 #4 & 1 #6	2 Elec	2.40	6.667		82	425		507	740
0750	3 #2 & 1 #4		2.20	7.273		128	465		593	850
0800	3 #1/0 & 1 #2		2	8		173	510		683	975
0850	3 #2/0 & 1 #1		1.80	8.889		202	565		767	1,100
0900	3 #4/0 & 1 #2/0		1.60	10		325	635		960	1,325
6500	Service entrance cap for copper SEU									
6600	100 amp	1 Elec	12	.667	Ea.	7	42.50		49.50	72.50
6700	150 amp		10	.800		12.95	51		63.95	93
6800	200 amp		8	1		19.65	63.50		83.15	120
9000	Minimum labor/equipment charge		4	2	Job		127		127	196

26 05 19.90 Wire

			Crew	Daily Output	Labor-Hours	Unit	Material	2021 Bare Costs Labor	Equipment	Total	Total Incl O&P
0010	**WIRE**, normal installation conditions in wireway, conduit, cable tray										
0020	600 volt, copper type THW, solid, #14	R260533-22	1 Elec	13	.615	C.L.F.	8.30	39		47.30	69
0030	#12			11	.727		10.25	46.50		56.75	82.50
0040	#10			10	.800		16.55	51		67.55	96.50
0050	Stranded, #14			13	.615		7.55	39		46.55	68.50
0100	#12			11	.727		17.45	46.50		63.95	90
0120	#10			10	.800		28	51		79	110
0140	#8			8	1		30.50	63.50		94	132
0160	#6			6.50	1.231		73	78.50		151.50	201
0180	#4		2 Elec	10.60	1.509		112	96		208	271
0200	#3			10	1.600		152	102		254	325
0220	#2			9	1.778		176	113		289	365
0240	#1			8	2		219	127		346	435
0260	1/0			6.60	2.424		285	154		439	550
0280	2/0			5.80	2.759		495	176		671	815
0300	3/0			5	3.200		350	204		554	700

For customer support on your Commercial Renovation Costs with RSMeans Data, call 800.448.8182.

495

26 05 19 – Low-Voltage Electrical Power Conductors and Cables

26 05 19.90 Wire		Crew	Daily Output	Labor-Hours	Unit	Material	2021 Bare Costs Labor	Equipment	Total	Total Incl O&P
0350	4/0	2 Elec	4.40	3.636	C.L.F.	570	232		802	985
0400	250 kcmil	3 Elec	6	4		620	255		875	1,075
0420	300 kcmil		5.70	4.211		890	268		1,158	1,400
0450	350 kcmil		5.40	4.444		585	283		868	1,075
0480	400 kcmil		5.10	4.706		1,525	300		1,825	2,125
0490	500 kcmil		4.80	5		1,250	320		1,570	1,875
0510	750 kcmil		3.30	7.273		1,050	465		1,515	1,875
0540	600 volt, aluminum type THHN, stranded, #6	1 Elec	8	1		16.25	63.50		79.75	116
0560	#4	2 Elec	13	1.231		20.50	78.50		99	143
0580	#2		10.60	1.509		28	96		124	179
0600	#1		9	1.778		39	113		152	217
0620	1/0		8	2		43.50	127		170.50	244
0640	2/0		7.20	2.222		53	142		195	276
0680	3/0		6.60	2.424		65.50	154		219.50	310
0700	4/0		6.20	2.581		70	164		234	330
0720	250 kcmil	3 Elec	8.70	2.759		84	176		260	365
0740	300 kcmil		8.10	2.963		122	189		311	425
0760	350 kcmil		7.50	3.200		121	204		325	450
0780	400 kcmil		6.90	3.478		146	222		368	500
0800	500 kcmil		6	4		157	255		412	565
0850	600 kcmil		5.70	4.211		209	268		477	640
0880	700 kcmil		5.10	4.706		238	300		538	720
9000	Minimum labor/equipment charge	1 Elec	4	2	Job		127		127	196

26 05 26 – Grounding and Bonding for Electrical Systems

26 05 26.80 Grounding

		Crew	Daily Output	Labor-Hours	Unit	Material	2021 Bare Costs Labor	Equipment	Total	Total Incl O&P
0010	**GROUNDING**									
0030	Rod, copper clad, 8' long, 1/2" diameter	1 Elec	5.50	1.455	Ea.	17	92.50		109.50	161
0040	5/8" diameter		5.50	1.455		15	92.50		107.50	159
0050	3/4" diameter		5.30	1.509		28.50	96		124.50	180
0080	10' long, 1/2" diameter		4.80	1.667		21	106		127	186
0090	5/8" diameter		4.60	1.739		20	111		131	192
0100	3/4" diameter		4.40	1.818		52	116		168	235
0130	15' long, 3/4" diameter		4	2		63	127		190	266
0260	Wire ground bare armored, #8-1 conductor		2	4	C.L.F.	77.50	255		332.50	475
0280	#4-1 conductor		1.60	5		148	320		468	655
0390	Bare copper wire, stranded, #8		11	.727		27.50	46.50		74	102
0400	#6		10	.800		43.50	51		94.50	127
0600	#2	2 Elec	10	1.600		112	102		214	280
0800	3/0		6.60	2.424		275	154		429	540
1000	4/0		5.70	2.807		305	179		484	610
1200	250 kcmil	3 Elec	7.20	3.333		375	212		587	740
1800	Water pipe ground clamps, heavy duty									
2000	Bronze, 1/2" to 1" diameter	1 Elec	8	1	Ea.	24	63.50		87.50	124
2100	1-1/4" to 2" diameter		8	1		44.50	63.50		108	147
2200	2-1/2" to 3" diameter		6	1.333		66	85		151	203
2400	Grounding, exothermic welding									
2500	Exothermic welding reusable mold, cable to cable, parallel, vertical	1 Elec	8	1	Ea.	425	63.50		488.50	570
2505	Splice single		8	1		360	63.50		423.50	500
2510	Splice single to double		8	1		425	63.50		488.50	570
2520	Cable to cable, termination, Tee		8	1		420	63.50		483.50	560
2530	Cable to rod, termination, Tee		8	1		330	63.50		393.50	460
2535	Cable to rebar over, termination, Tee		8	1		97.50	63.50		161	205

26 05 Common Work Results for Electrical

26 05 26 – Grounding and Bonding for Electrical Systems

26 05 26.80 Grounding

		Crew	Daily Output	Labor-Hours	Unit	Material	2021 Bare Costs Labor	Equipment	Total	Total Incl O&P
2540	Cable to rod, termination, 90 Deg	1 Elec	8	1	Ea.	355	63.50		418.50	490
2550	Cable to vertical flat steel		8	1		420	63.50		483.50	565
2555	Cable to 45 Deg		8	1		430	63.50		493.50	570
2730	Exothermic weld, 4/0 wire to 1" ground rod		7	1.143		2.74	73		75.74	115
2800	Brazed connections, #6 wire		12	.667		16	42.50		58.50	82.50
3000	#2 wire		10	.800		21.50	51		72.50	102
3100	3/0 wire		8	1		32	63.50		95.50	133
3200	4/0 wire		7	1.143		36.50	73		109.50	153
3400	250 kcmil wire		5	1.600		42.50	102		144.50	204
3600	500 kcmil wire		4	2		52.50	127		179.50	254
7000	Exothermic welding kit, multi vertical		8	1		715	63.50		778.50	885
9000	Minimum labor/equipment charge		4	2	Job		127		127	196

26 05 29 – Hangers and Supports for Electrical Systems

26 05 29.30 Fittings and Channel Support

		Crew	Daily Output	Labor-Hours	Unit	Material	2021 Bare Costs Labor	Equipment	Total	Total Incl O&P
0010	**FITTINGS & CHANNEL SUPPORT**									
0020	Rooftop channel support									
0200	2-7/8" L x 1-5/8" W, 12 ga. pre galv., dbl. base	1 Elec	46	.174	Ea.	3.76	11.10		14.86	21
0210	2-7/8" L x 1-5/8" W, 12 ga. hot dip galv., dbl. base		46	.174		2.19	11.10		13.29	19.40
0220	2-7/16" L x 1-5/8" W, 12 ga. pre galv., sngl base		48	.167		1.80	10.60		12.40	18.30
0230	2-7/16" L x 1-5/8" W, 12 ga. hot dip galv., sngl base		48	.167		6.40	10.60		17	23.50
0240	13/16" L x 1-5/8" W, 14 ga. pre galv., sngl base		56	.143		1.53	9.10		10.63	15.70
0250	13/16" L x 1-5/8" W, 14 ga. hot dip galv., sngl base		56	.143		6.30	9.10		15.40	21
0260	1-5/8" L x 1-5/8" W, 12 ga. pre galv., sngl base		54	.148		1.66	9.45		11.11	16.35
0270	1-5/8" L x 1-5/8" W, 12 ga. hot dip galv., sngl base		54	.148		1.76	9.45		11.21	16.45
0280	1-5/8" x 28", 12 ga. pre galv., dbl. H block base		42	.190		3.81	12.15		15.96	23
0290	1-5/8" x 36", 12 ga. pre galv., dbl. H block base		40	.200		4.03	12.75		16.78	24
0300	1-5/8" x 42", 12 ga. pre galv., dbl. H block base		34	.235		4.53	15		19.53	28
0310	1-5/8" x 50", 12 ga. pre galv., dbl. H block base		38	.211		4.83	13.40		18.23	26
0320	1-5/8" x 60", 12 ga. pre galv., dbl. H block base		36	.222		5.30	14.15		19.45	27.50
0330	13/16" x 8" H, threaded rod pre galv., H block base		35	.229		2.56	14.55		17.11	25.50
0340	13/16" x 12" H, threaded rod pre galv., H block base		34.60	.231		3.20	14.75		17.95	26
0350	1-5/8" x 16" H, threaded rod dbl. H block base		34	.235		4.77	15		19.77	28.50
0360	1-5/8" x 12" H, pre galv., dbl. H block base		34.60	.231		9.15	14.75		23.90	32.50
0370	1-5/8" x 24"H, pre galv., dbl. H block base		34	.235		9.65	15		24.65	33.50
0380	3-1/4" x 24" H, pre galv., dbl. H block base		32	.250		22	15.95		37.95	48.50
0390	3-1/4" x 36" H, pre galv., dbl. H block base		30	.267		24	17		41	52.50

26 05 33 – Raceway and Boxes for Electrical Systems

26 05 33.13 Conduit

		Crew	Daily Output	Labor-Hours	Unit	Material	2021 Bare Costs Labor	Equipment	Total	Total Incl O&P
0010	**CONDUIT** To 10' high, includes 2 terminations, 2 elbows, R260533-22									
0020	11 beam clamps, and 11 couplings per 100 L.F.									
1161	Field bends, 45° to 90°, 1/2" diameter	1 Elec	53	.151	Ea.		9.60		9.60	14.75
1162	3/4" diameter		47	.170			10.85		10.85	16.65
1163	1" diameter		44	.182			11.60		11.60	17.80
1164	1-1/4" diameter		23	.348			22		22	34
1165	1-1/2" diameter		21	.381			24.50		24.50	37.50
1166	2" diameter		16	.500			32		32	49
1691	See note on line 26 05 33.13 9995									
1750	Rigid galvanized steel, 1/2" diameter	1 Elec	90	.089	L.F.	2.62	5.65		8.27	11.60
1770	3/4" diameter		80	.100		2.92	6.35		9.27	13
1800	1" diameter		65	.123		4.18	7.85		12.03	16.65
1830	1-1/4" diameter		60	.133		5.95	8.50		14.45	19.55
1850	1-1/2" diameter		55	.145		6.80	9.25		16.05	22

497

26 05 33.13 Conduit

		Crew	Daily Output	Labor-Hours	Unit	Material	2021 Bare Costs Labor	2021 Bare Costs Equipment	Total	Total Incl O&P
1870	2" diameter	1 Elec	45	.178	L.F.	8.45	11.30		19.75	26.50
1991	Field bends, 45° to 90°, 1/2" diameter		44	.182	Ea.		11.60		11.60	17.80
1992	3/4" diameter		40	.200			12.75		12.75	19.55
1993	1" diameter		36	.222			14.15		14.15	21.50
1994	1-1/4" diameter		19	.421			27		27	41
1995	1-1/2" diameter		18	.444			28.50		28.50	43.50
1996	2" diameter		13	.615	↓		39		39	60
2500	Steel, intermediate conduit (IMC), 1/2" diameter		100	.080	L.F.	1.97	5.10		7.07	10
2530	3/4" diameter		90	.089		2.92	5.65		8.57	11.90
2550	1" diameter		70	.114		3.36	7.30		10.66	14.90
2570	1-1/4" diameter		65	.123		4.49	7.85		12.34	17
2600	1-1/2" diameter		60	.133		5.60	8.50		14.10	19.20
2630	2" diameter		50	.160		6.75	10.20		16.95	23
2650	2-1/2" diameter	↓	40	.200		12.10	12.75		24.85	33
2670	3" diameter	2 Elec	60	.267		16.10	17		33.10	43.50
2700	3-1/2" diameter		54	.296		20	18.85		38.85	51
2730	4" diameter	↓	50	.320	↓	23.50	20.50		44	57.50
2731	Field bends, 45° to 90°, 1/2" diameter	1 Elec	44	.182	Ea.		11.60		11.60	17.80
2732	3/4" diameter		40	.200			12.75		12.75	19.55
2733	1" diameter		36	.222			14.15		14.15	21.50
2734	1-1/4" diameter		19	.421			27		27	41
2735	1-1/2" diameter		18	.444			28.50		28.50	43.50
2736	2" diameter		13	.615	↓		39		39	60
5000	Electric metallic tubing (EMT), 1/2" diameter		170	.047	L.F.	.72	3		3.72	5.40
5020	3/4" diameter		130	.062		1.01	3.92		4.93	7.10
5040	1" diameter		115	.070		1.68	4.43		6.11	8.65
5060	1-1/4" diameter		100	.080		2.69	5.10		7.79	10.80
5080	1-1/2" diameter		90	.089		3.55	5.65		9.20	12.60
5100	2" diameter		80	.100		3.82	6.35		10.17	14
5120	2-1/2" diameter	↓	60	.133		5.50	8.50		14	19.10
5140	3" diameter	2 Elec	100	.160		8.85	10.20		19.05	25.50
5160	3-1/2" diameter		90	.178		10.25	11.30		21.55	28.50
5180	4" diameter	↓	80	.200	↓	11.65	12.75		24.40	32.50
5200	Field bends, 45° to 90°, 1/2" diameter	1 Elec	89	.090	Ea.		5.75		5.75	8.80
5220	3/4" diameter		80	.100			6.35		6.35	9.80
5240	1" diameter		73	.110			7		7	10.70
5260	1-1/4" diameter		38	.211			13.40		13.40	20.50
5280	1-1/2" diameter		36	.222			14.15		14.15	21.50
5300	2" diameter		26	.308			19.60		19.60	30
5320	Offsets, 1/2" diameter		65	.123			7.85		7.85	12.05
5340	3/4" diameter		62	.129			8.20		8.20	12.65
5360	1" diameter		53	.151			9.60		9.60	14.75
5380	1-1/4" diameter		30	.267			17		17	26
5400	1-1/2" diameter		28	.286	↓		18.20		18.20	28
9990	Minimum labor/equipment charge	↓	4	2	Job		127		127	196

26 05 33.16 Boxes for Electrical Systems

		Crew	Daily Output	Labor-Hours	Unit	Material	2021 Bare Costs Labor	2021 Bare Costs Equipment	Total	Total Incl O&P
0010	**BOXES FOR ELECTRICAL SYSTEMS**									
0020	Pressed steel, octagon, 4"	1 Elec	20	.400	Ea.	1.31	25.50		26.81	40.50
0060	Covers, blank		64	.125		.67	7.95		8.62	13
0100	Extension rings		40	.200		2.11	12.75		14.86	22
0150	Square, 4"		20	.400		2.20	25.50		27.70	41.50
0200	Extension rings	↓	40	.200	↓	2.68	12.75		15.43	22.50

26 05 33 – Raceway and Boxes for Electrical Systems

26 05 33.16 Boxes for Electrical Systems	Crew	Daily Output	Labor-Hours	Unit	Material	2021 Bare Costs Labor	Equipment	Total	Total Incl O&P
0250 Covers, blank	1 Elec	64	.125	Ea.	.98	7.95		8.93	13.35
0300 Plaster rings		64	.125		1.66	7.95		9.61	14.10
0652 Switchbox		24	.333		4.70	21		25.70	37.50
1100 Concrete, floor, 1 gang		5.30	1.509		118	96		214	278
9000 Minimum labor/equipment charge		4	2	Job		127		127	196

26 05 33.17 Outlet Boxes, Plastic

	Crew	Daily Output	Labor-Hours	Unit	Material	2021 Bare Costs Labor	Equipment	Total	Total Incl O&P
0010 **OUTLET BOXES, PLASTIC**									
0050 4" diameter, round with 2 mounting nails	1 Elec	25	.320	Ea.	2.49	20.50		22.99	34
0100 Bar hanger mounted		25	.320		4.83	20.50		25.33	37
0200 4", square with 2 mounting nails		25	.320		5.35	20.50		25.85	37.50
0300 Plaster ring		64	.125		2.38	7.95		10.33	14.85
0400 Switch box with 2 mounting nails, 1 gang		30	.267		2.04	17		19.04	28
0500 2 gang		25	.320		3.38	20.50		23.88	35
0600 3 gang		20	.400		5.70	25.50		31.20	45.50
0700 Old work box		30	.267		2.19	17		19.19	28.50
9000 Minimum labor/equipment charge		4	2	Job		127		127	196

26 05 33.18 Pull Boxes

	Crew	Daily Output	Labor-Hours	Unit	Material	2021 Bare Costs Labor	Equipment	Total	Total Incl O&P
0010 **PULL BOXES**									
0100 Steel, pull box, NEMA 1, type SC, 6" W x 6" H x 4" D	1 Elec	8	1	Ea.	9.50	63.50		73	108
0200 8" W x 8" H x 4" D		8	1		13.05	63.50		76.55	112
0300 10" W x 12" H x 6" D		5.30	1.509		32	96		128	184
0400 16" W x 20" H x 8" D		4	2		83	127		210	287
0500 20" W x 24" H x 8" D		3.20	2.500		97	159		256	350
0600 24" W x 36" H x 8" D		2.70	2.963		151	189		340	455
0650 Pull box, hinged, NEMA 1, 6" W x 6" H x 4" D		8	1		12.75	63.50		76.25	112
0800 12" W x 16" H x 6" D	1 Elec	4.70	1.702	Ea.	80	108		188	256
1000 20" W x 20" H x 6" D		3.60	2.222		216	142		358	455
1200 20" W x 20" H x 8" D		3.20	2.500		170	159		329	430
1400 24" W x 36" H x 8" D		2.70	2.963		250	189		439	565
1600 24" W x 42" H x 8" D		2	4		370	255		625	795

26 05 33.23 Wireway

	Crew	Daily Output	Labor-Hours	Unit	Material	2021 Bare Costs Labor	Equipment	Total	Total Incl O&P
0010 **WIREWAY** to 10' high									
0020 For higher elevations, see Section 26 05 36.40									
0100 NEMA 1, screw cover w/fittings and supports, 2-1/2" x 2-1/2"	1 Elec	45	.178	L.F.	15.85	11.30		27.15	35
0200 4" x 4"	"	40	.200		23	12.75		35.75	44.50
0400 6" x 6"	2 Elec	60	.267		31	17		48	60.50
0600 8" x 8"	"	40	.400		22	25.50		47.50	63
4475 NEMA 3R, screw cover w/fittings and supports, 4" x 4"	1 Elec	36	.222		17	14.15		31.15	40
4480 6" x 6"	2 Elec	55	.291		18.85	18.55		37.40	49.50
4485 8" x 8"		36	.444		33.50	28.50		62	80
4490 12" x 12"		18	.889		61	56.50		117.50	154

26 05 33.95 Cutting and Drilling

	Crew	Daily Output	Labor-Hours	Unit	Material	2021 Bare Costs Labor	Equipment	Total	Total Incl O&P
0010 **CUTTING AND DRILLING**									
0100 Hole drilling to 10' high, concrete wall									
0110 8" thick, 1/2" pipe size	R-31	12	.667	Ea.	.26	42.50	5.20	47.96	71
0120 3/4" pipe size		12	.667		.26	42.50	5.20	47.96	71
0130 1" pipe size		9.50	.842		.36	53.50	6.60	60.46	90
0140 1-1/4" pipe size		9.50	.842		.36	53.50	6.60	60.46	90
0150 1-1/2" pipe size		9.50	.842		.36	53.50	6.60	60.46	90
0160 2" pipe size		4.40	1.818		.55	116	14.25	130.80	194
0170 2-1/2" pipe size		4.40	1.818		.55	116	14.25	130.80	194
0180 3" pipe size		4.40	1.818		.55	116	14.25	130.80	194

26 05 33.95 Cutting and Drilling		Crew	Daily Output	Labor-Hours	Unit	Material	2021 Bare Costs Labor	Equipment	Total	Total Incl O&P
0190	3-1/2" pipe size	R-31	3.30	2.424	Ea.	.67	154	19	173.67	259
0200	4" pipe size		3.30	2.424		.67	154	19	173.67	259
0500	12" thick, 1/2" pipe size		9.40	.851		.39	54	6.65	61.04	91.50
0520	3/4" pipe size		9.40	.851		.39	54	6.65	61.04	91.50
0540	1" pipe size		7.30	1.096		.53	70	8.60	79.13	117
0560	1-1/4" pipe size		7.30	1.096		.53	70	8.60	79.13	117
0570	1-1/2" pipe size		7.30	1.096		.53	70	8.60	79.13	117
0580	2" pipe size		3.60	2.222		.83	142	17.40	160.23	237
0590	2-1/2" pipe size		3.60	2.222		.83	142	17.40	160.23	237
0600	3" pipe size		3.60	2.222		.83	142	17.40	160.23	237
0610	3-1/2" pipe size		2.80	2.857		1.01	182	22.50	205.51	305
0630	4" pipe size		2.50	3.200		1.01	204	25	230.01	345
0650	16" thick, 1/2" pipe size		7.60	1.053		.52	67	8.25	75.77	113
0670	3/4" pipe size		7	1.143		.52	73	8.95	82.47	122
0690	1" pipe size		6	1.333		.71	85	10.45	96.16	142
0710	1-1/4" pipe size		5.50	1.455		.71	92.50	11.40	104.61	155
0730	1-1/2" pipe size		5.50	1.455		.71	92.50	11.40	104.61	155
0750	2" pipe size		3	2.667		1.10	170	21	192.10	285
0770	2-1/2" pipe size		2.70	2.963		1.10	189	23	213.10	315
0790	3" pipe size		2.50	3.200		1.10	204	25	230.10	345
0810	3-1/2" pipe size		2.30	3.478		1.34	222	27	250.34	370
0830	4" pipe size		2	4		1.34	255	31.50	287.84	425
0850	20" thick, 1/2" pipe size		6.40	1.250		.65	79.50	9.80	89.95	133
0870	3/4" pipe size		6	1.333		.65	85	10.45	96.10	142
0890	1" pipe size		5	1.600		.89	102	12.55	115.44	172
0910	1-1/4" pipe size		4.80	1.667		.89	106	13.05	119.94	178
0930	1-1/2" pipe size		4.60	1.739		.89	111	13.60	125.49	186
0950	2" pipe size		2.70	2.963		1.38	189	23	213.38	315
0970	2-1/2" pipe size		2.40	3.333		1.38	212	26	239.38	355
0990	3" pipe size		2.20	3.636		1.38	232	28.50	261.88	390
1010	3-1/2" pipe size		2	4		1.68	255	31.50	288.18	425
1030	4" pipe size		1.70	4.706		1.68	300	37	338.68	500
1050	24" thick, 1/2" pipe size		5.50	1.455		.78	92.50	11.40	104.68	155
1070	3/4" pipe size		5.10	1.569		.78	100	12.30	113.08	167
1090	1" pipe size		4.30	1.860		1.07	119	14.55	134.62	199
1110	1-1/4" pipe size		4	2		1.07	127	15.65	143.72	214
1130	1-1/2" pipe size		4	2		1.07	127	15.65	143.72	214
1150	2" pipe size		2.40	3.333		1.65	212	26	239.65	355
1170	2-1/2" pipe size		2.20	3.636		1.65	232	28.50	262.15	390
1190	3" pipe size		2	4		1.65	255	31.50	288.15	425
1210	3-1/2" pipe size		1.80	4.444		2.02	283	35	320.02	475
1230	4" pipe size		1.50	5.333		2.02	340	42	384.02	570
1500	Brick wall, 8" thick, 1/2" pipe size		18	.444		.26	28.50	3.48	32.24	47.50
1520	3/4" pipe size		18	.444		.26	28.50	3.48	32.24	47.50
1540	1" pipe size		13.30	.602		.36	38.50	4.71	43.57	64.50
1560	1-1/4" pipe size		13.30	.602		.36	38.50	4.71	43.57	64.50
1580	1-1/2" pipe size		13.30	.602		.36	38.50	4.71	43.57	64.50
1600	2" pipe size		5.70	1.404		.55	89.50	11	101.05	150
1620	2-1/2" pipe size		5.70	1.404		.55	89.50	11	101.05	150
1640	3" pipe size		5.70	1.404		.55	89.50	11	101.05	150
1660	3-1/2" pipe size		4.40	1.818		.67	116	14.25	130.92	194
1680	4" pipe size		4	2		.67	127	15.65	143.32	214
1700	12" thick, 1/2" pipe size		14.50	.552		.39	35	4.32	39.71	59

For customer support on your Commercial Renovation Costs with RSMeans Data, call 800.448.8182.

26 05 33.95 Cutting and Drilling	Crew	Daily Output	Labor-Hours	Unit	Material	2021 Bare Costs Labor	Equipment	Total	Total Incl O&P
1720	3/4" pipe size	R-31	14.50	.552	Ea.	.39	35	4.32	39.71	59
1740	1" pipe size		11	.727		.53	46.50	5.70	52.73	78
1760	1-1/4" pipe size		11	.727		.53	46.50	5.70	52.73	78
1780	1-1/2" pipe size		11	.727		.53	46.50	5.70	52.73	78
1800	2" pipe size		5	1.600		.83	102	12.55	115.38	172
1820	2-1/2" pipe size		5	1.600		.83	102	12.55	115.38	172
1840	3" pipe size		5	1.600		.83	102	12.55	115.38	172
1860	3-1/2" pipe size		3.80	2.105		1.01	134	16.50	151.51	225
1880	4" pipe size		3.30	2.424		1.01	154	19	174.01	259
1900	16" thick, 1/2" pipe size		12.30	.650		.52	41.50	5.10	47.12	69.50
1920	3/4" pipe size		12.30	.650		.52	41.50	5.10	47.12	69.50
1940	1" pipe size		9.30	.860		.71	55	6.75	62.46	92
1960	1-1/4" pipe size		9.30	.860		.71	55	6.75	62.46	92
1980	1-1/2" pipe size		9.30	.860		.71	55	6.75	62.46	92
2000	2" pipe size		4.40	1.818		1.10	116	14.25	131.35	195
2010	2-1/2" pipe size		4.40	1.818		1.10	116	14.25	131.35	195
2030	3" pipe size		4.40	1.818		1.10	116	14.25	131.35	195
2050	3-1/2" pipe size		3.30	2.424		1.34	154	19	174.34	259
2070	4" pipe size		3	2.667		1.34	170	21	192.34	285
2090	20" thick, 1/2" pipe size		10.70	.748		.65	47.50	5.85	54	80
2110	3/4" pipe size		10.70	.748		.65	47.50	5.85	54	80
2130	1" pipe size		8	1		.89	63.50	7.85	72.24	108
2150	1-1/4" pipe size		8	1		.89	63.50	7.85	72.24	108
2170	1-1/2" pipe size		8	1		.89	63.50	7.85	72.24	108
2190	2" pipe size		4	2		1.38	127	15.65	144.03	215
2210	2-1/2" pipe size		4	2		1.38	127	15.65	144.03	215
2230	3" pipe size		4	2		1.38	127	15.65	144.03	215
2250	3-1/2" pipe size		3	2.667		1.68	170	21	192.68	286
2270	4" pipe size		2.70	2.963		1.68	189	23	213.68	315
2290	24" thick, 1/2" pipe size		9.40	.851		.78	54	6.65	61.43	91.50
2310	3/4" pipe size		9.40	.851		.78	54	6.65	61.43	91.50
2330	1" pipe size		7.10	1.127		1.07	72	8.80	81.87	121
2350	1-1/4" pipe size		7.10	1.127		1.07	72	8.80	81.87	121
2370	1-1/2" pipe size		7.10	1.127		1.07	72	8.80	81.87	121
2390	2" pipe size		3.60	2.222		1.65	142	17.40	161.05	238
2410	2-1/2" pipe size		3.60	2.222		1.65	142	17.40	161.05	238
2430	3" pipe size		3.60	2.222		1.65	142	17.40	161.05	238
2450	3-1/2" pipe size		2.80	2.857		2.02	182	22.50	206.52	305
2470	4" pipe size	▼	2.50	3.200	▼	2.02	204	25	231.02	345
3000	Knockouts to 8' high, metal boxes & enclosures									
3020	With hole saw, 1/2" pipe size	1 Elec	53	.151	Ea.		9.60		9.60	14.75
3040	3/4" pipe size		47	.170			10.85		10.85	16.65
3050	1" pipe size		40	.200			12.75		12.75	19.55
3060	1-1/4" pipe size		36	.222			14.15		14.15	21.50
3070	1-1/2" pipe size		32	.250			15.95		15.95	24.50
3080	2" pipe size		27	.296			18.85		18.85	29
3090	2-1/2" pipe size		20	.400			25.50		25.50	39
4010	3" pipe size		16	.500			32		32	49
4030	3-1/2" pipe size		13	.615			39		39	60
4050	4" pipe size	▼	11	.727	▼		46.50		46.50	71

26 05 39 – Underfloor Raceways for Electrical Systems

26 05 39.30 Conduit In Concrete Slab

		Crew	Daily Output	Labor-Hours	Unit	Material	2021 Bare Costs Labor	Equipment	Total	Total Incl O&P
0010	**CONDUIT IN CONCRETE SLAB** Including terminations,									
0020	fittings and supports									
3230	PVC, schedule 40, 1/2" diameter	1 Elec	270	.030	L.F.	.36	1.89		2.25	3.29
3250	3/4" diameter		230	.035		.37	2.22		2.59	3.81
3270	1" diameter		200	.040		.42	2.55		2.97	4.37
3300	1-1/4" diameter		170	.047		.73	3		3.73	5.40
3330	1-1/2" diameter		140	.057		.76	3.64		4.40	6.45
3350	2" diameter		120	.067		.90	4.25		5.15	7.50
4350	Rigid galvanized steel, 1/2" diameter		200	.040		2.32	2.55		4.87	6.45
4400	3/4" diameter		170	.047		2.48	3		5.48	7.30
4450	1" diameter		130	.062		3.73	3.92		7.65	10.10
4500	1-1/4" diameter		110	.073		5.25	4.63		9.88	12.90
4600	1-1/2" diameter		100	.080		5.90	5.10		11	14.35
4800	2" diameter		90	.089		7.20	5.65		12.85	16.65
9000	Minimum labor/equipment charge		4	2	Job		127		127	196

26 05 39.40 Conduit In Trench

		Crew	Daily Output	Labor-Hours	Unit	Material	2021 Bare Costs Labor	Equipment	Total	Total Incl O&P
0010	**CONDUIT IN TRENCH** Includes terminations and fittings									
0020	Does not include excavation or backfill, see Section 31 23 16									
0200	Rigid galvanized steel, 2" diameter	1 Elec	150	.053	L.F.	6.90	3.40		10.30	12.80
0400	2-1/2" diameter	"	100	.080		13.60	5.10		18.70	23
0600	3" diameter	2 Elec	160	.100		15.45	6.35		21.80	27
0800	3-1/2" diameter		140	.114		20.50	7.30		27.80	33.50
1000	4" diameter		100	.160		22	10.20		32.20	39.50
1200	5" diameter		80	.200		48	12.75		60.75	72.50
1400	6" diameter		60	.267		59.50	17		76.50	91.50
9000	Minimum labor/equipment charge	1 Elec	4	2	Job		127		127	196

26 05 83 – Wiring Connections

26 05 83.10 Motor Connections

		Crew	Daily Output	Labor-Hours	Unit	Material	2021 Bare Costs Labor	Equipment	Total	Total Incl O&P
0010	**MOTOR CONNECTIONS**									
0020	Flexible conduit and fittings, 115 volt, 1 phase, up to 1 HP motor	1 Elec	8	1	Ea.	6.20	63.50		69.70	105
9000	Minimum labor/equipment charge	"	4	2	Job		127		127	196

26 05 90 – Residential Applications

26 05 90.10 Residential Wiring

		Crew	Daily Output	Labor-Hours	Unit	Material	2021 Bare Costs Labor	Equipment	Total	Total Incl O&P
0010	**RESIDENTIAL WIRING**									
0020	20' avg. runs and #14/2 wiring incl. unless otherwise noted									
1000	Service & panel, includes 24' SE-AL cable, service eye, meter,									
1010	Socket, panel board, main bkr., ground rod, 15 or 20 amp									
1020	1-pole circuit breakers, and misc. hardware									
1100	100 amp, with 10 branch breakers	1 Elec	1.19	6.723	Ea.	400	430		830	1,100
1110	With PVC conduit and wire		.92	8.696		425	555		980	1,325
1120	With RGS conduit and wire		.73	10.959		600	700		1,300	1,725
1150	150 amp, with 14 branch breakers		1.03	7.767		650	495		1,145	1,475
1170	With PVC conduit and wire		.82	9.756		690	620		1,310	1,725
1180	With RGS conduit and wire		.67	11.940		1,050	760		1,810	2,325
1200	200 amp, with 18 branch breakers	2 Elec	1.80	8.889		1,075	565		1,640	2,075
1220	With PVC conduit and wire		1.46	10.959		1,125	700		1,825	2,325
1230	With RGS conduit and wire		1.24	12.903		1,550	820		2,370	3,000
1800	Lightning surge suppressor	1 Elec	32	.250		50.50	15.95		66.45	80.50
2000	Switch devices									
2100	Single pole, 15 amp, ivory, with a 1-gang box, cover plate,									
2110	Type NM (Romex) cable	1 Elec	17.10	.468	Ea.	6.85	30		36.85	53.50

26 05 90 – Residential Applications

26 05 90.10 Residential Wiring	Crew	Daily Output	Labor-Hours	Unit	Material	2021 Bare Costs Labor	Equipment	Total	Total Incl O&P
2120	Type MC cable	1 Elec	14.30	.559	Ea.	20.50	35.50		56	77
2130	EMT & wire		5.71	1.401		29	89.50		118.50	169
2150	3-way, #14/3, type NM cable		14.55	.550		8.50	35		43.50	63.50
2170	Type MC cable		12.31	.650		26	41.50		67.50	92
2180	EMT & wire		5	1.600		31	102		133	191
2200	4-way, #14/3, type NM cable		14.55	.550		15.40	35		50.40	71
2220	Type MC cable		12.31	.650		32.50	41.50		74	99.50
2230	EMT & wire		5	1.600		38	102		140	199
2250	S.P., 20 amp, #12/2, type NM cable		13.33	.600		11.85	38		49.85	71.50
2270	Type MC cable		11.43	.700		24.50	44.50		69	95.50
2280	EMT & wire		4.85	1.649		35.50	105		140.50	200
2290	S.P. rotary dimmer, 600 W, no wiring		17	.471		27.50	30		57.50	76
2300	S.P. rotary dimmer, 600 W, type NM cable		14.55	.550		31	35		66	88
2320	Type MC cable		12.31	.650		44.50	41.50		86	113
2330	EMT & wire		5	1.600		55	102		157	218
2350	3-way rotary dimmer, type NM cable		13.33	.600		22	38		60	83
2370	Type MC cable		11.43	.700		36	44.50		80.50	108
2380	EMT & wire		4.85	1.649		46	105		151	212
2400	Interval timer wall switch, 20 amp, 1-30 min., #12/2									
2410	Type NM cable	1 Elec	14.55	.550	Ea.	57.50	35		92.50	118
2420	Type MC cable		12.31	.650		64.50	41.50		106	135
2430	EMT & wire		5	1.600		81.50	102		183.50	247
2500	Decorator style									
2510	S.P., 15 amp, type NM cable	1 Elec	17.10	.468	Ea.	11.60	30		41.60	59
2520	Type MC cable		14.30	.559		25.50	35.50		61	82.50
2530	EMT & wire		5.71	1.401		34	89.50		123.50	174
2550	3-way, #14/3, type NM cable		14.55	.550		13.30	35		48.30	68.50
2570	Type MC cable		12.31	.650		30.50	41.50		72	97
2580	EMT & wire		5	1.600		36	102		138	197
2600	4-way, #14/3, type NM cable		14.55	.550		20	35		55	76
2620	Type MC cable		12.31	.650		37.50	41.50		79	105
2630	EMT & wire		5	1.600		42.50	102		144.50	204
2650	S.P., 20 amp, #12/2, type NM cable		13.33	.600		16.60	38		54.60	77
2670	Type MC cable		11.43	.700		29.50	44.50		74	101
2680	EMT & wire		4.85	1.649		40.50	105		145.50	206
2700	S.P., slide dimmer, type NM cable		17.10	.468		37	30		67	86.50
2720	Type MC cable		14.30	.559		50.50	35.50		86	110
2730	EMT & wire		5.71	1.401		60.50	89.50		150	204
2750	S.P., touch dimmer, type NM cable		17.10	.468		24.50	30		54.50	73
2770	Type MC cable		14.30	.559		38	35.50		73.50	96.50
2780	EMT & wire		5.71	1.401		48.50	89.50		138	190
2800	3-way touch dimmer, type NM cable		13.33	.600		48.50	38		86.50	112
2820	Type MC cable		11.43	.700		62.50	44.50		107	137
2830	EMT & wire		4.85	1.649		72.50	105		177.50	241
3000	Combination devices									
3100	S.P. switch/15 amp recpt., ivory, 1-gang box, plate									
3110	Type NM cable	1 Elec	11.43	.700	Ea.	19.50	44.50		64	90
3120	Type MC cable		10	.800		33	51		84	115
3130	EMT & wire		4.40	1.818		43.50	116		159.50	226
3150	S.P. switch/pilot light, type NM cable		11.43	.700		20.50	44.50		65	91
3170	Type MC cable		10	.800		34	51		85	116
3180	EMT & wire		4.43	1.806		44	115		159	226
3190	2-S.P. switches, 2-#14/2, no wiring		14	.571		5.20	36.50		41.70	61.50

503

26 05 90.10 Residential Wiring	Crew	Daily Output	Labor-Hours	Unit	Material	2021 Bare Costs Labor	2021 Bare Costs Equipment	Total	Total Incl O&P
3200	2-S.P. switches, 2-#14/2, type NM cables	1 Elec	10	.800	Ea.	21	51		72	102
3220	Type MC cable		8.89	.900		39	57.50		96.50	131
3230	EMT & wire		4.10	1.951		46.50	124		170.50	242
3250	3-way switch/15 amp recpt., #14/3, type NM cable		10	.800		26	51		77	107
3270	Type MC cable		8.89	.900		43.50	57.50		101	136
3280	EMT & wire		4.10	1.951		48.50	124		172.50	245
3300	2-3 way switches, 2-#14/3, type NM cables		8.89	.900		33	57.50		90.50	125
3320	Type MC cable		8	1		58	63.50		121.50	162
3330	EMT & wire		4	2		55.50	127		182.50	257
3350	S.P. switch/20 amp recpt., #12/2, type NM cable		10	.800		27	51		78	108
3370	Type MC cable		8.89	.900		34	57.50		91.50	125
3380	EMT & wire		4.10	1.951		50.50	124		174.50	247
3400	Decorator style									
3410	S.P. switch/15 amp recpt., type NM cable	1 Elec	11.43	.700	Ea.	24.50	44.50		69	95
3420	Type MC cable		10	.800		38	51		89	121
3430	EMT & wire		4.40	1.818		48	116		164	231
3450	S.P. switch/pilot light, type NM cable		11.43	.700		25	44.50		69.50	96
3470	Type MC cable		10	.800		39	51		90	121
3480	EMT & wire		4.40	1.818		49	116		165	232
3500	2-S.P. switches, 2-#14/2, type NM cables		10	.800		26	51		77	107
3520	Type MC cable		8.89	.900		44	57.50		101.50	136
3530	EMT & wire		4.10	1.951		51	124		175	247
3550	3-way/15 amp recpt., #14/3, type NM cable		10	.800		31	51		82	113
3570	Type MC cable		8.89	.900		48	57.50		105.50	141
3580	EMT & wire		4.10	1.951		53.50	124		177.50	250
3650	2-3 way switches, 2-#14/3, type NM cables		8.89	.900		38	57.50		95.50	130
3670	Type MC cable		8	1		63	63.50		126.50	167
3680	EMT & wire		4	2		60.50	127		187.50	263
3700	S.P. switch/20 amp recpt., #12/2, type NM cable		10	.800		31.50	51		82.50	114
3720	Type MC cable		8.89	.900		38.50	57.50		96	131
3730	EMT & wire		4.10	1.951		55.50	124		179.50	252
4000	Receptacle devices									
4010	Duplex outlet, 15 amp recpt., ivory, 1-gang box, plate									
4015	Type NM cable	1 Elec	14.55	.550	Ea.	7.30	35		42.30	62
4020	Type MC cable		12.31	.650		21	41.50		62.50	86.50
4030	EMT & wire		5.33	1.501		29.50	95.50		125	180
4050	With #12/2, type NM cable		12.31	.650		8.50	41.50		50	73
4070	Type MC cable		10.67	.750		21.50	48		69.50	97
4080	EMT & wire		4.71	1.699		32.50	108		140.50	202
4100	20 amp recpt., #12/2, type NM cable		12.31	.650		18.10	41.50		59.60	83.50
4120	Type MC cable		10.67	.750		31	48		79	108
4130	EMT & wire		4.71	1.699		42	108		150	212
4140	For GFI see Section 26 05 90.10 line 4300 below									
4150	Decorator style, 15 amp recpt., type NM cable	1 Elec	14.55	.550	Ea.	12.05	35		47.05	67.50
4170	Type MC cable		12.31	.650		26	41.50		67.50	92
4180	EMT & wire		5.33	1.501		34	95.50		129.50	185
4200	With #12/2, type NM cable		12.31	.650		13.25	41.50		54.75	78
4220	Type MC cable		10.67	.750		26	48		74	102
4230	EMT & wire		4.71	1.699		37	108		145	207
4250	20 amp recpt., #12/2, type NM cable		12.31	.650		23	41.50		64.50	88.50
4270	Type MC cable		10.67	.750		35.50	48		83.50	113
4280	EMT & wire		4.71	1.699		46.50	108		154.50	218
4300	GFI, 15 amp recpt., type NM cable		12.31	.650		21	41.50		62.50	86.50

26 05 90.10 Residential Wiring	Crew	Daily Output	Labor-Hours	Unit	Material	2021 Bare Costs Labor	2021 Bare Costs Equipment	Total	Total Incl O&P
4320 Type MC cable	1 Elec	10.67	.750	Ea.	34.50	48		82.50	112
4330 EMT & wire		4.71	1.699		43	108		151	214
4350 GFI with #12/2, type NM cable		10.67	.750		22	48		70	98
4370 Type MC cable		9.20	.870		35	55.50		90.50	124
4380 EMT & wire		4.21	1.900		46	121		167	237
4400 20 amp recpt., #12/2, type NM cable		10.67	.750		62.50	48		110.50	143
4420 Type MC cable		9.20	.870		75.50	55.50		131	168
4430 EMT & wire		4.21	1.900		86.50	121		207.50	281
4500 Weather-proof cover for above receptacles, add	▼	32	.250	▼	3.29	15.95		19.24	28
4550 Air conditioner outlet, 20 amp-240 volt recpt.									
4560 30' of #12/2, 2 pole circuit breaker									
4570 Type NM cable	1 Elec	10	.800	Ea.	59.50	51		110.50	144
4580 Type MC cable		9	.889		74	56.50		130.50	169
4590 EMT & wire		4	2		84	127		211	289
4600 Decorator style, type NM cable		10	.800		64.50	51		115.50	150
4620 Type MC cable		9	.889		79	56.50		135.50	174
4630 EMT & wire	▼	4	2	▼	89	127		216	294
4650 Dryer outlet, 30 amp-240 volt recpt., 20' of #10/3									
4660 2 pole circuit breaker									
4670 Type NM cable	1 Elec	6.41	1.248	Ea.	54	79.50		133.50	182
4680 Type MC cable		5.71	1.401		64	89.50		153.50	207
4690 EMT & wire	▼	3.48	2.299	▼	75	146		221	310
4700 Range outlet, 50 amp-240 volt recpt., 30' of #8/3									
4710 Type NM cable	1 Elec	4.21	1.900	Ea.	85.50	121		206.50	280
4720 Type MC cable		4	2		108	127		235	315
4730 EMT & wire		2.96	2.703		104	172		276	380
4750 Central vacuum outlet, type NM cable		6.40	1.250		61.50	79.50		141	190
4770 Type MC cable		5.71	1.401		74	89.50		163.50	219
4780 EMT & wire	▼	3.48	2.299		90.50	146		236.50	325
4800 30 amp-110 volt locking recpt., #10/2 circ. bkr.									
4810 Type NM cable	1 Elec	6.20	1.290	Ea.	74.50	82		156.50	208
4820 Type MC cable		5.40	1.481		89.50	94.50		184	244
4830 EMT & wire	▼	3.20	2.500	▼	106	159		265	360
4900 Low voltage outlets									
4910 Telephone recpt., 20' of 4/C phone wire	1 Elec	26	.308	Ea.	7.60	19.60		27.20	38.50
4920 TV recpt., 20' of RG59U coax wire, F type connector	"	16	.500	"	15.40	32		47.40	66
4950 Door bell chime, transformer, 2 buttons, 60' of bellwire									
4970 Economy model	1 Elec	11.50	.696	Ea.	55.50	44.50		100	129
4980 Custom model		11.50	.696		94	44.50		138.50	172
4990 Luxury model, 3 buttons	▼	9.50	.842	▼	193	53.50		246.50	295
6000 Lighting outlets									
6050 Wire only (for fixture), type NM cable	1 Elec	32	.250	Ea.	5.90	15.95		21.85	31
6070 Type MC cable		24	.333		12.25	21		33.25	46
6080 EMT & wire		10	.800		19.95	51		70.95	101
6100 Box (4") and wire (for fixture), type NM cable		25	.320		14.95	20.50		35.45	48
6120 Type MC cable		20	.400		21.50	25.50		47	62.50
6130 EMT & wire	▼	11	.727	▼	29	46.50		75.50	103
6200 Fixtures (use with line 6050 or 6100 above)									
6210 Canopy style, economy grade	1 Elec	40	.200	Ea.	22.50	12.75		35.25	44.50
6220 Custom grade		40	.200		53	12.75		65.75	78
6250 Dining room chandelier, economy grade		19	.421		80.50	27		107.50	130
6270 Luxury grade		15	.533		3,050	34		3,084	3,400
6310 Kitchen fixture (fluorescent), economy grade		30	.267		76	17		93	110

26 05 90.10 Residential Wiring	Crew	Daily Output	Labor-Hours	Unit	Material	2021 Bare Costs Labor	Equipment	Total	Total Incl O&P
6320 Custom grade	1 Elec	25	.320	Ea.	220	20.50		240.50	274
6350 Outdoor, wall mounted, economy grade		30	.267		38.50	17		55.50	68.50
6360 Custom grade		30	.267		116	17		133	154
6370 Luxury grade		25	.320		248	20.50		268.50	305
6410 Outdoor PAR floodlights, 1 lamp, 150 watt		20	.400		15.55	25.50		41.05	56
6420 2 lamp, 150 watt each		20	.400		28	25.50		53.50	70
6425 Motion sensing, 2 lamp, 150 watt each		20	.400		107	25.50		132.50	156
6430 For infrared security sensor, add		32	.250		102	15.95		117.95	137
6450 Outdoor, quartz-halogen, 300 watt flood		20	.400		40	25.50		65.50	83
6600 Recessed downlight, round, pre-wired, 50 or 75 watt trim		30	.267		46.50	17		63.50	77
6610 With shower light trim		30	.267		71.50	17		88.50	105
6620 With wall washer trim		28	.286		72	18.20		90.20	107
6630 With eye-ball trim		28	.286		31	18.20		49.20	62
6700 Porcelain lamp holder		40	.200		2.75	12.75		15.50	22.50
6710 With pull switch		40	.200		11.30	12.75		24.05	32
6750 Fluorescent strip, 2-20 watt tube, wrap around diffuser, 24"		24	.333		43.50	21		64.50	80.50
6760 1-34 watt tube, 48"		24	.333		113	21		134	157
6770 2-34 watt tubes, 48"		20	.400		148	25.50		173.50	201
6800 Bathroom heat lamp, 1-250 watt		28	.286		33.50	18.20		51.70	65
6810 2-250 watt lamps	▼	28	.286	▼	60	18.20		78.20	94
6820 For timer switch, see Section 26 05 90.10 line 2400									
6900 Outdoor post lamp, incl. post, fixture, 35' of #14/2									
6910 Type NM cable	1 Elec	3.50	2.286	Ea.	310	146		456	565
6920 Photo-eye, add		27	.296		15.20	18.85		34.05	45.50
6950 Clock dial time switch, 24 hr., w/enclosure, type NM cable		11.43	.700		71	44.50		115.50	147
6970 Type MC cable		11	.727		84.50	46.50		131	164
6980 EMT & wire	▼	4.85	1.649	▼	93	105		198	263
7000 Alarm systems									
7050 Smoke detectors, box, #14/3, type NM cable	1 Elec	14.55	.550	Ea.	31	35		66	88
7070 Type MC cable		12.31	.650		42.50	41.50		84	110
7080 EMT & wire	▼	5	1.600		47.50	102		149.50	209
7090 For relay output to security system, add				▼	9.80			9.80	10.80
8000 Residential equipment									
8050 Disposal hook-up, incl. switch, outlet box, 3' of flex									
8060 20 amp-1 pole circ. bkr., and 25' of #12/2									
8070 Type NM cable	1 Elec	10	.800	Ea.	25	51		76	106
8080 Type MC cable		8	1		38.50	63.50		102	141
8090 EMT & wire	▼	5	1.600	▼	52.50	102		154.50	215
8100 Trash compactor or dishwasher hook-up, incl. outlet box,									
8110 3' of flex, 15 amp-1 pole circ. bkr., and 25' of #14/2									
8120 Type NM cable	1 Elec	10	.800	Ea.	21.50	51		72.50	102
8130 Type MC cable		8	1		36.50	63.50		100	138
8140 EMT & wire	▼	5	1.600	▼	48	102		150	210
8150 Hot water sink dispenser hook-up, use line 8100									
8200 Vent/exhaust fan hook-up, type NM cable	1 Elec	32	.250	Ea.	5.90	15.95		21.85	31
8220 Type MC cable		24	.333		12.25	21		33.25	46
8230 EMT & wire	▼	10	.800	▼	19.95	51		70.95	101
8250 Bathroom vent fan, 50 CFM (use with above hook-up)									
8260 Economy model	1 Elec	15	.533	Ea.	18.60	34		52.60	72.50
8270 Low noise model		15	.533		50.50	34		84.50	108
8280 Custom model	▼	12	.667	▼	116	42.50		158.50	193
8300 Bathroom or kitchen vent fan, 110 CFM									
8310 Economy model	1 Elec	15	.533	Ea.	63	34		97	121

26 05 90.10 Residential Wiring	Crew	Daily Output	Labor-Hours	Unit	Material	2021 Bare Costs Labor	Equipment	Total	Total Incl O&P
8320	Low noise model	1 Elec	15	.533	Ea.	101	34		135	163
8350	Paddle fan, variable speed (w/o lights)									
8360	Economy model (AC motor)	1 Elec	10	.800	Ea.	132	51		183	224
8362	With light kit		10	.800		171	51		222	268
8370	Custom model (AC motor)		10	.800		320	51		371	435
8372	With light kit		10	.800		360	51		411	475
8380	Luxury model (DC motor)		8	1		279	63.50		342.50	405
8382	With light kit		8	1		320	63.50		383.50	450
8390	Remote speed switch for above, add		12	.667		40.50	42.50		83	110
8500	Whole house exhaust fan, ceiling mount, 36", variable speed									
8510	Remote switch, incl. shutters, 20 amp-1 pole circ. bkr.									
8520	30' of #12/2, type NM cable	1 Elec	4	2	Ea.	1,500	127		1,627	1,875
8530	Type MC cable		3.50	2.286		1,525	146		1,671	1,900
8540	EMT & wire		3	2.667		1,550	170		1,720	1,950
8600	Whirlpool tub hook-up, incl. timer switch, outlet box									
8610	3' of flex, 20 amp-1 pole GFI circ. bkr.									
8620	30' of #12/2, type NM cable	1 Elec	5	1.600	Ea.	241	102		343	420
8630	Type MC cable		4.20	1.905		252	121		373	465
8640	EMT & wire		3.40	2.353		263	150		413	520
8650	Hot water heater hook-up, incl. 1-2 pole circ. bkr., box;									
8660	3' of flex, 20' of #10/2, type NM cable	1 Elec	5	1.600	Ea.	29	102		131	189
8670	Type MC cable		4.20	1.905		43	121		164	234
8680	EMT & wire		3.40	2.353		48.50	150		198.50	284
9000	Heating/air conditioning									
9050	Furnace/boiler hook-up, incl. firestat, local on-off switch									
9060	Emergency switch, and 40' of type NM cable	1 Elec	4	2	Ea.	51	127		178	252
9070	Type MC cable		3.50	2.286		69	146		215	300
9080	EMT & wire		1.50	5.333		88	340		428	615
9100	Air conditioner hook-up, incl. local 60 amp disc. switch									
9110	3' sealtite, 40 amp, 2 pole circuit breaker									
9130	40' of #8/2, type NM cable	1 Elec	3.50	2.286	Ea.	149	146		295	390
9140	Type MC cable		3	2.667		180	170		350	460
9150	EMT & wire		1.30	6.154		186	390		576	805
9200	Heat pump hook-up, 1-40 & 1-100 amp 2 pole circ. bkr.									
9210	Local disconnect switch, 3' sealtite									
9220	40' of #8/2 & 30' of #3/2									
9230	Type NM cable	1 Elec	1.30	6.154	Ea.	425	390		815	1,075
9240	Type MC cable		1.08	7.407		510	470		980	1,275
9250	EMT & wire		.94	8.511		555	540		1,095	1,450
9500	Thermostat hook-up, using low voltage wire									
9520	Heating only, 25' of #18-3	1 Elec	24	.333	Ea.	6.15	21		27.15	39.50
9530	Heating/cooling, 25' of #18-4	"	20	.400	"	8.75	25.50		34.25	48.50

26 22 13 – Low-Voltage Distribution Transformers

26 22 13.10 Transformer, Dry-Type	Crew	Daily Output	Labor-Hours	Unit	Material	2021 Bare Costs Labor	Equipment	Total	Total Incl O&P
0010 **TRANSFORMER, DRY-TYPE**									
0050 Single phase, 240/480 volt primary, 120/240 volt secondary									
0100 1 kVA	1 Elec	2	4	Ea.	305	255		560	730
0300 2 kVA		1.60	5		610	320		930	1,175
0500 3 kVA		1.40	5.714		690	365		1,055	1,325
0700 5 kVA		1.20	6.667		840	425		1,265	1,575
0900 7.5 kVA	2 Elec	2.20	7.273		1,200	465		1,665	2,025
1100 10 kVA		1.60	10		1,550	635		2,185	2,675
1300 15 kVA		1.20	13.333		2,225	850		3,075	3,750
2300 3 phase, 480 volt primary, 120/208 volt secondary									
2310 Ventilated, 3 kVA	1 Elec	1	8	Ea.	1,250	510		1,760	2,150
2700 6 kVA		.80	10		1,325	635		1,960	2,425
2900 9 kVA		.70	11.429		1,550	730		2,280	2,825
3100 15 kVA	2 Elec	1.10	14.545		1,625	925		2,550	3,200
3300 30 kVA		.90	17.778		5,175	1,125		6,300	7,450
3500 45 kVA		.80	20		6,775	1,275		8,050	9,400
9000 Minimum labor/equipment charge	1 Elec	1	8	Job		510		510	785
9300 Energy efficient transformer 3 ph									
9400 15 kVA, 480VAC delta, 208Y/120VAC	R-3	1.50	13.333	Ea.	2,425	845	127	3,397	4,125
9405 30 kVA, 480VAC delta, 208Y/120VAC		1.50	13.333		2,550	845	127	3,522	4,250
9410 45 kVA, 480VAC delta, 208Y/120VAC		1.50	13.333		2,000	845	127	2,972	3,650
9415 75 kVA, 480VAC delta, 208Y/120VAC		1.25	16		2,425	1,025	153	3,603	4,425
9420 112 kVA, 480VAC delta, 208Y/120VAC		1.25	16		7,125	1,025	153	8,303	9,600
9425 150 kVA, 480VAC delta, 208Y/120VAC		1.25	16		4,500	1,025	153	5,678	6,700
9430 225 kVA, 480VAC delta, 208Y/120VAC		1.25	16		5,775	1,025	153	6,953	8,100
9435 300 kVA, 480VAC delta, 208Y/120VAC		1.25	16		8,050	1,025	153	9,228	10,600

26 24 Switchboards and Panelboards

26 24 16 – Panelboards

26 24 16.20 Panelboard and Load Center Circuit Breakers

26 24 16.20 Panelboard and Load Center Circuit Breakers	Crew	Daily Output	Labor-Hours	Unit	Material	2021 Bare Costs Labor	Equipment	Total	Total Incl O&P
0010 **PANELBOARD AND LOAD CENTER CIRCUIT BREAKERS**									
0050 Bolt-on, 10,000 amp I.C., 120 volt, 1 pole									
0100 15-50 amp	1 Elec	10	.800	Ea.	26.50	51		77.50	108
0200 60 amp		8	1		23	63.50		86.50	124
0300 70 amp		8	1		62	63.50		125.50	167
0350 240 volt, 2 pole									
0400 15-50 amp	1 Elec	8	1	Ea.	63.50	63.50		127	168
0500 60 amp		7.50	1.067		62	68		130	173
0600 80-100 amp		5	1.600		156	102		258	330
0700 3 pole, 15-60 amp		6.20	1.290		195	82		277	340
0800 70 amp		5	1.600		240	102		342	420
0900 80-100 amp		3.60	2.222		289	142		431	535
1000 22,000 amp I.C., 240 volt, 2 pole, 70-225 amp		2.70	2.963		175	189		364	480
1100 3 pole, 70-225 amp		2.30	3.478		430	222		652	810
1200 14,000 amp I.C., 277 volts, 1 pole, 15-30 amp		8	1		108	63.50		171.50	217
1300 22,000 amp I.C., 480 volts, 2 pole, 70-225 amp		2.70	2.963		134	189		323	435
1400 3 pole, 70-225 amp		2.30	3.478		1,100	222		1,322	1,575
2060 Plug-in tandem, 120/240 V, 2-15 A, 1 pole		11	.727		21.50	46.50		68	94.50
2070 1-15 A & 1-20 A		11	.727		24	46.50		70.50	97.50
2080 2-20 A		11	.727		22	46.50		68.50	95
2082 Arc fault circuit interrupter, 120/240 V, 1-15 A & 1-20 A, 1 pole		11	.727		50.50	46.50		97	127

26 24 Switchboards and Panelboards

26 24 16 – Panelboards

26 24 16.20 Panelboard and Load Center Circuit Breakers	Crew	Daily Output	Labor-Hours	Unit	Material	2021 Bare Costs Labor	Equipment	Total	Total Incl O&P
5110 NEMA 1 enclosure only, 600V, 3 p, 14k AIC, 100A	1 Elec	2.20	3.636	Ea.	147	232		379	515
9000 Minimum labor/equipment charge	↓	3	2.667	Job		170		170	261

26 24 16.30 Panelboards Commercial Applications

	Crew	Daily Output	Labor-Hours	Unit	Material	2021 Bare Costs Labor	Equipment	Total	Total Incl O&P
0010 **PANELBOARDS COMMERCIAL APPLICATIONS**									
0050 NQOD, w/20 amp 1 pole bolt-on circuit breakers									
0100 3 wire, 120/240 volts, 100 amp main lugs									
0150 10 circuits	1 Elec	1	8	Ea.	950	510		1,460	1,825
0200 14 circuits		.88	9.091		1,075	580		1,655	2,075
0250 18 circuits		.75	10.667		1,150	680		1,830	2,325
0300 20 circuits		.65	12.308		1,250	785		2,035	2,575
0600 4 wire, 120/208 volts, 100 amp main lugs, 12 circuits		1	8		1,025	510		1,535	1,900
0650 16 circuits		.75	10.667		1,150	680		1,830	2,300
0700 20 circuits		.65	12.308		1,325	785		2,110	2,650
0750 24 circuits		.60	13.333		415	850		1,265	1,750
0800 30 circuits	↓	.53	15.094		1,650	960		2,610	3,275
0850 225 amp main lugs, 32 circuits	2 Elec	.90	17.778		1,875	1,125		3,000	3,825
0900 34 circuits		.84	19.048		1,925	1,225		3,150	3,975
0950 36 circuits		.80	20		1,950	1,275		3,225	4,100
1000 42 circuits		.68	23.529		2,200	1,500		3,700	4,725
1010 400 amp main lugs, 42 circs	↓	.68	23.529	↓	2,200	1,500		3,700	4,725
1200 NEHB, w/20 amp, 1 pole bolt-on circuit breakers									
1250 4 wire, 277/480 volts, 100 amp main lugs, 12 circuits	1 Elec	.88	9.091	Ea.	1,500	580		2,080	2,550
1300 20 circuits	"	.60	13.333		2,150	850		3,000	3,675
1350 225 amp main lugs, 24 circuits	2 Elec	.90	17.778		2,525	1,125		3,650	4,525
1400 30 circuits		.80	20		3,025	1,275		4,300	5,275
1448 32 circuits		4.90	3.265		3,500	208		3,708	4,175
1450 36 circuits	↓	.72	22.222	↓	3,500	1,425		4,925	6,025
1600 NQOD panel, w/20 amp, 1 pole, circuit breakers									
2000 4 wire, 120/208 volts with main circuit breaker									
2050 100 amp main, 24 circuits	1 Elec	.47	17.021	Ea.	1,750	1,075		2,825	3,600
2100 30 circuits	"	.40	20		2,050	1,275		3,325	4,200
2200 225 amp main, 32 circuits	2 Elec	.72	22.222		3,400	1,425		4,825	5,925
2250 42 circuits	"	.56	28.571	↓	4,000	1,825		5,825	7,200
2300 400 amp main, 42 circuits	2 Elec	.48	33.333	Ea.	5,025	2,125		7,150	8,775
2350 600 amp main, 42 circuits	"	.40	40	"	7,400	2,550		9,950	12,100
2400 NEHB, with 20 amp, 1 pole circuit breaker									
2450 4 wire, 277/480 volts with main circuit breaker									
2500 100 amp main, 24 circuits	1 Elec	.42	19.048	Ea.	2,900	1,225		4,125	5,075
2550 30 circuits	"	.38	21.053		3,400	1,350		4,750	5,775
2600 225 amp main, 30 circuits	2 Elec	.72	22.222		4,250	1,425		5,675	6,875
2650 42 circuits	"	.56	28.571	↓	5,250	1,825		7,075	8,575
9000 Minimum labor/equipment charge	1 Elec	1	8	Job		510		510	785

26 27 Low-Voltage Distribution Equipment

26 27 13 – Electricity Metering

26 27 13.10 Meter Centers and Sockets

		Crew	Daily Output	Labor-Hours	Unit	Material	2021 Bare Costs Labor	2021 Bare Costs Equipment	Total	Total Incl O&P
0010	**METER CENTERS AND SOCKETS**									
0100	Sockets, single position, 4 terminal, 100 amp	1 Elec	3.20	2.500	Ea.	73	159		232	325
0200	150 amp		2.30	3.478		55	222		277	400
0300	200 amp		1.90	4.211		106	268		374	525
0400	Transformer rated, 20 amp		3.20	2.500		174	159		333	435
0500	Double position, 4 terminal, 100 amp		2.80	2.857		234	182		416	535
0600	150 amp		2.10	3.810		275	243		518	675
0700	200 amp		1.70	4.706		515	300		815	1,025
1100	Meter centers and sockets, three phase, single pos, 7 terminal, 100 amp		2.80	2.857		157	182		339	450
1200	200 amp		2.10	3.810		264	243		507	665
1400	400 amp		1.70	4.706		840	300		1,140	1,375
9000	Minimum labor/equipment charge		3	2.667	Job		170		170	261

26 27 16 – Electrical Cabinets and Enclosures

26 27 16.10 Cabinets

		Crew	Daily Output	Labor-Hours	Unit	Material	2021 Bare Costs Labor	2021 Bare Costs Equipment	Total	Total Incl O&P
0010	**CABINETS**									
7000	Cabinets, current transformer									
7050	Single door, 24" H x 24" W x 10" D	1 Elec	1.60	5	Ea.	143	320		463	650
7100	30" H x 24" W x 10" D		1.30	6.154		159	390		549	775
7150	36" H x 24" W x 10" D		1.10	7.273		405	465		870	1,150
7200	30" H x 30" W x 10" D		1	8		265	510		775	1,075
7250	36" H x 30" W x 10" D		.90	8.889		330	565		895	1,225
7300	36" H x 36" W x 10" D		.80	10		340	635		975	1,350
7500	Double door, 48" H x 36" W x 10" D		.60	13.333		630	850		1,480	2,000
7550	24" H x 24" W x 12" D		1	8		251	510		761	1,050
9990	Minimum labor/equipment charge		2	4	Job		255		255	390

26 27 23 – Indoor Service Poles

26 27 23.40 Surface Raceway

		Crew	Daily Output	Labor-Hours	Unit	Material	2021 Bare Costs Labor	2021 Bare Costs Equipment	Total	Total Incl O&P
0010	**SURFACE RACEWAY**									
0090	Metal, straight section									
0100	No. 500	1 Elec	100	.080	L.F.	1.13	5.10		6.23	9.10
0110	No. 700	1 Elec	100	.080	L.F.	1.26	5.10		6.36	9.25
0400	No. 1500, small pancake		90	.089		2.44	5.65		8.09	11.40
0600	No. 2000, base & cover, blank		90	.089		2.49	5.65		8.14	11.45
0800	No. 3000, base & cover, blank		75	.107		4.74	6.80		11.54	15.65
1000	No. 4000, base & cover, blank		65	.123		7.65	7.85		15.50	20.50
1200	No. 6000, base & cover, blank		50	.160		12.80	10.20		23	30
2400	Fittings, elbows, No. 500		40	.200	Ea.	2.06	12.75		14.81	22
2800	Elbow cover, No. 2000		40	.200		4.02	12.75		16.77	24
2880	Tee, No. 500		42	.190		3.98	12.15		16.13	23
2900	No. 2000		27	.296		13.40	18.85		32.25	44
3000	Switch box, No. 500		16	.500		11.95	32		43.95	62
3400	Telephone outlet, No. 1500		16	.500		16.10	32		48.10	66.50
3600	Junction box, No. 1500		16	.500		11.05	32		43.05	61
3800	Plugmold wired sections, No. 2000									
4000	1 circuit, 6 outlets, 3' long	1 Elec	8	1	Ea.	40.50	63.50		104	143
4100	2 circuits, 8 outlets, 6' long	"	5.30	1.509	"	60	96		156	214
9300	Non-metallic, straight section									
9310	7/16" x 7/8", base & cover, blank	1 Elec	160	.050	L.F.	2.17	3.19		5.36	7.30
9320	Base & cover w/adhesive		160	.050		1.49	3.19		4.68	6.55
9340	7/16" x 1-5/16", base & cover, blank		145	.055		2.38	3.51		5.89	8
9350	Base & cover w/adhesive		145	.055		2.27	3.51		5.78	7.90
9370	11/16" x 2-1/4", base & cover, blank		130	.062		2.88	3.92		6.80	9.15

510

26 27 23 – Indoor Service Poles

26 27 23.40 Surface Raceway	Crew	Daily Output	Labor-Hours	Unit	Material	2021 Bare Costs Labor	Equipment	Total	Total Incl O&P
9380	Base & cover w/adhesive	1 Elec	130	.062	L.F.	3.16	3.92		7.08	9.50
9385	1-11/16" x 5-1/4", two compartment base & cover w/screws		80	.100	↓	10.50	6.35		16.85	21.50
9400	Fittings, elbows, 7/16" x 7/8"		50	.160	Ea.	1.98	10.20		12.18	17.85
9410	7/16" x 1-5/16"		45	.178		2.05	11.30		13.35	19.65
9420	11/16" x 2-1/4"		40	.200		2.22	12.75		14.97	22
9425	1-11/16" x 5-1/4"		28	.286		11.65	18.20		29.85	41
9430	Tees, 7/16" x 7/8"		35	.229		2.56	14.55		17.11	25.50
9440	7/16" x 1-5/16"		32	.250		2.62	15.95		18.57	27.50
9450	11/16" x 2-1/4"		30	.267		2.70	17		19.70	29
9455	1-11/16" x 5-1/4"		24	.333		18.70	21		39.70	53
9460	Cover clip, 7/16" x 7/8"		80	.100		.52	6.35		6.87	10.35
9470	7/16" x 1-5/16"		72	.111		.47	7.10		7.57	11.35
9480	11/16" x 2-1/4"		64	.125		.79	7.95		8.74	13.10
9484	1-11/16" x 5-1/4"		42	.190		2.76	12.15		14.91	21.50
9486	Wire clip, 1-11/16" x 5-1/4"		68	.118		.48	7.50		7.98	12.05
9490	Blank end, 7/16" x 7/8"		50	.160		.72	10.20		10.92	16.45
9500	7/16" x 1-5/16"		45	.178		.83	11.30		12.13	18.30
9510	11/16" x 2-1/4"		40	.200	↓	1.23	12.75		13.98	21
9515	1-11/16" x 5-1/4"	1 Elec	38	.211	Ea.	5.75	13.40		19.15	27
9520	Round fixture box, 5.5" diam. x 1"		25	.320		11.70	20.50		32.20	44.50
9530	Device box, 1 gang		30	.267		5.30	17		22.30	32
9540	2 gang		25	.320	↓	7.65	20.50		28.15	40
9990	Minimum labor/equipment charge	↓	5	1.600	Job		102		102	157

26 27 26 – Wiring Devices

26 27 26.20 Wiring Devices Elements

		Crew	Daily Output	Labor-Hours	Unit	Material	2021 Bare Costs Labor	Equipment	Total	Total Incl O&P
0010	**WIRING DEVICES ELEMENTS**									
0200	Toggle switch, quiet type, single pole, 15 amp	1 Elec	40	.200	Ea.	.49	12.75		13.24	20
0500	20 amp		27	.296		4.72	18.85		23.57	34
0550	Rocker, 15 amp		40	.200		2.39	12.75		15.14	22
0560	20 amp		27	.296		6.35	18.85		25.20	36
0600	3 way, 15 amp		23	.348		1.49	22		23.49	35.50
0850	Rocker, 15 amp		23	.348		8.50	22		30.50	43.50
0860	20 amp		18	.444		8.30	28.50		36.80	52.50
0900	4 way, 15 amp		15	.533		11.35	34		45.35	64.50
1030	Rocker, 15 amp		15	.533		9.95	34		43.95	63
1040	20 amp		11	.727		17.90	46.50		64.40	90.50
1650	Dimmer switch, 120 volt, incandescent, 600 watt, 1 pole [G]		16	.500		20.50	32		52.50	71.50
2460	Receptacle, duplex, 120 volt, grounded, 15 amp		40	.200		1.38	12.75		14.13	21
2470	20 amp		27	.296		11	18.85		29.85	41
2480	Ground fault interrupting, 15 amp		27	.296		15.05	18.85		33.90	45.50
2482	20 amp		27	.296		55.50	18.85		74.35	90
2490	Dryer, 30 amp		15	.533		5.35	34		39.35	58
2500	Range, 50 amp		11	.727		14.55	46.50		61.05	87
2600	Wall plates, stainless steel, 1 gang		80	.100		1.41	6.35		7.76	11.35
2800	2 gang		53	.151		4.39	9.60		13.99	19.60
3200	Lampholder, keyless		26	.308		20	19.60		39.60	52
3400	Pullchain with receptacle		22	.364		21.50	23		44.50	59
4870	Receptacle box assembly, cast aluminum, 60A, 4P5W, 3P, 120/208V, NEMA PR4		22	.364		710	23		733	815
4875	100A, 4P5W, 3P, 347/600V, NEMA PR4		22	.364		815	23		838	930
4880	200A, 4P5W, 3P, 347/600V, NEMA PR4		22	.364		1,775	23		1,798	1,975
8100	Pin/sleeve, 20A, 480V, DSN1, male inlet, 3 pole, 3 phase, 4 wire		16	.500		45.50	32		77.50	99
8110	20A, 125V, DSN20, male inlet, 3 pole, 3 phase, 3 wire		16	.500		41.50	32		73.50	95

26 27 26 – Wiring Devices

26 27 26.20 Wiring Devices Elements	Crew	Daily Output	Labor-Hours	Unit	Material	2021 Bare Costs Labor	Equipment	Total	Total Incl O&P
8120	60A, 480V, DS60, male inlet, 3 pole, 3 phase, 3 wire	1 Elec	16	.500	Ea.	345	32		377	430
8130	Pin/sleeve, 20A, 480V, DSN10, female RCPT, 3P, 3 phase, 4 wire		16	.500		91.50	32		123.50	150
8140	30A, 480V, DS30, female RCPT, 3P, 3 phase, 4 wire		16	.500		320	32		352	400
8500	Wiring device terminal strip, 2 pole, 12-2AWG, 300VAC, CSA, 600VAC, 10A		36	.222		3.48	14.15		17.63	25.50
8505	4 pole, 12-2AWG, 300VAC, CSA, 600VAC, 10A		36	.222		4.92	14.15		19.07	27
8510	6 pole, 12-2AWG, 300VAC, CSA, 600VAC, 10A		36	.222		5.55	14.15		19.70	27.50
8515	8 pole, 12-2AWG, 300VAC, CSA, 600VAC, 10A		30	.267		6.20	17		23.20	33
8520	10 pole, 12-2AWG, 300VAC, CSA, 600VAC, 10A		30	.267		6.85	17		23.85	33.50
8525	12 pole, 12-2AWG, 300VAC, CSA, 600VAC, 10A		30	.267		7.55	17		24.55	34.50
8530	Wiring device terminal strip jumper, 10A		36	.222		.86	14.15		15.01	22.50
8535	Wiring device terminal strip, 2 pole, 12-22AWG, 300VAC, CSA, 600VAC, 20A		34	.235		3.78	15		18.78	27
8540	4 pole, 12-22AWG, 300VAC, CSA, 600VAC, 20A		34	.235		5.15	15		20.15	28.50
8545	6 pole, 12-22AWG, 300VAC, CSA, 600VAC, 20A		34	.235		5.90	15		20.90	29.50
8550	8 pole, 12-22AWG, 300VAC, CSA, 600VAC, 20A		30	.267		6.55	17		23.55	33
8555	10 pole, 12-22AWG, 300VAC, CSA, 600VAC, 20A		30	.267		7.25	17		24.25	34
8560	12 pole, 12-22AWG, 300VAC, CSA, 600VAC, 20A		30	.267		7.95	17		24.95	35
8565	Wiring device terminal strip jumper, 20A		34	.235		1.06	15		16.06	24
8570	Wiring device terminal shorting type no cover, 4 pole, 6-10AWG, 600VAC, 50A		30	.267		20	17		37	48
8575	6 pole, 6-10AWG, 600VAC, 50A		30	.267		26	17		43	54.50
8580	8 pole, 6-10AWG, 600VAC, 50A		30	.267		29.50	17		46.50	58.50
8585	12 pole, 6-10AWG, 600VAC, 50A		30	.267		33	17		50	62.50
8590	Wiring device terminal 2 shorting pin 1/2 cover, 2 pole, 8-4AWG, 600V, 45A		30	.267		21	17		38	49
8595	4 pole, 8-4AWG, 600V, 45A		30	.267		26	17		43	55
8600	6 pole, 8-4AWG, 600V, 45A		30	.267		32.50	17		49.50	61.50
8605	8 pole, 8-4AWG, 600V, 45A		30	.267		47	17		64	77.50
8610	12 pole, 8-4AWG, 600V, 45A		30	.267		56	17		73	87.50
8615	Wiring device terminal 2 shorting pin W/cover, 4 pole, 8-4AWG, 600V, 45A		30	.267		35.50	17		52.50	65
8620	6 pole, 8-4AWG, 600V, 45A		30	.267		42	17		59	72.50
8625	8 pole, 8-4AWG, 600V, 45A		30	.267		56.50	17		73.50	88
8630	12 pole, 8-4AWG, 600V, 45A		30	.267		67.50	17		84.50	100
9000	Minimum labor/equipment charge		4	2	Job		127		127	196

26 27 33 – Power Distribution Units

26 27 33.10 Power Distribution Unit Cabinet

		Crew	Daily Output	Labor-Hours	Unit	Material	Labor	Equipment	Total	Total Incl O&P
0010	**POWER DISTRIBUTION UNIT CABINET**									
0100	Power distribution unit, single cabinet, 50 kVA output, 480/208 input	1 Elec	2	4	Ea.	9,250	255		9,505	10,600
0110	75 kVA output, 480/208 input		2	4		10,000	255		10,255	11,400
0120	100 kVA output, 480/208 input		2	4		10,900	255		11,155	12,400
0130	125 kVA output, 480/208 input		1.50	5.333		12,000	340		12,340	13,700
0140	150 kVA output, 480/208 input		1.50	5.333		14,600	340		14,940	16,600
0150	200 kVA output, 480/208 input		1	8		16,100	510		16,610	18,500
0160	225 kVA output, 480/208 input		1	8		17,000	510		17,510	19,500

26 27 73 – Door Chimes

26 27 73.10 Doorbell System

		Crew	Daily Output	Labor-Hours	Unit	Material	Labor	Equipment	Total	Total Incl O&P
0010	**DOORBELL SYSTEM**, incl. transformer, button & signal									
0100	6" bell	1 Elec	4	2	Ea.	153	127		280	365
0200	Buzzer		4	2		123	127		250	330
1000	Door chimes, 2 notes		16	.500		32.50	32		64.50	84.50
1020	with ambient light		12	.667		120	42.50		162.50	197
1100	Tube type, 3 tube system		12	.667		236	42.50		278.50	325
1180	4 tube system		10	.800		475	51		526	605
1900	For transformer & button, add		5	1.600		16	102		118	175
3000	For push button only		24	.333		.92	21		21.92	33.50

26 27 Low-Voltage Distribution Equipment

26 27 73 – Door Chimes

26 27 73.10 Doorbell System	Crew	Daily Output	Labor-Hours	Unit	Material	2021 Bare Costs Labor	Equipment	Total	Total Incl O&P
3200	Bell transformer	1 Elec	16	.500	Ea.	26	32		58	77.50
9000	Minimum labor/equipment charge	▼	4	2	Job		127		127	196

26 28 Low-Voltage Circuit Protective Devices

26 28 16 – Enclosed Switches and Circuit Breakers

26 28 16.10 Circuit Breakers

		Crew	Daily Output	Labor-Hours	Unit	Material	2021 Bare Costs Labor	Equipment	Total	Total Incl O&P
0010	**CIRCUIT BREAKERS** (in enclosure)									
0100	Enclosed (NEMA 1), 600 volt, 3 pole, 30 amp	1 Elec	3.20	2.500	Ea.	550	159		709	850
0200	60 amp		2.80	2.857		700	182		882	1,050
0400	100 amp		2.30	3.478		800	222		1,022	1,225
0500	200 amp		1.50	5.333		1,900	340		2,240	2,600
0600	225 amp	▼	1.50	5.333		1,850	340		2,190	2,550
0700	400 amp	2 Elec	1.60	10	▼	3,150	635		3,785	4,450
9000	Minimum labor/equipment charge	1 Elec	4	2	Job		127		127	196

26 28 16.13 Circuit Breakers

		Crew	Daily Output	Labor-Hours	Unit	Material	2021 Bare Costs Labor	Equipment	Total	Total Incl O&P
0010	**CIRCUIT BREAKERS**									
0100	Circuit breaker current limiter, 225 to 400, ampere, 1 pole	1 Elec	48	.167	Ea.	810	10.60		820.60	905
0200	Circuit breaker current limiter, 200 to 500, ampere, 1 pole, thermal magnet		46	.174		490	11.10		501.10	555
0300	300 to 500, ampere, 1 pole, thermal magnet		46	.174		1,375	11.10		1,386.10	1,550
0400	Circuit breaker current limiter, 600 to 1000, ampere, 1 pole, tri pac	▼	46	.174	▼	1,775	11.10		1,786.10	1,975

26 28 16.20 Safety Switches

		Crew	Daily Output	Labor-Hours	Unit	Material	2021 Bare Costs Labor	Equipment	Total	Total Incl O&P
0010	**SAFETY SWITCHES**									
0100	General duty 240 volt, 3 pole NEMA 1, fusible, 30 amp	1 Elec	3.20	2.500	Ea.	65.50	159		224.50	315
0200	60 amp		2.30	3.478		111	222		333	460
0300	100 amp		1.90	4.211		193	268		461	620
0400	200 amp	▼	1.30	6.154		415	390		805	1,050
0500	400 amp	2 Elec	1.80	8.889	▼	1,050	565		1,615	2,025
1100	Heavy duty, 600 volt, 3 pole NEMA 1 nonfused									
1110	30 amp	1 Elec	3.20	2.500	Ea.	93.50	159		252.50	350
1500	60 amp		2.30	3.478		164	222		386	520
1700	100 amp		1.90	4.211		260	268		528	695
1900	200 amp	▼	1.30	6.154		400	390		790	1,050
2100	400 amp	2 Elec	1.80	8.889		920	565		1,485	1,875
2300	600 amp		1.20	13.333		1,800	850		2,650	3,275
2500	800 amp	▼	.94	17.021	▼	3,175	1,075		4,250	5,175
2700	1,200 amp	2 Elec	.80	20	Ea.	4,500	1,275		5,775	6,900
4350	600 volt, 3 pole, fusible, 30 amp	1 Elec	3.20	2.500		175	159		334	440
4380	60 amp		2.30	3.478		214	222		436	575
4400	100 amp		1.90	4.211		395	268		663	840
4420	200 amp	▼	1.30	6.154		570	390		960	1,225
4440	400 amp	2 Elec	1.80	8.889		1,475	565		2,040	2,500
4450	600 amp		1.20	13.333		2,600	850		3,450	4,150
4460	800 amp		.94	17.021		5,075	1,075		6,150	7,275
4480	1,200 amp	▼	.80	20		4,200	1,275		5,475	6,575
5510	Heavy duty, 600 volt, 3 pole 3 ph. NEMA 3R fusible, 30 amp	1 Elec	3.10	2.581		295	164		459	580
5520	60 amp		2.20	3.636		365	232		597	760
5530	100 amp		1.80	4.444		530	283		813	1,025
5540	200 amp		1.20	6.667		725	425		1,150	1,450
5550	400 amp	2 Elec	1.60	10	▼	1,800	635		2,435	2,950
5700	600 volt, 3 pole NEMA 3R nonfused									

26 28 Low-Voltage Circuit Protective Devices

26 28 16 – Enclosed Switches and Circuit Breakers

26 28 16.20 Safety Switches

		Crew	Daily Output	Labor-Hours	Unit	Material	2021 Bare Costs Labor	Equipment	Total	Total Incl O&P
5710	30 amp	1 Elec	3.10	2.581	Ea.	160	164		324	430
5900	60 amp		2.20	3.636		283	232		515	665
6100	100 amp		1.80	4.444		405	283		688	880
6300	200 amp		1.20	6.667		490	425		915	1,175
6500	400 amp	2 Elec	1.60	10		1,325	635		1,960	2,425
6700	600 amp	"	1	16		2,550	1,025		3,575	4,375
6900	600 volt, 6 pole NEMA 3R nonfused, 30 amp	1 Elec	2.70	2.963		1,150	189		1,339	1,575
7100	60 amp		2	4		1,450	255		1,705	2,000
7300	100 amp		1.50	5.333		1,900	340		2,240	2,625
7500	200 amp		1.20	6.667		4,050	425		4,475	5,100
9990	Minimum labor/equipment charge		3	2.667	Job		170		170	261

26 28 16.40 Time Switches

		Crew	Daily Output	Labor-Hours	Unit	Material	2021 Bare Costs Labor	Equipment	Total	Total Incl O&P
0010	**TIME SWITCHES**									
0100	Single pole, single throw, 24 hour dial	1 Elec	4	2	Ea.	144	127		271	355
0200	24 hour dial with reserve power		3.60	2.222		740	142		882	1,025
0300	Astronomic dial		3.60	2.222		257	142		399	500
0400	Astronomic dial with reserve power		3.30	2.424		960	154		1,114	1,275
0500	7 day calendar dial		3.30	2.424		214	154		368	470
0600	7 day calendar dial with reserve power		3.20	2.500		193	159		352	455
0700	Photo cell 2,000 watt		8	1		22	63.50		85.50	123
1080	Load management device, 4 loads		2	4		1,150	255		1,405	1,675
1100	8 loads		1	8		2,600	510		3,110	3,625
9000	Minimum labor/equipment charge		3.50	2.286	Job		146		146	224

26 28 16.50 Meter Socket Entry Hub

		Crew	Daily Output	Labor-Hours	Unit	Material	2021 Bare Costs Labor	Equipment	Total	Total Incl O&P
0010	**METER SOCKET ENTRY HUB**									
0100	Meter socket entry hub closing cap, 3/4 to 4 inch	1 Elec	40	.200	Ea.	2.63	12.75		15.38	22.50
0110	Meter socket entry hub, 3/4 conduit		40	.200		16.35	12.75		29.10	37.50
0120	1 inch conduit		40	.200		16.95	12.75		29.70	38
0130	1.25 inch conduit		40	.200		16.15	12.75		28.90	37.50
0140	1.50 inch conduit		40	.200		16.85	12.75		29.60	38
0150	2 inch conduit		40	.200		26	12.75		38.75	48
0160	2.50 inch conduit		40	.200		45.50	12.75		58.25	69.50
0170	3 inch conduit		40	.200		79	12.75		91.75	107

26 29 Low-Voltage Controllers

26 29 13 – Enclosed Controllers

26 29 13.40 Relays

		Crew	Daily Output	Labor-Hours	Unit	Material	2021 Bare Costs Labor	Equipment	Total	Total Incl O&P
0010	**RELAYS** Enclosed (NEMA 1)									
0050	600 volt AC, 1 pole, 12 amp	1 Elec	5.30	1.509	Ea.	90.50	96		186.50	248
0100	2 pole, 12 amp		5	1.600		90.50	102		192.50	257
0200	4 pole, 10 amp		4.50	1.778		121	113		234	305

26 29 23 – Variable-Frequency Motor Controllers

26 29 23.10 Variable Frequency Drives/Adj. Frequency Drives

			Crew	Daily Output	Labor-Hours	Unit	Material	2021 Bare Costs Labor	Equipment	Total	Total Incl O&P
0010	**VARIABLE FREQUENCY DRIVES/ADJ. FREQUENCY DRIVES**										
0100	Enclosed (NEMA 1), 460 volt, for 3 HP motor size	G	1 Elec	.80	10	Ea.	1,750	635		2,385	2,900
0110	5 HP motor size	G		.80	10		1,975	635		2,610	3,150
0120	7.5 HP motor size	G		.67	11.940		2,400	760		3,160	3,825
0130	10 HP motor size	G		.67	11.940		2,125	760		2,885	3,500
0140	15 HP motor size	G	2 Elec	.89	17.978		3,400	1,150		4,550	5,475
0150	20 HP motor size	G		.89	17.978		4,275	1,150		5,425	6,450

26 29 Low-Voltage Controllers

26 29 23 – Variable-Frequency Motor Controllers

26 29 23.10 Variable Frequency Drives/Adj. Frequency Drives		Crew	Daily Output	Labor-Hours	Unit	Material	2021 Bare Costs Labor	Equipment	Total	Total Incl O&P
0160	25 HP motor size	G 2 Elec	.67	23.881	Ea.	5,075	1,525		6,600	7,900
0170	30 HP motor size	G	.67	23.881		6,150	1,525		7,675	9,100
0180	40 HP motor size	G	.67	23.881		7,125	1,525		8,650	10,200
0190	50 HP motor size	G	.53	30.189		9,375	1,925		11,300	13,300
0200	60 HP motor size	G R-3	.56	35.714		11,800	2,275	340	14,415	16,800
0210	75 HP motor size	G	.56	35.714		13,600	2,275	340	16,215	18,800
0220	100 HP motor size	G	.50	40		16,400	2,550	380	19,330	22,400
0230	125 HP motor size	G	.50	40		17,800	2,550	380	20,730	23,900
0240	150 HP motor size	G	.50	40		23,300	2,550	380	26,230	29,900
0250	200 HP motor size	G	.42	47.619		32,900	3,025	455	36,380	41,400
1100	Custom-engineered, 460 volt, for 3 HP motor size	G 1 Elec	.56	14.286		3,575	910		4,485	5,325
1110	5 HP motor size	G	.56	14.286		3,625	910		4,535	5,400
1120	7.5 HP motor size	G	.47	17.021		3,075	1,075		4,150	5,050
1130	10 HP motor size	G	.47	17.021		3,250	1,075		4,325	5,275
1140	15 HP motor size	G 2 Elec	.62	25.806		4,925	1,650		6,575	7,950
1150	20 HP motor size	G	.62	25.806		2,650	1,650		4,300	5,450
1160	25 HP motor size	G	.47	34.043		5,675	2,175		7,850	9,550
1170	30 HP motor size	G	.47	34.043		6,575	2,175		8,750	10,600
1180	40 HP motor size	G	.47	34.043		7,425	2,175		9,600	11,500
1190	50 HP motor size	G	.37	43.243		11,000	2,750		13,750	16,300
1200	60 HP motor size	G R-3	.39	51.282		16,500	3,250	490	20,240	23,700
1210	75 HP motor size	G	.39	51.282		13,800	3,250	490	17,540	20,800
1220	100 HP motor size	G	.35	57.143		19,300	3,625	545	23,470	27,500
1230	125 HP motor size	G	.35	57.143		20,700	3,625	545	24,870	29,000
1240	150 HP motor size	G	.35	57.143		23,500	3,625	545	27,670	32,000
1250	200 HP motor size	G	.29	68.966		27,000	4,375	660	32,035	37,200
2000	For complex & special design systems to meet specific									
2010	requirements, obtain quote from vendor.									

26 32 Packaged Generator Assemblies

26 32 13 – Engine Generators

26 32 13.16 Gas-Engine-Driven Generator Sets

		Crew	Daily Output	Labor-Hours	Unit	Material	2021 Bare Costs Labor	Equipment	Total	Total Incl O&P
0010	**GAS-ENGINE-DRIVEN GENERATOR SETS**									
0020	Gas or gasoline operated, includes battery,									
0050	charger & muffler									
0200	7.5 kW	R-3	.83	24.096	Ea.	7,525	1,525	230	9,280	10,900
0300	11.5 kW		.71	28.169		10,700	1,775	269	12,744	14,800
0400	20 kW		.63	31.746		10,600	2,025	305	12,930	15,100
0500	35 kW		.55	36.364		15,000	2,300	345	17,645	20,400

26 33 Battery Equipment

26 33 19 – Battery Units

26 33 19.10 Battery Units

		Crew	Daily Output	Labor-Hours	Unit	Material	2021 Bare Costs Labor	2021 Bare Costs Equipment	Total	Total Incl O&P
0010	**BATTERY UNITS**									
0500	Salt water battery, 2.1 kWA, 24V, 750W, 30A, wired in crate, IP22 rated	1 Elec	16.25	.492	Ea.	940	31.50		971.50	1,075
0510	2.2 kWh, 48V, 800W, 17A, wired in crate, IP22 rated		16.25	.492		1,175	31.50		1,206.50	1,350
0520	25.9 kWh, 48V, 11700W, 240A, wired in crate, IP2X rated		16.25	.492		14,100	31.50		14,131.50	15,500
0700	Nickel iron nife battery, deep cycle, 100Ah, 12V, 10 cells		14.25	.561		1,000	36		1,036	1,150
0710	24V, 10 cells		14.25	.561		2,150	36		2,186	2,425
0720	48V, 10 cells		14.25	.561		4,025	36		4,061	4,500
0730	Nickel iron nife battery, deep cycle, 300Ah, 12V, 10 cells		12.25	.653		2,975	41.50		3,016.50	3,350
0740	24V, 10 cells		12.25	.653		5,950	41.50		5,991.50	6,625
0750	48V, 10 cells		12.25	.653		12,500	41.50		12,541.50	13,800

26 51 Interior Lighting

26 51 13 – Interior Lighting Fixtures, Lamps, and Ballasts

26 51 13.50 Interior Lighting Fixtures

		Crew	Daily Output	Labor-Hours	Unit	Material	2021 Bare Costs Labor	2021 Bare Costs Equipment	Total	Total Incl O&P
0010	**INTERIOR LIGHTING FIXTURES** Including lamps, mounting									
0030	hardware and connections									
0100	Fluorescent, C.W. lamps, troffer, recess mounted in grid, RS									
0130	Grid ceiling mount									
0200	Acrylic lens, 1' W x 4' L, two 40 watt	1 Elec	5.70	1.404	Ea.	54	89.50		143.50	197
0300	2' W x 2' L, two U40 watt		5.70	1.404		58	89.50		147.50	201
0600	2' W x 4' L, four 40 watt		4.70	1.702		65.50	108		173.50	240
0910	Acrylic lens, 1' W x 4' L, two 32 watt T8 **G**	1 Elec	5.70	1.404	Ea.	78	89.50		167.50	223
0930	2' W x 2' L, two U32 watt T8 **G**		5.70	1.404		131	89.50		220.50	281
0940	2' W x 4' L, two 32 watt T8 **G**		5.30	1.509		93	96		189	250
0950	2' W x 4' L, three 32 watt T8 **G**		5	1.600		82.50	102		184.50	248
0960	2' W x 4' L, four 32 watt T8 **G**		4.70	1.702		75.50	108		183.50	250
1000	Surface mounted, RS									
1030	Acrylic lens with hinged & latched door frame									
1100	1' W x 4' L, two 40 watt	1 Elec	7	1.143	Ea.	71	73		144	190
1200	2' W x 2' L, two U40 watt		7	1.143		76	73		149	196
1500	2' W x 4' L, four 40 watt		5.30	1.509		88	96		184	245
1501	2' W x 4' L, six 40 watt T8		5.20	1.538		88	98		186	248
2100	Strip fixture									
2130	Surface mounted									
2200	4' long, one 40 watt, RS	1 Elec	8.50	.941	Ea.	28	60		88	123
2300	4' long, two 40 watt, RS	"	8	1		50	63.50		113.50	153
2600	8' long, one 75 watt, SL	2 Elec	13.40	1.194		57.50	76		133.50	181
2700	8' long, two 75 watt, SL	"	12.40	1.290		72.50	82		154.50	206
3000	Strip, pendent mounted, industrial, white porcelain enamel									
3100	4' long, two 40 watt, RS	1 Elec	5.70	1.404	Ea.	53.50	89.50		143	196
3200	4' long, two 60 watt, HO	"	5	1.600		83	102		185	248
3300	8' long, two 75 watt, SL	2 Elec	8.80	1.818		101	116		217	289
4220	Metal halide, integral ballast, ceiling, recess mounted									
4230	prismatic glass lens, floating door									
4240	2' W x 2' L, 250 watt	1 Elec	3.20	2.500	Ea.	370	159		529	650
4250	2' W x 2' L, 400 watt	2 Elec	5.80	2.759		375	176		551	685
4260	Surface mounted, 2' W x 2' L, 250 watt	1 Elec	2.70	2.963		340	189		529	665
4270	400 watt	2 Elec	4.80	3.333		405	212		617	770
4280	High bay, aluminum reflector,									
4290	Single unit, 400 watt	2 Elec	4.60	3.478	Ea.	425	222		647	805
4300	Single unit, 1,000 watt		4	4		610	255		865	1,050

516

26 51 Interior Lighting

26 51 13 – Interior Lighting Fixtures, Lamps, and Ballasts

26 51 13.50 Interior Lighting Fixtures

		Crew	Daily Output	Labor-Hours	Unit	Material	2021 Bare Costs Labor	Equipment	Total	Total Incl O&P
4310	Twin unit, 400 watt	2 Elec	3.20	5	Ea.	825	320		1,145	1,400
4320	Low bay, aluminum reflector, 250W DX lamp	1 Elec	3.20	2.500	↓	365	159		524	645
4340	High pressure sodium integral ballast ceiling, recess mounted									
4350	prismatic glass lens, floating door									
4360	2' W x 2' L, 150 watt lamp	1 Elec	3.20	2.500	Ea.	400	159		559	685
4370	2' W x 2' L, 400 watt lamp	2 Elec	5.80	2.759		475	176		651	795
4380	Surface mounted, 2' W x 2' L, 150 watt lamp	1 Elec	2.70	2.963		485	189		674	825
4390	400 watt lamp	2 Elec	4.80	3.333	↓	545	212		757	920
4400	High bay, aluminum reflector,									
4410	Single unit, 400 watt lamp	2 Elec	4.60	3.478	Ea.	390	222		612	770
4430	Single unit, 1,000 watt lamp	"	4	4		545	255		800	990
4440	Low bay, aluminum reflector, 150 watt lamp	1 Elec	3.20	2.500	↓	335	159		494	615
4450	Incandescent, high hat can, round alzak reflector, prewired									
4470	100 watt	1 Elec	8	1	Ea.	68.50	63.50		132	173
4480	150 watt		8	1		104	63.50		167.50	213
4500	300 watt	↓	6.70	1.194	↓	241	76		317	380
4600	Square glass lens with metal trim, prewired									
4630	100 watt	1 Elec	6.70	1.194	Ea.	55	76		131	178
4680	150 watt		6.70	1.194		98	76		174	225
4700	200 watt		6.70	1.194		98	76		174	225
6010	Vapor tight, incandescent, ceiling mounted, 200 watt		6.20	1.290		70	82		152	203
6100	Fluorescent, surface mounted, 2 lamps, 4' L, RS, 40 watt		3.20	2.500		119	159		278	375
6850	Vandalproof, surface mounted, fluorescent, two 32 watt T8 [G]		3.20	2.500		268	159		427	540
6860	Incandescent, one 150 watt		8	1		90	63.50		153.50	197
6900	Mirror light, fluorescent, RS, acrylic enclosure, two 40 watt		8	1		117	63.50		180.50	226
6910	One 40 watt		8	1		100	63.50		163.50	209
6920	One 20 watt	↓	12	.667	↓	87.50	42.50		130	161
7500	Ballast replacement, by weight of ballast, to 15' high									
7520	Indoor fluorescent, less than 2 lb.	1 Elec	10	.800	Ea.	27	51		78	109
7540	Two 40W, watt reducer, 2 to 5 lb.		9.40	.851		68.50	54		122.50	159
7560	Two F96 slimline, over 5 lb.		8	1		109	63.50		172.50	217
7580	Vaportite ballast, less than 2 lb.		9.40	.851		27	54		81	114
7600	2 lb. to 5 lb.		8.90	.899		68.50	57.50		126	164
7620	Over 5 lb.		7.60	1.053		109	67		176	222
7630	Electronic ballast for two tubes		8	1		37.50	63.50		101	140
7640	Dimmable ballast one-lamp [G]		8	1		105	63.50		168.50	214
7650	Dimmable ballast two-lamp [G]		7.60	1.053		110	67		177	224
7660	Dimmable ballast three-lamp		6.90	1.159	↓	126	74		200	252
9000	Minimum labor/equipment charge	↓	3	2.667	Job		170		170	261

26 51 13.70 Residential Fixtures

		Crew	Daily Output	Labor-Hours	Unit	Material	2021 Bare Costs Labor	Equipment	Total	Total Incl O&P
0010	**RESIDENTIAL FIXTURES**									
0400	Fluorescent, interior, surface, circline, 32 watt & 40 watt	1 Elec	20	.400	Ea.	178	25.50		203.50	235
0500	2' x 2', two U-tube 32 watt T8		8	1		138	63.50		201.50	250
0700	Shallow under cabinet, two 20 watt		16	.500		65.50	32		97.50	121
0900	Wall mounted, 4' L, two 32 watt T8, with baffle		10	.800		163	51		214	258
2000	Incandescent, exterior lantern, wall mounted, 60 watt		16	.500		64.50	32		96.50	120
2100	Post light, 150 W, with 7' post		4	2		254	127		381	475
2500	Lamp holder, weatherproof with 150 W PAR		16	.500		35.50	32		67.50	88
2550	With reflector and guard		12	.667		63	42.50		105.50	134
2600	Interior pendent, globe with shade, 150 W		20	.400	↓	180	25.50		205.50	237
9000	Minimum labor/equipment charge	↓	4	2	Job		127		127	196

26 51 19 – LED Interior Lighting

26 51 19.10 LED Interior Lighting	Crew	Daily Output	Labor-Hours	Unit	Material	2021 Bare Costs Labor	Equipment	Total	Total Incl O&P
0010 **LED INTERIOR LIGHTING**									
7000 Interior LED fixts, tape rope kit, 120V, 3' strip, 4W, w/5' cord, 3,000K	1 Elec	24	.333	Ea.	20.50	21		41.50	55
7010 5,000K		24	.333		11.10	21		32.10	44.50
7020 Interior LED fixts, tape rope kit, 120V, 6' strip, 8W, w/5' cord, 3,000K		23	.348		15.20	22		37.20	50.50
7030 5,000K		23	.348		22.50	22		44.50	59
7040 Interior LED fixts, tape rope kit, 120V, 13' strip, 15W, w/5' cord, 3,000K		22	.364		26.50	23		49.50	64.50
7050 5,000K		22	.364		28	23		51	66.50
7060 Interior LED fixts, tape rope kit, 120V, 20' strip, 22W, w/5' cord, 3,000K		21	.381		40.50	24.50		65	82
7070 5,000K		21	.381		40.50	24.50		65	82
7080 Interior LED fixts, tape rope kit, 120V, 33' strip, 37W, w/5' cord, 3,000K		20	.400		60	25.50		85.50	105
7090 5,000K		20	.400		60	25.50		85.50	105
7100 Interior LED fixts, tape rope kit, 120V, 50' strip, 55W, w/5' cord, 3,000K		19	.421		90.50	27		117.50	141
7110 5,000K		19	.421		129	27		156	183
7120 Interior LED tape light, starstrand, dimmable, 12V, super star, 2", 3,000K		25	.320		11.15	20.50		31.65	44
7130 Ultra star, 2", 3,000K		25	.320		14.10	20.50		34.60	47
7140 Rainbow star, 2", 3,000K		25	.320		11.30	20.50		31.80	44
7150 Interior LED tape light, starstrand, dimmable, 24V, elite star, 4", 35K		25	.320		5.10	20.50		25.60	37
7160 12", 35K		25	.320		19.35	20.50		39.85	53
7170 60", 35K		23	.348		52.50	22		74.50	91.50
7180 120", 35K		21	.381		101	24.50		125.50	149
7190 240", 35K		18	.444		310	28.50		338.50	385
7200 Interior LED tape lighting channel , 36", AL flex, starstrand, elite star		48	.167		10.55	10.60		21.15	28

26 52 Safety Lighting

26 52 13 – Emergency and Exit Lighting

26 52 13.10 Emergency Lighting and Battery Units

	Crew	Daily Output	Labor-Hours	Unit	Material	2021 Bare Costs Labor	Equipment	Total	Total Incl O&P
0010 **EMERGENCY LIGHTING AND BATTERY UNITS**									
0300 Emergency light units, battery operated									
0350 Twin sealed beam light, 25 W, 6 V each									
0500 Lead battery operated	1 Elec	4	2	Ea.	139	127		266	350
0700 Nickel cadmium battery operated	"	4	2	"	340	127		467	570
9000 Minimum labor/equipment charge		4	2	Job		127		127	196

26 52 13.16 Exit Signs

	Crew	Daily Output	Labor-Hours	Unit	Material	2021 Bare Costs Labor	Equipment	Total	Total Incl O&P
0010 **EXIT SIGNS**									
0080 Exit light ceiling or wall mount, incandescent, single face	1 Elec	8	1	Ea.	72.50	63.50		136	178
0100 Double face	"	6.70	1.194		50.50	76		126.50	173
1500 Exit sign, 12 V, 1 face, remote end mounted (Type 602A1)	R-19	18	1.111		72.50	71		143.50	189
1780 With emergency battery, explosion proof	"	7.70	2.597		3,600	166		3,766	4,200
9000 Minimum labor/equipment charge	1 Elec	4	2	Job		127		127	196

26 55 Special Purpose Lighting

26 55 33 – Hazard Warning Lighting

26 55 33.10 Warning Beacons

		Crew	Daily Output	Labor-Hours	Unit	Material	2021 Bare Costs Labor	Equipment	Total	Total Incl O&P
0010	**WARNING BEACONS**									
0015	Surface mount with colored or clear lens									
0100	Rotating beacon									
0110	120V, 40 watt halogen	1 Elec	3.50	2.286	Ea.	233	146		379	480
0120	24V, 20 watt halogen	"	3.50	2.286	"	239	146		385	485
0200	Steady beacon									
0210	120V, 40 watt halogen	1 Elec	3.50	2.286	Ea.	236	146		382	485
0220	24V, 20 watt		3.50	2.286		236	146		382	485
0230	12V DC, incandescent		3.50	2.286		119	146		265	355
0300	Flashing beacon									
0310	120V, 40 watt halogen	1 Elec	3.50	2.286	Ea.	237	146		383	485
0320	24V, 20 watt halogen		3.50	2.286		222	146		368	470
0410	12V DC with two 6V lantern batteries		7	1.143		221	73		294	355

26 55 59 – Display Lighting

26 55 59.10 Track Lighting

		Crew	Daily Output	Labor-Hours	Unit	Material	2021 Bare Costs Labor	Equipment	Total	Total Incl O&P
0010	**TRACK LIGHTING**									
0080	Track, 1 circuit, 4' section	1 Elec	6.70	1.194	Ea.	50.50	76		126.50	173
0100	8' section	2 Elec	10.60	1.509		66	96		162	221
0300	3 circuits, 4' section	1 Elec	6.70	1.194		105	76		181	233
0400	8' section	2 Elec	10.60	1.509		131	96		227	292
9000	Minimum labor/equipment charge	1 Elec	3	2.667	Job		170		170	261

26 55 63 – Detention Lighting

26 55 63.10 Detention Lighting Fixtures

		Crew	Daily Output	Labor-Hours	Unit	Material	2021 Bare Costs Labor	Equipment	Total	Total Incl O&P
0010	**DETENTION LIGHTING FIXTURES**									
1000	Surface mounted, cold rolled steel, 14 ga., 1' x 4'	1 Elec	5.40	1.481	Ea.	855	94.50		949.50	1,100
1010	2' x 2'		5.40	1.481		825	94.50		919.50	1,050
1020	2' x 4'		5	1.600		1,000	102		1,102	1,250
1100	12 ga., 1' x 4'		5.40	1.481		1,075	94.50		1,169.50	1,325
1110	2' x 2'		5.40	1.481		825	94.50		919.50	1,050
1120	2' x 4'		5	1.600		1,075	102		1,177	1,325
3000	Fluorescent vandal resistant light fixture, high abuse troffer		5.40	1.481		222	94.50		316.50	390
3010	50" L x 8" W x 4" H		5.40	1.481		192	94.50		286.50	355
3020	32W		5.40	1.481		280	94.50		374.50	455
3030	T8/32W		5.40	1.481		203	94.50		297.50	370
3040	No lamp		5.60	1.429		335	91		426	510
3050	Lamp included		5.40	1.481		385	94.50		479.50	570
3060	16 ga. cold rolled steel		5.40	1.481		385	94.50		479.50	570
3070	16 ga. cold rolled steel, ceiling/wall mount		5.40	1.481		320	94.50		414.50	500
3080	w/lamp, white		5.40	1.481		251	94.50		345.50	420
3090	wo/lamp, bronze		5.60	1.429		223	91		314	385
3100	w/lamp 6" L x 4-1/2" W x 8-1/2" H		5.40	1.481		38	94.50		132.50	187
3110	w/lamp 5-7/8" W x 4-1/2" D x 8-1/2" H clear		5.40	1.481		65.50	94.50		160	217
3120	w/lamp white		5.40	1.481		320	94.50		414.50	495
3130	Tall wallpack, w/lamp, bronze		5.40	1.481		111	94.50		205.50	267
3140	Tall wallpack, w/lamp, white		5.40	1.481		110	94.50		204.50	266

26 56 Exterior Lighting

26 56 19 – LED Exterior Lighting

26 56 19.55 Roadway LED Luminaire

		Crew	Daily Output	Labor-Hours	Unit	Material	2021 Bare Costs Labor	2021 Bare Costs Equipment	Total	Total Incl O&P
0010	**ROADWAY LED LUMINAIRE**									
2000	Exterior fixtures, LED roadway, type 2, 50W, neutral slipfitter bronze	2 Elec	5.20	3.077	Ea.	755	196		951	1,125
2010	3, 50W, neutral slipfitter bronze		5.20	3.077		755	196		951	1,125
2020	4, 50W, neutral slipfitter bronze		5.20	3.077		755	196		951	1,125
2030	Exterior fixtures, LED roadway, type 2, 78W, neutral slipfitter bronze		5.20	3.077		725	196		921	1,100
2040	3, 78W, neutral slipfitter bronze		5.20	3.077		725	196		921	1,100
2050	4, 78W, neutral slipfitter bronze		5.20	3.077		740	196		936	1,125
2060	Exterior fixtures, LED roadway, type 2, 105W, neutral slipfitter bronze		5.20	3.077		865	196		1,061	1,250
2070	3, 105W, neutral slipfitter bronze		5.20	3.077		845	196		1,041	1,225
2080	4, 105W, neutral slipfitter bronze		5.20	3.077		865	196		1,061	1,250
2090	Exterior fixtures, LED roadway, type 2, 125W, neutral slipfitter bronze	▼	5.20	3.077	▼	840	196		1,036	1,225

26 56 23 – Area Lighting

26 56 23.10 Exterior Fixtures

		Crew	Daily Output	Labor-Hours	Unit	Material	2021 Bare Costs Labor	2021 Bare Costs Equipment	Total	Total Incl O&P
0010	**EXTERIOR FIXTURES** With lamps									
0200	Wall mounted, incandescent, 100 watt	1 Elec	8	1	Ea.	168	63.50		231.50	283
0400	Quartz, 500 watt		5.30	1.509		61.50	96		157.50	216
1100	Wall pack, low pressure sodium, 35 watt		4	2		207	127		334	425
1150	55 watt		4	2		243	127		370	465
1160	High pressure sodium, 70 watt		4	2		208	127		335	425
1170	150 watt		4	2		219	127		346	435
1175	High pressure sodium, 250 watt		4	2		219	127		346	435
1180	Metal halide, 175 watt		4	2		223	127		350	440
1190	250 watt		4	2		253	127		380	475
1195	400 watt		4	2		410	127		537	650
1250	Induction lamp, 40 watt		4	2		520	127		647	770
1260	80 watt		4	2		600	127		727	855
1278	LED, poly lens, 26 watt		4	2		1,675	127		1,802	2,050
1280	110 watt		4	2		820	127		947	1,100
1500	LED, glass lens, 13 watt	▼	4	2	▼	1,675	127		1,802	2,050

26 56 33 – Walkway Lighting

26 56 33.10 Walkway Luminaire

		Crew	Daily Output	Labor-Hours	Unit	Material	2021 Bare Costs Labor	2021 Bare Costs Equipment	Total	Total Incl O&P
0010	**WALKWAY LUMINAIRE**									
9000	Minimum labor/equipment charge	1 Elec	3.75	2.133	Job		136		136	209

26 56 33.55 Walkway LED Luminaire

		Crew	Daily Output	Labor-Hours	Unit	Material	2021 Bare Costs Labor	2021 Bare Costs Equipment	Total	Total Incl O&P
0010	**WALKWAY LED LUMINAIRE**									
0600	LED bollard pole, 6" diam. x 36" H, cast alum, surf. mtd., 8W, 6,000K, 120V	1 Elec	8.50	.941	Ea.	390	60		450	520
0610	6" diam. x 42" H, cast alum, surf. mtd., 12W, 5,100K, 120V		8.50	.941		625	60		685	775
0620	6" diam. x 42" H, cast alum, surf. mtd., 18W, 5,100K, 120V		8.50	.941		645	60		705	795
0630	6" diam. x 42" H, cast alum, surf. mtd., 24W, 5,100K, 120V		8.50	.941		660	60		720	815
0640	9.5" square x 42" H, cast resin, surf. mtd., A19, 9W, 120V		8.50	.941		615	60		675	770
0650	9.75" diam. x 42" H, cast resin, side mount, dome, 9W, 120V		8.50	.941		700	60		760	860
0660	9.75" diam. x 43" H, cast resin, burial base, dome, 9W, 120V		8.50	.941		700	60		760	860
0670	8" diam. x 36" H, concrete security base, w/rebar, 8W, HID		8	1		1,050	63.50		1,113.50	1,250
0680	10" diam. x 49" H, concrete base, w/Sch 40 Steel pipe, 8W		6.75	1.185		1,175	75.50		1,250.50	1,425
0690	12" diam. x 48" H, concrete base, w/Sch 40 Steel pipe, 8W		6.75	1.185		1,350	75.50		1,425.50	1,600
0700	8" diam. x 36" H, concrete, cast aluminum dome, 90 degree, 4.1W		7	1.143		1,050	73		1,123	1,250
0710	8" diam. x 36" H, concrete, cast aluminum dome, 180 degree, 4.1W		7	1.143		1,275	73		1,348	1,525
0720	8" diam. x 48" H, concrete, cast aluminum dome, 360 degree, 4.1W	▼	7	1.143	▼	1,575	73		1,648	1,850

26 56 Exterior Lighting

26 56 36 – Flood Lighting

26 56 36.20 Floodlights	Crew	Daily Output	Labor- Hours	Unit	Material	2021 Bare Costs Labor	Equipment	Total	Total Incl O&P
0010 **FLOODLIGHTS** with ballast and lamp,									
1290 floor mtd, mount with swivel bracket									
1400 Pole mounted, pole not included									
2250 Low pressure sodium, 55 watt	1 Elec	2.70	2.963	Ea.	420	189		609	755
2270 90 watt	"	2	4	"	600	255		855	1,050

26 61 Lighting Systems and Accessories

26 61 23 – Lamps Applications

26 61 23.10 Lamps		Crew	Daily Output	Labor- Hours	Unit	Material	2021 Bare Costs Labor	Equipment	Total	Total Incl O&P
0010 **LAMPS**										
0080 Fluorescent, rapid start, cool white, 2' long, 20 watt		1 Elec	1	8	C	283	510		793	1,100
0100 4' long, 40 watt			.90	8.889		297	565		862	1,200
0170 4' long, 34 watt energy saver	G		.90	8.889		3.99	565		568.99	875
0176 2' long, T8, 17 watt energy saver	G		1	8		365	510		875	1,175
0178 3' long, T8, 25 watt energy saver	G		.90	8.889		390	565		955	1,300
0180 4' long, T8, 32 watt energy saver	G		.90	8.889		710	565		1,275	1,650
0560 Twin tube compact lamp	G		.90	8.889		294	565		859	1,200
0570 Double twin tube compact lamp	G		.80	10		705	635		1,340	1,750
0600 Mercury vapor, mogul base, deluxe white, 100 watt			.30	26.667		14,600	1,700		16,300	18,700
0800 400 watt		1 Elec	.30	26.667	C	10,700	1,700		12,400	14,400
1000 Metal halide, mogul base, 175 watt			.30	26.667		1,125	1,700		2,825	3,825
1100 250 watt			.30	26.667		2,825	1,700		4,525	5,700
1350 High pressure sodium, 70 watt			.30	26.667		1,800	1,700		3,500	4,575
1370 150 watt			.30	26.667		1,650	1,700		3,350	4,400
1500 Low pressure sodium, 35 watt			.30	26.667		9,075	1,700		10,775	12,600
1600 90 watt			.30	26.667		14,100	1,700		15,800	18,100
1800 Incandescent, interior, A21, 100 watt			1.60	5		390	320		710	920
1900 A21, 150 watt			1.60	5		150	320		470	655
2300 R30, 75 watt			1.30	6.154		745	390		1,135	1,425
2500 Exterior, PAR 38, 75 watt			1.30	6.154		1,400	390		1,790	2,150
2600 PAR 38, 150 watt			1.30	6.154		2,150	390		2,540	2,950
9000 Minimum labor/equipment charge			4	2	Job		127		127	196

Division Notes

	CREW	DAILY OUTPUT	LABOR-HOURS	UNIT	BARE COSTS				TOTAL INCL O&P
					MAT.	LABOR	EQUIP.	TOTAL	

Estimating Tips

27 20 00 Data Communications
27 30 00 Voice Communications
27 40 00 Audio-Video Communications

- When estimating material costs for special systems, it is always prudent to obtain manufacturers' quotations for equipment prices and special installation requirements that may affect the total cost.

- For cost modifications for elevated tray installation, add the percentages to labor according to the height of the installation and only to the quantities exceeding the different height levels, not to the total tray quantities. Refer to subdivision 26 01 02.20 for labor adjustment factors.

- Do not overlook the costs for equipment used in the installation. If scissor lifts and boom lifts are available in the field, contractors may use them in lieu of the proposed ladders and rolling staging.

Reference Numbers

Reference numbers are shown at the beginning of some major classifications. These numbers refer to related items in the Reference Section. The reference information may be an estimating procedure, an alternate pricing method, or technical information.

Note: Not all subdivisions listed here necessarily appear. ■

Same Data. Simplified.

Enjoy the convenience and efficiency of accessing your costs anywhere:

- **Skip the multiplier** by setting your location
- **Quickly search,** edit, favorite and share costs
- **Stay on top of price changes** with automatic updates

Discover more at rsmeans.com/online

27 13 23.13 Communications Optical Fiber	Crew	Daily Output	Labor-Hours	Unit	Material	2021 Bare Costs Labor	Equipment	Total	Total Incl O&P
0010 **COMMUNICATIONS OPTICAL FIBER**									
0040 Specialized tools & techniques cause installation costs to vary.									
0070 Fiber optic, cable, bulk simplex, single mode	1 Elec	8	1	C.L.F.	22.50	63.50		86	123
0080 Multi mode		8	1		29.50	63.50		93	131
0090 4 strand, single mode		7.34	1.090		38	69.50		107.50	149
0095 Multi mode		7.34	1.090		50.50	69.50		120	163
0100 12 strand, single mode		6.67	1.199		72	76.50		148.50	196
0105 Multi mode		6.67	1.199		96.50	76.50		173	223
0150 Jumper				Ea.	31.50			31.50	34.50
0200 Pigtail					38			38	41.50
0300 Connector	1 Elec	24	.333		25	21		46	60.50
0350 Finger splice		32	.250		37.50	15.95		53.45	65.50
0400 Transceiver (low cost bi-directional)		8	1		440	63.50		503.50	585
0450 Rack housing, 4 rack spaces, 12 panels (144 fibers)		2	4		1,075	255		1,330	1,575
0500 Patch panel, 12 ports		6	1.333		315	85		400	475
0600 Cable connector panel, 6 fiber		40	.200		55.50	12.75		68.25	80.50
1000 Cable, 62.5 microns, direct burial, 4 fiber	R-15	1200	.040	L.F.	.77	2.50	.23	3.50	4.95
1020 Indoor, 2 fiber	R-19	1000	.020		.49	1.28		1.77	2.50
1040 Outdoor, aerial/duct	"	1670	.012		.64	.76		1.40	1.87
1060 50 microns, direct burial, 8 fiber	R-22	4000	.009		1.28	.54		1.82	2.24
1080 12 fiber		4000	.009		1.52	.54		2.06	2.50
1100 Indoor, 12 fiber		759	.049		1.38	2.86		4.24	5.90
1120 Connectors, 62.5 micron cable, transmission	R-19	40	.500	Ea.	14.90	32		46.90	65.50
1140 Cable splice	"	40	.500	"	17.35	32		49.35	68
1160 125 micron cable, transmission	R-19	16	1.250	Ea.	15.65	80		95.65	140
1180 Receiver, 1.2 mile range		20	1		340	64		404	475
1200 1.9 mile range		20	1		226	64		290	345
1220 6.2 mile range		5	4		375	255		630	805
1240 Transmitter, 1.2 mile range		20	1		375	64		439	515
1260 1.9 mile range		20	1		284	64		348	410
1280 6.2 mile range		5	4		1,175	255		1,430	1,700
1300 Modem, 1.2 mile range		5	4		200	255		455	610
1320 6.2 mile range		5	4		375	255		630	805
1340 1.9 mile range, 12 channel		5	4		2,100	255		2,355	2,725
1360 Repeater, 1.2 mile range		10	2		340	128		468	570
1380 1.9 mile range		10	2		500	128		628	745
1400 6.2 mile range		5	4		905	255		1,160	1,375
1420 1.2 mile range, digital		5	4		460	255		715	895
1510 1/4", black, non-metallic, flexible, tube, liquid tight		8	2.500	C.L.F.	221	160		381	490
1520 3/8"		8	2.500		130	160		290	390
1530 1/2"		7.34	2.725		129	174		303	410
1540 3/4"		7.34	2.725		234	174		408	525
1550 1"		7.34	2.725		365	174		539	665
1560 1-1/4"		6.67	2.999		600	191		791	955
1570 1-1/2"		6.67	2.999		695	191		886	1,050
1580 2"		6.35	3.150		1,100	201		1,301	1,525
2040 Fiber optic cable, 48 strand, single mode, armored, Plenum		1.67	11.976	M.L.F.	5,050	765		5,815	6,725
2042 72 strand		1.50	13.307		7,050	850		7,900	9,050
2044 96 strand		1.42	14.085		12,500	900		13,400	15,100
2046 144 strand		1.50	13.307		18,800	850		19,650	22,000

27 15 Communications Horizontal Cabling

27 15 01 – Communications Horizontal Cabling Applications

27 15 01.23 Audio-Video Communications Horizontal Cabling	Crew	Daily Output	Labor-Hours	Unit	Material	2021 Bare Costs Labor	Equipment	Total	Total Incl O&P
0010 **AUDIO-VIDEO COMMUNICATIONS HORIZONTAL CABLING**									
3000 S-video cable, MD4, 1M Pro interconnects, 3'	1 Elec	38	.211	Ea.	21	13.40		34.40	43.50
3010 6'		38	.211		29.50	13.40		42.90	53
3020 9'		38	.211		30	13.40		43.40	53.50
3030 16'		36	.222		50	14.15		64.15	76.50
3040 22'		36	.222		58.50	14.15		72.65	86
3300 HDMI cable, coaxial, HDMI/HDMI, Non plenum, 3', 24 kt terminations		40	.200		13.55	12.75		26.30	34.50
3310 6', 24 kt terminations		40	.200		15.80	12.75		28.55	37
3320 HDMI cable, coaxial, HDMI/HDMI metal, Non plenum, 3', 24 kt terminations		40	.200		16.40	12.75		29.15	37.50
3330 6', 24 kt terminations		40	.200		19.05	12.75		31.80	40.50
3340 15', 24 kt terminations		38	.211		26.50	13.40		39.90	49.50
3350 25', 24 kt terminations		38	.211		34.50	13.40		47.90	58.50

27 15 10 – Special Communications Cabling

27 15 10.23 Sound and Video Cables and Fittings

	Crew	Daily Output	Labor-Hours	Unit	Material	2021 Bare Costs Labor	Equipment	Total	Total Incl O&P
0010 **SOUND AND VIDEO CABLES & FITTINGS**									
1250 Nonshielded, #22-2 conductor	1 Elec	10	.800	C.L.F.	14.25	51		65.25	94

27 15 13 – Communications Copper Horizontal Cabling

27 15 13.13 Communication Cables and Fittings

	Crew	Daily Output	Labor-Hours	Unit	Material	2021 Bare Costs Labor	Equipment	Total	Total Incl O&P
0010 **COMMUNICATION CABLES AND FITTINGS**									
5000 High performance unshielded twisted pair (UTP)									
7000 Category 5, #24, 4 pair solid, PVC jacket	1 Elec	7	1.143	C.L.F.	57	73		130	175
7100 4 pair solid, plenum		7	1.143		17.05	73		90.05	131
7200 4 pair stranded, PVC jacket		7	1.143		27.50	73		100.50	142
7210 Category 5e, #24, 4 pair solid, PVC jacket		7	1.143		25	73		98	140
7212 4 pair solid, plenum		7	1.143		45.50	73		118.50	162
7214 4 pair stranded, PVC jacket		7	1.143		19.30	73		92.30	134
7240 Category 6, #24, 4 pair solid, PVC jacket		7	1.143		52.50	73		125.50	170
7242 4 pair solid, plenum		7	1.143		12.05	73		85.05	125
7244 4 pair stranded, PVC jacket		7	1.143		19.90	73		92.90	134
7300 Connector, RJ45, category 5		80	.100	Ea.	.75	6.35		7.10	10.65
7302 Shielded RJ45, category 5		72	.111		1.59	7.10		8.69	12.60
7310 Jack, UTP RJ45, category 3		72	.111		.54	7.10		7.64	11.45
7312 Category 5		65	.123		6.45	7.85		14.30	19.15
7314 Category 5e		65	.123		5.15	7.85		13	17.70
7316 Category 6		65	.123		3.80	7.85		11.65	16.25
7402 RJ45, green		16.50	.485		6.40	31		37.40	54.50
7408 RJ45, orange		16.50	.485		6.50	31		37.50	54.50
7420 5M, RJ45, 568B, cord set		16.50	.485		46.50	31		77.50	99
7422 3M, RJ45, 568B, cord set		16.50	.485		31.50	31		62.50	82.50
7424 1M, RJ45, 568B, cord set		16.50	.485		29.50	31		60.50	80
7450 M12 TO RJ45, 2M ultra-loch D-mode, cord set		15.50	.516		45	33		78	100

27 32 Voice Communications Terminal Equipment

27 32 26 – Ring-Down Emergency Telephones

27 32 26.10 Emergency Phone Stations

27 32 26.10 Emergency Phone Stations	Crew	Daily Output	Labor-Hours	Unit	Material	2021 Bare Costs Labor	Equipment	Total	Total Incl O&P
0010 **EMERGENCY PHONE STATIONS**									
0100 Hands free emergency speaker phone w/LED strobe	1 Elec	7.22	1.108	Ea.	860	70.50		930.50	1,050
0110 Wall/pole mounted emergency phone station w/LED strobe	"	4.60	1.739		1,525	111		1,636	1,850
2000 Call station, 30", pole mounted emergency phone, 120V/ VoIP, blue beacon	2 Elec	5	3.200		1,100	204		1,304	1,525
2010 Call station, 36"		5	3.200		1,575	204		1,779	2,050
2020 Call station, 5'-8"		4.80	3.333		2,650	212		2,862	3,250
2030 Call station, 9'		4.80	3.333		2,675	212		2,887	3,275

27 41 Audio-Video Systems

27 41 33 – Master Antenna Television Systems

27 41 33.10 TV Systems

	Crew	Daily Output	Labor-Hours	Unit	Material	2021 Bare Costs Labor	Equipment	Total	Total Incl O&P
0010 **TV SYSTEMS,** not including rough-in wires, cables & conduits									
5000 Antenna, small	1 Elec	6	1.333	Ea.	55	85		140	191
5100 Large	"	4	2	"	230	127		357	450

27 51 Distributed Audio-Video Communications Systems

27 51 16 – Public Address Systems

27 51 16.10 Public Address System

	Crew	Daily Output	Labor-Hours	Unit	Material	2021 Bare Costs Labor	Equipment	Total	Total Incl O&P
0010 **PUBLIC ADDRESS SYSTEM**									
0100 Conventional, office	1 Elec	5.33	1.501	Speaker	167	95.50		262.50	330
0200 Industrial		2.70	2.963	"	325	189		514	645
9000 Minimum labor/equipment charge		3.50	2.286	Job		146		146	224

27 51 19 – Sound Masking Systems

27 51 19.10 Sound System

	Crew	Daily Output	Labor-Hours	Unit	Material	2021 Bare Costs Labor	Equipment	Total	Total Incl O&P
0010 **SOUND SYSTEM,** not including rough-in wires, cables & conduits									
2000 Intercom, 30 station capacity, master station	2 Elec	2	8	Ea.	2,275	510		2,785	3,275
2020 10 station capacity	"	4	4		1,425	255		1,680	1,975
3600 House telephone, talking station	1 Elec	1.60	5		545	320		865	1,100
3800 Press to talk, release to listen	"	5.30	1.509		145	96		241	305
4000 System-on button					76			76	84
4200 Door release	1 Elec	4	2		136	127		263	345
4400 Combination speaker and microphone		8	1		232	63.50		295.50	355
4600 Termination box		3.20	2.500		73	159		232	325
4800 Amplifier or power supply		5.30	1.509		840	96		936	1,075
5000 Vestibule door unit		16	.500	Name	161	32		193	226
5200 Strip cabinet		27	.296	Ea.	291	18.85		309.85	350
5400 Directory		16	.500	"	156	32		188	220
9000 Minimum labor/equipment charge		3.50	2.286	Job		146		146	224

Estimating Tips

- When estimating material costs for electronic safety and security systems, it is always prudent to obtain manufacturers' quotations for equipment prices and special installation requirements that may affect the total cost.

- Fire alarm systems consist of control panels, annunciator panels, batteries with rack, charger, and fire alarm actuating and indicating devices. Some fire alarm systems include speakers, telephone lines, door closer controls, and other components. Be careful not to overlook the costs related to installation for these items. Also be aware of costs for integrated automation instrumentation and terminal devices, control equipment, control wiring, and programming. Insurance underwriters may have specific requirements for the type of materials to be installed or design requirements based on the hazard to be protected. Local jurisdictions may have requirements not covered by code. It is advisable to be aware of any special conditions.

- Security equipment includes items such as CCTV, access control, and other detection and identification systems to perform alert and alarm functions. Be sure to consider the costs related to installation for this security equipment, such as for integrated automation instrumentation and terminal devices, control equipment, control wiring, and programming.

Reference Numbers

Reference numbers are shown at the beginning of some major classifications. These numbers refer to related items in the Reference Section. The reference information may be an estimating procedure, an alternate pricing method, or technical information.

Same Data. Simplified.

Enjoy the convenience and efficiency of accessing your costs anywhere:

- **Skip the multiplier** by setting your location
- **Quickly search,** edit, favorite and share costs
- **Stay on top of price changes** with automatic updates

Discover more at rsmeans.com/online

---

28 15 Access Control Hardware Devices

28 15 11 – Integrated Credential Readers and Field Entry Management

28 15 11.11 Standard Card Readers

		Crew	Daily Output	Labor-Hours	Unit	Material	2021 Bare Costs Labor	2021 Bare Costs Equipment	Total	Total Incl O&P
0010	**STANDARD CARD READERS**									
0015	Card key access									
0020	Computerized system, processor, proximity reader and cards									
0030	Does not include door hardware, lockset or wiring									
0040	Card key system for 1 door				Ea.	1,675			1,675	1,825
0060	Card key system for 2 doors					2,400			2,400	2,625
0080	Card key system for 4 doors					2,600			2,600	2,875
0100	Processor for card key access system					1,225			1,225	1,350
0160	Magnetic lock for electric access, 600 pound holding force					192			192	212
0170	Magnetic lock for electric access, 1200 pound holding force					192			192	212
0200	Proximity card reader					197			197	217

28 15 11.15 Biometric Identity Devices

		Crew	Daily Output	Labor-Hours	Unit	Material	2021 Bare Costs Labor	2021 Bare Costs Equipment	Total	Total Incl O&P
0010	**BIOMETRIC IDENTITY DEVICES**									
0220	Hand geometry scanner, mem of 512 users, excl. striker/power	1 Elec	3	2.667	Ea.	2,450	170		2,620	2,950
0230	Memory upgrade for, adds 9,700 user profiles		8	1		440	63.50		503.50	580
0240	Adds 32,500 user profiles		8	1		580	63.50		643.50	740
0250	Prison type, memory of 256 users, excl. striker/power		3	2.667		2,800	170		2,970	3,325
0260	Memory upgrade for, adds 3,300 user profiles		8	1		410	63.50		473.50	550
0270	Adds 9,700 user profiles		8	1		505	63.50		568.50	655
0280	Adds 27,900 user profiles		8	1		770	63.50		833.50	950
0290	All weather, mem of 512 users, excl. striker/power		3	2.667		4,425	170		4,595	5,100
0300	Facial & fingerprint scanner, combination unit, excl. striker/power		3	2.667		940	170		1,110	1,275
0310	Access for, for initial setup, excl. striker/power		3	2.667		1,375	170		1,545	1,775

28 18 Security Access Detection Equipment

28 18 11 – Security Access Metal Detectors

28 18 11.13 Security Access Metal Detectors

		Crew	Daily Output	Labor-Hours	Unit	Material	2021 Bare Costs Labor	2021 Bare Costs Equipment	Total	Total Incl O&P
0010	**SECURITY ACCESS METAL DETECTORS**									
0240	Metal detector, hand-held, wand type, unit only				Ea.	129			129	138
0250	Metal detector, walk through portal type, single zone	1 Elec	2	4		3,850	255		4,105	4,625
0260	Multi-zone	"	2	4		4,975	255		5,230	5,875

28 18 13 – Security Access X-Ray Equipment

28 18 13.16 Security Access X-Ray Equipment

		Crew	Daily Output	Labor-Hours	Unit	Material	2021 Bare Costs Labor	2021 Bare Costs Equipment	Total	Total Incl O&P
0010	**SECURITY ACCESS X-RAY EQUIPMENT**									
0290	X-ray machine, desk top, for mail/small packages/letters	1 Elec	4	2	Ea.	3,375	127		3,502	3,925
0300	Conveyor type, incl. monitor		2	4		15,600	255		15,855	17,600
0310	Includes additional features		2	4		27,800	255		28,055	31,000
0320	X-ray machine, large unit, for airports, incl. monitor	2 Elec	1	16		39,000	1,025		40,025	44,500
0330	Full console	"	.50	32		67,000	2,050		69,050	76,500

28 18 15 – Security Access Explosive Detection Equipment

28 18 15.23 Security Access Explosive Detection Equipment

		Crew	Daily Output	Labor-Hours	Unit	Material	2021 Bare Costs Labor	2021 Bare Costs Equipment	Total	Total Incl O&P
0010	**SECURITY ACCESS EXPLOSIVE DETECTION EQUIPMENT**									
0270	Explosives detector, walk through portal type	1 Elec	2	4	Ea.	57,500	255		57,755	64,000
0280	Hand-held, battery operated				"				25,500	28,100

528

For customer support on your Commercial Renovation Costs with RSMeans Data, call 800.448.8182.

28 23 Video Management System

28 23 23 – Video Surveillance Systems Infrastructure

28 23 23.50 Video Surveillance Equipments	Crew	Daily Output	Labor-Hours	Unit	Material	2021 Bare Costs Labor	Equipment	Total	Total Incl O&P
0010 **VIDEO SURVEILLANCE EQUIPMENTS**									
0200 Video cameras, wireless, hidden in exit signs, clocks, etc., incl. receiver	1 Elec	3	2.667	Ea.	176	170		346	455
0210 Accessories for video recorder, single camera		3	2.667		177	170		347	455
0220 For multiple cameras		3	2.667		1,700	170		1,870	2,125
0230 Video cameras, wireless, for under vehicle searching, complete	↓	2	4	↓	15,200	255		15,455	17,200

28 31 Intrusion Detection

28 31 16 – Intrusion Detection Systems Infrastructure

28 31 16.50 Intrusion Detection

	Crew	Daily Output	Labor-Hours	Unit	Material	Labor	Equipment	Total	Total Incl O&P
0010 **INTRUSION DETECTION**, not including wires & conduits									
0100 Burglar alarm, battery operated, mechanical trigger	1 Elec	4	2	Ea.	289	127		416	515
0200 Electrical trigger		4	2		345	127		472	575
0400 For outside key control, add		8	1		89	63.50		152.50	196
0600 For remote signaling circuitry, add		8	1		147	63.50		210.50	260
0800 Card reader, flush type, standard		2.70	2.963		680	189		869	1,050
1000 Multi-code		2.70	2.963		1,250	189		1,439	1,675
1200 Door switches, hinge switch		5.30	1.509		65.50	96		161.50	221
1400 Magnetic switch		5.30	1.509		109	96		205	268
1600 Exit control locks, horn alarm		4	2		219	127		346	435
1800 Flashing light alarm		4	2		248	127		375	470
2000 Indicating panels, 1 channel	↓	2.70	2.963		222	189		411	535
2200 10 channel	2 Elec	3.20	5		1,175	320		1,495	1,800
2400 20 channel		2	8		2,500	510		3,010	3,525
2600 40 channel	↓	1.14	14.035		4,650	895		5,545	6,475
2800 Ultrasonic motion detector, 12 V	1 Elec	2.30	3.478		193	222		415	550
3000 Infrared photoelectric detector		4	2		122	127		249	330
3200 Passive infrared detector		4	2		55	127		182	257
3400 Glass break alarm switch		8	1		81	63.50		144.50	187
3420 Switchmats, 30" x 5'		5.30	1.509		120	96		216	280
3440 30" x 25'		4	2		211	127		338	430
3460 Police connect panel		4	2		292	127		419	515
3480 Telephone dialer		5.30	1.509		400	96		496	590
3500 Alarm bell		4	2		101	127		228	305
3520 Siren		4	2		152	127		279	365
3540 Microwave detector, 10' to 200'		2	4		555	255		810	1,000
3560 10' to 350'	↓	2	4	↓	1,700	255		1,955	2,275

28 46 Fire Detection and Alarm

28 46 11 – Fire Sensors and Detectors

28 46 11.21 Carbon-Monoxide Detection Sensors

	Crew	Daily Output	Labor-Hours	Unit	Material	Labor	Equipment	Total	Total Incl O&P
0010 **CARBON-MONOXIDE DETECTION SENSORS**									
8400 Smoke and carbon monoxide alarm battery operated photoelectric low profile	1 Elec	24	.333	Ea.	54.50	21		75.50	92.50
8410 low profile photoelectric battery powered		24	.333		42.50	21		63.50	79.50
8420 photoelectric low profile sealed lithium		24	.333		74.50	21		95.50	115
8430 Photoelectric low profile sealed lithium smoke and CO with voice combo		24	.333		52	21		73	89.50
8500 Carbon monoxide sensor, wall mount 1Mod 1 relay output smoke & heat		24	.333		455	21		476	535
8700 Carbon monoxide detector, battery operated, wall mounted	↓	16	.500	↓	44	32		76	97.50
8710 Hardwired, wall and ceiling mounted	1 Elec	8	1	Ea.	73	63.50		136.50	179
8720 Duct mounted	"	8	1		320	63.50		383.50	455

28 46 11.21 Carbon-Monoxide Detection Sensors		Crew	Daily Output	Labor-Hours	Unit	Material	2021 Bare Costs Labor	Equipment	Total	Total Incl O&P
8730	Continuous air monitoring system, PC based, incl. remote sensors, min.	2 Elec	.25	64	Ea.	31,600	4,075		35,675	41,000
8740	Maximum	"	.10	160	↓	57,000	10,200		67,200	78,000

28 46 11.27 Other Sensors

		Crew	Daily Output	Labor-Hours	Unit	Material	2021 Bare Costs Labor	Equipment	Total	Total Incl O&P
0010	**OTHER SENSORS**									
5200	Smoke detector, ceiling type	1 Elec	6.20	1.290	Ea.	126	82		208	264
5240	Smoke detector, addressable type		6	1.333		214	85		299	365
5400	Duct type		3.20	2.500		295	159		454	570
5420	Duct addressable type		3.20	2.500		180	159		339	445
8300	Smoke alarm with integrated strobe light 120 V, 16DB 60 fpm flash rate		16	.500		126	32		158	187
8310	Photoelectric smoke detector with strobe 120 V, 90 DB ceiling mount		12	.667		240	42.50		282.50	330
8320	120 V, 90 DB wall mount	↓	12	.667	↓	214	42.50		256.50	300

28 46 11.50 Fire and Heat Detectors

		Crew	Daily Output	Labor-Hours	Unit	Material	2021 Bare Costs Labor	Equipment	Total	Total Incl O&P
0010	**FIRE & HEAT DETECTORS**									
5000	Detector, rate of rise	1 Elec	8	1	Ea.	29.50	63.50		93	131
5100	Fixed temp fire alarm		7	1.143		39	73		112	155
5200	125 V SRF MT-135F		7.25	1.103		35.50	70.50		106	147
5300	MT-195F	↓	7.25	1.103	↓	36	70.50		106.50	148

28 46 20 – Fire Alarm

28 46 20.50 Alarm Panels and Devices

		Crew	Daily Output	Labor-Hours	Unit	Material	2021 Bare Costs Labor	Equipment	Total	Total Incl O&P
0010	**ALARM PANELS AND DEVICES**, not including wires & conduits									
2200	Intercom remote station	1 Elec	8	1	Ea.	94.50	63.50		158	202
2400	Intercom outlet		8	1		66	63.50		129.50	171
2600	Sound system, intercom handset		8	1		515	63.50		578.50	665
3590	Signal device, beacon		8	1		207	63.50		270.50	325
3594	2 zone	↓	1.35	5.926		385	375		760	1,000
3598	3 zone	2 Elec	2	8		465	510		975	1,300
3600	4 zone		2	8		435	510		945	1,275
3700	5 zone		1.50	10.667		740	680		1,420	1,875
3800	8 zone		1	16		420	1,025		1,445	2,025
3900	10 zone		1.25	12.800		890	815		1,705	2,225
4000	12 zone	↓	.67	23.988		2,550	1,525		4,075	5,175
4020	Alarm device, tamper, flow	1 Elec	8	1		206	63.50		269.50	325
4025	Fire alarm, loop expander card		16	.500		810	32		842	940
4200	Battery and rack		4	2		465	127		592	705
4400	Automatic charger		8	1		525	63.50		588.50	680
4600	Signal bell		8	1		68	63.50		131.50	173
4610	Fire alarm signal bell 10" red 20-24 V P		8	1		153	63.50		216.50	267
4800	Trouble buzzer or manual station		8	1		75.50	63.50		139	181
5425	Duct smoke and heat detector 2 wire		8	1		155	63.50		218.50	268
5430	Fire alarm duct detector controller	↓	3	2.667	↓	183	170		353	460
5435	Fire alarm duct detector sensor kit	1 Elec	8	1	Ea.	77.50	63.50		141	183
5440	Remote test station for smoke detector duct type		5.30	1.509		54.50	96		150.50	208
5460	Remote fire alarm indicator light		5.30	1.509		28	96		124	179
5600	Strobe and horn		5.30	1.509		129	96		225	290
5610	Strobe and horn (ADA type)		5.30	1.509		155	96		251	320
5620	Visual alarm (ADA type)		6.70	1.194		103	76		179	230
5800	electric bell		6.70	1.194		50.50	76		126.50	173
6000	Door holder, electro-magnetic		4	2		92	127		219	297
6200	Combination holder and closer		3.20	2.500		157	159		316	415
6600	Drill switch		8	1		485	63.50		548.50	635
6800	Master box		2.70	2.963		5,850	189		6,039	6,750
7000	Break glass station	↓	8	1		62.50	63.50		126	167

530

28 46 Fire Detection and Alarm

28 46 20 – Fire Alarm

28 46 20.50 Alarm Panels and Devices	Crew	Daily Output	Labor-Hours	Unit	Material	2021 Bare Costs Labor	Equipment	Total	Total Incl O&P
7800	Remote annunciator, 8 zone lamp	1 Elec	1.80	4.444	Ea.	219	283		502	675
8000	12 zone lamp	2 Elec	2.60	6.154		430	390		820	1,075
8200	16 zone lamp	"	2.20	7.273		385	465		850	1,125

28 47 Mass Notification

28 47 12 – Notification Systems

28 47 12.10 Mass Notification System

		Crew	Daily Output	Labor-Hours	Unit	Material	2021 Bare Costs Labor	Equipment	Total	Total Incl O&P
0010	**MASS NOTIFICATION SYSTEM**									
0100	Wireless command center, 10,000 devices	2 Elec	1.33	12.030	Ea.	3,050	765		3,815	4,525
0200	Option, email notification					2,125			2,125	2,325
0210	Remote device supervision & monitor					2,650			2,650	2,925
0300	Antenna VHF or UHF, for medium range	1 Elec	4	2		93.50	127		220.50	299
0310	For high-power transmitter		2	4		1,125	255		1,380	1,650
0400	Transmitter, 25 watt		4	2		1,750	127		1,877	2,125
0410	40 watt		2.66	3.008		2,025	192		2,217	2,525
0420	100 watt		1.33	6.015		6,625	385		7,010	7,875
0500	Wireless receiver/control module for speaker		8	1		287	63.50		350.50	415
0600	Desktop paging controller, stand alone		4	2		855	127		982	1,125
0700	Auxiliary level inputs		8.36	.957		103	61		164	207

Division Notes

		CREW	DAILY OUTPUT	LABOR-HOURS	UNIT	BARE COSTS				TOTAL INCL O&P
						MAT.	LABOR	EQUIP.	TOTAL	

Estimating Tips
31 05 00 Common Work Results for Earthwork

- Estimating the actual cost of performing earthwork requires careful consideration of the variables involved. This includes items such as type of soil, whether water will be encountered, dewatering, whether banks need bracing, disposal of excavated earth, and length of haul to fill or spoil sites, etc. If the project has large quantities of cut or fill, consider raising or lowering the site to reduce costs, while paying close attention to the effect on site drainage and utilities.

- If the project has large quantities of fill, creating a borrow pit on the site can significantly lower the costs.

- It is very important to consider what time of year the project is scheduled for completion. Bad weather can create large cost overruns from dewatering, site repair, and lost productivity from cold weather.

Reference Numbers

Reference numbers are shown at the beginning of some major classifications. These numbers refer to related items in the Reference Section. The reference information may be an estimating procedure, an alternate pricing method, or technical information.

Note: Not all subdivisions listed here necessarily appear. ∎

Same Data. Simplified.

Enjoy the convenience and efficiency of accessing your costs anywhere:

- **Skip the multiplier** by setting your location
- **Quickly search,** edit, favorite and share costs
- **Stay on top of price changes** with automatic updates

Discover more at rsmeans.com/online

31 05 Common Work Results for Earthwork

31 05 13 – Soils for Earthwork

31 05 13.10 Borrow

31 05 13.10 Borrow	Crew	Daily Output	Labor-Hours	Unit	Material	2021 Bare Costs Labor	Equipment	Total	Total Incl O&P
0010 **BORROW**									
0020 Spread, 200 HP dozer, no compaction, 2 mile RT haul									
0200 Common borrow	B-15	600	.047	C.Y.	13.50	2.45	4.46	20.41	23.50
0700 Screened loam		600	.047		29.50	2.45	4.46	36.41	41.50
0800 Topsoil, weed free		600	.047		27	2.45	4.46	33.91	39
0900 For 5 mile haul, add	B-34B	200	.040			2.05	2.90	4.95	6.45

31 05 16 – Aggregates for Earthwork

31 05 16.10 Borrow

31 05 16.10 Borrow	Crew	Daily Output	Labor-Hours	Unit	Material	2021 Bare Costs Labor	Equipment	Total	Total Incl O&P
0010 **BORROW**									
0020 Spread, with 200 HP dozer, no compaction, 2 mile RT haul									
0100 Bank run gravel	B-15	600	.047	L.C.Y.	29.50	2.45	4.46	36.41	41.50
0300 Crushed stone (1.40 tons per C.Y.), 1-1/2"		600	.047		25.50	2.45	4.46	32.41	37.50
0320 3/4"		600	.047		25.50	2.45	4.46	32.41	37.50
0340 1/2"		600	.047		19.60	2.45	4.46	26.51	30.50
0360 3/8"		600	.047		33	2.45	4.46	39.91	45.50
0400 Sand, washed, concrete		600	.047		33	2.45	4.46	39.91	45
0500 Dead or bank sand		600	.047		18.75	2.45	4.46	25.66	29.50
0600 Select structural fill		600	.047		30	2.45	4.46	36.91	42
0900 For 5 mile haul, add	B-34B	200	.040			2.05	2.90	4.95	6.45

31 06 Schedules for Earthwork

31 06 60 – Schedules for Special Foundations and Load Bearing Elements

31 06 60.14 Piling Special Costs

31 06 60.14 Piling Special Costs	Crew	Daily Output	Labor-Hours	Unit	Material	2021 Bare Costs Labor	Equipment	Total	Total Incl O&P
0010 **PILING SPECIAL COSTS**									
0011 Piling special costs, pile caps, see Section 03 30 53.40									
0500 Cutoffs, concrete piles, plain	1 Pile	5.50	1.455	Ea.		81.50		81.50	130
0600 With steel thin shell, add		38	.211			11.75		11.75	18.80
0700 Steel pile or "H" piles		19	.421			23.50		23.50	37.50
0800 Wood piles		38	.211			11.75		11.75	18.80
1000 Testing, any type piles, test load is twice the design load									
1050 50 ton design load, 100 ton test				Ea.				14,000	15,500
1100 100 ton design load, 200 ton test								20,000	22,000
1200 200 ton design load, 400 ton test								28,000	31,000

31 06 60.15 Mobilization

31 06 60.15 Mobilization	Crew	Daily Output	Labor-Hours	Unit	Material	2021 Bare Costs Labor	Equipment	Total	Total Incl O&P
0010 **MOBILIZATION**									
0020 Set up & remove, air compressor, 600 CFM	A-5	3.30	5.455	Ea.		245	15.05	260.05	405
0100 1,200 CFM	"	2.20	8.182			365	22.50	387.50	610
0200 Crane, with pile leads and pile hammer, 75 ton	B-19	.60	107			6,100	3,400	9,500	13,400
0300 150 ton	"	.36	178			10,200	5,675	15,875	22,400

31 14 Earth Stripping and Stockpiling

31 14 13 − Soil Stripping and Stockpiling

31 14 13.23 Topsoil Stripping and Stockpiling

		Crew	Daily Output	Labor-Hours	Unit	Material	2021 Bare Costs Labor	Equipment	Total	Total Incl O&P
0010	**TOPSOIL STRIPPING AND STOCKPILING**									
1500	Loam or topsoil, remove/stockpile on site									
1510	By hand, 6" deep, 50' haul, less than 100 S.Y.	B-1	100	.240	S.Y.		10.80		10.80	17.20
1520	By skid steer, 6" deep, 100' haul, 101-500 S.Y.	B-62	500	.048			2.31	.36	2.67	4.05
1530	100' haul, 501-900 S.Y.	"	900	.027			1.28	.20	1.48	2.25
1540	200' haul, 901-1,100 S.Y.	B-63	1000	.040			1.86	.18	2.04	3.16
1550	By dozer, 200' haul, 1,101-4,000 S.Y.	B-10B	4000	.003			.16	.38	.54	.67

31 23 Excavation and Fill

31 23 16 − Excavation

31 23 16.13 Excavating, Trench

		Crew	Daily Output	Labor-Hours	Unit	Material	2021 Bare Costs Labor	Equipment	Total	Total Incl O&P
0010	**EXCAVATING, TRENCH**									
0011	Or continuous footing									
0020	Common earth with no sheeting or dewatering included									
1400	By hand with pick and shovel 2' to 6' deep, light soil	1 Clab	8	1	B.C.Y.		44.50		44.50	70.50
1500	Heavy soil	"	4	2	"		89		89	141
1700	For tamping backfilled trenches, air tamp, add	A-1G	100	.080	E.C.Y.		3.55	.55	4.10	6.25
1900	Vibrating plate, add	B-18	180	.133	"		6	.92	6.92	10.55
2100	Trim sides and bottom for concrete pours, common earth		1500	.016	S.F.		.72	.11	.83	1.27
2300	Hardpan		600	.040	"		1.80	.28	2.08	3.17
9000	Minimum labor/equipment charge	1 Clab	4	2	Job		89		89	141

31 23 16.14 Excavating, Utility Trench

		Crew	Daily Output	Labor-Hours	Unit	Material	2021 Bare Costs Labor	Equipment	Total	Total Incl O&P
0010	**EXCAVATING, UTILITY TRENCH**									
0011	Common earth									
0050	Trenching with chain trencher, 12 HP, operator walking									
0100	4" wide trench, 12" deep	B-53	800	.010	L.F.		.56	.20	.76	1.09
0150	18" deep		750	.011			.59	.21	.80	1.15
0200	24" deep		700	.011			.63	.23	.86	1.24
0300	6" wide trench, 12" deep		650	.012			.68	.24	.92	1.34
0350	18" deep		600	.013			.74	.26	1	1.44
0400	24" deep		550	.015			.81	.29	1.10	1.58
0450	36" deep		450	.018			.99	.35	1.34	1.93
0600	8" wide trench, 12" deep		475	.017			.93	.33	1.26	1.83
0650	18" deep		400	.020			1.11	.40	1.51	2.17
0700	24" deep		350	.023			1.27	.45	1.72	2.48
0750	36" deep		300	.027			1.48	.53	2.01	2.89
0900	Minimum labor/equipment charge		2	4	Job		222	79.50	301.50	430
1000	Backfill by hand including compaction, add									
1050	4" wide trench, 12" deep	A-1G	800	.010	L.F.		.44	.07	.51	.79
1100	18" deep		530	.015			.67	.10	.77	1.18
1150	24" deep		400	.020			.89	.14	1.03	1.56
1300	6" wide trench, 12" deep		540	.015			.66	.10	.76	1.16
1350	18" deep		405	.020			.88	.13	1.01	1.55
1400	24" deep		270	.030			1.32	.20	1.52	2.31
1450	36" deep		180	.044			1.97	.30	2.27	3.47
1600	8" wide trench, 12" deep		400	.020			.89	.14	1.03	1.56
1650	18" deep		265	.030			1.34	.21	1.55	2.36
1700	24" deep		200	.040			1.78	.27	2.05	3.13
1750	36" deep		135	.059			2.63	.40	3.03	4.64
2000	Chain trencher, 40 HP operator riding									
2050	6" wide trench and backfill, 12" deep	B-54	1200	.007	L.F.		.37	.38	.75	.99

31 23 Excavation and Fill

31 23 16 – Excavation

31 23 16.14 Excavating, Utility Trench

		Crew	Daily Output	Labor-Hours	Unit	Material	2021 Bare Costs Labor	2021 Bare Costs Equipment	Total	Total Incl O&P
2100	18" deep	B-54	1000	.008	L.F.		.44	.45	.89	1.19
2150	24" deep		975	.008			.46	.46	.92	1.22
2200	36" deep		900	.009			.49	.50	.99	1.32
2250	48" deep		750	.011			.59	.60	1.19	1.58
2300	60" deep		650	.012			.68	.69	1.37	1.83
2400	8" wide trench and backfill, 12" deep		1000	.008			.44	.45	.89	1.19
2450	18" deep		950	.008			.47	.47	.94	1.25
2500	24" deep		900	.009			.49	.50	.99	1.32
2550	36" deep		800	.010			.56	.56	1.12	1.49
2600	48" deep		650	.012			.68	.69	1.37	1.83
2700	12" wide trench and backfill, 12" deep		975	.008			.46	.46	.92	1.22
2750	18" deep		860	.009			.52	.52	1.04	1.39
2800	24" deep		800	.010			.56	.56	1.12	1.49
2850	36" deep		725	.011			.61	.62	1.23	1.64
3000	16" wide trench and backfill, 12" deep		835	.010			.53	.54	1.07	1.42
3050	18" deep		750	.011			.59	.60	1.19	1.58
3100	24" deep	▼	700	.011	▼		.63	.64	1.27	1.70
3200	Compaction with vibratory plate, add								35%	35%
5100	Hand excavate and trim for pipe bells after trench excavation									
5200	8" pipe	1 Clab	155	.052	L.F.		2.29		2.29	3.65
5300	18" pipe	"	130	.062	"		2.73		2.73	4.35
9000	Minimum labor/equipment charge	A-1G	4	2	Job		89	13.65	102.65	156

31 23 16.16 Structural Excavation for Minor Structures

		Crew	Daily Output	Labor-Hours	Unit	Material	2021 Bare Costs Labor	2021 Bare Costs Equipment	Total	Total Incl O&P
0010	**STRUCTURAL EXCAVATION FOR MINOR STRUCTURES**									
0015	Hand, pits to 6' deep, sandy soil	1 Clab	8	1	B.C.Y.		44.50		44.50	70.50
0100	Heavy soil or clay		4	2			89		89	141
0300	Pits 6' to 12' deep, sandy soil		5	1.600			71		71	113
0500	Heavy soil or clay		3	2.667			118		118	189
0700	Pits 12' to 18' deep, sandy soil		4	2			89		89	141
0900	Heavy soil or clay		2	4			178		178	283
1100	Hand loading trucks from stock pile, sandy soil		12	.667			29.50		29.50	47
1300	Heavy soil or clay	▼	8	1	▼		44.50		44.50	70.50
1500	For wet or muck hand excavation, add to above								50%	50%
6000	Machine excavation, for spread and mat footings, elevator pits,									
6001	and small building foundations									
6030	Common earth, hydraulic backhoe, 1/2 C.Y. bucket	B-12E	55	.291	B.C.Y.		15.40	8.30	23.70	33
6035	3/4 C.Y. bucket	B-12F	90	.178			9.40	7.80	17.20	23.50
6040	1 C.Y. bucket	B-12A	108	.148			7.85	7.70	15.55	21
6050	1-1/2 C.Y. bucket	B-12B	144	.111			5.90	4.83	10.73	14.55
6060	2 C.Y. bucket	B-12C	200	.080			4.23	4.71	8.94	11.85
6070	Sand and gravel, 3/4 C.Y. bucket	B-12F	100	.160			8.45	7	15.45	21
6080	1 C.Y. bucket	B-12A	120	.133	▼		7.05	6.95	14	18.75
6090	1-1/2 C.Y. bucket	B-12B	160	.100	B.C.Y.		5.30	4.35	9.65	13.15
6100	2 C.Y. bucket	B-12C	220	.073			3.85	4.29	8.14	10.75
6110	Clay, till, or blasted rock, 3/4 C.Y. bucket	B-12F	80	.200			10.60	8.75	19.35	26.50
6120	1 C.Y. bucket	B-12A	95	.168			8.90	8.75	17.65	23.50
6130	1-1/2 C.Y. bucket	B-12B	130	.123			6.50	5.35	11.85	16.15
6140	2 C.Y. bucket	B-12C	175	.091			4.84	5.40	10.24	13.55
6230	Sandy clay & loam, hydraulic backhoe, 1/2 C.Y. bucket	B-12E	60	.267			14.10	7.60	21.70	30.50
6235	3/4 C.Y. bucket	B-12F	98	.163			8.65	7.15	15.80	21.50
6240	1 C.Y. bucket	B-12A	116	.138			7.30	7.20	14.50	19.40
6250	1-1/2 C.Y. bucket	B-12B	156	.103	▼		5.45	4.46	9.91	13.45

31 23 Excavation and Fill

31 23 16 – Excavation

31 23 16.16 Structural Excavation for Minor Structures	Crew	Daily Output	Labor-Hours	Unit	Material	2021 Bare Costs Labor	2021 Bare Costs Equipment	Total	Total Incl O&P
9000 Minimum labor/equipment charge	1 Clab	4	2	Job		89		89	141

31 23 16.42 Excavating, Bulk Bank Measure

	Crew	Daily Output	Labor-Hours	Unit	Material	Labor	Equipment	Total	Total Incl O&P
0010 **EXCAVATING, BULK BANK MEASURE** R312316-40									
0011 Common earth piled									
0020 For loading onto trucks, add								15%	15%
0200 Excavator, hydraulic, crawler mtd., 1 C.Y. cap. = 100 C.Y./hr. R312316-45	B-12A	800	.020	B.C.Y.		1.06	1.04	2.10	2.81
1200 Front end loader, track mtd., 1-1/2 C.Y. cap. = 70 C.Y./hr.	B-10N	560	.021			1.16	1.02	2.18	2.94
1500 Wheel mounted, 3/4 C.Y. cap. = 45 C.Y./hr.	B-10R	360	.033			1.80	.85	2.65	3.76
5000 Excavating, bulk bank measure, sandy clay & loam piled									
5020 For loading onto trucks, add								15%	15%
5100 Excavator, hydraulic, crawler mtd., 1 C.Y. cap. = 120 C.Y./hr.	B-12A	960	.017	B.C.Y.		.88	.87	1.75	2.34
9000 Minimum labor/equipment charge	B-10L	2	6	Job		325	203	528	735

31 23 16.46 Excavating, Bulk, Dozer

	Crew	Daily Output	Labor-Hours	Unit	Material	Labor	Equipment	Total	Total Incl O&P
0010 **EXCAVATING, BULK, DOZER**									
0011 Open site									
2000 80 HP, 50' haul, sand & gravel	B-10L	460	.026	B.C.Y.		1.41	.88	2.29	3.19
2200 150' haul, sand & gravel		230	.052			2.82	1.76	4.58	6.35
2400 300' haul, sand & gravel		120	.100			5.40	3.38	8.78	12.20
3000 105 HP, 50' haul, sand & gravel	B-10W	700	.017			.93	.92	1.85	2.47
3200 150' haul, sand & gravel		310	.039			2.10	2.07	4.17	5.55
3300 300' haul, sand & gravel		140	.086			4.64	4.58	9.22	12.35
4000 200 HP, 50' haul, sand & gravel	B-10B	1400	.009			.46	1.09	1.55	1.92
4200 150' haul, sand & gravel		595	.020			1.09	2.55	3.64	4.52
4400 300' haul, sand & gravel		310	.039			2.10	4.90	7	8.70

31 23 23 – Fill

31 23 23.13 Backfill

	Crew	Daily Output	Labor-Hours	Unit	Material	Labor	Equipment	Total	Total Incl O&P
0010 **BACKFILL** R312323-30									
0015 By hand, no compaction, light soil	1 Clab	14	.571	L.C.Y.		25.50		25.50	40.50
0100 Heavy soil		11	.727	"		32.50		32.50	51.50
0300 Compaction in 6" layers, hand tamp, add to above		20.60	.388	E.C.Y.		17.25		17.25	27.50
0400 Roller compaction operator walking, add	B-10A	100	.120	"		6.50	1.67	8.17	12.05
0500 Air tamp, add	B-9D	190	.211	E.C.Y.		9.45	1.73	11.18	16.90
0600 Vibrating plate, add	A-1D	60	.133			5.90	.53	6.43	10.05
0800 Compaction in 12" layers, hand tamp, add to above	1 Clab	34	.235			10.45		10.45	16.65
0900 Roller compaction operator walking, add	B-10A	150	.080			4.33	1.11	5.44	8
1000 Air tamp, add	B-9	285	.140			6.30	1.25	7.55	11.35
1100 Vibrating plate, add	A-1E	90	.089			3.95	1.84	5.79	8.30
1200 Trench, dozer, no compaction, 60 HP	B-10L	425	.028	L.C.Y.		1.53	.95	2.48	3.45
1300 Dozer backfilling, bulk, up to 300' haul, no compaction	B-10B	1200	.010	"		.54	1.27	1.81	2.24
1400 Air tamped, add	B-11B	80	.200	E.C.Y.		10	5.35	15.35	21.50
1900 Dozer backfilling, trench, up to 300' haul, no compaction	B-10B	900	.013	L.C.Y.		.72	1.69	2.41	2.99
2000 Air tamped, add	B-11B	80	.200	E.C.Y.		10	5.35	15.35	21.50
2350 Spreading in 8" layers, small dozer	B-10B	1060	.011	L.C.Y.		.61	1.43	2.04	2.54
2450 Compacting with vibrating plate, 8" lifts	A-1D	73	.110	E.C.Y.		4.87	.44	5.31	8.25

31 23 23.14 Backfill, Structural

	Crew	Daily Output	Labor-Hours	Unit	Material	Labor	Equipment	Total	Total Incl O&P
0010 **BACKFILL, STRUCTURAL**									
0011 Dozer or F.E. loader									
0020 From existing stockpile, no compaction									
1000 55 HP wheeled loader, 50' haul, common earth	B-11C	200	.080	L.C.Y.		4.14	1.08	5.22	7.70
2000 80 HP, 50' haul, sand & gravel	B-10L	1100	.011			.59	.37	.96	1.34
2010 Sandy clay & loam		1070	.011			.61	.38	.99	1.37

For customer support on your Commercial Renovation Costs with RSMeans Data, call 800.448.8182.

537

31 23 Excavation and Fill

31 23 23 – Fill

31 23 23.14 Backfill, Structural

		Crew	Daily Output	Labor-Hours	Unit	Material	2021 Bare Costs Labor	2021 Bare Costs Equipment	Total	Total Incl O&P
2020	Common earth	B-10L	975	.012	L.C.Y.		.67	.42	1.09	1.51
2040	Clay		850	.014			.76	.48	1.24	1.73
2400	300' haul, sand & gravel		370	.032			1.76	1.10	2.86	3.96
2410	Sandy clay & loam		360	.033			1.80	1.13	2.93	4.07
2420	Common earth		330	.036			1.97	1.23	3.20	4.44
2440	Clay		290	.041			2.24	1.40	3.64	5.05
3000	105 HP, 50' haul, sand & gravel	B-10W	1350	.009			.48	.47	.95	1.27
3010	Sandy clay & loam		1325	.009			.49	.48	.97	1.30
3020	Common earth		1225	.010			.53	.52	1.05	1.41
3040	Clay		1100	.011			.59	.58	1.17	1.57
3300	300' haul, sand & gravel		465	.026			1.40	1.38	2.78	3.71
3310	Sandy clay & loam		455	.026			1.43	1.41	2.84	3.79
3320	Common earth		415	.029			1.57	1.54	3.11	4.16
3340	Clay		370	.032			1.76	1.73	3.49	4.66

31 23 23.16 Fill By Borrow and Utility Bedding

		Crew	Daily Output	Labor-Hours	Unit	Material	2021 Bare Costs Labor	2021 Bare Costs Equipment	Total	Total Incl O&P
0010	**FILL BY BORROW AND UTILITY BEDDING**									
0049	Utility bedding, for pipe & conduit, not incl. compaction									
0050	Crushed or screened bank run gravel	B-6	150	.160	L.C.Y.	33.50	7.70	1.44	42.64	50
0100	Crushed stone 3/4" to 1/2"		150	.160		25.50	7.70	1.44	34.64	42
0200	Sand, dead or bank		150	.160		18.75	7.70	1.44	27.89	34
0500	Compacting bedding in trench	A-1D	90	.089	E.C.Y.		3.95	.35	4.30	6.70
0600	If material source exceeds 2 miles, add for extra mileage.									
0610	See Section 31 23 23.20 for hauling mileage add.									

31 23 23.17 General Fill

		Crew	Daily Output	Labor-Hours	Unit	Material	2021 Bare Costs Labor	2021 Bare Costs Equipment	Total	Total Incl O&P
0010	**GENERAL FILL**									
0011	Spread dumped material, no compaction									
0020	By dozer	B-10B	1000	.012	L.C.Y.		.65	1.52	2.17	2.69
0100	By hand	1 Clab	12	.667	"		29.50		29.50	47
9000	Minimum labor/equipment charge	"	4	2	Job		89		89	141

31 23 23.20 Hauling

		Crew	Daily Output	Labor-Hours	Unit	Material	2021 Bare Costs Labor	2021 Bare Costs Equipment	Total	Total Incl O&P
0010	**HAULING**									
0011	Excavated or borrow, loose cubic yards									
0012	no loading equipment, including hauling, waiting, loading/dumping									
0013	time per cycle (wait, load, travel, unload or dump & return)									
0014	8 C.Y. truck, 15 MPH avg., cycle 0.5 miles, 10 min. wait/ld./uld.	B-34A	320	.025	L.C.Y.		1.28	1.27	2.55	3.45
0016	cycle 1 mile		272	.029			1.51	1.50	3.01	4.06
0018	cycle 2 miles		208	.038			1.97	1.96	3.93	5.30
0020	cycle 4 miles		144	.056			2.85	2.83	5.68	7.65
0022	cycle 6 miles		112	.071			3.66	3.64	7.30	9.85
0024	cycle 8 miles		88	.091			4.66	4.63	9.29	12.55
0026	20 MPH avg., cycle 0.5 mile		336	.024			1.22	1.21	2.43	3.28
0028	cycle 1 mile		296	.027			1.39	1.38	2.77	3.72
0030	cycle 2 miles		240	.033			1.71	1.70	3.41	4.60
0032	cycle 4 miles		176	.045			2.33	2.32	4.65	6.25
0034	cycle 6 miles		136	.059			3.02	3	6.02	8.10
0036	cycle 8 miles		112	.071			3.66	3.64	7.30	9.85
0044	25 MPH avg., cycle 4 miles		192	.042			2.14	2.12	4.26	5.75
0046	cycle 6 miles		160	.050			2.57	2.55	5.12	6.90
0048	cycle 8 miles		128	.063			3.21	3.18	6.39	8.60
0050	30 MPH avg., cycle 4 miles		216	.037			1.90	1.89	3.79	5.10
0052	cycle 6 miles		176	.045			2.33	2.32	4.65	6.25
0054	cycle 8 miles		144	.056			2.85	2.83	5.68	7.65

31 23 23 – Fill

31 23 23.20 Hauling	Crew	Daily Output	Labor-Hours	Unit	Material	2021 Bare Costs Labor	Equipment	Total	Total Incl O&P
0114	15 MPH avg., cycle 0.5 mile, 15 min. wait/ld./uld.	B-34A	224	.036	L.C.Y.		1.83	1.82	3.65	4.92
0116	cycle 1 mile		200	.040			2.05	2.04	4.09	5.50
0118	cycle 2 miles		168	.048			2.44	2.43	4.87	6.55
0120	cycle 4 miles		120	.067			3.42	3.40	6.82	9.20
0122	cycle 6 miles		96	.083			4.28	4.25	8.53	11.45
0124	cycle 8 miles		80	.100			5.15	5.10	10.25	13.80
0126	20 MPH avg., cycle 0.5 mile		232	.034			1.77	1.76	3.53	4.75
0128	cycle 1 mile		208	.038			1.97	1.96	3.93	5.30
0130	cycle 2 miles		184	.043			2.23	2.22	4.45	6
0132	cycle 4 miles		144	.056			2.85	2.83	5.68	7.65
0134	cycle 6 miles		112	.071			3.66	3.64	7.30	9.85
0136	cycle 8 miles	B-34A	96	.083	L.C.Y.		4.28	4.25	8.53	11.45
0144	25 MPH avg., cycle 4 miles		152	.053			2.70	2.68	5.38	7.25
0146	cycle 6 miles		128	.063			3.21	3.18	6.39	8.60
0148	cycle 8 miles		112	.071			3.66	3.64	7.30	9.85
0150	30 MPH avg., cycle 4 miles		168	.048			2.44	2.43	4.87	6.55
0152	cycle 6 miles		144	.056			2.85	2.83	5.68	7.65
0154	cycle 8 miles		120	.067			3.42	3.40	6.82	9.20
0214	15 MPH avg., cycle 0.5 mile, 20 min. wait/ld./uld.		176	.045			2.33	2.32	4.65	6.25
0216	cycle 1 mile		160	.050			2.57	2.55	5.12	6.90
0218	cycle 2 miles		136	.059			3.02	3	6.02	8.10
0220	cycle 4 miles		104	.077			3.95	3.92	7.87	10.60
0222	cycle 6 miles		88	.091			4.66	4.63	9.29	12.55
0224	cycle 8 miles		72	.111			5.70	5.65	11.35	15.35
0226	20 MPH avg., cycle 0.5 mile		176	.045			2.33	2.32	4.65	6.25
0228	cycle 1 mile		168	.048			2.44	2.43	4.87	6.55
0230	cycle 2 miles		144	.056			2.85	2.83	5.68	7.65
0232	cycle 4 miles		120	.067			3.42	3.40	6.82	9.20
0234	cycle 6 miles		96	.083			4.28	4.25	8.53	11.45
0236	cycle 8 miles		88	.091			4.66	4.63	9.29	12.55
0244	25 MPH avg., cycle 4 miles		128	.063			3.21	3.18	6.39	8.60
0246	cycle 6 miles		112	.071			3.66	3.64	7.30	9.85
0248	cycle 8 miles		96	.083			4.28	4.25	8.53	11.45
0250	30 MPH avg., cycle 4 miles		136	.059			3.02	3	6.02	8.10
0252	cycle 6 miles		120	.067			3.42	3.40	6.82	9.20
0254	cycle 8 miles		104	.077			3.95	3.92	7.87	10.60
0314	15 MPH avg., cycle 0.5 mile, 25 min. wait/ld./uld.		144	.056			2.85	2.83	5.68	7.65
0316	cycle 1 mile		128	.063			3.21	3.18	6.39	8.60
0318	cycle 2 miles		112	.071			3.66	3.64	7.30	9.85
0320	cycle 4 miles		96	.083			4.28	4.25	8.53	11.45
0322	cycle 6 miles		80	.100			5.15	5.10	10.25	13.80
0324	cycle 8 miles		64	.125			6.40	6.35	12.75	17.25
0326	20 MPH avg., cycle 0.5 mile		144	.056			2.85	2.83	5.68	7.65
0328	cycle 1 mile		136	.059			3.02	3	6.02	8.10
0330	cycle 2 miles		120	.067			3.42	3.40	6.82	9.20
0332	cycle 4 miles		104	.077			3.95	3.92	7.87	10.60
0334	cycle 6 miles		88	.091			4.66	4.63	9.29	12.55
0336	cycle 8 miles		80	.100			5.15	5.10	10.25	13.80
0344	25 MPH avg., cycle 4 miles		112	.071			3.66	3.64	7.30	9.85
0346	cycle 6 miles		96	.083			4.28	4.25	8.53	11.45
0348	cycle 8 miles		88	.091			4.66	4.63	9.29	12.55
0350	30 MPH avg., cycle 4 miles		112	.071			3.66	3.64	7.30	9.85
0352	cycle 6 miles	B-34A	104	.077	L.C.Y.		3.95	3.92	7.87	10.60

539

31 23 23.20 Hauling	Crew	Daily Output	Labor-Hours	Unit	Material	2021 Bare Costs Labor	2021 Bare Costs Equipment	Total	Total Incl O&P
0354 cycle 8 miles	B-34A	96	.083	L.C.Y.		4.28	4.25	8.53	11.45
0414 15 MPH avg., cycle 0.5 mile, 30 min. wait/ld./uld.		120	.067			3.42	3.40	6.82	9.20
0416 cycle 1 mile		112	.071			3.66	3.64	7.30	9.85
0418 cycle 2 miles		96	.083			4.28	4.25	8.53	11.45
0420 cycle 4 miles		80	.100			5.15	5.10	10.25	13.80
0422 cycle 6 miles		72	.111			5.70	5.65	11.35	15.35
0424 cycle 8 miles		64	.125			6.40	6.35	12.75	17.25
0426 20 MPH avg., cycle 0.5 mile		120	.067			3.42	3.40	6.82	9.20
0428 cycle 1 mile		112	.071			3.66	3.64	7.30	9.85
0430 cycle 2 miles		104	.077			3.95	3.92	7.87	10.60
0432 cycle 4 miles		88	.091			4.66	4.63	9.29	12.55
0434 cycle 6 miles		80	.100			5.15	5.10	10.25	13.80
0436 cycle 8 miles		72	.111			5.70	5.65	11.35	15.35
0444 25 MPH avg., cycle 4 miles		96	.083			4.28	4.25	8.53	11.45
0446 cycle 6 miles		88	.091			4.66	4.63	9.29	12.55
0448 cycle 8 miles		80	.100			5.15	5.10	10.25	13.80
0450 30 MPH avg., cycle 4 miles		96	.083			4.28	4.25	8.53	11.45
0452 cycle 6 miles		88	.091			4.66	4.63	9.29	12.55
0454 cycle 8 miles		80	.100			5.15	5.10	10.25	13.80
0514 15 MPH avg., cycle 0.5 mile, 35 min. wait/ld./uld.		104	.077			3.95	3.92	7.87	10.60
0516 cycle 1 mile		96	.083			4.28	4.25	8.53	11.45
0518 cycle 2 miles		88	.091			4.66	4.63	9.29	12.55
0520 cycle 4 miles		72	.111			5.70	5.65	11.35	15.35
0522 cycle 6 miles		64	.125			6.40	6.35	12.75	17.25
0524 cycle 8 miles		56	.143			7.35	7.30	14.65	19.70
0526 20 MPH avg., cycle 0.5 mile		104	.077			3.95	3.92	7.87	10.60
0528 cycle 1 mile		96	.083			4.28	4.25	8.53	11.45
0530 cycle 2 miles		96	.083			4.28	4.25	8.53	11.45
0532 cycle 4 miles		80	.100			5.15	5.10	10.25	13.80
0534 cycle 6 miles		72	.111			5.70	5.65	11.35	15.35
0536 cycle 8 miles		64	.125			6.40	6.35	12.75	17.25
0544 25 MPH avg., cycle 4 miles		88	.091			4.66	4.63	9.29	12.55
0546 cycle 6 miles		80	.100			5.15	5.10	10.25	13.80
0548 cycle 8 miles		72	.111			5.70	5.65	11.35	15.35
0550 30 MPH avg., cycle 4 miles		88	.091			4.66	4.63	9.29	12.55
0552 cycle 6 miles		80	.100			5.15	5.10	10.25	13.80
0554 cycle 8 miles	▼	72	.111			5.70	5.65	11.35	15.35
1014 12 C.Y. truck, cycle 0.5 mile, 15 MPH avg., 15 min. wait/ld./uld.	B-34B	336	.024			1.22	1.72	2.94	3.85
1016 cycle 1 mile		300	.027			1.37	1.93	3.30	4.30
1018 cycle 2 miles	▼	252	.032	▼		1.63	2.30	3.93	5.15
1020 cycle 4 miles	B-34B	180	.044	L.C.Y.		2.28	3.22	5.50	7.20
1022 cycle 6 miles		144	.056			2.85	4.02	6.87	9
1024 cycle 8 miles		120	.067			3.42	4.83	8.25	10.75
1025 cycle 10 miles		96	.083			4.28	6.05	10.33	13.45
1026 20 MPH avg., cycle 0.5 mile		348	.023			1.18	1.66	2.84	3.71
1028 cycle 1 mile		312	.026			1.32	1.86	3.18	4.14
1030 cycle 2 miles		276	.029			1.49	2.10	3.59	4.68
1032 cycle 4 miles		216	.037			1.90	2.68	4.58	6
1034 cycle 6 miles		168	.048			2.44	3.45	5.89	7.70
1036 cycle 8 miles		144	.056			2.85	4.02	6.87	9
1038 cycle 10 miles		120	.067			3.42	4.83	8.25	10.75
1040 25 MPH avg., cycle 4 miles		228	.035			1.80	2.54	4.34	5.65
1042 cycle 6 miles	▼	192	.042	▼		2.14	3.02	5.16	6.75

For customer support on your Commercial Renovation Costs with RSMeans Data, call 800.448.8182.

31 23 23.20 Hauling		Crew	Daily Output	Labor-Hours	Unit	Material	2021 Bare Costs Labor	Equipment	Total	Total Incl O&P
1044	cycle 8 miles	B-34B	168	.048	L.C.Y.		2.44	3.45	5.89	7.70
1046	cycle 10 miles		144	.056			2.85	4.02	6.87	9
1050	30 MPH avg., cycle 4 miles		252	.032			1.63	2.30	3.93	5.15
1052	cycle 6 miles		216	.037			1.90	2.68	4.58	6
1054	cycle 8 miles		180	.044			2.28	3.22	5.50	7.20
1056	cycle 10 miles		156	.051			2.63	3.71	6.34	8.30
1060	35 MPH avg., cycle 4 miles		264	.030			1.55	2.19	3.74	4.89
1062	cycle 6 miles		228	.035			1.80	2.54	4.34	5.65
1064	cycle 8 miles		204	.039			2.01	2.84	4.85	6.35
1066	cycle 10 miles		180	.044			2.28	3.22	5.50	7.20
1068	cycle 20 miles		120	.067			3.42	4.83	8.25	10.75
1069	cycle 30 miles		84	.095			4.89	6.90	11.79	15.40
1070	cycle 40 miles		72	.111			5.70	8.05	13.75	17.95
1072	40 MPH avg., cycle 6 miles		240	.033			1.71	2.41	4.12	5.40
1074	cycle 8 miles		216	.037			1.90	2.68	4.58	6
1076	cycle 10 miles		192	.042			2.14	3.02	5.16	6.75
1078	cycle 20 miles		120	.067			3.42	4.83	8.25	10.75
1080	cycle 30 miles		96	.083			4.28	6.05	10.33	13.45
1082	cycle 40 miles		72	.111			5.70	8.05	13.75	17.95
1084	cycle 50 miles		60	.133			6.85	9.65	16.50	21.50
1094	45 MPH avg., cycle 8 miles		216	.037			1.90	2.68	4.58	6
1096	cycle 10 miles		204	.039			2.01	2.84	4.85	6.35
1098	cycle 20 miles		132	.061			3.11	4.39	7.50	9.80
1100	cycle 30 miles		108	.074			3.80	5.35	9.15	11.95
1102	cycle 40 miles		84	.095			4.89	6.90	11.79	15.40
1104	cycle 50 miles		72	.111			5.70	8.05	13.75	17.95
1106	50 MPH avg., cycle 10 miles		216	.037			1.90	2.68	4.58	6
1108	cycle 20 miles		144	.056			2.85	4.02	6.87	9
1110	cycle 30 miles	▼	108	.074	▼		3.80	5.35	9.15	11.95
1112	cycle 40 miles	B-34B	84	.095	L.C.Y.		4.89	6.90	11.79	15.40
1114	cycle 50 miles		72	.111			5.70	8.05	13.75	17.95
1214	15 MPH avg., cycle 0.5 mile, 20 min. wait/ld./uld.		264	.030			1.55	2.19	3.74	4.89
1216	cycle 1 mile		240	.033			1.71	2.41	4.12	5.40
1218	cycle 2 miles		204	.039			2.01	2.84	4.85	6.35
1220	cycle 4 miles		156	.051			2.63	3.71	6.34	8.30
1222	cycle 6 miles		132	.061			3.11	4.39	7.50	9.80
1224	cycle 8 miles		108	.074			3.80	5.35	9.15	11.95
1225	cycle 10 miles		96	.083			4.28	6.05	10.33	13.45
1226	20 MPH avg., cycle 0.5 mile		264	.030			1.55	2.19	3.74	4.89
1228	cycle 1 mile		252	.032			1.63	2.30	3.93	5.15
1230	cycle 2 miles		216	.037			1.90	2.68	4.58	6
1232	cycle 4 miles		180	.044			2.28	3.22	5.50	7.20
1234	cycle 6 miles		144	.056			2.85	4.02	6.87	9
1236	cycle 8 miles		132	.061			3.11	4.39	7.50	9.80
1238	cycle 10 miles		108	.074			3.80	5.35	9.15	11.95
1240	25 MPH avg., cycle 4 miles		192	.042			2.14	3.02	5.16	6.75
1242	cycle 6 miles		168	.048			2.44	3.45	5.89	7.70
1244	cycle 8 miles		144	.056			2.85	4.02	6.87	9
1246	cycle 10 miles		132	.061			3.11	4.39	7.50	9.80
1250	30 MPH avg., cycle 4 miles		204	.039			2.01	2.84	4.85	6.35
1252	cycle 6 miles		180	.044			2.28	3.22	5.50	7.20
1254	cycle 8 miles		156	.051			2.63	3.71	6.34	8.30
1256	cycle 10 miles	▼	144	.056	▼		2.85	4.02	6.87	9

31 23 23.20 Hauling		Crew	Daily Output	Labor-Hours	Unit	Material	2021 Bare Costs Labor	2021 Bare Costs Equipment	Total	Total Incl O&P
1260	35 MPH avg., cycle 4 miles	B-34B	216	.037	L.C.Y.		1.90	2.68	4.58	6
1262	cycle 6 miles		192	.042			2.14	3.02	5.16	6.75
1264	cycle 8 miles		168	.048			2.44	3.45	5.89	7.70
1266	cycle 10 miles		156	.051			2.63	3.71	6.34	8.30
1268	cycle 20 miles		108	.074			3.80	5.35	9.15	11.95
1269	cycle 30 miles		72	.111			5.70	8.05	13.75	17.95
1270	cycle 40 miles		60	.133			6.85	9.65	16.50	21.50
1272	40 MPH avg., cycle 6 miles		192	.042			2.14	3.02	5.16	6.75
1274	cycle 8 miles		180	.044			2.28	3.22	5.50	7.20
1276	cycle 10 miles		156	.051			2.63	3.71	6.34	8.30
1278	cycle 20 miles		108	.074			3.80	5.35	9.15	11.95
1280	cycle 30 miles		84	.095			4.89	6.90	11.79	15.40
1282	cycle 40 miles		72	.111			5.70	8.05	13.75	17.95
1284	cycle 50 miles		60	.133			6.85	9.65	16.50	21.50
1294	45 MPH avg., cycle 8 miles		180	.044			2.28	3.22	5.50	7.20
1296	cycle 10 miles		168	.048			2.44	3.45	5.89	7.70
1298	cycle 20 miles		120	.067			3.42	4.83	8.25	10.75
1300	cycle 30 miles	B-34B	96	.083	L.C.Y.		4.28	6.05	10.33	13.45
1302	cycle 40 miles		72	.111			5.70	8.05	13.75	17.95
1304	cycle 50 miles		60	.133			6.85	9.65	16.50	21.50
1306	50 MPH avg., cycle 10 miles		180	.044			2.28	3.22	5.50	7.20
1308	cycle 20 miles		132	.061			3.11	4.39	7.50	9.80
1310	cycle 30 miles		96	.083			4.28	6.05	10.33	13.45
1312	cycle 40 miles		84	.095			4.89	6.90	11.79	15.40
1314	cycle 50 miles		72	.111			5.70	8.05	13.75	17.95
1414	15 MPH avg., cycle 0.5 mile, 25 min. wait/ld./uld.		204	.039			2.01	2.84	4.85	6.35
1416	cycle 1 mile		192	.042			2.14	3.02	5.16	6.75
1418	cycle 2 miles		168	.048			2.44	3.45	5.89	7.70
1420	cycle 4 miles		132	.061			3.11	4.39	7.50	9.80
1422	cycle 6 miles		120	.067			3.42	4.83	8.25	10.75
1424	cycle 8 miles		96	.083			4.28	6.05	10.33	13.45
1425	cycle 10 miles		84	.095			4.89	6.90	11.79	15.40
1426	20 MPH avg., cycle 0.5 mile		216	.037			1.90	2.68	4.58	6
1428	cycle 1 mile		204	.039			2.01	2.84	4.85	6.35
1430	cycle 2 miles		180	.044			2.28	3.22	5.50	7.20
1432	cycle 4 miles		156	.051			2.63	3.71	6.34	8.30
1434	cycle 6 miles		132	.061			3.11	4.39	7.50	9.80
1436	cycle 8 miles		120	.067			3.42	4.83	8.25	10.75
1438	cycle 10 miles		96	.083			4.28	6.05	10.33	13.45
1440	25 MPH avg., cycle 4 miles		168	.048			2.44	3.45	5.89	7.70
1442	cycle 6 miles		144	.056			2.85	4.02	6.87	9
1444	cycle 8 miles		132	.061			3.11	4.39	7.50	9.80
1446	cycle 10 miles		108	.074			3.80	5.35	9.15	11.95
1450	30 MPH avg., cycle 4 miles		168	.048			2.44	3.45	5.89	7.70
1452	cycle 6 miles		156	.051			2.63	3.71	6.34	8.30
1454	cycle 8 miles		132	.061			3.11	4.39	7.50	9.80
1456	cycle 10 miles		120	.067			3.42	4.83	8.25	10.75
1460	35 MPH avg., cycle 4 miles		180	.044			2.28	3.22	5.50	7.20
1462	cycle 6 miles		156	.051			2.63	3.71	6.34	8.30
1464	cycle 8 miles		144	.056			2.85	4.02	6.87	9
1466	cycle 10 miles		132	.061			3.11	4.39	7.50	9.80
1468	cycle 20 miles		96	.083			4.28	6.05	10.33	13.45
1469	cycle 30 miles		72	.111			5.70	8.05	13.75	17.95

31 23 23 – Fill

31 23 23.20 Hauling		Crew	Daily Output	Labor-Hours	Unit	Material	2021 Bare Costs Labor	Equipment	Total	Total Incl O&P
1470	cycle 40 miles	B-34B	60	.133	L.C.Y.		6.85	9.65	16.50	21.50
1472	40 MPH avg., cycle 6 miles		168	.048			2.44	3.45	5.89	7.70
1474	cycle 8 miles		156	.051			2.63	3.71	6.34	8.30
1476	cycle 10 miles		144	.056			2.85	4.02	6.87	9
1478	cycle 20 miles		96	.083			4.28	6.05	10.33	13.45
1480	cycle 30 miles	B-34B	84	.095	L.C.Y.		4.89	6.90	11.79	15.40
1482	cycle 40 miles		60	.133			6.85	9.65	16.50	21.50
1484	cycle 50 miles		60	.133			6.85	9.65	16.50	21.50
1494	45 MPH avg., cycle 8 miles		156	.051			2.63	3.71	6.34	8.30
1496	cycle 10 miles		144	.056			2.85	4.02	6.87	9
1498	cycle 20 miles		108	.074			3.80	5.35	9.15	11.95
1500	cycle 30 miles		84	.095			4.89	6.90	11.79	15.40
1502	cycle 40 miles		72	.111			5.70	8.05	13.75	17.95
1504	cycle 50 miles		60	.133			6.85	9.65	16.50	21.50
1506	50 MPH avg., cycle 10 miles		156	.051			2.63	3.71	6.34	8.30
1508	cycle 20 miles		120	.067			3.42	4.83	8.25	10.75
1510	cycle 30 miles		96	.083			4.28	6.05	10.33	13.45
1512	cycle 40 miles		72	.111			5.70	8.05	13.75	17.95
1514	cycle 50 miles		60	.133			6.85	9.65	16.50	21.50
1614	15 MPH avg., cycle 0.5 mile, 30 min. wait/ld./uld.		180	.044			2.28	3.22	5.50	7.20
1616	cycle 1 mile		168	.048			2.44	3.45	5.89	7.70
1618	cycle 2 miles		144	.056			2.85	4.02	6.87	9
1620	cycle 4 miles		120	.067			3.42	4.83	8.25	10.75
1622	cycle 6 miles		108	.074			3.80	5.35	9.15	11.95
1624	cycle 8 miles		84	.095			4.89	6.90	11.79	15.40
1625	cycle 10 miles		84	.095			4.89	6.90	11.79	15.40
1626	20 MPH avg., cycle 0.5 mile		180	.044			2.28	3.22	5.50	7.20
1628	cycle 1 mile		168	.048			2.44	3.45	5.89	7.70
1630	cycle 2 miles		156	.051			2.63	3.71	6.34	8.30
1632	cycle 4 miles		132	.061			3.11	4.39	7.50	9.80
1634	cycle 6 miles		120	.067			3.42	4.83	8.25	10.75
1636	cycle 8 miles		108	.074			3.80	5.35	9.15	11.95
1638	cycle 10 miles		96	.083			4.28	6.05	10.33	13.45
1640	25 MPH avg., cycle 4 miles		144	.056			2.85	4.02	6.87	9
1642	cycle 6 miles		132	.061			3.11	4.39	7.50	9.80
1644	cycle 8 miles		108	.074			3.80	5.35	9.15	11.95
1646	cycle 10 miles		108	.074			3.80	5.35	9.15	11.95
1650	30 MPH avg., cycle 4 miles		144	.056			2.85	4.02	6.87	9
1652	cycle 6 miles		132	.061			3.11	4.39	7.50	9.80
1654	cycle 8 miles		120	.067			3.42	4.83	8.25	10.75
1656	cycle 10 miles		108	.074			3.80	5.35	9.15	11.95
1660	35 MPH avg., cycle 4 miles		156	.051			2.63	3.71	6.34	8.30
1662	cycle 6 miles		144	.056			2.85	4.02	6.87	9
1664	cycle 8 miles		132	.061			3.11	4.39	7.50	9.80
1666	cycle 10 miles		120	.067			3.42	4.83	8.25	10.75
1668	cycle 20 miles		84	.095			4.89	6.90	11.79	15.40
1669	cycle 30 miles	B-34B	72	.111	L.C.Y.		5.70	8.05	13.75	17.95
1670	cycle 40 miles		60	.133			6.85	9.65	16.50	21.50
1672	40 MPH avg., cycle 6 miles		144	.056			2.85	4.02	6.87	9
1674	cycle 8 miles		132	.061			3.11	4.39	7.50	9.80
1676	cycle 10 miles		120	.067			3.42	4.83	8.25	10.75
1678	cycle 20 miles		96	.083			4.28	6.05	10.33	13.45
1680	cycle 30 miles		72	.111			5.70	8.05	13.75	17.95

543

31 23 23.20 Hauling	Crew	Daily Output	Labor-Hours	Unit	Material	2021 Bare Costs Labor	Equipment	Total	Total Incl O&P
1682	cycle 40 miles	B-34B	60	.133	L.C.Y.		6.85	9.65	16.50	21.50
1684	cycle 50 miles		48	.167			8.55	12.05	20.60	27
1694	45 MPH avg., cycle 8 miles		144	.056			2.85	4.02	6.87	9
1696	cycle 10 miles		132	.061			3.11	4.39	7.50	9.80
1698	cycle 20 miles		96	.083			4.28	6.05	10.33	13.45
1700	cycle 30 miles		84	.095			4.89	6.90	11.79	15.40
1702	cycle 40 miles		60	.133			6.85	9.65	16.50	21.50
1704	cycle 50 miles		60	.133			6.85	9.65	16.50	21.50
1706	50 MPH avg., cycle 10 miles		132	.061			3.11	4.39	7.50	9.80
1708	cycle 20 miles		108	.074			3.80	5.35	9.15	11.95
1710	cycle 30 miles		84	.095			4.89	6.90	11.79	15.40
1712	cycle 40 miles		72	.111			5.70	8.05	13.75	17.95
1714	cycle 50 miles		60	.133			6.85	9.65	16.50	21.50
2000	Hauling, 8 C.Y. truck, small project cost per hour	B-34A	8	1	Hr.		51.50	51	102.50	138
2100	12 C.Y. truck	B-34B	8	1			51.50	72.50	124	162
2150	16.5 C.Y. truck	B-34C	8	1			51.50	79.50	131	170
2175	18 C.Y. 8 wheel truck	B-34I	8	1			51.50	94	145.50	186
2200	20 C.Y. truck	B-34D	8	1			51.50	81.50	133	172
2300	Grading at dump, or embankment if required, by dozer	B-10B	1000	.012	L.C.Y.		.65	1.52	2.17	2.69
2310	Spotter at fill or cut, if required	1 Clab	8	1	Hr.		44.50		44.50	70.50
9014	18 C.Y. truck, 8 wheels,15 min. wait/ld./uld.,15 MPH, cycle 0.5 mi.	B-34I	504	.016	L.C.Y.		.81	1.50	2.31	2.94
9016	cycle 1 mile		450	.018			.91	1.67	2.58	3.30
9018	cycle 2 miles		378	.021			1.09	1.99	3.08	3.92
9020	cycle 4 miles		270	.030			1.52	2.79	4.31	5.50
9022	cycle 6 miles		216	.037			1.90	3.49	5.39	6.85
9024	cycle 8 miles		180	.044			2.28	4.19	6.47	8.25
9025	cycle 10 miles		144	.056			2.85	5.25	8.10	10.30
9026	20 MPH avg., cycle 0.5 mile		522	.015			.79	1.44	2.23	2.84
9028	cycle 1 mile		468	.017			.88	1.61	2.49	3.17
9030	cycle 2 miles		414	.019			.99	1.82	2.81	3.58
9032	cycle 4 miles		324	.025			1.27	2.33	3.60	4.58
9034	cycle 6 miles		252	.032			1.63	2.99	4.62	5.90
9036	cycle 8 miles		216	.037			1.90	3.49	5.39	6.85
9038	cycle 10 miles		180	.044			2.28	4.19	6.47	8.25
9040	25 MPH avg., cycle 4 miles	B-34I	342	.023	L.C.Y.		1.20	2.20	3.40	4.33
9042	cycle 6 miles		288	.028			1.43	2.62	4.05	5.15
9044	cycle 8 miles		252	.032			1.63	2.99	4.62	5.90
9046	cycle 10 miles		216	.037			1.90	3.49	5.39	6.85
9050	30 MPH avg., cycle 4 miles		378	.021			1.09	1.99	3.08	3.92
9052	cycle 6 miles		324	.025			1.27	2.33	3.60	4.58
9054	cycle 8 miles		270	.030			1.52	2.79	4.31	5.50
9056	cycle 10 miles		234	.034			1.75	3.22	4.97	6.35
9060	35 MPH avg., cycle 4 miles		396	.020			1.04	1.90	2.94	3.74
9062	cycle 6 miles		342	.023			1.20	2.20	3.40	4.33
9064	cycle 8 miles		288	.028			1.43	2.62	4.05	5.15
9066	cycle 10 miles		270	.030			1.52	2.79	4.31	5.50
9068	cycle 20 miles		162	.049			2.53	4.65	7.18	9.15
9070	cycle 30 miles		126	.063			3.26	6	9.26	11.80
9072	cycle 40 miles		90	.089			4.56	8.35	12.91	16.50
9074	40 MPH avg., cycle 6 miles		360	.022			1.14	2.09	3.23	4.12
9076	cycle 8 miles		324	.025			1.27	2.33	3.60	4.58
9078	cycle 10 miles		288	.028			1.43	2.62	4.05	5.15
9080	cycle 20 miles		180	.044			2.28	4.19	6.47	8.25

544

31 23 23 – Fill

		Crew	Daily Output	Labor-Hours	Unit	Material	2021 Bare Costs Labor	Equipment	Total	Total Incl O&P
31 23 23.20 Hauling										
9082	cycle 30 miles	B-34I	144	.056	L.C.Y.		2.85	5.25	8.10	10.30
9084	cycle 40 miles		108	.074			3.80	7	10.80	13.75
9086	cycle 50 miles		90	.089			4.56	8.35	12.91	16.50
9094	45 MPH avg., cycle 8 miles		324	.025			1.27	2.33	3.60	4.58
9096	cycle 10 miles		306	.026			1.34	2.46	3.80	4.85
9098	cycle 20 miles		198	.040			2.07	3.81	5.88	7.50
9100	cycle 30 miles		144	.056			2.85	5.25	8.10	10.30
9102	cycle 40 miles		126	.063			3.26	6	9.26	11.80
9104	cycle 50 miles		108	.074			3.80	7	10.80	13.75
9106	50 MPH avg., cycle 10 miles		324	.025			1.27	2.33	3.60	4.58
9108	cycle 20 miles		216	.037			1.90	3.49	5.39	6.85
9110	cycle 30 miles		162	.049			2.53	4.65	7.18	9.15
9112	cycle 40 miles		126	.063			3.26	6	9.26	11.80
9114	cycle 50 miles		108	.074			3.80	7	10.80	13.75
9214	20 min. wait/ld./uld.,15 MPH, cycle 0.5 mi.		396	.020			1.04	1.90	2.94	3.74
9216	cycle 1 mile		360	.022			1.14	2.09	3.23	4.12
9218	cycle 2 miles		306	.026			1.34	2.46	3.80	4.85
9220	cycle 4 miles		234	.034			1.75	3.22	4.97	6.35
9222	cycle 6 miles		198	.040			2.07	3.81	5.88	7.50
9224	cycle 8 miles		162	.049			2.53	4.65	7.18	9.15
9225	cycle 10 miles		144	.056			2.85	5.25	8.10	10.30
9226	20 MPH avg., cycle 0.5 mile		396	.020			1.04	1.90	2.94	3.74
9228	cycle 1 mile	B-34I	378	.021	L.C.Y.		1.09	1.99	3.08	3.92
9230	cycle 2 miles		324	.025			1.27	2.33	3.60	4.58
9232	cycle 4 miles		270	.030			1.52	2.79	4.31	5.50
9234	cycle 6 miles		216	.037			1.90	3.49	5.39	6.85
9236	cycle 8 miles		198	.040			2.07	3.81	5.88	7.50
9238	cycle 10 miles		162	.049			2.53	4.65	7.18	9.15
9240	25 MPH avg., cycle 4 miles		288	.028			1.43	2.62	4.05	5.15
9242	cycle 6 miles		252	.032			1.63	2.99	4.62	5.90
9244	cycle 8 miles		216	.037			1.90	3.49	5.39	6.85
9246	cycle 10 miles		198	.040			2.07	3.81	5.88	7.50
9250	30 MPH avg., cycle 4 miles		306	.026			1.34	2.46	3.80	4.85
9252	cycle 6 miles		270	.030			1.52	2.79	4.31	5.50
9254	cycle 8 miles		234	.034			1.75	3.22	4.97	6.35
9256	cycle 10 miles		216	.037			1.90	3.49	5.39	6.85
9260	35 MPH avg., cycle 4 miles		324	.025			1.27	2.33	3.60	4.58
9262	cycle 6 miles		288	.028			1.43	2.62	4.05	5.15
9264	cycle 8 miles		252	.032			1.63	2.99	4.62	5.90
9266	cycle 10 miles		234	.034			1.75	3.22	4.97	6.35
9268	cycle 20 miles		162	.049			2.53	4.65	7.18	9.15
9270	cycle 30 miles		108	.074			3.80	7	10.80	13.75
9272	cycle 40 miles		90	.089			4.56	8.35	12.91	16.50
9274	40 MPH avg., cycle 6 miles		288	.028			1.43	2.62	4.05	5.15
9276	cycle 8 miles		270	.030			1.52	2.79	4.31	5.50
9278	cycle 10 miles		234	.034			1.75	3.22	4.97	6.35
9280	cycle 20 miles		162	.049			2.53	4.65	7.18	9.15
9282	cycle 30 miles		126	.063			3.26	6	9.26	11.80
9284	cycle 40 miles		108	.074			3.80	7	10.80	13.75
9286	cycle 50 miles		90	.089			4.56	8.35	12.91	16.50
9294	45 MPH avg., cycle 8 miles		270	.030			1.52	2.79	4.31	5.50
9296	cycle 10 miles		252	.032			1.63	2.99	4.62	5.90
9298	cycle 20 miles		180	.044			2.28	4.19	6.47	8.25

545

31 23 23.20 Hauling	Crew	Daily Output	Labor-Hours	Unit	Material	2021 Bare Costs Labor	2021 Bare Costs Equipment	Total	Total Incl O&P
9300	cycle 30 miles	B-34I	144	.056	L.C.Y.		2.85	5.25	8.10	10.30
9302	cycle 40 miles		108	.074			3.80	7	10.80	13.75
9304	cycle 50 miles		90	.089			4.56	8.35	12.91	16.50
9306	50 MPH avg., cycle 10 miles		270	.030			1.52	2.79	4.31	5.50
9308	cycle 20 miles		198	.040			2.07	3.81	5.88	7.50
9310	cycle 30 miles		144	.056			2.85	5.25	8.10	10.30
9312	cycle 40 miles		126	.063			3.26	6	9.26	11.80
9314	cycle 50 miles		108	.074			3.80	7	10.80	13.75
9414	25 min. wait/ld./uld.,15 MPH, cycle 0.5 mi.		306	.026			1.34	2.46	3.80	4.85
9416	cycle 1 mile		288	.028			1.43	2.62	4.05	5.15
9418	cycle 2 miles	B-34I	252	.032	L.C.Y.		1.63	2.99	4.62	5.90
9420	cycle 4 miles		198	.040			2.07	3.81	5.88	7.50
9422	cycle 6 miles		180	.044			2.28	4.19	6.47	8.25
9424	cycle 8 miles		144	.056			2.85	5.25	8.10	10.30
9425	cycle 10 miles		126	.063			3.26	6	9.26	11.80
9426	20 MPH avg., cycle 0.5 mile		324	.025			1.27	2.33	3.60	4.58
9428	cycle 1 mile		306	.026			1.34	2.46	3.80	4.85
9430	cycle 2 miles		270	.030			1.52	2.79	4.31	5.50
9432	cycle 4 miles		234	.034			1.75	3.22	4.97	6.35
9434	cycle 6 miles		198	.040			2.07	3.81	5.88	7.50
9436	cycle 8 miles		180	.044			2.28	4.19	6.47	8.25
9438	cycle 10 miles		144	.056			2.85	5.25	8.10	10.30
9440	25 MPH avg., cycle 4 miles		252	.032			1.63	2.99	4.62	5.90
9442	cycle 6 miles		216	.037			1.90	3.49	5.39	6.85
9444	cycle 8 miles		198	.040			2.07	3.81	5.88	7.50
9446	cycle 10 miles		180	.044			2.28	4.19	6.47	8.25
9450	30 MPH avg., cycle 4 miles		252	.032			1.63	2.99	4.62	5.90
9452	cycle 6 miles		234	.034			1.75	3.22	4.97	6.35
9454	cycle 8 miles		198	.040			2.07	3.81	5.88	7.50
9456	cycle 10 miles		180	.044			2.28	4.19	6.47	8.25
9460	35 MPH avg., cycle 4 miles		270	.030			1.52	2.79	4.31	5.50
9462	cycle 6 miles		234	.034			1.75	3.22	4.97	6.35
9464	cycle 8 miles		216	.037			1.90	3.49	5.39	6.85
9466	cycle 10 miles		198	.040			2.07	3.81	5.88	7.50
9468	cycle 20 miles		144	.056			2.85	5.25	8.10	10.30
9470	cycle 30 miles		108	.074			3.80	7	10.80	13.75
9472	cycle 40 miles		90	.089			4.56	8.35	12.91	16.50
9474	40 MPH avg., cycle 6 miles		252	.032			1.63	2.99	4.62	5.90
9476	cycle 8 miles		234	.034			1.75	3.22	4.97	6.35
9478	cycle 10 miles		216	.037			1.90	3.49	5.39	6.85
9480	cycle 20 miles		144	.056			2.85	5.25	8.10	10.30
9482	cycle 30 miles		126	.063			3.26	6	9.26	11.80
9484	cycle 40 miles		90	.089			4.56	8.35	12.91	16.50
9486	cycle 50 miles		90	.089			4.56	8.35	12.91	16.50
9494	45 MPH avg., cycle 8 miles		234	.034			1.75	3.22	4.97	6.35
9496	cycle 10 miles		216	.037			1.90	3.49	5.39	6.85
9498	cycle 20 miles		162	.049			2.53	4.65	7.18	9.15
9500	cycle 30 miles		126	.063			3.26	6	9.26	11.80
9502	cycle 40 miles		108	.074			3.80	7	10.80	13.75
9504	cycle 50 miles		90	.089			4.56	8.35	12.91	16.50
9506	50 MPH avg., cycle 10 miles		234	.034			1.75	3.22	4.97	6.35
9508	cycle 20 miles		180	.044			2.28	4.19	6.47	8.25
9510	cycle 30 miles	B-34I	144	.056	L.C.Y.		2.85	5.25	8.10	10.30

546

31 23 23 – Fill

31 23 23.20 Hauling		Crew	Daily Output	Labor-Hours	Unit	Material	2021 Bare Costs Labor	2021 Bare Costs Equipment	Total	Total Incl O&P
9512	cycle 40 miles	B-34I	108	.074	L.C.Y.		3.80	7	10.80	13.75
9514	cycle 50 miles		90	.089			4.56	8.35	12.91	16.50
9614	30 min. wait/ld./uld.,15 MPH, cycle 0.5 mi.		270	.030			1.52	2.79	4.31	5.50
9616	cycle 1 mile		252	.032			1.63	2.99	4.62	5.90
9618	cycle 2 miles		216	.037			1.90	3.49	5.39	6.85
9620	cycle 4 miles		180	.044			2.28	4.19	6.47	8.25
9622	cycle 6 miles		162	.049			2.53	4.65	7.18	9.15
9624	cycle 8 miles		126	.063			3.26	6	9.26	11.80
9625	cycle 10 miles		126	.063			3.26	6	9.26	11.80
9626	20 MPH avg., cycle 0.5 mile		270	.030			1.52	2.79	4.31	5.50
9628	cycle 1 mile		252	.032			1.63	2.99	4.62	5.90
9630	cycle 2 miles		234	.034			1.75	3.22	4.97	6.35
9632	cycle 4 miles		198	.040			2.07	3.81	5.88	7.50
9634	cycle 6 miles		180	.044			2.28	4.19	6.47	8.25
9636	cycle 8 miles		162	.049			2.53	4.65	7.18	9.15
9638	cycle 10 miles		144	.056			2.85	5.25	8.10	10.30
9640	25 MPH avg., cycle 4 miles		216	.037			1.90	3.49	5.39	6.85
9642	cycle 6 miles		198	.040			2.07	3.81	5.88	7.50
9644	cycle 8 miles		180	.044			2.28	4.19	6.47	8.25
9646	cycle 10 miles		162	.049			2.53	4.65	7.18	9.15
9650	30 MPH avg., cycle 4 miles		216	.037			1.90	3.49	5.39	6.85
9652	cycle 6 miles		198	.040			2.07	3.81	5.88	7.50
9654	cycle 8 miles		180	.044			2.28	4.19	6.47	8.25
9656	cycle 10 miles		162	.049			2.53	4.65	7.18	9.15
9660	35 MPH avg., cycle 4 miles		234	.034			1.75	3.22	4.97	6.35
9662	cycle 6 miles		216	.037			1.90	3.49	5.39	6.85
9664	cycle 8 miles		198	.040			2.07	3.81	5.88	7.50
9666	cycle 10 miles		180	.044			2.28	4.19	6.47	8.25
9668	cycle 20 miles		126	.063			3.26	6	9.26	11.80
9670	cycle 30 miles		108	.074			3.80	7	10.80	13.75
9672	cycle 40 miles		90	.089			4.56	8.35	12.91	16.50
9674	40 MPH avg., cycle 6 miles		216	.037			1.90	3.49	5.39	6.85
9676	cycle 8 miles		198	.040			2.07	3.81	5.88	7.50
9678	cycle 10 miles		180	.044			2.28	4.19	6.47	8.25
9680	cycle 20 miles		144	.056			2.85	5.25	8.10	10.30
9682	cycle 30 miles		108	.074			3.80	7	10.80	13.75
9684	cycle 40 miles		90	.089			4.56	8.35	12.91	16.50
9686	cycle 50 miles		72	.111			5.70	10.45	16.15	20.50
9694	45 MPH avg., cycle 8 miles		216	.037			1.90	3.49	5.39	6.85
9696	cycle 10 miles		198	.040			2.07	3.81	5.88	7.50
9698	cycle 20 miles	B-34I	144	.056	L.C.Y.		2.85	5.25	8.10	10.30
9700	cycle 30 miles		126	.063			3.26	6	9.26	11.80
9702	cycle 40 miles		108	.074			3.80	7	10.80	13.75
9704	cycle 50 miles		90	.089			4.56	8.35	12.91	16.50
9706	50 MPH avg., cycle 10 miles		198	.040			2.07	3.81	5.88	7.50
9708	cycle 20 miles		162	.049			2.53	4.65	7.18	9.15
9710	cycle 30 miles		126	.063			3.26	6	9.26	11.80
9712	cycle 40 miles		108	.074			3.80	7	10.80	13.75
9714	cycle 50 miles		90	.089			4.56	8.35	12.91	16.50

31 23 23 – Fill

31 23 23.23 Compaction	Crew	Daily Output	Labor-Hours	Unit	Material	2021 Bare Costs Labor	2021 Bare Costs Equipment	Total	Total Incl O&P
0010 COMPACTION									
5000 Riding, vibrating roller, 6" lifts, 2 passes	B-10Y	3000	.004	E.C.Y.		.22	.19	.41	.55
5020 3 passes		2300	.005			.28	.25	.53	.72
5040 4 passes		1900	.006			.34	.31	.65	.88
5050 8" lifts, 2 passes		4100	.003			.16	.14	.30	.41
5060 12" lifts, 2 passes		5200	.002			.12	.11	.23	.32
5080 3 passes		3500	.003			.19	.17	.36	.47
5100 4 passes		2600	.005			.25	.22	.47	.64
5600 Sheepsfoot or wobbly wheel roller, 6" lifts, 2 passes	B-10G	2400	.005			.27	.57	.84	1.04
5620 3 passes		1735	.007			.37	.79	1.16	1.45
5640 4 passes		1300	.009			.50	1.05	1.55	1.93
5680 12" lifts, 2 passes		5200	.002			.12	.26	.38	.49
5700 3 passes		3500	.003			.19	.39	.58	.72
5720 4 passes		2600	.005			.25	.52	.77	.97
7000 Walk behind, vibrating plate 18" wide, 6" lifts, 2 passes	A-1D	200	.040			1.78	.16	1.94	3.01
7020 3 passes		185	.043			1.92	.17	2.09	3.25
7040 4 passes		140	.057			2.54	.23	2.77	4.29
7200 12" lifts, 2 passes, 21" wide	A-1E	560	.014			.63	.30	.93	1.34
7220 3 passes		375	.021			.95	.44	1.39	2
7240 4 passes		280	.029			1.27	.59	1.86	2.67
7500 Vibrating roller 24" wide, 6" lifts, 2 passes	B-10A	420	.029			1.55	.40	1.95	2.87
7520 3 passes		280	.043			2.32	.60	2.92	4.30
7540 4 passes		210	.057			3.09	.79	3.88	5.70
7600 12" lifts, 2 passes		840	.014			.77	.20	.97	1.43
7620 3 passes		560	.021			1.16	.30	1.46	2.15
7640 4 passes		420	.029			1.55	.40	1.95	2.87
8000 Rammer tamper, 6" to 11", 4" lifts, 2 passes	A-1F	130	.062			2.73	.36	3.09	4.75
8050 3 passes		97	.082			3.66	.48	4.14	6.40
8100 4 passes		65	.123			5.45	.72	6.17	9.50
8200 8" lifts, 2 passes		260	.031			1.37	.18	1.55	2.38
8250 3 passes		195	.041			1.82	.24	2.06	3.17
8300 4 passes		130	.062			2.73	.36	3.09	4.75
8400 13" to 18", 4" lifts, 2 passes	A-1G	390	.021	E.C.Y.		.91	.14	1.05	1.60
8450 3 passes		290	.028			1.22	.19	1.41	2.16
8500 4 passes		195	.041			1.82	.28	2.10	3.21
8600 8" lifts, 2 passes		780	.010			.46	.07	.53	.81
8650 3 passes		585	.014			.61	.09	.70	1.07
8700 4 passes		390	.021			.91	.14	1.05	1.60
9000 Water, 3,000 gal. truck, 3 mile haul	B-45	1888	.008		1.23	.47	.44	2.14	2.58
9010 6 mile haul		1444	.011		1.23	.61	.58	2.42	2.94
9020 12 mile haul		1000	.016		1.23	.88	.83	2.94	3.66
9030 6,000 gal. wagon, 3 mile haul	B-59	2000	.004		1.23	.21	.23	1.67	1.94
9040 6 mile haul	"	1600	.005		1.23	.26	.29	1.78	2.08

31 23 23.24 Compaction, Structural

0010 COMPACTION, STRUCTURAL R312323-30									
0020 Steel wheel tandem roller, 5 tons	B-10E	8	1.500	Hr.		81	32.50	113.50	163
0050 Air tamp, 6" to 8" lifts, common fill	B-9	250	.160	E.C.Y.		7.15	1.42	8.57	12.95
0060 Select fill	"	300	.133			5.95	1.19	7.14	10.80
0600 Vibratory plate, 8" lifts, common fill	A-1D	200	.040			1.78	.16	1.94	3.01
0700 Select fill	"	216	.037			1.64	.15	1.79	2.78
9000 Minimum labor/equipment charge	1 Clab	4	2	Job		89		89	141

31 25 Erosion and Sedimentation Controls

31 25 14 – Stabilization Measures for Erosion and Sedimentation Control

31 25 14.16 Rolled Erosion Control Mats and Blankets

31 25 14.16 Rolled Erosion Control Mats and Blankets		Crew	Daily Output	Labor-Hours	Unit	Material	2021 Bare Costs Labor	2021 Bare Costs Equipment	Total	Total Incl O&P
0010	**ROLLED EROSION CONTROL MATS AND BLANKETS**									
0020	Jute mesh, 100 S.Y. per roll, 4' wide, stapled	G	B-80A	2400	.010	S.Y.	1	.44	.35	1.79	2.20
0070	Paper biodegradable mesh	G	B-1	2500	.010		.14	.43		.57	.84
0100	Plastic netting, stapled, 2" x 1" mesh, 20 mil	G	"	2500	.010		.31	.43		.74	1.03
0120	Revegetation mat, webbed	G	2 Clab	1000	.016		2.72	.71		3.43	4.12
0200	Polypropylene mesh, stapled, 6.5 oz./S.Y.	G	B-1	2500	.010		1.83	.43		2.26	2.70
0300	Tobacco netting, or jute mesh #2, stapled	G	"	2500	.010		.26	.43		.69	.98
0600	Straw in polymeric netting, biodegradable log		A-2	1000	.024	L.F.	2.25	1.10	.20	3.55	4.45
0705	Sediment Log, Filter Sock, 9"			1000	.024		2.25	1.10	.20	3.55	4.45
0710	Sediment Log, Filter Sock, 12"			1000	.024		3.50	1.10	.20	4.80	5.80
1000	Silt fence, install and remove	G	B-62	650	.037		.44	1.78	.28	2.50	3.59

31 31 Soil Treatment

31 31 16 – Termite Control

31 31 16.13 Chemical Termite Control

31 31 16.13 Chemical Termite Control		Crew	Daily Output	Labor-Hours	Unit	Material	2021 Bare Costs Labor	2021 Bare Costs Equipment	Total	Total Incl O&P
0010	**CHEMICAL TERMITE CONTROL**									
0020	Slab and walls, residential	1 Skwk	1200	.007	SF Flr.	.32	.38		.70	.95
0030	SS mesh, no chemicals, avg 1,400 S.F. home, min	G		1000	.008		.32	.46		.78	1.07
0100	Commercial, minimum		2496	.003		.33	.18		.51	.65
0200	Maximum		1645	.005		.50	.28		.78	.98
0390	Minimum labor/equipment charge	1 Skwk	4	2	Job		114		114	181
0400	Insecticides for termite control, minimum		14	.571	Gal.	2.14	32.50		34.64	54
0500	Maximum		11	.727	"	4.29	41.50		45.79	70.50

31 41 Shoring

31 41 13 – Timber Shoring

31 41 13.10 Building Shoring

31 41 13.10 Building Shoring		Crew	Daily Output	Labor-Hours	Unit	Material	2021 Bare Costs Labor	2021 Bare Costs Equipment	Total	Total Incl O&P
0010	**BUILDING SHORING**									
0020	Shoring, existing building, with timber, no salvage allowance	B-51	2.20	21.818	M.B.F.	1,250	990	90.50	2,330.50	3,050
1000	On cribbing with 35 ton screw jacks, per box and jack		3.60	13.333	Jack	65	605	55	725	1,100
1090	Minimum labor/equipment charge		2	24	Ea.		1,100	99.50	1,199.50	1,850
1100	Masonry openings in walls, see Section 02 41 19.16									

31 41 16 – Sheet Piling

31 41 16.10 Sheet Piling Systems

31 41 16.10 Sheet Piling Systems		Crew	Daily Output	Labor-Hours	Unit	Material	2021 Bare Costs Labor	2021 Bare Costs Equipment	Total	Total Incl O&P
0010	**SHEET PILING SYSTEMS**									
0020	Sheet piling, 50,000 psi steel, not incl. wales, 22 psf, left in place	B-40	10.81	5.920	Ton	1,900	340	325	2,565	2,975
0100	Drive, extract & salvage	R314116-40	6	10.667	"	530	610	590	1,730	2,200
1200	15' deep excavation, 22 psf, left in place		983	.065	S.F.	22	3.72	3.59	29.31	34
1300	Drive, extract & salvage		545	.117	"	5.95	6.70	6.50	19.15	24.50
2100	Rent steel sheet piling and wales, first month				Ton	330			330	365
2200	Per added month				"	33			33	36.50
3900	Wood, solid sheeting, incl. wales, braces and spacers,									
3910	drive, extract & salvage, 8' deep excavation	B-31	330	.121	S.F.	3.34	5.70	.78	9.82	13.60
4520	Left in place, 8' deep, 55 S.F./hr.		440	.091	"	6	4.26	.58	10.84	14.05
4990	Minimum labor/equipment charge		2	20	Job		935	128	1,063	1,650
5000	For treated lumber add cost of treatment to lumber									

31 43 Concrete Raising

31 43 13 – Pressure Grouting

31 43 13.13 Concrete Pressure Grouting	Crew	Daily Output	Labor-Hours	Unit	Material	2021 Bare Costs Labor	Equipment	Total	Total Incl O&P
0010 **CONCRETE PRESSURE GROUTING**									
0020 Grouting, pressure, cement & sand, 1:1 mix, minimum	B-61	124	.323	Bag	21.50	15.15	2.62	39.27	50.50
0100 Maximum		51	.784	"	21.50	37	6.35	64.85	89
0200 Cement and sand, 1:1 mix, minimum		250	.160	C.F.	43	7.50	1.30	51.80	61
0300 Maximum		100	.400		64.50	18.80	3.25	86.55	105
0400 Epoxy cement grout, minimum		137	.292		805	13.75	2.37	821.12	910
0500 Maximum		57	.702		805	33	5.70	843.70	945
0700 Alternate pricing method: (Add for materials)									
0710 5 person crew and equipment	B-61	1	40	Day		1,875	325	2,200	3,325

31 46 Needle Beams

31 46 13 – Cantilever Needle Beams

31 46 13.10 Needle Beams

	Crew	Daily Output	Labor-Hours	Unit	Material	2021 Bare Costs Labor	Equipment	Total	Total Incl O&P
0010 **NEEDLE BEAMS**									
0011 Incl. wood shoring 10' x 10' opening									
0400 Block, concrete, 8" thick	B-9	7.10	5.634	Ea.	67.50	252	50	369.50	530
0420 12" thick		6.70	5.970		73	267	53	393	565
0800 Brick, 4" thick with 8" backup block		5.70	7.018		73	315	62.50	450.50	650
1000 Brick, solid, 8" thick		6.20	6.452		67.50	289	57.50	414	600
1040 12" thick		4.90	8.163		73	365	72.50	510.50	740
1080 16" thick		4.50	8.889		84	400	79	563	815
2000 Add for additional floors of shoring	B-1	6	4		67.50	180		247.50	360
9000 Minimum labor/equipment charge	"	2	12	Job		540		540	860

31 48 Underpinning

31 48 13 – Underpinning Piers

31 48 13.10 Underpinning Foundations

	Crew	Daily Output	Labor-Hours	Unit	Material	2021 Bare Costs Labor	Equipment	Total	Total Incl O&P
0010 **UNDERPINNING FOUNDATIONS**									
0011 Including excavation,									
0020 forming, reinforcing, concrete and equipment									
0100 5' to 16' below grade, 100 to 500 C.Y.	B-52	2.30	24.348	C.Y.	345	1,225	249	1,819	2,600
0200 Over 500 C.Y.		2.50	22.400		310	1,125	229	1,664	2,400
0400 16' to 25' below grade, 100 to 500 C.Y.		2	28		380	1,425	286	2,091	2,975
0500 Over 500 C.Y.		2.10	26.667		355	1,350	273	1,978	2,850
0700 26' to 40' below grade, 100 to 500 C.Y.		1.60	35		410	1,775	360	2,545	3,675
0800 Over 500 C.Y.		1.80	31.111		380	1,575	320	2,275	3,275
0900 For under 50 C.Y., add					10%	40%			

550

For customer support on your Commercial Renovation Costs with RSMeans Data, call 800.448.8182.

31 62 Driven Piles

31 62 13 – Concrete Piles

31 62 13.23 Prestressed Concrete Piles

	Crew	Daily Output	Labor-Hours	Unit	Material	2021 Bare Costs Labor	2021 Bare Costs Equipment	Total	Total Incl O&P
0010 **PRESTRESSED CONCRETE PILES**, 200 piles									
0020 Unless specified otherwise, not incl. pile caps or mobilization									
3100 Precast, prestressed, 40' long, 10" thick, square	B-19	700	.091	V.L.F.	14.50	5.20	2.92	22.62	27.50
3200 12" thick, square		680	.094		26	5.40	3.01	34.41	40.50
3400 14" thick, square	↓	600	.107	↓	29.50	6.10	3.41	39.01	46

31 62 16 – Steel Piles

31 62 16.13 Steel Piles

	Crew	Daily Output	Labor-Hours	Unit	Material	2021 Bare Costs Labor	2021 Bare Costs Equipment	Total	Total Incl O&P
0010 **STEEL PILES**									
0100 Step tapered, round, concrete filled									
0110 8" tip, 12" butt, 60 ton capacity, 30' depth	B-19	760	.084	V.L.F.	17.75	4.81	2.69	25.25	30
0120 60' depth with extension		740	.086		37.50	4.94	2.76	45.20	52.50
0130 80' depth with extensions		700	.091		58.50	5.20	2.92	66.62	76
0250 "H" Sections, 50' long, HP8 x 36	↓	640	.100	↓	17.75	5.70	3.20	26.65	32

31 62 19 – Timber Piles

31 62 19.10 Wood Piles

	Crew	Daily Output	Labor-Hours	Unit	Material	2021 Bare Costs Labor	2021 Bare Costs Equipment	Total	Total Incl O&P
0010 **WOOD PILES**									
0011 Friction or end bearing, not including									
0050 mobilization or demobilization									
0100 ACZA treated piles, 1.0 lb/C.F., up to 30' long, 12" butts, 8" points	B-19	625	.102	V.L.F.	15.05	5.85	3.27	24.17	29.50
0200 30' to 39' long, 12" butts, 8" points	B-19	700	.091	V.L.F.	15.90	5.20	2.92	24.02	29
0800 ACZA Treated piles, 1.5 lb./C.F.									
0810 friction or end bearing, ASTM class B									
1000 Up to 30' long, 12" butts, 8" points	B-19	625	.102	V.L.F.	17.85	5.85	3.27	26.97	32.50
1100 30' to 39' long, 12" butts, 8" points		700	.091		18.50	5.20	2.92	26.62	32
2700 Mobilization for 10,000 L.F. pile job, add		3300	.019			1.11	.62	1.73	2.43
2800 25,000 L.F. pile job, add	↓	8500	.008	↓		.43	.24	.67	.94

31 62 23 – Composite Piles

31 62 23.13 Concrete-Filled Steel Piles

	Crew	Daily Output	Labor-Hours	Unit	Material	2021 Bare Costs Labor	2021 Bare Costs Equipment	Total	Total Incl O&P
0010 **CONCRETE-FILLED STEEL PILES** no mobilization or demobilization									
2600 Pipe piles, 50' L, 8" diam., 29 lb./L.F., no concrete	B-19	500	.128	V.L.F.	21	7.30	4.09	32.39	39
2700 Concrete filled		460	.139		26	7.95	4.45	38.40	46
2900 10" diameter, 34 lb./L.F., no concrete		500	.128		25.50	7.30	4.09	36.89	44
3000 Concrete filled		450	.142		31	8.10	4.54	43.64	52
3200 12" diameter, 44 lb./L.F., no concrete		475	.135		33	7.70	4.30	45	53.50
3300 Concrete filled		415	.154		37.50	8.80	4.93	51.23	60.50
3500 14" diameter, 46 lb./L.F., no concrete		430	.149		35.50	8.50	4.76	48.76	57.50
3600 Concrete filled	↓	355	.180	↓	40.50	10.30	5.75	56.55	67

31 63 Bored Piles

31 63 26 – Drilled Caissons

31 63 26.13 Fixed End Caisson Piles

	Crew	Daily Output	Labor-Hours	Unit	Material	2021 Bare Costs Labor	2021 Bare Costs Equipment	Total	Total Incl O&P
0010 **FIXED END CAISSON PILES**									
0015 Including excavation, concrete, 50 lb. reinforcing									
0020 per C.Y., not incl. mobilization, boulder removal, disposal									
0100 Open style, machine drilled, to 50' deep, in stable ground, no R316326-60									
0110 casings or ground water, 18" diam., 0.065 C.Y./L.F.	B-43	200	.240	V.L.F.	10.10	11.75	3.84	25.69	34
0200 24" diameter, 0.116 C.Y./L.F.		190	.253	"	18.05	12.35	4.04	34.44	44
0210 4' bell diameter, add		20	2.400	Ea.	275	117	38.50	430.50	530
0500 48" diameter, 0.465 C.Y./L.F.	↓	100	.480	V.L.F.	72.50	23.50	7.70	103.70	125

31 63 Bored Piles

31 63 26 – Drilled Caissons

31 63 26.13 Fixed End Caisson Piles

		Crew	Daily Output	Labor-Hours	Unit	Material	2021 Bare Costs Labor	Equipment	Total	Total Incl O&P
0510	9' bell diameter, add	B-43	2	24	Ea.	1,575	1,175	385	3,135	4,025
1200	Open style, machine drilled, to 50' deep, in wet ground, pulled									
1220	casing and pumping									
1400	24" diameter, 0.116 C.Y./L.F.	B-48	125	.448	V.L.F.	18.05	22.50	8.40	48.95	64
1410	4' bell diameter, add	"	19.80	2.828	Ea.	275	141	53	469	580
1700	48" diameter, 0.465 C.Y./L.F.	B-49	55	1.600	V.L.F.	72.50	83.50	29.50	185.50	244
1710	9' bell diameter, add	"	3.30	26.667	Ea.	1,575	1,400	495	3,470	4,500
2300	Open style, machine drilled, to 50' deep, in soft rocks and									
2320	medium hard shales									
2500	24" diameter, 0.116 C.Y./L.F.	B-49	30	2.933	V.L.F.	18.05	153	54.50	225.55	320
2510	4' bell diameter, add	B-49	10.90	8.073	Ea.	275	420	150	845	1,125
2800	48" diameter, 0.465 C.Y./L.F.		10	8.800	V.L.F.	72.50	460	163	695.50	985
2810	9' bell diameter, add		1.10	80	Ea.	1,175	4,175	1,475	6,825	9,525
3600	For rock excavation, sockets, add, minimum		120	.733	C.F.		38.50	13.60	52.10	75.50
3650	Average		95	.926			48.50	17.20	65.70	95.50
3700	Maximum		48	1.833			96	34	130	189
3900	For 50' to 100' deep, add				V.L.F.				7%	7%
4000	For 100' to 150' deep, add								25%	25%
4100	For 150' to 200' deep, add								30%	30%
4200	For casings left in place, add				Lb.	1.39			1.39	1.53
4300	For other than 50 lb. reinf. per C.Y., add or deduct				"	1.31			1.31	1.44
4400	For steel I-beam cores, add	B-49	8.30	10.602	Ton	2,100	555	197	2,852	3,425
4500	Load and haul excess excavation, 2 miles	B-34B	178	.045	L.C.Y.		2.31	3.25	5.56	7.25
4600	For mobilization, 50 mile radius, rig to 36"	B-43	2	24	Ea.		1,175	385	1,560	2,275
4650	Rig to 84"	B-48	1.75	32			1,600	600	2,200	3,175
4700	For low headroom, add								50%	50%
4750	For difficult access, add								25%	25%

31 63 29 – Drilled Concrete Piers and Shafts

31 63 29.13 Uncased Drilled Concrete Piers

		Crew	Daily Output	Labor-Hours	Unit	Material	2021 Bare Costs Labor	Equipment	Total	Total Incl O&P
0010	**UNCASED DRILLED CONCRETE PIERS**									
0020	Unless specified otherwise, not incl. pile caps or mobilization									
0800	Cast in place friction pile, 50' long, fluted,									
0810	tapered steel, 4,000 psi concrete, no reinforcing									
0900	12" diameter, 7 ga.	B-19	600	.107	V.L.F.	29.50	6.10	3.41	39.01	46
1200	18" diameter, 7 ga.	"	480	.133	"	44	7.60	4.26	55.86	65
1300	End bearing, fluted, constant diameter,									
1320	4,000 psi concrete, no reinforcing									
1340	12" diameter, 7 ga.	B-19	600	.107	V.L.F.	31	6.10	3.41	40.51	47.50
1400	18" diameter, 7 ga.	"	480	.133	"	49.50	7.60	4.26	61.36	71

Estimating Tips

32 01 00 Operations and Maintenance of Exterior Improvements

- Recycling of asphalt pavement is becoming very popular and is an alternative to removal and replacement. It can be a good value engineering proposal if removed pavement can be recycled, either at the project site or at another site that is reasonably close to the project site. Sections on repair of flexible and rigid pavement are included.

32 10 00 Bases, Ballasts, and Paving

- When estimating paving, keep in mind the project schedule. Also note that prices for asphalt and concrete are generally higher in the cold seasons. Lines for pavement markings, including tactile warning systems and fence lines, are included.

32 90 00 Planting

- The timing of planting and guarantee specifications often dictate the costs for establishing tree and shrub growth and a stand of grass or ground cover. Establish the work performance schedule to coincide with the local planting season. Maintenance and growth guarantees can add 20–100% to the total landscaping cost and can be contractually cumbersome. The cost to replace trees and shrubs can be as high as 5% of the total cost, depending on the planting zone, soil conditions, and time of year.

Reference Numbers

Reference numbers are shown at the beginning of some major classifications. These numbers refer to related items in the Reference Section. The reference information may be an estimating procedure, an alternate pricing method, or technical information.

Note: Not all subdivisions listed here necessarily appear. ■

Same Data. Simplified.

Enjoy the convenience and efficiency of accessing your costs anywhere:

- **Skip the multiplier** by setting your location
- **Quickly search,** edit, favorite and share costs
- **Stay on top of price changes** with automatic updates

Discover more at rsmeans.com/online

32 01 11 – Paving Cleaning

32 01 11.51 Rubber and Paint Removal From Paving

		Crew	Daily Output	Labor-Hours	Unit	Material	2021 Bare Costs Labor	2021 Bare Costs Equipment	Total	Total Incl O&P
0010	**RUBBER AND PAINT REMOVAL FROM PAVING**									
0015	Does not include traffic control costs									
0020	See other items in Section 32 17 23.13									
0100	Remove permanent painted traffic lines and markings	B-78A	500	.016	C.L.F.		.89	1.99	2.88	3.58
0200	Temporary traffic line tape	2 Clab	1500	.011	L.F.		.47		.47	.75
0300	Thermoplastic traffic lines and markings	B-79A	500	.024	C.L.F.		1.33	3.02	4.35	5.40
0400	Painted pavement markings	B-78B	500	.036	S.F.		1.64	.82	2.46	3.51
1000	Prepare and clean, 25,000 S.Y.	B-91A	7.36	2.174	Mile		106	114	220	294
2000	Minimum labor/equipment charge	1 Clab	2	4	Job		178		178	283

32 01 13 – Flexible Paving Surface Treatment

32 01 13.61 Slurry Seal (Latex Modified)

		Crew	Daily Output	Labor-Hours	Unit	Material	2021 Bare Costs Labor	2021 Bare Costs Equipment	Total	Total Incl O&P
0010	**SLURRY SEAL (LATEX MODIFIED)**									
0011	Chip seal, slurry seal, and microsurfacing, see section 32 12 36									

32 01 13.62 Asphalt Surface Treatment

		Crew	Daily Output	Labor-Hours	Unit	Material	2021 Bare Costs Labor	2021 Bare Costs Equipment	Total	Total Incl O&P
0010	**ASPHALT SURFACE TREATMENT**									
3000	Pavement overlay, polypropylene									
3040	6 oz./S.Y., ideal conditions	B-63	10000	.004	S.Y.	.83	.19	.02	1.04	1.23
3190	Prime coat, bituminous, 0.28 gallon/S.Y.	B-45	2400	.007	C.S.F.	8.80	.37	.35	9.52	10.65
3200	Tack coat, emulsion, 0.05 gal./S.Y., 1,000 S.Y.		2500	.006	S.Y.	.35	.35	.33	1.03	1.32
3240	10,000 S.Y.		10000	.002		.28	.09	.08	.45	.54
3275	10,000 S.Y.		10000	.002		.53	.09	.08	.70	.81
3280	0.15 gal./S.Y., 1,000 S.Y.		2500	.006		.96	.35	.33	1.64	1.99
3320	10,000 S.Y.		10000	.002		.77	.09	.08	.94	1.08
3780	Rubberized asphalt (latex) seal	↓	5000	.003		1.44	.18	.17	1.79	2.04
5400	Thermoplastic coal-tar, Type I, small or irregular area	B-90	2400	.027		4.23	1.31	.93	6.47	7.75
5450	Roadway or large area		8000	.008		4.23	.39	.28	4.90	5.60
5500	Type II, small or irregular area	↓	2400	.027	↓	5.40	1.31	.93	7.64	9.05
6000	Gravel surfacing on asphalt, screened and rolled	B-11L	160	.100	C.Y.	23.50	5.15	6.70	35.35	41.50
7000	For subbase treatment, see Section 31 32 13									

32 01 13.64 Sand Seal

		Crew	Daily Output	Labor-Hours	Unit	Material	2021 Bare Costs Labor	2021 Bare Costs Equipment	Total	Total Incl O&P
0010	**SAND SEAL**									
2080	Sand sealing, sharp sand, asphalt emulsion, small area	B-91	10000	.006	S.Y.	1.57	.34	.23	2.14	2.51
2120	Roadway or large area	"	18000	.004	"	1.35	.19	.13	1.67	1.93
3000	Sealing random cracks, min. 1/2" wide, to 1-1/2", 1,000 L.F.	B-77	2800	.014	L.F.	1.56	.65	.38	2.59	3.18
3040	10,000 L.F.		4000	.010	"	1.08	.46	.27	1.81	2.22
3080	Alternate method, 1,000 L.F.		200	.200	Gal.	40	9.15	5.35	54.50	64
3120	10,000 L.F.	↓	325	.123	"	32	5.60	3.30	40.90	47.50
3200	Multi-cracks (flooding), 1 coat, small area	B-92	460	.070	S.Y.	2.45	3.12	2.64	8.21	10.55
3320	Large area		1425	.022	"	15.25	1.01	.85	17.11	19.35
3360	Alternate method, small area		115	.278	Gal.	14.95	12.50	10.55	38	48
3400	Large area	↓	715	.045	"	14.05	2.01	1.70	17.76	20.50

32 01 13.66 Fog Seal

		Crew	Daily Output	Labor-Hours	Unit	Material	2021 Bare Costs Labor	2021 Bare Costs Equipment	Total	Total Incl O&P
0010	**FOG SEAL**									
0012	Sealcoating, 2 coat coal tar pitch emulsion over 10,000 S.Y.	B-45	5000	.003	S.Y.	.83	.18	.17	1.18	1.37
0030	1,000 to 10,000 S.Y.	"	3000	.005		.83	.29	.28	1.40	1.68
0100	Under 1,000 S.Y.	B-1	1050	.023		.83	1.03		1.86	2.55
0300	Petroleum resistant, over 10,000 S.Y.	B-45	5000	.003		1.52	.18	.17	1.87	2.13
0320	1,000 to 10,000 S.Y.	"	3000	.005	↓	1.52	.29	.28	2.09	2.44
0400	Under 1,000 S.Y.	B-1	1050	.023	S.Y.	1.52	1.03		2.55	3.31
0600	Non-skid pavement renewal, over 10,000 S.Y.	B-45	5000	.003		1.31	.18	.17	1.66	1.90
0620	1,000 to 10,000 S.Y.	"	3000	.005	↓	1.31	.29	.28	1.88	2.21

32 01 13 – Flexible Paving Surface Treatment

32 01 13.66 Fog Seal

		Crew	Daily Output	Labor-Hours	Unit	Material	2021 Bare Costs Labor	Equipment	Total	Total Incl O&P
0700	Under 1,000 S.Y.	B-1	1050	.023	S.Y.	1.31	1.03		2.34	3.08
0800	Prepare and clean surface for above	A-2	8545	.003	↓		.13	.02	.15	.24
1000	Hand seal asphalt curbing	B-1	4420	.005	L.F.	.64	.24		.88	1.09
1900	Asphalt surface treatment, single course, small area									
1901	0.30 gal./S.Y. asphalt material, 20 lb./S.Y. aggregate	B-91	5000	.013	S.Y.	1.48	.68	.46	2.62	3.19
1910	Roadway or large area		10000	.006		1.36	.34	.23	1.93	2.28
1950	Asphalt surface treatment, dbl. course for small area		3000	.021		2.53	1.13	.76	4.42	5.40
1960	Roadway or large area		6000	.011		2.28	.56	.38	3.22	3.81
1980	Asphalt surface treatment, single course, for shoulders	↓	7500	.009	↓	1.23	.45	.31	1.99	2.40

32 01 16 – Flexible Paving Rehabilitation

32 01 16.71 Cold Milling Asphalt Paving

		Crew	Daily Output	Labor-Hours	Unit	Material	2021 Bare Costs Labor	Equipment	Total	Total Incl O&P
0010	**COLD MILLING ASPHALT PAVING**									
5200	Cold planing & cleaning, 1" to 3" asphalt pavmt., over 25,000 S.Y.	B-71	6000	.009	S.Y.		.48	.77	1.25	1.60
5280	5,000 S.Y. to 10,000 S.Y.		4000	.014	"		.71	1.16	1.87	2.40
5285	5,000 S.Y. to 10,000 S.Y.	↓	36000	.002	S.F.		.08	.13	.21	.26
5300	Asphalt pavement removal from conc. base, no haul									
5320	Rip, load & sweep 1" to 3"	B-70	8000	.007	S.Y.		.36	.29	.65	.88
5330	3" to 6" deep	"	5000	.011			.57	.47	1.04	1.41
5340	Profile grooving, asphalt pavement load & sweep, 1" deep	B-71	12500	.004			.23	.37	.60	.77
5350	3" deep		9000	.006			.32	.52	.84	1.07
5360	6" deep	↓	5000	.011	↓		.57	.93	1.50	1.92

32 01 16.73 In Place Cold Reused Asphalt Paving

		Crew	Daily Output	Labor-Hours	Unit	Material	2021 Bare Costs Labor	Equipment	Total	Total Incl O&P
0010	**IN PLACE COLD REUSED ASPHALT PAVING**									
5000	Reclamation, pulverizing and blending with existing base									
5040	Aggregate base, 4" thick pavement, over 15,000 S.Y.	B-73	2400	.027	S.Y.		1.43	1.90	3.33	4.34
5080	5,000 S.Y. to 15,000 S.Y.		2200	.029			1.56	2.07	3.63	4.73
5120	8" thick pavement, over 15,000 S.Y.		2200	.029			1.56	2.07	3.63	4.73
5160	5,000 S.Y. to 15,000 S.Y.	↓	2000	.032	↓		1.72	2.27	3.99	5.20

32 01 16.74 In Place Hot Reused Asphalt Paving

			Crew	Daily Output	Labor-Hours	Unit	Material	2021 Bare Costs Labor	Equipment	Total	Total Incl O&P
0010	**IN PLACE HOT REUSED ASPHALT PAVING**										
5500	Recycle asphalt pavement at site										
5520	Remove, rejuvenate and spread 4" deep	G	B-72	2500	.026	S.Y.	4.61	1.33	3.33	9.27	10.80
5521	6" deep	G	"	2000	.032	"	6.75	1.66	4.16	12.57	14.65

32 01 17 – Flexible Paving Repair

32 01 17.10 Repair of Asphalt Pavement Holes

		Crew	Daily Output	Labor-Hours	Unit	Material	2021 Bare Costs Labor	Equipment	Total	Total Incl O&P
0010	**REPAIR OF ASPHALT PAVEMENT HOLES** (cold patch)									
0100	Flexible pavement repair holes, roadway, light traffic, 1 C.F. size each	B-37A	24	1	Ea.	9.50	46	14.80	70.30	100
0150	Group of two, 1 C.F. size each	B-37A	16	1.500	Set	19	69	22	110	156
0200	Group of three, 1 C.F. size each	"	12	2	"	28.50	91.50	29.50	149.50	210
0300	Medium traffic, 1 C.F. size each	B-37B	24	1.333	Ea.	9.50	60.50	14.80	84.80	123
0350	Group of two, 1 C.F. size each		16	2	Set	19	91	22	132	191
0400	Group of three, 1 C.F. size each	↓	12	2.667	"	28.50	121	29.50	179	257
0500	Highway/heavy traffic, 1 C.F. size each	B-37C	24	1.333	Ea.	9.50	62	23	94.50	135
0550	Group of two, 1 C.F. size each		16	2	Set	19	93	34.50	146.50	208
0600	Group of three, 1 C.F. size each	↓	12	2.667	"	28.50	124	46	198.50	281
0700	Add police officer and car for traffic control				Hr.	56.50			56.50	62
1000	Flexible pavement repair holes, parking lot, bag material, 1 C.F. size each	B-37D	18	.889	Ea.	60	41.50	6.25	107.75	139
1010	Economy bag material, 1 C.F. size each		18	.889	"	37.50	41.50	6.25	85.25	114
1100	Group of two, 1 C.F. size each		12	1.333	Set	120	62	9.35	191.35	241
1110	Group of two, economy bag, 1 C.F. size each		12	1.333		75.50	62	9.35	146.85	192
1200	Group of three, 1 C.F. size each	↓	10	1.600	↓	180	74.50	11.20	265.70	330

32 01 17.10 Repair of Asphalt Pavement Holes

		Crew	Daily Output	Labor-Hours	Unit	Material	2021 Bare Costs Labor	Equipment	Total	Total Incl O&P
1210	Economy bag, group of three, 1 C.F. size each	B-37D	10	1.600	Set	113	74.50	11.20	198.70	255
1300	Flexible pavement repair holes, parking lot, 1 C.F. single hole	A-3A	4	2	Ea.	60	111	44	215	288
1310	Economy material, 1 C.F. single hole		4	2	"	37.50	111	44	192.50	263
1400	Flexible pavement repair holes, parking lot, 1 C.F. four holes		2	4	Set	240	222	88.50	550.50	705
1410	Economy material, 1 C.F. four holes		2	4	"	151	222	88.50	461.50	610
1500	Flexible pavement repair holes, large parking lot, bulk matl, 1 C.F. size	B-37A	32	.750	Ea.	9.50	34.50	11.10	55.10	77.50

32 01 17.20 Repair of Asphalt Pavement Patches

		Crew	Daily Output	Labor-Hours	Unit	Material	2021 Bare Costs Labor	Equipment	Total	Total Incl O&P
0010	**REPAIR OF ASPHALT PAVEMENT PATCHES**									
0100	Flexible pavement patches, roadway, light traffic, sawcut 10-25 S.F.	B-89	12	1.333	Ea.		69.50	89	158.50	208
0150	Sawcut 26-60 S.F.		10	1.600			83.50	107	190.50	249
0200	Sawcut 61-100 S.F.		8	2			104	133	237	310
0210	Sawcut groups of small size patches		24	.667			35	44.50	79.50	104
0220	Large size patches		16	1			52	66.50	118.50	156
0300	Flexible pavement patches, roadway, light traffic, digout 10-25 S.F.	B-6	16	1.500			72	13.50	85.50	129
0350	Digout 26-60 S.F.		12	2			96	18	114	172
0400	Digout 61-100 S.F.		8	3			144	27	171	258
0450	Add 8 C.Y. truck, small project debris haulaway	B-34A	8	1	Hr.		51.50	51	102.50	138
0460	Add 12 C.Y. truck, small project debris haulaway	B-34B	8	1			51.50	72.50	124	162
0480	Add flagger for non-intersection medium traffic	1 Clab	8	1			44.50		44.50	70.50
0490	Add flasher truck for intersection medium traffic or heavy traffic	A-2B	8	1			49	25	74	106
0500	Flexible pavement patches, roadway, repave, cold, 15 S.F., 4" D	B-37	8	6	Ea.	45	279	32.50	356.50	530
0510	6" depth		8	6		68	279	32.50	379.50	555
0520	Repave, cold, 20 S.F., 4" depth		8	6		60	279	32.50	371.50	545
0530	6" depth		8	6		90.50	279	32.50	402	580
0540	Repave, cold, 25 S.F., 4" depth		8	6		74.50	279	32.50	386	565
0550	6" depth		8	6		113	279	32.50	424.50	605
0600	Repave, cold, 30 S.F., 4" depth		8	6		89.50	279	32.50	401	580
0610	6" depth		8	6		136	279	32.50	447.50	630
0640	Repave, cold, 40 S.F., 4" depth		8	6		119	279	32.50	430.50	610
0650	6" depth		8	6		181	279	32.50	492.50	680
0680	Repave, cold, 50 S.F., 4" depth		8	6		149	279	32.50	460.50	645
0690	6" depth		8	6		226	279	32.50	537.50	730
0720	Repave, cold, 60 S.F., 4" depth		8	6		179	279	32.50	490.50	680
0730	6" depth		8	6		272	279	32.50	583.50	780
0800	Repave, cold, 70 S.F., 4" depth		7	6.857		209	320	37	566	775
0810	6" depth		7	6.857		315	320	37	672	895
0820	Repave, cold, 75 S.F., 4" depth		7	6.857		223	320	37	580	790
0830	6" depth		7	6.857		340	320	37	697	920
0900	Repave, cold, 80 S.F., 4" depth		6	8		239	375	43	657	900
0910	6" depth		6	8		360	375	43	778	1,050
0940	Repave, cold, 90 S.F., 4" depth		6	8		269	375	43	687	935
0950	6" depth		6	8		410	375	43	828	1,100
0980	Repave, cold, 100 S.F., 4" depth		6	8		299	375	43	717	970
0990	6" depth		6	8		450	375	43	868	1,125
1000	Add flasher truck for paving operations in medium or heavy traffic	A-2B	8	1	Hr.		49	25	74	106
1100	Prime coat for repair 15-40 S.F., 25% overspray	B-37A	48	.500	Ea.	8.75	23	7.40	39.15	54.50
1150	41-60 S.F.		40	.600		17.50	27.50	8.90	53.90	73
1175	61-80 S.F.		32	.750		23.50	34.50	11.10	69.10	93
1200	81-100 S.F.		32	.750		29	34.50	11.10	74.60	99
1210	Groups of patches w/25% overspray		3600	.007	S.F.	.29	.31	.10	.70	.92
1300	Flexible pavement repair patches, street repave, hot, 60 S.F., 4" D	B-37	8	6	Ea.	97.50	279	32.50	409	590
1310	6" depth		8	6		148	279	32.50	459.50	645

556

32 01 17 – Flexible Paving Repair

32 01 17.20 Repair of Asphalt Pavement Patches

		Crew	Daily Output	Labor-Hours	Unit	Material	2021 Bare Costs Labor	Equipment	Total	Total Incl O&P
1320	Repave, hot, 70 S.F., 4" depth	B-37	7	6.857	Ea.	114	320	37	471	670
1330	6" depth		7	6.857		172	320	37	529	735
1340	Repave, hot, 75 S.F., 4" depth		7	6.857		121	320	37	478	680
1350	6" depth		7	6.857		184	320	37	541	750
1360	Repave, hot, 80 S.F., 4" depth		6	8		130	375	43	548	780
1370	6" depth		6	8		197	375	43	615	855
1380	Repave, hot, 90 S.F., 4" depth		6	8		146	375	43	564	800
1390	6" depth		6	8		222	375	43	640	880
1400	Repave, hot, 100 S.F., 4" depth		6	8		162	375	43	580	815
1410	6" depth		6	8		246	375	43	664	910
1420	Pave hot groups of patches, 4" depth		900	.053	S.F.	1.62	2.48	.29	4.39	6.05
1430	6" depth		900	.053	"	2.46	2.48	.29	5.23	6.95
1500	Add 8 C.Y. truck for hot asphalt paving operations	B-34A	1	8	Day		410	410	820	1,100
1550	Add flasher truck for hot paving operations in med/heavy traffic	A-2B	8	1	Hr.		49	25	74	106
2000	Add police officer and car for traffic control				"	56.50			56.50	62

32 01 17.61 Sealing Cracks In Asphalt Paving

		Crew	Daily Output	Labor-Hours	Unit	Material	2021 Bare Costs Labor	Equipment	Total	Total Incl O&P
0010	**SEALING CRACKS IN ASPHALT PAVING**									
0100	Sealing cracks in asphalt paving, 1/8" wide x 1/2" depth, slow set	B-37A	2000	.012	L.F.	.18	.55	.18	.91	1.27
0110	1/4" wide x 1/2" depth		2000	.012		.20	.55	.18	.93	1.30
0130	3/8" wide x 1/2" depth		2000	.012		.23	.55	.18	.96	1.33
0140	1/2" wide x 1/2" depth		1800	.013		.26	.61	.20	1.07	1.48
0150	3/4" wide x 1/2" depth		1600	.015		.32	.69	.22	1.23	1.69
0160	1" wide x 1/2" depth		1600	.015		.38	.69	.22	1.29	1.76
0165	1/8" wide x 1" depth, rapid set	B-37F	2000	.016		.19	.73	.27	1.19	1.66
0170	1/4" wide x 1" depth		2000	.016		.37	.73	.27	1.37	1.86
0175	3/8" wide x 1" depth		1800	.018		.56	.81	.30	1.67	2.22
0180	1/2" wide x 1" depth		1800	.018		.74	.81	.30	1.85	2.43
0185	5/8" wide x 1" depth		1800	.018		.93	.81	.30	2.04	2.63
0190	3/4" wide x 1" depth		1600	.020		1.11	.91	.33	2.35	3.04
0195	1" wide x 1" depth		1600	.020		1.48	.91	.33	2.72	3.45
0200	1/8" wide x 2" depth, rapid set		2000	.016		.37	.73	.27	1.37	1.86
0210	1/4" wide x 2" depth		2000	.016		.74	.73	.27	1.74	2.27
0220	3/8" wide x 2" depth		1800	.018		1.11	.81	.30	2.22	2.83
0230	1/2" wide x 2" depth		1800	.018		1.48	.81	.30	2.59	3.24
0240	5/8" wide x 2" depth		1800	.018		1.85	.81	.30	2.96	3.65
0250	3/4" wide x 2" depth		1600	.020		2.23	.91	.33	3.47	4.27
0260	1" wide x 2" depth		1600	.020		2.97	.91	.33	4.21	5.10
0300	Add flagger for non-intersection medium traffic	1 Clab	8	1	Hr.		44.50		44.50	70.50
0400	Add flasher truck for intersection medium traffic or heavy traffic	A-2B	8	1	"		49	25	74	106

32 01 19 – Rigid Paving Surface Treatment

32 01 19.61 Sealing of Joints In Rigid Paving

		Crew	Daily Output	Labor-Hours	Unit	Material	2021 Bare Costs Labor	Equipment	Total	Total Incl O&P
0010	**SEALING OF JOINTS IN RIGID PAVING**									
1000	Sealing, concrete pavement, preformed elastomeric	B-77	500	.080	L.F.	3.58	3.65	2.15	9.38	12.10
2000	Waterproofing, membrane, tar and fabric, small area	B-63	233	.172	S.Y.	15.10	8	.77	23.87	30
2500	Large area		1435	.028		13.30	1.30	.13	14.73	16.85
3000	Preformed rubberized asphalt, small area		100	.400		19.50	18.65	1.80	39.95	53
3500	Large area		367	.109		17.75	5.10	.49	23.34	28
9000	Minimum labor/equipment charge	1 Clab	2	4	Job		178		178	283

32 01 29.61 Partial Depth Patching of Rigid Pavement

		Crew	Daily Output	Labor-Hours	Unit	Material	2021 Bare Costs Labor	2021 Bare Costs Equipment	Total	Total Incl O&P
0010	**PARTIAL DEPTH PATCHING OF RIGID PAVEMENT**									
0100	Rigid pavement repair, roadway, light traffic, 25% pitting 15 S.F.	B-37F	16	2	Ea.	24	91	33	148	208
0110	Pitting 20 S.F.	"	16	2	"	32.50	91	33	156.50	217
0120	Pitting 25 S.F.	B-37F	16	2	Ea.	40.50	91	33	164.50	226
0130	Pitting 30 S.F.		16	2		48.50	91	33	172.50	235
0140	Pitting 40 S.F.		16	2		64.50	91	33	188.50	253
0150	Pitting 50 S.F.		12	2.667		80.50	121	44.50	246	330
0160	Pitting 60 S.F.		12	2.667		97	121	44.50	262.50	350
0170	Pitting 70 S.F.		12	2.667		113	121	44.50	278.50	365
0180	Pitting 75 S.F.		12	2.667		121	121	44.50	286.50	375
0190	Pitting 80 S.F.		8	4		129	182	66.50	377.50	505
0200	Pitting 90 S.F.		8	4		145	182	66.50	393.50	525
0210	Pitting 100 S.F.		8	4		161	182	66.50	409.50	540
0300	Roadway, light traffic, 50% pitting 15 S.F.		12	2.667		48.50	121	44.50	214	295
0310	Pitting 20 S.F.		12	2.667		64.50	121	44.50	230	315
0320	Pitting 25 S.F.		12	2.667		80.50	121	44.50	246	330
0330	Pitting 30 S.F.		12	2.667		97	121	44.50	262.50	350
0340	Pitting 40 S.F.		12	2.667		129	121	44.50	294.50	385
0350	Pitting 50 S.F.		8	4		161	182	66.50	409.50	540
0360	Pitting 60 S.F.		8	4		194	182	66.50	442.50	575
0370	Pitting 70 S.F.		8	4		226	182	66.50	474.50	610
0380	Pitting 75 S.F.		8	4		242	182	66.50	490.50	630
0390	Pitting 80 S.F.		6	5.333		258	243	88.50	589.50	765
0400	Pitting 90 S.F.		6	5.333		290	243	88.50	621.50	805
0410	Pitting 100 S.F.		6	5.333		325	243	88.50	656.50	840
1000	Rigid pavement repair, light traffic, surface patch, 2" deep, 15 S.F.		8	4		97	182	66.50	345.50	470
1010	Surface patch, 2" deep, 20 S.F.		8	4		129	182	66.50	377.50	505
1020	Surface patch, 2" deep, 25 S.F.		8	4		161	182	66.50	409.50	540
1030	Surface patch, 2" deep, 30 S.F.		8	4		194	182	66.50	442.50	575
1040	Surface patch, 2" deep, 40 S.F.		8	4		258	182	66.50	506.50	645
1050	Surface patch, 2" deep, 50 S.F.		6	5.333		325	243	88.50	656.50	840
1060	Surface patch, 2" deep, 60 S.F.		6	5.333		385	243	88.50	716.50	910
1070	Surface patch, 2" deep, 70 S.F.		6	5.333		450	243	88.50	781.50	980
1080	Surface patch, 2" deep, 75 S.F.		6	5.333		485	243	88.50	816.50	1,025
1090	Surface patch, 2" deep, 80 S.F.		5	6.400		515	291	106	912	1,150
1100	Surface patch, 2" deep, 90 S.F.		5	6.400		580	291	106	977	1,225
1200	Surface patch, 2" deep, 100 S.F.	▼	5	6.400	▼	645	291	106	1,042	1,300
2000	Add flagger for non-intersection medium traffic	1 Clab	8	1	Hr.		44.50		44.50	70.50
2100	Add flasher truck for intersection medium traffic or heavy traffic	A-2B	8	1			49	25	74	106
2200	Add police officer and car for traffic control				▼	56.50			56.50	62

32 01 29.70 Full Depth Patching of Rigid Pavement

		Crew	Daily Output	Labor-Hours	Unit	Material	2021 Bare Costs Labor	2021 Bare Costs Equipment	Total	Total Incl O&P
0010	**FULL DEPTH PATCHING OF RIGID PAVEMENT**									
0015	Pavement preparation includes sawcut, remove pavement and replace									
0020	6" of subbase and one layer of reinforcement									
0030	Trucking and haulaway of debris excluded									
0100	Rigid pavement replace, light traffic, prep, 15 S.F., 6" depth	B-37E	16	3.500	Ea.	24	173	58.50	255.50	365
0110	Replacement preparation 20 S.F., 6" depth		16	3.500		32	173	58.50	263.50	370
0115	25 S.F., 6" depth		16	3.500		40	173	58.50	271.50	380
0120	30 S.F., 6" depth		14	4		48	197	66.50	311.50	435
0125	35 S.F., 6" depth		14	4		56	197	66.50	319.50	445
0130	40 S.F., 6" depth		14	4		63.50	197	66.50	327	455
0135	45 S.F., 6" depth		14	4		71.50	197	66.50	335	465

32 01 29.70 Full Depth Patching of Rigid Pavement	Crew	Daily Output	Labor-Hours	Unit	Material	2021 Bare Costs Labor	Equipment	Total	Total Incl O&P
0140　　　　50 S.F., 6" depth	B-37E	12	4.667	Ea.	79.50	230	78	387.50	540
0145　　　　55 S.F., 6" depth		12	4.667		87.50	230	78	395.50	545
0150　　　　60 S.F., 6" depth		10	5.600		95.50	276	93.50	465	645
0155　　　　65 S.F., 6" depth		10	5.600		104	276	93.50	473.50	650
0160　　　　70 S.F., 6" depth		10	5.600		112	276	93.50	481.50	660
0165　　　　75 S.F., 6" depth		10	5.600		120	276	93.50	489.50	670
0170　　　　80 S.F., 6" depth		8	7		127	345	117	589	815
0175　　　　85 S.F., 6" depth		8	7		135	345	117	597	820
0180　　　　90 S.F., 6" depth		8	7		143	345	117	605	830
0185　　　　95 S.F., 6" depth		8	7		151	345	117	613	840
0190　　　　100 S.F., 6" depth		8	7		159	345	117	621	850
0200　Pavement preparation for 8" rigid paving same as 6"									
0290　Pavement preparation for 9" rigid paving same as 10"									
0300　Rigid pavement replace, light traffic, prep, 15 S.F., 10" depth	B-37E	16	3.500	Ea.	40.50	173	58.50	272	380
0302　Pavement preparation 10" includes sawcut, remove pavement and									
0304　　　replace 6" of subbase and two layers of reinforcement									
0306　Trucking and haulaway of debris excluded									
0310　　　Replacement preparation 20 S.F., 10" depth	B-37E	16	3.500	Ea.	54	173	58.50	285.50	395
0315　　　25 S.F., 10" depth		16	3.500		68	173	58.50	299.50	410
0320　　　30 S.F., 10" depth		14	4		81.50	197	66.50	345	475
0325　　　35 S.F., 10" depth		14	4		61.50	197	66.50	325	450
0330　　　40 S.F., 10" depth		14	4		108	197	66.50	371.50	505
0335　　　45 S.F., 10" depth		12	4.667		122	230	78	430	585
0340　　　50 S.F., 10" depth		12	4.667		136	230	78	444	600
0345　　　55 S.F., 10" depth		12	4.667		149	230	78	457	615
0350　　　60 S.F., 10" depth		10	5.600		163	276	93.50	532.50	715
0355　　　65 S.F., 10" depth		10	5.600		176	276	93.50	545.50	730
0360　　　70 S.F., 10" depth		10	5.600		190	276	93.50	559.50	745
0365　　　75 S.F., 10" depth		8	7		203	345	117	665	895
0370　　　80 S.F., 10" depth		8	7		217	345	117	679	910
0375　　　85 S.F., 10" depth	B-37E	8	7	Ea.	230	345	117	692	925
0380　　　90 S.F., 10" depth		8	7		244	345	117	706	940
0385　　　95 S.F., 10" depth		8	7		258	345	117	720	955
0390　　　100 S.F., 10" depth		8	7		271	345	117	733	970
0500　Add 8 C.Y. truck, small project debris haulaway	B-34A	8	1	Hr.		51.50	51	102.50	138
0550　Add 12 C.Y. truck, small project debris haulaway	B-34B	8	1			51.50	72.50	124	162
0600　Add flagger for non-intersection medium traffic	1 Clab	8	1			44.50		44.50	70.50
0650　Add flasher truck for intersection medium traffic or heavy traffic	A-2B	8	1			49	25	74	106
0700　Add police officer and car for traffic control					56.50			56.50	62
0990　Concrete will be replaced using quick set mix with aggregate									
1000　Rigid pavement replace, light traffic, repour, 15 S.F., 6" depth	B-37F	18	1.778	Ea.	173	81	29.50	283.50	350
1010　　　20 S.F., 6" depth		18	1.778		230	81	29.50	340.50	415
1015　　　25 S.F., 6" depth		18	1.778		288	81	29.50	398.50	475
1020　　　30 S.F., 6" depth		16	2		345	91	33	469	560
1025　　　35 S.F., 6" depth		16	2		405	91	33	529	625
1030　　　40 S.F., 6" depth		16	2		460	91	33	584	685
1035　　　45 S.F., 6" depth		16	2		520	91	33	644	750
1040　　　50 S.F., 6" depth		16	2		575	91	33	699	815
1045　　　55 S.F., 6" depth		12	2.667		635	121	44.50	800.50	935
1050　　　60 S.F., 6" depth		12	2.667		690	121	44.50	855.50	1,000
1055　　　65 S.F., 6" depth		12	2.667		750	121	44.50	915.50	1,075
1060　　　70 S.F., 6" depth		12	2.667		805	121	44.50	970.50	1,125
1065　　　75 S.F., 6" depth		10	3.200		865	146	53	1,064	1,250

32 01 29 – Rigid Paving Repair

32 01 29.70 Full Depth Patching of Rigid Pavement	Crew	Daily Output	Labor-Hours	Unit	Material	2021 Bare Costs Labor	Equipment	Total	Total Incl O&P
1070	80 S.F., 6" depth	B-37F	10	3.200	Ea.	920	146	53	1,119	1,300
1075	85 S.F., 6" depth		8	4		980	182	66.50	1,228.50	1,450
1080	90 S.F., 6" depth		8	4		1,025	182	66.50	1,273.50	1,525
1085	95 S.F., 6" depth		8	4		1,100	182	66.50	1,348.50	1,575
1090	100 S.F., 6" depth		8	4		1,150	182	66.50	1,398.50	1,650
1099	Concrete will be replaced using 4,500 psi concrete ready mix									
1100	Rigid pavement replace, light traffic, repour, 15 S.F., 6" depth	A-2	12	2	Ea.	360	91.50	16.55	468.05	560
1110	20-50 S.F., 6" depth		12	2		360	91.50	16.55	468.05	560
1120	55-65 S.F., 6" depth		10	2.400		400	110	19.85	529.85	635
1130	70-80 S.F., 6" depth		10	2.400		430	110	19.85	559.85	670
1140	85-100 S.F., 6" depth		8	3		505	138	25	668	800
1200	Repour 15-40 S.F., 8" depth		12	2		360	91.50	16.55	468.05	560
1210	45-50 S.F., 8" depth		12	2		400	91.50	16.55	508.05	605
1220	55-60 S.F., 8" depth		10	2.400		430	110	19.85	559.85	670
1230	65-80 S.F., 8" depth		10	2.400		505	110	19.85	634.85	750
1240	85-90 S.F., 8" depth		8	3		540	138	25	703	840
1250	95-100 S.F., 8" depth	A-2	8	3	Ea.	575	138	25	738	880
1300	Repour 15-30 S.F., 10" depth		12	2		360	91.50	16.55	468.05	560
1310	35-40 S.F., 10" depth		12	2		400	91.50	16.55	508.05	605
1320	45 S.F., 10" depth		12	2		430	91.50	16.55	538.05	640
1330	50-65 S.F., 10" depth		10	2.400		505	110	19.85	634.85	750
1340	70 S.F., 10" depth		10	2.400		540	110	19.85	669.85	790
1350	75-80 S.F., 10" depth		10	2.400		575	110	19.85	704.85	830
1360	85-95 S.F., 10" depth		8	3		645	138	25	808	955
1370	100 S.F., 10" depth		8	3		685	138	25	848	1,000

32 06 Schedules for Exterior Improvements

32 06 10 – Schedules for Bases, Ballasts, and Paving

32 06 10.10 Sidewalks, Driveways and Patios

32 06 10.10 Sidewalks, Driveways and Patios	Crew	Daily Output	Labor-Hours	Unit	Material	2021 Bare Costs Labor	Equipment	Total	Total Incl O&P
0010	**SIDEWALKS, DRIVEWAYS AND PATIOS** No base									
0020	Asphaltic concrete, 2" thick	B-37	720	.067	S.Y.	7.25	3.11	.36	10.72	13.30
0100	2-1/2" thick	"	660	.073	"	9.15	3.39	.39	12.93	15.90
0110	Bedding for brick or stone, mortar, 1" thick	D-1	300	.053	S.F.	.91	2.60		3.51	5.20
0120	2" thick	"	200	.080		2.29	3.90		6.19	8.75
0130	Sand, 2" thick	B-18	8000	.003		.33	.14	.02	.49	.60
0140	4" thick	"	4000	.006		.66	.27	.04	.97	1.21
0300	Concrete, 3,000 psi, CIP, 6 x 6 - W1.4 x W1.4 mesh,									
0310	broomed finish, no base, 4" thick	B-24	600	.040	S.F.	2.22	2.01		4.23	5.65
0350	5" thick		545	.044		2.77	2.22		4.99	6.55
0400	6" thick		510	.047		3.23	2.37		5.60	7.30
0450	For bank run gravel base, 4" thick, add	B-18	2500	.010		.68	.43	.07	1.18	1.51
0520	8" thick, add	"	1600	.015		1.38	.68	.10	2.16	2.71
0550	Exposed aggregate finish, add to above, minimum	B-24	1875	.013		.13	.64		.77	1.16
0600	Maximum	"	455	.053		.44	2.65		3.09	4.68
0950	Concrete tree grate, 5' square	B-6	25	.960	Ea.	425	46	8.65	479.65	555
0955	Tree well & cover, concrete, 3' square		25	.960		246	46	8.65	300.65	355
0960	Cast iron tree grate with frame, 2 piece, round, 5' diameter		25	.960		1,125	46	8.65	1,179.65	1,325
0980	Square, 5' side		25	.960		1,125	46	8.65	1,179.65	1,325
1700	Redwood, prefabricated, 4' x 4' sections	2 Carp	316	.051	S.F.	7.35	2.77		10.12	12.50
1750	Redwood planks, 1" thick, on sleepers	"	240	.067	"	7.35	3.65		11	13.90
2250	Stone dust, 4" thick	B-62	900	.027	S.Y.	5.95	1.28	.20	7.43	8.80

For customer support on your Commercial Renovation Costs with RSMeans Data, call 800.448.8182.

32 06 Schedules for Exterior Improvements

32 06 10 - Schedules for Bases, Ballasts, and Paving

32 06 10.10 Sidewalks, Driveways and Patios

		Crew	Daily Output	Labor-Hours	Unit	Material	2021 Bare Costs Labor	2021 Bare Costs Equipment	Total	Total Incl O&P
9000	Minimum labor/equipment charge	D-1	2	8	Job		390		390	625

32 06 10.20 Steps

		Crew	Daily Output	Labor-Hours	Unit	Material	Labor	Equipment	Total	Total Incl O&P
0010	**STEPS**									
0011	Incl. excav., borrow & concrete base as required									
0100	Brick steps	B-24	35	.686	LF Riser	18.55	34.50		53.05	75
0200	Railroad ties	2 Clab	25	.640	"	3.86	28.50		32.36	50
0300	Bluestone treads, 12" x 2" or 12" x 1-1/2"	B-24	30	.800	LF Riser	45	40		85	113
0490	Minimum labor/equipment charge	D-1	2	8	Job		390		390	625
0500	Concrete, cast in place, see Section 03 30 53.40									
0600	Precast concrete, see Section 03 41 23.50									

32 11 Base Courses

32 11 23 - Aggregate Base Courses

32 11 23.23 Base Course Drainage Layers

		Crew	Daily Output	Labor-Hours	Unit	Material	Labor	Equipment	Total	Total Incl O&P
0010	**BASE COURSE DRAINAGE LAYERS**									
0011	For Soil Stabilization, see Section 31 32									
0012	For roadways and large areas									
0050	Crushed 3/4" stone base, compacted, 3" deep	B-36C	5200	.008	S.Y.	2.50	.42	.83	3.75	4.32
0100	6" deep		5000	.008		5	.44	.86	6.30	7.15
0200	9" deep		4600	.009		7.50	.48	.93	8.91	10.05
0300	12" deep		4200	.010		10	.52	1.02	11.54	13
0301	Crushed 1-1/2" stone base, compacted to 4" deep	B-36B	6000	.011		5.90	.56	.83	7.29	8.30
0302	6" deep		5400	.012		8.85	.63	.92	10.40	11.75
0303	8" deep		4500	.014		11.85	.75	1.10	13.70	15.40
0304	12" deep		3800	.017		17.75	.89	1.31	19.95	22.50
0310	Minimum labor/equipment charge	B-36	1	40	Job		2,025	1,825	3,850	5,200
0350	Bank run gravel, spread and compacted									
0370	6" deep	B-32	6000	.005	S.Y.	5.70	.30	.47	6.47	7.30
0390	9" deep		4900	.007		8.55	.36	.58	9.49	10.65
0400	12" deep		4200	.008		11.45	.42	.68	12.55	14
6900	For small and irregular areas, add						50%	50%		
6990	Minimum labor/equipment charge	1 Clab	3	2.667	Job		118		118	189
7000	Prepare and roll sub-base, small areas to 2,500 S.Y.	B-32A	1500	.016	S.Y.		.87	1.16	2.03	2.64
8000	Large areas over 2,500 S.Y.	"	3500	.007			.37	.50	.87	1.13
8050	For roadways	B-32	4000	.008			.44	.71	1.15	1.47
9000	Minimum labor/equipment charge	1 Clab	4	2	Job		89		89	141

32 12 Flexible Paving

32 12 16 - Asphalt Paving

32 12 16.13 Plant-Mix Asphalt Paving

		Crew	Daily Output	Labor-Hours	Unit	Material	Labor	Equipment	Total	Total Incl O&P
0010	**PLANT-MIX ASPHALT PAVING**									
0080	Binder course, 1-1/2" thick	B-25	7725	.011	S.Y.	5.60	.55	.35	6.50	7.40
0120	2" thick		6345	.014		7.45	.67	.43	8.55	9.70
0130	2-1/2" thick		5620	.016		9.30	.76	.49	10.55	11.95
0160	3" thick		4905	.018		11.15	.87	.56	12.58	14.25
0170	3-1/2" thick		4520	.019		13	.95	.61	14.56	16.45
0300	Wearing course, 1" thick	B-25B	10575	.009		3.68	.45	.28	4.41	5.05
0340	1-1/2" thick		7725	.012		6.20	.61	.39	7.20	8.20
0380	2" thick		6345	.015		8.30	.75	.47	9.52	10.85

561

32 12 16.13 Plant-Mix Asphalt Paving

		Crew	Daily Output	Labor-Hours	Unit	Material	2021 Bare Costs Labor	Equipment	Total	Total Incl O&P
0420	2-1/2" thick	B-25B	5480	.018	S.Y.	10.25	.87	.54	11.66	13.25
0460	3" thick		4900	.020		12.25	.97	.61	13.83	15.65
0470	3-1/2" thick		4520	.021		14.35	1.05	.66	16.06	18.20
0480	4" thick		4140	.023		16.40	1.15	.72	18.27	20.50
0600	Porous pavement, 1-1/2" open graded friction course	B-25	7725	.011		15.30	.55	.35	16.20	18.10
0800	Alternate method of figuring paving costs									
0810	Binder course, 1-1/2" thick	B-25	630	.140	Ton	68.50	6.80	4.35	79.65	90.50
0811	2" thick		690	.128		68.50	6.20	3.97	78.67	89
0812	3" thick		800	.110		68.50	5.35	3.42	77.27	87
0813	4" thick		900	.098		68.50	4.75	3.04	76.29	86
0850	Wearing course, 1" thick	B-25B	575	.167		72	8.25	5.20	85.45	98.50
0851	1-1/2" thick		630	.152		72	7.55	4.74	84.29	96.50
0852	2" thick		690	.139		72	6.90	4.33	83.23	95
0853	2-1/2" thick		765	.125		72	6.20	3.90	82.10	93.50
0854	3" thick		800	.120		72	5.95	3.73	81.68	93
1000	Pavement replacement over trench, 2" thick	B-21	90	.311	S.Y.	7.65	15.90	2.12	25.67	36
1050	4" thick		70	.400		15.15	20.50	2.73	38.38	52
1080	6" thick		55	.509		24	26	3.47	53.47	71.50
3000	Prime coat, emulsion, 0.30 gal./S.Y., 1000 S.Y.	B-45	2500	.006		2.10	.35	.33	2.78	3.24
3100	Tack coat, emulsion, 0.10 gal./S.Y., 1000 S.Y.	"	2500	.006		.70	.35	.33	1.38	1.70

32 12 16.14 Asphaltic Concrete Paving

		Crew	Daily Output	Labor-Hours	Unit	Material	2021 Bare Costs Labor	Equipment	Total	Total Incl O&P
0011	**ASPHALTIC CONCRETE PAVING**, parking lots & driveways									
0015	No asphalt hauling included									
0020	6" stone base, 2" binder course, 1" topping	B-25C	9000	.005	S.F.	2.06	.26	.27	2.59	2.97
0025	2" binder course, 2" topping		9000	.005		2.49	.26	.27	3.02	3.45
0030	3" binder course, 2" topping		9000	.005		2.91	.26	.27	3.44	3.91
0035	4" binder course, 2" topping		9000	.005		3.32	.26	.27	3.85	4.36
0040	1-1/2" binder course, 1" topping		9000	.005		1.85	.26	.27	2.38	2.75
0042	3" binder course, 1" topping		9000	.005		2.47	.26	.27	3	3.43
0045	3" binder course, 3" topping		9000	.005		3.35	.26	.27	3.88	4.40
0050	4" binder course, 3" topping		9000	.005		3.76	.26	.27	4.29	4.85
0055	4" binder course, 4" topping		9000	.005		4.19	.26	.27	4.72	5.30
0300	Binder course, 1-1/2" thick		35000	.001		.62	.07	.07	.76	.87
0400	2" thick		25000	.002		.80	.10	.10	1	1.14
0500	3" thick		15000	.003		1.24	.16	.16	1.56	1.80
0600	4" thick		10800	.004		1.63	.22	.22	2.07	2.38
0800	Sand finish course, 3/4" thick		41000	.001		.31	.06	.06	.43	.49
0900	1" thick		34000	.001		.38	.07	.07	.52	.61
1000	Fill pot holes, hot mix, 2" thick	B-16	4200	.008		.82	.36	.14	1.32	1.62
1100	4" thick		3500	.009		1.20	.43	.17	1.80	2.18
1120	6" thick		3100	.010		1.61	.48	.19	2.28	2.75
1140	Cold patch, 2" thick	B-51	3000	.016	S.F.	1.24	.73	.07	2.04	2.60
1160	4" thick		2700	.018		2.37	.81	.07	3.25	3.98
1180	6" thick		1900	.025		3.69	1.15	.10	4.94	6
3000	Prime coat, emulsion, 0.30 gal./S.Y., 1000 S.Y.	B-45	2500	.006	S.Y.	2.10	.35	.33	2.78	3.24
3100	Tack coat, emulsion, 0.10 gal./S.Y., 1000 S.Y.	"	2500	.006	"	.70	.35	.33	1.38	1.70

32 12 Flexible Paving

32 12 36 – Seal Coats

32 12 36.13 Chip Seal

		Crew	Daily Output	Labor-Hours	Unit	Material	2021 Bare Costs Labor	2021 Bare Costs Equipment	Total	Total Incl O&P
0010	**CHIP SEAL**									
0011	Excludes crack repair and flush coat									
1000	Fine - PMCRS-2h (20lbs/sy, 1/4" (No.10), .30gal/sy app. rate)									
1010	Small, irregular areas	B-91	5000	.013	S.Y.	1.10	.68	.46	2.24	2.77
1020	Parking Lot	B-91D	15000	.007		1.10	.36	.23	1.69	2.02
1030	Roadway	"	30000	.003		1.10	.18	.11	1.39	1.61
1090	For Each .5% Latex Additive, Add					.23			.23	.25
1100	Medium Fine - PMCRS-2h (25lbs/sy, 5/16" (No.8), .35gal/sy app. rate)									
1110	Small, irregular areas	B-91	4000	.016	S.Y.	1.29	.84	.57	2.70	3.37
1120	Parking Lot	B-91D	12000	.009		1.29	.44	.28	2.01	2.42
1130	Roadway	"	24000	.004		1.29	.22	.14	1.65	1.92
1190	For Each .5% Latex Additive, Add					.28			.28	.31
1200	Medium - PMCRS-2h (30lbs/sy, 3/8" (No.6), .40gal/sy app. rate)									
1210	Small, irregular areas	B-91	3330	.019	S.Y.	1.41	1.01	.69	3.11	3.91
1220	Parking Lot	B-91D	10000	.010		1.41	.53	.34	2.28	2.76
1230	Roadway	"	20000	.005		1.41	.27	.17	1.85	2.16
1290	For Each .5% Latex Additive, Add					.34			.34	.37
1300	Course - PMCRS-2h (30lbs/sy, 1/2" (No.4), .40gal/sy app. rate)									
1310	Small, irregular areas	B-91	2500	.026	S.Y.	1.39	1.35	.92	3.66	4.66
1320	Parking Lot	B-91D	7500	.014		1.39	.71	.45	2.55	3.14
1330	Roadway	"	15000	.007		1.39	.36	.23	1.98	2.33
1390	For Each .5% Latex Additive, Add					.34			.34	.37
1400	Double - PMCRS-2h (Course Base with Fine Top)									
1410	Small, irregular areas	B-91	2000	.032	S.Y.	2.36	1.69	1.14	5.19	6.50
1420	Parking Lot	B-91D	6000	.017		2.36	.89	.57	3.82	4.62
1430	Roadway	"	12000	.009		2.36	.44	.28	3.08	3.61
1490	For Each .5% Latex Additive, Add					.56			.56	.62

32 12 36.14 Flush Coat

		Crew	Daily Output	Labor-Hours	Unit	Material	2021 Bare Costs Labor	2021 Bare Costs Equipment	Total	Total Incl O&P
0010	**FLUSH COAT**									
0011	Fog Seal with Sand Cover (18gal/sy, 6lbs/sy)									
1010	Small, irregular areas	B-91	2000	.032	S.Y.	.52	1.69	1.14	3.35	4.50
1020	Parking lot		6000	.011		.52	.56	.38	1.46	1.89
1030	Roadway		12000	.005		.52	.28	.19	.99	1.23

32 12 36.33 Slurry Seal

		Crew	Daily Output	Labor-Hours	Unit	Material	2021 Bare Costs Labor	2021 Bare Costs Equipment	Total	Total Incl O&P
0010	**SLURRY SEAL**									
0011	Includes sweeping and cleaning of area									
1000	Type I-PMCQS-1h-EAS (12lbs/sy, 1/8", 20% asphalt emulsion)									
1010	Small, irregular areas	B-90	8000	.008	S.Y.	.98	.39	.28	1.65	2
1020	Parking Lot		25000	.003		.98	.13	.09	1.20	1.37
1030	Roadway		50000	.001		.98	.06	.04	1.08	1.22
1090	For Each .5% Latex Additive, Add					.15			.15	.16
1100	Type II-PMCQS-1h-EAS (15lbs/sy, 1/4", 18% asphalt emulsion)									
1110	Small, irregular areas	B-90	6000	.011	S.Y.	1.10	.52	.37	1.99	2.45
1120	Parking lot		20000	.003		1.10	.16	.11	1.37	1.58
1130	Roadway		40000	.002		1.10	.08	.06	1.24	1.39
1190	For Each .5% Latex Additive, Add					.18			.18	.20
1200	Type III-PMCQS-1h-EAS (25lbs/sy, 3/8", 15% asphalt emulsion)									
1210	Small, irregular areas	B-90	4000	.016	S.Y.	1.51	.79	.56	2.86	3.52
1220	Parking lot		12000	.005		1.51	.26	.19	1.96	2.28
1230	Roadway		24000	.003		1.51	.13	.09	1.73	1.97
1290	For Each .5% Latex Additive, Add					.29			.29	.32

563

32 12 Flexible Paving

32 12 36 – Seal Coats

32 12 36.36 Microsurfacing

		Crew	Daily Output	Labor-Hours	Unit	Material	2021 Bare Costs Labor	2021 Bare Costs Equipment	Total	Total Incl O&P
0010	**MICROSURFACING**									
1100	Type II-MSE (20lbs/sy, 1/4", 18% microsurfacing emulsion)									
1110	Small, irregular areas	B-90	5000	.013	S.Y.	1.65	.63	.44	2.72	3.30
1120	Parking lot		15000	.004		1.65	.21	.15	2.01	2.30
1130	Roadway	↓	30000	.002	↓	1.65	.10	.07	1.82	2.06
1200	Type IIIa-MSE (32lbs/sy, 3/8", 15% microsurfacing emulsion)									
1210	Small, irregular areas	B-90	3000	.021	S.Y.	2.29	1.05	.74	4.08	4.99
1220	Parking lot		9000	.007		2.29	.35	.25	2.89	3.34
1230	Roadway	↓	18000	.004	↓	2.29	.17	.12	2.58	2.94

32 13 Rigid Paving

32 13 13 – Concrete Paving

32 13 13.25 Concrete Pavement, Highways

		Crew	Daily Output	Labor-Hours	Unit	Material	2021 Bare Costs Labor	2021 Bare Costs Equipment	Total	Total Incl O&P
0010	**CONCRETE PAVEMENT, HIGHWAYS**									
0015	Including joints, finishing and curing									
0020	Fixed form, 12' pass, unreinforced, 6" thick	B-26	3000	.029	S.Y.	26.50	1.44	1.19	29.13	33
0100	8" thick		2750	.032		36.50	1.58	1.30	39.38	44
0400	12" thick		1800	.049		52	2.41	1.99	56.40	63.50
0410	Conc. pavement, w/jt., fnsh.& curing, fix form, 24' pass, unreinforced, 6"T		6000	.015		25	.72	.60	26.32	29.50
0430	8" thick		5500	.016		34.50	.79	.65	35.94	40
0460	12" thick	↓	3600	.024		50	1.20	.99	52.19	58
0520	Welded wire fabric, sheets for rigid paving 2.33 lb./S.Y.	2 Rodm	389	.041		1.34	2.42		3.76	5.25
0530	Reinforcing steel for rigid paving 12 lb./S.Y.	"	666	.024	↓	7.85	1.42		9.27	10.80
0540	Reinforcing steel for rigid paving 18 lb./S.Y.	2 Rodm	444	.036	S.Y.	11.75	2.12		13.87	16.20
0620	Slip form, 12' pass, unreinforced, 6" thick	B-26A	5600	.016		25.50	.77	.66	26.93	30
0624	8" thick		5300	.017		35	.82	.70	36.52	40.50
0630	12" thick		3470	.025		50.50	1.25	1.06	52.81	58.50
0632	15" thick		2890	.030		64	1.50	1.28	66.78	74
0640	Slip form, 24' pass, unreinforced, 6" thick		11200	.008		25	.39	.33	25.72	28.50
0644	8" thick		10600	.008		33.50	.41	.35	34.26	38
0650	12" thick		6940	.013		49	.62	.53	50.15	55.50
0652	15" thick	↓	5780	.015		61.50	.75	.64	62.89	70
0700	Finishing, broom finish small areas	2 Cefi	120	.133			6.90		6.90	10.80
1000	Curing, with sprayed membrane by hand	2 Clab	1500	.011	↓	1.10	.47		1.57	1.96

32 14 Unit Paving

32 14 13 – Precast Concrete Unit Paving

32 14 13.18 Precast Concrete Plantable Pavers

		Crew	Daily Output	Labor-Hours	Unit	Material	2021 Bare Costs Labor	2021 Bare Costs Equipment	Total	Total Incl O&P
0010	**PRECAST CONCRETE PLANTABLE PAVERS** (50% grass)									
0300	3/4" crushed stone base for plantable pavers, 6" depth	B-62	1000	.024	S.Y.	3.90	1.15	.18	5.23	6.30
0400	8" depth		900	.027		5.20	1.28	.20	6.68	7.95
0500	10" depth		800	.030		6.50	1.44	.22	8.16	9.70
0600	12" depth	↓	700	.034	↓	7.80	1.65	.26	9.71	11.45
0700	Hydro seeding plantable pavers	B-81A	20	.800	M.S.F.	9.75	37.50	48.50	95.75	124
0800	Apply fertilizer and seed to plantable pavers	1 Clab	8	1	"	48	44.50		92.50	124

564

32 14 Unit Paving

32 14 16 – Brick Unit Paving

32 14 16.10 Brick Paving

		Crew	Daily Output	Labor-Hours	Unit	Material	2021 Bare Costs Labor	Equipment	Total	Total Incl O&P
0010	**BRICK PAVING**									
0012	4" x 8" x 1-1/2", without joints (4.5 bricks/S.F.)	D-1	110	.145	S.F.	2.74	7.10		9.84	14.40
0100	Grouted, 3/8" joint (3.9 bricks/S.F.)		90	.178		2.26	8.65		10.91	16.40
0200	4" x 8" x 2-1/4", without joints (4.5 bricks/S.F.)		110	.145		2.61	7.10		9.71	14.25
0300	Grouted, 3/8" joint (3.9 bricks/S.F.)		90	.178		2.26	8.65		10.91	16.40
0455	Pervious brick paving, 4" x 8" x 3-1/4", without joints (4.5 bricks/S.F.)		110	.145		2.83	7.10		9.93	14.50
0500	Bedding, asphalt, 3/4" thick	B-25	5130	.017		.68	.83	.53	2.04	2.66
0540	Course washed sand bed, 1" thick	B-18	5000	.005		.37	.22	.03	.62	.79
0580	Mortar, 1" thick	D-1	300	.053		.76	2.60		3.36	5
0620	2" thick		200	.080		1.52	3.90		5.42	7.95
1500	Brick on 1" thick sand bed laid flat, 4.5/S.F.		100	.160		3.08	7.80		10.88	15.90
2000	Brick pavers, laid on edge, 7.2/S.F.		70	.229		4.74	11.15		15.89	23
2500	For 4" thick concrete bed and joints, add		595	.027		1.34	1.31		2.65	3.57
2800	For steam cleaning, add	A-1H	950	.008		.10	.37	.08	.55	.80
9000	Minimum labor/equipment charge	1 Bric	2	4	Job		215		215	345

32 14 23 – Asphalt Unit Paving

32 14 23.10 Asphalt Blocks

		Crew	Daily Output	Labor-Hours	Unit	Material	2021 Bare Costs Labor	Equipment	Total	Total Incl O&P
0010	**ASPHALT BLOCKS**									
0020	Rectangular, 6" x 12" x 1-1/4", w/bed & neopr. adhesive	D-1	135	.119	S.F.	10.10	5.75		15.85	20.50
0100	3" thick		130	.123		14.15	6		20.15	25
0300	Hexagonal tile, 8" wide, 1-1/4" thick		135	.119		10.10	5.75		15.85	20.50
0400	2" thick		130	.123		14.15	6		20.15	25
0500	Square, 8" x 8", 1-1/4" thick		135	.119		10.10	5.75		15.85	20.50
0600	2" thick		130	.123		14.15	6		20.15	25
9000	Minimum labor/equipment charge	1 Bric	2	4	Job		215		215	345

32 14 40 – Stone Paving

32 14 40.10 Stone Pavers

		Crew	Daily Output	Labor-Hours	Unit	Material	2021 Bare Costs Labor	Equipment	Total	Total Incl O&P
0010	**STONE PAVERS**									
1300	Slate, natural cleft, irregular, 3/4" thick	D-1	92	.174	S.F.	9.70	8.45		18.15	24.50
1350	Random rectangular, gauged, 1/2" thick		105	.152		21	7.40		28.40	35
1400	Random rectangular, butt joint, gauged, 1/4" thick		150	.107		22.50	5.20		27.70	33.50
1450	For sand rubbed finish, add					10			10	11
1550	Granite blocks, 3-1/2" x 3-1/2" x 3-1/2"	D-1	92	.174		21.50	8.45		29.95	37.50
1600	4" to 12" long, 3" to 5" wide, 3" to 5" thick	"	98	.163		18.05	7.95		26	32.50

32 16 Curbs, Gutters, Sidewalks, and Driveways

32 16 13 – Curbs and Gutters

32 16 13.13 Cast-in-Place Concrete Curbs and Gutters

		Crew	Daily Output	Labor-Hours	Unit	Material	2021 Bare Costs Labor	Equipment	Total	Total Incl O&P
0010	**CAST-IN-PLACE CONCRETE CURBS AND GUTTERS**									
0290	Forms only, no concrete									
0300	Concrete, wood forms, 6" x 18", straight	C-2	500	.096	L.F.	4	5.10		9.10	12.55
0400	6" x 18", radius	"	200	.240		4.21	12.80		17.01	25
0404	Concrete, wood forms, 6" x 18", straight & concrete	C-2A	500	.096		7.50	5.05		12.55	16.30
0406	6" x 18", radius	"	200	.240		7.70	12.70		20.40	28.50
0415	Machine formed, 6" x 18", straight	B-69A	2000	.024		5.15	1.16	.62	6.93	8.15
0416	6" x 18", radius	"	900	.053		5.20	2.58	1.37	9.15	11.35

For customer support on your Commercial Renovation Costs with RSMeans Data, call 800.448.8182.

32 16 Curbs, Gutters, Sidewalks, and Driveways

32 16 13 – Curbs and Gutters

32 16 13.23 Precast Concrete Curbs and Gutters

		Crew	Daily Output	Labor-Hours	Unit	Material	2021 Bare Costs Labor	Equipment	Total	Total Incl O&P
0010	**PRECAST CONCRETE CURBS AND GUTTERS**									
0550	Precast, 6" x 18", straight	B-29	700	.080	L.F.	9.05	3.86	1.22	14.13	17.40
0600	6" x 18", radius	"	325	.172	"	9.65	8.30	2.62	20.57	26.50

32 16 13.33 Asphalt Curbs

		Crew	Daily Output	Labor-Hours	Unit	Material	2021 Bare Costs Labor	Equipment	Total	Total Incl O&P
0010	**ASPHALT CURBS**									
0012	Curbs, asphaltic, machine formed, 8" wide, 6" high, 40 L.F./ton	B-27	1000	.032	L.F.	1.84	1.44	.25	3.53	4.60
0100	8" wide, 8" high, 30 L.F./ton		900	.036		2.46	1.60	.28	4.34	5.55
0150	Asphaltic berm, 12" W, 3" to 6" H, 35 L.F./ton, before pavement	↓	700	.046		.04	2.05	.36	2.45	3.72
0200	12" W, 1-1/2" to 4" H, 60 L.F./ton, laid with pavement	B-2	1050	.038	↓	.02	1.71		1.73	2.75

32 16 13.43 Stone Curbs

		Crew	Daily Output	Labor-Hours	Unit	Material	2021 Bare Costs Labor	Equipment	Total	Total Incl O&P
0010	**STONE CURBS**									
1000	Granite, split face, straight, 5" x 16"	D-13	275	.175	L.F.	16.45	9.10	1.57	27.12	34.50
1100	6" x 18"	"	250	.192	"	21.50	10	1.73	33.23	42
1300	Radius curbing, 6" x 18", over 10' radius	B-29	260	.215	L.F.	26.50	10.40	3.27	40.17	49
1400	Corners, 2' radius	"	80	.700	Ea.	89	34	10.65	133.65	163
1600	Edging, 4-1/2" x 12", straight	D-13	300	.160	L.F.	8.25	8.35	1.44	18.04	24
1800	Curb inlets (guttermouth) straight	B-29	41	1.366	Ea.	198	66	21	285	345
2000	Indian granite (Belgian block)									
2100	Jumbo, 10-1/2" x 7-1/2" x 4", grey	D-1	150	.107	L.F.	8.30	5.20		13.50	17.45
2150	Pink		150	.107		9.80	5.20		15	19.10
2200	Regular, 9" x 4-1/2" x 4-1/2", grey		160	.100		4.77	4.87		9.64	13.10
2250	Pink		160	.100		6.50	4.87		11.37	15
2300	Cubes, 4" x 4" x 4", grey		175	.091		2.62	4.45		7.07	10.05
2350	Pink		175	.091		4.04	4.45		8.49	11.60
2400	6" x 6" x 6", pink	↓	155	.103	↓	13.50	5.05		18.55	23
2500	Alternate pricing method for Indian granite									
2550	Jumbo, 10-1/2" x 7-1/2" x 4" (30 lb.), grey				Ton	470			470	520
2600	Pink					570			570	625
2650	Regular, 9" x 4-1/2" x 4-1/2" (20 lb.), grey					335			335	370
2700	Pink					455			455	500
2750	Cubes, 4" x 4" x 4" (5 lb.), grey					320			320	350
2800	Pink					520			520	570
2850	6" x 6" x 6" (25 lb.), pink					525			525	580
2900	For pallets, add				↓	22			22	24

32 17 Paving Specialties

32 17 13 – Parking Bumpers

32 17 13.13 Metal Parking Bumpers

		Crew	Daily Output	Labor-Hours	Unit	Material	2021 Bare Costs Labor	Equipment	Total	Total Incl O&P
0010	**METAL PARKING BUMPERS**									
0015	Bumper rails for garages, 12 ga. rail, 6" wide, with steel									
0020	posts 12'-6" OC, minimum	E-4	190	.168	L.F.	19.75	10.25	.78	30.78	39
0030	Average		165	.194		24.50	11.80	.90	37.20	47
0100	Maximum	↓	140	.229	↓	29.50	13.90	1.06	44.46	56
1300	Pipe bollards, conc. filled/paint, 8' L x 4' D hole, 6" diam.	B-6	20	1.200	Ea.	680	57.50	10.80	748.30	850
1400	8" diam.		15	1.600		745	77	14.40	836.40	960
1500	12" diam.	↓	12	2		1,025	96	18	1,139	1,300
8900	Security bollards, SS, lighted, hyd., incl. controls, group of 3	L-7	.06	509		47,400	27,000		74,400	94,500
8910	Group of 5	"	.04	683	↓	83,000	36,200		119,200	149,000
9000	Minimum labor/equipment charge	E-4	2	16	Job		975	74.50	1,049.50	1,650

32 17 Paving Specialties

32 17 13 – Parking Bumpers

32 17 13.16 Plastic Parking Bumpers

		Crew	Daily Output	Labor-Hours	Unit	Material	2021 Bare Costs Labor	Equipment	Total	Total Incl O&P
0010	**PLASTIC PARKING BUMPERS**									
1600	Bollards, recycled plastic, 5" x 5" x 5' L ▣G	B-6	20	1.200	Ea.	145	57.50	10.80	213.30	262
1610	Wheel stops, recycled plastic, yellow, 4" x 6" x 6' L ▣G		20	1.200		55	57.50	10.80	123.30	163
1620	Speed bumps, recycled plastic, yellow, 3" H x 10" W x 6' L ▣G		20	1.200		191	57.50	10.80	259.30	315
1630	3" H x 10" W x 9' L ▣G		20	1.200		276	57.50	10.80	344.30	410

32 17 13.19 Precast Concrete Parking Bumpers

		Crew	Daily Output	Labor-Hours	Unit	Material	2021 Bare Costs Labor	Equipment	Total	Total Incl O&P
0010	**PRECAST CONCRETE PARKING BUMPERS**									
1000	Wheel stops, precast concrete incl. dowels, 6" x 10" x 6'-0"	B-2	120	.333	Ea.	58	14.95		72.95	87.50
1100	8" x 13" x 6'-0"	"	120	.333		86.50	14.95		101.45	119
1540	Bollards, precast concrete, free standing, round, 15" diam. x 36" H	B-6	14	1.714		232	82.50	15.45	329.95	400
1550	16" diam. x 30" H		14	1.714		298	82.50	15.45	395.95	475
1560	18" diam. x 34" H		14	1.714		465	82.50	15.45	562.95	655
1570	18" diam. x 72" H		14	1.714		735	82.50	15.45	832.95	955
1572	26" diam. x 18" H		12	2		420	96	18	534	630
1574	26" diam. x 30" H		12	2		695	96	18	809	935
1578	26" diam. x 50" H		12	2		940	96	18	1,054	1,200
1580	Square, 22" x 32" H		12	2		445	96	18	559	655
1585	22" x 41" H		12	2		655	96	18	769	890
1590	Hexagon, 26" x 41" H		10	2.400		545	115	21.50	681.50	805

32 17 13.26 Wood Parking Bumpers

		Crew	Daily Output	Labor-Hours	Unit	Material	2021 Bare Costs Labor	Equipment	Total	Total Incl O&P
0010	**WOOD PARKING BUMPERS**									
0020	Parking barriers, timber w/saddles, treated type									
0100	4" x 4" for cars	B-2	520	.077	L.F.	3.91	3.45		7.36	9.80
0200	6" x 6" for trucks	"	520	.077	"	8.20	3.45		11.65	14.50

32 17 23 – Pavement Markings

32 17 23.13 Painted Pavement Markings

		Crew	Daily Output	Labor-Hours	Unit	Material	2021 Bare Costs Labor	Equipment	Total	Total Incl O&P
0010	**PAINTED PAVEMENT MARKINGS**									
0020	Acrylic waterborne, white or yellow, 4" wide, less than 3,000 L.F.	B-78	20000	.002	L.F.	.12	.11	.05	.28	.36
0200	6" wide, less than 3,000 L.F.		11000	.004		.18	.20	.10	.48	.62
0500	8" wide, less than 3,000 L.F.		10000	.005		.23	.22	.11	.56	.73
0600	12" wide, less than 3,000 L.F.		4000	.012		.35	.55	.27	1.17	1.56
0620	Arrows or gore lines		2300	.021	S.F.	.21	.95	.47	1.63	2.26
0640	Temporary paint, white or yellow, less than 3,000 L.F.		15000	.003	L.F.	.13	.15	.07	.35	.45
0660	Removal	1 Clab	300	.027			1.18		1.18	1.89
0680	Temporary tape	2 Clab	1500	.011		.52	.47		.99	1.32
0710	Thermoplastic, white or yellow, 4" wide, less than 6,000 L.F.	B-79	15000	.003		.45	.12	.12	.69	.82
0730	6" wide, less than 6,000 L.F.		14000	.003		.68	.13	.13	.94	1.09
0740	8" wide, less than 6,000 L.F.		12000	.003		.90	.15	.15	1.20	1.39
0750	12" wide, less than 6,000 L.F.		6000	.007		1.33	.30	.30	1.93	2.28
0760	Arrows		660	.061	S.F.	.74	2.77	2.70	6.21	8.20
0770	Gore lines		2500	.016		.74	.73	.71	2.18	2.76
0780	Letters		660	.061		.75	2.77	2.70	6.22	8.20

32 17 23.14 Pavement Parking Markings

		Crew	Daily Output	Labor-Hours	Unit	Material	2021 Bare Costs Labor	Equipment	Total	Total Incl O&P
0010	**PAVEMENT PARKING MARKINGS**									
0790	Layout of pavement marking	A-2	25000	.001	L.F.		.04	.01	.05	.08
0800	Lines on pvmt., parking stall, paint, white, 4" wide	B-78B	400	.045	Stall	5.20	2.05	1.02	8.27	10.10
0825	Parking stall, small quantities	2 Pord	80	.200	"	10.40	9.30		19.70	26
0830	Lines on pvmt., parking stall, thermoplastic, white, 4" wide	B-79	300	.133	Stall	21.50	6.10	5.95	33.55	40
1000	Street letters and numbers	B-78B	1600	.011	S.F.	.88	.51	.25	1.64	2.07

32 31 13.20 Fence, Chain Link Industrial

32 31 13.20 Fence, Chain Link Industrial	Crew	Daily Output	Labor-Hours	Unit	Material	2021 Bare Costs Labor	Equipment	Total	Total Incl O&P
0010 **FENCE, CHAIN LINK INDUSTRIAL**									
0011 Schedule 40, including concrete									
0020 3 strands barb wire, 2" post @ 10' OC, set in concrete, 6' H									
0200 9 ga. wire, galv. steel, in concrete	B-80C	240	.100	L.F.	19.40	4.59	1.05	25.04	30
0248 Fence, add for vinyl coated fabric				S.F.	.71			.71	.78
0300 Aluminized steel	B-80C	240	.100	L.F.	26.50	4.59	1.05	32.14	37.50
0301 Fence, wrought iron		240	.100		43	4.59	1.05	48.64	55.50
0303 Fence, commercial 4' high		240	.100		32.50	4.59	1.05	38.14	44
0304 Fence, commercial 6' high		240	.100		58	4.59	1.05	63.64	72.50
0500 6 ga. wire, galv. steel		240	.100		25	4.59	1.05	30.64	36
0600 Aluminized steel		240	.100		34	4.59	1.05	39.64	46
0800 6 ga. wire, 6' high but omit barbed wire, galv. steel		250	.096		22.50	4.40	1.01	27.91	32.50
0900 Aluminized steel, in concrete		250	.096		30.50	4.40	1.01	35.91	42
0920 8' H, 6 ga. wire, 2-1/2" line post, galv. steel, in concrete		180	.133		35.50	6.10	1.41	43.01	50.50
0940 Aluminized steel, in concrete		180	.133		43.50	6.10	1.41	51.01	59.50
1100 Add for corner posts, 3" diam., galv. steel, in concrete		40	.600	Ea.	89	27.50	6.30	122.80	149
1200 Aluminized steel, in concrete		40	.600		99.50	27.50	6.30	133.30	160
1300 Add for braces, galv. steel		80	.300		49	13.75	3.16	65.91	79.50
1350 Aluminized steel		80	.300		52	13.75	3.16	68.91	83
1400 Gate for 6' high fence, 1-5/8" frame, 3' wide, galv. steel		10	2.400		208	110	25.50	343.50	430
1500 Aluminized steel, in concrete		10	2.400		224	110	25.50	359.50	450
2000 5'-0" high fence, 9 ga., no barbed wire, 2" line post, in concrete									
2010 10' OC, 1-5/8" top rail, in concrete									
2100 Galvanized steel, in concrete	B-80C	300	.080	L.F.	20.50	3.67	.84	25.01	29.50
2200 Aluminized steel, in concrete		300	.080	"	20	3.67	.84	24.51	29
2400 Gate, 4' wide, 5' high, 2" frame, galv. steel, in concrete		10	2.400	Ea.	229	110	25.50	364.50	455
2500 Aluminized steel, in concrete		10	2.400	"	183	110	25.50	318.50	405
3100 Overhead slide gate, chain link, 6' high, to 18' wide, in concrete		38	.632	L.F.	97	29	6.65	132.65	160
3110 Cantilever type, in concrete	B-80	48	.667		189	32.50	22	243.50	284
3120 8' high, in concrete		24	1.333		182	65	44	291	350
3130 10' high, in concrete		18	1.778		211	86.50	58.50	356	435
9000 Minimum labor/equipment charge		2	16	Job		780	525	1,305	1,800

32 31 13.25 Fence, Chain Link Residential

32 31 13.25 Fence, Chain Link Residential	Crew	Daily Output	Labor-Hours	Unit	Material	2021 Bare Costs Labor	Equipment	Total	Total Incl O&P
0010 **FENCE, CHAIN LINK RESIDENTIAL**									
0011 Schedule 20, 11 ga. wire, 1-5/8" post									
3300 Residential, 11 ga. wire, 1-5/8" line post @ 10' OC									
3310 1-3/8" top rail									
3350 4' high, galvanized steel	B-80C	475	.051	L.F.	12.90	2.32	.53	15.75	18.50
3400 Aluminized		475	.051	"	17.15	2.32	.53	20	23
3600 Gate, 3' wide, 1-3/8" frame, galv. steel		10	2.400	Ea.	111	110	25.50	246.50	325
3700 Aluminized		10	2.400	"	169	110	25.50	304.50	390
3900 3' high, galvanized steel		620	.039	L.F.	134	1.78	.41	136.19	151
4000 Aluminized		620	.039	"	122	1.78	.41	124.19	137
4200 Gate, 3' wide, 1-3/8" frame, galv. steel		12	2	Ea.	94	91.50	21	206.50	272
4300 Aluminized		12	2	"	99	91.50	21	211.50	278
9000 Minimum labor/equipment charge	B-1	2	12	Job		540		540	860

32 31 13.26 Tennis Court Fences and Gates

32 31 13.26 Tennis Court Fences and Gates	Crew	Daily Output	Labor-Hours	Unit	Material	2021 Bare Costs Labor	Equipment	Total	Total Incl O&P
0010 **TENNIS COURT FENCES AND GATES**									
3150 Tennis courts, 11 ga. wire, 1-3/4" mesh, 2-1/2" line posts									
3170 1-5/8" top rail, 3" corner and gate posts									
3190 10' high, galvanized steel	B-80	190	.168	L.F.	31.50	8.20	5.55	45.25	53.50
3210 Vinyl covered 9 ga. wire	"	190	.168	"	39.50	8.20	5.55	53.25	62.50

For customer support on your Commercial Renovation Costs with RSMeans Data, call 800.448.8182.

32 31 Fences and Gates

32 31 13 – Chain Link Fences and Gates

32 31 13.40 Fence, Fabric and Accessories

	32 31 13.40 Fence, Fabric and Accessories	Crew	Daily Output	Labor-Hours	Unit	Material	2021 Bare Costs Labor	2021 Bare Costs Equipment	Total	Total Incl O&P
0010	**FENCE, FABRIC & ACCESSORIES**									
8000	Components only, fabric, galvanized, 4' high	B-80C	585	.041	L.F.	3.64	1.88	.43	5.95	7.50
8040	6' high		430	.056	"	5.95	2.56	.59	9.10	11.30
8200	Posts, line post, 4' high		34	.706	Ea.	15.95	32.50	7.45	55.90	77.50
8240	6' high		26	.923	"	19.80	42.50	9.75	72.05	100
8400	Top rails, 1-3/8" OD		1000	.024	L.F.	1.72	1.10	.25	3.07	3.92
8440	1-5/8" OD		1000	.024	"	2.83	1.10	.25	4.18	5.15

32 31 13.80 Residential Chain Link Gate

	32 31 13.80 Residential Chain Link Gate	Crew	Daily Output	Labor-Hours	Unit	Material	Labor	Equipment	Total	Total Incl O&P
0010	**RESIDENTIAL CHAIN LINK GATE**									
0110	Residential 4' gate, single incl. hardware and concrete	B-80C	10	2.400	Ea.	191	110	25.50	326.50	415
0120	5'		10	2.400		201	110	25.50	336.50	425
0130	6'		10	2.400		213	110	25.50	348.50	435
0510	Residential 4' gate, double incl. hardware and concrete		10	2.400		290	110	25.50	425.50	525
0520	5'		10	2.400		305	110	25.50	440.50	540
0530	6'		10	2.400		350	110	25.50	485.50	590

32 31 13.82 Internal Chain Link Gate

	32 31 13.82 Internal Chain Link Gate	Crew	Daily Output	Labor-Hours	Unit	Material	Labor	Equipment	Total	Total Incl O&P
0010	**INTERNAL CHAIN LINK GATE**									
0110	Internal 6' gate, single incl. post flange, hardware and concrete	B-80C	10	2.400	Ea.	340	110	25.50	475.50	580
0120	8'		10	2.400		380	110	25.50	515.50	620
0130	10'		10	2.400		505	110	25.50	640.50	760
0510	Internal 6' gate, double incl. post flange, hardware and concrete		10	2.400		575	110	25.50	710.50	835
0520	8'		10	2.400		645	110	25.50	780.50	915
0530	10'		10	2.400		805	110	25.50	940.50	1,100

32 31 13.84 Industrial Chain Link Gate

	32 31 13.84 Industrial Chain Link Gate	Crew	Daily Output	Labor-Hours	Unit	Material	Labor	Equipment	Total	Total Incl O&P
0010	**INDUSTRIAL CHAIN LINK GATE**									
0110	Industrial 8' gate, single incl. hardware and concrete	B-80C	10	2.400	Ea.	525	110	25.50	660.50	780
0120	10'		10	2.400		590	110	25.50	725.50	855
0510	Industrial 8' gate, double incl. hardware and concrete		10	2.400		790	110	25.50	925.50	1,075
0520	10'		10	2.400		895	110	25.50	1,030.50	1,175

32 31 13.88 Chain Link Transom

	32 31 13.88 Chain Link Transom	Crew	Daily Output	Labor-Hours	Unit	Material	Labor	Equipment	Total	Total Incl O&P
0010	**CHAIN LINK TRANSOM**									
0110	Add for, single transom, 3' wide, incl. components & hardware	B-80C	10	2.400	Ea.	119	110	25.50	254.50	335
0120	Add for, double transom, 6' wide, incl. components & hardware	"	10	2.400	"	127	110	25.50	262.50	345

32 31 19 – Decorative Metal Fences and Gates

32 31 19.10 Decorative Fence

	32 31 19.10 Decorative Fence	Crew	Daily Output	Labor-Hours	Unit	Material	Labor	Equipment	Total	Total Incl O&P
0010	**DECORATIVE FENCE**									
5300	Tubular picket, steel, 6' sections, 1-9/16" posts, 4' high	B-80C	300	.080	L.F.	38.50	3.67	.84	43.01	49.50
5400	2" posts, 5' high		240	.100		44	4.59	1.05	49.64	56.50
5600	2" posts, 6' high		200	.120		53	5.50	1.26	59.76	68.50
5700	Staggered picket, 1-9/16" posts, 4' high		300	.080		45	3.67	.84	49.51	56.50
5800	2" posts, 5' high		240	.100		45	4.59	1.05	50.64	58
5900	2" posts, 6' high		200	.120		59	5.50	1.26	65.76	75
6200	Gates, 4' high, 3' wide	B-1	10	2.400	Ea.	335	108		443	540
6300	5' high, 3' wide		10	2.400		246	108		354	445
6400	6' high, 3' wide		10	2.400		370	108		478	575
6500	4' wide		10	2.400		330	108		438	530

For customer support on your Commercial Renovation Costs with RSMeans Data, call 800.448.8182.

569

32 31 Fences and Gates

32 31 23 – Plastic Fences and Gates

32 31 23.20 Fence, Recycled Plastic

		Crew	Daily Output	Labor-Hours	Unit	Material	2021 Bare Costs Labor	Equipment	Total	Total Incl O&P
0010	**FENCE, RECYCLED PLASTIC**									
9015	Fence rail, made from recycled plastic, various colors, 2 rail [G]	B-1	150	.160	L.F.	11.35	7.20		18.55	24
9018	3 rail [G]		150	.160		10.25	7.20		17.45	23
9020	4 rail [G]		150	.160		15.85	7.20		23.05	29
9030	Fence pole, made from recycled plastic, various colors, 7' [G]		96	.250	Ea.	60.50	11.25		71.75	84.50
9040	Stockade fence, made from recycled plastic, various colors, 4' high [G]	B-80C	160	.150	L.F.	32.50	6.90	1.58	40.98	48.50
9050	6' high [G]		160	.150	"	37	6.90	1.58	45.48	53
9060	6' pole [G]		96	.250	Ea.	43	11.45	2.64	57.09	68
9070	9' pole [G]		96	.250	"	42.50	11.45	2.64	56.59	67.50
9080	Picket fence, made from recycled plastic, various colors, 3' high [G]		160	.150	L.F.	23.50	6.90	1.58	31.98	38.50
9090	4' high [G]		160	.150	"	35	6.90	1.58	43.48	51
9100	3' high gate [G]		8	3	Ea.	24.50	138	31.50	194	281
9110	4' high gate [G]		8	3		31.50	138	31.50	201	289
9120	5' high pole [G]		96	.250		45.50	11.45	2.64	59.59	71.50
9130	6' high pole [G]		96	.250		42	11.45	2.64	56.09	67.50
9140	Pole cap only [G]					4.89			4.89	5.40
9150	Keeper pins only [G]					.41			.41	.45

32 31 26 – Wire Fences and Gates

32 31 26.10 Fences, Misc. Metal

		Crew	Daily Output	Labor-Hours	Unit	Material	2021 Bare Costs Labor	Equipment	Total	Total Incl O&P
0010	**FENCES, MISC. METAL**									
0012	Chicken wire, posts @ 4', 1" mesh, 4' high	B-80C	410	.059	L.F.	4.32	2.69	.62	7.63	9.70
0100	2" mesh, 6' high		350	.069		3.75	3.15	.72	7.62	9.90
0200	Galv. steel, 12 ga., 2" x 4" mesh, posts 5' OC, 3' high		300	.080		2.79	3.67	.84	7.30	9.85
0300	5' high		300	.080		3.30	3.67	.84	7.81	10.40
0400	14 ga., 1" x 2" mesh, 3' high		300	.080		3.10	3.67	.84	7.61	10.20
0500	5' high		300	.080		4.34	3.67	.84	8.85	11.55
1000	Kennel fencing, 1-1/2" mesh, 6' long, 3'-6" wide, 6'-2" high	2 Clab	4	4	Ea.	505	178		683	840
1050	12' long		4	4		715	178		893	1,075
1200	Top covers, 1-1/2" mesh, 6' long		15	1.067		146	47.50		193.50	237
1250	12' long		12	1.333		219	59		278	335
4492	Security fence, prison grade, barbed wire, set in concrete, 10' high	B-80	22	1.455	L.F.	64.50	71	48	183.50	236
4494	Security fence, prison grade, razor wire, set in concrete, 10' high		18	1.778		65.50	86.50	58.50	210.50	274
4500	Security fence, prison grade, set in concrete, 12' high		25	1.280		56	62.50	42	160.50	207
4600	16' high		20	1.600		87	78	52.50	217.50	278
4990	Security fence, prison grade, set in concrete, 10' high		25	1.280		56	62.50	42	160.50	207

32 31 29 – Wood Fences and Gates

32 31 29.20 Fence, Wood Rail

		Crew	Daily Output	Labor-Hours	Unit	Material	2021 Bare Costs Labor	Equipment	Total	Total Incl O&P
0010	**FENCE, WOOD RAIL**									
0012	Picket, No. 2 cedar, Gothic, 2 rail, 3' high	B-1	160	.150	L.F.	8.60	6.75		15.35	20
0050	Gate, 3'-6" wide	B-80C	9	2.667	Ea.	95	122	28	245	330
0600	Open rail, rustic, No. 1 cedar, 2 rail, 3' high		160	.150	L.F.	6.20	6.90	1.58	14.68	19.55
0650	Gate, 3' wide		9	2.667	Ea.	90	122	28	240	325
1200	Stockade, No. 2 cedar, treated wood rails, 6' high		160	.150	L.F.	9.10	6.90	1.58	17.58	22.50
1250	Gate, 3' wide		9	2.667	Ea.	140	122	28	290	380
3300	Board, shadow box, 1" x 6", treated pine, 6' high		160	.150	L.F.	13.35	6.90	1.58	21.83	27.50
3400	No. 1 cedar, 6' high		150	.160		24	7.35	1.69	33.04	39.50
3900	Basket weave, No. 1 cedar, 6' high		160	.150		38	6.90	1.58	46.48	54
3950	Gate, 3'-6" wide	B-1	8	3	Ea.	284	135		419	525
4000	Treated pine, 6' high		150	.160	L.F.	17.50	7.20		24.70	31
4200	Gate, 3'-6" wide		9	2.667	Ea.	194	120		314	405
8000	Posts only, 4' high		40	.600		13.55	27		40.55	58

32 31 Fences and Gates

32 31 29 – Wood Fences and Gates

32 31 29.20 Fence, Wood Rail	Crew	Daily Output	Labor-Hours	Unit	Material	2021 Bare Costs Labor	Equipment	Total	Total Incl O&P
8040	6' high	B-1	30	.800	Ea.	17.90	36		53.90	77
9000	Minimum labor/equipment charge	1 Clab	2	4	Job		178		178	283

32 32 Retaining Walls

32 32 29 – Timber Retaining Walls

32 32 29.10 Landscape Timber Retaining Walls

		Crew	Daily Output	Labor-Hours	Unit	Material	2021 Bare Costs Labor	Equipment	Total	Total Incl O&P
0010	**LANDSCAPE TIMBER RETAINING WALLS**									
0100	Treated timbers, 6" x 6"	1 Clab	265	.030	L.F.	2.84	1.34		4.18	5.25
0110	6" x 8'	"	200	.040	"	7.40	1.78		9.18	11
0120	Drilling holes in timbers for fastening, 1/2"	1 Carp	450	.018	Inch		.97		.97	1.55
0130	5/8"	1 Carp	450	.018	Inch		.97		.97	1.55
0140	Reinforcing rods for fastening, 1/2"	1 Clab	312	.026	L.F.	.46	1.14		1.60	2.31
0150	5/8"	"	312	.026	"	.71	1.14		1.85	2.59
0160	Reinforcing fabric	2 Clab	2500	.006	S.Y.	2.25	.28		2.53	2.93
0170	Gravel backfill		28	.571	C.Y.	26.50	25.50		52	70
0180	Perforated pipe, 4" diameter with silt sock		1200	.013	L.F.	1.26	.59		1.85	2.33
0190	Galvanized 60d common nails	1 Clab	625	.013	Ea.	.15	.57		.72	1.06
0200	20d common nails	"	3800	.002	"	.03	.09		.12	.19

32 32 53 – Stone Retaining Walls

32 32 53.10 Retaining Walls, Stone

		Crew	Daily Output	Labor-Hours	Unit	Material	2021 Bare Costs Labor	Equipment	Total	Total Incl O&P
0010	**RETAINING WALLS, STONE**									
0015	Including excavation, concrete footing and									
0020	stone 3' below grade. Price is exposed face area.									
0200	Decorative random stone, to 6' high, 1'-6" thick, dry set	D-1	35	.457	S.F.	71	22.50		93.50	114
0300	Mortar set		40	.400		73	19.50		92.50	112
0500	Cut stone, to 6' high, 1'-6" thick, dry set		35	.457		73.50	22.50		96	117
0600	Mortar set		40	.400		74	19.50		93.50	113
0800	Random stone, 6' to 10' high, 2' thick, dry set		45	.356		75.50	17.30		92.80	111
0900	Mortar set		50	.320		90	15.60		105.60	124
1100	Cut stone, 6' to 10' high, 2' thick, dry set		45	.356		75.50	17.30		92.80	111
1200	Mortar set		50	.320		90	15.60		105.60	124
9000	Minimum labor/equipment charge		2	8	Job		390		390	625

32 33 Site Furnishings

32 33 33 – Site Manufactured Planters

32 33 33.10 Planters

		Crew	Daily Output	Labor-Hours	Unit	Material	2021 Bare Costs Labor	Equipment	Total	Total Incl O&P
0010	**PLANTERS**									
0012	Concrete, sandblasted, precast, 48" diameter, 24" high	2 Clab	15	1.067	Ea.	665	47.50		712.50	805
0100	Fluted, precast, 7' diameter, 36" high		10	1.600		1,700	71		1,771	1,975
0300	Fiberglass, circular, 36" diameter, 24" high		15	1.067		755	47.50		802.50	905
0400	60" diameter, 24" high		10	1.600		1,200	71		1,271	1,450
9000	Minimum labor/equipment charge	1 Clab	2	4	Job		178		178	283

32 33 43 – Site Seating and Tables

32 33 43.13 Site Seating

		Crew	Daily Output	Labor-Hours	Unit	Material	2021 Bare Costs Labor	Equipment	Total	Total Incl O&P
0010	**SITE SEATING**									
0012	Seating, benches, park, precast conc., w/backs, wood rails, 4' long	2 Clab	5	3.200	Ea.	670	142		812	965
0100	8' long		4	4		1,175	178		1,353	1,550
0300	Fiberglass, without back, one piece, 4' long		10	1.600		1,700	71		1,771	2,000

571

32 33 Site Furnishings
32 33 43 – Site Seating and Tables

32 33 43.13 Site Seating	Crew	Daily Output	Labor-Hours	Unit	Material	2021 Bare Costs Labor	2021 Bare Costs Equipment	Total	Total Incl O&P
0400	8' long	2 Clab	7	2.286	Ea.	365	101		466	565
0500	Steel barstock pedestals w/backs, 2" x 3" wood rails, 4' long		10	1.600		1,325	71		1,396	1,575
0510	8' long		7	2.286		1,850	101		1,951	2,175
0515	Powder coated steel, 4" x 4" plastic slats, 6' long		8	2		550	89		639	745
0520	3" x 8" wood plank, 4' long	2 Clab	10	1.600	Ea.	1,500	71		1,571	1,775
0530	8' long		7	2.286		1,425	101		1,526	1,725
0540	Backless, 4" x 4" wood plank, 4' square		10	1.600		1,275	71		1,346	1,525
0550	8' long		7	2.286		1,250	101		1,351	1,525
0560	Powder coated steel, with back and 2 anti-vagrant dividers, 6' long		8	2		1,250	89		1,339	1,525
0600	Aluminum pedestals, with backs, aluminum slats, 8' long		8	2		505	89		594	700
0610	15' long		5	3.200		1,025	142		1,167	1,350
0620	Portable, aluminum slats, 8' long		8	2		450	89		539	635
0630	15' long		5	3.200		655	142		797	945
0800	Cast iron pedestals, back & arms, wood slats, 4' long		8	2		465	89		554	650
0820	8' long		5	3.200		1,050	142		1,192	1,375
0840	Backless, wood slats, 4' long		8	2		760	89		849	975
0860	8' long		5	3.200		1,300	142		1,442	1,675
1700	Steel frame, fir seat, 10' long		10	1.600		405	71		476	560
2000	Benches, park, with back, galv. stl. frame, 4" x 4" plastic slats, 6' long		7	2.286		555	101		656	770
9000	Minimum labor/equipment charge		2	8	Job		355		355	565

32 84 Planting Irrigation
32 84 23 – Underground Sprinklers
32 84 23.10 Sprinkler Irrigation System

		Crew	Daily Output	Labor-Hours	Unit	Material	2021 Bare Costs Labor	2021 Bare Costs Equipment	Total	Total Incl O&P
0010	**SPRINKLER IRRIGATION SYSTEM**									
0011	For lawns									
0800	Residential system, custom, 1" supply	B-20	2000	.012	S.F.	.28	.59		.87	1.25
0900	1-1/2" supply	"	1800	.013	"	.42	.66		1.08	1.50

32 91 Planting Preparation
32 91 13 – Soil Preparation
32 91 13.23 Structural Soil Mixing

		Crew	Daily Output	Labor-Hours	Unit	Material	2021 Bare Costs Labor	2021 Bare Costs Equipment	Total	Total Incl O&P
0010	**STRUCTURAL SOIL MIXING**									
0100	Rake topsoil, site material, harley rock rake, ideal	B-6	33	.727	M.S.F.	35	6.55		41.55	62.50
0200	Adverse	"	7	3.429			165	31	196	295
0300	Screened loam, york rake and finish, ideal	B-62	24	1			48	7.50	55.50	84.50
0400	Adverse	"	20	1.200			57.50	9	66.50	101
1000	Remove topsoil & stock pile on site, 75 HP dozer, 6" deep, 50' haul	B-10L	30	.400			21.50	13.55	35.05	49
1050	300' haul		6.10	1.967			106	66.50	172.50	240
1100	12" deep, 50' haul		15.50	.774			42	26	68	95
1150	300' haul		3.10	3.871			210	131	341	475
1200	200 HP dozer, 6" deep, 50' haul	B-10B	125	.096			5.20	12.15	17.35	21.50
1250	300' haul		30.70	.391			21	49.50	70.50	87.50
1300	12" deep, 50' haul		62	.194			10.50	24.50	35	43.50
1350	300' haul		15.40	.779			42	98.50	140.50	175
1400	Alternate method, 75 HP dozer, 50' haul	B-10L	860	.014	C.Y.		.76	.47	1.23	1.71
1450	300' haul	B-10L	114	.105	C.Y.		5.70	3.56	9.26	12.85
1500	200 HP dozer, 50' haul	B-10B	2660	.005			.24	.57	.81	1.01
1600	300' haul	"	570	.021			1.14	2.67	3.81	4.72

32 91 Planting Preparation

32 91 13 – Soil Preparation

32 91 13.23 Structural Soil Mixing

		Crew	Daily Output	Labor-Hours	Unit	Material	2021 Bare Costs Labor	Equipment	Total	Total Incl O&P
1800	Rolling topsoil, hand push roller	1 Clab	3200	.003	S.F.		.11		.11	.18
1850	Tractor drawn roller	B-66	10666	.001	"		.04	.03	.07	.09
2000	Root raking and loading, residential, no boulders	B-6	53.30	.450	M.S.F.		21.50	4.06	25.56	38.50
2100	With boulders		32	.750			36	6.75	42.75	64.50
2200	Municipal, no boulders		200	.120			5.75	1.08	6.83	10.30
2300	With boulders		120	.200			9.60	1.80	11.40	17.20
2400	Large commercial, no boulders	B-10B	400	.030			1.62	3.80	5.42	6.75
2500	With boulders	"	240	.050			2.71	6.35	9.06	11.20
3000	Scarify subsoil, residential, skid steer loader w/scarifiers, 50 HP	B-66	32	.250			13.90	8.35	22.25	30.50
3050	Municipal, skid steer loader w/scarifiers, 50 HP	"	120	.067			3.70	2.23	5.93	8.20
3100	Large commercial, 75 HP, dozer w/scarifier	B-10L	240	.050			2.71	1.69	4.40	6.10
3500	Screen topsoil from stockpile, vibrating screen, wet material (organic)	B-10P	200	.060	C.Y.		3.25	5.75	9	11.40
3550	Dry material	"	300	.040			2.17	3.82	5.99	7.60
3600	Mixing with conditioners, manure and peat	B-10R	550	.022			1.18	.56	1.74	2.46
3650	Mobilization add for 2 days or less operation	B-34K	3	2.667	Job		137	289	426	540
3800	Spread conditioned topsoil, 6" deep, by hand	B-1	360	.067	S.Y.	6	3		9	11.40
3850	300 HP dozer	B-10M	27	.444	M.S.F.	650	24	66	740	825
4000	Spread soil conditioners, alum. sulfate, 1#/S.Y., hand push spreader	1 Clab	17500	.001	S.Y.	19.20	.02		19.22	21
4050	Tractor spreader	B-66	700	.011	M.S.F.	2,125	.63	.38	2,126.01	2,350
4100	Fertilizer, 0.2#/S.Y., push spreader	1 Clab	17500	.001	S.Y.	.08	.02		.10	.12
4150	Tractor spreader	B-66	700	.011	M.S.F.	9.35	.63	.38	10.36	11.65
4200	Ground limestone, 1#/S.Y., push spreader	1 Clab	17500	.001	S.Y.	.16	.02		.18	.21
4250	Tractor spreader	B-66	700	.011	M.S.F.	17.80	.63	.38	18.81	21
4400	Manure, 18#/S.Y., push spreader	1 Clab	2500	.003	S.Y.	7.40	.14		7.54	8.40
4450	Tractor spreader	B-66	280	.029	M.S.F.	825	1.59	.96	827.55	910
4500	Perlite, 1" deep, push spreader	1 Clab	17500	.001	S.Y.	11.05	.02		11.07	12.20
4550	Tractor spreader	B-66	700	.011	M.S.F.	1,225	.63	.38	1,226.01	1,350
4600	Vermiculite, push spreader	1 Clab	17500	.001	S.Y.	8.45	.02		8.47	9.35
4650	Tractor spreader	B-66	700	.011	M.S.F.	940	.63	.38	941.01	1,025
5000	Spread topsoil, skid steer loader and hand dress	B-62	270	.089	C.Y.	27	4.28	.66	31.94	37.50
5100	Articulated loader and hand dress	B-100	320	.038		27	2.03	2.89	31.92	36.50
5200	Articulated loader and 75 HP dozer	B-10M	500	.024		27	1.30	3.57	31.87	36
5300	Road grader and hand dress	B-11L	1000	.016		27	.83	1.07	28.90	32.50
6000	Tilling topsoil, 20 HP tractor, disk harrow, 2" deep	B-66	450	.018	M.S.F.		.99	.59	1.58	2.19
6050	4" deep	"	360	.022	"		1.23	.74	1.97	2.74
6100	6" deep	B-66	270	.030	M.S.F.		1.64	.99	2.63	3.66
6150	26" rototiller, 2" deep	A-1J	1250	.006	S.Y.		.28	.04	.32	.50
6200	4" deep		1000	.008			.36	.05	.41	.63
6250	6" deep		750	.011			.47	.07	.54	.83

32 91 13.26 Planting Beds

		Crew	Daily Output	Labor-Hours	Unit	Material	2021 Bare Costs Labor	Equipment	Total	Total Incl O&P
0010	**PLANTING BEDS**									
0100	Backfill planting pit, by hand, on site topsoil	2 Clab	18	.889	C.Y.		39.50		39.50	63
0200	Prepared planting mix, by hand	"	24	.667			29.50		29.50	47
0300	Skid steer loader, on site topsoil	B-62	340	.071			3.40	.53	3.93	5.95
0400	Prepared planting mix	"	410	.059			2.82	.44	3.26	4.93
1000	Excavate planting pit, by hand, sandy soil	2 Clab	16	1			44.50		44.50	70.50
1100	Heavy soil or clay	"	8	2			89		89	141
1200	1/2 C.Y. backhoe, sandy soil	B-11C	150	.107			5.50	1.44	6.94	10.30
1300	Heavy soil or clay	"	115	.139			7.20	1.88	9.08	13.35
2000	Mix planting soil, incl. loam, manure, peat, by hand	2 Clab	60	.267		47.50	11.85		59.35	71
2100	Skid steer loader	B-62	150	.160		47.50	7.70	1.20	56.40	65.50
3000	Pile sod, skid steer loader	"	2800	.009	S.Y.		.41	.06	.47	.72

32 91 Planting Preparation

32 91 13 – Soil Preparation

32 91 13.26 Planting Beds

		Crew	Daily Output	Labor-Hours	Unit	Material	2021 Bare Costs Labor	2021 Bare Costs Equipment	Total	Total Incl O&P
3100	By hand	2 Clab	400	.040	S.Y.		1.78		1.78	2.83
4000	Remove sod, F.E. loader	B-10S	2000	.006			.32	.22	.54	.75
4100	Sod cutter	B-12K	3200	.005			.26	.31	.57	.76
4200	By hand	2 Clab	240	.067	↓		2.96		2.96	4.71

32 91 19 – Landscape Grading

32 91 19.13 Topsoil Placement and Grading

		Crew	Daily Output	Labor-Hours	Unit	Material	2021 Bare Costs Labor	2021 Bare Costs Equipment	Total	Total Incl O&P
0010	**TOPSOIL PLACEMENT AND GRADING**									
0700	Furnish and place, truck dumped, screened, 4" deep	B-10S	1300	.009	S.Y.	3.69	.50	.34	4.53	5.20
0800	6" deep		820	.015	"	4.73	.79	.54	6.06	7.05
0810	Minimum labor/equipment charge	↓	2	6	Ea.		325	221	546	755
0900	Fine grading and seeding, incl. lime, fertilizer & seed,									
1000	With equipment	B-14	1000	.048	S.Y.	.49	2.24	.22	2.95	4.33
2000	Minimum labor/equipment charge	1 Clab	4	2	Job		89		89	141

32 92 Turf and Grasses

32 92 19 – Seeding

32 92 19.13 Mechanical Seeding

		Crew	Daily Output	Labor-Hours	Unit	Material	2021 Bare Costs Labor	2021 Bare Costs Equipment	Total	Total Incl O&P
0010	**MECHANICAL SEEDING**									
0020	Mechanical seeding, 215 lb./acre	B-66	1.50	5.333	Acre	585	296	178	1,059	1,300
0100	44 lb./M.S.Y.	"	2500	.003	S.Y.	.24	.18	.11	.53	.66
0101	44 lb./M.S.Y.	1 Clab	13950	.001	S.F.	.03	.03		.06	.07
0300	Fine grading and seeding incl. lime, fertilizer & seed,									
0310	with equipment	B-14	1000	.048	S.Y.	.47	2.24	.22	2.93	4.31
0400	Fertilizer hand push spreader, 35 lb./M.S.F.	1 Clab	200	.040	M.S.F.	9	1.78		10.78	12.75
0600	Limestone hand push spreader, 50 lb./M.S.F.	"	180	.044	"	5.45	1.97		7.42	9.15
0800	Grass seed hand push spreader, 4.5 lb./M.S.F.	1 Clab	180	.044	M.S.F.	26.50	1.97		28.47	32.50
1000	Hydro or air seeding for large areas, incl. seed and fertilizer	B-81	8900	.003	S.Y.	.60	.14	.06	.80	.95
1100	With wood fiber mulch added	"	8900	.003	"	2.20	.14	.06	2.40	2.71
1300	Seed only, over 100 lb., field seed, minimum				Lb.	2.52			2.52	2.77
1400	Maximum					1.67			1.67	1.84
1500	Lawn seed, minimum					1.34			1.34	1.47
1600	Maximum				↓	3.01			3.01	3.31
1800	Aerial operations, seeding only, field seed	B-58	50	.480	Acre	820	23	46.50	889.50	990
1900	Lawn seed		50	.480		435	23	46.50	504.50	570
2100	Seed and liquid fertilizer, field seed		50	.480		905	23	46.50	974.50	1,075
2200	Lawn seed	↓	50	.480	↓	520	23	46.50	589.50	660
9000	Minimum labor/equipment charge	1 Clab	4	2	Job		89		89	141

32 92 23 – Sodding

32 92 23.10 Sodding Systems

		Crew	Daily Output	Labor-Hours	Unit	Material	2021 Bare Costs Labor	2021 Bare Costs Equipment	Total	Total Incl O&P
0010	**SODDING SYSTEMS**									
0020	Sodding, 1" deep, bluegrass sod, on level ground, over 8 M.S.F.	B-63	22	1.818	M.S.F.	212	85	8.15	305.15	375
0200	4 M.S.F.		17	2.353		400	110	10.55	520.55	625
0300	1,000 S.F.		13.50	2.963		335	138	13.30	486.30	605
0500	Sloped ground, over 8 M.S.F.		6	6.667		212	310	30	552	760
0600	4 M.S.F.		5	8		400	375	36	811	1,075
0700	1,000 S.F.		4	10		335	465	45	845	1,150
1000	Bent grass sod, on level ground, over 6 M.S.F.		20	2		330	93	9	432	525
1100	3 M.S.F.		18	2.222		360	104	10	474	570
1200	Sodding 1,000 S.F. or less		14	2.857		380	133	12.85	525.85	645
1500	Sloped ground, over 6 M.S.F.	↓	15	2.667	↓	330	124	11.95	465.95	575

32 92 Turf and Grasses

32 92 23 – Sodding

32 92 23.10 Sodding Systems	Crew	Daily Output	Labor-Hours	Unit	Material	2021 Bare Costs Labor	Equipment	Total	Total Incl O&P
1600	3 M.S.F.	B-63	13.50	2.963	M.S.F.	360	138	13.30	511.30	630
1700	1,000 S.F.	▼	12	3.333	▼	380	155	14.95	549.95	680

32 93 Plants

32 93 13 – Ground Covers

32 93 13.10 Ground Cover Plants

		Crew	Daily Output	Labor-Hours	Unit	Material	Labor	Equipment	Total	Total Incl O&P
0010	**GROUND COVER PLANTS**									
0012	Plants, pachysandra, in prepared beds	B-1	15	1.600	C	84.50	72		156.50	208
0200	Vinca minor, 1 yr., bare root, in prepared beds		12	2	"	99	90		189	253
0600	Stone chips, in 50 lb. bags, Georgia marble		520	.046	Bag	5.15	2.08		7.23	8.95
0700	Onyx gemstone		260	.092		40	4.16		44.16	50.50
0800	Quartz		260	.092	▼	17.15	4.16		21.31	25.50
0900	Pea gravel, truckload lots	▼	28	.857	Ton	18.70	38.50		57.20	82

32 93 33 – Shrubs

32 93 33.10 Shrubs and Trees

		Crew	Daily Output	Labor-Hours	Unit	Material	Labor	Equipment	Total	Total Incl O&P
0010	**SHRUBS AND TREES**									
0011	Evergreen, in prepared beds, B&B									
0100	Arborvitae pyramidal, 4'-5'	B-17	30	1.067	Ea.	101	52	21	174	217
0150	Globe, 12"-15"	B-1	96	.250	Ea.	17.70	11.25		28.95	37.50
0300	Cedar, blue, 8'-10'	B-17	18	1.778		248	87	34.50	369.50	450
0500	Hemlock, Canadian, 2-1/2'-3'	B-1	36	.667		32.50	30		62.50	83.50
0550	Holly, Savannah, 8'-10' H		9.68	2.479		254	112		366	460
0600	Juniper, andorra, 18"-24"		80	.300		55.50	13.50		69	82.50
0620	Wiltoni, 15"-18"	▼	80	.300		29.50	13.50		43	54
0640	Skyrocket, 4-1/2'-5'	B-17	55	.582		99	28.50	11.35	138.85	166
0660	Blue pfitzer, 2'-2-1/2'	B-1	44	.545		39.50	24.50		64	82.50
0680	Ketleerie, 2-1/2'-3'		50	.480		59	21.50		80.50	99.50
0700	Pine, black, 2-1/2'-3'		50	.480		56	21.50		77.50	96
0720	Mugo, 18"-24"	▼	60	.400		49	18.05		67.05	82
0740	White, 4'-5'	B-17	75	.427		51.50	21	8.30	80.80	99
0800	Spruce, blue, 18"-24"	B-1	60	.400		65.50	18.05		83.55	101
0840	Norway, 4'-5'	B-17	75	.427		80	21	8.30	109.30	130
0900	Yew, denisforma, 12"-15"	B-1	60	.400		36.50	18.05		54.55	69
1000	Capitata, 18"-24"		30	.800		32.50	36		68.50	93
1100	Hicksi, 2'-2-1/2'	▼	30	.800	▼	93	36		129	160

32 93 33.20 Shrubs

		Crew	Daily Output	Labor-Hours	Unit	Material	Labor	Equipment	Total	Total Incl O&P
0010	**SHRUBS**									
0011	Broadleaf Evergreen, planted in prepared beds									
0100	Andromeda, 15"-18", cont	B-1	96	.250	Ea.	38.50	11.25		49.75	60.50
0200	Azalea, 15"-18", cont		96	.250		37.50	11.25		48.75	59
0300	Barberry, 9"-12", cont		130	.185		22.50	8.30		30.80	38
0400	Boxwood, 15"-18", B&B		96	.250		49	11.25		60.25	71.50
0500	Euonymus, emerald gaiety, 12"-15", cont		115	.209		27	9.40		36.40	45
0600	Holly, 15"-18", B&B		96	.250		41.50	11.25		52.75	63.50
0900	Mount laurel, 18"-24", B&B		80	.300		77.50	13.50		91	107
1000	Paxistema, 9"-12" H		130	.185		23	8.30		31.30	39
1100	Rhododendron, 18"-24", cont		48	.500		44.50	22.50		67	85
1200	Rosemary, 1 gal. cont		600	.040		18.55	1.80		20.35	23.50
2000	Deciduous, planted in prepared beds, amelanchier, 2'-3', B&B		57	.421		134	19		153	178
2100	Azalea, 15"-18", B&B	▼	96	.250	▼	34.50	11.25		45.75	56

32 93 33 – Shrubs

32 93 33.20 Shrubs		Crew	Daily Output	Labor-Hours	Unit	Material	2021 Bare Costs Labor	Equipment	Total	Total Incl O&P
2300	Bayberry, 2'-3', B&B	B-1	57	.421	Ea.	33	19		52	66.50
2600	Cotoneaster, 15"-18", B&B	↓	80	.300		33.50	13.50		47	58.50
2800	Dogwood, 3'-4', B&B	B-17	40	.800		36	39	15.60	90.60	119
2900	Euonymus, alatus compacta, 15"-18", cont	B-1	80	.300		28	13.50		41.50	52.50
3200	Forsythia, 2'-3', cont	"	60	.400		15.40	18.05		33.45	45.50
3300	Hibiscus, 3'-4', B&B	B-17	75	.427		45.50	21	8.30	74.80	92
3400	Honeysuckle, 3'-4', B&B	B-1	60	.400		65	18.05		83.05	100
3500	Hydrangea, 2'-3', B&B	"	57	.421		30	19		49	63
3600	Lilac, 3'-4', B&B	B-17	40	.800	↓	22.50	39	15.60	77.10	104
3900	Privet, bare root, 18"-24"	B-1	80	.300	Ea.	18.15	13.50		31.65	41.50
4100	Quince, 2'-3', B&B	"	57	.421		31.50	19		50.50	64.50
4200	Russian olive, 3'-4', B&B	B-17	75	.427		26	21	8.30	55.30	70.50
4400	Spirea, 3'-4', B&B	B-1	70	.343		21	15.45		36.45	47.50
4500	Viburnum, 3'-4', B&B	B-17	40	.800	↓	27.50	39	15.60	82.10	110

32 93 43 – Trees

32 93 43.20 Trees

32 93 43.20 Trees			Crew	Daily Output	Labor-Hours	Unit	Material	2021 Bare Costs Labor	Equipment	Total	Total Incl O&P
0010	**TREES**										
0011	Deciduous, in prep. beds, balled & burlapped (B&B)										
0100	Ash, 2" caliper	G	B-17	8	4	Ea.	197	196	78	471	615
0200	Beech, 5'-6'	G		50	.640		216	31.50	12.45	259.95	300
0300	Birch, 6'-8', 3 stems	G		20	1.600		65	78	31	174	230
0500	Crabapple, 6'-8'	G		20	1.600		132	78	31	241	305
0600	Dogwood, 4'-5'	G		40	.800		131	39	15.60	185.60	223
0700	Eastern redbud, 4'-5'	G		40	.800		149	39	15.60	203.60	243
0800	Elm, 8'-10'	G		20	1.600		375	78	31	484	570
0900	Ginkgo, 6'-7'	G		24	1.333		163	65	26	254	310
1000	Hawthorn, 8'-10', 1" caliper	G		20	1.600		156	78	31	265	330
1100	Honeylocust, 10'-12', 1-1/2" caliper	G		10	3.200		229	156	62.50	447.50	570
1300	Larch, 8'	G		32	1		139	49	19.50	207.50	252
1400	Linden, 8'-10', 1" caliper	G		20	1.600		161	78	31	270	335
1500	Magnolia, 4'-5'	G		20	1.600		140	78	31	249	315
1600	Maple, red, 8'-10', 1-1/2" caliper	G		10	3.200		182	156	62.50	400.50	520
1700	Mountain ash, 8'-10', 1" caliper	G		16	2		202	98	39	339	420
1800	Oak, 2-1/2"-3" caliper	G		6	5.333		365	261	104	730	930
2100	Planetree, 9'-11', 1-1/4" caliper	G		10	3.200		300	156	62.50	518.50	645
2200	Plum, 6'-8', 1" caliper	G		20	1.600		120	78	31	229	291
2300	Poplar, 9'-11', 1-1/4" caliper	G		10	3.200		99	156	62.50	317.50	425
2500	Sumac, 2'-3'	G		75	.427		45.50	21	8.30	74.80	92
2700	Tulip, 5'-6'	G		40	.800		54.50	39	15.60	109.10	139
2800	Willow, 6'-8', 1" caliper	G	↓	20	1.600	↓	122	78	31	231	293
9000	Minimum labor/equipment charge		1 Clab	4	2	Job		89		89	141

32 94 Planting Accessories

32 94 13 – Landscape Edging

32 94 13.20 Edging		Crew	Daily Output	Labor-Hours	Unit	Material	2021 Bare Costs Labor	Equipment	Total	Total Incl O&P
0010	**EDGING**									
0050	Aluminum alloy, including stakes, 1/8" x 4", mill finish	B-1	390	.062	L.F.	2.27	2.77		5.04	6.90
0051	Black paint		390	.062		2.63	2.77		5.40	7.30
0052	Black anodized		390	.062		3.04	2.77		5.81	7.75
0100	Brick, set horizontally, 1-1/2 bricks per L.F.	D-1	370	.043		1.75	2.11		3.86	5.30
0150	Set vertically, 3 bricks per L.F.	"	135	.119		4.04	5.75		9.79	13.70
0200	Corrugated aluminum, roll, 4" wide	1 Carp	650	.012		2.08	.67		2.75	3.36
0250	6" wide	1 Carp	550	.015	L.F.	2.60	.80		3.40	4.13
0600	Railroad ties, 6" x 8"	2 Carp	170	.094		2.74	5.15		7.89	11.20
0650	7" x 9"		136	.118		3.04	6.45		9.49	13.60
0750	Redwood 2" x 4"		330	.048		5.45	2.65		8.10	10.20
0800	Steel edge strips, incl. stakes, 1/4" x 5"	B-1	390	.062		4.68	2.77		7.45	9.55
0850	3/16" x 4"	"	390	.062		3.70	2.77		6.47	8.50
9000	Minimum labor/equipment charge	1 Carp	4	2	Job		109		109	174

32 94 50 – Tree Guying

32 94 50.10 Tree Guying Systems		Crew	Daily Output	Labor-Hours	Unit	Material	2021 Bare Costs Labor	Equipment	Total	Total Incl O&P
0010	**TREE GUYING SYSTEMS**									
0015	Tree guying including stakes, guy wire and wrap									
0100	Less than 3" caliper, 2 stakes	2 Clab	35	.457	Ea.	14.50	20.50		35	48.50
0200	3" to 4" caliper, 3 stakes	"	21	.762	"	21	34		55	77
1000	Including arrowhead anchor, cable, turnbuckles and wrap									
1100	Less than 3" caliper, 3" anchors	2 Clab	20	.800	Ea.	25.50	35.50		61	84.50
1200	3" to 6" caliper, 4" anchors		15	1.067		35	47.50		82.50	114
1300	6" caliper, 6" anchors		12	1.333		29	59		88	127
1400	8" caliper, 8" anchors		9	1.778		125	79		204	263

32 96 Transplanting

32 96 23 – Plant and Bulb Transplanting

32 96 23.23 Planting

32 96 23.23 Planting		Crew	Daily Output	Labor-Hours	Unit	Material	2021 Bare Costs Labor	Equipment	Total	Total Incl O&P
0010	**PLANTING**									
0012	Moving shrubs on site, 12" ball	B-62	28	.857	Ea.		41	6.40	47.40	72
0100	24" ball	"	22	1.091	"		52.50	8.15	60.65	92

32 96 23.43 Moving Trees

32 96 23.43 Moving Trees		Crew	Daily Output	Labor-Hours	Unit	Material	2021 Bare Costs Labor	Equipment	Total	Total Incl O&P
0010	**MOVING TREES**, On site									
0300	Moving trees on site, 36" ball	B-6	3.75	6.400	Ea.		310	57.50	367.50	550
0400	60" ball	"	1	24	"		1,150	216	1,366	2,075

32 96 43 – Tree Transplanting

32 96 43.20 Tree Removal

32 96 43.20 Tree Removal		Crew	Daily Output	Labor-Hours	Unit	Material	2021 Bare Costs Labor	Equipment	Total	Total Incl O&P
0010	**TREE REMOVAL**									
0100	Dig & lace, shrubs, broadleaf evergreen, 18"-24" high	B-1	55	.436	Ea.		19.65		19.65	31.50
0200	2'-3'	"	35	.686			31		31	49
0300	3'-4'	B-6	30	.800			38.50	7.20	45.70	69
0400	4'-5'	"	20	1.200			57.50	10.80	68.30	103
1000	Deciduous, 12"-15"	B-1	110	.218			9.85		9.85	15.65
1100	18"-24"		65	.369			16.65		16.65	26.50
1200	2'-3'		55	.436			19.65		19.65	31.50
1300	3'-4'	B-6	50	.480			23	4.32	27.32	41.50
2000	Evergreen, 18"-24"	B-1	55	.436			19.65		19.65	31.50
2100	2' to 2.5'		50	.480			21.50		21.50	34.50
2200	2.5' to 3'		35	.686			31		31	49

32 96 43.20 Tree Removal		Crew	Daily Output	Labor-Hours	Unit	Material	2021 Bare Costs Labor	Equipment	Total	Total Incl O&P
2300	3' to 3.5'	B-1	20	1.200	Ea.		54		54	86
3000	Trees, deciduous, small, 2'-3'	↓	55	.436	↓		19.65		19.65	31.50
3100	3'-4'	B-6	50	.480	Ea.		23	4.32	27.32	41.50
3200	4'-5'		35	.686			33	6.20	39.20	59
3300	5'-6'		30	.800			38.50	7.20	45.70	69
4000	Shade, 5'-6'		50	.480			23	4.32	27.32	41.50
4100	6'-8'		35	.686			33	6.20	39.20	59
4200	8'-10'		25	.960			46	8.65	54.65	82.50
4300	2" caliper		12	2			96	18	114	172
5000	Evergreen, 4'-5'		35	.686			33	6.20	39.20	59
5100	5'-6'		25	.960			46	8.65	54.65	82.50
5200	6'-7'		19	1.263			61	11.40	72.40	109
5300	7'-8'		15	1.600			77	14.40	91.40	138
5400	8'-10'	↓	11	2.182	↓		105	19.65	124.65	188

Estimating Tips
33 10 00 Water Utilities
33 30 00 Sanitary Sewerage Utilities
33 40 00 Storm Drainage Utilities

- Never assume that the water, sewer, and drainage lines will go in at the early stages of the project. Consider the site access needs before dividing the site in half with open trenches, loose pipe, and machinery obstructions. Always inspect the site to establish that the site drawings are complete. Check off all existing utilities on your drawings as you locate them. Be especially careful with underground utilities because appurtenances are sometimes buried during regrading or repaving operations. If you find any discrepancies, mark up the site plan for further research. Differing site conditions can be very costly if discovered later in the project.

- See also Section 33 01 00 for restoration of pipe where removal/replacement may be undesirable. Use of new types of piping materials can reduce the overall project cost. Owners/design engineers should consider the installing contractor as a valuable source of current information on utility products and local conditions that could lead to significant cost savings.

Reference Numbers

Reference numbers are shown at the beginning of some major classifications. These numbers refer to related items in the Reference Section. The reference information may be an estimating procedure, an alternate pricing method, or technical information.

Note: Not all subdivisions listed here necessarily appear. ■

Same Data. Simplified.

Enjoy the convenience and efficiency of accessing your costs anywhere:

- **Skip the multiplier** by setting your location
- **Quickly search,** edit, favorite and share costs
- **Stay on top of price changes** with automatic updates

Discover more at rsmeans.com/online

33 01 30.11 Television Inspection of Sewers

		Daily Output	Labor-Hours	Unit	Material	2021 Bare Costs Labor	Equipment	Total	Total Incl O&P
		Crew							
0010	**TELEVISION INSPECTION OF SEWERS**								
0100	Pipe internal cleaning & inspection, cleaning, pressure pipe systems								
0120	Pig method, lengths 1000' to 10,000'								
0140	4" diameter thru 24" diameter, minimum			L.F.				3.60	4.14
0160	Maximum			"				18	21
6000	Sewage/sanitary systems								
6100	Power rodder with header & cutters								
6110	Mobilization charge, minimum			Total				695	800
6120	Mobilization charge, maximum			"				9,125	10,600
6140	Cleaning 4"-12" diameter			L.F.				6	6.60
6190	14"-24" diameter							8	8.80
6240	30" diameter							9.60	10.50
6250	36" diameter							12	13.20
6260	48" diameter							16	17.60
6270	60" diameter							24	26.50
6280	72" diameter							48	53
9000	Inspection, television camera with video								
9060	up to 500 linear feet			Total				715	820

33 01 30.72 Cured-In-Place Pipe Lining

		Crew	Daily Output	Labor-Hours	Unit	Material	2021 Bare Costs Labor	Equipment	Total	Total Incl O&P
0010	**CURED-IN-PLACE PIPE LINING**									
0011	Not incl. bypass or cleaning									
0020	Less than 10,000 L.F., urban, 6" to 10"	C-17E	130	.615	L.F.	9.45	35.50	.28	45.23	66.50
0050	10" to 12"		125	.640		11.65	37	.29	48.94	71.50
0070	12" to 16"		115	.696		11.95	40	.32	52.27	77
0100	16" to 20"		95	.842		14	48.50	.39	62.89	92.50
0200	24" to 36"		90	.889		15.40	51	.41	66.81	98.50
0300	48" to 72"		80	1		24	57.50	.46	81.96	118
0500	Rural, 6" to 10"	C-17E	180	.444	L.F.	9.45	25.50	.20	35.15	51
0550	10" to 12"		175	.457		11.65	26.50	.21	38.36	54.50
0570	12" to 16"		160	.500		12.45	29	.23	41.68	59.50
0600	16" to 20"		135	.593		14.45	34	.27	48.72	70
0700	24" to 36"		125	.640		15.75	37	.29	53.04	76
0800	48" to 72"		100	.800		24	46	.37	70.37	100
1000	Greater than 10,000 L.F., urban, 6" to 10"		160	.500		9.45	29	.23	38.68	56
1050	10" to 12"		155	.516		11.65	29.50	.24	41.39	60
1070	12" to 16"		140	.571		11.95	33	.26	45.21	65.50
1100	16" to 20"		120	.667		14.45	38.50	.31	53.26	77
1200	24" to 36"		115	.696		15.75	40	.32	56.07	81
1300	48" to 72"		95	.842		24	48.50	.39	72.89	103
1500	Rural, 6" to 10"		215	.372		9.45	21.50	.17	31.12	44.50
1550	10" to 12"		210	.381		11.65	22	.18	33.83	47.50
1570	12" to 16"		185	.432		11.95	25	.20	37.15	53
1600	16" to 20"		150	.533		14.45	30.50	.25	45.20	64.50
1700	24" to 36"		140	.571		15.75	33	.26	49.01	69.50
1800	48" to 72"		120	.667		25	38.50	.31	63.81	89
2000	Cured in place pipe, non-pressure, flexible felt resin									
2100	6" diameter				L.F.				25.50	28
2200	8" diameter								26.50	29
2300	10" diameter								29	32
2400	12" diameter								34.50	38
2500	15" diameter								59	65
2600	18" diameter								82	90

33 01 30.72 Cured-In-Place Pipe Lining

33 01 30.72 Cured-In-Place Pipe Lining		Crew	Daily Output	Labor-Hours	Unit	Material	2021 Bare Costs		Total	Total Incl O&P
							Labor	Equipment		
2700	21" diameter				L.F.				105	115
2800	24" diameter								177	195
2900	30" diameter								195	215
3000	36" diameter								205	225
3100	48" diameter				↓				218	240

33 01 30.75 Cured-In-place Pipe Lining

33 01 30.75 Cured-In-place Pipe Lining		Crew	Daily Output	Labor-Hours	Unit	Material	2021 Bare Costs		Total	Total Incl O&P
							Labor	Equipment		
0010	**CURED-IN-PLACE PIPE LINING**									
0050	Bypass, Sewer	C-17E	500	.160	L.F.		9.20	.07	9.27	14.70
0090	Mobilize - (Any Size) 1 Mob per Job.		1	80	Job		4,600	37	4,637	7,350
0091	Relocate - (°12") (8hrs, 3 per day)		3	26.667	Ea.		1,525	12.25	1,537.25	2,450
0092	Relocate - (12" to 24") (8hrs, 2 per day)		2	40			2,300	18.40	2,318.40	3,675
0093	Relocate - (24" to 42") (8hrs, 1.5 per day)		1.50	53.333			3,075	24.50	3,099.50	4,875
0094	Relocate - (42") (8hrs, 1.5 per day)		1	80	↓		4,600	37	4,637	7,350
0095	Additional Travel, $2.00 per Mile		3362.50	.024	Mile		1.37	.01	1.38	2.18
0096	Demobilize - (Any Size) 1 Demob per Job.		1	80	Job		4,600	37	4,637	7,350
0097	Refrigeration Truck (12", Cost per Each Location)	↓	3	26.667	Ea.		1,525	12.25	1,537.25	2,450
0098	Refrigeration Truck (12"-24", Each Pipe / Location)	C-17E	3	26.667	Ea.		1,525	12.25	1,537.25	2,450
0099	Refrigeration Truck (24"-42", Each Pipe / Location)		1.50	53.333			3,075	24.50	3,099.50	4,875
0100	Refrigeration Truck (42", Each Pipe / Location)		1	80	↓		4,600	37	4,637	7,350
1210	12", 6mm, Epoxy, 250k, 2ft		6	13.333	L.F.	48	765	6.15	819.15	1,275
1211	12", 6mm, Epoxy, 250k, 4ft		12	6.667		48	385	3.07	436.07	665
1212	12", 6mm, Epoxy, 250k, 10ft		30	2.667		48	153	1.23	202.23	297
1213	12", 6mm, Epoxy, 250k, 20ft		60	1.333		48	76.50	.61	125.11	176
1214	12", 6mm, Epoxy, 250k, 50ft		150	.533		48	30.50	.25	78.75	102
1215	12", 6mm, Epoxy, 250k, 100ft		300	.267		48	15.35	.12	63.47	77.50
1216	12", 6mm, Epoxy, 250k, 300ft		900	.089		48	5.10	.04	53.14	61
1217	12", 6mm, Epoxy, 250k, 328ft	↓	1000	.080		48	4.60	.04	52.64	60.50
1218	12", 6mm, Epoxy, 250k, Transitions, add					20%				
1219	12", 6mm, Epoxy, 250k, 45 and 90 deg, add					50%				
1230	12", 5mm, Epoxy, 400k, 2ft	C-17E	6	13.333		40	765	6.15	811.15	1,275
1231	12", 5mm, Epoxy, 400k, 4ft		12	6.667		40	385	3.07	428.07	655
1232	12", 5mm, Epoxy, 400k, 10ft		30	2.667		40	153	1.23	194.23	288
1233	12", 5mm, Epoxy, 400k, 20ft		60	1.333		40	76.50	.61	117.11	167
1234	12", 5mm, Epoxy, 400k, 50ft		150	.533		40	30.50	.25	70.75	93
1235	12", 5mm, Epoxy, 400k, 100ft		300	.267		40	15.35	.12	55.47	68.50
1236	12", 5mm, Epoxy, 400k, 300ft		900	.089		40	5.10	.04	45.14	52
1237	12", 5mm, Epoxy, 400k, 328ft	↓	1000	.080		40	4.60	.04	44.64	51.50
1238	12", 5mm, Epoxy, 400k, Transitions, add					20%				
1239	12", 5mm, Epoxy, 400k, 45 and 90 deg, add					50%				
2410	24", 12mm, Epoxy, 250k, 2ft	C-17E	4	20		192	1,150	9.20	1,351.20	2,050
2411	24", 12mm, Epoxy, 250k, 4ft		8	10		192	575	4.60	771.60	1,125
2412	24", 12mm, Epoxy, 250k, 10ft		20	4		192	230	1.84	423.84	580
2413	24", 12mm, Epoxy, 250k, 20ft		40	2		192	115	.92	307.92	395
2414	24", 12mm, Epoxy, 250k, 50ft		100	.800		192	46	.37	238.37	285
2415	24", 12mm, Epoxy, 250k, 100ft		200	.400		192	23	.18	215.18	249
2416	24", 12mm, Epoxy, 250k, 300ft		600	.133		192	7.65	.06	199.71	224
2417	24", 12mm, Epoxy, 250k, 328ft	↓	656	.122		192	7	.06	199.06	223
2418	24", 12mm, Epoxy, 250k, Transitions, add					20%				
2419	24", 12mm, Epoxy, 250k, 45 and 90 deg, add					50%				
2430	24", 10.5mm, Epoxy, 400k, 2ft	C-17E	4	20		168	1,150	9.20	1,327.20	2,025
2431	24", 10.5mm, Epoxy, 400k, 4ft		8	10		168	575	4.60	747.60	1,100
2432	24", 10.5mm, Epoxy, 400k, 10ft	↓	20	4	↓	168	230	1.84	399.84	550

33 01 Operation and Maintenance of Utilities

33 01 30 – Operation and Maintenance of Sewer Utilities

	33 01 30.75 Cured-In-place Pipe Lining	Crew	Daily Output	Labor-Hours	Unit	Material	2021 Bare Costs Labor	Equipment	Total	Total Incl O&P
2433	24", 10.5mm, Epoxy, 400k, 20ft	C-17E	40	2	L.F.	168	115	.92	283.92	370
2434	24", 10.5mm, Epoxy, 400k, 50ft		100	.800		168	46	.37	214.37	258
2435	24", 10.5mm, Epoxy, 400k, 100ft		200	.400		168	23	.18	191.18	222
2436	24", 10.5mm, Epoxy, 400k, 300ft		600	.133		168	7.65	.06	175.71	197
2437	24", 10.5mm, Epoxy, 400k, 328ft	↓	656	.122		168	7	.06	175.06	196
2438	24", 10.5mm, Epoxy, 400k, Transitions, add					20%				
2439	24", 10.5mm, Epoxy, 400k, 45 and 90 deg, add				↓	50%				

33 05 Common Work Results for Utilities

33 05 07 – Trenchless Installation of Utility Piping

33 05 07.23 Utility Boring and Jacking

		Crew	Daily Output	Labor-Hours	Unit	Material	2021 Bare Costs Labor	Equipment	Total	Total Incl O&P
0010	**UTILITY BORING AND JACKING**									
0011	Casing only, 100' minimum,									
0020	not incl. jacking pits or dewatering									
0100	Roadwork, 1/2" thick wall, 24" diameter casing	B-42	20	3.200	L.F.	142	159	53.50	354.50	470
0200	36" diameter		16	4		219	199	66.50	484.50	630
0300	48" diameter		15	4.267		305	212	71	588	750
0500	Railroad work, 24" diameter		15	4.267		142	212	71	425	570
0600	36" diameter		14	4.571		219	228	76	523	685
0700	48" diameter	↓	12	5.333		305	265	89	659	855
0900	For ledge, add								20%	20%
1000	Small diameter boring, 3", sandy soil	B-82	900	.018		23	.89	.20	24.09	27
1040	Rocky soil	"	500	.032	↓	23	1.60	.37	24.97	28.50

33 05 07.36 Microtunneling

		Crew	Daily Output	Labor-Hours	Unit	Material	2021 Bare Costs Labor	Equipment	Total	Total Incl O&P
0010	**MICROTUNNELING**									
0011	Not including excavation, backfill, shoring,									
0020	or dewatering, average 50'/day, slurry method									
0100	24" to 48" outside diameter, minimum				L.F.				965	965
0110	Adverse conditions, add				"				500	500
1000	Rent microtunneling machine, average monthly lease				Month				97,500	107,000
1010	Operating technician				Day				630	705
1100	Mobilization and demobilization, minimum				Job				41,200	45,900
1110	Maximum				"				445,500	490,500

33 05 61 – Concrete Manholes

33 05 61.10 Storm Drainage Manholes, Frames and Covers

		Crew	Daily Output	Labor-Hours	Unit	Material	2021 Bare Costs Labor	Equipment	Total	Total Incl O&P
0010	**STORM DRAINAGE MANHOLES, FRAMES & COVERS**									
0020	Excludes footing, excavation, backfill (See line items for frame & cover)									
0050	Brick, 4' inside diameter, 4' deep	D-1	1	16	Ea.	700	780		1,480	2,025
0100	6' deep		.70	22.857		1,000	1,125		2,125	2,900
0150	8' deep		.50	32	↓	1,300	1,550		2,850	3,925
0200	Add for 1' depth increase	↓	4	4	V.L.F.	176	195		371	510
1110	Precast, 4' ID, 4' deep	B-22	4.10	7.317	Ea.	905	380	70	1,355	1,675
1120	6' deep		3	10		1,075	515	95.50	1,685.50	2,100
1130	8' deep		2	15	↓	1,900	775	143	2,818	3,475
1140	For depths over 8', add	↓	16	1.875	V.L.F.	107	97	17.90	221.90	291

33 05 Common Work Results for Utilities

33 05 63 – Concrete Vaults and Chambers

33 05 63.13 Precast Concrete Utility Structures	Crew	Daily Output	Labor-Hours	Unit	Material	2021 Bare Costs Labor	Equipment	Total	Total Incl O&P
0010 **PRECAST CONCRETE UTILITY STRUCTURES**, 6" thick									
0050　5' x 10' x 6' high, ID	B-13	2	28	Ea.	1,825	1,350	293	3,468	4,475
0350　Hand hole, precast concrete, 1-1/2" thick									
0400　　1'-0" x 2'-0" x 1'-9", ID, light duty	B-1	4	6	Ea.	525	270		795	1,000
0450　　4'-6" x 3'-2" x 2'-0", OD, heavy duty	B-6	3	8	"	1,600	385	72	2,057	2,450

33 05 97 – Identification and Signage for Utilities

33 05 97.05 Utility Connection

	Crew	Daily Output	Labor-Hours	Unit	Material	2021 Bare Costs Labor	Equipment	Total	Total Incl O&P
0010 **UTILITY CONNECTION**									
0020　Water, sanitary, stormwater, gas, single connection	B-14	1	48	Ea.	4,925	2,225	216	7,366	9,225
0030　Telecommunication	"	3	16	"	420	745	72	1,237	1,725

33 05 97.10 Utility Accessories

		Crew	Daily Output	Labor-Hours	Unit	Material	2021 Bare Costs Labor	Equipment	Total	Total Incl O&P
0010 **UTILITY ACCESSORIES**	R312316-40									
0400　Underground tape, detectable, reinforced, alum. foil core, 2"		1 Clab	150	.053	C.L.F.	3	2.37		5.37	7.05
9000　Minimum labor/equipment charge		"	4	2	Job		89		89	141

33 11 Groundwater Sources

33 11 13 – Potable Water Supply Wells

33 11 13.10 Wells and Accessories

		Crew	Daily Output	Labor-Hours	Unit	Material	2021 Bare Costs Labor	Equipment	Total	Total Incl O&P
0010 **WELLS & ACCESSORIES**	R221113-50									
0011　Domestic										
0100　Drilled, 4" to 6" diameter		B-23	120	.333	L.F.		14.95	13.50	28.45	39
0200　　8" diameter		"	95.20	.420	"		18.80	17	35.80	48.50
1400　Remove & reset pump, minimum		B-21	4	7	Ea.		355	47.50	402.50	620
1420　　Maximum		"	2	14	"		715	95.50	810.50	1,225
1500　Pumps, installed in wells to 100' deep, 4" submersible										
1510　　1/2 HP		Q-1	3.22	4.969	Ea.	580	305		885	1,100
1520　　3/4 HP			2.66	6.015		980	365		1,345	1,650
1600　　1 HP			2.29	6.987		805	425		1,230	1,550
1700　　1-1/2 HP		Q-22	1.60	10		2,250	610	297	3,157	3,750
1800　　2 HP			1.33	12.030		1,800	735	360	2,895	3,525
1900　　3 HP			1.14	14.035		2,050	855	415	3,320	4,050
2000　　5 HP			1.14	14.035		2,575	855	415	3,845	4,600
2050　Remove and install motor only, 4 HP			1.14	14.035		1,525	855	415	2,795	3,450
5000　Wells to 180' deep, 4" submersible, 1 HP		B-21	1.10	25.455		2,425	1,300	174	3,899	4,900
5500　　2 HP			1.10	25.455		3,825	1,300	174	5,299	6,450
6000　　3 HP			1	28		4,450	1,425	191	6,066	7,350
7000　　5 HP			.90	31.111		6,425	1,600	212	8,237	9,800
9000　Minimum labor/equipment charge			1.80	15.556	Job		795	106	901	1,375

583

33 14 Water Utility Transmission and Distribution

33 14 13 – Public Water Utility Distribution Piping

33 14 13.15 Water Supply, Ductile Iron Pipe

		Crew	Daily Output	Labor-Hours	Unit	Material	2021 Bare Costs Labor	2021 Bare Costs Equipment	Total	Total Incl O&P
0010	**WATER SUPPLY, DUCTILE IRON PIPE** R331113-80									
0020	Not including excavation or backfill									
2000	Pipe, class 50 water piping, 18' lengths									
2020	Mechanical joint, 4" diameter	B-21	200	.140	L.F.	47	7.15	.95	55.10	64
3000	Push-on joint, 4" diameter		400	.070		24	3.57	.48	28.05	32.50
3020	6" diameter	↓	333.33	.084	↓	23.50	4.29	.57	28.36	33.50
8000	Piping, fittings, mechanical joint, AWWA C110									
8006	90° bend, 4" diameter	B-20	16	1.500	Ea.	173	74		247	310
8020	6" diameter		12.80	1.875		256	92.50		348.50	430
8200	Wye or tee, 4" diameter		10.67	2.249		475	111		586	695
8220	6" diameter		8.53	2.814		645	139		784	930
8398	45° bend, 4" diameter		16	1.500		201	74		275	340
8400	6" diameter	↓	12.80	1.875		279	92.50		371.50	450
8405	8" diameter	B-21	10.67	2.624		410	134	17.90	561.90	680
8450	Decreaser, 6" x 4" diameter	B-20	14.22	1.688		231	83		314	385
8460	8" x 6" diameter	B-21	11.64	2.406	↓	345	123	16.40	484.40	595
8550	Piping, butterfly valves, cast iron									
8560	4" diameter	B-20	6	4	Ea.	565	197		762	940
9600	Steel sleeve with tap, 4" diameter		3	8		550	395		945	1,225
9620	6" diameter	↓	2	12		590	590		1,180	1,600

33 14 17 – Site Water Utility Service Laterals

33 14 17.15 Tapping, Crosses and Sleeves

		Crew	Daily Output	Labor-Hours	Unit	Material	2021 Bare Costs Labor	2021 Bare Costs Equipment	Total	Total Incl O&P
0010	**TAPPING, CROSSES AND SLEEVES**									
4000	Drill and tap pressurized main (labor only)									
4100	6" main, 1" to 2" service	Q-1	3	5.333	Ea.		325		325	500
4150	8" main, 1" to 2" service	"	2.75	5.818	"		355		355	545
4500	Tap and insert gate valve									
4600	8" main, 4" branch	B-21	3.20	8.750	Ea.		445	59.50	504.50	775
4650	6" branch		2.70	10.370			530	70.50	600.50	920
4800	12" main, 6" branch	↓	2.35	11.915	↓		610	81.50	691.50	1,050

33 31 Sanitary Sewerage Piping

33 31 11 – Public Sanitary Sewerage Gravity Piping

33 31 11.25 Sewage Collection, Polyvinyl Chloride Pipe

		Crew	Daily Output	Labor-Hours	Unit	Material	2021 Bare Costs Labor	2021 Bare Costs Equipment	Total	Total Incl O&P
0010	**SEWAGE COLLECTION, POLYVINYL CHLORIDE PIPE**									
0020	Not including excavation or backfill									
2000	20' lengths, SDR 35, B&S, 4" diameter	B-20	375	.064	L.F.	2.21	3.16		5.37	7.45
2040	6" diameter		350	.069		4.26	3.38		7.64	10.05
2080	13' lengths, SDR 35, B&S, 8" diameter	↓	335	.072	↓	8.80	3.53		12.33	15.30
2120	10" diameter	B-21	330	.085	L.F.	12.05	4.33	.58	16.96	20.50
4000	Piping, DWV PVC, no exc./bkfill., 10' L, Sch 40, 4" diameter	B-20	375	.064		4.78	3.16		7.94	10.25
4010	6" diameter		350	.069		11.20	3.38		14.58	17.65
4020	8" diameter	↓	335	.072	↓	17.85	3.53		21.38	25.50

584

For customer support on your Commercial Renovation Costs with RSMeans Data, call 800.448.8182.

33 34 Onsite Wastewater Disposal

33 34 13 – Septic Tanks

33 34 13.13 Concrete Septic Tanks

		Crew	Daily Output	Labor-Hours	Unit	Material	2021 Bare Costs Labor	Equipment	Total	Total Incl O&P
0010	**CONCRETE SEPTIC TANKS**									
0011	Not including excavation or piping									
0015	Septic tanks, precast, 1,000 gallon	B-21	8	3.500	Ea.	1,000	179	24	1,203	1,400
0020	1,250 gallon		8	3.500		1,500	179	24	1,703	1,950
0060	1,500 gallon		7	4		1,675	204	27.50	1,906.50	2,200
0100	2,000 gallon		5	5.600		2,550	286	38	2,874	3,325
0140	2,500 gallon		5	5.600		2,375	286	38	2,699	3,125
0180	4,000 gallon		4	7		6,825	355	47.50	7,227.50	8,150
0220	5,000 gallon, 4 piece	B-13	3	18.667		9,000	900	196	10,096	11,500
0300	15,000 gallon, 4 piece	B-13B	1.70	32.941		25,700	1,600	580	27,880	31,500
0400	25,000 gallon, 4 piece		1.10	50.909		48,400	2,450	900	51,750	58,000
0500	40,000 gallon, 4 piece		.80	70		53,500	3,375	1,250	58,125	65,500
0520	50,000 gallon, 5 piece	B-13C	.60	93.333		61,500	4,500	3,850	69,850	79,500
0640	75,000 gallon, cast in place	C-14C	.25	448		75,000	23,400	108	98,508	120,000
0660	100,000 gallon	"	.15	747		93,000	39,000	179	132,179	164,000
1150	Leaching field chambers, 13' x 3'-7" x 1'-4", standard	B-13	16	3.500		475	169	36.50	680.50	835
1200	Heavy duty, 8' x 4' x 1'-6"		14	4		325	193	42	560	705
1300	13' x 3'-9" x 1'-6"		12	4.667		1,400	225	49	1,674	1,925
1350	20' x 4' x 1'-6"		5	11.200		1,250	540	117	1,907	2,350
1420	Leaching pit, precast concrete, 6' diameter, 3' deep	B-21	4.70	5.957		1,100	305	40.50	1,445.50	1,750
1600	Leaching pit, 6'-6" diameter, 6' deep		5	5.600		880	286	38	1,204	1,450
1620	8' deep		4	7		1,150	355	47.50	1,552.50	1,900
1700	8' diameter, H-20 load, 6' deep		4	7		1,400	355	47.50	1,802.50	2,150
1720	8' deep		3	9.333		2,575	475	63.50	3,113.50	3,650
2000	Velocity reducing pit, precast conc., 6' diameter, 3' deep		4.70	5.957		1,875	305	40.50	2,220.50	2,575

33 34 13.33 Polyethylene Septic Tanks

		Crew	Daily Output	Labor-Hours	Unit	Material	2021 Bare Costs Labor	Equipment	Total	Total Incl O&P
0010	**POLYETHYLENE SEPTIC TANKS**									
0015	High density polyethylene, 1,000 gallon	B-21	8	3.500	Ea.	1,600	179	24	1,803	2,050
0020	1,250 gallon		8	3.500		1,325	179	24	1,528	1,750
0025	1,500 gallon		7	4		1,400	204	27.50	1,631.50	1,875

33 34 16 – Septic Tank Effluent Filters

33 34 16.13 Septic Tank Gravity Effluent Filters

		Crew	Daily Output	Labor-Hours	Unit	Material	2021 Bare Costs Labor	Equipment	Total	Total Incl O&P
0010	**SEPTIC TANK GRAVITY EFFLUENT FILTERS**									
3000	Effluent filter, 4" diameter	1 Skwk	8	1	Ea.	36	57		93	130
3020	6" diameter	"	7	1.143	"	43.50	65.50		109	151
3030	8" diameter	1 Skwk	7	1.143	Ea.	254	65.50		319.50	380
3040	8" diameter, very fine		7	1.143		505	65.50		570.50	660
3050	10" diameter, very fine		6	1.333		240	76		316	385
3060	10" diameter		6	1.333		274	76		350	420
3080	12" diameter		6	1.333		660	76		736	845
3090	15" diameter		5	1.600		1,100	91.50		1,191.50	1,350

33 34 51 – Drainage Field Systems

33 34 51.10 Drainage Field Excavation and Fill

		Crew	Daily Output	Labor-Hours	Unit	Material	2021 Bare Costs Labor	Equipment	Total	Total Incl O&P
0010	**DRAINAGE FIELD EXCAVATION AND FILL**									
2200	Septic tank & drainage field excavation with 3/4 C.Y. backhoe	B-12F	145	.110	C.Y.		5.85	4.84	10.69	14.50
2400	4' trench for disposal field, 3/4 C.Y. backhoe	"	335	.048	L.F.		2.53	2.09	4.62	6.30
2600	Gravel fill, run of bank	B-6	150	.160	C.Y.	26.50	7.70	1.44	35.64	43
2800	Crushed stone, 3/4"	"	150	.160	"	41	7.70	1.44	50.14	58.50

585

For customer support on your Commercial Renovation Costs with RSMeans Data, call 800.448.8182.

33 34 Onsite Wastewater Disposal

33 34 51 – Drainage Field Systems

33 34 51.13 Utility Septic Tank Tile Drainage Field

		Crew	Daily Output	Labor-Hours	Unit	Material	2021 Bare Costs Labor	Equipment	Total	Total Incl O&P
0010	**UTILITY SEPTIC TANK TILE DRAINAGE FIELD**									
0015	Distribution box, concrete, 5 outlets	2 Clab	20	.800	Ea.	75	35.50		110.50	139
0020	7 outlets		16	1		93.50	44.50		138	174
0025	9 outlets		8	2		530	89		619	725
0115	Distribution boxes, HDPE, 5 outlets		20	.800		75	35.50		110.50	139
0117	6 outlets		15	1.067		75	47.50		122.50	158
0118	7 outlets		15	1.067		74.50	47.50		122	158
0120	8 outlets		10	1.600		79	71		150	200
0240	Distribution boxes, outlet flow leveler	1 Clab	50	.160		2.99	7.10		10.09	14.60
0300	Precast concrete, galley, 4' x 4' x 4'	B-21	16	1.750		246	89.50	11.95	347.45	425
0350	HDPE infiltration chamber 12" H x 15" W	2 Clab	300	.053	L.F.	7.05	2.37		9.42	11.50
0351	12" H x 15" W end cap	1 Clab	32	.250	Ea.	19.05	11.10		30.15	38.50
0355	chamber 12" H x 22" W	2 Clab	300	.053	L.F.	6.70	2.37		9.07	11.10
0356	12" H x 22" W end cap	1 Clab	32	.250	Ea.	23.50	11.10		34.60	43
0360	chamber 13" H x 34" W	2 Clab	300	.053	L.F.	14.45	2.37		16.82	19.60
0361	13" H x 34" W end cap	1 Clab	32	.250	Ea.	51	11.10		62.10	73.50
0365	chamber 16" H x 34" W	2 Clab	300	.053	L.F.	19.40	2.37		21.77	25.50
0366	16" H x 34" W end cap	1 Clab	32	.250	Ea.	16.90	11.10		28	36.50
0370	chamber 8" H x 16" W	2 Clab	300	.053	L.F.	10	2.37		12.37	14.75
0371	8" H x 16" W end cap	1 Clab	32	.250	Ea.	11.20	11.10		22.30	30

33 41 16.10 Piping, Subdrainage, Vitrified Clay

		Crew	Daily Output	Labor-Hours	Unit	Material	2021 Bare Costs Labor	Equipment	Total	Total Incl O&P
0010	**PIPING, SUBDRAINAGE, VITRIFIED CLAY**									
0020	Not including excavation and backfill									
1000	Foundation drain, 6" diameter	B-14	315	.152	L.F.	6.35	7.10	.69	14.14	19
1010	8" diameter		290	.166		9.50	7.70	.74	17.94	23.50
1020	12" diameter		275	.175		20.50	8.15	.79	29.44	36.50
3000	Perforated, 5' lengths, C700, 4" diameter		400	.120		5.30	5.60	.54	11.44	15.30
3020	6" diameter	B-14	315	.152	L.F.	6.85	7.10	.69	14.64	19.55
3040	8" diameter		290	.166		7.55	7.70	.74	15.99	21.50
3060	12" diameter		275	.175		20.50	8.15	.79	29.44	37

33 42 Stormwater Conveyance

33 42 11 – Stormwater Gravity Piping

33 42 11.40 Piping, Storm Drainage, Corrugated Metal

		Crew	Daily Output	Labor-Hours	Unit	Material	2021 Bare Costs Labor	Equipment	Total	Total Incl O&P
0010	**PIPING, STORM DRAINAGE, CORRUGATED METAL** R221113-50									
0020	Not including excavation or backfill									
2000	Corrugated metal pipe, galvanized									
2020	Bituminous coated with paved invert, 20' lengths									
2040	8" diameter, 16 ga.	B-14	330	.145	L.F.	7.25	6.80	.65	14.70	19.45
2080	12" diameter, 16 ga.	"	210	.229	"	12	10.65	1.03	23.68	31

33 42 11.60 Sewage/Drainage Collection, Concrete Pipe

		Crew	Daily Output	Labor-Hours	Unit	Material	2021 Bare Costs Labor	Equipment	Total	Total Incl O&P
0010	**SEWAGE/DRAINAGE COLLECTION, CONCRETE PIPE**									
0020	Not including excavation or backfill									
1000	Non-reinforced pipe, extra strength, B&S or T&G joints									
1010	6" diameter	B-14	265.04	.181	L.F.	8.05	8.45	.81	17.31	23
1020	8" diameter		224	.214		8.90	10	.96	19.86	26.50
1040	12" diameter		200	.240		10.25	11.20	1.08	22.53	30
2000	Reinforced culvert, class 3, no gaskets									
2010	12" diameter	B-14	150	.320	L.F.	16.50	14.90	1.44	32.84	43
2020	15" diameter	"	150	.320	"	20.50	14.90	1.44	36.84	47.50

33 42 Stormwater Conveyance

33 42 33 – Stormwater Curbside Drains and Inlets

33 42 33.13 Catch Basins

		Daily Output	Labor-Hours	Unit	Material	2021 Bare Costs Labor	Equipment	Total	Total Incl O&P
0010	**CATCH BASINS**								
0011	Not including footing & excavation								
1600	Frames & grates, C.I., 24" square, 500 lb.	B-6	7.80	3.077	Ea.	395	148	27.50	570.50	700
1700	26" D shape, 600 lb.	"	7	3.429	"	735	165	31	931	1,100
3320	Frames and covers, existing, raised for paving, 2", including									
3340	row of brick, concrete collar, up to 12" wide frame	B-6	18	1.333	Ea.	52.50	64	12	128.50	172

33 42 33.15 Remove and Replace Catch Basin Cover

		Crew	Daily Output	Labor-Hours	Unit	Material	2021 Bare Costs Labor	Equipment	Total	Total Incl O&P
0010	**REMOVE AND REPLACE CATCH BASIN COVER**									
0023	Remove catch basin cover	1 Clab	80	.100	Ea.		4.44		4.44	7.05
0033	Replace catch basin cover	"	80	.100	"		4.44		4.44	7.05

33 52 Hydrocarbon Transmission and Distribution

33 52 13 – Liquid Hydrocarbon Piping

33 52 13.16 Gasoline Piping

		Crew	Daily Output	Labor-Hours	Unit	Material	2021 Bare Costs Labor	Equipment	Total	Total Incl O&P
0010	**GASOLINE PIPING**									
0020	Primary containment pipe, fiberglass-reinforced									
0030	Plastic pipe 15' & 30' lengths									
0040	2" diameter	Q-6	425	.056	L.F.	7.25	3.60		10.85	13.55
0050	3" diameter		400	.060		9.85	3.83		13.68	16.70
0060	4" diameter		375	.064		12.75	4.08		16.83	20.50
0100	Fittings									
0110	Elbows, 90° & 45°, bell ends, 2"	Q-6	24	1	Ea.	43.50	64		107.50	147
0120	3" diameter		22	1.091		57	69.50		126.50	170
0130	4" diameter		20	1.200		71.50	76.50		148	197
0200	Tees, bell ends, 2"		21	1.143		57.50	73		130.50	175
0210	3" diameter		18	1.333		65.50	85		150.50	203
0220	4" diameter		15	1.600		80.50	102		182.50	246
0230	Flanges bell ends, 2"		24	1		31	64		95	133
0240	3" diameter		22	1.091		38.50	69.50		108	150
0250	4" diameter		20	1.200		43	76.50		119.50	165
0260	Sleeve couplings, 2"		21	1.143		11.95	73		84.95	125
0270	3" diameter		18	1.333		18.80	85		103.80	152
0280	4" diameter		15	1.600		21	102		123	180
0290	Threaded adapters, 2"		21	1.143		19.55	73		92.55	134
0300	3" diameter		18	1.333		25.50	85		110.50	159
0310	4" diameter		15	1.600		36.50	102		138.50	197
0320	Reducers, 2"		27	.889		26.50	56.50		83	117
0330	3" diameter		22	1.091		26.50	69.50		96	136
0340	4" diameter		20	1.200		35.50	76.50		112	157
1010	Gas station product line for secondary containment (double wall)									
1100	Fiberglass reinforced plastic pipe 25' lengths									
1120	Pipe, plain end, 3" diameter	Q-6	375	.064	L.F.	36	4.08		40.08	46
1130	4" diameter		350	.069		29.50	4.37		33.87	39
1140	5" diameter		325	.074		33	4.71		37.71	43.50
1150	6" diameter		300	.080		36.50	5.10		41.60	48
1200	Fittings									
1230	Elbows, 90° & 45°, 3" diameter	Q-6	18	1.333	Ea.	152	85		237	298
1240	4" diameter		16	1.500		183	95.50		278.50	350
1250	5" diameter		14	1.714		175	109		284	360
1260	6" diameter		12	2		218	128		346	435

33 52 Hydrocarbon Transmission and Distribution

33 52 13 – Liquid Hydrocarbon Piping

33 52 13.16 Gasoline Piping

		Crew	Daily Output	Labor-Hours	Unit	Material	2021 Bare Costs Labor	2021 Bare Costs Equipment	Total	Total Incl O&P
1270	Tees, 3" diameter	Q-6	15	1.600	Ea.	189	102		291	365
1280	4" diameter		12	2		180	128		308	395
1290	5" diameter		9	2.667		235	170		405	520
1300	6" diameter		6	4		330	255		585	760
1310	Couplings, 3" diameter		18	1.333		53.50	85		138.50	190
1320	4" diameter		16	1.500		122	95.50		217.50	282
1330	5" diameter		14	1.714		217	109		326	410
1340	6" diameter		12	2		340	128		468	570
1350	Cross-over nipples, 3" diameter		18	1.333		10.20	85		95.20	142
1360	4" diameter		16	1.500		12.40	95.50		107.90	162
1370	5" diameter		14	1.714		17.20	109		126.20	188
1380	6" diameter	Q-6	12	2	Ea.	18	128		146	217
1400	Telescoping, reducers, concentric 4" x 3"		18	1.333		43.50	85		128.50	179
1410	5" x 4"		17	1.412		93	90		183	241
1420	6" x 5"		16	1.500		261	95.50		356.50	435

33 52 16 – Gas Hydrocarbon Piping

33 52 16.20 Piping, Gas Service and Distribution, P.E.

		Crew	Daily Output	Labor-Hours	Unit	Material	2021 Bare Costs Labor	2021 Bare Costs Equipment	Total	Total Incl O&P
0010	**PIPING, GAS SERVICE AND DISTRIBUTION, POLYETHYLENE**									
0020	Not including excavation or backfill									
1000	60 psi coils, compression coupling @ 100', 1/2" diameter, SDR 11	B-20A	608	.053	L.F.	.56	2.80		3.36	5
1010	1" diameter, SDR 11		544	.059		1.13	3.13		4.26	6.15
1040	1-1/4" diameter, SDR 11		544	.059		1.61	3.13		4.74	6.65
1100	2" diameter, SDR 11		488	.066		2.84	3.49		6.33	8.55
1160	3" diameter, SDR 11		408	.078		5.05	4.17		9.22	12.05
1500	60 psi 40' joints with coupling, 3" diameter, SDR 11	B-21A	408	.098		7.50	5.35	1.06	13.91	17.80
1540	4" diameter, SDR 11		352	.114		14.50	6.25	1.23	21.98	27
1600	6" diameter, SDR 11		328	.122		37	6.70	1.32	45.02	52.50
1640	8" diameter, SDR 11		272	.147		62	8.05	1.59	71.64	82.50
9000	Minimum labor/equipment charge	B-20	2	12	Job		590		590	940

33 71 Electrical Utility Transmission and Distribution

33 71 19 – Electrical Underground Ducts and Manholes

33 71 19.17 Electric and Telephone Underground

		Crew	Daily Output	Labor-Hours	Unit	Material	2021 Bare Costs Labor	2021 Bare Costs Equipment	Total	Total Incl O&P
0010	**ELECTRIC AND TELEPHONE UNDERGROUND**									
0011	Not including excavation									
0200	backfill and cast in place concrete									
4200	Underground duct, banks ready for concrete fill, min. of 7.5"									
4400	between conduits, center to center									
4600	2 @ 2" diameter	2 Elec	240	.067	L.F.	1.91	4.25		6.16	8.60
4800	4 @ 2" diameter		120	.133		3.83	8.50		12.33	17.25
5600	4 @ 4" diameter		80	.200		6.30	12.75		19.05	26.50
6200	Rigid galvanized steel, 2 @ 2" diameter		180	.089		13.95	5.65		19.60	24
6400	4 @ 2" diameter		90	.178		28	11.30		39.30	48
7400	4 @ 4" diameter		34	.471		86.50	30		116.50	141
9990	Minimum labor/equipment charge	1 Elec	3.50	2.286	Job		146		146	224

Estimating Tips
34 11 00 Rail Tracks
This subdivision includes items that may involve either repair of existing or construction of new railroad tracks. Additional preparation work, such as the roadbed earthwork, would be found in Division 31. Additional new construction siding and turnouts are found in Subdivision 34 72. Maintenance of railroads is found under 34 01 23 Operation and Maintenance of Railways.

34 40 00 Traffic Signals
This subdivision includes traffic signal systems. Other traffic control devices such as traffic signs are found in Subdivision 10 14 53 Traffic Signage.

34 70 00 Vehicle Barriers
This subdivision includes security vehicle barriers, guide and guard rails, crash barriers, and delineators. The actual maintenance and construction of concrete and asphalt pavement are found in Division 32.

Reference Numbers
Reference numbers are shown at the beginning of some major classifications. These numbers refer to related items in the Reference Section. The reference information may be an estimating procedure, an alternate pricing method, or technical information.

Note: Not all subdivisions listed here necessarily appear. ■

Same Data. Simplified.

Enjoy the convenience and efficiency of accessing your costs anywhere:

- **Skip the multiplier** by setting your location
- **Quickly search,** edit, favorite and share costs
- **Stay on top of price changes** with automatic updates

Discover more at rsmeans.com/online

34 71 13.26 Vehicle Guide Rails	Crew	Daily Output	Labor-Hours	Unit	Material	2021 Bare Costs Labor	Equipment	Total	Total Incl O&P
0010 **VEHICLE GUIDE RAILS**									
0012 Corrugated stl., galv. stl. posts, 6'-3" OC - "W-Beam Guiderail"	B-80	850	.038	L.F.	12.10	1.84	1.24	15.18	17.55
0200 End sections, galvanized, flared		50	.640	Ea.	98.50	31	21	150.50	181
0300 Wrap around end		50	.640		141	31	21	193	228
0325 W-Beam, stl., galv. Terminal Connector, Concrete Mounted		8	4		249	195	132	576	730
0410 Thrie Beam, stl., galv. stl. posts, 6'-3" OC	B-80	850	.038	L.F.	23.50	1.84	1.24	26.58	30
0440 Thrie Beam, stl., galv, Wrap Around End Section		50	.640	Ea.	90	31	21	142	171
0450 Thrie Beam, stl., galv. Terminal Connector, Concrete Mounted		8	4	"	274	195	132	601	755
0510 Rub Rail, stl, galv, steel posts (not included), 6'-3" OC		850	.038	L.F.	1.70	1.84	1.24	4.78	6.15
1650 Timber guide rail, 4" x 8" with 6" x 8" wood posts, treated		960	.033	"	13.90	1.63	1.10	16.63	19.10

Estimating Tips

Products such as conveyors, material handling cranes and hoists, and other items specified in this division require trained installers. The general contractor may not have any choice as to who will perform the installation or when it will be performed. Long lead times are often required for these products, making early decisions in purchasing and scheduling necessary. The installation of this type of equipment may require the embedment of mounting hardware during the construction of floors, structural walls, or interior walls/partitions. Electrical connections will require coordination with the electrical contractor.

Reference Numbers

Reference numbers are shown at the beginning of some major classifications. These numbers refer to related items in the Reference Section. The reference information may be an estimating procedure, an alternate pricing method, or technical information.

Same Data. Simplified.

Enjoy the convenience and efficiency of accessing your costs anywhere:

- **Skip the multiplier** by setting your location
- **Quickly search,** edit, favorite and share costs
- **Stay on top of price changes** with automatic updates

Discover more at rsmeans.com/online

41 22 Cranes and Hoists

41 22 13 – Cranes

41 22 13.10 Crane Rail	Crew	Daily Output	Labor-Hours	Unit	Material	2021 Bare Costs Labor	Equipment	Total	Total Incl O&P
0010 **CRANE RAIL**									
0020 Box beam bridge, no equipment included	E-4	3400	.009	Lb.	1.32	.57	.04	1.93	2.43
0200 Running track only, 104 lb. per yard		5600	.006	"	.66	.35	.03	1.04	1.32
0210 Running track only, 104 lb. per yard, 20' piece	↓	160	.200	L.F.	23	12.15	.93	36.08	46

41 22 13.13 Bridge Cranes

	Crew	Daily Output	Labor-Hours	Unit	Material	Labor	Equipment	Total	Total Incl O&P
0010 **BRIDGE CRANES**									
0100 1 girder, 20' span, 3 ton	M-3	1	34	Ea.	22,000	2,300	143	24,443	28,000
0125 5 ton		1	34		23,600	2,300	143	26,043	29,700
0150 7.5 ton		1	34		28,700	2,300	143	31,143	35,200
0175 10 ton		.80	42.500		37,200	2,875	179	40,254	45,500
0176 12.5 ton		.80	42.500		44,600	2,875	179	47,654	53,500
0200 15 ton		.80	42.500		62,500	2,875	179	65,554	73,000
0225 30' span, 3 ton		1	34		22,900	2,300	143	25,343	28,900
0250 5 ton		1	34		24,700	2,300	143	27,143	30,900
0275 7.5 ton		1	34		30,200	2,300	143	32,643	36,900
0300 10 ton		.80	42.500		38,500	2,875	179	41,554	47,000
0325 15 ton	↓	.80	42.500		65,000	2,875	179	68,054	76,000
0350 2 girder, 40' span, 3 ton	M-4	.50	72		38,300	4,825	380	43,505	50,000
0375 5 ton		.50	72		40,200	4,825	380	45,405	52,000
0400 7.5 ton		.50	72		44,200	4,825	380	49,405	56,500
0425 10 ton		.40	90		41,000	6,025	475	47,500	55,000
0450 15 ton		.40	90		69,000	6,025	475	75,500	86,000
0475 25 ton		.30	120		83,000	8,025	635	91,660	104,500
0500 50' span, 3 ton		.50	72		41,700	4,825	380	46,905	54,000
0525 5 ton		.50	72		44,600	4,825	380	49,805	57,000
0550 7.5 ton		.50	72		47,800	4,825	380	53,005	60,500
0575 10 ton		.40	90		55,500	6,025	475	62,000	71,000
0600 15 ton		.40	90		72,500	6,025	475	79,000	89,500
0625 25 ton	↓	.30	120	↓	87,000	8,025	635	95,660	109,000

41 22 13.19 Jib Cranes

	Crew	Daily Output	Labor-Hours	Unit	Material	Labor	Equipment	Total	Total Incl O&P
0010 **JIB CRANES**									
0020 Jib crane, wall cantilever, 500 lb. capacity, 8' span	2 Mill	1	16	Ea.	1,375	940		2,315	2,950
0040 12' span		1	16		1,500	940		2,440	3,100
0060 16' span		1	16		1,750	940		2,690	3,375
0080 20' span	↓	1	16	↓	2,100	940		3,040	3,750
0100 1,000 lb. capacity, 8' span	2 Mill	1	16	Ea.	1,475	940		2,415	3,050
0120 12' span		1	16		1,600	940		2,540	3,200
0130 16' span		1	16		2,225	940		3,165	3,875
0150 20' span	↓	1	16	↓	2,475	940		3,415	4,175

Estimating Tips

- When estimating costs for the installation of electrical power generation equipment, factors to review include access to the job site, access and setting up at the installation site, required connections, uncrating pads, anchors, leveling, final assembly of the components, and temporary protection from physical damage, such as environmental exposure.

- Be aware of the costs of equipment supports, concrete pads, and vibration isolators. Cross-reference them against other trades' specifications. Also, review site and structural drawings for items that must be included in the estimates.

- It is important to include items that are not documented in the plans and specifications but must be priced. These items include, but are not limited to, testing, dust protection, roof penetration, core drilling concrete floors and walls, patching, cleanup, and final adjustments. Add a contingency or allowance for utility company fees for power hookups, if needed.

- The project size and scope of electrical power generation equipment will have a significant impact on cost. The intent of RSMeans cost data is to provide a benchmark cost so that owners, engineers, and electrical contractors will have a comfortable number with which to start a project. Additionally, there are many websites available to use for research and to obtain a vendor's quote to finalize costs.

Reference Numbers

Reference numbers are shown at the beginning of some major classifications. These numbers refer to related items in the Reference Section. The reference information may be an estimating procedure, an alternate pricing method, or technical information.

Same Data. Simplified.

Enjoy the convenience and efficiency of accessing your costs anywhere:

- **Skip the multiplier** by setting your location
- **Quickly search,** edit, favorite and share costs
- **Stay on top of price changes** with automatic updates

Discover more at rsmeans.com/online

---

48 13 Hydroelectric Plant Electrical Power Generation Equipment

48 13 13 – Hydroelectric Power Plant Water Turbines

48 13 13.10 Water Turbines and Components	Crew	Daily Output	Labor-Hours	Unit	Material	2021 Bare Costs Labor	Equipment	Total	Total Incl O&P
0010 **WATER TURBINES & COMPONENTS**									
0500 Water turbine, 100-500 watt, 3P, 48VDC, 54"W x 24"H, 260LBS	2 Elec	2	8	Ea.	10,300	510		10,810	12,100
0600 Water turbine, 1500-2000 watt, 3P, 130-200VAC to DC, 95LBS	2 Elec	3	5.333	Ea.	3,900	340		4,240	4,825

Assemblies Section

Table of Contents

Table of Contents

RSMeans data: Assemblies— How They Work

Assemblies estimating provides a fast and reasonably accurate way to develop construction costs. An assembly is the grouping of individual work items—with appropriate quantities—to provide a cost for a major construction component in a convenient unit of measure.

An assemblies estimate is often used during early stages of design development to compare the cost impact of various design alternatives on total building cost.

Assemblies estimates are also used as an efficient tool to verify construction estimates.

Assemblies estimates do not require a completed design or detailed drawings. Instead, they are based on the general size of the structure and other known parameters of the project. The degree of accuracy of an assemblies estimate is generally within +/- 15%.

Most assemblies consist of three major elements: a graphic, the system components, and the cost data itself. The **Graphic** is a visual representation showing the typical appearance of the assembly

❶ Unique 12-character Identifier

Our assemblies are identified by a **unique 12-character identifier**. The assemblies are numbered using UNIFORMAT II, ASTM Standard E1557. The first 5 characters represent this system to Level 3. The last 7 characters represent further breakdown in order to arrange items in understandable groups of similar tasks. Line numbers are consistent across all of our publications, so a line number in any assemblies data set will always refer to the same item.

❷ Narrative Descriptions

Our assemblies descriptions appear in two formats: narrative and table. **Narrative descriptions** are shown in a hierarchical structure to make them readable. In order to read a complete description, read up through the indents to the top of the section. Include everything that is above and to the left that is not contradicted by information below.

A10 Foundations

A1010 Standard Foundations

This page illustrates and describes a spread footing system including concrete, forms, reinforcing and anchor bolts. Lines within System Components give the unit price and total price on a cost each basis for this system. Prices for spread footing systems are on Line Items A1010 230 1200 thru 2300. Both material quantities and labor costs have been adjusted for the system listed.

Factors: To adjust for job conditions other than normal working situations use Lines A1010 230 2900 thru 4000.

Example: You are to install the system in an existing occupied building. Access to the site and protection to the building are mandatory. Go to Lines A1010 230 3600 and 3800 and apply these percentages to the appropriate MAT. and INST. costs.

System Components	QUANTITY	UNIT	COST EACH MAT.	COST EACH INST.	COST EACH TOTAL
Interior column footing, 3' square 1' thick, 2000 psi concrete, Including forms, reinforcing, and anchor bolts.					
Concrete, 2000 psi	.330	C.Y.	39.60		39.60
Placing concrete	.330	C.Y.		21.32	21.32
Forms, footing, 4 uses	12.000	SFCA	14.04	76.80	90.84
Reinforcing	11.000	Lb.	7.48	7.70	15.18
Anchor bolts, 3/4" diameter	2.000	Ea.	3.40	10.50	13.90
TOTAL			64.52	116.32	180.84

A1010 230	Spread Footing	COST EACH MAT.	COST EACH INST.	COST EACH TOTAL
1200	3' square x 1' thick, 2000 psi concrete	64.50	116	180.50
1300	3000 psi concrete	70	116	186
1400	4000 psi concrete	72	116	188
1600	For alternate footing systems:			
1700	4' square x 1' thick, 2000 psi concrete	106	165	271
1800	3000 psi concrete	116	165	281
1900	4000 psi concrete	119	165	284
2100	5' square x 1'-3" thick, 2000 psi concrete	198	272	470
2200	3000 psi concrete	217	272	489
2300	4000 psi concrete	223	272	495
2600				
2700				
2900	Cut & patch to match existing construction, add, minimum	2%	3%	
3000	Maximum	5%	9%	
3100	Dust protection, add, minimum	1%	2%	
3200	Maximum	4%	11%	
3300	Equipment usage curtailment, add, minimum	1%	1%	
3400	Maximum	3%	10%	
3500	Material handling & storage limitation, add, minimum	1%	1%	
3600	Maximum	6%	7%	
3700	Protection of existing work, add, minimum	2%	2%	
3800	Maximum	5%	7%	
3900	Shift work requirements, add, minimum		5%	
4000	Maximum		30%	

For supplemental customizable square foot estimating forms, visit: **RSMeans.com/2021books** in question. It is frequently accompanied by additional explanatory technical information describing the class of items. The **System Components** is a listing of the individual tasks that make up the assembly, including the quantity and unit of measure for each item, along with the cost of material and installation. The **Assemblies**

data below lists prices for other similar systems with dimensional and/or size variations.

All of our assemblies costs represent the cost for the installing contractor. An allowance for profit has been added to all material, labor, and equipment rental costs. A markup for labor burdens, including workers' compensation, fixed

overhead, and business overhead, is included with installation costs.

The information in RSMeans cost data represents a "national average" cost. This data should be modified to the project location using the **City Cost Indexes** or **Location Factors** tables found in the Reference Section.

A10 Foundations

A1010 Standard Foundations

This page illustrates and describes a spread footing system including concrete, forms, reinforcing and anchor bolts. Lines within System Components give the unit price and total price on a cost each basis for this system. Prices for spread footing systems are on Line Items A1010 230 1200 thru 2300. Both material quantities and labor costs have been adjusted for the system listed.

Factors: To adjust for job conditions other than normal working situations use Lines A1010 230 2900 thru 4000.

Example: You are to install the system in an existing occupied building. Access to the site and protection to the building are mandatory. Go to Lines A1010 230 3600 and 3800 and apply these percentages to the appropriate MAT. and INST. costs.

❸ Unit of Measure

All RSMeans data: Assemblies include a typical **Unit of Measure** used for estimating that item. For instance, for continuous footings or foundation walls the unit is linear feet (L.F.). For spread footings the unit is each (Ea.). The estimator needs to take special care that the unit in the data matches the unit in the takeoff. Abbreviations and unit conversions can be found in the Reference Section.

❹ System Components

System components are listed separately to detail what is included in the development of the total system price.

System Components ❹	QUANTITY	UNIT	COST EACH MAT.	COST EACH INST.	COST EACH TOTAL
Interior column footing, 3' square 1' thick, 2000 psi concrete, Including forms, reinforcing, and anchor bolts.					
Concrete, 2000 psi	.330	C.Y.	39.60		39.60
Placing concrete	.330	C.Y.		21.32	21.32
Forms, footing, 4 uses	12.000	SFCA	14.04	76.80	90.84
Reinforcing	11.000	Lb.	7.48	7.70	15.18
Anchor bolts, 3/4" diameter	2.000	Ea.	3.40	10.50	13.90
TOTAL			64.52	116.32	180.84

A1010 230	Spread Footing	COST EACH MAT.	COST EACH INST.	COST EACH TOTAL
1200	3' square x 1' thick, 2000 psi concrete	64.50	116	180.50
1300	3000 psi concrete	70	116	186
1400	4000 psi concrete	72	116	188
1600	For alternate footing systems:			
1700	4' square x 1' thick, 2000 psi concrete	106	165	271
1800	3000 psi concrete	116	165	281
1900	4000 psi concrete	119	165	284
2100	5' square x 1'-3" thick, 2000 psi concrete	198	272	470
2200	3000 psi concrete	217	272	489
2300	4000 psi concrete	223	272	495
2600				
2700				
2900	Cut & patch to match existing construction, add, minimum	2%	3%	
3000	Maximum	5%	9%	
3100	Dust protection, add, minimum	1%	2%	
3200	Maximum	4%	11%	
3300	Equipment usage curtailment, add, minimum	1%	1%	
3400	Maximum	3%	10%	
3500	Material handling & storage limitation, add, minimum	1%	1%	
3600	Maximum	6%	7%	
3700	Protection of existing work, add, minimum	2%	2%	
3800	Maximum	5%	7%	
3900	Shift work requirements, add, minimum		5%	
4000	Maximum		30%	

RSMeans data: Assemblies— How They Work (Continued)

Sample Estimate

This sample demonstrates the elements of an estimate, including a tally of the RSMeans data lines. Published assemblies costs include all markups for labor burden and profit for the installing contractor. This estimate adds a summary of the markups applied by a general contractor on the installing contractor's work. These figures represent the total cost to the owner. The location factor with RSMeans data is applied at the bottom of the estimate to adjust the cost of the work to a specific location.

Project Name:	Interior Fit-out, ABC Office				
Location:	**Anywhere, USA**		Date: 1/1/2021		R+R
Assembly Number	**Description**	**Qty.**	**Unit**		**Subtotal**
❶ C1010 124 1200	Wood partition, 2 x 4 @ 16" OC w/5/8" FR gypsum board	560	S.F.		$3,332.00
C1020 114 1800	Metal door & frame, flush hollow core, 3'-0" x 7'-0"	2	Ea.		$2,640.00
C3010 230 0080	Painting, brushwork, primer & 2 coats	1,120	S.F.		$1,601.60
C3020 410 0140	Carpet, tufted, nylon, roll goods, 12' wide, 26 oz	240	S.F.		$698.40
C3030 210 6000	Acoustic ceilings, 24" x 48" tile, tee grid suspension	200	S.F.		$1,836.00
D5020 125 0560	Receptacles incl plate, box, conduit, wire, 20 A duplex	8	Ea.		$2,520.00
D5020 125 0720	Light switch incl plate, box, conduit, wire, 20 A single pole	2	Ea.		$616.00
D5020 210 0560	Fluorescent fixtures, recess mounted, 20 per 1000 SF	200	S.F.		$2,484.00
	Assembly Subtotal				**$15,728.00**
	Sales Tax @ ❷	5 %			$ 393.20
	General Requirements @ ❸	7 %			$ 1,100.96
	Subtotal A				**$17,222.16**
	GC Overhead @ ❹	5 %			$ 861.11
	Subtotal B				**$18,083.27**
	GC Profit @ ❺	5 %			$ 904.16
	Subtotal C				**$18,987.43**
	Adjusted by Location Factor ❻	114.2			$ 21,683.65
	Architects Fee @ ❼	8 %			$ 1,734.69
	Contingency @ ❽	15 %			$ 3,252.55
	Project Total Cost				**$ 26,670.89**

This estimate is based on an interactive spreadsheet. You are free to download it and adjust it to your methodology. A copy of this spreadsheet is available at **RSMeans.com/2021books**.

① Work Performed

The body of the estimate shows the RSMeans data selected, including line numbers, a brief description of each item, its takeoff quantity and unit, and the total installed cost, including the installing contractor's overhead and profit.

② Sales Tax

If the work is subject to state or local sales taxes, the amount must be added to the estimate. In a conceptual estimate it can be assumed that one half of the total represents material costs. Therefore, apply the sales tax rate to 50% of the assembly subtotal.

③ General Requirements

This item covers project-wide needs provided by the general contractor. These items vary by project but may include temporary facilities and utilities, security, testing, project cleanup, etc. In assemblies estimates a percentage is used—typically between 5% and 15% of project cost.

④ General Contractor Overhead

This entry represents the general contractor's markup on all work to cover project administration costs.

⑤ General Contractor Profit

This entry represents the GC's profit on all work performed. The value included here can vary widely by project and is influenced by the GC's perception of the project's financial risk and market conditions.

⑥ Location Factor

RSMeans published data are based on national average costs. If necessary, adjust the total cost of the project using a location factor from the "Location Factor" table or the "City Cost Indexes" table found in the Reference Section. Use location factors if the work is general, covering the work of multiple trades. If the work is by a single trade (e.g., masonry) use the more specific data found in the City Cost Indexes.

To adjust costs by location factors, multiply the base cost by the factor and divide by 100.

⑦ Architect's Fee

If appropriate, add the design cost to the project estimate. These fees vary based on project complexity and size. Typical design and engineering fees can be found in the Reference Section.

⑧ Contingency

A factor for contingency may be added to any estimate to represent the cost of unknowns that may occur between the time that the estimate is performed and the time the project is constructed. The amount of the allowance will depend on the stage of design at which the estimate is done, as well as the contractor's assessment of the risk involved.

A1010 Standard Foundations

This page illustrates and describes a strip footing system including concrete, forms, reinforcing, keyway and dowels. Lines within System Components give the unit price and total price per linear foot for this system. Prices for strip footing systems are on Line Items A1010 120 1400 thru 2500. Both material quantities and labor costs have been adjusted for the system listed.

Factors: To adjust for job conditions other than normal working situations use Lines A1010 120 2900 thru 4000.

Example: You are to install this footing, and due to a lack of accessibility, only hand tools can be used. Material handling is also a problem. Go to Lines A1010 120 3400 and 3600 and apply these percentages to the appropriate MAT. and INST. costs.

System Components	QUANTITY	UNIT	COST PER L.F. MAT.	COST PER L.F. INST.	COST PER L.F. TOTAL
Strip footing, 2'-0" wide x 1'-0" thick, 2000 psi concrete including forms Reinforcing, keyway, and dowels.					
Concrete, 2000 psi	.074	C.Y.	8.88		8.88
Placing concrete	.074	C.Y.		2.19	2.19
Forms, footing, 4 uses	2.000	S.F.	5.64	11	16.64
Reinforcing	3.170	Lb.	2.16	2.22	4.38
Keyway, 2" x 4", 4 uses	1.000	L.F.	.41	1.31	1.72
Dowels, #4 bars, 2' long, 24" O.C.	.500	Ea.	.50	1.54	2.04
TOTAL			17.59	18.26	35.85

A1010 120	Strip Footing	COST PER L.F. MAT.	COST PER L.F. INST.	COST PER L.F. TOTAL
1400	2'-0" wide x 1' thick, 2000 psi concrete	17.60	18.25	35.85
1500	3000 psi concrete	18.85	18.25	37.10
1600	4000 psi concrete	19.20	18.25	37.45
1800	For alternate footing systems:			
1900	2'-6" wide x 1' thick, 2000 psi concrete	20.50	19.40	39.90
2000	3000 psi concrete	22	19.40	41.40
2100	4000 psi concrete	22.50	19.40	41.90
2300	3'-0" wide x 1' thick, 2000 psi concrete	23	20.50	43.50
2400	3000 psi concrete	25	20.50	45.50
2500	4000 psi concrete	25.50	20.50	46
2700				
2800				
2900	Cut & patch to match existing construction, add, minimum	2%	3%	
3000	Maximum	5%	9%	
3100	Dust protection, add, minimum	1%	2%	
3200	Maximum	4%	11%	
3300	Equipment usage curtailment, add, minimum	1%	1%	
3400	Maximum	3%	10%	
3500	Material handling & storage limitation, add, minimum	1%	1%	
3600	Maximum	6%	7%	
3700	Protection of existing work, add, minimum	2%	2%	
3800	Maximum	5%	7%	
3900	Shift work requirements, add, minimum		5%	
4000	Maximum		30%	

A10 Foundations

A1010 Standard Foundations

This page illustrates and describes a spread footing system including concrete, forms, reinforcing and anchor bolts. Lines within System Components give the unit price and total price on a cost each basis for this system. Prices for spread footing systems are on Line Items A1010 230 1200 thru 2300. Both material quantities and labor costs have been adjusted for the system listed.

Factors: To adjust for job conditions other than normal working situations use Lines A1010 230 2900 thru 4000.

Example: You are to install the system in an existing occupied building. Access to the site and protection to the building are mandatory. Go to Lines A1010 230 3600 and 3800 and apply these percentages to the appropriate MAT. and INST. costs.

System Components	QUANTITY	UNIT	COST EACH MAT.	COST EACH INST.	COST EACH TOTAL
Interior column footing, 3' square 1' thick, 2000 psi concrete, Including forms, reinforcing, and anchor bolts.					
Concrete, 2000 psi	.330	C.Y.	39.60		39.60
Placing concrete	.330	C.Y.		21.32	21.32
Forms, footing, 4 uses	12.000	SFCA	14.04	76.80	90.84
Reinforcing	11.000	Lb.	7.48	7.70	15.18
Anchor bolts, 3/4" diameter	2.000	Ea.	3.40	10.50	13.90
TOTAL			64.52	116.32	180.84

A1010 230	Spread Footing	COST EACH MAT.	COST EACH INST.	COST EACH TOTAL
1200	3' square x 1' thick, 2000 psi concrete	64.50	116	180.50
1300	3000 psi concrete	70	116	186
1400	4000 psi concrete	72	116	188
1600	For alternate footing systems:			
1700	4' square x 1' thick, 2000 psi concrete	106	165	271
1800	3000 psi concrete	116	165	281
1900	4000 psi concrete	119	165	284
2100	5' square x 1'-3" thick, 2000 psi concrete	198	272	470
2200	3000 psi concrete	217	272	489
2300	4000 psi concrete	223	272	495
2600				
2700				
2900	Cut & patch to match existing construction, add, minimum	2%	3%	
3000	Maximum	5%	9%	
3100	Dust protection, add, minimum	1%	2%	
3200	Maximum	4%	11%	
3300	Equipment usage curtailment, add, minimum	1%	1%	
3400	Maximum	3%	10%	
3500	Material handling & storage limitation, add, minimum	1%	1%	
3600	Maximum	6%	7%	
3700	Protection of existing work, add, minimum	2%	2%	
3800	Maximum	5%	7%	
3900	Shift work requirements, add, minimum		5%	
4000	Maximum		30%	

A1010 Standard Foundations

General: Footing drains can be placed either inside or outside of foundation walls depending upon the source of water to be intercepted. If the source of subsurface water is principally from grade or a subsurface stream above the bottom of the footing, outside drains should be used. For high water tables, use inside drains or both inside and outside.

The effectiveness of underdrains depends on good waterproofing. This must be carefully installed and protected during construction.

Costs below include the labor and materials for the pipe and 6″ of crushed stone around pipe. Excavation and backfill are not included.

System Components	QUANTITY	UNIT	COST PER L.F.		
			MAT.	INST.	TOTAL
Foundation underdrain, outside, PVC 4″ diameter.					
PVC pipe 4″ diam. S.D.R. 35	1.000	L.F.	2.43	5	7.43
Crushed stone 3/4″ to 1/2″	.070	C.Y.	2	.96	2.96
TOTAL			4.43	5.96	10.39

A1010 310	Foundation Underdrain	COST PER L.F.		
		MAT.	INST.	TOTAL
1000	Foundation underdrain, outside only, PVC, 4″ diameter	4.43	5.95	10.38
1100	6″ diameter	7.25	6.60	13.85
1400	Perforated HDPE, 6″ diameter	4.57	2.50	7.07
1450	8″ diameter	8.95	3.13	12.08
1500	12″ diameter	11.30	7.75	19.05
1600	Corrugated metal, 16 ga. asphalt coated, 6″ diameter	9.55	6.20	15.75
1650	8″ diameter	14.65	6.60	21.25
1700	10″ diameter	18.05	8.65	26.70
3000	Outside and inside, PVC, 4″ diameter	8.85	11.90	20.75
3100	6″ diameter	14.50	13.15	27.65
3400	Perforated HDPE, 6″ diameter	9.15	4.99	14.14
3450	8″ diameter	17.85	6.25	24.10
3500	12″ diameter	22.50	15.50	38
3600	Corrugated metal, 16 ga., asphalt coated, 6″ diameter	19.15	12.35	31.50
3650	8″ diameter	29.50	13.20	42.70
3700	10″ diameter	36	17.35	53.35
4700	Cut & patch to match existing construction, add, minimum	2%	3%	
4800	Maximum	5%	9%	
4900	Dust protection, add, minimum	1%	2%	
5000	Maximum	4%	11%	
5100	Equipment usage curtailment, add, minimum	1%	1%	
5200	Maximum	3%	10%	
5300	Material handling & storage limitation, add, minimum	1%	1%	
5400	Maximum	6%	7%	
5500	Protection of existing work, add, minimum	2%	2%	
5600	Maximum	5%	7%	
5700	Shift work requirements, add, minimum		5%	
5800	Maximum		30%	
5900	Temporary shoring and bracing, add, minimum	2%	5%	
6000	Maximum	5%	12%	

A10 Foundations

A1030 Slab on Grade

This page illustrates and describes a slab on grade system including slab, bank run gravel, bulkhead forms, placing concrete, welded wire fabric, vapor barrier, steel trowel finish and curing paper. Lines within System Components give the unit price and total price per square foot for this system. Prices for slab on grade systems are on Line Items A1030 110 1400 thru 2600. Both material quantities and labor costs have been adjusted for the system listed.

Factors: To adjust for job conditions other than normal working situations use Lines A1030 110 2900 thru 4000.

Example: You are to install the system at a site where protection of the existing building is required. Go to Line A1030 110 3800 and apply these percentages to the appropriate MAT. and INST. costs.

System Components	QUANTITY	UNIT	COST PER S.F. MAT.	COST PER S.F. INST.	COST PER S.F. TOTAL
Ground slab, 4″ thick, 3000 psi concrete, 4″ granular base, vapor barrier					
Welded wire fabric, screed and steel trowel finish.					
Concrete, 4″ thick, 3000 psi concrete	.012	C.Y.	1.64		1.64
Bank run gravel, 4″ deep	.012	C.Y.	.39	.11	.50
Polyethylene vapor barrier, 10 mil.	.011	C.S.F.	.10	.21	.31
Bulkhead forms, expansion material	.100	L.F.	.05	.44	.49
Welded wire fabric, 6 x 6 - #10/10	.011	C.S.F.	.18	.46	.64
Place concrete	.012	C.Y.		.39	.39
Screed & steel trowel finish	1.000	S.F.		1.15	1.15
TOTAL			2.36	2.76	5.12

A1030 110	Interior Slab on Grade	COST PER S.F. MAT.	COST PER S.F. INST.	COST PER S.F. TOTAL
1400	4″ thick slab, 3000 PSI concrete, 4″ deep bank run gravel	2.36	2.76	5.12
1500	6″ deep bank run gravel	2.09	2.67	4.76
1600	12″ deep bank run gravel	2.44	2.70	5.14
1700				
1800				
1900				
2000	For alternate slab systems:			
2100	5″ thick slab, 3000 psi concrete, 6″ deep bank run gravel	3.08	2.86	5.94
2200	12″ deep bank run gravel	3.77	2.90	6.67
2300				
2400				
2500	6″ thick slab, 3000 psi concrete, 6″ deep bank run gravel	3.62	2.99	6.61
2600	12″ deep bank run gravel	4.32	3.03	7.35
2700				
2900	Cut & patch to match existing construction, add, minimum	2%	3%	
3000	Maximum	5%	9%	
3100	Dust protection, add, minimum	1%	2%	
3200	Maximum	4%	11%	
3300	Equipment usage curtailment, add, minimum	1%	1%	
3400	Maximum	3%	10%	
3500	Material handling & storage limitation, add, minimum	1%	1%	
3600	Maximum	6%	7%	
3700	Protection of existing work, add, minimum	2%	2%	
3800	Maximum	5%	7%	
3900	Shift work requirements, add, minimum		5%	
4000	Maximum		30%	

607

A2010 Basement Excavation

This page illustrates and describes continuous footing and trench excavation systems including a wheel mounted backhoe, operator, equipment rental, fuel, oil, mobilization, no hauling or backfill. Lines within System Components give the unit price and total price per linear foot for this system. Prices for alternate continuous footing and trench excavation systems are on Line Items A2010 130 1200 thru 2700. Both material quantities and labor costs have been adjusted for the system listed.

Factors: To adjust for job conditions other than normal working situations use Lines A2010 130 2900 thru 4000.

Example: You are to install the system and supply dust protection. Go to Line A2010 130 3000 and apply this percentage to the appropriate TOTAL costs.

System Components	QUANTITY	UNIT	COST PER L.F.		
			EQUIP.	LABOR	TOTAL
Continuous footing or trench excav. w/ 3/4 C.Y. wheel mntd backhoe Including operator, equipment rental, fuel, oil and mobilization. Trench Is 4' wide at bottom, 3' deep w/sides sloped 1 to 2 and 200' long. Costs Are based on a production rate of 240 C.Y./day. No hauling/backfill incl.					
3/4 C.Y. backhoe, wheel mounted	.500	Day	120	650	770
Mobilization	1.000	Ea.	1,100	2,125	3,225
TOTAL			1,220	2,775	3,995
COST PER L.F.		L.F.	6.10	13.88	19.98

A2010 130	Excavation, Footings or Trench	COST PER L.F.		
		EQUIP.	LABOR	TOTAL
1200	For alternate trench sizes, 4' bottom with sloped sides:			
1300	2' deep, 50' long	7.30	26.50	33.80
1400	100' long	3.66	13.25	16.91
1500	300' long	4.07	9.25	13.32
1600	4' deep, 50' long	7.30	26.50	33.80
1700	100' long	3.66	13.25	16.91
1800	300' long	1.64	6.70	8.34
1900	6' deep, 50' long	7.30	26.50	33.80
2000	100' long	4.02	15.20	19.22
2100	300' long	2.42	10.90	13.32
2200	8' deep, 50' long	7.30	26.50	33.80
2300	100' long	4.86	19.75	24.61
2400	300' long	3.22	15.25	18.47
2500	10' deep, 50' long	8.50	33	41.50
2600	100' long	5.90	25.50	31.40
2700	300' long	4.18	20.50	24.68
2800				
2900	Dust protection, add, minimum		2%	2%
3000	Maximum		11%	11%
3100	Equipment usage curtailment, add, minimum		1%	1%
3200	Maximum		10%	10%
3300	Material handling & storage limitation, add, minimum		1%	1%
3400	Maximum		7%	7%
3500	Protection of existing work, add, minimum		2%	2%
3600	Maximum		7%	7%
3700	Shift work requirements, add, minimum		5%	5%
3800	Maximum		30%	30%
3900	Temporary shoring and bracing, add, minimum		5%	5%
4000	Maximum		12%	12%

A2010 Basement Excavation

This page illustrates and describes foundation excavation systems including a backhoe-loader, operator, equipment rental, fuel, oil, mobilization, hauling material, no backfilling. Lines within System Components give the unit price and total price per cubic yard for this system. Prices for foundation excavation systems are on Line Items A2010 140 1600 thru 2200. Both material quantities and labor costs have been adjusted for the system listed.

Factors: To adjust for job conditions other than normal working situations use Lines A2010 140 2900 thru 4000.

Example: You are to install the system with the use of temporary shoring and bracing. Go to Line A2010 140 3900 and apply these percentages to the appropriate TOTAL costs.

System Components	QUANTITY	UNIT	COST PER C.Y.		
			EQUIP.	LABOR	TOTAL
Foundation excav w/ 3/4 C.Y. backhoe-loader, incl. operator, equip Rental, fuel, oil, and mobilization. Hauling of excavated material is Included. Prices based on one day production of 360 C.Y. in medium soil Without backfilling.					
Backhoe-loader, 3/4 C.Y.	1.000	Day	240	1,300	1,540
Hauling, 12 C.Y. trucks, 1 mile round trip	360.000	L.C.Y.	910.80	936	1,846.80
Mobilization	1.000	Ea.	246	675	921
TOTAL			1,396.80	2,911	4,307.80
COST PER C.Y.		C.Y.	3.87	8.06	11.97

A2010 140	Excavation, Foundation	COST PER C.Y.		
		EQUIP.	LABOR	TOTAL
1600	100 C.Y.	5.65	13	18.65
1700	200 C.Y.	4.43	9.60	14.03
1800	300 C.Y.	4.05	8.65	12.70
1850	360 C.Y.	3.89	8.06	11.94
1900	400 C.Y.	3.81	7.85	11.66
2000	500 C.Y.	3.69	7.60	11.29
2100	600 C.Y.	3.58	7.20	10.78
2200	700 C.Y.	3.53	7.10	10.63
2300				
2400				
2900	Dust protection, add, minimum		2%	2%
3000	Maximum		11%	11%
3100	Equipment usage curtailment, add, minimum		1%	1%
3200	Maximum		10%	10%
3300	Material handling & storage limitation, add, minimum		1%	1%
3400	Maximum		7%	7%
3500	Protection of existing work, add, minimum		2%	2%
3600	Maximum		7%	7%
3700	Shift work requirements, add, minimum		5%	5%
3800	Maximum		30%	30%
3900	Temporary shoring and bracing, add, minimum		5%	5%
4000	Maximum		12%	12%

A2020 Basement Walls

This page illustrates and describes a concrete wall system including concrete, placing concrete, forms, reinforcing, insulation, waterproofing and anchor bolts. Lines within System Components give the unit price and total price per linear foot for this system. Prices for concrete wall systems are on Line Items A2020 120 1400 thru 2600. Both material quantities and labor costs have been adjusted for the system listed.

Factors: To adjust for job conditions other than normal working situations use Lines A2020 120 2900 thru 4000.

Example: You are to install this wall system where delivery of material is difficult. Go to Line A2020 120 3600 and apply these percentages to the appropriate MAT. and INST. costs.

System Components	QUANTITY	UNIT	MAT.	INST.	TOTAL
Cast in place concrete foundation wall, 8″ thick, 3′ high, 2500 psi					
Concrete including forms, reinforcing, waterproofing, and anchor bolts.					
Concrete, 2500 psi, 8″ thick, 3′ high	.070	C.Y.	8.68		8.68
Forms, wall, 4 uses	6.000	S.F.	10.62	48.30	58.92
Reinforcing	6.000	Lb.	4.08	2.94	7.02
Placing concrete	.070	C.Y.		3.73	3.73
Waterproofing	3.000	S.F.	1.44	3.99	5.43
Rigid insulaton, 1″ polystyrene	3.000	S.F.	1.05	3.06	4.11
Anchor bolts, 1/2″ diameter, 4′ O.C.	.250	Ea.	.47	1.31	1.78
TOTAL			26.34	63.33	89.67

A2020 120	Concrete Wall	MAT.	INST.	TOTAL
1400	8″ thick, 2500 psi concrete, 3′ high	26.50	63.50	90
1500	4′ high	36	84.50	120.50
1600	6′ high	53.50	126	179.50
1700	8′ high	71	167	238
1800	3500 psi concrete, 4′ high	37.50	84.50	122
1900	6′ high	55.50	126	181.50
2000	8′ high	74	167	241
2100	12″ thick, 2500 psi concrete, 4′ high	44.50	89	133.50
2200	6′ high	65	132	197
2300	8′ high	86	175	261
2400	3500 psi concrete, 4′ high	47	89	136
2500	8′ high	90.50	175	265.50
2600	10′ high	113	218	331
2700				
2900	Cut & patch to match existing construction, add, minimum	2%	3%	
3000	Maximum	5%	9%	
3100	Dust protection, add, minimum	1%	2%	
3200	Maximum	4%	11%	
3300	Equipment usage curtailment, add, minimum	1%	1%	
3400	Maximum	3%	10%	
3500	Material handling & storage limitation, add, minimum	1%	1%	
3600	Maximum	6%	7%	
3700	Protection of existing work, add, minimum	2%	2%	
3800	Maximum	5%	7%	
3900	Shift work requirements, add, minimum		5%	
4000	Maximum		30%	

A20 Basement Construction

A2020 Basement Walls

This page illustrates and describes a concrete block wall system including concrete block, masonry reinforcing, parging, waterproofing insulation and anchor bolts. Lines within System Components give the unit price and total price per linear foot for this system. Prices for concrete block wall systems are on Line Items A2020 140 1200 thru 2600. Both material quantities and labor costs have been adjusted for the system listed.

Factors: To adjust for job conditions other than normal working situations use Lines A2020 140 2900 thru 4000.

Example: You are to install the system to match an existing foundation wall. Go to Line A2020 140 3000 and apply these percentages to the appropriate MAT. and INST. costs.

System Components	QUANTITY	UNIT	COST PER L.F.		
			MAT.	INST.	TOTAL
Concrete block, 8″ thick, masonry reinforcing, parged and Waterproofed, insulation and anchor bolts, wall 2'-8″ high.					
Concrete block, 8″ x 8″ x 16″	2.670	S.F.	9.77	20.03	29.80
Masonry reinforcing	2.000	L.F.	.58	.46	1.04
Parging	.193	S.Y.	.91	3.05	3.96
Waterproofing	2.670	S.F.	1.28	3.55	4.83
Insulation, 1″ rigid polystyrene	2.670	S.F.	.93	2.72	3.65
Anchor bolts, 1/2″ diameter, 4' O.C.	.250	Ea.	.47	1.31	1.78
TOTAL			13.94	31.12	45.06

A2020 140	Concrete Block Wall	COST PER L.F.		
		MAT.	INST.	TOTAL
1200	8″ thick block, 2'-8″ high	13.95	31	44.95
1300	4' high	20.50	46	66.50
1400	6' high	30.50	68.50	99
1500	8' high	41	91	132
1600	Grouted solid, 4' high	27	62	89
1700	6' high	40	92.50	132.50
1800	8' high	53.50	123	176.50
2100	12″ thick block, 4' high	29.50	66.50	96
2200	6' high	44	99	143
2300	8' high	59	131	190
2400	Grouted solid, 4' high	40	83.50	123.50
2500	6' high	59.50	124	183.50
2600	8' high	79.50	165	244.50
2700				
2900	Cut & patch to match existing construction, add, minimum	2%	3%	
3000	Maximum	5%	9%	
3100	Dust protection, add, minimum	1%	2%	
3200	Maximum	4%	11%	
3300	Equipment usage curtailment, add, minimum	1%	1%	
3400	Maximum	3%	10%	
3500	Material handling & storage limitation, add, minimum	1%	1%	
3600	Maximum	6%	7%	
3700	Protection of existing work, add, minimum	2%	2%	
3800	Maximum	5%	7%	
3900	Shift work requirements, add, minimum		5%	
4000	Maximum		30%	

B1010 Floor Construction

This page illustrates and describes a reinforced concrete slab system including concrete, placing concrete, formwork, reinforcing, steel trowel finish, V.A. floor tile and acoustical spray ceiling finish. Lines within System Components give the unit price per square foot for this system. Prices for reinforced concrete slab systems are on Line Items B1010 225 1400 thru 1800. Both material quantities and labor costs have been adjusted for the system listed.

Factors: To adjust for job conditions other than normal working situations use Lines B1010 225 2700 thru 4000.

Example: You are to install the system to match an existing floor system. Go to Line B1010 225 2800 and apply these percentages to the appropriate MAT. and INST. costs.

System Components	QUANTITY	UNIT	COST PER S.F.		
			MAT.	INST.	TOTAL
Flat slab system, reinforced concrete with vinyl tile floor, sprayed					
Acoustical ceiling finish, not including columns.					
Concrete, 4000 psi, 6" thick	.020	C.Y.	2.84		2.84
Placing concrete	.020	C.Y.		.77	.77
Formwork, 4 use	1.000	S.F.	2.06	7.30	9.36
Edge form	.120	S.F.	.11	1.42	1.53
Reinforcing steel	3.000	Lb.	2.04	2.10	4.14
Steel trowel finish	1.000	S.F.		1.47	1.47
Vinyl floor tile, 1/8" thick, premium colors/patterns	1.000	S.F.	7.35	1.29	8.64
Acoustical spray ceiling finish, solvent based	1.000	S.F.	.31	.20	.51
TOTAL			14.71	14.55	29.26

B1010 225	Floor - Ceiling, Concrete Slab	COST PER S.F.		
		MAT.	INST.	TOTAL
1400	Concrete, 4000 psi, 6" thick	14.70	14.55	29.25
1500	7" thick	15.30	15.15	30.45
1600	8" thick	16.15	15.90	32.05
1700	9" thick	16.85	16.55	33.40
1800	10" thick	17.70	17.30	35
1900				
2000				
2100				
2200				
2300				
2400				
2500				
2700	Cut & patch to match existing construction, add, minimum	2%	3%	
2800	Maximum	5%	9%	
2900	Dust protection, add, minimum	1%	2%	
3000	Maximum	4%	11%	
3100	Equipment usage curtailment, add, minimum	1%	1%	
3200	Maximum	3%	10%	
3300	Material handling & storage limitation, add, minimum	1%	1%	
3400	Maximum	6%	7%	
3500	Protection of existing work, add, minimum	2%	2%	
3600	Maximum	5%	7%	
3700	Shift work requirements, add, minimum		5%	
3800	Maximum		30%	
3900	Temporary shoring and bracing, add, minimum	2%	5%	
4000	Maximum	5%	12%	

B1010 Floor Construction

This page illustrates and describes a hollow core prestressed concrete panel system including hollow core slab, grout, carpet, carpet padding, and sprayed ceiling. Lines within System Components give the unit price and total price per square foot for this system. Prices for hollow core prestressed concrete panel systems are on Line Items B1010 232 1200 thru 1600. Both material and labor costs have been adjusted for the system listed.

Factors: To adjust for job conditions other than normal working situations use Lines B1010 232 2700 thru 3800.

Example: You are to install the system where dust control is a major concern. Go to Line B1010 232 2800 and apply these percentages to the appropriate MAT. and INST. costs.

System Components	QUANTITY	UNIT	COST PER S.F.		
			MAT.	INST.	TOTAL
Precast hollow core plank with carpeted floors, padding					
And sprayed textured ceiling.					
Hollow core plank, 4" thick with grout topping	1.000	S.F.	9.25	3.95	13.20
Nylon carpet, 26 oz. medium traffic	.110	S.Y.	5.01	.95	5.96
Carpet padding, 20 oz./sq. yd.	.110	S.Y.	.59	.47	1.06
Sprayed texture ceiling	1.000	S.F.	.04	.84	.88
TOTAL			14.89	6.21	21.10

B1010 232	Floor - Ceiling, Conc. Panel	COST PER S.F.		
		MAT.	INST.	TOTAL
1200	Hollow core concrete plank, with grout, 4" thick	14.90	6.20	21.10
1300	6" thick	15.05	5.65	20.70
1400	8" thick	19.10	5.20	24.30
1500	10" thick	16.55	4.89	21.44
1600	12" thick	18.50	4.63	23.13
1700				
1800				
1900				
2000				
2100				
2200				
2300				
2400				
2500				
2700	Dust protection, add, minimum	1%	2%	
2800	Maximum	4%	11%	
2900	Equipment usage curtailment, add, minimum	1%	1%	
3000	Maximum	3%	10%	
3100	Material handling & storage limitation, add, minimum	1%	1%	
3200	Maximum	6%	7%	
3300	Protection of existing work, add, minimum	2%	2%	
3400	Maximum	5%	7%	
3500	Shift work requirements, add, minimum		5%	
3600	Maximum		30%	
3700	Temporary shoring and bracing, add, minimum	2%	5%	
3800	Maximum	5%	12%	

B1010 Floor Construction

This page illustrates and describes a hollow core plank system including precast hollow core plank, concrete topping, V.A. tile, concealed suspension system and ceiling tile. Lines within System Components give the unit price and total price per square foot for this system. Prices for hollow core plank systems are on Line Items B1010 233 1400 thru 1800. Both material quantities and labor costs have been adjusted for the system listed.

Factors: To adjust for job conditions other than normal working situations use Lines B1010 233 2900 thru 4000.

Example: You are to install the system where equipment usage is a problem. Go to Line B1010 233 3200 and apply these percentages to the appropriate MAT. and INST. costs.

System Components	QUANTITY	UNIT	COST PER S.F. MAT.	INST.	TOTAL
Precast hollow core plank with 2″ topping, vinyl composition floor tile					
And suspended acoustical tile ceiling.					
Hollow core concrete plank, 4″ thick	1.000	S.F.	9.25	3.95	13.20
Finishing floors, granolithic topping, 1:1:1-1/2 mix, 2″ thick	1.000	S.F.	2.17	6.74	8.91
Vinyl composition tile, .08″ thick	1.000	S.F.	1.67	1.29	2.96
Concealed Z bar suspension system, 12″ module	1.000	S.F.	1.05	1.34	2.39
Ceiling tile, mineral fiber, 3/4″ thick	1.000	S.F.	4.15	2.32	6.47
TOTAL			18.29	15.64	33.93

B1010 233	Floor - Ceiling, Conc. Plank	COST PER S.F. MAT.	INST.	TOTAL
1400	Hollow core concrete plank, 4″ thick	18.30	15.65	33.95
1500	6″ thick	18.45	15.05	33.50
1600	8″ thick	22.50	14.65	37.15
1700	10″ thick	19.95	14.30	34.25
1800	12″ thick	22	14.05	36.05
1900				
2000				
2100				
2200				
2300				
2400				
2500				
2600				
2700				
2900	Dust protection, add, minimum	1%	2%	
3000	Maximum	4%	11%	
3100	Equipment usage curtailment, add, minimum	1%	1%	
3200	Maximum	3%	10%	
3300	Material handling & storage limitation, add, minimum	1%	1%	
3400	Maximum	6%	7%	
3500	Protection of existing work, add, minimum	2%	2%	
3600	Maximum	5%	7%	
3700	Shift work requirements, add, minimum		5%	
3800	Maximum		30%	
3900	Temporary shoring and bracing, add, minimum	2%	5%	
4000	Maximum	5%	12%	

B1010 Floor Construction

This page illustrates and describes a structural steel w/metal decking and concrete system including steel beams, steel decking, shear studs, concrete, placing concrete, edge form, steel trowel finish, curing, wire fabric, fireproofing, beams and decking, tile floor, and suspended ceiling. Lines within System Components give the unit price and total price per square foot for this system. Prices for alternate structural steel w/metal decking and concrete systems are on Line Items B1010 243 1900 thru 2400. Both material quantities and labor costs have been adjusted for the system listed.

Factors: To adjust for job conditions other than normal working situations use Lines B1010 243 2900 thru 4000.

Example: You are to install the system where material handling and storage are a problem. Go to Line B1010 243 3400 and apply these percentages to the appropriate MAT. and INST. costs.

System Components	QUANTITY	UNIT	COST PER S.F. MAT.	COST PER S.F. INST.	COST PER S.F. TOTAL
Composite structural beams, 20 ga. 3″ deep steel decking, shear studs, 3000 psi, concrete, placing concrete, edge forms, steel trowel finish, Welded wire fabric fireproofing, tile floor and suspended ceiling.					
Steel deck, 20 gage, 3″ deep, galvanized	1.000	S.F.	4.15	1.10	5.25
Structural steel framing	.004	Ton	11.60	3.04	14.64
Shear studs, 3/4″	.250	Ea.	.15	.65	.80
Concrete, 3000 psi, 5″ thick	.015	C.Y.	2.06		2.06
Place concrete	.015	C.Y.		.58	.58
Edge form	.140	L.F.	.05	.74	.79
Steel trowel finish	1.000	S.F.		1.47	1.47
Welded wire fabric 6 x 6 - #10/10	.011	C.S.F.	.18	.46	.64
Fireproofing, sprayed	1.880	S.F.	1.22	2.48	3.70
Vinyl composition floor tile	1.000	S.F.	.97	1.29	2.26
Suspended acoustical ceiling	1.000	S.F.	6.85	1.83	8.68
TOTAL			27.23	13.64	40.87

B1010 243	Floor - Ceiling, Struc. Steel	COST PER S.F. MAT.	COST PER S.F. INST.	COST PER S.F. TOTAL
1900	Composite deck, galvanized, 3″ deep, 22 ga.	26	13.55	39.55
1950	20 ga.	27	13.65	40.65
2000	18 ga.	26.50	13.70	40.20
2100	16 ga.	29	13.75	42.75
2200	Composite deck, galvanized, 2″ deep, 22 ga.	25.50	13.40	38.90
2300	20 ga.	27	13.45	40.45
2400	16 ga.	28.50	13.55	42.05
2500				
2900	Dust protection, add, minimum	1%	2%	
3000	Maximum	4%	11%	
3100	Equipment usage curtailment, add, minimum	1%	1%	
3200	Maximum	3%	10%	
3300	Material handling & storage limitation, add, minimum	1%	1%	
3400	Maximum	6%	7%	
3500	Protection of existing work, add, minimum	2%	2%	
3600	Maximum	5%	7%	
3700	Shift work requirements, add, minimum		5%	
3800	Maximum		30%	
3900	Temporary shoring and bracing, add, minimum	2%	5%	
4000	Maximum	5%	12%	

B10 Superstructure

B1010 Floor Construction

This page illustrates and describes an open web joist and steel slab-form system including open web steel joist, slab form, concrete, placing concrete, wire fabric, steel trowel finish, tile floor and plasterboard. Lines within System Components give the unit price and total price per square foot for this system. Prices for open web joists and steel slab-form systems are on Line Items B1010 245 1800 thru 2200. Both material quantities and labor costs have been adjusted for the systems listed.

Factors: To adjust for job conditions other than normal working situations use Lines B1010 245 2900 thru 4000.

Example: You are to install the system in a congested commercial area and most work will be done at night. Go to Line B1010 245 3800 and apply this percentage to the appropriate INST. cost.

System Components	QUANTITY	UNIT	COST PER S.F. MAT.	COST PER S.F. INST.	COST PER S.F. TOTAL
Open web steel joists, 24″ O.C., slab form, 3000 psi concrete, welded					
Wire fabric, steel trowel finish, floor tile, 5/8″ drywall ceiling.					
Open web steel joists, 12″ deep, 5.2#/L.F., 24″ O.C.	2.630	Lb.	3.10	1.32	4.42
Slab form, 28 gage, 9/16″ deep, galvanized	1.000	S.F.	1.86	.83	2.69
Concrete, 3000 psi, 2-1/2″ thick	.008	C.Y.	1.10		1.10
Placing concrete	.008	C.Y.		.31	.31
Welded wire fabric 6 x 6 - #10/10	.011	C.S.F.	.18	.46	.64
Steel trowel finish	1.000	S.F.		1.47	1.47
Vinyl composition floor tile	1.000	S.F.	.97	1.29	2.26
Ceiling furring, 3/4″ channels, 24″ O.C.	1.000	S.F.	.28	1.61	1.89
Gypsum drywall, 5/8″ thick, finished	1.000	S.F.	.45	1.82	2.27
Paint ceiling	1.000	S.F.	.23	.77	1
TOTAL			8.17	9.88	18.05

B1010 245	Floor - Ceiling, Steel Joists	COST PER S.F. MAT.	COST PER S.F. INST.	COST PER S.F. TOTAL
1800	Open web joists, 12″ deep, 5.2#/L.F.	8.15	9.85	18
1900	16″ deep, 6.6#/L.F.	8.95	10.20	19.15
2000	20″ deep, 8.4# L.F.	10	10.65	20.65
2100	24″ deep, 11.5# L.F.	11.85	11.45	23.30
2200	26″ deep, 12.8# L.F.	12.60	11.75	24.35
2300				
2400				
2500				
2600				
2700				
2900	Dust protection, add, minimum	1%	2%	
3000	Maximum	4%	11%	
3100	Equipment usage curtailment, add, minimum	1%	1%	
3200	Maximum	3%	10%	
3300	Material handling & storage limitation, add, minimum	1%	1%	
3400	Maximum	6%	7%	
3500	Protection of existing work, add, minimum	2%	2%	
3600	Maximum	5%	7%	
3700	Shift work requirements, add, minimum		5%	
3800	Maximum		30%	
3900	Temporary shoring and bracing, add, minimum	2%	5%	
4000	Maximum	5%	12%	

B1010 Floor Construction

Joist Hangers

Existing Girder

Fiberglass Insulation, Kraft Faced, 9" Thick

Steel Cross Bridging

12" Open Web Wood Joists, 2'-0" o.c.

3/4" CDX Plywood Sub-floor

Existing Foundation

2" x 8" Sill

Anchor Bolts, 1/2" x 8"

Open Web Wood Joists

This page illustrates and describes floor framing systems including sills, joist hangers, joists, bridging, subfloor and insulation. Lines within systems components give the unit price and total price per square foot for this system. Prices for alternate joist types are on line items B1010 262 0100 through 0700. Both material quantities and labor costs have been adjusted for the system listed.

Factors: To adjust for job conditions other than normal working situations use lines B1010 262 2700 through 4000.

Example: You are to install the system where delivery of material is difficult. Go to line B1010 262 3400 and apply these percentages to the appropriate MAT. and INST. costs.

System Components	QUANTITY	UNIT	COST PER S.F. MAT.	COST PER S.F. INST.	COST PER S.F. TOTAL
Open web wood joists, sill, butt to girder, 5/8" subfloor, insul.					
Anchor bolt, hook type, with nut and washer, 1/2" x 8" long	.025	Ea.	.04	.13	.17
Joist & beam hanger, 18 ga. galvanized	.075	Ea.	.12	.32	.44
Bridging, steel, compression, for 2" x 12" joists	.010	C.Pr.	.01	.03	.04
Framing, open web wood joists, 12" deep	.001	M.L.F.	6.18	1.58	7.76
Framing, sills, 2" x 8"	.001	M.B.F.	1.33	2.08	3.41
Sub-floor, plywood, CDX, 3/4" thick	1.000	S.F.	1.27	1.03	2.30
Insulation, fiberglass, kraft face, 9" thick x 23" wide	1.000	S.F.	1	.52	1.52
TOTAL			9.95	5.69	15.64

B1010 262	Open Web Wood Joists	COST PER S.F. MAT.	COST PER S.F. INST.	COST PER S.F. TOTAL
0100	12" open web joists, sill, butt to girder, 5/8" sub-floor	9.95	5.70	15.65
0200	14" joists	10.75	5.80	16.55
0300	16" joists	10.30	5.90	16.20
0400	18" joists	9.40	6	15.40
0450	Wood struc." I " joists, 24" O.C., to 24' span, 50 PSF LL, butt to girder	6.60	5.30	11.90
0500	55 PSF LL	7	5.35	12.35
0600	to 30' span, 45 PSF LL	7.85	5.20	13.05
0700	55 PSF LL	10.35	5.30	15.65
2700	Cut & patch to match existing construction, add, minimum	2%	3%	
2800	Maximum	5%	9%	
2900	Dust protection, add, minimum	1%	2%	
3000	Maximum	4%	11%	
3100	Equipment usage curtailment, add, minimum	1%	1%	
3200	Maximum	3%	10%	
3300	Material handling & storage limitation, add, minimum	1%	1%	
3400	Maximum	6%	7%	
3500	Protection of existing work, add, minimum	2%	2%	
3600	Maximum	5%	7%	
3700	Shift work requirements, add, minimum		5%	
3800	Maximum		30%	
3900	Temporary shoring and bracing, add, minimum	2%	5%	
4000	Maximum	5%	12%	

619

B1010 Floor Construction

This page illustrates and describes a wood joist floor system including wood joist, oak floor, sub-floor, bridging, sand and finish floor, furring, plasterboard, taped, finished and painted ceiling. Lines within System Components give the unit price and total price per square foot for this system. Prices for wood joist floor systems are on Line Items B1010 263 1700 thru 2300. Both material quantities and labor costs have been adjusted for the system listed.

Factors: To adjust for job conditions other than normal working situations use Lines B1010 263 2700 thru 4000.

Example: You are to install the system during off peak hours, 6 P.M. to 2 A.M. Go to Line B1010 263 3800 and apply this percentage to the appropriate INST. costs.

System Components	QUANTITY	UNIT	COST PER S.F. MAT.	COST PER S.F. INST.	COST PER S.F. TOTAL
Wood joists, 2″ x 8″, 16″ O.C.,oak floor (sanded & finished),1/2″ sub					
Floor, 1″x 3″ bridging, furring, 5/8″ drywall, taped, finished and painted.					
Wood joists, 2″ x 8″, 16″ O.C.	1.000	L.F.	1.78	1.27	3.05
Subfloor plywood CDX 1/2″	1.000	SF Flr.	1.10	.93	2.03
Bridging 1″ x 3″	.150	Pr.	.16	.80	.96
Oak flooring, No. 1 common	1.000	S.F.	4.66	4.10	8.76
Sand and finish floor	1.000	S.F.	.25	1.92	2.17
Wood furring, 1″ x 3″, 16″ O.C.	1.000	L.F.	.68	1.99	2.67
Gypsum drywall, 5/8″ thick	1.000	S.F.	.40	.77	1.17
Tape and finishing	1.000	S.F.	.05	.70	.75
Paint ceiling	1.000	S.F.	.23	.77	1
TOTAL			9.31	13.25	22.56

B1010 263	Floor-Ceiling, Wood Joists		COST PER S.F. MAT.	COST PER S.F. INST.	COST PER S.F. TOTAL
1700	16″ on center, 2″ x 6″ joists		8.70	13.10	21.80
1750	2″ x 8″		9.30	13.25	22.55
1800	2″ x 10″		9.75	13.55	23.30
1900	2″ x 12″		10.30	13.55	23.85
2000	2″ x 14″		11.45	13.80	25.25
2100	12″ on center, 2″ x 10″ joists		10.35	13.90	24.25
2200	2″ x 12″		11	13.95	24.95
2300	2″ x 14″		12.45	14.25	26.70
2400					
2500					
2700	Cut & patch to match existing construction, add, minimum		2%	3%	
2800	Maximum		5%	9%	
2900	Dust protection, add, minimum		1%	2%	
3000	Maximum		4%	11%	
3100	Equipment usage curtailment, add, minimum		1%	1%	
3200	Maximum		3%	10%	
3300	Material handling & storage limitation, add, minimum		1%	1%	
3400	Maximum		6%	7%	
3500	Protection of existing work, add, minimum		2%	2%	
3600	Maximum		5%	7%	
3700	Shift work requirements, add, minimum			5%	
3800	Maximum			30%	
3900	Temporary shoring and bracing, add, minimum		2%	5%	
4000	Maximum		5%	12%	

B1020 Roof Construction

This page illustrates and describes a flat roof bar joist system including asphalt and gravel roof, lightweight concrete, metal decking, open web joists, and suspended acoustic ceiling. Lines within System Components give the unit price and total price per square foot for this system. Prices for flat roof bar joist systems are on Line Items B1020 106 1200 thru 1600. Both material quantities and labor costs have been adjusted for the system listed.

Factors: To adjust for job conditions other than normal working situations use Lines B1020 106 2700 thru 4000.

Example: You are to install the system where material handling and storage present some problem. Go to Line B1020 106 3300 and apply these percentages to the appropriate MAT. and INST. costs.

System Components			COST PER S.F.		
	QUANTITY	UNIT	MAT.	INST.	TOTAL
Three ply asphalt and gravel roof on lightweight concrete, metal decking on Web joist 4' O.C., with suspended acoustic ceiling.					
Open web joist, 10" deep, 4' O.C.	1.250	Lb.	1.48	.63	2.11
Metal decking, 1-1/2" deep, 22 Ga., galvanized	1.000	S.F.	2.99	.74	3.73
Gravel roofing, 3 ply asphalt	.010	C.S.F.	1.63	2.26	3.89
Perlite roof fill, 3" thick	1.000	S.F.	3.05	.62	3.67
Suspended ceiling, 3/4" mineral fiber on "Z" bar suspension	1.000	S.F.	2.61	4.65	7.26
TOTAL			11.76	8.90	20.66

B1020 106	Steel Joist Roof & Ceiling	COST PER S.F.		
		MAT.	INST.	TOTAL
1200	Open web bar joist, 10" deep, 5#/L.F.	11.75	8.90	20.65
1300	16" deep, 5#/L.F.	12.25	9.10	21.35
1400	18" deep, 6.6#/L.F.	12.65	9.25	21.90
1500	20" deep, 9.6#/L.F.	13.10	9.45	22.55
1600	24" deep, 11.52#/L.F.	13.70	9.70	23.40
1700				
1800				
1900				
2000				
2100				
2200				
2300				
2400				
2500				
2700	Cut & patch to match existing construction, add, minimum	2%	3%	
2800	Maximum	5%	9%	
2900	Dust protection, add, minimum	1%	2%	
3000	Maximum	4%	11%	
3100	Equipment usage curtailment, add, minimum	1%	1%	
3200	Maximum	3%	10%	
3300	Material handling & storage limitation, add, minimum	1%	1%	
3400	Maximum	6%	7%	
3500	Protection of existing work, add, minimum	2%	2%	
3600	Maximum	5%	7%	
3700	Shift work requirements, add, minimum		5%	
3800	Maximum		30%	
3900	Temporary shoring and bracing, add, minimum	2%	5%	
4000	Maximum	5%	12%	

B1020 Roof Construction

This page illustrates and describes a wood frame roof system including rafters, ceiling joists, sheathing, building paper, asphalt shingles, roof trim, furring, insulation, plaster and paint. Lines within System Components give the unit price and total price per square foot for this system. Prices for wood frame roof systems are on Line Items B1020 202 1800 thru 2700. Both material quantities and labor costs have been adjusted for the system listed.

Factors: To adjust for job conditions other than normal working situations use Lines B1020 202 3300 thru 4000.

Example: You are to install the system while protecting existing work. Go to Line B1020 202 3800 and apply these percentages to the appropriate MAT. and INST. costs.

System Components	QUANTITY	UNIT	COST PER S.F.		
			MAT.	INST.	TOTAL
Wood frame roof system, 4 in 12 pitch, including rafters, sheathing,					
Shingles, insulation, drywall, thin coat plaster, and painting.					
Rafters, 2" x 6", 16" O.C., 4 in 12 pitch	1.080	L.F.	1.26	1.50	2.76
Ceiling joists, 2" x 6", 16 O.C.	1.000	L.F.	1.17	1.11	2.28
Sheathing, 1/2" CDX	1.080	S.F.	1.19	1.08	2.27
Building paper, 15# felt	.011	C.S.F.	.07	.21	.28
Asphalt, standard strip shingles, inorganic, class A, 30 year	.011	C.S.F.	1.39	1.46	2.85
Roof trim	.100	L.F.	.22	.30	.52
Furring, 1" x 3", 16" O.C.	1.000	L.F.	.68	1.99	2.67
Fiberglass insulation, 6" batts	1.000	S.F.	.84	.52	1.36
Gypsum board, 1/2" thick	1.000	S.F.	.36	.77	1.13
Thin coat plaster	1.000	S.F.	.11	.87	.98
Paint, roller, 2 coats	1.000	S.F.	.23	.77	1
TOTAL			7.52	10.58	18.10

B1020 202	Wood Frame Roof & Ceiling	COST PER S.F.		
		MAT.	INST.	TOTAL
1800	Rafters 16" O.C., 2" x 6"	7.50	10.60	18.10
1900	2" x 8"	8.20	10.65	18.85
2000	2" x 10"	8.65	11.45	20.10
2100	2" x 12"	9.25	11.70	20.95
2200	Rafters 24" O.C., 2" x 6"	6.90	9.90	16.80
2300	2" x 8"	7.40	10	17.40
2400	2" x 10"	7.75	10.55	18.30
2500	2" x 12"	8.20	10.75	18.95
2600	Roof pitch, 6 in 12, add	3%	10%	
2700	8 in 12, add	5%	12%	
2800				
2900				
3000				
3100				
3300	Cut & patch to match existing construction, add, minimum	2%	3%	
3400	Maximum	5%	9%	
3500	Material handling & storage limitation, add, minimum	1%	1%	
3600	Maximum	6%	7%	
3700	Protection of existing work, add, minimum	2%	2%	
3800	Maximum	5%	7%	
3900	Shift work requirements, add, minimum		5%	
4000	Maximum		30%	

B1020 Roof Construction

5/8" Plywood Sheathing

Truss Anchors

Roof Trusses at 2'-0" o.c.

9" Fiberglass Insulation

1" x 8" Fascia Board

Wood Fabricated Roof Truss System

This page illustrates and describes a sloped roof truss framing system including trusses, truss anchors, sheathing, fascia and insulation. Lines B1020 210 1500 through 1800 are for a flat roof framing system including flat roof trusses, truss anchors, sheathing, fascia and insulation. Lines within system components give the unit price and total price per square foot for this system. Prices for alternate sizes and systems are given in line items B1020 210 0100 through 1800. Both material quantities and labor costs have been adjusted for the system listed.

Factors: To adjust for job conditions other than normal working situations use lines B1020 210 2700 through 4000.

Example: You are to install the system where crane placement and movement will be difficult. Go to line B1020 210 3200 and apply these percentages to the appropriate MAT. and INST. costs.

System Components	QUANTITY	UNIT	COST PER S.F.		
			MAT.	INST.	TOTAL
Roof truss, 2' O.C., 4/12 slope, 12' span, 5/8 sheath., fascia, insul.					
Timber connectors, rafter anchors, galv., 1 1/2" x 5 1/4"	.084	Ea.	.04	.40	.44
Plywood sheathing, CDX, 5/8" thick	1.054	S.F.	1.34	1.13	2.47
Truss, 2' O.C., metal plate connectors, 12' span	.042	Ea.	3.36	2.14	5.50
Molding, fascia trim, 1" x 8"	.167	L.F.	.37	.51	.88
Insulation, fiberglass, kraft face, 9" thick x 23" wide	1.000	S.F.	1	.52	1.52
TOTAL			6.11	4.70	10.81

B1020 210	Roof Truss, Wood	COST PER S.F.		
		MAT.	INST.	TOTAL
0100	Roof truss, 2'O.C., 4/12p, 1'ovhg, 12'span, 5/8"sheath, fascia, insul.	6.10	4.70	10.80
0200	20' span	5.45	3.89	9.34
0300	24' span	5.10	3.64	8.74
0400	26' span	7.20	4.84	12.04
0500	28' span	5.40	3.56	8.96
0600	30' span	5.40	3.53	8.93
0700	32' span	5.30	3.47	8.77
0800	34' span	4.88	3.42	8.30
0890	60' span	6.70	3.45	10.15
0900	8/12 pitch, 20' span	6.40	4.19	10.59
1000	24' span	6.25	3.92	10.17
1100	26' span	5.85	3.84	9.69
1200	28' span	6.50	3.82	10.32
1300	32' span	6.20	3.76	9.96
1400	36' span	6.30	3.69	9.99
1500	Flat roof frame, fab.strl.joists, 2' O.C., 15'-24' span, 50 PSF LL	5.35	3.19	8.54
1600	55 PSF LL	5.75	3.27	9.02
1700	24'-30'span, 45 PSF LL	6.55	3.05	9.60
1800	55 PSF LL	9.10	3.14	12.24
1900				
2700	Cut & patch to match existing construction, add, minimum	2%	3%	
2800	Maximum	5%	9%	
2900	Dust protection, add, minimum	1%	2%	
3000	Maximum	4%	11%	
3100	Equipment usage curtailment, add, minimum	1%	1%	
3200	Maximum	3%	10%	
3300	Material handling & storage limitation, add, minimum	1%	1%	
3400	Maximum	6%	7%	

B1020 Roof Construction

B1020 210	Roof Truss, Wood	COST PER S.F.		
		MAT.	INST.	TOTAL
3500	Protection of existing work, add, minimum	2%	2%	
3600	Maximum	5%	7%	
3700	Shift work requirements, add, minimum		5%	
3800	Maximum		30%	
3900	Temporary shoring and bracing, add, minimum	2%	5%	
4000	Maximum	5%	12%	

This page illustrates and describes a masonry concrete block system including concrete block wall, pointed, reinforcing, waterproofing, gypsum plaster on gypsum lath, and furring. Lines within System Components give the unit price and total price per square foot for this system. Prices for masonry concrete block wall systems are on Line Items B2010 108 1400 thru 2200. Both material quantities and labor costs have been adjusted for the system listed.

Factors: To adjust for job conditions other than normal working situations use Lines B2010 108 2700 thru 4000.

Example: You are to install the system and match existing construction at several locations. Go to Line B2010 108 2800 and apply these percentages to the appropriate MAT. and INST. costs.

System Components	QUANTITY	UNIT	COST PER S.F.		
			MAT.	INST.	TOTAL
Concrete block, 8″ thick, reinforced every 2 courses, waterproofing gypsum					
Plaster over gypsum lath on 1″ x 3″ furring, interior painting & baseboard.					
Concrete block, 8″ x 8″ x 16″, reinforced	1.000	S.F.	5.35	7.70	13.05
Silicone waterproofing, 2 coats	1.000	S.F.	.89	.22	1.11
Bituminous coating, 1/16″ thick	1.000	S.F.	.46	1.33	1.79
On concrete	1.000	L.F.	.74	2.68	3.42
Gypsum lath, 3/8″ thick	.110	S.Y.	.40	.87	1.27
Gypsum plaster, 2 coats	.110	S.Y.	.49	3.32	3.81
Painting, 2 coats	1.000	S.F.	.15	.60	.75
Baseboard wood, 9/16″ x 2-5/8″	.100	L.F.	.31	.29	.60
Paint baseboard, primer + 1 coat enamel	.100	L.F.	.02	.07	.09
TOTAL			8.81	17.08	25.89

B2010 108	Masonry Wall, Concrete Block	COST PER S.F.		
		MAT.	INST.	TOTAL
1400	8″ thick block, regular	8.80	17.10	25.90
1500	Fluted 2 sides	9.25	20	29.25
1600	Deep grooved	8.45	20	28.45
1700	Slump block	13.90	17.50	31.40
1800	Split rib	11	20	31
1900				
2000	12″ thick block, regular	8.55	20.50	29.05
2100	Slump block	18.25	19.05	37.30
2200	Split rib	12.05	23	35.05
2300				
2700	Cut & patch to match existing construction, add, minimum	2%	3%	
2800	Maximum	5%	9%	
2900	Dust protection, add, minimum	1%	2%	
3000	Maximum	4%	11%	
3100	Equipment usage curtailment, add, minimum	1%	1%	
3200	Maximum	3%	10%	
3300	Material handling & storage limitation, add, minimum	1%	1%	
3400	Maximum	6%	7%	
3500	Protection of existing work, add, minimum	2%	2%	
3600	Maximum	5%	7%	
3700	Shift work requirements, add, minimum		5%	
3800	Maximum		30%	
3900	Temporary shoring and bracing, add, minimum	2%	5%	
4000	Maximum	5%	12%	

For customer support on your Commercial Renovation Costs with RSMeans Data, call 800.448.8182.

B2010 Exterior Walls

This page illustrates and describes a masonry wall, brick-stone system including brick, concrete block, durawall, insulation, plasterboard, taped and finished, furring, baseboard and painting interior. Lines within System Components give the unit price and total price per square foot for this system. Prices for masonry wall, brick-stone systems are on Line Item B2010 129 1400 thru 2500. Both material quantities and labor costs have been adjusted for the system listed.

Factors: To adjust for job conditions other than normal working situations use Lines B2010 129 3100 thru 4200.

Example: You are to install the system without damaging the existing work. Go to Line B2010 129 3900 and apply these percentages to the appropriate MAT. and INST. costs.

System Components	QUANTITY	UNIT	COST PER S.F.		
			MAT.	INST.	TOTAL
Face brick, 4"thick, concrete block back-up, reinforce every second course, 3/4"insulation, furring, 1/2"drywall, taped, finish, and painted, baseboard					
Face brick, 4" brick	1.000	S.F.	5.60	14.50	20.10
Concrete back-up block, reinforced 8" thick	1.000	S.F.	3.24	8.10	11.34
3/4" rigid polystyrene insulation	1.000	S.F.	.67	.87	1.54
Furring, 1" x 3", wood, 16" O.C.	1.000	L.F.	.74	1.41	2.15
Drywall, 1/2" thick	1.000	S.F.	.36	.70	1.06
Taping & finishing	1.000	S.F.	.05	.70	.75
Painting, 2 coats	1.000	S.F.	.23	.77	1
Baseboard, wood, 9/16" x 2-5/8"	.100	L.F.	.31	.29	.60
Paint baseboard, primer + 1 coat enamel	.100	L.F.	.02	.07	.09
TOTAL			11.22	27.41	38.63

B2010 131	Masonry Wall, Brick - Stone	COST PER S.F.		
		MAT.	INST.	TOTAL
1400	Face brick, standard, red, 4" x 2-2/3" x 8" (6.75/S.F.)	11.20	27.50	38.70
1500	Norman, 4" x 2-2/3" x 12" (4.5 per S.F.)	13	23	36
1600	Roman, 4" x 2" x 12" (6.0 per S.F.)	14.55	25.50	40.05
1700	Engineer, 4" x 3-1/5" x 8" (5.63 per S.F.)	10	25	35
1800	S.C.R., 6" x 2-2/3" x 12" (4.5 per S.F.)	12.35	23	35.35
1900	Jumbo, 6" x 4" x 12" (3.0 per S.F.)	11.55	20.50	32.05
2000	Norwegian, 6" x 3-1/5" x 12" (3.75 per S.F.)	14.05	21.50	35.55
2100				
2200				
2300	Stone, veneer, fieldstone, 6" thick	13.75	23.50	37.25
2400	Marble, 2" thick	58.50	40	98.50
2500	Limestone, 2" thick	39.50	24.50	64
3100	Cut & patch to match existing construction, add, minimum	2%	3%	
3200	Maximum	5%	9%	
3300	Dust protection, add, minimum	1%	2%	
3400	Maximum	4%	11%	
3500	Equipment usage curtailment, add, minimum	1%	1%	
3600	Maximum	3%	10%	
3700	Material handling & storage limitation, add, minimum	1%	1%	
3800	Maximum	6%	7%	
3900	Protection of existing work, add, minimum	2%	2%	
4000	Maximum	5%	7%	
4100	Shift work requirements, add, minimum		5%	
4200	Maximum		30%	

B2010 Exterior Walls

This page illustrates and describes a parapet wall system including wall, coping, flashing and cant strip. Lines within System Components give the unit price and cost per lineal foot for this system. Prices for parapet wall systems are on Lines B2010 142 1300 thru 2400. Both material quantities and labor costs have been adjusted for the system listed.

Factors: To adjust for job conditions other than normal working situations, use Lines B2010 142 2900 thru 4000.

Example: You are to install the system without damaging the adjacent property. Go to Line B2010 142 3600 and apply these percentages to the appropriate MAT. and INST. costs.

System Components	QUANTITY	UNIT	COST PER L.F.		
			MAT.	INST.	TOTAL
Concrete block parapet, incl. reinf., coping, flashing & cant strip, 2′high					
8″ concrete block	2.000	S.F.	10.70	15.40	26.10
Masonry reinforcing	1.000	L.F.	.29	.23	.52
Coping, precast	1.000	L.F.	29.50	17.90	47.40
Roof cant	1.000	L.F.	.63	2.04	2.67
16 oz.	1.000	L.F.	8.05	3.10	11.15
Cap flashing	1.000	L.F.	8.70	5.75	14.45
TOTAL			**57.87**	**44.42**	**102.29**

B2010 142	Parapet Wall	COST PER L.F.		
		MAT.	INST.	TOTAL
1300	Concrete block, 8″ thick	58	44.50	102.50
1400	12″ thick	78	58	136
1500	Split rib block, 8″ thick	62.50	50.50	113
1600	12″ thick	78	59.50	137.50
1700	Brick, common brick, 8″ wall	69	76	145
1800	12″ wall	80	96	176
1900	4″ brick, 4″ backup block	63.50	73	136.50
2000	8″ backup block	64.50	75	139.50
2100	Stucco on masonry	58	49.50	107.50
2200	On wood frame	22.50	40.50	63
2300	Wood, T 1-11 siding	41	48.50	89.50
2400	Boards, 1″ x 6″ cedar	50.50	53.50	104
2500				
2600				
2900	Dust protection, add, minimum	1%	2%	
3000	Maximum	4%	11%	
3100	Equipment usage curtailment, add, minimum	1%	1%	
3200	Maximum	3%	10%	
3300	Material handling & storage limitation, add, minimum	1%	1%	
3400	Maximum	6%	7%	
3500	Protection of existing work, add, minimum	2%	2%	
3600	Maximum	5%	7%	
3700	Shift work requirements, add, minimum		5%	
3800	Maximum		30%	
3900	Temporary shoring and bracing, add, minimum	2%	5%	
4000	Maximum	5%	12%	

627

B2010 Exterior Walls

This page illustrates and describes a masonry cleaning and restoration system, including staging, cleaning, repointing. Lines within System Components give the unit price and cost per square foot for this system. Prices for systems are on Lines B2010 143 1300 thru 2500. Both material quantities and labor costs have been adjusted for the system listed.

Factors: To adjust for conditions other than normal working situations, use Lines B2010 143 2900 thru 4200.

Example: You are to clean a wall and be concerned about dust control. Go to line B2010 143 3100 and apply these percentages to the appropriate MAT. and INST. costs.

System Components	QUANTITY	UNIT	COST PER S.F. MAT.	COST PER S.F. INST.	COST PER S.F. TOTAL
Repoint existing building, brick, common bond, high pressure cleaning,					
Water only, soft old mortar.					
Scaffolding,steel tubular,bldg ext wall face, labor only, 1 to 5 stories	.010	C.S.F.		2.61	2.61
Cleaning masonry, high pressure wash, water only, avg soil, bio staining	1.100	S.F.		1.98	1.98
Repoint, brick common bond	1.000	S.F.	.63	7.20	7.83
TOTAL			.63	11.79	12.42

B2010 143		Masonry Restoration - Cleaning	COST PER S.F. MAT.	COST PER S.F. INST.	COST PER S.F. TOTAL
1300		Brick, common bond	.63	11.80	12.43
1400		Flemish bond	.66	12.25	12.91
1500		English bond	.66	13	13.66
1700		Stone, 2' x 2' blocks	.84	8.90	9.74
1800		2' x 4' blocks	.63	7.85	8.48
2000		Add to above prices for alternate cleaning systems:			
2100		Chemical brush and wash	.06	.79	.85
2200		High pressure chemical and water	.10	.68	.78
2300		Sandblasting, wet system	.43	1.88	2.31
2400		Dry system	.40	1.11	1.51
2500		Steam cleaning		.52	.52
2600					
2900		Cut & patch to match existing construction, add, minimum	2%	3%	
3000		Maximum	5%	9%	
3100		Dust protection, add, minimum	1%	2%	
3200		Maximum	4%	11%	
3300		Equipment usage curtailment, add, minimum	1%	1%	
3400		Maximum	3%	10%	
3500		Material handling & storage limitation, add, minimum	1%	1%	
3600		Maximum	6%	7%	
3900		Protection of existing work, add, minimum	2%	2%	
4000		Maximum	5%	7%	
4100		Shift work requirements, add, minimum		5%	
4200		Maximum		30%	

B2010 Exterior Walls

This page illustrates and describes a wood frame exterior wall system including wood studs, sheathing, felt, insulation, plasterboard, taped and finished, baseboard and painted interior. Lines within System Components give the unit price and total price per square foot for this system. Prices for wood frame exterior wall systems are on Line Items B2010 150 1600 thru 2700. Both material quantities and labor costs have been adjusted for the system listed.

Factors: To adjust for job conditions other than normal working situations use Lines B2010 150 3100 thru 4200.

Example: You are to install the system with need for complete temporary bracing. Go to Line B2010 150 4200 and apply these percentages to the appropriate MAT. and INST. costs.

System Components			COST PER S.F.		
	QUANTITY	UNIT	MAT.	INST.	TOTAL
Wood stud wall, cedar shingle siding, building paper, plywood sheathing, Insulation, 5/8″ drywall, taped, finished and painted, baseboard.					
2″ x 4″ wood studs, 16″ O.C.	.100	L.F.	.81	1.40	2.21
1/2″ CDX sheathing	1.000	S.F.	1.10	1	2.10
18″ No. 1 red cedar shingles, 7-1/2″ exposure	.008	C.S.F.	1.93	2.48	4.41
15# felt paper	.010	C.S.F.	.06	.19	.25
3-1/2″ fiberglass insulation	1.000	S.F.	.61	.52	1.13
5/8″ drywall	1.000	S.F.	.40	.70	1.10
Drywall finishing adder	1.000	S.F.	.05	.70	.75
Baseboard trim, stock pine, 9/16″ x 3-1/2″, painted	.100	L.F.	.31	.29	.60
Paint baseboard, primer + 1 coat enamel	.100	L.F.	.02	.07	.09
Paint, 2 coats, interior	1.000	S.F.	.23	.77	1
TOTAL			5.52	8.12	13.64

B2010 150	Wood Frame Exterior Wall	COST PER S.F.		
		MAT.	INST.	TOTAL
1600	Cedar shingle siding, painted	5.50	8.10	13.60
1700	Aluminum siding, horizontal clapboard	7.05	8.35	15.40
1800	Cedar bevel siding, 1/2″ x 6″, vertical , painted	8.45	8	16.45
1900	Redwood siding 1″ x 4″ to 1″ x 6″ vertical, T & G	9.10	8.80	17.90
2000	Board and batten	8.90	7.30	16.20
2100	Ship lap siding	9.40	7.75	17.15
2200	Plywood, grooved (T1-11) fir	5.10	7.70	12.80
2300	Redwood	6.45	7.70	14.15
2400	Southern yellow pine	5.15	7.70	12.85
2500	Masonry on stud wall, stucco, wire and plaster	4.52	12.35	16.87
2600	Stone veneer	11.70	16.15	27.85
2700	Brick veneer, brick $275 per M	9.20	20	29.20
3100	Cut & patch to match existing construction, add, minimum	2%	3%	
3200	Maximum	5%	9%	
3300	Dust protection, add, minimum	1%	2%	
3400	Maximum	4%	11%	
3500	Material handling & storage limitation, add, minimum	1%	1%	
3600	Maximum	6%	7%	
3700	Protection of existing work, add, minimum	2%	2%	
3800	Maximum	5%	7%	
3900	Shift work requirements, add, minimum		5%	
4000	Maximum		30%	
4100	Temporary shoring and bracing, add, minimum	2%	5%	
4200	Maximum	5%	12%	

B2010 Exterior Walls

B2010 190	Selective Price Sheet	COST PER S.F.		
		MAT.	INST.	TOTAL
0100	Exterior surface, masonry, concrete block, standard 4" thick	2.30	7.45	9.75
0200	6" thick	2.98	8	10.98
0300	8" thick	2.90	8.50	11.40
0400	12" thick	5.10	11.05	16.15
0500	Split rib, 4" thick	5.40	9.25	14.65
0600	8" thick	7.55	10.65	18.20
0700	Brick running bond, standard size, 6.75/S.F.	5.60	14.50	20.10
0800	Buff, 6.75/S.F.	5.90	14.50	20.40
0900	Stucco, on frame	.94	6.75	7.69
1000	On masonry	.38	5.25	5.63
1100	Metal, aluminum, horizontal, plain	3.44	2.71	6.15
1200	Insulated	3.32	2.71	6.03
1300	Vertical, plain	2.89	2.71	5.60
1400	Insulated	3.01	2.83	5.84
1500	Wood, beveled siding, "A" grade cedar, 1/2" x 6"	4.86	2.36	7.22
1600	1/2" x 8"	8.25	2.11	10.36
1700	Shingles, 16" #1 red, 7-1/2" exposure	2.41	3.40	5.81
1800	18" perfections, 7-1/2" exposure	2.43	2.84	5.27
1900	Handsplit, 10" exposure	4.15	2.79	6.94
2000	White cedar, 7-1/2" exposure	1.33	3.50	4.83
2100	Vertical, board & batten, redwood	5.50	3.17	8.67
2200	White pine	2.65	2.11	4.76
2300	T. & G. boards, redwood, 1" x 4"	7.70	3.17	10.87
2400	1' x 8"	9.90	2.11	12.01
2500				
2600	Interior surface, drywall, taped & finished, standard, 1/2"	.42	1.44	1.86
2700	5/8" thick	.45	1.44	1.89
2800	Fire resistant, 1/2" thick	.46	1.44	1.90
2900	5/8" thick	.47	1.44	1.91
3000	Moisture resistant, 1/2" thick	.46	1.44	1.90
3100	5/8" thick	.53	1.44	1.97
3200	Core board, 1" thick	.90	2.90	3.80
3300	Plaster, gypsum, 2 coats	.49	3.35	3.84
3400	3 coats	.69	4.04	4.73
3500	Perlite or vermiculite, 2 coats	.73	3.81	4.54
3600	3 coats	.82	4.74	5.56
3700	Gypsum lath, standard, 3/8" thick	.41	.88	1.29
3800	1/2" thick	.35	.94	1.29
3900	Fire resistant, 3/8" thick	.30	1.07	1.37
4000	1/2" thick	.34	1.15	1.49
4100	Metal lath, diamond, 2.5 lb.	.67	.88	1.55
4200	Rib, 3.4 lb.	.59	1.07	1.66
4300	Framing metal studs including top and bottom			
4400	Runners, walls 10' high			
4500	24" O.C., non load bearing 20 gauge, 2-1/2" wide	.30	1.16	1.46
4600	3-5/8" wide	.54	1.18	1.72
4700	4" wide	.58	1.18	1.76
4800	6" wide	.70	1.20	1.90
4900	Load bearing 18 gauge, 2-1/2" wide	.66	1.30	1.96
5000	3-5/8" wide	.79	1.33	2.12
5100	4" wide	.86	1.69	2.55
5200	6" wide	1.06	1.38	2.44
5300	16" O.C., non load bearing 20 gauge, 2-1/2" wide	.40	1.79	2.19
5400	3-5/8" wide	.71	1.81	2.52
5500	4" wide	.78	1.83	2.61
5600	6" wide	.95	1.86	2.81
5700	Load bearing 18 gauge, 2-1/2" wide	.89	1.81	2.70
5800	3-5/8" wide	1.06	1.84	2.90

B2010 Exterior Walls

B2010 190	Selective Price Sheet	COST PER S.F.		
		MAT.	INST.	TOTAL
5900	4" wide	1.08	1.89	2.97
6000	6" wide	1.41	1.91	3.32
6100	Framing wood studs incl. double top plate and			
6200	Single bottom plate, walls 10' high			
6300	24" O.C., 2" x 4"	.62	1.12	1.74
6400	2" x 6"	.94	1.21	2.15
6500	16" O.C., 2" x 4"	.81	1.40	2.21
6600	2" x 6"	1.23	1.55	2.78
6700	Sheathing, boards, 1" x 6"	2.74	2.14	4.88
6800	1" x 8"	2.96	1.82	4.78
6900	Plywood, 3/8" thick	.86	1.16	2.02
7000	1/2" thick	1.10	1.24	2.34
7100	5/8" thick	1.27	1.33	2.60
7200	3/4" thick	1.52	1.43	2.95
7300	Wood fiber, 5/8" thick	.75	1.16	1.91
7400	Gypsum weatherproof, 1/2" thick	.51	1.24	1.75
7500	Insulation, fiberglass batts, 3-1/2" thick, R13	.54	1	1.54
7600	6" thick, R19	.66	1.16	1.82
7700	Poured 4" thick, fiberglass wool, R4/inch	.61	3.48	4.09
7800	Mineral wool, R3/inch	.62	3.48	4.10
7900	Polystyrene, R4/inch	1.72	3.48	5.20
8000	Perlite or vermiculite, R2.7/inch	7.25	3.48	10.73
8100	Rigid, fiberglass, R4.3/inch, 1" thick	.63	.70	1.33
8200	R8.7/inch, 2" thick	1.38	.78	2.16

B2020 Exterior Windows

B2020 108			Vinyl Clad Windows				
	MATERIAL	TYPE	GLAZING	SIZE	DETAIL	COST PER UNIT	
						MAT.	INST.	TOTAL

	MATERIAL	TYPE	GLAZING	SIZE	DETAIL	MAT.	INST.	TOTAL
1000	Vinyl clad	casement	insul. glass	1'-4" x 4'-0"		465	161	626
1010				2'-0" x 3'-0"		370	161	531
1020				2'-0" x 4'-0"		450	169	619
1030				2'-0" x 5'-0"		470	178	648
1040				2'-0" x 6'-0"		650	178	828
1050				2'-4" x 4'-0"		470	178	648
1060				2'-6" x 5'-0"		595	173	768
1070				3'-0" x 5'-0"		795	178	973
1080				4'-0" x 3'-0"		930	178	1,108
1090				4'-0" x 4'-0"		755	178	933
1100				4'-8" x 4'-0"		825	178	1,003
1110				4'-8" x 5'-0"		1,000	207	1,207
1120				4'-8" x 6'-0"		1,075	207	1,282
1130				6'-0" x 4'-0"		950	207	1,157
1140				6'-0" x 5'-0"		1,275	207	1,482
3000		double-hung	insul. glass	2'-0" x 4'-0"		625	155	780
3050				2'-0" x 5'-0"		555	155	710
3100				2'-4" x 4'-0"		650	161	811
3150				2'-4" x 4'-8"		570	161	731
3200				2'-4" x 6'-0"		550	161	711
3250				2'-6" x 4'-0"		565	161	726
3300				2'-8" x 4'-0"		690	207	897
3350				2'-8" x 5'-0"		665	178	843
3375				2'-8" x 6'-0"		580	178	758
3400				3'-0 x 3'-6"		475	161	636
3450				3'-0 x 4'-0"		525	169	694
3500				3'-0 x 4'-8"		585	169	754
3550				3'-0 x 5'-0"		585	178	763
3600				3'-0 x 6'-0"		525	178	703
3700				4'-0 x 5'-0"		690	191	881
3800				4'-0 x 6'-0"		825	191	1,016

B2020 Exterior Windows

This page illustrates and describes an aluminum window system including double hung aluminum window, exterior and interior trim, hardware and insulating glass. Lines within System Components give the unit price and total price on a cost each basis for this system. Prices for aluminum window systems are on Line Items B2020 110 1000 thru 2400. Both material quantities and labor costs have been adjusted for the system listed.

Factors: To adjust for job conditions other than normal working situations use Lines B2020 110 3100 thru 4000.

Example: You are to install the system and cut and patch to match existing construction. Go to Line B2020 110 3200 and apply these percentages to the appropriate MAT. and INST. costs.

System Components	QUANTITY	UNIT	COST EACH MAT.	COST EACH INST.	COST EACH TOTAL
Double hung aluminum window, 2'-4" x 2'-6" exterior and interior trim, Hardware glazed with insulating glass.					
Double-hung aluminum window, 2'-4" x 2' 6"	1.000	Ea.	271.10	45.47	316.57
Interior Trim	1.000	Set	35.50	53.50	89
Hardware	1.000	Set	6.75	29	35.75
TOTAL			313.35	127.97	441.32

B2020 110	Windows - Aluminum	MAT.	INST.	TOTAL
1000	Aluminum, double hung, 2'-4" x 2'-6"	315	128	443
1100	2'-8" x 4'-6"	605	192	797
1200	3'- 4" x 5'-6"	945	288	1,233
1300	Casement, 3'-6" x 2'-4"	430	146	576
1400	4'-6" x 2'-4"	545	180	725
1500	5'-6" x 2'-4"	705	245	950
1600	Projected window, 2'-1" x 3'-0"	310	131	441
1700	3'-5" x 3'-0"	485	178	663
1800	4'-0" x 4'-0"	775	270	1,045
1900	Horizontal sliding 3'-0" x 3'-0"	276	153	429
2000	3'-6" x 4'-0"	410	208	618
2100	5'-0" x 6'-0"	875	380	1,255
2200	Picture window, 3'-8" x 3'-1"	290	170	460
2300	4'-4" x 4'-5"	470	248	718
2400	5'-8" x 4'-9"	685	355	1,040
2500				
2600				
2700				
2800				
2900				
3100	Cut & patch to match existing construction, add, minimum	2%	3%	
3200	Maximum	5%	9%	
3300	Dust protection, add, minimum	1%	2%	
3400	Maximum	4%	11%	
3500	Material handling & storage limitation, add, minimum	1%	1%	
3600	Maximum	6%	7%	
3700	Protection of existing work, add, minimum	2%	2%	
3800	Maximum	5%	7%	
3900	Shift work requirements, add, minimum		5%	
4000	Maximum		30%	

B2020 Exterior Windows

This page illustrates and describes a wood window system including wood picture window, exterior and interior trim, hardware and insulating glass. Lines within System Components give the unit price and total price on a cost each basis for this system. Prices for wood window systems are on Line Items B2020 112 1200 thru 2900. Both material quantities and labor costs have been adjusted for the system listed.

Factors: To adjust for job conditions other than normal working situations use Lines B2020 112 3100 thru 4000.

Example: You are to install the above system where dust control is vital. Go to Line B2020 112 3400 and apply these percentages to the appropriate MAT. and INST. costs.

System Components	QUANTITY	UNIT	COST EACH MAT.	COST EACH INST.	COST EACH TOTAL
Wood picture window 4'-0" x 4'-6", exterior and interior trim, hardware,					
Glazed with insulating glass.					
4'-0" x 4'-6" wood picture window with insulating glass	1.000	Ea.	675	127	802
Oak interior trim	1.000	Set	87.50	116	203.50
Hardware	1.000	Set	6.75	29	35.75
TOTAL			769.25	272	1,041.25

B2020 112	Windows - Wood	COST EACH MAT.	COST EACH INST.	COST EACH TOTAL
1200	Picture window, 4'-0" x 4'-6"	770	272	1,042
1300	5'-0" x 4'-0"	765	272	1,037
1400	6'-0" x 4'-6"	795	284	1,079
1500	Bow, bay window, 8'-0" x 5'-0", standard	2,725	284	3,009
1600	Deluxe	2,025	284	2,309
1700	Bow, bay window, 12'-0" x 6'-0" standard	3,975	375	4,350
1800	Deluxe	2,800	375	3,175
1900	Double hung, 3'-0" x 4'-0"	380	176	556
2000	4'-0" x 4'-6"	480	232	712
2100	Casement 2'-0" x 3'-0"	355	152	507
2200	2 leaf, 4'-0" x 4'-0"	1,025	254	1,279
2300	3 leaf, 6'-0" x 6'-0"	1,625	405	2,030
2400	Awning, 2'-10" x 1'-10"	400	152	552
2500	3'-6" x 2'-4"	435	176	611
2600	4'-0" x 3'-0"	650	232	882
2700	Horizontal sliding 3'-0" x 2'-0"	385	152	537
2800	4'-0" x 3'-6"	525	176	701
2900	6'-0" x 5'-0"	755	232	987
3100	Cut & patch to match existing construction, add, minimum	2%	3%	
3200	Maximum	5%	9%	
3300	Dust protection, add, minimum	1%	2%	
3400	Maximum	4%	11%	
3500	Material handling & storage limitation, add, minimum	1%	1%	
3600	Maximum	6%	7%	
3700	Protection of existing work, add, minimum	2%	2%	
3800	Maximum	5%	7%	
3900	Shift work requirements, add, minimum		5%	
4000	Maximum		30%	

B20 Exterior Enclosure

B2020 Exterior Windows

This page illustrates and describes storm window and door systems based on a cost each price. Prices for storm window and door systems are on Line Items B2020 116 0400 thru 2400. Both material quantities and labor costs have been adjusted for the system listed.

Factors: To adjust for job conditions other than normal working situations use Lines B2020 116 3100 thru 4000 and apply these percentages to the appropriate MAT and INST. costs.

Example: You are to install the system and protect all existing construction. Go to Line B2020 116 3800 and apply these percentages to the appropriate MAT. and INST. costs.

B2020 116	Storm Windows & Doors	MAT.	INST.	TOTAL
0350	Storm Window and door systems, single glazed			
0400	Window, custom aluminum anodized, 2'-0" x 3'-5"	103	46.50	149.50
0500	2'-6" x 5'-0"	150	50	200
0600	4'-0" x 6'-0"	266	55.50	321.50
0700	White painted aluminum, 2'-0" x 3'-5"	126	46.50	172.50
0800	2'-6" x 5'-0"	214	50	264
0900	4'-0" x 6'-0"	325	55.50	380.50
1000	Average quality aluminum, anodized, 2'-0" x 3'-5"	114	46.50	160.50
1100	2'-6" x 5'-0"	142	50	192
1200	4'-0" x 6'-0"	146	55.50	201.50
1300	White painted aluminum, 2'-0" x 3'-5"	118	46.50	164.50
1400	2'-6" x 5'-0"	201	50	251
1500	4'-0" x 6'-0"	133	55.50	188.50
1600	Mill finish, 2'-0" x 3'-5"	95	46.50	141.50
1700	2'-6" x 5'-0"	119	50	169
1800	4'-0" x 6'-0"	151	55.50	206.50
1900				
2000	Door, aluminum anodized 3'-0" x 6'-8"	211	99.50	310.50
2100	White painted aluminum, 3'-0" x 6'-8"	320	99.50	419.50
2200	Mill finish, 3'-0" x 6'-8"	293	99.50	392.50
2300				
2400	Wood, storm and screen, painted	375	155	530
2500				
2600				
2700				
2800				
3100	Cut & patch to match existing construction, add, minimum	2%	3%	
3200	Maximum	5%	9%	
3300	Dust protection, add, minimum	1%	2%	
3400	Maximum	4%	11%	
3500	Material handling & storage limitation, add, minimum	1%	1%	
3600	Maximum	6%	7%	
3700	Protection of existing work, add, minimum	2%	2%	
3800	Maximum	5%	7%	
3900	Shift work requirements, add, minimum		5%	
4000	Maximum		30%	

B2020 Exterior Windows

This page illustrates and describes a window wall system including aluminum tube framing, caulking, and glass. Lines within System Components give the unit price and total price per square foot for this system. Prices for window wall systems are on Line Items B2020 124 1300 thru 2600. Both material quantities and labor costs have been adjusted for the system listed.

Factors: To adjust for job conditions other than normal working situations use Lines B2020 124 3100 thru 4000.

Example: You are to install the system with need for complete dust protection. Go to Line B2020 124 3400 and apply these percentages to the appropriate MAT. and INST. costs.

System Components	QUANTITY	UNIT	COST PER S.F.		
			MAT.	INST.	TOTAL
Window wall, including aluminum header, sill, mullions, caulking And glass					
Header, mill finish, 1-3/4" x 4-1/2" deep	.167	L.F.	3.84	2.70	6.54
Sill, mill finish, 1-3/4" x 4-1/2" deep	.167	L.F.	3.84	2.63	6.47
Vertical mullion, 1-3/4" x 4-1/2" deep, 6' O.C.	.191	L.F.	9.85	6.57	16.42
Caulking	.381	L.F.	.12	.91	1.03
Glass, 1/4" plate	1.000	S.F.	9.15	11.15	20.30
TOTAL			26.80	23.96	50.76

B2020 124	Aluminum Frame, Window Wall	COST PER S.F.		
		MAT.	INST.	TOTAL
1300	Mill finish, 1-3/4" x 4-1/2" deep, 1/4" plate glass	27	24	51
1400	2" x 4-1/2" deep, insulating glass	43.50	27.50	71
1500	Thermo-break, 2-1/4" x 4-1/2" deep, insulating glass	44.50	28	72.50
1600	Bronze finish, 1-3/4" x 4-1/2" deep, 1/4" plate glass	30	26	56
1700	2" x 4-1/2" deep, insulating glass	45.50	30	75.50
1800	Thermo-break, 2-1/4" x 4-1/2" deep, insulating glass	47	30.50	77.50
2000	Black finish, 1-3/4" x 4-1/2" deep, 1/4" plate glass	31.50	27	58.50
2100	2" x 4-1/2" deep, insulating glass	46.50	31	77.50
2200	Thermo-break, 2-1/4" x 4-1/2" deep, insulating glass	48	31.50	79.50
2300				
2400	Stainless steel, 1-3/4" x 4-1/2" deep, 1/4" plate glass	40	33	73
2500	2" x 4-1/2" deep, insulating glass	53	37.50	90.50
2600	Thermo-break, 2-1/4" x 4-1/2" deep, insulating glass	55	38	93
2700				
3100	Cut & patch to match existing construction, add, minimum	2%	3%	
3200	Maximum	5%	9%	
3300	Dust protection, add, minimum	1%	2%	
3400	Maximum	4%	11%	
3500	Material handling, & storage limitation, add, minimum	1%	1%	
3600	Maximum	6%	7%	
3700	Protection of existing work, add, minimum	2%	2%	
3800	Maximum	5%	7%	
3900	Shift work requirements, add, minimum		5%	
4000	Maximum		30%	

B2030 Exterior Doors

This page illustrates and describes a commercial metal door system, including a single aluminum and glass door, narrow stiles, jamb, hardware weatherstripping, panic hardware and closer. Lines within System Components give the unit price and total price on a cost each basis for this system. Prices for commercial metal door systems are on Line Items B2030 125 1200 thru 2500. Both material quantities and labor costs have been adjusted for the system listed.

Factors: To adjust for job conditions other than normal working situations, use Lines B2030 125 3100 thru 4000.

Example: You are to install the system and cut and patch to match existing construction. Go to Line B2030 125 3200 and apply these percentages to the appropriate MAT. and INST. costs.

System Components			COST EACH		
	QUANTITY	UNIT	MAT.	INST.	TOTAL
Single aluminum and glass door, 3'-0"x7'-0", with narrow stiles, ext. jamb.					
Weatherstripping, 1/2" tempered insul. glass, panic hardware, and closer.					
Aluminum door, 3'-0" x 7'-0" x 1-3/4", narrow stiles	1.000	Ea.	1,200	780	1,980
Tempered insulating glass, 1/2" thick	20.000	S.F.	700	490	1,190
Panic hardware	1.000	Set	885	116	1,001
Surface mount regular arm	1.000	Ea.	170	107	277
TOTAL			2,955	1,493	4,448

B2030 125	Doors, Metal - Commercial	COST EACH		
		MAT.	INST.	TOTAL
1200	Single aluminum and glass, 3'-0" x 7'-0"	2,950	1,500	4,450
1300	With transom, 3'-0" x 10'-0"	3,500	1,775	5,275
1400	Anodized aluminum and glass, 3'-0" x 7'-0"	3,375	1,775	5,150
1500	With transom, 3'-0" x 10'-0"	4,025	2,075	6,100
1600	Steel, deluxe, hollow metal 3'-0" x 7'-0"	2,250	580	2,830
1700	With transom 3'-0" x 10'-0"	2,775	660	3,435
1800	Fire door, "A" label, 3'-0" x 7'-0"	2,225	580	2,805
1900	Double, aluminum and glass, 6'-0" x 7'-0"	4,300	2,575	6,875
2000	With transom, 6'-0" x 10'-0"	5,425	3,175	8,600
2100	Anodized aluminum and glass 6'-0" x 7'-0"	4,700	3,000	7,700
2200	With transom, 6'-0" x 10'-0"	6,050	3,675	9,725
2300	Steel, deluxe, hollow metal, 6'-0" x 7'-0"	4,000	1,000	5,000
2400	With transom, 6'-0" x 10'-0"	5,050	1,175	6,225
2500	Fire door, "A" label, 6'-0" x 7'-0"	3,825	1,000	4,825
2800				
2900				
3100	Cut & patch to match existing construction, add, minimum	2%	3%	
3200	Maximum	5%	9%	
3300	Dust protection, add, minimum	1%	2%	
3400	Maximum	4%	11%	
3500	Material handling & storage limitation, add, minimum	1%	1%	
3600	Maximum	6%	7%	
3700	Protection of existing work, add, minimum	2%	2%	
3800	Maximum	5%	7%	
3900	Shift work requirements, add, minimum		5%	
4000	Maximum		30%	

B2030 Exterior Doors

This page illustrates and describes sliding door systems including a sliding door, frame, interior and exterior trim with exterior staining. Lines within System Components give the unit price and total price on a cost each basis for this system. Prices for sliding door systems are on Line Items B2030 150 1000 thru 2400. Both material quantities and labor costs have been adjusted for the system listed.

Factors: To adjust for job conditions other than normal working situations use Lines B2030 150 2700 thru 4000.

Example: You are to install the system with temporary shoring and bracing. Go to Line B2030 150 3900 and apply these percentages to the appropriate MAT. and INST. costs.

System Components	QUANTITY	UNIT	COST EACH		
			MAT.	INST.	TOTAL
Sliding wood door, 6'-0" x 6'-8", with wood frame, interior and exterior					
Trim and exterior staining.					
Sliding wood door, standard, 6'-0" x 6'-8", insulated glass	1.000	Ea.	1,350	350	1,700
Interior & exterior trim	1.000	Set	30.40	55.80	86.20
Stain door	1.000	Ea.	4.79	34.50	39.29
Stain trim	1.000	Ea.	3.20	14.80	18
TOTAL			1,388.39	455.10	1,843.49

B2030 150	Doors, Sliding - Patio	COST EACH		
		MAT.	INST.	TOTAL
1000	Wood, standard, 6'-0" x 6'-8", insulated glass	1,400	455	1,855
1100	8'-0" x 6'-8"	1,650	575	2,225
1200	12'-0" x 6'-8"	2,975	700	3,675
1300	Vinyl coated, 6'-0" x 6'-8"	1,750	455	2,205
1400	8'-0" x 6'-8"	2,225	575	2,800
1500	12'-0" x 6'-8"	3,575	700	4,275
1700	Aluminum, standard, 6'-0" x 6'-8", insulated glass	1,050	490	1,540
1800	8'-0" x 6'-8"	1,800	620	2,420
1900	12'-0" x 6'-8"	1,825	735	2,560
2000	Anodized, 6'-0" x 6'-8"	1,775	490	2,265
2100	8'-0" x 6'-8"	2,025	620	2,645
2200	12'-0" x 6'-8"	3,500	735	4,235
2300				
2400	Deduct for single glazing	114		114
2500				
2700	Cut & patch to match existing construction, add, minimum	2%	3%	
2800	Maximum	5%	9%	
2900	Dust protection, add, minimum	1%	2%	
3000	Maximum	4%	11%	
3100	Equipment usage curtailment, add, minimum	1%	1%	
3200	Maximum	3%	10%	
3300	Material handling & storage limitation, add, minimum	1%	1%	
3400	Maximum	6%	7%	
3500	Protection of existing, work, add, minimum	2%	2%	
3600	Maximum	5%	7%	
3700	Shift work requirements, add, minimum		5%	
3800	Maximum		30%	
3900	Temporary shoring and bracing, add, minimum	2%	5%	
4000	Maximum	5%	12%	

B2030 Exterior Doors

This page illustrates and describes residential door systems including a door, frame, trim, hardware, weatherstripping, stained and finished. Lines within System Components give the unit price and total price on a cost each basis for this system. Prices for residential door systems are on Line Items B2030 240 1500 thru 2200. Both material quantities and labor costs have been adjusted for the system listed.

Factors: To adjust for job conditions other than normal working situations use Lines B2030 240 3100 thru 4000.

Example: You are to install the system with a material handling and storage limitation. Go to Line B2030 240 3600 and apply these percentages to the appropriate MAT. and INST. costs.

System Components	QUANTITY	UNIT	COST EACH MAT.	COST EACH INST.	COST EACH TOTAL
Single solid wood colonial door, 3' x 6'-8" with wood frame and trim,					
Stained and finished, including hardware and weatherstripping.					
Solid wood colonial door, fir 3' x 6'-8" x 1-3/4", hinges	1.000	Ea.	680	93	773
Hinges, ball bearing	1.000	Ea.	41.50		41.50
Exterior frame, with trim	1.000	Set	155.55	63.24	218.79
Interior trim	1.000	Set	25.50	58	83.50
Average quality	1.000	Set	54.50	50	104.50
Sill, oak 8" deep	3.500	Ea.	82.25	48.83	131.08
Weatherstripping	1.000	Set	24.50	91.50	116
Stained and finished	1.000	Ea.	10.25	73.50	83.75
TOTAL			1,074.05	478.07	1,552.12

B2030 240	Doors, Residential - Exterior	COST EACH MAT.	COST EACH INST.	COST EACH TOTAL
1400	Single doors, 3' x 6'-8"			
1500	Solid colonial door, fir	1,075	480	1,555
1600	Hollow metal exterior door, plain	875	445	1,320
1700	Solid core wood door, plain	565	460	1,025
1800				
1900	Double doors, 6' x 6'-8"			
2000	Solid colonial double doors, fir	1,850	635	2,485
2100	Hollow metal exterior doors, plain	1,450	595	2,045
2200	Hollow core wood doors, plain	835	615	1,450
2300				
2800				
2900				
3100	Cut & patch to match existing construction, add, minimum	2%	3%	
3200	Maximum	5%	9%	
3300	Dust protection, add, minimum	1%	2%	
3400	Maximum	4%	11%	
3500	Material handling & storage limitation, add, minimum	1%	1%	
3600	Maximum	6%	7%	
3700	Protection of existing work, add, minimum	2%	2%	
3800	Maximum	5%	7%	
3900	Shift work requirements, add, minimum		5%	
4000	Maximum		30%	

B2030 Exterior Doors

This page illustrates and describes overhead door systems including an overhead door, track, hardware, trim, and electric door opener. Lines within System Components give the unit price and total price on a cost each basis for this system. Prices for overhead door systems are on Line Items B2030 410 1200 thru 2300. Both material quantities and labor costs have been adjusted for the system listed.

Factors: To adjust for job conditions other than normal working situations use Lines B2030 410 3100 thru 4000.

Example: You are to install the system and match existing construction. Go to Line B2030 410 3200 and apply these percentages to the appropriate MAT. and INST. costs.

System Components	QUANTITY	UNIT	COST EACH		
			MAT.	INST.	TOTAL
Wood overhead door, commercial sectional, including track, door, hardware					
And trim, electrically operated.					
Commercial, heavy duty wood door, 8' x 8' x 1-3/4" thick	1.000	Ea.	1,325	695	2,020
Frame, 2 x 8, pressure treated	1.000	Set	50.40	88.08	138.48
Wood trim	1.000	Set	73.92	69.60	143.52
Painting, two coats	1.000	Ea.	33.12	234.72	267.84
Electric trolley operator	1.000	Ea.	1,250	350	1,600
TOTAL			2,732.44	1,437.40	4,169.84

B2030 410	Doors, Overhead	COST EACH		
		MAT.	INST.	TOTAL
1200	Commercial, wood, 1-3/4" thick, 8' x 8'	2,725	1,425	4,150
1300	12' x 12'	4,575	2,025	6,600
1400	14' x 14'	5,750	2,375	8,125
1500	Fiberglass & aluminum 12' x 12'	5,275	2,075	7,350
1600	20' x 20'	10,400	5,225	15,625
1700	Residential, wood, 9' x 7'	1,275	545	1,820
1800	16' x 7'	4,075	815	4,890
1900	Hardboard faced, 9' x 7'	940	545	1,485
2000	16' x 7'	1,700	815	2,515
2100	Fiberglass & aluminum, 9' x 7'	1,175	650	1,825
2200	16' x 7'	2,625	815	3,440
2300	For residential electric opener, add	710	87	797
2400				
2500				
2600				
2700				
2800				
2900				
3100	Cut & patch to match existing construction, add, minumum	2%	3%	
3200	Maximum	5%	9%	
3300	Dust protection, add, minimum	1%	2%	
3400	Maximum	4%	11%	
3500	Material handling & storage limitation, add, minimum	1%	1%	
3600	Maximum	6%	7%	
3700	Protection of existing work, add, minimum	2%	2%	
3800	Maximum	5%	7%	
3900	Shift work requirements, add, minimum		5%	
4000	Maximum		30%	

B2030 Exterior Doors

B2030 810	Selective Price Sheet	COST EACH		
		MAT.	INST.	TOTAL
0100	Door closer, rack and pinion	141	116	257
0200	Backcheck and adjustable power	141	116	257
0300	Regular, hinge face mount, all sizes, regular arm	141	116	257
0400	Hold open arm	112	116	228
0500	Top jamb mount, all sizes, regular arm	141	116	257
0600	Hold open arm	112	116	228
0700	Stop face mount, all sizes, regular arm	141	116	257
0800	Hold open arm	112	116	228
0900	Fusible link, hinge face mount, all sizes, regular arm	208	116	324
1000	Hold open arm	179	116	295
1100	Top jamb mount, all sizes, regular arm	208	116	324
1200	Hold open arm	179	116	295
1300	Stop face mount, all sizes, regular arm	208	116	324
1400	Hold open arm	179	116	295
1500				
1600				
1700	Door stops			
1800				
1900	Holder & bumper, floor or wall	39.50	22	61.50
2000	Wall bumper	15.20	22	37.20
2100	Floor bumper	4.93	22	26.93
2200	Plunger type, door mounted	36.50	22	58.50
2300	Hinges, full mortise, material only, per pair			
2400	Low frequency, 4-1/2" x 4-1/2", steel base, USP	27.50		27.50
2500	Brass base, US10	103		103
2600	Stainless steel base, US32	88		88
2700	Average frequency, 4-1/2" x 4-1/2", steel base, USP	53		53
2800	Brass base, US10	99.50		99.50
2900	Stainless steel base, US32	79.50		79.50
3000	High frequency, 4-1/2" x 4-1/2", steel base, USP	71.50		71.50
3100	Brass base, US10	66.50		66.50
3200	Stainless steel base, US32	118		118
3300	Kick plate			
3400				
3500	6" high, for 3'-0" door, aluminum	39	46.50	85.50
3600	Bronze	74	46.50	120.50
3700	Panic device			
3800				
3900	For rim locks, single door, exit	885	116	1,001
4000	Outside key and pull	1,475	139	1,614
4100	Bar and vertical rod, exit only	1,050	139	1,189
4200	Outside key and pull	1,375	174	1,549
4300	Lockset			
4400				
4500	Heavy duty, cylindrical, passage doors	131	58	189
4600	Classroom	715	87	802
4700	Bedroom, bathroom, and inner office doors	170	58	228
4800	Apartment, office, and corridor doors	305	69.50	374.50
4900	Standard duty, cylindrical, exit doors	215	69.50	284.50
5100	Passage doors	83	58	141
5200	Public restroom, classroom, & office doors	228	87	315
5300	Deadlock, mortise, heavy duty	212	77.50	289.50
5400	Double cylinder	211	77.50	288.50
5500	Entrance lock, cylinder, deadlocking latch	197	77.50	274.50
5600	Deadbolt	254	87	341
5700	Commercial, mortise, wrought knob, keyed, minimum	216	87	303
5800	Maximum	590	99.50	689.50
5900	Cast knob, keyed, minimum	300	77.50	377.50

B2030 Exterior Doors

B2030 810	Selective Price Sheet	COST EACH		
		MAT.	INST.	TOTAL
6000	Maximum	610	77.50	687.50
6100	Push-pull			
6200				
6300	Aluminum	10.35	58	68.35
6400	Bronze	28.50	58	86.50
6500	Door pull, designer style, minimum	29	58	87
6600	Maximum	276	127	403
6700	Threshold			
6800				
6900	3'-0" long door saddles, aluminum, minimum	21.50	14.50	36
7000	Maximum	70.50	58	128.50
7100	Bronze, minimum	46.50	11.60	58.10
7200	Maximum	174	58	232
7300	Rubber, 1/2" thick, 5-1/2" wide	45	35	80
7400	2-3/4" wide	54.50	35	89.50
7500	Weatherstripping, per set			
7600				
7700	Doors, wood frame, interlocking for 3' x 7' door, zinc	59	232	291
7800	Bronze	76	232	308
7900	Wood frame, spring type for 3' x 7' door, bronze	24.50	91.50	116
8000	Metal frame, spring type for 3' x 7' door, bronze	56	232	288
8100	For stainless steel, spring type, add	133%		
8200				
8300	Metal frame, extruded sections, 3' x 7' door, aluminum	36.50	232	268.50
8400	Bronze	105	232	337

B3010 Roof Coverings

B3010 160	Selective Price Sheet	COST PER S.F.		
		MAT.	INST.	TOTAL
0100	Roofing, built-up, asphalt roll roof, 3 ply organic/mineral surface	.82	1.85	2.67
0200	3 plies glass fiber felt type iv, 1 ply mineral surface	1.45	1.98	3.43
0300	Cold applied, 3 ply		.64	.64
0400	Coal tar pitch, 4 ply tarred felt	2.26	2.36	4.62
0500	Mopped, 3 ply glass fiber	1.86	2.61	4.47
0600	4 ply organic felt	2.26	2.36	4.62
0700	Elastomeric, hypalon, neoprene unreinforced	1.45	4.38	5.83
0800	Polyester reinforced	1.58	5.20	6.78
0900	Neoprene, 5 coats 60 mils	4.84	15.30	20.14
1000	Over 10,000 S.F.	4.84	7.95	12.79
1600				
1700	Asphalt, strip, 210-235#/sq.	.83	1.21	2.04
1800	235-240#/sq.	1.26	1.33	2.59
1900	Class A laminated	1.25	1.47	2.72
2000	Class C laminated	1.25	1.66	2.91
2100	Slate, buckingham, 3/16" thick	6.10	3.80	9.90
2200	Black, 1/4" thick	6.10	3.80	9.90
2300	Wood, shingles, 16" no. 1, 5" exp.	3.60	2.79	6.39
2400	Red cedar, 18" perfections	3.25	2.53	5.78
2500	Shakes, 24", 10" exposure	4.15	2.79	6.94
2600	18", 8-1/2" exposure	2.74	3.50	6.24
2700	Insulation, ceiling batts, fiberglass, 3-1/2" thick, R13	.39	.52	.91
2800	6" thick, R19	.51	.52	1.03
2900	9" thick, R30	1	.61	1.61
3000	12" thick, R38	1.24	.61	1.85
3100	Mineral fiber, 3-1/2" thick, R13	.92	.44	1.36
3200	6" thick, R19	1.45	.44	1.89
3300	Roof deck, fiberboard, 1" thick, R2.78	.96	.85	1.81
3400	Mineral fiber, 2" thick, R5.26	1.33	.85	2.18
3500	Perlite boards, 3/4" thick, R2.08	.55	.85	1.40
3600	2" thick, R5.26	1.30	.94	2.24
3700	Polystyrene extruded, 15 PSI, 1" thick, R5.26	.69	.55	1.24
3800	2" thick, R10	.86	.62	1.48
3900	40 PSI, 1" thick, R5	1.07	.55	1.62
4000	Tapered for drainage	1.08	.57	1.65
4300	Ceiling, plaster, gypsum, 2 coats	.49	3.81	4.30
4400	3 coats	.69	4.51	5.20
4500	Perlite or vermiculite, 2 coats	.73	4.45	5.18
4600	3 coats	.82	5.55	6.37
4700	Gypsum lath, plain 3/8" thick	.41	.88	1.29
4800	1/2" thick	.35	.94	1.29
4900	Firestop, 3/8" thick	.30	1.07	1.37
5000	1/2" thick	.34	1.15	1.49
5100	Metal lath, rib, 2.75 lb.	.47	1	1.47
5200	3.40 lb.	.59	1.07	1.66
5300	Diamond, 2.50 lb.	.67	1	1.67
5400	3.40 lb.	.50	1.25	1.75
5500	Drywall, taped and finished standard, 1/2" thick	.42	1.82	2.24
5600	5/8" thick	.45	1.82	2.27
5700	Fire resistant, 1/2" thick	.46	1.82	2.28
5800	5/8" thick	.47	1.82	2.29
5900	Water resist., 1/2" thick	.46	1.82	2.28
6000	5/8" thick	.53	1.82	2.35
6100	Finish, instead of taping			
6200	For thin coat plaster, add	.11	.87	.98
6300	Finish textured spray, add	.04	.84	.88
6500	Tile, stapled glued, mineral fiber plastic coated, 5/8" thick	2.42	2.32	4.74
6600	3/4" thick	4.15	2.32	6.47

643

B3010 Roof Coverings

B3010 160	Selective Price Sheet	COST PER S.F.		
		MAT.	INST.	TOTAL
6700	Wood fiber, 1/2" thick	1.46	1.74	3.20
6800	3/4" thick	1.43	1.74	3.17
6900	Suspended, fiberglass film faced, 5/8" thick	2.50	1.11	3.61
7000	3" thick	4.39	1.55	5.94
7100	Mineral fiber 5/8" thick, standard face	1.24	1.03	2.27
7200	Aluminum faced	4.18	1.16	5.34
7500	Ceiling suspension systems, for tile, "T" bar, class "A", 2' x 4' grid	1.43	.87	2.30
7600	2' x 2' grid	1.75	1.07	2.82
7700	Concealed "Z" bar, 12" module	1.05	1.34	2.39
7800				
7900	Plaster/drywall, 3/4" channels, steel furring, 16" O.C.	.42	2.33	2.75
8000	24" O.C.	.28	1.61	1.89
8100	1-1/2" channels, 16" O.C.	.56	2.60	3.16
8200	24" O.C.	.38	1.73	2.11
8300	Ceiling framing, 2" x 4" studs, 16" O.C.	.58	1.16	1.74
8400	24" O.C.	.39	.78	1.17

B30 Roofing

B3020 Roof Openings

This page illustrates and describes a roof hatch system. Lines within System Components give the unit price and total cost each for this system. Prices for systems are on Lines B3020 230 1400 thru 2400. Both material quantities and labor costs have been adjusted for the system listed.

Factors: To adjust for job conditions other than normal working situations use Lines B3020 230 2900 thru 4000.

Example: You are to install the system and cut and patch to match existing construction. Use line B3020 230 3000 and apply these percentages to the appropriate MAT. and INST. costs.

System Components			COST EACH		
	QUANTITY	UNIT	MAT.	INST.	TOTAL
Roof hatch, 2'-6" x 3'-0", aluminum, curb and cover included.					
Through steel construction.					
Roof hatch, 2'-6" x 3'-0", aluminum	1.000	Ea.	985	276	1,261
Cutout decking	11.000	L.F.	12.65	28.16	40.81
Frame opening	44.000	L.F.	38.28	421.08	459.36
Flashing, 16 oz. copper	16.000	S.F.	139.20	92	231.20
Cant strip, 4 x 4, mineral fiber	12.000	L.F.	7.56	24.48	32.04
TOTAL			1,182.69	841.72	2,024.41

B3020 230	Roof Hatches, Skylights	COST EACH		
		MAT.	INST.	TOTAL
1400	Roof hatch, aluminum, curb and cover, 2'-6" x 3'-0"	1,175	840	2,015
1500	2'-6" x 4'-6"	1,525	1,025	2,550
1600	2'-6" x 8'-0"	3,000	1,475	4,475
1700	Skylight, plexiglass dome, with curb mounting, 30" x 32"	705	795	1,500
1800	30" x 45"	840	995	1,835
1900	40" x 45"	995	1,325	2,320
2000	Smoke hatch, unlabeled, 2'-6" x 3'-0"	1,425	910	2,335
2100	2'-6" x 4'-6"	1,850	1,100	2,950
2200	2'-6" x 8'-0"	3,250	1,525	4,775
2300				
2400	For steel ladder, add per vertical linear foot	46	39	85
2450				
2600				
2700				
2900	Cut & patch to match existing construction, add, minimum	2%	3%	
3000	Maximum	5%	9%	
3100	Dust protection, add, minimum	1%	2%	
3200	Maximum	4%	11%	
3300	Equipment usage curtailment, add, minimum	1%	1%	
3400	Maximum	3%	10%	
3500	Material handling & storage limitation, add, minimum	1%	1%	
3600	Maximum	6%	7%	
3700	Protection of existing work, add, minimum	2%	2%	
3800	Maximum	5%	7%	
3900	Shift work requirements, add, minimum		5%	
4000	Maximum		30%	

645

B3020 Roof Openings

B3020 240	Selective Price Sheet	COST PER UNIT		
		MAT.	INST.	TOTAL
0100	Downspouts per L.F., aluminum, enameled .024" thick, 2" x 3"	2.45	4.53	6.98
0200	3" x 4"	2.53	5.85	8.38
0300	Round .025" thick, 3" diam.	2.12	4.29	6.41
0400	4" diam.	7.55	5.85	13.40
0500	Copper, round 16 oz. stock, 2" diam.	8.35	4.29	12.64
0600	3" diam.	11.45	4.29	15.74
0700	4" diam.	11.95	5.65	17.60
0800	5" diam.	15.85	6.30	22.15
0900	Rectangular, 2" x 3"	11.45	4.29	15.74
1000	3" x 4"	17.25	5.65	22.90
1100	Lead coated copper, round, 2" diam.	29.50	4.29	33.79
1200	3" diam.	28.50	4.29	32.79
1300	4" diam.	31.50	5.65	37.15
1400	5" diam.	29	6.30	35.30
1500	Rectangular, 2" x 3"	29	4.29	33.29
1600	3" x 4"	46.50	5.65	52.15
1700	Steel galvanized, round 28 gauge, 3" diam.	3.10	4.29	7.39
1800	4" diam.	2.30	5.65	7.95
1900	5" diam.	4.26	6.30	10.56
2000	6" diam.	4.11	7.75	11.86
2100	Rectangular, 2" x 3"	4.38	4.29	8.67
2200	3" x 4"	5.60	5.65	11.25
2300	Elbows, aluminum, round, 3" diam.	3.27	8.15	11.42
2400	4" diam.	7.95	8.15	16.10
2500	Rectangular, 2" x 3"	1.19	8.15	9.34
2600	3" x 4"	3.21	8.15	11.36
2700	Copper, round 16 oz., 2" diam.	11.95	8.15	20.10
2800	3" diam.	13.55	8.15	21.70
2900	4" diam.	14.40	8.15	22.55
3100	Rectangular, 2" x 3"	13.10	8.15	21.25
3200	3" x 4"	16.85	8.15	25
3300	Drip edge per L.F., aluminum, 5" wide	.88	1.74	2.62
3400	8" wide	1.54	1.74	3.28
3500	28" wide	8.75	6.95	15.70
3600				
3700	Steel galvanized, 5" wide	.76	1.74	2.50
3800	8" wide	.98	1.74	2.72
3900				
4000				
4100				
4200				
4300	Flashing 12" wide per S.F., aluminum, mill finish, .013" thick	1.43	4.58	6.01
4400	.019" thick	1.72	4.58	6.30
4500	.040" thick	2.62	4.58	7.20
4600	.050" thick	3.58	4.58	8.16
4700	Copper, mill finish, 16 oz.	8.70	5.75	14.45
4800	20 oz.	12.35	6.05	18.40
4900	24 oz.	17.15	6.30	23.45
5000	32 oz.	25	6.65	31.65
5100	Lead, 2.5 lb./S.F., 12" wide	6.95	4.92	11.87
5200	Over 12" wide	4.77	4.92	9.69
5900	Polyvinyl chloride, black, .010" thick	.30	2.33	2.63
6000	.020" thick	.29	2.33	2.62
6100	.030" thick	.36	2.33	2.69
6200	.056" thick	.97	2.33	3.30
6300	Steel, galvanized, 20 gauge	1.36	5.10	6.46
6400	30 gauge	.68	4.15	4.83
6500	Stainless, 32 gauge, .010" thick	3.91	4.28	8.19

B3020 Roof Openings

B3020 240	Selective Price Sheet	COST PER UNIT		
		MAT.	INST.	TOTAL
6600	28 gauge, .015" thick	5.60	4.28	9.88
6700	26 gauge, .018" thick	5.75	4.28	10.03
6800	24 gauge, .025" thick	6.55	4.28	10.83
6900	Gutters per L.F., aluminum, 5" box, .027" thick	3.36	6.75	10.11
7000	.032" thick	3.80	6.55	10.35
7100	Copper, half round, 4" wide	10.55	6.55	17.10
7200	6" wide	11.45	6.90	18.35
7300	Steel, 26 gauge galvanized, 5" wide	2.86	6.55	9.41
7400	6" wide	2.92	6.55	9.47
7500	Wood, treated hem-fir, 3" x 4"	11.95	6.95	18.90
7600	4" x 5"	26.50	6.95	33.45
7700	Reglet per L.F., aluminum, .025" thick	2.09	3.10	5.19
7800	Copper, 10 oz.	8.05	3.10	11.15
7900	Steel, galvanized, 24 gauge	1.66	3.10	4.76
8000	Stainless, .020" thick	4.53	3.10	7.63
8100	Counter flashing 12" wide per L.F., aluminum, .032" thick	2.21	5.45	7.66
8200	Aluminum, .025" thick	2.21	5.45	7.66
8300	Steel, galvanized, 24 gauge	1.54	5.45	6.99
8400	Stainless, .020" thick	7.55	5.45	13

For customer support on your Commercial Renovation Costs with RSMeans Data, call 800.448.8182.

C1010 Partitions

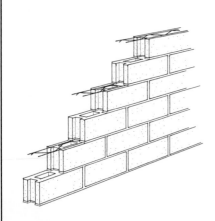

This page illustrates and describes a concrete block wall system including concrete block, horizontal reinforcing alternate courses, mortar, and tooled joints both sides. Lines within System Components give the unit price and total price per square foot for this system. Prices for concrete block wall systems are on Line Items C1010 106 0800 thru 2100. Both material quantities and labor costs have been adjusted for the system listed.

Factors: To adjust for job conditions other than normal working situations use Lines C1010 106 2900 thru 4000.

Example: You are to install the system and protect all existing work. Go to Line C1010 106 3600 and apply these percentages to the appropriate MAT. and INST. costs.

System Components	QUANTITY	UNIT	COST PER S.F.		
			MAT.	INST.	TOTAL
Concrete block partition, including horizontal reinforcing every second Course, mortar, tooled joints both sides.					
8" x 16" concrete block, normal weight, 4" thick	1.000	S.F.	2.30	7.45	9.75
Horizontal reinforcing every second course	.750	L.F.	.22	.17	.39
TOTAL			2.52	7.62	10.14

C1010 106	Partitions, Concrete Block	COST PER S.F.		
		MAT.	INST.	TOTAL
0800	8" x 16" concrete block, normal weight, 4" thick	2.52	7.60	10.12
0900	6" thick	3.20	8.15	11.35
1000	8" thick	3.12	8.65	11.77
1100	10" thick	3.73	9	12.73
1200	12" thick	5.30	11.20	16.50
1300	8" x 16" concrete block, lightweight, 4" thick	2.42	7.40	9.82
1400	6" thick	3.63	7.95	11.58
1500	8" thick	4.59	8.45	13.04
1600	10" thick	4.86	8.80	13.66
1700	12" thick	4.64	10.90	15.54
1800	8" x 16" glazed concrete block, 4" thick	14.05	9.40	23.45
1900	8" thick	18.40	10.45	28.85
2000	Structural facing tile, 6T series, glazed 2 sides, 4" thick	19.40	16.55	35.95
2100	6" thick	24.50	17.40	41.90
2200				
2300				
2400				
2500				
2600				
2900	Cut & patch to match existing construction, add, minimum	2%	3%	
3000	Maximum	5%	9%	
3100	Dust protection, add, minimum	1%	2%	
3200	Maximum	4%	11%	
3300	Material handling & storage limitation, add, minimum	1%	1%	
3400	Maximum	6%	7%	
3500	Protection of existing work, add, minimum	2%	2%	
3600	Maximum	5%	7%	
3700	Shift work requirements, add, minimum		5%	
3800	Maximum		30%	
3900	Temporary shoring and bracing, add, minimum	2%	5%	
4000	Maximum	5%	12%	

This page illustrates and describes a wood stud partition system including wood studs with plates, gypsum plasterboard – taped and finished, insulation, baseboard and painting. Lines within System Components give the unit price and total price per square foot for this system. Prices for wood stud partition systems are on Line Items C1010 132 1300 thru 2700. Both material quantities and labor costs have been adjusted for the system listed.

Factors: To adjust for job conditions other than normal working situations use Lines C1010 132 2900 thru 4000.

Example: You are to install the system where material handling and storage present a serious problem. Go to Line C1010 132 3400 and apply these percentages to the appropriate MAT. and INST. costs.

System Components	QUANTITY	UNIT	COST PER S.F.		
			MAT.	INST.	TOTAL
Wood stud wall, 2"x4", 16" O.C., dbl. top plate, sngl bot. plate, 5/8" dwl.					
taped, finished and painted on 2 faces, insulation, baseboard, wall 8' high.					
Wood studs, 2" x 4", 16" O.C., 8' high	1.000	S.F.	.86	1.74	2.60
Gypsum drywall, 5/8" thick	2.000	S.F.	.80	1.40	2.20
Taping and finishing	2.000	S.F.	.10	1.40	1.50
Insulation, 3-1/2" fiberglass batts	1.000	S.F.	.61	.52	1.13
Baseboard	.250	L.F.	.77	.73	1.50
Paint baseboard, primer + 2 coats	.250	L.F.	.06	.43	.49
Painting, roller, 2 coats	2.000	S.F.	.46	1.54	2
TOTAL			3.66	7.76	11.42

C1010 132	Partitions, Wood Stud	COST PER S.F.		
		MAT.	INST.	TOTAL
1300	2" x 3" studs, 8' high, 16" O.C.	3.61	7.65	11.26
1400	24" O.C.	3.43	7.35	10.78
1500	10' high, 16" O.C.	3.57	7.35	10.92
1600	24" O.C.	3.38	7.10	10.48
1650	2" x 4" studs, 8' high, 16" O.C.	3.66	7.75	11.41
1700	24" O.C.	3.47	7.40	10.87
1800	10' high, 16" O.C.	3.61	7.40	11.01
1900	24" O.C.	3.41	7.15	10.56
2000	12' high, 16" O.C.	3.54	7.40	10.94
2100	24" O.C.	3.35	7.15	10.50
2200	2" x 6" studs, 8' high, 16" O.C.	4.12	7.95	12.07
2300	24" O.C.	3.82	7.55	11.37
2400	10' high, 16" O.C.	4.03	7.55	11.58
2500	24" O.C.	3.73	7.25	10.98
2600	12' high, 16" O.C.	3.92	7.60	11.52
2700	24" O.C.	3.64	7.25	10.89
2900	Cut & patch to match existing construction, add, minimum	2%	3%	
3000	Maximum	5%	9%	
3100	Dust protection, add, minimum	1%	2%	
3200	Maximum	4%	11%	
3300	Material handling & storage limitation, add, minimum	1%	1%	
3400	Maximum	6%	7%	
3500	Protection of existing work, add, minimum	2%	2%	
3600	Maximum	5%	7%	
3700	Shift work requirements, add, minimum		5%	
3800	Maximum		30%	
3900	Temporary shoring and bracing, add, minimum	2%	5%	
4000	Maximum	5%	12%	

C1010 Partitions

This page illustrates and describes a non-load bearing metal stud partition system including metal studs with runners, gypsum plasterboard, taped and finished, insulation, baseboard and painting. Lines within System Components give the unit price and total price per square foot for this system. Prices for non-load bearing metal stud partition systems are on Line Items C1010 134 1300 thru 2300. Both material quantities and labor costs have been adjusted for the system listed.

Factors: To adjust for job conditions other than normal working situations use Lines C1010 134 2900 thru 4000.

Example: You are to install the system and cut and patch to match existing construction. Go to Line C1010 134 3000 and apply these percentages to the appropriate MAT. and INST. costs.

System Components	QUANTITY	UNIT	COST PER S.F.		
			MAT.	INST.	TOTAL
Non-load bearing metal studs, including top & bottom runners, 5/8″ drywall,					
Taped, finished and painted 2 faces, insulation, painted baseboard.					
Metal studs, 25 ga., 3-5/8″ wide, 24″ O.C.	1.000	S.F.	.39	.94	1.33
Gypsum drywall, 5/8″ thick	2.000	S.F.	.80	1.40	2.20
Taping & finishing	2.000	S.F.	.10	1.40	1.50
Insulation, 3-1/2″ fiberglass batts	1.000	S.F.	.61	.52	1.13
Baseboard	.200	L.F.	.62	.58	1.20
Paint baseboard, primer + 2 coats	.200	L.F.	.05	.35	.40
Painting, roller work, 2 coats	2.000	S.F.	.46	1.54	2
TOTAL			3.03	6.73	9.76

C1010 134	Partitions, Metal Stud, NLB	COST PER S.F.		
		MAT.	INST.	TOTAL
1300	Non-load bearing, 25 ga., 24″ O.C., 2-1/2″ wide	2.95	6.70	9.65
1350	3-5/8″ wide	3.02	6.75	9.77
1400	6″ wide	3.09	6.75	9.84
1500	16″ O.C., 2-1/2″ wide	3.03	6.95	9.98
1600	3-5/8″ wide	3.12	6.95	10.07
1700	6″ wide	3.21	7	10.21
1800	20 ga., 24″ O.C., 2-1/2″ wide	2.93	6.95	9.88
1900	3-5/8″ wide	3.21	6.95	10.16
2000	6″ wide	3.33	7	10.33
2100	16″ O.C., 2-1/2″ wide	3.01	7.25	10.26
2200	3-5/8″ wide	3.36	7.25	10.61
2300	6″ wide	3.51	7.30	10.81
2400				
2500				
2600				
2700				
2900	Cut & patch to match existing construction, add, minimum	2%	3%	
3000	Maximum	5%	9%	
3100	Dust protection, add, minimum	1%	2%	
3200	Maximum	4%	11%	
3300	Material handling & storage limitation, add, minimum	1%	1%	
3400	Maximum	6%	7%	
3500	Protection of existing work, add, minimum	2%	2%	
3600	Maximum	5%	7%	
3700	Shift work requirements, add, minimum		5%	
3800	Maximum		30%	
3900	Temporary shoring and bracing, add, minimum	2%	5%	
4000	Maximum	5%	12%	

C1010 Partitions

This page illustrates and describes a drywall system including gypsum plasterboard, taped and finished, metal studs with runners, insulation, baseboard and painting. Lines within System Components give the unit price and total price per square foot for this system. Prices for drywall systems are on Line Items C1010 136 1200 thru 1900. Both material quantities and labor costs have been adjusted for the system listed.

Factors: To adjust for job conditions other than normal working situations use Lines C1010 136 2900 thru 4000.

Example: You are to install the system and control dust in the work area. Go to Line C1010 136 3100 and apply these percentages to the appropriate MAT. and INST. costs.

System Components	QUANTITY	UNIT	COST PER S.F. MAT.	COST PER S.F. INST.	COST PER S.F. TOTAL
Gypsum drywall, taped, finished and painted 2 faces, galvanized metal studs					
Including top & bottom runners, insulation, painted baseboard, wall 10'high					
Gypsum drywall, 5/8" thick, standard	2.000	S.F.	.80	1.40	2.20
Taping and finishing	2.000	S.F.	.10	1.40	1.50
Metal studs, 20 ga., 3-5/8" wide, 24" O.C.	1.000	S.F.	.54	1.18	1.72
Insulation, 3-1/2" fiberglass batts	1.000	S.F.	.61	.52	1.13
Baseboard	.200	L.F.	.62	.58	1.20
Paint baseboard, primer + 2 coats	.200	L.F.	.03	.15	.18
Painting, roller 2 coats	2.000	S.F.	.36	1.48	1.84
TOTAL			3.06	6.71	9.77

C1010 136	Partitions, Drywall	COST PER S.F. MAT.	COST PER S.F. INST.	COST PER S.F. TOTAL
1200	Gypsum drywall, 5/8" thick, standard	3.06	6.70	9.76
1300	Fire resistant	2.95	6.45	9.40
1400	Water resistant	3.05	6.45	9.50
1500	1/2" thick, standard	2.83	6.45	9.28
1600	Fire resistant	2.93	6.45	9.38
1700	Water resistant	2.93	6.45	9.38
1800	3/8" thick, vinyl faced, standard	3.27	8.15	11.42
1900	5/8" thick, vinyl faced, fire resistant	4.55	8.15	12.70
2000				
2100				
2200				
2300				
2400				
2500				
2600				
2700				
2900	Cut & patch to match existing construction, add, minimum	2%	3%	
3000	Maximum	5%	9%	
3100	Dust protection, add, minimum	1%	2%	
3200	Maximum	4%	11%	
3300	Material handling & storage limitation, add, minimum	1%	1%	
3400	Maximum	6%	7%	
3500	Protection of existing work, add, minimum	2%	2%	
3600	Maximum	5%	7%	
3700	Shift work requirements, add, minimum		5%	
3800	Maximum		30%	
3900	Temporary shoring and bracing, add, minimum	2%	5%	
4000	Maximum	5%	12%	

C1010 Partitions

This page illustrates and describes a load bearing metal stud wall system including metal studs, sheetrock–taped and finished, insulation, baseboard and painting. Lines within System Components give the unit price and total price per square foot for this system. Prices for load bearing metal stud wall systems are on Line Items C1010 138 1500 thru 2500. Both material quantities and labor costs have been adjusted for the system listed.

Factors: To adjust for job conditions other than normal working situations use Lines C1010 138 2900 thru 4000.

Example: You are to install the system using temporary shoring and bracing. Go to Line C1010 138 3900 and apply these percentages to the appropriate MAT. and INST. costs.

System Components	QUANTITY	UNIT	COST PER S.F. MAT.	COST PER S.F. INST.	COST PER S.F. TOTAL
Load bearing, 18 ga., 3-5/8″, galvanized metal studs, 24″ O.C.,including Top and bottom runners, 1/2″ drywall, taped, finished and painted 2 Faces, 3″ insulation, and painted baseboard, wall 10′ high.					
Metal studs, 24″ OC, 18 ga., 3-5/8″ wide, galvanized	1.000	S.F.	.79	1.33	2.12
Gypsum drywall, 1/2″ thick	2.000	S.F.	.72	1.40	2.12
Taping and finishing	2.000	S.F.	.10	1.40	1.50
Insulation, 3-1/2″ fiberglass batts	1.000	S.F.	.61	.52	1.13
Baseboard	.200	L.F.	.62	.58	1.20
Paint baseboard, primer + 2 coats	.200	L.F.	.05	.35	.40
Painting, roller 2 coats	2.000	S.F.	.46	1.54	2
TOTAL			3.35	7.12	10.47

C1010 138	Partitions, Metal Stud, LB	COST PER S.F. MAT.	COST PER S.F. INST.	COST PER S.F. TOTAL
1500	Load bearing, 18 ga., 24″ O.C., 2-1/2″ wide	3.21	7.10	10.31
1550	3-5/8″ wide	3.34	7.10	10.44
1600	6″ wide	3.61	7.15	10.76
1700	16″ O.C. 2-1/2″ wide	3.38	7.40	10.78
1800	3-5/8″ wide	3.54	7.45	10.99
1900	6″ wide	3.87	7.50	11.37
2000	16 ga., 24″ O.C., 2-1/2″ wide	3.40	7.25	10.65
2100	3-5/8″ wide	3.52	7.30	10.82
2200	6″ wide	3.88	7.35	11.23
2300	16″ O.C., 2-1/2″ wide	3.61	7.65	11.26
2400	3-5/8″ wide	3.76	7.70	11.46
2500	6″ wide	4.21	7.75	11.96
2600				
2700				
2900	Cut & patch to match existing construction, add, minimum	2%	3%	
3000	Maximum	5%	9%	
3100	Dust protection, add, minumum	1%	2%	
3200	Maximum	4%	11%	
3300	Material handling & storage limitation, add, minimum	1%	1%	
3400	Maximum	6%	7%	
3500	Protection of existing work, add, minimum	2%	2%	
3600	Maximum	5%	7%	
3700	Shift work requirements, add, minimum		5%	
3800	Maximum		30%	
3900	Temporary shoring and bracing, add, minimum	2%	5%	
4000	Maximum	5%	12%	

This page illustrates and describes a plaster and lath system including gypsum plaster, gypsum lath, wood studs with plates, insulation, baseboard and painting. Lines within System Components give the unit price and price per square foot for this system. Prices for plaster and lath systems are on Line Items C1010 148 1400 thru 2500. Both material quantities and labor costs have been adjusted for the system listed.

Factors: To adjust for job conditions other than normal working situations use Lines C1010 148 2900 thru 4000.

Example: You are to install the system during evening hours only. Go to Line C1010 148 3800 and apply this percentage to the appropriate INST. costs.

System Components	QUANTITY	UNIT	COST PER S.F.		
			MAT.	INST.	TOTAL
Gypsum plaster, 2 coats, over 3/8" lath, 2 faces, 2" x 4" wood Stud partition, 24"O.C. incl. double top plate, single bottom plate, 3-1/2" Insulation, baseboard and painting.					
Gypsum plaster, 2 coats	.220	S.Y.	.97	6.64	7.61
Lath, gypsum, 3/8" thick	.220	S.Y.	.81	1.75	2.56
Wood studs, 2" x 4", 24" O.C.	1.000	S.F.	.62	1.12	1.74
Insulation, 3-1/2" fiberglass batts	1.000	S.F.	.61	.52	1.13
Baseboard, 9/16" x 3-1/2"	.200	L.F.	.62	.58	1.20
Paint baseboard, primer + 2 coats	.200	L.F.	.03	.15	.18
Paint, 2 coats	2.000	S.F.	.46	1.54	2
TOTAL			4.12	12.30	16.42

C1010 148	Partitions, Plaster & Lath	COST PER S.F.		
		MAT.	INST.	TOTAL
1400	Gypsum plaster, 2 coats	4.11	12.30	16.41
1500	3 coats	4.51	13.65	18.16
1600	Perlite plaster, 2 coats	4.58	13.20	17.78
1700	3 coats	4.76	15.05	19.81
1800				
1900				
2000	For alternate lath systems:			
2100	Gypsum lath, 1/2" thick	4.47	13.30	17.77
2200	Foil back, 3/8" thick	4.58	13.45	18.03
2300	1/2" thick	4.56	13.60	18.16
2400	Metal lath, 2.5 Lb. diamond	5.10	13.20	18.30
2500	3.4 Lb. diamond	4.76	13.45	18.21
2600				
2700				
2900	Cut & patch to match existing construction, add, minimum	2%	3%	
3000	Maximum	5%	9%	
3100	Dust protection, add, minimum	1%	2%	
3200	Maximum	4%	11%	
3300	Material handling & storage limitation, add, minimum	1%	1%	
3400	Maximum	6%	7%	
3500	Protection of existing work, add, minimum	2%	2%	
3600	Maximum	5%	7%	
3700	Shift work requirements, add, minimum		5%	
3800	Maximum		30%	
3900	Temporary shoring and bracing, add, minimum	2%	5%	
4000	Maximum	5%	12%	

655

C1010 Partitions

C1010 170	Selective Price Sheet	COST PER S.F.		
		MAT.	**INST.**	**TOTAL**
0100	Studs			
0200				
0300	24" O.C. metal, 10' high wall, including			
0400	Top and bottom runners			
0500	Non load bearing, galvanized 25 Ga., 1-5/8" wide	.37	.92	1.29
0600	2-1/2" wide	.32	.93	1.25
0700	3-5/8" wide	.39	.94	1.33
0800	4" wide	.35	.94	1.29
0900	6" wide	.46	.96	1.42
1000				
1100	Galvanized 20 Ga., 2-1/2" wide	.30	1.16	1.46
1200	3-5/8" wide	.54	1.18	1.72
1300	4" wide	.58	1.18	1.76
1400	6" wide	.70	1.20	1.90
1500	Load bearing, painted 18 Ga., 2-1/2" wide	.66	1.30	1.96
1600	3-5/8" wide	.79	1.33	2.12
1700	4" wide	.69	1.36	2.05
1800	6" wide	1.06	1.38	2.44
1900	Galvanized 18 Ga., 2-1/2" wide	.66	1.30	1.96
2000	3-5/8" wide	.79	1.33	2.12
2100	4" wide	.69	1.36	2.05
2200	6" wide	1.06	1.38	2.44
2300	Galvanized 16 Ga., 2-1/2" wide	.85	1.49	2.34
2400	3-5/8" wide	.97	1.52	2.49
2500	4" wide	1.04	1.55	2.59
2600	6" wide	1.33	1.59	2.92
2700				
2800				
2900	24" O.C. wood, 10' high wall, including			
3000	Double top plate and shoe			
3100	2" x 4"	.73	1.32	2.05
3200	2" x 6"	1.08	1.40	2.48
3300				
3400				
3500	Furring, 24" O.C. 10' high wall, metal, 3/4" channels	.28	1.61	1.89
3600	1-1/2" channels	.38	1.73	2.11
3700	Wood, on wood, 1" x 2" strips	.23	.64	.87
3800	1" x 3" strips	.34	.64	.98
3900	On masonry, 1" x 2" strips	.25	.71	.96
4000	1" x 3" strips	.37	.71	1.08
4100	On concrete, 1" x 2" strips	.25	1.34	1.59
4200	1" x 3" strips	.37	1.34	1.71
4300	Studs			
4400				
4500	16" O.C., metal, 10' high wall, including			
4600	Top and bottom runners			
4700	Non load bearing, galvanized 25 Ga., 1-5/8" wide	.51	1.41	1.92
4800	2-1/2" wide	.43	1.42	1.85
4900	3-5/8" wide	.52	1.45	1.97
5000	4" wide	.47	1.47	1.94
5100	6" wide	.62	1.48	2.10
5200				
5300	Galvanized 20 Ga., 2-1/2" wide	.40	1.79	2.19
5400	3-5/8" wide	.71	1.81	2.52
5500	4" wide	.78	1.83	2.61
5600	6" wide	.95	1.86	2.81
5700	Load bearing, painted 18 Ga., 2-1/2" wide	.89	1.81	2.70
5800	3-5/8" wide	1.06	1.84	2.90

C1010 Partitions

C1010 170	Selective Price Sheet	COST PER S.F.		
		MAT.	INST.	TOTAL
5900	4" wide	1.08	1.89	2.97
6000	6" wide	1.41	1.91	3.32
6100	Galvanized 18 Ga., 2-1/2" wide	.89	1.81	2.70
6200	3-5/8" wide	1.06	1.84	2.90
6300	4" wide	1.08	1.89	2.97
6400	6" wide	1.41	1.91	3.32
6500	Galvanized 16 Ga., 2-1/2" wide	1.16	2.05	3.21
6600	3-5/8" wide	1.34	2.10	3.44
6700	4" wide	1.44	2.15	3.59
6800	6" wide	1.82	2.20	4.02
6900				
7000				
7100	16" O.C., wood, 10' high wall, including			
7200	Double top plate and shoe			
7300	2" x 4"	.88	1.52	2.40
7400	2" x 6"	1.08	1.36	2.44
7500				
7600				
7700	Furring, 16" O.C. 10' high wall, metal, 3/4" channels	.42	2.33	2.75
7800	1-1/2" channels	.56	2.60	3.16
7900	Wood, on wood, 1" x 2" strips	.35	.95	1.30
8000	1" x 3" strips	.51	.95	1.46
8100	On masonry, 1" x 2" strips	.38	1.06	1.44
8200	1" x 3" strips	.56	1.06	1.62
8300	On concrete, 1" x 2" strips	.38	2.01	2.39
8400	1" x 3" strips	.56	2.01	2.57

C1010 Partitions

C1010 180	Selective Price Sheet, Drywall & Plaster	COST PER S.F.		
		MAT.	INST.	TOTAL
0100	Lath, gypsum perforated			
0200				
0300	Lath, gypsum, perforated, regular, 3/8" thick	.41	.88	1.29
0400	1/2" thick	.35	1.07	1.42
0500	Fire resistant, 3/8" thick	.30	1.07	1.37
0600	1/2" thick	.34	1.15	1.49
0700	Foil back, 3/8" thick	.41	1	1.41
0800	1/2" thick	.40	1.07	1.47
0900				
1000	Metal lath			
1100	Metal lath, diamond, painted, 2.5 lb.	.67	.88	1.55
1200	3.4 lb.	.50	1	1.50
1300	Rib painted, 2.75 lb	.47	1	1.47
1400	3.40 lb	.59	1.07	1.66
1500				
1600				
1700	Plaster, gypsum, 2 coats	.49	3.35	3.84
1800	3 coats	.69	4.04	4.73
1900	Perlite/vermiculite, 2 coats	.73	3.81	4.54
2000	3 coats	.82	4.74	5.56
2100	Bondcrete, 1 coat	.53	1.76	2.29
2200				
2500				
2600				
2700	Drywall, standard, 3/8" thick, no finish included	.39	.70	1.09
2800	1/2" thick, no finish included	.36	.70	1.06
2900	Taped and finished	.42	1.44	1.86
3000	5/8" thick, no finish included	.40	.70	1.10
3100	Taped and finished	.45	1.44	1.89
3200	Fire resistant, 1/2" thick, no finish included	.41	.70	1.11
3300	Taped and finished	.46	1.44	1.90
3400	5/8" thick, no finish included	.42	.70	1.12
3500	Taped and finished	.47	1.44	1.91
3600	Water resistant, 1/2" thick, no finish included	.41	.70	1.11
3700	Taped and finished	.46	1.44	1.90
3800	5/8" thick, no finish included	.47	.70	1.17
3900	Taped and finished	.53	1.44	1.97
4000	Finish, instead of taping			
4100	For thin coat plaster, add	.11	.87	.98
4200	Finish, textured spray, add	.04	.84	.88

C1010 Partitions

This page illustrates and describes a movable office partition system including demountable office partitions of various styles and sizes based on a cost per square foot basis. Prices for movable office partition systems are on Line Items C1010 210 0300 thru 3600.

C1010 210	Partitions, Movable Office	COST PER S.F.		
		MAT.	INST.	TOTAL
0250	Office partition, demountable, no deduction for door opening, add for doors			
0300	Air wall, cork finish, semi acoustic, 1-5/8" thick, minimum	46	4.29	50.29
0400	Maximum	55	7.35	62.35
0500	Acoustic, 2" thick, minimum	47	4.57	51.57
0600	Maximum	65	6.20	71.20
0700				
0800				
0900	In-plant modular office system, w/prehung steel door			
1000	3" thick honeycomb core panels			
1100	12' x 12', 2 wall	10.05	.60	10.65
1200	4 wall	21	1.79	22.79
1300				
1400				
1500	Gypsum, demountable, 3" to 3-3/4" thick x 9' high, vinyl clad	7.50	3.22	10.72
1600	Fabric clad	18.65	3.50	22.15
1700	1.75 system, vinyl clad hardboard, paper honeycomb core panel			
1800	1-3/4" to 2-1/2" thick x 9' high	12.55	3.22	15.77
1900	Unitized gypsum panel, 2" to 2-1/2" thick x 9' high, vinyl clad	16.20	3.22	19.42
2000	Fabric clad	26.50	3.50	30
2100				
2200				
2300	Unitized mineral fiber panel system, 2-1/4" thick x 9' high			
2400	Vinyl clad mineral fiber	16.10	3.22	19.32
2500	Fabric clad mineral fiber	24	3.50	27.50
2600				
2700	Movable steel walls, modular system			
2800	Unitized panels, 48" wide x 9' high			
2900	Baked enamel, pre-finished	18.20	2.55	20.75
3000	Fabric clad	26.50	2.78	29.28
3100	For acoustical partitions, add, unsealed	2.62		2.62
3200	Sealed	12.20		12.20
3300				
3400				
3500	Note: For door prices, see Div. 08			
3600	For door hardware prices, see Div. 08			
3700				
3800				
3900				
4000				

C1020 Interior Doors

This page illustrates and describes flush interior door systems including hollow core door, jamb, header and trim with hardware. Lines within System Components give the unit price and total price on a cost each basis for this system. Prices for flush interior door systems are on Line items C1020 106 1000 thru 2400. Both material quantities and labor costs have been adjusted for the system listed.

Factors: To adjust for job conditions other than normal working situations use Lines C1020 106 2700 thru 3800.

Example: You are to install the system in an area where dust protection is vital. Go to Line C1020 106 3000 and apply these percentages to the appropriate MAT. and INST. costs.

System Components	QUANTITY	UNIT	COST EACH MAT.	COST EACH INST.	COST EACH TOTAL
Single hollow core door, include jamb, header, trim and hardware, painted.					
Hollow core Lauan, 1-3/8" thick, 2'-0" x 6'-8", painted	1.000	Ea.	65	77.50	142.50
Paint door and frame, 1 coat	1.000	Ea.	4.79	34.50	39.29
Wood jamb, 4-9/16" deep	1.000	Set	112.80	59.52	172.32
Trim, casing	1.000	Set	48.64	89.28	137.92
Hardware, hinges	1.000	Set	27.50		27.50
Hardware, lockset	1.000	Set	48.50	43.50	92
TOTAL			307.23	304.30	611.53

C1020 106	Doors, Interior Flush, Wood	MAT.	INST.	TOTAL
1000	Lauan (Mahogany) hollow core, 1-3/8" x 2'-0" x 6'-8"	305	305	610
1100	2'-6" x 6'-8"	320	310	630
1200	2'-8" x 6'-8"	330	315	645
1300	3'-0" x 6'-8"	350	325	675
1500	Birch, hollow core, 1-3/8" x 2'-0" x 6'-8"	345	305	650
1600	2'-6" x 6'-8"	345	310	655
1700	2'-8" x 6'-8"	360	315	675
1800	3'-0" x 6'-8"	360	325	685
1900	Solid core, pre-hung, 1-3/8" x 2'-6" x 6'-8"	645	315	960
2000	2'-8" x 6'-8"	490	315	805
2100	3'-0" x 6'-8"	650	330	980
2200				
2400	For metal frame instead of wood, add	50%	20%	
2600				
2700	Cut & patch to match existing construction, add, minimum	2%	3%	
2800	Maximum	5%	9%	
2900	Dust protection, add, minimum	1%	2%	
3000	Maximum	4%	11%	
3100	Equipment usage curtailment, add, minimum	1%	1%	
3200	Maximum	3%	10%	
3300	Material handling & storage limitation, add, minimum	1%	1%	
3400	Maximum	6%	7%	
3500	Protection of existing work, add, minimum	2%	2%	
3600	Maximum	5%	7%	
3700	Shift work requirements, add, minimum		5%	
3800	Maximum		30%	
3900				

C1020 Interior Doors

This page illustrates and describes interior, solid and louvered door systems including a pine panel door, wood jambs, header, and trim with hardware. Lines within System Components give the unit price and total price on a cost each basis for this system. Prices for interior, solid and louvered systems are on Line Items C1020 108 1100 thru 2400. Both material quantities and labor costs have been adjusted for the system listed.

Factors: To adjust for job conditions other than normal working situations use Lines C1020 108 2900 thru 4000.

Example: You are to install the system during night hours only. Go to Line C1020 108 4000 and apply these percentages to the appropriate INST. costs.

System Components	QUANTITY	UNIT	COST EACH		
			MAT.	INST.	TOTAL
Single interior door, including jamb, header, trim and hardware, painted.					
Solid pine panel door, 1-3/8″ thick, 2′-0″ x 6′-8″	1.000	Ea.	282	77.50	359.50
Paint door and frame, 1 coat	1.000	Ea.	4.79	34.50	39.29
Wooden jamb, 4-5/8″ deep	1.000	Set	112.80	59.52	172.32
Trim, casing	1.000	Set	48.64	89.28	137.92
Hardware, hinges	1.000	Set	27.50		27.50
Hardware, lockset	1.000	Set	48.50	43.50	92
TOTAL			524.23	304.30	828.53

C1020 108	Doors, Interior Solid & Louvered	COST EACH		
		MAT.	INST.	TOTAL
1000				
1100	Solid pine, painted raised panel, 1-3/8″ x 2′-0″ x 6′-8″	525	305	830
1200	2′-6″ x 6′-8″	550	310	860
1300	2′-8″ x 6′-8″	550	315	865
1400	3′-0″ x 6′-8″	580	325	905
1500				
1600	Louvered pine, painted 1′-6″ x 6′-8″	420	300	720
1700	2′-0″ x 6′-8″	445	310	755
1800	2′-6″ x 6′-8″	460	315	775
1900	3′-0″ x 6′-8″	510	325	835
2200	For prehung door, deduct	5%	30%	
2300				
2400	For metal frame instead of wood, add	50%	20%	
2500				
2900	Cut & patch to match existing construction, add, minimum	2%	3%	
3000	Maximum	5%	9%	
3100	Dust protection, add, minimum	1%	2%	
3200	Maximum	4%	11%	
3300	Equipment usage curtailment, add, minimum	1%	1%	
3400	Maximum	3%	10%	
3500	Material handling & storage limitation, add, minimum	1%	1%	
3600	Maximum	6%	7%	
3700	Protection of existing work, add, minimum	2%	2%	
3800	Maximum	5%	7%	
3900	Shift work requirements, add, minimum		5%	
4000	Maximum		30%	

661

C1020 Interior Doors

This page illustrates and describes interior metal door systems including a metal door, metal frame and hardware. Lines within System Components give the unit price and total price on a cost each for this system. Prices for interior metal door systems are on Line Items C1020 110 1000 thru 2000. Both material quantities and labor costs have been adjusted for the system listed.

Factors: To adjust for job conditions other than normal working situations use Lines C1020 110 2900 thru 4000.

Example: You are to install the system while protecting existing construction. Go to Line C1020 110 3700 and apply these percentages to the appropriate MAT. and INST. costs.

System Components	QUANTITY	UNIT	COST EACH		
			MAT.	INST.	TOTAL
Single metal door, including frame and hardware.					
Hollow metal door, 1-3/8" thick, 2'-6" x 6'-8", painted	1.000	Ea.	630	93	723
Metal frame, 6-1/4" deep	1.000	Set	206	87	293
Paint door and frame, 1 coat	1.000	Ea.	4.79	34.50	39.29
Hardware, hinges	1.000	Set	64.50		64.50
Hardware, passage lockset	1.000	Set	48.50	43.50	92
TOTAL			953.79	258	1,211.79

C1020 110	Doors, Interior Flush, Metal	COST EACH		
		MAT.	INST.	TOTAL
1000	Hollow metal doors, 1-3/8" thick, 2'-6" x 6'-8"	955	258	1,213
1100	1-3/8" thick, 2'-8" x 6'-8"	955	258	1,213
1200	3'-0" x 7'-0"	955	271	1,226
1300				
1400	Interior fire door, 1-3/8" thick, 2'-6" x 6'-8"	1,025	258	1,283
1500	2'-8" x 6'-8"	955	258	1,213
1600	3'-0" x 7'-0"	1,025	271	1,296
1700				
1800	Add to fire doors:			
1900	Baked enamel finish	30%	15%	
2000	Galvanizing	20%		
2200				
2300				
2400				
2900	Cut & patch to match existing construction, add, minimum	2%	3%	
3000	Maximum	5%	9%	
3100	Dust protection, add, minimum	1%	2%	
3200	Maximum	4%	11%	
3300	Equipment usage curtailment, add, minimum	1%	1%	
3400	Maximum	3%	10%	
3500	Material handling & storage limitation, add, minimum	1%	1%	
3600	Maximum	6%	7%	
3700	Protection of existing work, add, minimum	2%	2%	
3800	Maximum	5%	7%	
3900	Shift work requirements, add, minimum		5%	
4000	Maximum		30%	

This page illustrates and describes an interior closet door system including an interior closet door, painted, with trim and hardware. Prices for interior closet door systems are on Line Items C1020 112 0500 thru 2200. Both material quantities and labor costs have been adjusted for the system listed.

Factors: To adjust for job conditions other than normal working situations use Lines C1020 112 2900 thru 4000.

Example: You are to install the system and match the existing construction. Go to Line C1020 112 2900 and apply these percentages to the appropriate MAT. and INST. costs.

C1020 112	Doors, Closet	COST PER SET		
		MAT.	INST.	TOTAL
0350	Interior closet door painted, including frame, trim and hardware, prehung.			
0400	Bi-fold doors			
0500	Pine paneled, 3'-0" x 6'-8"	470	252	722
0600	6'-0" x 6'-8"	675	340	1,015
0700	Birch, hollow core, 3'-0" x 6'-8"	236	274	510
0800	6'-0" x 6'-8"	345	365	710
0900	Lauan, hollow core, 3'-0" x 6'-8"	219	252	471
1000	6'-0" x 6'-8"	315	340	655
1100	Louvered pine, 3'-0" x 6'-8"	460	252	712
1200	6'-0" x 6'-8"	535	340	875
1300				
1400	Sliding, bi-passing closet doors			
1500	Pine paneled, 4'-0" x 6'-8"	695	268	963
1600	6'-0" x 6'-8"	1,175	340	1,515
1700	Birch, hollow core, 4'-0" x 6'-8"	430	268	698
1800	6'-0" x 6'-8"	580	340	920
1900	Lauan, hollow core, 4'-0" x 6'-8"	335	268	603
2000	6'-0" x 6'-8"	355	340	695
2100	Louvered pine, 4'-0" x 6'-8"	720	268	988
2200	6'-0" x 6'-8"	1,025	340	1,365
2300				
2400				
2500				
2600				
2700				
2800				
2900	Cut & patch to match existing construction, add, minimum	2%	3%	
3000	Maximum	5%	9%	
3100	Dust protection, add, minimum	1%	2%	
3200	Maximum	4%	11%	
3300	Equipment usage curtailment, add, minimum	1%	1%	
3400	Maximum	3%	10%	
3500	Material handling & storage limitation, add, minimum	1%	1%	
3600	Maximum	6%	7%	
3700	Protection of existing work, add, minimum	2%	2%	
3800	Maximum	5%	7%	
3900	Shift work requirements, add, minimum		5%	
4000	Maximum		30%	

C2010 Stair Construction

This page illustrates and describes a stair system based on a cost per flight price. Prices for various stair systems are on Line Items C2010 130 0700 thru 3200. Both material quantities and labor costs have been adjusted for the system listed.

Factors: To adjust for job conditions other than normal working situations use Lines C2010 130 3500 thru 4200.

Example: You are to install the system during evenings only. Go to Line C2010 130 4200 and apply this percentage to the appropriate MAT. and INST. costs.

System Components	QUANTITY	UNIT	COST PER FLIGHT		
			MAT.	INST.	TOTAL
Below are various stair systems based on cost per flight of stairs, no side Walls. Stairs are 4'-0" wide, railings are included unless otherwise noted.					

C2010 130	Stairs	COST PER FLIGHT		
		MAT.	INST.	TOTAL
0700	Concrete, cast in place, no nosings, no railings, 12 risers	400	2,375	2,775
0800	24 risers	800	4,750	5,550
0900	Add for 1 intermediate landing	214	660	874
1000	Concrete, cast in place, with nosings, no railings, 12 risers	1,375	2,625	4,000
1100	24 risers	2,775	5,250	8,025
1200	Add for 1 intermediate landing	296	680	976
1300	Steel, grating tread, safety nosing, 12 risers	5,675	1,250	6,925
1400	24 risers	11,400	2,500	13,900
1500	Add for intermediate landing	1,400	330	1,730
1600	Steel, metal pan tread for concrete in-fill, 12 risers	8,700	1,250	9,950
1700	24 risers	17,400	2,500	19,900
1800	Add for intermediate landing	1,400	330	1,730
1900	Spiral, industrial, 4' - 6" diameter, 12 risers	8,050	885	8,935
2000	24 risers	16,100	1,775	17,875
2100	Wood, box stairs, oak treads, 12 risers	2,400	790	3,190
2200	24 risers	4,775	1,575	6,350
2300	Add for 1 intermediate landing	252	178	430
2400	Wood, basement stairs, no risers, 12 steps	830	325	1,155
2500	24 steps	1,650	650	2,300
2600	Add for 1 intermediate landing	71	34	105
2700	Wood, open, rough sawn cedar, 12 steps	2,700	640	3,340
2800	24 steps	5,400	1,275	6,675
2900	Add for 1 intermediate landing	38	35	73
3000	Wood, residential, oak treads, 12 risers	2,650	3,500	6,150
3100	24 risers	5,275	6,975	12,250
3200	Add for 1 intermediate landing	217	161	378
3500	Dust protection, add, minimum	1%	2%	
3600	Maximum	4%	11%	
3700	Material handling & storage limitation, add, minimum	1%	1%	
3800	Maximum	6%	7%	
3900	Protection of existing work, add, minimum	2%	2%	
4000	Maximum	5%	7%	
4100	Shift work requirements, add, minimum		5%	
4200	Maximum		30%	

C3010 210	Selective Price Sheet	COST PER S.F.		
		MAT.	INST.	TOTAL
0100	Painting, on plaster or drywall, brushwork, primer and 1 ct.	.18	.87	1.05
0200	Primer and 2 ct.	.25	1.38	1.63
0300	Rollerwork, primer and 1 ct.	.18	.74	.92
0400	Primer and 2 ct.	.25	1.18	1.43
0500	Woodwork incl. puttying, brushwork, primer and 1 ct.	.15	1.31	1.46
0600	Primer and 2 ct.	.23	1.73	1.96
0700	Wood trim to 6" wide, enamel, primer and 1 ct.	.16	.74	.90
0800	Primer and 2 ct.	.24	.93	1.17
0900	Cabinets and casework, enamel, primer and 1 ct.	.16	1.47	1.63
1000	Primer and 2 ct.	.24	1.81	2.05
1100	On masonry or concrete, latex, brushwork, primer and 1 ct.	.29	1.23	1.52
1200	Primer and 2 ct.	.40	1.76	2.16
1300	For block filler, add	.20	1.51	1.71
1400				
1500	Varnish, wood trim, sealer 1 ct., sanding, puttying, quality work	.14	2.65	2.79
1600	Medium work	.11	2.06	2.17
1700	Without sanding	.17	.31	.48
1800				
1900	Wall coverings, wall paper, at low price per double roll, avg workmanship	.66	.92	1.58
2000	At medium price per double roll, average workmanship	1.49	1.10	2.59
2100	At high price per double roll, quality workmanship	2.57	1.35	3.92
2200				
2300	Grass cloths with lining paper	1.39	1.47	2.86
2400	Premium texture/color	3.82	1.68	5.50
2500	Vinyl, fabric backed, light weight	1.73	.92	2.65
2600	Medium weight	1.35	1.22	2.57
2700	Heavy weight	2.06	1.35	3.41
2800				
2900	Cork tiles, 12" x 12", 3/16" thick	4.44	2.44	6.88
3000	5/16" thick	3.72	2.50	6.22
3100	Granular surface, 12" x 36", 1/2" thick	1.49	1.52	3.01
3200	1" thick	1.94	1.58	3.52
3300	Aluminum foil	1.18	2.13	3.31
3400				
3500	Tile, ceramic, adhesive set, 4-1/4" x 4-1/4"	3.32	6.15	9.47
3600	6" x 6"	4.64	6.65	11.29
3700	Decorated, 4-1/4" x 4-1/4", minimum	4.10	4.32	8.42
3800	Color group 4	75.50	6.45	81.95
3900	For epoxy grout, add	.47	1.46	1.93
4000	Pregrouted sheets	6.40	4.86	11.26
4100	Glass mosaics, 3/4" tile on 12" sheets, minimum	21.50	15.95	37.45
4200	Color group 8	7.25	18.20	25.45
4300	Metal, tile pattern, 4' x 4' sheet, 24 ga., nailed			
4400	Stainless steel	31	2.72	33.72
4500	Aluminized steel	22.50	2.72	25.22
4600				
4700	Brick, interior veneer, 4" face brick, running bond, minimum	5.50	14.85	20.35
4800	Maximum	5.50	14.85	20.35
4900	Simulated, urethane pieces, set in mastic	9.10	4.65	13.75
5000	Fiberglass panels	10.90	3.48	14.38
5100	Wall coating, on drywall, thin coat, plain	.11	.87	.98
5200	Stipple	.11	.87	.98
5300	Textured spray	.04	.84	.88
5400				
5500	Paneling not incl. furring or trim, hardboard, tempered, 1/8" thick	.56	2.79	3.35
5600	1/4" thick	.61	2.79	3.40
5700	Plastic faced, 1/8" thick	.47	2.79	3.26
5800	1/4" thick	.91	2.79	3.70

C30 Interior Finishes

C3010 Wall Finishes

C3010 210	Selective Price Sheet	COST PER S.F.		
		MAT.	INST.	TOTAL
5900	Woodgrained, 1/4" thick, minimum	.78	2.79	3.57
6000	Maximum	1.47	3.28	4.75
6100	Plywood, 4' x 8' shts. 1/4" thick, prefin., birch faced, min.	2.17	2.79	4.96
6200	Maximum	1.13	3.98	5.11
6300	Walnut, minimum	2.94	2.79	5.73
6400	Maximum	2.72	3.48	6.20
6500	Mahogany, African	2.82	3.48	6.30
6600	Philippine	.46	2.79	3.25
6700	Chestnut	5.95	3.72	9.67
6800	Pecan	2.78	3.48	6.26
6900	Rosewood	3.41	4.36	7.77
7000	Teak	3.62	3.48	7.10
7100	Aromatic cedar, plywood	2.55	3.48	6.03
7200	Particle board	1.29	3.48	4.77
7300	Wood board, 3/4" thick, knotty pine	2.64	4.65	7.29
7400	Rough sawn cedar	4.90	4.65	9.55
7500	Redwood, clear	14	4.65	18.65
7600	Aromatic cedar	2.56	5.05	7.61

C3020 Floor Finishes

C3020 430	Selective Price Sheet	COST PER S.F.		
		MAT.	INST.	TOTAL
0100	Flooring, carpet, acrylic, 26 oz. light traffic	1.95	.95	2.90
0200	35 oz. heavy traffic	6.90	.95	7.85
0300	Nylon anti-static, 15 oz. light traffic	2.89	.95	3.84
0400	22 oz. medium traffic	4.72	.95	5.67
0500	26 oz. heavy traffic	6.60	1.02	7.62
0600	28 oz. heavy traffic	3.94	1.02	4.96
0700	Tile, foamed back, needle punch	4.69	1.13	5.82
0800	Tufted loop	3.88	1.13	5.01
0900	Wool, 36 oz. medium traffic	12.20	1.02	13.22
1000	42 oz. heavy traffic	13.45	1.02	14.47
1100	Composition, epoxy, with colored chips, minimum	3.65	5.30	8.95
1200	Maximum	4.90	7.25	12.15
1300	Trowelled, minimum	3.92	6.35	10.27
1400	Maximum	6.80	7.40	14.20
1500	Terrazzo, 1/4" thick, chemical resistant, minimum	6.65	7.35	14
1600	Maximum	11.30	9.85	21.15
1700	Resilient, asphalt tile, 1/8" thick	1.63	1.61	3.24
1800	Conductive flooring, rubber, 1/8" thick	8.40	2.05	10.45
1900	Cork tile 1/8" thick, standard finish	5.35	2.05	7.40
2000	Urethane finish	6.90	2.05	8.95
2100	PVC sheet goods for gyms, 1/4" thick	7.40	8.05	15.45
2200	3/8" thick	10.80	10.75	21.55
2300	Vinyl composition 12" x 12" tile, plain, 1/16" thick	.97	1.29	2.26
2400	1/8" thick	1.99	1.29	3.28
2500	Vinyl tile, 12" x 12" x 1/8" thick, minimum	7.35	1.29	8.64
2600	Premium colors/pattern	9.10	1.29	10.39
2700	Vinyl sheet goods, backed, .080" thick, plain pattern/colors	4.59	2.81	7.40
2900	Slate, random rectangular, 1/4" thick	25	8.35	33.35
3000	1/2" thick	23	11.90	34.90
3100	Natural cleft, irregular, 3/4" thick	10.70	13.60	24.30
3200	For sand rubbed finish, add	11		11
3300	Terrazzo, cast in place, bonded 1-3/4" thick, gray cement	4.25	19.65	23.90
3400	White cement	5	19.65	24.65
3500	Not bonded 3" thick, gray cement	5.25	21	26.25
3600	White cement	6.05	21	27.05
3700	Precast, 12" x 12" x 1" thick	30	21	51
3800	1-1/4" thick	28	21	49
3900	16" x 16" x 1-1/4" thick	31	25	56
4000	1-1/2" thick	28.50	28	56.50
4100	Marble travertine, standard, 12" x 12" x 3/4" thick	10.50	19.40	29.90
4200				
4300	Tile, ceramic, natural clay, thin set	5.45	6.35	11.80
4400	Porcelain, thin set	4.44	6.15	10.59
4500	Specialty, decorator finish	15.15	6.35	21.50
4600				
4700	Quarry, red, mud set, 4" x 4" x 1/2" thick	10.30	9.70	20
4800	6" x 6" x 1/2" thick	11.15	8.30	19.45
4900	Brown, imported, 6" x 6" x 3/4" thick	8.55	9.70	18.25
5000	8" x 8" x 1" thick	11.25	10.60	21.85
5100	Slate, Vermont, thin set, 6" x 6" x 1/4" thick	8.45	6.45	14.90
5200				
5300	Wood, maple strip, 25/32" x 2-1/4", finished, select	5.25	4.41	9.66
5400	2nd and better	6.90	4.41	11.31
5500	Oak, 25/32" x 2-1/4" finished, clear	4.82	4.41	9.23
5600	No. 1 common	4.83	4.41	9.24
5700	Parquet, standard, 5/16", finished, minimum	6.70	4.67	11.37
5800	Maximum	13.15	7.25	20.40
5900	Custom, finished, plain pattern	19.80	6.95	26.75

C3020 Floor Finishes

C3020 430	Selective Price Sheet	COST PER S.F.		
		MAT.	INST.	TOTAL
6000	Intricate pattern	27	13.95	40.95
6100	Prefinished, oak, 2-1/4" wide	7.35	4.10	11.45
6200	Ranch plank	7.60	4.81	12.41
6300	Sleepers on concrete, treated, 24" O.C., 1" x 2"	.38	.46	.84
6400	1" x 3"	.69	.59	1.28
6500	2" x 4"	.93	.82	1.75
6600	2" x 6"	1.36	.94	2.30
6700	Refinish old floors, soft wood	.25	1.41	1.66
6800	Hard wood	.25	4.35	4.60
6900	Subflooring, plywood, CDX, 1/2" thick	1.10	.93	2.03
7000	5/8" thick	1.27	1.03	2.30
7100	3/4" thick	1.52	1.11	2.63
7200				
7300	1" x 10" boards, S4S, laid regular	3.93	1.27	5.20
7400	Laid diagonal	3.93	1.55	5.48
7500	1" x 8" boards, S4S, laid regular	2.96	1.39	4.35
7600	Laid diagonal	2.96	1.64	4.60
7700	Underlayment, plywood, underlayment grade, 3/8" thick	1.63	.93	2.56
7800	1/2" thick	2.02	.96	2.98
7900	5/8" thick	2.37	1	3.37
8000	3/4" thick	2.43	1.07	3.50
8100	Particle board, 3/8" thick	.49	.93	1.42
8200	1/2" thick	.79	.96	1.75
8300	5/8" thick	1.08	1	2.08
8400	3/4" thick	1.17	1.07	2.24

C3030 Ceiling Finishes

This page illustrates suspended acoustical board systems including acoustic ceiling board, hangers, and T bar suspension. Lines within System Components give the unit price and total price per square foot for this system. Prices for suspended acoustical board systems are on Line Items C3030 210 1000 thru 2400. Both material quantities and labor costs have been adjusted for the system listed.

Factors: To adjust for job conditions other than normal working situations use Lines C3030 210 2900 thru 4000.

Example: You are to install the system and protect existing construction. Go to Line C3030 210 3800 and apply these percentages to the appropriate MAT. and INST. costs.

System Components	QUANTITY	UNIT	COST PER S.F. MAT.	COST PER S.F. INST.	COST PER S.F. TOTAL
Suspended acoustical ceiling board installed on exposed grid system.					
Fiberglass boards, film faced, 2' x 4', 5/8" thick	1.000	S.F.	2.50	1.11	3.61
Hangers, #12 wire	1.000	S.F.	.17	.67	.84
T bar suspension system, 2' x 4' grid	1.000	S.F.	1.43	.87	2.30
TOTAL			4.10	2.65	6.75

C3030 210	Ceiling, Suspended Acoustical	COST PER S.F. MAT.	COST PER S.F. INST.	COST PER S.F. TOTAL
1000	2' x 4' grid, fiberglass board, film faced, 5/8" thick	4.10	2.65	6.75
1100	Mineral fiber board, aluminum faced	6.10	2.61	8.71
1200	Standard faced	2.84	2.57	5.41
1300	Plastic faced	4.97	3.28	8.25
1400	Fiberglass, film faced, 3" thick, R11	6	3.09	9.09
1500	Grass cloth faced, 3/4" thick	5.30	2.93	8.23
1600	1" thick	5.30	2.98	8.28
1700	1-1/2" thick, nubby face	4.70	3.01	7.71
2200				
2300				
2400	Add for 2' x 2' grid system	.35	.21	.56
2500				
2600				
2700				
2900	Cut & patch to match existing construction, add, minimum	2%	3%	
3000	Maximum	5%	9%	
3100	Dust protection, add, minimum	1%	2%	
3200	Maximum	4%	11%	
3300	Equipment usage curtailment, add, minimum	1%	1%	
3400	Maximum	3%	10%	
3500	Material handling & storage limitation, add, minimum	1%	1%	
3600	Maximum	6%	7%	
3700	Protection of existing work, add, minimum	2%	2%	
3800	Maximum	5%	7%	
3900	Shift work requirements, add, minimum		5%	
4000	Maximum		30%	

C3030 Ceiling Finishes

This page illustrates and describes suspended gypsum board systems including gypsum board, metal furring, taping, finished and painted. Lines within System Components give the unit price and total price per square foot for this system. Prices for suspended gypsum board systems are on Line Items C3030 220 1400 thru 1700. Both material quantities and labor costs have been adjusted for the system listed.

Factors: To adjust for job conditions other than normal working situations use Lines C3030 220 2900 thru 4000.

Example: You are to install the system and control dust in the working area. Go to Line C3030 220 3200 and apply these percentages to the appropriate MAT. and INST. costs.

System Components	QUANTITY	UNIT	COST PER S.F.		
			MAT.	INST.	TOTAL
Suspended ceiling gypsum board,4' x 8' x 5/8" thick,					
On metal furring, taped, finished and painted.					
Gypsum drywall, 4' x 8', 5/8" thick, screwed	1.000	S.F.	.40	.77	1.17
Main runners, 1-1/2" C.R.C., 4' O.C.	.500	S.F.	.19	.87	1.06
25 ga., channels, 2' O.C.	1.000	S.F.	.28	1.61	1.89
Taped and finished	1.000	S.F.	.05	.70	.75
Paint, 2 coats, roller work	1.000	S.F.	.23	.77	1
TOTAL			1.15	4.72	5.87

C3030 220	Ceilings, Suspended Gypsum Board	COST PER S.F.		
		MAT.	INST.	TOTAL
1400	4' x 8' x 5/8", gypsum drywall, 2 coats paint	1.15	4.72	5.87
1500	Thin coat plaster, 2 coats paint	.98	4.12	5.10
1600	Spray-on sand finish, no paint	.91	4.09	5
1700	12" x 12" x 3/4" acoustical wood fiber tile	2.30	4.99	7.29
1800				
1900				
2000				
2100				
2200				
2300				
2400				
2500				
2600				
2700				
2900	Cut & patch to match existing construction, add, minimum	2%	3%	
3000	Maximum	5%	9%	
3100	Dust protection, add, minimum	1%	2%	
3200	Maximum	4%	11%	
3300	Equipment usage curtailment, add, minimum	1%	1%	
3400	Maximum	3%	10%	
3500	Material handling & storage limitation, add, minimum	1%	1%	
3600	Maximum	6%	7%	
3700	Protection of existing work, add, minimum	2%	2%	
3800	Maximum	5%	7%	
3900	Shift work requirements, add, minimum		5%	
4000	Maximum		30%	

C3030 Ceiling Finishes

This page illustrates and describes suspended plaster and lath systems including gypsum plaster, lath, furring and runners with ceiling painted. Lines within System Components give the unit price and total price per square foot for this system. Prices for suspended plaster and lath systems are on Line Items C3030 230 1200 thru 2300. Both material quantities and labor costs have been adjusted for the system listed.

Factors: To adjust for job conditions other than normal working situations use Lines C3030 230 2900 thru 4000.

Example: You are to install the system to match existing construction. Go to Line C3030 230 2900 and apply these percentages to the appropriate MAT. and INST. costs.

System Components	QUANTITY	UNIT	COST PER S.F.		
			MAT.	INST.	TOTAL
Gypsum plaster, 3 coats, on 3.4# rib lath, on 3/4″ C.R.C. furring on 1-1/2″ main runners, ceiling painted.	.110	S.Y.	.69	4.47	5.16
Gypsum plaster, 3 coats	.110	S.Y.	.58	1.06	1.64
3.4# rib lath	1.000	S.F.	.38	1.73	2.11
Main runners, 1-1/2″ C.R.C. 24″ O.C.	1.000	S.F.	.42	2.33	2.75
Furring 3/4″ C.R.C. 16″ OC	1.000	S.F.	.23	.77	1
Painting, 2 coats, roller work					
TOTAL			2.30	10.36	12.66

C3030 230	Ceiling, Suspended Plaster	COST PER S.F.		
		MAT.	INST.	TOTAL
1200	Gypsum plaster, 3 coats, on 3.4# rib lath	2.30	10.35	12.65
1300	On 2.5# diamond lath	2.38	10.30	12.68
1400	On 3/8″ gypsum lath	2.12	10.30	12.42
1500	2 coats, on 3.4# rib lath	2.07	10.35	12.42
1600	On 2.5# diamond lath	2.18	9.60	11.78
1700	On 3/8″ gypsum lath	1.92	9.60	11.52
1800	Perlite plaster, 3 coats, on 3.4# rib lath	2.42	11.40	13.82
1900	On 2.5# diamond lath	2.50	11.30	13.80
2000	On 3/8″ gypsum lath	2.24	11.30	13.54
2100	2 coats, on 3.4# rib lath	2.33	10.30	12.63
2200	On 2.5# diamond lath	2.41	10.25	12.66
2300	On 3/8″ gypsum lath	2.15	10.25	12.40
2400				
2500				
2600				
2700				
2900	Cut & patch to match existing construction, add, minimum	2%	3%	
3000	Maximum	5%	9%	
3100	Dust protection, add, minimum	1%	2%	
3200	Maximum	4%	11%	
3300	Equipment usage curtailment, add, minimum	1%	1%	
3400	Maximum	3%	10%	
3500	Material handling & storage limitation, add, minimum	1%	1%	
3600	Maximum	6%	7%	
3700	Protection of existing work, add, minimum	2%	2%	
3800	Maximum	5%	7%	
3900	Shift work requirements, add, minimum		5%	
4000	Maximum		30%	

C30 Interior Finishes

C3030 Ceiling Finishes

C3030 250	Selective Price Sheet	COST PER S.F.		
		MAT.	INST.	TOTAL
0100	Ceiling, plaster, gypsum, 2 coats	.49	3.81	4.30
0200	3 coats	.69	4.51	5.20
0300	Perlite or vermiculite, 2 coats	.73	4.45	5.18
0400	3 coats	.82	5.55	6.37
0500	Gypsum lath, plain, 3/8" thick	.41	.88	1.29
0600	1/2" thick	.35	.94	1.29
0700	Firestop, 3/8" thick	.30	1.07	1.37
0800	1/2" thick	.34	1.15	1.49
0900	Metal lath, rib, 2.75 lb.	.47	1	1.47
1000	3.40 lb.	.59	1.07	1.66
1100	Diamond, 2.50 lb.	.67	1	1.67
1200	3.40 lb.	.50	1.25	1.75
1500	Drywall, standard, 1/2" thick, no finish included	.36	.77	1.13
1600	Taped and finished	.42	1.82	2.24
1700	5/8" thick, no finish included	.40	.77	1.17
1800	Taped and finished	.45	1.82	2.27
1900	Fire resistant, 1/2" thick, no finish included	.41	.77	1.18
2000	Taped and finished	.46	1.82	2.28
2100	5/8" thick, no finish included	.42	.77	1.19
2200	Taped and finished	.47	1.82	2.29
2300	Water resistant, 1/2" thick, no finish included	.41	.77	1.18
2400	Taped and finished	.46	1.82	2.28
2500	5/8" thick, no finish included	.47	.77	1.24
2600	Taped and finished	.53	1.82	2.35
2700	Finish, instead of taping			
2800	For thin coat plaster, add	.11	.87	.98
2900	Finish, textured spray, add	.04	.84	.88
3000				
3100				
3200				
3300	Tile, stapled or glued, mineral fiber plastic coated, 5/8" thick	2.42	2.32	4.74
3400	3/4" thick	4.15	2.32	6.47
3500	Wood fiber, 1/2" thick	1.46	1.74	3.20
3600	3/4" thick	1.43	1.74	3.17
3700	Suspended, fiberglass, film faced, 5/8" thick	2.50	1.11	3.61
3800	3" thick	4.39	1.55	5.94
3900	Mineral fiber, 5/8" thick, standard	1.24	1.03	2.27
4000	Aluminum	4.18	1.16	5.34
4100	Wood fiber, reveal edge, 1" thick			
4200	3" thick			
4300	Framing, metal furring, 3/4" channels, 12" O.C.	.47	3.22	3.69
4400	16" O.C.	.42	2.33	2.75
4500	24" O.C.	.28	1.61	1.89
4600				
4700	1-1/2" channels, 12" O.C.	.63	3.56	4.19
4800	16" O.C.	.56	2.60	3.16
4900	24" O.C.	.38	1.73	2.11
5000				
5100				
5200				
5300	Ceiling suspension systems, for tile,			
5400	Concealed "Z" bar, 12" module	1.05	1.34	2.39
5500	Class A, "T" bar 2'-0" x 4'-0" grid	1.43	.87	2.30
5600	2'-0" x 2'-0" grid	1.75	1.07	2.82
5700	Carrier channels for lighting fixtures, add	.26	1.51	1.77
5800				
5900				
6000				

C3030 Ceiling Finishes

C3030 250	Selective Price Sheet	COST PER S.F.		
		MAT.	INST.	TOTAL
6100	Wood, furring 1" x 3", on wood, 12" O.C.	.68	1.99	2.67
6200	16" O.C..	.51	1.49	2
6300	24" O.C.	.34	1	1.34
6400				
6500	On concrete, 12" O.C.	.74	3.32	4.06
6600	16" O.C.	.56	2.49	3.05
6700	24" O.C.	.37	1.66	2.03
6800				
6900	Joists, 2" x 4", 12" O.C.	.99	1.43	2.42
7000	16" O.C.	.83	1.20	2.03
7100	24" O.C.	.67	.97	1.64
7200	32" O.C.	.51	.74	1.25
7300	2" x 6", 12" O.C.	1.53	1.45	2.98
7400	16" O.C.	1.27	1.20	2.47
7500	24" O.C.	1.01	.96	1.97
7600	32" O.C.	.75	.71	1.46

For customer support on your Commercial Renovation Costs with RSMeans Data, call 800.448.8182.

Same Data. Simplified.

Enjoy the convenience and efficiency of accessing your costs anywhere:

■ **Skip the multiplier** by setting your location

■ **Quickly search,** edit, favorite and share costs

■ **Stay on top of price changes** with automatic updates

Discover more at rsmeans.com/online

This page illustrates and describes oil hydraulic elevator systems. Prices for various oil hydraulic elevator systems are on Line Items D1010 130 0400 thru 2000. Both material quantities and labor costs have been adjusted for the system listed.

Factors: To adjust for job conditions other than normal working situations use Lines D1010 130 3100 thru 4000.

Example: You are to install the system with limited equipment usage. Go to Line D1010 130 3400 and apply this percentage to the appropriate TOTAL costs.

D1010 130	Oil Hydraulic Elevators	COST EACH		
		MAT.	INST.	TOTAL
0310	Oil-hydraulic elevator systems			
0320	Including piston and piston shaft			
0330				
0400	1500 Lb. passenger, 2 floors	44,800	22,100	66,900
0500	3 floors	55,500	34,200	89,700
0600	4 floors	80,500	48,100	128,600
0700	5 floors	90,500	60,000	150,500
0800	6 floors	100,500	71,500	172,000
0900				
1000	2500 Lb. passenger, 2 floors	48,300	22,100	70,400
1100	3 floors	59,000	34,200	93,200
1200	4 floors	79,000	48,100	127,100
1300	5 floors	89,500	60,000	149,500
1400	6 floors	104,000	71,500	175,500
1500				
1600	4000 Lb. passenger, 2 floors	55,000	22,100	77,100
1700	3 floors	67,000	34,200	101,200
1800	4 floors	90,500	48,100	138,600
1900	5 floors	101,000	60,000	161,000
2000	6 floors	110,500	71,500	182,000
2100				
2200				
2300				
2400				
2500				
2600				
2700				
2800				
2900				
3000				
3100	Dust protection, add, minimum	1%	2%	
3200	Maximum	4%	11%	
3300	Equipment usage curtailment, add, minimum	1%	1%	
3400	Maximum	3%	10%	
3500	Material handling & storage limitation, add, minimum	1%	1%	
3600	Maximum	6%	7%	
3700	Protection of existing work, add, minimum	2%	2%	
3800	Maximum	5%	7%	
3900	Shift work requirements, add, minimum		5%	
4000	Maximum		30%	

For customer support on your Commercial Renovation Costs with RSMeans Data, call 800.448.8182.

D2010 Plumbing Fixtures

This page illustrates and describes a women's public restroom system including a water closet, lavatory, accessories, and service piping. Lines within System Components give the unit price and total price on a cost each basis for this system. Prices for women's public restroom systems are on Line Items D2010 956 1300 thru 1700. Both material quantities and labor costs have been adjusted for the system listed.

Factors: To adjust for job conditions other than normal working situations use Lines D2010 956 2900 thru 4000.

Example: You are to install the system and protect surrounding area from dust. Go to Line D2010 956 3100 and apply these percentages to the MAT. and INST. costs.

System Components	QUANTITY	UNIT	COST EACH MAT.	COST EACH INST.	COST EACH TOTAL
Public women's restroom incl. water closet, lavatory, accessories and Necessary service piping to install this system in one wall.					
Water closet, wall mounted, one piece	1.000	Ea.	1,050	259	1,309
Rough-in waste and vent for water closet	1.000	Set	1,600	585	2,185
Lavatory, 20" x 18" P.E. cast iron with accessories	1.000	Ea.	335	188	523
Rough-in waste and vent for lavatory	1.000	Set	730	905	1,635
Toilet partition, painted metal between walls, floor mounted	1.000	Ea.	385	232	617
For handicap unit, add	1.000	Ea.	360		360
Grab bar, 36" long	2.000	Ea.	29.30	70	99.30
Mirror, 18" x 24", with stainless steel shelf	1.000	Ea.	91	35	126
Napkin/tampon dispenser, recessed	1.000	Ea.	700	46.50	746.50
Soap Dispenser, chrome, surface mounted, liquid	1.000	Ea.	74.50	35	109.50
Toilet tissue dispenser, surface mounted, stainless steel	1.000	Ea.	17.60	23	40.60
Towel dispenser, surface mounted, stainless steel	1.000	Ea.	42.50	43.50	86
TOTAL			5,414.90	2,422	7,836.90

D2010 956	Plumbing - Public Restroom	COST EACH MAT.	COST EACH INST.	COST EACH TOTAL
1200				
1300	Public women's restroom, one water closet, one lavatory	5,425	2,425	7,850
1400	Two water closets, two lavatories	11,000	4,625	15,625
1500				
1600	For each additional water closet over 2, add	3,700	1,075	4,775
1700	For each additional lavatory over 2, add	1,225	1,175	2,400
1800				
1900				
2400	NOTE: PLUMBING APPROXIMATIONS			
2500	WATER CONTROL: water meter, backflow preventer,			
2600	Shock absorbers, vacuum breakers, mixer....10 to 15% of fixtures			
2700	PIPE AND FITTINGS: .... 30 to 60% of fixtures			
2800				
2900	Cut & patch to match existing construction, add, minimum	2%	3%	
3000	Maximum	5%	9%	
3100	Dust protection, add, minimum	1%	2%	
3200	Maximum	4%	11%	
3300	Equipment usage curtailment, add, minimum	1%	1%	
3400	Maximum	3%	10%	
3500	Material handling & storage limitation, add, minimum	1%	1%	

D20 Plumbing

D2010 Plumbing Fixtures

D2010 956	Plumbing - Public Restroom	COST EACH		
		MAT.	INST.	TOTAL
3600	Maximum	6%	7%	
3700	Protection of existing work, add, minimum	2%	2%	
3800	Maximum	5%	7%	
3900	Shift work requirements, add, minimum		5%	
4000	Maximum		30%	

D20 Plumbing

D2010 Plumbing Fixtures

This page illustrates and describes a men's public restroom system including a water closet, urinal, lavatory, accessories and service piping. Lines within System Components give the unit price and total price on a cost each basis for this system. Prices for men's public restroom systems are on Line Items D2010 957 1700 thru 2200. Both material quantities and labor costs have been adjusted for the system listed.

Factors: To adjust for job conditions other than normal working situations use Lines D2010 957 2900 thru 4000.

Example: You are to install the system and match existing construction. Go to Line D2010 957 3000 and apply these percentages to the appropriate MAT. and INST. costs.

System Components	QUANTITY	UNIT	COST EACH		
			MAT.	INST.	TOTAL
Public men's restroom incl. water closet, urinal, lavatory, accessories,					
And necessary service piping to install this system in one wall.					
Water closet, wall mounted, one piece	1.000	Ea.	1,050	259	1,309
Rough-in waste & vent for water closet	1.000	Set	1,600	585	2,185
Urinal, wall hung	1.000	Ea.	370	500	870
Rough-in waste & vent for urinal	1.000	Set	920	530	1,450
Lavatory, 20" x 18", P.E. cast iron with accessories	1.000	Ea.	335	188	523
Rough-in waste & vent for lavatory	1.000	Set	730	905	1,635
Partition, painted mtl., between walls, floor mntd.	1.000	Ea.	385	232	617
For handicap unit, add	1.000	Ea.	360		360
Grab bars	2.000	Ea.	29.30	70	99.30
Urinal screen, powder coated steel, wall mounted	1.000	Ea.	114	139	253
Mirror, 18" x 24" with stainless steel shelf	1.000	Ea.	91	35	126
Soap dispenser, surface mounted, liquid	1.000	Ea.	74.50	35	109.50
Toilet tissue dispenser, surface mounted, stainless steel	1.000	Ea.	17.60	23	40.60
Towel dispenser, surface mounted, stainless steel	1.000	Ea.	42.50	43.50	86
TOTAL			6,118.90	3,544.50	9,663.40

D2010 957	Plumbing - Public Restroom	COST EACH		
		MAT.	INST.	TOTAL
1600				
1700	Public men's restroom, one water closet, one urinal, one lavatory	6,125	3,550	9,675
1800	Two water closets, two urinals, two lavatories	13,100	6,900	20,000
1900				
2000	For each additional water closet over 2, add	3,700	1,075	4,775
2100	For each additional urinal over 2, add	1,400	1,175	2,575
2200	For each additional lavatory over 2, add	1,225	1,175	2,400
2300				
2400	NOTE: PLUMBING APPROXIMATIONS			
2500	WATER CONTROL: water meter, backflow preventer,			
2600	Shock absorbers, vacuum breakers, mixer....10 to 15% of fixtures			
2700	PIPE AND FITTINGS: .... 30 to 60% of fixtures			
2800				
2900	Cut & patch to match existing construction, add, minimum	2%	3%	
3000	Maximum	5%	9%	
3100	Dust protection, add, minimum	1%	2%	
3200	Maximum	4%	11%	
3300	Equipment usage curtailment, add, minimum	1%	1%	

D20 Plumbing

D2010 Plumbing Fixtures

D2010 957	Plumbing - Public Restroom	COST EACH		
		MAT.	**INST.**	**TOTAL**
3400	Maximum	3%	10%	
3500	Material handling, & storage limitation, add, minimum	1%	1%	
3600	Maximum	6%	7%	
3700	Protection of existing work, add, minimum	2%	2%	
3800	Maximum	5%	7%	
3900	Shift work requirements, add, minimum		5%	
4000	Maximum		30%	

D2010 Plumbing Fixtures

This page illustrates and describes a two fixture lavatory system including a water closet, lavatory, accessories and all service piping. Lines within System Components give the unit price and total price on a cost each basis for this system. Prices for two fixture lavatory systems are on Line Items D2010 958 1800 and 1900. Both material quantities and labor costs have been adjusted for the system listed.

Factors: To adjust for job conditions other than normal working situations use Lines D2010 958 2900 thru 4000.

Example: You are to install the system while controlling dust in the work area. Go to Line D2010 958 3200 and apply these percentages to the appropriate MAT. and INST. costs.

System Components	QUANTITY	UNIT	COST EACH MAT.	COST EACH INST.	COST EACH TOTAL
Two fixture bathroom incl. water closet, lavatory, accessories and					
Necessary service piping to install this system in 2 walls.					
Water closet, floor mounted, 2 piece, close coupled	1.000	Ea.	211	284	495
Rough in waste & vent for water closet	1.000	Set	380	495	875
Lavatory, 20" x 18", P.E. cast iron with accessories	1.000	Ea.	335	188	523
Rough in waste & vent for lavatory	1.000	Set	730	905	1,635
Additional service piping					
1/2" copper pipe with sweat solder joints	10.000	L.F.	27.60	103	130.60
2" black schedule 40 steel pipe with threaded couplings	12.000	L.F.	100.20	282	382.20
4" cast iron soil pipe with lead and oakum joints	7.000	L.F.	185.50	192.50	378
Accessories					
Toilet tissue dispenser, chrome, single roll	1.000	Ea.	17.60	23	40.60
18" long stainless steel towel bar	1.000	Ea.	42	30.50	72.50
Medicine cabinet with mirror, 20" x 16", unlighted	1.000	Ea.	112	50	162
TOTAL			2,140.90	2,553	4,693.90

D2010 958	Plumbing - Two Fixture Bathroom	COST EACH MAT.	COST EACH INST.	COST EACH TOTAL
1800	Water closet, lavavatory & accessories, in 2 walls with all service piping	2,150	2,550	4,700
1900	Above system installed in 1 wall	2,025	2,225	4,250
2000				
2400	NOTE: PLUMBING APPROXIMATIONS			
2500	WATER CONTROL: water meter, backflow preventer,			
2600	Shock absorbers, vacuum breakers, mixer....10 to 15% of fixtures			
2700	PIPE AND FITTINGS: .... 30 to 60% of fixtures			
2800				
2900	Cut & patch to match existing construction, add, minimum	2%	3%	
3000	Maximum	5%	9%	
3100	Dust protection, add, minimum	1%	2%	
3200	Maximum	4%	11%	
3300	Equipment usage curtailment, add, minimum	1%	1%	
3400	Maximum	3%	10%	
3500	Material handling & storage limitation, add, minimum	1%	1%	
3600	Maximum	6%	7%	
3700	Protection of existing work, add, minimum	2%	2%	
3800	Maximum	5%	7%	
3900	Shift work requirements, add, minimum		5%	
4000	Maximum		30%	

D2010 Plumbing Fixtures

This page illustrates and describes a three fixture bathroom system including a water closet, tub, lavatory, accessories and service piping. Lines within System Components give the unit price and total price on a cost each basis for this system. Prices for a three fixture bathroom system is on Line Item D2010 959 1700. Both material quantities and labor costs have been adjusted for the system listed.

Factors: To adjust for job conditions other than normal working situations use Lines D2010 959 2900 thru 4000.

Example: You are to install the system and protect all existing work. Go to Line D2010 959 3800 and apply these percentages to the appropriate MAT. and INST. costs.

System Components	QUANTITY	UNIT	COST EACH MAT.	COST EACH INST.	COST EACH TOTAL
Three fixture bathroom incl. water closet, bathtub, lavatory, accessories,					
And necessary service piping to install this system in 1 wall.					
Water closet, tank vit china flr mtd close cpld, 2 pc. w/seat supply & stop	1.000	Ea.	211	284	495
For rough in, supply, waste and vent	.930	Set	353.40	460.35	813.75
Bath tub, P.E., cast iron, 5' long, with accessories	1.000	Ea.	1,250	340	1,590
Rough-in, supply, waste and vent, for all above tubs, add	.950	Set	479.75	688.75	1,168.50
Lavatory, 20" x 18" P.E. cast iron with accessories	1.000	Ea.	335	188	523
Rough-in, supply, waste and vent for above lavatories	1.000	Set	730	905	1,635
Accessories					
Toilet tissue dispenser, surface mounted, SS, single roll	1.000	Ea.	17.60	23	40.60
Towel bar, stainless steel, 18" long	2.000	Ea.	84	61	145
Medicine cabinets with mirror, sst. frame, 16" x 22", unlighted	1.000	Ea.	112	50	162
TOTAL			3,572.75	3,000.10	6,572.85

D2010 959	Plumbing - Three Fixture Bathroom	MAT.	INST.	TOTAL
1600				
1700	Water closet, lavatory, bathtub, in 1 wall with all service piping	3,575	3,000	6,575
2400	NOTE: PLUMBING APPROXIMATIONS			
2500	WATER CONTROL: water meter, backflow preventer,			
2600	Shock absorbers, vacuum breakers, mixer....10 to 15% of fixtures			
2700	PIPE AND FITTINGS: .... 30 to 60% of fixtures			
2800				
2900	Cut & patch to match existing construction, add, minimum	2%	3%	
3000	Maximum	5%	9%	
3100	Dust protection, add, minimum	1%	2%	
3200	Maximum	4%	11%	
3300	Equipment usage curtailment, add, minimum	1%	1%	
3400	Maximum	3%	10%	
3500	Material handling & storage limitation, add, minimum	1%	1%	
3600	Maximum	6%	7%	
3700	Protection of existing work, add, minimum	2%	2%	
3800	Maximum	5%	7%	
3900	Shift work requirements, add, minimum		5%	
4000	Maximum		30%	

D2010 Plumbing Fixtures

This page illustrates and describes a three fixture bathroom system including a water closet, tub, lavatory, accessories, and service piping. Lines within System Components give the unit price and total price on a cost each basis for this system. Prices for a three fixture bathroom system are on Line Item D2010 960 1900 thru 2000. Both material quantities and labor costs have been adjusted for the system listed.

Factors: To adjust for job conditions other than normal working situations use Lines D2010 960 2900 thru 4000.

Example: You are to install the system and protect the surrounding area from dust. Go to Line D2010 960 3100 and apply these percentages to the appropriate MAT. and INST. costs.

System Components	QUANTITY	UNIT	COST EACH		
			MAT.	INST.	TOTAL
Three fixture bathroom incl. water closet, bathtub, lavatory, accessories,					
And necessary service piping to install this system in 2 walls.					
Water closet, floor mounted, 2 piece, close coupled	1.000	Ea.	211	284	495
Rough-in waste & vent for water closet	1.000	Set	380	495	875
Bathtub, P.E. cast iron, 5' long with accessories	1.000	Ea.	1,250	340	1,590
Rough-in waste & vent for bathtub	1.000	Set	505	725	1,230
Lavatory, 20" x 18" P.E. cast iron with accessories	1.000	Ea.	335	188	523
Rough-in waste & vent for lavatory	1.000	Set	730	905	1,635
Additional service piping					
1-1/4" copper DWV type tubing, sweat solder joints	6.000	L.F.	72	83.40	155.40
2" black schedule 40 steel pipe with threaded couplings	12.000	L.F.	100.20	282	382.20
Accessories					
Toilet tissue dispenser, chrome, single roll	1.000	Ea.	17.60	23	40.60
18" long stainless steel towel bar	2.000	Ea.	84	61	145
Medicine cabinet with mirror, 20" x 16", unlighted	1.000	Ea.	112	50	162
TOTAL			3,796.80	3,436.40	7,233.20

D2010 960	Plumbing - Three Fixture Bathroom	COST EACH		
		MAT.	INST.	TOTAL
1900	Water closet, lavatory, bathtub, in 2 walls with all service piping	3,800	3,425	7,225
2000	Above system with corner contour tub, P.E. cast iron	5,775	3,425	9,200
2300				
2400	NOTE: PLUMBING APPROXIMATIONS			
2500	WATER CONTROL: water meter, backflow preventer,			
2600	Shock absorbers, vacuum breakers, mixer....10 to 15% of fixtures			
2700	PIPE AND FITTINGS: .... 30 to 60% of fixtures			
2800				
2900	Cut & patch to match existing construction, add, minimum	2%	3%	
3000	Maximum	5%	9%	
3100	Dust protection, add, minimum	1%	2%	
3200	Maximum	4%	11%	
3300	Equipment usage curtailment, add, minimum	1%	1%	
3400	Maximum	3%	10%	
3500	Material handling & storage limitation, add, minimum	1%	1%	
3600	Maximum	6%	7%	
3700	Protection of existing work, add, minimum	2%	2%	
3800	Maximum	5%	7%	
3900	Shift work requirements, add, minimum		5%	
4000	Maximum		30%	

D2010 Plumbing Fixtures

This page illustrates and describes a three fixture bathroom system including a water closet, shower, lavatory, accessories, and service piping, Lines within System Components give the unit price and total price on a cost each basis for this system. Prices for three fixture bathroom systems are on Line Items D2010 961 2100 thru 2200. Both material quantities and labor costs have been adjusted for the system listed.

Factors: To adjust for job conditions other than normal working situations use Lines D2010 961 2900 thru 4000.

Example: You are to install the system and protect existing construction. Go to Line D2010 961 3700 and apply these percentages to the appropriate MAT. and INST. costs.

System Components	QUANTITY	UNIT	COST EACH		
			MAT.	INST.	TOTAL
Three fixture bathroom incl. water closet, shower, lavatory, accessories, And necessary service piping to install this system in 2 walls.					
Water closet, floor mounted, 2 piece, close coupled	1.000	Ea.	211	284	495
Rough-in waste & vent for water closet	1.000	Set	380	495	875
32" shower, enameled stall, molded stone receptor	1.000	Ea.	1,275	300	1,575
Rough-in waste & vent for shower	1.000	Set	490	735	1,225
Lavatory, 20" x 18" P.E. cast iron with accessories	1.000	Ea.	335	188	523
Rough-in waste & vent for lavatory	1.000	Set	730	905	1,635
Additional service piping					
1-1/4" Copper DWV type tubing, sweat solder joints	4.000	L.F.	48	55.60	103.60
2" black schedule 40 steel pipe with threaded couplings	6.000	L.F.	50.10	141	191.10
4" cast iron soil with lead and oakum joints	7.000	L.F.	185.50	192.50	378
Accessories					
Toilet tissue dispenser, chrome, single roll	1.000	Ea.	17.60	23	40.60
18" long stainless steel towel bar	2.000	Ea.	84	61	145
Medicine cabinet with mirror, 20" x 16", unlighted	1.000	Ea.	112	50	162
TOTAL			3,918.20	3,430.10	7,348.30

D2010 961	Plumbing - Three Fixture Bathroom	COST EACH		
		MAT.	INST.	TOTAL
2100	Water closet, 32" shower, lavavatory, in 2 walls with all service piping	3,925	3,425	7,350
2200	Above system with 36" corner angle shower	5,075	3,450	8,525
2300				
2400	NOTE: PLUMBING APPROXIMATIONS			
2500	WATER CONTROL: water meter, backflow preventer,			
2600	Shock absorbers, vacuum breakers, mixer....10 to 15% of fixtures			
2700	PIPE AND FITTINGS: .... 30 to 60% of fixtures			
2800				
2900	Cut & patch to match existing construction, add, minimum	2%	3%	
3000	Maximum	5%	9%	
3100	Dust protection, add, minimum	1%	2%	
3200	Maximum	4%	11%	
3300	Equipment usage curtailment, add, minimum	1%	1%	
3400	Maximum	3%	10%	
3500	Material handling & storage limitation, add, minimum	1%	1%	
3600	Maximum	6%	7%	
3700	Protection of existing work, add, minimum	2%	2%	
3800	Maximum	5%	7%	
3900	Shift work requirements, add, minimum		5%	
4000	Maximum		30%	

D2010 Plumbing Fixtures

This page illustrates and describes a four fixture bathroom system including a water closet, shower, bathtub, lavatory, accessories, and service piping. Lines within System Components give the unit price and total price on a cost each basis for this system. Prices for four fixture bathroom systems are on Line Items D2010 962 1800 thru 1900. Both material quantities and labor costs have been adjusted for the system listed.

Factors: To adjust for job conditions other than normal working situations use Lines D2010 962 2900 thru 4000.

Example: You are to install the system during weekends and evenings. Go to Line D2010 962 4000 and apply these percentages to the appropriate MAT. and INST. costs.

System Components	QUANTITY	UNIT	COST EACH MAT.	COST EACH INST.	COST EACH TOTAL
Four fixture bathroom incl. water closet, shower, bathtub, lavatory					
Accessories and necessary service piping to install this system in 2 walls.					
Water closet, floor mounted, 2 piece, close coupled	1.000	Ea.	211	284	495
Rough-in waste & vent for water closet	1.000	Set	380	495	875
32" shower, enameled steel stall, molded stone receptor	1.000	Ea.	1,275	300	1,575
Rough-in waste & vent for shower	1.000	Set	490	735	1,225
Bathtub, P.E. cast iron, 5' long with accessories	1.000	Ea.	1,250	340	1,590
Rough-in waste & vent for bathtub	1.000	Set	505	725	1,230
Lavatory, 20" x 18" P.E. cast iron with accessories	1.000	Ea.	335	188	523
Rough-in waste & vent for lavatory	1.000	Set	730	905	1,635
Accessories					
Toilet tissue dispenser, chrome, single roll	1.000	Ea.	17.60	23	40.60
18" long stainless steel towel bar	2.000	Ea.	84	61	145
Medicine cabinet with mirror, 20" x 16" unlighted	1.000	Ea.	112	50	162
TOTAL			5,389.60	4,106	9,495.60

D2010 962	Plumbing - Four Fixture Bathroom	COST EACH MAT.	COST EACH INST.	COST EACH TOTAL
1800	Water closet, 32" shower, bathtub, lavavatory, in 2 walls	5,400	4,100	9,500
1900	Above system with 36" corner angle shower, plumbing in 3 walls	6,675	4,500	11,175
2000				
2300				
2400	NOTE: PLUMBING APPROXIMATIONS			
2500	WATER CONTROL: water meter, backflow preventer,			
2600	Shock absorbers, vacuum breakers, mixer....10 to 15% of fixtures			
2700	PIPE AND FITTINGS: .... 30 to 60% of fixtures			
2800				
2900	Cut & patch to match existing construction, add, minimum	2%	3%	
3000	Maximum	5%	9%	
3100	Dust protection, add, minimum	1%	2%	
3200	Maximum	4%	11%	
3300	Equipment usage curtailment, add, minimum	1%	1%	
3400	Maximum	3%	10%	
3500	Material handling & storage limitation, add, minimum	1%	1%	
3600	Maximum	6%	7%	
3700	Protection of existing work, add, minimum	2%	2%	
3800	Maximum	5%	7%	
3900	Shift work requirements, add, minimum		5%	
4000	Maximum		30%	

685

D2010 Plumbing Fixtures

This page illustrates and describes a five fixture bathroom system including a water closet, shower, bathtub, lavatories, accessories, and service piping. Lines within System Components give the unit price and total price on a cost each basis for this system. Prices for five fixture bathroom systems are on Line Items D2010 963 1800 thru 1900. Both material quantities and labor costs have been adjusted for the system listed.

Factors: To adjust for job conditions other than normal working situations use Lines D2010 963 2900 thru 4000.

Example: You are to install and match any existing construction. Go to Line D2010 963 2900 and apply these percentages to the appropriate MAT. and INST. costs.

System Components	QUANTITY	UNIT	COST EACH MAT.	COST EACH INST.	COST EACH TOTAL
Five fixture bathroom incl. water closet, shower, bathtub, 2 lavatories, Accessories and necessary service piping to install this system in 1 wall.					
Water closet, floor mounted, 2 piece, close coupled	1.000	Ea.	211	284	495
Rough-in waste & vent for water closet	1.000	Set	380	495	875
36" corner angle shower, enameled steel stall, molded stone receptor	1.000	Ea.	2,425	315	2,740
Rough-in waste & vent for shower	1.000	Set	490	735	1,225
Bathtub, P.E. cast iron, 5' long with accessories	1.000	Ea.	1,250	340	1,590
Rough-in waste & vent for bathtub	1.000	Set	505	725	1,230
Countertop, plastic laminate, with backsplash	2.000	Ea.	240	92	332
Vanity base, 2 doors, 24" wide	2.000	Ea.	840	254	1,094
Lavatories, 20" x 18" cabinet mntd, PECI	2.000	Ea.	880	470	1,350
Rough-in waste & vent for lavatories	1.600	Set	712	1,048	1,760
Accessories					
Toilet tissue dispenser, chrome, single roll	1.000	Ea.	17.60	23	40.60
18" long stainless steel towel bars	2.000	Ea.	84	61	145
Medicine cabinet with mirror, 20" x 16", unlighted	2.000	Ea.	224	100	324
TOTAL			8,258.60	4,942	13,200.60

D2010 963	Plumbing - Five Fixture Bathroom	COST EACH MAT.	COST EACH INST.	COST EACH TOTAL
1800	Water closet, shower, bathtub, 2 lavatories, on 1 wall	8,250	4,950	13,200
1900	Above system installed in 2 walls with all necessary service piping	8,325	5,100	13,425
2000				
2100				
2200				
2300				
2400	NOTE: PLUMBING APPROXIMATIONS			
2500	WATER CONTROL: water meter, backflow preventer,			
2600	Shock absorbers, vacuum breakers, mixer....10 to 15% of fixtures			
2700	PIPE AND FITTINGS: .... 30 to 60% of fixtures			
2800				
2900	Cut & patch to match existing construction, add, minimum	2%	3%	
3000	Maximum	5%	9%	
3100	Dust protection, add, minimum	1%	2%	
3200	Maximum	4%	11%	
3300	Equipment usage curtailment, add, minimum	1%	1%	
3400	Maximum	3%	10%	
3500	Material handling & storage limitation, add, minimum	1%	1%	

686

D2010 Plumbing Fixtures

D2010 963		Plumbing - Five Fixture Bathroom	COST EACH		
			MAT.	INST.	TOTAL
3600		Maximum	6%	7%	
3700		Protection of existing work, add, minimum	2%	2%	
3800		Maximum	5%	7%	
3900		Shift work requirements, add, minimum		5%	
4000		Maximum		30%	

687

D3010 Energy Supply

This page illustrates and describes an oil fired hot water baseboard system including an oil fired boiler, fin tube radiation and all fittings and piping. Lines within System Components give the unit price and total price per square foot for this system. Prices for oil fired hot water baseboard systems are on Line Items D3010 540 1600 thru 2100. Both material quantities and labor costs have been adjusted for the system listed.

Factors: To adjust for job conditions other than normal working situations use Lines D3010 540 2900 thru 4000.

Example: You are to install the system while protecting all existing work. Go to Line D3010 540 3700 and apply these percentages to the appropriate MAT. and INST. costs.

System Components	QUANTITY	UNIT	COST PER S.F.		
			MAT.	INST.	TOTAL
Oil fired hot water baseboard system including boiler, fin tube radiation					
And all necessary fittings and piping.					
Area to 800 S.F.					
Boiler, cast iron, w/oil piping, 97 MBH	1.000	Ea.	3,062.50	1,843.75	4,906.25
Copper piping	130.000	L.F.	1,534	1,872	3,406
Fin tube radiation	68.000	L.F.	584.80	1,768	2,352.80
Circulator	1.000	Ea.	745	251	996
Oil tank	1.000	Ea.	525	305	830
Expansion tank, ASME	1.000	Ea.	800	108	908
TOTAL			7,251.30	6,147.75	13,399.05
COST PER S.F.		S.F.	9.06	7.68	16.74

D3010 540	Heating - Oil Fired Hot Water	COST PER S.F.		
		MAT.	INST.	TOTAL
1600	Cast iron boiler, area to 800 S.F.	9.06	7.69	16.75
1700	To 1000 S.F.	7.85	6.60	14.45
1800	To 1200 S.F.	6.75	6	12.75
1900	To 1600 S.F.	8.50	6	14.50
2000	To 2000 S.F.	7.05	5.30	12.35
2100	To 3000 S.F.	5.15	4.48	9.63
2200				
2300				
2700				
2800				
2900	Cut & patch to match existing construction, add, minimum	2%	3%	
3000	Maximum	5%	9%	
3100	Dust protection, add, minimum	1%	2%	
3200	Maximum	4%	11%	
3300	Equipment usage curtailment, add, minimum	1%	1%	
3400	Maximum	3%	10%	
3500	Material handling & storage limitation, add, minimum	1%	1%	
3600	Maximum	6%	7%	
3700	Protection of existing work, add, minimum	2%	2%	
3800	Maximum	5%	7%	
3900	Shift work requirements, add, minimum		5%	
4000	Maximum		30%	

D3010 Energy Supply

This page illustrates and describes a gas fired hot water baseboard system including a gas fired boiler, fin tube radiation, fittings and piping. Lines within System Components give the unit price and total price per square foot for this system. Prices for gas fired hot water baseboard systems are on Line Items D3010 550 1400 thru 1900. Both material quantities and labor costs have been adjusted for the system listed.

Factors: To adjust for job conditions other than normal working situations use Lines D3010 550 2900 thru 4000.

Example: You are to install the system with minimal equipment usage. Go to Line D3010 550 3400 and apply these percentages to the appropriate MAT. and INST. costs.

System Components	QUANTITY	UNIT	COST PER S.F. MAT.	COST PER S.F. INST.	COST PER S.F. TOTAL
Gas fired hot water baseboard system including boiler, fin tube radiation,					
All necessary fittings and piping.					
Area to 800 S.F.					
Cast iron boiler, insulating jacket, gas piping, 80 MBH	1.000	Ea.	3,937.50	2,750	6,687.50
Copper piping	130.000	L.F.	1,534	1,872	3,406
Fin tube radiation	68.000	L.F.	584.80	1,768	2,352.80
Circulator, flange connection	1.000	Ea.	745	251	996
Expansion tank, ASME	1.000	Ea.	800	108	908
TOTAL			7,601.30	6,749	14,350.30
COST PER S.F.		S.F.	9.50	8.44	17.94

D3010 550	Heating - Gas Fired Hot Water	COST PER S.F. MAT.	COST PER S.F. INST.	COST PER S.F. TOTAL
1400	Cast iron boiler, area to 800 S.F.	9.50	8.44	17.94
1500	To 1000 S.F.	7.90	7.40	15.30
1600	To 1200 S.F.	6.80	6.65	13.45
1700	To 1600 S.F.	8.55	6.50	15.05
1800	To 2000 S.F.	7.05	5.80	12.85
1900	To 3000 S.F.	5.45	5.10	10.55
2000				
2100				
2900	Cut & patch to match existing construction, add, minimum	2%	3%	
3000	Maximum	5%	9%	
3100	Dust protection, add, minimum	1%	2%	
3200	Maximum	4%	11%	
3300	Equipment usage curtailment, add, minimum	1%	1%	
3400	Maximum	3%	10%	
3500	Material handling & storage limitation, add, minimum	1%	1%	
3600	Maximum	6%	7%	
3700	Protection of existing work, add, minimum	2%	2%	
3800	Maximum	5%	7%	
3900	Shift work requirements, add, minimum		5%	
4000	Maximum		30%	

D3010 Energy Supply

This page illustrates and describes a wood burning stove system including a free standing stove, preformed hearth, masonry chimney, and necessary piping and fittings. Lines within System Components give the unit price and total price on a cost each basis for this system. Prices for wood burning stove systems are on Line Items D3010 991 1400 thru 2000. Both material quantities and labor costs have been adjusted for the system listed.

Factors: To adjust for job conditions other than normal working situations use Lines D3010 991 2900 thru 4000.

Example: You are to install the system and protect existing construction. Go to Line D3010 991 3500 and apply these percentages to the appropriate MAT., and INST. costs.

System Components

System Components	QUANTITY	UNIT	COST EACH MAT.	COST EACH INST.	COST EACH TOTAL
Cast iron, free standing, wood burning stove with preformed hearth, masonry Chimney, and all necessary piping and fittings to install in chimney.					
Cast iron wood burning stove, stove pipe	1.000	Ea.	1,600	1,075	2,675
Wall panel or hearth, non-combustible	1.000	Ea.	1,026	197.60	1,223.60
16" x 16" brick chimney, 8" x 8" flue	20.000	V.L.F.	725	1,725	2,450
Concrete in place, chimney foundations	.500	C.Y.	103	144.47	247.47
TOTAL			3,454	3,142.07	6,596.07

D3010 991 — Wood Burning Stoves

		COST EACH MAT.	COST EACH INST.	COST EACH TOTAL
1400	Cast iron, free standing with preformed hearth	3,450	3,150	6,600
1500				
1600	Installed in existing fireplace	2,150	1,175	3,325
1700				
1800	Installed with insulated metal chimney			
1900	System including ceiling package,			
2000	Metal chimney to 10 L.F.	3,875	1,675	5,550
2100				
2200				
2300				
2400				
2500				
2600				
2700				
2900	Dust protection, add, minimum	1%	2%	
3000	Maximum	4%	11%	
3100	Equipment usage curtailment, add, minimum	1%	1%	
3200	Maximum	3%	10%	
3300	Material handling & storage limitation, add, minimum	1%	1%	
3400	Maximum	6%	7%	
3500	Protection of existing work, add, minimum	2%	2%	
3600	Maximum	5%	7%	
3700	Shift work requirements, add, minimum		5%	
3800	Maximum		30%	
3900	Temporary shoring and bracing, add, minimum	2%	5%	
4000	Maximum	5%	12%	

D3010 Energy Supply

This page illustrates and describes masonry fireplace systems including a brick fireplace, footing, foundation, hearth, firebox, chimney and flue. Lines within System Components give the unit price and total price on a cost each basis for this system. Prices for masonry fireplace systems are on Line Items D3010 992 1200 thru 1500. Both material quantities and labor costs have been adjusted for the system listed.

Factors: To adjust for job conditions other than normal working situations use Lines D3010 992 2700 thru 4000.

Example: You are to install the system with some temporary shoring and bracing. Go to Line D3010 992 3900 and apply these percentages to the appropriate MAT. and INST. costs.

System Components	QUANTITY	UNIT	COST EACH MAT.	INST.	TOTAL
Brick masonry fireplace, including footing, foundation, hearth, firebox, Chimney and flue, chimney 12′ above firebox.					
Footing, 4′ x 7′ x 12″ thick, 3000 psi concrete	1.040	C.Y.	238.16	344.32	582.48
Foundation, 12″ concrete block	180.000	S.F.	1,062	2,250	3,312
Fireplace, brick faced, 6′-0″ wide x 5′-0″ high	1.000	Ea.	690	3,125	3,815
Hearth	1.000	Ea.	236	625	861
Chimney, 20″ x 20″, one 12″ x 12″ flue, 12 V.L.F.	12.000	V.L.F.	684	1,098	1,782
Mantle, wood	7.000	L.F.	84.70	135.45	220.15
TOTAL			2,994.86	7,577.77	10,572.63

D3010 992	Masonry Fireplace	COST EACH MAT.	INST.	TOTAL
1200	Chimney, 20″ x 20″, one 12″ x 12″ flue, 12 V.L.F.	3,000	7,575	10,575
1300	18 V.L.F.	3,325	8,125	11,450
1400	24 V.L.F.	3,675	8,675	12,350
1500	Fieldstone face instead of brick	3,300	7,575	10,875
1600				
1700				
1800				
1900				
2000				
2100				
2200				
2300				
2700	Cut & patch to match existing construction, add, minimum	2%	3%	
2800	Maximum	5%	9%	
2900	Dust protection, add, minimum	1%	2%	
3000	Maximum	4%	11%	
3100	Equipment usage curtailment, add, minimum	1%	1%	
3200	Maximum	3%	10%	
3300	Material handling & storage limitation, add, minimum	1%	1%	
3400	Maximum	6%	7%	
3500	Protection of existing work, add, minimum	2%	2%	
3600	Maximum	5%	7%	
3700	Shift work requirements, add, minimum		5%	
3800	Maximum		30%	
3900	Temporary shoring and bracing, add, minimum	2%	5%	
4000	Maximum	5%	12%	

691

D3020   Heat Generating Systems

This page illustrates and describes an oil fired forced air system including an oil fired furnace, ductwork, registers and hookups. Lines within System Components give the unit price and total price per square foot for this system. Prices for oil fired forced air systems are on Line Items D3020 122 1600 thru 2800. Both material quantities and labor costs have been adjusted for the system listed.

Factors: To adjust for job conditions other than normal working situations use Lines D3020 122 3100 thru 4200.

Example: You are to install the system during evenings and weekends. Go to Line D3020 122 4200 and apply this percentage to the appropriate INST. cost.

System Components	QUANTITY	UNIT	COST PER S.F.		
			MAT.	INST.	TOTAL
Oil fired hot air heating system including furnace, ductwork, registers					
And all necessary hookups.					
Area to 800 S.F., heat only					
Furnace, oil, atomizing gun type burner, w/oil piping	1.000	Ea.	4,656.25	512.50	5,168.75
Oil tank, steel, 275 gallon	1.000	Ea.	525	305	830
Duct, galvanized, steel	312.000	Lb.	202.80	3,026.40	3,229.20
Insulation, blanket type, duct work	270.000	S.F.	1,323	4,455	5,778
Flexible duct, 6" diameter, insulated	100.000	L.F.	419	565	984
Registers, baseboard, gravity, 12" x 6"	8.000	Ea.	108	284	392
Return, damper, 18" x 36"	1.000	Ea.	184	81.50	265.50
TOTAL			7,418.05	9,229.40	16,647.45
COST PER S.F.		S.F.	9.27	11.54	20.81

D3020 122	Heating-Cooling, Oil, Forced Air	COST PER S.F.		
		MAT.	INST.	TOTAL
1600	Oil fired, area to 800 S.F.	9.28	11.53	20.81
1700	To 1000 S.F.	7.45	9.30	16.75
1800	To 1200 S.F.	6.40	8.05	14.45
1900	To 1600 S.F.	5.10	6.50	11.60
2000	To 2000 S.F.	5.25	8.85	14.10
2100	To 3000 S.F.	3.98	7.05	11.03
2200	For combined heating and cooling systems:			
2300	Oil fired, heating and cooling, area to 800 S.F.	12.25	12.90	25.15
2400	To 1000 S.F.	9.60	10.45	20.05
2500	To 1200 S.F.	8.20	9	17.20
2600	To 1600 S.F.	6.90	7.25	14.15
2700	To 2000 S.F.	6.65	9.55	16.20
2800	To 3000 S.F.	4.94	7.55	12.49
3100	Cut & patch to match existing construction, add, minimum	2%	3%	
3200	Maximum	5%	9%	
3300	Dust protection, add, minimum	1%	2%	
3400	Maximum	4%	11%	
3500	Equipment usage curtailment, add, minimum	1%	1%	
3600	Maximum	3%	10%	
3700	Material handling & storage limitation, add, minimum	1%	1%	
3800	Maximum	6%	7%	
3900	Protection of existing work, add, minimum	2%	2%	
4000	Maximum	5%	7%	
4100	Shift work requirements, add, minimum		5%	
4200	Maximum		30%	

D3020 Heat Generating Systems

This page illustrates and describes a gas fired forced air system including a gas fired furnace, ductwork, registers and hookups. Lines within System Components give the unit price and total price per square foot for this system. Prices for gas fired forced air systems are on Line Items D3020 124 1400 thru 2600. Both material quantities and labor costs have been adjusted for the system listed.

Factors: To adjust for job conditions other than normal working situations use Lines D3020 124 2900 thru 4000.

Example: You are to install the system with material handling and storage limitations. Go to Line D3020 124 3500 and apply these percentages to the appropriate MAT. and INST. costs.

System Components	QUANTITY	UNIT	COST PER S.F. MAT.	COST PER S.F. INST.	COST PER S.F. TOTAL
Gas fired hot air heating system including furnace, ductwork, registers					
And all necessary hookups.					
Area to 800 S.F., heat only					
Furnace, gas, AGA certified, direct drive, w/gas piping, 44 MBH	1.000	Ea.	987.50	456.25	1,443.75
Duct, galvanized steel	312.000	Lb.	202.80	3,026.40	3,229.20
Insulation, blanket type, ductwork	270.000	S.F.	1,323	4,455	5,778
Flexible duct, 6″ diameter, insulated	100.000	L.F.	419	565	984
Registers, baseboard, gravity, 12″ x 6″	8.000	Ea.	108	284	392
Return, damper, 18″ x 36″	1.000	Ea.	184	81.50	265.50
TOTAL			3,224.30	8,868.15	12,092.45
COST PER S.F.		S.F.	4.03	11.09	15.12

D3020 124	Heating-Cooling, Gas, Forced Air	COST PER S.F. MAT.	COST PER S.F. INST.	COST PER S.F. TOTAL
1400	Gas fired, area to 800 S.F.	4.03	11.09	15.12
1500	To 1000 S.F.	3.17	8.95	12.12
1600	To 1200 S.F.	2.84	7.75	10.59
1700	To 1600 S.F.	2.85	7.05	9.90
1800	To 2000 S.F.	3.11	8.70	11.81
1900	To 3000 S.F.	2.45	6.95	9.40
2000	For combined heating and cooling systems:			
2100	Gas fired, heating and cooling, area to 800 S.F.	7	12.45	19.45
2200	To 1000 S.F.	5.30	10.10	15.40
2300	To 1200 S.F.	4.61	8.70	13.31
2400	To 1600 S.F.	4.67	7.85	12.52
2500	To 2000 S.F.	4.50	9.40	13.90
2600	To 3000 S.F.	3.41	7.45	10.86
2900	Cut & patch to match existing construction, add, minimum	2%	3%	
3000	Maximum	5%	9%	
3100	Dust protection, add, minimum	1%	2%	
3200	Maximum	4%	11%	
3300	Equipment usage curtailment, add, minimum	1%	1%	
3400	Maximum	3%	10%	
3500	Material handling & storage limitation, add, minimum	1%	1%	
3600	Maximum	6%	7%	
3700	Protection of existing work, add, minimum	2%	2%	
3800	Maximum	5%	7%	
3900	Shift work requirements, add, minimum		5%	
4000	Maximum		30%	

D3020 Heat Generating Systems

Cast Iron Boiler, Hot Water, Gas Fired

This page illustrates and describes boilers including expansion tank, circulating pump and all service piping. Lines within Systems Components give the material and installation price on a cost each basis for the components. Prices for boiler systems are on Line Items D3020 130 1010 thru D3020 138 1040. Material quantities and labor costs have been adjusted for the system listed.

Factors: To adjust for job conditions other than normal working situations use Line D3020 138 2700 thru 4000.

System Components	QUANTITY	UNIT	COST EACH MAT.	COST EACH INST.	COST EACH TOTAL
Cast iron boiler, including piping, chimney and accessories					
Boilers, gas fired, std controls, CI, insulated, HW, gross output 100 MBH	1.000	Ea.	3,075	2,375	5,450
Pipe, black steel, Sch 40, threaded, W/coupling & hangers, 10' OC, 3/4" dia	22.500	L.F.	81	308.25	389.25
Pipe, black steel, Sch 40, threaded, W/coupling & hangers, 10' OC, 1" dia.	21.500	L.F.	88.15	338.63	426.78
Elbow, 90°, black, straight, 3/4" dia.	9.000	Ea.	51.75	535.50	587.25
Elbow, 90°, black, straight, 1" dia.	6.000	Ea.	60	387	447
Tee, black, straight, 3/4" dia.	2.000	Ea.	18.30	186	204.30
Tee, black, straight, 1" dia.	2.000	Ea.	31.20	208	239.20
Tee, black, reducing, 1" dia.	2.000	Ea.	35.60	208	243.60
Pipe cap, black, 3/4" dia.	1.000	Ea.	6.50	26	32.50
Union, black with brass seat, 3/4" dia.	2.000	Ea.	41	129	170
Union, black with brass seat, 1" dia.	2.000	Ea.	54	139	193
Pipe nipples, black, 3/4" dia.	5.000	Ea.	9	34.25	43.25
Pipe nipples, black, 1" dia.	3.000	Ea.	6.15	23.63	29.78
Valves, bronze, gate, N.R.S., threaded, class 150, 3/4" size	2.000	Ea.	242	84	326
Valves, bronze, gate, N.R.S., threaded, class 150, 1" size	2.000	Ea.	304	88	392
Gas cock, brass, 3/4" size	1.000	Ea.	27	38	65
Thermometer, stem type, 9" case, 8" stem, 3/4" NPT	2.000	Ea.	422	60	482
Tank, steel, liquid expansion, ASME, painted, 15 gallon capacity	1.000	Ea.	970	89.50	1,059.50
Pump, circulating, bronze, flange connection, 3/4" to 1-1/2" size, 1/8 HP	1.000	Ea.	1,350	251	1,601
Vent chimney, all fuel, pressure tight, double wall, SS, 6" dia.	20.000	L.F.	1,420	490	1,910
Vent chimney, elbow, 90° fixed, 6" dia.	2.000	Ea.	900	98	998
Vent chimney, Tee, 6" dia.	2.000	Ea.	544	122	666
Vent chimney, ventilated roof thimble, 6" dia.	1.000	Ea.	291	56.50	347.50
Vent chimney, flat roof flashing, 6" dia.	1.000	Ea.	127	49	176
Vent chimney, stack cap, 6" diameter	1.000	Ea.	325	32	357
Insulation, fiberglass pipe covering, 1" wall, 1" IPS	20.000	L.F.	25	126	151
TOTAL			10,504.65	6,482.26	16,986.91

D3020 130	Boiler, Cast Iron, Hot Water, Gas	COST EACH MAT.	COST EACH INST.	COST EACH TOTAL
1000	For alternate boilers:			
1010	Boiler, cast iron, gas, hot water, 100 MBH	10,500	6,475	16,975
1020	200 MBH	12,300	7,400	19,700
1030	320 MBH	10,700	8,100	18,800
1040	440 MBH	15,300	10,400	25,700
1050	544 MBH	25,300	16,400	41,700
1060	765 MBH	30,700	17,600	48,300
1070	1088 MBH	34,700	18,800	53,500

D30 HVAC

D3020 Heat Generating Systems

D3020 130	Boiler, Cast Iron, Hot Water, Gas	COST EACH		
		MAT.	INST.	TOTAL
1080	1530 MBH	43,300	24,900	68,200
1090	2312 MBH	53,000	29,300	82,300

D3020 134	Boiler, Cast Iron, Steam, Gas	COST EACH		
		MAT.	INST.	TOTAL
1010	Boiler, cast iron, gas, steam, 100 MBH	8,175	6,275	14,450
1020	200 MBH	20,000	11,800	31,800
1030	320 MBH	21,100	13,300	34,400
1040	544 MBH	29,100	20,800	49,900
1050	765 MBH	33,300	21,700	55,000
1060	1275 MBH	40,200	23,900	64,100
1070	2675 MBH	53,500	34,800	88,300

D3020 136	Boiler, Cast Iron, Hot Water, Gas/Oil	COST EACH		
		MAT.	INST.	TOTAL
1010	Boiler, cast iron, gas & oil, hot water, 584 MBH	32,200	18,200	50,400
1020	876 MBH	42,500	19,500	62,000
1030	1168 MBH	50,000	22,000	72,000
1040	1460 MBH	55,000	26,500	81,500
1050	2044 MBH	67,000	27,500	94,500
1060	2628 MBH	83,500	33,300	116,800

D3020 138	Boiler, Cast Iron, Steam, Gas/Oil	COST EACH		
		MAT.	INST.	TOTAL
1010	Boiler, cast iron, gas & oil, steam, 810 MBH	38,500	23,600	62,100
1020	1360 MBH	44,000	25,100	69,100
1030	2040 MBH	50,500	32,400	82,900
1040	2700 MBH	53,500	35,900	89,400
2700	Cut & patch to match existing construction, add, minimum	2%	3%	
2800	Maximum	5%	9%	
2900	Dust protection, add, minimum	1%	2%	
3000	Maximum	4%	11%	
3100	Equipment usage curtailment, add, minimum	1%	1%	
3200	Maximum	3%	10%	
3300	Material handling & storage limitation, add, minimum	1%	1%	
3400	Maximum	6%	7%	
3500	Protection of existing work, add, minimum	2%	2%	
3600	Maximum	5%	7%	
3700	Shift work requirements, add, minimum		5%	
3800	Maximum		30%	
3900	Temporary shoring and bracing, add, minimum	2%	5%	
4000	Maximum	5%	12%	

D3020 Heat Generating Systems

Flow ↓   ↑ Flow

Gate Valve →
Y Strainer →
Pressure Gauge →
Expansion Joint →

← Combination Valve
← Pressure Gauge
← Expansion Joint

Base Mounted End-Suction Pump

This page illustrates and describes end suction base mounted circulating pumps, including strainer and piping. Lines within system components give the material and installation price on a cost each basis for the components. Prices for circulating pumps are on Line items D3020 330 1010 thru 1050. Material quantities and labor costs have been adjusting for the system listed.

Factors: To adjust for job conditions other than normal working situations uses Lines D3020 330 2700 thru 4000.

System Components			COST EACH		
	QUANTITY	UNIT	MAT.	INST.	TOTAL
Circulator pump, end suction, including pipe,					
Valves, fittings and guages.					
Pump, circulating, CI, base mounted, 2-1/2" size, 3 HP, to 150 GPM	1.000	Ea.	10,900	835	11,735
Pipe, black steel, Sch. 40, on yoke & roll hangers, 10' O.C., 2-1/2" dia	12.000	L.F.	166.80	414.24	581.04
Elbow, 90°, weld joint, steel, 2-1/2" pipe size	1.000	Ea.	36	202.80	238.80
Flange, weld neck, 150 LB, 2-1/2" pipe size	8.000	Ea.	312	811.20	1,123.20
Valve, iron body, gate, 125 lb., N.R.S., flanged, 2-1/2" size	1.000	Ea.	1,025	300	1,325
Strainer, Y type, iron body, flanged, 125 lb., 2-1/2" pipe size	1.000	Ea.	182	305	487
Multipurpose valve, CI body, 2" size	1.000	Ea.	655	105	760
Expansion joint, flanged spool, 6" F to F, 2-1/2" dia	2.000	Ea.	780	244	1,024
T-O-L, weld joint, socket, 1/4" pipe size, nozzle	2.000	Ea.	18	141.30	159.30
T-O-L, weld joint, socket, 1/2" pipe size, nozzle	2.000	Ea.	18	147.70	165.70
Control gauges, pressure or vacuum, 3-1/2" diameter dial	2.000	Ea.	48	53	101
Pressure/temperature relief plug, 316 SS, 3/4" OD, 7-1/2" insertion	2.000	Ea.	156	53	209
Insulation, fiberglass pipe covering, 1-1/2" wall, 2-1/2" IPS	12.000	L.F.	48.72	92.40	141.12
Water flow	1.000	Ea.	1,425	1,275	2,700
Pump balancing	1.000	Ea.		370	370
TOTAL			15,770.52	5,349.64	21,120.16

D3020 330	Circulating Pump Systems, End Suction	COST EACH		
		MAT.	INST.	TOTAL
1000	For alternate pump sizes:			
1010	Pump, base mtd with motor, end-suction, 2-1/2" size, 3 HP, to 150 GPM	15,800	5,350	21,150
1020	3" size, 5 HP, to 225 GPM	19,100	6,050	25,150
1030	4" size, 7-1/2 HP, to 350 GPM	20,000	7,175	27,175
1040	5" size, 15 HP, to 1000 GPM	27,800	10,600	38,400
1050	6" size, 25 HP, to 1550 GPM	38,700	12,800	51,500
2700	Cut & patch to match existing construction, add, minimum	2%	3%	
2800	Maximum	5%	9%	
2900	Dust protection, add, minimum	1%	2%	
3000	Maximum	4%	11%	
3100	Equipment usage curtailment, add, minimum	1%	1%	
3200	Maximum	3%	10%	
3300	Material handling & storage limitation, add, minimum	1%	1%	
3400	Maximum	6%	7%	
3500	Protection of existing work, add, minimum	2%	2%	
3600	Maximum	5%	7%	
3700	Shift work requirements, add, minimum		5%	
3800	Maximum		30%	

For customer support on your Commercial Renovation Costs with RSMeans Data, call 800.448.8182.

D3020 Heat Generating Systems

D3020 330	Circulating Pump Systems, End Suction	COST EACH		
		MAT.	INST.	TOTAL
3900	Temporary shoring and bracing, add, minimum	2%	5%	
4000	Maximum	5%	12%	

D3040 Distribution Systems

Galvanized Steel Duct

Fiberglass Duct Insulation

Flexible
Fiberglass
Duct

Horizontal Fan Coil Air Conditioning System

This page illustrates and describes fan coil air conditioners including duct work, duct installation, piping and diffusers. Lines within Systems Components give the material and installation price on a cost each basis for the components. Prices for fan coil A/C unit systems are on Line Items D3040 118 1010 thru D3040 126 1120. Material quantities and labor costs have been adjusted for the system listed.

Factors: To adjust for job conditions other than normal working situations use Lines D3040 126 2700 thru 4000.

System Components	QUANTITY	UNIT	COST EACH		
			MAT.	INST.	TOTAL
Fan coil A/C system, horizontal, with housing, controls, 2 pipe, 1/2 ton					
Fan coil A/C, horizontal housing, filters, chilled water, 1/2 ton cooling	1.000	Ea.	1,600	190	1,790
Pipe, black steel, Sch 40, threaded, W/cplgs & hangers, 10' OC, 3/4" dia.	21.500	L.F.	77.40	294.55	371.95
Elbow, 90°, black, straight, 3/4" dia.	6.000	Ea.	34.50	357	391.50
Tee, black, straight, 3/4" dia.	2.000	Ea.	18.30	186	204.30
Union, black with brass seat, 3/4" dia.	2.000	Ea.	41	129	170
Pipe nipples, 3/4" diam	3.000	Ea.	5.40	20.55	25.95
Valves, bronze, gate, N.R.S., threaded, class 150, 3/4" size	2.000	Ea.	242	84	326
Circuit setter, balance valve, bronze body, threaded, 3/4" pipe size	1.000	Ea.	99.50	42	141.50
Insulation, fiberglass pipe covering, 1" wall, 3/4" IPS	20.000	L.F.	22.60	121	143.60
Ductwork, 12" x 8" fabricated, galvanized steel, 12 LF	55.000	Lb.	35.75	533.50	569.25
Insulation, ductwork, blanket type, fiberglass, 1" thk, 1-1/2 LB density	40.000	S.F.	196	660	856
Diffusers, aluminum, OB damper, ceiling, perf, 24"x24" panel size, 6"x6"	2.000	Ea.	404	102	506
Ductwork, flexible, fiberglass fabric, insulated, 1"thk, PE jacket, 6" dia	16.000	L.F.	67.04	90.40	157.44
Round volume control damper 6" dia.	2.000	Ea.	84	74	158
Fan coil unit balancing	1.000	Ea.		106	106
Re-heat coil balancing	1.000	Ea.		156	156
Diffuser/register, high, balancing	2.000	Ea.		294	294
TOTAL			2,927.49	3,440	6,367.49

D3040 118	Fan Coil A/C Unit, Two Pipe	COST EACH		
		MAT.	INST.	TOTAL
1000	For alternate A/C units:			
1010	Fan coil A/C system, cabinet mounted, controls, 2 pipe, 1/2 Ton	1,300	1,575	2,875
1020	1 Ton	1,550	1,750	3,300
1030	1-1/2 Ton	1,600	1,775	3,375
1040	2 Ton	2,700	2,175	4,875
1050	3 Ton	4,100	2,250	6,350

D3040 120	Fan Coil A/C Unit, Two Pipe, Electric Heat	COST EACH		
		MAT.	INST.	TOTAL
1010	Fan coil A/C system, cabinet mntd, elect. ht, controls, 2 pipe, 1/2 Ton	2,250	1,575	3,825
1020	1 Ton	2,700	1,750	4,450
1030	1-1/2 Ton	3,100	1,775	4,875
1040	2 Ton	4,675	2,200	6,875
1050	3 Ton	7,550	2,250	9,800

D3040 Distribution Systems

D3040 122	Fan Coil A/C Unit, Four Pipe	COST EACH		
		MAT.	INST.	TOTAL
1010	Fan coil A/C system, cabinet mounted, controls, 4 pipe, 1/2 Ton	2,150	3,050	5,200
1020	1 Ton	2,450	3,250	5,700
1030	1-1/2 Ton	2,500	3,275	5,775
1040	2 Ton	3,250	3,400	6,650
1050	3 Ton	5,375	3,650	9,025

D3040 124	Fan Coil A/C, Horizontal, Duct Mount, 2 Pipe	COST EACH		
		MAT.	INST.	TOTAL
1010	Fan coil A/C system, horizontal w/housing, controls, 2 pipe, 1/2 Ton	2,925	3,450	6,375
1020	1 Ton	3,925	5,475	9,400
1030	1-1/2 Ton	4,375	7,350	11,725
1040	2 Ton	5,825	9,125	14,950
1050	3 Ton	7,425	12,700	20,125
1060	3-1/2 Ton	7,800	13,000	20,800
1070	4 Ton	8,250	15,100	23,350
1080	5 Ton	9,800	19,400	29,200
1090	6 Ton	10,000	20,700	30,700
1100	7 Ton	10,200	21,300	31,500
1110	8 Ton	10,200	21,600	31,800
1120	10 Ton	10,900	22,100	33,000

D3040 126	Fan Coil A/C, Horiz., Duct Mount, 2 Pipe, Elec. Ht.	COST EACH		
		MAT.	INST.	TOTAL
1010	Fan coil A/C system, horiz. hsng, elect. ht, ctrls, 2 pipe, 1/2 Ton	3,200	3,450	6,650
1020	1 Ton	4,375	5,475	9,850
1030	1-1/2 Ton	4,850	7,350	12,200
1040	2 Ton	6,525	9,125	15,650
1050	3 Ton	10,200	13,000	23,200
1060	3-1/2 Ton	11,300	13,000	24,300
1070	4 Ton	12,800	15,200	28,000
1080	5 Ton	15,200	19,400	34,600
1090	6 Ton	15,400	20,800	36,200
1100	7 Ton	16,300	21,500	37,800
1110	8 Ton	16,600	22,200	38,800
1120	10 Ton	17,800	21,300	39,100
2700	Cut & patch to match existing construction, add, minimum	2%	3%	
2800	Maximum	5%	9%	
2900	Dust protection, add, minimum	1%	2%	
3000	Maximum	4%	11%	
3100	Equipment usage curtailment, add, minimum	1%	1%	
3200	Maximum	3%	10%	
3300	Material handling & storage limitation, add, minimum	1%	1%	
3400	Maximum	6%	7%	
3500	Protection of existing work, add, minimum	2%	2%	
3600	Maximum	5%	7%	
3700	Shift work requirements, add, minimum		5%	
3800	Maximum		30%	
3900	Temporary shoring and bracing, add, minimum	2%	5%	
4000	Maximum	5%	12%	

699

D3040 Distribution Systems

Plate Heat Exchanger   Shell and Tube Heat Exchanger

This page illustrates and describes plate and shell and tube type heat exchangers including pressure gauge and exchanger control system. Lines within Systems Components give the material and installation price on a cost each basis for the components. Prices for heat exchanger systems are on Line Items D3040 610 1010 thru D3040 620 1040. Material quantities and labor costs have been adjusted for the system listed.

Factors To adjust for job conditions other then normal working situations use Lines D3040 620 2700 thru 4000.

System Components	QUANTITY	UNIT	COST EACH		
			MAT.	INST.	TOTAL
Shell and tube type heat exchanger with related valves and piping.					
Heat exchanger, 4 pass, 3/4″ O.D. copper tubes, by steam at 10 psi, 40 GPM	1.000	Ea.	7,200	380	7,580
Pipe, black steel, Sch 40, threaded, w/couplings & hangers, 10′ OC, 2″ dia.	50.500	L.F.	421.68	1,186.75	1,608.43
Elbow, 90°, straight, 2″ dia.	16.000	Ea.	600	1,336	1,936
Reducer, black steel, concentric, 2″ dia.	2.000	Ea.	63	143	206
Valves, bronze, gate, rising stem, threaded, class 150, 2″ size	6.000	Ea.	2,010	456	2,466
Valves, bronze, globe, class 150, rising stem, threaded, 2″ size	2.000	Ea.	1,730	152	1,882
Strainers, Y type, bronze body, screwed, 125 lb, 2″ pipe size	2.000	Ea.	516	130	646
Valve, electric motor actuated, brass, 2 way, screwed, 1-1/2″ pipe size	1.000	Ea.	465	64.50	529.50
Union, black with brass seat, 2″ dia.	8.000	Ea.	552	708	1,260
Tee, black, straight, 2″ dia.	9.000	Ea.	423	1,233	1,656
Tee, black, reducing run and outlet, 2″ dia.	4.000	Ea.	226	548	774
Pipe nipple, black, 2″ dia.	2.000	Ea.	4.18	11.75	15.93
Thermometers, stem type, 9″ case, 8″ stem, 3/4″ NPT	2.000	Ea.	422	60	482
Gauges, pressure or vacuum, 3-1/2″ diameter dial	2.000	Ea.	48	53	101
Insulation, fiberglass pipe covering, 1″ wall, 2″ IPS	40.000	L.F.	96.40	278	374.40
Heat exchanger control system	1.000	Ea.	3,075	2,675	5,750
Coil balancing	1.000	Ea.		156	156
TOTAL			17,852.26	9,571	27,423.26

D3040 610	Heat Exchanger, Plate Type	COST EACH		
		MAT.	INST.	TOTAL
1000	For alternate heat exchangers:			
1010	Plate heat exchanger, 400 GPM	71,000	19,600	90,600
1020	800 GPM	116,000	24,800	140,800
1030	1200 GPM	176,500	33,600	210,100
1040	1800 GPM	238,500	41,800	280,300

D3040 620	Heat Exchanger, Shell & Tube	COST EACH		
		MAT.	INST.	TOTAL
1010	Shell & tube heat exchanger, 40 GPM	17,900	9,575	27,475
1020	96 GPM	69,500	17,200	86,700
1030	240 GPM	158,500	23,300	181,800
1040	600 GPM	116,500	32,700	149,200
2700	Cut & patch to match existing construction, add, minimum	2%	3%	
2800	Maximum	5%	9%	
2900	Dust protection, add, minimum	1%	2%	
3000	Maximum	4%	11%	

D3040 Distribution Systems

D3040 620	Heat Exchanger, Shell & Tube	COST EACH		
		MAT.	INST.	TOTAL
3100	Equipment usage curtailment, add, minimum	1%	1%	
3200	Maximum	3%	10%	
3300	Material handling & storage limitation, add, minimum	1%	1%	
3400	Maximum	6%	7%	
3500	Protection of existing work, add, minimum	2%	2%	
3600	Maximum	5%	7%	
3700	Shift work requirements, add, minimum		5%	
3800	Maximum		30%	
3900	Temporary shoring and bracing, add, minimum	2%	5%	
4000	Maximum	5%	12%	

D3050 Terminal & Package Units

Weather Cap Roof Vent →

Roof Flashing →

Vent Pipe

90° Elbow

Tee

Tee Cap

Gas Cock →

Gas Unit Heater

Hydronic Unit Heater

This page illustrates and describes unit heaters including piping and vents. Lines within Systems Components give the material and installation price on a cost each basis for this system. Prices for unit heater systems are on Line items D3050 120 1010 thru D3050 140 1040. Material quantities and labor costs have been adjusted for the system listed.

Factors: To adjust for job conditions other than normal working situations use Lines D3050 140 2700 thru 4000.

System Components	QUANTITY	UNIT	COST EACH		
			MAT.	INST.	TOTAL
Unit heaters complete with piping, thermostats and chimney as required					
Space heater, propeller fan, 20 MBH output	1.000	Ea.	855	179	1,034
Pipe, black steel, Sch 40, threaded, w/coupling & hangers, 10' OC, 3/4" dia	21.000	L.F.	75.60	287.70	363.30
Elbow, 90°, black steel, straight, 1/2" dia.	3.000	Ea.	14.25	166.50	180.75
Elbow, 90°, black steel, straight, 3/4" dia.	3.000	Ea.	17.25	178.50	195.75
Tee, black steel, reducing, 3/4" dia.	1.000	Ea.	14.10	93	107.10
Union, black with brass seat, 3/4" dia.	1.000	Ea.	20.50	64.50	85
Pipe nipples, black, 1/2" dia	2.000	Ea.	5.92	26.50	32.42
Pipe nipples, black, 3/4" dia	2.000	Ea.	3.60	13.70	17.30
Pipe cap, black, 3/4" dia.	1.000	Ea.	3.60	13.70	17.30
Gas cock, brass, 3/4" size	1.000	Ea.	6.50	26	32.50
Thermostat, 1 set back, electric, timed	1.000	Ea.	27	38	65
Wiring, thermostat hook-up,25' of #18-3	1.000	Ea.	109	105	214
Vent chimney, prefab metal, U.L. listed, gas, double wall, galv. st, 4" dia	1.000	Ea.	6.75	32.50	39.25
Vent chimney, gas, double wall, galv. steel, elbow 90°, 4" dia.	12.000	L.F.	127.80	258	385.80
Vent chimney, gas, double wall, galv. steel, Tee, 4" dia.	2.000	Ea.	62	86	148
Vent chimney, gas, double wall, galv. steel, T cap, 4" dia.	1.000	Ea.	43.50	56.50	100
Vent chimney, gas, double wall, galv. steel, roof flashing, 4" dia.	1.000	Ea.	2.89	35	37.89
Vent chimney, gas, double wall, galv. steel, top, 4" dia.	1.000	Ea.	10.50	43	53.50
	1.000	Ea.	21.50	33.50	55
TOTAL			**1,423.66**	**1,722.90**	**3,146.56**

D3050 120	Unit Heaters, Gas	COST EACH		
		MAT.	INST.	TOTAL
1000	For alternate unit heaters:			
1010	Space heater, suspended, gas fired, propeller fan, 20 MBH	1,425	1,725	3,150
1020	60 MBH	2,650	1,800	4,450
1030	100 MBH	1,975	1,900	3,875
1040	160 MBH	2,325	2,075	4,400
1050	200 MBH	2,900	2,325	5,225
1060	280 MBH	5,175	2,475	7,650
1070	320 MBH	4,900	2,800	7,700

D3050 130	Unit Heaters, Hydronic	COST EACH		
		MAT.	INST.	TOTAL
1010	Space heater, suspended, horiz. mount, HW, prop. fan, 20 MBH	2,350	1,850	4,200
1020	60 MBH	2,925	2,075	5,000
1030	100 MBH	3,575	2,275	5,850
1040	150 MBH	3,775	2,500	6,275

D3050 Terminal & Package Units

D3050 130	Unit Heaters, Hydronic	COST EACH		
		MAT.	INST.	TOTAL
1050	200 MBH	4,475	2,800	7,275
1060	300 MBH	4,650	3,300	7,950

D3050 140	Cabinet Unit Heaters, Hydronic	COST EACH		
		MAT.	INST.	TOTAL
1010	Unit heater, cabinet type, horizontal blower, hot water, 20 MBH	2,975	1,800	4,775
1020	60 MBH	4,150	1,975	6,125
1030	100 MBH	4,500	2,200	6,700
1040	120 MBH	4,500	2,275	6,775
2700	Cut & patch to match existing construction, add, minimum	2%	3%	
2800	Maximum	5%	9%	
2900	Dust protection, add, minimum	1%	2%	
3000	Maximum	4%	11%	
3100	Equipment usage curtailment, add, minimum	1%	1%	
3200	Maximum	3%	10%	
3300	Material handling & storage limitation, add, minimum	1%	1%	
3400	Maximum	6%	7%	
3500	Protection of existing work, add, minimum	2%	2%	
3600	Maximum	5%	7%	
3700	Shift work requirements, add, minimum		5%	
3800	Maximum		30%	
3900	Temporary shoring and bracing, add, minimum	2%	5%	
4000	Maximum	5%	12%	

D3050 Terminal & Package Units

Gas Cock ←

Roof Curb ←

Rooftop Air Conditioner Unit

Gas Piping

This page illustrates and describes packaged air conditioners including rooftop DX unit, pipes and fittings. Lines within Systems Components give the material and installation price on a cost each basis for the components. Prices for A/C systems are on Line Items D3050 175 1010 thru D3050 180 1060. Material quantities and labor costs have been adjusted for the system listed.

Factors: To adjust for job conditions other than normal working situations use Lines D3050 175 2700 thru 4000.

System Components	QUANTITY	UNIT	COST EACH		
			MAT.	INST.	TOTAL
Packaged rooftop DX unit, gas, with related piping and valves.					
Roof top A/C, curb, economizer, sgl zone, elec cool, gas ht, 5 ton, 112 MBH	1.000	Ea.	4,650	2,700	7,350
Pipe, black steel, Sch 40, threaded, w/cplgs & hangers, 10' OC, 1" dia	21.000	L.F.	86.10	330.75	416.85
Elbow, 90°, black, straight, 3/4" dia.	3.000	Ea.	17.25	178.50	195.75
Elbow, 90°, black, straight, 1" dia.	3.000	Ea.	30	193.50	223.50
Tee, black, reducing, 1" dia.	1.000	Ea.	17.80	104	121.80
Union, black with brass seat, 1" dia.	1.000	Ea.	27	69.50	96.50
Pipe nipple, black, 3/4" dia	2.000	Ea.	7.20	27.40	34.60
Pipe nipple, black, 1" dia	2.000	Ea.	4.10	15.75	19.85
Cap, black, 1" dia.	1.000	Ea.	7.90	28	35.90
Gas cock, brass, 1" size	1.000	Ea.	46	44	90
Control system, pneumatic, Rooftop A/C unit	1.000	Ea.	3,175	2,975	6,150
Rooftop unit heat/cool balancing	1.000	Ea.		570	570
TOTAL			8,068.35	7,236.40	15,304.75

D3050 175	Rooftop Air Conditioner, Const. Volume	COST EACH		
		MAT.	INST.	TOTAL
1000	For alternate A/C systems:			
1010	A/C, Rooftop, DX cool, gas heat, curb, economizer, fltrs, 5 Ton	8,075	7,225	15,300
1020	7-1/2 Ton	9,825	7,525	17,350
1030	12-1/2 Ton	14,300	8,275	22,575
1040	18 Ton	19,100	9,075	28,175
1050	25 Ton	43,600	10,300	53,900
1060	40 Ton	53,000	13,800	66,800

D3050 180	Rooftop Air Conditioner, Variable Air Volume	COST EACH		
		MAT.	INST.	TOTAL
1010	A/C, Rooftop, DX cool, gas heat, curb, ecmizr, fltrs, VAV, 12-1/2 Ton	16,700	8,850	25,550
1020	18 Ton	22,800	9,750	32,550
1030	25 Ton	48,300	11,200	59,500
1040	40 Ton	60,500	15,100	75,600
1050	60 Ton	86,000	20,600	106,600
1060	80 Ton	109,500	25,800	135,300
2700	Cut & patch to match existing construction, add, minimum	2%	3%	
2800	Maximum	5%	9%	
2900	Dust protection, add, minimum	1%	2%	
3000	Maximum	4%	11%	
3100	Equipment usage curtailment, add, minimum	1%	1%	
3200	Maximum	3%	10%	

D3050 Terminal & Package Units

D3050 180	Rooftop Air Conditioner, Variable Air Volume	COST EACH		
		MAT.	INST.	TOTAL
3300	Material handling & storage limitation, add, minimum	1%	1%	
3400	Maximum	6%	7%	
3500	Protection of existing work, add, minimum	2%	2%	
3600	Maximum	5%	7%	
3700	Shift work requirements, add, minimum		5%	
3800	Maximum		30%	
3900	Temporary shoring and bracing, add, minimum	2%	5%	
4000	Maximum	5%	12%	

D3050 Terminal & Package Units

Condenser Supply Water Piping

Condenser Return Water Piping

Self–Contained Air Conditioner Unit, Water Cooled

This page illustrates and describes packaged electrical air conditioning units including multizone control and related piping. Lines within Systems Components give the material and installation price on a cost each basis for the components. Prices for packaged electrical A/C unit systems are on Line Items D3050 210 1010 thru D3050 225 1060. Material quantities and labor costs have been adjusted for the system listed.

Factors: To adjust for job conditions other than normal working situations use Lines D3050 225 2700 thru 4000.

System Components	QUANTITY	UNIT	COST EACH MAT.	COST EACH INST.	COST EACH TOTAL
Packaged water cooled electric air conditioning unit					
with related piping and valves.					
Self-contained, water cooled, elect. heat, not inc tower, 5 ton, const vol	1.000	Ea.	9,500	1,975	11,475
Pipe, black steel, Sch 40, threaded, w/cplg & hangers, 10' OC, 1-1/4" dia.	21.500	L.F.	113.95	363.35	477.30
Elbow, 90°, black, straight, 1-1/4" dia.	6.000	Ea.	98.70	411	509.70
Tee, black, straight, 1-1/4" dia.	2.000	Ea.	51	214	265
Tee, black, reducing, 1-1/4" dia.	2.000	Ea.	62	214	276
Thermometers, stem type, 9" case, 8" stem, 3/4" NPT	2.000	Ea.	422	60	482
Union, black with brass seat, 1-1/4" dia.	2.000	Ea.	95	143	238
Pipe nipple, black, 1-1/4" dia	3.000	Ea.	7.95	25.35	33.30
Valves, bronze, gate, N.R.S., threaded, class 150, 1-1/4" size	2.000	Ea.	412	111	523
Circuit setter, bal valve, bronze body, threaded, 1-1/4" pipe size	1.000	Ea.	189	56	245
Insulation, fiberglass pipe covering, 1" wall, 1-1/4" IPS	20.000	L.F.	26.80	132	158.80
Control system, pneumatic, A/C Unit with heat	1.000	Ea.	3,175	2,975	6,150
Re-heat coil balancing	1.000	Ea.		156	156
Rooftop unit heat/cool balancing	1.000	Ea.		570	570
TOTAL			14,153.40	7,405.70	21,559.10

D3050 210	AC Unit, Package, Elec. Ht., Water Cooled		MAT.	INST.	TOTAL
1000	For alternate A/C systems:				
1010	A/C, Self contained, single pkg., water cooled, elect. heat, 5 Ton		14,200	7,400	21,600
1020	10 Ton		21,400	9,100	30,500
1030	20 Ton		28,600	11,700	40,300
1040	30 Ton		50,500	12,700	63,200
1050	40 Ton		65,500	14,000	79,500
1060	50 Ton		86,000	16,800	102,800

D3050 215	AC Unit, Package, Elec. Ht., Water Cooled, VAV		MAT.	INST.	TOTAL
1010	A/C, Self contained, single pkg., water cooled, elect. ht, VAV, 10 Ton		24,000	9,100	33,100
1020	20 Ton		33,300	11,700	45,000
1030	30 Ton		57,500	12,700	70,200
1040	40 Ton		73,000	14,000	87,000
1050	50 Ton		96,500	16,800	113,300
1060	60 Ton		109,000	20,500	129,500

D3050 Terminal & Package Units

D3050 220	AC Unit, Package, Hot Water Coil, Water Cooled	COST EACH		
		MAT.	INST.	TOTAL
1010	A/C, Self contn'd, single pkg., water cool, H/W ht, const. vol, 5 Ton	12,500	10,200	22,700
1020	10 Ton	17,600	12,200	29,800
1030	20 Ton	41,800	15,400	57,200
1040	30 Ton	54,000	19,600	73,600
1050	40 Ton	63,500	23,800	87,300
1060	50 Ton	74,000	29,200	103,200

D3050 225	AC Unit, Package, HW Coil, Water Cooled, VAV	COST EACH		
		MAT.	INST.	TOTAL
1010	A/C, Self contn'd, single pkg., water cool, H/W ht, VAV, 10 Ton	20,200	11,500	31,700
1020	20 Ton	29,300	15,000	44,300
1030	30 Ton	50,500	18,800	69,300
1040	40 Ton	66,500	20,400	86,900
1050	50 Ton	86,500	24,100	110,600
1060	60 Ton	100,500	26,200	126,700
2700	Cut & patch to match existing construction, add, minimum	2%	3%	
2800	Maximum	5%	9%	
2900	Dust protection, add, minimum	1%	2%	
3000	Maximum	4%	11%	
3100	Equipment usage curtailment, add, minimum	1%	1%	
3200	Maximum	3%	10%	
3300	Material handling & storage limitation, add, minimum	1%	1%	
3400	Maximum	6%	7%	
3500	Protection of existing work, add, minimum	2%	2%	
3600	Maximum	5%	7%	
3700	Shift work requirements, add, minimum		5%	
3800	Maximum		30%	
3900	Temporary shoring and bracing, add, minimum	2%	5%	
4000	Maximum	5%	12%	

D3050   Terminal & Package Units

Thru—Wall Heat Pump

Thru—Wall Air Conditioner

This page illustrates and describes thru-wall air conditioning and heat pump units. Lines within Systems Components give the material and installation price on a cost each basis for the components. Prices for thru-wall unit systems are on Line Items D3050 255 1010 thru D3050 260 1050. Material quantities and labor costs have been adjusted for the system listed.

Factors: To adjust for job conditions other than normal working conditions use Lines D3050 255 2700 thru 4000.

System Components	QUANTITY	UNIT	COST EACH		
			MAT.	INST.	TOTAL
Electric thru-wall air conditioning unit.					
A/C unit, thru-wall, electric heat., cabinet, louver, 1/2 ton	1.000	Ea.	1,025	253	1,278
TOTAL			1,025	253	1,278

D3050 255	Thru-Wall A/C Unit	COST EACH		
		MAT.	INST.	TOTAL
1000	For alternate A/C units:			
1010	A/C Unit, thru-the-wall, supplemental elect. heat, cabinet, louver, 1/2 Ton	1,025	253	1,278
1020	3/4 Ton	1,075	305	1,380
1030	1 Ton	1,350	380	1,730
1040	1-1/2 Ton	2,725	635	3,360
1050	2 Ton	3,000	800	3,800

D3050 260	Thru-Wall Heat Pump	COST EACH		
		MAT.	INST.	TOTAL
1010	Heat pump, thru-the-wall, cabinet, louver, 1/2 Ton	2,975	190	3,165
1020	3/4 Ton	3,125	253	3,378
1030	1 Ton	4,125	380	4,505
1040	Supplemental. elect. heat, 1-1/2 Ton	3,625	980	4,605
1050	2 Ton	3,675	1,000	4,675
2700	Cut & patch to match existing construction, add, minimum	2%	3%	
2800	Maximum	5%	9%	
2900	Dust protection, add, minimum	1%	2%	
3000	Maximum	4%	11%	
3100	Equipment usage curtailment, add, minimum	1%	1%	
3200	Maximum	3%	10%	
3300	Material handling & storage limitation, add, minimum	1%	1%	
3400	Maximum	6%	7%	
3500	Protection of existing work, add, minimum	2%	2%	
3600	Maximum	5%	7%	
3700	Shift work requirements, add, minimum		5%	
3800	Maximum		30%	
3900	Temporary shoring and bracing, add, minimum	2%	5%	
4000	Maximum	5%	12%	

For customer support on your Commercial Renovation Costs with RSMeans Data, call 800.448.8182.

D4010 Sprinklers

This page illustrates and describes a dry type fire sprinkler system. Lines within System Components give the unit cost of a system for a 2,000 square foot building. Lines D4010 320 1700 thru 2400 give the square foot costs for alternate systems. Both material quantities and labor costs have been adjusted for the system listed.

Factors: To adjust for conditions other than normal working conditions, use Lines D4010 320 2900 thru 4000.

System Components	QUANTITY	UNIT	COST EACH		
			MAT.	INST.	TOTAL
Pre-action fire sprinkler system, ordinary hazard, open area to 2000 S.F.					
On one floor.					
Air compressor	1.000	Ea.	1,550	630	2,180
4″ OS & Y valve	1.000	Ea.	1,100	500	1,600
Pipe riser 4″ diameter	10.000	L.F.	240	437	677
Sprinkler head supply piping, 1″	163.000	L.F.	307.50	1,181.25	1,488.75
Sprinkler head supply piping, 1-1/4″	163.000	L.F.	291.50	929.50	1,221
Sprinkler head supply piping, 2-1/2″	163.000	L.F.	596.60	1,140	1,736.60
Pipe fittings, 1″	8.000	Ea.	124.80	832	956.80
Pipe fittings, 1-1/4″	8.000	Ea.	204	856	1,060
Pipe fittings, 2-1/2″	4.000	Ea.	464	668	1,132
Pipe fittings, 4″	2.000	Ea.	820	750	1,570
Detectors	2.000	Ea.	1,920	103	2,023
Sprinkler heads	16.000	Ea.	262.40	824	1,086.40
Fire department connection	1.000	Ea.	645	296	941
TOTAL			8,525.80	9,146.75	17,672.55

D4010 320	Fire Sprinkler Systems, Dry	COST PER S.F.		
		MAT.	INST.	TOTAL
1700	Ordinary hazard, one floor, area to 2000 S.F./floor	4.26	4.57	8.83
1800	For each additional floor, add per floor	1.66	3.81	5.47
1900	Area to 3200 S.F./ floor	3.27	4.54	7.81
2000	For each additional floor, add per floor	1.64	4.06	5.70
2100	Area to 5000 S.F./ floor	3.72	4.62	8.34
2200	For each additional floor, add per floor	2.29	4.29	6.58
2300	Area to 8000 S.F./ floor	2.75	3.89	6.64
2400	For each additional floor, add per floor	1.74	3.68	5.42
2500				
2600				
2700				
2800				
2900	Cut & patch to match existing construction, add, minimum	2%	3%	
3000	Maximum	5%	9%	
3100	Dust protection, add, minimum	1%	2%	
3200	Maximum	4%	11%	
3300	Equipment usage curtailment, add, minimum	1%	1%	
3400	Maximum	3%	10%	

D4010 Sprinklers

D4010 320	Fire Sprinkler Systems, Dry	COST PER S.F.		
		MAT.	INST.	TOTAL
3500	Material handling & storage limitation, add, minimum	1%	1%	
3600	Maximum	6%	7%	
3700	Protection of existing work, add, minimum	2%	2%	
3800	Maximum	5%	7%	
3900	Shift work requirements, add, minimum		5%	
4000	Maximum		30%	

D4010 Sprinklers

This page illustrates and describes a wet type fire sprinkler system. Lines within System Components give the unit cost of a system for a 2,000 square foot building. Lines D4010 420 1900 thru 2600 give the square foot costs for alternate systems. Both material quantities and labor costs have been adjusted for the system listed.

Factors: To adjust for conditions other than normal working conditions, use Lines D4010 420 2900 thru 4000.

System Components	QUANTITY	UNIT	COST EACH MAT.	COST EACH INST.	COST EACH TOTAL
Wet pipe fire sprinkler system, ordinary hazard, open area to 2000 S.F.					
On one floor.					
4" OS & Y valve	1.000	Ea.	1,100	500	1,600
Wet pipe alarm valve	1.000	Ea.	2,725	740	3,465
Water motor alarm	1.000	Ea.	355	205	560
3" check valve	1.000	Ea.	480	495	975
Pipe riser, 4" diameter	10.000	L.F.	240	437	677
Water gauges and trim	1.000	Set	3,775	1,475	5,250
Electric fire horn	1.000	Ea.	55.50	117	172.50
Sprinkler head supply piping, 1" diameter	168.000	L.F.	307.50	1,181.25	1,488.75
Sprinkler head supply piping, 1-1/2" diameter	168.000	L.F.	341	1,034	1,375
Sprinkler head supply piping, 2-1/2" diameter	168.000	L.F.	596.60	1,140	1,736.60
Pipe fittings, 1"	8.000	Ea.	124.80	832	956.80
Pipe fittings, 1-1/2"	8.000	Ea.	252	928	1,180
Pipe fittings, 2-1/2"	4.000	Ea.	464	668	1,132
Pipe fittings, 4"	2.000	Ea.	820	750	1,570
Sprinkler heads	16.000	Ea.	262.40	824	1,086.40
Fire department connection	1.000	Ea.	645	296	941
TOTAL			12,543.80	11,622.25	24,166.05

D4010 420	Fire Sprinkler Systems, Wet	COST PER S.F. MAT.	COST PER S.F. INST.	COST PER S.F. TOTAL
1900	Ordinary hazard, one floor, area to 2000 S.F./floor	8.75	6.15	14.90
2000	For each additional floor, add per floor	1.70	3.90	5.60
2100	Area to 3200 S.F./floor	6.05	5.45	11.50
2200	For each additional floor, add per floor	1.64	4.06	5.70
2300	Area to 5000 S.F./ floor	5.10	5.20	10.30
2400	For each additional floor, add per floor	2.29	4.29	6.58
2500	Area to 8000 S.F./ floor	3.50	4.24	7.74
2600	For each additional floor, add per floor	1.87	3.90	5.77
2700				
2900	Cut & patch to match existing construction, add minimum	2%	3%	
3000	Maximum	5%	9%	
3100	Dust protection, add, minimum	1%	2%	
3200	Maximum	4%	11%	
3300	Equipment usage curtailment, add, minimum	1%	1%	
3400	Maximum	3%	10%	
3500	Material handling & storage limitation, add, minimum	1%	1%	
3600	Maximum	6%	7%	

711

D5010 Electrical Service/Distribution

This page illustrates and describes commercial service systems including a meter socket, service head and cable, entrance switch, steel conduit, copper wire, panel board, ground rod, wire, and conduit. Lines within System Components give the unit price and total price on a cost each basis for this system. Prices for commercial service systems are on Line Items D5010 210 1500 thru 1700. Both material quantities and labor costs have been adjusted for the system listed.

Factors: To adjust for job conditions other than normal working situations use Lines D5010 210 2900 thru 4000.

Example: You are to install the system with maximum equipment usage curtailment. Go to Line D5010 210 3400 and apply these percentages to the appropriate MAT. and INST. costs.

System Components	QUANTITY	UNIT	COST EACH		
			MAT.	INST.	TOTAL
Commercial electric service including service breakers, metering 120/208 Volt, 3 phase, 4 wire, feeder, and panel board.					
100 Amp Service					
Meter socket	1.000	Ea.	80.50	245	325.50
Service head	.200	C.L.F.	28.20	157	185.20
Service entrance switch	1.000	Ea.	212	410	622
Rigid steel conduit	20.000	L.F.	98.80	241	339.80
600 volt copper wire #3	1.000	C.L.F.	168	157	325
Panel board, 24 circuits, 20 Amp breakers	1.000	Ea.	455	1,300	1,755
Ground rod	1.000	Ea.	18.70	142	160.70
TOTAL			1,061.20	2,652	3,713.20

D5010 210	Commercial Service - 3 Phase	COST EACH		
		MAT.	INST.	TOTAL
1500	120/208 Volt, 3 phase, 4 wire service, 60 Amp	1,500	2,050	3,550
1550	100 Amp	1,050	2,650	3,700
1600	200 Amp	4,125	4,675	8,800
1700	400 Amp	6,875	5,850	12,725
1800				
1900				
2000				
2100				
2200				
2300				
2400				
2500				
2900	Cut & patch to match existing construction, add, minimum	2%	3%	
3000	Maximum	5%	9%	
3100	Dust protection, add, minimum	1%	2%	
3200	Maximum	4%	11%	
3300	Equipment usage curtailment, add, minimum	1%	1%	
3400	Maximum	3%	10%	
3500	Material handling & storage limitation, add, minimum	1%	1%	
3600	Maximum	6%	7%	
3700	Protection of existing work, add, minimum	2%	2%	
3800	Maximum	5%	7%	
3900	Shift work requirements, add, minimum		5%	
4000	Maximum		30%	

D5010 Electrical Service/Distribution

This page illustrates and describes a residential, single phase system including a weather cap, service entrance cable, meter socket, entrance switch, ground rod, ground cable, EMT, and panelboard. Lines within System Components give the unit price and total price on a cost each basis for this system. Prices for a residential, single phase system are also given. Both material quantities and labor costs have been adjusted for the system listed.

Factors: To adjust for job conditions other than normal working situations use Lines D5010 220 2900 thru 4000.

Example: You are to install the system with a minimum equipment usage curtailment. Go to Line D5010 220 3300 and apply these percentages to the appropriate MAT. and INST. costs.

System Components	QUANTITY	UNIT	COST EACH MAT.	COST EACH INST.	COST EACH TOTAL
100 Amp Service, single phase					
Weathercap	1.000	Ea.	7.70	65	72.70
Service entrance cable	.200	C.L.F.	28.20	157	185.20
Meter socket	1.000	Ea.	80.50	245	325.50
Entrance disconnect switch	1.000	Ea.	212	410	622
Ground rod, with clamp	1.000	Ea.	23	163	186
Ground cable	.100	C.L.F.	16.30	49	65.30
Panelboard, 12 circuit	1.000	Ea.	238	650	888
TOTAL			605.70	1,739	2,344.70
200 Amp Service , single phase					
Weathercap	1.000	Ea.	21.50	98	119.50
Service entrance cable	.200	C.L.F.	71	225	296
Meter socket	1.000	Ea.	116	410	526
Entrance disconnect switch	1.000	Ea.	455	600	1,055
Ground rod, with clamp	1.000	Ea.	57	178	235
Ground cable	.100	C.L.F.	54.50	27	81.50
3/4" EMT	10.000	L.F.	11.10	60	71.10
Panelboard, 24 circuit	1.000	Ea.	485	1,075	1,560
TOTAL			1,271.10	2,673	3,944.10

D5010 220	Residential Service - Single Phase	COST EACH MAT.	COST EACH INST.	COST EACH TOTAL
2800				
2810	100 Amp Service, single phase	605	1,750	2,355
2820	200 Amp Service, single phase	1,275	2,675	3,950
2900	Cut & patch to match existing construction, add, minimum	2%	3%	
3000	Maximum	5%	9%	
3100	Dust protection, add, minimum	1%	2%	
3200	Maximum	4%	11%	
3300	Equipment usage curtailment, add, minimum	1%	1%	
3400	Maximum	3%	10%	
3500	Material handling & storage limitation, add, minimum	1%	1%	
3600	Maximum	6%	7%	
3700	Protection of existing work, add, minimum	2%	2%	
3800	Maximum	5%	7%	
3900	Shift work requirements, add, minimum		5%	
4000	Maximum		30%	

D5020 Lighting and Branch Wiring

D5020 180	Selective Price Sheet	COST EACH		
		MAT.	INST.	TOTAL
0100	Using non-metallic sheathed, cable, air conditioning receptacle	19.90	78.50	98.40
0200	Disposal wiring	10.30	87	97.30
0300	Dryer circuit	30.50	142	172.50
0400	Duplex receptacle	19.90	60	79.90
0500	Fire alarm or smoke detector	84.50	78.50	163
0600	Furnace circuit & switch	26.50	130	156.50
0700	Ground fault receptacle	72	98	170
0800	Heater circuit	18.70	98	116.70
0900	Lighting wiring	22	49	71
1000	Range circuit	82	196	278
1100	Switches single pole	13	49	62
1200	3-way	16.85	65	81.85
1300	Water heater circuit	28.50	157	185.50
1400	Weatherproof receptacle	272	130	402
1500	Using BX cable, air conditioning receptacle	34	94.50	128.50
1600	Disposal wiring	24	104	128
1700	Dryer circuit	41	170	211
1800	Duplex receptacle	34	72.50	106.50
1900	Fire alarm or smoke detector	84.50	78.50	163
2000	Furnace circuit & switch	41.50	157	198.50
2100	Ground fault receptacle	82	119	201
2200	Heater circuit	26.50	119	145.50
2300	Lighting wiring	27	59	86
2400	Range circuit	102	237	339
2500	Switches, single pole	27	59	86
2600	3-way	23	78.50	101.50
2700	Water heater circuit	44	186	230
2800	Weatherproof receptacle	280	157	437
2900	Using EMT conduit, air conditioning receptacle	46	117	163
3000	Disposal wiring	37.50	130	167.50
3100	Dryer circuit	53.50	212	265.50
3200	Duplex receptacle	46	90	136
3300	Fire alarm or smoke detector	104	117	221
3400	Furnace circuit & switch	51	196	247
3500	Ground fault receptacle	100	145	245
3600	Heater circuit	38.50	145	183.50
3700	Lighting wiring	37.50	73	110.50
3800	Range circuit	96.50	290	386.50
3900	Switches, single pole	39	73	112
4000	3-way	33.50	98	131.50
4100	Water heater circuit	50	230	280
4200	Weatherproof receptacle	292	196	488
4300	Using aluminum conduit, air conditioning receptacle	47	157	204
4400	Disposal wiring	39	174	213
4500	Dryer circuit	61.50	280	341.50
4600	Duplex receptacle	45	120	165
4700	Fire alarm or smoke detector	110	157	267
4800	Furnace circuit & switch	55	261	316
4900	Ground fault receptacle	104	196	300
5000	Heater circuit	39	196	235
5100	Lighting wiring	46.50	98	144.50
5200	Range circuit	112	390	502
5300	Switches, single pole	48	98	146
5400	3-way	42.50	130	172.50
5500	Water heater circuit	57	315	372
5600	Weatherproof receptacle	310	261	571
5700	Using galvanized steel conduit	53.50	167	220.50
5800	Disposal wiring	41.50	186	227.50

D5020 Lighting and Branch Wiring

D5020 180	Selective Price Sheet	COST EACH		
		MAT.	INST.	TOTAL
5900	Dryer circuit	74	300	374
6000	Duplex receptacle	53.50	128	181.50
6100	Fire alarm or smoke detector	112	167	279
6200	Furnace circuit & switch	57	280	337
6300	Ground fault receptacle	114	206	320
6400	Heater circuit	46	206	252
6500	Lighting wiring	50	104	154
6600	Range circuit	122	410	532
6700	Switches, single pole	58	104	162
6800	3-way	53.50	135	188.50
6900	Water heater circuit	67	325	392
7000	Weatherproof receptacle	310	280	590
7100				
7200				
7300				
7400				
7500				
7600				
7700				
7800				
7900				
8000				
8100				
8200				
8300				
8400				

This page illustrates and describes fluorescent lighting systems including a fixture, lamp, outlet box and wiring. Lines within System Components give the unit price and total price on a cost each basis for this system. Prices for fluorescent lighting systems are on Line Items D5020 248 1200 thru 1500. Both material quantities and labor costs have been adjusted for the system listed.

Factors: To adjust for job conditions other than normal working situations use Lines D5020 248 2900 thru 4000.

Example: You are to install the system during evening hours. Go to Line D5020 248 3900 and apply this percentage to the appropriate INST. cost.

System Components	QUANTITY	UNIT	COST EACH MAT.	COST EACH INST.	COST EACH TOTAL
Fluorescent lighting, including fixture, lamp, outlet box and wiring.					
Recessed lighting fixture, on suspended system	1.000	Ea.	72.50	167	239.50
Outlet box	1.000	Ea.	2.42	43.50	45.92
#12 wire	.660	C.L.F.	7	46.86	53.86
Conduit, EMT, 1/2" conduit	20.000	L.F.	15.80	92	107.80
TOTAL			97.72	349.36	447.08

D5020 248	Lighting, Fluorescent	COST EACH MAT.	COST EACH INST.	COST EACH TOTAL
1200	Recessed lighting fixture, on suspended system	97.50	350	447.50
1300	Surface mounted, 2' x 4', acrylic prismatic diffuser	121	365	486
1400	Strip fixture, 8' long, two 8' lamps	104	345	449
1500	Pendant mounted, industrial, 8' long, with reflectors	136	395	531
1600				
1700				
1800				
1900				
2000				
2100				
2200				
2300				
2900	Cut & patch to match existing construction, add, minimum	2%	3%	
3000	Maximum	5%	9%	
3100	Dust protection, add, minimum	1%	2%	
3200	Maximum	4%	11%	
3300	Equipment usage curtailment, add, minimum	1%	1%	
3400	Maximum	3%	10%	
3500	Material handling & storage limitation, add, minimum	1%	1%	
3600	Maximum	6%	7%	
3700	Protection of existing work, add, minimum	2%	2%	
3800	Maximum	5%	7%	
3900	Shift work requirements, add, minimum		5%	
4000	Maximum		30%	

D50 Electrical

D5020 Lighting and Branch Wiring

This page illustrates and describes incandescent lighting systems including a fixture, lamp, outlet box, conduit and wiring. Lines within System Components give the unit price and total price on a cost each basis for this system. Prices for an alternate incandescent lighting system are also given. Both material quantities and labor costs have been adjusted for the system listed.

Factors: To adjust for conditions other than normal working situations use Lines D5020 250 2900 thru 4000.

Example: You are to install the system and cut and match existing construction. Go to Line D5020 250 3000 and apply this percentage to the appropriate INST. costs.

System Components			COST EACH		
	QUANTITY	UNIT	MAT.	INST.	TOTAL
Incandescent light fixture, including lamp, outlet box, conduit and wiring.					
Recessed wide reflector with flat glass lens	1.000	Ea.	108	117	225
Outlet box	1.000	Ea.	2.42	43.50	45.92
Armored cable, 3 wire	.200	C.L.F.	13.90	71	84.90
TOTAL			124.32	231.50	355.82
Recessed flood light fixture, including lamp, outlet box, conduit & wire					
Recessed, R-40 flood lamp with refractor	1.000	Ea.	115	98	213
150 watt R-40 flood lamp	.010	Ea.	19.75	6	25.75
Outlet box	1.000	Ea.	2.42	43.50	45.92
Outlet box cover	1.000	Ea.	1.08	12.25	13.33
Romex, 12-2 with ground	.200	C.L.F.	5.20	71	76.20
Conduit, 1/2" EMT	20.000	L.F.	15.80	92	107.80
TOTAL			159.25	322.75	482

D5020 250	Lighting, Incandescent	COST EACH		
		MAT.	INST.	TOTAL
2810	Incandescent light fixture, including lamp, outlet box, conduit and wiring.	124	232	356
2820	Recessed flood light fixture, including lamp, outlet box, conduit & wire	159	325	484
2900	Cut & patch to match existing construction, add, minimum	2%	3%	
3000	Maximum	5%	9%	
3100	Dust protection, add, minimum	1%	2%	
3200	Maximum	4%	11%	
3300	Equipment usage curtailment, add, minimum	1%	1%	
3400	Maximum	3%	10%	
3500	Material handling & storage limitation, add, minimum	1%	1%	
3600	Maximum	6%	7%	
3700	Protection of existing work, add, minimum	2%	2%	
3800	Maximum	5%	7%	
3900	Shift work requirements, add, minimum		5%	
4000	Maximum		30%	

D5020 Lighting and Branch Wiring

This page illustrates and describes high intensity lighting systems including a lamp, EMT conduit, EMT "T" fitting with cover, and wire. Lines within System Components give the unit price and total price on a cost each basis for this system. Prices for high intensity lighting systems are on Line Items D5020 252 1100 thru 1800. Both material quantities and labor costs have been adjusted for the system listed.

Factors: To adjust for job conditions other than normal working situations use Lines D5020 252 2900 thru 4000.

Example: You are to install the system and protect existing construction. Go to Line D5020 252 3700 and apply these percentages to the appropriate MAT. and INST. costs.

System Components			COST EACH		
	QUANTITY	UNIT	MAT.	INST.	TOTAL
High intensity lighting system consisting of 400 watt metal halide fixture And lamp with 1/2" EMT conduit and fittings using #12 wire, high bay.					
Single unit, 400 watt	1.000	Ea.	465	340	805
Electric metallic tubing (EMT), 1/2" diameter	30.000	Ea.	23.70	138	161.70
1/2" EMT "T" fitting with cover	1.000	Ea.	6.25	49	55.25
Wire, 600 volt, type THWN-THHN, copper, solid, #12	.600	Ea.	6.36	42.60	48.96
TOTAL			501.31	569.60	1,070.91

D5020 252	Lighting, High Intensity	COST EACH		
		MAT.	INST.	TOTAL
1100	High bay: 400 watt, metal halide fixture and lamp	500	570	1,070
1200	High pressure sodium and lamp	465	570	1,035
1400	1000 watt, metal halide	705	620	1,325
1500	High pressure sodium	635	620	1,255
1700	Low bay: 250 watt, metal halide	435	475	910
1800	150 watt, high pressure sodium	405	475	880
1900				
2000				
2100				
2200				
2300				
2400				
2500				
2600				
2700				
2900	Cut & patch to match existing construction, add, minimum	2%	3%	
3000	Maximum	5%	9%	
3100	Dust protection, add, minimum	1%	2%	
3200	Maximum	4%	11%	
3300	Equipment usage curtailment, add, minimum	1%	1%	
3400	Maximum	3%	10%	
3500	Material handling & storage limitation, add, minimum	1%	1%	
3600	Maximum	6%	7%	
3700	Protection of existing work, add, minimum	2%	2%	
3800	Maximum	5%	7%	
3900	Shift work requirements, add, minimum		5%	
4000	Maximum		30%	

For customer support on your Commercial Renovation Costs with RSMeans Data, call 800.448.8182.

D5090 Other Electrical Systems

This page illustrates and describes baseboard heat systems including a thermostat, outlet box, breaker, and feed. Lines within System Components give the unit price and total price on a cost each basis for this system. Prices for baseboard heat systems are on Line Items D5090 430 1500 thru 1800. Both material quantities and labor costs have been adjusted for the system listed.

Factors: To adjust for job conditions other than normal working situations use Lines D5090 430 2900 thru 4000.

Example: You are to install the system during evenings and weekends only. Go to Line D5090 430 4000 and apply this percentage to the appropriate INST. cost.

System Components	QUANTITY	UNIT	COST EACH MAT.	COST EACH INST.	COST EACH TOTAL
Baseboard heat including thermostat, outlet box, breaker and feed.					
Electric baseboard heater, 4' long	1.000	Ea.	42.50	117	159.50
Thermostat, integral	1.000	Ea.	32	49	81
Romex, 12-3 with ground	.400	C.L.F.	16.40	156	172.40
Panel board breaker, 15-50 amp	1.000	Ea.	70	98	168
Outlet box	1.000	Ea.	2.42	43.50	45.92
TOTAL			163.32	463.50	626.82

D5090 530	Heat, Baseboard	COST EACH MAT.	COST EACH INST.	COST EACH TOTAL
1400	Heat, electric baseboard			
1500	2' long	155	445	600
1550	4' long	163	465	628
1600	6' long	179	505	684
1700	8' long	201	545	746
1800	10' long	228	585	813
1900				
2000				
2100				
2200				
2300				
2400				
2500				
2600				
2700				
2900	Cut & patch to match existing construction, add, minimum	2%	3%	
3000	Maximum	5%	9%	
3100	Dust protection, add, minimum	1%	2%	
3200	Maximum	4%	11%	
3300	Equipment usage curtailment, add, minimum	1%	1%	
3400	Maximum	3%	10%	
3500	Material handling & storage limitation, add, minimum	1%	1%	
3600	Maximum	6%	7%	
3700	Protection of existing work, add, minimum	2%	2%	
3800	Maximum	5%	7%	
3900	Shift work requirements, add, minimum		5%	
4000	Maximum		30%	

719

Same Data. Simplified.

Enjoy the convenience and efficiency of accessing your costs anywhere:

- **Skip the multiplier** by setting your location
- **Quickly search,** edit, favorite and share costs
- **Stay on top of price changes** with automatic updates

Discover more at rsmeans.com/online

This page illustrates and describes kitchen systems including top and bottom cabinets, custom laminated plastic top, single bowl sink, and appliances. Lines within System Components give the unit price and total price on a cost each basis for this system. Prices for kitchen systems are on Line Items E1090 310 1400 thru 1600. Both material quantities and labor costs have been adjusted for the system listed.

Factors: To adjust for job conditions other than normal working situations use Lines E1090 310 2900 thru 4000.

Example: You are to install the system and protect the work area from dust. Go to Line E1090 310 3200 and apply these percentages to the appropriate MAT. and INST. costs.

System Components	QUANTITY	UNIT	COST EACH		
			MAT.	INST.	TOTAL
Kitchen cabinets including wall and base cabinets, custom laminated					
Plastic top,sink & appliances,no plumbing or electrical rough-in included.					
Prefinished wood cabinets, average quality, wall and base	20.000	L.F.	4,460	930	5,390
Custom laminated plastic counter top	20.000	L.F.	369	460	829
Stainless steel sink, 22" x 25"	1.000	Ea.	765	268	1,033
Faucet, top mount	1.000	Ea.	91.50	83.50	175
Dishwasher, built-in	1.000	Ea.	340	405	745
Compactor, built-in	1.000	Ea.	775	139	914
Range hood, vented, 30" wide	1.000	Ea.	117	299	416
TOTAL			6,917.50	2,584.50	9,502

E1090 310	Kitchens	COST EACH		
		MAT.	INST.	TOTAL
1400	Prefinished wood cabinets, average quality	6,925	2,575	9,500
1500	Prefinished wood cabinets, high quality	13,800	2,975	16,775
1600	Custom cabinets, built in place, high quality	18,300	3,400	21,700
1700				
1800				
1900				
2000				
2100				
2200	NOTE: No plumbing or electric rough-ins are included in the above			
2300	Prices, for plumbing see Division 22, for electric see Division 26.			
2400				
2500				
2600				
2700				
2900	Cut & patch to match existing construction, add, minimum	2%	3%	
3000	Maximum	5%	9%	
3100	Dust protection, add, minimum	1%	2%	
3200	Maximum	4%	11%	
3300	Equipment usage curtailment, add, minimum	1%	1%	
3400	Maximum	3%	10%	
3500	Material handling & storage limitation, add, minimum	1%	1%	
3600	Maximum	6%	7%	
3700	Protection of existing work, add, minimum	2%	2%	
3800	Maximum	5%	7%	
3900	Shift work requirements, add, minimum		5%	
4000	Maximum		30%	

E1090 Other Equipment

E1090 315	Selective Price Sheet	COST EACH		
		MAT.	INST.	TOTAL
0100	Cabinets standard wood, base, one drawer one door, 12" wide	350	56	406
0200	15" wide	405	58	463
0300	18" wide	237	60	297
0400	21" wide	335	61.50	396.50
0500	24" wide	460	62.50	522.50
0600				
0700	Two drawers two doors, 27" wide	480	63.50	543.50
0800	30" wide	465	65	530
0900	33" wide	315	66.50	381.50
1000	36" wide	535	68.50	603.50
1100	42" wide	520	70.50	590.50
1200	48" wide	640	73.50	713.50
1300	Drawer base (4 drawers), 12" wide	325	56	381
1400	15" wide	325	58	383
1500	18" wide	360	60	420
1600	24" wide	415	62.50	477.50
1700	Sink or range base, 30" wide	400	65	465
1800	33" wide	450	66.50	516.50
1900	36" wide	460	68.50	528.50
2000	42" wide	505	70.50	575.50
2100	Corner base, 36" wide	625	77.50	702.50
2200	Lazy Susan with revolving door	465	84.50	549.50
2300	Cabinets standard wood, wall two doors, 12" high, 30" wide	237	56	293
2500	15" high, 30" high	350	58	408
2600	36" wide	166	61.50	227.50
2700	24" high, 30" wide	177	60	237
2800	36" wide	510	61.50	571.50
2900	30" high, 30" wide	430	72	502
3000	36" wide	410	74	484
3100	42" wide	485	75.50	560.50
3200	48" wide	830	75.50	905.50
3300	One door, 30" high, 12" wide	116	63.50	179.50
3400	15" wide	262	65	327
3500	18" wide	289	66.50	355.50
3600	24" wide	335	68.50	403.50
3700	Corner, 30" high, 24" wide	215	77.50	292.50
3800	36" wide	500	84.50	584.50
3900	Broom, 84" high, 24" deep, 18" wide	870	139	1,009
4000	Oven, 84" high, 24" deep, 27" wide	1,125	174	1,299
4100	Valance board, 4' long	104	14.10	118.10
4200	6" long	156	21	177
4300	Counter tops, laminated plastic, stock 25" wide w/backsplash, min.	18.45	23	41.45
4400	Maximum	44.50	28	72.50
4500	Custom, 7/8" thick, no splash	41	23	64
4600	Cove splash	52	23	75
4700	1-1/4" thick, no splash	45	25	70
4800	Square splash	50	25	75
4900	Post formed	13.90	23	36.90
5000				
5100	Maple laminated 1-1/2" thick, no splash	98	25	123
5200	Square splash	116	25	141
5300				
5400				
5500	Appliances, range, free standing, minimum	505	113	618
5600	Maximum	2,525	283	2,808
5700	Built-in, minimum	955	130	1,085
5800	Maximum	2,200	695	2,895
5900	Counter top range 4 burner, minimum	370	130	500

723

E1090 Other Equipment

E1090 315	Selective Price Sheet	COST EACH		
		MAT.	INST.	TOTAL
6000	Maximum	2,050	261	2,311
6100	Compactor, built-in, minimum	775	139	914
6200	Maximum	1,250	232	1,482
6300	Dishwasher, built-in, minimum	340	405	745
6400	Maximum	490	810	1,300
6500	Garbage disposer, sink-pipe, minimum	120	162	282
6600	Maximum	266	162	428
6700	Range hood, 30" wide, 2 speed, minimum	117	299	416
6800	Maximum	1,150	500	1,650
6900	Refrigerator, no frost, 12 cu. ft.	500	113	613
7000	20 cu. ft.	620	189	809
7100	Plumb. not incl. rough-ins, sinks porc. C.I., single bowl, 21" x 24"	335	268	603
7200	21" x 30"	1,000	268	1,268
7300	Double bowl, 20" x 32"	430	315	745
7400				
7500	Stainless steel, single bowl, 19" x 18"	690	268	958
7600	22" x 25"	765	268	1,033
7700				
7800				
7900				
8000				
8100				
8200				
8300				
8400				

For customer support on your Commercial Renovation Costs with RSMeans Data, call 800.448.8182.

Same Data. Simplified.

Enjoy the convenience and efficiency of accessing your costs anywhere:

- **Skip the multiplier** by setting your location
- **Quickly search,** edit, favorite and share costs
- **Stay on top of price changes** with automatic updates

Discover more at rsmeans.com/online

G1030 Site Earthwork

This page illustrates and describes utility trench excavation with backfill systems including a trench, backfill, excavated material, utility pipe or ductwork not included. Lines within System Components give the unit price and total price on a linear foot basis for this system. Prices for utility trench excavation with backfill systems are on Line Items G1030 810 1100 thru 2500. Both material quantities and labor costs have been adjusted for the system listed.

Factors: To adjust for job conditions other than normal working situations use Lines G1030 810 3100 thru 4000.

Example: You are to install the system and protect existing construction. Go to Line G1030 810 3600 and apply these percentages to the appropriate TOTAL cost.

System Components	QUANTITY	UNIT	COST PER L.F.		
			MAT.	INST.	TOTAL
Cont. utility trench excav. with 3/8 C.Y. wheel mtd. backhoe, Backfilling with excav. mat. and comp. in 12" lifts. Trench is 2' wide by 4' dp. Cost of utility piping or ductwork is not incl. Cost based on an Excavation production rate of 150 C.Y. daily no hauling included.					
Machine excavate trench, 2' wide by 4' deep	.296	C.Y.	.47	2.58	3.05
Dozer backfill with excavated material	.296	C.Y.	.31	.71	1.02
Compact in 12" lifts, vibrating plate	.296	C.Y.	.30	2.83	3.13
TOTAL			1.08	6.12	7.20

G1030 810	Excavation, Utility Trench	COST PER L.F.		
		MAT.	INST.	TOTAL
1100	2' wide x 2' deep	.54	3.06	3.60
1200	3' deep	.81	4.58	5.40
1250	4' deep	1.08	6.10	7.20
1300	With sloping sides, 2' wide x 5' deep	3.04	17.20	20
1400	6' deep	4.06	23	27
1500	7' deep	5.20	29.50	34.50
1600	8' deep	6.50	36.50	43
1700	9' deep	7.90	45	52.50
1800	10' deep	9.45	53.50	63
2000	For hauling excavated material up to 2 miles & backfilling w/ gravel			
2100	Gravel, 2' wide by 2' deep, add	5.40	4.08	9.48
2200	4' deep, add	10.80	8.15	18.95
2300	6' deep, add	40.50	30.50	71
2400	8' deep, add	65	49	114
2500	10' deep, add	94.50	71.50	166
2600				
2700				
2800	For shallow, hand excavated trenches, no backfill, to 6' dp, light soil		70.50	70.50
2900	Heavy soil		141	141
3000				
3100	Equipment usage curtailment, add, minimum	1%	1%	
3200	Maximum	3%	10%	
3300	Material handling & storage limitation, add, minimum	1%	1%	
3400	Maximum	6%	7%	
3500	Protection of existing work, add, minimum	2%	2%	
3600	Maximum	5%	7%	
3700	Shift work requirements, add, minimum		5%	
3800	Maximum		30%	
3900	Temporary shoring and bracing, add, minimum	2%	5%	
4000	Maximum	5%	12%	

G2010 Roadways

This page illustrates and describes driveway systems including concrete slab, gravel base, broom finish, compaction, and joists. Lines within System Components give the unit price and total price on a cost each basis for this system. Prices for driveway systems are on Line Items G2010 240 1200 thru 2700. Both material quantities and labor costs have been adjusted for the system listed.

Factors: To adjust for job conditions other than normal working situations use Lines G2010 240 2900 thru 4000.

Example: You are to install the system and match existing construction. Go to Line G2010 240 3000 and apply these percentages to the appropriate MAT. and INST. costs.

System Components	QUANTITY	UNIT	COST EACH MAT.	COST EACH INST.	COST EACH TOTAL
Complete driveway, 10' x 30', 4" concrete slab, 3000 psi, on 6" crushed Stone base, concrete broom finished and cured with 10' x 10' joints.					
Grade and compact subgrade	33.000	S.Y.		87.12	87.12
Place and compact 6" crushed stone base	33.000	S.Y.	181.50	53.79	235.29
Place and remove edgeforms (4 use)	80.000	L.F.	40	354.40	394.40
Concrete for slab, 3000 psi	4.000	C.Y.	548		548
Place concrete, 4" thick slab, direct chute	4.000	C.Y.		130.16	130.16
Screed, float and broom finish slab	4.000	C.Y.		303	303
Spray on membrane curing compound	3.000	S.F.	40.20	35.70	75.90
Saw cut 3" deep joints	60.000	L.F.	7.80	152.40	160.20
TOTAL			817.50	1,116.57	1,934.07

G2010 240	Driveways	COST EACH MAT.	COST EACH INST.	COST EACH TOTAL
1200	Concrete: 10' x 30' with 4" concrete on 6" crushed stone	820	1,125	1,945
1300	With 6" concrete	1,100	1,175	2,275
1400	20' x 30' with 4" concrete	1,625	2,200	3,825
1500	With 6" concrete	2,175	2,325	4,500
1600	10' wide, for each additional 10' length over 30', 4" thick, add	260	365	625
1700	6" thick, add	360	390	750
1800	20' wide, for each additional 10' length over 30', 4" thick, add	515	720	1,235
1900	6" thick, add	685	760	1,445
2000	Asphalt: 10' x 30' with 2" binder, 1" topping on 6" cr. st., sealed	645	1,225	1,870
2100	3" binder, 1" topping	780	1,225	2,005
2200	20' x 30' with 2" binder, 1" topping	1,300	1,500	2,800
2300	3" binding, 1" topping	1,575	1,525	3,100
2400	10' wide for each add'l. 10' length over 30', 2" b + 1" t, add	215	97	312
2500	3" binder, 1" topping	261	102	363
2600	20' wide for each add'l. 10' length over 30', 2" b + 1" t, add	430	194	624
2700	3" binder, 1" topping	520	204	724
2800				
2900	Cut & patch to match existing construction, add, minimum	2%	3%	
3000	Maximum	5%	9%	
3100	Dust protection, add, minimum	1%	2%	
3200	Maximum	4%	11%	
3300	Equipment usage curtailment, add, minimum	1%	1%	
3400	Maximum	3%	10%	
3500	Material handling & storage limitation, add, minimum	1%	1%	
3600	Maximum	6%	7%	
3700	Protection of existing work, add, minimum	2%	2%	
3800	Maximum	5%	7%	
3900	Shift work requirements, add, minimum		5%	
4000	Maximum		30%	

G20 Site Improvements

G2020 Parking Lots

This page illustrates and describes asphalt parking lot systems including asphalt binder, topping, crushed stone base, painted parking stripes and concrete parking blocks. Lines within System Components give the unit price and total price per square yard for this system. Prices for asphalt parking lot systems are on Line Items G2020 230 1400 thru 2500. Both material quantities and labor costs have been adjusted for the system listed.

Factors: To adjust for job conditions other than normal working situations use Lines G2020 230 2900 thru 4000.

Example: You are to install the system and match existing construction. Go to Line G2020 230 2900 and apply these percentages to the appropriate MAT. and INST. costs.

System Components	QUANTITY	UNIT	COST PER S.Y.		
			MAT.	INST.	TOTAL
Parking lot consisting of 2″ asphalt binder and 1″ topping on 6″					
Crushed stone base with painted parking stripes and concrete parking blocks.					
Fine grade and compact subgrade	1.000	S.Y.		2.64	2.64
6″ crushed stone base, stone	.320	Ton	1.76	.52	2.28
2″ asphalt binder	1.000	S.Y.	8.15	1.55	9.70
1″ asphalt topping	1.000	S.Y.	4.05	1.02	5.07
Paint parking stripes	.500	L.F.	.07	.12	.19
6″ x 10″ x 6′ precast concrete parking blocks	.020	Ea.	1.27	.48	1.75
Mobilization of equipment	.005	Ea.		4.61	4.61
TOTAL			15.30	10.94	26.24

G2020 230	Parking Lots, Asphalt	COST PER S.Y.		
		MAT.	INST.	TOTAL
1400	2″ binder plus 1″ topping on 6″ crushed stone	15.30	10.95	26.25
1500	9″ crushed stone	22	12.20	34.20
1600	12″ crushed stone	24.50	12.35	36.85
1700	On bank run gravel, 6″ deep	19.85	11.40	31.25
1800	9″ deep	23	11.60	34.60
1900	12″ deep	26	11.80	37.80
2000	3″ binder plus 1″ topping on 6″ crushed stone	23	12.50	35.50
2100	9″ deep crushed stone	26	12.65	38.65
2200	12″ deep crushed stone	28.50	12.80	41.30
2300	On bank run gravel, 6″ deep	24	11.85	35.85
2400	9″ deep	27	12.05	39.05
2500	12″ deep	30	12.25	42.25
2600				
2700				
2900	Cut & patch to match existing construction, add, minimum	2%	3%	
3000	Maximum	5%	9%	
3100	Dust protection, add, minimum	1%	2%	
3200	Maximum	4%	11%	
3300	Equipment usage curtailment, add, minimum	1%	1%	
3400	Maximum	3%	10%	
3500	Material handling & storage limitation, add, minimum	1%	1%	
3600	Maximum	6%	7%	
3700	Protection of existing work, add, minimum	2%	2%	
3800	Maximum	5%	7%	
3900	Shift work requirements, add, minimum		5%	
4000	Maximum		30%	

G2020 Parking Lots

This page illustrates and describes concrete parking lot systems including a concrete slab, unreinforced, compacted gravel base, broom finished, cured, joints, painted parking stripes, and parking blocks. Lines within System Components give the unit price and total price per square yard for this system. Prices for concrete parking lot systems are on Line Items G2020 240 1500 thru 2500. All materials have been adjusted according to the system listed.

Factors: To adjust for job conditions other than normal working situations use Lines G2020 240 2900 thru 4000.

Example: You are to install the system and protect existing construction. Go to Line G2020 240 3800 and apply these percentages to the appropriate MAT. and INST. costs.

System Components	QUANTITY	UNIT	COST PER S.Y.		
			MAT.	INST.	TOTAL
Parking lot, 4" slab, 3000 psi concrete on 6" crushed stone base					
With 12' x 12' joints, painted parking strips, conc. parking blocks.					
Grade and compact subgrade	1.000	S.Y.		2.64	2.64
6" crushed stone base, stone	.320	Ton	1.76	.52	2.28
Place and remove edge forms, 4 use	.250	L.F.	.13	1.11	1.24
Concrete for slab, 3000 psi	1.000	S.F.	15.07		15.07
Place concrete, 4" thick slab, direct chute	1.000	S.F.		3.58	3.58
Spray on membrane curing compound	1.000	S.F.	1.21	.75	1.96
Saw cut 3" deep joints	2.250	L.F.	.29	5.72	6.01
Paint parking stripes	.500	L.F.	.07	.12	.19
6" x 10" x 6' precast concrete parking blocks	.020	Ea.	1.27	.48	1.75
TOTAL			19.80	14.92	34.72

G2020 240	Parking Lots, Concrete	COST PER S.Y.		
		MAT.	INST.	TOTAL
1500	4" concrete slab, 3000 psi on 3" crushed stone	21	15.95	36.95
1550	6" crushed stone	19.80	14.90	34.70
1600	9" crushed stone	26.50	16.15	42.65
1700	On bank run gravel, 6" deep	24.50	15.35	39.85
1800	9" deep	27.50	15.60	43.10
1900	On compacted subgrade only	18.05	14.40	32.45
2000	6" concrete slab, 3000 psi on 3" crushed stone	28	17.70	45.70
2100	6" crushed stone	31	17.75	48.75
2200	9" crushed stone	33.50	17.90	51.40
2300	On bank run gravel, 6" deep	31.50	17.10	48.60
2400	9" deep	34.50	17.30	51.80
2500	On compacted subgrade only	31	17.75	48.75
2600				
2700				
2900	Cut & patch to match existing construction, add, minimum	2%	3%	
3000	Maximum	5%	9%	
3100	Dust protection, add, minimum	1%	2%	
3200	Maximum	4%	11%	
3300	Equipment usage curtailment, add, minimum	1%	1%	
3400	Maximum	3%	10%	
3500	Material handling & storage limitation, add, minimum	1%	1%	
3600	Maximum	6%	7%	
3700	Protection of existing work, add, minimum	2%	2%	
3800	Maximum	5%	7%	
3900	Shift work requirements, add, minimum		5%	
4000	Maximum		30%	

729

G2030 Pedestrian Paving

This page illustrates and describes sidewalk systems including concrete, welded wire and broom finish. Lines within System Components give unit price and total price per square foot for this system. Prices for sidewalk systems are on Line Items G2030 105 1700 thru 1900. Both material quantities and labor costs have been adjusted for the system listed.

Factors: To adjust for job conditions other than normal working situations use Lines G2030 105 2900 thru 4000.

Example: You are to install the system and match existing construction. Go to Line G2030 105 2900 and apply these percentages to the appropriate MAT. and INST. costs.

System Components	QUANTITY	UNIT	COST PER S.F. MAT.	COST PER S.F. INST.	COST PER S.F. TOTAL
4" thick concrete sidewalk with welded wire fabric					
3000 psi air entrained concrete, broom finish.					
Gravel fill, 4" deep	.012	C.Y.	.39	.11	.50
Compact fill	.012	C.Y.		.03	.03
Hand grade	1.000	S.F.		2.72	2.72
Edge form	.250	L.F.	.13	1.11	1.24
Welded wire fabric	.011	S.F.	.18	.46	.64
Concrete, 3000 psi air entrained	.012	C.Y.	1.64		1.64
Place concrete	.012	C.Y.		.39	.39
Broom finish	1.000	S.F.		1.01	1.01
TOTAL			2.34	5.83	8.17

G2030 105	Sidewalks	COST PER S.F. MAT.	COST PER S.F. INST.	COST PER S.F. TOTAL
1700	Asphalt (bituminous), 2" thick	1.26	3.45	4.71
1750	4" thick	2.34	5.85	8.19
1800	Brick, on sand, bed, 4.5 brick per S.F.	3.39	12.50	15.89
1900	Flagstone, slate, 1" thick, rectangular	16.10	13.60	29.70
2000				
2100				
2200				
2300				
2400				
2500				
2600				
2700				
2900	Cut & patch to match existing construction, add, minimum	2%	3%	
3000	Maximum	5%	9%	
3100	Dust protection, add, minimum	1%	2%	
3200	Maximum	4%	11%	
3300	Equipment usage curtailment, add, minimum	1%	1%	
3400	Maximum	3%	10%	
3500	Material handling & storage limitation, add, minimum	1%	1%	
3600	Maximum	6%	7%	
3700	Protection of existing work, add, minimum	2%	2%	
3800	Maximum	5%	7%	
3900	Shift work requirements, add, minimum		5%	
4000	Maximum		30%	

G2050 Landscaping

This page describes landscaping—lawn establishment systems including loam, lime, fertilizer, side and top mulching. Lines within system components give the unit price and total price per square yard for this system. Prices for landscaping—lawn establishment systems are on Line Items G2050 420 1000 thru 1200. Both material quantities and labor costs have been adjusted for the system listed.

Factors: To adjust for job conditions other than normal working situations use Lines G2050 420 2900 thru 4000.

Example: You are to install the system and provide dust protection. Go to Line G2050 420 3200 and apply these percentages to the appropriate MAT. and INST. costs.

System Components	QUANTITY	UNIT	COST PER S.Y. MAT.	COST PER S.Y. INST.	COST PER S.Y. TOTAL
Establishing lawns with loam, lime, fertilizer, seed and top mulching					
On rough graded areas.					
Furnish and place loam 4" deep	.110	C.Y.	4.06	1.15	5.21
Fine grade, lime, fertilize and seed	1.000	S.Y.	.54	3.79	4.33
Hay mulch, 1 bale/M.S.F.	1.000	S.Y.	.29	.69	.98
Rolling with hand roller	1.000	S.Y.		1.20	1.20
TOTAL			4.89	6.83	11.72

G2050 420	Landscaping - Lawn Establishment	COST PER S.Y. MAT.	COST PER S.Y. INST.	COST PER S.Y. TOTAL
1000	Establishing lawns with loam, lime, fertilizer, seed and hay mulch	4.89	6.85	11.74
1100	Above system with jute mesh in place of hay mulch	5.70	7.25	12.95
1200	Above system with sod in place of seed	6.30	2.49	8.79
1300				
1400				
1500				
1600				
1700				
1800				
1900				
2000				
2100				
2200				
2300				
2400				
2500				
2600				
2700				
2900	Cut & patch to match existing construction, add, minimum	2%	3%	
3000	Maximum	5%	9%	
3100	Dust protection, add, minimum	1%	2%	
3200	Maximum	4%	11%	
3300	Equipment usage curtailment, add, minimum	1%	1%	
3400	Maximum	3%	10%	
3500	Material handling & storage limitation, add, minimum	1%	1%	
3600	Maximum	6%	7%	
3700	Protection of existing work, add, minimum	2%	2%	
3800	Maximum	5%	7%	
3900	Shift work requirements, add, minimum		5%	
4000	Maximum		30%	

G2050 Landscaping

G2050 510	Selective Price Sheet	COST PER UNIT		
		MAT.	INST.	TOTAL
0100	Shrubs and trees, Evergreen, in prepared beds			
0200				
0300	Arborvitae pyramidal, 4'-5'	111	106	217
0400	Globe, 12"-15"	19.50	17.95	37.45
0500	Cedar, blue, 8'-10'	273	176	449
0600	Hemlock, Canadian, 2-1/2'-3'	35.50	48	83.50
0700	Juniper, andora, 18"-24"	61	21.50	82.50
0800	Wiltoni, 15"-18"	32.50	21.50	54
0900	Skyrocket, 4-1/2'-5'	109	57.50	166.50
1000	Blue pfitzer, 2'-2-1/2'	43.50	39	82.50
1100	Ketleerie, 2-1/2'-3'	65	34.50	99.50
1200	Pine, black, 2-1/2'-3'	61.50	34.50	96
1300	Mugo, 18"-24"	53.50	28.50	82
1400	White, 4'-5'	57	42	99
1500	Spruce, blue, 18"-24"	72	28.50	100.50
1600	Norway, 4'-5'	88	42	130
1700	Yew, denisforma, 12"-15"	40.50	28.50	69
1800	Capitata, 18"-24"	35.50	57.50	93
1900	Hicksi, 2'-2-1/2'	102	57.50	159.50
2000				
2100	Trees, Deciduous, in prepared beds			
2200				
2300	Beech, 5'-6'	238	63	301
2400	Dogwood, 4'-5'	144	79	223
2500	Elm, 8'-10'	410	159	569
2600	Magnolia, 4'-5'	154	159	313
2700	Maple, red, 8'-10', 1-1/2" caliper	201	315	516
2800	Oak, 2-1/2"-3" caliper	400	530	930
2900	Willow, 6'-8', 1" caliper	134	159	293
3000				
3100	Shrubs, Broadleaf Evergreen, in prepared beds			
3200				
3300	Andromeda, 15"-18", container	42.50	17.95	60.45
3400	Azalea, 15"-18", container	41	17.95	58.95
3500	Barberry, 9"-12", container	24.50	13.25	37.75
3600	Boxwood, 15"-18", B & B	53.50	17.95	71.45
3700	Euonymus, emerald gaiety, 12"-15", container	30	15	45
3800	Holly, 15"-18", B & B	45.50	17.95	63.45
3900	Mount laurel, 18"-24", B & B	85.50	21.50	107
4000	Privet, 18"-24", B & B	25.50	13.25	38.75
4100	Rhododendron, 18"-24", container	49	36	85
4200	Rosemary, 1 gal. container	20.50	2.87	23.37
4300	Deciduous, amalanchier, 2'-3', B & B	148	30	178
4400	Azalea, 15"-18", B & B	38	17.95	55.95
4500	Bayberry, 2'-3', B & B	36.50	30	66.50
4600	Cotoneaster, 15"-18", B & B	37	21.50	58.50
4700	Dogwood, 3'-4', B & B	40	79	119
4800	Euonymus, alatus compacta, 15"-18", container	31	21.50	52.50
4900	Forsythia, 2'-3', container	16.95	28.50	45.45
5000	Honeysuckle, 3'-4', B & B	71.50	28.50	100
5100	Hydrangea, 2'-3', B & B	33	30	63
5200	Lilac, 3'-4', B & B	25	79	104
5300	Quince, 2'-3', B & B	34.50	30	64.50
5400				
5500	Ground Cover			
5600				
5700	Plants, Pachysandra, prepared beds, per hundred	93	115	208
5800	Vinca minor or English ivy, per hundred	109	144	253

G2050 Landscaping

G2050 510	Selective Price Sheet	COST PER UNIT		
		MAT.	INST.	TOTAL
5900	Plant bed prep. 18" dp., mach., per square foot	2.90	.51	3.41
6000	By hand, per square foot	2.90	3.65	6.55
6100	Stone chips, Georgia marble, 50# bags, per bag	5.65	3.31	8.96
6200	Onyx gemstone, per bag	44	6.60	50.60
6300	Quartz, per bag	18.85	6.60	25.45
6400	Pea gravel, truck load lots, per cubic yard	20.50	61.50	82
6500	Mulch, polyethylene mulch, per square yard	.67	.57	1.24
6600	Wood chips, 2" deep, per square yard	1.46	2.57	4.03
6700	Peat moss, 1" deep, per square yard	5.65	.63	6.28
6800	Erosion control per square yard, Jute mesh stapled	1.10	1.10	2.20
6900	Plastic netting, stapled	.34	.69	1.03
7000	Polypropylene mesh, stapled	2.01	.69	2.70
7100	Tobacco netting, stapled	.29	.69	.98
7200	Lawns per square yard, seeding incl. fine grade, limestone, fert. and seed	.54	3.79	4.33
7300	Sodding, incl. fine grade, level ground	2.10	1.29	3.39
7500	Edging, per linear foot, Redwood, untreated, 1" x 4"	3.27	2.79	6.06
7600	2" x 4"	6	4.22	10.22
7700	Stl. edge strips, 1/4" x 5" inc. stakes	4.53	4.22	8.75
7800	3/16" x 4"	4.48	4.22	8.70
7900	Brick edging, set on edge	3.86	9.25	13.11
8000	Set flat	1.93	3.38	5.31
8100				
8200				
8300				
8400				

G3020 Sanitary Sewer

This page illustrates and describes drainage and utilities – septic system including a tank, distribution box, excavation, piping, crushed stone and backfill. Lines within System Components give the unit price and total price on a cost each basis for this system. Prices for drainage and utilities – septic systems are on Line Items G3020 610 1600 thru 2000. Both material quantities and labor costs have been adjusted for the system listed.

Factors: To adjust for job conditions other than normal working situations use Lines G3020 610 3100 thru 4000.

Example: You are to install the system with a material handling limitation. Go to Line G3020 610 3500 and apply these percentages to the appropriate MAT. and INST. costs.

System Components	QUANTITY	UNIT	COST EACH		
			MAT.	INST.	TOTAL
Septic system including tank, distribution box, excavation, piping, crushed Stone and backfill for a 1000 S.F. leaching field.					
Precast concrete septic tank, 1000 gal. capacity	1.000	Ea.	1,100	309.50	1,409.50
Concrete distribution box	1.000	Ea.	103	70.50	173.50
Sch. 40 sewer pipe, PVC, 4″ diameter	170.000	L.F.	413.10	850	1,263.10
Sch. 40 sewer pipe fittings, PVC, 4″ diameter	8.000	Ea.	168	724	892
Excavation for septic tank	18.000	C.Y.		261	261
Excavation for disposal field	101.000	C.Y.		634.28	634.28
Crushed stone backfill	76.000	C.Y.	2,242	1,044.24	3,286.24
Backfill with excavated material	26.000	C.Y.		58.24	58.24
Asphalt felt paper, #15	6.000	Sq.	37.50	113.10	150.60
TOTAL			4,063.60	4,064.86	8,128.46

G3020 610	Septic Systems	COST EACH		
		MAT.	INST.	TOTAL
1600	1000 gal. tank with 1000 S.F. field	4,075	4,075	8,150
1700	1000 gal. tank with 2000 S.F. field	5,950	5,975	11,925
1800	With leaching pits	3,825	2,475	6,300
1900	2000 gal. tank with 2000 S.F. field	7,675	6,150	13,825
2000	With leaching pits	5,675	3,150	8,825
2100				
2200				
2300				
2400				
2500				
2600				
2700				
2800				
2900				
3100	Dust protection, add, minimum	1%	2%	
3200	Maximum	4%	11%	
3300	Equipment usage curtailment, add, minimum	1%	1%	
3400	Maximum	3%	10%	
3500	Material handling & storage limitation, add, minimum	1%	1%	
3600	Maximum	6%	7%	
3700	Protection of existing work, add, minimum	2%	2%	
3800	Maximum	5%	7%	
3900	Shift work requirements, add, minimum		5%	
4000	Maximum		30%	

G3060 Fuel Distribution

2" Vent Cap

3" Fill Box

Manway Extension

Supply

Return

Pea Gravel Backfill

Hold Down Straps

Concrete Slab w/ Reinforcing Steel

Fiberglass Underground Fuel Storage Tank, Single Wall

This page illustrates and describes fiberglass tanks including hold down slab, reinforcing steel and peastone gravel. Lines within Systems Components give the material and installation price on a cost each basis for the components. Prices for fiberglass tank systems are on Line Items G3060 305 1010 thru G3060 310 1100. Material quantities and labor costs have been adjusted for the system listed.

Factors: To adjust for job conditions other than normal working situations use Lines G3060 310 2700 thru 4000.

System Components	QUANTITY	UNIT	COST EACH		
			MAT.	INST.	TOTAL
Double wall underground fiberglass storage tanks including, excavation,					
hold down slab, backfill, manway extension and related piping.					
Excavating trench, no sheeting or dewatering, 6'-10 ' D, 3/4 CY hyd backhoe	18.000	Ea.		167.94	167.94
Corrosion resistance wrap & coat, small diam. pipe, 1" diam., add	120.000	Ea.	105.60		105.60
Fiberglass-reinforced elbows 90 & 45 degree, 2 inch	7.000	Ea.	336	689.50	1,025.50
Fiberglass-reinforced threaded adapters, 2 inch	2.000	Ea.	43	224	267
Remote tank gauging system, 30', 5" pointer travel	1.000	Ea.	6,450	420	6,870
Corrosion resistance wrap & coat, small diam. pipe. 4 " diam. add	30.000	Ea.	82.50		82.50
Pea gravel	9.000	Ea.	1,080	553.50	1,633.50
Reinforcing in hold down pad, #3 to #7	120.000	Ea.	81.60	84	165.60
Tank leak detector system, 8 channel, external monitoring	1.000	Ea.	1,375		1,375
Tubbing copper, type L, 3/4" diam.	120.000	Ea.	402	1,320	1,722
Elbow 90 degree ,wrought cu x cu 3/4 " diam.	4.000	Ea.	17.12	176	193.12
Pipe, black steel welded, sch 40, 3" diam.	30.000	Ea.	511.50	1,132.50	1,644
Elbow 90 degree, steel welded butt joint, 3 " diam.	3.000	Ea.	76.50	695.70	772.20
Fiberglass-reinforced plastic pipe, 2 inch	156.000	Ea.	1,248	865.80	2,113.80
Fiberglass-reinforced plastic pipe, 3 inch	36.000	Ea.	1,422	226.80	1,648.80
Fiberglass-reinforced elbows 90 & 45 degree, 3 inch	3.000	Ea.	501	393	894
Tank leak detection system, probes, well monitoring, liquid phase detection	2.000	Ea.	1,490		1,490
Double wall fiberglass annular space	1.000	Ea.	465		465
Flange, weld neck, 150 lb, 3 " pipe size	1.000	Ea.	43	115.45	158.45
Vent protector/breather, 2" diam.	1.000	Ea.	60	26.50	86.50
Corrosion resistance wrap & coat, small diam. pipe, 2" diam., add	36.000	Ea.	43.56		43.56
Concrete hold down pad, 6' x 5' x 8" thick	30.000	Ea.	108.90	56.10	165
600 gallon capacity, underground storage tank, double wall fiberglass	1.000	Ea.	10,200	625	10,825
Tank, for hold-downs 500-4,000 gal., add	1.000	Ea.	500	201	701
Foot valve, single poppet, 3/4" diam.	1.000	Ea.	132	47	179
Fuel fill box, locking inner cover, 3" diam.	1.000	Ea.	129	169	298
Fuel oil specialties, valve,ball chk, globe type fusible, 3/4" diam.	2.000	Ea.	374	84	458
TOTAL			27,277.28	8,272.79	35,550.07

G3060 310	Fiberglass Fuel Tank, Double Wall	COST EACH		
		MAT.	INST.	TOTAL
1010	Storage tank, fuel, underground, double wall fiberglass, 600 Gal.	27,300	8,275	35,575
1020	2500 Gal.	41,300	11,400	52,700
1030	4000 Gal.	49,900	14,100	64,000
1040	6000 Gal.	57,000	15,900	72,900
1050	8000 Gal.	70,500	19,200	89,700
1060	10,000 Gal.	82,000	22,100	104,100

735

G3060 Fuel Distribution

G3060 310	Fiberglass Fuel Tank, Double Wall	COST EACH		
		MAT.	INST.	TOTAL
1070	15,000 Gal.	122,500	25,400	147,900
1080	20,000 Gal.	139,500	30,700	170,200
1090	25,000 Gal.	160,500	35,700	196,200
1100	30,000 Gal.	183,500	37,800	221,300
2700	Cut & patch to match existing construction, add, minimum	2%	3%	
2800	Maximum	5%	9%	
2900	Dust protection, add, minimum	1%	2%	
3000	Maximum	4%	11%	
3100	Equipment usage curtailment, add, minimum	1%	1%	
3200	Maximum	3%	10%	
3300	Material handling & storage limitation, add, minimum	1%	1%	
3400	Maximum	6%	7%	
3500	Protection of existing work, add, minimum	2%	2%	
3600	Maximum	5%	7%	
3700	Shift work requirements, add, minimum		5%	
3800	Maximum		30%	
3900	Temporary shoring and bracing, add, minimum	2%	5%	
4000	Maximum	5%	12%	

736

G3060 Fuel Distribution

Steel Underground Fuel Storage Tank, Double Wall

System Components	QUANTITY	UNIT	COST EACH		
			MAT.	INST.	TOTAL
Double wall underground steel storage tank including, excavation, hold down slab, backfill, manway extension and related piping.					
Fiberglass-reinforced plastic pipe, 2 inch	156.000	Ea.	1,248	865.80	2,113.80
Fiberglass-reinforced plastic pipe, 3 inch	36.000	Ea.	1,422	226.80	1,648.80
Fiberglass-reinforced elbows 90 & 45 degree, 3 inch	3.000	Ea.	501	393	894
Tank leak detection system, probes, well monitoring, liquid phase detection	2.000	Ea.	1,490		1,490
Flange, weld neck, 150 lb, 3" pipe size	1.000	Ea.	43	115.45	158.45
Vent protector/breather, 2" diam.	1.000	Ea.	60	26.50	86.50
Concrete hold down pad, 6' x 5' x 8" thick	30.000	Ea.	108.90	56.10	165
Tanks, for hold-downs, 500-2,000 gal, add	1.000	Ea.	236	201	437
Double wall steel tank annular space	1.000	Ea.	281		281
Foot valve, single poppet, 3/4" diam.	1.000	Ea.	132	47	179
Fuel fill box, locking inner cover, 3" diam.	1.000	Ea.	129	169	298
Fuel oil specialties, valve, ball chk, globe type fusible, 3/4" diam.	2.000	Ea.	374	84	458
Pea gravel	9.000	Ea.	1,080	553.50	1,633.50
Reinforcing in hold down pad, #3 to #7	120.000	Ea.	81.60	84	165.60
Tank, for manways, add	1.000	Ea.	1,325		1,325
Tank leak detector system, 8 channel, external monitoring	1.000	Ea.	1,375		1,375
Tubbing copper, type L, 3/4" diam.	120.000	Ea.	402	1,320	1,722
Elbow 90 degree, wrought cu x cu 3/4" diam.	8.000	Ea.	34.24	352	386.24
Pipe, black steel welded, sch 40, 3" diam.	30.000	Ea.	511.50	1,132.50	1,644
Elbow 90 degree, steel welded butt joint, 3" diam.	3.000	Ea.	76.50	695.70	772.20
Excavating trench, no sheeting or dewatering, 6'-10' D, 3/4 CY hyd backhoe	18.000	Ea.		167.94	167.94
Fiberglass-reinforced elbows 90 & 45 degree, 2 inch	7.000	Ea.	336	689.50	1,025.50
Fiberglass-reinforced threaded adapters, 2 inch	2.000	Ea.	43	224	267
500 gallon capacity, underground storage tank, double wall steel	1.000	Ea.	3,925	635	4,560
Remote tank gauging system, 30', 5" pointer travel	1.000	Ea.	6,450	420	6,870
TOTAL			21,664.74	8,458.79	30,123.53

G3060 320	Steel Fuel Tank, Double Wall	COST EACH		
		MAT.	INST.	TOTAL
1010	Storage tank, fuel, underground, double wall steel, 500 Gal.	21,700	8,450	30,150
1020	2000 Gal.	27,400	11,000	38,400
1030	4000 Gal.	37,500	14,300	51,800
1040	6000 Gal.	42,500	15,800	58,300
1050	8000 Gal.	48,400	19,500	67,900
1060	10,000 Gal.	55,500	22,400	77,900

G3060 Fuel Distribution

G3060 320	Steel Fuel Tank, Double Wall	COST EACH		
		MAT.	INST.	TOTAL
1070	15,000 Gal.	67,500	26,000	93,500
1080	20,000 Gal.	94,000	31,200	125,200
1090	25,000 Gal.	107,000	35,700	142,700
1100	30,000 Gal.	107,000	38,000	145,000
1110	40,000 Gal.	223,000	47,700	270,700
1120	50,000 Gal.	270,000	57,000	327,000

G3060 325	Steel Tank, Above Ground, Single Wall	COST EACH		
		MAT.	INST.	TOTAL
1010	Storage tank, fuel, above ground, single wall steel, 550 Gal.	16,700	5,975	22,675
1020	2000 Gal.	24,400	6,100	30,500
1030	5000 Gal.	36,700	9,400	46,100
1040	10,000 Gal.	61,500	10,000	71,500
1050	15,000 Gal.	68,500	10,300	78,800
1060	20,000 Gal.	84,000	10,600	94,600
1070	25,000 Gal.	98,000	10,900	108,900
1080	30,000 Gal.	111,000	11,300	122,300

G3060 330	Steel Tank, Above Ground, Double Wall	COST EACH		
		MAT.	INST.	TOTAL
1010	Storage tank, fuel, above ground, double wall steel, 500 Gal.	15,300	6,050	21,350
1020	2000 Gal.	20,400	6,175	26,575
1030	4000 Gal.	29,200	6,300	35,500
1040	6000 Gal.	33,300	9,750	43,050
1050	8000 Gal.	37,000	10,000	47,000
1060	10,000 Gal.	49,300	10,200	59,500
1070	15,000 Gal.	60,500	10,500	71,000
1080	20,000 Gal.	68,500	10,900	79,400
1090	25,000 Gal.	81,000	11,200	92,200
1100	30,000 Gal.	89,000	11,600	100,600
2700	Cut & patch to match existing construction, add, minimum	2%	3%	
2800	Maximum	5%	9%	
2900	Dust protection, add, minimum	1%	2%	
3000	Maximum	4%	11%	
3100	Equipment usage curtailment, add, minimum	1%	1%	
3200	Maximum	3%	10%	
3300	Material handling & storage limitation, add, minimum	1%	1%	
3400	Maximum	6%	7%	
3500	Protection of existing work, add, minimum	2%	2%	
3600	Maximum	5%	7%	
3700	Shift work requirements, add, minimum		5%	
3800	Maximum		30%	
3900	Temporary shoring and bracing, add, minimum	2%	5%	
4000	Maximum	5%	12%	

G4020 Site Lighting

This page illustrates and describes light poles for parking or walkway area lighting. Included are aluminum or steel light poles, single or multiple fixture bracket arms, excavation, concrete footing, backfill and compaction. Lines within system components give the unit price and total price on a cost per each basis. Prices for systems are shown on lines G4020 210 0200 through 1440. Both material quantities and labor costs have been adjusted for the system listed.

Factors: To adjust for job conditions other than normal working situations use lines G4020 210 2700 through 3600.

Example: You are to install the system where you must be careful not to damage existing walks or parking areas. Go to line G4020 210 330 and apply these percentages to the appropriate MAT. and INST. costs.

System Components	QUANTITY	UNIT	COST EACH MAT.	COST EACH INST.	COST EACH TOTAL
Light poles, aluminum, 20' high, 1 arm bracket					
Aluminum light pole, 20', no concrete base	1.000	Ea.	665	747.50	1,412.50
Bracket arm for aluminum light pole	1.000	Ea.	151	98	249
Excavation by hand, pits to 6' deep, heavy soil or clay	2.368	C.Y.		333.89	333.89
Footing, concrete incl forms, reinforcing, spread, under 1 C.Y.	.465	C.Y.	106.49	153.95	260.44
Backfill by hand	1.903	C.Y.		98	98
Compaction vibrating plate	1.903	C.Y.		15.83	15.83
TOTAL			922.49	1,447.17	2,369.66

G4020 210	Light Pole (Installed)	COST EACH MAT.	COST EACH INST.	COST EACH TOTAL
0200	Light pole, aluminum, 20' high, 1 arm bracket	920	1,450	2,370
0240	2 arm brackets	1,075	1,450	2,525
0280	3 arm brackets	1,225	1,500	2,725
0320	4 arm brackets	1,375	1,500	2,875
0360	30' high, 1 arm bracket	2,000	1,825	3,825
0400	2 arm brackets	2,150	1,825	3,975
0440	3 arm brackets	2,300	1,875	4,175
0480	4 arm brackets	2,475	1,900	4,375
0680	40' high, 1 arm bracket	2,500	2,450	4,950
0720	2 arm brackets	2,650	2,450	5,100
0760	3 arm brackets	2,800	2,500	5,300
0800	4 arm brackets	2,975	2,500	5,475
0840	Steel, 20' high, 1 arm bracket	1,600	1,525	3,125
0880	2 arm brackets	1,675	1,525	3,200
0920	3 arm brackets	1,625	1,575	3,200
0960	4 arm brackets	1,725	1,575	3,300
1000	30' high, 1 arm bracket	1,775	1,950	3,725
1040	2 arm brackets	1,850	1,950	3,800
1080	3 arm brackets	1,800	2,000	3,800
1120	4 arm brackets	1,900	2,000	3,900
1320	40' high, 1 arm bracket	2,050	2,625	4,675
1360	2 arm brackets	2,125	2,625	4,750
1400	3 arm brackets	2,075	2,675	4,750
1440	4 arm brackets	2,175	2,675	4,850
2700	Cut & patch to match existing construction, add, minimum	2%	3%	
2800	Maximum	5%	9%	
2900	Equipment usage curtailment, add, minimum	1%	1%	
3000	Maximum	3%	10%	

G4020 Site Lighting

G4020 210	Light Pole (Installed)	COST EACH		
		MAT.	INST.	TOTAL
3100	Material handling & storage limitation, add, minimum	1%	1%	
3200	Maximum	6%	7%	
3300	Protection of existing work, add, minimum	2%	2%	
3400	Maximum	5%	7%	
3500	Shift work requirements, add, minimum		5%	
3600	Maximum		30%	

Reference Section

All the reference information is in one section, making it easy to find what you need to know ... and easy to use the data set on a daily basis. This section is visually identified by a vertical black bar on the page edges.

In this Reference Section, we've included Equipment Rental Costs, a listing of rental and operating costs; Crew Listings, a full listing of all crews and equipment, and their costs; Historical Cost Indexes for cost comparisons over time; City Cost Indexes and Location Factors for adjusting costs to the region you are in; Reference Tables, where you will find explanations, estimating information and procedures, or technical data; Change Orders, information on pricing changes to contract documents; and an explanation of all the Abbreviations in the data set.

Table of Contents

Estimating Tips

- This section contains the average costs to rent and operate hundreds of pieces of construction equipment. This is useful information when one is estimating the time and material requirements of any particular operation in order to establish a unit or total cost. Bare equipment costs shown on a unit cost line include, not only rental, but also operating costs for equipment under normal use.

Rental Costs

- Equipment rental rates are obtained from the following industry sources throughout North America: contractors, suppliers, dealers, manufacturers, and distributors.
- Rental rates vary throughout the country, with larger cities generally having lower rates. Lease plans for new equipment are available for periods in excess of six months, with a percentage of payments applying toward purchase.
- Monthly rental rates vary from 2% to 5% of the purchase price of the equipment depending on the anticipated life of the equipment and its wearing parts.
- Weekly rental rates are about 1/3 of the monthly rates, and daily rental rates are about 1/3 of the weekly rate.
- Rental rates can also be treated as reimbursement costs for contractor-owned equipment. Owned equipment costs include depreciation, loan payments, interest, taxes, insurance, storage, and major repairs.

Operating Costs

- The operating costs include parts and labor for routine servicing, such as the repair and replacement of pumps, filters, and worn lines. Normal operating expendables, such as fuel, lubricants, tires, and electricity (where applicable), are also included.
- Extraordinary operating expendables with highly variable wear patterns, such as diamond bits and blades, are excluded. These costs can be found as material costs in the Unit Price section.
- The hourly operating costs listed do not include the operator's wages.

Equipment Cost/Day

- Any power equipment required by a crew is shown in the Crew Listings with a daily cost.
- This daily cost of equipment needed by a crew includes both the rental cost and the operating cost and is based on dividing the weekly rental rate by 5 (the number of working days in the week), then adding the hourly operating cost multiplied by 8 (the number of hours in a day). This "Equipment Cost/Day" is shown in the far right column of the Equipment Rental section.
- If equipment is needed for only one or two days, it is best to develop your own cost by including components for daily rent and hourly operating costs. This is important when the listed Crew for a task does not contain the equipment needed, such as a crane for lifting mechanical heating/cooling equipment up onto a roof.

- If the quantity of work is less than the crew's Daily Output shown for a Unit Price line item that includes a bare unit equipment cost, the recommendation is to estimate one day's rental cost and operating cost for equipment shown in the Crew Listing for that line item.
- Please note, in some cases the equipment description in the crew is followed by a time period in parenthesis. For example: (daily) or (monthly). In these cases the equipment cost/day is calculated by adding the rental cost per time period to the hourly operating cost multiplied by 8.

Mobilization, Demobilization Costs

- The cost to move construction equipment from an equipment yard or rental company to the job site and back again is not included in equipment rental costs listed in the Reference Section. It is also not included in the bare equipment cost of any Unit Price line item or in any equipment costs shown in the Crew Listings.
- Mobilization (to the site) and demobilization (from the site) costs can be found in the Unit Price section.
- If a piece of equipment is already at the job site, it is not appropriate to utilize mobilization or demobilization costs again in an estimate. ∎

Same Data. Simplified.

Enjoy the convenience and efficiency of accessing your costs anywhere:

- **Skip the multiplier** by setting your location
- **Quickly search**, edit, favorite and share costs
- **Stay on top of price changes** with automatic updates

Discover more at rsmeans.com/online

---

01 54 33 | Equipment Rental

		UNIT	HOURLY OPER. COST	RENT PER DAY	RENT PER WEEK	RENT PER MONTH	EQUIPMENT COST/DAY	
10	**0010**	**CONCRETE EQUIPMENT RENTAL** without operators R015433 -10							**10**
	0200	Bucket, concrete lightweight, 1/2 C.Y.	Ea.	.76	38.50	116.16	350	29.30	
	0300	1 C.Y.		.99	63.50	190	570	45.90	
	0400	1-1/2 C.Y.		1.29	60.50	181.82	545	46.70	
	0500	2 C.Y.		1.40	74	222.23	665	55.65	
	0580	8 C.Y.		7.01	94.50	282.83	850	112.65	
	0600	Cart, concrete, self-propelled, operator walking, 10 C.F.		2.88	157	469.50	1,400	116.95	
	0700	Operator riding, 18 C.F.		4.85	167	500.50	1,500	138.95	
	0800	Conveyer for concrete, portable, gas, 16" wide, 26' long		10.71	162	484.86	1,450	182.70	
	0900	46' long		11.09	177	530.31	1,600	194.80	
	1000	56' long		11.26	194	580.82	1,750	206.25	
	1100	Core drill, electric, 2-1/2 H.P., 1" to 8" bit diameter		1.58	83.50	250	750	62.65	
	1150	11 H.P., 8" to 18" cores		5.45	130	390	1,175	121.60	
	1200	Finisher, concrete floor, gas, riding trowel, 96" wide		9.75	155	465.11	1,400	171.05	
	1300	Gas, walk-behind, 3 blade, 36" trowel		2.06	101	302.50	910	76.95	
	1400	4 blade, 48" trowel		3.10	116	347.50	1,050	94.30	
	1500	Float, hand-operated (Bull float), 48" wide		.08	13	39	117	8.45	
	1570	Curb builder, 14 H.P., gas, single screw		14.14	256	767.69	2,300	266.65	
	1590	Double screw		12.50	256	767.69	2,300	253.55	
	1600	Floor grinder, concrete and terrazzo, electric, 22" path		3.07	119	357.50	1,075	96.05	
	1700	Edger, concrete, electric, 7" path		1.19	57.50	172.50	520	44.05	
	1750	Vacuum pick-up system for floor grinders, wet/dry		1.63	103	309.38	930	74.95	
	1800	Mixer, powered, mortar and concrete, gas, 6 C.F., 18 H.P.		7.48	89	267.50	805	113.35	
	1900	10 C.F., 25 H.P.		9.08	124	372.50	1,125	147.15	
	2000	16 C.F.		9.44	146	436.88	1,300	162.90	
	2100	Concrete, stationary, tilt drum, 2 C.Y.		7.98	81	242.43	725	112.35	
	2120	Pump, concrete, truck mounted, 4" line, 80' boom		31.24	290	868.70	2,600	423.65	
	2140	5" line, 110' boom		39.99	290	868.70	2,600	493.65	
	2160	Mud jack, 50 C.F. per hr.		6.50	231	691.93	2,075	190.35	
	2180	225 C.F. per hr.		8.61	296	888.91	2,675	246.65	
	2190	Shotcrete pump rig, 12 C.Y./hr.		16.73	226	676.78	2,025	269.20	
	2200	35 C.Y./hr.		15.91	290	868.70	2,600	301.05	
	2600	Saw, concrete, manual, gas, 18 H.P.		5.59	115	345	1,025	113.70	
	2650	Self-propelled, gas, 30 H.P.		7.97	82	245.62	735	112.85	
	2675	V-groove crack chaser, manual, gas, 6 H.P.		1.65	100	300	900	73.25	
	2700	Vibrators, concrete, electric, 60 cycle, 2 H.P.		.47	70	210	630	45.80	
	2800	3 H.P.		.57	81	242.50	730	53.05	
	2900	Gas engine, 5 H.P.		1.56	17.05	51.21	154	22.70	
	3000	8 H.P.		2.10	17.25	51.74	155	27.15	
	3050	Vibrating screed, gas engine, 8 H.P.		2.83	108	325	975	87.65	
	3120	Concrete transit mixer, 6 x 4, 250 H.P., 8 C.Y., rear discharge		51.08	70.50	212.13	635	451.10	
	3200	Front discharge		59.30	136	409.10	1,225	556.20	
	3300	6 x 6, 285 H.P., 12 C.Y., rear discharge		58.56	167	500	1,500	568.50	
	3400	Front discharge		61.02	167	500	1,500	588.15	
20	**0010**	**EARTHWORK EQUIPMENT RENTAL** without operators R015433 -10							**20**
	0040	Aggregate spreader, push type, 8' to 12' wide	Ea.	2.62	65	195	585	59.95	
	0045	Tailgate type, 8' wide		2.56	64	191.92	575	58.90	
	0055	Earth auger, truck mounted, for fence & sign posts, utility poles		13.95	152	454.55	1,375	202.55	
	0060	For borings and monitoring wells		42.95	84	252.53	760	394.15	
	0070	Portable, trailer mounted		2.32	92.50	277.50	835	74.05	
	0075	Truck mounted, for caissons, water wells		86.00	134	402	1,200	768.40	
	0080	Horizontal boring machine, 12" to 36" diameter, 45 H.P.		22.93	110	330	990	249.45	
	0090	12" to 48" diameter, 65 H.P.		31.47	130	390	1,175	329.75	
	0095	Auger, for fence posts, gas engine, hand held		.45	84.50	254.05	760	54.40	
	0100	Excavator, diesel hydraulic, crawler mounted, 1/2 C.Y. cap.		21.92	470	1,408.39	4,225	457	
	0120	5/8 C.Y. capacity		29.30	615	1,851.77	5,550	604.75	
	0140	3/4 C.Y. capacity		32.95	730	2,190.83	6,575	701.80	
	0150	1 C.Y. capacity		41.58	835	2,500	7,500	832.65	

01 54 33 | Equipment Rental

		UNIT	HOURLY OPER. COST	RENT PER DAY	RENT PER WEEK	RENT PER MONTH	EQUIPMENT COST/DAY
0200	1-1/2 C.Y. capacity	Ea.	49.02	505	1,518	4,550	695.80
0300	2 C.Y. capacity		57.09	810	2,430	7,300	942.70
0320	2-1/2 C.Y. capacity		83.38	1,500	4,500	13,500	1,567
0325	3-1/2 C.Y. capacity		121.20	2,025	6,072	18,200	2,184
0330	4-1/2 C.Y. capacity		152.99	3,700	11,132	33,400	3,450
0335	6 C.Y. capacity		194.11	3,250	9,765.80	29,300	3,506
0340	7 C.Y. capacity		176.77	3,425	10,303.22	30,900	3,475
0342	Excavator attachments, bucket thumbs		3.43	261	783.90	2,350	184.20
0345	Grapples		3.17	225	674.15	2,025	160.20
0346	Hydraulic hammer for boom mounting, 4000 ft lb.		13.60	900	2,702.04	8,100	649.20
0347	5000 ft lb.		16.10	960	2,884.20	8,650	705.60
0348	8000 ft lb.		23.75	1,225	3,643.20	10,900	918.65
0349	12,000 ft lb.		25.95	1,125	3,373	10,100	882.20
0350	Gradall type, truck mounted, 3 ton @ 15' radius, 5/8 C.Y.		43.83	835	2,500	7,500	850.65
0370	1 C.Y. capacity		59.93	840	2,525.30	7,575	984.55
0400	Backhoe-loader, 40 to 45 H.P., 5/8 C.Y. capacity		12.01	286	857.50	2,575	267.55
0450	45 H.P. to 60 H.P., 3/4 C.Y. capacity		18.19	118	353.54	1,050	216.20
0460	80 H.P., 1-1/4 C.Y. capacity		20.54	118	353.54	1,050	235.05
0470	112 H.P., 1-1/2 C.Y. capacity		33.28	620	1,855.22	5,575	637.30
0482	Backhoe-loader attachment, compactor, 20,000 lb.		6.50	157	470.34	1,400	146.05
0485	Hydraulic hammer, 750 ft lb.		3.71	108	324.01	970	94.50
0486	Hydraulic hammer, 1200 ft lb.		6.61	207	621.89	1,875	177.25
0500	Brush chipper, gas engine, 6" cutter head, 35 H.P.		9.25	212	635	1,900	201
0550	Diesel engine, 12" cutter head, 130 H.P.		23.88	292	875	2,625	366.05
0600	15" cutter head, 165 H.P.		26.82	515	1,545	4,625	523.60
0750	Bucket, clamshell, general purpose, 3/8 C.Y.		1.41	92.50	277.78	835	66.85
0800	1/2 C.Y.		1.53	92.50	277.78	835	67.80
0850	3/4 C.Y.		1.65	92.50	277.78	835	68.80
0900	1 C.Y.		1.71	92.50	277.78	835	69.25
0950	1-1/2 C.Y.		2.81	92.50	277.78	835	78.05
1000	2 C.Y.		2.94	95	285	855	80.55
1010	Bucket, dragline, medium duty, 1/2 C.Y.		.83	92.50	277.78	835	62.15
1020	3/4 C.Y.		.79	92.50	277.78	835	61.85
1030	1 C.Y.		.80	92.50	277.78	835	62
1040	1-1/2 C.Y.		1.27	92.50	277.78	835	65.70
1050	2 C.Y.		1.30	92.50	277.78	835	65.95
1070	3 C.Y.		2.10	92.50	277.78	835	72.30
1200	Compactor, manually guided 2-drum vibratory smooth roller, 7.5 H.P.		7.28	181	542.50	1,625	166.75
1250	Rammer/tamper, gas, 8"		2.23	48.50	146.06	440	47
1260	15"		2.65	55.50	167.23	500	54.65
1300	Vibratory plate, gas, 18" plate, 3000 lb. blow		2.15	24.50	73.55	221	31.90
1350	21" plate, 5000 lb. blow		2.64	241	722.50	2,175	165.60
1370	Curb builder/extruder, 14 H.P., gas, single screw		12.50	256	767.69	2,300	253.55
1390	Double screw		12.50	256	767.69	2,300	253.55
1500	Disc harrow attachment, for tractor		.48	83.50	249.80	750	53.80
1810	Feller buncher, shearing & accumulating trees, 100 H.P.		43.75	465	1,393.97	4,175	628.80
1860	Grader, self-propelled, 25,000 lb.		33.65	1,125	3,366.73	10,100	942.50
1910	30,000 lb.		33.15	1,350	4,040.48	12,100	1,073
1920	40,000 lb.		52.35	1,325	4,000	12,000	1,219
1930	55,000 lb.		67.53	1,675	5,000	15,000	1,540
1950	Hammer, pavement breaker, self-propelled, diesel, 1000 to 1250 lb.		28.65	600	1,800	5,400	589.25
2000	1300 to 1500 lb.		43.19	1,025	3,051.51	9,150	955.80
2050	Pile driving hammer, steam or air, 4150 ft lb. @ 225 bpm		12.26	505	1,518	4,550	401.70
2100	8750 ft lb. @ 145 bpm		14.47	710	2,125.20	6,375	540.85
2150	15,000 ft lb. @ 60 bpm		14.81	845	2,530	7,600	624.45
2200	24,450 ft lb. @ 111 bpm		15.83	980	2,934.80	8,800	713.60
2250	Leads, 60' high for pile driving hammers up to 20,000 ft lb.		3.70	305	910.80	2,725	211.80
2300	90' high for hammers over 20,000 ft lb.		5.50	545	1,639.44	4,925	371.85

01 54 33	Equipment Rental	UNIT	HOURLY OPER. COST	RENT PER DAY	RENT PER WEEK	RENT PER MONTH	EQUIPMENT COST/DAY	
20	2350	Diesel type hammer, 22,400 ft lb.	Ea.	17.98	495	1,489.40	4,475	441.70	20
	2400	41,300 ft lb.		25.91	625	1,881.35	5,650	583.55	
	2450	141,000 ft lb.		41.70	995	2,978.80	8,925	929.35	
	2500	Vib. elec. hammer/extractor, 200 kW diesel generator, 34 H.P.		41.74	725	2,168.78	6,500	767.70	
	2550	80 H.P.		73.68	1,050	3,135.58	9,400	1,217	
	2600	150 H.P.		136.35	2,000	6,035.99	18,100	2,298	
	2700	Hydro excavator w/ext boom 12 C.Y., 1200 gallons		38.15	1,625	4,857.60	14,600	1,277	
	2800	Log chipper, up to 22" diameter, 600 H.P.		45.00	325	975	2,925	555	
	2850	Logger, for skidding & stacking logs, 150 H.P.		43.84	940	2,813.18	8,450	913.35	
	2860	Mulcher, diesel powered, trailer mounted		20.06	310	924.26	2,775	345.35	
	2900	Rake, spring tooth, with tractor		14.85	375	1,123.58	3,375	343.50	
	3000	Roller, vibratory, tandem, smooth drum, 20 H.P.		7.88	325	978.60	2,925	258.75	
	3050	35 H.P.		10.22	275	825	2,475	246.80	
	3100	Towed type vibratory compactor, smooth drum, 50 H.P.		25.50	525	1,581.85	4,750	520.35	
	3150	Sheepsfoot, 50 H.P.		25.87	365	1,100	3,300	426.95	
	3170	Landfill compactor, 220 H.P.		70.51	1,675	5,035.45	15,100	1,571	
	3200	Pneumatic tire roller, 80 H.P.		13.04	410	1,228.10	3,675	349.90	
	3250	120 H.P.		19.56	565	1,700	5,100	496.50	
	3300	Sheepsfoot vibratory roller, 240 H.P.		62.77	1,425	4,303.41	12,900	1,363	
	3320	340 H.P.		84.58	2,200	6,565.78	19,700	1,990	
	3350	Smooth drum vibratory roller, 75 H.P.		23.55	660	1,982.18	5,950	584.80	
	3400	125 H.P.		27.86	750	2,247.17	6,750	672.35	
	3410	Rotary mower, brush, 60", with tractor		18.96	365	1,097.45	3,300	371.15	
	3420	Rototiller, walk-behind, gas, 5 H.P.		2.15	60	180	540	53.25	
	3422	8 H.P.		2.84	116	347.50	1,050	92.20	
	3440	Scrapers, towed type, 7 C.Y. capacity		6.50	129	386.72	1,150	129.30	
	3450	10 C.Y. capacity		7.27	172	517.37	1,550	161.65	
	3500	15 C.Y. capacity		7.46	199	595.76	1,775	178.85	
	3525	Self-propelled, single engine, 14 C.Y. capacity		134.48	2,250	6,739.92	20,200	2,424	
	3550	Dual engine, 21 C.Y. capacity		142.64	2,525	7,575.90	22,700	2,656	
	3600	31 C.Y. capacity		189.53	3,650	10,954.15	32,900	3,707	
	3640	44 C.Y. capacity		234.76	4,700	14,083.90	42,300	4,695	
	3650	Elevating type, single engine, 11 C.Y. capacity		62.42	935	2,800	8,400	1,059	
	3700	22 C.Y. capacity		115.65	1,625	4,850	14,600	1,895	
	3710	Screening plant, 110 H.P. w/5' x 10' screen		21.32	650	1,952.56	5,850	561.10	
	3720	5' x 16' screen		26.92	1,225	3,666.67	11,000	948.65	
	3850	Shovel, crawler-mounted, front-loading, 7 C.Y. capacity		220.61	3,975	11,915.21	35,700	4,148	
	3855	12 C.Y. capacity		339.92	5,500	16,514.06	49,500	6,022	
	3860	Shovel/backhoe bucket, 1/2 C.Y.		2.72	74	221.58	665	66.05	
	3870	3/4 C.Y.		2.69	83	248.76	745	71.25	
	3880	1 C.Y.		2.78	92	275.42	825	77.35	
	3890	1-1/2 C.Y.		2.98	108	324.01	970	88.65	
	3910	3 C.Y.		3.47	146	438.98	1,325	115.55	
	3950	Stump chipper, 18" deep, 30 H.P.		6.95	223	668	2,000	189.20	
	4110	Dozer, crawler, torque converter, diesel 80 H.P.		25.48	335	1,010.12	3,025	405.85	
	4150	105 H.P.		34.64	605	1,818.22	5,450	640.80	
	4200	140 H.P.		41.64	660	1,980	5,950	729.15	
	4260	200 H.P.		63.72	1,675	5,050.60	15,200	1,520	
	4310	300 H.P.		81.45	1,900	5,667.20	17,000	1,785	
	4360	410 H.P.		107.70	3,250	9,727.46	29,200	2,807	
	4370	500 H.P.		134.58	3,925	11,788.10	35,400	3,434	
	4380	700 H.P.		232.23	5,525	16,587.71	49,800	5,175	
	4400	Loader, crawler, torque conv., diesel, 1-1/2 C.Y., 80 H.P.		29.80	555	1,667.71	5,000	572	
	4450	1-1/2 to 1-3/4 C.Y., 95 H.P.		30.54	705	2,120	6,350	668.35	
	4510	1-3/4 to 2-1/4 C.Y., 130 H.P.		48.19	900	2,700	8,100	925.50	
	4530	2-1/2 to 3-1/4 C.Y., 190 H.P.		58.30	1,125	3,400	10,200	1,146	
	4560	3-1/2 to 5 C.Y., 275 H.P.		72.05	1,475	4,400	13,200	1,456	
	4610	Front end loader, 4WD, articulated frame, diesel, 1 to 1-1/4 C.Y., 70 H.P.		16.77	286	857.06	2,575	305.60	

01 54 33 | Equipment Rental

			UNIT	HOURLY OPER. COST	RENT PER DAY	RENT PER WEEK	RENT PER MONTH	EQUIPMENT COST/DAY	
20	4620	1-1/2 to 1-3/4 C.Y., 95 H.P.	Ea.	20.17	465	1,400	4,200	441.40	20
	4650	1-3/4 to 2 C.Y., 130 H.P.		21.26	380	1,140	3,425	398.05	
	4710	2-1/2 to 3-1/2 C.Y., 145 H.P.		29.79	665	2,000	6,000	638.30	
	4730	3 to 4-1/2 C.Y., 185 H.P.		32.38	835	2,500	7,500	759	
	4760	5-1/4 to 5-3/4 C.Y., 270 H.P.		53.67	900	2,692.98	8,075	967.95	
	4810	7 to 9 C.Y., 475 H.P.		91.99	2,575	7,744.59	23,200	2,285	
	4870	9 to 11 C.Y., 620 H.P.		133.10	2,725	8,204.77	24,600	2,706	
	4880	Skid-steer loader, wheeled, 10 C.F., 30 H.P. gas		9.66	170	511.19	1,525	179.50	
	4890	1 C.Y., 78 H.P., diesel		18.60	495	1,487.50	4,475	446.30	
	4892	Skid-steer attachment, auger		.75	149	447.50	1,350	95.50	
	4893	Backhoe		.75	123	370.35	1,100	80.05	
	4894	Broom		.71	136	407.50	1,225	87.20	
	4895	Forks		.16	32	96	288	20.45	
	4896	Grapple		.73	89.50	267.93	805	59.45	
	4897	Concrete hammer		1.06	183	550	1,650	118.50	
	4898	Tree spade		.61	105	313.56	940	67.55	
	4899	Trencher		.66	102	305	915	66.25	
	4900	Trencher, chain, boom type, gas, operator walking, 12 H.P.		4.21	208	624.51	1,875	158.60	
	4910	Operator riding, 40 H.P.		16.83	525	1,580	4,750	450.70	
	5000	Wheel type, diesel, 4' deep, 12" wide		69.33	975	2,926.54	8,775	1,140	
	5100	6' deep, 20" wide		88.37	885	2,655	7,975	1,238	
	5150	Chain type, diesel, 5' deep, 8" wide		16.45	365	1,097.45	3,300	351.05	
	5200	Diesel, 8' deep, 16" wide		90.47	1,950	5,853.08	17,600	1,894	
	5202	Rock trencher, wheel type, 6" wide x 18" deep		47.46	91	272.73	820	434.20	
	5206	Chain type, 18" wide x 7' deep		105.44	286	858.60	2,575	1,015	
	5210	Tree spade, self-propelled		13.79	232	696.98	2,100	249.70	
	5250	Truck, dump, 2-axle, 12 ton, 8 C.Y. payload, 220 H.P.		24.16	355	1,071.50	3,225	407.60	
	5300	Three axle dump, 16 ton, 12 C.Y. payload, 400 H.P.		45.03	365	1,095.41	3,275	579.35	
	5310	Four axle dump, 25 ton, 18 C.Y. payload, 450 H.P.		50.45	585	1,750	5,250	753.60	
	5350	Dump trailer only, rear dump, 16-1/2 C.Y.		5.80	153	459.89	1,375	138.35	
	5400	20 C.Y.		6.26	172	517.37	1,550	153.55	
	5450	Flatbed, single axle, 1-1/2 ton rating		19.22	74.50	223.67	670	198.50	
	5500	3 ton rating		25.95	1,075	3,212.18	9,625	850.05	
	5550	Off highway rear dump, 25 ton capacity		63.42	1,525	4,600	13,800	1,427	
	5600	35 ton capacity		67.70	675	2,020.24	6,050	945.65	
	5610	50 ton capacity		84.88	2,175	6,500	19,500	1,979	
	5620	65 ton capacity		90.64	2,025	6,062.12	18,200	1,938	
	5630	100 ton capacity		122.70	2,975	8,936.41	26,800	2,769	
	6000	Vibratory plow, 25 H.P., walking		6.84	305	909.11	2,725	236.55	
40	0010	**GENERAL EQUIPMENT RENTAL** without operators							40
	0020	Aerial lift, scissor type, to 20' high, 1200 lb. capacity, electric	Ea.	3.52	154	462.50	1,400	120.70	
	0030	To 30' high, 1200 lb. capacity		3.82	235	705	2,125	171.55	
	0040	Over 30' high, 1500 lb. capacity		5.20	288	865	2,600	214.55	
	0070	Articulating boom, to 45' high, 500 lb. capacity, diesel		10.04	250	750	2,250	230.30	
	0075	To 60' high, 500 lb. capacity		13.82	305	909.11	2,725	292.40	
	0080	To 80' high, 500 lb. capacity		16.24	995	2,987.50	8,975	727.45	
	0085	To 125' high, 500 lb. capacity		18.56	1,675	4,997.50	15,000	1,148	
	0100	Telescoping boom to 40' high, 500 lb. capacity, diesel		11.37	320	954.56	2,875	281.90	
	0105	To 45' high, 500 lb. capacity		12.66	325	974.77	2,925	296.20	
	0110	To 60' high, 500 lb. capacity		16.56	267	800	2,400	292.45	
	0115	To 80' high, 500 lb. capacity		21.52	360	1,077.80	3,225	387.75	
	0120	To 100' high, 500 lb. capacity		29.06	985	2,950	8,850	822.45	
	0125	To 120' high, 500 lb. capacity		29.51	1,525	4,545	13,600	1,145	
	0195	Air compressor, portable, 6.5 CFM, electric		.91	53.50	160	480	39.30	
	0196	Gasoline		.66	56	167.50	505	38.80	
	0200	Towed type, gas engine, 60 CFM		9.55	129	387.50	1,175	153.85	
	0300	160 CFM		10.60	213	637.50	1,925	212.30	

Note: R015433-10 reference at row 0020; R015433-15 reference at row 0070.

For customer support on your Commercial Renovation Costs with RSMeans Data, call 800.448.8182.

747

01 54 33 | Equipment Rental

		UNIT	HOURLY OPER. COST	RENT PER DAY	RENT PER WEEK	RENT PER MONTH	EQUIPMENT COST/DAY	
0400	Diesel engine, rotary screw, 250 CFM	Ea.	12.23	175	525	1,575	202.85	40
0500	365 CFM		16.19	355	1,070	3,200	343.55	
0550	450 CFM		20.19	300	900	2,700	341.50	
0600	600 CFM		34.51	251	752.54	2,250	426.55	
0700	750 CFM		35.04	525	1,580	4,750	596.30	
0930	Air tools, breaker, pavement, 60 lb.		.57	81.50	245	735	53.60	
0940	80 lb.		.57	81	242.50	730	53.05	
0950	Drills, hand (jackhammer), 65 lb.		.68	64	192.50	580	43.90	
0960	Track or wagon, swing boom, 4" drifter		62.50	1,050	3,135.41	9,400	1,127	
0970	5" drifter		62.50	1,050	3,135.41	9,400	1,127	
0975	Track mounted quarry drill, 6" diameter drill		102.94	1,900	5,722.33	17,200	1,968	
0980	Dust control per drill		1.05	25.50	76.26	229	23.65	
0990	Hammer, chipping, 12 lb.		.61	46.50	140	420	32.85	
1000	Hose, air with couplings, 50' long, 3/4" diameter		.07	11	33	99	7.15	
1100	1" diameter		.08	12.35	37	111	8.05	
1200	1-1/2" diameter		.22	35	105	315	22.80	
1300	2" diameter		.24	45	135	405	28.95	
1400	2-1/2" diameter		.36	57.50	172.50	520	37.40	
1410	3" diameter		.42	58.50	175	525	38.35	
1450	Drill, steel, 7/8" x 2'		.09	13.05	39.12	117	8.50	
1460	7/8" x 6'		.12	19.85	59.58	179	12.85	
1520	Moil points		.03	7	21	63	4.40	
1525	Pneumatic nailer w/accessories		.48	39.50	119.19	360	27.70	
1530	Sheeting driver for 60 lb. breaker		.04	7.85	23.52	70.50	5.05	
1540	For 90 lb. breaker		.13	10.65	31.88	95.50	7.45	
1550	Spade, 25 lb.		.51	7.50	22.47	67.50	8.55	
1560	Tamper, single, 35 lb.		.60	58.50	175	525	39.75	
1570	Triple, 140 lb.		.90	62.50	187.09	560	44.60	
1580	Wrenches, impact, air powered, up to 3/4" bolt		.43	50	149.75	450	33.40	
1590	Up to 1-1/4" bolt		.58	67.50	202.50	610	45.15	
1600	Barricades, barrels, reflectorized, 1 to 99 barrels		.03	4	12	36	2.65	
1610	100 to 200 barrels		.03	4.46	13.38	40	2.90	
1620	Barrels with flashers, 1 to 99 barrels		.04	6.45	19.39	58	4.15	
1630	100 to 200 barrels		.03	5.15	15.52	46.50	3.35	
1640	Barrels with steady burn type C lights		.05	8.55	25.61	77	5.50	
1650	Illuminated board, trailer mounted, with generator		3.32	141	423.30	1,275	111.20	
1670	Portable barricade, stock, with flashers, 1 to 6 units		.04	6.45	19.34	58	4.15	
1680	25 to 50 units		.03	6	18.03	54	3.85	
1685	Butt fusion machine, wheeled, 1.5 HP electric, 2" - 8" diameter pipe		2.66	190	568.50	1,700	134.95	
1690	Tracked, 20 HP diesel, 4" - 12" diameter pipe		10.04	565	1,702.05	5,100	420.75	
1695	83 HP diesel, 8" - 24" diameter pipe		51.84	1,125	3,358.65	10,100	1,086	
1700	Carts, brick, gas engine, 1000 lb. capacity		2.98	65.50	196.97	590	63.20	
1800	1500 lb., 7-1/2' lift		2.95	70	210.50	630	65.70	
1822	Dehumidifier, medium, 6 lb./hr., 150 CFM		1.20	77.50	232.03	695	56.05	
1824	Large, 18 lb./hr., 600 CFM		2.22	585	1,750	5,250	367.75	
1830	Distributor, asphalt, trailer mounted, 2000 gal., 38 H.P. diesel		11.13	355	1,071.32	3,225	303.25	
1840	3000 gal., 38 H.P. diesel		13.02	385	1,149.71	3,450	334.10	
1850	Drill, rotary hammer, electric		1.13	72	216.17	650	52.25	
1860	Carbide bit, 1-1/2" diameter, add to electric rotary hammer		.03	41.50	125	375	25.25	
1865	Rotary, crawler, 250 H.P.		137.40	2,450	7,375	22,100	2,574	
1870	Emulsion sprayer, 65 gal., 5 H.P. gas engine		2.80	108	324.01	970	87.20	
1880	200 gal., 5 H.P. engine		7.31	181	543.50	1,625	167.20	
1900	Floor auto-scrubbing machine, walk-behind, 28" path		5.69	223	670	2,000	179.55	
1930	Floodlight, mercury vapor, or quartz, on tripod, 1000 watt		.46	36.50	110	330	25.70	
1940	2000 watt		.60	28.50	85.55	257	21.90	
1950	Floodlights, trailer mounted with generator, 1 - 300 watt light		3.59	79.50	238.30	715	76.35	
1960	2 - 1000 watt lights		4.54	88.50	265.48	795	89.40	
2000	4 - 300 watt lights		4.29	101	303.11	910	95	

01 54 33 | Equipment Rental

		UNIT	HOURLY OPER. COST	RENT PER DAY	RENT PER WEEK	RENT PER MONTH	EQUIPMENT COST/DAY	
40	2005	Foam spray rig, incl. box trailer, compressor, generator, proportioner	Ea.	25.76	540	1,620.05	4,850	530.15	40
	2015	Forklift, pneumatic tire, rough terr, straight mast, 5000 lb, 12' lift, gas		18.82	221	663.70	2,000	283.30	
	2025	8000 lb, 12' lift		22.95	365	1,097.45	3,300	403.10	
	2030	5000 lb, 12' lift, diesel		15.59	247	740.71	2,225	272.85	
	2035	8000 lb, 12' lift, diesel		16.90	280	839.82	2,525	303.15	
	2045	All terrain, telescoping boom, diesel, 5000 lb, 10' reach, 19' lift		17.40	200	600	1,800	259.25	
	2055	6600 lb, 29' reach, 42' lift		21.29	236	707.08	2,125	311.75	
	2065	10,000 lb, 31' reach, 45' lift		23.30	267	800	2,400	346.45	
	2070	Cushion tire, smooth floor, gas, 5000 lb capacity		8.32	262	785	2,350	223.60	
	2075	8000 lb capacity		11.47	320	957.50	2,875	283.25	
	2085	Diesel, 5000 lb capacity		7.82	257	770	2,300	216.55	
	2090	12,000 lb capacity		12.16	400	1,206.59	3,625	338.60	
	2095	20,000 lb capacity		17.41	800	2,405	7,225	620.30	
	2100	Generator, electric, gas engine, 1.5 kW to 3 kW		2.60	46.50	140	420	48.80	
	2200	5 kW		3.25	91	272.50	820	80.45	
	2300	10 kW		5.98	108	322.50	970	112.35	
	2400	25 kW		7.47	405	1,222.25	3,675	304.25	
	2500	Diesel engine, 20 kW		9.29	222	665	2,000	207.35	
	2600	50 kW		16.09	360	1,085	3,250	345.70	
	2700	100 kW		28.85	485	1,455	4,375	521.85	
	2800	250 kW		54.84	825	2,472.50	7,425	933.25	
	2850	Hammer, hydraulic, for mounting on boom, to 500 ft lb.		2.93	94.50	283.25	850	80.05	
	2860	1000 ft lb.		4.65	141	423.30	1,275	121.85	
	2900	Heaters, space, oil or electric, 50 MBH		1.48	47	141.42	425	40.10	
	3000	100 MBH		2.74	47	141.42	425	50.25	
	3100	300 MBH		8.00	136	409.10	1,225	145.80	
	3150	500 MBH		13.28	208	625	1,875	231.20	
	3200	Hose, water, suction with coupling, 20' long, 2" diameter		.02	5.65	17	51	3.55	
	3210	3" diameter		.03	14.15	42.50	128	8.75	
	3220	4" diameter		.03	28.50	85.01	255	17.25	
	3230	6" diameter		.11	41	122.96	370	25.50	
	3240	8" diameter		.27	54	161.92	485	34.55	
	3250	Discharge hose with coupling, 50' long, 2" diameter		.01	6.50	19.50	58.50	4	
	3260	3" diameter		.01	7.35	22	66	4.50	
	3270	4" diameter		.02	21	63.25	190	12.80	
	3280	6" diameter		.06	29.50	88.04	264	18.10	
	3290	8" diameter		.24	38	113.85	340	24.70	
	3295	Insulation blower		.84	118	353.54	1,050	77.40	
	3300	Ladders, extension type, 16' to 36' long		.18	41.50	125	375	26.45	
	3400	40' to 60' long		.64	123	368	1,100	78.75	
	3405	Lance for cutting concrete		2.22	66.50	200	600	57.80	
	3407	Lawn mower, rotary, 22", 5 H.P.		1.06	38.50	115	345	31.50	
	3408	48" self-propelled		2.92	160	481	1,450	119.60	
	3410	Level, electronic, automatic, with tripod and leveling rod		1.06	43	129.50	390	34.40	
	3430	Laser type, for pipe and sewer line and grade		2.20	117	350	1,050	87.55	
	3440	Rotating beam for interior control		.91	64	192.50	580	45.80	
	3460	Builder's optical transit, with tripod and rod		.10	43	129.50	390	26.70	
	3500	Light towers, towable, with diesel generator, 2000 watt		4.31	102	307.29	920	95.90	
	3600	4000 watt		4.55	142	425	1,275	121.40	
	3700	Mixer, powered, plaster and mortar, 6 C.F., 7 H.P.		2.08	86.50	260	780	68.60	
	3800	10 C.F., 9 H.P.		2.26	124	372.50	1,125	92.60	
	3850	Nailer, pneumatic		.48	30.50	91	273	22.10	
	3900	Paint sprayers complete, 8 CFM		.86	62.50	186.74	560	44.20	
	4000	17 CFM		1.61	111	333.84	1,000	79.70	
	4020	Pavers, bituminous, rubber tires, 8' wide, 50 H.P., diesel		32.31	575	1,724.57	5,175	603.40	
	4030	10' wide, 150 H.P.		96.77	1,975	5,905.34	17,700	1,955	
	4050	Crawler, 8' wide, 100 H.P., diesel		88.64	2,075	6,245.03	18,700	1,958	
	4060	10' wide, 150 H.P.		105.21	2,375	7,133.45	21,400	2,268	

01 54 33 | Equipment Rental

		UNIT	HOURLY OPER. COST	RENT PER DAY	RENT PER WEEK	RENT PER MONTH	EQUIPMENT COST/DAY
4070	Concrete paver, 12' to 24' wide, 250 H.P.	Ea.	88.67	1,700	5,121.45	15,400	1,734
4080	Placer-spreader-trimmer, 24' wide, 300 H.P.		118.92	2,575	7,760.56	23,300	2,503
4100	Pump, centrifugal gas pump, 1-1/2" diam., 65 GPM		3.97	54.50	164.10	490	64.60
4200	2" diameter, 130 GPM		5.04	44	132.50	400	66.80
4300	3" diameter, 250 GPM		5.18	55	165	495	74.40
4400	6" diameter, 1500 GPM		22.46	92.50	277.78	835	235.25
4500	Submersible electric pump, 1-1/4" diameter, 55 GPM		.40	35	105	315	24.25
4600	1-1/2" diameter, 83 GPM		.45	43.50	130	390	29.60
4700	2" diameter, 120 GPM		1.66	55	165	495	46.30
4800	3" diameter, 300 GPM		3.07	109	327.50	985	90.05
4900	4" diameter, 560 GPM		14.90	76	227.50	685	164.70
5000	6" diameter, 1590 GPM		22.30	65.50	196.97	590	217.80
5100	Diaphragm pump, gas, single, 1-1/2" diameter		1.14	38.50	116.16	350	32.35
5200	2" diameter		4.02	92.50	277.78	835	87.70
5300	3" diameter		4.09	86	258	775	84.35
5400	Double, 4" diameter		6.10	96	287.88	865	106.35
5450	Pressure washer 5 GPM, 3000 psi		3.92	110	330	990	97.35
5460	7 GPM, 3000 psi		5.00	90	270	810	93.95
5470	High pressure water jet 10 KSI		40.02	730	2,185.92	6,550	757.40
5480	40 KSI		28.21	990	2,975.28	8,925	820.75
5500	Trash pump, self-priming, gas, 2" diameter		3.85	109	328.29	985	96.50
5600	Diesel, 4" diameter		6.74	163	489.91	1,475	151.95
5650	Diesel, 6" diameter		17.02	163	489.91	1,475	234.15
5655	Grout pump		18.92	284	851.83	2,550	321.75
5700	Salamanders, L.P. gas fired, 100,000 Btu		2.92	57.50	172.50	520	57.85
5705	50,000 Btu		1.68	23	69	207	27.25
5720	Sandblaster, portable, open top, 3 C.F. capacity		.61	132	395	1,175	83.85
5730	6 C.F. capacity		1.02	133	399	1,200	87.90
5740	Accessories for above		.14	24	72.12	216	15.55
5750	Sander, floor		.78	73.50	220	660	50.20
5760	Edger		.52	35	105	315	25.20
5800	Saw, chain, gas engine, 18" long		1.77	63.50	190	570	52.20
5900	Hydraulic powered, 36" long		.79	59	176.77	530	41.65
5950	60" long		.79	65.50	196.97	590	45.70
6000	Masonry, table mounted, 14" diameter, 5 H.P.		1.34	76.50	230	690	56.70
6050	Portable cut-off, 8 H.P.		1.83	72.50	217.50	655	58.15
6100	Circular, hand held, electric, 7-1/4" diameter		.23	13.85	41.50	125	10.15
6200	12" diameter		.24	41	122.50	370	26.40
6250	Wall saw, w/hydraulic power, 10 H.P.		3.32	99.50	299	895	86.40
6275	Shot blaster, walk-behind, 20" wide		4.79	284	851.83	2,550	208.70
6280	Sidewalk broom, walk-behind		2.27	83.50	250.85	755	68.30
6300	Steam cleaner, 100 gallons per hour		3.38	83.50	250.85	755	77.20
6310	200 gallons per hour		4.38	101	303.11	910	95.70
6340	Tar kettle/pot, 400 gallons		16.65	128	383.85	1,150	209.95
6350	Torch, cutting, acetylene-oxygen, 150' hose, excludes gases		.46	15.50	46.51	140	12.95
6360	Hourly operating cost includes tips and gas		21.17	7.10	21.23	63.50	173.60
6410	Toilet, portable chemical		.13	23.50	70.03	210	15.05
6420	Recycle flush type		.16	29	86.75	260	18.65
6430	Toilet, fresh water flush, garden hose,		.20	34.50	103.47	310	22.25
6440	Hoisted, non-flush, for high rise		.16	28	84.66	254	18.20
6465	Tractor, farm with attachment		17.57	395	1,184.37	3,550	377.45
6480	Trailers, platform, flush deck, 2 axle, 3 ton capacity		1.71	96	287.41	860	71.15
6500	25 ton capacity		6.31	145	433.76	1,300	137.20
6600	40 ton capacity		8.14	206	616.66	1,850	188.45
6700	3 axle, 50 ton capacity		8.82	228	683.33	2,050	207.25
6800	75 ton capacity		11.21	305	907.63	2,725	271.20
6810	Trailer mounted cable reel for high voltage line work		5.94	28.50	85.86	258	64.75
6820	Trailer mounted cable tensioning rig		11.79	28.50	85.86	258	111.50

01 54 33 | Equipment Rental

		UNIT	HOURLY OPER. COST	RENT PER DAY	RENT PER WEEK	RENT PER MONTH	EQUIPMENT COST/DAY	
40	6830	Cable pulling rig	Ea.	74.51	28.50	85.86	258	613.30	**40**
	6850	Portable cable/wire puller, 8000 lb max pulling capacity		3.74	121	363.64	1,100	102.65	
	6900	Water tank trailer, engine driven discharge, 5000 gallons		7.24	160	480.79	1,450	154.10	
	6925	10,000 gallons		9.87	218	653.25	1,950	209.60	
	6950	Water truck, off highway, 6000 gallons		72.61	845	2,529.89	7,600	1,087	
	7010	Tram car for high voltage line work, powered, 2 conductor		6.95	29.50	88	264	73.20	
	7020	Transit (builder's level) with tripod		.10	17.75	53.30	160	11.45	
	7030	Trench box, 3000 lb., 6' x 8'		.57	97.50	293.15	880	63.15	
	7040	7200 lb., 6' x 20'		.73	189	566.72	1,700	119.15	
	7050	8000 lb., 8' x 16'		1.09	205	615	1,850	131.70	
	7060	9500 lb., 8' x 20'		1.22	235	705.51	2,125	150.85	
	7065	11,000 lb., 8' x 24'		1.28	221	663.70	2,000	142.95	
	7070	12,000 lb., 10' x 20'		1.51	267	799.57	2,400	172	
	7100	Truck, pickup, 3/4 ton, 2 wheel drive		9.35	62.50	187.09	560	112.20	
	7200	4 wheel drive		9.60	167	500	1,500	176.75	
	7250	Crew carrier, 9 passenger		12.81	109	328.29	985	168.15	
	7290	Flat bed truck, 20,000 lb. GVW		15.44	134	402.40	1,200	204.05	
	7300	Tractor, 4 x 2, 220 H.P.		22.52	218	653.25	1,950	310.80	
	7410	330 H.P.		32.72	298	893.64	2,675	440.45	
	7500	6 x 4, 380 H.P.		36.53	345	1,034.74	3,100	499.15	
	7600	450 H.P.		44.76	420	1,254.23	3,775	608.95	
	7610	Tractor, with A frame, boom and winch, 225 H.P.		25.04	296	886.76	2,650	377.65	
	7620	Vacuum truck, hazardous material, 2500 gallons		12.94	315	940.67	2,825	291.70	
	7625	5,000 gallons		13.18	445	1,332.62	4,000	371.95	
	7650	Vacuum, HEPA, 16 gallon, wet/dry		.86	122	365	1,100	79.90	
	7655	55 gallon, wet/dry		.79	25.50	76.50	230	21.60	
	7660	Water tank, portable		.74	162	485.11	1,450	102.90	
	7690	Sewer/catch basin vacuum, 14 C.Y., 1500 gallons		17.52	670	2,012	6,025	542.60	
	7700	Welder, electric, 200 amp		3.86	33.50	101.01	305	51.10	
	7800	300 amp		5.61	104	313.14	940	107.55	
	7900	Gas engine, 200 amp		9.05	59	176.77	530	107.80	
	8000	300 amp		10.25	111	333.96	1,000	148.75	
	8100	Wheelbarrow, any size		.07	11.15	33.50	101	7.20	
	8200	Wrecking ball, 4000 lb.		2.53	62	186	560	57.45	
50	0010	**HIGHWAY EQUIPMENT RENTAL** without operators							**50**
	0050	Asphalt batch plant, portable drum mixer, 100 ton/hr.	Ea.	89.47	1,550	4,677.24	14,000	1,651	
	0060	200 ton/hr.		103.21	1,675	4,990.80	15,000	1,824	
	0070	300 ton/hr.		121.29	1,950	5,853.08	17,600	2,141	
	0100	Backhoe attachment, long stick, up to 185 H.P., 10-1/2' long		.37	26	77.34	232	18.45	
	0140	Up to 250 H.P., 12' long		.42	29	86.75	260	20.70	
	0180	Over 250 H.P., 15' long		.57	39.50	118.11	355	28.20	
	0200	Special dipper arm, up to 100 H.P., 32' long		1.17	80.50	241.44	725	57.65	
	0240	Over 100 H.P., 33' long		1.46	101	303.11	910	72.30	
	0280	Catch basin/sewer cleaning truck, 3 ton, 9 C.Y., 1000 gal.		35.81	425	1,280.36	3,850	542.60	
	0300	Concrete batch plant, portable, electric, 200 C.Y./hr.		24.47	565	1,698.44	5,100	535.50	
	0520	Grader/dozer attachment, ripper/scarifier, rear mounted, up to 135 H.P.		3.19	64	192.32	575	63.95	
	0540	Up to 180 H.P.		4.18	97	290.56	870	91.60	
	0580	Up to 250 H.P.		5.92	155	465.11	1,400	140.40	
	0700	Pvmt. removal bucket, for hyd. excavator, up to 90 H.P.		2.18	59	176.64	530	52.80	
	0740	Up to 200 H.P.		2.33	75.50	225.76	675	63.80	
	0780	Over 200 H.P.		2.55	92.50	276.98	830	75.80	
	0900	Aggregate spreader, self-propelled, 187 H.P.		51.21	750	2,247.17	6,750	859.10	
	1000	Chemical spreader, 3 C.Y.		3.20	99.50	299	895	85.40	
	1900	Hammermill, traveling, 250 H.P.		68.03	520	1,565.69	4,700	857.40	
	2000	Horizontal borer, 3" diameter, 13 H.P. gas driven		5.47	234	702.03	2,100	184.15	
	2150	Horizontal directional drill, 20,000 lb. thrust, 78 H.P. diesel		27.86	535	1,606.09	4,825	544.10	
	2160	30,000 lb. thrust, 115 H.P.		34.24	625	1,868.72	5,600	647.65	
	2170	50,000 lb. thrust, 170 H.P.		49.09	720	2,156.61	6,475	824.05	

Note: Row 0050 area includes reference box **R015433 -10**.

		UNIT	HOURLY OPER. COST	RENT PER DAY	RENT PER WEEK	RENT PER MONTH	EQUIPMENT COST/DAY	
50	2190	Mud trailer for HDD, 1500 gallons, 175 H.P., gas	Ea.	25.76	177	530.31	1,600	312.15	**50**
	2200	Hydromulcher, diesel, 3000 gallon, for truck mounting		17.61	193	580	1,750	256.85	
	2300	Gas, 600 gallon		7.57	96	287.88	865	118.15	
	2400	Joint & crack cleaner, walk behind, 25 H.P.		3.19	46	137.38	410	53	
	2500	Filler, trailer mounted, 400 gallons, 20 H.P.		8.43	173	517.50	1,550	170.95	
	3000	Paint striper, self-propelled, 40 gallon, 22 H.P.		6.83	123	368.69	1,100	128.35	
	3100	120 gallon, 120 H.P.		17.83	385	1,151.54	3,450	372.90	
	3200	Post drivers, 6" I-Beam frame, for truck mounting		12.54	325	969.72	2,900	294.25	
	3400	Road sweeper, self-propelled, 8' wide, 90 H.P.		36.34	720	2,164.75	6,500	723.65	
	3450	Road sweeper, vacuum assisted, 4 C.Y., 220 gallons		58.98	680	2,038.13	6,125	879.45	
	4000	Road mixer, self-propelled, 130 H.P.		46.79	835	2,508.46	7,525	876	
	4100	310 H.P.		75.91	2,175	6,558.59	19,700	1,919	
	4220	Cold mix paver, incl. pug mill and bitumen tank, 165 H.P.		96.11	2,350	7,015.87	21,000	2,172	
	4240	Pavement brush, towed		3.47	101	303.11	910	88.35	
	4250	Paver, asphalt, wheel or crawler, 130 H.P., diesel		95.37	2,300	6,898.28	20,700	2,143	
	4300	Paver, road widener, gas, 1' to 6', 67 H.P.		47.22	985	2,952.67	8,850	968.30	
	4400	Diesel, 2' to 14', 88 H.P.		57.06	1,175	3,501.40	10,500	1,157	
	4600	Slipform pavers, curb and gutter, 2 track, 75 H.P.		58.53	1,275	3,814.96	11,400	1,231	
	4700	4 track, 165 H.P.		36.11	855	2,560.72	7,675	801.05	
	4800	Median barrier, 215 H.P.		59.13	1,350	4,076.26	12,200	1,288	
	4901	Trailer, low bed, 75 ton capacity		10.84	286	857.06	2,575	258.10	
	5000	Road planer, walk behind, 10" cutting width, 10 H.P.		2.48	258	773	2,325	174.45	
	5100	Self-propelled, 12" cutting width, 64 H.P.		8.34	192	575.77	1,725	181.85	
	5120	Traffic line remover, metal ball blaster, truck mounted, 115 H.P.		47.03	1,025	3,100	9,300	996.25	
	5140	Grinder, truck mounted, 115 H.P.		51.41	1,025	3,100	9,300	1,031	
	5160	Walk-behind, 11 H.P.		3.59	143	430	1,300	114.75	
	5200	Pavement profiler, 4' to 6' wide, 450 H.P.		218.77	1,275	3,838.46	11,500	2,518	
	5300	8' to 10' wide, 750 H.P.		334.94	1,350	4,015.23	12,000	3,483	
	5400	Roadway plate, steel, 1" x 8' x 20'		.09	61.50	184.69	555	37.65	
	5600	Stabilizer, self-propelled, 150 H.P.		41.55	1,050	3,131.37	9,400	958.70	
	5700	310 H.P.		76.95	1,325	3,939.47	11,800	1,404	
	5800	Striper, truck mounted, 120 gallon paint, 460 H.P.		49.24	350	1,046	3,150	603.10	
	5900	Thermal paint heating kettle, 115 gallons		7.78	75	225	675	107.25	
	6000	Tar kettle, 330 gallon, trailer mounted		12.40	96	287.50	865	156.70	
	7000	Tunnel locomotive, diesel, 8 to 12 ton		30.11	625	1,881.35	5,650	617.20	
	7005	Electric, 10 ton		29.60	715	2,142.65	6,425	665.30	
	7010	Muck cars, 1/2 C.Y. capacity		2.33	27	81	243	34.80	
	7020	1 C.Y. capacity		2.54	35	105.56	315	41.45	
	7030	2 C.Y. capacity		2.69	39.50	118.11	355	45.10	
	7040	Side dump, 2 C.Y. capacity		2.90	49	146.33	440	52.50	
	7050	3 C.Y. capacity		3.90	53.50	160.96	485	63.40	
	7060	5 C.Y. capacity		5.69	69.50	207.99	625	87.10	
	7100	Ventilating blower for tunnel, 7-1/2 H.P.		2.16	53.50	159.91	480	49.30	
	7110	10 H.P.		2.45	55.50	167.23	500	53.05	
	7120	20 H.P.		3.58	72.50	217.40	650	72.15	
	7140	40 H.P.		6.21	96	287.43	860	107.20	
	7160	60 H.P.		8.79	103	308.33	925	132	
	7175	75 H.P.		10.49	160	480.79	1,450	180.10	
	7180	200 H.P.		21.03	315	945.90	2,850	357.45	
	7800	Windrow loader, elevating		54.49	1,700	5,125	15,400	1,461	
60	0010	**LIFTING AND HOISTING EQUIPMENT RENTAL** without operators							**60**
	0150	Crane, flatbed mounted, 3 ton capacity	Ea.	14.58	203	610.30	1,825	238.75	
	0200	Crane, climbing, 106' jib, 6000 lb. capacity, 410 fpm		40.13	2,675	8,034	24,100	1,928	
	0300	101' jib, 10,250 lb. capacity, 270 fpm		46.90	2,300	6,868.82	20,600	1,749	
	0500	Tower, static, 130' high, 106' jib, 6200 lb. capacity at 400 fpm		45.61	2,300	6,916	20,700	1,748	
	0520	Mini crawler spider crane, up to 24" wide, 1990 lb. lifting capacity		12.65	555	1,672.31	5,025	435.70	
	0525	Up to 30" wide, 6450 lb. lifting capacity		14.70	660	1,985.87	5,950	514.75	
	0530	Up to 52" wide, 6680 lb. lifting capacity		23.38	810	2,430.08	7,300	673.05	

Reference notes in table: R015433-10 (row 0010/0150), R312316-45 (row 0200/0300)

01 54 33 | Equipment Rental

		UNIT	HOURLY OPER. COST	RENT PER DAY	RENT PER WEEK	RENT PER MONTH	EQUIPMENT COST/DAY	
60	0535	Up to 55" wide, 8920 lb. lifting capacity	Ea.	26.10	895	2,691.37	8,075	747.10	60
	0540	Up to 66" wide, 13,350 lb. lifting capacity		35.34	1,400	4,180.77	12,500	1,119	
	0600	Crawler mounted, lattice boom, 1/2 C.Y., 15 tons at 12' radius		37.34	855	2,558	7,675	810.30	
	0700	3/4 C.Y., 20 tons at 12' radius		54.72	960	2,874	8,625	1,013	
	0800	1 C.Y., 25 tons at 12' radius		68.10	1,025	3,038	9,125	1,152	
	0900	1-1/2 C.Y., 40 tons at 12' radius		66.98	1,150	3,476	10,400	1,231	
	1000	2 C.Y., 50 tons at 12' radius		89.66	1,375	4,120	12,400	1,541	
	1100	3 C.Y., 75 tons at 12' radius		65.65	2,400	7,210	21,600	1,967	
	1200	100 ton capacity, 60' boom		86.78	2,700	8,080.96	24,200	2,310	
	1300	165 ton capacity, 60' boom		107.18	3,025	9,091.08	27,300	2,676	
	1400	200 ton capacity, 70' boom		139.61	3,875	11,616.38	34,800	3,440	
	1500	350 ton capacity, 80' boom		184.04	4,200	12,626.50	37,900	3,998	
	1600	Truck mounted, lattice boom, 6 x 4, 20 tons at 10' radius		40.16	2,000	6,026	18,100	1,526	
	1700	25 tons at 10' radius		43.16	2,400	7,210	21,600	1,787	
	1800	8 x 4, 30 tons at 10' radius		54.71	2,575	7,725	23,200	1,983	
	1900	40 tons at 12' radius		54.71	2,825	8,446	25,300	2,127	
	2000	60 tons at 15' radius		54.23	1,675	5,000.09	15,000	1,434	
	2050	82 tons at 15' radius		60.03	1,800	5,404.14	16,200	1,561	
	2100	90 tons at 15' radius		67.06	1,950	5,883.95	17,700	1,713	
	2200	115 tons at 15' radius		75.66	2,200	6,591.03	19,800	1,923	
	2300	150 tons at 18' radius		81.91	2,775	8,343	25,000	2,324	
	2350	165 tons at 18' radius		87.91	2,825	8,498	25,500	2,403	
	2400	Truck mounted, hydraulic, 12 ton capacity		29.80	395	1,186.89	3,550	475.80	
	2500	25 ton capacity		36.72	490	1,464.67	4,400	586.70	
	2550	33 ton capacity		54.71	910	2,727.32	8,175	983.15	
	2560	40 ton capacity		54.71	910	2,727.32	8,175	983.15	
	2600	55 ton capacity		54.32	925	2,777.83	8,325	990.15	
	2700	80 ton capacity		71.13	1,475	4,444.53	13,300	1,458	
	2720	100 ton capacity		75.72	1,575	4,722.31	14,200	1,550	
	2740	120 ton capacity		106.34	1,850	5,555.66	16,700	1,962	
	2760	150 ton capacity		113.80	2,050	6,186.99	18,600	2,148	
	2800	Self-propelled, 4 x 4, with telescoping boom, 5 ton		15.29	435	1,298	3,900	381.95	
	2900	12-1/2 ton capacity		21.63	435	1,298	3,900	432.65	
	3000	15 ton capacity		34.77	455	1,363.66	4,100	550.85	
	3050	20 ton capacity		22.55	655	1,969.73	5,900	574.35	
	3100	25 ton capacity		37.06	1,425	4,293.01	12,900	1,155	
	3150	40 ton capacity		45.35	665	1,994.99	5,975	761.80	
	3200	Derricks, guy, 20 ton capacity, 60' boom, 75' mast		22.97	1,450	4,378	13,100	1,059	
	3300	100' boom, 115' mast		36.41	2,050	6,180	18,500	1,527	
	3400	Stiffleg, 20 ton capacity, 70' boom, 37' mast		25.66	635	1,906	5,725	586.50	
	3500	100' boom, 47' mast		39.71	685	2,060	6,175	729.70	
	3550	Helicopter, small, lift to 1250 lb. maximum, w/pilot		100.14	2,175	6,500.12	19,500	2,101	
	3600	Hoists, chain type, overhead, manual, 3/4 ton		.15	10.25	30.70	92	7.30	
	3900	10 ton		.80	6.25	18.78	56.50	10.10	
	4000	Hoist and tower, 5000 lb. cap., portable electric, 40' high		4.82	146	438	1,325	126.15	
	4100	For each added 10' section, add		.12	32.50	98	294	20.55	
	4200	Hoist and single tubular tower, 5000 lb. electric, 100' high		7.03	108	325	975	121.20	
	4300	For each added 6'-6" section, add		.21	39.50	118	355	25.25	
	4400	Hoist and double tubular tower, 5000 lb., 100' high		7.65	108	325	975	126.20	
	4500	For each added 6'-6" section, add		.23	42.50	128	385	27.45	
	4550	Hoist and tower, mast type, 6000 lb., 100' high		9.32	95.50	286.87	860	131.95	
	4570	For each added 10' section, add		.13	32.50	98	294	20.65	
	4600	Hoist and tower, personnel, electric, 2000 lb., 100' @ 125 fpm		17.67	25.50	75.76	227	156.55	
	4700	3000 lb., 100' @ 200 fpm		20.22	25.50	75.76	227	176.95	
	4800	3000 lb., 150' @ 300 fpm		22.44	25.50	75.76	227	194.70	
	4900	4000 lb., 100' @ 300 fpm		23.22	25.50	75.76	227	200.90	
	5000	6000 lb., 100' @ 275 fpm	▼	24.95	25.50	75.76	227	214.75	
	5100	For added heights up to 500', add	L.F.	.01	3.33	10	30	2.10	

01 54 33 | Equipment Rental

		UNIT	HOURLY OPER. COST	RENT PER DAY	RENT PER WEEK	RENT PER MONTH	EQUIPMENT COST/DAY	
60	**5200**	Jacks, hydraulic, 20 ton	Ea.	.05	19.85	59.60	179	12.30	**60**
	5500	100 ton		.40	26	78.50	236	18.95	
	6100	Jacks, hydraulic, climbing w/50' jackrods, control console, 30 ton cap.		2.19	32	96	288	36.70	
	6150	For each added 10' jackrod section, add		.05	5	15	45	3.40	
	6300	50 ton capacity		3.52	34.50	103	310	48.75	
	6350	For each added 10' jackrod section, add		.06	5	15	45	3.50	
	6500	125 ton capacity		9.20	53.50	160	480	105.55	
	6550	For each added 10' jackrod section, add		.62	5	15	45	8	
	6600	Cable jack, 10 ton capacity with 200' cable		1.60	36.50	110	330	34.80	
	6650	For each added 50' of cable, add		.22	15.35	46	138	11	
70	**0010**	**WELLPOINT EQUIPMENT RENTAL** without operators							**70**
	0020	Based on 2 months rental							
	0100	Combination jetting & wellpoint pump, 60 H.P. diesel	Ea.	15.83	305	922	2,775	311.05	
	0200	High pressure gas jet pump, 200 H.P., 300 psi	"	34.17	278	833.35	2,500	440.05	
	0300	Discharge pipe, 8" diameter	L.F.	.01	1.41	4.24	12.75	.90	
	0350	12" diameter		.01	2.09	6.26	18.80	1.35	
	0400	Header pipe, flows up to 150 GPM, 4" diameter		.01	.74	2.22	6.65	.50	
	0500	400 GPM, 6" diameter		.01	1.08	3.23	9.70	.70	
	0600	800 GPM, 8" diameter		.01	1.41	4.24	12.75	.95	
	0700	1500 GPM, 10" diameter		.01	1.75	5.25	15.75	1.15	
	0800	2500 GPM, 12" diameter		.03	2.09	6.26	18.80	1.45	
	0900	4500 GPM, 16" diameter		.03	2.42	7.27	22	1.70	
	0950	For quick coupling aluminum and plastic pipe, add		.03	9.60	28.84	86.50	6.05	
	1100	Wellpoint, 25' long, with fittings & riser pipe, 1-1/2" or 2" diameter	Ea.	.07	133	399	1,200	80.35	
	1200	Wellpoint pump, diesel powered, 4" suction, 20 H.P.		7.07	152	454.55	1,375	147.50	
	1300	6" suction, 30 H.P.		9.49	140	420	1,250	159.90	
	1400	8" suction, 40 H.P.		12.85	253	757.59	2,275	254.35	
	1500	10" suction, 75 H.P.		18.75	268	803.05	2,400	310.60	
	1600	12" suction, 100 H.P.		27.51	300	904.06	2,700	400.90	
	1700	12" suction, 175 H.P.		39.37	320	959.61	2,875	506.90	
80	**0010**	**MARINE EQUIPMENT RENTAL** without operators	Ea.						**80**
	0200	Barge, 400 Ton, 30' wide x 90' long		17.84	1,200	3,632.05	10,900	869.15	
	0240	800 Ton, 45' wide x 90' long		22.41	1,500	4,468.20	13,400	1,073	
	2000	Tugboat, diesel, 100 H.P.		29.92	240	721.18	2,175	383.65	
	2040	250 H.P.		58.10	435	1,306.49	3,925	726.10	
	2080	380 H.P.		126.49	1,300	3,919.48	11,800	1,796	
	3000	Small work boat, gas, 16-foot, 50 H.P.		11.48	48.50	145.01	435	120.85	
	4000	Large, diesel, 48-foot, 200 H.P.		75.57	1,375	4,154.64	12,500	1,436	

R015433 -10

Crew No.	Bare Costs		Incl. Subs O&P		Cost Per Labor-Hour	

Crew A-1	Hr.	Daily	Hr.	Daily	Bare Costs	Incl. O&P
1 Building Laborer	$44.40	$355.20	$70.70	$565.60	$44.40	$70.70
1 Concrete Saw, Gas Manual		113.70		125.07	14.21	15.63
8 L.H., Daily Totals		$468.90		$690.67	$58.61	$86.33

Crew A-1A	Hr.	Daily	Hr.	Daily	Bare Costs	Incl. O&P
1 Skilled Worker	$57.10	$456.80	$90.50	$724.00	$57.10	$90.50
1 Shot Blaster, 20"		208.70		229.57	26.09	28.70
8 L.H., Daily Totals		$665.50		$953.57	$83.19	$119.20

Crew A-1B	Hr.	Daily	Hr.	Daily	Bare Costs	Incl. O&P
1 Building Laborer	$44.40	$355.20	$70.70	$565.60	$44.40	$70.70
1 Concrete Saw		112.85		124.14	14.11	15.52
8 L.H., Daily Totals		$468.05		$689.74	$58.51	$86.22

Crew A-1C	Hr.	Daily	Hr.	Daily	Bare Costs	Incl. O&P
1 Building Laborer	$44.40	$355.20	$70.70	$565.60	$44.40	$70.70
1 Chain Saw, Gas, 18"		52.20		57.42	6.53	7.18
8 L.H., Daily Totals		$407.40		$623.02	$50.92	$77.88

Crew A-1D	Hr.	Daily	Hr.	Daily	Bare Costs	Incl. O&P
1 Building Laborer	$44.40	$355.20	$70.70	$565.60	$44.40	$70.70
1 Vibrating Plate, Gas, 18"		31.90		35.09	3.99	4.39
8 L.H., Daily Totals		$387.10		$600.69	$48.39	$75.09

Crew A-1E	Hr.	Daily	Hr.	Daily	Bare Costs	Incl. O&P
1 Building Laborer	$44.40	$355.20	$70.70	$565.60	$44.40	$70.70
1 Vibrating Plate, Gas, 21"		165.60		182.16	20.70	22.77
8 L.H., Daily Totals		$520.80		$747.76	$65.10	$93.47

Crew A-1F	Hr.	Daily	Hr.	Daily	Bare Costs	Incl. O&P
1 Building Laborer	$44.40	$355.20	$70.70	$565.60	$44.40	$70.70
1 Rammer/Tamper, Gas, 8"		47.00		51.70	5.88	6.46
8 L.H., Daily Totals		$402.20		$617.30	$50.27	$77.16

Crew A-1G	Hr.	Daily	Hr.	Daily	Bare Costs	Incl. O&P
1 Building Laborer	$44.40	$355.20	$70.70	$565.60	$44.40	$70.70
1 Rammer/Tamper, Gas, 15"		54.65		60.12	6.83	7.51
8 L.H., Daily Totals		$409.85		$625.72	$51.23	$78.21

Crew A-1H	Hr.	Daily	Hr.	Daily	Bare Costs	Incl. O&P
1 Building Laborer	$44.40	$355.20	$70.70	$565.60	$44.40	$70.70
1 Exterior Steam Cleaner		77.20		84.92	9.65	10.62
8 L.H., Daily Totals		$432.40		$650.52	$54.05	$81.31

Crew A-1J	Hr.	Daily	Hr.	Daily	Bare Costs	Incl. O&P
1 Building Laborer	$44.40	$355.20	$70.70	$565.60	$44.40	$70.70
1 Cultivator, Walk-Behind, 5 H.P.		53.25		58.58	6.66	7.32
8 L.H., Daily Totals		$408.45		$624.17	$51.06	$78.02

Crew A-1K	Hr.	Daily	Hr.	Daily	Bare Costs	Incl. O&P
1 Building Laborer	$44.40	$355.20	$70.70	$565.60	$44.40	$70.70
1 Cultivator, Walk-Behind, 8 H.P.		92.20		101.42	11.53	12.68
8 L.H., Daily Totals		$447.40		$667.02	$55.92	$83.38

Crew A-1M	Hr.	Daily	Hr.	Daily	Bare Costs	Incl. O&P
1 Building Laborer	$44.40	$355.20	$70.70	$565.60	$44.40	$70.70
1 Snow Blower, Walk-Behind		68.30		75.13	8.54	9.39
8 L.H., Daily Totals		$423.50		$640.73	$52.94	$80.09

Crew A-2	Hr.	Daily	Hr.	Daily	Bare Costs	Incl. O&P
2 Laborers	$44.40	$710.40	$70.70	$1131.20	$45.87	$73.08
1 Truck Driver (light)	48.80	390.40	77.85	622.80		
1 Flatbed Truck, Gas, 1.5 Ton		198.50		218.35	8.27	9.10
24 L.H., Daily Totals		$1299.30		$1972.35	$54.14	$82.18

Crew A-2A	Hr.	Daily	Hr.	Daily	Bare Costs	Incl. O&P
2 Laborers	$44.40	$710.40	$70.70	$1131.20	$45.87	$73.08
1 Truck Driver (light)	48.80	390.40	77.85	622.80		
1 Flatbed Truck, Gas, 1.5 Ton		198.50		218.35		
1 Concrete Saw		112.85		124.14	12.97	14.27
24 L.H., Daily Totals		$1412.15		$2096.49	$58.84	$87.35

Crew A-2B	Hr.	Daily	Hr.	Daily	Bare Costs	Incl. O&P
1 Truck Driver (light)	$48.80	$390.40	$77.85	$622.80	$48.80	$77.85
1 Flatbed Truck, Gas, 1.5 Ton		198.50		218.35	24.81	27.29
8 L.H., Daily Totals		$588.90		$841.15	$73.61	$105.14

Crew A-3A	Hr.	Daily	Hr.	Daily	Bare Costs	Incl. O&P
1 Equip. Oper. (light)	$55.50	$444.00	$86.60	$692.80	$55.50	$86.60
1 Pickup Truck, 4x4, 3/4 Ton		176.75		194.43	22.09	24.30
8 L.H., Daily Totals		$620.75		$887.23	$77.59	$110.90

Crew A-3B	Hr.	Daily	Hr.	Daily	Bare Costs	Incl. O&P
1 Equip. Oper. (medium)	$59.00	$472.00	$92.05	$736.40	$55.15	$86.95
1 Truck Driver (heavy)	51.30	410.40	81.85	654.80		
1 Dump Truck, 12 C.Y., 400 H.P.		579.35		637.28		
1 F.E. Loader, W.M., 2.5 C.Y.		638.30		702.13	76.10	83.71
16 L.H., Daily Totals		$2100.05		$2730.61	$131.25	$170.66

Crew A-3C	Hr.	Daily	Hr.	Daily	Bare Costs	Incl. O&P
1 Equip. Oper. (light)	$55.50	$444.00	$86.60	$692.80	$55.50	$86.60
1 Loader, Skid Steer, 78 H.P.		446.30		490.93	55.79	61.37
8 L.H., Daily Totals		$890.30		$1183.73	$111.29	$147.97

Crew A-3D	Hr.	Daily	Hr.	Daily	Bare Costs	Incl. O&P
1 Truck Driver (light)	$48.80	$390.40	$77.85	$622.80	$48.80	$77.85
1 Pickup Truck, 4x4, 3/4 Ton		176.75		194.43		
1 Flatbed Trailer, 25 Ton		137.20		150.92	39.24	43.17
8 L.H., Daily Totals		$704.35		$968.14	$88.04	$121.02

Crew A-3E	Hr.	Daily	Hr.	Daily	Bare Costs	Incl. O&P
1 Equip. Oper. (crane)	$61.45	$491.60	$95.85	$766.80	$56.38	$88.85
1 Truck Driver (heavy)	51.30	410.40	81.85	654.80		
1 Pickup Truck, 4x4, 3/4 Ton		176.75		194.43	11.05	12.15
16 L.H., Daily Totals		$1078.75		$1616.03	$67.42	$101.00

Crew A-3F	Hr.	Daily	Hr.	Daily	Bare Costs	Incl. O&P
1 Equip. Oper. (crane)	$61.45	$491.60	$95.85	$766.80	$56.38	$88.85
1 Truck Driver (heavy)	51.30	410.40	81.85	654.80		
1 Pickup Truck, 4x4, 3/4 Ton		176.75		194.43		
1 Truck Tractor, 6x4, 380 H.P.		499.15		549.07		
1 Lowbed Trailer, 75 Ton		258.10		283.91	58.38	64.21
16 L.H., Daily Totals		$1836.00		$2449.00	$114.75	$153.06

For customer support on your Commercial Renovation Costs with RSMeans Data, call 800.448.8182.

755

Crew No.	Bare Costs		Incl. Subs O&P		Cost Per Labor-Hour	
Crew A-3G	Hr.	Daily	Hr.	Daily	Bare Costs	Incl. O&P
1 Equip. Oper. (crane)	$61.45	$491.60	$95.85	$766.80	$56.38	$88.85
1 Truck Driver (heavy)	51.30	410.40	81.85	654.80		
1 Pickup Truck, 4x4, 3/4 Ton		176.75		194.43		
1 Truck Tractor, 6x4, 450 H.P.		608.95		669.85		
1 Lowbed Trailer, 75 Ton		258.10		283.91	65.24	71.76
16 L.H., Daily Totals		$1945.80		$2569.78	$121.61	$160.61
Crew A-3H	Hr.	Daily	Hr.	Daily	Bare Costs	Incl. O&P
1 Equip. Oper. (crane)	$61.45	$491.60	$95.85	$766.80	$61.45	$95.85
1 Hyd. Crane, 12 Ton (Daily)		733.15		806.47	91.64	100.81
8 L.H., Daily Totals		$1224.75		$1573.27	$153.09	$196.66
Crew A-3I	Hr.	Daily	Hr.	Daily	Bare Costs	Incl. O&P
1 Equip. Oper. (crane)	$61.45	$491.60	$95.85	$766.80	$61.45	$95.85
1 Hyd. Crane, 25 Ton (Daily)		810.50		891.55	101.31	111.44
8 L.H., Daily Totals		$1302.10		$1658.35	$162.76	$207.29
Crew A-3J	Hr.	Daily	Hr.	Daily	Bare Costs	Incl. O&P
1 Equip. Oper. (crane)	$61.45	$491.60	$95.85	$766.80	$61.45	$95.85
1 Hyd. Crane, 40 Ton (Daily)		1287.00		1415.70	160.88	176.96
8 L.H., Daily Totals		$1778.60		$2182.50	$222.32	$272.81
Crew A-3K	Hr.	Daily	Hr.	Daily	Bare Costs	Incl. O&P
1 Equip. Oper. (crane)	$61.45	$491.60	$95.85	$766.80	$56.98	$88.88
1 Equip. Oper. (oiler)	52.50	420.00	81.90	655.20		
1 Hyd. Crane, 55 Ton (Daily)		1377.00		1514.70		
1 P/U Truck, 3/4 Ton (Daily)		143.85		158.24	95.05	104.56
16 L.H., Daily Totals		$2432.45		$3094.93	$152.03	$193.43
Crew A-3L	Hr.	Daily	Hr.	Daily	Bare Costs	Incl. O&P
1 Equip. Oper. (crane)	$61.45	$491.60	$95.85	$766.80	$56.98	$88.88
1 Equip. Oper. (oiler)	52.50	420.00	81.90	655.20		
1 Hyd. Crane, 80 Ton (Daily)		2058.00		2263.80		
1 P/U Truck, 3/4 Ton (Daily)		143.85		158.24	137.62	151.38
16 L.H., Daily Totals		$3113.45		$3844.03	$194.59	$240.25
Crew A-3M	Hr.	Daily	Hr.	Daily	Bare Costs	Incl. O&P
1 Equip. Oper. (crane)	$61.45	$491.60	$95.85	$766.80	$56.98	$88.88
1 Equip. Oper. (oiler)	52.50	420.00	81.90	655.20		
1 Hyd. Crane, 100 Ton (Daily)		2253.00		2478.30		
1 P/U Truck, 3/4 Ton (Daily)		143.85		158.24	149.80	164.78
16 L.H., Daily Totals		$3308.45		$4058.53	$206.78	$253.66
Crew A-3N	Hr.	Daily	Hr.	Daily	Bare Costs	Incl. O&P
1 Equip. Oper. (crane)	$61.45	$491.60	$95.85	$766.80	$61.45	$95.85
1 Tower Crane (monthly)		1737.00		1910.70	217.13	238.84
8 L.H., Daily Totals		$2228.60		$2677.50	$278.57	$334.69
Crew A-3P	Hr.	Daily	Hr.	Daily	Bare Costs	Incl. O&P
1 Equip. Oper. (light)	$55.50	$444.00	$86.60	$692.80	$55.50	$86.60
1 A.T. Forklift, 31' reach, 45' lift		346.45		381.10	43.31	47.64
8 L.H., Daily Totals		$790.45		$1073.90	$98.81	$134.24
Crew A-3Q	Hr.	Daily	Hr.	Daily	Bare Costs	Incl. O&P
1 Equip. Oper. (light)	$55.50	$444.00	$86.60	$692.80	$55.50	$86.60
1 Pickup Truck, 4x4, 3/4 Ton		176.75		194.43		
1 Flatbed Trailer, 3 Ton		71.15		78.27	30.99	34.09
8 L.H., Daily Totals		$691.90		$965.49	$86.49	$120.69

Crew No.	Bare Costs		Incl. Subs O&P		Cost Per Labor-Hour	
Crew A-3R	Hr.	Daily	Hr.	Daily	Bare Costs	Incl. O&P
1 Equip. Oper. (light)	$55.50	$444.00	$86.60	$692.80	$55.50	$86.60
1 Forklift, Smooth Floor, 8,000 Lb.		283.25		311.57	35.41	38.95
8 L.H., Daily Totals		$727.25		$1004.38	$90.91	$125.55
Crew A-4	Hr.	Daily	Hr.	Daily	Bare Costs	Incl. O&P
2 Carpenters	$54.70	$875.20	$87.10	$1393.60	$51.95	$82.58
1 Painter, Ordinary	46.45	371.60	73.55	588.40		
24 L.H., Daily Totals		$1246.80		$1982.00	$51.95	$82.58
Crew A-5	Hr.	Daily	Hr.	Daily	Bare Costs	Incl. O&P
2 Laborers	$44.40	$710.40	$70.70	$1131.20	$44.89	$71.49
.25 Truck Driver (light)	48.80	97.60	77.85	155.70		
.25 Flatbed Truck, Gas, 1.5 Ton		49.63		54.59	2.76	3.03
18 L.H., Daily Totals		$857.63		$1341.49	$47.65	$74.53
Crew A-6	Hr.	Daily	Hr.	Daily	Bare Costs	Incl. O&P
1 Instrument Man	$57.10	$456.80	$90.50	$724.00	$54.88	$87.10
1 Rodman/Chainman	52.65	421.20	83.70	669.60		
1 Level, Electronic		34.40		37.84	2.15	2.37
16 L.H., Daily Totals		$912.40		$1431.44	$57.02	$89.47
Crew A-7	Hr.	Daily	Hr.	Daily	Bare Costs	Incl. O&P
1 Chief of Party	$68.50	$548.00	$105.80	$846.40	$59.42	$93.33
1 Instrument Man	57.10	456.80	90.50	724.00		
1 Rodman/Chainman	52.65	421.20	83.70	669.60		
1 Level, Electronic		34.40		37.84	1.43	1.58
24 L.H., Daily Totals		$1460.40		$2277.84	$60.85	$94.91
Crew A-8	Hr.	Daily	Hr.	Daily	Bare Costs	Incl. O&P
1 Chief of Party	$68.50	$548.00	$105.80	$846.40	$57.73	$90.92
1 Instrument Man	57.10	456.80	90.50	724.00		
2 Rodmen/Chainmen	52.65	842.40	83.70	1339.20		
1 Level, Electronic		34.40		37.84	1.08	1.18
32 L.H., Daily Totals		$1881.60		$2947.44	$58.80	$92.11
Crew A-9	Hr.	Daily	Hr.	Daily	Bare Costs	Incl. O&P
1 Asbestos Foreman	$61.45	$491.60	$97.05	$776.40	$61.01	$96.39
7 Asbestos Workers	60.95	3413.20	96.30	5392.80		
64 L.H., Daily Totals		$3904.80		$6169.20	$61.01	$96.39
Crew A-10A	Hr.	Daily	Hr.	Daily	Bare Costs	Incl. O&P
1 Asbestos Foreman	$61.45	$491.60	$97.05	$776.40	$61.12	$96.55
2 Asbestos Workers	60.95	975.20	96.30	1540.80		
24 L.H., Daily Totals		$1466.80		$2317.20	$61.12	$96.55
Crew A-10B	Hr.	Daily	Hr.	Daily	Bare Costs	Incl. O&P
1 Asbestos Foreman	$61.45	$491.60	$97.05	$776.40	$61.08	$96.49
3 Asbestos Workers	60.95	1462.80	96.30	2311.20		
32 L.H., Daily Totals		$1954.40		$3087.60	$61.08	$96.49
Crew A-10C	Hr.	Daily	Hr.	Daily	Bare Costs	Incl. O&P
3 Asbestos Workers	$60.95	$1462.80	$96.30	$2311.20	$60.95	$96.30
1 Flatbed Truck, Gas, 1.5 Ton		198.50		218.35	8.27	9.10
24 L.H., Daily Totals		$1661.30		$2529.55	$69.22	$105.40

For customer support on your Commercial Renovation Costs with RSMeans Data, call 800.448.8182.

Crew No.	Bare Costs Hr.	Daily	Incl. Subs O&P Hr.	Daily	Cost Per Labor-Hour Bare Costs	Incl. O&P
Crew A-10D	Hr.	Daily	Hr.	Daily	Bare Costs	Incl. O&P
2 Asbestos Workers	$60.95	$975.20	$96.30	$1540.80	$58.96	$92.59
1 Equip. Oper. (crane)	61.45	491.60	95.85	766.80		
1 Equip. Oper. (oiler)	52.50	420.00	81.90	655.20		
1 Hydraulic Crane, 33 Ton		983.15		1081.46	30.72	33.80
32 L.H., Daily Totals		$2869.95		$4044.26	$89.69	$126.38

Crew No.	Bare Costs Hr.	Daily	Incl. Subs O&P Hr.	Daily	Cost Per Labor-Hour Bare Costs	Incl. O&P
Crew A-11	Hr.	Daily	Hr.	Daily	Bare Costs	Incl. O&P
1 Asbestos Foreman	$61.45	$491.60	$97.05	$776.40	$61.01	$96.39
7 Asbestos Workers	60.95	3413.20	96.30	5392.80		
2 Chip. Hammers, 12 Lb., Elec.		65.70		72.27	1.03	1.13
64 L.H., Daily Totals		$3970.50		$6241.47	$62.04	$97.52

Crew No.	Bare Costs Hr.	Daily	Incl. Subs O&P Hr.	Daily	Cost Per Labor-Hour Bare Costs	Incl. O&P
Crew A-12	Hr.	Daily	Hr.	Daily	Bare Costs	Incl. O&P
1 Asbestos Foreman	$61.45	$491.60	$97.05	$776.40	$61.01	$96.39
7 Asbestos Workers	60.95	3413.20	96.30	5392.80		
1 Trk-Mtd Vac, 14 CY, 1500 Gal.		542.60		596.86		
1 Flatbed Truck, 20,000 GVW		204.05		224.46	11.67	12.83
64 L.H., Daily Totals		$4651.45		$6990.52	$72.68	$109.23

Crew No.	Bare Costs Hr.	Daily	Incl. Subs O&P Hr.	Daily	Cost Per Labor-Hour Bare Costs	Incl. O&P
Crew A-13	Hr.	Daily	Hr.	Daily	Bare Costs	Incl. O&P
1 Equip. Oper. (light)	$55.50	$444.00	$86.60	$692.80	$55.50	$86.60
1 Trk-Mtd Vac, 14 CY, 1500 Gal.		542.60		596.86		
1 Flatbed Truck, 20,000 GVW		204.05		224.46	93.33	102.66
8 L.H., Daily Totals		$1190.65		$1514.12	$148.83	$189.26

Crew No.	Bare Costs Hr.	Daily	Incl. Subs O&P Hr.	Daily	Cost Per Labor-Hour Bare Costs	Incl. O&P
Crew B-1	Hr.	Daily	Hr.	Daily	Bare Costs	Incl. O&P
1 Labor Foreman (outside)	$46.40	$371.20	$73.90	$591.20	$45.07	$71.77
2 Laborers	44.40	710.40	70.70	1131.20		
24 L.H., Daily Totals		$1081.60		$1722.40	$45.07	$71.77

Crew No.	Bare Costs Hr.	Daily	Incl. Subs O&P Hr.	Daily	Cost Per Labor-Hour Bare Costs	Incl. O&P
Crew B-1A	Hr.	Daily	Hr.	Daily	Bare Costs	Incl. O&P
1 Labor Foreman (outside)	$46.40	$371.20	$73.90	$591.20	$45.07	$71.77
2 Laborers	44.40	710.40	70.70	1131.20		
2 Cutting Torches		25.90		28.49		
2 Sets of Gases		347.20		381.92	15.55	17.10
24 L.H., Daily Totals		$1454.70		$2132.81	$60.61	$88.87

Crew No.	Bare Costs Hr.	Daily	Incl. Subs O&P Hr.	Daily	Cost Per Labor-Hour Bare Costs	Incl. O&P
Crew B-1B	Hr.	Daily	Hr.	Daily	Bare Costs	Incl. O&P
1 Labor Foreman (outside)	$46.40	$371.20	$73.90	$591.20	$49.16	$77.79
2 Laborers	44.40	710.40	70.70	1131.20		
1 Equip. Oper. (crane)	61.45	491.60	95.85	766.80		
2 Cutting Torches		25.90		28.49		
2 Sets of Gases		347.20		381.92		
1 Hyd. Crane, 12 Ton		475.80		523.38	26.53	29.18
32 L.H., Daily Totals		$2422.10		$3422.99	$75.69	$106.97

Crew No.	Bare Costs Hr.	Daily	Incl. Subs O&P Hr.	Daily	Cost Per Labor-Hour Bare Costs	Incl. O&P
Crew B-1C	Hr.	Daily	Hr.	Daily	Bare Costs	Incl. O&P
1 Labor Foreman (outside)	$46.40	$371.20	$73.90	$591.20	$45.07	$71.77
2 Laborers	44.40	710.40	70.70	1131.20		
1 Telescoping Boom Lift, to 60'		292.45		321.69	12.19	13.40
24 L.H., Daily Totals		$1374.05		$2044.10	$57.25	$85.17

Crew No.	Bare Costs Hr.	Daily	Incl. Subs O&P Hr.	Daily	Cost Per Labor-Hour Bare Costs	Incl. O&P
Crew B-1D	Hr.	Daily	Hr.	Daily	Bare Costs	Incl. O&P
2 Laborers	$44.40	$710.40	$70.70	$1131.20	$44.40	$70.70
1 Small Work Boat, Gas, 50 H.P.		120.85		132.94		
1 Pressure Washer, 7 GPM		93.95		103.35	13.43	14.77
16 L.H., Daily Totals		$925.20		$1367.48	$57.83	$85.47

Crew No.	Bare Costs Hr.	Daily	Incl. Subs O&P Hr.	Daily	Cost Per Labor-Hour Bare Costs	Incl. O&P
Crew B-1E	Hr.	Daily	Hr.	Daily	Bare Costs	Incl. O&P
1 Labor Foreman (outside)	$46.40	$371.20	$73.90	$591.20	$44.90	$71.50
3 Laborers	44.40	1065.60	70.70	1696.80		
1 Work Boat, Diesel, 200 H.P.		1436.00		1579.60		
2 Pressure Washers, 7 GPM		187.90		206.69	50.75	55.82
32 L.H., Daily Totals		$3060.70		$4074.29	$95.65	$127.32

Crew No.	Bare Costs Hr.	Daily	Incl. Subs O&P Hr.	Daily	Cost Per Labor-Hour Bare Costs	Incl. O&P
Crew B-1F	Hr.	Daily	Hr.	Daily	Bare Costs	Incl. O&P
2 Skilled Workers	$57.10	$913.60	$90.50	$1448.00	$52.87	$83.90
1 Laborer	44.40	355.20	70.70	565.60		
1 Small Work Boat, Gas, 50 H.P.		120.85		132.94		
1 Pressure Washer, 7 GPM		93.95		103.35	8.95	9.85
24 L.H., Daily Totals		$1483.60		$2249.88	$61.82	$93.75

Crew No.	Bare Costs Hr.	Daily	Incl. Subs O&P Hr.	Daily	Cost Per Labor-Hour Bare Costs	Incl. O&P
Crew B-1G	Hr.	Daily	Hr.	Daily	Bare Costs	Incl. O&P
2 Laborers	$44.40	$710.40	$70.70	$1131.20	$44.40	$70.70
1 Small Work Boat, Gas, 50 H.P.		120.85		132.94	7.55	8.31
16 L.H., Daily Totals		$831.25		$1264.14	$51.95	$79.01

Crew No.	Bare Costs Hr.	Daily	Incl. Subs O&P Hr.	Daily	Cost Per Labor-Hour Bare Costs	Incl. O&P
Crew B-1H	Hr.	Daily	Hr.	Daily	Bare Costs	Incl. O&P
2 Skilled Workers	$57.10	$913.60	$90.50	$1448.00	$52.87	$83.90
1 Laborer	44.40	355.20	70.70	565.60		
1 Small Work Boat, Gas, 50 H.P.		120.85		132.94	5.04	5.54
24 L.H., Daily Totals		$1389.65		$2146.53	$57.90	$89.44

Crew No.	Bare Costs Hr.	Daily	Incl. Subs O&P Hr.	Daily	Cost Per Labor-Hour Bare Costs	Incl. O&P
Crew B-1J	Hr.	Daily	Hr.	Daily	Bare Costs	Incl. O&P
1 Labor Foreman (inside)	$44.90	$359.20	$71.50	$572.00	$44.65	$71.10
1 Laborer	44.40	355.20	70.70	565.60		
16 L.H., Daily Totals		$714.40		$1137.60	$44.65	$71.10

Crew No.	Bare Costs Hr.	Daily	Incl. Subs O&P Hr.	Daily	Cost Per Labor-Hour Bare Costs	Incl. O&P
Crew B-1K	Hr.	Daily	Hr.	Daily	Bare Costs	Incl. O&P
1 Carpenter Foreman (inside)	$55.20	$441.60	$87.90	$703.20	$54.95	$87.50
1 Carpenter	54.70	437.60	87.10	696.80		
16 L.H., Daily Totals		$879.20		$1400.00	$54.95	$87.50

Crew No.	Bare Costs Hr.	Daily	Incl. Subs O&P Hr.	Daily	Cost Per Labor-Hour Bare Costs	Incl. O&P
Crew B-2	Hr.	Daily	Hr.	Daily	Bare Costs	Incl. O&P
1 Labor Foreman (outside)	$46.40	$371.20	$73.90	$591.20	$44.80	$71.34
4 Laborers	44.40	1420.80	70.70	2262.40		
40 L.H., Daily Totals		$1792.00		$2853.60	$44.80	$71.34

Crew No.	Bare Costs Hr.	Daily	Incl. Subs O&P Hr.	Daily	Cost Per Labor-Hour Bare Costs	Incl. O&P
Crew B-2A	Hr.	Daily	Hr.	Daily	Bare Costs	Incl. O&P
1 Labor Foreman (outside)	$46.40	$371.20	$73.90	$591.20	$45.07	$71.77
2 Laborers	44.40	710.40	70.70	1131.20		
1 Telescoping Boom Lift, to 60'		292.45		321.69	12.19	13.40
24 L.H., Daily Totals		$1374.05		$2044.10	$57.25	$85.17

Crew No.	Bare Costs Hr.	Daily	Incl. Subs O&P Hr.	Daily	Cost Per Labor-Hour Bare Costs	Incl. O&P
Crew B-3	Hr.	Daily	Hr.	Daily	Bare Costs	Incl. O&P
1 Labor Foreman (outside)	$46.40	$371.20	$73.90	$591.20	$49.47	$78.51
2 Laborers	44.40	710.40	70.70	1131.20		
1 Equip. Oper. (medium)	59.00	472.00	92.05	736.40		
2 Truck Drivers (heavy)	51.30	820.80	81.85	1309.60		
1 Crawler Loader, 3 C.Y.		1146.00		1260.60		
2 Dump Trucks, 12 C.Y., 400 H.P.		1158.70		1274.57	48.01	52.82
48 L.H., Daily Totals		$4679.10		$6303.57	$97.48	$131.32

Crew No.	Bare Costs Hr.	Daily	Incl. Subs O&P Hr.	Daily	Cost Per Labor-Hour Bare Costs	Incl. O&P
Crew B-3A	Hr.	Daily	Hr.	Daily	Bare Costs	Incl. O&P
4 Laborers	$44.40	$1420.80	$70.70	$2262.40	$47.32	$74.97
1 Equip. Oper. (medium)	59.00	472.00	92.05	736.40		
1 Hyd. Excavator, 1.5 C.Y.		695.80		765.38	17.40	19.13
40 L.H., Daily Totals		$2588.60		$3764.18	$64.72	$94.10

Crew No.	Bare Costs		Incl. Subs O&P		Cost Per Labor-Hour	

Crew B-3B	Hr.	Daily	Hr.	Daily	Bare Costs	Incl. O&P
2 Laborers	$44.40	$710.40	$70.70	$1131.20	$49.77	$78.83
1 Equip. Oper. (medium)	59.00	472.00	92.05	736.40		
1 Truck Driver (heavy)	51.30	410.40	81.85	654.80		
1 Backhoe Loader, 80 H.P.		235.05		258.56		
1 Dump Truck, 12 C.Y., 400 H.P.		579.35		637.28	25.45	28.00
32 L.H., Daily Totals		$2407.20		$3418.24	$75.22	$106.82

Crew B-3C	Hr.	Daily	Hr.	Daily	Bare Costs	Incl. O&P
3 Laborers	$44.40	$1065.60	$70.70	$1696.80	$48.05	$76.04
1 Equip. Oper. (medium)	59.00	472.00	92.05	736.40		
1 Crawler Loader, 4 C.Y.		1456.00		1601.60	45.50	50.05
32 L.H., Daily Totals		$2993.60		$4034.80	$93.55	$126.09

Crew B-4	Hr.	Daily	Hr.	Daily	Bare Costs	Incl. O&P
1 Labor Foreman (outside)	$46.40	$371.20	$73.90	$591.20	$45.88	$73.09
4 Laborers	44.40	1420.80	70.70	2262.40		
1 Truck Driver (heavy)	51.30	410.40	81.85	654.80		
1 Truck Tractor, 220 H.P.		310.80		341.88		
1 Flatbed Trailer, 40 Ton		188.45		207.29	10.40	11.44
48 L.H., Daily Totals		$2701.65		$4057.57	$56.28	$84.53

Crew B-5	Hr.	Daily	Hr.	Daily	Bare Costs	Incl. O&P
1 Labor Foreman (outside)	$46.40	$371.20	$73.90	$591.20	$48.86	$77.26
4 Laborers	44.40	1420.80	70.70	2262.40		
2 Equip. Oper. (medium)	59.00	944.00	92.05	1472.80		
1 Air Compressor, 250 cfm		202.85		223.13		
2 Breakers, Pavement, 60 lb.		107.20		117.92		
2 -50' Air Hoses, 1.5"		45.60		50.16		
1 Crawler Loader, 3 C.Y.		1146.00		1260.60	26.82	29.50
56 L.H., Daily Totals		$4237.65		$5978.22	$75.67	$106.75

Crew B-5D	Hr.	Daily	Hr.	Daily	Bare Costs	Incl. O&P
1 Labor Foreman (outside)	$46.40	$371.20	$73.90	$591.20	$49.16	$77.83
4 Laborers	44.40	1420.80	70.70	2262.40		
2 Equip. Oper. (medium)	59.00	944.00	92.05	1472.80		
1 Truck Driver (heavy)	51.30	410.40	81.85	654.80		
1 Air Compressor, 250 cfm		202.85		223.13		
2 Breakers, Pavement, 60 lb.		107.20		117.92		
2 -50' Air Hoses, 1.5"		45.60		50.16		
1 Crawler Loader, 3 C.Y.		1146.00		1260.60		
1 Dump Truck, 12 C.Y., 400 H.P.		579.35		637.28	32.52	35.77
64 L.H., Daily Totals		$5227.40		$7270.30	$81.68	$113.60

Crew B-5E	Hr.	Daily	Hr.	Daily	Bare Costs	Incl. O&P
1 Labor Foreman (outside)	$46.40	$371.20	$73.90	$591.20	$49.16	$77.83
4 Laborers	44.40	1420.80	70.70	2262.40		
2 Equip. Oper. (medium)	59.00	944.00	92.05	1472.80		
1 Truck Driver (heavy)	51.30	410.40	81.85	654.80		
1 Water Tank Trailer, 5000 Gal.		154.10		169.51		
1 High Pressure Water Jet 40 KSI		820.75		902.83		
2 -50' Air Hoses, 1.5"		45.60		50.16		
1 Crawler Loader, 3 C.Y.		1146.00		1260.60		
1 Dump Truck, 12 C.Y., 400 H.P.		579.35		637.28	42.90	47.19
64 L.H., Daily Totals		$5892.20		$8001.58	$92.07	$125.02

Crew B-6	Hr.	Daily	Hr.	Daily	Bare Costs	Incl. O&P
2 Laborers	$44.40	$710.40	$70.70	$1131.20	$48.10	$76.00
1 Equip. Oper. (light)	55.50	444.00	86.60	692.80		
1 Backhoe Loader, 48 H.P.		216.20		237.82	9.01	9.91
24 L.H., Daily Totals		$1370.60		$2061.82	$57.11	$85.91

Crew B-6A	Hr.	Daily	Hr.	Daily	Bare Costs	Incl. O&P
.5 Labor Foreman (outside)	$46.40	$185.60	$73.90	$295.60	$50.64	$79.88
1 Laborer	44.40	355.20	70.70	565.60		
1 Equip. Oper. (medium)	59.00	472.00	92.05	736.40		
1 Vacuum Truck, 5000 Gal.		371.95		409.14	18.60	20.46
20 L.H., Daily Totals		$1384.75		$2006.74	$69.24	$100.34

Crew B-6B	Hr.	Daily	Hr.	Daily	Bare Costs	Incl. O&P
2 Labor Foremen (outside)	$46.40	$742.40	$73.90	$1182.40	$45.07	$71.77
4 Laborers	44.40	1420.80	70.70	2262.40		
1 S.P. Crane, 4x4, 5 Ton		381.95		420.14		
1 Flatbed Truck, Gas, 1.5 Ton		198.50		218.35		
1 Butt Fusion Mach., 4"-12" diam.		420.75		462.82	20.86	22.94
48 L.H., Daily Totals		$3164.40		$4546.12	$65.92	$94.71

Crew B-6C	Hr.	Daily	Hr.	Daily	Bare Costs	Incl. O&P
2 Labor Foremen (outside)	$46.40	$742.40	$73.90	$1182.40	$45.07	$71.77
4 Laborers	44.40	1420.80	70.70	2262.40		
1 S.P. Crane, 4x4, 12 Ton		432.65		475.92		
1 Flatbed Truck, Gas, 3 Ton		850.05		935.05		
1 Butt Fusion Mach., 8"-24" diam.		1086.00		1194.60	49.35	54.28
48 L.H., Daily Totals		$4531.90		$6050.37	$94.41	$126.05

Crew B-6D	Hr.	Daily	Hr.	Daily	Bare Costs	Incl. O&P
.5 Labor Foreman (outside)	$46.40	$185.60	$73.90	$295.60	$50.64	$79.88
1 Laborer	44.40	355.20	70.70	565.60		
1 Equip. Oper. (medium)	59.00	472.00	92.05	736.40		
1 Hydro Excavator, 12 C.Y.		1277.00		1404.70	63.85	70.23
20 L.H., Daily Totals		$2289.80		$3002.30	$114.49	$150.12

Crew B-7	Hr.	Daily	Hr.	Daily	Bare Costs	Incl. O&P
1 Labor Foreman (outside)	$46.40	$371.20	$73.90	$591.20	$47.17	$74.79
4 Laborers	44.40	1420.80	70.70	2262.40		
1 Equip. Oper. (medium)	59.00	472.00	92.05	736.40		
1 Brush Chipper, 12", 130 H.P.		366.05		402.65		
1 Crawler Loader, 3 C.Y.		1146.00		1260.60		
2 Chain Saws, Gas, 36" Long		83.30		91.63	33.24	36.56
48 L.H., Daily Totals		$3859.35		$5344.89	$80.40	$111.35

Crew B-7A	Hr.	Daily	Hr.	Daily	Bare Costs	Incl. O&P
2 Laborers	$44.40	$710.40	$70.70	$1131.20	$48.10	$76.00
1 Equip. Oper. (light)	55.50	444.00	86.60	692.80		
1 Rake w/Tractor		343.50		377.85		
2 Chain Saws, Gas, 18"		104.40		114.84	18.66	20.53
24 L.H., Daily Totals		$1602.30		$2316.69	$66.76	$96.53

Crew B-7B	Hr.	Daily	Hr.	Daily	Bare Costs	Incl. O&P
1 Labor Foreman (outside)	$46.40	$371.20	$73.90	$591.20	$47.76	$75.80
4 Laborers	44.40	1420.80	70.70	2262.40		
1 Equip. Oper. (medium)	59.00	472.00	92.05	736.40		
1 Truck Driver (heavy)	51.30	410.40	81.85	654.80		
1 Brush Chipper, 12", 130 H.P.		366.05		402.65		
1 Crawler Loader, 3 C.Y.		1146.00		1260.60		
2 Chain Saws, Gas, 36" Long		83.30		91.63		
1 Dump Truck, 8 C.Y., 220 H.P.		407.60		448.36	35.77	39.34
56 L.H., Daily Totals		$4677.35		$6448.05	$83.52	$115.14

For customer support on your Commercial Renovation Costs with RSMeans Data, call 800.448.8182.

Crew No.	Bare Costs		Incl. Subs O&P		Cost Per Labor-Hour	
Crew B-7C	Hr.	Daily	Hr.	Daily	Bare Costs	Incl. O&P
1 Labor Foreman (outside)	$46.40	$371.20	$73.90	$591.20	$47.76	$75.80
4 Laborers	44.40	1420.80	70.70	2262.40		
1 Equip. Oper. (medium)	59.00	472.00	92.05	736.40		
1 Truck Driver (heavy)	51.30	410.40	81.85	654.80		
1 Brush Chipper, 12", 130 H.P.		366.05		402.65		
1 Crawler Loader, 3 C.Y.		1146.00		1260.60		
2 Chain Saws, Gas, 36" Long		83.30		91.63		
1 Dump Truck, 12 C.Y., 400 H.P.		579.35		637.28	38.83	42.72
56 L.H., Daily Totals		$4849.10		$6636.97	$86.59	$118.52

Crew No.	Bare Costs		Incl. Subs O&P		Cost Per Labor-Hour	
Crew B-8	Hr.	Daily	Hr.	Daily	Bare Costs	Incl. O&P
1 Labor Foreman (outside)	$46.40	$371.20	$73.90	$591.20	$51.04	$80.63
2 Laborers	44.40	710.40	70.70	1131.20		
2 Equip. Oper. (medium)	59.00	944.00	92.05	1472.80		
1 Equip. Oper. (oiler)	52.50	420.00	81.90	655.20		
2 Truck Drivers (heavy)	51.30	820.80	81.85	1309.60		
1 Hyd. Crane, 25 Ton		586.70		645.37		
1 Crawler Loader, 3 C.Y.		1146.00		1260.60		
2 Dump Trucks, 12 C.Y., 400 H.P.		1158.70		1274.57	45.18	49.70
64 L.H., Daily Totals		$6157.80		$8340.54	$96.22	$130.32

Crew No.	Bare Costs		Incl. Subs O&P		Cost Per Labor-Hour	
Crew B-9	Hr.	Daily	Hr.	Daily	Bare Costs	Incl. O&P
1 Labor Foreman (outside)	$46.40	$371.20	$73.90	$591.20	$44.80	$71.34
4 Laborers	44.40	1420.80	70.70	2262.40		
1 Air Compressor, 250 cfm		202.85		223.13		
2 Breakers, Pavement, 60 lb.		107.20		117.92		
2 -50' Air Hoses, 1.5"		45.60		50.16	8.89	9.78
40 L.H., Daily Totals		$2147.65		$3244.82	$53.69	$81.12

Crew No.	Bare Costs		Incl. Subs O&P		Cost Per Labor-Hour	
Crew B-9A	Hr.	Daily	Hr.	Daily	Bare Costs	Incl. O&P
2 Laborers	$44.40	$710.40	$70.70	$1131.20	$46.70	$74.42
1 Truck Driver (heavy)	51.30	410.40	81.85	654.80		
1 Water Tank Trailer, 5000 Gal.		154.10		169.51		
1 Truck Tractor, 220 H.P.		310.80		341.88		
2 -50' Discharge Hoses, 3"		9.00		9.90	19.75	21.72
24 L.H., Daily Totals		$1594.70		$2307.29	$66.45	$96.14

Crew No.	Bare Costs		Incl. Subs O&P		Cost Per Labor-Hour	
Crew B-9B	Hr.	Daily	Hr.	Daily	Bare Costs	Incl. O&P
2 Laborers	$44.40	$710.40	$70.70	$1131.20	$46.70	$74.42
1 Truck Driver (heavy)	51.30	410.40	81.85	654.80		
2 -50' Discharge Hoses, 3"		9.00		9.90		
1 Water Tank Trailer, 5000 Gal.		154.10		169.51		
1 Truck Tractor, 220 H.P.		310.80		341.88		
1 Pressure Washer		97.35		107.08	23.80	26.18
24 L.H., Daily Totals		$1692.05		$2414.38	$70.50	$100.60

Crew No.	Bare Costs		Incl. Subs O&P		Cost Per Labor-Hour	
Crew B-9D	Hr.	Daily	Hr.	Daily	Bare Costs	Incl. O&P
1 Labor Foreman (outside)	$46.40	$371.20	$73.90	$591.20	$44.80	$71.34
4 Common Laborers	44.40	1420.80	70.70	2262.40		
1 Air Compressor, 250 cfm		202.85		223.13		
2 -50' Air Hoses, 1.5"		45.60		50.16		
2 Air Powered Tampers		79.50		87.45	8.20	9.02
40 L.H., Daily Totals		$2119.95		$3214.34	$53.00	$80.36

Crew No.	Bare Costs		Incl. Subs O&P		Cost Per Labor-Hour	
Crew B-9E	Hr.	Daily	Hr.	Daily	Bare Costs	Incl. O&P
1 Cement Finisher	$51.80	$414.40	$81.10	$648.80	$48.10	$75.90
1 Laborer	44.40	355.20	70.70	565.60		
1 Chip. Hammers, 12 Lb., Elec.		32.85		36.13	2.05	2.26
16 L.H., Daily Totals		$802.45		$1250.54	$50.15	$78.16

Crew No.	Bare Costs		Incl. Subs O&P		Cost Per Labor-Hour	
Crew B-10	Hr.	Daily	Hr.	Daily	Bare Costs	Incl. O&P
1 Equip. Oper. (medium)	$59.00	$472.00	$92.05	$736.40	$54.13	$84.93
.5 Laborer	44.40	177.60	70.70	282.80		
12 L.H., Daily Totals		$649.60		$1019.20	$54.13	$84.93

Crew No.	Bare Costs		Incl. Subs O&P		Cost Per Labor-Hour	
Crew B-10A	Hr.	Daily	Hr.	Daily	Bare Costs	Incl. O&P
1 Equip. Oper. (medium)	$59.00	$472.00	$92.05	$736.40	$54.13	$84.93
.5 Laborer	44.40	177.60	70.70	282.80		
1 Roller, 2-Drum, W.B., 7.5 H.P.		166.75		183.43	13.90	15.29
12 L.H., Daily Totals		$816.35		$1202.63	$68.03	$100.22

Crew No.	Bare Costs		Incl. Subs O&P		Cost Per Labor-Hour	
Crew B-10B	Hr.	Daily	Hr.	Daily	Bare Costs	Incl. O&P
1 Equip. Oper. (medium)	$59.00	$472.00	$92.05	$736.40	$54.13	$84.93
.5 Laborer	44.40	177.60	70.70	282.80		
1 Dozer, 200 H.P.		1520.00		1672.00	126.67	139.33
12 L.H., Daily Totals		$2169.60		$2691.20	$180.80	$224.27

Crew No.	Bare Costs		Incl. Subs O&P		Cost Per Labor-Hour	
Crew B-10C	Hr.	Daily	Hr.	Daily	Bare Costs	Incl. O&P
1 Equip. Oper. (medium)	$59.00	$472.00	$92.05	$736.40	$54.13	$84.93
.5 Laborer	44.40	177.60	70.70	282.80		
1 Dozer, 200 H.P.		1520.00		1672.00		
1 Vibratory Roller, Towed, 23 Ton		520.35		572.38	170.03	187.03
12 L.H., Daily Totals		$2689.95		$3263.59	$224.16	$271.97

Crew No.	Bare Costs		Incl. Subs O&P		Cost Per Labor-Hour	
Crew B-10D	Hr.	Daily	Hr.	Daily	Bare Costs	Incl. O&P
1 Equip. Oper. (medium)	$59.00	$472.00	$92.05	$736.40	$54.13	$84.93
.5 Laborer	44.40	177.60	70.70	282.80		
1 Dozer, 200 H.P.		1520.00		1672.00		
1 Sheepsft. Roller, Towed		426.95		469.64	162.25	178.47
12 L.H., Daily Totals		$2596.55		$3160.84	$216.38	$263.40

Crew No.	Bare Costs		Incl. Subs O&P		Cost Per Labor-Hour	
Crew B-10E	Hr.	Daily	Hr.	Daily	Bare Costs	Incl. O&P
1 Equip. Oper. (medium)	$59.00	$472.00	$92.05	$736.40	$54.13	$84.93
.5 Laborer	44.40	177.60	70.70	282.80		
1 Tandem Roller, 5 Ton		258.75		284.63	21.56	23.72
12 L.H., Daily Totals		$908.35		$1303.83	$75.70	$108.65

Crew No.	Bare Costs		Incl. Subs O&P		Cost Per Labor-Hour	
Crew B-10F	Hr.	Daily	Hr.	Daily	Bare Costs	Incl. O&P
1 Equip. Oper. (medium)	$59.00	$472.00	$92.05	$736.40	$54.13	$84.93
.5 Laborer	44.40	177.60	70.70	282.80		
1 Tandem Roller, 10 Ton		246.80		271.48	20.57	22.62
12 L.H., Daily Totals		$896.40		$1290.68	$74.70	$107.56

Crew No.	Bare Costs		Incl. Subs O&P		Cost Per Labor-Hour	
Crew B-10G	Hr.	Daily	Hr.	Daily	Bare Costs	Incl. O&P
1 Equip. Oper. (medium)	$59.00	$472.00	$92.05	$736.40	$54.13	$84.93
.5 Laborer	44.40	177.60	70.70	282.80		
1 Sheepsfoot Roller, 240 H.P.		1363.00		1499.30	113.58	124.94
12 L.H., Daily Totals		$2012.60		$2518.50	$167.72	$209.88

Crew No.	Bare Costs		Incl. Subs O&P		Cost Per Labor-Hour	
Crew B-10H	Hr.	Daily	Hr.	Daily	Bare Costs	Incl. O&P
1 Equip. Oper. (medium)	$59.00	$472.00	$92.05	$736.40	$54.13	$84.93
.5 Laborer	44.40	177.60	70.70	282.80		
1 Diaphragm Water Pump, 2"		87.70		96.47		
1 -20' Suction Hose, 2"		3.55		3.90		
2 -50' Discharge Hoses, 2"		8.00		8.80	8.27	9.10
12 L.H., Daily Totals		$748.85		$1128.38	$62.40	$94.03

For customer support on your Commercial Renovation Costs with RSMeans Data, call 800.448.8182.

759

Crew No.		Bare Costs	Incl. Subs O&P		Cost Per Labor-Hour	

Crew B-10I	Hr.	Daily	Hr.	Daily	Bare Costs	Incl. O&P
1 Equip. Oper. (medium)	$59.00	$472.00	$92.05	$736.40	$54.13	$84.93
.5 Laborer	44.40	177.60	70.70	282.80		
1 Diaphragm Water Pump, 4"		106.35		116.99		
1 -20' Suction Hose, 4"		17.25		18.98		
2 -50' Discharge Hoses, 4"		25.60		28.16	12.43	13.68
12 L.H., Daily Totals		$798.80		$1183.32	$66.57	$98.61

Crew B-10J	Hr.	Daily	Hr.	Daily	Bare Costs	Incl. O&P
1 Equip. Oper. (medium)	$59.00	$472.00	$92.05	$736.40	$54.13	$84.93
.5 Laborer	44.40	177.60	70.70	282.80		
1 Centrifugal Water Pump, 3"		74.40		81.84		
1 -20' Suction Hose, 3"		8.75		9.63		
2 -50' Discharge Hoses, 3"		9.00		9.90	7.68	8.45
12 L.H., Daily Totals		$741.75		$1120.57	$61.81	$93.38

Crew B-10K	Hr.	Daily	Hr.	Daily	Bare Costs	Incl. O&P
1 Equip. Oper. (medium)	$59.00	$472.00	$92.05	$736.40	$54.13	$84.93
.5 Laborer	44.40	177.60	70.70	282.80		
1 Centr. Water Pump, 6"		235.25		258.77		
1 -20' Suction Hose, 6"		25.50		28.05		
2 -50' Discharge Hoses, 6"		36.20		39.82	24.75	27.22
12 L.H., Daily Totals		$946.55		$1345.85	$78.88	$112.15

Crew B-10L	Hr.	Daily	Hr.	Daily	Bare Costs	Incl. O&P
1 Equip. Oper. (medium)	$59.00	$472.00	$92.05	$736.40	$54.13	$84.93
.5 Laborer	44.40	177.60	70.70	282.80		
1 Dozer, 80 H.P.		405.85		446.44	33.82	37.20
12 L.H., Daily Totals		$1055.45		$1465.64	$87.95	$122.14

Crew B-10M	Hr.	Daily	Hr.	Daily	Bare Costs	Incl. O&P
1 Equip. Oper. (medium)	$59.00	$472.00	$92.05	$736.40	$54.13	$84.93
.5 Laborer	44.40	177.60	70.70	282.80		
1 Dozer, 300 H.P.		1785.00		1963.50	148.75	163.63
12 L.H., Daily Totals		$2434.60		$2982.70	$202.88	$248.56

Crew B-10N	Hr.	Daily	Hr.	Daily	Bare Costs	Incl. O&P
1 Equip. Oper. (medium)	$59.00	$472.00	$92.05	$736.40	$54.13	$84.93
.5 Laborer	44.40	177.60	70.70	282.80		
1 F.E. Loader, T.M., 1.5 C.Y.		572.00		629.20	47.67	52.43
12 L.H., Daily Totals		$1221.60		$1648.40	$101.80	$137.37

Crew B-10O	Hr.	Daily	Hr.	Daily	Bare Costs	Incl. O&P
1 Equip. Oper. (medium)	$59.00	$472.00	$92.05	$736.40	$54.13	$84.93
.5 Laborer	44.40	177.60	70.70	282.80		
1 F.E. Loader, T.M., 2.25 C.Y.		925.50		1018.05	77.13	84.84
12 L.H., Daily Totals		$1575.10		$2037.25	$131.26	$169.77

Crew B-10P	Hr.	Daily	Hr.	Daily	Bare Costs	Incl. O&P
1 Equip. Oper. (medium)	$59.00	$472.00	$92.05	$736.40	$54.13	$84.93
.5 Laborer	44.40	177.60	70.70	282.80		
1 Crawler Loader, 3 C.Y.		1146.00		1260.60	95.50	105.05
12 L.H., Daily Totals		$1795.60		$2279.80	$149.63	$189.98

Crew B-10Q	Hr.	Daily	Hr.	Daily	Bare Costs	Incl. O&P
1 Equip. Oper. (medium)	$59.00	$472.00	$92.05	$736.40	$54.13	$84.93
.5 Laborer	44.40	177.60	70.70	282.80		
1 Crawler Loader, 4 C.Y.		1456.00		1601.60	121.33	133.47
12 L.H., Daily Totals		$2105.60		$2620.80	$175.47	$218.40

Crew B-10R	Hr.	Daily	Hr.	Daily	Bare Costs	Incl. O&P
1 Equip. Oper. (medium)	$59.00	$472.00	$92.05	$736.40	$54.13	$84.93
.5 Laborer	44.40	177.60	70.70	282.80		
1 F.E. Loader, W.M., 1 C.Y.		305.60		336.16	25.47	28.01
12 L.H., Daily Totals		$955.20		$1355.36	$79.60	$112.95

Crew B-10S	Hr.	Daily	Hr.	Daily	Bare Costs	Incl. O&P
1 Equip. Oper. (medium)	$59.00	$472.00	$92.05	$736.40	$54.13	$84.93
.5 Laborer	44.40	177.60	70.70	282.80		
1 F.E. Loader, W.M., 1.5 C.Y.		441.40		485.54	36.78	40.46
12 L.H., Daily Totals		$1091.00		$1504.74	$90.92	$125.40

Crew B-10T	Hr.	Daily	Hr.	Daily	Bare Costs	Incl. O&P
1 Equip. Oper. (medium)	$59.00	$472.00	$92.05	$736.40	$54.13	$84.93
.5 Laborer	44.40	177.60	70.70	282.80		
1 F.E. Loader, W.M., 2.5 C.Y.		638.30		702.13	53.19	58.51
12 L.H., Daily Totals		$1287.90		$1721.33	$107.33	$143.44

Crew B-10U	Hr.	Daily	Hr.	Daily	Bare Costs	Incl. O&P
1 Equip. Oper. (medium)	$59.00	$472.00	$92.05	$736.40	$54.13	$84.93
.5 Laborer	44.40	177.60	70.70	282.80		
1 F.E. Loader, W.M., 5.5 C.Y.		967.95		1064.74	80.66	88.73
12 L.H., Daily Totals		$1617.55		$2083.95	$134.80	$173.66

Crew B-10V	Hr.	Daily	Hr.	Daily	Bare Costs	Incl. O&P
1 Equip. Oper. (medium)	$59.00	$472.00	$92.05	$736.40	$54.13	$84.93
.5 Laborer	44.40	177.60	70.70	282.80		
1 Dozer, 700 H.P.		5175.00		5692.50	431.25	474.38
12 L.H., Daily Totals		$5824.60		$6711.70	$485.38	$559.31

Crew B-10W	Hr.	Daily	Hr.	Daily	Bare Costs	Incl. O&P
1 Equip. Oper. (medium)	$59.00	$472.00	$92.05	$736.40	$54.13	$84.93
.5 Laborer	44.40	177.60	70.70	282.80		
1 Dozer, 105 H.P.		640.80		704.88	53.40	58.74
12 L.H., Daily Totals		$1290.40		$1724.08	$107.53	$143.67

Crew B-10X	Hr.	Daily	Hr.	Daily	Bare Costs	Incl. O&P
1 Equip. Oper. (medium)	$59.00	$472.00	$92.05	$736.40	$54.13	$84.93
.5 Laborer	44.40	177.60	70.70	282.80		
1 Dozer, 410 H.P.		2807.00		3087.70	233.92	257.31
12 L.H., Daily Totals		$3456.60		$4106.90	$288.05	$342.24

Crew B-10Y	Hr.	Daily	Hr.	Daily	Bare Costs	Incl. O&P
1 Equip. Oper. (medium)	$59.00	$472.00	$92.05	$736.40	$54.13	$84.93
.5 Laborer	44.40	177.60	70.70	282.80		
1 Vibr. Roller, Towed, 12 Ton		584.80		643.28	48.73	53.61
12 L.H., Daily Totals		$1234.40		$1662.48	$102.87	$138.54

Crew B-11A	Hr.	Daily	Hr.	Daily	Bare Costs	Incl. O&P
1 Equipment Oper. (med.)	$59.00	$472.00	$92.05	$736.40	$51.70	$81.38
1 Laborer	44.40	355.20	70.70	565.60		
1 Dozer, 200 H.P.		1520.00		1672.00	95.00	104.50
16 L.H., Daily Totals		$2347.20		$2974.00	$146.70	$185.88

Crew B-11B	Hr.	Daily	Hr.	Daily	Bare Costs	Incl. O&P
1 Equipment Oper. (light)	$55.50	$444.00	$86.60	$692.80	$49.95	$78.65
1 Laborer	44.40	355.20	70.70	565.60		
1 Air Powered Tamper		39.75		43.73		
1 Air Compressor, 365 cfm		343.55		377.90		
2 -50' Air Hoses, 1.5"		45.60		50.16	26.81	29.49
16 L.H., Daily Totals		$1228.10		$1730.19	$76.76	$108.14

Crew No.	Bare Costs		Incl. Subs O&P		Cost Per Labor-Hour	

Crew B-11C

	Hr.	Daily	Hr.	Daily	Bare Costs	Incl. O&P
1 Equipment Oper. (med.)	$59.00	$472.00	$92.05	$736.40	$51.70	$81.38
1 Laborer	44.40	355.20	70.70	565.60		
1 Backhoe Loader, 48 H.P.		216.20		237.82	13.51	14.86
16 L.H., Daily Totals		$1043.40		$1539.82	$65.21	$96.24

Crew B-11K

	Hr.	Daily	Hr.	Daily	Bare Costs	Incl. O&P
1 Equipment Oper. (med.)	$59.00	$472.00	$92.05	$736.40	$51.70	$81.38
1 Laborer	44.40	355.20	70.70	565.60		
1 Trencher, Chain Type, 8' D		1894.00		2083.40	118.38	130.21
16 L.H., Daily Totals		$2721.20		$3385.40	$170.07	$211.59

Crew B-11L

	Hr.	Daily	Hr.	Daily	Bare Costs	Incl. O&P
1 Equipment Oper. (med.)	$59.00	$472.00	$92.05	$736.40	$51.70	$81.38
1 Laborer	44.40	355.20	70.70	565.60		
1 Grader, 30,000 Lbs.		1073.00		1180.30	67.06	73.77
16 L.H., Daily Totals		$1900.20		$2482.30	$118.76	$155.14

Crew B-11M

	Hr.	Daily	Hr.	Daily	Bare Costs	Incl. O&P
1 Equipment Oper. (med.)	$59.00	$472.00	$92.05	$736.40	$51.70	$81.38
1 Laborer	44.40	355.20	70.70	565.60		
1 Backhoe Loader, 80 H.P.		235.05		258.56	14.69	16.16
16 L.H., Daily Totals		$1062.25		$1560.56	$66.39	$97.53

Crew B-11S

	Hr.	Daily	Hr.	Daily	Bare Costs	Incl. O&P
1 Equipment Operator (med.)	$59.00	$472.00	$92.05	$736.40	$54.13	$84.93
.5 Laborer	44.40	177.60	70.70	282.80		
1 Dozer, 300 H.P.		1785.00		1963.50		
1 Ripper, Beam & 1 Shank		91.60		100.76	156.38	172.02
12 L.H., Daily Totals		$2526.20		$3083.46	$210.52	$256.95

Crew B-11T

	Hr.	Daily	Hr.	Daily	Bare Costs	Incl. O&P
1 Equipment Operator (med.)	$59.00	$472.00	$92.05	$736.40	$54.13	$84.93
.5 Laborer	44.40	177.60	70.70	282.80		
1 Dozer, 410 H.P.		2807.00		3087.70		
1 Ripper, Beam & 2 Shanks		140.40		154.44	245.62	270.18
12 L.H., Daily Totals		$3597.00		$4261.34	$299.75	$355.11

Crew B-11W

	Hr.	Daily	Hr.	Daily	Bare Costs	Incl. O&P
1 Equipment Operator (med.)	$59.00	$472.00	$92.05	$736.40	$51.37	$81.77
1 Common Laborer	44.40	355.20	70.70	565.60		
10 Truck Drivers (heavy)	51.30	4104.00	81.85	6548.00		
1 Dozer, 200 H.P.		1520.00		1672.00		
1 Vibratory Roller, Towed, 23 Ton		520.35		572.38		
10 Dump Trucks, 8 C.Y., 220 H.P.		4076.00		4483.60	63.71	70.08
96 L.H., Daily Totals		$11047.55		$14577.99	$115.08	$151.85

Crew B-11Y

	Hr.	Daily	Hr.	Daily	Bare Costs	Incl. O&P
1 Labor Foreman (outside)	$46.40	$371.20	$73.90	$591.20	$49.49	$78.17
5 Common Laborers	44.40	1776.00	70.70	2828.00		
3 Equipment Operators (med.)	59.00	1416.00	92.05	2209.20		
1 Dozer, 80 H.P.		405.85		446.44		
2 Rollers, 2-Drum, W.B., 7.5 H.P.		333.50		366.85		
4 Vibrating Plates, Gas, 21"		662.40		728.64	19.47	21.42
72 L.H., Daily Totals		$4964.95		$7170.32	$68.96	$99.59

Crew B-12A

	Hr.	Daily	Hr.	Daily	Bare Costs	Incl. O&P
1 Equip. Oper. (crane)	$61.45	$491.60	$95.85	$766.80	$52.92	$83.28
1 Laborer	44.40	355.20	70.70	565.60		
1 Hyd. Excavator, 1 C.Y.		832.65		915.91	52.04	57.24
16 L.H., Daily Totals		$1679.45		$2248.32	$104.97	$140.52

Crew B-12B

	Hr.	Daily	Hr.	Daily	Bare Costs	Incl. O&P
1 Equip. Oper. (crane)	$61.45	$491.60	$95.85	$766.80	$52.92	$83.28
1 Laborer	44.40	355.20	70.70	565.60		
1 Hyd. Excavator, 1.5 C.Y.		695.80		765.38	43.49	47.84
16 L.H., Daily Totals		$1542.60		$2097.78	$96.41	$131.11

Crew B-12C

	Hr.	Daily	Hr.	Daily	Bare Costs	Incl. O&P
1 Equip. Oper. (crane)	$61.45	$491.60	$95.85	$766.80	$52.92	$83.28
1 Laborer	44.40	355.20	70.70	565.60		
1 Hyd. Excavator, 2 C.Y.		942.70		1036.97	58.92	64.81
16 L.H., Daily Totals		$1789.50		$2369.37	$111.84	$148.09

Crew B-12D

	Hr.	Daily	Hr.	Daily	Bare Costs	Incl. O&P
1 Equip. Oper. (crane)	$61.45	$491.60	$95.85	$766.80	$52.92	$83.28
1 Laborer	44.40	355.20	70.70	565.60		
1 Hyd. Excavator, 3.5 C.Y.		2184.00		2402.40	136.50	150.15
16 L.H., Daily Totals		$3030.80		$3734.80	$189.43	$233.43

Crew B-12E

	Hr.	Daily	Hr.	Daily	Bare Costs	Incl. O&P
1 Equip. Oper. (crane)	$61.45	$491.60	$95.85	$766.80	$52.92	$83.28
1 Laborer	44.40	355.20	70.70	565.60		
1 Hyd. Excavator, .5 C.Y.		457.00		502.70	28.56	31.42
16 L.H., Daily Totals		$1303.80		$1835.10	$81.49	$114.69

Crew B-12F

	Hr.	Daily	Hr.	Daily	Bare Costs	Incl. O&P
1 Equip. Oper. (crane)	$61.45	$491.60	$95.85	$766.80	$52.92	$83.28
1 Laborer	44.40	355.20	70.70	565.60		
1 Hyd. Excavator, .75 C.Y.		701.80		771.98	43.86	48.25
16 L.H., Daily Totals		$1548.60		$2104.38	$96.79	$131.52

Crew B-12G

	Hr.	Daily	Hr.	Daily	Bare Costs	Incl. O&P
1 Equip. Oper. (crane)	$61.45	$491.60	$95.85	$766.80	$52.92	$83.28
1 Laborer	44.40	355.20	70.70	565.60		
1 Crawler Crane, 15 Ton		810.30		891.33		
1 Clamshell Bucket, .5 C.Y.		67.80		74.58	54.88	60.37
16 L.H., Daily Totals		$1724.90		$2298.31	$107.81	$143.64

Crew B-12H

	Hr.	Daily	Hr.	Daily	Bare Costs	Incl. O&P
1 Equip. Oper. (crane)	$61.45	$491.60	$95.85	$766.80	$52.92	$83.28
1 Laborer	44.40	355.20	70.70	565.60		
1 Crawler Crane, 25 Ton		1152.00		1267.20		
1 Clamshell Bucket, 1 C.Y.		69.25		76.17	76.33	83.96
16 L.H., Daily Totals		$2068.05		$2675.78	$129.25	$167.24

Crew B-12I

	Hr.	Daily	Hr.	Daily	Bare Costs	Incl. O&P
1 Equip. Oper. (crane)	$61.45	$491.60	$95.85	$766.80	$52.92	$83.28
1 Laborer	44.40	355.20	70.70	565.60		
1 Crawler Crane, 20 Ton		1013.00		1114.30		
1 Dragline Bucket, .75 C.Y.		61.85		68.03	67.18	73.90
16 L.H., Daily Totals		$1921.65		$2514.74	$120.10	$157.17

Crew B-12J

	Hr.	Daily	Hr.	Daily	Bare Costs	Incl. O&P
1 Equip. Oper. (crane)	$61.45	$491.60	$95.85	$766.80	$52.92	$83.28
1 Laborer	44.40	355.20	70.70	565.60		
1 Gradall, 5/8 C.Y.		850.65		935.72	53.17	58.48
16 L.H., Daily Totals		$1697.45		$2268.11	$106.09	$141.76

For customer support on your Commercial Renovation Costs with RSMeans Data, call 800.448.8182.

761

Crew No.	Bare Costs		Incl. Subs O&P		Cost Per Labor-Hour	
Crew B-12K	Hr.	Daily	Hr.	Daily	Bare Costs	Incl. O&P
1 Equip. Oper. (crane)	$61.45	$491.60	$95.85	$766.80	$52.92	$83.28
1 Laborer	44.40	355.20	70.70	565.60		
1 Gradall, 3 Ton, 1 C.Y.		984.55		1083.01	61.53	67.69
16 L.H., Daily Totals		$1831.35		$2415.41	$114.46	$150.96
Crew B-12L	Hr.	Daily	Hr.	Daily	Bare Costs	Incl. O&P
1 Equip. Oper. (crane)	$61.45	$491.60	$95.85	$766.80	$52.92	$83.28
1 Laborer	44.40	355.20	70.70	565.60		
1 Crawler Crane, 15 Ton		810.30		891.33		
1 F.E. Attachment, .5 C.Y.		66.05		72.66	54.77	60.25
16 L.H., Daily Totals		$1723.15		$2296.39	$107.70	$143.52
Crew B-12M	Hr.	Daily	Hr.	Daily	Bare Costs	Incl. O&P
1 Equip. Oper. (crane)	$61.45	$491.60	$95.85	$766.80	$52.92	$83.28
1 Laborer	44.40	355.20	70.70	565.60		
1 Crawler Crane, 20 Ton		1013.00		1114.30		
1 F.E. Attachment, .75 C.Y.		71.25		78.38	67.77	74.54
16 L.H., Daily Totals		$1931.05		$2525.07	$120.69	$157.82
Crew B-12N	Hr.	Daily	Hr.	Daily	Bare Costs	Incl. O&P
1 Equip. Oper. (crane)	$61.45	$491.60	$95.85	$766.80	$52.92	$83.28
1 Laborer	44.40	355.20	70.70	565.60		
1 Crawler Crane, 25 Ton		1152.00		1267.20		
1 F.E. Attachment, 1 C.Y.		77.35		85.08	76.83	84.52
16 L.H., Daily Totals		$2076.15		$2684.68	$129.76	$167.79
Crew B-12O	Hr.	Daily	Hr.	Daily	Bare Costs	Incl. O&P
1 Equip. Oper. (crane)	$61.45	$491.60	$95.85	$766.80	$52.92	$83.28
1 Laborer	44.40	355.20	70.70	565.60		
1 Crawler Crane, 40 Ton		1231.00		1354.10		
1 F.E. Attachment, 1.5 C.Y.		88.65		97.52	82.48	90.73
16 L.H., Daily Totals		$2166.45		$2784.01	$135.40	$174.00
Crew B-12P	Hr.	Daily	Hr.	Daily	Bare Costs	Incl. O&P
1 Equip. Oper. (crane)	$61.45	$491.60	$95.85	$766.80	$52.92	$83.28
1 Laborer	44.40	355.20	70.70	565.60		
1 Crawler Crane, 40 Ton		1231.00		1354.10		
1 Dragline Bucket, 1.5 C.Y.		65.70		72.27	81.04	89.15
16 L.H., Daily Totals		$2143.50		$2758.77	$133.97	$172.42
Crew B-12Q	Hr.	Daily	Hr.	Daily	Bare Costs	Incl. O&P
1 Equip. Oper. (crane)	$61.45	$491.60	$95.85	$766.80	$52.92	$83.28
1 Laborer	44.40	355.20	70.70	565.60		
1 Hyd. Excavator, 5/8 C.Y.		604.75		665.23	37.80	41.58
16 L.H., Daily Totals		$1451.55		$1997.63	$90.72	$124.85
Crew B-12S	Hr.	Daily	Hr.	Daily	Bare Costs	Incl. O&P
1 Equip. Oper. (crane)	$61.45	$491.60	$95.85	$766.80	$52.92	$83.28
1 Laborer	44.40	355.20	70.70	565.60		
1 Hyd. Excavator, 2.5 C.Y.		1567.00		1723.70	97.94	107.73
16 L.H., Daily Totals		$2413.80		$3056.10	$150.86	$191.01
Crew B-12T	Hr.	Daily	Hr.	Daily	Bare Costs	Incl. O&P
1 Equip. Oper. (crane)	$61.45	$491.60	$95.85	$766.80	$52.92	$83.28
1 Laborer	44.40	355.20	70.70	565.60		
1 Crawler Crane, 75 Ton		1967.00		2163.70		
1 F.E. Attachment, 3 C.Y.		115.55		127.11	130.16	143.18
16 L.H., Daily Totals		$2929.35		$3623.20	$183.08	$226.45

Crew No.	Bare Costs		Incl. Subs O&P		Cost Per Labor-Hour	
Crew B-12V	Hr.	Daily	Hr.	Daily	Bare Costs	Incl. O&P
1 Equip. Oper. (crane)	$61.45	$491.60	$95.85	$766.80	$52.92	$83.28
1 Laborer	44.40	355.20	70.70	565.60		
1 Crawler Crane, 75 Ton		1967.00		2163.70		
1 Dragline Bucket, 3 C.Y.		72.30		79.53	127.46	140.20
16 L.H., Daily Totals		$2886.10		$3575.63	$180.38	$223.48
Crew B-12Y	Hr.	Daily	Hr.	Daily	Bare Costs	Incl. O&P
1 Equip. Oper. (crane)	$61.45	$491.60	$95.85	$766.80	$50.08	$79.08
2 Laborers	44.40	710.40	70.70	1131.20		
1 Hyd. Excavator, 3.5 C.Y.		2184.00		2402.40	91.00	100.10
24 L.H., Daily Totals		$3386.00		$4300.40	$141.08	$179.18
Crew B-12Z	Hr.	Daily	Hr.	Daily	Bare Costs	Incl. O&P
1 Equip. Oper. (crane)	$61.45	$491.60	$95.85	$766.80	$50.08	$79.08
2 Laborers	44.40	710.40	70.70	1131.20		
1 Hyd. Excavator, 2.5 C.Y.		1567.00		1723.70	65.29	71.82
24 L.H., Daily Totals		$2769.00		$3621.70	$115.38	$150.90
Crew B-13	Hr.	Daily	Hr.	Daily	Bare Costs	Incl. O&P
1 Labor Foreman (outside)	$46.40	$371.20	$73.90	$591.20	$48.28	$76.35
4 Laborers	44.40	1420.80	70.70	2262.40		
1 Equip. Oper. (crane)	61.45	491.60	95.85	766.80		
1 Equip. Oper. (oiler)	52.50	420.00	81.90	655.20		
1 Hyd. Crane, 25 Ton		586.70		645.37	10.48	11.52
56 L.H., Daily Totals		$3290.30		$4920.97	$58.76	$87.87
Crew B-13A	Hr.	Daily	Hr.	Daily	Bare Costs	Incl. O&P
1 Labor Foreman (outside)	$46.40	$371.20	$73.90	$591.20	$50.83	$80.44
2 Laborers	44.40	710.40	70.70	1131.20		
2 Equipment Operators (med.)	59.00	944.00	92.05	1472.80		
2 Truck Drivers (heavy)	51.30	820.80	81.85	1309.60		
1 Crawler Crane, 75 Ton		1967.00		2163.70		
1 Crawler Loader, 4 C.Y.		1456.00		1601.60		
2 Dump Trucks, 8 C.Y., 220 H.P.		815.20		896.72	75.68	83.25
56 L.H., Daily Totals		$7084.60		$9166.82	$126.51	$163.69
Crew B-13B	Hr.	Daily	Hr.	Daily	Bare Costs	Incl. O&P
1 Labor Foreman (outside)	$46.40	$371.20	$73.90	$591.20	$48.28	$76.35
4 Laborers	44.40	1420.80	70.70	2262.40		
1 Equip. Oper. (crane)	61.45	491.60	95.85	766.80		
1 Equip. Oper. (oiler)	52.50	420.00	81.90	655.20		
1 Hyd. Crane, 55 Ton		990.15		1089.17	17.68	19.45
56 L.H., Daily Totals		$3693.75		$5364.77	$65.96	$95.80
Crew B-13C	Hr.	Daily	Hr.	Daily	Bare Costs	Incl. O&P
1 Labor Foreman (outside)	$46.40	$371.20	$73.90	$591.20	$48.28	$76.35
4 Laborers	44.40	1420.80	70.70	2262.40		
1 Equip. Oper. (crane)	61.45	491.60	95.85	766.80		
1 Equip. Oper. (oiler)	52.50	420.00	81.90	655.20		
1 Crawler Crane, 100 Ton		2310.00		2541.00	41.25	45.38
56 L.H., Daily Totals		$5013.60		$6816.60	$89.53	$121.72
Crew B-13D	Hr.	Daily	Hr.	Daily	Bare Costs	Incl. O&P
1 Laborer	$44.40	$355.20	$70.70	$565.60	$52.92	$83.28
1 Equip. Oper. (crane)	61.45	491.60	95.85	766.80		
1 Hyd. Excavator, 1 C.Y.		832.65		915.91		
1 Trench Box		119.15		131.07	59.49	65.44
16 L.H., Daily Totals		$1798.60		$2379.38	$112.41	$148.71

For customer support on your Commercial Renovation Costs with RSMeans Data, call 800.448.8182.

Crew No.	Bare Costs Hr.	Daily	Incl. Subs O&P Hr.	Daily	Cost Per Labor-Hour Bare Costs	Incl. O&P
Crew B-13E	Hr.	Daily	Hr.	Daily	Bare Costs	Incl. O&P
1 Laborer	$44.40	$355.20	$70.70	$565.60	$52.92	$83.28
1 Equip. Oper. (crane)	61.45	491.60	95.85	766.80		
1 Hyd. Excavator, 1.5 C.Y.		695.80		765.38		
1 Trench Box		119.15		131.07	50.93	56.03
16 L.H., Daily Totals		$1661.75		$2228.84	$103.86	$139.30

Crew No.	Bare Costs Hr.	Daily	Incl. Subs O&P Hr.	Daily	Cost Per Labor-Hour Bare Costs	Incl. O&P
Crew B-13F	Hr.	Daily	Hr.	Daily	Bare Costs	Incl. O&P
1 Laborer	$44.40	$355.20	$70.70	$565.60	$52.92	$83.28
1 Equip. Oper. (crane)	61.45	491.60	95.85	766.80		
1 Hyd. Excavator, 3.5 C.Y.		2184.00		2402.40		
1 Trench Box		119.15		131.07	143.95	158.34
16 L.H., Daily Totals		$3149.95		$3865.86	$196.87	$241.62

Crew No.	Bare Costs Hr.	Daily	Incl. Subs O&P Hr.	Daily	Cost Per Labor-Hour Bare Costs	Incl. O&P
Crew B-13G	Hr.	Daily	Hr.	Daily	Bare Costs	Incl. O&P
1 Laborer	$44.40	$355.20	$70.70	$565.60	$52.92	$83.28
1 Equip. Oper. (crane)	61.45	491.60	95.85	766.80		
1 Hyd. Excavator, .75 C.Y.		701.80		771.98		
1 Trench Box		119.15		131.07	51.31	56.44
16 L.H., Daily Totals		$1667.75		$2235.45	$104.23	$139.72

Crew No.	Bare Costs Hr.	Daily	Incl. Subs O&P Hr.	Daily	Cost Per Labor-Hour Bare Costs	Incl. O&P
Crew B-13H	Hr.	Daily	Hr.	Daily	Bare Costs	Incl. O&P
1 Laborer	$44.40	$355.20	$70.70	$565.60	$52.92	$83.28
1 Equip. Oper. (crane)	61.45	491.60	95.85	766.80		
1 Gradall, 5/8 C.Y.		850.65		935.72		
1 Trench Box		119.15		131.07	60.61	66.67
16 L.H., Daily Totals		$1816.60		$2399.18	$113.54	$149.95

Crew No.	Bare Costs Hr.	Daily	Incl. Subs O&P Hr.	Daily	Cost Per Labor-Hour Bare Costs	Incl. O&P
Crew B-13I	Hr.	Daily	Hr.	Daily	Bare Costs	Incl. O&P
1 Laborer	$44.40	$355.20	$70.70	$565.60	$52.92	$83.28
1 Equip. Oper. (crane)	61.45	491.60	95.85	766.80		
1 Gradall, 3 Ton, 1 C.Y.		984.55		1083.01		
1 Trench Box		119.15		131.07	68.98	75.88
16 L.H., Daily Totals		$1950.50		$2546.47	$121.91	$159.15

Crew No.	Bare Costs Hr.	Daily	Incl. Subs O&P Hr.	Daily	Cost Per Labor-Hour Bare Costs	Incl. O&P
Crew B-13J	Hr.	Daily	Hr.	Daily	Bare Costs	Incl. O&P
1 Laborer	$44.40	$355.20	$70.70	$565.60	$52.92	$83.28
1 Equip. Oper. (crane)	61.45	491.60	95.85	766.80		
1 Hyd. Excavator, 2.5 C.Y.		1567.00		1723.70		
1 Trench Box		119.15		131.07	105.38	115.92
16 L.H., Daily Totals		$2532.95		$3187.17	$158.31	$199.20

Crew No.	Bare Costs Hr.	Daily	Incl. Subs O&P Hr.	Daily	Cost Per Labor-Hour Bare Costs	Incl. O&P
Crew B-13K	Hr.	Daily	Hr.	Daily	Bare Costs	Incl. O&P
2 Equip. Opers. (crane)	$61.45	$983.20	$95.85	$1533.60	$61.45	$95.85
1 Hyd. Excavator, .75 C.Y.		701.80		771.98		
1 Hyd. Hammer, 4000 ft-lb		649.20		714.12		
1 Hyd. Excavator, .75 C.Y.		701.80		771.98	128.30	141.13
16 L.H., Daily Totals		$3036.00		$3791.68	$189.75	$236.98

Crew No.	Bare Costs Hr.	Daily	Incl. Subs O&P Hr.	Daily	Cost Per Labor-Hour Bare Costs	Incl. O&P
Crew B-13L	Hr.	Daily	Hr.	Daily	Bare Costs	Incl. O&P
2 Equip. Opers. (crane)	$61.45	$983.20	$95.85	$1533.60	$61.45	$95.85
1 Hyd. Excavator, 1.5 C.Y.		695.80		765.38		
1 Hyd. Hammer, 5000 ft-lb		705.60		776.16		
1 Hyd. Excavator, .75 C.Y.		701.80		771.98	131.45	144.60
16 L.H., Daily Totals		$3086.40		$3847.12	$192.90	$240.44

Crew No.	Bare Costs Hr.	Daily	Incl. Subs O&P Hr.	Daily	Cost Per Labor-Hour Bare Costs	Incl. O&P
Crew B-13M	Hr.	Daily	Hr.	Daily	Bare Costs	Incl. O&P
2 Equip. Opers. (crane)	$61.45	$983.20	$95.85	$1533.60	$61.45	$95.85
1 Hyd. Excavator, 2.5 C.Y.		1567.00		1723.70		
1 Hyd. Hammer, 8000 ft-lb		918.65		1010.52		
1 Hyd. Excavator, 1.5 C.Y.		695.80		765.38	198.84	218.72
16 L.H., Daily Totals		$4164.65		$5033.19	$260.29	$314.57

Crew No.	Bare Costs Hr.	Daily	Incl. Subs O&P Hr.	Daily	Cost Per Labor-Hour Bare Costs	Incl. O&P
Crew B-13N	Hr.	Daily	Hr.	Daily	Bare Costs	Incl. O&P
2 Equip. Opers. (crane)	$61.45	$983.20	$95.85	$1533.60	$61.45	$95.85
1 Hyd. Excavator, 3.5 C.Y.		2184.00		2402.40		
1 Hyd. Hammer, 12,000 ft-lb		882.20		970.42		
1 Hyd. Excavator, 1.5 C.Y.		695.80		765.38	235.13	258.64
16 L.H., Daily Totals		$4745.20		$5671.80	$296.57	$354.49

Crew No.	Bare Costs Hr.	Daily	Incl. Subs O&P Hr.	Daily	Cost Per Labor-Hour Bare Costs	Incl. O&P
Crew B-14	Hr.	Daily	Hr.	Daily	Bare Costs	Incl. O&P
1 Labor Foreman (outside)	$46.40	$371.20	$73.90	$591.20	$46.58	$73.88
4 Laborers	44.40	1420.80	70.70	2262.40		
1 Equip. Oper. (light)	55.50	444.00	86.60	692.80		
1 Backhoe Loader, 48 H.P.		216.20		237.82	4.50	4.95
48 L.H., Daily Totals		$2452.20		$3784.22	$51.09	$78.84

Crew No.	Bare Costs Hr.	Daily	Incl. Subs O&P Hr.	Daily	Cost Per Labor-Hour Bare Costs	Incl. O&P
Crew B-14A	Hr.	Daily	Hr.	Daily	Bare Costs	Incl. O&P
1 Equip. Oper. (crane)	$61.45	$491.60	$95.85	$766.80	$55.77	$87.47
.5 Laborer	44.40	177.60	70.70	282.80		
1 Hyd. Excavator, 4.5 C.Y.		3450.00		3795.00	287.50	316.25
12 L.H., Daily Totals		$4119.20		$4844.60	$343.27	$403.72

Crew No.	Bare Costs Hr.	Daily	Incl. Subs O&P Hr.	Daily	Cost Per Labor-Hour Bare Costs	Incl. O&P
Crew B-14B	Hr.	Daily	Hr.	Daily	Bare Costs	Incl. O&P
1 Equip. Oper. (crane)	$61.45	$491.60	$95.85	$766.80	$55.77	$87.47
.5 Laborer	44.40	177.60	70.70	282.80		
1 Hyd. Excavator, 6 C.Y.		3506.00		3856.60	292.17	321.38
12 L.H., Daily Totals		$4175.20		$4906.20	$347.93	$408.85

Crew No.	Bare Costs Hr.	Daily	Incl. Subs O&P Hr.	Daily	Cost Per Labor-Hour Bare Costs	Incl. O&P
Crew B-14C	Hr.	Daily	Hr.	Daily	Bare Costs	Incl. O&P
1 Equip. Oper. (crane)	$61.45	$491.60	$95.85	$766.80	$55.77	$87.47
.5 Laborer	44.40	177.60	70.70	282.80		
1 Hyd. Excavator, 7 C.Y.		3475.00		3822.50	289.58	318.54
12 L.H., Daily Totals		$4144.20		$4872.10	$345.35	$406.01

Crew No.	Bare Costs Hr.	Daily	Incl. Subs O&P Hr.	Daily	Cost Per Labor-Hour Bare Costs	Incl. O&P
Crew B-14F	Hr.	Daily	Hr.	Daily	Bare Costs	Incl. O&P
1 Equip. Oper. (crane)	$61.45	$491.60	$95.85	$766.80	$55.77	$87.47
.5 Laborer	44.40	177.60	70.70	282.80		
1 Hyd. Shovel, 7 C.Y.		4148.00		4562.80	345.67	380.23
12 L.H., Daily Totals		$4817.20		$5612.40	$401.43	$467.70

Crew No.	Bare Costs Hr.	Daily	Incl. Subs O&P Hr.	Daily	Cost Per Labor-Hour Bare Costs	Incl. O&P
Crew B-14G	Hr.	Daily	Hr.	Daily	Bare Costs	Incl. O&P
1 Equip. Oper. (crane)	$61.45	$491.60	$95.85	$766.80	$55.77	$87.47
.5 Laborer	44.40	177.60	70.70	282.80		
1 Hyd. Shovel, 12 C.Y.		6022.00		6624.20	501.83	552.02
12 L.H., Daily Totals		$6691.20		$7673.80	$557.60	$639.48

Crew No.	Bare Costs Hr.	Daily	Incl. Subs O&P Hr.	Daily	Cost Per Labor-Hour Bare Costs	Incl. O&P
Crew B-14J	Hr.	Daily	Hr.	Daily	Bare Costs	Incl. O&P
1 Equip. Oper. (medium)	$59.00	$472.00	$92.05	$736.40	$54.13	$84.93
.5 Laborer	44.40	177.60	70.70	282.80		
1 F.E. Loader, 8 C.Y.		2285.00		2513.50	190.42	209.46
12 L.H., Daily Totals		$2934.60		$3532.70	$244.55	$294.39

Crew No.	Bare Costs		Incl. Subs O&P		Cost Per Labor-Hour	

Crew B-14K	Hr.	Daily	Hr.	Daily	Bare Costs	Incl. O&P
1 Equip. Oper. (medium)	$59.00	$472.00	$92.05	$736.40	$54.13	$84.93
.5 Laborer	44.40	177.60	70.70	282.80		
1 F.E. Loader, 10 C.Y.		2706.00		2976.60	225.50	248.05
12 L.H., Daily Totals		$3355.60		$3995.80	$279.63	$332.98

Crew B-15	Hr.	Daily	Hr.	Daily	Bare Costs	Incl. O&P
1 Equipment Oper. (med.)	$59.00	$472.00	$92.05	$736.40	$52.51	$83.17
.5 Laborer	44.40	177.60	70.70	282.80		
2 Truck Drivers (heavy)	51.30	820.80	81.85	1309.60		
2 Dump Trucks, 12 C.Y., 400 H.P.		1158.70		1274.57		
1 Dozer, 200 H.P.		1520.00		1672.00	95.67	105.23
28 L.H., Daily Totals		$4149.10		$5275.37	$148.18	$188.41

Crew B-16	Hr.	Daily	Hr.	Daily	Bare Costs	Incl. O&P
1 Labor Foreman (outside)	$46.40	$371.20	$73.90	$591.20	$46.63	$74.29
2 Laborers	44.40	710.40	70.70	1131.20		
1 Truck Driver (heavy)	51.30	410.40	81.85	654.80		
1 Dump Truck, 12 C.Y., 400 H.P.		579.35		637.28	18.10	19.92
32 L.H., Daily Totals		$2071.35		$3014.49	$64.73	$94.20

Crew B-17	Hr.	Daily	Hr.	Daily	Bare Costs	Incl. O&P
2 Laborers	$44.40	$710.40	$70.70	$1131.20	$48.90	$77.46
1 Equip. Oper. (light)	55.50	444.00	86.60	692.80		
1 Truck Driver (heavy)	51.30	410.40	81.85	654.80		
1 Backhoe Loader, 48 H.P.		216.20		237.82		
1 Dump Truck, 8 C.Y., 220 H.P.		407.60		448.36	19.49	21.44
32 L.H., Daily Totals		$2188.60		$3164.98	$68.39	$98.91

Crew B-17A	Hr.	Daily	Hr.	Daily	Bare Costs	Incl. O&P
2 Labor Foremen (outside)	$46.40	$742.40	$73.90	$1182.40	$47.54	$75.61
6 Laborers	44.40	2131.20	70.70	3393.60		
1 Skilled Worker Foreman (out)	59.10	472.80	93.65	749.20		
1 Skilled Worker	57.10	456.80	90.50	724.00		
80 L.H., Daily Totals		$3803.20		$6049.20	$47.54	$75.61

Crew B-17B	Hr.	Daily	Hr.	Daily	Bare Costs	Incl. O&P
2 Laborers	$44.40	$710.40	$70.70	$1131.20	$48.90	$77.46
1 Equip. Oper. (light)	55.50	444.00	86.60	692.80		
1 Truck Driver (heavy)	51.30	410.40	81.85	654.80		
1 Backhoe Loader, 48 H.P.		216.20		237.82		
1 Dump Truck, 12 C.Y., 400 H.P.		579.35		637.28	24.86	27.35
32 L.H., Daily Totals		$2360.35		$3353.91	$73.76	$104.81

Crew B-18	Hr.	Daily	Hr.	Daily	Bare Costs	Incl. O&P
1 Labor Foreman (outside)	$46.40	$371.20	$73.90	$591.20	$45.07	$71.77
2 Laborers	44.40	710.40	70.70	1131.20		
1 Vibrating Plate, Gas, 21"		165.60		182.16	6.90	7.59
24 L.H., Daily Totals		$1247.20		$1904.56	$51.97	$79.36

Crew B-19	Hr.	Daily	Hr.	Daily	Bare Costs	Incl. O&P
1 Pile Driver Foreman (outside)	$57.90	$463.20	$92.40	$739.20	$57.11	$90.35
4 Pile Drivers	55.90	1788.80	89.20	2854.40		
2 Equip. Oper. (crane)	61.45	983.20	95.85	1533.60		
1 Equip. Oper. (oiler)	52.50	420.00	81.90	655.20		
1 Crawler Crane, 40 Ton		1231.00		1354.10		
1 Lead, 90' High		371.85		409.04		
1 Hammer, Diesel, 22k ft-lb		441.70		485.87	31.95	35.14
64 L.H., Daily Totals		$5699.75		$8031.40	$89.06	$125.49

Crew B-19A	Hr.	Daily	Hr.	Daily	Bare Costs	Incl. O&P
1 Pile Driver Foreman (outside)	$57.90	$463.20	$92.40	$739.20	$57.11	$90.35
4 Pile Drivers	55.90	1788.80	89.20	2854.40		
2 Equip. Oper. (crane)	61.45	983.20	95.85	1533.60		
1 Equip. Oper. (oiler)	52.50	420.00	81.90	655.20		
1 Crawler Crane, 75 Ton		1967.00		2163.70		
1 Lead, 90' High		371.85		409.04		
1 Hammer, Diesel, 41k ft-lb		583.55		641.90	45.66	50.23
64 L.H., Daily Totals		$6577.60		$8997.04	$102.78	$140.58

Crew B-19B	Hr.	Daily	Hr.	Daily	Bare Costs	Incl. O&P
1 Pile Driver Foreman (outside)	$57.90	$463.20	$92.40	$739.20	$57.11	$90.35
4 Pile Drivers	55.90	1788.80	89.20	2854.40		
2 Equip. Oper. (crane)	61.45	983.20	95.85	1533.60		
1 Equip. Oper. (oiler)	52.50	420.00	81.90	655.20		
1 Crawler Crane, 40 Ton		1231.00		1354.10		
1 Lead, 90' High		371.85		409.04		
1 Hammer, Diesel, 22k ft-lb		441.70		485.87		
1 Barge, 400 Ton		869.15		956.07	45.53	50.08
64 L.H., Daily Totals		$6568.90		$8987.47	$102.64	$140.43

Crew B-19C	Hr.	Daily	Hr.	Daily	Bare Costs	Incl. O&P
1 Pile Driver Foreman (outside)	$57.90	$463.20	$92.40	$739.20	$57.11	$90.35
4 Pile Drivers	55.90	1788.80	89.20	2854.40		
2 Equip. Oper. (crane)	61.45	983.20	95.85	1533.60		
1 Equip. Oper. (oiler)	52.50	420.00	81.90	655.20		
1 Crawler Crane, 75 Ton		1967.00		2163.70		
1 Lead, 90' High		371.85		409.04		
1 Hammer, Diesel, 41k ft-lb		583.55		641.90		
1 Barge, 400 Ton		869.15		956.07	59.24	65.17
64 L.H., Daily Totals		$7446.75		$9953.10	$116.36	$155.52

Crew B-20	Hr.	Daily	Hr.	Daily	Bare Costs	Incl. O&P
1 Labor Foreman (outside)	$46.40	$371.20	$73.90	$591.20	$49.30	$78.37
1 Skilled Worker	57.10	456.80	90.50	724.00		
1 Laborer	44.40	355.20	70.70	565.60		
24 L.H., Daily Totals		$1183.20		$1880.80	$49.30	$78.37

Crew B-20A	Hr.	Daily	Hr.	Daily	Bare Costs	Incl. O&P
1 Labor Foreman (outside)	$46.40	$371.20	$73.90	$591.20	$53.16	$83.13
1 Laborer	44.40	355.20	70.70	565.60		
1 Plumber	67.70	541.60	104.40	835.20		
1 Plumber Apprentice	54.15	433.20	83.50	668.00		
32 L.H., Daily Totals		$1701.20		$2660.00	$53.16	$83.13

Crew B-21	Hr.	Daily	Hr.	Daily	Bare Costs	Incl. O&P
1 Labor Foreman (outside)	$46.40	$371.20	$73.90	$591.20	$51.04	$80.86
1 Skilled Worker	57.10	456.80	90.50	724.00		
1 Laborer	44.40	355.20	70.70	565.60		
.5 Equip. Oper. (crane)	61.45	245.80	95.85	383.40		
.5 S.P. Crane, 4x4, 5 Ton		190.97		210.07	6.82	7.50
28 L.H., Daily Totals		$1619.97		$2474.27	$57.86	$88.37

Crew B-21A	Hr.	Daily	Hr.	Daily	Bare Costs	Incl. O&P
1 Labor Foreman (outside)	$46.40	$371.20	$73.90	$591.20	$54.82	$85.67
1 Laborer	44.40	355.20	70.70	565.60		
1 Plumber	67.70	541.60	104.40	835.20		
1 Plumber Apprentice	54.15	433.20	83.50	668.00		
1 Equip. Oper. (crane)	61.45	491.60	95.85	766.80		
1 S.P. Crane, 4x4, 12 Ton		432.65		475.92	10.82	11.90
40 L.H., Daily Totals		$2625.45		$3902.72	$65.64	$97.57

For customer support on your Commercial Renovation Costs with RSMeans Data, call 800.448.8182.

Crew No.	Bare Costs		Incl. Subs O&P		Cost Per Labor-Hour	

Crew B-21B	Hr.	Daily	Hr.	Daily	Bare Costs	Incl. O&P
1 Labor Foreman (outside)	$46.40	$371.20	$73.90	$591.20	$48.21	$76.37
3 Laborers	44.40	1065.60	70.70	1696.80		
1 Equip. Oper. (crane)	61.45	491.60	95.85	766.80		
1 Hyd. Crane, 12 Ton		475.80		523.38	11.90	13.08
40 L.H., Daily Totals		$2404.20		$3578.18	$60.10	$89.45

Crew B-21C	Hr.	Daily	Hr.	Daily	Bare Costs	Incl. O&P
1 Labor Foreman (outside)	$46.40	$371.20	$73.90	$591.20	$48.28	$76.35
4 Laborers	44.40	1420.80	70.70	2262.40		
1 Equip. Oper. (crane)	61.45	491.60	95.85	766.80		
1 Equip. Oper. (oiler)	52.50	420.00	81.90	655.20		
2 Cutting Torches		25.90		28.49		
2 Sets of Gases		347.20		381.92		
1 Lattice Boom Crane, 90 Ton		1713.00		1884.30	37.25	40.98
56 L.H., Daily Totals		$4789.70		$6570.31	$85.53	$117.33

Crew B-22	Hr.	Daily	Hr.	Daily	Bare Costs	Incl. O&P
1 Labor Foreman (outside)	$46.40	$371.20	$73.90	$591.20	$51.73	$81.86
1 Skilled Worker	57.10	456.80	90.50	724.00		
1 Laborer	44.40	355.20	70.70	565.60		
.75 Equip. Oper. (crane)	61.45	368.70	95.85	575.10		
.75 S.P. Crane, 4x4, 5 Ton		286.46		315.11	9.55	10.50
30 L.H., Daily Totals		$1838.36		$2771.01	$61.28	$92.37

Crew B-22A	Hr.	Daily	Hr.	Daily	Bare Costs	Incl. O&P
1 Labor Foreman (outside)	$46.40	$371.20	$73.90	$591.20	$50.75	$80.33
1 Skilled Worker	57.10	456.80	90.50	724.00		
2 Laborers	44.40	710.40	70.70	1131.20		
1 Equipment Operator, Crane	61.45	491.60	95.85	766.80		
1 S.P. Crane, 4x4, 5 Ton		381.95		420.14		
1 Butt Fusion Mach., 4"-12" diam.		420.75		462.82	20.07	22.07
40 L.H., Daily Totals		$2832.70		$4096.17	$70.82	$102.40

Crew B-22B	Hr.	Daily	Hr.	Daily	Bare Costs	Incl. O&P
1 Labor Foreman (outside)	$46.40	$371.20	$73.90	$591.20	$50.75	$80.33
1 Skilled Worker	57.10	456.80	90.50	724.00		
2 Laborers	44.40	710.40	70.70	1131.20		
1 Equip. Oper. (crane)	61.45	491.60	95.85	766.80		
1 S.P. Crane, 4x4, 5 Ton		381.95		420.14		
1 Butt Fusion Mach., 8"-24" diam.		1086.00		1194.60	36.70	40.37
40 L.H., Daily Totals		$3497.95		$4827.94	$87.45	$120.70

Crew B-22C	Hr.	Daily	Hr.	Daily	Bare Costs	Incl. O&P
1 Skilled Worker	$57.10	$456.80	$90.50	$724.00	$50.75	$80.60
1 Laborer	44.40	355.20	70.70	565.60		
1 Butt Fusion Mach., 2"-8" diam.		134.95		148.44	8.43	9.28
16 L.H., Daily Totals		$946.95		$1438.05	$59.18	$89.88

Crew B-23	Hr.	Daily	Hr.	Daily	Bare Costs	Incl. O&P
1 Labor Foreman (outside)	$46.40	$371.20	$73.90	$591.20	$44.80	$71.34
4 Laborers	44.40	1420.80	70.70	2262.40		
1 Drill Rig, Truck-Mounted		768.40		845.24		
1 Flatbed Truck, Gas, 3 Ton		850.05		935.05	40.46	44.51
40 L.H., Daily Totals		$3410.45		$4633.90	$85.26	$115.85

Crew B-23A	Hr.	Daily	Hr.	Daily	Bare Costs	Incl. O&P
1 Labor Foreman (outside)	$46.40	$371.20	$73.90	$591.20	$49.93	$78.88
1 Laborer	44.40	355.20	70.70	565.60		
1 Equip. Oper. (medium)	59.00	472.00	92.05	736.40		
1 Drill Rig, Truck-Mounted		768.40		845.24		
1 Pickup Truck, 3/4 Ton		112.20		123.42	36.69	40.36
24 L.H., Daily Totals		$2079.00		$2861.86	$86.63	$119.24

Crew B-23B	Hr.	Daily	Hr.	Daily	Bare Costs	Incl. O&P
1 Labor Foreman (outside)	$46.40	$371.20	$73.90	$591.20	$49.93	$78.88
1 Laborer	44.40	355.20	70.70	565.60		
1 Equip. Oper. (medium)	59.00	472.00	92.05	736.40		
1 Drill Rig, Truck-Mounted		768.40		845.24		
1 Pickup Truck, 3/4 Ton		112.20		123.42		
1 Centr. Water Pump, 6"		235.25		258.77	46.49	51.14
24 L.H., Daily Totals		$2314.25		$3120.64	$96.43	$130.03

Crew B-24	Hr.	Daily	Hr.	Daily	Bare Costs	Incl. O&P
1 Cement Finisher	$51.80	$414.40	$81.10	$648.80	$50.30	$79.63
1 Laborer	44.40	355.20	70.70	565.60		
1 Carpenter	54.70	437.60	87.10	696.80		
24 L.H., Daily Totals		$1207.20		$1911.20	$50.30	$79.63

Crew B-25	Hr.	Daily	Hr.	Daily	Bare Costs	Incl. O&P
1 Labor Foreman (outside)	$46.40	$371.20	$73.90	$591.20	$48.56	$76.81
7 Laborers	44.40	2486.40	70.70	3959.20		
3 Equip. Oper. (medium)	59.00	1416.00	92.05	2209.20		
1 Asphalt Paver, 130 H.P.		2143.00		2357.30		
1 Tandem Roller, 10 Ton		246.80		271.48		
1 Roller, Pneum. Whl., 12 Ton		349.90		384.89	31.13	34.25
88 L.H., Daily Totals		$7013.30		$9773.27	$79.70	$111.06

Crew B-25B	Hr.	Daily	Hr.	Daily	Bare Costs	Incl. O&P
1 Labor Foreman (outside)	$46.40	$371.20	$73.90	$591.20	$49.43	$78.08
7 Laborers	44.40	2486.40	70.70	3959.20		
4 Equip. Oper. (medium)	59.00	1888.00	92.05	2945.60		
1 Asphalt Paver, 130 H.P.		2143.00		2357.30		
2 Tandem Rollers, 10 Ton		493.60		542.96		
1 Roller, Pneum. Whl., 12 Ton		349.90		384.89	31.11	34.22
96 L.H., Daily Totals		$7732.10		$10781.15	$80.54	$112.30

Crew B-25C	Hr.	Daily	Hr.	Daily	Bare Costs	Incl. O&P
1 Labor Foreman (outside)	$46.40	$371.20	$73.90	$591.20	$49.60	$78.35
3 Laborers	44.40	1065.60	70.70	1696.80		
2 Equip. Oper. (medium)	59.00	944.00	92.05	1472.80		
1 Asphalt Paver, 130 H.P.		2143.00		2357.30		
1 Tandem Roller, 10 Ton		246.80		271.48	49.79	54.77
48 L.H., Daily Totals		$4770.60		$6389.58	$99.39	$133.12

Crew B-25D	Hr.	Daily	Hr.	Daily	Bare Costs	Incl. O&P
1 Labor Foreman (outside)	$46.40	$371.20	$73.90	$591.20	$49.82	$78.69
3 Laborers	44.40	1065.60	70.70	1696.80		
2.125 Equip. Oper. (medium)	59.00	1003.00	92.05	1564.85		
.125 Truck Driver (heavy)	51.30	51.30	81.85	81.85		
.125 Truck Tractor, 6x4, 380 H.P.		62.39		68.63		
.125 Dist. Tanker, 3000 Gallon		41.76		45.94		
1 Asphalt Paver, 130 H.P.		2143.00		2357.30		
1 Tandem Roller, 10 Ton		246.80		271.48	49.88	54.87
50 L.H., Daily Totals		$4985.06		$6678.05	$99.70	$133.56

For customer support on your Commercial Renovation Costs with RSMeans Data, call 800.448.8182.

765

Crew No.	Bare Costs		Incl. Subs O&P		Cost Per Labor-Hour	
Crew B-25E	Hr.	Daily	Hr.	Daily	Bare Costs	Incl. O&P
1 Labor Foreman (outside)	$46.40	$371.20	$73.90	$591.20	$50.03	$79.01
3 Laborers	44.40	1065.60	70.70	1696.80		
2.250 Equip. Oper. (medium)	59.00	1062.00	92.05	1656.90		
.25 Truck Driver (heavy)	51.30	102.60	81.85	163.70		
.25 Truck Tractor, 6x4, 380 H.P.		124.79		137.27		
.25 Dist. Tanker, 3000 Gallon		83.53		91.88		
1 Asphalt Paver, 130 H.P.		2143.00		2357.30		
1 Tandem Roller, 10 Ton		246.80		271.48	49.96	54.96
52 L.H., Daily Totals		$5199.51		$6966.52	$99.99	$133.97

Crew B-26	Hr.	Daily	Hr.	Daily	Bare Costs	Incl. O&P
1 Labor Foreman (outside)	$46.40	$371.20	$73.90	$591.20	$49.23	$77.77
6 Laborers	44.40	2131.20	70.70	3393.60		
2 Equip. Oper. (medium)	59.00	944.00	92.05	1472.80		
1 Rodman (reinf.)	58.90	471.20	92.20	737.60		
1 Cement Finisher	51.80	414.40	81.10	648.80		
1 Grader, 30,000 Lbs.		1073.00		1180.30		
1 Paving Mach. & Equip.		2503.00		2753.30	40.64	44.70
88 L.H., Daily Totals		$7908.00		$10777.60	$89.86	$122.47

Crew B-26A	Hr.	Daily	Hr.	Daily	Bare Costs	Incl. O&P
1 Labor Foreman (outside)	$46.40	$371.20	$73.90	$591.20	$49.23	$77.77
6 Laborers	44.40	2131.20	70.70	3393.60		
2 Equip. Oper. (medium)	59.00	944.00	92.05	1472.80		
1 Rodman (reinf.)	58.90	471.20	92.20	737.60		
1 Cement Finisher	51.80	414.40	81.10	648.80		
1 Grader, 30,000 Lbs.		1073.00		1180.30		
1 Paving Mach. & Equip.		2503.00		2753.30		
1 Concrete Saw		112.85		124.14	41.92	46.11
88 L.H., Daily Totals		$8020.85		$10901.74	$91.15	$123.88

Crew B-26B	Hr.	Daily	Hr.	Daily	Bare Costs	Incl. O&P
1 Labor Foreman (outside)	$46.40	$371.20	$73.90	$591.20	$50.04	$78.96
6 Laborers	44.40	2131.20	70.70	3393.60		
3 Equip. Oper. (medium)	59.00	1416.00	92.05	2209.20		
1 Rodman (reinf.)	58.90	471.20	92.20	737.60		
1 Cement Finisher	51.80	414.40	81.10	648.80		
1 Grader, 30,000 Lbs.		1073.00		1180.30		
1 Paving Mach. & Equip.		2503.00		2753.30		
1 Concrete Pump, 110' Boom		493.65		543.01	42.39	46.63
96 L.H., Daily Totals		$8873.65		$12057.01	$92.43	$125.59

Crew B-26C	Hr.	Daily	Hr.	Daily	Bare Costs	Incl. O&P
1 Labor Foreman (outside)	$46.40	$371.20	$73.90	$591.20	$48.25	$76.34
6 Laborers	44.40	2131.20	70.70	3393.60		
1 Equip. Oper. (medium)	59.00	472.00	92.05	736.40		
1 Rodman (reinf.)	58.90	471.20	92.20	737.60		
1 Cement Finisher	51.80	414.40	81.10	648.80		
1 Paving Mach. & Equip.		2503.00		2753.30		
1 Concrete Saw		112.85		124.14	32.70	35.97
80 L.H., Daily Totals		$6475.85		$8985.03	$80.95	$112.31

Crew B-27	Hr.	Daily	Hr.	Daily	Bare Costs	Incl. O&P
1 Labor Foreman (outside)	$46.40	$371.20	$73.90	$591.20	$44.90	$71.50
3 Laborers	44.40	1065.60	70.70	1696.80		
1 Berm Machine		253.55		278.90	7.92	8.72
32 L.H., Daily Totals		$1690.35		$2566.91	$52.82	$80.22

Crew B-28	Hr.	Daily	Hr.	Daily	Bare Costs	Incl. O&P
2 Carpenters	$54.70	$875.20	$87.10	$1393.60	$51.27	$81.63
1 Laborer	44.40	355.20	70.70	565.60		
24 L.H., Daily Totals		$1230.40		$1959.20	$51.27	$81.63

Crew B-29	Hr.	Daily	Hr.	Daily	Bare Costs	Incl. O&P
1 Labor Foreman (outside)	$46.40	$371.20	$73.90	$591.20	$48.28	$76.35
4 Laborers	44.40	1420.80	70.70	2262.40		
1 Equip. Oper. (crane)	61.45	491.60	95.85	766.80		
1 Equip. Oper. (oiler)	52.50	420.00	81.90	655.20		
1 Gradall, 5/8 C.Y.		850.65		935.72	15.19	16.71
56 L.H., Daily Totals		$3554.25		$5211.31	$63.47	$93.06

Crew B-30	Hr.	Daily	Hr.	Daily	Bare Costs	Incl. O&P
1 Equip. Oper. (medium)	$59.00	$472.00	$92.05	$736.40	$53.87	$85.25
2 Truck Drivers (heavy)	51.30	820.80	81.85	1309.60		
1 Hyd. Excavator, 1.5 C.Y.		695.80		765.38		
2 Dump Trucks, 12 C.Y., 400 H.P.		1158.70		1274.57	77.27	85.00
24 L.H., Daily Totals		$3147.30		$4085.95	$131.14	$170.25

Crew B-31	Hr.	Daily	Hr.	Daily	Bare Costs	Incl. O&P
1 Labor Foreman (outside)	$46.40	$371.20	$73.90	$591.20	$46.86	$74.62
3 Laborers	44.40	1065.60	70.70	1696.80		
1 Carpenter	54.70	437.60	87.10	696.80		
1 Air Compressor, 250 cfm		202.85		223.13		
1 Sheeting Driver		7.45		8.20		
2 -50' Air Hoses, 1.5"		45.60		50.16	6.40	7.04
40 L.H., Daily Totals		$2130.30		$3266.29	$53.26	$81.66

Crew B-32	Hr.	Daily	Hr.	Daily	Bare Costs	Incl. O&P
1 Laborer	$44.40	$355.20	$70.70	$565.60	$55.35	$86.71
3 Equip. Oper. (medium)	59.00	1416.00	92.05	2209.20		
1 Grader, 30,000 Lbs.		1073.00		1180.30		
1 Tandem Roller, 10 Ton		246.80		271.48		
1 Dozer, 200 H.P.		1520.00		1672.00	88.74	97.62
32 L.H., Daily Totals		$4611.00		$5898.58	$144.09	$184.33

Crew B-32A	Hr.	Daily	Hr.	Daily	Bare Costs	Incl. O&P
1 Laborer	$44.40	$355.20	$70.70	$565.60	$54.13	$84.93
2 Equip. Oper. (medium)	59.00	944.00	92.05	1472.80		
1 Grader, 30,000 Lbs.		1073.00		1180.30		
1 Roller, Vibratory, 25 Ton		672.35		739.59	72.72	80.00
24 L.H., Daily Totals		$3044.55		$3958.28	$126.86	$164.93

Crew B-32B	Hr.	Daily	Hr.	Daily	Bare Costs	Incl. O&P
1 Laborer	$44.40	$355.20	$70.70	$565.60	$54.13	$84.93
2 Equip. Oper. (medium)	59.00	944.00	92.05	1472.80		
1 Dozer, 200 H.P.		1520.00		1672.00		
1 Roller, Vibratory, 25 Ton		672.35		739.59	91.35	100.48
24 L.H., Daily Totals		$3491.55		$4449.98	$145.48	$185.42

Crew B-32C	Hr.	Daily	Hr.	Daily	Bare Costs	Incl. O&P
1 Labor Foreman (outside)	$46.40	$371.20	$73.90	$591.20	$52.03	$81.91
2 Laborers	44.40	710.40	70.70	1131.20		
3 Equip. Oper. (medium)	59.00	1416.00	92.05	2209.20		
1 Grader, 30,000 Lbs.		1073.00		1180.30		
1 Tandem Roller, 10 Ton		246.80		271.48		
1 Dozer, 200 H.P.		1520.00		1672.00	59.16	65.08
48 L.H., Daily Totals		$5337.40		$7055.38	$111.20	$146.99

For customer support on your Commercial Renovation Costs with RSMeans Data, call 800.448.8182.

Crew No.	Bare Costs		Incl. Subs O&P		Cost Per Labor-Hour	

Crew B-33A	Hr.	Daily	Hr.	Daily	Bare Costs	Incl. O&P
1 Equip. Oper. (medium)	$59.00	$472.00	$92.05	$736.40	$54.83	$85.95
.5 Laborer	44.40	177.60	70.70	282.80		
.25 Equip. Oper. (medium)	59.00	118.00	92.05	184.10		
1 Scraper, Towed, 7 C.Y.		129.30		142.23		
1.25 Dozers, 300 H.P.		2231.25		2454.38	168.61	185.47
14 L.H., Daily Totals		$3128.15		$3799.91	$223.44	$271.42

Crew B-33B	Hr.	Daily	Hr.	Daily	Bare Costs	Incl. O&P
1 Equip. Oper. (medium)	$59.00	$472.00	$92.05	$736.40	$54.83	$85.95
.5 Laborer	44.40	177.60	70.70	282.80		
.25 Equip. Oper. (medium)	59.00	118.00	92.05	184.10		
1 Scraper, Towed, 10 C.Y.		161.65		177.82		
1.25 Dozers, 300 H.P.		2231.25		2454.38	170.92	188.01
14 L.H., Daily Totals		$3160.50		$3835.49	$225.75	$273.96

Crew B-33C	Hr.	Daily	Hr.	Daily	Bare Costs	Incl. O&P
1 Equip. Oper. (medium)	$59.00	$472.00	$92.05	$736.40	$54.83	$85.95
.5 Laborer	44.40	177.60	70.70	282.80		
.25 Equip. Oper. (medium)	59.00	118.00	92.05	184.10		
1 Scraper, Towed, 15 C.Y.		178.85		196.74		
1.25 Dozers, 300 H.P.		2231.25		2454.38	172.15	189.37
14 L.H., Daily Totals		$3177.70		$3854.41	$226.98	$275.32

Crew B-33D	Hr.	Daily	Hr.	Daily	Bare Costs	Incl. O&P
1 Equip. Oper. (medium)	$59.00	$472.00	$92.05	$736.40	$54.83	$85.95
.5 Laborer	44.40	177.60	70.70	282.80		
.25 Equip. Oper. (medium)	59.00	118.00	92.05	184.10		
1 S.P. Scraper, 14 C.Y.		2424.00		2666.40		
.25 Dozer, 300 H.P.		446.25		490.88	205.02	225.52
14 L.H., Daily Totals		$3637.85		$4360.57	$259.85	$311.47

Crew B-33E	Hr.	Daily	Hr.	Daily	Bare Costs	Incl. O&P
1 Equip. Oper. (medium)	$59.00	$472.00	$92.05	$736.40	$54.83	$85.95
.5 Laborer	44.40	177.60	70.70	282.80		
.25 Equip. Oper. (medium)	59.00	118.00	92.05	184.10		
1 S.P. Scraper, 21 C.Y.		2656.00		2921.60		
.25 Dozer, 300 H.P.		446.25		490.88	221.59	243.75
14 L.H., Daily Totals		$3869.85		$4615.77	$276.42	$329.70

Crew B-33F	Hr.	Daily	Hr.	Daily	Bare Costs	Incl. O&P
1 Equip. Oper. (medium)	$59.00	$472.00	$92.05	$736.40	$54.83	$85.95
.5 Laborer	44.40	177.60	70.70	282.80		
.25 Equip. Oper. (medium)	59.00	118.00	92.05	184.10		
1 Elev. Scraper, 11 C.Y.		1059.00		1164.90		
.25 Dozer, 300 H.P.		446.25		490.88	107.52	118.27
14 L.H., Daily Totals		$2272.85		$2859.07	$162.35	$204.22

Crew B-33G	Hr.	Daily	Hr.	Daily	Bare Costs	Incl. O&P
1 Equip. Oper. (medium)	$59.00	$472.00	$92.05	$736.40	$54.83	$85.95
.5 Laborer	44.40	177.60	70.70	282.80		
.25 Equip. Oper. (medium)	59.00	118.00	92.05	184.10		
1 Elev. Scraper, 22 C.Y.		1895.00		2084.50		
.25 Dozer, 300 H.P.		446.25		490.88	167.23	183.96
14 L.H., Daily Totals		$3108.85		$3778.68	$222.06	$269.91

Crew B-33K	Hr.	Daily	Hr.	Daily	Bare Costs	Incl. O&P
1 Equipment Operator (med.)	$59.00	$472.00	$92.05	$736.40	$54.83	$85.95
.25 Equipment Operator (med.)	59.00	118.00	92.05	184.10		
.5 Laborer	44.40	177.60	70.70	282.80		
1 S.P. Scraper, 31 C.Y.		3707.00		4077.70		
.25 Dozer, 410 H.P.		701.75		771.92	314.91	346.40
14 L.H., Daily Totals		$5176.35		$6052.93	$369.74	$432.35

Crew B-34A	Hr.	Daily	Hr.	Daily	Bare Costs	Incl. O&P
1 Truck Driver (heavy)	$51.30	$410.40	$81.85	$654.80	$51.30	$81.85
1 Dump Truck, 8 C.Y., 220 H.P.		407.60		448.36	50.95	56.05
8 L.H., Daily Totals		$818.00		$1103.16	$102.25	$137.90

Crew B-34B	Hr.	Daily	Hr.	Daily	Bare Costs	Incl. O&P
1 Truck Driver (heavy)	$51.30	$410.40	$81.85	$654.80	$51.30	$81.85
1 Dump Truck, 12 C.Y., 400 H.P.		579.35		637.28	72.42	79.66
8 L.H., Daily Totals		$989.75		$1292.09	$123.72	$161.51

Crew B-34C	Hr.	Daily	Hr.	Daily	Bare Costs	Incl. O&P
1 Truck Driver (heavy)	$51.30	$410.40	$81.85	$654.80	$51.30	$81.85
1 Truck Tractor, 6x4, 380 H.P.		499.15		549.07		
1 Dump Trailer, 16.5 C.Y.		138.35		152.19	79.69	87.66
8 L.H., Daily Totals		$1047.90		$1356.05	$130.99	$169.51

Crew B-34D	Hr.	Daily	Hr.	Daily	Bare Costs	Incl. O&P
1 Truck Driver (heavy)	$51.30	$410.40	$81.85	$654.80	$51.30	$81.85
1 Truck Tractor, 6x4, 380 H.P.		499.15		549.07		
1 Dump Trailer, 20 C.Y.		153.55		168.91	81.59	89.75
8 L.H., Daily Totals		$1063.10		$1372.77	$132.89	$171.60

Crew B-34E	Hr.	Daily	Hr.	Daily	Bare Costs	Incl. O&P
1 Truck Driver (heavy)	$51.30	$410.40	$81.85	$654.80	$51.30	$81.85
1 Dump Truck, Off Hwy., 25 Ton		1427.00		1569.70	178.38	196.21
8 L.H., Daily Totals		$1837.40		$2224.50	$229.68	$278.06

Crew B-34F	Hr.	Daily	Hr.	Daily	Bare Costs	Incl. O&P
1 Truck Driver (heavy)	$51.30	$410.40	$81.85	$654.80	$51.30	$81.85
1 Dump Truck, Off Hwy., 35 Ton		945.65		1040.21	118.21	130.03
8 L.H., Daily Totals		$1356.05		$1695.02	$169.51	$211.88

Crew B-34G	Hr.	Daily	Hr.	Daily	Bare Costs	Incl. O&P
1 Truck Driver (heavy)	$51.30	$410.40	$81.85	$654.80	$51.30	$81.85
1 Dump Truck, Off Hwy., 50 Ton		1979.00		2176.90	247.38	272.11
8 L.H., Daily Totals		$2389.40		$2831.70	$298.68	$353.96

Crew B-34H	Hr.	Daily	Hr.	Daily	Bare Costs	Incl. O&P
1 Truck Driver (heavy)	$51.30	$410.40	$81.85	$654.80	$51.30	$81.85
1 Dump Truck, Off Hwy., 65 Ton		1938.00		2131.80	242.25	266.48
8 L.H., Daily Totals		$2348.40		$2786.60	$293.55	$348.32

Crew B-34I	Hr.	Daily	Hr.	Daily	Bare Costs	Incl. O&P
1 Truck Driver (heavy)	$51.30	$410.40	$81.85	$654.80	$51.30	$81.85
1 Dump Truck, 18 C.Y., 450 H.P.		753.60		828.96	94.20	103.62
8 L.H., Daily Totals		$1164.00		$1483.76	$145.50	$185.47

Crew B-34J	Hr.	Daily	Hr.	Daily	Bare Costs	Incl. O&P
1 Truck Driver (heavy)	$51.30	$410.40	$81.85	$654.80	$51.30	$81.85
1 Dump Truck, Off Hwy., 100 Ton		2769.00		3045.90	346.13	380.74
8 L.H., Daily Totals		$3179.40		$3700.70	$397.43	$462.59

Crew No.	Bare Costs		Incl. Subs O&P		Cost Per Labor-Hour	
	Hr.	Daily	Hr.	Daily	Bare Costs	Incl. O&P
Crew B-34K					Bare Costs	Incl. O&P
1 Truck Driver (heavy)	$51.30	$410.40	$81.85	$654.80	$51.30	$81.85
1 Truck Tractor, 6x4, 450 H.P.		608.95		669.85		
1 Lowbed Trailer, 75 Ton		258.10		283.91	108.38	119.22
8 L.H., Daily Totals		$1277.45		$1608.56	$159.68	$201.07
Crew B-34L	Hr.	Daily	Hr.	Daily	Bare Costs	Incl. O&P
1 Equip. Oper. (light)	$55.50	$444.00	$86.60	$692.80	$55.50	$86.60
1 Flatbed Truck, Gas, 1.5 Ton		198.50		218.35	24.81	27.29
8 L.H., Daily Totals		$642.50		$911.15	$80.31	$113.89
Crew B-34M	Hr.	Daily	Hr.	Daily	Bare Costs	Incl. O&P
1 Equip. Oper. (light)	$55.50	$444.00	$86.60	$692.80	$55.50	$86.60
1 Flatbed Truck, Gas, 3 Ton		850.05		935.05	106.26	116.88
8 L.H., Daily Totals		$1294.05		$1627.86	$161.76	$203.48
Crew B-34N	Hr.	Daily	Hr.	Daily	Bare Costs	Incl. O&P
1 Truck Driver (heavy)	$51.30	$410.40	$81.85	$654.80	$55.15	$86.95
1 Equip. Oper. (medium)	59.00	472.00	92.05	736.40		
1 Truck Tractor, 6x4, 380 H.P.		499.15		549.07		
1 Flatbed Trailer, 40 Ton		188.45		207.29	42.98	47.27
16 L.H., Daily Totals		$1570.00		$2147.56	$98.13	$134.22
Crew B-34P	Hr.	Daily	Hr.	Daily	Bare Costs	Incl. O&P
1 Pipe Fitter	$68.35	$546.80	$105.40	$843.20	$58.72	$91.77
1 Truck Driver (light)	48.80	390.40	77.85	622.80		
1 Equip. Oper. (medium)	59.00	472.00	92.05	736.40		
1 Flatbed Truck, Gas, 3 Ton		850.05		935.05		
1 Backhoe Loader, 48 H.P.		216.20		237.82	44.43	48.87
24 L.H., Daily Totals		$2475.45		$3375.28	$103.14	$140.64
Crew B-34Q	Hr.	Daily	Hr.	Daily	Bare Costs	Incl. O&P
1 Pipe Fitter	$68.35	$546.80	$105.40	$843.20	$59.53	$93.03
1 Truck Driver (light)	48.80	390.40	77.85	622.80		
1 Equip. Oper. (crane)	61.45	491.60	95.85	766.80		
1 Flatbed Trailer, 25 Ton		137.20		150.92		
1 Dump Truck, 8 C.Y., 220 H.P.		407.60		448.36		
1 Hyd. Crane, 25 Ton		586.70		645.37	47.15	51.86
24 L.H., Daily Totals		$2560.30		$3477.45	$106.68	$144.89
Crew B-34R	Hr.	Daily	Hr.	Daily	Bare Costs	Incl. O&P
1 Pipe Fitter	$68.35	$546.80	$105.40	$843.20	$59.53	$93.03
1 Truck Driver (light)	48.80	390.40	77.85	622.80		
1 Equip. Oper. (crane)	61.45	491.60	95.85	766.80		
1 Flatbed Trailer, 25 Ton		137.20		150.92		
1 Dump Truck, 8 C.Y., 220 H.P.		407.60		448.36		
1 Hyd. Crane, 25 Ton		586.70		645.37		
1 Hyd. Excavator, 1 C.Y.		832.65		915.91	81.84	90.02
24 L.H., Daily Totals		$3392.95		$4393.36	$141.37	$183.06
Crew B-34S	Hr.	Daily	Hr.	Daily	Bare Costs	Incl. O&P
2 Pipe Fitters	$68.35	$1093.60	$105.40	$1686.40	$62.36	$97.13
1 Truck Driver (heavy)	51.30	410.40	81.85	654.80		
1 Equip. Oper. (crane)	61.45	491.60	95.85	766.80		
1 Flatbed Trailer, 40 Ton		188.45		207.29		
1 Truck Tractor, 6x4, 380 H.P.		499.15		549.07		
1 Hyd. Crane, 80 Ton		1458.00		1603.80		
1 Hyd. Excavator, 2 C.Y.		942.70		1036.97	96.51	106.16
32 L.H., Daily Totals		$5083.90		$6505.13	$158.87	$203.29
Crew B-34T	Hr.	Daily	Hr.	Daily	Bare Costs	Incl. O&P
2 Pipe Fitters	$68.35	$1093.60	$105.40	$1686.40	$62.36	$97.13
1 Truck Driver (heavy)	51.30	410.40	81.85	654.80		
1 Equip. Oper. (crane)	61.45	491.60	95.85	766.80		
1 Flatbed Trailer, 40 Ton		188.45		207.29		
1 Truck Tractor, 6x4, 380 H.P.		499.15		549.07		
1 Hyd. Crane, 80 Ton		1458.00		1603.80	67.05	73.75
32 L.H., Daily Totals		$4141.20		$5468.16	$129.41	$170.88
Crew B-34U	Hr.	Daily	Hr.	Daily	Bare Costs	Incl. O&P
1 Truck Driver (heavy)	$51.30	$410.40	$81.85	$654.80	$53.40	$84.22
1 Equip. Oper. (light)	55.50	444.00	86.60	692.80		
1 Truck Tractor, 220 H.P.		310.80		341.88		
1 Flatbed Trailer, 25 Ton		137.20		150.92	28.00	30.80
16 L.H., Daily Totals		$1302.40		$1840.40	$81.40	$115.03
Crew B-34V	Hr.	Daily	Hr.	Daily	Bare Costs	Incl. O&P
1 Truck Driver (heavy)	$51.30	$410.40	$81.85	$654.80	$56.08	$88.10
1 Equip. Oper. (crane)	61.45	491.60	95.85	766.80		
1 Equip. Oper. (light)	55.50	444.00	86.60	692.80		
1 Truck Tractor, 6x4, 450 H.P.		608.95		669.85		
1 Equipment Trailer, 50 Ton		207.25		227.97		
1 Pickup Truck, 4x4, 3/4 Ton		176.75		194.43	41.37	45.51
24 L.H., Daily Totals		$2338.95		$3206.65	$97.46	$133.61
Crew B-34W	Hr.	Daily	Hr.	Daily	Bare Costs	Incl. O&P
5 Truck Drivers (heavy)	$51.30	$2052.00	$81.85	$3274.00	$53.92	$85.29
2 Equip. Opers. (crane)	61.45	983.20	95.85	1533.60		
1 Equip. Oper. (mechanic)	61.50	492.00	95.95	767.60		
1 Laborer	44.40	355.20	70.70	565.60		
4 Truck Tractors, 6x4, 380 H.P.		1996.60		2196.26		
2 Equipment Trailers, 50 Ton		414.50		455.95		
2 Flatbed Trailers, 40 Ton		376.90		414.59		
1 Pickup Truck, 4x4, 3/4 Ton		176.75		194.43		
1 S.P. Crane, 4x4, 20 Ton		574.35		631.78	49.15	54.07
72 L.H., Daily Totals		$7421.50		$10033.81	$103.08	$139.36
Crew B-35	Hr.	Daily	Hr.	Daily	Bare Costs	Incl. O&P
1 Labor Foreman (outside)	$46.40	$371.20	$73.90	$591.20	$56.75	$88.81
1 Skilled Worker	57.10	456.80	90.50	724.00		
2 Welders	67.70	1083.20	104.40	1670.40		
1 Laborer	44.40	355.20	70.70	565.60		
1 Equip. Oper. (crane)	61.45	491.60	95.85	766.80		
1 Equip. Oper. (oiler)	52.50	420.00	81.90	655.20		
2 Welder, Electric, 300 amp		215.10		236.61		
1 Hyd. Excavator, .75 C.Y.		701.80		771.98	16.37	18.01
56 L.H., Daily Totals		$4094.90		$5981.79	$73.12	$106.82
Crew B-35A	Hr.	Daily	Hr.	Daily	Bare Costs	Incl. O&P
1 Labor Foreman (outside)	$46.40	$371.20	$73.90	$591.20	$53.42	$83.99
2 Laborers	44.40	710.40	70.70	1131.20		
1 Skilled Worker	57.10	456.80	90.50	724.00		
1 Welder (plumber)	67.70	541.60	104.40	835.20		
1 Equip. Oper. (crane)	61.45	491.60	95.85	766.80		
1 Equip. Oper. (oiler)	52.50	420.00	81.90	655.20		
1 Welder, Gas Engine, 300 amp		148.75		163.63		
1 Crawler Crane, 75 Ton		1967.00		2163.70	37.78	41.56
56 L.H., Daily Totals		$5107.35		$7030.93	$91.20	$125.55

For customer support on your Commercial Renovation Costs with RSMeans Data, call 800.448.8182.

Crew No.	Bare Costs		Incl. Subs O&P		Cost Per Labor-Hour	
Crew B-36	Hr.	Daily	Hr.	Daily	Bare Costs	Incl. O&P
1 Labor Foreman (outside)	$46.40	$371.20	$73.90	$591.20	$50.64	$79.88
2 Laborers	44.40	710.40	70.70	1131.20		
2 Equip. Oper. (medium)	59.00	944.00	92.05	1472.80		
1 Dozer, 200 H.P.		1520.00		1672.00		
1 Aggregate Spreader		59.95		65.94		
1 Tandem Roller, 10 Ton		246.80		271.48	45.67	50.24
40 L.H., Daily Totals		$3852.35		$5204.63	$96.31	$130.12

Crew No.	Bare Costs		Incl. Subs O&P		Cost Per Labor-Hour	
Crew B-36A	Hr.	Daily	Hr.	Daily	Bare Costs	Incl. O&P
1 Labor Foreman (outside)	$46.40	$371.20	$73.90	$591.20	$53.03	$83.36
2 Laborers	44.40	710.40	70.70	1131.20		
4 Equip. Oper. (medium)	59.00	1888.00	92.05	2945.60		
1 Dozer, 200 H.P.		1520.00		1672.00		
1 Aggregate Spreader		59.95		65.94		
1 Tandem Roller, 10 Ton		246.80		271.48		
1 Roller, Pneum. Whl., 12 Ton		349.90		384.89	38.87	42.76
56 L.H., Daily Totals		$5146.25		$7062.31	$91.90	$126.11

Crew No.	Bare Costs		Incl. Subs O&P		Cost Per Labor-Hour	
Crew B-36B	Hr.	Daily	Hr.	Daily	Bare Costs	Incl. O&P
1 Labor Foreman (outside)	$46.40	$371.20	$73.90	$591.20	$52.81	$83.17
2 Laborers	44.40	710.40	70.70	1131.20		
4 Equip. Oper. (medium)	59.00	1888.00	92.05	2945.60		
1 Truck Driver (heavy)	51.30	410.40	81.85	654.80		
1 Grader, 30,000 Lbs.		1073.00		1180.30		
1 F.E. Loader, Crl, 1.5 C.Y.		668.35		735.18		
1 Dozer, 300 H.P.		1785.00		1963.50		
1 Roller, Vibratory, 25 Ton		672.35		739.59		
1 Truck Tractor, 6x4, 450 H.P.		608.95		669.85		
1 Water Tank Trailer, 5000 Gal.		154.10		169.51	77.53	85.28
64 L.H., Daily Totals		$8341.75		$10780.73	$130.34	$168.45

Crew No.	Bare Costs		Incl. Subs O&P		Cost Per Labor-Hour	
Crew B-36C	Hr.	Daily	Hr.	Daily	Bare Costs	Incl. O&P
1 Labor Foreman (outside)	$46.40	$371.20	$73.90	$591.20	$54.94	$86.38
3 Equip. Oper. (medium)	59.00	1416.00	92.05	2209.20		
1 Truck Driver (heavy)	51.30	410.40	81.85	654.80		
1 Grader, 30,000 Lbs.		1073.00		1180.30		
1 Dozer, 300 H.P.		1785.00		1963.50		
1 Roller, Vibratory, 25 Ton		672.35		739.59		
1 Truck Tractor, 6x4, 450 H.P.		608.95		669.85		
1 Water Tank Trailer, 5000 Gal.		154.10		169.51	107.33	118.07
40 L.H., Daily Totals		$6491.00		$8177.94	$162.28	$204.45

Crew No.	Bare Costs		Incl. Subs O&P		Cost Per Labor-Hour	
Crew B-37	Hr.	Daily	Hr.	Daily	Bare Costs	Incl. O&P
1 Labor Foreman (outside)	$46.40	$371.20	$73.90	$591.20	$46.58	$73.88
4 Laborers	44.40	1420.80	70.70	2262.40		
1 Equip. Oper. (light)	55.50	444.00	86.60	692.80		
1 Tandem Roller, 5 Ton		258.75		284.63	5.39	5.93
48 L.H., Daily Totals		$2494.75		$3831.03	$51.97	$79.81

Crew No.	Bare Costs		Incl. Subs O&P		Cost Per Labor-Hour	
Crew B-37A	Hr.	Daily	Hr.	Daily	Bare Costs	Incl. O&P
2 Laborers	$44.40	$710.40	$70.70	$1131.20	$45.87	$73.08
1 Truck Driver (light)	48.80	390.40	77.85	622.80		
1 Flatbed Truck, Gas, 1.5 Ton		198.50		218.35		
1 Tar Kettle, T.M.		156.70		172.37	14.80	16.28
24 L.H., Daily Totals		$1456.00		$2144.72	$60.67	$89.36

Crew No.	Bare Costs		Incl. Subs O&P		Cost Per Labor-Hour	
Crew B-37B	Hr.	Daily	Hr.	Daily	Bare Costs	Incl. O&P
3 Laborers	$44.40	$1065.60	$70.70	$1696.80	$45.50	$72.49
1 Truck Driver (light)	48.80	390.40	77.85	622.80		
1 Flatbed Truck, Gas, 1.5 Ton		198.50		218.35		
1 Tar Kettle, T.M.		156.70		172.37	11.10	12.21
32 L.H., Daily Totals		$1811.20		$2710.32	$56.60	$84.70

Crew No.	Bare Costs		Incl. Subs O&P		Cost Per Labor-Hour	
Crew B-37C	Hr.	Daily	Hr.	Daily	Bare Costs	Incl. O&P
2 Laborers	$44.40	$710.40	$70.70	$1131.20	$46.60	$74.28
2 Truck Drivers (light)	48.80	780.80	77.85	1245.60		
2 Flatbed Trucks, Gas, 1.5 Ton		397.00		436.70		
1 Tar Kettle, T.M.		156.70		172.37	17.30	19.03
32 L.H., Daily Totals		$2044.90		$2985.87	$63.90	$93.31

Crew No.	Bare Costs		Incl. Subs O&P		Cost Per Labor-Hour	
Crew B-37D	Hr.	Daily	Hr.	Daily	Bare Costs	Incl. O&P
1 Laborer	$44.40	$355.20	$70.70	$565.60	$46.60	$74.28
1 Truck Driver (light)	48.80	390.40	77.85	622.80		
1 Pickup Truck, 3/4 Ton		112.20		123.42	7.01	7.71
16 L.H., Daily Totals		$857.80		$1311.82	$53.61	$81.99

Crew No.	Bare Costs		Incl. Subs O&P		Cost Per Labor-Hour	
Crew B-37E	Hr.	Daily	Hr.	Daily	Bare Costs	Incl. O&P
3 Laborers	$44.40	$1065.60	$70.70	$1696.80	$49.33	$78.06
1 Equip. Oper. (light)	55.50	444.00	86.60	692.80		
1 Equip. Oper. (medium)	59.00	472.00	92.05	736.40		
2 Truck Drivers (light)	48.80	780.80	77.85	1245.60		
4 Barrels w/ Flasher		16.60		18.26		
1 Concrete Saw		112.85		124.14		
1 Rotary Hammer Drill		52.25		57.48		
1 Hammer Drill Bit		25.25		27.77		
1 Loader, Skid Steer, 30 H.P.		179.50		197.45		
1 Conc. Hammer Attach.		118.50		130.35		
1 Vibrating Plate, Gas, 18"		31.90		35.09		
2 Flatbed Trucks, Gas, 1.5 Ton		397.00		436.70	16.68	18.34
56 L.H., Daily Totals		$3696.25		$5398.84	$66.00	$96.41

Crew No.	Bare Costs		Incl. Subs O&P		Cost Per Labor-Hour	
Crew B-37F	Hr.	Daily	Hr.	Daily	Bare Costs	Incl. O&P
3 Laborers	$44.40	$1065.60	$70.70	$1696.80	$45.50	$72.49
1 Truck Driver (light)	48.80	390.40	77.85	622.80		
4 Barrels w/ Flasher		16.60		18.26		
1 Concrete Mixer, 10 C.F.		147.15		161.87		
1 Air Compressor, 60 cfm		153.85		169.24		
1 -50' Air Hose, 3/4"		7.15		7.87		
1 Spade (Chipper)		8.55		9.40		
1 Flatbed Truck, Gas, 1.5 Ton		198.50		218.35	16.62	18.28
32 L.H., Daily Totals		$1987.80		$2904.58	$62.12	$90.77

Crew No.	Bare Costs		Incl. Subs O&P		Cost Per Labor-Hour	
Crew B-37G	Hr.	Daily	Hr.	Daily	Bare Costs	Incl. O&P
1 Labor Foreman (outside)	$46.40	$371.20	$73.90	$591.20	$46.58	$73.88
4 Laborers	44.40	1420.80	70.70	2262.40		
1 Equip. Oper. (light)	55.50	444.00	86.60	692.80		
1 Berm Machine		253.55		278.90		
1 Tandem Roller, 5 Ton		258.75		284.63	10.67	11.74
48 L.H., Daily Totals		$2748.30		$4109.93	$57.26	$85.62

Crew No.	Bare Costs		Incl. Subs O&P		Cost Per Labor-Hour	
Crew B-37H	Hr.	Daily	Hr.	Daily	Bare Costs	Incl. O&P
1 Labor Foreman (outside)	$46.40	$371.20	$73.90	$591.20	$46.58	$73.88
4 Laborers	44.40	1420.80	70.70	2262.40		
1 Equip. Oper. (light)	55.50	444.00	86.60	692.80		
1 Tandem Roller, 5 Ton		258.75		284.63		
1 Flatbed Truck, Gas, 1.5 Ton		198.50		218.35		
1 Tar Kettle, T.M.		156.70		172.37	12.79	14.07
48 L.H., Daily Totals		$2849.95		$4221.74	$59.37	$87.95

Crew No.	Bare Costs		Incl. Subs O&P		Cost Per Labor-Hour	
Crew B-37I	**Hr.**	**Daily**	**Hr.**	**Daily**	**Bare Costs**	**Incl. O&P**
3 Laborers	$44.40	$1065.60	$70.70	$1696.80	$49.33	$78.06
1 Equip. Oper. (light)	55.50	444.00	86.60	692.80		
1 Equip. Oper. (medium)	59.00	472.00	92.05	736.40		
2 Truck Drivers (light)	48.80	780.80	77.85	1245.60		
4 Barrels w/ Flasher		16.60		18.26		
1 Concrete Saw		112.85		124.14		
1 Rotary Hammer Drill		52.25		57.48		
1 Hammer Drill Bit		25.25		27.77		
1 Air Compressor, 60 cfm		153.85		169.24		
1 -50' Air Hose, 3/4"		7.15		7.87		
1 Spade (Chipper)		8.55		9.40		
1 Loader, Skid Steer, 30 H.P.		179.50		197.45		
1 Conc. Hammer Attach.		118.50		130.35		
1 Concrete Mixer, 10 C.F.		147.15		161.87		
1 Vibrating Plate, Gas, 18"		31.90		35.09		
2 Flatbed Trucks, Gas, 1.5 Ton		397.00		436.70	22.33	24.56
56 L.H., Daily Totals		**$4012.95**		**$5747.20**	**$71.66**	**$102.63**
Crew B-37J	**Hr.**	**Daily**	**Hr.**	**Daily**	**Bare Costs**	**Incl. O&P**
1 Labor Foreman (outside)	$46.40	$371.20	$73.90	$591.20	$46.58	$73.88
4 Laborers	44.40	1420.80	70.70	2262.40		
1 Equip. Oper. (light)	55.50	444.00	86.60	692.80		
1 Air Compressor, 60 cfm		153.85		169.24		
1 -50' Air Hose, 3/4"		7.15		7.87		
2 Concrete Mixers, 10 C.F.		294.30		323.73		
2 Flatbed Trucks, Gas, 1.5 Ton		397.00		436.70		
1 Shot Blaster, 20"		208.70		229.57	22.10	24.31
48 L.H., Daily Totals		**$3297.00**		**$4713.50**	**$68.69**	**$98.20**
Crew B-37K	**Hr.**	**Daily**	**Hr.**	**Daily**	**Bare Costs**	**Incl. O&P**
1 Labor Foreman (outside)	$46.40	$371.20	$73.90	$591.20	$46.58	$73.88
4 Laborers	44.40	1420.80	70.70	2262.40		
1 Equip. Oper. (light)	55.50	444.00	86.60	692.80		
1 Air Compressor, 60 cfm		153.85		169.24		
1 -50' Air Hose, 3/4"		7.15		7.87		
2 Flatbed Trucks, Gas, 1.5 Ton		397.00		436.70		
1 Shot Blaster, 20"		208.70		229.57	15.97	17.57
48 L.H., Daily Totals		**$3002.70**		**$4389.77**	**$62.56**	**$91.45**
Crew B-38	**Hr.**	**Daily**	**Hr.**	**Daily**	**Bare Costs**	**Incl. O&P**
1 Labor Foreman (outside)	$46.40	$371.20	$73.90	$591.20	$49.94	$78.79
2 Laborers	44.40	710.40	70.70	1131.20		
1 Equip. Oper. (light)	55.50	444.00	86.60	692.80		
1 Equip. Oper. (medium)	59.00	472.00	92.05	736.40		
1 Backhoe Loader, 48 H.P.		216.20		237.82		
1 Hyd. Hammer (1200 lb.)		177.25		194.97		
1 F.E. Loader, W.M., 4 C.Y.		759.00		834.90		
1 Pvmt. Rem. Bucket		63.80		70.18	30.41	33.45
40 L.H., Daily Totals		**$3213.85**		**$4489.48**	**$80.35**	**$112.24**
Crew B-39	**Hr.**	**Daily**	**Hr.**	**Daily**	**Bare Costs**	**Incl. O&P**
1 Labor Foreman (outside)	$46.40	$371.20	$73.90	$591.20	$46.58	$73.88
4 Laborers	44.40	1420.80	70.70	2262.40		
1 Equip. Oper. (light)	55.50	444.00	86.60	692.80		
1 Air Compressor, 250 cfm		202.85		223.13		
2 Breakers, Pavement, 60 lb.		107.20		117.92		
2 -50' Air Hoses, 1.5"		45.60		50.16	7.41	8.15
48 L.H., Daily Totals		**$2591.65**		**$3937.61**	**$53.99**	**$82.03**

Crew No.	Bare Costs		Incl. Subs O&P		Cost Per Labor-Hour	
Crew B-40	**Hr.**	**Daily**	**Hr.**	**Daily**	**Bare Costs**	**Incl. O&P**
1 Pile Driver Foreman (outside)	$57.90	$463.20	$92.40	$739.20	$57.11	$90.35
4 Pile Drivers	55.90	1788.80	89.20	2854.40		
2 Equip. Oper. (crane)	61.45	983.20	95.85	1533.60		
1 Equip. Oper. (oiler)	52.50	420.00	81.90	655.20		
1 Crawler Crane, 40 Ton		1231.00		1354.10		
1 Vibratory Hammer & Gen.		2298.00		2527.80	55.14	60.65
64 L.H., Daily Totals		**$7184.20**		**$9664.30**	**$112.25**	**$151.00**
Crew B-40B	**Hr.**	**Daily**	**Hr.**	**Daily**	**Bare Costs**	**Incl. O&P**
1 Labor Foreman (outside)	$46.40	$371.20	$73.90	$591.20	$48.92	$77.29
3 Laborers	44.40	1065.60	70.70	1696.80		
1 Equip. Oper. (crane)	61.45	491.60	95.85	766.80		
1 Equip. Oper. (oiler)	52.50	420.00	81.90	655.20		
1 Lattice Boom Crane, 40 Ton		2127.00		2339.70	44.31	48.74
48 L.H., Daily Totals		**$4475.40**		**$6049.70**	**$93.24**	**$126.04**
Crew B-41	**Hr.**	**Daily**	**Hr.**	**Daily**	**Bare Costs**	**Incl. O&P**
1 Labor Foreman (outside)	$46.40	$371.20	$73.90	$591.20	$45.91	$72.93
4 Laborers	44.40	1420.80	70.70	2262.40		
.25 Equip. Oper. (crane)	61.45	122.90	95.85	191.70		
.25 Equip. Oper. (oiler)	52.50	105.00	81.90	163.80		
.25 Crawler Crane, 40 Ton		307.75		338.52	6.99	7.69
44 L.H., Daily Totals		**$2327.65**		**$3547.63**	**$52.90**	**$80.63**
Crew B-42	**Hr.**	**Daily**	**Hr.**	**Daily**	**Bare Costs**	**Incl. O&P**
1 Labor Foreman (outside)	$46.40	$371.20	$73.90	$591.20	$49.78	$79.00
4 Laborers	44.40	1420.80	70.70	2262.40		
1 Equip. Oper. (crane)	61.45	491.60	95.85	766.80		
1 Equip. Oper. (oiler)	52.50	420.00	81.90	655.20		
1 Welder	60.30	482.40	97.55	780.40		
1 Hyd. Crane, 25 Ton		586.70		645.37		
1 Welder, Gas Engine, 300 amp		148.75		163.63		
1 Horz. Boring Csg. Mch.		329.75		362.73	16.64	18.31
64 L.H., Daily Totals		**$4251.20**		**$6227.72**	**$66.43**	**$97.31**
Crew B-43	**Hr.**	**Daily**	**Hr.**	**Daily**	**Bare Costs**	**Incl. O&P**
1 Labor Foreman (outside)	$46.40	$371.20	$73.90	$591.20	$48.92	$77.29
3 Laborers	44.40	1065.60	70.70	1696.80		
1 Equip. Oper. (crane)	61.45	491.60	95.85	766.80		
1 Equip. Oper. (oiler)	52.50	420.00	81.90	655.20		
1 Drill Rig, Truck-Mounted		768.40		845.24	16.01	17.61
48 L.H., Daily Totals		**$3116.80**		**$4555.24**	**$64.93**	**$94.90**
Crew B-44	**Hr.**	**Daily**	**Hr.**	**Daily**	**Bare Costs**	**Incl. O&P**
1 Pile Driver Foreman (outside)	$57.90	$463.20	$92.40	$739.20	$56.10	$88.95
4 Pile Drivers	55.90	1788.80	89.20	2854.40		
2 Equip. Oper. (crane)	61.45	983.20	95.85	1533.60		
1 Laborer	44.40	355.20	70.70	565.60		
1 Crawler Crane, 40 Ton		1231.00		1354.10		
1 Lead, 60' High		211.80		232.98		
1 Hammer, Diesel, 15K ft.-lbs.		624.45		686.89	32.30	35.53
64 L.H., Daily Totals		**$5657.65**		**$7966.77**	**$88.40**	**$124.48**
Crew B-45	**Hr.**	**Daily**	**Hr.**	**Daily**	**Bare Costs**	**Incl. O&P**
1 Equip. Oper. (medium)	$59.00	$472.00	$92.05	$736.40	$55.15	$86.95
1 Truck Driver (heavy)	51.30	410.40	81.85	654.80		
1 Dist. Tanker, 3000 Gallon		334.10		367.51		
1 Truck Tractor, 6x4, 380 H.P.		499.15		549.07	52.08	57.29
16 L.H., Daily Totals		**$1715.65**		**$2307.78**	**$107.23**	**$144.24**

For customer support on your Commercial Renovation Costs with RSMeans Data, call 800.448.8182.

Crew No.	Bare Costs		Incl. Subs O&P		Cost Per Labor-Hour	
Crew B-46	Hr.	Daily	Hr.	Daily	Bare Costs	Incl. O&P
1 Pile Driver Foreman (outside)	$57.90	$463.20	$92.40	$739.20	$50.48	$80.48
2 Pile Drivers	55.90	894.40	89.20	1427.20		
3 Laborers	44.40	1065.60	70.70	1696.80		
1 Chain Saw, Gas, 36" Long		41.65		45.81	.87	.95
48 L.H., Daily Totals		$2464.85		$3909.01	$51.35	$81.44

Crew No.	Bare Costs		Incl. Subs O&P		Cost Per Labor-Hour	
Crew B-47	Hr.	Daily	Hr.	Daily	Bare Costs	Incl. O&P
1 Blast Foreman (outside)	$46.40	$371.20	$73.90	$591.20	$48.77	$77.07
1 Driller	44.40	355.20	70.70	565.60		
1 Equip. Oper. (light)	55.50	444.00	86.60	692.80		
1 Air Track Drill, 4"		1127.00		1239.70		
1 Air Compressor, 600 cfm		426.55		469.20		
2 -50' Air Hoses, 3"		76.70		84.37	67.93	74.72
24 L.H., Daily Totals		$2800.65		$3642.88	$116.69	$151.79

Crew No.	Bare Costs		Incl. Subs O&P		Cost Per Labor-Hour	
Crew B-47A	Hr.	Daily	Hr.	Daily	Bare Costs	Incl. O&P
1 Drilling Foreman (outside)	$46.40	$371.20	$73.90	$591.20	$53.45	$83.88
1 Equip. Oper. (heavy)	61.45	491.60	95.85	766.80		
1 Equip. Oper. (oiler)	52.50	420.00	81.90	655.20		
1 Air Track Drill, 5"		1127.00		1239.70	46.96	51.65
24 L.H., Daily Totals		$2409.80		$3252.90	$100.41	$135.54

Crew No.	Bare Costs		Incl. Subs O&P		Cost Per Labor-Hour	
Crew B-47C	Hr.	Daily	Hr.	Daily	Bare Costs	Incl. O&P
1 Laborer	$44.40	$355.20	$70.70	$565.60	$49.95	$78.65
1 Equip. Oper. (light)	55.50	444.00	86.60	692.80		
1 Air Compressor, 750 cfm		596.30		655.93		
2 -50' Air Hoses, 3"		76.70		84.37		
1 Air Track Drill, 4"		1127.00		1239.70	112.50	123.75
16 L.H., Daily Totals		$2599.20		$3238.40	$162.45	$202.40

Crew No.	Bare Costs		Incl. Subs O&P		Cost Per Labor-Hour	
Crew B-47E	Hr.	Daily	Hr.	Daily	Bare Costs	Incl. O&P
1 Labor Foreman (outside)	$46.40	$371.20	$73.90	$591.20	$44.90	$71.50
3 Laborers	44.40	1065.60	70.70	1696.80		
1 Flatbed Truck, Gas, 3 Ton		850.05		935.05	26.56	29.22
32 L.H., Daily Totals		$2286.85		$3223.05	$71.46	$100.72

Crew No.	Bare Costs		Incl. Subs O&P		Cost Per Labor-Hour	
Crew B-47G	Hr.	Daily	Hr.	Daily	Bare Costs	Incl. O&P
1 Labor Foreman (outside)	$46.40	$371.20	$73.90	$591.20	$47.67	$75.47
2 Laborers	44.40	710.40	70.70	1131.20		
1 Equip. Oper. (light)	55.50	444.00	86.60	692.80		
1 Air Track Drill, 4"		1127.00		1239.70		
1 Air Compressor, 600 cfm		426.55		469.20		
2 -50' Air Hoses, 3"		76.70		84.37		
1 Gunite Pump Rig		321.75		353.93	61.00	67.10
32 L.H., Daily Totals		$3477.60		$4562.40	$108.68	$142.57

Crew No.	Bare Costs		Incl. Subs O&P		Cost Per Labor-Hour	
Crew B-47H	Hr.	Daily	Hr.	Daily	Bare Costs	Incl. O&P
1 Skilled Worker Foreman (out)	$59.10	$472.80	$93.65	$749.20	$57.60	$91.29
3 Skilled Workers	57.10	1370.40	90.50	2172.00		
1 Flatbed Truck, Gas, 3 Ton		850.05		935.05	26.56	29.22
32 L.H., Daily Totals		$2693.25		$3856.26	$84.16	$120.51

Crew No.	Bare Costs		Incl. Subs O&P		Cost Per Labor-Hour	
Crew B-48	Hr.	Daily	Hr.	Daily	Bare Costs	Incl. O&P
1 Labor Foreman (outside)	$46.40	$371.20	$73.90	$591.20	$49.86	$78.62
3 Laborers	44.40	1065.60	70.70	1696.80		
1 Equip. Oper. (crane)	61.45	491.60	95.85	766.80		
1 Equip. Oper. (oiler)	52.50	420.00	81.90	655.20		
1 Equip. Oper. (light)	55.50	444.00	86.60	692.80		
1 Centr. Water Pump, 6"		235.25		258.77		
1 -20' Suction Hose, 6"		25.50		28.05		
1 -50' Discharge Hose, 6"		18.10		19.91		
1 Drill Rig, Truck-Mounted		768.40		845.24	18.70	20.57
56 L.H., Daily Totals		$3839.65		$5554.77	$68.57	$99.19

Crew No.	Bare Costs		Incl. Subs O&P		Cost Per Labor-Hour	
Crew B-49	Hr.	Daily	Hr.	Daily	Bare Costs	Incl. O&P
1 Labor Foreman (outside)	$46.40	$371.20	$73.90	$591.20	$52.25	$82.41
3 Laborers	44.40	1065.60	70.70	1696.80		
2 Equip. Oper. (crane)	61.45	983.20	95.85	1533.60		
2 Equip. Oper. (oilers)	52.50	840.00	81.90	1310.40		
1 Equip. Oper. (light)	55.50	444.00	86.60	692.80		
2 Pile Drivers	55.90	894.40	89.20	1427.20		
1 Hyd. Crane, 25 Ton		586.70		645.37		
1 Centr. Water Pump, 6"		235.25		258.77		
1 -20' Suction Hose, 6"		25.50		28.05		
1 -50' Discharge Hose, 6"		18.10		19.91		
1 Drill Rig, Truck-Mounted		768.40		845.24	18.57	20.42
88 L.H., Daily Totals		$6232.35		$9049.34	$70.82	$102.83

Crew No.	Bare Costs		Incl. Subs O&P		Cost Per Labor-Hour	
Crew B-50	Hr.	Daily	Hr.	Daily	Bare Costs	Incl. O&P
2 Pile Driver Foremen (outside)	$57.90	$926.40	$92.40	$1478.40	$54.27	$86.12
6 Pile Drivers	55.90	2683.20	89.20	4281.60		
2 Equip. Oper. (crane)	61.45	983.20	95.85	1533.60		
1 Equip. Oper. (oiler)	52.50	420.00	81.90	655.20		
3 Laborers	44.40	1065.60	70.70	1696.80		
1 Crawler Crane, 40 Ton		1231.00		1354.10		
1 Lead, 60' High		211.80		232.98		
1 Hammer, Diesel, 15K ft.-lbs.		624.45		686.89		
1 Air Compressor, 600 cfm		426.55		469.20		
2 -50' Air Hoses, 3"		76.70		84.37		
1 Chain Saw, Gas, 36" Long		41.65		45.81	23.32	25.66
112 L.H., Daily Totals		$8690.55		$12518.97	$77.59	$111.78

Crew No.	Bare Costs		Incl. Subs O&P		Cost Per Labor-Hour	
Crew B-51	Hr.	Daily	Hr.	Daily	Bare Costs	Incl. O&P
1 Labor Foreman (outside)	$46.40	$371.20	$73.90	$591.20	$45.47	$72.42
4 Laborers	44.40	1420.80	70.70	2262.40		
1 Truck Driver (light)	48.80	390.40	77.85	622.80		
1 Flatbed Truck, Gas, 1.5 Ton		198.50		218.35	4.14	4.55
48 L.H., Daily Totals		$2380.90		$3694.75	$49.60	$76.97

Crew No.	Bare Costs		Incl. Subs O&P		Cost Per Labor-Hour	
Crew B-52	Hr.	Daily	Hr.	Daily	Bare Costs	Incl. O&P
1 Carpenter Foreman (outside)	$56.70	$453.60	$90.30	$722.40	$50.76	$80.39
1 Carpenter	54.70	437.60	87.10	696.80		
3 Laborers	44.40	1065.60	70.70	1696.80		
1 Cement Finisher	51.80	414.40	81.10	648.80		
.5 Rodman (reinf.)	58.90	235.60	92.20	368.80		
.5 Equip. Oper. (medium)	59.00	236.00	92.05	368.20		
.5 Crawler Loader, 3 C.Y.		573.00		630.30	10.23	11.26
56 L.H., Daily Totals		$3415.80		$5132.10	$61.00	$91.64

Crew No.	Bare Costs		Incl. Subs O&P		Cost Per Labor-Hour	
Crew B-53	Hr.	Daily	Hr.	Daily	Bare Costs	Incl. O&P
1 Equip. Oper. (light)	$55.50	$444.00	$86.60	$692.80	$55.50	$86.60
1 Trencher, Chain, 12 H.P.		158.60		174.46	19.82	21.81
8 L.H., Daily Totals		$602.60		$867.26	$75.33	$108.41

Crews - Renovation

Crew No.	Bare Costs		Incl. Subs O&P		Cost Per Labor-Hour	

Crew B-54	Hr.	Daily	Hr.	Daily	Bare Costs	Incl. O&P
1 Equip. Oper. (light)	$55.50	$444.00	$86.60	$692.80	$55.50	$86.60
1 Trencher, Chain, 40 H.P.		450.70		495.77	56.34	61.97
8 L.H., Daily Totals		$894.70		$1188.57	$111.84	$148.57

Crew B-54A	Hr.	Daily	Hr.	Daily	Bare Costs	Incl. O&P
.17 Labor Foreman (outside)	$46.40	$63.10	$73.90	$100.50	$57.17	$89.41
1 Equipment Operator (med.)	59.00	472.00	92.05	736.40		
1 Wheel Trencher, 67 H.P.		1140.00		1254.00	121.79	133.97
9.36 L.H., Daily Totals		$1675.10		$2090.90	$178.96	$223.39

Crew B-54B	Hr.	Daily	Hr.	Daily	Bare Costs	Incl. O&P
.25 Labor Foreman (outside)	$46.40	$92.80	$73.90	$147.80	$56.48	$88.42
1 Equipment Operator (med.)	59.00	472.00	92.05	736.40		
1 Wheel Trencher, 150 H.P.		1238.00		1361.80	123.80	136.18
10 L.H., Daily Totals		$1802.80		$2246.00	$180.28	$224.60

Crew B-54D	Hr.	Daily	Hr.	Daily	Bare Costs	Incl. O&P
1 Laborer	$44.40	$355.20	$70.70	$565.60	$51.70	$81.38
1 Equipment Operator (med.)	59.00	472.00	92.05	736.40		
1 Rock Trencher, 6" Width		434.20		477.62	27.14	29.85
16 L.H., Daily Totals		$1261.40		$1779.62	$78.84	$111.23

Crew B-54E	Hr.	Daily	Hr.	Daily	Bare Costs	Incl. O&P
1 Laborer	$44.40	$355.20	$70.70	$565.60	$51.70	$81.38
1 Equipment Operator (med.)	59.00	472.00	92.05	736.40		
1 Rock Trencher, 18" Width		1015.00		1116.50	63.44	69.78
16 L.H., Daily Totals		$1842.20		$2418.50	$115.14	$151.16

Crew B-55	Hr.	Daily	Hr.	Daily	Bare Costs	Incl. O&P
2 Laborers	$44.40	$710.40	$70.70	$1131.20	$45.87	$73.08
1 Truck Driver (light)	48.80	390.40	77.85	622.80		
1 Truck-Mounted Earth Auger		394.15		433.57		
1 Flatbed Truck, Gas, 3 Ton		850.05		935.05	51.84	57.03
24 L.H., Daily Totals		$2345.00		$3122.62	$97.71	$130.11

Crew B-56	Hr.	Daily	Hr.	Daily	Bare Costs	Incl. O&P
1 Laborer	$44.40	$355.20	$70.70	$565.60	$49.95	$78.65
1 Equip. Oper. (light)	55.50	444.00	86.60	692.80		
1 Air Track Drill, 4"		1127.00		1239.70		
1 Air Compressor, 600 cfm		426.55		469.20		
1 -50' Air Hose, 3"		38.35		42.19	99.49	109.44
16 L.H., Daily Totals		$2391.10		$3009.49	$149.44	$188.09

Crew B-57	Hr.	Daily	Hr.	Daily	Bare Costs	Incl. O&P
1 Labor Foreman (outside)	$46.40	$371.20	$73.90	$591.20	$50.77	$79.94
2 Laborers	44.40	710.40	70.70	1131.20		
1 Equip. Oper. (crane)	61.45	491.60	95.85	766.80		
1 Equip. Oper. (light)	55.50	444.00	86.60	692.80		
1 Equip. Oper. (oiler)	52.50	420.00	81.90	655.20		
1 Crawler Crane, 25 Ton		1152.00		1267.20		
1 Clamshell Bucket, 1 C.Y.		69.25		76.17		
1 Centr. Water Pump, 6"		235.25		258.77		
1 -20' Suction Hose, 6"		25.50		28.05		
20 -50' Discharge Hoses, 6"		362.00		398.20	38.42	42.26
48 L.H., Daily Totals		$4281.20		$5865.60	$89.19	$122.20

Crew B-58	Hr.	Daily	Hr.	Daily	Bare Costs	Incl. O&P
2 Laborers	$44.40	$710.40	$70.70	$1131.20	$48.10	$76.00
1 Equip. Oper. (light)	55.50	444.00	86.60	692.80		
1 Backhoe Loader, 48 H.P.		216.20		237.82		
1 Small Helicopter, w/ Pilot		2101.00		2311.10	96.55	106.21
24 L.H., Daily Totals		$3471.60		$4372.92	$144.65	$182.21

Crew B-59	Hr.	Daily	Hr.	Daily	Bare Costs	Incl. O&P
1 Truck Driver (heavy)	$51.30	$410.40	$81.85	$654.80	$51.30	$81.85
1 Truck Tractor, 220 H.P.		310.80		341.88		
1 Water Tank Trailer, 5000 Gal.		154.10		169.51	58.11	63.92
8 L.H., Daily Totals		$875.30		$1166.19	$109.41	$145.77

Crew B-59A	Hr.	Daily	Hr.	Daily	Bare Costs	Incl. O&P
2 Laborers	$44.40	$710.40	$70.70	$1131.20	$46.70	$74.42
1 Truck Driver (heavy)	51.30	410.40	81.85	654.80		
1 Water Tank Trailer, 5000 Gal.		154.10		169.51		
1 Truck Tractor, 220 H.P.		310.80		341.88	19.37	21.31
24 L.H., Daily Totals		$1585.70		$2297.39	$66.07	$95.72

Crew B-60	Hr.	Daily	Hr.	Daily	Bare Costs	Incl. O&P
1 Labor Foreman (outside)	$46.40	$371.20	$73.90	$591.20	$51.45	$80.89
2 Laborers	44.40	710.40	70.70	1131.20		
1 Equip. Oper. (crane)	61.45	491.60	95.85	766.80		
2 Equip. Oper. (light)	55.50	888.00	86.60	1385.60		
1 Equip. Oper. (oiler)	52.50	420.00	81.90	655.20		
1 Crawler Crane, 40 Ton		1231.00		1354.10		
1 Lead, 60' High		211.80		232.98		
1 Hammer, Diesel, 15K ft.-lbs.		624.45		686.89		
1 Backhoe Loader, 48 H.P.		216.20		237.82	40.78	44.85
56 L.H., Daily Totals		$5164.65		$7041.80	$92.23	$125.75

Crew B-61	Hr.	Daily	Hr.	Daily	Bare Costs	Incl. O&P
1 Labor Foreman (outside)	$46.40	$371.20	$73.90	$591.20	$47.02	$74.52
3 Laborers	44.40	1065.60	70.70	1696.80		
1 Equip. Oper. (light)	55.50	444.00	86.60	692.80		
1 Cement Mixer, 2 C.Y.		112.35		123.58		
1 Air Compressor, 160 cfm		212.30		233.53	8.12	8.93
40 L.H., Daily Totals		$2205.45		$3337.92	$55.14	$83.45

Crew B-62	Hr.	Daily	Hr.	Daily	Bare Costs	Incl. O&P
2 Laborers	$44.40	$710.40	$70.70	$1131.20	$48.10	$76.00
1 Equip. Oper. (light)	55.50	444.00	86.60	692.80		
1 Loader, Skid Steer, 30 H.P.		179.50		197.45	7.48	8.23
24 L.H., Daily Totals		$1333.90		$2021.45	$55.58	$84.23

Crew B-62A	Hr.	Daily	Hr.	Daily	Bare Costs	Incl. O&P
2 Laborers	$44.40	$710.40	$70.70	$1131.20	$48.10	$76.00
1 Equip. Oper. (light)	55.50	444.00	86.60	692.80		
1 Loader, Skid Steer, 30 H.P.		179.50		197.45		
1 Trencher Attachment		66.25		72.88	10.24	11.26
24 L.H., Daily Totals		$1400.15		$2094.32	$58.34	$87.26

Crew B-63	Hr.	Daily	Hr.	Daily	Bare Costs	Incl. O&P
4 Laborers	$44.40	$1420.80	$70.70	$2262.40	$46.62	$73.88
1 Equip. Oper. (light)	55.50	444.00	86.60	692.80		
1 Loader, Skid Steer, 30 H.P.		179.50		197.45	4.49	4.94
40 L.H., Daily Totals		$2044.30		$3152.65	$51.11	$78.82

For customer support on your Commercial Renovation Costs with RSMeans Data, call 800.448.8182.

Crew No.	Bare Costs Hr.	Daily	Incl. Subs O&P Hr.	Daily	Cost Per Labor-Hour Bare Costs	Incl. O&P
Crew B-63B	**Hr.**	**Daily**	**Hr.**	**Daily**	**Bare Costs**	**Incl. O&P**
1 Labor Foreman (inside)	$44.90	$359.20	$71.50	$572.00	$47.30	$74.88
2 Laborers	44.40	710.40	70.70	1131.20		
1 Equip. Oper. (light)	55.50	444.00	86.60	692.80		
1 Loader, Skid Steer, 78 H.P.		446.30		490.93	13.95	15.34
32 L.H., Daily Totals		$1959.90		$2886.93	$61.25	$90.22
Crew B-64	**Hr.**	**Daily**	**Hr.**	**Daily**	**Bare Costs**	**Incl. O&P**
1 Laborer	$44.40	$355.20	$70.70	$565.60	$46.60	$74.28
1 Truck Driver (light)	48.80	390.40	77.85	622.80		
1 Power Mulcher (small)		201.00		221.10		
1 Flatbed Truck, Gas, 1.5 Ton		198.50		218.35	24.97	27.47
16 L.H., Daily Totals		$1145.10		$1627.85	$71.57	$101.74
Crew B-65	**Hr.**	**Daily**	**Hr.**	**Daily**	**Bare Costs**	**Incl. O&P**
1 Laborer	$44.40	$355.20	$70.70	$565.60	$46.60	$74.28
1 Truck Driver (light)	48.80	390.40	77.85	622.80		
1 Power Mulcher (Large)		345.35		379.88		
1 Flatbed Truck, Gas, 1.5 Ton		198.50		218.35	33.99	37.39
16 L.H., Daily Totals		$1289.45		$1786.64	$80.59	$111.66
Crew B-66	**Hr.**	**Daily**	**Hr.**	**Daily**	**Bare Costs**	**Incl. O&P**
1 Equip. Oper. (light)	$55.50	$444.00	$86.60	$692.80	$55.50	$86.60
1 Loader-Backhoe, 40 H.P.		267.55		294.31	33.44	36.79
8 L.H., Daily Totals		$711.55		$987.11	$88.94	$123.39
Crew B-67	**Hr.**	**Daily**	**Hr.**	**Daily**	**Bare Costs**	**Incl. O&P**
1 Millwright	$58.75	$470.00	$90.75	$726.00	$57.13	$88.67
1 Equip. Oper. (light)	55.50	444.00	86.60	692.80		
1 R.T. Forklift, 5,000 Lb., diesel		272.85		300.13	17.05	18.76
16 L.H., Daily Totals		$1186.85		$1718.93	$74.18	$107.43
Crew B-67B	**Hr.**	**Daily**	**Hr.**	**Daily**	**Bare Costs**	**Incl. O&P**
1 Millwright Foreman (inside)	$59.25	$474.00	$91.55	$732.40	$59.00	$91.15
1 Millwright	58.75	470.00	90.75	726.00		
16 L.H., Daily Totals		$944.00		$1458.40	$59.00	$91.15
Crew B-68	**Hr.**	**Daily**	**Hr.**	**Daily**	**Bare Costs**	**Incl. O&P**
2 Millwrights	$58.75	$940.00	$90.75	$1452.00	$57.67	$89.37
1 Equip. Oper. (light)	55.50	444.00	86.60	692.80		
1 R.T. Forklift, 5,000 Lb., diesel		272.85		300.13	11.37	12.51
24 L.H., Daily Totals		$1656.85		$2444.93	$69.04	$101.87
Crew B-68A	**Hr.**	**Daily**	**Hr.**	**Daily**	**Bare Costs**	**Incl. O&P**
1 Millwright Foreman (inside)	$59.25	$474.00	$91.55	$732.40	$58.92	$91.02
2 Millwrights	58.75	940.00	90.75	1452.00		
1 Forklift, Smooth Floor, 8,000 Lb.		283.25		311.57	11.80	12.98
24 L.H., Daily Totals		$1697.25		$2495.97	$70.72	$104.00
Crew B-68B	**Hr.**	**Daily**	**Hr.**	**Daily**	**Bare Costs**	**Incl. O&P**
1 Millwright Foreman (inside)	$59.25	$474.00	$91.55	$732.40	$62.79	$96.79
2 Millwrights	58.75	940.00	90.75	1452.00		
2 Electricians	63.70	1019.20	97.85	1565.60		
2 Plumbers	67.70	1083.20	104.40	1670.40		
1 R.T. Forklift, 5,000 Lb., gas		283.30		311.63	5.06	5.56
56 L.H., Daily Totals		$3799.70		$5732.03	$67.85	$102.36

Crew No.	Bare Costs Hr.	Daily	Incl. Subs O&P Hr.	Daily	Cost Per Labor-Hour Bare Costs	Incl. O&P
Crew B-68C	**Hr.**	**Daily**	**Hr.**	**Daily**	**Bare Costs**	**Incl. O&P**
1 Millwright Foreman (inside)	$59.25	$474.00	$91.55	$732.40	$62.35	$96.14
1 Millwright	58.75	470.00	90.75	726.00		
1 Electrician	63.70	509.60	97.85	782.80		
1 Plumber	67.70	541.60	104.40	835.20		
1 R.T. Forklift, 5,000 Lb., gas		283.30		311.63	8.85	9.74
32 L.H., Daily Totals		$2278.50		$3388.03	$71.20	$105.88
Crew B-68D	**Hr.**	**Daily**	**Hr.**	**Daily**	**Bare Costs**	**Incl. O&P**
1 Labor Foreman (inside)	$44.90	$359.20	$71.50	$572.00	$48.27	$76.27
1 Laborer	44.40	355.20	70.70	565.60		
1 Equip. Oper. (light)	55.50	444.00	86.60	692.80		
1 R.T. Forklift, 5,000 Lb., gas		283.30		311.63	11.80	12.98
24 L.H., Daily Totals		$1441.70		$2142.03	$60.07	$89.25
Crew B-68E	**Hr.**	**Daily**	**Hr.**	**Daily**	**Bare Costs**	**Incl. O&P**
1 Struc. Steel Foreman (inside)	$60.80	$486.40	$98.35	$786.80	$60.40	$97.71
3 Struc. Steel Workers	60.30	1447.20	97.55	2341.20		
1 Welder	60.30	482.40	97.55	780.40		
1 Forklift, Smooth Floor, 8,000 Lb.		283.25		311.57	7.08	7.79
40 L.H., Daily Totals		$2699.25		$4219.98	$67.48	$105.50
Crew B-68F	**Hr.**	**Daily**	**Hr.**	**Daily**	**Bare Costs**	**Incl. O&P**
1 Skilled Worker Foreman (out)	$59.10	$472.80	$93.65	$749.20	$57.77	$91.55
2 Skilled Workers	57.10	913.60	90.50	1448.00		
1 R.T. Forklift, 5,000 Lb., gas		283.30		311.63	11.80	12.98
24 L.H., Daily Totals		$1669.70		$2508.83	$69.57	$104.53
Crew B-68G	**Hr.**	**Daily**	**Hr.**	**Daily**	**Bare Costs**	**Incl. O&P**
2 Structural Steel Workers	$60.30	$964.80	$97.55	$1560.80	$60.30	$97.55
1 R.T. Forklift, 5,000 Lb., gas		283.30		311.63	17.71	19.48
16 L.H., Daily Totals		$1248.10		$1872.43	$78.01	$117.03
Crew B-69	**Hr.**	**Daily**	**Hr.**	**Daily**	**Bare Costs**	**Incl. O&P**
1 Labor Foreman (outside)	$46.40	$371.20	$73.90	$591.20	$48.92	$77.29
3 Laborers	44.40	1065.60	70.70	1696.80		
1 Equip. Oper. (crane)	61.45	491.60	95.85	766.80		
1 Equip. Oper. (oiler)	52.50	420.00	81.90	655.20		
1 Hyd. Crane, 80 Ton		1458.00		1603.80	30.38	33.41
48 L.H., Daily Totals		$3806.40		$5313.80	$79.30	$110.70
Crew B-69A	**Hr.**	**Daily**	**Hr.**	**Daily**	**Bare Costs**	**Incl. O&P**
1 Labor Foreman (outside)	$46.40	$371.20	$73.90	$591.20	$48.40	$76.53
3 Laborers	44.40	1065.60	70.70	1696.80		
1 Equip. Oper. (medium)	59.00	472.00	92.05	736.40		
1 Concrete Finisher	51.80	414.40	81.10	648.80		
1 Curb/Gutter Paver, 2-Track		1231.00		1354.10	25.65	28.21
48 L.H., Daily Totals		$3554.20		$5027.30	$74.05	$104.74
Crew B-69B	**Hr.**	**Daily**	**Hr.**	**Daily**	**Bare Costs**	**Incl. O&P**
1 Labor Foreman (outside)	$46.40	$371.20	$73.90	$591.20	$48.40	$76.53
3 Laborers	44.40	1065.60	70.70	1696.80		
1 Equip. Oper. (medium)	59.00	472.00	92.05	736.40		
1 Cement Finisher	51.80	414.40	81.10	648.80		
1 Curb/Gutter Paver, 4-Track		801.05		881.15	16.69	18.36
48 L.H., Daily Totals		$3124.25		$4554.35	$65.09	$94.88

For customer support on your Commercial Renovation Costs with RSMeans Data, call 800.448.8182.

773

Crew No.	Bare Costs		Incl. Subs O&P		Cost Per Labor-Hour	
Crew B-70	Hr.	Daily	Hr.	Daily	Bare Costs	Incl. O&P
1 Labor Foreman (outside)	$46.40	$371.20	$73.90	$591.20	$50.94	$80.31
3 Laborers	44.40	1065.60	70.70	1696.80		
3 Equip. Oper. (medium)	59.00	1416.00	92.05	2209.20		
1 Grader, 30,000 Lbs.		1073.00		1180.30		
1 Ripper, Beam & 1 Shank		91.60		100.76		
1 Road Sweeper, S.P., 8' wide		723.65		796.01		
1 F.E. Loader, W.M., 1.5 C.Y.		441.40		485.54	41.60	45.76
56 L.H., Daily Totals		$5182.45		$7059.81	$92.54	$126.07

Crew No.	Bare Costs		Incl. Subs O&P		Cost Per Labor-Hour	
Crew B-71	Hr.	Daily	Hr.	Daily	Bare Costs	Incl. O&P
1 Labor Foreman (outside)	$46.40	$371.20	$73.90	$591.20	$50.94	$80.31
3 Laborers	44.40	1065.60	70.70	1696.80		
3 Equip. Oper. (medium)	59.00	1416.00	92.05	2209.20		
1 Pvmt. Profiler, 750 H.P.		3483.00		3831.30		
1 Road Sweeper, S.P., 8' wide		723.65		796.01		
1 F.E. Loader, W.M., 1.5 C.Y.		441.40		485.54	83.00	91.30
56 L.H., Daily Totals		$7500.85		$9610.06	$133.94	$171.61

Crew No.	Bare Costs		Incl. Subs O&P		Cost Per Labor-Hour	
Crew B-72	Hr.	Daily	Hr.	Daily	Bare Costs	Incl. O&P
1 Labor Foreman (outside)	$46.40	$371.20	$73.90	$591.20	$51.95	$81.78
3 Laborers	44.40	1065.60	70.70	1696.80		
4 Equip. Oper. (medium)	59.00	1888.00	92.05	2945.60		
1 Pvmt. Profiler, 750 H.P.		3483.00		3831.30		
1 Hammermill, 250 H.P.		857.40		943.14		
1 Windrow Loader		1461.00		1607.10		
1 Mix Paver, 165 H.P.		2172.00		2389.20		
1 Roller, Pneum. Whl., 12 Ton		349.90		384.89	130.05	143.06
64 L.H., Daily Totals		$11648.10		$14389.23	$182.00	$224.83

Crew No.	Bare Costs		Incl. Subs O&P		Cost Per Labor-Hour	
Crew B-73	Hr.	Daily	Hr.	Daily	Bare Costs	Incl. O&P
1 Labor Foreman (outside)	$46.40	$371.20	$73.90	$591.20	$53.77	$84.44
2 Laborers	44.40	710.40	70.70	1131.20		
5 Equip. Oper. (medium)	59.00	2360.00	92.05	3682.00		
1 Road Mixer, 310 H.P.		1919.00		2110.90		
1 Tandem Roller, 10 Ton		246.80		271.48		
1 Hammermill, 250 H.P.		857.40		943.14		
1 Grader, 30,000 Lbs.		1073.00		1180.30		
.5 F.E. Loader, W.M., 1.5 C.Y.		220.70		242.77		
.5 Truck Tractor, 220 H.P.		155.40		170.94		
.5 Water Tank Trailer, 5000 Gal.		77.05		84.75	71.08	78.19
64 L.H., Daily Totals		$7990.95		$10408.68	$124.86	$162.64

Crew No.	Bare Costs		Incl. Subs O&P		Cost Per Labor-Hour	
Crew B-74	Hr.	Daily	Hr.	Daily	Bare Costs	Incl. O&P
1 Labor Foreman (outside)	$46.40	$371.20	$73.90	$591.20	$53.67	$84.56
1 Laborer	44.40	355.20	70.70	565.60		
4 Equip. Oper. (medium)	59.00	1888.00	92.05	2945.60		
2 Truck Drivers (heavy)	51.30	820.80	81.85	1309.60		
1 Grader, 30,000 Lbs.		1073.00		1180.30		
1 Ripper, Beam & 1 Shank		91.60		100.76		
2 Stabilizers, 310 H.P.		2808.00		3088.80		
1 Flatbed Truck, Gas, 3 Ton		850.05		935.05		
1 Chem. Spreader, Towed		85.40		93.94		
1 Roller, Vibratory, 25 Ton		672.35		739.59		
1 Water Tank Trailer, 5000 Gal.		154.10		169.51		
1 Truck Tractor, 220 H.P.		310.80		341.88	94.46	103.90
64 L.H., Daily Totals		$9480.50		$12061.83	$148.13	$188.47

Crew No.	Bare Costs		Incl. Subs O&P		Cost Per Labor-Hour	
Crew B-75	Hr.	Daily	Hr.	Daily	Bare Costs	Incl. O&P
1 Labor Foreman (outside)	$46.40	$371.20	$73.90	$591.20	$54.01	$84.95
1 Laborer	44.40	355.20	70.70	565.60		
4 Equip. Oper. (medium)	59.00	1888.00	92.05	2945.60		
1 Truck Driver (heavy)	51.30	410.40	81.85	654.80		
1 Grader, 30,000 Lbs.		1073.00		1180.30		
1 Ripper, Beam & 1 Shank		91.60		100.76		
2 Stabilizers, 310 H.P.		2808.00		3088.80		
1 Dist. Tanker, 3000 Gallon		334.10		367.51		
1 Truck Tractor, 6x4, 380 H.P.		499.15		549.07		
1 Roller, Vibratory, 25 Ton		672.35		739.59	97.83	107.61
56 L.H., Daily Totals		$8503.00		$10783.22	$151.84	$192.56

Crew No.	Bare Costs		Incl. Subs O&P		Cost Per Labor-Hour	
Crew B-76	Hr.	Daily	Hr.	Daily	Bare Costs	Incl. O&P
1 Dock Builder Foreman (outside)	$57.90	$463.20	$92.40	$739.20	$56.98	$90.22
5 Dock Builders	55.90	2236.00	89.20	3568.00		
2 Equip. Oper. (crane)	61.45	983.20	95.85	1533.60		
1 Equip. Oper. (oiler)	52.50	420.00	81.90	655.20		
1 Crawler Crane, 50 Ton		1541.00		1695.10		
1 Barge, 400 Ton		869.15		956.07		
1 Hammer, Diesel, 15K ft.-lbs.		624.45		686.89		
1 Lead, 60' High		211.80		232.98		
1 Air Compressor, 600 cfm		426.55		469.20		
2 -50' Air Hoses, 3"		76.70		84.37	52.08	57.29
72 L.H., Daily Totals		$7852.05		$10620.62	$109.06	$147.51

Crew No.	Bare Costs		Incl. Subs O&P		Cost Per Labor-Hour	
Crew B-76A	Hr.	Daily	Hr.	Daily	Bare Costs	Incl. O&P
1 Labor Foreman (outside)	$46.40	$371.20	$73.90	$591.20	$47.79	$75.64
5 Laborers	44.40	1776.00	70.70	2828.00		
1 Equip. Oper. (crane)	61.45	491.60	95.85	766.80		
1 Equip. Oper. (oiler)	52.50	420.00	81.90	655.20		
1 Crawler Crane, 50 Ton		1541.00		1695.10		
1 Barge, 400 Ton		869.15		956.07	37.66	41.42
64 L.H., Daily Totals		$5468.95		$7492.36	$85.45	$117.07

Crew No.	Bare Costs		Incl. Subs O&P		Cost Per Labor-Hour	
Crew B-77	Hr.	Daily	Hr.	Daily	Bare Costs	Incl. O&P
1 Labor Foreman (outside)	$46.40	$371.20	$73.90	$591.20	$45.68	$72.77
3 Laborers	44.40	1065.60	70.70	1696.80		
1 Truck Driver (light)	48.80	390.40	77.85	622.80		
1 Crack Cleaner, 25 H.P.		53.00		58.30		
1 Crack Filler, Trailer Mtd.		170.95		188.04		
1 Flatbed Truck, Gas, 3 Ton		850.05		935.05	26.85	29.54
40 L.H., Daily Totals		$2901.20		$4092.20	$72.53	$102.31

Crew No.	Bare Costs		Incl. Subs O&P		Cost Per Labor-Hour	
Crew B-78	Hr.	Daily	Hr.	Daily	Bare Costs	Incl. O&P
1 Labor Foreman (outside)	$46.40	$371.20	$73.90	$591.20	$45.47	$72.42
4 Laborers	44.40	1420.80	70.70	2262.40		
1 Truck Driver (light)	48.80	390.40	77.85	622.80		
1 Paint Striper, S.P., 40 Gallon		128.35		141.19		
1 Flatbed Truck, Gas, 3 Ton		850.05		935.05		
1 Pickup Truck, 3/4 Ton		112.20		123.42	22.72	24.99
48 L.H., Daily Totals		$3273.00		$4676.06	$68.19	$97.42

Crew No.	Bare Costs		Incl. Subs O&P		Cost Per Labor-Hour	
Crew B-78A	Hr.	Daily	Hr.	Daily	Bare Costs	Incl. O&P
1 Equip. Oper. (light)	$55.50	$444.00	$86.60	$692.80	$55.50	$86.60
1 Line Rem. (Metal Balls) 115 H.P.		996.25		1095.88	124.53	136.98
8 L.H., Daily Totals		$1440.25		$1788.68	$180.03	$223.58

For customer support on your Commercial Renovation Costs with RSMeans Data, call 800.448.8182.

Crew No.	Bare Costs		Incl. Subs O&P		Cost Per Labor-Hour	
Crew B-78B	Hr.	Daily	Hr.	Daily	Bare Costs	Incl. O&P
2 Laborers	$44.40	$710.40	$70.70	$1131.20	$45.63	$72.47
.25 Equip. Oper. (light)	55.50	111.00	86.60	173.20		
1 Pickup Truck, 3/4 Ton		112.20		123.42		
1 Line Rem.,11 H.P.,Walk Behind		114.75		126.22		
.25 Road Sweeper, S.P., 8' wide		180.91		199.00	22.66	24.92
18 L.H., Daily Totals		$1229.26		$1753.05	$68.29	$97.39
Crew B-78C	Hr.	Daily	Hr.	Daily	Bare Costs	Incl. O&P
1 Labor Foreman (outside)	$46.40	$371.20	$73.90	$591.20	$45.47	$72.42
4 Laborers	44.40	1420.80	70.70	2262.40		
1 Truck Driver (light)	48.80	390.40	77.85	622.80		
1 Paint Striper, T.M., 120 Gal.		603.10		663.41		
1 Flatbed Truck, Gas, 3 Ton		850.05		935.05		
1 Pickup Truck, 3/4 Ton		112.20		123.42	32.61	35.87
48 L.H., Daily Totals		$3747.75		$5198.28	$78.08	$108.30
Crew B-78D	Hr.	Daily	Hr.	Daily	Bare Costs	Incl. O&P
2 Labor Foremen (outside)	$46.40	$742.40	$73.90	$1182.40	$45.24	$72.06
7 Laborers	44.40	2486.40	70.70	3959.20		
1 Truck Driver (light)	48.80	390.40	77.85	622.80		
1 Paint Striper, T.M., 120 Gal.		603.10		663.41		
1 Flatbed Truck, Gas, 3 Ton		850.05		935.05		
3 Pickup Trucks, 3/4 Ton		336.60		370.26		
1 Air Compressor, 60 cfm		153.85		169.24		
1 -50' Air Hose, 3/4"		7.15		7.87		
1 Breaker, Pavement, 60 lb.		53.60		58.96	25.05	27.56
80 L.H., Daily Totals		$5623.55		$7969.19	$70.29	$99.61
Crew B-78E	Hr.	Daily	Hr.	Daily	Bare Costs	Incl. O&P
2 Labor Foremen (outside)	$46.40	$742.40	$73.90	$1182.40	$45.10	$71.83
9 Laborers	44.40	3196.80	70.70	5090.40		
1 Truck Driver (light)	48.80	390.40	77.85	622.80		
1 Paint Striper, T.M., 120 Gal.		603.10		663.41		
1 Flatbed Truck, Gas, 3 Ton		850.05		935.05		
4 Pickup Trucks, 3/4 Ton		448.80		493.68		
2 Air Compressors, 60 cfm		307.70		338.47		
2 -50' Air Hoses, 3/4"		14.30		15.73		
2 Breakers, Pavement, 60 lb.		107.20		117.92	24.28	26.71
96 L.H., Daily Totals		$6660.75		$9459.86	$69.38	$98.54
Crew B-78F	Hr.	Daily	Hr.	Daily	Bare Costs	Incl. O&P
2 Labor Foremen (outside)	$46.40	$742.40	$73.90	$1182.40	$45.00	$71.67
11 Laborers	44.40	3907.20	70.70	6221.60		
1 Truck Driver (light)	48.80	390.40	77.85	622.80		
1 Paint Striper, T.M., 120 Gal.		603.10		663.41		
1 Flatbed Truck, Gas, 3 Ton		850.05		935.05		
7 Pickup Trucks, 3/4 Ton		785.40		863.94		
3 Air Compressors, 60 cfm		461.55		507.70		
3 -50' Air Hoses, 3/4"		21.45		23.59		
3 Breakers, Pavement, 60 lb.		160.80		176.88	25.74	28.31
112 L.H., Daily Totals		$7922.35		$11197.39	$70.74	$99.98
Crew B-79	Hr.	Daily	Hr.	Daily	Bare Costs	Incl. O&P
1 Labor Foreman (outside)	$46.40	$371.20	$73.90	$591.20	$45.68	$72.77
3 Laborers	44.40	1065.60	70.70	1696.80		
1 Truck Driver (light)	48.80	390.40	77.85	622.80		
1 Paint Striper, T.M., 120 Gal.		603.10		663.41		
1 Heating Kettle, 115 Gallon		107.25		117.97		
1 Flatbed Truck, Gas, 3 Ton		850.05		935.05		
2 Pickup Trucks, 3/4 Ton		224.40		246.84	44.62	49.08
40 L.H., Daily Totals		$3612.00		$4874.08	$90.30	$121.85

Crew No.	Bare Costs		Incl. Subs O&P		Cost Per Labor-Hour	
Crew B-79A	Hr.	Daily	Hr.	Daily	Bare Costs	Incl. O&P
1.5 Equip. Oper. (light)	$55.50	$666.00	$86.60	$1039.20	$55.50	$86.60
.5 Line Remov. (Grinder) 115 H.P.		515.50		567.05		
1 Line Rem. (Metal Balls) 115 H.P.		996.25		1095.88	125.98	138.58
12 L.H., Daily Totals		$2177.75		$2702.13	$181.48	$225.18
Crew B-79B	Hr.	Daily	Hr.	Daily	Bare Costs	Incl. O&P
1 Laborer	$44.40	$355.20	$70.70	$565.60	$44.40	$70.70
1 Set of Gases		173.60		190.96	21.70	23.87
8 L.H., Daily Totals		$528.80		$756.56	$66.10	$94.57
Crew B-79C	Hr.	Daily	Hr.	Daily	Bare Costs	Incl. O&P
1 Labor Foreman (outside)	$46.40	$371.20	$73.90	$591.20	$45.31	$72.18
5 Laborers	44.40	1776.00	70.70	2828.00		
1 Truck Driver (light)	48.80	390.40	77.85	622.80		
1 Paint Striper, T.M., 120 Gal.		603.10		663.41		
1 Heating Kettle, 115 Gallon		107.25		117.97		
1 Flatbed Truck, Gas, 3 Ton		850.05		935.05		
3 Pickup Trucks, 3/4 Ton		336.60		370.26		
1 Air Compressor, 60 cfm		153.85		169.24		
1 -50' Air Hose, 3/4"		7.15		7.87		
1 Breaker, Pavement, 60 lb.		53.60		58.96	37.71	41.48
56 L.H., Daily Totals		$4649.20		$6364.76	$83.02	$113.66
Crew B-79D	Hr.	Daily	Hr.	Daily	Bare Costs	Incl. O&P
2 Labor Foremen (outside)	$46.40	$742.40	$73.90	$1182.40	$45.45	$72.39
5 Laborers	44.40	1776.00	70.70	2828.00		
1 Truck Driver (light)	48.80	390.40	77.85	622.80		
1 Paint Striper, T.M., 120 Gal.		603.10		663.41		
1 Heating Kettle, 115 Gallon		107.25		117.97		
1 Flatbed Truck, Gas, 3 Ton		850.05		935.05		
4 Pickup Trucks, 3/4 Ton		448.80		493.68		
1 Air Compressor, 60 cfm		153.85		169.24		
1 -50' Air Hose, 3/4"		7.15		7.87		
1 Breaker, Pavement, 60 lb.		53.60		58.96	34.75	38.22
64 L.H., Daily Totals		$5132.60		$7079.38	$80.20	$110.62
Crew B-79E	Hr.	Daily	Hr.	Daily	Bare Costs	Incl. O&P
2 Labor Foremen (outside)	$46.40	$742.40	$73.90	$1182.40	$45.24	$72.06
7 Laborers	44.40	2486.40	70.70	3959.20		
1 Truck Driver (light)	48.80	390.40	77.85	622.80		
1 Paint Striper, T.M., 120 Gal.		603.10		663.41		
1 Heating Kettle, 115 Gallon		107.25		117.97		
1 Flatbed Truck, Gas, 3 Ton		850.05		935.05		
5 Pickup Trucks, 3/4 Ton		561.00		617.10		
2 Air Compressors, 60 cfm		307.70		338.47		
2 -50' Air Hoses, 3/4"		14.30		15.73		
2 Breakers, Pavement, 60 lb.		107.20		117.92	31.88	35.07
80 L.H., Daily Totals		$6169.80		$8570.06	$77.12	$107.13
Crew B-80	Hr.	Daily	Hr.	Daily	Bare Costs	Incl. O&P
1 Labor Foreman (outside)	$46.40	$371.20	$73.90	$591.20	$48.77	$77.26
1 Laborer	44.40	355.20	70.70	565.60		
1 Truck Driver (light)	48.80	390.40	77.85	622.80		
1 Equip. Oper. (light)	55.50	444.00	86.60	692.80		
1 Flatbed Truck, Gas, 3 Ton		850.05		935.05		
1 Earth Auger, Truck-Mtd.		202.55		222.81	32.89	36.18
32 L.H., Daily Totals		$2613.40		$3630.26	$81.67	$113.45

For customer support on your **Commercial Renovation Costs** with RSMeans Data, call 800.448.8182.

775

Crew No.	Bare Costs		Incl. Subs O&P		Cost Per Labor-Hour	

Crew B-80A	Hr.	Daily	Hr.	Daily	Bare Costs	Incl. O&P
3 Laborers	$44.40	$1065.60	$70.70	$1696.80	$44.40	$70.70
1 Flatbed Truck, Gas, 3 Ton		850.05		935.05	35.42	38.96
24 L.H., Daily Totals		$1915.65		$2631.86	$79.82	$109.66

Crew B-80B	Hr.	Daily	Hr.	Daily	Bare Costs	Incl. O&P
3 Laborers	$44.40	$1065.60	$70.70	$1696.80	$47.17	$74.67
1 Equip. Oper. (light)	55.50	444.00	86.60	692.80		
1 Crane, Flatbed Mounted, 3 Ton		238.75		262.63	7.46	8.21
32 L.H., Daily Totals		$1748.35		$2652.22	$54.64	$82.88

Crew B-80C	Hr.	Daily	Hr.	Daily	Bare Costs	Incl. O&P
2 Laborers	$44.40	$710.40	$70.70	$1131.20	$45.87	$73.08
1 Truck Driver (light)	48.80	390.40	77.85	622.80		
1 Flatbed Truck, Gas, 1.5 Ton		198.50		218.35		
1 Manual Fence Post Auger, Gas		54.40		59.84	10.54	11.59
24 L.H., Daily Totals		$1353.70		$2032.19	$56.40	$84.67

Crew B-81	Hr.	Daily	Hr.	Daily	Bare Costs	Incl. O&P
1 Laborer	$44.40	$355.20	$70.70	$565.60	$51.57	$81.53
1 Equip. Oper. (medium)	59.00	472.00	92.05	736.40		
1 Truck Driver (heavy)	51.30	410.40	81.85	654.80		
1 Hydromulcher, T.M., 3000 Gal.		256.85		282.54		
1 Truck Tractor, 220 H.P.		310.80		341.88	23.65	26.02
24 L.H., Daily Totals		$1805.25		$2581.22	$75.22	$107.55

Crew B-81A	Hr.	Daily	Hr.	Daily	Bare Costs	Incl. O&P
1 Laborer	$44.40	$355.20	$70.70	$565.60	$46.60	$74.28
1 Truck Driver (light)	48.80	390.40	77.85	622.80		
1 Hydromulcher, T.M., 600 Gal.		118.15		129.97		
1 Flatbed Truck, Gas, 3 Ton		850.05		935.05	60.51	66.56
16 L.H., Daily Totals		$1713.80		$2253.42	$107.11	$140.84

Crew B-82	Hr.	Daily	Hr.	Daily	Bare Costs	Incl. O&P
1 Laborer	$44.40	$355.20	$70.70	$565.60	$49.95	$78.65
1 Equip. Oper. (light)	55.50	444.00	86.60	692.80		
1 Horiz. Borer, 6 H.P.		184.15		202.57	11.51	12.66
16 L.H., Daily Totals		$983.35		$1460.96	$61.46	$91.31

Crew B-82A	Hr.	Daily	Hr.	Daily	Bare Costs	Incl. O&P
2 Laborers	$44.40	$710.40	$70.70	$1131.20	$49.95	$78.65
2 Equip. Opers. (light)	55.50	888.00	86.60	1385.60		
2 Dump Truck, 8 C.Y., 220 H.P.		815.20		896.72		
1 Flatbed Trailer, 25 Ton		137.20		150.92		
1 Horiz. Dir. Drill, 20k lb. Thrust		544.10		598.51		
1 Mud Trailer for HDD, 1500 Gal.		312.15		343.37		
1 Pickup Truck, 4x4, 3/4 Ton		176.75		194.43		
1 Flatbed Trailer, 3 Ton		71.15		78.27		
1 Loader, Skid Steer, 78 H.P.		446.30		490.93	78.21	86.04
32 L.H., Daily Totals		$4101.25		$5269.94	$128.16	$164.69

Crew B-82B	Hr.	Daily	Hr.	Daily	Bare Costs	Incl. O&P
2 Laborers	$44.40	$710.40	$70.70	$1131.20	$49.95	$78.65
2 Equip. Opers. (light)	55.50	888.00	86.60	1385.60		
2 Dump Truck, 8 C.Y., 220 H.P.		815.20		896.72		
1 Flatbed Trailer, 25 Ton		137.20		150.92		
1 Horiz. Dir. Drill, 30k lb. Thrust		647.65		712.41		
1 Mud Trailer for HDD, 1500 Gal.		312.15		343.37		
1 Pickup Truck, 4x4, 3/4 Ton		176.75		194.43		
1 Flatbed Trailer, 3 Ton		71.15		78.27		
1 Loader, Skid Steer, 78 H.P.		446.30		490.93	81.45	89.59
32 L.H., Daily Totals		$4204.80		$5383.84	$131.40	$168.25

Crew B-82C	Hr.	Daily	Hr.	Daily	Bare Costs	Incl. O&P
2 Laborers	$44.40	$710.40	$70.70	$1131.20	$49.95	$78.65
2 Equip. Opers. (light)	55.50	888.00	86.60	1385.60		
2 Dump Truck, 8 C.Y., 220 H.P.		815.20		896.72		
1 Flatbed Trailer, 25 Ton		137.20		150.92		
1 Horiz. Dir. Drill, 50k lb. Thrust		824.05		906.46		
1 Mud Trailer for HDD, 1500 Gal.		312.15		343.37		
1 Pickup Truck, 4x4, 3/4 Ton		176.75		194.43		
1 Flatbed Trailer, 3 Ton		71.15		78.27		
1 Loader, Skid Steer, 78 H.P.		446.30		490.93	86.96	95.66
32 L.H., Daily Totals		$4381.20		$5577.88	$136.91	$174.31

Crew B-82D	Hr.	Daily	Hr.	Daily	Bare Costs	Incl. O&P
1 Equip. Oper. (light)	$55.50	$444.00	$86.60	$692.80	$55.50	$86.60
1 Mud Trailer for HDD, 1500 Gal.		312.15		343.37	39.02	42.92
8 L.H., Daily Totals		$756.15		$1036.17	$94.52	$129.52

Crew B-83	Hr.	Daily	Hr.	Daily	Bare Costs	Incl. O&P
1 Tugboat Captain	$59.00	$472.00	$92.05	$736.40	$51.70	$81.38
1 Tugboat Hand	44.40	355.20	70.70	565.60		
1 Tugboat, 250 H.P.		726.10		798.71	45.38	49.92
16 L.H., Daily Totals		$1553.30		$2100.71	$97.08	$131.29

Crew B-84	Hr.	Daily	Hr.	Daily	Bare Costs	Incl. O&P
1 Equip. Oper. (medium)	$59.00	$472.00	$92.05	$736.40	$59.00	$92.05
1 Rotary Mower/Tractor		371.15		408.26	46.39	51.03
8 L.H., Daily Totals		$843.15		$1144.67	$105.39	$143.08

Crew B-85	Hr.	Daily	Hr.	Daily	Bare Costs	Incl. O&P
3 Laborers	$44.40	$1065.60	$70.70	$1696.80	$48.70	$77.20
1 Equip. Oper. (medium)	59.00	472.00	92.05	736.40		
1 Truck Driver (heavy)	51.30	410.40	81.85	654.80		
1 Telescoping Boom Lift, to 80'		387.75		426.52		
1 Brush Chipper, 12", 130 H.P.		366.05		402.65		
1 Pruning Saw, Rotary		26.40		29.04	19.50	21.46
40 L.H., Daily Totals		$2728.20		$3946.22	$68.20	$98.66

Crew B-86	Hr.	Daily	Hr.	Daily	Bare Costs	Incl. O&P
1 Equip. Oper. (medium)	$59.00	$472.00	$92.05	$736.40	$59.00	$92.05
1 Stump Chipper, S.P.		189.20		208.12	23.65	26.02
8 L.H., Daily Totals		$661.20		$944.52	$82.65	$118.07

Crew B-86A	Hr.	Daily	Hr.	Daily	Bare Costs	Incl. O&P
1 Equip. Oper. (medium)	$59.00	$472.00	$92.05	$736.40	$59.00	$92.05
1 Grader, 30,000 Lbs.		1073.00		1180.30	134.13	147.54
8 L.H., Daily Totals		$1545.00		$1916.70	$193.13	$239.59

Crew B-86B	Hr.	Daily	Hr.	Daily	Bare Costs	Incl. O&P
1 Equip. Oper. (medium)	$59.00	$472.00	$92.05	$736.40	$59.00	$92.05
1 Dozer, 200 H.P.		1520.00		1672.00	190.00	209.00
8 L.H., Daily Totals		$1992.00		$2408.40	$249.00	$301.05

Crew B-87	Hr.	Daily	Hr.	Daily	Bare Costs	Incl. O&P
1 Laborer	$44.40	$355.20	$70.70	$565.60	$56.08	$87.78
4 Equip. Oper. (medium)	59.00	1888.00	92.05	2945.60		
2 Feller Bunchers, 100 H.P.		1257.60		1383.36		
1 Log Chipper, 22" Tree		555.00		610.50		
1 Dozer, 105 H.P.		640.80		704.88		
1 Chain Saw, Gas, 36" Long		41.65		45.81	62.38	68.61
40 L.H., Daily Totals		$4738.25		$6255.76	$118.46	$156.39

For customer support on your Commercial Renovation Costs with RSMeans Data, call 800.448.8182.

Crew B-88

Crew B-88	Hr.	Daily	Hr.	Daily	Bare Costs	Incl. O&P
1 Laborer	$44.40	$355.20	$70.70	$565.60	$56.91	$89.00
6 Equip. Oper. (medium)	59.00	2832.00	92.05	4418.40		
2 Feller Bunchers, 100 H.P.		1257.60		1383.36		
1 Log Chipper, 22" Tree		555.00		610.50		
2 Log Skidders, 50 H.P.		1826.70		2009.37		
1 Dozer, 105 H.P.		640.80		704.88		
1 Chain Saw, Gas, 36" Long		41.65		45.81	77.17	84.89
56 L.H., Daily Totals		$7508.95		$9737.92	$134.09	$173.89

Crew B-89

Crew B-89	Hr.	Daily	Hr.	Daily	Bare Costs	Incl. O&P
1 Equip. Oper. (light)	$55.50	$444.00	$86.60	$692.80	$52.15	$82.22
1 Truck Driver (light)	48.80	390.40	77.85	622.80		
1 Flatbed Truck, Gas, 3 Ton		850.05		935.05		
1 Concrete Saw		112.85		124.14		
1 Water Tank, 65 Gal.		102.90		113.19	66.61	73.27
16 L.H., Daily Totals		$1900.20		$2487.98	$118.76	$155.50

Crew B-89A

Crew B-89A	Hr.	Daily	Hr.	Daily	Bare Costs	Incl. O&P
1 Skilled Worker	$57.10	$456.80	$90.50	$724.00	$50.75	$80.60
1 Laborer	44.40	355.20	70.70	565.60		
1 Core Drill (Large)		121.60		133.76	7.60	8.36
16 L.H., Daily Totals		$933.60		$1423.36	$58.35	$88.96

Crew B-89B

Crew B-89B	Hr.	Daily	Hr.	Daily	Bare Costs	Incl. O&P
1 Equip. Oper. (light)	$55.50	$444.00	$86.60	$692.80	$52.15	$82.22
1 Truck Driver (light)	48.80	390.40	77.85	622.80		
1 Wall Saw, Hydraulic, 10 H.P.		86.40		95.04		
1 Generator, Diesel, 100 kW		521.85		574.03		
1 Water Tank, 65 Gal.		102.90		113.19		
1 Flatbed Truck, Gas, 3 Ton		850.05		935.05	97.58	107.33
16 L.H., Daily Totals		$2395.60		$3032.92	$149.72	$189.56

Crew B-89C

Crew B-89C	Hr.	Daily	Hr.	Daily	Bare Costs	Incl. O&P
1 Cement Finisher	$51.80	$414.40	$81.10	$648.80	$51.80	$81.10
1 Masonry cut-off saw, gas		58.15		63.97	7.27	8.00
8 L.H., Daily Totals		$472.55		$712.76	$59.07	$89.10

Crew B-90

Crew B-90	Hr.	Daily	Hr.	Daily	Bare Costs	Incl. O&P
1 Labor Foreman (outside)	$46.40	$371.20	$73.90	$591.20	$49.15	$77.86
3 Laborers	44.40	1065.60	70.70	1696.80		
2 Equip. Oper. (light)	55.50	888.00	86.60	1385.60		
2 Truck Drivers (heavy)	51.30	820.80	81.85	1309.60		
1 Road Mixer, 310 H.P.		1919.00		2110.90		
1 Dist. Truck, 2000 Gal.		303.25		333.57	34.72	38.19
64 L.H., Daily Totals		$5367.85		$7427.68	$83.87	$116.06

Crew B-90A

Crew B-90A	Hr.	Daily	Hr.	Daily	Bare Costs	Incl. O&P
1 Labor Foreman (outside)	$46.40	$371.20	$73.90	$591.20	$53.03	$83.36
2 Laborers	44.40	710.40	70.70	1131.20		
4 Equip. Oper. (medium)	59.00	1888.00	92.05	2945.60		
2 Graders, 30,000 Lbs.		2146.00		2360.60		
1 Tandem Roller, 10 Ton		246.80		271.48		
1 Roller, Pneum. Whl., 12 Ton		349.90		384.89	48.98	53.87
56 L.H., Daily Totals		$5712.30		$7684.97	$102.01	$137.23

Crew B-90B

Crew B-90B	Hr.	Daily	Hr.	Daily	Bare Costs	Incl. O&P
1 Labor Foreman (outside)	$46.40	$371.20	$73.90	$591.20	$52.03	$81.91
2 Laborers	44.40	710.40	70.70	1131.20		
3 Equip. Oper. (medium)	59.00	1416.00	92.05	2209.20		
1 Roller, Pneum. Whl., 12 Ton		349.90		384.89		
1 Road Mixer, 310 H.P.		1919.00		2110.90	47.27	52.00
48 L.H., Daily Totals		$4766.50		$6427.39	$99.30	$133.90

Crew B-90C

Crew B-90C	Hr.	Daily	Hr.	Daily	Bare Costs	Incl. O&P
1 Labor Foreman (outside)	$46.40	$371.20	$73.90	$591.20	$50.45	$79.85
4 Laborers	44.40	1420.80	70.70	2262.40		
3 Equip. Oper. (medium)	59.00	1416.00	92.05	2209.20		
3 Truck Drivers (heavy)	51.30	1231.20	81.85	1964.40		
3 Road Mixers, 310 H.P.		5757.00		6332.70	65.42	71.96
88 L.H., Daily Totals		$10196.20		$13359.90	$115.87	$151.82

Crew B-90D

Crew B-90D	Hr.	Daily	Hr.	Daily	Bare Costs	Incl. O&P
1 Labor Foreman (outside)	$46.40	$371.20	$73.90	$591.20	$49.52	$78.45
6 Laborers	44.40	2131.20	70.70	3393.60		
3 Equip. Oper. (medium)	59.00	1416.00	92.05	2209.20		
3 Truck Drivers (heavy)	51.30	1231.20	81.85	1964.40		
3 Road Mixers, 310 H.P.		5757.00		6332.70	55.36	60.89
104 L.H., Daily Totals		$10906.60		$14491.10	$104.87	$139.34

Crew B-90E

Crew B-90E	Hr.	Daily	Hr.	Daily	Bare Costs	Incl. O&P
1 Labor Foreman (outside)	$46.40	$371.20	$73.90	$591.20	$50.26	$79.41
4 Laborers	44.40	1420.80	70.70	2262.40		
3 Equip. Oper. (medium)	59.00	1416.00	92.05	2209.20		
1 Truck Driver (heavy)	51.30	410.40	81.85	654.80		
1 Road Mixer, 310 H.P.		1919.00		2110.90	26.65	29.32
72 L.H., Daily Totals		$5537.40		$7828.50	$76.91	$108.73

Crew B-91

Crew B-91	Hr.	Daily	Hr.	Daily	Bare Costs	Incl. O&P
1 Labor Foreman (outside)	$46.40	$371.20	$73.90	$591.20	$52.81	$83.17
2 Laborers	44.40	710.40	70.70	1131.20		
4 Equip. Oper. (medium)	59.00	1888.00	92.05	2945.60		
1 Truck Driver (heavy)	51.30	410.40	81.85	654.80		
1 Dist. Tanker, 3000 Gallon		334.10		367.51		
1 Truck Tractor, 6x4, 380 H.P.		499.15		549.07		
1 Aggreg. Spreader, S.P.		859.10		945.01		
1 Roller, Pneum. Whl., 12 Ton		349.90		384.89		
1 Tandem Roller, 10 Ton		246.80		271.48	35.77	39.34
64 L.H., Daily Totals		$5669.05		$7840.76	$88.58	$122.51

Crew B-91A

Crew B-91A	Hr.	Daily	Hr.	Daily	Bare Costs	Incl. O&P
2 Truck Drivers (light)	$48.80	$780.80	$77.85	$1245.60	$48.80	$77.85
1 Road Sweeper, S.P., 8' wide		723.65		796.01		
1 Pickup Truck, 3/4 Ton		112.20		123.42	52.24	57.46
16 L.H., Daily Totals		$1616.65		$2165.03	$101.04	$135.31

Crew B-91B

Crew B-91B	Hr.	Daily	Hr.	Daily	Bare Costs	Incl. O&P
1 Laborer	$44.40	$355.20	$70.70	$565.60	$51.70	$81.38
1 Equipment Oper. (med.)	59.00	472.00	92.05	736.40		
1 Road Sweeper, Vac. Assist.		879.45		967.39	54.97	60.46
16 L.H., Daily Totals		$1706.65		$2269.40	$106.67	$141.84

Crew B-91C

Crew B-91C	Hr.	Daily	Hr.	Daily	Bare Costs	Incl. O&P
1 Laborer	$44.40	$355.20	$70.70	$565.60	$46.60	$74.28
1 Truck Driver (light)	48.80	390.40	77.85	622.80		
1 Catch Basin Cleaning Truck		542.60		596.86	33.91	37.30
16 L.H., Daily Totals		$1288.20		$1785.26	$80.51	$111.58

For customer support on your Commercial Renovation Costs with RSMeans Data, call 800.448.8182.

Crew No.	Bare Costs		Incl. Subs O&P		Cost Per Labor-Hour	

Crew B-91D	Hr.	Daily	Hr.	Daily	Bare Costs	Incl. O&P
1 Labor Foreman (outside)	$46.40	$371.20	$73.90	$591.20	$51.23	$80.87
5 Laborers	44.40	1776.00	70.70	2828.00		
5 Equip. Oper. (medium)	59.00	2360.00	92.05	3682.00		
2 Truck Drivers (heavy)	51.30	820.80	81.85	1309.60		
1 Aggreg. Spreader, S.P.		859.10		945.01		
2 Truck Tractors, 6x4, 380 H.P.		998.30		1098.13		
2 Dist. Tankers, 3000 Gallon		668.20		735.02		
2 Pavement Brushes, Towed		176.70		194.37		
2 Rollers Pneum. Whl., 12 Ton		699.80		769.78	32.71	35.98
104 L.H., Daily Totals		$8730.10		$12153.11	$83.94	$116.86

Crew B-92	Hr.	Daily	Hr.	Daily	Bare Costs	Incl. O&P
1 Labor Foreman (outside)	$46.40	$371.20	$73.90	$591.20	$44.90	$71.50
3 Laborers	44.40	1065.60	70.70	1696.80		
1 Crack Cleaner, 25 H.P.		53.00		58.30		
1 Air Compressor, 60 cfm		153.85		169.24		
1 Tar Kettle, T.M.		156.70		172.37		
1 Flatbed Truck, Gas, 3 Ton		850.05		935.05	37.92	41.72
32 L.H., Daily Totals		$2650.40		$3622.96	$82.83	$113.22

Crew B-93	Hr.	Daily	Hr.	Daily	Bare Costs	Incl. O&P
1 Equip. Oper. (medium)	$59.00	$472.00	$92.05	$736.40	$59.00	$92.05
1 Feller Buncher, 100 H.P.		628.80		691.68	78.60	86.46
8 L.H., Daily Totals		$1100.80		$1428.08	$137.60	$178.51

Crew B-94A	Hr.	Daily	Hr.	Daily	Bare Costs	Incl. O&P
1 Laborer	$44.40	$355.20	$70.70	$565.60	$44.40	$70.70
1 Diaphragm Water Pump, 2"		87.70		96.47		
1 -20' Suction Hose, 2"		3.55		3.90		
2 -50' Discharge Hoses, 2"		8.00		8.80	12.41	13.65
8 L.H., Daily Totals		$454.45		$674.77	$56.81	$84.35

Crew B-94B	Hr.	Daily	Hr.	Daily	Bare Costs	Incl. O&P
1 Laborer	$44.40	$355.20	$70.70	$565.60	$44.40	$70.70
1 Diaphragm Water Pump, 4"		106.35		116.99		
1 -20' Suction Hose, 4"		17.25		18.98		
2 -50' Discharge Hoses, 4"		25.60		28.16	18.65	20.52
8 L.H., Daily Totals		$504.40		$729.72	$63.05	$91.22

Crew B-94C	Hr.	Daily	Hr.	Daily	Bare Costs	Incl. O&P
1 Laborer	$44.40	$355.20	$70.70	$565.60	$44.40	$70.70
1 Centrifugal Water Pump, 3"		74.40		81.84		
1 -20' Suction Hose, 3"		8.75		9.63		
2 -50' Discharge Hoses, 3"		9.00		9.90	11.52	12.67
8 L.H., Daily Totals		$447.35		$666.97	$55.92	$83.37

Crew B-94D	Hr.	Daily	Hr.	Daily	Bare Costs	Incl. O&P
1 Laborer	$44.40	$355.20	$70.70	$565.60	$44.40	$70.70
1 Centr. Water Pump, 6"		235.25		258.77		
1 -20' Suction Hose, 6"		25.50		28.05		
2 -50' Discharge Hoses, 6"		36.20		39.82	37.12	40.83
8 L.H., Daily Totals		$652.15		$892.25	$81.52	$111.53

Crew C-1	Hr.	Daily	Hr.	Daily	Bare Costs	Incl. O&P
3 Carpenters	$54.70	$1312.80	$87.10	$2090.40	$52.13	$83.00
1 Laborer	44.40	355.20	70.70	565.60		
32 L.H., Daily Totals		$1668.00		$2656.00	$52.13	$83.00

Crew C-2	Hr.	Daily	Hr.	Daily	Bare Costs	Incl. O&P
1 Carpenter Foreman (outside)	$56.70	$453.60	$90.30	$722.40	$53.32	$84.90
4 Carpenters	54.70	1750.40	87.10	2787.20		
1 Laborer	44.40	355.20	70.70	565.60		
48 L.H., Daily Totals		$2559.20		$4075.20	$53.32	$84.90

Crew C-2A	Hr.	Daily	Hr.	Daily	Bare Costs	Incl. O&P
1 Carpenter Foreman (outside)	$56.70	$453.60	$90.30	$722.40	$52.83	$83.90
3 Carpenters	54.70	1312.80	87.10	2090.40		
1 Cement Finisher	51.80	414.40	81.10	648.80		
1 Laborer	44.40	355.20	70.70	565.60		
48 L.H., Daily Totals		$2536.00		$4027.20	$52.83	$83.90

Crew C-3	Hr.	Daily	Hr.	Daily	Bare Costs	Incl. O&P
1 Rodman Foreman (outside)	$60.90	$487.20	$95.30	$762.40	$55.10	$86.51
4 Rodmen (reinf.)	58.90	1884.80	92.20	2950.40		
1 Equip. Oper. (light)	55.50	444.00	86.60	692.80		
2 Laborers	44.40	710.40	70.70	1131.20		
3 Stressing Equipment		56.85		62.53		
.5 Grouting Equipment		123.33		135.66	2.82	3.10
64 L.H., Daily Totals		$3706.57		$5734.99	$57.92	$89.61

Crew C-4	Hr.	Daily	Hr.	Daily	Bare Costs	Incl. O&P
1 Rodman Foreman (outside)	$60.90	$487.20	$95.30	$762.40	$59.40	$92.97
3 Rodmen (reinf.)	58.90	1413.60	92.20	2212.80		
3 Stressing Equipment		56.85		62.53	1.78	1.95
32 L.H., Daily Totals		$1957.65		$3037.74	$61.18	$94.93

Crew C-4A	Hr.	Daily	Hr.	Daily	Bare Costs	Incl. O&P
2 Rodmen (reinf.)	$58.90	$942.40	$92.20	$1475.20	$58.90	$92.20
4 Stressing Equipment		75.80		83.38	4.74	5.21
16 L.H., Daily Totals		$1018.20		$1558.58	$63.64	$97.41

Crew C-5	Hr.	Daily	Hr.	Daily	Bare Costs	Incl. O&P
1 Rodman Foreman (outside)	$60.90	$487.20	$95.30	$762.40	$58.64	$91.69
4 Rodmen (reinf.)	58.90	1884.80	92.20	2950.40		
1 Equip. Oper. (crane)	61.45	491.60	95.85	766.80		
1 Equip. Oper. (oiler)	52.50	420.00	81.90	655.20		
1 Hyd. Crane, 25 Ton		586.70		645.37	10.48	11.52
56 L.H., Daily Totals		$3870.30		$5780.17	$69.11	$103.22

Crew C-6	Hr.	Daily	Hr.	Daily	Bare Costs	Incl. O&P
1 Labor Foreman (outside)	$46.40	$371.20	$73.90	$591.20	$45.97	$72.97
4 Laborers	44.40	1420.80	70.70	2262.40		
1 Cement Finisher	51.80	414.40	81.10	648.80		
2 Gas Engine Vibrators		54.30		59.73	1.13	1.24
48 L.H., Daily Totals		$2260.70		$3562.13	$47.10	$74.21

Crew C-6A	Hr.	Daily	Hr.	Daily	Bare Costs	Incl. O&P
2 Cement Finishers	$51.80	$828.80	$81.10	$1297.60	$51.80	$81.10
1 Concrete Vibrator, Elec, 2 HP		45.80		50.38	2.86	3.15
16 L.H., Daily Totals		$874.60		$1347.98	$54.66	$84.25

Crew No.	Bare Costs		Incl. Subs O&P		Cost Per Labor-Hour	
Crew C-7	Hr.	Daily	Hr.	Daily	Bare Costs	Incl. O&P
1 Labor Foreman (outside)	$46.40	$371.20	$73.90	$591.20	$47.97	$75.83
5 Laborers	44.40	1776.00	70.70	2828.00		
1 Cement Finisher	51.80	414.40	81.10	648.80		
1 Equip. Oper. (medium)	59.00	472.00	92.05	736.40		
1 Equip. Oper. (oiler)	52.50	420.00	81.90	655.20		
2 Gas Engine Vibrators		54.30		59.73		
1 Concrete Bucket, 1 C.Y.		45.90		50.49		
1 Hyd. Crane, 55 Ton		990.15		1089.17	15.14	16.66
72 L.H., Daily Totals		$4543.95		$6658.98	$63.11	$92.49
Crew C-8	Hr.	Daily	Hr.	Daily	Bare Costs	Incl. O&P
1 Labor Foreman (outside)	$46.40	$371.20	$73.90	$591.20	$48.89	$77.18
3 Laborers	44.40	1065.60	70.70	1696.80		
2 Cement Finishers	51.80	828.80	81.10	1297.60		
1 Equip. Oper. (medium)	59.00	472.00	92.05	736.40		
1 Concrete Pump (Small)		423.65		466.01	7.57	8.32
56 L.H., Daily Totals		$3161.25		$4788.02	$56.45	$85.50
Crew C-8A	Hr.	Daily	Hr.	Daily	Bare Costs	Incl. O&P
1 Labor Foreman (outside)	$46.40	$371.20	$73.90	$591.20	$47.20	$74.70
3 Laborers	44.40	1065.60	70.70	1696.80		
2 Cement Finishers	51.80	828.80	81.10	1297.60		
48 L.H., Daily Totals		$2265.60		$3585.60	$47.20	$74.70
Crew C-8B	Hr.	Daily	Hr.	Daily	Bare Costs	Incl. O&P
1 Labor Foreman (outside)	$46.40	$371.20	$73.90	$591.20	$47.72	$75.61
3 Laborers	44.40	1065.60	70.70	1696.80		
1 Equip. Oper. (medium)	59.00	472.00	92.05	736.40		
1 Vibrating Power Screed		87.65		96.42		
1 Roller, Vibratory, 25 Ton		672.35		739.59		
1 Dozer, 200 H.P.		1520.00		1672.00	57.00	62.70
40 L.H., Daily Totals		$4188.80		$5532.40	$104.72	$138.31
Crew C-8C	Hr.	Daily	Hr.	Daily	Bare Costs	Incl. O&P
1 Labor Foreman (outside)	$46.40	$371.20	$73.90	$591.20	$48.40	$76.53
3 Laborers	44.40	1065.60	70.70	1696.80		
1 Cement Finisher	51.80	414.40	81.10	648.80		
1 Equip. Oper. (medium)	59.00	472.00	92.05	736.40		
1 Shotcrete Rig, 12 C.Y./hr		269.20		296.12		
1 Air Compressor, 160 cfm		212.30		233.53		
4 -50' Air Hoses, 1"		32.20		35.42		
4 -50' Air Hoses, 2"		115.80		127.38	13.11	14.43
48 L.H., Daily Totals		$2952.70		$4365.65	$61.51	$90.95
Crew C-8D	Hr.	Daily	Hr.	Daily	Bare Costs	Incl. O&P
1 Labor Foreman (outside)	$46.40	$371.20	$73.90	$591.20	$49.52	$78.08
1 Laborer	44.40	355.20	70.70	565.60		
1 Cement Finisher	51.80	414.40	81.10	648.80		
1 Equipment Oper. (light)	55.50	444.00	86.60	692.80		
1 Air Compressor, 250 cfm		202.85		223.13		
2 -50' Air Hoses, 1"		16.10		17.71	6.84	7.53
32 L.H., Daily Totals		$1803.75		$2739.24	$56.37	$85.60

Crew No.	Bare Costs		Incl. Subs O&P		Cost Per Labor-Hour	
Crew C-8E	Hr.	Daily	Hr.	Daily	Bare Costs	Incl. O&P
1 Labor Foreman (outside)	$46.40	$371.20	$73.90	$591.20	$47.82	$75.62
3 Laborers	44.40	1065.60	70.70	1696.80		
1 Cement Finisher	51.80	414.40	81.10	648.80		
1 Equipment Oper. (light)	55.50	444.00	86.60	692.80		
1 Shotcrete Rig, 35 C.Y./hr.		301.05		331.15		
1 Air Compressor, 250 cfm		202.85		223.13		
4 -50' Air Hoses, 1"		32.20		35.42		
4 -50' Air Hoses, 2"		115.80		127.38	13.58	14.94
48 L.H., Daily Totals		$2947.10		$4346.69	$61.40	$90.56
Crew C-9	Hr.	Daily	Hr.	Daily	Bare Costs	Incl. O&P
1 Cement Finisher	$51.80	$414.40	$81.10	$648.80	$49.02	$77.28
2 Laborers	44.40	710.40	70.70	1131.20		
1 Equipment Oper. (light)	55.50	444.00	86.60	692.80		
1 Grout Pump, 50 C.F./hr.		190.35		209.38		
1 Air Compressor, 160 cfm		212.30		233.53		
2 -50' Air Hoses, 1"		16.10		17.71		
2 -50' Air Hoses, 2"		57.90		63.69	14.90	16.38
32 L.H., Daily Totals		$2045.45		$2997.11	$63.92	$93.66
Crew C-10	Hr.	Daily	Hr.	Daily	Bare Costs	Incl. O&P
1 Laborer	$44.40	$355.20	$70.70	$565.60	$49.33	$77.63
2 Cement Finishers	51.80	828.80	81.10	1297.60		
24 L.H., Daily Totals		$1184.00		$1863.20	$49.33	$77.63
Crew C-10B	Hr.	Daily	Hr.	Daily	Bare Costs	Incl. O&P
3 Laborers	$44.40	$1065.60	$70.70	$1696.80	$47.36	$74.86
2 Cement Finishers	51.80	828.80	81.10	1297.60		
1 Concrete Mixer, 10 C.F.		147.15		161.87		
2 Trowels, 48" Walk-Behind		188.60		207.46	8.39	9.23
40 L.H., Daily Totals		$2230.15		$3363.72	$55.75	$84.09
Crew C-10C	Hr.	Daily	Hr.	Daily	Bare Costs	Incl. O&P
1 Laborer	$44.40	$355.20	$70.70	$565.60	$49.33	$77.63
2 Cement Finishers	51.80	828.80	81.10	1297.60		
1 Trowel, 48" Walk-Behind		94.30		103.73	3.93	4.32
24 L.H., Daily Totals		$1278.30		$1966.93	$53.26	$81.96
Crew C-10D	Hr.	Daily	Hr.	Daily	Bare Costs	Incl. O&P
1 Laborer	$44.40	$355.20	$70.70	$565.60	$49.33	$77.63
2 Cement Finishers	51.80	828.80	81.10	1297.60		
1 Vibrating Power Screed		87.65		96.42		
1 Trowel, 48" Walk-Behind		94.30		103.73	7.58	8.34
24 L.H., Daily Totals		$1365.95		$2063.34	$56.91	$85.97
Crew C-10E	Hr.	Daily	Hr.	Daily	Bare Costs	Incl. O&P
1 Laborer	$44.40	$355.20	$70.70	$565.60	$49.33	$77.63
2 Cement Finishers	51.80	828.80	81.10	1297.60		
1 Vibrating Power Screed		87.65		96.42		
1 Cement Trowel, 96" Ride-On		171.05		188.16	10.78	11.86
24 L.H., Daily Totals		$1442.70		$2147.77	$60.11	$89.49
Crew C-10F	Hr.	Daily	Hr.	Daily	Bare Costs	Incl. O&P
1 Laborer	$44.40	$355.20	$70.70	$565.60	$49.33	$77.63
2 Cement Finishers	51.80	828.80	81.10	1297.60		
1 Telescoping Boom Lift, to 60'		292.45		321.69	12.19	13.40
24 L.H., Daily Totals		$1476.45		$2184.90	$61.52	$91.04

Crews - Renovation

Crew No.	Bare Costs		Incl. Subs O&P		Cost Per Labor-Hour	
Crew C-11	Hr.	Daily	Hr.	Daily	Bare Costs	Incl. O&P
1 Struc. Steel Foreman (outside)	$62.30	$498.40	$100.75	$806.00	$59.78	$95.98
6 Struc. Steel Workers	60.30	2894.40	97.55	4682.40		
1 Equip. Oper. (crane)	61.45	491.60	95.85	766.80		
1 Equip. Oper. (oiler)	52.50	420.00	81.90	655.20		
1 Lattice Boom Crane, 150 Ton		2324.00		2556.40	32.28	35.51
72 L.H., Daily Totals		$6628.40		$9466.80	$92.06	$131.48
Crew C-12	Hr.	Daily	Hr.	Daily	Bare Costs	Incl. O&P
1 Carpenter Foreman (outside)	$56.70	$453.60	$90.30	$722.40	$54.44	$86.36
3 Carpenters	54.70	1312.80	87.10	2090.40		
1 Laborer	44.40	355.20	70.70	565.60		
1 Equip. Oper. (crane)	61.45	491.60	95.85	766.80		
1 Hyd. Crane, 12 Ton		475.80		523.38	9.91	10.90
48 L.H., Daily Totals		$3089.00		$4668.58	$64.35	$97.26
Crew C-13	Hr.	Daily	Hr.	Daily	Bare Costs	Incl. O&P
1 Struc. Steel Worker	$60.30	$482.40	$97.55	$780.40	$58.43	$94.07
1 Welder	60.30	482.40	97.55	780.40		
1 Carpenter	54.70	437.60	87.10	696.80		
1 Welder, Gas Engine, 300 amp		148.75		163.63	6.20	6.82
24 L.H., Daily Totals		$1551.15		$2421.22	$64.63	$100.88
Crew C-14	Hr.	Daily	Hr.	Daily	Bare Costs	Incl. O&P
1 Carpenter Foreman (outside)	$56.70	$453.60	$90.30	$722.40	$53.39	$84.30
5 Carpenters	54.70	2188.00	87.10	3484.00		
4 Laborers	44.40	1420.80	70.70	2262.40		
4 Rodmen (reinf.)	58.90	1884.80	92.20	2950.40		
2 Cement Finishers	51.80	828.80	81.10	1297.60		
1 Equip. Oper. (crane)	61.45	491.60	95.85	766.80		
1 Equip. Oper. (oiler)	52.50	420.00	81.90	655.20		
1 Hyd. Crane, 80 Ton		1458.00		1603.80	10.13	11.14
144 L.H., Daily Totals		$9145.60		$13742.60	$63.51	$95.43
Crew C-14A	Hr.	Daily	Hr.	Daily	Bare Costs	Incl. O&P
1 Carpenter Foreman (outside)	$56.70	$453.60	$90.30	$722.40	$54.68	$86.69
16 Carpenters	54.70	7001.60	87.10	11148.80		
4 Rodmen (reinf.)	58.90	1884.80	92.20	2950.40		
2 Laborers	44.40	710.40	70.70	1131.20		
1 Cement Finisher	51.80	414.40	81.10	648.80		
1 Equip. Oper. (medium)	59.00	472.00	92.05	736.40		
1 Gas Engine Vibrator		27.15		29.86		
1 Concrete Pump (Small)		423.65		466.01	2.25	2.48
200 L.H., Daily Totals		$11387.60		$17833.88	$56.94	$89.17
Crew C-14B	Hr.	Daily	Hr.	Daily	Bare Costs	Incl. O&P
1 Carpenter Foreman (outside)	$56.70	$453.60	$90.30	$722.40	$54.57	$86.47
16 Carpenters	54.70	7001.60	87.10	11148.80		
4 Rodmen (reinf.)	58.90	1884.80	92.20	2950.40		
2 Laborers	44.40	710.40	70.70	1131.20		
2 Cement Finishers	51.80	828.80	81.10	1297.60		
1 Equip. Oper. (medium)	59.00	472.00	92.05	736.40		
1 Gas Engine Vibrator		27.15		29.86		
1 Concrete Pump (Small)		423.65		466.01	2.17	2.38
208 L.H., Daily Totals		$11802.00		$18482.68	$56.74	$88.86
Crew C-14C	Hr.	Daily	Hr.	Daily	Bare Costs	Incl. O&P
1 Carpenter Foreman (outside)	$56.70	$453.60	$90.30	$722.40	$52.29	$82.94
6 Carpenters	54.70	2625.60	87.10	4180.80		
2 Rodmen (reinf.)	58.90	942.40	92.20	1475.20		
4 Laborers	44.40	1420.80	70.70	2262.40		
1 Cement Finisher	51.80	414.40	81.10	648.80		
1 Gas Engine Vibrator		27.15		29.86	.24	.27
112 L.H., Daily Totals		$5883.95		$9319.47	$52.54	$83.21
Crew C-14D	Hr.	Daily	Hr.	Daily	Bare Costs	Incl. O&P
1 Carpenter Foreman (outside)	$56.70	$453.60	$90.30	$722.40	$54.35	$86.28
18 Carpenters	54.70	7876.80	87.10	12542.40		
2 Rodmen (reinf.)	58.90	942.40	92.20	1475.20		
2 Laborers	44.40	710.40	70.70	1131.20		
1 Cement Finisher	51.80	414.40	81.10	648.80		
1 Equip. Oper. (medium)	59.00	472.00	92.05	736.40		
1 Gas Engine Vibrator		27.15		29.86		
1 Concrete Pump (Small)		423.65		466.01	2.25	2.48
200 L.H., Daily Totals		$11320.40		$17752.28	$56.60	$88.76
Crew C-14E	Hr.	Daily	Hr.	Daily	Bare Costs	Incl. O&P
1 Carpenter Foreman (outside)	$56.70	$453.60	$90.30	$722.40	$53.34	$84.23
2 Carpenters	54.70	875.20	87.10	1393.60		
4 Rodmen (reinf.)	58.90	1884.80	92.20	2950.40		
3 Laborers	44.40	1065.60	70.70	1696.80		
1 Cement Finisher	51.80	414.40	81.10	648.80		
1 Gas Engine Vibrator		27.15		29.86	.31	.34
88 L.H., Daily Totals		$4720.75		$7441.86	$53.64	$84.57
Crew C-14F	Hr.	Daily	Hr.	Daily	Bare Costs	Incl. O&P
1 Labor Foreman (outside)	$46.40	$371.20	$73.90	$591.20	$49.56	$77.99
2 Laborers	44.40	710.40	70.70	1131.20		
6 Cement Finishers	51.80	2486.40	81.10	3892.80		
1 Gas Engine Vibrator		27.15		29.86	.38	.41
72 L.H., Daily Totals		$3595.15		$5645.06	$49.93	$78.40
Crew C-14G	Hr.	Daily	Hr.	Daily	Bare Costs	Incl. O&P
1 Labor Foreman (outside)	$46.40	$371.20	$73.90	$591.20	$48.91	$77.10
2 Laborers	44.40	710.40	70.70	1131.20		
4 Cement Finishers	51.80	1657.60	81.10	2595.20		
1 Gas Engine Vibrator		27.15		29.86	.48	.53
56 L.H., Daily Totals		$2766.35		$4347.47	$49.40	$77.63
Crew C-14H	Hr.	Daily	Hr.	Daily	Bare Costs	Incl. O&P
1 Carpenter Foreman (outside)	$56.70	$453.60	$90.30	$722.40	$53.53	$84.75
2 Carpenters	54.70	875.20	87.10	1393.60		
1 Rodman (reinf.)	58.90	471.20	92.20	737.60		
1 Laborer	44.40	355.20	70.70	565.60		
1 Cement Finisher	51.80	414.40	81.10	648.80		
1 Gas Engine Vibrator		27.15		29.86	.57	.62
48 L.H., Daily Totals		$2596.75		$4097.86	$54.10	$85.37
Crew C-14L	Hr.	Daily	Hr.	Daily	Bare Costs	Incl. O&P
1 Carpenter Foreman (outside)	$56.70	$453.60	$90.30	$722.40	$51.19	$81.40
6 Carpenters	54.70	2625.60	87.10	4180.80		
4 Laborers	44.40	1420.80	70.70	2262.40		
1 Cement Finisher	51.80	414.40	81.10	648.80		
1 Gas Engine Vibrator		27.15		29.86	.28	.31
96 L.H., Daily Totals		$4941.55		$7844.27	$51.47	$81.71

For customer support on your Commercial Renovation Costs with RSMeans Data, call 800.448.8182.

Crew No.	Bare Costs		Incl. Subs O&P		Cost Per Labor-Hour	

Crew C-14M	Hr.	Daily	Hr.	Daily	Bare Costs	Incl. O&P
1 Carpenter Foreman (outside)	$56.70	$453.60	$90.30	$722.40	$53.08	$83.91
2 Carpenters	54.70	875.20	87.10	1393.60		
1 Rodman (reinf.)	58.90	471.20	92.20	737.60		
2 Laborers	44.40	710.40	70.70	1131.20		
1 Cement Finisher	51.80	414.40	81.10	648.80		
1 Equip. Oper. (medium)	59.00	472.00	92.05	736.40		
1 Gas Engine Vibrator		27.15		29.86		
1 Concrete Pump (Small)		423.65		466.01	7.04	7.75
64 L.H., Daily Totals		$3847.60		$5865.88	$60.12	$91.65

Crew C-15	Hr.	Daily	Hr.	Daily	Bare Costs	Incl. O&P
1 Carpenter Foreman (outside)	$56.70	$453.60	$90.30	$722.40	$51.31	$81.22
2 Carpenters	54.70	875.20	87.10	1393.60		
3 Laborers	44.40	1065.60	70.70	1696.80		
2 Cement Finishers	51.80	828.80	81.10	1297.60		
1 Rodman (reinf.)	58.90	471.20	92.20	737.60		
72 L.H., Daily Totals		$3694.40		$5848.00	$51.31	$81.22

Crew C-16	Hr.	Daily	Hr.	Daily	Bare Costs	Incl. O&P
1 Labor Foreman (outside)	$46.40	$371.20	$73.90	$591.20	$48.89	$77.18
3 Laborers	44.40	1065.60	70.70	1696.80		
2 Cement Finishers	51.80	828.80	81.10	1297.60		
1 Equip. Oper. (medium)	59.00	472.00	92.05	736.40		
1 Gunite Pump Rig		321.75		353.93		
2 -50' Air Hoses, 3/4"		14.30		15.73		
2 -50' Air Hoses, 2"		57.90		63.69	7.03	7.74
56 L.H., Daily Totals		$3131.55		$4755.35	$55.92	$84.92

Crew C-16A	Hr.	Daily	Hr.	Daily	Bare Costs	Incl. O&P
1 Laborer	$44.40	$355.20	$70.70	$565.60	$51.75	$81.24
2 Cement Finishers	51.80	828.80	81.10	1297.60		
1 Equip. Oper. (medium)	59.00	472.00	92.05	736.40		
1 Gunite Pump Rig		321.75		353.93		
2 -50' Air Hoses, 3/4"		14.30		15.73		
2 -50' Air Hoses, 2"		57.90		63.69		
1 Telescoping Boom Lift, to 60'		292.45		321.69	21.45	23.59
32 L.H., Daily Totals		$2342.40		$3354.64	$73.20	$104.83

Crew C-17	Hr.	Daily	Hr.	Daily	Bare Costs	Incl. O&P
2 Skilled Worker Foremen (out)	$59.10	$945.60	$93.65	$1498.40	$57.50	$91.13
8 Skilled Workers	57.10	3654.40	90.50	5792.00		
80 L.H., Daily Totals		$4600.00		$7290.40	$57.50	$91.13

Crew C-17A	Hr.	Daily	Hr.	Daily	Bare Costs	Incl. O&P
2 Skilled Worker Foremen (out)	$59.10	$945.60	$93.65	$1498.40	$57.55	$91.19
8 Skilled Workers	57.10	3654.40	90.50	5792.00		
.125 Equip. Oper. (crane)	61.45	61.45	95.85	95.85		
.125 Hyd. Crane, 80 Ton		182.25		200.47	2.25	2.48
81 L.H., Daily Totals		$4843.70		$7586.73	$59.80	$93.66

Crew C-17B	Hr.	Daily	Hr.	Daily	Bare Costs	Incl. O&P
2 Skilled Worker Foremen (out)	$59.10	$945.60	$93.65	$1498.40	$57.60	$91.25
8 Skilled Workers	57.10	3654.40	90.50	5792.00		
.25 Equip. Oper. (crane)	61.45	122.90	95.85	191.70		
.25 Hyd. Crane, 80 Ton		364.50		400.95		
.25 Trowel, 48" Walk-Behind		23.57		25.93	4.73	5.21
82 L.H., Daily Totals		$5110.98		$7908.98	$62.33	$96.45

Crew C-17C	Hr.	Daily	Hr.	Daily	Bare Costs	Incl. O&P
2 Skilled Worker Foremen (out)	$59.10	$945.60	$93.65	$1498.40	$57.64	$91.30
8 Skilled Workers	57.10	3654.40	90.50	5792.00		
.375 Equip. Oper. (crane)	61.45	184.35	95.85	287.55		
.375 Hyd. Crane, 80 Ton		546.75		601.42	6.59	7.25
83 L.H., Daily Totals		$5331.10		$8179.38	$64.23	$98.55

Crew C-17D	Hr.	Daily	Hr.	Daily	Bare Costs	Incl. O&P
2 Skilled Worker Foremen (out)	$59.10	$945.60	$93.65	$1498.40	$57.69	$91.35
8 Skilled Workers	57.10	3654.40	90.50	5792.00		
.5 Equip. Oper. (crane)	61.45	245.80	95.85	383.40		
.5 Hyd. Crane, 80 Ton		729.00		801.90	8.68	9.55
84 L.H., Daily Totals		$5574.80		$8475.70	$66.37	$100.90

Crew C-17E	Hr.	Daily	Hr.	Daily	Bare Costs	Incl. O&P
2 Skilled Worker Foremen (out)	$59.10	$945.60	$93.65	$1498.40	$57.50	$91.13
8 Skilled Workers	57.10	3654.40	90.50	5792.00		
1 Hyd. Jack with Rods		36.70		40.37	.46	.50
80 L.H., Daily Totals		$4636.70		$7330.77	$57.96	$91.63

Crew C-18	Hr.	Daily	Hr.	Daily	Bare Costs	Incl. O&P
.125 Labor Foreman (outside)	$46.40	$46.40	$73.90	$73.90	$44.62	$71.06
1 Laborer	44.40	355.20	70.70	565.60		
1 Concrete Cart, 10 C.F.		116.95		128.65	12.99	14.29
9 L.H., Daily Totals		$518.55		$768.14	$57.62	$85.35

Crew C-19	Hr.	Daily	Hr.	Daily	Bare Costs	Incl. O&P
.125 Labor Foreman (outside)	$46.40	$46.40	$73.90	$73.90	$44.62	$71.06
1 Laborer	44.40	355.20	70.70	565.60		
1 Concrete Cart, 18 C.F.		138.95		152.85	15.44	16.98
9 L.H., Daily Totals		$540.55		$792.35	$60.06	$88.04

Crew C-20	Hr.	Daily	Hr.	Daily	Bare Costs	Incl. O&P
1 Labor Foreman (outside)	$46.40	$371.20	$73.90	$591.20	$47.40	$75.07
5 Laborers	44.40	1776.00	70.70	2828.00		
1 Cement Finisher	51.80	414.40	81.10	648.80		
1 Equip. Oper. (medium)	59.00	472.00	92.05	736.40		
2 Gas Engine Vibrators		54.30		59.73		
1 Concrete Pump (Small)		423.65		466.01	7.47	8.21
64 L.H., Daily Totals		$3511.55		$5330.15	$54.87	$83.28

Crew C-21	Hr.	Daily	Hr.	Daily	Bare Costs	Incl. O&P
1 Labor Foreman (outside)	$46.40	$371.20	$73.90	$591.20	$47.40	$75.07
5 Laborers	44.40	1776.00	70.70	2828.00		
1 Cement Finisher	51.80	414.40	81.10	648.80		
1 Equip. Oper. (medium)	59.00	472.00	92.05	736.40		
2 Gas Engine Vibrators		54.30		59.73		
1 Concrete Conveyer		206.25		226.88	4.07	4.48
64 L.H., Daily Totals		$3294.15		$5091.01	$51.47	$79.55

Crew C-22	Hr.	Daily	Hr.	Daily	Bare Costs	Incl. O&P
1 Rodman Foreman (outside)	$60.90	$487.20	$95.30	$762.40	$59.19	$92.63
4 Rodmen (reinf.)	58.90	1884.80	92.20	2950.40		
.125 Equip. Oper. (crane)	61.45	61.45	95.85	95.85		
.125 Equip. Oper. (oiler)	52.50	52.50	81.90	81.90		
.125 Hyd. Crane, 25 Ton		73.34		80.67	1.75	1.92
42 L.H., Daily Totals		$2559.29		$3971.22	$60.94	$94.55

Crew No.	Bare Costs		Incl. Subs O&P		Cost Per Labor-Hour	

Crew C-23

	Hr.	Daily	Hr.	Daily	Bare Costs	Incl. O&P
2 Skilled Worker Foremen (out)	$59.10	$945.60	$93.65	$1498.40	$57.48	$90.81
6 Skilled Workers	57.10	2740.80	90.50	4344.00		
1 Equip. Oper. (crane)	61.45	491.60	95.85	766.80		
1 Equip. Oper. (oiler)	52.50	420.00	81.90	655.20		
1 Lattice Boom Crane, 90 Ton		1713.00		1884.30	21.41	23.55
80 L.H., Daily Totals		$6311.00		$9148.70	$78.89	$114.36

Crew C-24

	Hr.	Daily	Hr.	Daily	Bare Costs	Incl. O&P
2 Skilled Worker Foremen (out)	$59.10	$945.60	$93.65	$1498.40	$57.48	$90.81
6 Skilled Workers	57.10	2740.80	90.50	4344.00		
1 Equip. Oper. (crane)	61.45	491.60	95.85	766.80		
1 Equip. Oper. (oiler)	52.50	420.00	81.90	655.20		
1 Lattice Boom Crane, 150 Ton		2324.00		2556.40	29.05	31.95
80 L.H., Daily Totals		$6922.00		$9820.80	$86.53	$122.76

Crew C-25

	Hr.	Daily	Hr.	Daily	Bare Costs	Incl. O&P
2 Rodmen (reinf.)	$58.90	$942.40	$92.20	$1475.20	$47.42	$77.03
2 Rodmen Helpers	35.95	575.20	61.85	989.60		
32 L.H., Daily Totals		$1517.60		$2464.80	$47.42	$77.03

Crew C-27

	Hr.	Daily	Hr.	Daily	Bare Costs	Incl. O&P
2 Cement Finishers	$51.80	$828.80	$81.10	$1297.60	$51.80	$81.10
1 Concrete Saw		112.85		124.14	7.05	7.76
16 L.H., Daily Totals		$941.65		$1421.73	$58.85	$88.86

Crew C-28

	Hr.	Daily	Hr.	Daily	Bare Costs	Incl. O&P
1 Cement Finisher	$51.80	$414.40	$81.10	$648.80	$51.80	$81.10
1 Portable Air Compressor, Gas		38.80		42.68	4.85	5.34
8 L.H., Daily Totals		$453.20		$691.48	$56.65	$86.44

Crew C-29

	Hr.	Daily	Hr.	Daily	Bare Costs	Incl. O&P
1 Laborer	$44.40	$355.20	$70.70	$565.60	$44.40	$70.70
1 Pressure Washer		97.35		107.08	12.17	13.39
8 L.H., Daily Totals		$452.55		$672.68	$56.57	$84.09

Crew C-30

	Hr.	Daily	Hr.	Daily	Bare Costs	Incl. O&P
1 Laborer	$44.40	$355.20	$70.70	$565.60	$44.40	$70.70
1 Concrete Mixer, 10 C.F.		147.15		161.87	18.39	20.23
8 L.H., Daily Totals		$502.35		$727.47	$62.79	$90.93

Crew C-31

	Hr.	Daily	Hr.	Daily	Bare Costs	Incl. O&P
1 Cement Finisher	$51.80	$414.40	$81.10	$648.80	$51.80	$81.10
1 Grout Pump		321.75		353.93	40.22	44.24
8 L.H., Daily Totals		$736.15		$1002.73	$92.02	$125.34

Crew C-32

	Hr.	Daily	Hr.	Daily	Bare Costs	Incl. O&P
1 Cement Finisher	$51.80	$414.40	$81.10	$648.80	$48.10	$75.90
1 Laborer	44.40	355.20	70.70	565.60		
1 Crack Chaser Saw, Gas, 6 H.P.		73.25		80.58		
1 Vacuum Pick-Up System		74.95		82.44	9.26	10.19
16 L.H., Daily Totals		$917.80		$1377.42	$57.36	$86.09

Crew D-1

	Hr.	Daily	Hr.	Daily	Bare Costs	Incl. O&P
1 Bricklayer	$53.70	$429.60	$86.30	$690.40	$48.70	$78.25
1 Bricklayer Helper	43.70	349.60	70.20	561.60		
16 L.H., Daily Totals		$779.20		$1252.00	$48.70	$78.25

Crew D-2

	Hr.	Daily	Hr.	Daily	Bare Costs	Incl. O&P
3 Bricklayers	$53.70	$1288.80	$86.30	$2071.20	$50.15	$80.52
2 Bricklayer Helpers	43.70	699.20	70.20	1123.20		
.5 Carpenter	54.70	218.80	87.10	348.40		
44 L.H., Daily Totals		$2206.80		$3542.80	$50.15	$80.52

Crew D-3

	Hr.	Daily	Hr.	Daily	Bare Costs	Incl. O&P
3 Bricklayers	$53.70	$1288.80	$86.30	$2071.20	$49.94	$80.20
2 Bricklayer Helpers	43.70	699.20	70.20	1123.20		
.25 Carpenter	54.70	109.40	87.10	174.20		
42 L.H., Daily Totals		$2097.40		$3368.60	$49.94	$80.20

Crew D-4

	Hr.	Daily	Hr.	Daily	Bare Costs	Incl. O&P
1 Bricklayer	$53.70	$429.60	$86.30	$690.40	$49.15	$78.33
2 Bricklayer Helpers	43.70	699.20	70.20	1123.20		
1 Equip. Oper. (light)	55.50	444.00	86.60	692.80		
1 Grout Pump, 50 C.F./hr.		190.35		209.38	5.95	6.54
32 L.H., Daily Totals		$1763.15		$2715.78	$55.10	$84.87

Crew D-5

	Hr.	Daily	Hr.	Daily	Bare Costs	Incl. O&P
1 Bricklayer	53.70	429.60	86.30	690.40	53.70	86.30
8 L.H., Daily Totals		$429.60		$690.40	$53.70	$86.30

Crew D-6

	Hr.	Daily	Hr.	Daily	Bare Costs	Incl. O&P
3 Bricklayers	$53.70	$1288.80	$86.30	$2071.20	$48.94	$78.60
3 Bricklayer Helpers	43.70	1048.80	70.20	1684.80		
.25 Carpenter	54.70	109.40	87.10	174.20		
50 L.H., Daily Totals		$2447.00		$3930.20	$48.94	$78.60

Crew D-7

	Hr.	Daily	Hr.	Daily	Bare Costs	Incl. O&P
1 Tile Layer	$51.70	$413.60	$80.70	$645.60	$46.65	$72.83
1 Tile Layer Helper	41.60	332.80	64.95	519.60		
16 L.H., Daily Totals		$746.40		$1165.20	$46.65	$72.83

Crew D-8

	Hr.	Daily	Hr.	Daily	Bare Costs	Incl. O&P
3 Bricklayers	$53.70	$1288.80	$86.30	$2071.20	$49.70	$79.86
2 Bricklayer Helpers	43.70	699.20	70.20	1123.20		
40 L.H., Daily Totals		$1988.00		$3194.40	$49.70	$79.86

Crew D-9

	Hr.	Daily	Hr.	Daily	Bare Costs	Incl. O&P
3 Bricklayers	$53.70	$1288.80	$86.30	$2071.20	$48.70	$78.25
3 Bricklayer Helpers	43.70	1048.80	70.20	1684.80		
48 L.H., Daily Totals		$2337.60		$3756.00	$48.70	$78.25

Crew D-10

	Hr.	Daily	Hr.	Daily	Bare Costs	Incl. O&P
1 Bricklayer Foreman (outside)	$55.70	$445.60	$89.50	$716.00	$53.64	$85.46
1 Bricklayer	53.70	429.60	86.30	690.40		
1 Bricklayer Helper	43.70	349.60	70.20	561.60		
1 Equip. Oper. (crane)	61.45	491.60	95.85	766.80		
1 S.P. Crane, 4x4, 12 Ton		432.65		475.92	13.52	14.87
32 L.H., Daily Totals		$2149.05		$3210.72	$67.16	$100.33

Crew D-11

	Hr.	Daily	Hr.	Daily	Bare Costs	Incl. O&P
1 Bricklayer Foreman (outside)	$55.70	$445.60	$89.50	$716.00	$51.03	$82.00
1 Bricklayer	53.70	429.60	86.30	690.40		
1 Bricklayer Helper	43.70	349.60	70.20	561.60		
24 L.H., Daily Totals		$1224.80		$1968.00	$51.03	$82.00

Crew No.	Bare Costs		Incl. Subs O&P		Cost Per Labor-Hour	
Crew D-12	**Hr.**	**Daily**	**Hr.**	**Daily**	**Bare Costs**	**Incl. O&P**
1 Bricklayer Foreman (outside)	$55.70	$445.60	$89.50	$716.00	$49.20	$79.05
1 Bricklayer	53.70	429.60	86.30	690.40		
2 Bricklayer Helpers	43.70	699.20	70.20	1123.20		
32 L.H., Daily Totals		$1574.40		$2529.60	$49.20	$79.05
Crew D-13	**Hr.**	**Daily**	**Hr.**	**Daily**	**Bare Costs**	**Incl. O&P**
1 Bricklayer Foreman (outside)	$55.70	$445.60	$89.50	$716.00	$52.16	$83.19
1 Bricklayer	53.70	429.60	86.30	690.40		
2 Bricklayer Helpers	43.70	699.20	70.20	1123.20		
1 Carpenter	54.70	437.60	87.10	696.80		
1 Equip. Oper. (crane)	61.45	491.60	95.85	766.80		
1 S.P. Crane, 4x4, 12 Ton		432.65		475.92	9.01	9.91
48 L.H., Daily Totals		$2936.25		$4469.11	$61.17	$93.11
Crew D-14	**Hr.**	**Daily**	**Hr.**	**Daily**	**Bare Costs**	**Incl. O&P**
3 Bricklayers	$53.70	$1288.80	$86.30	$2071.20	$51.20	$82.28
1 Bricklayer Helper	43.70	349.60	70.20	561.60		
32 L.H., Daily Totals		$1638.40		$2632.80	$51.20	$82.28
Crew E-1	**Hr.**	**Daily**	**Hr.**	**Daily**	**Bare Costs**	**Incl. O&P**
1 Welder Foreman (outside)	$62.30	$498.40	$100.75	$806.00	$59.37	$94.97
1 Welder	60.30	482.40	97.55	780.40		
1 Equip. Oper. (light)	55.50	444.00	86.60	692.80		
1 Welder, Gas Engine, 300 amp		148.75		163.63	6.20	6.82
24 L.H., Daily Totals		$1573.55		$2442.82	$65.56	$101.78
Crew E-2	**Hr.**	**Daily**	**Hr.**	**Daily**	**Bare Costs**	**Incl. O&P**
1 Struc. Steel Foreman (outside)	$62.30	$498.40	$100.75	$806.00	$59.64	$95.53
4 Struc. Steel Workers	60.30	1929.60	97.55	3121.60		
1 Equip. Oper. (crane)	61.45	491.60	95.85	766.80		
1 Equip. Oper. (oiler)	52.50	420.00	81.90	655.20		
1 Lattice Boom Crane, 90 Ton		1713.00		1884.30	30.59	33.65
56 L.H., Daily Totals		$5052.60		$7233.90	$90.22	$129.18
Crew E-3	**Hr.**	**Daily**	**Hr.**	**Daily**	**Bare Costs**	**Incl. O&P**
1 Struc. Steel Foreman (outside)	$62.30	$498.40	$100.75	$806.00	$60.97	$98.62
1 Struc. Steel Worker	60.30	482.40	97.55	780.40		
1 Welder	60.30	482.40	97.55	780.40		
1 Welder, Gas Engine, 300 amp		148.75		163.63	6.20	6.82
24 L.H., Daily Totals		$1611.95		$2530.43	$67.16	$105.43
Crew E-3A	**Hr.**	**Daily**	**Hr.**	**Daily**	**Bare Costs**	**Incl. O&P**
1 Struc. Steel Foreman (outside)	$62.30	$498.40	$100.75	$806.00	$60.97	$98.62
1 Struc. Steel Worker	60.30	482.40	97.55	780.40		
1 Welder	60.30	482.40	97.55	780.40		
1 Welder, Gas Engine, 300 amp		148.75		163.63		
1 Telescoping Boom Lift, to 40'		281.90		310.09	17.94	19.74
24 L.H., Daily Totals		$1893.85		$2840.51	$78.91	$118.35
Crew E-4	**Hr.**	**Daily**	**Hr.**	**Daily**	**Bare Costs**	**Incl. O&P**
1 Struc. Steel Foreman (outside)	$62.30	$498.40	$100.75	$806.00	$60.80	$98.35
3 Struc. Steel Workers	60.30	1447.20	97.55	2341.20		
1 Welder, Gas Engine, 300 amp		148.75		163.63	4.65	5.11
32 L.H., Daily Totals		$2094.35		$3310.82	$65.45	$103.46

Crew No.	Bare Costs		Incl. Subs O&P		Cost Per Labor-Hour	
Crew E-5	**Hr.**	**Daily**	**Hr.**	**Daily**	**Bare Costs**	**Incl. O&P**
2 Struc. Steel Foremen (outside)	$62.30	$996.80	$100.75	$1612.00	$60.03	$96.45
5 Struc. Steel Workers	60.30	2412.00	97.55	3902.00		
1 Equip. Oper. (crane)	61.45	491.60	95.85	766.80		
1 Welder	60.30	482.40	97.55	780.40		
1 Equip. Oper. (oiler)	52.50	420.00	81.90	655.20		
1 Lattice Boom Crane, 90 Ton		1713.00		1884.30		
1 Welder, Gas Engine, 300 amp		148.75		163.63	23.27	25.60
80 L.H., Daily Totals		$6664.55		$9764.33	$83.31	$122.05
Crew E-6	**Hr.**	**Daily**	**Hr.**	**Daily**	**Bare Costs**	**Incl. O&P**
3 Struc. Steel Foremen (outside)	$62.30	$1495.20	$100.75	$2418.00	$59.96	$96.38
9 Struc. Steel Workers	60.30	4341.60	97.55	7023.60		
1 Equip. Oper. (crane)	61.45	491.60	95.85	766.80		
1 Welder	60.30	482.40	97.55	780.40		
1 Equip. Oper. (oiler)	52.50	420.00	81.90	655.20		
1 Equip. Oper. (light)	55.50	444.00	86.60	692.80		
1 Lattice Boom Crane, 90 Ton		1713.00		1884.30		
1 Welder, Gas Engine, 300 amp		148.75		163.63		
1 Air Compressor, 160 cfm		212.30		233.53		
2 Impact Wrenches		90.30		99.33	16.91	18.60
128 L.H., Daily Totals		$9839.15		$14717.58	$76.87	$114.98
Crew E-7	**Hr.**	**Daily**	**Hr.**	**Daily**	**Bare Costs**	**Incl. O&P**
1 Struc. Steel Foreman (outside)	$62.30	$498.40	$100.75	$806.00	$60.03	$96.45
4 Struc. Steel Workers	60.30	1929.60	97.55	3121.60		
1 Equip. Oper. (crane)	61.45	491.60	95.85	766.80		
1 Equip. Oper. (oiler)	52.50	420.00	81.90	655.20		
1 Welder Foreman (outside)	62.30	498.40	100.75	806.00		
2 Welders	60.30	964.80	97.55	1560.80		
1 Lattice Boom Crane, 90 Ton		1713.00		1884.30		
2 Welder, Gas Engine, 300 amp		297.50		327.25	25.13	27.64
80 L.H., Daily Totals		$6813.30		$9927.95	$85.17	$124.10
Crew E-8	**Hr.**	**Daily**	**Hr.**	**Daily**	**Bare Costs**	**Incl. O&P**
1 Struc. Steel Foreman (outside)	$62.30	$498.40	$100.75	$806.00	$59.73	$95.87
4 Struc. Steel Workers	60.30	1929.60	97.55	3121.60		
1 Welder Foreman (outside)	62.30	498.40	100.75	806.00		
4 Welders	60.30	1929.60	97.55	3121.60		
1 Equip. Oper. (crane)	61.45	491.60	95.85	766.80		
1 Equip. Oper. (oiler)	52.50	420.00	81.90	655.20		
1 Equip. Oper. (light)	55.50	444.00	86.60	692.80		
1 Lattice Boom Crane, 90 Ton		1713.00		1884.30		
4 Welder, Gas Engine, 300 amp		595.00		654.50	22.19	24.41
104 L.H., Daily Totals		$8519.60		$12508.80	$81.92	$120.28
Crew E-9	**Hr.**	**Daily**	**Hr.**	**Daily**	**Bare Costs**	**Incl. O&P**
2 Struc. Steel Foremen (outside)	$62.30	$996.80	$100.75	$1612.00	$59.96	$96.38
5 Struc. Steel Workers	60.30	2412.00	97.55	3902.00		
1 Welder Foreman (outside)	62.30	498.40	100.75	806.00		
5 Welders	60.30	2412.00	97.55	3902.00		
1 Equip. Oper. (crane)	61.45	491.60	95.85	766.80		
1 Equip. Oper. (oiler)	52.50	420.00	81.90	655.20		
1 Equip. Oper. (light)	55.50	444.00	86.60	692.80		
1 Lattice Boom Crane, 90 Ton		1713.00		1884.30		
5 Welder, Gas Engine, 300 amp		743.75		818.13	19.19	21.11
128 L.H., Daily Totals		$10131.55		$15039.23	$79.15	$117.49

Crew No.	Bare Costs		Incl. Subs O&P		Cost Per Labor-Hour	

Crew E-10	Hr.	Daily	Hr.	Daily	Bare Costs	Incl. O&P
1 Welder Foreman (outside)	$62.30	$498.40	$100.75	$806.00	$61.30	$99.15
1 Welder	60.30	482.40	97.55	780.40		
1 Welder, Gas Engine, 300 amp		148.75		163.63		
1 Flatbed Truck, Gas, 3 Ton		850.05		935.05	62.42	68.67
16 L.H., Daily Totals		$1979.60		$2685.08	$123.72	$167.82

Crew E-11	Hr.	Daily	Hr.	Daily	Bare Costs	Incl. O&P
2 Painters, Struc. Steel	$47.20	$755.20	$80.50	$1288.00	$48.58	$79.58
1 Building Laborer	44.40	355.20	70.70	565.60		
1 Equip. Oper. (light)	55.50	444.00	86.60	692.80		
1 Air Compressor, 250 cfm		202.85		223.13		
1 Sandblaster, Portable, 3 C.F.		83.85		92.23		
1 Set Sand Blasting Accessories		15.55		17.11	9.45	10.39
32 L.H., Daily Totals		$1856.65		$2878.88	$58.02	$89.96

Crew E-11A	Hr.	Daily	Hr.	Daily	Bare Costs	Incl. O&P
2 Painters, Struc. Steel	$47.20	$755.20	$80.50	$1288.00	$48.58	$79.58
1 Building Laborer	44.40	355.20	70.70	565.60		
1 Equip. Oper. (light)	55.50	444.00	86.60	692.80		
1 Air Compressor, 250 cfm		202.85		223.13		
1 Sandblaster, Portable, 3 C.F.		83.85		92.23		
1 Set Sand Blasting Accessories		15.55		17.11		
1 Telescoping Boom Lift, to 60'		292.45		321.69	18.58	20.44
32 L.H., Daily Totals		$2149.10		$3200.57	$67.16	$100.02

Crew E-11B	Hr.	Daily	Hr.	Daily	Bare Costs	Incl. O&P
2 Painters, Struc. Steel	$47.20	$755.20	$80.50	$1288.00	$46.27	$77.23
1 Building Laborer	44.40	355.20	70.70	565.60		
2 Paint Sprayer, 8 C.F.M.		88.40		97.24		
1 Telescoping Boom Lift, to 60'		292.45		321.69	15.87	17.46
24 L.H., Daily Totals		$1491.25		$2272.53	$62.14	$94.69

Crew E-12	Hr.	Daily	Hr.	Daily	Bare Costs	Incl. O&P
1 Welder Foreman (outside)	$62.30	$498.40	$100.75	$806.00	$58.90	$93.67
1 Equip. Oper. (light)	55.50	444.00	86.60	692.80		
1 Welder, Gas Engine, 300 amp		148.75		163.63	9.30	10.23
16 L.H., Daily Totals		$1091.15		$1662.43	$68.20	$103.90

Crew E-13	Hr.	Daily	Hr.	Daily	Bare Costs	Incl. O&P
1 Welder Foreman (outside)	$62.30	$498.40	$100.75	$806.00	$60.03	$96.03
.5 Equip. Oper. (light)	55.50	222.00	86.60	346.40		
1 Welder, Gas Engine, 300 amp		148.75		163.63	12.40	13.64
12 L.H., Daily Totals		$869.15		$1316.03	$72.43	$109.67

Crew E-14	Hr.	Daily	Hr.	Daily	Bare Costs	Incl. O&P
1 Welder Foreman (outside)	$62.30	$498.40	$100.75	$806.00	$62.30	$100.75
1 Welder, Gas Engine, 300 amp		148.75		163.63	18.59	20.45
8 L.H., Daily Totals		$647.15		$969.63	$80.89	$121.20

Crew E-16	Hr.	Daily	Hr.	Daily	Bare Costs	Incl. O&P
1 Welder Foreman (outside)	$62.30	$498.40	$100.75	$806.00	$61.30	$99.15
1 Welder	60.30	482.40	97.55	780.40		
1 Welder, Gas Engine, 300 amp		148.75		163.63	9.30	10.23
16 L.H., Daily Totals		$1129.55		$1750.03	$70.60	$109.38

Crew E-17	Hr.	Daily	Hr.	Daily	Bare Costs	Incl. O&P
1 Struc. Steel Foreman (outside)	$62.30	$498.40	$100.75	$806.00	$61.30	$99.15
1 Structural Steel Worker	60.30	482.40	97.55	780.40		
16 L.H., Daily Totals		$980.80		$1586.40	$61.30	$99.15

Crew E-18	Hr.	Daily	Hr.	Daily	Bare Costs	Incl. O&P
1 Struc. Steel Foreman (outside)	$62.30	$498.40	$100.75	$806.00	$60.44	$97.09
3 Structural Steel Workers	60.30	1447.20	97.55	2341.20		
1 Equipment Operator (med.)	59.00	472.00	92.05	736.40		
1 Lattice Boom Crane, 20 Ton		1526.00		1678.60	38.15	41.97
40 L.H., Daily Totals		$3943.60		$5562.20	$98.59	$139.06

Crew E-19	Hr.	Daily	Hr.	Daily	Bare Costs	Incl. O&P
1 Struc. Steel Foreman (outside)	$62.30	$498.40	$100.75	$806.00	$59.37	$94.97
1 Structural Steel Worker	60.30	482.40	97.55	780.40		
1 Equip. Oper. (light)	55.50	444.00	86.60	692.80		
1 Lattice Boom Crane, 20 Ton		1526.00		1678.60	63.58	69.94
24 L.H., Daily Totals		$2950.80		$3957.80	$122.95	$164.91

Crew E-20	Hr.	Daily	Hr.	Daily	Bare Costs	Incl. O&P
1 Struc. Steel Foreman (outside)	$62.30	$498.40	$100.75	$806.00	$59.72	$95.78
5 Structural Steel Workers	60.30	2412.00	97.55	3902.00		
1 Equip. Oper. (crane)	61.45	491.60	95.85	766.80		
1 Equip. Oper. (oiler)	52.50	420.00	81.90	655.20		
1 Lattice Boom Crane, 40 Ton		2127.00		2339.70	33.23	36.56
64 L.H., Daily Totals		$5949.00		$8469.70	$92.95	$132.34

Crew E-22	Hr.	Daily	Hr.	Daily	Bare Costs	Incl. O&P
1 Skilled Worker Foreman (out)	$59.10	$472.80	$93.65	$749.20	$57.77	$91.55
2 Skilled Workers	57.10	913.60	90.50	1448.00		
24 L.H., Daily Totals		$1386.40		$2197.20	$57.77	$91.55

Crew E-24	Hr.	Daily	Hr.	Daily	Bare Costs	Incl. O&P
3 Structural Steel Workers	$60.30	$1447.20	$97.55	$2341.20	$59.98	$96.17
1 Equipment Operator (med.)	59.00	472.00	92.05	736.40		
1 Hyd. Crane, 25 Ton		586.70		645.37	18.33	20.17
32 L.H., Daily Totals		$2505.90		$3722.97	$78.31	$116.34

Crew E-25	Hr.	Daily	Hr.	Daily	Bare Costs	Incl. O&P
1 Welder Foreman (outside)	$62.30	$498.40	$100.75	$806.00	$62.30	$100.75
1 Cutting Torch		12.95		14.24	1.62	1.78
8 L.H., Daily Totals		$511.35		$820.25	$63.92	$102.53

Crew E-26	Hr.	Daily	Hr.	Daily	Bare Costs	Incl. O&P
1 Struc. Steel Foreman (outside)	$62.30	$498.40	$100.75	$806.00	$61.64	$98.97
1 Struc. Steel Worker	60.30	482.40	97.55	780.40		
1 Welder	60.30	482.40	97.55	780.40		
.25 Electrician	63.70	127.40	97.85	195.70		
.25 Plumber	67.70	135.40	104.40	208.80		
1 Welder, Gas Engine, 300 amp		148.75		163.63	5.31	5.84
28 L.H., Daily Totals		$1874.75		$2934.93	$66.96	$104.82

Crew E-27	Hr.	Daily	Hr.	Daily	Bare Costs	Incl. O&P
1 Struc. Steel Foreman (outside)	$62.30	$498.40	$100.75	$806.00	$59.72	$95.78
5 Struc. Steel Workers	60.30	2412.00	97.55	3902.00		
1 Equip. Oper. (crane)	61.45	491.60	95.85	766.80		
1 Equip. Oper. (oiler)	52.50	420.00	81.90	655.20		
1 Hyd. Crane, 12 Ton		475.80		523.38		
1 Hyd. Crane, 80 Ton		1458.00		1603.80	30.22	33.24
64 L.H., Daily Totals		$5755.80		$8257.18	$89.93	$129.02

Crew F-3	Hr.	Daily	Hr.	Daily	Bare Costs	Incl. O&P
4 Carpenters	$54.70	$1750.40	$87.10	$2787.20	$56.05	$88.85
1 Equip. Oper. (crane)	61.45	491.60	95.85	766.80		
1 Hyd. Crane, 12 Ton		475.80		523.38	11.90	13.08
40 L.H., Daily Totals		$2717.80		$4077.38	$67.94	$101.93

Crew No.	Bare Costs		Incl. Subs O&P		Cost Per Labor-Hour	

Crew F-4	Hr.	Daily	Hr.	Daily	Bare Costs	Incl. O&P
4 Carpenters	$54.70	$1750.40	$87.10	$2787.20	$55.46	$87.69
1 Equip. Oper. (crane)	61.45	491.60	95.85	766.80		
1 Equip. Oper. (oiler)	52.50	420.00	81.90	655.20		
1 Hyd. Crane, 55 Ton		990.15		1089.17	20.63	22.69
48 L.H., Daily Totals		$3652.15		$5298.36	$76.09	$110.38

Crew F-5	Hr.	Daily	Hr.	Daily	Bare Costs	Incl. O&P
1 Carpenter Foreman (outside)	$56.70	$453.60	$90.30	$722.40	$55.20	$87.90
3 Carpenters	54.70	1312.80	87.10	2090.40		
32 L.H., Daily Totals		$1766.40		$2812.80	$55.20	$87.90

Crew F-6	Hr.	Daily	Hr.	Daily	Bare Costs	Incl. O&P
2 Carpenters	$54.70	$875.20	$87.10	$1393.60	$51.93	$82.29
2 Building Laborers	44.40	710.40	70.70	1131.20		
1 Equip. Oper. (crane)	61.45	491.60	95.85	766.80		
1 Hyd. Crane, 12 Ton		475.80		523.38	11.90	13.08
40 L.H., Daily Totals		$2553.00		$3814.98	$63.83	$95.37

Crew F-7	Hr.	Daily	Hr.	Daily	Bare Costs	Incl. O&P
2 Carpenters	$54.70	$875.20	$87.10	$1393.60	$49.55	$78.90
2 Building Laborers	44.40	710.40	70.70	1131.20		
32 L.H., Daily Totals		$1585.60		$2524.80	$49.55	$78.90

Crew G-1	Hr.	Daily	Hr.	Daily	Bare Costs	Incl. O&P
1 Roofer Foreman (outside)	$50.20	$401.60	$86.40	$691.20	$44.99	$77.41
4 Roofers Composition	48.20	1542.40	82.95	2654.40		
2 Roofer Helpers	35.95	575.20	61.85	989.60		
1 Application Equipment		194.80		214.28		
1 Tar Kettle/Pot		209.95		230.94		
1 Crew Truck		168.15		184.97	10.23	11.25
56 L.H., Daily Totals		$3092.10		$4965.39	$55.22	$88.67

Crew G-2	Hr.	Daily	Hr.	Daily	Bare Costs	Incl. O&P
1 Plasterer	$49.85	$398.80	$79.25	$634.00	$46.27	$73.60
1 Plasterer Helper	44.55	356.40	70.85	566.80		
1 Building Laborer	44.40	355.20	70.70	565.60		
1 Grout Pump, 50 C.F./hr.		190.35		209.38	7.93	8.72
24 L.H., Daily Totals		$1300.75		$1975.79	$54.20	$82.32

Crew G-2A	Hr.	Daily	Hr.	Daily	Bare Costs	Incl. O&P
1 Roofer Composition	$48.20	$385.60	$82.95	$663.60	$42.85	$71.83
1 Roofer Helper	35.95	287.60	61.85	494.80		
1 Building Laborer	44.40	355.20	70.70	565.60		
1 Foam Spray Rig, Trailer-Mtd.		530.15		583.16		
1 Pickup Truck, 3/4 Ton		112.20		123.42	26.76	29.44
24 L.H., Daily Totals		$1670.75		$2430.59	$69.61	$101.27

Crew G-3	Hr.	Daily	Hr.	Daily	Bare Costs	Incl. O&P
2 Sheet Metal Workers	$65.45	$1047.20	$102.00	$1632.00	$54.92	$86.35
2 Building Laborers	44.40	710.40	70.70	1131.20		
32 L.H., Daily Totals		$1757.60		$2763.20	$54.92	$86.35

Crew G-4	Hr.	Daily	Hr.	Daily	Bare Costs	Incl. O&P
1 Labor Foreman (outside)	$46.40	$371.20	$73.90	$591.20	$45.07	$71.77
2 Building Laborers	44.40	710.40	70.70	1131.20		
1 Flatbed Truck, Gas, 1.5 Ton		198.50		218.35		
1 Air Compressor, 160 cfm		212.30		233.53	17.12	18.83
24 L.H., Daily Totals		$1492.40		$2174.28	$62.18	$90.59

Crew G-5	Hr.	Daily	Hr.	Daily	Bare Costs	Incl. O&P
1 Roofer Foreman (outside)	$50.20	$401.60	$86.40	$691.20	$43.70	$75.20
2 Roofers Composition	48.20	771.20	82.95	1327.20		
2 Roofer Helpers	35.95	575.20	61.85	989.60		
1 Application Equipment		194.80		214.28	4.87	5.36
40 L.H., Daily Totals		$1942.80		$3222.28	$48.57	$80.56

Crew G-6A	Hr.	Daily	Hr.	Daily	Bare Costs	Incl. O&P
2 Roofers Composition	$48.20	$771.20	$82.95	$1327.20	$48.20	$82.95
1 Small Compressor, Electric		39.30		43.23		
2 Pneumatic Nailers		55.40		60.94	5.92	6.51
16 L.H., Daily Totals		$865.90		$1431.37	$54.12	$89.46

Crew G-7	Hr.	Daily	Hr.	Daily	Bare Costs	Incl. O&P
1 Carpenter	$54.70	$437.60	$87.10	$696.80	$54.70	$87.10
1 Small Compressor, Electric		39.30		43.23		
1 Pneumatic Nailer		27.70		30.47	8.38	9.21
8 L.H., Daily Totals		$504.60		$770.50	$63.08	$96.31

Crew H-1	Hr.	Daily	Hr.	Daily	Bare Costs	Incl. O&P
2 Glaziers	$52.65	$842.40	$83.70	$1339.20	$56.48	$90.63
2 Struc. Steel Workers	60.30	964.80	97.55	1560.80		
32 L.H., Daily Totals		$1807.20		$2900.00	$56.48	$90.63

Crew H-2	Hr.	Daily	Hr.	Daily	Bare Costs	Incl. O&P
2 Glaziers	$52.65	$842.40	$83.70	$1339.20	$49.90	$79.37
1 Building Laborer	44.40	355.20	70.70	565.60		
24 L.H., Daily Totals		$1197.60		$1904.80	$49.90	$79.37

Crew H-3	Hr.	Daily	Hr.	Daily	Bare Costs	Incl. O&P
1 Glazier	$52.65	$421.20	$83.70	$669.60	$47.35	$75.80
1 Helper	42.05	336.40	67.90	543.20		
16 L.H., Daily Totals		$757.60		$1212.80	$47.35	$75.80

Crew H-4	Hr.	Daily	Hr.	Daily	Bare Costs	Incl. O&P
1 Carpenter	$54.70	$437.60	$87.10	$696.80	$51.44	$81.57
1 Carpenter Helper	42.05	336.40	67.90	543.20		
.5 Electrician	63.70	254.80	97.85	391.40		
20 L.H., Daily Totals		$1028.80		$1631.40	$51.44	$81.57

Crew J-1	Hr.	Daily	Hr.	Daily	Bare Costs	Incl. O&P
3 Plasterers	$49.85	$1196.40	$79.25	$1902.00	$47.73	$75.89
2 Plasterer Helpers	44.55	712.80	70.85	1133.60		
1 Mixing Machine, 6 C.F.		113.35		124.69	2.83	3.12
40 L.H., Daily Totals		$2022.55		$3160.28	$50.56	$79.01

Crew J-2	Hr.	Daily	Hr.	Daily	Bare Costs	Incl. O&P
3 Plasterers	$49.85	$1196.40	$79.25	$1902.00	$48.79	$77.32
2 Plasterer Helpers	44.55	712.80	70.85	1133.60		
1 Lather	54.10	432.80	84.45	675.60		
1 Mixing Machine, 6 C.F.		113.35		124.69	2.36	2.60
48 L.H., Daily Totals		$2455.35		$3835.89	$51.15	$79.91

Crew J-3	Hr.	Daily	Hr.	Daily	Bare Costs	Incl. O&P
1 Terrazzo Worker	$51.75	$414.00	$80.80	$646.40	$47.65	$74.40
1 Terrazzo Helper	43.55	348.40	68.00	544.00		
1 Floor Grinder, 22" Path		96.05		105.66		
1 Terrazzo Mixer		162.90		179.19	16.18	17.80
16 L.H., Daily Totals		$1021.35		$1475.24	$63.83	$92.20

Crew No.	Bare Costs		Incl. Subs O&P		Cost Per Labor-Hour	

Crew J-4

	Hr.	Daily	Hr.	Daily	Bare Costs	Incl. O&P
2 Cement Finishers	$51.80	$828.80	$81.10	$1297.60	$49.33	$77.63
1 Laborer	44.40	355.20	70.70	565.60		
1 Floor Grinder, 22" Path		96.05		105.66		
1 Floor Edger, 7" Path		44.05		48.45		
1 Vacuum Pick-Up System		74.95		82.44	8.96	9.86
24 L.H., Daily Totals		$1399.05		$2099.76	$58.29	$87.49

Crew J-4A

	Hr.	Daily	Hr.	Daily	Bare Costs	Incl. O&P
2 Cement Finishers	$51.80	$828.80	$81.10	$1297.60	$48.10	$75.90
2 Laborers	44.40	710.40	70.70	1131.20		
1 Floor Grinder, 22" Path		96.05		105.66		
1 Floor Edger, 7" Path		44.05		48.45		
1 Vacuum Pick-Up System		74.95		82.44		
1 Floor Auto Scrubber		179.55		197.51	12.33	13.56
32 L.H., Daily Totals		$1933.80		$2862.86	$60.43	$89.46

Crew J-4B

	Hr.	Daily	Hr.	Daily	Bare Costs	Incl. O&P
1 Laborer	$44.40	$355.20	$70.70	$565.60	$44.40	$70.70
1 Floor Auto Scrubber		179.55		197.51	22.44	24.69
8 L.H., Daily Totals		$534.75		$763.11	$66.84	$95.39

Crew J-6

	Hr.	Daily	Hr.	Daily	Bare Costs	Incl. O&P
2 Painters	$46.45	$743.20	$73.55	$1176.80	$48.20	$76.10
1 Building Laborer	44.40	355.20	70.70	565.60		
1 Equip. Oper. (light)	55.50	444.00	86.60	692.80		
1 Air Compressor, 250 cfm		202.85		223.13		
1 Sandblaster, Portable, 3 C.F.		83.85		92.23		
1 Set Sand Blasting Accessories		15.55		17.11	9.45	10.39
32 L.H., Daily Totals		$1844.65		$2767.68	$57.65	$86.49

Crew J-7

	Hr.	Daily	Hr.	Daily	Bare Costs	Incl. O&P
2 Painters	$46.45	$743.20	$73.55	$1176.80	$46.45	$73.55
1 Floor Belt Sander		50.20		55.22		
1 Floor Sanding Edger		25.20		27.72	4.71	5.18
16 L.H., Daily Totals		$818.60		$1259.74	$51.16	$78.73

Crew K-1

	Hr.	Daily	Hr.	Daily	Bare Costs	Incl. O&P
1 Carpenter	$54.70	$437.60	$87.10	$696.80	$51.75	$82.47
1 Truck Driver (light)	48.80	390.40	77.85	622.80		
1 Flatbed Truck, Gas, 3 Ton		850.05		935.05	53.13	58.44
16 L.H., Daily Totals		$1678.05		$2254.66	$104.88	$140.92

Crew K-2

	Hr.	Daily	Hr.	Daily	Bare Costs	Incl. O&P
1 Struc. Steel Foreman (outside)	$62.30	$498.40	$100.75	$806.00	$57.13	$92.05
1 Struc. Steel Worker	60.30	482.40	97.55	780.40		
1 Truck Driver (light)	48.80	390.40	77.85	622.80		
1 Flatbed Truck, Gas, 3 Ton		850.05		935.05	35.42	38.96
24 L.H., Daily Totals		$2221.25		$3144.26	$92.55	$131.01

Crew L-1

	Hr.	Daily	Hr.	Daily	Bare Costs	Incl. O&P
1 Electrician	$63.70	$509.60	$97.85	$782.80	$65.70	$101.13
1 Plumber	67.70	541.60	104.40	835.20		
16 L.H., Daily Totals		$1051.20		$1618.00	$65.70	$101.13

Crew L-2

	Hr.	Daily	Hr.	Daily	Bare Costs	Incl. O&P
1 Carpenter	$54.70	$437.60	$87.10	$696.80	$48.38	$77.50
1 Carpenter Helper	42.05	336.40	67.90	543.20		
16 L.H., Daily Totals		$774.00		$1240.00	$48.38	$77.50

Crew L-3

	Hr.	Daily	Hr.	Daily	Bare Costs	Incl. O&P
1 Carpenter	$54.70	$437.60	$87.10	$696.80	$59.64	$93.51
.5 Electrician	63.70	254.80	97.85	391.40		
.5 Sheet Metal Worker	65.45	261.80	102.00	408.00		
16 L.H., Daily Totals		$954.20		$1496.20	$59.64	$93.51

Crew L-3A

	Hr.	Daily	Hr.	Daily	Bare Costs	Incl. O&P
1 Carpenter Foreman (outside)	$56.70	$453.60	$90.30	$722.40	$59.62	$94.20
.5 Sheet Metal Worker	65.45	261.80	102.00	408.00		
12 L.H., Daily Totals		$715.40		$1130.40	$59.62	$94.20

Crew L-4

	Hr.	Daily	Hr.	Daily	Bare Costs	Incl. O&P
2 Skilled Workers	$57.10	$913.60	$90.50	$1448.00	$52.08	$82.97
1 Helper	42.05	336.40	67.90	543.20		
24 L.H., Daily Totals		$1250.00		$1991.20	$52.08	$82.97

Crew L-5

	Hr.	Daily	Hr.	Daily	Bare Costs	Incl. O&P
1 Struc. Steel Foreman (outside)	$62.30	$498.40	$100.75	$806.00	$60.75	$97.76
5 Struc. Steel Workers	60.30	2412.00	97.55	3902.00		
1 Equip. Oper. (crane)	61.45	491.60	95.85	766.80		
1 Hyd. Crane, 25 Ton		586.70		645.37	10.48	11.52
56 L.H., Daily Totals		$3988.70		$6120.17	$71.23	$109.29

Crew L-5A

	Hr.	Daily	Hr.	Daily	Bare Costs	Incl. O&P
1 Struc. Steel Foreman (outside)	$62.30	$498.40	$100.75	$806.00	$61.09	$97.92
2 Structural Steel Workers	60.30	964.80	97.55	1560.80		
1 Equip. Oper. (crane)	61.45	491.60	95.85	766.80		
1 S.P. Crane, 4x4, 25 Ton		1155.00		1270.50	36.09	39.70
32 L.H., Daily Totals		$3109.80		$4404.10	$97.18	$137.63

Crew L-5B

	Hr.	Daily	Hr.	Daily	Bare Costs	Incl. O&P
1 Struc. Steel Foreman (outside)	$62.30	$498.40	$100.75	$806.00	$62.33	$97.79
2 Structural Steel Workers	60.30	964.80	97.55	1560.80		
2 Electricians	63.70	1019.20	97.85	1565.60		
2 Steamfitters/Pipefitters	68.35	1093.60	105.40	1686.40		
1 Equip. Oper. (crane)	61.45	491.60	95.85	766.80		
1 Equip. Oper. (oiler)	52.50	420.00	81.90	655.20		
1 Hyd. Crane, 80 Ton		1458.00		1603.80	20.25	22.27
72 L.H., Daily Totals		$5945.60		$8644.60	$82.58	$120.06

Crew L-6

	Hr.	Daily	Hr.	Daily	Bare Costs	Incl. O&P
1 Plumber	$67.70	$541.60	$104.40	$835.20	$66.37	$102.22
.5 Electrician	63.70	254.80	97.85	391.40		
12 L.H., Daily Totals		$796.40		$1226.60	$66.37	$102.22

Crew L-7

	Hr.	Daily	Hr.	Daily	Bare Costs	Incl. O&P
2 Carpenters	$54.70	$875.20	$87.10	$1393.60	$53.04	$83.95
1 Building Laborer	44.40	355.20	70.70	565.60		
.5 Electrician	63.70	254.80	97.85	391.40		
28 L.H., Daily Totals		$1485.20		$2350.60	$53.04	$83.95

Crew L-8

	Hr.	Daily	Hr.	Daily	Bare Costs	Incl. O&P
2 Carpenters	$54.70	$875.20	$87.10	$1393.60	$57.30	$90.56
.5 Plumber	67.70	270.80	104.40	417.60		
20 L.H., Daily Totals		$1146.00		$1811.20	$57.30	$90.56

For customer support on your Commercial Renovation Costs with RSMeans Data, call 800.448.8182.

Crew No.	Bare Costs		Incl. Subs O&P		Cost Per Labor-Hour	

Crew L-9

Crew L-9	Hr.	Daily	Hr.	Daily	Bare Costs	Incl. O&P
1 Labor Foreman (inside)	$44.90	$359.20	$71.50	$572.00	$50.19	$79.86
2 Building Laborers	44.40	710.40	70.70	1131.20		
1 Struc. Steel Worker	60.30	482.40	97.55	780.40		
.5 Electrician	63.70	254.80	97.85	391.40		
36 L.H., Daily Totals		$1806.80		$2875.00	$50.19	$79.86

Crew L-10

Crew L-10	Hr.	Daily	Hr.	Daily	Bare Costs	Incl. O&P
1 Struc. Steel Foreman (outside)	$62.30	$498.40	$100.75	$806.00	$61.35	$98.05
1 Structural Steel Worker	60.30	482.40	97.55	780.40		
1 Equip. Oper. (crane)	61.45	491.60	95.85	766.80		
1 Hyd. Crane, 12 Ton		475.80		523.38	19.82	21.81
24 L.H., Daily Totals		$1948.20		$2876.58	$81.17	$119.86

Crew L-11

Crew L-11	Hr.	Daily	Hr.	Daily	Bare Costs	Incl. O&P
2 Wreckers	$44.40	$710.40	$71.85	$1149.60	$51.44	$81.54
1 Equip. Oper. (crane)	61.45	491.60	95.85	766.80		
1 Equip. Oper. (light)	55.50	444.00	86.60	692.80		
1 Hyd. Excavator, 2.5 C.Y.		1567.00		1723.70		
1 Loader, Skid Steer, 78 H.P.		446.30		490.93	62.92	69.21
32 L.H., Daily Totals		$3659.30		$4823.83	$114.35	$150.74

Crew M-1

Crew M-1	Hr.	Daily	Hr.	Daily	Bare Costs	Incl. O&P
3 Elevator Constructors	$90.30	$2167.20	$138.35	$3320.40	$85.79	$131.44
1 Elevator Apprentice	72.25	578.00	110.70	885.60		
5 Hand Tools		50.50		55.55	1.58	1.74
32 L.H., Daily Totals		$2795.70		$4261.55	$87.37	$133.17

Crew M-3

Crew M-3	Hr.	Daily	Hr.	Daily	Bare Costs	Incl. O&P
1 Electrician Foreman (outside)	$65.70	$525.60	$100.90	$807.20	$67.62	$104.39
1 Common Laborer	44.40	355.20	70.70	565.60		
.25 Equipment Operator (med.)	59.00	118.00	92.05	184.10		
1 Elevator Constructor	90.30	722.40	138.35	1106.80		
1 Elevator Apprentice	72.25	578.00	110.70	885.60		
.25 S.P. Crane, 4x4, 20 Ton		143.59		157.95	4.22	4.65
34 L.H., Daily Totals		$2442.79		$3707.25	$71.85	$109.04

Crew M-4

Crew M-4	Hr.	Daily	Hr.	Daily	Bare Costs	Incl. O&P
1 Electrician Foreman (outside)	$65.70	$525.60	$100.90	$807.20	$66.92	$103.35
1 Common Laborer	44.40	355.20	70.70	565.60		
.25 Equipment Operator, Crane	61.45	122.90	95.85	191.70		
.25 Equip. Oper. (oiler)	52.50	105.00	81.90	163.80		
1 Elevator Constructor	90.30	722.40	138.35	1106.80		
1 Elevator Apprentice	72.25	578.00	110.70	885.60		
.25 S.P. Crane, 4x4, 40 Ton		190.45		209.50	5.29	5.82
36 L.H., Daily Totals		$2599.55		$3930.20	$72.21	$109.17

Crew Q-1

Crew Q-1	Hr.	Daily	Hr.	Daily	Bare Costs	Incl. O&P
1 Plumber	$67.70	$541.60	$104.40	$835.20	$60.92	$93.95
1 Plumber Apprentice	54.15	433.20	83.50	668.00		
16 L.H., Daily Totals		$974.80		$1503.20	$60.92	$93.95

Crew Q-1A

Crew Q-1A	Hr.	Daily	Hr.	Daily	Bare Costs	Incl. O&P
.25 Plumber Foreman (outside)	$69.70	$139.40	$107.50	$215.00	$68.10	$105.02
1 Plumber	67.70	541.60	104.40	835.20		
10 L.H., Daily Totals		$681.00		$1050.20	$68.10	$105.02

Crew Q-1C

Crew Q-1C	Hr.	Daily	Hr.	Daily	Bare Costs	Incl. O&P
1 Plumber	$67.70	$541.60	$104.40	$835.20	$60.28	$93.32
1 Plumber Apprentice	54.15	433.20	83.50	668.00		
1 Equip. Oper. (medium)	59.00	472.00	92.05	736.40		
1 Trencher, Chain Type, 8' D		1894.00		2083.40	78.92	86.81
24 L.H., Daily Totals		$3340.80		$4323.00	$139.20	$180.13

Crew Q-2

Crew Q-2	Hr.	Daily	Hr.	Daily	Bare Costs	Incl. O&P
2 Plumbers	$67.70	$1083.20	$104.40	$1670.40	$63.18	$97.43
1 Plumber Apprentice	54.15	433.20	83.50	668.00		
24 L.H., Daily Totals		$1516.40		$2338.40	$63.18	$97.43

Crew Q-3

Crew Q-3	Hr.	Daily	Hr.	Daily	Bare Costs	Incl. O&P
1 Plumber Foreman (inside)	$68.20	$545.60	$105.20	$841.60	$64.44	$99.38
2 Plumbers	67.70	1083.20	104.40	1670.40		
1 Plumber Apprentice	54.15	433.20	83.50	668.00		
32 L.H., Daily Totals		$2062.00		$3180.00	$64.44	$99.38

Crew Q-4

Crew Q-4	Hr.	Daily	Hr.	Daily	Bare Costs	Incl. O&P
1 Plumber Foreman (inside)	$68.20	$545.60	$105.20	$841.60	$64.44	$99.38
1 Plumber	67.70	541.60	104.40	835.20		
1 Welder (plumber)	67.70	541.60	104.40	835.20		
1 Plumber Apprentice	54.15	433.20	83.50	668.00		
1 Welder, Electric, 300 amp		107.55		118.31	3.36	3.70
32 L.H., Daily Totals		$2169.55		$3298.30	$67.80	$103.07

Crew Q-5

Crew Q-5	Hr.	Daily	Hr.	Daily	Bare Costs	Incl. O&P
1 Steamfitter	$68.35	$546.80	$105.40	$843.20	$61.52	$94.88
1 Steamfitter Apprentice	54.70	437.60	84.35	674.80		
16 L.H., Daily Totals		$984.40		$1518.00	$61.52	$94.88

Crew Q-6

Crew Q-6	Hr.	Daily	Hr.	Daily	Bare Costs	Incl. O&P
2 Steamfitters	$68.35	$1093.60	$105.40	$1686.40	$63.80	$98.38
1 Steamfitter Apprentice	54.70	437.60	84.35	674.80		
24 L.H., Daily Totals		$1531.20		$2361.20	$63.80	$98.38

Crew Q-7

Crew Q-7	Hr.	Daily	Hr.	Daily	Bare Costs	Incl. O&P
1 Steamfitter Foreman (inside)	$68.85	$550.80	$106.20	$849.60	$65.06	$100.34
2 Steamfitters	68.35	1093.60	105.40	1686.40		
1 Steamfitter Apprentice	54.70	437.60	84.35	674.80		
32 L.H., Daily Totals		$2082.00		$3210.80	$65.06	$100.34

Crew Q-8

Crew Q-8	Hr.	Daily	Hr.	Daily	Bare Costs	Incl. O&P
1 Steamfitter Foreman (inside)	$68.85	$550.80	$106.20	$849.60	$65.06	$100.34
1 Steamfitter	68.35	546.80	105.40	843.20		
1 Welder (steamfitter)	68.35	546.80	105.40	843.20		
1 Steamfitter Apprentice	54.70	437.60	84.35	674.80		
1 Welder, Electric, 300 amp		107.55		118.31	3.36	3.70
32 L.H., Daily Totals		$2189.55		$3329.11	$68.42	$104.03

Crew Q-9

Crew Q-9	Hr.	Daily	Hr.	Daily	Bare Costs	Incl. O&P
1 Sheet Metal Worker	$65.45	$523.60	$102.00	$816.00	$58.90	$91.80
1 Sheet Metal Apprentice	52.35	418.80	81.60	652.80		
16 L.H., Daily Totals		$942.40		$1468.80	$58.90	$91.80

Crew Q-10

Crew Q-10	Hr.	Daily	Hr.	Daily	Bare Costs	Incl. O&P
2 Sheet Metal Workers	$65.45	$1047.20	$102.00	$1632.00	$61.08	$95.20
1 Sheet Metal Apprentice	52.35	418.80	81.60	652.80		
24 L.H., Daily Totals		$1466.00		$2284.80	$61.08	$95.20

Crew No.	Bare Costs		Incl. Subs O&P		Cost Per Labor-Hour	
Crew Q-11	Hr.	Daily	Hr.	Daily	Bare Costs	Incl. O&P
1 Sheet Metal Foreman (inside)	$65.95	$527.60	$102.80	$822.40	$62.30	$97.10
2 Sheet Metal Workers	65.45	1047.20	102.00	1632.00		
1 Sheet Metal Apprentice	52.35	418.80	81.60	652.80		
32 L.H., Daily Totals		$1993.60		$3107.20	$62.30	$97.10
Crew Q-12	Hr.	Daily	Hr.	Daily	Bare Costs	Incl. O&P
1 Sprinkler Installer	$66.50	$532.00	$102.70	$821.60	$59.85	$92.42
1 Sprinkler Apprentice	53.20	425.60	82.15	657.20		
16 L.H., Daily Totals		$957.60		$1478.80	$59.85	$92.42
Crew Q-13	Hr.	Daily	Hr.	Daily	Bare Costs	Incl. O&P
1 Sprinkler Foreman (inside)	$67.00	$536.00	$103.45	$827.60	$63.30	$97.75
2 Sprinkler Installers	66.50	1064.00	102.70	1643.20		
1 Sprinkler Apprentice	53.20	425.60	82.15	657.20		
32 L.H., Daily Totals		$2025.60		$3128.00	$63.30	$97.75
Crew Q-14	Hr.	Daily	Hr.	Daily	Bare Costs	Incl. O&P
1 Asbestos Worker	$60.95	$487.60	$96.30	$770.40	$54.85	$86.65
1 Asbestos Apprentice	48.75	390.00	77.00	616.00		
16 L.H., Daily Totals		$877.60		$1386.40	$54.85	$86.65
Crew Q-15	Hr.	Daily	Hr.	Daily	Bare Costs	Incl. O&P
1 Plumber	$67.70	$541.60	$104.40	$835.20	$60.92	$93.95
1 Plumber Apprentice	54.15	433.20	83.50	668.00		
1 Welder, Electric, 300 amp		107.55		118.31	6.72	7.39
16 L.H., Daily Totals		$1082.35		$1621.51	$67.65	$101.34
Crew Q-16	Hr.	Daily	Hr.	Daily	Bare Costs	Incl. O&P
2 Plumbers	$67.70	$1083.20	$104.40	$1670.40	$63.18	$97.43
1 Plumber Apprentice	54.15	433.20	83.50	668.00		
1 Welder, Electric, 300 amp		107.55		118.31	4.48	4.93
24 L.H., Daily Totals		$1623.95		$2456.70	$67.66	$102.36
Crew Q-17	Hr.	Daily	Hr.	Daily	Bare Costs	Incl. O&P
1 Steamfitter	$68.35	$546.80	$105.40	$843.20	$61.52	$94.88
1 Steamfitter Apprentice	54.70	437.60	84.35	674.80		
1 Welder, Electric, 300 amp		107.55		118.31	6.72	7.39
16 L.H., Daily Totals		$1091.95		$1636.31	$68.25	$102.27
Crew Q-17A	Hr.	Daily	Hr.	Daily	Bare Costs	Incl. O&P
1 Steamfitter	$68.35	$546.80	$105.40	$843.20	$61.50	$95.20
1 Steamfitter Apprentice	54.70	437.60	84.35	674.80		
1 Equip. Oper. (crane)	61.45	491.60	95.85	766.80		
1 Hyd. Crane, 12 Ton		475.80		523.38		
1 Welder, Electric, 300 amp		107.55		118.31	24.31	26.74
24 L.H., Daily Totals		$2059.35		$2926.49	$85.81	$121.94
Crew Q-18	Hr.	Daily	Hr.	Daily	Bare Costs	Incl. O&P
2 Steamfitters	$68.35	$1093.60	$105.40	$1686.40	$63.80	$98.38
1 Steamfitter Apprentice	54.70	437.60	84.35	674.80		
1 Welder, Electric, 300 amp		107.55		118.31	4.48	4.93
24 L.H., Daily Totals		$1638.75		$2479.51	$68.28	$103.31
Crew Q-19	Hr.	Daily	Hr.	Daily	Bare Costs	Incl. O&P
1 Steamfitter	$68.35	$546.80	$105.40	$843.20	$62.25	$95.87
1 Steamfitter Apprentice	54.70	437.60	84.35	674.80		
1 Electrician	63.70	509.60	97.85	782.80		
24 L.H., Daily Totals		$1494.00		$2300.80	$62.25	$95.87

Crew No.	Bare Costs		Incl. Subs O&P		Cost Per Labor-Hour	
Crew Q-20	Hr.	Daily	Hr.	Daily	Bare Costs	Incl. O&P
1 Sheet Metal Worker	$65.45	$523.60	$102.00	$816.00	$59.86	$93.01
1 Sheet Metal Apprentice	52.35	418.80	81.60	652.80		
.5 Electrician	63.70	254.80	97.85	391.40		
20 L.H., Daily Totals		$1197.20		$1860.20	$59.86	$93.01
Crew Q-21	Hr.	Daily	Hr.	Daily	Bare Costs	Incl. O&P
2 Steamfitters	$68.35	$1093.60	$105.40	$1686.40	$63.77	$98.25
1 Steamfitter Apprentice	54.70	437.60	84.35	674.80		
1 Electrician	63.70	509.60	97.85	782.80		
32 L.H., Daily Totals		$2040.80		$3144.00	$63.77	$98.25
Crew Q-22	Hr.	Daily	Hr.	Daily	Bare Costs	Incl. O&P
1 Plumber	$67.70	$541.60	$104.40	$835.20	$60.92	$93.95
1 Plumber Apprentice	54.15	433.20	83.50	668.00		
1 Hyd. Crane, 12 Ton		475.80		523.38	29.74	32.71
16 L.H., Daily Totals		$1450.60		$2026.58	$90.66	$126.66
Crew Q-22A	Hr.	Daily	Hr.	Daily	Bare Costs	Incl. O&P
1 Plumber	$67.70	$541.60	$104.40	$835.20	$56.92	$88.61
1 Plumber Apprentice	54.15	433.20	83.50	668.00		
1 Laborer	44.40	355.20	70.70	565.60		
1 Equip. Oper. (crane)	61.45	491.60	95.85	766.80		
1 Hyd. Crane, 12 Ton		475.80		523.38	14.87	16.36
32 L.H., Daily Totals		$2297.40		$3358.98	$71.79	$104.97
Crew Q-23	Hr.	Daily	Hr.	Daily	Bare Costs	Incl. O&P
1 Plumber Foreman (outside)	$69.70	$557.60	$107.50	$860.00	$65.47	$101.32
1 Plumber	67.70	541.60	104.40	835.20		
1 Equip. Oper. (medium)	59.00	472.00	92.05	736.40		
1 Lattice Boom Crane, 20 Ton		1526.00		1678.60	63.58	69.94
24 L.H., Daily Totals		$3097.20		$4110.20	$129.05	$171.26
Crew R-1	Hr.	Daily	Hr.	Daily	Bare Costs	Incl. O&P
1 Electrician Foreman	$64.20	$513.60	$98.60	$788.80	$59.53	$91.44
3 Electricians	63.70	1528.80	97.85	2348.40		
2 Electrician Apprentices	50.95	815.20	78.25	1252.00		
48 L.H., Daily Totals		$2857.60		$4389.20	$59.53	$91.44
Crew R-1A	Hr.	Daily	Hr.	Daily	Bare Costs	Incl. O&P
1 Electrician	$63.70	$509.60	$97.85	$782.80	$57.33	$88.05
1 Electrician Apprentice	50.95	407.60	78.25	626.00		
16 L.H., Daily Totals		$917.20		$1408.80	$57.33	$88.05
Crew R-1B	Hr.	Daily	Hr.	Daily	Bare Costs	Incl. O&P
1 Electrician	$63.70	$509.60	$97.85	$782.80	$55.20	$84.78
2 Electrician Apprentices	50.95	815.20	78.25	1252.00		
24 L.H., Daily Totals		$1324.80		$2034.80	$55.20	$84.78
Crew R-1C	Hr.	Daily	Hr.	Daily	Bare Costs	Incl. O&P
2 Electricians	$63.70	$1019.20	$97.85	$1565.60	$57.33	$88.05
2 Electrician Apprentices	50.95	815.20	78.25	1252.00		
1 Portable cable puller, 8000 lb.		102.65		112.92	3.21	3.53
32 L.H., Daily Totals		$1937.05		$2930.51	$60.53	$91.58

For customer support on your Commercial Renovation Costs with RSMeans Data, call 800.448.8182.

Crew No.	Bare Costs		Incl. Subs O&P		Cost Per Labor-Hour	

Crew R-2	Hr.	Daily	Hr.	Daily	Bare Costs	Incl. O&P
1 Electrician Foreman	$64.20	$513.60	$98.60	$788.80	$59.81	$92.07
3 Electricians	63.70	1528.80	97.85	2348.40		
2 Electrician Apprentices	50.95	815.20	78.25	1252.00		
1 Equip. Oper. (crane)	61.45	491.60	95.85	766.80		
1 S.P. Crane, 4x4, 5 Ton		381.95		420.14	6.82	7.50
56 L.H., Daily Totals		$3731.15		$5576.15	$66.63	$99.57

Crew R-3	Hr.	Daily	Hr.	Daily	Bare Costs	Incl. O&P
1 Electrician Foreman	$64.20	$513.60	$98.60	$788.80	$63.45	$97.75
1 Electrician	63.70	509.60	97.85	782.80		
.5 Equip. Oper. (crane)	61.45	245.80	95.85	383.40		
.5 S.P. Crane, 4x4, 5 Ton		190.97		210.07	9.55	10.50
20 L.H., Daily Totals		$1459.97		$2165.07	$73.00	$108.25

Crew R-4	Hr.	Daily	Hr.	Daily	Bare Costs	Incl. O&P
1 Struc. Steel Foreman (outside)	$62.30	$498.40	$100.75	$806.00	$61.38	$98.25
3 Struc. Steel Workers	60.30	1447.20	97.55	2341.20		
1 Electrician	63.70	509.60	97.85	782.80		
1 Welder, Gas Engine, 300 amp		148.75		163.63	3.72	4.09
40 L.H., Daily Totals		$2603.95		$4093.63	$65.10	$102.34

Crew R-5	Hr.	Daily	Hr.	Daily	Bare Costs	Incl. O&P
1 Electrician Foreman	$64.20	$513.60	$98.60	$788.80	$55.87	$87.03
4 Electrician Linemen	63.70	2038.40	97.85	3131.20		
2 Electrician Operators	63.70	1019.20	97.85	1565.60		
4 Electrician Groundmen	42.05	1345.60	67.90	2172.80		
1 Crew Truck		168.15		184.97		
1 Flatbed Truck, 20,000 GVW		204.05		224.46		
1 Pickup Truck, 3/4 Ton		112.20		123.42		
.2 Hyd. Crane, 55 Ton		198.03		217.83		
.2 Hyd. Crane, 12 Ton		95.16		104.68		
.2 Earth Auger, Truck-Mtd.		40.51		44.56		
1 Tractor w/Winch		377.65		415.42	13.59	14.95
88 L.H., Daily Totals		$6112.55		$8973.73	$69.46	$101.97

Crew R-6	Hr.	Daily	Hr.	Daily	Bare Costs	Incl. O&P
1 Electrician Foreman	$64.20	$513.60	$98.60	$788.80	$55.87	$87.03
4 Electrician Linemen	63.70	2038.40	97.85	3131.20		
2 Electrician Operators	63.70	1019.20	97.85	1565.60		
4 Electrician Groundmen	42.05	1345.60	67.90	2172.80		
1 Crew Truck		168.15		184.97		
1 Flatbed Truck, 20,000 GVW		204.05		224.46		
1 Pickup Truck, 3/4 Ton		112.20		123.42		
.2 Hyd. Crane, 55 Ton		198.03		217.83		
.2 Hyd. Crane, 12 Ton		95.16		104.68		
.2 Earth Auger, Truck-Mtd.		40.51		44.56		
1 Tractor w/Winch		377.65		415.42		
3 Cable Trailers		194.25		213.68		
.5 Tensioning Rig		55.75		61.33		
.5 Cable Pulling Rig		306.65		337.32	19.91	21.91
88 L.H., Daily Totals		$6669.20		$9586.04	$75.79	$108.93

Crew R-7	Hr.	Daily	Hr.	Daily	Bare Costs	Incl. O&P
1 Electrician Foreman	$64.20	$513.60	$98.60	$788.80	$45.74	$73.02
5 Electrician Groundmen	42.05	1682.00	67.90	2716.00		
1 Crew Truck		168.15		184.97	3.50	3.85
48 L.H., Daily Totals		$2363.75		$3689.76	$49.24	$76.87

Crew R-8	Hr.	Daily	Hr.	Daily	Bare Costs	Incl. O&P
1 Electrician Foreman	$64.20	$513.60	$98.60	$788.80	$56.57	$87.99
3 Electrician Linemen	63.70	1528.80	97.85	2348.40		
2 Electrician Groundmen	42.05	672.80	67.90	1086.40		
1 Pickup Truck, 3/4 Ton		112.20		123.42		
1 Crew Truck		168.15		184.97	5.84	6.42
48 L.H., Daily Totals		$2995.55		$4531.98	$62.41	$94.42

Crew R-9	Hr.	Daily	Hr.	Daily	Bare Costs	Incl. O&P
1 Electrician Foreman	$64.20	$513.60	$98.60	$788.80	$52.94	$82.97
1 Electrician Lineman	63.70	509.60	97.85	782.80		
2 Electrician Operators	63.70	1019.20	97.85	1565.60		
4 Electrician Groundmen	42.05	1345.60	67.90	2172.80		
1 Pickup Truck, 3/4 Ton		112.20		123.42		
1 Crew Truck		168.15		184.97	4.38	4.82
64 L.H., Daily Totals		$3668.35		$5618.39	$57.32	$87.79

Crew R-10	Hr.	Daily	Hr.	Daily	Bare Costs	Incl. O&P
1 Electrician Foreman	$64.20	$513.60	$98.60	$788.80	$60.17	$92.98
4 Electrician Linemen	63.70	2038.40	97.85	3131.20		
1 Electrician Groundman	42.05	336.40	67.90	543.20		
1 Crew Truck		168.15		184.97		
3 Tram Cars		219.60		241.56	8.08	8.89
48 L.H., Daily Totals		$3276.15		$4889.73	$68.25	$101.87

Crew R-11	Hr.	Daily	Hr.	Daily	Bare Costs	Incl. O&P
1 Electrician Foreman	$64.20	$513.60	$98.60	$788.80	$60.69	$93.79
4 Electricians	63.70	2038.40	97.85	3131.20		
1 Equip. Oper. (crane)	61.45	491.60	95.85	766.80		
1 Common Laborer	44.40	355.20	70.70	565.60		
1 Crew Truck		168.15		184.97		
1 Hyd. Crane, 12 Ton		475.80		523.38	11.50	12.65
56 L.H., Daily Totals		$4042.75		$5960.74	$72.19	$106.44

Crew R-12	Hr.	Daily	Hr.	Daily	Bare Costs	Incl. O&P
1 Carpenter Foreman (inside)	$55.20	$441.60	$87.90	$703.20	$51.90	$82.61
4 Carpenters	54.70	1750.40	87.10	2787.20		
4 Common Laborers	44.40	1420.80	70.70	2262.40		
1 Equip. Oper. (medium)	59.00	472.00	92.05	736.40		
1 Steel Worker	60.30	482.40	97.55	780.40		
1 Dozer, 200 H.P.		1520.00		1672.00		
1 Pickup Truck, 3/4 Ton		112.20		123.42	18.55	20.40
88 L.H., Daily Totals		$6199.40		$9065.02	$70.45	$103.01

Crew R-13	Hr.	Daily	Hr.	Daily	Bare Costs	Incl. O&P
1 Electrician Foreman	$64.20	$513.60	$98.60	$788.80	$61.55	$94.86
3 Electricians	63.70	1528.80	97.85	2348.40		
.25 Equip. Oper. (crane)	61.45	122.90	95.85	191.70		
1 Equipment Oiler	52.50	420.00	81.90	655.20		
.25 Hydraulic Crane, 33 Ton		245.79		270.37	5.85	6.44
42 L.H., Daily Totals		$2831.09		$4254.47	$67.41	$101.30

Crew R-15	Hr.	Daily	Hr.	Daily	Bare Costs	Incl. O&P
1 Electrician Foreman	$64.20	$513.60	$98.60	$788.80	$62.42	$96.10
4 Electricians	63.70	2038.40	97.85	3131.20		
1 Equipment Oper. (light)	55.50	444.00	86.60	692.80		
1 Telescoping Boom Lift, to 40'		281.90		310.09	5.87	6.46
48 L.H., Daily Totals		$3277.90		$4922.89	$68.29	$102.56

For customer support on your Commercial Renovation Costs with RSMeans Data, call 800.448.8182.

789

Crew No.	Bare Costs		Incl. Subs O & P		Cost Per Labor-Hour	

Crew R-18	Hr.	Daily	Hr.	Daily	Bare Costs	Incl. O&P
.25 Electrician Foreman	$64.20	$128.40	$98.60	$197.20	$55.89	$85.85
1 Electrician	63.70	509.60	97.85	782.80		
2 Electrician Apprentices	50.95	815.20	78.25	1252.00		
26 L.H., Daily Totals		$1453.20		$2232.00	$55.89	$85.85

Crew R-19	Hr.	Daily	Hr.	Daily	Bare Costs	Incl. O&P
.5 Electrician Foreman	$64.20	$256.80	$98.60	$394.40	$63.80	$98.00
2 Electricians	63.70	1019.20	97.85	1565.60		
20 L.H., Daily Totals		$1276.00		$1960.00	$63.80	$98.00

Crew R-21	Hr.	Daily	Hr.	Daily	Bare Costs	Incl. O&P
1 Electrician Foreman	$64.20	$513.60	$98.60	$788.80	$63.71	$97.89
3 Electricians	63.70	1528.80	97.85	2348.40		
.1 Equip. Oper. (medium)	59.00	47.20	92.05	73.64		
.1 S.P. Crane, 4x4, 25 Ton		115.50		127.05	3.52	3.87
32.8 L.H., Daily Totals		$2205.10		$3337.89	$67.23	$101.77

Crew R-22	Hr.	Daily	Hr.	Daily	Bare Costs	Incl. O&P
.66 Electrician Foreman	$64.20	$338.98	$98.60	$520.61	$58.30	$89.54
2 Electricians	63.70	1019.20	97.85	1565.60		
2 Electrician Apprentices	50.95	815.20	78.25	1252.00		
37.28 L.H., Daily Totals		$2173.38		$3338.21	$58.30	$89.54

Crew R-30	Hr.	Daily	Hr.	Daily	Bare Costs	Incl. O&P
.25 Electrician Foreman (outside)	$65.70	$131.40	$100.90	$201.80	$51.98	$81.38
1 Electrician	63.70	509.60	97.85	782.80		
2 Laborers (Semi-Skilled)	44.40	710.40	70.70	1131.20		
26 L.H., Daily Totals		$1351.40		$2115.80	$51.98	$81.38

Crew R-31	Hr.	Daily	Hr.	Daily	Bare Costs	Incl. O&P
1 Electrician	$63.70	$509.60	$97.85	$782.80	$63.70	$97.85
1 Core Drill, Electric, 2.5 H.P.		62.65		68.92	7.83	8.61
8 L.H., Daily Totals		$572.25		$851.72	$71.53	$106.46

For customer support on your Commercial Renovation Costs with RSMeans Data, call 800.448.8182.

Historical Cost Indexes

The table below lists both the RSMeans® historical cost index based on Jan. 1, 1993 = 100 as well as the computed value of an index based on Jan. 1, 2021 costs. Since the Jan. 1, 2021 figure is estimated, space is left to write in the actual index figures as they become available through the quarterly *RSMeans Construction Cost Indexes*.

To compute the actual index based on Jan. 1, 2021 = 100, divide the historical cost index for a particular year by the actual Jan. 1, 2021 construction cost index. Space has been left to advance the index figures as the year progresses.

Year	Historical Cost Index Jan. 1, 1993 = 100		Current Index Based on Jan. 1, 2021 = 100		Year	Historical Cost Index Jan. 1, 1993 = 100	Current Index Based on Jan. 1, 2021 = 100		Year	Historical Cost Index Jan. 1, 1993 = 100	Current Index Based on Jan. 1, 2021 = 100	
	Est.	Actual	Est.	Actual		Actual	Est.	Actual		Actual	Est.	Actual
Oct 2021					July 2004	143.7	60.7		1984	82.0	34.6	
July 2021					2003	132.0	55.8		1983	80.2	33.9	
April 2021					2002	128.7	54.4		1982	76.1	32.2	
Jan 2021	236.7		100.0		2001	125.1	52.9		1981	70.0	29.6	
2020		234.6	99.1		2000	120.9	51.1		1980	62.9	26.6	
2019		232.2	98.1		1999	117.6	49.7		1979	57.8	24.4	
2018		222.9	94.2		1998	115.1	48.6		1978	53.5	22.6	
2017		213.6	90.2		1997	112.8	47.7		1977	49.5	20.9	
2016		207.3	87.6		1996	110.2	46.6		1976	46.9	19.8	
2015		206.2	87.1		1995	107.6	45.5		1975	44.8	18.9	
2014		204.9	86.6		1994	104.4	44.1		1974	41.4	17.5	
2013		201.2	85.0		1993	101.7	43.0		1973	37.7	15.9	
2012		194.6	82.2		1992	99.4	42.0		1972	34.8	14.7	
2011		191.2	80.8		1991	96.8	40.9		1971	32.1	13.6	
2010		183.5	77.5		1990	94.3	39.8		1970	28.7	12.1	
2009		180.1	76.1		1989	92.1	38.9		1969	26.9	11.4	
2008		180.4	76.2		1988	89.9	38.0					
2007		169.4	71.6		1987	87.7	37.0					
2006		162.0	68.4		1986	84.2	35.6					
2005		151.6	64.0		1985	82.6	34.9					

Adjustments to Costs

The "Historical Cost Index" can be used to convert national average building costs at a particular time to the approximate building costs for some other time.

Example:

Estimate and compare construction costs for different years in the same city.

To estimate the national average construction cost of a building in 1970, knowing that it cost $900,000 in 2021:

INDEX in 1970 = 28.7

INDEX in 2021 = 236.7

Note: The city cost indexes for Canada can be used to convert U.S. national averages to local costs in Canadian dollars.

Example:

To estimate and compare the cost of a building in Toronto, ON in 2021 with the known cost of $600,000 (US$) in New York, NY in 2021:

INDEX Toronto = 112.5

INDEX New York = 134.1

$$\frac{\text{INDEX Toronto}}{\text{INDEX New York}} \times \text{Cost New York} = \text{Cost Toronto}$$

$$\frac{112.5}{134.1} \times \$600,000 = .839 \times \$600,000 = \$503,400$$

The construction cost of the building in Toronto is $503,400 (CN$).

Time Adjustment Using the Historical Cost Indexes:

$$\frac{\text{Index for Year A}}{\text{Index for Year B}} \times \text{Cost in Year B} = \text{Cost in Year A}$$

$$\frac{\text{INDEX 1970}}{\text{INDEX 2021}} \times \text{Cost 2021} = \text{Cost 1970}$$

$$\frac{28.7}{236.7} \times \$900,000 = .121 \times \$900,000 = \$108,900$$

The construction cost of the building in 1970 was $108,900.

*Historical Cost Index updates and other resources are provided on the following website:
rsmeans.com/2021books

How to Use the City Cost Indexes

What you should know before you begin

RSMeans City Cost Indexes (CCI) are an extremely useful tool for when you want to compare costs from city to city and region to region.

This publication contains average construction cost indexes for 731 U.S. and Canadian cities covering over 930 three-digit zip code locations, as listed directly under each city.

Keep in mind that a City Cost Index number is a percentage ratio of a specific city's cost to the national average cost of the same item at a stated time period.

In other words, these index figures represent relative construction factors (or, if you prefer, multipliers) for material and installation costs, as well as the weighted average for Total In Place costs for each CSI MasterFormat division. Installation costs include both labor and equipment rental costs. When estimating equipment rental rates only for a specific location, use 01 54 33 EQUIPMENT RENTAL COSTS in the Reference Section.

The 30 City Average Index is the average of 30 major U.S. cities and serves as a national average. This national average represents the baseline from which all locations are calculated.

Index figures for both material and installation are based on the 30 major city average of 100 and represent the cost relationship as of July 1, 2020. The index for each division is computed from representative material and labor quantities for that division. The weighted average for each city is a weighted total of the components listed above it. It does not include relative productivity between trades or cities.

As changes occur in local material prices, labor rates, and equipment rental rates (including fuel costs), the impact of these changes should be accurately measured by the change in the City Cost Index for each particular city (as compared to the 30 city average).

Therefore, if you know (or have estimated) building costs in one city today, you can easily convert those costs to expected building costs in another city.

In addition, by using the Historical Cost Index, you can easily convert national average building costs at a particular time to the approximate building costs for some other time. The City Cost Indexes can then be applied to calculate the costs for a particular city.

Quick calculations

Location Adjustment Using the City Cost Indexes:

$$\frac{\text{Index for City A}}{\text{Index for City B}} \times \text{Cost in City B} = \text{Cost in City A}$$

Time Adjustment for the National Average Using the Historical Cost Index:

$$\frac{\text{Index for Year A}}{\text{Index for Year B}} \times \text{Cost in Year B} = \text{Cost in Year A}$$

Adjustment from the National Average:

$$\frac{\text{Index for City A}}{100} \times \text{National Average Cost} = \text{Cost in City A}$$

Since each of the other RSMeans data sets contains many different items, any *one* item multiplied by the particular city index may give incorrect results. However, the larger the number of items compiled, the closer the results should be to actual costs for that particular city.

The City Cost Indexes for Canadian cities are calculated using Canadian material and equipment prices and labor rates in Canadian dollars. When compared to the baseline mentioned previously, the resulting index enables the user to convert baseline costs to local costs. Therefore, indexes for Canadian cities can be used to convert U.S. national average prices to local costs in Canadian dollars.

How to use this section

1. Compare costs from city to city.

In using the RSMeans Indexes, remember that an index number is not a fixed number but a ratio: It's a percentage ratio of a building component's cost at any stated time to the national average cost of that same component at the same time period. Put in the form of an equation:

$$\frac{\text{Specific City Cost}}{\text{National Average Cost}} \times 100 = \text{City Index Number}$$

Therefore, when making cost comparisons between cities, do not subtract one city's index number from the index number of another city and read the result as a percentage difference. Instead, divide one city's index number by that of the other city. The resulting number may then be used as a multiplier to calculate cost differences from city to city.

The formula used to find cost differences between cities for the purpose of comparison is as follows:

$$\frac{\text{City A Index}}{\text{City B Index}} \times \text{City B Cost (Known)} = \text{City A Cost (Unknown)}$$

In addition, you can use RSMeans CCI to calculate and compare costs division by division between cities using the same basic formula. (Just be sure that you're comparing similar divisions.)

2. Compare a specific city's construction costs with the national average.

When you're studying construction location feasibility, it's advisable to compare a prospective project's cost index with an index of the national average cost.

For example, divide the weighted average index of construction costs of a specific city by that of the 30 City Average, which = 100.

$$\frac{\text{City Index}}{100} = \text{\% of National Average}$$

As a result, you get a ratio that indicates the relative cost of construction in that city in comparison with the national average.

3. Convert U.S. national average to actual costs in Canadian City.

$$\frac{\text{Index for Canadian City}}{100} \times \text{National Average Cost} = \text{Cost in Canadian City in \$ CAN}$$

4. Adjust construction cost data based on a national average.

When you use a source of construction cost data which is based on a national average (such as RSMeans cost data), it is necessary to adjust those costs to a specific location.

$$\frac{\text{City Index}}{100} \times \frac{\text{Cost Based on}}{\text{National Average Costs}} = \frac{\text{City Cost}}{\text{(Unknown)}}$$

5. When applying the City Cost Indexes to demolition projects, use the appropriate division installation index.
For example, for removal of existing doors and windows, use the Division 8 (Openings) index.

What you might like to know about how we developed the Indexes

The information presented in the CCI is organized according to the Construction Specifications Institute (CSI) MasterFormat 2018 classification system.

To create a reliable index, RSMeans researched the building type most often constructed in the United States and Canada. Because it was concluded that no one type of building completely represented the building construction industry, nine different types of buildings were combined to create a composite model.

The exact material, labor, and equipment quantities are based on detailed analyses of these nine building types, and then each quantity is weighted in proportion to expected usage. These various material items, labor hours, and equipment rental rates are thus combined to form a composite building representing as closely as possible the actual usage of materials, labor, and equipment in the North American building construction industry.

The following structures were chosen to make up that composite model:

1. Factory, 1 story
2. Office, 2–4 stories
3. Store, Retail
4. Town Hall, 2–3 stories
5. High School, 2–3 stories
6. Hospital, 4–8 stories
7. Garage, Parking
8. Apartment, 1–3 stories
9. Hotel/Motel, 2–3 stories

For the purposes of ensuring the timeliness of the data, the components of the index for the composite model have been streamlined. They currently consist of:

- specific quantities of 66 commonly used construction materials;
- specific labor-hours for 21 building construction trades; and
- specific days of equipment rental for 6 types of construction equipment (normally used to install the 66 material items by the 21 trades.) Fuel costs and routine maintenance costs are included in the equipment cost.

Material and equipment price quotations are gathered quarterly from cities in the United States and Canada. These prices and the latest negotiated labor wage rates for 21 different building trades are used to compile the quarterly update of the City Cost Index.

The 30 major U.S. cities used to calculate the national average are:

Atlanta, GA	Memphis, TN
Baltimore, MD	Milwaukee, WI
Boston, MA	Minneapolis, MN
Buffalo, NY	Nashville, TN
Chicago, IL	New Orleans, LA
Cincinnati, OH	New York, NY
Cleveland, OH	Philadelphia, PA
Columbus, OH	Phoenix, AZ
Dallas, TX	Pittsburgh, PA
Denver, CO	St. Louis, MO
Detroit, MI	San Antonio, TX
Houston, TX	San Diego, CA
Indianapolis, IN	San Francisco, CA
Kansas City, MO	Seattle, WA
Los Angeles, CA	Washington, DC

What the CCI does not indicate

The weighted average for each city is a total of the divisional components weighted to reflect typical usage. It does not include the productivity variations between trades or cities.

In addition, the CCI does not take into consideration factors such as the following:

- managerial efficiency
- competitive conditions
- automation
- restrictive union practices
- unique local requirements
- regional variations due to specific building codes

ALABAMA

DIVISION		UNITED STATES 30 CITY AVERAGE			ANNISTON 362			BIRMINGHAM 350 - 352			BUTLER 369			DECATUR 356			DOTHAN 363		
		MAT.	INST.	TOTAL	MAT.	INST.	TOTAL	MAT.	INST.	TOTAL	MAT.	INST.	TOTAL	MAT.	INST.	TOTAL	MAT.	INST.	TOTAL
015433	CONTRACTOR EQUIPMENT		100.0	100.0		100.5	100.5		102.7	102.7		98.2	98.2		100.5	100.5		98.2	98.2
0241, 31 - 34	SITE & INFRASTRUCTURE, DEMOLITION	100.0	100.0	100.0	89.8	87.0	87.9	88.9	90.9	90.3	101.8	83.2	89.0	81.6	86.8	85.1	99.5	83.1	88.2
0310	Concrete Forming & Accessories	100.0	100.0	100.0	88.9	66.7	70.0	94.3	67.4	71.4	84.5	69.1	71.4	94.5	65.7	70.0	94.3	67.3	71.3
0320	Concrete Reinforcing	100.0	100.0	100.0	88.6	69.5	79.4	92.3	69.5	81.3	93.7	71.5	83.0	86.4	68.4	77.7	93.7	69.6	82.1
0330	Cast-in-Place Concrete	100.0	100.0	100.0	83.3	67.0	77.1	101.0	68.9	88.9	81.2	67.3	76.0	93.8	67.0	83.7	81.2	67.3	76.0
03	CONCRETE	100.0	100.0	100.0	91.4	69.0	81.4	97.0	69.8	84.8	92.5	70.5	82.6	90.0	68.3	80.3	91.8	69.4	81.7
04	MASONRY	100.0	100.0	100.0	96.7	62.5	75.8	89.8	62.6	73.2	101.5	62.5	77.7	87.1	62.6	72.1	102.9	62.5	78.3
05	METALS	100.0	100.0	100.0	101.4	94.5	99.3	98.7	92.8	96.8	100.3	95.9	99.0	100.9	93.8	98.7	100.4	95.1	98.8
06	WOOD, PLASTICS & COMPOSITES	100.0	100.0	100.0	85.3	67.5	76.0	93.4	67.7	80.0	79.6	70.1	74.6	97.4	65.6	80.7	92.5	67.5	79.4
07	THERMAL & MOISTURE PROTECTION	100.0	100.0	100.0	95.0	62.3	80.9	94.7	66.1	82.4	95.0	65.7	82.4	93.6	65.3	81.5	95.0	64.7	82.0
08	OPENINGS	100.0	100.0	100.0	94.4	68.2	88.0	102.3	68.7	94.1	94.5	70.6	88.6	106.7	67.3	97.1	94.5	68.6	88.2
0920	Plaster & Gypsum Board	100.0	100.0	100.0	86.0	67.2	73.7	97.5	67.2	77.6	84.0	69.9	74.8	96.9	65.2	76.1	93.4	67.2	76.2
0950, 0980	Ceilings & Acoustic Treatment	100.0	100.0	100.0	85.8	67.2	74.1	96.2	67.2	78.0	85.8	69.9	75.8	94.3	65.2	76.1	85.8	67.2	74.1
0960	Flooring	100.0	100.0	100.0	83.5	68.4	79.1	95.8	68.4	87.8	88.2	68.4	82.4	89.0	68.4	83.0	93.3	68.4	86.0
0970, 0990	Wall Finishes & Painting/Coating	100.0	100.0	100.0	87.4	53.7	67.2	91.6	53.7	68.9	87.4	45.5	62.3	81.3	60.5	68.8	87.4	78.1	81.8
09	FINISHES	100.0	100.0	100.0	84.4	65.7	74.3	94.3	65.9	78.9	87.4	66.3	76.0	89.3	65.3	76.3	89.9	68.4	78.3
COVERS	DIVS. 10 - 14, 25, 28, 41, 43, 44, 46	100.0	100.0	100.0	100.0	84.1	96.2	100.0	84.7	96.4	100.0	85.2	96.5	100.0	83.9	96.2	100.0	85.0	96.5
21, 22, 23	FIRE SUPPRESSION, PLUMBING & HVAC	100.0	100.0	100.0	102.3	51.9	82.0	100.8	65.0	86.4	98.5	64.4	84.7	100.8	64.9	86.4	98.5	63.8	84.5
26, 27, 3370	ELECTRICAL, COMMUNICATIONS & UTIL.	100.0	100.0	100.0	99.2	57.7	78.6	98.6	65.4	82.2	101.0	57.8	79.7	94.2	65.2	79.9	99.9	73.3	86.7
MF2018	WEIGHTED AVERAGE	100.0	100.0	100.0	97.1	66.6	83.9	98.2	70.9	86.4	97.1	69.7	85.2	96.9	70.2	85.4	97.2	71.6	86.1

ALABAMA

DIVISION		EVERGREEN 364			GADSDEN 359			HUNTSVILLE 357 - 358			JASPER 355			MOBILE 365 - 366			MONTGOMERY 360 - 361		
		MAT.	INST.	TOTAL	MAT.	INST.	TOTAL	MAT.	INST.	TOTAL	MAT.	INST.	TOTAL	MAT.	INST.	TOTAL	MAT.	INST.	TOTAL
015433	CONTRACTOR EQUIPMENT		98.2	98.2		100.5	100.5		100.5	100.5		100.5	100.5		98.2	98.2		103.5	103.5
0241, 31 - 34	SITE & INFRASTRUCTURE, DEMOLITION	102.3	83.2	89.1	87.5	87.0	87.2	81.3	86.7	85.0	87.4	87.0	87.1	95.1	83.2	86.9	93.5	92.2	92.6
0310	Concrete Forming & Accessories	80.9	67.0	69.0	86.0	67.2	70.0	94.5	65.3	69.7	91.7	67.1	70.8	93.6	67.0	71.0	97.8	67.4	71.9
0320	Concrete Reinforcing	93.8	71.5	83.0	91.6	69.5	80.9	86.4	72.8	79.8	86.4	69.5	78.2	91.3	71.5	81.8	99.5	69.6	85.1
0330	Cast-in-Place Concrete	81.2	67.2	75.9	93.8	67.2	83.8	91.3	66.7	82.1	104.0	67.2	90.1	85.2	67.2	78.4	85.1	68.9	79.0
03	CONCRETE	92.9	69.5	82.4	94.3	69.3	83.1	88.8	68.8	79.9	97.7	69.3	84.9	87.9	69.6	79.7	90.8	69.8	81.4
04	MASONRY	101.5	62.5	77.7	85.8	62.5	71.6	88.5	62.0	72.3	83.5	62.5	70.7	100.5	62.5	77.3	99.7	62.6	77.0
05	METALS	100.4	95.7	98.9	98.7	95.0	97.5	100.9	95.5	99.2	98.6	94.9	97.5	102.5	95.8	100.4	101.5	92.8	98.8
06	WOOD, PLASTICS & COMPOSITES	75.5	67.5	71.3	87.0	67.5	76.8	97.4	65.3	80.6	94.4	67.5	80.3	91.1	67.5	78.8	95.4	67.7	80.9
07	THERMAL & MOISTURE PROTECTION	95.0	64.7	82.0	93.9	65.4	81.6	93.6	65.0	81.3	93.9	65.2	81.6	94.6	64.7	81.7	93.3	66.5	81.8
08	OPENINGS	94.5	69.1	88.3	103.4	68.6	94.9	106.4	68.2	97.1	103.3	68.6	94.9	97.0	69.1	90.2	95.7	68.7	89.2
0920	Plaster & Gypsum Board	83.5	67.2	72.8	89.3	67.2	74.8	96.9	65.0	76.0	93.3	67.2	76.2	90.6	67.2	75.3	93.0	67.2	76.1
0950, 0980	Ceilings & Acoustic Treatment	85.8	67.2	74.1	87.9	67.2	74.9	95.7	65.0	76.4	87.9	67.2	74.9	92.9	67.2	76.8	94.8	67.2	77.5
0960	Flooring	86.1	68.4	80.9	85.3	68.4	80.4	89.0	68.4	83.0	87.4	68.4	81.9	92.8	68.4	85.7	92.5	68.4	85.5
0970, 0990	Wall Finishes & Painting/Coating	87.4	45.5	62.3	81.3	53.7	64.7	81.3	60.6	68.9	81.3	53.7	64.7	90.5	45.5	63.5	92.1	53.7	69.1
09	FINISHES	86.7	64.8	74.9	86.0	65.7	75.0	89.6	65.1	76.3	87.1	65.7	75.5	90.7	64.8	76.7	93.1	65.9	78.3
COVERS	DIVS. 10 - 14, 25, 28, 41, 43, 44, 46	100.0	84.9	96.4	100.0	84.1	96.3	100.0	83.7	96.2	100.0	84.1	96.2	100.0	84.9	96.5	100.0	84.7	96.4
21, 22, 23	FIRE SUPPRESSION, PLUMBING & HVAC	98.5	58.1	82.2	103.7	65.5	88.3	100.8	63.8	85.9	103.7	64.9	88.1	100.6	60.4	84.4	100.6	63.3	85.6
26, 27, 3370	ELECTRICAL, COMMUNICATIONS & UTIL.	98.6	57.8	78.4	94.2	65.4	80.0	95.0	63.4	79.4	93.9	57.7	76.0	101.6	60.3	81.2	101.5	73.2	87.5
MF2018	WEIGHTED AVERAGE	96.8	67.8	84.3	97.2	70.8	85.8	96.9	69.9	85.2	97.6	69.6	85.5	97.8	68.7	85.2	98.0	71.8	86.6

ALABAMA / ALASKA

DIVISION		PHENIX CITY 368			SELMA 367			TUSCALOOSA 354			ANCHORAGE 995 - 996			FAIRBANKS 997			JUNEAU 998		
		MAT.	INST.	TOTAL	MAT.	INST.	TOTAL	MAT.	INST.	TOTAL	MAT.	INST.	TOTAL	MAT.	INST.	TOTAL	MAT.	INST.	TOTAL
015433	CONTRACTOR EQUIPMENT		98.2	98.2		98.2	98.2		100.5	100.5		107.6	107.6		111.2	111.2		107.6	107.6
0241, 31 - 34	SITE & INFRASTRUCTURE, DEMOLITION	105.9	83.1	90.2	99.3	83.1	88.2	81.8	87.0	85.4	124.0	121.8	122.5	123.2	121.2	121.8	139.7	115.8	123.3
0310	Concrete Forming & Accessories	88.7	68.6	71.6	85.9	67.2	70.0	94.4	67.1	71.2	112.9	113.4	113.3	120.7	112.7	113.9	124.0	113.4	115.0
0320	Concrete Reinforcing	93.7	68.2	81.4	93.7	69.5	82.1	86.4	69.5	78.2	143.1	118.2	131.1	145.0	118.2	132.1	146.2	118.2	132.7
0330	Cast-in-Place Concrete	81.2	66.9	75.8	81.2	67.2	75.9	95.1	67.2	84.6	117.0	114.9	116.2	116.0	112.4	114.6	116.9	114.9	116.2
03	CONCRETE	95.8	69.6	84.1	91.3	69.3	81.4	90.6	69.3	81.0	112.6	114.1	113.3	105.1	113.1	108.7	119.7	114.1	117.2
04	MASONRY	101.5	61.9	77.3	105.9	62.5	79.4	87.4	62.5	72.2	185.2	116.7	143.4	187.5	115.0	143.3	171.5	116.7	138.0
05	METALS	100.3	94.6	98.5	100.3	95.0	98.7	100.1	95.0	98.5	126.3	103.8	119.3	122.7	105.5	117.3	121.4	103.8	115.9
06	WOOD, PLASTICS & COMPOSITES	84.9	69.6	76.9	81.5	67.5	74.2	97.4	67.5	81.7	102.3	111.1	106.9	117.3	111.0	114.0	117.0	111.1	113.9
07	THERMAL & MOISTURE PROTECTION	95.4	65.4	82.5	94.8	65.3	82.0	93.7	65.4	81.5	177.8	113.6	150.2	188.1	112.6	155.7	191.7	113.6	158.2
08	OPENINGS	94.4	69.4	88.3	94.4	68.6	88.1	106.4	68.6	97.2	131.0	113.6	126.8	133.2	112.9	128.2	131.7	113.6	127.3
0920	Plaster & Gypsum Board	87.4	69.4	75.6	85.4	67.2	73.5	96.9	67.2	77.4	139.6	111.3	121.0	165.5	111.3	129.9	141.3	111.3	121.6
0950, 0980	Ceilings & Acoustic Treatment	85.8	69.4	75.5	85.8	67.2	74.1	95.7	67.2	77.8	120.9	111.3	114.9	107.6	111.3	109.9	116.2	111.3	113.1
0960	Flooring	90.0	68.4	83.7	88.5	68.4	82.7	89.0	68.4	83.0	114.5	116.7	115.1	117.5	116.7	117.3	126.7	116.7	123.8
0970, 0990	Wall Finishes & Painting/Coating	87.4	79.8	82.8	87.4	53.7	67.2	81.3	53.7	64.7	108.6	117.6	114.0	109.0	113.4	111.6	108.9	117.6	114.1
09	FINISHES	88.8	69.6	78.4	87.4	65.7	75.6	89.6	65.7	76.7	123.7	114.5	118.7	125.5	113.5	119.0	125.5	114.5	119.5
COVERS	DIVS. 10 - 14, 25, 28, 41, 43, 44, 46	100.0	84.8	96.4	100.0	84.1	96.3	100.0	84.1	96.3	100.0	110.1	102.4	100.0	109.3	102.2	100.0	110.1	102.4
21, 22, 23	FIRE SUPPRESSION, PLUMBING & HVAC	98.5	63.0	84.2	98.5	63.8	84.5	100.8	64.3	86.1	101.2	104.1	102.4	101.2	106.5	103.3	101.8	104.1	102.7
26, 27, 3370	ELECTRICAL, COMMUNICATIONS & UTIL.	100.5	68.2	84.5	99.6	73.3	86.6	94.7	65.4	80.2	115.8	107.9	111.9	125.5	107.9	116.8	109.2	107.9	108.5
MF2018	WEIGHTED AVERAGE	97.7	70.8	86.1	96.9	71.2	85.8	96.9	70.5	85.5	119.7	110.4	115.7	120.2	111.0	116.2	119.7	110.4	115.7

		ALASKA			ARIZONA														
		KETCHIKAN			CHAMBERS			FLAGSTAFF			GLOBE			KINGMAN			MESA/TEMPE		
DIVISION		999			865			860			855			864			852		
		MAT.	INST.	TOTAL	MAT.	INST.	TOTAL	MAT.	INST.	TOTAL	MAT.	INST.	TOTAL	MAT.	INST.	TOTAL	MAT.	INST.	TOTAL
015433	CONTRACTOR EQUIPMENT		111.2	111.2		85.9	85.9		85.9	85.9		87.5	87.5		85.9	85.9		87.5	87.5
0241, 31 - 34	SITE & INFRASTRUCTURE, DEMOLITION	176.0	121.3	138.4	69.8	87.5	82.0	89.3	87.6	88.2	103.1	88.7	93.2	69.8	87.6	82.0	93.2	88.9	90.2
0310	Concrete Forming & Accessories	112.4	113.3	113.2	96.6	68.8	73.0	102.5	68.9	74.0	86.2	69.0	71.6	94.7	64.9	69.3	89.0	70.1	72.9
0320	Concrete Reinforcing	108.2	118.2	113.0	99.5	73.6	87.0	99.3	73.6	86.9	104.3	73.6	89.5	99.6	73.6	87.0	105.0	73.6	89.8
0330	Cast-in-Place Concrete	234.6	113.8	189.0	88.4	67.8	80.6	88.5	67.8	80.7	79.6	67.4	75.0	88.1	67.8	80.4	80.3	67.6	75.5
03	CONCRETE	177.3	113.8	148.8	93.4	69.2	82.6	114.1	69.3	94.0	98.0	69.3	85.1	93.1	67.5	81.6	89.7	69.9	80.8
04	MASONRY	194.7	116.6	147.1	94.9	60.0	73.6	95.0	60.7	74.1	100.5	59.9	75.8	94.9	60.1	73.6	100.7	60.0	75.9
05	METALS	122.8	105.4	117.4	101.2	70.2	91.5	101.7	70.7	92.1	102.3	71.4	92.7	101.9	70.6	92.2	102.7	71.8	93.0
06	WOOD, PLASTICS & COMPOSITES	107.8	111.0	109.5	97.3	70.4	83.3	104.0	70.4	86.4	80.2	70.6	75.2	92.3	65.0	78.0	83.8	72.0	77.6
07	THERMAL & MOISTURE PROTECTION	194.3	112.5	159.2	99.3	70.0	86.7	101.2	70.4	88.0	99.0	68.7	86.0	99.2	69.6	86.5	98.7	69.1	86.0
08	OPENINGS	128.9	113.5	125.2	105.8	66.8	96.3	105.9	70.0	97.2	95.3	66.9	88.4	106.0	65.4	96.1	95.4	67.9	88.7
0920	Plaster & Gypsum Board	148.8	111.3	124.2	94.9	70.0	78.6	98.4	70.0	79.8	85.5	70.0	75.3	87.4	64.4	72.3	88.9	71.5	77.5
0950, 0980	Ceilings & Acoustic Treatment	101.1	111.3	107.5	116.5	70.0	87.4	117.1	70.0	87.6	103.7	70.0	82.6	117.1	64.4	84.1	103.7	71.5	83.5
0960	Flooring	117.2	116.7	117.1	89.1	61.5	81.1	91.6	62.7	83.2	100.0	61.5	88.7	87.7	61.5	80.1	101.7	62.7	90.3
0970, 0990	Wall Finishes & Painting/Coating	109.0	117.6	114.1	84.7	56.9	68.0	84.7	56.9	68.0	86.9	56.9	68.9	84.7	56.9	68.0	86.9	56.9	68.9
09	FINISHES	125.7	114.4	119.6	95.2	66.0	79.4	98.3	66.3	80.9	98.1	66.2	80.8	93.8	62.8	77.1	97.9	67.2	81.3
COVERS	DIVS. 10 - 14, 25, 28, 41, 43, 44, 46	100.0	109.8	102.3	100.0	84.0	96.2	100.0	84.0	96.2	100.0	84.4	96.3	100.0	83.4	96.1	100.0	84.5	96.4
21, 22, 23	FIRE SUPPRESSION, PLUMBING & HVAC	98.6	104.1	100.8	98.5	75.7	89.3	101.4	76.1	91.2	96.3	75.8	88.0	98.5	76.1	89.5	101.5	75.8	91.1
26, 27, 3370	ELECTRICAL, COMMUNICATIONS & UTIL.	125.5	107.9	116.8	101.5	64.1	83.0	100.6	60.8	80.9	97.3	64.6	81.1	101.5	64.6	83.2	95.4	64.6	80.1
MF2018	WEIGHTED AVERAGE	130.1	110.9	121.8	98.2	70.4	86.2	102.4	70.4	88.6	98.2	70.7	86.3	98.2	69.8	85.9	98.0	71.1	86.4

		ARIZONA												ARKANSAS					
		PHOENIX			PRESCOTT			SHOW LOW			TUCSON			BATESVILLE			CAMDEN		
DIVISION		850,853			863			859			856 - 857			725			717		
		MAT.	INST.	TOTAL	MAT.	INST.	TOTAL	MAT.	INST.	TOTAL	MAT.	INST.	TOTAL	MAT.	INST.	TOTAL	MAT.	INST.	TOTAL
015433	CONTRACTOR EQUIPMENT		94.1	94.1		85.9	85.9		87.5	87.5		87.5	87.5		86.8	86.8		86.8	86.8
0241, 31 - 34	SITE & INFRASTRUCTURE, DEMOLITION	92.8	96.1	95.0	77.0	87.7	84.3	105.4	88.7	94.0	88.6	88.9	88.8	71.6	82.8	79.3	77.4	82.8	81.1
0310	Concrete Forming & Accessories	92.8	70.9	74.1	98.4	71.0	75.1	92.7	69.0	72.6	89.4	77.4	79.2	85.5	59.9	63.7	77.9	60.4	63.0
0320	Concrete Reinforcing	99.7	73.6	87.1	99.3	73.6	86.9	105.2	73.6	89.8	87.1	73.7	80.6	83.0	67.4	75.5	91.3	68.5	80.3
0330	Cast-in-Place Concrete	84.5	69.9	79.0	88.4	67.9	80.7	79.7	67.4	75.0	82.6	67.6	77.0	68.7	73.2	70.4	74.9	73.3	74.3
03	CONCRETE	92.6	71.0	82.9	99.0	70.3	86.1	100.2	69.3	86.3	88.2	73.2	81.5	71.2	66.6	69.1	75.0	67.0	71.4
04	MASONRY	93.7	60.6	73.5	95.0	60.1	73.7	100.6	59.9	75.8	87.5	60.6	71.1	96.4	61.0	74.8	102.4	61.8	77.6
05	METALS	104.2	72.4	94.3	101.8	70.9	92.2	102.1	71.4	92.5	103.4	72.0	93.6	95.5	75.7	89.4	101.7	76.0	93.7
06	WOOD, PLASTICS & COMPOSITES	88.5	72.3	80.0	98.9	73.2	85.5	87.8	70.6	78.8	84.0	81.8	82.9	92.8	61.3	76.3	86.3	61.8	73.5
07	THERMAL & MOISTURE PROTECTION	99.2	69.9	86.6	99.8	70.5	87.3	99.3	68.7	86.1	99.6	70.3	87.0	104.4	61.1	85.8	98.1	61.8	82.5
08	OPENINGS	101.2	70.6	93.7	105.9	68.3	96.7	94.7	68.5	88.3	92.0	76.3	88.1	99.8	59.4	90.0	103.1	60.0	92.6
0920	Plaster & Gypsum Board	96.9	71.5	80.2	95.1	72.9	80.5	90.9	70.0	77.2	93.2	81.6	85.6	89.0	60.8	70.5	82.7	61.3	68.7
0950, 0980	Ceilings & Acoustic Treatment	116.2	71.5	88.2	115.2	72.9	88.7	103.7	70.0	82.6	105.0	81.6	90.3	92.4	60.8	72.6	90.9	61.3	72.4
0960	Flooring	102.6	62.7	90.9	90.1	64.0	82.5	103.6	61.5	91.3	93.3	65.3	85.1	84.9	70.4	80.7	85.7	71.8	81.7
0970, 0990	Wall Finishes & Painting/Coating	94.3	56.7	71.8	84.7	56.9	68.0	86.9	56.9	68.9	87.5	56.9	69.2	88.6	52.4	66.9	92.7	54.0	69.5
09	FINISHES	102.7	67.7	83.8	95.6	68.2	80.8	100.1	66.2	81.8	96.3	73.5	84.0	83.6	60.9	71.3	83.8	61.7	71.8
COVERS	DIVS. 10 - 14, 25, 28, 41, 43, 44, 46	100.0	85.6	96.6	100.0	84.3	96.3	100.0	84.4	96.3	100.0	85.6	96.6	100.0	78.8	95.0	100.0	78.8	95.0
21, 22, 23	FIRE SUPPRESSION, PLUMBING & HVAC	96.7	76.9	88.7	101.4	75.8	91.1	96.3	75.8	88.0	101.4	73.7	90.2	97.3	51.7	78.9	97.1	57.2	81.0
26, 27, 3370	ELECTRICAL, COMMUNICATIONS & UTIL.	97.6	62.7	80.3	100.4	64.6	82.7	95.1	64.6	80.0	97.2	59.7	78.6	96.2	58.2	77.4	94.1	56.0	75.2
MF2018	WEIGHTED AVERAGE	98.4	72.1	87.0	99.9	71.1	87.5	98.4	70.8	86.5	96.9	71.8	86.1	92.4	62.9	79.6	94.1	64.1	81.1

		ARKANSAS																	
		FAYETTEVILLE			FORT SMITH			HARRISON			HOT SPRINGS			JONESBORO			LITTLE ROCK		
DIVISION		727			729			726			719			724			720 - 722		
		MAT.	INST.	TOTAL	MAT.	INST.	TOTAL	MAT.	INST.	TOTAL	MAT.	INST.	TOTAL	MAT.	INST.	TOTAL	MAT.	INST.	TOTAL
015433	CONTRACTOR EQUIPMENT		86.8	86.8		86.8	86.8		86.8	86.8		86.8	86.8		109.9	109.9		92.6	92.6
0241, 31 - 34	SITE & INFRASTRUCTURE, DEMOLITION	71.1	82.8	79.1	76.5	82.6	80.7	76.1	82.8	80.7	80.3	82.7	81.9	95.2	99.4	98.1	83.6	91.3	88.9
0310	Concrete Forming & Accessories	80.5	60.4	63.4	103.6	59.9	66.4	90.9	60.3	64.8	75.1	59.8	62.1	89.2	60.3	64.6	101.6	60.5	66.6
0320	Concrete Reinforcing	83.0	65.1	74.4	84.0	64.9	74.8	82.6	68.5	75.8	89.6	68.4	79.4	80.2	66.9	73.8	85.2	67.5	76.7
0330	Cast-in-Place Concrete	68.7	73.3	70.4	78.5	74.2	76.9	76.2	73.2	75.0	76.6	72.6	75.1	74.8	74.2	74.6	76.2	76.1	76.2
03	CONCRETE	70.9	66.4	68.9	78.2	66.5	72.9	77.5	66.9	72.7	78.0	66.5	72.8	75.2	68.0	71.9	79.7	67.7	74.3
04	MASONRY	87.2	61.8	71.7	94.6	60.6	73.9	96.5	61.8	75.3	76.7	58.3	65.5	88.6	60.6	71.5	91.6	61.5	73.3
05	METALS	95.5	74.8	89.1	97.9	74.2	90.5	96.7	75.8	90.2	101.7	75.5	93.5	92.4	90.0	91.7	98.0	74.2	90.6
06	WOOD, PLASTICS & COMPOSITES	88.2	61.8	74.4	114.0	61.8	86.7	99.8	61.8	79.9	83.2	61.8	72.0	97.2	62.1	78.8	107.3	62.0	83.6
07	THERMAL & MOISTURE PROTECTION	105.2	61.8	86.6	105.8	60.9	86.5	104.7	61.8	86.3	98.3	59.8	81.8	110.5	60.9	89.2	100.1	62.4	83.9
08	OPENINGS	99.8	59.1	89.9	101.8	58.6	91.3	100.6	59.9	90.7	103.1	59.5	92.4	105.4	59.8	94.3	95.5	59.1	86.6
0920	Plaster & Gypsum Board	88.4	61.8	70.7	94.6	61.3	72.8	93.7	61.3	72.5	81.0	61.3	68.1	102.6	61.3	75.5	109.0	61.3	77.7
0950, 0980	Ceilings & Acoustic Treatment	92.4	61.3	72.9	94.3	61.3	73.7	94.3	61.3	73.7	90.9	61.3	72.4	97.0	61.3	74.7	97.4	61.3	74.8
0960	Flooring	81.9	71.8	79.0	91.9	70.4	85.6	87.5	71.8	82.9	84.6	67.7	79.7	61.3	70.4	64.0	93.8	80.4	89.9
0970, 0990	Wall Finishes & Painting/Coating	88.6	54.0	67.9	88.6	51.6	66.5	88.6	54.0	67.9	92.7	51.9	68.3	78.4	53.0	63.2	91.9	53.0	68.6
09	FINISHES	82.8	61.7	71.3	87.1	61.0	73.0	85.8	61.7	72.7	83.6	60.3	71.0	82.4	61.4	71.0	95.4	63.4	78.1
COVERS	DIVS. 10 - 14, 25, 28, 41, 43, 44, 46	100.0	78.8	95.0	100.0	78.7	95.0	100.0	78.8	95.0	100.0	78.5	94.9	100.0	77.6	94.7	100.0	79.4	95.1
21, 22, 23	FIRE SUPPRESSION, PLUMBING & HVAC	97.4	61.6	83.0	101.2	48.3	79.9	97.3	49.5	78.0	97.1	49.0	77.7	101.8	50.8	81.2	100.8	48.4	79.7
26, 27, 3370	ELECTRICAL, COMMUNICATIONS & UTIL.	91.2	51.9	71.7	94.0	58.3	76.3	95.0	55.4	75.4	95.8	62.8	79.5	99.8	60.0	80.1	102.3	59.6	81.2
MF2018	WEIGHTED AVERAGE	91.3	64.2	79.6	95.1	61.9	80.8	93.7	62.3	80.1	93.4	62.5	80.1	94.8	65.7	82.2	96.0	63.5	81.9

For customer support on your Commercial Renovation Costs with RSMeans Data, call 800.448.8182.

795

DIVISION		ARKANSAS											CALIFORNIA					
		PINE BLUFF 716			RUSSELLVILLE 728			TEXARKANA 718			WEST MEMPHIS 723			ALHAMBRA 917 - 918			ANAHEIM 928		
		MAT.	INST.	TOTAL	MAT.	INST.	TOTAL	MAT.	INST.	TOTAL	MAT.	INST.	TOTAL	MAT.	INST.	TOTAL	MAT.	INST.	TOTAL
015433	CONTRACTOR EQUIPMENT		86.8	86.8		86.8	86.8		88.0	88.0		109.9	109.9		92.5	92.5		97.5	97.5
0241, 31 - 34	SITE & INFRASTRUCTURE, DEMOLITION	82.8	82.6	82.7	72.8	82.6	79.5	93.4	84.6	87.4	101.1	99.2	99.8	99.1	101.8	101.0	100.7	102.6	102.1
0310	Concrete Forming & Accessories	74.8	60.1	62.3	86.6	59.6	63.6	81.6	60.2	63.4	95.6	60.1	65.4	115.5	135.6	132.6	102.0	139.4	133.8
0320	Concrete Reinforcing	91.2	67.4	79.7	83.5	68.3	76.2	90.8	68.5	80.0	80.2	60.4	70.6	98.9	133.2	115.4	93.2	133.1	112.5
0330	Cast-in-Place Concrete	76.6	73.0	75.2	71.9	72.5	72.2	83.5	73.0	79.6	78.4	73.7	76.6	82.3	126.4	99.0	88.3	129.7	104.0
03	CONCRETE	78.7	66.6	73.3	74.0	66.3	70.5	77.3	66.8	72.6	81.4	66.5	74.7	92.6	130.8	109.8	96.9	133.8	113.5
04	MASONRY	109.6	60.6	79.7	93.9	58.3	72.2	89.2	60.8	71.9	77.5	58.2	65.8	113.1	137.8	128.2	77.2	136.8	113.6
05	METALS	102.5	75.3	94.0	95.5	75.4	89.2	94.8	75.8	88.9	91.4	87.4	90.2	79.2	113.8	90.0	107.0	115.7	109.7
06	WOOD, PLASTICS & COMPOSITES	82.8	61.8	71.8	94.6	61.8	77.4	91.2	61.8	75.8	104.4	62.1	82.3	101.6	133.4	118.2	96.3	138.5	118.4
07	THERMAL & MOISTURE PROTECTION	98.4	61.2	82.4	105.5	59.8	85.9	99.1	61.0	82.8	111.1	59.8	89.1	103.4	128.8	114.3	109.4	132.9	119.5
08	OPENINGS	104.3	59.2	93.3	99.8	59.5	90.0	109.1	60.0	97.1	102.9	57.8	91.9	88.1	133.7	99.2	103.9	136.2	111.8
0920	Plaster & Gypsum Board	80.7	61.3	68.0	89.0	61.3	70.8	84.2	61.3	69.2	104.6	61.3	76.2	96.7	134.7	121.6	111.5	139.7	130.0
0950, 0980	Ceilings & Acoustic Treatment	90.9	61.3	72.4	92.4	61.3	72.9	96.6	61.3	74.5	95.6	61.3	74.1	115.1	134.7	127.4	106.8	139.7	127.4
0960	Flooring	84.4	71.8	80.7	84.4	67.7	79.5	86.5	70.8	81.9	63.8	67.7	65.0	104.7	119.9	109.1	98.9	119.9	105.0
0970, 0990	Wall Finishes & Painting/Coating	92.7	51.6	68.1	88.6	51.9	66.6	92.7	54.0	69.5	78.4	51.9	62.5	102.8	119.8	113.0	90.1	119.8	107.9
09	FINISHES	83.5	61.3	71.5	83.7	60.3	71.1	86.5	61.4	72.9	83.7	60.6	71.2	103.9	130.6	118.4	100.3	133.8	118.4
COVERS	DIVS. 10 - 14, 25, 28, 41, 43, 44, 46	100.0	78.7	95.0	100.0	78.5	94.9	100.0	78.7	95.0	100.0	79.3	95.1	100.0	116.6	103.9	100.0	117.7	104.2
21, 22, 23	FIRE SUPPRESSION, PLUMBING & HVAC	101.0	51.7	81.1	97.4	48.5	77.7	101.0	54.9	82.4	98.0	61.5	83.3	96.9	128.4	109.6	100.8	128.4	112.0
26, 27, 3370	ELECTRICAL, COMMUNICATIONS & UTIL.	94.2	57.6	76.1	94.0	55.4	74.9	95.7	57.6	76.9	101.1	62.8	82.1	121.7	131.9	126.7	91.7	114.7	103.1
MF2018	WEIGHTED AVERAGE	96.2	62.8	81.8	92.5	61.3	79.0	95.1	63.7	81.5	94.2	67.5	82.7	97.4	127.0	110.2	99.5	125.9	110.9

DIVISION		CALIFORNIA																
		BAKERSFIELD 932 - 933			BERKELEY 947			EUREKA 955			FRESNO 936 - 938			INGLEWOOD 903 - 905			LONG BEACH 906 - 908		
		MAT.	INST.	TOTAL	MAT.	INST.	TOTAL	MAT.	INST.	TOTAL	MAT.	INST.	TOTAL	MAT.	INST.	TOTAL	MAT.	INST.	TOTAL
015433	CONTRACTOR EQUIPMENT		100.3	100.3		98.8	98.8		95.3	95.3		95.5	95.5		95.6	95.6		95.6	95.6
0241, 31 - 34	SITE & INFRASTRUCTURE, DEMOLITION	100.0	108.7	106.0	115.4	103.4	107.2	112.0	100.4	104.1	102.2	100.9	101.3	87.5	99.7	95.9	94.5	99.7	98.1
0310	Concrete Forming & Accessories	100.8	138.4	132.8	115.8	173.3	164.7	111.2	160.0	152.7	100.3	159.0	150.3	105.7	136.0	131.5	99.8	136.0	130.6
0320	Concrete Reinforcing	95.8	133.1	113.8	87.7	134.6	110.4	101.3	134.7	117.4	80.1	133.7	106.0	99.7	133.2	115.9	98.9	133.2	115.5
0330	Cast-in-Place Concrete	84.7	130.1	101.9	110.8	136.5	120.5	95.9	132.9	109.9	91.7	132.5	107.1	81.8	129.1	99.6	93.1	129.1	106.7
03	CONCRETE	90.9	133.4	109.9	107.8	151.7	127.5	107.5	144.6	124.1	93.6	143.8	116.1	90.5	132.1	109.2	99.9	132.1	114.3
04	MASONRY	91.2	136.3	118.7	120.6	152.8	140.3	99.7	157.9	135.2	98.9	143.6	126.2	72.7	137.9	112.5	81.7	137.9	116.0
05	METALS	101.8	113.6	105.5	107.1	119.4	110.9	106.7	119.1	110.6	102.1	117.7	107.0	87.5	116.3	96.5	87.4	116.3	96.4
06	WOOD, PLASTICS & COMPOSITES	92.7	137.5	116.1	109.5	180.0	146.4	110.1	164.4	138.5	98.7	164.4	133.1	99.0	133.8	117.2	91.7	133.8	113.7
07	THERMAL & MOISTURE PROTECTION	108.0	123.5	114.6	114.9	155.3	132.2	112.9	153.3	130.3	98.9	133.7	113.8	105.2	129.9	115.8	105.5	129.9	116.0
08	OPENINGS	93.1	135.2	103.4	90.7	165.9	109.0	103.3	146.8	113.9	95.0	150.6	108.6	87.5	134.0	98.8	87.4	134.0	98.8
0920	Plaster & Gypsum Board	97.5	138.4	124.4	110.3	181.9	157.3	116.3	166.2	149.0	92.4	166.2	140.8	103.4	134.7	123.9	98.6	134.7	122.3
0950, 0980	Ceilings & Acoustic Treatment	109.8	138.4	127.7	105.7	181.9	153.4	111.3	166.2	145.7	102.6	166.2	142.4	102.5	134.7	122.7	102.5	134.7	122.7
0960	Flooring	99.9	119.9	105.7	121.5	149.0	129.5	102.5	149.0	116.1	99.3	122.1	106.0	113.7	119.9	115.5	110.4	119.9	113.2
0970, 0990	Wall Finishes & Painting/Coating	92.6	108.4	102.1	108.4	168.0	144.1	91.7	135.6	118.0	97.9	122.3	112.5	108.9	119.8	115.4	108.9	119.8	115.4
09	FINISHES	100.1	131.9	117.3	109.8	169.8	142.3	105.5	157.2	133.5	97.8	150.6	126.4	105.1	131.0	119.1	104.0	131.0	118.6
COVERS	DIVS. 10 - 14, 25, 28, 41, 43, 44, 46	100.0	127.0	106.4	100.0	132.5	107.7	100.0	129.6	107.0	100.0	129.5	107.0	100.0	117.6	104.1	100.0	117.6	104.1
21, 22, 23	FIRE SUPPRESSION, PLUMBING & HVAC	100.9	127.1	111.5	97.6	167.8	125.9	97.1	133.2	111.7	100.9	131.4	113.2	96.1	128.5	109.2	96.1	127.5	108.8
26, 27, 3370	ELECTRICAL, COMMUNICATIONS & UTIL.	105.5	109.7	107.6	100.1	155.8	127.7	98.1	130.8	114.3	96.2	108.6	102.3	99.5	131.9	115.5	99.3	131.9	115.4
MF2018	WEIGHTED AVERAGE	99.0	124.8	110.1	103.3	151.9	124.3	102.6	137.6	117.7	98.6	131.2	112.7	93.6	127.4	108.2	95.2	127.2	109.0

DIVISION		CALIFORNIA																
		LOS ANGELES 900 - 902			MARYSVILLE 959			MODESTO 953			MOJAVE 935			OAKLAND 946			OXNARD 930		
		MAT.	INST.	TOTAL	MAT.	INST.	TOTAL	MAT.	INST.	TOTAL	MAT.	INST.	TOTAL	MAT.	INST.	TOTAL	MAT.	INST.	TOTAL
015433	CONTRACTOR EQUIPMENT		104.7	104.7		95.3	95.3		95.3	95.3		95.5	95.5		98.8	98.8		94.5	94.5
0241, 31 - 34	SITE & INFRASTRUCTURE, DEMOLITION	94.0	110.4	105.3	108.4	100.0	102.6	103.7	100.4	101.4	94.8	100.6	98.8	121.8	103.4	109.2	102.4	98.8	99.9
0310	Concrete Forming & Accessories	103.2	140.0	134.5	100.4	159.0	150.2	96.1	159.3	149.9	109.7	138.1	133.8	103.2	173.4	162.9	101.9	139.4	133.8
0320	Concrete Reinforcing	98.1	131.4	114.2	101.3	133.5	116.9	105.0	133.8	118.9	96.3	133.0	114.0	89.7	137.2	112.6	94.5	133.1	113.3
0330	Cast-in-Place Concrete	84.6	131.3	102.2	107.1	132.4	116.7	95.9	132.6	109.7	79.3	128.8	98.0	105.1	136.5	117.0	92.4	129.1	106.2
03	CONCRETE	96.2	134.4	113.3	108.1	143.7	124.0	99.1	143.9	119.2	85.7	132.9	106.9	108.9	152.2	128.3	93.5	133.6	111.5
04	MASONRY	88.1	138.0	118.5	100.7	141.0	125.3	99.7	144.6	127.1	97.0	136.3	120.9	129.2	152.8	143.6	99.8	136.5	122.2
05	METALS	95.8	116.4	102.2	106.2	116.6	109.4	102.9	117.2	107.4	99.4	114.8	104.2	102.5	120.4	108.1	97.3	115.4	103.0
06	WOOD, PLASTICS & COMPOSITES	107.3	139.0	123.9	96.4	164.4	132.0	91.2	164.4	129.5	98.7	137.4	118.9	95.6	180.0	139.8	93.3	138.6	117.0
07	THERMAL & MOISTURE PROTECTION	101.7	132.5	114.9	112.3	137.8	123.3	111.9	139.8	123.9	104.8	121.6	112.1	113.2	155.3	131.3	108.1	131.7	118.2
08	OPENINGS	98.2	137.6	107.8	102.6	148.4	113.8	101.4	150.9	113.4	90.3	135.2	101.3	90.8	166.5	109.2	92.8	136.3	103.4
0920	Plaster & Gypsum Board	98.4	139.7	125.5	109.4	166.2	146.6	111.5	166.2	147.4	102.3	138.4	125.9	104.5	181.9	155.3	96.0	139.7	124.7
0950, 0980	Ceilings & Acoustic Treatment	109.6	139.7	128.5	110.6	166.2	145.4	106.8	166.2	144.0	104.9	138.4	125.9	107.7	181.9	154.2	104.7	139.7	126.6
0960	Flooring	110.9	119.9	113.5	98.1	114.0	102.7	98.4	133.0	108.5	100.7	115.5	105.0	114.4	149.0	124.5	93.1	119.9	100.9
0970, 0990	Wall Finishes & Painting/Coating	108.1	119.8	115.1	91.7	135.6	118.0	91.7	135.6	118.0	84.5	108.6	99.0	108.4	168.0	144.1	84.5	114.6	102.5
09	FINISHES	105.8	134.1	121.2	102.6	150.7	128.6	101.6	154.0	129.9	98.7	131.1	116.3	107.9	170.0	141.5	95.9	133.3	116.1
COVERS	DIVS. 10 - 14, 25, 28, 41, 43, 44, 46	100.0	118.9	104.5	100.0	129.6	107.0	100.0	129.6	107.0	100.0	113.0	103.1	100.0	132.5	107.7	100.0	117.9	104.2
21, 22, 23	FIRE SUPPRESSION, PLUMBING & HVAC	100.4	128.6	111.8	97.1	126.1	108.8	100.8	131.4	113.2	97.0	127.0	109.1	101.3	167.9	128.2	100.8	128.5	112.0
26, 27, 3370	ELECTRICAL, COMMUNICATIONS & UTIL.	98.1	131.9	114.8	95.3	116.3	105.6	97.4	110.2	103.7	94.8	109.7	102.2	99.4	163.0	130.9	100.6	117.4	108.9
MF2018	WEIGHTED AVERAGE	98.4	129.3	111.8	101.8	130.7	114.3	100.8	132.1	114.3	95.6	123.6	107.7	103.9	153.2	125.2	98.2	125.8	110.1

CALIFORNIA

DIVISION		PALM SPRINGS 922			PALO ALTO 943			PASADENA 910 - 912			REDDING 960			RICHMOND 948			RIVERSIDE 925		
		MAT.	INST.	TOTAL	MAT.	INST.	TOTAL	MAT.	INST.	TOTAL	MAT.	INST.	TOTAL	MAT.	INST.	TOTAL	MAT.	INST.	TOTAL
015433	CONTRACTOR EQUIPMENT		96.3	96.3		98.8	98.8		92.5	92.5		95.3	95.3		98.8	98.8		96.3	96.3
0241, 31 - 34	SITE & INFRASTRUCTURE, DEMOLITION	91.8	100.8	98.0	111.6	103.4	106.0	95.9	101.8	100.0	124.5	100.2	107.8	120.5	103.4	108.8	99.2	100.8	100.3
0310	Concrete Forming & Accessories	98.4	132.1	127.0	101.1	163.9	154.6	102.8	135.6	130.7	102.8	155.1	147.3	119.0	173.0	165.0	102.4	139.4	133.9
0320	Concrete Reinforcing	106.5	132.9	119.3	87.7	136.9	111.4	99.8	133.2	115.9	134.5	133.8	134.2	87.7	136.9	111.4	103.5	133.1	117.8
0330	Cast-in-Place Concrete	84.2	129.6	101.3	93.7	136.4	109.8	78.1	126.4	96.3	117.0	132.6	122.9	107.8	136.4	118.6	91.6	129.7	106.0
03	CONCRETE	91.3	130.4	108.9	97.6	147.8	120.2	88.3	130.8	107.4	118.6	142.0	129.1	110.9	152.0	129.3	97.3	133.8	113.7
04	MASONRY	75.2	134.1	111.1	102.6	149.2	131.0	99.1	137.8	122.7	120.2	144.6	135.1	120.4	149.2	138.0	76.2	136.5	113.0
05	METALS	107.6	115.2	110.0	100.0	119.7	106.2	79.2	113.8	90.0	103.2	117.3	107.6	100.1	119.8	106.2	107.0	115.7	109.7
06	WOOD, PLASTICS & COMPOSITES	90.6	129.0	110.7	92.8	167.7	132.0	86.2	133.4	110.9	104.7	158.6	132.9	113.2	180.0	148.2	96.3	138.5	118.4
07	THERMAL & MOISTURE PROTECTION	108.7	129.3	117.5	112.4	149.9	128.5	102.9	128.3	113.8	126.7	139.9	132.4	113.1	152.8	130.1	109.6	132.8	119.6
08	OPENINGS	100.0	131.3	107.6	90.8	155.9	106.7	88.1	133.7	99.2	116.4	147.7	124.0	90.8	165.8	109.1	102.6	136.2	110.8
0920	Plaster & Gypsum Board	106.1	129.9	121.7	102.8	169.2	146.4	91.4	134.7	119.8	120.9	160.2	146.7	110.7	181.9	157.4	110.6	139.7	129.7
0950, 0980	Ceilings & Acoustic Treatment	103.6	129.9	120.1	106.3	169.2	145.7	115.1	134.7	127.4	138.0	160.2	151.9	106.3	181.9	153.7	111.5	139.7	129.1
0960	Flooring	101.0	115.5	105.2	112.9	149.0	123.5	98.2	119.9	104.6	88.8	126.7	99.9	124.1	149.0	131.3	102.4	119.9	107.5
0970, 0990	Wall Finishes & Painting/Coating	88.7	114.6	104.2	108.4	168.0	144.1	102.8	119.8	113.0	101.6	135.6	122.0	108.4	168.0	144.1	88.7	119.8	107.3
09	FINISHES	98.7	126.7	113.9	106.3	162.6	136.7	101.1	130.7	117.1	108.9	149.7	131.0	111.2	169.8	142.9	102.0	133.8	119.2
COVERS	DIVS. 10 - 14, 25, 28, 41, 43, 44, 46	100.0	116.7	103.9	100.0	131.2	107.4	100.0	116.6	103.9	100.0	129.0	106.8	100.0	132.5	107.6	100.0	120.4	104.8
21, 22, 23	FIRE SUPPRESSION, PLUMBING & HVAC	97.1	125.9	108.7	97.6	164.8	124.7	96.9	128.4	109.6	101.1	131.4	113.3	97.6	162.4	123.7	100.8	128.4	112.0
26, 27, 3370	ELECTRICAL, COMMUNICATIONS & UTIL.	94.4	106.7	100.5	99.3	170.0	134.3	118.7	131.9	125.2	99.5	121.7	110.5	99.8	140.3	119.9	91.5	116.0	103.6
MF2018	WEIGHTED AVERAGE	97.3	121.9	108.0	99.4	150.8	121.6	95.4	127.0	109.0	107.8	132.6	118.6	102.8	148.2	122.4	99.4	126.0	110.9

CALIFORNIA

DIVISION		SACRAMENTO 942,956 - 958			SALINAS 939			SAN BERNARDINO 923 - 924			SAN DIEGO 919 - 921			SAN FRANCISCO 940 - 941			SAN JOSE 951		
		MAT.	INST.	TOTAL	MAT.	INST.	TOTAL	MAT.	INST.	TOTAL	MAT.	INST.	TOTAL	MAT.	INST.	TOTAL	MAT.	INST.	TOTAL
015433	CONTRACTOR EQUIPMENT		97.6	97.6		95.5	95.5		96.3	96.3		103.0	103.0		111.2	111.2		97.0	97.0
0241, 31 - 34	SITE & INFRASTRUCTURE, DEMOLITION	99.6	108.5	105.8	115.6	100.9	105.5	78.2	100.8	93.7	104.5	109.5	107.9	121.3	114.9	116.9	135.4	95.7	108.2
0310	Concrete Forming & Accessories	101.9	157.3	149.1	105.9	157.9	150.1	106.5	139.4	134.5	104.0	129.4	125.6	107.4	168.5	159.4	102.5	173.6	163.0
0320	Concrete Reinforcing	83.3	133.8	107.7	95.0	134.3	113.9	103.5	133.1	117.8	96.0	131.1	112.9	100.8	135.5	117.5	92.7	137.4	114.3
0330	Cast-in-Place Concrete	88.9	133.2	105.6	91.2	132.9	107.0	63.3	129.7	88.4	89.5	125.9	103.3	117.4	133.9	123.6	111.4	135.9	120.6
03	CONCRETE	100.4	143.0	119.5	103.5	143.5	121.5	72.1	133.8	99.8	97.1	127.6	110.8	120.5	149.1	133.3	105.0	152.5	126.3
04	MASONRY	103.5	144.6	128.6	96.9	150.1	129.4	83.0	136.5	115.6	90.7	132.6	116.3	136.8	153.9	147.2	128.4	156.1	145.3
05	METALS	98.2	112.2	102.6	102.0	118.3	107.1	107.0	115.7	109.7	95.8	115.0	101.8	107.2	124.6	112.7	101.1	125.7	108.8
06	WOOD, PLASTICS & COMPOSITES	90.8	161.6	127.8	98.7	161.4	131.5	100.4	138.5	120.3	95.5	126.9	111.9	101.5	174.3	139.5	103.1	179.8	143.2
07	THERMAL & MOISTURE PROTECTION	122.2	141.7	130.6	105.5	145.8	122.8	107.8	132.8	118.6	108.1	120.8	113.5	119.0	156.1	134.9	108.5	160.6	130.8
08	OPENINGS	103.6	149.3	114.7	93.9	155.7	108.9	100.0	137.8	109.2	100.1	127.5	106.8	99.4	162.2	114.7	92.6	166.4	110.6
0920	Plaster & Gypsum Board	98.9	163.1	141.0	97.2	163.1	140.4	112.3	139.7	130.3	97.5	127.4	117.1	100.4	175.9	149.9	107.1	181.9	156.2
0950, 0980	Ceilings & Acoustic Treatment	106.3	163.1	141.9	104.9	163.1	141.4	106.8	139.7	127.4	126.2	127.4	127.0	119.8	175.9	154.9	107.6	181.9	154.2
0960	Flooring	112.7	126.7	116.8	95.0	142.9	109.0	104.6	119.9	109.0	108.0	119.9	111.5	111.2	142.9	120.5	92.1	149.0	108.7
0970, 0990	Wall Finishes & Painting/Coating	105.2	135.6	123.4	85.4	168.0	134.9	88.7	119.8	107.3	103.5	119.8	113.3	109.1	176.8	149.7	92.0	168.0	137.5
09	FINISHES	104.6	151.3	129.9	98.3	157.7	130.4	100.2	133.8	118.4	108.8	126.5	118.4	109.1	165.9	139.8	100.9	170.0	138.3
COVERS	DIVS. 10 - 14, 25, 28, 41, 43, 44, 46	100.0	129.8	107.0	100.0	129.3	106.9	100.0	117.7	104.2	100.0	116.1	103.8	100.0	128.9	106.8	100.0	132.1	107.6
21, 22, 23	FIRE SUPPRESSION, PLUMBING & HVAC	101.2	130.5	113.0	97.0	136.9	113.1	97.1	128.4	109.7	100.7	127.0	111.3	101.2	179.4	132.8	100.8	173.4	130.1
26, 27, 3370	ELECTRICAL, COMMUNICATIONS & UTIL.	95.3	121.6	108.3	95.7	132.6	114.0	94.4	111.9	103.0	105.3	103.3	104.3	100.0	186.3	142.7	100.5	188.8	144.2
MF2018	WEIGHTED AVERAGE	101.0	133.1	114.9	99.2	137.8	115.9	95.0	125.4	108.1	100.3	121.3	109.4	107.6	159.0	129.8	103.0	158.4	126.9

CALIFORNIA

DIVISION		SAN LUIS OBISPO 934			SAN MATEO 944			SAN RAFAEL 949			SANTA ANA 926 - 927			SANTA BARBARA 931			SANTA CRUZ 950		
		MAT.	INST.	TOTAL	MAT.	INST.	TOTAL	MAT.	INST.	TOTAL	MAT.	INST.	TOTAL	MAT.	INST.	TOTAL	MAT.	INST.	TOTAL
015433	CONTRACTOR EQUIPMENT		95.5	95.5		98.8	98.8		98.0	98.0		96.3	96.3		95.5	95.5		97.0	97.0
0241, 31 - 34	SITE & INFRASTRUCTURE, DEMOLITION	107.4	100.6	102.7	118.1	103.5	108.1	111.8	108.2	109.4	90.2	100.8	97.5	102.4	100.6	101.2	135.0	95.6	107.9
0310	Concrete Forming & Accessories	111.6	139.5	135.4	108.1	173.7	163.9	114.7	173.0	164.3	107.0	139.4	134.5	102.5	139.4	133.9	102.6	158.0	149.7
0320	Concrete Reinforcing	96.3	133.1	114.0	87.7	137.6	111.8	88.4	134.8	110.8	107.1	133.1	119.7	94.5	133.1	113.1	114.4	134.3	124.0
0330	Cast-in-Place Concrete	98.1	129.0	109.8	104.3	136.7	116.5	121.4	135.5	126.7	80.8	129.7	99.3	92.0	129.0	106.0	110.7	134.6	119.7
03	CONCRETE	101.1	133.6	115.7	107.3	152.5	127.6	131.4	151.0	140.2	88.9	133.8	109.1	93.4	133.6	111.4	107.6	144.4	124.1
04	MASONRY	98.5	135.2	120.9	120.1	159.0	143.8	97.5	155.8	133.0	72.5	136.8	111.8	97.2	135.2	120.4	132.2	150.2	143.2
05	METALS	100.1	115.4	104.9	99.9	121.2	106.5	101.2	115.6	105.7	107.0	115.7	109.7	97.9	115.3	103.3	108.6	122.6	112.9
06	WOOD, PLASTICS & COMPOSITES	100.9	138.6	120.6	101.7	180.0	142.7	101.5	179.8	142.5	102.4	138.5	121.3	93.3	138.6	117.0	103.1	161.5	133.6
07	THERMAL & MOISTURE PROTECTION	105.6	130.8	116.4	112.8	160.0	133.1	117.9	155.8	134.2	109.1	132.9	119.3	105.0	131.0	116.2	108.0	148.5	125.4
08	OPENINGS	92.1	135.8	102.8	90.8	166.5	109.2	101.3	165.1	116.8	99.3	136.2	108.3	93.6	137.8	104.4	93.9	155.7	108.9
0920	Plaster & Gypsum Board	102.8	139.7	127.0	108.7	181.9	156.7	111.1	181.9	157.5	113.4	139.7	130.7	96.0	139.7	124.7	114.8	163.1	146.5
0950, 0980	Ceilings & Acoustic Treatment	104.9	139.7	126.7	106.3	181.9	153.7	115.9	181.9	157.3	106.8	139.7	127.4	104.7	139.7	126.6	108.1	163.1	142.6
0960	Flooring	101.5	119.9	106.9	116.9	149.0	126.3	129.4	142.9	133.3	105.2	119.9	109.5	94.1	119.9	101.7	96.2	142.9	109.8
0970, 0990	Wall Finishes & Painting/Coating	84.5	114.6	102.5	108.4	168.0	144.1	104.6	168.0	142.6	88.7	119.9	107.4	84.5	114.6	102.5	92.0	168.0	137.5
09	FINISHES	100.1	133.3	118.0	108.7	170.2	142.0	113.0	168.6	143.1	101.7	133.8	119.0	96.3	133.3	116.3	103.3	157.8	132.8
COVERS	DIVS. 10 - 14, 25, 28, 41, 43, 44, 46	100.0	126.9	106.3	100.0	132.6	107.7	100.0	131.9	107.5	100.0	117.7	104.2	100.0	117.9	104.2	100.0	129.6	107.0
21, 22, 23	FIRE SUPPRESSION, PLUMBING & HVAC	97.0	128.5	109.7	97.6	169.3	126.5	97.6	179.6	130.6	97.1	128.4	109.7	100.8	128.5	112.0	100.8	137.0	115.4
26, 27, 3370	ELECTRICAL, COMMUNICATIONS & UTIL.	94.8	113.6	104.1	99.3	178.8	138.6	96.2	125.7	110.8	94.4	114.7	104.4	93.9	111.7	102.7	99.7	132.6	116.0
MF2018	WEIGHTED AVERAGE	98.4	125.5	110.1	101.9	156.6	125.5	105.0	150.3	124.6	97.1	125.7	109.5	97.4	125.0	109.4	104.9	138.1	119.2

For customer support on your Commercial Renovation Costs with RSMeans Data, call 800.448.8182.

797

CALIFORNIA / COLORADO

DIVISION		SANTA ROSA 954			STOCKTON 952			SUSANVILLE 961			VALLEJO 945			VAN NUYS 913 - 916			ALAMOSA 811		
		MAT.	INST.	TOTAL	MAT.	INST.	TOTAL	MAT.	INST.	TOTAL	MAT.	INST.	TOTAL	MAT.	INST.	TOTAL	MAT.	INST.	TOTAL
015433	CONTRACTOR EQUIPMENT		95.8	95.8		95.3	95.3		95.3	95.3		98.0	98.0		92.5	92.5		87.8	87.8
0241, 31 - 34	SITE & INFRASTRUCTURE, DEMOLITION	104.1	100.3	101.5	103.4	100.4	101.3	131.6	100.0	109.9	99.9	108.5	105.8	113.5	101.8	105.5	135.8	81.2	98.3
0310	Concrete Forming & Accessories	99.5	172.4	161.5	100.8	157.1	148.7	104.3	150.2	143.4	103.8	172.1	161.9	110.3	135.6	131.8	100.4	68.5	73.3
0320	Concrete Reinforcing	102.2	134.8	117.9	105.0	133.8	118.9	134.5	133.6	134.1	89.5	134.7	111.3	99.8	133.2	115.9	107.3	66.6	87.7
0330	Cast-in-Place Concrete	105.2	134.1	116.1	93.4	132.5	108.2	106.4	132.5	116.3	96.7	134.7	111.1	82.4	126.4	99.0	98.6	70.9	88.1
03	CONCRETE	108.0	150.6	127.1	98.2	142.9	118.3	120.9	139.8	129.4	105.0	150.3	125.3	102.2	130.8	115.0	112.5	69.6	91.7
04	MASONRY	98.8	154.8	132.9	99.7	144.6	127.1	118.6	141.0	132.3	75.8	154.8	124.0	113.1	137.8	128.2	128.3	58.5	85.7
05	METALS	107.4	120.3	111.4	103.2	117.2	107.5	102.2	116.7	106.7	101.1	115.1	105.5	78.4	113.8	89.4	103.2	77.4	95.2
06	WOOD, PLASTICS & COMPOSITES	92.3	179.6	138.0	97.4	161.4	130.9	106.6	152.4	130.6	89.3	179.8	136.6	95.6	133.4	115.4	93.9	71.0	81.9
07	THERMAL & MOISTURE PROTECTION	109.5	155.2	129.1	112.3	138.9	123.7	129.0	136.4	132.2	115.8	155.0	132.6	104.2	128.3	114.5	108.4	66.6	90.5
08	OPENINGS	100.9	165.7	116.7	101.4	149.2	113.0	117.4	141.8	123.3	102.9	164.4	117.9	88.0	133.7	99.1	96.4	69.9	89.9
0920	Plaster & Gypsum Board	108.3	181.9	156.6	111.5	163.1	145.4	121.4	153.8	142.7	105.4	181.9	155.6	94.9	134.7	121.0	83.8	70.3	75.0
0950, 0980	Ceilings & Acoustic Treatment	106.8	181.9	153.9	114.7	163.1	145.0	128.9	153.8	144.5	117.3	181.9	157.8	112.6	134.7	126.4	112.0	70.3	85.9
0960	Flooring	101.5	136.3	111.7	98.4	133.0	108.5	89.3	126.7	100.2	122.7	149.0	130.4	101.6	119.9	107.0	106.8	66.9	95.1
0970, 0990	Wall Finishes & Painting/Coating	88.7	168.0	136.2	91.7	132.0	115.8	101.6	135.6	122.0	105.6	168.0	143.0	102.8	119.9	113.0	97.5	76.1	84.7
09	FINISHES	100.9	167.2	136.8	103.4	151.9	129.6	108.0	145.9	128.5	109.3	169.6	141.9	103.3	130.7	118.1	102.6	69.0	84.4
COVERS	DIVS. 10 - 14, 25, 28, 41, 43, 44, 46	100.0	131.1	107.3	100.0	129.2	106.9	100.0	128.3	106.7	100.0	131.6	107.4	100.0	116.6	103.9	100.0	84.8	96.4
21, 22, 23	FIRE SUPPRESSION, PLUMBING & HVAC	97.1	179.8	130.4	100.8	131.4	113.2	97.3	131.4	111.0	101.3	147.6	120.0	96.9	128.4	109.6	97.1	69.8	86.1
26, 27, 3370	ELECTRICAL, COMMUNICATIONS & UTIL.	94.7	125.7	110.0	97.4	112.8	105.0	99.8	121.7	110.6	92.9	126.7	109.6	118.6	131.9	125.2	95.5	60.9	78.4
MF2018	WEIGHTED AVERAGE	101.2	149.8	122.2	101.0	131.9	114.3	107.3	131.0	117.5	100.5	143.4	119.0	98.5	127.0	110.8	103.3	69.2	88.6

COLORADO

DIVISION		BOULDER 803			COLORADO SPRINGS 808 - 809			DENVER 800 - 802			DURANGO 813			FORT COLLINS 805			FORT MORGAN 807		
		MAT.	INST.	TOTAL	MAT.	INST.	TOTAL	MAT.	INST.	TOTAL	MAT.	INST.	TOTAL	MAT.	INST.	TOTAL	MAT.	INST.	TOTAL
015433	CONTRACTOR EQUIPMENT		89.3	89.3		87.3	87.3		96.4	96.4		87.8	87.8		89.3	89.3		89.3	89.3
0241, 31 - 34	SITE & INFRASTRUCTURE, DEMOLITION	94.7	87.8	90.0	97.2	83.4	87.7	101.1	97.6	98.7	129.1	81.3	96.3	107.4	87.9	94.1	97.0	87.3	90.4
0310	Concrete Forming & Accessories	103.7	69.1	74.3	93.6	68.5	72.3	102.6	68.7	73.8	106.6	68.3	74.0	101.6	68.4	73.4	104.2	68.3	73.6
0320	Concrete Reinforcing	102.3	66.8	85.2	101.5	69.1	85.8	100.3	69.1	85.2	107.3	66.6	87.7	102.4	66.8	85.2	102.6	66.8	85.3
0330	Cast-in-Place Concrete	113.5	71.4	97.6	116.6	71.5	99.6	118.1	72.5	100.9	113.4	70.9	97.4	128.2	71.0	106.6	111.4	70.2	95.8
03	CONCRETE	107.8	70.0	90.8	111.4	70.2	92.9	115.2	70.6	95.1	114.2	69.5	94.1	119.2	69.6	96.9	106.2	69.2	89.6
04	MASONRY	104.1	63.9	79.6	107.5	62.1	79.8	113.5	62.7	82.5	116.0	59.5	81.6	120.9	61.9	84.9	119.6	63.7	85.5
05	METALS	93.9	76.5	88.5	97.1	77.6	91.0	99.6	77.3	92.6	103.2	77.4	95.2	95.2	76.5	89.4	93.6	76.6	88.3
06	WOOD, PLASTICS & COMPOSITES	102.9	70.5	86.0	91.8	70.7	80.7	107.2	70.7	88.1	103.4	71.0	86.4	100.2	70.5	84.7	102.9	70.5	86.0
07	THERMAL & MOISTURE PROTECTION	106.3	71.0	91.2	107.3	70.2	91.4	104.5	72.1	90.6	108.3	66.9	90.5	106.7	70.2	91.0	106.3	69.3	90.4
08	OPENINGS	95.0	69.7	88.8	99.1	70.3	92.1	102.4	70.3	94.6	102.9	69.9	94.9	95.0	69.7	88.8	95.0	69.7	88.8
0920	Plaster & Gypsum Board	123.3	70.3	88.5	108.5	70.3	83.5	119.1	70.3	87.1	97.1	70.3	79.5	116.5	70.3	86.2	123.3	70.3	88.5
0950, 0980	Ceilings & Acoustic Treatment	97.4	70.3	80.4	107.6	70.3	84.2	111.0	70.3	85.5	112.0	70.3	85.9	97.4	70.3	80.4	97.4	70.3	80.4
0960	Flooring	107.5	73.2	97.5	98.7	71.3	90.7	103.3	74.7	95.0	112.3	65.8	98.7	103.5	74.7	95.1	108.0	73.2	97.8
0970, 0990	Wall Finishes & Painting/Coating	96.2	76.1	84.2	95.9	80.7	86.8	105.9	74.7	87.2	97.5	76.1	84.7	96.2	76.1	84.2	96.2	76.1	84.2
09	FINISHES	102.3	70.4	85.1	100.2	70.1	83.9	104.6	70.2	86.0	105.2	68.9	85.6	102.4	69.9	84.8	102.4	69.9	84.8
COVERS	DIVS. 10 - 14, 25, 28, 41, 43, 44, 46	100.0	84.5	96.4	100.0	84.1	96.3	100.0	84.3	96.3	100.0	84.7	96.4	100.0	83.8	96.2	100.0	83.8	96.2
21, 22, 23	FIRE SUPPRESSION, PLUMBING & HVAC	97.2	71.9	87.0	101.1	71.5	89.2	100.9	74.7	90.4	97.1	56.8	80.9	101.0	71.5	89.1	97.2	70.9	86.6
26, 27, 3370	ELECTRICAL, COMMUNICATIONS & UTIL.	99.0	79.8	89.5	102.1	73.9	88.1	103.7	79.8	91.8	95.1	53.5	74.5	99.0	79.8	89.5	99.3	75.2	87.4
MF2018	WEIGHTED AVERAGE	99.2	73.7	88.2	101.8	72.3	89.1	104.0	75.1	91.5	103.6	65.5	87.1	102.8	73.2	90.0	99.9	72.5	88.0

COLORADO

DIVISION		GLENWOOD SPRINGS 816			GOLDEN 804			GRAND JUNCTION 815			GREELEY 806			MONTROSE 814			PUEBLO 810		
		MAT.	INST.	TOTAL	MAT.	INST.	TOTAL	MAT.	INST.	TOTAL	MAT.	INST.	TOTAL	MAT.	INST.	TOTAL	MAT.	INST.	TOTAL
015433	CONTRACTOR EQUIPMENT		90.9	90.9		89.3	89.3		90.9	90.9		89.3	89.3		89.4	89.4		87.8	87.8
0241, 31 - 34	SITE & INFRASTRUCTURE, DEMOLITION	145.1	88.7	106.3	107.6	87.6	93.9	129.0	89.1	101.6	94.0	87.8	89.8	138.5	85.4	102.0	120.8	81.5	93.8
0310	Concrete Forming & Accessories	97.0	68.1	72.4	96.3	68.7	72.8	105.4	69.3	74.7	99.2	68.4	73.0	96.6	68.3	72.5	102.8	68.8	73.9
0320	Concrete Reinforcing	106.1	66.8	87.2	102.6	66.8	85.3	106.5	66.6	87.3	102.3	66.8	85.2	106.0	66.8	87.1	102.6	69.0	86.4
0330	Cast-in-Place Concrete	98.6	69.9	87.7	111.5	71.4	96.4	109.2	72.2	95.2	107.1	71.0	93.5	98.6	70.4	88.0	97.9	71.9	88.1
03	CONCRETE	117.9	69.1	96.0	117.2	69.8	95.9	110.7	70.4	92.6	102.7	69.6	87.9	108.4	69.3	90.9	100.8	70.5	87.2
04	MASONRY	102.2	63.6	78.6	122.7	62.7	86.1	136.2	61.7	90.8	113.7	63.2	82.9	108.4	63.6	81.1	98.7	60.9	75.6
05	METALS	102.9	77.3	94.9	93.8	76.5	88.4	104.6	77.3	96.1	95.2	76.6	89.4	102.0	77.6	94.4	106.4	78.9	97.8
06	WOOD, PLASTICS & COMPOSITES	88.9	70.7	79.4	94.1	70.5	81.8	101.4	70.7	85.3	97.2	70.5	83.2	89.7	70.8	79.8	96.6	71.0	83.2
07	THERMAL & MOISTURE PROTECTION	108.2	68.0	90.9	107.5	70.6	91.6	107.2	69.3	90.9	105.9	70.5	90.7	108.4	68.0	91.0	106.8	68.6	90.4
08	OPENINGS	101.9	69.8	94.1	95.0	69.7	88.8	102.6	69.8	94.6	95.0	69.7	88.8	103.1	69.9	95.0	98.1	70.5	91.4
0920	Plaster & Gypsum Board	125.3	70.3	89.3	113.4	70.3	85.1	138.6	70.3	93.8	115.1	70.3	85.7	82.7	70.3	74.6	88.1	70.3	76.4
0950, 0980	Ceilings & Acoustic Treatment	111.4	70.3	85.7	97.4	70.3	80.4	111.4	70.3	85.7	97.4	70.3	80.4	112.0	70.3	85.9	119.7	70.3	88.8
0960	Flooring	105.8	73.2	96.3	101.0	73.2	92.9	111.6	69.6	99.3	102.3	72.6	93.7	109.5	69.8	97.9	108.1	72.6	97.7
0970, 0990	Wall Finishes & Painting/Coating	97.4	76.1	84.7	96.2	76.1	84.2	97.4	76.1	84.7	96.2	76.1	84.2	97.5	76.1	84.7	97.5	76.1	84.7
09	FINISHES	108.5	70.1	87.7	100.3	70.1	84.0	110.2	69.9	88.4	99.4	69.8	83.4	103.2	69.5	85.0	103.6	70.1	85.5
COVERS	DIVS. 10 - 14, 25, 28, 41, 43, 44, 46	100.0	84.1	96.3	100.0	84.1	96.3	100.0	84.9	96.5	100.0	83.8	96.2	100.0	84.4	96.3	100.0	84.8	96.4
21, 22, 23	FIRE SUPPRESSION, PLUMBING & HVAC	97.1	56.7	80.8	97.2	71.3	86.7	100.9	73.3	89.8	101.0	71.4	89.1	97.1	56.8	80.9	100.9	71.6	89.1
26, 27, 3370	ELECTRICAL, COMMUNICATIONS & UTIL.	93.0	53.5	73.4	99.3	77.6	88.5	94.8	52.5	73.9	99.0	79.8	89.5	94.8	53.5	74.3	95.5	63.3	79.6
MF2018	WEIGHTED AVERAGE	103.6	66.6	87.6	101.5	73.0	89.2	105.6	70.0	90.2	99.9	73.3	88.4	102.3	66.3	86.7	101.6	70.7	88.2

For customer support on your Commercial Renovation Costs with RSMeans Data, call 800.448.8182.

		COLORADO			CONNECTICUT														
	DIVISION	SALIDA			BRIDGEPORT			BRISTOL			HARTFORD			MERIDEN			NEW BRITAIN		
		812			066			060			061			064			060		
		MAT.	INST.	TOTAL	MAT.	INST.	TOTAL	MAT.	INST.	TOTAL	MAT.	INST.	TOTAL	MAT.	INST.	TOTAL	MAT.	INST.	TOTAL
015433	CONTRACTOR EQUIPMENT		89.4	89.4		94.2	94.2		94.2	94.2		100.4	100.4		94.6	94.6		94.2	94.2
0241, 31 - 34	SITE & INFRASTRUCTURE, DEMOLITION	128.8	85.0	98.7	107.9	95.6	99.4	107.1	95.6	99.2	102.3	104.7	103.9	104.5	96.2	98.8	107.2	95.6	99.2
0310	Concrete Forming & Accessories	105.4	68.4	74.0	104.3	115.7	114.0	104.3	117.0	115.1	102.9	117.4	115.2	104.0	117.0	115.0	104.8	117.0	115.2
0320	Concrete Reinforcing	105.7	66.6	86.9	116.0	144.7	129.8	116.0	144.7	129.8	110.4	144.7	126.9	116.0	144.7	129.8	116.0	144.7	129.8
0330	Cast-in-Place Concrete	113.0	70.5	97.0	99.8	126.6	110.0	93.5	126.6	106.0	95.3	128.4	107.8	90.0	126.6	103.8	95.1	126.6	107.0
03	CONCRETE	109.0	69.4	91.3	101.9	123.9	111.8	99.0	124.5	110.5	100.2	125.2	111.4	97.3	124.5	109.5	99.7	124.5	110.9
04	MASONRY	136.5	59.5	89.5	111.8	130.5	123.2	103.7	130.5	120.1	107.3	130.6	121.5	103.4	130.5	119.9	105.6	130.5	120.8
05	METALS	101.7	77.5	94.1	103.5	118.7	108.2	103.5	118.7	108.2	108.2	117.6	111.1	100.3	118.6	106.0	99.4	118.7	105.4
06	WOOD, PLASTICS & COMPOSITES	97.9	70.8	83.8	106.2	112.7	109.6	106.2	114.5	110.6	97.6	114.8	106.6	106.2	114.5	110.6	106.2	114.5	110.6
07	THERMAL & MOISTURE PROTECTION	107.2	66.9	89.9	98.9	121.6	108.7	99.1	120.6	108.3	103.5	121.6	111.3	99.1	120.6	108.3	99.1	120.5	108.3
08	OPENINGS	96.4	69.9	90.0	96.6	119.9	102.3	96.6	121.6	102.7	98.3	121.7	104.0	98.8	120.9	104.2	96.6	121.6	102.7
0920	Plaster & Gypsum Board	82.9	70.3	74.7	122.9	112.9	116.3	122.9	114.8	117.6	110.1	114.8	113.2	125.4	114.8	118.4	122.9	114.8	117.6
0950, 0980	Ceilings & Acoustic Treatment	112.0	70.3	85.9	102.9	112.9	109.2	102.9	114.8	110.3	103.3	114.8	110.5	111.0	114.8	113.4	102.9	114.8	110.3
0960	Flooring	115.2	66.7	101.1	96.3	123.6	104.2	96.3	123.6	104.2	100.3	126.0	107.8	96.3	118.8	102.8	96.3	123.6	104.2
0970, 0990	Wall Finishes & Painting/Coating	97.5	76.1	84.7	90.7	126.8	112.3	90.7	130.8	114.7	98.9	130.8	118.0	90.7	126.8	112.3	90.7	130.8	114.7
09	FINISHES	103.9	68.8	84.9	100.3	117.7	109.8	100.4	119.2	110.6	101.8	119.9	111.6	102.7	118.0	111.0	100.4	119.2	110.6
COVERS	DIVS. 10 - 14, 25, 28, 41, 43, 44, 46	100.0	84.5	96.3	100.0	112.2	102.9	100.0	112.4	102.9	100.0	113.1	103.1	100.0	112.4	102.9	100.0	112.4	102.9
21, 22, 23	FIRE SUPPRESSION, PLUMBING & HVAC	97.1	70.9	86.6	101.0	119.0	108.3	101.0	119.0	108.3	100.9	119.0	108.2	97.2	119.0	106.0	101.0	119.0	108.3
26, 27, 3370	ELECTRICAL, COMMUNICATIONS & UTIL.	95.0	60.9	78.1	94.1	105.2	99.6	94.1	103.9	99.0	97.1	110.8	103.9	94.1	101.9	98.0	94.2	103.9	99.0
MF2018	WEIGHTED AVERAGE	102.9	69.8	88.6	100.9	116.7	107.7	100.1	116.9	107.4	101.7	118.7	109.0	98.8	116.4	106.4	99.7	116.9	107.1

		CONNECTICUT																	
	DIVISION	NEW HAVEN			NEW LONDON			NORWALK			STAMFORD			WATERBURY			WILLIMANTIC		
		065			063			068			069			067			062		
		MAT.	INST.	TOTAL	MAT.	INST.	TOTAL	MAT.	INST.	TOTAL	MAT.	INST.	TOTAL	MAT.	INST.	TOTAL	MAT.	INST.	TOTAL
015433	CONTRACTOR EQUIPMENT		94.6	94.6		94.6	94.6		94.2	94.2		94.2	94.2		94.2	94.2		94.2	94.2
0241, 31 - 34	SITE & INFRASTRUCTURE, DEMOLITION	107.2	96.2	99.7	98.6	96.2	97.0	107.7	95.6	99.4	108.4	95.6	99.6	107.6	95.6	99.3	107.8	95.4	99.3
0310	Concrete Forming & Accessories	104.0	116.9	115.0	104.0	116.9	115.0	104.3	115.7	114.0	104.3	116.0	114.3	104.3	117.0	115.1	104.3	116.9	115.0
0320	Concrete Reinforcing	116.0	144.7	129.8	90.9	144.7	116.9	116.0	144.6	129.8	116.0	144.7	129.8	116.0	144.7	129.8	116.0	144.7	129.8
0330	Cast-in-Place Concrete	96.7	124.7	107.3	82.4	124.7	98.3	98.2	125.8	108.7	99.8	126.0	109.7	99.8	126.6	110.0	93.2	124.7	105.1
03	CONCRETE	114.3	123.8	118.6	87.5	123.8	103.8	101.2	123.7	111.3	101.9	123.9	111.8	101.9	124.5	112.1	98.9	123.8	110.0
04	MASONRY	104.0	130.5	120.2	102.3	130.5	119.5	103.4	129.7	119.4	104.3	129.7	119.9	104.3	130.5	120.3	103.6	130.5	120.0
05	METALS	99.7	118.6	105.6	99.3	118.6	105.3	103.5	118.6	108.2	103.5	119.0	108.3	103.5	118.6	108.2	103.2	118.5	108.0
06	WOOD, PLASTICS & COMPOSITES	106.2	114.5	110.6	106.2	114.5	110.6	106.2	112.7	109.6	106.2	112.7	109.6	106.2	114.5	110.6	106.2	114.5	110.6
07	THERMAL & MOISTURE PROTECTION	99.2	119.4	107.9	99.0	120.3	108.1	99.2	121.3	108.7	99.1	121.3	108.6	99.1	119.6	107.9	99.3	120.0	108.2
08	OPENINGS	96.6	120.9	102.5	98.9	120.1	104.1	96.6	119.9	102.3	96.6	119.9	102.3	96.6	120.9	102.5	98.9	120.9	104.3
0920	Plaster & Gypsum Board	122.9	114.8	117.6	122.9	114.8	117.6	122.9	112.9	116.3	122.9	112.9	116.3	122.9	114.8	117.6	122.9	114.8	117.6
0950, 0980	Ceilings & Acoustic Treatment	102.9	114.8	110.3	101.4	114.8	109.8	102.9	112.9	109.2	102.9	112.9	109.2	102.9	114.8	110.3	101.4	114.8	109.8
0960	Flooring	96.3	123.6	104.2	96.3	121.2	103.5	96.3	123.6	104.2	96.3	123.6	104.2	96.3	123.6	104.2	96.3	115.5	101.9
0970, 0990	Wall Finishes & Painting/Coating	90.7	124.4	110.9	90.7	126.8	112.3	90.7	126.8	112.3	90.7	126.8	112.3	90.7	126.8	112.3	90.7	126.8	112.3
09	FINISHES	100.4	118.5	110.2	99.5	118.4	109.7	100.4	117.7	109.8	100.5	117.7	109.8	100.3	118.8	110.3	100.2	117.4	109.5
COVERS	DIVS. 10 - 14, 25, 28, 41, 43, 44, 46	100.0	112.4	102.9	100.0	112.4	102.9	100.0	112.2	102.9	100.0	112.4	102.9	100.0	112.4	102.9	100.0	112.4	102.9
21, 22, 23	FIRE SUPPRESSION, PLUMBING & HVAC	101.0	119.0	108.3	97.2	119.0	106.0	101.0	119.0	108.3	101.0	119.0	108.3	101.0	119.0	108.3	101.0	118.9	108.2
26, 27, 3370	ELECTRICAL, COMMUNICATIONS & UTIL.	94.1	106.2	100.1	91.9	103.9	97.8	94.1	107.1	100.6	94.1	156.4	125.0	93.8	107.3	100.4	94.1	107.1	100.5
MF2018	WEIGHTED AVERAGE	101.5	117.0	108.2	96.7	116.6	105.3	100.4	116.9	107.5	100.5	123.9	110.6	100.5	117.2	107.7	100.3	116.9	107.5

		D.C.			DELAWARE									FLORIDA					
	DIVISION	WASHINGTON			DOVER			NEWARK			WILMINGTON			DAYTONA BEACH			FORT LAUDERDALE		
		200 - 205			199			197			198			321			333		
		MAT.	INST.	TOTAL	MAT.	INST.	TOTAL	MAT.	INST.	TOTAL	MAT.	INST.	TOTAL	MAT.	INST.	TOTAL	MAT.	INST.	TOTAL
015433	CONTRACTOR EQUIPMENT		104.0	104.0		115.6	115.6		117.0	117.0		115.8	115.8		98.2	98.2		91.4	91.4
0241, 31 - 34	SITE & INFRASTRUCTURE, DEMOLITION	105.0	95.4	98.4	102.7	103.6	103.3	102.0	106.0	104.7	98.9	104.5	102.8	117.0	82.6	93.4	94.9	72.4	79.4
0310	Concrete Forming & Accessories	100.6	73.0	77.1	100.0	100.2	100.2	96.2	99.9	99.4	98.0	100.4	100.0	98.6	61.6	67.1	93.8	67.7	71.6
0320	Concrete Reinforcing	115.4	90.3	103.3	100.7	116.3	108.2	98.8	116.3	107.2	96.7	116.4	106.2	90.5	61.5	76.5	88.7	57.6	73.7
0330	Cast-in-Place Concrete	98.3	79.7	91.3	106.2	108.3	107.0	88.6	106.3	95.3	103.4	108.3	105.2	85.5	65.0	77.8	88.0	72.3	82.1
03	CONCRETE	103.7	79.2	92.7	98.6	106.8	102.3	92.1	106.2	98.4	96.6	106.9	101.2	88.7	64.5	77.8	90.6	69.1	81.0
04	MASONRY	99.1	87.1	91.8	103.9	101.7	102.5	102.6	101.6	102.0	99.8	101.7	100.9	87.0	60.0	70.5	90.9	72.8	79.9
05	METALS	104.5	96.2	101.9	105.3	123.5	111.0	107.4	125.8	113.1	105.5	123.6	111.2	100.0	89.2	96.7	96.8	87.7	94.0
06	WOOD, PLASTICS & COMPOSITES	100.7	71.4	85.4	96.1	98.1	97.1	91.0	97.8	94.6	89.5	98.1	94.0	93.5	60.3	76.1	79.1	62.3	70.3
07	THERMAL & MOISTURE PROTECTION	100.8	86.0	94.5	105.4	112.4	108.4	109.9	111.2	110.5	104.9	112.4	108.1	100.8	64.5	85.2	105.5	68.1	89.5
08	OPENINGS	100.8	74.8	94.5	91.1	109.9	95.7	91.1	109.7	95.6	88.6	109.9	93.8	92.9	59.8	84.8	94.3	59.9	85.9
0920	Plaster & Gypsum Board	118.0	70.6	86.9	103.9	97.9	100.0	99.8	97.9	98.6	100.9	97.9	99.0	90.9	59.8	70.5	111.4	61.8	78.9
0950, 0980	Ceilings & Acoustic Treatment	108.3	70.6	84.7	105.5	97.9	100.7	100.8	97.9	99.0	101.9	97.9	99.4	86.1	59.8	69.7	94.4	61.8	74.0
0960	Flooring	94.7	76.7	89.5	91.5	106.8	95.9	91.9	106.8	96.2	99.5	106.8	101.6	104.4	61.0	91.8	102.1	61.0	90.1
0970, 0990	Wall Finishes & Painting/Coating	101.5	70.0	82.6	91.4	117.9	107.3	84.1	117.9	104.4	94.5	117.9	108.5	99.0	62.1	76.9	90.6	58.4	71.3
09	FINISHES	101.8	72.3	85.8	97.1	102.3	99.9	92.3	102.1	97.6	98.6	102.3	100.6	94.9	60.9	76.5	97.1	64.8	79.6
COVERS	DIVS. 10 - 14, 25, 28, 41, 43, 44, 46	100.0	95.5	99.0	100.0	100.0	100.0	100.0	99.3	99.8	100.0	106.3	101.5	100.0	82.0	95.8	100.0	88.0	97.2
21, 22, 23	FIRE SUPPRESSION, PLUMBING & HVAC	100.8	89.4	96.2	99.4	120.9	108.0	99.6	120.8	108.2	100.7	120.9	108.9	100.7	73.7	89.8	100.7	76.0	90.7
26, 27, 3370	ELECTRICAL, COMMUNICATIONS & UTIL.	96.4	103.9	100.1	98.2	109.9	104.0	98.0	109.9	103.9	97.4	109.9	103.6	98.0	61.4	79.9	95.1	67.5	81.4
MF2018	WEIGHTED AVERAGE	101.2	88.0	95.5	99.5	110.4	104.2	98.7	110.6	103.8	99.1	110.7	104.1	97.2	69.0	85.0	96.5	72.2	86.0

FLORIDA

DIVISION		FORT MYERS 339,341			GAINESVILLE 326,344			JACKSONVILLE 320,322			LAKELAND 338			MELBOURNE 329			MIAMI 330 - 332,340		
		MAT.	INST.	TOTAL	MAT.	INST.	TOTAL	MAT.	INST.	TOTAL	MAT.	INST.	TOTAL	MAT.	INST.	TOTAL	MAT.	INST.	TOTAL
015433	CONTRACTOR EQUIPMENT		98.2	98.2		98.2	98.2		98.2	98.2		98.2	98.2		98.2	98.2		95.7	95.7
0241, 31 - 34	SITE & INFRASTRUCTURE, DEMOLITION	106.7	82.7	90.2	125.2	82.4	95.8	117.0	82.2	93.1	108.7	82.7	90.9	124.6	82.6	95.7	94.8	80.7	85.1
0310	Concrete Forming & Accessories	89.8	62.0	66.1	93.1	55.5	61.1	98.5	61.8	67.3	86.0	63.2	66.6	94.6	62.3	67.1	101.3	64.3	69.9
0320	Concrete Reinforcing	89.7	75.0	82.6	96.0	59.8	78.5	90.5	59.7	75.7	91.9	75.1	83.8	91.5	67.1	79.7	98.0	57.6	78.5
0330	Cast-in-Place Concrete	91.8	64.1	81.3	98.2	64.6	85.5	86.4	64.5	78.1	93.9	65.1	83.0	103.1	65.0	88.7	85.2	69.6	79.4
03	CONCRETE	91.0	66.7	80.1	99.5	61.3	82.3	89.1	64.0	77.9	92.6	67.6	81.4	99.7	65.8	84.5	89.6	66.5	79.2
04	MASONRY	85.7	57.3	68.4	99.7	59.2	75.0	86.8	57.7	69.1	101.1	60.0	76.0	84.3	60.0	69.4	94.7	62.6	75.1
05	METALS	98.8	93.7	97.2	98.8	87.5	95.3	98.6	87.0	95.0	98.7	94.2	97.3	108.4	91.4	103.1	97.0	85.5	93.4
06	WOOD, PLASTICS & COMPOSITES	75.9	61.0	68.1	86.6	52.7	68.9	93.5	61.3	76.6	71.2	62.7	66.8	88.6	61.0	74.2	94.1	62.5	77.6
07	THERMAL & MOISTURE PROTECTION	105.3	61.3	86.4	101.2	61.8	84.3	101.1	62.3	84.5	105.2	62.2	86.7	101.3	63.3	85.0	104.8	64.8	87.6
08	OPENINGS	95.7	63.3	87.8	92.6	55.2	83.5	92.9	59.9	84.9	95.6	64.2	88.0	92.2	61.5	84.7	96.7	60.1	87.8
0920	Plaster & Gypsum Board	107.3	60.6	76.6	87.9	52.0	64.4	90.9	60.8	71.1	105.0	62.3	77.0	87.9	60.6	70.0	98.6	61.8	74.5
0950, 0980	Ceilings & Acoustic Treatment	89.2	60.6	71.3	80.5	52.0	62.7	86.1	60.8	70.3	89.2	62.3	72.3	85.5	60.6	69.9	95.1	61.8	74.2
0960	Flooring	99.0	73.7	91.6	101.9	59.8	89.6	104.4	61.0	91.8	97.1	61.0	86.6	102.2	61.0	90.2	100.9	61.6	89.5
0970, 0990	Wall Finishes & Painting/Coating	95.1	62.5	75.5	99.0	62.5	77.1	99.0	62.5	77.1	95.1	62.5	75.5	99.0	79.5	87.3	96.8	58.4	73.8
09	FINISHES	96.6	63.9	78.9	93.4	56.2	73.3	95.0	61.5	76.8	95.8	62.3	77.7	94.4	63.2	77.5	96.9	62.6	78.3
COVERS	DIVS. 10 - 14, 25, 28, 41, 43, 44, 46	100.0	80.8	95.5	100.0	79.3	95.1	100.0	80.3	95.4	100.0	82.2	95.8	100.0	82.1	95.8	100.0	85.2	96.5
21, 22, 23	FIRE SUPPRESSION, PLUMBING & HVAC	97.6	57.0	81.2	97.8	60.6	82.8	100.7	64.0	85.9	97.6	58.4	81.8	100.7	74.1	90.0	96.3	67.9	84.8
26, 27, 3370	ELECTRICAL, COMMUNICATIONS & UTIL.	96.3	64.2	80.4	98.0	56.4	77.4	97.7	64.4	81.3	94.8	56.4	75.8	99.0	66.5	82.9	97.7	77.1	87.5
MF2018	WEIGHTED AVERAGE	96.4	66.7	83.6	98.3	63.8	83.4	97.0	66.9	84.0	97.1	66.3	83.8	99.9	70.6	87.2	96.2	70.4	85.1

FLORIDA

DIVISION		ORLANDO 327 - 328,347			PANAMA CITY 324			PENSACOLA 325			SARASOTA 342			ST. PETERSBURG 337			TALLAHASSEE 323		
		MAT.	INST.	TOTAL	MAT.	INST.	TOTAL	MAT.	INST.	TOTAL	MAT.	INST.	TOTAL	MAT.	INST.	TOTAL	MAT.	INST.	TOTAL
015433	CONTRACTOR EQUIPMENT		103.5	103.5		98.2	98.2		98.2	98.2		98.2	98.2		98.2	98.2		103.5	103.5
0241, 31 - 34	SITE & INFRASTRUCTURE, DEMOLITION	114.1	91.6	98.7	130.1	82.6	97.5	129.4	82.6	97.4	116.8	82.7	93.4	110.9	82.6	91.5	110.1	91.9	97.6
0310	Concrete Forming & Accessories	103.9	61.6	67.9	97.6	64.3	69.2	95.6	61.5	66.6	95.9	63.0	67.9	93.8	61.1	66.0	101.8	62.1	68.1
0320	Concrete Reinforcing	96.5	65.0	81.3	94.5	65.8	80.7	96.9	66.7	82.4	92.3	72.6	82.8	91.9	75.0	83.8	103.1	59.9	82.3
0330	Cast-in-Place Concrete	105.9	66.6	91.0	90.8	64.9	81.0	112.2	64.7	94.3	101.2	65.0	87.5	94.9	64.9	83.6	88.6	66.5	80.3
03	CONCRETE	99.2	65.5	84.0	98.1	66.4	83.9	107.2	65.3	88.4	94.9	67.0	82.4	94.3	66.6	81.8	93.2	64.8	80.4
04	MASONRY	93.2	60.0	73.0	91.2	59.4	71.8	109.8	58.9	78.7	91.4	60.0	72.3	139.1	57.9	89.6	91.5	59.5	72.0
05	METALS	95.2	88.0	93.0	99.7	90.5	96.8	100.7	90.7	97.6	102.2	92.5	99.1	99.6	93.8	97.8	99.9	86.1	95.6
06	WOOD, PLASTICS & COMPOSITES	94.0	60.6	76.5	92.2	64.3	77.6	90.7	61.5	75.3	93.5	62.7	77.4	80.7	60.0	69.9	98.6	61.5	79.2
07	THERMAL & MOISTURE PROTECTION	108.2	65.6	89.9	101.5	62.5	84.7	101.4	61.6	84.3	100.3	62.2	83.9	105.5	61.1	86.4	96.2	63.0	82.0
08	OPENINGS	97.7	60.7	88.7	91.1	63.0	84.2	91.1	61.5	83.9	97.7	63.2	89.3	94.5	62.7	86.7	97.8	60.1	88.6
0920	Plaster & Gypsum Board	92.7	59.8	71.1	90.1	63.9	72.9	98.9	60.8	73.9	98.3	62.3	74.6	109.3	59.6	76.6	101.5	60.8	74.8
0950, 0980	Ceilings & Acoustic Treatment	99.0	59.8	74.5	85.5	63.9	72.0	85.5	60.8	70.0	91.1	62.3	73.0	90.6	59.6	71.1	99.9	60.8	75.4
0960	Flooring	98.8	61.0	87.7	104.1	72.0	94.7	99.9	61.0	88.6	105.4	52.4	89.9	101.0	57.4	88.3	101.4	60.1	89.4
0970, 0990	Wall Finishes & Painting/Coating	94.5	62.1	75.1	99.0	62.5	77.1	99.0	62.5	77.1	95.1	62.5	75.6	95.1	62.5	75.5	99.7	62.5	77.4
09	FINISHES	98.0	61.1	78.0	96.0	65.1	79.3	95.7	61.2	77.0	98.5	60.8	78.1	98.0	60.0	77.4	100.0	61.4	79.1
COVERS	DIVS. 10 - 14, 25, 28, 41, 43, 44, 46	100.0	82.6	95.9	100.0	79.4	95.2	100.0	78.7	95.0	100.0	80.9	95.5	100.0	78.4	95.0	100.0	78.9	95.0
21, 22, 23	FIRE SUPPRESSION, PLUMBING & HVAC	96.4	56.0	80.1	100.7	64.2	86.0	100.7	62.0	85.1	100.3	56.2	82.5	100.7	58.4	83.6	96.5	64.2	83.5
26, 27, 3370	ELECTRICAL, COMMUNICATIONS & UTIL.	98.3	63.2	80.9	96.9	56.4	76.9	100.2	52.3	76.5	97.5	56.4	77.1	95.3	59.6	77.6	103.3	56.4	80.1
MF2018	WEIGHTED AVERAGE	97.9	66.3	84.3	98.7	67.2	85.1	101.2	65.3	85.7	99.1	65.3	84.5	100.4	65.8	85.4	98.1	66.7	84.6

FLORIDA / GEORGIA

DIVISION		TAMPA 335 - 336,346			WEST PALM BEACH 334,349			ALBANY 317,398			ATHENS 306			ATLANTA 300 - 303,399			AUGUSTA 308 - 309		
		MAT.	INST.	TOTAL	MAT.	INST.	TOTAL	MAT.	INST.	TOTAL	MAT.	INST.	TOTAL	MAT.	INST.	TOTAL	MAT.	INST.	TOTAL
015433	CONTRACTOR EQUIPMENT		98.2	98.2		91.4	91.4		92.3	92.3		91.0	91.0		97.5	97.5		91.0	91.0
0241, 31 - 34	SITE & INFRASTRUCTURE, DEMOLITION	111.3	84.6	93.0	91.1	72.4	78.2	104.9	74.5	84.0	101.9	88.8	92.9	99.5	97.2	97.9	95.4	89.4	91.3
0310	Concrete Forming & Accessories	97.0	63.4	68.4	97.7	67.4	71.9	90.2	67.7	71.0	91.9	42.0	49.5	97.2	75.8	78.9	93.1	73.6	76.5
0320	Concrete Reinforcing	88.7	75.2	82.2	91.2	54.8	73.7	91.8	71.8	82.1	92.3	66.2	79.8	94.5	71.8	83.6	92.6	70.8	82.1
0330	Cast-in-Place Concrete	92.8	65.2	82.4	83.7	72.2	79.4	86.0	69.7	79.8	103.9	69.5	90.9	107.3	72.4	94.1	98.1	70.8	87.8
03	CONCRETE	93.0	67.8	81.7	87.5	68.5	79.0	84.0	70.8	78.1	97.8	56.1	79.1	100.4	74.4	88.7	90.9	72.8	82.8
04	MASONRY	91.4	60.0	72.3	90.4	70.0	78.0	98.5	69.3	80.7	80.0	68.2	72.8	92.1	70.3	78.8	94.1	69.4	79.0
05	METALS	98.8	94.4	97.4	95.7	86.3	92.8	105.9	98.3	103.5	95.3	77.3	89.7	96.3	83.2	92.2	95.0	83.4	91.4
06	WOOD, PLASTICS & COMPOSITES	85.1	62.7	73.4	84.1	62.3	72.7	80.5	68.0	73.9	94.7	34.5	63.2	104.6	78.6	91.0	96.2	76.2	85.7
07	THERMAL & MOISTURE PROTECTION	105.8	62.2	87.1	105.2	68.0	89.2	99.5	68.8	86.3	94.3	65.5	82.0	95.9	73.9	86.5	94.0	71.0	84.1
08	OPENINGS	95.6	64.2	88.0	93.8	59.3	85.4	85.4	70.3	81.7	91.4	49.5	81.2	99.6	76.7	94.0	91.4	75.1	87.5
0920	Plaster & Gypsum Board	112.2	62.3	79.5	114.9	61.8	80.1	101.3	67.7	79.3	92.6	33.3	53.7	99.3	78.3	85.5	103.4	76.1	82.0
0950, 0980	Ceilings & Acoustic Treatment	94.4	62.3	74.3	89.2	61.8	72.0	83.6	67.7	73.7	102.6	33.3	59.2	93.9	78.3	84.1	103.3	76.1	86.2
0960	Flooring	102.1	61.0	90.1	104.0	52.4	88.9	103.5	71.1	94.1	93.9	71.1	87.3	95.6	72.5	88.9	94.1	71.1	87.4
0970, 0990	Wall Finishes & Painting/Coating	95.1	62.5	75.5	90.6	57.8	70.9	85.1	95.0	91.0	89.7	89.5	89.6	96.3	96.7	96.5	89.7	88.4	89.0
09	FINISHES	99.7	62.3	79.5	96.6	63.0	78.4	94.1	70.7	81.4	95.5	50.2	71.0	95.4	77.5	85.7	95.2	74.8	84.1
COVERS	DIVS. 10 - 14, 25, 28, 41, 43, 44, 46	100.0	81.0	95.5	100.0	87.9	97.2	100.0	84.3	96.3	100.0	80.5	95.4	100.0	86.3	96.8	100.0	85.4	96.6
21, 22, 23	FIRE SUPPRESSION, PLUMBING & HVAC	100.7	58.5	83.6	97.6	66.1	84.9	100.7	69.4	88.1	97.2	63.1	83.5	101.0	72.0	89.3	101.0	66.4	87.1
26, 27, 3370	ELECTRICAL, COMMUNICATIONS & UTIL.	95.0	63.2	79.3	95.6	64.9	80.4	95.2	62.0	78.8	99.2	61.5	80.6	98.9	72.7	85.9	99.7	68.5	84.3
MF2018	WEIGHTED AVERAGE	98.0	67.4	84.8	95.1	69.0	83.8	96.5	72.2	86.0	95.9	63.8	82.0	98.7	76.7	89.2	96.4	73.5	86.5

		GEORGIA																
DIVISION		COLUMBUS			DALTON			GAINESVILLE			MACON			SAVANNAH			STATESBORO		
		318 - 319			307			305			310 - 312			313 - 314			304		
		MAT.	INST.	TOTAL	MAT.	INST.	TOTAL	MAT.	INST.	TOTAL	MAT.	INST.	TOTAL	MAT.	INST.	TOTAL	MAT.	INST.	TOTAL
015433	CONTRACTOR EQUIPMENT		92.3	92.3		106.1	106.1		91.0	91.0		101.8	101.8		97.7	97.7		94.0	94.0
0241, 31 - 34	SITE & INFRASTRUCTURE, DEMOLITION	104.8	74.8	84.2	101.6	93.8	96.2	101.6	88.7	92.7	106.3	88.8	94.3	106.2	83.8	90.8	102.8	75.0	83.7
0310	Concrete Forming & Accessories	90.1	70.3	73.3	83.5	62.2	65.4	96.0	39.9	48.3	89.8	69.2	72.3	96.2	71.6	75.3	78.1	50.2	54.3
0320	Concrete Reinforcing	91.7	71.8	82.1	91.8	57.7	75.3	92.1	60.2	76.7	92.9	71.8	82.7	97.8	70.9	84.8	91.4	67.5	79.8
0330	Cast-in-Place Concrete	85.7	70.0	79.8	100.9	68.0	88.5	109.2	69.0	94.0	84.5	70.9	79.4	86.8	71.3	81.0	103.7	68.6	90.5
03	CONCRETE	83.8	72.2	78.6	97.0	65.3	82.8	99.6	55.0	79.6	83.4	72.0	78.3	84.5	72.8	79.2	97.2	61.9	81.4
04	MASONRY	98.2	69.4	80.7	81.1	66.6	72.2	88.5	67.3	75.5	112.4	68.8	85.8	92.4	70.3	78.9	83.2	68.1	74.0
05	METALS	105.5	99.0	103.5	96.2	92.0	94.9	94.5	77.3	89.2	100.7	99.2	100.2	101.8	95.8	99.9	99.8	96.9	98.9
06	WOOD, PLASTICS & COMPOSITES	80.5	71.4	75.7	77.4	63.0	69.9	99.0	32.0	64.0	85.3	70.1	77.3	92.0	73.3	82.2	70.3	45.5	57.4
07	THERMAL & MOISTURE PROTECTION	99.4	70.1	86.8	96.1	65.7	83.0	94.3	63.4	81.0	98.0	71.2	86.5	97.4	70.6	85.9	94.6	62.8	81.0
08	OPENINGS	85.4	72.8	82.3	92.9	64.5	86.0	91.4	48.1	80.8	85.2	72.0	82.0	94.9	73.6	89.7	93.9	57.4	85.0
0920	Plaster & Gypsum Board	101.3	71.3	81.6	82.3	62.5	69.3	94.3	30.7	52.6	104.0	69.9	81.6	103.3	72.9	83.4	84.2	44.6	58.2
0950, 0980	Ceilings & Acoustic Treatment	83.6	71.3	75.9	118.4	62.5	83.4	102.6	30.7	57.6	79.2	69.9	73.4	98.2	72.9	82.4	113.4	44.6	70.3
0960	Flooring	103.5	71.1	94.1	94.8	71.1	87.9	95.5	71.1	88.4	81.4	71.1	78.4	97.9	72.5	90.5	111.4	71.1	99.6
0970, 0990	Wall Finishes & Painting/Coating	85.1	85.9	85.6	84.2	66.7	72.2	89.7	89.5	89.6	87.1	95.0	91.8	86.9	85.3	85.9	88.1	66.7	75.3
09	FINISHES	94.0	71.7	81.9	104.6	64.0	82.7	96.0	48.6	70.3	83.4	71.8	77.1	96.9	73.1	84.0	107.4	54.2	78.6
COVERS	DIVS. 10 - 14, 25, 28, 41, 43, 44, 46	100.0	84.7	96.4	100.0	81.0	95.5	100.0	80.0	95.3	100.0	84.4	96.3	100.0	85.4	96.6	100.0	81.6	95.7
21, 22, 23	FIRE SUPPRESSION, PLUMBING & HVAC	100.7	66.3	86.8	97.3	57.9	81.4	97.2	62.6	83.3	100.7	71.9	89.1	100.8	64.3	86.1	98.0	66.3	85.2
26, 27, 3370	ELECTRICAL, COMMUNICATIONS & UTIL.	95.3	70.8	83.2	108.0	59.9	84.2	99.2	70.7	85.1	93.9	62.8	78.5	99.2	63.8	81.7	99.7	63.4	81.7
MF2018	WEIGHTED AVERAGE	96.4	73.4	86.4	97.7	68.0	84.9	96.5	64.3	82.6	95.3	74.4	86.3	97.3	72.8	86.7	97.9	67.1	84.6

		GEORGIA						HAWAII									IDAHO		
DIVISION		VALDOSTA			WAYCROSS			HILO			HONOLULU			STATES & POSS., GUAM			BOISE		
		316			315			967			968			969			836 - 837		
		MAT.	INST.	TOTAL	MAT.	INST.	TOTAL	MAT.	INST.	TOTAL	MAT.	INST.	TOTAL	MAT.	INST.	TOTAL	MAT.	INST.	TOTAL
015433	CONTRACTOR EQUIPMENT		92.3	92.3		92.3	92.3		96.0	96.0		100.7	100.7		159.0	159.0		92.3	92.3
0241, 31 - 34	SITE & INFRASTRUCTURE, DEMOLITION	114.8	74.5	87.1	111.1	74.1	85.7	146.1	100.4	114.7	153.4	108.5	122.6	192.5	97.3	127.1	84.6	88.8	87.5
0310	Concrete Forming & Accessories	80.0	44.0	49.4	81.9	61.5	64.5	106.1	124.1	121.4	123.2	124.2	124.0	108.4	63.3	70.1	80.4	82.9
0320	Concrete Reinforcing	93.8	60.7	77.8	93.8	56.6	75.8	140.0	122.0	131.3	160.1	122.1	141.7	251.6	43.8	151.4	101.2	76.0	89.0
0330	Cast-in-Place Concrete	84.2	69.5	78.7	95.1	68.5	85.1	196.9	125.0	169.7	153.7	126.2	143.3	170.6	104.3	145.6	89.2	93.5	90.8
03	CONCRETE	88.2	58.1	74.7	91.1	64.8	79.3	153.2	123.2	139.7	146.4	123.5	136.1	156.1	74.9	119.7	97.6	84.3	91.7
04	MASONRY	104.4	69.3	83.0	105.1	68.1	82.5	140.4	124.4	130.7	132.4	124.5	127.6	200.5	48.8	108.0	125.0	84.2	100.1
05	METALS	105.0	93.4	101.4	104.0	86.6	98.7	110.0	107.8	109.4	124.9	106.4	119.1	141.9	81.4	123.0	107.7	80.4	99.2
06	WOOD, PLASTICS & COMPOSITES	67.9	36.3	51.3	69.6	61.2	65.2	108.3	124.7	116.9	135.2	124.8	129.8	119.7	63.6	90.4	89.5	79.4	84.2
07	THERMAL & MOISTURE PROTECTION	99.7	65.4	84.9	99.4	64.9	84.6	127.6	118.4	123.6	144.9	119.6	134.1	151.5	70.0	116.5	99.1	85.4	93.2
08	OPENINGS	82.2	50.5	74.5	82.5	60.7	77.2	114.7	122.3	116.6	128.7	122.4	127.2	118.9	52.8	102.8	97.1	74.0	91.4
0920	Plaster & Gypsum Board	94.2	35.1	55.4	94.2	60.8	72.3	122.9	125.3	124.5	169.2	125.3	140.4	249.0	53.0	120.4	94.3	79.1	84.3
0950, 0980	Ceilings & Acoustic Treatment	82.3	35.1	52.7	80.9	60.8	68.3	125.1	125.3	125.2	132.7	125.3	128.1	231.1	53.0	119.5	112.5	79.1	91.6
0960	Flooring	96.9	71.1	89.4	98.3	71.1	90.4	105.9	140.4	116.0	122.8	140.4	127.9	123.3	47.3	101.1	91.8	86.5	90.3
0970, 0990	Wall Finishes & Painting/Coating	85.1	93.1	89.9	85.1	66.7	74.1	96.1	137.0	120.6	109.0	137.0	125.8	101.6	39.6	64.5	89.3	45.7	63.2
09	FINISHES	91.8	51.9	70.2	91.4	63.4	76.3	111.9	129.0	121.1	126.8	129.0	128.0	186.6	58.2	117.1	96.1	77.7	86.1
COVERS	DIVS. 10 - 14, 25, 28, 41, 43, 44, 46	100.0	80.9	95.5	100.0	81.2	95.6	100.0	112.1	102.9	100.0	112.3	102.9	100.0	92.9	98.3	100.0	86.9	96.9
21, 22, 23	FIRE SUPPRESSION, PLUMBING & HVAC	100.7	69.1	88.0	98.4	60.5	83.1	101.1	108.9	104.2	101.1	108.9	104.3	104.1	52.5	83.2	101.2	71.5	89.2
26, 27, 3370	ELECTRICAL, COMMUNICATIONS & UTIL.	93.8	56.7	75.5	97.1	63.4	80.4	105.8	122.0	113.8	106.7	122.0	114.3	150.4	49.2	100.3	97.2	69.3	83.4
MF2018	WEIGHTED AVERAGE	96.7	65.4	83.2	96.6	66.9	83.8	115.6	117.2	116.3	120.3	117.8	119.2	138.0	63.4	105.7	100.9	78.2	91.1

		IDAHO															ILLINOIS		
DIVISION		COEUR D'ALENE			IDAHO FALLS			LEWISTON			POCATELLO			TWIN FALLS			BLOOMINGTON		
		838			834			835			832			833			617		
		MAT.	INST.	TOTAL	MAT.	INST.	TOTAL	MAT.	INST.	TOTAL	MAT.	INST.	TOTAL	MAT.	INST.	TOTAL	MAT.	INST.	TOTAL
015433	CONTRACTOR EQUIPMENT		87.3	87.3		92.3	92.3		87.3	87.3		92.3	92.3		92.3	92.3		98.8	98.8
0241, 31 - 34	SITE & INFRASTRUCTURE, DEMOLITION	84.4	83.1	83.5	83.3	88.6	86.9	91.3	84.4	86.5	86.1	88.8	87.9	93.0	88.6	90.0	94.4	92.2	92.9
0310	Concrete Forming & Accessories	103.6	78.3	82.1	90.7	75.5	77.8	109.1	81.3	85.5	97.4	79.9	82.5	98.5	79.8	82.6	80.3	111.8	107.1
0320	Concrete Reinforcing	109.3	99.8	104.7	103.2	74.2	89.2	109.3	100.0	104.8	101.6	76.1	89.3	103.6	77.0	90.7	95.4	97.5	96.4
0330	Cast-in-Place Concrete	96.5	81.5	90.8	85.0	80.1	83.1	100.3	81.5	93.4	91.7	93.3	92.3	94.2	81.6	89.4	94.4	109.3	100.0
03	CONCRETE	103.8	83.3	94.6	89.5	77.2	84.0	107.5	84.8	97.3	96.7	84.0	91.0	104.6	80.1	93.6	91.5	109.3	99.5
04	MASONRY	126.7	83.6	100.4	120.2	84.2	98.2	127.1	85.3	101.6	122.5	87.2	101.0	125.4	81.8	98.8	115.8	116.7	116.4
05	METALS	101.4	88.5	97.4	115.9	79.1	104.4	100.8	89.8	97.4	116.0	80.1	104.8	116.0	80.0	104.8	93.0	121.1	101.7
06	WOOD, PLASTICS & COMPOSITES	90.7	77.1	83.6	82.7	73.6	78.0	96.9	80.1	88.1	89.5	79.4	84.2	90.7	79.4	84.7	76.5	109.2	93.6
07	THERMAL & MOISTURE PROTECTION	152.8	80.4	121.7	97.7	72.2	86.8	153.1	82.5	122.8	98.3	74.7	88.2	99.2	80.7	91.2	95.5	107.9	100.8
08	OPENINGS	114.1	75.0	104.6	99.8	68.6	92.2	107.0	81.1	100.7	97.7	68.2	90.5	100.5	64.3	91.7	91.4	114.3	96.9
0920	Plaster & Gypsum Board	170.7	76.8	109.1	83.2	73.2	76.6	172.4	79.9	111.7	85.0	79.1	81.1	87.2	79.1	81.9	85.4	109.8	101.4
0950, 0980	Ceilings & Acoustic Treatment	146.0	76.8	102.6	113.9	73.2	88.4	146.0	79.9	104.6	119.7	79.1	94.2	116.5	79.1	93.0	84.1	109.8	100.2
0960	Flooring	129.9	73.5	113.5	91.5	73.5	86.2	133.5	78.0	117.3	95.2	78.0	90.2	96.5	82.3	92.3	85.5	115.3	94.2
0970, 0990	Wall Finishes & Painting/Coating	106.4	70.3	84.7	89.3	37.6	58.3	106.4	72.1	85.8	89.2	40.7	60.2	89.3	39.9	59.7	85.6	128.3	111.2
09	FINISHES	159.1	76.3	114.3	94.3	71.2	81.8	160.4	79.2	116.5	97.5	75.8	85.7	98.1	76.6	86.4	86.1	114.2	101.3
COVERS	DIVS. 10 - 14, 25, 28, 41, 43, 44, 46	100.0	88.2	97.2	100.0	86.3	96.8	100.0	97.4	99.4	100.0	86.9	96.9	100.0	86.9	96.9	100.0	105.5	101.3
21, 22, 23	FIRE SUPPRESSION, PLUMBING & HVAC	100.6	81.5	92.9	102.3	78.2	92.6	102.4	85.6	95.6	100.9	73.0	89.7	100.9	71.5	89.0	96.8	100.4	98.2
26, 27, 3370	ELECTRICAL, COMMUNICATIONS & UTIL.	88.4	78.5	83.5	88.0	68.2	78.2	86.8	84.7	85.8	94.3	67.5	81.1	89.1	70.7	80.0	94.7	87.0	90.8
MF2018	WEIGHTED AVERAGE	108.2	81.5	96.6	100.2	76.9	90.1	108.4	84.9	98.2	101.8	77.8	91.4	102.9	76.9	91.7	94.8	105.4	99.4

For customer support on your Commercial Renovation Costs with RSMeans Data, call 800.448.8182.

801

ILLINOIS

DIVISION		CARBONDALE 629			CENTRALIA 628			CHAMPAIGN 618 - 619			CHICAGO 606 - 608			DECATUR 625			EAST ST. LOUIS 620 - 622		
		MAT.	INST.	TOTAL	MAT.	INST.	TOTAL	MAT.	INST.	TOTAL	MAT.	INST.	TOTAL	MAT.	INST.	TOTAL	MAT.	INST.	TOTAL
015433	CONTRACTOR EQUIPMENT		107.2	107.2		107.2	107.2		99.7	99.7		97.0	97.0		99.7	99.7		107.2	107.2
0241, 31 - 34	SITE & INFRASTRUCTURE, DEMOLITION	103.0	92.8	96.0	103.4	94.7	97.4	103.2	93.2	96.3	105.6	100.0	101.7	96.7	93.4	94.4	106.0	94.3	97.9
0310	Concrete Forming & Accessories	88.6	104.8	102.4	90.5	109.9	107.0	86.9	112.4	108.6	100.5	159.0	150.3	91.5	116.6	112.9	86.0	112.8	108.8
0320	Concrete Reinforcing	90.9	98.5	94.6	90.9	99.1	94.8	95.4	96.8	96.1	104.8	151.0	127.1	89.1	96.9	92.9	90.8	105.1	97.7
0330	Cast-in-Place Concrete	92.1	98.4	94.5	92.5	114.7	100.9	109.5	106.3	108.3	120.6	154.2	133.3	100.5	113.9	105.6	94.1	117.7	103.0
03	CONCRETE	82.4	103.1	91.7	82.9	111.3	95.6	103.5	108.3	105.7	112.4	155.4	131.7	92.9	112.9	101.9	83.9	114.7	97.7
04	MASONRY	81.6	106.8	97.0	81.6	117.8	103.7	141.6	120.2	128.5	102.7	164.4	140.4	77.1	120.0	103.3	81.9	121.5	106.1
05	METALS	98.5	127.7	107.6	98.6	130.8	108.7	93.0	117.8	100.7	95.7	147.2	111.8	102.5	119.3	107.8	99.7	134.5	110.6
06	WOOD, PLASTICS & COMPOSITES	85.6	102.2	94.3	88.4	107.4	98.3	83.4	110.1	97.4	108.0	158.3	134.3	87.7	116.1	102.6	82.6	109.5	96.6
07	THERMAL & MOISTURE PROTECTION	90.6	96.7	93.2	90.7	106.5	97.5	96.3	110.4	102.3	94.7	149.2	118.1	95.9	111.7	102.7	90.7	109.1	98.6
08	OPENINGS	87.1	112.0	93.2	87.1	114.8	93.9	91.9	113.4	97.1	101.2	168.5	117.6	97.9	116.8	102.5	87.2	115.3	94.0
0920	Plaster & Gypsum Board	86.8	102.6	97.2	87.9	107.9	101.0	87.1	110.7	102.6	96.0	159.9	138.0	89.8	116.9	107.6	85.4	110.1	101.6
0950, 0980	Ceilings & Acoustic Treatment	79.6	102.6	94.0	79.6	107.9	97.4	84.1	110.7	100.8	89.6	159.9	133.7	86.6	116.9	105.6	79.6	110.1	98.7
0960	Flooring	116.3	111.3	114.8	117.3	111.3	115.5	88.9	111.3	95.4	95.4	161.3	114.6	104.6	114.0	107.3	115.2	113.5	114.7
0970, 0990	Wall Finishes & Painting/Coating	99.9	99.9	99.9	99.9	106.3	103.7	85.6	109.4	99.9	93.9	162.4	135.0	92.8	112.2	104.4	91.0	112.6	102.7
09	FINISHES	91.4	106.2	99.4	91.9	109.8	101.6	88.0	112.2	101.1	94.8	160.6	130.4	92.0	116.2	105.1	91.0	112.6	102.7
COVERS	DIVS. 10 - 14, 25, 28, 41, 43, 44, 46	100.0	101.3	100.3	100.0	100.2	100.1	100.0	106.1	101.4	100.0	127.2	106.4	100.0	106.2	101.5	100.0	107.4	101.8
21, 22, 23	FIRE SUPPRESSION, PLUMBING & HVAC	96.7	101.7	98.7	96.7	93.2	95.3	96.8	101.9	98.9	100.4	134.9	114.3	100.5	95.9	98.7	100.5	96.7	99.0
26, 27, 3370	ELECTRICAL, COMMUNICATIONS & UTIL.	91.9	103.8	97.8	92.9	102.3	97.6	97.4	91.9	94.6	96.1	136.4	116.0	94.9	99.9	97.4	92.6	105.6	99.1
MF2018	WEIGHTED AVERAGE	92.7	105.2	98.1	92.9	106.7	98.9	98.4	106.3	101.8	100.5	144.4	119.5	96.7	107.6	101.4	94.1	109.8	100.9

ILLINOIS

DIVISION		EFFINGHAM 624			GALESBURG 614			JOLIET 604			KANKAKEE 609			LA SALLE 613			NORTH SUBURBAN 600 - 603		
		MAT.	INST.	TOTAL	MAT.	INST.	TOTAL	MAT.	INST.	TOTAL	MAT.	INST.	TOTAL	MAT.	INST.	TOTAL	MAT.	INST.	TOTAL
015433	CONTRACTOR EQUIPMENT		99.7	99.7		98.8	98.8		90.4	90.4		90.4	90.4		98.8	98.8		90.4	90.4
0241, 31 - 34	SITE & INFRASTRUCTURE, DEMOLITION	101.0	92.0	94.8	96.7	92.1	93.5	101.7	94.8	97.0	94.9	93.1	93.7	96.1	92.9	93.9	101.0	93.9	96.1
0310	Concrete Forming & Accessories	95.9	110.3	108.2	86.8	112.8	108.9	100.1	157.8	149.2	92.5	139.9	132.8	101.1	120.7	117.7	99.2	152.6	144.6
0320	Concrete Reinforcing	92.1	89.3	90.7	94.9	100.6	97.6	111.9	138.0	124.5	112.6	128.7	120.4	95.1	126.5	110.3	111.9	142.0	126.4
0330	Cast-in-Place Concrete	100.2	105.4	102.1	97.2	104.9	100.1	109.6	146.6	123.5	102.1	130.3	112.8	97.1	116.9	104.6	109.6	146.8	123.7
03	CONCRETE	93.6	105.6	99.0	94.4	108.8	100.9	104.1	149.8	124.6	97.9	134.3	114.2	95.3	121.3	106.9	104.1	148.2	123.9
04	MASONRY	85.7	110.5	100.8	116.0	117.4	116.8	101.8	159.4	136.7	98.3	140.9	124.3	116.0	124.3	121.1	98.8	152.2	131.4
05	METALS	99.5	112.4	103.6	93.0	123.5	102.5	93.8	138.8	107.8	93.8	130.6	105.3	93.0	139.3	107.5	94.8	138.9	108.6
06	WOOD, PLASTICS & COMPOSITES	90.3	110.1	100.6	83.3	111.4	98.0	101.1	158.5	131.1	93.0	139.8	117.5	99.4	119.0	109.6	100.0	152.9	127.7
07	THERMAL & MOISTURE PROTECTION	95.4	104.2	99.2	95.7	105.1	99.7	99.3	145.0	118.9	98.4	133.2	113.3	95.9	117.0	105.0	99.7	141.0	117.4
08	OPENINGS	93.1	110.3	97.3	91.4	112.9	96.6	99.2	163.6	114.9	92.5	150.5	106.6	91.4	128.4	100.4	99.2	161.8	114.5
0920	Plaster & Gypsum Board	89.3	110.7	103.3	87.1	112.1	103.5	90.0	160.4	136.2	89.4	141.2	123.4	94.5	119.9	111.1	94.3	154.7	133.9
0950, 0980	Ceilings & Acoustic Treatment	79.6	110.7	99.1	84.1	112.1	101.6	89.3	160.4	133.9	89.3	141.2	121.9	84.1	119.9	106.5	89.3	154.7	130.3
0960	Flooring	105.9	111.3	107.4	88.6	115.3	96.4	94.2	154.7	111.9	91.1	143.8	106.5	95.5	119.1	102.4	94.6	152.2	111.4
0970, 0990	Wall Finishes & Painting/Coating	92.8	104.1	99.6	85.6	93.8	90.5	83.9	169.6	135.2	83.9	129.2	111.0	85.6	126.7	110.2	85.4	156.1	127.8
09	FINISHES	90.6	110.5	101.4	87.3	111.7	100.5	91.4	159.6	128.3	90.1	140.1	117.1	90.3	119.9	106.3	92.1	153.8	125.5
COVERS	DIVS. 10 - 14, 25, 28, 41, 43, 44, 46	100.0	104.7	101.1	100.0	105.7	101.3	100.0	123.1	105.4	100.0	111.6	102.7	100.0	106.7	101.6	100.0	117.7	104.2
21, 22, 23	FIRE SUPPRESSION, PLUMBING & HVAC	96.7	99.5	97.9	96.8	100.7	98.4	100.4	131.4	112.9	96.6	123.4	107.4	96.8	122.3	107.1	100.3	129.7	112.2
26, 27, 3370	ELECTRICAL, COMMUNICATIONS & UTIL.	93.3	103.7	98.5	95.4	83.2	89.4	95.2	133.5	114.2	95.1	131.9	113.3	93.0	131.9	112.3	95.1	130.5	112.6
MF2018	WEIGHTED AVERAGE	95.3	104.9	99.4	95.5	104.8	99.5	98.5	140.3	116.5	95.1	129.5	110.0	95.8	122.2	107.2	98.5	137.3	115.3

ILLINOIS

DIVISION		PEORIA 615 - 616			QUINCY 623			ROCK ISLAND 612			ROCKFORD 610 - 611			SOUTH SUBURBAN 605			SPRINGFIELD 626 - 627		
		MAT.	INST.	TOTAL	MAT.	INST.	TOTAL	MAT.	INST.	TOTAL	MAT.	INST.	TOTAL	MAT.	INST.	TOTAL	MAT.	INST.	TOTAL
015433	CONTRACTOR EQUIPMENT		98.8	98.8		99.7	99.7		98.8	98.8		98.8	98.8		90.4	90.4		105.4	105.4
0241, 31 - 34	SITE & INFRASTRUCTURE, DEMOLITION	97.7	93.0	94.5	99.7	92.4	94.7	94.9	90.9	92.1	97.1	94.1	95.0	101.0	93.8	96.0	102.5	102.4	102.4
0310	Concrete Forming & Accessories	90.6	117.9	113.8	93.5	113.8	110.8	88.4	95.3	94.3	94.3	130.0	124.6	99.2	152.6	144.6	90.9	119.9	115.6
0320	Concrete Reinforcing	92.5	103.8	98.0	91.7	82.1	87.0	94.9	94.2	94.6	87.6	132.9	109.4	111.9	142.0	126.4	88.3	101.9	94.9
0330	Cast-in-Place Concrete	94.3	114.9	102.1	100.4	102.1	101.0	95.1	94.2	94.8	96.5	129.2	109.6	109.6	146.7	123.6	95.6	113.1	102.2
03	CONCRETE	91.7	115.2	102.3	93.2	104.9	98.5	92.3	96.0	94.0	92.3	130.9	109.6	104.1	148.1	123.8	89.3	114.9	100.8
04	MASONRY	114.6	122.5	119.4	109.1	108.5	108.7	115.8	96.1	103.8	89.4	142.0	121.5	98.8	152.1	131.3	89.0	124.3	110.5
05	METALS	95.7	125.1	104.8	99.6	110.1	102.9	93.0	115.9	100.2	95.7	144.2	110.8	94.8	138.7	108.5	100.1	120.3	106.4
06	WOOD, PLASTICS & COMPOSITES	91.8	115.9	104.4	87.5	116.1	102.5	85.0	94.2	89.8	91.7	126.6	110.0	100.0	152.9	127.7	87.1	119.0	103.8
07	THERMAL & MOISTURE PROTECTION	96.4	112.7	103.4	95.4	102.6	98.5	95.7	93.9	94.9	99.1	129.3	112.1	99.7	141.0	117.4	98.1	116.3	105.9
08	OPENINGS	96.6	120.9	102.5	93.8	111.3	98.1	91.4	99.5	93.3	96.6	138.1	106.7	99.2	161.8	114.5	98.8	119.9	103.9
0920	Plaster & Gypsum Board	91.9	116.6	108.1	87.9	116.9	106.9	87.1	94.4	91.9	91.9	127.6	115.3	94.3	154.7	133.9	90.4	119.5	109.5
0950, 0980	Ceilings & Acoustic Treatment	89.2	116.6	106.4	79.6	116.9	103.0	84.1	94.4	90.5	89.2	127.6	113.3	89.3	154.7	130.3	88.5	119.5	108.0
0960	Flooring	92.3	121.0	100.7	104.6	108.4	105.7	89.8	90.3	89.9	92.3	124.0	101.5	94.6	152.2	111.4	105.3	128.7	112.1
0970, 0990	Wall Finishes & Painting/Coating	85.6	140.2	118.3	92.8	106.2	100.8	85.6	91.0	88.8	85.6	144.2	120.8	85.4	156.1	127.8	98.4	114.0	107.7
09	FINISHES	90.3	121.0	106.9	90.0	113.2	102.5	87.6	93.7	90.9	90.3	130.0	111.8	92.1	153.8	125.5	95.5	121.7	109.7
COVERS	DIVS. 10 - 14, 25, 28, 41, 43, 44, 46	100.0	106.8	101.6	100.0	104.5	101.1	100.0	92.3	98.2	100.0	113.3	103.1	100.0	117.7	104.2	100.0	107.9	101.9
21, 22, 23	FIRE SUPPRESSION, PLUMBING & HVAC	100.6	102.2	101.2	96.7	98.0	97.2	96.8	92.9	95.2	100.7	115.5	106.7	100.3	129.7	112.2	100.5	102.5	101.3
26, 27, 3370	ELECTRICAL, COMMUNICATIONS & UTIL.	96.3	93.5	94.9	91.5	79.2	85.4	88.9	89.7	89.3	96.5	128.3	112.3	95.1	130.5	112.6	100.9	91.4	96.2
MF2018	WEIGHTED AVERAGE	97.4	109.9	102.8	96.1	101.1	98.3	94.5	95.5	94.9	96.3	126.3	109.3	98.5	137.2	115.3	97.8	110.4	103.2

INDIANA

DIVISION		ANDERSON 460			BLOOMINGTON 474			COLUMBUS 472			EVANSVILLE 476 - 477			FORT WAYNE 467 - 468			GARY 463 - 464		
		MAT.	INST.	TOTAL	MAT.	INST.	TOTAL	MAT.	INST.	TOTAL	MAT.	INST.	TOTAL	MAT.	INST.	TOTAL	MAT.	INST.	TOTAL
015433	CONTRACTOR EQUIPMENT		92.7	92.7		79.6	79.6		79.6	79.6		108.7	108.7		92.7	92.7		92.7	92.7
0241, 31 - 34	SITE & INFRASTRUCTURE, DEMOLITION	99.7	86.5	90.6	87.3	85.1	85.8	83.6	85.1	84.6	92.8	112.6	106.4	100.5	86.4	90.9	100.3	90.2	93.4
0310	Concrete Forming & Accessories	94.5	78.9	81.3	103.2	80.5	83.9	96.8	79.5	82.1	95.6	80.1	82.4	92.9	73.8	76.7	94.6	109.1	107.0
0320	Concrete Reinforcing	102.1	83.9	93.3	89.6	84.9	87.3	90.0	84.9	87.6	98.1	80.4	89.6	102.1	77.2	90.1	102.1	112.5	107.1
0330	Cast-in-Place Concrete	102.2	75.4	92.1	99.8	75.1	90.5	99.4	73.3	89.5	95.4	83.2	90.8	108.6	74.3	95.7	106.8	109.8	107.9
03	CONCRETE	94.9	79.2	87.8	99.2	79.1	90.2	98.4	78.0	89.2	99.3	81.5	91.3	97.8	75.3	87.7	97.0	109.9	102.8
04	MASONRY	83.7	75.3	78.6	90.4	73.1	79.8	90.2	73.1	79.7	85.8	78.7	81.4	87.6	71.2	77.6	85.1	107.9	99.0
05	METALS	102.9	89.4	98.7	100.8	75.6	92.9	100.8	75.1	92.8	93.4	84.6	90.7	102.9	86.8	97.8	102.9	106.8	104.1
06	WOOD, PLASTICS & COMPOSITES	92.8	79.4	85.8	110.3	81.3	95.1	104.0	80.1	91.5	91.9	79.4	85.4	92.7	74.1	83.0	90.6	107.5	99.4
07	THERMAL & MOISTURE PROTECTION	107.9	75.3	93.9	95.3	77.3	87.6	94.7	76.6	86.9	99.8	81.7	92.0	107.6	76.0	94.0	106.5	103.5	105.2
08	OPENINGS	92.9	77.4	89.1	97.3	78.9	92.8	93.8	78.2	90.0	91.7	76.9	88.1	92.9	71.3	87.6	92.9	112.6	97.7
0920	Plaster & Gypsum Board	100.4	79.3	86.5	95.4	81.7	86.4	92.3	80.4	84.5	90.9	78.7	82.9	99.8	73.8	82.8	94.2	108.2	103.3
0950, 0980	Ceilings & Acoustic Treatment	96.0	79.3	85.5	78.5	81.7	80.5	78.5	80.4	79.7	82.1	78.7	80.0	96.0	73.8	82.1	96.0	108.2	103.6
0960	Flooring	94.1	76.0	88.8	100.5	81.8	95.1	95.2	81.8	91.3	95.4	72.3	88.6	94.1	70.6	87.2	94.1	107.6	98.0
0970, 0990	Wall Finishes & Painting/Coating	92.0	67.9	77.5	84.7	79.7	81.7	84.7	79.7	81.7	90.9	83.6	86.5	92.0	71.0	79.4	92.0	117.3	107.1
09	FINISHES	92.8	77.2	84.3	91.3	80.8	85.6	89.3	80.1	84.3	90.2	78.8	84.1	92.6	72.9	81.9	91.8	109.5	101.4
COVERS	DIVS. 10 - 14, 25, 28, 41, 43, 44, 46	100.0	89.9	97.6	100.0	89.3	97.5	100.0	89.2	97.5	100.0	91.6	98.0	100.0	90.2	97.7	100.0	102.9	100.7
21, 22, 23	FIRE SUPPRESSION, PLUMBING & HVAC	100.6	76.7	90.9	100.2	77.6	91.1	96.4	76.9	88.6	100.5	78.2	91.5	100.6	71.6	88.9	100.6	103.1	101.6
26, 27, 3370	ELECTRICAL, COMMUNICATIONS & UTIL.	89.4	84.9	87.2	100.2	85.0	92.7	99.6	87.1	93.4	96.2	83.2	89.7	90.1	75.0	82.6	100.6	108.1	104.3
MF2018	WEIGHTED AVERAGE	96.8	80.4	89.7	98.2	79.7	90.2	96.5	79.4	89.1	96.0	83.2	90.5	97.5	75.9	88.2	98.3	105.8	101.5

INDIANA

DIVISION		INDIANAPOLIS 461 - 462			KOKOMO 469			LAFAYETTE 479			LAWRENCEBURG 470			MUNCIE 473			NEW ALBANY 471		
		MAT.	INST.	TOTAL	MAT.	INST.	TOTAL	MAT.	INST.	TOTAL	MAT.	INST.	TOTAL	MAT.	INST.	TOTAL	MAT.	INST.	TOTAL
015433	CONTRACTOR EQUIPMENT		87.9	87.9		92.7	92.7		79.6	79.6		98.9	98.9		91.0	91.0		88.9	88.9
0241, 31 - 34	SITE & INFRASTRUCTURE, DEMOLITION	99.5	93.7	95.5	95.4	86.5	89.3	84.3	85.1	84.8	81.6	99.5	93.8	87.1	85.5	86.0	79.5	87.3	84.9
0310	Concrete Forming & Accessories	97.4	85.0	86.8	97.6	77.3	80.3	93.8	81.9	83.7	92.7	77.6	79.8	93.1	78.5	80.7	92.1	76.2	78.6
0320	Concrete Reinforcing	101.0	85.3	93.4	92.3	85.2	88.9	89.6	85.2	87.4	88.8	78.0	83.6	99.1	83.8	91.7	90.2	80.4	85.5
0330	Cast-in-Place Concrete	99.3	85.1	94.0	101.2	81.0	93.6	99.9	78.4	91.8	93.5	72.7	85.6	104.9	74.6	93.5	96.5	72.6	87.5
03	CONCRETE	97.7	84.5	91.8	91.7	80.6	86.7	98.6	80.9	90.6	91.8	76.6	85.0	97.5	78.7	89.0	97.2	76.1	87.3
04	MASONRY	94.5	79.0	85.0	83.4	76.0	78.8	95.3	76.2	83.6	75.0	73.0	73.8	92.1	75.4	81.9	81.4	70.3	74.7
05	METALS	99.7	75.4	92.1	99.2	89.9	96.3	99.1	75.7	91.8	95.5	86.1	92.6	102.5	89.3	98.4	97.6	82.0	92.8
06	WOOD, PLASTICS & COMPOSITES	96.5	85.6	90.8	96.0	76.7	85.9	100.7	83.1	91.5	89.9	77.9	83.6	101.3	79.0	89.6	92.5	77.1	84.4
07	THERMAL & MOISTURE PROTECTION	97.6	81.3	90.6	106.9	76.4	93.8	94.7	78.9	87.9	99.7	75.9	89.5	97.7	76.7	88.7	88.3	73.3	81.9
08	OPENINGS	103.5	81.2	98.1	88.2	76.3	85.3	92.3	79.9	89.3	93.5	74.1	88.8	90.8	77.2	87.4	91.3	73.2	86.9
0920	Plaster & Gypsum Board	87.9	85.3	86.2	105.1	76.5	86.3	89.9	83.6	85.7	70.8	78.0	75.5	90.9	79.3	83.3	88.3	77.1	81.0
0950, 0980	Ceilings & Acoustic Treatment	97.6	85.3	89.9	96.6	76.5	84.0	74.7	83.6	80.2	86.5	78.0	81.2	78.5	79.3	79.0	82.1	77.1	79.0
0960	Flooring	95.9	81.8	91.8	98.2	87.9	95.2	94.0	81.8	90.4	69.7	81.8	73.2	94.8	76.0	89.3	92.6	51.5	80.6
0970, 0990	Wall Finishes & Painting/Coating	99.6	79.7	87.7	92.0	67.2	77.2	84.7	82.3	83.2	85.5	71.8	77.3	84.7	67.9	74.6	90.9	64.4	75.0
09	FINISHES	94.1	84.1	88.7	94.6	78.1	85.6	87.7	82.1	84.7	80.7	77.9	79.2	88.7	76.8	82.3	89.3	70.1	78.9
COVERS	DIVS. 10 - 14, 25, 28, 41, 43, 44, 46	100.0	92.3	98.2	100.0	89.7	97.6	100.0	89.4	97.5	100.0	88.8	97.4	100.0	88.9	97.4	100.0	88.6	97.3
21, 22, 23	FIRE SUPPRESSION, PLUMBING & HVAC	100.5	78.5	91.6	96.8	77.1	88.9	96.4	77.9	88.9	97.4	73.8	87.9	100.2	76.5	90.7	96.7	76.7	88.7
26, 27, 3370	ELECTRICAL, COMMUNICATIONS & UTIL.	102.5	87.1	94.9	93.3	77.5	85.5	99.1	80.3	89.8	94.2	73.9	84.2	92.2	75.7	84.0	94.9	74.8	84.9
MF2018	WEIGHTED AVERAGE	99.5	82.9	92.3	94.9	79.8	88.4	96.2	79.9	89.1	93.0	78.3	86.6	96.6	78.9	89.0	94.3	76.3	86.5

INDIANA / IOWA

DIVISION		SOUTH BEND 465 - 466			TERRE HAUTE 478			WASHINGTON 475			BURLINGTON 526			CARROLL 514			CEDAR RAPIDS 522 - 524		
		MAT.	INST.	TOTAL	MAT.	INST.	TOTAL	MAT.	INST.	TOTAL	MAT.	INST.	TOTAL	MAT.	INST.	TOTAL	MAT.	INST.	TOTAL
015433	CONTRACTOR EQUIPMENT		108.4	108.4		108.7	108.7		108.7	108.7		97.0	97.0		97.0	97.0		93.8	93.8
0241, 31 - 34	SITE & INFRASTRUCTURE, DEMOLITION	100.0	95.3	96.8	94.3	112.8	107.0	93.8	113.1	107.1	98.1	88.8	91.7	86.9	89.6	88.7	100.4	89.1	92.6
0310	Concrete Forming & Accessories	96.2	77.8	80.5	96.8	78.6	81.3	97.7	81.6	84.0	98.0	90.7	91.8	83.6	80.8	81.2	104.8	85.9	88.7
0320	Concrete Reinforcing	100.3	82.9	91.9	98.1	84.0	91.3	90.8	85.0	88.0	94.3	93.2	93.8	95.0	81.8	88.6	95.0	99.3	97.1
0330	Cast-in-Place Concrete	102.3	80.1	93.9	92.4	78.9	87.3	100.5	85.8	95.0	105.0	52.5	85.2	105.0	78.9	95.1	105.3	84.4	97.4
03	CONCRETE	101.5	80.9	92.2	102.3	80.0	92.3	108.0	83.9	97.2	96.3	78.4	88.2	94.9	81.1	88.7	96.4	88.4	92.8
04	MASONRY	81.6	75.4	77.8	93.6	75.4	82.5	85.9	78.3	81.2	101.6	68.9	81.7	103.1	69.4	82.6	107.5	82.7	92.4
05	METALS	106.8	102.5	105.5	94.1	86.1	91.6	88.6	87.1	88.1	86.0	97.3	89.5	86.1	93.7	88.5	88.5	102.5	92.8
06	WOOD, PLASTICS & COMPOSITES	97.3	77.1	86.7	94.5	78.4	86.1	94.8	80.9	87.5	93.7	94.8	94.3	77.6	85.4	81.7	102.0	85.6	93.4
07	THERMAL & MOISTURE PROTECTION	102.2	79.6	92.5	99.9	79.9	91.3	99.8	82.7	92.5	104.1	73.6	91.0	104.3	75.3	91.9	105.2	81.4	95.0
08	OPENINGS	92.9	76.5	88.9	92.2	76.6	88.4	89.1	80.3	87.0	94.3	92.3	93.8	98.5	81.1	94.3	99.0	87.8	96.3
0920	Plaster & Gypsum Board	88.4	76.6	80.7	90.9	77.7	82.2	90.9	80.2	83.9	99.9	95.0	96.7	95.9	85.4	89.0	105.4	85.7	92.5
0950, 0980	Ceilings & Acoustic Treatment	96.2	76.6	83.9	82.1	77.7	79.3	77.6	80.2	79.2	103.1	95.0	98.0	103.1	85.4	92.0	105.6	85.7	93.1
0960	Flooring	92.3	87.3	90.8	95.4	76.6	89.9	96.4	81.8	92.1	95.3	65.9	86.7	89.3	75.9	85.4	109.7	87.9	103.3
0970, 0990	Wall Finishes & Painting/Coating	95.9	83.4	88.4	90.9	79.1	83.8	90.9	86.5	88.3	88.7	80.6	83.8	88.7	80.6	83.8	90.6	72.3	79.6
09	FINISHES	93.1	79.8	85.9	90.2	78.0	83.6	89.7	81.9	85.5	95.9	86.2	90.7	92.1	80.3	85.7	101.4	84.7	92.4
COVERS	DIVS. 10 - 14, 25, 28, 41, 43, 44, 46	100.0	92.0	98.1	100.0	91.1	97.9	100.0	91.8	98.1	100.0	93.8	98.5	100.0	89.7	97.6	100.0	93.1	98.4
21, 22, 23	FIRE SUPPRESSION, PLUMBING & HVAC	100.5	75.9	90.6	100.5	78.1	91.5	96.7	80.0	90.0	96.9	79.8	90.0	96.9	74.1	87.7	100.7	84.4	94.1
26, 27, 3370	ELECTRICAL, COMMUNICATIONS & UTIL.	103.5	88.8	96.2	94.7	85.3	90.0	95.1	83.8	89.5	99.7	69.6	84.8	100.3	74.1	87.3	97.7	80.2	89.1
MF2018	WEIGHTED AVERAGE	99.6	83.4	92.6	96.9	82.9	90.8	95.1	84.8	90.7	95.8	81.1	89.5	95.5	79.3	88.5	98.3	86.6	93.2

For customer support on your Commercial Renovation Costs with RSMeans Data, call 800.448.8182.

803

IOWA

DIVISION		COUNCIL BLUFFS 515			CRESTON 508			DAVENPORT 527 - 528			DECORAH 521			DES MOINES 500 - 503,509			DUBUQUE 520		
		MAT.	INST.	TOTAL	MAT.	INST.	TOTAL	MAT.	INST.	TOTAL	MAT.	INST.	TOTAL	MAT.	INST.	TOTAL	MAT.	INST.	TOTAL
015433	CONTRACTOR EQUIPMENT		93.2	93.2		97.0	97.0		97.0	97.0		97.0	97.0		103.5	103.5		92.7	92.7
0241, 31 - 34	SITE & INFRASTRUCTURE, DEMOLITION	105.5	86.2	92.2	92.0	90.6	91.0	98.8	92.7	94.6	96.7	88.6	91.1	98.1	101.0	100.1	98.2	86.5	90.2
0310	Concrete Forming & Accessories	82.7	76.4	77.3	77.2	83.5	82.6	104.3	100.5	101.1	95.1	69.1	73.0	96.1	87.1	88.4	84.5	83.2	83.4
0320	Concrete Reinforcing	96.9	84.5	91.0	94.0	81.8	88.1	95.0	99.9	97.4	94.3	81.2	88.0	103.3	99.6	101.5	93.7	80.8	87.5
0330	Cast-in-Place Concrete	109.4	88.2	101.4	107.8	82.2	98.1	101.4	96.5	99.6	102.2	72.9	91.1	91.9	92.7	92.2	103.1	83.8	95.8
03	CONCRETE	98.3	82.9	91.4	95.5	83.5	90.1	94.6	99.6	96.8	94.2	73.7	85.0	95.4	91.5	93.6	93.0	83.8	88.9
04	MASONRY	107.3	78.0	89.4	104.4	76.5	87.4	103.1	95.1	98.3	123.7	66.3	88.7	89.6	87.2	88.1	108.6	70.0	85.1
05	METALS	93.1	96.0	94.0	86.4	94.3	88.9	88.5	108.4	94.7	86.1	92.5	88.1	92.0	96.4	93.4	87.1	93.8	89.2
06	WOOD, PLASTICS & COMPOSITES	76.1	75.1	75.6	69.3	85.4	77.7	102.0	100.3	101.1	90.1	68.5	78.8	90.4	85.6	87.9	78.2	84.2	81.3
07	THERMAL & MOISTURE PROTECTION	104.4	75.6	92.1	106.0	78.5	94.2	104.6	93.8	100.0	104.3	66.8	88.2	98.6	85.5	92.9	104.7	76.9	92.8
08	OPENINGS	98.1	80.0	93.7	107.9	82.8	101.8	99.0	101.9	99.7	97.2	75.0	91.8	100.6	87.8	97.5	98.1	84.2	94.7
0920	Plaster & Gypsum Board	95.9	75.0	82.2	91.6	85.4	87.5	105.4	100.7	102.3	98.8	67.9	78.5	84.6	85.4	85.1	95.9	84.3	88.3
0950, 0980	Ceilings & Acoustic Treatment	103.1	75.0	85.5	88.4	85.4	86.5	105.6	100.7	102.5	103.1	67.9	81.0	91.0	85.4	87.5	103.1	84.3	91.3
0960	Flooring	88.3	85.3	87.4	79.3	65.9	75.4	97.8	90.6	95.7	94.8	65.9	86.4	92.2	92.0	92.2	100.5	71.3	92.0
0970, 0990	Wall Finishes & Painting/Coating	85.6	57.0	68.4	82.2	80.6	81.2	88.7	93.0	91.3	88.7	80.6	83.8	93.1	82.9	87.0	89.8	79.3	83.5
09	FINISHES	93.0	75.6	83.6	83.5	80.1	81.7	98.0	97.9	97.9	95.6	69.5	81.5	90.3	87.4	88.7	96.7	80.6	88.0
COVERS	DIVS. 10 - 14, 25, 28, 41, 43, 44, 46	100.0	90.5	97.8	100.0	92.1	98.1	100.0	97.3	99.4	100.0	89.2	97.5	100.0	94.7	98.8	100.0	91.9	98.1
21, 22, 23	FIRE SUPPRESSION, PLUMBING & HVAC	100.7	79.6	92.2	96.8	76.0	88.4	100.7	95.6	98.6	96.9	70.9	86.4	100.5	85.0	94.3	100.7	76.3	90.9
26, 27, 3370	ELECTRICAL, COMMUNICATIONS & UTIL.	102.1	82.0	92.2	93.3	74.1	83.8	95.9	88.1	92.0	97.7	45.5	71.9	103.8	91.4	97.6	100.9	77.0	89.1
MF2018	WEIGHTED AVERAGE	98.8	81.9	91.5	95.2	81.1	89.1	97.3	96.6	97.0	96.7	71.0	85.6	97.3	90.0	94.2	97.3	80.6	90.1

IOWA

DIVISION		FORT DODGE 505			MASON CITY 504			OTTUMWA 525			SHENANDOAH 516			SIBLEY 512			SIOUX CITY 510 - 511		
		MAT.	INST.	TOTAL	MAT.	INST.	TOTAL	MAT.	INST.	TOTAL	MAT.	INST.	TOTAL	MAT.	INST.	TOTAL	MAT.	INST.	TOTAL
015433	CONTRACTOR EQUIPMENT		97.0	97.0		97.0	97.0		92.7	92.7		93.2	93.2		97.0	97.0		97.0	97.0
0241, 31 - 34	SITE & INFRASTRUCTURE, DEMOLITION	99.5	87.5	91.2	99.5	88.5	92.0	98.1	83.7	88.2	103.2	85.6	91.1	108.1	88.5	94.7	110.2	90.7	96.8
0310	Concrete Forming & Accessories	77.8	75.1	75.5	82.2	68.7	70.7	92.7	83.9	85.2	84.7	74.3	75.8	85.4	36.4	43.7	104.8	75.2	79.6
0320	Concrete Reinforcing	94.0	81.1	87.8	93.9	81.2	87.8	94.3	93.3	93.8	96.9	81.7	89.6	96.9	81.1	89.3	95.0	97.9	96.4
0330	Cast-in-Place Concrete	101.2	39.9	78.0	101.2	68.3	88.7	105.7	61.2	88.9	105.8	79.3	95.8	103.7	53.1	84.6	104.4	87.3	97.9
03	CONCRETE	91.1	64.9	79.4	91.4	71.9	82.6	95.7	78.4	88.0	95.9	78.3	88.0	94.9	51.9	75.6	95.8	84.3	90.7
04	MASONRY	103.4	49.8	70.7	117.4	65.1	85.5	104.3	52.9	73.0	106.8	71.4	85.2	126.4	50.0	79.8	101.1	68.4	81.2
05	METALS	86.5	92.2	88.3	86.5	92.5	88.4	86.0	98.0	89.7	92.1	93.8	92.6	86.2	91.6	87.9	88.5	101.2	92.4
06	WOOD, PLASTICS & COMPOSITES	69.8	85.4	78.0	74.4	68.5	71.3	86.8	94.6	90.9	78.2	75.8	77.0	79.2	32.9	55.0	102.0	74.6	87.7
07	THERMAL & MOISTURE PROTECTION	105.4	62.2	86.8	104.9	67.3	88.8	104.9	67.1	88.7	103.7	70.0	89.3	104.0	52.7	82.0	104.6	71.6	90.4
08	OPENINGS	102.1	73.1	95.1	94.6	75.0	89.9	98.5	89.4	96.3	90.2	73.6	86.2	95.4	44.3	82.9	99.0	80.9	94.6
0920	Plaster & Gypsum Board	91.6	85.4	87.5	91.6	67.9	76.1	96.2	95.0	95.4	95.9	75.7	82.6	95.9	31.4	53.6	105.4	74.2	84.9
0950, 0980	Ceilings & Acoustic Treatment	88.4	85.4	86.5	88.4	67.9	75.6	103.1	95.0	98.0	103.1	75.7	85.9	103.1	31.4	58.2	105.6	74.2	86.0
0960	Flooring	80.5	65.9	76.3	82.6	65.9	77.7	103.7	65.9	92.7	89.1	70.5	83.7	90.2	65.9	83.2	97.9	71.3	90.1
0970, 0990	Wall Finishes & Painting/Coating	82.2	77.4	79.3	82.2	80.6	81.2	89.8	80.6	84.3	85.6	80.6	82.6	88.7	79.6	83.2	88.7	67.8	76.1
09	FINISHES	85.3	74.9	79.7	85.8	69.2	76.8	97.7	81.5	88.9	93.1	74.0	82.8	95.4	44.5	67.8	99.5	73.1	85.2
COVERS	DIVS. 10 - 14, 25, 28, 41, 43, 44, 46	100.0	85.6	96.6	100.0	88.9	97.4	100.0	87.3	97.0	100.0	83.7	96.2	100.0	80.0	95.3	100.0	90.5	97.8
21, 22, 23	FIRE SUPPRESSION, PLUMBING & HVAC	96.8	68.0	85.2	96.8	74.0	87.6	96.9	69.6	85.9	96.9	79.9	90.0	96.9	68.1	85.3	100.7	78.8	91.9
26, 27, 3370	ELECTRICAL, COMMUNICATIONS & UTIL.	98.7	66.8	82.9	97.9	45.5	72.0	99.6	68.3	84.1	97.7	75.5	86.7	97.7	45.5	71.9	97.7	74.1	86.0
MF2018	WEIGHTED AVERAGE	95.0	71.0	84.6	95.0	71.2	84.7	96.4	75.7	87.5	96.1	78.7	88.6	96.9	60.1	81.0	98.0	80.3	90.3

DIVISION		IOWA						KANSAS											
		SPENCER 513			WATERLOO 506 - 507			BELLEVILLE 669			COLBY 677			DODGE CITY 678			EMPORIA 668		
		MAT.	INST.	TOTAL	MAT.	INST.	TOTAL	MAT.	INST.	TOTAL	MAT.	INST.	TOTAL	MAT.	INST.	TOTAL	MAT.	INST.	TOTAL
015433	CONTRACTOR EQUIPMENT		97.0	97.0		97.0	97.0		100.6	100.6		100.6	100.6		100.6	100.6		98.8	98.8
0241, 31 - 34	SITE & INFRASTRUCTURE, DEMOLITION	108.1	87.3	93.9	105.1	90.4	95.0	110.4	88.3	95.2	103.8	88.8	93.5	106.8	89.1	94.6	102.5	85.9	91.1
0310	Concrete Forming & Accessories	92.5	36.1	44.5	94.4	70.1	73.7	90.9	53.1	58.7	93.9	59.1	64.3	87.1	60.8	64.7	82.1	67.5	69.7
0320	Concrete Reinforcing	96.9	81.0	89.3	94.6	81.1	88.1	96.1	99.7	97.8	96.5	99.7	98.0	94.0	99.5	96.7	94.8	99.8	97.2
0330	Cast-in-Place Concrete	103.7	62.8	88.3	108.4	84.8	99.5	113.4	81.1	101.2	108.8	83.5	99.2	110.7	87.9	102.1	109.6	83.6	99.8
03	CONCRETE	95.4	55.1	77.3	97.0	78.2	88.6	107.6	72.4	91.6	104.3	76.0	91.6	105.5	78.2	93.3	100.3	79.8	91.1
04	MASONRY	126.4	50.0	79.8	104.1	76.4	87.2	91.2	58.0	70.9	97.2	61.9	75.6	111.1	59.3	79.8	96.9	63.3	76.4
05	METALS	86.2	91.3	87.8	88.8	94.5	90.6	93.4	97.6	94.7	90.6	98.3	93.1	92.2	98.0	94.0	93.1	98.6	94.8
06	WOOD, PLASTICS & COMPOSITES	86.7	32.9	58.6	89.3	65.8	77.0	85.8	49.3	66.7	92.8	55.3	73.2	84.5	56.2	69.7	76.8	66.4	71.4
07	THERMAL & MOISTURE PROTECTION	104.9	53.2	82.7	105.2	77.7	93.4	91.3	60.3	78.0	97.1	62.5	82.2	97.0	66.5	83.9	89.4	71.9	81.9
08	OPENINGS	106.1	44.3	91.0	95.1	74.5	90.0	93.7	60.6	85.6	97.5	63.9	89.3	97.5	64.3	89.4	91.8	72.0	87.0
0920	Plaster & Gypsum Board	96.2	31.4	53.7	99.5	65.2	77.0	85.3	48.3	61.0	97.1	54.4	69.1	90.6	55.3	67.4	82.2	65.9	71.5
0950, 0980	Ceilings & Acoustic Treatment	103.1	31.4	58.2	89.7	65.2	74.3	82.2	48.3	60.9	79.7	54.4	63.9	79.7	55.3	64.4	82.2	65.9	72.2
0960	Flooring	93.1	65.9	85.2	87.9	79.9	85.5	86.9	65.2	82.5	87.7	65.2	81.1	83.8	66.5	78.7	85.0	65.2	79.2
0970, 0990	Wall Finishes & Painting/Coating	88.7	54.5	68.2	82.2	84.6	83.6	89.0	98.1	94.4	94.9	98.1	96.8	94.9	98.1	96.8	89.0	98.1	94.4
09	FINISHES	96.2	40.6	66.1	89.0	72.4	80.0	86.5	58.6	71.4	86.6	63.1	73.9	84.6	64.3	73.7	83.7	69.6	76.1
COVERS	DIVS. 10 - 14, 25, 28, 41, 43, 44, 46	100.0	80.0	95.3	100.0	90.6	97.8	100.0	79.5	95.2	100.0	81.5	95.6	100.0	85.5	96.6	100.0	81.3	95.6
21, 22, 23	FIRE SUPPRESSION, PLUMBING & HVAC	96.9	68.1	85.3	100.6	81.0	92.7	96.7	68.0	85.1	96.7	67.7	85.0	100.5	76.3	90.7	96.7	70.0	85.9
26, 27, 3370	ELECTRICAL, COMMUNICATIONS & UTIL.	99.2	45.5	72.6	94.9	64.2	79.7	99.8	60.4	80.3	96.8	65.4	81.2	94.6	71.2	83.0	97.4	64.4	81.1
MF2018	WEIGHTED AVERAGE	98.3	59.9	81.7	96.5	78.4	88.7	96.9	69.1	84.9	96.5	71.7	85.7	98.0	74.8	88.0	95.2	74.0	86.0

KANSAS

DIVISION		FORT SCOTT 667			HAYS 676			HUTCHINSON 675			INDEPENDENCE 673			KANSAS CITY 660 - 662			LIBERAL 679		
		MAT.	INST.	TOTAL	MAT.	INST.	TOTAL	MAT.	INST.	TOTAL	MAT.	INST.	TOTAL	MAT.	INST.	TOTAL	MAT.	INST.	TOTAL
015433	CONTRACTOR EQUIPMENT		99.7	99.7		100.6	100.6		100.6	100.6		100.6	100.6		97.3	97.3		100.6	100.6
0241, 31 - 34	SITE & INFRASTRUCTURE, DEMOLITION	99.2	86.0	90.1	108.8	88.6	94.9	88.5	88.9	88.8	107.2	89.0	94.7	94.0	86.8	89.1	108.5	88.6	94.8
0310	Concrete Forming & Accessories	99.4	79.8	82.7	91.4	57.6	62.7	81.6	54.7	58.7	102.5	65.4	70.9	95.9	99.3	98.8	87.5	56.8	61.4
0320	Concrete Reinforcing	94.1	96.0	95.0	94.0	99.7	96.8	94.0	99.7	96.7	93.5	96.0	94.7	91.3	104.3	97.6	95.4	99.4	97.3
0330	Cast-in-Place Concrete	101.6	78.9	93.1	85.8	81.5	84.2	79.5	83.4	81.0	111.1	83.5	100.7	87.1	96.7	90.7	85.8	80.1	83.7
03	CONCRETE	96.0	83.1	90.2	96.7	74.6	86.8	80.2	73.9	77.4	106.7	78.2	93.9	88.6	99.7	93.6	98.5	73.7	87.4
04	MASONRY	99.4	53.8	71.6	108.5	58.1	77.8	97.2	61.9	75.6	94.6	61.9	74.6	98.5	98.1	98.3	109.0	51.4	73.8
05	METALS	93.1	96.9	94.3	90.2	98.3	92.8	90.0	97.8	92.5	89.9	97.0	92.1	101.0	105.9	102.5	90.5	97.5	92.7
06	WOOD, PLASTICS & COMPOSITES	96.1	87.4	91.5	89.8	55.3	71.8	79.2	49.5	63.7	103.9	63.5	82.8	92.1	99.8	96.1	85.1	55.3	69.5
07	THERMAL & MOISTURE PROTECTION	90.5	69.5	81.5	97.4	61.0	81.8	95.8	61.6	81.1	97.2	71.5	86.1	89.9	97.8	93.3	97.6	58.2	80.7
08	OPENINGS	91.8	82.7	89.6	97.4	63.9	89.3	97.4	60.7	88.5	95.6	67.5	88.7	93.0	98.5	94.3	97.5	63.9	89.3
0920	Plaster & Gypsum Board	87.9	87.4	87.5	94.9	54.4	68.3	89.2	48.5	62.5	106.2	62.9	77.8	81.9	100.1	93.9	91.5	54.4	67.2
0950, 0980	Ceilings & Acoustic Treatment	82.2	87.4	85.4	79.7	54.4	63.9	79.7	48.5	60.1	79.7	62.9	69.2	82.2	100.1	93.4	79.7	54.4	63.9
0960	Flooring	99.0	63.8	88.7	86.4	65.2	80.2	80.7	65.2	76.2	92.1	63.8	83.9	79.7	96.0	84.4	84.0	65.2	78.5
0970, 0990	Wall Finishes & Painting/Coating	90.5	73.8	80.5	94.9	98.1	96.8	94.9	98.1	96.8	94.9	98.1	96.8	96.5	104.4	101.2	94.9	98.1	96.8
09	FINISHES	88.7	77.4	82.6	86.4	62.1	73.3	82.1	59.7	70.0	89.3	68.1	77.8	84.2	99.3	92.3	85.5	61.6	72.6
COVERS	DIVS. 10 - 14, 25, 28, 41, 43, 44, 46	100.0	82.2	95.8	100.0	80.2	95.3	100.0	80.8	95.5	100.0	82.4	95.9	100.0	95.6	99.0	100.0	79.6	95.2
21, 22, 23	FIRE SUPPRESSION, PLUMBING & HVAC	96.7	64.2	83.6	96.7	65.3	84.0	96.7	67.7	85.0	96.7	67.1	84.8	100.4	103.0	101.4	96.7	66.0	84.3
26, 27, 3370	ELECTRICAL, COMMUNICATIONS & UTIL.	96.7	65.4	81.2	96.0	65.4	80.9	92.5	58.7	75.7	94.0	70.1	82.2	101.6	97.1	99.4	94.6	65.4	80.1
MF2018	WEIGHTED AVERAGE	95.2	73.9	86.0	96.0	70.4	84.9	91.9	69.7	82.3	96.4	73.6	86.6	96.4	99.2	97.6	96.0	69.5	84.6

DIVISION		KANSAS SALINA 674			KANSAS TOPEKA 664 - 666			KANSAS WICHITA 670 - 672			KENTUCKY ASHLAND 411 - 412			KENTUCKY BOWLING GREEN 421 - 422			KENTUCKY CAMPTON 413 - 414		
		MAT.	INST.	TOTAL	MAT.	INST.	TOTAL	MAT.	INST.	TOTAL	MAT.	INST.	TOTAL	MAT.	INST.	TOTAL	MAT.	INST.	TOTAL
015433	CONTRACTOR EQUIPMENT		100.6	100.6		104.7	104.7		106.1	106.1		95.7	95.7		88.9	88.9		95.0	95.0
0241, 31 - 34	SITE & INFRASTRUCTURE, DEMOLITION	97.5	88.7	91.5	98.6	96.9	97.4	95.0	98.7	97.6	112.9	76.0	87.6	80.0	87.5	85.1	87.9	87.8	87.8
0310	Concrete Forming & Accessories	83.6	61.6	64.9	91.8	68.0	71.6	91.0	57.3	62.3	87.2	86.3	86.5	87.8	81.4	82.3	89.9	78.4	80.1
0320	Concrete Reinforcing	93.5	99.6	96.4	92.7	101.1	96.8	87.4	101.4	94.1	91.8	91.3	91.5	89.0	79.2	84.3	89.8	88.7	89.2
0330	Cast-in-Place Concrete	96.1	84.2	91.6	96.5	88.8	93.6	89.9	78.8	85.7	87.8	90.9	89.0	87.1	68.3	80.0	97.2	65.4	85.2
03	CONCRETE	93.3	77.4	86.1	91.4	82.0	87.2	87.6	73.6	81.3	92.9	90.0	91.6	91.5	76.9	84.9	94.9	76.1	86.5
04	MASONRY	126.2	55.1	82.8	94.2	65.1	76.4	95.3	53.2	69.6	92.5	88.2	89.9	94.7	69.6	79.4	91.3	52.4	67.6
05	METALS	92.0	98.6	94.1	97.3	98.6	97.7	94.1	96.6	94.9	96.7	106.5	99.8	98.4	85.1	94.2	97.6	88.6	94.8
06	WOOD, PLASTICS & COMPOSITES	80.8	60.7	70.3	90.6	66.7	78.1	96.3	55.1	74.8	74.3	84.7	79.7	86.3	83.8	85.0	84.5	85.3	84.9
07	THERMAL & MOISTURE PROTECTION	96.5	61.8	81.6	94.1	84.1	89.8	95.3	60.9	80.5	90.5	85.1	88.2	88.3	77.9	83.9	99.9	64.8	84.8
08	OPENINGS	97.4	67.3	90.1	103.2	74.8	96.3	100.9	64.6	92.1	90.5	86.1	89.4	91.3	79.5	88.4	92.5	84.1	90.4
0920	Plaster & Gypsum Board	89.2	59.9	70.0	94.3	65.8	75.6	93.9	53.9	67.7	56.9	84.7	75.2	84.1	84.0	84.0	84.1	84.7	84.5
0950, 0980	Ceilings & Acoustic Treatment	79.7	59.9	67.3	95.5	65.8	76.9	91.9	53.9	68.1	77.8	84.7	82.1	82.1	84.0	83.3	82.1	84.7	83.7
0960	Flooring	82.1	66.5	77.6	92.3	66.5	84.7	93.0	66.9	85.4	74.8	77.5	75.6	90.4	60.3	81.6	92.8	61.4	83.6
0970, 0990	Wall Finishes & Painting/Coating	94.9	98.1	96.8	95.3	98.1	97.0	96.4	57.3	73.0	91.6	86.1	88.3	90.9	66.9	76.5	90.9	51.3	67.2
09	FINISHES	83.3	65.3	73.6	93.6	69.5	80.6	92.7	57.6	73.7	76.9	84.6	81.1	88.0	76.0	81.5	88.8	72.9	80.2
COVERS	DIVS. 10 - 14, 25, 28, 41, 43, 44, 46	100.0	85.5	96.6	100.0	85.9	96.7	100.0	85.4	96.6	100.0	83.6	96.1	100.0	89.3	97.5	100.0	87.6	97.1
21, 22, 23	FIRE SUPPRESSION, PLUMBING & HVAC	100.5	69.4	87.9	100.6	74.5	90.1	100.3	70.6	88.3	96.4	81.0	90.2	100.5	77.9	91.4	96.7	72.8	87.1
26, 27, 3370	ELECTRICAL, COMMUNICATIONS & UTIL.	94.4	73.4	84.0	102.0	72.0	87.2	99.5	73.4	86.6	92.7	85.0	88.9	95.2	76.5	85.9	92.9	84.9	89.0
MF2018	WEIGHTED AVERAGE	96.8	73.3	86.6	98.1	78.0	89.4	96.5	72.2	86.0	93.6	86.3	90.5	95.2	78.4	87.9	94.9	76.3	86.9

KENTUCKY

DIVISION		CORBIN 407 - 409			COVINGTON 410			ELIZABETHTOWN 427			FRANKFORT 406			HAZARD 417 - 418			HENDERSON 424		
		MAT.	INST.	TOTAL	MAT.	INST.	TOTAL	MAT.	INST.	TOTAL	MAT.	INST.	TOTAL	MAT.	INST.	TOTAL	MAT.	INST.	TOTAL
015433	CONTRACTOR EQUIPMENT		95.0	95.0		98.9	98.9		88.9	88.9		99.9	99.9		95.0	95.0		108.7	108.7
0241, 31 - 34	SITE & INFRASTRUCTURE, DEMOLITION	92.1	88.2	89.4	83.1	98.8	93.9	73.8	86.5	82.5	90.3	97.6	95.3	85.7	88.8	87.8	82.2	111.3	102.2
0310	Concrete Forming & Accessories	86.2	73.1	75.1	85.1	68.2	70.7	81.8	69.7	71.5	101.4	75.9	79.7	85.9	79.3	80.3	93.3	74.2	77.0
0320	Concrete Reinforcing	87.1	87.9	87.5	88.5	74.9	81.9	89.4	79.8	84.7	103.5	80.1	92.2	90.2	88.2	89.2	89.1	80.3	84.9
0330	Cast-in-Place Concrete	90.3	69.6	82.5	93.0	78.2	87.4	78.7	64.9	73.5	90.5	74.1	84.3	93.4	66.9	83.4	77.0	81.1	78.5
03	CONCRETE	84.3	75.0	80.2	93.2	73.8	84.5	83.5	70.4	77.6	88.8	76.4	83.2	91.7	76.9	85.1	88.6	78.1	83.9
04	MASONRY	90.6	57.5	70.4	105.7	69.0	83.3	78.7	59.2	66.8	88.7	71.2	78.0	90.2	54.4	68.4	98.1	75.5	84.3
05	METALS	92.5	87.7	91.0	95.5	86.5	92.7	97.5	84.0	93.3	95.2	84.2	91.7	97.6	88.6	94.8	88.3	85.3	87.4
06	WOOD, PLASTICS & COMPOSITES	74.1	74.9	74.5	81.5	65.9	73.3	80.3	71.4	75.7	100.2	75.0	87.0	80.7	85.3	83.1	89.1	73.2	80.8
07	THERMAL & MOISTURE PROTECTION	104.5	66.4	88.1	99.9	66.5	86.9	87.7	65.4	78.2	102.1	72.9	89.5	99.8	66.4	85.4	99.1	79.0	90.5
08	OPENINGS	87.1	67.9	82.4	94.3	69.5	88.2	91.3	70.3	86.2	98.7	75.7	93.1	92.8	84.0	90.6	89.5	75.6	86.1
0920	Plaster & Gypsum Board	93.4	74.1	80.7	67.4	65.8	66.3	83.0	71.2	75.3	99.3	74.1	82.7	83.0	84.7	84.1	87.2	72.3	77.4
0950, 0980	Ceilings & Acoustic Treatment	79.5	74.1	76.1	86.5	65.8	73.5	82.1	71.2	75.3	96.5	74.1	82.4	82.1	84.7	83.7	77.6	72.3	74.3
0960	Flooring	88.3	61.4	80.4	67.0	79.6	70.7	87.5	70.4	82.5	95.8	75.0	89.7	90.9	61.4	82.3	94.4	72.6	88.0
0970, 0990	Wall Finishes & Painting/Coating	85.4	59.0	69.6	85.5	67.7	74.8	90.9	64.9	75.3	96.1	87.4	90.9	90.9	51.3	67.2	90.9	82.6	85.9
09	FINISHES	85.6	69.6	76.9	79.6	69.8	74.3	86.6	68.7	76.9	95.7	77.0	85.6	88.0	73.4	80.1	87.9	74.8	80.8
COVERS	DIVS. 10 - 14, 25, 28, 41, 43, 44, 46	100.0	89.0	97.4	100.0	87.8	97.1	100.0	86.1	96.7	100.0	91.1	97.9	100.0	88.3	97.2	100.0	52.8	88.9
21, 22, 23	FIRE SUPPRESSION, PLUMBING & HVAC	97.2	70.6	86.5	97.5	72.4	87.3	96.9	74.3	87.8	100.7	77.2	91.2	96.8	73.9	87.5	96.9	75.6	88.3
26, 27, 3370	ELECTRICAL, COMMUNICATIONS & UTIL.	89.3	84.9	87.1	96.0	69.8	83.0	92.8	75.0	84.0	100.1	75.0	87.7	92.9	84.9	89.0	94.7	75.0	84.9
MF2018	WEIGHTED AVERAGE	91.8	75.0	84.6	94.9	75.0	86.3	91.8	73.4	83.8	96.8	78.6	88.9	94.3	77.1	86.9	92.7	78.7	86.7

For customer support on your Commercial Renovation Costs with RSMeans Data, call 800.448.8182.

805

		KENTUCKY																
	DIVISION	LEXINGTON			LOUISVILLE			OWENSBORO			PADUCAH			PIKEVILLE			SOMERSET		
		403 - 405			400 - 402			423			420			415 - 416			425 - 426		
		MAT.	INST.	TOTAL	MAT.	INST.	TOTAL	MAT.	INST.	TOTAL	MAT.	INST.	TOTAL	MAT.	INST.	TOTAL	MAT.	INST.	TOTAL
015433	CONTRACTOR EQUIPMENT		95.0	95.0		95.3	95.3		108.7	108.7		108.7	108.7		95.7	95.7		95.0	95.0
0241, 31 - 34	SITE & INFRASTRUCTURE, DEMOLITION	94.6	90.7	91.9	88.4	97.1	94.4	92.7	112.7	106.5	84.9	111.5	103.2	124.2	75.3	90.6	78.2	88.3	85.1
0310	Concrete Forming & Accessories	100.4	75.0	78.8	101.1	81.5	84.4	91.5	78.4	80.4	89.1	78.2	79.8	97.5	82.4	84.7	87.1	73.8	75.8
0320	Concrete Reinforcing	95.5	86.8	91.3	87.9	86.7	87.3	89.1	79.8	84.6	89.6	78.5	84.3	92.3	91.2	91.8	89.4	88.1	88.8
0330	Cast-in-Place Concrete	92.4	83.9	89.2	87.0	69.7	80.5	89.8	83.9	87.6	82.0	77.8	80.4	96.5	85.8	92.5	77.0	84.4	79.8
03	CONCRETE	87.4	80.6	84.3	85.1	78.5	82.1	100.4	81.0	91.7	93.0	78.5	86.5	106.4	86.5	97.5	78.7	80.5	79.5
04	MASONRY	89.2	71.8	78.6	87.3	70.6	77.1	90.5	80.0	84.1	93.3	75.9	82.7	89.7	79.8	83.6	84.9	60.6	70.1
05	METALS	94.9	88.7	93.0	97.1	86.1	93.6	89.9	87.3	89.1	86.8	85.8	86.5	96.6	106.3	99.6	97.5	88.3	94.7
06	WOOD, PLASTICS & COMPOSITES	92.1	73.2	82.2	97.8	84.1	90.6	86.7	77.5	81.9	84.1	78.7	81.3	85.8	84.7	85.2	81.4	74.9	78.0
07	THERMAL & MOISTURE PROTECTION	104.9	78.0	93.4	101.8	75.1	90.3	99.8	82.0	92.2	99.2	78.9	90.5	91.4	76.9	85.2	99.1	68.7	86.0
08	OPENINGS	87.4	76.6	84.8	88.7	79.1	86.3	89.5	78.4	86.8	88.8	78.1	86.2	91.0	82.9	89.1	91.9	74.2	87.6
0920	Plaster & Gypsum Board	102.8	72.3	82.8	94.0	84.0	87.4	85.8	76.7	79.8	85.2	77.9	80.4	60.6	84.7	76.4	83.0	74.1	77.1
0950, 0980	Ceilings & Acoustic Treatment	82.0	72.3	75.9	99.0	84.0	89.6	77.6	76.7	77.1	77.6	77.9	77.8	77.8	84.7	82.1	82.1	74.1	77.1
0960	Flooring	93.4	63.8	84.8	96.2	60.1	85.6	93.8	60.3	84.0	92.7	72.6	86.8	79.2	61.4	74.0	91.2	61.4	82.5
0970, 0990	Wall Finishes & Painting/Coating	85.4	80.5	82.5	93.7	66.8	77.6	90.9	86.0	87.9	90.9	69.6	78.2	91.6	87.4	89.0	90.9	64.9	75.3
09	FINISHES	89.1	72.8	80.3	95.2	76.1	84.9	88.1	75.3	81.1	87.3	76.2	81.3	79.5	79.3	79.4	87.3	70.3	78.1
COVERS	DIVS. 10 - 14, 25, 28, 41, 43, 44, 46	100.0	90.7	97.8	100.0	89.8	97.6	100.0	93.4	98.4	100.0	86.4	96.8	100.0	45.6	87.2	100.0	89.2	97.5
21, 22, 23	FIRE SUPPRESSION, PLUMBING & HVAC	101.0	76.9	91.2	100.7	78.4	91.7	100.5	77.9	91.4	96.9	74.5	87.9	96.4	78.4	89.2	96.9	70.5	86.2
26, 27, 3370	ELECTRICAL, COMMUNICATIONS & UTIL.	91.5	76.3	84.0	95.5	76.5	86.1	94.7	73.8	84.4	96.7	74.4	85.7	95.3	85.0	90.2	93.3	84.9	89.2
MF2018	WEIGHTED AVERAGE	94.2	78.8	87.5	94.9	79.6	88.3	95.3	81.8	89.4	93.0	80.0	87.4	96.1	82.1	90.1	92.1	76.6	85.4

		LOUISIANA																
	DIVISION	ALEXANDRIA			BATON ROUGE			HAMMOND			LAFAYETTE			LAKE CHARLES			MONROE		
		713 - 714			707 - 708			704			705			706			712		
		MAT.	INST.	TOTAL	MAT.	INST.	TOTAL	MAT.	INST.	TOTAL	MAT.	INST.	TOTAL	MAT.	INST.	TOTAL	MAT.	INST.	TOTAL
015433	CONTRACTOR EQUIPMENT		88.0	88.0		92.5	92.5		86.6	86.6		86.6	86.6		86.1	86.1		88.0	88.0
0241, 31 - 34	SITE & INFRASTRUCTURE, DEMOLITION	99.2	84.7	89.2	101.4	92.8	95.5	99.5	82.6	87.9	101.1	84.5	89.7	101.8	84.2	89.7	99.2	84.6	89.2
0310	Concrete Forming & Accessories	77.1	61.1	63.5	94.2	71.5	74.9	74.4	54.3	57.3	91.4	65.5	69.4	91.9	69.1	72.5	76.7	60.6	63.0
0320	Concrete Reinforcing	92.6	53.8	73.9	88.3	53.8	71.7	86.3	51.8	69.6	87.5	51.8	70.3	87.5	53.8	71.3	91.6	53.8	73.3
0330	Cast-in-Place Concrete	87.2	67.1	79.6	92.0	69.5	83.6	87.0	65.6	78.9	86.6	67.5	79.3	91.1	68.4	82.6	87.2	66.3	79.3
03	CONCRETE	82.0	62.6	73.3	89.2	68.3	79.8	84.3	58.6	72.8	85.3	64.3	75.9	87.5	66.6	78.1	81.8	62.2	73.0
04	MASONRY	106.0	64.5	80.7	86.4	62.3	71.7	90.4	62.2	73.2	90.4	65.3	75.1	89.9	66.8	75.8	100.9	63.2	77.9
05	METALS	93.5	70.3	86.2	93.7	73.6	87.4	85.6	68.0	80.1	84.9	68.3	79.7	84.9	69.2	80.0	93.4	70.3	86.2
06	WOOD, PLASTICS & COMPOSITES	86.2	60.1	72.6	96.1	75.0	85.0	75.4	52.3	63.3	95.7	65.7	80.0	94.0	69.5	81.2	85.6	60.1	72.2
07	THERMAL & MOISTURE PROTECTION	99.6	66.2	85.2	98.0	68.1	85.2	97.3	63.3	82.7	97.9	67.1	84.7	97.6	68.2	85.0	99.5	65.6	85.0
08	OPENINGS	110.5	59.5	98.1	97.9	71.9	91.5	96.2	52.8	85.7	99.6	59.8	89.9	99.6	63.2	90.8	110.5	58.0	97.7
0920	Plaster & Gypsum Board	81.3	59.6	67.1	98.4	74.6	82.8	100.4	51.6	68.4	109.1	65.3	80.4	109.1	69.3	83.0	81.0	59.6	67.0
0950, 0980	Ceilings & Acoustic Treatment	92.2	59.6	71.8	100.3	74.6	84.2	103.0	51.6	70.8	99.8	65.3	78.2	100.5	69.3	81.0	92.2	59.6	71.8
0960	Flooring	84.6	67.6	79.6	93.6	67.6	86.0	92.7	67.6	85.3	101.3	67.6	91.5	101.3	67.6	91.5	84.2	67.6	79.4
0970, 0990	Wall Finishes & Painting/Coating	92.7	60.4	73.3	95.8	60.4	74.6	94.9	57.0	72.2	94.9	56.9	72.1	94.9	60.4	74.2	92.7	62.2	74.4
09	FINISHES	84.9	61.9	72.5	96.6	70.1	82.3	95.8	56.4	74.4	98.7	65.0	80.5	98.9	68.1	82.2	84.8	61.8	72.3
COVERS	DIVS. 10 - 14, 25, 28, 41, 43, 44, 46	100.0	82.7	95.9	100.0	85.3	96.5	100.0	82.3	95.8	100.0	84.8	96.4	100.0	85.7	96.6	100.0	82.3	95.8
21, 22, 23	FIRE SUPPRESSION, PLUMBING & HVAC	101.0	63.0	85.6	100.9	61.5	85.0	97.3	59.8	82.2	101.1	61.2	85.0	101.1	64.2	86.2	101.0	61.9	85.2
26, 27, 3370	ELECTRICAL, COMMUNICATIONS & UTIL.	92.7	54.9	74.0	103.0	59.3	81.4	95.9	67.0	81.6	96.8	61.4	79.3	96.5	66.1	81.4	94.2	58.5	76.5
MF2018	WEIGHTED AVERAGE	96.1	64.7	82.5	97.0	68.3	84.6	93.2	63.4	80.3	95.1	65.9	82.5	95.3	68.4	83.7	96.0	64.6	82.4

		LOUISIANA									MAINE							
	DIVISION	NEW ORLEANS			SHREVEPORT			THIBODAUX			AUGUSTA			BANGOR			BATH		
		700 - 701			710 - 711			703			043			044			045		
		MAT.	INST.	TOTAL	MAT.	INST.	TOTAL	MAT.	INST.	TOTAL	MAT.	INST.	TOTAL	MAT.	INST.	TOTAL	MAT.	INST.	TOTAL
015433	CONTRACTOR EQUIPMENT		88.6	88.6		94.3	94.3		86.6	86.6		99.3	99.3		94.2	94.2		94.2	94.2
0241, 31 - 34	SITE & INFRASTRUCTURE, DEMOLITION	102.0	93.6	96.2	101.8	94.0	96.5	101.9	84.3	89.8	90.4	101.7	98.2	92.8	93.0	92.9	90.1	91.5	91.1
0310	Concrete Forming & Accessories	94.6	69.8	73.5	93.5	62.4	67.1	85.2	62.3	65.7	101.2	78.6	82.0	95.5	78.4	81.0	90.7	78.1	80.0
0320	Concrete Reinforcing	87.7	53.2	71.0	93.0	52.8	73.6	86.3	51.8	69.7	97.7	80.9	89.6	88.8	80.9	85.0	87.9	80.6	84.4
0330	Cast-in-Place Concrete	83.7	71.5	79.1	90.3	68.3	82.0	93.5	65.4	82.9	90.3	111.5	98.3	71.4	110.3	86.1	71.4	110.4	86.1
03	CONCRETE	88.9	67.4	79.3	88.9	63.4	77.4	89.0	62.2	77.0	96.1	90.9	93.8	87.5	90.5	88.9	87.5	90.4	88.8
04	MASONRY	92.9	63.2	74.8	88.9	63.6	73.5	112.5	61.7	81.5	93.0	93.1	93.1	108.3	92.7	98.7	114.5	90.6	99.9
05	METALS	95.9	61.8	85.2	95.5	68.7	87.1	85.6	68.2	80.2	115.0	92.3	107.9	98.7	93.8	97.2	97.0	93.2	95.8
06	WOOD, PLASTICS & COMPOSITES	94.8	72.6	83.2	97.7	62.3	79.2	82.7	63.3	72.5	98.5	76.0	86.7	94.6	75.9	84.8	87.7	75.9	81.5
07	THERMAL & MOISTURE PROTECTION	96.1	68.6	84.3	98.0	67.1	84.8	97.5	64.3	83.3	110.2	100.7	106.1	107.7	100.0	104.4	110.9	99.0	103.9
08	OPENINGS	98.7	65.0	90.5	104.5	57.6	93.1	100.5	54.8	89.4	102.9	79.1	97.1	96.6	82.0	93.1	96.6	78.1	92.1
0920	Plaster & Gypsum Board	101.0	72.1	82.0	97.8	61.6	74.0	102.1	62.9	76.3	107.5	75.0	86.2	110.3	75.0	87.2	105.2	75.0	85.4
0950, 0980	Ceilings & Acoustic Treatment	95.8	72.1	80.9	100.9	61.6	76.3	103.0	62.9	77.9	97.1	75.0	83.3	80.2	75.0	77.0	79.5	75.0	76.7
0960	Flooring	107.9	68.9	96.6	92.9	67.6	85.5	98.4	67.0	89.3	91.0	109.8	96.5	84.2	109.8	91.7	82.6	105.5	89.3
0970, 0990	Wall Finishes & Painting/Coating	99.9	59.9	75.9	94.6	62.2	75.2	95.3	57.0	72.4	95.3	89.8	92.0	89.1	98.4	94.7	89.1	84.7	86.5
09	FINISHES	100.1	69.0	83.3	95.3	63.2	77.9	97.8	62.6	78.8	96.6	84.6	90.1	88.9	85.4	87.0	87.5	83.1	85.1
COVERS	DIVS. 10 - 14, 25, 28, 41, 43, 44, 46	100.0	85.4	96.6	100.0	83.2	96.0	100.0	83.3	96.1	100.0	100.7	100.2	100.0	100.3	100.1	100.0	96.2	99.1
21, 22, 23	FIRE SUPPRESSION, PLUMBING & HVAC	101.0	61.4	85.0	100.8	62.5	85.3	97.3	59.6	82.1	101.2	75.0	90.6	101.3	74.7	90.6	97.5	74.9	88.4
26, 27, 3370	ELECTRICAL, COMMUNICATIONS & UTIL.	100.3	69.4	85.0	101.4	66.6	83.7	94.8	67.0	81.0	99.2	78.4	88.9	98.4	70.9	84.8	96.8	78.4	87.7
MF2018	WEIGHTED AVERAGE	97.7	68.3	85.0	97.7	66.8	84.4	95.5	65.0	82.3	101.6	86.0	94.9	97.5	84.5	91.9	96.2	84.6	91.2

MAINE

DIVISION		HOULTON 047 MAT.	INST.	TOTAL	KITTERY 039 MAT.	INST.	TOTAL	LEWISTON 042 MAT.	INST.	TOTAL	MACHIAS 046 MAT.	INST.	TOTAL	PORTLAND 040-041 MAT.	INST.	TOTAL	ROCKLAND 048 MAT.	INST.	TOTAL
015433	CONTRACTOR EQUIPMENT		94.2	94.2		94.2	94.2		94.2	94.2		94.2	94.2		99.4	99.4		94.2	94.2
0241, 31-34	SITE & INFRASTRUCTURE, DEMOLITION	91.9	91.5	91.7	81.0	91.5	88.2	90.1	93.0	92.1	91.3	91.5	91.5	90.6	102.0	98.4	87.7	91.5	90.3
0310	Concrete Forming & Accessories	100.0	78.1	81.3	89.1	78.3	79.9	101.8	78.4	81.9	96.4	78.1	80.8	102.8	78.5	82.2	97.8	78.1	81.1
0320	Concrete Reinforcing	88.8	80.6	84.9	87.5	80.7	84.2	109.1	80.9	95.5	88.8	80.6	84.9	108.3	80.9	95.1	88.8	80.6	84.9
0330	Cast-in-Place Concrete	71.4	109.3	85.7	71.9	110.4	86.4	72.9	110.3	87.0	71.4	110.3	86.1	86.7	111.4	96.0	72.9	110.4	87.0
03	CONCRETE	88.6	90.0	89.2	81.9	90.5	85.8	87.5	90.5	88.9	88.0	90.4	89.1	95.9	90.8	93.6	85.3	90.4	87.6
04	MASONRY	91.5	90.6	91.0	111.6	90.6	98.8	92.2	92.7	92.5	91.5	90.6	91.0	98.3	92.7	94.9	86.2	90.6	88.9
05	METALS	97.3	93.1	96.0	91.6	93.4	92.1	102.4	93.9	99.8	97.3	93.2	96.0	108.6	92.4	103.5	97.1	93.2	95.9
06	WOOD, PLASTICS & COMPOSITES	99.2	75.9	87.0	89.7	75.9	82.4	101.5	75.9	88.1	95.6	75.9	85.3	101.7	76.0	88.2	96.7	75.9	85.8
07	THERMAL & MOISTURE PROTECTION	107.8	99.0	104.0	108.1	99.0	104.2	107.5	100.0	104.3	107.7	99.0	104.0	110.7	101.0	106.6	107.4	99.0	103.8
08	OPENINGS	96.7	78.1	92.2	96.6	81.5	92.9	99.6	82.0	95.3	96.7	78.1	92.2	97.0	82.0	93.4	96.6	78.1	92.1
0920	Plaster & Gypsum Board	112.0	75.0	87.8	100.3	75.0	83.7	115.4	75.0	88.9	110.9	75.0	87.4	107.5	75.0	86.2	110.9	75.0	87.4
0950, 0980	Ceilings & Acoustic Treatment	79.5	75.0	76.7	90.6	75.0	80.9	89.7	75.0	80.5	79.5	75.0	76.7	96.1	75.0	82.9	79.5	75.0	76.7
0960	Flooring	85.3	105.5	91.2	89.0	105.5	93.8	86.8	109.8	93.5	84.7	105.5	90.8	90.7	109.8	96.3	85.0	105.5	91.2
0970, 0990	Wall Finishes & Painting/Coating	89.1	84.7	86.5	79.1	97.3	90.0	89.1	98.4	94.7	89.1	84.7	86.5	95.4	98.4	97.2	89.1	84.7	86.5
09	FINISHES	89.3	83.1	86.0	91.9	84.5	87.9	92.3	85.4	88.5	88.9	83.1	85.8	95.2	85.5	89.9	88.7	83.1	85.7
COVERS	DIVS. 10-14, 25, 28, 41, 43, 44, 46	100.0	93.1	98.4	100.0	96.2	99.1	100.0	100.3	100.1	100.0	93.1	98.4	100.0	100.6	100.1	100.0	96.2	99.1
21, 22, 23	FIRE SUPPRESSION, PLUMBING & HVAC	97.5	74.9	88.4	97.6	74.9	88.4	101.3	74.8	90.6	97.5	74.9	88.4	100.7	74.8	90.3	97.5	74.9	88.4
26, 27, 3370	ELECTRICAL, COMMUNICATIONS & UTIL.	100.1	78.4	89.4	93.6	78.4	86.1	100.1	73.9	87.1	100.1	78.4	89.4	103.2	73.8	88.7	100.0	78.4	89.3
MF2018	WEIGHTED AVERAGE	95.9	84.4	91.0	94.4	84.9	90.3	98.0	84.9	92.4	95.8	84.5	90.9	100.6	85.6	94.1	95.0	84.6	90.5

MAINE / MARYLAND

DIVISION		WATERVILLE 049 MAT.	INST.	TOTAL	ANNAPOLIS 214 MAT.	INST.	TOTAL	BALTIMORE 210-212 MAT.	INST.	TOTAL	COLLEGE PARK 207-208 MAT.	INST.	TOTAL	CUMBERLAND 215 MAT.	INST.	TOTAL	EASTON 216 MAT.	INST.	TOTAL
015433	CONTRACTOR EQUIPMENT		94.2	94.2		105.0	105.0		102.6	102.6		106.0	106.0		100.4	100.4		100.4	100.4
0241, 31-34	SITE & INFRASTRUCTURE, DEMOLITION	91.8	91.5	91.6	98.7	94.3	95.7	99.8	95.2	96.7	101.7	90.8	94.2	89.9	87.3	88.1	96.5	85.6	89.0
0310	Concrete Forming & Accessories	90.1	78.1	79.9	102.6	75.4	79.5	103.2	75.3	79.5	84.9	73.6	75.3	94.6	80.5	82.6	92.3	71.4	74.5
0320	Concrete Reinforcing	88.8	80.6	84.9	104.8	90.4	97.9	113.9	90.5	102.6	110.6	97.8	104.4	95.2	87.4	91.4	94.3	83.1	88.9
0330	Cast-in-Place Concrete	71.4	110.3	86.1	114.5	77.1	100.3	123.0	78.2	106.1	103.7	73.5	92.3	95.0	85.2	91.3	105.5	63.2	89.5
03	CONCRETE	89.1	90.4	89.6	103.1	79.9	92.7	113.1	79.9	98.2	103.1	79.4	92.5	90.3	84.6	87.8	98.3	72.1	86.5
04	MASONRY	101.8	90.6	95.0	103.3	73.4	85.1	106.0	75.2	87.2	110.9	71.6	86.9	103.3	88.9	94.5	118.6	56.5	80.7
05	METALS	97.2	93.2	96.0	104.8	102.3	104.0	103.2	97.6	101.4	90.7	110.5	96.9	100.2	103.7	101.3	100.4	100.9	100.6
06	WOOD, PLASTICS & COMPOSITES	87.0	75.9	81.2	99.2	74.7	86.4	105.6	74.5	89.4	76.9	73.1	74.9	89.3	78.2	83.5	86.6	78.1	82.1
07	THERMAL & MOISTURE PROTECTION	107.7	99.0	104.0	99.9	80.7	91.7	99.6	81.4	91.7	101.9	79.2	92.2	97.6	81.2	90.6	97.8	71.7	86.6
08	OPENINGS	96.7	78.1	92.2	102.1	81.0	97.0	100.8	81.0	96.0	92.7	82.1	90.1	97.7	82.5	94.0	96.1	81.0	92.4
0920	Plaster & Gypsum Board	105.2	75.0	85.4	102.9	74.2	84.1	104.9	73.9	84.6	109.2	72.6	85.2	108.0	78.1	88.4	108.0	77.9	88.3
0950, 0980	Ceilings & Acoustic Treatment	79.5	75.0	76.7	94.3	74.2	81.7	106.9	73.9	86.2	113.8	72.6	88.0	107.3	78.1	89.0	107.3	77.9	88.9
0960	Flooring	82.3	105.5	89.1	94.7	75.9	89.2	96.5	75.9	90.5	88.1	74.8	84.2	89.4	94.3	90.8	88.5	72.2	83.7
0970, 0990	Wall Finishes & Painting/Coating	89.1	84.7	86.5	96.0	69.4	80.1	97.7	69.4	80.7	100.3	69.4	81.8	91.8	80.6	85.1	91.8	69.4	78.4
09	FINISHES	87.6	83.1	85.2	95.2	74.3	83.9	99.8	74.2	85.9	99.0	72.6	84.7	98.5	83.0	90.1	98.7	71.5	83.9
COVERS	DIVS. 10-14, 25, 28, 41, 43, 44, 46	100.0	93.0	98.3	100.0	89.5	97.5	100.0	89.7	97.6	100.0	82.3	95.8	100.0	91.4	98.0	100.0	83.8	96.2
21, 22, 23	FIRE SUPPRESSION, PLUMBING & HVAC	97.5	74.9	88.4	100.8	82.9	93.6	100.8	81.6	93.0	97.1	83.4	91.6	96.8	70.7	86.3	96.8	69.4	85.8
26, 27, 3370	ELECTRICAL, COMMUNICATIONS & UTIL.	100.1	78.4	89.4	101.1	85.6	93.4	96.7	86.9	91.8	96.4	103.9	100.1	97.2	79.9	88.6	96.7	60.0	78.6
MF2018	WEIGHTED AVERAGE	96.2	84.5	91.2	101.4	83.4	93.6	102.3	83.1	94.0	97.5	85.8	92.5	97.1	83.1	91.0	98.9	72.6	87.5

MARYLAND / MASSACHUSETTS

DIVISION		ELKTON 219 MAT.	INST.	TOTAL	HAGERSTOWN 217 MAT.	INST.	TOTAL	SALISBURY 218 MAT.	INST.	TOTAL	SILVER SPRING 209 MAT.	INST.	TOTAL	WALDORF 206 MAT.	INST.	TOTAL	BOSTON 020-022, 024 MAT.	INST.	TOTAL
015433	CONTRACTOR EQUIPMENT		100.4	100.4		100.4	100.4		100.4	100.4		97.9	97.9		97.9	97.9		105.8	105.8
0241, 31-34	SITE & INFRASTRUCTURE, DEMOLITION	84.4	86.1	85.6	88.9	87.5	87.9	96.4	85.3	88.8	88.8	82.9	84.8	95.0	82.8	86.6	92.0	105.4	101.2
0310	Concrete Forming & Accessories	99.5	89.0	90.6	93.6	78.0	80.3	109.4	48.7	57.8	93.3	72.9	75.9	101.4	72.8	77.1	104.7	136.7	131.9
0320	Concrete Reinforcing	94.3	111.8	102.8	95.2	87.4	91.4	94.3	62.1	78.8	109.4	97.7	103.7	110.1	97.7	104.1	117.1	149.5	132.8
0330	Cast-in-Place Concrete	85.4	69.5	79.4	90.5	85.3	88.5	105.5	61.1	88.7	106.2	73.9	94.0	118.9	73.7	101.8	98.2	142.4	114.9
03	CONCRETE	83.2	87.3	85.0	86.8	83.5	85.3	99.4	57.5	80.6	100.9	79.0	91.1	111.1	78.9	96.6	103.4	140.4	120.0
04	MASONRY	103.2	64.3	79.5	109.7	88.9	97.0	117.7	53.1	78.3	110.2	71.9	86.9	94.8	71.9	80.8	107.9	144.2	130.1
05	METALS	100.5	112.7	104.3	100.3	103.9	101.4	100.5	91.6	97.7	95.2	106.5	98.7	95.2	106.0	98.6	102.4	134.6	112.4
06	WOOD, PLASTICS & COMPOSITES	95.0	97.9	96.5	88.2	74.4	81.0	107.9	49.6	77.4	84.3	72.4	78.1	92.4	72.4	81.9	104.9	135.7	121.0
07	THERMAL & MOISTURE PROTECTION	97.3	77.5	88.8	97.8	84.5	92.1	98.2	67.0	84.8	104.3	83.9	95.5	104.9	83.9	95.9	108.4	138.0	121.1
08	OPENINGS	96.1	99.5	96.9	96.1	79.8	92.1	96.3	59.6	87.4	84.8	81.7	84.1	85.4	81.7	84.5	100.3	143.3	110.8
0920	Plaster & Gypsum Board	110.0	98.3	102.3	108.0	74.2	85.8	118.0	48.7	72.5	116.0	72.6	87.5	118.5	72.6	88.4	105.5	136.3	125.7
0950, 0980	Ceilings & Acoustic Treatment	107.3	98.3	101.6	110.1	74.2	87.6	107.3	48.7	70.6	121.4	72.6	90.9	121.4	72.6	90.9	94.0	136.3	120.5
0960	Flooring	90.9	72.2	85.4	88.9	94.3	90.5	94.5	72.2	88.0	94.0	74.8	88.4	97.6	74.8	91.0	96.8	162.4	115.9
0970, 0990	Wall Finishes & Painting/Coating	91.8	69.4	78.4	91.8	69.4	78.4	91.8	69.4	78.4	108.0	69.4	84.9	108.0	69.4	84.9	101.6	154.9	133.5
09	FINISHES	98.8	85.0	91.4	98.8	79.5	88.4	101.8	53.9	75.9	99.6	72.0	84.7	101.3	72.1	85.5	98.5	143.8	123.0
COVERS	DIVS. 10-14, 25, 28, 41, 43, 44, 46	100.0	53.7	89.1	100.0	91.0	97.9	100.0	79.6	95.2	100.0	80.8	95.5	100.0	78.8	95.0	100.0	117.4	104.1
21, 22, 23	FIRE SUPPRESSION, PLUMBING & HVAC	96.8	75.1	88.0	100.6	88.3	95.6	96.8	67.5	85.0	97.1	83.8	91.7	97.1	83.8	91.7	96.7	127.1	109.0
26, 27, 3370	ELECTRICAL, COMMUNICATIONS & UTIL.	98.2	83.7	91.0	97.0	79.9	88.5	95.7	57.9	77.0	93.8	103.9	98.8	91.8	103.9	97.8	98.3	129.3	113.7
MF2018	WEIGHTED AVERAGE	96.1	83.0	90.5	97.7	86.2	92.8	99.3	65.0	84.5	96.7	84.9	91.6	97.4	84.8	92.0	100.2	132.9	114.3

MASSACHUSETTS

DIVISION		BROCKTON 023			BUZZARDS BAY 025			FALL RIVER 027			FITCHBURG 014			FRAMINGHAM 017			GREENFIELD 013		
		MAT.	INST.	TOTAL	MAT.	INST.	TOTAL	MAT.	INST.	TOTAL	MAT.	INST.	TOTAL	MAT.	INST.	TOTAL	MAT.	INST.	TOTAL
015433	CONTRACTOR EQUIPMENT		97.1	97.1		97.1	97.1		98.0	98.0		94.2	94.2		95.9	95.9		94.2	94.2
0241, 31 - 34	SITE & INFRASTRUCTURE, DEMOLITION	91.0	96.4	94.7	80.6	96.0	91.2	90.2	96.5	94.5	82.8	95.9	91.8	79.6	95.4	90.4	86.5	94.5	92.0
0310	Concrete Forming & Accessories	101.9	122.8	119.7	99.4	122.0	118.7	101.9	122.5	119.4	94.5	115.2	112.1	102.5	122.7	119.7	92.6	119.0	115.1
0320	Concrete Reinforcing	105.1	144.2	124.0	84.3	121.3	102.1	105.1	121.3	112.9	85.2	135.8	109.6	85.2	144.0	113.5	88.8	120.8	104.3
0330	Cast-in-Place Concrete	84.5	133.0	102.8	70.2	132.7	93.8	81.7	133.3	101.2	77.1	132.7	98.1	77.1	132.9	98.2	79.2	117.8	93.8
03	CONCRETE	91.6	129.8	108.7	77.7	125.4	99.1	91.9	125.8	105.5	78.6	124.6	99.2	81.3	129.7	103.0	82.0	118.5	98.4
04	MASONRY	104.3	136.3	123.8	96.8	136.3	120.9	104.9	136.2	124.0	98.7	128.1	116.7	105.3	132.1	121.7	103.3	118.6	112.6
05	METALS	100.9	129.6	109.9	95.5	119.2	102.9	100.9	119.8	106.8	100.3	121.8	107.0	100.3	128.7	109.2	102.9	113.1	106.1
06	WOOD, PLASTICS & COMPOSITES	101.3	122.2	112.2	97.4	122.2	110.4	101.3	122.4	112.4	95.2	112.6	104.3	102.9	122.0	112.9	92.8	122.6	108.4
07	THERMAL & MOISTURE PROTECTION	105.5	127.4	114.9	104.4	124.1	112.8	105.4	124.0	113.3	102.4	120.0	109.9	102.6	125.1	112.2	102.5	109.9	105.7
08	OPENINGS	97.8	130.5	105.8	93.9	119.4	100.1	97.8	120.5	103.4	99.6	123.0	105.3	90.2	130.3	99.9	99.8	121.0	105.0
0920	Plaster & Gypsum Board	92.9	122.6	112.4	87.2	122.6	110.5	92.9	122.6	112.4	108.7	112.8	111.4	112.4	122.6	119.1	109.7	123.0	118.4
0950, 0980	Ceilings & Acoustic Treatment	98.6	122.6	113.7	80.4	122.6	106.9	98.6	122.6	113.7	86.4	112.8	103.0	86.4	122.6	109.1	94.7	123.0	112.5
0960	Flooring	89.9	158.8	110.0	87.4	158.8	108.2	88.7	158.8	109.4	89.1	158.8	109.4	90.7	158.8	110.6	88.3	136.7	102.5
0970, 0990	Wall Finishes & Painting/Coating	87.0	138.9	118.1	87.0	138.9	118.1	87.0	138.9	118.1	84.9	138.9	117.2	85.8	138.9	117.6	84.9	114.9	102.9
09	FINISHES	92.8	131.5	113.7	86.3	131.5	110.7	92.5	131.6	113.7	90.1	125.8	109.5	90.9	131.3	112.8	92.4	123.0	108.9
COVERS	DIVS. 10 - 14, 25, 28, 41, 43, 44, 46	100.0	112.8	103.0	100.0	112.8	103.0	100.0	113.3	103.1	100.0	106.3	101.5	100.0	112.4	102.9	100.0	105.4	101.3
21, 22, 23	FIRE SUPPRESSION, PLUMBING & HVAC	101.7	104.3	102.8	97.2	99.5	98.1	101.7	104.0	102.6	97.6	100.5	98.8	97.6	116.3	105.2	97.6	98.2	97.8
26, 27, 3370	ELECTRICAL, COMMUNICATIONS & UTIL.	100.8	99.8	100.3	98.0	96.4	97.2	100.6	101.2	100.9	99.8	100.2	100.0	97.0	119.8	108.3	99.8	101.5	100.7
MF2018	WEIGHTED AVERAGE	98.9	117.8	107.1	93.3	114.1	102.3	98.5	116.0	106.1	95.4	113.2	103.1	94.9	122.4	106.8	96.7	109.5	102.3

MASSACHUSETTS

DIVISION		HYANNIS 026			LAWRENCE 019			LOWELL 018			NEW BEDFORD 027			PITTSFIELD 012			SPRINGFIELD 010 - 011		
		MAT.	INST.	TOTAL	MAT.	INST.	TOTAL	MAT.	INST.	TOTAL	MAT.	INST.	TOTAL	MAT.	INST.	TOTAL	MAT.	INST.	TOTAL
015433	CONTRACTOR EQUIPMENT		97.1	97.1		96.6	96.6		94.2	94.2		98.0	98.0		94.2	94.2		94.2	94.2
0241, 31 - 34	SITE & INFRASTRUCTURE, DEMOLITION	87.4	96.2	93.4	92.0	96.1	94.8	91.1	96.3	94.7	88.6	96.5	94.0	92.1	94.6	93.8	91.5	95.0	93.9
0310	Concrete Forming & Accessories	92.7	122.5	118.0	103.8	123.2	120.3	100.4	125.2	121.5	101.9	122.6	119.5	100.3	106.5	105.6	100.6	120.3	117.4
0320	Concrete Reinforcing	84.3	121.3	102.2	105.5	141.8	123.0	106.3	141.6	123.3	105.1	121.3	112.9	88.2	117.5	102.3	106.3	120.9	113.3
0330	Cast-in-Place Concrete	77.1	132.9	98.1	89.2	133.1	105.8	81.1	136.1	101.8	72.0	133.3	95.1	88.5	116.0	98.9	84.4	119.7	97.7
03	CONCRETE	83.3	125.6	102.3	94.9	129.6	110.5	87.0	131.3	106.9	85.8	125.6	103.8	88.0	111.6	98.6	88.6	119.7	102.6
04	MASONRY	103.4	136.3	123.4	110.9	138.8	127.9	97.8	139.9	123.5	103.3	134.8	122.5	98.4	113.3	107.5	98.1	121.9	112.6
05	METALS	97.1	119.7	104.1	103.3	128.9	111.3	103.3	125.6	110.2	100.9	119.8	106.8	103.0	111.3	105.6	106.3	113.2	108.4
06	WOOD, PLASTICS & COMPOSITES	89.7	122.2	106.7	103.7	122.2	113.4	102.6	122.2	112.8	101.3	122.4	112.4	102.6	107.1	104.9	102.6	122.6	113.3
07	THERMAL & MOISTURE PROTECTION	104.9	126.5	114.2	103.2	128.1	113.9	102.9	129.3	114.2	105.3	123.6	113.1	103.0	106.3	104.4	102.9	111.4	106.6
08	OPENINGS	94.4	120.6	100.8	93.7	129.7	102.5	100.0	129.7	107.9	97.8	124.9	104.4	100.9	111.6	103.5	100.9	121.0	105.8
0920	Plaster & Gypsum Board	83.2	122.6	109.1	115.1	122.6	120.0	115.1	122.6	120.0	92.9	122.6	112.4	115.1	107.1	109.8	115.1	123.0	120.3
0950, 0980	Ceilings & Acoustic Treatment	89.7	122.6	110.3	96.2	122.6	112.8	96.2	122.6	112.8	98.6	122.6	113.7	96.2	107.1	103.0	96.2	123.0	113.0
0960	Flooring	84.8	158.8	106.4	91.3	158.8	111.0	91.3	158.8	111.0	88.7	158.8	109.1	91.6	131.5	103.3	90.7	136.7	104.1
0970, 0990	Wall Finishes & Painting/Coating	87.0	138.9	118.1	85.0	138.9	117.3	84.9	138.9	117.2	87.0	138.9	118.1	84.9	114.9	102.9	85.7	114.9	103.2
09	FINISHES	87.6	131.5	111.3	94.5	131.4	114.5	94.5	132.8	115.2	92.4	131.6	113.6	94.6	112.2	104.1	94.4	123.8	110.3
COVERS	DIVS. 10 - 14, 25, 28, 41, 43, 44, 46	100.0	112.8	103.0	100.0	112.9	103.0	100.0	114.6	103.4	100.0	113.3	103.1	100.0	102.9	100.7	100.0	106.5	101.5
21, 22, 23	FIRE SUPPRESSION, PLUMBING & HVAC	101.7	108.6	104.5	101.0	120.3	108.8	101.0	126.6	111.3	101.7	104.0	102.6	101.0	95.2	98.7	101.0	99.9	100.6
26, 27, 3370	ELECTRICAL, COMMUNICATIONS & UTIL.	98.8	101.2	100.0	99.0	125.6	112.2	99.4	124.4	111.8	101.6	103.2	102.4	99.4	101.5	100.5	99.5	102.9	101.2
MF2018	WEIGHTED AVERAGE	96.0	116.9	105.1	99.4	124.9	110.4	98.4	126.4	110.5	98.1	116.3	106.0	98.6	105.1	101.4	99.1	110.8	104.2

DIVISION		MASSACHUSETTS WORCESTER 015 - 016			MICHIGAN ANN ARBOR 481			BATTLE CREEK 490			BAY CITY 487			DEARBORN 481			DETROIT 482		
		MAT.	INST.	TOTAL	MAT.	INST.	TOTAL	MAT.	INST.	TOTAL	MAT.	INST.	TOTAL	MAT.	INST.	TOTAL	MAT.	INST.	TOTAL
015433	CONTRACTOR EQUIPMENT		94.2	94.2		106.7	106.7		94.4	94.4		106.7	106.7		106.7	106.7		97.5	97.5
0241, 31 - 34	SITE & INFRASTRUCTURE, DEMOLITION	91.4	95.9	94.5	82.0	90.0	87.5	93.9	79.0	83.7	72.7	88.3	83.4	81.7	90.2	87.5	100.3	102.4	101.8
0310	Concrete Forming & Accessories	101.0	124.0	120.5	98.0	105.4	104.3	100.4	79.4	82.6	98.1	79.2	82.0	97.9	106.1	104.9	98.6	104.1	103.3
0320	Concrete Reinforcing	106.3	148.7	126.8	99.0	104.8	101.8	90.4	81.5	86.1	99.0	103.9	101.3	99.0	106.1	102.4	99.6	102.4	99.6
0330	Cast-in-Place Concrete	83.9	132.8	102.4	85.2	99.1	90.5	83.2	93.3	87.0	81.6	85.2	83.0	83.4	100.0	89.7	101.5	99.3	100.7
03	CONCRETE	88.4	130.9	107.4	88.0	104.0	95.2	84.4	84.4	84.4	86.3	87.1	86.7	87.1	104.9	95.1	99.6	101.5	100.5
04	MASONRY	97.7	134.7	120.3	96.0	100.1	98.5	93.9	77.4	83.8	95.6	79.8	85.9	95.9	123.5	112.7	101.0	98.3	99.4
05	METALS	106.3	127.9	113.1	99.3	118.8	105.4	105.5	83.2	98.5	99.9	114.3	104.4	99.4	120.7	106.1	100.8	93.0	98.4
06	WOOD, PLASTICS & COMPOSITES	103.1	124.1	114.1	90.9	107.1	99.3	93.0	78.6	85.5	90.9	79.1	84.7	90.9	107.1	99.3	93.7	106.0	100.1
07	THERMAL & MOISTURE PROTECTION	102.9	123.0	111.6	104.6	100.1	102.6	96.3	78.8	88.8	102.1	82.1	93.5	103.1	107.8	105.1	102.2	103.7	102.9
08	OPENINGS	100.9	132.7	108.6	94.1	102.1	96.1	86.7	75.5	84.0	94.1	83.6	91.6	94.1	102.1	96.1	97.0	100.9	97.9
0920	Plaster & Gypsum Board	115.1	124.6	121.3	93.9	107.1	102.6	87.6	75.0	79.3	93.9	78.4	83.7	93.9	107.1	102.6	91.5	106.1	101.1
0950, 0980	Ceilings & Acoustic Treatment	96.2	124.6	114.0	80.2	107.1	97.1	87.9	75.0	79.8	80.9	78.4	79.3	80.2	107.1	97.1	84.2	106.1	97.9
0960	Flooring	91.3	160.0	111.4	92.8	110.9	98.0	92.1	67.2	84.8	92.8	85.5	90.7	92.0	104.9	95.8	94.9	104.8	97.8
0970, 0990	Wall Finishes & Painting/Coating	84.9	138.9	117.2	83.4	100.4	93.6	86.5	72.6	78.2	83.4	82.4	82.8	83.4	98.6	92.5	94.8	97.3	96.3
09	FINISHES	94.5	132.8	115.2	90.5	106.0	98.9	88.7	76.0	81.8	90.2	80.0	84.7	90.3	105.2	98.3	94.0	103.9	99.4
COVERS	DIVS. 10 - 14, 25, 28, 41, 43, 44, 46	100.0	107.6	101.8	100.0	101.8	100.4	100.0	90.2	97.7	100.0	95.1	98.8	100.0	102.2	100.5	100.0	101.0	100.2
21, 22, 23	FIRE SUPPRESSION, PLUMBING & HVAC	101.0	105.1	102.7	100.7	94.1	98.0	100.6	79.9	92.3	100.7	80.2	92.4	100.7	103.1	101.6	100.5	101.5	100.9
26, 27, 3370	ELECTRICAL, COMMUNICATIONS & UTIL.	99.5	106.4	102.9	97.2	105.3	101.2	95.6	74.8	85.3	96.3	79.4	88.0	97.2	100.1	98.6	99.8	101.6	100.7
MF2018	WEIGHTED AVERAGE	99.1	118.6	107.5	96.3	102.0	98.7	95.7	79.4	88.6	95.7	85.3	91.2	96.1	106.0	100.4	99.5	100.9	100.1

City Cost Indexes - V2

MICHIGAN

DIVISION		FLINT 484 - 485			GAYLORD 497			GRAND RAPIDS 493,495			IRON MOUNTAIN 498 - 499			JACKSON 492			KALAMAZOO 491		
		MAT.	INST.	TOTAL	MAT.	INST.	TOTAL	MAT.	INST.	TOTAL	MAT.	INST.	TOTAL	MAT.	INST.	TOTAL	MAT.	INST.	TOTAL
015433	CONTRACTOR EQUIPMENT		106.7	106.7		101.9	101.9		98.3	98.3		89.3	89.3		101.9	101.9		94.4	94.4
0241, 31 - 34	SITE & INFRASTRUCTURE, DEMOLITION	70.3	88.9	83.1	87.6	76.8	80.2	93.3	87.7	89.5	96.4	85.6	89.0	111.0	78.4	88.6	94.2	79.0	83.8
0310	Concrete Forming & Accessories	101.5	84.9	87.4	98.6	71.4	75.4	97.0	78.1	80.9	92.0	75.8	78.3	95.2	79.4	81.8	100.4	79.8	82.8
0320	Concrete Reinforcing	99.0	104.2	101.5	84.3	88.6	86.4	93.6	81.5	87.8	84.3	80.9	82.7	81.9	103.9	92.5	90.4	80.0	85.4
0330	Cast-in-Place Concrete	85.8	88.7	86.9	83.0	75.3	80.1	87.5	91.2	88.9	98.3	64.9	85.7	82.9	90.0	85.6	84.9	93.4	88.1
03	CONCRETE	88.5	90.9	89.6	81.2	77.5	79.5	93.3	83.1	88.7	89.5	73.7	82.4	76.6	88.7	82.0	87.4	84.4	86.1
04	MASONRY	96.1	89.0	91.7	104.3	69.7	83.2	92.1	76.8	82.8	90.8	76.4	82.0	84.9	83.3	83.9	92.6	81.2	85.6
05	METALS	99.4	114.9	104.2	107.2	109.2	107.9	102.5	82.5	96.2	106.6	89.5	101.3	107.5	112.5	109.0	105.5	83.1	98.5
06	WOOD, PLASTICS & COMPOSITES	94.9	84.1	89.2	86.6	70.9	78.4	93.2	76.7	84.6	84.7	76.2	80.3	85.3	77.1	81.0	93.0	78.6	85.5
07	THERMAL & MOISTURE PROTECTION	102.4	86.2	95.4	95.0	71.4	84.9	98.0	73.4	87.4	98.4	71.4	86.8	94.3	84.6	90.1	96.3	80.1	89.4
08	OPENINGS	94.1	85.5	92.0	86.0	75.8	83.5	101.8	76.2	95.6	92.5	67.3	86.3	85.3	83.7	84.9	86.7	76.7	84.3
0920	Plaster & Gypsum Board	95.6	83.5	87.7	87.1	69.4	75.5	95.3	73.4	80.9	52.9	76.3	68.3	85.7	75.7	79.1	87.6	75.0	79.3
0950, 0980	Ceilings & Acoustic Treatment	80.2	83.5	82.3	86.0	69.4	75.6	98.2	73.4	82.7	85.8	76.3	79.9	86.0	75.7	79.5	87.9	75.0	79.8
0960	Flooring	92.8	88.8	91.6	85.5	81.7	84.4	94.8	78.6	90.0	102.3	86.8	97.8	84.0	75.1	81.4	92.1	73.5	86.6
0970, 0990	Wall Finishes & Painting/Coating	83.4	80.7	81.8	82.6	77.7	79.7	97.1	77.7	85.5	99.1	66.3	79.4	82.6	89.3	86.6	86.5	78.7	81.8
09	FINISHES	89.8	84.6	87.0	87.8	73.2	79.9	95.4	77.6	85.8	88.8	76.8	82.3	88.9	78.6	83.4	88.7	78.0	82.9
COVERS	DIVS. 10 - 14, 25, 28, 41, 43, 44, 46	100.0	97.2	99.3	100.0	94.4	98.7	100.0	100.9	100.2	100.0	93.0	98.4	100.0	97.7	99.5	100.0	101.8	100.4
21, 22, 23	FIRE SUPPRESSION, PLUMBING & HVAC	100.7	85.9	94.7	97.0	75.3	88.3	100.6	81.9	93.0	96.9	80.0	90.1	97.0	81.7	90.9	100.6	80.5	92.5
26, 27, 3370	ELECTRICAL, COMMUNICATIONS & UTIL.	97.2	92.8	95.0	93.8	73.5	83.7	100.7	86.4	93.6	99.4	77.7	88.7	97.2	99.2	98.2	95.4	77.2	86.4
MF2018	WEIGHTED AVERAGE	95.9	90.9	93.8	94.6	78.1	87.5	98.9	82.2	91.6	96.5	78.9	88.9	94.1	88.0	91.4	96.0	80.9	89.5

MICHIGAN / MINNESOTA

DIVISION		LANSING 488 - 489			MUSKEGON 494			ROYAL OAK 480,483			SAGINAW 486			TRAVERSE CITY 496			BEMIDJI 566		
		MAT.	INST.	TOTAL	MAT.	INST.	TOTAL	MAT.	INST.	TOTAL	MAT.	INST.	TOTAL	MAT.	INST.	TOTAL	MAT.	INST.	TOTAL
015433	CONTRACTOR EQUIPMENT		110.1	110.1		94.4	94.4		87.2	87.2		106.7	106.7		89.3	89.3		95.8	95.8
0241, 31 - 34	SITE & INFRASTRUCTURE, DEMOLITION	92.8	97.2	95.8	92.0	78.9	83.0	86.2	88.5	87.8	73.7	88.3	83.7	81.9	84.6	83.8	95.2	92.3	93.2
0310	Concrete Forming & Accessories	95.5	82.2	84.2	101.0	78.5	81.9	93.7	104.1	102.6	98.0	82.3	84.6	92.0	69.8	73.1	87.3	84.8	85.1
0320	Concrete Reinforcing	100.8	103.9	102.3	91.1	81.6	86.5	89.7	99.8	94.5	99.0	103.9	101.3	85.6	76.5	81.2	97.1	100.7	98.8
0330	Cast-in-Place Concrete	94.0	88.4	91.9	82.9	91.6	86.2	74.6	93.5	81.7	84.2	85.2	84.6	76.7	77.0	76.8	99.7	95.0	97.9
03	CONCRETE	94.2	89.4	92.0	82.9	83.5	83.2	75.7	99.0	86.1	87.5	88.5	88.0	73.8	74.4	74.0	91.3	92.5	91.9
04	MASONRY	89.3	85.2	86.8	91.4	78.0	83.2	89.6	95.4	93.2	97.6	79.8	86.7	89.1	71.7	78.5	100.5	103.0	102.0
05	METALS	98.3	111.4	102.4	103.1	83.5	97.0	102.8	90.0	98.8	99.4	114.1	104.0	106.5	87.3	100.5	89.2	118.8	98.4
06	WOOD, PLASTICS & COMPOSITES	89.7	80.4	84.8	90.9	77.4	83.8	86.4	107.1	97.2	88.0	83.5	85.6	84.7	69.5	76.7	70.2	80.1	75.4
07	THERMAL & MOISTURE PROTECTION	101.0	85.2	94.2	95.3	72.9	85.7	100.6	96.8	99.0	103.2	82.6	94.3	97.4	71.1	86.1	106.3	90.0	99.3
08	OPENINGS	102.5	83.4	97.9	86.0	76.7	83.8	93.9	99.8	95.4	92.3	86.0	90.8	92.5	63.0	85.3	99.4	103.7	100.4
0920	Plaster & Gypsum Board	86.8	79.8	82.2	69.5	73.8	72.3	91.0	107.1	101.6	93.9	82.9	86.7	52.9	69.4	63.7	98.3	80.1	86.3
0950, 0980	Ceilings & Acoustic Treatment	84.2	79.8	81.5	87.9	73.8	79.0	79.6	107.1	96.9	80.2	82.9	81.9	85.8	69.4	75.5	131.6	80.1	99.3
0960	Flooring	97.2	93.6	96.2	90.7	84.1	88.8	89.8	101.8	93.3	92.8	85.5	90.7	102.3	81.7	96.3	92.2	104.1	95.7
0970, 0990	Wall Finishes & Painting/Coating	96.8	77.7	85.4	84.7	79.0	81.3	84.9	89.3	87.6	83.4	82.4	82.8	99.1	37.6	62.2	85.6	96.5	92.1
09	FINISHES	91.9	83.3	87.2	85.2	79.2	82.0	89.4	102.5	96.5	90.2	82.6	86.1	87.7	68.2	77.1	101.1	89.0	94.5
COVERS	DIVS. 10 - 14, 25, 28, 41, 43, 44, 46	100.0	97.0	99.3	100.0	101.5	100.3	100.0	95.3	98.9	100.0	95.6	99.0	100.0	91.6	98.0	100.0	97.5	99.4
21, 22, 23	FIRE SUPPRESSION, PLUMBING & HVAC	100.5	85.9	94.6	100.4	81.5	92.8	97.1	96.6	96.9	100.7	79.7	92.2	96.9	75.2	88.1	97.1	83.6	91.7
26, 27, 3370	ELECTRICAL, COMMUNICATIONS & UTIL.	101.3	94.5	97.9	95.8	70.2	83.2	99.0	99.3	99.2	95.6	85.2	90.5	95.3	73.4	84.5	103.3	96.9	100.2
MF2018	WEIGHTED AVERAGE	98.1	90.5	94.8	94.5	79.7	88.1	94.1	96.9	95.3	95.6	86.7	91.8	93.5	75.2	85.6	96.8	94.6	95.8

MINNESOTA

DIVISION		BRAINERD 564			DETROIT LAKES 565			DULUTH 556 - 558			MANKATO 560			MINNEAPOLIS 553 - 555			ROCHESTER 559		
		MAT.	INST.	TOTAL	MAT.	INST.	TOTAL	MAT.	INST.	TOTAL	MAT.	INST.	TOTAL	MAT.	INST.	TOTAL	MAT.	INST.	TOTAL
015433	CONTRACTOR EQUIPMENT		98.4	98.4		95.8	95.8		104.2	104.2		98.4	98.4		108.3	108.3		99.5	99.5
0241, 31 - 34	SITE & INFRASTRUCTURE, DEMOLITION	95.6	97.0	96.6	93.4	92.5	92.8	100.9	103.6	102.8	92.2	97.9	96.1	96.0	107.6	104.0	98.0	96.1	96.7
0310	Concrete Forming & Accessories	88.3	85.4	85.8	83.8	80.4	81.0	100.3	104.6	104.0	97.8	101.7	101.1	102.4	116.9	114.7	104.9	105.9	105.7
0320	Concrete Reinforcing	95.9	100.8	98.3	97.1	100.6	98.8	108.1	101.2	104.7	95.8	110.2	102.7	94.0	108.1	100.8	100.3	110.5	105.2
0330	Cast-in-Place Concrete	108.4	99.7	105.1	96.8	98.1	97.3	102.2	103.0	102.5	99.8	110.4	103.8	90.7	118.0	101.0	97.5	102.5	99.4
03	CONCRETE	95.1	94.4	94.8	88.9	91.6	90.1	102.0	102.3	102.1	90.9	107.2	98.2	95.9	116.4	105.1	93.3	106.5	99.2
04	MASONRY	127.3	112.3	118.2	126.0	108.9	115.6	100.4	112.3	107.7	114.9	125.9	121.6	109.4	121.3	116.7	101.3	114.3	109.2
05	METALS	90.2	118.7	99.1	89.2	118.0	98.2	97.5	120.1	104.6	90.1	123.8	100.6	96.1	125.4	105.2	96.6	126.9	106.1
06	WOOD, PLASTICS & COMPOSITES	84.4	77.1	80.6	66.9	71.9	69.5	97.6	97.7	97.6	95.5	92.2	93.7	103.6	114.6	109.3	104.7	104.6	104.6
07	THERMAL & MOISTURE PROTECTION	103.8	98.2	101.4	106.1	94.0	100.9	98.0	106.4	101.6	104.3	100.4	102.6	100.1	118.1	107.8	104.4	98.6	101.9
08	OPENINGS	86.3	102.0	90.1	99.4	99.2	99.3	103.1	108.5	104.4	90.7	111.8	95.8	98.9	123.7	104.9	98.2	120.4	103.6
0920	Plaster & Gypsum Board	85.8	77.2	80.2	98.0	71.7	80.7	93.3	98.0	96.4	89.8	92.7	91.7	96.1	115.0	108.5	101.9	105.3	104.1
0950, 0980	Ceilings & Acoustic Treatment	60.2	77.2	70.8	131.6	71.7	94.0	92.0	98.0	95.8	60.2	92.7	80.5	101.2	115.0	109.8	94.5	105.3	101.3
0960	Flooring	90.9	104.1	94.8	90.9	81.6	88.2	94.8	103.5	97.4	93.0	88.8	91.8	101.9	121.5	107.6	94.9	88.8	93.1
0970, 0990	Wall Finishes & Painting/Coating	80.1	96.5	90.0	85.6	79.4	81.9	91.4	108.9	101.9	90.8	98.8	95.6	99.5	127.3	116.2	84.4	102.5	95.2
09	FINISHES	81.5	89.0	85.6	100.6	79.8	89.3	93.3	101.5	97.8	83.2	98.8	91.6	98.7	118.8	109.6	92.8	102.8	98.2
COVERS	DIVS. 10 - 14, 25, 28, 41, 43, 44, 46	100.0	99.3	99.8	100.0	98.5	99.6	100.0	101.2	100.3	100.0	105.4	101.3	100.0	106.7	101.6	100.0	102.0	100.5
21, 22, 23	FIRE SUPPRESSION, PLUMBING & HVAC	96.2	87.4	92.7	97.1	86.4	92.8	100.5	96.4	98.8	96.2	95.6	96.0	100.6	113.3	105.7	100.6	101.7	101.1
26, 27, 3370	ELECTRICAL, COMMUNICATIONS & UTIL.	101.4	101.9	101.7	103.1	68.4	86.0	101.4	101.9	101.7	106.9	99.5	103.2	104.2	115.4	109.8	101.4	99.5	100.4
MF2018	WEIGHTED AVERAGE	95.5	97.9	96.5	97.6	90.3	94.4	99.9	103.9	101.6	95.5	105.0	99.6	99.7	116.6	107.0	98.4	106.1	101.7

For customer support on your Commercial Renovation Costs with RSMeans Data, call 800.448.8182.

809

MINNESOTA / MISSISSIPPI

		SAINT PAUL 550-551			ST. CLOUD 563			THIEF RIVER FALLS 567			WILLMAR 562			WINDOM 561			BILOXI 395		
DIVISION		MAT.	INST.	TOTAL	MAT.	INST.	TOTAL	MAT.	INST.	TOTAL	MAT.	INST.	TOTAL	MAT.	INST.	TOTAL	MAT.	INST.	TOTAL
015433	CONTRACTOR EQUIPMENT		104.2	104.2		98.4	98.4		95.8	95.8		98.4	98.4		98.4	98.4		98.7	98.7
0241, 31-34	SITE & INFRASTRUCTURE, DEMOLITION	97.6	104.3	102.2	91.4	98.3	96.2	94.1	92.1	92.8	90.3	97.0	94.9	84.1	96.2	92.4	101.8	83.7	89.4
0310	Concrete Forming & Accessories	102.2	120.4	117.7	85.2	119.7	114.5	88.3	84.1	84.7	84.9	85.2	85.2	89.9	81.2	82.5	86.3	68.0	70.7
0320	Concrete Reinforcing	105.3	110.8	108.0	95.9	110.5	102.9	97.4	100.6	98.9	95.5	110.1	102.6	95.5	109.2	102.1	89.6	64.5	77.5
0330	Cast-in-Place Concrete	102.5	119.7	109.0	95.6	117.5	103.9	98.8	92.0	96.2	97.1	78.5	90.1	84.0	82.8	83.5	121.8	66.9	101.0
03	CONCRETE	101.6	119.0	109.4	86.8	117.9	100.7	90.1	91.1	90.5	86.7	88.7	87.6	78.0	88.2	82.6	98.8	68.6	85.3
04	MASONRY	98.7	123.0	113.5	110.2	119.5	115.8	100.5	103.0	102.0	113.7	112.3	112.9	125.9	89.1	103.5	90.1	63.9	74.1
05	METALS	97.4	126.6	106.5	90.9	125.4	101.7	89.3	117.8	98.2	90.0	123.4	100.4	89.9	121.5	99.8	90.1	90.6	90.3
06	WOOD, PLASTICS & COMPOSITES	101.3	118.2	110.1	81.1	117.8	100.3	71.6	80.1	76.1	80.7	77.2	78.9	86.1	77.2	81.4	83.0	68.6	75.5
07	THERMAL & MOISTURE PROTECTION	101.5	119.2	109.1	104.0	112.7	107.7	107.2	87.9	98.9	103.7	97.9	101.2	103.7	81.5	94.2	101.4	63.5	85.1
08	OPENINGS	97.9	127.9	105.2	91.1	127.7	100.0	99.4	103.7	100.4	88.2	101.1	91.3	91.6	101.1	93.9	96.1	63.1	88.1
0920	Plaster & Gypsum Board	92.5	119.0	109.9	85.8	119.0	107.6	98.3	80.1	86.3	85.8	77.3	80.2	85.8	77.3	80.2	112.4	68.3	83.5
0950, 0980	Ceilings & Acoustic Treatment	93.3	119.0	109.4	60.2	119.0	97.1	131.6	80.1	99.3	60.2	77.3	70.9	60.2	77.3	70.9	90.4	68.3	76.6
0960	Flooring	93.5	121.5	101.7	87.7	119.6	97.0	91.9	81.6	88.9	89.2	81.6	87.0	91.8	81.6	88.8	93.8	64.3	85.2
0970, 0990	Wall Finishes & Painting/Coating	90.6	131.3	115.0	90.8	127.3	112.7	85.6	79.4	81.9	85.6	79.4	81.9	85.6	98.6	93.5	80.7	48.3	61.3
09	FINISHES	93.1	121.8	108.6	80.9	120.7	102.5	101.0	83.2	91.3	81.0	83.3	82.2	81.2	82.6	82.0	92.5	65.4	77.8
COVERS	DIVS. 10-14, 25, 28, 41, 43, 44, 46	100.0	107.0	101.6	100.0	106.1	101.4	100.0	97.4	99.4	100.0	99.2	99.8	100.0	96.2	99.1	100.0	82.9	96.0
21, 22, 23	FIRE SUPPRESSION, PLUMBING & HVAC	100.5	117.9	107.5	100.0	109.3	103.8	97.1	83.3	91.5	96.2	100.6	98.0	96.2	80.7	90.0	101.1	56.6	83.1
26, 27, 3370	ELECTRICAL, COMMUNICATIONS & UTIL.	102.1	119.9	110.9	101.4	119.9	110.6	100.9	68.4	84.8	101.4	80.6	91.1	106.9	99.4	103.2	100.2	56.1	78.4
MF2018	WEIGHTED AVERAGE	99.3	119.2	107.9	94.9	115.9	104.0	96.4	89.4	93.4	93.7	96.6	94.9	94.0	91.7	93.0	97.1	66.6	83.9

MISSISSIPPI

		CLARKSDALE 386			COLUMBUS 397			GREENVILLE 387			GREENWOOD 389			JACKSON 390-392			LAUREL 394		
DIVISION		MAT.	INST.	TOTAL	MAT.	INST.	TOTAL	MAT.	INST.	TOTAL	MAT.	INST.	TOTAL	MAT.	INST.	TOTAL	MAT.	INST.	TOTAL
015433	CONTRACTOR EQUIPMENT		98.7	98.7		98.7	98.7		98.7	98.7		98.7	98.7		103.9	103.9		98.7	98.7
0241, 31-34	SITE & INFRASTRUCTURE, DEMOLITION	99.6	82.0	87.5	100.1	82.8	88.2	105.8	83.7	90.6	102.6	81.8	88.3	97.2	92.6	94.0	105.3	82.1	89.4
0310	Concrete Forming & Accessories	83.7	42.6	48.7	75.8	44.7	49.4	80.2	62.0	64.7	94.0	42.7	50.4	87.3	67.4	70.3	75.9	57.6	60.4
0320	Concrete Reinforcing	99.6	63.4	82.1	96.0	63.5	80.3	100.1	63.7	82.5	99.6	63.5	82.1	100.9	64.6	83.4	96.6	31.2	65.1
0330	Cast-in-Place Concrete	100.3	57.4	84.1	124.2	59.6	99.8	103.3	66.1	89.2	107.8	57.2	88.7	105.0	67.6	90.9	121.5	58.8	97.8
03	CONCRETE	93.3	53.5	75.4	100.3	55.3	80.1	98.6	65.5	83.8	99.6	53.5	78.9	96.4	68.4	83.8	102.3	55.3	81.2
04	MASONRY	91.9	50.0	66.4	114.1	52.7	76.6	137.0	63.5	92.2	92.5	49.9	66.5	97.1	64.3	77.1	110.5	52.6	75.2
05	METALS	92.3	86.7	90.6	87.2	89.2	87.8	93.4	90.3	92.4	92.3	86.5	90.5	96.6	88.6	94.1	87.3	75.3	83.6
06	WOOD, PLASTICS & COMPOSITES	78.7	42.9	59.9	69.9	43.7	56.2	74.8	60.7	67.4	92.4	42.9	66.5	88.3	67.8	77.5	70.9	62.3	66.4
07	THERMAL & MOISTURE PROTECTION	97.7	50.1	77.3	101.3	53.5	80.8	98.1	62.7	83.0	98.2	52.3	78.4	99.8	64.7	84.7	101.5	55.6	81.8
08	OPENINGS	94.9	46.4	83.1	95.8	46.8	83.8	94.6	56.3	85.3	94.9	46.4	83.1	99.2	60.4	89.8	93.1	49.9	82.6
0920	Plaster & Gypsum Board	91.5	41.9	59.0	101.8	42.7	63.0	91.0	60.2	70.8	101.7	41.9	62.5	95.9	67.2	77.1	101.8	61.9	75.6
0950, 0980	Ceilings & Acoustic Treatment	89.2	41.9	59.5	84.0	42.7	58.1	91.3	60.2	71.8	89.2	41.9	59.5	93.6	67.2	77.1	84.0	61.9	70.1
0960	Flooring	99.6	64.3	89.3	87.5	64.3	80.7	98.0	64.3	88.2	105.4	64.3	93.4	91.5	65.6	84.0	86.2	64.3	79.9
0970, 0990	Wall Finishes & Painting/Coating	90.9	44.7	63.2	80.7	44.7	59.1	90.9	56.2	70.1	90.9	44.7	63.2	87.0	56.2	68.5	80.7	44.7	59.1
09	FINISHES	92.8	46.5	67.7	87.7	47.6	66.0	93.3	61.6	76.1	96.2	46.5	69.3	92.5	66.0	78.2	87.9	58.6	72.0
COVERS	DIVS. 10-14, 25, 28, 41, 43, 44, 46	100.0	45.4	87.2	100.0	46.4	87.4	100.0	81.9	95.7	100.0	45.4	87.1	100.0	83.2	96.0	100.0	32.4	84.1
21, 22, 23	FIRE SUPPRESSION, PLUMBING & HVAC	99.1	48.9	78.8	99.0	50.8	79.5	100.9	56.5	83.0	99.1	49.2	79.0	101.2	60.0	84.6	99.0	45.4	77.4
26, 27, 3370	ELECTRICAL, COMMUNICATIONS & UTIL.	97.4	40.1	69.0	98.0	53.0	75.7	97.4	56.1	77.0	97.4	37.5	67.8	102.6	56.1	79.6	99.3	55.8	77.8
MF2018	WEIGHTED AVERAGE	95.7	53.8	77.6	96.7	57.1	79.6	99.4	65.2	84.6	97.0	53.5	78.2	98.6	67.9	85.3	96.8	56.4	79.4

MISSISSIPPI / MISSOURI

		MCCOMB 396			MERIDIAN 393			TUPELO 388			BOWLING GREEN 633			CAPE GIRARDEAU 637			CHILLICOTHE 646		
DIVISION		MAT.	INST.	TOTAL	MAT.	INST.	TOTAL	MAT.	INST.	TOTAL	MAT.	INST.	TOTAL	MAT.	INST.	TOTAL	MAT.	INST.	TOTAL
015433	CONTRACTOR EQUIPMENT		98.7	98.7		98.7	98.7		98.7	98.7		104.0	104.0		104.0	104.0		98.6	98.6
0241, 31-34	SITE & INFRASTRUCTURE, DEMOLITION	93.0	81.9	85.4	97.4	83.8	88.0	97.2	81.9	86.7	88.2	86.1	86.8	90.2	86.3	87.5	101.7	84.8	90.1
0310	Concrete Forming & Accessories	75.8	43.9	48.6	73.3	68.1	68.9	80.7	44.1	49.6	91.6	90.7	90.8	84.2	81.4	81.8	77.7	92.3	90.1
0320	Concrete Reinforcing	97.2	32.7	66.1	96.0	64.6	80.8	97.4	63.3	81.0	100.6	92.0	96.4	101.9	84.9	93.7	100.7	97.8	99.3
0330	Cast-in-Place Concrete	107.8	57.0	88.6	115.3	66.9	97.1	100.3	58.3	84.4	82.7	91.5	86.0	81.8	84.7	82.9	97.5	82.5	91.8
03	CONCRETE	89.9	48.6	71.3	94.4	68.7	82.8	92.8	54.5	75.6	87.9	92.7	90.1	87.1	85.0	86.1	96.7	90.7	94.0
04	MASONRY	115.5	49.5	75.3	89.7	63.9	74.0	126.1	51.0	80.3	111.9	93.8	100.9	109.4	80.1	91.5	101.3	88.4	93.4
05	METALS	87.4	74.3	83.3	88.4	90.8	89.1	92.2	86.2	90.4	88.9	115.3	97.1	90.0	111.5	96.7	82.3	108.6	90.5
06	WOOD, PLASTICS & COMPOSITES	69.9	45.2	57.0	67.6	68.6	68.1	75.4	43.7	58.8	90.7	91.2	90.9	83.0	79.6	81.2	80.0	93.7	87.2
07	THERMAL & MOISTURE PROTECTION	100.8	52.3	80.0	101.0	63.9	85.1	97.7	50.5	77.4	96.6	94.7	95.8	96.1	83.0	90.5	94.1	88.5	91.7
08	OPENINGS	95.8	39.7	82.1	95.5	64.2	87.9	94.9	49.5	83.8	98.2	94.6	97.3	98.2	78.0	93.3	86.4	92.6	87.9
0920	Plaster & Gypsum Board	101.8	44.3	64.1	101.8	68.3	79.8	91.0	42.7	59.3	95.9	91.5	93.0	95.4	79.6	85.0	103.7	93.7	97.1
0950, 0980	Ceilings & Acoustic Treatment	84.0	44.3	59.1	84.7	68.3	74.4	89.2	42.7	60.1	89.6	91.5	90.8	89.6	79.6	83.3	88.5	93.7	91.7
0960	Flooring	87.5	64.3	80.7	86.1	64.3	79.8	98.2	64.3	88.3	95.6	92.4	94.7	92.1	84.6	89.9	92.4	95.1	93.2
0970, 0990	Wall Finishes & Painting/Coating	80.7	44.7	59.1	80.7	56.2	66.0	90.9	43.1	62.3	95.1	101.2	98.7	95.1	76.7	84.1	90.4	98.3	95.1
09	FINISHES	87.1	47.7	65.8	87.2	66.3	75.9	92.3	47.4	68.0	96.5	91.9	94.0	95.4	80.8	87.5	95.9	93.9	94.7
COVERS	DIVS. 10-14, 25, 28, 41, 43, 44, 46	100.0	48.2	87.8	100.0	82.9	96.0	100.0	46.4	87.4	100.0	96.8	99.3	100.0	95.6	99.0	100.0	96.6	99.2
21, 22, 23	FIRE SUPPRESSION, PLUMBING & HVAC	99.0	48.3	78.6	101.1	60.1	84.5	99.2	50.2	79.4	96.9	94.8	96.1	100.7	98.2	99.7	97.3	95.2	96.4
26, 27, 3370	ELECTRICAL, COMMUNICATIONS & UTIL.	96.8	53.9	75.6	99.3	58.1	78.9	97.2	53.0	75.3	99.4	74.0	86.8	99.4	94.8	97.1	94.2	73.0	83.7
MF2018	WEIGHTED AVERAGE	95.1	53.8	77.3	95.4	67.8	83.5	97.2	56.4	79.6	95.6	92.3	94.2	96.4	90.6	93.9	93.7	90.7	92.4

MISSOURI

DIVISION		COLUMBIA 652			FLAT RIVER 636			HANNIBAL 634			HARRISONVILLE 647			JEFFERSON CITY 650 - 651			JOPLIN 648		
		MAT.	INST.	TOTAL	MAT.	INST.	TOTAL	MAT.	INST.	TOTAL	MAT.	INST.	TOTAL	MAT.	INST.	TOTAL	MAT.	INST.	TOTAL
015433	CONTRACTOR EQUIPMENT		107.2	107.2		104.0	104.0		104.0	104.0		98.6	98.6		112.3	112.3		102.0	102.0
0241, 31 - 34	SITE & INFRASTRUCTURE, DEMOLITION	93.0	91.4	91.9	90.7	85.8	87.3	86.0	85.9	85.9	93.6	85.9	88.3	93.0	99.4	97.4	103.0	89.7	93.9
0310	Concrete Forming & Accessories	77.7	89.1	87.4	98.4	85.0	87.0	89.9	79.6	81.1	75.1	95.5	92.5	90.8	78.5	80.4	88.0	75.8	77.6
0320	Concrete Reinforcing	87.1	100.0	93.3	101.9	98.0	100.0	100.1	91.9	96.1	100.3	105.7	102.9	90.9	89.1	90.0	104.0	102.3	103.2
0330	Cast-in-Place Concrete	83.7	87.0	84.9	85.3	86.7	85.8	78.3	89.5	82.5	99.9	95.7	98.3	88.8	85.6	87.6	105.6	74.7	93.9
03	CONCRETE	79.4	91.9	85.0	90.8	89.5	90.2	84.5	87.0	85.6	93.2	98.2	95.4	85.6	84.6	85.2	96.3	81.2	89.5
04	MASONRY	146.8	91.4	113.0	109.3	73.2	87.3	102.9	90.6	95.4	95.6	95.3	95.4	99.2	91.5	94.5	94.5	80.6	86.0
05	METALS	94.1	119.5	102.0	88.8	116.6	97.5	88.9	114.8	97.0	82.7	113.1	92.2	93.6	112.1	99.4	85.2	106.6	91.8
06	WOOD, PLASTICS & COMPOSITES	71.4	86.7	79.4	99.3	86.5	92.6	88.8	78.3	83.3	77.0	95.4	86.6	89.1	73.1	80.7	90.2	75.2	82.3
07	THERMAL & MOISTURE PROTECTION	92.1	89.7	91.0	96.8	86.8	92.5	96.4	89.2	93.3	93.3	96.2	94.5	99.2	88.9	94.8	93.7	80.7	88.2
08	OPENINGS	94.3	88.7	93.0	98.2	93.8	97.1	98.2	80.6	93.9	86.1	99.4	89.4	94.5	78.4	90.6	87.4	78.5	85.2
0920	Plaster & Gypsum Board	72.2	86.7	81.7	101.9	86.7	91.9	95.7	78.3	84.3	100.0	95.3	96.9	80.3	72.4	75.1	110.4	74.6	86.9
0950, 0980	Ceilings & Acoustic Treatment	79.8	86.7	84.1	89.6	86.7	87.8	89.6	78.3	82.5	88.5	95.3	92.8	84.2	72.4	76.8	89.1	74.6	80.0
0960	Flooring	87.3	95.9	89.8	98.9	80.7	93.6	94.8	92.4	94.1	87.9	96.1	90.3	93.5	68.2	86.1	116.9	71.5	103.6
0970, 0990	Wall Finishes & Painting/Coating	94.1	82.0	86.8	95.1	68.3	79.1	95.1	89.6	91.8	94.6	102.7	99.4	95.1	82.0	87.2	90.0	72.2	79.3
09	FINISHES	80.9	89.1	85.4	98.4	81.7	89.4	96.1	81.9	88.4	93.7	96.2	95.1	88.4	76.1	81.7	102.5	74.7	87.4
COVERS	DIVS. 10 - 14, 25, 28, 41, 43, 44, 46	100.0	96.9	99.3	100.0	94.7	98.8	100.0	94.4	98.7	100.0	98.0	99.5	100.0	97.6	99.4	100.0	93.4	98.4
21, 22, 23	FIRE SUPPRESSION, PLUMBING & HVAC	100.9	101.2	101.0	96.9	92.2	95.0	96.9	93.1	95.4	97.2	96.1	96.7	101.0	95.6	98.8	101.1	69.9	88.5
26, 27, 3370	ELECTRICAL, COMMUNICATIONS & UTIL.	94.2	82.5	88.4	103.4	94.8	99.1	98.3	74.0	86.3	99.9	97.5	98.7	100.9	82.5	91.8	92.5	74.7	83.7
MF2018	WEIGHTED AVERAGE	95.7	94.5	95.2	96.6	90.6	94.0	94.5	88.5	91.9	93.2	97.4	95.0	95.7	90.0	93.2	95.2	80.1	88.7

MISSOURI

DIVISION		KANSAS CITY 640 - 641			KIRKSVILLE 635			POPLAR BLUFF 639			ROLLA 654 - 655			SEDALIA 653			SIKESTON 638		
		MAT.	INST.	TOTAL	MAT.	INST.	TOTAL	MAT.	INST.	TOTAL	MAT.	INST.	TOTAL	MAT.	INST.	TOTAL	MAT.	INST.	TOTAL
015433	CONTRACTOR EQUIPMENT		105.1	105.1		94.1	94.1		96.4	96.4		107.2	107.2		97.3	97.3		96.4	96.4
0241, 31 - 34	SITE & INFRASTRUCTURE, DEMOLITION	95.5	100.3	98.8	89.1	81.1	83.6	76.6	84.6	82.1	91.4	90.2	90.6	89.1	85.5	86.6	79.9	85.2	83.6
0310	Concrete Forming & Accessories	87.2	100.5	98.5	82.2	75.6	76.6	82.5	76.3	77.2	84.9	90.5	89.6	82.9	75.4	76.5	83.2	76.4	77.4
0320	Concrete Reinforcing	92.2	107.0	99.4	100.8	79.4	90.5	103.9	72.4	88.7	87.6	88.9	88.2	86.4	105.1	95.4	103.2	72.4	88.4
0330	Cast-in-Place Concrete	103.6	100.5	102.4	85.3	78.7	82.8	65.6	79.8	71.0	85.6	90.6	87.5	89.5	77.9	85.1	70.1	79.8	73.8
03	CONCRETE	94.3	102.1	97.8	102.5	78.6	91.8	78.9	78.1	78.6	81.1	91.8	85.9	94.1	82.7	89.0	82.3	78.2	80.4
04	MASONRY	101.2	100.7	100.9	116.0	81.3	94.8	107.7	71.0	85.3	117.2	81.8	95.6	124.2	78.2	96.2	107.1	71.0	85.1
05	METALS	91.9	110.2	97.6	88.6	98.7	91.7	89.1	95.3	91.0	93.4	113.4	99.7	92.4	110.5	98.1	89.5	95.4	91.3
06	WOOD, PLASTICS & COMPOSITES	85.1	100.4	93.1	76.7	73.9	75.2	75.7	77.2	76.5	78.4	92.3	85.7	73.0	74.1	73.5	77.4	77.2	77.3
07	THERMAL & MOISTURE PROTECTION	92.2	103.0	96.9	101.9	86.7	95.4	100.5	79.2	91.4	92.3	88.7	90.8	96.9	83.8	91.3	100.7	78.4	91.1
08	OPENINGS	96.9	100.2	97.7	103.7	75.9	96.9	104.6	72.4	96.8	94.3	87.7	92.7	99.4	83.8	95.6	104.6	72.4	96.8
0920	Plaster & Gypsum Board	110.5	100.5	103.9	91.1	73.7	79.7	91.4	77.2	82.1	73.6	92.4	86.0	68.5	73.7	71.9	93.4	77.2	82.7
0950, 0980	Ceilings & Acoustic Treatment	90.7	100.5	96.9	88.4	73.7	79.2	89.6	77.2	81.8	79.8	92.4	87.7	79.8	73.7	76.0	89.6	77.2	81.8
0960	Flooring	93.1	99.9	95.1	73.8	92.0	79.1	87.1	80.7	85.2	91.0	92.0	91.3	71.2	69.6	70.7	87.7	80.7	85.6
0970, 0990	Wall Finishes & Painting/Coating	96.2	111.5	105.4	90.9	74.6	81.1	90.2	63.9	74.4	94.1	85.9	89.2	94.1	98.3	96.6	90.2	63.9	74.4
09	FINISHES	96.8	101.5	99.4	95.1	77.8	85.7	94.9	75.2	84.2	82.3	90.3	86.6	80.2	75.9	77.9	95.6	75.5	84.7
COVERS	DIVS. 10 - 14, 25, 28, 41, 43, 44, 46	100.0	99.4	99.9	100.0	93.9	98.6	100.0	93.0	98.4	100.0	96.8	99.2	100.0	92.0	98.1	100.0	93.0	98.4
21, 22, 23	FIRE SUPPRESSION, PLUMBING & HVAC	100.8	101.4	101.0	97.0	92.5	95.2	97.0	90.4	94.3	97.1	93.9	95.8	97.0	90.4	94.4	97.0	90.4	94.3
26, 27, 3370	ELECTRICAL, COMMUNICATIONS & UTIL.	101.7	99.7	100.7	98.3	73.9	86.2	98.6	94.8	96.7	93.0	76.4	84.8	93.8	97.4	95.6	97.8	94.8	96.3
MF2018	WEIGHTED AVERAGE	97.5	101.8	99.3	98.0	83.6	91.8	94.4	84.3	90.0	93.4	90.6	92.2	95.6	88.1	92.4	94.9	84.4	90.4

MISSOURI / MONTANA

DIVISION		SPRINGFIELD 656 - 658			ST. JOSEPH 644 - 645			ST. LOUIS 630 - 631			BILLINGS 590 - 591			BUTTE 597			GREAT FALLS 594		
		MAT.	INST.	TOTAL	MAT.	INST.	TOTAL	MAT.	INST.	TOTAL	MAT.	INST.	TOTAL	MAT.	INST.	TOTAL	MAT.	INST.	TOTAL
015433	CONTRACTOR EQUIPMENT		99.7	99.7		98.6	98.6		107.9	107.9		96.1	96.1		95.8	95.8		95.8	95.8
0241, 31 - 34	SITE & INFRASTRUCTURE, DEMOLITION	91.6	89.3	90.0	97.5	84.2	88.3	96.9	98.8	98.2	91.9	90.6	91.0	98.2	90.3	92.8	101.9	90.6	94.1
0310	Concrete Forming & Accessories	90.7	78.3	80.2	86.8	93.4	92.4	94.6	102.5	101.3	99.1	69.4	73.8	85.7	69.1	71.6	99.4	69.2	73.7
0320	Concrete Reinforcing	84.0	99.6	91.5	97.5	108.6	102.9	91.4	112.4	101.5	90.1	81.0	85.7	97.7	80.6	89.4	90.1	81.0	85.7
0330	Cast-in-Place Concrete	90.9	77.0	85.7	98.1	96.5	97.5	92.4	104.8	97.1	112.2	70.3	96.4	123.5	70.1	103.3	130.5	71.1	108.1
03	CONCRETE	90.5	82.7	87.0	92.3	98.0	94.8	92.0	106.0	98.3	95.7	72.6	85.4	99.0	72.3	87.1	104.1	72.8	90.0
04	MASONRY	92.8	84.5	87.7	96.4	93.2	94.5	87.6	107.7	99.8	131.1	83.1	101.8	126.6	76.1	95.8	131.5	80.3	100.3
05	METALS	98.5	104.9	100.5	88.5	113.9	96.4	94.3	121.5	102.8	104.6	89.4	99.9	98.9	88.5	95.7	102.1	89.4	98.1
06	WOOD, PLASTICS & COMPOSITES	80.0	76.3	78.1	90.2	93.5	91.9	92.1	100.1	96.3	89.4	66.5	77.5	75.0	66.5	70.6	90.9	66.5	78.1
07	THERMAL & MOISTURE PROTECTION	96.1	78.5	88.6	93.6	92.7	93.2	95.3	105.6	99.7	108.7	72.2	93.0	108.3	69.8	91.7	108.9	72.1	93.1
08	OPENINGS	101.7	88.2	98.4	90.3	99.9	92.7	100.0	106.8	101.7	98.4	66.4	90.6	96.8	66.4	89.4	99.6	71.0	92.6
0920	Plaster & Gypsum Board	74.8	76.0	75.6	111.9	93.4	99.8	98.1	100.5	99.7	117.1	66.1	83.7	116.3	66.1	83.4	126.4	66.1	86.9
0950, 0980	Ceilings & Acoustic Treatment	79.8	76.0	77.4	94.9	93.4	94.0	86.3	100.5	95.2	110.0	66.1	82.5	117.4	66.1	85.3	119.3	66.1	86.0
0960	Flooring	90.3	71.5	84.8	97.5	99.9	98.2	97.5	97.1	97.4	90.6	77.1	86.7	87.9	79.4	85.4	94.6	77.1	89.5
0970, 0990	Wall Finishes & Painting/Coating	88.5	104.5	98.1	90.4	103.0	97.9	99.3	106.8	103.8	87.6	88.4	88.1	86.1	64.5	73.2	86.1	91.1	89.1
09	FINISHES	84.4	80.0	82.0	99.4	95.7	97.4	97.2	101.5	99.5	97.8	72.2	84.0	98.3	70.1	83.0	102.4	72.3	86.1
COVERS	DIVS. 10 - 14, 25, 28, 41, 43, 44, 46	100.0	94.3	98.7	100.0	97.3	99.4	100.0	101.8	100.4	100.0	92.3	98.2	100.0	92.3	98.2	100.0	92.0	98.1
21, 22, 23	FIRE SUPPRESSION, PLUMBING & HVAC	100.9	74.0	90.1	101.1	90.6	96.8	100.7	104.5	102.3	100.8	74.4	90.2	100.8	68.3	87.7	100.8	74.0	90.0
26, 27, 3370	ELECTRICAL, COMMUNICATIONS & UTIL.	97.4	67.1	82.4	99.9	80.0	90.1	103.2	96.7	100.0	99.2	72.4	85.9	105.1	70.5	88.0	98.7	73.4	86.1
MF2018	WEIGHTED AVERAGE	96.5	81.4	90.0	96.0	93.3	94.9	97.5	104.6	100.6	101.4	77.2	90.9	101.3	74.4	89.7	102.8	77.1	91.7

For customer support on your Commercial Renovation Costs with RSMeans Data, call 800.448.8182.

811

MONTANA

DIVISION		HAVRE 595 MAT.	INST.	TOTAL	HELENA 596 MAT.	INST.	TOTAL	KALISPELL 599 MAT.	INST.	TOTAL	MILES CITY 593 MAT.	INST.	TOTAL	MISSOULA 598 MAT.	INST.	TOTAL	WOLF POINT 592 MAT.	INST.	TOTAL
015433	CONTRACTOR EQUIPMENT		95.8	95.8		100.9	100.9		95.8	95.8		95.8	95.8		95.8	95.8		95.8	95.8
0241, 31 - 34	SITE & INFRASTRUCTURE, DEMOLITION	105.1	90.1	94.8	92.1	98.1	96.2	87.8	90.2	89.5	94.1	90.1	91.4	81.0	90.3	87.4	111.6	90.1	96.8
0310	Concrete Forming & Accessories	78.0	67.7	69.2	92.1	68.9	74.0	89.3	68.8	71.9	97.3	67.9	72.3	89.3	69.3	72.2	89.4	67.8	71.1
0320	Concrete Reinforcing	98.5	75.1	87.2	101.6	80.6	91.5	100.3	84.7	92.8	98.1	75.2	87.0	99.4	84.5	92.2	99.5	74.5	87.5
0330	Cast-in-Place Concrete	133.0	68.3	108.6	95.2	71.2	86.1	107.2	69.5	93.0	117.4	68.3	98.9	91.0	70.1	83.1	131.5	67.3	107.3
03	CONCRETE	106.6	70.2	90.3	96.9	72.5	86.0	89.2	72.7	81.8	96.1	70.3	84.5	78.4	73.1	76.0	110.1	69.8	92.0
04	MASONRY	127.6	77.2	96.9	115.5	75.1	90.9	125.5	79.2	97.3	133.6	77.2	99.2	153.1	75.9	106.0	134.8	77.2	99.7
05	METALS	95.1	86.4	92.4	100.9	86.7	96.5	95.0	90.2	93.5	94.3	86.6	91.9	95.5	89.9	93.8	94.4	86.3	91.9
06	WOOD, PLASTICS & COMPOSITES	65.7	66.5	66.2	94.6	66.8	80.1	78.8	66.5	72.4	87.5	66.5	76.5	78.8	66.5	72.4	77.9	66.5	72.0
07	THERMAL & MOISTURE PROTECTION	108.6	63.7	89.3	104.8	70.4	90.0	107.8	70.7	91.9	108.2	65.0	89.7	107.4	71.9	92.2	109.2	64.9	90.2
08	OPENINGS	97.3	65.1	89.5	96.1	66.6	88.9	97.3	67.3	90.0	96.8	65.1	89.1	96.8	67.3	89.6	96.8	65.0	89.1
0920	Plaster & Gypsum Board	111.7	66.1	81.8	114.1	66.1	82.6	116.3	66.1	83.4	124.7	66.1	86.3	116.3	66.1	83.4	118.8	66.1	84.2
0950, 0980	Ceilings & Acoustic Treatment	117.4	66.1	85.3	122.3	66.1	87.1	117.4	66.1	85.3	112.9	66.1	83.6	117.4	66.1	85.3	112.9	66.1	83.6
0960	Flooring	85.5	86.8	85.9	97.2	79.4	92.0	89.6	86.8	88.8	94.5	86.8	92.2	89.6	79.4	86.7	91.0	86.8	89.8
0970, 0990	Wall Finishes & Painting/Coating	86.1	64.5	73.2	93.9	57.2	71.9	86.1	86.7	86.4	86.1	64.5	73.2	86.1	86.7	86.4	86.1	64.5	73.2
09	FINISHES	97.7	70.9	83.2	105.0	69.3	85.6	98.2	73.8	85.0	100.2	70.9	84.3	97.7	72.5	84.1	99.7	70.9	84.1
COVERS	DIVS. 10 - 14, 25, 28, 41, 43, 44, 46	100.0	84.7	96.4	100.0	92.6	98.2	100.0	85.4	96.6	100.0	84.7	96.4	100.0	92.2	98.2	100.0	84.7	96.4
21, 22, 23	FIRE SUPPRESSION, PLUMBING & HVAC	97.0	65.6	84.3	100.9	67.9	87.6	97.0	66.1	84.5	97.0	71.7	86.8	100.8	68.4	87.7	97.0	71.6	86.8
26, 27, 3370	ELECTRICAL, COMMUNICATIONS & UTIL.	98.7	66.7	82.8	105.3	70.9	88.3	102.3	65.6	84.1	98.7	71.7	85.3	103.2	69.1	86.3	98.7	71.7	85.3
MF2018	WEIGHTED AVERAGE	100.2	72.5	88.3	101.2	74.7	89.7	98.0	74.1	87.7	99.1	74.6	88.5	98.8	74.8	88.5	101.3	74.5	89.7

NEBRASKA

DIVISION		ALLIANCE 693 MAT.	INST.	TOTAL	COLUMBUS 686 MAT.	INST.	TOTAL	GRAND ISLAND 688 MAT.	INST.	TOTAL	HASTINGS 689 MAT.	INST.	TOTAL	LINCOLN 683 - 685 MAT.	INST.	TOTAL	MCCOOK 690 MAT.	INST.	TOTAL
015433	CONTRACTOR EQUIPMENT		92.2	92.2		98.8	98.8		98.8	98.8		98.8	98.8		104.9	104.9		98.8	98.8
0241, 31 - 34	SITE & INFRASTRUCTURE, DEMOLITION	97.6	92.3	94.0	100.4	86.8	91.1	105.6	87.3	93.0	103.8	86.8	92.1	95.3	96.6	96.2	100.5	86.8	91.1
0310	Concrete Forming & Accessories	83.9	53.2	57.8	95.7	72.2	75.7	95.2	67.4	71.5	98.9	69.9	74.2	96.8	73.9	77.3	89.2	53.8	59.0
0320	Concrete Reinforcing	110.1	83.2	97.1	100.7	81.9	91.6	100.1	80.4	90.6	100.1	72.5	86.8	95.3	80.7	88.3	102.7	72.7	88.2
0330	Cast-in-Place Concrete	104.2	78.3	94.4	104.6	78.3	94.7	110.7	75.6	97.5	110.7	70.8	95.6	87.5	81.9	85.4	112.6	70.7	96.8
03	CONCRETE	113.9	68.0	93.3	100.2	77.2	89.9	104.8	73.8	90.9	105.1	71.9	90.2	93.8	78.9	87.1	103.2	64.6	85.9
04	MASONRY	109.4	74.2	87.9	115.8	74.2	90.4	109.1	73.0	87.1	119.3	70.9	89.8	98.4	77.2	85.5	103.6	74.2	85.7
05	METALS	96.7	83.0	92.4	89.2	95.8	91.3	91.0	94.8	92.2	91.7	91.5	91.6	98.6	93.4	97.0	91.7	91.5	91.6
06	WOOD, PLASTICS & COMPOSITES	78.2	47.0	61.9	92.3	72.3	81.8	91.4	64.0	77.0	95.5	69.6	81.9	100.2	72.6	85.7	85.8	47.9	66.0
07	THERMAL & MOISTURE PROTECTION	101.1	72.3	88.7	101.4	76.7	90.8	101.5	77.2	91.1	101.6	74.1	89.8	100.3	79.0	91.1	96.1	72.6	86.0
08	OPENINGS	90.9	55.9	82.4	91.7	71.0	86.7	91.7	66.3	85.5	91.7	67.2	85.8	105.8	68.3	96.7	91.3	53.7	82.1
0920	Plaster & Gypsum Board	75.5	45.9	56.1	84.4	71.8	76.2	83.5	63.3	70.3	85.2	69.1	74.6	90.3	71.8	78.2	87.1	46.8	60.7
0950, 0980	Ceilings & Acoustic Treatment	88.5	45.9	61.8	84.8	71.8	76.7	84.8	63.3	71.3	84.8	69.1	74.9	106.7	71.8	84.8	84.1	46.8	60.7
0960	Flooring	90.9	48.2	78.8	83.7	83.3	83.6	83.4	75.3	81.1	84.8	77.5	82.7	94.9	82.6	91.3	89.6	83.3	87.8
0970, 0990	Wall Finishes & Painting/Coating	156.3	49.2	92.1	74.2	57.2	64.0	74.2	59.5	65.4	74.2	57.2	64.0	94.7	75.6	83.2	85.6	42.9	60.0
09	FINISHES	91.0	56.8	72.5	85.1	72.5	78.3	85.1	67.1	75.4	85.8	69.7	77.1	96.5	75.4	85.1	88.2	56.6	71.1
COVERS	DIVS. 10 - 14, 25, 28, 41, 43, 44, 46	100.0	85.1	96.5	100.0	88.0	97.2	100.0	88.1	97.2	100.0	87.7	97.1	100.0	89.7	97.6	100.0	85.3	96.5
21, 22, 23	FIRE SUPPRESSION, PLUMBING & HVAC	97.1	71.4	86.7	97.0	71.7	86.8	100.8	80.2	92.5	97.0	71.0	86.5	100.6	80.3	92.4	96.8	71.7	86.7
26, 27, 3370	ELECTRICAL, COMMUNICATIONS & UTIL.	92.7	63.2	78.1	96.1	78.4	87.3	94.9	66.0	80.6	94.3	76.5	85.5	105.4	66.0	85.9	94.4	63.3	79.0
MF2018	WEIGHTED AVERAGE	98.4	70.4	86.3	96.0	77.7	88.1	97.4	76.2	88.2	97.1	75.2	87.7	99.8	79.3	90.9	95.9	70.2	84.8

NEBRASKA / NEVADA

DIVISION		NORFOLK 687 MAT.	INST.	TOTAL	NORTH PLATTE 691 MAT.	INST.	TOTAL	OMAHA 680 - 681 MAT.	INST.	TOTAL	VALENTINE 692 MAT.	INST.	TOTAL	CARSON CITY 897 MAT.	INST.	TOTAL	ELKO 898 MAT.	INST.	TOTAL
015433	CONTRACTOR EQUIPMENT		88.5	88.5		98.8	98.8		94.2	94.2		91.9	91.9		95.7	95.7		92.3	92.3
0241, 31 - 34	SITE & INFRASTRUCTURE, DEMOLITION	82.6	85.9	84.9	102.3	86.6	91.5	89.9	95.2	93.6	85.3	91.3	89.4	80.5	93.7	89.6	65.3	88.0	80.9
0310	Concrete Forming & Accessories	82.3	71.4	73.0	91.8	72.8	75.6	92.3	78.3	80.4	80.6	51.4	55.7	100.8	84.7	87.1	105.3	92.7	94.6
0320	Concrete Reinforcing	100.8	63.3	82.7	102.2	77.7	90.4	103.7	81.0	92.7	102.7	63.2	83.6	116.4	117.8	117.1	117.6	109.1	113.5
0330	Cast-in-Place Concrete	105.3	68.1	91.2	112.6	58.0	92.0	86.7	79.6	84.0	99.4	51.6	81.4	97.1	84.3	92.3	90.9	69.3	82.8
03	CONCRETE	98.8	69.4	85.6	103.3	69.7	88.2	94.2	79.5	87.6	100.8	54.7	80.1	106.5	90.2	99.2	97.2	87.2	92.7
04	MASONRY	123.9	74.1	93.5	93.1	73.7	81.2	96.8	81.6	87.5	104.8	73.6	85.8	116.8	67.2	86.5	123.4	64.9	87.7
05	METALS	92.5	78.1	88.0	91.0	93.4	91.8	98.6	84.8	94.3	102.5	77.5	94.7	109.5	95.4	105.1	113.1	91.6	106.4
06	WOOD, PLASTICS & COMPOSITES	76.8	71.8	74.2	87.9	74.2	80.8	92.6	78.0	85.0	73.6	45.7	59.0	87.1	85.6	86.3	96.2	98.8	97.6
07	THERMAL & MOISTURE PROTECTION	100.3	74.7	89.3	96.0	74.8	86.9	95.9	82.2	90.0	95.8	70.6	85.0	114.3	80.1	99.6	110.4	70.2	93.2
08	OPENINGS	92.9	66.1	86.4	90.7	70.6	85.8	99.7	79.0	94.7	92.7	51.4	82.7	101.1	83.8	96.9	102.4	89.2	99.2
0920	Plaster & Gypsum Board	84.4	71.8	76.1	87.1	73.9	78.4	92.2	77.9	82.9	88.7	45.0	60.0	98.5	85.2	89.7	103.2	99.0	100.5
0950, 0980	Ceilings & Acoustic Treatment	99.4	71.8	82.1	84.1	73.9	77.7	93.6	77.9	83.8	98.4	45.0	64.9	107.1	85.2	93.3	105.3	99.0	101.4
0960	Flooring	102.7	83.3	97.0	90.8	75.3	86.3	95.0	90.4	93.7	115.7	83.3	106.3	99.0	64.1	88.8	99.1	64.1	88.9
0970, 0990	Wall Finishes & Painting/Coating	133.5	57.2	87.8	85.6	56.1	67.9	103.2	58.1	76.1	154.7	58.1	96.8	91.6	85.1	87.7	87.4	75.6	80.3
09	FINISHES	101.7	72.2	85.7	88.6	71.7	79.5	96.7	78.2	86.7	108.1	56.8	80.4	98.3	80.4	88.7	95.4	86.8	90.8
COVERS	DIVS. 10 - 14, 25, 28, 41, 43, 44, 46	100.0	86.9	96.9	100.0	87.9	97.2	100.0	89.3	97.5	100.0	83.9	96.2	100.0	106.3	101.5	100.0	86.2	96.8
21, 22, 23	FIRE SUPPRESSION, PLUMBING & HVAC	96.6	70.8	86.2	100.6	73.4	89.6	100.6	84.3	93.9	96.3	70.1	85.7	100.9	81.4	93.0	99.0	73.3	88.6
26, 27, 3370	ELECTRICAL, COMMUNICATIONS & UTIL.	95.2	78.4	86.8	92.9	66.0	79.6	102.9	83.1	93.1	90.6	63.2	77.1	100.9	92.0	96.5	97.5	85.0	91.3
MF2018	WEIGHTED AVERAGE	97.4	74.5	87.5	96.1	74.9	86.9	98.6	83.0	91.9	98.0	67.4	84.8	103.2	85.6	95.6	101.5	81.8	93.0

NEVADA / NEW HAMPSHIRE

| DIVISION | | NEVADA | | | | | | | | | NEW HAMPSHIRE | | | | | | | | |
| | | ELY 893 | | | LAS VEGAS 889 - 891 | | | RENO 894 - 895 | | | CHARLESTON 036 | | | CLAREMONT 037 | | | CONCORD 032 - 033 | | |
		MAT.	INST.	TOTAL	MAT.	INST.	TOTAL	MAT.	INST.	TOTAL	MAT.	INST.	TOTAL	MAT.	INST.	TOTAL	MAT.	INST.	TOTAL
015433	CONTRACTOR EQUIPMENT		92.3	92.3		92.3	92.3		92.3	92.3		94.2	94.2		94.2	94.2		98.6	98.6
0241, 31 - 34	SITE & INFRASTRUCTURE, DEMOLITION	70.7	89.1	83.3	73.8	92.7	86.8	70.9	89.8	83.9	81.6	93.4	89.7	75.5	93.4	87.8	91.0	102.3	98.7
0310	Concrete Forming & Accessories	97.9	97.8	97.8	99.1	112.9	110.9	94.5	84.5	85.9	86.6	82.9	83.5	93.3	83.0	84.5	99.3	94.4	95.1
0320	Concrete Reinforcing	116.2	109.3	112.9	107.1	125.5	116.0	109.9	123.8	116.6	87.5	84.4	86.0	87.5	84.4	86.0	93.9	84.5	89.4
0330	Cast-in-Place Concrete	97.6	91.3	95.2	94.5	109.4	100.1	103.2	82.2	95.3	86.0	111.8	95.7	79.0	111.8	91.4	104.7	113.8	108.2
03	CONCRETE	106.0	97.3	102.1	101.0	113.3	106.5	105.1	90.5	98.6	90.3	93.5	91.7	83.0	93.5	87.7	102.2	99.3	100.9
04	MASONRY	129.0	71.7	94.1	115.5	105.1	109.2	122.4	67.1	88.7	92.3	95.4	94.2	93.1	95.4	94.5	105.2	98.9	101.4
05	METALS	113.0	94.4	107.2	122.1	105.3	116.8	114.8	98.7	109.8	99.2	90.8	96.6	99.2	90.8	96.6	106.7	89.8	101.4
06	WOOD, PLASTICS & COMPOSITES	86.5	100.9	94.0	85.0	112.0	99.1	81.0	85.3	83.2	87.6	80.6	83.3	94.7	80.6	87.3	98.2	94.4	96.2
07	THERMAL & MOISTURE PROTECTION	110.9	88.0	101.1	125.1	103.2	115.7	110.4	78.5	96.7	107.6	102.2	105.3	107.4	102.2	105.2	111.3	107.1	109.5
08	OPENINGS	102.3	90.3	99.4	101.3	115.0	104.7	100.2	84.9	96.5	96.6	79.8	92.5	97.6	79.8	93.3	97.4	87.4	95.0
0920	Plaster & Gypsum Board	98.1	101.1	100.1	92.4	112.5	105.6	87.8	85.2	86.1	100.3	79.9	86.9	100.8	79.9	87.1	100.2	94.1	96.2
0950, 0980	Ceilings & Acoustic Treatment	105.3	101.1	102.7	111.7	112.5	112.2	108.5	85.2	93.9	90.6	79.9	83.9	90.6	79.9	83.9	92.6	94.1	93.5
0960	Flooring	96.6	64.1	87.1	88.5	102.0	92.4	93.3	64.1	84.8	87.5	107.9	93.4	90.0	107.9	95.2	96.0	107.9	99.5
0970, 0990	Wall Finishes & Painting/Coating	87.4	106.7	99.0	89.9	118.5	107.1	87.4	85.1	86.1	79.1	85.7	83.1	79.1	85.7	83.1	95.0	85.7	89.4
09	FINISHES	94.4	93.1	93.7	93.0	112.2	103.4	92.6	80.2	85.9	90.6	87.2	88.8	90.9	87.2	88.9	95.0	95.9	95.5
COVERS	DIVS. 10 - 14, 25, 28, 41, 43, 44, 46	100.0	67.0	92.2	100.0	106.4	101.5	100.0	105.6	101.3	100.0	84.4	96.3	100.0	84.4	96.3	100.0	106.8	101.6
21, 22, 23	FIRE SUPPRESSION, PLUMBING & HVAC	99.0	91.6	96.0	101.1	103.6	102.1	100.9	81.4	93.0	97.6	78.3	89.8	97.6	78.3	89.8	101.2	94.5	94.8
26, 27, 3370	ELECTRICAL, COMMUNICATIONS & UTIL.	97.8	89.5	93.7	102.1	109.9	105.9	98.1	92.0	95.1	94.6	72.4	83.7	94.6	72.4	83.7	95.9	72.4	84.3
MF2018	WEIGHTED AVERAGE	102.9	89.5	97.1	104.2	107.0	105.4	102.8	85.6	95.4	95.6	85.6	91.3	94.8	85.7	90.8	100.8	91.3	96.7

NEW HAMPSHIRE / NEW JERSEY

| DIVISION | | NEW HAMPSHIRE | | | | | | | | | | | | | | | NEW JERSEY | | |
| | | KEENE 034 | | | LITTLETON 035 | | | MANCHESTER 031 | | | NASHUA 030 | | | PORTSMOUTH 038 | | | ATLANTIC CITY 082,084 | | |
		MAT.	INST.	TOTAL	MAT.	INST.	TOTAL	MAT.	INST.	TOTAL	MAT.	INST.	TOTAL	MAT.	INST.	TOTAL	MAT.	INST.	TOTAL
015433	CONTRACTOR EQUIPMENT		94.2	94.2		94.2	94.2		99.1	99.1		94.2	94.2		94.2	94.2		91.8	91.8
0241, 31 - 34	SITE & INFRASTRUCTURE, DEMOLITION	89.5	93.4	92.2	75.6	92.4	87.1	87.4	102.4	97.7	91.2	93.6	92.8	85.0	94.3	91.4	92.0	97.2	95.6
0310	Concrete Forming & Accessories	91.8	83.2	84.5	103.8	77.3	81.3	100.2	94.8	95.6	100.8	94.7	95.7	88.2	94.3	93.4	112.0	145.5	140.5
0320	Concrete Reinforcing	87.5	84.4	86.0	88.2	84.4	86.4	110.5	84.6	98.0	109.3	84.6	97.4	87.5	84.6	86.1	82.0	137.7	108.9
0330	Cast-in-Place Concrete	86.4	111.9	96.0	77.5	103.4	87.3	98.7	114.3	104.6	81.8	113.4	93.7	77.5	113.6	91.2	79.8	135.5	100.8
03	CONCRETE	90.0	93.7	91.6	82.4	88.0	85.0	101.6	99.7	100.7	90.3	99.4	94.4	82.5	99.4	90.1	87.8	139.2	110.9
04	MASONRY	94.9	95.4	95.2	105.1	80.6	90.1	100.7	98.9	99.6	97.7	98.8	98.4	93.2	98.3	96.3	106.3	140.0	126.9
05	METALS	99.9	91.2	97.2	99.9	90.8	97.1	107.4	90.3	102.0	106.0	91.8	101.5	101.5	93.2	98.9	104.0	116.0	107.7
06	WOOD, PLASTICS & COMPOSITES	93.0	80.6	86.5	105.7	80.6	92.6	100.4	94.5	97.3	104.3	94.4	99.1	89.1	94.4	91.9	117.5	147.1	133.0
07	THERMAL & MOISTURE PROTECTION	108.2	102.2	105.6	107.6	95.5	102.4	112.5	107.1	110.2	108.7	106.1	107.6	108.1	105.7	107.1	100.7	136.4	116.0
08	OPENINGS	95.3	83.1	92.3	98.5	79.8	94.0	96.3	92.5	95.4	99.5	90.1	97.2	100.0	81.1	95.4	96.7	142.0	107.7
0920	Plaster & Gypsum Board	100.5	79.9	87.0	115.8	79.9	92.2	99.2	94.1	95.8	110.7	94.1	99.8	100.3	94.1	96.2	109.6	148.2	134.9
0950, 0980	Ceilings & Acoustic Treatment	90.6	79.9	83.9	90.6	79.9	83.9	89.3	94.1	92.3	100.9	94.1	96.6	91.4	94.1	93.1	111.8	148.2	126.1
0960	Flooring	89.6	107.9	95.0	100.4	107.9	102.6	95.0	110.1	99.4	93.3	110.1	98.2	87.7	110.1	94.2	102.7	158.5	119.4
0970, 0990	Wall Finishes & Painting/Coating	79.1	98.4	90.7	79.1	85.7	83.1	96.3	118.0	109.3	79.1	118.0	102.4	79.1	97.4	90.1	77.1	147.2	119.1
09	FINISHES	92.5	88.6	90.4	96.0	83.5	89.2	94.3	99.9	97.3	97.4	99.9	98.7	91.5	97.4	94.7	94.4	149.6	124.3
COVERS	DIVS. 10 - 14, 25, 28, 41, 43, 44, 46	100.0	90.3	97.7	100.0	91.1	97.9	100.0	107.0	101.7	100.0	106.9	101.6	100.0	106.6	101.5	100.0	116.5	103.9
21, 22, 23	FIRE SUPPRESSION, PLUMBING & HVAC	97.6	78.4	89.8	97.6	70.6	86.7	101.2	85.5	94.9	101.4	85.5	95.0	101.4	85.0	94.8	100.4	130.0	112.4
26, 27, 3370	ELECTRICAL, COMMUNICATIONS & UTIL.	94.6	72.4	83.7	95.4	47.3	71.6	97.4	78.5	88.0	96.3	78.5	87.5	95.0	77.6	86.4	92.2	138.0	114.9
MF2018	WEIGHTED AVERAGE	96.1	86.2	91.8	96.1	77.7	88.1	100.5	93.0	97.3	99.3	92.3	96.3	96.5	91.5	94.3	97.8	132.6	112.8

NEW JERSEY

| DIVISION | | CAMDEN 081 | | | DOVER 078 | | | ELIZABETH 072 | | | HACKENSACK 076 | | | JERSEY CITY 073 | | | LONG BRANCH 077 | | |
		MAT.	INST.	TOTAL	MAT.	INST.	TOTAL	MAT.	INST.	TOTAL	MAT.	INST.	TOTAL	MAT.	INST.	TOTAL	MAT.	INST.	TOTAL
015433	CONTRACTOR EQUIPMENT		91.8	91.8		94.2	94.2		94.2	94.2		94.2	94.2		91.8	91.8		91.4	91.4
0241, 31 - 34	SITE & INFRASTRUCTURE, DEMOLITION	92.8	97.5	96.0	107.6	98.9	101.6	113.1	98.9	103.3	108.7	98.9	101.9	98.3	98.8	98.6	103.5	97.6	99.5
0310	Concrete Forming & Accessories	101.8	146.6	139.9	96.0	148.3	140.5	109.0	148.4	142.5	96.0	148.1	140.3	100.0	148.3	141.1	100.6	134.2	129.2
0320	Concrete Reinforcing	107.7	137.9	122.2	72.3	152.4	110.9	72.3	152.4	110.9	72.3	152.4	110.9	93.8	152.4	122.1	72.3	151.9	110.7
0330	Cast-in-Place Concrete	77.4	135.7	99.4	82.7	135.9	102.8	71.1	137.8	96.2	80.8	137.7	102.3	64.5	135.9	91.5	71.7	128.9	93.3
03	CONCRETE	88.5	139.8	111.5	85.9	143.3	111.7	83.0	144.0	110.4	84.3	143.8	111.0	80.6	143.2	108.7	84.6	134.2	106.9
04	MASONRY	95.5	138.5	121.8	98.0	141.7	124.6	115.6	141.7	131.5	102.2	141.7	126.3	92.5	141.7	122.5	107.4	129.8	121.1
05	METALS	110.3	116.4	112.2	101.5	125.2	108.9	103.1	125.3	110.0	101.6	125.0	108.9	107.9	122.3	112.4	101.6	120.6	108.0
06	WOOD, PLASTICS & COMPOSITES	104.7	149.8	128.3	93.1	149.8	122.8	109.1	149.8	130.4	93.1	149.8	122.8	94.2	149.8	123.3	95.4	134.7	115.9
07	THERMAL & MOISTURE PROTECTION	100.5	135.9	115.7	102.2	140.6	118.6	102.5	140.8	118.9	102.0	133.1	115.3	101.7	140.6	118.4	101.8	126.2	112.3
08	OPENINGS	98.8	143.4	109.7	102.7	144.8	113.0	101.2	144.8	111.8	100.6	144.8	111.4	99.2	144.8	110.3	95.3	134.1	104.7
0920	Plaster & Gypsum Board	105.8	151.0	135.4	118.5	151.0	139.8	125.5	151.0	142.2	118.5	151.0	139.8	121.8	151.0	140.9	120.4	135.5	130.3
0950, 0980	Ceilings & Acoustic Treatment	98.1	151.0	131.2	83.8	151.0	125.9	85.2	151.0	126.4	83.8	151.0	125.9	93.5	151.0	129.5	83.8	135.5	116.2
0960	Flooring	98.9	158.5	116.3	82.4	183.6	112.0	87.4	183.6	115.5	82.4	183.6	112.0	83.3	183.6	112.6	83.6	168.8	108.5
0970, 0990	Wall Finishes & Painting/Coating	77.1	147.2	119.1	82.6	142.9	118.7	82.6	142.9	118.7	82.6	142.9	118.7	82.7	142.9	118.8	82.7	147.2	121.4
09	FINISHES	94.8	150.4	124.9	90.5	154.7	125.2	93.8	155.2	127.1	90.4	154.7	125.2	93.3	155.2	126.9	91.5	142.7	119.2
COVERS	DIVS. 10 - 14, 25, 28, 41, 43, 44, 46	100.0	115.8	103.7	100.0	125.9	106.1	100.0	125.9	106.1	100.0	125.9	106.1	100.0	125.9	106.1	100.0	113.2	103.1
21, 22, 23	FIRE SUPPRESSION, PLUMBING & HVAC	100.7	132.5	113.6	100.4	135.3	114.5	100.7	135.2	114.6	100.4	135.0	114.5	100.7	135.0	114.7	100.4	125.0	110.3
26, 27, 3370	ELECTRICAL, COMMUNICATIONS & UTIL.	95.8	141.5	118.4	94.7	144.4	119.3	95.2	144.4	119.5	94.7	143.1	118.7	98.7	143.1	120.6	94.5	125.1	109.6
MF2018	WEIGHTED AVERAGE	98.9	133.7	114.0	97.6	137.5	114.8	98.9	137.6	115.6	97.4	137.1	114.6	97.7	137.1	114.7	97.1	126.6	109.9

For customer support on your Commercial Renovation Costs with RSMeans Data, call 800.448.8182.

813

NEW JERSEY

DIVISION		NEW BRUNSWICK 088 - 089			NEWARK 070 - 071			PATERSON 074 - 075			POINT PLEASANT 087			SUMMIT 079			TRENTON 085 - 086		
		MAT.	INST.	TOTAL	MAT.	INST.	TOTAL	MAT.	INST.	TOTAL	MAT.	INST.	TOTAL	MAT.	INST.	TOTAL	MAT.	INST.	TOTAL
015433	CONTRACTOR EQUIPMENT		91.4	91.4		99.5	99.5		94.2	94.2		91.4	91.4		94.2	94.2		95.5	95.5
0241, 31 - 34	SITE & INFRASTRUCTURE, DEMOLITION	105.8	98.0	100.4	114.9	107.9	110.1	110.5	98.8	102.5	107.2	97.6	100.6	110.3	98.9	102.4	90.8	103.4	99.5
0310	Concrete Forming & Accessories	105.4	147.6	141.3	99.8	148.4	141.2	98.2	148.0	140.6	99.0	134.2	129.0	98.9	148.2	140.9	98.9	133.8	128.6
0320	Concrete Reinforcing	83.0	151.9	116.2	97.7	152.4	124.1	93.8	152.4	122.1	83.0	151.9	116.2	72.3	152.4	110.9	107.7	110.7	109.2
0330	Cast-in-Place Concrete	98.6	132.0	111.3	93.5	138.9	110.6	82.2	137.6	103.1	98.6	130.8	110.8	68.7	137.7	94.8	94.5	129.4	107.7
03	CONCRETE	104.2	141.4	120.9	93.7	144.3	116.4	88.7	143.8	113.4	103.8	134.9	117.7	80.2	143.9	108.8	95.4	127.2	109.7
04	MASONRY	104.4	135.2	123.2	102.5	141.7	126.4	98.5	141.7	124.8	92.6	129.8	115.3	101.0	141.7	125.8	102.5	129.9	119.2
05	METALS	104.1	121.0	109.3	109.6	123.9	114.0	102.9	125.0	109.8	104.1	120.7	109.2	101.5	125.2	108.9	109.6	106.4	108.6
06	WOOD, PLASTICS & COMPOSITES	110.5	149.7	131.0	96.4	149.9	124.4	95.9	149.8	124.1	101.9	134.7	119.1	97.2	149.8	124.7	96.7	134.7	116.6
07	THERMAL & MOISTURE PROTECTION	101.0	136.2	116.1	104.1	141.9	120.3	102.3	133.1	115.5	101.0	128.4	112.8	102.5	140.8	119.0	100.9	132.6	114.5
08	OPENINGS	92.1	144.8	104.9	101.2	144.9	111.8	105.7	144.8	115.2	93.8	138.5	104.7	106.6	144.8	115.9	97.2	125.2	104.1
0920	Plaster & Gypsum Board	108.2	151.0	136.3	112.0	151.0	137.6	121.8	151.0	140.9	102.8	135.5	124.3	120.4	151.0	140.5	97.7	135.5	122.5
0950, 0980	Ceilings & Acoustic Treatment	89.1	151.0	127.9	92.7	151.0	129.2	93.5	151.0	129.5	89.1	135.5	118.2	83.8	151.0	125.9	92.7	135.5	119.5
0960	Flooring	100.3	183.6	124.6	97.6	183.6	122.7	83.3	183.6	112.6	97.6	158.5	115.4	83.6	183.6	112.8	104.3	168.8	123.2
0970, 0990	Wall Finishes & Painting/Coating	77.1	142.9	116.5	96.1	142.9	124.2	82.6	142.9	118.7	77.1	147.2	119.1	82.6	142.9	118.7	92.4	147.2	125.3
09	FINISHES	94.8	154.7	127.2	98.0	155.3	129.0	93.4	154.7	126.6	93.3	141.0	119.1	91.5	154.7	125.7	97.0	142.8	121.8
COVERS	DIVS. 10 - 14, 25, 28, 41, 43, 44, 46	100.0	125.7	106.1	100.0	126.2	106.2	100.0	125.9	106.1	100.0	105.9	101.4	100.0	125.9	106.1	100.0	113.3	103.1
21, 22, 23	FIRE SUPPRESSION, PLUMBING & HVAC	100.4	127.8	111.5	100.8	134.2	114.3	100.7	135.1	114.6	100.4	125.0	110.3	100.4	132.3	113.3	100.8	124.6	110.4
26, 27, 3370	ELECTRICAL, COMMUNICATIONS & UTIL.	92.7	128.7	110.5	103.3	143.1	123.0	98.7	144.4	121.3	92.2	125.1	108.5	95.2	144.4	119.5	100.5	123.9	112.1
MF2018	WEIGHTED AVERAGE	99.7	132.2	113.8	101.8	137.9	117.4	99.3	137.3	115.7	99.1	126.5	110.9	97.7	136.9	114.6	100.5	124.4	110.8

| DIVISION | | NEW JERSEY VINELAND 080,083 | | | NEW MEXICO | | | | | | | | | | | | | | |
					ALBUQUERQUE 870 - 872			CARRIZOZO 883			CLOVIS 881			FARMINGTON 874			GALLUP 873		
		MAT.	INST.	TOTAL	MAT.	INST.	TOTAL	MAT.	INST.	TOTAL	MAT.	INST.	TOTAL	MAT.	INST.	TOTAL	MAT.	INST.	TOTAL
015433	CONTRACTOR EQUIPMENT		91.8	91.8		105.2	105.2		105.2	105.2		105.2	105.2		105.2	105.2		105.2	105.2
0241, 31 - 34	SITE & INFRASTRUCTURE, DEMOLITION	96.7	97.3	97.1	88.4	94.4	92.5	106.6	94.4	98.2	93.9	94.4	94.2	94.6	94.4	94.5	103.2	94.4	97.1
0310	Concrete Forming & Accessories	95.7	133.1	127.5	103.1	65.0	70.7	95.9	65.0	69.6	95.9	65.0	69.5	103.2	65.0	70.7	103.2	65.0	70.7
0320	Concrete Reinforcing	82.0	125.1	102.8	96.3	69.5	83.4	109.2	69.5	90.0	110.4	69.5	90.7	105.3	69.5	88.1	100.8	69.5	85.7
0330	Cast-in-Place Concrete	86.0	129.1	102.3	90.4	68.9	82.2	93.4	68.9	84.1	93.3	68.8	84.1	91.2	68.9	82.8	85.8	68.9	79.4
03	CONCRETE	92.2	129.2	108.8	94.1	68.5	82.6	116.6	68.5	95.0	104.3	68.4	88.2	97.6	68.5	84.5	104.4	68.5	88.3
04	MASONRY	94.6	129.9	116.1	105.7	62.4	79.3	107.4	62.4	80.0	107.4	62.4	80.0	112.7	62.4	82.0	101.8	62.4	77.8
05	METALS	103.9	111.5	106.3	106.2	89.8	101.1	104.7	89.8	100.0	104.3	89.6	99.7	103.9	89.8	99.5	103.0	89.8	98.9
06	WOOD, PLASTICS & COMPOSITES	98.3	133.3	116.6	104.5	64.9	83.8	89.5	64.9	76.6	89.5	64.9	76.6	104.6	64.9	83.8	104.6	64.9	83.8
07	THERMAL & MOISTURE PROTECTION	100.4	131.1	113.6	98.9	72.0	87.4	103.9	72.0	90.2	102.5	72.0	89.4	99.1	72.0	87.4	100.3	72.0	88.1
08	OPENINGS	93.4	131.8	102.8	98.9	65.1	90.7	96.5	65.1	88.9	96.7	65.1	88.9	96.7	65.1	92.3	101.1	65.1	92.3
0920	Plaster & Gypsum Board	101.1	134.0	122.7	126.4	63.9	85.4	82.9	63.9	70.5	82.9	63.9	70.5	111.9	63.9	80.4	111.9	63.9	80.4
0950, 0980	Ceilings & Acoustic Treatment	89.1	134.0	117.3	111.4	63.9	81.6	112.0	63.9	81.9	112.0	63.9	81.9	108.5	63.9	80.6	108.5	63.9	80.6
0960	Flooring	96.9	158.5	114.9	88.4	65.0	81.6	96.5	65.0	87.3	96.5	65.0	87.3	89.8	65.0	82.6	89.8	65.0	82.6
0970, 0990	Wall Finishes & Painting/Coating	77.1	147.2	119.1	92.0	51.1	67.5	89.3	51.1	66.4	89.3	51.1	66.4	86.4	51.1	65.3	86.4	51.1	65.3
09	FINISHES	92.0	140.1	118.0	99.0	63.3	79.6	97.5	63.3	78.9	96.2	63.3	78.4	96.6	63.3	78.5	97.8	63.3	79.1
COVERS	DIVS. 10 - 14, 25, 28, 41, 43, 44, 46	100.0	113.2	103.1	100.0	86.4	96.8	100.0	86.4	96.8	100.0	86.4	96.8	100.0	86.4	96.8	100.0	86.4	96.8
21, 22, 23	FIRE SUPPRESSION, PLUMBING & HVAC	100.4	124.7	110.2	101.4	68.1	87.9	98.7	68.1	86.3	98.7	67.7	86.2	101.2	68.1	87.9	98.9	68.1	86.5
26, 27, 3370	ELECTRICAL, COMMUNICATIONS & UTIL.	92.2	138.0	114.9	86.6	69.5	78.1	89.7	69.5	79.7	87.8	69.5	78.7	84.9	69.5	77.3	84.3	69.5	77.0
MF2018	WEIGHTED AVERAGE	97.2	126.6	109.9	98.9	71.6	87.1	101.4	71.6	88.5	99.1	71.5	87.2	99.3	71.6	87.3	99.2	71.6	87.3

NEW MEXICO

DIVISION		LAS CRUCES 880			LAS VEGAS 877			ROSWELL 882			SANTA FE 875			SOCORRO 878			TRUTH/CONSEQUENCES 879		
		MAT.	INST.	TOTAL	MAT.	INST.	TOTAL	MAT.	INST.	TOTAL	MAT.	INST.	TOTAL	MAT.	INST.	TOTAL	MAT.	INST.	TOTAL
015433	CONTRACTOR EQUIPMENT		81.1	81.1		105.2	105.2		105.2	105.2		109.7	109.7		105.2	105.2		81.1	81.1
0241, 31 - 34	SITE & INFRASTRUCTURE, DEMOLITION	95.6	74.7	81.2	93.4	94.4	94.1	96.5	94.4	95.0	95.6	102.8	100.5	90.0	94.4	93.0	108.8	74.7	85.4
0310	Concrete Forming & Accessories	92.5	64.0	68.3	103.2	65.0	70.7	95.9	65.0	69.6	101.9	65.1	70.6	103.2	65.0	70.7	100.9	64.0	69.5
0320	Concrete Reinforcing	106.0	69.3	88.3	102.6	69.5	86.6	110.4	69.5	90.7	102.1	69.5	86.4	104.5	69.5	87.6	98.0	69.4	84.2
0330	Cast-in-Place Concrete	88.2	62.1	78.3	88.7	68.9	81.2	93.3	68.9	84.1	94.2	70.0	85.1	86.9	68.9	80.1	95.3	62.1	82.7
03	CONCRETE	82.0	65.2	74.5	95.2	68.5	83.2	105.2	68.5	88.7	99.1	68.8	85.5	94.1	68.5	82.6	87.3	65.2	77.4
04	MASONRY	103.0	60.7	77.2	102.1	62.4	77.9	118.1	62.4	84.2	98.8	62.5	76.7	102.0	62.4	77.9	98.6	62.1	76.4
05	METALS	103.1	82.1	96.5	102.7	89.8	98.7	105.6	89.8	100.7	101.0	87.7	96.8	103.0	89.8	98.9	102.5	82.1	96.2
06	WOOD, PLASTICS & COMPOSITES	79.9	63.9	71.5	104.6	64.9	83.8	89.5	64.9	76.6	101.3	64.9	82.3	104.6	64.9	83.8	96.9	63.9	79.6
07	THERMAL & MOISTURE PROTECTION	91.8	67.2	81.2	98.6	72.0	87.2	102.7	72.0	89.5	100.5	73.4	88.8	98.6	72.0	87.2	89.2	67.6	79.9
08	OPENINGS	92.1	64.6	85.4	97.7	65.1	89.8	96.5	65.1	88.8	99.1	65.2	90.9	97.6	65.1	89.7	91.1	64.6	84.6
0920	Plaster & Gypsum Board	82.0	63.9	70.1	111.9	63.9	80.4	82.9	63.9	70.5	125.2	63.9	85.0	111.9	63.9	80.4	113.4	63.9	80.9
0950, 0980	Ceilings & Acoustic Treatment	96.8	63.9	76.2	108.5	63.9	80.6	112.0	63.9	81.9	109.4	63.9	80.9	108.5	63.9	80.6	98.5	63.9	76.9
0960	Flooring	127.1	65.0	109.0	89.8	65.0	82.6	96.5	65.0	87.3	98.9	65.0	89.0	89.8	65.0	82.6	102.8	65.0	91.8
0970, 0990	Wall Finishes & Painting/Coating	78.5	51.1	62.1	86.4	51.1	65.3	89.3	51.1	66.4	96.7	51.1	69.4	86.4	51.1	65.3	79.1	51.1	62.3
09	FINISHES	104.6	62.5	81.8	96.4	63.3	78.5	96.3	63.3	78.4	103.8	63.3	81.9	96.3	63.3	78.4	107.6	62.5	83.2
COVERS	DIVS. 10 - 14, 25, 28, 41, 43, 44, 46	100.0	84.1	96.3	100.0	86.4	96.8	100.0	86.4	96.8	100.0	86.5	96.8	100.0	86.4	96.8	100.0	84.1	96.2
21, 22, 23	FIRE SUPPRESSION, PLUMBING & HVAC	101.7	67.8	88.0	98.9	68.1	86.5	100.9	68.1	87.7	101.2	68.1	87.8	98.9	68.1	86.5	98.8	67.8	86.3
26, 27, 3370	ELECTRICAL, COMMUNICATIONS & UTIL.	90.0	80.7	85.4	86.2	69.5	77.9	88.9	69.5	79.3	100.7	69.4	85.2	84.7	69.5	77.2	88.1	69.4	78.9
MF2018	WEIGHTED AVERAGE	96.8	70.0	85.2	97.4	71.6	86.3	100.7	71.6	88.1	100.5	72.2	88.3	97.1	71.6	86.1	96.8	68.6	84.6

City Cost Indexes - V2

DIVISION		NEW MEXICO TUCUMCARI 884			NEW YORK														
					ALBANY 120 - 122			BINGHAMTON 137 - 139			BRONX 104			BROOKLYN 112			BUFFALO 140 - 142		
		MAT.	INST.	TOTAL	MAT.	INST.	TOTAL	MAT.	INST.	TOTAL	MAT.	INST.	TOTAL	MAT.	INST.	TOTAL	MAT.	INST.	TOTAL
015433	CONTRACTOR EQUIPMENT		105.2	105.2		117.0	117.0		115.8	115.8		102.9	102.9		107.5	107.5		101.3	101.3
0241, 31 - 34	SITE & INFRASTRUCTURE, DEMOLITION	93.5	94.4	94.1	78.5	107.4	98.4	94.1	87.0	89.2	98.8	183.3	170.1	106.9	183.3	171.9	99.2	103.8	102.4
0310	Concrete Forming & Accessories	95.9	64.9	69.5	99.3	104.4	103.7	100.5	92.2	93.4	94.9	183.3	170.1	97.9	183.3	169.7	103.4	115.3	113.6
0320	Concrete Reinforcing	108.2	69.5	89.5	99.2	112.1	105.4	96.5	106.7	101.4	95.6	173.5	133.2	97.9	231.2	162.2	95.5	113.4	104.2
0330	Cast-in-Place Concrete	93.3	68.8	84.1	76.1	115.5	91.0	105.2	103.6	104.6	82.3	167.1	114.3	107.2	165.5	129.2	117.3	122.0	119.1
03	CONCRETE	103.6	68.4	87.8	89.6	110.6	99.0	94.1	101.0	97.2	86.3	175.5	126.3	105.6	183.9	140.7	107.1	116.9	111.5
04	MASONRY	119.1	62.4	84.5	97.2	116.3	108.9	109.7	103.6	106.0	88.5	184.6	147.1	119.1	184.6	159.1	110.8	123.5	118.5
05	METALS	104.3	89.6	99.7	103.1	123.4	109.4	96.0	135.0	108.2	92.9	173.2	117.9	103.9	173.4	125.6	93.9	107.2	98.0
06	WOOD, PLASTICS & COMPOSITES	89.5	64.9	76.6	97.4	100.6	99.1	102.1	89.0	95.2	93.5	182.4	140.0	105.3	182.1	145.5	100.9	114.0	107.8
07	THERMAL & MOISTURE PROTECTION	102.5	72.0	89.4	107.7	109.7	108.6	110.1	93.1	102.8	102.8	163.8	129.0	110.3	163.2	133.0	104.0	111.2	107.1
08	OPENINGS	96.4	65.1	88.3	96.4	101.9	97.7	90.5	94.1	91.4	92.5	191.0	116.5	87.9	190.9	113.0	100.0	109.5	102.3
0920	Plaster & Gypsum Board	82.9	63.9	70.5	105.1	100.4	102.0	113.6	88.6	97.2	92.2	184.6	152.8	106.0	184.6	157.6	119.9	114.3	116.2
0950, 0980	Ceilings & Acoustic Treatment	112.0	63.9	81.9	97.3	100.4	99.3	102.4	88.6	93.8	80.3	184.6	145.7	96.1	184.6	151.6	102.1	114.3	109.7
0960	Flooring	96.5	65.0	87.3	92.6	107.6	97.0	103.0	97.8	101.5	99.4	182.7	123.7	110.0	182.7	131.2	94.1	119.1	101.4
0970, 0990	Wall Finishes & Painting/Coating	89.3	51.1	66.4	93.9	99.4	97.2	84.5	101.3	94.5	104.7	164.2	140.4	109.6	164.2	142.3	93.8	114.4	106.1
09	FINISHES	96.1	63.3	78.3	94.8	103.9	99.7	96.7	93.4	94.9	93.8	181.8	141.4	107.8	181.6	147.7	101.6	116.4	109.6
COVERS	DIVS. 10 - 14, 25, 28, 41, 43, 44, 46	100.0	86.4	96.8	100.0	100.9	100.2	100.0	98.0	99.5	100.0	137.9	108.9	100.0	137.2	108.8	100.0	106.3	101.5
21, 22, 23	FIRE SUPPRESSION, PLUMBING & HVAC	98.7	67.7	86.2	101.2	108.0	103.9	101.6	98.6	100.4	100.4	171.5	129.1	100.5	171.5	129.1	100.9	102.0	101.3
26, 27, 3370	ELECTRICAL, COMMUNICATIONS & UTIL.	89.7	69.5	79.7	101.8	107.2	104.5	100.2	97.3	98.7	93.6	183.2	137.9	100.2	183.2	141.3	99.9	104.3	102.1
MF2018	WEIGHTED AVERAGE	99.7	71.5	87.6	98.3	109.4	103.1	98.5	100.5	99.3	94.8	171.2	127.8	102.7	173.1	133.1	100.0	109.8	104.8

DIVISION		NEW YORK																	
		ELMIRA 148 - 149			FAR ROCKAWAY 116			FLUSHING 113			GLENS FALLS 128			HICKSVILLE 115,117,118			JAMAICA 114		
		MAT.	INST.	TOTAL	MAT.	INST.	TOTAL	MAT.	INST.	TOTAL	MAT.	INST.	TOTAL	MAT.	INST.	TOTAL	MAT.	INST.	TOTAL
015433	CONTRACTOR EQUIPMENT		118.7	118.7		107.5	107.5		107.5	107.5		110.8	110.8		107.5	107.5		107.5	107.5
0241, 31 - 34	SITE & INFRASTRUCTURE, DEMOLITION	97.9	87.0	90.4	120.6	117.5	118.5	120.6	117.5	118.5	70.0	96.4	88.1	111.0	116.0	114.4	114.9	117.5	116.7
0310	Concrete Forming & Accessories	83.6	94.4	92.8	92.0	183.3	169.7	96.2	183.3	170.3	82.5	96.2	94.2	88.1	153.9	144.1	96.2	183.3	170.3
0320	Concrete Reinforcing	97.6	105.5	101.4	97.9	231.2	162.2	99.6	231.2	163.1	95.2	107.4	101.1	97.9	169.7	132.5	97.9	231.2	162.2
0330	Cast-in-Place Concrete	103.2	102.9	103.1	116.0	165.5	134.7	116.0	165.5	134.7	73.3	107.2	86.1	98.4	155.6	120.0	107.2	165.5	129.2
03	CONCRETE	92.8	101.7	96.8	111.7	183.9	144.1	112.2	183.9	144.4	78.7	103.3	89.7	97.4	156.5	123.9	104.9	183.9	140.4
04	MASONRY	106.2	103.1	104.3	124.2	184.6	161.0	117.4	184.6	158.4	102.3	107.9	105.7	114.2	168.6	147.4	122.3	184.6	160.3
05	METALS	92.4	136.3	106.1	104.0	173.4	125.6	104.0	173.4	125.6	96.4	122.5	104.6	105.5	147.9	118.7	104.0	173.4	125.6
06	WOOD, PLASTICS & COMPOSITES	81.9	92.7	87.6	86.5	182.1	136.5	91.8	182.1	139.0	83.3	92.9	88.3	82.6	151.9	118.9	91.8	182.1	139.0
07	THERMAL & MOISTURE PROTECTION	110.8	92.9	103.1	110.2	163.4	133.0	110.2	163.4	133.1	101.0	102.3	101.5	109.8	155.1	129.2	110.0	163.4	132.9
08	OPENINGS	96.8	95.8	96.6	86.7	190.9	112.1	86.7	190.9	112.1	90.2	95.4	91.5	87.1	159.2	104.7	86.7	190.9	112.1
0920	Plaster & Gypsum Board	110.2	92.6	98.6	92.7	184.6	153.0	95.6	184.6	154.0	90.9	92.8	92.1	92.5	153.6	132.6	95.6	184.6	154.0
0950, 0980	Ceilings & Acoustic Treatment	109.1	92.6	98.7	83.4	184.6	146.8	83.4	184.6	146.8	88.5	92.8	91.2	82.6	153.6	127.1	83.4	184.6	146.8
0960	Flooring	86.8	97.8	90.0	104.5	182.7	127.3	106.1	182.7	128.4	83.8	102.8	89.4	103.4	180.3	125.9	106.1	182.7	128.4
0970, 0990	Wall Finishes & Painting/Coating	90.7	88.9	89.6	109.6	164.2	142.3	109.6	164.2	142.3	85.3	99.4	93.8	109.6	164.2	142.3	109.6	164.2	142.3
09	FINISHES	97.2	94.1	95.5	101.7	181.6	144.9	102.5	181.6	145.3	86.6	97.1	92.3	100.3	159.5	132.3	102.1	181.6	145.1
COVERS	DIVS. 10 - 14, 25, 28, 41, 43, 44, 46	100.0	98.2	99.6	100.0	137.2	108.8	100.0	137.2	108.8	100.0	91.9	98.1	100.0	131.3	107.4	100.0	137.2	108.8
21, 22, 23	FIRE SUPPRESSION, PLUMBING & HVAC	97.3	94.4	96.1	96.6	171.5	126.8	96.6	171.5	126.8	97.5	104.1	100.1	100.5	155.3	122.6	96.6	171.5	126.8
26, 27, 3370	ELECTRICAL, COMMUNICATIONS & UTIL.	97.6	100.3	98.9	106.4	183.2	144.4	106.4	183.2	144.4	95.8	101.1	98.4	99.7	145.8	122.5	98.9	183.2	140.6
MF2018	WEIGHTED AVERAGE	96.9	100.4	98.4	102.7	173.1	133.1	102.5	173.1	133.0	92.8	103.2	97.3	100.5	151.7	122.6	100.9	173.1	132.1

DIVISION		NEW YORK																	
		JAMESTOWN 147			KINGSTON 124			LONG ISLAND CITY 111			MONTICELLO 127			MOUNT VERNON 105			NEW ROCHELLE 108		
		MAT.	INST.	TOTAL	MAT.	INST.	TOTAL	MAT.	INST.	TOTAL	MAT.	INST.	TOTAL	MAT.	INST.	TOTAL	MAT.	INST.	TOTAL
015433	CONTRACTOR EQUIPMENT		89.2	89.2		107.5	107.5		107.5	107.5		107.5	107.5		102.9	102.9		102.9	102.9
0241, 31 - 34	SITE & INFRASTRUCTURE, DEMOLITION	99.4	87.2	91.0	137.1	112.4	120.1	119.0	117.5	118.0	132.6	112.4	118.7	104.6	102.9	103.5	103.8	102.9	103.2
0310	Concrete Forming & Accessories	83.7	86.7	86.2	83.9	126.0	119.7	101.0	183.3	171.0	92.0	126.0	120.9	84.9	134.4	127.0	99.9	131.0	126.4
0320	Concrete Reinforcing	97.8	105.9	101.7	95.6	152.8	123.2	97.9	231.2	162.2	94.8	152.8	122.8	94.5	172.2	132.0	94.6	172.1	132.0
0330	Cast-in-Place Concrete	107.1	101.1	104.8	99.0	140.7	114.7	110.6	165.5	131.3	92.6	140.8	110.8	91.9	145.2	112.1	91.9	144.9	111.9
03	CONCRETE	95.9	95.3	95.6	98.8	135.6	115.3	108.1	183.9	142.1	94.1	135.6	112.7	95.0	145.5	117.7	94.5	143.8	116.6
04	MASONRY	116.0	100.3	106.4	118.4	148.9	137.0	115.6	184.6	157.7	110.9	148.9	134.1	94.5	151.4	129.2	94.5	151.4	129.2
05	METALS	90.0	100.3	93.2	104.8	133.9	113.9	103.9	173.4	125.6	104.8	134.0	113.9	92.7	168.0	116.1	92.9	167.2	116.1
06	WOOD, PLASTICS & COMPOSITES	80.7	83.3	82.1	84.7	119.9	103.1	98.6	182.1	142.3	92.9	119.8	107.0	83.3	128.0	106.7	100.9	124.4	113.2
07	THERMAL & MOISTURE PROTECTION	110.3	91.1	102.0	123.0	139.6	130.2	110.2	163.4	133.0	122.7	139.6	130.0	103.6	141.4	119.8	103.8	139.0	118.9
08	OPENINGS	96.7	89.6	95.0	92.4	137.8	103.5	86.7	190.9	112.1	88.5	137.7	100.5	92.5	162.7	109.6	92.5	160.7	109.2
0920	Plaster & Gypsum Board	98.3	82.9	88.2	91.3	120.7	110.6	100.9	184.6	155.8	91.9	120.7	110.8	88.3	128.7	114.8	100.4	125.0	116.5
0950, 0980	Ceilings & Acoustic Treatment	104.6	82.9	91.0	75.7	120.7	103.9	83.4	184.6	146.8	75.7	120.7	103.9	78.4	128.7	109.9	78.4	125.0	107.6
0960	Flooring	89.2	97.8	91.7	100.9	153.2	116.2	107.9	182.7	129.7	103.4	153.2	117.9	90.8	171.4	114.3	98.9	156.5	115.7
0970, 0990	Wall Finishes & Painting/Coating	92.2	95.8	94.4	111.2	124.7	119.3	109.6	164.2	142.3	111.2	117.5	115.0	103.2	164.2	139.8	103.2	164.2	139.8
09	FINISHES	95.3	88.9	91.8	98.5	130.1	115.6	103.6	181.6	145.8	98.9	129.2	115.3	90.9	143.1	119.1	94.8	137.9	118.1
COVERS	DIVS. 10 - 14, 25, 28, 41, 43, 44, 46	100.0	91.9	98.1	100.0	115.1	103.6	100.0	137.2	108.8	100.0	115.1	103.6	100.0	123.2	105.5	100.0	107.8	101.8
21, 22, 23	FIRE SUPPRESSION, PLUMBING & HVAC	97.1	89.2	93.9	97.4	128.3	109.8	100.5	171.5	129.1	97.4	132.3	111.5	96.8	146.3	116.7	96.8	146.0	116.6
26, 27, 3370	ELECTRICAL, COMMUNICATIONS & UTIL.	96.6	91.4	94.0	97.2	112.0	104.5	99.3	183.2	140.8	97.2	112.0	104.5	91.7	164.2	127.6	91.7	136.4	113.9
MF2018	WEIGHTED AVERAGE	97.1	92.3	95.0	101.2	128.8	113.1	102.2	173.1	132.8	99.8	129.5	112.7	94.9	147.0	117.4	95.3	141.4	115.2

For customer support on your Commercial Renovation Costs with RSMeans Data, call 800.448.8182.

815

		NEW YORK																
DIVISION		NEW YORK 100 - 102			NIAGARA FALLS 143			PLATTSBURGH 129			POUGHKEEPSIE 125 - 126			QUEENS 110			RIVERHEAD 119		
		MAT.	INST.	TOTAL	MAT.	INST.	TOTAL	MAT.	INST.	TOTAL	MAT.	INST.	TOTAL	MAT.	INST.	TOTAL	MAT.	INST.	TOTAL
015433	CONTRACTOR EQUIPMENT		105.6	105.6		89.2	89.2		92.5	92.5		107.5	107.5		107.5	107.5		107.5	107.5
0241, 31 - 34	SITE & INFRASTRUCTURE, DEMOLITION	106.6	112.7	110.8	101.5	88.1	92.3	107.7	93.5	98.0	133.6	111.4	118.3	114.2	117.5	116.5	112.0	115.7	114.6
0310	Concrete Forming & Accessories	103.6	186.4	174.0	83.6	108.9	105.1	88.0	88.0	88.0	83.9	161.5	149.9	88.3	183.3	169.1	93.3	154.5	145.4
0320	Concrete Reinforcing	107.3	175.3	140.1	96.5	106.7	101.4	99.5	106.7	103.0	95.6	152.4	123.0	99.6	231.2	163.1	99.7	241.9	168.3
0330	Cast-in-Place Concrete	95.3	168.6	123.0	110.8	119.4	114.0	89.6	96.6	92.2	95.8	130.2	108.8	101.8	165.5	125.9	98.0	169.2	130.0
03	CONCRETE	99.3	177.7	134.5	98.2	111.8	104.3	91.1	94.2	92.5	96.2	147.9	119.4	100.6	183.9	138.0	98.0	169.2	130.0
04	MASONRY	99.0	186.2	152.2	123.1	119.9	121.1	96.8	93.3	94.7	110.0	133.7	124.5	109.4	184.6	155.3	119.6	168.6	149.5
05	METALS	105.2	174.6	126.8	92.5	100.8	95.0	100.7	97.6	99.7	104.9	133.6	113.8	103.9	173.4	125.6	106.0	174.2	127.3
06	WOOD, PLASTICS & COMPOSITES	99.3	186.1	144.7	80.6	104.6	93.2	90.5	85.2	87.7	84.7	173.3	131.1	82.8	182.1	134.7	88.5	151.9	121.7
07	THERMAL & MOISTURE PROTECTION	105.0	168.7	132.3	110.4	105.1	108.1	116.3	92.8	106.2	123.0	138.8	129.8	109.7	163.4	132.7	111.3	154.4	129.8
08	OPENINGS	96.2	197.3	120.8	96.7	100.8	97.7	97.6	91.2	96.0	92.5	164.3	110.0	86.7	190.9	112.1	87.1	177.0	109.0
0920	Plaster & Gypsum Board	99.4	188.3	157.7	98.3	104.8	102.6	111.6	84.5	93.8	91.3	175.6	146.6	92.5	184.6	152.9	93.8	153.6	133.1
0950, 0980	Ceilings & Acoustic Treatment	98.1	188.3	154.6	104.6	104.8	104.7	104.9	84.5	92.1	75.7	175.6	138.3	83.4	184.6	146.8	83.3	153.6	127.8
0960	Flooring	99.8	184.4	124.5	89.2	106.9	94.4	106.3	100.2	104.5	100.9	151.4	115.6	103.4	182.7	126.6	104.5	180.3	126.6
0970, 0990	Wall Finishes & Painting/Coating	104.0	169.0	143.0	92.2	106.1	100.6	106.7	90.6	97.1	111.2	117.5	115.0	109.6	164.2	142.3	109.6	164.2	142.3
09	FINISHES	99.6	185.1	145.9	95.4	108.5	102.5	98.2	89.8	93.7	98.3	157.8	130.5	100.6	181.6	144.4	100.9	159.5	132.6
COVERS	DIVS. 10 - 14, 25, 28, 41, 43, 44, 46	100.0	141.5	109.8	100.0	97.5	99.4	100.0	87.9	97.2	100.0	110.5	102.5	100.0	137.2	108.8	100.0	128.5	106.7
21, 22, 23	FIRE SUPPRESSION, PLUMBING & HVAC	99.4	174.6	129.7	97.1	100.2	98.4	97.4	95.6	96.6	97.4	114.6	104.3	100.5	171.5	129.1	100.8	157.4	123.6
26, 27, 3370	ELECTRICAL, COMMUNICATIONS & UTIL.	99.3	185.3	141.9	95.4	95.2	95.3	93.4	88.3	90.9	97.2	116.7	106.9	99.7	183.2	141.0	101.1	146.0	123.3
MF2018	WEIGHTED AVERAGE	100.3	174.2	132.2	98.0	103.4	100.3	97.7	92.8	95.6	100.3	131.7	113.9	100.5	173.1	131.8	101.3	156.9	125.3

		NEW YORK																
DIVISION		ROCHESTER 144 - 146			SCHENECTADY 123			STATEN ISLAND 103			SUFFERN 109			SYRACUSE 130 - 132			UTICA 133 - 135		
		MAT.	INST.	TOTAL	MAT.	INST.	TOTAL	MAT.	INST.	TOTAL	MAT.	INST.	TOTAL	MAT.	INST.	TOTAL	MAT.	INST.	TOTAL
015433	CONTRACTOR EQUIPMENT		117.0	117.0		110.8	110.8		102.9	102.9		102.9	102.9		110.8	110.8		110.8	110.8
0241, 31 - 34	SITE & INFRASTRUCTURE, DEMOLITION	89.1	104.7	99.8	79.9	97.2	91.8	109.3	107.1	107.8	100.9	100.7	100.8	92.9	96.5	95.4	71.8	96.0	88.4
0310	Concrete Forming & Accessories	103.4	98.9	99.6	100.9	104.1	103.6	84.3	183.5	168.7	93.5	135.6	129.4	99.6	93.3	94.2	100.5	92.8	93.9
0320	Concrete Reinforcing	96.8	105.7	101.1	94.2	112.0	102.8	95.6	205.9	148.8	94.6	144.8	118.8	97.5	106.8	102.0	97.5	105.7	101.5
0330	Cast-in-Place Concrete	99.2	107.0	102.1	88.5	113.5	97.9	91.9	167.1	120.3	89.1	133.7	106.0	97.7	107.2	101.3	89.3	106.6	95.9
03	CONCRETE	98.0	104.1	100.8	91.9	109.9	100.0	96.7	180.6	134.4	91.9	136.1	111.7	97.0	101.9	99.2	95.1	101.2	97.8
04	MASONRY	104.3	106.9	105.9	99.6	116.2	109.7	100.6	184.6	151.8	94.3	136.2	119.9	100.9	108.0	105.2	92.5	107.5	101.6
05	METALS	101.2	121.3	107.4	100.7	125.7	108.5	91.0	173.5	116.7	91.0	130.1	103.2	98.9	121.8	106.5	98.9	88.4	93.4
06	WOOD, PLASTICS & COMPOSITES	104.7	97.0	100.6	104.9	100.3	102.5	82.0	182.4	134.5	93.0	137.7	116.4	98.9	89.3	93.9	94.9	99.1	96.7
07	THERMAL & MOISTURE PROTECTION	117.5	102.1	110.9	102.6	108.5	105.2	103.1	163.8	129.2	103.6	135.8	117.4	105.3	99.4	102.8	94.9	99.1	96.7
08	OPENINGS	100.6	98.4	100.1	95.3	101.7	96.9	92.5	191.0	116.5	92.5	145.7	105.5	92.5	92.4	92.5	95.1	91.7	94.2
0920	Plaster & Gypsum Board	106.4	96.8	100.1	102.1	100.4	101.0	88.5	184.6	151.5	91.4	138.7	122.4	102.4	89.1	93.7	102.4	88.1	93.0
0950, 0980	Ceilings & Acoustic Treatment	104.8	96.8	99.8	99.2	100.4	100.0	80.3	184.6	145.7	78.4	138.7	116.2	102.4	89.1	94.0	102.4	88.1	93.5
0960	Flooring	92.2	107.5	96.7	91.4	107.6	96.2	94.5	182.7	120.2	94.6	163.7	114.8	91.9	96.7	93.3	89.6	98.1	92.1
0970, 0990	Wall Finishes & Painting/Coating	93.7	99.7	97.3	85.3	99.4	93.8	104.7	164.2	140.4	103.2	120.9	113.8	87.1	102.6	96.4	81.0	102.6	93.9
09	FINISHES	98.8	100.4	99.7	93.6	103.7	99.0	92.8	181.8	141.0	92.1	140.1	118.1	95.3	94.1	94.6	93.6	93.9	93.8
COVERS	DIVS. 10 - 14, 25, 28, 41, 43, 44, 46	100.0	99.7	99.9	100.0	100.1	100.0	100.0	137.9	108.9	100.0	108.9	102.1	100.0	98.5	99.7	100.0	98.0	99.5
21, 22, 23	FIRE SUPPRESSION, PLUMBING & HVAC	100.9	91.1	96.6	101.2	107.9	103.9	100.4	171.5	129.1	96.8	117.8	105.2	101.2	95.8	99.0	101.2	101.7	101.4
26, 27, 3370	ELECTRICAL, COMMUNICATIONS & UTIL.	102.6	96.5	99.6	99.8	107.2	103.5	93.6	183.2	137.9	97.3	110.4	103.8	100.1	106.8	103.4	98.0	106.8	102.4
MF2018	WEIGHTED AVERAGE	100.8	101.2	101.0	97.9	108.6	102.5	96.5	171.9	129.1	94.9	125.4	108.1	98.7	101.6	99.9	96.8	102.5	99.3

		NEW YORK									NORTH CAROLINA								
DIVISION		WATERTOWN 136			WHITE PLAINS 106			YONKERS 107			ASHEVILLE 287 - 288			CHARLOTTE 281 - 282			DURHAM 277		
		MAT.	INST.	TOTAL	MAT.	INST.	TOTAL	MAT.	INST.	TOTAL	MAT.	INST.	TOTAL	MAT.	INST.	TOTAL	MAT.	INST.	TOTAL
015433	CONTRACTOR EQUIPMENT		110.8	110.8		102.9	102.9		102.9	102.9		97.0	97.0		100.2	100.2		102.2	102.2
0241, 31 - 34	SITE & INFRASTRUCTURE, DEMOLITION	79.3	96.3	90.9	99.0	103.0	101.7	106.6	103.0	104.1	96.5	76.0	82.4	99.5	83.5	88.5	101.8	84.5	89.9
0310	Concrete Forming & Accessories	83.9	96.0	94.2	98.8	145.4	138.4	99.0	145.5	138.6	93.8	59.6	64.7	98.3	64.7	69.8	101.5	59.6	65.9
0320	Concrete Reinforcing	98.2	106.8	102.3	94.6	172.3	132.1	98.4	172.3	134.0	93.6	62.9	78.8	96.6	68.3	83.0	95.4	63.6	80.1
0330	Cast-in-Place Concrete	104.1	109.4	106.1	81.6	145.4	105.7	91.3	145.4	111.8	102.3	70.1	90.1	104.4	72.6	92.4	115.8	70.1	98.5
03	CONCRETE	107.5	103.8	105.8	86.0	150.5	114.9	94.6	150.6	119.7	89.6	65.6	78.8	92.6	69.5	82.3	99.0	65.7	84.1
04	MASONRY	93.7	111.5	104.6	93.9	151.4	129.0	96.6	151.4	130.0	85.3	62.6	71.4	87.3	65.2	73.8	86.9	62.6	72.0
05	METALS	97.7	121.6	105.1	92.5	168.3	116.2	101.3	168.4	122.3	101.0	89.1	97.3	101.9	89.1	97.9	117.4	89.3	108.6
06	WOOD, PLASTICS & COMPOSITES	78.9	91.9	85.7	98.9	142.5	121.7	98.8	142.5	121.7	91.4	57.0	73.4	92.6	63.2	77.3	96.9	57.0	76.1
07	THERMAL & MOISTURE PROTECTION	95.2	101.2	97.8	103.4	144.0	120.8	103.8	144.6	121.3	98.7	61.5	82.7	91.9	66.2	80.9	101.7	61.4	84.4
08	OPENINGS	95.1	95.4	95.1	92.5	170.7	111.6	96.2	170.7	114.3	89.3	57.6	81.6	99.2	61.5	90.1	95.9	57.8	86.6
0920	Plaster & Gypsum Board	93.1	91.7	92.2	94.8	143.6	126.8	99.7	143.6	128.5	108.8	55.9	74.1	100.2	62.3	75.3	95.8	55.9	69.6
0950, 0980	Ceilings & Acoustic Treatment	102.4	91.7	95.7	78.4	143.6	119.3	98.1	143.6	126.6	89.6	55.9	68.5	92.1	62.3	73.4	98.1	55.9	71.6
0960	Flooring	82.6	98.1	87.1	97.1	180.3	121.4	96.5	182.7	121.7	98.2	63.7	88.1	97.2	68.4	88.8	104.5	63.7	92.6
0970, 0990	Wall Finishes & Painting/Coating	81.0	101.0	93.0	103.2	164.2	139.8	103.2	164.2	139.8	98.6	54.4	72.1	95.0	66.8	78.1	100.1	54.4	72.7
09	FINISHES	90.9	96.4	93.9	92.9	153.5	125.7	98.7	153.9	128.6	96.0	59.3	76.2	95.1	65.3	79.0	96.7	59.3	76.5
COVERS	DIVS. 10 - 14, 25, 28, 41, 43, 44, 46	100.0	99.6	99.9	100.0	127.5	106.5	100.0	134.8	108.2	100.0	84.2	96.3	100.0	85.0	96.5	100.0	84.2	96.3
21, 22, 23	FIRE SUPPRESSION, PLUMBING & HVAC	101.2	89.9	96.6	101.0	146.4	119.3	101.0	146.4	119.3	101.8	58.8	84.5	101.0	63.4	85.8	101.8	58.8	84.5
26, 27, 3370	ELECTRICAL, COMMUNICATIONS & UTIL.	100.0	88.7	94.4	91.7	164.2	127.6	97.4	164.2	130.5	99.9	55.7	78.0	101.6	84.2	93.0	95.4	55.2	75.5
MF2018	WEIGHTED AVERAGE	98.4	98.9	98.6	94.9	149.8	118.6	99.1	150.1	121.1	97.0	64.6	83.0	98.4	72.1	87.0	101.2	65.2	85.6

NORTH CAROLINA

DIVISION		ELIZABETH CITY 279			FAYETTEVILLE 283			GASTONIA 280			GREENSBORO 270,272-274			HICKORY 286			KINSTON 285		
		MAT.	INST.	TOTAL	MAT.	INST.	TOTAL	MAT.	INST.	TOTAL	MAT.	INST.	TOTAL	MAT.	INST.	TOTAL	MAT.	INST.	TOTAL
015433	CONTRACTOR EQUIPMENT		106.2	106.2		102.2	102.2		97.0	97.0		102.2	102.2		102.2	102.2		102.2	102.2
0241, 31 - 34	SITE & INFRASTRUCTURE, DEMOLITION	105.9	85.8	92.1	95.7	84.2	87.8	96.2	76.0	82.3	101.6	84.5	89.8	94.9	84.2	87.6	93.7	84.2	87.2
0310	Concrete Forming & Accessories	84.6	61.5	64.9	93.0	58.1	63.3	101.1	59.8	66.0	101.3	59.6	65.8	89.0	59.7	64.1	84.9	58.1	62.1
0320	Concrete Reinforcing	93.4	66.5	80.4	97.2	63.6	81.0	93.9	63.7	79.3	94.2	63.6	79.5	93.6	63.6	79.1	93.1	63.6	78.9
0330	Cast-in-Place Concrete	116.0	70.1	98.7	107.3	68.4	92.6	100.0	70.8	89.0	114.9	70.7	98.2	102.3	69.9	90.1	98.8	67.8	87.0
03	CONCRETE	98.8	67.0	84.5	91.6	64.4	79.4	88.5	66.0	78.4	98.4	65.9	83.8	89.3	65.6	78.7	86.5	64.2	76.5
04	MASONRY	99.1	58.5	74.4	88.2	58.5	70.1	89.4	62.6	73.0	83.7	62.6	70.8	75.0	62.5	67.4	81.5	58.5	67.5
05	METALS	103.3	91.4	99.6	121.6	89.4	111.5	101.7	89.5	97.9	109.7	89.4	103.4	101.1	88.7	97.2	99.9	89.3	96.6
06	WOOD, PLASTICS & COMPOSITES	77.7	61.2	69.1	89.8	57.0	72.7	100.8	57.0	77.9	96.6	57.0	75.9	84.9	57.0	70.3	81.0	57.0	68.4
07	THERMAL & MOISTURE PROTECTION	100.7	59.8	83.2	98.0	59.6	81.5	99.0	61.4	82.9	101.3	61.4	84.2	99.1	61.4	82.9	98.9	59.6	82.0
08	OPENINGS	93.3	60.7	85.4	89.4	57.8	81.7	92.5	57.8	84.1	95.9	57.8	86.6	89.4	57.8	81.7	89.4	57.8	81.7
0920	Plaster & Gypsum Board	89.7	59.6	70.0	112.1	55.9	75.2	114.7	55.9	76.1	96.8	55.9	70.0	108.8	55.9	74.1	108.7	55.9	74.1
0950, 0980	Ceilings & Acoustic Treatment	98.1	59.6	74.0	92.8	55.9	69.7	94.7	55.9	70.4	98.1	55.9	71.6	89.6	55.9	68.5	94.7	55.9	70.4
0960	Flooring	96.3	63.7	86.8	98.3	63.7	88.2	101.6	63.7	90.5	104.5	63.7	92.6	98.1	63.7	88.1	95.2	63.7	86.0
0970, 0990	Wall Finishes & Painting/Coating	100.1	54.4	72.7	98.6	54.4	72.1	98.6	54.4	72.1	100.1	54.4	72.7	98.6	54.4	72.1	98.6	54.4	72.1
09	FINISHES	93.8	60.9	76.0	97.2	58.3	76.2	98.9	59.3	77.5	96.8	59.3	76.6	96.2	59.3	76.2	96.3	58.3	75.8
COVERS	DIVS. 10 - 14, 25, 28, 41, 43, 44, 46	100.0	85.3	96.5	100.0	82.9	96.0	100.0	84.3	96.3	100.0	84.2	96.3	100.0	84.3	96.3	100.0	82.8	96.0
21, 22, 23	FIRE SUPPRESSION, PLUMBING & HVAC	97.9	55.8	80.9	101.6	56.7	83.5	101.8	57.8	84.1	101.7	58.8	84.4	98.0	57.8	81.8	98.0	55.6	80.9
26, 27, 3370	ELECTRICAL, COMMUNICATIONS & UTIL.	95.1	66.6	81.0	99.9	55.2	77.8	99.4	84.2	91.9	94.7	55.7	75.4	97.9	84.2	91.1	97.7	53.7	76.0
MF2018	WEIGHTED AVERAGE	98.1	66.6	84.5	100.5	64.0	84.7	97.7	68.5	85.1	99.7	65.3	84.8	95.2	69.0	83.9	94.9	63.5	81.3

NORTH CAROLINA / NORTH DAKOTA

DIVISION		MURPHY 289			RALEIGH 275 - 276			ROCKY MOUNT 278			WILMINGTON 284			WINSTON-SALEM 271			BISMARCK 585		
		MAT.	INST.	TOTAL	MAT.	INST.	TOTAL	MAT.	INST.	TOTAL	MAT.	INST.	TOTAL	MAT.	INST.	TOTAL	MAT.	INST.	TOTAL
015433	CONTRACTOR EQUIPMENT		97.0	97.0		106.9	106.9		102.2	102.2		97.0	97.0		102.2	102.2		100.9	100.9
0241, 31 - 34	SITE & INFRASTRUCTURE, DEMOLITION	97.1	75.7	82.4	101.1	93.2	95.7	103.8	84.1	90.3	97.5	75.7	82.5	102.0	84.5	90.0	101.0	99.3	99.8
0310	Concrete Forming & Accessories	101.9	57.9	64.5	99.4	59.3	65.3	92.2	59.0	64.0	95.3	58.1	63.6	103.1	59.6	66.1	107.3	78.9	83.1
0320	Concrete Reinforcing	93.1	61.3	77.8	95.7	63.6	80.2	93.4	63.6	79.0	94.3	63.6	79.5	94.2	63.6	79.5	94.4	100.2	97.2
0330	Cast-in-Place Concrete	105.9	67.7	91.5	119.2	70.7	100.9	113.5	69.1	96.7	101.9	67.8	89.0	117.8	70.1	99.8	99.4	89.1	95.5
03	CONCRETE	92.5	63.7	79.6	99.4	65.6	84.2	99.3	65.0	83.9	89.6	64.2	78.2	99.9	65.7	84.5	96.5	86.6	92.1
04	MASONRY	78.1	58.5	66.2	85.1	61.4	70.6	76.9	61.2	67.3	75.8	58.5	65.2	84.0	62.6	70.9	102.0	84.1	91.1
05	METALS	98.8	88.5	95.6	102.0	87.0	97.3	102.5	88.5	98.1	100.6	89.3	97.1	106.8	89.3	101.4	93.6	94.5	93.8
06	WOOD, PLASTICS & COMPOSITES	101.6	56.9	78.2	93.7	57.2	74.6	86.5	57.0	71.1	93.2	57.0	74.3	96.6	57.0	75.9	103.2	75.4	88.6
07	THERMAL & MOISTURE PROTECTION	99.0	59.6	82.1	94.6	62.0	80.6	101.3	60.9	83.9	98.7	59.6	81.9	101.3	61.4	84.2	107.9	86.4	98.6
08	OPENINGS	89.3	57.2	81.5	98.0	57.9	88.2	92.6	57.8	84.1	89.5	57.8	81.7	95.9	57.8	86.6	103.4	84.6	98.8
0920	Plaster & Gypsum Board	113.3	55.7	75.5	91.5	55.9	68.1	90.9	55.9	67.9	110.2	55.9	74.6	96.8	55.9	70.0	91.1	74.9	80.5
0950, 0980	Ceilings & Acoustic Treatment	89.6	55.7	68.4	97.4	55.9	71.4	94.9	55.9	70.5	92.8	55.9	69.7	98.1	55.9	71.6	113.7	74.9	89.4
0960	Flooring	101.9	63.7	90.8	96.2	63.7	86.7	99.9	63.7	89.4	98.9	63.7	88.6	104.5	63.7	92.6	88.9	54.9	79.0
0970, 0990	Wall Finishes & Painting/Coating	98.6	54.4	72.1	96.7	54.4	71.4	100.1	54.4	72.7	98.6	54.4	72.1	100.1	54.4	72.7	89.1	61.0	72.3
09	FINISHES	97.9	58.2	76.4	95.1	59.1	75.6	94.4	59.0	75.3	97.1	58.3	76.1	96.8	59.3	76.6	97.5	73.2	84.3
COVERS	DIVS. 10 - 14, 25, 28, 41, 43, 44, 46	100.0	82.8	96.0	100.0	84.1	96.3	100.0	83.8	96.2	100.0	82.8	96.0	100.0	84.2	96.3	100.0	94.3	98.6
21, 22, 23	FIRE SUPPRESSION, PLUMBING & HVAC	98.0	55.6	80.9	100.9	58.2	83.7	97.9	57.0	81.4	101.8	56.7	83.6	101.7	58.8	84.4	100.5	78.1	91.5
26, 27, 3370	ELECTRICAL, COMMUNICATIONS & UTIL.	100.6	53.7	77.4	100.4	54.2	77.5	96.7	55.2	76.2	100.2	53.7	77.2	94.7	55.7	75.4	98.8	76.2	87.6
MF2018	WEIGHTED AVERAGE	96.1	62.6	81.6	99.0	65.3	84.4	97.1	64.5	83.0	96.6	63.1	82.1	99.5	65.3	84.7	99.1	83.1	92.2

NORTH DAKOTA

DIVISION		DEVILS LAKE 583			DICKINSON 586			FARGO 580 - 581			GRAND FORKS 582			JAMESTOWN 584			MINOT 587		
		MAT.	INST.	TOTAL	MAT.	INST.	TOTAL	MAT.	INST.	TOTAL	MAT.	INST.	TOTAL	MAT.	INST.	TOTAL	MAT.	INST.	TOTAL
015433	CONTRACTOR EQUIPMENT		95.8	95.8		95.8	95.8		100.9	100.9		95.8	95.8		95.8	95.8		95.8	95.8
0241, 31 - 34	SITE & INFRASTRUCTURE, DEMOLITION	104.7	90.7	95.1	113.1	90.6	97.7	101.2	99.0	99.7	109.3	91.4	97.0	103.8	90.7	94.8	106.7	91.2	96.0
0310	Concrete Forming & Accessories	106.2	69.3	74.8	93.9	69.1	72.8	96.6	69.8	73.8	98.7	76.2	79.6	95.8	69.2	73.2	93.5	69.4	73.0
0320	Concrete Reinforcing	97.8	100.2	99.0	98.7	100.1	99.4	97.4	100.8	99.1	96.3	100.3	98.2	98.4	100.7	99.5	99.7	100.1	99.9
0330	Cast-in-Place Concrete	116.6	78.4	102.2	105.5	78.3	95.3	96.6	84.9	92.2	105.5	84.9	97.7	115.2	78.4	101.3	105.5	78.6	95.4
03	CONCRETE	101.9	78.6	91.5	101.2	78.5	91.0	98.7	81.1	90.8	98.7	84.1	92.1	100.4	78.7	90.7	97.2	78.8	88.9
04	MASONRY	115.6	82.4	95.4	116.1	79.8	93.9	98.2	91.2	94.0	109.1	85.6	94.7	128.5	91.1	105.7	109.0	79.1	90.8
05	METALS	92.6	95.3	93.5	92.5	94.7	93.2	98.0	95.1	97.1	92.6	96.1	93.7	92.6	95.9	93.6	92.9	95.6	93.7
06	WOOD, PLASTICS & COMPOSITES	98.8	65.2	81.2	84.4	65.2	74.4	91.8	65.5	78.0	89.6	72.6	80.7	86.6	65.2	75.4	84.0	65.2	74.2
07	THERMAL & MOISTURE PROTECTION	106.3	81.0	95.4	106.9	81.4	95.9	104.3	85.9	96.4	106.5	84.2	96.9	106.1	83.8	96.5	106.2	81.9	95.8
08	OPENINGS	100.6	79.0	95.3	100.6	79.0	95.3	100.8	79.1	95.5	99.2	83.0	95.3	100.6	79.0	95.3	99.4	79.0	94.4
0920	Plaster & Gypsum Board	102.2	64.8	77.6	93.1	64.8	74.5	87.1	64.8	72.4	94.3	72.3	79.9	94.0	64.8	74.8	93.1	64.8	74.5
0950, 0980	Ceilings & Acoustic Treatment	109.4	64.8	81.4	109.4	64.8	81.4	97.9	64.8	77.1	109.4	72.3	86.2	109.4	64.8	81.4	109.4	64.8	81.4
0960	Flooring	96.3	54.9	84.2	89.3	54.9	79.2	100.3	54.9	87.1	91.2	54.9	80.6	90.0	54.9	79.8	89.0	54.9	79.0
0970, 0990	Wall Finishes & Painting/Coating	83.5	53.7	65.6	83.5	53.7	65.6	94.7	64.1	76.4	83.5	62.9	71.2	83.5	53.7	65.6	83.5	54.7	66.3
09	FINISHES	97.1	65.2	79.8	94.7	65.2	78.7	97.0	66.6	80.5	94.9	71.3	82.1	94.1	65.2	78.5	93.8	65.3	78.4
COVERS	DIVS. 10 - 14, 25, 28, 41, 43, 44, 46	100.0	82.8	95.9	100.0	82.7	95.9	100.0	91.8	98.1	100.0	90.8	97.8	100.0	82.7	95.9	100.0	91.2	97.9
21, 22, 23	FIRE SUPPRESSION, PLUMBING & HVAC	96.9	74.8	88.0	96.9	69.9	86.0	100.6	70.7	88.5	100.7	78.0	91.5	96.9	68.9	85.6	100.7	68.8	87.8
26, 27, 3370	ELECTRICAL, COMMUNICATIONS & UTIL.	94.2	70.0	82.2	100.7	65.3	83.2	100.6	66.8	83.9	96.8	70.1	83.6	94.2	65.7	80.1	99.2	69.9	84.7
MF2018	WEIGHTED AVERAGE	98.6	77.8	89.6	99.2	75.7	89.0	99.6	78.9	90.6	98.8	81.0	91.1	98.7	76.9	89.3	98.7	76.5	89.1

For customer support on your Commercial Renovation Costs with RSMeans Data, call 800.448.8182.

817

OHIO (and NORTH DAKOTA)

DIVISION		WILLISTON 588 MAT.	INST.	TOTAL	AKRON 442-443 MAT.	INST.	TOTAL	ATHENS 457 MAT.	INST.	TOTAL	CANTON 446-447 MAT.	INST.	TOTAL	CHILLICOTHE 456 MAT.	INST.	TOTAL	CINCINNATI 451-452 MAT.	INST.	TOTAL
015433	CONTRACTOR EQUIPMENT		95.8	95.8		87.0	87.0		83.5	83.5		87.0	87.0		94.0	94.0		96.6	96.6
0241, 31 - 34	SITE & INFRASTRUCTURE, DEMOLITION	106.6	91.5	96.2	96.4	91.2	92.8	108.7	82.7	90.8	96.5	91.1	92.8	95.0	92.1	93.0	93.3	98.5	96.9
0310	Concrete Forming & Accessories	100.7	76.2	79.8	102.6	79.9	83.3	94.3	76.1	78.8	102.6	72.0	76.6	97.7	78.6	81.4	100.4	74.9	78.7
0320	Concrete Reinforcing	100.6	100.1	100.4	92.2	87.2	89.8	93.1	85.0	89.2	92.2	71.5	82.2	90.0	84.6	87.4	95.5	83.7	89.8
0330	Cast-in-Place Concrete	105.5	84.9	97.7	107.2	85.5	99.0	111.7	92.1	104.3	108.2	84.5	99.2	101.4	88.8	96.6	96.8	76.5	89.1
03	CONCRETE	98.6	84.0	92.1	101.3	82.8	93.0	103.4	83.1	94.3	101.8	76.2	90.3	97.9	83.7	91.5	96.0	77.2	87.6
04	MASONRY	103.2	82.2	90.4	99.8	85.7	91.2	79.7	93.2	87.9	100.5	78.5	87.1	86.3	86.7	86.6	91.7	79.2	84.1
05	METALS	92.7	95.8	93.7	94.3	79.0	89.5	102.3	79.1	95.1	94.3	72.4	87.5	94.1	88.4	92.3	96.3	81.9	91.8
06	WOOD, PLASTICS & COMPOSITES	91.5	72.6	81.6	104.1	78.3	90.6	87.1	71.5	78.9	104.5	69.3	86.1	99.3	75.3	86.8	101.3	73.6	86.8
07	THERMAL & MOISTURE PROTECTION	106.5	83.3	96.5	104.1	88.1	97.2	102.9	87.4	96.2	105.0	84.9	96.4	104.3	84.4	95.8	102.1	80.9	93.0
08	OPENINGS	100.7	83.0	96.4	107.3	80.0	100.6	97.4	73.0	91.5	101.1	67.2	92.9	88.4	75.2	85.2	98.6	73.2	92.4
0920	Plaster & Gypsum Board	94.3	72.3	79.9	98.1	77.9	84.8	94.2	70.8	78.8	99.2	68.6	79.1	97.5	75.2	82.9	96.5	73.2	81.2
0950, 0980	Ceilings & Acoustic Treatment	109.4	72.3	86.2	93.2	77.9	83.6	102.8	70.8	82.7	93.2	68.6	77.8	95.0	75.2	82.6	88.5	73.2	78.9
0960	Flooring	92.3	54.9	81.4	91.6	79.5	88.1	124.9	71.8	109.4	91.8	70.7	85.6	101.5	71.8	92.8	101.5	79.3	95.0
0970, 0990	Wall Finishes & Painting/Coating	83.5	60.4	69.7	97.9	86.5	91.1	104.9	84.9	92.9	97.9	77.3	85.5	102.2	84.9	91.8	101.5	73.0	84.4
09	FINISHES	95.1	71.2	82.1	95.5	80.3	87.3	102.9	75.6	88.1	95.7	71.5	82.6	99.4	77.2	87.4	97.2	75.1	85.3
COVERS	DIVS. 10 - 14, 25, 28, 41, 43, 44, 46	100.0	93.0	98.4	100.0	91.6	98.0	100.0	90.3	97.7	100.0	90.2	97.7	100.0	89.4	97.5	100.0	88.1	97.2
21, 22, 23	FIRE SUPPRESSION, PLUMBING & HVAC	96.9	74.7	87.9	100.5	83.7	93.7	96.7	77.3	88.9	100.5	75.5	90.4	97.2	88.0	93.5	100.4	78.1	91.4
26, 27, 3370	ELECTRICAL, COMMUNICATIONS & UTIL.	97.1	75.1	86.2	99.2	78.8	89.1	100.7	86.4	93.6	98.5	81.9	90.3	99.8	86.4	93.2	99.1	74.0	86.7
MF2018	WEIGHTED AVERAGE	97.8	80.7	90.4	99.7	83.0	92.5	99.2	81.8	91.7	99.2	77.5	89.8	96.3	85.3	91.5	98.0	79.2	89.9

OHIO

DIVISION		CLEVELAND 441 MAT.	INST.	TOTAL	COLUMBUS 430-432 MAT.	INST.	TOTAL	DAYTON 453-454 MAT.	INST.	TOTAL	HAMILTON 450 MAT.	INST.	TOTAL	LIMA 458 MAT.	INST.	TOTAL	LORAIN 440 MAT.	INST.	TOTAL
015433	CONTRACTOR EQUIPMENT		91.8	91.8		93.7	93.7		87.4	87.4		94.0	94.0		86.8	86.8		87.0	87.0
0241, 31 - 34	SITE & INFRASTRUCTURE, DEMOLITION	95.5	95.2	95.3	102.9	94.1	96.8	91.4	91.5	91.5	91.4	91.7	91.6	102.5	82.4	88.7	95.7	91.4	92.7
0310	Concrete Forming & Accessories	101.3	89.9	91.6	100.6	76.3	80.0	99.8	74.8	78.5	99.9	74.6	78.4	94.3	73.7	76.7	102.6	72.0	76.6
0320	Concrete Reinforcing	92.8	90.7	91.8	104.6	78.8	92.2	95.5	75.4	85.8	95.5	73.6	84.9	93.1	75.6	84.6	92.2	87.5	89.9
0330	Cast-in-Place Concrete	106.9	95.9	102.7	102.9	86.0	96.5	86.8	77.9	83.5	93.1	79.5	87.9	102.6	86.1	96.4	102.0	88.3	96.8
03	CONCRETE	101.7	91.7	97.2	100.1	80.2	91.2	89.3	76.0	83.4	92.2	76.8	85.3	96.3	78.5	88.3	98.9	80.3	90.5
04	MASONRY	104.6	97.3	100.1	87.9	83.9	85.5	85.4	74.1	78.5	85.7	77.8	80.9	108.9	74.9	88.2	96.4	91.0	93.1
05	METALS	95.9	83.8	92.1	103.2	79.6	95.9	95.6	75.5	89.4	95.7	84.3	92.1	102.3	79.0	95.1	94.9	79.8	90.2
06	WOOD, PLASTICS & COMPOSITES	98.4	87.9	92.9	100.3	75.1	87.1	103.2	74.5	88.2	102.2	74.5	87.7	87.0	72.6	79.4	104.1	67.7	85.0
07	THERMAL & MOISTURE PROTECTION	100.9	96.7	99.1	92.7	83.8	88.9	108.1	76.3	94.4	104.4	76.8	92.5	102.5	79.9	92.8	104.9	87.6	97.5
08	OPENINGS	100.2	86.0	96.7	97.8	73.6	91.9	95.1	72.2	89.5	93.0	72.3	87.9	97.4	71.0	91.0	101.1	74.2	94.6
0920	Plaster & Gypsum Board	98.1	87.5	91.1	92.7	74.5	80.7	99.0	74.3	82.8	99.0	74.3	82.8	94.2	71.9	79.5	98.1	66.9	77.6
0950, 0980	Ceilings & Acoustic Treatment	91.2	87.5	88.8	88.8	74.5	79.8	95.7	74.3	82.3	95.0	74.3	82.1	102.8	71.9	83.4	93.2	66.9	76.7
0960	Flooring	92.5	87.3	91.0	96.0	75.3	89.9	105.2	69.0	94.7	102.5	74.9	94.5	124.2	73.6	109.4	91.8	87.3	90.5
0970, 0990	Wall Finishes & Painting/Coating	101.9	89.3	94.3	99.5	77.7	86.4	102.2	67.4	81.3	102.2	67.9	81.6	104.9	72.8	85.7	97.9	86.5	91.1
09	FINISHES	95.4	89.2	92.0	93.2	75.7	83.8	100.6	72.6	85.5	99.6	73.9	85.7	102.3	73.0	86.4	95.6	75.4	84.7
COVERS	DIVS. 10 - 14, 25, 28, 41, 43, 44, 46	100.0	97.4	99.4	100.0	88.7	97.3	100.0	87.2	97.0	100.0	87.1	97.0	100.0	91.6	98.0	100.0	86.5	96.8
21, 22, 23	FIRE SUPPRESSION, PLUMBING & HVAC	100.5	91.3	96.8	100.4	89.8	96.1	101.2	77.8	91.7	101.0	71.4	89.1	96.7	87.4	93.0	100.5	84.6	94.1
26, 27, 3370	ELECTRICAL, COMMUNICATIONS & UTIL.	99.1	92.0	95.6	101.2	81.0	91.2	97.6	72.9	85.4	97.8	72.7	85.4	101.0	72.9	87.1	98.6	75.3	87.1
MF2018	WEIGHTED AVERAGE	99.4	91.5	96.0	99.2	83.2	92.3	96.9	76.6	88.1	96.9	76.7	88.2	99.6	79.0	90.7	98.7	81.8	91.4

OHIO

DIVISION		MANSFIELD 448-449 MAT.	INST.	TOTAL	MARION 433 MAT.	INST.	TOTAL	SPRINGFIELD 455 MAT.	INST.	TOTAL	STEUBENVILLE 439 MAT.	INST.	TOTAL	TOLEDO 434-436 MAT.	INST.	TOTAL	YOUNGSTOWN 444-445 MAT.	INST.	TOTAL
015433	CONTRACTOR EQUIPMENT		87.0	87.0		87.4	87.4		87.4	87.4		91.1	91.1		91.1	91.1		87.0	87.0
0241, 31 - 34	SITE & INFRASTRUCTURE, DEMOLITION	91.5	91.1	91.2	94.9	88.0	90.2	91.7	91.5	91.6	139.6	95.6	109.4	99.5	88.3	91.8	96.2	91.0	92.6
0310	Concrete Forming & Accessories	90.5	70.9	73.8	98.8	76.5	79.8	99.8	74.8	78.5	98.9	76.3	79.7	103.1	81.7	84.9	102.6	74.0	78.3
0320	Concrete Reinforcing	84.0	75.3	79.8	98.8	75.3	87.4	95.5	75.4	85.8	96.1	95.6	95.9	107.2	78.9	93.5	92.2	84.2	88.3
0330	Cast-in-Place Concrete	99.2	84.3	93.6	89.0	84.7	87.4	89.2	77.9	84.9	96.7	86.0	92.7	97.5	86.8	93.4	106.1	83.9	97.7
03	CONCRETE	93.3	76.3	85.7	88.1	79.1	84.1	90.4	76.0	84.0	91.9	82.8	87.8	96.1	83.2	90.3	100.8	79.1	91.0
04	MASONRY	98.8	85.2	90.5	90.5	86.1	87.8	85.6	74.1	78.5	78.7	87.3	83.9	96.6	88.0	91.3	100.2	83.0	89.7
05	METALS	95.2	74.6	88.7	102.2	77.5	94.5	95.6	75.5	89.3	98.4	82.6	93.5	103.0	84.2	97.1	94.4	77.6	89.2
06	WOOD, PLASTICS & COMPOSITES	89.7	67.7	78.2	95.0	74.9	84.5	104.4	74.5	88.8	89.8	73.8	81.5	99.8	81.1	90.0	104.1	72.1	87.3
07	THERMAL & MOISTURE PROTECTION	103.3	85.4	95.7	90.0	84.9	87.8	108.0	76.3	94.4	101.6	84.2	94.2	90.8	87.1	89.3	105.2	84.8	96.4
08	OPENINGS	102.3	68.4	94.1	90.9	72.4	86.4	93.4	71.9	88.2	91.1	76.7	87.6	93.4	78.3	89.7	101.1	77.7	95.4
0920	Plaster & Gypsum Board	91.2	66.9	75.2	91.7	74.5	80.4	99.0	74.3	82.8	89.5	72.8	78.6	93.7	80.8	85.2	98.1	71.4	80.6
0950, 0980	Ceilings & Acoustic Treatment	93.9	66.9	77.0	95.0	74.5	82.1	95.7	74.3	82.3	92.0	72.8	80.0	95.0	80.8	86.1	93.2	71.4	79.6
0960	Flooring	86.4	88.4	87.0	96.4	88.4	94.1	105.2	69.0	94.7	125.5	90.8	115.4	96.9	90.1	94.9	91.8	85.9	90.1
0970, 0990	Wall Finishes & Painting/Coating	97.9	74.1	83.6	101.1	74.1	84.9	102.2	67.4	81.3	113.5	82.9	95.2	101.1	82.9	90.2	97.9	76.2	84.9
09	FINISHES	92.9	73.6	82.4	95.1	78.0	85.9	100.6	72.6	85.5	111.6	78.8	93.9	95.9	83.1	89.0	95.6	75.7	84.8
COVERS	DIVS. 10 - 14, 25, 28, 41, 43, 44, 46	100.0	89.8	97.6	100.0	88.4	97.3	100.0	87.1	97.0	100.0	91.8	98.1	100.0	86.7	96.9	100.0	90.0	97.6
21, 22, 23	FIRE SUPPRESSION, PLUMBING & HVAC	96.7	83.3	91.3	96.6	88.7	93.4	101.2	77.8	91.7	97.1	87.1	93.1	100.4	89.2	95.9	100.5	81.7	92.9
26, 27, 3370	ELECTRICAL, COMMUNICATIONS & UTIL.	96.6	87.4	92.0	97.1	87.4	92.3	97.6	77.2	87.5	91.5	102.5	96.9	102.3	99.6	101.0	98.6	71.6	85.2
MF2018	WEIGHTED AVERAGE	96.6	81.1	89.9	95.5	83.6	90.3	96.9	77.2	88.4	97.1	87.4	92.9	98.9	87.7	94.1	99.1	79.7	90.7

		OHIO			OKLAHOMA														
DIVISION		ZANESVILLE			ARDMORE			CLINTON			DURANT			ENID			GUYMON		
		437 - 438			734			736			747			737			739		
		MAT.	INST.	TOTAL	MAT.	INST.	TOTAL	MAT.	INST.	TOTAL	MAT.	INST.	TOTAL	MAT.	INST.	TOTAL	MAT.	INST.	TOTAL
015433	CONTRACTOR EQUIPMENT		87.4	87.4		77.7	77.7		76.9	76.9		76.9	76.9		76.9	76.9		76.9	76.9
0241, 31 - 34	SITE & INFRASTRUCTURE, DEMOLITION	98.0	87.9	91.0	94.4	89.1	90.8	95.7	87.7	90.2	92.9	85.0	87.5	97.9	87.7	90.9	99.9	86.8	90.9
0310	Concrete Forming & Accessories	95.2	75.5	78.5	90.9	53.4	59.0	89.5	53.4	58.8	83.1	52.9	57.4	93.5	53.6	59.6	97.0	53.1	59.7
0320	Concrete Reinforcing	98.2	88.2	93.4	88.2	66.3	77.6	88.7	66.3	77.9	92.1	60.6	76.9	88.1	66.3	77.6	88.7	60.3	75.0
0330	Cast-in-Place Concrete	93.8	83.1	89.8	93.3	67.5	83.5	90.1	67.5	81.6	90.3	67.3	81.6	90.1	67.6	81.6	90.2	66.9	81.4
03	CONCRETE	91.7	80.5	86.7	86.4	60.5	74.8	85.9	60.5	74.5	85.3	59.2	73.6	86.4	60.7	74.8	89.0	59.1	75.6
04	MASONRY	88.9	83.0	85.3	96.7	56.7	72.3	123.2	56.7	82.7	88.5	59.6	70.8	104.6	56.7	75.4	98.4	53.1	70.8
05	METALS	103.7	84.7	97.8	98.0	61.0	86.5	98.1	61.0	86.6	93.2	58.6	82.4	99.6	61.2	87.6	98.6	57.3	85.8
06	WOOD, PLASTICS & COMPOSITES	89.7	74.9	82.0	99.0	52.5	74.7	98.0	52.5	74.2	86.4	52.5	68.7	101.9	52.5	76.1	106.1	52.5	78.1
07	THERMAL & MOISTURE PROTECTION	90.1	81.2	86.3	101.2	62.7	84.7	101.3	62.7	84.7	97.7	62.0	82.4	101.5	62.7	84.8	101.8	59.4	83.6
08	OPENINGS	90.9	77.3	87.6	103.4	53.3	91.2	103.4	53.3	91.2	95.9	51.9	85.2	104.6	54.5	92.4	103.6	51.9	91.0
0920	Plaster & Gypsum Board	87.5	74.5	78.9	89.9	51.9	65.0	89.6	51.9	64.9	79.2	51.9	61.3	90.5	51.9	65.2	90.5	51.9	65.2
0950, 0980	Ceilings & Acoustic Treatment	95.0	74.5	82.1	93.5	51.9	67.5	93.5	51.9	67.5	87.1	51.9	65.1	93.5	51.9	67.5	93.5	51.9	67.5
0960	Flooring	94.5	71.8	87.9	86.9	54.7	77.5	85.8	54.7	76.7	92.3	49.3	79.8	87.5	71.6	82.9	89.1	54.7	79.0
0970, 0990	Wall Finishes & Painting/Coating	101.1	84.9	91.4	84.3	43.5	59.9	84.3	43.5	59.9	91.5	43.5	62.8	84.3	43.5	59.9	84.3	40.2	57.9
09	FINISHES	94.2	75.1	83.9	88.1	51.4	68.2	88.0	51.4	68.2	87.3	50.3	67.3	88.6	54.9	70.4	89.5	52.3	69.3
COVERS	DIVS. 10 - 14, 25, 28, 41, 43, 44, 46	100.0	87.5	97.1	100.0	77.4	94.7	100.0	77.4	94.7	100.0	77.4	94.7	100.0	77.4	94.7	100.0	77.4	94.7
21, 22, 23	FIRE SUPPRESSION, PLUMBING & HVAC	96.6	85.6	92.2	97.3	65.5	84.4	97.3	65.5	84.4	97.4	65.4	84.5	101.1	65.5	86.7	97.3	65.2	84.3
26, 27, 3370	ELECTRICAL, COMMUNICATIONS & UTIL.	97.3	86.4	91.9	95.2	70.0	82.7	96.1	70.0	83.2	97.6	66.6	82.3	96.1	70.0	83.2	97.5	63.2	80.5
MF2018	WEIGHTED AVERAGE	96.1	83.0	90.4	95.9	63.8	82.0	97.3	63.7	82.8	93.8	62.7	80.4	97.8	64.2	83.3	97.0	61.7	81.7

		OKLAHOMA																	
DIVISION		LAWTON			MCALESTER			MIAMI			MUSKOGEE			OKLAHOMA CITY			PONCA CITY		
		735			745			743			744			730 - 731			746		
		MAT.	INST.	TOTAL	MAT.	INST.	TOTAL	MAT.	INST.	TOTAL	MAT.	INST.	TOTAL	MAT.	INST.	TOTAL	MAT.	INST.	TOTAL
015433	CONTRACTOR EQUIPMENT		77.7	77.7		76.9	76.9		88.0	88.0		88.0	88.0		87.3	87.3		76.9	76.9
0241, 31 - 34	SITE & INFRASTRUCTURE, DEMOLITION	94.0	89.1	90.7	86.3	86.9	86.7	87.6	84.5	85.5	88.7	84.5	85.8	94.3	99.0	97.5	93.5	87.3	89.2
0310	Concrete Forming & Accessories	96.7	53.6	60.0	81.0	39.8	46.0	95.5	53.1	59.5	100.4	56.6	63.1	99.6	62.7	68.2	90.4	53.3	58.8
0320	Concrete Reinforcing	88.3	66.3	77.7	91.7	60.4	76.6	90.3	66.3	78.7	91.2	65.7	78.9	98.4	66.4	83.0	91.2	66.3	79.2
0330	Cast-in-Place Concrete	87.2	67.5	79.7	79.2	66.8	74.5	83.0	68.3	77.4	84.0	68.9	78.3	91.6	72.8	84.5	92.8	67.4	83.2
03	CONCRETE	83.0	60.6	73.0	76.5	53.1	66.0	81.0	61.7	72.3	82.6	63.3	74.0	90.4	66.7	79.8	87.4	60.5	75.3
04	MASONRY	98.4	56.7	73.0	106.3	56.6	76.0	91.3	56.8	70.2	108.2	48.3	71.6	98.8	57.5	73.6	83.8	56.7	67.3
05	METALS	103.4	61.2	90.3	93.2	57.4	82.0	93.1	75.7	87.7	94.7	74.9	88.5	96.6	64.0	86.4	93.1	60.9	83.1
06	WOOD, PLASTICS & COMPOSITES	105.0	52.5	77.6	83.7	34.9	58.2	100.4	52.6	75.4	105.5	56.8	80.0	100.9	64.2	81.7	95.3	52.5	72.9
07	THERMAL & MOISTURE PROTECTION	101.2	62.7	84.7	97.3	59.3	81.0	97.8	62.0	82.5	98.1	59.9	81.7	93.9	65.8	81.8	97.9	62.9	82.9
08	OPENINGS	106.4	54.5	93.8	95.9	42.2	82.8	95.9	53.4	85.5	97.1	55.7	87.0	99.8	60.9	90.3	95.9	53.3	85.5
0920	Plaster & Gypsum Board	92.8	51.9	66.0	78.7	33.8	49.2	84.9	51.9	63.3	87.8	56.2	67.1	93.9	63.7	74.1	83.8	51.9	62.9
0950, 0980	Ceilings & Acoustic Treatment	102.4	51.9	70.8	87.1	33.8	53.7	87.1	51.9	65.1	97.9	56.2	71.8	90.7	63.7	73.8	87.1	51.9	65.1
0960	Flooring	89.3	71.6	84.2	91.2	54.7	80.5	98.6	49.3	84.2	101.2	31.7	80.9	89.6	71.6	84.3	95.6	54.7	83.7
0970, 0990	Wall Finishes & Painting/Coating	84.3	43.5	59.9	91.5	40.2	60.8	91.5	40.2	60.8	91.5	40.2	60.8	90.1	43.5	62.2	91.5	43.5	62.8
09	FINISHES	91.1	54.9	71.5	86.4	40.7	61.6	89.3	50.0	68.0	93.0	49.5	69.5	91.1	61.9	75.3	89.0	52.3	69.1
COVERS	DIVS. 10 - 14, 25, 28, 41, 43, 44, 46	100.0	77.4	94.7	100.0	75.5	94.2	100.0	77.8	94.8	100.0	78.7	95.0	100.0	79.5	95.2	100.0	77.4	94.7
21, 22, 23	FIRE SUPPRESSION, PLUMBING & HVAC	101.1	65.5	86.7	97.4	60.5	82.5	97.4	60.6	82.5	101.2	62.1	85.4	101.0	65.9	86.8	97.4	60.6	82.5
26, 27, 3370	ELECTRICAL, COMMUNICATIONS & UTIL.	97.5	66.6	82.2	96.2	64.5	80.5	97.4	64.5	81.2	95.9	86.7	91.4	102.4	70.0	86.3	95.9	63.2	79.7
MF2018	WEIGHTED AVERAGE	98.1	63.9	83.3	93.2	58.3	78.1	93.5	63.0	80.3	96.1	65.7	83.0	97.6	67.8	84.8	93.9	61.8	80.0

		OKLAHOMA												OREGON					
DIVISION		POTEAU			SHAWNEE			TULSA			WOODWARD			BEND			EUGENE		
		749			748			740 - 741			738			977			974		
		MAT.	INST.	TOTAL	MAT.	INST.	TOTAL	MAT.	INST.	TOTAL	MAT.	INST.	TOTAL	MAT.	INST.	TOTAL	MAT.	INST.	TOTAL
015433	CONTRACTOR EQUIPMENT		86.8	86.8		76.9	76.9		88.0	88.0		76.9	76.9		95.5	95.5		95.5	95.5
0241, 31 - 34	SITE & INFRASTRUCTURE, DEMOLITION	74.4	80.3	78.4	96.4	87.3	90.1	94.7	84.2	87.5	96.1	87.7	90.3	112.5	95.7	100.9	102.9	95.7	98.0
0310	Concrete Forming & Accessories	88.1	52.9	58.1	83.0	53.3	57.7	100.2	57.0	63.4	89.6	42.1	49.2	103.1	101.8	102.0	99.5	102.1	101.7
0320	Concrete Reinforcing	92.2	66.2	79.7	91.2	62.4	77.3	91.4	66.3	79.3	88.1	66.3	77.6	90.5	114.3	102.0	94.4	114.4	104.0
0330	Cast-in-Place Concrete	83.0	68.2	77.4	95.7	67.4	85.0	91.5	70.8	83.7	90.1	67.4	81.6	118.3	101.0	111.8	114.6	101.1	109.5
03	CONCRETE	83.0	61.5	73.3	88.9	59.8	75.9	87.6	64.3	77.1	86.1	55.4	72.3	110.9	103.2	107.5	102.0	103.4	102.6
04	MASONRY	91.5	56.8	70.3	107.7	56.7	76.6	92.2	57.8	71.2	92.3	56.7	70.6	101.7	102.7	102.3	98.6	102.7	101.1
05	METALS	93.2	75.4	87.6	93.1	59.5	82.6	97.9	76.1	91.1	98.2	60.9	86.6	107.3	98.1	104.4	108.0	98.5	105.0
06	WOOD, PLASTICS & COMPOSITES	91.5	52.6	71.2	86.2	52.5	68.6	104.6	56.8	79.6	98.1	37.3	66.3	95.9	102.1	99.1	91.5	102.1	97.0
07	THERMAL & MOISTURE PROTECTION	97.9	62.0	82.5	97.9	61.6	82.3	98.1	64.9	83.9	101.4	61.2	84.2	118.6	102.9	111.8	117.7	105.5	112.5
08	OPENINGS	95.9	53.4	85.5	95.9	52.4	85.3	98.8	56.9	88.6	103.4	44.9	89.2	97.0	105.0	99.0	97.3	105.0	99.2
0920	Plaster & Gypsum Board	81.8	51.9	62.2	79.2	51.9	61.3	87.8	56.2	67.0	89.6	36.3	54.6	116.1	102.2	107.0	113.9	102.2	106.2
0950, 0980	Ceilings & Acoustic Treatment	87.1	51.9	65.1	87.1	51.9	65.1	97.9	56.2	71.8	93.5	36.3	57.7	90.0	102.2	97.6	90.7	102.2	97.9
0960	Flooring	94.9	49.3	81.6	92.3	54.7	81.3	100.0	62.4	89.0	85.8	52.1	76.0	101.3	106.4	102.8	99.7	106.4	101.7
0970, 0990	Wall Finishes & Painting/Coating	91.5	43.5	62.8	91.5	40.2	60.8	91.5	54.9	69.6	84.3	43.5	59.9	93.9	76.5	83.5	93.9	66.6	77.6
09	FINISHES	87.0	50.4	67.2	87.5	51.0	67.8	92.9	57.0	73.5	88.0	42.0	63.1	100.7	99.9	100.3	99.0	98.8	98.9
COVERS	DIVS. 10 - 14, 25, 28, 41, 43, 44, 46	100.0	77.7	94.7	100.0	77.4	94.7	100.0	78.6	95.0	100.0	75.8	94.3	100.0	102.5	100.6	100.0	102.5	100.6
21, 22, 23	FIRE SUPPRESSION, PLUMBING & HVAC	97.4	60.6	82.5	97.4	65.4	84.5	101.2	61.3	85.1	97.3	65.5	84.4	97.2	102.3	99.3	101.2	117.3	107.7
26, 27, 3370	ELECTRICAL, COMMUNICATIONS & UTIL.	96.0	64.5	80.4	97.6	70.0	83.9	97.6	69.9	83.9	97.4	70.0	83.8	98.3	96.1	97.2	97.6	96.1	96.8
MF2018	WEIGHTED AVERAGE	93.0	62.6	79.9	95.3	63.3	81.5	96.9	65.5	83.3	95.9	61.2	80.9	102.3	100.5	101.5	101.6	103.7	102.5

For customer support on your Commercial Renovation Costs with RSMeans Data, call 800.448.8182.

819

OREGON

DIVISION		KLAMATH FALLS 976			MEDFORD 975			PENDLETON 978			PORTLAND 970 - 972			SALEM 973			VALE 979		
		MAT.	INST.	TOTAL	MAT.	INST.	TOTAL	MAT.	INST.	TOTAL	MAT.	INST.	TOTAL	MAT.	INST.	TOTAL	MAT.	INST.	TOTAL
015433	CONTRACTOR EQUIPMENT		95.5	95.5		95.5	95.5		93.1	93.1		95.5	95.5		101.4	101.4		93.1	93.1
0241, 31 - 34	SITE & INFRASTRUCTURE, DEMOLITION	116.6	95.7	102.2	110.8	95.7	100.4	110.6	89.2	95.9	105.6	95.7	98.8	98.5	104.0	102.3	97.5	89.2	91.8
0310	Concrete Forming & Accessories	96.1	101.4	100.6	94.9	101.6	100.6	97.2	101.7	101.1	101.0	102.0	101.9	99.2	102.2	101.7	103.8	100.4	100.9
0320	Concrete Reinforcing	90.5	114.2	101.9	92.0	114.3	102.7	89.7	114.4	101.6	95.1	114.5	104.4	102.2	114.5	108.1	87.6	114.2	100.4
0330	Cast-in-Place Concrete	118.4	95.7	109.8	118.3	100.9	111.7	119.2	96.2	110.5	117.8	101.1	111.5	108.3	103.0	106.3	93.7	96.4	94.7
03	CONCRETE	113.8	101.2	108.1	108.3	103.1	106.0	94.7	101.5	97.8	103.7	103.4	103.6	99.7	104.0	101.6	78.8	100.9	88.7
04	MASONRY	114.7	102.7	107.4	95.2	102.7	99.8	105.6	104.9	105.2	100.5	104.8	103.1	101.3	104.9	103.5	103.5	104.9	104.3
05	METALS	107.3	97.8	104.3	107.6	97.9	104.5	115.5	98.7	110.2	109.4	98.6	106.0	116.4	97.2	110.4	115.3	97.2	109.7
06	WOOD, PLASTICS & COMPOSITES	86.2	102.1	94.5	85.1	102.1	94.0	88.7	102.2	95.8	92.7	102.1	97.6	85.5	102.4	94.3	97.7	102.2	100.1
07	THERMAL & MOISTURE PROTECTION	118.8	98.4	110.0	118.4	98.0	109.7	110.5	95.7	104.1	117.7	103.4	111.6	114.0	104.8	110.0	109.9	91.8	102.1
08	OPENINGS	97.0	105.0	99.0	99.6	105.0	100.9	93.1	105.0	96.0	95.3	105.0	97.7	102.7	105.1	103.3	93.1	92.3	92.9
0920	Plaster & Gypsum Board	110.2	102.2	104.9	109.6	102.2	104.7	96.3	102.2	100.2	113.5	102.2	106.1	109.9	102.2	104.8	103.4	102.2	102.6
0950, 0980	Ceilings & Acoustic Treatment	97.6	102.2	100.5	102.6	102.2	102.3	65.8	102.2	88.6	92.8	102.2	98.7	105.1	102.2	103.3	65.8	102.2	88.6
0960	Flooring	98.3	106.4	100.7	97.8	106.4	100.3	67.3	106.4	78.7	97.6	106.4	100.1	100.0	106.4	101.9	69.4	106.4	80.2
0970, 0990	Wall Finishes & Painting/Coating	93.9	76.8	83.7	93.9	76.8	83.7	85.0	76.8	80.1	93.8	76.8	83.6	95.5	74.5	82.9	85.0	76.5	79.9
09	FINISHES	101.3	99.9	100.6	101.4	99.9	100.6	71.3	100.6	86.8	98.9	99.9	99.5	100.9	99.6	100.2	71.9	100.0	87.1
COVERS	DIVS. 10 - 14, 25, 28, 41, 43, 44, 46	100.0	102.4	100.6	100.0	102.4	100.6	100.0	105.2	101.2	100.0	102.6	100.6	100.0	103.1	100.7	100.0	102.6	100.6
21, 22, 23	FIRE SUPPRESSION, PLUMBING & HVAC	97.2	102.3	99.2	101.2	108.1	104.0	100.0	108.8	103.5	101.2	111.1	105.2	100.2	108.2	103.4	100.0	69.8	87.8
26, 27, 3370	ELECTRICAL, COMMUNICATIONS & UTIL.	97.2	79.6	88.5	100.6	79.6	90.2	90.3	93.3	91.8	97.9	110.5	104.1	101.5	96.1	98.8	90.3	65.5	78.0
MF2018	WEIGHTED AVERAGE	103.3	97.8	100.9	103.1	99.3	101.4	98.3	101.0	99.5	102.0	104.7	103.2	103.3	102.8	103.0	96.0	87.8	92.4

PENNSYLVANIA

DIVISION		ALLENTOWN 181			ALTOONA 166			BEDFORD 155			BRADFORD 167			BUTLER 160			CHAMBERSBURG 172		
		MAT.	INST.	TOTAL	MAT.	INST.	TOTAL	MAT.	INST.	TOTAL	MAT.	INST.	TOTAL	MAT.	INST.	TOTAL	MAT.	INST.	TOTAL
015433	CONTRACTOR EQUIPMENT		110.8	110.8		110.8	110.8		108.0	108.0		110.8	110.8		110.8	110.8		110.0	110.0
0241, 31 - 34	SITE & INFRASTRUCTURE, DEMOLITION	91.6	95.0	93.9	94.9	95.1	95.0	102.8	91.4	94.8	90.1	92.9	92.0	85.8	94.8	92.0	87.6	92.2	90.7
0310	Concrete Forming & Accessories	98.9	109.3	107.7	82.5	87.6	86.9	81.1	78.4	78.8	85.0	93.1	91.9	83.9	92.0	90.8	87.4	74.0	76.0
0320	Concrete Reinforcing	97.5	112.8	104.9	94.5	109.6	101.8	93.7	100.7	97.1	96.5	100.8	98.6	95.1	121.2	107.7	98.9	106.9	102.7
0330	Cast-in-Place Concrete	88.5	103.6	94.2	98.6	90.6	95.6	107.2	82.3	97.8	94.2	87.1	91.5	87.1	97.0	90.8	93.8	90.0	92.4
03	CONCRETE	91.8	108.8	99.4	87.2	93.8	90.2	100.9	85.1	93.8	93.5	93.4	93.5	79.4	100.0	88.6	97.4	87.0	92.7
04	MASONRY	96.6	98.8	97.9	100.3	89.8	93.9	119.2	76.9	93.4	97.9	78.6	86.1	102.6	91.8	96.0	98.3	75.8	84.6
05	METALS	99.8	122.3	106.8	93.8	118.2	101.4	98.8	111.2	102.7	97.5	111.5	101.9	93.5	123.0	102.7	99.2	116.8	104.7
06	WOOD, PLASTICS & COMPOSITES	98.2	110.8	104.8	75.5	86.2	81.1	78.1	78.9	78.5	81.3	97.8	90.0	77.1	90.9	84.3	83.6	73.0	78.0
07	THERMAL & MOISTURE PROTECTION	105.3	112.6	108.4	103.3	95.0	99.7	99.6	85.1	93.4	105.1	85.4	96.6	102.7	94.0	99.0	96.8	79.3	89.3
08	OPENINGS	92.5	108.5	96.4	86.3	89.6	87.1	93.6	83.6	91.1	92.4	93.0	92.5	86.3	99.3	89.4	89.0	78.4	86.4
0920	Plaster & Gypsum Board	99.7	111.2	107.3	91.2	85.9	87.7	98.2	78.5	85.3	91.6	97.9	95.7	91.2	90.7	90.9	115.9	72.4	87.3
0950, 0980	Ceilings & Acoustic Treatment	92.2	111.2	104.1	95.4	85.9	89.4	113.3	78.5	91.5	95.5	97.9	97.0	96.1	90.7	92.7	96.9	72.4	81.5
0960	Flooring	91.9	97.7	93.6	85.4	103.7	90.7	92.2	98.9	94.1	86.1	98.9	89.8	86.4	100.5	90.5	91.4	74.4	86.4
0970, 0990	Wall Finishes & Painting/Coating	87.1	108.2	99.7	82.8	112.0	100.3	89.7	100.6	96.2	87.1	100.6	95.2	82.8	101.1	93.8	87.1	95.9	92.4
09	FINISHES	92.4	107.2	100.4	90.4	92.5	91.5	101.6	83.9	92.0	90.4	95.4	93.1	90.3	94.0	92.3	93.8	75.7	84.0
COVERS	DIVS. 10 - 14, 25, 28, 41, 43, 44, 46	100.0	103.5	100.8	100.0	98.6	99.7	100.0	94.9	98.8	100.0	97.7	99.5	100.0	99.7	99.9	100.0	92.6	98.3
21, 22, 23	FIRE SUPPRESSION, PLUMBING & HVAC	101.2	116.9	107.5	100.5	90.6	96.5	93.1	80.5	88.0	97.4	86.3	92.9	96.7	92.5	95.0	95.1	85.5	91.2
26, 27, 3370	ELECTRICAL, COMMUNICATIONS & UTIL.	99.4	99.5	99.5	90.7	113.3	101.9	94.7	113.3	103.9	94.1	113.3	103.6	91.2	103.7	97.4	93.6	82.7	88.2
MF2018	WEIGHTED AVERAGE	97.5	108.1	102.1	94.2	97.5	95.6	98.0	90.1	94.6	95.6	95.0	95.3	92.1	98.7	95.0	95.4	86.0	91.3

PENNSYLVANIA

DIVISION		DOYLESTOWN 189			DUBOIS 158			ERIE 164 - 165			GREENSBURG 156			HARRISBURG 170 - 171			HAZLETON 182		
		MAT.	INST.	TOTAL	MAT.	INST.	TOTAL	MAT.	INST.	TOTAL	MAT.	INST.	TOTAL	MAT.	INST.	TOTAL	MAT.	INST.	TOTAL
015433	CONTRACTOR EQUIPMENT		89.5	89.5		108.0	108.0		110.8	110.8		108.0	108.0		112.2	112.2		110.8	110.8
0241, 31 - 34	SITE & INFRASTRUCTURE, DEMOLITION	104.5	82.9	89.6	107.6	91.5	96.6	91.8	94.9	93.9	98.7	93.2	94.9	88.7	98.1	95.1	84.6	93.9	91.0
0310	Concrete Forming & Accessories	81.3	122.1	116.1	80.5	80.8	80.7	98.1	87.4	89.0	87.9	86.7	86.8	100.5	89.6	91.2	79.0	85.6	84.6
0320	Concrete Reinforcing	94.3	143.7	118.1	93.0	113.7	103.0	96.5	109.3	102.7	93.0	120.9	106.5	103.9	112.4	108.0	94.7	107.7	101.0
0330	Cast-in-Place Concrete	83.6	124.7	99.1	103.3	93.2	99.5	96.9	91.0	94.7	99.4	96.4	98.3	92.5	102.4	96.2	83.6	91.8	86.7
03	CONCRETE	87.3	126.3	104.8	102.9	92.2	98.1	86.4	93.8	89.7	96.1	97.3	96.6	92.6	99.2	95.6	84.4	93.0	88.3
04	MASONRY	100.2	129.3	117.9	120.1	89.5	101.4	88.2	90.7	89.7	130.3	88.0	104.5	93.7	92.9	93.2	110.1	85.8	95.3
05	METALS	97.3	121.9	105.0	98.8	116.7	104.4	94.0	117.3	101.3	98.8	120.7	105.6	106.0	118.6	109.9	99.5	117.8	105.2
06	WOOD, PLASTICS & COMPOSITES	76.8	121.6	100.2	76.9	78.9	78.0	94.9	85.7	90.1	85.0	84.1	84.5	100.3	86.4	93.0	75.1	84.1	79.8
07	THERMAL & MOISTURE PROTECTION	101.9	125.8	112.2	99.9	91.2	96.2	103.6	90.7	98.1	99.5	91.8	96.2	98.9	108.2	102.9	104.5	96.2	100.9
08	OPENINGS	94.9	128.3	103.0	93.6	86.6	91.9	86.4	90.7	87.5	93.5	95.6	94.0	100.6	87.8	97.4	93.0	89.9	92.2
0920	Plaster & Gypsum Board	90.1	122.2	111.2	96.8	78.5	84.8	99.7	85.4	90.3	98.6	83.9	89.0	122.6	85.9	98.5	90.4	83.7	86.0
0950, 0980	Ceilings & Acoustic Treatment	91.6	122.2	110.8	113.3	78.5	91.5	92.2	85.4	87.9	112.7	83.9	94.6	102.3	85.9	92.0	92.9	83.7	87.2
0960	Flooring	75.9	132.9	92.6	91.9	98.9	93.9	92.5	103.7	95.8	95.9	74.4	89.6	95.2	94.1	94.9	83.1	85.2	83.7
0970, 0990	Wall Finishes & Painting/Coating	86.7	136.8	116.7	89.7	101.1	96.6	92.9	93.5	93.3	89.7	101.1	96.6	93.7	89.5	91.2	87.1	99.1	94.3
09	FINISHES	84.1	125.4	106.4	101.8	85.3	92.9	93.5	90.5	91.9	102.3	85.2	93.0	99.2	89.6	94.0	88.3	86.6	87.4
COVERS	DIVS. 10 - 14, 25, 28, 41, 43, 44, 46	100.0	114.9	103.5	100.0	96.7	99.2	100.0	98.9	99.7	100.0	98.8	99.7	100.0	98.6	99.7	100.0	98.5	99.6
21, 22, 23	FIRE SUPPRESSION, PLUMBING & HVAC	96.7	126.3	108.7	93.1	83.4	89.2	100.5	93.8	97.8	93.1	84.7	89.7	96.4	98.5	97.2	97.4	91.0	94.8
26, 27, 3370	ELECTRICAL, COMMUNICATIONS & UTIL.	93.5	127.5	110.3	95.2	113.3	104.1	92.2	90.5	91.4	95.2	113.3	104.2	100.6	89.0	94.9	94.8	86.7	90.8
MF2018	WEIGHTED AVERAGE	94.7	122.6	106.7	98.5	94.0	96.6	94.0	94.6	94.3	98.0	95.8	97.1	98.5	97.0	97.9	95.1	92.5	94.0

PENNSYLVANIA

DIVISION		INDIANA 157			JOHNSTOWN 159			KITTANNING 162			LANCASTER 175 - 176			LEHIGH VALLEY 180			MONTROSE 188		
		MAT.	INST.	TOTAL	MAT.	INST.	TOTAL	MAT.	INST.	TOTAL	MAT.	INST.	TOTAL	MAT.	INST.	TOTAL	MAT.	INST.	TOTAL
015433	CONTRACTOR EQUIPMENT		108.0	108.0		108.0	108.0		110.8	110.8		110.0	110.0		110.8	110.8		110.8	110.8
0241, 31 - 34	SITE & INFRASTRUCTURE, DEMOLITION	96.9	92.1	93.6	103.3	93.4	96.5	88.3	94.7	92.7	79.7	94.3	89.7	88.4	94.3	92.4	87.2	94.0	91.8
0310	Concrete Forming & Accessories	81.7	89.8	88.6	80.5	87.3	86.3	83.9	86.8	86.3	89.5	87.8	88.0	91.8	106.3	104.1	80.0	86.1	85.2
0320	Concrete Reinforcing	92.4	121.2	106.3	93.7	121.1	106.9	95.1	121.0	107.6	98.5	112.5	105.2	94.7	102.9	98.6	99.2	110.7	104.7
0330	Cast-in-Place Concrete	97.5	93.6	96.0	108.1	89.8	101.2	90.5	94.0	91.8	79.8	96.7	86.2	90.5	95.3	92.3	88.7	89.3	88.9
03	CONCRETE	93.7	97.7	95.5	101.9	95.3	98.9	81.9	96.6	88.5	85.7	96.6	90.6	90.8	102.8	96.2	89.4	92.9	91.0
04	MASONRY	115.5	89.6	99.7	116.4	89.2	99.8	104.9	85.1	92.8	104.0	89.5	95.1	96.6	92.5	94.1	96.5	87.7	91.1
05	METALS	98.9	120.5	105.6	98.9	120.6	105.6	93.6	121.8	102.4	99.2	121.3	106.1	99.4	117.1	104.9	97.6	118.5	104.1
06	WOOD, PLASTICS & COMPOSITES	78.9	90.8	85.1	76.9	86.1	81.7	77.1	86.3	81.9	86.8	86.2	86.5	88.8	108.8	99.2	76.0	84.1	80.2
07	THERMAL & MOISTURE PROTECTION	99.3	92.5	96.4	99.6	91.9	96.3	102.9	90.9	97.7	96.1	105.1	100.0	105.0	106.9	105.8	104.6	86.9	97.0
08	OPENINGS	93.6	94.9	93.9	93.5	92.3	93.2	86.3	96.8	88.8	89.0	87.8	88.7	92.9	102.7	95.3	89.8	88.5	89.5
0920	Plaster & Gypsum Board	98.5	90.7	93.4	96.7	85.9	89.6	91.2	86.0	87.8	118.7	85.9	97.2	92.4	109.1	103.3	91.1	83.7	86.3
0950, 0980	Ceilings & Acoustic Treatment	113.3	90.7	99.1	112.7	85.9	95.9	96.1	86.0	89.8	96.9	85.9	90.0	92.9	109.1	103.1	95.5	83.7	88.1
0960	Flooring	92.7	98.9	94.5	91.9	103.7	95.3	86.4	98.9	90.0	92.5	94.4	93.0	88.8	89.8	89.1	83.8	98.9	88.2
0970, 0990	Wall Finishes & Painting/Coating	89.7	101.1	96.6	89.7	112.0	103.1	82.8	101.1	93.8	87.1	89.5	88.6	87.1	96.2	92.5	87.1	99.1	94.3
09	FINISHES	101.4	92.2	96.4	101.2	92.2	96.4	90.5	89.6	90.0	93.9	88.3	90.9	90.5	102.3	96.9	89.5	89.0	89.2
COVERS	DIVS. 10 - 14, 25, 28, 41, 43, 44, 46	100.0	98.0	99.6	100.0	98.2	99.6	100.0	97.8	99.5	100.0	96.5	99.2	100.0	102.7	100.6	100.0	99.1	99.8
21, 22, 23	FIRE SUPPRESSION, PLUMBING & HVAC	93.1	81.5	88.4	93.1	88.3	91.1	96.7	86.6	92.6	95.1	96.0	95.5	97.4	112.0	103.3	97.4	90.3	94.5
26, 27, 3370	ELECTRICAL, COMMUNICATIONS & UTIL.	95.2	113.3	104.2	95.2	113.3	104.2	90.7	113.3	101.9	94.8	96.4	95.6	94.8	124.6	109.5	94.1	91.7	92.9
MF2018	WEIGHTED AVERAGE	96.8	96.2	96.5	98.0	97.2	97.7	92.6	96.6	94.3	94.2	96.5	95.2	95.6	107.5	100.8	94.6	93.3	94.0

PENNSYLVANIA

DIVISION		NEW CASTLE 161			NORRISTOWN 194			OIL CITY 163			PHILADELPHIA 190 - 191			PITTSBURGH 150 - 152			POTTSVILLE 179		
		MAT.	INST.	TOTAL	MAT.	INST.	TOTAL	MAT.	INST.	TOTAL	MAT.	INST.	TOTAL	MAT.	INST.	TOTAL	MAT.	INST.	TOTAL
015433	CONTRACTOR EQUIPMENT		110.8	110.8		95.3	95.3		110.8	110.8		100.4	100.4		96.9	96.9		110.0	110.0
0241, 31 - 34	SITE & INFRASTRUCTURE, DEMOLITION	86.2	95.3	92.5	94.7	91.9	92.8	84.8	93.1	90.5	97.9	102.1	100.8	102.7	93.3	96.3	82.6	92.7	89.5
0310	Concrete Forming & Accessories	83.9	95.9	94.1	81.6	120.7	114.9	83.9	89.4	88.6	95.6	140.6	133.9	98.9	99.1	99.1	80.3	84.4	83.8
0320	Concrete Reinforcing	94.1	99.4	96.6	94.7	143.7	119.7	95.1	90.4	92.8	107.8	151.5	128.9	92.6	125.8	108.6	97.7	108.7	103.0
0330	Cast-in-Place Concrete	87.9	96.9	91.3	83.6	122.4	98.3	85.4	90.6	87.4	93.9	133.6	108.9	106.7	100.9	104.5	85.0	92.5	87.8
03	CONCRETE	79.7	98.0	87.9	86.6	124.8	103.8	78.2	91.5	84.1	100.3	138.9	117.6	101.5	104.3	102.8	89.1	93.0	90.8
04	MASONRY	102.7	94.8	97.9	111.2	125.7	120.0	101.0	85.0	91.3	99.9	135.2	121.4	105.5	100.7	102.6	97.5	83.9	89.2
05	METALS	93.6	115.1	100.3	102.9	122.0	108.8	93.6	112.7	99.5	106.5	124.1	112.0	100.3	108.9	103.0	99.4	118.5	105.4
06	WOOD, PLASTICS & COMPOSITES	77.1	96.4	87.2	74.4	121.5	99.1	77.1	90.9	84.3	93.8	142.0	119.0	100.3	99.2	99.7	75.6	82.3	79.1
07	THERMAL & MOISTURE PROTECTION	102.8	93.4	98.8	108.8	124.3	115.4	102.6	87.6	96.2	103.7	137.2	118.1	99.9	99.6	99.8	96.3	95.2	95.8
08	OPENINGS	86.3	97.0	88.9	86.6	128.3	96.7	86.3	92.3	87.7	97.7	143.0	108.8	97.0	104.7	98.9	89.0	89.6	89.2
0920	Plaster & Gypsum Board	91.2	96.4	94.6	87.2	122.2	110.2	91.2	90.7	90.9	103.0	143.0	129.3	100.4	99.1	99.5	113.3	81.9	92.7
0950, 0980	Ceilings & Acoustic Treatment	96.1	96.4	95.3	95.1	122.2	112.1	96.1	90.7	92.7	99.7	143.0	126.9	106.5	99.1	101.9	96.9	81.9	87.5
0960	Flooring	86.4	103.4	91.3	87.1	132.9	100.5	86.4	98.9	90.0	98.4	149.0	113.2	99.8	106.7	101.8	88.3	98.9	91.4
0970, 0990	Wall Finishes & Painting/Coating	82.8	109.2	98.6	85.2	136.8	116.2	82.8	101.1	93.8	95.6	168.5	139.3	95.3	110.6	104.5	87.1	99.1	94.3
09	FINISHES	90.4	98.7	94.9	87.4	124.4	107.4	90.2	92.3	91.4	98.7	145.6	124.1	102.5	101.3	101.9	92.1	87.7	89.7
COVERS	DIVS. 10 - 14, 25, 28, 41, 43, 44, 46	100.0	100.3	100.1	100.0	113.6	103.2	100.0	98.2	99.6	100.0	119.6	104.6	100.0	101.4	100.3	100.0	98.3	99.6
21, 22, 23	FIRE SUPPRESSION, PLUMBING & HVAC	96.7	96.0	96.4	95.7	124.4	107.3	96.7	89.6	93.9	96.4	136.9	112.7	96.5	98.6	97.3	95.1	91.2	93.5
26, 27, 3370	ELECTRICAL, COMMUNICATIONS & UTIL.	91.2	100.8	95.9	93.4	136.7	114.9	92.7	103.7	98.2	99.2	157.6	128.1	96.8	111.5	104.1	93.2	87.9	90.6
MF2018	WEIGHTED AVERAGE	92.2	98.9	95.1	95.1	123.4	107.3	92.1	94.4	93.1	99.7	136.9	115.8	99.3	102.6	100.7	94.0	92.5	93.3

PENNSYLVANIA

DIVISION		READING 195 - 196			SCRANTON 184 - 185			STATE COLLEGE 168			STROUDSBURG 183			SUNBURY 178			UNIONTOWN 154		
		MAT.	INST.	TOTAL	MAT.	INST.	TOTAL	MAT.	INST.	TOTAL	MAT.	INST.	TOTAL	MAT.	INST.	TOTAL	MAT.	INST.	TOTAL
015433	CONTRACTOR EQUIPMENT		117.0	117.0		110.8	110.8		110.0	110.0		110.8	110.8		110.8	110.8		108.0	108.0
0241, 31 - 34	SITE & INFRASTRUCTURE, DEMOLITION	99.9	105.0	103.4	92.1	94.6	93.8	82.5	93.8	90.3	86.2	94.2	91.7	94.0	93.5	93.7	97.4	93.0	94.4
0310	Concrete Forming & Accessories	98.4	89.2	90.6	99.0	89.9	91.3	84.0	88.0	87.4	85.8	87.2	87.0	92.3	81.7	83.2	74.5	91.8	89.2
0320	Concrete Reinforcing	98.8	147.8	122.4	97.5	112.5	104.8	95.8	109.9	102.6	97.8	110.9	104.1	100.5	106.9	103.6	93.0	121.3	106.7
0330	Cast-in-Place Concrete	74.9	97.3	83.4	92.4	93.5	92.8	89.1	90.8	89.7	87.1	90.8	88.5	92.9	91.1	92.2	97.5	96.3	97.0
03	CONCRETE	85.9	103.6	93.8	93.6	96.4	94.9	93.7	94.1	93.9	88.2	94.0	90.8	92.7	90.9	91.9	93.3	99.6	96.1
04	MASONRY	99.8	92.0	95.0	97.0	94.1	95.2	102.4	92.9	96.6	94.4	92.3	93.1	98.1	77.6	85.6	133.8	94.1	109.6
05	METALS	103.2	135.3	113.2	101.9	120.9	107.8	97.4	118.8	104.1	99.5	119.0	105.6	99.2	117.4	104.9	98.6	121.2	105.6
06	WOOD, PLASTICS & COMPOSITES	94.0	86.1	89.9	98.2	88.5	93.1	83.4	86.2	84.8	82.3	84.1	83.2	84.9	82.3	83.5	70.4	90.8	81.1
07	THERMAL & MOISTURE PROTECTION	109.9	106.2	108.3	105.1	92.7	99.8	104.3	105.9	105.0	104.7	87.2	97.2	97.3	88.5	93.6	99.2	94.6	97.2
08	OPENINGS	90.8	103.1	93.8	92.5	92.7	92.5	89.7	89.6	89.6	93.0	91.1	92.5	89.1	86.6	88.5	93.5	97.4	94.4
0920	Plaster & Gypsum Board	96.3	85.9	89.5	102.4	88.2	93.1	92.2	85.9	88.1	90.9	83.7	86.2	112.5	81.9	92.4	94.7	90.7	92.1
0950, 0980	Ceilings & Acoustic Treatment	86.7	85.9	86.2	102.4	88.2	93.5	92.3	85.9	88.3	91.6	83.7	86.7	93.7	81.9	86.3	112.7	90.7	98.9
0960	Flooring	91.9	94.4	92.6	91.9	96.9	93.4	89.3	96.4	91.4	86.6	89.8	87.5	89.2	98.9	92.1	88.7	105.0	93.5
0970, 0990	Wall Finishes & Painting/Coating	84.1	108.2	98.5	87.1	108.2	99.7	87.1	112.0	102.0	87.1	99.1	94.3	87.1	99.1	94.3	89.7	109.2	101.4
09	FINISHES	88.4	90.7	89.7	95.2	92.6	93.8	89.7	91.3	90.6	89.2	88.2	88.7	92.9	86.1	89.2	99.6	95.3	97.3
COVERS	DIVS. 10 - 14, 25, 28, 41, 43, 44, 46	100.0	98.9	99.8	100.0	99.6	99.9	100.0	96.9	99.3	100.0	100.1	100.0	100.0	94.2	98.6	100.0	99.5	99.9
21, 22, 23	FIRE SUPPRESSION, PLUMBING & HVAC	101.1	114.5	106.5	101.2	96.0	99.1	97.4	96.3	97.0	97.4	93.3	95.7	95.1	83.2	90.3	93.1	88.9	91.4
26, 27, 3370	ELECTRICAL, COMMUNICATIONS & UTIL.	99.5	96.5	98.0	99.5	92.0	95.8	93.3	113.3	103.2	94.8	136.5	115.4	93.6	88.5	91.1	92.7	113.3	102.9
MF2018	WEIGHTED AVERAGE	97.3	104.7	100.5	98.3	96.8	97.7	95.2	99.1	96.9	95.0	100.9	97.6	95.0	89.3	92.5	97.1	99.2	98.0

For customer support on your Commercial Renovation Costs with RSMeans Data, call 800.448.8182.

821

PENNSYLVANIA

DIVISION		WASHINGTON 153			WELLSBORO 169			WESTCHESTER 193			WILKES-BARRE 186 - 187			WILLIAMSPORT 177			YORK 173 - 174		
		MAT.	INST.	TOTAL	MAT.	INST.	TOTAL	MAT.	INST.	TOTAL	MAT.	INST.	TOTAL	MAT.	INST.	TOTAL	MAT.	INST.	TOTAL
015433	CONTRACTOR EQUIPMENT		108.0	108.0		110.8	110.8		95.3	95.3		110.8	110.8		110.8	110.8		110.0	110.0
0241, 31 - 34	SITE & INFRASTRUCTURE, DEMOLITION	97.5	93.9	95.0	93.5	93.4	93.5	100.7	92.8	95.3	84.2	94.6	91.4	85.2	94.4	91.5	84.7	94.3	91.3
0310	Concrete Forming & Accessories	81.9	98.1	95.7	84.2	81.0	81.5	88.6	121.9	117.0	88.8	89.9	89.7	88.6	86.9	87.2	84.0	87.3	86.8
0320	Concrete Reinforcing	93.0	121.5	106.8	95.8	110.6	102.9	96.5	143.7	119.3	96.5	112.5	104.2	99.7	112.5	105.9	100.5	112.5	106.3
0330	Cast-in-Place Concrete	97.5	96.9	97.3	93.4	84.6	90.1	92.7	124.5	104.7	83.6	93.4	87.3	78.6	91.6	83.5	85.4	96.6	89.6
03	CONCRETE	93.8	102.8	97.8	95.9	89.0	92.8	94.2	126.1	108.5	85.3	96.4	90.3	80.9	94.5	87.0	90.3	96.4	93.0
04	MASONRY	114.3	96.7	103.5	102.6	78.0	87.6	105.6	129.3	120.1	110.4	94.1	100.5	89.7	91.9	91.1	99.5	89.5	93.4
05	METALS	98.6	122.4	106.0	97.5	117.7	103.8	102.9	122.0	108.8	97.6	120.9	104.9	99.2	120.9	106.0	100.9	121.2	107.2
06	WOOD, PLASTICS & COMPOSITES	79.0	98.8	89.3	80.7	81.5	81.1	81.9	121.5	102.6	85.1	88.5	86.8	81.0	86.2	83.7	79.4	85.7	82.7
07	THERMAL & MOISTURE PROTECTION	99.3	96.9	98.3	105.4	82.1	95.4	109.3	121.4	114.5	104.5	92.7	99.4	96.6	91.3	94.3	96.4	105.0	100.1
08	OPENINGS	93.5	103.6	95.9	92.3	86.7	91.0	86.6	128.3	96.7	89.8	92.7	90.5	89.1	91.4	89.7	89.0	87.4	88.6
0920	Plaster & Gypsum Board	98.4	98.9	98.7	90.8	81.1	84.4	87.8	122.2	110.4	91.9	88.2	89.5	113.3	85.9	95.3	113.3	85.4	95.0
0950, 0980	Ceilings & Acoustic Treatment	112.7	98.9	104.1	92.3	81.1	85.3	95.1	122.2	112.1	95.5	88.2	90.9	96.9	85.9	90.0	96.2	85.4	89.4
0960	Flooring	92.8	105.0	96.4	85.6	98.9	89.5	90.0	132.9	102.5	87.5	96.9	90.3	88.1	94.4	89.9	89.8	94.4	91.1
0970, 0990	Wall Finishes & Painting/Coating	89.7	109.2	101.4	87.1	99.1	94.3	85.2	136.8	116.2	87.1	108.2	99.7	87.1	108.2	99.7	87.1	89.5	88.6
09	FINISHES	101.3	100.3	100.7	89.8	86.1	87.8	88.8	125.3	108.6	90.4	92.6	91.6	92.7	90.0	91.2	92.4	88.0	90.0
COVERS	DIVS. 10 - 14, 25, 28, 41, 43, 44, 46	100.0	100.4	100.1	100.0	94.1	98.6	100.0	114.7	103.5	100.0	99.6	99.9	100.0	96.3	99.1	100.0	96.4	99.2
21, 22, 23	FIRE SUPPRESSION, PLUMBING & HVAC	93.1	92.5	92.8	97.4	83.2	91.7	95.7	126.3	108.0	97.4	96.0	96.8	95.1	93.4	94.4	100.9	96.0	98.9
26, 27, 3370	ELECTRICAL, COMMUNICATIONS & UTIL.	94.7	113.3	103.9	94.1	91.7	92.9	93.3	127.5	110.2	94.8	92.0	93.4	94.1	85.1	89.6	95.6	83.9	89.8
MF2018	WEIGHTED AVERAGE	96.6	101.9	98.9	96.2	89.3	93.2	96.1	123.2	107.8	94.9	96.8	95.7	92.8	94.2	93.4	96.2	94.6	95.5

DIVISION		PUERTO RICO SAN JUAN 009			RHODE ISLAND NEWPORT 028			PROVIDENCE 029			SOUTH CAROLINA AIKEN 298			BEAUFORT 299			CHARLESTON 294		
		MAT.	INST.	TOTAL	MAT.	INST.	TOTAL	MAT.	INST.	TOTAL	MAT.	INST.	TOTAL	MAT.	INST.	TOTAL	MAT.	INST.	TOTAL
015433	CONTRACTOR EQUIPMENT		83.7	83.7		96.9	96.9		102.4	102.4		101.8	101.8		101.8	101.8		101.8	101.8
0241, 31 - 34	SITE & INFRASTRUCTURE, DEMOLITION	129.7	83.1	97.7	86.0	95.7	92.7	89.3	104.5	99.8	133.3	83.3	99.0	128.2	83.3	97.4	112.3	83.5	92.5
0310	Concrete Forming & Accessories	86.6	21.0	30.8	101.8	120.8	118.0	101.6	121.0	118.1	97.7	62.5	67.8	96.2	38.1	46.8	95.2	62.9	67.7
0320	Concrete Reinforcing	179.7	18.0	101.7	105.1	121.3	112.9	99.2	121.3	109.9	94.4	57.8	76.8	93.5	63.6	79.1	93.4	64.2	79.3
0330	Cast-in-Place Concrete	97.1	31.1	72.2	68.7	116.8	86.9	94.2	118.3	103.3	89.6	65.3	80.4	89.5	65.3	80.4	105.1	65.6	90.2
03	CONCRETE	97.5	25.0	65.0	84.3	119.2	99.9	98.1	119.7	107.8	99.0	64.3	83.4	96.5	54.3	77.5	93.4	65.7	81.0
04	MASONRY	87.9	23.3	48.5	97.5	121.7	112.3	106.6	121.8	115.9	85.7	62.1	71.3	102.1	62.1	77.7	103.5	65.6	80.4
05	METALS	127.6	38.7	99.9	100.9	115.9	105.6	105.3	114.5	108.2	102.4	87.6	97.8	102.4	89.7	98.4	104.4	90.8	100.2
06	WOOD, PLASTICS & COMPOSITES	84.8	20.0	50.9	101.2	119.8	110.9	103.2	120.1	112.0	94.6	64.5	78.9	92.7	31.3	60.6	91.2	64.5	77.3
07	THERMAL & MOISTURE PROTECTION	136.9	26.4	89.5	105.0	117.2	110.2	109.2	118.1	113.1	96.7	61.5	81.6	96.3	58.5	80.1	95.3	63.5	81.7
08	OPENINGS	155.1	18.6	121.8	97.8	120.6	103.4	99.4	120.7	104.6	94.1	60.2	85.8	94.1	43.2	81.7	97.6	62.7	89.1
0920	Plaster & Gypsum Board	157.4	18.0	66.0	91.6	120.2	110.4	105.2	120.2	115.0	91.2	63.6	73.1	95.8	29.5	52.3	96.5	63.6	74.9
0950, 0980	Ceilings & Acoustic Treatment	226.3	18.0	95.8	84.2	120.2	106.8	96.9	120.2	111.5	80.1	63.6	69.8	85.8	29.5	50.5	85.8	63.6	71.9
0960	Flooring	212.2	28.4	158.5	88.7	125.9	99.5	89.1	125.9	99.8	101.5	85.7	96.9	103.0	72.3	94.1	102.7	80.0	96.1
0970, 0990	Wall Finishes & Painting/Coating	183.5	23.4	87.6	87.0	116.8	104.9	93.8	116.8	107.6	90.9	62.9	74.1	90.9	54.2	68.9	90.9	68.0	77.1
09	FINISHES	201.9	22.4	104.8	88.9	121.8	106.7	95.1	122.0	109.6	93.4	66.9	79.0	95.4	44.1	67.6	93.4	66.7	78.9
COVERS	DIVS. 10 - 14, 25, 28, 41, 43, 44, 46	99.3	25.1	81.8	100.0	107.8	101.8	100.0	108.3	101.9	100.0	83.3	96.1	100.0	76.1	94.4	100.0	83.3	96.1
21, 22, 23	FIRE SUPPRESSION, PLUMBING & HVAC	117.7	18.0	77.5	100.9	111.4	105.1	101.2	111.4	105.3	97.8	52.6	79.6	97.8	58.2	81.8	101.6	58.3	84.1
26, 27, 3370	ELECTRICAL, COMMUNICATIONS & UTIL.	117.5	24.1	71.3	101.3	97.9	99.4	102.1	97.6	99.8	95.2	59.6	77.6	98.3	63.9	81.3	96.9	56.8	77.1
MF2018	WEIGHTED AVERAGE	124.9	28.3	83.1	97.0	112.6	103.8	101.0	113.3	106.3	98.1	65.2	83.9	98.9	61.5	82.7	99.4	67.0	85.4

DIVISION		SOUTH CAROLINA COLUMBIA 290 - 292			FLORENCE 295			GREENVILLE 296			ROCK HILL 297			SPARTANBURG 293			SOUTH DAKOTA ABERDEEN 574		
		MAT.	INST.	TOTAL	MAT.	INST.	TOTAL	MAT.	INST.	TOTAL	MAT.	INST.	TOTAL	MAT.	INST.	TOTAL	MAT.	INST.	TOTAL
015433	CONTRACTOR EQUIPMENT		105.4	105.4		101.8	101.8		101.8	101.8		101.8	101.8		101.8	101.8		95.8	95.8
0241, 31 - 34	SITE & INFRASTRUCTURE, DEMOLITION	112.3	92.3	98.6	122.4	83.2	95.5	117.2	83.5	94.0	114.1	83.2	92.9	117.0	83.5	94.0	100.1	90.9	93.8
0310	Concrete Forming & Accessories	98.1	62.9	68.1	81.5	62.9	65.6	95.0	63.0	67.7	92.9	62.2	66.8	98.3	63.0	68.2	100.0	66.5	71.5
0320	Concrete Reinforcing	95.8	64.2	80.6	93.0	64.2	79.1	92.9	61.5	77.7	93.7	62.7	78.8	92.9	64.2	79.1	92.9	70.1	81.9
0330	Cast-in-Place Concrete	106.4	65.3	90.9	89.5	65.3	80.4	89.5	65.6	80.5	89.5	64.8	80.2	89.5	65.6	80.5	111.2	75.5	97.7
03	CONCRETE	93.4	65.7	81.0	91.4	65.5	79.8	90.4	65.3	79.2	88.6	64.9	78.0	90.6	65.8	79.5	100.0	71.2	87.1
04	MASONRY	95.8	65.6	77.4	85.8	65.6	73.4	83.5	65.6	72.5	109.3	61.3	80.0	85.8	65.6	73.5	118.1	69.8	88.6
05	METALS	101.5	88.3	97.4	103.2	90.0	99.1	103.2	90.5	99.3	102.4	89.3	98.3	103.2	90.9	99.4	91.7	81.5	88.6
06	WOOD, PLASTICS & COMPOSITES	95.4	64.5	79.3	75.9	64.5	69.9	91.1	64.5	77.2	89.2	64.5	76.3	95.5	64.5	79.3	98.9	62.7	79.9
07	THERMAL & MOISTURE PROTECTION	91.1	64.3	79.6	95.6	63.5	81.9	95.6	63.5	81.8	95.5	56.7	78.8	95.7	63.5	81.9	102.9	73.5	90.3
08	OPENINGS	99.8	62.7	90.7	94.2	62.7	86.5	94.1	62.4	86.4	94.1	61.3	86.1	94.1	62.7	86.5	96.9	55.4	86.8
0920	Plaster & Gypsum Board	94.4	63.6	74.2	85.3	63.6	71.1	89.8	63.6	72.6	89.2	63.6	72.4	92.0	63.6	73.4	99.6	62.2	75.1
0950, 0980	Ceilings & Acoustic Treatment	88.5	63.6	72.9	81.4	63.6	70.2	80.1	63.6	69.8	80.1	63.6	69.8	80.1	63.6	69.8	103.6	62.2	77.7
0960	Flooring	92.8	80.0	89.1	93.7	80.0	89.7	100.3	80.0	94.4	99.2	72.3	91.3	101.7	80.0	95.4	93.9	41.8	78.7
0970, 0990	Wall Finishes & Painting/Coating	92.4	68.0	77.8	90.9	68.0	77.1	90.9	68.0	77.1	90.9	62.9	74.1	90.9	68.0	77.1	83.5	31.8	52.5
09	FINISHES	92.2	66.7	78.4	89.4	66.7	77.1	91.2	66.7	77.9	90.5	64.4	76.4	91.9	66.7	78.3	94.1	59.6	75.4
COVERS	DIVS. 10 - 14, 25, 28, 41, 43, 44, 46	100.0	83.3	96.1	100.0	83.3	96.1	100.0	83.4	96.1	100.0	83.1	96.0	100.0	83.4	96.1	100.0	86.0	96.7
21, 22, 23	FIRE SUPPRESSION, PLUMBING & HVAC	101.0	56.5	83.0	101.6	56.5	83.4	101.6	55.5	83.0	97.8	52.3	79.4	101.6	55.5	83.0	100.6	51.4	80.8
26, 27, 3370	ELECTRICAL, COMMUNICATIONS & UTIL.	100.3	60.4	80.5	95.2	60.4	78.0	97.0	83.5	90.3	97.0	83.5	90.3	97.0	83.5	90.3	99.8	64.8	82.4
MF2018	WEIGHTED AVERAGE	98.8	67.6	85.3	97.4	67.0	84.2	97.5	70.0	85.6	97.3	68.2	84.7	97.7	70.2	85.8	99.1	66.7	85.1

SOUTH DAKOTA

DIVISION		MITCHELL 573			MOBRIDGE 576			PIERRE 575			RAPID CITY 577			SIOUX FALLS 570 - 571			WATERTOWN 572		
		MAT.	INST.	TOTAL	MAT.	INST.	TOTAL	MAT.	INST.	TOTAL	MAT.	INST.	TOTAL	MAT.	INST.	TOTAL	MAT.	INST.	TOTAL
015433	CONTRACTOR EQUIPMENT		95.8	95.8		95.8	95.8		100.9	100.9		95.8	95.8		102.0	102.0		95.8	95.8
0241, 31 - 34	SITE & INFRASTRUCTURE, DEMOLITION	96.3	90.2	92.1	96.3	90.2	92.1	99.5	98.4	98.7	98.4	90.5	92.9	93.7	101.1	98.8	96.2	90.2	92.1
0310	Concrete Forming & Accessories	98.9	42.8	51.2	87.8	43.2	49.9	99.0	46.2	54.1	108.3	55.8	63.7	101.8	84.8	87.4	83.7	72.1	73.8
0320	Concrete Reinforcing	92.4	67.3	80.3	94.8	67.2	81.5	94.8	97.5	96.1	86.9	97.6	92.0	97.3	99.8	98.5	89.8	67.2	78.9
0330	Cast-in-Place Concrete	108.0	50.1	86.2	108.0	74.2	95.3	112.8	77.8	99.6	107.2	75.4	95.2	88.9	82.6	86.5	108.0	74.1	95.2
03	CONCRETE	97.7	51.2	76.8	97.3	59.7	80.4	102.3	67.3	86.6	97.2	70.9	85.4	94.3	86.9	91.0	96.4	72.7	85.8
04	MASONRY	104.5	74.2	86.1	112.4	69.8	86.4	105.3	74.4	86.5	111.7	72.7	87.9	106.1	79.2	89.7	139.5	70.6	97.5
05	METALS	90.7	80.4	87.5	90.8	80.7	87.6	94.2	88.6	92.4	93.4	90.7	92.6	94.8	93.0	94.3	90.7	80.6	87.6
06	WOOD, PLASTICS & COMPOSITES	97.7	32.5	63.6	84.4	32.2	57.1	103.8	33.0	66.8	104.4	46.3	74.0	98.4	84.0	90.9	79.7	71.7	75.5
07	THERMAL & MOISTURE PROTECTION	102.7	70.1	88.7	102.6	71.7	89.4	103.0	72.7	90.0	103.3	77.2	92.1	103.3	86.1	95.9	102.4	73.2	89.8
08	OPENINGS	96.1	38.2	82.0	98.6	37.5	83.7	99.5	62.4	90.5	100.6	69.8	93.1	103.8	90.9	100.6	96.1	59.2	87.1
0920	Plaster & Gypsum Board	98.5	31.1	54.3	92.8	30.9	52.2	95.0	31.4	53.3	99.4	45.3	63.9	91.5	83.8	86.5	90.8	71.4	78.1
0950, 0980	Ceilings & Acoustic Treatment	100.4	31.1	57.0	103.6	30.9	58.0	109.6	31.4	60.6	104.9	45.3	67.6	105.6	83.8	91.9	100.4	71.4	82.3
0960	Flooring	93.5	74.8	88.1	89.0	44.3	76.0	99.9	29.8	79.4	93.2	69.2	86.2	95.8	77.2	90.3	87.6	41.8	74.2
0970, 0990	Wall Finishes & Painting/Coating	83.5	34.8	54.3	83.5	35.7	54.8	93.6	118.9	108.8	83.5	118.9	104.7	96.5	118.9	109.90	83.5	31.8	52.5
09	FINISHES	92.9	46.2	67.6	91.6	42.0	64.8	99.8	50.0	72.8	94.1	64.2	77.9	97.4	87.0	91.8	90.1	64.3	76.1
COVERS	DIVS. 10 - 14, 25, 28, 41, 43, 44, 46	100.0	77.8	94.8	100.0	77.7	94.7	100.0	85.8	96.7	100.0	86.1	96.7	100.0	91.4	98.0	100.0	81.4	95.6
21, 22, 23	FIRE SUPPRESSION, PLUMBING & HVAC	96.8	46.7	76.6	96.8	65.8	84.3	100.6	75.1	90.3	100.6	74.2	90.0	100.5	72.2	89.1	96.8	49.6	77.8
26, 27, 3370	ELECTRICAL, COMMUNICATIONS & UTIL.	98.3	61.0	79.9	99.8	38.3	69.3	102.9	45.9	74.7	96.7	45.9	71.6	102.6	66.7	84.8	97.6	61.0	79.5
MF2018	WEIGHTED AVERAGE	96.6	59.6	80.6	97.1	60.7	81.4	100.2	68.9	86.7	98.7	71.0	86.7	99.3	82.1	91.8	97.7	66.6	84.3

TENNESSEE

DIVISION		CHATTANOOGA 373 - 374			COLUMBIA 384			COOKEVILLE 385			JACKSON 383			JOHNSON CITY 376			KNOXVILLE 377 - 379		
		MAT.	INST.	TOTAL	MAT.	INST.	TOTAL	MAT.	INST.	TOTAL	MAT.	INST.	TOTAL	MAT.	INST.	TOTAL	MAT.	INST.	TOTAL
015433	CONTRACTOR EQUIPMENT		103.4	103.4		98.2	98.2		98.2	98.2		104.2	104.2		97.4	97.4		97.4	97.4
0241, 31 - 34	SITE & INFRASTRUCTURE, DEMOLITION	106.1	91.8	96.3	89.8	82.3	84.6	95.5	79.0	84.2	99.4	91.6	94.0	112.8	81.3	91.2	91.9	82.1	85.2
0310	Concrete Forming & Accessories	101.1	62.0	67.9	80.5	60.3	63.3	80.7	31.2	38.6	87.8	39.7	46.9	85.3	59.8	63.6	99.9	63.1	68.6
0320	Concrete Reinforcing	96.3	68.3	82.8	87.4	61.9	75.1	87.4	61.6	75.0	87.4	69.2	78.6	96.9	64.3	81.2	96.3	67.8	82.5
0330	Cast-in-Place Concrete	98.2	63.8	85.2	89.4	63.2	79.5	101.3	56.2	84.3	98.9	57.7	83.3	79.0	58.0	71.1	92.3	65.1	82.0
03	CONCRETE	95.3	65.4	81.9	91.9	63.3	79.1	101.8	47.6	77.5	92.6	53.3	75.0	104.8	61.7	85.5	92.8	66.2	80.9
04	MASONRY	100.5	56.9	73.9	115.5	53.2	77.5	110.6	37.8	66.2	115.8	43.8	71.9	115.3	43.9	71.7	78.5	56.3	65.0
05	METALS	94.4	90.6	93.2	93.2	87.6	91.5	93.3	86.6	91.2	95.8	90.3	94.1	91.7	88.7	90.8	94.8	90.0	93.3
06	WOOD, PLASTICS & COMPOSITES	108.9	62.7	84.7	67.8	61.2	64.4	68.0	28.4	47.3	82.4	38.3	59.3	80.3	65.3	72.4	97.4	63.3	79.5
07	THERMAL & MOISTURE PROTECTION	100.7	62.2	84.2	95.5	60.3	80.4	96.0	50.5	76.5	97.7	52.4	78.2	95.9	57.1	79.3	93.8	62.6	80.4
08	OPENINGS	101.7	61.3	91.8	91.0	51.3	81.3	91.0	34.2	77.1	97.4	43.8	84.3	98.0	62.1	89.3	95.5	56.6	86.0
0920	Plaster & Gypsum Board	81.2	62.3	68.8	88.2	60.8	70.2	88.2	27.1	48.1	90.2	37.2	55.4	101.5	65.0	77.5	108.9	62.9	78.7
0950, 0980	Ceilings & Acoustic Treatment	99.0	62.3	76.0	80.9	60.8	68.3	80.9	27.1	47.1	88.7	37.2	56.4	94.8	65.0	76.1	96.1	62.9	75.3
0960	Flooring	98.9	61.4	87.9	83.3	51.5	74.0	83.4	47.7	73.0	82.7	56.1	75.0	92.8	56.1	82.1	98.3	56.1	86.0
0970, 0990	Wall Finishes & Painting/Coating	92.3	56.7	71.0	81.9	53.8	65.0	81.9	53.8	65.0	84.0	60.6	70.0	89.1	56.7	69.7	89.1	56.7	69.7
09	FINISHES	94.8	61.0	76.5	87.8	57.7	71.5	88.3	34.7	59.3	88.1	43.5	64.0	98.2	58.9	76.9	92.3	60.9	75.3
COVERS	DIVS. 10 - 14, 25, 28, 41, 43, 44, 46	100.0	81.3	95.6	100.0	81.0	95.5	100.0	73.5	93.8	100.0	75.0	94.1	100.0	78.1	94.9	100.0	82.1	95.8
21, 22, 23	FIRE SUPPRESSION, PLUMBING & HVAC	101.1	56.7	83.2	98.3	70.2	87.0	98.3	63.9	84.5	101.0	58.6	83.9	100.8	53.0	81.5	100.8	62.9	85.5
26, 27, 3370	ELECTRICAL, COMMUNICATIONS & UTIL.	101.8	79.8	90.9	94.9	48.7	72.0	96.3	59.0	77.8	101.1	63.3	82.4	92.9	50.5	72.0	97.9	57.7	78.0
MF2018	WEIGHTED AVERAGE	99.0	68.7	85.9	95.2	64.3	82.0	96.6	56.1	79.1	98.1	60.0	81.6	99.2	60.5	82.5	95.7	66.0	82.9

TENNESSEE / TEXAS

DIVISION		MCKENZIE 382			MEMPHIS 375,380 - 381			NASHVILLE 370 - 372			ABILENE 795 - 796			AMARILLO 790 - 791			AUSTIN 786 - 787		
		MAT.	INST.	TOTAL	MAT.	INST.	TOTAL	MAT.	INST.	TOTAL	MAT.	INST.	TOTAL	MAT.	INST.	TOTAL	MAT.	INST.	TOTAL
015433	CONTRACTOR EQUIPMENT		98.2	98.2		99.2	99.2		106.9	106.9		88.0	88.0		94.3	94.3		92.5	92.5
0241, 31 - 34	SITE & INFRASTRUCTURE, DEMOLITION	95.1	79.2	84.2	90.0	91.3	90.9	103.9	99.0	100.5	93.8	84.9	87.7	93.9	94.4	94.2	97.0	91.0	92.9
0310	Concrete Forming & Accessories	89.4	33.4	41.8	98.0	72.1	76.0	102.1	68.6	73.6	99.0	60.4	66.2	99.3	54.0	60.8	98.7	54.8	61.3
0320	Concrete Reinforcing	87.6	62.5	75.5	104.0	71.5	88.3	102.4	69.0	86.3	97.1	50.5	74.6	97.8	50.4	74.9	92.2	45.2	69.5
0330	Cast-in-Place Concrete	99.1	57.3	83.3	93.9	69.8	84.8	91.2	70.5	83.4	86.0	66.6	78.7	83.9	67.6	77.8	93.6	67.1	83.6
03	CONCRETE	100.6	49.2	77.5	96.7	72.0	85.6	94.5	70.4	83.7	86.5	61.6	75.3	93.3	59.0	77.9	89.2	58.2	75.3
04	MASONRY	114.4	43.2	71.0	93.6	60.0	73.1	92.1	63.9	74.9	101.3	63.3	78.1	103.5	61.7	76.0	98.8	63.3	77.1
05	METALS	93.3	87.2	91.4	84.5	83.1	84.1	107.0	85.7	100.4	108.3	70.2	96.4	103.5	68.6	92.6	102.5	65.4	91.0
06	WOOD, PLASTICS & COMPOSITES	77.8	30.6	53.1	100.2	73.0	86.0	105.3	67.9	85.7	103.7	62.1	81.9	107.1	53.2	78.9	96.9	54.8	74.9
07	THERMAL & MOISTURE PROTECTION	96.1	51.4	76.9	93.2	67.3	82.1	99.2	68.9	86.2	100.4	64.7	85.1	102.3	63.8	85.8	96.3	65.1	82.9
08	OPENINGS	91.0	35.8	77.5	99.2	66.9	91.4	100.8	66.5	92.5	101.7	57.9	91.0	103.1	53.0	90.9	101.3	49.9	88.8
0920	Plaster & Gypsum Board	91.0	29.3	50.5	95.7	72.4	80.4	90.3	66.9	75.0	101.7	61.6	75.4	113.8	52.3	73.4	88.4	53.9	65.8
0950, 0980	Ceilings & Acoustic Treatment	80.9	29.3	48.6	101.5	72.4	83.2	89.3	66.9	75.3	97.3	61.6	74.9	95.4	52.3	68.4	99.4	53.9	70.9
0960	Flooring	86.2	51.5	76.1	98.1	56.1	85.8	96.8	64.1	87.2	96.5	71.2	89.1	96.3	67.8	88.0	94.5	60.9	84.7
0970, 0990	Wall Finishes & Painting/Coating	81.9	42.0	58.0	90.6	74.8	81.1	97.7	67.9	79.9	91.5	57.0	70.8	94.5	54.3	70.4	94.3	42.1	63.0
09	FINISHES	89.4	36.0	60.5	97.5	69.4	82.3	95.3	67.3	80.1	92.2	62.1	75.9	97.7	56.1	75.2	94.8	54.1	72.8
COVERS	DIVS. 10 - 14, 25, 28, 41, 43, 44, 46	100.0	74.3	94.0	100.0	85.3	96.5	100.0	85.1	96.5	100.0	79.6	95.2	100.0	79.4	95.1	100.0	79.1	95.1
21, 22, 23	FIRE SUPPRESSION, PLUMBING & HVAC	98.3	55.8	81.2	100.7	74.2	90.0	101.1	82.1	93.4	101.2	50.4	80.7	100.9	52.4	81.4	100.8	59.6	84.2
26, 27, 3370	ELECTRICAL, COMMUNICATIONS & UTIL.	96.1	53.3	74.9	101.6	64.8	83.4	99.6	62.0	81.0	97.6	55.2	76.6	100.3	60.6	80.6	99.4	59.4	79.6
MF2018	WEIGHTED AVERAGE	96.7	54.7	78.6	96.5	72.5	86.1	100.0	74.5	89.0	99.0	61.6	82.9	99.9	61.8	83.5	98.6	62.3	82.9

For customer support on your Commercial Renovation Costs with RSMeans Data, call 800.448.8182.

823

TEXAS

DIVISION		BEAUMONT 776 - 777			BROWNWOOD 768			BRYAN 778			CHILDRESS 792			CORPUS CHRISTI 783 - 784			DALLAS 752 - 753		
		MAT.	INST.	TOTAL	MAT.	INST.	TOTAL	MAT.	INST.	TOTAL	MAT.	INST.	TOTAL	MAT.	INST.	TOTAL	MAT.	INST.	TOTAL
015433	CONTRACTOR EQUIPMENT		92.1	92.1		88.0	88.0		92.1	92.1		88.0	88.0		97.9	97.9		107.6	107.6
0241, 31 - 34	SITE & INFRASTRUCTURE, DEMOLITION	90.8	88.8	89.4	102.1	84.8	90.3	81.3	88.6	86.3	104.0	84.9	90.8	141.2	79.9	99.1	108.1	99.1	101.9
0310	Concrete Forming & Accessories	102.4	57.2	63.9	97.6	58.5	64.4	80.7	56.1	59.8	97.5	58.7	64.5	106.9	54.1	62.0	96.4	62.9	67.9
0320	Concrete Reinforcing	93.9	59.3	77.2	84.6	48.7	67.3	96.3	48.7	73.3	97.3	48.7	73.8	82.4	50.1	66.8	103.3	51.8	78.5
0330	Cast-in-Place Concrete	98.6	63.2	85.2	107.8	56.3	88.3	79.0	56.8	70.6	88.2	56.4	76.2	114.2	67.3	96.5	100.1	68.9	88.7
03	CONCRETE	95.5	60.6	79.8	95.6	56.9	78.3	79.7	56.1	69.1	95.0	57.0	77.9	96.2	59.9	79.9	97.2	64.7	82.6
04	MASONRY	99.2	63.9	77.7	126.7	58.2	84.9	139.3	58.1	89.8	104.9	57.8	76.1	81.5	61.7	69.4	102.7	60.0	76.6
05	METALS	101.1	74.6	92.8	103.1	68.7	92.3	101.0	69.9	91.3	105.5	68.8	94.0	98.8	83.9	94.1	103.4	80.7	96.3
06	WOOD, PLASTICS & COMPOSITES	111.4	57.3	83.1	102.8	60.9	80.9	76.5	57.2	66.4	103.0	60.9	81.0	121.9	53.6	86.1	97.9	64.6	80.5
07	THERMAL & MOISTURE PROTECTION	92.8	64.7	80.7	92.9	61.1	79.2	86.2	62.1	75.8	101.0	61.0	83.8	99.3	63.1	83.8	90.4	67.8	80.7
08	OPENINGS	94.6	55.8	85.1	100.6	55.7	89.6	95.4	53.9	85.3	96.1	55.7	86.3	101.5	51.5	89.3	96.6	60.0	87.7
0920	Plaster & Gypsum Board	106.6	56.8	73.9	96.9	60.5	73.0	91.6	56.7	68.7	100.7	60.5	74.3	90.6	52.8	65.8	93.3	63.6	73.8
0950, 0980	Ceilings & Acoustic Treatment	116.3	56.8	79.0	92.5	60.5	72.4	105.7	56.7	75.0	93.5	60.5	72.8	103.2	52.8	71.6	109.9	63.6	80.9
0960	Flooring	120.5	77.1	107.8	76.0	52.1	69.0	89.7	63.9	82.1	94.7	52.1	82.3	111.3	73.7	100.3	97.7	69.8	89.5
0970, 0990	Wall Finishes & Painting/Coating	95.9	51.3	69.2	89.5	49.2	65.3	93.2	54.7	70.1	91.5	49.2	66.1	106.8	45.6	70.1	102.3	54.3	73.5
09	FINISHES	104.1	60.0	80.3	84.0	56.4	69.1	90.6	56.9	72.4	91.9	56.4	72.7	105.7	56.4	79.0	101.5	63.2	80.8
COVERS	DIVS. 10 - 14, 25, 28, 41, 43, 44, 46	100.0	80.3	95.4	100.0	79.1	95.1	100.0	79.8	95.2	100.0	79.1	95.1	100.0	78.4	94.9	100.0	81.3	95.6
21, 22, 23	FIRE SUPPRESSION, PLUMBING & HVAC	101.0	60.8	84.8	97.0	47.3	77.0	97.2	59.5	82.0	97.4	51.6	78.9	101.0	60.8	84.8	101.0	62.1	85.3
26, 27, 3370	ELECTRICAL, COMMUNICATIONS & UTIL.	99.6	61.0	80.5	99.4	44.4	72.2	98.0	61.0	79.7	97.6	53.7	75.9	96.7	66.1	81.6	98.2	60.5	79.5
MF2018	WEIGHTED AVERAGE	99.2	64.9	84.4	99.1	57.2	81.0	96.3	62.4	81.7	98.7	59.4	81.7	100.2	64.6	84.8	100.1	67.4	86.0

TEXAS

DIVISION		DEL RIO 788			DENTON 762			EASTLAND 764			EL PASO 798 - 799,885			FORT WORTH 760 - 761			GALVESTON 775		
		MAT.	INST.	TOTAL	MAT.	INST.	TOTAL	MAT.	INST.	TOTAL	MAT.	INST.	TOTAL	MAT.	INST.	TOTAL	MAT.	INST.	TOTAL
015433	CONTRACTOR EQUIPMENT		87.2	87.2		97.4	97.4		88.0	88.0		94.3	94.3		94.3	94.3		104.9	104.9
0241, 31 - 34	SITE & INFRASTRUCTURE, DEMOLITION	120.3	82.8	94.6	101.8	79.3	86.3	104.9	84.8	91.1	89.7	94.0	92.7	97.7	94.4	95.4	108.5	87.3	93.9
0310	Concrete Forming & Accessories	102.5	49.3	57.2	106.3	58.9	65.9	98.7	58.6	64.6	100.5	57.8	64.2	97.1	60.7	66.1	89.8	53.8	59.2
0320	Concrete Reinforcing	83.0	45.3	64.8	86.1	48.9	68.1	84.9	48.7	67.4	99.2	50.5	75.7	93.6	51.4	73.2	95.7	56.9	77.0
0330	Cast-in-Place Concrete	122.9	55.8	97.6	83.3	57.5	73.5	113.9	56.4	92.2	76.1	66.4	72.4	96.9	67.4	85.7	104.9	57.8	87.1
03	CONCRETE	115.8	51.8	87.1	75.3	58.5	67.8	100.3	57.0	80.9	90.0	60.2	76.6	89.5	62.1	77.2	96.9	57.8	79.3
04	MASONRY	95.8	58.0	72.8	135.3	60.1	89.4	95.1	58.2	72.6	96.8	63.3	76.3	93.8	61.2	73.9	96.0	58.2	72.9
05	METALS	98.6	65.4	88.2	102.6	84.1	96.9	102.9	68.7	92.2	103.7	67.8	92.5	105.4	69.1	94.1	102.6	89.1	98.4
06	WOOD, PLASTICS & COMPOSITES	103.1	48.6	74.6	115.7	61.0	87.1	109.5	60.9	84.1	94.1	58.4	75.4	96.5	62.3	78.6	92.5	53.9	72.3
07	THERMAL & MOISTURE PROTECTION	96.4	60.4	81.0	91.4	62.4	79.0	93.2	61.1	79.4	94.6	64.6	81.7	90.9	65.1	79.8	85.6	62.4	75.7
08	OPENINGS	96.7	46.0	84.4	118.8	55.4	103.4	70.6	55.7	67.0	96.3	53.7	85.9	101.0	58.2	90.6	99.4	53.9	88.3
0920	Plaster & Gypsum Board	86.9	47.8	61.3	101.8	60.5	74.7	96.9	60.5	73.0	112.0	57.6	76.3	106.5	61.6	77.1	99.1	53.2	69.0
0950, 0980	Ceilings & Acoustic Treatment	98.6	47.8	66.8	96.3	60.5	73.8	92.5	60.5	72.4	92.4	57.6	70.6	100.9	61.6	76.3	110.8	53.2	74.7
0960	Flooring	95.7	52.1	83.0	72.0	52.1	66.2	95.3	52.1	82.7	97.6	70.6	89.7	95.3	61.5	85.5	106.0	63.9	93.7
0970, 0990	Wall Finishes & Painting/Coating	95.1	40.5	62.4	100.1	46.5	68.0	90.8	49.2	65.9	98.2	49.2	68.8	100.8	54.4	73.0	106.0	51.1	73.1
09	FINISHES	98.2	48.2	71.2	82.8	56.2	68.4	89.9	56.4	71.8	96.8	59.1	76.4	97.6	60.0	77.3	101.0	54.6	75.9
COVERS	DIVS. 10 - 14, 25, 28, 41, 43, 44, 46	100.0	77.2	94.6	100.0	79.4	95.1	100.0	79.1	95.1	100.0	78.9	95.0	100.0	80.2	95.3	100.0	79.7	95.2
21, 22, 23	FIRE SUPPRESSION, PLUMBING & HVAC	97.1	56.6	80.8	97.0	53.3	79.4	97.0	47.3	77.0	100.9	63.7	85.9	100.8	56.8	83.1	97.1	59.6	82.0
26, 27, 3370	ELECTRICAL, COMMUNICATIONS & UTIL.	98.6	57.6	78.3	101.7	56.5	79.3	99.3	56.4	78.1	100.4	49.6	75.3	100.8	59.8	80.5	99.5	61.1	80.5
MF2018	WEIGHTED AVERAGE	100.7	58.2	82.3	98.9	61.6	82.8	95.7	58.9	79.8	98.3	63.3	83.2	99.0	63.9	83.9	98.8	64.0	83.8

TEXAS

DIVISION		GIDDINGS 789			GREENVILLE 754			HOUSTON 770 - 772			HUNTSVILLE 773			LAREDO 780			LONGVIEW 756		
		MAT.	INST.	TOTAL	MAT.	INST.	TOTAL	MAT.	INST.	TOTAL	MAT.	INST.	TOTAL	MAT.	INST.	TOTAL	MAT.	INST.	TOTAL
015433	CONTRACTOR EQUIPMENT		87.2	87.2		97.5	97.5		102.2	102.2		92.1	92.1		87.2	87.2		88.8	88.8
0241, 31 - 34	SITE & INFRASTRUCTURE, DEMOLITION	105.5	83.1	90.1	97.4	81.8	86.7	107.7	94.9	98.9	97.2	88.6	91.3	99.5	83.3	88.3	95.1	87.7	90.0
0310	Concrete Forming & Accessories	100.2	49.4	57.0	87.0	55.5	60.2	94.8	59.5	64.8	88.7	52.2	57.7	102.5	53.7	61.0	82.4	58.7	62.2
0320	Concrete Reinforcing	83.7	46.2	65.6	103.8	48.9	77.3	95.5	50.6	73.8	96.5	48.7	73.5	83.0	50.1	67.1	102.7	48.4	76.5
0330	Cast-in-Place Concrete	104.1	56.1	86.0	102.5	57.4	85.5	97.5	68.6	86.6	108.7	56.7	89.1	87.8	62.8	78.4	119.3	56.3	95.5
03	CONCRETE	93.4	52.2	74.9	93.2	56.8	76.9	94.9	62.2	80.2	103.4	54.3	81.3	87.8	57.2	74.1	109.5	56.8	85.9
04	MASONRY	103.2	58.1	75.7	163.6	60.1	100.5	94.2	63.7	75.6	136.9	58.1	88.8	89.3	62.4	72.9	159.1	56.8	96.7
05	METALS	98.0	66.4	88.2	100.8	82.5	95.1	105.6	75.0	96.1	100.9	69.8	91.2	101.2	68.5	91.0	93.6	67.4	85.4
06	WOOD, PLASTICS & COMPOSITES	102.4	48.6	74.3	84.2	56.4	69.7	98.0	59.8	78.0	86.5	52.1	68.5	103.1	53.5	77.1	77.8	61.0	69.0
07	THERMAL & MOISTURE PROTECTION	96.9	61.1	81.5	89.4	61.3	77.3	87.8	68.4	79.5	87.3	61.5	76.2	95.7	63.2	81.7	90.6	60.2	77.6
08	OPENINGS	95.9	46.8	83.9	92.3	53.2	82.8	106.3	56.9	94.3	95.4	50.5	84.4	96.9	50.4	85.5	83.5	55.3	76.6
0920	Plaster & Gypsum Board	86.1	47.8	61.0	79.3	55.6	63.7	100.1	58.7	72.9	95.5	51.5	66.6	88.1	52.8	64.9	77.5	60.5	66.3
0950, 0980	Ceilings & Acoustic Treatment	98.6	47.8	66.8	103.3	55.6	73.4	103.7	58.7	75.5	105.7	51.5	71.7	103.0	52.8	71.6	96.3	60.5	73.8
0960	Flooring	95.4	52.1	82.8	92.3	52.1	80.5	105.4	75.2	96.6	94.3	52.1	82.0	95.6	60.9	85.5	98.2	52.1	84.7
0970, 0990	Wall Finishes & Painting/Coating	95.1	42.1	63.3	93.4	49.2	66.9	106.5	58.0	77.4	93.2	54.7	70.1	95.1	42.1	63.3	84.8	46.5	61.9
09	FINISHES	96.9	48.4	70.6	95.3	53.8	72.9	105.6	62.0	82.0	93.3	51.9	70.9	97.7	53.3	73.7	99.7	56.2	76.1
COVERS	DIVS. 10 - 14, 25, 28, 41, 43, 44, 46	100.0	77.5	94.7	100.0	79.0	95.1	100.0	82.1	95.8	100.0	79.2	95.1	100.0	78.0	94.8	100.0	75.1	94.1
21, 22, 23	FIRE SUPPRESSION, PLUMBING & HVAC	97.2	58.5	81.6	97.3	55.4	80.4	101.0	63.8	85.9	97.2	59.5	82.0	100.9	56.8	83.1	97.2	55.0	80.2
26, 27, 3370	ELECTRICAL, COMMUNICATIONS & UTIL.	95.9	53.5	74.9	95.4	56.5	76.1	102.0	65.1	83.7	98.0	56.9	77.7	98.7	59.1	79.1	95.5	49.0	72.5
MF2018	WEIGHTED AVERAGE	97.2	58.3	80.5	99.6	61.3	83.1	101.3	67.3	86.6	99.9	60.7	83.0	97.6	60.9	81.8	99.8	59.2	82.3

		TEXAS																
	DIVISION	**LUBBOCK**			**LUFKIN**			**MCALLEN**			**MCKINNEY**			**MIDLAND**			**ODESSA**		
		793 - 794			759			785			750			797			797		
		MAT.	INST.	TOTAL	MAT.	INST.	TOTAL	MAT.	INST.	TOTAL	MAT.	INST.	TOTAL	MAT.	INST.	TOTAL	MAT.	INST.	TOTAL
015433	CONTRACTOR EQUIPMENT		100.1	100.1		88.8	88.8		98.0	98.0		97.5	97.5		100.1	100.1		88.0	88.0
0241, 31 - 34	SITE & INFRASTRUCTURE, DEMOLITION	118.4	83.8	94.7	90.0	87.1	88.0	145.4	80.0	100.5	93.8	81.8	85.5	120.9	83.8	95.4	94.0	85.0	87.8
0310	Concrete Forming & Accessories	99.3	53.1	60.0	85.7	52.4	57.4	106.9	49.9	58.4	86.0	55.5	60.1	103.8	60.4	66.9	99.0	60.4	66.2
0320	Concrete Reinforcing	98.3	50.4	75.2	104.6	63.6	84.8	82.6	50.0	66.9	103.8	48.9	77.3	99.4	50.2	75.7	97.1	50.3	74.5
0330	Cast-in-Place Concrete	86.2	66.7	78.8	106.7	55.7	87.4	124.0	57.8	99.0	96.1	57.4	81.5	91.6	64.3	81.3	86.0	65.7	78.4
03	CONCRETE	85.3	59.3	73.6	101.4	56.3	81.2	103.7	54.6	81.7	88.3	56.8	74.2	89.4	61.7	77.0	86.5	61.3	75.2
04	MASONRY	100.6	61.6	76.9	120.5	58.0	82.3	96.1	61.7	75.1	176.4	60.1	105.4	117.8	59.2	82.1	101.3	61.6	77.1
05	METALS	112.0	84.9	103.5	100.9	71.4	91.7	98.5	83.2	93.7	100.7	82.5	95.0	110.1	84.7	102.1	107.5	69.9	95.8
06	WOOD, PLASTICS & COMPOSITES	104.8	52.1	77.2	84.8	52.2	67.8	120.0	48.7	82.8	83.0	56.4	69.1	110.2	62.2	85.1	103.7	62.1	81.9
07	THERMAL & MOISTURE PROTECTION	90.3	63.4	78.7	90.4	59.6	77.2	99.2	62.4	83.4	89.2	61.3	77.2	90.5	63.4	78.9	100.4	63.6	84.6
08	OPENINGS	108.7	52.0	94.9	63.4	54.4	61.2	100.6	47.1	87.6	92.3	53.2	82.8	110.2	57.8	97.4	101.7	57.7	91.0
0920	Plaster & Gypsum Board	101.9	51.3	68.7	76.1	51.5	60.0	91.5	47.8	62.9	79.3	55.6	63.7	103.0	61.6	75.9	101.7	61.6	75.4
0950, 0980	Ceilings & Acoustic Treatment	97.9	51.3	68.7	91.2	51.5	66.3	103.0	47.8	68.4	103.3	55.6	73.4	94.7	61.6	74.0	97.3	61.6	74.9
0960	Flooring	90.1	67.7	83.6	131.5	52.1	108.3	110.8	73.7	100.0	91.7	52.1	80.2	91.3	60.9	82.4	96.5	66.5	87.7
0970, 0990	Wall Finishes & Painting/Coating	102.4	52.2	72.3	84.8	49.2	63.5	106.8	40.5	67.1	93.4	49.2	66.9	102.4	49.2	70.5	91.5	52.2	67.9
09	FINISHES	94.7	55.2	73.3	107.5	51.7	77.3	106.1	53.0	77.4	95.0	53.8	72.7	94.8	59.3	75.6	92.2	60.7	75.2
COVERS	DIVS. 10 - 14, 25, 28, 41, 43, 44, 46	100.0	78.9	95.0	100.0	79.5	95.2	100.0	77.7	94.7	100.0	79.0	95.1	100.0	80.0	95.3	100.0	79.7	95.2
21, 22, 23	FIRE SUPPRESSION, PLUMBING & HVAC	100.5	50.6	80.4	97.2	55.5	80.4	97.2	47.1	77.0	97.3	57.5	81.2	96.7	50.6	78.1	101.2	50.6	80.8
26, 27, 3370	ELECTRICAL, COMMUNICATIONS & UTIL.	96.4	52.2	74.5	96.7	59.1	78.1	96.5	30.4	63.8	95.5	56.5	76.2	96.4	52.6	74.7	97.7	52.5	75.3
MF2018	WEIGHTED AVERAGE	100.4	60.7	83.3	96.6	60.5	81.0	100.9	55.2	81.2	99.5	61.8	83.2	100.9	61.7	84.0	98.9	60.8	82.5

		TEXAS																
	DIVISION	**PALESTINE**			**SAN ANGELO**			**SAN ANTONIO**			**TEMPLE**			**TEXARKANA**			**TYLER**		
		758			769			781 - 782			765			755			757		
		MAT.	INST.	TOTAL	MAT.	INST.	TOTAL	MAT.	INST.	TOTAL	MAT.	INST.	TOTAL	MAT.	INST.	TOTAL	MAT.	INST.	TOTAL
015433	CONTRACTOR EQUIPMENT		88.8	88.8		88.0	88.0		95.0	95.0		88.0	88.0		88.8	88.8		88.8	88.8
0241, 31 - 34	SITE & INFRASTRUCTURE, DEMOLITION	95.3	87.7	90.1	98.5	84.9	89.1	97.5	95.9	96.4	87.3	84.4	85.3	84.7	87.8	86.8	94.3	87.8	89.8
0310	Concrete Forming & Accessories	76.2	58.7	61.3	97.9	49.7	56.9	101.3	55.7	62.5	101.9	49.3	57.2	93.9	59.0	64.2	87.6	58.8	63.1
0320	Concrete Reinforcing	102.0	48.7	76.3	84.5	50.2	67.9	89.1	47.3	68.9	84.7	48.4	67.2	101.8	51.2	77.4	102.7	51.2	77.9
0330	Cast-in-Place Concrete	97.5	56.3	81.9	101.7	62.4	86.8	83.8	69.4	78.3	83.3	55.9	72.9	98.2	64.8	85.6	117.2	56.3	94.2
03	CONCRETE	102.4	56.9	82.0	91.2	55.3	75.1	86.3	59.6	74.3	77.8	52.5	66.5	94.1	60.4	79.0	109.0	57.4	85.8
04	MASONRY	115.0	56.8	79.5	123.2	58.2	83.6	89.0	63.4	73.4	134.3	58.1	87.8	181.0	56.8	105.3	169.7	56.8	100.9
05	METALS	100.7	67.5	90.3	103.3	69.4	92.7	103.4	64.2	91.2	102.9	67.2	91.8	93.5	68.8	85.8	100.5	68.5	90.5
06	WOOD, PLASTICS & COMPOSITES	75.0	61.0	67.7	103.1	48.6	74.6	103.8	54.9	78.2	112.5	48.6	79.1	91.3	61.0	75.5	86.7	61.0	73.3
07	THERMAL & MOISTURE PROTECTION	90.9	60.4	77.8	92.7	60.9	79.1	88.7	67.6	79.6	92.3	60.1	78.5	90.3	61.9	78.1	90.8	60.4	77.7
08	OPENINGS	63.3	55.7	61.5	100.6	49.3	88.1	102.3	51.7	90.0	67.3	48.9	62.8	83.5	56.3	76.8	63.3	56.3	61.6
0920	Plaster & Gypsum Board	75.6	60.5	65.7	96.9	47.8	64.7	92.1	53.9	67.0	96.9	47.8	64.7	82.6	60.5	68.1	76.1	60.5	65.9
0950, 0980	Ceilings & Acoustic Treatment	91.2	60.5	71.9	92.5	47.8	64.5	93.2	53.9	68.5	92.5	47.8	64.5	96.3	60.5	73.8	91.2	60.5	71.9
0960	Flooring	121.6	52.1	101.3	76.1	52.1	69.1	105.7	73.9	96.4	96.9	52.1	83.8	107.4	60.9	93.8	134.1	52.1	110.2
0970, 0990	Wall Finishes & Painting/Coating	84.8	49.2	63.5	89.5	49.2	65.3	101.1	45.6	67.8	90.8	42.1	61.6	84.8	49.2	63.5	84.8	49.2	63.5
09	FINISHES	105.2	56.5	78.8	83.8	49.2	65.1	106.7	57.7	80.2	89.1	48.4	67.1	102.5	58.2	78.5	108.6	56.5	80.4
COVERS	DIVS. 10 - 14, 25, 28, 41, 43, 44, 46	100.0	79.2	95.1	100.0	77.5	94.7	100.0	79.7	95.2	100.0	77.8	94.8	100.0	79.2	95.1	100.0	79.2	95.1
21, 22, 23	FIRE SUPPRESSION, PLUMBING & HVAC	97.2	54.1	79.9	97.0	49.8	78.0	100.9	60.0	84.4	97.0	50.6	78.3	97.2	53.0	79.4	97.2	55.6	80.4
26, 27, 3370	ELECTRICAL, COMMUNICATIONS & UTIL.	93.3	46.6	70.2	103.2	52.6	78.1	100.2	61.5	81.0	100.4	50.1	75.5	96.6	52.5	74.8	95.5	51.7	73.9
MF2018	WEIGHTED AVERAGE	95.9	58.9	80.0	98.7	57.3	80.8	98.9	63.8	83.7	94.1	56.4	77.8	99.1	60.4	82.4	100.1	60.2	82.8

		TEXAS															UTAH		
	DIVISION	**VICTORIA**			**WACO**			**WAXAHACHIE**			**WHARTON**			**WICHITA FALLS**			**LOGAN**		
		779			766 - 767			751			774			763			843		
		MAT.	INST.	TOTAL	MAT.	INST.	TOTAL	MAT.	INST.	TOTAL	MAT.	INST.	TOTAL	MAT.	INST.	TOTAL	MAT.	INST.	TOTAL
015433	CONTRACTOR EQUIPMENT		103.5	103.5		88.0	88.0		97.5	97.5		104.9	104.9		88.0	88.0		91.5	91.5
0241, 31 - 34	SITE & INFRASTRUCTURE, DEMOLITION	113.6	85.0	93.9	96.0	84.6	88.2	95.6	81.7	86.0	119.1	87.0	97.1	96.8	85.0	88.7	97.2	86.8	90.0
0310	Concrete Forming & Accessories	89.8	49.6	55.6	100.2	60.4	66.3	86.0	58.9	63.0	84.0	51.3	56.1	100.2	60.5	66.4	100.2	70.4	74.9
0320	Concrete Reinforcing	91.9	48.5	71.0	84.3	47.7	66.7	103.8	48.9	77.3	95.6	48.5	72.9	84.3	50.4	68.0	101.6	87.8	94.9
0330	Cast-in-Place Concrete	117.8	57.7	95.1	90.1	65.7	80.9	101.5	57.2	84.8	121.1	56.8	96.8	96.3	71.1	86.7	85.1	75.0	81.3
03	CONCRETE	104.9	54.4	82.2	84.2	60.8	73.7	92.2	58.3	77.0	108.9	54.8	84.6	87.1	63.2	76.4	105.3	75.5	91.9
04	MASONRY	113.9	58.2	79.9	93.2	62.4	74.4	164.3	60.0	100.7	97.3	58.1	73.4	93.7	62.0	74.4	108.5	63.1	80.8
05	METALS	101.0	85.7	96.2	105.4	68.1	93.7	100.8	82.2	95.0	102.6	85.2	97.2	105.3	70.2	94.4	107.5	84.3	100.3
06	WOOD, PLASTICS & COMPOSITES	95.1	48.7	70.9	110.5	62.1	85.2	83.0	61.1	71.6	85.0	50.8	67.1	110.5	62.1	85.2	83.0	70.7	76.5
07	THERMAL & MOISTURE PROTECTION	88.6	59.5	76.1	93.1	64.9	81.0	89.3	61.8	77.5	85.9	62.0	75.7	93.1	63.8	80.5	100.9	68.7	87.0
08	OPENINGS	99.2	49.1	87.0	78.2	57.5	73.2	92.3	55.8	83.4	99.4	50.3	87.4	78.2	57.9	73.2	92.4	73.1	87.7
0920	Plaster & Gypsum Board	94.7	47.8	63.9	97.0	61.6	73.8	79.8	60.5	67.1	94.3	50.0	65.2	97.0	61.6	73.8	83.4	70.2	74.7
0950, 0980	Ceilings & Acoustic Treatment	112.1	47.8	71.8	93.1	61.6	73.4	105.2	60.5	77.2	110.8	50.0	72.7	93.1	61.6	73.4	113.9	70.2	86.5
0960	Flooring	104.3	52.1	89.0	96.3	69.9	88.6	91.7	52.1	80.2	103.0	63.5	91.5	96.9	74.3	90.3	96.5	62.9	86.7
0970, 0990	Wall Finishes & Painting/Coating	106.2	54.7	75.3	90.8	57.0	70.6	93.4	49.2	66.9	106.0	53.1	74.3	92.8	54.3	69.7	89.3	60.8	72.3
09	FINISHES	98.3	49.9	72.1	89.5	61.8	74.5	95.7	56.6	74.5	100.1	52.9	74.6	89.9	62.4	75.0	97.3	67.7	81.3
COVERS	DIVS. 10 - 14, 25, 28, 41, 43, 44, 46	100.0	79.1	95.1	100.0	79.6	95.2	100.0	79.5	95.2	100.0	79.4	95.1	100.0	79.6	95.2	100.0	85.6	96.6
21, 22, 23	FIRE SUPPRESSION, PLUMBING & HVAC	97.2	59.4	81.9	100.8	59.7	84.2	97.3	55.4	80.4	97.1	59.0	81.7	100.8	54.3	82.1	100.9	74.6	90.3
26, 27, 3370	ELECTRICAL, COMMUNICATIONS & UTIL.	104.2	52.0	78.3	103.3	55.4	79.6	95.4	56.5	76.2	103.2	57.0	80.3	105.3	54.7	80.3	94.6	70.9	82.9
MF2018	WEIGHTED AVERAGE	101.0	60.8	83.6	95.8	63.2	81.7	99.5	62.0	83.3	101.0	62.1	84.2	96.5	62.5	81.8	100.7	74.0	89.2

For customer support on your Commercial Renovation Costs with RSMeans Data, call 800.448.8182.

825

		UTAH												VERMONT					
		OGDEN			PRICE			PROVO			SALT LAKE CITY			BELLOWS FALLS			BENNINGTON		
	DIVISION	842,844			845			846 - 847			840 - 841			051			052		
		MAT.	INST.	TOTAL	MAT.	INST.	TOTAL	MAT.	INST.	TOTAL	MAT.	INST.	TOTAL	MAT.	INST.	TOTAL	MAT.	INST.	TOTAL
015433	CONTRACTOR EQUIPMENT		91.5	91.5		90.6	90.6		90.6	90.6		91.5	91.5		94.2	94.2		94.2	94.2
0241, 31 - 34	SITE & INFRASTRUCTURE, DEMOLITION	85.1	86.8	86.3	94.4	84.4	87.5	93.5	85.3	87.9	84.9	86.7	86.1	86.8	93.3	91.3	86.2	93.3	91.1
0310	Concrete Forming & Accessories	100.2	70.4	74.9	102.9	63.3	69.2	102.1	70.4	75.2	102.6	70.4	75.2	99.6	98.4	98.6	96.7	98.2	98.0
0320	Concrete Reinforcing	101.3	87.8	94.7	108.9	80.0	95.0	109.9	87.8	99.2	103.4	87.8	95.9	82.2	82.8	82.5	82.2	82.8	82.5
0330	Cast-in-Place Concrete	86.4	75.0	82.1	85.2	70.8	79.7	85.2	75.0	81.3	94.4	75.0	87.1	86.5	111.7	96.0	86.5	111.6	96.0
03	CONCRETE	94.3	75.5	85.9	106.5	69.4	89.9	104.9	75.5	91.7	114.1	75.5	96.8	86.9	100.2	92.9	86.7	100.1	92.7
04	MASONRY	102.8	63.1	78.6	114.0	66.3	84.9	114.2	63.1	83.0	116.8	63.1	84.0	98.8	91.9	94.6	108.9	91.9	98.5
05	METALS	108.1	84.3	100.7	104.6	80.6	97.1	105.6	84.3	98.9	111.8	84.3	103.2	100.2	90.8	97.2	100.1	90.6	97.2
06	WOOD, PLASTICS & COMPOSITES	83.0	70.7	76.5	86.4	61.8	73.6	84.7	70.7	77.4	85.0	70.7	77.5	106.0	103.2	104.5	102.9	103.2	103.1
07	THERMAL & MOISTURE PROTECTION	99.7	68.7	86.4	102.7	67.0	87.4	102.8	68.7	88.1	107.6	68.7	90.9	101.4	87.7	95.5	101.3	87.7	95.5
08	OPENINGS	92.4	73.1	87.7	96.1	73.4	90.5	96.1	73.1	90.5	94.2	73.1	89.0	100.6	94.0	99.0	100.6	94.0	99.0
0920	Plaster & Gypsum Board	83.4	70.2	74.7	86.6	61.0	69.8	84.3	70.2	75.0	92.8	70.2	77.9	113.9	103.1	106.8	113.0	103.1	106.5
0950, 0980	Ceilings & Acoustic Treatment	113.9	70.2	86.5	113.9	61.0	80.8	113.9	70.2	86.5	106.8	70.2	83.8	81.4	103.1	95.0	81.4	103.1	95.0
0960	Flooring	94.6	62.9	85.3	97.7	53.5	84.8	97.5	62.9	87.4	98.6	62.9	88.1	90.3	110.7	96.3	89.4	110.7	95.6
0970, 0990	Wall Finishes & Painting/Coating	89.3	60.8	72.3	89.3	57.8	70.4	89.3	60.8	72.3	92.4	62.1	74.2	82.8	96.7	91.1	82.8	96.7	91.1
09	FINISHES	95.5	67.7	80.4	98.5	60.2	77.8	98.0	67.7	81.6	97.0	67.8	81.2	88.5	101.2	95.4	88.1	101.2	95.2
COVERS	DIVS. 10 - 14, 25, 28, 41, 43, 44, 46	100.0	85.6	96.6	100.0	84.6	96.4	100.0	85.6	96.6	100.0	85.6	96.6	100.0	100.5	100.1	100.0	100.4	100.1
21, 22, 23	FIRE SUPPRESSION, PLUMBING & HVAC	100.9	74.6	90.3	98.9	62.2	84.1	100.9	74.6	90.3	101.2	74.6	90.5	97.6	83.5	91.9	97.6	83.5	91.9
26, 27, 3370	ELECTRICAL, COMMUNICATIONS & UTIL.	95.0	70.9	83.1	98.9	66.2	82.7	95.1	71.6	83.4	97.5	68.6	83.2	104.3	76.6	90.6	104.3	52.4	78.6
MF2018	WEIGHTED AVERAGE	98.6	74.0	88.0	101.1	68.6	87.1	101.1	74.0	89.4	103.3	73.7	90.5	97.1	90.6	94.3	97.5	87.2	93.0

		VERMONT																	
		BRATTLEBORO			BURLINGTON			GUILDHALL			MONTPELIER			RUTLAND			ST. JOHNSBURY		
	DIVISION	053			054			059			056			057			058		
		MAT.	INST.	TOTAL	MAT.	INST.	TOTAL	MAT.	INST.	TOTAL	MAT.	INST.	TOTAL	MAT.	INST.	TOTAL	MAT.	INST.	TOTAL
015433	CONTRACTOR EQUIPMENT		94.2	94.2		99.4	99.4		94.2	94.2		99.4	99.4		94.2	94.2		94.2	94.2
0241, 31 - 34	SITE & INFRASTRUCTURE, DEMOLITION	87.7	93.3	91.6	91.6	103.4	99.7	85.8	93.0	90.7	90.0	102.5	98.6	90.6	94.3	93.2	85.8	92.4	90.3
0310	Concrete Forming & Accessories	99.9	98.4	98.6	98.8	81.9	84.5	96.0	92.7	93.2	101.0	99.2	99.5	100.2	81.8	84.6	93.2	92.7	92.8
0320	Concrete Reinforcing	81.3	82.8	82.0	105.8	82.8	94.7	82.9	82.8	82.8	96.1	82.8	89.7	102.8	82.8	93.1	81.3	82.7	82.0
0330	Cast-in-Place Concrete	89.3	111.7	97.7	107.3	113.9	109.8	83.9	103.5	91.3	99.6	113.9	105.0	84.8	112.9	95.4	83.9	103.5	91.3
03	CONCRETE	88.9	100.2	93.9	104.9	93.4	99.8	84.4	94.8	89.1	100.2	101.2	100.6	89.7	93.1	91.2	84.0	94.8	88.8
04	MASONRY	108.6	91.9	98.4	109.6	94.2	100.2	108.9	77.8	90.0	108.7	94.2	99.9	89.7	94.1	92.4	136.9	77.8	100.9
05	METALS	100.1	90.8	97.2	108.3	89.2	102.4	100.2	90.5	97.1	106.7	89.2	101.3	106.5	90.6	101.6	100.2	90.4	97.1
06	WOOD, PLASTICS & COMPOSITES	106.4	103.2	104.7	99.7	79.8	89.3	101.1	103.2	102.2	101.3	103.3	102.3	106.6	79.7	92.5	95.1	103.2	99.4
07	THERMAL & MOISTURE PROTECTION	101.5	87.7	95.6	107.3	96.8	102.8	101.2	81.3	92.6	110.1	99.3	105.5	101.6	95.8	99.1	101.0	81.3	92.6
08	OPENINGS	100.6	94.2	99.1	103.1	81.5	97.9	100.6	94.4	99.1	103.6	94.5	101.4	103.7	81.5	98.3	100.6	94.4	99.1
0920	Plaster & Gypsum Board	113.9	103.1	106.8	115.4	79.0	91.5	120.4	103.1	109.0	117.5	103.1	108.0	114.4	79.0	91.1	122.4	103.1	109.7
0950, 0980	Ceilings & Acoustic Treatment	81.4	103.1	95.0	93.3	79.0	84.3	81.4	103.1	95.0	83.9	103.1	95.9	87.6	79.0	82.2	81.4	103.1	95.0
0960	Flooring	90.6	110.7	96.4	96.9	110.7	100.9	93.4	110.7	98.4	97.4	110.7	101.3	90.3	110.7	96.3	96.7	110.7	100.8
0970, 0990	Wall Finishes & Painting/Coating	82.8	96.7	91.1	95.3	104.6	100.9	82.8	104.6	95.9	98.3	104.6	102.1	82.8	104.6	95.9	82.8	104.6	95.9
09	FINISHES	88.7	101.2	95.4	97.0	88.9	92.6	90.2	98.5	94.7	95.4	102.7	99.4	90.2	88.8	89.4	91.5	98.5	95.3
COVERS	DIVS. 10 - 14, 25, 28, 41, 43, 44, 46	100.0	100.5	100.1	100.0	98.9	99.7	100.0	95.7	99.0	100.0	101.4	100.3	100.0	98.6	99.7	100.0	95.7	99.0
21, 22, 23	FIRE SUPPRESSION, PLUMBING & HVAC	97.6	83.5	91.9	101.2	68.1	87.8	97.6	59.5	82.3	97.4	68.1	85.6	101.4	68.0	87.9	97.6	59.5	82.3
26, 27, 3370	ELECTRICAL, COMMUNICATIONS & UTIL.	104.3	76.6	90.6	104.0	56.4	80.5	104.3	52.4	78.6	102.8	56.4	79.9	104.3	56.4	80.6	104.3	52.4	78.6
MF2018	WEIGHTED AVERAGE	97.8	90.4	94.7	103.1	82.3	94.1	97.3	79.2	89.5	101.2	80.0	94.6	99.4	81.6	91.7	98.7	79.2	90.3

		VERMONT			VIRGINIA														
		WHITE RIVER JCT.			ALEXANDRIA			ARLINGTON			BRISTOL			CHARLOTTESVILLE			CULPEPER		
	DIVISION	050			223			222			242			229			227		
		MAT.	INST.	TOTAL	MAT.	INST.	TOTAL	MAT.	INST.	TOTAL	MAT.	INST.	TOTAL	MAT.	INST.	TOTAL	MAT.	INST.	TOTAL
015433	CONTRACTOR EQUIPMENT		94.2	94.2		103.4	103.4		102.2	102.2		102.2	102.2		106.2	106.2		102.2	102.2
0241, 31 - 34	SITE & INFRASTRUCTURE, DEMOLITION	90.2	92.5	91.8	114.5	86.0	94.9	125.0	83.8	96.7	108.9	84.0	91.8	113.5	85.7	94.4	112.0	83.5	92.4
0310	Concrete Forming & Accessories	93.7	93.2	93.2	94.9	72.8	76.1	93.1	61.8	66.5	87.7	59.2	63.4	85.9	60.9	64.6	82.9	68.7	70.8
0320	Concrete Reinforcing	82.2	82.7	82.5	86.9	88.5	87.7	98.0	86.2	92.3	98.0	70.4	84.7	97.4	71.9	85.1	98.0	88.5	93.4
0330	Cast-in-Place Concrete	89.3	104.3	94.9	107.2	76.7	95.7	104.3	75.2	93.3	103.9	49.2	83.3	108.1	73.3	94.9	106.8	74.5	94.6
03	CONCRETE	90.5	95.2	92.6	99.4	78.3	89.9	103.8	72.4	89.7	100.2	59.5	81.9	120.7	68.9	86.5	107.9	75.6	87.9
04	MASONRY	122.1	79.1	95.9	91.9	72.8	80.3	105.7	71.4	84.8	95.2	55.0	70.7	120.7	55.3	80.8	108.0	70.4	85.0
05	METALS	100.2	90.4	97.1	105.9	102.2	104.8	104.4	101.1	103.4	103.2	93.4	100.2	103.5	96.4	101.3	103.6	101.1	102.8
06	WOOD, PLASTICS & COMPOSITES	99.2	103.2	101.3	94.6	71.7	82.6	90.7	57.7	73.4	81.8	57.7	69.2	79.9	58.3	68.6	78.3	67.7	72.8
07	THERMAL & MOISTURE PROTECTION	101.5	81.9	93.1	100.6	79.9	91.7	102.5	77.4	91.8	101.8	60.7	84.1	101.4	68.1	87.1	101.6	77.6	91.3
08	OPENINGS	100.6	94.4	99.1	95.5	76.0	90.8	93.8	67.8	87.4	96.4	64.7	88.7	94.8	65.3	87.6	95.1	73.8	89.9
0920	Plaster & Gypsum Board	111.6	103.1	106.0	101.6	71.0	81.5	98.1	56.6	70.9	94.2	56.6	69.5	94.2	56.6	69.5	94.4	66.9	76.3
0950, 0980	Ceilings & Acoustic Treatment	81.4	103.1	95.0	99.0	71.0	81.5	96.5	56.6	71.5	95.8	56.6	71.2	95.8	56.6	71.2	96.5	66.9	77.9
0960	Flooring	88.4	110.7	94.9	99.0	75.3	92.1	97.5	74.8	90.8	94.1	55.0	82.7	92.5	55.0	81.6	92.5	71.0	86.2
0970, 0990	Wall Finishes & Painting/Coating	82.8	104.6	95.9	111.0	70.1	86.5	111.0	70.1	86.5	97.7	75.0	84.1	97.7	59.1	74.6	111.0	69.8	86.4
09	FINISHES	88.0	98.8	93.9	98.3	72.2	84.2	98.0	63.5	79.3	94.7	59.8	75.8	94.3	58.2	74.8	94.9	68.4	80.6
COVERS	DIVS. 10 - 14, 25, 28, 41, 43, 44, 46	100.0	96.1	99.1	100.0	88.8	97.4	100.0	79.8	95.2	100.0	82.2	95.8	100.0	83.8	96.2	100.0	87.6	97.1
21, 22, 23	FIRE SUPPRESSION, PLUMBING & HVAC	97.6	60.2	82.5	101.5	86.1	95.3	101.5	85.6	95.1	97.7	49.7	78.3	97.7	86.0	93.0	97.7	83.2	91.8
26, 27, 3370	ELECTRICAL, COMMUNICATIONS & UTIL.	104.3	52.3	78.6	95.4	100.8	98.0	93.4	102.2	97.7	95.0	35.0	65.3	95.0	72.0	83.6	97.0	97.3	97.1
MF2018	WEIGHTED AVERAGE	98.7	79.6	90.4	100.0	84.8	93.4	101.0	81.6	92.6	98.5	59.4	81.6	99.8	74.2	88.7	99.1	82.0	91.7

For customer support on your Commercial Renovation Costs with RSMeans Data, call 800.448.8182.

VIRGINIA

DIVISION		FAIRFAX 220-221			FARMVILLE 239			FREDERICKSBURG 224-225			GRUNDY 246			HARRISONBURG 228			LYNCHBURG 245		
		MAT.	INST.	TOTAL	MAT.	INST.	TOTAL	MAT.	INST.	TOTAL	MAT.	INST.	TOTAL	MAT.	INST.	TOTAL	MAT.	INST.	TOTAL
015433	CONTRACTOR EQUIPMENT		102.2	102.2		106.2	106.2		102.2	102.2		102.2	102.2		102.2	102.2		102.2	102.2
0241, 31 - 34	SITE & INFRASTRUCTURE, DEMOLITION	123.2	83.8	96.1	109.8	84.6	92.5	111.5	83.4	92.2	106.7	82.5	90.1	120.0	83.8	95.2	107.5	84.0	91.4
0310	Concrete Forming & Accessories	86.3	61.6	65.3	102.5	60.2	66.5	86.3	63.3	66.7	91.5	58.3	63.2	81.7	61.2	64.3	87.7	60.6	64.6
0320	Concrete Reinforcing	98.0	88.5	93.4	95.4	64.9	80.7	98.7	77.4	88.5	96.7	42.2	70.4	98.0	80.5	89.6	97.4	70.7	84.5
0330	Cast-in-Place Concrete	104.3	75.3	93.4	107.7	83.8	98.6	105.9	72.5	93.3	103.9	49.3	83.3	104.3	75.5	93.4	103.9	74.1	92.7
03	CONCRETE	103.4	72.7	89.6	102.5	70.9	88.3	97.6	70.6	85.5	98.8	53.9	78.7	101.2	71.3	87.8	98.7	68.9	85.3
04	MASONRY	105.6	72.0	85.1	105.6	56.1	75.4	106.9	81.6	91.5	97.1	56.0	72.0	104.5	71.4	84.3	112.5	55.3	77.6
05	METALS	103.7	101.8	103.1	101.3	89.6	97.7	103.6	97.9	101.8	103.2	74.2	94.2	103.5	99.0	102.1	103.4	95.4	100.9
06	WOOD, PLASTICS & COMPOSITES	81.8	57.1	68.9	98.5	58.3	77.4	81.8	62.4	71.7	85.4	57.7	70.9	77.0	57.9	67.0	81.8	57.9	69.3
07	THERMAL & MOISTURE PROTECTION	102.2	77.2	91.5	103.8	69.7	89.1	101.6	79.5	92.1	101.8	59.1	83.4	102.0	76.8	91.2	101.6	68.2	87.3
08	OPENINGS	93.8	68.0	87.5	94.3	53.2	84.3	94.8	67.5	88.2	96.4	44.4	83.7	95.1	67.0	88.2	95.1	64.8	87.7
0920	Plaster & Gypsum Board	94.4	56.0	69.2	106.4	56.6	73.7	94.4	61.5	72.8	94.2	56.6	69.5	94.2	56.8	69.6	94.2	56.8	69.6
0950, 0980	Ceilings & Acoustic Treatment	96.5	56.0	71.1	90.1	56.6	69.1	96.5	61.5	74.5	95.8	56.6	71.2	95.8	56.8	71.4	95.8	56.8	71.4
0960	Flooring	94.3	76.4	89.1	95.3	55.0	83.5	94.3	69.1	86.9	95.3	55.0	83.5	92.3	76.9	87.8	94.1	55.0	82.7
0970, 0990	Wall Finishes & Painting/Coating	111.0	70.1	86.5	93.3	54.5	70.0	111.0	54.3	77.0	97.7	29.7	57.0	111.0	58.9	79.8	97.7	75.0	84.1
09	FINISHES	96.5	63.6	78.7	94.4	58.0	74.7	95.4	62.4	77.5	94.9	54.9	73.2	95.4	62.6	77.7	94.5	59.6	75.6
COVERS	DIVS. 10 - 14, 25, 28, 41, 43, 44, 46	100.0	87.0	96.9	100.0	83.7	96.2	100.0	74.3	93.9	100.0	82.4	95.8	100.0	82.4	95.9	100.0	82.0	95.8
21, 22, 23	FIRE SUPPRESSION, PLUMBING & HVAC	97.7	84.0	92.2	98.0	52.9	79.8	97.7	76.1	89.0	97.7	63.1	83.7	97.7	75.0	88.5	97.7	86.0	93.0
26, 27, 3370	ELECTRICAL, COMMUNICATIONS & UTIL.	95.9	102.2	99.0	89.8	69.9	79.9	93.6	102.1	97.8	95.0	69.8	82.6	95.2	78.3	86.8	95.9	65.3	80.7
MF2018	WEIGHTED AVERAGE	99.9	81.7	92.0	98.5	66.0	84.5	98.6	79.9	90.5	98.4	63.2	83.2	99.4	75.6	89.1	99.1	73.1	87.9

VIRGINIA

DIVISION		NEWPORT NEWS 236			NORFOLK 233-235			PETERSBURG 238			PORTSMOUTH 237			PULASKI 243			RICHMOND 230-232		
		MAT.	INST.	TOTAL	MAT.	INST.	TOTAL	MAT.	INST.	TOTAL	MAT.	INST.	TOTAL	MAT.	INST.	TOTAL	MAT.	INST.	TOTAL
015433	CONTRACTOR EQUIPMENT		106.3	106.3		107.9	107.9		106.2	106.2		106.2	106.2		102.2	102.2		108.0	108.0
0241, 31 - 34	SITE & INFRASTRUCTURE, DEMOLITION	109.0	85.8	93.0	109.6	92.1	97.6	112.4	85.7	94.1	107.7	85.4	92.4	106.1	82.8	90.1	101.8	92.3	95.3
0310	Concrete Forming & Accessories	101.5	60.5	66.6	103.2	60.2	66.6	93.8	61.0	65.9	89.0	60.2	64.5	91.5	59.0	63.8	96.3	60.9	66.2
0320	Concrete Reinforcing	95.3	66.8	81.5	105.0	66.8	86.5	94.9	71.9	83.8	94.9	66.8	81.3	96.7	80.0	88.6	104.0	71.9	88.5
0330	Cast-in-Place Concrete	104.6	76.1	93.8	116.2	77.0	101.4	111.1	76.3	98.0	103.6	64.7	88.9	103.9	80.3	95.0	88.4	77.3	84.2
03	CONCRETE	99.6	68.5	85.8	104.1	68.8	88.2	105.0	70.0	89.3	98.2	64.7	83.2	98.8	71.7	86.7	90.5	70.1	81.4
04	MASONRY	100.4	55.3	72.9	104.4	55.3	74.5	114.9	55.3	78.5	106.6	55.2	75.3	91.0	53.5	68.1	103.3	55.3	74.1
05	METALS	103.8	94.0	100.7	105.2	91.0	100.8	101.4	96.4	99.9	102.7	93.6	99.9	103.3	94.8	100.6	105.9	93.6	102.0
06	WOOD, PLASTICS & COMPOSITES	96.9	58.5	76.8	99.2	58.3	77.8	86.7	58.3	71.8	82.2	58.3	69.7	85.4	57.7	70.9	96.5	58.3	76.5
07	THERMAL & MOISTURE PROTECTION	103.7	66.3	87.6	101.2	68.8	87.3	103.7	68.5	88.6	103.7	66.6	87.8	101.8	68.6	87.5	98.8	69.7	86.3
08	OPENINGS	94.6	63.8	87.1	95.3	64.2	87.7	94.0	65.3	87.0	94.7	63.7	87.2	96.4	56.0	86.6	101.6	65.3	92.8
0920	Plaster & Gypsum Board	107.9	56.8	74.3	99.3	56.8	71.4	100.4	56.6	71.6	101.4	56.6	72.0	94.2	56.6	69.5	98.9	56.8	71.3
0950, 0980	Ceilings & Acoustic Treatment	95.8	56.8	71.4	90.7	56.8	69.4	92.0	56.6	69.8	95.8	56.6	71.2	95.8	56.6	71.2	93.6	56.8	70.5
0960	Flooring	95.3	55.0	83.5	95.9	55.0	83.9	91.3	55.0	80.7	88.3	55.0	78.6	95.3	55.0	83.5	91.1	55.0	80.6
0970, 0990	Wall Finishes & Painting/Coating	93.3	58.0	72.1	97.7	58.9	74.4	93.3	59.1	72.8	93.3	58.9	72.7	97.7	46.0	66.7	92.5	59.1	72.5
09	FINISHES	95.8	58.2	75.4	94.5	58.2	74.9	93.0	58.2	74.2	92.9	58.2	74.1	94.9	56.0	73.8	93.7	58.2	74.5
COVERS	DIVS. 10 - 14, 25, 28, 41, 43, 44, 46	100.0	83.8	96.2	100.0	83.4	96.1	100.0	83.8	96.2	100.0	81.5	95.6	100.0	81.5	95.7	100.0	83.5	96.1
21, 22, 23	FIRE SUPPRESSION, PLUMBING & HVAC	101.8	65.5	87.1	101.1	65.0	86.5	98.0	86.0	93.1	101.8	65.5	87.1	97.7	63.9	84.1	101.1	86.0	95.0
26, 27, 3370	ELECTRICAL, COMMUNICATIONS & UTIL.	91.9	72.1	82.1	95.9	62.3	79.3	92.0	72.0	82.1	90.6	62.3	76.6	95.0	80.3	87.7	98.6	72.0	85.5
MF2018	WEIGHTED AVERAGE	99.5	69.5	86.5	100.7	68.3	86.7	99.4	74.4	88.6	99.0	67.4	85.3	98.1	69.7	85.8	99.6	74.6	88.8

VIRGINIA / WASHINGTON

DIVISION		ROANOKE 240-241			STAUNTON 244			WINCHESTER 226			CLARKSTON 994			EVERETT 982			OLYMPIA 985		
		MAT.	INST.	TOTAL	MAT.	INST.	TOTAL	MAT.	INST.	TOTAL	MAT.	INST.	TOTAL	MAT.	INST.	TOTAL	MAT.	INST.	TOTAL
015433	CONTRACTOR EQUIPMENT		102.2	102.2		106.2	106.2		102.2	102.2		88.4	88.4		97.7	97.7		100.2	100.2
0241, 31 - 34	SITE & INFRASTRUCTURE, DEMOLITION	106.9	84.1	91.2	110.1	84.4	92.5	118.7	83.5	94.5	102.2	83.1	89.1	93.2	103.3	100.1	94.8	105.9	102.4
0310	Concrete Forming & Accessories	99.0	60.8	66.5	91.1	60.1	64.8	84.5	63.9	67.0	102.5	62.0	68.0	114.5	105.5	106.8	98.0	106.2	105.0
0320	Concrete Reinforcing	97.7	70.7	84.7	97.4	69.3	83.8	97.4	76.4	87.3	106.6	99.4	103.1	107.5	113.7	110.5	120.0	113.5	116.9
0330	Cast-in-Place Concrete	118.0	83.8	105.1	108.1	83.6	98.8	104.3	61.2	88.0	87.0	79.6	84.2	100.9	111.7	105.0	88.8	114.3	98.5
03	CONCRETE	103.3	72.4	89.4	100.2	71.5	87.3	100.7	66.8	85.5	90.0	75.1	83.3	94.5	108.6	100.8	93.0	109.7	100.5
04	MASONRY	98.4	57.1	73.2	107.8	56.0	76.2	101.8	66.8	80.5	100.6	87.8	92.8	109.1	104.3	106.2	106.7	102.2	104.0
05	METALS	105.7	95.3	102.4	103.5	90.8	99.5	103.6	96.1	101.3	92.9	88.6	91.6	116.2	99.5	111.0	117.2	98.0	111.2
06	WOOD, PLASTICS & COMPOSITES	95.7	57.9	75.9	85.4	58.3	71.2	79.9	62.6	70.9	95.5	56.5	75.1	112.1	105.1	108.4	92.4	105.4	99.2
07	THERMAL & MOISTURE PROTECTION	101.6	70.4	88.2	101.4	69.6	87.7	102.1	74.2	90.1	157.0	79.1	123.6	113.5	107.0	110.7	109.4	107.5	108.6
08	OPENINGS	95.5	64.8	88.0	95.1	54.1	85.1	96.5	65.8	89.0	116.6	64.4	103.9	105.6	107.8	106.2	109.3	108.0	109.0
0920	Plaster & Gypsum Board	101.6	56.8	72.2	94.2	56.6	69.5	94.4	61.6	72.9	143.7	55.3	85.7	109.5	105.5	106.9	105.5	105.5	105.5
0950, 0980	Ceilings & Acoustic Treatment	99.0	56.8	72.5	95.8	56.6	71.2	96.5	61.6	74.6	91.6	55.3	68.9	104.4	105.5	105.1	114.7	105.5	108.9
0960	Flooring	99.0	56.7	86.7	94.9	55.0	83.2	93.7	71.0	87.1	85.8	73.9	82.3	109.8	101.4	107.4	99.9	91.6	97.5
0970, 0990	Wall Finishes & Painting/Coating	97.7	75.0	84.1	97.7	29.2	56.7	111.0	73.9	88.8	83.5	68.7	74.6	95.4	92.6	93.7	97.0	92.6	94.3
09	FINISHES	97.3	60.3	77.2	94.7	55.3	73.4	95.9	65.6	79.5	105.6	63.1	82.6	106.2	103.0	104.5	102.6	101.5	102.0
COVERS	DIVS. 10 - 14, 25, 28, 41, 43, 44, 46	100.0	82.3	95.8	100.0	83.8	96.2	100.0	78.4	94.9	100.0	93.5	98.5	100.0	102.5	100.6	100.0	103.6	100.9
21, 22, 23	FIRE SUPPRESSION, PLUMBING & HVAC	101.5	67.2	87.6	97.7	56.8	81.2	97.7	77.5	89.6	97.5	76.8	89.2	101.2	102.5	101.7	101.1	108.1	103.9
26, 27, 3370	ELECTRICAL, COMMUNICATIONS & UTIL.	95.0	59.7	77.5	94.1	77.8	86.0	93.9	88.4	91.2	91.6	92.1	91.8	105.7	105.9	105.8	103.5	105.6	104.6
MF2018	WEIGHTED AVERAGE	100.5	69.1	87.0	99.0	67.8	85.5	99.2	76.4	89.3	99.9	79.5	91.1	104.5	104.2	104.4	103.9	105.2	104.5

For customer support on your Commercial Renovation Costs with RSMeans Data, call 800.448.8182.

827

City Cost Indexes - V2

DIVISION		WASHINGTON																	
		RICHLAND 993			SEATTLE 980 - 981,987			SPOKANE 990 - 992			TACOMA 983 - 984			VANCOUVER 986			WENATCHEE 988		
		MAT.	INST.	TOTAL	MAT.	INST.	TOTAL	MAT.	INST.	TOTAL	MAT.	INST.	TOTAL	MAT.	INST.	TOTAL	MAT.	INST.	TOTAL
015433	CONTRACTOR EQUIPMENT		88.4	88.4		100.2	100.2		88.4	88.4		97.7	97.7		94.1	94.1		97.7	97.7
0241, 31 - 34	SITE & INFRASTRUCTURE, DEMOLITION	104.7	83.9	90.5	97.8	105.0	102.8	104.2	83.9	90.3	96.3	103.3	101.1	106.3	90.8	95.6	104.9	99.9	101.4
0310	Concrete Forming & Accessories	102.7	82.2	85.2	110.0	105.8	106.4	108.0	81.7	85.7	104.5	106.2	105.9	104.9	97.1	98.3	105.9	73.7	78.5
0320	Concrete Reinforcing	102.2	99.3	100.8	103.9	111.8	107.7	102.8	100.1	101.5	106.2	113.7	109.8	107.1	113.7	110.3	107.1	91.4	99.5
0330	Cast-in-Place Concrete	87.2	85.3	86.5	106.5	110.7	108.1	90.6	85.1	88.5	103.8	112.4	107.0	115.9	100.3	110.0	105.9	87.4	98.9
03	CONCRETE	89.5	86.2	88.0	101.4	108.1	104.4	91.7	86.1	89.2	96.3	109.1	102.0	106.2	100.9	103.8	104.1	81.8	94.1
04	MASONRY	101.7	85.8	92.0	114.3	102.8	107.3	102.3	100.4	101.2	108.1	105.5	106.5	108.7	104.8	106.3	111.4	90.2	98.5
05	METALS	93.3	89.6	92.2	115.8	100.2	110.9	95.4	90.3	93.8	118.1	99.5	112.3	115.6	100.0	110.7	115.4	86.1	106.2
06	WOOD, PLASTICS & COMPOSITES	95.7	80.5	87.8	108.8	105.2	106.9	104.2	80.5	91.8	100.4	105.1	102.9	93.9	96.2	95.1	102.3	70.8	85.8
07	THERMAL & MOISTURE PROTECTION	158.7	87.1	128.0	112.4	105.5	109.9	155.4	88.0	126.5	113.3	108.5	111.2	113.7	100.9	108.2	112.9	84.3	100.6
08	OPENINGS	114.7	76.6	105.4	105.7	106.7	106.0	115.3	77.4	106.0	106.4	107.8	106.7	102.5	101.2	102.2	105.9	70.3	97.2
0920	Plaster & Gypsum Board	143.7	80.0	101.9	104.6	105.5	105.2	132.3	80.0	97.9	106.7	105.5	105.9	104.8	96.6	99.4	110.3	70.3	84.1
0950, 0980	Ceilings & Acoustic Treatment	98.0	80.0	86.7	107.2	105.5	106.1	94.4	80.0	85.4	107.6	105.5	106.3	106.2	96.6	100.2	100.4	70.3	81.5
0960	Flooring	86.1	81.8	84.9	106.6	98.7	104.3	85.3	101.4	90.0	102.0	101.4	101.9	107.8	102.2	106.2	105.2	73.9	96.1
0970, 0990	Wall Finishes & Painting/Coating	83.5	75.8	78.9	108.1	91.2	98.0	83.7	79.1	80.9	95.4	92.6	93.7	97.6	79.1	86.5	95.4	67.5	78.7
09	FINISHES	107.4	80.5	92.8	106.5	102.3	104.2	104.8	84.8	94.0	104.5	103.3	103.8	103.0	95.7	99.1	105.2	71.9	87.2
COVERS	DIVS. 10 - 14, 25, 28, 41, 43, 44, 46	100.0	97.3	99.4	100.0	102.7	100.6	100.0	97.2	99.3	100.0	102.9	100.7	100.0	98.1	99.5	100.0	95.3	98.9
21, 22, 23	FIRE SUPPRESSION, PLUMBING & HVAC	101.4	111.4	105.4	101.2	117.2	107.6	101.3	84.2	94.4	101.3	108.1	104.0	101.3	107.5	103.8	97.5	87.6	93.5
26, 27, 3370	ELECTRICAL, COMMUNICATIONS & UTIL.	89.1	101.3	95.2	103.1	114.5	108.7	88.0	83.5	85.8	105.5	105.7	105.6	111.7	109.1	110.4	106.3	91.0	98.7
MF2018	WEIGHTED AVERAGE	100.7	93.0	97.4	105.4	108.4	106.7	101.0	86.8	94.9	104.9	105.9	105.2	106.2	102.2	104.4	105.0	85.6	96.6

DIVISION		WASHINGTON			WEST VIRGINIA														
		YAKIMA 989			BECKLEY 258 - 259			BLUEFIELD 247 - 248			BUCKHANNON 262			CHARLESTON 250 - 253			CLARKSBURG 263 - 264		
		MAT.	INST.	TOTAL	MAT.	INST.	TOTAL	MAT.	INST.	TOTAL	MAT.	INST.	TOTAL	MAT.	INST.	TOTAL	MAT.	INST.	TOTAL
015433	CONTRACTOR EQUIPMENT		97.7	97.7		102.2	102.2		102.2	102.2		102.2	102.2		107.1	107.1		102.2	102.2
0241, 31 - 34	SITE & INFRASTRUCTURE, DEMOLITION	98.7	101.7	100.8	100.1	84.5	89.4	100.6	84.5	89.5	106.8	84.7	91.6	99.1	94.7	96.1	107.5	84.7	91.8
0310	Concrete Forming & Accessories	104.8	102.1	102.5	85.3	84.3	84.5	88.1	84.1	84.7	87.4	87.3	87.3	97.9	88.8	90.2	84.5	87.4	87.0
0320	Concrete Reinforcing	106.7	100.2	103.6	95.7	85.1	90.6	96.3	78.6	87.8	96.9	86.3	91.8	101.7	85.3	93.8	96.9	96.2	96.6
0330	Cast-in-Place Concrete	110.8	87.3	101.9	100.0	89.1	95.9	101.6	89.0	96.8	101.3	90.8	97.3	96.8	90.5	94.4	111.0	86.6	101.8
03	CONCRETE	100.9	96.1	98.7	91.9	87.3	89.8	94.8	86.0	90.8	97.6	89.4	94.0	91.2	89.7	90.5	101.6	89.7	96.3
04	MASONRY	101.5	102.1	101.9	91.7	85.4	87.8	92.6	85.4	88.2	104.0	88.3	94.5	87.2	90.5	89.2	108.0	88.3	96.0
05	METALS	116.1	91.1	108.3	98.0	102.6	99.5	103.5	100.1	102.4	103.7	103.3	103.6	96.8	100.9	98.1	103.7	106.8	104.7
06	WOOD, PLASTICS & COMPOSITES	100.8	105.1	103.0	80.4	84.4	82.5	84.0	84.4	84.2	83.1	86.8	85.0	92.2	88.6	90.3	79.1	86.8	83.1
07	THERMAL & MOISTURE PROTECTION	113.5	95.8	105.9	102.6	83.1	94.2	101.5	83.1	93.6	101.8	86.7	95.3	97.9	87.6	93.5	101.7	86.4	95.2
08	OPENINGS	105.9	100.8	104.6	94.4	83.1	91.7	96.9	81.6	93.2	96.9	84.7	94.0	95.0	85.4	92.6	96.9	86.9	94.5
0920	Plaster & Gypsum Board	106.3	105.5	105.8	86.6	84.1	85.0	93.3	84.1	87.2	93.9	86.5	89.1	92.6	88.1	89.7	91.7	86.5	88.3
0950, 0980	Ceilings & Acoustic Treatment	101.9	105.5	104.1	81.0	84.1	82.9	93.3	84.1	87.5	95.8	86.5	90.0	90.6	88.1	89.1	95.8	86.5	90.0
0960	Flooring	103.1	81.8	96.9	89.1	98.4	91.8	91.7	98.4	93.7	91.4	94.7	92.4	93.4	98.4	94.9	90.3	94.7	91.6
0970, 0990	Wall Finishes & Painting/Coating	95.4	79.1	85.6	90.2	87.7	88.7	97.7	85.5	90.4	97.7	89.0	92.5	92.3	89.2	90.4	97.7	89.0	92.5
09	FINISHES	103.6	95.9	99.4	86.6	87.4	87.0	92.7	87.2	89.7	93.8	88.9	91.2	92.6	90.7	91.6	93.1	88.9	90.8
COVERS	DIVS. 10 - 14, 25, 28, 41, 43, 44, 46	100.0	100.4	100.1	100.0	91.6	98.0	100.0	91.6	98.0	100.0	92.9	98.3	100.0	93.4	98.5	100.0	92.9	98.3
21, 22, 23	FIRE SUPPRESSION, PLUMBING & HVAC	101.3	113.5	106.2	98.0	89.4	94.6	97.7	80.1	90.6	97.7	91.1	95.0	101.1	91.3	97.1	97.7	91.3	95.1
26, 27, 3370	ELECTRICAL, COMMUNICATIONS & UTIL.	108.2	101.3	104.8	91.9	81.0	86.5	94.2	81.0	87.7	95.3	90.7	93.0	98.9	86.4	92.7	95.3	90.7	93.0
MF2018	WEIGHTED AVERAGE	105.0	101.6	103.5	95.2	87.7	91.9	97.3	85.2	92.1	98.7	90.5	95.1	96.6	91.1	94.2	99.3	91.0	95.7

DIVISION		WEST VIRGINIA																	
		GASSAWAY 266			HUNTINGTON 255 - 257			LEWISBURG 249			MARTINSBURG 254			MORGANTOWN 265			PARKERSBURG 261		
		MAT.	INST.	TOTAL	MAT.	INST.	TOTAL	MAT.	INST.	TOTAL	MAT.	INST.	TOTAL	MAT.	INST.	TOTAL	MAT.	INST.	TOTAL
015433	CONTRACTOR EQUIPMENT		102.2	102.2		102.2	102.2		102.2	102.2		102.2	102.2		102.2	102.2		102.2	102.2
0241, 31 - 34	SITE & INFRASTRUCTURE, DEMOLITION	104.2	84.7	90.8	105.0	85.9	91.9	116.4	84.5	94.5	103.8	84.9	90.9	101.5	85.8	90.7	110.4	85.8	93.5
0310	Concrete Forming & Accessories	86.7	84.5	84.8	98.2	88.2	89.7	84.8	84.1	84.2	85.3	76.5	77.8	84.8	87.7	87.3	89.4	89.3	89.3
0320	Concrete Reinforcing	96.9	87.3	92.3	97.1	90.3	93.8	96.9	78.6	88.1	95.7	90.6	93.2	96.9	96.3	96.6	96.3	86.2	91.4
0330	Cast-in-Place Concrete	106.0	88.9	99.6	109.0	94.2	103.4	101.7	89.0	96.9	104.7	84.8	97.2	101.3	92.9	98.1	103.5	92.8	99.5
03	CONCRETE	98.1	87.6	93.4	97.1	91.7	94.7	104.2	86.0	96.0	95.3	83.1	89.8	94.5	92.0	93.4	99.7	91.0	95.8
04	MASONRY	108.8	88.2	96.2	89.7	92.1	91.2	96.4	85.4	89.7	92.9	80.5	85.3	126.8	88.3	103.4	82.5	85.5	84.3
05	METALS	103.6	103.3	103.5	100.5	105.2	102.0	103.6	100.2	102.5	98.4	102.5	99.7	103.7	107.1	104.8	104.4	103.2	104.0
06	WOOD, PLASTICS & COMPOSITES	81.9	83.2	82.6	92.8	87.3	89.9	79.5	84.4	82.1	80.4	75.9	78.1	79.5	86.8	83.3	83.5	89.2	86.5
07	THERMAL & MOISTURE PROTECTION	101.5	86.3	95.0	102.8	86.9	96.0	102.5	83.1	94.2	102.8	77.1	91.8	101.5	87.3	95.4	101.8	86.8	95.4
08	OPENINGS	95.3	82.9	92.3	93.7	85.8	91.8	96.9	81.6	93.2	96.1	71.1	90.0	98.0	86.9	95.3	95.9	83.4	92.8
0920	Plaster & Gypsum Board	92.8	82.8	86.3	92.0	87.0	88.7	91.7	84.1	86.7	86.6	75.3	79.2	91.7	86.5	88.3	94.2	89.0	90.8
0950, 0980	Ceilings & Acoustic Treatment	95.8	82.8	87.7	81.0	87.0	84.8	95.8	84.1	88.5	81.0	75.3	77.4	95.8	86.5	90.0	95.8	89.0	91.5
0960	Flooring	91.3	98.4	93.3	96.7	100.4	97.8	90.5	98.4	92.8	89.1	94.4	90.6	90.5	94.7	91.7	94.4	98.4	95.6
0970, 0990	Wall Finishes & Painting/Coating	97.7	89.2	92.6	90.2	89.2	89.6	97.7	61.4	75.9	90.2	79.7	83.9	97.7	89.0	92.5	97.7	89.2	92.6
09	FINISHES	93.3	87.6	90.2	89.5	90.5	90.1	94.2	84.5	89.0	86.8	79.7	83.0	92.7	88.9	90.7	94.8	91.1	92.8
COVERS	DIVS. 10 - 14, 25, 28, 41, 43, 44, 46	100.0	92.5	98.2	100.0	93.1	98.4	100.0	91.6	98.0	100.0	77.1	94.6	100.0	92.9	98.3	100.0	93.1	98.4
21, 22, 23	FIRE SUPPRESSION, PLUMBING & HVAC	97.7	81.5	91.2	101.8	91.6	97.7	97.7	89.4	94.3	98.0	77.1	89.6	97.7	91.3	95.1	101.4	91.2	97.3
26, 27, 3370	ELECTRICAL, COMMUNICATIONS & UTIL.	95.3	86.4	90.9	95.0	89.8	92.4	92.1	81.0	86.6	96.8	74.0	85.5	95.5	90.7	93.1	95.4	85.0	90.3
MF2018	WEIGHTED AVERAGE	98.7	87.3	93.7	97.8	91.7	95.1	99.0	86.8	93.8	96.5	80.8	89.7	99.3	91.4	95.9	98.9	90.0	95.1

DIVISION		WEST VIRGINIA									WISCONSIN								
		PETERSBURG 268			ROMNEY 267			WHEELING 260			BELOIT 535			EAU CLAIRE 547			GREEN BAY 541 - 543		
		MAT.	INST.	TOTAL	MAT.	INST.	TOTAL	MAT.	INST.	TOTAL	MAT.	INST.	TOTAL	MAT.	INST.	TOTAL	MAT.	INST.	TOTAL
015433	CONTRACTOR EQUIPMENT		102.2	102.2		102.2	102.2		102.2	102.2		96.6	96.6		98.1	98.1		95.8	95.8
0241, 31 - 34	SITE & INFRASTRUCTURE, DEMOLITION	100.8	85.6	90.4	103.7	85.6	91.3	111.0	85.7	93.7	96.1	98.3	97.6	96.4	97.7	97.3	99.9	93.7	95.7
0310	Concrete Forming & Accessories	88.6	85.3	85.8	84.0	85.2	85.0	91.5	87.1	87.7	99.0	93.1	94.0	100.2	98.1	98.4	112.4	102.8	104.3
0320	Concrete Reinforcing	96.3	87.3	91.9	96.9	90.9	94.0	95.7	96.3	96.0	102.7	136.5	119.0	93.7	113.9	103.5	91.9	113.9	102.5
0330	Cast-in-Place Concrete	101.3	90.7	97.3	106.0	82.4	97.1	103.5	93.0	99.6	101.9	96.5	99.8	97.8	103.1	99.8	101.1	103.0	101.8
03	CONCRETE	94.6	88.6	91.9	97.9	86.3	92.7	99.7	91.8	96.2	98.8	102.0	100.2	93.6	102.8	97.7	96.9	105.0	100.5
04	MASONRY	99.1	88.3	92.5	96.8	88.2	91.6	107.0	86.8	94.7	95.1	109.8	104.0	92.1	111.9	104.1	122.5	111.0	115.5
05	METALS	103.8	103.3	103.6	103.8	104.4	104.0	104.5	107.1	105.3	96.7	113.6	102.0	91.7	108.5	96.9	94.3	108.9	98.8
06	WOOD, PLASTICS & COMPOSITES	84.3	84.4	84.4	78.5	84.4	81.6	85.4	86.8	86.1	95.7	90.9	93.2	101.8	96.5	99.0	112.8	102.9	107.6
07	THERMAL & MOISTURE PROTECTION	101.6	81.9	93.2	101.7	81.8	93.1	102.0	86.0	95.1	105.0	93.5	100.1	103.7	105.4	104.5	106.2	105.9	106.0
08	OPENINGS	98.0	80.5	93.8	98.0	81.4	93.9	96.6	86.9	94.2	97.9	105.5	99.7	103.0	96.1	101.3	98.8	105.8	100.5
0920	Plaster & Gypsum Board	93.9	84.1	87.5	91.4	84.1	86.6	94.2	86.5	89.2	90.9	91.3	91.2	99.7	96.9	97.9	98.3	103.5	101.7
0950, 0980	Ceilings & Acoustic Treatment	95.8	84.1	88.5	95.8	84.1	88.5	95.8	86.5	90.0	87.4	91.3	89.9	94.1	96.9	95.9	86.3	103.5	97.1
0960	Flooring	92.3	94.7	93.0	90.2	94.4	91.5	95.3	94.7	95.1	95.4	113.0	100.6	82.3	117.9	92.7	97.7	117.9	103.6
0970, 0990	Wall Finishes & Painting/Coating	97.7	89.0	92.5	97.7	89.0	92.5	97.7	89.0	92.5	88.9	98.7	94.7	80.1	83.6	82.2	87.9	106.2	98.9
09	FINISHES	93.5	87.5	90.3	92.8	87.5	89.9	95.0	88.5	91.5	92.5	96.7	94.8	89.0	99.9	94.9	92.4	106.0	99.7
COVERS	DIVS. 10 - 14, 25, 28, 41, 43, 44, 46	100.0	92.5	98.2	100.0	92.5	98.2	100.0	91.6	98.0	100.0	98.2	99.6	100.0	96.7	99.2	100.0	98.1	99.5
21, 22, 23	FIRE SUPPRESSION, PLUMBING & HVAC	97.7	81.9	91.3	97.7	81.6	91.2	101.5	90.5	97.1	98.4	91.2	95.5	100.6	90.3	96.5	100.9	97.4	99.5
26, 27, 3370	ELECTRICAL, COMMUNICATIONS & UTIL.	98.0	74.0	86.1	97.4	74.0	85.8	93.2	90.7	91.9	98.8	81.1	90.1	102.4	102.9	102.6	97.9	102.9	100.4
MF2018	WEIGHTED AVERAGE	98.3	85.6	92.8	98.5	85.3	92.8	100.1	90.9	96.1	97.8	97.3	97.6	97.4	100.4	98.7	99.4	103.1	101.0

DIVISION		WISCONSIN																	
		KENOSHA 531			LA CROSSE 546			LANCASTER 538			MADISON 537			MILWAUKEE 530,532			NEW RICHMOND 540		
		MAT.	INST.	TOTAL	MAT.	INST.	TOTAL	MAT.	INST.	TOTAL	MAT.	INST.	TOTAL	MAT.	INST.	TOTAL	MAT.	INST.	TOTAL
015433	CONTRACTOR EQUIPMENT		94.6	94.6		98.1	98.1		96.6	96.6		101.6	101.6		90.8	90.8		98.4	98.4
0241, 31 - 34	SITE & INFRASTRUCTURE, DEMOLITION	103.0	96.2	98.3	90.1	97.7	95.3	94.8	97.7	96.8	94.9	107.7	103.7	91.0	97.2	95.3	93.0	96.7	95.5
0310	Concrete Forming & Accessories	110.2	113.8	113.3	84.6	98.0	96.0	98.2	92.0	92.9	103.6	106.2	105.8	102.6	116.9	114.8	93.8	89.2	89.9
0320	Concrete Reinforcing	102.5	113.9	108.0	93.5	113.9	103.3	103.9	102.1	103.0	103.4	114.0	108.5	102.0	116.7	109.1	91.0	109.7	100.0
0330	Cast-in-Place Concrete	110.9	105.5	108.8	87.9	103.0	93.6	101.3	96.0	99.3	93.7	108.9	99.4	94.5	118.1	103.4	101.8	99.0	100.7
03	CONCRETE	103.7	110.6	106.8	85.0	102.8	92.9	98.5	95.4	97.1	99.9	108.2	103.6	100.0	116.2	107.3	91.1	96.6	93.6
04	MASONRY	93.0	115.8	106.9	91.3	111.9	103.8	95.1	109.8	104.1	94.5	119.5	109.7	97.4	121.2	111.9	117.0	109.9	112.7
05	METALS	97.8	106.1	100.4	91.6	108.4	96.8	94.0	100.2	95.9	100.8	104.6	102.0	95.7	98.9	96.7	92.0	105.5	96.2
06	WOOD, PLASTICS & COMPOSITES	105.7	115.2	110.7	84.0	96.5	90.5	94.8	90.9	92.8	98.9	103.0	101.0	102.3	115.6	109.3	89.7	87.1	88.4
07	THERMAL & MOISTURE PROTECTION	105.7	109.9	107.5	103.1	104.8	103.8	104.8	93.2	99.8	106.0	111.2	108.2	104.4	115.3	109.1	104.0	100.1	102.3
08	OPENINGS	92.2	112.7	97.2	102.9	96.2	101.3	93.8	88.5	92.6	101.5	105.7	102.5	102.5	113.5	105.2	88.5	87.0	88.1
0920	Plaster & Gypsum Board	78.8	116.3	103.4	95.1	96.9	96.3	89.5	91.3	90.7	100.5	103.5	102.5	92.9	116.0	108.0	85.8	87.5	86.9
0950, 0980	Ceilings & Acoustic Treatment	87.4	116.3	105.5	92.9	96.9	95.4	82.3	91.3	88.0	95.7	103.5	100.6	88.8	116.0	105.8	58.9	87.5	76.8
0960	Flooring	113.2	117.9	114.6	76.3	117.9	88.4	95.1	105.9	98.2	95.0	117.9	101.7	101.5	118.1	106.4	91.2	109.3	96.5
0970, 0990	Wall Finishes & Painting/Coating	99.7	122.4	113.3	80.1	81.1	80.7	88.9	93.0	91.4	90.9	104.7	99.2	100.8	122.2	113.6	90.8	77.8	83.0
09	FINISHES	96.7	116.2	107.2	85.9	99.7	93.4	90.9	94.5	92.9	95.8	108.0	102.4	96.3	118.1	108.1	81.9	91.4	87.0
COVERS	DIVS. 10 - 14, 25, 28, 41, 43, 44, 46	100.0	99.8	100.0	100.0	96.7	99.2	100.0	98.2	99.6	100.0	103.4	100.8	100.0	104.2	101.0	100.0	90.3	97.7
21, 22, 23	FIRE SUPPRESSION, PLUMBING & HVAC	100.8	97.3	99.4	100.6	90.3	96.5	94.7	85.0	90.8	96.1	100.6	97.9	96.1	117.1	104.6	96.2	84.8	91.6
26, 27, 3370	ELECTRICAL, COMMUNICATIONS & UTIL.	100.0	102.9	101.4	102.7	102.9	102.8	98.5	81.7	90.2	98.9	102.8	100.9	97.8	102.8	100.3	100.8	81.7	91.4
MF2018	WEIGHTED AVERAGE	99.2	106.2	102.2	95.7	100.3	97.7	95.8	92.9	94.6	98.6	106.4	101.9	97.9	111.7	103.9	94.9	92.9	94.0

DIVISION		WISCONSIN																	
		OSHKOSH 549			PORTAGE 539			RACINE 534			RHINELANDER 545			SUPERIOR 548			WAUSAU 544		
		MAT.	INST.	TOTAL	MAT.	INST.	TOTAL	MAT.	INST.	TOTAL	MAT.	INST.	TOTAL	MAT.	INST.	TOTAL	MAT.	INST.	TOTAL
015433	CONTRACTOR EQUIPMENT		95.8	95.8		96.6	96.6		96.6	96.6		95.8	95.8		98.4	98.4		95.8	95.8
0241, 31 - 34	SITE & INFRASTRUCTURE, DEMOLITION	91.1	92.9	92.3	85.7	98.6	94.5	96.7	99.8	98.8	103.5	92.4	95.9	89.7	96.4	94.3	86.9	93.5	91.5
0310	Concrete Forming & Accessories	92.5	92.4	92.4	89.7	92.7	92.2	99.4	110.8	109.1	89.7	91.5	91.2	91.6	85.6	86.5	91.9	94.3	94.0
0320	Concrete Reinforcing	92.0	105.5	98.6	104.0	102.2	103.1	102.7	113.9	108.1	92.2	102.3	97.1	91.0	109.5	99.9	92.2	113.7	102.6
0330	Cast-in-Place Concrete	93.7	95.7	94.5	86.9	94.7	89.9	99.9	105.3	101.9	106.3	94.4	101.8	95.9	94.7	95.5	87.4	89.2	88.1
03	CONCRETE	86.7	96.2	91.0	86.2	95.5	90.4	97.9	109.2	103.0	98.4	94.8	96.8	85.9	93.5	89.3	81.9	96.3	88.4
04	MASONRY	105.8	111.3	109.1	94.0	109.8	103.6	95.3	115.8	107.8	125.9	109.1	115.7	116.3	107.6	111.0	105.3	109.1	107.6
05	METALS	92.2	104.0	95.9	94.7	104.3	97.7	98.3	106.1	100.8	92.0	102.8	95.4	93.0	105.8	97.0	91.9	107.6	96.8
06	WOOD, PLASTICS & COMPOSITES	89.0	91.0	90.1	83.8	90.9	87.5	96.1	111.2	104.0	85.9	91.0	88.6	87.8	82.9	85.2	88.4	93.9	91.3
07	THERMAL & MOISTURE PROTECTION	105.1	85.3	96.6	104.2	99.4	102.1	105.1	109.1	106.8	106.0	83.6	96.4	103.7	94.7	99.8	104.9	95.2	100.7
08	OPENINGS	95.4	92.8	94.8	94.0	96.0	94.5	97.9	110.4	101.0	95.4	88.5	93.8	87.9	91.0	88.6	95.6	96.3	95.8
0920	Plaster & Gypsum Board	86.1	91.3	89.5	83.2	91.3	88.5	90.9	112.1	104.8	86.1	91.3	89.5	86.1	83.1	84.1	86.1	94.3	91.5
0950, 0980	Ceilings & Acoustic Treatment	86.3	91.3	89.4	86.1	91.3	89.4	87.4	112.1	102.9	86.3	91.3	89.4	60.2	83.1	74.5	86.1	94.3	91.3
0960	Flooring	89.2	109.5	95.1	91.1	109.3	96.4	95.4	117.9	102.0	88.5	109.3	94.6	92.3	117.4	99.7	89.1	109.3	95.0
0970, 0990	Wall Finishes & Painting/Coating	85.6	106.2	98.0	88.9	93.0	91.4	88.8	121.1	108.2	85.6	76.2	79.9	80.1	100.7	92.5	85.6	79.3	81.8
09	FINISHES	87.5	97.3	92.8	89.2	95.4	92.6	92.5	113.6	103.9	88.3	93.4	91.1	81.5	92.5	87.5	87.2	95.4	91.7
COVERS	DIVS. 10 - 14, 25, 28, 41, 43, 44, 46	100.0	81.9	95.7	100.0	98.3	99.6	100.0	99.4	99.9	100.0	96.0	99.1	100.0	89.1	97.4	100.0	96.4	99.1
21, 22, 23	FIRE SUPPRESSION, PLUMBING & HVAC	97.1	84.4	92.0	94.7	91.3	93.4	100.6	99.2	100.0	97.1	84.4	92.0	96.2	86.8	92.5	97.1	85.0	92.2
26, 27, 3370	ELECTRICAL, COMMUNICATIONS & UTIL.	101.3	81.2	91.3	101.8	90.5	96.2	99.6	102.9	101.2	100.8	75.2	88.1	105.0	94.9	100.0	102.3	75.2	88.9
MF2018	WEIGHTED AVERAGE	95.2	92.7	94.1	94.2	96.5	95.2	98.5	106.2	101.8	98.0	91.1	95.0	94.6	94.6	94.6	94.5	92.9	93.8

For customer support on your Commercial Renovation Costs with RSMeans Data, call 800.448.8182.

829

WYOMING

DIVISION		CASPER 826			CHEYENNE 820			NEWCASTLE 827			RAWLINS 823			RIVERTON 825			ROCK SPRINGS 829 - 831		
		MAT.	INST.	TOTAL	MAT.	INST.	TOTAL	MAT.	INST.	TOTAL	MAT.	INST.	TOTAL	MAT.	INST.	TOTAL	MAT.	INST.	TOTAL
015433	CONTRACTOR EQUIPMENT		98.5	98.5		92.3	92.3		92.3	92.3		92.3	92.3		92.3	92.3		92.3	92.3
0241, 31 - 34	SITE & INFRASTRUCTURE, DEMOLITION	98.8	96.8	97.4	91.5	87.2	88.5	83.3	87.2	86.0	97.1	87.2	90.3	90.7	87.2	88.3	87.0	87.2	87.1
0310	Concrete Forming & Accessories	99.6	64.4	69.7	103.4	63.8	69.7	92.8	64.2	68.5	97.4	64.2	69.1	91.6	64.1	68.2	99.2	64.0	69.2
0320	Concrete Reinforcing	105.9	81.4	94.1	97.3	81.5	89.7	104.5	81.6	93.5	104.2	81.6	93.3	105.2	81.6	93.8	105.2	80.9	93.5
0330	Cast-in-Place Concrete	104.7	79.4	95.1	98.7	78.0	90.9	99.7	78.0	91.5	99.8	78.0	91.5	99.7	77.9	91.5	99.7	77.9	91.4
03	CONCRETE	109.2	73.0	93.0	101.7	72.3	88.6	101.8	72.5	88.7	116.4	72.5	96.9	110.7	72.4	93.5	102.3	72.3	88.8
04	MASONRY	96.8	65.0	77.4	98.5	66.5	79.0	95.5	68.0	78.7	95.5	68.0	78.7	95.5	68.0	78.7	152.7	61.1	96.8
05	METALS	101.5	79.0	94.5	103.8	80.4	96.5	100.0	80.5	93.9	100.0	80.5	93.9	100.1	80.3	93.9	100.9	79.7	94.2
06	WOOD, PLASTICS & COMPOSITES	94.7	62.2	77.7	94.4	61.4	77.2	83.1	61.9	72.0	87.7	61.9	74.2	81.8	61.9	71.4	92.4	61.9	76.4
07	THERMAL & MOISTURE PROTECTION	109.2	67.6	91.3	105.5	67.5	89.2	106.5	67.2	89.6	108.0	67.2	90.5	107.4	70.4	91.5	106.6	68.3	90.2
08	OPENINGS	109.2	67.1	99.0	107.0	66.7	97.2	111.0	65.5	99.9	110.6	65.5	99.6	110.8	65.5	99.8	111.4	66.3	100.4
0920	Plaster & Gypsum Board	96.9	61.1	73.4	85.8	60.6	69.3	82.8	61.1	68.6	83.1	61.1	68.7	82.8	61.1	68.6	94.7	61.1	72.7
0950, 0980	Ceilings & Acoustic Treatment	119.8	61.1	83.0	107.8	60.6	78.2	110.7	61.1	79.6	110.7	61.1	79.6	110.7	61.1	79.6	110.7	61.1	79.6
0960	Flooring	103.9	72.8	94.8	102.9	72.8	94.1	96.6	67.6	88.1	99.7	67.6	90.4	95.9	67.6	87.7	102.4	58.3	89.5
0970, 0990	Wall Finishes & Painting/Coating	98.3	58.1	74.2	97.7	58.1	74.0	94.3	74.6	82.5	94.3	74.6	82.5	94.3	57.6	72.3	94.3	74.6	82.5
09	FINISHES	103.7	64.6	82.6	99.4	64.1	80.3	95.0	65.2	78.9	97.2	65.2	79.9	95.6	63.3	78.1	98.3	63.3	79.4
COVERS	DIVS. 10 - 14, 25, 28, 41, 43, 44, 46	100.0	88.6	97.3	100.0	87.9	97.2	100.0	98.4	99.6	100.0	98.4	99.6	100.0	87.2	97.0	100.0	85.0	96.5
21, 22, 23	FIRE SUPPRESSION, PLUMBING & HVAC	100.9	74.6	90.3	101.2	74.6	90.5	99.1	71.8	88.1	99.1	71.8	88.1	99.1	71.8	88.1	101.1	71.8	89.3
26, 27, 3370	ELECTRICAL, COMMUNICATIONS & UTIL.	97.0	62.0	79.7	95.2	67.7	81.6	94.0	60.0	77.2	94.0	60.0	77.2	94.0	64.2	79.3	92.7	64.8	78.9
MF2018	WEIGHTED AVERAGE	102.5	72.3	89.4	101.0	72.4	88.7	99.4	71.3	87.3	101.9	71.3	88.7	100.8	71.4	88.1	103.3	70.6	89.1

WYOMING / CANADA

DIVISION		SHERIDAN 828			WHEATLAND 822			WORLAND 824			YELLOWSTONE NAT'L PA 821			BARRIE, ONTARIO			BATHURST, NEW BRUNSWICK		
		MAT.	INST.	TOTAL	MAT.	INST.	TOTAL	MAT.	INST.	TOTAL	MAT.	INST.	TOTAL	MAT.	INST.	TOTAL	MAT.	INST.	TOTAL
015433	CONTRACTOR EQUIPMENT		92.3	92.3		92.3	92.3		92.3	92.3		92.3	92.3		97.9	97.9		97.4	97.4
0241, 31 - 34	SITE & INFRASTRUCTURE, DEMOLITION	91.1	87.2	88.4	87.6	87.2	87.3	84.9	87.2	86.5	85.0	87.2	86.5	116.9	91.6	99.5	104.6	87.4	92.8
0310	Concrete Forming & Accessories	101.8	64.0	69.7	94.8	63.9	68.5	94.9	64.2	68.8	95.0	64.1	68.7	125.7	80.9	87.6	104.9	56.8	63.9
0320	Concrete Reinforcing	105.2	81.6	93.8	104.5	80.8	93.1	105.2	81.6	93.8	107.0	79.0	93.5	171.9	84.6	129.8	137.9	56.0	98.4
0330	Cast-in-Place Concrete	103.1	77.9	93.6	104.1	77.9	94.2	99.7	78.0	91.5	99.7	77.9	91.5	155.6	80.1	127.1	112.6	54.5	90.6
03	CONCRETE	110.8	72.4	93.6	107.0	72.2	91.4	102.0	72.6	88.8	102.2	72.0	88.7	141.5	81.8	114.7	116.6	56.9	89.8
04	MASONRY	95.8	68.0	78.9	95.8	56.3	71.7	95.5	68.0	78.7	95.5	68.0	78.8	169.0	87.4	119.2	164.9	56.2	98.6
05	METALS	103.6	80.3	96.4	99.9	79.3	93.5	100.1	80.5	94.0	100.7	79.0	94.0	111.5	90.6	105.0	113.4	72.2	100.6
06	WOOD, PLASTICS & COMPOSITES	96.1	61.9	78.2	85.2	61.9	73.0	85.3	61.9	73.1	85.3	61.9	73.1	116.6	79.7	97.3	95.0	57.0	75.2
07	THERMAL & MOISTURE PROTECTION	107.9	68.0	90.8	106.7	63.9	88.3	106.5	67.2	89.6	106.0	67.2	89.3	114.2	83.8	101.1	110.4	56.8	87.4
08	OPENINGS	111.6	65.5	100.4	109.6	65.5	98.8	111.2	65.5	100.1	104.3	64.6	94.6	90.4	79.6	87.8	84.2	50.3	75.9
0920	Plaster & Gypsum Board	104.5	61.1	76.0	82.8	61.1	68.6	82.8	61.1	68.6	83.0	61.1	68.6	151.7	79.4	104.2	119.4	56.0	77.8
0950, 0980	Ceilings & Acoustic Treatment	114.9	61.1	81.2	110.7	61.1	79.6	110.7	61.1	79.6	111.3	61.1	79.8	95.4	79.4	85.3	115.9	56.0	78.4
0960	Flooring	101.0	57.5	88.3	98.2	67.6	89.3	98.2	67.6	89.3	98.2	67.6	89.3	121.5	85.1	110.8	101.2	40.5	83.4
0970, 0990	Wall Finishes & Painting/Coating	96.6	58.1	73.5	94.3	55.6	71.1	94.3	74.6	82.5	94.3	57.6	72.3	102.9	81.6	90.1	108.6	46.6	71.5
09	FINISHES	103.0	61.8	80.7	95.7	63.1	78.1	95.4	65.2	79.1	95.6	63.3	78.1	112.9	81.9	96.1	109.1	52.7	78.6
COVERS	DIVS. 10 - 14, 25, 28, 41, 43, 44, 46	100.0	87.9	97.2	100.0	87.9	97.2	100.0	98.4	99.6	100.0	98.4	99.6	139.3	62.8	121.3	131.2	55.8	113.4
21, 22, 23	FIRE SUPPRESSION, PLUMBING & HVAC	99.1	71.8	88.1	99.1	71.8	88.1	99.1	71.8	88.1	99.1	71.8	88.1	106.2	91.1	100.1	106.2	63.0	88.8
26, 27, 3370	ELECTRICAL, COMMUNICATIONS & UTIL.	95.9	64.2	80.2	94.0	64.2	79.3	94.0	64.2	79.3	93.2	64.2	78.9	117.2	81.6	99.6	117.0	55.0	86.3
MF2018	WEIGHTED AVERAGE	102.4	71.1	88.9	100.1	69.9	87.0	99.5	71.9	87.6	98.9	71.4	87.0	118.0	85.2	103.8	112.7	60.8	90.3

CANADA

DIVISION		BRANDON, MANITOBA			BRANTFORD, ONTARIO			BRIDGEWATER, NOVA SCOTIA			CALGARY, ALBERTA			CAP-DE-LA-MADELEINE, QUEBEC			CHARLESBOURG, QUEBEC		
		MAT.	INST.	TOTAL	MAT.	INST.	TOTAL	MAT.	INST.	TOTAL	MAT.	INST.	TOTAL	MAT.	INST.	TOTAL	MAT.	INST.	TOTAL
015433	CONTRACTOR EQUIPMENT		99.6	99.6		97.9	97.9		97.3	97.3		122.7	122.7		98.1	98.1		98.1	98.1
0241, 31 - 34	SITE & INFRASTRUCTURE, DEMOLITION	124.4	89.9	100.7	116.1	91.9	99.4	100.2	89.3	92.7	126.4	112.5	116.9	96.2	90.6	92.4	96.2	90.6	92.4
0310	Concrete Forming & Accessories	149.5	64.9	77.5	129.0	87.6	93.8	97.4	66.7	71.3	125.3	92.0	96.9	136.3	77.8	86.5	136.3	77.8	86.5
0320	Concrete Reinforcing	167.0	53.1	112.0	160.7	83.3	123.3	136.9	46.4	93.2	133.9	78.7	107.3	136.9	70.5	104.8	136.9	70.5	104.8
0330	Cast-in-Place Concrete	108.0	68.4	93.0	129.0	99.6	117.9	132.5	65.1	107.0	134.5	101.3	122.0	104.1	85.4	97.0	104.1	85.4	97.0
03	CONCRETE	125.5	65.1	98.4	128.4	91.4	111.8	126.4	63.6	98.3	136.8	93.7	117.5	114.1	79.8	98.7	114.1	79.8	98.7
04	MASONRY	221.4	59.4	122.6	173.0	91.5	123.3	167.9	64.2	104.6	231.4	86.3	142.9	169.1	74.8	111.5	169.1	74.8	111.5
05	METALS	126.0	77.2	110.8	111.4	91.2	105.1	110.6	74.9	99.5	128.4	98.4	119.1	109.7	84.0	101.7	109.7	84.0	101.7
06	WOOD, PLASTICS & COMPOSITES	150.7	65.9	106.3	120.4	86.6	102.7	86.3	66.5	75.9	99.8	91.9	95.7	132.8	78.0	104.1	132.8	78.0	104.1
07	THERMAL & MOISTURE PROTECTION	126.8	66.8	101.0	120.1	89.0	106.8	114.8	65.8	93.8	136.1	94.8	118.4	113.9	81.9	100.2	113.9	81.9	100.2
08	OPENINGS	100.2	58.9	90.1	87.7	85.3	87.1	82.7	60.3	77.3	82.4	81.2	82.1	89.2	71.1	84.8	89.2	71.1	84.8
0920	Plaster & Gypsum Board	114.9	64.8	82.0	114.7	86.4	96.1	120.4	65.7	84.5	132.6	90.9	105.2	143.9	77.4	100.3	143.9	77.4	100.3
0950, 0980	Ceilings & Acoustic Treatment	122.1	64.8	86.2	102.7	86.4	92.5	102.7	65.7	79.5	146.1	90.9	111.5	102.7	77.4	86.9	102.7	77.4	86.9
0960	Flooring	133.4	60.4	112.1	115.3	85.1	106.5	96.7	57.4	85.2	124.1	81.3	111.6	105.3	83.8	106.1	105.3	83.8	106.1
0970, 0990	Wall Finishes & Painting/Coating	114.4	52.6	77.4	107.2	89.5	96.6	107.2	57.7	77.6	107.8	101.3	103.9	107.2	81.4	91.7	107.2	81.4	91.7
09	FINISHES	124.1	63.1	91.1	109.7	87.6	97.8	104.5	64.3	82.8	127.8	91.5	108.2	113.1	79.6	95.0	113.1	79.6	95.0
COVERS	DIVS. 10 - 14, 25, 28, 41, 43, 44, 46	131.2	58.0	114.0	131.2	64.7	115.5	131.2	58.7	114.1	131.2	88.4	121.1	131.2	73.1	117.5	131.2	73.1	117.5
21, 22, 23	FIRE SUPPRESSION, PLUMBING & HVAC	106.5	76.8	94.5	106.2	93.8	101.2	106.2	77.4	94.6	106.6	86.0	98.3	106.6	82.0	96.7	106.6	82.0	96.7
26, 27, 3370	ELECTRICAL, COMMUNICATIONS & UTIL.	116.4	62.1	89.5	114.5	81.1	98.0	118.8	57.7	88.6	114.0	90.4	102.3	113.2	64.2	89.0	113.2	64.2	89.0
MF2018	WEIGHTED AVERAGE	122.8	68.9	99.5	115.2	88.8	103.8	113.3	68.7	94.0	123.3	91.7	109.7	112.8	78.2	97.8	112.8	78.2	97.8

For customer support on your Commercial Renovation Costs with RSMeans Data, call 800.448.8182.

CANADA

DIVISION		CHARLOTTETOWN, PRINCE EDWARD ISLAND			CHICOUTIMI, QUEBEC			CORNER BROOK, NEWFOUNDLAND			CORNWALL, ONTARIO			DALHOUSIE, NEW BRUNSWICK			DARTMOUTH, NOVA SCOTIA		
		MAT.	INST.	TOTAL	MAT.	INST.	TOTAL	MAT.	INST.	TOTAL	MAT.	INST.	TOTAL	MAT.	INST.	TOTAL	MAT.	INST.	TOTAL
015433	CONTRACTOR EQUIPMENT		115.1	115.1		97.9	97.9		97.7	97.7		97.9	97.9		98.0	98.0		96.8	96.8
0241, 31 - 34	SITE & INFRASTRUCTURE, DEMOLITION	132.1	101.4	111.0	101.7	90.5	94.0	127.6	87.3	99.9	114.3	91.4	98.6	99.9	87.8	91.6	116.0	88.9	97.4
0310	Concrete Forming & Accessories	125.5	52.1	63.0	138.3	89.8	97.0	122.9	74.1	81.4	126.3	81.0	87.8	104.3	57.0	64.0	111.1	66.7	73.3
0320	Concrete Reinforcing	153.8	45.3	101.4	104.0	93.7	99.0	153.1	47.5	102.1	160.7	83.0	123.2	140.0	56.0	99.5	160.1	46.4	105.2
0330	Cast-in-Place Concrete	150.0	56.3	114.6	106.0	93.8	101.4	125.7	61.8	101.5	116.0	90.7	106.4	109.0	54.7	88.5	121.3	65.0	100.0
03	CONCRETE	147.7	54.0	105.6	107.7	92.0	100.6	160.3	65.9	118.0	122.2	85.2	105.6	120.0	57.1	91.7	140.5	63.6	106.0
04	MASONRY	189.1	53.8	106.6	167.5	89.8	120.1	216.3	72.6	128.7	172.0	83.8	118.2	170.1	56.2	100.6	231.9	64.2	129.6
05	METALS	133.6	75.8	115.6	113.2	91.5	106.4	126.4	73.9	110.0	111.2	89.9	104.6	104.6	72.8	94.6	126.8	74.6	110.5
06	WOOD, PLASTICS & COMPOSITES	102.2	51.7	75.8	133.0	90.7	110.9	125.7	80.1	101.8	118.2	80.5	98.5	93.2	57.1	74.3	112.1	66.4	88.2
07	THERMAL & MOISTURE PROTECTION	135.4	56.3	101.5	111.0	95.6	104.4	130.8	64.1	102.2	119.9	83.9	104.5	119.6	56.8	92.6	128.1	65.8	101.3
08	OPENINGS	85.8	45.0	75.9	87.9	76.3	85.1	106.0	65.3	96.1	89.2	79.3	86.8	85.7	50.3	77.1	91.6	60.3	83.9
0920	Plaster & Gypsum Board	126.1	49.9	76.1	138.9	90.5	107.2	148.0	79.7	103.1	172.0	80.2	111.7	127.7	56.0	80.6	144.2	65.7	92.7
0950, 0980	Ceilings & Acoustic Treatment	131.4	49.9	80.3	114.6	90.5	99.5	122.7	79.7	95.7	106.6	80.2	90.0	105.0	56.0	74.3	131.0	65.7	90.1
0960	Flooring	119.9	53.9	100.6	117.8	83.8	107.9	114.3	48.3	95.0	115.3	83.9	106.1	103.2	61.9	91.1	108.9	57.4	93.9
0970, 0990	Wall Finishes & Painting/Coating	108.7	38.8	66.8	108.6	103.3	105.4	114.3	55.0	78.7	107.2	83.4	93.0	110.3	46.6	72.2	114.3	57.7	80.4
09	FINISHES	123.0	51.3	84.2	116.0	91.2	102.6	123.8	68.7	94.0	118.7	82.0	98.8	109.3	57.1	81.1	122.4	64.2	90.9
COVERS	DIVS. 10 - 14, 25, 28, 41, 43, 44, 46	131.2	56.4	113.6	131.2	78.2	118.7	131.2	58.4	114.1	131.2	62.4	115.0	131.2	55.9	113.4	131.2	58.7	114.1
21, 22, 23	FIRE SUPPRESSION, PLUMBING & HVAC	106.5	57.4	86.7	106.2	80.3	95.7	106.5	64.1	89.4	106.6	91.8	100.7	106.3	63.1	88.8	106.5	77.4	94.7
26, 27, 3370	ELECTRICAL, COMMUNICATIONS & UTIL.	119.8	46.7	83.6	113.5	84.3	99.0	113.5	50.7	82.4	114.6	82.1	98.5	117.6	52.0	85.1	117.6	57.7	88.0
MF2018	WEIGHTED AVERAGE	124.1	58.7	95.8	112.5	86.9	101.4	127.3	66.7	101.1	115.3	85.5	102.4	112.4	61.0	90.2	124.0	68.6	100.1

CANADA

DIVISION		EDMONTON, ALBERTA			FORT MCMURRAY, ALBERTA			FREDERICTON, NEW BRUNSWICK			GATINEAU, QUEBEC			GRANBY, QUEBEC			HALIFAX, NOVA SCOTIA		
		MAT.	INST.	TOTAL	MAT.	INST.	TOTAL	MAT.	INST.	TOTAL	MAT.	INST.	TOTAL	MAT.	INST.	TOTAL	MAT.	INST.	TOTAL
015433	CONTRACTOR EQUIPMENT		123.6	123.6		100.2	100.2		115.0	115.0		98.1	98.1		98.1	98.1		112.9	112.9
0241, 31 - 34	SITE & INFRASTRUCTURE, DEMOLITION	126.6	114.2	118.1	120.8	92.5	101.4	116.1	102.5	106.8	96.0	90.6	92.3	96.4	90.6	92.4	101.8	102.3	102.1
0310	Concrete Forming & Accessories	128.7	92.0	97.4	126.2	85.9	91.9	128.1	57.7	68.2	136.3	77.7	86.4	136.3	77.6	86.4	125.4	85.1	91.1
0320	Concrete Reinforcing	133.5	78.7	107.1	148.7	78.6	114.9	138.3	56.2	98.7	144.7	70.5	108.9	144.7	70.5	108.9	153.3	75.3	115.7
0330	Cast-in-Place Concrete	143.8	101.3	127.8	171.2	95.1	142.5	111.1	56.8	90.6	102.7	85.4	96.1	106.1	85.3	98.3	97.9	82.3	92.0
03	CONCRETE	141.3	93.7	119.9	146.3	88.3	120.3	127.7	58.6	96.7	114.5	79.7	98.9	116.1	79.7	99.7	122.5	83.4	105.0
04	MASONRY	230.8	86.3	142.7	214.4	83.6	134.6	187.9	57.6	108.4	168.9	74.8	111.5	169.2	74.8	111.6	184.4	83.7	122.9
05	METALS	131.9	98.3	121.4	138.5	89.6	123.2	133.8	81.1	117.4	109.7	83.9	101.6	109.9	83.8	101.8	133.5	95.0	121.5
06	WOOD, PLASTICS & COMPOSITES	99.1	91.9	95.3	114.1	85.6	99.1	104.4	57.5	79.8	132.8	78.0	104.1	132.8	78.0	104.1	101.6	85.6	93.2
07	THERMAL & MOISTURE PROTECTION	141.3	94.8	121.3	128.8	89.4	111.9	135.5	58.3	102.4	113.9	81.9	100.2	113.9	80.5	99.5	137.8	85.9	115.5
08	OPENINGS	81.1	81.2	81.1	89.2	77.7	86.4	88.1	49.5	78.7	89.2	66.8	83.8	89.2	66.8	83.8	89.1	77.0	86.2
0920	Plaster & Gypsum Board	131.9	90.9	105.0	115.2	85.0	95.4	128.8	56.0	81.0	114.1	77.4	90.0	114.1	77.4	90.0	122.7	85.0	97.9
0950, 0980	Ceilings & Acoustic Treatment	141.7	90.9	109.9	111.7	85.0	95.0	132.7	56.0	84.6	102.7	77.4	86.9	102.7	77.4	86.9	129.1	85.0	101.4
0960	Flooring	126.3	81.3	113.1	115.3	81.3	105.4	120.7	64.6	104.3	115.3	83.8	106.1	115.3	83.8	106.1	113.9	78.1	103.4
0970, 0990	Wall Finishes & Painting/Coating	110.4	101.3	105.0	107.3	85.9	94.5	109.9	59.4	79.7	107.2	81.4	91.7	107.2	81.4	91.7	109.5	87.5	96.3
09	FINISHES	126.8	91.5	107.7	113.1	85.4	98.1	126.3	59.3	90.1	108.9	79.6	93.0	108.9	79.6	93.0	121.1	84.7	101.4
COVERS	DIVS. 10 - 14, 25, 28, 41, 43, 44, 46	131.2	89.6	121.4	131.2	86.5	120.7	131.2	56.8	113.7	131.2	73.1	117.5	131.2	73.1	117.5	131.2	65.6	115.7
21, 22, 23	FIRE SUPPRESSION, PLUMBING & HVAC	106.4	86.0	98.2	106.7	90.7	100.2	106.7	71.8	92.6	106.6	82.0	96.7	106.2	82.0	96.4	106.5	78.9	95.4
26, 27, 3370	ELECTRICAL, COMMUNICATIONS & UTIL.	123.4	90.4	107.1	108.5	76.3	92.6	122.1	68.5	95.6	113.2	64.2	89.0	113.9	64.2	89.3	123.7	87.2	105.6
MF2018	WEIGHTED AVERAGE	125.3	91.9	110.9	123.9	86.2	107.6	121.9	67.8	98.5	112.5	78.0	97.6	112.7	77.9	97.7	120.5	84.9	105.1

CANADA

DIVISION		HAMILTON, ONTARIO			HULL, QUEBEC			JOLIETTE, QUEBEC			KAMLOOPS, BRITISH COLUMBIA			KINGSTON, ONTARIO			KITCHENER, ONTARIO		
		MAT.	INST.	TOTAL	MAT.	INST.	TOTAL	MAT.	INST.	TOTAL	MAT.	INST.	TOTAL	MAT.	INST.	TOTAL	MAT.	INST.	TOTAL
015433	CONTRACTOR EQUIPMENT		110.1	110.1		98.1	98.1		98.1	98.1		101.3	101.3		100.1	100.1		99.4	99.4
0241, 31 - 34	SITE & INFRASTRUCTURE, DEMOLITION	105.3	100.8	102.2	96.0	90.6	92.3	96.5	90.6	92.5	119.7	94.1	102.1	114.3	95.0	101.0	93.6	96.1	95.3
0310	Concrete Forming & Accessories	132.8	91.0	97.2	136.3	77.7	86.4	136.3	77.8	86.5	123.0	80.8	87.1	126.5	81.1	87.8	118.4	84.7	89.7
0320	Concrete Reinforcing	134.4	98.4	117.0	144.7	70.5	108.9	136.9	70.5	104.8	107.3	73.8	91.1	160.7	83.0	123.2	94.7	98.2	96.4
0330	Cast-in-Place Concrete	106.1	97.9	103.0	102.7	85.4	96.1	106.9	85.4	98.8	92.6	89.6	91.5	116.0	90.7	106.4	110.3	93.3	103.9
03	CONCRETE	120.1	95.2	108.9	114.5	79.7	98.9	115.4	79.8	99.4	126.4	83.2	107.0	124.3	85.3	106.8	102.2	90.4	96.9
04	MASONRY	189.2	96.3	132.5	168.9	74.8	111.5	169.3	74.8	111.6	177.4	82.2	119.3	179.6	83.9	121.2	150.0	94.5	116.1
05	METALS	132.0	102.4	122.7	109.9	83.9	101.8	109.9	84.0	101.8	111.9	86.2	103.9	112.6	89.9	105.6	122.8	96.1	114.5
06	WOOD, PLASTICS & COMPOSITES	110.7	90.7	100.2	132.8	78.0	104.1	132.8	78.0	104.1	100.5	79.6	89.5	118.2	80.6	98.6	109.9	83.0	95.8
07	THERMAL & MOISTURE PROTECTION	132.8	97.6	117.7	113.9	81.9	100.1	113.9	81.9	100.1	129.8	79.5	108.2	119.9	85.0	104.9	115.3	94.3	106.3
08	OPENINGS	85.5	89.8	86.5	89.2	66.8	83.8	89.2	71.1	84.8	86.3	77.2	84.1	89.2	79.0	86.7	80.3	83.9	81.2
0920	Plaster & Gypsum Board	124.4	90.0	101.7	114.1	77.4	90.0	143.9	77.4	100.3	100.1	78.8	86.2	175.4	80.3	113.0	104.8	82.7	90.3
0950, 0980	Ceilings & Acoustic Treatment	137.4	90.0	107.7	102.7	77.4	86.9	102.7	77.4	86.9	102.7	78.8	87.8	119.3	80.3	94.8	106.7	82.7	91.7
0960	Flooring	121.1	91.0	112.3	115.3	83.8	106.1	115.3	83.8	106.1	114.8	48.0	95.3	115.3	83.9	106.1	102.9	91.1	99.5
0970, 0990	Wall Finishes & Painting/Coating	107.2	99.2	102.4	107.2	81.4	91.7	107.2	81.4	91.7	107.2	74.4	87.5	107.2	77.1	89.2	101.7	89.8	94.6
09	FINISHES	123.1	91.8	106.1	108.9	79.6	93.0	113.1	79.6	95.0	109.3	74.3	90.4	122.2	81.3	100.1	103.2	85.9	93.9
COVERS	DIVS. 10 - 14, 25, 28, 41, 43, 44, 46	131.2	87.6	120.9	131.2	73.1	117.5	131.2	73.1	117.5	131.2	81.6	119.5	131.2	62.4	115.0	131.2	85.2	120.3
21, 22, 23	FIRE SUPPRESSION, PLUMBING & HVAC	106.8	87.7	99.1	106.2	82.0	96.4	106.2	82.0	96.4	106.2	85.2	97.7	106.6	91.9	100.7	105.5	87.7	98.3
26, 27, 3370	ELECTRICAL, COMMUNICATIONS & UTIL.	115.5	99.8	107.7	115.3	64.2	90.0	113.9	64.2	89.3	117.3	73.0	95.4	114.6	80.8	97.9	111.3	97.5	104.5
MF2018	WEIGHTED AVERAGE	119.2	94.5	108.6	112.6	78.0	97.7	113.0	78.2	97.9	115.6	81.6	100.9	116.5	85.5	103.1	109.9	91.2	101.8

For customer support on your Commercial Renovation Costs with RSMeans Data, call 800.448.8182.

831

CANADA

DIVISION		LAVAL, QUEBEC			LETHBRIDGE, ALBERTA			LLOYDMINSTER, ALBERTA			LONDON, ONTARIO			MEDICINE HAT, ALBERTA			MONCTON, NEW BRUNSWICK		
		MAT.	INST.	TOTAL	MAT.	INST.	TOTAL	MAT.	INST.	TOTAL	MAT.	INST.	TOTAL	MAT.	INST.	TOTAL	MAT.	INST.	TOTAL
015433	CONTRACTOR EQUIPMENT		98.1	98.1		100.2	100.2		100.2	100.2		113.0	113.0		100.2	100.2		97.4	97.4
0241, 31 - 34	SITE & INFRASTRUCTURE, DEMOLITION	96.4	90.7	92.5	113.8	93.1	99.6	113.8	92.5	99.2	104.4	101.5	102.4	112.6	92.6	98.8	104.1	89.2	93.8
0310	Concrete Forming & Accessories	136.5	78.4	87.0	128.1	86.0	92.3	125.8	76.9	84.2	131.9	85.8	92.7	128.0	76.8	84.4	104.9	63.8	70.0
0320	Concrete Reinforcing	144.7	71.2	109.2	148.7	78.6	114.9	148.7	78.5	114.9	124.4	97.2	111.3	148.7	78.5	114.9	137.9	66.7	103.6
0330	Cast-in-Place Concrete	106.1	86.0	98.5	128.4	95.1	115.9	119.2	91.7	108.8	118.5	97.3	110.5	119.2	91.6	108.8	108.4	65.3	92.1
03	CONCRETE	116.1	80.3	100.0	126.5	88.3	109.4	122.0	83.0	104.5	124.5	92.6	110.1	122.2	82.9	104.6	114.6	66.0	92.8
04	MASONRY	169.2	75.4	112.0	188.1	83.6	124.4	168.3	77.6	113.0	200.2	95.4	136.3	168.3	77.6	113.0	164.6	71.7	107.9
05	METALS	109.8	84.2	101.8	132.7	89.6	119.3	111.9	89.4	104.9	132.0	104.1	123.3	112.1	89.4	105.0	113.4	85.0	104.5
06	WOOD, PLASTICS & COMPOSITES	132.9	78.6	104.5	117.9	85.6	101.0	114.1	76.4	94.4	113.7	83.9	98.1	117.9	76.4	96.2	95.0	63.0	78.3
07	THERMAL & MOISTURE PROTECTION	114.7	82.4	100.8	126.1	89.4	110.4	122.4	85.2	106.4	128.3	95.3	114.1	128.5	85.2	109.9	114.6	69.3	95.2
08	OPENINGS	89.2	67.5	83.9	89.2	77.7	86.4	89.2	72.7	85.2	81.8	85.0	82.6	89.2	72.7	85.2	84.2	59.8	78.3
0920	Plaster & Gypsum Board	114.4	78.1	90.6	106.2	85.0	92.3	101.6	75.6	84.6	128.9	82.9	98.7	103.9	75.6	85.4	119.4	62.2	81.9
0950, 0980	Ceilings & Acoustic Treatment	102.7	78.1	87.3	111.7	85.0	95.0	102.7	75.6	85.8	135.0	82.9	102.4	102.7	75.6	85.8	115.9	62.2	82.2
0960	Flooring	115.3	84.7	106.4	115.3	81.3	105.4	115.3	81.3	105.4	115.7	91.1	108.5	115.3	81.3	105.4	101.2	63.3	90.1
0970, 0990	Wall Finishes & Painting/Coating	107.2	82.1	92.2	107.2	93.8	99.2	107.3	73.1	86.8	109.6	96.1	101.5	107.2	73.1	86.8	108.6	79.9	91.4
09	FINISHES	108.9	80.2	93.4	111.0	86.3	97.6	108.3	77.1	91.4	123.5	87.5	104.0	108.5	77.1	91.5	109.1	65.1	85.3
COVERS	DIVS. 10 - 14, 25, 28, 41, 43, 44, 46	131.2	73.7	117.6	131.2	85.3	120.4	131.2	83.5	119.9	131.2	87.1	120.8	131.2	82.3	119.7	131.2	58.2	114.0
21, 22, 23	FIRE SUPPRESSION, PLUMBING & HVAC	105.8	82.7	96.5	106.6	87.6	98.9	106.6	87.6	98.9	106.6	86.3	98.4	106.2	84.5	97.4	106.2	70.5	91.8
26, 27, 3370	ELECTRICAL, COMMUNICATIONS & UTIL.	114.3	64.8	89.8	110.1	76.3	93.4	107.7	76.3	92.2	110.1	98.2	104.2	107.7	76.3	92.2	120.9	85.5	103.4
MF2018	WEIGHTED AVERAGE	112.6	78.6	97.9	118.9	85.7	104.6	113.6	82.6	100.2	119.2	92.9	107.8	113.7	81.9	100.0	113.0	73.2	95.8

CANADA

DIVISION		MONTREAL, QUEBEC			MOOSE JAW, SASKATCHEWAN			NEW GLASGOW, NOVA SCOTIA			NEWCASTLE, NEW BRUNSWICK			NORTH BAY, ONTARIO			OSHAWA, ONTARIO		
		MAT.	INST.	TOTAL	MAT.	INST.	TOTAL	MAT.	INST.	TOTAL	MAT.	INST.	TOTAL	MAT.	INST.	TOTAL	MAT.	INST.	TOTAL
015433	CONTRACTOR EQUIPMENT		116.2	116.2		96.5	96.5		96.8	96.8		97.4	97.4		97.4	97.4		99.4	99.4
0241, 31 - 34	SITE & INFRASTRUCTURE, DEMOLITION	114.3	102.2	106.0	113.6	87.1	95.4	109.9	88.9	95.5	104.6	87.4	92.8	125.0	90.6	101.4	103.8	96.3	98.6
0310	Concrete Forming & Accessories	132.9	90.6	96.9	108.5	53.6	61.8	111.0	66.7	73.3	104.9	57.0	64.1	150.5	78.6	89.4	124.9	88.2	93.6
0320	Concrete Reinforcing	124.6	93.8	109.8	105.0	59.8	83.1	153.1	46.4	101.6	137.9	56.0	98.4	181.1	82.6	133.5	150.0	99.2	125.5
0330	Cast-in-Place Concrete	122.7	97.5	113.2	115.6	62.0	95.4	121.3	65.0	100.0	112.6	54.6	90.7	116.8	78.2	102.3	127.5	102.1	118.0
03	CONCRETE	126.5	94.2	112.0	106.5	58.7	85.0	139.6	63.6	105.5	116.6	57.0	89.8	141.6	79.8	113.8	124.7	95.2	111.5
04	MASONRY	185.6	89.9	127.2	166.9	55.2	98.8	215.9	64.2	123.4	164.9	56.2	98.6	222.8	80.2	135.8	152.8	97.7	119.2
05	METALS	142.7	101.5	129.8	108.6	74.0	97.8	124.4	74.6	108.8	113.4	72.5	100.6	125.2	89.4	114.0	113.1	96.8	108.0
06	WOOD, PLASTICS & COMPOSITES	109.7	91.4	100.1	97.1	52.4	73.7	112.1	66.4	88.2	95.0	57.0	75.2	154.0	79.2	114.8	118.1	86.7	101.7
07	THERMAL & MOISTURE PROTECTION	126.1	97.1	113.7	113.1	58.4	89.6	128.1	65.8	101.3	114.6	56.8	89.8	134.2	80.5	111.1	116.2	99.6	109.1
08	OPENINGS	87.6	78.6	85.4	85.5	50.5	77.0	91.6	60.3	83.9	84.2	50.3	75.9	98.4	77.0	93.2	85.1	87.2	85.6
0920	Plaster & Gypsum Board	126.6	90.5	102.9	97.5	51.3	67.2	141.8	65.7	91.9	119.4	56.0	77.8	138.3	78.8	99.3	108.5	86.6	94.1
0950, 0980	Ceilings & Acoustic Treatment	139.7	90.5	108.9	102.7	51.3	70.5	122.1	65.7	86.8	115.9	56.0	78.4	122.1	78.8	95.0	102.2	86.6	92.4
0960	Flooring	118.9	86.8	109.5	105.0	52.3	89.6	108.9	57.4	93.9	101.2	61.9	89.7	133.4	83.9	119.0	105.9	93.2	102.2
0970, 0990	Wall Finishes & Painting/Coating	115.1	103.3	108.0	107.2	59.4	78.6	114.3	57.7	80.4	108.6	46.6	71.5	114.3	82.8	95.4	101.7	103.0	102.5
09	FINISHES	124.7	92.2	107.1	104.6	53.7	77.0	119.9	64.2	89.8	109.1	57.1	80.9	127.2	80.4	101.8	104.1	90.2	96.6
COVERS	DIVS. 10 - 14, 25, 28, 41, 43, 44, 46	131.2	79.8	119.1	131.2	55.5	113.4	131.2	58.7	114.1	131.2	55.8	113.4	131.2	61.2	114.7	131.2	85.9	120.5
21, 22, 23	FIRE SUPPRESSION, PLUMBING & HVAC	106.8	80.5	96.2	106.6	68.4	91.2	106.5	77.4	94.7	106.2	63.0	88.8	106.5	90.0	99.8	105.5	87.8	98.3
26, 27, 3370	ELECTRICAL, COMMUNICATIONS & UTIL.	118.0	84.3	101.4	116.3	55.4	86.2	113.9	57.7	86.1	116.4	55.0	86.0	115.2	82.0	98.7	112.2	100.3	106.3
MF2018	WEIGHTED AVERAGE	122.1	89.4	108.0	111.0	62.4	90.0	121.9	68.6	98.9	112.8	61.4	90.5	125.0	83.4	107.0	112.4	93.6	104.3

CANADA

DIVISION		OTTAWA, ONTARIO			OWEN SOUND, ONTARIO			PETERBOROUGH, ONTARIO			PORTAGE LA PRAIRIE, MANITOBA			PRINCE ALBERT, SASKATCHEWAN			PRINCE GEORGE, BRITISH COLUMBIA		
		MAT.	INST.	TOTAL	MAT.	INST.	TOTAL	MAT.	INST.	TOTAL	MAT.	INST.	TOTAL	MAT.	INST.	TOTAL	MAT.	INST.	TOTAL
015433	CONTRACTOR EQUIPMENT		112.7	112.7		97.9	97.9		97.9	97.9		100.1	100.1		96.5	96.5		101.3	101.3
0241, 31 - 34	SITE & INFRASTRUCTURE, DEMOLITION	108.0	101.0	103.2	116.9	91.5	99.4	116.1	91.3	99.1	114.6	90.2	97.8	109.6	87.2	94.2	122.9	94.1	103.1
0310	Concrete Forming & Accessories	126.4	85.0	91.2	125.7	77.3	84.5	129.0	79.6	87.0	128.2	64.5	74.0	108.5	53.4	61.6	114.6	76.1	81.9
0320	Concrete Reinforcing	134.5	97.2	116.5	171.9	84.6	129.7	160.7	83.1	123.2	148.7	53.1	102.6	109.6	59.7	85.5	107.3	73.8	91.1
0330	Cast-in-Place Concrete	120.6	99.1	112.5	155.6	74.7	125.1	129.0	79.8	110.4	119.2	67.9	99.8	104.8	61.9	88.6	136.7	81.1	111.8
03	CONCRETE	126.4	92.9	111.4	141.5	78.3	113.1	128.4	80.9	107.1	115.1	64.8	92.5	102.1	58.5	82.5	136.7	81.1	111.8
04	MASONRY	189.1	95.7	132.1	169.0	85.3	117.9	173.0	86.2	120.1	171.7	58.5	102.7	166.2	55.2	98.5	179.3	82.2	120.1
05	METALS	132.7	105.0	124.0	111.5	90.4	104.9	111.4	90.1	104.7	112.1	77.4	101.3	108.7	73.8	97.8	111.9	86.3	103.9
06	WOOD, PLASTICS & COMPOSITES	106.3	82.7	94.0	116.6	76.2	95.5	120.4	78.0	98.2	117.9	65.9	90.7	97.1	52.4	73.7	100.5	73.1	86.1
07	THERMAL & MOISTURE PROTECTION	141.4	95.9	121.9	114.2	81.3	100.0	120.1	85.8	105.4	113.6	66.4	93.4	113.0	57.4	89.1	123.8	78.8	104.5
08	OPENINGS	90.5	84.3	89.0	90.4	76.4	87.0	87.7	78.6	85.5	89.2	58.9	81.8	84.5	50.5	76.2	86.3	73.7	83.2
0920	Plaster & Gypsum Board	124.0	81.8	96.3	151.7	75.8	101.9	114.7	77.6	90.4	103.6	64.8	78.2	97.5	51.3	67.2	100.1	72.2	81.8
0950, 0980	Ceilings & Acoustic Treatment	145.7	81.8	105.6	95.4	75.8	83.1	102.7	77.6	87.0	102.7	64.8	79.0	102.7	51.3	70.5	102.7	65.7	79.5
0960	Flooring	114.4	86.8	106.4	121.5	85.1	110.8	115.3	83.9	106.1	115.3	60.4	99.3	105.0	52.3	89.6	110.5	65.7	97.4
0970, 0990	Wall Finishes & Painting/Coating	109.6	91.3	98.6	102.9	81.6	90.1	107.2	84.8	93.8	107.3	52.6	74.6	107.2	50.6	73.3	107.2	74.4	87.5
09	FINISHES	126.0	85.5	104.1	112.9	79.3	94.7	109.7	81.0	94.2	108.4	62.8	83.8	104.6	52.7	76.5	108.1	73.6	89.4
COVERS	DIVS. 10 - 14, 25, 28, 41, 43, 44, 46	131.2	85.0	120.3	139.3	61.7	121.0	131.2	62.6	115.0	131.2	57.7	113.9	131.2	55.5	113.4	131.2	80.9	119.3
21, 22, 23	FIRE SUPPRESSION, PLUMBING & HVAC	106.7	87.9	99.1	106.2	89.9	99.7	106.2	93.2	101.0	106.2	76.3	94.1	106.6	61.7	88.5	106.2	85.2	97.7
26, 27, 3370	ELECTRICAL, COMMUNICATIONS & UTIL.	111.3	98.1	104.8	118.5	80.8	99.9	114.5	81.6	98.2	116.0	53.6	85.1	116.3	55.4	86.2	114.5	73.0	94.0
MF2018	WEIGHTED AVERAGE	120.6	93.0	108.7	118.1	83.5	103.2	115.2	85.2	102.3	113.6	67.5	93.6	110.2	60.7	88.8	116.5	81.0	101.2

CANADA

DIVISION		QUEBEC CITY, QUEBEC			RED DEER, ALBERTA			REGINA, SASKATCHEWAN			RIMOUSKI, QUEBEC			ROUYN-NORANDA, QUEBEC			SAINT HYACINTHE, QUEBEC		
		MAT.	INST.	TOTAL	MAT.	INST.	TOTAL	MAT.	INST.	TOTAL	MAT.	INST.	TOTAL	MAT.	INST.	TOTAL	MAT.	INST.	TOTAL
015433	CONTRACTOR EQUIPMENT		117.5	117.5		100.2	100.2		124.0	124.0		98.1	98.1		98.1	98.1		98.1	98.1
0241, 31 - 34	SITE & INFRASTRUCTURE, DEMOLITION	113.4	102.2	105.7	112.6	92.6	98.8	126.0	113.0	117.1	96.3	90.6	92.4	96.0	90.6	92.3	96.4	90.6	92.4
0310	Concrete Forming & Accessories	129.6	90.7	96.5	140.1	76.8	86.3	131.7	88.2	94.7	136.3	89.8	96.8	136.3	77.7	86.4	136.3	77.7	86.4
0320	Concrete Reinforcing	121.9	93.9	108.4	148.7	78.5	114.9	121.5	84.6	103.7	103.3	93.7	98.7	144.7	70.5	108.9	144.7	70.5	108.9
0330	Cast-in-Place Concrete	129.8	97.6	117.6	119.2	91.6	108.8	151.6	94.6	130.1	108.0	93.8	102.7	102.7	85.4	96.1	106.1	85.4	98.3
03	CONCRETE	129.2	94.4	113.6	122.9	82.9	105.0	141.3	90.6	118.5	111.4	92.0	102.7	114.5	79.7	98.9	116.1	79.7	99.8
04	MASONRY	189.0	89.9	128.5	168.3	77.6	113.0	221.2	83.1	137.0	168.7	89.8	120.6	168.9	74.8	111.5	169.2	74.8	111.6
05	METALS	136.5	103.3	126.2	112.1	89.4	105.0	137.4	97.7	125.0	109.4	91.6	103.9	109.9	83.9	101.8	109.9	83.9	101.8
06	WOOD, PLASTICS & COMPOSITES	112.5	91.4	101.4	117.9	76.4	96.2	105.7	89.2	97.1	132.8	90.7	110.8	132.8	78.0	104.1	132.8	78.0	104.1
07	THERMAL & MOISTURE PROTECTION	123.0	97.1	111.9	138.2	85.2	115.5	148.5	83.4	120.5	113.9	95.6	106.0	113.9	81.9	100.1	114.4	81.9	100.4
08	OPENINGS	88.5	85.7	87.9	89.2	72.7	85.2	88.2	77.0	85.4	88.8	76.3	85.8	89.2	66.8	83.8	89.2	66.8	83.8
0920	Plaster & Gypsum Board	123.0	90.5	101.7	103.9	75.6	85.4	142.0	88.1	106.6	143.7	90.5	108.8	113.9	77.4	90.0	113.9	77.4	90.0
0950, 0980	Ceilings & Acoustic Treatment	134.2	90.5	106.8	102.7	75.6	85.8	156.4	88.1	113.6	102.1	90.5	94.8	102.1	77.4	86.6	102.1	77.4	86.6
0960	Flooring	115.6	86.8	107.2	118.3	81.3	107.5	125.4	91.1	115.4	116.4	83.8	106.9	115.3	83.8	106.1	115.3	83.8	106.1
0970, 0990	Wall Finishes & Painting/Coating	115.9	103.3	108.3	107.2	73.1	86.8	109.6	85.1	95.0	110.0	103.3	105.9	107.2	81.4	91.7	107.2	81.4	91.7
09	FINISHES	123.5	92.2	106.6	109.4	77.1	91.9	134.2	89.4	109.9	113.5	91.2	101.4	108.7	79.6	93.0	108.7	79.6	93.0
COVERS	DIVS. 10 - 14, 25, 28, 41, 43, 44, 46	131.2	79.8	119.1	131.2	82.3	119.7	131.2	67.6	116.2	131.2	78.3	118.7	131.2	73.1	117.5	131.2	73.1	117.5
21, 22, 23	FIRE SUPPRESSION, PLUMBING & HVAC	106.5	80.5	96.0	106.2	84.5	97.4	106.4	82.8	96.9	106.2	80.3	95.8	106.2	82.0	96.4	102.4	82.0	94.1
26, 27, 3370	ELECTRICAL, COMMUNICATIONS & UTIL.	124.2	84.3	104.5	107.7	76.3	92.2	121.0	88.1	104.7	113.9	84.3	99.2	113.9	64.2	89.3	114.5	64.2	89.6
MF2018	WEIGHTED AVERAGE	122.2	89.9	108.2	114.2	81.9	100.2	127.0	88.5	110.3	112.3	86.9	101.4	112.4	78.0	97.6	111.8	78.0	97.2

CANADA

DIVISION		SAINT JOHN, NEW BRUNSWICK			SARNIA, ONTARIO			SASKATOON, SASKATCHEWAN			SAULT STE MARIE, ONTARIO			SHERBROOKE, QUEBEC			SOREL, QUEBEC		
		MAT.	INST.	TOTAL	MAT.	INST.	TOTAL	MAT.	INST.	TOTAL	MAT.	INST.	TOTAL	MAT.	INST.	TOTAL	MAT.	INST.	TOTAL
015433	CONTRACTOR EQUIPMENT		97.4	97.4		97.9	97.9		97.0	97.0		97.9	97.9		98.1	98.1		98.1	98.1
0241, 31 - 34	SITE & INFRASTRUCTURE, DEMOLITION	104.8	89.2	94.1	114.7	91.4	98.7	111.8	90.0	96.8	105.1	91.0	95.4	96.4	90.6	92.4	96.5	90.6	92.5
0310	Concrete Forming & Accessories	128.1	61.3	71.2	127.5	86.1	92.3	108.3	87.3	90.4	115.0	83.4	88.1	136.3	77.7	86.4	136.3	77.8	86.5
0320	Concrete Reinforcing	137.9	66.7	103.6	114.6	84.4	100.1	112.2	84.4	98.8	103.1	83.2	93.5	144.7	70.5	108.9	136.9	70.5	104.8
0330	Cast-in-Place Concrete	110.9	65.3	93.7	119.0	92.1	108.8	112.0	89.5	103.5	106.9	78.5	96.2	106.1	85.4	98.3	106.9	85.4	98.8
03	CONCRETE	117.5	64.8	93.8	117.5	88.3	104.4	106.6	87.8	98.1	102.6	82.3	93.5	116.1	79.7	99.8	115.4	79.8	99.4
04	MASONRY	188.3	71.5	117.1	185.9	88.7	126.6	179.7	83.0	120.7	169.9	89.2	120.7	169.2	74.8	111.6	169.3	74.8	111.6
05	METALS	113.3	84.9	104.5	111.3	90.5	104.8	105.4	87.1	99.7	110.5	93.9	105.3	109.7	83.9	101.6	109.9	84.0	101.8
06	WOOD, PLASTICS & COMPOSITES	120.8	59.5	88.7	119.3	85.6	101.7	93.8	88.3	90.9	105.1	84.9	94.5	132.8	78.0	104.1	132.8	78.0	104.1
07	THERMAL & MOISTURE PROTECTION	115.0	69.2	95.3	120.2	89.1	106.8	116.7	81.0	101.4	118.9	85.1	104.4	113.9	81.9	100.2	113.9	81.9	100.1
08	OPENINGS	84.1	56.9	77.5	90.7	82.2	88.6	85.5	76.5	83.3	82.5	83.3	82.7	89.2	66.8	83.8	89.2	71.1	84.8
0920	Plaster & Gypsum Board	133.4	58.5	84.3	139.9	85.4	104.1	116.5	88.1	97.9	106.4	84.7	92.1	113.9	77.4	90.0	143.7	77.4	100.2
0950, 0980	Ceilings & Acoustic Treatment	120.4	58.5	81.6	107.8	85.4	93.8	124.2	88.1	101.6	102.7	84.7	91.4	102.1	77.4	86.6	102.1	77.4	86.6
0960	Flooring	113.3	63.3	98.7	115.3	91.8	108.4	108.0	91.1	103.1	107.9	88.7	102.3	115.3	83.8	106.1	115.3	83.8	106.1
0970, 0990	Wall Finishes & Painting/Coating	108.6	79.9	91.4	107.2	95.9	100.4	110.3	85.1	95.2	107.2	88.9	96.2	107.2	81.4	91.7	107.2	81.4	91.7
09	FINISHES	115.6	63.1	87.2	114.5	88.5	100.5	114.4	88.7	100.5	105.7	85.3	94.6	108.7	79.6	93.0	112.9	79.6	94.9
COVERS	DIVS. 10 - 14, 25, 28, 41, 43, 44, 46	131.2	58.1	114.0	131.2	63.9	115.3	131.2	65.9	115.8	131.2	83.8	120.0	131.2	73.1	117.5	131.2	73.1	117.5
21, 22, 23	FIRE SUPPRESSION, PLUMBING & HVAC	106.2	72.0	92.4	106.2	98.9	103.2	106.0	82.6	96.5	106.2	87.8	98.8	106.6	82.0	96.7	106.2	82.0	96.4
26, 27, 3370	ELECTRICAL, COMMUNICATIONS & UTIL.	123.5	85.5	104.7	117.6	83.6	100.8	118.6	88.0	103.5	116.2	82.0	99.3	113.9	64.2	89.3	113.9	64.2	89.3
MF2018	WEIGHTED AVERAGE	115.6	72.9	97.1	115.5	89.4	104.2	112.0	85.1	100.4	110.6	86.4	100.1	112.8	78.0	97.7	112.9	78.2	97.9

CANADA

DIVISION		ST. CATHARINES, ONTARIO			ST JEROME, QUEBEC			ST. JOHN'S, NEWFOUNDLAND			SUDBURY, ONTARIO			SUMMERSIDE, PRINCE EDWARD ISLAND			SYDNEY, NOVA SCOTIA		
		MAT.	INST.	TOTAL	MAT.	INST.	TOTAL	MAT.	INST.	TOTAL	MAT.	INST.	TOTAL	MAT.	INST.	TOTAL	MAT.	INST.	TOTAL
015433	CONTRACTOR EQUIPMENT		97.3	97.3		98.1	98.1		120.5	120.5		97.3	97.3		96.8	96.8		96.8	96.8
0241, 31 - 34	SITE & INFRASTRUCTURE, DEMOLITION	93.8	92.8	93.1	96.0	90.6	92.3	113.1	111.0	111.6	93.9	92.4	92.8	119.4	86.4	96.7	106.1	88.9	94.3
0310	Concrete Forming & Accessories	116.1	90.6	94.4	136.3	77.7	86.4	127.4	83.3	89.8	111.3	85.6	89.4	111.3	51.5	60.4	111.0	66.7	73.3
0320	Concrete Reinforcing	95.5	98.3	96.8	144.7	70.5	108.9	168.7	80.9	126.4	96.2	96.8	96.5	151.1	45.2	100.0	153.1	46.4	101.6
0330	Cast-in-Place Concrete	105.3	95.0	101.4	102.7	85.4	96.1	135.2	96.3	120.5	106.2	91.5	100.7	115.2	53.1	91.7	93.6	65.0	82.8
03	CONCRETE	99.9	93.7	97.1	114.5	79.7	98.9	143.9	88.2	118.9	100.1	89.9	95.5	149.5	52.1	105.8	126.7	63.6	98.4
04	MASONRY	149.6	97.0	117.5	168.9	74.8	111.5	212.1	88.8	136.9	149.7	93.2	115.2	215.2	53.7	116.7	213.6	64.2	122.5
05	METALS	113.0	96.0	107.7	109.9	83.9	101.8	139.5	95.2	125.7	112.5	95.4	107.1	124.4	68.0	106.8	124.4	74.6	108.8
06	WOOD, PLASTICS & COMPOSITES	107.2	90.4	98.4	132.8	78.0	104.1	105.1	81.6	92.8	102.4	84.9	93.2	112.8	51.1	80.5	112.1	66.4	88.2
07	THERMAL & MOISTURE PROTECTION	115.3	97.6	107.7	113.9	81.9	100.1	137.8	92.7	118.5	114.5	92.6	105.1	127.6	55.6	96.7	128.1	65.8	101.3
08	OPENINGS	79.8	88.3	81.9	89.2	66.8	83.8	85.6	72.7	82.5	80.4	83.7	81.2	102.3	44.7	88.3	91.6	60.3	83.9
0920	Plaster & Gypsum Board	98.8	90.3	93.2	113.9	77.4	90.0	146.3	80.6	103.2	97.6	84.7	89.1	143.0	49.9	81.9	141.8	65.7	91.9
0950, 0980	Ceilings & Acoustic Treatment	102.2	90.3	94.8	102.1	77.4	86.6	144.0	80.6	104.3	97.1	84.7	89.3	122.1	49.9	76.9	122.1	65.7	86.8
0960	Flooring	101.7	88.1	97.8	115.3	83.8	106.1	118.7	50.0	98.7	99.9	88.7	96.6	109.0	53.9	93.9	102.9	57.3	93.9
0970, 0990	Wall Finishes & Painting/Coating	101.7	99.6	100.4	107.2	81.4	91.7	111.8	96.7	102.8	101.7	90.7	95.2	114.3	38.8	69.0	114.3	57.7	80.4
09	FINISHES	101.0	91.2	95.7	108.7	79.6	93.0	131.0	78.4	102.5	99.0	86.5	92.2	121.1	50.8	83.1	119.9	64.2	89.8
COVERS	DIVS. 10 - 14, 25, 28, 41, 43, 44, 46	131.2	66.2	115.9	131.2	73.1	117.5	131.2	65.4	115.7	131.2	85.1	120.3	131.2	55.0	113.2	131.2	58.7	114.1
21, 22, 23	FIRE SUPPRESSION, PLUMBING & HVAC	105.5	87.1	98.1	106.2	82.0	96.4	106.3	81.4	96.3	105.9	85.4	97.7	106.5	57.3	86.6	106.5	77.4	94.7
26, 27, 3370	ELECTRICAL, COMMUNICATIONS & UTIL.	112.8	98.5	105.7	114.6	64.2	89.6	122.9	78.3	100.8	110.7	98.3	104.6	112.8	46.7	80.1	113.9	57.7	86.1
MF2018	WEIGHTED AVERAGE	108.0	92.1	101.1	112.5	78.0	97.6	126.2	85.2	108.5	107.7	90.3	100.2	124.4	56.4	95.0	120.1	68.6	97.9

For customer support on your Commercial Renovation Costs with RSMeans Data, call 800.448.8182.

833

		CANADA																
DIVISION		THUNDER BAY, ONTARIO			TIMMINS, ONTARIO			TORONTO, ONTARIO			TROIS RIVIERES, QUEBEC			TRURO, NOVA SCOTIA			VANCOUVER, BRITISH COLUMBIA		
		MAT.	INST.	TOTAL	MAT.	INST.	TOTAL	MAT.	INST.	TOTAL	MAT.	INST.	TOTAL	MAT.	INST.	TOTAL	MAT.	INST.	TOTAL
015433	CONTRACTOR EQUIPMENT		97.3	97.3		97.9	97.9		112.2	112.2		97.6	97.6		97.3	97.3		130.1	130.1
0241, 31 - 34	SITE & INFRASTRUCTURE, DEMOLITION	98.3	92.7	94.4	116.1	91.0	98.8	107.8	101.6	103.6	106.3	90.3	95.3	100.4	89.3	92.7	115.2	116.1	115.8
0310	Concrete Forming & Accessories	124.9	89.4	94.7	129.0	78.7	86.2	131.3	98.5	103.4	160.0	77.8	90.1	97.4	66.7	71.3	132.4	89.3	95.7
0320	Concrete Reinforcing	85.3	98.1	91.5	160.7	82.6	123.0	134.4	100.9	118.2	153.1	70.5	113.3	136.9	46.4	93.2	130.1	91.4	111.4
0330	Cast-in-Place Concrete	116.0	94.3	107.8	129.0	78.3	109.8	106.9	109.2	107.8	96.9	85.3	92.5	133.9	65.1	107.9	121.8	98.7	113.0
03	CONCRETE	107.6	92.8	101.0	128.4	79.8	106.6	120.3	103.1	112.6	129.2	79.7	107.0	127.1	63.6	98.6	130.1	94.0	113.9
04	MASONRY	150.2	96.7	117.6	173.0	80.2	116.4	189.2	104.2	137.4	218.0	74.8	130.6	168.0	64.2	104.7	180.3	93.1	127.1
05	METALS	113.0	95.2	107.5	111.4	89.7	104.6	132.5	106.3	124.3	123.6	83.7	111.2	110.6	74.9	99.5	132.2	106.9	124.3
06	WOOD, PLASTICS & COMPOSITES	118.1	88.6	102.7	120.4	79.2	98.9	113.5	97.0	104.9	170.1	77.9	121.9	86.3	66.5	75.9	108.4	88.8	98.1
07	THERMAL & MOISTURE PROTECTION	115.6	95.2	106.9	120.1	80.5	103.1	139.9	105.1	125.0	127.1	81.9	107.7	114.8	65.8	93.8	145.8	90.2	121.9
08	OPENINGS	79.1	86.9	81.0	87.7	77.0	85.1	84.1	95.7	87.0	100.2	71.1	93.1	82.7	60.3	77.3	84.7	86.0	85.0
0920	Plaster & Gypsum Board	123.9	88.5	100.7	114.7	78.8	91.2	123.9	96.6	106.0	171.2	77.4	109.7	120.4	65.7	84.5	135.1	87.5	103.9
0950, 0980	Ceilings & Acoustic Treatment	97.1	88.5	91.7	102.7	78.8	87.8	143.2	96.6	114.0	120.8	77.4	93.6	102.7	65.7	79.5	147.4	87.5	109.9
0960	Flooring	105.9	94.3	102.5	115.3	83.9	106.1	115.9	96.6	110.3	133.4	83.8	118.9	96.7	57.4	85.2	130.0	84.8	116.8
0970, 0990	Wall Finishes & Painting/Coating	101.7	92.0	95.9	107.2	82.8	92.6	106.0	103.0	104.2	114.3	81.4	94.5	107.2	57.7	77.6	105.8	95.5	99.6
09	FINISHES	104.8	90.5	97.1	109.7	80.4	93.8	123.6	98.7	110.1	130.7	79.6	103.0	104.5	64.3	82.8	131.7	89.0	108.6
COVERS	DIVS. 10 - 14, 25, 28, 41, 43, 44, 46	131.2	66.3	115.9	131.2	61.3	114.7	131.2	90.4	121.6	131.2	73.0	117.5	131.2	58.7	114.1	131.2	86.6	120.7
21, 22, 23	FIRE SUPPRESSION, PLUMBING & HVAC	105.5	87.2	98.1	106.2	90.0	99.7	106.8	95.0	102.0	106.5	82.0	96.6	106.2	77.4	94.6	106.8	86.4	98.6
26, 27, 3370	ELECTRICAL, COMMUNICATIONS & UTIL.	111.3	97.8	104.6	116.2	82.0	99.3	113.7	100.0	106.9	114.2	64.2	89.5	113.6	57.7	85.9	118.8	79.5	99.4
MF2018	WEIGHTED AVERAGE	109.3	91.6	101.6	115.4	83.4	101.6	119.3	99.9	111.0	122.8	78.1	103.5	112.8	68.7	93.7	121.7	91.7	108.7

		CANADA																	
DIVISION		VICTORIA, BRITISH COLUMBIA			WHITEHORSE, YUKON			WINDSOR, ONTARIO			WINNIPEG, MANITOBA			YARMOUTH, NOVA SCOTIA			YELLOWKNIFE, NWT		
		MAT.	INST.	TOTAL	MAT.	INST.	TOTAL	MAT.	INST.	TOTAL	MAT.	INST.	TOTAL	MAT.	INST.	TOTAL	MAT.	INST.	TOTAL
015433	CONTRACTOR EQUIPMENT		104.8	104.8		139.6	139.6		97.3	97.3		122.1	122.1		96.8	96.8		126.5	126.5
0241, 31 - 34	SITE & INFRASTRUCTURE, DEMOLITION	123.6	99.0	106.7	134.2	120.5	124.8	90.4	92.7	92.0	112.7	109.6	110.6	111.0	66.7	73.3	138.9	74.7	84.3
0310	Concrete Forming & Accessories	114.0	86.6	90.7	136.3	56.3	68.2	124.9	87.2	92.8	131.3	63.3	73.4	153.1	46.4	101.6	136.0	64.1	101.3
0320	Concrete Reinforcing	109.7	91.2	100.8	164.6	61.7	114.9	93.6	97.1	95.3	125.0	58.4	92.9	153.1	46.4	101.6	136.0	64.1	101.3
0330	Cast-in-Place Concrete	116.0	94.5	107.9	150.5	72.3	121.0	107.9	95.7	103.3	141.6	70.1	114.6	120.0	65.0	99.2	172.6	86.5	140.1
03	CONCRETE	138.8	90.4	117.1	156.1	64.8	115.2	101.4	92.2	97.2	137.1	66.2	105.3	139.0	63.6	105.1	162.8	78.1	124.8
04	MASONRY	182.8	93.0	128.0	256.6	56.9	134.8	149.8	96.0	116.9	211.7	62.8	120.9	215.8	64.2	123.3	244.6	67.6	136.6
05	METALS	106.7	91.9	102.1	143.5	93.1	127.8	113.0	95.5	107.6	139.8	84.8	122.7	124.4	74.6	108.8	138.9	90.0	123.7
06	WOOD, PLASTICS & COMPOSITES	98.2	86.1	91.9	120.5	55.1	86.3	118.1	86.0	101.3	105.3	64.1	83.8	112.1	66.4	88.2	129.9	76.2	101.8
07	THERMAL & MOISTURE PROTECTION	126.3	87.3	109.6	143.9	62.8	109.1	115.4	94.7	106.5	140.8	67.9	109.5	128.1	65.8	101.3	142.4	77.5	114.5
08	OPENINGS	87.0	80.6	85.4	99.6	52.4	88.1	79.0	85.8	80.6	87.8	57.8	80.5	91.6	60.3	83.9	94.1	64.6	86.9
0920	Plaster & Gypsum Board	109.6	85.5	93.8	180.6	52.4	96.5	109.0	85.8	93.8	138.3	62.3	88.4	141.8	65.7	91.9	189.8	74.7	114.2
0950, 0980	Ceilings & Acoustic Treatment	104.4	85.5	92.6	170.0	52.4	96.3	97.1	85.8	90.0	148.7	62.3	94.6	122.1	65.7	86.8	168.4	74.7	109.6
0960	Flooring	113.6	65.7	99.7	124.5	53.8	103.8	105.9	91.8	101.8	121.0	65.6	104.8	108.9	57.4	93.9	124.3	80.9	111.6
0970, 0990	Wall Finishes & Painting/Coating	110.3	95.5	101.5	117.1	51.7	77.9	101.7	92.5	96.2	106.0	51.1	73.1	114.3	57.7	80.4	120.3	73.9	92.5
09	FINISHES	111.8	84.1	96.8	148.1	55.1	97.7	102.4	88.5	94.9	131.5	62.8	94.3	119.9	64.2	89.8	150.9	75.3	110.0
COVERS	DIVS. 10 - 14, 25, 28, 41, 43, 44, 46	131.2	63.2	115.2	131.2	59.8	114.4	131.2	65.7	115.8	131.2	61.0	114.6	131.2	58.7	114.1	131.2	62.0	114.9
21, 22, 23	FIRE SUPPRESSION, PLUMBING & HVAC	106.3	84.7	97.6	106.9	69.9	92.0	105.5	87.6	98.3	106.8	61.8	88.7	106.5	77.4	94.7	107.6	86.2	99.0
26, 27, 3370	ELECTRICAL, COMMUNICATIONS & UTIL.	116.5	79.0	98.0	138.0	56.2	97.5	116.0	98.1	107.1	123.2	61.6	92.7	113.9	57.7	86.1	132.2	76.3	104.5
MF2018	WEIGHTED AVERAGE	116.8	86.4	103.7	136.0	68.6	106.9	108.6	91.2	101.1	125.8	68.4	101.0	121.9	68.6	98.8	135.0	81.1	111.7

For customer support on your Commercial Renovation Costs with RSMeans Data, call 800.448.8182.

Costs shown in RSMeans cost data publications are based on national averages for materials and installation. To adjust these costs to a specific location, simply multiply the base cost by the factor and divide by 100 for that city. The data is arranged alphabetically by state and postal zip code numbers. For a city not listed, use the factor for a nearby city with similar economic characteristics.

STATE/ZIP	CITY	MAT.	INST.	TOTAL
ALABAMA				
350-352	Birmingham	98.2	70.9	86.4
354	Tuscaloosa	96.9	70.5	85.5
355	Jasper	97.6	69.6	85.5
356	Decatur	96.9	70.2	85.4
357-358	Huntsville	96.9	69.9	85.2
359	Gadsden	97.2	70.8	85.8
360-361	Montgomery	98.0	71.8	86.6
362	Anniston	97.1	66.6	83.9
363	Dothan	97.2	71.6	86.1
364	Evergreen	96.8	67.8	84.3
365-366	Mobile	97.8	68.7	85.2
367	Selma	96.9	71.2	85.8
368	Phenix City	97.7	70.8	86.1
369	Butler	97.1	69.7	85.2
ALASKA				
995-996	Anchorage	119.7	110.4	115.7
997	Fairbanks	120.2	111.0	116.2
998	Juneau	119.7	110.4	115.7
999	Ketchikan	130.1	110.9	121.8
ARIZONA				
850,853	Phoenix	98.4	72.1	87.0
851,852	Mesa/Tempe	98.0	71.1	86.4
855	Globe	98.2	70.7	86.3
856-857	Tucson	96.9	71.8	86.1
859	Show Low	98.4	70.8	86.5
860	Flagstaff	102.4	70.4	88.6
863	Prescott	99.9	71.1	87.5
864	Kingman	98.2	69.8	85.9
865	Chambers	98.2	70.4	86.2
ARKANSAS				
716	Pine Bluff	96.2	62.8	81.8
717	Camden	94.1	64.1	81.1
718	Texarkana	95.1	63.7	81.5
719	Hot Springs	93.4	62.5	80.1
720-722	Little Rock	96.0	63.5	81.9
723	West Memphis	94.2	67.5	82.7
724	Jonesboro	94.8	65.7	82.2
725	Batesville	92.4	62.9	79.6
726	Harrison	93.7	62.3	80.1
727	Fayetteville	91.3	64.2	79.6
728	Russellville	92.5	61.3	79.0
729	Fort Smith	95.1	61.9	80.8
CALIFORNIA				
900-902	Los Angeles	98.4	129.3	111.8
903-905	Inglewood	93.6	127.4	108.2
906-908	Long Beach	95.2	127.2	109.0
910-912	Pasadena	95.4	127.0	109.0
913-916	Van Nuys	98.5	127.0	110.8
917-918	Alhambra	97.4	127.0	110.2
919-921	San Diego	100.3	121.3	109.4
922	Palm Springs	97.3	121.9	108.0
923-924	San Bernardino	95.0	125.4	108.1
925	Riverside	99.4	126.0	110.9
926-927	Santa Ana	97.1	125.7	109.5
928	Anaheim	99.5	125.9	110.9
930	Oxnard	98.2	125.8	110.1
931	Santa Barbara	97.4	125.0	109.4
932-933	Bakersfield	99.0	124.8	110.1
934	San Luis Obispo	98.4	125.5	110.1
935	Mojave	95.6	123.6	107.7
936-938	Fresno	98.6	131.2	112.7
939	Salinas	99.2	137.8	115.9
940-941	San Francisco	107.6	159.0	129.8
942,956-958	Sacramento	101.0	133.1	114.9
943	Palo Alto	99.4	150.8	121.6
944	San Mateo	101.9	156.6	125.5
945	Vallejo	100.5	143.4	119.0
946	Oakland	103.9	153.2	125.2
947	Berkeley	103.3	151.9	124.3
948	Richmond	102.8	148.2	122.4
949	San Rafael	105.0	150.3	124.6
950	Santa Cruz	104.9	138.1	119.2

STATE/ZIP	CITY	MAT.	INST.	TOTAL
CALIFORNIA (CONT'D)				
951	San Jose	103.0	158.4	126.9
952	Stockton	101.0	131.9	114.3
953	Modesto	100.8	132.1	114.3
954	Santa Rosa	101.2	149.8	122.2
955	Eureka	102.6	137.6	117.7
959	Marysville	101.8	130.7	114.3
960	Redding	107.8	132.6	118.6
961	Susanville	107.3	131.0	117.5
COLORADO				
800-802	Denver	104.0	75.1	91.5
803	Boulder	99.2	73.7	88.2
804	Golden	101.5	73.0	89.2
805	Fort Collins	102.8	73.2	90.0
806	Greeley	99.9	73.3	88.4
807	Fort Morgan	99.9	72.5	88.0
808-809	Colorado Springs	101.8	72.3	89.1
810	Pueblo	101.6	70.7	88.2
811	Alamosa	103.3	69.2	88.6
812	Salida	102.9	69.8	88.6
813	Durango	103.6	65.5	87.1
814	Montrose	102.3	66.3	86.7
815	Grand Junction	105.6	70.0	90.2
816	Glenwood Springs	103.6	66.6	87.6
CONNECTICUT				
060	New Britain	99.7	116.9	107.1
061	Hartford	101.7	118.7	109.0
062	Willimantic	100.3	116.9	107.5
063	New London	96.7	116.6	105.3
064	Meriden	98.8	116.4	106.4
065	New Haven	101.5	117.0	108.2
066	Bridgeport	100.9	116.7	107.7
067	Waterbury	100.5	117.2	107.7
068	Norwalk	100.4	116.9	107.5
069	Stamford	100.5	123.9	110.6
D.C.				
200-205	Washington	101.3	88.0	95.5
DELAWARE				
197	Newark	98.7	110.6	103.8
198	Wilmington	99.1	110.7	104.1
199	Dover	99.5	110.4	104.2
FLORIDA				
320,322	Jacksonville	97.0	66.9	84.0
321	Daytona Beach	97.2	69.0	85.0
323	Tallahassee	98.1	66.7	84.6
324	Panama City	98.7	67.2	85.1
325	Pensacola	101.2	65.3	85.7
326,344	Gainesville	98.3	63.8	83.4
327-328,347	Orlando	97.9	66.3	84.3
329	Melbourne	99.9	70.6	87.2
330-332,340	Miami	96.2	70.4	85.1
333	Fort Lauderdale	96.5	72.2	86.0
334,349	West Palm Beach	95.1	69.0	83.8
335-336,346	Tampa	98.0	67.4	84.8
337	St. Petersburg	100.4	65.8	85.4
338	Lakeland	97.1	66.3	83.8
339,341	Fort Myers	96.4	66.7	83.6
342	Sarasota	99.1	65.3	84.5
GEORGIA				
300-303,399	Atlanta	98.7	76.7	89.2
304	Statesboro	97.9	67.1	84.6
305	Gainesville	96.5	64.3	82.6
306	Athens	95.9	63.8	82.0
307	Dalton	97.7	68.0	84.9
308-309	Augusta	96.4	73.5	86.5
310-312	Macon	95.3	74.4	86.3
313-314	Savannah	97.3	72.8	86.7
315	Waycross	96.6	66.9	83.8
316	Valdosta	96.7	65.4	83.2
317,398	Albany	96.5	72.2	86.0
318-319	Columbus	96.4	73.4	86.4

STATE/ZIP	CITY	MAT.	INST.	TOTAL
HAWAII				
967	Hilo	115.6	117.2	116.3
968	Honolulu	120.3	117.8	119.2
STATES & POSS.				
969	Guam	138.0	63.4	105.7
IDAHO				
832	Pocatello	101.8	77.8	91.4
833	Twin Falls	102.9	76.9	91.7
834	Idaho Falls	100.2	76.9	90.1
835	Lewiston	108.4	84.9	98.2
836-837	Boise	100.9	78.2	91.1
838	Coeur d'Alene	108.1	81.5	96.6
ILLINOIS				
600-603	North Suburban	98.5	137.3	115.3
604	Joliet	98.5	140.3	116.5
605	South Suburban	98.5	137.2	115.3
606-608	Chicago	100.5	144.4	119.5
609	Kankakee	95.1	129.5	110.0
610-611	Rockford	96.3	126.3	109.3
612	Rock Island	94.5	95.5	94.9
613	La Salle	95.8	122.2	107.2
614	Galesburg	95.5	104.8	99.5
615-616	Peoria	97.4	109.9	102.8
617	Bloomington	94.8	105.4	99.4
618-619	Champaign	98.4	106.3	101.8
620-622	East St. Louis	94.1	109.8	100.9
623	Quincy	96.1	101.1	98.3
624	Effingham	95.3	104.9	99.4
625	Decatur	96.7	107.6	101.4
626-627	Springfield	97.8	110.4	103.2
628	Centralia	92.9	106.7	98.9
629	Carbondale	92.7	105.2	98.1
INDIANA				
460	Anderson	96.8	80.4	89.7
461-462	Indianapolis	99.5	82.9	92.3
463-464	Gary	98.3	105.8	101.5
465-466	South Bend	99.6	83.4	92.6
467-468	Fort Wayne	97.5	75.9	88.2
469	Kokomo	94.9	79.8	88.4
470	Lawrenceburg	93.0	78.3	86.6
471	New Albany	94.3	76.3	86.5
472	Columbus	96.5	79.4	89.1
473	Muncie	96.6	78.9	89.0
474	Bloomington	98.2	79.7	90.2
475	Washington	95.1	84.8	90.7
476-477	Evansville	96.0	83.2	90.5
478	Terre Haute	96.9	82.9	90.8
479	Lafayette	96.2	79.9	89.1
IOWA				
500-503,509	Des Moines	97.3	90.0	94.2
504	Mason City	95.0	71.2	84.7
505	Fort Dodge	95.0	71.0	84.6
506-507	Waterloo	96.5	78.4	88.7
508	Creston	95.2	81.1	89.1
510-511	Sioux City	98.0	80.3	90.3
512	Sibley	96.9	60.1	81.0
513	Spencer	98.3	59.9	81.7
514	Carroll	95.5	79.3	88.5
515	Council Bluffs	98.8	81.9	91.5
516	Shenandoah	96.1	78.7	88.6
520	Dubuque	97.3	80.6	90.1
521	Decorah	96.7	71.0	85.6
522-524	Cedar Rapids	98.3	86.6	93.2
525	Ottumwa	96.4	75.7	87.5
526	Burlington	95.8	81.1	89.5
527-528	Davenport	97.3	96.6	97.0
KANSAS				
660-662	Kansas City	96.4	99.2	97.6
664-666	Topeka	98.1	78.0	89.4
667	Fort Scott	95.2	73.9	86.0
668	Emporia	95.2	74.0	86.0
669	Belleville	96.9	69.1	84.9
670-672	Wichita	96.5	72.2	86.0
673	Independence	96.4	73.6	86.6
674	Salina	96.8	73.3	86.6
675	Hutchinson	91.9	69.7	82.3
676	Hays	96.0	70.4	84.9
677	Colby	96.5	71.7	85.7

STATE/ZIP	CITY	MAT.	INST.	TOTAL
KANSAS (CONT'D)				
678	Dodge City	98.0	74.8	88.0
679	Liberal	96.0	69.5	84.6
KENTUCKY				
400-402	Louisville	94.9	79.6	88.3
403-405	Lexington	94.2	78.8	87.5
406	Frankfort	96.8	78.6	88.9
407-409	Corbin	91.8	75.0	84.6
410	Covington	94.9	75.0	86.3
411-412	Ashland	93.6	86.3	90.5
413-414	Campton	94.9	76.3	86.9
415-416	Pikeville	96.1	82.1	90.1
417-418	Hazard	94.3	77.1	86.9
420	Paducah	93.0	80.0	87.4
421-422	Bowling Green	95.2	78.4	87.9
423	Owensboro	95.3	81.8	89.4
424	Henderson	92.7	78.7	86.7
425-426	Somerset	92.1	76.6	85.4
427	Elizabethtown	91.8	73.4	83.8
LOUISIANA				
700-701	New Orleans	97.7	68.3	85.0
703	Thibodaux	95.5	65.0	82.3
704	Hammond	93.2	63.4	80.3
705	Lafayette	95.1	65.9	82.5
706	Lake Charles	95.3	68.4	83.7
707-708	Baton Rouge	97.0	68.3	84.6
710-711	Shreveport	97.7	66.8	84.4
712	Monroe	96.0	64.6	82.4
713-714	Alexandria	96.1	64.7	82.5
MAINE				
039	Kittery	94.4	84.9	90.3
040-041	Portland	100.6	85.6	94.1
042	Lewiston	98.0	84.9	92.4
043	Augusta	101.6	86.0	94.9
044	Bangor	97.5	84.5	91.9
045	Bath	96.2	84.6	91.2
046	Machias	95.8	84.5	90.9
047	Houlton	95.9	84.4	91.0
048	Rockland	95.0	84.6	90.5
049	Waterville	96.2	84.5	91.2
MARYLAND				
206	Waldorf	97.4	84.8	92.0
207-208	College Park	97.5	85.8	92.5
209	Silver Spring	96.7	84.9	91.6
210-212	Baltimore	102.3	83.1	94.0
214	Annapolis	101.4	83.4	93.6
215	Cumberland	97.1	83.1	91.0
216	Easton	98.9	72.6	87.5
217	Hagerstown	97.7	86.2	92.8
218	Salisbury	99.3	65.0	84.5
219	Elkton	96.1	83.0	90.5
MASSACHUSETTS				
010-011	Springfield	99.1	110.8	104.2
012	Pittsfield	98.6	105.1	101.4
013	Greenfield	96.7	109.5	102.3
014	Fitchburg	95.4	113.2	103.1
015-016	Worcester	99.1	118.6	107.5
017	Framingham	94.9	122.4	106.8
018	Lowell	98.4	126.4	110.5
019	Lawrence	99.4	124.9	110.4
020-022, 024	Boston	100.2	132.9	114.3
023	Brockton	98.9	117.8	107.1
025	Buzzards Bay	93.3	114.1	102.3
026	Hyannis	96.0	116.9	105.1
027	New Bedford	98.1	116.3	106.0
MICHIGAN				
480,483	Royal Oak	94.1	96.9	95.3
481	Ann Arbor	96.3	102.0	98.7
482	Detroit	99.5	100.9	100.1
484-485	Flint	95.9	90.9	93.8
486	Saginaw	95.6	86.7	91.8
487	Bay City	95.7	85.3	91.2
488-489	Lansing	98.1	90.5	94.8
490	Battle Creek	95.7	79.4	88.6
491	Kalamazoo	96.0	80.9	89.5
492	Jackson	94.1	88.0	91.4
493,495	Grand Rapids	98.9	82.2	91.6
494	Muskegon	94.5	79.7	88.1

For customer support on your Commercial Renovation Costs with RSMeans Data, call 800.448.8182.

STATE/ZIP	CITY	MAT.	INST.	TOTAL
MICHIGAN (CONT'D)				
496	Traverse City	93.5	75.2	85.6
497	Gaylord	94.6	78.1	87.5
498-499	Iron Mountain	96.5	78.9	88.9
MINNESOTA				
550-551	Saint Paul	99.3	119.2	107.9
553-555	Minneapolis	99.7	116.6	107.0
556-558	Duluth	99.9	103.9	101.6
559	Rochester	98.4	106.1	101.7
560	Mankato	95.5	105.0	99.6
561	Windom	94.0	91.7	93.0
562	Willmar	93.7	96.6	94.9
563	St. Cloud	94.9	115.9	104.0
564	Brainerd	95.5	97.9	96.5
565	Detroit Lakes	97.6	90.3	94.4
566	Bemidji	96.8	94.6	95.8
567	Thief River Falls	96.4	89.4	93.4
MISSISSIPPI				
386	Clarksdale	95.7	53.8	77.6
387	Greenville	99.4	65.2	84.6
388	Tupelo	97.2	56.4	79.6
389	Greenwood	97.0	53.5	78.2
390-392	Jackson	98.6	67.9	85.3
393	Meridian	95.4	67.8	83.5
394	Laurel	96.8	56.4	79.4
395	Biloxi	97.1	66.6	83.9
396	McComb	95.1	53.8	77.3
397	Columbus	96.7	57.1	79.6
MISSOURI				
630-631	St. Louis	97.5	104.6	100.6
633	Bowling Green	95.6	92.3	94.2
634	Hannibal	94.5	88.5	91.9
635	Kirksville	98.0	83.6	91.8
636	Flat River	96.6	90.6	94.0
637	Cape Girardeau	96.4	90.6	93.9
638	Sikeston	94.9	84.4	90.4
639	Poplar Bluff	94.4	84.3	90.0
640-641	Kansas City	97.5	101.8	99.3
644-645	St. Joseph	96.0	93.3	94.9
646	Chillicothe	93.7	90.7	92.4
647	Harrisonville	93.2	97.4	95.0
648	Joplin	95.2	80.1	88.7
650-651	Jefferson City	95.7	90.0	93.2
652	Columbia	95.7	94.5	95.2
653	Sedalia	95.6	88.1	92.4
654-655	Rolla	93.4	90.6	92.2
656-658	Springfield	96.5	81.4	90.0
MONTANA				
590-591	Billings	101.4	77.2	90.9
592	Wolf Point	101.3	74.5	89.7
593	Miles City	99.1	74.6	88.5
594	Great Falls	102.8	77.1	91.7
595	Havre	100.2	72.5	88.3
596	Helena	101.2	74.7	89.7
597	Butte	101.3	74.4	89.7
598	Missoula	98.8	74.8	88.5
599	Kalispell	98.0	74.1	87.7
NEBRASKA				
680-681	Omaha	98.6	83.0	91.9
683-685	Lincoln	99.8	79.3	90.9
686	Columbus	96.0	77.7	88.1
687	Norfolk	97.4	74.5	87.5
688	Grand Island	97.4	76.2	88.2
689	Hastings	97.1	75.2	87.7
690	McCook	95.9	70.2	84.8
691	North Platte	96.1	74.9	86.9
692	Valentine	98.0	67.4	84.8
693	Alliance	98.4	70.4	86.3
NEVADA				
889-891	Las Vegas	104.2	107.0	105.4
893	Ely	102.9	89.5	97.1
894-895	Reno	102.8	85.6	95.4
897	Carson City	103.2	85.6	95.6
898	Elko	101.5	81.8	93.0
NEW HAMPSHIRE				
030	Nashua	99.3	92.3	96.3
031	Manchester	100.5	93.0	97.3

STATE/ZIP	CITY	MAT.	INST.	TOTAL
NEW HAMPSHIRE (CONT'D)				
032-033	Concord	100.8	91.3	96.7
034	Keene	96.1	86.2	91.8
035	Littleton	96.1	77.7	88.1
036	Charleston	95.6	85.6	91.3
037	Claremont	94.8	85.7	90.8
038	Portsmouth	96.5	91.5	94.3
NEW JERSEY				
070-071	Newark	101.8	137.9	117.4
072	Elizabeth	98.9	137.6	115.6
073	Jersey City	97.7	137.1	114.7
074-075	Paterson	99.3	137.3	115.7
076	Hackensack	97.4	137.1	114.6
077	Long Branch	97.1	126.6	109.9
078	Dover	97.6	137.5	114.8
079	Summit	97.7	136.9	114.6
080,083	Vineland	97.2	126.6	109.9
081	Camden	98.9	133.7	114.0
082,084	Atlantic City	97.8	132.6	112.8
085-086	Trenton	100.5	124.4	110.8
087	Point Pleasant	99.1	126.5	110.9
088-089	New Brunswick	99.7	132.2	113.8
NEW MEXICO				
870-872	Albuquerque	98.9	71.6	87.1
873	Gallup	99.2	71.6	87.3
874	Farmington	99.3	71.6	87.3
875	Santa Fe	100.5	72.2	88.3
877	Las Vegas	97.4	71.6	86.3
878	Socorro	97.1	71.6	86.1
879	Truth/Consequences	96.8	68.6	84.6
880	Las Cruces	96.8	70.0	85.2
881	Clovis	99.1	71.5	87.2
882	Roswell	100.7	71.6	88.1
883	Carrizozo	101.4	71.6	88.5
884	Tucumcari	99.7	71.5	87.6
NEW YORK				
100-102	New York	100.3	174.2	132.2
103	Staten Island	96.5	171.9	129.1
104	Bronx	94.8	171.2	127.8
105	Mount Vernon	94.9	147.0	117.4
106	White Plains	94.9	149.8	118.6
107	Yonkers	99.1	150.1	121.1
108	New Rochelle	95.3	141.4	115.2
109	Suffern	94.9	125.4	108.1
110	Queens	100.5	173.1	131.8
111	Long Island City	102.2	173.1	132.8
112	Brooklyn	102.7	173.1	133.1
113	Flushing	102.5	173.1	133.0
114	Jamaica	100.9	173.1	132.1
115,117,118	Hicksville	100.5	151.7	122.6
116	Far Rockaway	102.7	173.1	133.1
119	Riverhead	101.3	156.9	125.3
120-122	Albany	98.3	109.4	103.1
123	Schenectady	97.9	108.6	102.5
124	Kingston	101.2	128.8	113.1
125-126	Poughkeepsie	100.3	131.7	113.9
127	Monticello	99.8	129.5	112.7
128	Glens Falls	92.8	103.2	97.3
129	Plattsburgh	97.7	92.8	95.6
130-132	Syracuse	98.7	101.6	99.9
133-135	Utica	96.8	102.5	99.3
136	Watertown	98.4	98.9	98.6
137-139	Binghamton	98.5	100.5	99.3
140-142	Buffalo	100.9	109.8	104.8
143	Niagara Falls	98.0	103.4	100.3
144-146	Rochester	100.8	101.2	101.0
147	Jamestown	97.1	92.3	95.0
148-149	Elmira	96.9	100.4	98.4
NORTH CAROLINA				
270,272-274	Greensboro	99.7	65.3	84.8
271	Winston-Salem	99.5	65.3	84.7
275-276	Raleigh	99.0	65.3	84.4
277	Durham	101.2	65.2	85.6
278	Rocky Mount	97.1	64.5	83.0
279	Elizabeth City	98.1	66.6	84.5
280	Gastonia	97.7	68.5	85.1
281-282	Charlotte	98.4	72.1	87.0
283	Fayetteville	100.5	64.0	84.7
284	Wilmington	96.6	63.1	82.1
285	Kinston	94.9	63.5	81.3

STATE/ZIP	CITY	MAT.	INST.	TOTAL
NORTH CAROLINA (CONT'D)				
286	Hickory	95.2	69.0	83.9
287-288	Asheville	97.0	64.6	83.0
289	Murphy	96.1	62.6	81.6
NORTH DAKOTA				
580-581	Fargo	99.6	78.9	90.6
582	Grand Forks	98.8	81.0	91.1
583	Devils Lake	98.6	77.8	89.6
584	Jamestown	98.7	76.9	89.3
585	Bismarck	99.1	83.1	92.2
586	Dickinson	99.2	75.7	89.0
587	Minot	98.7	76.5	89.1
588	Williston	97.8	80.7	90.4
OHIO				
430-432	Columbus	99.2	83.2	92.3
433	Marion	95.5	83.6	90.3
434-436	Toledo	98.9	87.7	94.1
437-438	Zanesville	96.1	83.0	90.4
439	Steubenville	97.1	87.4	92.9
440	Lorain	98.7	81.8	91.4
441	Cleveland	99.4	91.5	96.0
442-443	Akron	99.7	83.0	92.5
444-445	Youngstown	99.1	79.7	90.7
446-447	Canton	99.2	77.5	89.8
448-449	Mansfield	96.6	81.1	89.9
450	Hamilton	96.9	76.7	88.2
451-452	Cincinnati	98.0	79.2	89.9
453-454	Dayton	96.9	76.6	88.1
455	Springfield	96.9	77.2	88.4
456	Chillicothe	96.3	85.3	91.5
457	Athens	99.2	81.8	91.7
458	Lima	99.6	79.0	90.7
OKLAHOMA				
730-731	Oklahoma City	97.6	67.8	84.8
734	Ardmore	95.9	63.8	82.0
735	Lawton	98.1	63.9	83.3
736	Clinton	97.3	63.7	82.8
737	Enid	97.8	64.2	83.3
738	Woodward	95.9	61.2	80.9
739	Guymon	97.0	61.7	81.7
740-741	Tulsa	96.9	65.5	83.3
743	Miami	93.5	63.0	80.3
744	Muskogee	96.1	65.7	83.0
745	McAlester	93.2	58.3	78.1
746	Ponca City	93.9	61.8	80.0
747	Durant	93.8	62.7	80.4
748	Shawnee	95.3	63.3	81.5
749	Poteau	93.0	62.6	79.9
OREGON				
970-972	Portland	102.0	104.7	103.2
973	Salem	103.3	102.8	103.0
974	Eugene	101.6	103.7	102.5
975	Medford	103.1	99.3	101.4
976	Klamath Falls	103.3	97.8	100.9
977	Bend	102.3	100.5	101.5
978	Pendleton	98.3	101.0	99.5
979	Vale	96.0	87.8	92.4
PENNSYLVANIA				
150-152	Pittsburgh	99.3	102.6	100.7
153	Washington	96.6	101.9	98.9
154	Uniontown	97.1	99.2	98.0
155	Bedford	98.0	90.1	94.6
156	Greensburg	98.0	95.8	97.1
157	Indiana	96.8	96.2	96.5
158	Dubois	98.5	94.0	96.6
159	Johnstown	98.0	97.2	97.7
160	Butler	92.1	98.7	95.0
161	New Castle	92.2	98.9	95.1
162	Kittanning	92.6	96.6	94.3
163	Oil City	92.1	94.4	93.1
164-165	Erie	94.0	94.6	94.3
166	Altoona	94.2	97.5	95.6
167	Bradford	95.6	95.0	95.3
168	State College	95.2	99.1	96.9
169	Wellsboro	96.2	89.3	93.2
170-171	Harrisburg	98.5	97.0	97.9
172	Chambersburg	95.4	86.0	91.3
173-174	York	96.2	94.6	95.5
175-176	Lancaster	94.2	96.5	95.2

STATE/ZIP	CITY	MAT.	INST.	TOTAL
PENNSYLVANIA (CONT'D)				
177	Williamsport	92.8	94.2	93.4
178	Sunbury	95.0	89.3	92.5
179	Pottsville	94.0	92.5	93.3
180	Lehigh Valley	95.6	107.5	100.8
181	Allentown	97.5	108.1	102.1
182	Hazleton	95.1	92.5	94.0
183	Stroudsburg	95.0	100.9	97.6
184-185	Scranton	98.3	96.8	97.7
186-187	Wilkes-Barre	94.9	96.8	95.7
188	Montrose	94.6	93.3	94.0
189	Doylestown	94.7	122.6	106.7
190-191	Philadelphia	99.7	136.9	115.8
193	Westchester	96.1	123.2	107.8
194	Norristown	95.1	123.4	107.3
195-196	Reading	97.3	104.7	100.5
PUERTO RICO				
009	San Juan	124.9	28.3	83.1
RHODE ISLAND				
028	Newport	97.0	112.6	103.8
029	Providence	101.0	113.3	106.3
SOUTH CAROLINA				
290-292	Columbia	98.8	67.6	85.3
293	Spartanburg	97.7	70.2	85.8
294	Charleston	99.4	67.0	85.4
295	Florence	97.4	67.0	84.2
296	Greenville	97.5	70.0	85.6
297	Rock Hill	97.3	68.2	84.7
298	Aiken	98.1	65.2	83.9
299	Beaufort	98.9	61.5	82.7
SOUTH DAKOTA				
570-571	Sioux Falls	99.3	82.1	91.8
572	Watertown	97.7	66.6	84.3
573	Mitchell	96.6	59.6	80.6
574	Aberdeen	99.1	66.7	85.1
575	Pierre	100.2	68.9	86.7
576	Mobridge	97.1	60.7	81.4
577	Rapid City	98.7	71.0	86.7
TENNESSEE				
370-372	Nashville	100.0	74.5	89.0
373-374	Chattanooga	99.0	68.7	85.9
375,380-381	Memphis	96.5	72.5	86.1
376	Johnson City	99.2	60.5	82.5
377-379	Knoxville	95.7	66.0	82.9
382	McKenzie	96.7	54.7	78.6
383	Jackson	98.1	60.0	81.6
384	Columbia	95.2	64.6	82.0
385	Cookeville	96.6	56.1	79.1
TEXAS				
750	McKinney	99.5	61.8	83.2
751	Waxahachie	99.5	62.0	83.3
752-753	Dallas	100.1	67.4	86.0
754	Greenville	99.6	61.3	83.1
755	Texarkana	99.1	60.4	82.4
756	Longview	99.8	59.2	82.3
757	Tyler	100.1	60.2	82.8
758	Palestine	95.9	58.9	80.0
759	Lufkin	96.6	60.5	81.0
760-761	Fort Worth	99.0	63.9	83.9
762	Denton	98.9	61.6	82.8
763	Wichita Falls	96.5	62.5	81.8
764	Eastland	95.7	58.9	79.8
765	Temple	94.1	56.4	77.8
766-767	Waco	95.8	63.2	81.7
768	Brownwood	99.1	57.2	81.0
769	San Angelo	98.7	57.3	80.8
770-772	Houston	101.3	67.3	86.6
773	Huntsville	99.9	60.7	83.0
774	Wharton	101.0	62.1	84.2
775	Galveston	98.8	64.0	83.8
776-777	Beaumont	99.2	64.9	84.4
778	Bryan	96.3	62.4	81.7
779	Victoria	101.0	60.8	83.6
780	Laredo	97.6	60.9	81.8
781-782	San Antonio	98.9	63.8	83.7
783-784	Corpus Christi	100.2	64.6	84.8
785	McAllen	100.9	55.2	81.2
786-787	Austin	98.6	62.3	82.9

For customer support on your Commercial Renovation Costs with RSMeans Data, call 800.448.8182.

Location Factors - Commercial - V2

STATE/ZIP	CITY	MAT.	INST.	TOTAL
TEXAS (CONT'D)				
788	Del Rio	100.7	58.2	82.3
789	Giddings	97.3	58.3	80.5
790-791	Amarillo	99.9	61.8	83.5
792	Childress	98.7	59.4	81.7
793-794	Lubbock	100.4	60.7	83.3
795-796	Abilene	99.0	61.6	82.9
797	Midland	100.9	61.7	84.0
798-799,885	El Paso	98.3	63.3	83.2
UTAH				
840-841	Salt Lake City	103.3	73.7	90.5
842,844	Ogden	98.6	74.0	88.0
843	Logan	100.7	74.0	89.2
845	Price	101.1	68.6	87.1
846-847	Provo	101.1	74.0	89.4
VERMONT				
050	White River Jct.	98.7	79.6	90.4
051	Bellows Falls	97.1	90.6	94.3
052	Bennington	97.5	87.2	93.0
053	Brattleboro	97.8	90.7	94.7
054	Burlington	103.1	82.3	94.1
056	Montpelier	101.2	86.0	94.6
057	Rutland	99.4	81.6	91.7
058	St. Johnsbury	98.7	79.2	90.3
059	Guildhall	97.3	79.2	89.5
VIRGINIA				
220-221	Fairfax	99.9	81.7	92.0
222	Arlington	101.0	81.6	92.6
223	Alexandria	100.0	84.8	93.4
224-225	Fredericksburg	98.6	79.9	90.5
226	Winchester	99.2	76.4	89.3
227	Culpeper	99.1	82.0	91.7
228	Harrisonburg	99.4	75.6	89.1
229	Charlottesville	99.8	74.2	88.7
230-232	Richmond	99.6	74.6	88.8
233-235	Norfolk	100.7	68.3	86.7
236	Newport News	99.5	69.5	86.5
237	Portsmouth	99.0	67.4	85.3
238	Petersburg	99.4	74.4	88.6
239	Farmville	98.5	66.0	84.5
240-241	Roanoke	100.5	69.1	87.0
242	Bristol	98.5	59.4	81.6
243	Pulaski	98.1	69.7	85.8
244	Staunton	99.0	67.8	85.5
245	Lynchburg	99.1	73.1	87.9
246	Grundy	98.4	63.2	83.2
WASHINGTON				
980-981,987	Seattle	105.4	108.4	106.7
982	Everett	104.5	104.2	104.4
983-984	Tacoma	104.9	105.7	105.2
985	Olympia	103.9	105.2	104.5
986	Vancouver	106.2	102.2	104.4
988	Wenatchee	105.0	85.6	96.6
989	Yakima	105.0	101.6	103.5
990-992	Spokane	101.0	86.8	94.9
993	Richland	100.7	93.0	97.4
994	Clarkston	99.9	79.5	91.1
WEST VIRGINIA				
247-248	Bluefield	97.3	85.2	92.1
249	Lewisburg	99.0	86.8	93.8
250-253	Charleston	96.6	91.1	94.2
254	Martinsburg	96.5	80.8	89.7
255-257	Huntington	97.8	91.7	95.1
258-259	Beckley	95.2	87.7	91.9
260	Wheeling	100.1	90.9	96.1
261	Parkersburg	98.9	90.0	95.1
262	Buckhannon	98.7	90.5	95.1
263-264	Clarksburg	99.3	91.0	95.7
265	Morgantown	99.3	91.4	95.9
266	Gassaway	98.7	87.3	93.7
267	Romney	98.5	85.3	92.8
268	Petersburg	98.3	85.6	92.8
WISCONSIN				
530,532	Milwaukee	97.9	111.7	103.9
531	Kenosha	99.2	106.2	102.2
534	Racine	98.5	106.2	101.8
535	Beloit	97.8	97.3	97.6
537	Madison	98.6	106.4	101.9

STATE/ZIP	CITY	MAT.	INST.	TOTAL
WISCONSIN (CONT'D)				
538	Lancaster	95.8	92.9	94.6
539	Portage	94.2	96.5	95.2
540	New Richmond	94.9	92.9	94.0
541-543	Green Bay	99.4	103.1	101.0
544	Wausau	94.5	92.9	93.8
545	Rhinelander	98.0	91.1	95.0
546	La Crosse	95.7	100.3	97.7
547	Eau Claire	97.4	100.4	98.7
548	Superior	94.6	94.6	94.6
549	Oshkosh	95.2	92.7	94.1
WYOMING				
820	Cheyenne	101.0	72.4	88.7
821	Yellowstone Nat'l Park	98.9	71.4	87.0
822	Wheatland	100.1	69.9	87.0
823	Rawlins	101.9	71.3	88.7
824	Worland	99.5	71.9	87.6
825	Riverton	100.8	71.4	88.1
826	Casper	102.5	72.3	89.4
827	Newcastle	99.4	71.3	87.3
828	Sheridan	102.4	71.1	88.9
829-831	Rock Springs	103.3	70.6	89.1
CANADIAN FACTORS (reflect Canadian currency)				
ALBERTA				
	Calgary	123.3	91.7	109.7
	Edmonton	125.3	91.9	110.9
	Fort McMurray	123.9	86.2	107.6
	Lethbridge	118.9	85.7	104.6
	Lloydminster	113.6	82.6	100.2
	Medicine Hat	113.7	81.9	100.0
	Red Deer	114.2	81.9	100.2
BRITISH COLUMBIA				
	Kamloops	115.6	81.6	100.9
	Prince George	116.5	81.0	101.2
	Vancouver	121.7	91.7	108.7
	Victoria	116.8	86.4	103.7
MANITOBA				
	Brandon	122.8	68.9	99.5
	Portage la Prairie	113.6	67.5	93.6
	Winnipeg	125.8	68.4	101.0
NEW BRUNSWICK				
	Bathurst	112.7	60.8	90.3
	Dalhousie	112.4	61.0	90.2
	Fredericton	121.9	67.8	98.5
	Moncton	113.0	73.2	95.8
	Newcastle	112.8	61.4	90.5
	St. John	115.6	72.9	97.1
NEWFOUNDLAND				
	Corner Brook	127.3	66.7	101.1
	St. Johns	126.2	85.2	108.5
NORTHWEST TERRITORIES				
	Yellowknife	135.0	81.1	111.7
NOVA SCOTIA				
	Bridgewater	113.3	68.7	94.0
	Dartmouth	124.0	68.6	100.1
	Halifax	120.5	84.9	105.1
	New Glasgow	121.9	68.6	98.9
	Sydney	120.1	68.6	97.9
	Truro	112.8	68.7	93.7
	Yarmouth	121.9	68.6	98.8
ONTARIO				
	Barrie	118.0	85.2	103.8
	Brantford	115.2	88.8	103.8
	Cornwall	115.3	85.5	102.4
	Hamilton	119.2	94.5	108.6
	Kingston	116.5	85.5	103.1
	Kitchener	109.9	91.2	101.8
	London	119.2	92.9	107.8
	North Bay	125.0	83.4	107.0
	Oshawa	112.4	93.6	104.3
	Ottawa	120.6	93.0	108.7
	Owen Sound	118.1	83.5	103.2
	Peterborough	115.2	85.2	102.3
	Sarnia	115.5	89.4	104.2

STATE/ZIP	CITY	MAT.	INST.	TOTAL
ONTARIO (CONT'D)				
	Sault Ste. Marie	110.6	86.4	100.1
	St. Catharines	108.0	92.1	101.1
	Sudbury	107.7	90.3	100.2
	Thunder Bay	109.3	91.6	101.6
	Timmins	115.4	83.4	101.6
	Toronto	119.3	99.9	111.0
	Windsor	108.6	91.2	101.1
PRINCE EDWARD ISLAND				
	Charlottetown	124.1	58.7	95.8
	Summerside	124.4	56.4	95.0
QUEBEC				
	Cap-de-la-Madeleine	112.8	78.2	97.8
	Charlesbourg	112.8	78.2	97.8
	Chicoutimi	112.5	86.9	101.4
	Gatineau	112.5	78.0	97.6
	Granby	112.7	77.9	97.7
	Hull	112.6	78.0	97.7
	Joliette	113.0	78.2	97.9
	Laval	112.6	78.6	97.9
	Montreal	122.1	89.4	108.0
	Quebec City	122.2	89.9	108.2
	Rimouski	112.3	86.9	101.4
	Rouyn-Noranda	112.4	78.0	97.6
	Saint-Hyacinthe	111.8	78.0	97.2
	Sherbrooke	112.8	78.0	97.7
	Sorel	112.9	78.2	97.9
	Saint-Jerome	112.5	78.0	97.6
	Trois-Rivieres	122.8	78.1	103.5
SASKATCHEWAN				
	Moose Jaw	111.0	62.4	90.0
	Prince Albert	110.2	60.7	88.8
	Regina	127.0	88.5	110.3
	Saskatoon	112.0	85.1	100.4
YUKON				
	Whitehorse	136.0	68.6	106.9

For customer support on your Commercial Renovation Costs with RSMeans Data, call 800.448.8182.

R011105-05 Tips for Accurate Estimating

1. Use pre-printed or columnar forms for orderly sequence of dimensions and locations and for recording telephone quotations.

2. Use only the front side of each paper or form except for certain pre-printed summary forms.

3. Be consistent in listing dimensions: For example, length x width x height. This helps in rechecking to ensure that, the total length of partitions is appropriate for the building area.

4. Use printed (rather than measured) dimensions where given.

5. Add up multiple printed dimensions for a single entry where possible.

6. Measure all other dimensions carefully.

7. Use each set of dimensions to calculate multiple related quantities.

8. Convert foot and inch measurements to decimal feet when listing. Memorize decimal equivalents to .01 parts of a foot (1/8″ equals approximately .01′).

9. Do not "round off" quantities until the final summary.

10. Mark drawings with different colors as items are taken off.

11. Keep similar items together, different items separate.

12. Identify location and drawing numbers to aid in future checking for completeness.

13. Measure or list everything on the drawings or mentioned in the specifications.

14. It may be necessary to list items not called for to make the job complete.

15. Be alert for: Notes on plans such as N.T.S. (not to scale); changes in scale throughout the drawings; reduced size drawings; discrepancies between the specifications and the drawings.

16. Develop a consistent pattern of performing an estimate. For example:
    a. Start the quantity takeoff at the lower floor and move to the next higher floor.
    b. Proceed from the main section of the building to the wings.
    c. Proceed from south to north or vice versa, clockwise or counterclockwise.
    d. Take off floor plan quantities first, elevations next, then detail drawings.

17. List all gross dimensions that can be either used again for different quantities, or used as a rough check of other quantities for verification (exterior perimeter, gross floor area, individual floor areas, etc.).

18. Utilize design symmetry or repetition (repetitive floors, repetitive wings, symmetrical design around a center line, similar room layouts, etc.). Note: Extreme caution is needed here so as not to omit or duplicate an area.

19. Do not convert units until the final total is obtained. For instance, when estimating concrete work, keep all units to the nearest cubic foot, then summarize and convert to cubic yards.

20. When figuring alternatives, it is best to total all items involved in the basic system, then total all items involved in the alternates. Therefore you work with positive numbers in all cases. When adds and deducts are used, it is often confusing whether to add or subtract a portion of an item; especially on a complicated or involved alternate.

R011105-50   Metric Conversion Factors

Description: This table is primarily for converting customary U.S. units in the left hand column to SI metric units in the right hand column. In addition, conversion factors for some commonly encountered Canadian and non-SI metric units are included.

If You Know		Multiply By		To Find
Length	Inches	x	25.4[a]	=	Millimeters
	Feet	x	0.3048[a]	=	Meters
	Yards	x	0.9144[a]	=	Meters
	Miles (statute)	x	1.609	=	Kilometers
Area	Square inches	x	645.2	=	Square millimeters
	Square feet	x	0.0929	=	Square meters
	Square yards	x	0.8361	=	Square meters
Volume (Capacity)	Cubic inches	x	16,387	=	Cubic millimeters
	Cubic feet	x	0.02832	=	Cubic meters
	Cubic yards	x	0.7646	=	Cubic meters
	Gallons (U.S. liquids)[b]	x	0.003785	=	Cubic meters[c]
	Gallons (Canadian liquid)[b]	x	0.004546	=	Cubic meters[c]
	Ounces (U.S. liquid)[b]	x	29.57	=	Milliliters[c, d]
	Quarts (U.S. liquid)[b]	x	0.9464	=	Liters[c, d]
	Gallons (U.S. liquid)[b]	x	3.785	=	Liters[c, d]
Force	Kilograms force[d]	x	9.807	=	Newtons
	Pounds force	x	4.448	=	Newtons
	Pounds force	x	0.4536	=	Kilograms force[d]
	Kips	x	4448	=	Newtons
	Kips	x	453.6	=	Kilograms force[d]
Pressure, Stress, Strength (Force per unit area)	Kilograms force per square centimeter[d]	x	0.09807	=	Megapascals
	Pounds force per square inch (psi)	x	0.006895	=	Megapascals
	Kips per square inch	x	6.895	=	Megapascals
	Pounds force per square inch (psi)	x	0.07031	=	Kilograms force per square centimeter[d]
	Pounds force per square foot	x	47.88	=	Pascals
	Pounds force per square foot	x	4.882	=	Kilograms force per square meter[d]
Flow	Cubic feet per minute	x	0.4719	=	Liters per second
	Gallons per minute	x	0.0631	=	Liters per second
	Gallons per hour	x	1.05	=	Milliliters per second
Bending Moment Or Torque	Inch-pounds force	x	0.01152	=	Meter-kilograms force[d]
	Inch-pounds force	x	0.1130	=	Newton-meters
	Foot-pounds force	x	0.1383	=	Meter-kilograms force[d]
	Foot-pounds force	x	1.356	=	Newton-meters
	Meter-kilograms force[d]	x	9.807	=	Newton-meters
Mass	Ounces (avoirdupois)	x	28.35	=	Grams
	Pounds (avoirdupois)	x	0.4536	=	Kilograms
	Tons (metric)	x	1000	=	Kilograms
	Tons, short (2000 pounds)	x	907.2	=	Kilograms
	Tons, short (2000 pounds)	x	0.9072	=	Megagrams[e]
Mass per Unit Volume	Pounds mass per cubic foot	x	16.02	=	Kilograms per cubic meter
	Pounds mass per cubic yard	x	0.5933	=	Kilograms per cubic meter
	Pounds mass per gallon (U.S. liquid)[b]	x	119.8	=	Kilograms per cubic meter
	Pounds mass per gallon (Canadian liquid)[b]	x	99.78	=	Kilograms per cubic meter
Temperature	Degrees Fahrenheit	(F-32)/1.8		=	Degrees Celsius
	Degrees Fahrenheit	(F+459.67)/1.8		=	Degrees Kelvin
	Degrees Celsius	C+273.15		=	Degrees Kelvin

[a]The factor given is exact
[b]One U.S. gallon = 0.8327 Canadian gallon
[c]1 liter = 1000 milliliters = 1000 cubic centimeters
   1 cubic decimeter = 0.001 cubic meter

[d]Metric but not SI unit
[e]Called "tonne" in England and
   "metric ton" in other metric countries

R011105-60   Weights and Measures

Measures of Length
1 Mile = 1760 Yards = 5280 Feet
1 Yard = 3 Feet = 36 inches
1 Foot = 12 Inches
1 Mil = 0.001 Inch
1 Fathom = 2 Yards = 6 Feet
1 Rod = 5.5 Yards = 16.5 Feet
1 Hand = 4 Inches
1 Span = 9 Inches
1 Micro-inch = One Millionth Inch or 0.000001 Inch
1 Micron = One Millionth Meter + 0.00003937 Inch

Surveyor's Measure
1 Mile = 8 Furlongs = 80 Chains
1 Furlong = 10 Chains = 220 Yards
1 Chain = 4 Rods = 22 Yards = 66 Feet = 100 Links
1 Link = 7.92 Inches

Square Measure
1 Square Mile = 640 Acres = 6400 Square Chains
1 Acre = 10 Square Chains = 4840 Square Yards =
    43,560 Sq. Ft.
1 Square Chain = 16 Square Rods = 484 Square Yards =
    4356 Sq. Ft.
1 Square Rod = 30.25 Square Yards = 272.25 Square Feet = 625 Square
    Lines
1 Square Yard = 9 Square Feet
1 Square Foot = 144 Square Inches
An Acre equals a Square 208.7 Feet per Side

Cubic Measure
1 Cubic Yard = 27 Cubic Feet
1 Cubic Foot = 1728 Cubic Inches
1 Cord of Wood = 4 x 4 x 8 Feet = 128 Cubic Feet
1 Perch of Masonry = 16½ x 1½ x 1 Foot = 24.75 Cubic Feet

Avoirdupois or Commercial Weight
1 Gross or Long Ton = 2240 Pounds
1 Net or Short Ton = 2000 Pounds
1 Pound = 16 Ounces = 7000 Grains
1 Ounce = 16 Drachms = 437.5 Grains
1 Stone = 14 Pounds

Power
1 British Thermal Unit per Hour = 0.2931 Watts
1 Ton (Refrigeration) = 3.517 Kilowatts
1 Horsepower (Boiler) = 9.81 Kilowatts
1 Horsepower (550 ft-lb/s) = 0.746 Kilowatts

Shipping Measure
For Measuring Internal Capacity of a Vessel:
    1 Register Ton = 100 Cubic Feet

For Measurement of Cargo:
    Approximately 40 Cubic Feet of Merchandise is considered a Shipping
    Ton, unless that bulk would weigh more than 2000 Pounds, in which case
    Freight Charge may be based upon weight.

40 Cubic Feet = 32.143 U.S. Bushels = 31.16 Imp. Bushels

Liquid Measure
1 Imperial Gallon = 1.2009 U.S. Gallon = 277.42 Cu. In.
1 Cubic Foot = 7.48 U.S. Gallons

R011110-10   Architectural Fees

Tabulated below are typical percentage fees by project size, for good professional architectural service. Fees may vary from those listed depending upon degree of design difficulty and economic conditions in any particular area.

Rates can be interpolated horizontally and vertically. Various portions of the same project requiring different rates should be adjusted proportionately. For alterations, add 50% to the fee for the first $500,000 of project cost and add 25% to the fee for project cost over $500,000.

Architectural fees tabulated below include Structural, Mechanical and Electrical Engineering Fees. They do not include the fees for special consultants such as kitchen planning, security, acoustical, interior design, etc.

Civil Engineering fees are included in the Architectural fee for project sites requiring minimal design such as city sites. However, separate Civil Engineering fees must be added when utility connections require design, drainage calculations are needed, stepped foundations are required, or provisions are required to protect adjacent wetlands.

Building Types	Total Project Size in Thousands of Dollars						
	100	250	500	1,000	5,000	10,000	50,000
Factories, garages, warehouses, repetitive housing	9.0%	8.0%	7.0%	6.2%	5.3%	4.9%	4.5%
Apartments, banks, schools, libraries, offices, municipal buildings	12.2	12.3	9.2	8.0	7.0	6.6	6.2
Churches, hospitals, homes, laboratories, museums, research	15.0	13.6	12.7	11.9	9.5	8.8	8.0
Memorials, monumental work, decorative furnishings	—	16.0	14.5	13.1	10.0	9.0	8.3

R011110-30   Engineering Fees

Typical **Structural Engineering Fees** based on type of construction and total project size. These fees are included in Architectural Fees.

Type of Construction	Total Project Size (in thousands of dollars)			
	$500	$500-$1,000	$1,000-$5,000	Over $5000
Industrial buildings, factories & warehouses	Technical payroll times 2.0 to 2.5	1.60%	1.25%	1.00%
Hotels, apartments, offices, dormitories, hospitals, public buildings, food stores		2.00%	1.70%	1.20%
Museums, banks, churches and cathedrals		2.00%	1.75%	1.25%
Thin shells, prestressed concrete, earthquake resistive		2.00%	1.75%	1.50%
Parking ramps, auditoriums, stadiums, convention halls, hangars & boiler houses		2.50%	2.00%	1.75%
Special buildings, major alterations, underpinning & future expansion		Add to above 0.5%	Add to above 0.5%	Add to above 0.5%

For complex reinforced concrete or unusually complicated structures, add 20% to 50%.

Typical **Mechanical and Electrical Engineering Fees** are based on the size of the subcontract. The fee structure for both is shown below. These fees are included in Architectural Fees.

Type of Construction	Subcontract Size							
	$25,000	$50,000	$100,000	$225,000	$350,000	$500,000	$750,000	$1,000,000
Simple structures	6.4%	5.7%	4.8%	4.5%	4.4%	4.3%	4.2%	4.1%
Intermediate structures	8.0	7.3	6.5	5.6	5.1	5.0	4.9	4.8
Complex structures	10.1	9.0	9.0	8.0	7.5	7.5	7.0	7.0

For renovations, add 15% to 25% to applicable fee.

R012153-10   Repair and Remodeling

Cost figures are based on new construction utilizing the most cost-effective combination of labor, equipment and material with the work scheduled in proper sequence to allow the various trades to accomplish their work in an efficient manner.

The costs for repair and remodeling work must be modified due to the following factors that may be present in any given repair and remodeling project.

1. Equipment usage curtailment due to the physical limitations of the project, with only hand-operated equipment being used.

2. Increased requirement for shoring and bracing to hold up the building while structural changes are being made and to allow for temporary storage of construction materials on above-grade floors.

3. Material handling becomes more costly due to having to move within the confines of an enclosed building. For multi-story construction, low capacity elevators and stairwells may be the only access to the upper floors.

4. Large amount of cutting and patching and attempting to match the existing construction is required. It is often more economical to remove entire walls rather than create many new door and window openings. This sort of trade-off has to be carefully analyzed.

5. Cost of protection of completed work is increased since the usual sequence of construction usually cannot be accomplished.

6. Economies of scale usually associated with new construction may not be present. If small quantities of components must be custom fabricated due to job requirements, unit costs will naturally increase. Also, if only small work areas are available at a given time, job scheduling between trades becomes difficult and subcontractor quotations may reflect the excessive start-up and shut-down phases of the job.

7. Work may have to be done on other than normal shifts and may have to be done around an existing production facility which has to stay in production during the course of the repair and remodeling.

8. Dust and noise protection of adjoining non-construction areas can involve substantial special protection and alter usual construction methods.

9. Job may be delayed due to unexpected conditions discovered during demolition or removal. These delays ultimately increase construction costs.

10. Piping and ductwork runs may not be as simple as for new construction. Wiring may have to be snaked through walls and floors.

11. Matching "existing construction" may be impossible because materials may no longer be manufactured. Substitutions may be expensive.

12. Weather protection of existing structure requires additional temporary structures to protect building at openings.

13. On small projects, because of local conditions, it may be necessary to pay a tradesman for a minimum of four hours for a task that is completed in one hour.

All of the above areas can contribute to increased costs for a repair and remodeling project. Each of the above factors should be considered in the planning, bidding and construction stage in order to minimize the increased costs associated with repair and remodeling jobs.

R012153-60   Security Factors

Contractors entering, working in, and exiting secure facilities often lose productive time during a normal workday. The recommended allowances in this section are intended to provide for the loss of productivity by increasing labor costs. Note that different costs are associated with searches upon entry only and searches upon entry and exit. Time spent in a queue is unpredictable and not part of these allowances. Contractors should plan ahead for this situation.

Security checkpoints are designed to reflect the level of security required to gain access or egress. An extreme example is when contractors, along with any materials, tools, equipment, and vehicles, must be physically searched and have all materials, tools, equipment, and vehicles inventoried and documented prior to both entry and exit.

Physical searches without going through the documentation process represent the next level and take up less time.

Electronic searches—passing through a detector or x-ray machine with no documentation of materials, tools, equipment, and vehicles—take less time than physical searches.

Visual searches of materials, tools, equipment, and vehicles represent the next level of security.

Finally, access by means of an ID card or displayed sticker takes the least amount of time.

Another consideration is if the searches described above are performed each and every day, or if they are performed only on the first day with access granted by ID card or displayed sticker for the remainder of the project. The figures for this situation have been calculated to represent the initial check-in as described and subsequent entry by ID card or displayed sticker for up to 20 days on site. For the situation described above, where the time period is beyond 20 days, the impact on labor cost is negligible.

There are situations where tradespeople must be accompanied by an escort and observed during the work day. The loss of freedom of movement will slow down productivity for the tradesperson. Costs for the observer have not been included. Those costs are normally born by the owner.

R012153-65   Infectious Disease Precautions

Contractors entering and working on job sites may be required to take precautions to prevent the spread of infectious diseases. Those precautions will reduce the amount of productive time during a normal workday. The recommended allowances in this section are intended to provide for the loss of productive time by increasing labor costs.

The estimator should be aware that one, many or none of these precautions apply to the line items in an estimate. Job site requirements and sound judgement must be applied.

Labor cost implications are based upon:

Temperature checks	Once per day
Donning/doffing masks and gloves	Four times per day (2 times each)
Washing hands (additional over normal)	Eight times per day
Informational meetings	Once per day
Maintaining social distance	Throughout the day
Disinfecting tools or equipment	Throughout the day

R012157-20   Construction Time Requirements

The table below lists the construction durations for various building types along with their respective project sizes and project values. Design time runs 25% to 40% of construction time.

Building Type	Size S.F.	Project Value	Construction Duration
Industrial/Warehouse	100,000	$8,000,000	14 months
	500,000	$32,000,000	19 months
	1,000,000	$75,000,000	21 months
Offices/Retail	50,000	$7,000,000	15 months
	250,000	$28,000,000	23 months
	500,000	$58,000,000	34 months
Institutional/Hospitals/Laboratory	200,000	$45,000,000	31 months
	500,000	$110,000,000	52 months
	750,000	$160,000,000	55 months
	1,000,000	$210,000,000	60 months

R012909-80   Sales Tax by State

State sales tax on materials is tabulated below (5 states have no sales tax). Many states allow local jurisdictions, such as a county or city, to levy additional sales tax.

Some projects may be sales tax exempt, particularly those constructed with public funds.

State	Tax (%)	State	Tax (%)	State	Tax (%)	State	Tax (%)
Alabama	4	Illinois	6.25	Montana	0	Rhode Island	7
Alaska	0	Indiana	7	Nebraska	5.5	South Carolina	6
Arizona	5.6	Iowa	6	Nevada	6.85	South Dakota	4.5
Arkansas	6.5	Kansas	6.5	New Hampshire	0	Tennessee	7
California	7.25	Kentucky	6	New Jersey	6.625	Texas	6.25
Colorado	2.9	Louisiana	4.45	New Mexico	5.125	Utah	6.10
Connecticut	6.35	Maine	5.5	New York	4	Vermont	6
Delaware	0	Maryland	6	North Carolina	4.75	Virginia	5.3
District of Columbia	6	Massachusetts	6.25	North Dakota	5	Washington	6.5
Florida	6	Michigan	6	Ohio	5.75	West Virginia	6
Georgia	4	Minnesota	6.875	Oklahoma	4.5	Wisconsin	5
Hawaii	4	Mississippi	7	Oregon	0	Wyoming	4
Idaho	6	Missouri	4.225	Pennsylvania	6	Average	5.11%

Sales Tax by Province (Canada)

GST - a value-added tax, which the government imposes on most goods and services provided in or imported into Canada. PST - a retail sales tax, which five of the provinces impose on the prices of most goods and some services. QST - a value-added tax, similar to the federal GST, which Quebec imposes. HST - Three provinces have combined their retail sales taxes with the federal GST into one harmonized tax.

Province	PST (%)	QST(%)	GST(%)	HST(%)
Alberta	7	0	5	0
British Columbia	7	0	5	0
Manitoba	7	0	5	0
New Brunswick	0	0	0	15
Newfoundland	0	0	0	15
Northwest Territories	0	0	5	0
Nova Scotia	0	0	0	15
Ontario	0	0	0	13
Prince Edward Island	0	0	0	15
Quebec	9.975	0	5	0
Saskatchewan	6	0	5	0
Yukon	0	0	5	0

R012909-85   Unemployment Taxes and Social Security Taxes

State unemployment tax rates vary not only from state to state, but also with the experience rating of the contractor. The federal unemployment tax rate is 6.2% of the first $7,000 of wages. This is reduced by a credit of up to 5.4% for timely payment to the state. The minimum federal unemployment tax is 0.6% after all credits.

Social security (FICA) for 2021 is estimated at time of publication to be 7.65% of wages up to $137,700.

R012909-86   Unemployment Tax by State

Information is from the U.S. Department of Labor, state unemployment tax rates.

State	Tax (%)	State	Tax (%)	State	Tax (%)	State	Tax (%)
Alabama	6.80	Illinois	6.93	Montana	6.30	Rhode Island	9.49
Alaska	5.4	Indiana	7.4	Nebraska	5.4	South Carolina	5.46
Arizona	12.76	Iowa	7.5	Nevada	5.4	South Dakota	9.35
Arkansas	14.3	Kansas	7.6	New Hampshire	7.5	Tennessee	10.0
California	6.2	Kentucky	9.3	New Jersey	5.8	Texas	6.5
Colorado	8.15	Louisiana	6.2	New Mexico	5.4	Utah	7.1
Connecticut	6.8	Maine	5.46	New York	9.1	Vermont	7.7
Delaware	8.20	Maryland	7.50	North Carolina	5.76	Virginia	6.21
District of Columbia	7	Massachusetts	12.65	North Dakota	10.74	Washington	7.73
Florida	5.4	Michigan	10.3	Ohio	9	West Virginia	8.5
Georgia	5.4	Minnesota	9.1	Oklahoma	5.5	Wisconsin	12.0
Hawaii	5.6	Mississippi	5.6	Oregon	5.4	Wyoming	8.8
Idaho	5.4	Missouri	8.37	Pennsylvania	11.03	Median	7.40%

R012909-90   Overtime

One way to improve the completion date of a project or eliminate negative float from a schedule is to compress activity duration times. This can be achieved by increasing the crew size or working overtime with the proposed crew.

To determine the costs of working overtime to compress activity duration times, consider the following examples. Below is an overtime efficiency and cost chart based on a five, six, or seven day week with an eight through twelve hour day. Payroll percentage increases for time and one half and double times are shown for the various working days.

Days per Week	Hours per Day	Production Efficiency					Payroll Cost Factors	
		1st Week	2nd Week	3rd Week	4th Week	Average 4 Weeks	@ 1-1/2 Times	@ 2 Times
	8	100%	100%	100%	100%	100%	1.000	1.000
	9	100	100	95	90	96	1.056	1.111
5	10	100	95	90	85	93	1.100	1.200
	11	95	90	75	65	81	1.136	1.273
	12	90	85	70	60	76	1.167	1.333
	8	100	100	95	90	96	1.083	1.167
	9	100	95	90	85	93	1.130	1.259
6	10	95	90	85	80	88	1.167	1.333
	11	95	85	70	65	79	1.197	1.394
	12	90	80	65	60	74	1.222	1.444
	8	100	95	85	75	89	1.143	1.286
	9	95	90	80	70	84	1.183	1.365
7	10	90	85	75	65	79	1.214	1.429
	11	85	80	65	60	73	1.240	1.481
	12	85	75	60	55	69	1.262	1.524

R013113-40 Builder's Risk Insurance

Builder's risk insurance is insurance on a building during construction. Premiums are paid by the owner or the contractor. Blasting, collapse and underground insurance would raise total insurance costs.

R013113-50 General Contractor's Overhead

There are two distinct types of overhead on a construction project: Project overhead and main office overhead. Project overhead includes those costs at a construction site not directly associated with the installation of construction materials. Examples of project overhead costs include the following:

1. Superintendent
2. Construction office and storage trailers
3. Temporary sanitary facilities
4. Temporary utilities
5. Security fencing
6. Photographs
7. Cleanup
8. Performance and payment bonds

The above project overhead items are also referred to as general requirements and therefore are estimated in Division 1. Division 1 is the first division listed in the CSI MasterFormat but it is usually the last division estimated. The sum of the costs in Divisions 1 through 49 is referred to as the sum of the direct costs.

All construction projects also include indirect costs. The primary components of indirect costs are the contractor's main office overhead and profit. The amount of the main office overhead expense varies depending on the following:

1. Owner's compensation
2. Project managers' and estimators' wages
3. Clerical support wages
4. Office rent and utilities
5. Corporate legal and accounting costs
6. Advertising
7. Automobile expenses
8. Association dues
9. Travel and entertainment expenses

These costs are usually calculated as a percentage of annual sales volume. This percentage can range from 35% for a small contractor doing less than $500,000 to 5% for a large contractor with sales in excess of $100 million.

R013113-55 Installing Contractor's Overhead

Installing contractors (subcontractors) also incur costs for general requirements and main office overhead.

Included within the total incl. overhead and profit costs is a percent mark-up for overhead that includes:

1. Compensation and benefits for office staff and project managers
2. Office rent, utilities, business equipment, and maintenance
3. Corporate legal and accounting costs

4. Advertising
5. Vehicle expenses (for office staff and project managers)
6. Association dues
7. Travel, entertainment
8. Insurance
9. Small tools and equipment

R013113-60 Workers' Compensation Insurance Rates by Trade

The table below tabulates the national averages for workers' compensation insurance rates by trade and type of building. The average "Insurance Rate" is multiplied by the "% of Building Cost" for each trade. This produces the "Workers' Compensation" cost by % of total labor cost, to be added for each trade by building type to determine the weighted average workers' compensation rate for the building types analyzed.

Trade	Insurance Rate (% Labor Cost) Range		Insurance Rate (% Labor Cost) Average	% of Building Cost Office Bldgs.	% of Building Cost Schools & Apts.	% of Building Cost Mfg.	Workers' Compensation Office Bldgs.	Workers' Compensation Schools & Apts.	Workers' Compensation Mfg.
Excavation, Grading, etc.	2.1 % to	16.5%	9.3%	4.8%	4.9%	4.5%	0.45%	0.46%	0.42%
Piles & Foundations	3.1 to	28.0	15.6	7.1	5.2	8.7	1.11	0.81	1.36
Concrete	2.5 to	24.9	13.7	5.0	14.8	3.7	0.69	2.03	0.51
Masonry	3.2 to	53.5	28.4	6.9	7.5	1.9	1.96	2.13	0.54
Structural Steel	3.3 to	30.8	17.1	10.7	3.9	17.6	1.83	0.67	3.01
Miscellaneous & Ornamental Metals	2.3 to	22.8	12.5	2.8	4.0	3.6	0.35	0.50	0.45
Carpentry & Millwork	2.6 to	29.1	15.8	3.7	4.0	0.5	0.58	0.63	0.08
Metal or Composition Siding	4.2 to	135.3	69.7	2.3	0.3	4.3	1.60	0.21	3.00
Roofing	4.2 to	105.1	54.7	2.3	2.6	3.1	1.26	1.42	1.70
Doors & Hardware	3.1 to	29.1	16.1	0.9	1.4	0.4	0.14	0.23	0.06
Sash & Glazing	2.6 to	20.5	11.5	3.5	4.0	1.0	0.40	0.46	0.12
Lath & Plaster	2.1 to	33.7	17.9	3.3	6.9	0.8	0.59	1.24	0.14
Tile, Marble & Floors	1.9 to	17.4	9.7	2.6	3.0	0.5	0.25	0.29	0.05
Acoustical Ceilings	1.7 to	21.5	11.6	2.4	0.2	0.3	0.28	0.02	0.03
Painting	2.8 to	37.7	20.3	1.5	1.6	1.6	0.30	0.32	0.32
Interior Partitions	2.6 to	29.1	15.8	3.9	4.3	4.4	0.62	0.68	0.70
Miscellaneous Items	1.6 to	97.7	49.7	5.2	3.7	9.7	2.58	1.84	4.82
Elevators	1.0 to	8.9	5.0	2.1	1.1	2.2	0.11	0.06	0.11
Sprinklers	1.4 to	14.2	7.8	0.5	—	2.0	0.04	—	0.16
Plumbing	1.4 to	13.2	7.3	4.9	7.2	5.2	0.36	0.53	0.38
Heat., Vent., Air Conditioning	2.8 to	15.8	9.3	13.5	11.0	12.9	1.26	1.02	1.20
Electrical	1.2 to	10.6	5.9	10.1	8.4	11.1	0.60	0.50	0.65
Total	1.0 % to	135.3%	68.1	100.0%	100.0%	100.0%	17.36%	16.05%	19.81%
Overall Weighted Average							17.74%		

Workers' Compensation Insurance Rates by States

The table below lists the weighted average Workers' Compensation base rate for each state with a factor comparing this with the national average of 15.1%.

State	Weighted Average	Factor	State	Weighted Average	Factor	State	Weighted Average	Factor
Alabama	17.2%	159	Kentucky	17.5%	162	North Dakota	8.3%	77
Alaska	11.9	110	Louisiana	25.5	236	Ohio	6.4	59
Arizona	11.1	103	Maine	13.5	125	Oklahoma	13.3	123
Arkansas	7.3	68	Maryland	15.2	141	Oregon	10.8	100
California	32.6	302	Massachusetts	12.0	111	Pennsylvania	26.1	242
Colorado	8.1	75	Michigan	5.8	54	Rhode Island	11.0	102
Connecticut	19.8	183	Minnesota	20.7	192	South Carolina	24.6	228
Delaware	18.2	169	Mississippi	13.7	127	South Dakota	11.3	105
District of Columbia	11.3	105	Missouri	21.4	198	Tennessee	8.8	81
Florida	12.5	116	Montana	12.2	113	Texas	8.0	74
Georgia	50.0	463	Nebraska	16.6	154	Utah	9.2	85
Hawaii	14.0	130	Nevada	12.0	111	Vermont	13.6	126
Idaho	12.9	119	New Hampshire	12.7	118	Virginia	9.9	92
Illinois	27.9	258	New Jersey	21.5	199	Washington	10.4	96
Indiana	4.9	45	New Mexico	20.7	192	West Virginia	6.7	62
Iowa	15.1	140	New York	25.9	240	Wisconsin	16.0	148
Kansas	8.8	81	North Carolina	17.4	161	Wyoming	7.0	65
Weighted Average for U.S. is			15.1% of payroll = 100%					

The weighted average skilled worker rate for 35 trades is 17.74%. For bidding purposes, apply the full value of Workers' Compensation directly to total labor costs, or if labor is 38%, materials 42% and overhead and profit 20% of total cost, carry 38/80 x 17.74% = 8.43% of cost (before overhead and profit) into overhead. Rates vary not only from state to state but also with the experience rating of the contractor.

Rates are the most current available at the time of publication.

R013113-80   Performance Bond

This table shows the cost of a Performance Bond for a construction job scheduled to be completed in 12 months. Add 1% of the premium cost per month for jobs requiring more than 12 months to complete. The rates are "standard" rates offered to contractors that the bonding company considers financially sound and capable of doing the work. Preferred rates are offered by some bonding companies based upon financial strength of the contractor. Actual rates vary from contractor to contractor and from bonding company to bonding company. Contractors should prequalify through a bonding agency before submitting a bid on a contract that requires a bond.

| Contract Amount | Building Construction Class B Projects | | | Highways & Bridges | | | | | |
				Class A New Construction			Class A-1 Highway Resurfacing		
First $ 100,000 bid	$25.00 per M			$15.00 per M			$9.40 per M		
Next 400,000 bid	$ 2,500	plus $15.00	per M	$ 1,500	plus $10.00	per M	$ 940	plus $7.20	per M
Next 2,000,000 bid	8,500	plus 10.00	per M	5,500	plus 7.00	per M	3,820	plus 5.00	per M
Next 2,500,000 bid	28,500	plus 7.50	per M	19,500	plus 5.50	per M	15,820	plus 4.50	per M
Next 2,500,000 bid	47,250	plus 7.00	per M	33,250	plus 5.00	per M	28,320	plus 4.50	per M
Over 7,500,000 bid	64,750	plus 6.00	per M	45,750	plus 4.50	per M	39,570	plus 4.00	per M

R015423-10   Steel Tubular Scaffolding

On new construction, tubular scaffolding is efficient up to 60' high or five stories. Above this it is usually better to use a hung scaffolding if construction permits. Swing scaffolding operations may interfere with tenants. In this case, the tubular is more practical at all heights.

In repairing or cleaning the front of an existing building the cost of tubular scaffolding per S.F. of building front increases as the height increases above the first tier. The first tier cost is relatively high due to leveling and alignment.

The minimum efficient crew for erecting and dismantling is three workers. They can set up and remove 18 frame sections per day up to 5 stories high. For 6 to 12 stories high, a crew of four is most efficient. Use two or more on top and two on the bottom for handing up or hoisting. They can also set up and remove 18 frame sections per day. At 7' horizontal spacing, this will run about 800 S.F. per day of erecting and dismantling. Time for placing and removing planks must be added to the above. A crew of three can place and remove 72 planks per day up to 5 stories. For over 5 stories, a crew of four can place and remove 80 planks per day.

The table below shows the number of pieces required to erect tubular steel scaffolding for 1000 S.F. of building frontage. This area is made up of a scaffolding system that is 12 frames (11 bays) long by 2 frames high.

For jobs under twenty-five frames, add 50% to rental cost. Rental rates will be lower for jobs over three months duration. Large quantities for long periods can reduce rental rates by 20%.

Description of Component	Number of Pieces for 1000 S.F. of Building Front	Unit
5' Wide Standard Frame, 6'-4" High	24	Ea.
Leveling Jack & Plate	24	
Cross Brace	44	
Side Arm Bracket, 21"	12	
Guardrail Post	12	
Guardrail, 7' section	22	
Stairway Section	2	
Stairway Starter Bar	1	
Stairway Inside Handrail	2	
Stairway Outside Handrail	2	
Walk-Thru Frame Guardrail	2	

Scaffolding is often used as falsework over 15' high during construction of cast-in-place concrete beams and slabs. Two foot wide scaffolding is generally used for heavy beam construction. The span between frames depends upon the load to be carried with a maximum span of 5'.

Heavy duty shoring frames with a capacity of 10,000#/leg can be spaced up to 10' O.C. depending upon form support design and loading.

Scaffolding used as horizontal shoring requires less than half the material required with conventional shoring.

On new construction, erection is done by carpenters.

Rolling towers supporting horizontal shores can reduce labor and speed the job. For maintenance work, catwalks with spans up to 70' can be supported by the rolling towers.

R015423-20 Pump Staging

Pump staging is generally not available for rent. The table below shows the number of pieces required to erect pump staging for 2400 S.F. of building frontage. This area is made up of a pump jack system that is 3 poles (2 bays) wide by 2 poles high.

Item	Number of Pieces for 2400 S.F. of Building Front	Unit
Aluminum pole section, 24' long	6	Ea.
Aluminum splice joint, 6' long	3	
Aluminum foldable brace	3	
Aluminum pump jack	3	
Aluminum support for workbench/back safety rail	3	
Aluminum scaffold plank/workbench, 14" wide x 24' long	4	
Safety net, 22' long	2	
Aluminum plank end safety rail	2	↓

The cost in place for this 2400 S.F. will depend on how many uses are realized during the life of the equipment.

R015433-10   Contractor Equipment

Rental Rates shown elsewhere in the data set pertain to late model high quality machines in excellent working condition, rented from equipment dealers. Rental rates from contractors may be substantially lower than the rental rates from equipment dealers depending upon economic conditions; for older, less productive machines, reduce rates by a maximum of 15%. Any overtime must be added to the base rates. For shift work, rates are lower. Usual rule of thumb is 150% of one shift rate for two shifts; 200% for three shifts.

For periods of less than one week, operated equipment is usually more economical to rent than renting bare equipment and hiring an operator.

Costs to move equipment to a job site (mobilization) or from a job site (demobilization) are not included in rental rates, nor in any Equipment costs on any Unit Price line items or crew listings. These costs can be found elsewhere. If a piece of equipment is already at a job site, it is not appropriate to utilize mob/demob costs in an estimate again.

Rental rates vary throughout the country with larger cities generally having lower rates. Lease plans for new equipment are available for periods in excess of six months with a percentage of payments applying toward purchase.

Rental rates can also be treated as reimbursement costs for contractor-owned equipment. Owned equipment costs include depreciation, loan payments, interest, taxes, insurance, storage, and major repairs.

Monthly rental rates vary from 2% to 5% of the cost of the equipment depending on the anticipated life of the equipment and its wearing parts. Weekly rates are about 1/3 the monthly rates and daily rental rates are about 1/3 the weekly rates.

The hourly operating costs for each piece of equipment include costs to the user such as fuel, oil, lubrication, normal expendables for the equipment, and a percentage of the mechanic's wages chargeable to maintenance. The hourly operating costs listed do not include the operator's wages.

The daily cost for equipment used in the standard crews is figured by dividing the weekly rate by five, then adding eight times the hourly operating cost to give the total daily equipment cost, not including the operator. This figure is in the right hand column of the Equipment listings under Equipment Cost/Day.

Pile Driving rates shown for the pile hammer and extractor do not include leads, cranes, boilers or compressors. Vibratory pile driving requires an added field specialist during set-up and pile driving operation for the electric model. The hydraulic model requires a field specialist for set-up only. Up to 125 reuses of sheet piling are possible using vibratory drivers. For normal conditions, crane capacity for hammer type and size is as follows.

Crane Capacity	Hammer Type and Size		
	Air or Steam	Diesel	Vibratory
25 ton	to 8,750 ft.-lb.		70 H.P.
40 ton	15,000 ft.-lb.	to 32,000 ft.-lb.	170 H.P.
60 ton	25,000 ft.-lb.		300 H.P.
100 ton		112,000 ft.-lb.	

Cranes should be specified for the job by size, building and site characteristics, availability, performance characteristics, and duration of time required.

Backhoes & Shovels rent for about the same as equivalent size cranes but maintenance and operating expenses are higher. The crane operator's rate must be adjusted for high boom heights. Average adjustments: for 150' boom add 2% per hour; over 185', add 4% per hour; over 210', add 6% per hour; over 250', add 8% per hour and over 295', add 12% per hour.

Tower Cranes of the climbing or static type have jibs from 50' to 200' and capacities at maximum reach range from 4,000 to 14,000 pounds. Lifting capacities increase up to maximum load as the hook radius decreases.

Typical rental rates, based on purchase price, are about 2% to 3% per month.

Erection and dismantling run between 500 and 2000 labor hours. Climbing operation takes 10 labor hours per 20' climb. Crane dead time is about 5 hours per 40' climb. If crane is bolted to side of the building add cost of ties and extra mast sections. Climbing cranes have from 80' to 180' of mast while static cranes have 80' to 800' of mast.

Truck Cranes can be converted to tower cranes by using tower attachments. Mast heights over 400' have been used.

A single 100' high material **Hoist and Tower** can be erected and dismantled in about 400 labor hours; a double 100' high hoist and tower in about 600 labor hours. Erection times for additional heights are 3 and 4 labor hours per vertical foot respectively up to 150', and 4 to 5 labor hours per vertical foot over 150' high. A 40' high portable Buck hoist takes about 160 labor hours to erect and dismantle. Additional heights take 2 labor hours per vertical foot to 80' and 3 labor hours per vertical foot for the next 100'. Most material hoists do not meet local code requirements for carrying personnel.

A 150' high **Personnel Hoist** requires about 500 to 800 labor hours to erect and dismantle. Budget erection time at 5 labor hours per vertical foot for all trades. Local code requirements or labor scarcity requiring overtime can add up to 50% to any of the above erection costs.

Earthmoving Equipment: The selection of earthmoving equipment depends upon the type and quantity of material, moisture content, haul distance, haul road, time available, and equipment available. Short haul cut and fill operations may require dozers only, while another operation may require excavators, a fleet of trucks, and spreading and compaction equipment. Stockpiled material and granular material are easily excavated with front end loaders. Scrapers are most economically used with hauls between 300' and 1-1/2 miles if adequate haul roads can be maintained. Shovels are often used for blasted rock and any material where a vertical face of 8' or more can be excavated. Special conditions may dictate the use of draglines, clamshells, or backhoes. Spreading and compaction equipment must be matched to the soil characteristics, the compaction required and the rate the fill is being supplied.

R015433-15   Heavy Lifting

Hydraulic Climbing Jacks

The use of hydraulic heavy lift systems is an alternative to conventional type crane equipment. The lifting, lowering, pushing, or pulling mechanism is a hydraulic climbing jack moving on a square steel jackrod from 1-5/8" to 4" square, or a steel cable. The jackrod or cable can be vertical or horizontal, stationary or movable, depending on the individual application. When the jackrod is stationary, the climbing jack will climb the rod and push or pull the load along with itself. When the climbing jack is stationary, the jackrod is movable with the load attached to the end and the climbing jack will lift or lower the jackrod with the attached load. The heavy lift system is normally operated by a single control lever located at the hydraulic pump.

The system is flexible in that one or more climbing jacks can be applied wherever a load support point is required, and the rate of lift synchronized.

Economic benefits have been demonstrated on projects such as: erection of ground assembled roofs and floors, complete bridge spans, girders and trusses, towers, chimney liners and steel vessels, storage tanks, and heavy machinery. Other uses are raising and lowering offshore work platforms, caissons, tunnel sections and pipelines.

R015436-50 Mobilization

Costs to move rented construction equipment to a job site from an equipment dealer's or contractor's yard (mobilization) or off the job site (demobilization) are not included in the rental or operating rates, nor in the equipment cost on a unit price line or in a crew listing. These costs can be found consolidated in the Mobilization section of the data and elsewhere in particular site work sections. If a piece of equipment is already on the job site, it is not appropriate to include mob/demob costs in a new estimate that requires use of that equipment. The following table identifies approximate sizes of rented construction equipment that would be hauled on a towed trailer. Because this listing is not all-encompassing, the user can infer as to what size trailer might be required for a piece of equipment not listed.

3-ton Trailer	20-ton Trailer	40-ton Trailer	50-ton Trailer
20 H.P. Excavator	110 H.P. Excavator	200 H.P. Excavator	270 H.P. Excavator
50 H.P. Skid Steer	165 H.P. Dozer	300 H.P. Dozer	Small Crawler Crane
35 H.P. Roller	150 H.P. Roller	400 H.P. Scraper	500 H.P. Scraper
40 H.P. Trencher	Backhoe	450 H.P. Art. Dump Truck	500 H.P. Art. Dump Truck

For customer support on your Commercial Renovation Costs with RSMeans Data, call 800.448.8182.

853

R024119-10   Demolition Defined

Whole Building Demolition - Demolition of the whole building with no concern for any particular building element, component, or material type being demolished. This type of demolition is accomplished with large pieces of construction equipment that break up the structure, load it into trucks and haul it to a disposal site, but disposal or dump fees are not included. Demolition of below-grade foundation elements, such as footings, foundation walls, grade beams, slabs on grade, etc., is not included. Certain mechanical equipment containing flammable liquids or ozone-depleting refrigerants, electric lighting elements, communication equipment components, and other building elements may contain hazardous waste, and must be removed, either selectively or carefully, as hazardous waste before the building can be demolished.

Foundation Demolition - Demolition of below-grade foundation footings, foundation walls, grade beams, and slabs on grade. This type of demolition is accomplished by hand or pneumatic hand tools, and does not include saw cutting, or handling, loading, hauling, or disposal of the debris.

Gutting - Removal of building interior finishes and electrical/mechanical systems down to the load-bearing and sub-floor elements of the rough building frame, with no concern for any particular building element, component, or material type being demolished. This type of demolition is accomplished by hand or pneumatic hand tools, and includes loading into trucks, but not hauling, disposal or dump fees, scaffolding, or shoring. Certain mechanical equipment containing flammable liquids or ozone-depleting refrigerants, electric lighting elements, communication equipment components, and other building elements may contain hazardous waste, and must be removed, either selectively or carefully, as hazardous waste, before the building is gutted.

Selective Demolition - Demolition of a selected building element, component, or finish, with some concern for surrounding or adjacent elements, components, or finishes (see the first Subdivision (s) at the beginning of appropriate Divisions). This type of demolition is accomplished by hand or pneumatic hand tools, and does not include handling, loading,

storing, hauling, or disposal of the debris, scaffolding, or shoring. "Gutting" methods may be used in order to save time, but damage that is caused to surrounding or adjacent elements, components, or finishes may have to be repaired at a later time.

Careful Removal - Removal of a piece of service equipment, building element or component, or material type, with great concern for both the removed item and surrounding or adjacent elements, components, or finishes. The purpose of careful removal may be to protect the removed item for later re-use, preserve a higher salvage value of the removed item, or replace an item while taking care to protect surrounding or adjacent elements, components, connections, or finishes from cosmetic and/or structural damage. An approximation of the time required to perform this type of removal is 1/3 to 1/2 the time it would take to install a new item of like kind (see Reference Number R220105-10). This type of removal is accomplished by hand or pneumatic hand tools, and does not include loading, hauling, or storing the removed item, scaffolding, shoring, or lifting equipment.

Cutout Demolition - Demolition of a small quantity of floor, wall, roof, or other assembly, with concern for the appearance and structural integrity of the surrounding materials. This type of demolition is accomplished by hand or pneumatic hand tools, and does not include saw cutting, handling, loading, hauling, or disposal of debris, scaffolding, or shoring.

Rubbish Handling - Work activities that involve handling, loading or hauling of debris. Generally, the cost of rubbish handling must be added to the cost of all types of demolition, with the exception of whole building demolition.

Minor Site Demolition - Demolition of site elements outside the footprint of a building. This type of demolition is accomplished by hand or pneumatic hand tools, or with larger pieces of construction equipment, and may include loading a removed item onto a truck (check the Crew for equipment used). It does not include saw cutting, hauling or disposal of debris, and, sometimes, handling or loading.

R024119-20   Dumpsters

Dumpster rental costs on construction sites are presented in two ways.

The cost per week rental includes the delivery of the dumpster; its pulling or emptying once per week, and its final removal. The assumption is made that the dumpster contractor could choose to empty a dumpster by simply bringing in an empty unit and removing the full one. These costs also include the disposal of the materials in the dumpster.

The Alternate Pricing can be used when actual planned conditions are not approximated by the weekly numbers. For example, these lines can be used when a dumpster is needed for 4 weeks and will need to be emptied 2 or 3 times per week. Conversely the Alternate Pricing lines can be used when a dumpster will be rented for several weeks or months but needs to be emptied only a few times over this period.

R024119-30   Rubbish Handling Chutes

To correctly estimate the cost of rubbish handling chute systems, the individual components must be priced separately. First choose the size of the system; a 30-inch diameter chute is quite common, but the sizes range from 18 to 36 inches in diameter. The 30-inch chute comes in a standard weight and two thinner weights. The thinner weight chutes are sometimes chosen for cost savings, but they are more easily damaged.

There are several types of major chute pieces that make up the chute system. The first component to consider is the top chute section (top intake hopper) where the material is dropped into the chute at the highest point. After determining the top chute, the intermediate chute pieces called the regular chute sections are priced. Next, the number of chute control door sections (intermediate intake hoppers) must be determined. In the more complex systems, a chute control door section is provided at each floor level. The last major component to consider is bolt down frames; these are usually provided at every other floor level.

There are a number of accessories to consider for safe operation and control. There are covers for the top chute and the chute door sections. The top

chute can have a trough that allows for better loading of the chute. For the safest operation, a chute warning light system can be added that will warn the other chute intake locations not to load while another is being used. There are dust control devices that spray a water mist to keep down the dust as the debris is loaded into a Dumpster. There are special breakaway cords that are used to prevent damage to the chute if the dumpster is removed without disconnecting from the chute. There are chute liners that can be installed to protect the chute structure from physical damage from rough abrasive materials. Warning signs can be posted at each floor level that is provided with a chute control door section.

In summary, a complete rubbish handling chute system will include one top section, several intermediate regular sections, several intermediate control door (intake hopper) sections and bolt down frames at every other floor level starting with the top floor. If so desired, the system can also include covers and a light warning system for a safer operation. The bottom of the chute should always be above the Dumpster and should be tied off with a breakaway cord to the Dumpster.

R026510-20   Underground Storage Tank Removal

Underground storage tank removal can be divided into two categories: non-leaking and leaking. Prior to removing an underground storage tank, tests should be made, with the proper authorities present, to determine whether a tank has been leaking or the surrounding soil has been contaminated.

To safely remove liquid underground storage tanks:
1. Excavate to the top of the tank.
2. Disconnect all piping.
3. Open all tank vents and access ports.
4. Remove all liquids and/or sludge.
5. Purge the tank with an inert gas.
6. Provide access to the inside of the tank and clean out the interior using proper personal protective equipment (PPE).
7. Excavate soil surrounding the tank using proper PPE for on-site personnel.
8. Pull and properly dispose of the tank.
9. Clean up the site of all contaminated material.
10. Install new tanks or close the excavation.

R028213-20   Asbestos Removal Process

Asbestos removal is accomplished by a specialty contractor who understands the federal and state regulations regarding the handling and disposal of the material. The process of asbestos removal is divided into many individual steps. An accurate estimate can be calculated only after all the steps have been priced.

The steps are generally as follows:
1. Obtain an asbestos abatement plan from an industrial hygienist.
2. Monitor the air quality in and around the removal area and along the path of travel between the removal area and transport area. This establishes the background contamination.
3. Construct a two part decontamination chamber at entrance to removal area.
4. Install a HEPA filter to create a negative pressure in the removal area.
5. Install wall, floor and ceiling protection as required by the plan, usually 2 layers of fireproof 6 mil polyethylene.
6. Industrial hygienist visually inspects work area to verify compliance with plan.
7. Provide temporary supports for conduit and piping affected by the removal process.
8. Proceed with asbestos removal and bagging process. Monitor air quality as described in Step #2. Discontinue operations when contaminate levels exceed applicable standards.
9. Document the legal disposal of materials in accordance with EPA standards.
10. Thoroughly clean removal area including all ledges, crevices and surfaces.
11. Post abatement inspection by industrial hygienist to verify plan compliance.
12. Provide a certificate from a licensed industrial hygienist attesting that contaminate levels are within acceptable standards before returning area to regular use.

R028319-60   Lead Paint Remediation Methods

Lead paint remediation can be accomplished by the following methods.

1. Abrasive blast
2. Chemical stripping
3. Power tool cleaning with vacuum collection system
4. Encapsulation
5. Remove and replace
6. Enclosure

Each of these methods has strengths and weaknesses depending on the specific circumstances of the project. The following is an overview of each method.

1. **Abrasive blasting** is usually accomplished with sand or recyclable metallic blast. Before work can begin, the area must be contained to ensure the blast material with lead does not escape to the atmosphere. The use of vacuum blast greatly reduces the containment requirements. Lead abatement equipment that may be associated with this work includes a negative air machine. In addition, it is necessary to have an industrial hygienist monitor the project on a continual basis. When the work is complete, the spent blast sand with lead must be disposed of as a hazardous material. If metallic shot was used, the lead is separated from the shot and disposed of as hazardous material. Worker protection includes disposable clothing and respiratory protection.

2. **Chemical stripping** requires strong chemicals to be applied to the surface to remove the lead paint. Before the work can begin, the area under/adjacent to the work area must be covered to catch the chemical and removed lead. After the chemical is applied to the painted surface it is usually covered with paper. The chemical is left in place for the specified period, then the paper with lead paint is pulled or scraped off. The process may require several chemical applications. The paper with chemicals and lead paint adhered to it, plus the containment and loose scrapings collected by a HEPA (High Efficiency Particulate Air Filter) vac, must be disposed of as hazardous material. The chemical stripping process usually requires a neutralizing agent and several wash downs after the paint is removed. Worker protection includes a neoprene or other compatible protective clothing and respiratory

protection with face shield. An industrial hygienist is required intermittently during the process.

3. **Power tool cleaning** is accomplished using shrouded needle blasting guns. The shrouding with different end configurations is held up against the surface to be cleaned. The area is blasted with hardened needles and the shroud captures the lead with a HEPA vac and deposits it in a holding tank. An industrial hygienist monitors the project. Protective clothing and a respirator are required until air samples prove otherwise. When the work is complete the lead must be disposed of as hazardous material.

4. **Encapsulation** is a method that leaves the well bonded lead paint in place after the peeling paint has been removed. Before the work can begin, the area under/adjacent to the work must be covered to catch the scrapings. The scraped surface is then washed with a detergent and rinsed. The prepared surface is covered with approximately 10 mils of paint. A reinforcing fabric can also be embedded in the paint covering. The scraped paint and containment must be disposed of as hazardous material. Workers must wear protective clothing and respirators.

5. **Removing and replacing** are effective ways to remove lead paint from windows, gypsum walls and concrete masonry surfaces. The painted materials are removed and new materials are installed. Workers should wear a respirator and tyvek suit. The demolished materials must be disposed of as hazardous waste if it fails the TCLP (Toxicity Characteristic Leachate Process) test.

6. **Enclosure** is the process that permanently seals lead painted materials in place. This process has many applications such as covering lead painted drywall with new drywall, covering exterior construction with tyvek paper then residing, or covering lead painted structural members with aluminum or plastic. The seams on all enclosing materials must be securely sealed. An industrial hygienist monitors the project, and protective clothing and a respirator are required until air samples prove otherwise.

All the processes require clearance monitoring and wipe testing as required by the hygienist.

R031113-10   Wall Form Materials

Aluminum Forms

Approximate weight is 3 lbs. per S.F.C.A. Standard widths are available from 4″ to 36″ with 36″ most common. Standard lengths of 2′, 4′, 6′ to 8′ are available. Forms are lightweight and fewer ties are needed with the wider widths. The form face is either smooth or textured.

Metal Framed Plywood Forms

Manufacturers claim over 75 reuses of plywood and over 300 reuses of steel frames. Many specials such as corners, fillers, pilasters, etc. are available. Monthly rental is generally about 15% of purchase price for first month and 9% per month thereafter with 90% of rental applied to purchase for the first month and decreasing percentages thereafter. Aluminum framed forms cost 25% to 30% more than steel framed.

After the first month, extra days may be prorated from the monthly charge. Rental rates do not include ties, accessories, cleaning, loss of hardware or freight in and out. Approximate weight is 5 lbs. per S.F. for steel; 3 lbs. per S.F. for aluminum.

Forms can be rented with option to buy.

Plywood Forms, Job Fabricated

There are two types of plywood used for concrete forms.
1. Exterior plyform which is completely waterproof. This is face oiled to facilitate stripping. Ten reuses can be expected with this type with 25 reuses possible.
2. An overlaid type consists of a resin fiber fused to exterior plyform. No oiling is required except to facilitate cleaning. This is available in both high density (HDO) and medium density overlaid (MDO). Using HDO, 50 reuses can be expected with 200 possible.

Plyform is available in 5/8″ and 3/4″ thickness. High density overlaid is available in 3/8″, 1/2″, 5/8″ and 3/4″ thickness.

5/8″ thick is sufficient for most building forms, while 3/4″ is best on heavy construction.

Plywood Forms, Modular, Prefabricated

There are many plywood forming systems without frames. Most of these are manufactured from 1-1/8″ (HDO) plywood and have some hardware attached. These are used principally for foundation walls 8′ or less high. With care and maintenance, 100 reuses can be attained with decreasing quality of surface finish.

Steel Forms

Approximate weight is 6-1/2 lbs. per S.F.C.A. including accessories. Standard widths are available from 2″ to 24″, with 24″ most common. Standard lengths are from 2′ to 8′, with 4′ the most common. Forms are easily ganged into modular units.

Forms are usually leased for 15% of the purchase price per month prorated daily over 30 days.

Rental may be applied to sale price, and usually rental forms are bought. With careful handling and cleaning 200 to 400 reuses are possible.

Straight wall gang forms up to 12′ x 20′ or 8′ x 30′ can be fabricated. These crane handled forms usually lease for approx. 9% per month.

Individual job analysis is available from the manufacturer at no charge.

R031113-40   Forms for Reinforced Concrete

Design Economy

Avoid many sizes in proportioning beams and columns.

From story to story avoid changing column dimensions. Gain strength by adding steel or using a richer mix. If a change in size of column is necessary, vary one dimension only to minimize form alterations. Keep beams and columns the same width.

From floor to floor in a multi-story building vary beam depth, not width, as that will leave the slab panel form unchanged. It is cheaper to vary the strength of a beam from floor to floor by means of a steel area than by 2″ changes in either width or depth.

Cost Factors

Material includes the cost of lumber, cost of rent for metal pans or forms if used, nails, form ties, form oil, bolts and accessories.

Labor includes the cost of carpenters to make up, erect, remove and repair, plus common labor to clean and move. Having carpenters remove forms minimizes repairs.

Improper alignment and condition of forms will increase finishing cost. When forms are heavily oiled, concrete surfaces must be neutralized before finishing. Special curing compounds will cause spillages to spall off in first frost. Gang forming methods will reduce costs on large projects.

Materials Used

Boards are seldom used unless their architectural finish is required. Generally, steel, fiberglass and plywood are used for contact surfaces. Labor on plywood is 10% less than with boards. The plywood is backed up with

2 x 4′s at 12″ to 32″ O.C. Walers are generally 2 - 2 x 4′s. Column forms are held together with steel yokes or bands. Shoring is with adjustable shoring or scaffolding for high ceilings.

Reuse

Floor and column forms can be reused four or possibly five times without excessive repair. Remember to allow for 10% waste on each reuse.

When modular sized wall forms are made, up to twenty uses can be expected with exterior plyform.

When forms are reused, the cost to erect, strip, clean and move will not be affected. 10% replacement of lumber should be included and about one hour of carpenter time for repairs on each reuse per 100 S.F.

The reuse cost for certain accessory items normally rented on a monthly basis will be lower than the cost for the first use.

After the fifth use, new material required plus time needed for repair prevent the form cost from dropping further; it may go up. Much depends on care in stripping, the number of special bays, changes in beam or column sizes and other factors.

Costs for multiple use of formwork may be developed as follows:

2 Uses	3 Uses	4 Uses
$\dfrac{\text{(1st Use + Reuse)}}{2} = \text{avg. cost/2 uses}$	$\dfrac{\text{(1st Use + 2 Reuses)}}{3} = \text{avg. cost/3 uses}$	$\dfrac{\text{(1st use + 3 Reuses)}}{4} = \text{avg. cost/4 uses}$

R031113-60 Formwork Labor-Hours

Item	Unit	Hours Required			Total Hours	Multiple Use		
		Fabricate	Erect & Strip	Clean & Move	1 Use	2 Use	3 Use	4 Use
Beam and Girder, interior beams, 12" wide	100 S.F.	6.4	8.3	1.3	16.0	13.3	12.4	12.0
Hung from steel beams		5.8	7.7	1.3	14.8	12.4	11.6	11.2
Beam sides only, 36" high		5.8	7.2	1.3	14.3	11.9	11.1	10.7
Beam bottoms only, 24" wide		6.6	13.0	1.3	20.9	18.1	17.2	16.7
Box out for openings		9.9	10.0	1.1	21.0	16.6	15.1	14.3
Buttress forms, to 8' high		6.0	6.5	1.2	13.7	11.2	10.4	10.0
Centering, steel, 3/4" rib lath			1.0		1.0			
3/8" rib lath or slab form	↓		0.9		0.9			
Chamfer strip or keyway	100 L.F.		1.5		1.5	1.5	1.5	1.5
Columns, fiber tube 8" diameter			20.6		20.6			
12"			21.3		21.3			
16"			22.9		22.9			
20"			23.7		23.7			
24"			24.6		24.6			
30"	↓		25.6		25.6			
Columns, round steel, 12" diameter			22.0		22.0	22.0	22.0	22.0
16"			25.6		25.6	25.6	25.6	25.6
20"			30.5		30.5	30.5	30.5	30.5
24"	↓		37.7		37.7	37.7	37.7	37.7
Columns, plywood 8" x 8"	100 S.F.	7.0	11.0	1.2	19.2	16.2	15.2	14.7
12" x 12"		6.0	10.5	1.2	17.7	15.2	14.4	14.0
16" x 16"		5.9	10.0	1.2	17.1	14.7	13.8	13.4
24" x 24"		5.8	9.8	1.2	16.8	14.4	13.6	13.2
Columns, steel framed plywood 8" x 8"			10.0	1.0	11.0	11.0	11.0	11.0
12" x 12"			9.3	1.0	10.3	10.3	10.3	10.3
16" x 16"			8.5	1.0	9.5	9.5	9.5	9.5
24" x 24"			7.8	1.0	8.8	8.8	8.8	8.8
Drop head forms, plywood		9.0	12.5	1.5	23.0	19.0	17.7	17.0
Coping forms		8.5	15.0	1.5	25.0	21.3	20.0	19.4
Culvert, box			14.5	4.3	18.8	18.8	18.8	18.8
Curb forms, 6" to 12" high, on grade		5.0	8.5	1.2	14.7	12.7	12.1	11.7
On elevated slabs		6.0	10.8	1.2	18.0	15.5	14.7	14.3
Edge forms to 6" high, on grade	100 L.F.	2.0	3.5	0.6	6.1	5.6	5.4	5.3
7" to 12" high	100 S.F.	2.5	5.0	1.0	8.5	7.8	7.5	7.4
Equipment foundations		10.0	18.0	2.0	30.0	25.5	24.0	23.3
Flat slabs, including drops		3.5	6.0	1.2	10.7	9.5	9.0	8.8
Hung from steel		3.0	5.5	1.2	9.7	8.7	8.4	8.2
Closed deck for domes		3.0	5.8	1.2	10.0	9.0	8.7	8.5
Open deck for pans		2.2	5.3	1.0	8.5	7.9	7.7	7.6
Footings, continuous, 12" high		3.5	3.5	1.5	8.5	7.3	6.8	6.6
Spread, 12" high		4.7	4.2	1.6	10.5	8.7	8.0	7.7
Pile caps, square or rectangular		4.5	5.0	1.5	11.0	9.3	8.7	8.4
Grade beams, 24" deep		2.5	5.3	1.2	9.0	8.3	8.0	7.9
Lintel or Sill forms		8.0	17.0	2.0	27.0	23.5	22.3	21.8
Spandrel beams, 12" wide		9.0	11.2	1.3	21.5	17.5	16.2	15.5
Stairs			25.0	4.0	29.0	29.0	29.0	29.0
Trench forms in floor		4.5	14.0	1.5	20.0	18.3	17.7	17.4
Walls, Plywood, at grade, to 8' high		5.0	6.5	1.5	13.0	11.0	9.7	9.5
8' to 16'		7.5	8.0	1.5	17.0	13.8	12.7	12.1
16' to 20'		9.0	10.0	1.5	20.5	16.5	15.2	14.5
Foundation walls, to 8' high		4.5	6.5	1.0	12.0	10.3	9.7	9.4
8' to 16' high		5.5	7.5	1.0	14.0	11.8	11.0	10.6
Retaining wall to 12' high, battered		6.0	8.5	1.5	16.0	13.5	12.7	12.3
Radial walls to 12' high, smooth		8.0	9.5	2.0	19.5	16.0	14.8	14.3
2' chords		7.0	8.0	1.5	16.5	13.5	12.5	12.0
Prefabricated modular, to 8' high		—	4.3	1.0	5.3	5.3	5.3	5.3
Steel, to 8' high		—	6.8	1.2	8.0	8.0	8.0	8.0
8' to 16' high		—	9.1	1.5	10.6	10.3	10.2	10.2
Steel framed plywood to 8' high		—	6.8	1.2	8.0	7.5	7.3	7.2
8' to 16' high	↓	—	9.3	1.2	10.5	9.5	9.2	9.0

R032110-10   Reinforcing Steel Weights and Measures

Bar Designation No.**	Nominal Weight Lb./Ft.	U.S. Customary Units Nominal Dimensions*			SI Units Nominal Dimensions*			
		Diameter in.	Cross Sectional Area, in.²	Perimeter in.	Nominal Weight kg/m	Diameter mm	Cross Sectional Area, cm²	Perimeter mm
3	.376	.375	.11	1.178	.560	9.52	.71	29.9
4	.668	.500	.20	1.571	.994	12.70	1.29	39.9
5	1.043	.625	.31	1.963	1.552	15.88	2.00	49.9
6	1.502	.750	.44	2.356	2.235	19.05	2.84	59.8
7	2.044	.875	.60	2.749	3.042	22.22	3.87	69.8
8	2.670	1.000	.79	3.142	3.973	25.40	5.10	79.8
9	3.400	1.128	1.00	3.544	5.059	28.65	6.45	90.0
10	4.303	1.270	1.27	3.990	6.403	32.26	8.19	101.4
11	5.313	1.410	1.56	4.430	7.906	35.81	10.06	112.5
14	7.650	1.693	2.25	5.320	11.384	43.00	14.52	135.1
18	13.600	2.257	4.00	7.090	20.238	57.33	25.81	180.1

* The nominal dimensions of a deformed bar are equivalent to those of a plain round bar having the same weight per foot as the deformed bar.
** Bar numbers are based on the number of eighths of an inch included in the nominal diameter of the bars.

R032110-80   Shop-Fabricated Reinforcing Steel

The material prices for reinforcing, shown in the unit cost sections of the data set, are for 50 tons or more of shop-fabricated reinforcing steel and include:

1. Mill base price of reinforcing steel
2. Mill grade/size/length extras
3. Mill delivery to the fabrication shop
4. Shop storage and handling
5. Shop drafting/detailing
6. Shop shearing and bending
7. Shop listing
8. Shop delivery to the job site

Both material and installation costs can be considerably higher for small jobs consisting primarily of smaller bars, while material costs may be slightly lower for larger jobs.

R032205-30   Common Stock Styles of Welded Wire Fabric

This table provides some of the basic specifications, sizes, and weights of welded wire fabric used for reinforcing concrete.

New Designation	Old Designation		Steel Area per Foot				Approximate Weight per 100 S.F.	
Spacing — Cross Sectional Area (in.) — (Sq. in. 100)	Spacing — Wire Gauge (in.) — (AS & W)		Longitudinal		Transverse			
			in.	cm	in.	cm	lbs	kg
Rolls								
6 x 6 — W1.4 x W1.4	6 x 6 — 10 x 10		.028	.071	.028	.071	21	9.53
6 x 6 — W2.0 x W2.0	6 x 6 — 8 x 8	1	.040	.102	.040	.102	29	13.15
6 x 6 — W2.9 x W2.9	6 x 6 — 6 x 6		.058	.147	.058	.147	42	19.05
6 x 6 — W4.0 x W4.0	6 x 6 — 4 x 4		.080	.203	.080	.203	58	26.91
4 x 4 — W1.4 x W1.4	4 x 4 — 10 x 10		.042	.107	.042	.107	31	14.06
4 x 4 — W2.0 x W2.0	4 x 4 — 8 x 8	1	.060	.152	.060	.152	43	19.50
4 x 4 — W2.9 x W2.9	4 x 4 — 6 x 6		.087	.227	.087	.227	62	28.12
4 x 4 — W4.0 x W4.0	4 x 4 — 4 x 4		.120	.305	.120	.305	85	38.56
Sheets								
6 x 6 — W2.9 x W2.9	6 x 6 — 6 x 6		.058	.147	.058	.147	42	19.05
6 x 6 — W4.0 x W4.0	6 x 6 — 4 x 4		.080	.203	.080	.203	58	26.31
6 x 6 — W5.5 x W5.5	6 x 6 — 2 x 2	2	.110	.279	.110	.279	80	36.29
4 x 4 — W1.4 x W1.4	4 x 4 — 4 x 4		.120	.305	.120	.305	85	38.56

NOTES: 1. Exact W—number size for 8 gauge is W2.1
      2. Exact W—number size for 2 gauge is W5.4

The above table was compiled with the following excerpts from the WRI Manual of Standard Practices, 7th Edition, Copyright 2006. Reproduced with permission of the Wire Reinforcement Institute, Inc.:

1. Chapter 3, page 7, Table 1 Common Styles of Metric Wire Reinforcement (WWR) With Equivalent US Customary Units
2. Chapter 6, page 19, Table 5 Customary Units
3. Chapter 6, Page 23, Table 7 Customary Units (in.) Welded Plain Wire Reinforcement
4. Chapter 6, Page 25 Table 8 Wire Size Comparison
5. Chapter 9, Page 30, Table 9 Weight of Longitudinal Wires Weight (Mass) Estimating Tables
6. Chapter 9, Page 31, Table 9M Weight of Longitudinal Wires Weight (Mass) Estimating Tables
7. Chapter 9, Page 32, Table 10 Weight of Transverse Wires Based on 62″ lengths of transverse wire (60″ width plus 1″ overhand each side)
8. Chapter 9, Page 33, Table 10M Weight of Transverse Wires

R033105-10 Proportionate Quantities

The tables below show both quantities per S.F. of floor areas as well as form and reinforcing quantities per C.Y. Unusual structural requirements would increase the ratios below. High strength reinforcing would reduce the steel weights. Figures are for 3000 psi concrete and 60,000 psi reinforcing unless specified otherwise.

Type of Construction	Live Load	Span	Per S.F. of Floor Area				Per C.Y. of Concrete		
			Concrete	Forms	Reinf.	Pans	Forms	Reinf.	Pans
Flat Plate	50 psf	15 Ft.	.46 C.F.	1.06 S.F.	1.71 lb.		62 S.F.	101 lb.	
		20	.63	1.02	2.40		44	104	
		25	.79	1.02	3.03		35	104	
	100	15	.46	1.04	2.14		61	126	
		20	.71	1.02	2.72		39	104	
		25	.83	1.01	3.47		33	113	
Flat Plate (waffle construction) 20" domes	50	20	.43	1.00	2.10	.84 S.F.	63	135	53 S.F.
		25	.52	1.00	2.90	.89	52	150	46
		30	.64	1.00	3.70	.87	42	155	37
	100	20	.51	1.00	2.30	.84	53	125	45
		25	.64	1.00	3.20	.83	42	135	35
		30	.76	1.00	4.40	.81	36	160	29
Waffle Construction 30" domes	50	25	.69	1.06	1.83	.68	42	72	40
		30	.74	1.06	2.39	.69	39	87	39
		35	.86	1.05	2.71	.69	33	85	39
		40	.78	1.00	4.80	.68	35	165	40
Flat Slab (two way with drop panels)	50	20	.62	1.03	2.34		45	102	
		25	.77	1.03	2.99		36	105	
		30	.95	1.03	4.09		29	116	
	100	20	.64	1.03	2.83		43	119	
		25	.79	1.03	3.88		35	133	
		30	.96	1.03	4.66		29	131	
	200	20	.73	1.03	3.03		38	112	
		25	.86	1.03	4.23		32	133	
		30	1.06	1.03	5.30		26	135	
One Way Joists 20" Pans	50	15	.36	1.04	1.40	.93	78	105	70
		20	.42	1.05	1.80	.94	67	120	60
		25	.47	1.05	2.60	.94	60	150	54
	100	15	.38	1.07	1.90	.93	77	140	66
		20	.44	1.08	2.40	.94	67	150	58
		25	.52	1.07	3.50	.94	55	185	49
One Way Joists 8" x 16" filler blocks	50	15	.34	1.06	1.80	.81 Ea.	84	145	64 Ea.
		20	.40	1.08	2.20	.82	73	145	55
		25	.46	1.07	3.20	.83	63	190	49
	100	15	.39	1.07	1.90	.81	74	130	56
		20	.46	1.09	2.80	.82	64	160	48
		25	.53	1.10	3.60	.83	56	190	42
One Way Beam & Slab	50	15	.42	1.30	1.73		84	111	
		20	.51	1.28	2.61		68	138	
		25	.64	1.25	2.78		53	117	
	100	15	.42	1.30	1.90		84	122	
		20	.54	1.35	2.69		68	154	
		25	.69	1.37	3.93		54	145	
	200	15	.44	1.31	2.24		80	137	
		20	.58	1.40	3.30		65	163	
		25	.69	1.42	4.89		53	183	
Two Way Beam & Slab	100	15	.47	1.20	2.26		69	130	
		20	.63	1.29	3.06		55	131	
		25	.83	1.33	3.79		43	123	
	200	15	.49	1.25	2.70		41	149	
		20	.66	1.32	4.04		54	165	
		25	.88	1.32	6.08		41	187	

R033105-10   Proportionate Quantities (cont.)

Item	Size	Forms	Reinforcing	Minimum	Maximum
	4000 psi Concrete and 60,000 psi Reinforcing—Form and Reinforcing Quantities per C.Y.				
	10″ x 10″	130 S.F.C.A.	#5 to #11	220 lbs.	875 lbs.
	12″ x 12″	108	#6 to #14	200	955
	14″ x 14″	92	#7 to #14	190	900
	16″ x 16″	81	#6 to #14	187	1082
	18″ x 18″	72	#6 to #14	170	906
	20″ x 20″	65	#7 to #18	150	1080
Columns	22″ x 22″	59	#8 to #18	153	902
(square tied)	24″ x 24″	54	#8 to #18	164	884
	26″ x 26″	50	#9 to #18	169	994
	28″ x 28″	46	#9 to #18	147	864
	30″ x 30″	43	#10 to #18	146	983
	32″ x 32″	40	#10 to #18	175	866
	34″ x 34″	38	#10 to #18	157	772
	36″ x 36″	36	#10 to #18	175	852
	38″ x 38″	34	#10 to #18	158	765
	40″ x 40″	32	#10 to #18	143	692

Item	Size	Form	Spiral	Reinforcing	Minimum	Maximum
	12″ diameter	34.5 L.F.	190 lbs.	#4 to #11	165 lbs.	1505 lb.
		34.5	190	#14 & #18	—	1100
	14″	25	170	#4 to #11	150	970
		25	170	#14 & #18	800	1000
	16″	19	160	#4 to #11	160	950
		19	160	#14 & #18	605	1080
	18″	15	150	#4 to #11	160	915
		15	150	#14 & #18	480	1075
	20″	12	130	#4 to #11	155	865
		12	130	#14 & #18	385	1020
	22″	10	125	#4 to #11	165	775
		10	125	#14 & #18	320	995
	24″	9	120	#4 to #11	195	800
		9	120	#14 & #18	290	1150
Columns	26″	7.3	100	#4 to #11	200	729
(spirally reinforced)		7.3	100	#14 & #18	235	1035
	28″	6.3	95	#4 to #11	175	700
		6.3	95	#14 & #18	200	1075
	30″	5.5	90	#4 to #11	180	670
		5.5	90	#14 & #18	175	1015
	32″	4.8	85	#4 to #11	185	615
		4.8	85	#14 & #18	155	955
	34″	4.3	80	#4 to #11	180	600
		4.3	80	#14 & #18	170	855
	36″	3.8	75	#4 to #11	165	570
		3.8	75	#14 & #18	155	865
	40″	3.0	70	#4 to #11	165	500
		3.0	70	#14 & #18	145	765

    For customer support on your Commercial Renovation Costs with RSMeans Data, call 800.448.8182.

R033105-10 Proportionate Quantities (cont.)

				3000 psi Concrete and 60,000 psi Reinforcing—Form and Reinforcing Quantities per C.Y.		
Item	**Type**	**Loading**	**Height**	**C.Y./L.F.**	**Forms/C.Y.**	**Reinf./C.Y.**
Retaining Walls	Cantilever	Level Backfill	4 Ft.	0.2 C.Y.	49 S.F.	35 lbs.
			8	0.5	42	45
			12	0.8	35	70
			16	1.1	32	85
			20	1.6	28	105
		Highway Surcharge	4	0.3	41	35
			8	0.5	36	55
			12	0.8	33	90
			16	1.2	30	120
			20	1.7	27	155
		Railroad Surcharge	4	0.4	28	45
			8	0.8	25	65
			12	1.3	22	90
			16	1.9	20	100
			20	2.6	18	120
	Gravity, with Vertical Face	Level Backfill	4	0.4	37	None
			7	0.6	27	
			10	1.2	20	
		Sloping Backfill	4	0.3	31	↓
			7	0.8	21	
			10	1.6	15	

		Live Load in Kips per Linear Foot						
	Span	**Under 1 Kip**		**2 to 3 Kips**		**4 to 5 Kips**		**6 to 7 Kips**	
		Forms	**Reinf.**	**Forms**	**Reinf.**	**Forms**	**Reinf.**	**Forms**	**Reinf.**
Beams	10 Ft.	—	—	90 S.F.	170 #	85 S.F.	175 #	75 S.F.	185 #
	16	130 S.F.	165 #	85	180	75	180	65	225
	20	110	170	75	185	62	200	51	200
	26	90	170	65	215	62	215	—	—
	30	85	175	60	200	—	—	—	—

Item	**Size**	**Type**	**Forms per C.Y.**	**Reinforcing per C.Y.**
Spread Footings	Under 1 C.Y.	1,000 psf soil	24 S.F.	44 lbs.
		5,000	24	42
		10,000	24	52
	1 C.Y. to 5 C.Y.	1,000	14	49
		5,000	14	50
		10,000	14	50
	Over 5 C.Y.	1,000	9	54
		5,000	9	52
		10,000	9	56
Pile Caps (30 Ton Concrete Piles)	Under 5 C.Y.	shallow caps	20	65
		medium	20	50
		deep	20	40
	5 C.Y. to 10 C.Y.	shallow	14	55
		medium	15	45
		deep	15	40
	10 C.Y. to 20 C.Y.	shallow	11	60
		medium	11	45
		deep	12	35
	Over 20 C.Y.	shallow	9	60
		medium	9	45
		deep	10	40

For customer support on your Commercial Renovation Costs with RSMeans Data, call 800.448.8182.

863

R033105-10   Proportionate Quantities (cont.)

Item	Size		Pile Spacing	50 T Pile	100 T Pile	50 T Pile	100 T Pile
Pile Caps (Steel H Piles)	Under 5 C.Y.		24″ O.C.	24 S.F.	24 S.F.	75 lbs.	90 lbs.
			30″	25	25	80	100
			36″	24	24	80	110
	5 C.Y. to 10 C.Y.		24″	15	15	80	110
			30″	15	15	85	110
			36″	15	15	75	90
	Over 10 C.Y.		24″	13	13	85	90
			30″	11	11	85	95
			36″	10	10	85	90

3000 psi Concrete and 60,000 psi Reinforcing — Form and Reinforcing Quantities per C.Y.

		8″ Thick		10″ Thick		12″ Thick		15″ Thick	
	Height	Forms	Reinf.	Forms	Reinf.	Forms	Reinf.	Forms	Reinf.
Basement Walls	7 Ft.	81 S.F.	44 lbs.	65 S.F.	45 lbs.	54 S.F.	44 lbs.	41 S.F.	43 lbs.
	8	44			45		44		43
	9	46			45		44		43
	10	57			45		44		43
	12	83			50		52		43
	14	116			65		64		51
	16				86		90		65
	18						106		70

R033105-20   Materials for One C.Y. of Concrete

This is an approximate method of figuring quantities of cement, sand and coarse aggregate for a field mix with waste allowance included.

With crushed gravel as coarse aggregate, to determine barrels of cement required, divide 10 by total mix; that is, for 1:2:4 mix, 10 divided by 7 = 1-3/7 barrels.

If the coarse aggregate is crushed stone, use 10-1/2 instead of 10 as given for gravel.

To determine tons of sand required, multiply barrels of cement by parts of sand and then by 0.2; that is, for the 1:2:4 mix, as above, 1-3/7 x 2 x .2 = .57 tons.

Tons of crushed gravel are in the same ratio to tons of sand as parts in the mix, or 4/2 x .57 = 1.14 tons.

1 bag cement = 94#	1 C.Y. sand or crushed gravel = 2700#	1 C.Y. crushed stone = 2575#
4 bags = 1 barrel	1 ton sand or crushed gravel = 20 C.F.	1 ton crushed stone = 21 C.F.

Average carload of cement is 692 bags; of sand or gravel is 56 tons.

Do not stack stored cement over 10 bags high.

R033105-70   Placing Ready-Mixed Concrete

For ground pours allow for 5% waste when figuring quantities.

Prices in the front of the data set assume normal deliveries. If deliveries are made before 8 A.M. or after 5 P.M. or on Saturday afternoons add 30%. Negotiated discounts for large volumes are not included in prices in front of the data set.

For the lower floors without truck access, concrete may be wheeled in rubber-tired buggies, conveyer handled, crane handled or pumped. Pumping is economical if there is top steel. Conveyers are more efficient for thick slabs.

At higher floors the rubber-tired buggies may be hoisted by a hoisting tower and wheeled to the location. Placement by a conveyer is limited to three floors and is best for high-volume pours. Pumped concrete is best when the building has no crane access. Concrete may be pumped directly as high as thirty-six stories using special pumping techniques. Normal maximum height is about fifteen stories.

The best pumping aggregate is screened and graded bank gravel rather than crushed stone.

Pumping downward is more difficult than pumping upward. The horizontal distance from pump to pour may increase preparation time prior to pour. Placing by cranes, either mobile, climbing or tower types, continues as the most efficient method for high-rise concrete buildings.

R033105-85   Lift Slabs

The cost advantage of the lift slab method is due to placing all concrete, reinforcing steel, inserts and electrical conduit at ground level and in reduction of formwork. Minimum economical project size is about 30,000 S.F. Slabs may be tilted for parking garage ramps.

It is now used in all types of buildings and has gone up to 22 stories high in apartment buildings. Current trend is to use post-tensioned flat plate slabs with spans from 22' to 35'. Cylindrical void forms are used when deep slabs are required. One pound of prestressing steel is about equal to seven pounds of conventional reinforcing.

To be considered cured for stressing and lifting, a slab must have attained 75% of design strength. Seven days are usually sufficient with four to five days possible if high early strength cement is used. Slabs can be stacked using two coats of a non-bonding agent to insure that slabs do not stick to each other. Lifting is done by companies specializing in this work. Lift rate is 5' to 15' per hour with an average of 10' per hour. Total areas up to 33,000 S.F. have been lifted at one time. 24 to 36 jacking columns are common. Most economical bay sizes are 24' to 28' with four to fourteen stories most efficient. Continuous design reduces reinforcing steel cost. Use of post-tensioned slabs allows larger bay sizes.

R033543-10   Polished Concrete Floors

A polished concrete floor has a glossy mirror-like appearance and is created by grinding the concrete floor with finer and finer diamond grits, similar to sanding wood, until the desired level of reflective clarity and sheen are achieved. The technical term for this type of polished concrete is bonded abrasive polished concrete. The basic piece of equipment used in the polishing process is a walk-behind planetary grinder for working large floor areas. This grinder drives diamond-impregnated abrasive discs, which progress from coarse- to fine-grit discs.

The process begins with the use of very coarse diamond segments or discs bonded in a metallic matrix. These segments are coarse enough to allow the removal of pits, blemishes, stains, and light coatings from the floor surface in preparation for final smoothing. The condition of the original concrete surface will dictate the grit coarseness of the initial grinding step which will generally end up being a three- to four-step process using ever finer grits. The purpose of this initial grinding step is to remove surface coatings and blemishes and to cut down into the cream for very fine aggregate exposure, or deeper into the fine aggregate layer just below the cream layer, or even deeper into the coarse aggregate layer. These initial grinding steps will progress up to the 100/120 grit. If wet grinding is done, a waste slurry is produced that must be removed between grit changes and disposed of properly. If dry grinding is done, a high performance vacuum will pick up the dust during grinding and collect it in bags which must be disposed of properly.

The process continues with honing the floor in a series of steps that progresses from 100-grit to 400-grit diamond abrasive discs embedded in a plastic or resin matrix. At some point during, or just prior to, the honing step, one or two coats of stain or dye can be sprayed onto the surface to give color to the concrete, and two coats of densifier/hardener must be applied to the floor surface and allowed to dry. This sprayed-on densifier/hardener will penetrate about 1/8″ into the concrete to make the surface harder, denser and more abrasion-resistant.

The process ends with polishing the floor surface in a series of steps that progresses from resin-impregnated 800-grit (medium polish) to 1500-grit (high polish) to 3000-grit (very high polish), depending on the desired level of reflective clarity and sheen.

The Concrete Polishing Association of America (CPAA) has defined the flooring options available when processing concrete to a desired finish. The first category is aggregate exposure, the grinding of a concrete surface with bonded abrasives, in as many abrasive grits necessary, to achieve one of the following classes:

A. Cream – very little surface cut depth; little aggregate exposure

B. Fine aggregate (salt and pepper) – surface cut depth of 1/16″; fine aggregate exposure with little or no medium aggregate exposure at random locations

C. Medium aggregate – surface cut depth of 1/8″; medium aggregate exposure with little or no large aggregate exposure at random locations

D. Large aggregate – surface cut depth of 1/4″; large aggregate exposure with little or no fine aggregate exposure at random locations

The second CPAA defined category is reflective clarity and sheen, the polishing of a concrete surface with the minimum number of bonded abrasives as indicated to achieve one of the following levels:

1. Ground – flat appearance with none to very slight diffused reflection; none to very low reflective sheen; using a minimum total of 4 grit levels up to 100-grit

2. Honed – matte appearance with or without slight diffused reflection; low to medium reflective sheen; using a minimum total of 5 grit levels up to 400-grit

3. Semi-polished – objects being reflected are not quite sharp and crisp but can be easily identified; medium to high reflective sheen; using a minimum total of 6 grit levels up to 800-grit

4. Highly-polished – objects being reflected are sharp and crisp as would be seen in a mirror-like reflection; high to highest reflective sheen; using a minimum total of up to 8 grit levels up to 1500-grit or 3000-grit

The CPAA defines reflective clarity as the degree of sharpness and crispness of the reflection of overhead objects when viewed 5' above and perpendicular to the floor surface. Reflective sheen is the degree of gloss reflected from a surface when viewed at least 20' from and at an angle to the floor surface. These terms are relatively subjective. The final outcome depends on the internal makeup and surface condition of the original concrete floor, the experience of the floor polishing crew, and the expectations of the owner. Before the grinding, honing, and polishing work commences on the main floor area, it might be beneficial to do a mock-up panel in the same floor but in an out of the way place to demonstrate the sequence of steps with increasingly fine abrasive grits and to demonstrate the final reflective clarity and reflective sheen. This mock-up panel will be within the area of, and part of, the final work.

©Concrete Polishing Association of America ″Glossary.″

R034105-30   Prestressed Precast Concrete Structural Units

Type	Location	Depth	Span in Ft.		Live Load Lb. per S.F.
Double Tee 8' to 10'	Floor	28" to 34"	60 to 80		50 to 80
	Roof	12" to 24"	30 to 50		40
	Wall	Width 8'	Up to 55' high		Wind
Multiple Tee 8'	Roof	8" to 12"	15 to 40		40
	Floor	8" to 12"	15 to 30		100
Plank ⦿⦿⦿ or ▱▱▱	Roof or Floor		Roof	Floor	40 for Roof 100 for Floor
		4"	13	12	
		6"	22	18	
		8"	26	25	
		10"	33	29	
		12"	42	32	
Single Tee 8' to 10'	Roof	28" 32" 36" 48"	40 80 100 120		40
AASHO Girder	Bridges	Type 4 5 6	100 110 125		Highway
Box Beam 4'	Bridges	15" 27" 33"	40 to 100		Highway

The majority of precast projects today utilize double tees rather than single tees because of speed and ease of installation. As a result casting beds at manufacturing plants are normally formed for double tees. Single tee projects will therefore require an initial set up charge to be spread over the individual single tee costs.

For floors, a 2" to 3" topping is field cast over the shapes. For roofs, insulating concrete or rigid insulation is placed over the shapes.

Member lengths up to 40' are standard haul, 40' to 60' require special permits and lengths over 60' must be escorted. Excessive width and/or length can add up to 100% on hauling costs.

Large heavy members may require two cranes for lifting which would increase erection costs by about 45%. An eight man crew can install 12 to 20 double tees, or 45 to 70 quad tees or planks per day.

Grouting of connections must also be included.

Several system buildings utilizing precast members are available. Heights can go up to 22 stories for apartment buildings. The optimum design ratio is 3 S.F. of surface to 1 S.F. of floor area.

R035216-10 Lightweight Concrete

Lightweight aggregate concrete is usually purchased ready mixed, but it can also be field mixed.

Vermiculite or Perlite comes in bags of 4 C.F. under various trade names. The weight is about 8 lbs. per C.F. For insulating roof fill use 1:6 mix. For a structural deck use 1:4 mix over gypsum boards, steeltex, steel centering, etc., supported by closely spaced joists or bulb trees. For structural slabs use 1:3:2 vermiculite sand concrete over steeltex, metal lath, steel centering, etc., on joists spaced 2'-0" O.C. for maximum L.L. of 80 P.S.F. Use same mix for slab base fill over steel flooring or regular reinforced concrete slab when tile, terrazzo or other finish is to be laid over.

For slabs on grade use 1:3:2 mix when tile, etc., finish is to be laid over. If radiant heating units are installed use a 1:6 mix for a base. After coils are in place, cover with a regular granolithic finish (mix 1:3:2) to a minimum depth of 1-1/2" over top of units.

Reinforce all slabs with 6 x 6 or 10 x 10 welded wire mesh.

R042110-20 Common and Face Brick

Common building brick manufactured according to ASTM C62 and facing brick manufactured according to ASTM C216 are the two standard bricks available for general building use.

Building brick is made in three grades: SW, where high resistance to damage caused by cyclic freezing is required; MW, where moderate resistance to cyclic freezing is needed; and NW, where little resistance to cyclic freezing is needed. Facing brick is made in only the two grades SW and MW. Additionally, facing brick is available in three types: FBS, for general use; FBX, for general use where a higher degree of precision and lower permissible variation in size than FBS are needed; and FBA, for general use to produce characteristic architectural effects resulting from non-uniformity in size and texture of the units.

In figuring the material cost of brickwork, an allowance of 25% mortar waste and 3% brick breakage was included. If bricks are delivered palletized with 280 to 300 per pallet, or packaged, allow only 1-1/2% for breakage. Packaged or palletized delivery is practical when a job is big enough to have a crane or other equipment available to handle a package of brick. This is so on all industrial work but not always true on small commercial buildings.

The use of buff and gray face is increasing, and there is a continuing trend to the Norman, Roman, Jumbo and SCR brick.

Common red clay brick for backup is not used that often. Concrete block is the most usual backup material with occasional use of sand lime or cement brick. Building brick is commonly used in solid walls for strength and as a fire stop.

Brick panels built on the ground and then crane erected to the upper floors have proven to be economical. This allows the work to be done under cover and without scaffolding.

R042110-50 Brick, Block & Mortar Quantities

	Running Bond					For Other Bonds Standard Size Add to S.F. Quantities in Table to Left		
	Number of Brick per S.F. of Wall - Single Wythe with 3/8" Joints				C.F. of Mortar per M Bricks, Waste Included			
Type Brick	Nominal Size (incl. mortar) L H W	Modular Coursing	Number of Brick per S.F.	3/8" Joint	1/2" Joint	Bond Type	Description	Factor
Standard	8 x 2-2/3 x 4	3C=8"	6.75	8.1	10.3	Common	full header every fifth course	+20%
Economy	8 x 4 x 4	1C=4"	4.50	9.1	11.6		full header every sixth course	+16.7%
Engineer	8 x 3-1/5 x 4	5C=16"	5.63	8.5	10.8	English	full header every second course	+50%
Fire	9 x 2-1/2 x 4-1/2	2C=5"	6.40	550 # Fireclay	—	Flemish	alternate headers every course	+33.3%
Jumbo	12 x 4 x 6 or 8	1C=4"	3.00	22.5	29.2		every sixth course	+5.6%
Norman	12 x 2-2/3 x 4	3C=8"	4.50	11.2	14.3	Header = W x H exposed		+100%
Norwegian	12 x 3-1/5 x 4	5C=16"	3.75	11.7	14.9	Rowlock = H x W exposed		+100%
Roman	12 x 2 x 4	2C=4"	6.00	10.7	13.7	Rowlock stretcher = L x W exposed		+33.3%
SCR	12 x 2-2/3 x 6	3C=8"	4.50	21.8	28.0	Soldier = H x L exposed		—
Utility	12 x 4 x 4	1C=4"	3.00	12.3	15.7	Sailor = W x L exposed		-33.3%

Concrete Blocks Nominal Size		Approximate Weight per S.F.		Blocks per 100 S.F.	Mortar per M block, waste included	
		Standard	Lightweight		Partitions	Back up
2"	x 8" x 16"	20 PSF	15 PSF	113	27 C.F.	36 C.F.
4"		30	20		41	51
6"		42	30		56	66
8"		55	38		72	82
10"		70	47		87	97
12"		85	55		102	112

Brick & Mortar Quantities
©Brick Industry Association. 2009 Feb. Technical Notes on
Brick Construction 10:
  Dimensioning and Estimating Brick Masonry. Reston (VA): BIA. Table 1
  Modular Brick Sizes and Table 4 Quantity Estimates for Brick Masonry.

R050521-20 Welded Structural Steel

Usual weight reductions with welded design run 10% to 20% compared with bolted or riveted connections. This amounts to about the same total cost compared with bolted structures since field welding is more expensive than bolts. For normal spans of 18' to 24' figure 6 to 7 connections per ton.

Trusses — For welded trusses add 4% to weight of main members for connections. Up to 15% less steel can be expected in a welded truss compared to one that is shop bolted. Cost of erection is the same whether shop bolted or welded.

General — Typical electrodes for structural steel welding are E6010, E6011, E60T and E70T. Typical buildings vary between 2# to 8# of weld rod per

ton of steel. Buildings utilizing continuous design require about three times as much welding as conventional welded structures. In estimating field erection by welding, it is best to use the average linear feet of weld per ton to arrive at the welding cost per ton. The type, size and position of the weld will have a direct bearing on the cost per linear foot. A typical field welder will deposit 1.8# to 2# of weld rod per hour manually. Using semiautomatic methods can increase production by as much as 50% to 75%.

R051223-10 Structural Steel

The bare material prices for structural steel, shown in the unit cost sections of the data set, are for 100 tons of shop-fabricated structural steel and include:

1. Mill base price of structural steel
2. Mill scrap/grade/size/length extras
3. Mill delivery to a metals service center (warehouse)
4. Service center storage and handling
5. Service center delivery to a fabrication shop
6. Shop storage and handling
7. Shop drafting/detailing
8. Shop fabrication
9. Shop coat of primer paint
10. Shop listing
11. Shop delivery to the job site

In unit cost sections of the data set that contain items for field fabrication of steel components, the bare material cost of steel includes:

1. Mill base price of structural steel
2. Mill scrap/grade/size/length extras
3. Mill delivery to a metals service center (warehouse)
4. Service center storage and handling
5. Service center delivery to the job site

R051223-15 Structural Steel Estimating for Repair and Remodeling Projects

The correct approach to estimating structural steel is dependent upon the amount of steel required for the particular project. If the project is a sizable addition to a building, the data can be used directly from the data set. This is not the case however if the project requires a small amount of steel, for instance to reinforce existing roof or floor structural systems. To better understand this, please refer to the unit price line for a W16x31 beam with bolted connections. Assume your project requires the reinforcement of the structural members of the roof system of a 3 story building required for the installation a new piece of HVAC equipment. The project will need 4 pieces of W16x31 that are 30' long. After pricing this 120 L.F. job using the unit prices for a W16x31, an analysis will reveal that the price is wholly inadequate.

The first problem is apparent if you examine the amount of steel that can be installed per day. The unit price line indicates 900 linear feet per day can be installed by a 5 man crew with an 90 ton crane. This productivity is correct for new construction but certainly is not for repair and remodeling work. Installation of new structural steel considers that each member is installed from the foundation to the roof of the structure in a planned and systematic manner with a crane having unrestricted access to all parts of the project. Additionally each connection is planned and detailed with full field access for fit-up and final bolting. The erection is planned and progresses such that interferences and conflicts with other structural members are minimized, if not completely eliminated. All of these assumptions are clearly not the case with a repair and remodeling job, and a significant decrease in the stated productivity will be observed.

A crane will certainly be needed to lift the members into the general area of the project but in most cases will not be able to place the beams into

their final position. An opening in the existing roof may not be large enough to permit the beams to pass through and it may be necessary to bring them into the building through existing windows or doors. Moving the beams to the actual area where they will be installed may involve hand labor and the use of dollies. Finally, hoists and/or jacks may be needed for final positioning.

The connection of new members to existing can often be accomplished by field bolting with accurate field measurements and good planning but in many cases access to both sides of existing members is not possible and field welding becomes the only alternative. In addition to the cost of the actual welding, protection of existing finishes, systems and structure and fire protection must be considered.

New beams can never be installed tight to the existing decks or floors which they must support and the use of shims and tack welding becomes necessary. Additionally, further planning and cost are involved in assuring that existing loads are minimized during the installation and shimming process.

It is apparent that installation of structural steel as part of a repair and remodeling project involves more than simply installing the members and estimating the cost in the same manner as new construction. The best procedure for estimating the total cost are adequate planning and coordination of each process and activity that will be needed. Unit costs for the materials, labor and equipment can then be attached to each needed activity and a final, complete price can be determined.

R053100-10 Decking Descriptions

General - All Deck Products

A steel deck is made by cold forming structural grade sheet steel into a repeating pattern of parallel ribs. The strength and stiffness of the panels are the result of the ribs and the material properties of the steel. Deck lengths can be varied to suit job conditions, but because of shipping considerations, are usually less than 40 feet. Standard deck width varies with the product used but full sheets are usually 12″, 18″, 24″, 30″, or 36″. The deck is typically furnished in a standard width with the ends cut square. Any cutting for width, such as at openings or for angular fit, is done at the job site.

The deck is typically attached to the building frame with arc puddle welds, self-drilling screws, or powder or pneumatically driven pins. Sheet to sheet fastening is done with screws, button punching (crimping), or welds.

Composite Floor Deck

After installation and adequate fastening, a floor deck serves several purposes. It (a) acts as a working platform, (b) stabilizes the frame, (c) serves as a concrete form for the slab, and (d) reinforces the slab to carry the design loads applied during the life of the building. Composite decks are distinguished by the presence of shear connector devices as part of the deck. These devices are designed to mechanically lock the concrete and deck together so that the concrete and the deck work together to carry subsequent floor loads. These shear connector devices can be rolled-in embossments, lugs, holes, or wires welded to the panels. The deck profile can also be used to interlock concrete and steel.

Composite deck finishes are either galvanized (zinc coated) or phosphatized/painted. Galvanized deck has a zinc coating on both the top and bottom surfaces. The phosphatized/painted deck has a bare (phosphatized) top surface that will come into contact with the concrete. This bare top surface can be expected to develop rust before the concrete is placed. The bottom side of the deck has a primer coat of paint.

A composite floor deck is normally installed so the panel ends do not overlap on the supporting beams. Shear lugs or panel profile shapes often prevent a tight metal to metal fit if the panel ends overlap; the air gap caused by overlapping will prevent proper fusion with the structural steel supports when the panel end laps are shear stud welded.

Adequate end bearing of the deck must be obtained as shown on the drawings. If bearing is actually less in the field than shown on the drawings, further investigation is required.

Roof Deck

A roof deck is not designed to act compositely with other materials. A roof deck acts alone in transferring horizontal and vertical loads into the building frame. Roof deck rib openings are usually narrower than floor deck rib openings. This provides adequate support of the rigid thermal insulation board.

A roof deck is typically installed to endlap approximately 2″ over supports. However, it can be butted (or lapped more than 2″) to solve field fit problems. Since designers frequently use the installed deck system as part of the horizontal bracing system (the deck as a diaphragm), any fastening substitution or change should be approved by the designer. Continuous perimeter support of the deck is necessary to limit edge deflection in the finished roof and may be required for diaphragm shear transfer.

Standard roof deck finishes are galvanized or primer painted. The standard factory applied paint for roof decks is a primer paint and is not intended to weather for extended periods of time. Field painting or touching up of abrasions and deterioration of the primer coat or other protective finishes is the responsibility of the contractor.

Cellular Deck

A cellular deck is made by attaching a bottom steel sheet to a roof deck or composite floor deck panel. A cellular deck can be used in the same manner as a floor deck. Electrical, telephone, and data wires are easily run through the chase created between the deck panel and the bottom sheet.

When used as part of the electrical distribution system, the cellular deck must be installed so that the ribs line up and create a smooth cell transition at abutting ends. The joint that occurs at butting cell ends must be taped or otherwise sealed to prevent wet concrete from seeping into the cell. Cell interiors must be free of welding burrs, or other sharp intrusions, to prevent damage to wires.

When used as a roof deck, the bottom flat plate is usually left exposed to view. Care must be maintained during erection to keep good alignment and prevent damage.

A cellular deck is sometimes used with the flat plate on the top side to provide a flat working surface. Installation of the deck for this purpose requires special methods for attachment to the frame because the flat plate, now on the top, can prevent direct access to the deck material that is bearing on the structural steel. It may be advisable to treat the flat top surface to prevent slipping.

A cellular deck is always furnished galvanized or painted over galvanized.

Form Deck

A form deck can be any floor or roof deck product used as a concrete form. Connections to the frame are by the same methods used to anchor floor and roof decks. Welding washers are recommended when welding a deck that is less than 20 gauge thickness.

A form deck is furnished galvanized, prime painted, or uncoated. Galvanized deck must be used for those roof deck systems where a form deck is used to carry a lightweight insulating concrete fill.

R061110-30   Lumber Product Material Prices

The price of forest products fluctuates widely from location to location and from season to season depending upon economic conditions. The bare material prices in the unit cost sections of the data set show the National Average material prices in effect Jan. 1 of this data year. It must be noted that lumber prices in general may change significantly during the year.

Availability of certain items depends upon geographic location and must be checked prior to firm-price bidding.

R061636-20   Plywood

There are two types of plywood used in construction: interior, which is moisture-resistant but not waterproofed, and exterior, which is waterproofed.

The grade of the exterior surface of the plywood sheets is designated by the first letter: A, for smooth surface with patches allowed; B, for solid surface with patches and plugs allowed; C, which may be surface plugged or may have knot holes up to 1″ wide; and D, which is used only for interior type plywood and may have knot holes up to 2-1/2″ wide. "Structural Grade" is specifically designed for engineered applications such as box beams. All CC & DD grades have roof and floor spans marked on them.

Underlayment-grade plywood runs from 1/4″ to 1-1/4″ thick. Thicknesses 5/8″ and over have optional tongue and groove joints which eliminate the need for blocking the edges. Underlayment 19/32″ and over may be referred to as Sturd-i-Floor.

The price of plywood can fluctuate widely due to geographic and economic conditions.

Typical uses for various plywood grades are as follows:

AA-AD Interior — cupboards, shelving, paneling, furniture

BB Plyform — concrete form plywood

CDX — wall and roof sheathing

Structural — box beams, girders, stressed skin panels

AA-AC Exterior — fences, signs, siding, soffits, etc.

Underlayment — base for resilient floor coverings

Overlaid HDO — high density for concrete forms & highway signs

Overlaid MDO — medium density for painting, siding, soffits & signs

303 Siding — exterior siding, textured, striated, embossed, etc.

R075113-20   Built-Up Roofing

Asphalt is available in kegs of 100 lbs. each; coal tar pitch in 560 lb. kegs. Prepared roofing felts are available in a wide range of sizes, weights and characteristics. However, the most commonly used are #15 (432 S.F. per roll, 13 lbs. per square) and #30 (216 S.F. per roll, 27 lbs. per square).

Inter-ply bitumen varies from 24 lbs. per sq. (asphalt) to 30 lbs. per sq. (coal tar) per ply, MF4@ 25%. Flood coat bitumen also varies from 60 lbs. per sq. (asphalt) to 75 lbs. per sq. (coal tar), MF4@ 25%. Expendable equipment (mops, brooms, screeds, etc.) runs about 16% of the bitumen cost. For new, inexperienced crews this factor may be much higher.

A rigid insulation board is typically applied in two layers. The first is mechanically attached to nailable decks or spot or solid mopped to non-nailable decks; the second layer is then spot or solid mopped to the first layer. Membrane application follows the insulation, except in protected membrane roofs, where the membrane goes down first and the insulation on top, followed with ballast (stone or concrete pavers). Insulation and related labor costs are NOT included in prices for built-up roofing.

R075213-30　Modified Bitumen Roofing

The cost of modified bitumen roofing is highly dependent on the type of installation that is planned. Installation is based on the type of modifier used in the bitumen. The two most popular modifiers are atactic polypropylene (APP) and styrene butadiene styrene (SBS). The modifiers are added to heated bitumen during the manufacturing process to change its characteristics. A polyethylene, polyester or fiberglass reinforcing sheet is then sandwiched between layers of this bitumen. When completed, the result is a pre-assembled, built-up roof that has increased elasticity and weatherability. Some manufacturers include a surfacing material such as ceramic or mineral granules, metal particles or sand.

The preferred method of adhering SBS-modified bitumen roofing to the substrate is with hot-mopped asphalt (much the same as built-up roofing). This installation method requires a tar kettle/pot to heat the asphalt, as well as the labor, tools and equipment necessary to distribute and spread the hot asphalt.

The alternative method for applying APP and SBS modified bitumen is as follows. A skilled installer uses a torch to melt a small pool of bitumen off the membrane. This pool must form across the entire roll for proper adhesion. The installer must unroll the roofing at a pace slow enough to melt the bitumen, but fast enough to prevent damage to the rest of the membrane.

Modified bitumen roofing provides the advantages of both built-up and single-ply roofing. Labor costs are reduced over those of built-up roofing because only a single ply is necessary. The elasticity of single-ply roofing is attained with the reinforcing sheet and polymer modifiers. Modifieds have some self-healing characteristics and because of their multi-layer construction, they offer the reliability and safety of built-up roofing.

R078413-30　Firestopping

Firestopping is the sealing of structural, mechanical, electrical and other penetrations through fire-rated assemblies. The basic components of firestop systems are safing insulation and firestop sealant on both sides of wall penetrations and the top side of floor penetrations.

Pipe penetrations are assumed to be through concrete, grout, or joint compound and can be sleeved or unsleeved. Costs for the penetrations and sleeves are not included. An annular space of 1″ is assumed. Escutcheons are not included.

A metallic pipe is assumed to be copper, aluminum, cast iron or similar metallic material. An insulated metallic pipe is assumed to be covered with a thermal insulating jacket of varying thickness and materials.

A non-metallic pipe is assumed to be PVC, CPVC, FR Polypropylene or similar plastic piping material. Intumescent firestop sealants or wrap strips are included. Collars on both sides of wall penetrations and a sheet metal plate on the underside of floor penetrations are included.

Ductwork is assumed to be sheet metal, stainless steel or similar metallic material. Duct penetrations are assumed to be through concrete, grout or joint compound. Costs for penetrations and sleeves are not included. An annular space of 1/2″ is assumed.

Multi-trade openings include costs for sheet metal forms, firestop mortar, wrap strips, collars and sealants as necessary.

Structural penetrations joints are assumed to be 1/2″ or less. CMU walls are assumed to be within 1-1/2″ of the metal deck. Drywall walls are assumed to be tight to the underside of metal decking.

Metal panel, glass or curtain wall systems include a spandrel area of 5′ filled with mineral wool foil-faced insulation. Fasteners and stiffeners are included.

R081313-20 Steel Door Selection Guide

Standard steel doors are classified into four levels, as recommended by the Steel Door Institute in the chart below. Each of the four levels offers a range of construction models and designs to meet architectural requirements for preference and appearance, including full flush, seamless, and stile & rail. Recommended minimum gauge requirements are also included.

For complete standard steel door construction specifications and available sizes, refer to the Steel Door Institute Technical Data Series, ANSI A250.8-98 (SDI-100), and ANSI A250.4-94 Test Procedure and Acceptance Criteria for Physical Endurance of Steel Door and Hardware Reinforcements.

Level		Model	Construction	For Full Flush or Seamless		
				Min. Gauge	Thickness (in)	Thickness (mm)
I	Standard Duty	1	Full Flush	20	0.032	0.8
		2	Seamless			
II	Heavy Duty	1	Full Flush	18	0.042	1.0
		2	Seamless			
III	Extra Heavy Duty	1	Full Flush	16	0.053	1.3
		2	Seamless			
		3	*Stile & Rail			
IV	Maximum Duty	1	Full Flush	14	0.067	1.6
		2	Seamless			

*Stiles & rails are 16 gauge; flush panels, when specified, are 18 gauge

R085123-10 Steel Sash

An ironworker crew will erect 25 S.F. or 1.3 sash unit per hour, whichever is less.

A mechanic will point 30 L.F. per hour.

A painter will paint 90 S.F. per coat per hour.

A Glazier production depends on light size.

Allow 1 lb. special steel sash putty per 16″ x 20″ light.

R085313-20 Replacement Windows

Replacement windows are typically measured per United Inch.

United Inches are calculated by rounding the width and height of the window opening up to the nearest inch, then adding the two figures.

The labor cost for replacement windows includes removal of sash, existing sash balance or weights, parting bead where necessary and installation of new window.

Debris hauling and dump fees are not included.

R087110-10   Hardware Finishes

This table describes hardware finishes used throughout the industry. It also shows the base metal and the respective symbols in the three predominate systems of identification. Many of these are used in pricing descriptions in Division Eight.

US″	BMHA*	CDN	Base	Description
US P	600	CP	Steel	Primed for Painting
US 1B	601	C1B	Steel	Bright Black Japanned
US 2C	602	C2C	Steel	Zinc Plated
US 2G	603	C2G	Steel	Zinc Plated
US 3	605	C3	Brass	Bright Brass, Clear Coated
US 4	606	C4	Brass	Satin Brass, Clear Coated
US 5	609	C5	Brass	Satin Brass, Blackened, Satin Relieved, Clear Coated
US 7	610	C7	Brass	Satin Brass, Blackened, Bright Relieved, Clear Coated
US 9	611	C9	Bronze	Bright Bronze, Clear Coated
US 10	612	C10	Bronze	Satin Bronze, Clear Coated
US 10A	641	C10A	Steel	Antiqued Bronze, Oiled and Lacquered
US 10B	613	C10B	Bronze	Antiqued Bronze, Oiled
US 11	616	C11	Bronze	Satin Bronze, Blackened, Satin Relieved, Clear Coated
US 14	618	C14	Brass/Bronze	Bright Nickel Plated, Clear Coated
US 15	619	C15	Brass/Bronze	Satin Nickel, Clear Coated
US 15A	620	C15A	Brass/Bronze	Satin Nickel Plated, Blackened, Satin Relieved, Clear Coated
US 17A	621	C17A	Brass/Bronze	Nickel Plated, Blackened, Relieved, Clear Coated
US 19	622	C19	Brass/Bronze	Flat Black Coated
US 20	623	C20	Brass/Bronze	Statuary Bronze, Light
US 20A	624	C20A	Brass/Bronze	Statuary Bronze, Dark
US 26	625	C26	Brass/Bronze	Bright Chromium
US 26D	626	C26D	Brass/Bronze	Satin Chromium
US 20	627	C27	Aluminum	Satin Aluminum Clear
US 28	628	C28	Aluminum	Anodized Dull Aluminum
US 32	629	C32	Stainless Steel	Bright Stainless Steel
US 32D	630	C32D	Stainless Steel	Stainless Steel
US 3	632	C3	Steel	Bright Brass Plated, Clear Coated
US 4	633	C4	Steel	Satin Brass, Clear Coated
US 7	636	C7	Steel	Satin Brass Plated, Blackened, Bright Relieved, Clear Coated
US 9	637	C9	Steel	Bright Bronze Plated, Clear Coated
US 5	638	C5	Steel	Satin Brass Plated, Blackened, Bright Relieved, Clear Coated
US 10	639	C10	Steel	Satin Bronze Plated, Clear Coated
US 10B	640	C10B	Steel	Antique Bronze, Oiled
US 10A	641	C10A	Steel	Antiqued Bronze, Oiled and Lacquered
US 11	643	C11	Steel	Satin Bronze Plated, Blackened, Bright Relieved, Clear Coated
US 14	645	C14	Steel	Bright Nickel Plated, Clear Coated
US 15	646	C15	Steel	Satin Nickel Plated, Clear Coated
US 15A	647	C15A	Steel	Nickel Plated, Blackened, Bright Relieved, Clear Coated
US 17A	648	C17A	Steel	Nickel Plated, Blackened, Relieved, Clear Coated
US 20	649	C20	Steel	Statuary Bronze, Light
US 20A	650	C20A	Steel	Statuary Bronze, Dark
US 26	651	C26	Steel	Bright Chromium Plated
US 26D	652	C26D	Steel	Satin Chromium Plated

* - BMHA Builders Hardware Manufacturing Association

″ - US Equivalent

^ - Canadian Equivalent

Japanning is imitating Asian lacquer work

R088110-10   Glazing Productivity

Some glass sizes are estimated by the "united inch" (height + width). The table below shows the number of lights glazed in an eight-hour period by the crew size indicated, for glass up to 1/4" thick. Square or nearly square lights are more economical on a S.F. basis. Long slender lights will have a high S.F. installation cost. For insulated glass reduce production by 33%. For 1/2" float glass reduce production by 50%. Production time for glazing with two glaziers per day averages: 1/4" float glass 120 S.F.; 1/2" float glass 55 S.F.; 1/2" insulated glass 95 S.F.; 3/4" insulated glass 75 S.F.

Glazing Method	United Inches per Light							
	40″	60″	80″	100″	135″	165″	200″	240″
Number of Men in Crew	1	1	1	1	2	3	3	4
Industrial sash, putty	60	45	24	15	18	—	—	—
With stops, putty bed	50	36	21	12	16	8	4	3
Wood stops, rubber	40	27	15	9	11	6	3	2
Metal stops, rubber	30	24	14	9	9	6	3	2
Structural glass	10	7	4	3	—	—	—	—
Corrugated glass	12	9	7	4	4	4	3	—
Storefronts	16	15	13	11	7	6	4	4
Skylights, putty glass	60	36	21	12	16	—	—	—
Thiokol set	15	15	11	9	9	6	3	2
Vinyl set, snap on	18	18	13	12	12	7	5	4
Maximum area per light	2.8 S.F.	6.3 S.F.	11.1 S.F.	17.4 S.F.	31.6 S.F.	47 S.F.	69 S.F.	100 S.F.

R092000-50   Lath, Plaster and Gypsum Board

Gypsum board lath is available in 3/8" thick x 16" wide x 4' long sheets as a base material for multi-layer plaster applications. It is also available as a base for either multi-layer or veneer plaster applications in 1/2" and 5/8" thick–4' wide x 8', 10' or 12' long sheets. Fasteners are screws or blued ring shank nails for wood framing and screws for metal framing.

Metal lath is available in diamond mesh patterns with flat or self-furring profiles. Paper backing is available for applications where excessive plaster waste needs to be avoided. A slotted mesh ribbed lath should be used in areas where the span between structural supports is greater than normal. Most metal lath comes in 27" x 96" sheets. Diamond mesh weighs 1.75, 2.5 or 3.4 pounds per square yard, slotted mesh lath weighs 2.75 or 3.4 pounds per square yard. Metal lath can be nailed, screwed or tied in place.

Many **accessories** are available. Corner beads, flat reinforcing strips, casing beads, control and expansion joints, furring brackets and channels are some examples. Note that accessories are not included in plaster or stucco line items.

Plaster is defined as a material or combination of materials that when mixed with a suitable amount of water, forms a plastic mass or paste. When applied to a surface, the paste adheres to it and subsequently hardens, preserving in a rigid state the form or texture imposed during the period of elasticity.

Gypsum plaster is made from ground calcined gypsum. It is mixed with aggregates and water for use as a base coat plaster.

Vermiculite plaster is a fire-retardant plaster covering used on steel beams, concrete slabs and other heavy construction materials. Vermiculite is a group name for certain clay minerals, hydrous silicates or aluminum, magnesium and iron that have been expanded by heat.

Perlite plaster is a plaster using perlite as an aggregate instead of sand. Perlite is a volcanic glass that has been expanded by heat.

Gauging plaster is a mix of gypsum plaster and lime putty that when applied produces a quick drying finish coat.

Veneer plaster is a one or two component gypsum plaster used as a thin finish coat over special gypsum board.

Keenes cement is a white cementitious material manufactured from gypsum that has been burned at a high temperature and ground to a fine powder. Alum is added to accelerate the set. The resulting plaster is hard and strong and accepts and maintains a high polish, hence it is used as a finishing plaster.

Stucco is a Portland cement based plaster used primarily as an exterior finish.

Plaster is used on both interior and exterior surfaces. Generally it is applied in multiple-coat systems. A three-coat system uses the terms scratch, brown and finish to identify each coat. A two-coat system uses base and finish to describe each coat. Each type of plaster and application system has attributes that are chosen by the designer to best fit the intended use.

Gypsum Plaster Quantities for 100 S.Y.	2 Coat, 5/8" Thick		3 Coat, 3/4" Thick		
	Base	Finish	Scratch	Brown	Finish
	1:3 Mix	2:1 Mix	1:2 Mix	1:3 Mix	2:1 Mix
Gypsum plaster	1,300 lb.		1,350 lb.	650 lb.	
Sand	1.75 C.Y.		1.85 C.Y.	1.35 C.Y.	
Finish hydrated lime		340 lb.			340 lb.
Gauging plaster		170 lb.			170 lb.

Vermiculite or Perlite Plaster Quantities for 100 S.Y.	2 Coat, 5/8" Thick		3 Coat, 3/4" Thick		
	Base	Finish	Scratch	Brown	Finish
Gypsum plaster	1,250 lb.		1,450 lb.	800 lb.	
Vermiculite or perlite	7.8 bags		8.0 bags	3.3 bags	
Finish hydrated lime		340 lb.			340 lb.
Gauging plaster		170 lb.			170 lb.

Stucco–Three-Coat System Quantities for 100 S.Y.	On Wood Frame	On Masonry
Portland cement	29 bags	21 bags
Sand	2.6 C.Y.	2.0 C.Y.
Hydrated lime	180 lb.	120 lb.

R092910-10   Levels of Gypsum Board Finish

In the past, contract documents often used phrases such as "industry standard" and "workmanlike finish" to specify the expected quality of gypsum board wall and ceiling installations. The vagueness of these descriptions led to unacceptable work and disputes.

In order to resolve this problem, four major trade associations concerned with the manufacture, erection, finish and decoration of gypsum board wall and ceiling systems developed an industry-wide *Recommended Levels of Gypsum Board Finish.*

The finish of gypsum board walls and ceilings for specific final decoration is dependent on a number of factors. A primary consideration is the location of the surface and the degree of decorative treatment desired. Painted and unpainted surfaces in warehouses and other areas where appearance is normally not critical may simply require the taping of wallboard joints and 'spotting' of fastener heads. Blemish-free, smooth, monolithic surfaces often intended for painted and decorated walls and ceilings in habitated structures, ranging from single-family dwellings through monumental buildings, require additional finishing prior to the application of the final decoration.

Other factors to be considered in determining the level of finish of the gypsum board surface are (1) the type of angle of surface illumination (both natural and artificial lighting), and (2) the paint and method of application or the type and finish of wallcovering specified as the final decoration. Critical lighting conditions, gloss paints, and thin wall coverings require a higher level of gypsum board finish than heavily textured surfaces which are subsequently painted or surfaces which are to be decorated with heavy grade wall coverings.

The following descriptions were developed by the Association of the Wall and Ceiling Industries-International (AWCI), Ceiling & Interior Systems Construction Association (CISCA), Gypsum Association (GA), and Painting and Decorating Contractors of America (PDCA) as a guide.

Level 0: Used in temporary construction or wherever the final decoration has not been determined. Unfinished. No taping, finishing or corner beads are required. Also could be used where non-predecorated panels will be used in demountable-type partitions that are to be painted as a final finish.

Level 1: Frequently used in plenum areas above ceilings, in attics, in areas where the assembly would generally be concealed, or in building service corridors and other areas not normally open to public view. Some degree of sound and smoke control is provided; in some geographic areas, this level is referred to as "fire-taping," although this level of finish does not typically meet fire-resistant assembly requirements. Where a fire resistance rating is required for the gypsum board assembly, details of construction should be in accordance with reports of fire tests of assemblies that have met the requirements of the fire rating acceptable.

All joints and interior angles shall have tape embedded in joint compound. Accessories are optional at specifier discretion in corridors and other areas with pedestrian traffic. Tape and fastener heads need not be covered with joint compound. Surface shall be free of excess joint compound. Tool marks and ridges are acceptable.

Level 2: It may be specified for standard gypsum board surfaces in garages, warehouse storage, or other similar areas where surface appearance is not of primary importance.

All joints and interior angles shall have tape embedded in joint compound and shall be immediately wiped with a joint knife or trowel, leaving a thin coating of joint compound over all joints and interior angles. Fastener heads and accessories shall be covered with a coat of joint compound. Surface shall be free of excess joint compound. Tool marks and ridges are acceptable.

Level 3: Typically used in areas receiving heavy texture (spray or hand applied) finishes before final painting, or where commercial-grade (heavy duty) wall coverings are to be applied as the final decoration. This level of finish should not be used where smooth painted surfaces or where lighter weight wall coverings are specified. The prepared surface shall be coated with a drywall primer prior to the application of final finishes.

All joints and interior angles shall have tape embedded in joint compound and shall be immediately wiped with a joint knife or trowel, leaving a thin coating of joint compound over all joints and interior angles. One additional coat of joint compound shall be applied over all joints and interior angles. Fastener heads and accessories shall be covered with two separate coats of joint compound. All joint compounds shall be smooth and free of tool marks and ridges. The prepared surface shall be covered with a drywall primer prior to the application of the final decoration.

Level 4: This level should be used where residential grade (light duty) wall coverings, flat paints, or light textures are to be applied. The prepared surface shall be coated with a drywall primer prior to the application of final finishes. Release agents for wall coverings are specifically formulated to minimize damage if coverings are subsequently removed.

The weight, texture, and sheen level of the wall covering material selected should be taken into consideration when specifying wall coverings over this level of drywall treatment. Joints and fasteners must be sufficiently concealed if the wall covering material is lightweight, contains limited pattern, has a glossy finish, or has any combination of these features. In critical lighting areas, flat paints applied over light textures tend to reduce joint photographing. Gloss, semi-gloss, and enamel paints are not recommended over this level of finish.

All joints and interior angles shall have tape embedded in joint compound and shall be immediately wiped with a joint knife or trowel, leaving a thin coating of joint compound over all joints and interior angles. In addition, two separate coats of joint compound shall be applied over all flat joints and one separate coat of joint compound applied over interior angles. Fastener heads and accessories shall be covered with three separate coats of joint compound. All joint compounds shall be smooth and free of tool marks and ridges. The prepared surface shall be covered with a drywall primer like Sheetrock first coat prior to the application of the final decoration.

Level 5: The highest quality finish is the most effective method to provide a uniform surface and minimize the possibility of joint photographing and of fasteners showing through the final decoration. This level of finish is required where gloss, semi-gloss, or enamel is specified; when flat joints are specified over an untextured surface; or where critical lighting conditions occur. The prepared surface shall be coated with a drywall primer prior to the application of the final decoration.

All joints and interior angles shall have tape embedded in joint compound and be immediately wiped with a joint knife or trowel, leaving a thin coating of joint compound over all joints and interior angles. Two separate coats of joint compound shall be applied over all flat joints and one separate coat of joint compound applied over interior angles. Fastener heads and accessories shall be covered with three separate coats of joint compound.

A thin skim coat of joint compound shall be trowel applied to the entire surface. Excess compound is immediately troweled off, leaving a film or skim coating of compound completely covering the paper. As an alternative to a skim coat, a material manufactured especially for this purpose may be applied such as Sheetrock Tuff-Hide primer surfacer. The surface must be smooth and free of tool marks and ridges. The prepared surface shall be covered with a drywall primer prior to the application of the final decoration.

R142000-10   Freight Elevators

Capacities run from 2,000 lbs. to over 100,000 lbs. with 3,000 lbs. to 10,000 lbs. most common. Travel speeds are generally lower and the control is less intricate than on passenger elevators. Costs in the unit price sections are for hydraulic and geared elevators.

R142000-20   Elevator Selective Costs   See R142000-40 for cost development.

A. Base Unit	Passenger		Freight		Hospital	
	Hydraulic	Electric	Hydraulic	Electric	Hydraulic	Electric
Capacity	1,500 lb.	2,000 lb.	2,000 lb.	4,000 lb.	4,000 lb.	4,000 lb.
Speed	100 F.P.M.	200 F.P.M.	100 F.P.M.	200 F.P.M.	100 F.P.M.	200 F.P.M.
#Stops/Travel Ft.	2/12	4/40	2/20	4/40	2/20	4/40
Push Button Oper.	Yes	Yes	Yes	Yes	Yes	Yes
Telephone Box & Wire	"	"	"	"	"	"
Emergency Lighting	"	"	No	No	"	"
Cab	Plastic Lam. Walls	Plastic Lam. Walls	Painted Steel	Painted Steel	Plastic Lam. Walls	Plastic Lam. Walls
Cove Lighting	Yes	Yes	No	No	Yes	Yes
Floor	V.C.T.	V.C.T.	Wood w/Safety Treads	Wood w/Safety Treads	V.C.T.	V.C.T.
Doors, & Speedside Slide	Yes	Yes	Yes	Yes	Yes	Yes
Gates, Manual	No	No	No	No	No	No
Signals, Lighted Buttons	Car and Hall	Car and Hall	Car and Hall	Car and Hall	Car and Hall	Car and Hall
O.H. Geared Machine	N.A.	Yes	N.A.	Yes	N.A.	Yes
Variable Voltage Contr.	"	"	N.A.	"	"	"
Emergency Alarm	Yes	"	Yes	"	Yes	"
Class "A" Loading	N.A.	N.A.	"	"	N.A.	N.A.

R142000-30   Passenger Elevators

Electric elevators are used generally but hydraulic elevators can be used for lifts up to 70' and where large capacities are required. Hydraulic speeds are limited to 200 F.P.M. but cars are self leveling at the stops. On low rises, hydraulic installation runs about 15% less than standard electric types but on higher rises this installation cost advantage is reduced. Maintenance of hydraulic elevators is about the same as the electric type but the underground portion is not included in the maintenance contract.

In electric elevators there are several control systems available, the choice of which will be based upon elevator use, size, speed and cost criteria. The two types of drives are geared for low speeds and gearless for 450 F.P.M. and over.

The tables on the preceding pages illustrate typical installed costs of the various types of elevators available.

R142000-40   Elevator Cost Development

To price a new car or truck from the factory, you must start with the manufacturer's basic model, then add or exchange optional equipment and features. The same is true for pricing elevators.

Example:

Requirement: One-passenger elevator, five-story hydraulic, 2,500 lb. capacity, 12' floor to floor, speed 150 F.P.M., emergency power switching and maintenance contract.

Description	Adjustment
A. Base Elevator: Hydraulic Passenger, 1500 lb. Capacity, 100 fpm, 2 Stops, Standard Finish	1 Ea.
B. Capacity Adjustment (2,500 lb.)	1 Ea.
C. Excess Travel Adjustment: 48' Total Travel (4 x 12') minus 12' Base Unit Travel =	36 V.L.F.
D. Stops Adjustment: 5 Total Stops minus 2 Stops (Base Unit) =	3 Stops
E. Speed Adjustment (150 F.P.M.)	1 Ea.
F. Options:	
1. Intercom Service	1 Ea.
2. Emergency Power Switching, Automatic	1 Ea.
3. Stainless Steel Entrance Doors	5 Ea.
4. Maintenance Contract (12 Months)	1 Ea.
5. Position Indicator for main floor level (none indicated in Base Unit)	1 Ea.

R220102-20   Labor Adjustment Factors

Labor Adjustment Factors are provided for Divisions 21, 22, and 23 to assist the mechanical estimator account for the various complexities and special conditions of any particular project. While a single percentage has been entered on each line of Division 22 01 02.20, it should be understood that these are just suggested midpoints of ranges of values commonly used by mechanical estimators. They may be increased or decreased depending on the severity of the special conditions.

The group for "existing occupied buildings" has been the subject of requests for explanation. Actually there are two stages to this group: buildings that are existing and "finished" but unoccupied, and those that also are occupied. Buildings that are "finished" may result in higher labor costs due to the

workers having to be more careful not to damage finished walls, ceilings, floors, etc. and may necessitate special protective coverings and barriers. Also corridor bends and doorways may not accommodate long pieces of pipe or larger pieces of equipment. Work above an already hung ceiling can be very time consuming. The addition of occupants may force the work to be done on premium time (nights and/or weekends), eliminate the possible use of some preferred tools such as pneumatic drivers, powder charged drivers, etc. The estimator should evaluate the access to the work area and just how the work is going to be accomplished to arrive at an increase in labor costs over "normal" new construction productivity.

R220105-10   Demolition (Selective vs. Removal for Replacement)

Demolition can be divided into two basic categories.

One type of demolition involves the removal of material with no concern for its replacement. The labor-hours to estimate this work are found under "Selective Demolition" in the Fire Protection, Plumbing and HVAC Divisions. It is selective in that individual items or all the material installed as a system or trade grouping such as plumbing or heating systems are removed. This may be accomplished by the easiest way possible, such as sawing, torch cutting, or sledge hammering as well as simple unbolting.

The second type of demolition is the removal of some items for repair or replacement. This removal may involve careful draining, opening of unions,

disconnecting and tagging electrical connections, capping pipes/ducts to prevent entry of debris or leakage of the material contained as well as transporting the item away from its in-place location to a truck/dumpster. An approximation of the time required to accomplish this type of demolition is to use half of the time indicated as necessary to install a new unit. For example: installation of a new pump might be listed as requiring 6 labor-hours so if we had to estimate the removal of the old pump we would allow an additional 3 hours for a total of 9 hours. That is, the complete replacement of a defective pump with a new pump would be estimated to take 9 labor-hours.

R221113-50   Pipe Material Considerations

1. Malleable fittings should be used for gas service.
2. Malleable fittings are used where there are stresses/strains due to expansion and vibration.
3. Cast fittings may be broken as an aid to disassembling heating lines frozen by long use, temperature and minerals.
4. A cast iron pipe is extensively used for underground and submerged service.
5. Type M (light wall) copper tubing is available in hard temper only and is used for nonpressure and less severe applications than K and L.

6. Type L (medium wall) copper tubing, available hard or soft for interior service.
7. Type K (heavy wall) copper tubing, available in hard or soft temper for use where conditions are severe. For underground and interior service.
8. Hard drawn tubing requires fewer hangers or supports but should not be bent. Silver brazed fittings are recommended, but soft solder is normally used.
9. Type DMV (very light wall) copper tubing designed for drainage, waste and vent plus other non-critical pressure services.

Domestic/Imported Pipe and Fittings Costs

The prices shown in this publication for steel/cast iron pipe and steel, cast iron, and malleable iron fittings are based on domestic production sold at the normal trade discounts. The above listed items of foreign manufacture may be available at prices 1/3 to 1/2 of those shown. Some imported items after minor machining or finishing operations are being sold as domestic to further complicate the system.

Caution: Most pipe prices in this data set also include a coupling and pipe hangers which for the larger sizes can add significantly to the per foot cost and should be taken into account when comparing "book cost" with the quoted supplier's cost.

R260105-30 Electrical Demolition (Removal for Replacement)

The purpose of this reference number is to provide a guide to users for electrical "removal for replacement" by applying the rule of thumb: 1/3 of new installation time (typical range from 20% to 50%) for removal. Remember to use reasonable judgment when applying the suggested percentage factor. For example:

Contractors have been requested to remove an existing fluorescent lighting fixture and replace with a new fixture utilizing energy saver lamps and electronic ballast:

In order to fully understand the extent of the project, contractors should visit the job site and estimate the time to perform the renovation work in accordance with applicable national, state and local regulation and codes.

The contractor may need to add extra labor hours to his estimate if he discovers unknown concealed conditions such as: contaminated asbestos ceiling, broken acoustical ceiling tile and need to repair, patch and touch-up paint all the damaged or disturbed areas, tasks normally assigned to general contractors. In addition, the owner could request that the contractors salvage the materials removed and turn over the materials to the owner or dispose of the materials to a reclamation station. The normal removal item is 0.5 labor hour for a lighting fixture and 1.5 labor-hours for new installation time. Revise the estimate times from 2 labor-hours work up to a minimum 4 labor-hours work for just fluorescent lighting fixture.

For removal of large concentrations of lighting fixtures in the same area, apply an "economy of scale" to reduce estimating labor hours.

Electrical

R2605 Common Work Results for Electrical

R260519-90 Wire

Wire quantities are taken off by either measuring each cable run or by extending the conduit and raceway quantities times the number of conductors in the raceway. Ten percent should be added for waste and tie-ins. Keep in mind that the unit of measure of wire is C.L.F. not L.F. as in raceways so the formula would read:

$$\frac{\text{(L.F. Raceway x No. of Conductors) x 1.10}}{100} = \text{C.L.F.}$$

Price per C.L.F. of wire includes:
1. Setting up wire coils or spools on racks
2. Attaching wire to pull in means
3. Measuring and cutting wire
4. Pulling wire into a raceway
5. Identifying and tagging

Price does not include:
1. Connections to breakers, panelboards, or equipment
2. Splices

Job Conditions: Productivity is based on new construction to a height of 15′ using rolling staging in an unobstructed area. Material staging is assumed to be within 100′ of work being performed.

Economy of Scale: If more than three wires at a time are being pulled, deduct the following percentages from the labor of that grouping:

4-5 wires	25%
6-10 wires	30%
11-15 wires	35%
over 15	40%

If a wire pull is less than 100′ in length and is interrupted several times by boxes, lighting outlets, etc., it may be necessary to add the following lengths to each wire being pulled:

Junction box to junction box	2 L.F.
Lighting panel to junction box	6 L.F.
Distribution panel to sub panel	8 L.F.
Switchboard to distribution panel	12 L.F.
Switchboard to motor control center	20 L.F.
Switchboard to cable tray	40 L.F.

Measure of Drops and Riser: It is important when taking off wire quantities to include the wire for drops to electrical equipment. If heights of electrical equipment are not clearly stated, use the following guide:

	Bottom A.F.F.	Top A.F.F.	Inside Cabinet
Safety switch to 100A	5′	6′	2′
Safety switch 400 to 600A	4′	6′	3′
100A panel 12 to 30 circuit	4′	6′	3′
42 circuit panel	3′	6′	4′
Switch box	3′	3′6″	1′
Switchgear	0′	8′	8′
Motor control centers	0′	8′	8′
Transformers - wall mount	4′	8′	2′
Transformers - floor mount	0′	12′	4′

For customer support on your Commercial Renovation Costs with RSMeans Data, call 800.448.8182.

879

R260533-22   Conductors in Conduit

The table below lists the maximum number of conductors for various sized conduit using THW, TW or THWN insulations.

Copper Wire Size	1/2"			3/4"			1"			1-1/4"			1-1/2"			2"			2-1/2"			3"		3-1/2"		4"	
	TW	THW	THWN	TW	THW	THWN	TW	THW	THWN	TW	THW	THWN	TW	THW	THWN	TW	THW	THWN	TW	THW	THWN	THW	THWN	THW	THWN	THW	THWN
#14	9	6	13	15	10	24	25	16	39	44	29	69	60	40	94	99	65	154	142	93		143		192			
#12	7	4	10	12	8	18	19	13	29	35	24	51	47	32	70	78	53	114	111	76	164	117		157			
#10	5	4	6	9	6	11	15	11	18	26	19	32	36	26	44	60	43	73	85	61	104	95	160	127		163	
#8	2	1	3	4	3	5	7	5	9	12	10	16	17	13	22	28	22	36	40	32	51	49	79	66	106	85	136
#6		1	1		2	4		4	6		7	11		10	15		16	26		23	37	36	57	48	76	62	98
#4		1	1		1	2		3	4		5	7		7	9		12	16		17	22	27	35	36	47	47	60
#3		1	1		1	1		2	3		4	6		6	8		10	13		15	19	23	29	31	39	40	51
#2		1	1		1	1		2	3		4	5		5	7		9	11		13	16	20	25	27	33	34	43
#1					1	1		1	1		3	3		4	5		6	8		9	12	14	18	19	25	25	32
1/0					1	1		1	1		2	3		3	4		5	7		8	10	12	15	16	21	21	27
2/0					1	1		1	1		1	2		3	3		5	6		7	8	10	13	14	17	18	22
3/0					1	1		1	1		1	1		2	3		4	5		6	7	9	11	12	14	15	18
4/0						1		1	1		1	1		1	2		3	4		5	6	7	9	10	12	13	15
250 kcmil								1	1		1	1		1	1		2	3		4	4	6	7	8	10	10	12
300								1	1		1	1		1	1		2	3		3	4	5	6	7	8	9	11
350									1		1	1		1	1		1	2		3	3	4	5	6	7	8	9
400											1	1		1	1		1	1		2	3	4	5	5	6	7	8
500											1	1		1	1		1	1		1	2	3	4	4	5	6	7
600												1		1	1		1	1		1	1	3	3	4	4	5	5
700														1	1		1	1		1	1	2	3	3	4	4	5
750														1	1		1	1		1	1	2	2	3	3	4	4

For customer support on your Commercial Renovation Costs with RSMeans Data, call 800.448.8182.

R312316-40   Excavating

The selection of equipment used for structural excavation and bulk excavation or for grading is determined by the following factors.

1. Quantity of material
2. Type of material
3. Depth or height of cut
4. Length of haul
5. Condition of haul road
6. Accessibility of site
7. Moisture content and dewatering requirements
8. Availability of excavating and hauling equipment

Some additional costs must be allowed for hand trimming the sides and bottom of concrete pours and other excavation below the general excavation.

Number of B.C.Y. per truck = 1.5 C.Y. bucket x 8 passes = 12 loose C.Y.

$$= 12 \text{ x } \frac{100}{118} = 10.2 \text{ B.C.Y. per truck}$$

Truck Haul Cycle:

Load truck, 8 passes	=	4 minutes
Haul distance, 1 mile	=	9 minutes
Dump time	=	2 minutes
Return, 1 mile	=	7 minutes
Spot under machine	=	1 minute
		23 minute cycle

Add the mobilization and demobilization costs to the total excavation costs. When equipment is rented for more than three days, there is often no mobilization charge by the equipment dealer. On larger jobs outside of urban areas, scrapers can move earth economically provided a dump site or fill area and adequate haul roads are available. Excavation within sheeting bracing or cofferdam bracing is usually done with a clamshell and production

When planning excavation and fill, the following should also be considered.

1. Swell factor
2. Compaction factor
3. Moisture content
4. Density requirements

A typical example for scheduling and estimating the cost of excavation of a 15′ deep basement on a dry site when the material must be hauled off the site is outlined below.

Assumptions:

1. Swell factor, 18%
2. No mobilization or demobilization
3. Allowance included for idle time and moving on job
4. No dewatering, sheeting, or bracing
5. No truck spotter or hand trimming

Fleet Haul Production per day in B.C.Y.

$$4 \text{ trucks x } \frac{50 \text{ min. hour}}{23 \text{ min. haul cycle}} \text{ x 8 hrs. x 10.2 B.C.Y.}$$

$$= 4 \text{ x } 2.2 \text{ x } 8 \text{ x } 10.2 = 718 \text{ B.C.Y./day}$$

is low, since the clamshell may have to be guided by hand between the bracing. When excavating or filling an area enclosed with a wellpoint system, add 10% to 15% to the cost to allow for restricted access. When estimating earth excavation quantities for structures, allow work space outside the building footprint for construction of the foundation and a slope of 1:1 unless sheeting is used.

R312316-45 Excavating Equipment

The table below lists theoretical hourly production in C.Y./hr. bank measure for some typical excavation equipment. Figures assume 50 minute hours, 83% job efficiency, 100% operator efficiency, 90° swing and properly sized hauling units, which must be modified for adverse digging and loading conditions. Actual production costs in the front of the data set average about 50% of the theoretical values listed here.

Equipment	Soil Type	B.C.Y. Weight	% Swell	1 C.Y.	1-1/2 C.Y.	2 C.Y.	2-1/2 C.Y.	3 C.Y.	3-1/2 C.Y.	4 C.Y.
Hydraulic Excavator	Moist loam, sandy clay	3400 lb.	40%	165	195	200	275	330	385	440
"Backhoe"	Sand and gravel	3100	18	140	170	225	240	285	330	380
15' Deep Cut	Common earth	2800	30	150	180	230	250	300	350	400
	Clay, hard, dense	3000	33	120	140	190	200	240	260	320
Power Shovel Optimum Cut (Ft.)	Moist loam, sandy clay	3400	40	170 (6.0)	245 (7.0)	295 (7.8)	335 (8.4)	385 (8.8)	435 (9.1)	475 (9.4)
	Sand and gravel	3100	18	165 (6.0)	225 (7.0)	275 (7.8)	325 (8.4)	375 (8.8)	420 (9.1)	460 (9.4)
	Common earth	2800	30	145 (7.8)	200 (9.2)	250 (10.2)	295 (11.2)	335 (12.1)	375 (13.0)	425 (13.8)
	Clay, hard, dense	3000	33	120 (9.0)	175 (10.7)	220 (12.2)	255 (13.3)	300 (14.2)	335 (15.1)	375 (16.0)
Drag Line Optimum Cut (Ft.)	Moist loam, sandy clay	3400	40	130 (6.6)	180 (7.4)	220 (8.0)	250 (8.5)	290 (9.0)	325 (9.5)	385 (10.0)
	Sand and gravel	3100	18	130 (6.6)	175 (7.4)	210 (8.0)	245 (8.5)	280 (9.0)	315 (9.5)	375 (10.0)
	Common earth	2800	30	110 (8.0)	160 (9.0)	190 (9.9)	220 (10.5)	250 (11.0)	280 (11.5)	310 (12.0)
	Clay, hard, dense	3000	33	90 (9.3)	130 (10.7)	160 (11.8)	190 (12.3)	225 (12.8)	250 (13.3)	280 (12.0)

				Wheel Loaders				Track Loaders		
				3 C.Y.	4 C.Y.	6 C.Y.	8 C.Y.	2-1/4 C.Y.	3 C.Y.	4 C.Y.
Loading Tractors	Moist loam, sandy clay	3400	40	260	340	510	690	135	180	250
	Sand and gravel	3100	18	245	320	480	650	130	170	235
	Common earth	2800	30	230	300	460	620	120	155	220
	Clay, hard, dense	3000	33	200	270	415	560	110	145	200
	Rock, well-blasted	4000	50	180	245	380	520	100	130	180

R312323-30 Compacting Backfill

Compaction of fill in embankments, around structures, in trenches, and under slabs is important to control settlement. Factors affecting compaction are:

1. Soil gradation
2. Moisture content
3. Equipment used
4. Depth of fill per lift
5. Density required

Production Rate:

$$\frac{1.75' \text{ plate width} \times 50 \text{ F.P.M.} \times 50 \text{ min./hr.} \times .67' \text{ lift}}{27 \text{ C.F. per C.Y.}} = 108.5 \text{ C.Y./hr.}$$

Production Rate for 4 Passes:

$$\frac{108.5 \text{ C.Y.}}{4 \text{ passes}} = 27.125 \text{ C.Y./hr.} \times 8 \text{ hrs.} = 217 \text{ C.Y./day}$$

Example:

Compact granular fill around a building foundation using a 21″ wide x 24″ vibratory plate in 8″ lifts. Operator moves at 50 F.P.M. working a 50 minute hour to develop 95% Modified Proctor Density with 4 passes.

R314116-40   Wood Sheet Piling

Wood sheet piling may be used for depths to 20' where there is no ground water. If moderate ground water is encountered Tongue & Groove sheeting will help to keep it out. When considerable ground water is present, steel sheeting must be used.

For estimating purposes on trench excavation, sizes are as follows:

Depth	Sheeting	Wales	Braces	B.F. per S.F.
To 8'	3 x 12's	6 x 8's, 2 line	6 x 8's, @ 10'	4.0 @ 8'
8' x 12'	3 x 12's	10 x 10's, 2 line	10 x 10's, @ 9'	5.0 average
12' to 20'	3 x 12's	12 x 12's, 3 line	12 x 12's, @ 8'	7.0 average

Sheeting to be toed in at least 2' depending upon soil conditions. A five person crew with an air compressor and sheeting driver can drive and brace 440 SF/day at 8' deep, 360 SF/day at 12' deep, and 320 SF/day at 16' deep.

For normal soils, piling can be pulled in 1/3 the time to install. Pulling difficulty increases with the time in the ground. Production can be increased by high pressure jetting.

R316326-60   Caissons

The three principal types of caissons are:

(1) Belled Caissons, which except for shallow depths and poor soil conditions, are generally recommended. They provide more bearing than shaft area. Because of its conical shape, no horizontal reinforcement of the bell is required.

(2) Straight Shaft Caissons are used where relatively light loads are to be supported by caissons that rest on high value bearing strata. While the shaft is larger in diameter than for belled types this is more than offset by the savings in time and labor.

(3) Keyed Caissons are used when extremely heavy loads are to be carried. A keyed or socketed caisson transfers its load into rock by a combination of end-bearing and shear reinforcing of the shaft. The most economical shaft often consists of a steel casing, a steel wide flange core and concrete. Allowable compressive stresses of .225 f'c for concrete, 16,000 psi for the wide flange core, and 9,000 psi for the steel casing are commonly used. The usual range of shaft diameter is 18" to 84". The number of sizes specified for any one project should be limited due to the problems of casing and auger storage. When hand work is to be performed, shaft diameters should not be less than 32". When inspection of borings is required a minimum shaft diameter of 30" is recommended. Concrete caissons are intended to be poured against earth excavation so permanent forms, which add to cost, should not be used if the excavation is clean and the earth is sufficiently impervious to prevent excessive loss of concrete.

Soil Conditions for Belling		
Good	Requires Handwork	Not Recommended
Clay	Hard Shale	Silt
Sandy Clay	Limestone	Sand
Silty Clay	Sandstone	Gravel
Clayey Silt	Weathered Mica	Igneous Rock
Hard-pan		
Soft Shale		
Decomposed Rock		

R329219-50   Seeding

The type of grass is determined by light, shade and moisture content of soil plus intended use. Fertilizer should be disked 4" before seeding. For steep slopes disk five tons of mulch and lay two tons of hay or straw on surface per acre after seeding. Surface mulch can be staked, lightly disked or tar emulsion sprayed. Material for mulch can be wood chips, peat moss, partially rotted hay or straw, wood fibers and sprayed emulsions. Hemp seed blankets with fertilizer are also available. For spring seeding, watering is necessary. Late fall seeding may have to be reseeded in the spring. Hydraulic seeding, power mulching, and aerial seeding can be used on large areas.

R331113-80   Piping Designations

There are several systems currently in use to describe pipe and fittings. The following paragraphs will help to identify and clarify classifications of piping systems used for water distribution.

Piping may be classified by schedule. Piping schedules include 5S, 10S, 10, 20, 30, Standard, 40, 60, Extra Strong, 80, 100, 120, 140, 160 and Double Extra Strong. These schedules are dependent upon the pipe wall thickness. The wall thickness of a particular schedule may vary with pipe size.

Ductile iron pipe for water distribution is classified by Pressure Classes such as Class 150, 200, 250, 300 and 350. These classes are actually the rated water working pressure of the pipe in pounds per square inch (psi). The pipe in these pressure classes is designed to withstand the rated water working pressure plus a surge allowance of 100 psi.

The American Water Works Association (AWWA) provides standards for various types of **plastic pipe.** C-900 is the specification for polyvinyl chloride (PVC) piping used for water distribution in sizes ranging from 4″ through 12″. C-901 is the specification for polyethylene (PE) pressure pipe, tubing and fittings used for water distribution in sizes ranging from 1/2″ through 3″. C-905 is the specification for PVC piping sizes 14″ and greater.

PVC pressure-rated pipe is identified using the standard dimensional ratio (SDR) method. This method is defined by the American Society for Testing and Materials (ASTM) Standard D 2241. This pipe is available in SDR numbers 64, 41, 32.5, 26, 21, 17, and 13.5. A pipe with an SDR of 64 will have the thinnest wall while a pipe with an SDR of 13.5 will have the thickest wall. When the pressure rating (PR) of a pipe is given in psi, it is based on a line supplying water at 73 degrees F.

The National Sanitation Foundation (NSF) seal of approval is applied to products that can be used with potable water. These products have been tested to ANSI/NSF Standard 14.

Valves and strainers are classified by American National Standards Institute (ANSI) Classes. These Classes are 125, 150, 200, 250, 300, 400, 600, 900, 1500 and 2500. Within each class there is an operating pressure range dependent upon temperature. Design parameters should be compared to the appropriate material dependent, pressure-temperature rating chart for accurate valve selection.

Change Orders

Change Order Considerations

A change order is a written document usually prepared by the design professional and signed by the owner, the architect/engineer, and the contractor. A change order states the agreement of the parties to: an addition, deletion, or revision in the work; an adjustment in the contract sum, if any; or an adjustment in the contract time, if any. Change orders, or "extras", in the construction process occur after execution of the construction contract and impact architects/engineers, contractors, and owners.

Change orders that are properly recognized and managed can ensure orderly, professional, and profitable progress for everyone involved in the project. There are many causes for change orders and change order requests. In all cases, change orders or change order requests should be addressed promptly and in a precise and prescribed manner. The following paragraphs include information regarding change order pricing and procedures.

The Causes of Change Orders

Reasons for issuing change orders include:

- Unforeseen field conditions that require a change in the work
- Correction of design discrepancies, errors, or omissions in the contract documents
- Owner-requested changes, either by design criteria, scope of work, or project objectives
- Completion date changes for reasons unrelated to the construction process
- Changes in building code interpretations, or other public authority requirements that require a change in the work
- Changes in availability of existing or new materials and products

Procedures

Properly written contract documents must include the correct change order procedures for all parties—owners, design professionals, and contractors—to follow in order to avoid costly delays and litigation.

Being "in the right" is not always a sufficient or acceptable defense. The contract provisions requiring notification and documentation must be adhered to within a defined or reasonable time frame.

The appropriate method of handling change orders is by a written proposal and acceptance by all parties involved. Prior to starting work on a project, all parties should identify their

authorized agents who may sign and accept change orders, as well as any limits placed on their authority.

Time may be a critical factor when the need for a change arises. For such cases, the contractor might be directed to proceed on a "time and materials" basis, rather than wait for all paperwork to be processed—a delay that could impede progress. In this situation, the contractor must still follow the prescribed change order procedures including, but not limited to, notification and documentation.

Lack of documentation can be very costly, especially if legal judgments are to be made, and if certain field personnel are no longer available. For time and material change orders, the contractor should keep accurate daily records of all labor and material allocated to the change.

Owners or awarding authorities who do considerable and continual building construction (such as the federal government) realize the inevitability of change orders for numerous reasons, both predictable and unpredictable. As a result, the federal government, the American Institute of Architects (AIA), the Engineers Joint Contract Documents Committee (EJCDC), and other contractor, legal, and technical organizations have developed standards and procedures to be followed by all parties to achieve contract continuance and timely completion, while being financially fair to all concerned.

Pricing Change Orders

When pricing change orders, regardless of their cause, the most significant factor is when the change occurs. The need for a change may be perceived in the field or requested by the architect/engineer *before* any of the actual installation has begun, or may evolve or appear *during* construction when the item of work in question is partially installed. In the latter cases, the original sequence of construction is disrupted, along with all contiguous and supporting systems. Change orders cause the greatest impact when they occur *after* the installation has been completed and must be uncovered, or even replaced. Post-completion changes may be caused by necessary design changes, product failure, or changes in the owner's requirements that are not discovered until the building or the systems begin to function.

Specified procedures of notification and record keeping must be adhered to and enforced regardless of the stage of construction: *before, during,* or *after* installation. Some bidding documents anticipate change orders by requiring that unit prices including overhead and profit percentages—for additional as well as deductible changes—be listed. Generally these unit prices do not fully take into account the ripple effect, or impact on other trades, and should be used for general guidance only.

When pricing change orders, it is important to classify the time frame in which the change occurs. There are two basic time frames for change orders: *pre-installation change orders,* which occur before the start of construction, and *post-installation change orders,* which involve reworking after the original installation. Change orders that occur between these stages may be priced according to the extent of work completed using a combination of techniques developed for pricing *pre-* and *post-installation* changes.

Factors To Consider When Pricing Change Orders

As an estimator begins to prepare a change order, the following questions should be reviewed to determine their impact on the final price.

General

- *Is the change order work* pre-installation *or* post-installation?

  Change order work costs vary according to how much of the installation has been completed. Once workers have the project scoped in their minds, even though they have not started, it can be difficult to refocus. Consequently they may spend more than the normal amount of time understanding the change. Also, modifications to work in place, such as trimming or refitting, usually take more time than was initially estimated. The greater the amount of work in place, the more reluctant workers are to change it. Psychologically they may resent the change and as a result the rework takes longer than normal. Post-installation change order estimates must include demolition of existing work as required to accomplish the change. If the work is performed at a later time, additional obstacles, such as building finishes, may be present which must be protected. Regardless of whether the change occurs pre-installation or post-installation, attempt to isolate the identifiable factors and price them separately. For example, add shipping costs that may be required pre-installation or any demolition required post-installation. Then analyze the potential impact on productivity of psychological and/or learning curve factors and adjust the output rates accordingly. One approach is to break down the typical workday into segments and quantify the impact on each segment.

Change Order Installation Efficiency

The labor-hours expressed (for new construction) are based on average installation time, using an efficiency level. For change order situations, adjustments to this efficiency level should reflect the daily labor-hour allocation for that particular occurrence.

- *Will the change substantially delay the original completion date?*

A significant change in the project may cause the original completion date to be extended. The extended schedule may subject the contractor to new wage rates dictated by relevant labor contracts. Project supervision and other project overhead must also be extended beyond the original completion date. The schedule extension may also put installation into a new weather season. For example, underground piping scheduled for October installation was delayed until January. As a result, frost penetrated the trench area, thereby changing the degree of difficulty of the task. Changes and delays may have a ripple effect throughout the project. This effect must be analyzed and negotiated with the owner.

- *What is the net effect of a deduct change order?*

In most cases, change orders resulting in a deduction or credit reflect only bare costs. The contractor may retain the overhead and profit based on the original bid.

Materials

- *Will you have to pay more or less for the new material, required by the change order, than you paid for the original purchase?*

The same material prices or discounts will usually apply to materials purchased for change orders as new construction. In some instances, however, the contractor may forfeit the advantages of competitive pricing for change orders. Consider the following example:

A contractor purchased over $20,000 worth of fan coil units for an installation and obtained the maximum discount. Some time later it was determined the project required an additional matching unit. The contractor has to purchase this unit from the original supplier to ensure a match. The supplier at this time may not discount the unit because of the small quantity, and he is no longer in a competitive situation. The impact of quantity on purchase can add between 0% and 25% to material prices and/or subcontractor quotes.

- *If materials have been ordered or delivered to the job site, will they be subject to a cancellation charge or restocking fee?*

Check with the supplier to determine if ordered materials are subject to a cancellation charge. Delivered materials not used as a result of a change order may be subject to a restocking fee if returned to the supplier. Common restocking charges run between 20% and 40%. Also, delivery charges to return the goods to the supplier must be added.

Labor

- *How efficient is the existing crew at the actual installation?*

Is the same crew that performed the initial work going to do the change order? Possibly the change consists of the installation of a unit identical to one already installed; therefore, the change should take less time. Be sure to consider this potential productivity increase and modify the productivity rates accordingly.

- *If the crew size is increased, what impact will that have on supervision requirements?*

Under most bargaining agreements or management practices, there is a point at which a working foreman is replaced by a nonworking foreman. This replacement increases project overhead by adding a nonproductive worker. If additional workers are added to accelerate the project or to perform changes while maintaining the schedule, be sure to add additional supervision time if warranted. Calculate the hours involved and the additional cost directly if possible.

- *What are the other impacts of increased crew size?*

The larger the crew, the greater the potential for productivity to decrease. Some of the factors that cause this productivity loss are: overcrowding (producing restrictive conditions in the working space) and possibly a shortage of any special tools and equipment required. Such factors affect not only the crew working on the elements directly involved in the change order, but other crews whose movements may also be hampered. As the crew increases, check its basic composition for changes by the addition or deletion of apprentices or nonworking foreman, and quantify the potential effects of equipment shortages or other logistical factors.

- *As new crews, unfamiliar with the project, are brought onto the site, how long will it take them to become oriented to the project requirements?*

The orientation time for a new crew to become 100% effective varies with the site and type of project. Orientation is easiest at a new construction site and most difficult at existing, very restrictive renovation sites. The type of work also affects orientation time. When all elements of the work are exposed, such as concrete or masonry work, orientation is decreased. When the work is concealed or less visible, such as existing electrical systems, orientation takes longer. Usually orientation can be accomplished in one day or less. Costs for added orientation should be itemized and added to the total estimated cost.

- *How much actual production can be gained by working overtime?*

Short term overtime can be used effectively to accomplish more work in a day. However, as overtime is scheduled to run beyond several weeks, studies have shown marked decreases in output. The following chart shows the effect of long term overtime on worker efficiency. If the anticipated change requires extended overtime to keep the job on schedule, these factors can be used as a guide to predict the impact on time and cost. Add project overhead, particularly supervision, that may also be incurred.

For customer support on your Commercial Renovation Costs with RSMeans Data, call 800.448.8182.

Days per Week	Hours per Day	Production Efficiency					Payroll Cost Factors	
		1st Week	2nd Week	3rd Week	4th Week	Average 4 Weeks	@ 1-1/2 Times	@ 2 Times
5	8	100%	100%	100%	100%	100%	100%	100%
	9	100	100	95	90	96	1.056	1.111
	10	100	95	90	85	93	1.100	1.200
	11	95	90	75	65	81	1.136	1.273
	12	90	85	70	60	76	1.167	1.333
6	8	100	100	95	90	96	1.083	1.167
	9	100	95	90	85	93	1.130	1.259
	10	95	90	85	80	88	1.167	1.333
	11	95	85	70	65	79	1.197	1.394
	12	90	80	65	60	74	1.222	1.444
7	8	100	95	85	75	89	1.143	1.286
	9	95	90	80	70	84	1.183	1.365
	10	90	85	75	65	79	1.214	1.429
	11	85	80	65	60	73	1.240	1.481
	12	85	75	60	55	69	1.262	1.524

Effects of Overtime

Caution: Under many labor agreements, Sundays and holidays are paid at a higher premium than the normal overtime rate.

The use of long-term overtime is counterproductive on almost any construction job; that is, the longer the period of overtime, the lower the actual production rate. Numerous studies have been conducted, and while they have resulted in slightly different numbers, all reach the same conclusion. The figure above tabulates the effects of overtime work on efficiency.

As illustrated, there can be a difference between the *actual* payroll cost per hour and the *effective* cost per hour for overtime work. This is due to the reduced production efficiency with the increase in weekly hours beyond 40. This difference between actual and effective cost results from overtime work over a prolonged period. Short-term overtime work does not result in as great a reduction in efficiency and, in such cases, effective cost may not vary significantly from the actual payroll cost. As the total hours per week are increased on a regular basis, more time is lost due to fatigue, lowered morale, and an increased accident rate.

As an example, assume a project where workers are working 6 days a week, 10 hours per day. From the figure above (based on productivity studies), the average effective productive hours over a 4-week period are:

$$0.875 \times 60 = 52.5$$

Depending upon the locale and day of week, overtime hours may be paid at time and a half or double time. For time and a half, the overall (average) *actual* payroll cost (including regular and overtime hours) is determined as follows:

$$\frac{40 \text{ reg. hrs.} + (20 \text{ overtime hrs.} \times 1.5)}{60 \text{ hrs.}} = 1.167$$

Based on 60 hours, the payroll cost per hour will be 116.7% of the normal rate at 40 hours per week. However, because the effective production (efficiency) for 60 hours is reduced to the equivalent of 52.5 hours, the effective cost of overtime is calculated as follows:

For time and a half:

$$\frac{40 \text{ reg. hrs.} + (20 \text{ overtime hrs.} \times 1.5)}{52.5 \text{ hrs.}} = 1.33$$

The installed cost will be 133% of the normal rate (for labor).

Thus, when figuring overtime, the actual cost per unit of work will be higher than the apparent overtime payroll dollar increase, due to the reduced productivity of the longer work week. These efficiency calculations are true only for those cost factors determined by hours worked. Costs that are applied weekly or monthly, such as equipment rentals, will not be similarly affected.

Equipment

- *What equipment is required to complete the change order?*

Change orders may require extending the rental period of equipment already on the job site, or the addition of special equipment brought in to accomplish the change work. In either case, the additional rental charges and operator labor charges must be added.

Summary

The preceding considerations and others you deem appropriate should be analyzed and applied to a change order estimate. The impact of each should be quantified and listed on the estimate to form an audit trail.

Change orders that are properly identified, documented, and managed help to ensure the orderly, professional, and profitable progress of the work. They also minimize potential claims or disputes at the end of the project.

Back by customer demand!

You asked and we listened. For customer convenience and estimating ease, we have made the 2021 Project Costs available for download at **RSMeans.com/2021books**. You will also find sample estimates, an RSMeans data overview video, and a book registration form to receive quarterly data updates throughout 2021.

Estimating Tips

- The cost figures available in the download were derived from hundreds of projects contained in the RSMeans database of completed construction projects. They include the contractor's overhead and profit. The figures have been adjusted to January of the current year.

- These projects were located throughout the U.S. and reflect a tremendous variation in square foot (S.F.) costs. This is due to differences, not only in labor and material costs, but also in individual owners' requirements. For instance, a bank in a large city would have different features than one in a rural area. This is true of all the different types of buildings analyzed. Therefore, caution should be exercised when using these Project Costs. For example, for courthouses, costs in the database are local courthouse costs and will not apply to the larger, more elaborate federal courthouses.

- None of the figures "go with" any others. All individual cost items were computed and tabulated separately. Thus, the sum of the median figures for plumbing, HVAC, and electrical will not normally total up to the total mechanical and electrical costs arrived at by separate analysis and tabulation of the projects.

- Each building was analyzed as to total and component costs and percentages. The figures were arranged in ascending order with the results tabulated as shown. The 1/4 column shows that 25% of the projects had lower costs and 75% had higher. The 3/4 column shows that 75% of the projects had lower costs and 25% had higher. The median column shows that 50% of the projects had lower costs and 50% had higher.

- Project Costs are useful in the conceptual stage when no details are available. As soon as details become available in the project design, the square foot approach should be discontinued and the project should be priced as to its particular components. When more precision is required, or for estimating the replacement cost of specific buildings, the current edition of *Square Foot Costs with RSMeans data* should be used.

- In using the figures in this section, it is recommended that the median column be used for preliminary figures if no additional information is available. The median figures, when multiplied by the total city construction cost index figures (see City Cost Indexes) and then multiplied by the project size modifier at the end of this section, should present a fairly accurate base figure, which would then have to be adjusted in view of the estimator's experience, local economic conditions, code requirements, and the owner's particular requirements. There is no need to factor in the percentage figures, as these should remain constant from city to city.

- The editors of this data would greatly appreciate receiving cost figures on one or more of your recent projects, which would then be included in the averages for next year. All cost figures received will be kept confidential, except that they will be averaged with other similar projects to arrive at square foot cost figures for next year.

See the website above for details and the discount available for submitting one or more of your projects.

Same Data. Simplified.

Enjoy the convenience and efficiency of accessing your costs anywhere:

- **Skip the multiplier** by setting your location
- **Quickly search,** edit, favorite and share costs
- **Stay on top of price changes** with automatic updates

Discover more at rsmeans.com/online

50 17 00 | Project Costs

			UNIT	UNIT COSTS			% OF TOTAL			
				1/4	MEDIAN	3/4	1/4	MEDIAN	3/4	
01	0000	**Auto Sales with Repair**	S.F.							01
	0100	Architectural		107	120	130	58%	64%	67%	
	0200	Plumbing		8.95	9.40	12.55	4.84%	5.20%	6.80%	
	0300	Mechanical		12	16.10	17.75	6.40%	8.70%	10.15%	
	0400	Electrical		18.45	23	28.50	9.05%	11.70%	15.90%	
	0500	Total Project Costs		180	187	193				
02	0000	**Banking Institutions**	S.F.							02
	0100	Architectural		161	198	241	59%	65%	69%	
	0200	Plumbing		6.50	9.10	12.60	2.12%	3.39%	4.19%	
	0300	Mechanical		12.95	17.90	21	4.41%	5.10%	10.75%	
	0400	Electrical		31.50	38	58.50	10.45%	13.05%	15.90%	
	0500	Total Project Costs		268	300	370				
03	0000	**Court House**	S.F.							03
	0100	Architectural		85	167	167	54.50%	58.50%	58.50%	
	0200	Plumbing		3.22	3.22	3.22	2.07%	2.07%	2.07%	
	0300	Mechanical		20	20	20	12.95%	12.95%	12.95%	
	0400	Electrical		26	26	26	16.60%	16.60%	16.60%	
	0500	Total Project Costs		155	286	286				
04	0000	**Data Centers**	S.F.							04
	0100	Architectural		192	192	192	68%	68%	68%	
	0200	Plumbing		10.55	10.55	10.55	3.71%	3.71%	3.71%	
	0300	Mechanical		27	27	27	9.45%	9.45%	9.45%	
	0400	Electrical		25.50	25.50	25.50	9%	9%	9%	
	0500	Total Project Costs		284	284	284				
05	0000	**Detention Centers**	S.F.							05
	0100	Architectural		178	188	200	52%	53%	60.50%	
	0200	Plumbing		18.80	23	27.50	5.15%	7.10%	7.25%	
	0300	Mechanical		24	34	41	7.55%	9.50%	13.80%	
	0400	Electrical		39	46.50	60.50	10.90%	14.85%	17.95%	
	0500	Total Project Costs		300	320	375				
06	0000	**Fire Stations**	S.F.							06
	0100	Architectural		101	130	190	46%	54.50%	61.50%	
	0200	Plumbing		10.25	14.05	18.35	4.59%	5.60%	6.30%	
	0300	Mechanical		15.10	22	29.50	6.05%	8.25%	10.20%	
	0400	Electrical		23.50	30	40	10.75%	12.55%	14.95%	
	0500	Total Project Costs		211	241	335				
07	0000	**Gymnasium**	S.F.							07
	0100	Architectural		89	118	118	57%	64.50%	64.50%	
	0200	Plumbing		2.19	7.15	7.15	1.58%	3.48%	3.48%	
	0300	Mechanical		3.35	30	30	2.42%	14.65%	14.65%	
	0400	Electrical		11	21.50	21.50	7.95%	10.35%	10.35%	
	0500	Total Project Costs		139	206	206				
08	0000	**Hospitals**	S.F.							08
	0100	Architectural		108	178	193	43%	47.50%	48%	
	0200	Plumbing		7.95	15.15	33	6%	7.45%	7.65%	
	0300	Mechanical		52.50	59	77	14.20%	17.95%	23.50%	
	0400	Electrical		24	48	62	10.95%	13.75%	16.85%	
	0500	Total Project Costs		253	375	405				
09	0000	**Industrial Buildings**	S.F.							09
	0100	Architectural		59	116	235	52%	56%	82%	
	0200	Plumbing		1.70	4.62	13.35	1.57%	2.11%	6.30%	
	0300	Mechanical		4.86	9.25	44	4.77%	5.55%	14.80%	
	0400	Electrical		7.40	8.45	70.50	7.85%	13.55%	16.20%	
	0500	Total Project Costs		88	133	435				
10	0000	**Medical Clinics & Offices**	S.F.							10
	0100	Architectural		90.50	122	163	48.50%	55.50%	62.50%	
	0200	Plumbing		9	13.50	21.50	4.47%	6.65%	8.75%	
	0300	Mechanical		14.60	22.50	41.50	7.80%	10.90%	16.05%	
	0400	Electrical		19.90	24.50	38	8.90%	11.40%	14.05%	
	0500	Total Project Costs		168	224	295				

50 17 00 | Project Costs

			UNIT	UNIT COSTS			% OF TOTAL			
				1/4	MEDIAN	3/4	1/4	MEDIAN	3/4	
11	0000	**Mixed Use**	S.F.							11
	0100	Architectural		92	130	198	45.50%	52.50%	61.50%	
	0200	Plumbing		6.25	11.45	12.15	3.31%	3.47%	4.18%	
	0300	Mechanical		15.25	25	46	4.68%	13.60%	17.05%	
	0400	Electrical		16.15	36	53.50	8.30%	11.40%	15.65%	
	0500	Total Project Costs		190	335	340				
12	0000	**Multi-Family Housing**	S.F.							12
	0100	Architectural		77.50	105	155	54.50%	61.50%	66.50%	
	0200	Plumbing		6.90	11.10	15.10	5.30%	6.85%	8%	
	0300	Mechanical		7.15	9.55	27.50	4.92%	6.90%	10.40%	
	0400	Electrical		10	15.70	22.50	6.20%	8%	10.25%	
	0500	Total Project Costs		128	210	253				
13	0000	**Nursing Home & Assisted Living**	S.F.							13
	0100	Architectural		72.50	94.50	119	51.50%	55.50%	63.50%	
	0200	Plumbing		7.80	11.75	12.90	6.25%	7.40%	8.80%	
	0300	Mechanical		6.40	9.45	18.50	4.04%	6.70%	9.55%	
	0400	Electrical		10.60	16.70	23.50	7%	10.75%	13.10%	
	0500	Total Project Costs		123	161	194				
14	0000	**Office Buildings**	S.F.							14
	0100	Architectural		93	130	179	54.50%	61%	69%	
	0200	Plumbing		5.15	8.10	15.15	2.70%	3.78%	5.85%	
	0300	Mechanical		10.10	17.15	26.50	5.60%	8.20%	11.10%	
	0400	Electrical		12.90	22	34	7.75%	10%	12.70%	
	0500	Total Project Costs		159	202	285				
15	0000	**Parking Garage**	S.F.							15
	0100	Architectural		32	39	40.50	70%	79%	88%	
	0200	Plumbing		1.05	1.10	2.06	2.05%	2.70%	2.83%	
	0300	Mechanical		.82	1.25	4.76	2.11%	3.62%	3.81%	
	0400	Electrical		2.80	3.08	6.45	5.30%	6.35%	7.95%	
	0500	Total Project Costs		39	47.50	51.50				
16	0000	**Parking Garage/Mixed Use**	S.F.							16
	0100	Architectural		103	113	115	61%	62%	65.50%	
	0200	Plumbing		3.33	4.36	6.65	2.47%	2.72%	3.66%	
	0300	Mechanical		14.20	16	23	7.80%	13.10%	13.60%	
	0400	Electrical		14.90	21.50	22	8.20%	12.65%	18.15%	
	0500	Total Project Costs		169	176	182				
17	0000	**Police Stations**	S.F.							17
	0100	Architectural		117	131	165	49%	56.50%	61%	
	0200	Plumbing		15.45	18.55	18.65	5.05%	5.55%	9.05%	
	0300	Mechanical		35	48.50	50.50	13%	14.55%	16.55%	
	0400	Electrical		26.50	29	30.50	9.15%	12.10%	14%	
	0500	Total Project Costs		219	270	305				
18	0000	**Police/Fire**	S.F.							18
	0100	Architectural		114	114	350	55.50%	66%	68%	
	0200	Plumbing		9.15	9.45	35	5.45%	5.50%	5.55%	
	0300	Mechanical		13.95	22	80	8.35%	12.70%	12.80%	
	0400	Electrical		15.90	20.50	91.50	9.50%	11.75%	14.55%	
	0500	Total Project Costs		167	173	630				
19	0000	**Public Assembly Buildings**	S.F.							19
	0100	Architectural		116	160	240	57.50%	61.50%	66%	
	0200	Plumbing		6.15	9	13.50	2.60%	3.36%	4.79%	
	0300	Mechanical		12.95	23	35.50	6.55%	8.95%	12.45%	
	0400	Electrical		19.15	26	41.50	8.60%	10.75%	13%	
	0500	Total Project Costs		186	255	375				
20	0000	**Recreational**	S.F.							20
	0100	Architectural		111	176	238	49.50%	59%	65.50%	
	0200	Plumbing		8.60	14.85	25.50	3.12%	4.76%	7.40%	
	0300	Mechanical		13.25	20	32	4.98%	6.60%	11.55%	
	0400	Electrical		19.15	27.50	40	7.15%	8.95%	10.75%	
	0500	Total Project Costs		198	296	445				

50 17 00	Project Costs	UNIT	UNIT COSTS			% OF TOTAL			
			1/4	MEDIAN	3/4	1/4	MEDIAN	3/4	
21 0000	**Restaurants**	S.F.							21
0100	Architectural		127	195	250	59%	60%	63.50%	
0200	Plumbing		13.90	32	40.50	7.35%	7.75%	8.95%	
0300	Mechanical		15	17.70	37.50	6.50%	8.15%	11.15%	
0400	Electrical		14.90	24.50	48.50	7.10%	10.30%	11.60%	
0500	Total Project Costs		210	305	420				
22 0000	**Retail**	S.F.							22
0100	Architectural		56.50	88.50	182	54.50%	60%	64.50%	
0200	Plumbing		7.05	9.85	12.25	4.18%	5.45%	8.45%	
0300	Mechanical		6.60	9.35	17.20	4.98%	6.15%	7.05%	
0400	Electrical		10.50	21.50	31.50	7.90%	11.25%	12.45%	
0500	Total Project Costs		86	153	292				
23 0000	**Schools**	S.F.							23
0100	Architectural		98	127	173	51%	56%	60.50%	
0200	Plumbing		7.75	10.70	15.65	3.66%	4.59%	7%	
0300	Mechanical		18.70	26.50	38.50	8.95%	12%	14.55%	
0400	Electrical		18.45	25.50	33	9.45%	11.25%	13.30%	
0500	Total Project Costs		178	227	310				
24 0000	**University, College & Private School Classroom & Admin Buildings**	S.F.							24
0100	Architectural		126	160	194	50.50%	55%	59.50%	
0200	Plumbing		8.50	11.50	19.90	3.02%	4.46%	7.35%	
0300	Mechanical		25	37.50	46.50	9.95%	11.95%	14.70%	
0400	Electrical		20	28.50	35.50	7.70%	9.90%	11.55%	
0500	Total Project Costs		219	293	380				
25 0000	**University, College & Private School Dormitories**	S.F.							25
0100	Architectural		81.50	143	152	53%	61.50%	68%	
0200	Plumbing		10.80	15.25	17.50	6.35%	6.65%	8.95%	
0300	Mechanical		4.84	20.50	32.50	4.13%	9%	11.80%	
0400	Electrical		5.75	19.95	30.50	4.75%	7.35%	10.60%	
0500	Total Project Costs		121	229	271				
26 0000	**University, College & Private School Science, Eng. & Lab Buildings**	S.F.							26
0100	Architectural		140	166	195	48.50%	56.50%	58%	
0200	Plumbing		9.65	14.65	24.50	3.29%	3.77%	5%	
0300	Mechanical		37	69	71	11.70%	19.40%	23.50%	
0400	Electrical		29	38	39	9%	12.05%	13.15%	
0500	Total Project Costs		293	315	370				
27 0000	**University, College & Private School Student Union Buildings**	S.F.							27
0100	Architectural		111	292	292	54.50%	54.50%	59.50%	
0200	Plumbing		16.80	16.80	25	3.13%	4.27%	11.45%	
0300	Mechanical		32	51.50	51.50	9.60%	9.60%	14.55%	
0400	Electrical		28	48.50	48.50	9.05%	12.80%	13.15%	
0500	Total Project Costs		219	535	535				
28 0000	**Warehouses**	S.F.							28
0100	Architectural		47.50	72.50	132	60.50%	67%	71.50%	
0200	Plumbing		2.48	5.30	10.20	2.82%	3.72%	5%	
0300	Mechanical		2.93	16.70	26	4.56%	8.15%	10.70%	
0400	Electrical		6.15	20	33.50	7.75%	10.10%	18.30%	
0500	Total Project Costs		71	113	228				

Square Foot Project Size Modifier

One factor that affects the S.F. cost of a particular building is the size. In general, for buildings built to the same specifications in the same locality, the larger building will have the lower S.F. cost. This is due mainly to the decreasing contribution of the exterior walls plus the economy of scale usually achievable in larger buildings. The Area Conversion Scale shown below will give a factor to convert costs for the typical size building to an adjusted cost for the particular project.

The Square Foot Base Size lists the median costs, most typical project size in our accumulated data, and the range in size of the projects.

The Size Factor for your project is determined by dividing your project area in S.F. by the typical project size for the particular Building Type. With this factor, enter the Area Conversion Scale at the appropriate Size Factor and determine the appropriate Cost Multiplier for your building size.

Example: Determine the cost per S.F. for a 107,200 S.F. Multi-family housing.

$$\frac{\text{Proposed building area} = 107{,}200 \text{ S.F.}}{\text{Typical size from below} = 53{,}600 \text{ S.F.}} = 2.00$$

Enter Area Conversion Scale at 2.0, intersect curve, read horizontally the appropriate cost multiplier of .94. Size adjusted cost becomes .94 x $210.00 = $197.40 based on national average costs.

Note: For Size Factors less than .50, the Cost Multiplier is 1.1
For Size Factors greater than 3.5, the Cost Multiplier is .90

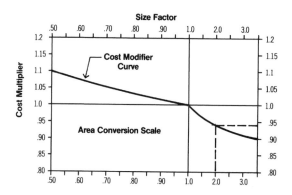

System	Median Cost (Total Project Costs)	Typical Size Gross S.F. (Median of Projects)	Typical Range (Low – High) (Projects)	
Auto Sales with Repair	$187.00	24,900	4,680 –	29,253
Banking Institutions	300.00	9,300	3,267 –	38,148
Court House	286.00	47,600	24,680 –	70,500
Data Centers	284.00	14,400	14,369 –	14,369
Detention Centers	320.00	37,800	12,257 –	183,339
Fire Stations	241.00	12,500	6,300 –	49,577
Gymnasium	206.00	52,400	22,844 –	81,992
Hospitals	375.00	87,100	22,428 –	410,273
Industrial Buildings	133.00	21,100	5,146 –	200,625
Medical Clinics & Offices	224.00	22,500	2,273 –	327,000
Mixed Use	335.00	28,500	7,200 –	188,944
Multi-Family Housing	210.00	53,600	2,472 –	1,161,450
Nursing Home & Assisted Living	161.00	38,200	1,515 –	242,555
Office Buildings	202.00	20,600	1,115 –	930,000
Parking Garage	47.5	151,800	99,884 –	287,040
Parking Garage/Mixed Use	176.00	254,200	5,302 –	318,033
Police Stations	270.00	28,500	15,377 –	88,625
Police/Fire	173.00	44,300	8,600 –	50,279
Public Assembly Buildings	255.00	21,000	2,232 –	235,301
Recreational	296.00	27,700	1,000 –	223,787
Restaurants	305.00	6,000	3,975 –	42,000
Retail	153.00	26,000	4,035 –	84,270
Schools	227.00	69,900	1,344 –	410,848
University, College & Private School Classroom & Admin Buildings	293.00	48,200	8,806 –	196,187
University, College & Private School Dormitories	229.00	39,600	1,500 –	126,889
University, College & Private School Science, Eng. & Lab Buildings	315.00	60,000	5,311 –	117,643
University, College & Private School Student Union Buildings	535.00	47,300	42,075 –	50,000
Warehouses	113.00	11,100	640 –	303,750

A	Area Square Feet; Ampere	Brk., brk	Brick	Csc	Cosecant
AAFES	Army and Air Force Exchange Service	brkt	Bracket	C.S.F.	Hundred Square Feet
		Brs.	Brass	CSI	Construction Specifications Institute
ABS	Acrylonitrile Butadiene Stryrene; Asbestos Bonded Steel	Brz.	Bronze		
		Bsn.	Basin	CT	Current Transformer
A.C., AC	Alternating Current; Air-Conditioning; Asbestos Cement; Plywood Grade A & C	Btr.	Better	CTS	Copper Tube Size
		BTU	British Thermal Unit	Cu	Copper, Cubic
		BTUH	BTU per Hour	Cu. Ft.	Cubic Foot
		Bu.	Bushels	cw	Continuous Wave
ACI	American Concrete Institute	BUR	Built-up Roofing	C.W.	Cool White; Cold Water
ACR	Air Conditioning Refrigeration	BX	Interlocked Armored Cable	Cwt.	100 Pounds
ADA	Americans with Disabilities Act	°C	Degree Centigrade	C.W.X.	Cool White Deluxe
AD	Plywood, Grade A & D	c	Conductivity, Copper Sweat	C.Y.	Cubic Yard (27 cubic feet)
Addit.	Additional	C	Hundred; Centigrade	C.Y./Hr.	Cubic Yard per Hour
Adh.	Adhesive	C/C	Center to Center, Cedar on Cedar	Cyl.	Cylinder
Adj.	Adjustable	C-C	Center to Center	d	Penny (nail size)
af	Audio-frequency	Cab	Cabinet	D	Deep; Depth; Discharge
AFFF	Aqueous Film Forming Foam	Cair.	Air Tool Laborer	Dis., Disch.	Discharge
AFUE	Annual Fuel Utilization Efficiency	Cal.	Caliper	Db	Decibel
AGA	American Gas Association	Calc	Calculated	Dbl.	Double
Agg.	Aggregate	Cap.	Capacity	DC	Direct Current
A.H., Ah	Ampere Hours	Carp.	Carpenter	DDC	Direct Digital Control
A hr.	Ampere-hour	C.B.	Circuit Breaker	Demob.	Demobilization
A.H.U., AHU	Air Handling Unit	C.C.A.	Chromate Copper Arsenate	d.f.t.	Dry Film Thickness
A.I.A.	American Institute of Architects	C.C.F.	Hundred Cubic Feet	d.f.u.	Drainage Fixture Units
AIC	Ampere Interrupting Capacity	cd	Candela	D.H.	Double Hung
Allow.	Allowance	cd/sf	Candela per Square Foot	DHW	Domestic Hot Water
alt., alt	Alternate	CD	Grade of Plywood Face & Back	DI	Ductile Iron
Alum.	Aluminum	CDX	Plywood, Grade C & D, exterior glue	Diag.	Diagonal
a.m.	Ante Meridiem			Diam., Dia	Diameter
Amp.	Ampere	Cefi.	Cement Finisher	Distrib.	Distribution
Anod.	Anodized	Cem.	Cement	Div.	Division
ANSI	American National Standards Institute	CF	Hundred Feet	Dk.	Deck
		C.F.	Cubic Feet	D.L.	Dead Load; Diesel
APA	American Plywood Association	CFM	Cubic Feet per Minute	DLH	Deep Long Span Bar Joist
Approx.	Approximate	CFRP	Carbon Fiber Reinforced Plastic	dlx	Deluxe
Apt.	Apartment	c.g.	Center of Gravity	Do.	Ditto
Asb.	Asbestos	CHW	Chilled Water; Commercial Hot Water	DOP	Dioctyl Phthalate Penetration Test (Air Filters)
A.S.B.C.	American Standard Building Code				
Asbe.	Asbestos Worker	C.I., CI	Cast Iron	Dp., dp	Depth
ASCE	American Society of Civil Engineers	C.I.P., CIP	Cast in Place	D.P.S.T.	Double Pole, Single Throw
A.S.H.R.A.E.	American Society of Heating, Refrig. & AC Engineers	Circ.	Circuit	Dr.	Drive
		C.L.	Carload Lot	DR	Dimension Ratio
ASME	American Society of Mechanical Engineers	CL	Chain Link	Drink.	Drinking
		Clab.	Common Laborer	D.S.	Double Strength
ASTM	American Society for Testing and Materials	Clam	Common Maintenance Laborer	D.S.A.	Double Strength A Grade
		C.L.F.	Hundred Linear Feet	D.S.B.	Double Strength B Grade
Attchmt.	Attachment	CLF	Current Limiting Fuse	Dty.	Duty
Avg., Ave.	Average	CLP	Cross Linked Polyethylene	DWV	Drain Waste Vent
AWG	American Wire Gauge	cm	Centimeter	DX	Deluxe White, Direct Expansion
AWWA	American Water Works Assoc.	CMP	Corr. Metal Pipe	dyn	Dyne
Bbl.	Barrel	CMU	Concrete Masonry Unit	e	Eccentricity
B&B, BB	Grade B and Better; Balled & Burlapped	CN	Change Notice	E	Equipment Only; East; Emissivity
		Col.	Column	Ea.	Each
B&S	Bell and Spigot	CO₂	Carbon Dioxide	EB	Encased Burial
B.&W.	Black and White	Comb.	Combination	Econ.	Economy
b.c.c.	Body-centered Cubic	comm.	Commercial, Communication	E.C.Y	Embankment Cubic Yards
B.C.Y.	Bank Cubic Yards	Compr.	Compressor	EDP	Electronic Data Processing
BE	Bevel End	Conc.	Concrete	EIFS	Exterior Insulation Finish System
B.F.	Board Feet	Cont., cont	Continuous; Continued, Container	E.D.R.	Equiv. Direct Radiation
Bg. cem.	Bag of Cement	Corkbd.	Cork Board	Eq.	Equation
BHP	Boiler Horsepower; Brake Horsepower	Corr.	Corrugated	EL	Elevation
		Cos	Cosine	Elec.	Electrician; Electrical
B.I.	Black Iron	Cot	Cotangent	Elev.	Elevator; Elevating
bidir.	bidirectional	Cov.	Cover	EMT	Electrical Metallic Conduit; Thin Wall Conduit
Bit., Bitum.	Bituminous	C/P	Cedar on Paneling		
Bit., Conc.	Bituminous Concrete	CPA	Control Point Adjustment	Eng.	Engine, Engineered
Bk.	Backed	Cplg.	Coupling	EPDM	Ethylene Propylene Diene Monomer
Bkrs.	Breakers	CPM	Critical Path Method		
Bldg., bldg	Building	CPVC	Chlorinated Polyvinyl Chloride	EPS	Expanded Polystyrene
Blk.	Block	C.Pr.	Hundred Pair	Eqhv.	Equip. Oper., Heavy
Bm.	Beam	CRC	Cold Rolled Channel	Eqlt.	Equip. Oper., Light
Boil.	Boilermaker	Creos.	Creosote	Eqmd.	Equip. Oper., Medium
bpm	Blows per Minute	Crpt.	Carpet & Linoleum Layer	Eqmm.	Equip. Oper., Master Mechanic
BR	Bedroom	CRT	Cathode-ray Tube	Eqol.	Equip. Oper., Oilers
Brg., brng.	Bearing	CS	Carbon Steel, Constant Shear Bar Joist	Equip.	Equipment
Brhe.	Bricklayer Helper			ERW	Electric Resistance Welded
Bric.	Bricklayer				

Abbreviations

E.S.	Energy Saver	H	High Henry	Lath.	Lather	
Est.	Estimated	HC	High Capacity	Lav.	Lavatory	
esu	Electrostatic Units	H.D., HD	Heavy Duty; High Density	lb.; #	Pound	
E.W.	Each Way	H.D.O.	High Density Overlaid	L.B., LB	Load Bearing; L Conduit Body	
EWT	Entering Water Temperature	HDPE	High Density Polyethylene Plastic	L. & E.	Labor & Equipment	
Excav.	Excavation	Hdr.	Header	lb./hr.	Pounds per Hour	
excl	Excluding	Hdwe.	Hardware	lb./L.F.	Pounds per Linear Foot	
Exp., exp	Expansion, Exposure	H.I.D., HID	High Intensity Discharge	lbf/sq.in.	Pound-force per Square Inch	
Ext., ext	Exterior; Extension	Help.	Helper Average	L.C.L.	Less than Carload Lot	
Extru.	Extrusion	HEPA	High Efficiency Particulate Air	L.C.Y.	Loose Cubic Yard	
f.	Fiber Stress		Filter	Ld.	Load	
F	Fahrenheit; Female; Fill	Hg	Mercury	LE	Lead Equivalent	
Fab., fab	Fabricated; Fabric	HIC	High Interrupting Capacity	LED	Light Emitting Diode	
FBGS	Fiberglass	HM	Hollow Metal	L.F.	Linear Foot	
F.C.	Footcandles	HMWPE	High Molecular Weight	L.F. Hdr	Linear Feet of Header	
f.c.c.	Face-centered Cubic		Polyethylene	L.F. Nose	Linear Foot of Stair Nosing	
f'c.	Compressive Stress in Concrete;	HO	High Output	L.F. Rsr	Linear Foot of Stair Riser	
	Extreme Compressive Stress	Horiz.	Horizontal	Lg.	Long; Length; Large	
F.E.	Front End	H.P., HP	Horsepower; High Pressure	L & H	Light and Heat	
FEP	Fluorinated Ethylene Propylene	H.P.F.	High Power Factor	LH	Long Span Bar Joist	
	(Teflon)	Hr.	Hour	L.H.	Labor Hours	
F.G.	Flat Grain	Hrs./Day	Hours per Day	L.L., LL	Live Load	
F.H.A.	Federal Housing Administration	HSC	High Short Circuit	L.L.D.	Lamp Lumen Depreciation	
Fig.	Figure	Ht.	Height	lm	Lumen	
Fin.	Finished	Htg.	Heating	lm/sf	Lumen per Square Foot	
FIPS	Female Iron Pipe Size	Htrs.	Heaters	lm/W	Lumen per Watt	
Fixt.	Fixture	HVAC	Heating, Ventilation & Air-	LOA	Length Over All	
FJP	Finger jointed and primed		Conditioning	log	Logarithm	
Fl. Oz.	Fluid Ounces	Hvy.	Heavy	L-O-L	Lateralolet	
Flr.	Floor	HW	Hot Water	long.	Longitude	
Flrs.	Floors	Hyd.; Hydr.	Hydraulic	L.P., LP	Liquefied Petroleum; Low Pressure	
FM	Frequency Modulation;	Hz	Hertz (cycles)	L.P.F.	Low Power Factor	
	Factory Mutual	I.	Moment of Inertia	LR	Long Radius	
Fmg.	Framing	IBC	International Building Code	L.S.	Lump Sum	
FM/UL	Factory Mutual/Underwriters Labs	I.C.	Interrupting Capacity	Lt.	Light	
Fdn.	Foundation	ID	Inside Diameter	Lt. Ga.	Light Gauge	
FNPT	Female National Pipe Thread	I.D.	Inside Dimension; Identification	L.T.L.	Less than Truckload Lot	
Fori.	Foreman, Inside	I.F.	Inside Frosted	Lt. Wt.	Lightweight	
Foro.	Foreman, Outside	I.M.C.	Intermediate Metal Conduit	L.V.	Low Voltage	
Fount.	Fountain	In.	Inch	M	Thousand; Material; Male;	
fpm	Feet per Minute	Incan.	Incandescent		Light Wall Copper Tubing	
FPT	Female Pipe Thread	Incl.	Included; Including	M²CA	Meters Squared Contact Area	
Fr	Frame	Int.	Interior	m/hr.; M.H.	Man-hour	
F.R.	Fire Rating	Inst.	Installation	mA	Milliampere	
FRK	Foil Reinforced Kraft	Insul., insul	Insulation/Insulated	Mach.	Machine	
FSK	Foil/Scrim/Kraft	I.P.	Iron Pipe	Mag. Str.	Magnetic Starter	
FRP	Fiberglass Reinforced Plastic	I.P.S., IPS	Iron Pipe Size	Maint.	Maintenance	
FS	Forged Steel	IPT	Iron Pipe Threaded	Marb.	Marble Setter	
FSC	Cast Body; Cast Switch Box	I.W.	Indirect Waste	Mat; Mat'l.	Material	
Ft., ft	Foot; Feet	J	Joule	Max.	Maximum	
Ftng.	Fitting	J.I.C.	Joint Industrial Council	MBF	Thousand Board Feet	
Ftg.	Footing	K	Thousand; Thousand Pounds;	MBH	Thousand BTU's per hr.	
Ft lb.	Foot Pound		Heavy Wall Copper Tubing, Kelvin	MC	Metal Clad Cable	
Furn.	Furniture	K.A.H.	Thousand Amp. Hours	MCC	Motor Control Center	
FVNR	Full Voltage Non-Reversing	kcmil	Thousand Circular Mils	M.C.F.	Thousand Cubic Feet	
FVR	Full Voltage Reversing	KD	Knock Down	MCFM	Thousand Cubic Feet per Minute	
FXM	Female by Male	K.D.A.T.	Kiln Dried After Treatment	M.C.M.	Thousand Circular Mils	
Fy.	Minimum Yield Stress of Steel	kg	Kilogram	MCP	Motor Circuit Protector	
g	Gram	kG	Kilogauss	MD	Medium Duty	
G	Gauss	kgf	Kilogram Force	MDF	Medium-density fibreboard	
Ga.	Gauge	kHz	Kilohertz	M.D.O.	Medium Density Overlaid	
Gal., gal.	Gallon	Kip	1000 Pounds	Med.	Medium	
Galv., galv	Galvanized	KJ	Kilojoule	MF	Thousand Feet	
GC/MS	Gas Chromatograph/Mass	K.L.	Effective Length Factor	M.F.B.M.	Thousand Feet Board Measure	
	Spectrometer	K.L.F.	Kips per Linear Foot	Mfg.	Manufacturing	
Gen.	General	Km	Kilometer	Mfrs.	Manufacturers	
GFI	Ground Fault Interrupter	KO	Knock Out	mg	Milligram	
GFRC	Glass Fiber Reinforced Concrete	K.S.F.	Kips per Square Foot	MGD	Million Gallons per Day	
Glaz.	Glazier	K.S.I.	Kips per Square Inch	MGPH	Million Gallons per Hour	
GPD	Gallons per Day	kV	Kilovolt	MH, M.H.	Manhole; Metal Halide; Man-Hour	
gpf	Gallon per Flush	kVA	Kilovolt Ampere	MHz	Megahertz	
GPH	Gallons per Hour	kVAR	Kilovar (Reactance)	Mi.	Mile	
gpm, GPM	Gallons per Minute	KW	Kilowatt	MI	Malleable Iron; Mineral Insulated	
GR	Grade	KWh	Kilowatt-hour	MIPS	Male Iron Pipe Size	
Gran.	Granular	L	Labor Only; Length; Long;	mj	Mechanical Joint	
Grnd.	Ground		Medium Wall Copper Tubing	m	Meter	
GVW	Gross Vehicle Weight	Lab.	Labor	mm	Millimeter	
GWB	Gypsum Wall Board	lat	Latitude	Mill.	Millwright	
				Min., min.	Minimum, Minute	

Misc.	Miscellaneous	PCM	Phase Contrast Microscopy	SBS	Styrene Butadiere Styrene
ml	Milliliter, Mainline	PDCA	Painting and Decorating	SC	Screw Cover
M.L.F.	Thousand Linear Feet		Contractors of America	SCFM	Standard Cubic Feet per Minute
Mo.	Month	P.E., PE	Professional Engineer;	Scaf.	Scaffold
Mobil.	Mobilization		Porcelain Enamel;	Sch., Sched.	Schedule
Mog.	Mogul Base		Polyethylene; Plain End	S.C.R.	Modular Brick
MPH	Miles per Hour	P.E.C.I.	Porcelain Enamel on Cast Iron	S.D.	Sound Deadening
MPT	Male Pipe Thread	Perf.	Perforated	SDR	Standard Dimension Ratio
MRGWB	Moisture Resistant Gypsum	PEX	Cross Linked Polyethylene	S.E.	Surfaced Edge
	Wallboard	Ph.	Phase	Sel.	Select
MRT	Mile Round Trip	P.I.	Pressure Injected	SER, SEU	Service Entrance Cable
ms	Millisecond	Pile.	Pile Driver	S.F.	Square Foot
M.S.F.	Thousand Square Feet	Pkg.	Package	S.F.C.A.	Square Foot Contact Area
Mstz.	Mosaic & Terrazzo Worker	Pl.	Plate	S.F. Flr.	Square Foot of Floor
M.S.Y.	Thousand Square Yards	Plah.	Plasterer Helper	S.F.G.	Square Foot of Ground
Mtd., mtd., mtd	Mounted	Plas.	Plasterer	S.F. Hor.	Square Foot Horizontal
Mthe.	Mosaic & Terrazzo Helper	plf	Pounds Per Linear Foot	SFR	Square Feet of Radiation
Mtng.	Mounting	Pluh.	Plumber Helper	S.F. Shlf.	Square Foot of Shelf
Mult.	Multi; Multiply	Plum.	Plumber	S4S	Surface 4 Sides
MUTCD	Manual on Uniform Traffic Control	Ply.	Plywood	Shee.	Sheet Metal Worker
	Devices	p.m.	Post Meridiem	Sin.	Sine
M.V.A.	Million Volt Amperes	Pntd.	Painted	Skwk.	Skilled Worker
M.V.A.R.	Million Volt Amperes Reactance	Pord.	Painter, Ordinary	SL	Saran Lined
MV	Megavolt	pp	Pages	S.L.	Slimline
MW	Megawatt	PP, PPL	Polypropylene	Sldr.	Solder
MXM	Male by Male	P.P.M.	Parts per Million	SLH	Super Long Span Bar Joist
MYD	Thousand Yards	Pr.	Pair	S.N.	Solid Neutral
N	Natural; North	P.E.S.B.	Pre-engineered Steel Building	SO	Stranded with oil resistant inside
nA	Nanoampere	Prefab.	Prefabricated		insulation
NA	Not Available; Not Applicable	Prefin.	Prefinished	S-O-L	Socketolet
N.B.C.	National Building Code	Prop.	Propelled	sp	Standpipe
NC	Normally Closed	PSF, psf	Pounds per Square Foot	S.P.	Static Pressure; Single Pole; Self-
NEMA	National Electrical Manufacturers	PSI, psi	Pounds per Square Inch		Propelled
	Assoc.	PSIG	Pounds per Square Inch Gauge	Spri.	Sprinkler Installer
NEHB	Bolted Circuit Breaker to 600V.	PSP	Plastic Sewer Pipe	spwg	Static Pressure Water Gauge
NFPA	National Fire Protection Association	Pspr.	Painter, Spray	S.P.D.T.	Single Pole, Double Throw
NLB	Non-Load-Bearing	Psst.	Painter, Structural Steel	SPF	Spruce Pine Fir; Sprayed
NM	Non-Metallic Cable	P.T.	Potential Transformer		Polyurethane Foam
nm	Nanometer	P. & T.	Pressure & Temperature	S.P.S.T.	Single Pole, Single Throw
No.	Number	Ptd.	Painted	SPT	Standard Pipe Thread
NO	Normally Open	Ptns.	Partitions	Sq.	Square; 100 Square Feet
N.O.C.	Not Otherwise Classified	Pu	Ultimate Load	Sq. Hd.	Square Head
Nose.	Nosing	PVC	Polyvinyl Chloride	Sq. In.	Square Inch
NPT	National Pipe Thread	Pvmt.	Pavement	S.S.	Single Strength; Stainless Steel
NQOD	Combination Plug-on/Bolt on	PRV	Pressure Relief Valve	S.S.B.	Single Strength B Grade
	Circuit Breaker to 240V.	Pwr.	Power	sst, ss	Stainless Steel
N.R.C., NRC	Noise Reduction Coefficient/	Q	Quantity Heat Flow	Sswk.	Structural Steel Worker
	Nuclear Regulator Commission	Qt.	Quart	Sswl.	Structural Steel Welder
N.R.S.	Non Rising Stem	Quan., Qty.	Quantity	St.; Stl.	Steel
ns	Nanosecond	Q.C.	Quick Coupling	STC	Sound Transmission Coefficient
NTP	Notice to Proceed	r	Radius of Gyration	Std.	Standard
nW	Nanowatt	R	Resistance	Stg.	Staging
OB	Opposing Blade	R.C.P.	Reinforced Concrete Pipe	STK	Select Tight Knot
OC	On Center	Rect.	Rectangle	STP	Standard Temperature & Pressure
OD	Outside Diameter	recpt.	Receptacle	Stpi.	Steamfitter, Pipefitter
O.D.	Outside Dimension	Reg.	Regular	Str.	Strength; Starter; Straight
ODS	Overhead Distribution System	Reinf.	Reinforced	Strd.	Stranded
O.G.	Ogee	Req'd.	Required	Struct.	Structural
O.H.	Overhead	Res.	Resistant	Sty.	Story
O&P	Overhead and Profit	Resi.	Residential	Subj.	Subject
Oper.	Operator	RF	Radio Frequency	Subs.	Subcontractors
Opng.	Opening	RFID	Radio-frequency Identification	Surf.	Surface
Orna.	Ornamental	Rgh.	Rough	Sw.	Switch
OSB	Oriented Strand Board	RGS	Rigid Galvanized Steel	Swbd.	Switchboard
OS&Y	Outside Screw and Yoke	RHW	Rubber, Heat & Water Resistant;	S.Y.	Square Yard
OSHA	Occupational Safety and Health		Residential Hot Water	Syn.	Synthetic
	Act	rms	Root Mean Square	S.Y.P.	Southern Yellow Pine
Ovhd.	Overhead	Rnd.	Round	Sys.	System
OWG	Oil, Water or Gas	Rodm.	Rodman	t.	Thickness
Oz.	Ounce	Rofc.	Roofer, Composition	T	Temperature; Ton
P.	Pole; Applied Load; Projection	Rofp.	Roofer, Precast	Tan	Tangent
p.	Page	Rohe.	Roofer Helpers (Composition)	T.C.	Terra Cotta
Pape.	Paperhanger	Rots.	Roofer, Tile & Slate	T & C	Threaded and Coupled
P.A.P.R.	Powered Air Purifying Respirator	R.O.W.	Right of Way	T.D.	Temperature Difference
PAR	Parabolic Reflector	RPM	Revolutions per Minute	TDD	Telecommunications Device for
P.B., PB	Push Button	R.S.	Rapid Start		the Deaf
Pc., Pcs.	Piece, Pieces	Rsr	Riser	T.E.M.	Transmission Electron Microscopy
P.C.	Portland Cement; Power Connector	RT	Round Trip	temp	Temperature, Tempered, Temporary
P.C.F.	Pounds per Cubic Foot	S.	Suction; Single Entrance; South	TFFN	Nylon Jacketed Wire

For customer support on your Commercial Renovation Costs with RSMeans Data, call 800.448.8182.

Abbreviations

TFE	Tetrafluoroethylene (Teflon)	U.L., UL	Underwriters Laboratory	w/	With
T. & G.	Tongue & Groove;	Uld.	Unloading	W.C., WC	Water Column; Water Closet
	Tar & Gravel	Unfin.	Unfinished	W.F.	Wide Flange
Th., Thk.	Thick	UPS	Uninterruptible Power Supply	W.G.	Water Gauge
Thn.	Thin	URD	Underground Residential	Wldg.	Welding
Thrded	Threaded		Distribution	W. Mile	Wire Mile
Tilf.	Tile Layer, Floor	US	United States	W-O-L	Weldolet
Tilh.	Tile Layer, Helper	USGBC	U.S. Green Building Council	W.R.	Water Resistant
THHN	Nylon Jacketed Wire	USP	United States Primed	Wrck.	Wrecker
THW.	Insulated Strand Wire	UTMCD	Uniform Traffic Manual For Control	WSFU	Water Supply Fixture Unit
THWN	Nylon Jacketed Wire		Devices	W.S.P.	Water, Steam, Petroleum
T.L., TL	Truckload	UTP	Unshielded Twisted Pair	WT., Wt.	Weight
T.M.	Track Mounted	V	Volt	WWF	Welded Wire Fabric
Tot.	Total	VA	Volt Amperes	XFER	Transfer
T-O-L	Threadolet	VAT	Vinyl Asbestos Tile	XFMR	Transformer
tmpd	Tempered	V.C.T.	Vinyl Composition Tile	XHD	Extra Heavy Duty
TPO	Thermoplastic Polyolefin	VAV	Variable Air Volume	XHHW	Cross-Linked Polyethylene Wire
T.S.	Trigger Start	VC	Veneer Core	XLPE	Insulation
Tr.	Trade	VDC	Volts Direct Current	XLP	Cross-linked Polyethylene
Transf.	Transformer	Vent.	Ventilation	Xport	Transport
Trhv.	Truck Driver, Heavy	Vert.	Vertical	Y	Wye
Trlr	Trailer	V.F.	Vinyl Faced	yd	Yard
Trlt.	Truck Driver, Light	V.G.	Vertical Grain	yr	Year
TTY	Teletypewriter	VHF	Very High Frequency	Δ	Delta
TV	Television	VHO	Very High Output	%	Percent
T.W.	Thermoplastic Water Resistant	Vib.	Vibrating	~	Approximately
	Wire	VLF	Vertical Linear Foot	Ø	Phase; diameter
UCI	Uniform Construction Index	VOC	Volatile Organic Compound	@	At
UF	Underground Feeder	Vol.	Volume	#	Pound; Number
UGND	Underground Feeder	VRP	Vinyl Reinforced Polyester	<	Less Than
UHF	Ultra High Frequency	W	Wire; Watt; Wide; West	>	Greater Than
U.I.	United Inch			Z	Zone

Index

For customer support on your Commercial Renovation Costs with RSMeans Data, call 800.448.8182.

Index

For customer support on your Commercial Renovation Costs with RSMeans Data, call 800.448.8182.

For customer support on your Commercial Renovation Costs with RSMeans Data, call 800.448.8182.

For customer support on your Commercial Renovation Costs with RSMeans Data, call 800.448.8182.

Index

For customer support on your Commercial Renovation Costs with RSMeans Data, call 800.448.8182.

Index

Index

For customer support on your Commercial Renovation Costs with RSMeans Data, call 800.448.8182.

Index

For customer support on your Commercial Renovation Costs with RSMeans Data, call 800.448.8182.

For customer support on your Commercial Renovation Costs with RSMeans Data, call 800.448.8182.

921

Index

Division Notes

	CREW	DAILY OUTPUT	LABOR-HOURS	UNIT	BARE COSTS				TOTAL INCL O&P
					MAT.	LABOR	EQUIP.	TOTAL	

Division Notes

	CREW	DAILY OUTPUT	LABOR-HOURS	UNIT	BARE COSTS				TOTAL INCL O&P
					MAT.	LABOR	EQUIP.	TOTAL	

Unit prices according to the latest MasterFormat®

ost Data Selection Guide

e following table provides definitive information on the content of each cost data publication. The number of lines of data provided in each unit price or emblies division, as well as the number of crews, is listed for each data set. The presence of other elements such as reference tables, square foot models, uipment rental costs, historical cost indexes, and city cost indexes, is also indicated. You can use the table to help select the RSMeans data set that has e quantity and type of information you most need in your work.

nit Cost visions	Building Construction	Mechanical	Electrical	Commercial Renovation	Square Foot	Site Work Landsc.	Green Building	Interior	Concrete Masonry	Open Shop	Heavy Construction	Light Commercial	Facilities Construction	Plumbing	Residential
1	621	454	475	576	0	543	200	377	505	620	560	322	1092	462	219
2	825	347	157	781	0	972	181	466	221	824	739	550	1268	355	340
3	1745	341	232	1265	0	1537	1043	355	2274	1745	1930	538	2028	317	445
4	960	22	0	920	0	724	180	613	1158	928	614	532	1175	0	446
5	1895	158	155	1099	0	858	1793	1106	735	1895	1031	985	1912	204	752
6	2468	18	18	2127	0	110	589	1544	281	2464	123	2157	2141	22	2677
7	1593	215	128	1633	0	580	761	532	523	1590	26	1326	1693	227	1046
8	2140	80	3	2733	0	255	1138	1813	105	2142	0	2328	2966	0	1552
9	2125	86	45	1943	0	313	464	2216	424	2062	15	1779	2379	54	1544
10	1088	17	10	684	0	232	32	898	136	1088	34	588	1179	237	224
11	1095	199	166	539	0	135	56	923	29	1062	0	229	1115	162	108
12	539	0	2	297	0	219	147	1552	14	506	0	272	1571	23	216
13	745	149	157	252	0	370	124	250	77	721	271	109	761	115	103
14	273	36	0	223	0	0	0	257	0	273	0	12	293	16	6
21	127	0	41	37	0	0	0	293	0	127	0	121	665	685	259
22	1165	7543	160	1226	0	2010	1061	849	20	1154	2109	875	7505	9400	719
23	1263	7010	639	1033	0	250	864	775	38	1246	191	980	5244	2011	579
25	0	0	14	14	0	0	0	0	0	0	0	0	0	0	0
26	1513	491	10473	1293	0	859	645	1160	55	1439	648	1360	10254	399	636
27	95	0	467	105	0	0	0	71	0	95	45	67	389	0	56
28	143	79	223	124	0	0	28	97	0	127	0	70	209	57	41
31	1511	733	610	807	0	3261	284	7	1216	1456	3280	607	1568	660	616
32	906	49	8	944	0	4523	419	418	367	877	1942	496	1803	140	544
33	1319	1100	563	310	0	3142	33	0	253	596	3277	135	1790	2165	161
34	112	0	47	9	0	195	0	0	36	67	226	0	141	0	0
35	18	0	0	0	0	327	0	0	0	18	442	0	84	0	0
41	63	0	0	34	0	8	0	22	0	62	31	0	69	14	0
44	75	79	0	0	0	0	0	0	0	0	264	0	75	75	0
46	23	16	0	0	0	274	261	0	0	23	0	0	33	33	0
48	8	0	36	2	0	0	21	0	0	8	15	8	21	0	8
Totals	26453	19222	14829	21010	0	21697	10324	16594	8467	25215	17813	16446	51423	17833	13297

Assem Div	Building Construction	Mechanical	Elec-trical	Commercial Renovation	Square Foot	Site Work Landscape	Assemblies	Green Building	Interior	Concrete Masonry	Heavy Construction	Light Commercial	Facilities Construction	Plumbing	Asm Div	Residential
A		15	0	190	165	579	599	0	0	537	573	155	24	0	1	378
B		0	0	848	2567	0	5673	56	329	1979	368	2108	174	0	2	211
C		0	0	647	956	0	1338	0	1645	146	0	846	251	0	3	591
D		1060	945	712	1866	72	2544	330	827	0	0	1353	1108	1090	4	851
E		0	0	85	262	0	302	0	5	0	0	259	5	0	5	391
F		0	0	0	114	0	143	0	0	0	0	114	0	0	6	357
G		527	447	318	312	3378	792	0	0	535	1349	205	293	677	7	307
															8	760
															9	80
															10	0
															11	0
															12	0
Totals		1602	1392	2800	6242	4029	11391	386	2806	3197	2290	5040	1855	1767		3926

Reference Section	Building Construction Costs	Mechanical	Electrical	Commercial Renovation	Square Foot	Site Work Landscape	Assem.	Green Building	Interior	Concrete Masonry	Open Shop	Heavy Construction	Light Commercial	Facilities Construction	Plumbing	Resi.
Reference Tables	yes	yes	yes	yes	no	yes	yes	yes	yes	yes	yes	yes	yes	yes	yes	yes
Models					111			25					50			28
Crews	584	584	584	564		584	584	584	584	584	562	584	562	564	584	560
Equipment Rental Costs	yes	yes	yes	yes		yes		yes	yes	yes	yes	yes	yes	yes	yes	yes
Historical Cost Indexes	yes	yes	yes	yes	yes	yes	yes	yes	yes	yes	yes	yes	yes	yes	yes	no
City Cost Indexes	yes	yes	yes	yes	yes	yes	yes	yes	yes	yes	yes	yes	yes	yes	yes	yes

Virtual classes offered monthly, call for the schedule—see note for details

2021 Training Class Options   📞 877.620.6245

RSMeans offers training classes in a variety of formats—eLearning training modules, instructor-led classes, on-site training, and virtual training classes. Due to our current Covid-19 situation, RSMeans has adjusted our current training offerings and instructor-led classes for 2021 have not been determined. Please visit our website at https://www.rsmeans.com/products/training/seminars for our current schedule of class offerings.

Training classes that are offered in one or more of our formats are: Building Systems and the Construction Process; Construction Cost Estimating: Concepts & Practice; Introduction to Estimating; Scope of Work for Facilities Estimating; Facilities Construction Estimating; Maintenance & Repair Estimating for Facilities; Mechanical & Electrical Estimating; RSMeans CostWorks CD; and RSMeans Online Training.

If you have any questions or you would like to register for any class or purchase a training module, call us at 877-620.6245.

Facilities Construction Estimating

In this two-day course, professionals working in facilities management can get help with their daily challenges to establish budgets for all phases of a project.

Some of what you'll learn:
- Determining the full scope of a project
- Understanding of RSMeans data and what is included in prices
- Identifying appropriate factors to be included in your estimate
- Creative solutions to estimating issues
- Organizing estimates for presentation and discussion
- Special estimating techniques for repair/remodel and maintenance projects
- Appropriate use of contingency, city cost indexes, and reference notes
- Techniques to get to the correct estimate quickly

Who should attend: facility managers, engineers, contractors, facility tradespeople, planners, and project managers.

Construction Cost Estimating: Concepts and Practice

This one- or two-day introductory course to improve estimating skills and effectiveness starts with the details of interpreting bid documents and ends with the summary of the estimate and bid submission.

Some of what you'll learn:
- Using the plans and specifications to create estimates
- The takeoff process—deriving all tasks with correct quantities
- Developing pricing using various sources; how subcontractor pricing fits in
- Summarizing the estimate to arrive at the final number
- Formulas for area and cubic measure, adding waste and adjusting productivity to specific projects
- Evaluating subcontractors' proposals and prices
- Adding insurance and bonds
- Understanding how labor costs are calculated
- Submitting bids and proposals

Who should attend: project managers, architects, engineers, owners' representatives, contractors, and anyone who's responsible for budgeting or estimating construction projects.

Assessing Scope of Work for Facilities Construction Estimating

This two-day practical training program addresses the vital importance of understanding the scope of projects in order to produce accurate cost estimates for facilities repair and remodeling.

Some of what you'll learn:
- Discussions of site visits, plans/specs, record drawings of facilities, and site-specific lists
- Review of CSI divisions, including means, methods, materials, and the challenges of scoping each topic
- Exercises in scope identification and scope writing for accurate estimating of projects
- Hands-on exercises that require scope, takeoff, and pricing

Who should attend: corporate and government estimators, planners, facility managers, and others who need to produce accurate project estimates.

Maintenance & Repair Estimating for Facilities

This two-day course teaches attendees how to plan, budget, and estimate the cost of ongoing and preventive maintenance and repair for existing buildings and grounds.

Some of what you'll learn:
- The most financially favorable maintenance, repair, and replacement scheduling and estimating
- Auditing and value engineering facilities
- Preventive planning and facilities upgrading
- Determining both in-house and contract-out service costs
- Annual, asset-protecting M&R plan

Who should attend: facility managers, maintenance supervisors, buildings and grounds superintendents, plant managers, planners, estimators, and others involved in facilities planning and budgeting.

Mechanical & Electrical Estimating

This two-day course teaches attendees how to prepare more accurate and complete mechanical/electrical estimates, avoid the pitfalls of omission and double-counting, and understand the composition and rationale within the RSMeans mechanical/electrical database.

Some of what you'll learn:
- The unique way mechanical and electrical systems are interrelated
- M&E estimates—conceptual, planning, budgeting, and bidding stages
- Order of magnitude, square foot, assemblies, and unit price estimating
- Comparative cost analysis of equipment and design alternatives

Who should attend: architects, engineers, facilities managers, mechanical and electrical contractors, and others who need a highly reliable method for developing, understanding, and evaluating mechanical and electrical contracts.

Building Systems and the Construction Process

This one-day course was written to assist novices and those outside the industry in obtaining a solid understanding of the construction process - from both a building systems and construction administration approach.

Some of what you'll learn:
- Various systems used and how components come together to create a building
- Start with foundation and end with the physical systems of the structure such as HVAC and Electrical
- Focus on the process from start of design through project closeout

This training session requires you to bring a laptop computer to class.

Who should attend: building professionals or novices to help make the crossover to the construction industry; suited for anyone responsible for providing high level oversight on construction projects.

Practical Project Management for Construction Professionals

In this two-day course you will acquire the essential knowledge and develop the skills to effectively and efficiently execute the day-to-day responsibilities of the construction project manager.

Some of what you'll learn:
- General conditions of the construction contract
- Contract modifications: change orders and construction change directives
- Negotiations with subcontractors and vendors
- Effective writing: notification and communications
- Dispute resolution: claims and liens

Who should attend: architects, engineers, owners' representatives, and project managers.

RSMeans data Training

for our Online
Solution

...ating is vital to the decision-making process at each state of
...line solution works the way you do. It's systematic, flexible,
...s one-day class you will see how you can estimate any phase
...aster and better.

...what you'll learn:
...stomizing our online estimating solution
...Making the most of RSMeans "Circle Reference" numbers
- How to integrate your cost data
- Generating reports, exporting estimates to MS Excel, sharing, collaborating, and more

Also offered as a self-paced or on-site training program!

Training for our CD Estimating Solution

This one-day course helps users become more familiar with the functionality of the CD. Each menu, icon, screen, and function found in the program is explained in depth. Time is devoted to hands-on estimating exercises.

Some of what you'll learn:
- Searching the database using all navigation methods
- Exporting RSMeans data to your preferred spreadsheet format
- Viewing crews, assembly components, and much more
- Automatically regionalizing the database

This training session requires you to bring a laptop computer to class.

When you register for this course you will receive an outline for your laptop requirements.

Also offered as a self-paced or on-site training program!

Site Work Estimating with RSMeans data

This one-day program focuses directly on site work costs. Accurately scoping, quantifying, and pricing site preparation, underground utility work, and improvements to exterior site elements are often the most difficult estimating tasks on any project. Some of what you'll learn:
- Evaluation of site work and understanding site scope including: site clearing, grading, excavation, disposal and trucking of materials, backfill and compaction, underground utilities, paving, sidewalks, and seeding & planting.
- Unit price site work estimates—Correct use of RSMeans site work cost data to develop a cost estimate.
- Using and modifying assemblies—Save valuable time when estimating site work activities using custom assemblies.

Who should attend: engineers, contractors, estimators, project managers, owners' representatives, and others who are concerned with the proper preparation and/or evaluation of site work estimates.

Please bring a laptop with capability to access the internet.

Facilities Estimating Using the CD

This two-day class combines hands-on skill-building with best estimating practices and real-life problems. You will learn key concepts, tips, pointers, and guidelines to save time and avoid cost oversights and errors.

Some of what you'll learn:
- Estimating process concepts
- Customizing and adapting RSMeans cost data
- Establishing scope of work to account for all known variables
- Budget estimating: when, why, and how
- Site visits: what to look for and what you can't afford to overlook
- How to estimate repair and remodeling variables

This training session requires you to bring a laptop computer to class.

Who should attend: facility managers, architects, engineers, contractors, facility tradespeople, planners, project managers, and anyone involved with JOC, SABRE, or IDIQ.

Registration Information

RSMeans data Training

How to register
By Phone
Register by phone at 877.620.6245

Online
Register online at
RSMeans.com/products/services/training

Note: Purchase Orders or Credits Cards are required to register.

Two-day seminar registration fee - $1,300*.

One-Day Construction Cost Estimating or Building Systems and the Construction Process - $825*.

Two-day virtual training classes - $825*.

Three-day virtual training classes - $995*.

Instructor-led Government pricing
All federal government employees save off the regular seminar price. Other promotional discounts cannot be combined with the government discount. Call 781.422.5115 for government pricing.

CANCELLATION POLICY FOR INSTRUCTOR-LED CLASSES:
If you are unable to attend a seminar, substitutions may be made at any time before the session starts by notifying the seminar registrar at 781.422.5115 or your sales representative.
If you cancel twenty-one (21) days or more prior to the seminar, there will be no penalty and your registration fees will be refunded. These cancellations must be received by the seminar registrar or your sales representative and will be confirmed to be eligible for cancellation.
If you cancel fewer than twenty-one (21) days prior to the seminar, you will forfeit the registration fee.
In the unfortunate event of an RSMeans cancellation, RSMeans will work with you to reschedule your attendance in the same seminar at a later date or will

fully refund your registration fee. RSMeans cannot be responsible for any non-refundable travel expenses incurred by you or another as a result of your registration, attendance at, or cancellation of an RSMeans seminar.
Any on-demand training modules are not eligible for cancellation, substitution, transfer, return, or refund.

AACE approved courses
Many seminars described and offered here have been approved for 14 hours (1.4 recertification credits) of credit by the AACE International Certification Board toward meeting the continuing education requirements for recertification as a Certified Cost Engineer/Certified Cost Consultant.

AIA Continuing Education
We are registered with the AIA Continuing Education System (AIA/CES) and are committed to developing quality learning activities in accordance with the CES criteria. Many seminars meet the AIA/CES criteria for Quality Level 2. AIA members may receive 14 learning units (LUs) for each two-day RSMeans course.

Daily course schedule
The first day of each seminar session begins at 8:30 a.m. and ends at 4:30 p.m. The second day begins at 8:00 a.m. and ends at 4:00 p.m. Participants are urged to bring a hand-held calculator since many actual problems will be worked out in each session.

Continental breakfast
Your registration includes the cost of a continental breakfast and a morning and afternoon refreshment break. These informal segments allow you to discuss topics of mutual interest with other seminar attendees. (You are free to make your own lunch and dinner arrangements.)

Hotel/transportation arrangements
We arrange to hold a block of rooms at most host hotels. To take advantage of special group rates when making your reservation, be sure to mention that you are attending the RSMeans Institute data seminar. You are, of course, free to stay at the lodging place of your choice. (Hotel reservations and transportation arrangements should be made directly by seminar attendees.)

Important
Class sizes are limited, so please register as soon as possible.

*Note: Pricing subject to change.